Beginning and Intermediate Algebra

Julie Miller
Molly O'Neill

Daytona Beach Community College

Boston Burr Ridge, IL Dubuque, IA Madison, WI New York San Francisco St. Louis
Bangkok Bogotá Caracas Kuala Lumpur Lisbon London Madrid Mexico City
Milan Montreal New Delhi Santiago Seoul Singapore Sydney Taipei Toronto

BEGINNING AND INTERMEDIATE ALGEBRA

Published by McGraw-Hill, a business unit of The McGraw-Hill Companies, Inc., 1221 Avenue of the Americas, New York, NY 10020.

Some ancillaries, including electronic and print components, may not be available to customers outside the United States.

This book is printed on acid-free paper.

4 5 6 7 8 9 0 QPD/QPD 0 9 8 7
1 2 3 4 5 6 7 8 9 0 QPD/QPD 0 9 8 7 6 5 4

ISBN 978-0-07-296533-9
MHID 0-07-296533-9 (Student Edition)
ISBN 978-0-07-296534-6
MHID 0-07-296534-7 (Annotated Instructor's Edition)

Publisher, Mathematics and Statistics: *William K. Barter*
Publisher, Developmental Mathematics: *Elizabeth J. Haefele*
Senior Developmental Editor: *Erin Brown*
Executive Marketing Manager: *Michael Weitz*
Marketing Manager: *Steven R. Stembridge*
Project Coordinator: *April R. Southwood*
Senior Production Supervisor: *Sherry L. Kane*
Media Technology Producer: *Amber M. Huebner*
Designer: *Laurie B. Janssen*
(USE) Cover Image: *Robert W. Madden/Gettyimages*
Lead Photo Research Coordinator: *Carrie K. Burger*
Supplement Producer: *Brenda A. Ernzen*
Compositor: *The GTS Companies/York, PA Campus*
Typeface: *10/12 Times Ten*
Printer: *Quebecor World Dubuque*

Photo Credits: Page 1: © Vol. 38/Corbis; p. 32: © PhotoDisc/Getty R-F website; p. 33: © Vol. 154/Corbis; p. 58: © Elena Rooraid/PhotoEdit; p. 65: © Vol. 44/Corbis; p. 75: © Corbis R-F Website; p. 97: © Corbis R-F Website; p. 131: © Vol. 44/Corbis; p. 137: © Judy Griesdieck/Corbis; p. 146: © Corbis R-F Website; p. 155: © Corbis R-F Website; p. 169: © Robert Brenner/PhotoEdit; p. 183: © Vol. 188/Corbis; p. 190: © David Young-Wolff/PhotoEdit; p. 201: © Tony Freeman/PhotoEdit; p. 212: © Michael Newman/PhotoEdit; p. 214: © Vol. 51/PhotoDisc/Getty; p. 224: © Vol. 107/Corbis; p. 233: © EyeWire/Getty website; p. 245: © Corbis R-F Website; p. 272: © Vol. 26/Corbis; p. 305: © Corbis R-F Website; p. 309: © Vol. 1/PhotoDisc; p. 375: © Vol. 62/Corbis; p. 383: © Paul Morris/Corbis; p. 411: © EyeWire/Getty R-F website; p. 438: © PhotoDisc/Getty R-F website; p. 462: © Susan Van Etten/PhotoEdit; p. 470: © Vol. 20/Corbis; p. 500: NOAA; p. 511: © Vol. 59/Corbis; p. 517: © Vol. 247/Corbis; p. 534: ©. Vol. 132/Corbis; p. 543: © Linda Waymire; p. 552: © Jeff Greenberg/PhotoEdit; p. 562: © PhotoDisc/Getty R-F Website; p. 599: © Vol. 67/PhotoDisc; p. 607: © Quest/SPL/Photo Researchers, Inc.; p. 635: © The McGraw-Hill Companies, Inc./John Thoeming, Inc.; p. 652: © Corbis R-F Website; p. 657: © Vol. 44/Corbis; p. 699: © Tony Freeman/PhotoEdit; p. 707: © Nathan Benn/Corbis; p. 725: © Vol. 56/Corbis; p. 735: © Corbis R-F Website; p. 748: © Vol. 102/Corbis; p. 749: © Corbis R-F Website; p. 773: © Mary Kate Denny/PhotoEdit; p. 782: © Stone/Getty; p. 787: © Gary Conner/PhotoEdit; p. 808: © Corbis; p. 823: © PhotoDisc/Getty R-F Website; p. 863: © Jacana Scientific Control/Scott Berthoule/Photo Researchers, Inc.; p. 869: © Oliver Meckes/Photo Researchers, Inc.; p. 871: © PhotoDisc/Getty R-F Website; p. 921: © Corbis R-F Website; p. 957: © Corbis R-F Website.

Library of Congress Cataloging-in-Publication Data

Miller, Julie, 1962–
Beginning and intermediate algebra / Julie Miller, Molly O'Neill.—1st ed.
p. cm.
Includes index.
ISBN 0-07-296533-9 (acid-free paper)
1. Algebra—Textbooks. I. O'Neill, Molly, 1953–. II. Title.

QA152.3 .M57 2006
512.9—dc22

2004057890
CIP

www.mhhe.com

Table of Contents

Dedication

To my uncle, Dr. Don D. Miller, mathematician and WWII codebreaker

—Julie Miller

To my son and best friend, Stephen

—Molly O'Neill

About the Authors

JULIE MILLER

Julie Miller has been on the faculty of the Mathematics Department at Daytona Beach Community College for 16 years, where she has taught developmental and upper-level courses. Prior to her work at DBCC, she worked as a software engineer for General Electric in the area of flight and radar simulation. Julie earned a bachelor of science in applied mathematics from Union College in Schenectady, New York, and a master of science in mathematics from the University of Florida. In addition to this textbook, she has authored several course supplements for college algebra, trigonometry, and precalculus, as well as several short works of fiction and nonfiction for young readers.

"My father is a medical researcher, and I got hooked on math and science when I was young and would visit his laboratory. I can remember using graph paper to plot data points for his experiments and doing simple calculations. He would then tell me what the peaks and features in the graph meant in the context of his experiment. I think that applications and hands-on experience made math come alive for me and I'd like to see math come alive for my students."

—Julie Miller

MOLLY O'NEILL

Molly O'Neill is also from Daytona Beach Community College, where she has taught for 18 years in the Mathematics Department. She has taught a variety of courses from developmental mathematics to calculus. Before she came to Florida, Molly taught as an adjunct instructor at the University of Michigan–Dearborn, Eastern Michigan University, Wayne State University, and Oakland Community College. Molly earned a bachelor of science in mathematics and a master of arts and teaching from Western Michigan University in Kalamazoo, Michigan. Besides this textbook, she has authored several course supplements for college algebra, trigonometry, and precalculus and has reviewed texts for developmental mathematics.

"I differ from many of my colleagues in that math was not always easy for me. But in seventh grade I had a teacher who taught me that if I follow the rules of mathematics, even I could solve math problems. Once I understood this, I enjoyed math to the point of choosing it for my career. I now have the greatest job because I get to do math everyday and I have the opportunity to influence my students just as I was influenced. Authoring these texts has given me another avenue to reach even more students."

—Molly O'Neill

PREFACE

FROM THE AUTHORS

First and foremost, we would like to thank the students and colleagues who have helped us prepare this text. The content and organization of the text is based on an accumulation of our own notes and experiences as teachers. Every day in the classroom our students ask good, probing questions, and every day we seek better ways to convey mathematical concepts. These ideas have found their way from the classroom and into the book.

We originally embarked on this textbook project because we were seeing a lack of student success in courses *above* the level of Beginning and Intermediate Algebra. We wanted to build a better bridge between Intermediate Algebra and college level algebra. Our goal was to use pedagogical features to make the material *stick*. To accomplish this, we employed the following methods.

Active Learning

First, we believe students retain more of what they learn when they are actively engaged in the classroom. Consequently, as we wrote each section of text, we also wrote accompanying worksheets called Classroom Activities to foster accountability and to encourage classroom participation. Classroom Activities resemble the examples that students encounter in the textbook. The activities can be assigned to individual students, or to pairs or groups of students. Most of the activities have been tested in the classroom with our own students.

Critical Thinking

Another one of our core goals as teachers is to make students think and process what they learn. Therefore, we emphasized critical thinking and the interpretation of mathematical results. When students obtain an answer to an exercise, we want them to understand what the answer means in the context of the problem. Our examples and exercises often ask students to, "**Interpret the meaning in the context of the problem**." If students can explain the concepts, their understanding will increase.

Writing Style

We have each taught developmental mathematics for close to 20 years, and we know and understand our students very well. Our intent is to offer a compassionate voice to the students. Our writing style reflects the language and tone that we use daily within our classrooms. We have created special Tips and Avoiding Mistakes boxes that highlight points we emphasize in our own lectures. Therefore, students who use the book should feel very comfortable with the reading level.

Real World Applications

Another critical component of the text is the inclusion of contemporary real-world examples and applications. We based examples and applications on information students encounter daily when they turn on the news, read a magazine, or surf the World Wide Web. We incorporated data for students to create mathematical models in the form of functions, equations, and graphs. When students encounter facts or information that is meaningful to them, they will relate better to the material and remember more of what they learn.

Problem-Solving

Lastly, we crafted our exercise sets to help students solidify their understanding of the content and refine their problem-solving skills. At the beginning of the Practice Exercises, we inserted review problems so that concepts within a chapter could be continually reinforced. We balanced each set of Practice Exercises with both drill-and-practice and conceptual problems that require critical thinking or writing. Throughout the text, you will see exercises with multiple parts, such as (a), (b), (c), etc. These multistep problems are carefully crafted to build problem-solving skills. Each part leads in to the next until a final result is achieved.

Language of Mathematics

We place special emphasis on the skill of translating mathematical notation to English expressions and vice versa by writing "translating expressions" exercises. We also created challenge exercises in the Expanding Your Skills sections of the exercise sets. Graphing Calculator Exercises also appear. They can be found at the *end* of appropriate exercise sets so that instructors who do not encourage the use of a graphing calculator can skip them.

With these measures, we trust that students will achieve success and gain the foundation they need to move on to higher-level mathematics courses.

LISTENING TO STUDENTS' AND INSTRUCTORS' CONCERNS

Although this is a first edition text, the core material has effectively been revised from our Beginning Algebra text and from our Intermediate Algebra text. These two books have been on the market for over a year, and we have had extensive feedback from users. Some of the instructors who have used our texts have kept user diaries. This has helped us to continue to strengthen the content. In effect, this first edition text is based on second edition material. Our editorial staff has amassed the results of reviewer questionnaires, user diaries, focus groups, and symposia. We have read virtually thousands of pages of reviews. Furthermore, McGraw-Hill symposia have brought faculty together from across the United States to discuss issues and trends in developmental mathematics. These efforts have involved hundreds of faculty and have explored issues such as content, readability, and even the aesthetics of page layout.

WHAT SETS THIS BOOK APART?

While all core content is covered in this textbook, the organization is a departure from common practice in some ways. However, we feel that the aspects of the organization that make it unique are factors that will contribute directly to students' success.

Chapter R

Chapter R is a reference chapter. We designed it to help students reacquaint themselves with the fundamentals of fractions and geometry. This chapter also addresses study skills and helpful hints to use the resources provided in the text and supplements.

Graphing and Functions

The graphical interpretation of algebra is a critical skill that carries through to upper level mathematics courses. However, students often have difficulty with graphing. Therefore, we offer graphing on two levels. In Chapter 3, we present concepts related to the "beginning algebra" level of graphing. This includes an introduction to a rectangular coordinate system, identifying linear equations in two variables, and finding the slope of a line. We graph linear equations using a table of points, using the x- and y-intercepts, and using the slope-intercept form. At this level, we also find an equation of a line given its slope and y-intercept.

In Chapter 7, we review the "beginning algebra" level of graphing and then we present the "intermediate algebra" level of graphing. We give the point-slope formula to find an equation of a line and then follow up with applications of linear equations and modeling.

We devote the latter part of Chapter 7 to functions, beginning with the general concept of a relation. We introduce the graphs of six basic functions to initiate a "repertoire" of functions that students can carry with them to college-level algebra courses. After introducing functions in Chapter 7, we take a functional approach throughout the remainder of the text. This includes work with polynomial, quadratic, radical, rational, exponential, and logarithmic functions.

Factoring

After factoring out the greatest common factor, we present two methods for factoring trinomials—the grouping method and the trial-and-error method. In both cases, we teach the most general case first (that is, trinomials with a leading coefficient *not* equal to 1). The reason is that we always want the student to consider the leading coefficient and its factors whether it's a 1 or any other number. Thus, when students take the product of the inner terms and the product of the outer terms, the factors of the leading coefficient always come into play.

We recognize that our approach to factoring trinomials is unusual. However, because we do not present the easier case first, students will not default to the easier case when it doesn't apply. When students find a method that's easy—like finding two numbers that multiply to "c" and add up to "b"—students may tune out the fact that it doesn't work all the time. We want to prevent students from thinking that $6x^2 + 19x + 15$ is not factorable because no two factors of 15 add up to 19.

In Section 5.6, "General Factoring Summary," we first review and summarize the techniques learned up to that point. Then we present advanced techniques such as using substitution in factoring and grouping three terms with one term. These advanced techniques are denoted with the header "Part II" both in the text and in the exercises (see page 348, Concept 3, and page 352). The Practice Exercises in Section 5.6 are divided into two groups, Part I and Part II. This makes it easy for instructors to identify the level of the exercises. Those instructors who only want to cover the Beginning Algebra level of factoring can assign only the Part I exercises. The instructors who want to cover the Intermediate Algebra level of factoring can assign both Part I and Part II.

Inequalities

A student who completes Intermediate Algebra should be able to recognize and solve a variety of equations and inequalities. However, identifying different types of equations and inequalities and knowing what approach to take is a skill that is often overlooked. We designed Chapter 9 as a synthesis chapter for students to solve all kinds of equations and inequalities appropriate at this level. Our exercise sets involve a "mixture" of problem types for students to recognize and solve. While we present traditional methods for solving inequalities, we also show the test point method. All of the inequalities in this chapter can be solved by solving the related equation and testing regions on the number line.

Calculator Usage

The use of a scientific or graphing calculator often inspires great debate among faculty teaching developmental mathematics. Our Calculator Connections boxes offer screen shots and some keystrokes to support applications where a calculator might enhance learning. Our approach is to use a calculator as a verification tool *after* analytical methods have been applied. As you move into the Intermediate Algebra level of the text, and as more coverage is given to graphing and functions, you will see a gradual transition into more calculator coverage. Graphing calculator exercises are provided at the end of exercise sets in appropriate sections. However, the Calculator Connections boxes and graphing calculator exercises are self-contained units and may be employed or easily omitted at the recommendation of the instructor.

Beginning Algebra Review

For students who place into Intermediate Algebra, we have provided concise review sections at the back of the text. Each section of the review corresponds with a chapter from the first portion of the text, that is, Chapters 1–6. Each section provides exercises that target the main points of the corresponding chapter.

ACKNOWLEDGMENTS AND REVIEWERS

The development of this textbook would never have been possible without the creative ideas and constructive feedback offered by many reviewers. We are especially thankful to the following instructors for their valuable feedback and careful review of the manuscript.

Randall Allbritton, *Daytona Beach Community College*
Wesley L. Anderson, *Northwest Vista College*
Shirley M. Brown, *Weatherford College*
Nate Callender, *Dyersburg State Community College*
Marc Campbell, *Daytona Beach Community College*
James D. Chandler Jr., *Tidewater Community College*
Rhoderick Fleming, *Wake Technical Community College*
David French, *Tidewater Community College*
Becky Hubiak, *Tidewater Community College*
Laura Kalbaugh, *Wake Technical Community College*
Michael Kirby, *Tidewater Community College*
Nickos Lambros, *DeVry University–Dupage Campus*
Kirsten Lollis, *San Diego City College*
Jean P. Millen, *Georgia Perimeter College*
Marilyn Peacock, *Tidewater Community College*
Elizabeth Sievers, *College of DuPage*
Pamela S. Webster, *Texas A&M University–Commerce*
Marilyn A. Zopp, *McHenry County College*

In addition, we would also like to thank the following reviewers of *Beginning Algebra, Intermediate Algebra,* and *Algebra for College Students*. Their suggestions were also helpful in developing this manuscript:

Marwan Abu-Sawwa, *Florida Community College at Jacksonville*
Jannette Avery, *Monroe Community College*
Pam Baenziger, *Kirkwood Community College*
Jo Battaglia, *Pennsylvania State University*
Mary Kay Best, *Coastal Bend College*
Paul Blankenship, *Lexington Community College*
James C. Boyett, *Arkansas State University, Beebe*
Debra D. Bryant, *Tennessee Technological University*
Connie L. Buller, *Metropolitan Community College, Omaha*
Gerald F. Busald, *San Antonio College*
Jimmy Chang, *St. Petersburg College*
Oiyin Pauline Chow, *Harrisburg Area Community College*
Elizabeth Condon, *Queens Community College*
Pat C. Cook, *Weatherford College*
Jacqueline Coomes, *Eastern Washington University*
Julane Crabtree, *Johnson County Community College*
Cynthia M. Craig, *Augusta State University*
Katherine W. Creery, *University of Memphis*
Bettyann Daley, *University of Delaware*
Antonio David, *Del Mar College*
Andres Delgado, *Orange County Community College*
Irene Durancyzk, *Eastern Michigan University*
Dennis C. Ebersole, *Northampton Community College*
James M. Edmondson, *Santa Barbara City College*
Robert A. Farinelli, *Community College of Allegheny County, Boyce*
Pat Foard, *South Plains College*
Toni Fountain, *Chattanooga State Technical Community College*
Dr. Paul F. Foutz, *Temple College*
Chris J. Gardiner, *Eastern Michigan University*
Jacqueline B. Giles, *Houston County Community College*
Pamela Heard, *Langston University*
Celeste Hernandez, *Richland College*
Julie Hess, *Grand Rapids Community College*
Kayana Hoagland, *South Puget Sound Community College*
Bruce Hoelter, *Raritan Valley Community College*
Rosalie Hojegian, *Possaic County Community College*
Lori Holdren, *Manatee Community College*
Glenn Hunt, *Riverside Community College*
Sarah Jackman, *Richland College*
Donald L. James, *Montgomery College*

Nancy R. Johnson, *Manatee Community College*
Steven Kahn, *Anne Arundel Community College*
Richard Karwatka, *University of Wisconsin, Parkside*
Jane Keller, *Metropolitan Community College*
William A. Kincaid, Sr., *Wilmington College*
Jeff A. Koleno, *Lorain County Community College*
Kathryn Kozak, *Coconino Community College*
Betty J. Larson, *South Dakota State University*
Mary M. Leeseberg, *Manatee Community College*
Deann Leoni, *Edmonds Community College*
Deanna Li, *North Seattle Community College*
Pamela A. Lipka, *University of Wisconsin*
J. Robert Malena, *Community College of Allegheny County—South Campus*
Ana M. Mantilla, *Pima Community College*
Maria M. Maspons, *Miami Dade Community College*
Timothy McKenna, *University of Michigan, Dearborn*
Debbie K. Millard, *Florida Community College*
Jean P. Millen, *Georgia Perimeter College*
Mary Ann Misko, *Gadsden State Community College*
Jeffery Mock, *Diablo Valley College*
Stephen E. Mussack, *Moorpark College*
Cameron Neal, Jr., *Temple College*
Sue Nolen, *Mesa Community College*
Timothy Norfolk, *University of Akron*
Linda Padilla, *Joliet Junior College*
Don Piele, *University of Wisconsin, Parkside*
Bernard J. Pina, *Dona Ana Branch Community College—NMSU*
Rita Beth Pruitt, *South Plains College*
Nancy C. Ressler, *Oakton Community College*
Reynaldo Rivera, Jr., *Estrella Mountain Community College*
Mary Romans, *Kent State University*
Frances Rosamond, *National University*
Katalin Rozsa, *Mesa Community College*
Elizabeth Russell, *Glendale Community College*
Fred Safier, *City College of San Francisco*
Ned W. Schillow, *Lehigh Carbon Community College*
Mary Lee Seitz, *Erie Community College*
Cindy Shaber, *Boise State University*
Lisa Sheppard, *Lorain County Community College*
Donna Sherrill, *Arkansas Technical University*
Bruce Sisko, *Belleville Area College (Southwestern Illinois College)*
Carolyn R. Smith, *Washington State University*
Sandra L. Spain, *Thomas Nelson Community College*
Brian Starr, *National University*

Daryl Stephens, *East Tennessee State University*
Dr. Bryan Stewart, *Tarrant County College*
Elizabeth A. Swift, *Cerritos College*
Irving C. Tang, *Oklahoma State University, Oklahoma City*
Alexis Thurman, *County College of Morris*
Peggy Tibbs, *Arkansas Technical University*
Dr. Lee Topham, *Kingwood College*
Dr. Roy N. Tucker, *Palo Alto College*
Jeannine G. Vigerust, *New Mexico State University*
Sandra Vrem, *College of the Redwoods*
John C. Wenger, *Harold Washington College*
Charles M. Wheeler, *Montgomery College*
Ethel R. Wheland, *University of Akron*
Denise A. Widup, *University of Wisconsin, Parkside*
Thomas L. Wolters, *Muskegon Community College*

Special thanks go to David French of Tidewater Community College for preparing the *Instructor's Solutions Manual* and the *Student's Solutions Manual*, to Sherri Messersmith of College of DuPage for her appearance in and work on the video series, and to Lauri Semarne and Cindy Trimble for their work in ensuring accuracy in the content.

In addition we are forever grateful to the many people behind the scenes at McGraw-Hill who have made this project possible. Our sincerest thanks to Erin Brown, the best editor in the business, thanks for keeping the train on the track. To Liz Haefele, David Dietz, Bill Barter, and Michael Lange for believing in us and supporting us for many years. To Cindy Schmerbach and April Southwood for their patience during production and to Steve Stembridge for his creative ideas promoting our efforts. We further appreciate the hard work of Pat Steele.

Finally, we give special thanks to all the students and instructors who use *Beginning and Intermediate Algebra* in their classes.

Julie Miller and Molly O'Neill

A COMMITMENT TO ACCURACY

You have a right to expect an accurate textbook, and McGraw-Hill invests considerable time and effort to make sure that we deliver one. Listed below are the many steps we take to make sure this happens.

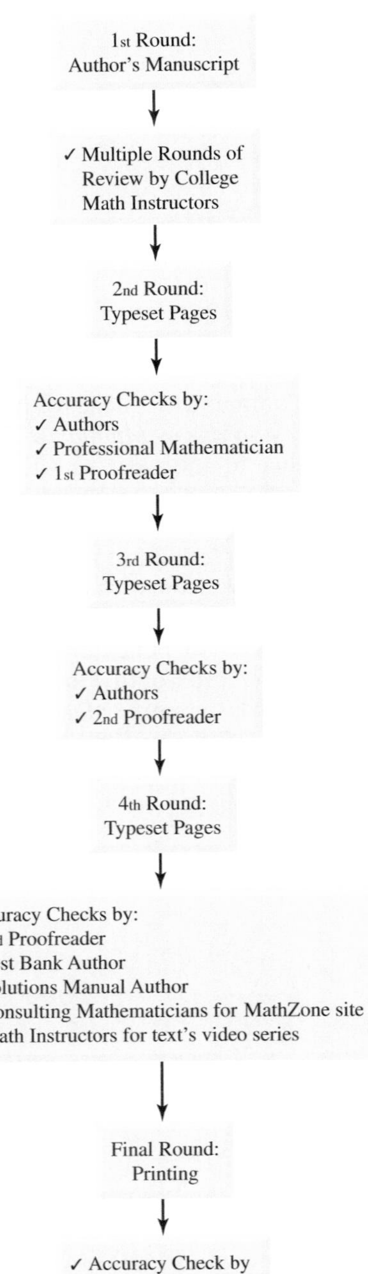

OUR ACCURACY VERIFICATION PROCESS

First Round

Step 1: Numerous **college math instructors** review the manuscript and report on any errors that they may find, and the authors make these corrections in their final manuscript.

Second Round

Step 2: Once the manuscript has been typeset, the **authors** check their manuscript against the first page proofs to ensure that all illustrations, graphs, examples, exercises, solutions, and answers have been correctly laid out on the pages, and that all notation is correctly used.

Step 3: An outside, **professional mathematician** works through every example and exercise in the page proofs to verify the accuracy of the answers.

Step 4: A **proofreader** adds a triple layer of accuracy assurance in the first pages by hunting for errors, then a second, corrected round of page proofs is produced.

Third Round

Step 5: The **author team** reviews the second round of page proofs for two reasons: 1) to make certain that any previous corrections were properly made, and 2) to look for any errors they might have missed on the first round.

Step 6: A **second proofreader** is added to the project to examine the new round of page proofs to double check the author team's work and to lend a fresh, critical eye to the book before the third round of paging.

Fourth Round

Step 7: A **third proofreader** inspects the third round of page proofs to verify that all previous corrections have been properly made and that there are no new or remaining errors.

Step 8: Meanwhile, in partnership with **independent mathematicians,** the text accuracy is verified from a variety of fresh perspectives:

- The **test bank author** checks for consistency and accuracy as they prepare the computerized test item file.
- The **solutions manual author** works every single exercise and verifies their answers, reporting any errors to the publisher.
- A **consulting group of mathematicians,** who write material for the text's MathZone site, notifies the publisher of any errors they encounter in the page proofs.
- A video production company employing **expert math instructors** for the text's videos will alert the publisher of any errors they might find in the page proofs.

Final Round

Step 9: The **project manager,** who has overseen the book from the beginning, performs a **fourth proofread** of the textbook during the printing process, providing a final accuracy review.

⇒ What results is a mathematics textbook that is as accurate and error-free as is humanly possible, and our authors and publishing staff are confident that our many layers of quality assurance have produced textbooks that are the leaders of the industry for their integrity and correctness.

GUIDED TOUR: FEATURES AND SUPPLEMENTS

Chapter Opener

Each chapter opens with an application relating to topics presented in the chapter. The Chapter Openers also contain references to Technology Connections—Internet activities found in the *Instructor's Resource Manual*—that further the scope of the application.

chapter

3

GRAPHING LINEAR EQUATIONS IN TWO VARIABLES

In Chapter 3 we investigate linear relationships and their graphs. In mathematical applications, a linear relationship may be given by a fixed value plus a constant rate of increase or decrease. For example:

Suppose 8 inches of snow is on the ground at 12:00 noon. A blizzard dumps more snow at a rate of $\frac{3}{4}$ in. per hour for nine hours. The depth of the snow, d (in inches), is given by:

$$d = \frac{3}{4}x + 8$$

where x is the number of hours after the storm began $(0 \le x \le 9)$.

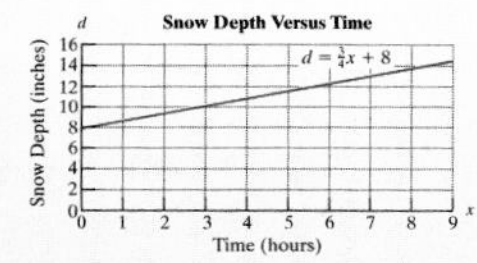

This and other linear equations may be analyzed and graphed using the techniques presented in Chapter 3. For more information see the Technology Connections in MathZone at

www.mhhe.com/miller_oneill

section

3.2 LINEAR EQUATIONS IN TWO VARIABLES

Concepts

1. Definition of a Linear Equation in Two Variables
2. Solutions to Linear Equations in Two Variables
3. Graphing Linear Equations in Two Variables by Plotting Points
4. Applications of Linear Equations in Two Variables

1. Definition of a Linear Equation in Two Variables

Recall that an equation in the form $ax + b = 0$, where $a \neq 0$, is called a linear equation in one variable. A solution to such an equation is a value of x that makes the equation a true statement. For example, $3x + 6 = 0$ has a solution of $x = -2$.

In this section we will look at linear equations in *two* variables.

Definition of a Linear Equation in Two Variables

Let a, b, and c be real numbers such that a and b are not both zero. Then, an equation that can be written in the form:

$$ax + by = c$$

is called a **linear equation in two variables.**

2. Solutions to Linear Equations in Two Variables

The equation $x + y = 4$ is a linear equation in two variables. A solution to such an equation is an ordered pair (x, y) that makes the equation a true statement. Several solutions to the equation $x + y = 4$ are listed here:

Solution:	Check:
(x, y)	$x + y = 4$
$(2, 2)$	$(2) + (2) = 4$ ✔
$(1, 3)$	$(1) + (3) = 4$ ✔
$(4, 0)$	$(4) + (0) = 4$ ✔
$(-1, 5)$	$(-1) + (5) = 4$ ✔

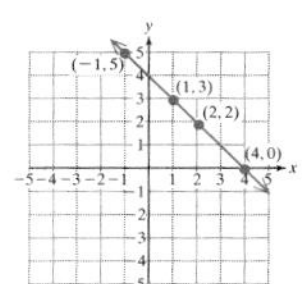

Figure 3-5

By graphing these ordered pairs, we see that the solution points line-up (Figure 3-5).

Notice that there are infinitely many solutions to the equation $x + y = 4$ so they cannot all be listed. Therefore, to visualize all solutions to the equation $x + y = 4$, we draw the line through the points in the graph. Every point on the line represents an ordered pair solution to the equation $x + y = 4$, and the line represents the set of *all* solutions to the equation.

example 1 Determining Solutions to a Linear Equation

For the linear equation, $4x - 5y = 8$, determine whether the given ordered pair is a solution.

a. $(2, 0)$ b. $(3, 1)$ c. $\left(1, -\frac{4}{5}\right)$

Solution:

a. $4x - 5y = 8$

$4(2) - 5(0) \stackrel{?}{=} 8$ Substitute $x = 2$ and $y = 0$.

$8 - 0 = 8$ ✔ (true) The ordered pair $(2, 0)$ is a solution.

Concepts

A list of important concepts is provided at the beginning of each section. Each concept corresponds to a heading within the section, making it easy for students to locate topics as they study or as they work through homework exercises.

Avoiding Mistakes

Through marginal notes labeled Avoiding Mistakes students are alerted to common errors and are shown methods to avoid them.

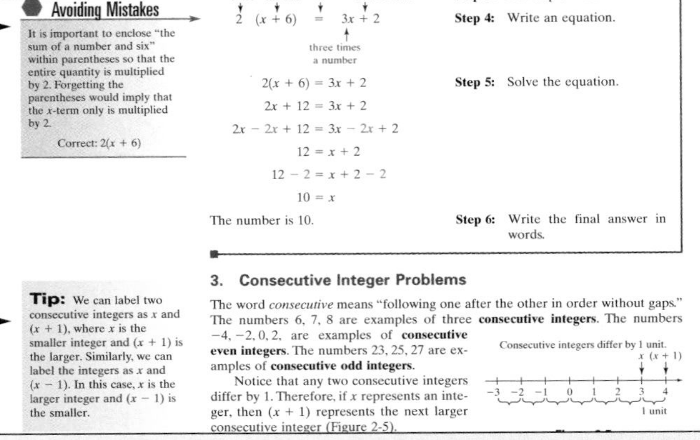

Avoiding Mistakes

It is important to enclose "the sum of a number and six" within parentheses so that the entire quantity is multiplied by 2. Forgetting the parentheses would imply that the x-term only is multiplied by 2.

Correct: $2(x + 6)$

$2 \quad (x + 6) \quad = \quad 3x + 2$ (three times a number) **Step 4:** Write an equation.

$2(x + 6) = 3x + 2$ **Step 5:** Solve the equation.

$2x + 12 = 3x + 2$

$2x - 2x + 12 = 3x - 2x + 2$

$12 = x + 2$

$12 - 2 = x + 2 - 2$

$10 = x$

The number is 10. **Step 6:** Write the final answer in words.

3. Consecutive Integer Problems

Tip: We can label two consecutive integers as x and $(x + 1)$, where x is the smaller integer and $(x + 1)$ is the larger. Similarly, we can label the integers as x and $(x - 1)$. In this case, x is the larger integer and $(x - 1)$ is the smaller.

The word *consecutive* means "following one after the other in order without gaps." The numbers 6, 7, 8 are examples of three **consecutive integers**. The numbers $-4, -2, 0, 2,$ are examples of **consecutive even integers**. The numbers 23, 25, 27 are examples of **consecutive odd integers**.

Notice that any two consecutive integers differ by 1. Therefore, if x represents an integer, then $(x + 1)$ represents the next larger consecutive integer (Figure 2-5).

Tips

Tip boxes appear throughout the text and offer helpful hints and insight.

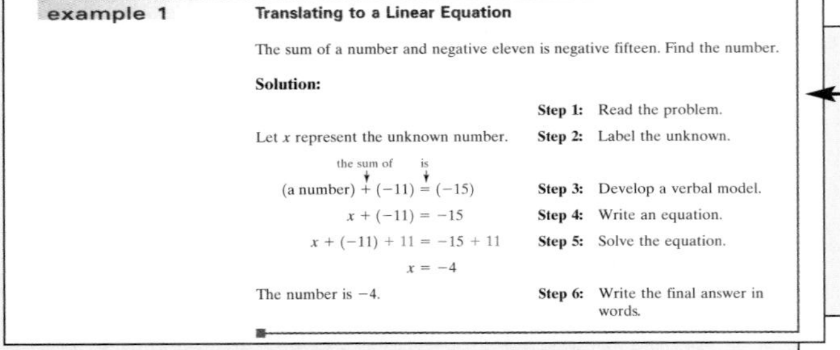

example 1 **Translating to a Linear Equation**

The sum of a number and negative eleven is negative fifteen. Find the number.

Solution:

Step 1: Read the problem.

Let x represent the unknown number. **Step 2:** Label the unknown.

(a number) + (−11) = (−15) (the sum of, is) **Step 3:** Develop a verbal model.

$x + (-11) = -15$ **Step 4:** Write an equation.

$x + (-11) + 11 = -15 + 11$ **Step 5:** Solve the equation.

$x = -4$

The number is −4. **Step 6:** Write the final answer in words.

Worked Examples

Examples are set off in boxes and organized so that students can easily follow the solutions. Explanations appear beside each step in columnar form, where appropriate.

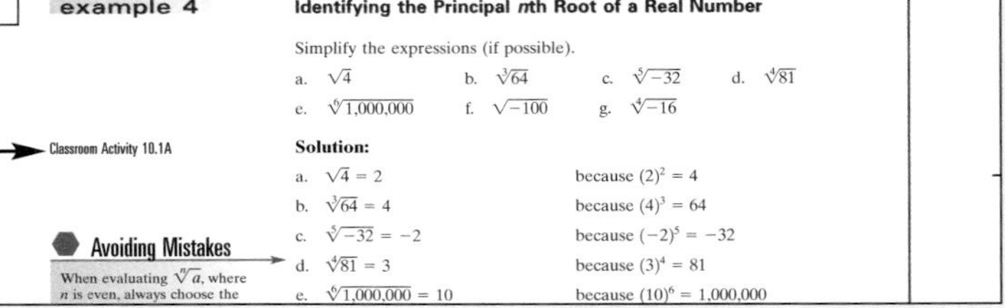

example 4 **Identifying the Principal *n*th Root of a Real Number**

Simplify the expressions (if possible).

a. $\sqrt{4}$ b. $\sqrt[3]{64}$ c. $\sqrt[5]{-32}$ d. $\sqrt[4]{81}$
e. $\sqrt[6]{1{,}000{,}000}$ f. $\sqrt{-100}$ g. $\sqrt[4]{-16}$

Classroom Activity 10.1A

Solution:

a. $\sqrt{4} = 2$ because $(2)^2 = 4$
b. $\sqrt[3]{64} = 4$ because $(4)^3 = 64$
c. $\sqrt[5]{-32} = -2$ because $(-2)^5 = -32$
d. $\sqrt[4]{81} = 3$ because $(3)^4 = 81$
e. $\sqrt[6]{1{,}000{,}000} = 10$ because $(10)^6 = 1{,}000{,}000$

Avoiding Mistakes

When evaluating $\sqrt[n]{a}$, where n is even, always choose the

References to Classroom Activities (AIE only)

Throughout each section of the Annotated Instructor's Edition (AIE), references are made to Classroom Activities. These references appear on the page where the activities might be introduced. The activities are given in the *Instructor's Resource Manual,* available at www.mhhe.com/miller_oneill.

Practice Exercises

The Practice Exercises comprise a variety of problem types.

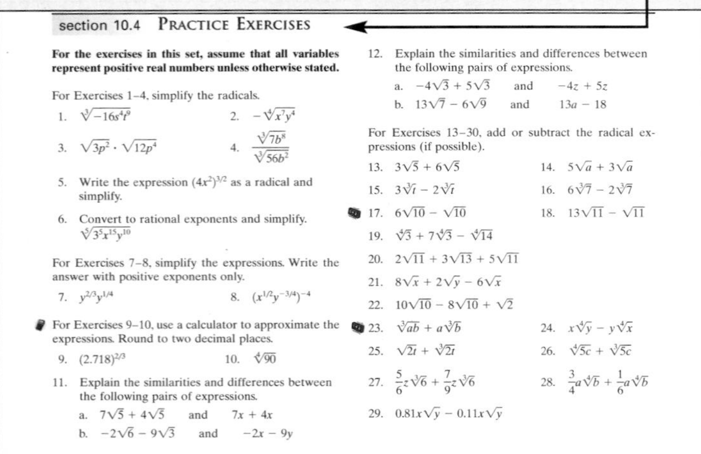

section 10.4 PRACTICE EXERCISES

For the exercises in this set, assume that all variables represent positive real numbers unless otherwise stated.

For Exercises 1–4, simplify the radicals.

1. $\sqrt[3]{-16s^4t^9}$ 2. $-\sqrt[4]{x^7y^4}$
3. $\sqrt{3p^2} \cdot \sqrt{12p^4}$ 4. $\dfrac{\sqrt[3]{7b^8}}{\sqrt[3]{56b^2}}$
5. Write the expression $(4x^2)^{3/2}$ as a radical and simplify.
6. Convert to rational exponents and simplify. $\sqrt[5]{3^5x^{15}y^{10}}$

For Exercises 7–8, simplify the expressions. Write the answer with positive exponents only.

7. $y^{2/3}y^{1/4}$ 8. $(x^{1/2}y^{-3/4})^{-4}$

For Exercises 9–10, use a calculator to approximate the expressions. Round to two decimal places.

9. $(2.718)^{2/3}$ 10. $\sqrt[4]{90}$

11. Explain the similarities and differences between the following pairs of expressions.
a. $7\sqrt{5} + 4\sqrt{5}$ and $7x + 4x$
b. $-2\sqrt{6} - 9\sqrt{3}$ and $-2x - 9y$

12. Explain the similarities and differences between the following pairs of expressions.
a. $-4\sqrt{3} + 5\sqrt{3}$ and $-4z + 5z$
b. $13\sqrt{7} - 6\sqrt{9}$ and $13a - 18$

For Exercises 13–30, add or subtract the radical expressions (if possible).

13. $3\sqrt{5} + 6\sqrt{5}$ 14. $5\sqrt{a} + 3\sqrt{a}$
15. $3\sqrt[3]{t} - 2\sqrt[3]{t}$ 16. $6\sqrt[3]{7} - 2\sqrt[3]{7}$
17. $6\sqrt{10} - \sqrt{10}$ 18. $13\sqrt{11} - \sqrt{11}$
19. $\sqrt[4]{3} + 7\sqrt[4]{3} - \sqrt[4]{14}$
20. $2\sqrt{11} + 3\sqrt{13} + 5\sqrt{11}$
21. $8\sqrt{x} + 2\sqrt{y} - 6\sqrt{x}$
22. $10\sqrt{10} - 8\sqrt{10} + \sqrt{2}$
23. $\sqrt[3]{ab} + a\sqrt[3]{b}$ 24. $x\sqrt[4]{y} - y\sqrt[4]{x}$
25. $\sqrt{2t} + \sqrt[3]{2t}$ 26. $\sqrt[4]{5c} + \sqrt[3]{5c}$
27. $\frac{5}{6}z\sqrt[3]{6} + \frac{7}{9}z\sqrt[3]{6}$ 28. $\frac{3}{4}a\sqrt[4]{b} + \frac{1}{6}a\sqrt[4]{b}$
29. $0.81x\sqrt{y} - 0.11x\sqrt{y}$

Review Problems appear within the Practice Exercises to help students retain their knowledge of concepts previously learned. Each review problem is labeled with a section number in the AIE, referencing the section where the problem type is introduced.

Exercises Keyed to Video are labeled with an icon to help students and instructors identify those exercises that appear in the video series that accompanies *Beginning and Intermediate Algebra.*

Calculator Exercises signify situations where a calculator would provide assistance for time-consuming calculations. These exercises were carefully designed to demonstrate the types of situations where a calculator is a handy tool rather than a "crutch." They are designed for use with either a scientific or a graphing calculator.

Step-by-Step Problem-Solving

Many of the exercises contain multiple parts that serve to step students through the process of solving a problem.

For Exercises 58–67, solve the geometry formulas for the indicated variables.

58. a. A rectangle has length l and width w. Write a formula for the area.
    b. Solve the formula for the width, w.
    c. The area of a rectangular volleyball court is 1740.5 ft^2 and the length is 59 ft. Find the width.

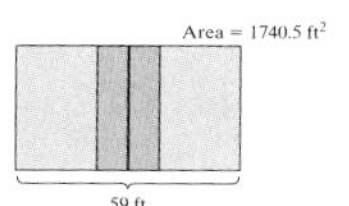

Figure for Exercise 58

In many of the multiple-part exercises, students are asked interpretative questions to increase their conceptual understanding.

6. The water bill charge for a certain utility company is \$4.20 per 1000 gallons used. The total cost, y, depends on the number of thousands of gallons of water, x, according to the equation

$$y = 4.20x \quad x \geq 0$$

   a. Determine the cost of using 3000 gallons. (*Hint:* $x = 3$.)
   b. Determine the cost of using 5000 gallons.
   c. What is the y-intercept of the equation? Interpret the meaning of the y-intercept in the context of this problem.
   d. What is the slope of the equation? Interpret the meaning of the slope in the context of this problem.
   e. Graph the equation.

Modeling

Many of the exercises contain data and graphs so that students can build models in the form of equations, graphs, and functions. Through this approach, students see the relevance of mathematics.

10. In a certain city, the time required to commute to work, y, (in minutes) by car is related linearly to the distance traveled, x (in miles).

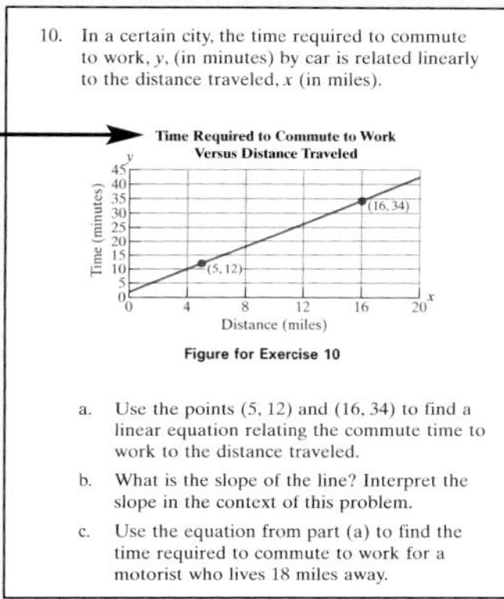

Figure for Exercise 10

   a. Use the points (5, 12) and (16, 34) to find a linear equation relating the commute time to work to the distance traveled.
   b. What is the slope of the line? Interpret the slope in the context of this problem.
   c. Use the equation from part (a) to find the time required to commute to work for a motorist who lives 18 miles away.

30. $7.5\sqrt{pq} - 6.3\sqrt{pq}$
31. Explain the process for adding the two radicals. $3\sqrt{2} + 7\sqrt{50}$.
32. Explain the process for adding the two radicals. $\sqrt{8} + \sqrt{32}$.

For Exercises 33–50, add or subtract the radical expressions as indicated.

33. $\sqrt{36} + \sqrt{81}$
34. $3\sqrt{80} - 5\sqrt{45}$
35. $2\sqrt{12} + \sqrt{48}$
36. $5\sqrt{32} + 2\sqrt{50}$
37. $4\sqrt{7} + \sqrt{63} - 2\sqrt{28}$
38. $8\sqrt{3} - 2\sqrt{27} + \sqrt{75}$
39. $3\sqrt{2a} - \sqrt{8a} - \sqrt{72a}$
40. $\sqrt{12t} - \sqrt{27t} + 5\sqrt{3t}$
41. $2s^2\sqrt[3]{s^2t^6} + 3t^2\sqrt[3]{8s^8}$
42. $4\sqrt[3]{x^4} - 2x\sqrt[3]{x}$
43. $7\sqrt[3]{x^4} - x\sqrt[3]{x}$
44. $6\sqrt[3]{y^{10}} - 3y^2\sqrt[3]{y^4}$
45. $5p\sqrt{20p^2} + p^2\sqrt{80}$
46. $2q\sqrt{48q^2} - \sqrt{27q^4}$
47. $\frac{3}{2}ab\sqrt{24a^3} + \frac{4}{3}\sqrt{54a^5b^2} - a^2b\sqrt{150a}$
48. $mn\sqrt{72n} + \frac{2}{3}n\sqrt{8m^2n} - \frac{5}{6}\sqrt{50m^2n^3}$
49. $x\sqrt[3]{16} - 2\sqrt[3]{27x} + \sqrt[3]{54x^3}$
50. $5\sqrt[4]{y^5} - 2y\sqrt[4]{y} + \sqrt[4]{16y^7}$

For Exercises 51–56, answer true or false. If an answer is false, explain why.

51. $\sqrt{x} + \sqrt{y} = \sqrt{x + y}$
52. $\sqrt{x} + \sqrt{x} = 2\sqrt{x}$
53. $5\sqrt[3]{x} + 2\sqrt[3]{x} = 7\sqrt[3]{x}$
54. $6\sqrt{x} + 5\sqrt[3]{x} = 11\sqrt{x}$
55. $\sqrt{y} + \sqrt{y} = \sqrt{2y}$
56. $\sqrt{c^2 + d^2} = c + d$

For Exercises 57–60, translate the English phrase into an algebraic expression. Simplify each expression if possible.

57. The sum of the square root of 48 and the square root of 12.
58. The sum of the cube root of 16 and the cube root of 2.
59. The difference of 5 times the cube root of x^6 and the square of x.
60. The sum of the cube of y and the fourth root of y^{12}.

For Exercises 61–64, write an English phrase that translates the mathematical expression. (Answers may vary.)

61. $\sqrt{18} - 5^2$
62. $4^3 - \sqrt[3]{4}$
63. $\sqrt[4]{x} + y^3$
64. $a^4 + \sqrt{a}$

Expanding Your Skills

65. a. An irregularly shaped garden is shown in the figure. All distances are expressed in yards. Find the perimeter. *Hint:* Use the Pythagorean theorem to find the length of each side. Write the final answer in radical form.
    b. Approximate your answer to two decimal places.
    c. If edging costs \$1.49 per foot and sales tax is 6%, find the total cost of edging the garden.

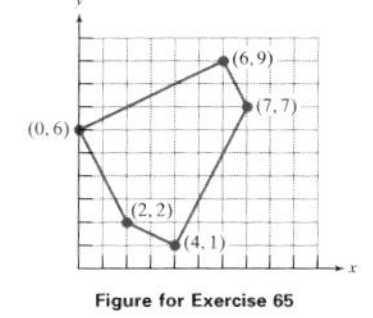

Figure for Exercise 65

Writing Exercises offer students an opportunity to conceptualize and communicate their understanding of algebra. These, along with the translating expressions exercises enable students to strengthen their command of mathematical language and notation and improve their reading skills and writing skills.

Geometry Exercises appear throughout the Practice Exercises and encourage students to review and apply geometry concepts.

The following icons appear in both the AIE and the student edition:

 Calculator Exercises

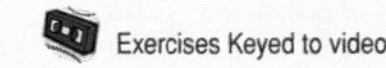 Exercises Keyed to video

Applications based on real-world facts and figures motivate students and enable them to hone their problem-solving skills.

Graphing Calculator Exercises, found at the end of the Practice Exercise sets, offer students and instructors a means of using the graphing calculator to explore concepts.

Expanding Your Skills, found near the end of most Practice Exercise sets, challenge students' knowledge of the concepts presented.

d. Use the equation to find the year in which the sales equaled \$14,470 million.

36. The net earning for the Emerson Electric Co. from 1994 to 1999 is given by the equation $y = 105x + 800$. The value of y is in millions of dollars and x is the time in years after 1994. That is, $x = 0$ corresponds to 1994, $x = 1$ corresponds to 1995, and so on.

 a. Find the net earnings for the year 1994.
 b. Find the net earnings for the year 1997.
 c. Use the equation to find the year in which the net earnings equaled 1010 million dollars.
 d. Use the equation to find the year in which the net earnings equaled 1325 million dollars.

EXPANDING YOUR SKILLS

37. Graph the lines on the same set of coordinate axes, and observe any similarities and differences.
 a. $y = 2x$
 b. $y = 2x + 3$
 c. $y = 2x - 2$

38. Graph the lines on the same set of coordinate axes, and observe any similarities and differences.
 a. $x + y = 0$
 b. $x + y = 3$
 c. $x + y = -1$

GRAPHING CALCULATOR EXERCISES

For Exercises 39–43, use a graphing calculator to graph the equations on the standard viewing window. Then graph the equation again on the suggested window.

39. $y = -3x - 2$
 a. Standard viewing window
 b. Window defined by: $-5 \le x \le 5$ and $-5 \le y \le 5$

40. $y = 0.33x - 0.25$
 a. Standard viewing window
 b. Window defined by: $-2 \le x \le 2$ and $-2 \le y \le 2$

41. $2x + y = 5$
 a. Standard viewing window
 b. Window defined by: $-1 \le x \le 8$ and $-4 \le y \le 8$

42. $3x - 2y = 32$
 a. Standard viewing window
 b. Window defined by: $-10 \le x \le 20$ and $-20 \le y \le 10$

43. $x + 2y = -50$
 a. Standard viewing window
 b. Window defined by $-60 \le x \le 20$ and $-30 \le y \le 20$

End-of-Chapter Summary and Exercises

The **Summary**, located at the end of each chapter, outlines key concepts for each section and illustrates those concepts with examples. The Summary also provides a list of important terms that mirror those appearing in the Vocabulary Worksheets of the Student Portfolio found in the *Instructor's Resource Manual*. With this list, students can quickly identify important ideas and vocabulary to be reviewed before quizzes or exams.

Following the Summary is a set of **Review Exercises** that are organized by section. A **Chapter Test** appears after each set of Review Exercises. Chapters 2–14 also include **Cumulative Reviews** that follow the Chapter Tests. These end-of-chapter materials provide students with ample opportunity to prepare for quizzes or exams.

chapter 5 SUMMARY

SECTION 5.1—GREATEST COMMON FACTOR AND FACTORING BY GROUPING

KEY CONCEPTS:	EXAMPLES:
The greatest common factor (GCF) is the greatest factor common to all terms of a polynomial. To factor out the GCF from a polynomial, use the distributive property.	**Factor out the GCF:** $3x^2(a + b) - 6x(a + b)$ $= 3x(a + b)x - 3x(a + b)(2)$ $= 3x(a + b)(x - 2)$
A four-term polynomial may be factorable by grouping.	
Steps to Factoring by Grouping 1. Identify and factor out the GCF from all four terms. 2. Factor out the GCF from the first pair of terms. Factor out the GCF or its opposite from the second pair of terms. 3. If the two terms share a common binomial factor, factor out the binomial factor.	**Factor by Grouping:** $60xa - 30xb - 80ya + 40yb$ $= 10[6xa - 3xb - 8ya + 4yb]$ $= 10[3x(2a - b) - 4y(2a - b)]$ $= 10[(2a - b)(3x - 4y)]$ $= 10(2a - b)(3x - 4y)$

chapter 5 REVIEW EXERCISES

Section 5.1

For Exercises 1–6, identify the greatest common factor between each pair of terms.

1. 24, 18
2. $16x^2, 20x^3$
3. $15a^2b^4, 22ab^5$
4. $3(x + 5), x(x + 5)$
5. $2c^3(3c - 5), 4c(3c - 5)$
6. $-2wyz, -4xyz$

For Exercises 7–14, factor out the greatest common factor.

7. $6x^2 + 2x^3 - 8x$
8. $11w^3 - 44w^2$
9. $32y^2 - 48$
10. $5a^3 + 9a^2 + 2a$

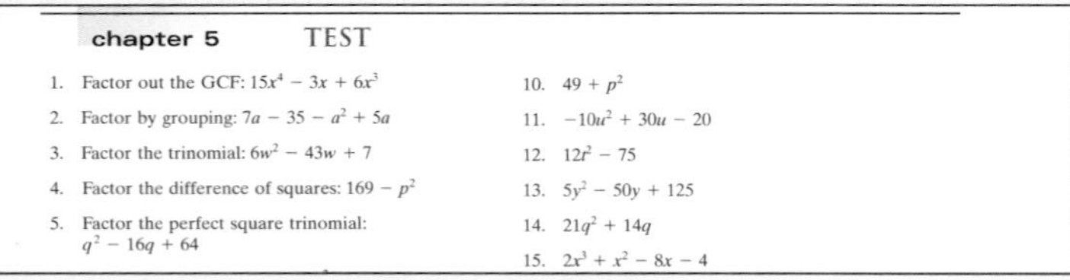

chapter 5 TEST

1. Factor out the GCF: $15x^4 - 3x + 6x^3$
2. Factor by grouping: $7a - 35 - a^2 + 5a$
3. Factor the trinomial: $6w^2 - 43w + 7$
4. Factor the difference of squares: $169 - p^2$
5. Factor the perfect square trinomial: $q^2 - 16q + 64$
10. $49 + p^2$
11. $-10u^2 + 30u - 20$
12. $12t^2 - 75$
13. $5y^2 - 50y + 125$
14. $21q^2 + 14q$
15. $2x^3 + x^2 - 8x - 4$

CUMULATIVE REVIEW EXERCISES, CHAPTERS 1–5

For Exercises 1–2, simplify completely.

1. $\dfrac{|4 - 25 \div (-5) \cdot 2|}{\sqrt{8^2 + 6^2}}$
2. $-\dfrac{2}{3} - \dfrac{1}{3}(3^2 + \sqrt{81})$
3. Solve for x: $-3.5 - 2.5x = 1.5(x - 3)$
4. Solve for t: $5 - 2(t + 4) = 3t + 12$
5. Solve for y: $3x - 2y = 8$
6. The circumference of a circular fountain is 50 ft. Find the radius of the fountain. Round to the nearest tenth of a foot.
11. $(2w - 7)^2$
12. Divide using long division: $(r^4 + 2r^3 - 5r + 1) \div (r - 3)$

For Exercises 13–15, simplify the expressions. Write the final answer using positive exponents only.

13. $\dfrac{c^{12}c^{-5}}{c^3}$
14. $\left(\dfrac{2a^2b^{-3}}{c}\right)^{-2}$
15. $\left(\dfrac{1}{2}\right)^0 - \left(\dfrac{1}{4}\right)^{-2}$
16. Divide. Write the final answer in scientific notation: $\dfrac{8.0 \times 10^{-3}}{5.0 \times 10^{-6}}$

SUPPLEMENTS

MULTIMEDIA SUPPLEMENTS

Easy, Free, Has it all . . .

McGraw-Hill's MathZone is a complete, online tutorial and course management system for mathematics and statistics, designed for greater ease of use than any other system available. Free upon adoption of a McGraw-Hill title, instructors can use the system to create and share courses and assignments with colleagues and adjuncts in a matter of a few clicks of the mouse. All assignments, questions, e-Professors, online tutoring, and video lectures are directly tied to text-specific materials in *Beginning and Intermediate Algebra*. MathZone courses are customized to your textbook, but you can edit questions and algorithms, import your own content, and create announcements and due dates for assignments. MathZone has automatic grading and reporting of easy-to-assign algorithmically generated homework, quizzing, and testing. All student activity within MathZone is automatically recorded and available to you through a fully integrated gradebook that can be downloaded to Excel.

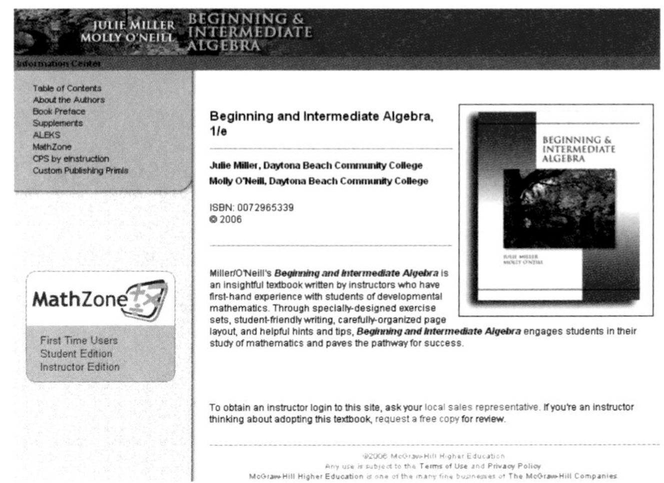

Instructor's Resource Manual (Instructors only)

The *Instructor's Resource Manual* (IRM), written by the authors, contains numerous classroom activities designed for group or individual work, a series of Internet Activities tied to the Chapter Openers in the textbook, and materials for a student portfolio. Several **Classroom Activities** are available for each section in the textbook to be used as a complement to lecture. The activities give students a chance to practice problems similar to the examples found within the sections. Each activity can be completed in 5 to 10 minutes. With increasing demands on faculty schedules, these ready-made lessons offer a convenient means for both full-time and adjunct faculty to promote active learning in the classroom.

In addition, Internet activities called **Technology Connections** are provided for each chapter to give students working applications of topics covered in the chapter. For instructors who want to use the Classroom Activities or Technology Connections as cooperative learning lessons, the IRM also provides strategies for successful implementation of cooperative learning.

The **Student Portfolio** is another significant feature of the IRM. The materials in the portfolio provide guidelines for the student to maintain an organized notebook with notes, homework, tests, test corrections, a calendar, a record of grades, and key terms. The portfolio is a vehicle to reinforce students' responsibility for their own learning.

This supplement is available online at www.mhhe.com/miller_oneill.

Section 6.6 Rational Equations

Activity 6.6A Solving Rational Equations

1. Given the equation $\frac{1}{2}y - \frac{3}{4} = 2 + \frac{5}{12}y$

   a. What number could be used to clear fractions? ____________________

   b. Solve the equation by clearing fractions first.

   ____

2. Given the equation: $\frac{12}{y^2 - 2y} = \frac{6}{y-2} + \frac{3}{y}$

   a. Are there any values of y for which the expressions in this equation are undefined?

ALEKS is an artificial intelligence-based system for individualized math learning, available for Higher Education from McGraw-Hill over the World Wide Web.

ALEKS delivers precise assessments of math knowledge, guides the student in the selection of appropriate new study material, and records student progress toward mastery of goals.

ALEKS interacts with a student much as a skilled human tutor would, moving between explanation and practice as needed, correcting and analyzing errors, defining terms and changing topics on request. By accurately assessing a student's knowledge, ALEKS can focus clearly on what the student is ready to learn next, helping to master the course content more quickly and easily.

ALEKS is:

- **A comprehensive course management system.** It tells the instructor exactly what students know and don't know.
- **Artificial intelligence.** It totally individualizes assessment and learning.
- **Customizable.** ALEKS can be set to cover the material in your course.
- **Web-based.** It uses a standard browser for easy Internet access.
- **Inexpensive.** There are no setup fees or site license fees.

ALEKS 2.0 adds the following new features:

- **Automatic Textbook Integration**
- **New Instructor Module**
- **Instructor-Created Quizzes**
- **New Message Center**

ALEKS maintains the features that have made it so popular including:

- **Web-Based Delivery** No complicated network or lab setup
- **Immediate Feedback** for students in learning mode
- **Integrated Tracking of Student Progress and Activity**
- **Individualized Instruction** which gives students problems they are ***Ready to Learn***

For more information please contact your McGraw-Hill Sales Representative or visit ALEKS at http://www.highedmath.aleks.com

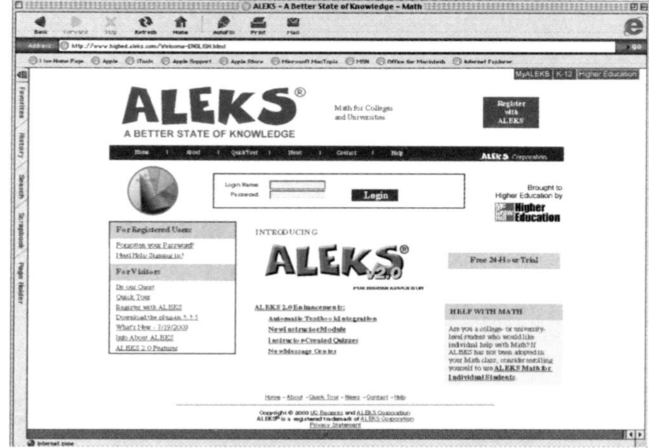

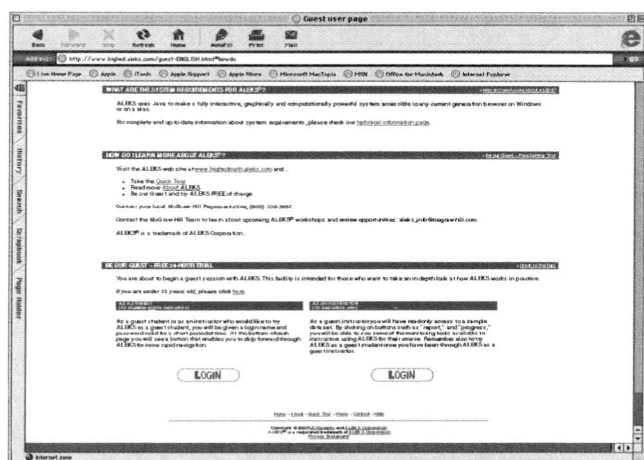

Instructor's Testing and Resource CD (Instructors only)

This cross-platform CD-ROM includes a computerized test bank utilizing Brownstone Diploma testing software that enables users to create customized exams quickly. The program enables instructors to search for questions by topic or format, to edit existing questions or to add new ones, and to scramble questions and answer keys for multiple versions of the same test.

NetTutor

NetTutor is a revolutionary system that enables students to interact with a live tutor over the World Wide Web by using NetTutor's Web-based, graphical chat capabilities. Students can also submit questions and receive answers, browse previously answered questions, and view previous live chat sessions.

NetTutor can be accessed through MathZone.

Miller/O'Neill Video Series (Videotapes or DVDs)

The video series is based on problems taken directly from the Practice Exercises. The Practice Exercises contain icons (in both the student and the instructor edition) that show which problems from the text appear in the video series. A mathematics instructor presents selected problems and works through them, following the solution methodology employed in the text. The videos series is available on videotapes or DVDs. The video series is close-captioned for the hearing impaired, ADA compliant, and subtitled in Spanish.

PRINTED SUPPLEMENTS

Annotated Instructor's Edition (Instructors only)

The Annotated Instructor's Edition (AIE) contains answers to all exercises and tests. The answers to most exercises are printed in green next to each problem (answers not appearing on the page are found in the back of the AIE in an appendix). This ancillary also provides valuable keys that serve as a useful guide to instructors as they assign homework problems and structure lessons. Icons and references are placed throughout the Practice Exercises so that instructors can easily identify problem types, such as writing exercises, geometry exercises, translating expressions exercises, review exercises, challenge problems (Expanding Your Skills), and calculator exercises. Students *do not see* all of these icons in the Student Edition of the text. Students see only the video icons, the calculator icons, and the references to Graphing Calculator Exercises and Expanding Your Skills.

Instructor's Solutions Manual (Instructors only)

The *Instructor's Solutions Manual* contains comprehensive, worked-out solutions to all exercises in the Practice Exercise sets, the end-of-chapter Review Exercises, the Chapter Tests, and the Cumulative Review Exercises.

Student's Solutions Manual

The *Student's Solutions Manual* contains comprehensive, worked-out solutions to the odd-numbered exercises in the Practice Exercise sets, the end-of-chapter Review Exercises, the Chapter Tests, and the Cumulative Review Exercises.

chapter

Reference: Fractions, Geometry, and Study Skills

In a recent year, the price per share of stock for Hershey Foods was $\$46\frac{3}{8}$, or, equivalently, \$46.375. A year later, the price per share rose to $\$61\frac{5}{8}$, or \$61.625. The net increase is given by

$$\$61\tfrac{5}{8} - \$46\tfrac{3}{8} = \$15\tfrac{1}{4}$$

or equivalently, \$15.25.

Therefore, if a stockholder bought 100 shares and sold the stock a year later, the increase in value is given by

$$100 \times \$15.25 = \$1525$$

The yield (or percent increase) during this 1-year period is found by dividing the difference in the selling price and original price by the original price.

$$\frac{\$15.25}{\$46.375} \approx 0.33, \text{ or } 33\%$$

Hence, the stockholder made approximately 33% growth on the investment.

This chapter focuses on operations on fractions as well as an introduction to geometry.

For further information see the Technology Connections in MathZone at

www.mhhe.com/miller_oneill

section

R.1 FRACTIONS

Concepts

1. Basic Definitions
2. Prime Factorization
3. Reducing Fractions to Lowest Terms
4. Multiplying Fractions
5. Dividing Fractions
6. Adding and Subtracting Fractions
7. Operations on Mixed Numbers

1. Basic Definitions

The study of algebra involves many of the operations and procedures used in arithmetic. Therefore, we begin this text by reviewing the basic operations of addition, subtraction, multiplication, and division on fractions and mixed numbers.

In day-to-day life, the numbers we use for counting are

the **natural numbers**: 1, 2, 3, 4, . . . and

the **whole numbers**: 0, 1, 2, 3, . . .

Whole numbers are used to count the number of whole units in a quantity. A fraction is used to express part of a whole unit. If a child gains $2\frac{1}{2}$ lb, the child has gained two whole pounds plus a portion of a pound. To express the additional half pound mathematically, we may use the fraction, $\frac{1}{2}$.

A Fraction and Its Parts

Fractions are numbers of the form $\frac{a}{b}$, where $\frac{a}{b} = a \div b$ and b does not equal zero.
In the fraction $\frac{a}{b}$, the **numerator** is a, and the **denominator** is b.

The denominator of a fraction indicates how many equal parts divide the whole. The numerator indicates how many parts are being represented. For instance, suppose Jack wants to plant carrots in $\frac{2}{5}$ of a rectangular garden. He can divide the garden into five equal parts and use two of the parts for carrots (Figure R-1).

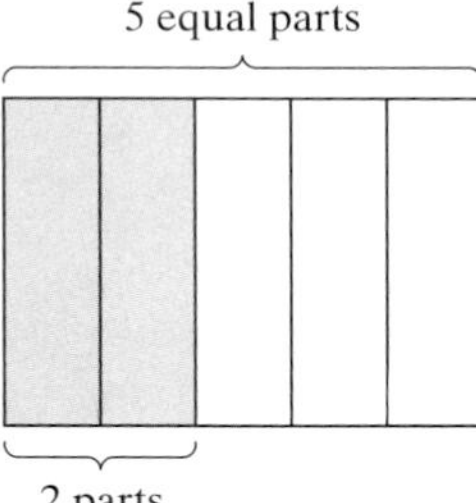

The shaded region represents $\frac{2}{5}$ of the garden.

Figure R-1

Definition of a Proper Fraction, an Improper Fraction, and a Mixed Number

1. If the numerator of a fraction is less than the denominator, the fraction is a **proper fraction**. A proper fraction represents a quantity that is less than a whole unit.
2. If the numerator of a fraction is greater than or equal to the denominator, then the fraction is an **improper fraction**. An improper fraction represents a quantity greater than or equal to a whole unit.
3. A **mixed number** is a whole number added to a fraction.

Proper Fractions: $\frac{3}{5}$ $\frac{1}{8}$

Improper Fractions: $\frac{7}{5}$ $\frac{8}{8}$

Mixed Numbers: $1\frac{1}{5}$ $2\frac{3}{8}$

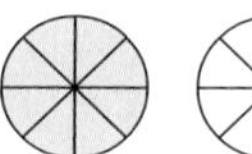

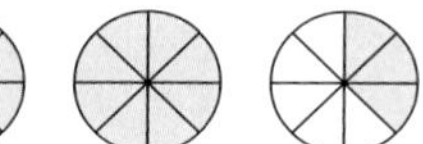

2. Prime Factorization

To perform operations on fractions it is important to understand the concept of a factor. For example, when the numbers 2 and 6 are multiplied, the result (called the **product**) is 12.

$$2 \times 6 = 12$$

(2 and 6: factors; 12: product)

The numbers 2 and 6 are said to be **factors*** of 12. The number 12 is said to be factored when it is written as the product of two or more natural numbers. For example, 12 can be factored in several ways:

$$12 = 1 \times 12 \qquad 12 = 2 \times 6 \qquad 12 = 3 \times 4 \qquad 12 = 2 \times 2 \times 3$$

A natural number greater than 1 whose only factors are 1 and itself is called a **prime number**. The first several prime numbers are 2, 3, 5, 7, 11, and 13. A natural number greater than 1 that is not prime is called a **composite number**. That is, a composite number has factors other than itself and 1. The first several composite numbers are 4, 6, 8, 9, 10, 12, 14, 15, and 16.

The number 1 is neither prime nor composite.

example 1 Writing a Natural Number as a Product of Prime Factors

Write each number as a product of prime factors.

a. 12 b. 30

Solution:

a. $12 = 2 \times 2 \times 3$

Divide 12 by prime numbers until the number 1 is obtained.

$$\begin{array}{r} 2\,\underline{|\,12} \\ 2\,\underline{|\,\;6} \\ 3 \end{array}$$

Or use a factor tree

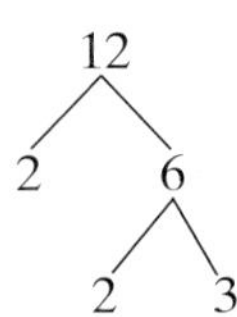

*In this context, we refer only to natural number factors.

b. $30 = 2 \times 3 \times 5$

$$\begin{array}{r|l} 2 & 30 \\ 3 & 15 \\ & 5 \end{array}$$

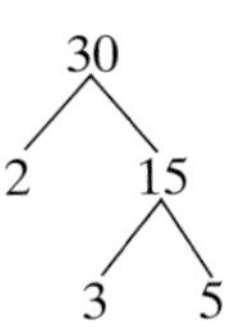

3. Reducing Fractions to Lowest Terms

The process of factoring numbers can be used to reduce fractions to lowest terms. A fractional portion of a whole can be represented by infinitely many fractions. For example, Figure R-2 shows that $\frac{1}{2}$ is equivalent to $\frac{2}{4}, \frac{3}{6}, \frac{4}{8}$, and so on.

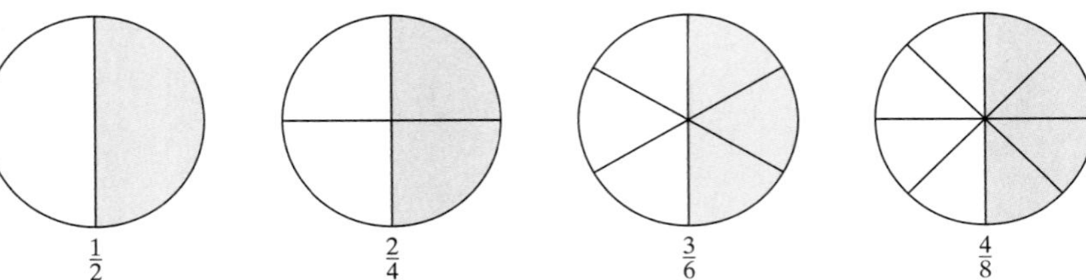

Figure R-2

The fraction $\frac{1}{2}$ is said to be in lowest terms because the numerator and denominator share no common factor other than 1. The fraction $\frac{2}{4}$ is not in lowest terms because the numerator and denominator are both divisible by 2.

To reduce a fraction to lowest terms, the goal is to "divide out" common factors from both the numerator and denominator.

example 2

Reducing a Fraction to Lowest Terms

Reduce the fraction $\frac{12}{30}$ to lowest terms.

Solution:

From Example 1, we have $12 = 2 \times 2 \times 3$ and $30 = 2 \times 3 \times 5$. Hence,

$$\frac{12}{30} = \frac{2 \times 2 \times 3}{2 \times 3 \times 5}$$

The common factors of 2 and 3 may be "divided out" of the numerator and denominator.

$$= \frac{\overset{1}{\cancel{2}} \times 2 \times \overset{1}{\cancel{3}}}{\underset{1}{\cancel{2}} \times \underset{1}{\cancel{3}} \times 5} = \frac{2}{5}$$

Multiply $1 \times 2 \times 1 = 2$.

Multiply $1 \times 1 \times 5 = 5$.

example 3

Avoiding Mistakes

In Example …the …ors of 2 and 7 divid…ominator, numerato… 1. It is leaving …ember to impor… of 1. The writ… the fraction re… i…

Reducing a Fraction to Lowest Terms

Reduce $\frac{14}{42}$ to lowest terms.

Solution:

$$\frac{14}{42} = \frac{2 \times 7}{2 \times 3 \times 7}$$ Factor the numerator and denominator.

$$= \frac{\overset{1}{\cancel{2}} \times \overset{1}{\cancel{7}}}{\underset{1}{\cancel{2}} \times 3 \times \underset{1}{\cancel{7}}} = \frac{1}{3}$$

Multiply $1 \times 1 = 1$.

Multiply $1 \times 3 \times 1 = 3$.

4. Multiplying Fractions

Multiplying Fractions

If b is not zero and d is not zero, then

$$\frac{a}{b} \times \frac{c}{d} = \frac{a \times c}{b \times d}$$

To multiply fractions, multiply the numerators and multiply the denominators.

example 4

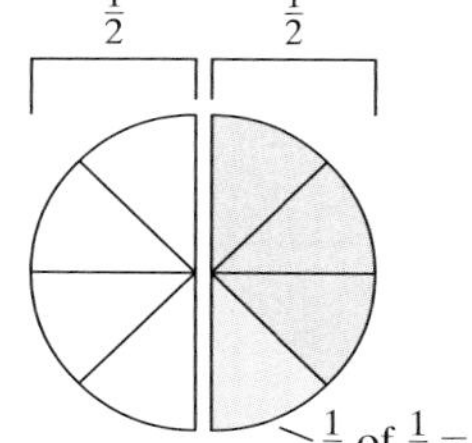

Figure R-3

Multiplying Fractions

Multiply the fractions: $\frac{1}{4} \times \frac{1}{2}$

Solution:

$$\frac{1}{4} \times \frac{1}{2} = \frac{1 \times 1}{4 \times 2} = \frac{1}{8}$$ Multiply the numerators. Multiply the denominators.

Notice that the product $\frac{1}{4} \times \frac{1}{2}$ represents a quantity that is $\frac{1}{4}$ of $\frac{1}{2}$. Taking $\frac{1}{4}$ of a quantity is equivalent to dividing the quantity by 4. One-half of a pie divided into four pieces leaves pieces that each represent $\frac{1}{8}$ of the pie (Figure R-3).

example 5

Multiplying Fractions

Multiply the fractions.

a. $\frac{7}{10} \times \frac{15}{14}$ b. $\frac{2}{13} \times \frac{13}{2}$ c. $5 \times \frac{1}{5}$

Solution:

a. $\dfrac{7}{10} \times \dfrac{15}{14} = \dfrac{7 \times 15}{10 \times 14}$

Multiply the nu multiply the dertors and

Divide out commtors. the numerator andrs in

$= \dfrac{\overset{1}{\cancel{7}} \times 3 \times \overset{1}{\cancel{5}}}{2 \times \underset{1}{\cancel{5}} \times 2 \times \underset{1}{\cancel{7}}} = \dfrac{3}{4}$

Multiply 1 × 3 × 1 =ator.

Multiply 2 × 1 × 2 × 1 =

b. $\dfrac{2}{13} \times \dfrac{13}{2} = \dfrac{2 \times 13}{13 \times 2} = \dfrac{\overset{1}{\cancel{2}} \times \overset{1}{\cancel{13}}}{\underset{1}{\cancel{13}} \times \underset{1}{\cancel{2}}} = \dfrac{1}{1} = 1$

Multiply 1 × 1 = 1.

Multiply 1 × 1 = 1.

c. $5 \times \dfrac{1}{5} = \dfrac{5}{1} \times \dfrac{1}{5}$

The whole number 5 can be written as $\frac{5}{1}$.

$= \dfrac{\overset{1}{\cancel{5}} \times 1}{1 \times \underset{1}{\cancel{5}}} = \dfrac{1}{1} = 1$

Multiply and reduce to lowest terms.

5. Dividing Fractions

Before we divide fractions, we need to know how to find the reciprocal of a fraction. Notice from Example 5 that $\frac{2}{13} \times \frac{13}{2} = 1$ and $5 \times \frac{1}{5} = 1$. The numbers $\frac{2}{13}$ and $\frac{13}{2}$ are said to be reciprocals because their product is 1. Likewise the numbers 5 and $\frac{1}{5}$ are reciprocals because their product is 1.

The Reciprocal of a Number

Two numbers are **reciprocals** of each other if their product is 1. Therefore the reciprocal of the fraction

$$\frac{a}{b} \text{ is } \frac{b}{a} \qquad \text{because} \qquad \frac{a}{b} \times \frac{b}{a} = 1, \text{ for } a, b \neq 0$$

Number	Reciprocal	Product
$\dfrac{2}{13}$	$\dfrac{13}{2}$	$\dfrac{2}{13} \times \dfrac{13}{2} = 1$
$\dfrac{1}{8}$	$\dfrac{8}{1}$ (or equivalently 8)	$\dfrac{1}{8} \times 8 = 1$
$6\left(\text{or equivalently } \dfrac{6}{1}\right)$	$\dfrac{1}{6}$	$6 \times \dfrac{1}{6} = 1$

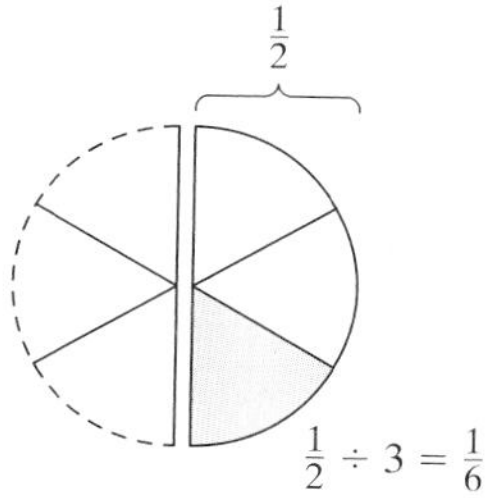

Figure R-4

To understand the concept of dividing fractions, consider a pie that is half-eaten. Suppose the remaining half must be divided among three people, that is: $\frac{1}{2} \div 3$. However, dividing by 3 is equivalent to taking $\frac{1}{3}$ of the remaining $\frac{1}{2}$ of the pie (Figure R-4).

$$\frac{1}{2} \div 3 = \frac{1}{2} \cdot \frac{1}{3} = \frac{1}{6}$$

This example illustrates that *dividing two numbers is equivalent to multiplying the first number by the reciprocal of the second number.*

Division of Fractions

Let a, b, c, and d be numbers such that b, c, and d are not zero. Then,

multiply

$$\frac{a}{b} \div \frac{c}{d} = \frac{a}{b} \times \frac{d}{c}$$

reciprocal

To divide fractions, multiply the first fraction by the reciprocal of the second fraction.

example 6 **Dividing Fractions**

Divide the fractions.

a. $\frac{8}{5} \div \frac{3}{10}$ b. $\frac{12}{13} \div 6$

Solution:

a. $\frac{8}{5} \div \frac{3}{10} = \frac{8}{5} \times \frac{10}{3}$ Multiply by the reciprocal of $\frac{3}{10}$, which is $\frac{10}{3}$.

$= \frac{8 \times \overset{2}{\cancel{10}}}{\underset{1}{\cancel{5}} \times 3} = \frac{16}{3}$ Multiply and reduce to lowest terms.

b. $\frac{12}{13} \div 6 = \frac{12}{13} \div \frac{6}{1}$ Write the whole number 6 as $\frac{6}{1}$.

$= \frac{12}{13} \times \frac{1}{6}$ Multiply by the reciprocal of $\frac{6}{1}$, which is $\frac{1}{6}$.

$\frac{\overset{2}{\cancel{12}} \times 1}{13 \times \underset{1}{\cancel{6}}} = \frac{2}{13}$ Multiply and reduce to lowest terms.

6. Adding and Subtracting Fractions

Adding and Subtracting Fractions

Two fractions can be added or subtracted only if they have a common denominator. Let a, b, and c, be numbers such that b does not equal zero. Then,

$$\frac{a}{b} + \frac{c}{b} = \frac{a + c}{b} \quad \text{and} \quad \frac{a}{b} - \frac{c}{b} = \frac{a - c}{b}$$

To add or subtract fractions with the same denominator, add or subtract the numerators and write the result over the common denominator.

example 7 **Adding and Subtracting Fractions with the Same Denominator**

Tip: $\frac{1}{12} + \frac{7}{12}$ can be visualized as the sum of the pink and blue sections in the figure shown here.

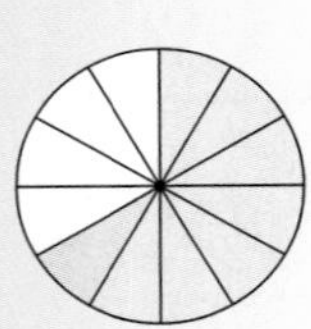

Add or subtract as indicated.

a. $\frac{1}{12} + \frac{7}{12}$ b. $\frac{13}{5} - \frac{3}{5}$

Solution:

a. $\frac{1}{12} + \frac{7}{12} = \frac{1 + 7}{12}$ Add the numerators.

$= \frac{8}{12}$ Simplify.

$= \frac{2}{3}$ Reduce to lowest terms.

b. $\frac{13}{5} - \frac{3}{5} = \frac{13 - 3}{5}$ Subtract the numerators.

$= \frac{10}{5}$ Simplify.

$= 2$ Reduce to lowest terms.

In Example 7, we added and subtracted fractions with the same denominators. To add or subtract fractions with different denominators we must first write them as equivalent fractions with a common denominator. A common denominator may be *any* common multiple of the denominators. We will use the least common multiple to form the **least common denominator (LCD)**.

Consider the fractions $\frac{1}{3}$ and $\frac{1}{2}$. A common denominator must be a multiple of 3 and a multiple of 2. Hence the product $3 \times 2 = 6$ can be used as a common denominator. Notice that the numbers 6, 12, 18, 24 and so on are all multiples of both 3 and 2. However, 6 is the *least* common multiple.

Consider the fractions $\frac{1}{4}$ and $\frac{1}{8}$. Notice that the product $4 \times 8 = 32$ is a common multiple of 4 and 8. However, the least common multiple of 4 and 8 is 8. To understand why, write each denominator as a product of prime factors.

$$4 = 2 \times 2 \quad \text{and} \quad 8 = 2 \times 2 \times 2$$

Both 4 and 8 are composed of repeated factors of 2. However, any number that is a multiple of $2 \times 2 \times 2 = 8$ is automatically a multiple of $2 \times 2 = 4$. Therefore, it is sufficient to use the factor 2 the maximum number of times it appears in the factorization of either denominator. The LCD of $\frac{1}{4}$ and $\frac{1}{8}$ is $\underbrace{2 \times 2 \times 2}_{\text{3 factors of 2}} = 8$.

Steps to Finding the Least Common Denominator of Two Fractions

1. Write each denominator as a product of prime factors.
2. The LCD is the product of unique prime factors from both denominators. If a factor is repeated in either denominator, use that factor the maximum number of times it appears in either denominator.

example 8

Finding the LCD of Fractions

Find the LCD of the given fractions.

a. $\frac{1}{9}$ and $\frac{1}{15}$ b. $\frac{3}{10}$, $\frac{1}{6}$, and $\frac{7}{8}$

Solution:

a. $9 = 3 \times 3$ and $15 = 3 \times 5$ — Factor the denominators.

$\text{LCD} = 3 \times 3 \times 5 = 45$ — The LCD is the product of the factors of 3 and 5, where 3 is repeated twice.

b. $10 = 2 \times 5$, $6 = 2 \times 3$, and $8 = 2 \times 2 \times 2$ — Factor the denominators.

$\text{LCD} = 2 \times 2 \times 2 \times 3 \times 5 = 120$ — The LCD is the product of the factors 2, 3, and 5, where 2 is repeated three times.

To add or subtract fractions, a common denominator must first be determined. Then it is necessary to convert each fraction to an equivalent fraction with the common denominator.

Writing Equivalent Fractions

To write a fraction as an equivalent fraction with a common denominator, multiply the numerator and denominator by the factors from the common denominator that are missing from the denominator of the original fraction.

Note: Multiplying the numerator and denominator by the *same* nonzero quantity will not change the value of the fraction.

example 9 **Writing Equivalent Fractions**

Write the fractions $\frac{1}{9}$ and $\frac{1}{15}$ as equivalent fractions with the LCD as the denominator.

Solution:

From Example 8(a), we know that the LCD for $\frac{1}{9}$ and $\frac{1}{15}$ is 45.

$\frac{1}{9} = \frac{1 \times 5}{9 \times 5} = \frac{5}{45}$ Multiply numerator and denominator by 5. This creates a denominator of 45.

$\frac{1}{15} = \frac{1 \times 3}{15 \times 3} = \frac{3}{45}$ Multiply numerator and denominator by 3. This creates a denominator of 45.

example 10 **Adding Fractions with Different Denominators**

Suppose Nakeysha ate $\frac{1}{2}$ of an ice-cream pie, and her friend Carla ate $\frac{1}{3}$ of the pie. How much of the ice-cream pie was eaten?

Solution:

$\frac{1}{2} + \frac{1}{3}$ The LCD is $3 \times 2 = 6$.

$= \frac{1 \times 3}{2 \times 3} + \frac{1 \times 2}{3 \times 2}$ Multiply numerator and denominator by the missing factors.

$= \frac{3}{6} + \frac{2}{6}$

$= \frac{5}{6}$ Add the fractions.

Together, Nakeysha and Carla ate $\frac{5}{6}$ of the ice-cream pie.

By converting the fractions $\frac{1}{2}$ and $\frac{1}{3}$ to the same denominator, we are able to add *like* size pieces of pie (Figure R-5).

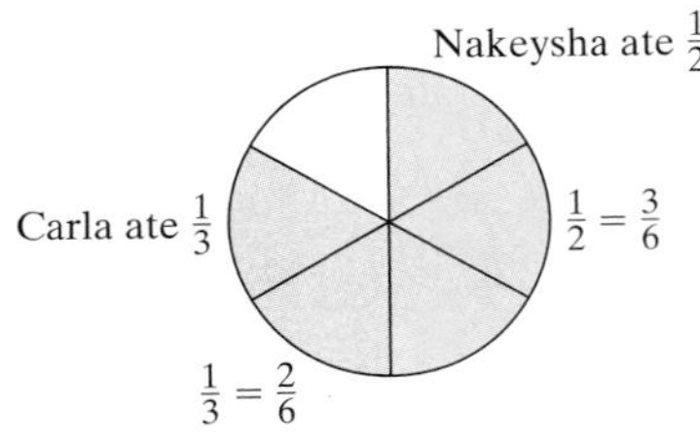

Figure R-5

example 11

Adding and Subtracting Fractions

Simplify: $\frac{1}{2} + \frac{5}{8} - \frac{11}{16}$.

Solution:

$\frac{1}{2} + \frac{5}{8} - \frac{11}{16}$

Factor the denominators: $2 = 2$, $8 = 2 \times 2 \times 2$, and $16 = 2 \times 2 \times 2 \times 2$.
The LCD is $2 \times 2 \times 2 \times 2 = 16$.
Next write the fractions as equivalent fractions with the LCD.

$= \frac{1 \times 2 \times 2 \times 2}{2 \times 2 \times 2 \times 2} + \frac{5 \times 2}{8 \times 2} - \frac{11}{16}$

Multiply numerators and denominators by the factors missing from each denominator.

$= \frac{8}{16} + \frac{10}{16} - \frac{11}{16}$

$= \frac{8 + 10 - 11}{16}$

Add and subtract the numerators.

$= \frac{7}{16}$

Simplify.

7. Operations on Mixed Numbers

Recall that a mixed number is a whole number added to a fraction. The number $3\frac{1}{2}$ represents the sum of three wholes plus a half. That is, $3\frac{1}{2} = 3 + \frac{1}{2}$. For this reason, any mixed number can be converted to an improper fraction by using addition.

$$3\tfrac{1}{2} = 3 + \frac{1}{2} = \frac{6}{2} + \frac{1}{2} = \frac{7}{2}$$

Tip: A shortcut to writing a mixed number as an improper fraction is to multiply the whole number by the denominator of the fraction. Then add this value to the numerator of the fraction and write the result over the denominator.

$3\frac{1}{2} \longrightarrow$ Multiply the whole number by the denominator: $3 \times 2 = 6$.

Add the numerator: $6 + 1 = 7$.

Write the result over the denominator: $\frac{7}{2}$.

To add, subtract, multiply, or divide mixed numbers, we will first write the mixed number as an improper fraction.

example 12 Operations on Mixed Numbers

Perform the indicated operations.

a. $5\frac{1}{3} - 2\frac{1}{4}$ b. $7\frac{1}{2} \div 3$

Solution:

a. $5\frac{1}{3} - 2\frac{1}{4}$

$= \dfrac{16}{3} - \dfrac{9}{4}$ Write the mixed numbers as improper fractions.

$= \dfrac{16 \times 4}{3 \times 4} - \dfrac{9 \times 3}{4 \times 3}$ The LCD is 12. Multiply numerators and denominators by the missing factors from the denominators.

$= \dfrac{64}{12} - \dfrac{27}{12}$

$= \dfrac{37}{12}$ or $3\frac{1}{12}$ Subtract the fractions.

Tip: An improper fraction can also be written as a mixed number. Both answers are acceptable. Note that

$$\frac{37}{12} = \frac{36}{12} + \frac{1}{12} = 3 + \frac{1}{12} \text{ or } 3\frac{1}{12}$$

b. $7\frac{1}{2} \div 3$

$= \dfrac{15}{2} \div \dfrac{3}{1}$ Write the mixed number and whole number as fractions.

$= \dfrac{\overset{5}{\cancel{15}}}{2} \times \dfrac{1}{\underset{1}{\cancel{3}}}$ Multiply by the reciprocal of $\frac{3}{1}$, which is $\frac{1}{3}$.

$= \dfrac{5}{2}$ or $2\frac{1}{2}$ The answer may be written as an improper fraction or as a mixed number.

Avoiding Mistakes

Remember that when dividing (or multiplying) fractions, a common denominator is not necessary.

section R.1 Practice Exercises

For Exercises 1–8, identify the numerator and denominator of the fraction. Then determine if the fraction is a proper fraction or an improper fraction.

1. $\dfrac{7}{8}$
2. $\dfrac{2}{3}$
3. $\dfrac{9}{5}$
4. $\dfrac{5}{2}$
5. $\dfrac{6}{6}$
6. $\dfrac{4}{4}$
7. $\dfrac{12}{1}$
8. $\dfrac{5}{1}$

For Exercises 9–16, write a proper or improper fraction associated with the shaded region of each figure.

9. 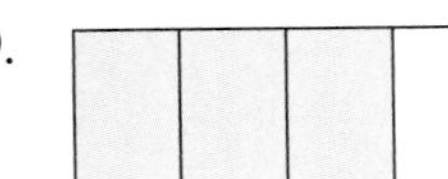

10. 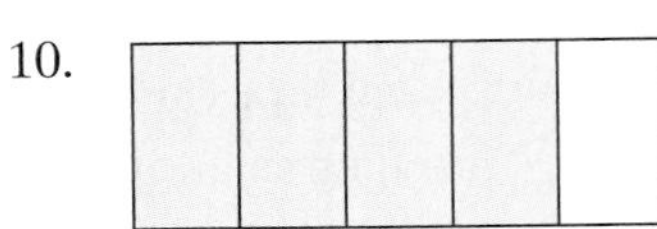

11. 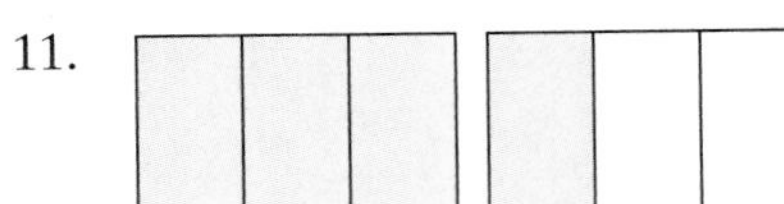

12. 

13. 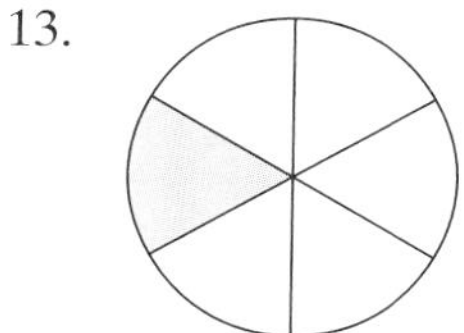

14. 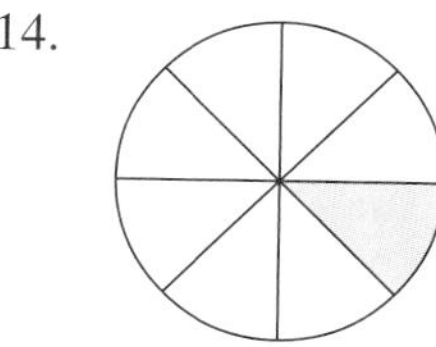

15. 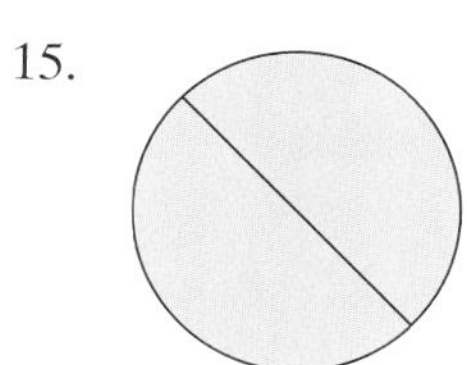

16. 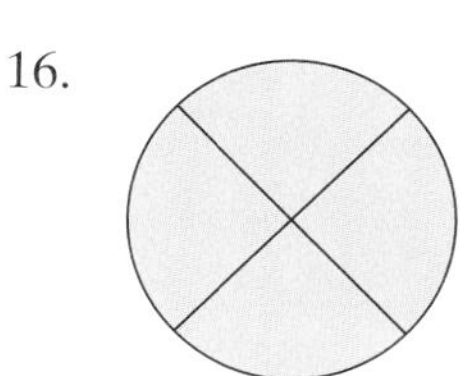

For Exercises 17–20, write both an improper fraction and a mixed number associated with the shaded region of each figure.

17. 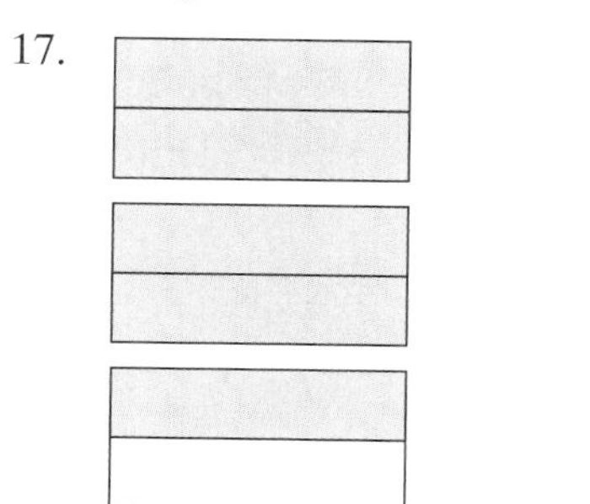

18. 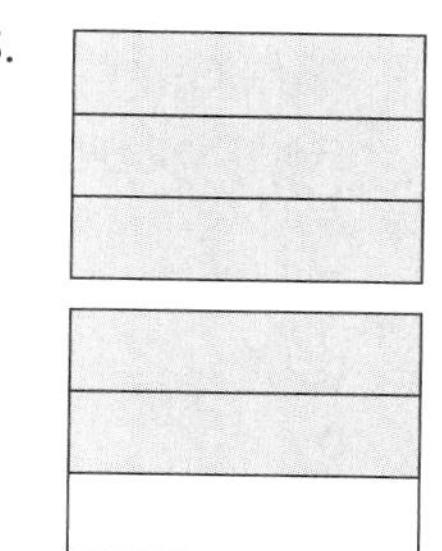

19. 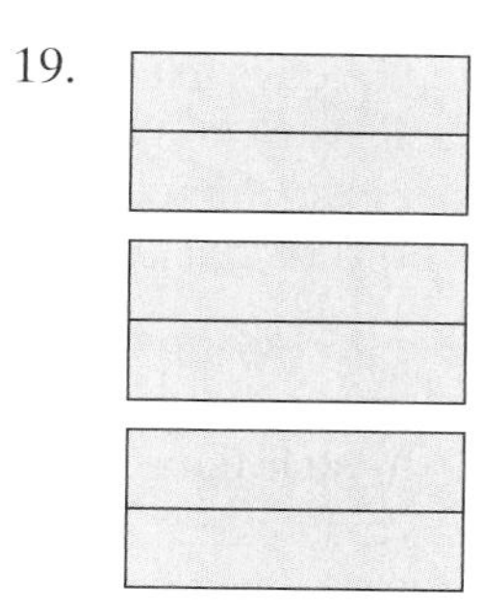

20. 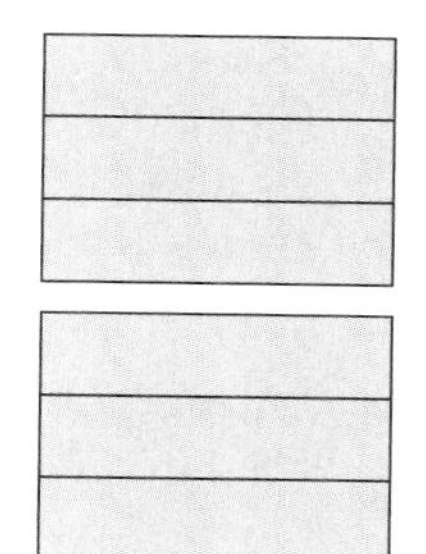

21. Explain the difference between the set of whole numbers and the set of natural numbers.

22. Explain the difference between a proper fraction and an improper function.

23. Write a fraction that reduces to $\frac{1}{2}$. (Answers may vary.)

24. Write a fraction that reduces to $\frac{1}{3}$. (Answers may vary.)

For Exercises 25–32, identify the number as either a prime number or a composite number.

25. 5
26. 9
27. 4
28. 2
29. 39
30. 23
31. 53
32. 51

For Exercises 33–40, write the number as a product of prime factors.

33. 36
34. 70
35. 42
36. 35
37. 110
38. 136
39. 135
40. 105

For Exercises 41–52, reduce each fraction to lowest terms. Use prime factorization if necessary.

41. $\dfrac{3}{15}$
42. $\dfrac{8}{12}$
43. $\dfrac{6}{16}$
44. $\dfrac{12}{20}$
45. $\dfrac{42}{48}$
46. $\dfrac{35}{80}$
47. $\dfrac{48}{64}$
48. $\dfrac{32}{48}$
49. $\dfrac{110}{176}$
50. $\dfrac{70}{120}$
51. $\dfrac{150}{200}$
52. $\dfrac{119}{210}$

For Exercises 53–54, identify if the statement is true or false. If it is false, rewrite as a true statement.

53. When multiplying or dividing fractions it is necessary to have a common denominator.

54. When dividing two fractions it is necessary to multiply the first fraction by the reciprocal of the second fraction.

For Exercises 55–58, convert the improper fraction to a mixed number.

55. $\dfrac{11}{9}$

56. $\dfrac{26}{6}$

57. $\dfrac{7}{2}$

58. $\dfrac{12}{5}$

For Exercises 59–62, convert the mixed number to an improper fraction.

59. $5\dfrac{2}{5}$

60. $2\dfrac{2}{3}$

61. $1\dfrac{7}{8}$

62. $4\dfrac{1}{2}$

For Exercises 63–74, multiply or divide as indicated.

63. $\dfrac{10}{13} \times \dfrac{26}{15}$

64. $\dfrac{15}{28} \times \dfrac{7}{9}$

65. $\dfrac{3}{7} \div \dfrac{9}{14}$

66. $\dfrac{7}{25} \div \dfrac{1}{5}$

67. $\dfrac{9}{10} \times 5$

68. $\dfrac{3}{7} \times 14$

69. $4\dfrac{3}{5} \div \dfrac{1}{10}$

70. $2\dfrac{4}{5} \div \dfrac{7}{11}$

71. $3\dfrac{1}{5} \times \dfrac{7}{8}$

72. $2\dfrac{1}{2} \times \dfrac{4}{5}$

73. $1\dfrac{2}{9} \div 6$

74. $2\dfrac{2}{5} \div \dfrac{2}{7}$

75. Stephen's take-home pay is \$1200 a month. If his rent is $\frac{1}{4}$ of his pay, how much is his rent?

76. Gus decides to save $\frac{1}{3}$ of his pay each month. If his monthly pay is \$2112, how much will he save each month?

77. A recipe for a casserole calls for $\frac{1}{3}$ of a dozen eggs. How many eggs are needed?

78. Shontell had time to print out only $\frac{3}{5}$ of her book report before school. If the report is 10 pages long, how many pages did she print out?

79. Gail buys 6 lb of mixed nuts to be divided into decorative jars that will each hold $\frac{3}{4}$ lb of nuts. How many jars will she be able to fill?

Figure for Exercise 79

80. Natalie has 4 yd of material with which she can make holiday aprons. If it takes $\frac{1}{2}$ yd of material per apron, how many aprons can she make?

81. There are 4 cups of oatmeal in a box. If each serving is $\frac{1}{3}$ of a cup, how many servings are contained in the box?

82. A board $26\frac{3}{8}$ in. long must be cut into three pieces of equal length. Find the length of each piece.

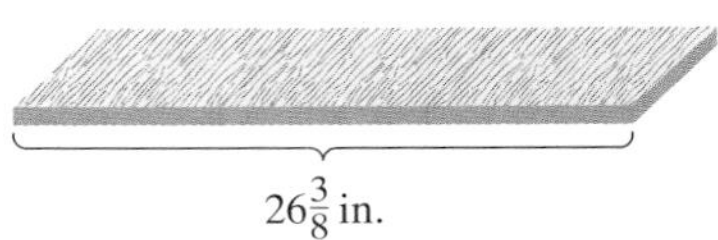

Figure for Exercise 82

83. Richard makes candy for the holidays. He requires $1\frac{3}{4}$ tsp of vanilla for each batch of candy. If a bottle of vanilla contains 21 tsp, how many batches can he make before he runs out of vanilla?

84. A piece of rope $52\frac{1}{2}$ in. long must be cut into eight equal pieces. Find the length of each piece.

For Exercises 85–90, find the least common denominator for each pair of fractions.

85. $\dfrac{1}{6}, \dfrac{5}{24}$

86. $\dfrac{1}{12}, \dfrac{11}{30}$

87. $\dfrac{9}{20}, \dfrac{3}{8}$

88. $\dfrac{13}{24}, \dfrac{7}{40}$

89. $\dfrac{7}{10}, \dfrac{11}{45}$

90. $\dfrac{1}{20}, \dfrac{1}{30}$

For Exercises 91–114, add or subtract as indicated.

91. $\dfrac{5}{14} + \dfrac{1}{14}$

92. $\dfrac{9}{5} + \dfrac{1}{5}$

93. $\dfrac{17}{24} - \dfrac{5}{24}$

94. $\dfrac{11}{18} - \dfrac{5}{18}$

95. $\dfrac{1}{8} + \dfrac{3}{4}$

96. $\dfrac{3}{16} + \dfrac{1}{2}$

97. $\dfrac{3}{8} - \dfrac{3}{10}$

98. $\dfrac{12}{35} - \dfrac{1}{10}$

99. $\dfrac{7}{26} - \dfrac{2}{13}$

100. $\dfrac{11}{24} - \dfrac{5}{16}$

101. $\dfrac{7}{18} + \dfrac{5}{12} - \dfrac{1}{6}$

102. $\dfrac{3}{16} + \dfrac{9}{20} - \dfrac{1}{4}$

103. $\dfrac{3}{4} - \dfrac{1}{20} + \dfrac{2}{5}$

104. $\dfrac{1}{6} - \dfrac{1}{24} + \dfrac{5}{12}$

105. $\dfrac{5}{12} + \dfrac{5}{16} - \dfrac{1}{3}$

106. $\dfrac{3}{25} + \dfrac{8}{35} - \dfrac{1}{5}$

107. $2\frac{1}{8} + 1\frac{3}{8}$

108. $1\frac{3}{14} + 1\frac{1}{14}$

109. $1\frac{5}{6} - \frac{7}{8}$

110. $2\frac{1}{3} - \frac{5}{6}$

111. $1\frac{1}{6} + 3\frac{3}{4}$

112. $4\frac{1}{2} + 2\frac{2}{3}$

113. $1 - \dfrac{7}{8}$

114. $2 - \dfrac{3}{7}$

115. A futon, when set up as a sofa, measures $3\frac{5}{6}$ ft wide. When it is opened to be used as a bed, the width is increased by $1\frac{3}{4}$ ft. What is the total width of this bed?

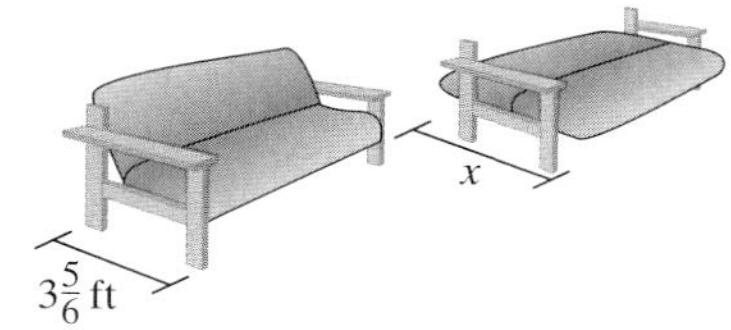

Figure for Exercise 115

116. If Sally adds a lace that is $\frac{7}{8}$ in. wide to a skirt that is $20\frac{1}{2}$ in. long, what will be the final length of the skirt?

117. Three children ate Cheerios for breakfast. Aaron had $\frac{1}{2}$ cup, Sean had $\frac{1}{3}$ cup and Sheila had $\frac{3}{4}$ cup. How much cereal was eaten that morning?

118. José orders two seafood platters for a party. One platter has $1\frac{1}{2}$ lb of shrimp and the other has $\frac{3}{4}$ lb of shrimp. How many pounds does he have altogether?

119. Ayako took a trip to the store $5\frac{1}{2}$ miles away. If she rode the bus for $4\frac{5}{6}$ miles and walked the rest of the way, how far did she have to walk?

120. Average rainfall in Tampa for the month of November is $2\frac{3}{4}$ in. One stormy weekend $3\frac{1}{8}$ in. of rain fell. How many inches of rain over the average is this?

121. A plane trip from Orlando to Detroit takes $2\frac{3}{4}$ hours. If the plane has traveled for $1\frac{1}{2}$ hours, how much time remains for the flight?

122. Pete started working out at the gym several months ago. His waist measured $38\frac{1}{2}$ in. when he began and is now $33\frac{3}{4}$ in. How many inches did he lose around his waist?

123. Find the perimeter (distance around) the parking lot.

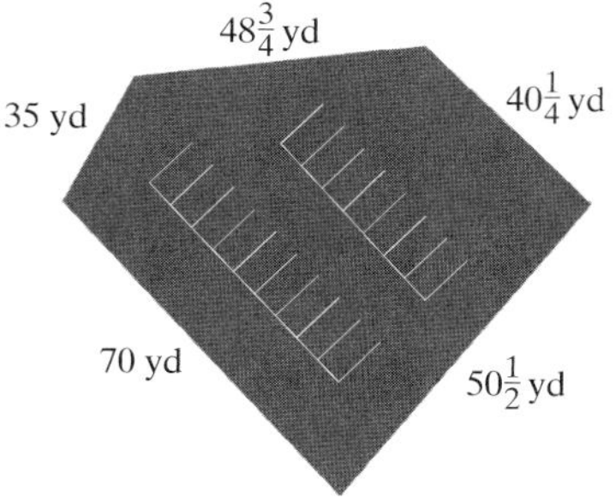

Figure for Exercise 123

124. A booth for display at a science fair is in the shape of a triangle. Find the distance around the triangle.

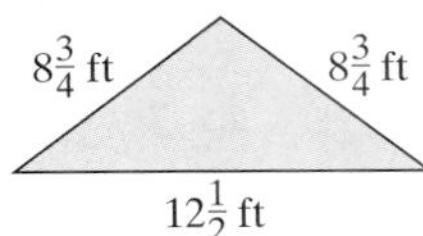

Figure for Exercise 124

For Exercises 125–128, approximate the sum by rounding each fraction to the nearest whole number.

125. $\frac{7}{8} + \frac{16}{15}$

126. $\frac{6}{5} + \frac{21}{22}$

127. $\frac{21}{4} + \frac{98}{100} + \frac{80}{41}$

128. $\frac{29}{5} + \frac{51}{10} + \frac{7}{8}$

Key Terms:

composite number
denominator
factor
fraction
improper fraction
least common denominator (LCD)
mixed number
natural numbers
numerator
prime number
product
proper fraction
reciprocal
whole numbers

Concepts

1. **Perimeter**
2. **Area**
3. **Volume**
4. **Angles**
5. **Triangles**

section

R.2 Introduction to Geometry

1. Perimeter

In this section we present several facts and formulas that may be used throughout the text in applications of geometry. One of the most important uses of geometry involves the measurement of objects of various shapes. We begin with an introduction to perimeter, area, and volume for several common shapes and objects.

Perimeter is defined as the distance around a figure. For example, if we were to put up a fence around a field, the perimeter would determine the amount of fencing. For a polygon (a closed figure constructed from line segments) the perimeter is the sum of the lengths of the sides. For a circle the distance around the outside is called the **circumference**.

Rectangle

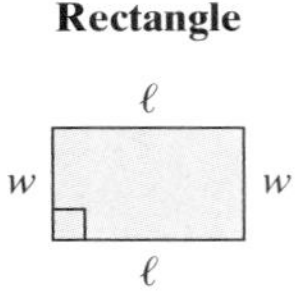

$P = 2\ell + 2w$

Square

s s s s

$P = 4s$

Triangle

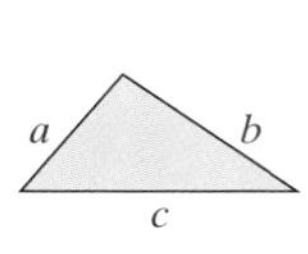

$P = a + b + c$

Circle

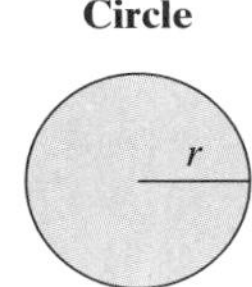

Circumference: $C = 2\pi r$

example 1 Finding Perimeter and Circumference

Find the perimeter or circumference as indicated. Use 3.14 for π.

a. Perimeter of the polygon

b. Perimeter of the rectangle

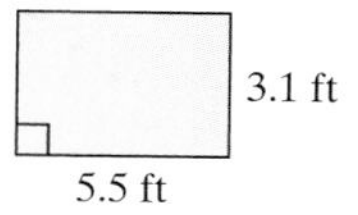

c. Circumference of the circle

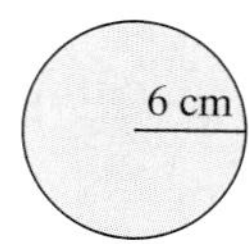

Solution:

a. $P = (8\text{ m}) + (2\text{ m}) + (6\text{ m}) + (4\text{ m}) + (3\text{ m})$ Add the lengths of the sides.

$= 23\text{ m}$ The perimeter is 23 m.

b. $P = 2\ell + 2w$

$= 2(5.5\text{ ft}) + 2(3.1\text{ ft})$ Substitute $\ell = 5.5$ ft and $w = 3.1$ ft.

$= 11\text{ ft} + 6.2\text{ ft}$

$= 17.2\text{ ft}$ The perimeter is 17.2 ft.

c. $C = 2\pi r$

$= 2(3.14)(6\text{ cm})$ Substitute 3.14 for π and $r = 6$ cm.

$= 6.28(6\text{ cm})$

$= 37.68\text{ cm}$ The circumference is 37.68 cm.

Tip: If a calculator is used to find the circumference of a circle, use the π key to get a more accurate answer.

2. Area

The **area** of a geometric figure is the number of square units that can be enclosed within the figure. For example, the rectangle shown in Figure R-6 encloses 6 square inches (6 in.^{2}). In applications, we would find the area of a region if we were laying carpet or putting down sod for a lawn.

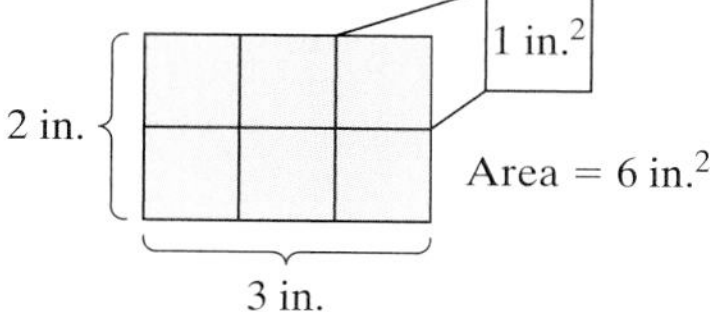

Figure R-6

The formulas used to compute the area for several common geometric shapes are given here:

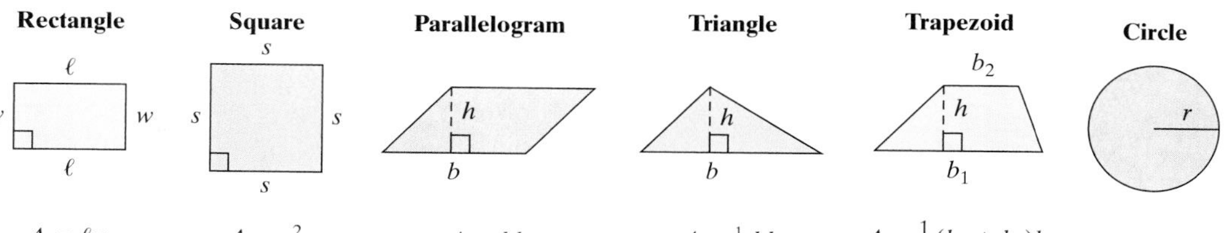

example 2 **Finding Area**

Find the area enclosed by each figure.

a. 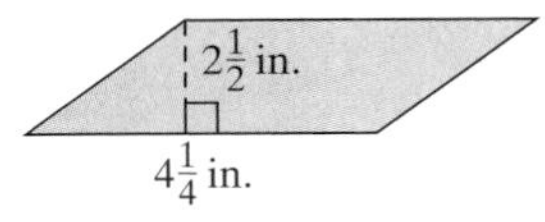

b. 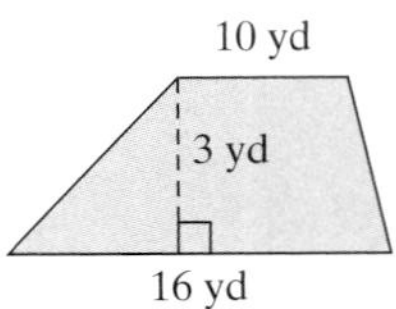

Solution:

Tip: Notice that the units of area are square units such as square inches (in.^2), square feet (ft^2), square yards (yd^2), square centimeters (cm^2), and so on.

a. $A = bh$ — The figure is a parallelogram.

$= (4\frac{1}{4}\text{ in.})(2\frac{1}{2}\text{ in.})$ — Substitute $b = 4\frac{1}{4}$ in. and $h = 2\frac{1}{2}$ in.

$= \left(\frac{17}{4}\text{ in.}\right)\left(\frac{5}{2}\text{ in.}\right)$

$= \frac{85}{8}\text{ in.}^2$ or $10\frac{5}{8}\text{ in.}^2$

b. $A = \frac{1}{2}(b_1 + b_2)h$ — The figure is a trapezoid.

$= \frac{1}{2}(16\text{ yd} + 10\text{ yd})(3\text{ yd})$ — Substitute $b_1 = 16$ yd, $b_2 = 10$ yd, and $h = 3$ yd.

$= \frac{1}{2}(26\text{ yd})(3\text{ yd})$

$= (13\text{ yd})(3\text{ yd})$

$= 39\text{ yd}^2$ — The area is 39 yd^2.

Tip: Notice that several of the formulas presented thus far involve multiple operations. The order in which we perform the arithmetic is called the order of operations and is covered in detail in Section 1.2. For now, we will follow these guidelines in the order given below:

1. Perform operations within parentheses first.
2. Evaluate expressions with exponents.
3. Perform multiplication or division in order from left to right.
4. Perform addition or subtraction in order from left to right.

example 3 **Finding Area of a Circle**

Find the area of a circular fountain if the diameter is 50 ft. Use 3.14 for π.

Solution:

We are given the diameter of the circle. To find the radius, take half of the diameter, $\frac{1}{2}(50) = 25$. Therefore, the radius is 25 ft.

$A = \pi r^2$	
$= (3.14)(25 \text{ ft})^2$	Substitute 3.14 for π and $r = 25$ ft.
$= (3.14)(625 \text{ ft}^2)$	Evaluate $(25 \text{ ft})^2 = 625 \text{ ft}^2$.
$= 1962.5 \text{ ft}^2$	The area of the fountain is 1962.5 ft^2.

3. Volume

The **volume** of a solid is the number of cubic units that can be enclosed within a solid. The solid shown in Figure R-7 contains 18 cubic inches (18 in.^{3}). In applications, volume might refer to the amount of water in a swimming pool.

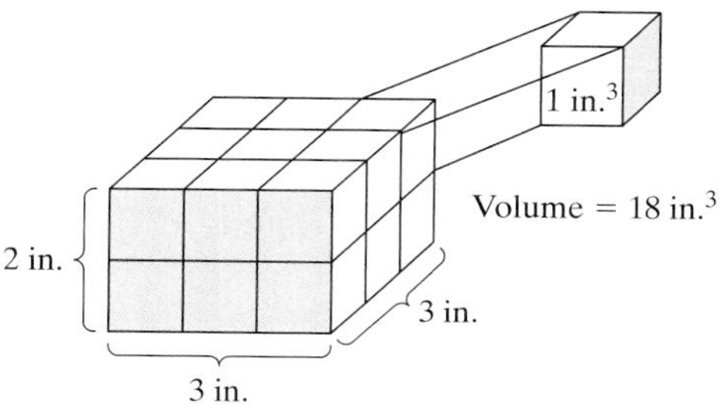

Figure R-7

The formulas used to compute the volume of several common solids are given here:

Rectangular Solid	**Cube**	**Right Circular Cylinder**
$V = \ell wh$	$V = s^3$	$V = \pi r^2 h$

Tip: Notice that the volume formulas for the three figures above are given by the product of the area of the base and the height of the figure:

$V = \ell wh$	$V = s \cdot s \cdot s$	$V = \pi r^2 h$
Area of Rectangular Base	Area of Square Base	Area of Circular Base

Right Circular Cone **Sphere**

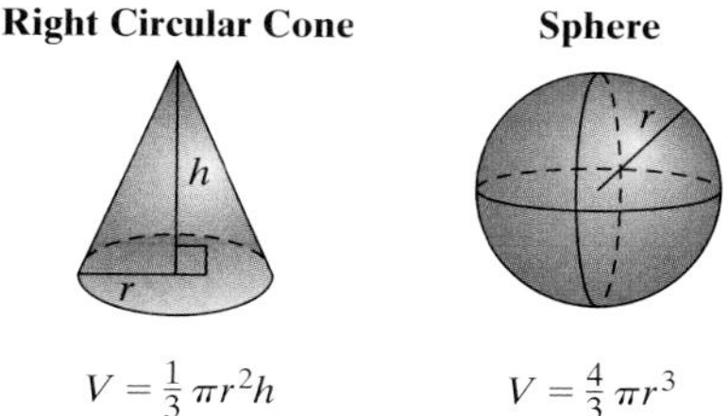

$V = \frac{1}{3}\pi r^2 h$ $V = \frac{4}{3}\pi r^3$

example 4 **Finding Volume**

Find the volume of each object.

a. 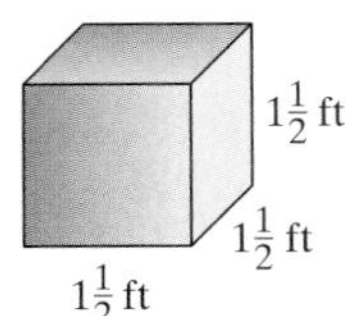

b. 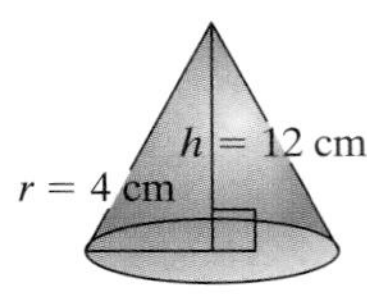

Solution:

a. $V = s^3$ — The object is a cube.

$= (1\frac{1}{2}\text{ ft})^3$ — Substitute $s = 1\frac{1}{2}$ ft.

$= \left(\frac{3}{2}\text{ ft}\right)^3$

$= \left(\frac{3}{2}\text{ ft}\right)\left(\frac{3}{2}\text{ ft}\right)\left(\frac{3}{2}\text{ ft}\right)$

$= \frac{27}{8}\text{ ft}^3$ or $3\frac{3}{8}\text{ ft}^3$

Tip: Notice that the units of volume are cubic units such as cubic inches (in.^{3}), cubic feet (ft^3), cubic yards (yd^3), cubic centimeters (cm^3), and so on.

b. $V = \frac{1}{3}\pi r^2 h$ — The object is a right circular cone.

$= \frac{1}{3}(3.14)(4\text{ cm})^2(12\text{ cm})$ — Substitute 3.14 for π, $r = 4$ cm, and $h = 12$ cm.

$= \frac{1}{3}(3.14)(16\text{ cm}^2)(12\text{ cm})$ — Evaluate $(4\text{ cm})^2 = 16\text{ cm}^2$.

$= 200.96\text{ cm}^3$

example 5 **Finding Volume in an Application**

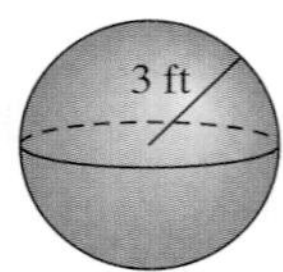

An underground gas tank is in the shape of a sphere with radius 3 ft.

a. Find the volume of the tank. Use 3.14 for π.
b. Find the cost to fill the tank with gasoline if gasoline costs \$9/ft^3.

Solution:

a. $V = \frac{4}{3}\pi r^3$

$= \frac{4}{3}(3.14)(3\text{ ft})^3$ — Substitute 3.14 for π and $r = 3$ ft.

$= \frac{4}{3}(3.14)(27\text{ ft}^3)$ — Evaluate $(3\text{ ft})^3 = 27\text{ ft}^3$.

$= 113.04\text{ ft}^3$ — The tank holds 113.04 ft^3 of gasoline.

b. Cost $= (\$9/\text{ft}^3)(113.04\ \text{ft}^3)$

$= \$1017.36$ It will cost \$1017.36 to fill the tank.

4. Angles

Applications involving angles and their measure come up often in the study of algebra, trigonometry, calculus, and applied sciences. The most common unit used to measure an angle is the degree (°). Several examples of angles and their corresponding degree measure are shown in Figure R-8.

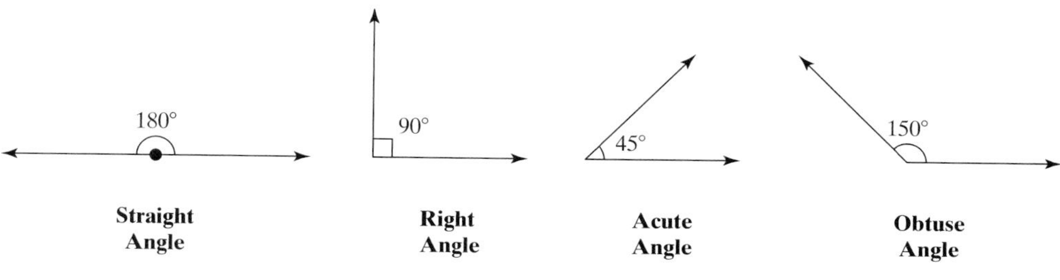

Figure R-8

- An angle that measures 180° is called a **straight angle**.
- An angle that measures 90° is a **right angle** (right angles are often marked at the vertex with a square or corner symbol, □).
- An angle that measures between 0° and 90° is called an **acute angle**.
- An angle that measures between 90° and 180° is called an **obtuse angle**.
- Two angles with the same measure are **equal angles** (or **congruent angles**).

The measure of an angle will be denoted by the symbol m written in front of the angle. Therefore, the measure of $\angle A$ is denoted $m(\angle A)$.

- Two nonnegative angles are said to be **complementary** if their sum is 90°.
- Two nonnegative angles are said to be **supplementary** if their sum is 180°.

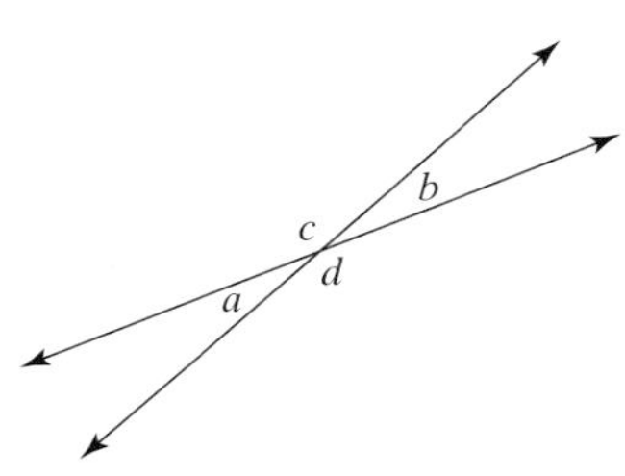

Figure R-9

When two lines intersect, four angles are formed. In Figure R-9, $\angle a$ and $\angle b$ are said to be a pair of **vertical angles**. Another set of vertical angles is the pair $\angle c$ and $\angle d$. An important property of vertical angles is that the measures of two vertical angles are *equal*. In the figure $m(\angle a) = m(\angle b)$ and $m(\angle c) = m(\angle d)$.

Parallel lines are lines that lie in the same plane and do not intersect. In Figure R-10, the lines L_1 and L_2 are parallel lines. If a line intersects two parallel lines, the line is called a **transversal**. In Figure R-10, the line m is a transversal and forms eight angles with the parallel lines L_1 and L_2.

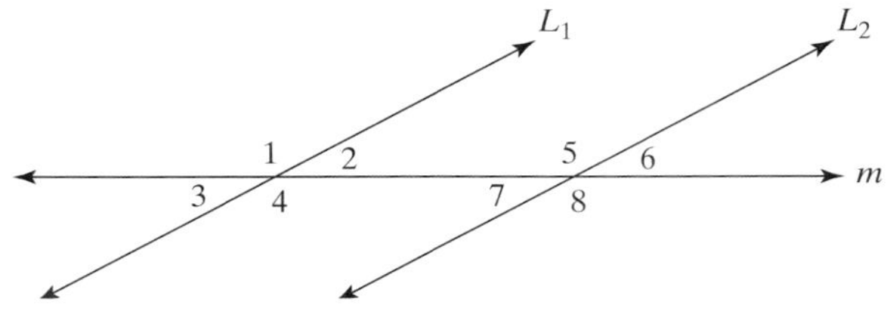

Figure R-10

The measures of angles 1–8 in Figure R-10 have the following special properties.

L_1 and L_2 are Parallel. Line m is a Transversal.	**Name of Angles**	**Property**
	The following pairs of angles are called **alternate interior angles**: $\angle 2$ and $\angle 7$ $\angle 4$ and $\angle 5$	Alternate interior angles are equal in measure. $m(\angle 2) = m(\angle 7)$ $m(\angle 4) = m(\angle 5)$
	The following pairs of angles are called **alternate exterior angles**: $\angle 1$ and $\angle 8$ $\angle 3$ and $\angle 6$	Alternate exterior angles are equal in measure. $m(\angle 1) = m(\angle 8)$ $m(\angle 3) = m(\angle 6)$
	The following pairs of angles are called **corresponding angles**: $\angle 1$ and $\angle 5$ $\angle 2$ and $\angle 6$ $\angle 3$ and $\angle 7$ $\angle 4$ and $\angle 8$	Corresponding angles are equal in measure. $m(\angle 1) = m(\angle 5)$ $m(\angle 2) = m(\angle 6)$ $m(\angle 3) = m(\angle 7)$ $m(\angle 4) = m(\angle 8)$

example 6 Finding Unknown Angles in a Diagram

Find the measure of each angle and explain how the angle is related to the given angle of 70°.

a. $\angle a$
b. $\angle b$
c. $\angle c$
d. $\angle d$

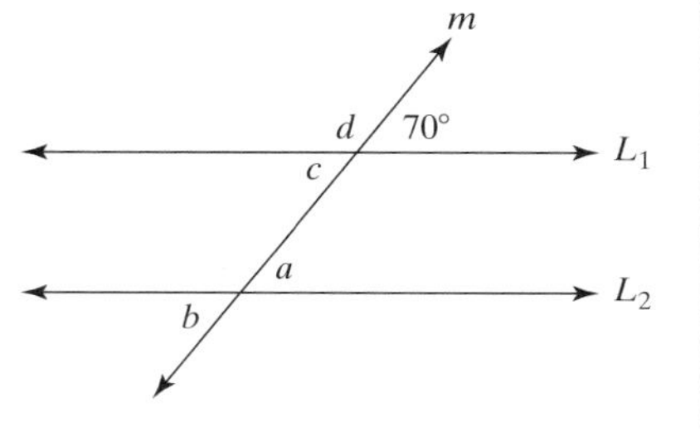

Solution:

a. $m(\angle a) = 70°$ — $\angle a$ is a corresponding angle to the given angle of 70°.

b. $m(\angle b) = 70°$ — $\angle b$ and the given angle of 70° are alternate exterior angles.

c. $m(\angle c) = 70°$ — $\angle c$ and the given angle of 70° are vertical angles.

d. $m(\angle d) = 110°$ — $\angle d$ is the supplement of the given angle of 70°.

5. Triangles

Triangles are categorized by the measures of the angles (Figure R-11) and by the number of equal sides or angles (Figure R-12).

- An **acute triangle** is a triangle in which all three angles are acute.
- A **right triangle** is a triangle in which one angle is a right angle.
- An **obtuse triangle** is a triangle in which one angle is obtuse.

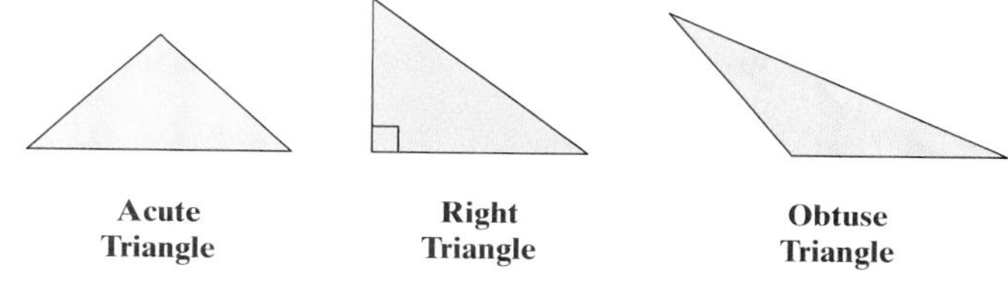

Figure R-11

- An **equilateral triangle** is a triangle in which all three sides (and all three angles) are equal.
- An **isosceles triangle** is a triangle in which two sides are equal (the angles opposite the equal sides are also equal).
- A **scalene triangle** is a triangle in which no sides (or angles) are equal.

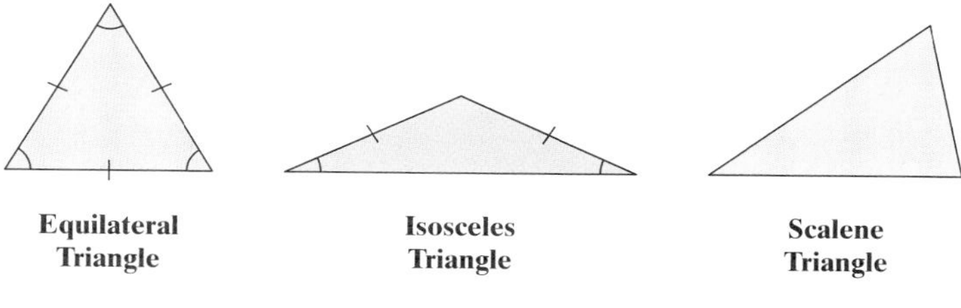

Figure R-12

The following important property is true for all triangles.

Sum of the Angles in a Triangle

The sum of the measures of the angles of a triangle is 180°.

example 7 Finding Unknown Angles in a Diagram

Find the measure of each angle in the figure.

a. $\angle a$
b. $\angle b$
c. $\angle c$
d. $\angle d$
e. $\angle e$

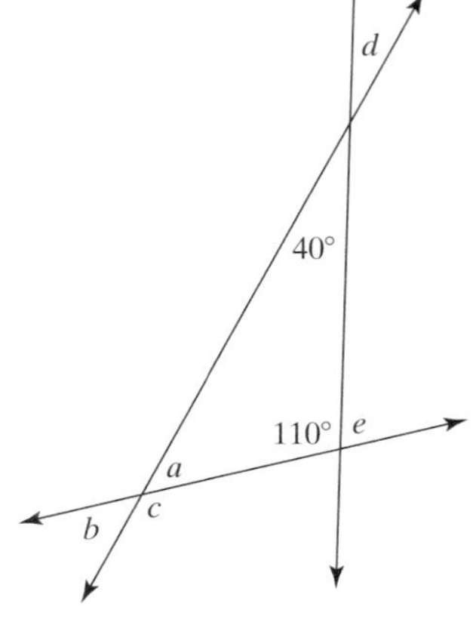

Solution:

a. $m(\angle a) = 30°$ The sum of the angles in a triangle is 180°.

b. $m(\angle b) = 30°$ $\angle a$ and $\angle b$ are vertical angles and are equal.

c. $m(\angle c) = 150°$ $\angle c$ and $\angle a$ are supplementary angles ($\angle c$ and $\angle b$ are also supplementary).

d. $m(\angle d) = 40°$ $\angle d$ and the given angle of 40° are vertical angles.

e. $m(\angle e) = 70°$ $\angle e$ and the given angle of 110° are supplementary angles.

section R.2 Practice Exercises

1. Identify which of the following units could be measures of perimeter.
   a. Square inches (in.^{2})
   b. Meters (m)
   c. Cubic feet (ft^3)
   d. Cubic meters (m^3)
   e. Miles (mi)
   f. Square centimeters (cm^2)
   g. Square yards (yd^2)
   h. Cubic inches (in.^{3})
   i. Kilometers (km)

For Exercises 2–5, find the perimeter and area of each figure.

2. 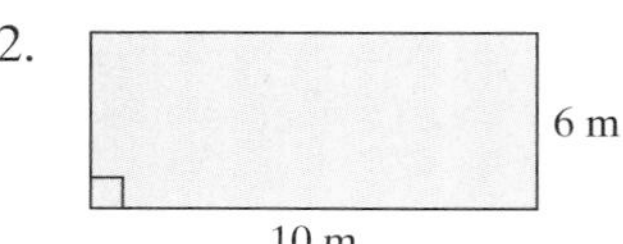

3. 

4. 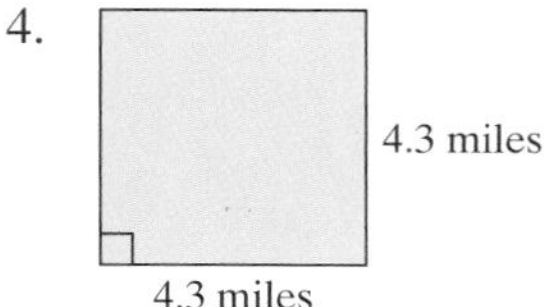

5. 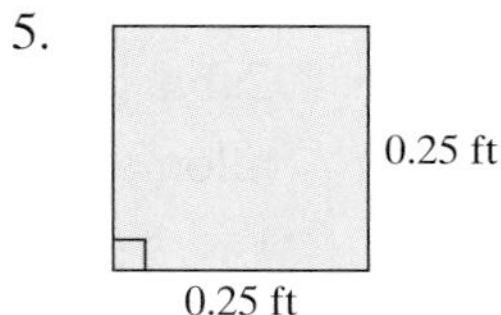

6. Identify which of the following units could be measures of circumference.
   a. Square inches ($in.^2$)
   b. Meters (m)
   c. Cubic feet (ft^3)
   d. Cubic meters (m^3)
   e. Miles (mi)
   f. Square centimeters (cm^2)
   g. Square yards (yd^2)
   h. Cubic inches ($in.^3$)
   i. Kilometers (km)

For Exercises 7–10, find the perimeter or circumference. Use 3.14 for π.

7. 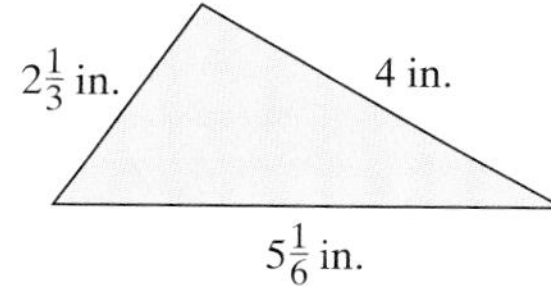

8. 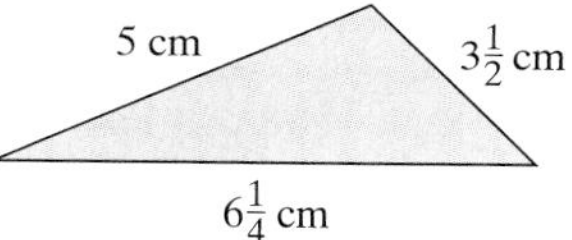

9. 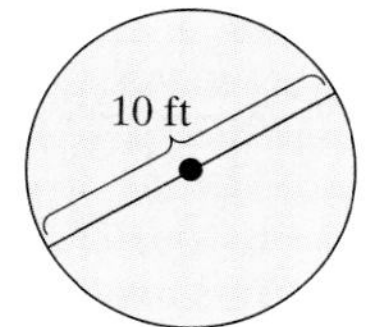

10. 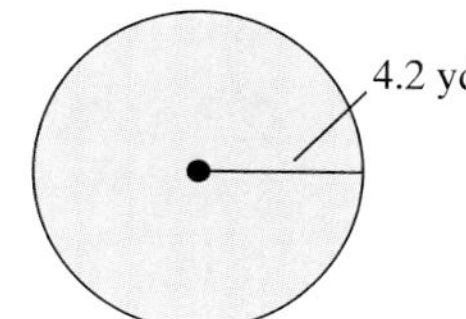

11. Identify which of the following units could be measures of area.
   a. Square inches ($in.^2$)
   b. Meters (m)
   c. Cubic feet (ft^3)
   d. Cubic meters (m^3)
   e. Miles (mi)
   f. Square centimeters (cm^2)
   g. Square yards (yd^2)
   h. Cubic inches ($in.^3$)
   i. Kilometers (km)

For Exercises 12–21, find the area. Use 3.14 for π.

12. 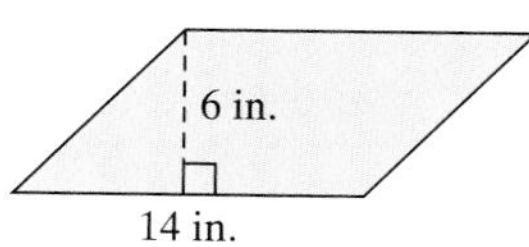

13. 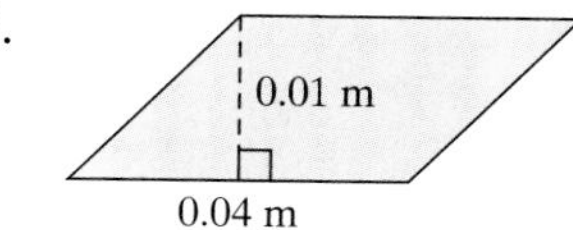

14. 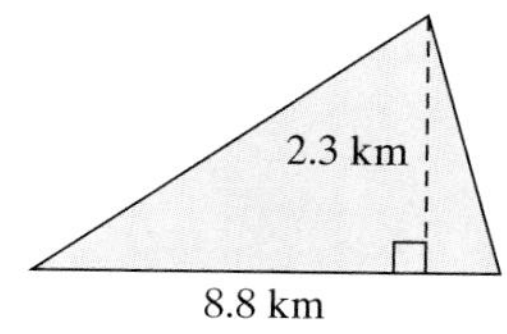

15. 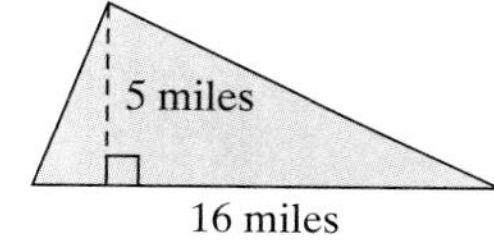

16. 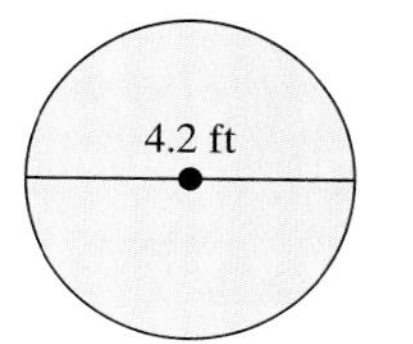

17. 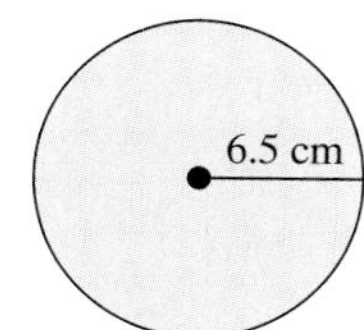

18. 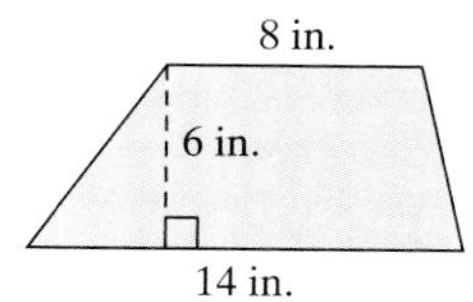

19. 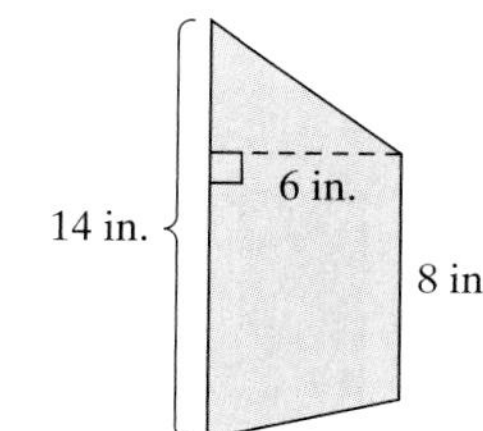

20. 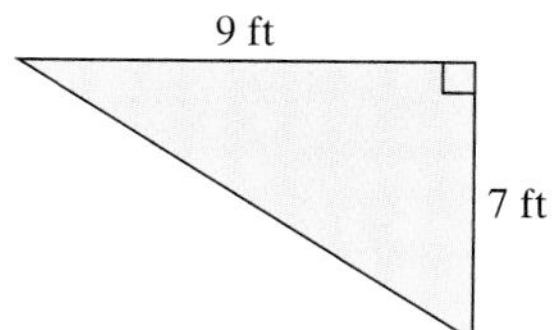

21. 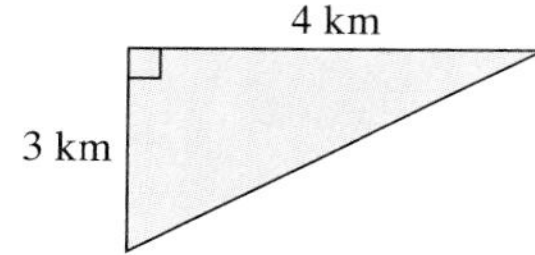

22. Identify which of the following units could be measures of volume.
    a. Square inches ($in.^2$)
    b. Meters (m)
    c. Cubic feet (ft^3)
    d. Cubic meters (m^3)
    e. Miles (mi)
    f. Square centimeters (cm^2)
    g. Square yards (yd^2)
    h. Cubic inches ($in.^3$)
    i. Kilometers (km)

For Exercises 23–26, find the volume of each figure. Use 3.14 for π.

23. 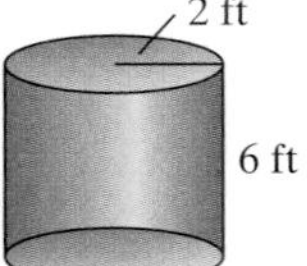

24. 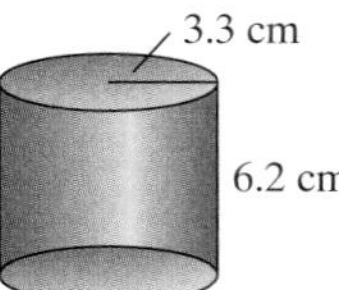

25. 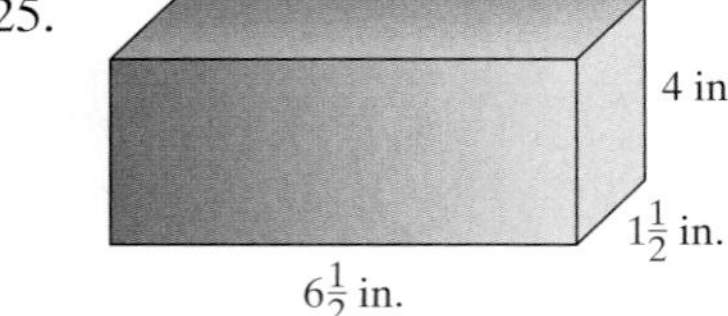

26. 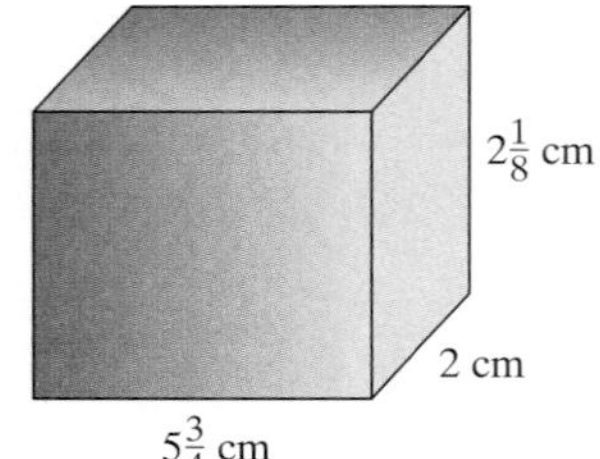

27. Find the volume of a spherical balloon whose radius is 9 in. Use 3.14 for π.

28. Find the volume of a spherical ball whose radius is 3 in. Use 3.14 for π.

29. Find the volume of a snow cone in the shape of a right circular cone whose radius is 3 cm and whose height is 12 cm. Use 3.14 for π.

30. Find the volume of a pile of gravel in the shape of a right circular cone whose radius is 10 yd and whose height is 18 yd. Use 3.14 for π.

31. Find the volume of a cube that is 3.2 ft on a side.

32. Find the volume of a cube that is 10.5 cm on a side.

For Exercises 33–34, find the perimeter.

33. 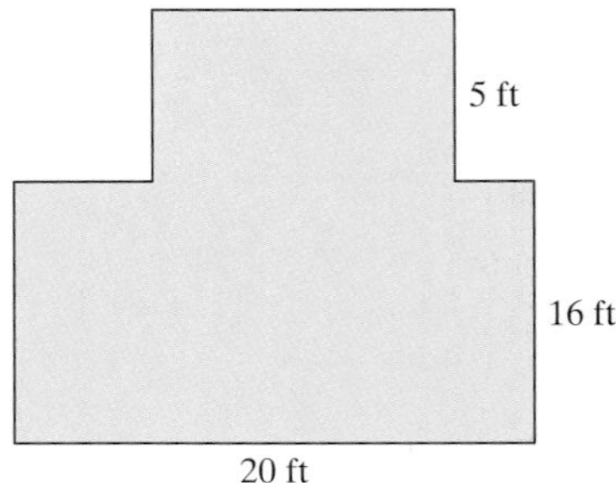

34. 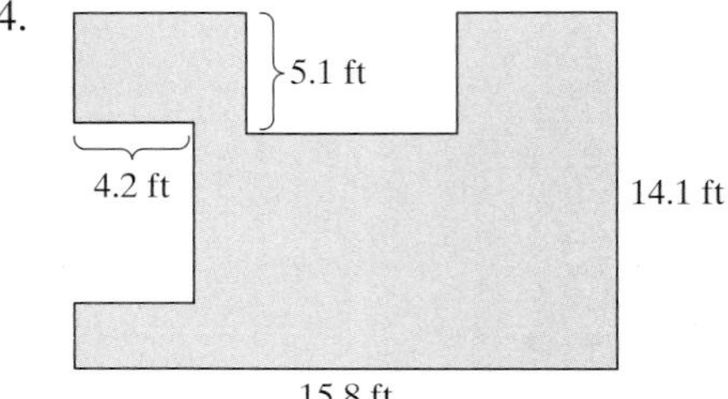

For Exercises 35–38, find the area of the shaded region. Use 3.14 for π.

35. 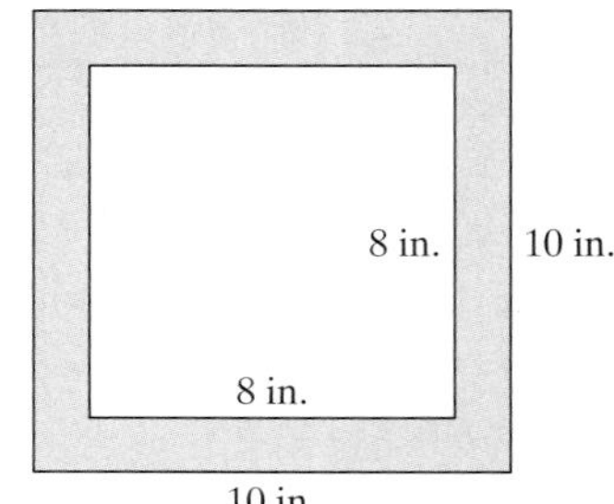

36. 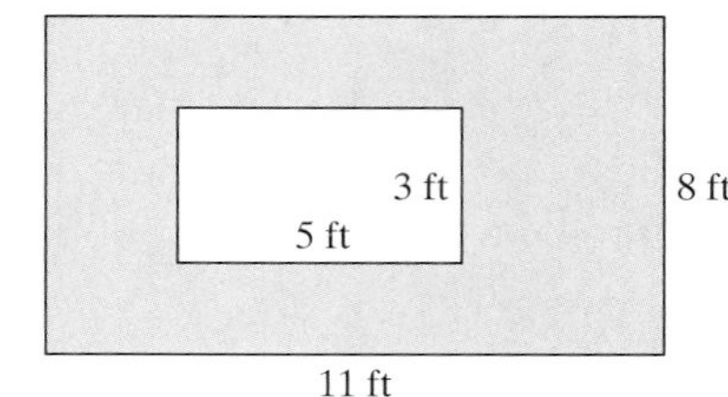

37.

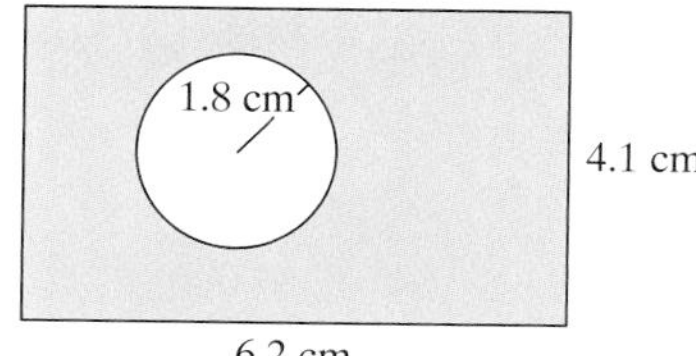

38.

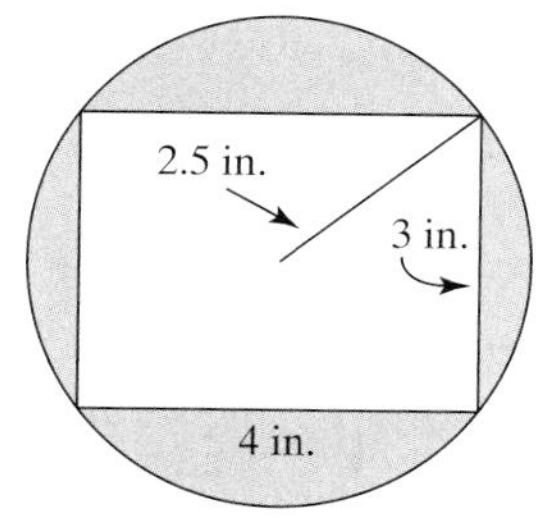

39. A wall measuring 20 ft by 8 ft can be painted for $50.

   a. What is the price per square foot? Round to the nearest cent.

   b. At this rate, how much would it cost to paint the remaining three walls that measure 20 ft by 8 ft, 16 ft by 8 ft, and 16 ft by 8 ft, respectively? Round to the nearest dollar.

40. Suppose it costs $320 to carpet a 16 ft by 12 ft room.

   a. What is the price per square foot? Round to the nearest cent.

   b. At this rate, how much would it cost to carpet a room that is 20 ft by 32 ft?

41. If you were to purchase fencing for a garden, would you measure the perimeter or area of the garden?

42. If you were to purchase sod (grass) for your front yard, would you measure the perimeter or area of the yard?

43. a. Find the area of a circular pizza that is 8 in. in diameter (the radius is 4 in.). Use 3.14 for π.

   b. Find the area of a circular pizza that is 12 in. in diameter (the radius is 6 in.).

   c. Assume that the 8-in. and 12-in. pizzas have the same thickness. Which would provide more pizza, two 8-in. pizzas or one 12-in. pizza?

44. Find the area of a circular stained glass window that is 16 in. in diameter. Use 3.14 for π.

45. Find the volume of a soup can in the shape of a right circular cylinder if its radius is 3.2 cm and its height is 9 cm. Use 3.14 for π.

46. Find the volume of a coffee mug in the shape of a right circular cylinder whose radius is 2.5 in. and whose height is 6 in. Use 3.14 for π.

For Exercises 47–54, answer True or False. If an answer is false, explain why.

47. The sum of the measures of two right angles equals the measure of a straight angle.

48. Two right angles are complementary.

49. Two right angles are supplementary.

50. Two acute angles cannot be supplementary.

51. Two obtuse angles cannot be supplementary.

52. An obtuse angle and an acute angle can be supplementary.

53. If a triangle is equilateral, then it is not scalene.

54. If a triangle is isosceles, then it is also scalene.

55. What angle is its own complement?

56. What angle is its own supplement?

57. If possible find two acute angles that are supplementary.

58. If possible find two acute angles that are complementary. Answers may vary.

59. If possible find an obtuse angle and an acute angle that are supplementary. Answers may vary.

60. If possible find two obtuse angles that are supplementary.

61. Refer to the figure at the top of page 28.

   a. State all pairs of vertical angles.

   b. State all pairs of supplementary angles.

c. If the measure of $\angle 4$ is 80°, find the measures of $\angle 1$, $\angle 2$, and $\angle 3$.

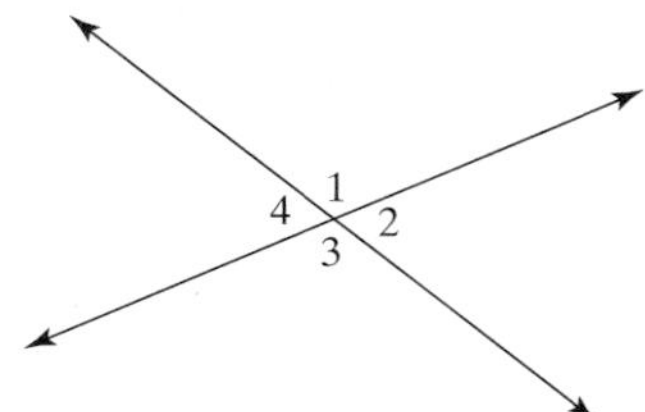

Figure for Exercise 61

62. Refer to the figure.
   a. State all pairs of vertical angles.
   b. State all pairs of supplementary angles.
   c. If the measure of $\angle a$ is 25°, find the measures of $\angle b$, $\angle c$, and $\angle d$.

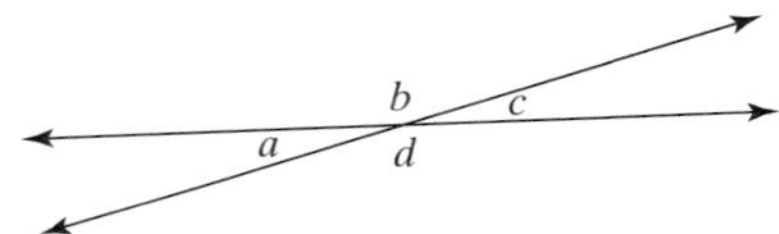

Figure for Exercise 62

For Exercises 63–70, find the complement of each angle.

63. 33°	64. 87°	65. 12°	66. 45°
67. 30°	68. 20°	69. 70°	70. 60°

For Exercises 71–78, find the supplement of each angle.

71. 33°	72. 87°	73. 122°	74. 90°
75. 45°	76. 150°	77. 135°	78. 30°

For Exercises 79–86, refer to the figure. Assume that L_1 and L_2 are parallel lines cut by the transversal, n.

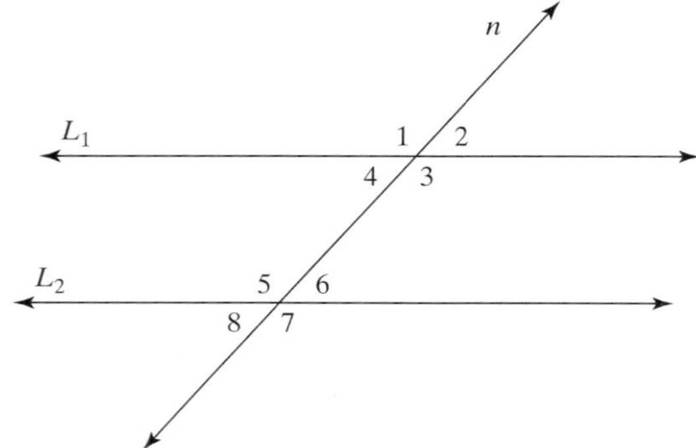

Figure for Exercises 79–86

79. $m(\angle 5) = m(\angle \_\_\_\_)$ Reason: Vertical angles are equal.

80. $m(\angle 5) = m(\angle \_\_\_\_)$ Reason: Alternate interior angles are equal.

81. $m(\angle 5) = m(\angle \_\_\_\_)$ Reason: Corresponding angles are equal.

82. $m(\angle 7) = m(\angle \_\_\_\_)$ Reason: Corresponding angles are equal.

83. $m(\angle 7) = m(\angle \_\_\_\_)$ Reason: Alternate exterior angles are equal.

84. $m(\angle 7) = m(\angle \_\_\_\_)$ Reason: Vertical angles are equal.

85. $m(\angle 3) = m(\angle \_\_\_\_)$ Reason: Alternate interior angles are equal.

86. $m(\angle 3) = m(\angle \_\_\_\_)$ Reason: Vertical angles are equal.

87. Find the measure of angles a–g in the figure. Assume that L_1 and L_2 are parallel and that n is a transversal.

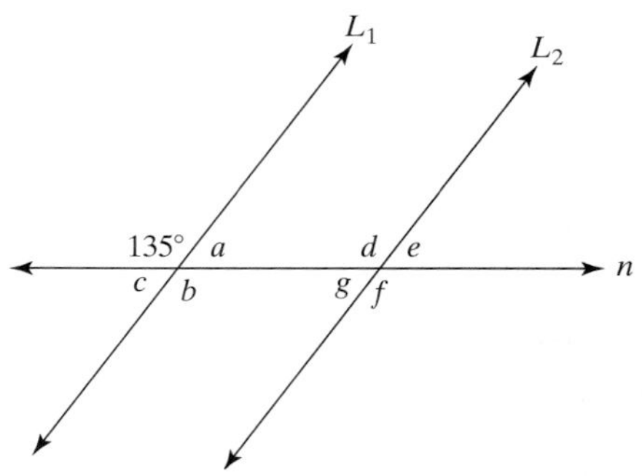

Figure for Exercise 87

88. Find the measure of angles a–g in the figure. Assume that L_1 and L_2 are parallel and that n is a transversal.

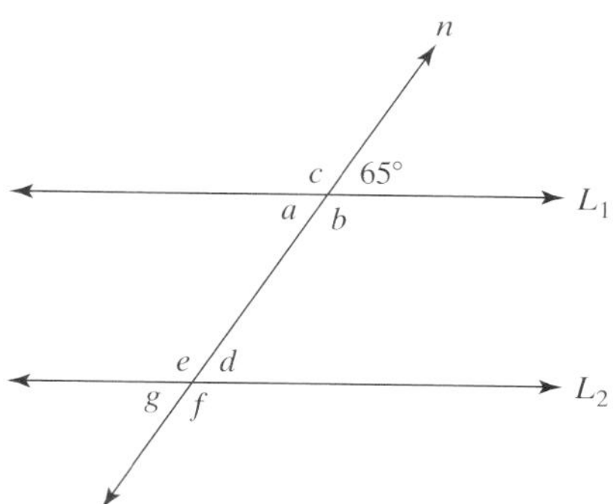

Figure for Exercise 88

For Exercises 89–92, identify the triangle as equilateral, isosceles, or scalene.

89.

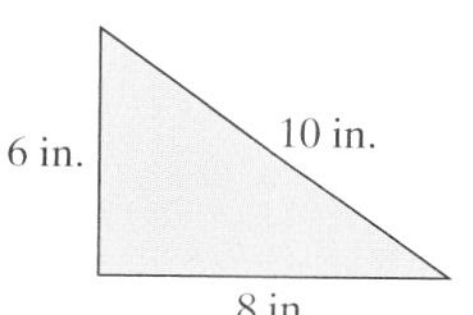

90.

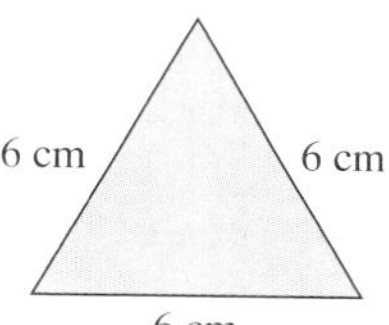

91.

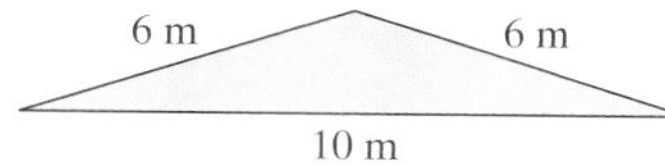

92.

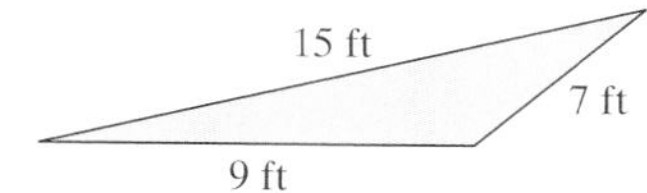

93. Can a triangle be both a right triangle and an obtuse triangle? Explain.

94. Can a triangle be both a right triangle and an isosceles triangle? Explain.

For Exercises 95–98, find the missing angles.

95.

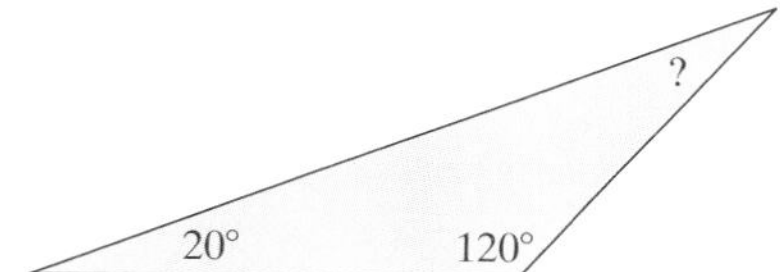

96.

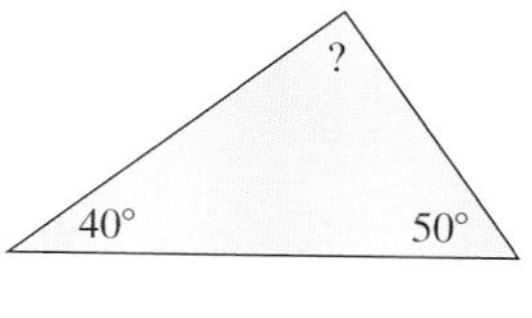

97.

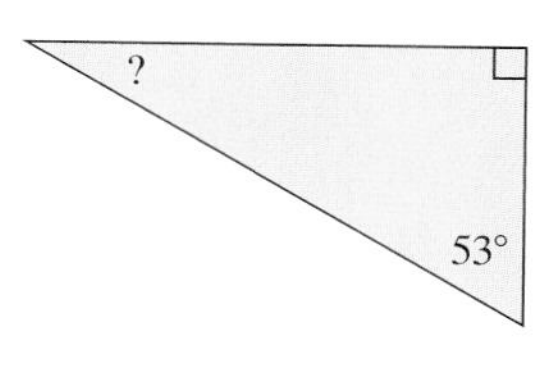

98.

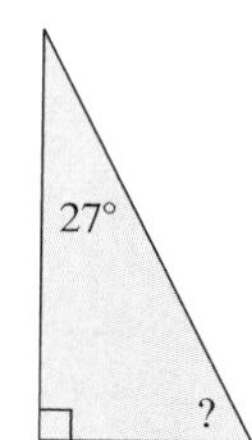

99. Refer to the figure. Find the measure of angles a–j.

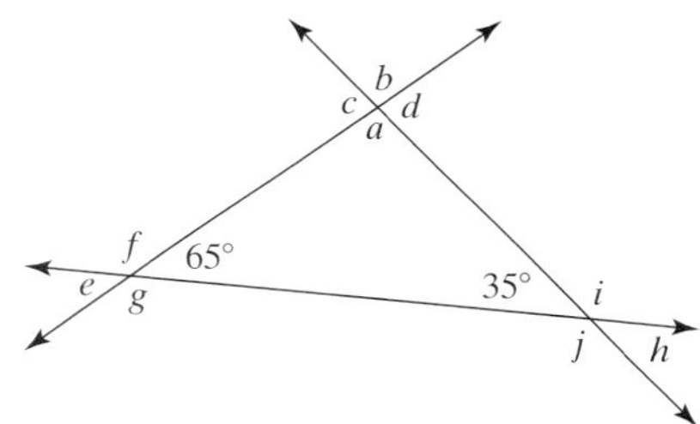

Figure for Exercise 99

100. Refer to the figure. Find the measure of angles a–j.

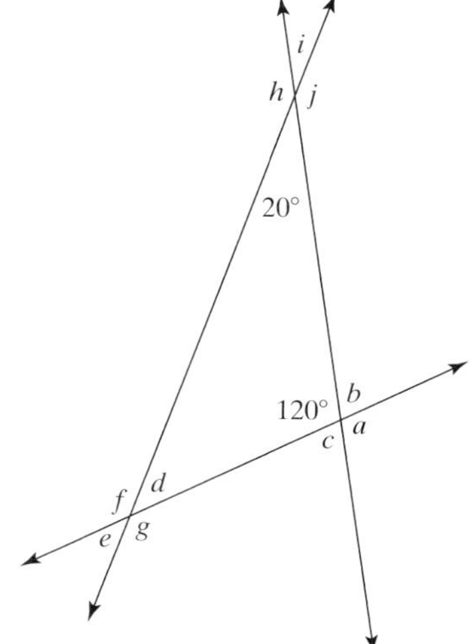

Figure for Exercise 100

101. Refer to the figure. Find the measure of angles a–k. Assume that L_1 and L_2 are parallel.

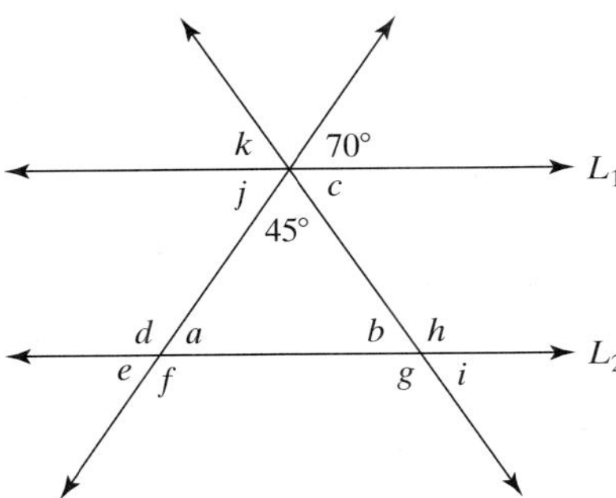

Figure for Exercise 101

102. Refer to the figure. Find the measure of angles a–k. Assume that L_1 and L_2 are parallel.

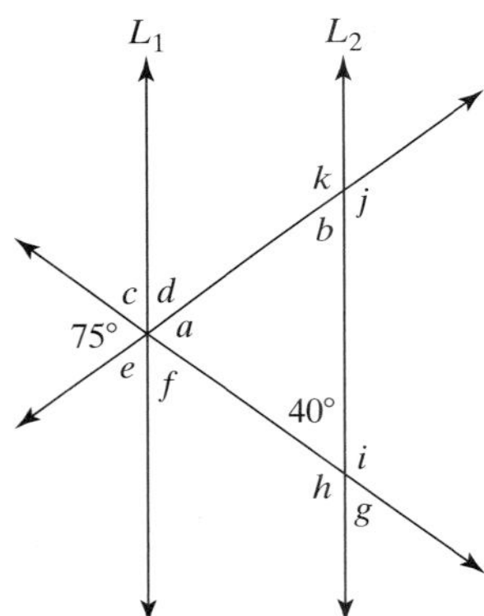

Figure for Exercise 102

Key Terms:

acute angle
acute triangle
alternate exterior angles
alternate interior angles
area of a circle
area of a parallelogram
area of a rectangle
area of a square
area of a trapezoid
area of a triangle
circumference
complementary angles
congruent angles
corresponding angles
equal angles
equilateral triangle
isosceles triangle
obtuse angle
obtuse triangle
parallel lines
perimeter of a polygon
perimeter of a rectangle
perimeter of a square
right angle
right triangle
scalene triangle
straight angle
supplementary angles
transversal
vertical angles
volume of a cube
volume of a rectangular solid
volume of a right circular cone
volume of a right circular cylinder
volume of a sphere

---

section

R.3 Study Tips

Concepts

1. **Before the Course**
2. **During the Course**
3. **Preparation for Exams**
4. **Where to Go for Help**

In taking a course in algebra, you are making a commitment to yourself, your instructor, and your classmates. Following some or all of the study tips discussed here can help you be successful in this endeavor. The features of this text that will assist you are printed in blue.

1. Before the Course

- Purchase the necessary materials for the course before the course begins or on the first day.

- Obtain a three-ring binder to keep and organize your notes, homework, tests, and any other materials acquired in the class. We call this type of notebook a portfolio.
- Arrange your schedule so that you have enough time to attend class and to do homework. A common rule of thumb is to set aside at least 2 hours for homework for every hour spent in class. That is, if you are taking a 4-credit-hour course, plan on at least 8 hours a week for homework. If you experience difficulty in mathematics, plan for more time. A 4-credit-hour course will then take *at least* 12 hours each week—about the same as a part-time job.
- Communicate with your employer and family members the importance of your success in this course so that they can support you.
- Be sure to find out the type of calculator (if any) that your instructor requires.

2. During the Course

- Read the section in the text *before* the lecture to familiarize yourself with the material and terminology.
- Attend every class and be on time.
- Take notes in class. Write down all of the examples that the instructor presents. Read the notes after class and add any comments to make your notes clearer to you. Use a tape recorder to record the lecture if the instructor permits recording of lectures.
- Ask questions in class.
- Read the section in the text *after* the lecture and pay special attention to the Tip boxes and Avoiding Mistakes boxes.
- Do homework every night. Even if your class does not meet everyday, you should still do some work every night to keep the material fresh in your mind.
- Check your homework with the answers that are supplied in the back of this text. Correct the exercises that do not match and circle or star the ones that you cannot correct yourself. This way you can easily find them and ask your instructor the next day.
- Write the definition and give an example of each Key Term usually found at the end of each Summary section at the end of the chapter.
- Form a study group with fellow students in your class, and exchange phone numbers. You will be surprised by how much you can learn by talking about mathematics with other students.
- If you use a calculator in your class, read the Calculator Connections boxes to learn how and when to use your calculator.

3. Preparation for Exams

- Look over your homework. Pay special attention to the exercises you have circled or starred to be sure that you have learned that concept.
- Read through the Summary at the end of the chapter. Be sure that you understand each concept and example. If not, go to the section in the text and reread that section.
- Give yourself enough time to take the Chapter Test uninterrupted. Then check the answers. For each problem you answered incorrectly, go to the Review Exercises and do all of the problems that are similar.

- To prepare for the final exam, complete the Cumulative Review Exercises at the end of each chapter, starting with Chapter 2. If you complete the cumulative reviews after finishing each chapter, then you will be preparing for the final exam throughout the course. The Cumulative Review Exercises are another excellent tool for helping you retain material.

4. Where to Go for Help

- At the first sign of trouble, see your instructor. Most instructors have specific office hours set aside to help students. Don't wait until after you have failed an exam to seek assistance.
- Get a tutor. Most colleges and universities have free tutoring available.
- When your instructor and tutor are unavailable, use the Student Solutions Manual for step-by-step solutions to the odd-numbered problems in the exercise sets.
- Work with another student from your class.
- Work on the computer. Many mathematics tutorial programs and websites are available on the Internet, including the one that accompanies this text: www.mhhe.com/miller_oneill

chapter

1

The Set of Real Numbers

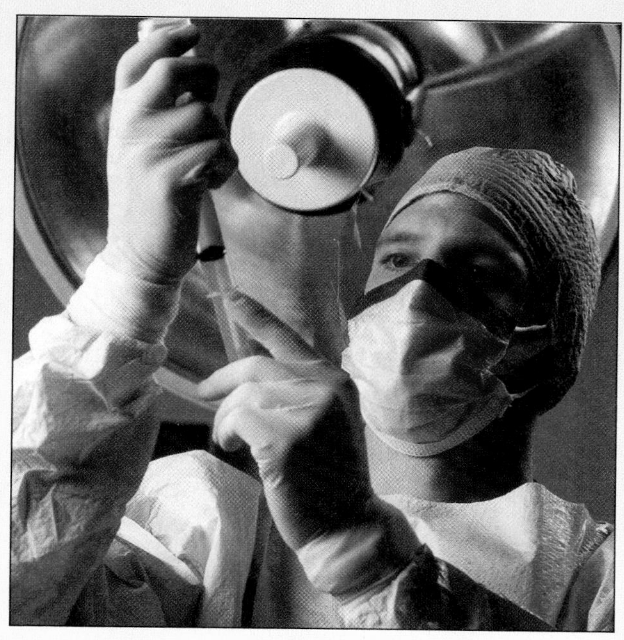

Many of the activities we perform every day are followed in a natural order. For example, we would not put on our shoes before pulling on our socks. Nor would a doctor begin surgery before giving an anesthetic.

In mathematics, it is also necessary to follow a prescribed order of operations to simplify an algebraic expression. For more information see the Technology Connections in MathZone at

www.mhhe.com/miller_oneill

For example, the expression $24 - 6 \times 2$ is properly simplified by performing multiplication before subtraction:

$$24 - 6 \times 2 = 24 - 12 = 12$$

When parentheses are present, the expressions within parentheses must be performed first:

$$(24 - 6) \times 2 = 18 \times 2 = 36$$

In this chapter, the order of operations for simplifying algebraic expressions is discussed at length.

section 1.1

Sets of Numbers and the Real Number Line

Concepts

1. Real Number Line
2. Plotting Points on the Number Line
3. Set of Real Numbers
4. Inequalities
5. Opposite of a Real Number
6. Absolute Value of a Real Number

1. Real Number Line

The numbers we work with on a day-to-day basis are all part of the set of **real numbers**. The real numbers encompass zero, all positive, and all negative numbers, including those represented by fractions and decimal numbers. The set of real numbers can be represented graphically on a horizontal number line with a point labeled as 0. Positive real numbers are graphed to the right of 0, and negative real numbers are graphed to the left. Zero is neither positive nor negative. Each point on the number line corresponds to exactly one real number. For this reason, this number line is called the **real number line** (Figure 1-1).

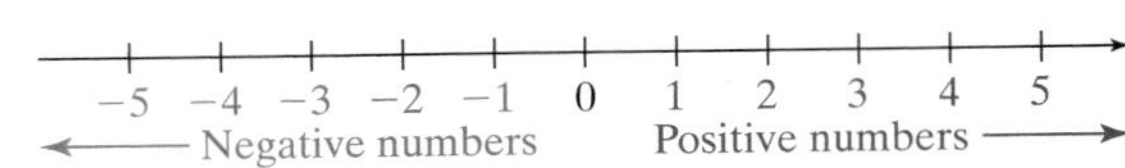

Figure 1-1

2. Plotting Points on the Number Line

example 1 **Plotting Points on the Real Number Line**

Plot the points on the real number line that represent the following real numbers.

a. -3 b. $\frac{3}{2}$ c. -4.7 d. $\frac{16}{5}$

Solution:

a. Because -3 is negative, it lies three units to the left of zero.
b. The fraction $\frac{3}{2}$ can be expressed as the mixed number $1\frac{1}{2}$ which lies half-way between 1 and 2 on the number line.
c. The negative number -4.7 lies $\frac{7}{10}$ units to the left of -4 on the number line.
d. The fraction $\frac{16}{5}$ can be expressed as the mixed number $3\frac{1}{5}$ which lies $\frac{1}{5}$ unit to the right of 3 on the number line.

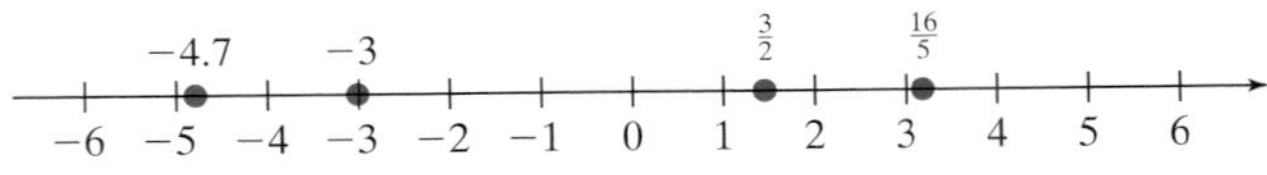

3. Set of Real Numbers

In mathematics, a well-defined collection of elements is called a set. The symbols { } are used to enclose the elements of the set. For example, the set {A, B, C, D, E} represents the set of the first five letters of the alphabet.

Several sets of numbers are used extensively in algebra that are subsets (or part) of the set of real numbers. These are the

set of natural numbers
set of whole numbers
set of integers
set of rational numbers
set of irrational numbers

Definition of the Natural Numbers, Whole Numbers, and Integers

The set of **natural numbers** is $\{1, 2, 3, \ldots\}$
The set of **whole numbers** is $\{0, 1, 2, 3, \ldots\}$
The set of **integers** is $\{\ldots -3, -2, -1, 0, 1, 2, 3, \ldots\}$

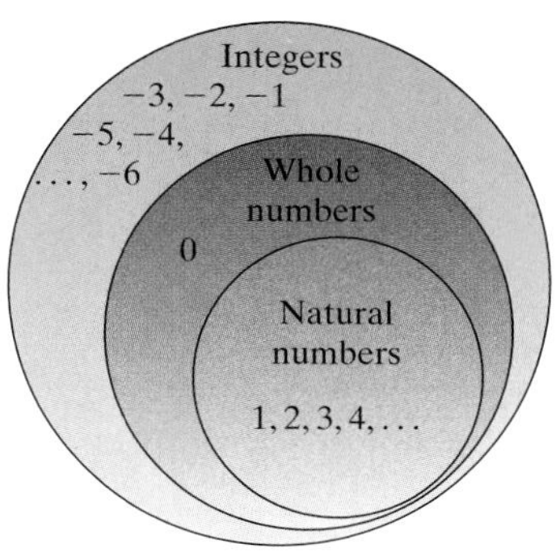

Figure 1-2

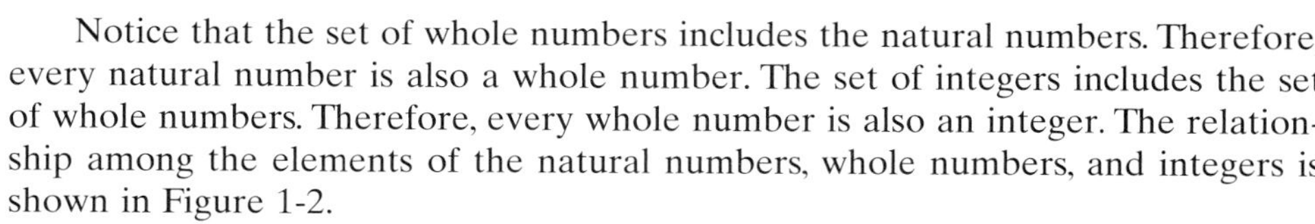

Notice that the set of whole numbers includes the natural numbers. Therefore, every natural number is also a whole number. The set of integers includes the set of whole numbers. Therefore, every whole number is also an integer. The relationship among the elements of the natural numbers, whole numbers, and integers is shown in Figure 1-2.

Fractions are also among the numbers we use frequently. A number that can be written as a fraction whose numerator is an integer and whose denominator is a nonzero integer is called a rational number.

Definition of the Rational Numbers

The set of **rational numbers** is the set of numbers that can be expressed in the form $\frac{p}{q}$, where both p and q are integers and q does not equal 0.

We also say that a rational number $\frac{p}{q}$ is a *ratio* of two integers, p and q, where q is not equal to zero.

example 2 **Identifying Rational Numbers**

Show that the following numbers are rational numbers by finding an equivalent ratio of two integers.

a. $\dfrac{-2}{3}$ b. -12 c. 0.5 d. $0.\overline{6}$

Solution:

a. The fraction $\frac{-2}{3}$ is a rational number because it can be expressed as the ratio of -2 and 3.
b. The number -12 is a rational number because it can be expressed as the ratio of -12 and 1. That is, $-12 = \frac{-12}{1}$. In this example, we see that an integer is also a rational number.

Tip: Any rational number can be represented by a terminating decimal or by a repeating decimal.

c. The terminating decimal 0.5 is a rational number because it can be expressed as the ratio of 5 and 10. That is, $0.5 = \frac{5}{10}$. In this example we see that a terminating decimal is also a rational number.

d. The repeating decimal $0.\overline{6}$ is a rational number because it can be expressed as the ratio of 2 and 3. That is, $0.\overline{6} = \frac{2}{3}$. In this example we see that a repeating decimal is also a rational number.

Some real numbers, such as the number π, cannot be represented by the ratio of two integers. In decimal form, an irrational number is a nonterminating, nonrepeating decimal. The value of π, for example, can be approximated as $\pi \approx 3.1415926535897932$. However, the decimal digits continue indefinitely with no repeated pattern. Other examples of irrational numbers are the square roots of nonperfect squares, such as $\sqrt{3}$ (read as "the positive square root of 3"). The expression $\sqrt{3}$ is a number that when multiplied by itself equals 3.

Definition of the Irrational Numbers

The set of **irrational numbers** is the set of real numbers that are not rational.

Note: An irrational number cannot be written as a terminating decimal or as a repeating decimal.

The set of real numbers consists of both the rational and the irrational numbers. The relationship among these important sets of numbers is illustrated in Figure 1-3:

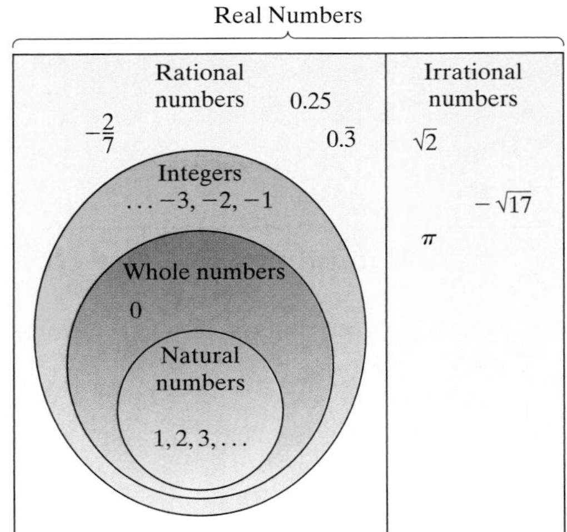

Figure 1-3

example 3

Classifying Numbers by Set

Check the set(s) to which each number belongs. The numbers may belong to more than one set.

	Natural Numbers	Whole Numbers	Integers	Rational Numbers	Irrational Numbers	Real Numbers
5						
$\frac{-47}{3}$						
1.48						
$\sqrt{7}$						
0						

Solution:

	Natural Numbers	Whole Numbers	Integers	Rational Numbers	Irrational Numbers	Real Numbers
5	✔	✔	✔	✔ (ratio of 5 and 1)		✔
$\frac{-47}{3}$				✔ (ratio of −47 and 3)		✔
1.48				✔ (ratio of 148 and 100)		✔
$\sqrt{7}$					✔	✔
0		✔	✔	✔ (ratio of 0 and 1)		✔

4. Inequalities

The relative size of two real numbers can be compared using the real number line. Suppose a and b represent two real numbers. We say that a is less than b, denoted $a < b$, if a lies to the left of b on the number line.

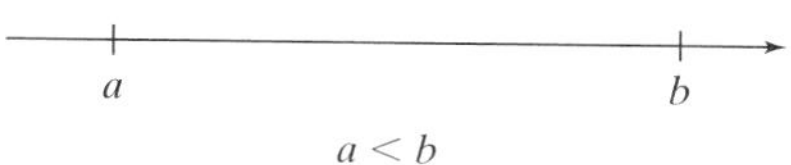

$a < b$

We say that a is greater than b, denoted $a > b$, if a lies to the right of b on the number line.

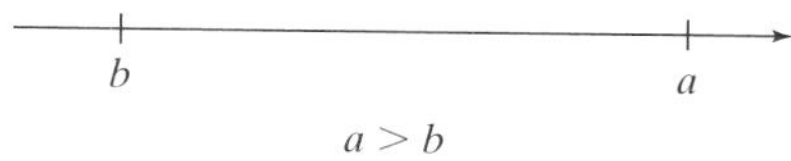

$a > b$

Table 1-1 summarizes the symbols that compare two real numbers a and b.

Table 1-1

Mathematical Expression	Translation	Example
$a < b$	a is less than b.	$2 < 3$
$a > b$	a is greater than b.	$5 > 1$
$a \leq b$	a is less than or equal to b.	$4 \leq 4$
$a \geq b$	a is greater than or equal to b.	$10 \geq 9$
$a = b$	a is equal to b.	$6 = 6$
$a \neq b$	a is not equal to b.	$7 \neq 0$
$a \approx b$	a is approximately equal to b.	$2.3 \approx 2$

The symbols $<$, $>$, $\leq$, $\geq$, and $\neq$ are called inequality signs, and the expressions $a < b$, $a > b$, $a \leq b$, $a \geq b$, and $a \neq b$ are called **inequalities**.

example 4 Ordering Real Numbers

The average temperatures (in degrees Celsius) for selected cities in the United States and Canada in January are shown in Table 1-2.

Table 1-2

City	Temp (°C)
Prince George, British Columbia	−12.1
Corpus Christi, Texas	13.4
Parkersburg, West Virginia	−0.9
San Jose, California	9.7
Juneau, Alaska	−5.7
New Bedford, Massachusetts	−0.2
Durham, North Carolina	4.2

a. Plot a point on the real number line representing the temperature of each city.
b. Then compare the temperatures between the following cities and fill in the blank with the appropriate inequality sign: $<$ or $>$.

Solution:

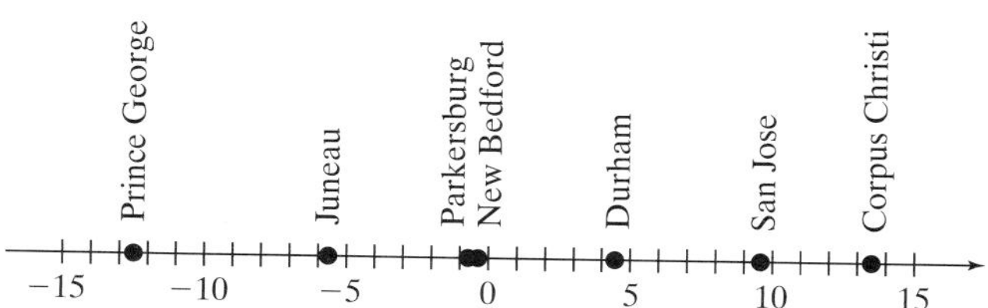

a. Temperature of San Jose $\boxed{<}$ temperature of Corpus Christi
b. Temperature of Juneau $\boxed{>}$ temperature of Prince George
c. Temperature of Parkersburg $\boxed{<}$ temperature of New Bedford
d. Temperature of Parkersburg $\boxed{>}$ temperature of Prince George

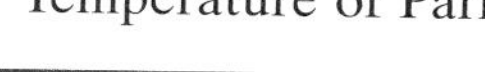

5. Opposite of a Real Number

To gain mastery of any algebraic skill, it is necessary to know the meaning of key definitions and key symbols. Two important definitions are the opposite of a real number and the absolute value of a real number.

Definition of the Opposite of a Real Number

Two numbers that are the same distance from 0 but on opposite sides of 0 on the number line are called **opposites** of each other. Symbolically, we denote the opposite of a real number a as $-a$.

example 5 Finding the Opposite of a Real Number

a. Find the opposite of 5.
b. Find the opposite of $-\frac{4}{7}$.
c. Evaluate $-(0.46)$.
d. Evaluate $-(-\frac{11}{3})$.

Solution:

a. The opposite of 5 is -5.

b. The opposite of $-\dfrac{4}{7}$ is $\dfrac{4}{7}$.

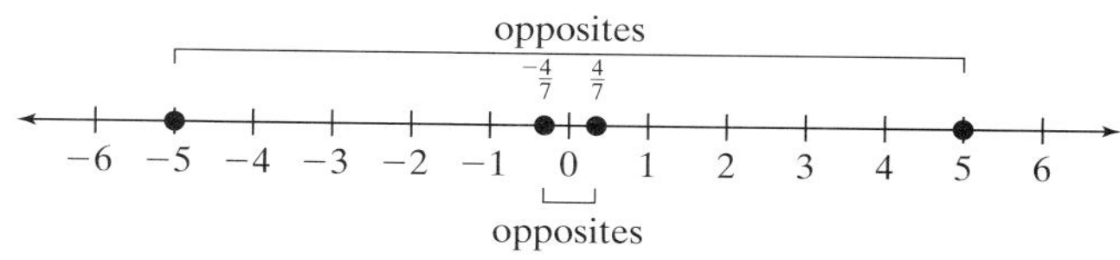

c. $-(0.46) = -0.46$ The expression $-(0.46)$ represents the opposite of 0.46.

d. $-\left(-\dfrac{11}{3}\right) = \dfrac{11}{3}$ The expression $-(-\frac{11}{3})$ represents the opposite of $-\frac{11}{3}$.

6. Absolute Value of a Real Number

The concept of absolute value will be used to define the addition of real numbers in Section 1.3.

Informal Definition of the Absolute Value of a Real Number

The **absolute value** of a real number a, denoted $|a|$, is the distance between a and 0 on the number line.

Note: The absolute value of any real number is nonnegative.

For example, $|3| = 3$ and $|-3| = 3$.

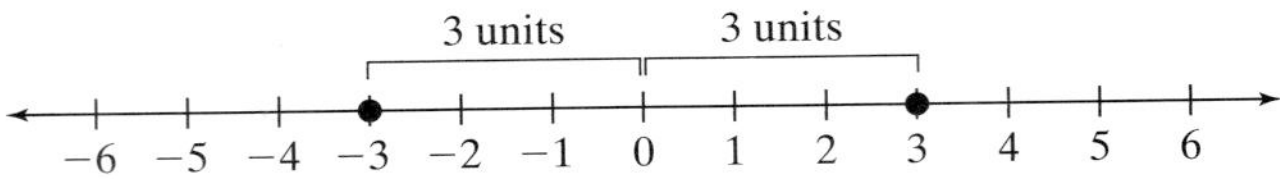

example 6 **Finding the Absolute Value of a Real Number**

Evaluate the absolute value expressions.

a. $|-4|$ b. $|\frac{1}{2}|$ c. $|-6.2|$ d. $|0|$

Solution:

a. $|-4| = 4$ -4 is 4 units from 0 on the number line.

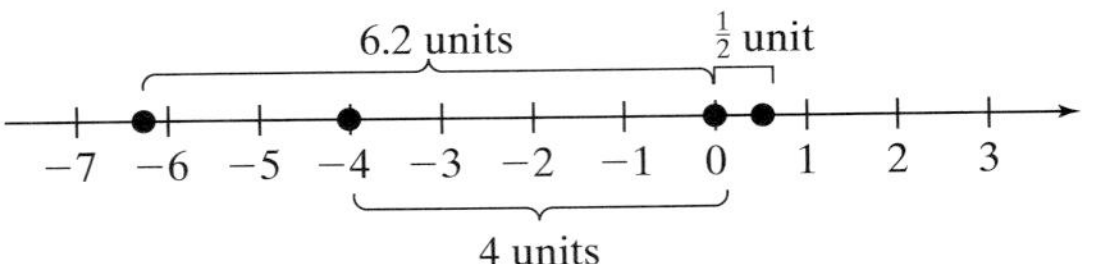

b. $|\frac{1}{2}| = \frac{1}{2}$ $\frac{1}{2}$ is $\frac{1}{2}$ unit from 0 on the number line.
c. $|-6.2| = 6.2$ -6.2 is 6.2 units from 0 on the number line.
d. $|0| = 0$ 0 is 0 units from 0 on the number line.

The absolute value of a number a is its distance from zero on the number line. The definition of $|a|$ may also be given symbolically depending on whether a is negative or nonnegative.

Definition of the Absolute Value of a Real Number

Let a be a real number. Then

1. If a is nonnegative (that is, $a \geq 0$), then $|a| = a$.
2. If a is negative (that is, $a < 0$), then $|a| = -a$.

This definition states that if a is a nonnegative number, then $|a|$ equals a itself. If a is a negative number, then $|a|$ equals the opposite of a. For example,

$|9| = 9$ Because 9 is positive, then $|9|$ equals the number 9 itself.

$|-7| = 7$ Because -7 is negative, then $|-7|$ equals the opposite of -7 which is 7.

example 7

Comparing Absolute Value Expressions

Determine if the statements are true or false.

a. $3 \leq 3$ b. $-|5| = |-5|$

Solution:

a. $3 \leq 3$ True. The symbol $\leq$ means "less then *or* equal to." Since 3 is equal to 3, the statement $3 \leq 3$ is true.

b. $-|5| = |-5|$ False. On the left-hand side, $-|5|$ is the opposite of $|5|$. Hence $-|5| = -5$. On the right-hand side, $|-5| = 5$. Therefore, the original statement simplifies to $-5 = 5$, which is false.

Calculator Connections

Scientific and graphing calculators approximate irrational numbers by using rational numbers in the form of terminating decimals. For example, consider approximating π and $\sqrt{3}$:

Scientific Calculator:

Enter: [π] (or [2nd] [π]) **Result:** 3.141592654

Enter: [3] [√] **Result:** 1.732050808

Graphing Calculator:

Enter: [2nd] [π] [ENTER]

Enter: [2nd] [√] [3] [ENTER]

π
3.141592654
√(3
1.732050808

Note that when writing approximations, we use the symbol, $\approx$.

$$\pi \approx 3.141592654$$
$$\sqrt{3} \approx 1.73205808$$

section 1.1 Practice Exercises

1. Plot the numbers on a real number line: $\{1, -2, -\pi, 0, -\frac{5}{2}, 5.1\}$

2. Plot the numbers on a real number line: $\{3, -4, \frac{1}{8}, -1.7, -\frac{4}{3}, 1.75\}$

For Exercises 3–22, describe each number as (a) a terminating decimal, (b) a repeating decimal, or (c) a nonterminating, nonrepeating decimal.

3. 0.29
4. 3.8
5. $\frac{1}{9}$
6. $\frac{1}{3}$
7. $\frac{1}{8}$
8. $\frac{1}{5}$
9. 5
10. 2
11. 2π
12. 3π
13. -0.125
14. -3.24
15. -3
16. -6
17. $\frac{7}{20}$
18. $\frac{5}{8}$
19. $0.\overline{2}$
20. $0.\overline{6}$
21. $\sqrt{6}$
22. $\sqrt{10}$

23. List all of the numbers from Exercises 3–22 that are rational numbers.

24. List all of the numbers from Exercises 3–22 that are irrational numbers.

25. Describe the set of natural numbers.

26. Describe the set of rational numbers.

27. Describe the set of irrational numbers.

28. Describe the set of whole numbers.

29. Describe the set of real numbers.

30. Describe the set of integers.

31. List three numbers that are real numbers but not rational numbers.

32. List three numbers that are real numbers but not irrational numbers.

33. List three numbers that are integers but not natural numbers.

34. List three numbers that are integers but not whole numbers.

35. List three numbers that are rational but not natural numbers.

36. List three numbers that are rational but not integers.

For Exercises 37–43, let $A = \{-\frac{3}{2}, \sqrt{11}, -4, 0.\overline{6}, \frac{0}{5}, \sqrt{7}, 1\}$

37. Are all of the numbers in set A real numbers?

38. List all of the rational numbers in set A.

39. List all of the whole numbers in set A.

40. List all of the natural numbers in set A.

41. List all of the irrational numbers in set A.

42. List all of the integers in set A.

43. Plot the real numbers of set A on a number line. (*Hint:* $\sqrt{11} \approx 3.3$ and $\sqrt{7} \approx 2.6$)

44. The elevations of selected cities in the United States are shown in the figure. Plot a point on the real number line representing the elevation of each city. Then compare the elevations and fill in the blank with the appropriate inequality sign: < or >. (A negative number indicates that the city is below sea level.)

Figure for Exercise 44

a. Elevation of Tucson ______ elevation of Cincinnati.

b. Elevation of New Orleans ______ elevation of Chicago.

c. Elevation of New Orleans ______ elevation of Houston.

d. Elevation of Chicago ______ elevation of Cincinnati.

45. The elevations of selected cities in the United States are given in the table. Plot a point on the real number line representing the elevation of each city. Then compare the elevations and fill in the blank with the appropriate inequality sign: < or >.

City	Elevation (in feet)*
Dallas, TX	390
Kansas City, MO	720
Long Beach, CA	−7
Denver, CO	5130
Philadelphia, PA	0

Table for Exercise 45

*A negative number indicates that the city is below sea level.

a. Elevation of Kansas City ______ elevation of Dallas.

b. Elevation of Philadelphia ______ elevation of Kansas City.

c. Elevation of Long Beach ______ elevation of Philadelphia.

d. Elevation of Dallas ______ elevation of Denver.

46. Several scores for the LPGA Samsung World Championship of women's golf are given in the table. Plot a point on the real number line representing the score of each golfer, then compare the scores and fill in the blank with the appropriate inequality sign: < or >.

LPGA Golfers	Final Score with Respect to Par
Annika Sorenstam	7
Laura Davies	−4
Lorie Kane	0
Cindy McCurdy	3
Se Ri Pak	−8

Table for Exercise 46

a. Kane's score ______ Pak's score.

b. Sorenstam's score ______ Davies' score.

c. Pak's score ______ McCurdy's score.

d. Kane's score ______ Davies' score.

47. Several scores for the LPGA Samsung World Championship of women's golf are given in the following table. Plot a point on the real number line representing the score of each golfer, then compare the scores and fill in the blank with the appropriate inequality sign: < or >.

LPGA Golfers	Final Score with Respect to Par
Akiko Fukushima	−6
Juli Inkster	2
Karrie Webb	−7
Lorie Kane	0
Meg Mallon	−3

Table for Exercise 47

a. Fukushima's score _______ Mallon's score.

b. Kane's score _______ Webb's score.

c. Inkster's score _______ Mallon's score.

d. Fukushima's _______ Webb's score.

For Exercises 48–55, write the opposite and the absolute value for each number.

48. 18 49. 2 50. −6.1 51. −2.5

52. $-\frac{5}{8}$ 53. $-\frac{1}{3}$ 54. $\frac{7}{3}$ 55. $\frac{1}{9}$

The opposite of a is denoted as $-a$. For Exercises 56–59, evaluate the opposites.

56. $-(-3)$ 57. $-(-5.1)$

58. $-\left(-\frac{7}{3}\right)$ 59. $-(-7)$

For Exercises 60–61, answer true or false. If a statement is false, explain why.

60. If n is positive, then $|n|$ is negative.

61. If m is negative, then $|m|$ is negative.

For Exercises 62–85, determine if the statements are true or false. Use the real number line to justify the answer.

62. $5 > 2$ 63. $8 < 10$

64. $6 < 6$ 65. $19 > 19$

66. $-7 \geq -7$

67. $-1 \leq -1$

68. $\frac{3}{2} \leq \frac{1}{6}$

69. $-\frac{1}{5} \geq 0$

70. $-5 > -2$

71. $6 < -10$

72. $8 \neq 8$

73. $10 \neq 10$

74. $|-2| \geq |-1|$

75. $|3| \leq |-1|$

76. $\left|-\frac{1}{9}\right| = \left|\frac{1}{9}\right|$

77. $\left|-\frac{1}{3}\right| = \left|\frac{1}{3}\right|$

78. $|7| \neq |-7|$

79. $|-13| \neq |13|$

80. $-1 < |-1|$

81. $-6 < |-6|$

82. $|-8| \geq |8|$

83. $|-11| \geq |11|$

84. $|-2| \leq |2|$

85. $|-21| \leq |21|$

Expanding Your Skills

86. For what numbers, a, is $-a$ positive?

87. For what numbers, a, is $|a| = a$?

Concepts

1. Variables and Expressions
2. Evaluating Algebraic Expressions
3. Exponential Expressions
4. Square Roots
5. Order of Operations
6. Translations

section 1.2 Order of Operations

1. Variables and Expressions

A **variable** is a symbol or letter such as x, y, and z, used to represent an unknown number. **Constants** are values that are not variable such as the numbers 3, -1.5, $\frac{2}{7}$, and π. An algebraic **expression** is a collection of variables and constants using algebraic operations. For example, $\frac{3}{x}$, $y + 7$, and $t - 1.4$ are algebraic expressions.

The symbols used to show the four basic operations of addition, subtraction, multiplication, and division are summarized in Table 1-3.

Table 1-3

Operation	Symbols	Translation
Addition	$a + b$	**sum** of a and b a plus b b added to a b more than a a increased by b the total of a and b
Subtraction	$a - b$	**difference** of a and b a minus b b subtracted from a a decreased by b b less than a
Multiplication	$a \times b, a \cdot b, a(b), (a)b, (a)(b), ab$ (*Note*: We rarely use the notation $a \times b$ because the symbol, $\times$, might be confused with the variable, x.)	**product** of a and b a times b a multiplied by b
Division	$a \div b, \frac{a}{b}, a/b, b\overline{)a}$	**quotient** of a and b a divided by b b divided into a ratio of a and b a over b a per b

2. Evaluating Algebraic Expressions

The value of an algebraic expression depends on the values of the variables within the expression.

example 1

Evaluating an Algebraic Expression

Evaluate the algebraic expression when $p = 4$ and $q = \frac{3}{4}$.

a. $100 - p$ b. $8q$

Solution:

a. $100 - p$

$100 - (\quad)$	When substituting a number for a variable, use parentheses.
$= 100 - (4)$	Substitute $p = 4$ in the parentheses.
$= 96$	Subtract.

b. $8q$

$= 8(\quad)$	When substituting a number for a variable, use parentheses.
$= 8\left(\frac{3}{4}\right)$	Substitute $q = \frac{3}{4}$ in the parentheses.
$= \frac{8}{1} \cdot \frac{3}{4}$	Write the whole number as a fraction.
$= \frac{24}{4}$	Multiply fractions.
$= 6$	Reduce.

3. Exponential Expressions

In algebra, repeated multiplication can be expressed using exponents. The expression, $4 \cdot 4 \cdot 4$ can be written as

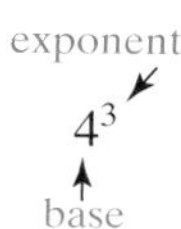

In the expression 4^3, the number 4 is the base, and 3 is the exponent, or power. The exponent indicates how many factors of the base to multiply. For example,

$$5^3 = 5 \cdot 5 \cdot 5 = 125 \quad \text{and} \quad 7^2 = 7 \cdot 7 = 49.$$

If an exponent is not explicitly written over a base, the exponent is assumed to be 1. For example,

$$4 = 4^1 \quad \text{and} \quad 2 = 2^1.$$

Definition of b^n

Let b represent any real number and n represent a positive integer. Then,

$$b^n = \underbrace{b \cdot b \cdot b \cdot b \ldots \cdot b}_{n \text{ factors of } b}$$

b^n is read as "b to the nth power."
b is called the **base** and n is called the **exponent**, or **power**.
b^2 is read as "b squared" and b^3 is read as "b cubed."

The exponent, n, is a count of the number of factors of the base.

example 2

Evaluating Exponential Expressions

Translate the expression into words and then evaluate the expression.

a. 2^5 b. 5^2 c. $\left(\frac{3}{4}\right)^3$ d. 1^6

Solution:

a. The expression 2^5 is read as "two to the fifth power."
$2^5 = (2)(2)(2)(2)(2) = 32$

b. The expression 5^2 is read as "five to the second power" or "five, squared."
$5^2 = (5)(5) = 25$

c. The expression $(\frac{3}{4})^3$ is read as "three-fourths to the third power" or "three-fourths, cubed."

$$\left(\frac{3}{4}\right)^3 = \left(\frac{3}{4}\right)\left(\frac{3}{4}\right)\left(\frac{3}{4}\right) = \frac{27}{64}$$

d. The expression 1^6 is read as "one to the sixth power."
$1^6 = (1)(1)(1)(1)(1)(1) = 1$

4. Square Roots

The reverse operation to squaring a number is to find its **square roots**. For example, finding a square root of 9 is equivalent to asking "what number(s) when squared equals 9?" The symbol, $\sqrt{\ }$, (called a radical sign) is used to find the *principal* square root of a number. By definition, the principal square root of a number is nonnegative. Therefore, $\sqrt{9}$, is the nonnegative number that when squared equals 9. Hence $\sqrt{9} = 3$ because 3 is nonnegative and $(3)^2 = 9$. Several more examples follow:

$\sqrt{64} = 8$ because $(8)^2 = 64$

$\sqrt{121} = 11$ because $(11)^2 = 121$

$\sqrt{0} = 0$ because $(0)^2 = 0$

Tip: To simplify square roots, it is advisable to become familiar with the following squares and square roots.

$0^2 = 0 \rightarrow \sqrt{0} = 0$	$7^2 = 49 \rightarrow \sqrt{49} = 7$
$1^2 = 1 \rightarrow \sqrt{1} = 1$	$8^2 = 64 \rightarrow \sqrt{64} = 8$
$2^2 = 4 \rightarrow \sqrt{4} = 2$	$9^2 = 81 \rightarrow \sqrt{81} = 9$
$3^2 = 9 \rightarrow \sqrt{9} = 3$	$10^2 = 100 \rightarrow \sqrt{100} = 10$
$4^2 = 16 \rightarrow \sqrt{16} = 4$	$11^2 = 121 \rightarrow \sqrt{121} = 11$
$5^2 = 25 \rightarrow \sqrt{25} = 5$	$12^2 = 144 \rightarrow \sqrt{144} = 12$
$6^2 = 36 \rightarrow \sqrt{36} = 6$	$13^2 = 169 \rightarrow \sqrt{169} = 13$

5. Order of Operations

When algebraic expressions contain numerous operations, it is important to evaluate the operations in the proper order. This is called the **order of operations**. Parentheses ( ), brackets [ ], and braces { } are used for grouping numbers and algebraic expressions. It is important to recognize that operations must be done within parentheses and other grouping symbols first. Other grouping symbols include absolute value bars, radical signs, and fraction bars.

Order of Operations

1. Simplify expressions within parentheses and other grouping symbols first. These include absolute value bars, fraction bars, and radicals. If imbedded parentheses are present, start with the innermost parenthesis.
2. Evaluate expressions involving exponents and radicals.
3. Perform multiplication or division in the order that they occur from left to right.
4. Perform addition or subtraction in the order that they occur from left to right.

example 3 **Applying the Order of Operations**

Simplify the expressions.

a. $\frac{1}{2}\left(\frac{5}{6} - \frac{3}{4}\right)$

b. $25 - 12 \div 3 \cdot 4$

c. $6.2 - |-2.1| + \sqrt{15 - 6}$

d. $28 - 2[(6 - 3)^2 + 4]$

Solution:

a. $\frac{1}{2}\left(\frac{5}{6} - \frac{3}{4}\right)$ Subtract fractions within the parentheses.

$= \frac{1}{2}\left(\frac{10}{12} - \frac{9}{12}\right)$ The least common denominator is 12.

$= \frac{1}{2}\left(\frac{1}{12}\right)$

$= \frac{1}{24}$ Multiply fractions.

b. $25 - 12 \div 3 \cdot 4$ Multiply or divide in order from left to right.

$= 25 - 4 \cdot 4$ Notice that the operation $12 \div 3$ is performed first (not $3 \cdot 4$).

$= 25 - 16$ Multiply $4 \cdot 4$ before subtracting.

$= 9$ Subtract.

c. $6.2 - |-2.1| + \sqrt{15 - 6}$

$= 6.2 - |-2.1| + \sqrt{9}$ Simplify within the square root.

$= 6.2 - (2.1) + 3$ Simplify the square root and absolute value.

$= 4.1 + 3$ Add or subtract from left to right.

$= 7.1$ Add.

d. $28 - 2[(6 - 3)^2 + 4]$

$= 28 - 2[(3)^2 + 4]$ Simplify within the inner parentheses first.

$= 28 - 2[(9) + 4]$ Simplify exponents.

$= 28 - 2[13]$ Add within the square brackets.

$= 28 - 26$ Multiply before subtracting.

$= 2$ Subtract.

Tip: Notice that after evaluating an absolute value, the absolute value bars are removed. Thus,

$$|-2.1| = 2.1$$

6. Translations

example 4 **Translating from English Form to Algebraic Form**

Translate each English phrase to an algebraic expression.

a. The quotient of x and 5
b. The difference of p and the square root of q
c. Seven less than n
d. Eight more than the absolute value of w

Avoiding Mistakes

Recall that "b less than a" is translated as $a - b$. Therefore, the statement "seven less than n" must be translated as $n - 7$, not $7 - n$.

Solution:

a. $\frac{x}{5}$ or $x \div 5$

b. $p - \sqrt{q}$

c. $n - 7$

d. $|w| + 8$

example 5

Translating from English Form to Algebraic Form

Translate each English phrase into an algebraic expression. Then evaluate the expression for $a = 6$, $b = 4$, and $c = 20$.

a. The product of a and the square root of b
b. Twice the sum of b and c
c. The difference of twice a and b

Solution:

a. $a\sqrt{b}$ Translate.

$= (\quad)\sqrt{(\quad)}$ Use parentheses to substitute a number for a variable.

$= (6)\sqrt{(4)}$ Substitute $a = 6$ and $b = 4$.

$= 6 \cdot 2$ Simplify the radical first.

$= 12$ Multiply.

Avoiding Mistakes

To compute "twice the sum of b and c" it is necessary to take the sum first and then multiply by 2. To ensure the proper order, the sum of b and c must be enclosed in parentheses. The proper translation is:

$2(b + c)$

b. $2(b + c)$ Translate.

$= 2((\quad) + (\quad))$ Use parentheses to substitute a number for a variable.

$= 2((4) + (20))$ Substitute $b = 4$ and $c = 20$.

$= 2(24)$ Simplify within the parentheses first.

$= 48$ Multiply.

c. $2a - b$ Translate.

$= 2(\quad) - (\quad)$ Use parentheses to substitute a number for a variable.

$= 2(6) - (4)$ Substitute $a = 6$ and $b = 4$.

$= 12 - 4$ Multiply first.

$= 8$ Subtract.

Calculator Connections

On a calculator, we enter exponents greater than the second power by using the key labeled [y^x] or [^]. For example, evaluate 2^4 and 10^6:

Scientific Calculator:

Enter: [2] [y^x] [4] [=] **Result:** 16

Enter: [10] [y^x] [6] [=] **Result:** 1000000

Graphing Calculator:

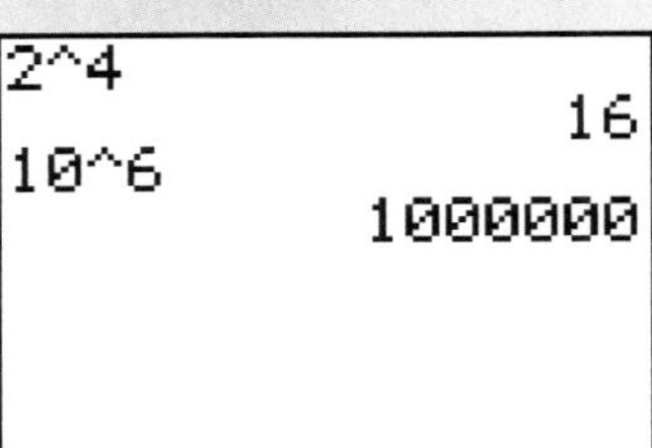

Most calculators also have the capability to enter several operations at once. However, it is important to note that fraction bars and radicals require user-defined parentheses to ensure that the proper order of operations is followed.

a. $130 - 2(5 - 1)^3$ b. $\dfrac{18 - 2}{11 - 9}$ c. $\sqrt{25 - 9}$

Scientific Calculator:

Enter: [130] [−] [2] [×] [(] [5] [−] [1] [)] [y^x] [3] [=] **Result:** 2

Enter: [(] [18] [−] [2] [)] [÷] [(] [11] [−] [9] [)] [=] **Result:** 8

Enter: [(] [25] [−] [9] [)] [√] **Result:** 4

Graphing Calculator:

```
130-2*(5-1)^3
                2
(18-2)/(11-9)
                8
√(25-9)
                4
```

section 1.2 Practice Exercises

For Exercises 1–4, let
$C = \{7, \frac{1}{3}, 0, -2, -\sqrt{9}, \pi, 3.25, -0.\overline{1}, \sqrt{5}\}$

1. Plot the integers from set C on a number line.

2. Plot the whole numbers from set C on a number line.

3. Plot the irrational numbers from set C on a number line. (*Hint:* $\sqrt{5} \approx 2.2$)

4. Plot the rational numbers from set C on a number line.

5. The high temperatures (in degrees Celsius, °C) in Montreal, Quebec, Canada, for a week in January are given in the table.

Day	High (°C)
Sunday	−8
Monday	−9
Tuesday	−12
Wednesday	−8
Thursday	−9
Friday	−8
Saturday	−7

Table for Exercise 5

Fill in the blank with the appropriate symbol (>, <, =).

a. Temperature on Tuesday ______ temperature on Friday.

b. Temperature on Wednesday ______ temperature on Monday.

c. Temperature on Saturday ______ temperature on Thursday.

d. Temperature on Tuesday ______ temperature on Saturday.

6. The high temperatures for Edmonton, Alberta, Canada, for a week in January are given in the table.

Day	High (°C)
Sunday	1
Monday	2
Tuesday	2
Wednesday	5
Thursday	−2
Friday	−3
Saturday	−3

Table for Exercise 6

Fill in the blank with the appropriate symbol (>, <, =).

a. Temperature on Tuesday ______ temperature on Friday.

b. Temperature on Wednesday ______ temperature on Monday.

c. Temperature on Saturday ______ temperature on Thursday.

d. Temperature on Friday ______ temperature on Saturday.

For Exercises 7–14, write each of the products using exponents.

7. $\frac{1}{6} \cdot \frac{1}{6} \cdot \frac{1}{6} \cdot \frac{1}{6}$

8. $10 \cdot 10 \cdot 10 \cdot 10 \cdot 10 \cdot 10$

9. $a \cdot a \cdot a \cdot b \cdot b$

10. $7 \cdot x \cdot x \cdot y \cdot y$

11. $5c \cdot 5c \cdot 5c \cdot 5c \cdot 5c$

12. $3 \cdot w \cdot z \cdot z \cdot z \cdot z$

13. $8 \cdot y \cdot x \cdot x \cdot x \cdot x \cdot x \cdot x$

14. $\frac{2}{3}t \cdot \frac{2}{3}t \cdot \frac{2}{3}t$

For Exercises 15–22, write each expression in expanded form using the definition of an exponent.

15. x^3 16. y^4 17. $(2b)^3$ 18. $(8c)^2$

19. $10y^5$ 20. x^2y^3 21. $2wz^2$ 22. $3a^3b$

For Exercises 23–30, simplify the expressions.

23. 5^2 24. 4^3 25. $\left(\frac{1}{7}\right)^2$ 26. $\left(\frac{1}{2}\right)^5$

27. $(0.25)^3$ 28. $(0.8)^2$ 29. 2^6 30. 13^2

For Exercises 31–38, simplify the square roots.

31. $\sqrt{81}$ 32. $\sqrt{64}$ 33. $\sqrt{4}$ 34. $\sqrt{9}$

35. $\sqrt{100}$ 36. $\sqrt{49}$ 37. $\sqrt{16}$ 38. $\sqrt{36}$

For Exercises 39–66, use the order of operations to simplify the expressions.

39. $8 + 2 \cdot 6$

40. $7 + 3 \cdot 4$

41. $(8 + 2)6$

42. $(7 + 3)4$

43. $4 + 2 \div 2 \cdot 3 + 1$

44. $5 + 6 \cdot 2 \div 4 - 1$

45. $\frac{1}{4} \cdot \frac{2}{3} - \frac{1}{6}$

46. $\frac{3}{4} \cdot \frac{2}{3} + \frac{2}{3}$

47. $\frac{9}{8} - \frac{1}{3} \cdot \frac{3}{4}$

48. $\frac{11}{6} - \frac{3}{8} \cdot \frac{4}{3}$

49. $3[5 + 2(8 - 3)]$

50. $2[4 + 3(6 - 4)]$

51. $10 + |-6|$

52. $18 + |-3|$

53. $21 - |8 - 2|$

54. $12 - |6 - 1|$

55. $2^2 + \sqrt{9} \cdot 5$

56. $3^2 + \sqrt{16} \cdot 2$

57. $\sqrt{9 + 16} - 2$

58. $\sqrt{36 + 13} - 5$

59. $\frac{7 + 3(8 - 2)}{(7 + 3)(8 - 2)}$

60. $\frac{16 - 8 \div 4}{4 + 8 \div 4 - 2}$

61. $\frac{15 - 5(3 \cdot 2 - 4)}{10 - 2(4 \cdot 5 - 16)}$

62. $\frac{5(7 - 3) + 8(6 - 4)}{4[7 + 3(2 \cdot 9 - 8)]}$

63. $[4^2 \cdot (6 - 4) \div 8] + [7 \cdot (8 - 3)]$

64. $(18 \div \sqrt{4}) \cdot \{[(9^2 - 1) \div 2] - 15\}$

65. $48 - 13 \cdot 3 + [(50 - 7 \cdot 5) + 2]$

66. $80 \div 16 \cdot 2 + (6^2 - |-2|)$

For Exercises 67–78, evaluate the expressions for the given substitutions.

67. $y - 3$ when $y = 18$

68. $3q$ when $q = 5$

69. $\frac{15}{t}$ when $t = 5$

70. $8 + w$ when $w = 12$

71. $2(c + 1) - 5$ when $c = 4$

72. $4(4x - 1)$ when $x = \frac{3}{4}$

73. $5 + 6d$ when $d = \frac{2}{3}$

74. $\frac{6}{5}h - 1$ when $h = 10$

75. $p^2 + \frac{2}{9}$ when $p = \frac{2}{3}$

76. $z^3 - \frac{2}{27}$ when $z = \frac{2}{3}$

77. $5(x + 2.3)$ when $x = 1.1$

78. $3(2.1 - y)$ when $y = 0.5$

79. The area of a rectangle may be computed as $A = \ell w$, where ℓ is the length of the rectangle and w is the width. Find the area for the rectangle.

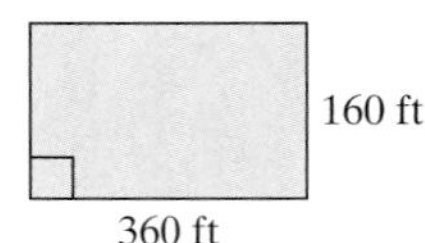

Figure for Exercise 79

80. The perimeter of the rectangular field from Exercise 79 can be computed as $P = 2\ell + 2w$. Find the perimeter.

81. The area of a trapezoid is given by $A = \frac{1}{2}(b_1 + b_2)h$, where b_1 and b_2 are the lengths of the two parallel sides and h is the height. Find the area of the window shown in the figure.

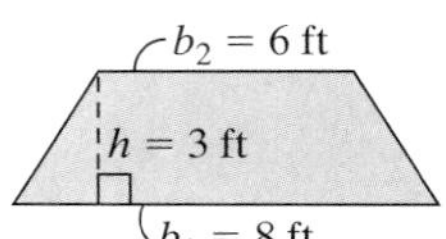

Figure for Exercise 81

82. The volume of a rectangular solid is given by $V = \ell wh$, where ℓ is the length of the solid, w is the width, and h is the height. Find the volume of the box shown in the figure.

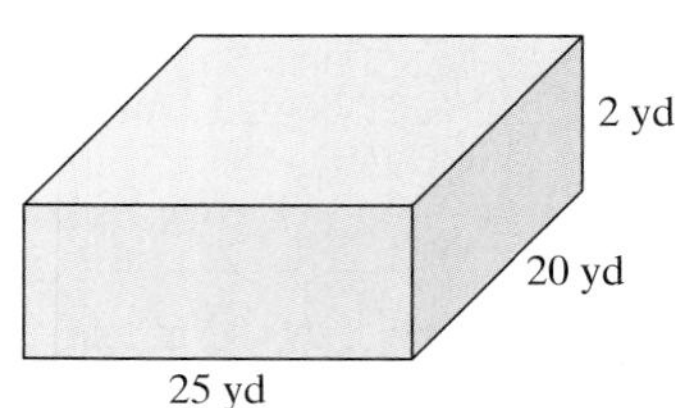

Figure for Exercise 82

83. a. For the expression $5x^3$, what is the base for the exponent 3?

b. Does 5 have an exponent? If so, what is it?

84. a. For the expression $2y^4$, what is the base for the exponent 4?
    b. Does 2 have an exponent? If so, what is it?

For Exercises 85–98, translate each English phrase into an algebraic expression.

85. The product of 3 and x
86. The sum of b and 6
87. The quotient of x and 7
88. Four divided by k
89. The difference of 2 and a
90. Three subtracted from t
91. x more than twice y
92. Nine decreased by the product of 3 and p
93. Four times the sum of x and 12
94. Twice the difference of x and 3
95. Twice x subtracted from 21
96. The quotient of twice x and 11
97. Fourteen less than t
98. Q less than 3

For Exercises 99–116, translate each algebraic expression into an English phrase. (Answers may vary.)

99. $5 + r$
100. $18 - x$
101. $s - 14$
102. $y + 12$
103. $\dfrac{5}{2p}$
104. xyz
105. $7x + 1$
106. $c - 2d$
107. 5^2
108. 6^3
109. $\sqrt{5}$
110. $\sqrt{10}$
111. 7^3
112. 10^2
113. $2 + x^2$
114. $z^2 + 16$
115. $3 + \sqrt{r}$
116. $21 - \sqrt{w}$
117. Some students use the following common memorization device (mnemonic) to help them remember the order of operations: the acronym PEMDAS or **P**lease **E**xcuse **M**y **D**ear **A**unt **S**ally to remember **P**arentheses, **E**xponents, **M**ultiplication, **D**ivision, **A**ddition, and **S**ubtraction. The problem with this mnemonic is that it suggests that multiplication is done before division and similarly, it suggests that addition is performed before subtraction. Explain why following this acronym may give the incorrect answer for the expressions:

    a. $36 \div 4 \cdot 3$ b. $36 - 4 + 3$

118. If you use the acronym **P**lease **E**xcuse **M**y **D**ear **A**unt **S**ally to remember the order of operations, what must you keep in mind about the last four steps?
119. Explain why the acronym **P**lease **E**xcuse **D**r. **M**ichael **S**mith's **A**unt could also be used as a memory device for the order of operations.

Expanding Your Skills

For Exercises 120–123, use the order of operations to simplify the expressions.

120. $\dfrac{\sqrt{\frac{1}{9}} + \frac{2}{3}}{\sqrt{\frac{4}{25}} + \frac{3}{5}}$
121. $\dfrac{5 - \sqrt{9}}{\sqrt{\frac{4}{9}} + \frac{1}{3}}$
122. $\dfrac{|-2|}{|-10| - |2|}$
123. $\dfrac{|-4|^2}{2^2 + \sqrt{144}}$

Graphing Calculator Exercises

For Exercises 124–132, simplify the expression without the use of a calculator. Then enter the expression into the calculator to verify your answer.

124. $\dfrac{4 + 6}{8 - 3}$
125. $110 - 5(2 + 1) - 4$
126. $100 - 2(5 - 3)^3$
127. $3 + (4 - 1)^2$
128. $(12 - 6 + 1)^2$
129. $3 \cdot 8 - \sqrt{32 + 2^2}$
130. $\sqrt{18 - 2}$
131. $(4 \cdot 3 - 3 \cdot 3)^3$
132. $\dfrac{20 - 3^2}{26 - 2^2}$

section

1.3 Addition of Real Numbers

Concepts

1. Addition of Real Numbers and the Number Line
2. Addition of Real Numbers
3. Translations
4. Applications Involving Addition of Real Numbers

1. Addition of Real Numbers and the Number Line

Adding real numbers can be visualized on the number line. To add a positive number, move to the right on the number line. To add a negative number, move to the left on the number line. The following example may help to illustrate the process.

On a winter day in Detroit, suppose the temperature starts out at 5 degrees Fahrenheit (5°F) at noon, and then drops 12° two hours later when a cold front passes through. The resulting temperature can be represented by the expression $5° + (-12°)$. On the number line, start at 5 and count 12 units to the left (Figure 1-4). The resulting temperature at 2:00 P.M. is -7°F.

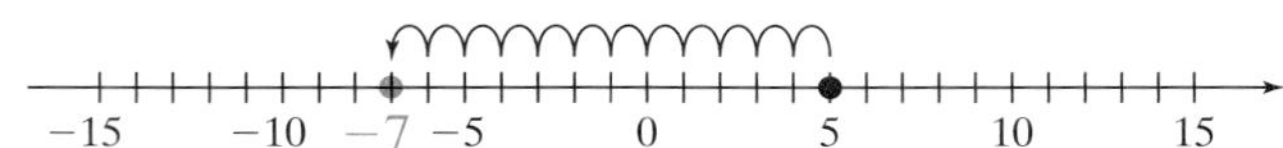

Figure 1-4

example 1 **Using the Number Line to Add Real Numbers**

Use the number line to add the numbers.

a. $-5 + 2$ b. $-1 + (-4)$ c. $4 + (-7)$

Solution:

a. $-5 + 2 = -3$

Start at -5 and count 2 units to the right.

Tip: Note that we move to the left when we add a negative number, and we move right when we add a positive number.

b. $-1 + (-4) = -5$

Start at -1 and count 4 units to the left.

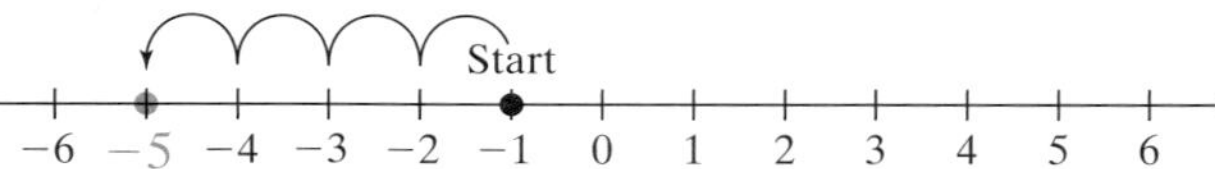

c. $4 + (-7) = -3$

Start at 4 and count 7 units to the left.

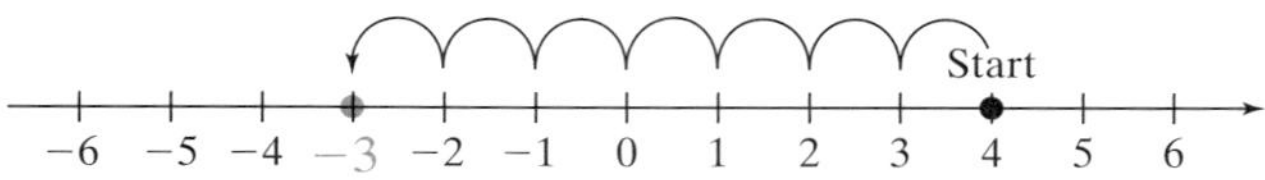

2. Addition of Real Numbers

When adding large numbers or numbers that involve fractions or decimals, counting units on the number line can be cumbersome. Study the following example to determine a pattern for adding two numbers with the *same* sign.

$1 + 4 = 5$

$-1 + (-4) = -5$

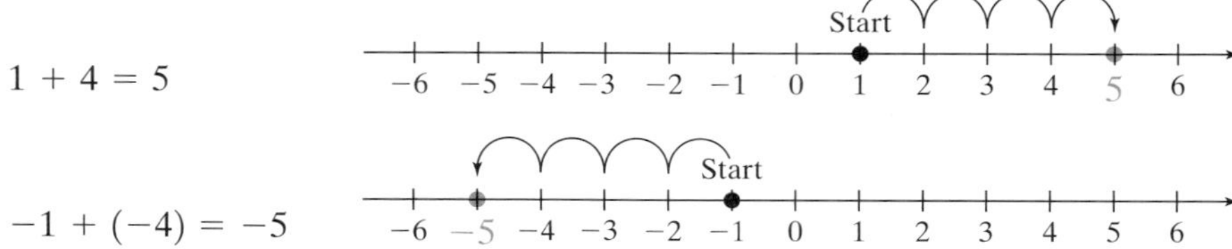

Adding Numbers with the *Same* Sign

To add two numbers with the *same* sign, add their absolute values and apply the common sign.

Study the following example to determine a pattern for adding two numbers with *different* signs.

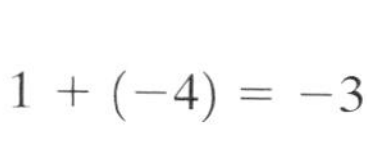

$1 + (-4) = -3$

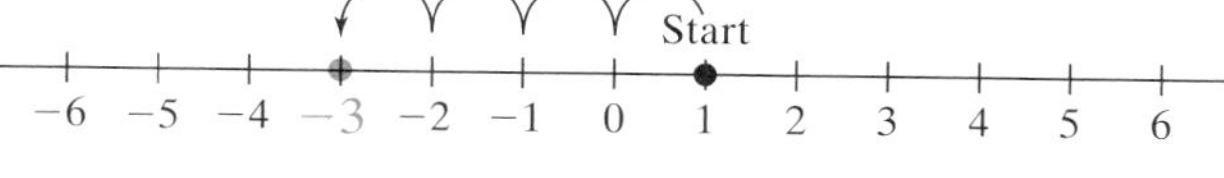

$-1 + 4 = 3$

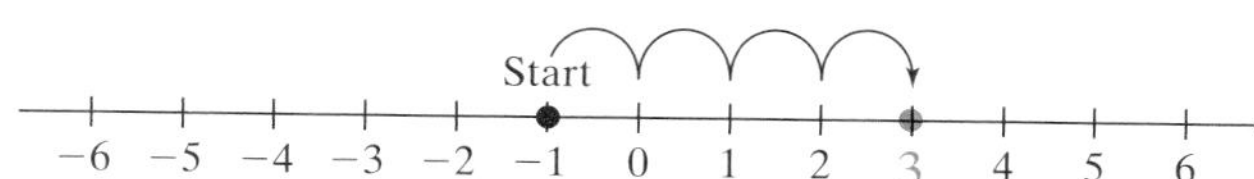

Adding Numbers with *Different* Signs

To add two numbers with *different* signs, subtract the smaller absolute value from the larger absolute value. Then apply the sign of the number having the larger absolute value.

example 2 **Adding Real Numbers**

Add the numbers.

a. $-12 + 15$ b. $7 + (-13)$ c. $-4.5 + (-2.1)$

Solution:

a. $-12 + 15$ Different signs

Subtract the absolute values. Apply the sign of the larger absolute value.

$$= +(|15| - |-12|) = +(15 - 12) = 3$$

Sign of the number with the larger absolute value (arrow to +)

Subtract the absolute values.

b. $7 + (-13)$ Different signs

Subtract the absolute values. Apply the sign of the larger absolute value.

$$= -(|-13| - |7|) = -(13 - 7) = -6$$

Sign of the number with the larger absolute value

Subtract the absolute values.

c. $-4.5 + (-2.1)$ Same signs. Add the absolute values and keep the common sign.

$$= -(|-4.5| + |-2.1|) = -(4.5 + 2.1) = -6.6$$

Keep the common sign. Add the absolute values.

3. Translations

example 3

Translating Expressions Involving the Addition of Real Numbers

Translate each English phrase into an algebraic expression. Then simplify the result.

a. The sum of -12, -8, -9, and -1
b. Negative three-tenths added to $-\frac{7}{8}$
c. The sum of -12 and its opposite

Solution:

a. $-12 + (-8) + (-9) + (-1)$ Translate.

$= -20 + (-9) + (-1)$ Add from left to right.

$= -29 + (-1)$

$= -30$

b. $-\frac{7}{8} + \left(-\frac{3}{10}\right)$ Translate.

$= -\frac{35}{40} + \left(-\frac{12}{40}\right)$ Get a common denominator.

$= -\frac{47}{40}$ The numbers have the same signs. Add their absolute values and keep the common sign. $-\left(\frac{35}{40} + \frac{12}{40}\right)$.

c. $-12 + (12)$ Translate.

$= 0$ Add.

Tip: The sum of any number and its opposite is 0.

4. Applications Involving Addition of Real Numbers

example 4

Adding Real Numbers in Applications

a. A running back on a football team gains 4 yards. On the next play, the quarterback is sacked and loses 13 yd. Write a mathematical expression to describe this situation and then simplify the result.

b. A student has \$120 in her checking account. After depositing her paycheck of \$215, she writes a check for \$255 to cover her portion of the rent and another check for \$294 to cover her car payment. Write a mathematical expression to describe this situation and then simplify the result.

Solution:

a. $4 + (-13)$ The loss of 13 yd can be interpreted as adding -13 yd.

$= -9$ The football team has a net loss of 9 yd.

b. $\underbrace{120 + 215} + (-255) + (-294)$ Writing a check is equivalent to adding a negative amount to the bank account.

$= \underbrace{335 + (-255)} + (-294)$ Use the order of operations. Add from left to right.

$= 80 + (-294)$

$= -214$ The student has overdrawn her account by \$214.

section 1.3 Practice Exercises

For Exercises 1–8, classify the numbers as rational or irrational.

1. $\frac{2}{9}$
2. $-\frac{2}{3}$
3. $-\frac{5}{8}$
4. $\frac{1}{2}$
5. π
6. $-\sqrt{11}$
7. -4
8. 3

Plot the points in set A on a number line. Then for Exercises 9–14 use the number line to place the appropriate inequality (<, >) between the expressions.

$$A = \left\{-\frac{1}{3}, 0, \sqrt{3}, -4, \frac{1}{8}, -2\tfrac{3}{4}, \sqrt{25}\right\}$$

9. 0 ______ $-\frac{1}{3}$
10. $\sqrt{3}$ ______ $\sqrt{25}$
11. $\sqrt{25}$ ______ $\sqrt{3}$
12. $\frac{1}{8}$ ______ $-\frac{1}{3}$
13. $-2\frac{3}{4}$ ______ -4
14. 0 ______ $\sqrt{3}$

For Exercises 15–46, add the integers.

15. $6 + (-3)$
16. $8 + (-2)$
17. $2 + (-5)$
18. $7 + (-3)$
19. $-19 + 2$
20. $-25 + 18$
21. $-4 + 11$
22. $-3 + 9$
23. $-16 + (-3)$
24. $-12 + (-23)$
25. $-2 + (-21)$
26. $-13 + (-1)$
27. $0 + (-5)$
28. $0 + (-4)$
29. $-3 + 0$
30. $-8 + 0$
31. $-16 + 16$
32. $11 + (-11)$
33. $41 + (-41)$
34. $-15 + 15$
35. $4 + (-9)$
36. $6 + (-9)$
37. $7 + (-2) + (-8)$
38. $2 + (-3) + (-6)$
39. $-17 + (-3) + 20$
40. $-9 + (-6) + 15$
41. $-3 + (-8) + (-12)$
42. $-8 + (-2) + (-13)$
43. $-42 + (-3) + 45 + (-6)$
44. $36 + (-3) + (-8) + (-25)$
45. $-5 + (-3) + (-7) + 4 + 8$
46. $-13 + (-1) + 5 + 2 + (-20)$

47. The temperature in Minneapolis, Minnesota, began at $-5°F$ (5° below zero) at 6:00 A.M. By noon the temperature had risen 13°, and by the end of the day, the temperature had dropped 11° from its noon time high. Write an expression using addition that describes the change in temperatures during the day. Then evaluate the expression to give the temperature at the end of the day.

48. The temperature in Toronto, Ontario, Canada, began at 4°F. A cold front went through at noon, and the temperature dropped 9°. By 4:00 P.M. the temperature had risen 2° from its noon time low. Write an expression using addition that describes the changes in temperature during the day. Then evaluate the expression to give the temperature at the end of the day.

49. During a football game, the University of Oklahoma's team gained 3 yd, lost 5 yd, and then gained 14 yd. Write an expression using addition that describes the team's total loss or gain and evaluate the expression.

50. During a football game, the Nebraska Cornhuskers lost 2 yd, gained 6 yd, and then lost 5 yd. Write an expression using addition that describes the team's total loss or gain and evaluate the expression.

51. State the rule for adding two numbers with different signs.

52. State the rule for adding two numbers with the same signs.

For Exercises 53–72, add the rational numbers.

53. $23.81 + (-2.51)$

54. $-9.23 + 10.53$

55. $-\frac{2}{7} + \frac{1}{14}$

56. $-\frac{1}{8} + \frac{5}{16}$

57. $-2.1 + \left(-\frac{3}{10}\right)$

58. $-8.3 + \left(-\frac{9}{10}\right)$

59. $\frac{3}{4} + (-0.5)$

60. $-\frac{3}{2} + 0.45$

61. $8.23 + (-8.23)$

62. $-7.5 + 7.5$

63. $-\frac{7}{8} + 0$

64. $0 + \left(-\frac{21}{22}\right)$

65. $-\frac{2}{3} + \left(-\frac{1}{9}\right) + 2$

66. $-\frac{1}{4} + \left(-\frac{3}{2}\right) + 2$

67. $-47.36 + 24.28$

68. $-0.015 + (0.0026)$

69. $516.816 + (-22.13)$

70. $87.02 + (-93.19)$

71. $-0.000617 + (-0.0015)$

72. $-5315.26 + (-314.89)$

73. Yoshima has \$52.23 in her checking account. She writes a check for groceries for \$52.95.

a. Write an addition problem that expresses Yoshima's transaction.

b. Is Yoshima's account overdrawn?

74. Mohammad has \$40.02 in his checking account. He writes a check for a pair of shoes for \$40.96.

a. Write an addition problem that expresses Mohammad's transaction.

b. Is Mohammad's account overdrawn?

75. In the game show Jeopardy a contestant responds to six questions with the following outcomes: +\$100, +\$200, −\$500, +\$300, +\$100, −\$200

a. Write an expression using addition to describe the contestant's scoring activity.

b. Evaluate the expression from part (a) to determine the contestant's final outcome.

76. A company that has been in business for 5 years has the following profit and loss record.

Year	Profit/Loss ($)
1	−50,000
2	−32,000
3	−5000
4	13,000
5	26,000

Table for Exercise 76

a. Write an expression using addition to describe the company's profit/loss activity.

b. Evaluate the expression from part (a) to determine the company's net profit or loss.

For Exercises 77–82, evaluate the expression for $x = -3$, $y = -2$, and $z = 16$.

77. $x + y + \sqrt{z}$

78. $2z + x + y$

79. $y + 3\sqrt{z}$

80. $-\sqrt{z} + y$

81. $|x| + |y|$

82. $z + x + |y|$

For Exercises 83–92, translate the English phrase into an algebraic expression. Then evaluate the expression.

83. The sum of -6 and -10

84. The sum of -3 and 5

85. Negative three increased by 8

86. Twenty-one increased by 4

87. Seventeen more than -21

88. Twenty-four more than -7

89. Three times the sum of -14 and 20

90. Two times the sum of 6 and -10

91. Five more than the sum of -7 and -2

92. Negative six more than the sum of 4 and -1

Concepts

1. **Subtraction of Real Numbers**
2. **Translations**
3. **Applications Involving Subtraction**
4. **Applying the Order of Operations**

section

1.4 Subtraction of Real Numbers

1. Subtraction of Real Numbers

In the previous section, we learned the rules for adding real numbers. Subtraction of real numbers is defined in terms of the addition process. For example, consider the following subtraction problem and the corresponding addition problem:

$$6 - 4 = 2 \quad \Leftrightarrow \quad 6 + (-4) = 2$$

Start

−6 −5 −4 −3 −2 −1 0 1 2 3 4 5 6

In each case, we start at 6 on the number line and move to the left 4 units. That is, adding the opposite of 4 produces the same result as subtracting 4. This is true in general. To subtract two real numbers, add the opposite of the second number to the first number.

Subtraction of Real Numbers

If a and b are real numbers, then $a - b = a + (-b)$.

$$\left.\begin{aligned} 10 - 4 &= 10 + (-4) = 6 \\ -10 - 4 &= -10 + (-4) = -14 \end{aligned}\right\} \quad \text{Subtracting 4 is the same as adding } -4.$$

$$\left.\begin{aligned} 10 - (-4) &= 10 + (4) = 14 \\ -10 - (-4) &= -10 + (4) = -6 \end{aligned}\right\} \quad \text{Subtracting } -4 \text{ is the same as adding 4.}$$

example 1 **Subtracting Real Numbers**

Subtract the numbers.

a. $4 - (-9)$ b. $-6 - 9$ c. $-11 - (-5)$ d. $7 - 10$

Solution:

a. $4 - (-9)$

$= 4 + (9) = 13$

Change subtraction to addition. Take the opposite of −9.

b. $-6 - 9$

$= -6 + (-9) = -15$

Change subtraction to addition. Take the opposite of 9.

c. $-11 - (-5)$

$= -11 + (5) = -6$

Change subtraction to addition. Take the opposite of −5.

d. $7 - 10$

$= 7 + (-10) = -3$

Change subtraction to addition. Take the opposite of 10.

2. Translations

example 2 **Translating Expressions Involving Subtraction**

Write an algebraic expression for each English phrase and then simplify the result.

a. The difference of −7 and −5
b. 12.4 subtracted from −4.7
c. −24 decreased by the sum of −10 and 13
d. Seven-fourths less than one-third

Solution:

a. $-7 - (-5)$	Translate.
$= -7 + (5)$	Rewrite subtraction in terms of addition.
$= -2$	Simplify.
b. $-4.7 - 12.4$	Translate.
$= -4.7 + (-12.4)$	Rewrite subtraction in terms of addition.
$= -17.1$	Simplify.
c. $-24 - (-10 + 13)$	Translate.
$= -24 - (3)$	Simplify inside parentheses.
$= -24 + (-3)$	Rewrite subtraction in terms of addition.
$= -27$	Simplify.

Tip: Recall that "b subtracted from a" is translated as $a - b$. Hence, −4.7 is written first and then 12.4 is subtracted.

Tip: Parentheses must be used around the sum of −10 and 13 so that −24 is decreased by the entire quantity $(-10 + 13)$.

d. $\frac{1}{3} - \frac{7}{4}$

$= \frac{1}{3} + \left(-\frac{7}{4}\right)$ Rewrite subtraction in terms of addition.

$= \frac{4}{12} + \left(-\frac{21}{12}\right)$ Get a common denominator.

$= -\frac{17}{12}$ or $-1\frac{5}{12}$

3. Applications Involving Subtraction

example 3 **Adding and Subtracting Real Numbers in an Application**

During one of his turns on *Jeopardy*, Harold selected the category "Show Tunes." He got the \$200, \$600, and \$1000 questions correct, but he got the \$400 and \$800 questions incorrect. Write an expression that determines Harold's score. Then simplify the expression to find his total winnings for that category.

Solution:

$200 - 400 + 600 - 800 + 1000$

$= 200 + (-400) + 600 + (-800) + 1000$ Rewrite subtraction in terms of addition.

$= -200 + 600 + (-800) + 1000$ Add from left to right.

$= 400 + (-800) + 1000$

$= -400 + 1000$

$= 600$ Harold won \$600 in that category.

example 4 **Using Subtraction of Real Numbers in an Application**

The highest recorded temperature in North America was 134°F, recorded on July 10, 1913, in Death Valley, California. The lowest temperature of −81°F was recorded on February 3, 1947, in Snag, Yukon, Canada.

Find the difference between the highest and lowest recorded temperatures in North America.

Solution:

$134 - (-81)$

$= 134 + (81)$ Rewrite subtraction in terms of addition.

$= 215$ Add.

The difference between the highest and lowest temperatures is 215°F.

4. Applying the Order of Operations

example 5

Applying the Order of Operations

Simplify the expressions.

a. $-6 + \{10 - [7 - (-4)]\}$

b. $5 - \sqrt{35 - (-14)} - 2$

c. $\left(-\frac{5}{8} - \frac{2}{3}\right) - \left(\frac{1}{8} + 2\right)$

d. $-6 - |7 - 11| + (-3 + 7)^2$

Solution:

a. $-6 + \{10 - [7 - (-4)]\}$ — Work inside the inner brackets first.

$= -6 + \{10 - [7 + (4)]\}$ — Rewrite subtraction in terms of addition.

$= -6 + [10 - (11)]$ — Simplify the expression inside brackets.

$= -6 + [10 + (-11)]$ — Rewrite subtraction in terms of addition.

$= -6 + (-1)$

$= -7$ — Add.

b. $5 - \sqrt{35 - (-14)} - 2$ — Work inside the radical first.

$= 5 - \sqrt{35 + (14)} - 2$ — Rewrite subtraction in terms of addition.

$= 5 - \sqrt{49} - 2$

$= 5 - 7 - 2$ — Simplify the radical.

$= 5 + (-7) + (-2)$ — Rewrite subtraction in terms of addition.

$= -2 + (-2)$ — Add from left to right.

$= -4$

c. $\left(-\frac{5}{8} - \frac{2}{3}\right) - \left(\frac{1}{8} + 2\right)$ — Work inside the parentheses first.

$= \left[-\frac{5}{8} + \left(-\frac{2}{3}\right)\right] - \left(\frac{1}{8} + 2\right)$ — Rewrite subtraction in terms of addition.

$= \left[-\frac{15}{24} + \left(-\frac{16}{24}\right)\right] - \left(\frac{1}{8} + \frac{16}{8}\right)$ — Get a common denominator in each parentheses.

$= \left(-\frac{31}{24}\right) - \left(\frac{17}{8}\right)$ — Add fractions in each parentheses.

$= \left(-\frac{31}{24}\right) + \left(-\frac{17}{8}\right)$ — Rewrite subtraction in terms of addition.

$= -\frac{31}{24} + \left(-\frac{51}{24}\right)$ — Get a common denominator.

$= -\frac{82}{24}$ — Add.

$= -\frac{41}{12}$ — Reduce to lowest terms.

d. $-6 - |7 - 11| + (-3 + 7)^2$ — Simplify within absolute value and parentheses first.

$= -6 - |7 + (-11)| + (-3 + 7)^2$ — Rewrite subtraction in terms of addition.

$= -6 - |-4| + (4)^2$

$= -6 - (4) + 16$ — Simplify absolute value and exponent.

$= -6 + (-4) + 16$ — Rewrite subtraction in terms of addition.

$= -10 + 16$ — Add from left to right.

$= 6$

Calculator Connections

Most calculators can add, subtract, multiply, and divide signed numbers. It is important to note, however, that the key used for the negative sign is different from the key used for subtraction. On a scientific calculator, the [+/−] key or [+○−] key is used to enter a negative number or to change the sign of an existing number. On a graphing calculator, the [(−)] key is used. These keys should not be confused with the [−] key which is used for subtraction. For example, try simplifying the following expressions.

a. $-7 + (-4) - 6$

b. $-3.1 - (-0.5) + 1.1$

Scientific Calculator:

Enter: [7] [+/−] [+] [(] [4] [+/−] [)] [−] [6] [=] Result: -17

Enter: [3] [.] [1] [+/−] [−] [(] [0] [.] [5] [+/−] [)] [+] [1] [.] [1] [=] Result: -1.5

Graphing Calculator:

```
-7+(-4)-6
                 -17
-3.1-(-0.5)+1.1
                -1.5
```

section 1.4 PRACTICE EXERCISES

1. List five whole numbers.
2. List five rational numbers.
3. List five irrational numbers.
4. List five integers.
5. List five natural numbers.
6. List five real numbers.

For Exercises 7–10, translate each English phrase into an algebraic expression.

7. The square root of 6
8. The square of x
9. Negative seven increased by 10
10. Two more than $-b$

For Exercises 11–18, fill in the blank to make each statement correct.

11. $5 - 3 = 5 + $ ______
12. $8 - 7 = 8 + $ ______
13. $-2 - 12 = -2 + $ ______
14. $-4 - 9 = -4 + $ ______
15. $7 - (-4) = 7 + $ ______
16. $13 - (-4) = 13 + $ ______
17. $-9 - (-3) = -9 + $ ______
18. $-15 - (-10) = -15 + $ ______

For Exercises 19–64, subtract the rational numbers.

19. $3 - 5$
20. $9 - 12$
21. $3 - (-5)$
22. $9 - (-12)$
23. $-3 - 5$
24. $-9 - 12$
25. $-3 - (-5)$
26. $-9 - (-5)$
27. $23 - 17$
28. $14 - 2$
29. $23 - (-17)$
30. $14 - (-2)$
31. $-23 - 17$
32. $-14 - 2$
33. $-23 - (-17)$
34. $-14 - (-2)$
35. $-6 - 14$
36. $-9 - 12$
37. $-7 - 17$
38. $-8 - 21$
39. $13 - (-12)$
40. $20 - (-5)$
41. $-14 - (-9)$
42. $-21 - (-17)$
43. $-\frac{6}{5} - \frac{3}{10}$
44. $-\frac{2}{9} - \frac{5}{3}$
45. $\frac{3}{8} - \left(-\frac{4}{3}\right)$
46. $\frac{7}{10} - \left(-\frac{5}{6}\right)$
47. $\frac{1}{2} - \frac{1}{10}$
48. $\frac{2}{7} - \frac{3}{14}$
49. $-\frac{11}{12} - \left(-\frac{1}{4}\right)$
50. $-\frac{7}{8} - \left(-\frac{1}{6}\right)$
51. $6.8 - (-2.4)$
52. $7.2 - (-1.9)$
53. $3.1 - 8.82$
54. $1.8 - 9.59$
55. $-4 - 3 - 2 - 1$
56. $-10 - 9 - 8 - 7$
57. $6 - 8 - 2 - 10$
58. $20 - 50 - 10 - 5$
59. $-36.75 - 14.25$
60. $-84.21 - 112.16$
61. $-112.846 + (-13.03) - 47.312$
62. $-96.473 + (-36.02) - 16.617$
63. $0.085 - (-3.14) + (0.018)$
64. $0.00061 - (-0.00057) + (0.0014)$

For Exercises 65–74, translate each English phrase into an algebraic expression. Then evaluate the expression.

65. Six minus -7
66. Eighteen minus -1
67. Eighteen subtracted from 3
68. Twenty-one subtracted from 8
69. The difference of -5 and -11
70. The difference of -2 and -18

71. Negative thirteen subtracted from -1

72. Negative thirty-one subtracted from -19

73. Twenty less than -32

74. Seven less than -3

For Exercises 75–88, perform the indicated operations. Remember to perform addition or subtraction as they occur from left to right.

75. $6 + 8 - (-2) - 4 + 1$

76. $-3 - (-4) + 1 - 2 - 5$

77. $-1 - 7 + (-3) - 8 + 10$

78. $13 - 7 + 4 - 3 - (-1)$

79. $-\frac{13}{10} + \frac{8}{15} - \left(-\frac{2}{5}\right)$

80. $\frac{11}{14} - \left(-\frac{9}{7}\right) - \frac{3}{2}$

81. $\frac{2}{3} + \frac{5}{9} - \frac{4}{3} - \left(-\frac{1}{6}\right)$

82. $-\frac{9}{8} - \frac{1}{4} - \left(-\frac{5}{6}\right) + \frac{1}{8}$

83. $2 - (-8) + 7 + 3 - 15$

84. $8 - (-13) + 1 - 9$

85. $-6 + (-1) + (-8) + (-10)$

86. $-8 + (-3) + (-5) + (-2)$

87. $-6 - 1 - 8 - 10$

88. $-8 - 3 - 5 - 2$

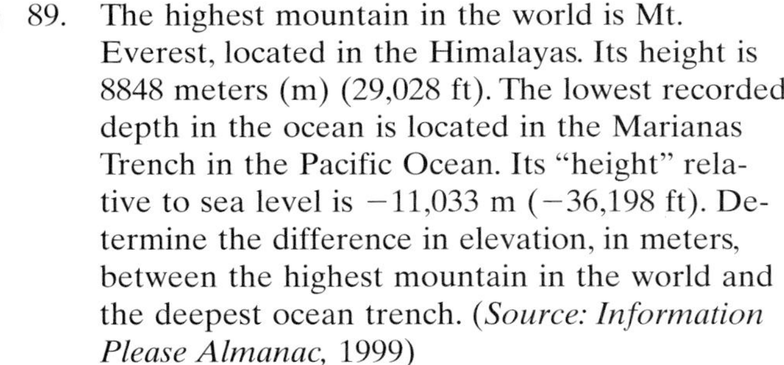

89. The highest mountain in the world is Mt. Everest, located in the Himalayas. Its height is 8848 meters (m) (29,028 ft). The lowest recorded depth in the ocean is located in the Marianas Trench in the Pacific Ocean. Its "height" relative to sea level is $-11{,}033$ m ($-36{,}198$ ft). Determine the difference in elevation, in meters, between the highest mountain in the world and the deepest ocean trench. (*Source: Information Please Almanac,* 1999)

90. The lowest point in North America is located in Death Valley, California, at an elevation of -282 ft (-86 m). The highest point in North America is Mt. McKinley, Alaska, at an elevation of 20,320 ft (6194 m). Find the difference in elevation, in feet, between the highest and lowest points in North America. (*Source: Information Please Almanac,* 1999)

91. On the game, *Jeopardy,* Jasper selected the category "The Last." He got the first four questions correct (worth \$200, \$400, \$600, and \$800) but then missed the last question (worth \$1000). Write an expression that determines Jasper's score. Then simplify the expression to find his total winnings for that category.

92. On Ethyl's turn in *Jeopardy*, she chose the category "Birds of a Feather." She already had \$1200 when she selected a Double Jeopardy question. She wagered \$500 but guessed incorrectly (therefore she lost \$500). On her next turn, she got the \$800 question correct. Write an expression that determines Ethyl's score. Then simplify the expression to find her total winnings for that category.

93. In Ohio, the highest temperature ever recorded was 113°F and the lowest was -39°F. Find the difference between the highest and lowest temperatures. (*Source: Information Please Almanac,* 1999)

94. In Mississippi, the highest temperature ever recorded was 115°F and the lowest was -19°F. Find the difference between the highest and lowest temperatures. (*Source: Information Please Almanac,* 1999)

For Exercises 95–102, evaluate the expressions for $a = -2$, $b = -6$, and $c = -1$.

95. $(a + b) - c$
96. $(a - b) + c$
97. $a - (b + c)$
98. $a + (b - c)$
99. $(a - b) - c$
100. $(a + b) + c$
101. $a - (b - c)$
102. $a + (b + c)$

For Exercises 103–108, evaluate the expression using the order of operations.

103. $\sqrt{29 + (-4)} - 7$
104. $8 - \sqrt{98 + (-3) + 5}$
105. $|10 + (-3)| - |-12 + (-6)|$
106. $|6 - 8| + |12 - 5|$
107. $\dfrac{3 - 4 + 5}{4 + (-2)}$
108. $\dfrac{12 - 14 + 6}{6 + (-2)}$

Graphing Calculator Exercises

For Exercises 109–116, simplify the expression without the use of a calculator. Then use the calculator to verify your answer.

109. $-8 + (-5)$
110. $4 + (-5) + (-1)$
111. $627 - (-84)$
112. $-0.06 - 0.12$
113. $-3.2 + (-14.5)$
114. $-472 + (-518)$
115. $-12 - 9 + 4$
116. $209 - 108 + (-63)$

Concepts
1. Multiplication of Real Numbers
2. Exponential Expressions
3. Division of Real Numbers
4. Division Involving Zero
5. Applying the Order of Operations

section

1.5 Multiplication and Division of Real Numbers

1. Multiplication of Real Numbers

Multiplication of real numbers can be interpreted as repeated addition.

example 1

Multiplying Real Numbers

Multiply the real numbers by writing the expressions as repeated addition.

a. $3(4)$
b. $3(-4)$

Solution:

a. $3(4) = 4 + 4 + 4 = 12$ Add 3 groups of 4.

b. $3(-4) = -4 + (-4) + (-4) = -12$ Add 3 groups of -4.

The results from Example 1 suggest that the product of two positive numbers is positive and the product of a positive number and a negative number is negative. Refer to Table 1-4 and observe the pattern across the bottom row.

Table 1-4

×	3	2	1	0	−1	−2	−3
2	6	4	2	0	−2	−4	−6

The product of two positive numbers is *positive*.

The product of a positive number and a negative number is *negative*.

The pattern along the bottom row shows that as 2 is multiplied by consecutively smaller integers, the product decreases by 2. As this pattern continues, notice that the product of a negative number and a positive number is negative.

To determine the sign of the product of two negative numbers, consider Table 1-5 and note the pattern across the bottom row.

Table 1-5

$\times$	3	2	1	0	−1	−2	−3
−2	−6	−4	−2	0	2	4	6

The product of a positive number and a negative number is *negative*. The product of two negative numbers is *positive*.

As −2 is multiplied by consecutively smaller integers, the product increases by 2. As this pattern continues, notice that the product of two negative numbers is positive.

Multiplication of Real Numbers

1. The product of two real numbers with the *same* sign is positive.
2. The product of two real numbers with *different* signs is negative.
3. The product of any real number and zero is zero.

example 2 **Multiplying Real Numbers**

Multiply the real numbers.

a. $-8(-4)$ b. $-2.5(-1.7)$ c. $-7(10)$

d. $\frac{1}{2}(-8)$ e. $0(-8.3)$ f. $-\frac{2}{7}\left(-\frac{7}{2}\right)$

Solution:

a. $-8(-4) = 32$

b. $-2.5(-1.7) = 4.25$

Same signs. Product is positive.

c. $-7(10) = -70$

d. $\frac{1}{2}(-8) = -4$

Different signs. Product is negative.

e. $0(-8.3)$

$= 0$ The product of any real number and zero is zero.

f. $-\frac{2}{7}\left(-\frac{7}{2}\right)$

$= \frac{14}{14}$ Multiply. *Same signs*. Product is positive.

$= 1$ Reduce to lowest terms.

Observe the pattern for repeated multiplications.

$$(-1)(-1) = 1$$

$$\underbrace{(-1)(-1)}(-1) = (1)(-1) = -1$$

$$\underbrace{(-1)(-1)}(-1)(-1) = \underbrace{(1)(-1)}(-1) = (-1)(-1) = 1$$

$$\underbrace{(-1)(-1)}(-1)(-1)(-1) = \underbrace{(1)(-1)}(-1)(-1) = \underbrace{(-1)(-1)}(-1) = (1)(-1) = -1$$

The pattern demonstrated in these examples indicates that

- The product of an even number of negative factors is positive.
- The product of an odd number of negative factors is negative.

2. Exponential Expressions

Recall that for any real number b and any positive integer, n:

$$b^n = \underbrace{b \cdot b \cdot b \cdot b \cdot \cdots \cdot b}_{n \text{ factors of } b}$$

Be particularly careful when evaluating exponential expressions involving negative numbers. An exponential expression with a negative base is written with parentheses around the base, such as $(-2)^4$.

To evaluate $(-2)^4$, the base -2 is multiplied four times:

$$(-2)^4 = (-2)(-2)(-2)(-2) = 16$$

If parentheses are *not* used, the expression -2^4 has a different meaning:

- The expression -2^4 has a base of 2 (not -2) and can be interpreted as $-1 \cdot 2^4$. Hence,

$$-2^4 = -1(2)(2)(2)(2) = -16$$

- The expression -2^4 can also be interpreted as the opposite of 2^4. Hence,

$$-2^4 = -(2 \cdot 2 \cdot 2 \cdot 2) = -16$$

example 3

Evaluating Exponential Expressions

Simplify.

a. $(-5)^2$ b. -5^2 c. $\left(-\frac{1}{2}\right)^3$ d. -0.4^3

Solution:

a. $(-5)^2 = (-5)(-5) = 25$ Multiply two factors of -5.

b. $-5^2 = -1(5)(5) = -25$ Multiply -1 by two factors of 5.

c. $\left(-\frac{1}{2}\right)^3 = \left(-\frac{1}{2}\right)\left(-\frac{1}{2}\right)\left(-\frac{1}{2}\right) = -\frac{1}{8}$ Multiply three factors of $-\frac{1}{2}$.

d. $-0.4^3 = -1(0.4)(0.4)(0.4) = -0.064$ Multiply -1 by three factors of 0.4.

3. Division of Real Numbers

Notice from Example 2(f) that $-\frac{2}{7}(-\frac{7}{2}) = 1$. Recall that two numbers are *reciprocals* if their product is 1. Symbolically, if a is a nonzero real number, then the reciprocal of a is $\frac{1}{a}$ because $a \cdot \frac{1}{a} = 1$. This definition also implies that a number and its reciprocal have the same sign.

The Reciprocal of a Real Number

Let a be a nonzero real number. Then, the **reciprocal** of a is $\frac{1}{a}$.

Recall that to subtract two real numbers, we add the opposite of the second number to the first number. In a similar way, division of real numbers is defined in terms of multiplication. To divide two real numbers, we multiply the first number by the reciprocal of the second number.

Division of Real Numbers

Let a and b be real numbers such that $b \neq 0$. Then, $a \div b = a \cdot \frac{1}{b}$.

Consider the quotient $10 \div 5$. The reciprocal of 5 is $\frac{1}{5}$, so we have

multiply

$$10 \div 5 = 2 \qquad \text{or equivalently,} \qquad 10 \cdot \frac{1}{5} = 2$$

reciprocal

Because division of real numbers can be expressed in terms of multiplication, then the sign rules that apply to multiplication also apply to division.

Division of Real Numbers

1. The quotient of two real numbers with the *same* sign is positive.
2. The quotient of two real numbers with *different* signs is negative.

example 4 **Dividing Real Numbers**

Divide the real numbers.

a. $200 \div (-10)$ b. $\dfrac{-48}{16}$ c. $\dfrac{-6.25}{-1.25}$ d. $\dfrac{-9}{-5}$

e. $15 \div -25$ f. $-\dfrac{3}{14} \div \dfrac{9}{7}$ g. $\dfrac{\frac{2}{5}}{-\frac{2}{5}}$

Solution:

a. $200 \div (-10) = -20$ *Different signs*. Quotient is negative.

b. $\frac{-48}{16} = -3$ *Different signs*. Quotient is negative.

c. $\frac{-6.25}{-1.25} = 5$ *Same signs*. Quotient is positive.

Tip: If the numerator and denominator of a fraction are both negative, then the quotient is positive. Therefore, $\frac{-9}{-5}$ can be simplified to $\frac{9}{5}$.

d. $\frac{-9}{-5} = \frac{9}{5}$ *Same signs*. Quotient is positive.

e. $15 \div -25$ *Different signs.* Quotient is negative.

$= \frac{15}{-25}$

$= -\frac{3}{5}$

Tip: If the numerator and denominator of a fraction have opposite signs then the quotient will be negative. Therefore, a fraction has the same value whether the negative sign is written in the numerator, in the denominator, or in front of the fraction.

$$\frac{-3}{5} = \frac{3}{-5} = -\frac{3}{5}$$

f. $-\frac{3}{14} \div \frac{9}{7}$ *Different signs.* Quotient is negative.

$= -\frac{3}{14} \cdot \frac{7}{9}$ Multiply by the reciprocal of $\frac{9}{7}$ which is $\frac{7}{9}$.

$= -\frac{21}{126}$ Multiply fractions.

$= -\frac{1}{6}$ Reduce to lowest terms.

g. $\frac{\frac{2}{5}}{-\frac{2}{5}}$ This is equivalent to $\frac{2}{5} \div (-\frac{2}{5})$.

$= \frac{2}{5}\left(-\frac{5}{2}\right)$ Multiply by the reciprocal of $-\frac{2}{5}$, which is $-\frac{5}{2}$.

$= -1$ Multiply fractions.

4. Division Involving Zero

Multiplication can be used to check any division problem. If $\frac{a}{b} = c$, then $bc = a$ (provided that $b \neq 0$). For example,

$$\frac{8}{-4} = -2 \quad \rightarrow \quad \underline{\text{Check}}: \quad (-4)(-2) = 8 \checkmark$$

This relationship between multiplication and division can be used to investigate division problems involving the number zero.

1. The quotient of 0 and any nonzero number is 0. For example,

$$\frac{0}{6} = 0 \qquad \text{because } 6 \cdot 0 = 0 \checkmark$$

2. The quotient of any nonzero number and 0 is undefined. For example,

$$\frac{6}{0} = ?$$

Finding the quotient $\frac{6}{0}$ is equivalent to asking, "what number times zero will equal 6?" That is, $(0)(?) = 6$. No real number satisfies this condition. Therefore, we say that division by zero is undefined.

3. The quotient of 0 and 0 cannot be determined. Evaluating an expression of the form $\frac{0}{0} = ?$ is equivalent to asking "what number times zero will equal 0?" That is, $(0)(?) = 0$. Any real number will satisfy this requirement; however, expressions involving $\frac{0}{0}$ are usually discussed in advanced mathematics courses.

Division Involving Zero

Let a represent a nonzero real number. Then,

1. $\frac{0}{a} = 0$ 2. $\frac{a}{0}$ is undefined

5. Applying the Order of Operations

example 5

Applying the Order of Operations

Simplify the expressions.

a. $-36 \div (-27) \div (-9)$

b. $\dfrac{4 + \sqrt{30 - 5}}{-5 - 1}$

c. $\dfrac{24 - 2[-3 + (5 - 8)]^2}{2|-12 + 3|}$

Solution:

a. $\underbrace{-36 \div (-27)} \div (-9)$

$= \dfrac{-36}{-27} \div -9$ Divide from left to right.

$= \dfrac{4}{3} \div -9$ Reduce to lowest terms.

$= \dfrac{4}{3}\left(-\dfrac{1}{9}\right)$ Multiply by the reciprocal of -9, which is $-\frac{1}{9}$.

$= -\dfrac{4}{27}$ The product of two numbers with different signs is negative.

b. $\dfrac{4 + \sqrt{30 - 5}}{-5 - 1}$ Simplify numerator and denominator separately.

$= \dfrac{4 + \sqrt{25}}{-6}$ Simplify within the radical and simplify the denominator.

$= \dfrac{4 + 5}{-6}$ Simplify the radical.

$= \dfrac{9}{-6}$

$= \dfrac{3}{-2}$ or $-\dfrac{3}{2}$ Reduce to lowest terms. The quotient of two numbers with different signs is negative.

c. $\dfrac{24 - 2[-3 + (5 - 8)]^2}{2|-12 + 3|}$ Simplify numerator and denominator separately.

$= \dfrac{24 - 2[-3 + (-3)]^2}{2|-9|}$ Simplify within the inner parentheses and absolute value.

$= \dfrac{24 - 2[-6]^2}{2(9)}$ Simplify within brackets, [ ]. Simplify the absolute value.

$= \dfrac{24 - 2(36)}{2(9)}$ Simplify exponents.

$= \dfrac{24 - 72}{18}$ Perform multiplication before subtraction.

$= \dfrac{-48}{18}$ or $-\dfrac{8}{3}$ Reduce to lowest terms.

example 6 Evaluating an Algebraic Expression

Given $y = -6$, evaluate the expressions.

a. y^2 b. $-y^2$

Solution:

a. y^2

$= (\ \ )^2$ When substituting a number for a variable, use parentheses.

$= (-6)^2$ Substitute $y = -6$.

$= 36$ Multiply $(-6)(-6) = 36$.

b. $-y^2$

$= -(\quad)^2$	When substituting a number for a variable, use parentheses.
$= -(-6)^2$	Substitute $y = -6$.
$= -(36)$	Evaluate $(-6)^2 = (-6)(-6) = 36$.
$= -36$	$-(36)$ is the same as $-1 \cdot (36)$.

Calculator Connections

Be particularly careful when raising a negative number to an even power on a calculator. For example, the expressions $(-4)^2$ and -4^2 have different values. That is, $(-4)^2 = 16$ and $-4^2 = -16$. Verify these expressions on a calculator.

Scientific Calculator:

To evaluate $(-4)^2$

Enter: (4 +/−) x^2 **Result:** 16

To evaluate -4^2 on a scientific calculator, it is important to square 4 first and then take its opposite.

Enter: 4 x^2 +/− **Result:** −16

Graphing Calculator:

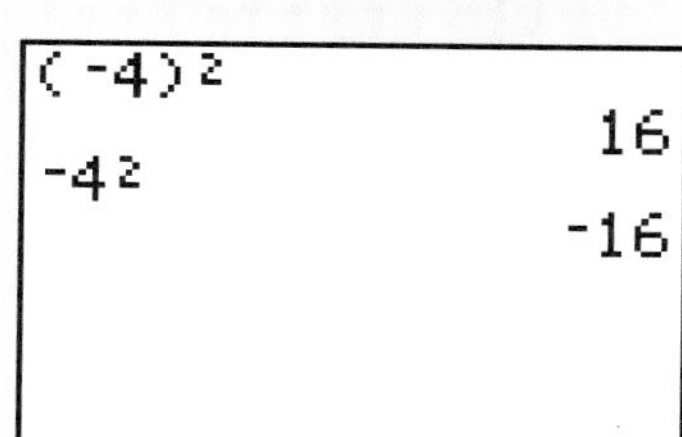

The graphing calculator allows for several methods of denoting the multiplication of two real numbers. For example, consider the product of −8 and 4.

```
-8*4
                -32
-8(4)
                -32
(-8)(4)
                -32
```

section 1.5 Practice Exercises

For Exercises 1–8, determine if the expression is true or false.

1. $6 + (-2) > -5 + 6$
2. $\sqrt{100} - (-4) \leq 7 - (-13)$
3. $-21 - 4 \geq -\sqrt{4} - 12$
4. $-8 - \sqrt{16} \geq 9 - \sqrt{16}$
5. $|-6| + |-14| \leq |-3| + |-17|$
6. $16 - |-2| \leq |-3| - (-11)$
7. $\sqrt{36} - |-6| > 0$
8. $\sqrt{9} + |-3| \leq 0$

Multiplication can be thought of as repeated addition. For each product in Exercises 9–12, write an equivalent addition problem.

9. $5(4)$
10. $2(6)$
11. $3(-2)$
12. $5(-6)$

For Exercises 13–18, show how multiplication can be used to check the division problems.

13. $\frac{14}{-2} = -7$
14. $\frac{-18}{-6} = 3$
15. $\frac{0}{-5} = 0$
16. $\frac{0}{-4} = 0$
17. $\frac{6}{0}$ is undefined
18. $\frac{-4}{0}$ is undefined

For Exercises 19–90, multiply or divide as indicated.

19. $2 \cdot 3$
20. $8 \cdot 6$
21. $2(-3)$
22. $8(-6)$
23. $(-2)3$
24. $(-8)6$
25. $(-2)(-3)$
26. $(-8)(-6)$
27. $24 \div 3$
28. $52 \div 2$
29. $24 \div (-3)$
30. $52 \div (-2)$
31. $(-24) \div 3$
32. $(-52) \div 2$
33. $(-24) \div (-3)$
34. $(-52) \div (-2)$
35. $-6 \cdot 0$
36. $-8 \cdot 0$
37. $-18 \div 0$
38. $-42 \div 0$
39. $0\left(-\frac{2}{5}\right)$
40. $0\left(-\frac{1}{8}\right)$
41. $0 \div \left(-\frac{1}{10}\right)$
42. $0 \div \left(\frac{4}{9}\right)$
43. $\frac{-14}{-7}$
44. $\frac{-21}{-3}$
45. $\frac{13}{-65}$
46. $\frac{7}{-77}$
47. $\frac{-9}{6}$
48. $\frac{-15}{10}$
49. $\frac{-30}{-100}$
50. $\frac{-250}{-1000}$
51. $\frac{26}{-13}$
52. $\frac{52}{-4}$
53. $1.72(-4.6)$
54. $361.3(-14.9)$
55. $-0.02(-4.6)$
56. $-0.06(-2.15)$
57. $\frac{14.4}{-2.4}$
58. $\frac{50.4}{-6.3}$
59. $\frac{-5.25}{-2.5}$
60. $\frac{-8.5}{-27.2}$
61. $(-3)^2$
62. $(-7)^2$
63. -3^2
64. -7^2
65. $\left(-\frac{2}{3}\right)^3$
66. $\left(-\frac{1}{5}\right)^3$
67. $(-0.2)^4$
68. $(-0.1)^4$
69. -0.2^4
70. -0.1^4
71. $-|-3|$
72. $-|-5|$
73. $-(-3)$
74. $-(-5)$
75. $-|7|$
76. $-|8|$
77. $|-7|$
78. $|-8|$
79. $(-2)(-5)(-3)$
80. $(-6)(-1)(-10)$

81. $(-8)(-4)(-1)(-3)$

82. $(-6)(-3)(-1)(-5)$

83. $100 \div (-10) \div (-5)$

84. $150 \div (-15) \div (-2)$

85. $\frac{2}{5} \cdot \frac{1}{3} \cdot \left(-\frac{10}{11}\right)$

86. $\left(-\frac{9}{8}\right) \cdot \left(-\frac{2}{3}\right) \cdot \left(1\frac{5}{12}\right)$

87. $\left(1\frac{1}{3}\right) \div 3 \div \left(-\frac{7}{9}\right)$

88. $-\frac{7}{8} \div \left(3\frac{1}{4}\right) \div (-2)$

89. $-12 \div (-6) \div (-2)$

90. $-36 \div (-2) \div 6$

91. For 3 weeks Jim pays \$2 a week for lottery tickets. Jim has one winning ticket for \$3; write an expression that describes his net gain or loss. How much money has Jim won or lost?

92. Stephanie pays \$2 a week for 6 weeks for lottery tickets. Stephanie has one winning ticket for \$5; write an expression that describes her net gain or loss. How much money has Stephanie won or lost?

For Exercises 93–108, multiply or divide as indicated.

93. $87 \div (-3)$

94. $96 \div (-6)$

95. $-4(-12)$

96. $(-5)(-11)$

97. $2.8(-5.1)$

98. $(7.21)(-0.3)$

99. $(-6.8) \div (-0.02)$

100. $(-12.3) \div (-0.03)$

101. $\left(-\frac{2}{15}\right)\left(\frac{25}{3}\right)$

102. $\left(-\frac{5}{16}\right)\left(\frac{4}{9}\right)$

103. $\left(-\frac{7}{8}\right) \div \left(-\frac{9}{16}\right)$

104. $\left(-\frac{22}{23}\right) \div \left(-\frac{11}{3}\right)$

105. $12 \div (-2)(4)$

106. $(-6) \cdot 7 \div (-2)$

107. $\left(-\frac{12}{5}\right) \div (-6) \cdot \left(-\frac{1}{8}\right)$

108. $10 \cdot \frac{1}{3} \div \frac{25}{6}$

109. Is the expression $\frac{10}{5x}$ equal to $10/5x$? Explain.

110. Is the expression $10/(5x)$ equal to $\frac{10}{5x}$? Explain.

For Exercises 111–118, translate the English phrase into an algebraic expression. Then evaluate the expression.

111. The product of -3.75 and 0.3

112. The product of -0.4 and -1.258

113. The quotient of $\frac{16}{5}$ and $\left(-\frac{8}{9}\right)$

114. The quotient of $\left(-\frac{3}{14}\right)$ and $\frac{1}{7}$

115. The number -0.4 plus the quantity 6 times -0.42

116. The number 0.5 plus the quantity -2 times 0.125

117. The number $-\frac{1}{4}$ minus the quantity 6 times $-\frac{1}{3}$

118. Negative five minus the quantity $\left(-\frac{5}{6}\right)$ times $\frac{3}{8}$

119. Evaluate the expressions in parts (a) and (b).
   a. $-4 - 3 - 2 - 1$
   b. $-4(-3)(-2)(-1)$
   c. Explain the difference between the operations in parts (a) and (b).

120. Evaluate the expressions in parts (a) and (b).
   a. $-10 - 9 - 8 - 7$
   b. $-10(-9)(-8)(-7)$
   c. Explain the difference between the operations in parts (a) and (b).

For Exercises 121–134, perform the indicated operations.

121. $8 - 2^3 \cdot 5 + 3 - (-6)$

122. $-14 \div (-7) - 8 \cdot 2 + 3^3$

123. $-(2 - 8)^2 \div (-6) \cdot 2$

124. $-(3 - 5)^2 \cdot 6 \div (-4)$

125. $\dfrac{6(-4) - 2(5 - 8)}{-6 - 3 - 5}$

126. $\dfrac{3(-4) - 5(9 - 11)}{-9 - 2 - 3}$

127. $\dfrac{-4+5}{(-2)\cdot 5+10}$

128. $\dfrac{-3+10}{2(-4)+8}$

129. $|-5|-|-7|$

130. $|-8|-|-2|$

131. $-|-1|-|5|$

132. $-|-10|-|6|$

133. $\dfrac{|2-9|-|5-7|}{10-15}$

134. $\dfrac{|-2+6|-|3-5|}{13-11}$

For Exercises 135–142, evaluate the expression for $x=-2$, $y=-4$, and $z=6$.

135. x^2-2y

136. $3y^2-z$

137. $4(2x-z)$

138. $6(3x+y)$

139. $\dfrac{3x+2y}{y}$

140. $\dfrac{2z-y}{x}$

141. $\dfrac{x+2y}{x-2y}$

142. $\dfrac{x-z}{x^2-z^2}$

143. Evaluate x^2+6 for $x=2$, and for $x=-2$

144. Evaluate y^3+6 for $y=2$ and for $y=-2$

Graphing Calculator Exercises

For Exercises 145–154, simplify the expression without the use of a calculator. Then use the calculator to verify your answer.

145. $-6(5)$

146. $\dfrac{-5.2}{2.6}$

147. $(-5)(-5)(-5)(-5)$

148. $(-5)^4$

149. -5^4

150. -2.4^2

151. $(-2.4)^2$

152. $(-1)(-1)(-1)$

153. $\dfrac{-8.4}{-2.1}$

154. $90 \div (-5)(2)$

Concepts

1. Commutative Properties of Real Numbers
2. Associative Properties of Real Numbers
3. Identity and Inverse Properties of Real Numbers
4. Distributive Property of Multiplication over Addition
5. Simplifying Algebraic Expressions
6. Clearing Parentheses and Combining *Like* Terms

section

1.6 Properties of Real Numbers and Simplifying Expressions

1. Commutative Properties of Real Numbers

When getting dressed in the morning, it makes no difference whether you put on your left shoe first and then your right shoe, or vice versa. This example illustrates a process in which the order does not affect the outcome. Such a process or operation is said to be commutative.

In algebra, the operations of addition and multiplication are commutative because the order in which we add or multiply two real numbers does not affect the result. For example,

$$10+5=5+10 \quad \text{and} \quad 10\cdot 5=5\cdot 10$$

Commutative Properties of Real Numbers

If a and b are real numbers, then

1. $a+b=b+a$ **commutative property of addition**
2. $ab=ba$ **commutative property of multiplication**

It is important to note that although the operations of addition and multiplication are commutative, subtraction and division are *not* commutative. For example,

$$\underbrace{10-5}_{5} \neq \underbrace{5-10}_{-5} \quad \text{and} \quad \underbrace{10\div 5}_{2} \neq \underbrace{5\div 10}_{\frac{1}{2}}$$

example 1 **Applying the Commutative Property of Addition**

Use the commutative property of addition to rewrite each expression.

a. $-3 + (-7)$ b. $3x^3 + 5x^4$

Solution:

a. $-3 + (-7) = -7 + (-3)$ Change the order of -3 and -7.
b. $3x^3 + 5x^4 = 5x^4 + 3x^3$ Change the order of $3x^3$ and $5x^4$.

Recall that subtraction is not a commutative operation. However, if we rewrite the difference of two numbers, $a - b$, as $a + (-b)$, we can apply the commutative property of addition. This is demonstrated in the next example.

example 2 **Applying the Commutative Property of Addition**

Rewrite the expression in terms of addition. Then apply the commutative property of addition.

a. $5a - 3b$ b. $z^2 - \frac{1}{4}$

Solution:

a. $5a - 3b$

$= 5a + (-3b)$ Rewrite subtraction as addition of $-3b$.

$= -3b + 5a$ Apply the commutative property of addition.

b. $z^2 - \frac{1}{4}$

$= z^2 + \left(-\frac{1}{4}\right)$ Rewrite subtraction as addition of $-\frac{1}{4}$.

$= -\frac{1}{4} + z^2$ Apply the commutative property of addition.

example 3 **Applying the Commutative Property of Multiplication**

Use the commutative property of multiplication to rewrite each expression.

a. $12(-6)$ b. $x \cdot 4$

Solution:

a. $12(-6) = -6(12)$ Change the order of 12 and -6.

b. $x \cdot 4 = 4 \cdot x$ (or simply $4x$) Change the order of x and 4.

2. Associative Properties of Real Numbers

The associative property of real numbers states that the manner in which three or more real numbers are grouped under addition or multiplication will not affect the outcome. For example,

$$(5 + 10) + 2 = 5 + (10 + 2) \quad \text{and} \quad (5 \cdot 10)2 = 5(10 \cdot 2)$$
$$15 + 2 = 5 + 12 \qquad\qquad (50)2 = 5(20)$$
$$17 = 17 \qquad\qquad 100 = 100$$

Associative Properties of Real Numbers

If a, b, and c represent real numbers, then

1. $(a + b) + c = a + (b + c)$ **associative property of addition**
2. $(ab)c = a(bc)$ **associative property of multiplication**

example 4

Applying the Associative Property of Multiplication

Use the associative property of multiplication to rewrite each expression. Then simplify the expression if possible.

a. $(5y)y$ b. $4(5z)$ c. $-\frac{3}{2}\left(-\frac{2}{3}w\right)$

Solution:

a. $(5y)y$

$= 5(y \cdot y)$ Apply the associative property of multiplication.

$= 5y^2$ Simplify.

b. $4(5z)$

$= (4 \cdot 5)z$ Apply the associative property of multiplication.

$= 20z$ Simplify.

c. $-\frac{3}{2}\left(-\frac{2}{3}w\right)$

$= \left[-\frac{3}{2}\left(-\frac{2}{3}\right)\right]w$ Apply the associative property of multiplication.

$= 1w$ Simplify.

$= w$

Note: In most cases, a detailed application of the associative property will not be shown when multiplying two expressions. Instead, the process will be written in one step, such as

$$(5y)y = 5y^2, \quad 4(5z) = 20z \quad \text{and} \quad -\frac{3}{2}\left(-\frac{2}{3}w\right) = w$$

3. Identity and Inverse Properties of Real Numbers

The number 0 has a special role under the operation of addition. Zero added to any real number does not change the number. Therefore, the number 0 is said to be the **additive identity** (also called the identity element of addition). For example,

$$-4 + 0 = -4 \qquad 0 + 5.7 = 5.7 \qquad 0 + \frac{3}{4} = \frac{3}{4}$$

The number 1 has a special role under the operation of multiplication. Any real number multiplied by 1 does not change the number. Therefore, the number 1 is said to be the **multiplicative identity** (also called the identity element of multiplication). For example,

$$(-8)1 = -8 \qquad 1(-2.85) = -2.85 \qquad 1\left(\frac{1}{5}\right) = \frac{1}{5}$$

Identity Properties of Real Numbers

If a is a real number, then

1. $a + 0 = 0 + a = a$ **identity property of addition**
2. $a \cdot 1 = 1 \cdot a = a$ **identity property of multiplication**

The sum of a number and its opposite equals 0. For example, $-12 + 12 = 0$. For any real number, a, the opposite of a (also called the **additive inverse** of a) is $-a$ and $a + (-a) = -a + a = 0$. The inverse property of addition states that the sum of any number and its additive inverse is the identity element of addition, 0. For example,

Number	Additive Inverse (opposite)	Sum
9	-9	$9 + (-9) = 0$
-21.6	21.6	$-21.6 + 21.6 = 0$
$\frac{2}{7}$	$-\frac{2}{7}$	$\frac{2}{7} + \left(-\frac{2}{7}\right) = 0$

If b is a nonzero real number, then the reciprocal of b (also called the **multiplicative inverse** of b) is $\frac{1}{b}$. The inverse property of multiplication states that the product of b and its multiplicative inverse is the identity element of multiplication, 1. Symbolically, we have $b \cdot \frac{1}{b} = \frac{1}{b} \cdot b = 1$. For example,

Number	Multiplicative Inverse (reciprocal)	Product
7	$\frac{1}{7}$	$7 \cdot \frac{1}{7} = 1$
3.14	$\frac{1}{3.14}$	$3.14\left(\frac{1}{3.14}\right) = 1$
$-\frac{3}{5}$	$-\frac{5}{3}$	$-\frac{3}{5}\left(-\frac{5}{3}\right) = 1$

Inverse Properties of Real Numbers

If a is a real number and b is a nonzero real number, then

1. $a + (-a) = -a + a = 0$ **inverse property of addition**
2. $b \cdot \frac{1}{b} = \frac{1}{b} \cdot b = 1$ **inverse property of multiplication**

4. Distributive Property of Multiplication over Addition

The operations of addition and multiplication are related by an important property called the **distributive property of multiplication over addition**. Consider the expression $6(2 + 3)$. The order of operations indicates that the sum $2 + 3$ is evaluated first, and then the result is multiplied by 6:

$$\begin{aligned} &6(2 + 3) \\ &= 6(5) \\ &= 30 \end{aligned}$$

Notice that the same result is obtained if the factor of 6 is multiplied by each of the numbers 2 and 3, and then their products are added:

$6(2 + 3)$ The factor of 6 is distributed to the numbers 2 and 3.

$$\begin{aligned} &= 6(2) + 6(3) \\ &= 12 + 18 \\ &= 30 \end{aligned}$$

Tip: The mathematical definition of the distributive property is consistent with the everyday meaning of the word *distribute*. To distribute means to "spread out from one to many." In the mathematical context, the factor a is distributed to both b and c in the parentheses.

The distributive property of multiplication over addition states that this is true in general.

Distributive Property of Multiplication over Addition

If a, b, and c are real numbers, then

$$a(b + c) = ab + ac \quad \text{and} \quad (b + c)a = ab + ac$$

example 5 Applying the Distributive Property

Apply the distributive property: $2(a + 6b + 7)$

Solution:

$2(a + 6b + 7)$

$= 2(a + 6b + 7)$

$= 2(a) + 2(6b) + 2(7)$ Apply the distributive property.

$= 2a + 12b + 14$ Simplify.

Tip: Notice that the parentheses are removed after the distributive property is applied. Sometimes this is referred to as *clearing parentheses.*

Because the difference of two expressions $a - b$ can be written in terms of addition as $a + (-b)$, the distributive property can be applied when the operation of subtraction is present within the parentheses. For example,

$5(y - 7)$

$= 5[y + (-7)]$ Rewrite subtraction as addition of -7.

$= 5[y + (-7)]$ Apply the distributive property.

$= 5(y) + 5(-7)$

$= 5y + (-35)$ or $5y - 35$ Simplify.

example 6 **Applying the Distributive Property**

Use the distributive property to rewrite each expression.

a. $-(-3a + 2b + 5c)$ b. $-6(2 - 4x)$

Solution:

a. $-(-3a + 2b + 5c)$

$= -1(-3a + 2b + 5c)$ The negative sign preceding the parentheses can be interpreted as taking the opposite of the quantity that follows or as $-1(-3a + 2b + 5c)$.

$= -1(-3a + 2b + 5c)$

$= -1(-3a) + (-1)(2b) + (-1)(5c)$ Apply the distributive property.

$= 3a + (-2b) + (-5c)$ Simplify.

$= 3a - 2b - 5c$

b. $-6(2 - 4x)$

$= -6[2 + (-4x)]$ Change subtraction to addition of $-4x$.

$= -6[2 + (-4x)]$

$= -6(2) + (-6)(-4x)$ Apply the distributive property. Notice that multiplying by -6 changes the signs of all terms to which it is applied.

$= -12 + 24x$ Simplify.

Tip: Notice that a negative factor preceding the parentheses changes the signs of all terms to which it is multiplied.

$-1(-3a + 2b + 5c)$

$= +3a - 2b - 5c$

Note: In most cases, the distributive property will be applied without as much detail as shown in Examples 5 and 6. Instead, the distributive property will be applied in one step.

$2(a + 6b + 7)$ 1 step $= 2a + 12b + 14$

$-(3a + 2b + 5c)$ 1 step $= -3a - 2b - 5c$

$-6(2 - 4x)$ 1 step $= -12 + 24x$

5. Simplifying Algebraic Expressions

An algebraic expression is the sum of one or more terms. A term is a constant or the product of a constant and one or more variables. For example, the expression

$$-7x^2 + xy - 100 \quad \text{or} \quad -7x^2 + xy + (-100)$$

consists of the terms $-7x^2$, xy, and -100. The terms $-7x^2$ and xy are **variable terms** and the term -100 is called a **constant term**. It is important to distinguish between a term and the factors within a term. For example, the quantity xy is one term, and the values x and y are factors within the term. The constant factor in a term is called the numerical coefficient (or simply **coefficient**) of the term. In the terms $-7x^2$, xy, and -100, the coefficients are -7, 1, and -100 respectively.

Terms are said to be ***like* terms** if they each have the same variables, and the corresponding variables are raised to the same powers. For example,

***Like* Terms**			***Unlike* Terms**			
$-3b$	and	$5b$	$-5c$	and	$7d$	(different variables)
$17xy$	and	$-4xy$	$6xy$	and	$3x$	(different variables)
$9p^2q^3$	and	p^2q^3	$4p^2q^3$	and	$8p^3q^2$	(different powers)
$5w$	and	$2w$	$5w$	and	2	(different variables)
7	and	10	7	and	$10a$	(different variables)

example 7

Identifying Terms, Factors, Coefficients and *Like* Terms

a. List the terms of the expression $5x^2 - 3x + 2$.
b. Identify the coefficient of the term $6yz^3$.
c. Which of the pairs are *like* terms: $8b, 3b^2$ or $4c^2d, -6c^2d$.

Solution:

a. The terms of the expression $5x^2 - 3x + 2$ are $5x^2$, $-3x$, and 2.
b. The coefficient of $6yz^3$ is 6.
c. $4c^2d$ and $-6c^2d$ are *like* terms.

6. Clearing Parentheses and Combining *Like* Terms

Two terms can be added or subtracted only if they are *like* terms. To add or subtract *like* terms, we use the distributive property as shown in Example 8.

example 8

Using the Distributive Property to Add and Subtract *Like* Terms

Add or subtract as indicated.

a. $7x + 2x$ b. $-2p + 3p - p$

Solution:

a. $7x + 2x$

$= (7 + 2)x$ Apply the distributive property.

$= 9x$ Simplify.

b. $-2p + 3p - p$

$= -2p + 3p - 1p$ Note that $-p$ equals $-1p$.

$= (-2 + 3 - 1)p$ Apply the distributive property.

$= (0)p$ Simplify.

$= 0$

Although the distributive property is used to add and subtract *like* terms, it is tedious to write each step. Observe that adding or subtracting *like* terms is a matter of combining the coefficients and leaving the variable factors unchanged. This can be shown in one step, a shortcut that we will use throughout the text. For example,

$$7x + 2x = 9x \qquad -2p + 3p - 1p = 0p = 0 \qquad -3a - 6a = -9a$$

example 9 **Using the Distributive Property to Add and Subtract *Like* Terms**

a. $3yz + 5 - 2yz + 9$ b. $1.2w^3 + 5.7w^3$

Solution:

a. $3yz + 5 - 2yz + 9$

$= 3yz - 2yz + 5 + 9$ Arrange *like* terms together.

$= 1yz + 14$ Combine *like* terms.

$= yz + 14$

b. $1.2w^3 + 5.7w^3$

$= 6.9w^3$ Combine *like* terms.

Tip: Notice that constants such as 5 and 9 are *like* terms.

Examples 10 and 11 illustrate how the distributive property is used to clear parentheses.

example 10 **Clearing Parentheses and Combining *Like* Terms**

Simplify by clearing parentheses and combining *like* terms: $5 - 2(3x + 7)$

Solution:

$5 - 2(3x + 7)$ The order of operations indicates that we must perform multiplication before subtraction.

It is important to understand that a factor of -2 (not 2) will be multiplied to all terms within the parentheses. To see why this is so, we may rewrite the subtraction in terms of addition.

$= 5 + (-2)(3x + 7)$	Change subtraction to addition.
$= 5 + (-2)(3x + 7)$	A factor of -2 is to be distributed to terms in the parentheses.
$= 5 + (-2)(3x) + (-2)(7)$	Apply the distributive property.
$= 5 + (-6x) + (-14)$	Simplify.
$= 5 + (-14) + (-6x)$	Arrange *like* terms together.
$= -9 + (-6x)$	Combine *like* terms.
$= -9 - 6x$	Simplify by writing addition of the opposite as subtraction.

example 11 Clearing Parentheses and Combining *Like* Terms

Simplify by clearing parentheses and combining *like* terms.

a. $10(5y + 2) - 6(y - 1)$

b. $\frac{1}{4}(4k + 2) - \frac{1}{2}(6k + 1)$

c. $-(4s - 6t) - (3t + 5s) - 2s$

Solution:

a. $10(5y + 2) - 6(y - 1)$

$= 50y + 20 - 6y + 6$	Apply the distributive property. Notice that a factor of -6 is distributed through the second parentheses and changes the signs.
$= 50y - 6y + 20 + 6$	Arrange *like* terms together.
$= 44y + 26$	Combine *like* terms.

b. $\frac{1}{4}(4k + 2) - \frac{1}{2}(6k + 1)$

$= \frac{4}{4}k + \frac{2}{4} - \frac{6}{2}k - \frac{1}{2}$	Apply the distributive property. Notice that a factor of $-\frac{1}{2}$ is distributed through the second parentheses and changes the signs.
$= k + \frac{1}{2} - 3k - \frac{1}{2}$	Simplify fractions.
$= k - 3k + \frac{1}{2} - \frac{1}{2}$	Arrange *like* terms together.
$= -2k + 0$	Combine *like* terms.
$= -2k$	

c. $-(4s - 6t) - (3t + 5s) - 2s$

$= -4s + 6t - 3t - 5s - 2s$ Apply the distributive property.

$= -4s - 5s - 2s + 6t - 3t$ Arrange *like* terms together.

$= -11s + 3t$ Combine *like* terms.

section 1.6 Practice Exercises

For Exercises 1–8, perform the indicated operations.

1. $-13 - (-5)$
2. $-1 - (-19)$
3. $18 \div (-4)$
4. $-27 \div 5$
5. $\frac{25}{21} - \frac{6}{7}$
6. $\frac{8}{9} - \frac{1}{3}$
7. $\left(-\frac{3}{5}\right)\left(\frac{4}{27}\right)$
8. $\left(\frac{1}{6}\right)\left(-\frac{8}{3}\right)$
9. What is another name for multiplicative inverse?
10. What is another name for additive inverse?
11. What is the additive identity?
12. What is the multiplicative identity?

For Exercises 13–22, match the statements with the properties of multiplication and addition.

13. $6 \cdot \frac{1}{6} = 1$
14. $7(4 \cdot 9) = (7 \cdot 4)9$
15. $2(3 + k) = 6 + 2k$
16. $3 \cdot 7 = 7 \cdot 3$
17. $5 + (-5) = 0$
18. $18 \cdot 1 = 18$
19. $(3 + 7) + 19 = 3 + (7 + 19)$
20. $23 + 6 = 6 + 23$
21. $3 + 0 = 3$
22. $-4(2a - b - 3c) = -8a + 4b + 12c$

a. Commutative property of addition
b. Inverse property of multiplication
c. Commutative property of multiplication
d. Associative property of addition
e. Identity property of multiplication
f. Associative property of multiplication
g. Inverse property of addition
h. Identity property of addition
i. Distributive property of multiplication over addition

For Exercises 23–44, use the distributive property to clear parentheses.

23. $6(5x + 1)$
24. $2(x + 7)$
25. $-2(a + 8)$
26. $-3(2z + 9)$
27. $3(5c - d)$
28. $4(w - 13z)$
29. $-7(y - 2)$
30. $-2(4x - 1)$
31. $\frac{1}{3}(m - 3)$
32. $\frac{2}{5}(n - 5)$
33. $\frac{3}{8}(4 + 8s)$
34. $\frac{4}{9}(3 - 9t)$
35. $-\frac{2}{3}(x - 6)$
36. $-\frac{1}{4}(2b - 8)$
37. $-(2p + 10)$
38. $-(7q + 1)$
39. $-(-3w - 5z)$
40. $-(-7a - b)$
41. $4(x + 2y - z)$
42. $-6(2a - b + c)$
43. $-(-6w + x - 3y)$
44. $-(-p - 5q - 10r)$

The distributive property can be used to simplify a product of two numbers by writing one of the factors as a sum or difference. For example,

$$2(98) = 2(100 - 2) = 200 - 4 = 196$$
$$5(27) = 5(20 + 7) = 100 + 35 = 135$$

For Exercises 45–48, rewrite the expression in parentheses as a sum or difference. Then use the distributive property to evaluate the products without the use of a calculator.

45. $4(92)$
46. $3(81)$
47. $4(902)$
48. $5(799)$

For Exercises 49–52, for each expression, list the terms and their coefficients.

49. $3xy - 6x^2 + y - 17$
50. $2x - y + 18xy + 5$
51. $x^4 - 10xy + 12 - y$
52. $-x + 8y - 9x^2y - 3$
53. Explain why $12x$ and $12x^2$ are not *like* terms.
54. Explain why $3x$ and $3xy$ are not *like* terms.
55. Explain why $7z$ and $\sqrt{13}z$ are *like* terms.
56. Explain why $2x$ and $8x$ are *like* terms.
57. Write three different *like* terms.
58. Write three terms that are not *like*.

For Exercises 59–66, simplify by combining *like* terms.

59. $5k - 10k - 12k + 16 + 7$
60. $-4p - 2p + 8p - 15 + 3$
61. $9x - 7y + 12x + 14y$
62. $2y - 8z + y - 5z - 3y$
63. $\frac{1}{4}a + b - \frac{3}{4}a - 5b$
64. $\frac{2}{5} + 2t - \frac{3}{5} + t - \frac{6}{5}$
65. $2.8z - 8.1z + 6 - 15.2$
66. $2.4 - 8.4w - 2w + 0.9$

For Exercises 67–90, simplify by clearing parentheses and combining *like* terms.

67. $-3(2x - 4) + 10$
68. $-2(4a + 3) - 14$
69. $4(w + 3) - 12$
70. $5(2r + 6) - 30$
71. $5 - 3(x - 4)$
72. $4 - 2(3x + 8)$
73. $-3(2t + 4) + 8(2t - 4)$
74. $-5(5y + 9) + 3(3y + 6)$
75. $2(w - 5) - (2w + 8)$
76. $6(x + 3) - (6x - 5)$
77. $-\frac{1}{3}(6t + 9) + 10$
78. $-\frac{3}{4}(8 + 4q) + 7$
79. $10(5.1a - 3.1) + 4$
80. $100(-3.14p - 1.05) + 212$
81. $-4m + 2(m - 3) + 2m$
82. $-3b + 4(b + 2) - 8b$
83. $\frac{1}{2}(10q - 2) + \frac{1}{3}(2 - 3q)$
84. $\frac{1}{5}(15 - 4p) - \frac{1}{10}(10p + 5)$
85. $7n - 2(n - 3) - 6 + n$
86. $8k - 4(k - 1) + 7 - k$
87. $6(x + 3) - 12 - 4(x - 3)$
88. $5(y - 4) + 3 - 6(y - 7)$
89. $6.1(5.3z - 4.1) - 5.8$
90. $-3.6(1.7q - 4.2) + 14.6$

For Exercises 91–102, name the property that justifies each statement.

91. $4(3a) = (4 \cdot 3)a$
92. $\frac{7}{8} \cdot \frac{8}{7} = 1$
93. $(-6 + 1) + 10 = -6 + (1 + 10)$
94. $8 + (-8) = 0$
95. $-3(b + 7) = -3b - 21$
96. $5(12) = 12(5)$
97. $0 + \left(-\frac{2}{3}\right) = -\frac{2}{3}$
98. $2 + (4 + (-1)) = (2 + 4) + (-1)$
99. $-25(1) = -25$
100. $5(2y + 9) = 10y + 45$
101. $-3.82 + 3.82 = 0$
102. $3 + \left(-\frac{1}{2} + 2\right) = \left(-\frac{1}{2} + 2\right) + 3$

Expanding Your Skills

For Exercises 103–110, determine if the expressions are equivalent. If two expressions are not equivalent, state why.

103. $3a + b, b + 3a$

104. $4y + 1, 1 + 4y$

105. $2c + 7, 9c$

106. $5z + 4, 9z$

107. $5x - 3, 3 - 5x$

108. $6d - 7, 7 - 6d$

109. $5x - 3, -3 + 5x$

110. $8 - 2x, -2x + 8$

111. Which grouping of terms is easier for computation,

$$(14\tfrac{2}{7} + 2\tfrac{1}{3}) + \tfrac{2}{3} \quad \text{or} \quad 14\tfrac{2}{7} + (2\tfrac{1}{3} + \tfrac{2}{3})?$$

112. Which grouping of terms is easier for computation,

$$(5\tfrac{1}{8} + 18\tfrac{2}{5}) + 1\tfrac{1}{5} \quad \text{or} \quad 5\tfrac{1}{8} + (18\tfrac{2}{5} + 1\tfrac{1}{5})?$$

113. As a small child in school, the great mathematician Karl Friedrich Gauss (1777–1855) was said to have found the sum of the integers from 1 to 100 mentally:

$$1 + 2 + 3 + 4 + \cdots + 99 + 100$$

Rather than adding the numbers sequentially, he added the numbers in pairs:

$$(1 + 99) + (2 + 98) + (3 + 97) + \cdots$$

a. Use this technique to add the integers from 1 to 10.

$$1 + 2 + 3 + 4 + 5 + 6 + 7 + 8 + 9 + 10$$

b. Use this technique to add the integers from 1 to 20.

---

chapter 1 Summary

Section 1.1—Sets of Numbers and the Real Number Line

Key Concepts:

Natural numbers: $\{1, 2, 3, \ldots\}$

Whole numbers: $\{0, 1, 2, 3, \ldots\}$

Integers: $\{\ldots -3, -2, -1, 0, 1, 2, 3, \ldots\}$

Rational numbers: The set of numbers that can be expressed in the form $\frac{p}{q}$, where p and q are integers and q does not equal 0.

Irrational numbers: The set of real numbers that are not rational.

Real numbers: The set of both the rational numbers and the irrational numbers.

$a < b$	"a is less than b."
$a > b$	"a is greater than b."
$a \leq b$	"a is less than or equal to b."
$a \geq b$	"a is greater than or equal to b."

Examples:

Some rational numbers: $\frac{1}{7}, -0.5, 0.\overline{3}$

Some irrational numbers: $\sqrt{7}, -\sqrt{2}, \pi$

The Real Number Line:

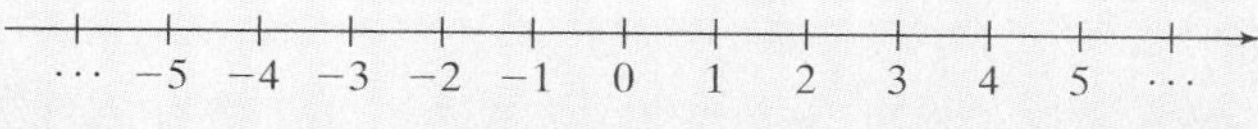

$5 < 7$	"5 is less than 7."
$-2 > -10$	"-2 is greater than -10."
$y \leq 3.4$	"y is less than or equal to 3.4."
$x \geq \frac{1}{2}$	"x is greater than or equal to $\frac{1}{2}$."

Two numbers that are the same distance from zero but on opposite sides of zero on the number line are called opposites. The opposite of a is denoted $-a$.

5 and -5 are opposites.

The absolute value of a real number, a, denoted $|a|$, is the distance between a and 0 on the number line.

If $a \geq 0$, $|a| = a$

If $a < 0$, $|a| = -a$

$|7| = 7$

$|-7| = 7$

Key Terms:

absolute value	opposite
inequality	rational numbers
integers	real number line
irrational numbers	real numbers
natural numbers	whole numbers

Section 1.2—Order of Operations

Key Concepts:

A variable is a symbol or letter used to represent an unknown number.

A constant is a value that is not variable.

An algebraic expression is a collection of variables and constants under algebraic operations.

$b^n = \underbrace{b \cdot b \cdot b \cdot b \ldots \cdot b}_{n \text{ factors of } b}$ b is the base, n is the exponent

$\sqrt{x}$ is the positive square root of x.

The Order of Operations:

1. Simplify expressions within parentheses and other grouping symbols first.
2. Evaluate expressions involving exponents and radicals.
3. Do multiplication or division in the order that they occur from left to right.
4. Do addition or subtraction in the order that they occur from left to right.

Key Terms:

base	power
constant	product
difference	quotient
exponent	square root
expression	sum
order of operations	variable

Examples:

Variables: x, y, z, a, b

Constants: $2, -3, \pi$

Expressions: $2x + 5, 3a + b^2$

$5^3 = 5 \cdot 5 \cdot 5 = 125$

$\sqrt{49} = 7$

Simplify:

$$10 + 5(3 - 1)^2 - \sqrt{5 - 1}$$

$$\begin{aligned} &= 10 + 5(2)^2 - \sqrt{4} \\ &= 10 + 5(4) - 2 \\ &= 10 + 20 - 2 \\ &= 30 - 2 \\ &= 28 \end{aligned}$$

Section 1.3—Addition of Real Numbers

Key Concepts:

Addition of Two Real Numbers:

Same Signs:
Add the absolute values of the numbers and apply the common sign to the sum.

Different Signs:
Subtract the smaller absolute value from the larger absolute value. Then apply the sign of the number having the larger absolute value.

Examples:

$-3 + (-4) = -7$

$-1.3 + (-9.1) = -10.4$

$-5 + 7 = 2$

$\frac{2}{3} + \left(-\frac{7}{3}\right) = -\frac{5}{3}$

Section 1.4—Subtraction of Real Numbers

Key Concepts:

Subtraction of Two Real Numbers:

Add the opposite of the second number to the first number. That is,

$$a - b = a + (-b)$$

Examples:

$7 - (-5) = 7 + (5) = 12$

$-3 - 5 = -3 + (-5) = -8$

$-11 - (-2) = -11 + (2) = -9$

Section 1.5—Multiplication and Division of Real Numbers

Key Concepts:

Multiplication and Division of Two Real Numbers:

Same Signs:
Product is positive.
Quotient is positive.

Different Signs:
Product is negative.
Quotient is negative.

The reciprocal of a number a is $\frac{1}{a}$.

Examples:

$(-5)(-2) = 10$ $\qquad$ $\frac{-20}{-4} = 5$

$(-3)(7) = -21$ $\qquad$ $\frac{-4}{8} = -\frac{1}{2}$

The reciprocal of -6 is $-\frac{1}{6}$.

Multiplication and Division Involving Zero:

The product of any real number and 0 is 0. $4 \cdot 0 = 0$

The quotient of 0 and any nonzero real number is 0. $0 \div 4 = 0$

The quotient of any nonzero real number and 0 is undefined. $4 \div 0$ is undefined

KEY TERM

reciprocal

SECTION 1.6—PROPERTIES OF REAL NUMBERS AND SIMPLIFYING EXPRESSIONS

KEY CONCEPTS: / EXAMPLES:

The Properties of Real Numbers:

Commutative Property of Addition:

$$a + b = b + a$$

Example: $-5 + (-7) = (-7) + (-5)$

Associative Property of Addition:

$$(a + b) + c = a + (b + c)$$

Example: $(2 + 3) + 10 = 2 + (3 + 10)$

Identity Property of Addition:
The number 0 is said to be the identity element for addition because:

$$0 + a = a \quad \text{and} \quad a + 0 = a$$

Example: $0 + \frac{3}{4} = \frac{3}{4}$ and $\frac{3}{4} + 0 = \frac{3}{4}$

Inverse Property of Addition:

$$a + (-a) = 0 \quad \text{and} \quad -a + a = 0$$

Example: $1.5 + (-1.5) = 0$ and $-1.5 + 1.5 = 0$

Commutative Property of Multiplication:

$$ab = ba$$

Example: $(-3)(-4) = (-4)(-3)$

Associative Property of Multiplication:

$$(ab)c = a(bc)$$

Example: $(2 \cdot 3)6 = 2(3 \cdot 6)$

Identity Property of Multiplication:
The number 1 is said to be the identity element for multiplication because:

$$1 \cdot a = a \quad \text{and} \quad a \cdot 1 = a$$

Example: $1 \cdot 5 = 5$ and $5 \cdot 1 = 5$

Inverse Property of Multiplication:

$$a \cdot \frac{1}{a} = 1 \quad \text{and} \quad \frac{1}{a} \cdot a = 1$$

$$6 \cdot \frac{1}{6} = 1 \quad \text{and} \quad \frac{1}{6} \cdot 6 = 1$$

The Distributive Property of Multiplication over Addition:

$$a(b + c) = ab + ac$$

Simplify using the distributive property:

$$2(x + 4y) = 2x + 8y$$

$$-(a + 6b - 5c) = -a - 6b + 5c$$

A term is a constant or the product of a constant and one or more variables.

The coefficient of a term is the numerical factor of the term.

Examples of Terms:

$-2x$, Coefficient is -2.

$5yz^2$ Coefficient is 5.

Like terms have the same variables, and the corresponding variables have the same powers.

Two terms can be added or subtracted if they are *like* terms. Sometimes it is necessary to clear parentheses before adding or subtracting *like* terms.

***Like* Terms:**

$3x$ and $-5x$, $4a^2b$ and $2a^2b$

Simplify:

$$-4d + 12d + d = 9d$$

Simplify:

$$\begin{aligned} &-2w - 4(w - 2) + 3 \\ &= -2w - 4w + 8 + 3 \\ &= -6w + 11 \end{aligned}$$

Key Terms

additive identity
additive inverse
associative property of addition
associative property of multiplication
coefficient
commutative property of addition
commutative property of multiplication
constant term
distributive property of multiplication over addition
identity property of addition
identity property of multiplication
inverse property of addition
inverse property of multiplication
like terms
multiplicative identity
multiplicative inverse
variable term

chapter 1 REVIEW EXERCISES

Section 1.1

1. Given the set $\{7, \frac{1}{3}, -4, 0, -\sqrt{3}, -0.\overline{2}, \pi, 1\}$
   a. List the natural numbers.
   b. List the integers.
   c. List the whole numbers.
   d. List the rational numbers.
   e. List the irrational numbers.
   f. List the real numbers.

For Exercises 2–5, determine the absolute values.

2. $\left|\frac{1}{2}\right|$ 3. $|-6|$ 4. $|-\sqrt{7}|$ 5. $|0|$

For Exercises 6–13, identify whether the inequality is true or false.

6. $-6 > -1$ 7. $0 < -5$ 8. $-10 \leq 0$

9. $5 \neq -5$ 10. $7 \geq 7$ 11. $7 \geq -7$

12. $0 \leq -3$ 13. $-\frac{2}{3} \leq -\frac{2}{3}$

Section 1.2

For Exercises 14–23, translate the English phrases into algebraic expressions.

14. The product of x and $\frac{2}{3}$
15. The quotient of 7 and y
16. The sum of 2 and $3b$
17. The difference of a and 5
18. Two more than $5k$
19. Seven less than $13z$
20. The quotient of 6 and x, decreased by 18
21. The product of y and 3, increased by 12
22. Three-eighths subtracted from z
23. Five subtracted from two times p

For Exercises 24–29, simplify the expressions.

24. 6^3 25. 15^2 26. $\sqrt{36}$

27. $\left(\frac{1}{4}\right)^2$ 28. $\frac{1}{\sqrt{100}}$ 29. $\left(\frac{3}{2}\right)^3$

For Exercises 30–33, perform the indicated operations.

30. $15 - 7 \cdot 2 + 12$
31. $|-11| + |5| - (7 - 2)$
32. $4^2 - (5 - 2)^2$ 33. $22 - 3(8 \div 4)^2$

Section 1.3

For Exercises 34–46, add the rational numbers.

34. $-6 + 8$ 35. $14 + (-10)$

36. $21 + (-6)$ 37. $-12 + (-5)$

38. $\frac{2}{7} + \left(-\frac{1}{9}\right)$ 39. $\left(-\frac{8}{11}\right) + \left(\frac{1}{2}\right)$

40. $\left(-\frac{1}{10}\right) + \left(-\frac{5}{6}\right)$ 41. $\left(-\frac{5}{2}\right) + \left(-\frac{1}{5}\right)$

42. $-8.17 + 6.02$ 43. $2.9 + (-7.18)$

44. $13 + (-2) + (-8)$ 45. $-5 + (-7) + 20$

46. $2 + 5 + (-8) + (-7) + 0 + 13 + (-1)$

47. Under what conditions will the expression $a + b$ be negative?

48. The high temperatures (in degrees Celsius) for the province of Alberta, Canada, during a week in January were $-8, -1, -4, -3, -4, 0$, and 7. What was the average high temperature for that week? Round to the nearest tenth of a degree.

Section 1.4

For Exercises 49–61, subtract the rational numbers.

49. $13 - 25$ 50. $31 - (-2)$

51. $-8 - (-7)$ 52. $-2 - 15$

53. $\left(-\frac{7}{9}\right) - \frac{5}{6}$ 54. $\frac{1}{3} - \frac{9}{8}$

55. $7 - 8.2$

56. $-1.05 - 3.2$

57. $-16.1 - (-5.9)$

58. $7.09 - (-5)$

59. $\frac{11}{2} - \left(-\frac{1}{6}\right) - \frac{7}{3}$

60. $-\frac{4}{5} - \frac{7}{10} - \left(-\frac{13}{20}\right)$

61. $6 - 14 - (-1) - 10 - (-21) - 5$

62. Under what conditions will the expression $a - b$ be negative?

For Exercises 63–67, write an algebraic expression and simplify.

63. -18 subtracted from -7

64. The difference of -6 and 41

65. Seven decreased by 13

66. Five subtracted from the difference of 20 and -7

67. The sum of 6 and -12, decreased by 21

68. In Nevada, the highest temperature ever recorded was 125°F and the lowest was -50°F. Find the difference between the highest and lowest temperatures. (*Source: Information Please Almanac* 1999)

Section 1.5

For Exercises 69–88, multiply or divide as indicated.

69. $10(-17)$

70. $(-7)13$

71. $(-52) \div 26$

72. $(-48) \div (-16)$

73. $\frac{7}{4} \div \left(-\frac{21}{2}\right)$

74. $\frac{2}{3}\left(-\frac{12}{11}\right)$

75. $-\frac{21}{5} \cdot 0$

76. $\frac{3}{4} \div 0$

77. $0 \div (-14)$

78. $\frac{0}{3} \cdot \frac{1}{8}$

79. $(-0.45)(-5)$

80. $(-2.1) \div (-0.07)$

81. $\frac{-21}{14}$

82. $\frac{-13}{-52}$

83. $(5)(-2)(3)$

84. $(-6)(-5)(15)$

85. $\left(-\frac{1}{2}\right)\left(\frac{7}{8}\right)\left(-\frac{4}{7}\right)$

86. $\left(\frac{12}{13}\right)\left(-\frac{1}{6}\right)\left(\frac{13}{14}\right)$

87. $40 \div 4 \div (-5)$

88. $\frac{10}{11} \div \frac{7}{11} \div \frac{5}{9}$

For Exercises 89–92, perform the indicated operations.

89. $9 - 4[-2(4 - 8) - 5(3 - 1)]$

90. $\frac{8(-3) - 6}{-7 - (-2)}$

91. $\frac{2}{3} - \left(\frac{3}{8} + \frac{5}{6}\right) \div \frac{5}{3}$

92. $5.4 - (0.3)^2 \div 0.09$

For Exercises 93–96, evaluate the expression with the given substitution.

93. $3(x + 2) \div y$ for $x = 4$ and $y = -9$

94. $a^2 - bc$ for $a = -6$, $b = 5$, and $c = 2$

95. $w + xy - \sqrt{z}$ for $w = 12$, $x = 6$, $y = -5$, and $z = 25$

96. $(u - v)^2 + (u^2 - v^2)$ for $u = 5$ and $v = -3$

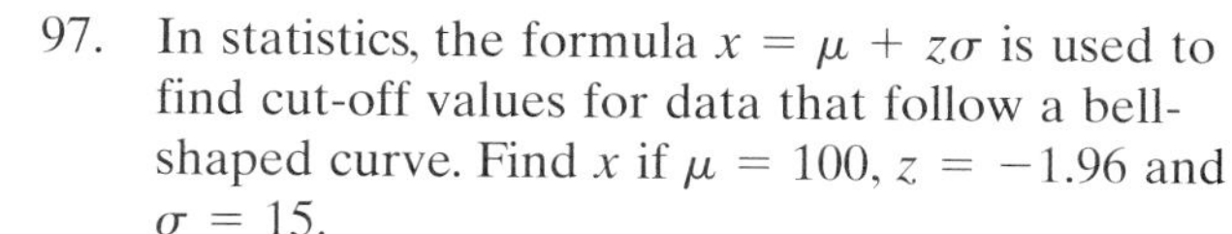

97. In statistics, the formula $x = \mu + z\sigma$ is used to find cut-off values for data that follow a bell-shaped curve. Find x if $\mu = 100$, $z = -1.96$ and $\sigma = 15$.

For Exercises 98–104, answer true or false. If a statement is false, explain why.

98. If n is positive, then $-n$ is negative.

99. If m is negative, then m^3 is negative.

100. If m is negative, then m^4 is negative.

101. If $m > 0$ and $n > 0$, then $mn > 0$.

102. If $p < 0$ and $q < 0$, then $pq < 0$.

103. A number and its reciprocal have the same signs.

104. A nonzero number and its opposite have different signs.

Section 1.6

For Exercises 105–112, answers may vary.

105. Give an example of the commutative property of addition.

106. Give an example of the associative property of addition.

107. Give an example of the inverse property of addition.

108. Give an example of the identity property of addition.

109. Give an example of the commutative property of multiplication.

110. Give an example of the associative property of multiplication.

111. Give an example of the inverse property of multiplication.

112. Give an example of the identity property of multiplication.

113. Explain why $5x - 2y$ is the same as $-2y + 5x$.

114. Explain why $3a - 9y$ is the same as $-9y + 3a$.

115. List the terms of the expression: $3y + 10x - 12 + xy$

116. Identify the coefficients for the terms listed in Exercise 115.

117. Simplify each expression by combining *like* terms.

   a. $3a + 3b - 4b + 5a - 10$

   b. $-6p + 2q + 9 - 13q - p + 7$

118. Use the distributive property to clear the parentheses.

   a. $-2(4z + 9)$ b. $5(4w - 8y + 1)$

For Exercises 119–124, simplify the expression.

119. $2p - (p + 5) + 3$

120. $6(h + 3) - 7h - 4$

121. $\frac{1}{2}(-6q) + q - 4\left(3q + \frac{1}{4}\right)$

122. $0.3b + 12(0.2 - 0.5b)$

123. $-4[2(x + 1) - (3x + 8)]$

124. $5[(7y - 3) + 3(y + 8)]$

chapter 1 TEST

1. Is $0.\overline{315}$ a rational number or irrational number? Explain your reasoning.

2. Plot the points on a number line: $|3|$, 0, -2, 0.5, $\left|-\frac{3}{2}\right|$, $\sqrt{16}$.

3. Use the number line in Exercise 2 to identify whether the statements are true or false.

   a. $|3| < -2$ b. $0 \leq \left|-\frac{3}{2}\right|$

   c. $-2 < 0.5$ d. $|3| \geq \left|-\frac{3}{2}\right|$

4. Use the definition of exponents to expand the expressions:

   a. $(4x)^3$ b. $4x^3$

5. a. Translate the expression into an English phrase: $2(a - b)$. (Answers may vary.)

   b. Translate the expression into an English phrase: $2a - b$. (Answers may vary.)

6. Translate the phrase into an algebraic expression: "The quotient of the square root of c and the square of d."

For Exercises 7–21, perform the indicated operations.

7. $18 + (-12)$ 8. $21 - (-7)$

9. $-\frac{1}{8} + \left(-\frac{3}{4}\right)$ 10. $-10.06 - (-14.72)$

11. $-14 + (-2) - 16$

12. $-84 \div 7$ 13. $38 \div 0$

14. $7(-4)$ 15. $-22 \cdot 0$

16. $(-16)(-2)(-1)(-3)$ 17. $\frac{2}{5} \div \left(-\frac{7}{10}\right) \cdot \left(-\frac{7}{6}\right)$

18. $(8 - 10)\frac{3}{2} + (-5)$

19. $8 - [(2 - 4) - (8 - 9)]$

20. $\dfrac{\sqrt{5^2 - 4^2}}{|-12 + 3|}$ 21. $\dfrac{|4 - 10|}{2 - 3(5 - 1)}$

22. Identify the property that justifies each statement.

   a. $6(-8) = (-8)6$

   b. $5 + 0 = 5$

   c. $(2 + 3) + 4 = 2 + (3 + 4)$

   d. $\frac{1}{7} \cdot 7 = 1$

   e. $8[7(-3)] = (8 \cdot 7)(-3)$

For Exercises 23–26, simplify the expression.

23. $3k - 20 + (-9k) + 12$

24. $4(p - 5) - (8p + 3)$

25. $\frac{1}{2}(12p - 4) + \frac{1}{3}(2 - 6p)$

26. $-3.2(1.6x - 4.1) + 1.4x$

For Exercises 27–28, evaluate the expression given the values $x = -6$, $y = 2$, and $z = -7$.

27. $-x^2 - 4y + z$ 28. $|y - z|$

For Exercises 29–30, simplify the expressions.

29. $(-8)(-6)(-4)(-3)$ 30. $-8 - 6 - 4 - 3$

For Exercises 31–32, translate the English statement to an algebraic expression. Then simplify the expression.

31. Subtract -4 from 12.

32. Find the difference of 6 and 8.

chapter

2

LINEAR EQUATIONS AND INEQUALITIES

The construction, from 1932 to 1937, of the Golden Gate Bridge in San Francisco is perhaps one of the greatest architectural achievements of the twentieth century. The total length of the bridge including the approaches spans 1.7 miles and has a total weight of 887,000 tons. The bridge has two main cables that pass over the tops of two large towers. Together, these cables contain 80,000 miles of steel wire—enough to circle the earth three times.

During the construction of the bridge, workers had to contend with foggy conditions, strong ocean currents, and powerful winds sweeping in from the Pacific Ocean. Eleven men lost their lives during construction.

The design and implementation of an engineering project of this magnitude requires heavy reliance on mathematics. In this chapter, we begin building a foundation of algebraic skills through the study of linear equations and related applications. For more information about applications of linear equations see the Technology Connections in MathZone at

www.mhhe.com/miller_oneill

Concepts

1. Definition of a Linear Equation in One Variable
2. Addition and Subtraction Properties of Equality
3. Multiplication and Division Properties of Equality
4. Translations
5. Applications of Linear Equations

section

2.1 Addition, Subtraction, Multiplication, and Division Properties of Equality

1. Definition of a Linear Equation in One Variable

An **equation** is a mathematical statement that indicates that two algebraic expressions are equal. Therefore, an equation must contain an equal sign, =. Some examples of equations are

$$x - 3 = 9 \qquad y^2 = 25 \qquad \frac{-3}{w} + \frac{1}{2} = 0$$

A **solution to an equation** is a value of the variable that makes the equation a true statement. That is, a solution to an equation makes the right-hand side of the equation equal to the left-hand side.

Equation:	Solution(s):	Check:	
$x - 3 = 9$	$x = 12$	$(12) - 3 \stackrel{?}{=} 9$ $9 = 9$ ✔	
$y^2 = 25$	$y = 5$ and $y = -5$	$(5)^2 \stackrel{?}{=} 25$ $25 = 25$ ✔	$(-5)^2 \stackrel{?}{=} 25$ $25 = 25$ ✔
$\frac{-3}{w} + \frac{1}{2} = 0$	$w = 6$	$\frac{-3}{(6)} + \frac{1}{2} = 0$ $-\frac{1}{2} + \frac{1}{2} \stackrel{?}{=} 0$ $0 = 0$ ✔	

The set of all solutions to an equation is called the **solution set**. Hence the solution set for $x - 3 = 9$ is $\{12\}$, the solution set for $y^2 = 25$ is $\{5, -5\}$ and the solution set for $\frac{-3}{w} + \frac{1}{2} = 0$ is $\{6\}$.

In the study of algebra, you will encounter a variety of equations. In this chapter, we will focus on a specific type of equation called a linear equation in one variable.

Definition of a Linear Equation in One Variable

Let a and b be real numbers such that $a \neq 0$. A **linear equation in one variable** is an equation that can be written in the form

$$ax + b = 0$$

Because the variable has an implied exponent of 1, a linear equation is sometimes called a first-degree equation. The following equations are linear equations in one variable.

$$2x + 3 = 0 \qquad \frac{1}{5}a - \frac{2}{7} = 0 \qquad -3.5z + 14 = 0$$

2. Addition and Subtraction Properties of Equality

If two equations have the same solution set, then the equations are said to be equivalent. For example, the following equations are equivalent because the solution set for each equation is $\{6\}$.

Equivalent Equations:	Check the Solution $x = 6$:
$2x - 5 = 7$	$2(6) - 5 = 7 \Rightarrow 12 - 5 = 7$ ✔
$2x = 12$	$2(6) = 12 \Rightarrow 12 = 12$ ✔
$x = 6$	$6 = 6 \Rightarrow 6 = 6$ ✔

To solve a linear equation, $ax + b = 0$, the goal is to find *all* values of x that make the equation true. One general strategy for solving an equation is to rewrite it as an equivalent but simpler equation. This process is repeated until the equation can be written in the form $x =$ number. The addition and subtraction properties of equality help us do this.

Addition and Subtraction Properties of Equality

Let a, b, and c represent algebraic expressions.

1. **Addition property of equality:** If $a = b$, then, $a + c = b + c$
2. **Subtraction property of equality:** If $a = b$, then, $a - c = b - c$

The addition and subtraction properties of equality indicate that adding or subtracting the same quantity to each side of an equation results in an equivalent equation. This is true because if two quantities are increased or decreased by the same amount, then the resulting quantities will also be equal (Figure 2-1).

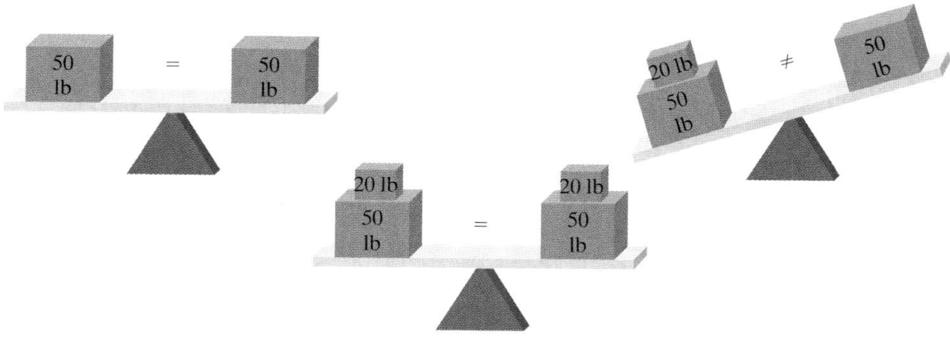

Figure 2-1

example 1

Applying the Addition and Subtraction Properties of Equality

Solve the equations.

a. $p - 4 = 11$ b. $w + 5 = -2$ c. $\frac{9}{4} = q - \frac{3}{4}$ d. $-1.2 + z = 4.6$

Solution:

In each equation, the goal is to isolate the variable on one side of the equation. To accomplish this, we use the fact that the sum of a number and its opposite is zero and the difference of a number and itself is zero.

a. $p - 4 = 11$

$p - 4 + 4 = 11 + 4$ To isolate p, add 4 to both sides $(-4 + 4 = 0)$.

$p + 0 = 15$ Simplify.

$p = 15$

Check: $p - 4 = 11$ Check the solution by substituting $p = 15$ back in the original equation.

$15 - 4 \stackrel{?}{=} 11$

$11 \stackrel{?}{=} 11$ ✔

b. $w + 5 = -2$

$w + 5 - 5 = -2 - 5$ To isolate w, subtract 5 from both sides. $(5 - 5 = 0)$.

$w + 0 = -7$ Simplify.

$w = -7$

Check: $w + 5 = -2$ Check the solution by substituting $w = -7$ back in the original equation.

$-7 + 5 \stackrel{?}{=} -2$

$-2 \stackrel{?}{=} -2$ ✔

Tip: The variable may be isolated on either side of the equation.

c. $\frac{9}{4} = q - \frac{3}{4}$

$\frac{9}{4} + \frac{3}{4} = q - \frac{3}{4} + \frac{3}{4}$ To isolate q, add $\frac{3}{4}$ to both sides $(-\frac{3}{4} + \frac{3}{4} = 0)$.

$\frac{12}{4} = q + 0$ Simplify.

$3 = q$ or equivalently $q = 3$

Check: $\frac{9}{4} = q - \frac{3}{4}$ Check the solution.

$\frac{9}{4} \stackrel{?}{=} 3 - \frac{3}{4}$ Substitute $q = 3$ in the original equation.

$\frac{9}{4} \stackrel{?}{=} \frac{12}{4} - \frac{3}{4}$ Get a common denominator.

$\frac{9}{4} \stackrel{?}{=} \frac{9}{4}$ ✔

d. $-1.2 + z = 4.6$

$-1.2 + 1.2 + z = 4.6 + 1.2$ To isolate z, add 1.2 to both sides.

$0 + z = 5.8$

$z = 5.8$

Check: $-1.2 + z = 4.6$ Check the equation.

$-1.2 + 5.8 \stackrel{?}{=} 4.6$ Substitute $z = 5.8$ in the original equation.

$4.6 \stackrel{?}{=} 4.6$ ✔

3. Multiplication and Division Properties of Equality

Adding or subtracting the same quantity to both sides of an equation results in an equivalent equation. In a similar way, multiplying or dividing both sides of an equation by the same nonzero quantity also results in an equivalent equation. This is stated formally as the multiplication and division properties of equality.

Multiplication and Division Properties of Equality

Let a, b, and c represent algebraic expressions.

1. **Multiplication property of equality**: If $a = b$ then, $ac = bc$

2. **Division property of equality**: If $a = b$ then, $\frac{a}{c} = \frac{b}{c}$ (provided $c \neq 0$)

To understand the multiplication property of equality, suppose we start with a true equation such as $10 = 10$. If both sides of the equation are multiplied by a constant such as 3, the result is also a true statement (Figure 2-2).

$10 = 10$

$3 \cdot 10 = 3 \cdot 10$

$30 = 30$

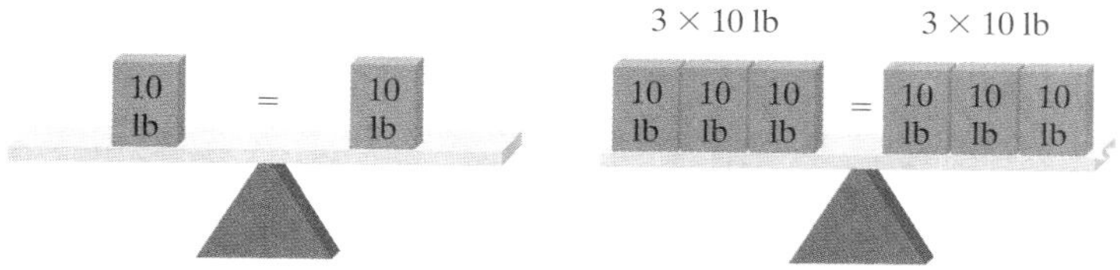

Figure 2-2

Similarly, if the equation is divided by a nonzero real number such as 2, the result is also a true statement (Figure 2-3).

$$10 = 10$$
$$\frac{10}{2} = \frac{10}{2}$$
$$5 = 5$$

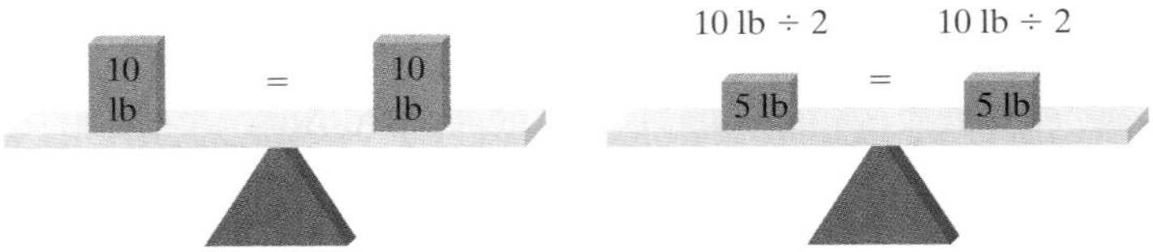

10 lb | = | 10 lb
10 lb ÷ 2 | 10 lb ÷ 2
5 lb | = | 5 lb

Figure 2-3

Tip: Recall that the product of a number and its reciprocal is 1. For example,

$$\frac{1}{5}(5) = 1$$
$$\frac{3}{2} \cdot \frac{2}{3} = 1,$$
$$-\frac{7}{2}\left(-\frac{2}{7}\right) = 1$$

To solve an equation in the variable x, the goal is to write the equation in the form $x =$ number. In particular, notice that we desire the coefficient of x to be 1. That is, we want to write the equation as $1x =$ number. Therefore, to solve an equation such as $5x = 15$, we can multiply both sides of the equation by the reciprocal of the x-term coefficient. In this case, multiply both sides by the reciprocal of 5, which is $\frac{1}{5}$.

$$5x = 15$$
$$\frac{1}{5}(5x) = \frac{1}{5}(15) \quad \text{Multiply by } \tfrac{1}{5}.$$
$$1x = 3 \quad \text{The coefficient of the } x\text{-term is now 1.}$$
$$x = 3$$

Tip: Recall that the quotient of a nonzero real number and itself is 1. For example,

$$\frac{5}{5} = 1$$
$$\frac{\frac{3}{2}}{\frac{3}{2}} = 1$$
$$\frac{-3.5}{-3.5} = 1$$

The division property of equality can also be used to solve the equation $5x = 15$ by dividing both sides by the coefficient of the x-term. In this case, divide both sides by 5 to make the coefficient of x equal to 1.

$$5x = 15$$
$$\frac{5x}{5} = \frac{15}{5} \quad \text{Divide by 5.}$$
$$1x = 3 \quad \text{The coefficient on the } x\text{-term is now 1.}$$
$$x = 3$$

example 2 **Applying the Multiplication and Division Properties of Equality**

Solve the equations using the multiplication or division properties of equality.

a. $12x = 60$ b. $-\frac{2}{9}q = \frac{1}{3}$ c. $-24.752 = -2.72z$

d. $\frac{d}{6} = -4$ e. $-x = 8$

Solution:

a. $12x = 60$

$\dfrac{12x}{12} = \dfrac{60}{12}$ To obtain a coefficient of 1 for the x-term, divide both sides by 12.

$1x = 5$ Simplify.

$x = 5$ Check: $12x = 60$

$12(5) \stackrel{?}{=} 60$

$60 \stackrel{?}{=} 60$ ✔

Tip: When applying the multiplication or division properties of equality to obtain a coefficient of 1 for the variable term, we will generally use the following convention:

- If the coefficient of the variable term is expressed as a fraction, we will usually multiply both sides by its reciprocal.
- Otherwise, we will divide both sides by the coefficient itself.

b. $-\dfrac{2}{9}q = \dfrac{1}{3}$

$\left(-\dfrac{9}{2}\right)\left(-\dfrac{2}{9}q\right) = \dfrac{1}{3}\left(-\dfrac{9}{2}\right)$ To obtain a coefficient of 1 for the q-term, multiply by the reciprocal of $-\frac{2}{9}$, which is $-\frac{9}{2}$.

$1q = -\dfrac{3}{2}$ Simplify. The product of a number and its reciprocal is 1.

$q = -\dfrac{3}{2}$ The solution checks in the original equation.

c. $-24.752 = -2.72z$

$\dfrac{-24.752}{-2.72} = \dfrac{-2.72z}{-2.72}$ To obtain a coefficient of 1 for the z-term, divide by -2.72.

$9.1 = 1z$ Simplify.

$9.1 = z$

$z = 9.1$ The solution checks.

d. $\dfrac{d}{6} = -4$

$\dfrac{1}{6}d = -4$ $\frac{d}{6}$ is equivalent to $\frac{1}{6}d$.

$\dfrac{6}{1} \cdot \dfrac{1}{6}d = -4 \cdot \dfrac{6}{1}$ To obtain a coefficient of 1 for the d-term, multiply by the reciprocal of $\frac{1}{6}$, which is $\frac{6}{1}$.

$1d = -24$ Simplify.

$d = -24$ The solution checks.

e. $-x = 8$ Note that $-x$ is equivalent to $-1 \cdot x$.

$-1x = 8$

$\frac{-1x}{-1} = \frac{8}{-1}$ To obtain a coefficient of 1 for the x-term, divide by -1.

$x = -8$ The solution checks.

It is important to distinguish between cases where the addition or subtraction properties of equality should be used to isolate a variable versus those in which the multiplication or division properties of equality should be used. Remember the goal is to isolate the variable term and obtain a coefficient of 1. Compare the equations:

$$5 + x = 20 \qquad \text{and} \qquad 5x = 20$$

In the first equation, the relationship between 5 and x is addition. Therefore, we want to reverse the process by subtracting 5 from both sides. In the second equation, the relationship between 5 and x is multiplication. To isolate x, we reverse the process by dividing by 5 or equivalently, multiplying by the reciprocal, $\frac{1}{5}$.

$$5 + x = 20 \qquad \text{and} \qquad 5x = 20$$

$$5 - 5 + x = 20 - 5 \qquad\qquad \frac{5x}{5} = \frac{20}{5}$$

$$x = 15 \qquad\qquad x = 4$$

4. Translations

example 3

Translating to a Linear Equation

Write an algebraic equation to represent each English sentence. Then solve the equation.

a. The quotient of a number and 4 is 6.
b. The product of a number and 4 is 6.
c. Negative twelve is equal to the sum of -5 and a number.
d. The value 1.4 subtracted from a number is 5.7.

Solution:

For each case we will let x represent the unknown number.

a. The quotient of a number and 4 is 6.

$\frac{x}{4} = 6$

$4 \cdot \frac{x}{4} = 4 \cdot 6$ Multiply both sides by 4.

$\frac{4}{1} \cdot \frac{x}{4} = 4 \cdot 6$

$x = 24$ Check: $\frac{24}{4} = 6$ ✔

b. The product of a number and 4 is 6.

$$4x = 6$$

$$\frac{4x}{4} = \frac{6}{4} \qquad \text{Divide both sides by 4.}$$

$$x = \frac{3}{2} \qquad \underline{\text{Check}}: 4(\tfrac{3}{2}) = 6 \checkmark$$

c. Negative twelve is equal to the sum of -5 and a number.

$$-12 = -5 + x$$

$$-12 + 5 = -5 + 5 + x \qquad \text{Add 5 to both sides.}$$

$$-7 = x \qquad \underline{\text{Check}}: -12 = -5 + (-7) \checkmark$$

d. The value 1.4 subtracted from a number is 5.7.

$$x - 1.4 = 5.7$$

$$x - 1.4 + 1.4 = 5.7 + 1.4 \qquad \text{Add 1.4 to both sides.}$$

$$x = 7.1 \qquad \underline{\text{Check}}: 7.1 - 1.4 = 5.7 \checkmark$$

5. Applications of Linear Equations

example 4 **Using Linear Equations in an Application**

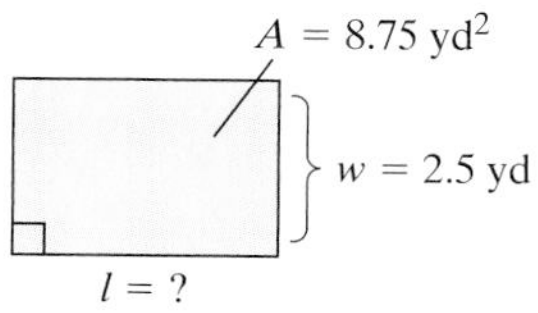

$A = 8.75 \text{ yd}^2$

$w = 2.5$ yd

$l = ?$

Figure 2-4

The area, A, of a rectangle is given by the formula $A = lw$, where l represents the length of the rectangle and w represents the width. Find the length of a rectangular tablecloth whose width is 2.5 yd and whose area is 8.75 yd^2 (Figure 2-4).

Solution:

$$A = lw$$

$$8.75 = l \cdot (2.5) \qquad \text{Substitute } A = 8.75 \text{ and } w = 2.5, \text{ and solve for } l.$$

$$\frac{8.75}{2.5} = \frac{l \cdot (2.5)}{2.5} \qquad \text{To obtain a coefficient of 1 for } l, \text{ divide both sides by 2.5.}$$

$$3.5 = l$$

The length of the tablecloth is 3.5 yd.

section 2.1 PRACTICE EXERCISES

For Exercises 1–8, identify the following as either an expression or an equation.

1. $5x + 3x$
2. $x - 4 + 5x$
3. $8x + 2 = 7$
4. $9 = 2x - 4$
5. $3x^2 + x = -3$
6. $7x^3 + x = 10$
7. $8 - x^3 + 12x$
8. $7x - 5 + x^2$
9. Explain how to determine if a number is a solution to an equation.
10. Is $x = 3$ a solution to the equation $x^2 - 6 = 5$?
11. Is $x = 3$ a solution to the equation $x^2 - 2 = 0$?
12. Is $x = -2$ a solution to the equation $3x + 9 = 3$?
13. Is $x = -4$ a solution to the equation $x + 5 = 1$?
14. Determine if $x = 2$ is a solution to any of the following equations:
    a. $2x + 1 = 5$ b. $x^2 - 2 = 3$
    c. $x - 5 = 0$ d. $\frac{2}{x} + 3 = 4$
15. Determine if $x = 0$ is a solution to any of the following equations:
    a. $5x + 2 = 6$ b. $4 = x + 4$
    c. $x^2 - 4 = 4$ d. $6x^2 + 1 = 1$
16. Explain how to check an equation.

For Exercises 17–36, solve the equation using the addition or subtraction property of equality. Be sure to check your answers.

17. $x + 6 = 5$
18. $x - 2 = 10$
19. $q - 14 = 6$
20. $w + 3 = -5$
21. $2 + m = -15$
22. $-6 + n = 10$
23. $-23 = y - 7$
24. $-9 = -21 + b$
25. $5 = z - \frac{1}{2}$
26. $-7 = p + \frac{2}{3}$
27. $x + \frac{5}{2} = \frac{1}{2}$
28. $x - \frac{2}{3} = \frac{7}{3}$
29. $4.1 = 2.8 + a$
30. $5.1 = -2.5 + y$
31. $4 + c = 4$
32. $-13 + b = -13$
33. $-6.02 + c = -8.15$
34. $p + 0.035 = -1.12$
35. $3.245 + t = -0.0225$
36. $-1.004 + k = 3.0589$

For Exercises 37–42, write an algebraic equation to represent each English sentence. (Let x represent the unknown number.) Then solve the equation.

37. The sum of negative eight and a number is forty-two.
38. The sum of thirty-one and a number is thirteen.
39. The difference of a number and negative six is eighteen.
40. The sum of negative twelve and a number is negative fifteen.
41. The sum of a number and $\frac{5}{8}$ is $\frac{13}{8}$.
42. The difference of a number and $\frac{2}{3}$ is $\frac{1}{3}$.

For Exercises 43–62, solve the equations using the multiplication or division property of equality. Be sure to check your answers.

43. $6x = 54$
44. $2w = 8$
45. $12 = -3p$
46. $6 = -2q$
47. $-5y = 0$
48. $-3k = 0$
49. $-\frac{y}{5} = 3$
50. $-\frac{z}{7} = 1$
51. $\frac{4}{5} = -t$
52. $-\frac{3}{7} = -h$
53. $\frac{2}{5}a = -4$
54. $\frac{3}{8}b = -9$
55. $-\frac{1}{5}b = -\frac{4}{5}$
56. $-\frac{3}{10}w = \frac{2}{5}$

57. $-41 = -x$

58. $32 = -y$

59. $3.81 = -0.03p$

60. $2.75 = -0.5q$

61. $5.82y = -15.132$

62. $-32.3x = -0.4522$

For Exercises 63–66, write an algebraic equation to represent each English sentence. (Let x represent the unknown number.) Then solve the equation.

63. The product of a number and seven is negative sixty-three.

64. The product of negative three and a number is 24.

65. The quotient of a number and 12 is one-third.

66. Eighteen is equal to the quotient of a number and two.

For Exercises 67–94, solve the equation using the appropriate property of equality.

67. $a - 9 = 1$

68. $b - 2 = -4$

69. $1 = -9x$

70. $-4 = -2k$

71. $-\frac{2}{3}h = 8$

72. $\frac{3}{4}p = 15$

73. $\frac{2}{3} + t = 8$

74. $\frac{3}{4} + y = 15$

75. $\frac{r}{3} = -12$

76. $\frac{d}{-4} = 5$

77. $k + 16 = 32$

78. $-18 = -9 + t$

79. $16k = 32$

80. $-18 = -9t$

81. $7 = -4q$

82. $-3s = 10$

83. $-4 + q = 7$

84. $s - 3 = 10$

85. $-\frac{1}{3}d = 12$

86. $-\frac{2}{5}m = 10$

87. $4 = \frac{1}{2} + z$

88. $3 = \frac{1}{4} + p$

89. $1.2y = 4.8$

90. $4.3w = 8.6$

91. $4.8 = 1.2 + y$

92. $8.6 = w - 4.3$

93. $0.0034 = y - 0.405$

94. $-0.98 = m + 1.0034$

The formula to find the area of a rectangle is given by $A = lw$ (area equals length times width). Use this formula to solve for the indicated variables in Exercises 95–98.

95. Find the length of a rectangle with width 23 cm and area 322 cm^2.

96. Find the length of a rectangle with width 10 in. and area 150 in^2.

97. Find the width of the rectangular table shown in the figure.

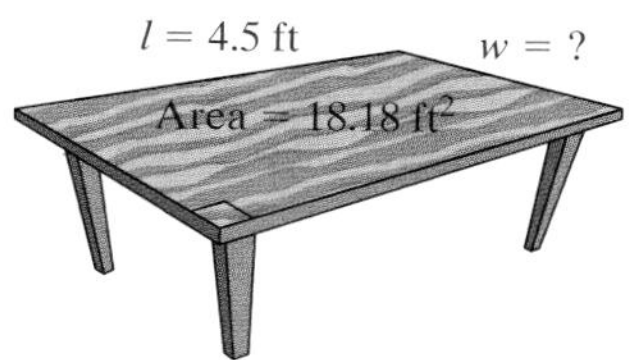

Figure for Exercise 97

98. Find the width of the rectangular garden shown in the figure.

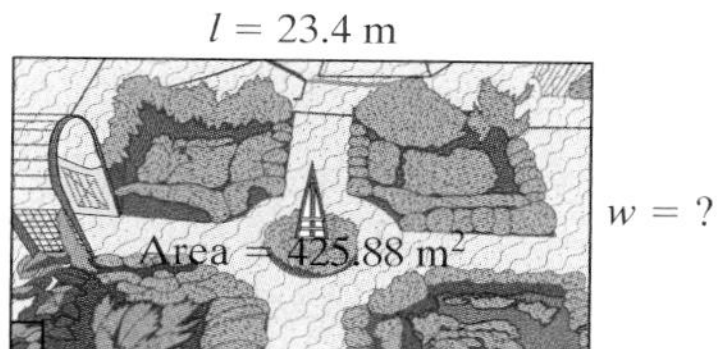

Figure for Exercise 98

Expanding Your Skills

A relationship among the variables distance (d), rate (r), and time (t) is given by the formula, $d = rt$ (distance equals rate times time). Use this formula to solve for the indicated variables in Exercises 99–102.

99. If Jonas runs at a rate of 4.5 mph, how long will it take him to run 6 miles?

100. If Tatyana walks at a rate of 3 mph, how long will it take her to walk 8 miles?

101. An in-line skater skated 6 miles in 45 min ($\frac{3}{4}$ hr). What was her average speed in mph?

102. A skate-boarder skated 3 miles in 30 min ($\frac{1}{2}$ hr). What was his average speed in mph?

section

2.2 Solving Linear Equations

Concepts

1. Solving Linear Equations Involving Multiple Steps
2. Steps to Solve a Linear Equation in One Variable
3. Conditional Equations, Identities, and Contradictions

1. Solving Linear Equations Involving Multiple Steps

In Section 2.1 we studied a one-step process to solve linear equations by using the addition, subtraction, multiplication, and division properties of equality. In this section, we will learn how to solve linear equations that require multiple steps. When solving an equation always keep in mind that the ultimate goal is to isolate the variable.

example 1

Solving a Linear Equation

Solve the equation: $-2w - 7 = 11$

Solution:

In the expression $-2w - 7$, w is first multiplied by -2 and then 7 is subtracted. To solve the equation $-2w - 7 = 11$, we must reverse these steps to isolate w. Therefore, we first add 7 and then divide by -2.

$$-2w - 7 = 11$$

$$-2w - 7 + 7 = 11 + 7$$ Add 7 to both sides of the equation. This isolates the w-term.

$$-2w = 18$$

$$\frac{-2w}{-2} = \frac{18}{-2}$$ Next, apply the division property of equality to obtain a coefficient of 1 for w. Divide by -2 on both sides.

$$w = -9$$

Check:

$$-2w - 7 = 11$$

$$-2(-9) - 7 \stackrel{?}{=} 11$$ Substitute $w = -9$ in the original equation.

$$18 - 7 \stackrel{?}{=} 11$$

$$11 = 11 \checkmark$$

In Example 2, the variable x appears on both sides of the equation. In this case, apply the addition or subtraction properties of equality to collect the variable terms on one side of the equation and the constant terms on the other side. Then use the multiplication or division properties of equality to get a coefficient of 1.

example 2

Solving a Linear Equation

Solve the equation: $6x - 4 = 2x - 8$

Solution:

$$6x - 4 = 2x - 8$$

$$6x - 2x - 4 = 2x - 2x - 8$$ Subtract $2x$ from both sides.

$4x - 4 = 0x - 8$	Simplify.
$4x - 4 = -8$	The x-terms have now been combined on one side of the equation.
$4x - 4 + 4 = -8 + 4$	Add 4 to both sides of the equation. This combines the constant terms on the other side of the equation.
$4x = -4$	
$\dfrac{4x}{4} = \dfrac{-4}{4}$	To obtain a coefficient of 1 for x, divide both sides of the equation by 4.
$x = -1$	The solution checks.

Tip: It is important to note that the variable may be isolated on either side of the equation. We will solve the equation from Example 2 again, this time isolating the variable on the right-hand side.

$6x - 4 = 2x - 8$	
$6x - 6x - 4 = 2x - 6x - 8$	Subtract $6x$ on both sides.
$0x - 4 = -4x - 8$	
$-4 = -4x - 8$	
$-4 + 8 = -4x - 8 + 8$	Add 8 to both sides.
$4 = -4x$	
$\dfrac{4}{-4} = \dfrac{-4x}{-4}$	Divide both sides by -4.
$-1 = x$ or equivalently $x = -1$	

2. Steps to Solve a Linear Equation in One Variable

In some cases it is necessary to simplify both sides of a linear equation before applying the properties of equality. Therefore, we offer the following steps to solve a linear equation in one variable.

Steps to Solve a Linear Equation in One Variable

1. Simplify both sides of the equation by clearing parentheses and combining *like* terms.
2. Use the addition and subtraction properties of equality to collect the variable terms on one side of the equation.
3. Use the addition and subtraction properties of equality to collect the constant terms on the other side of the equation.
4. Use the multiplication and division properties of equality to make the coefficient of the variable term equal to 1.
5. Check your answer.

example 3 Solving Linear Equations

Solve the equations:

a. $\frac{1}{5}x + 3 = 2$ b. $2.2y - 8.3 = 6.2y + 12.1$ c. $7 + 3 = 2(p - 3)$

Solution:

a. $\frac{1}{5}x + 3 = 2$

Step 1: The right- and left-hand sides are already simplified.

Step 2: All variable terms are already on one side of the equation (the left side).

$\frac{1}{5}x + 3 - 3 = 2 - 3$

$\frac{1}{5}x = -1$

Step 3: Use the subtraction property of equality to collect the constant terms on the other side of the equation (the right side).

$5\left(\frac{1}{5}x\right) = 5(-1)$

$1x = -5$

Step 4: Use the multiplication property of equality to make the coefficient of the x-term equal to 1.

$x = -5$

Step 5: Check: $\frac{1}{5}x + 3 = 2$

$\frac{1}{5}(-5) + 3 \stackrel{?}{=} 2$

$-1 + 3 \stackrel{?}{=} 2$

$2 = 2$ ✔

Tip: In Example 3(b), the variable terms were moved to the right side of the equation. Once that decision was made you have no choice but to move the constants to the other side, the left side.

b. $2.2y - 8.3 = 6.2y + 12.1$

Step 1: The right- and left-hand sides are already simplified.

$2.2y - 2.2y - 8.3 = 6.2y - 2.2y + 12.1$

$-8.3 = 4.0y + 12.1$

Step 2: Subtract $2.2y$ from both sides to collect the variable terms on one side of the equation.

$-8.3 - 12.1 = 4.0y + 12.1 - 12.1$

$-20.4 = 4.0y$

Step 3: Subtract 12.1 from both sides to collect the constant terms on the other side.

$\frac{-20.4}{4.0} = \frac{4.0y}{4.0}$

$-5.1 = y$

Step 4: To obtain a coefficient of 1 for the y-term, divide both sides of the equation by 4.0.

$y = -5.1$

Step 5: The check is left to the reader.

c. $7 + 3 = 2(p - 3)$

$10 = 2p - 6$ **Step 1:** Simplify both sides of the equation by clearing parentheses and combining *like* terms.

Step 2: The variable terms are already on one side of the equation.

$10 + 6 = 2p - 6 + 6$ **Step 3:** Add 6 to both sides to collect the constant terms on the *other* side.

$16 = 2p$

$\frac{16}{2} = \frac{2p}{2}$ **Step 4:** Divide both sides by 2 to obtain a coefficient of 1 for p.

$8 = p$ or $p = 8$ **Step 5:** The check is left to the reader.

Example 4 illustrates the procedure for solving linear equations when multiple steps of simplification are necessary.

example 4 Solving Linear Equations

Solve the equations:

a. $2 + 7x - 5 = 6(x + 3) + 2x$
b. $2[9 - (z - 3) + 4z] = 4z - 5(z + 2) - 8$

Solution:

a. $2 + 7x - 5 = 6(x + 3) + 2x$

$-3 + 7x = 6x + 18 + 2x$ **Step 1:** Add *like* terms on the left. Clear parentheses on the right.

$-3 + 7x = 8x + 18$ Combine *like* terms.

$-3 + 7x - 7x = 8x - 7x + 18$ **Step 2:** Subtract $7x$ from both sides.

$-3 = x + 18$ Simplify.

$-3 - 18 = x + 18 - 18$ **Step 3:** Subtract 18 from both sides.

$-21 = x$ **Step 4:** Because the coefficient of the x-term is already 1, there is no need to apply the multiplication or division property of equality.

$x = -21$

Step 5: The check is left to the reader.

b. $2[9 - (z - 3) + 4z] = 4z - 5(z + 2) - 8$

$2(9 - z + 3 + 4z) = 4z - 5z - 10 - 8$ **Step 1:** Clear parentheses.

$2(12 + 3z) = -z - 18$ Combine *like* terms.

$24 + 6z = -z - 18$ Clear parentheses.

Tip: In Example 4(b), the variable terms were collected on the left side of the equation. Therefore, the constants were collected on the right side.

$$24 + 6z + z = -z + z - 18$$ **Step 2:** Add z to both sides.

$$24 + 7z = -18$$

$$24 - 24 + 7z = -18 - 24$$ **Step 3:** Subtract 24 from both sides.

$$7z = -42$$

$$\frac{7z}{7} = \frac{-42}{7}$$ **Step 4:** Divide both sides by 7.

$$z = -6$$ **Step 5:** The check is left for the reader.

3. Conditional Equations, Identities, and Contradictions

The solutions to an equation are the values of x that make the equation a true statement. Some equations have one unique solution, while others may have no solution or infinitely many solutions.

I. Conditional Equations

An equation that is true for some values of the variable but false for other values is called a **conditional equation**. The equation $x + 4 = 6$, for example, is true on the condition that $x = 2$. For other values of x, the statement $x + 4 = 6$ is false.

II. Contradictions

Some equations have no solution, such as $x + 1 = x + 2$. There is no value of x, that when increased by 1 will equal the same value increased by 2. If we tried to solve the equation by subtracting x from both sides, we get the contradiction $1 = 2$. This indicates that the equation has no solution. An equation that has no solution is called a **contradiction**.

$$x + 1 = x + 2$$

$$x - x + 1 = x - x + 2$$

$$1 = 2 \quad \text{(contradiction)}$$ No solution.

III. Identities

An equation that has all real numbers as its solution set is called an **identity**. For example, consider the equation, $x + 4 = x + 4$. Because the left- and right-hand sides are identically equal, any real number substituted for x will result in equal quantities on both sides. If we subtract x from both sides of the equation, we get the identity $4 = 4$. In such a case, the solution is the set of all real numbers.

$$x + 4 = x + 4$$

$$x - x + 4 = x - x + 4$$

$$4 = 4 \quad \text{(identity)}$$ The solution is all real numbers.

example 5 **Identifying Conditional Equations, Contradictions, and Identities**

Identify each equation as a conditional equation, a contradiction, or an identity. Then describe the solution.

a. $4k - 5 = 2(2k - 3) + 1$ b. $2(b - 4) = 2b - 7$ c. $3x + 7 = 2x - 5$

Solution:

a.

$4k - 5 = 2(2k - 3) + 1$	
$4k - 5 = 4k - 6 + 1$	Clear parentheses.
$4k - 5 = 4k - 5$	Combine *like* terms.
$4k - 4k - 5 = 4k - 4k - 5$	Subtract $4k$ from both sides.
$-5 = -5$	

This is an identity. The solution is all real numbers.

b.

$2(b - 4) = 2b - 7$	
$2b - 8 = 2b - 7$	Clear parentheses.
$2b - 2b - 8 = 2b - 2b - 7$	Subtract $2b$ from both sides.
$-8 = -7$	

This is a contradiction. There is no solution.

c.

$3x + 7 = 2x - 5$	
$3x - 2x + 7 = 2x - 2x - 5$	Subtract $2x$ from both sides.
$x + 7 = -5$	Simplify.
$x + 7 - 7 = -5 - 7$	Add 7 to both sides.
$x = -12$	

This is a conditional equation. The solution is $x = -12$. (The equation is true only on the condition that $x = -12$.)

section 2.2 Practice Exercises

For Exercises 1–8, simplify the expressions by clearing parentheses and combining *like* terms.

1. $-3(4t) + 5t - 6$
2. $8(2x) - 3x + 9$
3. $5z + 2 - 7z - 3z$
4. $10 - 4w + 7w - 2 + w$
5. $-(-7p + 9) + (3p - 1)$
6. $8y - (2y + 3) - 19$
7. $5(3a) + 5(3 + a)$
8. $-2(6 + b) - 2(6b)$
9. Explain the difference between simplifying an expression and solving an equation.

For Exercises 10–17, solve the equations using the addition, subtraction, multiplication, or division property of equality.

10. $5w = -30$
11. $-7y = 21$
12. $x + 8 = -15$
13. $z - 23 = -28$

14. $6 = a - \frac{7}{8}$

15. $12 = b + \frac{1}{5}$

16. $-\frac{9}{8} = -\frac{3}{4}k$

17. $-\frac{3}{10} = -6h$

18. Which properties of equality would you apply to solve the equation $2x + 6 = 0$?

19. Which properties of equality would you apply to solve the equation $-2x - 6 = 0$?

For Exercises 20–45, solve the equations using the steps outlined in the text.

20. $6z + 1 = 13$

21. $5x + 2 = -13$

22. $3y - 4 = 14$

23. $-7w - 5 = -19$

24. $-2p + 8 = 3$

25. $4q + 5 = 2$

26. $6 = 7m - 1$

27. $-9 = 4n - 1$

28. $-\frac{1}{2} - 4x = 8$

29. $2b - \frac{1}{4} = 5$

30. $0.2x + 3.1 = -5.3$

31. $-1.8 + 2.4a = -6.6$

32. $\frac{5}{8} = \frac{1}{4} - \frac{1}{2}p$

33. $\frac{6}{7} = \frac{1}{7} + \frac{5}{3}r$

34. $7w - 6w + 1 = 10 - 4$

35. $5v - 3 - 4v = 13$

36. $11h - 8 - 9h = -16$

37. $6u - 5 - 8u = -7$

38. $3a + 7 = 2a - 19$

39. $6b - 20 = 14 + 5b$

40. $-4r - 28 = -78 - r$

41. $-6x - 7 = -3 - 8x$

42. $-2z - 8 = -z$

43. $-7t + 4 = -6t$

44. $\frac{5}{6}x + \frac{2}{3} = -\frac{1}{6}x - \frac{5}{3}$

45. $\frac{3}{7}x - \frac{1}{4} = -\frac{4}{7}x - \frac{5}{4}$

For Exercises 46–61, solve the equations using the steps outlined in the text.

46. $3(2p - 4) = 15$

47. $4(t + 15) = 20$

48. $6(3x + 2) - 10 = -4$

49. $4(2k + 1) - 1 = 5$

50. $2(y - 3) - y = 6$

51. $4(w - 5) - 3w = 2$

52. $17(s + 3) = 4(s - 10) + 13$

53. $5(4 + p) = 3(3p - 1) - 9$

54. $6(3t - 4) + 10 = 5(t - 2) - (3t + 4)$

55. $-5y + 2(2y + 1) = 2(5y - 1) - 7$

56. $-2[(4p + 1) - (3p - 1)] = 5(3 - p) - 9$

57. $5 - (6k + 1) = 2[(5k - 3) - (k - 2)]$

58. $0.2w - 0.47 = 0.53 - 0.2(2w - 13)$

59. $0.4z - 0.15 = 0.65 - 0.3(6 - 2z)$

60. $3(-0.9n + 0.5) = -3.5n + 1.3$

61. $7(0.4m - 0.1) = 5.2m + 0.86$

62. A conditional linear equation has (choose one): One solution, no solution, or infinitely many solutions.

63. An equation that is a contradiction has (choose one): One solution, no solution, or infinitely many solutions.

64. An equation that is an identity has (choose one): One solution, no solution, or infinitely many solutions.

For Exercises 65–74, identify as a conditional equation, a contradiction, or an identity. Then describe the solution.

65. $2(k - 7) = 2k - 13$

66. $5h + 4 = 5(h + 1) - 1$

67. $7x + 3 = 6(x - 2)$

68. $3y - 1 = 1 + 3y$

69. $3 - 5.2p = -5.2p + 3$

70. $2(q + 3) = 4q + q - 9$

71. $5(h - 1) - 1 = 2(h - 3) + 3h$

72. $2(5w - 3) - 5w = 5w - 6$

73. $2(5x + 7) - 2x = 2(7 - x)$

74. $6(2y - 1) - y = -(6 + y)$

Expanding Your Skills

75. Suppose $x = -5$ is a solution to the equation $x + a = 10$. Find the value of a.

76. Suppose $x = 6$ is a solution to the equation $x + a = -12$. Find the value of a.

77. Suppose $x = 3$ is a solution to the equation $ax = 12$. Find the value of a.

78. Suppose $x = 11$ is a solution to the equation $ax = 49.5$. Find the value of a.

For Exercises 79–80, identify as a conditional equation, a contradiction, or an identity.

79. $2(q + 1) - 1.5 = 5(0.4q + 0.16)$

80. $1.1(0.2k - 0.6) - 0.06 = 0.22(k + 1) + 0.05$

Concepts

1. Clearing Fractions and Decimals
2. Steps to Solving a Linear Equation
3. Solving Linear Equations with Fractions
4. Solving Linear Equations with Decimals

section

2.3 Linear Equations: Clearing Fractions and Decimals

1. Clearing Fractions and Decimals

Linear equations that contain fractions can be solved in different ways. The first procedure, illustrated here, uses the method outlined in Section 2.2.

$$\frac{5}{6}x - \frac{3}{4} = \frac{1}{3}$$

$$\frac{5}{6}x - \frac{3}{4} + \frac{3}{4} = \frac{1}{3} + \frac{3}{4}$$ To isolate the variable term, add $\frac{3}{4}$ to both sides.

$$\frac{5}{6}x = \frac{4}{12} + \frac{9}{12}$$ Find a common denominator on the right-hand side.

$$\frac{5}{6}x = \frac{13}{12}$$ Simplify.

$$\frac{6}{5}\left(\frac{5}{6}x\right) = \frac{\overset{1}{\cancel{6}}}{5}\left(\frac{13}{\underset{2}{\cancel{12}}}\right)$$ Multiply by the reciprocal of $\frac{5}{6}$, which is $\frac{6}{5}$.

$$x = \frac{13}{10}$$

Sometimes it is simpler to solve an equation with fractions by eliminating the fractions first using a process called **clearing fractions**. To clear fractions in the equation $\frac{5}{6}x - \frac{3}{4} = \frac{1}{3}$, we can multiply both sides of the equation by the least common denominator (LCD) of all terms in the equation. In this case, the LCD of $\frac{5}{6}x$, $\frac{3}{4}$, and $\frac{1}{3}$ is 12. Because each denominator in the equation is a factor of 12, we can reduce common factors to leave integer coefficients for each term.

example 1 **Solving a Linear Equation by Clearing Fractions**

Solve the equation $\frac{5}{6}x - \frac{3}{4} = \frac{1}{3}$ by clearing fractions first.

Solution:

Tip: Recall that the multiplication property of equality indicates that multiplying both sides of an equation by a nonzero constant results in an equivalent equation.

$$\frac{5}{6}x - \frac{3}{4} = \frac{1}{3}$$

$$12\left(\frac{5}{6}x - \frac{3}{4}\right) = 12\left(\frac{1}{3}\right)$$ Multiply both sides of the equation by the LCD, 12.

$$\frac{\overset{2}{\cancel{12}}}{1}\left(\frac{5}{\cancel{6}}x\right) - \frac{\overset{3}{\cancel{12}}}{1}\left(\frac{3}{\cancel{4}}\right) = \frac{\overset{4}{\cancel{12}}}{1}\left(\frac{1}{\cancel{3}}\right)$$ Apply the distributive property (recall that $12 = \frac{12}{1}$).

$$2(5x) - 3(3) = 4(1)$$ Reduce common factors to clear the fractions.

$$10x - 9 = 4$$

$$10x - 9 + 9 = 4 + 9$$ Add 9 to both sides.

$$10x = 13$$

$$\frac{10x}{10} = \frac{13}{10}$$ Divide both sides by 10.

$$x = \frac{13}{10}$$ Simplify.

Tip: The fractions in this equation can be eliminated by multiplying both sides of the equation by *any* common multiple of the denominators. For example, try multiplying both sides of the equation by 24:

$$24\left(\frac{5}{6}x - \frac{3}{4}\right) = 24\left(\frac{1}{3}\right)$$

$$\frac{\overset{4}{\cancel{24}}}{1}\left(\frac{5}{\cancel{6}}x\right) - \frac{\overset{6}{\cancel{24}}}{1}\left(\frac{3}{\cancel{4}}\right) = \frac{\overset{8}{\cancel{24}}}{1}\left(\frac{1}{\cancel{3}}\right)$$

$$20x - 18 = 8$$

$$20x = 26$$

$$\frac{20x}{20} = \frac{26}{20}$$

$$x = \frac{13}{10}$$

The same procedure used to clear fractions in an equation can be used to clear decimals. For example, consider the equation $0.05x + 0.25 = 0.2$. Because any terminating decimal can be written as a fraction, the equation can be interpreted as $\frac{5}{100}x + \frac{25}{100} = \frac{2}{10}$. A convenient common denominator for all terms in this equation is 100. Therefore, we can multiply the original equation by 100 to clear decimals.

example 2 Solving a Linear Equation by Clearing Decimals

Solve the equation $0.05x + 0.25 = 0.2$ by clearing decimals first.

Solution:

$0.05x + 0.25 = 0.2$	
$100(0.05x + 0.25) = 100(0.2)$	Multiply both sides of the equation by 100.
$100(0.05x) + 100(0.25) = 100(0.2)$	Apply the distributive property.
$5x + 25 = 20$	Simplify (decimals have been cleared).
$5x + 25 - 25 = 20 - 25$	Subtract 25 from both sides.
$5x = -5$	
$\frac{5x}{5} = \frac{-5}{5}$	Divide both sides by 5.
$x = -1$	Simplify.

This equation can be checked by hand or by using a calculator.

Tip: Notice that multiplying a decimal number by 100 has the effect of moving the decimal point two places to the right. Similarly, multiplying by 10 moves the decimal point one place to the right, multiplying by 1000 moves the decimal point three places to the right, and so on.

2. Steps to Solving a Linear Equation

In this section, we combine the process for clearing fractions and decimals with the general strategies for solving linear equations. To solve a linear equation, it is important to follow these steps.

Steps for Solving a Linear Equation in One Variable

1. Consider clearing fractions or decimals (if any are present) by multiplying both sides of the equation by a common denominator of all terms.
2. Simplify both sides of the equation by clearing parentheses and combining *like* terms.
3. Use the addition and subtraction properties of equality to collect the variable terms on one side of the equation.
4. Use the addition and subtraction properties of equality to collect the constant terms on the other side of the equation.
5. Use the multiplication and division properties of equality to make the coefficient of the variable term equal to 1.
6. Check your answer.

3. Solving Linear Equations with Fractions

example 3

Solving Linear Equations

a. $\frac{1}{6}x - \frac{2}{3} = \frac{1}{5}x - 1$ b. $\frac{1}{3}(x + 7) - \frac{1}{2}(x + 1) = 4$

Solution:

a.

$$\frac{1}{6}x - \frac{2}{3} = \frac{1}{5}x - 1$$ The LCD of $\frac{1}{6}x$, $\frac{2}{3}$, and $\frac{1}{5}x$ is 30.

$$30\left(\frac{1}{6}x - \frac{2}{3}\right) = 30\left(\frac{1}{5}x - 1\right)$$ Multiply by the LCD, 30.

$$\frac{\overset{5}{\cancel{30}}}{1} \cdot \frac{1}{\cancel{6}}x - \frac{\overset{10}{\cancel{30}}}{1} \cdot \frac{2}{\cancel{3}} = \frac{\overset{6}{\cancel{30}}}{1} \cdot \frac{1}{\cancel{5}}x - 30(1)$$ Apply the distributive property (recall $30 = \frac{30}{1}$).

$$5x - 20 = 6x - 30$$ Clear fractions.

$$5x - 6x - 20 = 6x - 6x - 30$$ Subtract $6x$ from both sides.

$$-x - 20 = -30$$

$$-x - 20 + 20 = -30 + 20$$ Add 20 to both sides.

$$-x = -10$$

$$\frac{-x}{-1} = \frac{-10}{-1}$$ Divide both sides by -1.

$$x = 10$$

b.

$$\frac{1}{3}(x + 7) - \frac{1}{2}(x + 1) = 4$$ The LCD of $\frac{1}{3}$ and $\frac{1}{2}$ is 6.

$$6\left[\frac{1}{3}(x + 7) - \frac{1}{2}(x + 1)\right] = 6(4)$$ Multiply both sides by 6.

$$\frac{6}{1}\left[\frac{1}{3}(x + 7)\right] - \frac{6}{1}\left[\frac{1}{2}(x + 1)\right] = 6(4)$$ Apply the distributive property.

The product is 2. The product is -3.

$$2(x + 7) - 3(x + 1) = 24$$

$$2x + 14 - 3x - 3 = 24$$ Clear parentheses.

$$-x + 11 = 24$$ Combine *like* terms.

$$-x + 11 - 11 = 24 - 11$$ Subtract 11.

$$-x = 13$$

$$\frac{-x}{-1} = \frac{13}{-1}$$ Divide by -1.

$$x = -13$$

Avoiding Mistakes

Notice that on the left-hand side of this equation, the product of 6 and $\frac{1}{3}$ is taken first, and then the result of 2 is distributed through the parentheses. Similarly, the product of 6 and $-\frac{1}{2}$ is taken first. The result of -3 is then distributed through the parentheses.

example 4 **Solving a Linear Equation with Fractions**

Solve $\frac{x-2}{5} - \frac{x-4}{2} = 2 + \frac{x+4}{10}$.

Solution:

$$\frac{x-2}{5} - \frac{x-4}{2} = \frac{2}{1} + \frac{x+4}{10}$$ The LCM of 5, 2, 1, and 10 is 10.

$$10\left(\frac{x-2}{5} - \frac{x-4}{2}\right) = 10\left(\frac{2}{1} + \frac{x+4}{10}\right)$$ Multiply both sides by 10.

$$\frac{\overset{2}{\cancel{10}}}{1}\cdot\left(\frac{x-2}{\cancel{5}}\right) - \frac{\overset{5}{\cancel{10}}}{1}\cdot\left(\frac{x-4}{\cancel{2}}\right) = \frac{10}{1}\cdot\left(\frac{2}{1}\right) + \frac{\overset{1}{\cancel{10}}}{1}\cdot\left(\frac{x+4}{\cancel{10}}\right)$$ Apply the distributive property.

$$2(x-2) - 5(x-4) = 20 + 1(x+4)$$ Clear fractions.

$$2x - 4 - 5x + 20 = 20 + x + 4$$ Apply the distributive property.

$$-3x + 16 = x + 24$$ Simplify both sides of the equation.

$$-3x - x + 16 = x - x + 24$$ Subtract x from both sides.

$$-4x + 16 = 24$$

$$-4x + 16 - 16 = 24 - 16$$ Subtract 16 from both sides.

$$-4x = 8$$

$$\frac{-4x}{-4} = \frac{8}{-4}$$ Divide both sides by -4.

$$x = -2$$

Avoiding Mistakes

In Example 4, several of the fractions in the equation have two terms in the numerator. It is important to enclose these fractions in parentheses when clearing fractions. In this way, we will remember to use the distributive property to multiply the factors shown in blue with both terms from the numerator of the fractions.

4. Solving Linear Equations with Decimals

The process of clearing decimals is similar to clearing fractions in a linear equation. In this case, we multiply both sides of the equation by a convenient power of 10. We find the coefficient of the term within the equation that has the greatest number of digits following the decimal point. If that coefficient is represented to the tenths-place, we multiply by 10. If the coefficient is represented to the hundredths-place, we multiply by 100 and so on.

example 5 Solving Linear Equations Containing Decimals

Solve the equations.

a. $2.5x + 3 = 1.7x - 6.6$ b. $0.20(x + 4) - 0.45(x + 9) = 12$

Solution:

a.

$$2.5x + 3 = 1.7x - 6.6$$

$$10(2.5x + 3) = 10(1.7x - 6.6)$$ Multiply both sides of the equation by 10.

$$25x + 30 = 17x - 66$$ Apply the distributive property.

$$25x - 17x + 30 = 17x - 17x - 66$$ Subtract $17x$ from both sides.

$$8x + 30 = -66$$

$$8x + 30 - 30 = -66 - 30$$ Subtract 30 from both sides.

$$8x = -96$$

$$\frac{8x}{8} = \frac{-96}{8}$$ Divide both sides by 8.

$$x = -12$$

Tip: Example 5(a) could also been solved without clearing decimals.

$$2.5x + 3 = 1.7x - 6.6$$

Subtracting $1.7x$ from both sides, we have

$$0.8x + 3 = -6.6$$

$$0.8x + 3 - 3 = -6.6 - 3$$

$$0.8x = -9.6$$

$$\frac{0.8x}{0.8} = \frac{-9.6}{0.8}$$

$$x = -12$$

b.

$$0.20(x + 4) - 0.45(x + 9) = 12$$

$$100[0.20(x + 4) - 0.45(x + 9)] = 100(12)$$ Multiply both sides by 100.

$$\underbrace{100[0.20}(x + 4)] - \underbrace{100[0.45}(x + 9)] = 100(12)$$ Apply the distributive property.

$$20(x + 4) - 45(x + 9) = 1200$$ Multiply factors from left to right.

$$20x + 80 - 45x - 405 = 1200$$ Apply the distributive property.

$$-25x - 325 = 1200$$ Simplify both sides.

$$-25x - 325 + 325 = 1200 + 325$$ Add 325 to both sides.

$$-25x = 1525$$

$$\frac{-25x}{-25} = \frac{1525}{-25}$$ Divide both sides by -25.

$$x = -61$$

section 2.3 PRACTICE EXERCISES

For Exercises 1–6, solve the equation.

1. $25x = -15$
2. $42 = 6y$
3. $34 = m - 12$
4. $-19 + n = 14$
5. $2 + 3(b - 6) - 2b = 6$
6. $6(z + 2) - 12 = 14$
7. Solve the equation and describe the solution set: $7x + 2 = 7(x - 12)$
8. Solve the equation and describe the solution set: $2(3x - 6) = 3(2x - 4)$

For Exercises 9–12, determine which of the values could be used to clear fractions or decimals in the given equation.

9. $\frac{2}{3}x - \frac{1}{6} = \frac{x}{9}$ Values: 6, 9, 12, 18, 24, 36
10. $\frac{1}{4}x - \frac{2}{7} = \frac{1}{2}x + 2$ Values: 4, 7, 14, 21, 28, 42
11. $0.02x + 0.5 = 0.35x + 1.2$ Values: 10; 100; 1000; 10,000
12. $0.003 - 0.002x = 0.1x$ Values: 10; 100; 1000; 10,000

For Exercises 13–32, solve the equation.

13. $\frac{1}{2}x + 3 = 5$
14. $\frac{1}{3}y - 4 = 9$
15. $\frac{1}{6}y + 2 = \frac{5}{12}$
16. $\frac{2}{15}z + 3 = \frac{7}{5}$
17. $\frac{1}{3}q + \frac{3}{5} = \frac{1}{15}q - \frac{2}{5}$
18. $\frac{3}{7}x - 5 = \frac{24}{7}x + 7$
19. $\frac{12}{5}w + 7 = 31 - \frac{3}{5}w$
20. $-\frac{1}{9}p - \frac{5}{18} = -\frac{1}{6}p + \frac{1}{3}$
21. $\frac{1}{4}(3m - 4) - \frac{1}{5} = \frac{1}{4}m + \frac{3}{10}$
22. $\frac{1}{25}(20 - t) = \frac{4}{25}t - \frac{3}{5}$
23. $\frac{1}{6}(5s + 3) = \frac{1}{2}(s + 11)$
24. $\frac{1}{12}(4n - 3) = \frac{1}{12}n - \frac{3}{4}$
25. $\frac{2}{3}x + 4 = \frac{2}{3}x - 6$
26. $-\frac{1}{9}a + \frac{2}{9} = \frac{1}{3} - \frac{1}{9}a$
27. $\frac{1}{6}(2c - 1) = \frac{1}{3}c - \frac{1}{6}$
28. $\frac{3}{4}b - \frac{1}{2} = \frac{1}{2}(2 - 3b)$
29. $\frac{2x + 1}{3} + \frac{x - 1}{3} = 5$
30. $\frac{4y - 2}{5} - \frac{y + 4}{5} = -3$
31. $\frac{3w - 2}{6} = 1 - \frac{w - 1}{3}$
32. $\frac{z - 7}{4} = \frac{6z - 1}{8} - 2$
33. Negative eight is half of a number. Find the number.
34. One third of a number equals five. Find the number.
35. The sum of $\frac{2}{5}$ and twice a number is the same as the sum of $\frac{11}{5}$ and the number. Find the number.
36. The difference of three times a number and $\frac{5}{9}$ is the same as the sum of twice the number and $\frac{1}{9}$. Find the number.
37. The sum of twice a number and $\frac{3}{4}$ is the same as the difference of four times the number and $\frac{1}{8}$. Find the number.

38. The difference of a number and $-\frac{11}{12}$ is the same as the difference of three times the number and $\frac{1}{6}$. Find the number.

For Exercises 39–50, solve the equation.

39. $9.2y - 4.3 = 50.9$

40. $-6.3x + 1.5 = -4.8$

41. $21.1w + 4.6 = 10.9w + 35.2$

42. $0.05z + 0.2 = 0.15z - 10.5$

43. $0.2p - 1.4 = 0.2(p - 7)$

44. $0.5(3q + 87) = 1.5q + 43.5$

45. $0.20x + 53.60 = x$

46. $z + 0.06z = 3816$

47. $0.15(90) + 0.05p = 0.10(90 + p)$

48. $0.25(60) + 0.10x = 0.15(60 + x)$

49. $0.40(y + 10) + 0.60y = 2$

50. $0.75(x - 2) + 0.25x = 0.5$

For Exercises 51–74, solve the equation.

51. $2b + 23 = 6b - 5$

52. $-x = 7$

53. $\frac{y}{4} = -2$

54. $10p - 9 + 2p - 3 = 8p - 18$

55. $0.5(2a - 3) - 0.1 = 0.4(6 + 2a)$

56. $-\frac{5}{9}w + \frac{11}{12} = \frac{23}{36}$

57. $-6x = 0$

58. $15.2q = -2.4q - 176$

59. $9.8h + 2 = 3.8h + 20$

60. $-k - 41 = 3 - k$

61. $\frac{1}{4}(x + 4) = \frac{1}{5}(2x + 3)$

62. $7y + 3(2y + 5) = 10y + 17$

63. $2z - 7 = 2(z - 13)$

64. $x - 17.8 = -21.3$

65. $\frac{4}{5}w = 10$

66. $5c + 25 = 20$

67. $4b - 8 - b = -3b + 2(3b - 4)$

68. $36 = 6z + 9$

69. $-3a + 1 = 19$

70. $-5(1 - x) + x = -(6 - 2x) + 6$

71. $3(4h - 2) - (5h - 8) = 8 - (2h + 3)$

72. $1.72w - 0.04w = 0.42$

73. $\frac{3}{8}t - \frac{5}{8} = \frac{1}{2}t + \frac{1}{8}$

74. $3(8x - 1) + 10 = 6(5 + 4x) - 23$

Expanding Your Skills

For Exercises 75–78, solve the equation.

75. $\frac{1}{2}a + 0.4 = -0.7 - \frac{3}{5}a$

76. $\frac{3}{4}c - 0.11 = 0.23(c - 5)$

77. $0.8 + \frac{7}{10}b = \frac{3}{2}b - 0.8$

78. $0.78 - \frac{1}{25}h = \frac{3}{5}h - 0.5$

Concepts

1. Problem-Solving Strategies
2. Translations Involving Linear Equations
3. Consecutive Integer Problems
4. Applications of Linear Equations

section

2.4 Applications of Linear Equations: Introduction to Problem Solving

1. Problem-Solving Strategies

Linear equations can be used to solve many real-world applications. However, with "word problems" students often do not know where to start. To help organize the problem-solving process, we offer the following **problem-solving flowchart**.

Problem-Solving Flowchart for Word Problems

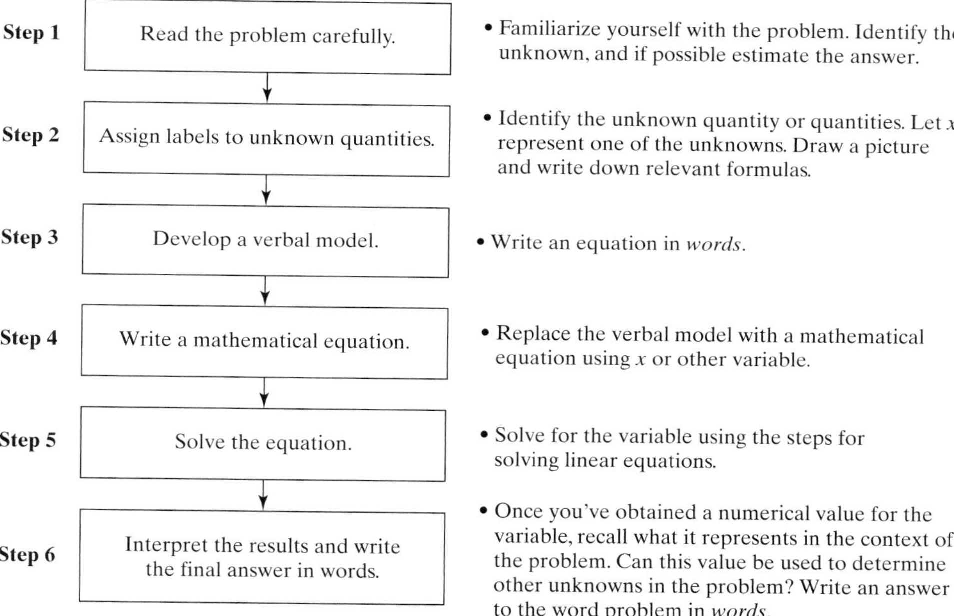

2. Translations Involving Linear Equations

We begin our work with word problems with practice translating between an English sentence and an algebraic equation. Recall from Section 1.2, that several key words translate to the algebraic operations of addition, subtraction, multiplication, and division.

Addition: $a + b$	**Subtraction:** $a - b$
The sum of a and b	The difference of a and b
a plus b	a minus b
b added to a	b subtracted from a
b more than a	a decreased by b
a increased by b	b less than a
the total of a and b	

Multiplication: $a \cdot b$	Division: $a \div b$
The product of a and b	The quotient of a and b
a times b	a divided by b
a multiplied by b	b divided into a
	The ratio of a and b
	a over b
	a per b

example 1

Translating to a Linear Equation

The sum of a number and negative eleven is negative fifteen. Find the number.

Solution:

Step 1: Read the problem.

Let x represent the unknown number. **Step 2:** Label the unknown.

$$\underset{\text{the sum of}}{}\quad\underset{\text{is}}{}$$
$$(\text{a number}) + (-11) = (-15)$$ **Step 3:** Develop a verbal model.

$$x + (-11) = -15$$ **Step 4:** Write an equation.

$$x + (-11) + 11 = -15 + 11$$ **Step 5:** Solve the equation.

$$x = -4$$

The number is -4. **Step 6:** Write the final answer in words.

example 2

Translating to a Linear Equation

Forty less than five times a number is fifty-two less than the number. Find the number.

Avoiding Mistakes

It is important to remember that subtraction is not a commutative operation. Therefore, the order in which two real numbers are subtracted affects the outcome. The expression "forty less than five times a number" must be translated as: $5x - 40$ (not $40 - 5x$). Similarly, "fifty-two less than the number" must be translated as: $x - 52$ (not $52 - x$).

Solution:

Step 1: Read the problem.

Let x represent the unknown number. **Step 2:** Label the unknown.

$$\begin{pmatrix}\text{5 times}\\ \text{a number}\end{pmatrix} - (40) = \begin{pmatrix}\text{the}\\ \text{number}\end{pmatrix} - (52)$$ **Step 3:** Develop a verbal model.

$$5x \quad - 40 \quad = \quad x \quad - 52$$ **Step 4:** Write an equation.

$$5x - 40 = x - 52$$ **Step 5:** Solve the equation.

$$5x - x - 40 = x - x - 52$$

$$4x - 40 = -52$$

$$4x - 40 + 40 = -52 + 40$$

$$4x = -12$$

$$\frac{4x}{4} = \frac{-12}{4}$$

$$x = -3$$

The number is -3. **Step 6:** Write the final answer in words.

example 3

Translating to a Linear Equation

Twice the sum of a number and six is two more than three times the number. Find the number.

Solution:

Step 1: Read the problem.

Let x represent the unknown number. **Step 2:** Label the unknown.

Step 3: Develop a verbal model.

twice → 2, the sum → $(x + 6)$, is → $=$, 2 more than → $+ 2$, three times a number → $3x$

$$2\ (x + 6) = 3x + 2$$

Step 4: Write an equation.

Avoiding Mistakes

It is important to enclose "the sum of a number and six" within parentheses so that the entire quantity is multiplied by 2. Forgetting the parentheses would imply that the x-term only is multiplied by 2.

Correct: $2(x + 6)$

$$2(x + 6) = 3x + 2$$

Step 5: Solve the equation.

$$2x + 12 = 3x + 2$$

$$2x - 2x + 12 = 3x - 2x + 2$$

$$12 = x + 2$$

$$12 - 2 = x + 2 - 2$$

$$10 = x$$

The number is 10. **Step 6:** Write the final answer in words.

3. Consecutive Integer Problems

Tip: We can label two consecutive integers as x and $(x + 1)$, where x is the smaller integer and $(x + 1)$ is the larger. Similarly, we can label the integers as x and $(x - 1)$. In this case, x is the larger integer and $(x - 1)$ is the smaller.

The word *consecutive* means "following one after the other in order without gaps." The numbers 6, 7, 8 are examples of three **consecutive integers**. The numbers $-4, -2, 0, 2,$ are examples of **consecutive even integers**. The numbers 23, 25, 27 are examples of **consecutive odd integers**.

Notice that any two consecutive integers differ by 1. Therefore, if x represents an integer, then $(x + 1)$ represents the next larger consecutive integer (Figure 2-5).

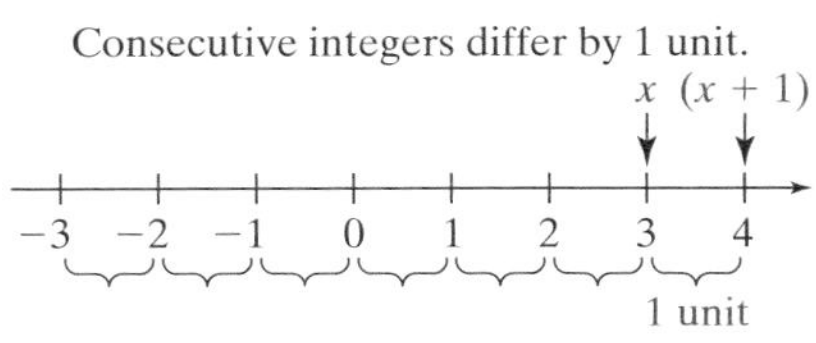

Figure 2-5

Any two consecutive even integers differ by 2. Therefore, if x represents an even integer, then $(x + 2)$ represents the next larger consecutive even integer (Figure 2-6).

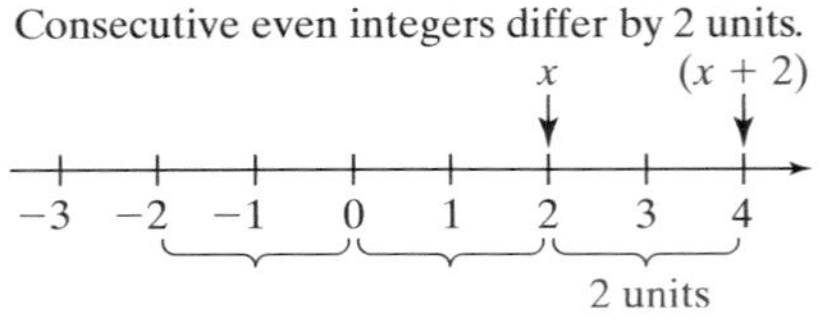

Consecutive even integers differ by 2 units.
x
(x + 2)
−3 −2 −1 0 1 2 3 4
2 units

Figure 2-6

Likewise, any two consecutive odd integers differ by 2. If x represents an odd integer, then $(x + 2)$ is the next larger odd integer (Figure 2-7).

Consecutive odd integers differ by 2 units.
x
(x + 2)
−3 −2 −1 0 1 2 3 4
2 units

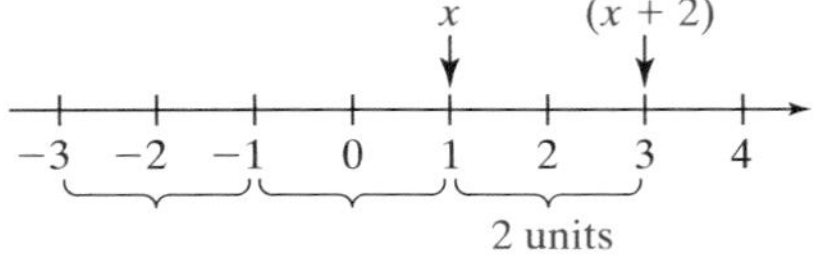

Figure 2-7

example 4

Solving an Application Involving Consecutive Integers

The sum of two consecutive odd integers is −188. Find the integers.

Solution:

In this example we have two unknown integers. We can let x represent either of the two unknowns.

Step 1: Read the problem.

Suppose x represents the first odd integer. **Step 2:** Label the variables.

Then $(x + 2)$ represents the second odd integer.

$$\begin{pmatrix}\text{First}\\\text{integer}\end{pmatrix} + \begin{pmatrix}\text{second}\\\text{integer}\end{pmatrix} = (\text{total})$$

Step 3: Write an equation in words.

$$x + (x + 2) = -188$$

Step 4: Write a mathematical equation.

$$x + (x + 2) = -188$$

$$2x + 2 = -188$$

Step 5: Solve for x.

$$2x + 2 - 2 = -188 - 2$$

$$2x = -190$$

$$\frac{2x}{2} = \frac{-190}{2}$$

$$x = -95$$

Tip: With word problems it is advisable to check that the answer is reasonable.

The numbers −95 and −93 are consecutive odd integers. Furthermore, their sum is −188 as desired.

The first integer is $x = -95$.

The second integer is $x + 2 = -95 + 2 = -93$.

The two integers are −95 and −93.

Step 6: Interpret the results and write the answer in words.

example 5

Solving an Application Involving Consecutive Integers

Ten times the smallest of three consecutive integers is twenty-two more than three times the sum of the integers. Find the integers.

Solution:

Step 1: Read the problem.

Let x represent the first integer.

$x + 1$ represents the second consecutive integer.

$x + 2$ represents the third consecutive integer.

Step 2: Label the variables.

$$\begin{pmatrix}\text{10 times}\\ \text{the first}\\ \text{integer}\end{pmatrix} = \begin{pmatrix}\text{3 times}\\ \text{the sum of}\\ \text{the integers}\end{pmatrix} + 22$$

Step 3: Write an equation in words.

$$\underset{\text{10 times the first integer}}{10x} \underset{\text{is}}{=} \underset{\text{3 times}}{3}\underbrace{[(x) + (x + 1) + (x + 2)]}_{\text{the sum of the integers}} \underset{\text{22 more than}}{+ 22}$$

Step 4: Write a mathematical equation.

$$10x = 3(x + x + 1 + x + 2) + 22$$

Step 5: Solve the equation.

$$10x = 3(3x + 3) + 22$$

$$10x = 9x + 9 + 22$$

Clear parentheses.

$$10x = 9x + 31$$

Combine *like* terms.

$$10x - 9x = 9x - 9x + 31$$

$$x = 31$$

Isolate the x-terms on one side.

The first integer is $x = 31$.

The second integer is $x + 1 = 31 + 1 = 32$.

The third integer is $x + 2 = 31 + 2 = 33$.

The three integers are 31, 32, and 33.

Step 6: Interpret the results and write the answer in words.

4. Applications of Linear Equations

example 6

Using a Linear Equation in an Application

As of June 6, 1998, the two films with the largest total box office revenues were *Titanic* and *Star Wars*. Together the two films grossed \$1043 million. If *Titanic* made \$121 million more than *Star Wars*, find the total box office revenue for each film. (*Source: Information Please Almanac,* 1999)

Solution:

In this example we have two unknowns. The variable x can represent *either* quantity. The revenue from *Titanic* is stated as \$121 million more than the revenue generated by *Star Wars*. Therefore, if x represents the revenue from *Star Wars*, the revenue for *Titanic* can be stated in terms of x as $(x + 121)$.

Step 1: Read the problem.

Let x represent the revenue from *Star Wars* (in \$millions).

$x + 121$ represents the revenue from *Titanic* (in \$millions).

Step 2: Label the unknowns.

$$\begin{pmatrix}\text{Revenue from}\\ \textit{Star Wars}\end{pmatrix} + \begin{pmatrix}\text{revenue from}\\ \textit{Titanic}\end{pmatrix} = \begin{pmatrix}\text{total}\\ \text{revenue}\end{pmatrix}$$

Step 3: Develop a verbal equation.

$$x \quad + \quad (x + 121) \quad = \quad 1043$$

Step 4: Write a mathematical equation.

$$x + (x + 121) = 1043$$

Step 5: Solve the equation.

$$2x + 121 = 1043$$

$$2x + 121 - 121 = 1043 - 121$$

$$2x = 922$$

$$\frac{2x}{2} = \frac{922}{2}$$

$$x = 461$$

Step 6: Interpret the results.

Revenue from *Star Wars*: $x = 461$.

Revenue from *Titanic*: $x + 121 = 461 + 121 = 582$.

As of June 6, 1998, the revenue made from *Star Wars* was $461 million and the revenue made from *Titanic* was $582 million.

section 2.4 PRACTICE EXERCISES

For Exercises 1–6, write an algebraic expression to represent the English sentence. Then solve the equation.

1. The sum of a number and sixteen is negative thirty-one. Find the number.
2. The sum of a number and negative twenty-one is fourteen. Find the number.
3. The difference of a number and six is negative three. Find the number.
4. The difference of a number and negative four is negative twelve. Find the number.
5. Sixteen less than a number is negative one. Find the number.
6. Ten less than a number is negative thirteen. Find the number.

For Exercises 7–12, identify as a conditional equation, a contradiction, or an identity. Then describe the solution.

7. $4(t - 2) = 1 + 4t$
8. $2x - 3 = 2(5 + x)$
9. $-5(y + 4) = 15$
10. $4(p - 1) + 8 = -10$
11. $7(2m - 2) - 5m = 9m - 14$
12. $-5n - 2(n + 4) = -7n - 8$

For Exercises 13–24, use the problem-solving flowchart (page 123) to solve the problems.

13. Twice the sum of a number and seven is eight. Find the number.
14. Twice the sum of a number and negative two is sixteen. Find the number.
15. Five times the difference of a number and three is four less than four times the number. Find the number.
16. Three times the difference of a number and seven is one less than twice the number. Find the number.
17. A number added to five is the same as twice the number. Find the number.
18. Three times a number is the same as the difference of twice the number and seven. Find the number.
19. The sum of six times a number and ten is equal to the difference of the number and fifteen. Find the number.
20. The difference of fourteen and three times a number is the same as the sum of the number and negative ten. Find the number.
21. If three is added to five times a number, the result is forty-three more than the number. Find the number.
22. If seven is added to three times a number, the result is thirty-one more than the number.
23. If the difference of a number and four is tripled, the result is six more than the number. Find the number.
24. Twice the sum of a number and eleven is twenty-two less than three times the number. Find the number.

For Exercises 25–30, use the problem-solving flowchart (page 123) to solve the problems.

25. In 1983, 104 more Democrats than Republicans were in the U.S. House of Representatives. If the total number of representatives in the House was 434, find the number of representatives from each party.

26. In 1995, 12 more Republicans than Democrats were in the U.S. House of Representatives. If the House had a total of 434 representatives from these two parties, find the number of Democrats and the number of Republicans.

27. A board is 86 cm in length and must be cut so that one piece is 20 cm longer than the other piece. Find the length of each piece.

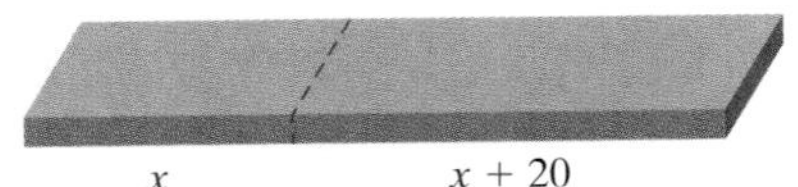

Figure for Exercise 27

28. A rope is 54 in. in length and must be cut into two pieces. If one piece must be twice as long as the other, find the length of each piece.

29. The longest river in Africa is the Nile. It is 2455 km longer than the Congo River, also in Africa. The sum of the lengths of these rivers is 11,195 km. What is the length of each river?

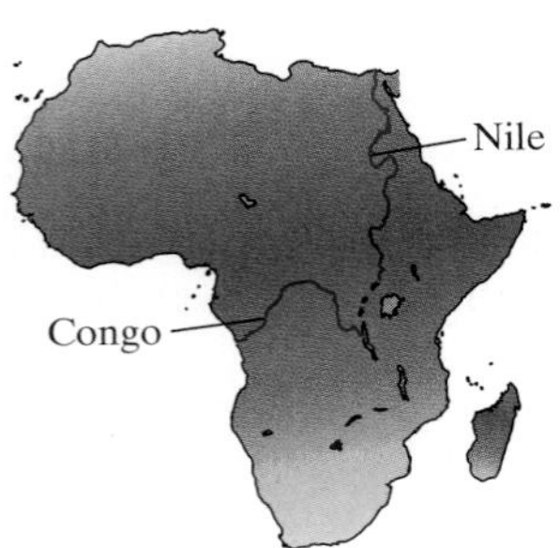

Figure for Exercise 29

30. The average depth of the Gulf of Mexico is three times the depth of the Red Sea. The difference between the average depths is 1078 m. What is the average depth of the Gulf of Mexico and the average depth of the Red Sea?

31. a. If x represents the smallest of three consecutive integers, write an expression to represent each of the next two consecutive integers.

    b. If x represents the largest of three consecutive integers, write an expression to represent each of the previous two consecutive integers.

32. a. If x represents the smallest of three consecutive even integers, write an expression to represent each of the next two larger consecutive even integers.

    b. If x represents the largest of three consecutive even integers, write an expression to represent each of the previous two consecutive even integers.

33. a. If x represents the smallest of three consecutive odd integers, write an expression to represent each of the next two larger consecutive odd integers.

    b. If x represents the largest of three consecutive odd integers, write an expression to represent each of the previous two consecutive odd integers.

34. Is it possible to find two consecutive integers that differ by 5? Explain.

35. The sum of the page numbers on two facing pages in a book is 941. What are the page numbers?

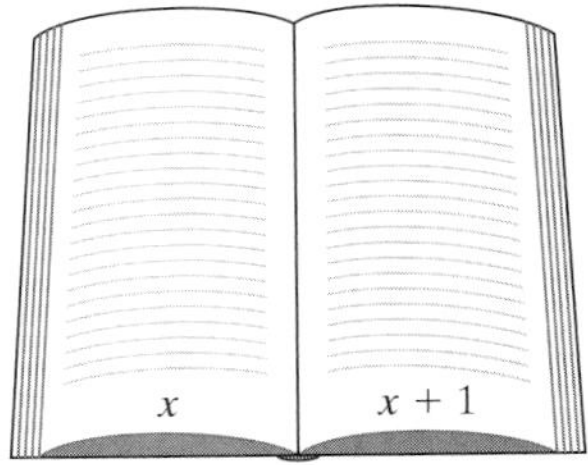

Figure for Exercise 35

36. Three raffle tickets are represented by three consecutive integers. If the sum of the three integers is 2,666,031, find the numbers.

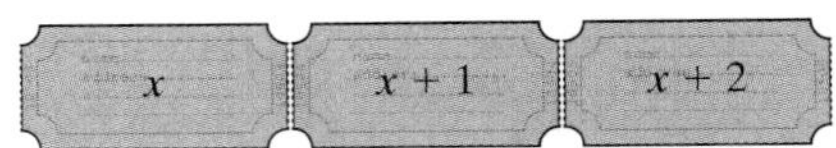

Figure for Exercise 36

37. Three consecutive odd integers are such that 3 times the smallest is 9 more than twice the largest. Find the three numbers.

38. Three consecutive even integers are such that the sum of the two larger integers is 232 more than three times the smallest integer. Find the three integers.

39. Three consecutive integers are such that three times the largest exceeds the sum of the two smaller integers by 47. Find the integers.

40. Four times the smallest of three consecutive odd integers is 236 more than the sum of the other two integers. Find the integers.

41. The perimeter of a triangle is 42 in. The lengths of the sides are represented by three consecutive integers. Find the lengths of the sides of the triangle.

42. The perimeter of a triangle is 54 m. The lengths of the sides are represented by three consecutive even integers. Find the lengths of the three sides.

43. The perimeter of a pentagon (a five-sided polygon) is 80 in. The five sides are represented by consecutive integers. Find the measures of the sides.

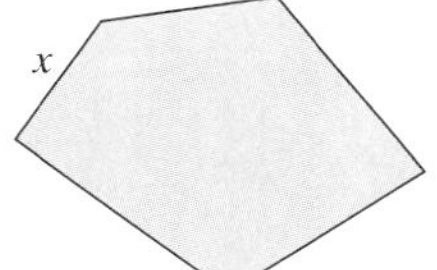

Figure for Exercise 43

44. The perimeter of a pentagon (a five-sided polygon) is 95 in. The five sides are represented by consecutive integers. Find the measures of the sides.

45. The area of Greenland is 201,900 km^2 less than three times the area of New Guinea. What is the area of New Guinea if the area of Greenland is 2,175,600 km^2?

46. The deepest point in the Pacific Ocean is 676 m more than twice the deepest point in the Arctic Ocean. If the deepest point in the Pacific is 10,920 m, how many meters is the deepest point in the Arctic Ocean?

47. Asia and Africa are the two largest continents in the world. The land area of Asia is approximately 14,514,000 km^2 larger than the land area of Africa. Together their total area is 74,644,000 km^2. Find the land area of Asia and the land area of Africa.

48. Mt. Everest, the highest mountain in the world is 2654 m higher than Mt. McKinley, the highest mountain in the United States. If the sum of their heights is 15,042 m, find the height of each mountain.

section

2.5 Solving Applications with Percent

Concepts

1. Conversions between Percents and Decimals
2. Solving Percent Problems
3. Applications Involving Commission
4. Applications Involving Sales Tax
5. Applications Involving Simple Interest

1. Conversions between Percents and Decimals

The concept of percent is widely used in a variety of mathematical applications.

- "A stock decreased by 12 percent for the year."
- "The sales tax in a certain state is 6 percent."
- "54 percent of graduating college seniors in the United States are women."

The word **percent** means "per hundred" and is denoted by the "%" symbol. It refers to the number of parts in 100 parts. A 6% sales tax, for example, means that there is a 6¢ tax for every 100¢ spent.

The % symbol implies "division by 100" or equivalently "multiplication by $\frac{1}{100}$ or 0.01."

Percent Notation

$p\%$ means $\dfrac{p}{100}$, or $p \times \dfrac{1}{100}$, or $p \times 0.01$

The preceding definition implies that to convert a percent to a decimal, remove the % symbol and multiply by 0.01.

example 1 **Converting Percents to Decimals**

Convert the percents to decimals:

a. 78% b. 412% c. 0.045%

Tip: Multiplying by 0.01 is equivalent to dividing by 100. This has the effect of moving the decimal point two places to the left.

Solution:

a. $78\% = 78 \times 0.01 = 0.78$
b. $412\% = 412 \times 0.01 = 4.12$
c. $0.045\% = 0.045 \times 0.01 = 0.00045$

To convert a decimal to a percent, we reverse the process just used. We multiply the number by 100 and attach the % symbol.

example 2 **Converting Decimals to Percents**

Convert the decimals to percents:

a. 0.92 b. 10.80 c. 0.005

Solution:

a. $0.92 = 0.92 \times 100\%$ Multiply by 100 and attach the % symbol.
$= 92\%$
b. $10.80 = 10.80 \times 100\%$ Multiply by 100 and attach the % symbol.
$= 1080\%$
c. $0.005 = 0.005 \times 100\%$ Multiply by 100 and attach the % symbol.
$= 0.5\%$

2. Solving Percent Problems

One method of solving a percent problem is to translate the English sentence into a linear equation.

example 3 **Solving Basic Percent Equations**

a. What percent of 60 is 25.2?
b. 8.2 is 125% of what number?
c. 2% of 1500 is what number?

Solution:

a. Let x represent the unknown percent.

Step 1: Read the problem.
Step 2: Label the variables.
Step 3: Create a verbal model.
Step 4: Write a mathematical equation.

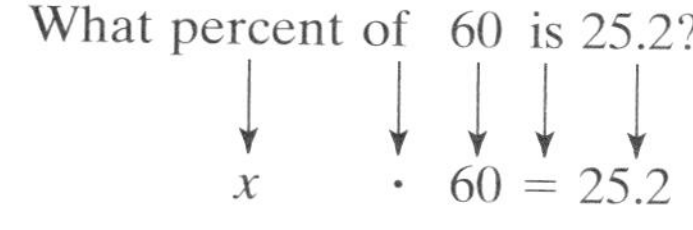

What percent of 60 is 25.2?

$x \cdot 60 = 25.2$

$$60x = 25.2$$

$$\frac{60x}{60} = \frac{25.2}{60}$$

$$x = 0.42 \text{ or } 42\%$$

Step 5: Solve the equation.

Step 6: Interpret the results and write the answer in words.

25.2 is 42% of 60.

b. Let x represent the unknown number.

Step 1: Read the problem.
Step 2: Label the variables.
Step 3: Create verbal model.
Step 4: Write a mathematical equation.

8.2 is 125% of what number?

$8.2 = 1.25 \cdot x$

$$8.2 = 1.25x$$

$$\frac{8.2}{1.25} = \frac{1.25x}{1.25}$$

$$6.56 = x$$

Step 5: Solve the equation.

Step 6: Interpret the results and write the answer in words.

8.2 is 125% of 6.56.

Avoiding Mistakes

Be sure to use the decimal or fraction form of a percent within an equation.

$125\% = 1.25$

c. Let x represent the unknown number.

Step 1: Read the problem.
Step 2: Label the variable.
Step 3: Create a verbal model.
Step 4: Write a mathematical equation.

2% of 1500 is what number?

$0.02 \cdot 1500 = x$

$$30 = x$$

Step 5: Solve the equation.

Step 6: Interpret the results.

2% of 1500 is 30.

3. Applications Involving Commission

Percents are often used to represent rates. For example, salespeople may receive all or part of their salary as a percentage of their sales. This is called a *commission*.

The *commission rate* is the percent of the sales that a salesperson receives in income.

$$\text{commission} = (\text{commission rate}) \cdot (\$ \text{ in merchandise sold})$$

example 4 **Using Percents to Find Commission**

Terrance works for a prestigious clothing shop and earns $10,000 per year plus 12.5% commission on sales. What is Terrance's total commission if his total in sales is $260,000?

Solution: — **Step 1:** Read the problem.

Let x represent the total commission. — **Step 2:** Label the variable.

$$\begin{pmatrix}\text{Total}\\ \text{commission}\end{pmatrix} = \begin{pmatrix}\text{commission}\\ \text{rate}\end{pmatrix}\begin{pmatrix}\text{total}\\ \text{sales}\end{pmatrix}$$
Step 3: Write an equation in words.

$$x = (0.125)(\$260{,}000)$$
Step 4: Write a mathematical equation.

$$x = \$32{,}500$$
Step 5: Solve the equation.

Terrance made $32,500 in commission. — **Step 6:** Interpret the results and write the answer in words.

4. Applications Involving Sales Tax

Another common use of percents is in computing sales tax.

$$\text{sales tax} = (\text{tax rate}) \cdot (\$ \text{ in merchandise})$$

example 5 **Applying Percents**

A video game is purchased for a total of $48.15 including sales tax. If the tax rate is 7%, find the original price of the video game before sales tax.

Solution: — **Step 1:** Read the problem.

Let x represent the price of the video game. — **Step 2:** Label variables.

$0.07x$ represents the amount of sales tax.

$$\begin{pmatrix}\text{Original}\\ \text{price}\end{pmatrix} + \begin{pmatrix}\text{sales}\\ \text{tax}\end{pmatrix} = \begin{pmatrix}\text{total}\\ \text{cost}\end{pmatrix}$$
Step 3: Write a verbal equation.

$$x + 0.07x = \$48.15$$
Step 4: Write a mathematical equation.

$$1.07x = 48.15$$
Step 5: Solve for x.

$$100(1.07x) = 100(48.15)$$
Multiply by 100 to clear decimals.

Tip: The equation in Example 5 could have been solved easily without clearing decimals.

$$x + 0.07x = 48.15$$
$$1.07x = 48.15$$
$$\frac{1.07x}{1.07} = \frac{48.15}{1.07}$$
$$x = 45$$

This illustrates the point that the decision to clear decimals (or fractions) is a matter of preference.

$$107x = 4815$$

$$\frac{107x}{107} = \frac{4815}{107}$$ Divide both sides by 107.

$$x = 45$$ **Step 6:** Interpret the results and write the answer in words.

The original price was \$45.

5. Applications Involving Simple Interest

One important application of percents is in computing simple interest on a loan or on an investment.

Banks hold large quantities of money for its customers. However, because all bank customers are unlikely to withdraw all of their money on a single day, a bank does not keep all the money in cash. Instead, it keeps some cash for day-to-day transactions but invests the remaining portion of the money. Because a bank uses its customer's money to make investments, and because it wants to attract more customers the bank pays interest on the money. **Simple interest** is interest that is earned on principal (the original amount of money invested in an account). The following formula is used to compute simple interest:

$$\begin{pmatrix}\text{Simple}\\\text{interest}\end{pmatrix} = \begin{pmatrix}\text{principal}\\\text{invested}\end{pmatrix}\begin{pmatrix}\text{annual}\\\text{interest rate}\end{pmatrix}\begin{pmatrix}\text{time}\\\text{in years}\end{pmatrix}$$

This formula is often written symbolically as $I = Prt$.

For example, to find the simple interest earned on \$2000 invested at 7.5% interest for 3 years, we have

$$I = P \cdot r \cdot t$$

$$\begin{aligned}\text{Interest} &= (\$2000)(0.075)(3)\\ &= \$450\end{aligned}$$

example 6

Applying Simple Interest

Jorge wants to save money for his daughter's college education. If Jorge needs to have \$4340 at the end of 4 years, how much money would he need to invest at a 6% simple interest rate?

Solution:

Step 1: Read the problem.

Let x represent the original amount invested. **Step 2:** Label the variables.

$$\begin{pmatrix}\text{Original}\\\text{principal}\end{pmatrix} + (\text{interest}) = (\text{total})$$

Step 3: Write an equation in words.

Recall that interest is computed by the formula $I = Prt$.

$$(P) \quad + \quad (Prt) \quad = (\text{total})$$

$$x \quad + \quad x(0.06)(4) = 4340$$

Step 4: Write a mathematical equation.

$$x + 0.24x = 4340$$

$$1.24x = 4340$$ **Step 5:** Solve the equation.

$$\frac{1.24x}{1.24} = \frac{4340}{1.24}$$

$$x = 3500$$

The original investment should be $3500. **Step 6:** Interpret the results and write the answer in words.

section 2.5 Practice Exercises

1. Twice the sum of two consecutive odd integers is -280. What are the integers?
2. A truck with a camper weighs 7650 pounds. If the truck weighs 2290 pounds more than the camper, what is the weight of the camper?

For Exercises 3 and 4, solve the equations.

3. $\frac{a}{15} + 6 = \frac{a}{5} + \frac{2}{3}$
4. $\frac{z}{3} + 2 = \frac{z}{6} - \frac{1}{12}$

For Exercises 5–8, convert the percents to decimal form.

5. 57%
6. 13%
7. 135%
8. 250%

For Exercises 9–12, convert the decimal numbers to percent form.

9. 0.69
10. 0.82
11. 0.006
12. 0.0003

For Exercises 13–24, find the missing values.

13. 45 is what percent of 360?
14. 338 is what percent of 520?
15. 544 is what percent of 640?
16. 576 is what percent of 800?
17. What is 0.5% of 150?
18. What is 9.5% of 616?
19. What is 42% of 740?
20. What is 56% of 280?
21. 177 is 20% of what number?
22. 126 is 15% of what number?
23. 275 is 12.5% of what number?
24. 594 is 45% of what number?

The number of AIDS cases acquired in the United States in 2002 by race or ethnic group is shown in the figure. Use this figure to answer Exercises 25–28. Round the answers to the nearest tenth of a percent. *(Source: Centers for Disease Control)*

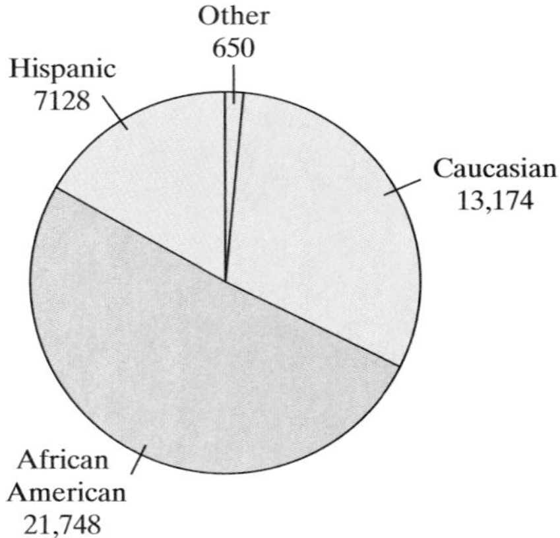

Figure for Exercises 25–28

25. What percent of reported AIDS cases is in the African American category?

26. What percent of reported AIDS cases is in the Hispanic category?

27. What percent of reported AIDS cases is in the Caucasian category?

28. What percent of reported AIDS cases is described as being in the other or unknown category?

For Exercises 29–42, solve for the unknown quantity.

29. Molly buys a golf outfit that costs \$74.95. If the sales tax rate is 7%, for how much should she write her check?

30. Patrick purchased golf shoes for \$85.98. If the sales tax rate is 6%, how much was charged to his Visa credit card?

31. The sales tax for a screwdriver set came to \$1.04. If the sales tax rate is 6.5%, what was the price of the screwdriver?

32. The sales tax for a picture frame came to \$1.32. If the sales tax rate is 5.5%, what was the price of the picture frame?

33. Sun Lei bought a laptop computer over the Internet for \$1800. The total cost, including tax, came to \$1890. What is the sales tax rate?

34. Jamie purchased a compact disc and paid \$18.26. If the disc price is \$16.99, what is the sales tax rate (rounded to the nearest tenth of a percent)?

35. A dress is marked at 30% off. If the sale price is \$20.97, what was the original price of the dress?

36. A jacket is on sale for \$53.60. If this represents a 20% discount, what was the original price of the jacket?

37. The local car dealership pays its sales personnel a commission of 25% of the dealer profit on each car. The dealer made a profit of \$18,250 on the cars Joëlle sold last month. What was her commission last month?

38. Dan sold a beachfront home for \$650,000. If his commission rate is 4%, what did he earn on the sale of that home?

39. A salesperson at You-Bought-It discount store earns 3% commission on all appliances that he sells. If Geoff's commission for the month was \$116.37, how much did he sell?

40. Anna makes a commission at an appliance store. In addition to her base salary, she earns a 2.5% commission on her sales. If Anna's commission for a month was \$260, how much did she sell?

41. For selling software, Tom received a bonus commission based on sales over \$500. If he received \$180 in commission for selling a total of \$2300 worth of software, what is his commission rate?

42. In addition to an hourly salary, Jessica earns a commission for selling ice cream bars at the beach. If she sells \$708 worth of ice cream and receives a commission of \$56.64, what is her commission rate?

For Exercises 43–50, solve the applications involving simple interest.

43. How much interest will Pam earn in 4 years if she invests \$3000 in an account that pays 3.5% simple interest?

44. How much interest will Roxanne have to pay if she borrows \$2000 for 2 years at a simple interest rate of 4%?

45. Bob borrowed some money for 1 year at 5% simple interest. If he had to pay back a total of \$1260, how much did he originally borrow?

46. Mike borrowed some money for 2 years at 6% simple interest. If he had to pay back a total of \$3640, how much did he originally borrow?

47. If \$1500 grows to \$1950 after 5 years, find the simple interest rate.

48. If \$9000 grows to \$10,440 in 2 years, find the simple interest rate.

49. A new bank offered simple interest loans at 11% for new customers. If a customer took out a loan for \$2000 to be paid back in 18 months ($\frac{3}{2}$ years), find
   a. the interest on the loan.
   b. the total amount that the customer owes (principal + interest).

50. Rafael has \$3000 saved for a future trip to Europe. If he invests in an account that pays 4% simple interest, how much will he have after $2\frac{1}{2}$ years?

Expanding Your Skills

51. Diane sells women's sportswear at a department store. She earns a regular salary and, as a bonus, she receives a commission of 4% on all merchandise sold over \$200. If Diane earned an extra \$25.80 last week in commission, how much merchandise did she sell over \$200?

52. Bob's position in men's formal wear pays by commission. He earns 5% for the first \$1000 he sells and 6% on sales thereafter. If Bob's commission last week was \$170, how much merchandise did he sell?

53. A sweater is on sale for 15% off. With a sales tax rate of 5%, the sales tax amounts to \$2.55. What was the original price of the sweater before the sale?

54. A set of golf clubs is on sale for 25% off. With a sales tax rate of 7%, the sales tax amounts to \$21.00. What was the original price of the clubs?

Concepts

1. Formulas
2. Geometry Applications: Perimeter
3. Geometry Applications: Complementary Angles
4. Geometry Applications: Angles Inscribed within a Triangle

section 2.6 Formulas and Applications of Geometry

1. Formulas

Formulas (or **literal equations**) are equations that contain several variables. For example, the perimeter of a triangle (distance around the triangle) can be found by the formula $P = a + b + c$, where a, b, and c are the lengths of the sides (Figure 2-8).

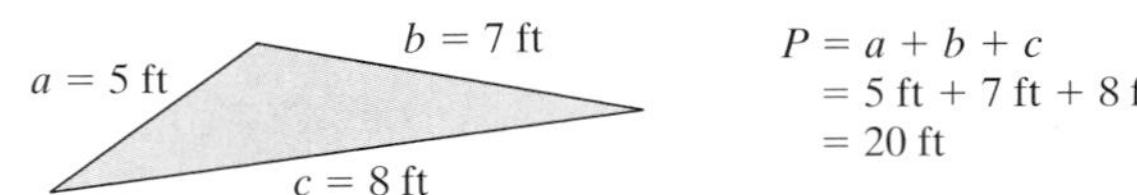

Figure 2-8

In this section, we will learn how to rewrite formulas to solve for a different variable within the formula. Suppose, for example, that the perimeter of a triangle is known and two of the sides are known (say sides a and b). Then the third side, c, can be found by subtracting the lengths of the known sides from the perimeter (Figure 2-9).

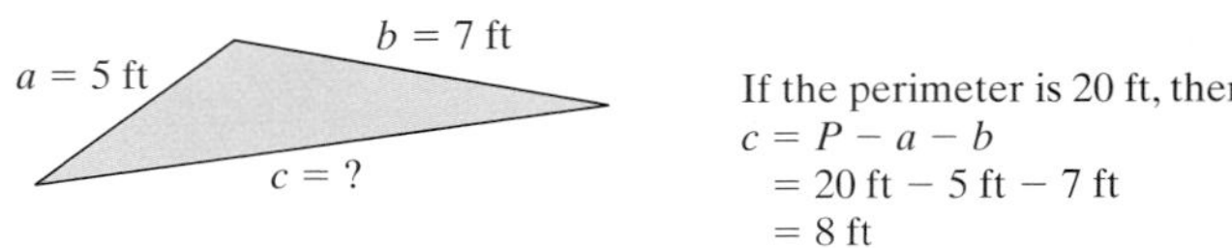

Figure 2-9

To solve a formula for a different variable, we use the same properties of equality outlined in the earlier sections of this chapter. For example, consider the two equations $2x + 3 = 11$ and $wx + y = z$. Suppose we want to solve for x in each case:

$2x + 3 = 11$		$wx + y = z$	
$2x + 3 - 3 = 11 - 3$	Subtract 3.	$wx + y - y = z - y$	Subtract y.
$2x = 8$		$wx = z - y$	
$\dfrac{2x}{2} = \dfrac{8}{2}$	Divide by 2.	$\dfrac{wx}{w} = \dfrac{z - y}{w}$	Divide by w.
$x = 4$		$x = \dfrac{z - y}{w}$	

The equation on the left has only one variable and we are able to simplify the equation to find a numerical value for x. The equation on the right has multiple variables. Because we do not know the values of w, y, and z, we are not able to simplify further. The value of x is left as a formula in terms of w, y, and z.

example 1 Solving Formulas for an Indicated Variable

Solve the formulas for the indicated variables.

a. $d = rt$ for t b. $5x + 2y = 12$ for y

Solution:

a. $d = rt$ for t The goal is to isolate the variable t.

$$\frac{d}{r} = \frac{rt}{r}$$

Because the relationship between r and t is multiplication, we reverse the process by dividing both sides by r.

$$\frac{d}{r} = t \text{ or equivalently } t = \frac{d}{r}$$

b. $5x + 2y = 12$ for y The goal is to solve for y.

$5x - 5x + 2y = 12 - 5x$ Subtract $5x$ from both sides to isolate the y-term.

$$2y = -5x + 12$$

$\dfrac{2y}{2} = \dfrac{-5x + 12}{2}$ Divide both sides by 2 to isolate y.

$$y = \frac{-5x + 12}{2}$$

Tip: The original equation $d = rt$ represents the distance traveled, d, in terms of the rate of speed, r, and the time of travel, t.

The equation $t = \frac{d}{r}$ represents the same relationship among the variables, however, the time of travel is expressed in terms of the distance and rate.

Tip: On the right-hand side we chose to write the variable term first $-5x + 12$ as is customary. However, it is also correct to write the expression as $12 - 5x$.

Tip: The expression $\dfrac{-5x + 12}{2}$

can also be written with the divisor 2 applied individually to each term in the numerator. Hence the answer may appear in several different forms. Each is correct:

$$y = \frac{-5x + 12}{2} \quad \text{or} \quad y = \frac{-5x}{2} + \frac{12}{2} \quad \Rightarrow \quad y = -\frac{5}{2}x + 6$$

example 2

Solving Formulas for an Indicated Variable

The formula $C = \frac{5}{9}(F - 32)$ is used to find the temperature, C, in degrees Celsius for a given temperature expressed in degrees Fahrenheit, F. Solve the formula $C = \frac{5}{9}(F - 32)$ for F.

Solution:

$$C = \frac{5}{9}(F - 32)$$

$$9C = 9 \cdot \frac{5}{9}(F - 32) \qquad \text{Multiply by 9 to clear fractions.}$$

$$9C = 5(F - 32) \qquad \text{Simplify.}$$

$$9C = 5F - 160 \qquad \text{Clear parentheses.}$$

$$9C + 160 = 5F - 160 + 160 \qquad \text{Add 160 to both sides.}$$

$$9C + 160 = 5F$$

$$\frac{9C + 160}{5} = \frac{5F}{5} \qquad \text{Divide both sides by 5 to isolate } F.$$

$$\frac{9C + 160}{5} = F$$

The answer can be written in several forms:

$$F = \frac{9C + 160}{5} \quad \text{or} \quad F = \frac{9C}{5} + \frac{160}{5} \quad \Rightarrow \quad F = \frac{9}{5}C + 32$$

2. Geometry Applications: Perimeter

In Section R.2, we presented numerous facts and formulas related to geometry. Sometimes these are needed to solve applications in geometry.

example 3

Solving a Geometry Application Involving Perimeter

The length of a rectangular lot is 1 m less than twice the width. If the perimeter is 190 m, find the length and width.

Solution:

Step 1: Read the problem.

Let x represent the width of the rectangle. **Step 2:** Label the variables.

Then $2x - 1$ represents the length.

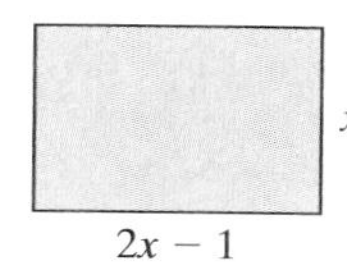

$$P = 2l + 2w$$ **Step 3:** Perimeter formula

$$190 = 2(2x - 1) + 2(x)$$ **Step 4:** Write an equation in terms of x.

$$190 = 4x - 2 + 2x$$ **Step 5:** Solve for x.

$$190 = 6x - 2$$

$$192 = 6x$$

$$\frac{192}{6} = \frac{6x}{6}$$

$$32 = x$$

The width is $x = 32$.

The length is $2x - 1 = 2(32) - 1 = 63$. **Step 6:** Interpret the results and write the answer in words.

The width of the rectangular lot is 32 m and the length is 63 m.

3. Geometry Applications: Complementary Angles

example 4

Solving a Geometry Application Involving Complementary Angles

Two complementary angles are drawn such that one angle is 4° more than seven times the other angle. Find the measure of each angle.

Solution: **Step 1:** Read the problem.

Let x represent the measure of one angle. **Step 2:** Label the variables.

Then $7x + 4$ represents the measure of the other angle.

The angles are complementary, so their sum must be 90°.

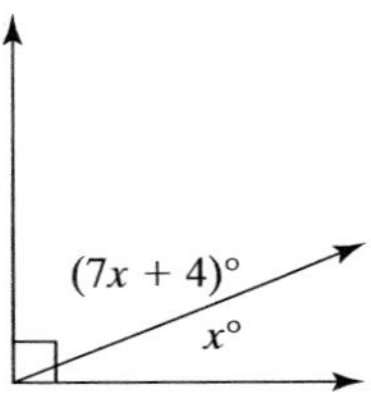

$$\begin{pmatrix}\text{Measure of}\\ \text{first angle}\end{pmatrix} + \begin{pmatrix}\text{measure of}\\ \text{second angle}\end{pmatrix} = 90°$$ **Step 3:** Create a verbal equation.

$$x + 7x + 4 = 90$$ **Step 4:** Write a mathematical equation.

$$8x + 4 = 90$$ **Step 5:** Solve for x.

$$8x = 86$$

$$\frac{8x}{8} = \frac{86}{8}$$

$$x = 10.75$$

Step 6: Interpret the results and write the answer in words.

One angle is $x = 10.75°$.

The other angle is $7x + 4 = 7(10.75°) + 4° = 79.25°$.

The angles are 10.75° and 79.25°.

4. Geometry Applications: Angles Inscribed within a Triangle

example 5

Solving a Geometry Application

One angle in a triangle is twice as large as the smallest angle. The third angle is 10° more than seven times the smallest angle. Find the measure of each angle.

Solution:

Step 1: Read the problem.

Let x represent the measure of the smallest angle. **Step 2:** Label the variables.

Then $2x$ and $7x + 10$ represent the measures of the other two angles.

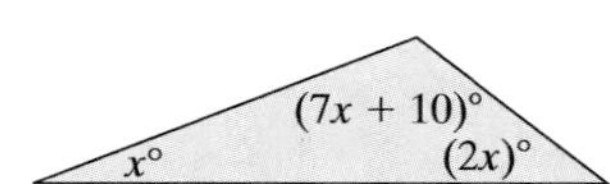

The sum of the angles must be 180°. **Step 3:** Create a verbal equation.

$x + (2x) + (7x + 10) = 180$ **Step 4:** Write a mathematical equation.

$$x + 2x + 7x + 10 = 180$$ **Step 5:** Solve for x.

$$10x + 10 = 180$$

$$10x = 170$$

$$x = 17$$

Step 6: Interpret the results and write the answer in words.

The smallest angle is $x = 17°$.

The other angles are $2x = 2(17°) = 34°$

$7x + 10 = 7(17°) + 10° = 129°$

The angles are 17°, 34°, and 129°.

section 2.6 Practice Exercises

For Exercises 1–6, solve the equation.

1. $3(2y + 3) - 4(-y + 1) = 7y - 10$
2. $-(3w + 4) + 5(w - 2) - 3(6w - 8) = 10$
3. $\frac{1}{2}(x - 3) + \frac{3}{4} = 3x - \frac{3}{4}$
4. $\frac{5}{6}x + \frac{1}{2} = \frac{1}{4}(x - 4)$
5. $0.5(y + 2) - 0.3 = 0.4y + 0.5$
6. $-0.02(1 - 4m) = 0.6 + 0.18m$

For Exercises 7–38, solve for the indicated variable.

7. $P = a + b + c$ for a
8. $P = a + b + c$ for b
9. $x = y - z$ for y
10. $c + d = e$ for d
11. $p = 250 + q$ for q
12. $y = 35 + x$ for x
13. $d = rt$ for t
14. $d = rt$ for r
15. $PV = nrt$ for t
16. $P_1V_1 = P_2V_2$ for V_1
17. $x - y = 5$ for x
18. $x + y = -2$ for y
19. $3x + y = -19$ for y
20. $x - 6y = -10$ for x
21. $2x + 3y = 6$ for y
22. $5x + 2y = 10$ for y
23. $-2x - y = 9$ for x
24. $3x - y = -13$ for x
25. $4x - 3y = 12$ for y
26. $6x - 3y = 4$ for y
27. $ax + by = c$ for y
28. $ax + by = c$ for x
29. $A = P(1 + rt)$ for t
30. $P = 2(L + w)$ for L
31. $a = 2(b + c)$ for c
32. $3(x + y) = z$ for x
33. $Q = \frac{x + y}{2}$ for y
34. $Q = \frac{a - b}{2}$ for a

35. $M = \frac{a}{S}$ for a

36. $A = \frac{1}{3}(a + b + c)$ for c

37. $P = I^2R$ for R

38. $F = \frac{GMm}{d^2}$ for m

For Exercises 39–57, use the problem-solving flowchart (page 123) from Section 2.4.

39. The perimeter of a rectangular garden is 24 ft. The length is 2 ft more than the width. Find the length and the width of the garden.

40. In a small rectangular wallet photo, the width is 7 cm less than the length. If the border (perimeter) of the photo is 34 cm, find the length and width.

41. A builder buys a rectangular lot of land such that the length is 5 m less than two times the width. If the perimeter is 590 m, find the length and the width.

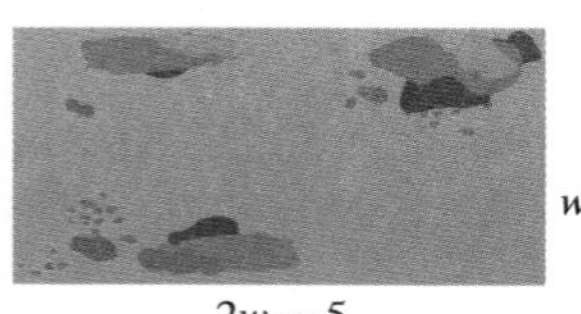

Figure for Exercise 41

42. The perimeter of a rectangular pool is 140 yd. If the length is 10 yd more than the width, find the length and the width.

43. The largest angle in a triangle is three times the smallest angle. The middle angle is two times the smallest angle. Given that the sum of the angles in a triangle is 180°, find the measure of each angle.

44. The smallest angle in a triangle is 90° less than the largest angle. The middle angle is 60° less than the largest angle. Find the measure of each angle.

45. The smallest angle in a triangle is half the largest angle. The middle angle is 30° less than the largest angle. Find the measure of each angle.

46. The largest angle of a triangle is three times the middle angle. The smallest angle is 10° less than the middle angle. Find the measure of each angle.

47. Find the value of x and the measure of each angle labeled in the figure.

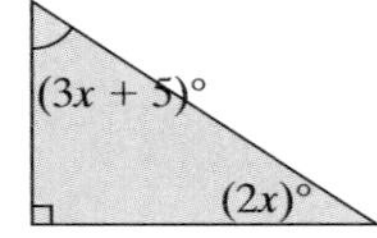

Figure for Exercise 47

48. Find the value of y and the measure of each angle labeled in the figure.

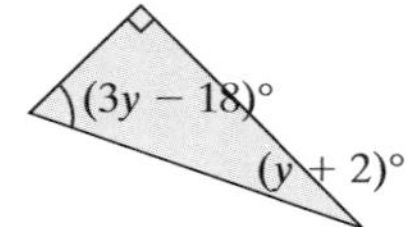

Figure for Exercise 48

49. Sometimes memory devices are helpful for remembering mathematical facts. Recall that the sum of two complementary angles is 90°. That is, two complementary angles when added together form a right angle or "corner." The words **C**omplementary and **C**orner both start with the letter **C**. Derive your own memory device for remembering that the sum of two supplementary angles is 180°.

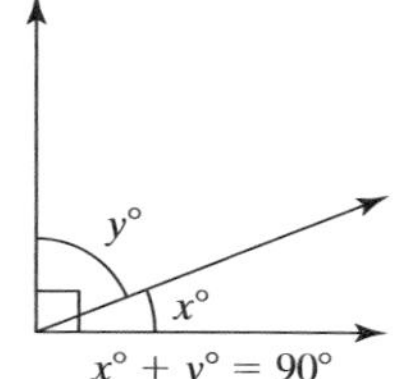

Complementary angles form a "corner"

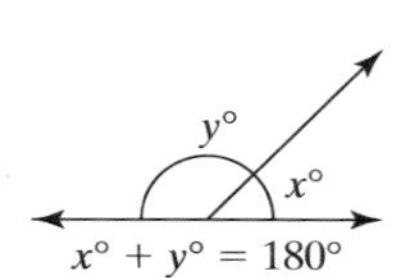

Supplementary angles . . .

50. Two angles are complementary. One angle is 20° less than the other angle. Find the measures of the angles.

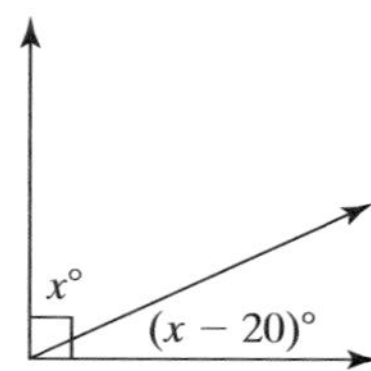

Figure for Exercise 50

51. Two angles are complementary. One angle is twice as large as the other angle. Find the measures of the angles.

52. Two angles are complementary. One angle is 4° less than three times the other angle. Find the measures of the angles.

53. Two angles are supplementary. One angle is three times as large as the other angle. Find the measures of the angles.

54. Two angles are supplementary. One angle is twice as large as the other angle. Find the measures of the angles.

55. Two angles are supplementary. One angle is 6° more than four times the other. Find the measures of the two angles.

56. Find the measures of the vertical angles labeled in the figure by first solving for x.

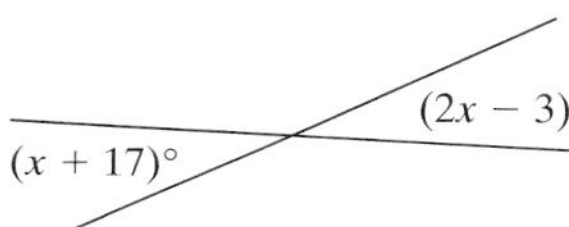

Figure for Exercise 56

57. Find the measures of the vertical angles labeled in the figure by first solving for y.

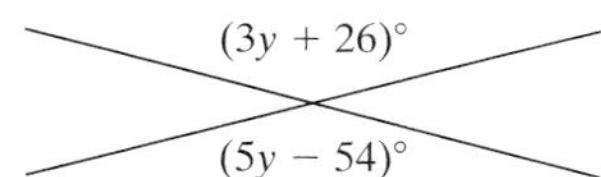

Figure for Exercise 57

For Exercises 58–67, solve the geometry formulas for the indicated variables.

58. a. A rectangle has length l and width w. Write a formula for the area.
    b. Solve the formula for the width, w.
    c. The area of a rectangular volleyball court is 1740.5 ft^2 and the length is 59 ft. Find the width.

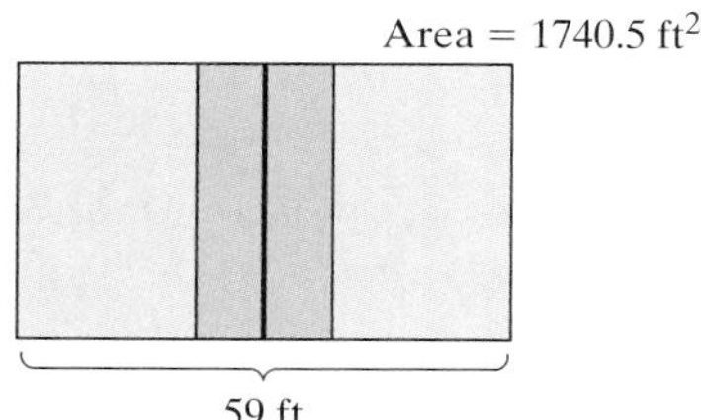

Figure for Exercise 58

59. a. A parallelogram has height h and base b. Write a formula for the area.
    b. Solve the formula for the base, b.
    c. Find the base of the parallelogram pictured if the area is 40 m^2.

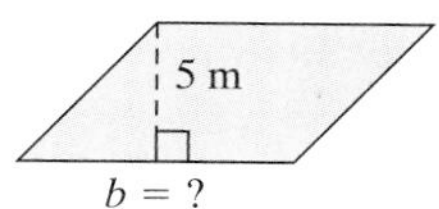

Figure for Exercise 59

60. a. A rectangle has length l and width w. Write a formula for the perimeter.
    b. Solve the formula for the length, l.
    c. The perimeter of the soccer field at Giants Stadium is 338 m. If the width is 66 m, find the length.

61. a. The length of each side of a square is s. Write a formula for the perimeter of the square.
    b. Solve the formula for the length of a side, s.
    c. The Pyramid of Khufu (known as the Great Pyramid) at Giza has a square base. If the distance around the bottom is 921.6 m, find the length of the sides at the bottom of the pyramid.

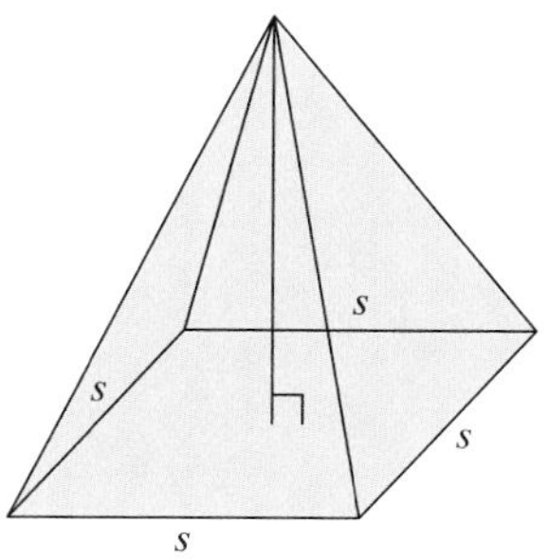

Figure for Exercise 61

62. a. A circle has a radius of r. Write a formula for the circumference.
    b. Solve the formula for the radius, r.
    c. The circumference of the circular Buckingham Fountain in Chicago is approximately

880 ft. Find the radius. Round to the nearest foot.

63. a. A rectangular solid has height h, width w, and length l. Write a formula for the volume of the solid.
    b. Solve the formula for the height, h.
    c. A rectangular box mounted on the back of a pick-up truck holds 45 ft^3 of cargo space. If the length and width of the box are 4.5 ft and 5.0 ft, respectively, find the height of the box.

64. The volume of a cylinder is given by $V = \pi r^2 h$. Solve for the height, h.

65. The volume of a sphere is given by $V = \frac{4}{3}\pi r^3$. Solve for r^3.

66. The area of a trapezoid is given by $A = \frac{1}{2}(B + b)h$. Solve for B.

67. The volume of a pyramid with a rectangular base is given by $V = \frac{1}{3}\,lwh$. Solve for the height, h.

Expanding Your Skills

For Exercises 68–72, find the indicated perimeters, areas, or volumes. Be sure to include the proper units. Use 3.14 for π where necessary.

68. a. Find the area of a circle with radius 10 m.
    b. Find the volume of a right circular cylinder with radius 10 m and height 25 m.

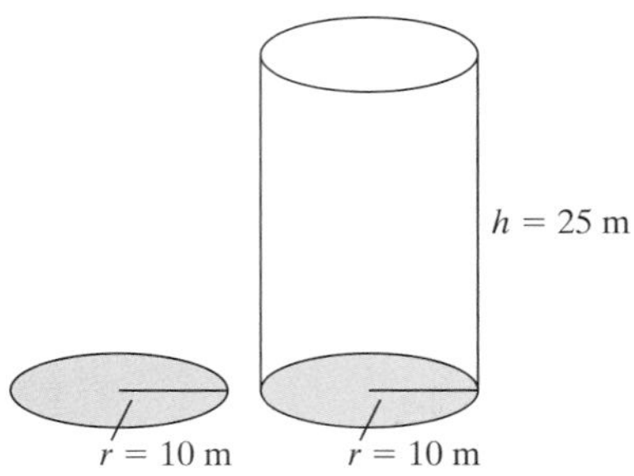

Figure for Exercise 68

69. a. Find the area of a circle with radius 20 ft.
    b. Find the volume of a right circular cylinder with radius 20 ft and height 8 ft.

70. a. Find the area of a parallelogram with base 30 in. and height 12 in.
    b. Find the area of a triangle with base 30 in. and height 12 in.
    c. Compare the areas found in parts (a) and (b).

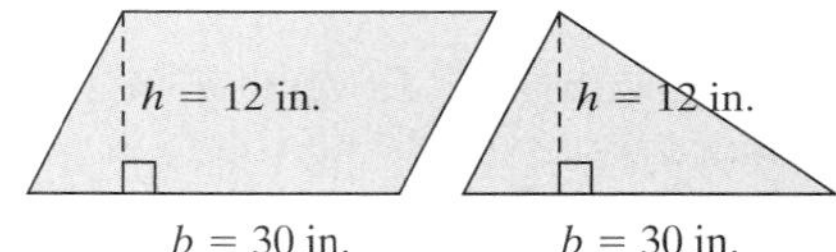

Figure for Exercise 70

71. a. Find the area of a parallelogram with base 7 m and height 4 m.
    b. Find the area of a triangle with base 7 m and height 4 m.
    c. Compare the areas found in parts (a) and (b).

72. a. Find the volume of a right circular cylinder with radius 6 ft and height 5 ft.
    b. Find the volume of a right circular cone with radius 6 ft and height 5 ft.
    c. Compare the volumes found in parts (a) and (b).

Graphing Calculator Exercises

For Exercises 73–76, approximate the expressions with a calculator. Round to three decimal places if necessary.

73. $\dfrac{880}{2\pi}$

74. $\dfrac{1600}{\pi(4)^2}$

75. $\dfrac{20}{(-0.05)(5)}$

76. $\dfrac{10}{0.5(6 + 4)}$

---

section

2.7 More Applications of Linear Equations

Concepts

1. **Applications Involving Ticket Sales**
2. **Applications Involving Mixtures**
3. **Applications Involving Principal and Interest**
4. **Applications Involving Distance, Rate, and Time**

1. Applications Involving Ticket Sales

Application problems that involve finding a combination of two or more quantities to meet specified constraints are called mixture problems. In the next example, we are asked to find the number of adults and the number of children that attended a play. The constraints are that exactly 282 people were present and that the total revenue was $1046. Algebra is used to find the right combination of the number of adults and number of children that attended the play.

example 1 **Solving an Application Involving Ticket Sales**

Two hundred eighty-two people attended a recent performance of *Cinderella*. Adult tickets sold for $5 and children's tickets sold for $3. Find the number of adults and the number of children that attended the play if the total revenue was $1046.

Solution: **Step 1:** Read the problem.

This example has two unknowns: the number of adults and the number of children.

We can let x represent either quantity. However, regardless of the choice of x, the other unknown will be represented as $282 - x$. For example,

if x represents the number of children, then

$$\begin{pmatrix}\text{Number}\\\text{of adults}\end{pmatrix} = \begin{pmatrix}\text{total number}\\\text{of people}\end{pmatrix} - \begin{pmatrix}\text{number of}\\\text{children}\end{pmatrix}$$

$$\text{Number of adults} = 282 - x$$

Let x represent the number of children. **Step 2:** Label the variables.

Then, $282 - x$ represents the number of adults.

The children's tickets are $3 each, so the value of all children's tickets is $3x$.

The adult tickets are $5 each, so the value of all adult tickets is $5(282 - x)$.

Sometimes it is helpful to organize the information in a chart to label the variables.

	Children's Tickets	Adult Tickets	Total
Number of tickets	x	$282 - x$	282
Value of tickets	$3x$	$5(282 - x)$	1046

The total revenue is equal to the sum of the revenue from the children's tickets and from the adult tickets. Using the bottom row of the chart, we have

$$\begin{pmatrix}\text{Value of}\\ \text{children's tickets}\end{pmatrix} + \begin{pmatrix}\text{value of}\\ \text{adult tickets}\end{pmatrix} = \begin{pmatrix}\text{total value}\\ \text{of tickets}\end{pmatrix}$$

Step 3: Create the verbal equation.

$$3x \quad + \quad 5(282 - x) \quad = \quad 1046$$

Step 4: Write a mathematical equation.

$$3x + 5(282 - x) = 1046$$

$$3x + 1410 - 5x = 1046$$

Step 5: Solve for x.

$$-2x + 1410 = 1046$$

$$-2x = -364$$

$$\frac{-2x}{-2} = \frac{-364}{-2}$$

$$x = 182$$

Tip: Check the answer. 182 children's tickets and 100 adult tickets make 282 total tickets sold.

Furthermore, 182 children's tickets at $3 each amounts to $546. One hundred adult tickets at $5 each amounts to $500. Therefore, the total revenue is $1046 as desired.

The number of children is 182.

Step 6: Interpret the results.

The number of adults is $282 - x = 282 - 182 = 100$.

There were 182 children and 100 adults that attended the performance.

2. Applications Involving Mixtures

example 2 **Solving a Problem Involving a Coffee Mixture**

Java Joe's sells a Kenyan coffee for $14 per lb. The house blend sells for $8 per lb. How many pounds of the Kenyan coffee should be mixed with the house blend to produce 24 pounds of a coffee worth $12.50 per lb?

Solution:

Step 1: Read the problem.

Let x represent the amount of the Kenyan coffee.

Then $24 - x$ is the amount of the house blend.

Step 2: Label the variables.

Tip: The choice to let x represent the amount of Kenyan coffee was arbitrary. We could have also worked the problem by letting x represent the amount of house blend, and $24 - x$ be the amount of Kenyan coffee.

	Kenyan coffee $14 per lb	**House blend $8 per lb**	**Total $12.50 per lb**
Amount	x	$24 - x$	24
Cost	$14x$	$8(24 - x)$	12.50(24)

The total cost is equal to the cost for the Kenyan coffee plus the cost for the house blend.

$$\begin{pmatrix}\text{Cost of}\\\text{Kenyan coffee}\end{pmatrix} + \begin{pmatrix}\text{cost of}\\\text{house blend}\end{pmatrix} = \begin{pmatrix}\text{total}\\\text{cost}\end{pmatrix}$$

Step 3: Create a verbal equation.

$$14x \quad + \quad 8(24 - x) \quad = 12.50(24)$$

Step 4: Write a mathematical equation.

$$14x + 192 - 8x = 300$$

Step 5: Solve for x.

$$6x + 192 = 300$$

$$6x = 108$$

$$\frac{6x}{6} = \frac{108}{6}$$

$$x = 18$$

If $x = 18$, then $24 - x = 6$.

Step 6: Interpret the results in words.

Therefore, Java Joe's requires 18 lb of the Kenyan coffee and 6 lb of the house blend.

3. Applications Involving Principal and Interest

example 3

Solving an Application Involving Principal and Interest

Shana invests some money in an account that earns 5% simple interest and three times that amount in an account that earns 7% simple interest. If the total interest is $390 at the end of 1 year, find the amount invested in each account.

Solution:

In this application, we are "mixing" money between two accounts to find the correct combination that yields exactly \$390 in interest. We can set up a chart to label the variables. Recall that simple interest is computed by the formula: $I = Prt$. The time of the investment is 1 year, so we have: $I = Pr$

Step 1: Read the problem.

Let x represent the amount invested at 5%. **Step 2:** Label the variables.

Then $3x$ represents the amount invested at 7%.

	5% Account	7% Account	Total
Principal	x	$3x$	
Interest	$0.05x$	$0.07(3x)$	390

The total interest is equal to the sum of the interest from the 5% account and from the 7% account. Using the bottom row of the chart, we have

$$\begin{pmatrix}\text{Interest earned}\\ \text{in 5\% account}\end{pmatrix} + \begin{pmatrix}\text{interest earned}\\ \text{in 7\% account}\end{pmatrix} = \begin{pmatrix}\text{total}\\ \text{interest}\end{pmatrix}$$

Step 3: Create a verbal equation.

$$0.05x \quad + \quad 0.07(3x) \quad = \quad 390$$

Step 4: Write a mathematical equation.

$$0.05x + 0.07(3x) = 390$$

$$0.05x + 0.21x = 390$$

Step 5: Solve for x.

$$0.26x = 390$$

$$\frac{0.26x}{0.26} = \frac{390}{0.26}$$

$$x = 1500$$

The amount invested at 5% is \$1500. **Step 6:** Interpret the results.

The amount invested at 7% is $3x = 3(\$1500) = \$4500$.

Shana invested \$1500 in the 5% account and \$4500 in the 7% account.

Tip: Check your answer. First notice that \$4500 is three times \$1500. Furthermore, if Shana invested \$1500 at 5%, the interest from that account is 0.05(\$1500) = \$75. Likewise \$4500 invested at 7% produces 0.07(\$4500) = \$315 in interest. Therefore, the total interest is \$75 + \$315 = \$390, as expected.

example 4 **Solving an Application Involving Principal and Interest**

Tillie has \$15,000 to invest. She puts part of the money in a savings account that pays 3% simple interest. She invests the rest at 6% simple interest. If her total interest at the end of 1 year is \$720, how much did she invest in the 3% account?

Solution:

Step 1: Read the problem.

Let x represent the amount invested at 3%. **Step 2:** Label the variables.

Then $15{,}000 - x$ represents the amount in the 6% account.

	3% account	6% account	Total
Principal	x	$15{,}000 - x$	15,000
Interest	$0.03x$	$0.06(15{,}000 - x)$	720

$$\begin{pmatrix}\text{Interest earned}\\ \text{in 3\% account}\end{pmatrix} + \begin{pmatrix}\text{interest earned}\\ \text{in 6\% account}\end{pmatrix} = \begin{pmatrix}\text{total}\\ \text{interest}\end{pmatrix}$$

Step 3: Write a verbal model.

$$0.03x + 0.06(15{,}000 - x) = 720$$

Step 4: Write a mathematical equation.

$$0.03x + 900 - 0.06x = 720$$

Step 5: Solve for x.

$$-0.03x + 900 = 720$$

Combine like terms.

$$-0.03x = -180$$

Subtract 900.

$$\frac{-0.03x}{-0.03} = \frac{-180}{-0.03}$$

$$x = 6000$$

Tillie invested \$6000 in the 3% account.

Step 6: Write the answer in words.

4. Applications Involving Distance, Rate, and Time

The formula: (distance) = (rate)(time) or simply, $d = rt$, relates the distance traveled to the rate of travel and the time of travel.

For example, if a car travels at 60 mph for 3 hours, then

$$\begin{aligned} d &= (60 \text{ mph})(3 \text{ hours}) \\ &= 180 \text{ miles} \end{aligned}$$

If a car travels at 60 mph for x hours, then

$$\begin{aligned} d &= (60 \text{ mph})(x \text{ hours}) \\ &= 60x \text{ miles} \end{aligned}$$

example 5

Understanding the Relationship Involving Distance, Rate, and Time

a. If a car travels 240 miles in 4 hours, what is the average speed?
b. If a bicyclist rides 18 miles per hour and covers 63 miles, for how long did he ride?
c. If a plane travels 900 kilometers per hour for 4 hours, how far did it travel?

Solution:

a. The relationship $d = rt$ can be solved for rate, r, by dividing both sides by t.

$$d = rt \quad \Rightarrow \quad \frac{d}{t} = \frac{rt}{t} \quad \Rightarrow \quad r = \frac{d}{t}$$

$r = \dfrac{240 \text{ mi}}{4 \text{ hr}}$ Substitute $d = 240$ mi, $t = 4$ hr.

$r = 60$ mph The car travels 60 mph.

b. The relationship $d = rt$ can be solved for time, t, by dividing both sides by r.

$$d = rt \quad \Rightarrow \quad \frac{d}{r} = \frac{rt}{r} \quad \Rightarrow \quad t = \frac{d}{r}$$

$t = \dfrac{63 \text{ mi}}{18 \text{ mph}}$ Substitute $d = 63$ mi, $r = 18$ mph.

$t = 3.5$ hr The bicyclist rode for 3.5 hr.

c. From $d = rt$, we have,

$d = (900 \text{ km/hr})(4 \text{ hr})$ Substitute $r = 900$ km/hr, $t = 4$ hr.

$d = 3600$ km The plane traveled 3600 km.

example 6 Solving an Application Involving Distance, Rate, and Time

One bicyclist rides 4 mph faster than another bicyclist. The faster rider takes 3 hr to complete a race while the slower rider takes 4 hr. Find the speed for each rider.

Solution: **Step 1:** Read the problem.

The problem is asking us to find the speed of each rider.

Let x represent the speed of the slower rider. Then $(x + 4)$ is the speed of the faster rider.

Step 2: Label the variables and organize the information given in the problem. A distance-rate-time chart may be helpful.

	Distance	Rate	Time
Faster rider	$3(x + 4)$	$x + 4$	3
Slower rider	$4(x)$	x	4

To complete the first column, we can use the relationship, $d = rt$.

The distance traveled by the faster rider is

(faster rider's rate)(faster rider's time) $= (x + 4)(3)$

The distance traveled by the slower rider is

(slower rider's rate)(slower rider's time) $= (x)(4)$

Tip: Check that the answer is reasonable. If the slower rider rides at 12 mph for 4 hr, he travels 48 mi. If the faster rider rides at 16 mph for 3 hr, he also travels 48 mi as expected.

Because the riders are riding in the same race, their distances are equal.

$$\begin{pmatrix}\text{Distance}\\ \text{by faster rider}\end{pmatrix} = \begin{pmatrix}\text{distance}\\ \text{by slower rider}\end{pmatrix}$$ **Step 3:** Set up a verbal model.

$$3(x + 4) = 4(x)$$ **Step 4:** Write a mathematical equation.

$$3x + 12 = 4x$$ **Step 5:** Solve the equation.

$$12 = x$$ Subtract $3x$ from both sides.

The variable x represents the slower rider's rate. The quantity $x + 4$ is the faster rider's rate. Thus, if $x = 12$, then $x + 4 = 16$.

The slower rider travels 12 mph and the faster rider travels 16 mph.

example 7 Solving an Application Involving Distance, Rate, and Time

Two families that live 270 miles apart plan to meet for an afternoon picnic. To share the driving, they want to meet somewhere between their two homes. Both families leave at 9:00 A.M., but one family averages 12 mph faster than the other family. If the families meet at the designated spot $2\frac{1}{2}$ hours later, determine

a. The average rate of speed for each family.
b. The distance each family traveled to the picnic.

Solution: **Step 1:** Read the problem and draw a sketch.

For simplicity, we will call the two families, Family A and Family B. Let Family A be the family that travels at the slower rate (Figure 2-10).

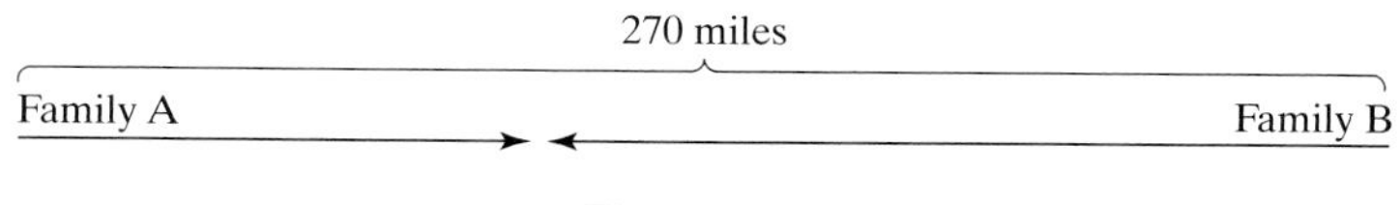

Figure 2-10

Let x represent the rate of Family A. Then $(x + 12)$ is the rate of Family B. **Step 2:** Label the variables.

	Distance	Rate	Time
Family A	$2.5x$	x	2.5
Family B	$2.5(x + 12)$	$x + 12$	2.5

To complete the first column, we can use the relationship $d = rt$.

The distance traveled by Family A is: $d_A = (x)(2.5)$.
The distance traveled by Family B is $d_B = (x + 12)(2.5)$.

To set up an equation, recall that the total distance between the two families is given as 270 miles.

$$\begin{pmatrix}\text{Distance}\\ \text{traveled by}\\ \text{Family A}\end{pmatrix} + \begin{pmatrix}\text{distance}\\ \text{traveled by}\\ \text{Family B}\end{pmatrix} = \begin{pmatrix}\text{total}\\ \text{distance}\end{pmatrix}$$

Step 3: Create a verbal equation.

$$2.5x \quad + \quad 2.5(x + 12) \quad = \quad 270$$

Step 4: Write a mathematical equation.

$$2.5x + 2.5(x + 12) = 270$$
$$2.5x + 2.5x + 30 = 270$$
$$5.0x + 30 = 270$$
$$5x = 240$$
$$\frac{5x}{5} = \frac{240}{5}$$
$$x = 48$$

Step 5: Solve for x.

a. Family A traveled 48 (mph).

Family B traveled $x + 12 = 48 + 12 = 60$ (mph).

Step 6: Interpret the results and write the answer in words.

b. To compute the distance each family traveled, use $d = rt$:

Family A traveled: (48 mph)(2.5 hr) = 120 miles

Family B traveled: (60 mph)(2.5 hr) = 150 miles

section 2.7 PRACTICE EXERCISES

For Exercises 1–8, solve for the indicated variable.

1. $3x + 7 = 13$ for x
2. $V = \pi r^2 h$ for r^2
3. $4x + 5y = 20$ for y
4. $4b - 2b + 8 = 3(b - 6) - 9$ for b
5. $0.3(a + 4) = 7.1a - 2.2$ for a
6. $\frac{2}{3}(p + 9) = \frac{5}{6}p$ for p
7. $-7 = -4r - 6$ for r
8. $V = \frac{1}{3}lwh$ for h
9. Adult tickets to a school play cost \$6 per ticket.
   a. If 200 adult tickets are sold, how much revenue is produced?
   b. If x adult tickets are sold, write an expression that represents the revenue.
   c. If $750 - x$ adult tickets are sold, write an expression that represents the revenue.
10. Children's tickets to a movie cost \$3.50 per ticket.
    a. If a theater sells 600 children's tickets, how much revenue is produced?
    b. If a theater sells x children's tickets, write an expression that represents the revenue.
    c. If a theater sells $2x$ children's tickets, write an expression that represents the revenue.

11. The local church had an ice cream social and sold tickets for $3 and $2. When the social was over, 81 tickets had been sold totaling $215. How many of each type of ticket did the church sell?

	$3 Tickets	$2 Tickets	Total
Number of tickets			
Value of tickets			

Table for Exercise 11

12. A high school had two raffles to raise funds to purchase books for the library. One raffle offered tickets at $0.50 to win a new portable radio and the other offered tickets for $2 to win a new bike. There were 224 tickets sold bringing in $251.50 for the library. How many of each type of ticket were sold?

13. Two raffles are being held at a potluck dinner fund-raiser. One raffle ticket costs $2.00 for a weekend vacation. The other costs $1.00 per ticket for free passes to a movie theater. If 208 tickets were sold and a total of $320 was received, how many of each type of ticket were sold?

14. In a performance of the play, "Company," 375 tickets were sold. The price of the orchestra level seats was $25, and the balcony seats sold for $21. If the total revenue was $8875.00, how many of each type of seat were sold?

15. A certain granola mixture is 10% peanuts.

   a. If a container has 20 lb of granola, how many pounds of peanuts are in it?

   b. If a container has x pounds of granola, write an expression that represents the number of pounds of peanuts in the granola.

   c. If a container has $x + 3$ pounds of granola, write an expression that represents the number of pounds of peanuts.

16. A certain blend of coffee sells for $9.00 per pound.

   a. If a container has 20 lb of coffee, how much will it cost?

   b. If a container has x pounds of coffee, write an expression that represents the cost.

   c. If a container has $40 - x$ pounds of this coffee, write an expression that represents the cost.

17. The Coffee Company wishes to mix coffee worth $12 per pound with coffee worth $8 per pound to produce 50 lb of coffee worth $8.80 per pound. How many pounds of the $12 coffee and how many pounds of the $8 coffee must be used?

	$12 Coffee	$8 Coffee	Total
Number of pounds			
Value of coffee			

Table for Exercise 17

18. The Nut House sells pecans worth $4 per pound and cashews worth $6 per pound. How many pounds of pecans and how many pounds of cashews must be mixed to form 16 lb of a nut mixture worth $4.50 per pound?

19. Sally wishes to mix raisins with granola to sell in decorated containers at a flea market. She can get raisins for $1.69 per pound and granola for $2.59 per pound. How many pounds of raisins and how many pounds of granola should she use to make 6 lb of a mixture worth $2.29?

20. Gina is going to a party and was asked to bring 3 lb of mixed nuts that includes cashews and peanuts. If the cashews cost $6.00 per pound and the peanuts cost $1.50 per pound, how many pounds of each did she buy if her total cost was $9.00?

21. Two different teas are mixed to make a blend that will be sold at a fair. Black tea sells for $2.20/lb and orange pekoe tea sells for $3.00/lb. How much of each should be used to obtain 4 lb of a blend selling for $2.50/lb?

22. A nut mixture consists of almonds and cashews. Almonds are $4.98/lb and cashews are $6.98/lb. How many pounds of each type of nut should be mixed to produce 16 lb selling for $5.73/lb?

23. A savings account earns 6% simple interest per year.

    a. If a woman invests $5000 in the account, how much interest will she earn after 1 year?

    b. If a woman has x dollars invested in the account, write an expression that represents the total interest earned after 1 year.

    c. If a woman has $\$20{,}000 - x$ invested in the account, write an expression that represents the total interest earned after 1 year.

24. A money market account earns 3.5% simple interest per year.

    a. If a man invests $4000 in the account, how much interest will he earn after 1 year?

    b. If a man has x dollars invested in the account, write an expression that represents the total interest earned after 1 year.

    c. If a man has $\$10{,}000 - x$ invested in the account, write an expression that represents the total interest earned after 1 year.

25. Nora has an account with her bank that pays 6% simple interest. She also has a savings account with her credit union that pays 8% simple interest. She received a total of $104 in interest in 1 year. If there is $800 more in the bank account than in the credit union, how much is in each account?

	6% Account	8% Account	Total
Principal			
Interest			

Table for Exercise 25

26. Bob has $8000 to invest. He puts part of the money in an account that pays 8% simple interest and invests the rest at 12% simple interest. If his total interest after 1 year was $840, how much did he invest at 12%?

27. Darrell has a total of $12,500 in two accounts. One account pays 8% simple interest and the other pays 12% simple interest. If he earned $1,160 in the first year, how much did he invest in each account?

28. Randall had a total of $15,000 invested in two accounts, one paying 9% simple interest and one paying 10% simple interest. How much was invested in each account if the interest after 1 year was $1432?

29. Ms. Walsh deposited some money in an account paying 5% simple interest and twice that amount in an account paying 6% simple interest. If the total interest from the two accounts is $765 for 1 year, how much was deposited into each account?

30. Mr. Campbell had some money in his bank earning 4.5% simple interest. He had $5000 more deposited in a credit union earning 6% simple interest. If his total interest for 1 year was $1140, how much did he have in each account?

31. How can $8750 be divided between a 6% account and an 8% account so that each account earns the same amount of simple interest after 1 year?

32. Iacco has a total of $20,000 to invest. She invests part in an account earning 5% simple interest and part in an account earning 3% simple interest. How much money is invested in each account if the interest from each account is the same at the end of 1 year?

33. a. If a car travels 60 mph for 5 hr, find the distance traveled.

    b. If a car travels at x miles per hour for 5 hr, write an expression that represents the distance traveled.

    c. If a car travels at $x + 12$ mph for 5 hr, write an expression that represents the distance traveled.

34. a. If a plane travels at 550 mph for 2.5 hr, find the distance traveled.

    b. If a plane travels at x miles per hour for 2.5 hr, write an expression that represents the distance traveled.

    c. If a plane travels at $x - 100$ mph for 2.5 hr, write an expression that represents the distance traveled.

35. A car travels 55 mph for 4 hr, and another car travels the same distance at 44 mph. For how many hours does the second car travel?

36. A car leaves a bus station and travels 45 mph. A bus leaves the same station, traveling 60 mph in

the same direction 1 hr later. How long will it take the bus to overtake the car?

37. Two cars are 192 miles apart and traveling toward each other on the same road. They meet in 2 hr. One car is traveling 4 mph faster than the other. What is the average speed of each car?

	Distance	Rate	Time
Faster car			
Slower car			

Table for Exercise 37

38. Two cars are 190 miles apart and traveling toward each other along the same road. They meet in 2 hr. One car is traveling 5 mph slower than the other. What is the average speed of each car?

39. A Piper Cub airplane has an average air speed that is 10 mph faster than a Cessna 150 airplane. If the combined distance traveled by these two small airplanes is 690 miles after 3 hr, what is the average speed of each plane?

40. The express train travels 25 mph faster than a cargo train. It takes the express train 6 hr to travel a route and it takes 9 hr for the cargo train to travel the same route. Find the speed of each train.

41. A woman can hike 1 mph faster down a trail to Archuletta Lake than she can on the return trip uphill. It takes her 3 hr to get to the lake and 6 hr to return. What is her speed hiking down to the lake?

42. A Cessna 182 airplane has an average air speed that is 50 mph slower than a Mooney airplane. If the combined distance traveled by these two small airplanes is 660 miles after 2 hr, what is the average speed of each plane?

43. Two boats travelling the same direction leave a harbor at noon. After 3 hr they are 60 miles apart. If one boat travels twice as fast as the other, find the average rate of each boat.

44. Two canoes travel down a river, starting at 9:00. One canoe travels twice as fast as the other. After 3.5 hr, the canoes are 5.25 miles apart. Find the average rate of each canoe.

Expanding Your Skills

45. Nickels are worth \$0.05 each.
    a. If a piggy bank contains 70 nickels, how much money is this?
    b. If a bank contains x nickels, write an expression that represents the total value.
    c. If a bank contains $30 + x$ nickels, write an expression that represents the total value.

46. Quarters are worth \$0.25 each.
    a. If a child has 12 quarters, how much money is this?
    b. If a child has x quarters, write an expression that represents the total value.
    c. If a child has $20 - x$ quarters, write an expression that represents the total value.

47. Jean-Paul has 12 coins in his pocket consisting of quarters and dimes. The total value of these coins is \$1.65. How many quarters and dimes does Jean-Paul have?

48. In Anna's purse there are 11 coins consisting of nickels and dimes. If the total value of these coins is \$0.95, how many of each type of coin are there?

49. A bank customer withdraws \$200 from her account in \$10 and \$20 bills. She requests twice as many \$20 bills as \$10 bills. How many of each kind of bill does she receive?

50. A woman withdraws \$250 from her account in \$5, \$10, and \$20 bills. If she receives four \$10 bills and twice as many \$5 bills as \$20 bills, how many of each type of bill does she receive?

section

2.8 Linear Inequalities

Concepts

1. Solutions to Inequalities
2. Graphing Linear Inequalities
3. Set-Builder Notation and Interval Notation
4. Addition and Subtraction Properties of Inequality
5. Multiplication and Division Properties of Inequality
6. Solving Inequalities of the Form $a < x < b$
7. Applications of Linear Inequalities

1. Solutions to Inequalities

Recall that $a < b$ (equivalently $b > a$) means that a lies to the left of b on the number line. The statement $a > b$ (equivalently $b < a$) means that a lies to the right of b on the number line. If $a = b$, then a and b are represented by the same point on the number line.

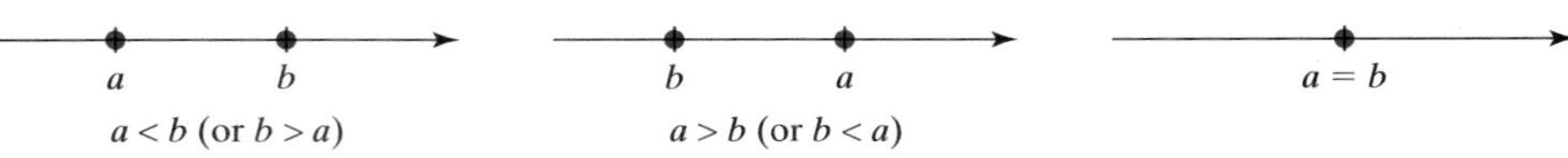

A Linear Inequality in One Variable

A **linear inequality in one variable**, x, is defined as any relationship of the form:

$$ax + b < 0,\ ax + b \le 0,\ ax + b > 0, \text{ or } ax + b \ge 0, \text{ where } a \ne 0.$$

The following inequalities are linear inequalities in one variable.

$$2x - 3 < 0 \qquad -4z - 3 > 0 \qquad a \le 4 \qquad 5.2y \ge 10.4$$

The number line is a useful tool to visualize the solution set of an equation or inequality. For example, the solution set to the equation $x = 2$ is $\{2\}$ and may be graphed as a single point on the number line.

$x = 2$ −6 −5 −4 −3 −2 −1 0 1 2 3 4 5 6

The solution set to an inequality is the set of real numbers that make the inequality a true statement. For example, the solution set to the inequality $x \ge 2$ is all real numbers 2 or greater. Because the solution set has an infinite number of values, we cannot list all of the individual solutions. However, we can graph the solution set on the number line.

$x \ge 2$ −6 −5 −4 −3 −2 −1 0 1 2 3 4 5 6

The square bracket symbol, [, is used on the graph to indicate that the point $x = 2$ is included in the solution set. By convention, square brackets, either [or], are used to *include* a point on a graph. Parentheses, (or), are used to *exclude* a point on a graph.

The solution set of the inequality $x > 2$ includes the real numbers greater than 2 but not including 2. Therefore, a (symbol is used on the graph to indicate that $x = 2$ is not included.

$x > 2$

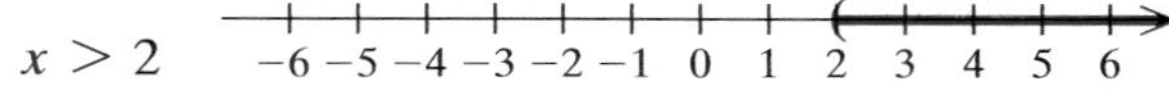

2. Graphing Linear Inequalities

example 1

Graphing Linear Inequalities

Graph the solution sets.

a. $x > -1$ b. $c \leq \dfrac{7}{3}$

Solution:

a. $x > -1$

−6 −5 −4 −3 −2 −1 0 1 2 3 4 5 6

The solution set is the set of all real numbers strictly greater than −1. Therefore, we graph the region on the number line to the right of −1. Because $x = -1$ is not included in the solution set, we use the (symbol at $x = -1$.

b. $c \leq \frac{7}{3}$ is equivalent to $c \leq 2\frac{1}{3}$.

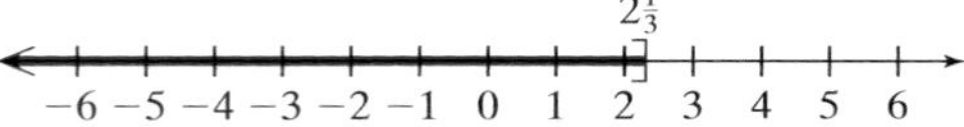

The solution set is the set of all real numbers less than or equal to $2\frac{1}{3}$. Therefore, graph the region on the number line to the left of and including $2\frac{1}{3}$. Use the symbol] to indicate that $c = 2\frac{1}{3}$ is included in the solution set.

Tip: Some textbooks use a closed circle or an open circle (● or ○) rather than a bracket or parenthesis to denote inclusion or exclusion of a value on the real number line. For example the solution sets for the inequalities $x > -1$ and $c \leq \frac{7}{3}$ are graphed here.

$x > -1$ −6 −5 −4 −3 −2 −1 0 1 2 3 4 5 6

$\frac{7}{3}$

$c \leq \frac{7}{3}$ −6 −5 −4 −3 −2 −1 0 1 2 3 4 5 6

A statement that involves more than one inequality is called a compound inequality. One type of compound inequality is used to indicate that one number is between two others. For example, the inequality $-2 < x < 5$ means that $-2 < x$ and $x < 5$. In words, this is easiest to understand if we read the variable first: x is greater than −2 and x is less than 5. The numbers satisfied by these two conditions are those between −2 and 5.

example 2 **Graphing a Compound Inequality**

Graph the solution set of the inequality: $-4.1 < y \leq -1.7$

Solution:

$-4.1 < y \leq -1.7$ means that

$-4.1 < y$ and $y \leq -1.7$

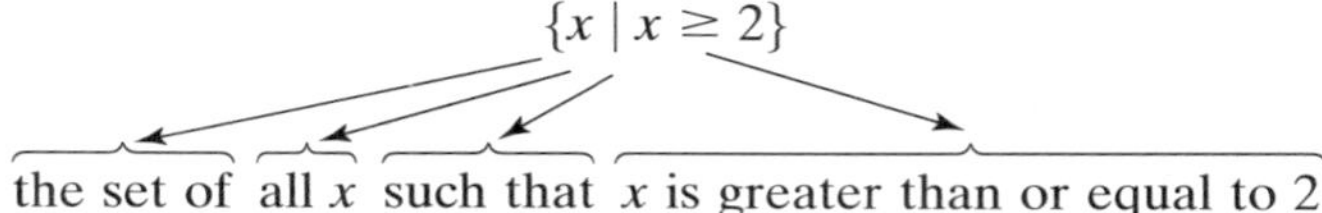

Shade the region of the number line greater than -4.1 and less than or equal to -1.7. These are the values of y between -4.1 and -1.7 (including $y = -1.7$).

3. Set-Builder Notation and Interval Notation

Graphing the solution set to an inequality is one way to define the set. Two other methods are to use **set-builder notation** or **interval notation.**

Set-Builder Notation

The solution to the inequality $x \geq 2$ can be expressed in set-builder notation as follows:

$$\{x \mid x \geq 2\}$$

the set of all x such that x is greater than or equal to 2

Interval Notation

To understand interval notation, first think of a number line extending infinitely far to the right and infinitely far to the left. Sometimes we use the infinity symbol, ∞, or negative infinity symbol, $-\infty$, to label the far right and far left ends of the number line (Figure 2-11).

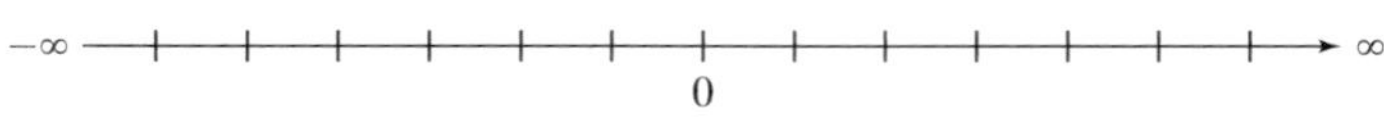

Figure 2-11

To express the solution set of an inequality in interval notation, sketch the graph first. Then use the endpoints to define the interval.

Inequality	Graph	Interval Notation
$x \geq 2$	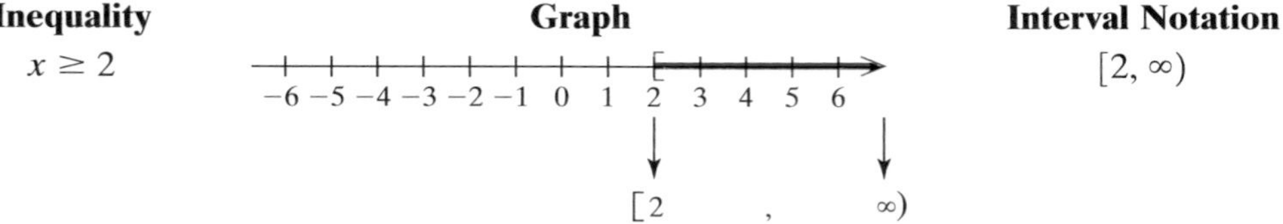	$[2, \infty)$

The graph of the solution set $x \geq 2$ begins at 2 and extends infinitely far to the right. The corresponding interval notation begins at 2 and extends to ∞. Notice that a square bracket [is used at 2 for both the graph and the interval notation. A parenthesis is always used at ∞ (and for $-\infty$), because there is no endpoint.

Using Interval Notation

- The endpoints used in interval notation are always written from left to right. That is, the smaller number is written first, followed by a comma, followed by the larger number.
- A parenthesis, (or), indicates that an endpoint is excluded from the set.
- A square bracket, [or], indicates that an endpoint is included in the set.
- Parentheses, (and), are always used with $-\infty$ and ∞.

example 3 **Using Set-Builder Notation and Interval Notation**

Complete the chart.

Set-Builder Notation	Graph	Interval Notation
	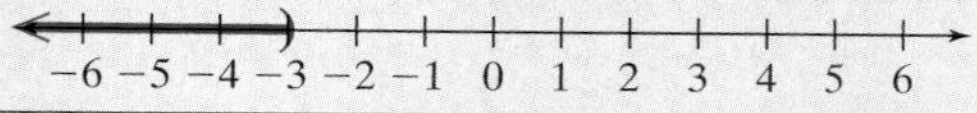	
		$[-\frac{1}{2}, \infty)$
$\{y \mid -2 \le y < 4\}$		

Solution:

Set-Builder Notation	Graph	Interval Notation
$\{x \mid x < -3\}$	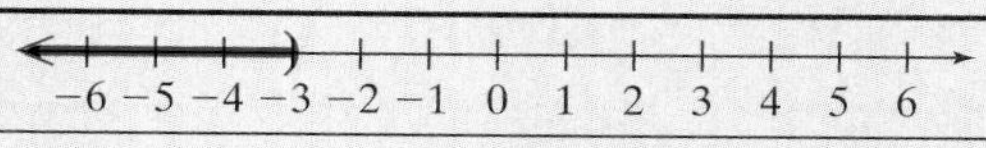	$(-\infty, -3)$
$\{x \mid x \ge -\frac{1}{2}\}$	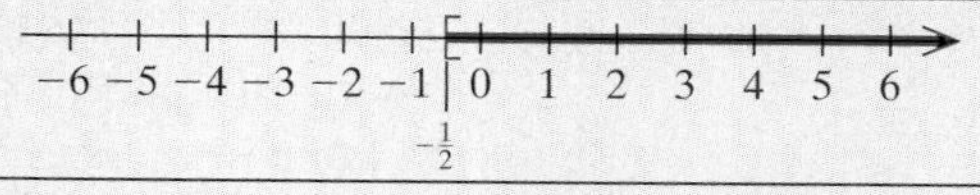	$[-\frac{1}{2}, \infty)$
$\{y \mid -2 \le y < 4\}$	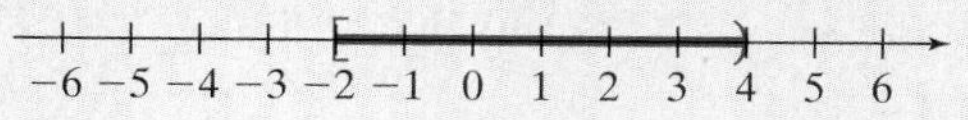	$[-2, 4)$

4. Addition and Subtraction Properties of Inequality

The process to solve a linear inequality is very similar to the method used to solve linear equations. Recall that adding or subtracting the same quantity to both sides of an equation results in an equivalent equation. The addition and subtraction properties of inequality state that the same is true for an inequality.

Addition and Subtraction Properties of Inequality

Let a, b, and c represent real numbers.

1. ***Addition Property of Inequality**: If $a < b$, then $a + c < b + c$
2. ***Subtraction Property of Inequality**: If $a < b$, then $a - c < b - c$

*These properties may also be stated for $a \leq b$, $a > b$, and $a \geq b$.

To illustrate the addition and subtraction properties of inequality, consider the inequality $10 < 20$. If we subtract or add a real number such as 5 to both sides, the left-hand side will still be less than the right-hand side (Figure 2-12).

$$10 < 20 \qquad\qquad 10 < 20$$
$$10 + 5 < 20 + 5 \qquad 10 - 5 < 20 - 5$$
$$15 < 25 \qquad\qquad 5 < 15$$

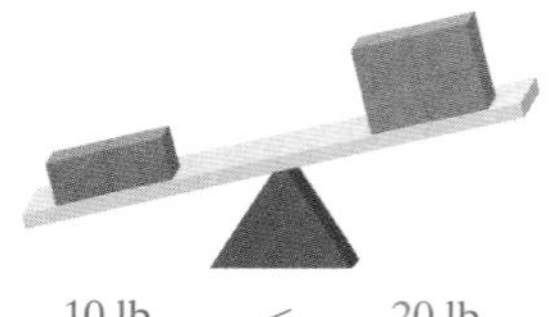

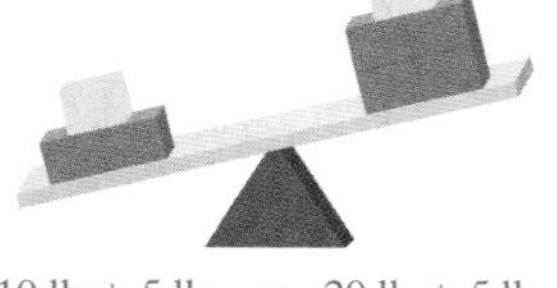

Figure 2-12

example 4

Solving a Linear Inequality

Solve the inequality and graph the solution set. Express the solution set in set-builder notation and in interval notation.

$$-2p + 5 < -3p + 6$$

Solution:

$-2p + 5 < -3p + 6$

$-2p + 3p + 5 < -3p + 3p + 6$ — Addition property of inequality (add $3p$ to both sides).

$p + 5 < 6$ — Simplify.

$p + 5 - 5 < 6 - 5$ — Subtraction property of inequality.

$p < 1$

Graph: −6 −5 −4 −3 −2 −1 0 1 2 3 4 5 6

Set-builder notation: $\{p \mid p < 1\}$

Interval notation: $(-\infty, 1)$

Tip: The solution to an inequality gives a set of values that make the original inequality true. Therefore, you can test your final answer by using *test points.* That is, pick a value in the proposed solution set and verify that it makes the original inequality true. Furthermore, any test point picked outside the solution set should make the original inequality false. For example,

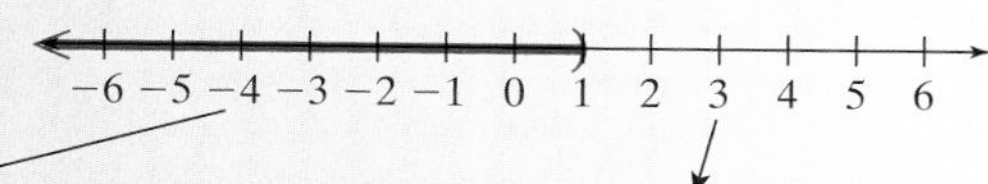

Pick $x = -4$ as an arbitrary test point within the proposed solution set.

$$-2p + 5 < -3p + 6$$
$$-2(-4) + 5 \overset{?}{<} -3(-4) + 6$$
$$8 + 5 \overset{?}{<} 12 + 6$$
$$13 < 18 \checkmark \quad \text{True}$$

Pick $x = 3$ as an arbitrary test point outside the proposed solution set.

$$-2p + 5 < -3p + 6$$
$$-2(3) + 5 \overset{?}{<} -3(3) + 6$$
$$-6 + 5 \overset{?}{<} -9 + 6$$
$$-1 \overset{?}{<} -3 \quad \text{False}$$

5. Multiplication and Division Properties of Inequality

Multiplying both sides of an equation by the same quantity results in an equivalent equation. However, the same is not always true for an inequality. If you multiply or divide an inequality by a negative quantity, the direction of the inequality symbol must be reversed.

For example, consider multiplying or dividing the inequality, $4 < 5$ by -1.

Multiply/Divide by -1: $4 < 5$ → $-4 > -5$

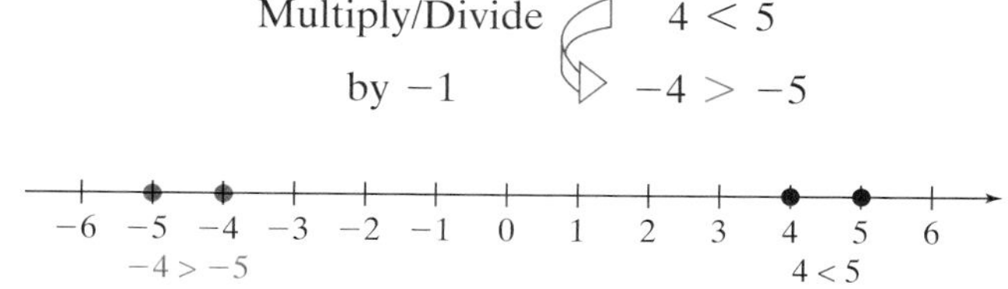

Figure 2-13

The number 4 lies to the left of 5 on the number line. However, −4 lies to the right of −5 (Figure 2-13). Changing the sign of two numbers changes their relative position on the number line. This is stated formally in the **multiplication** and **division properties of inequality**.

Multiplication and Division Properties of Inequality

Let a, b, and c represent real numbers. Then

*If c is positive and $a < b$, then

$$ac < bc \text{ and } \frac{a}{c} < \frac{b}{c}$$

*If c is negative and $a < b$, then

$$ac > bc \text{ and } \frac{a}{c} > \frac{b}{c}$$

The second statement indicates that if both sides of an inequality are multiplied or divided by a negative quantity, the inequality sign must be reversed.

*These properties may also be stated for $a \le b$, $a > b$, and $a \ge b$.

example 5 **Solving a Linear Inequality**

Solve the inequality $-5x - 3 \le 12$. Graph the solution set and write the answer in interval notation.

Solution:

$-5x - 3 \le 12$

$-5x - 3 + 3 \le 12 + 3$ Add 3 to both sides.

$-5x \le 15$

$\frac{-5x}{-5} \ge \frac{15}{-5}$ Divide by -5. Reverse the direction of the inequality sign.

$x \ge -3$

Interval notation: $[-3, \infty)$

−6 −5 −4 −3 −2 −1 0 1 2 3 4 5 6

Tip: The inequality $-5x - 3 \le 12$, could have been solved by isolating x on the right-hand side of the inequality. This would create a positive coefficient on the variable term and eliminate the need to divide by a negative number.

$-5x - 3 \le 12$

$-3 \le 5x + 12$

$-15 \le 5x$ Notice that the coefficient of x is positive.

$\frac{-15}{5} \le \frac{5x}{5}$ Do not reverse the inequality sign, because we are dividing by a positive number.

$-3 \le x$, or equivalently, $x \ge -3$

example 6 **Solving Linear Inequalities**

Solve the inequality $-\frac{1}{4}k + \frac{1}{6} \le 2 + \frac{2}{3}k$. Graph the solution set and write the answer in interval notation.

Solution:

$$-\frac{1}{4}k + \frac{1}{6} \le 2 + \frac{2}{3}k$$

$$12\left(-\frac{1}{4}k + \frac{1}{6}\right) \le 12\left(2 + \frac{2}{3}k\right)$$ Multiply both sides by 12 to clear fractions. (Because we multiplied by a positive number, the inequality sign is not reversed.)

$$\frac{12}{1}\left(-\frac{1}{4}k\right) + \frac{12}{1}\left(\frac{1}{6}\right) \le 12(2) + \frac{12}{1}\left(\frac{2}{3}k\right)$$ Apply the distributive property.

$$-3k + 2 \le 24 + 8k$$ Simplify.

$$-3k - 8k + 2 \le 24 + 8k - 8k$$ Subtract $8k$ from both sides.

$$-11k + 2 \le 24$$

$$-11k + 2 - 2 \le 24 - 2$$ Subtract 2 from both sides.

$$-11k \le 22$$

$$\frac{-11k}{-11} \ge \frac{22}{-11}$$ Divide both sides by -11. Reverse the inequality sign.

$$k \ge -2$$

Graph: −4 −3 −2 −1 0 1 2 3 4

Interval notation: $[-2, \infty)$

6. Solving Inequalities of the Form $a < x < b$

To solve a compound inequality of the form $a < x < b$ we can work with the inequality as a three-part inequality and isolate the variable, x, as demonstrated in Example 7.

example 7 **Solving a Compound Inequality of the Form $a < x < b$**

Solve the inequality: $-3 \le 2x + 1 < 7$. Graph the solution and write the answer in interval notation.

Solution:

To solve the compound inequality $-3 \le 2x + 1 < 7$ isolate the variable x in the middle. The operations performed on the middle portion of the inequality must also be performed on the left-hand side and right-hand side.

$$-3 \leq 2x + 1 < 7$$

$$-3 - 1 \leq 2x + 1 - 1 < 7 - 1$$ Subtract 1 from all three parts of the inequality.

$$-4 \leq 2x < 6$$ Simplify.

$$\frac{-4}{2} \leq \frac{2x}{2} < \frac{6}{2}$$ Divide by 2 in all three parts of the inequality.

$$-2 \leq x < 3$$

Graph: −6 −5 −4 −3 −2 −1 0 1 2 3 4 5 6

Interval notation: $[-2, 3)$

7. Applications of Linear Inequalities

Table 2-1 provides several commonly used translations to express inequalities.

Table 2-1

English Phrase	Mathematical Inequality
a is less than *b*	$a < b$
a is greater than *b* *a* exceeds *b*	$a > b$
a is less than or equal to *b* *a* is at most *b* *a* is no more than *b*	$a \leq b$
a is greater than or equal to *b* *a* is at least *b* *a* is no less than *b*	$a \geq b$

example 8 **Translating Expressions Involving Inequalities**

Translate the English phrases into mathematical inequalities:

a. Claude's annual salary, *s*, is no more than $40,000.
b. A citizen must be at least 18 years old to vote. (Let *x* represent a citizens' age.)
c. An amusement park ride has a height requirement between 48 in. and 70 in. (Let *h* represent height in inches.)

Solution:

a. $s \leq 40{,}000$
b. $x \geq 18$
c. $48 < h < 70$

Linear inequalities are found in a variety of applications. See how the next example can help you determine the minimum grade you need on an exam to get an A in your math course.

example 9 **Solving an Application with Linear Inequalities**

To earn an A in a math class, Alsha must average at least 90 on all of her tests. Suppose Alsha has scored 79, 86, 93, 90, and 95 on her first five math tests. Determine the minimum score she needs on her sixth test to get an A in the class.

Solution:

Let x represent the score on the sixth exam. — Label the variable.

$$\left(\begin{array}{c}\text{Average of}\\ \text{all tests}\end{array}\right) \geq 90$$ Create a verbal model.

$$\frac{79 + 86 + 93 + 90 + 95 + x}{6} \geq 90$$ The average score is found by taking the sum of the test scores and dividing by the number of scores.

$$\frac{443 + x}{6} \geq 90$$ Simplify.

$$6\left(\frac{443 + x}{6}\right) \geq (90)6$$ Multiply both sides by 6 to clear fractions.

$$443 + x \geq 540$$ Solve the inequality.

$$x \geq 540 - 443$$ Subtract 443 from both sides.

$$x \geq 97$$ Interpret the results.

Alsha must score at least 97 on her sixth test to receive an A in the course.

section 2.8 Practice Exercises

1. a. Simplify the expression: $3(x + 2) - (2x - 7)$
   b. Simplify the expression: $-(5x - 1) - 2(x + 6)$
   c. Solve the equation: $3(x + 2) - (2x - 7) = -(5x - 1) - 2(x + 6)$
2. a. Simplify the expression: $6 - 8(x + 3) + 5x$
   b. Simplify the expression: $5x - (2x - 5) + 13$
   c. Solve the equation: $6 - 8(x + 3) + 5x = 5x - (2x - 5) + 13$
3. The Ryder Cup is an annual golf tournament between the United States and Europe. In 1985 Europe won with five points more than the United States. If the total number of points scored was 28, how many points did each team score?
4. One angle of a triangle measures twice the smallest angle. The third angle is six times the measure of the smallest angle. Find the measures of each angle in the triangle.

For Exercises 5–10, graph each inequality and write the set in interval notation.

Set-Builder Notation	Graph	Interval Notation
5. $\{x \mid x \geq 6\}$		
6. $\left\{x \mid \frac{1}{2} < x \leq 4\right\}$		
7. $\{x \mid x \leq 2.1\}$		
8. $\left\{x \mid x > \frac{7}{3}\right\}$		
9. $\{x \mid -2 < x \leq 7\}$		
10. $\{x \mid x < -5\}$		

For Exercises 11–16, write each set in set-builder notation and in interval notation.

Set-Builder Notation	Graph	Interval Notation
11.	$\frac{3}{4}$	
12.	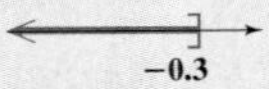 -0.3	
13.	-1 8	
14.	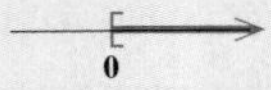 0	
15.	 -14	
16.	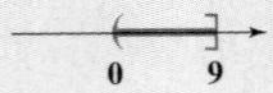 0 9	

For Exercises 17–22, graph each set and write the set in set-builder notation.

Set-Builder Notation	Graph	Interval Notation
17.		$[18, \infty)$
18.		$[-10, -2]$
19.		$(-\infty, -0.6)$
20.		$\left(-\infty, \frac{5}{3}\right)$
21.		$[-3.5, 7.1)$
22.		$[-10, \infty)$

For Exercises 23–30, solve the equation in part (a). For part (b), solve the inequality and graph the solution set.

23. a. $x + 3 = 6$
    b. $x + 3 > 6$

24. a. $y - 6 = 12$
    b. $y - 6 \geq 12$

25. a. $p - 4 = 9$
    b. $p - 4 \leq 9$

26. a. $k + 8 = 10$
    b. $k + 8 < 10$

27. a. $4c = -12$
    b. $4c < -12$

28. a. $5d = -35$
    b. $5d > -35$

29. a. $-10z = 15$
    b. $-10z \leq 15$

30. a. $-2w = 14$
    b. $-2w < 14$

For Exercises 31–34, determine whether the given number is a solution to the inequality.

31. $-2x + 5 < 4 \quad x = -2$

32. $-3y - 7 > 5 \quad y = 6$

33. $4(p + 7) - 1 > 2 + p \quad p = 1$

34. $3 - k < 2(-1 + k) \quad k = 4$

For Exercises 35–74, solve the inequality. Graph the solution set and write the set in interval notation.

35. $x + 5 \leq 6$

36. $y - 7 < 6$

37. $q - 7 > 3$

38. $r + 4 \geq -1$

39. $4 < 1 + z$

40. $3 > z - 6$

41. $2 \geq a - 6$

42. $7 \leq b + 12$

43. $3c > 6$

44. $4d \leq 12$

45. $-3c > 6$

46. $-4d \leq 12$

47. $-h \leq -14$

48. $-q > -7$

49. $12 \geq -\frac{x}{2}$

50. $6 < -\frac{m}{3}$

51. $-2 \leq p + 1 < 4$

52. $0 < k + 7 < 6$

53. $-3 < 6h - 3 < 12$

54. $-6 \leq 4a - 2 \leq 12$

55. $5 < \frac{1}{2}x < 6$

56. $-6 \leq 3x \leq 12$

57. $-5 \leq 4x - 1 < 15$

58. $-2 < \frac{1}{3}x - 2 \leq 2$

59. $0.6z \geq 54$

60. $-0.7w > 28$

61. $-\frac{2}{3}y < 6$

62. $\frac{3}{4}x \leq -12$

63. $-2x - 4 \leq 11$

64. $-3x + 1 > 0$

65. $-7b - 3 \leq 2b$

66. $3t \geq 7t - 35$

67. $4n + 2 < 6n + 8$

68. $2w - 1 \leq 5w + 8$

69. $8 - 6(x - 3) > -4x + 12$

70. $3 - 4(h - 2) > -5h + 6$

71. $\frac{7}{6}p + \frac{4}{3} \geq \frac{11}{6}p - \frac{7}{6}$

72. $\frac{1}{3}w - \frac{1}{2} \leq \frac{5}{6}w + \frac{1}{2}$

73. $-1.2a - 0.4 < -0.4a + 2$

74. $-0.4c + 1.2 > -2c - 0.4$

75. Let x represent a student's average in a math class. The grading scale is given here.

A	$93 \leq x \leq 100$
B+	$89 \leq x < 93$
B	$84 \leq x < 89$
C+	$80 \leq x < 84$
C	$75 \leq x < 80$
F	$0 \leq x < 75$

a. Write the range of scores corresponding to each letter grade in interval notation.

b. If Stephan's average is 84.01, what grade will he receive?

c. If Estella's average is 79.89, what grade will she receive?

76. Let x represent a student's average in a science class. The grading scale is given here.

A	$90 \leq x \leq 100$
B+	$86 \leq x < 90$
B	$80 \leq x < 86$
C+	$76 \leq x < 80$
C	$70 \leq x < 76$
D+	$66 \leq x < 70$
D	$60 \leq x < 66$
F	$0 \leq x < 60$

a. Write the range of scores corresponding to each letter grade in interval notation.

b. If Jacque's average is 89.99, what is her grade?

c. If Marc's average is 66.01, what is his grade?

For Exercises 77–88, translate the English phrase into a mathematical inequality.

77. The speed of a car, s, was at least 110 km/hr.

78. The length of a fish, L, was at least 10 in.

79. Tasha's average test score, t, exceeded 90.

80. The wind speed, w, exceeded 75 mph.

81. The height of a cave, h, was no more than 2 ft.

82. The temperature of the water in Blue Spring, t, is no more than 72°F.

83. The temperature on the tennis court, t, was no less than 100°F.

84. The length of the hike, L, was no less than 8 km.

85. The depth, d, of a certain pool was at most 10 ft.

86. The amount of rain, a, in a recent storm was at most 2 in.

87. The snowfall, h, in Monroe County is between 2 in. and 5 in.

88. The cost, c, of carpeting a room is between \$300 and \$400.

89. The average summer rainfall for Miami, Florida, for June, July, and August is 7.4 in. per month. If Miami receives 5.9 in. of rain in June and 6.1 in. in July, how much rain is required in August to exceed the 3-month summer average?

90. The average winter snowfall for Burlington, Vermont, for December, January, and February is 18.7 in. per month. If Burlington received 22 in. of snow in December and 24 in. in January, how much snow is required in February to exceed the 3-month winter average?

91. An artist paints wooden birdhouses. She buys the birdhouses for \$9 each. However, for large orders, the price per birdhouse is discounted by a percentage off the original price. Let x represent the number of birdhouses ordered. The corresponding discount is given in the table.

Size of Order	Discount
$x \le 49$	0%
$50 \le x \le 99$	5%
$100 \le x \le 199$	10%
$x \ge 200$	20%

Table for Exercise 91

a. If the artist makes an order for 190 birdhouses, compute the total cost.

b. Which costs more, 190 bird houses or 200 birdhouses? Explain your answer.

92. A wholesaler sells T-shirts to a surf shop at \$8 per shirt. However, for large orders, the price per shirt is discounted by a percentage off the original price. Let x represent the number of shirts ordered. The corresponding discount is given in the table.

Number of Shirts Ordered	Discount
$x \le 24$	0%
$25 \le x \le 49$	2%
$50 \le x \le 99$	4%
$100 \le x \le 149$	6%
$x \ge 150$	8%

Table for Exercise 92

a. If the surf shop orders 50 shirts, compute the total cost.

b. Which costs more, 148 shirts or 150 shirts? Explain your answer.

93. Maggie sells lemonade at an art show. She has a fixed cost of \$75 to cover the registration fee for the art show. In addition, her cost to produce each lemonade is \$0.17. If x represents the number of lemonades, then the total cost to produce x lemonades is given by

$$\text{Cost} = 75 + 0.17x$$

If Maggie sells each lemonade for \$2, then her revenue (the amount she brings in) for selling x lemonades is given by

$$\text{Revenue} = 2.00x$$

a. Write an inequality that expresses the number of lemonades, x, that Maggie must sell to make a profit. Profit is realized when the revenue is greater than the cost (Revenue > cost).

b. Solve the inequality in part (a).

94. Two rental car companies rent subcompact cars at a discount. Company A rents for \$14.95 per day plus 22 cents per mile. Company B rents for \$18.95 a day plus 18 cents per mile. Let x represent the number of miles driven in one day.

The cost to rent a subcompact car for one day from Company A is

$$\text{Cost}_A = 14.95 + 0.22x$$

The cost to rent a subcompact car for one day from Company B is

$$\text{Cost}_B = 18.95 + 0.18x$$

a. Write an inequality that expresses the number of miles, x, for which the daily cost to rent from Company A is less than the daily cost to rent from Company B

$$(\text{Cost}_A < \text{Cost}_B)$$

b. Solve the inequality in part (a).

Expanding Your Skills

For Exercises 95–100, solve the inequality. Graph the solution set and write the set in interval notation.

95. $3(x + 2) - (2x - 7) \le (5x - 1) - 2(x + 6)$

96. $6 - 8(y + 3) + 5y > 5y - (2y - 5) + 13$

97. $-2 - \frac{w}{4} \le \frac{1 + w}{3}$

98. $\frac{z - 3}{4} - 1 > \frac{z}{2}$

99. $-0.703 < 0.122p - 2.472$

100. $3.88 - 1.335t \ge 5.66$

chapter 2 SUMMARY

SECTION 2.1—ADDITION, SUBTRACTION, MULTIPLICATION, AND DIVISION PROPERTIES OF EQUALITY

KEY CONCEPTS:

An equation is an algebraic statement that indicates two expressions are equal. A solution to an equation is a value of the variable that makes the equation a true statement. The set of all solutions to an equation is the solution set of the equation.

A linear equation in one variable can be written in the form $ax + b = 0$, where $a \neq 0$.

EXAMPLES:

An example of an equation: $2x + 1 = 9$
The solution is $x = 4$.

The equation $5x + 20 = 0$ is a linear equation in one variable.

Addition Property of Equality:

If $a = b$, then $a + c = b + c$

Solve:

$$x - 5 = 12$$
$$x - 5 + 5 = 12 + 5$$
$$x = 17$$

Subtraction Property of Equality:

If $a = b$, then $a - c = b - c$

Solve:

$$z + 1.44 = 2.33$$
$$z + 1.44 - 1.44 = 2.33 - 1.44$$
$$z = 0.89$$

Multiplication Property of Equality:

If $a = b$, then $ac = bc$

Solve:

$$\frac{3}{4}x = 12$$
$$\frac{4}{3} \cdot \frac{3}{4}x = 12 \cdot \frac{4}{3}$$
$$x = 16$$

Division Property of Equality:

If $a = b$, then $\frac{a}{c} = \frac{b}{c}$ $(c \neq 0)$

Solve:

$$16 = 8y$$
$$\frac{16}{8} = \frac{8y}{8}$$
$$2 = y$$

KEY TERMS:

addition property of equality
division property of equality
equation
linear equation in one variable
multiplication property of equality
solution set
solution to an equation
subtraction property of equality

Section 2.2—Solving Linear Equations

Key Concepts:

Steps for Solving a Linear Equation in One Variable:

1. Simplify both sides of the equation by clearing parentheses and collecting *like* terms.
2. Use the addition and subtraction properties of equality to collect the variable terms on one side of the equation.
3. Use the addition and subtraction properties of equality to collect the constant terms on the other side of the equation.
4. Use the multiplication and division properties of equality to make the coefficient of the variable term equal to 1.
5. Check your answer.

Examples:

Solve the equation:

$5y + 7 = 3(y - 1) + 2$	
$5y + 7 = 3y - 3 + 2$	Clear parentheses.
$5y + 7 = 3y - 1$	Combine *like* terms.
$2y + 7 = -1$	Isolate variable term by subtracting $3y$ from both sides.
$2y = -8$	Isolate constant term by subtracting 7 from both sides.
$y = -4$	Divide both sides by 2.

Check:

$$5(-4) + 7 \stackrel{?}{=} 3((-4) - 1) + 2$$
$$-20 + 7 \stackrel{?}{=} 3(-5) + 2$$
$$-13 \stackrel{?}{=} -15 + 2$$
$$-13 = -13 \checkmark$$

A conditional equation is true for some values of the variable but is false for other values.

$x + 5 = 7$ is a conditional equation because it is true only on the condition that $x = 2$.

An equation that has all real numbers as its solution set is an identity.

Solve:

$$x + 4 = 2(x + 2) - x$$
$$x + 4 = 2x + 4 - x$$
$$x + 4 = x + 4$$
$$4 = 4 \quad \text{is an identity}$$

The solution is all real numbers.

An equation that has no solution is a contradiction.

Solve:

$$y - 5 = 2(y + 3) - y$$
$$y - 5 = 2y + 6 - y$$
$$y - 5 = y + 6$$
$$-5 = 6 \quad \text{is a contradiction}$$

There is no solution.

Key Terms:

conditional equation
contradiction
identity

Section 2.3—Linear Equations: Clearing Fractions and Decimals

Key Concepts:

To Clear Fractions (or Decimals) in an Equation

Multiply both sides of the equation by a common denominator of all the fractions.

Examples:

Solve the equation:

$-1.2x - 5.1 = 16.5$ — Multiply both sides by 10.

$10(-1.2x - 5.1) = 10(16.5)$

$-12x - 51 = 165$ — Apply the distributive property.

$-12x = 216$ — Add 51 to both sides.

$\frac{-12x}{-12} = \frac{216}{-12}$ — Divide by -12.

$x = -18$

Steps for Solving a Linear Equation in One Variable

1. Consider clearing fractions or decimals if any exist in the equation.
2. Simplify both sides of the equation by clearing parentheses and combining *like* terms.
3. Use the addition and subtraction properties of equality to collect the variable terms on one side of the equation.
4. Use the addition and subtraction properties of equality to collect the constant terms on the other side of the equation.
5. Use the multiplication and division properties of equality to make the coefficient of the variable term equal to 1.
6. Check your answer.

Solve the equation:

$\frac{1}{2}x - 2 - \frac{3}{4}x = \frac{7}{4}$

$\frac{4}{1}\left(\frac{1}{2}x - 2 - \frac{3}{4}x\right) = \frac{4}{1}\left(\frac{7}{4}\right)$ — Multiply by LCD.

$\frac{4}{1}\left(\frac{1}{2}x\right) - \frac{4}{1}(2) - \frac{4}{1}\left(\frac{3}{4}x\right) = \frac{4}{1}\left(\frac{7}{4}\right)$

$2x - 8 - 3x = 7$ — Apply distributive property.

$-x - 8 = 7$ — Combine *like* terms.

$-x = 15$ — Add 8 to both sides.

$x = -15$ — Divide by -1.

Key Term:

clearing fractions

Section 2.4—Applications of Linear Equations: Introduction to Problem Solving

Key Concepts:

Problem-Solving Steps for Word Problems

1. Read the problem carefully.
2. Assign labels to unknown quantities.
3. Develop a verbal model.
4. Write a mathematical equation.
5. Solve the equation.
6. Interpret the results and write the answer in words.

Key Terms:

consecutive even integers
consecutive integers
consecutive odd integers
problem-solving flow chart

Examples:

Consecutive integer problem:

The perimeter of a triangle is 54 m. The lengths of the sides are represented by three consecutive even integers. Find the lengths of the three sides.

1. Read the problem.
2. Let x represent one side, $x + 2$ represent the second side, and $x + 4$ represent the third side.

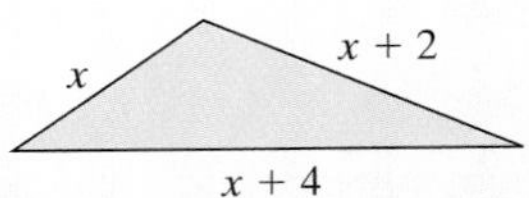

3. (First side) + (second side) + (third side) = perimeter
4. $x + (x + 2) + (x + 4) = 54$
5. $3x + 6 = 54$

$$3x = 48$$
$$x = 16$$
$$x + 2 = 18$$
$$x + 4 = 20$$

6. The lengths of the three sides are 16 m, 18 m, and 20 m.

Section 2.5—Solving Applications with Percent

Key Concepts:

Applications Involving Percents:

commission = (commission rate) ($ sold)

sales tax = (tax rate)($ in merchandise)

simple interest = Prt

Key Terms:

percent
simple interest

Examples:

Solve:
The cost of a CD plus tax is $16.43. If the tax rate is 6%, find the cost of the CD before tax.

Let x represent the pretax cost of the CD.

Then $0.06x$ represents the tax on the CD.

$$\begin{pmatrix}\text{Pretax}\\ \text{cost of CD}\end{pmatrix} + (\text{tax}) = \begin{pmatrix}\text{total}\\ \text{cost}\end{pmatrix}$$

$$x + 0.06x = 16.43$$
$$1.06x = 16.43$$
$$\frac{1.06x}{1.06} = \frac{16.43}{1.06}$$
$$x = 15.5$$

The CD cost $15.50 before tax.

Section 2.6—Formulas and Applications of Geometry

Key Concepts:

A literal equation is an equation that has more than one variable. Often such an equation can be manipulated to solve for different variables.

Formulas from Section R.2 can be used in applications involving geometry.

Key Term:

literal equations

Examples:

Given $P = 2a + b$ Solve for b.

$$P - 2a = 2a - 2a + b$$

$$P - 2a = b$$

or $b = P - 2a$

Solve:

Two complementary angles are drawn such that one angle is 5° more than three times the other angle. Find the measures of the two angles.

Let x represent one angle.

Let $3x + 5$ represent the other angle.

$$x + (3x + 5) = 90$$

$$4x + 5 = 90$$

$$4x = 85$$

$$x = 21.25$$

$$3x + 5 = 68.75$$

The two angles are 21.25° and 68.75°.

Section 2.7—More Applications of Linear Equations

Key Concepts:

The following types of word problems are introduced in Section 2.7.

1. Mixture problems
2. Applications involving distance, rate, and time
3. Principal and interest problems

Examples:

Principal and interest problem:

The amount Estella invested in a 6% savings account is \$3500 less than the amount she invested in a 10% savings account. At the end of one year, she received \$750 in simple interest. Find the amount Estella invested in each account.

Let x represent the amount at 10%.

Then, $x - 3500$ is the amount at 6%.

Distance, rate, and time problem:

A driver makes a trip in a rain storm in 6 hr. In good weather she increases her speed by 10 mph and the trip only takes 5 hr. Find the driver's speed during the rain.

	d	r	t
Rain	$6x$	x	6
No rain	$5(x+10)$	$x+10$	5

The distance traveled is the same.

$$6x = 5(x+10)$$
$$6x = 5x + 50$$
$$x = 50$$

The driver drove 50 mph in the rain.

$$\begin{pmatrix}\text{Interest from}\\ \text{10\% account}\end{pmatrix} + \begin{pmatrix}\text{interest from}\\ \text{6\% account}\end{pmatrix} = \begin{pmatrix}\text{total}\\ \text{interest}\end{pmatrix}$$

$0.10x + 0.06(x - 3500) = 750$ Multiply both sides of the equation by 100.

$$10x + 6(x - 3500) = 75{,}000$$
$$10x + 6x - 21{,}000 = 75{,}000$$
$$16x = 96{,}000$$
$$\frac{16x}{16} = \frac{96{,}000}{16}$$
$$x = 6000$$
$$x - 3500 = 2500$$

\$6000 was invested at 10% and \$2500 was invested at 6%.

SECTION 2.8—LINEAR INEQUALITIES

KEY CONCEPTS:

A linear inequality in one variable, x, is any relationship in the form: $ax + b < 0$, $ax + b > 0$, $ax + b \leq 0$, or $ax + b \geq 0$, where $a \neq 0$.

The solution set to an inequality can be expressed as a graph or in set-builder notation or in interval notation.

When graphing an inequality or when writing interval notation a parenthesis, (or), is used to denote that an endpoint is *not included* in a solution set. A square bracket, [or], is used to show that an endpoint *is included* in a solution set. Parenthesis (or) are always used with $-\infty$ and ∞.

The inequality $a < x < b$ is used to show that x is greater than a and less than b. That is, x is *between* a and b.

Multiplying or dividing an inequality by a negative quantity requires the direction of the inequality sign to be reversed.

KEY TERMS:

addition property of inequality
division property of inequality
interval notation
linear inequality in one variable
multiplication property of inequality
set-builder notation
subtraction property of inequality

EXAMPLES:

$3x - 1 < 11$ is a linear inequality.

Set-Builder Notation	Interval Notation	Graph
$\{x \mid x > a\}$	(a, ∞)	a
$\{x \mid x \geq a\}$	$[a, \infty)$	a
$\{x \mid x < a\}$	$(-\infty, a)$	a
$\{x \mid x \leq a\}$	$(-\infty, a]$	a
$\{x \mid a < x < b\}$	(a, b)	a b

Solve for *x*:

$$-2x + 6 \geq 14$$

$-2x + 6 - 6 \geq 14 - 6$ Subtract 6.

$-2x \geq 8$ Simplify.

$\dfrac{-2x}{-2} \leq \dfrac{8}{-2}$ Divide by -2. Reverse the inequality sign.

$$x \leq -4$$

Set-builder notation: $\{x \mid x \leq -4\}$

Graph: -4

Interval notation: $(-\infty, -4]$

chapter 2 REVIEW EXERCISES

Section 2.1

1. Label the following as either an expression or an equation:
   a. $3x + y = 10$ b. $9x + 10y - 2xy$
   c. $4(x + 3) = 12$ d. $-5x = 7$

2. Explain how to determine whether an equation is linear.

3. Identify which equations are linear.
   a. $4x^2 + 8 = -10$ b. $x + 18 = 72$
   c. $-3 + 2y^2 = 0$ d. $-4p - 5 = 6p$

4. For the equation, $4y + 9 = -3$, determine if the given numbers are solutions.
   a. $y = 3$ b. $y = 0$
   c. $y = -3$ d. $y = -2$

For Exercises 5–12, solve the equation using the addition, subtraction, multiplication, or division properties of equality.

5. $a + 6 = -2$ 6. $6 = z - 9$

7. $-\frac{3}{4} + k = \frac{9}{2}$ 8. $0.1r = 7$

9. $-5x = 21$ 10. $\frac{t}{3} = -20$

11. $-\frac{2}{5}k = \frac{4}{7}$ 12. $-m = -27$

13. The quotient of a number and negative six is equal to negative ten. Find the number.

14. The difference of a number and $-\frac{1}{8}$ is $\frac{5}{12}$. Find the number.

15. Four subtracted from a number is negative twelve. Find the number.

16. Six subtracted from a number is negative eight. Find the number.

Section 2.2

For Exercises 17–28, solve the equation.

17. $4d + 2 = 6$ 18. $5c - 6 = -9$

19. $-7c = -3c - 9$ 20. $-28 = 5w + 2$

21. $\frac{b}{3} + 1 = 0$ 22. $\frac{2}{3}h - 5 = 7$

23. $-3p + 7 = 5p + 1$

24. $4t - 6 = -12t + 16$

25. $4a - 9 = 3(a - 3)$

26. $3(2c + 5) = -2(c - 8)$

27. $7b + 3(b - 1) + 16 = 2(b + 8)$

28. $2 + (17 - x) + 2(x - 1) = 4(x + 2) - 8$

29. Explain the difference between an equation that is a contradiction and an equation that is an identity.

30. Label each equation as a conditional equation, a contradiction, or an identity.
    a. $x + 3 = 3 + x$ b. $3x - 19 = 2x + 1$
    c. $5x + 6 = 5x - 28$ d. $2x - 8 = 2(x - 4)$
    e. $-8x - 9 = -8(x - 9)$

Section 2.3

For Exercises 31–50, solve the equation.

31. $\frac{x}{8} - \frac{1}{4} = \frac{1}{2}$ 32. $\frac{y}{15} - \frac{2}{3} = \frac{4}{5}$

33. $\frac{4z + 7}{5} = z + 2$ 34. $\frac{5y - 3}{6} = 1 + y$

35. $\frac{1}{10}p - 3 = \frac{2}{5}p$ 36. $\frac{1}{4}y - \frac{3}{4} = \frac{1}{2}y + 1$

37. $-\frac{1}{4}(2 - 3t) = \frac{3}{4}$ 38. $\frac{2}{7}(w + 4) = \frac{1}{2}$

39. $17.3 - 2.7q = 10.55$

40. $4.9z + 4.6 = 3.2z - 2.2$

41. $5.74a + 9.28 = 2.24a - 5.42$

42. $62.84t - 123.66 = 4(2.36 + 2.4t)$

43. $0.05x + 0.10(24 - x) = 0.75(24)$

44. $0.20(x + 4) + 0.65x = 0.20(854)$

45. $100 - (t - 6) = -(t - 1)$

46. $3 - (x + 4) + 5 = 3x + 10 - 4x$

47. $5t - (2t + 14) = 3t - 14$

48. $9 - 6(2z + 1) = -3(4z - 1)$

49. $4w + 5[2(w + 4) - 3] = 2(w - 3) - 5$

50. $3r + 2[3(r - 1) + 2] = 2(3r + 4)$

Section 2.4

51. The sum of twice a number and four is forty. Find the number.

52. The difference of a number and −5 is −11. Find the number.

53. Twelve added to the sum of a number and two is forty-four. Find the number.

54. Twenty added to the sum of a number and six is thirty-seven. Find the number.

55. Three times a number is the same as the difference of twice the number and seven. Find the number.

56. Eight less than five times a number is forty-eight less than the number. Find the number.

57. Three times the largest of three consecutive even integers is 76 more than the sum of the other two integers. Find the integers.

58. Ten times the smallest of three consecutive integers is 213 more than the sum of the other two integers. Find the integers.

59. The perimeter of a triangle is 78 in. The lengths of the sides are represented by three consecutive integers. Find the lengths of the sides of the triangle.

60. The perimeter of a pentagon (a five-sided polygon) is 190 in. The five sides are represented by consecutive integers. Find the measures of the sides.

61. The minimum salary for a Major League baseball player in 1985 was $60,000. This was twice the minimum salary in 1980. What was the minimum salary in 1980?

62. The state of Indiana has approximately 2.1 million more people than Kentucky. Together their populations total 10.3 million. Approximately how many people are in each state?

Section 2.5

For Exercises 63–68, solve the problems involving percents.

63. What is 35% of 68?

64. What is 4% of 720?

65. 53.5 is what percent of 428?

66. 68.4 is what percent of 72?

67. 24 is 15% of what number?

68. 8.75 is 0.5% of what number?

69. A novel originally selling at $29.99 is on sale for 12% off. What is the sale price of the book? Round to the nearest cent.

70. What would be the total price (including tax) of a novel that sells for $26.39, if the sales tax rate is 7%?

71. In one week a salesperson receives $238 in commission. If this represents 14% of sales, how much did she sell?

72. Mara's yearly salary is $20,000 plus a 16% commission on sales. If her total salary was $44,000, how much money in merchandise did she sell?

73. Anna Tsao invested $3000 in an account paying 8% simple interest.
    a. How much interest will she earn in $3\frac{1}{2}$ years?
    b. What will her balance be at that time?

74. Eduardo invested money in an account earning 4% simple interest. At the end of 5 years, he had a total of $14,400. How much money did he originally invest?

Section 2.6

For Exercises 75–80, solve for the indicated variable.

75. $C = K - 273$ for K

76. $K = C + 273$ for C

77. $P = 4s$ for s

78. $y = mx + b$ for x

79. $2x + 5y = -2$ for y

80. $4(a + b) = Q$ for b

81. Pat is planning to extend the patio on the back of his house. He has drawn the shape shown in the figure, and he needs to know the width of the rectangular portion so that the total area will be approximately 85 ft^2. (Round the answer to the nearest foot.)

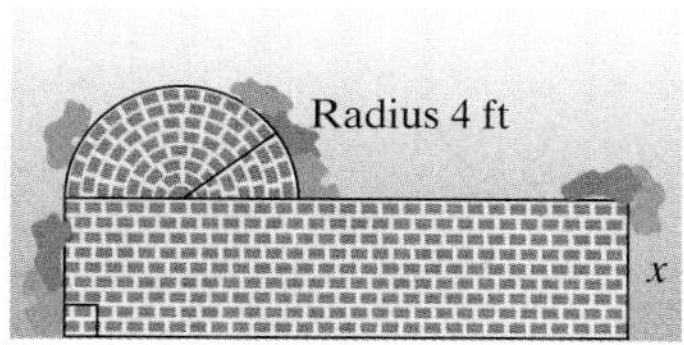

Figure for Exercise 81

82. The smallest angle of a triangle is 2° more than $\frac{1}{4}$ of the largest angle. The middle angle is 2° less than the largest angle. Find the measures of each angle.

83. One angle is 6° less than twice a second angle. If the two angles are complementary, what are their measures?

84. A rectangular window has width 1 ft less than its length. The perimeter is 18 ft. Find the length and the width of the window.

Section 2.7

85. Peggy invests some money in a savings account that pays 3.5% simple interest and buys some stock that yields 12%. Because of the volatility of the stock market, she invests half as much money in the stock as she invests in her savings account. If her total interest for the first year is $285.00, how much did Peggy invest in the savings account and how much did she invest in the stock?

86. Twenty pounds of nut mixture is to be made from a blend that costs $4/lb and another blend that costs $8/lb. How many pounds of each blend must be used to create 20 pounds of a nut mixture that costs $7/lb?

87. At a day care center, Suzanne wants to buy a treat for each of the 24 children. Ice cream on a stick costs $1.50 each and Popsicles cost $1.00 each. If Suzanne spends $29.00, how many of each type of treat does she buy?

88. A biker and jogger leave home at the same time, one traveling east and the other traveling west. If the biker maintains an average rate of 9 mph and the jogger maintains an average rate of 5 mph, how long will it take for them to be 10.5 miles apart?

89. The O'Neill family and Miller family leave at the same time for a reunion. The two families live 210 miles apart. The O'Neills travel in a small, four-cylinder car and drive an average of 5 mph slower than the Miller family. If it takes the families 2 hours to meet at a point between them, how fast does each family travel?

Section 2.8

90. Graph the inequalities and write the set in interval notation.

a. $\{x \mid x > -2\}$

b. $\left\{x \mid x \le \frac{1}{2}\right\}$

c. $\{x \mid -1 < x \le 4\}$

91. A landscaper buys potted geraniums from a nursery at a price of $5 per plant. However, for large orders, the price per plant is discounted by a percentage off the original price. Let x represent the number of potted plants ordered. The corresponding discount is given in the following table.

Number of Plants	Discount
$x \le 99$	0%
$100 \le x \le 199$	2%
$200 \le x \le 299$	4%
$x \ge 300$	6%

Table for Exercise 91

a. Find the cost to purchase 130 plants.

b. Which costs more, 300 plants or 295 plants? Explain your answer.

For Exercises 92–101, solve the inequality. Graph the solution set and express the answer in interval notation.

92. $c + 6 < 23$

93. $3w - 4 > -5$

94. $-2x - 7 \geq 5$

95. $5(y + 2) \leq -4$

96. $-\frac{3}{7}a \leq -21$

97. $1.3 > 0.4t - 12.5$

98. $4k + 23 < 7k - 31$

99. $\frac{6}{5}h - \frac{1}{5} \leq \frac{3}{10} + h$

100. $-6 < 2b \leq 14$

101. $-2 \leq z + 4 \leq 9$

102. The summer average rainfall for Bermuda for June, July, and August is 5.3 in. per month. If Bermuda receives 6.3 in. of rain in June and 7.1 in. in July, how much rain is required in August to exceed the 3-month summer average?

103. Reggie sells hot dogs at a ballpark. He has a fixed cost of \$33 to use the concession stand at the park. In addition, the cost for each hot dog is \$0.40. If x represents the number of hot dogs sold, then the total cost is given by

$$\text{Cost} = 33 + 0.40x$$

If Reggie sells each hot dog for \$1.50, then his revenue (the amount he brings in) for selling x hot dogs is given by

$$\text{Revenue} = 1.50x$$

a. Write an inequality that expresses the number of hot dogs, x, that Reggie must sell to make a profit. Profit is realized when the revenue is greater than the cost (revenue > cost).

b. Solve the inequality in part (a).

chapter 2 TEST

For Exercises 1–11, solve the equation.

1. $t + 3 = -13$

2. $8 = p - 4$

3. $\frac{t}{8} = -\frac{2}{9}$

4. $-3x + 5 = -2$

5. $2(p - 4) = p + 7$

6. $2 + d = 2 - 3(d - 5) - 2$

7. $\frac{3}{7} + \frac{2}{5}x = -\frac{1}{5}x + 1$

8. $3h + 1 = 3(h + 1)$

9. $\frac{4y - 3}{5} = y + 2$

10. $0.5c - 1.9 = 2.8 + 0.6c$

11. $-5(x + 2) + 8x = -2 + 3x - 8$

12. Solve the equation for y: $3x + y = -4$

13. Solve the equation for r: $C = 2\pi r$

14. $t = 4(x + y + z)$ for z

15. The perimeter of a pentagon (a five-sided polygon) is 315 in. The five sides are represented by consecutive integers. Find the measures of the sides.

16. A couple purchased two NHL hockey tickets and two NBA basketball tickets for \$153.92. A hockey ticket cost \$4.32 more than a basketball ticket. What were the prices of the individual tickets?

17. The total bill for a pair of basketball shoes (including sales tax) is \$87.74. If the tax rate is 7%, find the cost of the shoes before tax.

18. A couple has \$7200 to invest between two accounts, one that earns 8% simple interest and the other that earns 10% simple interest. If they earn \$656 in interest after 1 year, how much was invested in each account?

19. Two angles are complementary. One angle is 26° more than the other angle. What are the measures of the angles?

20. Two bikers leave the Harley-Davidson shop, traveling on the same road and in the same direction. One biker drives 5 mph faster than the other. The faster biker stops after $1\frac{1}{2}$ hr but the slower biker stops after 1 hr. If the two bikers are $32\frac{1}{2}$ miles apart, what was the speed of each biker?

21. The deadliest hurricane in the United States in the twentieth century occurred in 1900 off the coast of Galveston, Texas. The second deadliest hurricane occurred in 1928, making landfall in south Florida. Together the two storms accounted for 9836 deaths, and the Texas hurricane caused 6164 more deaths than the south Florida hurricane. Find the number of deaths caused by the Texas hurricane.

22. Graph the inequalities and write the set in interval notation.

    a. $\{x \mid x < 0\}$ b. $\{x \mid -2 \le x < 5\}$

For Exercises 23–25, solve the inequality. Graph the solution and write the solution set in interval notation.

23. $5x + 14 > -2x$

24. $2(3 - x) \ge 14$

25. $-13 \le 3p + 2 \le 5$

26. The average winter snowfall for Syracuse, New York, for December, January, and February is 27.5 in. per month. If Syracuse receives 24 in. of snow in December and 32 in. in January, how much snow is required in February to exceed the 3-month average?

CUMULATIVE REVIEW EXERCISES, CHAPTERS 1–2

For Exercises 1–8, perform the indicated operations.

1. $-12 + 5 - (-3)$

2. $\left|-\frac{1}{5} + \frac{7}{10}\right|$

3. $19.8 \div (-7.2)$

4. $5 - 2[3 - (4 - 7)]$

5. $-\frac{2}{3} + \left(\frac{1}{2}\right)^2$

6. $-3^2 + (-5)^2$

7. $\sqrt{5 - (-20) - 3^2}$

8. $\dfrac{-3.5 + 2.8 \div 0.4}{[-1 + 2(-8)] + 4(3 - 5)^2}$

9. Place the numbers in the appropriate location in the diagram:

$$\left\{\sqrt{7}, -2, 2, \frac{3}{5}, \pi, 0.\overline{7}, 0.25\right\}$$

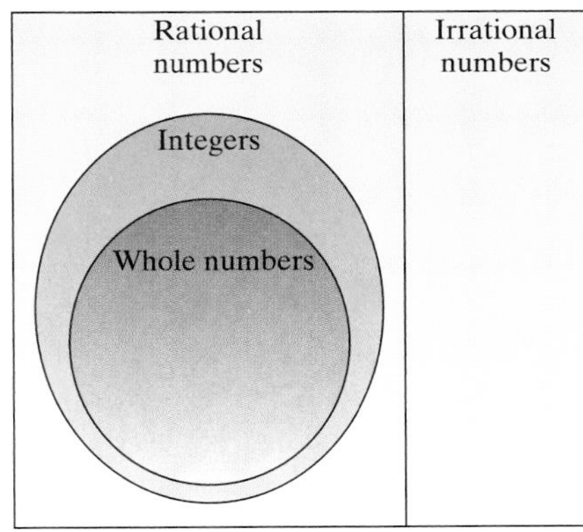

Figure for Exercise 9

10. Translate the mathematical expressions and simplify the results.

    a. One-third of three-fourths.

    b. The square root of the difference of five squared and nine.

11. Write the reciprocal of each number, if possible.

    a. $-\frac{2}{5}$ b. 3 c. 0

12. List the terms of the expression: $-7x^2y + 4xy - 6$

13. Identify the coefficient of the term: ab

14. Simplify: $-4[2x - 3(x + 4)] + 5(x - 7)$

15. Solve for x: $-2.5x - 5.2 = 12.8$

16. Solve for p: $-5(p - 3) + 2p = 3(5 - p)$

17. The sum of two consecutive odd integers is 156. Find the integers.

18. The total bill for a man's three-piece suit (including sales tax) is \$374.50. If the tax rate is 7%, find the cost of the suit before tax.

19. The area of a triangle is 41 cm^2. Find the height of the triangle if the base is 12 cm.

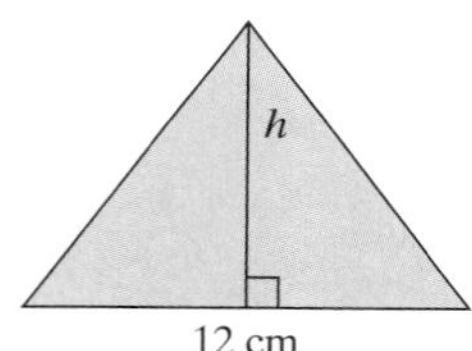

Figure for Exercise 19

20. Rachael invests twice as much money in an account earning 9% simple interest as she does in an account earning 4% simple interest. If the total interest is \$792 after 1 year, find the amount originally invested in each account.

21. An athlete bicycles an average of 12 mph faster than she runs. She runs 0.8 hr and bicycles for 1.2 hr for a total distance of 30.4 miles. Find her average rate running and her average rate bicycling.

22. Solve the inequality. Graph the solution set on a number line and express the solution in interval notation: $-2x - 3(x + 1) < 7$.

23. What number represents the opposite of the absolute value of -10?

chapter

3

Graphing Linear Equations in Two Variables

In Chapter 3 we investigate linear relationships and their graphs. In mathematical applications, a linear relationship may be given by a fixed value plus a constant rate of increase or decrease. For example:

Suppose 8 inches of snow is on the ground at 12:00 noon. A blizzard dumps more snow at a rate of $\frac{3}{4}$ in. per hour for nine hours. The depth of the snow, d (in inches), is given by:

$$d = \frac{3}{4}x + 8$$

where x is the number of hours after the storm began $(0 \le x \le 9)$.

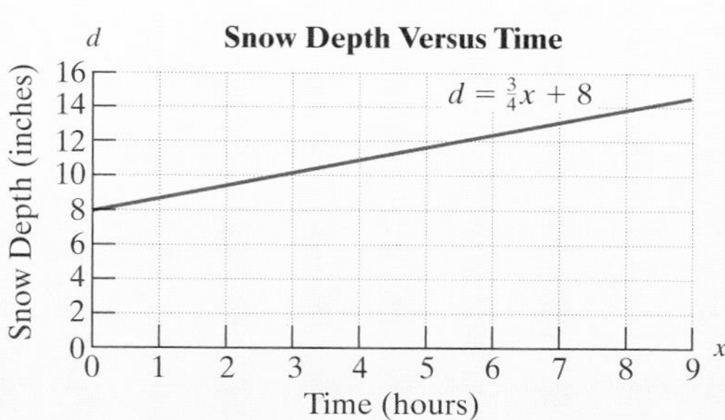

This and other linear equations may be analyzed and graphed using the techniques presented in Chapter 3. For more information see the Technology Connections in MathZone at

www.mhhe.com/miller_oneill

Concepts

1. Introduction to Graphing
2. Rectangular Coordinate System
3. Plotting Points in a Rectangular Coordinate System
4. Applications of Plotting Points

section

3.1 Rectangular Coordinate System

1. Introduction to Graphing

Mathematics is a powerful tool used by scientists and has directly contributed to the highly technical world we live in. Applications of mathematics have led to advances in the sciences, business, computer technology, and medicine.

One fundamental application of mathematics is the graphical representation of numerical information (or **data**). For example, Table 3-1 represents the number of clients admitted to a drug and alcohol rehabilitation program over a 12-month period.

Table 3-1

	Month	Number of Clients
Jan.	1	55
Feb.	2	62
March	3	64
April	4	60
May	5	70
June	6	73
July	7	77
Aug.	8	80
Sept.	9	80
Oct.	10	74
Nov.	11	85
Dec.	12	90

In table form, the information is difficult to picture and interpret. It appears that on a monthly basis, the number of clients fluctuates. However, when the data are represented in a graph, an upward trend is clear (Figure 3-1).

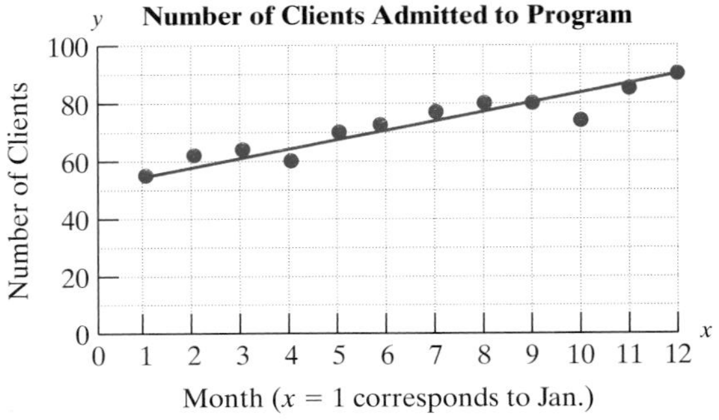

Figure 3-1

From the increase in clients depicted in this graph, management for the rehabilitation center might make plans for the future. If the trend continues, management might consider expanding its facilities and increasing its staff to accommodate the expected increase in clients.

example 1 **Interpreting a Graph**

Refer to Figure 3-1 and Table 3-1.

a. For which month was the number of clients the greatest?
b. How many clients were served in the first month (January)?
c. Which month corresponds to 60 clients served?
d. Between which two months did the number of patients decrease?
e. Between which two months did the number of clients remain the same?

Solution:

a. Month 12 (December) corresponds to the highest point on the graph, which represents the most clients.
b. In month 1 (January), there were 55 clients served.
c. Month 4 (April).
d. The number of clients decreased between months 3 and 4 and between months 9 and 10.
e. The number of clients remained the same between months 8 and 9.

2. Rectangular Coordinate System

In Example 1, two variables are represented, time and the number of clients. To picture two variables, we use a graph with two number lines drawn at right angles to each other (Figure 3-2). This forms a **rectangular coordinate system**. The horizontal line is called the ***x*-axis**, and the vertical line is called the ***y*-axis**. The point where the lines intersect is called the **origin**. On the x-axis, the numbers to the right of the origin are positive and the numbers to the left are negative. On the y-axis, the numbers above the origin are positive and the numbers below are negative. The x- and y-axes divide the graphing area into four regions called **quadrants**.

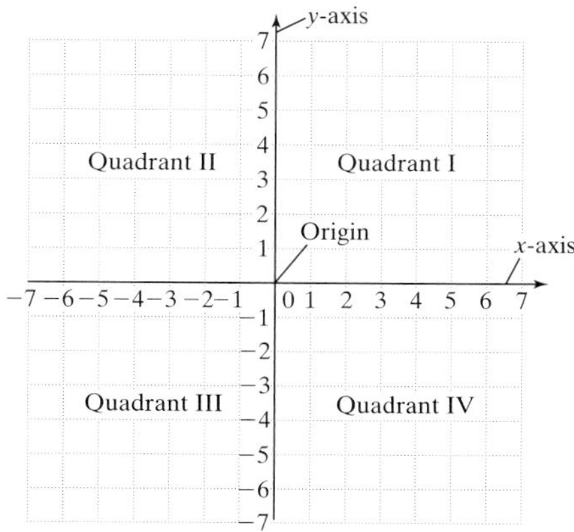

Figure 3-2

3. Plotting Points in a Rectangular Coordinate System

Points graphed in a rectangular coordinate system are defined by two numbers as an **ordered pair**, (x, y). The first number (called the first coordinate, or the abscissa) is the horizontal position from the origin. The second number (called the second coordinate, or the ordinate) is the vertical position from the origin. Example 2 shows how points are plotted in a rectangular coordinate system.

example 2 Plotting Points in a Rectangular Coordinate System

Plot the points.

a. $(4, 5)$ b. $(-4, -5)$ c. $(-1, 3)$ d. $(3, -1)$

e. $\left(\frac{1}{2}, -\frac{7}{3}\right)$ f. $(-2, 0)$ g. $(0, 0)$ h. $(\pi, 1.1)$

Solution:

See Figure 3-3.

a. The ordered pair $(4, 5)$ indicates that $x = 4$ and $y = 5$. Move 4 units in the positive x-direction (4 units to the right), and from there move 5 units in the positive y-direction (5 units up). Then plot the point. The point is in Quadrant I.
b. The ordered pair $(-4, -5)$ indicates that $x = -4$ and $y = -5$. Move 4 units in the negative x-direction (4 units to the left), and from there move 5 units in the negative y-direction (5 units down). Then plot the point. The point is in Quadrant III.
c. The ordered pair $(-1, 3)$ indicates that $x = -1$ and $y = 3$. Move 1 unit to the left and 3 units up. The point is in Quadrant II.
d. The ordered pair $(3, -1)$ indicates that $x = 3$ and $y = -1$. Move 3 units to the right and 1 unit down. The point is in Quadrant IV.
e. The improper fraction $-\frac{7}{3}$ can be written as the mixed number $-2\frac{1}{3}$. Therefore, to plot the point $(\frac{1}{2}, -\frac{7}{3})$ move to the right $\frac{1}{2}$ unit, and down $2\frac{1}{3}$ units. The point is in Quadrant IV.
f. The point $(-2, 0)$ indicates $y = 0$. Therefore, the point is on the x-axis.
g. The point $(0, 0)$ is at the origin.
h. The irrational number, π, can be approximated as 3.14. Thus, the point $(\pi, 1.1)$ is located approximately 3.14 units to the right and 1.1 units up. The point is in Quadrant I.

Figure 3-3

Tip: Notice that changing the order of the x- and y-coordinates changes the location of the point. The point $(-1, 3)$ for example is in Quadrant II, whereas $(3, -1)$ is in Quadrant IV (Figure 3-3). This is why points are represented by *ordered* pairs. The order of the coordinates is important.

4. Applications of Plotting Points

example 3

Plotting Points in an Application

The daily low temperatures (in degrees Fahrenheit) for one week in January for Sudbury, Ontario, Canada, are given in Table 3-2.

a. Write an ordered pair for each row in the table using the day number as the x-coordinate and the temperature as the y-coordinate.
b. Plot the ordered pairs from part (a) on a rectangular coordinate system.

Table 3-2

Day Number, x	Temperature, (°F), y
1	−3
2	−5
3	1
4	6
5	5
6	0
7	−4

Solution:

a. (1, −3)
(2, −5)
(3, 1)
(4, 6)
(5, 5)
(6, 0)
(7, −4)

Each ordered pair represents the day number and the corresponding low temperature for that day.

Tip: The graph in Example 3(b) shows only Quadrants I and IV because all of the x-coordinates are positive.

b.

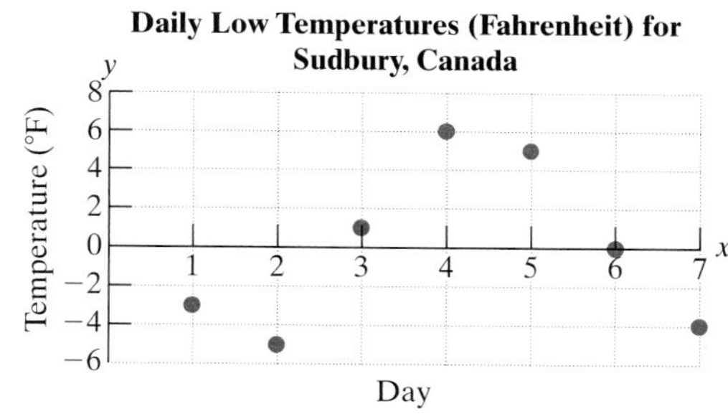

example 4

Plotting Points in an Application

The price per share of Nextel stock (in dollars) over a period of 5 days in January 2004 is shown in Figure 3-4.

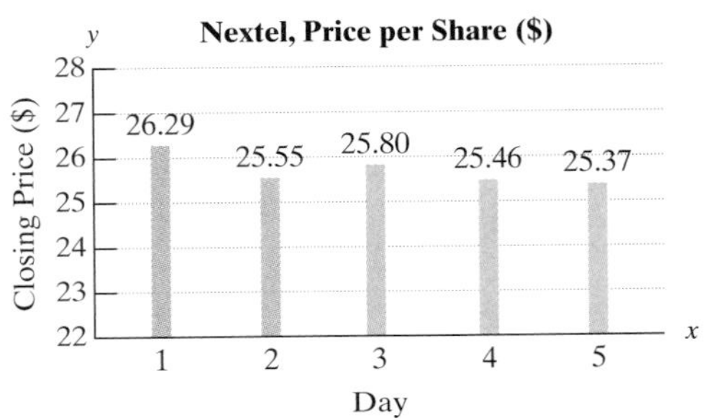

Figure 3-4

a. Use the information provided in the graph to write the ordered pairs representing the day number as x and the price as y. Interpret the meaning of the first ordered pair.
b. How much did the stock lose between the first day and the second day?
c. How much did the stock gain between day 2 and day 3?

Solution:

a. (1, 26.29), (2, 25.55), (3, 25.80), (4, 25.46), (5, 25.37);
(1, 26.29) means the price per share on day 1 was \$26.29.

b. We compute the daily gain or loss for a stock as follows:
Gain or loss = (current day's price) − (previous day's price)
For day 2, we have a loss: (day 2 price) − (day 1 price)

$$= \$25.55 - \$26.29$$
$$= -\$0.74 \quad \text{(Nextel lost \$0.74 per share)}$$

c. For day 3, we have a gain: (day 3 price) − (day 2 price)

$$= \$25.80 - \$25.55$$
$$= \$0.25 \quad \text{(Nextel gained \$0.25 per share)}$$

example 5

Determining Points from a Graph

A map of a national park is drawn so that the origin is placed at the ranger station. Four fire observation towers are located at points A, B, C, and D. Estimate the coordinates of the fire towers relative to the ranger station (all distances are in miles).

Solution:

Point A: $(-1, -2)$
Point B: $(2, -1)$
Point C: $\left(1\frac{1}{2}, \frac{1}{2}\right)$ or $\left(\frac{3}{2}, \frac{1}{2}\right)$ or $(1.5, 0.5)$
Point D: $\left(-\frac{1}{2}, 1\right)$ or $(-0.5, 1)$

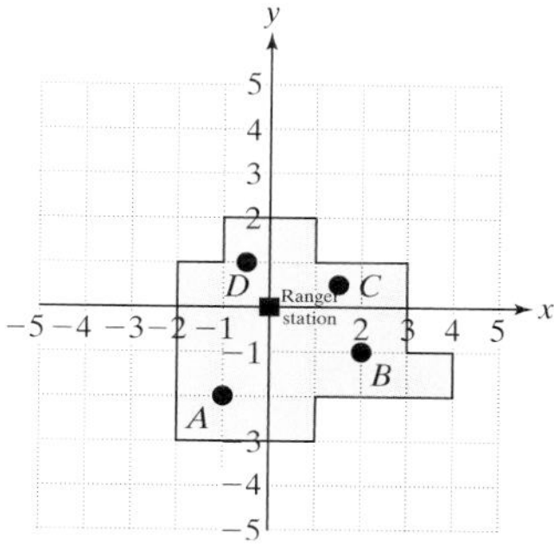

section 3.1 Practice Exercises

1. Plot the points on a rectangular coordinate system.
 a. $(2, 6)$ b. $(6, 2)$
 c. $(0, -3)$ d. $(-3, 0)$

2. Plot the points on a rectangular coordinate system.
 a. $(-1, 2)$ b. $(2, -1)$
 c. $(0, 7)$ d. $(7, 0)$

3. Plot the points on a rectangular coordinate system.
 a. $(4, 5)$ b. $(-4, 5)$
 c. $(4, -5)$ d. $(-4, -5)$

4. Plot the points on a rectangular coordinate system.
 a. $(2, 3)$ b. $(-2, 3)$
 c. $(2, -3)$ d. $(-2, -3)$

5. Plot the points on a rectangular coordinate system.
 a. $(-1, 5)$ b. $(0, 4)$ c. $\left(-2, -\frac{3}{2}\right)$
 d. $(2, -0.75)$ e. $(4, 2)$ f. $(-6.1, 0)$
 g. $(0, 0)$

6. Plot the points on a rectangular coordinate system.
 a. $(7, 0)$ b. $(-3, -1)$ c. $(0, 0)$
 d. $(0, 1.5)$ e. $(6, 1)$ f. $\left(-\frac{1}{4}, 4\right)$
 g. $\left(\frac{1}{4}, -4\right)$

For Exercises 7–14, identify the quadrant in which the given point is found.

7. $(13, -2)$
8. $(25, 16)$
9. $(-8, 14)$
10. $(-82, -71)$
11. $(-5, -19)$
12. $(-31, 6)$
13. $\left(\frac{5}{2}, \frac{7}{4}\right)$
14. $(9, -40)$

15. Explain why the point $(0, -5)$ is *not* located in Quadrant IV.

16. Explain why the point $(-1, 0)$ is *not* located in Quadrant II.

17. Explain where the point $(\frac{7}{8}, 0)$ is located.

18. Explain where the point $(0, \frac{6}{5})$ is located.

For Exercises 19 and 20, refer to the graph.

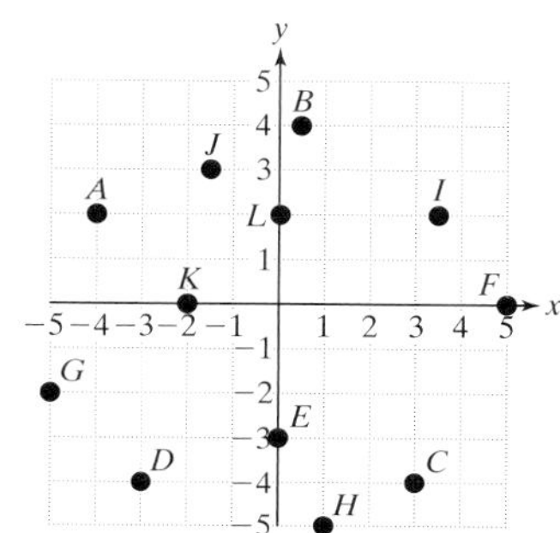

19. Estimate the coordinates of the points A, B, C, D, E, and F.

20. Estimate the coordinates of the points G, H, I, J, K, and L.

21. A movie theater has kept records of popcorn sales versus movie attendance.

a. Write the corresponding ordered pairs using the movie attendance as the x-variable and sales of popcorn as the y-variable. Interpret the meaning of the first ordered pair.

b. Plot the data points on a rectangular coordinate system.

Movie Attendance (number of people)	Sales of Popcorn ($)
250	225
175	193
315	330
220	209
450	570
400	480
190	185

Table for Exercise 21

22. The age and systolic blood pressure (in millimeters of mercury, mm Hg) for eight different women are given in the table.

a. Write the corresponding ordered pairs using the woman's age as the x-variable and the systolic blood pressure as the y-variable. Interpret the meaning of the first ordered pair.

b. Plot the data points on a rectangular coordinate system.

Age (years)	Systolic Blood Pressure (mm Hg)
57	149
41	120
71	158
36	115
64	151
25	110
40	118
77	165

Table for Exercise 22

23. The following ordered pairs give the population of the U.S. colonies from 1700 to 1770. Let x represent the year, where $x = 0$ corresponds to 1700, $x = 10$ corresponds to 1710, and so on. Let y represent the population of the colonies.

(0, 251000) (10, 332000) (20, 466000)
(30, 629000) (40, 906000) (50, 1171000)
(60, 1594000) (70, 2148000)

a. Interpret the meaning of the ordered pair (10, 332000).

b. Plot the points on a rectangular coordinate system.

(*Source: Information Please Almanac*)

24. The *poverty threshold* is defined by the federal government as an annual income at or below a certain value (defined yearly). The poverty threshold for selected years between 1960 and 1995 is given by the ordered pairs shown here. Let x represent the year where $x = 0$ corresponds to 1960, $x = 5$ corresponds to 1965 and so on. Let y represent the poverty threshold measured in dollars.

(0, 1490) (5, 1582) (10, 1954)
(20, 4190) (30, 6652) (35, 7763)

a. Interpret the meaning of the ordered pair (10, 1954).

b. Plot the points on a rectangular coordinate system.

(*Source: Information Please Almanac*)

25. The following table shows the average temperature in degrees Celsius for Montreal, Quebec, Canada, by month.

Month, (x)		Temperature °C, (y)
Jan.	1	−10.2
Feb.	2	−9.0
March	3	−2.5
April	4	5.7
May	5	13.0
June	6	18.3
July	7	20.9
Aug.	8	19.6
Sept.	9	14.8
Oct.	10	8.7
Nov.	11	2.0
Dec.	12	−6.9

Table for Exercise 25

a. Write the corresponding ordered pairs, letting $x = 1$ correspond to the month of January.

b. Plot the ordered pairs on a rectangular coordinate system.

26. The table shows the average temperature in degrees Fahrenheit for Fairbanks, Alaska, by month.

Month, (x)		Temperature °F, (y)
Jan.	1	−12.8
Feb.	2	−4.0
March	3	8.4
April	4	30.2
May	5	48.2
June	6	59.4
July	7	61.5
Aug.	8	56.7
Sept.	9	45.0
Oct.	10	25.0
Nov.	11	6.1
Dec.	12	−10.1

Table for Exercise 26

a. Write the corresponding ordered pairs, letting $x = 1$ correspond to the month of January.

b. Plot the ordered pairs on a rectangular coordinate system.

27. The number of patients served by a certain hospice care center is shown in the graph for the first 12 months after it opened.

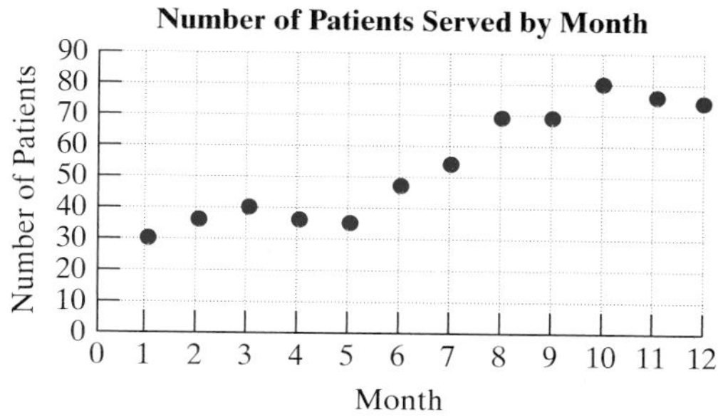

Figure for Exercise 27

a. For which month was the number of patients greatest?

b. How many patients did the center serve in the first month?

c. Between which months did the number of patients decrease?

d. Between which 2 months did the number of patients remain the same?

e. Which month corresponds to 40 patients served?

f. Approximately how many patients were served during the 10th month?

28. The number of housing permits (in thousands) issued by a certain county in Texas between 1994 and 2003 is shown in the graph.

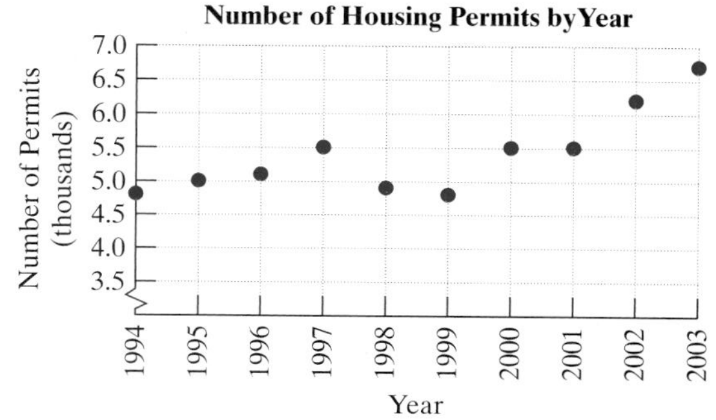

Figure for Exercise 28

a. For which year was the number of permits greatest?

b. How many permits did the county issue in 1997?

c. Between which years did the number of permits decrease?

d. Between which two years did the number of permits remain the same?

e. Which year corresponds to 5000 permits issued?

29. The price per share of a stock (in dollars) over a period of 5 days is shown in the graph.

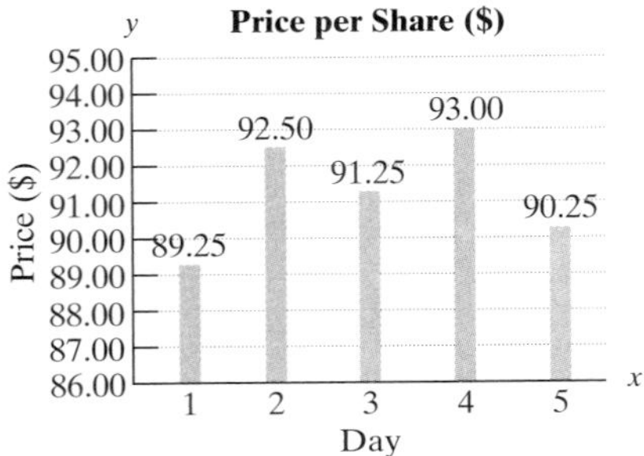

Figure for Exercise 29

a. Use the information provided in the graph to write the ordered pairs corresponding to the price of the stock each day. Interpret the meaning of the first ordered pair.

b. What was the gain in price between day 3 and day 4?

c. What was the loss in price between day 4 and day 5?

30. The price per share of a stock (in dollars) over a period of 5 days is shown in the graph.

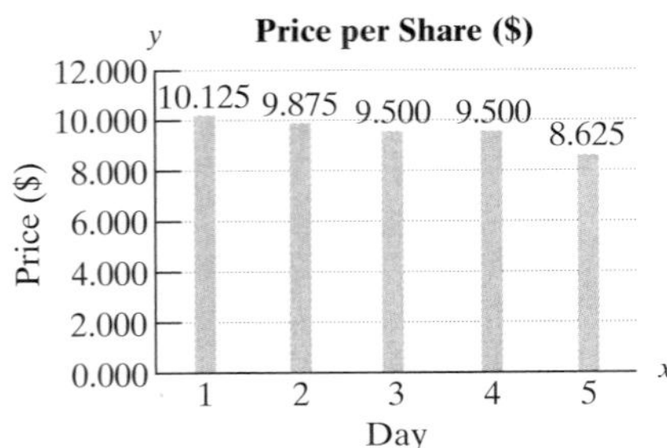

Figure for Exercise 30

a. Use the information provided in the graph to write the ordered pairs corresponding to the price of the stock each day. Interpret the meaning of the first ordered pair.

b. What was the loss between day 4 and day 5?

31. A map of a park is laid out with the visitor center located at the origin. Five visitors are in the park located at points A, B, C, D, and E. All distances are in meters.

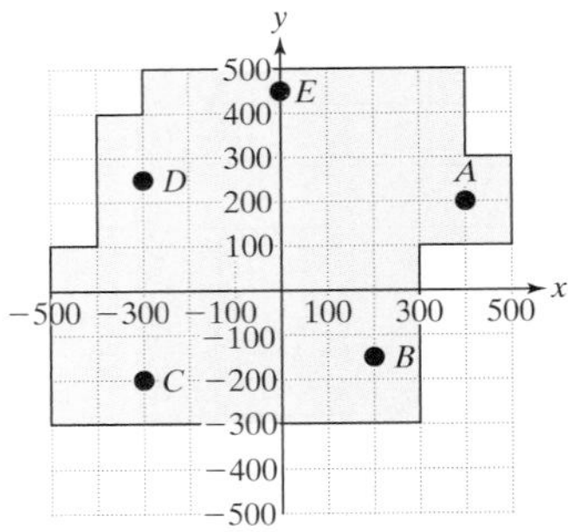

Figure for Exercise 31

a. Estimate the coordinates of each hiker.

b. How far apart are visitors C and D?

32. A townhouse has a sprinkler system in the backyard. With the water source at the origin, the sprinkler heads are located at points A, B, C, D, and E. All distances are in feet.

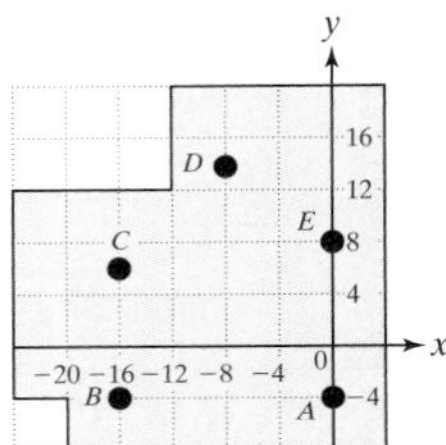

Figure for Exercise 32

a. Estimate the coordinates of each sprinkler head.

b. How far is the distance from sprinkler heads B and C?

section

3.2 Linear Equations in Two Variables

Concepts

1. Definition of a Linear Equation in Two Variables
2. Solutions to Linear Equations in Two Variables
3. Graphing Linear Equations in Two Variables by Plotting Points
4. Applications of Linear Equations in Two Variables

1. Definition of a Linear Equation in Two Variables

Recall that an equation in the form $ax + b = 0$, where $a \neq 0$, is called a linear equation in one variable. A solution to such an equation is a value of x that makes the equation a true statement. For example, $3x + 6 = 0$ has a solution of $x = -2$.

In this section we will look at linear equations in *two* variables.

Definition of a Linear Equation in Two Variables

Let a, b, and c be real numbers such that a and b are not both zero. Then, an equation that can be written in the form:

$$ax + by = c$$

is called a **linear equation in two variables.**

2. Solutions to Linear Equations in Two Variables

The equation $x + y = 4$ is a linear equation in two variables. A solution to such an equation is an ordered pair (x, y) that makes the equation a true statement. Several solutions to the equation $x + y = 4$ are listed here:

Solution:	Check:
(x, y)	$x + y = 4$
$(2, 2)$	$(2) + (2) = 4$ ✔
$(1, 3)$	$(1) + (3) = 4$ ✔
$(4, 0)$	$(4) + (0) = 4$ ✔
$(-1, 5)$	$(-1) + (5) = 4$ ✔

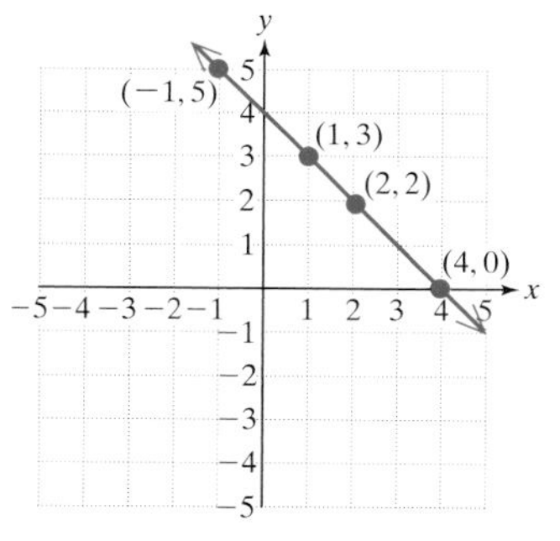

Figure 3-5

By graphing these ordered pairs, we see that the solution points line-up (Figure 3-5).

Notice that there are infinitely many solutions to the equation $x + y = 4$ so they cannot all be listed. Therefore, to visualize all solutions to the equation $x + y = 4$, we draw the line through the points in the graph. Every point on the line represents an ordered pair solution to the equation $x + y = 4$, and the line represents the set of *all* solutions to the equation.

example 1 **Determining Solutions to a Linear Equation**

For the linear equation, $4x - 5y = 8$, determine whether the given ordered pair is a solution.

a. $(2, 0)$ b. $(3, 1)$ c. $\left(1, -\frac{4}{5}\right)$

Solution:

a.
$$4x - 5y = 8$$
$$4(2) - 5(0) \stackrel{?}{=} 8 \qquad \text{Substitute } x = 2 \text{ and } y = 0.$$
$$8 - 0 = 8 \checkmark \text{ (true)} \qquad \text{The ordered pair } (2, 0) \text{ is a solution.}$$

b. $4x - 5y = 8$

$4(3) - 5(1) \stackrel{?}{=} 8$ Substitute $x = 3$ and $y = 1$.

$12 - 5 \neq 8$ The ordered pair (3, 1) is *not* a solution.

c. $4x - 5y = 8$

$4(1) - 5\left(-\frac{4}{5}\right) \stackrel{?}{=} 8$ Substitute $x = 1$ and $y = -\frac{4}{5}$.

$4 + 4 = 8$ ✔ (true) The ordered pair $\left(1, -\frac{4}{5}\right)$ is a solution.

3. Graphing Linear Equations in Two Variables by Plotting Points

The word *linear* means "relating to or resembling a line." It is not surprising then that the solution set for any linear equation in two variables forms a line in a rectangular coordinate system. Because two points determine a line, to graph a linear equation it is sufficient to find two solution points and draw the line between them. We will find three solution points and use the third point as a check point. This process is demonstrated in Example 2.

example 2

Graphing a Linear Equation

Graph the equation $x - 2y = 8$.

Solution:

We will find three ordered pairs that are solutions to $x - 2y = 8$. To find the ordered pairs, choose arbitrary values of x or y, such as those shown in the table. Then complete the table to find the corresponding ordered pairs.

x	y	
2		→ (2,)
	−1	→ (, −1)
0		→ (0,)

From the first row, substitute $x = 2$:

$$\begin{aligned} x - 2y &= 8 \\ (2) - 2y &= 8 \\ -2y &= 8 - 2 \\ -2y &= 6 \\ y &= -3 \end{aligned}$$

From the second row, substitute $y = -1$:

$$\begin{aligned} x - 2y &= 8 \\ x - 2(-1) &= 8 \\ x + 2 &= 8 \\ x &= 8 - 2 \\ x &= 6 \end{aligned}$$

From the third row, substitute $x = 0$:

$$\begin{aligned} x - 2y &= 8 \\ (0) - 2y &= 8 \\ -2y &= 8 \\ y &= -4 \end{aligned}$$

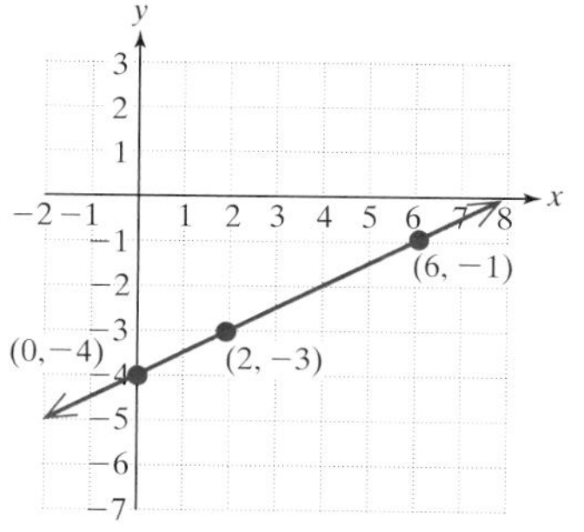

Figure 3-6

The completed table is shown below with the corresponding ordered pairs.

x	y	
2	−3	→ (2, −3)
6	−1	→ (6, −1)
0	−4	→ (0, −4)

To graph the equation, plot the three solutions and draw the line through the points (Figure 3-6).

Tip: Only two points are needed to graph a line. However, in Example 2, we found a third ordered pair, (0, −4). Notice that this point "lines up" with the other two points. If the three points do not "line up," then we know that a mistake was made in solving for at least one of the ordered pairs.

In Example 2, the original values for x and y given in the table were picked arbitrarily by the authors. It is important to note, however, that once you pick an arbitrary value for x, the corresponding y-value is determined by the equation. Similarly, once you pick an arbitrary value for y, the x-value is determined by the equation.

example 3 Graphing a Linear Equation

Graph the equation $4x + 3y = 15$.

Solution:

We will find three ordered pairs that are solutions to the equation $4x + 3y = 15$. In the next table, we have selected arbitrary values for x and y and must complete the ordered pairs. Notice that in this case, we are choosing zero for x and zero for y to illustrate that the resulting equation is often easy to solve.

x	y	
0		→ (0,)
	0	→ (, 0)
3		→ (3,)

From the first row, substitute $x = 0$:

$$\begin{aligned} 4x + 3y &= 15 \\ 4(0) + 3y &= 15 \\ 3y &= 15 \\ y &= 5 \end{aligned}$$

From the second row, substitute $y = 0$:

$$\begin{aligned} 4x + 3y &= 15 \\ 4x + 3(0) &= 15 \\ 4x &= 15 \\ x &= \frac{15}{4} \text{ or } 3\frac{3}{4} \end{aligned}$$

From the third row, substitute $x = 3$:

$$\begin{aligned} 4x + 3y &= 15 \\ 4(3) + 3y &= 15 \\ 12 + 3y &= 15 \\ 3y &= 3 \\ y &= 1 \end{aligned}$$

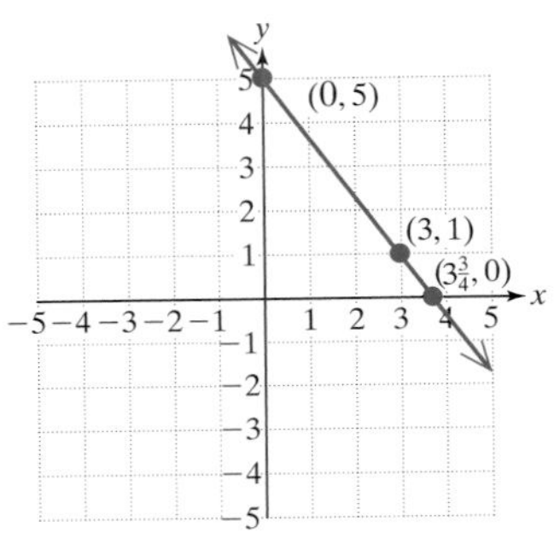

Figure 3-7

The completed table is shown with the corresponding ordered pairs.

x	y	
0	5	→ (0, 5)
$3\frac{3}{4}$	0	→ $(3\frac{3}{4}, 0)$
3	1	→ (3, 1)

To graph the equation, plot the three solutions and draw the line through the points (Figure 3-7).

example 4 Graphing a Linear Equation in Two Variables

Graph the line $y = -\frac{1}{3}x + 1$.

Solution:

Because the y-variable is isolated in the equation, it is easy to substitute a value for x and simplify the right-hand side to find y. Since any number for x can be picked, choose numbers that are multiples of 3 that will simplify easily when multiplied by $-\frac{1}{3}$.

x	y
3	
0	
−3	

$$y = -\frac{1}{3}x + 1$$

Let $x = 3$:

$$y = -\frac{1}{3}(3) + 1$$
$$y = -1 + 1$$
$$y = 0$$

Let $x = 0$:

$$y = -\frac{1}{3}(0) + 1$$
$$y = 0 + 1$$
$$y = 1$$

Let $x = -3$:

$$y = -\frac{1}{3}(-3) + 1$$
$$y = 1 + 1$$
$$y = 2$$

x	y
3	0
0	1
−3	2

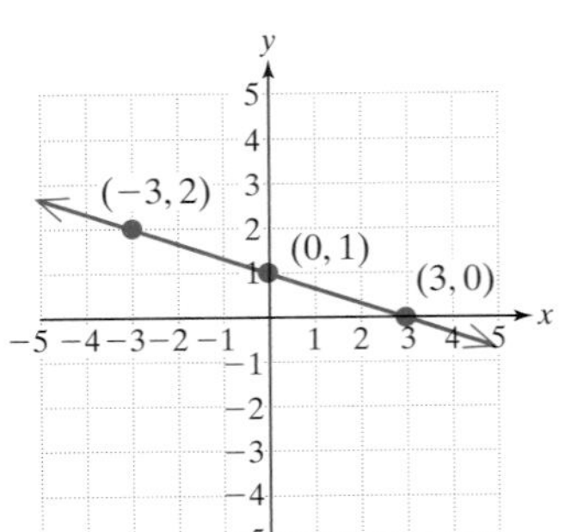

Figure 3-8

The line through the three ordered pairs (3, 0), (0, 1), and (−3, 2) is shown in Figure 3-8. The line represents the set of all solutions to the equation $y = -\frac{1}{3}x + 1$.

4. Applications of Linear Equations in Two Variables

example 5

Applying a Linear Equation in Two Variables

Tip: The condition $5 \le x \le 45$ indicates that the formula is valid for x values (skin fold measurements) between 5 mm and 45 mm, inclusive.

Exercise physiologists can estimate the percent of body fat for an individual by taking skin fold measurements from the abdomen, triceps, and thigh. Skin fold measurements are measured in millimeters with a pinching device called calipers. Based on experimental data, scientists have found that the following equation relates skin fold measurements, x, (in millimeters) to y% body fat.

$$y = 0.33x + 0.25 \qquad \text{for } 5 \le x \le 45$$

For parts (a)–(c), use the linear equation to find the percentage of body fat for the indicated skin fold measurements. Write the answers as ordered pairs and interpret the meaning of each ordered pair.

a. 15 mm ($x = 15$)
b. 25 mm ($x = 25$)
c. 40 mm ($x = 40$)
d. Graph the line using the ordered pairs from parts (a)–(c).
e. If a person has 10.15% body fat ($y = 10.15$), what is the corresponding skin fold measurement?

Solution:

a. The variable x represents skin fold, so substitute $x = 15$ into the equation and solve for y.

$$\begin{aligned} y &= 0.33x + 0.25 \\ &= 0.33(15) + 0.25 \\ &= 4.95 + 0.25 \\ &= 5.20 \end{aligned}$$

The ordered pair (15, 5.20) indicates that a person with a 15 mm skin fold will have approximately 5.2% body fat.

b. $y = 0.33x + 0.25$

$= 0.33(25) + 0.25$ Substitute $x = 25$ into the equation and solve for y.

$= 8.25 + 0.25$

$= 8.50$

The ordered pair (25, 8.50) indicates that a person with a 25-mm skin fold will have approximately 8.5% body fat.

c. $y = 0.33x + 0.25$

$= 0.33(40) + 0.25$ Substitute $x = 40$ into the equation and solve for y.

$= 13.2 + 0.25$

$= 13.45$

The ordered pair (40, 13.45) indicates that a person with a 40-mm skin fold will have approximately 13.45% body fat.

d. In application problems, a linear equation may have restrictions on the x-variable. It would not make sense, for instance, to have a negative skin fold measurement. In this case skin fold measurements are restricted to $5 \leq x \leq 45$. For this reason the line is drawn only for x-values between 5 and 45 mm (Figure 3-9).

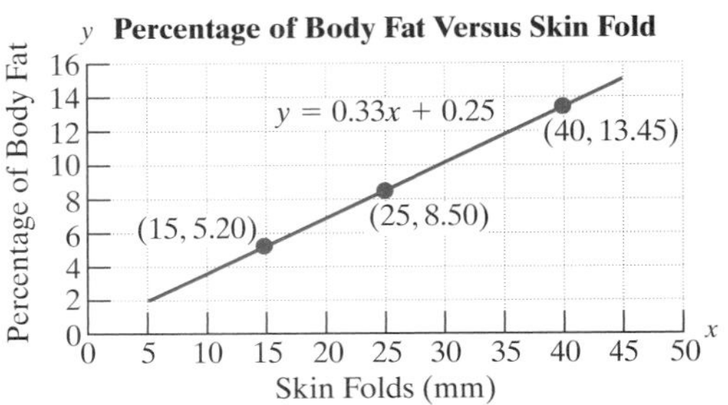

Figure 3-9

e. Because the variable y represents y% body fat, substitute $y = 10.15$ into the equation and solve for x.

$$y = 0.33x + 0.25$$

$$10.15 = 0.33x + 0.25 \qquad \text{Substitute } y = 10.15.$$

$$9.9 = 0.33x \qquad \text{Solve for } x.$$

$$\frac{9.9}{0.33} = \frac{0.33x}{0.33}$$

$$30 = x$$ A person with 10.15% body fat will have a 30-mm skin fold.

Calculator Connections

A viewing window of a graphing calculator shows a portion of a rectangular coordinate system. The standard viewing window for many calculators shows the x-axis between -10 and 10 and the y-axis between -10 and 10 (Figure 3-10). Furthermore, the scale defined by the "tic" marks on both the x- and y-axes is usually set to 1.

The "Standard Viewing Window"

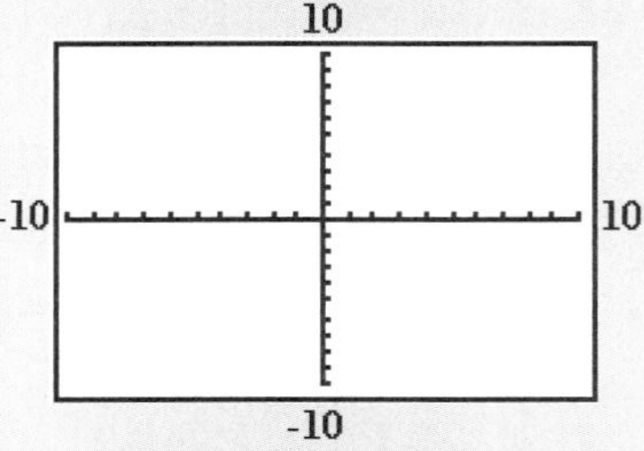

Figure 3-10

To graph an equation in x and y on a graphing calculator, the equation must be written with the y-variable isolated. Therefore, the equation $x + 3y = 3$ must first be written as $y = -\frac{1}{3}x + 1$ before it can be entered into a graphing calculator. To enter the equation $y = -\frac{1}{3}x + 1$ use parentheses around the fraction $\frac{1}{3}$. The *Graph* option displays the graph of the line.

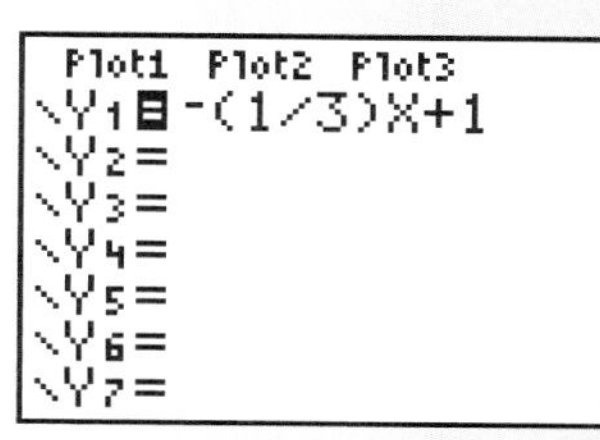

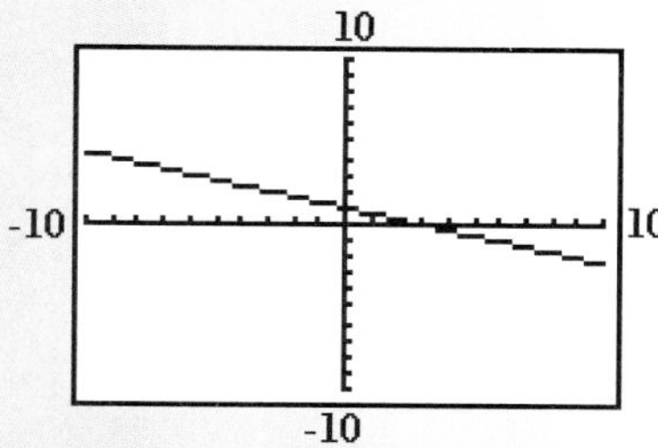

Sometimes the standard viewing window does not provide an adequate display for the graph of an equation. For example, the graph of $y = -x + 15$ is visible only in a small portion of the upper right corner of the standard viewing window.

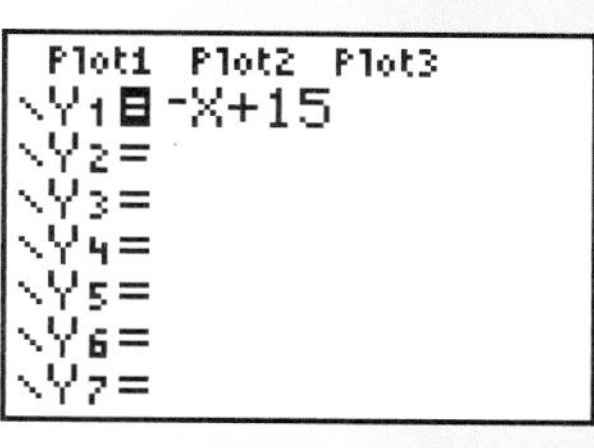

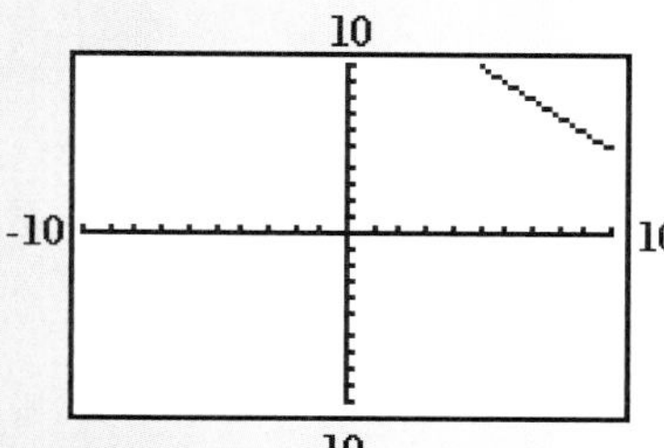

To see where this line crosses the x- and y-axes, we can change the viewing window to accommodate larger values of x and y. Most calculators have a *Range* feature or *Window* feature that allows the user to change the minimum and maximum x- and y-values.

To get a better picture of the equation $y = -x + 15$, change the minimum x-value to -10 and the maximum x-value to 20. Similarly, use a minimum y-value of -10 and a maximum y-value of 20.

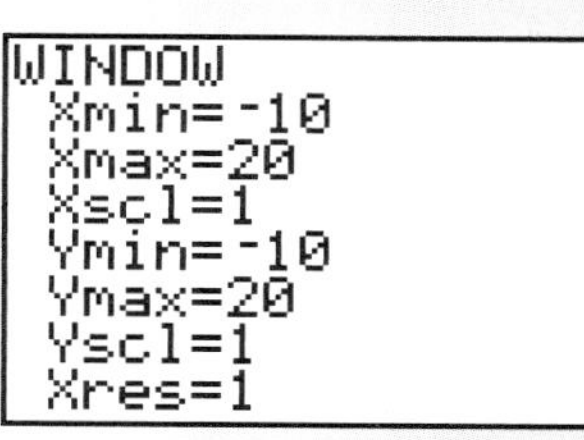

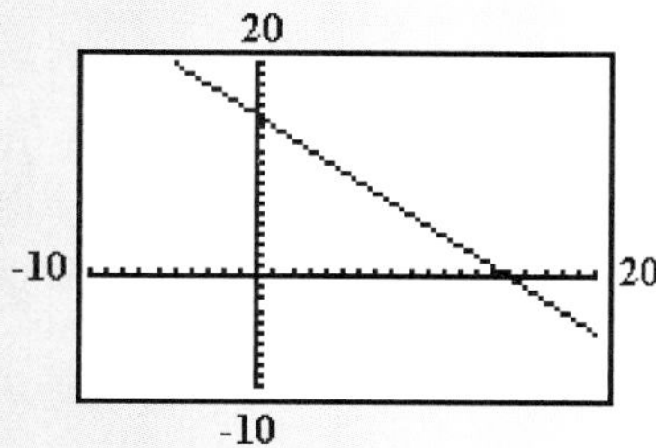

section 3.2 Practice Exercises

For Exercises 1–6, determine if the given ordered pair is a solution to the equation.

1. $x + y = -5$; $(-2, -3)$
2. $x - y = 6$; $(8, 2)$
3. $y = 3x - 2$; $(1, 1)$
4. $y = -\frac{1}{3}x + 3$; $(-3, 4)$
5. $y = -\frac{5}{2}x + 5$; $(-2, 0)$
6. $4x + 5y = 20$; $(-5, -4)$

For Exercises 7–20, complete the table, and graph the corresponding ordered pairs. Draw the line defined by the points to represent all solutions to the equation.

7. $x + y = 3$

x	y
2	
	3
−1	
	0

8. $x + y = -2$

x	y
1	
	0
−3	
	2

9. $x - y = 6$

x	y
3	
	−6
6	
	1

10. $x - y = -1$

x	y
3	
	0
1	
	3

11. $2x - 3y = 6$

x	y
0	
	0
2	

12. $4x + 2y = 8$

x	y
0	
	0
3	

13. $y = 5x + 1$

x	y
1	
	11
−1	

14. $y = -3x - 3$

x	y
−2	
	0
−4	

15. $y = \frac{2}{7}x - 5$

x	y
7	
−14	
0	

16. $y = -\frac{3}{5}x - 2$

x	y
0	
5	
10	

17. $5x + 3y = 12$

x	y
1	
	4
−1	

18. $4x - 3y = 6$

x	y
−2	
	4
0	

19. $y = -3.4x + 5.8$

x	y
0	
1	
2	

20. $y = -1.2x + 4.6$

x	y
0	
1	
2	

For Exercises 21–30, graph the equations.

21. $2x + y = 4$
22. $x + 3y = 3$
23. $x - 3y = 6$
24. $4x - y = 8$
25. $y = -5x - 5$
26. $y = 3x - 2$
27. $y = \frac{1}{2}x + 2$
28. $y = \frac{1}{3}x - 4$
29. $y = -1.2x - 1$
30. $y = -2.5x + 3$

31. The students in the ninth grade at Atlantic High School pick up aluminum cans to be recycled. The current value of aluminum is \$0.69 per pound. If the students pay \$20 to rent a truck to haul the cans, then the following equation expresses the amount of money that they earn, y, given the number of pounds of aluminum, x.

$$y = 0.69x - 20 \quad (x \geq 0)$$

a. Let $x = 55$ and solve for y.

b. Let $y = 80.05$ and solve for x.

c. Write the ordered pairs from parts (a) and (b) and interpret their meaning in the context of the problem.

d. Graph the ordered pairs and the line defined by the points.

32. The store "CDs R US" sells all compact discs for \$13.99. The following equation represents the revenue, y, (in dollars) generated by selling x CDs.

$$y = 13.99x \quad (x \geq 0)$$

a. Find y when $x = 13$.

b. Find x when $y = 279.80$.

c. Write the ordered pairs from parts (a) and (b) and interpret their meaning in the context of the problem.

d. Graph the ordered pairs and the line defined by the points.

33. The value of a car depreciates the minute that it is driven off of the lot. For a Hyundai Accent, the value of the car is given by the equation $y = -1531x + 11{,}599$ $(x \geq 0)$ where y is the value of the car in dollars, x years after its purchase.

a. Find y when $x = 1$.

b. Find x when $y = 7006$.

c. Write the ordered pairs from parts (a) and (b) and interpret their meaning in the context of the problem.

34. The enrollment in Catholic schools in the United States declined after 1970. This decline can be approximated by the linear equation

$$y = -94{,}378x + 4{,}363{,}000$$

where y is the total enrollment in Catholic schools and x is the number of years after 1970 (that is, $x = 0$ corresponds to 1970, $x = 1$ corresponds to 1971, and so on) (*Source: Information Please Almanac*)

a. Find y when x is 5.

b. Find x when y is 3,136,086.

c. Write the ordered pairs from parts (a) and (b) and interpret their meaning in the context of the problem.

35. The total sales for the Emerson Electric Co. from 1994 to 1999 is given by the equation $y = 1136x + 8790$. The value of y is in millions of dollars and x is the time in years after 1994. That is, $x = 0$ corresponds to 1994, $x = 1$ corresponds to 1995, and so on.

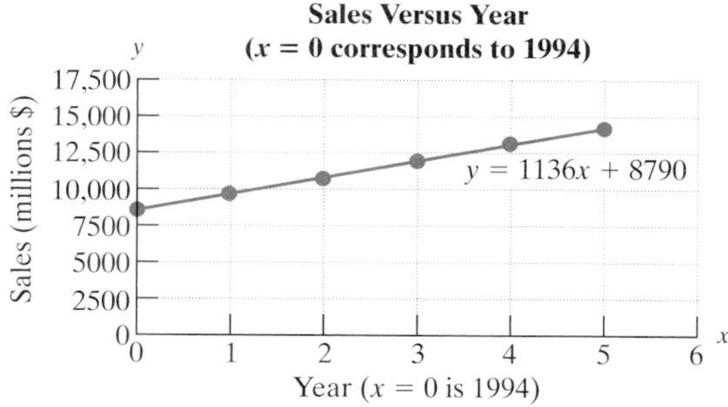

(*Source:* Emerson Electric Co., 1999 Annual Report.)

Figure for Exercise 35

a. Find the amount in sales reported for the year 1995.

b. Find the amount in sales reported for the year 1998.

c. Use the equation to find the year in which the sales equaled \$12,198 million.

d. Use the equation to find the year in which the sales equaled \$14,470 million.

36. The net earning for the Emerson Electric Co. from 1994 to 1999 is given by the equation $y = 105x + 800$. The value of y is in millions of dollars and x is the time in years after 1994. That is, $x = 0$ corresponds to 1994, $x = 1$ corresponds to 1995, and so on.

    a. Find the net earnings for the year 1994.

    b. Find the net earnings for the year 1997.

    c. Use the equation to find the year in which the net earnings equaled 1010 million dollars.

    d. Use the equation to find the year in which the net earnings equaled 1325 million dollars.

Expanding Your Skills

37. Graph the lines on the same set of coordinate axes, and observe any similarities and differences.

    a. $y = 2x$

    b. $y = 2x + 3$

    c. $y = 2x - 2$

38. Graph the lines on the same set of coordinate axes, and observe any similarities and differences.

    a. $x + y = 0$

    b. $x + y = 3$

    c. $x + y = -1$

Graphing Calculator Exercises

For Exercises 39–43, use a graphing calculator to graph the equations on the standard viewing window. Then graph the equation again on the suggested window.

39. $y = -3x - 2$

    a. Standard viewing window

    b. Window defined by: $-5 \le x \le 5$ and $-5 \le y \le 5$

40. $y = 0.33x - 0.25$

    a. Standard viewing window

    b. Window defined by: $-2 \le x \le 2$ and $-2 \le y \le 2$

41. $2x + y = 5$

    a. Standard viewing window

    b. Window defined by: $-1 \le x \le 8$ and $-4 \le y \le 8$

42. $3x - 2y = 32$

    a. Standard viewing window

    b. Window defined by: $-10 \le x \le 20$ and $-20 \le y \le 10$

43. $x + 2y = -50$

    a. Standard viewing window

    b. Window defined by $-60 \le x \le 20$ and $-30 \le y \le 20$

Concepts

1. Definition of x- and y-Intercepts
2. Determining x- and y-Intercepts from a Graph
3. Determining x- and y-Intercepts from an Equation
4. Applications of x- and y-Intercepts
5. Horizontal and Vertical Lines

section

3.3 x- and y-Intercepts, Horizontal and Vertical Lines

1. Definition of x- and y-Intercepts

For many applications of graphing it is advantageous to know the points where a graph intersects the x- or y-axis. These points are called the x- and y-intercepts.

In Figure 3-11, the x-intercept is at the point $(-3, 0)$ and the y-intercept is at $(0, 2)$. In Figure 3-12, there are three x-intercepts, one at $(-4, 0)$, one at $(1, 0)$, and one at $(5, 0)$. The y-intercept is at $(0, 20)$.

Notice that any point on the x-axis must have a y-coordinate of zero. Similarly, any point on the y-axis must have an x-coordinate of zero.

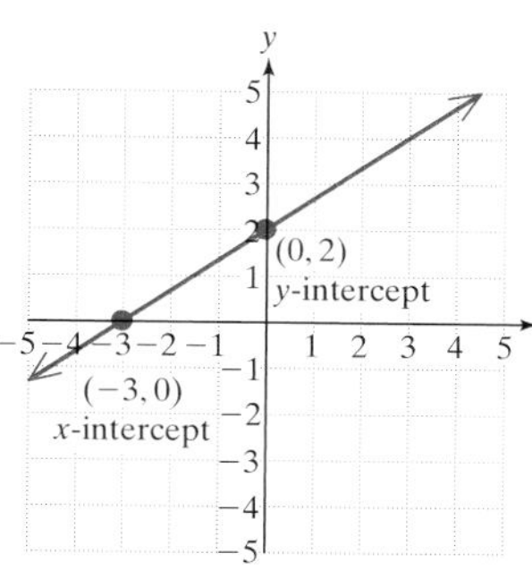

Figure 3-11

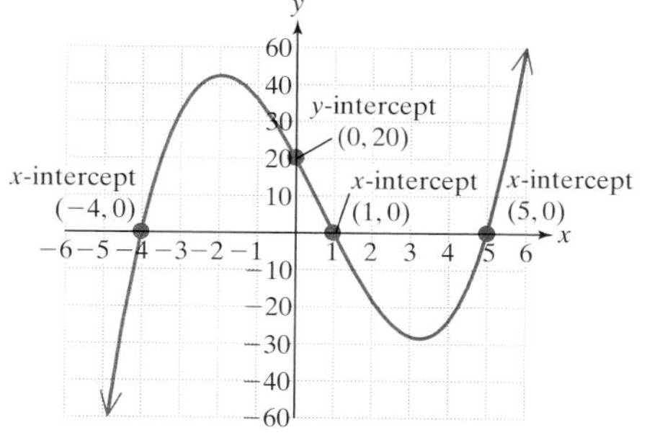

Figure 3-12

*Definition of *x*- and *y*-Intercepts

An ***x*-intercept** of an equation is a point $(a, 0)$ where the graph intersects the *x*-axis.

A ***y*-intercept** of an equation is a point $(0, b)$ where the graph intersects the *y*-axis.

*In some applications, an *x*-intercept is defined as the *x-coordinate* of a point of intersection that a graph makes with the *x*-axis. For example, if an *x*-intercept is at the point $(5, 0)$, it is sometimes stated simply as 5 (the *y*-coordinate is assumed to be zero). Similarly, a *y*-intercept is sometimes defined as the *y-coordinate* of a point of intersection that a graph makes with the *y*-axis. For example, if a *y*-intercept is at the point $(0, -3)$, it may be stated simply as -3 (the *x*-coordinate is assumed to be zero).

2. Determining *x*- and *y*-Intercepts from a Graph

example 1 **Determining the *x*- and *y*-Intercepts from a Graph**

Identify the *x*- and *y*-intercepts from the graphs.

a.

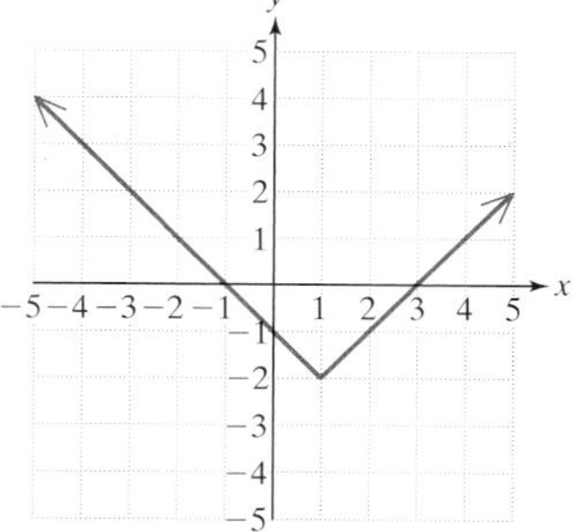

b.

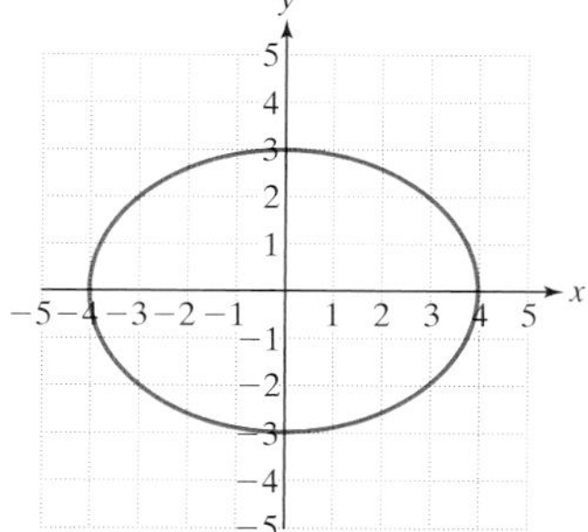

c.

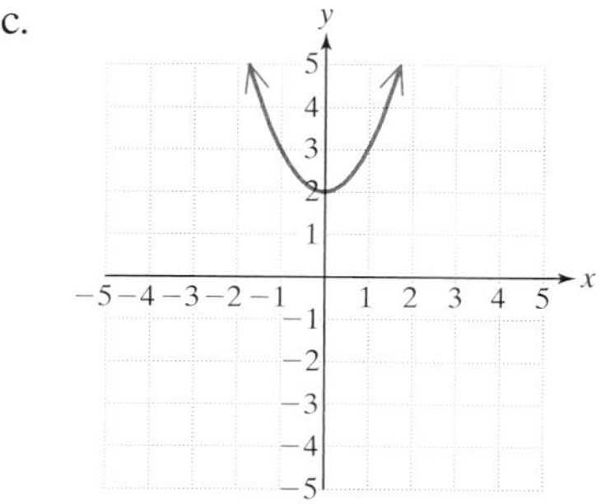

d.

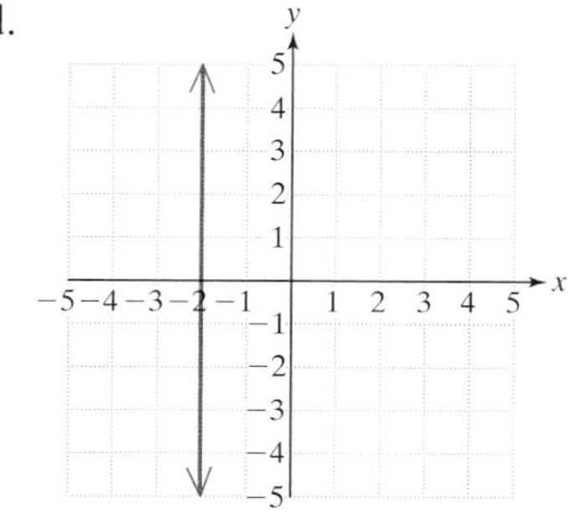

Solution:

a. x-intercepts: $(-1, 0)$ and $(3, 0)$
y-intercept: $(0, -1)$

b. x-intercepts: $(-4, 0)$ and $(4, 0)$
y-intercepts: $(0, -3)$ and $(0, 3)$

c. x-intercept: None
y-intercept: $(0, 2)$

d. x-intercept: $(-2, 0)$
y-intercept: None

3. Determining x- and y-Intercepts from an Equation

Although any two points may be used to graph a line, in some cases it is convenient to use the x- and y-intercepts of the line. To find the x- and y-intercepts of any two-variable equation in x and y, follow these steps:

Steps to Finding the x- and y-Intercepts from an Equation

Given an equation in x and y,

1. Find the x-intercept(s) by substituting $y = 0$ into the equation and solving for x.
2. Find the y-intercept(s) by substituting $x = 0$ into the equation and solving for y.

example 2

Finding the x- and y-Intercepts from an Equation

Given the equation $-3x + 2y = 8$,

a. Find the x-intercept.
b. Find the y-intercept.
c. Use the intercepts to graph the equation.

Tip: Finding the x- and y-intercepts is equivalent to completing the ordered pairs in the following table.

x	y
	0
0	

Solution:

a. To find the x-intercept, substitute $y = 0$.

$$-3x + 2y = 8$$
$$-3x + 2(0) = 8$$
$$-3x = 8$$
$$\frac{-3x}{-3} = \frac{8}{-3}$$
$$x = -\frac{8}{3}$$

The x-intercept is $(-\frac{8}{3}, 0)$.

b. To find the y-intercept, substitute $x = 0$.

$$-3x + 2y = 8$$
$$-3(0) + 2y = 8$$
$$2y = 8$$
$$y = 4$$

The y-intercept is $(0, 4)$.

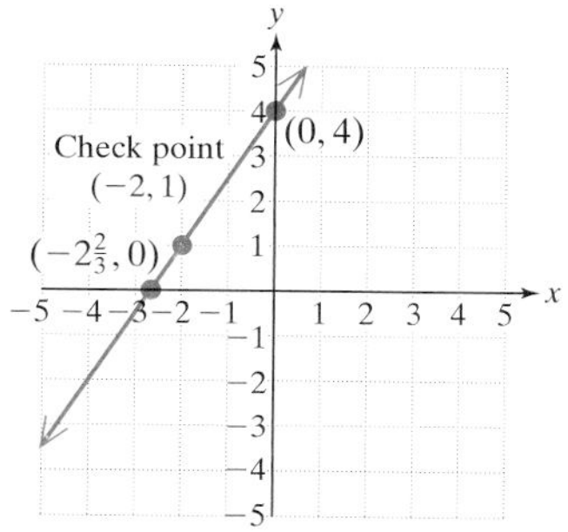

Figure 3-13

c. The line through the ordered pairs $(-\frac{8}{3}, 0)$ and $(0, 4)$ is shown in Figure 3-13. Note that the point $(-\frac{8}{3}, 0)$ can be written as $(-2\frac{2}{3}, 0)$.

The line represents the set of all solutions to the equation $-3x + 2y = 8$. Note: A third ordered pair can be found as a check point. If we choose an arbitrary value of x, such as $x = -2$, we have

$$-3x + 2y = 8$$
$$-3(-2) + 2y = 8$$
$$6 + 2y = 8$$
$$2y = 2$$
$$y = 1$$

Notice that the point $(-2, 1)$ "lines up" with the other two points. Therefore, we can be reasonably sure that we have graphed the line correctly.

example 3 Finding the x- and y-Intercepts from an Equation

Given the equation $y = -\frac{1}{2}x - 2$, find the x- and y-intercepts and graph the equation.

Solution:

To find the x-intercept, substitute $y = 0$.

$$y = -\frac{1}{2}x - 2$$
$$0 = -\frac{1}{2}x - 2$$
$$2 = -\frac{1}{2}x$$
$$-2(2) = -2\left(-\frac{1}{2}x\right) \quad \text{Clear fractions}$$
$$-4 = x$$

The x-intercept is $(-4, 0)$

To find the y-intercept, substitute $x = 0$.

$$y = -\frac{1}{2}x - 2$$
$$y = -\frac{1}{2}(0) - 2$$
$$y = -2$$

The y-intercept is $(0, -2)$

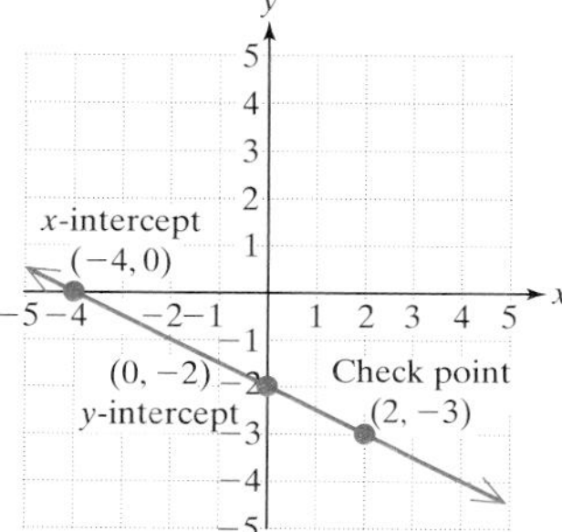

Figure 3-14

The graph of the line through $(-4, 0)$ and $(0, -2)$ is shown in Figure 3-14. We can verify the graph by finding a third ordered pair solution to $y = -\frac{1}{2}x - 2$. Letting x be an arbitary value such as $x = 2$, we have

$$y = -\frac{1}{2}(2) - 2$$
$$y = -1 - 2$$
$$y = -3$$

The third ordered pair, $(2, -3)$ "lines up" with the other two points.

example 4

Finding *x*- and *y*-Intercepts from an Equation

Given the equation $3x - 5y = 0$, find the x- and y-intercepts and graph the equation.

Solution:

To find the x-intercept, substitute $y = 0$.

$3x - 5(0) = 0$

$3x - 0 = 0$

$3x = 0$

$x = 0$

The x-intercept is $(0, 0)$

To find the y-intercept, substitute $x = 0$.

$3(0) - 5y = 0$

$0 - 5y = 0$

$-5y = 0$

$y = 0$

The y-intercept is $(0, 0)$

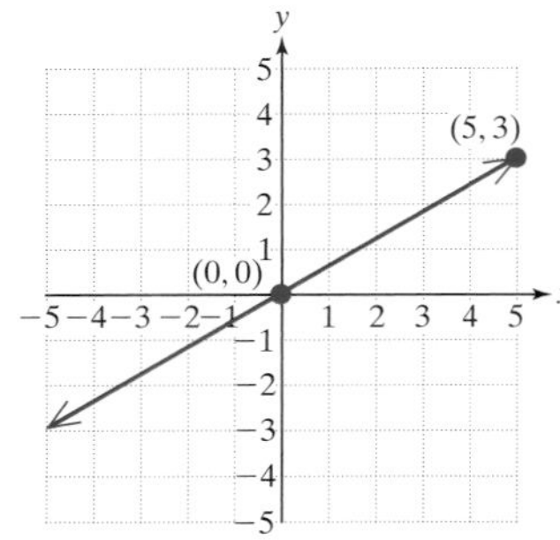

Figure 3-15

The equation $3x - 5y = 0$ has only one intercept $(0, 0)$ that is both the x-intercept and the y-intercept. In general, equations of the form $ax + by = 0$ will pass through the origin, yielding only one intercept. Because the x-intercept and the y-intercept are the same point, at least one more point is needed to graph the line. For example, if $x = 5$, then $y = 3$ and we have the point $(5, 3)$. See Figure 3-15.

4. Applications of *x*- and *y*-Intercepts

example 5

Interpreting the *x*- and *y*-Intercepts in an Application

The guarantee period for an \$80 truck battery is 4 years. If the battery does not last the entire 4 years, the manufacturer will pay the customer a refund for a portion of the loss. The amount the company will pay the customer, y, is given by the equation:

$y = 80 - 20x \quad 0 \le x \le 4$ where y is in dollars, and x is the time in years that the battery lasts.

a. If a battery lasts for only 1 year, how much will the company refund to the customer?
b. If the battery lasts for $3\frac{1}{2}$ years, how much will the company refund to the customer?
c. Find the y-intercept of this equation and interpret the meaning of the y-intercept in the context of this problem.
d. Find the x-intercept of this equation and interpret the meaning of the x-intercept in the context of this problem.

Solution:

a. $y = 80 - 20x$

$y = 80 - 20(1)$ Substitute $x = 1$.

$y = 60$

If a battery lasts only 1 year, the company will pay the customer \$60.

b. $y = 80 - 20x$

$y = 80 - 20(3.5)$ Substitute $x = 3.5$.

$y = 80 - 70$

$y = 10$

If a battery lasts only $3\frac{1}{2}$ years, the company will pay the customer \$10.

c. $y = 80 - 20x$

$y = 80 - 20(0)$ To find the y-intercept, substitute $x = 0$.

$y = 80$

The y-intercept is (0, 80). The y-intercept indicates that if a battery lasts 0 years, the company will give a full refund of \$80.

d. $y = 80 - 20x$

$0 = 80 - 20x$ To find the x-intercept, substitute $y = 0$.

$-80 = -20x$ Solve for x.

$4 = x$

The x-intercept is (4, 0). The x-intercept indicates that if a battery lasts 4 years, the company will pay the customer \$0. That is, if the battery lasts for the full warranty period, the company does not have to issue a refund.

The answers to parts (a)–(d) in Example 5 can be verified from the graph of the equation, $y = 80 - 20x$ (Figure 3-16). Notice that the longer a battery lasts, the less money the company must refund to the customer. After 4 years ($x = 4$), the company pays nothing ($y = 0$) to the customer. On the other hand, if a battery is dead at the time of purchase ($x = 0$), the company gives a full refund ($y = 80$).

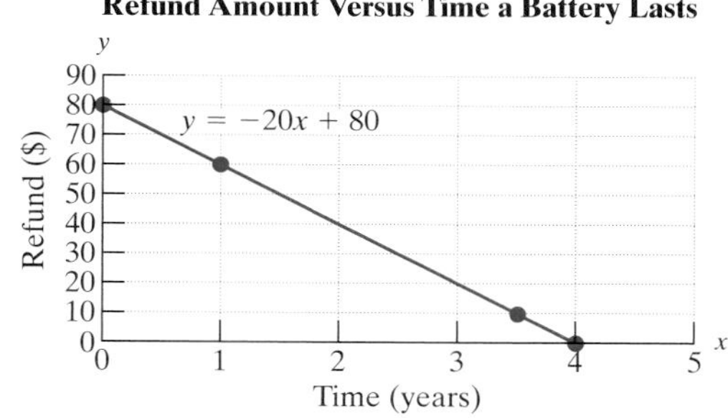

Figure 3-16

5. Horizontal and Vertical Lines

Recall that a linear equation can be written in the form of $ax + by = c$, where a and b are not both zero. However, if a or b is 0 then the line is either parallel to the x-axis (horizontal) or parallel to the y-axis (vertical).

Definitions of Vertical and Horizontal Lines

1. A **vertical line** is a line that can be written in the form, $x = k$, where k is a constant. (A vertical line is parallel to the y-axis.)
2. A **horizontal line** is a line that can be written in the form, $y = k$, where k is a constant. (A horizontal line is parallel to the x-axis.)

example 6 Graphing a Horizontal Line

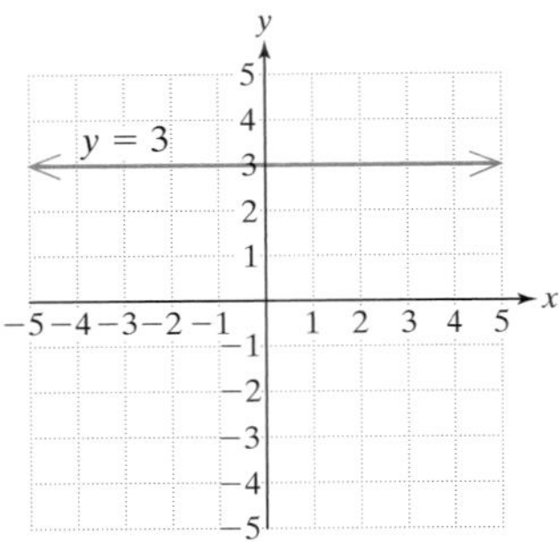

Figure 3-17

Graph the line $y = 3$.

Solution:

Because this equation is in the form $y = k$, the line is horizontal and must cross the y-axis at $y = 3$ (Figure 3-17).

Alternative Solution:

Create a table of values for the equation $y = 3$. The choice for the y-coordinate must be 3, but x can be any real number.

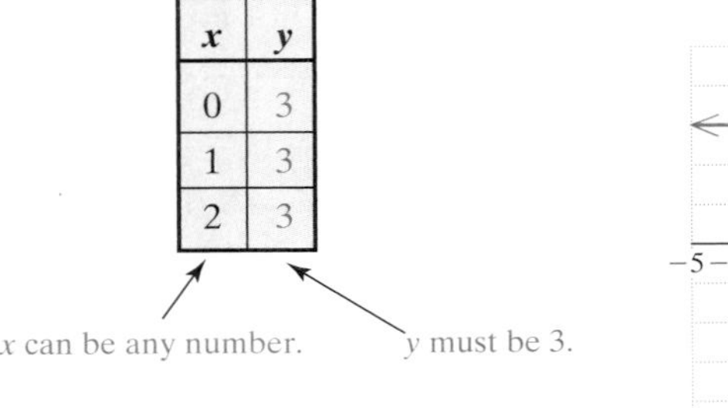

x	y
0	3
1	3
2	3

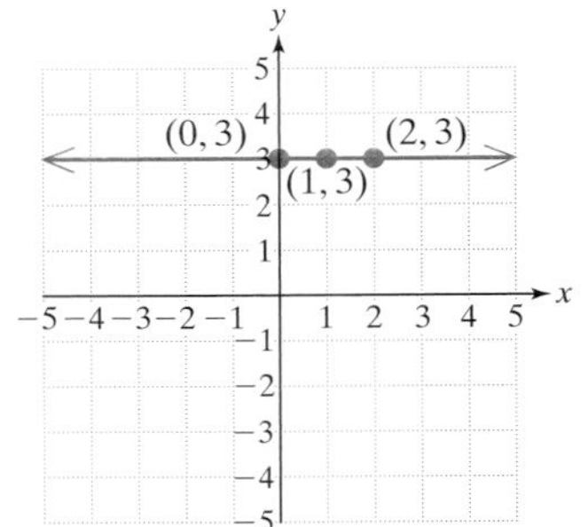

Tip: Notice that a horizontal line has a y-intercept, but does not have an x-intercept (unless the horizontal line is the x-axis itself).

example 7 Graphing a Vertical Line

Figure 3-18

Graph the line $4x = -8$.

Solution:

Because the equation does not have a y-variable, we can solve the equation for x.

$$4x = -8 \qquad \text{is equivalent to} \qquad x = -2$$

This equation is in the form $x = k$, indicating that the line is vertical and must cross the x-axis at $x = -2$ (Figure 3-18).

Alternative Solution:

Create a table of values for the equation $x = -2$. The choice for the x-coordinate must be -2 but y can be any real number.

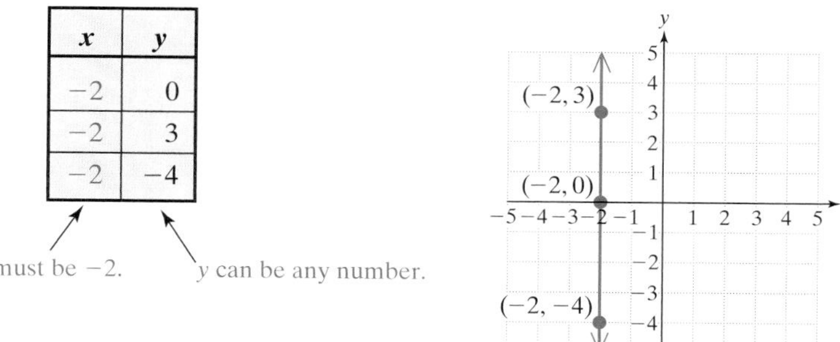

x	y
-2	0
-2	3
-2	-4

Tip: Notice that a vertical line has an x-intercept but does not have a y-intercept (unless the vertical line is the y-axis itself).

Calculator Connections

Algebraically, we can find that the x- and y-intercepts of the equation $y = \frac{2}{3}x - 2$ are the points (3, 0) and (0, −2), respectively. Students are urged to master the algebraic methods to study equations and their related graphs. However, a graphing calculator is an excellent verification tool to confirm the results.

Many calculators have a *Trace* feature that moves the cursor along the graph. The coordinates of the points along the graph are updated as the cursor moves.

To verify that the x-intercept is (3, 0), move the cursor to the point where the line crosses the x-axis. Depending on the viewing window, the coordinates will be *close* to $x = 3$ and $y = 0$. Similarly, to verify that the y-intercept is $(0, -2)$, move the cursor to the point where the line crosses the y-axis. Depending on the viewing window, the coordinates will be *close* to $x = 0$ and $y = -2$.

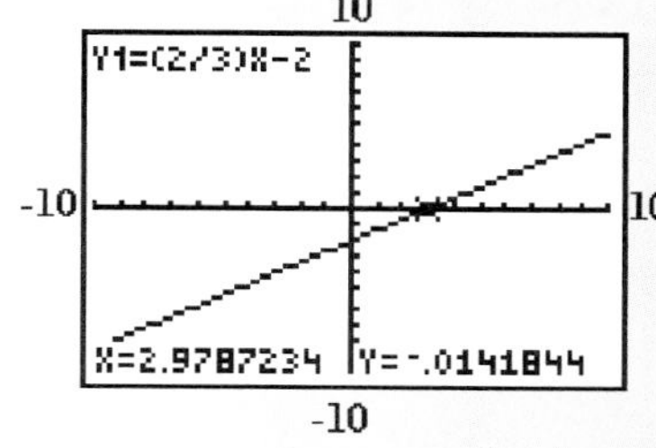

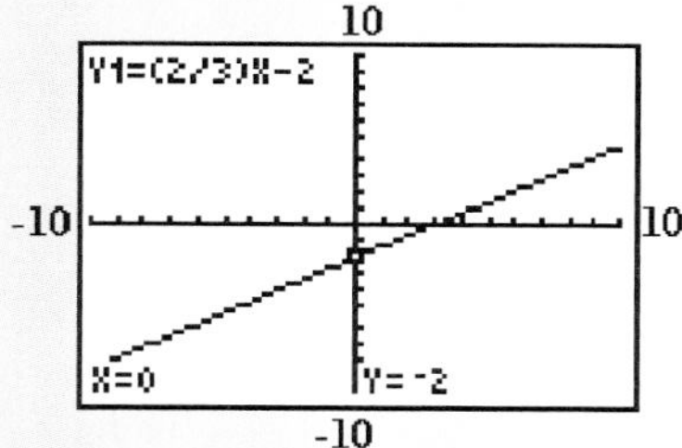

Tip: A "friendly" window on most calculators may give the exact coordinates of the x-intercept of a graph (Figure 3-19). A friendly window is set so that the coordinates of each pixel are terminating decimals. A friendly window can be defined by using a *ZDecimal* option or a comparable feature given in your user's manual.

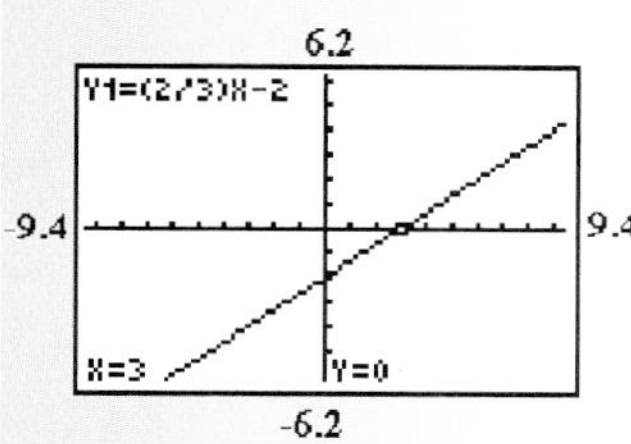

Figure 3-19

section 3.3 Practice Exercises

For Exercises 1–4, identify the quadrant in which the given point is found.

1. $(-2, 7)$
2. $(4, -15)$
3. $\left(-\frac{4}{3}, -\frac{1}{2}\right)$
4. $(3.7, 5.9)$

5. Determine whether the ordered pairs are solutions to the equation $x + 2y = 6$.

   a. $(0, 3)$ b. $(1, 2)$

   c. $(-4, 5)$ d. $(8, -1)$

6. Determine whether the ordered pairs are solutions to the equation $y = -\frac{3}{4}x - \frac{1}{4}$.

   a. $\left(0, -\frac{1}{4}\right)$ b. $(1, -1)$

   c. $(4, -3)$ d. $\left(-\frac{1}{3}, 0\right)$

7. State the definition of an x-intercept.

8. Given a two-variable equation in x and y, explain how to find an x-intercept.

9. Given a two-variable equation in x and y, explain how to find a y-intercept.

10. State the definition of a y-intercept.

For Exercises 11–16, identify the x- and y-intercepts from the graphs.

11.

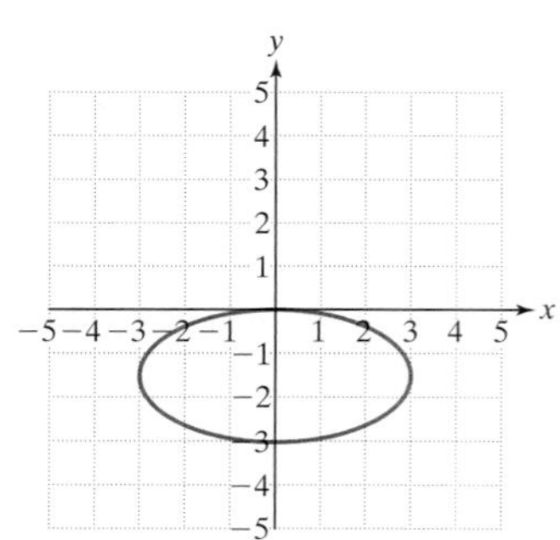

12.

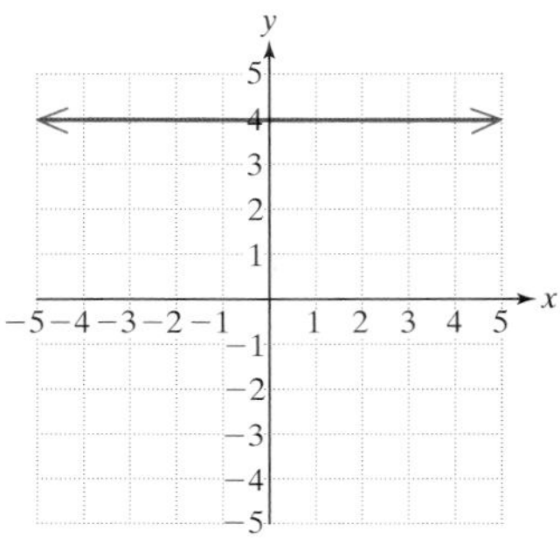

13.

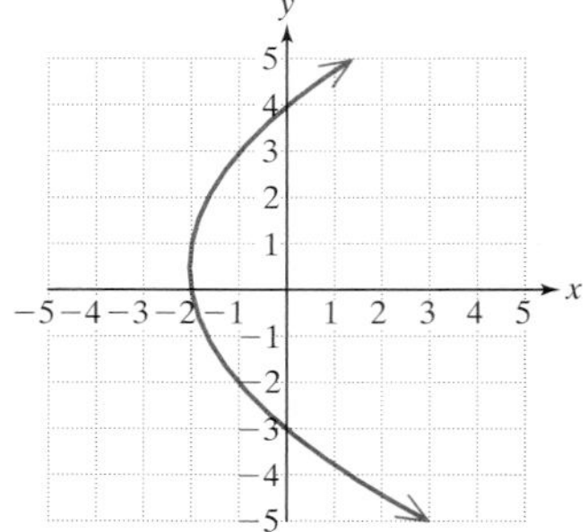

14.

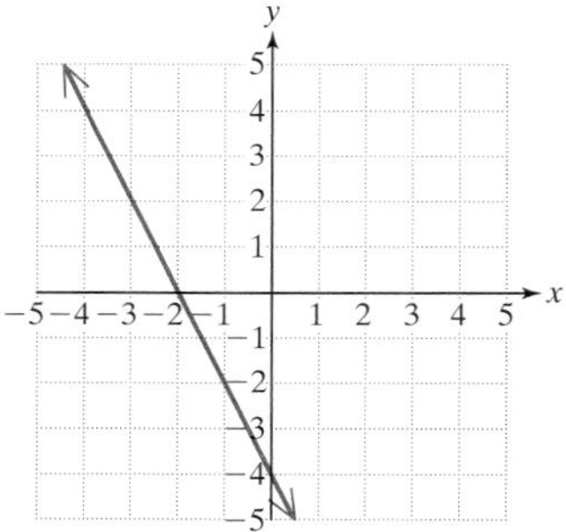

15.

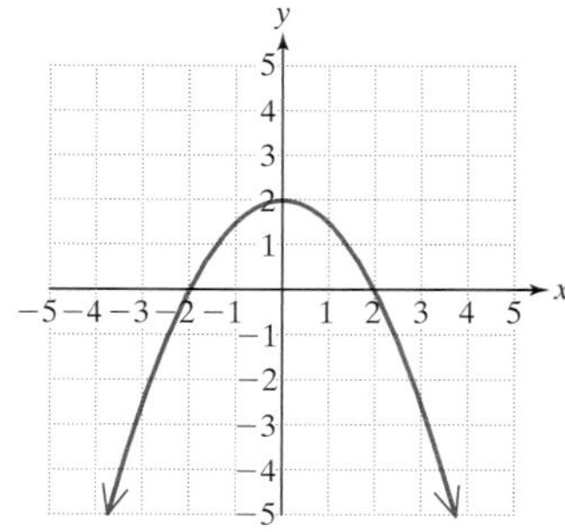

16.

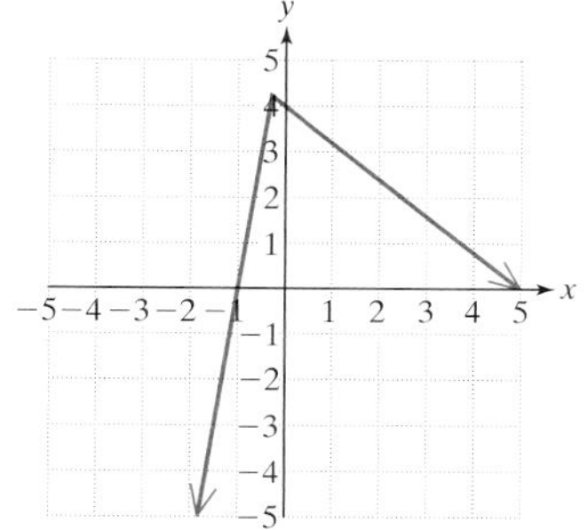

For Exercises 17–28, find the x- and y-intercepts.

17. $2x - 4y = 8$
18. $6x + 3y = 12$
19. $y = 5x + 10$
20. $y = 4x - 12$
21. $x = y + 2$
22. $x - y = 4$
23. $y = 4x$
24. $y = -2x$
25. $y = -\frac{1}{2}x + 3$
26. $y = \frac{1}{4}x - 2$
27. $4x - 7y = 9$
28. $2x + 8y = 7$
29. True or False. If the statement is false, rewrite it to be true.
    a. The line $x = 3$ is horizontal.
    b. The line $y = -4$ is horizontal.
30. True or False. If the statement is false, rewrite it to be true.
    a. A line parallel to the y-axis is vertical.
    b. A line perpendicular to the x-axis is vertical.

For Exercises 31–42, identify the equation as representing a vertical line or a horizontal line. Then graph the line.

31. $x = 3$
32. $y = -1$
33. $-2y = 8$
34. $5x = 20$
35. $x + 3 = 7$
36. $y - 8 = -13$
37. $3y = 0$
38. $5x = 0$
39. $2x + 7 = 10$
40. $-3y + 2 = 9$
41. $9 = 3 + 4y$
42. $7 = -2x - 5$
43. Explain why some lines may not have both an x- and a y-intercept.
44. The x-intercept is on which axis?
45. The y-intercept is on which axis?
46. Which of the lines will have only one intercept?
    a. $2x - 3y = 6$ b. $x = 5$
    c. $2y = 8$ d. $-x + y = 0$
47. Which of the lines will have only one intercept?
    a. $y = 2$ b. $x + y = 0$
    c. $2x - 10 = 2$ d. $x + 4y = 8$

For Exercises 48–69, find the x- and y-intercepts (if they exist), and graph the line.

48. $5x + y = 15$
49. $x - 3y = -9$
50. $y = \frac{2}{3}x - 1$
51. $y = -\frac{3}{4}x + 2$
52. $x - 3 = y$
53. $2x + 8 = y$
54. $-3x + y = 0$
55. $2x - 2 = 0$
56. $5y = 8$
57. $y + 9 = 7$
58. $x - 3 = -2$
59. $4x = 5$
60. $25y = 10x + 100$
61. $20x = -40y + 200$
62. $1.2x - 2.4y = 3.6$
63. $-8.1x - 10.8y = 16.2$
64. $x = 2y$
65. $x = -5y$
66. $2 = 4 + x$
67. $3 = 7 + y$
68. $10 = 5y$
69. $-6 = -6x$
70. The guarantee period for a car air-conditioning unit is 3 years. If the air-conditioner does not last the entire 3 years, the dealer will pay the customer a refund for a portion of the loss. The amount, y, that the dealer will refund to the customer is given by the equation

    $y = -500x + 1500 \quad 0 \le x \le 3$ where y is in dollars, and x is the time in years that the air-conditioner lasts.

    The graph of the equation is shown on page 212.

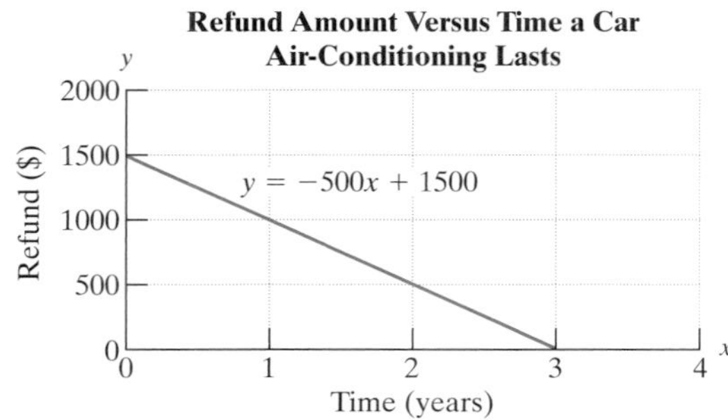

Figure for Exercise 70

a. If an air-conditioner lasts for only 1 year, how much will the dealer refund to the customer? Verify your answer from the graph.

b. If the air-conditioner lasts for 2 years, how much will the dealer refund to the customer?

c. Find the y-intercept of this equation and interpret the meaning of the y-intercept in the context of this problem.

d. Find the x-intercept of this equation and interpret the meaning of the x-intercept in the context of this problem.

71. The guarantee period for a stereo system is 2 years. If the stereo does not last the entire 2 years, the store will pay the customer a refund for a portion of the loss. The amount, y, that the store will refund to the customer is given by the equation

$y = -400x + 800 \quad 0 \le x \le 2$ where y is in dollars, and x is the time in years that the stereo lasts.

The graph of the equation is shown.

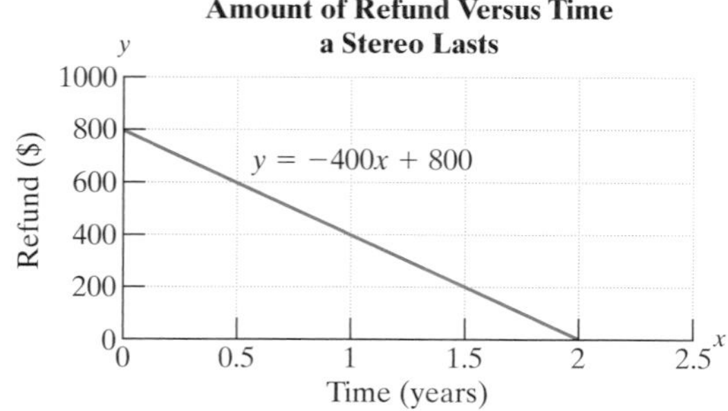

Figure for Exercise 71

a. If a stereo lasts for only 1 year, how much will the store refund to the customer? Verify your answer from the graph.

b. If the stereo lasts for $1\frac{1}{2}$ years, how much will the store refund to the customer?

c. Find the y-intercept of this equation and interpret the meaning of the y-intercept in the context of this problem.

d. Find the x-intercept of this equation and interpret the meaning of the x-intercept in the context of this problem.

72. A stadium holds 75,000 seats. If a concert promoter charges x dollars per ticket for a concert, then the total number of tickets sold, N, can be expressed as

$N = 75{,}000 - 500x \quad (x \ge 0)$ where x is the price per ticket.

The graph of the equation is shown. The variable N is used in place of y.

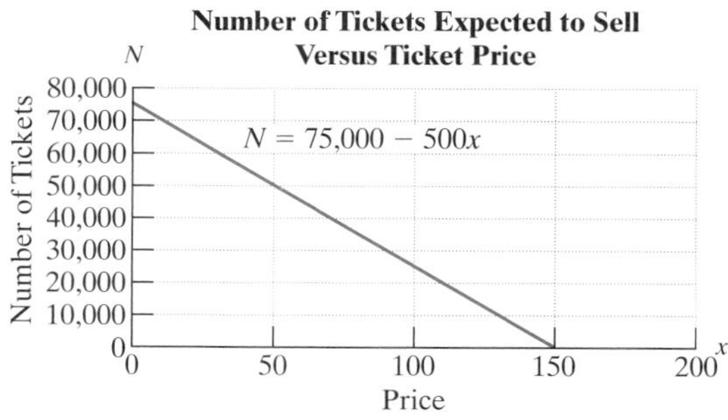

Figure for Exercise 72

a. If the ticket price is \$20, how many tickets can the promoter expect to sell? Verify your answer from the graph.

b. If the ticket price is \$50, how many tickets can the promoter expect to sell?

c. Find the N-intercept of this equation and interpret the meaning of the N-intercept in the context of this problem.

d. Find the x-intercept of this equation and interpret the meaning of the x-intercept in the context of this problem.

73. A certain sports arena holds 20,000 seats. If a sports franchise charges x dollars per ticket for a game, then the total number of tickets sold, N, can be approximated by

$N = 20{,}000 - 400x \quad (x \geq 0)$ where x is the price per ticket.

The graph of the equation is shown. The variable N is used in place of y.

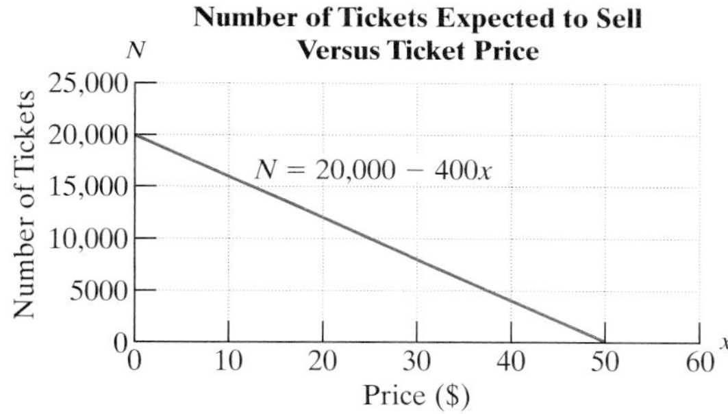

Figure for Exercise 73

a. If the ticket price is \$10, how many tickets can the promoter expect to sell? Verify your answer from the graph.

b. If the ticket price is \$40, how many tickets can the promoter expect to sell?

c. Find the N-intercept of this equation and interpret the meaning of the N-intercept in the context of this problem.

d. Find the x-intercept of this equation and interpret the meaning of the x-intercept in the context of this problem.

Expanding Your Skills

74. Write an equation representing the x-axis.

75. Write an equation representing the y-axis.

76. Write an equation of the line parallel to the x-axis and passing through the point $(0, 2)$.

77. Write an equation of the line parallel to the y-axis and passing through the point $(4, 0)$. (*Hint:* Sketch the line first.)

78. Write an equation of the line perpendicular to the x-axis and passing through the point $(3, -2)$.

79. Write an equation of the line perpendicular to the y-axis and passing through the point $(-1, 5)$.

80. Under what conditions will the x- and y-intercepts of the equation $ax + by = c$ be at the origin?

Graphing Calculator Exercises

For Exercises 81–85, use algebraic methods to find the x- and y-intercepts. Then use a graphing calculator to verify your results.

81. $y = 2x - 4$

82. $y = -3x + 6$

83. $3x + 4y = 6$

84. $-2x + 5y = -4$

85. $y = x - 15$

Concepts

1. Introduction to Slope
2. Slope Formula
3. Positive, Negative, Zero, and Undefined Slopes
4. Parallel and Perpendicular Lines
5. Applications of Slope

section

3.4 Slope of a Line

1. Introduction to Slope

The x- and y-intercepts represent the points where a line crosses the x- and y-axes. Another important feature of a line is its slope. Geometrically, the slope of a line measures the "steepness" of the line. For example, two ski runs are depicted by the lines in Figure 3-20.

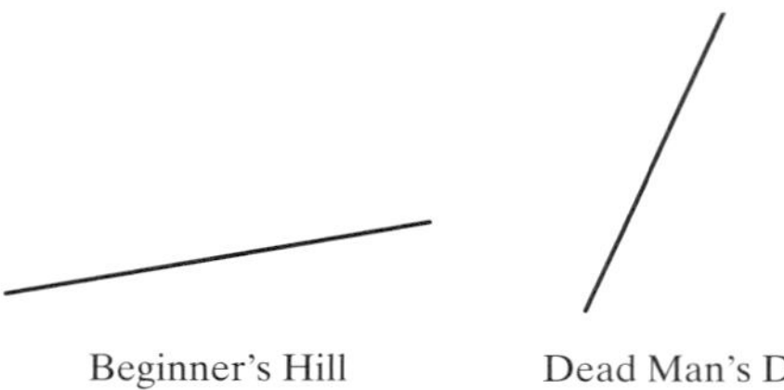

Figure 3-20

By visual inspection, Dead Man's Drop is "steeper" than Beginner's Hill. To measure the slope of a line quantitatively, consider two points on the line. The slope of the line is the ratio of the vertical change (change in y) between the two points and the horizontal change (change in x). As a memory device, we might think of the slope of a line as "rise over run." See Figure 3-21.

$$\text{Slope} = \frac{\text{change in } y}{\text{change in } x} = \frac{\text{rise}}{\text{run}}$$

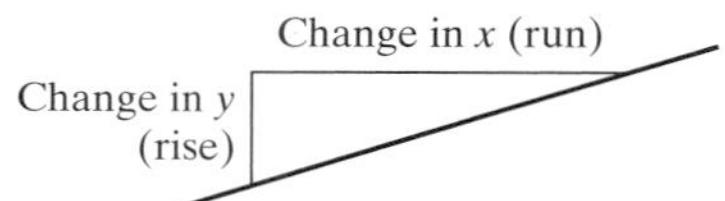

Figure 3-21

To move from point A to point B on Beginner's Hill, rise 2 ft and move to the right 6 ft (Figure 3-22).

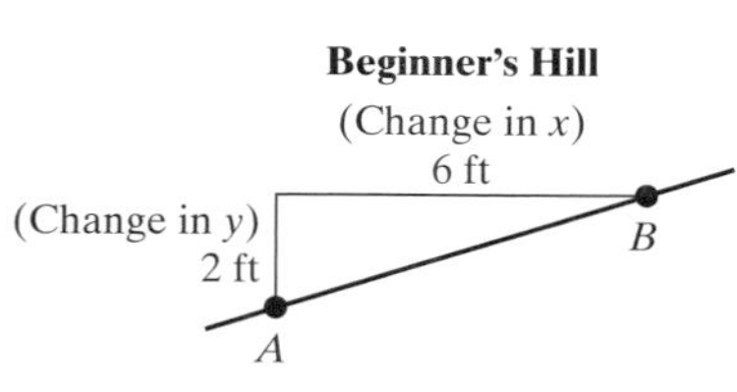

Figure 3-22

$$\text{Slope} = \frac{\text{change in } y}{\text{change in } x} = \frac{2 \text{ ft}}{6 \text{ ft}} = \frac{1}{3}$$

To move from point A to point B on Dead Man's Drop, rise 12 ft and move to the right 6 ft (Figure 3-23).

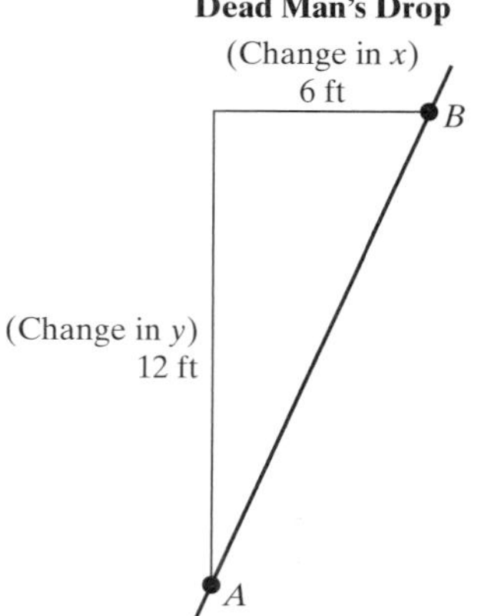

Figure 3-23

$$\text{Slope} = \frac{\text{change in } y}{\text{change in } x} = \frac{12 \text{ ft}}{6 \text{ ft}} = \frac{2}{1} = 2$$

The slope of Dead Man's Drop is greater than the slope of Beginner's Hill, confirming the observation that Dead Man's Drop is steeper. On Dead Man's Drop, there is a 12-ft change in elevation for every 6 ft of horizontal distance (a 2:1 ratio). On Beginner's Hill there is only a 2-ft change in elevation for every 6 ft of horizontal distance (a 1:3 ratio).

example 1 **Finding Slope in an Application**

Find the slope of the ramp up the stairs.

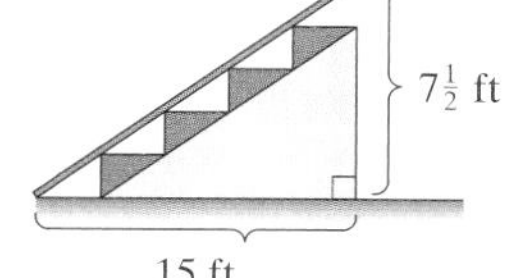

Solution:

$$\text{Slope} = \frac{\text{change in } y}{\text{change in } x} = \frac{7\frac{1}{2}\text{ ft}}{15\text{ ft}}$$

$$\frac{7\frac{1}{2}}{15} = \frac{\frac{15}{2}}{\frac{15}{1}}$$ Write the mixed number as an improper fraction.

$$\frac{15}{2} \cdot \frac{1}{15} = \frac{1}{2}$$ Multiply by the reciprocal and simplify.

The slope is $\frac{1}{2}$.

2. Slope Formula

The slope of a line may be found using any two points on the line—call these points (x_1, y_1) and (x_2, y_2). The change in y between the points can be found by taking the difference of the y-values: $y_2 - y_1$. The change in x can be found by taking the difference of the x-values in the same order: $x_2 - x_1$ (Figure 3-24).

The slope of a line is often symbolized by the letter m and is given by the following formula.

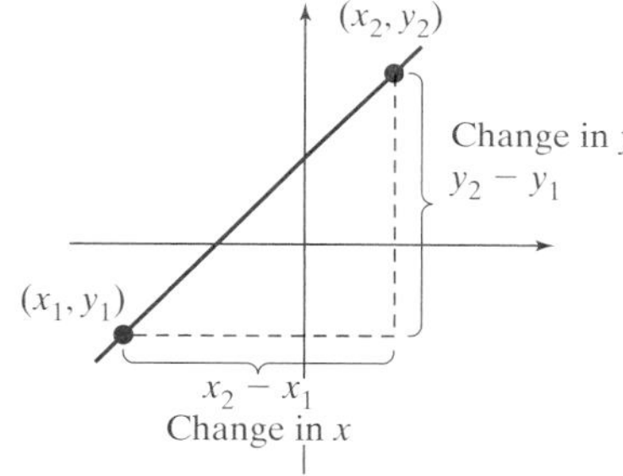

Figure 3-24

Definition of the Slope of a Line

The **slope** of a line passing through the distinct points (x_1, y_1) and (x_2, y_2) is

$$m = \frac{y_2 - y_1}{x_2 - x_1} \quad \text{provided } x_2 - x_1 \neq 0$$

example 2 **Finding the Slope of a Line Given Two Points**

Find the slope of the line through the points $(-1, 3)$ and $(-4, -2)$.

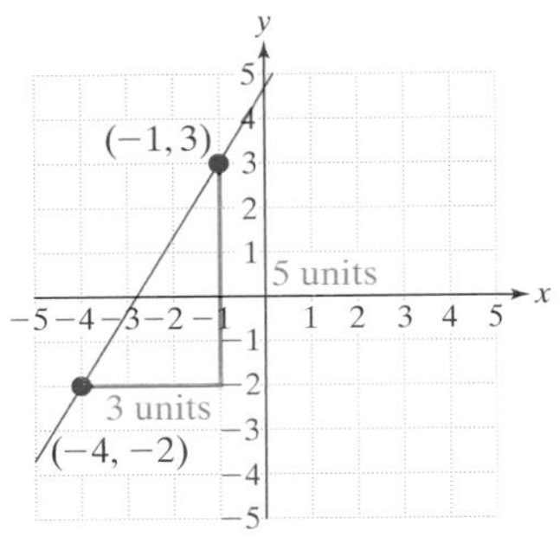

Figure 3-25

Solution:

To use the slope formula, first label the coordinates of each point and then substitute the coordinates into the slope formula.

$$\begin{matrix} (-1, 3) & \text{and} & (-4, -2) \\ (x_1, y_1) & & (x_2, y_2) \end{matrix}$$ Label the points.

$$m = \frac{y_2 - y_1}{x_2 - x_1} = \frac{(-2) - (3)}{(-4) - (-1)}$$ Apply the slope formula.

$$= \frac{-5}{-3}; \quad \text{hence, } m = \frac{5}{3}$$ Reduce to lowest terms.

The slope of the line can be verified from the graph (Figure 3-25).

Tip: The slope formula is not dependent on which point is labeled (x_1, y_1) and which point is labeled (x_2, y_2). In Example 2, reversing the order in which the points are labeled results in the same slope.

$$\begin{matrix} (-1, 3) & \text{and} & (-4, -2) \\ (x_2, y_2) & & (x_1, y_1) \end{matrix}$$ Label the points.

then

$$m = \frac{(3) - (-2)}{(-1) - (-4)} = \frac{5}{3}$$ Apply the slope formula.

To help you apply the slope formula correctly, keep this note in mind. The first terms in both the numerator and the denominator must be from the same ordered pair. For example:

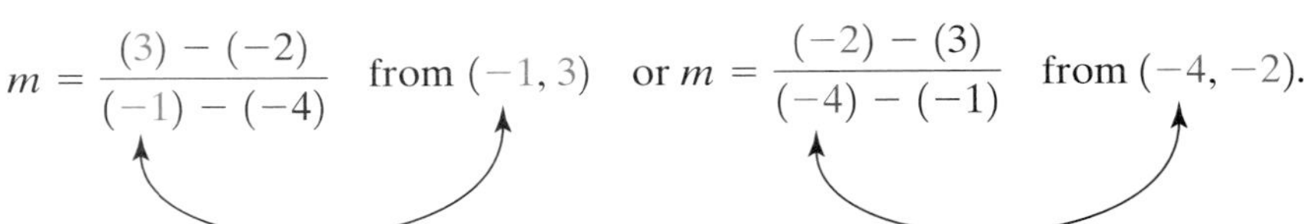

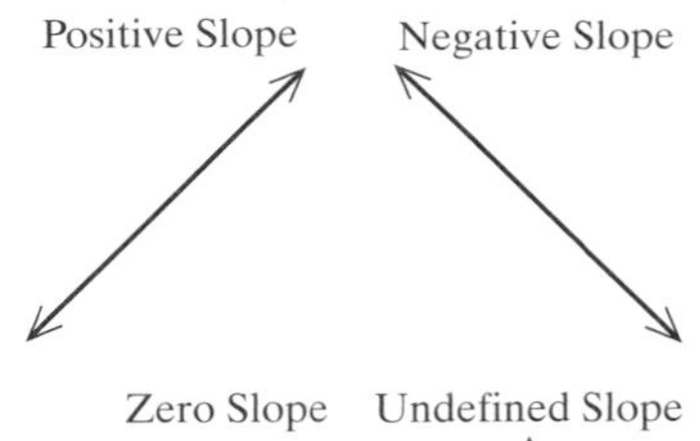

3. Positive, Negative, Zero, and Undefined Slopes

The value of the slope of a line may be positive, negative, zero, or undefined.

Lines that increase, or rise, from left to right have a positive slope.
Lines that decrease, or fall, from left to right have a negative slope.
Horizontal lines have a slope of zero.
Vertical lines have an undefined slope.

example 3 **Finding the Slope of a Line Given Two Points**

Find the slope of the line passing through the points $(-5, 0)$ and $(2, -3)$.

Solution:

$(-5, 0)$ and $(2, -3)$ Label the points.
(x_1, y_1) (x_2, y_2)

$$m = \frac{y_2 - y_1}{x_2 - x_1} = \frac{(-3) - (0)}{(2) - (-5)} \quad \text{Apply the slope formula.}$$

$$= \frac{-3}{7} \quad \text{or} \quad -\frac{3}{7} \quad \text{Simplify.}$$

By graphing the points $(-5, 0)$ and $(2, -3)$, we can verify that the slope is $-\frac{3}{7}$ (Figure 3-26). Notice that the line slopes downward from left to right.

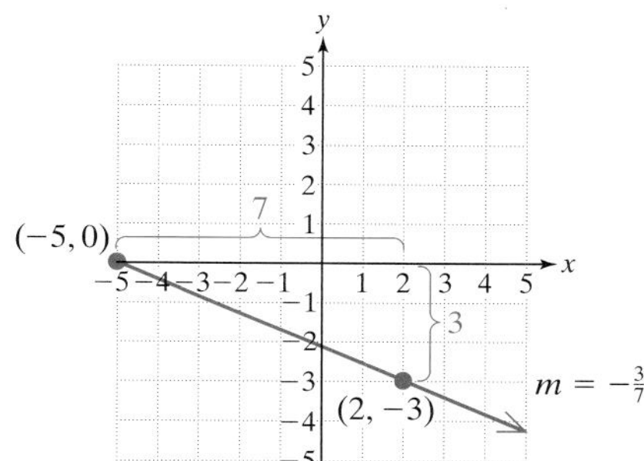

Figure 3-26

example 4

Determining the Slope of a Horizontal and Vertical Line

a. Find the slope of the line passing through the points $(2, -1)$ and $(2, 4)$.
b. Find the slope of the line passing through the points $(3, -2)$ and $(-4, -2)$.

Solution:

a. $(2, -1)$ and $(2, 4)$ Label the points.
(x_1, y_1) (x_2, y_2)

$$m = \frac{y_2 - y_1}{x_2 - x_1} = \frac{(4) - (-1)}{(2) - (2)} \quad \text{Apply the slope formula.}$$

$$m = \frac{5}{0} \quad \text{Undefined}$$

Because the slope, m, is undefined, we expect the points to form a vertical line as shown in Figure 3-27.

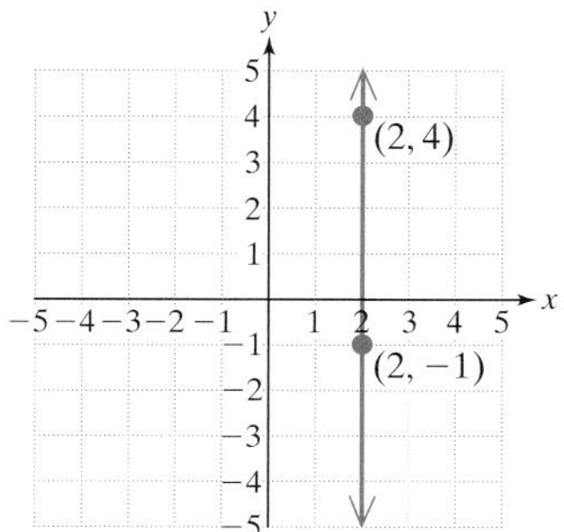

Figure 3-27

b. $(3, -2)$ and $(-4, -2)$ Label the points.
(x_1, y_1) (x_2, y_2)

$$m = \frac{y_2 - y_1}{x_2 - x_1} = \frac{(-2) - (-2)}{(-4) - (3)} \quad \text{Apply the slope formula.}$$

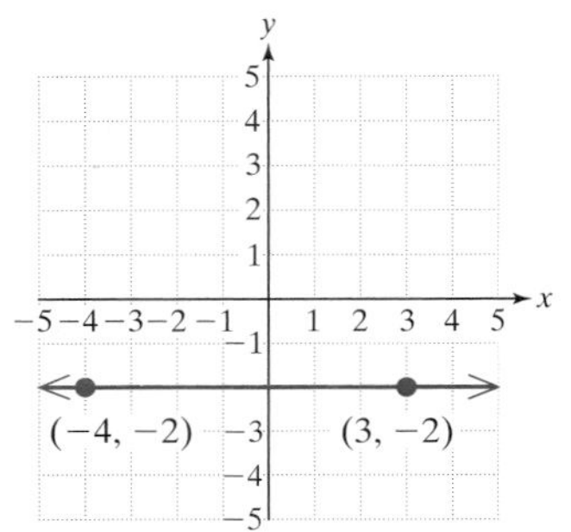

Figure 3-28

$$m = \frac{-2 + 2}{-4 - 3} = \frac{0}{-7} = 0$$

Because the slope is 0, we expect the points to form a horizontal line, as shown in Figure 3-28.

4. Parallel and Perpendicular Lines

Lines that do not intersect are called *parallel lines*. Parallel lines have the same slope and different *y*-intercepts (Figure 3-29).

Lines that intersect at a right angle are *perpendicular lines*. If two lines are perpendicular then the slope of one line is the opposite of the reciprocal of the slope of the other line (provided neither line is vertical) (Figure 3-30).

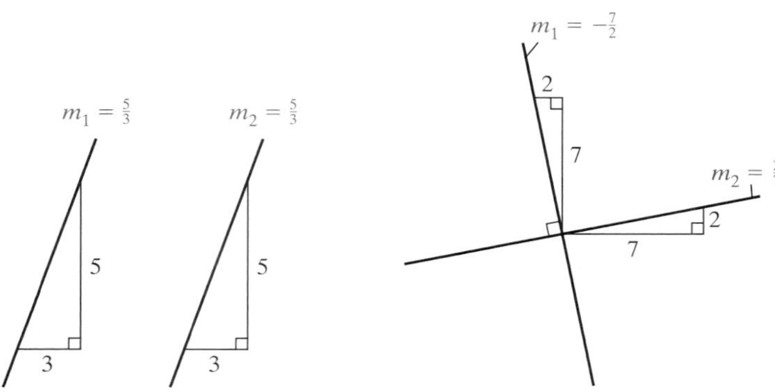

Figure 3-29

Figure 3-30

Slopes of Parallel Lines

If m_1 and m_2 represent the slopes of two parallel (nonvertical) lines, then

$$m_1 = m_2.$$

See Figure 3-29.

Slopes of Perpendicular Lines

If $m_1 \neq 0$ and $m_2 \neq 0$ represent the slopes of two perpendicular lines, then

$$m_1 = -\frac{1}{m_2} \quad \text{or} \quad m_2 = -\frac{1}{m_1}$$

or equivalently, $m_1 m_2 = -1$. See Figure 3-30.

example 5 **Determining the Slope of Parallel and Perpendicular Lines**

Suppose a given line has a slope of $-\frac{1}{4}$.

a. Find the slope of a line parallel to the given line.
b. Find the slope of a line perpendicular to the given line.

Solution:

a. Parallel lines must have the same slope. The slope of a line parallel to the given line is: $m = -\frac{1}{4}$.

b. Perpendicular lines must have opposite and reciprocal slopes. The slope of a line perpendicular to the given line is: $m = +\frac{4}{1}$ or simply, $m = 4$.

5. Applications of Slope

In many applications, the interpretation of slope refers to the *rate of change* of the y-variable to the x-variable.

example 6 **Interpreting Slope in an Application**

Mario earns \$10.00/hr working for a landscaping company. Shannelle earns \$15.00/hr working for an in-home nursing agency. Figure 3-31 shows their total earnings versus the number of hours they work.

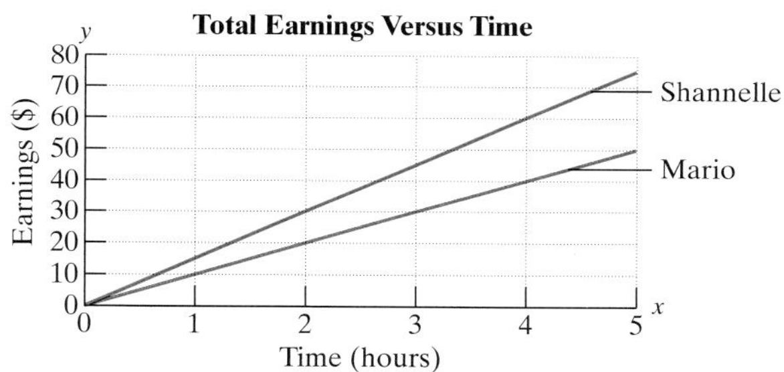

Figure 3-31

a. Find the slope of the line representing Mario's earnings.
b. Find the slope of the line representing Shannelle's earnings.

Tip: To find the slope for Example 6(a) we can pick points from the graph, such as (1, 10) and (2, 20). Then use the slope formula to calculate the slope.

$m = \frac{y_2 - y_1}{x_2 - x_1}$ becomes

$m = \frac{20 - 10}{2 - 1} = \frac{10}{1} = 10.$

Solution:

a. After 1 hr, Mario earns \$10. After 2 hr, he earns \$20, and so on. For each 1-hr change in time, there is a \$10 increase in wages. The slope of the line representing Mario's earnings is \$10/hr.
b. After 1 hr, Shannelle earns \$15. After 2 hr, she earns \$30, and so on. For each 1-hr change in time, there is a \$15 increase in wages. The slope of the line representing Shannelle's earnings is \$15/hr.

example 7 **Interpreting Slope in an Application**

Figure 3-32 depicts the annual median income for males in the United States between 1996 and 2001. The trend is approximately linear. Find the slope of the line and interpret the meaning of the slope in the context of this problem.

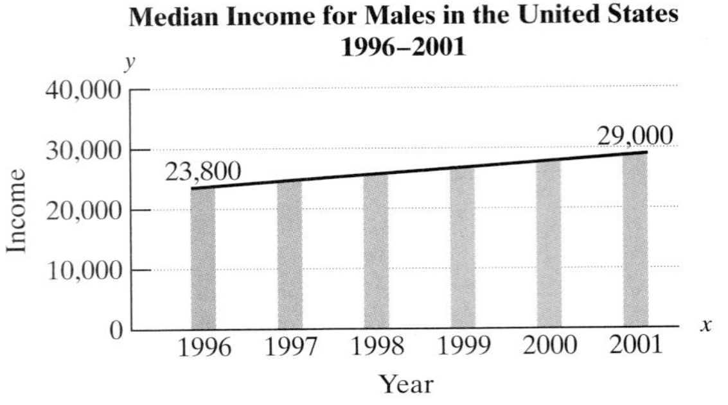

Figure 3-32

Source: U.S. Department of the Census.

Solution:

To determine the slope we need to know two points on the line. From the graph, the median income for males in 1996 was $23,800. This corresponds to the ordered pair (1996, 23,800). In 2001, the median income was $29,000. This corresponds to the ordered pair (2001, 29,000).

$$\underset{(x_1, y_1)}{(1996, 23{,}800)} \quad \text{and} \quad \underset{(x_2, y_2)}{(2001, 29{,}000)}$$ Label the points.

$$m = \frac{y_2 - y_1}{x_2 - x_1} = \frac{(29{,}000) - (23{,}800)}{(2001) - (1996)}$$ Apply the slope formula.

$$m = \frac{5200}{5} = 1040$$ Simplify.

The slope indicates that the median income for males in the United States increased at a rate of approximately $1040 per year between 1996 and 2001.

section 3.4 Practice Exercises

For Exercises 1–8, find the x- and y-intercepts (if they exist). Then graph the lines.

1. $x - 3y = 6$
2. $y = -3$
3. $x - 5 = 2$
4. $y = \frac{2}{3}x$
5. $2y - 3 = 0$
6. $4x + y = 8$
7. $2x = 4y$
8. $-1 = 2x + 3$

For Exercises 9–12, fill in the blank with the appropriate term: zero, negative, positive, or undefined.

9. The slope of a line parallel to the y-axis is ______.
10. The slope of a horizontal line is ______.
11. The slope of a line that rises from left to right is ______.
12. The slope of a line that falls from left to right is ______.

For Exercises 13–20, label the lines as having a positive, negative, zero, or undefined slope.

13.

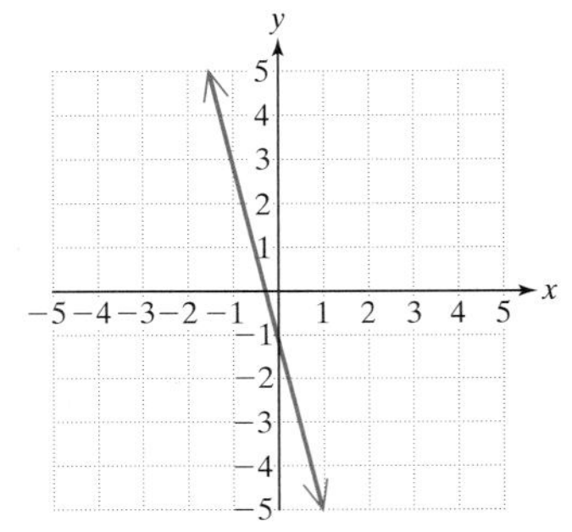

14.

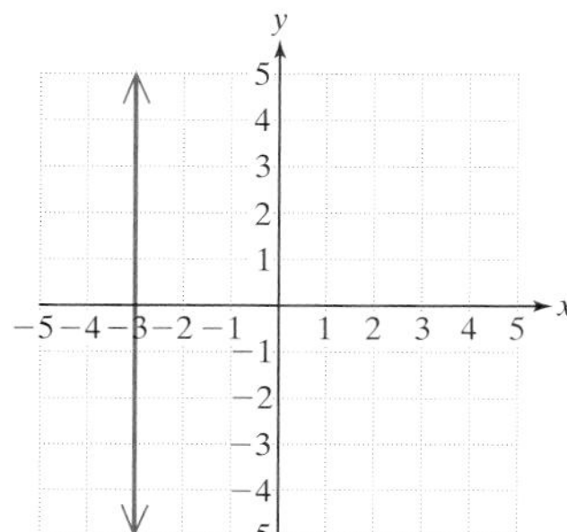

15.

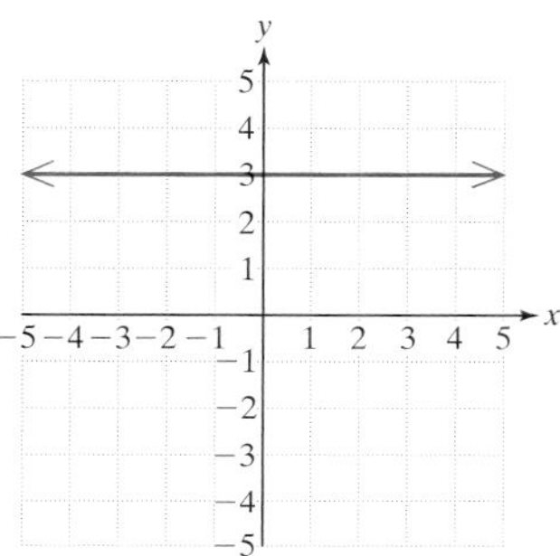

16.

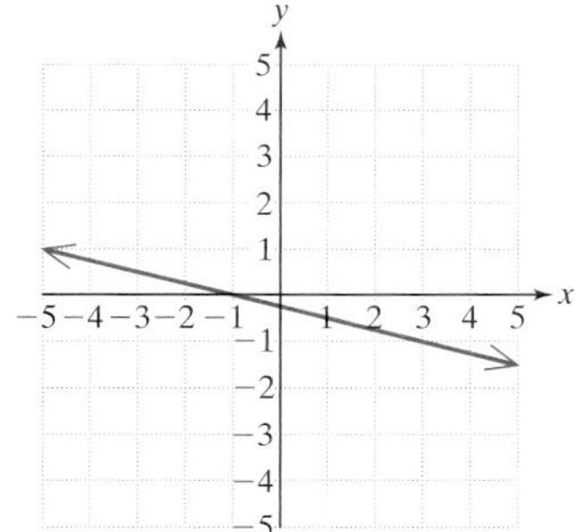

17.

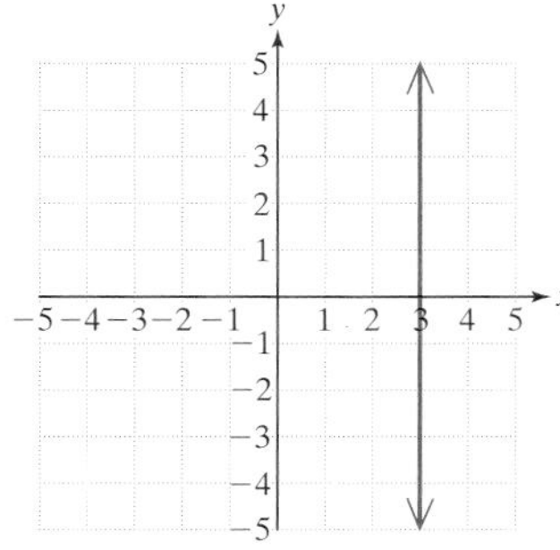

18.

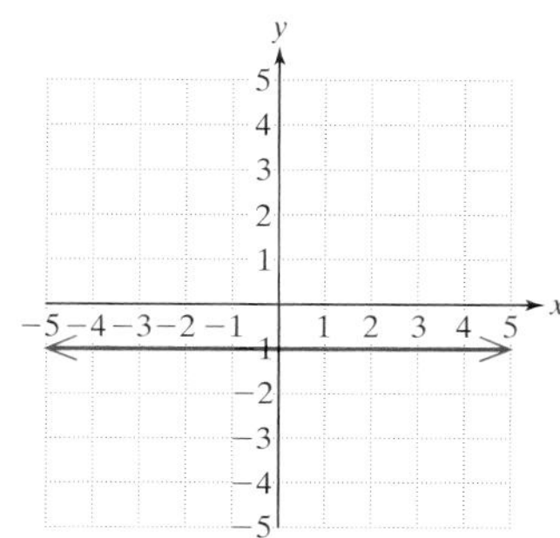

19.

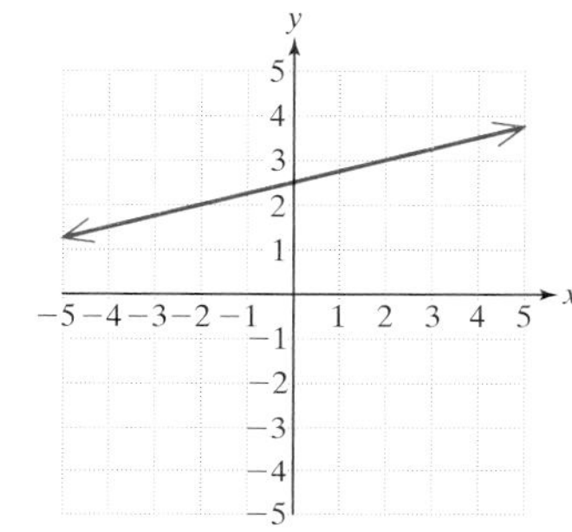

20.

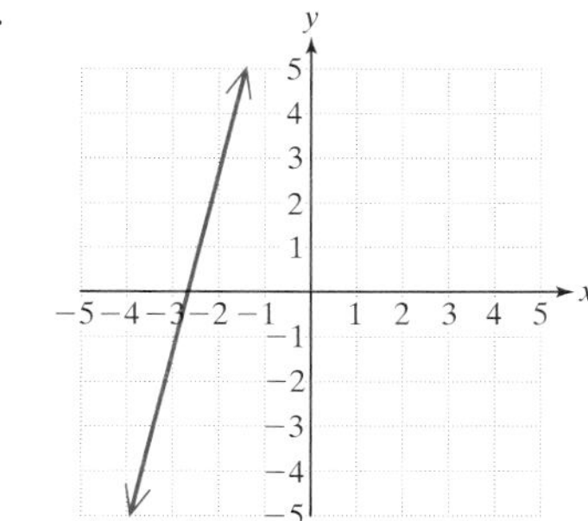

For Exercises 21–38, find the slope of the line that passes through the two points.

21. (2, 4) and (−1, −1)

22. (0, 4) and (3, 0)

23. (−2, 3) and (−1, 0)

24. (−3, −4) and (1, −5)

25. (5, 3) and (−2, 3)

26. (0, −1) and (−4, −1)

27. (2, −7) and (2, 5)

28. (−4, 3) and (−4, −4)

29. $\left(\frac{1}{2}, \frac{3}{5}\right)$ and $\left(\frac{1}{4}, -\frac{4}{5}\right)$

30. $\left(-\frac{2}{7}, \frac{1}{3}\right)$ and $\left(\frac{8}{7}, -\frac{5}{6}\right)$

31. $(3\frac{3}{4}, -1\frac{1}{4})$ and $(-5, 6\frac{1}{2})$

32. $(-6\frac{7}{8}, 5\frac{2}{5})$ and $(-10, 4\frac{1}{10})$

33. (6.8, −3.4) and (−3.2, 1.1)

34. (−3.15, 8.25) and (6.85, −4.25)

35. (−5.50, 1.75) and (−1.50, −4.80)

36. (11.2, 8.4) and (−3.8, −11.6)

37. (1994, 35,000) and (2000, 24,000)

38. (1988, 4.65) and (1998, 9.25)

39. Point *A* is located 3 units up and 4 units to the right of point *B*. What is the slope of the line passing through the points?

40. Point *A* is located 2 units up and 5 units to the left of point *B*. What is the slope of the line passing through the points?

41. Point *A* is located 3 units up and 3 units to the left of point *B*. What is the slope of the line passing through the points?

42. Point *A* is located 2 units down and 2 units to the left of point *B*. What is the slope of the line passing through the points?

43. Point *A* is located 5 units to the right of point *B*. What is the slope of the line passing through the points?

44. Point *A* is located 3 units down from point *B*. What is the slope of the line passing through the points?

45. In 1980, there were 304 thousand male inmates in federal and state prisons. By 1996, the number increased to 1069 thousand.

Let *x* represent the year and let *y* represent the number of male inmates in federal and state prisons (in thousands).

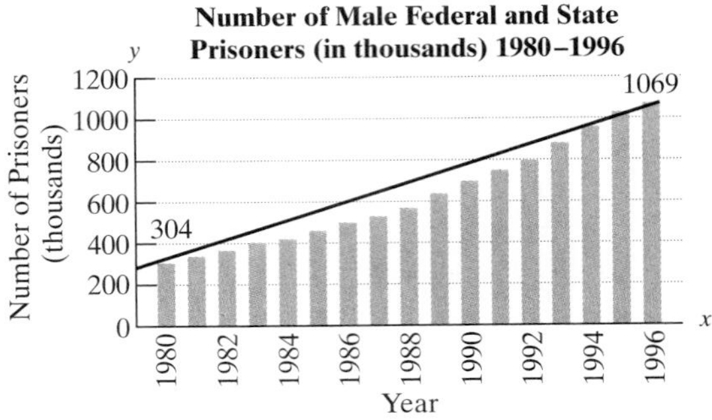

Figure for Exercise 45

(*Source:* U.S. Bureau of Statistics)

a. Using the ordered pairs (1980, 304) and (1996, 1069), find the slope of the line.

b. Interpret the meaning of the slope in the context of this problem.

46. In 1980, there were 12 thousand female inmates in federal and state prisons. By 1996, the number increased to 70 thousand.

Let *x* represent the year and let *y* represent the number of female inmates in federal and state prisons (in thousands).

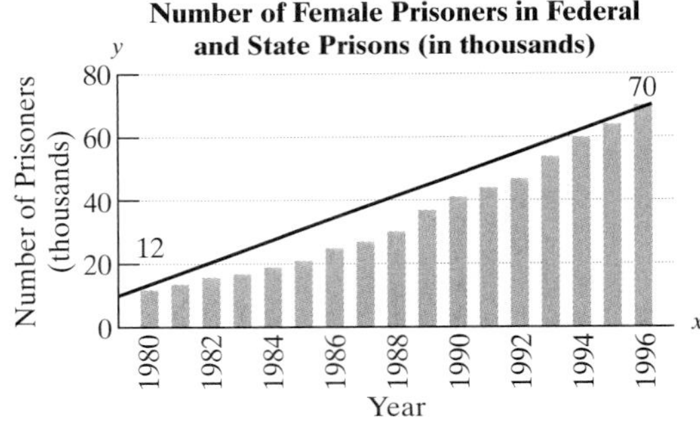

Figure for Exercise 46

(*Source:* U.S. Bureau of Statistics)

a. Using the ordered pairs (1980, 12) and (1996, 70), find the slope of the line.

b. Interpret the meaning of the slope in the context of this problem.

47. The following graph depicts the median income for females in the United States between 1990 and 1996.

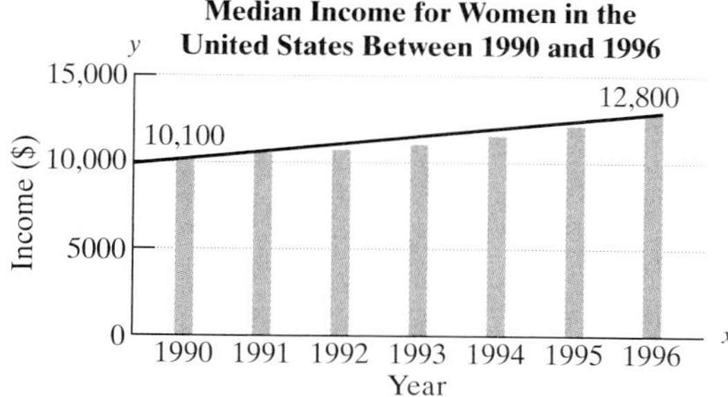

Figure for Exercise 47

(*Source:* U.S. Bureau of the Census)

a. Find the slope of the line and interpret the meaning of the slope in the context of this problem.

b. Compare the slopes for the rise in median income per year for women and for men (see Example 7). Based on the slopes of the two lines, will the median income for women ever catch-up to the median income for men if these linear trends continue? Explain.

For Exercises 48–55, draw a line as indicated. Answers may vary.

48. Draw a line with a positive slope and a positive y-intercept.

49. Draw a line with a positive slope and a negative y-intercept.

50. Draw a line with a negative slope and a negative y-intercept.

51. Draw a line with a negative slope and positive y-intercept.

52. Draw a line with a zero slope and a positive y-intercept.

53. Draw a line with a zero slope and a negative y-intercept.

54. Draw a line with undefined slope and a negative x-intercept.

55. Draw a line with undefined slope and a positive x-intercept.

For Exercises 56–61, answers may vary.

56. a. Draw a line with a positive slope.
    b. Draw a line parallel to the line in part (a).

57. a. Draw a line with a negative slope.
    b. Draw a line parallel to the line in part (a).

58. a. Draw a line with a negative slope.
    b. Draw a line perpendicular to the line in part (a).

59. a. Draw a line with a positive slope.
    b. Draw a line perpendicular to the line in part (a).

60. a. Draw a line with an undefined slope.
    b. Draw a line perpendicular to the line in part (a).

61. a. Draw a line with a slope of 0.
    b. Draw a line parallel to the line in part (a).

62. Suppose a given line has a slope of -2.
    a. What is the slope of a line parallel to the given line?
    b. What is the slope of a line perpendicular to the given line?

63. Suppose a given line has a slope of $\frac{2}{3}$.
    a. What is the slope of a line parallel to the given line?
    b. What is the slope of a line perpendicular to the given line?

64. Suppose a given line has a slope of 0.
    a. What is the slope of a line parallel to the given line?
    b. What is the slope of a line perpendicular to the given line?

65. Suppose a line has an undefined slope.
    a. What is the slope of a line parallel to the given line?
    b. What is the slope of a line perpendicular to the given line?

For Exercises 66–73, find the slopes of the lines l_1 and l_2 determined by the two given points. Then identify whether l_1 and l_2 are parallel, perpendicular, or neither.

66. l_1: $(2, 4)$ and $(-1, -2)$
    l_2: $(1, 7)$ and $(0, 5)$

67. l_1: $(0, 0)$ and $(-2, 4)$
    l_2: $(1, -5)$ and $(-1, -1)$

68. l_1: $(1, 9)$ and $(0, 4)$
    l_2: $(5, 2)$ and $(10, 1)$

69. l_1: $(3, -4)$ and $(-1, -8)$
    l_2: $(5, -5)$ and $(-2, 2)$

70. l_1: $(4, 4)$ and $(0, 3)$
    l_2: $(1, 7)$ and $(-1, -1)$

71. l_1: $(3, 5)$ and $(-2, -5)$
    l_2: $(2, 0)$ and $(-4, -3)$

72. l_1: $(3.1, 6.3)$ and $(3.1, -5.7)$
    l_2: $(1.2, 4.7)$ and $(1.2, -5.3)$

73. l_1: $(4.5, -6.7)$ and $(-2.3, -6.7)$
    l_2: $(-2.2, -6.7)$ and $(-1.4, -6.7)$

74. Determine the pitch (slope) of the roof.

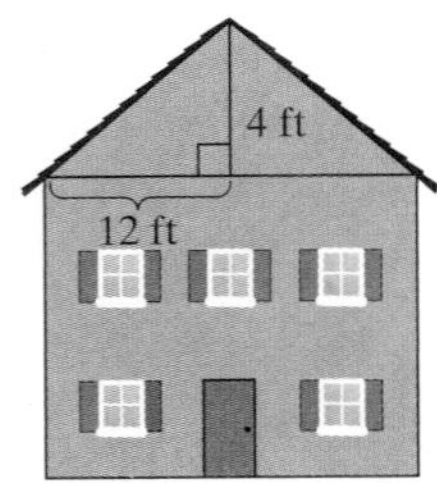

Figure for Exercise 74

75. Find the height, y, so that the pitch (slope) of the garage roof is $\frac{1}{4}$.

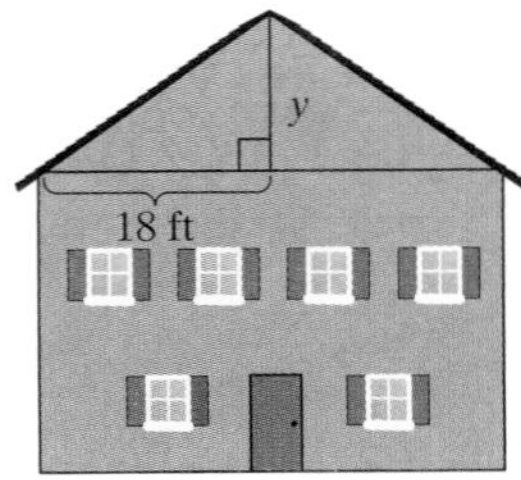

Figure for Exercise 75

76. The distance, d, (in miles) between a lightning strike and an observer is given by the equation:

    $d = 0.2t$ where t is the time (in seconds) between seeing lightning and hearing thunder.

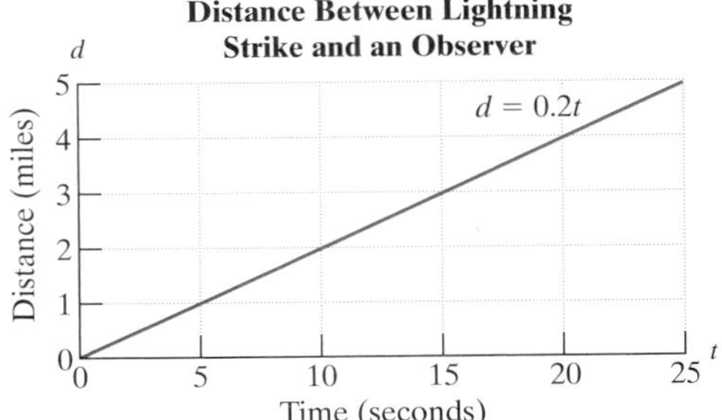

Figure for Exercise 76

a. If an observer counts 5 sec between seeing lightning and hearing thunder, how far away was the lightning strike?

b. If an observer counts 10 sec between seeing lightning and hearing thunder, how far away was the lightning strike?

c. If an observer counts 15 sec between seeing lightning and hearing thunder, how far away was the lightning strike?

d. What is the slope of the line? Interpret the meaning of the slope in the context of this problem.

77. Jorge is paid by the hour according to the equation

$P = 11.50x$ P is his total pay (in dollars) and x is the number of hours worked

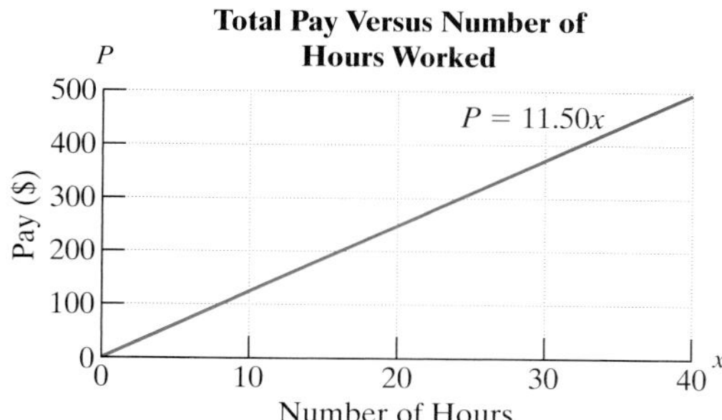

Figure for Exercise 77

a. How much money will Jorge earn if he works 20 hr?

b. How much money will Jorge earn if he works 21 hr?

c. How much money will Jorge earn if he works 22 hr?

d. What is the slope of the line? Interpret the meaning of the slope in the context of this problem.

Expanding Your Skills

78. Find the slope between the points $(a + b, 4m - n)$ and $(a - b, m + 2n)$.

79. Find the slope between the points $(3c - d, s + t)$ and $(c - 2d, s - t)$.

80. Find the x-intercept of the line $ax + by = c$.

81. Find the y-intercept of the line $ax + by = c$.

82. Find another point on the line that contains the point $(2, -1)$ and has a slope of $\frac{2}{5}$.

83. Find another point on the line that contains the point $(-3, 4)$ and has a slope of $\frac{1}{4}$.

---

section

3.5 Equations of a Line: Slope-Intercept Form

Concepts

1. Slope-Intercept Form of a Line
2. Identifying the Slope and y-Intercept of a Line
3. Graphing a Line from Its Slope and y-Intercept
4. Determining Whether Two Lines Are Parallel, Perpendicular, or Neither
5. Writing an Equation of a Line Given Its Slope and y-Intercept
6. Different Forms of Linear Equations: A Summary

1. Slope-Intercept Form of a Line

In Section 3.2, we learned that an equation of the form: $ax + by = c$ (where a and b are not both zero) represents a line in a rectangular coordinate system. An equation of a line written in this way is said to be in **standard form**. In this section, we will learn a new form, called **slope-intercept form**, which is useful in determining the slope and y-intercept of a line.

Let $(0, b)$ represent the y-intercept of a line. Let (x, y) represent any other point on the line. Then the slope of the line can be found as follows:

Let $(0, b)$ represent (x_1, y_1) and let (x, y) represent (x_2, y_2). Apply the slope formula.

$$m = \frac{(y_2 - y_1)}{(x_2 - x_1)} \rightarrow m = \frac{y - b}{x - 0} \qquad \text{Apply the slope formula.}$$

$$m = \frac{y - b}{x} \qquad \text{Simplify.}$$

$$mx = \left(\frac{y - b}{x}\right)x$$ Multiply by x to clear fractions.

$$mx = y - b$$

$$mx + b = y - b + b$$ To isolate y, add b to both sides.

$$mx + b = y \quad \text{or} \quad y = mx + b$$ The equation is in slope-intercept form.

Slope-Intercept Form of a Line

$y = mx + b$ is the slope-intercept form of a line.

m is the slope and the point $(0, b)$ is the y-intercept.

example 1

Identifying the Slope and *y*-Intercept of a Line

For each equation, identify the slope and y-intercept.

a. $y = 3x - 1$

b. $y = \frac{4}{3}x + \frac{5}{2}$

c. $y = -2.7x + 5$

d. $y = 4x$

Solution:

Each equation is written in slope-intercept form, $y = mx + b$. The slope is the coefficient on x, and the y-intercept is the constant term.

a. $y = 3x - 1$; The slope is 3. The y-intercept is $(0, -1)$.

b. $y = \frac{4}{3}x + \frac{5}{2}$; The slope is $\frac{4}{3}$. The y-intercept is $\left(0, \frac{5}{2}\right)$.

c. $y = -2.7x + 5$; The slope is -2.7. The y-intercept is $(0, 5)$.

d. The line $y = 4x$ can be written as $y = 4x + 0$. Therefore,
The slope is 4. The y-intercept is $(0, 0)$.

2. Identifying the Slope and *y*-Intercept of a Line

Given an equation of a line, we can write the equation in slope-intercept form by solving the equation for the y-variable. This is demonstrated in Example 2.

example 2

Identifying the Slope and *y*-Intercept of a Line

Given $-5x - 2y = 6$,

a. Write the slope-intercept form of the line.
b. Identify the slope and y-intercept.

Solution:

a. Write the equation in slope-intercept form, $y = mx + b$, by solving for y.

$$-5x - 2y = 6$$

$$-2y = 5x + 6 \quad \text{Add } 5x \text{ to both sides.}$$

$$\frac{-2y}{-2} = \frac{5x + 6}{-2} \quad \text{Divide both sides by } -2.$$

$$y = \frac{5x}{-2} + \frac{6}{-2}$$

$$y = -\frac{5}{2}x - 3 \quad \text{Slope-intercept form.}$$

b. The slope is $-\frac{5}{2}$ and the y-intercept is $(0, -3)$.

3. Graphing a Line from Its Slope and *y*-Intercept

Slope-intercept form is a useful tool to graph a line. The y-intercept is a known point on the line. The slope indicates the direction of the line and can be used to find a second point. Using slope-intercept form to graph a line is demonstrated in Example 3.

example 3

Graphing a Line Using the Slope and *y*-Intercept

Graph the line $y = -\frac{5}{2}x - 3$ by using the slope and y-intercept.

Solution:

First plot the y-intercept, $(0, -3)$.
The slope, $m = -\frac{5}{2}$ can be written as

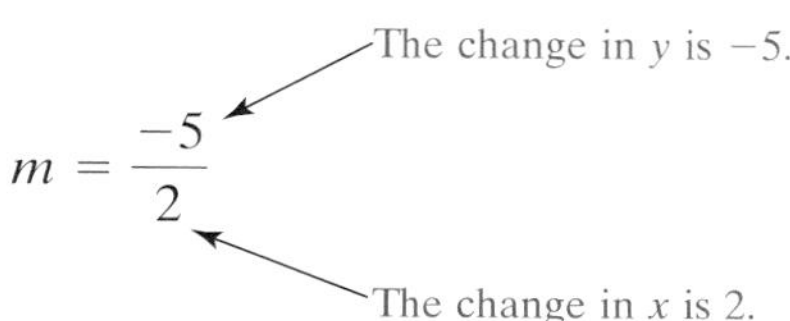

To find a second point on the line, start at the y-intercept and move down 5 units and to the right 2 units. Then draw the line through the two points (Figure 3-33).

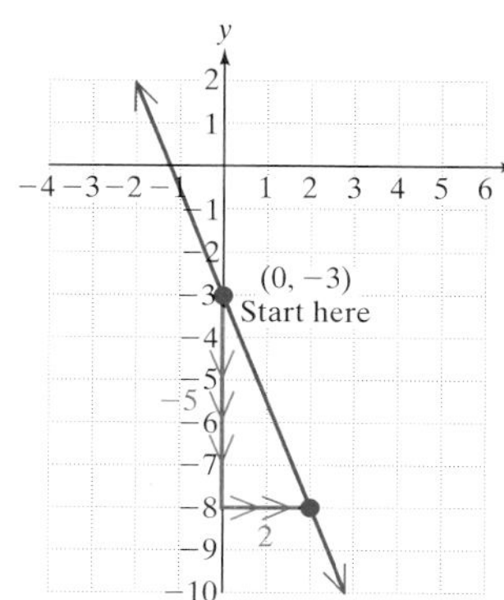

Figure 3-33

Tip: Example 3 illustrates that for a line with a slope of $-\frac{5}{2}$ we can start at the y-intercept and either move down 5 units and to the right 2 units or up 5 units and to the left 2 units to plot a second point.

Similarly, the slope can be written as

$$m = \frac{5}{-2}$$

The change in y is 5.

The change in x is -2.

To find a second point, start at the y-intercept and move up 5 units and to the left 2 units. Then draw the line through the two points (Figure 3-34).

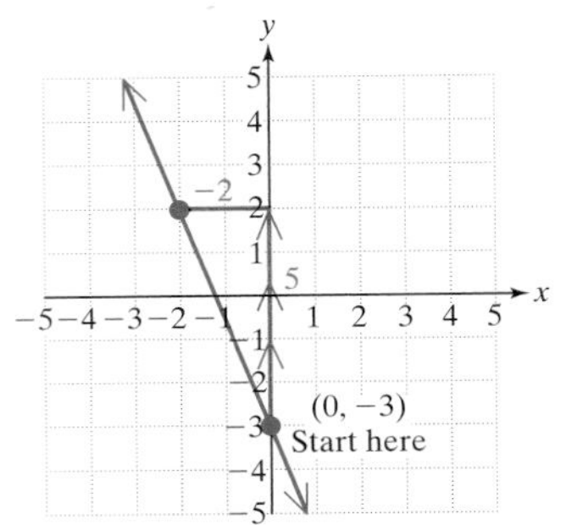

Figure 3-34

example 4

Graphing a Line from Its Slope and y-Intercept

Graph the line $y = 4x$ by using the slope and y-intercept.

Solution:

The line can be written as $y = 4x + 0$. Therefore, we can plot the y-intercept at (0, 0). The slope $m = 4$ can be written as

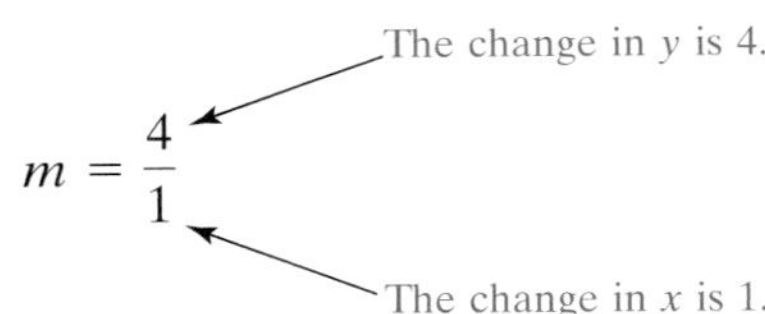

$$m = \frac{4}{1}$$

To find a second point on the line, start at the y-intercept and move up 4 units and to the right 1 unit. Then draw the line through the two points (Figure 3-35).

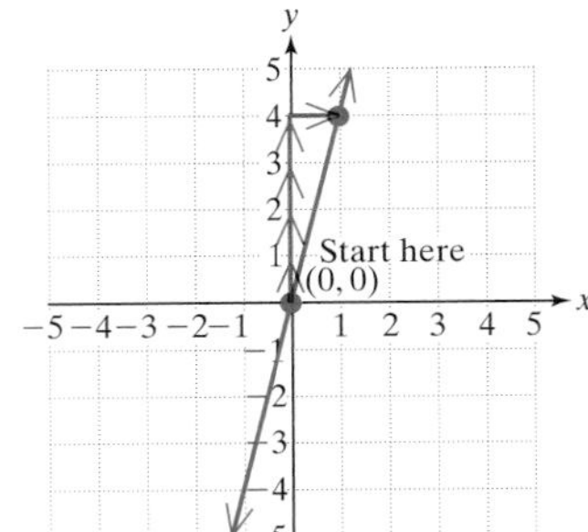

Figure 3-35

4. Determining Whether Two Lines Are Parallel, Perpendicular, or Neither

Slope intercept-form provides a means to find the slope of a line by inspection. Furthermore, if the slopes of two lines are known, then we can compare the slopes to determine if the lines are parallel, perpendicular, or neither parallel nor perpendicular. (Recall that two distinct nonvertical lines are parallel if their slopes are equal. The lines are perpendicular if the slope of one line is the opposite of the reciprocal of the slope of the other line.)

example 5 **Determining If Two Lines Are Parallel, Perpendicular, or Neither**

For each pair of lines determine if they are parallel, perpendicular, or neither.

a. l_1: $y = 3x - 5$
   l_2: $y = 3x + 1$

b. l_1: $y = \frac{1}{2}x + 3$
   l_2: $y = -2x + 4$

c. $-3x + 2y = 4$
   $2x - 3y = -3$

d. l_1: $x = 2$
   l_2: $2y = 8$

Solution:

a. The slope of l_1 is 3.

The slope of l_2 is 3.

Because the slopes are the same, the lines are parallel.

b. The slope of l_1 is $\frac{1}{2}$.

The slope of l_2 is -2.

The slope $\frac{1}{2}$ is the opposite of the reciprocal of -2. Therefore the lines are perpendicular.

c. Solve the equations for y.

$$\begin{aligned} -3x + 2y &= 4 \\ 2y &= 3x + 4 \\ y &= \frac{3}{2}x + 2 \end{aligned}$$

Slope of l_1 is $\frac{3}{2}$

$$\begin{aligned} 2x - 3y &= -3 \\ -3y &= -2x - 3 \\ y &= \frac{2}{3}x + 1 \end{aligned}$$

Slope of l_2 is $\frac{2}{3}$

The slopes are not the same. Therefore, the lines are not parallel. The values of the slopes are reciprocals, but they are not opposite in sign. Therefore, the lines are not perpendicular. The lines are neither parallel nor perpendicular.

d. The equation $x = 2$ represents a vertical line because the equation is in the form $x = k$. The equation $2y = 8$ can be simplified to $y = 4$ which represents a horizontal line.

In this example, we do not need to analyze the slopes because vertical lines and horizontal lines are perpendicular.

5. Writing an Equation of a Line Given Its Slope and *y*-Intercept

The slope-intercept form of a line can be used to write an equation of a line when the slope is known and the y-intercept is known.

example 6

Writing an Equation of a Line Using Slope-Intercept Form

Write an equation of the line whose slope is $\frac{2}{3}$ and whose y-intercept is (0, 8).

Solution:

The slope is given as $m = \frac{2}{3}$ and the y-intercept $(0, b)$ is given as (0, 8). Substitute the values $m = \frac{2}{3}$ and $b = 8$, into the slope-intercept form of a line.

$$y = mx + b$$

$$y = \frac{2}{3}x + 8$$

example 7

Writing an Equation of a Line Given Its Slope and *y*-Intercept

Write an equation of the line that passes through the origin and whose slope is -2.

Solution:

The slope is given as $m = -2$. Furthermore, the line passes through the origin. Therefore, the y-intercept is (0, 0) and the corresponding value of b is 0. The slope-intercept form of the line becomes:

$$y = mx + b$$

$$y = -2x + 0 \text{ or equivalently } y = -2x$$

6. Different Forms of Linear Equations: A Summary

In this chapter we have seen that linear equations can be written in several different forms. These are summarized in Table 3-3.

Table 3-3

Form	Example	Comments
Standard Form $ax + by = c$	$4x + 2y = 8$	a and b must not *both* be zero.
Horizontal Line $y = k$ (k is constant)	$y = 4$	The slope is zero and the y-intercept is $(0, k)$.
Vertical Line $x = k$ (k is constant)	$x = -1$	The slope is undefined and the x-intercept is $(k, 0)$.
Slope-Intercept Form $y = mx + b$ Slope $= m$ y-intercept is $(0, b)$	$y = -3x + 7$ Slope $= -3$ y-intercept is (0, 7)	Solving a linear equation for y results in slope-intercept form. The coefficient of the x-term is the slope, and the constant defines the location of the y-intercept.

Calculator Connections

In Example 5(b) we found that the lines $y = \frac{1}{2}x + 3$ and $y = -2x + 4$ are perpendicular. We can verify our results by graphing the lines on a graphing calculator.

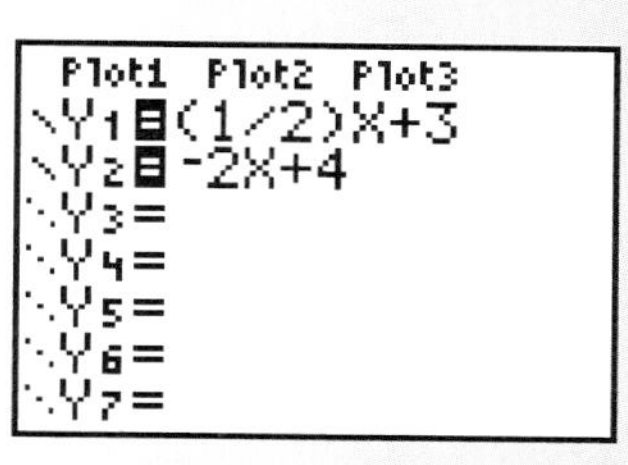

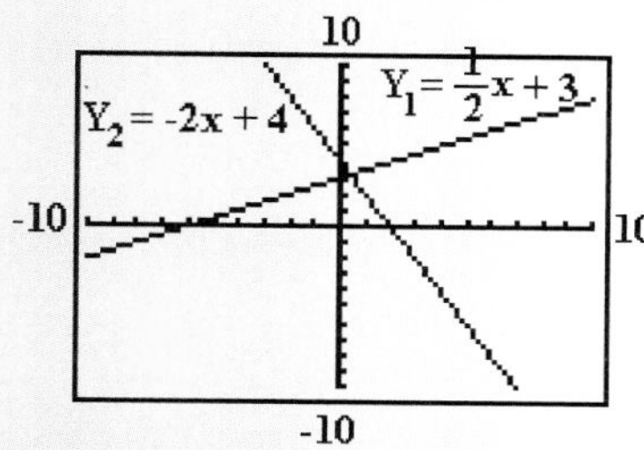

Notice that the lines do not appear perpendicular in the calculator display. That is, they do not appear to form a right angle at the point of intersection. Because many calculators have a rectangular screen, the standard viewing window is elongated in the horizontal direction. To eliminate this distortion, try using a *ZSquare* option. This feature will set the viewing window so that equal distances on the display denote an equal number of units on the graph.

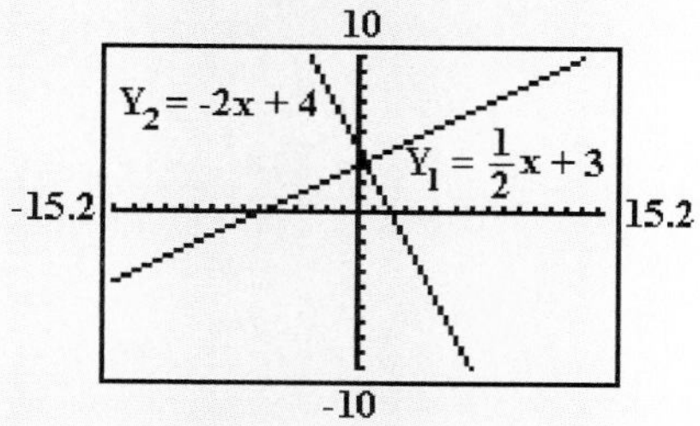

section 3.5 Practice Exercises

For Exercises 1–8 find the x- and y-intercepts (if they exist).

1. $x - 5y = 10$
2. $3x + y = -12$
3. $3y = -9$
4. $2 + y = 5$
5. $-4x = 6y$
6. $2x - 2y = 3$
7. $-x + 3 = 8$
8. $5x = 20$

For Exercises 9–20, write each equation in slope-intercept form, if possible. Then identify the slope and y-intercept, if they exist.

9. $2x - 5y = 4$
10. $3x + 2y = 9$
11. $3x - y = 5$
12. $7x - 3y = -6$
13. $x + y = 6$
14. $x - y = 1$
15. $x + 6 = 8$
16. $-4 + x = 1$
17. $-8y = 2$
18. $1 - y = 9$
19. $3y - 2x = 0$
20. $5x = 6y$
21. Graph the line through the point (0, 2), having a slope of -4.
22. Graph the line through the point (0, -1), having a slope of -3.
23. Graph the line through the point (0, -5), having a slope of $\frac{3}{2}$.
24. Graph the line through the point (0, 3), having a slope of $-\frac{1}{4}$.

For Exercises 25–44, write each equation in slope-intercept form (if possible) and graph the line.

25. $x - 2y = 6$
26. $5x - 2y = 2$
27. $2x + y = 9$
28. $-6x + y = 8$

29. $3y + 6x = -1$
30. $-4y - 2x = -7$
31. $2x = -4y + 6$
32. $3x = y - 7$
33. $x + y = 0$
34. $x - y = 0$
35. $0.2x - 0.5y = 0.1$
36. $1.5x + 2.5y = 5.0$
37. $5y = 9x$
38. $-2x = 5y$
39. $3y + 2 = 0$
40. $1 + 5y = 6$
41. $3x + 1 = 7$
42. $-2x - 5 = 1$
43. $\frac{1}{2}x + \frac{1}{4}y = \frac{1}{2}$
44. $\frac{1}{3}x - \frac{1}{6}y = \frac{1}{2}$

For Exercises 45–50, let m_1 and m_2 represent the slopes of two lines. Determine if the lines are parallel, perpendicular, or neither.

45. $m_1 = -2,\ m_2 = \frac{1}{2}$
46. $m_1 = \frac{2}{3},\ m_2 = \frac{3}{2}$
47. $m_1 = 1,\ m_2 = \frac{4}{4}$
48. $m_1 = \frac{3}{4},\ m_2 = -\frac{8}{6}$
49. $m_1 = \frac{2}{7},\ m_2 = -\frac{2}{7}$
50. $m_1 = 5,\ m_2 = 5$

For Exercises 51–66, determine if the lines l_1 and l_2 are parallel, perpendicular, or neither.

51. l_1: $y = -2x - 3$
    l_2: $y = \frac{1}{2}x + 4$
52. l_1: $y = \frac{4}{3}x - 2$
    l_2: $y = -\frac{3}{4}x + 6$
53. l_1: $y = \frac{4}{5}x - \frac{1}{2}$
    l_2: $y = \frac{5}{4}x - \frac{2}{3}$
54. l_1: $y = \frac{1}{5}x + 1$
    l_2: $y = 5x - 3$
55. l_1: $y = -9x + 6$
    l_2: $y = -9x - 1$
56. l_1: $y = 4x - 1$
    l_2: $y = 4x + \frac{1}{2}$
57. l_1: $x = 3$
    l_2: $y = \frac{7}{4}$
58. l_1: $y = \frac{2}{3}$
    l_2: $x = 6$
59. l_1: $2x = 4$
    l_2: $6 = x$
60. l_1: $2y = 7$
    l_2: $y = 4$
61. l_1: $2x + 3y = 6$
    l_2: $3x - 2y = 12$
62. l_1: $4x + 5y = 20$
    l_2: $5x - 4y = 60$
63. l_1: $4x + 2y = 6$
    l_2: $4x + 8y = 16$
64. l_1: $3x + y = 5$
    l_2: $x + 3y = 18$
65. l_1: $y = \frac{1}{5}x - 3$
    l_2: $2x - 10y = 20$
66. l_1: $y = \frac{1}{3}x + 2$
    l_2: $-x + 3y = 12$

For Exercises 67–72, write an equation of the line given the following information. Write the answer in slope-intercept form if possible.

67. The slope is $-\frac{1}{3}$ and the y-intercept is $(0, 2)$.
68. The slope is $\frac{2}{3}$ and the y-intercept is $(0, -1)$.
69. The slope is 5 and the line passes through the origin.
70. The slope is -3 and the line passes through the origin.
71. The slope is 6 and the line passes through the point $(0, -2)$.
72. The slope is -4 and the line passes through the point $(0, -3)$.
73. The cost for a rental car is \$49.95 per day plus a flat fee of \$31.95 for insurance. The equation, $C = 49.95x + 31.95$ represents the total cost, C, (in dollars) to rent the car for x days.
    a. Identify the slope. Interpret the meaning of the slope in the context of this problem.
    b. Identify the C-intercept. Interpret the meaning of the C-intercept in the context of this problem.
    c. Use the equation to determine how much it would cost to rent the car for 1 week.

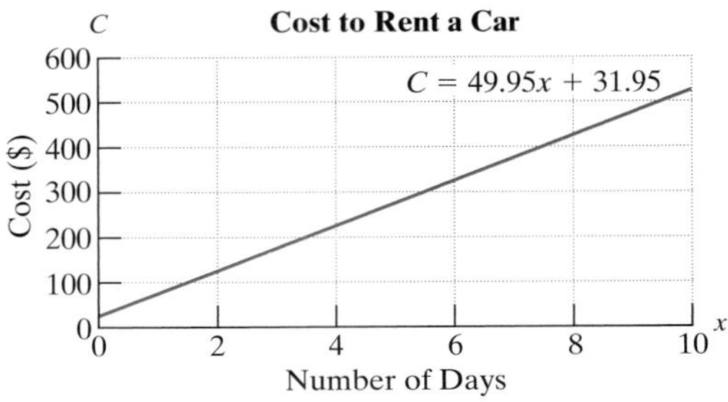

Figure for Exercise 73

74. A phone bill is determined each month by a \$16.95 flat fee plus \$0.10/min of long distance. The equation $C = 0.10x + 16.95$ represents the total monthly cost, C, for x minutes of long distance.

    a. Identify the slope. Interpret the meaning of the slope in the context of this problem.

    b. Identify the C-intercept. Interpret the meaning of the C-intercept in the context of this problem.

    c. Use the equation to determine the total cost of 234 min of long distance.

Phone Bill Cost Versus Number of Minutes of Long Distance

$C = 0.10x + 16.95$

Cost (\$) vs. Number of Minutes

Figure for Exercise 74

75. The total cost, y (in dollars), to get a car towed x miles using Buddy's Towing on a weekday is given by the equation $y = 2.5x + 45$.

    a. How much would it cost to tow a car 12 miles on a weekday?

    b. What is the slope and what does it mean in the context of the problem?

    c. What is the y-intercept and what does it mean in the context of the problem?

76. The monthly cost, C (in dollars), for the use of a particular cellular phone is \$25 plus \$0.03 per minute for all calls (incoming and outgoing) including long distance. The monthly cost is given by the equation $C = 0.03x + 25$, where x is the number of minutes used.

    a. How much would it cost if Luis talked on his cellular phone for 75 minutes in one month?

    b. What is the slope and what does it mean in the context of the problem?

    c. What is the y-intercept and what does it mean in the context of the problem?

Expanding Your Skills

For Exercises 77–80, graph the lines in parts a, b, and c on the same coordinate axes and state any similarities or differences.

77. a. $y = 2x + 1$
    b. $y = 3x + 1$
    c. $y = 4x + 1$

78. a. $y = \frac{3}{4}x - 1$
    b. $y = \frac{1}{2}x - 1$
    c. $y = \frac{1}{4}x - 1$

79. a. $y = -\frac{2}{3}x$
    b. $y = -\frac{2}{3}x + 3$
    c. $y = -\frac{2}{3}x - 1$

80. a. $y = 4x$
    b. $y = 4x - 3$
    c. $y = 4x + 3$

81. A linear equation is said to be written in standard form if it can be written as $ax + by = c$, where a and b are not both zero. Write the equation $ax + by = c$ in slope-intercept form to show that the slope is given by the ratio, $-\frac{a}{b}$.

For Exercises 82–85 use the result of Exercise 81 to find the slope of the line.

82. $2x + 5y = 8$

83. $6x + 7y = -9$

84. $4x - 3y = -5$

85. $11x - 8y = 4$

Graphing Calculator Exercises

For each pair of lines in Exercises 86–88, determine if the lines are parallel, perpendicular, or neither. Then use a square viewing window to graph the lines on a graphing calculator to verify your results.

86. $x + y = 1$
    $x - y = -3$

87. $3x + y = -2$
    $6x + 2y = 6$

88. $2x - y = 4$
    $3x + 2y = 4$

89. Graph the lines: $y = x + 1$ and $y = 0.99x + 3$. Are these lines parallel? Explain.

90. Graph the lines $y = -2x - 1$ and $y = -2x - 0.99$. Are these lines the same? Explain.

---

chapter 3

Summary

Section 3.1—Rectangular Coordinate System

Key Concepts:

Graphical representation of numerical data is often helpful to study problems in real-world applications.

A rectangular coordinate system is made up of a horizontal line called the *x*-axis and a vertical line called the *y*-axis. The point where the lines meet is the origin. The four regions of the plane are called quadrants.

The point (x, y) is an ordered pair. The first element in the ordered pair is the point's horizontal position from the origin. The second element in the ordered pair is the point's vertical position from the origin.

Key Terms:

data
ordered pair
origin
quadrant
rectangular coordinate system
x-axis
y-axis

Examples:

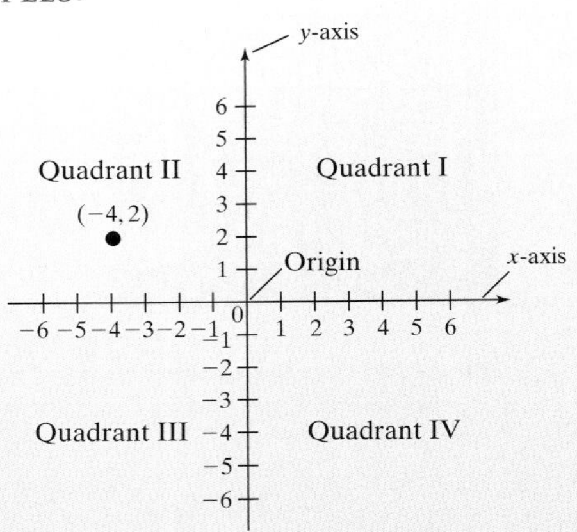

Section 3.2—Linear Equations in Two Variables

Key Concepts:

An equation written in the form $ax + by = c$ (where a and b are not both zero) is a linear equation in two variables.

A solution to a linear equation in x and y is an ordered pair (x, y) that makes the equation a true statement. The graph of the set of all solutions of a

Examples:

The equation $2x + y = 2$ is a linear equation in two variables.

linear equation in two variables is a line in a rectangular coordinate system.

A linear equation can be graphed by finding at least two solutions and graphing the line through the points.

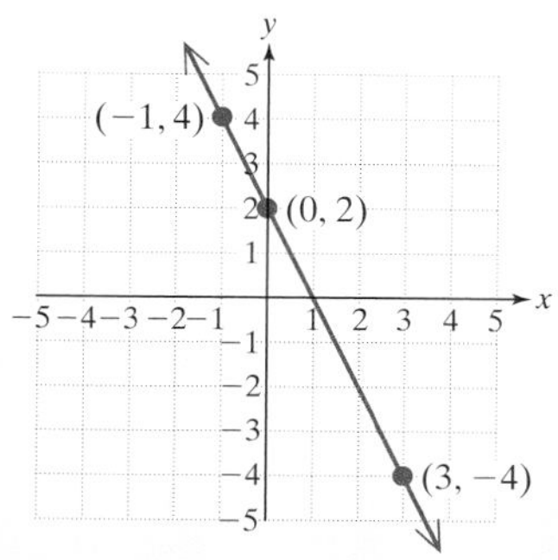

KEY TERM:

linear equation in two variables

Graph the equation $2x + y = 2$:

Select arbitrary values of x or y such as those shown in the table. Then complete the table to find the corresponding ordered pairs.

x	y
0	
	4
3	

$$2x + y = 2 \qquad 2x + y = 2 \qquad 2x + y = 2$$
$$2(0) + y = 2 \qquad 2x + (4) = 2 \qquad 2(3) + y = 2$$
$$0 + y = 2 \qquad 2x = -2 \qquad 6 + y = 2$$
$$y = 2 \qquad x = -1 \qquad y = -4$$

x	y	
0	2	→ (0, 2)
−1	4	→ (−1, 4)
3	−4	→ (3, −4)

SECTION 3.3—x- AND y-INTERCEPTS, HORIZONTAL AND VERTICAL LINES

KEY CONCEPTS:

An x-intercept of a graph is a point $(a, 0)$ where the graph intersects the x-axis.

A y-intercept of a graph is a point $(0, b)$ where the graph intersects the y-axis.

Finding x- and y-Intercepts from an Equation

1. To find an x-intercept from an equation, substitute $y = 0$ and solve for x.

EXAMPLES:

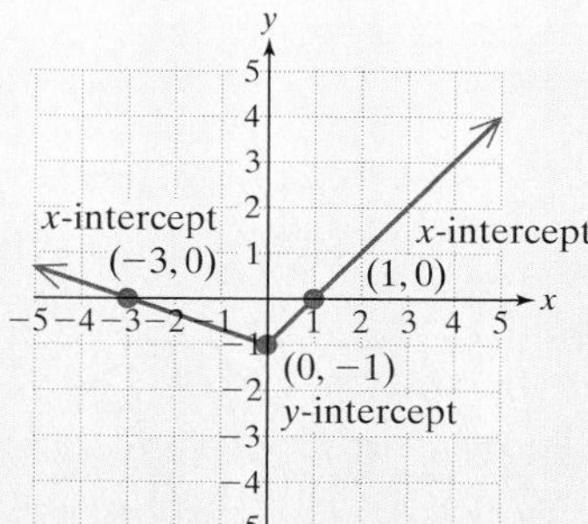

Find the x- and y-intercepts of $3x - y = 3$:

1. To find the x-intercept, substitute $y = 0$

$$3x - y = 3$$
$$3x - (0) = 3$$
$$3x = 3$$
$$x = 1 \qquad \text{The } x\text{-intercept is } (1, 0)$$

2. To find a y-intercept from an equation, substitute $x = 0$ and solve for y.

2. To find the y-intercept, substitute $x = 0$

$$3x - y = 3$$
$$3(0) - y = 3$$
$$0 - y = 3$$
$$-y = 3$$
$$y = -3$$ The y-intercept is $(0, -3)$

A vertical line can be written in the form $x = k$.
A horizontal line can be written in the form $y = k$.

Key Terms:

x-intercept
y-intercept
horizontal line
vertical line

$x = 3$ is a Vertical Line

$y = 3$ is a Horizontal Line

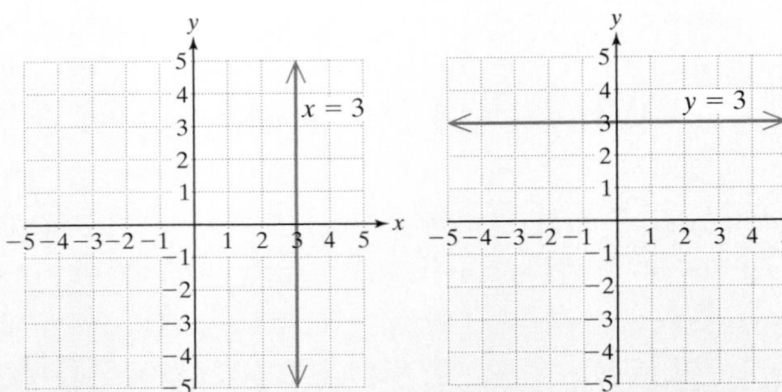

Section 3.4—Slope of a Line

Key Concepts:

The slope, m, of a line between two points (x_1, y_1) and (x_2, y_2) is given by

$$m = \frac{y_2 - y_1}{x_2 - x_1} \quad \text{or} \quad \frac{\text{change in } y}{\text{change in } x}$$

The slope of a line may be positive, negative, zero, or undefined.

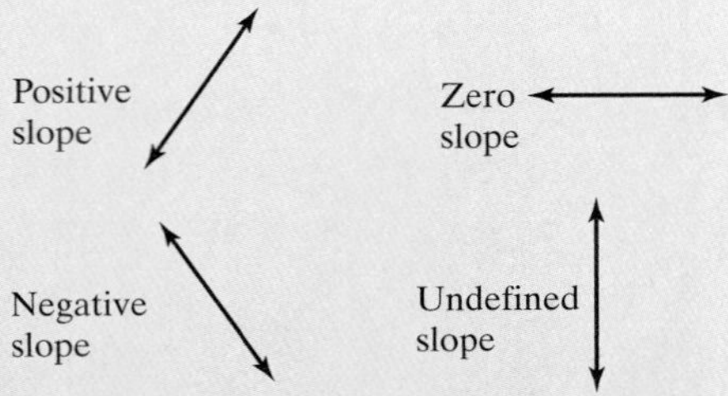

If m_1 and m_2 represent the slopes of two parallel (nonvertical) lines, then $m_1 = m_2$.

Examples:

Find the slope of the line between (1, −3) and (−3, 7)

$$m = \frac{7 - (-3)}{-3 - 1} = \frac{10}{-4} = -\frac{5}{2}$$

Find the slope of the line $y = -2$

$m = 0$ because the line is horizontal.

Find the slope of the line $x = 4$

m is undefined because the line is vertical.

If $m_1 \neq 0$ and $m_2 \neq 0$ represent the slopes of two perpendicular lines, then

$$m_1 = -\frac{1}{m_2} \qquad \text{or} \qquad m_2 = -\frac{1}{m_1}$$

or equivalently, $m_1 m_2 = -1$

Key Term:

slope

The slopes of two lines are given. Determine whether the lines are parallel, perpendicular, or neither.

a. $m_1 = -7$ and $m_2 = -7$ Parallel

b. $m_1 = -\frac{1}{5}$ and $m_2 = 5$ Perpendicular

c. $m_1 = -\frac{3}{2}$ and $m_2 = -\frac{2}{3}$ Neither

Section 3.5—Equations of a Line: Slope-Intercept Form

Key Concepts:

The slope-intercept form of a line is

$$y = mx + b$$

where m is the slope of the line and $(0, b)$ is the y-intercept.

Slope-intercept form is used to identify the slope and y-intercept of a line when the equation is given.

Slope-intercept form can also be used to graph a line.

Slope-intercept form can be used to construct an equation of a line given the y-intercept and the slope of the line.

Key Terms:

slope-intercept form of a line
standard form of a line

Examples:

Find the slope and y-intercept:

$$7x - 2y = 4$$

$$-2y = -7x + 4 \qquad \text{Solve for } y.$$

$$y = \frac{7}{2}x - 2$$

The slope is $\frac{7}{2}$. The y-intercept is $(0, -2)$.

Graph the line:

$$y = \frac{7}{2}x - 2$$

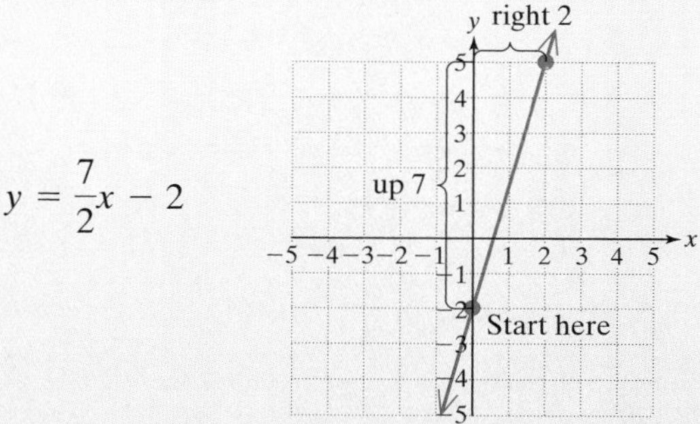

Find an equation of the line having y-intercept $(0, 5)$ and slope -4.

Using slope-intercept form:

$$y = mx + b$$

$$y = -4x + 5$$

chapter 3 REVIEW EXERCISES

Section 3.1

1. Graph the points on a rectangular coordinate system

   a. $\left(\frac{1}{2}, 5\right)$ b. $(-1, 4)$ c. $(2, -1)$

   d. $(0, 3)$ e. $(0, 0)$ f. $\left(-\frac{8}{5}, 0\right)$

   g. $(-2, -5)$ h. $(3, 1)$

2. Estimate the coordinates of the points A, B, C, D, E, and F.

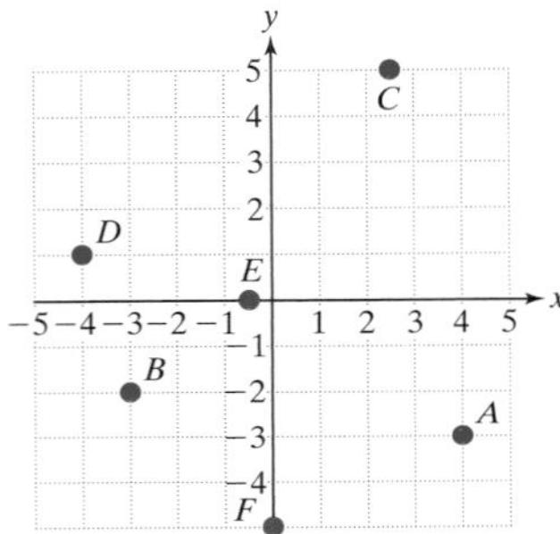

For Exercises 3–8, determine the quadrant in which the given point is found.

3. $(-2, -10)$
4. $(-4, 6)$
5. $(3, -5)$
6. $\left(\frac{1}{2}, \frac{7}{5}\right)$
7. $(\pi, -2.7)$
8. $(-1.2, -6.8)$
9. On which axis is the point $(2, 0)$ found?
10. On which axis is the point $(0, -3)$ found?
11. The price per share of the stock (in dollars) over a period of 5 days is shown in the graph.

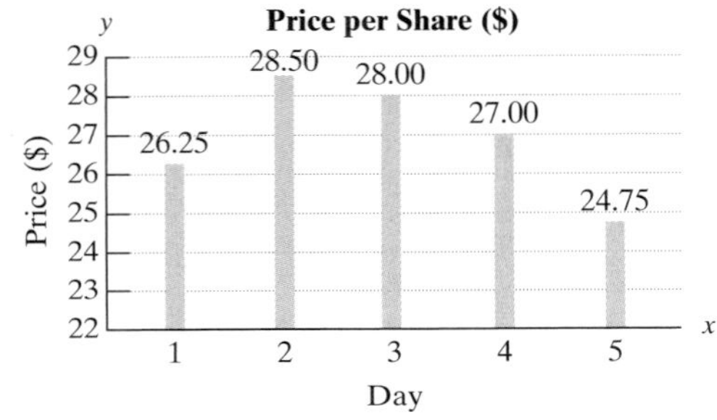

   a. Use the information provided in the graph to write the ordered pairs corresponding to the price of the stock each day. Interpret the meaning of the first ordered pair.

   b. On which day was the price the highest?

   c. What was the increase in price between day 1 and day 2?

12. The number of people living below the poverty level (in millions) for selected years between 1965 and 1995 is given by the ordered pairs shown here. Let x represent the year, where $x = 0$ corresponds to 1965, $x = 5$ corresponds to 1970, and so on. Let y represent the number of people (in millions) living below the poverty level. (*Source: U.S. Bureau of the Census*)

   $(0, 25.9)$ $(5, 29.3)$ $(10, 33.1)$

   $(25, 33.6)$ $(30, 36.4)$

   a. Interpret the meaning of the ordered pair $(25, 33.6)$.

   b. Plot the points on a rectangular coordinate system.

Section 3.2

For Exercises 13–16, determine if the given ordered pair is a solution to the equation.

13. $5x - 3y = 12;\ (0, 4)$
14. $2x - 4y = -6;\ (3, 0)$
15. $y = \frac{1}{3}x - 2;\ (9, 1)$
16. $y = -\frac{2}{5}x + 1;\ (-10, 5)$

For Exercises 17–20, complete the table and graph the corresponding ordered pairs. Graph the line through the points to represent all solutions to the equation.

17. $3x - y = 5$
18. $\frac{1}{2}x + 3y = 6$

x	y
2	
	4
1	

x	y
	2
−2	
	3

19. $y = \frac{2}{3}x - 1$

x	y
0	
3	
−6	

20. $y = -2x - 3$

x	y
0	
−3	
1	

For Exercises 21–24, graph the line.

21. $x + 2y = 4$

22. $x - y = 5$

23. $y = 3x - 2$

24. $y = \frac{1}{4}x$

25. At a certain gas station in the late 1990's, gasoline cost \$1.54 per gallon. If x gallons of gasoline was purchased, the total cost, y, (in dollars) is given by the equation

$$y = 1.54x \quad (x \geq 0)$$

a. Use the equation to find the cost of purchasing 12.4 gallons of gas.

b. Use the equation to complete the table.

Number of Gallons x	Cost (\$) y
5.0	
7.5	
10.0	
12.5	
15.0	

Table for Exercise 25

c. Write the table values from part (b) as ordered pairs and graph the ordered pairs on a rectangular coordinate system.

d. How many gallons did a person purchase if the cost was \$10.01?

e. How many gallons did a person purchase if the cost was \$26.18?

26. A phone company charges 6 cents per minute (\$0.06/min) of long distance time. If x represents the length of a call in minutes, then the cost, y, (in dollars) is given by the equation

$$y = 0.06x \quad (x \geq 0)$$

a. Use the equation to find the cost of 22 min of long distance.

b. Use the equation to complete the table.

Number of Minutes of Long Distance x	Cost (\$) y
5	
15	
25	
35	
45	

Table for Exercise 26

c. Write the table values from part (b) as ordered pairs and graph the ordered pairs on a rectangular coordinate system.

d. How many minutes did a person talk if the long distance charge is \$2.22?

e. How many minutes did a person talk if the long distance charge is \$4.68?

Section 3.3

For Exercises 27–30, identify the line as horizontal or vertical. Then graph the line.

27. $x = 4$

28. $x = -\frac{3}{2}$

29. $6y + 1 = 13$

30. $5y - 1 = 14$

For Exercises 31–38, find the x- and y-intercepts if they exist.

31. $-4x + 8y = 12$

32. $2x + y = 6$

33. $y = 8x$

34. $5x - y = 0$

35. $6y = -24$

36. $2y - 3 = 1$

37. $2x + 5 = 0$

38. $-3x + 1 = 0$

39. An automobile depreciates immediately after purchase. The equation $V = -5000n + 30{,}000$ determines the value of a certain automobile over the first 5 years of ownership, where V is the value of the automobile (in dollars) and n is the number of years of ownership. See figure.

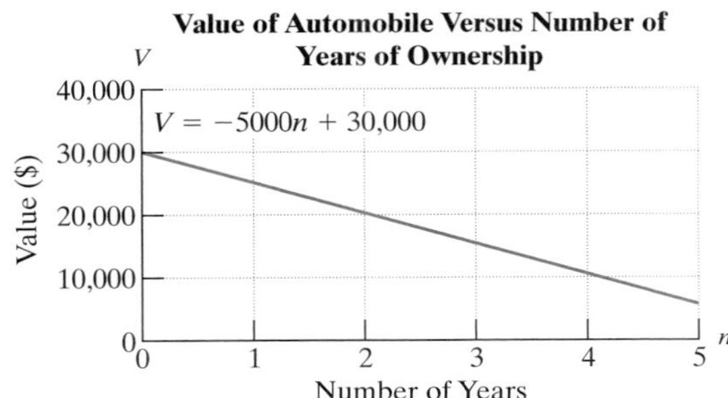

Figure for Exercise 39

a. Find the V-intercept for this equation.

b. Interpret the meaning of the V-intercept in terms of the value of the automobile and the number of years since its purchase.

c. Find the value of the automobile after 3 years.

Section 3.4

40. Draw a line with a positive slope.

41. Draw a line with a negative slope.

42. Draw a line with the slope of 0.

43. Draw a line with an undefined slope.

44. What is the slope of this ladder leaning up against a wall?

45. Point A is located 4 units down and 2 units to the right of point B. What is the slope of the line through points A and B?

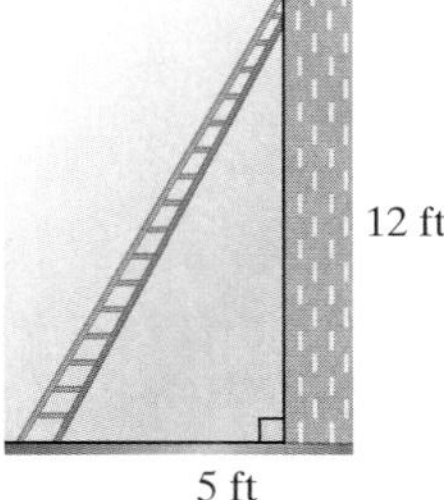

Figure for Exercise 44

46. Determine the slope of the line that passes through the points $(7, -9)$ and $(-5, -1)$.

47. Determine the slope of the line that has x- and y-intercepts of $(-1, 0)$ and $(0, 8)$.

48. Determine the slope of the line that passes through the points $(3, 0)$ and $(3, -7)$.

49. Determine the slope of the horizontal line $y = -1$.

50. A given line has a slope of -5.

a. What is the slope of a line parallel to the given line?

b. What is the slope of a line perpendicular to the given line?

51. A given line has a slope of 0.

a. What is the slope of a line parallel to the given line?

b. What is the slope of a line perpendicular to the given line?

For Exercises 52–55, find the slopes of the lines l_1 and l_2 from the two given points. Then determine whether l_1 and l_2 are parallel, perpendicular or neither.

52. l_1: $(3, 7)$ and $(0, 5)$
l_2: $(6, 3)$ and $(-3, -3)$

53. l_1: $(-2, 1)$ and $(-1, 9)$
l_2: $(0, -6)$ and $(2, 10)$

54. l_1: $(0, \frac{5}{6})$ and $(2, 0)$
l_2: $(0, \frac{6}{5})$ and $(-\frac{1}{2}, 0)$

55. l_1: $(1, 1)$ and $(1, -8)$
l_2: $(4, -5)$ and $(7, -5)$

Section 3.5

For Exercises 56–63, write each equation in slope-intercept form. Identify the slope and the y-intercept, and graph the line.

56. $5x - 2y = 10$

57. $3x + 4y = 12$

58. $2x + y = -3$

59. $3x - y = 4$

60. $x - 3y = 0$

61. $5y - 8 = 4$

62. $2y = -5$

63. $y - x = 0$

For Exercises 64–67, determine whether the lines l_1 and l_2 are parallel, perpendicular, or neither.

64. l_1: $y = \frac{3}{5}x + 3$
l_2: $y = \frac{5}{3}x + 1$

65. l_1: $2x - 5y = 10$
l_2: $5x + 2y = 20$

66. l_1: $x = 5$
l_2: $2y = 6$

67. l_1 $y = \frac{1}{3}x - 4$
l_2 $-x + 3y = 6$

68. Write an equation of the line whose slope is $-\frac{4}{3}$ and whose y-intercept is $(0, -1)$.

69. Write an equation of the line whose slope is 0 and whose y-intercept is $(0, 2)$.

70. Write an equation of the line passing through the origin and having a slope of -8.

chapter 3 TEST

1. In which quadrant is the given point found?
   a. $\left(-\frac{7}{2}, 4\right)$ b. $(4.6, -2)$ c. $(-37, -45)$
2. What is the y-coordinate for a point on the x-axis?
3. What is the x-coordinate for a point on the y-axis?
4. The following table depicts a boy's height versus his age. Let x represent the boy's age and y represent his height.

Age (years), x	Height (inches), y
5	46
7	50
9	55
11	60

Table for Exercise 4

   a. Write the data as ordered pairs and interpret the meaning of the first ordered pair.
   b. Graph the ordered pairs on a rectangular coordinate system.
   c. From the graph estimate the boy's height at age 10.
   d. The data appear to follow an upward trend up to the boy's teenage years. Do you think this trend will continue? Would it be reasonable to use these data to predict the boy's height at age 25?
5. Determine whether the ordered pair is a solution to the equation $2x - y = 6$
   a. $(0, 6)$ b. $(4, 2)$
   c. $(3, 0)$ d. $\left(\frac{9}{2}, 3\right)$
6. Given the equation $y = \frac{1}{4}x - 2$, complete the table. Plot the ordered pairs and graph the line through the points to represent the set of all solutions to the equation.

x	y
0	
4	
6	

7. If x represents an adult's age, then the person's maximum recommended heart rate, y, during exercise is approximated by the equation

$$y = 220 - x \quad (x \geq 18)$$

   a. Use the equation to find the maximum recommended heart rate for a person who is 18 years old.
   b. Use the equation to complete the following ordered pairs: (20,), (30,), (40,), (50,), (60,).

For Exercises 8–9, determine whether the equation represents a horizontal or vertical line. Then graph the line.

8. $-6y = 18$
9. $5x + 1 = 8$
10. Find the x-intercept and the y-intercept of the line $-4x + 3y = 6$.
11. What is the average slope of the hill?

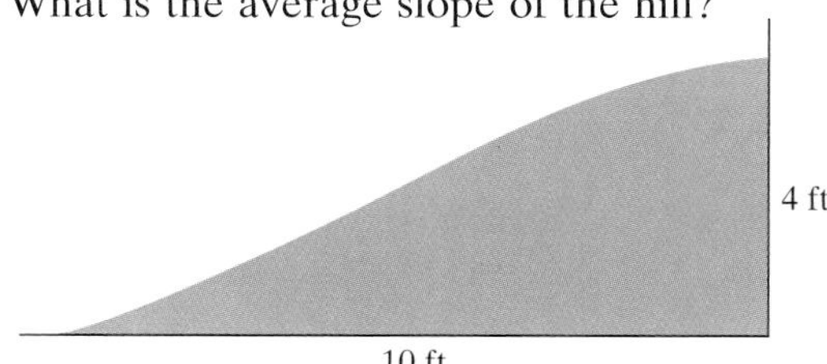

Figure for Exercise 11

12. a. Find the slope of the line that passes through the points $(-2, 0)$ and $(-5, -1)$.
    b. Find the slope of the line $4x - 3y = 9$.

13. a. What is the slope of a line parallel to the line $x + 4y = -16$?
    b. What is the slope of a line perpendicular to the line $x + 4y = -16$?

14. a. What is the slope of the line $x = 5$?
    b. What is the slope of the line $y = -3$?

For Exercises 15–18, find the x- and y-intercepts if they exist, and graph the lines.

15. $y = 8x + 2$
16. $2x + 9y = 0$
17. $x - 3 = 0$
18. $12y = -4$

19. Determine whether the lines l_1 and l_2 are parallel, perpendicular, or neither.

    l_1: $2y = 3x - 3$ $\quad$ l_2: $4x = -6y + 1$

20. Write an equation of the line that has y-intercept $(0, \frac{1}{2})$ and slope $\frac{1}{4}$. Write the equation in slope-intercept form.

For Exercises 21–22, find the slope and y-intercept of the line. Then use the slope and y-intercept to graph the line.

21. $2x + 3y = -9$
22. $5y - 4x = 0$

23. Girard's weekly salary is \$400 plus 15% commission on sales. His total salary y, (in dollars), can be represented by the equation $y = 0.15x + 400$, where x represents his sales in dollars.
    a. What is Girard's weekly salary if he sells \$1400 in merchandise?
    b. What was the amount in sales if Girard's salary for a certain week was \$730?
    c. What is the y-intercept of the graph of the equation $y = 0.15x + 400$? What does the y-intercept mean in the context of this problem?
    d. What is the slope of the graph of the equation $y = 0.15x + 400$? What does the slope mean in the context of this problem?

CUMULATIVE REVIEW EXERCISES FOR CHAPTERS 1–3

1. Identify the numbers as rational or irrational.

   a. -3 $\quad$ b. $\dfrac{5}{4}$ $\quad$ c. $\sqrt{10}$ $\quad$ d. 0

2. Write the opposite and the absolute value for each number.

   a. $-\dfrac{2}{3}$ $\quad$ b. 5.3

3. Simplify the expression using the order of operations: $32 \div 2 \cdot 4 + 5$

4. Add: $3 + (-8) + 2 + (-10)$

5. Subtract: $16 - 5 - (-7)$

For Exercises 6–7, translate the English phrase into an algebraic expression. Then evaluate the expression.

6. The quotient of $\dfrac{3}{4}$ and $-\dfrac{7}{8}$.

7. The product of -2.1 and -6.

8. Name the property that is illustrated by the following statement. $6 + (8 + 2) = (6 + 8) + 2$

For Exercises 9–12, solve the equation.

9. $6x - 10 = 14$

10. $3(m + 2) - 3 = 2m + 8$

11. $\dfrac{2}{3}y - \dfrac{1}{6} = y + \dfrac{4}{3}$

12. $1.7z + 2 = -2(0.3z + 1.3)$

13. The area of Texas is 267,277 mi^2. If this is 712 mi^2 less than 29 times the area of Maine, find the area of Maine.

14. For the formula $3a + b = c$, solve for a.

15. Plot the following points on a rectangular coordinate system.

    a. (3, 2) b. (−1, 2) c. (0, −4) d. (2, 0)

16. Estimate the coordinates of the points A, B, C, and D.

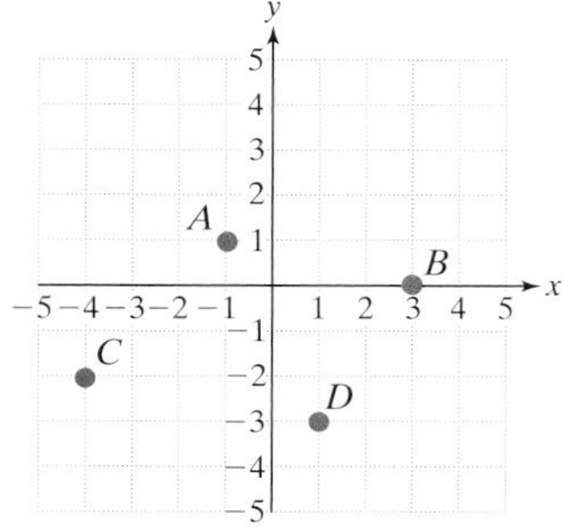

17. Find the x- and y-intercepts of $-2x + 4y = 4$.

18. Complete the ordered pairs for the equation $2y + x = 7$. $(1, \quad)$ $(\quad, -3)$

19. Write the equation in slope-intercept form. Then identify the slope and the y-intercept.
$3x + 2y = -12$

20. Write an equation of the line with slope $\frac{1}{2}$ and y-intercept $(0, -5)$.

21. Explain why the line $2x + 3 = 5$ has only one intercept.

22. Explain why the line $-6x + 2y = 0$ has only one intercept.

chapter

4

Polynomials and Properties of Exponents

The Pythagorean theorem states that the sum of the squares of the legs of a right triangle equals the square of the hypotenuse:

$$a^2 + b^2 = c^2$$

b (leg), c (hypotenuse), a (leg)

The applications of the Pythagorean theorem extend to many different fields such as construction, navigation, engineering, and physics.

For example, suppose two cables must be used to support an antenna tower at points 30 ft and 75 ft up the tower. Each cable is fastened at a point 40 ft from the base of the tower. The Pythagorean theorem can be used to show that the lengths of the cables are 50 ft and 85 ft.

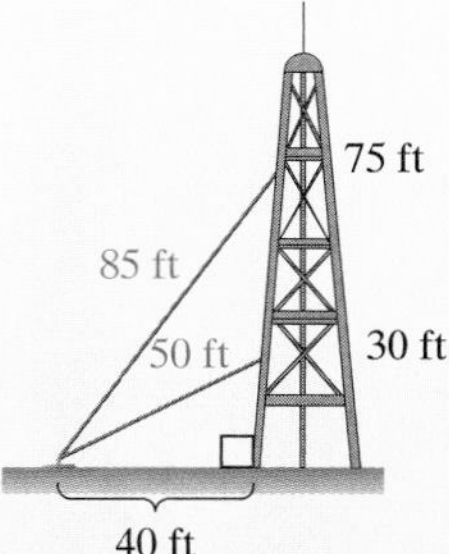

$$(40)^2 + (30)^2 = (50)^2 \qquad (40)^2 + (75)^2 = (85)^2$$
$$1600 + 900 = 2500 \qquad 1600 + 5625 = 7225$$
$$2500 = 2500 \checkmark \qquad 7225 = 7225 \checkmark$$

For more information see the Technology Connections in MathZone at

www.mhhe.com/miller_oneill

Concepts

1. Review of Exponential Notation
2. Evaluating Expressions with Exponents
3. Multiplying and Dividing Common Bases
4. Simplifying Expressions with Exponents
5. Applications of Exponents

section

4.1 Exponents: Multiplying and Dividing Common Bases

1. Review of Exponential Notation

Recall that an **exponent** is used to show repeated multiplication of the **base**.

Definition of b^n

Let b represent any real number and n represent a positive integer. Then,

$$b^n = \underbrace{b \cdot b \cdot b \cdot b \cdot \cdots \cdot b}_{n \text{ factors of } b}$$

example 1 **Evaluating Expressions with Exponents**

For each expression, identify the exponent and base. Then evaluate the expression.

a. 6^2 b. $\left(-\frac{1}{2}\right)^3$ c. 0.8^4

Solution:

Expression	Base	Exponent	Result
a. 6^2	6	2	$(6)(6) = 36$
b. $\left(-\frac{1}{2}\right)^3$	$-\frac{1}{2}$	3	$\left(-\frac{1}{2}\right)\left(-\frac{1}{2}\right)\left(-\frac{1}{2}\right) = -\frac{1}{8}$
c. 0.8^4	0.8	4	$(0.8)(0.8)(0.8)(0.8) = 0.4096$

Note that if no exponent is explicitly written for an expression, then the expression has an implied exponent of 1. For example,

$$x = x^1$$
$$y = y^1$$
$$5 = 5^1$$

2. Evaluating Expressions with Exponents

Particular care must be taken when evaluating exponential expressions involving negative numbers. An exponential expression with a negative base is written with parentheses around the base, such as $(-3)^2$.

To evaluate $(-3)^2$, we have: $(-3)^2 = (-3)(-3) = 9$

If no parentheses are present, the expression -3^2, is the *opposite* of 3^2, or equivalently, $-1 \cdot 3^2$.

Hence: $-3^2 = -1(3^2) = -1(3)(3) = -9$

example 2

Evaluating Expressions with Exponents

Evaluate each expression.

a. -5^4 b. $(-5)^4$ c. $(-0.2)^3$ d. -0.2^3

Solution:

a. -5^4

$= -1 \cdot 5^4$ 5 is the base with exponent 4.

$= -1 \cdot 5 \cdot 5 \cdot 5 \cdot 5$ Multiply -1 times 4 factors of 5.

$= -625$

b. $(-5)^4$ Parentheses indicate that -5 is the base with exponent 4.

$= (-5)(-5)(-5)(-5)$ Multiply 4 factors of -5.

$= 625$

c. $(-0.2)^3$ Parentheses indicate that -0.2 is the base with exponent 3.

$= (-0.2)(-0.2)(-0.2)$ Multiply 3 factors of -0.2.

$= -0.008$

d. -0.2^3

$= -1 \cdot 0.2^3$ 0.2 is the base with exponent 3.

$= -1 \cdot 0.2 \cdot 0.2 \cdot 0.2$ Multiply -1 times 3 factors of 0.2.

$= -0.008$

example 3

Evaluating Expressions with Exponents

Evaluate each expression for $a = 2$ and $b = -3$

a. $5a^2$ b. $(5a)^2$ c. $5ab^2$ d. $(b + a)^2$

Solution:

a. $5a^2$

$= 5(\ )^2$ Use parentheses to substitute a number for a variable.

$= 5(2)^2$ Substitute $a = 2$.

$= 5(4)$ Simplify.

$= 20$

Tip: In the expression $5ab^2$, the exponent, 2, applies only to the variable b. The constant, 5, and the variable, a, both have an implied exponent of 1.

Avoiding Mistakes

Be sure to follow the order of operations. In Example 3(d), it would not be correct to square the terms within the parentheses before adding.

b. $(5a)^2$

$= (5(\ ))^2$ Use parentheses to substitute a number for a variable.

$= (5(2))^2$ Substitute $a = 2$.

$= (10)^2$ Simplify inside the parentheses first.

$= 100$

c. $5ab^2$

$= 5(2)(-3)^2$ Substitute $a = 2, b = -3$.

$= 5(2)(9)$ Simplify.

$= 90$ Multiply.

d. $(b + a)^2$

$= ((-3) + (2))^2$ Substitute $b = -3$ and $a = 2$.

$= (-1)^2$ Simplify within the parentheses first.

$= 1$

3. Multiplying and Dividing Common Bases

In this section, we investigate the effect of multiplying or dividing two quantities with the same base. For example, consider the expressions: x^5x^2 and $\frac{x^5}{x^2}$. Simplifying each expression, we have:

$$x^5x^2 = (x \cdot x \cdot x \cdot x \cdot x)(x \cdot x) = \overbrace{x \cdot x \cdot x \cdot x \cdot x \cdot x \cdot x}^{\text{7 factors of } x} = x^7$$

$$\frac{x^5}{x^2} = \frac{x \cdot x \cdot x \cdot \cancel{x} \cdot \cancel{x}}{\cancel{x} \cdot \cancel{x}} = \frac{x \cdot x \cdot x}{1} = x^3$$

These examples suggest that to multiply two quantities with the same base, we add the exponents. To divide two quantities with the same base, we subtract the exponent in the denominator from the exponent in the numerator. These rules are stated formally as Properties 1 and 2 of exponents.

Multiplication of Like Bases

Assume that $b \neq 0$ is a real number and that m and n represent positive integers. Then,

$$\text{Property 1:} \quad b^m b^n = b^{m+n}$$

Division of Like Bases

Assume that $b \neq 0$ is a real number and that m and n represent positive integers such that $m > n$. Then,

$$\text{Property 2:} \quad \frac{b^m}{b^n} = b^{m-n}$$

4. Simplifying Expressions with Exponents

example 4 Simplifying Expressions with Exponents

Simplify the expressions.

a. w^3w^4 b. 2^32^4 c. $\dfrac{t^6}{t^4}$ d. $\dfrac{5^6}{5^4}$ e. $\dfrac{z^4z^5}{z^3}$ f. $\dfrac{10^7}{10^2 \cdot 10}$

Solution:

a. w^3w^4 $(w \cdot w \cdot w)(w \cdot w \cdot w \cdot w)$

$= w^{3+4}$ Add the exponents (the base is unchanged).

$= w^7$

Avoiding Mistakes

When we use Property 1 to add exponents, the base does not change. In Example 4(b), we have $2^32^4 = 2^7$.

b. 2^32^4 $(2 \cdot 2 \cdot 2)(2 \cdot 2 \cdot 2 \cdot 2)$

$= 2^{3+4}$ Add the exponents (the base is unchanged).

$= 2^7$ or 128

c. $\dfrac{t^6}{t^4}$ $\dfrac{\cancel{t} \cdot \cancel{t} \cdot \cancel{t} \cdot \cancel{t} \cdot t \cdot t}{\cancel{t} \cdot \cancel{t} \cdot \cancel{t} \cdot \cancel{t}}$

$= t^{6-4}$ Subtract the exponents (the base is unchanged).

$= t^2$

d. $\dfrac{5^6}{5^4}$ $\dfrac{\cancel{5} \cdot \cancel{5} \cdot \cancel{5} \cdot \cancel{5} \cdot 5 \cdot 5}{\cancel{5} \cdot \cancel{5} \cdot \cancel{5} \cdot \cancel{5}}$

$= 5^{6-4}$ Subtract the exponents (the base is unchanged).

$= 5^2$ or 25

e. $\dfrac{z^4z^5}{z^3}$

$= \dfrac{z^{4+5}}{z^3}$ Add the exponents in the numerator (the base is unchanged).

$= \dfrac{z^9}{z^3}$

$= z^{9-3}$ Subtract the exponents (the base is unchanged).

$= z^6$

f. $\dfrac{10^7}{10^2 \cdot 10}$

$= \dfrac{10^7}{10^2 \cdot 10^1}$ Note that 10 is equivalent to 10^1.

$= \dfrac{10^7}{10^{2+1}}$ Add the exponents in the denominator (the base is unchanged).

$= \dfrac{10^7}{10^3}$

$= 10^{7-3}$ Subtract the exponents (the base is unchanged).

$= 10^4$ or 10,000 Simplify.

example 5

Simplifying Expressions with Exponents

Use the commutative and associative properties of real numbers and the properties of exponents to simplify the expressions.

a. $(3p^2q^4)(2pq^5)$

b. $\dfrac{16w^9z^3}{3w^8z}$

Solution:

a. $(3p^2q^4)(2pq^5)$

$= (3 \cdot 2)(p^2p)(q^4q^5)$ Apply the associative and commutative properties of multiplication to group coefficients and like bases.

$= (3 \cdot 2)p^{2+1}q^{4+5}$ Add the exponents when multiplying like bases.

$= 6p^3q^9$ Simplify.

b. $\dfrac{16w^9z^3}{3w^8z}$

$= \left(\dfrac{16}{3}\right)\left(\dfrac{w^9}{w^8}\right)\left(\dfrac{z^3}{z}\right)$ Group like coefficients and factors.

$= \left(\dfrac{16}{3}\right)w^{9-8}z^{3-1}$ Subtract the exponents when dividing like bases.

$= \left(\dfrac{16}{3}\right)wz^2$ or $\dfrac{16wz^2}{3}$ Simplify.

5. Applications of Exponents

Recall that simple interest on an investment or loan is computed by the formula $I = Prt$, where P is the amount of principal, r is the interest rate, and t is the time in years. Simple interest is based only on the original principal. However, in most day-to-day applications, the interest computed on money invested or borrowed is compound interest. **Compound interest** is computed on the original principal and on the interest already accrued.

Suppose $1000 is invested at 8% interest for 3 years. Compare the total amount in the account if the money earns simple interest versus if the interest is compounded annually.

Simple Interest

The simple interest earned is given by $I = Prt$

$$= (1000)(0.08)(3)$$
$$= \$240$$

Thus, the total amount in the account after 3 years is \$1240.

Compound Interest (annual)

Table 4-1 outlines the process to compute interest compounded annually over 3 years. The table shows that the interest earned the first year is based on the original principal only. The interest earned the second year is based on the original principal and on the first year interest. The interest earned the third year is based on the original principal and on the interest earned the first and second years.

Table 4-1

Year	Interest Earned $I = Prt$	Total Amount in the Account
First year	$I = (\$1000)(0.08)(1) = \$80$	\$1000 + \$80 = \$1080
Second year	$I = (\$1080)(0.08)(1) = \$86.40$	\$1080 + \$86.40 = \$1166.40
Third year	$I = (\$1166.40)(0.08)(1) = \$93.31$	\$1166.40 + \$93.31 = \$1259.71

The difference in the account balance for interest compounded annually versus for simple interest is \$1259.71 − \$1240 = \$19.71.

The total amount, A, in an account earning annual compound interest may be computed quickly using the following formula:

$A = P(1 + r)^t$ where P is the amount of principal, r is the annual interest rate (expressed in decimal form), and t is the number of years.

For example, for \$1000 invested at 8% interest compounded annually for 3 years, we have

$$P = 1000$$
$$r = 0.08$$
$$t = 3$$
$$A = P(1 + r)^t$$
$$\begin{aligned} A &= 1000(1 + 0.08)^3 \\ &= 1000(1.08)^3 \\ &= 1000(1.259712) \\ &= 1259.712 \end{aligned}$$

Rounding to the nearest cent, we have $A = \$1259.71$, as expected.

example 6

Using Exponents in an Application

Find the amount in an account after 8 years if the initial investment is \$7000, invested at 2.25% interest compounded annually.

Solution:

Identify the values for each variable.

$P = 7000$

$r = 0.0225$ Note that the decimal form of a percent is used for calculations.

$t = 8$

$A = P(1 + r)^t$

$= 7000(1 + 0.0225)^8$ Substitute.

$= 7000(1.0225)^8$ Simplify inside the parentheses.

$= 7000(1.194831142)$ Approximate $(1.0225)^8$.

$= 8363.82$ Multiply (round to the nearest cent).

The amount in the account after 8 years is \$8363.82.

Calculator Connections

In Example 6, it was necessary to evaluate the expression $(1.0225)^8$. Recall that the [^] or [y^x] key may be used to enter expressions with exponents.

Scientific Calculator

Enter:  **Result:** 1.194831142

Graphing Calculator

```
1.0225^8
         1.194831142
```

section 4.1 PRACTICE EXERCISES

For this exercise set, assume all variables represent nonzero real numbers.

For Exercises 1–8, identify the base and the exponent.

1. r^4
2. c^3
3. 5^2
4. 3^5
5. $(-4)^8$
6. $(-1)^4$
7. x
8. q
9. What base corresponds to the exponent 5 in the expression $x^3y^5z^2$?
10. What base corresponds to the exponent 2 in the expression w^3v^2?
11. What base corresponds to the exponent 6 in the expression $4x^6$?
12. What base corresponds to the exponent 3 in the expression $2y^3$?
13. Evaluate the two expressions and compare the answers. Do the expressions have the same value? Why or why not? -5^2 and $(-5)^2$
14. Evaluate the two expressions and compare the answers. Do the expressions have the same value? Why or why not? -3^4 and $(-3)^4$
15. Evaluate the two expressions and compare the answers. Do the expressions have the same value? Why or why not? -2^5 and $(-2)^5$
16. Evaluate the two expressions and compare the answers. Do the expressions have the same value? Why or why not? -5^3 and $(-5)^3$
17. Evaluate the two expressions and compare the answers: $(\frac{1}{2})^3$ and $\frac{1}{2^3}$.
18. Evaluate the two expressions and compare the answers: $(\frac{1}{5})^2$ and $\frac{1}{5^2}$.
19. Evaluate the two expressions and compare the answers: $(\frac{3}{10})^2$ and $(0.3)^2$.
20. Evaluate the two expressions and compare the answers: $(\frac{7}{10})^3$ and $(0.7)^3$.
21. Expand the following expressions first. Then simplify using exponents.
    a. $x^4 \cdot x^3$ b. $5^4 \cdot 5^3$
22. Expand the following expressions first. Then simplify using exponents.
    a. $y^2 \cdot y^4$ b. $3^2 \cdot 3^4$

For Exercises 23–32, simplify the expressions. Write the answers in exponent form.

23. z^5z^3
24. w^4w^7
25. $a \cdot a^8$
26. p^4p
27. $4^5 \cdot 4^9$
28. $6^7 \cdot 6^5$
29. $9^4 \cdot 9$
30. $12 \cdot 12^6$
31. $c^5c^2c^7$
32. $b^7b^2b^8$
33. Expand the following expressions. Then reduce to lowest terms.
    a. $\frac{p^8}{p^3}$ b. $\frac{8^8}{8^3}$
34. Expand the following expressions. Then reduce to lowest terms.
    a. $\frac{w^5}{w^2}$ b. $\frac{4^5}{4^2}$

For Exercises 35–44, simplify the expressions. Write the answers in exponent form.

35. $\frac{x^8}{x^6}$
36. $\frac{z^5}{z^4}$
37. $\frac{a^{10}}{a}$
38. $\frac{b^{12}}{b}$
39. $\frac{7^{13}}{7^6}$
40. $\frac{2^6}{2^4}$
41. $\frac{5^8}{5}$
42. $\frac{3^5}{3}$
43. $\frac{y^{13}}{y^{12}}$
44. $\frac{w^7}{w^6}$

For Exercises 45–58, simplify the expressions. Write the answers in exponent form.

45. $\frac{h^3h^8}{h^7}$
46. $\frac{n^5n^4}{n^2}$

47. $\dfrac{x^9x}{x^5}$

48. $\dfrac{kk^3}{k^2}$

49. $\dfrac{7^2 \cdot 7^6}{7}$

50. $\dfrac{5^3 \cdot 5^8}{5}$

51. $\dfrac{x^{13}}{x^3x^4}$

52. $\dfrac{t^{10}}{t^5t^3}$

53. $\dfrac{10^{20}}{10^3 \cdot 10^8}$

54. $\dfrac{3^{15}}{3^2 \cdot 3^{10}}$

55. $\dfrac{6^8 \cdot 6^5}{6^2 \cdot 6}$

56. $\dfrac{2^{14} \cdot 2}{2^3 \cdot 2^6}$

57. $\dfrac{z^3z^{11}}{z^4z^6}$

58. $\dfrac{w^{12}w^2}{w^4w^5}$

For Exercises 59–70, use the commutative and associative properties of real numbers and the properties of exponents to simplify the expressions.

59. $(5a^2b)(8a^3b^4)$

60. $(10xy^3)(3x^4y)$

61. $(r^6s^4)(13r^2s)$

62. $(6p^2q^8)(7p^5q^3)$

63. $\left(\frac{2}{3}m^{13}n^8\right)(24m^7n^2)$

64. $\left(\frac{1}{4}c^6d^6\right)(28c^2d^7)$

65. $\dfrac{14c^4d^5}{7c^3d}$

66. $\dfrac{36h^5k^2}{9h^3k}$

67. $\dfrac{2x^3y^5}{8xy^3}$

68. $\dfrac{13w^8z^3}{26w^2z}$

69. $\dfrac{25h^3jk^5}{12h^2k}$

70. $\dfrac{15m^5np^{12}}{4mp^9}$

For Exercises 71–76, use the geometry formulas found in Section R.2. Use 3.14 for π where applicable.

71. Find the area of the pizza shown in the figure. Round to the nearest square inch.

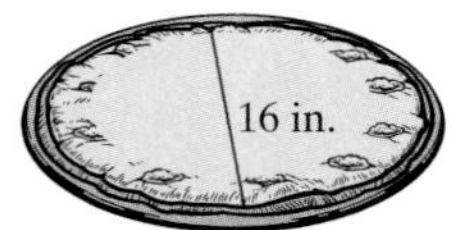

Figure for Exercise 71

72. Find the area of a circular pool 50 ft in diameter. Round to the nearest square foot.

73. Find the volume of the sphere shown in the figure. Round to the nearest cubic centimeter.

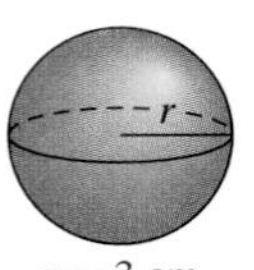

Figure for Exercise 73

74. Find the volume of a spherical balloon that is 10 in. in diameter. Round to the nearest cubic inch.

75. The employees at a craft shop make square napkins out of decorative holiday material. To make one napkin, a square piece of material 24 in. on a side is required. How many square inches of material are required to make 60 napkins? How many square feet is this if 1 ft^2 = 144 in.^{2}?

76. A construction company must pave a square parking lot. Find the area if the lot is 120 ft on a side. What is the area of the lot in square yards if 1 yd^2 = 9 ft^2?

Use the formula $A = P(1 + r)^t$ for Exercises 77–80.

77. Find the amount in an account after 2 years if the initial investment is \$5000, invested at 7% interest compounded annually.

78. Find the amount in an account after 5 years if the initial investment is \$2000, invested at 4% interest compounded annually.

79. Find the amount in an account after 3 years if the initial investment is \$4000, invested at 6% interest compounded annually.

80. Find the amount in an account after 4 years if the initial investment is \$10,000, invested at 5% interest compounded annually.

Expanding Your Skills

For Exercises 81–88, simplify the expressions using the addition or subtraction rules of exponents. Assume that a, b, m, and n represent positive integers.

81. $x^n x^{n+1}$
82. $y^a y^{2a}$
83. $p^{3m+5}p^{-m-2}$
84. $q^{4b-3}q^{-4b+4}$
85. $\dfrac{z^{b+1}}{z^b}$
86. $\dfrac{w^{5n+3}}{w^{2n}}$
87. $\dfrac{r^{3a+3}}{r^{3a}}$
88. $\dfrac{t^{3+2m}}{t^{2m}}$

Graphing Calculator Exercises

For Exercises 89–94, use a calculator to evaluate the expressions.

89. $(1.06)^5$
90. $(1.02)^{40}$
91. $5000(1.06)^5$
92. $2000(1.02)^{40}$
93. $3000(1 + 0.06)^2$
94. $1000(1 + 0.05)^3$

Concepts

1. Power Rule for Exponents
2. The Properties $(ab)^m = a^m b^m$ and $\left(\dfrac{a}{b}\right)^m = \dfrac{a^m}{b^m}$
3. Simplifying Expressions with Exponents

section

4.2 More Properties of Exponents

1. Power Rule for Exponents

The expression $(x^2)^3$ indicates that the quantity x^2 is cubed.

$$(x^2)^3 = (x^2)(x^2)(x^2) = (x \cdot x)(x \cdot x)(x \cdot x) = x^6$$

From this example, it appears that to raise a base to successive powers, we multiply the exponents and leave the base unchanged. This is stated formally as the **power rule for exponents**.

Power Rule for Exponents

Assume that $b \neq 0$ is a real number and that m and n represent positive integers. Then,

$$\text{Property 3:}\quad (b^m)^n = b^{m \cdot n}$$

example 1

Simplifying Expressions with Exponents

Simplify the expressions.

a. $(s^4)^2$ b. $(3^4)^2$ c. $(x^2x^5)^4$

Solution:

a. $(s^4)^2$

$= s^{4 \cdot 2}$ Multiply exponents (the base is unchanged).

$= s^8$

b. $(3^4)^2$

$= 3^{4 \cdot 2}$ Multiply exponents (the base is unchanged).

$= 3^8$ or 6561

c. $(x^2x^5)^4$

$= (x^7)^4$ Simplify inside the parentheses by adding exponents.

$= x^{7 \cdot 4}$ Multiply exponents (the base is unchanged).

$= x^{28}$

2. The Properties $(ab)^m = a^m b^m$ and $\left(\frac{a}{b}\right)^m = \frac{a^m}{b^m}$

Consider the following expressions and their simplified forms:

$$(xy)^3 = (xy)(xy)(xy) = (x \cdot x \cdot x)(y \cdot y \cdot y) = x^3y^3$$

$$\left(\frac{x}{y}\right)^3 = \left(\frac{x}{y}\right)\left(\frac{x}{y}\right)\left(\frac{x}{y}\right) = \left(\frac{x \cdot x \cdot x}{y \cdot y \cdot y}\right) = \frac{x^3}{y^3}$$

The expressions were simplified using the commutative and associative properties of multiplication. The simplified forms for each expression could have been reached in one step by applying the exponent to each factor inside the parentheses.

Power of a Product and Power of a Quotient

Assume that a and b are real numbers such that $a \neq 0, b \neq 0$. Let m represent a positive integer. Then,

Property 4: $(ab)^m = a^m b^m$

Property 5: $\left(\frac{a}{b}\right)^m = \frac{a^m}{b^m}$

Avoiding Mistakes

The power rule of exponents can be applied to a product of bases but in general cannot be applied to a sum or difference of bases.

$(ab)^n = a^n b^n$
Power rule can be applied.

$(a + b)^n \neq a^n + b^n$
Power rule *cannot* be applied.

Applying these properties of exponents, we have

$$(xy)^3 = x^3y^3 \quad \text{and} \quad \left(\frac{x}{y}\right)^3 = \frac{x^3}{y^3}$$

3. Simplifying Expressions with Exponents

example 2 **Simplifying Expressions with Exponents**

Simplify the expressions.

a. $(-2xyz)^4$ b. $(5x^2y^7)^3$ c. $\left(\frac{1}{3xy^4}\right)^2$

Solution:

a. $(-2xyz)^4$

$= (-2)^4x^4y^4z^4$ or $16x^4y^4z^4$ Raise each factor within parentheses to the fourth power.

b. $(5x^2y^7)^3$

$= 5^3(x^2)^3(y^7)^3$ Raise each factor within parentheses to the third power.

$= 125x^6y^{21}$ Multiply exponents and simplify.

c. $\left(\frac{1}{3xy^4}\right)^2$

$= \frac{1^2}{3^2x^2(y^4)^2}$ Square each factor within parentheses.

$= \frac{1}{9x^2y^8}$ Multiply exponents and simplify.

The properties of exponents can be used along with the properties of real numbers to simplify complicated expressions.

example 3 **Simplifying Expressions with Exponents**

Simplify the expressions.

a. $\frac{(x^2)^6(x^3)}{(x^7)^2}$ b. $(3cd^2)(2cd^3)^3$ c. $\left(\frac{x^7yz^4}{8xz^3}\right)^2$

Solution:

a. $\frac{(x^2)^6(x^3)}{(x^7)^2}$ Clear parentheses by applying the power rule.

$= \frac{x^{2\cdot 6}x^3}{x^{7\cdot 2}}$ Multiply exponents.

$$= \frac{x^{12}x^3}{x^{14}}$$

$$= \frac{x^{12+3}}{x^{14}}$$ Add exponents in the numerator.

$$= \frac{x^{15}}{x^{14}}$$

$= x^{15-14}$ Subtract exponents.

$= x$ Simplify.

b. $(3cd^2)(2cd^3)^3$ Clear parentheses by applying the power rule.

$= 3cd^2 \cdot 2^3c^3d^9$ Raise each factor in the second parentheses to the third power.

$= 3 \cdot 2^3cc^3d^2d^9$ Group like factors.

$= 3 \cdot 8c^{1+3}d^{2+9}$ Add exponents from like factors.

$= 24c^4d^{11}$ Simplify.

c. $\left(\frac{x^7yz^4}{8xz^3}\right)^2$

$$= \left(\frac{x^{7-1}yz^{4-3}}{8}\right)^2$$ Simplify inside the first parentheses by subtracting exponents from like factors.

$$= \left(\frac{x^6yz}{8}\right)^2$$

$$= \frac{(x^6)^2y^2z^2}{8^2}$$ Apply the power rule of exponents.

$$= \frac{x^{12}y^2z^2}{64}$$

section 4.2 Practice Exercises

For this exercise set assume all variables represent nonzero real numbers.

For Exercises 1–8, simplify.

1. $4^2 \cdot 4^7$
2. $5^8 \cdot 5^3 \cdot 5$
3. $a^{13} \cdot a \cdot a^6$
4. $y^{14}y^3$
5. $\frac{d^{13}d}{d^5}$
6. $\frac{3^8 \cdot 3}{3^2}$
7. $\frac{7^{11}}{7^5}$
8. $\frac{z^4}{z^3}$

9. Explain when to add exponents versus when to multiply exponents.

10. Explain when to add exponents versus when to subtract exponents.

For Exercises 11–22, simplify and write the answers in exponent form.

11. $(5^3)^4$ 12. $(2^8)^7$ 13. $(12^3)^2$

14. $(6^4)^4$ 15. $(y^7)^2$ 16. $(z^6)^4$

17. $(w^5)^5$ 18. $(t^3)^6$ 19. $(a^2a^4)^6$

20. $(z \cdot z^3)^2$ 21. $(y^3y^4)^2$ 22. $(w^5w)^4$

For Exercises 23–26, simplify the expressions and compare the properties used.

23. a. $\frac{x^7}{x^5}$ b. $x^7 \cdot x^5$ c. $(x^7)^5$

24. a. $\frac{w^4}{w^2}$ b. $w^4 \cdot w^2$ c. $(w^4)^2$

25. a. $\left(\frac{y^2}{z^3}\right)^4$ b. $(y^2z^3)^4$

26. a. $\left(\frac{c}{d^5}\right)^3$ b. $(cd^5)^3$

For Exercises 27–70, simplify the expression.

27. $\left(\frac{2}{3}\right)^3$ 28. $\left(\frac{1}{6}\right)^2$ 29. $\left(\frac{1}{4}\right)^2$

30. $\left(\frac{3}{4}\right)^3$ 31. $\left(\frac{x}{y}\right)^5$ 32. $\left(\frac{w}{z}\right)^7$

33. $\left(\frac{1}{t}\right)^4$ 34. $\left(\frac{2}{r}\right)^4$ 35. $(-3a)^4$

36. $(-2x)^5$ 37. $(-3abc)^3$ 38. $(-5xyz)^2$

39. $\frac{9u^5}{3u^2}$ 40. $\frac{45p^8}{9p^3}$

41. $(2xy^2)(3x^2y)$ 42. $(4a^3b^4)(2ab^2)$

43. $\frac{a^4b^7}{a^2b}$ 44. $\frac{s^6t^4}{st^3}$

45. $\frac{(4m)^5}{(4m)^4}$ 46. $\frac{(2q)^6}{(2q)^5}$

47. $\frac{(2a+b)^7}{(2a+b)^5}$ 48. $\frac{(u+v)^5}{(u+v)^3}$

49. $(6u^2v^4)^3$ 50. $(3a^5b^2)^6$

51. $5(x^2y)^4$ 52. $18(u^3v^4)^2$

53. $\left(\frac{4}{rs^4}\right)^5$ 54. $\left(\frac{2}{h^7k}\right)^3$

55. $\left(\frac{3p}{q^3}\right)^5$ 56. $\left(\frac{5x^2}{y^3}\right)^4$

57. $\frac{y^8(y^3)^4}{(y^2)^3}$ 58. $\frac{(w^3)^2(w^4)^5}{(w^4)^2}$

59. $(x^2)^5(x^3)^7$ 60. $(y^3)^4(y^2)^5$

61. $(a^2b)^3(a^4b^3)^5$ 62. $(c^3d^5)^2(cd^3)^3$

63. $\frac{(5a^3b)^4(a^2b)^4}{(5ab)^2}$ 64. $\frac{(6s^3)^2(s^4t^5)^2}{(3s^4t^2)^2}$

65. $\frac{(21x^5y)(2x^8y^4)}{14xy}$ 66. $\frac{(4u^3v^3)(9u^4v)}{12u^5v^2}$

67. $\left(\frac{2c^3d^4}{3c^2d}\right)^2$ 68. $\left(\frac{x^3y^5z}{5xy^2}\right)^2$

69. $(2c^3d^2)^5\left(\frac{c^6d^8}{4c^2d}\right)^3$ 70. $\left(\frac{s^5t^6}{2s^2t}\right)^2(10s^3t^3)^2$

Expanding Your Skills

For Exercises 71–78, simplify the expressions using the addition or subtraction properties of exponents. Assume that a, b, m, and n represent positive integers.

71. $(x^m)^2$ 72. $(y^3)^n$ 73. $(5a^{2n})^3$

74. $(3b^4)^m$ 75. $\left(\frac{m^2}{n^3}\right)^b$ 76. $\left(\frac{x^5}{y^3}\right)^m$

77. $\left(\frac{3a^3}{5b^4}\right)^n$ 78. $\left(\frac{4m^6}{3n^2}\right)^b$

79. Evaluate the two expressions and compare the answers: $(2^2)^3$ and $(2^3)^2$.

80. Evaluate the two expressions and compare the answers: $(4^4)^2$ and $(4^2)^4$.

81. Evaluate the two expressions and compare the answers. Which expression is greater? Why?

$$2^{(2^4)} \quad \text{and} \quad (2^2)^4$$

82. Evaluate the two expressions and compare the answers. Which expression is greater? Why?

$$3^{(2^4)} \quad \text{and} \quad (3^2)^4$$

---

section

4.3 Definitions of b^0 and b^{-n}

Concepts

1. **Definitions of b^0 and b^{-n}**
2. **Properties of Integer Exponents: A Summary**
3. **Simplifying Expressions with Exponents**

1. Definitions of b^0 and b^{-n}

In Sections 4.1 and 4.2, we learned several rules that enable us to manipulate expressions containing *positive* integer exponents. In this section, we present two definitions that can be used to simplify expressions with negative exponents or with an exponent of zero.

Let m and n be positive integers. Recall that Property 2 of exponents states that

$$\frac{b^m}{b^n} = b^{m-n}$$

provided that the exponent in the numerator is greater than the exponent in the denominator, $m > n$. If the condition $m > n$ is not imposed, then the difference of exponents $m - n$ may be zero or negative. Therefore, we want to define the expressions b^0 and b^{-n} so that the properties of exponents can be extended to include zero and negative exponents. For example, we know that

$$1 = \frac{5}{5} \qquad 1 = \frac{5^2}{5^2} \qquad 1 = \frac{5^3}{5^3}$$

and so on. If we subtract exponents in any of these expressions, the result is 5^0.

Subtract exponents.

$$1 = \frac{5^3}{5^3} = 5^{3-3} = 5^0. \qquad \text{Therefore, we will define } 5^0 \text{ as } 1.$$

The same logic may be applied to any nonzero base, b, so we define $b^0 = 1$.

Definition of b^0

Let b be a real number such that $b \neq 0$. Then, $b^0 = 1$.

Tip: The expression $b^0 = 1$ provided b is *not* zero. Therefore, the expression 0^0 cannot be simplified by this rule.

For $x \neq 0$ and $y \neq 0$, the following expressions all equal 1:

$$4^0 = 1 \qquad \left(-\frac{1}{2}\right)^0 = 1 \qquad x^0 = 1 \qquad (xy)^0 = 1$$

Next, for a positive integer n, we want to define the expression b^{-n} so that the properties of exponents can be extended to include negative exponents.

Consider the expression $\dfrac{x^4}{x^7} = \dfrac{\not{x} \cdot \not{x} \cdot \not{x} \cdot \not{x}}{\not{x} \cdot \not{x} \cdot \not{x} \cdot \not{x} \cdot x \cdot x \cdot x} = \dfrac{1}{x^3}$

Subtract exponents.

By subtracting exponents, we have $\dfrac{x^4}{x^7} = x^{4-7} = x^{-3}$

Hence, $x^{-3} = \dfrac{1}{x^3}$

This example illustrates the following rule.

Definition of b^{-n}

Let n be an integer and b be a real number such that $b \neq 0$. Then,

$$b^{-n} = \left(\frac{1}{b}\right)^n = \frac{1}{b^n}$$

Tip: Observe that $\dfrac{1}{x^{-2}} = \dfrac{1}{\frac{1}{x^2}} = 1 \div \dfrac{1}{x^2} = 1 \cdot \dfrac{x^2}{1} = x^2$. Therefore, $\dfrac{1}{x^{-2}} = x^2$. In general, $\dfrac{1}{b^{-n}} = b^n$ for $b \neq 0$.

This definition indicates that to evaluate b^{-n}, we must take the *reciprocal of the base* and change the sign of the exponent. For example,

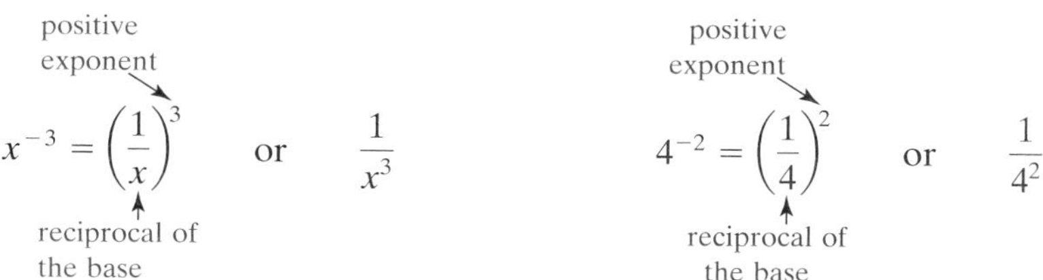

example 1 Simplifying Expressions with Negative and Zero Exponents

Simplify the expressions. Write the answers with positive exponents only. Assume all variables represent nonzero real numbers.

a. c^{-3} b. $\left(\dfrac{3}{4}\right)^{-2}$ c. $(50)^0 - 2^{-2}$ d. $\dfrac{p^4}{p^6}$

Solution:

a. c^{-3}

$= \left(\frac{1}{c}\right)^3$ Take the reciprocal of the base and make the exponent positive.

$= \frac{1}{c^3}$ Simplify.

b. $\left(\frac{3}{4}\right)^{-2}$

$= \left(\frac{4}{3}\right)^2$ Take the reciprocal of the base and make the exponent positive.

$= \frac{4^2}{3^2}$ Simplify.

$= \frac{16}{9}$

c. $(50)^0 - 2^{-2}$ Note that $(50)^0 = 1$.

$= 1 - \left(\frac{1}{2}\right)^2$ Take the reciprocal of the base, and make the exponent positive.

$= 1 - \frac{1}{4}$ Evaluate $\left(\frac{1}{2}\right)^2 = \frac{1}{4}$.

$= \frac{4}{4} - \frac{1}{4}$ Get a common denominator.

$= \frac{3}{4}$ Simplify.

d. $\frac{p^4}{p^6}$

$= p^{4-6}$ Subtract exponents.

$= p^{-2}$ Simplify.

$= \left(\frac{1}{p}\right)^2$ Take the reciprocal of the base and make the exponent positive.

$= \frac{1}{p^2}$

Tip: Example 1(d) can also be simplified by reducing common factors:

$$\frac{p^4}{p^6} = \frac{\overset{1}{\cancel{p}} \cdot \overset{1}{\cancel{p}} \cdot \overset{1}{\cancel{p}} \cdot \overset{1}{\cancel{p}}}{\cancel{p} \cdot \cancel{p} \cdot \cancel{p} \cdot \cancel{p} \cdot p \cdot p} = \frac{1}{p \cdot p} = \frac{1}{p^2}.$$

2. Properties of Integer Exponents: A Summary

Table 4-2 summarizes the properties and definitions of integer exponents.

Table 4-2

Properties of Integer Exponents

Assume that a and b are real numbers ($a \neq 0, b \neq 0$) and that m and n represent integers.

Property	Example	Details/Notes
Multiplication of Like Bases 1. $b^m b^n = b^{m+n}$	$b^2 b^4 = b^{2+4} = b^6$	$b^2 b^4 = (b \cdot b)(b \cdot b \cdot b \cdot b) = b^6$
Division of Like Bases 2. $\dfrac{b^m}{b^n} = b^{m-n}$	$\dfrac{b^5}{b^2} = b^{5-2} = b^3$	$\dfrac{b^5}{b^2} = \dfrac{\not{b} \cdot \not{b} \cdot b \cdot b \cdot b}{\not{b} \cdot \not{b}} = b^3$
The Power Rule 3. $(b^m)^n = b^{m \cdot n}$	$(b^4)^2 = b^{4 \cdot 2} = b^8$	$(b^4)^2 = (b \cdot b \cdot b \cdot b)(b \cdot b \cdot b \cdot b) = b^8$
Power of a Product 4. $(ab)^m = a^m b^m$	$(ab)^3 = a^3 b^3$	$(ab)^3 = (ab)(ab)(ab)$ $= (a \cdot a \cdot a)(b \cdot b \cdot b) = a^3 b^3$
Power of a Quotient 5. $\left(\dfrac{a}{b}\right)^m = \dfrac{a^m}{b^m}$	$\left(\dfrac{a}{b}\right)^3 = \dfrac{a^3}{b^3}$	$\left(\dfrac{a}{b}\right)^3 = \left(\dfrac{a}{b}\right)\left(\dfrac{a}{b}\right)\left(\dfrac{a}{b}\right) = \dfrac{a \cdot a \cdot a}{b \cdot b \cdot b} = \dfrac{a^3}{b^3}$

Definitions

Assume that b is a real number ($b \neq 0$) and that n represents an integer.

Definition	Example	Details/Notes
$b^0 = 1$	$(4)^0 = 1$	Any nonzero quantity raised to the zero power equals 1.
$b^{-n} = \left(\dfrac{1}{b}\right)^n = \dfrac{1}{b^n}$	$b^{-5} = \left(\dfrac{1}{b}\right)^5 = \dfrac{1}{b^5}$	To simplify a negative exponent, take the reciprocal of the base and make the exponent positive.

3. Simplifying Expressions with Exponents

example 2 **Simplifying Expressions with Exponents**

Simplify the following expressions. Write the answers with positive exponents only. Assume all variables are nonzero.

a. $\dfrac{z^2}{w^{-4}w^4z^{-8}}$ b. $(-4ab^{-2})^{-3}$ c. $\left(\dfrac{2p^{-4}q^3}{5p^2q}\right)^{-1}$

Solution:

a. $\dfrac{z^2}{w^{-4}w^4z^{-8}}$

$= \dfrac{z^2}{w^{-4+4}z^{-8}}$ Add the exponents in the denominator.

$= \dfrac{z^2}{w^0z^{-8}}$

$= \dfrac{z^2}{(1)z^{-8}}$ Recall that $w^0 = 1$.

$= z^{2-(-8)}$ Subtract the exponents.

$= z^{10}$ Simplify.

b. $(-4ab^{-2})^{-3}$

$= (-4)^{-3}a^{-3}(b^{-2})^{-3}$ Apply the power rule of exponents.

$= (-4)^{-3}a^{-3}b^6$

$= \left(\dfrac{1}{-4}\right)^3\left(\dfrac{1}{a}\right)^3 b^6$ Simplify the negative exponents.

$= \dfrac{1}{-64} \cdot \dfrac{1}{a^3} \cdot b^6$ Simplify.

$= -\dfrac{b^6}{64a^3}$ Multiply fractions.

c. $\left(\dfrac{2p^{-4}q^3}{5p^2q}\right)^{-1}$ The negative exponent outside the parentheses can be eliminated by taking the reciprocal of the quantity within the parentheses.

$= \left(\dfrac{5p^2q}{2p^{-4}q^3}\right)^1$ Take the reciprocal of the base and make the exponent positive.

$= \dfrac{5p^2q}{2p^{-4}q^3}$

$= \dfrac{5p^{2-(-4)}q^{1-3}}{2}$ Subtract the exponents.

$= \dfrac{5p^6q^{-2}}{2}$ Simplify.

$= \dfrac{5p^6}{2} \cdot \dfrac{1}{q^2}$ Simplify the negative exponent.

$= \dfrac{5p^6}{2q^2}$ Simplify.

example 3 **Simplifying an Expression with Exponents**

Simplify the expression $2^{-1} + 3^{-1} + 5^0$. Write the answer with positive exponents only.

Solution:

$2^{-1} + 3^{-1} + 5^0$

$= \frac{1}{2} + \frac{1}{3} + 1$ Simplify negative exponents. Simplify $5^0 = 1$.

$= \frac{3}{6} + \frac{2}{6} + \frac{6}{6}$ Get a common denominator.

$= \frac{11}{6}$ Simplify.

section 4.3 Practice Exercises

For this set of exercises, assume all variables represent nonzero real numbers.

For Exercises 1–10, simplify the expressions.

1. b^3b^8
2. c^7c^2
3. $\frac{x^6}{x^2}$
4. $\frac{y^9}{y^8}$
5. $\frac{9^4 \cdot 9^8}{9}$
6. $\frac{3^{14}}{3^3 \cdot 3^5}$
7. $(6ab^3c^2)^5$
8. $(7w^7z^2)^4$
9. $\left(\frac{s^2t^5}{4}\right)^3$
10. $\left(\frac{5k^3}{h^7}\right)^2$
11. Simplify.

    a. 8^0 b. $\frac{8^4}{8^4}$
12. Simplify.

    a. d^0 b. $\frac{d^3}{d^3}$

For Exercises 13–26, simplify the expression.

13. p^0
14. k^0
15. 5^0
16. 2^0
17. -4^0
18. -1^0
19. $(-6)^0$
20. $(-2)^0$
21. $(8x)^0$
22. $(-3y^3)^0$
23. $-7x^0$
24. $6y^0$
25. ab^0
26. pq^0
27. Simplify and write the answers with positive exponents.

    a. t^{-5} b. $\frac{t^3}{t^8}$
28. Simplify and write the answers with positive exponents.

    a. 4^{-3} b. $\frac{4^2}{4^5}$
29. Explain what is wrong with the following logic.
    $\frac{x^4}{x^{-6}} = x^{4-6} = x^{-2}$
30. Explain what is wrong with the following logic.
    $\frac{y^5}{y^{-3}} = y^{5-3} = y^{-2}$
31. Explain what is wrong with the following logic.
    $2a^{-3} = \frac{1}{2a^3}$
32. Explain what is wrong with the following logic.
    $5b^{-2} = \frac{1}{5b^2}$

For Exercises 33–76, simplify the expression. Write the answer with positive exponents only.

33. $\left(\frac{2}{7}\right)^{-3}$

34. $\left(\frac{5}{4}\right)^{-1}$

35. $\left(-\frac{1}{5}\right)^{-2}$

36. $\left(-\frac{1}{3}\right)^{-3}$

37. a^{-3}

38. c^{-5}

39. 12^{-1}

40. 4^{-2}

41. $(4b)^{-2}$

42. $(3z)^{-1}$

43. $6x^{-2}$

44. $7y^{-1}$

45. $w^{-4}w^{-2}$

46. $z^{-3}z^{-1}$

47. $x^{-8}x^{4}$

48. $s^{5}s^{-6}$

49. $a^{-8}a^{8}$

50. $q^{3}q^{-3}$

51. $y^{17}y^{-13}$

52. $b^{20}b^{-14}$

53. $(m^{-6}n^{9})^{3}$

54. $(c^{4}d^{-5})^{-2}$

55. $(-3j^{-5}k^{6})^{4}$

56. $(6xy^{-11})^{-3}$

57. $\frac{p^3}{p^9}$

58. $\frac{q^2}{q^{10}}$

59. $\frac{r^{-5}}{r^{-2}}$

60. $\frac{s^{-4}}{s^3}$

61. $\frac{7^3}{7^2 \cdot 7^8}$

62. $\frac{3^4 \cdot 3}{3^7}$

63. $\frac{a^{-1}b^2}{a^3b^8}$

64. $\frac{k^{-4}h^{-1}}{k^6h}$

65. $\frac{w^{-8}(w^2)^{-5}}{w^3}$

66. $\frac{p^2p^{-7}}{(p^2)^3}$

67. $(-8y^{-12})(2y^{16}z^{-2})$

68. $(5p^{-2}q^{5})(-2p^{-4}q^{-1})$

69. $\frac{-18a^{10}b^6}{108a^{-2}b^6}$

70. $\frac{-35x^{-4}y^{-3}}{-21x^2y^{-3}}$

71. $\frac{(-4c^{12}d^7)^2}{(5c^{-3}d^{10})^{-1}}$

72. $\frac{(s^3t^{-2})^4}{(3s^{-4}t^6)^{-2}}$

73. $\left(\frac{2}{p^6p^3}\right)^{-3}$

74. $\left(\frac{5x}{x^7}\right)^{-2}$

75. $\left(\frac{5cd^{-3}}{10d^5}\right)^{-1}$

76. $\left(\frac{4m^{10}n^4}{2m^{12}n^{-2}}\right)^{-1}$

Mixed Review

Exercises 77–98 use all of the properties and definitions presented in this section and Sections 4.1 and 4.2. Simplify each expression and write the answers with positive exponents only. Assume that all variables represent nonzero real numbers.

77. $\left(\frac{1}{2}\right)^{-1} + \left(\frac{1}{3}\right)^{0}$

78. $\left(\frac{1}{4}\right)^{0} - \left(\frac{1}{5}\right)^{-1}$

79. $(2^5b^{-3})^{-3}$

80. $(3^{-2}y^3)^{-2}$

81. $\left(\frac{3x}{2y}\right)^{-4}$

82. $\left(\frac{6c}{5d^3}\right)^{-2}$

83. $(3ab^2)(a^2b)^3$

84. $(4x^2y^3)^3(xy^2)$

85. $\left(\frac{xy^2}{x^3y}\right)^4$

86. $\left(\frac{a^3b}{a^5b^3}\right)^5$

87. $\frac{(t^{-2})^3}{t^{-4}}$

88. $\frac{(p^3)^{-4}}{p^{-5}}$

89. $\left(\frac{2w^2x^3}{3y^0}\right)^3$

90. $\left(\frac{5a^0b^4}{4c^3}\right)^2$

91. $\frac{q^3r^{-2}}{s^{-1}t^5}$

92. $\frac{n^{-3}m^2}{p^{-3}q^{-1}}$

93. $\frac{(y^{-3})^2(y^5)}{(y^{-3})^{-4}}$

94. $\frac{(w^2)^{-4}(w^{-2})}{(w^5)^{-4}}$

95. $\left(\frac{-2a^2b^{-3}}{a^{-4}b^{-5}}\right)^{-3}$

96. $\left(\frac{-3x^{-4}y^3}{2x^5y^{-2}}\right)^{-2}$

97. $(5h^{-2}k^0)^3(5k^{-2})^{-4}$

98. $(6m^3n^{-5})^{-4}(6m^0n^{-2})^5$

Expanding Your Skills

For Exercises 99–104, simplify the expression.

99. $5^{-1} + 2^{-2}$

100. $4^{-2} + 8^{-1}$

101. $10^0 - 10^{-1}$

102. $3^0 - 3^{-2}$

103. $\frac{4^{-1} + 3^{-2}}{1 + 2^{-3}}$

104. $\frac{2^{-3} + 4^{-1}}{5^{-1} + 1}$

Concepts

1. **Introduction to Scientific Notation**
2. **Writing Numbers in Scientific Notation**
3. **Writing Numbers without Scientific Notation**
4. **Multiplying and Dividing Numbers in Scientific Notation**
5. **Applications of Scientific Notation**

section

4.4 Scientific Notation

1. Introduction to Scientific Notation

In many applications in mathematics, it is necessary to work with very large or very small numbers. For example, the number of movie tickets sold in the United States and Canada for a recent year is estimated to be 1,680,000,000. The weight of a flea is approximately 0.00066 lb. To avoid writing numerous zeros in very large or small numbers, scientific notation was devised as a shortcut. Scientific notation is useful when performing calculations and when comparing the relative sizes of very large or very small numbers.

The principle behind scientific notation is to use a power of 10 to express the magnitude of the number. Consider the following powers of 10:

$$10^0 = 1$$

$$10^1 = 10 \qquad 10^{-1} = \frac{1}{10^1} = \frac{1}{10} = 0.1$$

$$10^2 = 100 \qquad 10^{-2} = \frac{1}{10^2} = \frac{1}{100} = 0.01$$

$$10^3 = 1000 \qquad 10^{-3} = \frac{1}{10^3} = \frac{1}{1000} = 0.001$$

$$10^4 = 10{,}000 \qquad 10^{-4} = \frac{1}{10^4} = \frac{1}{10{,}000} = 0.0001$$

In the base-10 numbering system, each place value to the left and right of the decimal point represents a different power of 10 (Figure 4-1).

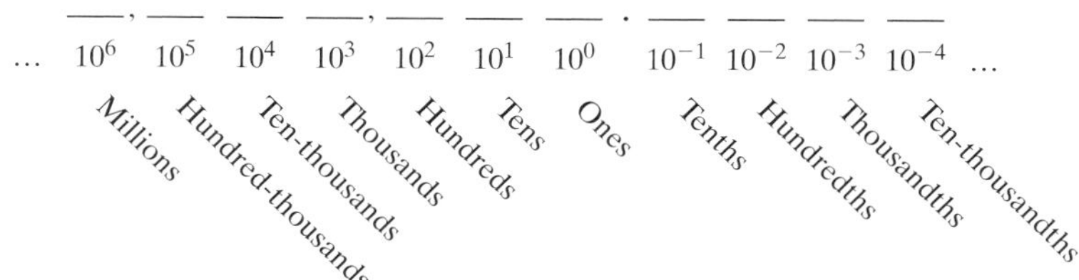

Figure 4-1

Therefore, a number such as 4000 can be written as 4.0×1000, or equivalently, 4.0×10^3. Similarly, the number 0.07 can be written as $7.0 \times \frac{1}{100}$, or equivalently, 7.0×10^{-2}.

Definition of a Number Written in Scientific Notation

A number expressed in the form: $a \times 10^n$, where $1 \leq |a| < 10$ and n is an integer is said to be written in **scientific notation**.

The numbers 4.0×10^3 and 7.0×10^{-2} are both expressed in scientific notation. To write a positive number in scientific notation, we apply the following guidelines:

1. Move the decimal point so that its new location is to the right of the first nonzero digit. The number should now be greater than or equal to 1 but less than 10. Count the number of places that the decimal point is moved.
2. If the original number is *large* (greater than or equal to 10), use the number of places the decimal point was moved as a *positive* power of 10.

$$\underbrace{450{,}000}_{5 \text{ places}} = 4.5 \times 100{,}000 = 4.5 \times 10^5$$

3. If the original number is *small* (between 0 and 1), use the number of places the decimal point was moved as a *negative* power of 10.

$$\underbrace{0.0002}_{4 \text{ places}} = 2.0 \times 0.0001 = 2.0 \times 10^{-4}$$

4. If the original number is greater than or equal to 1 but less than 10, use 0 as the power of 10.

$7.592 = 7.592 \times 10^0$ *Note*: A number between 1 and 10 is seldom written in scientific notation.

2. Writing Numbers in Scientific Notation

example 1 **Writing Numbers in Scientific Notation**

Write the numbers in scientific notation.

a. 53,000 b. 0.00053

Solution:

a. $53{,}000. = 5.3 \times 10^4$ To write 53,000 in scientific notation, the decimal point must be moved four places to the left. Because 53,000 is larger than 10, a *positive* power of 10 is used.

b. $0.00053 = 5.3 \times 10^{-4}$ To write 0.00053 in scientific notation, the decimal point must be moved four places to the right. Because 0.00053 is less than 1, a *negative* power of 10 is used.

example 2 **Writing Numbers in Scientific Notation**

Write the numerical values in scientific notation.

a. The number of movie tickets sold in the United States and Canada for a recent year is estimated to be 1,680,000,000.
b. The weight of a flea is approximately 0.00066 lb.
c. The temperature on a January day in Fargo dropped to −43°F.
d. A bench is 8.2 ft long.

Solution:

a. $1{,}680{,}000{,}000 = 1.68 \times 10^9$
b. $0.00066 \text{ lb} = 6.6 \times 10^{-4} \text{ lb}$
c. $-43°\text{F} = -4.3 \times 10^1 \text{ °F}$
d. $8.2 \text{ ft} = 8.2 \times 10^0 \text{ ft}$

Tip: For a number written in scientific notation, the power of 10 is sometimes called the **order of magnitude** (or simply the magnitude) of the number.

- The order of magnitude of the number of movie tickets is $\$10^9$ (billions of dollars).
- The mass of a flea is on the order of 10^{-4} lb (ten-thousandths of a pound).

3. Writing Numbers without Scientific Notation

example 3 **Writing Numbers without Scientific Notation**

Write the numerical values without scientific notation.

a. The mass of a proton is approximately 1.67×10^{-24} g.
b. The "nearby" star Vega is approximately 1.552×10^{14} miles from earth.

Solution:

a. $$\begin{aligned} 1.67 \times 10^{-24} \text{ g} &= (1.67 \times 0.000\,000\,000\,000\,000\,000\,000\,001) \text{ g} \\ &= 0.000\,000\,000\,000\,000\,000\,000\,001\,67 \text{ g} \end{aligned}$$

Because the power of 10 is negative, the value of 1.67×10^{-24} is a decimal number between 0 and 1. Move the decimal point 24 places to the *left.*

b. $$\begin{aligned} 1.552 \times 10^{14} \text{ miles} &= (1.552 \times 100{,}000{,}000{,}000{,}000) \text{ miles} \\ &= 155{,}200{,}000{,}000{,}000 \text{ miles} \end{aligned}$$

Because the power of 10 is a positive integer, the value of 1.552×10^{14} is a large number greater than 10. Move the decimal point 14 places to the *right.*

4. Multiplying and Dividing Numbers in Scientific Notation

To multiply or divide two numbers in scientific notation, use the commutative and associative properties of multiplication to group the powers of 10. For example,

$$400 \times 2000 = (4 \times 10^2)(2 \times 10^3) = (4 \cdot 2) \times (10^2 \cdot 10^3) = 8 \times 10^5$$

$$\frac{0.00054}{150} = \frac{5.4 \times 10^{-4}}{1.5 \times 10^2} = \left(\frac{5.4}{1.5}\right) \times \left(\frac{10^{-4}}{10^2}\right) = 3.6 \times 10^{-6}$$

example 4 **Multiplying and Dividing Numbers in Scientific Notation**

a. $(8.7 \times 10^4)(2.5 \times 10^{-12})$ b. $\dfrac{4.25 \times 10^{13}}{8.5 \times 10^{-2}}$

Solution:

a. $(8.7 \times 10^4)(2.5 \times 10^{-12})$

$= (8.7 \cdot 2.5) \times (10^4 \cdot 10^{-12})$	Commutative and associative properties of multiplication
$= 21.75 \times 10^{-8}$	This number is not in proper scientific notation because 21.75 is not between 1 and 10.
$= (2.175 \times 10^1) \times 10^{-8}$	Rewrite 21.75 as 2.175×10^1.
$= 2.175 \times (10^1 \times 10^{-8})$	Associative property of multiplication
$= 2.175 \times 10^{-7}$	Simplify.

b. $\dfrac{4.25 \times 10^{13}}{8.5 \times 10^{-2}}$

$= \left(\dfrac{4.25}{8.5}\right) \times \left(\dfrac{10^{13}}{10^{-2}}\right)$	Commutative and associative properties
$= 0.5 \times 10^{15}$	The number 0.5×10^{15} is not in proper scientific notation because 0.5 is not between 1 and 10.
$= (5.0 \times 10^{-1}) \times 10^{15}$	Rewrite 0.5 as 5.0×10^{-1}.
$= 5.0 \times (10^{-1} \times 10^{15})$	Associative property of multiplication
$= 5.0 \times 10^{14}$	Simplify.

5. Applications of Scientific Notation

example 5

Applying Scientific Notation

If a spacecraft travels at 1.6×10^4 mph, how long will it take the craft to travel to Mars if the distance is approximately 8.0×10^7 miles?

Solution:

Since $d = rt$, then $t = \dfrac{d}{r}$

$$t = \frac{8.0 \times 10^7 \text{ miles}}{1.6 \times 10^4 \text{ miles/hour}}$$

$$= \left(\frac{8.0}{1.6}\right) \times \left(\frac{10^7}{10^4}\right) \text{hr}$$

$$= 5.0 \times 10^3 \text{ hr}$$

The time required to travel to Mars is approximately 5.0×10^3 hr $= 5000$ hr, or 208 days.

Calculator Connections

Both scientific and graphing calculators can perform calculations involving numbers written in scientific notation. Most calculators use an [EE] key or an [EXP] key to enter the power of 10. Try using your calculator to evaluate:

a. 2.7×10^5 b. 7.1×10^{-3}

Scientific Calculator

Enter: [2] [.] [7] [EE] (or [EXP]) [5] [=] **Result**: 270000

Enter: [7] [.] [1] [EE] (or [EXP]) [3] [+/−] [=] **Result**: 0.0071

Graphing Calculator

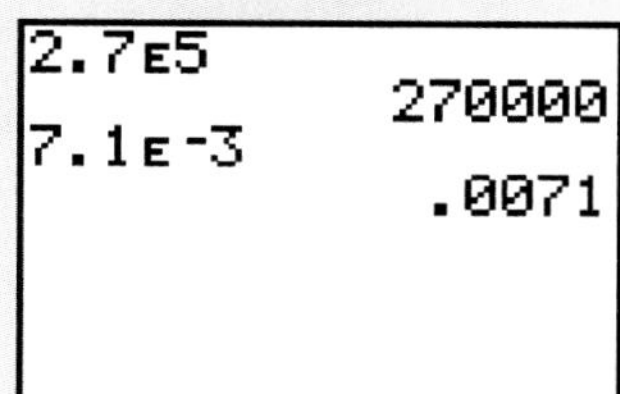

We recommend that you use parentheses to enclose each number written in scientific notation when performing calculations. Try using your calculator to perform the calculations from Example 4.

a. $(8.7 \times 10^4)(2.5 \times 10^{-12})$ b. $\dfrac{4.25 \times 10^{13}}{8.5 \times 10^{-2}}$

Scientific Calculator

Enter: [(] [8] [.] [7] [EE] [4] [)] [×] [(] [2] [.] [5] [EE] [1] [2] [+/−] [)] [=] **Result**: 0.000000218

Enter: [(] [4] [.] [2] [5] [EE] [1] [3] [)] [÷] [(] [8] [.] [5] [EE] [2] [+/−] [)] [=] **Result**: 5E14

Notice that the answers are shown on the calculator in scientific notation. The calculator does not have room to display enough zeros.

Graphing Calculator

```
(8.7E4)*(2.5E-12
)
          2.175E-7
(4.25E13)/(8.5E-
2)
              5E14
```

section 4.4 Practice Exercises

For Exercises 1–12, simplify the expressions. Assume all variables represent nonzero real numbers.

1. a^3a^{-4}
2. b^5b^8
3. $10^3 \cdot 10^{-4}$
4. $10^5 \cdot 10^8$
5. $\dfrac{x^3}{x^6}$
6. $\dfrac{y^2}{y^7}$
7. $\dfrac{10^3}{10^6}$
8. $\dfrac{10^2}{10^7}$
9. $\dfrac{z^9z^4}{z^3}$
10. $\dfrac{w^{-2}w^5}{w^{-1}}$
11. $\dfrac{10^9 \cdot 10^4}{10^3}$
12. $\dfrac{10^{-2} \cdot 10^5}{10^{-1}}$
13. Explain how you would write the number 0.000 000 000 23 in scientific notation.
14. Explain how you would write the number 23,000,000,000,000 in scientific notation.
15. Write the numerical values in scientific notation: In the world's largest tanker disaster, Amoco Cadiz spilled 68,000,000 gal of oil off Portsall, France, causing widespread environmental damage over 100 miles of Brittany coast.
16. Write the numerical values in scientific notation: The human heart pumps about 1400 L of blood per day. That would mean that it pumps approximately 10,000,000 L per year.

For Exercises 17–24, write the numbers in scientific notation.

17. 420,000,000
18. 750,000
19. 0.000008
20. 0.000125
21. The mass of a proton is approximately 0.000 000 000 000 000 000 000 0017 g.
22. The estimated wealth of Bill Gates in 1999 was $130,150,000,000.
23. The number of shares of Microsoft Corporation owned by Bill Gates in 1999 was 141,159,990.
24. One gram is equivalent to 0.0035 oz.
25. Explain how you would write the number 3.1×10^{-9} without scientific notation.
26. Explain how you would write the number 3.1×10^{9} without scientific notation.

For Exercises 27–34, write the numbers without scientific notation.

27. 5×10^{-5}
28. 2×10^{-7}
29. 2.8×10^{3}
30. 9.1×10^{6}
31. One picogram (pg) is equal to 1×10^{-12} g.
32. A nanometer (nm) is approximately 3.94×10^{-8} in.
33. A normal diet contains between 1.6×10^3 Cal and 2.8×10^3 Cal per day.

34. The total land area of Texas is approximately 2.62×10^5 square miles.

For Exercises 35–54, multiply or divide as indicated. Write the answers in scientific notation.

35. $(2.5 \times 10^6)(2.0 \times 10^{-2})$
36. $(2.0 \times 10^{-7})(3.0 \times 10^{13})$
37. $(1.2 \times 10^4)(3 \times 10^7)$
38. $(3.2 \times 10^{-3})(2.5 \times 10^8)$
39. $\dfrac{7.7 \times 10^6}{3.5 \times 10^2}$
40. $\dfrac{9.5 \times 10^{11}}{1.9 \times 10^3}$
41. $\dfrac{9.0 \times 10^{-6}}{4.0 \times 10^7}$
42. $\dfrac{7.0 \times 10^{-2}}{5.0 \times 10^9}$
43. $(8.0 \times 10^{10})(4.0 \times 10^3)$

44. $(6.0 \times 10^{-4})(3.0 \times 10^{-2})$

45. $(3.2 \times 10^{-4})(7.6 \times 10^{-7})$

46. $(5.9 \times 10^{12})(3.6 \times 10^{9})$

47. $\dfrac{2.1 \times 10^{11}}{7.0 \times 10^{-3}}$

48. $\dfrac{1.6 \times 10^{14}}{8.0 \times 10^{-5}}$

49. $\dfrac{5.7 \times 10^{-2}}{9.5 \times 10^{-8}}$

50. $\dfrac{2.72 \times 10^{-6}}{6.8 \times 10^{-4}}$

51. $6{,}000{,}000{,}000 \times 0.0000000023$

52. $0.000055 \times 40{,}000$

53. $\dfrac{0.0000000003}{6000}$

54. $\dfrac{420{,}000}{0.0000021}$

55. If a piece of paper is 3.0×10^{-3} in. thick, how thick is a stack of 1.25×10^{3} pieces of paper?

56. A box of staples contains 5.0×10^{3} staples and weighs 15 oz. How much does one staple weigh? Write your answer in scientific notation.

57. In the year 2000, \$$6.0 \times 10^{8}$ was spent on 350,000 30-second television commercials for campaign ads for political candidates in the United States. Based on these figures, determine the average cost of a 30-second television commercial. (*Source:* Television Bureau of Advertising)

58. A state lottery had a jackpot of \$$5.2 \times 10^{7}$. This week the winner was a group of office employees that included 13 people. How much would each person receive?

59. Dinosaurs became extinct about 65 million years ago.
    a. Write the number 65 million in scientific notation.
    b. How many days is 65 million years?
    c. How many hours is 65 million years?
    d. How many seconds is 65 million years?

60. The earth is 111,600,000 km from the sun.
    a. Write the number 111,600,000 in scientific notation.
    b. If there are 1000 m in a kilometer, how many meters is the earth from the sun?
    c. If there are 100 cm in a meter, how many centimeters is the earth from the sun?

61. Of all the workers in the United States who earn minimum wage, the pie chart shows a breakdown by educational level. (*Source:* U.S. Department of Labor)

Minimum-Wage Work Force by Educational Level

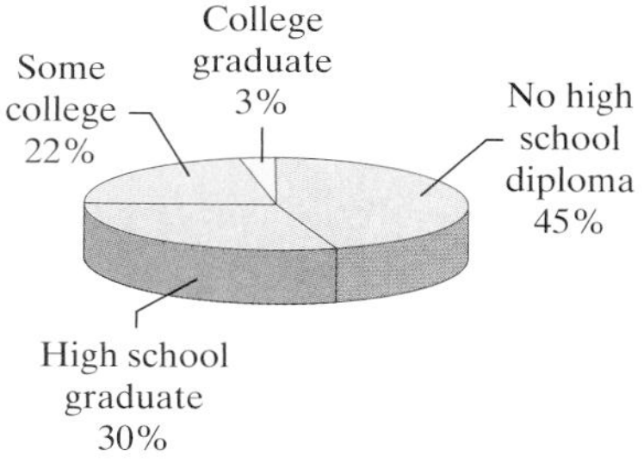

Figure for Exercise 61

    a. In a group of 1.5×10^{6} minimum-wage workers, how many would you expect to have no high school diploma?
    b. In a group of 1.5×10^{6} minimum-wage workers, how many would you expect to be college graduates?

62. In a recent year, the Hershey Foods Corporation reported \$$2.87 \times 10^{9}$ in annual revenue. This was a 10% loss from the figure reported 9 months earlier. What was the revenue 9 months earlier?

63. In a recent year, the IBM Corporation reported \$$6.337 \times 10^{10}$ in annual revenue. This was a 12% increase from the figure reported 9 months earlier. What was the revenue 9 months earlier?

Graphing Calculator Exercises

For Exercises 64–69, use a calculator to perform the indicated operations:

64. $(5.2 \times 10^{6})(4.6 \times 10^{-3})$

65. $(2.19 \times 10^{-8})(7.84 \times 10^{-4})$

66. $\dfrac{4.76 \times 10^{-5}}{2.38 \times 10^{9}}$

67. $\dfrac{8.5 \times 10^{4}}{4.0 \times 10^{-1}}$

68. $\dfrac{(9.6 \times 10^{7})(4.0 \times 10^{-3})}{2.0 \times 10^{-2}}$

69. $\dfrac{(5.0 \times 10^{-12})(6.4 \times 10^{-5})}{(1.6 \times 10^{-8})(4.0 \times 10^{2})}$

---

section

4.5 Addition and Subtraction of Polynomials

Concepts

1. **Introduction to Polynomials**
2. **Applications of Polynomials**
3. **Addition of Polynomials**
4. **Subtraction of Polynomials**
5. **Polynomials and Applications to Geometry**

1. Introduction to Polynomials

One commonly used algebraic expression is called a polynomial. A **polynomial** in one variable, x, is defined as a sum of terms of the form ax^n, where a is a real number and the exponent, n, is a nonnegative integer. For each term, a is called the **coefficient**, and n is called the **degree of the term**. For example:

Term (expressed in the form ax^n)	Coefficient	Degree
$-12z^7$	-12	7
$x^3 \rightarrow$ rewrite as $1x^3$	1	3
$10w \rightarrow$ rewrite as $10w^1$	10	1
$7 \rightarrow$ rewrite as $7x^0$	7	0

If a polynomial has exactly one term, it is categorized as a **monomial**. A two-term polynomial is called a **binomial**, and a three-term polynomial is called a **trinomial**. Usually the terms of a polynomial are written in descending order according to degree. The term with highest degree is called the **leading term**, and its coefficient is called the **leading coefficient**. The **degree of a polynomial** is the largest degree of all of its terms. Thus, the leading term determines the degree of the polynomial.

	Expression	Descending Order	Leading Coefficient	Degree of Polynomial
Monomials	$-3x^4$	$-3x^4$	-3	4
	17	17	17	0
Binomials	$4y^3 - 6y^5$	$-6y^5 + 4y^3$	-6	5
	$\frac{1}{2} - \frac{1}{4}c$	$-\frac{1}{4}c + \frac{1}{2}$	$-\frac{1}{4}$	1
Trinomials	$4p - 3p^3 + 8p^6$	$8p^6 - 3p^3 + 4p$	8	6
	$7a^4 - 1.2a^8 + 3a^3$	$-1.2a^8 + 7a^4 + 3a^3$	-1.2	8

example 1

Identifying the Parts of a Polynomial

Given: $4.5a - 2.7a^{10} + 1.6 - 3.7a^5$

a. List the terms of the polynomial, and state the coefficient and degree of each term.
b. Write the polynomial in descending order.
c. State the degree of the polynomial and the leading coefficient.

Solution:

a.

term:	coefficient:	degree:
$4.5a$	4.5	1
$-2.7a^{10}$	-2.7	10
1.6	1.6	0
$-3.7a^5$	-3.7	5

b. $-2.7a^{10} - 3.7a^5 + 4.5a + 1.6$

c. The degree of the polynomial is 10 and the leading coefficient is -2.7.

Polynomials may have more than one variable. In such a case, the degree of a term is the sum of the exponents of the variables contained in the term. For example, the term, $32x^2y^5z$, has degree 8 because the exponents applied to x, y, and z are 2, 5, and 1, respectively. The following polynomial has a degree of 11 because the highest degree of its terms is 11.

$$32x^2y^5z - 2x^3y + 2x^2yz^8 + 7$$

$32x^2y^5z$	$2x^3y$	$2x^2yz^8$	7
degree 8	degree 4	degree 11	degree 0

2. Applications of Polynomials

example 2

Using Polynomials in an Application

A child throws a ball upward and the height of the ball, h (in feet), can be computed by the following equation:

$h = -32t^2 + 64t + 2$ where t is the time (in seconds) after the ball is released.

a. Find the height of the ball after 0.5 sec, 1 sec, and 1.5 sec.
b. Find the height of the ball at the time of release.

Solution:

a. $h = -32t^2 + 64t + 2$

$= -32(0.5)^2 + 64(0.5) + 2$ Substitute $t = 0.5$.

$= -32(0.25) + 32 + 2$

$= -8 + 32 + 2$

$= 26$ The height of the ball after 0.5 sec is 26 ft.

$h = -32t^2 + 64t + 2$

$= -32(1)^2 + 64(1) + 2$ Substitute $t = 1$.

$= -32(1) + 64 + 2$

$= -32 + 64 + 2$

$= 34$ The height of the ball after 1 sec is 34 ft.

$h = -32t^2 + 64t + 2$

$= -32(1.5)^2 + 64(1.5) + 2$ Substitute $t = 1.5$.

$= -32(2.25) + 96 + 2$

$= -72 + 96 + 2$

$= 26$ The height of the ball after 1.5 sec is 26 ft.

b. $h = -32t^2 + 64t + 2$

$= -32(0)^2 + 64(0) + 2$ At the time of release, $t = 0$.

$= 0 + 0 + 2$

$= 2$ The height of the ball at the time of release is 2 ft.

3. Addition of Polynomials

Recall that two terms are said to be ***like* terms** if they each have the same variables, and the corresponding variables are raised to the same powers.

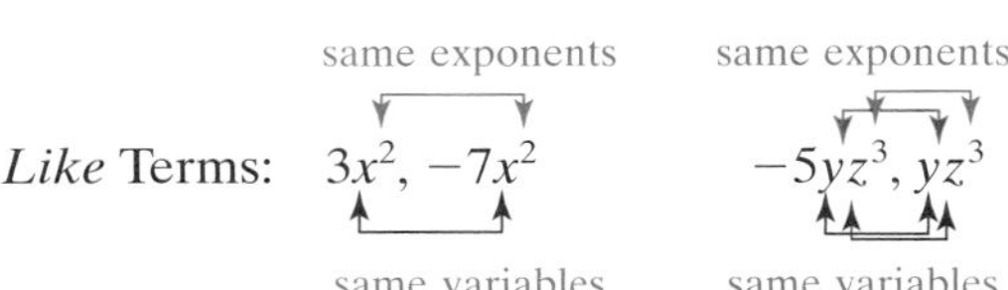

different exponents

Unlike Terms: $9z^2, 12z^6$ $\frac{1}{3}w^6, \frac{2}{5}p^6$ $4y, 7$

different variables different variables

Recall that the distributive property is used to add or subtract *like* terms. For example,

$3x^2 + 9x^2 - 2x^2$

$= (3 + 9 - 2)x^2$ Apply the distributive property.

$= (10)x^2$ Simplify.

$= 10x^2$

example 3 **Adding Polynomials**

Add the polynomials.

a. $3x^2y + 5x^2y$ b. $(-3c^3 + 5c^2 - 7c) + (11c^3 + 6c^2 + 3)$

c. $\left(\frac{1}{4}w^2 - \frac{2}{3}w\right) + \left(\frac{3}{4}w^2 + \frac{1}{6}w - \frac{1}{2}\right)$

Solution:

a. $3x^2y + 5x^2y$

$= (3 + 5)x^2y$ Apply the distributive property.

$= (8)x^2y$

$= 8x^2y$ Simplify.

Tip: Although the distributive property is used to combine *like* terms, the process is simplified by combining the coefficients of *like* terms.

b. $(-3c^3 + 5c^2 - 7c) + (11c^3 + 6c^2 + 3)$

$= -3c^3 + 11c^3 + 5c^2 + 6c^2 - 7c + 3$ Clear parentheses and group *like* terms.

$= 8c^3 + 11c^2 - 7c + 3$ Combine *like* terms.

Tip: Polynomials can also be added by combining *like* terms in columns. The sum of the polynomials from Example 3(b) is shown here.

$$\begin{array}{rrrrr} & -3c^3 + & 5c^2 & - 7c & + 0 \\ + & 11c^3 + & 6c^2 & + 0c & + 3 \\ \hline & 8c^3 + & 11c^2 & - 7c & + 3 \end{array}$$

Place holders such as 0 and $0c$ may be used to help line-up *like* terms.

Avoiding Mistakes

Example 3(c) is an expression, not an equation. Therefore, we cannot clear the fractions.

c. $\left(\frac{1}{4}w^2 - \frac{2}{3}w\right) + \left(\frac{3}{4}w^2 + \frac{1}{6}w - \frac{1}{2}\right)$

$= \frac{1}{4}w^2 + \frac{3}{4}w^2 - \frac{2}{3}w + \frac{1}{6}w - \frac{1}{2}$ Clear parentheses and group *like* terms.

$= \frac{1}{4}w^2 + \frac{3}{4}w^2 - \frac{4}{6}w + \frac{1}{6}w - \frac{1}{2}$ Get common denominators for *like* terms.

$= \frac{4}{4}w^2 - \frac{3}{6}w - \frac{1}{2}$ Add *like* terms.

$= w^2 - \frac{1}{2}w - \frac{1}{2}$ Simplify.

4. Subtraction of Polynomials

The opposite (or additive inverse) of a real number a is $-a$. Similarly, if A is a polynomial, then $-A$ is its opposite.

example 4

Finding the Opposite of a Polynomial

Find the opposite of the polynomials.

a. $5x$ b. $3a - 4b - c$ c. $5.5y^4 - 2.4y^3 + 1.1y - 3$

Solution:

Tip: Notice that the sign of each term is changed when finding the opposite of a polynomial.

a. The opposite of $5x$ is $-(5x)$, or $-5x$.
b. The opposite of $3a - 4b - c$ is $-(3a - 4b - c)$ or $-3a + 4b + c$.
c. The opposite of $5.5y^4 - 2.4y^3 + 1.1y - 3$ is $-(5.5y^4 - 2.4y^3 + 1.1y - 3)$, or equivalently, $-5.5y^4 + 2.4y^3 - 1.1y + 3$.

Subtraction of two polynomials is similar to subtracting real numbers. Add the opposite of the second polynomial to the first polynomial.

Definition of Subtraction of Polynomials

If A and B are polynomials, then $A - B = A + (-B)$.

example 5

Subtracting Polynomials

Subtract the polynomials.

a. $(-4p^4 + 5p^2 - 3) - (11p^2 + 4p - 6)$
b. $(a^2 - 2ab + 7b^2) - (-8a^2 - 6ab + 2b^2)$

Solution:

a. $(-4p^4 + 5p^2 - 3) - (11p^2 + 4p - 6)$

$= (-4p^4 + 5p^2 - 3) + (-11p^2 - 4p + 6)$ Add the opposite of the second polynomial.

$= -4p^4 + 5p^2 - 11p^2 - 4p - 3 + 6$ Group *like* terms.

$= -4p^4 - 6p^2 - 4p + 3$ Combine *like* terms.

Tip: Two polynomials can also be subtracted in columns by adding the opposite of the second polynomial to the first polynomial. Place holders (shown in red) may be used to help line up *like* terms.

$$\begin{array}{r} -4p^4 + 0p^3 + 5p^2 + 0p - 3 \\ -(0p^4 + 0p^3 + 11p^2 + 4p - 6) \end{array} \xrightarrow{\text{add the opposite}} \begin{array}{r} -4p^4 + 0p^3 + 5p^2 + 0p - 3 \\ + \underline{-0p^4 - 0p^3 - 11p^2 - 4p + 6} \\ -4p^4 \qquad - 6p^2 - 4p + 3 \end{array}$$

Hence the difference of the polynomials is $-4p^4 - 6p^2 - 4p + 3$.

b. $(a^2 - 2ab + 7b^2) - (-8a^2 - 6ab + 2b^2)$

$= (a^2 - 2ab + 7b^2) + (8a^2 + 6ab - 2b^2)$ Add the opposite of the second polynomial.

$= a^2 + 8a^2 - 2ab + 6ab + 7b^2 - 2b^2$ Group *like* terms.

$= 9a^2 + 4ab + 5b^2$ Combine *like* terms.

Tip: Recall that $a - b = a + (-b)$, or equivalently, $a + -1b$. Therefore, subtraction of polynomials can be simplified by applying the distributive property to clear parentheses.

$(a^2 - 2ab + 7b^2) - (-8a^2 - 6ab + 2b^2)$

$= 1(a^2 - 2ab + 7b^2) - 1(-8a^2 - 6ab + 2b^2)$

$= a^2 - 2ab + 7b^2 + 8a^2 + 6ab - 2b^2$ Apply the distributive property.

$= a^2 + 8a^2 - 2ab + 6ab + 7b^2 - 2b^2$ Group *like* terms.

$= 9a^2 + 4ab + 5b^2$ Combine *like* terms.

example 6 Subtracting Polynomials

Subtract: $\frac{1}{3}t^4 + \frac{1}{2}t^2$ from $t^2 - 4$ and simplify the result.

Solution:

To subtract a from b, we write $b - a$. Thus, to subtract $\overbrace{\frac{1}{3}t^4 + \frac{1}{2}t^2}^{a}$ from $\overbrace{t^2 - 4}^{b}$, we have

$$\overbrace{(t^2 - 4)}^{b} \overbrace{-}^{-} \overbrace{\left(\frac{1}{3}t^4 + \frac{1}{2}t^2\right)}^{a}$$

$= t^2 - 4 - \frac{1}{3}t^4 - \frac{1}{2}t^2$ Apply the distributive property.

$= -\frac{1}{3}t^4 + t^2 - \frac{1}{2}t^2 - 4$ Group *like* terms in descending order.

$= -\frac{1}{3}t^4 + \frac{2}{2}t^2 - \frac{1}{2}t^2 - 4$ The t^2-terms are the only *like* terms.

Get a common denominator for the t^2-terms.

$= -\frac{1}{3}t^4 + \frac{1}{2}t^2 - 4$ Add *like* terms.

5. Polynomials and Applications to Geometry

example 7 **Adding Polynomials in Geometry**

$2x^3 + 1$

x

$4x^3 - 2x^2$

$x^2 + 100$

Figure 4-2

Find a polynomial that represents the perimeter of the polygon in Figure 4-2.

Solution:

The perimeter of a polygon is the sum of the lengths of the sides.

$$P = (x) + (2x^3 + 1) + (4x^3 - 2x^2) + (x^2 + 100)$$

$= x + 2x^3 + 1 + 4x^3 - 2x^2 + x^2 + 100$ Clear parentheses.

$= 2x^3 + 4x^3 - 2x^2 + x^2 + x + 1 + 100$ Group *like* terms.

$= 6x^3 - x^2 + x + 101$ Combine *like* terms.

The polynomial $6x^3 - x^2 + x + 101$ represents the perimeter of the figure.

section 4.5 Practice Exercises

For Exercises 1–6, simplify the expressions. Assume all variables represent nonzero real numbers.

1. $\dfrac{p^3 \cdot 4p}{p^2}$
2. $(3x)(5x^{-4})$
3. $(6y^{-3})(2y^9)$
4. $\dfrac{8t^{-6}}{4t^{-2}}$
5. $\dfrac{8^3 \cdot 8^{-4}}{8^{-2} \cdot 8^6}$
6. $\dfrac{3^4 \cdot 3^{-8}}{3^{12} \cdot 3^{-4}}$
7. Explain the difference between 3.0×10^7 and 3^7.
8. Explain the difference between 4.0×10^{-2} and 4^{-2}.
9. Write the polynomial in descending order:
$$6 + 7x^2 - 7x^4 + 9x$$
10. Write the polynomial in descending order:
$$\frac{1}{2}y + y^2 - 12y^4 + y^3 - 6$$

For Exercises 11–22, categorize the expression as a monomial, a binomial, or a trinomial. Then identify the coefficient and degree of the leading term.

11. $10a^2 + 5a$
12. $7z + 13z^2 - 15$
13. $6x^2$
14. 9
15. $2t - t^4 - 5t$
16. $7x + 2$
17. $12y^4 - 3y + 1$
18. 23
19. $5bc^2$
20. $4 - 2c$
21. $w^4 - w^2$
22. $-32xyz$
23. Explain why the terms $3x$ and $3x^2$ are not *like* terms.
24. Explain why the terms $4w^3$ and $4z^3$ are not *like* terms.

For Exercises 25–40, add the polynomials.

25. $23x^2y + 12x^2y$
26. $-5ab^3 + 17ab^3$
27. $(6y + 3x) + (4y - 3x)$
28. $(2z - 5h) + (-3z + h)$
29. $3b^2 + (5b^2 - 9)$

30. $4c + (3 - 10c)$

31. $(7y^2 + 2y - 9) + (-3y^2 - y)$

32. $(-3w^2 + 4w - 6) + (5w^2 + 2)$

33. $(6a + 2b - 5c) + (-2a - 2b - 3c)$

34. $(-13x + 5y + 10z) + (-3x - 3y + 2z)$

35. $\left(\frac{2}{5}a + \frac{1}{4}b - \frac{5}{6}\right) + \left(\frac{3}{5}a - \frac{3}{4}b - \frac{7}{6}\right)$

36. $\left(\frac{5}{9}x + \frac{1}{10}y\right) + \left(-\frac{4}{9}x + \frac{3}{10}y\right)$

37. $\left(z - \frac{8}{3}\right) + \left(\frac{4}{3}z^2 - z + 1\right)$

38. $\left(-\frac{7}{5}r + 1\right) + \left(-\frac{3}{5}r^2 + \frac{7}{5}r + 1\right)$

39. $(7.9t^3 + 2.6t - 1.1) + (-3.4t^2 + 3.4t - 3.1)$

40. $(0.34y^2 + 1.23) + (3.42y - 7.56)$

41. Find a polynomial that represents the perimeter of the figure.

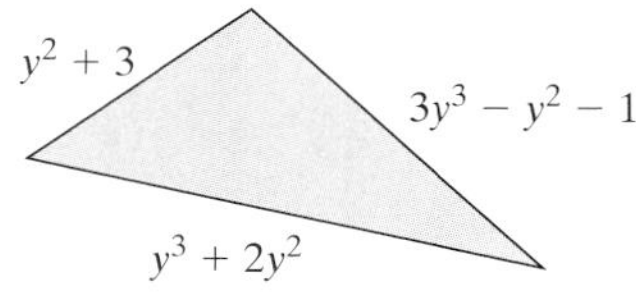

Figure for Exercise 41

42. Find a polynomial that represents the perimeter of the figure.

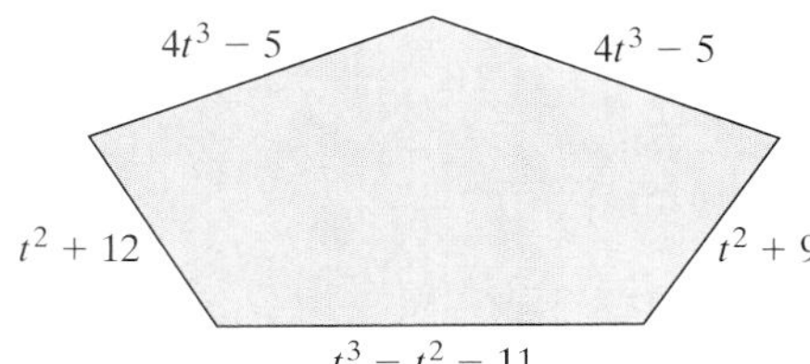

Figure for Exercise 42

43. A ball is dropped off a building and the height of the ball, h (in feet), can be computed by the equation:

$h = -16t^2 + 150$ where t is the time (in seconds) after the ball is released.

a. Find the height of the ball after 1 sec, 1.5 sec, and 2 sec.

b. Find the height of the building by determining the height at the time of release.

44. An object is dropped off a building and the height of the object, h (in meters), can be computed by the equation:

$h = -4.9t^2 + 45$ where t is the time (in seconds) after the object is released.

a. Find the height of the object after 1 sec, 1.5 sec, and 2 sec.

b. Find the height of the building by determining the height at the time of release.

For Exercises 45–52, find the opposite of each polynomial.

45. $4h - 5$

46. $5k - 12$

47. $-2m^2 + 3m - 15$

48. $-n^2 - 6n + 9$

49. $3v^3 + 5v^2 + 10v + 22$

50. $7u^4 + 3v^2 + 17$

51. $-9t^4 - 8t - 39$

52. $-5r^5 - 3r^3 - r - 23$

For Exercises 53–72, subtract the polynomials.

53. $4a^3b^2 - 12a^3b^2$

54. $5yz^4 - 14yz^4$

55. $-32x^3 - 21x^3$

56. $-23c^5 - 12c^5$

57. $(7a - 7) - (12a - 4)$

58. $(4x + 3v) - (-3x + v)$

59. $(4k + 3) - (-12k - 6)$

60. $(3h - 15) - (8h - 13)$

61. $25s - (23s - 14)$

62. $3x^2 - (-x^2 - 12)$

63. $(5t^2 - 3t - 2) - (2t^2 + t + 1)$

64. $(k^2 + 2k + 1) - (3k^2 - 6k + 2)$

65. $(10r - 6s + 2t) - (12r - 3s - t)$

66. $(a - 14b + 7c) - (-3a - 8b + 2c)$

67. $\left(\frac{7}{8}x + \frac{2}{3}y - \frac{3}{10}\right) - \left(\frac{1}{8}x + \frac{1}{3}y\right)$

68. $\left(r - \frac{1}{12}s\right) - \left(\frac{1}{2}r - \frac{5}{12}s - \frac{4}{11}\right)$

69. $\left(\frac{2}{3}h^2 - \frac{1}{5}h - \frac{3}{4}\right) - \left(\frac{4}{3}h^2 - \frac{4}{5}h + \frac{7}{4}\right)$

70. $\left(\frac{3}{8}p^3 - \frac{5}{7}p^2 - \frac{2}{5}\right) - \left(\frac{5}{8}p^3 - \frac{2}{7}p^2 + \frac{7}{5}\right)$

71. $(4.5x^4 - 3.1x^2 - 6.7) - (2.1x^4 + 4.4x)$

72. $(1.3c^3 + 4.8) - (4.3c^2 - 2c - 2.2)$

73. If the perimeter of the figure can be represented by the polynomial $5a^2 - 2a + 1$, find a polynomial that represents the length of the missing side.

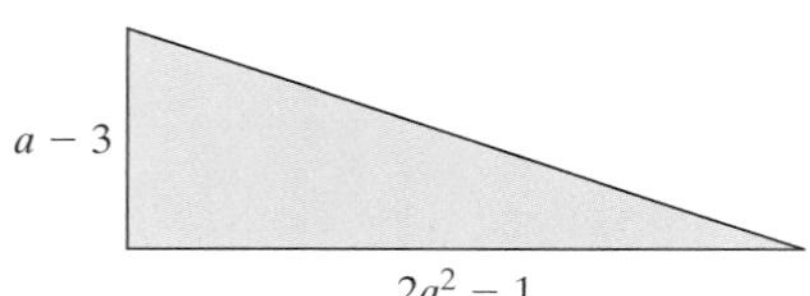

Figure for Exercise 73

74. If the perimeter of the figure can be represented by the polynomial $6w^3 - 2w - 3$, find a polynomial that represents the length of the missing side.

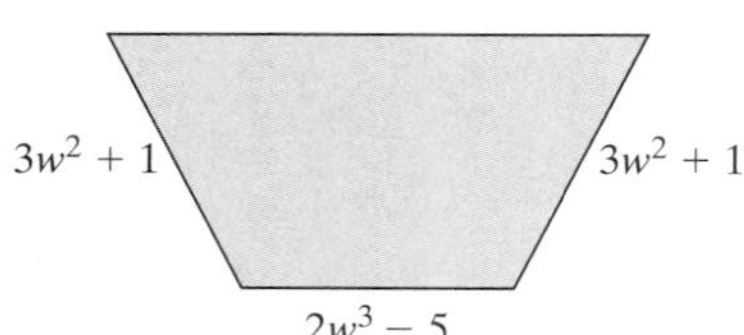

Figure for Exercise 74

75. Subtract $3x^3 - 5x + 10$ from $-2x^2 + 6x - 21$.

76. Subtract $7a^5 - 2a^3 - 5a$ from $3a^5 - 9a^2 + 3a - 8$.

77. Find the difference of $4b^3 + 6b - 7$ and $-12b^2 + 11b + 5$.

78. Find the difference of $-5y^2 + 3y - 21$ and $-4y^2 - 5y + 23$.

For Exercises 79–88, perform the indicated operation.

79. $(2ab^2 + 9a^2b) + (7ab^2 - 3ab + 7a^2b)$

80. $(8x^2y - 3xy - 6xy^2) + (3x^2y - 12xy)$

81. $(4z^5 + z^3 - 3z + 13) - (-z^4 - 8z^3 + 15)$

82. $(-15t^4 - 23t^2 + 16t) - (21t^3 + 18t^2 + t)$

83. $(9x^4 + 2x^3 - x + 5) + (9x^3 - 3x^2 + 8x + 3) - (7x^4 - x + 12)$

84. $(-6y^3 - 9y^2 + 23) - (7y^2 + 2y - 11) + (3y^3 - 25)$

85. $(5w^2 - 3w + 2) + (-4w + 6) - (7w^2 - 10)$

86. $(10u^3 - 5u^2 + 4) - (2u^3 + 5u^2 + u) - (u^3 - 3u + 9)$

87. $(7p^2q - 3pq^2) - (8p^2q + pq) + (4pq - pq^2)$

88. $(12c^2d - 2cd + 8cd^2) - (-c^2d + 4cd) - (5cd - 2cd^2)$

89. Write a binomial of degree 3. (Answers may vary.)

90. Write a trinomial of degree 6. (Answers may vary.)

91. Write a monomial of degree 5. (Answers may vary.)

92. Write a monomial of degree 1. (Answers may vary.)

93. Write a trinomial with leading coefficient −6. (Answers may vary.)

94. Write a binomial with leading coefficient 13. (Answers may vary.)

section

4.6 Multiplication of Polynomials

Concepts

1. Multiplication of Monomials
2. Multiplication of Polynomials
3. Special Case Products: Difference of Squares and Perfect Square Trinomials
4. Multiplication of Polynomials and Applications to Geometry

1. Multiplication of Monomials

The properties of exponents covered in Sections 4.1–4.3 can be used to simplify many algebraic expressions including the multiplication of monomials. To multiply monomials, first use the associative and commutative properties of multiplication to group coefficients and like bases. Then simplify the result by using the properties of exponents.

example 1

Multiplying Monomials

Multiply the monomials.

a. $(3x^4)(4x^2)$ b. $(-4c^5d)(2c^2d^3e)$ c. $\left(\frac{1}{3}a^4b^3\right)\left(\frac{3}{4}b^7\right)$

Solution:

a. $(3x^4)(4x^2)$

$= (3 \cdot 4)(x^4x^2)$ Group coefficients and like bases.

$= 12x^6$ Add the exponents and simplify.

b. $(-4c^5d)(2c^2d^3e)$

$= (-4 \cdot 2)(c^5c^2)(dd^3)(e)$ Group coefficients and like bases.

$= -8c^7d^4e$ Simplify.

c. $\left(\frac{1}{3}a^4b^3\right)\left(\frac{3}{4}b^7\right)$

$= \left(\frac{1}{3} \cdot \frac{3}{4}\right)(a^4)(b^3b^7)$ Group coefficients and like bases.

$= \frac{1}{4}a^4b^{10}$ Simplify.

2. Multiplication of Polynomials

The distributive property is used to multiply polynomials: $a(b + c) = ab + ac$.

example 2 **Multiplying a Polynomial by a Monomial**

Multiply the polynomials

a. $2t(4t - 3)$ b. $-3a^2\left(-4a^2 + 2a - \frac{1}{3}\right)$

Solution:

a. $2t(4t - 3)$ Multiply each term of the polynomial by $2t$.

$= (2t)(4t) + 2t(-3)$ Apply the distributive property.

$= 8t^2 - 6t$ Simplify each term.

b. $-3a^2\left(-4a^2 + 2a - \frac{1}{3}\right)$ Multiply each term of the polynomial by $-3a^2$.

$= (-3a^2)(-4a^2) + (-3a^2)(2a) + (-3a^2)\left(-\frac{1}{3}\right)$ Apply the distributive property.

$= 12a^4 - 6a^3 + a^2$ Simplify each term.

Thus far, we have illustrated polynomial multiplication involving monomials. Next, the distributive property will be used to multiply polynomials with more than one term.

$(x + 3)(x + 5) = (x + 3)x + (x + 3)5$ Apply the distributive property.

$= (x + 3)x + (x + 3)5$ Apply the distributive property again.

$= x \cdot x + 3 \cdot x + x \cdot 5 + 3 \cdot 5$

$= x^2 + 3x + 5x + 15$

$= x^2 + 8x + 15$ Combine *like* terms.

Note: Using the distributive property results in multiplying each term of the first polynomial by each term of the second polynomial.

$$\begin{aligned}(x + 3)(x + 5) &= x \cdot x + x \cdot 5 + 3 \cdot x + 3 \cdot 5 \\ &= x^2 + 5x + 3x + 15 \\ &= x^2 + 8x + 15\end{aligned}$$

example 3 **Multiplying a Polynomial by a Polynomial**

Multiply the polynomials.

a. $(c - 7)(c + 2)$ b. $(10x + 3y)(2x - 4y)$ c. $(y - 2)(3y^2 + y - 5)$

Solution:

a. $(c - 7)(c + 2)$ — Multiply each term in the first polynomial by each term in the second.

$= (c)(c) + (c)(2) + (-7)(c) + (-7)(2)$ — Apply the distributive property.

$= c^2 + 2c - 7c - 14$ — Simplify.

$= c^2 - 5c - 14$ — Combine *like* terms.

Tip: Notice that the product of two *binomials* equals the sum of the products of the **F**irst terms, the **O**uter terms, the **I**nner terms and the **L**ast terms. The acronym, **FOIL** (First Outer Inner Last) can be used as a memory device to multiply two binomials.

Outer terms, First terms, Inner terms, Last terms

	First	Outer	Inner	Last
$(c - 7)(c + 2) =$	$(c)(c)$ +	$(c)(2)$ +	$(-7)(c)$ +	$(-7)(2)$
$=$	c^2 +	$2c$ −	$7c$ −	14
$= c^2 - 5c - 14$				

b. $(10x + 3y)(2x - 4y)$ — Multiply each term in the first polynomial by each term in the second.

$= (10x)(2x) + (10x)(-4y) + (3y)(2x) + (3y)(-4y)$ — Apply the distributive property.

$= 20x^2 - 40xy + 6xy - 12y^2$ — Simplify each term.

$= 20x^2 - 34xy - 12y^2$ — Combine *like* terms.

Avoiding Mistakes

It is important to note that the acronym FOIL does not apply to Example 3(c) because the product does not involve two binomials.

c. $(y - 2)(3y^2 + y - 5)$ — Multiply each term in the first polynomial by each term in the second.

$= (y)(3y^2) + (y)(y) + (y)(-5) + (-2)(3y^2) + (-2)(y) + (-2)(-5)$

$= 3y^3 + y^2 - 5y - 6y^2 - 2y + 10$ — Simplify each term.

$= 3y^3 - 5y^2 - 7y + 10$ — Combine *like* terms.

Tip: Multiplication of polynomials can be performed vertically by a process similar to column multiplication of real numbers. For example,

$$\begin{array}{r} 235 \\ \times\ 21 \\ \hline 235 \\ 4700 \\ \hline 4935 \end{array} \qquad \begin{array}{r} 3y^2 + y - 5 \\ \times \qquad y - 2 \\ \hline -6y^2 - 2y + 10 \\ 3y^3 + y^2 - 5y + 0 \\ \hline 3y^3 - 5y^2 - 7y + 10 \end{array}$$

Note: When multiplying by the column method, it is important to *align like* terms vertically before adding terms.

3. Special Case Products: Difference of Squares and Perfect Square Trinomials

In some cases the product of two binomials takes on a special pattern.

I. The first special case occurs when multiplying the sum and difference of the same two terms. For example:

$$(2x + 3)(2x - 3)$$
$$= 4x^2 - 6x + 6x - 9$$
$$= 4x^2 - 9$$

Notice that the middle terms are opposites. This leaves only the difference between the square of the first term and the square of the second term. For this reason, the product is called a difference of squares.

Note: The sum and difference of the same two terms are called **conjugates**. Thus, the expressions $2x + 3$ and $2x - 3$ are conjugates of each other.

II. The second special case involves the square of a binomial. For example:

$$(3x + 7)^2$$
$$= (3x + 7)(3x + 7)$$
$$= 9x^2 + 21x + 21x + 49$$
$$= 9x^2 + 42x + 49$$
$$= (3x)^2 + 2(3x)(7) + (7)^2$$

When squaring a binomial, the product will be a trinomial called a perfect square trinomial. The first and third terms are formed by squaring the terms of the binomial. The middle term equals twice the product of the terms in the binomial.

Note: The expression $(3x - 7)^2$ also expands to a perfect square trinomial, but the middle term will be negative:

$$(3x - 7)(3x - 7) = 9x^2 - 21x - 21x + 49 = 9x^2 - 42x + 49$$

Special Case Product Formulas

1. $(a + b)(a - b) = a^2 - b^2$ The product is called a **difference of squares**.
2. $\left.\begin{array}{l}(a + b)^2 = a^2 + 2ab + b^2 \\ (a - b)^2 = a^2 - 2ab + b^2\end{array}\right\}$ The product is called a **perfect square trinomial**.

You should become familiar with these special case products because they will be used again in Chapter 5 to factor polynomials.

example 4 **Finding Special Products**

Use the special product formulas to multiply the polynomials.

a. $(x - 9)(x + 9)$ b. $\left(\frac{1}{2}p - 6\right)\left(\frac{1}{2}p + 6\right)$

Tip: The product of two conjugates can be checked by applying the distributive property:

$(x - 9)(x + 9)$
$= x^2 + 9x - 9x - 81$
$= x^2 - 81$

Solution:

a. $(x - 9)(x + 9)$ Apply the formula: $(a + b)(a - b) = a^2 - b^2$.

$a^2 - b^2$

$= (x)^2 - (9)^2$ Substitute $a = x$ and $b = 9$.

$= x^2 - 81$

b. $\left(\frac{1}{2}p - 6\right)\left(\frac{1}{2}p + 6\right)$ Apply the formula: $(a + b)(a - b) = a^2 - b^2$.

$a^2 - b^2$

$= \left(\frac{1}{2}p\right)^2 - (6)^2$ Substitute $a = \frac{1}{2}p$ and $b = 6$.

$= \frac{1}{4}p^2 - 36$ Simplify each term.

example 5 **Finding Special Products**

Use the special product formulas to multiply the polynomials.

a. $(3w - 4)^2$ b. $(5x^2 + 2)^2$

Solution:

a. $(3w - 4)^2$ Apply the formula: $(a - b)^2 = a^2 - 2ab + b^2$.

$a^2 - 2ab + b^2$

$= (3w)^2 - 2(3w)(4) + (4)^2$ Substitute $a = 3w$ and $b = 4$.

$= 9w^2 - 24w + 16$ Simplify each term.

Tip: The square of a binomial can be checked by explicitly writing the product of the two binomials and applying the distributive property:

$$(3w - 4)^2 = (3w - 4)(3w - 4) = 9w^2 - 12w - 12w + 16$$
$$= 9w^2 - 24w + 16$$

Avoiding Mistakes

The property for squaring two factors is different than the property for squaring two terms.

$(ab)^2 = a^2 b^2$ but

$(a + b)^2 = a^2 + 2ab + b^2$

b. $(5x^2 + 2)^2$ — Apply the formula: $(a + b)^2 = a^2 + 2ab + b^2$.

$a^2 + 2ab + b^2$

$= (5x^2)^2 + 2(5x^2)(2) + (2)^2$ — Substitute $a = 5x^2$ and $b = 2$.

$= 25x^4 + 20x^2 + 4$ — Simplify each term.

4. Multiplication of Polynomials and Applications to Geometry

example 6 **Using Special Case Products in an Application of Geometry**

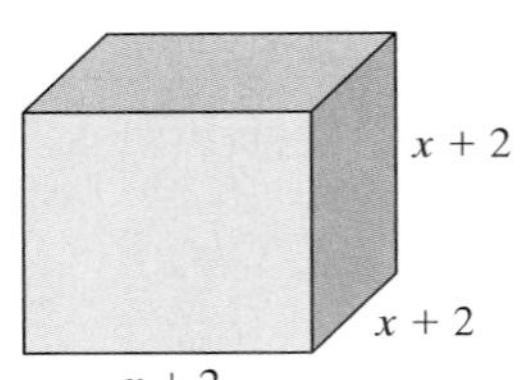

Figure 4-3

Find a polynomial that represents the volume of the cube (Figure 4-3).

Solution:

Volume = (length)(width)(height)

$V = (x + 2)(x + 2)(x + 2)$ or $V = (x + 2)^3$

To expand $(x + 2)(x + 2)(x + 2)$, multiply the first two factors. Then multiply the result by the last factor.

$V = (x + 2)(x + 2)(x + 2)$

$= (x^2 + 4x + 4)(x + 2)$

Tip: $(x + 2)(x + 2) = (x + 2)^2$ and results in a perfect square trinomial.

$$(x + 2)^2 = (x)^2 + 2(x)(2) + (2)^2$$
$$= x^2 + 4x + 4$$

$= (x^2)(x) + (x^2)(2) + (4x)(x) + (4x)(2) + (4)(x) + (4)(2)$ — Apply the distributive property.

$= x^3 + 2x^2 + 4x^2 + 8x + 4x + 8$ — Group *like* terms.

$= x^3 + 6x^2 + 12x + 8$ — Combine *like* terms.

The volume of the cube can be represented by

$V = (x + 2)^3 = x^3 + 6x^2 + 12x + 8.$

section 4.6 Practice Exercises

For Exercises 1–12, simplify the expressions (if possible).

1. $4x + 5x$
2. $2y^2 - 4y^2$
3. $(4x)(5x)$
4. $(2y^2)(-4y^2)$
5. $-5a^3b - 2a^3b$
6. $7uvw^2 + uvw^2$
7. $(-5a^3b)(-2a^3b)$
8. $(7uvw^2)(uvw^2)$
9. $-c + 4c^2$
10. $3t + 3t^3$
11. $(-c)(4c^2)$
12. $(3t)(3t^3)$

For Exercises 13–16, perform the indicated operations. Write the answers in scientific notation.

13. $(4.3 \times 10^6)(2.3 \times 10^6)$
14. $(3.9 \times 10^{-5})(1.5 \times 10^{-5})$
15. $(-2.1 \times 10^{-12})(9.3 \times 10^{-12})$
16. $(4.4 \times 10^9)(5.4 \times 10^9)$

For Exercises 17–52, multiply the expressions.

17. $8(4x)$
18. $-2(6y)$
19. $-10(5z)$
20. $7(3p)$
21. $(x^{10})(4x^3)$
22. $(a^{13}b^4)(12ab^4)$
23. $(4m^3n^7)(-3m^6n)$
24. $(2c^7d)(-c^3d^{11})$
25. $8pq(2pq - 3p + 5q)$
26. $5ab(2ab + 6a - 3b)$
27. $(k^2 - 13k - 6)(-4k)$
28. $(h^2 + 5h - 12)(-2h)$
29. $-15pq(3p^2 + p^3q^2 - 2q)$
30. $-4u^2v(2u - 5uv^3 + v)$
31. $(y - 10)(y + 9)$
32. $(x + 5)(x - 6)$
33. $(m - 12)(m - 2)$
34. $(n - 7)(n - 2)$
35. $(p - 2)(p + 1)$
36. $(q + 11)(q - 5)$
37. $(w + 8)(w + 3)$
38. $(z + 10)(z + 4)$
39. $(p - 3)(p - 11)$
40. $(y - 7)(y - 10)$
41. $(6x - 1)(2x + 5)$
42. $(3x + 7)(x - 8)$
43. $(4a - 9)(2a - 1)$
44. $(3b + 5)(b - 5)$
45. $(3t - 7)(3t + 1)$
46. $(5w - 2)(2w - 5)$
47. $(3x + 4)(x + 8)$
48. $(7y + 1)(3y + 5)$
49. $(5s + 3)(s^2 + s - 2)$
50. $(t - 4)(2t^2 - t + 6)$
51. $(3w - 2)(9w^2 + 6w + 4)$
52. $(z + 5)(z^2 - 5z + 25)$

For Exercises 53–60, multiply the expressions.

53. $(3a - 4b)(3a + 4b)$
54. $(5y + 7x)(5y - 7x)$
55. $(9k + 6)(9k - 6)$
56. $(2h - 5)(2h + 5)$
57. $\left(\frac{1}{2} - t\right)\left(\frac{1}{2} + t\right)$
58. $\left(r + \frac{1}{4}\right)\left(r - \frac{1}{4}\right)$
59. $(u^3 + 5v)(u^3 - 5v)$
60. $(8w^2 - x)(8w^2 + x)$

For Exercises 61–68, square the binomials.

61. $(a + b)^2$
62. $(a - b)^2$
63. $(x - y)^2$
64. $(x + y)^2$
65. $(2c + 5)^2$
66. $(5d - 9)^2$
67. $(3t^2 - 4s)^2$
68. $(u^2 + 4v)^2$

69. a. Evaluate $(2 + 4)^2$ by working within the parentheses first.
    b. Evaluate $2^2 + 4^2$.
    c. Compare the answers to parts (a) and (b) and make a conjecture about $(a + b)^2$ and $a^2 + b^2$.

70. a. Evaluate $(6 - 5)^2$ by working within the parentheses first.
    b. Evaluate $6^2 - 5^2$.
    c. Compare the answers to parts (a) and (b) and make a conjecture about $(a - b)^2$ and $a^2 - b^2$.

71. Find a polynomial expression that represents the area of the rectangle shown in the figure.

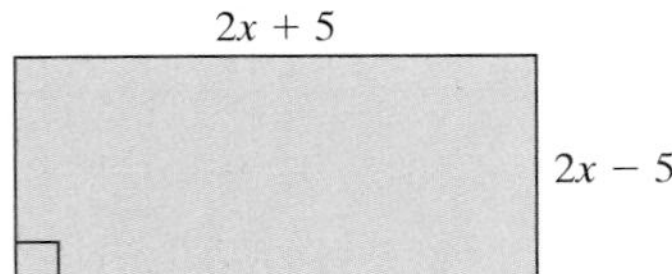

Figure for Exercise 71

72. Find a polynomial expression that represents the area of the rectangle shown in the figure.

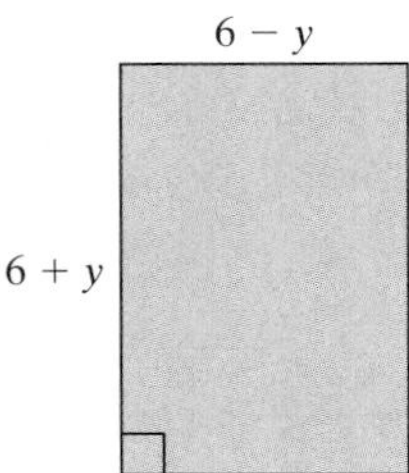

Figure for Exercise 72

73. Find a polynomial expression that represents the area of the square shown in the figure.

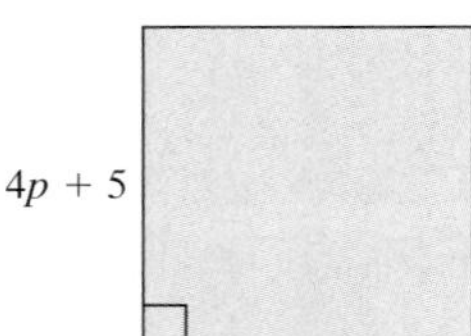

Figure for Exercise 73

74. Find a polynomial expression that represents the area of the square shown in the figure.

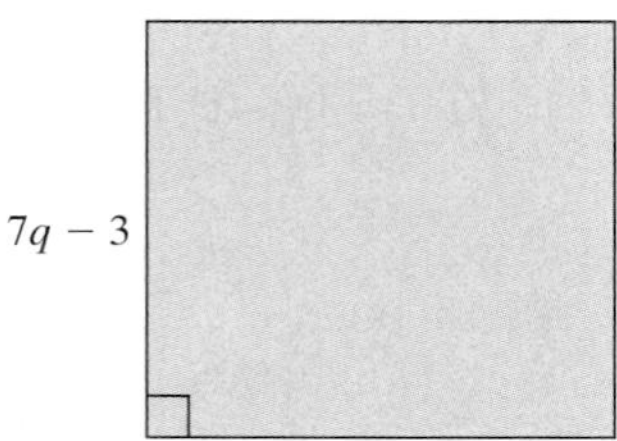

Figure for Exercise 74

For Exercises 75–96, multiply the expressions.

75. $(7x + y)(7x - y)$

76. $(9w - 4z)(9w + 4z)$

77. $(5s + 3t)^2$

78. $(5s - 3t)^2$

79. $(7x - 3y)(3x - 8y)$

80. $(5a - 4b)(2a - b)$

81. $\left(\frac{2}{3}t + 2\right)(3t + 4)$

82. $\left(\frac{1}{5}s + 6\right)(5s - 3)$

83. $(5z + 3)(z^2 + 4z - 1)$

84. $(2k - 5)(2k^2 + 3k + 5)$

85. $\left(\frac{1}{3}m - n\right)^2$

86. $\left(\frac{2}{5}p - q\right)^2$

87. $6w^2(7w - 14)$

88. $4v^3(v + 12)$

89. $(4y - 8.1)(4y + 8.1)$

90. $(2h + 2.7)(2h - 2.7)$

91. $(3c^2 + 4)(7c^2 - 8)$

92. $(5k^3 - 9)(k^3 - 2)$

93. $(3.1x + 4.5)^2$

94. $(2.5y + 1.1)^2$

95. $(k - 4)^3$

96. $(h + 3)^3$

97. Find a polynomial that represents the area of the triangle shown in the figure.
(Recall: $A = \frac{1}{2}bh$.)

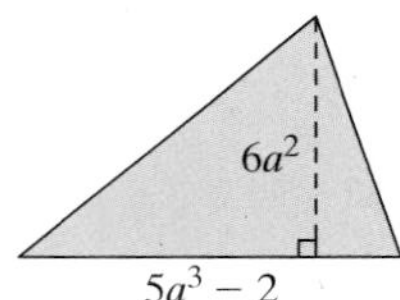

Figure for Exercise 97

98. Find a polynomial that represents the area of the triangle shown in the figure.

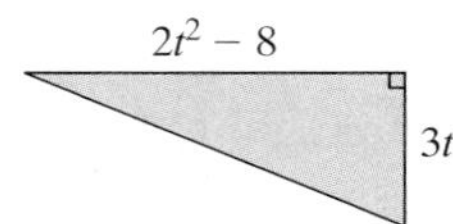

Figure for Exercise 98

99. Find a polynomial that represents the volume of the cube shown in the figure.

(Recall: $V = s^3$)

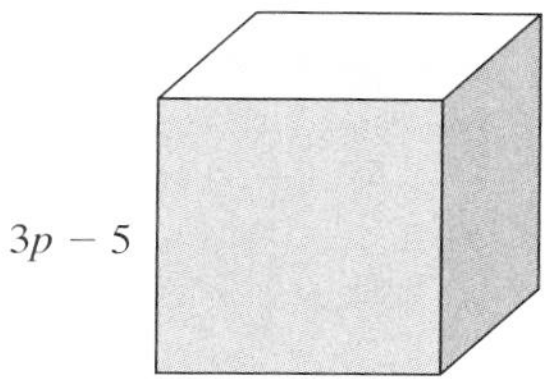

Figure for Exercise 99

100. Find a polynomial that represents the volume of the rectangular solid shown in the figure.

(Recall: $V = lwh$.)

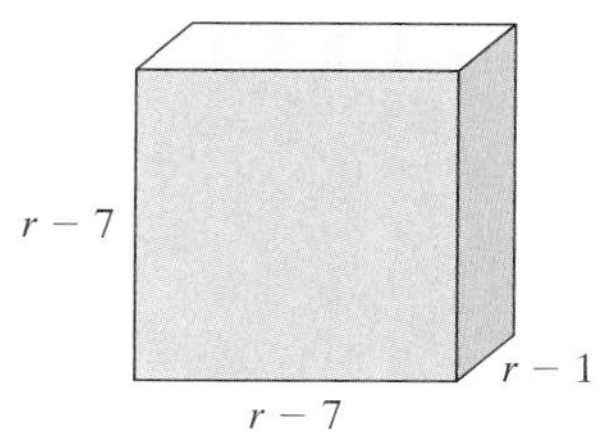

Figure for Exercise 100

Expanding Your Skills

For Exercises 101–103, multiply the expressions containing more than two factors. (*Hint*: First multiply two of the factors, then multiply that product by the third factor.)

101. $2a(3a - 4)(a + 5)$

102. $5x(x + 2)(6x - 1)$

103. $(x - 3)(2x + 1)(x - 4)$

104. $(y - 2)(2y - 3)(y + 3)$

105. What binomial when multiplied by $(3x + 5)$ will produce a product of $6x^2 - 11x - 35$? [*Hint*: Let the quantity $(a + b)$ represent the unknown binomial.] Then find a and b such that $(3x + 5)(a + b) = 6x^2 - 11x - 35$.

106. What binomial when multiplied by $(2x - 4)$ will produce a product of $2x^2 + 8x - 24$?

Concepts

1. Division by a Monomial
2. Long Division

section

4.7 Division of Polynomials

1. Division by a Monomial

Division of polynomials will be presented in this section as two separate cases: The first case illustrates **division by a monomial** divisor. The second case illustrates division by a polynomial with two or more terms.

To divide a polynomial by a monomial, divide each individual term in the polynomial by the divisor and simplify the result.

Dividing a Polynomial by a Monomial

If a, b, and c are polynomials such that $c \neq 0$, then

$$\frac{a + b}{c} = \frac{a}{c} + \frac{b}{c} \qquad \text{Similarly,} \qquad \frac{a - b}{c} = \frac{a}{c} - \frac{b}{c}$$

example 1 **Dividing a Polynomial by a Monomial**

Divide the polynomials.

a. $\dfrac{5a^3 - 10a^2 + 20a}{5a}$ b. $(12y^2z^3 - 15yz^2 + 6y^2z) \div (-6y^2z)$

Solution:

a. $\dfrac{5a^3 - 10a^2 + 20a}{5a}$

$= \dfrac{5a^3}{5a} - \dfrac{10a^2}{5a} + \dfrac{20a}{5a}$ Divide each term in the numerator by $5a$.

$= a^2 - 2a + 4$ Simplify each term using the properties of exponents.

b. $(12y^2z^3 - 15yz^2 + 6y^2z) \div (-6y^2z)$

$= \dfrac{12y^2z^3 - 15yz^2 + 6y^2z}{-6y^2z}$

$= \dfrac{12y^2z^3}{-6y^2z} - \dfrac{15yz^2}{-6y^2z} + \dfrac{6y^2z}{-6y^2z}$ Divide each term by $-6y^2z$.

$= -2z^2 + \dfrac{5z}{2y} - 1$ Simplify each term.

2. Long Division

If the divisor has two or more terms, a **long division** process similar to the division of real numbers is used.

example 2 **Using Long Division to Divide Polynomials**

Divide the polynomials using long division: $(2x^2 - x + 3) \div (x - 3)$

Solution:

$x - 3\overline{)2x^2 - x + 3}$ Divide the leading term in the dividend by the leading term in the divisor.

$\dfrac{2x^2}{x} = 2x$

This is the first term in the quotient.

$$\begin{array}{r} 2x \\ x - 3\overline{)2x^2 - x + 3} \\ -(2x^2 - 6x) \end{array}$$

Multiply $2x$ by the divisor.

$2x(x - 3) = 2x^2 - 6x$

and subtract the result.

$$\begin{array}{r} 2x \phantom{{}-x+3} \\ x-3\overline{)2x^2-x+3} \\ \underline{-2x^2+6x} \phantom{{}+3} \\ 5x \phantom{{}+3} \end{array}$$

Subtract the quantity $2x^2 - 6x$. To do this, add the opposite.

$$\begin{array}{r} 2x+5 \\ x-3\overline{)2x^2-x+3} \\ \underline{-2x^2+6x} \downarrow \\ 5x+3 \end{array}$$

Bring down the next column and repeat the process.

Divide the leading term by x: $(5x)/x = 5$. Place 5 in the quotient.

$$\begin{array}{r} 2x+5 \\ x-3\overline{)2x^2-x+3} \\ \underline{-2x^2+6x} \phantom{{}+3} \\ 5x+3 \\ -(5x-15) \end{array}$$

Multiply the divisor by 5: $5(x - 3) = 5x - 15$ and subtract the result.

$$\begin{array}{r} 2x+5 \\ x-3\overline{)2x^2-x+3} \\ \underline{-2x^2+6x} \phantom{{}+3} \\ 5x+3 \\ \underline{-5x+15} \\ 18 \end{array}$$

Subtract the quantity $5x - 15$ by adding the opposite.

The remainder is 18.

Summary:

The quotient is	$2x + 5$
The remainder is	18
The divisor is	$x - 3$
The dividend is	$2x^2 - x + 3$

The solution to a long division problem is usually written in the form:

$$\text{Quotient} + \frac{\text{remainder}}{\text{divisor}}$$

Hence

$$(2x^2 - x + 3) \div (x - 3) = 2x + 5 + \frac{18}{x - 3}$$

The division of polynomials can be checked in the same fashion as the division of real numbers. To check, we have

$$\text{Dividend} = (\text{divisor})(\text{quotient}) + \text{remainder}$$

$$\begin{aligned} 2x^2 - x + 3 &\stackrel{?}{=} (x - 3)(2x + 5) + (18) \\ &\stackrel{?}{=} 2x^2 - 6x + 5x - 15 + (18) \\ &\stackrel{?}{=} 2x^2 - x + 3 \checkmark \end{aligned}$$

example 3

Using Long Division to Divide Polynomials

Divide the polynomials using long division: $(2w^3 + 8w^2 - 16) \div (2w + 4)$

Solution:

First note that the dividend has a missing power of w and can be written as $2w^3 + 8w^2 + 0w - 16$. The term $0w$ is a place holder for the missing term. It is helpful to use the place holder to keep the powers of w lined up.

$$\begin{array}{r} w^2 \\ 2w + 4 \overline{) 2w^3 + 8w^2 + 0w - 16} \\ -(2w^3 + 4w^2) \end{array}$$

Divide $2w^3 \div 2w = w^2$. This is the first term of the quotient.
Then multiply $w^2(2w + 4) = 2w^3 + 4w^2$.

$$\begin{array}{r} w^2 \\ 2w + 4 \overline{) 2w^3 + 8w^2 + 0w - 16} \\ \underline{-2w^3 - 4w^2} \\ 4w^2 + 0w \end{array}$$

Subtract by adding the opposite.
Bring down the next column and repeat the process.

$$\begin{array}{r} w^2 + 2w \\ 2w + 4 \overline{) 2w^3 + 8w^2 + 0w - 16} \\ \underline{-2w^3 - 4w^2} \\ 4w^2 + 0w \\ -(4w^2 + 8w) \end{array}$$

Divide $4w^2$ by the leading term in the divisor. $4w^2 \div 2w = 2w$. Place $2w$ in the quotient.
Multiply $2w(2w + 4) = 4w^2 + 8w$.

$$\begin{array}{r} w^2 + 2w \\ 2w + 4 \overline{) 2w^3 + 8w^2 + 0w - 16} \\ \underline{-2w^3 - 4w^2} \\ 4w^2 + 0w \\ \underline{-4w^2 - 8w} \\ -8w - 16 \end{array}$$

Subtract by adding the opposite.
Bring down the next column and repeat.

$$\begin{array}{r} w^2 + 2w - 4 \\ 2w + 4 \overline{) 2w^3 + 8w^2 + 0w - 16} \\ \underline{-2w^3 - 4w^2} \\ 4w^2 + 0w \\ \underline{-4w^2 - 8w} \\ -8w - 16 \\ -(-8w - 16) \end{array}$$

Divide $-8w$ by the leading term in the divisor. $-8w \div 2w = -4$. Place -4 in the quotient.
Multiply $-4(2w + 4) = -8w - 16$.

$$\begin{array}{r} w^2 + 2w - 4 \\ 2w + 4 \overline{) 2w^3 + 8w^2 + 0w - 16} \\ \underline{-2w^3 - 4w^2} \\ 4w^2 + 0w \\ \underline{-4w^2 - 8w} \\ -8w - 16 \\ \underline{8w + 16} \\ 0 \end{array}$$

Subtract by adding the opposite.
The remainder is 0.

The quotient is $w^2 + 2w - 4$ and the remainder is 0.

In Example 3 the remainder is zero. Therefore, we say that $2w + 4$ divides *evenly* into $2w^3 + 8w^2 - 16$. For this reason, the divisor and quotient are factors of $2w^3 + 8w^2 - 16$. To check, we have

$$\text{Dividend} = (\text{divisor})(\text{quotient}) + \text{remainder}$$

$$\begin{aligned} 2w^3 + 8w^2 - 16 &\stackrel{?}{=} (2w + 4)(w^2 + 2w - 4) + 0 \\ &\stackrel{?}{=} 2w^3 + 4w^2 - 8w + 4w^2 + 8w - 16 \\ &\stackrel{?}{=} 2w^3 + 8w^2 - 16 \checkmark \end{aligned}$$

example 4

Using Long Division to Divide Polynomials

Divide the polynomials using long division.

$$\frac{2y + y^4 - 5}{1 + y^2}$$

Solution:

First note that both the dividend and divisor should be written in descending order:

$$\frac{y^4 + 2y - 5}{y^2 + 1}$$

Also note that the dividend and the divisor have missing powers of y. Use place holders.

$$y^2 + 0y + 1 \overline{)y^4 + 0y^3 + 0y^2 + 2y - 5}$$

$$\begin{array}{r} y^2 \\ y^2 + 0y + 1 \overline{)y^4 + 0y^3 + 0y^2 + 2y - 5} \\ -(y^4 + 0y^3 + y^2) \end{array}$$

Divide $y^4 \div y^2 = y^2$. This is the first term of the quotient.

Multiply $y^2(y^2 + 0y + 1) = y^4 + 0y^3 + y^2$

$$\begin{array}{r} y^2 \\ y^2 + 0y + 1 \overline{)y^4 + 0y^3 + 0y^2 + 2y - 5} \\ \underline{-y^4 - 0y^3 - y^2} \\ -y^2 + 2y - 5 \end{array}$$

Subtract by adding the opposite.

Bring down the next columns.

$$\begin{array}{r} y^2 - 1 \\ y^2 + 0y + 1 \overline{)y^4 + 0y^3 + 0y^2 + 2y - 5} \\ \underline{-y^4 - 0y^3 - y^2} \\ -y^2 + 2y - 5 \\ -(\underline{-y^2 - 0y - 1}) \end{array}$$

Divide $-y^2 \div y^2 = -1$.

Multiply $-1(y^2 + 0y + 1) = -y^2 - 0y - 1$.

$$
\begin{array}{r}
y^2 \qquad\qquad -1 \\
y^2 + 0y + 1 \overline{\big) y^4 + 0y^3 + 0y^2 + 2y - 5} \\
\underline{-y^4 - 0y^3 - y^2} \qquad\qquad \\
-y^2 + 2y - 5 \\
\underline{y^2 + 0y + 1} \\
2y - 4
\end{array}
$$

Subtract by adding the opposite.

Remainder

Therefore,

$$\frac{y^4 + 2y - 5}{y^2 + 1} = y^2 - 1 + \frac{2y - 4}{y^2 + 1}$$

Tip: Recall that

- Long division is used when the divisor has *two or more terms.*
- If the divisor has *one term* (when the divisor is a monomial) then divide each term in the dividend by the monomial divisor.

example 5

Determining Whether Long Division Is Necessary

Determine whether long division is necessary for each division of polynomials.

a. $\dfrac{2p^5 - 8p^4 + 4p - 16}{p^2 - 2p + 1}$

b. $\dfrac{2p^5 - 8p^4 + 4p - 16}{2p^2}$

c. $(3z^3 - 5z^2 + 10) \div (15z^3)$

d. $(3z^3 - 5z^2 + 10) \div (3z + 1)$

Solution:

a. $\dfrac{2p^5 - 8p^4 + 4p - 16}{p^2 - 2p + 1}$ The divisor has three terms. Use long division.

b. $\dfrac{2p^5 - 8p^4 + 4p - 16}{2p^2}$ The divisor has one term. No long division.

c. $(3z^3 - 5z^2 + 10) \div (15z^3)$ The divisor has one term. No long division.

d. $(3z^3 - 5z^2 + 10) \div (3z + 1)$ The divisor has two terms. Use long division.

section 4.7 PRACTICE EXERCISES

For Exercises 1–8, perform the indicated operations.

1. $(6z^5 - 2z^3 + z - 6) - (10z^4 + 2z^3 + z^2 + z)$
2. $(7a^2 + a - 6) + (2a^2 + 5a + 11)$
3. $(10x + y)(x - 3y)$
4. $8b^2(2b^2 - 5b + 12)$
5. $(2w^3 + 5)^2$
6. $\left(\frac{4}{3}y^2 - \frac{1}{2}y + \frac{3}{8}\right) - \left(\frac{1}{3}y^2 + \frac{1}{4}y - \frac{1}{8}\right)$
7. $\left(\frac{7}{8}w - 1\right)\left(\frac{7}{8}w + 1\right)$
8. $(2x + 1)(5x - 3)$
9. There are two methods for dividing polynomials. Explain when long division is used.
10. Explain how to check a polynomial division problem.
11. a. Divide

    $$\frac{15t^3 + 18t^2}{3t}$$

    b. Check by multiplying the quotient by the divisor.
12. a. Divide: $(-9y^4 + 6y^2 - y) \div (3y)$

    b. Check by multiplying the quotient by the divisor.

For Exercises 13–28, divide the polynomials.

13. $(6a^2 + 4a - 14) \div (2)$
14. $\frac{4b^2 + 16b - 12}{4}$
15. $\frac{-5x^2 - 20x + 5}{-5}$
16. $\frac{-3y^3 + 12y - 6}{-3}$
17. $\frac{3p^3 - p^2}{p}$
18. $(7q^4 + 5q^2) \div q$
19. $(4m^2 + 8m) \div 4m^2$
20. $\frac{n^2 - 8}{n}$
21. $\frac{14y^4 - 7y^3 + 21y^2}{-7y^2}$
22. $(25a^5 - 5a^4 + 15a^3 - 5a) \div (-5a)$
23. $(4x^3 - 24x^2 - x + 8) \div (4x)$
24. $\frac{20w^3 + 15w^2 - w + 5}{10w}$
25. $\frac{-a^3b^2 + a^2b^2 - ab^3}{-a^2b^2}$
26. $(3x^4y^3 - x^2y^2 - xy^3) \div (-x^2y^2)$
27. $(6t^4 - 2t^3 + 3t^2 - t + 4) \div (2t^3)$
28. $\frac{2y^3 - 2y^2 + 3y - 9}{2y^2}$
29. a. Divide: $(z^2 + 7z + 11) \div (z + 5)$

    b. Check by multiplying the quotient by the divisor and adding the remainder.
30. a. Divide

    $$\frac{2w^2 - 7w + 3}{w - 4}$$

    b. Check by multiplying the quotient by the divisor and adding the remainder.

For Exercises 31–50, divide the polynomials.

31. $\frac{t^2 + 4t + 3}{t + 1}$
32. $(3x^2 + 8x + 4) \div (x + 2)$
33. $(7b^2 - 3b - 4) \div (b - 1)$
34. $\frac{w^2 - w - 2}{w - 2}$
35. $\frac{5k^2 - 29k - 6}{5k + 1}$
36. $(4y^2 + 25y - 21) \div (4y - 3)$
37. $(4p^3 + 12p^2 + p - 12) \div (2p + 3)$
38. $\frac{12a^3 - 2a^2 - 17a - 5}{3a + 1}$
39. $\frac{-k - 6 + k^2}{1 + k}$
40. $(1 + h^2 + 3h) \div (2 + h)$

41. $(4x^3 - 8x^2 + 15x - 16) \div (2x - 3)$

42. $\dfrac{3b^3 + b^2 + 17b - 49}{3b - 5}$

43. $\dfrac{9 + a^2}{a + 3}$

44. $(3 + m^2) \div (m + 3)$

45. $(w^4 + 5w^3 - 5w^2 - 15w + 7) \div (w^2 - 3)$

46. $\dfrac{p^4 - p^3 - 4p^2 - 2p - 15}{p^2 + 2}$

47. $\dfrac{2n^4 + 5n^3 - 11n^2 - 20n + 12}{2n^2 + 3n - 2}$

48. $(6y^4 - 5y^3 - 8y^2 + 16y - 8) \div (2y^2 - 3y + 2)$

49. $(5x^3 - 4x - 9) \div (5x^2 + 5x + 1)$

50. $\dfrac{3a^3 - 5a + 16}{3a^2 - 6a + 7}$

51. Explain why $(x^3 - 8) \div (x - 2)$ is *not* $(x^2 + 4)$.

52. Explain why $(y^3 + 27) \div (y + 3)$ is *not* $(y^2 + 9)$.

For Exercises 53–64, determine which method to use to divide the polynomials: monomial division or long division. Then use that method to divide the polynomials.

53. $\dfrac{9a^3 + 12a^2}{3a}$

54. $\dfrac{3y^2 + 17y - 12}{y + 6}$

55. $(p^3 + p^2 - 4p - 4) \div (p^2 - p - 2)$

56. $(q^3 + 1) \div (q + 1)$

57. $\dfrac{t^4 + t^2 - 16}{t + 2}$

58. $\dfrac{-8m^5 - 4m^3 + 4m^2}{-2m^2}$

59. $(w^4 + w^2 - 5) \div (w^2 - 2)$

60. $(2k^2 + 9k + 7) \div (k + 1)$

61. $\dfrac{n^3 - 64}{n - 4}$

62. $\dfrac{15s^2 + 34s + 28}{5s + 3}$

63. $(9r^3 - 12r^2 + 9) \div (-3r^2)$

64. $(6x^4 - 16x^3 + 15x^2 - 5x + 10) \div (3x + 1)$

Expanding Your Skills

For Exercises 65–72, divide the polynomials and note any patterns.

65. $(x^2 - 1) \div (x - 1)$

66. $(x^3 - 1) \div (x - 1)$

67. $(x^4 - 1) \div (x - 1)$

68. $(x^5 - 1) \div (x - 1)$

69. $x^2 \div (x - 1)$

70. $x^3 \div (x - 1)$

71. $x^4 \div (x - 1)$

72. $x^5 \div (x - 1)$

chapter 4 SUMMARY

SECTION 4.1—EXPONENTS: MULTIPLYING AND DIVIDING COMMON BASES

KEY CONCEPTS:

Definition

$$b^n = \underbrace{b \cdot b \cdot b \cdot b \cdot \cdots \cdot b}_{n \text{ factors of } b}$$ b is the base, n is the exponent

Multiplying Common Bases

$b^m b^n = b^{m+n}$ (m, n positive integers)

Dividing Common Bases

$\dfrac{b^m}{b^n} = b^{m-n}$ ($b \neq 0$, m, n, positive integers)

KEY TERMS:

base
compound interest
exponent

EXAMPLES:

$3^4 = 3 \cdot 3 \cdot 3 \cdot 3 = 81$ 3 is the base
4 is the exponent

Compare: $(-5)^2$ versus -5^2

$(-5)^2 = (-5)(-5) = 25$
versus
$-5^2 = -1(5^2) = -1(5)(5) = -25$

Simplify: $x^3 \cdot x^4 \cdot x^2 \cdot x = x^{3+4+2+1} = x^{10}$

Simplify: $\dfrac{c^4 d^{10}}{cd^5} = c^{4-1}d^{10-5} = c^3 d^5$

SECTION 4.2—MORE PROPERTIES OF EXPONENTS

KEY CONCEPTS:

Power Rule for Exponents

$(b^m)^n = b^{mn}$ ($b \neq 0$, m, n positive integers)

Power of a Product and Power of a Quotient

Assume m and n are positive integers and a and b are real numbers where $a \neq 0, b \neq 0$.
Then,

$(ab)^m = a^m b^m$ and $\left(\dfrac{a}{b}\right)^m = \dfrac{a^m}{b^m}$

KEY TERMS:

power rule for exponents

EXAMPLES:

Simplify: $(x^4)^5 = x^{20}$

Simplify: $(4uv^2)^3 = 4^3u^3(v^2)^3 = 64u^3v^6$

Simplify: $\left(\dfrac{p^5q^3}{5pq^2}\right)^2 = \left(\dfrac{p^{5-1}q^{3-2}}{5}\right)^2 = \left(\dfrac{p^4q}{5}\right)^2$

$= \dfrac{p^8q^2}{25}$

Section 4.3—Definitions of b^0 and b^{-n}

Key Concepts:

Definitions

If b is a real number such that $b \neq 0$ and n is an integer, then:

1. $b^0 = 1$

2. $b^{-n} = \left(\frac{1}{b}\right)^n = \frac{1}{b^n}$

Examples:

Simplify: $4^0 = 1$

Simplify: $y^{-7} = \frac{1}{y^7}$

Simplify:

$$\left(\frac{2a^3b}{a^{-2}c^{-4}}\right)^{-2}$$

$$= \left(\frac{2a^{3-(-2)}b}{c^{-4}}\right)^{-2}$$

$$= \left(\frac{2a^5b}{c^{-4}}\right)^{-2} = \frac{2^{-2}a^{-10}b^{-2}}{c^8}$$

$$= \frac{1}{2^2a^{10}b^2c^8}$$

$$= \frac{1}{4a^{10}b^2c^8}$$

Section 4.4—Scientific Notation

Key Concepts:

A number written in scientific notation is expressed in the form:

$a \times 10^n$ where $1 \leq |a| < 10$ and n is an integer. The value 10^n is sometimes called the order of magnitude or simply the magnitude of the number.

Key Terms:

order of magnitude
scientific notation

Examples:

Write the numbers in scientific notation:

$$35{,}000 = 3.5 \times 10^4$$

$$0.000\,000\,548 = 5.48 \times 10^{-7}$$

Multiply: $(3.5 \times 10^4)(2.0 \times 10^{-6})$
$= 7.0 \times 10^{-2}$

Divide:

$$\frac{8.4 \times 10^{-9}}{2.1 \times 10^3} = 4.0 \times 10^{-9-3} = 4.0 \times 10^{-12}$$

Section 4.5—Addition and Subtraction of Polynomials

Key Concepts:

A polynomial in one variable is a finite sum of terms of the form ax^n, where a is a real number and the exponent, n is a nonnegative integer. For each term, a is called the coefficient of the term and n is the degree of the term. The term with highest degree is the leading term and its coefficient is called the leading coefficient. The degree of the polynomial is the largest degree of all its terms.

Examples:

Given: $4x^5 - 8x^3 + 9x - 5$

Coefficients of each term:	4, −8, 9, −5
Degree of each term:	5, 3, 1, 0
Leading term:	$4x^5$
Leading coefficient:	4
Degree of polynomial:	5

To add or subtract polynomials, add or subtract *like* terms.

Perform the indicated operations:

$$(2x^4 - 5x^3 + 1) - (x^4 + 3) + (x^3 - 4x - 7)$$
$$= 2x^4 - 5x^3 + 1 - x^4 - 3 + x^3 - 4x - 7$$
$$= 2x^4 - x^4 - 5x^3 + x^3 - 4x + 1 - 3 - 7$$
$$= x^4 - 4x^3 - 4x - 9$$

KEY TERMS:

binomial
coefficient
degree of a polynomial
degree of a term
leading coefficient
leading term
like terms
monomial
polynomial
trinomial

SECTION 4.6—MULTIPLICATION OF POLYNOMIALS

KEY CONCEPTS:

Multiplying Monomials

Use the commutative and associative properties of multiplication to group coefficients and like bases.

EXAMPLES:

Multiply: $(5a^2b)(-2ab^3)$
$$= (5 \cdot -2)(a^2a)(bb^3)$$
$$= -10a^3b^4$$

Multiplying Polynomials

Multiply each term in the first polynomial by each term in the second.

Multiply: $-3ab^2(2a^2 - 5ab)$
$$= (-3ab^2)(2a^2) + (-3ab^2)(-5ab)$$
$$= -6a^3b^2 + 15a^2b^3$$

Multiply: $(y + 3)(2y - 5)$
$$= 2y^2 - 5y + 6y - 15$$
$$= 2y^2 + y - 15$$

Multiply: $(x - 2)(3x^2 - 4x + 11)$
$$= 3x^3 - 4x^2 + 11x - 6x^2 + 8x - 22$$
$$= 3x^3 - 10x^2 + 19x - 22$$

Product of Conjugates

Results in a difference of squares

$$(a + b)(a - b) = a^2 - b^2$$

Multiply: $(3w - 4v)(3w + 4v)$
$$= (3w)^2 - (4v)^2$$
$$= 9w^2 - 16v^2$$

Square of a Binomial

Results in a perfect square trinomial

$$(a + b)^2 = a^2 + 2ab + b^2$$
$$(a - b)^2 = a^2 - 2ab + b^2$$

Multiply: $(5c - 8d)^2$
$$= (5c)^2 - 2(5c)(8d) + (8d)^2$$
$$= 25c^2 - 80cd + 64d^2$$

KEY TERMS:

conjugates
difference of squares
perfect square trinomial

Section 4.7—Division of Polynomials

Key Concepts:

Division of Polynomials

1. Division by a monomial, use the properties:

$$\frac{a+b}{c} = \frac{a}{c} + \frac{b}{c} \quad \text{and} \quad \frac{a-b}{c} = \frac{a}{c} - \frac{b}{c}$$

2. If the divisor has more than one term, use long division.

Key Terms:

division by a monomial
long division

Examples:

Divide: $\dfrac{-3x^2 - 6x + 9}{-3x}$

$$= \frac{-3x^2}{-3x} - \frac{6x}{-3x} + \frac{9}{-3x}$$

$$= x + 2 - \frac{3}{x}$$

Divide: $(3x^2 - 5x + 1) \div (x + 2)$

$$\begin{array}{r} 3x - 11 \\ x + 2 \overline{)\, 3x^2 - 5x + 1} \\ -(3x^2 + 6x) \\ -11x + 1 \\ -(-11x - 22) \\ 23 \end{array}$$

$$3x - 11 + \frac{23}{x + 2}$$

chapter 4 REVIEW EXERCISES

Section 4.1

For Exercises 1–4, identify the base and the exponent.

1. 5^3 2. x^4 3. $(-2)^0$ 4. y

5. Evaluate the expressions.
   a. 6^2 b. $(-6)^2$ c. -6^2

6. Evaluate the expressions.
   a. 4^3 b. $(-4)^3$ c. -4^3

For Exercises 7–18, simplify and write the answers in exponent form. Assume that all variables represent nonzero real numbers.

7. $5^3 \cdot 5^{10}$
8. a^7a^4
9. $x \cdot x^6 \cdot x^2$
10. $6^3 \cdot 6 \cdot 6^5$
11. $\dfrac{10^7}{10^4}$
12. $\dfrac{y^{14}}{y^8}$
13. $\dfrac{b^9}{b}$
14. $\dfrac{7^8}{7}$
15. $\dfrac{k^2k^3}{k^4}$
16. $\dfrac{8^4 \cdot 8^7}{8^{11}}$
17. $\dfrac{2^8 \cdot 2^{10}}{2^3 \cdot 2^7}$
18. $\dfrac{q^3q^{12}}{qq^8}$
19. Explain why $2^2 \cdot 4^4$ does *not* equal 8^6.
20. Explain why $\dfrac{10^5}{5^2}$ does *not* equal 2^3.

For Exercises 21–22, use the formula

$$A = P(1 + r)^t$$

21. Find the amount in an account after 3 years if the initial investment is \$6000, invested at 6% interest compounded annually.
22. Find the amount in an account after 2 years if the initial investment is \$20,000, invested at 5% interest compounded annually.

Section 4.2

For Exercises 23–40, simplify the expressions. Write the answers in exponent form. Assume all variables represent nonzero real numbers.

23. $(7^3)^4$

24. $(c^2)^6$

25. $(p^4p^2)^3$

26. $(9^5 \cdot 9^2)^4$

27. $\left(\frac{a}{b}\right)^2$

28. $\left(\frac{1}{3}\right)^4$

29. $\left(\frac{5}{c^2d^5}\right)^2$

30. $\left(-\frac{m^2}{4n^6}\right)^5$

31. $(2ab^2)^4$

32. $(-x^7y)^2$

33. $\left(\frac{-3x^3}{5y^2z}\right)^3$

34. $\left(\frac{r^3}{s^2t^6}\right)^5$

35. $\frac{a^4(a^2)^8}{(a^3)^3}$

36. $\frac{(8^3)^4 \cdot 8^{10}}{(8^4)^5}$

37. $\frac{(4h^2k)^2(h^3k)^4}{(2hk^3)^2}$

38. $\frac{(p^3q)^3(2p^2q^4)^4}{(8p)(pq^3)^2}$

39. $\left(\frac{2x^4y^3}{4xy^2}\right)^2$

40. $\left(\frac{a^4b^6}{ab^4}\right)^3$

Section 4.3

For Exercises 41–62, simplify the expressions. Assume all variables represent nonzero real numbers.

41. 8^0

42. $(-b)^0$

43. 1^0

44. $-x^0$

45. $2y^0$

46. $(2y)^0$

47. z^{-5}

48. 10^{-4}

49. $(6a)^{-2}$

50. $6a^{-2}$

51. $4^0 + 4^{-2}$

52. $9^{-1} + 9^0$

53. $t^{-6}t^{-2}$

54. r^8r^{-9}

55. $\frac{12x^{-2}y^3}{6x^4y^{-4}}$

56. $\frac{8ab^{-3}c^0}{10a^{-5}b^{-4}c^{-1}}$

57. $(-2m^2n^{-4})^{-4}$

58. $(3u^{-5}v^2)^{-3}$

59. $\frac{(k^{-6})^{-2}(k^3)}{5k^{-6}k^0}$

60. $\frac{(3h)^{-2}(h^{-5})^{-3}}{h^{-4}h^8}$

61. $\frac{7^0}{3^{-1} - 6^{-1}}$

62. $\frac{2^0 - 2^{-1}}{2^{-1} - 2^{-2}}$

Section 4.4

63. Write the numbers in scientific notation.
   a. The federal debt was estimated to be \$5,914,800,000,000 in the year 2001.
   b. The width of a piece of paper is 0.0042 in.
   c. The area of the Pacific Ocean is 166,241,000 km^2.

64. Write the numbers without scientific notation.
   a. The pH of 10 means the hydrogen ion concentration is 1×10^{-10}.
   b. The Social Security income was estimated to be \$$5.811 \times 10^{11}$ in the year 2002.
   c. A fund-raising event for neurospinal research raised \$$2.56 \times 10^5$.

For Exercises 65–68, perform the indicated operations. Write the answers in scientific notation.

65. $(4.1 \times 10^{-6})(2.3 \times 10^{11})$

66. $\frac{9.3 \times 10^3}{6.0 \times 10^{-7}}$

67. $\frac{2000}{0.000008}$

68. $(0.000078)(21{,}000{,}000)$

69. Use your calculator to evaluate 5^{20}. Why is scientific notation necessary on your calculator to express the answer?

70. Use your calculator to evaluate $(0.4)^{30}$. Why is scientific notation necessary on your calculator to express the answer?

71. The average distance between the earth and sun is 9.3×10^7 miles.

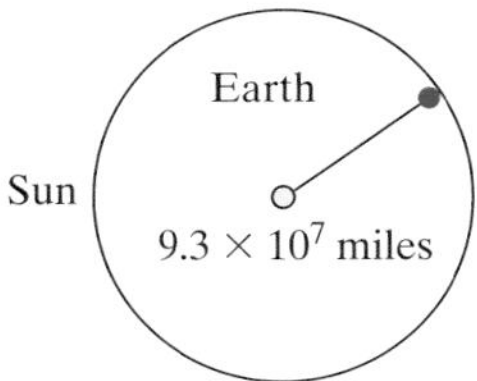

Figure for Exercise 71

   a. If the earth's orbit is approximated by a circle, find the total distance the earth travels around the sun in one orbit. (*Hint*: The circumference of a circle is given by $C = 2\pi r$.) Express the answer in scientific notation.

b. If the earth makes one complete trip around the sun in 1 year (365 days = 8.76×10^3 hr), find the average speed that the earth travels around the sun in miles per hour. Express the answer in scientific notation.

72. The average distance between the planet Mercury and the sun is 3.6×10^7 miles.

a. If Mercury's orbit is approximated by a circle, find the total distance Mercury travels around the sun in one orbit. (*Hint*: The circumference of a circle is given by $C = 2\pi r$.) Express the answer in scientific notation.

b. If Mercury makes one complete trip around the sun in 88 days (2.112×10^3 hr), find the average speed that Mercury travels around the sun in miles per hour. Express the answer in scientific notation.

Section 4.5

73. For the polynomial $7x^4 - x + 6$

a. Classify as a monomial, a binomial, or a trinomial.

b. Identify the degree of the polynomial.

c. Identify the leading coefficient.

74. For the polynomial $2y^3 - 5y^7$

a. Classify as a monomial, a binomial, or a trinomial.

b. Identify the degree of the polynomial.

c. Identify the leading coefficient.

For Exercises 75–80, add or subtract as indicated.

75. $(4x + 2) + (3x - 5)$

76. $(7y^2 - 11y - 6) - (8y^2 + 3y - 4)$

77. $(9a^2 - 6) - (-5a^2 + 2a)$

78. $(8w^4 - 6w + 3) + (2w^4 + 2w^3 - w + 1)$

79. $\left(5x^3 - \frac{1}{4}x^2 + \frac{5}{8}x + 2\right) + \left(\frac{5}{2}x^3 + \frac{1}{2}x^2 - \frac{1}{8}x\right)$

80. $(-0.02b^5 + b^4 - 0.7b + 0.3) + (0.03b^5 - 0.1b^3 + b + 0.03)$

81. Subtract $(9x^2 + 4x + 6)$ from $(7x^2 - 5x)$.

82. Find the difference of $(x^2 - 5x - 3)$ and $(6x^2 + 4x + 9)$.

83. Write a trinomial of degree 2 with a leading coefficient of -5. (Answers may vary.)

84. Write a binomial of degree 6 with leading coefficient 6. (Answers may vary.)

85. Find a polynomial that represents the perimeter of the given rectangle.

2w + 3

w

Figure for Exercise 85

Section 4.6

For Exercises 86–101, multiply the expressions.

86. $(25x^4y^3)(-3x^2y)$

87. $(9a^6)(2a^2b^4)$

88. $5c(3c^3 - 7c + 5)$

89. $(x^2 + 5x - 3)(-2x)$

90. $(5k - 4)(k + 1)$

91. $(4t - 1)(5t + 2)$

92. $(q + 8)(6q - 1)$

93. $(2a - 6)(a + 5)$

94. $\left(7a + \frac{1}{2}\right)\left(7a + \frac{1}{2}\right)$

95. $(b - 4)^2$

96. $(4p^2 + 6p + 9)(2p - 3)$

97. $(2w - 1)(-w^2 - 3w - 4)$

98. $(b - 4)(b + 4)$

99. $\left(\frac{1}{3}r^4 - s^2\right)\left(\frac{1}{3}r^4 + s^2\right)$

100. $(-7z^2 + 6)^2$

101. $(2h + 3)(h^4 - h^3 + h^2 - h + 1)$

102. Find a polynomial that represents the area of the given rectangle.

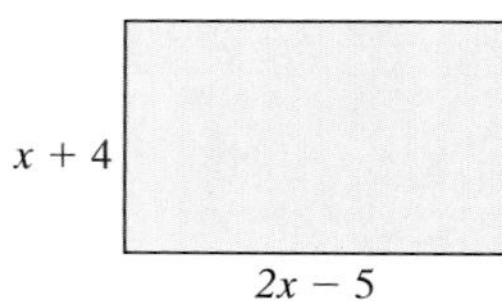

Figure for Exercise 102

Section 4.7

For Exercises 103–118, divide the polynomials.

103. $\dfrac{20y^3 - 10y^2}{5y}$

104. $(18a^3b^2 - 9a^2b - 27ab^2) \div 9ab$

105. $(12x^4 - 8x^3 + 4x^2) \div (-4x^2)$

106. $\dfrac{10z^7w^4 - 15z^3w^2 - 20zw}{-20z^2w}$

107. $\dfrac{x^2 + 7x + 10}{x + 5}$

108. $(2t^2 + t - 10) \div (t - 2)$

109. $(2p^2 + p - 16) \div (2p + 7)$

110. $\dfrac{5a^2 + 27a - 22}{5a - 3}$

111. $\dfrac{b^3 - 125}{b - 5}$

112. $(z^3 + 4z^2 + 5z + 20) \div (5 + z^2)$

113. $(y^4 - 4y^3 + 5y^2 - 3y + 2) \div (y^2 + 3)$

114. $\dfrac{8x^3 + 27}{2x + 3}$

115. $\dfrac{5 - 5x^2 + 6x^3}{3x + 2}$

116. $(4y^2 + 6y) \div (2y - 1)$

117. $(3t^4 - 8t^3 + t^2 - 4t - 5) \div (3t^2 + t + 1)$

118. $\dfrac{2w^4 + w^3 + 4w - 3}{2w^2 - w + 3}$

---

chapter 4 TEST

Assume all variables represent nonzero real numbers.

1. Expand the expression using the definition of exponents, then simplify: $\dfrac{3^4 \cdot 3^3}{3^6}$

For Exercises 2–11, simplify the expression. Write the answer with positive exponents only.

2. $9^5 \cdot 9$

3. $\dfrac{q^{10}}{q^2}$

4. $(3a^2b)^3$

5. $\left(\dfrac{2x}{y^3}\right)^4$

6. $(-7)^0$

7. c^{-3}

8. $\dfrac{14^3 \cdot 14^9}{14^{10} \cdot 14}$

9. $\dfrac{(s^2t)^3(7s^4t)^4}{(7s^2t^3)^2}$

10. $(2a^0b^{-6})^2$

11. $\left(\dfrac{6a^{-5}b}{8ab^{-2}}\right)^{-2}$

12. a. Write the number in scientific notation: 43,000,000,000

    b. Write the number without scientific notation: 5.6×10^{-6}

13. The average amount of water flowing over Niagara Falls is 1.68×10^5 m^3/min.

    a. How many cubic meters of water flow over the falls in one day?

    b. How many cubic meters of water flow over the falls in one year?

14. Write the polynomial in descending order: $4x + 5x^3 - 7x^2 + 11$.

    a. Identify the degree of the polynomial.

    b. Identify the leading coefficient of the polynomial.

15. Perform the indicated operations.

    $(7w^2 - 11w - 6) + (8w^2 + 3w + 4) - (-9w^2 - 5w + 2)$

For Exercises 16–20, multiply the polynomials.

16. $-2x^3(5x^2 + x - 15)$ 17. $(4a - 3)(2a - 1)$

18. $(4y - 5)(y^2 - 5y + 3)$

19. $(2 + 3b)(2 - 3b)$ 20. $(5z - 6)^2$

21. Find the perimeter and the area of the rectangle shown in the figure.

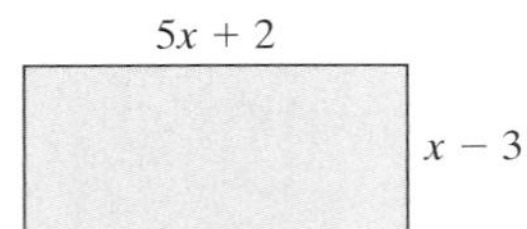

Figure for Exercise 21

22. Divide:

a. $(-12x^8 + x^6 - 8x^3) \div (4x^2)$

b. $\dfrac{2y^2 - 13y + 21}{y - 3}$

CUMULATIVE REVIEW EXERCISES, CHAPTERS 1–4

For Exercises 1–3, simplify completely.

1. $-5 - \frac{1}{2}[4 - 3(-7)]$ 2. $|-3^2 + 5|$

3. $\dfrac{-3 - \sqrt{14 - (-2) + 3^2}}{-3.44 + 1.2^2}$

4. Translate the phrase into a mathematical expression and simplify:

   The difference of the square of five and the square root of four.

5. Which of the following are rational numbers? $\left\{-7, \frac{0}{4}, 2, 0.8, \sqrt{100}, \sqrt{101}\right\}$

6. Solve for x: $\frac{1}{2}(x - 6) + \frac{2}{3} = \frac{1}{4}x$

7. Solve for y: $-2y - 3 = -5(y - 1) + 3y$

8. For a point in a rectangular coordinate system, in which quadrant are both the x- and y-coordinates negative?

9. For a point in a rectangular coordinate system, on which axis is the x-coordinate zero and the y-coordinate nonzero?

10. In a triangle, one angle is 23° more than the smallest angle. The third angle is 10° more than the sum of the other two angles. Find the measure of each angle.

11. A salesperson makes 3% commission on the sale of merchandise. If his total commission at the end of a week is \$360, what was the value of the merchandise he sold?

12. A farmer wants to enclose a rectangular lot that is five times as long as it is wide. One of the shorter sides of the lot is adjacent to a barn and does not require fencing. The farmer plans to fence the other three sides. If the farmer has 264 ft of fencing available, what should the dimensions of the lot be?

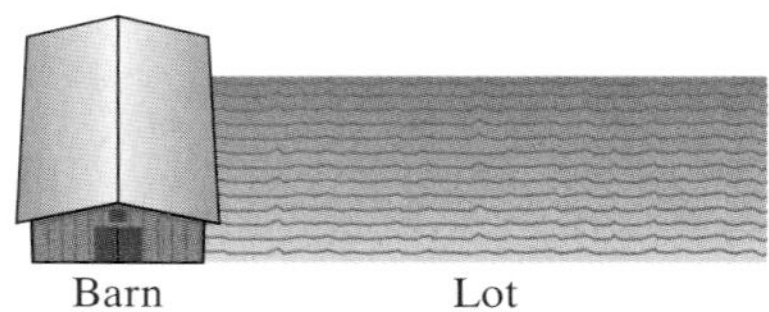

Figure for Exercise 12

13. A snow storm lasts for 9 hr and dumps snow at a rate of $1\frac{1}{2}$ in./hr. If there was already 6 in. of snow on the ground before the storm, the snow depth is given by the equation:

    $y = \frac{3}{2}x + 6$ where y is the snow depth in inches and x is the time in hours.

a. Find the snow depth after 4 hr.
b. Find the snow depth at the end of the storm.
c. How long had it snowed when the total depth of snow was $14\frac{1}{4}$ in.?
d. Complete the table and graph the corresponding ordered pairs.

Time (hr) x	Snow Depth (in.) y
0	
2	
4	
6	
8	
9	

Table for Exercise 13

14. Solve the inequality. Graph the solution set on the real number line and express the solution in interval notation. $2 - 3(2x + 4) \leq -2x - (x - 5)$

For Exercises 15–19, perform the indicated operations.

15. $(2x^2 + 3x - 7) - (-3x^2 + 12x + 8)$

16. $(2y + 3z)(-y - 5z)$

17. $(4t - 3)^2$

18. $\left(\frac{2}{5}a + \frac{1}{3}\right)\left(\frac{2}{5}a - \frac{1}{3}\right)$

19. $(7x - 8) - (x + 3)^2$

For Exercises 20–21, divide the polynomials.

20. $(12a^4b^3 - 6a^2b^2 + 3ab) \div (-3ab)$

21. $\dfrac{4m^3 - 5m + 2}{m - 2}$

For Exercises 22–24, use the properties of exponents to simplify the expressions. Write the answers with positive exponents only. Assume all variables represent nonzero real numbers.

22. $\dfrac{(x^2)^3(x^4x^{-6})}{x^5}$

23. $\left(\dfrac{2c^2d^4}{8cd^6}\right)^2$

24. $\dfrac{10a^{-2}b^{-3}}{5a^0b^{-6}}$

25. Write the following numbers in scientific notation.

a. 407,100,000 b. 0.000 004 071

26. Write the following numbers without scientific notation.

a. 3.89×10^{-4} b. 4.5×10^{12}

27. Perform the indicated operations and write the final answer in scientific notation.

$$\frac{(8.2 \times 10^{-2})(6.8 \times 10^{-6})}{2.0 \times 10^{-5}}$$

chapter

5 FACTORING POLYNOMIALS

Ignoring air resistance, the distance, d (in feet), that a skydiver falls in t seconds is approximated by the formula:

$$d = 16t^2$$

After 1 second, the skydiver will have fallen 16 ft. After 2 seconds the skydiver will have fallen 64 ft, and after 3 seconds, the skydiver will have fallen 144 ft. Notice that each consecutive 1-second interval results in a larger increase in the distance fallen as shown in the graph.

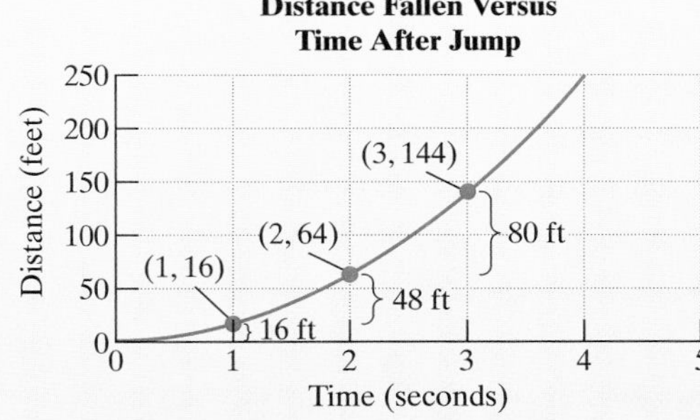

To graph this and other quadratic relationships, see the Technology Connections in MathZone at

www.mhhe.com/miller_oneill

Concepts

1. Introduction to Factoring
2. Greatest Common Factor of Two Integers
3. GCF of Two or More Monomials
4. Factoring out the Greatest Common Factor
5. Factoring out a Negative Factor
6. Factoring out a Binomial Factor
7. Factoring by Grouping

section

5.1 Greatest Common Factor and Factoring by Grouping

1. Introduction to Factoring

Chapter 5 is devoted to a mathematical operation called **factoring**. To factor an integer means to write the integer as a product of two or more integers. To factor a polynomial means to express the polynomial as a product of two or more polynomials.

In the product $2 \cdot 5 = 10$, for example, 2 and 5 are factors of 10.

In the product $(3x + 4)(2x - 1) = 6x^2 + 5x - 4$, the quantities $(3x + 4)$ and $(2x - 1)$ are factors of $6x^2 + 5x - 4$.

2. Greatest Common Factor of Two Integers

We begin our study of factoring by factoring integers. The number 20 for example can be factored as $1 \cdot 20$, $2 \cdot 10$, $4 \cdot 5$, or $2 \cdot 2 \cdot 5$. The product $2 \cdot 2 \cdot 5$ (or equivalently $2^2 \cdot 5$) consists only of prime numbers and is called the **prime factorization**.

The **greatest common factor** (denoted **GCF**) of two or more integers is the greatest factor common to each integer. To find the greatest common factor of two integers, it is often helpful to express the numbers as a product of prime factors as shown in Example 1.

example 1 Identifying the GCF of Two Integers

Find the greatest common factor of each pair of integers.

a. 24 and 36 b. 105 and 40

Solution:

First find the prime factorization of each number. Then find the product of common factors.

a.
$$\begin{array}{r|l} 2 & 24 \\ \hline 2 & 12 \\ \hline 2 & 6 \\ \hline & 3 \end{array} \qquad \begin{array}{r|l} 2 & 36 \\ \hline 2 & 18 \\ \hline 3 & 9 \\ \hline & 3 \end{array}$$

Factors of 24 = $2 \cdot 2 \cdot 2 \cdot 3$

Factors of 36 = $2 \cdot 2 \cdot 3 \cdot 3$

The numbers 24 and 36 share two factors of 2 and one factor of 3. Therefore, the greatest common factor is $2 \cdot 2 \cdot 3 = 12$.

b.
$$\begin{array}{r|l} 5 & 105 \\ \hline 3 & 21 \\ \hline & 7 \end{array} \qquad \begin{array}{r|l} 5 & 40 \\ \hline 2 & 8 \\ \hline 2 & 4 \\ \hline & 2 \end{array}$$

Factors of 105 = $3 \cdot 7 \cdot 5$

Factors of 40 = $2 \cdot 2 \cdot 2 \cdot 5$

The greatest common factor is 5.

3. GCF of Two or More Monomials

example 2 **Identifying the Greatest Common Factor of Two or More Polynomials**

Find the GCF among each group of terms.

a. $7x^3, 14x^2, 21x^4$ b. $15a^4b, 25a^3b^2$ c. $8c^2d^7e, 6c^3d^4$

Solution:

List the factors of each term.

a.
$$7x^3 = 7 \cdot x \cdot x \cdot x$$
$$14x^2 = 2 \cdot 7 \cdot x \cdot x$$
$$21x^4 = 3 \cdot 7 \cdot x \cdot x \cdot x \cdot x$$

The GCF is $7x^2$.

b.
$$15a^4b = 3 \cdot 5 \cdot a \cdot a \cdot a \cdot a \cdot b$$
$$25a^3b^2 = 5 \cdot 5 \cdot a \cdot a \cdot a \cdot b \cdot b$$

The GCF is $5a^3b$.

Tip: Notice that the expressions $15a^4b$ and $25a^3b^2$ share factors of 5, a, and b. The GCF is the product of the common factors, where each factor is raised to the lowest power to which it occurs in the original expressions.

$$15a^4b = 3 \cdot 5a^4b$$
$$25a^3b^2 = 5^2a^3b^2$$

Lowest power of 5 is 1: 5^1

Lowest power of a is 3: a^3

Lowest power of b is 1: b^1

The GCF is $5a^3b$.

c.
$$8c^2d^7e = 2^3c^2d^7e$$
$$6c^3d^4 = 2 \cdot 3c^3d^4$$

The common factors are 2, c, and d.

The lowest power of 2 is 1: 2^1

The lowest power of c is 2: c^2

The lowest power of d is 4: d^4

The GCF is $2c^2d^4$.

Sometimes two polynomials share a common binomial factor as shown in Example 3.

example 3

Finding the Greatest Common Binomial Factor

Find the greatest common factor between the terms: $3x(a + b)$ and $2y(a + b)$

Solution:

$$\left.\begin{array}{l} 3x(a+b) \\ 2y(a+b) \end{array}\right\}$$ The only common factor is the binomial $(a + b)$. The GCF is $(a + b)$.

4. Factoring out the Greatest Common Factor

The process of factoring a polynomial is the reverse process of multiplying polynomials. Both operations use the distributive property: $ab + ac = a(b + c)$.

Multiply

$$\begin{aligned} 5y(y^2 + 3y + 1) &= 5y(y^2) + 5y(3y) + 5y(1) \\ &= 5y^3 + 15y^2 + 5y \end{aligned}$$

Factor

$$\begin{aligned} 5y^3 + 15y^2 + 5y &= 5y(y^2) + 5y(3y) + 5y(1) \\ &= 5y(y^2 + 3y + 1) \end{aligned}$$

Steps to Removing the Greatest Common Factor

1. Identify the GCF of all terms of the polynomial.
2. Write each term as the product of the GCF and another factor.
3. Use the distributive property to remove the GCF.

Note: To check the factorization, multiply the polynomials to remove parentheses.

example 4

Factoring out the Greatest Common Factor

Factor out the GCF.

a. $4x - 20$ b. $6w^2 + 3w$ c. $15y^3 + 12y^4$ d. $9a^4b - 18a^5b + 27a^6b$

Solution:

a. $4x - 20$ The GCF is 4.

$= 4(x) - 4(5)$ Write each term as the product of the GCF and another factor.

$= 4(x - 5)$ Use the distributive property to factor out the GCF.

Tip: Any factoring problem can be checked by multiplying the factors:

Check: $4(x - 5) = 4x - 20$ ✔

Avoiding Mistakes

In Example 4(b) the GCF, $3w$, is equal to one of the terms of the polynomial. In such a case, you must leave a 1 in place of that term after the GCF is factored out.

b. $6w^2 + 3w$ — The GCF is $3w$.

$= 3w(2w) + 3w(1)$ — Write each term as the product of $3w$ and another factor.

$= 3w(2w + 1)$ — Use the distributive property to factor out the GCF.

c. $15y^3 + 12y^4$ — The GCF is $3y^3$.

$= 3y^3(5) + 3y^3(4y)$ — Write each term as the product of $3y^3$ and another factor.

$= 3y^3(5 + 4y)$ — Use the distributive property to factor out the GCF.

d. $9a^4b - 18a^5b + 27a^6b$ — The GCF is $9a^4b$.

$= 9a^4b(1) - 9a^4b(2a) + 9a^4b(3a^2)$ — Write each term as the product of $9a^4b$ and another factor.

$= 9a^4b(1 - 2a + 3a^2)$ — Use the distributive property to factor out the GCF.

5. Factoring out a Negative Factor

Sometimes it is advantageous to factor out the *opposite* of the GCF when the leading coefficient of the polynomial is negative. This is demonstrated in Example 5. Notice that this *changes the signs* of the remaining terms inside the parentheses.

example 5

Factoring out a Negative Factor

Factor out the quantity $-4pq$ from the polynomial $-12p^3q - 8p^2q^2 + 4pq^3$

Solution:

$-12p^3q - 8p^2q^2 + 4pq^3$ — The GCF is $4pq$. However, in this case, we will factor out the *opposite* of the GCF, $-4pq$.

$= -4pq(3p^2) + (-4pq)(2pq) + (-4pq)(-q^2)$ — Write each term as the product of $-4pq$ and another factor.

$= -4pq[3p^2 + 2pq + (-q^2)]$ — Factor out $-4pq$.

$= -4pq(3p^2 + 2pq - q^2)$ — Notice that each sign within the trinomial has changed.

Tip: To verify that this is the correct factorization and that the signs are correct, multiply the factors.

6. Factoring out a Binomial Factor

The distributive property may also be used to factor out a common factor that consists of more than one term as shown in Example 6.

example 6

Factoring out a Binomial Factor

Factor out the greatest common factor: $2w(x + 3) - 5(x + 3)$

Solution:

$2w(x + 3) - 5(x + 3)$	The greatest common factor is the quantity $(x + 3)$.
$= (x + 3)(2w) - (x + 3)(5)$	Write each term as the product of $(x + 3)$ and another factor.
$= (x + 3)(2w - 5)$	Use the distributive property to factor out the GCF.

7. Factoring by Grouping

When two binomials are multiplied, the product before simplifying contains four terms. For example:

$$(x + 4)(3a + 2b) = (x + 4)(3a) + (x + 4)(2b)$$
$$= (x + 4)(3a) + (x + 4)(2b)$$
$$= 3ax + 12a + 2bx + 8b$$

In Example 7, we learn how to reverse this process. That is, given a four-term polynomial, we will factor it as a product of two binomials. The process is called **factoring by grouping**.

example 7

Factoring by Grouping

Factor by grouping: $3ax + 12a + 2bx + 8b$

Solution:

$3ax + 12a + 2bx + 8b$	**Step 1:**	Identify and factor out the GCF from all four terms. In this case, the GCF is 1.
$= 3ax + 12a \mid + 2bx + 8b$		Group the first pair of terms and the second pair of terms.

$= 3a(x + 4) + 2b(x + 4)$ **Step 2:** Factor out the GCF from each pair of terms. *Note:* The two terms now share a common binomial factor of $(x + 4)$.

$= (x + 4)(3a + 2b)$ **Step 3:** Factor out the common binomial factor.

Check: $(x + 4)(3a + 2b) = 3ax + 2bx + 12a + 8b$ ✔

Note: Step 2 results in two terms with a common binomial factor. If the two binomials are different, step 3 cannot be performed. In such a case, the original polynomial may not be factorable by grouping.

Tip: One frequently asked question when factoring is whether the order can be switched between the factors. The answer is yes. Because multiplication is commutative, the order in which the factors are written does not matter.

$$(x + 4)(3a + 2b) = (3a + 2b)(x + 4)$$

Steps to Factoring by Grouping

To factor a four-term polynomial by grouping:

1. Identify and factor out the GCF from all four terms.
2. Factor out the GCF from the first pair of terms. Factor out the GCF from the second pair of terms. (Sometimes it is necessary to factor out the opposite of the GCF.)
3. If the two terms share a common binomial factor, factor out the binomial factor.

example 8 **Factoring by Grouping**

Factor the polynomials by grouping.

a. $ax + ay - bx - by$ b. $16w^4 - 40w^3 - 12w^2 + 30w$

Solution:

a. $ax + ay - bx - by$ **Step 1:** Identify and factor out the GCF from all four terms. In this case, the GCF is 1.

$= ax + ay \;\vdots\; - bx - by$ Group the first pair of terms and the second pair of terms.

Avoiding Mistakes

In step 2, the expression $a(x + y) - b(x + y)$ is not yet factored because it is a *difference*, not a product. To factor the expression, you must carry it one step further.

$$a(x + y) - b(x + y)$$
$$= (x + y)(a - b)$$

The factored form must be represented as a product.

$= a(x + y) - b(x + y)$ **Step 2:** Factor out a from the first pair of terms.

Factor out $-b$ from the second pair of terms. (This causes sign changes within the second parentheses.)

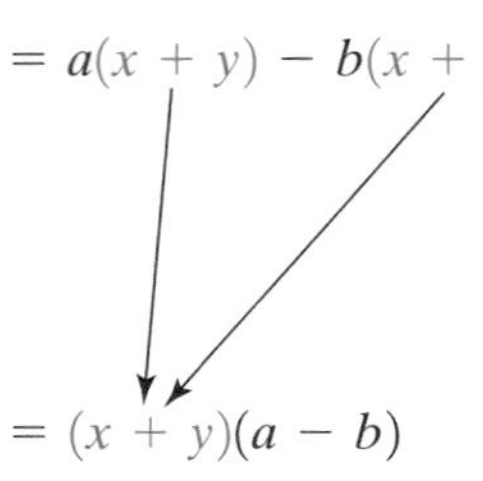

$= (x + y)(a - b)$ **Step 3:** Factor out the common binomial factor.

b. $16w^4 - 40w^3 - 12w^2 + 30w$ **Step 1:** Identify and factor out the GCF from all four terms. In this case, the GCF is $2w$.

$= 2w(8w^3 - 20w^2 - 6w + 15)$

$= 2w[8w^3 - 20w^2 \mid - 6w + 15]$ Group the first pair of terms and the second pair of terms.

$= 2w[4w^2(2w - 5) - 3(2w - 5)]$ **Step 2:** Factor out $4w^2$ from the first pair of terms.

Factor out -3 from the second pair of terms. (This causes sign changes within the second parentheses.)

$= 2w[(2w - 5)(4w^2 - 3)]$ **Step 3:** Factor out the common binomial factor.

section 5.1 PRACTICE EXERCISES

For Exercises 1–12, identify the greatest common factor between each pair of terms.

1. 28, 63
2. 24, 40
3. 42, 30
4. 18, 52
5. $2a^2b, 3ab^2$
6. $3x^3y^2, 5xy^4$
7. $12w^3z, 16w^2z$
8. $20cd, 15c^3d$
9. $7(x - y), 9(x - y)$
10. $(2a - b), 3(2a - b)$
11. $14(3x + 1)^2, 7(3x + 1)$
12. $a^2(w + z), a^3(w + z)^2$
13. a. Use the distributive property to multiply $3(x - 2y)$.
    b. Use the distributive property to factor $3x - 6y$.
14. a. Use the distributive property to multiply $a^2(5a + b)$.
    b. Use the distributive property to factor $5a^3 + a^2b$.

For Exercises 15–38, factor out the GCF.

15. $4p + 12$
16. $3q - 15$
17. $5c^2 - 10c$
18. $24d + 16d^3$
19. $x^5 + x^3$
20. $y^2 - y^4$
21. $t^4 - 4t$
22. $7r^3 - r^5$
23. $2ab + 4a^3b$
24. $5u^3v^2 - 5uv$
25. $38x^2y - 19x^2y^4$
26. $100a^5b^3 + 16a^2b$
27. $42p^3q^2 + 14pq^2 - 7p^4q^4$

28. $8m^2n^3 - 24m^2n^2 + 4m^3n$

29. $t^5 + 2rt^3 - 3t^4 + 4r^2t^2$

30. $u^2v + 5u^3v^2 - 2u^2 + 8uv$

31. $13(a + 6) - 4b(a + 6)$

32. $7(x^2 + 1) - y(x^2 + 1)$

33. $8v(w^2 - 2) + (w^2 - 2)$

34. $t(r + 2) + (r + 2)$

35. $21x(x + 3) + 7x^2(x + 3)$

36. $5y^3(y - 2) - 20y(y - 2)$

37. $6(z - 1)^3 + 7z(z - 1)^2 - (z - 1)$

38. $4(q + 5)^2 + 5q(q + 5) - (q + 5)$

39. For the polynomial $-2x^3 - 4x^2 + 8x$
    a. Factor out $-2x$.
    b. Factor out $2x$.

40. For the polynomial $-9y^5 + 3y^3 - 12y$
    a. Factor out $-3y$.
    b. Factor out $3y$.

41. Factor out -1 from the polynomial $-8t^2 - 9t - 2$.

42. Factor out -1 from the polynomial $-6x^3 - 2x - 5$.

43. Factor out -1 from the polynomial $-4y^3 + 5y - 7$.

44. Factor out -1 from the polynomial $-w^2 + w - 5$.

For Exercises 45–52, factor out the opposite of the greatest common factor.

45. $-15p^3 - 30p^2$

46. $-24m^3 - 12m^4$

47. $-q^4 + 2q^2 - 9q$

48. $-r^3 + 9r^2 - 5r$

49. $-7x - 6y - 2z$

50. $-4a + 5b - c$

51. $-3(2c + 5) - 4c(2c + 5)$

52. $-6n(4n - 1) - 7(4n - 1)$

For Exercises 53–68, factor by grouping.

53. $8a^2 - 4ab + 6ac - 3bc$

54. $4x^3 + 3x^2y + 4xy^2 + 3y^3$

55. $3q + 3p + qr + pr$

56. $xy - xz + 7y - 7z$

57. $6x^2 + 3x + 4x + 2$

58. $4y^2 + 8y + 7y + 14$

59. $2t^2 + 6t - 5t - 15$

60. $2p^2 - p - 6p + 3$

61. $6y^2 - 2y - 9y + 3$

62. $5a^2 + 30a - 2a - 12$

63. $b^4 + b^3 - 4b - 4$

64. $8w^5 + 12w^2 - 10w^3 - 15$

65. $3j^2k + 15k + j^2 + 5$

66. $2ab^2 - 6ac + b^2 - 3c$

67. $14w^6x^6 + 7w^6 - 2x^6 - 1$

68. $18p^4q - 9p^5 - 2q + p$

For Exercises 69–74, factor out the GCF first. Then factor by grouping.

69. $15x^4 + 15x^2y^2 + 10x^3y + 10xy^3$

70. $2a^3b - 4a^2b + 32ab - 64b$

71. $4abx - 4b^2x - 4ab + 4b^2$

72. $p^2q - pq^2 - rp^2q + rpq^2$

73. $6st^2 - 18st - 6t^4 + 18t^3$

74. $15j^3 - 10j^2k - 15j^2k^2 + 10jk^3$

75. The formula $P = 2l + 2w$ represents the perimeter, P, of a rectangle given the length, l, and the width, w. Factor out the GCF and write an equivalent formula in factored form.

76. The formula $P = 2a + 2b$ represents the perimeter, P, of a parallelogram given the base, b, and an adjacent side, a. Factor out the GCF and write an equivalent formula in factored form.

77. The formula $S = 2\pi r^2 + 2\pi rh$ represents the surface area, S, of a cylinder with radius, r, and height, h. Factor out the GCF and write an equivalent formula in factored form.

78. The formula $A = P + Prt$ represents the total amount of money, A, in an account that earns simple interest at a rate, r, for t years. Factor out the GCF and write an equivalent formula in factored form.

Expanding Your Skills

79. Factor out $\frac{1}{7}$ from $\frac{1}{7}x^2 + \frac{3}{7}x - \frac{5}{7}$.

80. Factor out $\frac{1}{5}$ from $\frac{6}{5}y^2 - \frac{4}{5}y + \frac{1}{5}$.

81. Factor out $\frac{1}{4}$ from $\frac{5}{4}w^2 + \frac{3}{4}w + \frac{9}{4}$.

82. Factor out $\frac{1}{6}$ from $\frac{1}{6}p^2 - \frac{3}{6}p + \frac{5}{6}$.

83. Factor out $\frac{1}{12}$ from $\frac{1}{12}z^2 + \frac{1}{3}z + \frac{1}{2}$.
(*Hint:* Write each coefficient as an equivalent fraction with a common denominator of 12.)

84. Factor out $\frac{1}{15}$ from $\frac{1}{5}t^2 - \frac{2}{3}t - \frac{4}{15}$.

85. Factor out $\frac{1}{6}$ from $\frac{5}{6}q^2 + \frac{1}{3}q - 2$.

86. Factor out $\frac{1}{8}$ from $\frac{3}{8}x^2 - \frac{1}{2}x - 1$

87. Write a polynomial that has a GCF of $3x$. (Answers may vary.)

88. Write a polynomial that has a GCF of $7y$. (Answers may vary.)

89. Write a polynomial that has a GCF of $4p^2q$. (Answers may vary.)

90. Write a polynomial that has a GCF of $2ab^2$. (Answers may vary.)

Concepts

1. **Grouping Method to Factor Trinomials**
2. **Factoring Trinomials with a Leading Coefficient of 1**
3. **Prime Polynomials**

section

5.2 Factoring Trinomials: Grouping Method

We have already learned how to factor out the GCF from a polynomial and how to factor a four-term polynomial by grouping. As we work through this chapter, we will expand our knowledge of factoring by learning how to factor trinomials and binomials.

There are two commonly used methods for factoring trinomials. The grouping method (or "ac" method) is presented here and the trial-and-error method is presented in the next section.

1. Grouping Method to Factor Trinomials

The product of two binomials results in a four-term expression that can sometimes be simplified to a trinomial. To factor the trinomial, we want to reverse the process.

Multiply

Multiply the binomials. Add the middle terms.

$$(2x + 3)(x + 2) = \longrightarrow 2x^2 + 4x + 3x + 6 = \longrightarrow 2x^2 + 7x + 6$$

Factor

$$2x^2 + 7x + 6 = \longrightarrow 2x^2 + 4x + 3x + 6 = \longrightarrow (2x + 3)(x + 2)$$

Rewrite the middle term as a sum or difference of terms. Factor by grouping.

To factor a trinomial, $ax^2 + bx + c$, by the grouping method, we rewrite the middle term, bx, as a sum or difference of terms. The goal is to produce a four-term polynomial that can be factored by grouping. The process is outlined in the following box.

Grouping Method Factor $ax^2 + bx + c$ $(a \neq 0)$

1. Multiply the coefficients of the first and last terms (ac).
2. Find two integers whose product is ac and whose sum is b. (If no pair of integers can be found, then the trinomial cannot be factored further and is called a **prime polynomial**.)
3. Rewrite the middle term bx as the sum of two terms whose coefficients are the integers found in step 2.
4. Factor by grouping.

The grouping method for factoring trinomials is illustrated in Example 1. However, before we begin, keep these two important guidelines in mind:

- For any factoring problem you encounter, always factor out the GCF from all terms first.
- To factor a trinomial, write the trinomial in descending order, $ax^2 + bx + c$.

example 1 **Factoring a Trinomial by the Grouping Method**

Factor the trinomial by the grouping method: $2x^2 + 7x + 6$

Solution:

$2x^2 + 7x + 6$ — Factor out the GCF from all terms. In this case, the GCF is 1.

$2x^2 + 7x + 6$ — **Step 1:** The trinomial is written in the form $ax^2 + bx + c$.

$a = 2, b = 7, c = 6$ — Find the product $ac = (2)(6) = 12$.

Tip: Note that the negative factors of 12 do not need to be considered in this case because we are trying to form a sum of positive 7.

12	12
$1 \cdot 12$	$(-1)(-12)$
$2 \cdot 6$	$(-2)(-6)$
$3 \cdot 4$	$(-3)(-4)$

Step 2: List all the factors of *ac* and search for the pair whose sum equals the value of *b*.

That is, list the factors of 12 and find the pair whose sum equals 7.

The numbers 3 and 4 satisfy both conditions: $3 \cdot 4 = 12$ and $3 + 4 = 7$.

$2x^2 + 7x + 6$

$= 2x^2 + 3x + 4x + 6$

Step 3: Write the middle term of the trinomial as the sum of two terms whose coefficients are the selected pair of numbers: 3 and 4.

$= 2x^2 + 3x \mid + 4x + 6$

Step 4: Factor by grouping.

$= x(2x + 3) + 2(2x + 3)$

$= (2x + 3)(x + 2)$

Check: $(2x + 3)(x + 2) = 2x^2 + 4x + 3x + 6$

$= 2x^2 + 7x + 6$ ✔

Tip: One frequently asked question is whether the order matters when we rewrite the middle term of the trinomial as two terms (step 3). The answer is no. From the previous example, the two middle terms in step 3 could have been reversed to obtain the same result:

$$\begin{aligned} &2x^2 + 7x + 6 \\ &= 2x^2 + 4x + 3x + 6 \\ &= 2x(x + 2) + 3(x + 2) \\ &= (x + 2)(2x + 3) \end{aligned}$$

This example also points out that the order in which two factors are written does not matter. The expression $(x + 2)(2x + 3)$ is equivalent to $(2x + 3)(x + 2)$ because multiplication is a commutative operation.

example 2 Factoring Trinomials by the Grouping Method

Factor the trinomial by the grouping method: $-2x + 8x^2 - 3$

Solution:

$-2x + 8x^2 - 3$

First rewrite the polynomial in the form $ax^2 + bx + c$.

$= 8x^2 - 2x - 3$

The GCF is 1.

$a = 8, b = -2, c = -3$ **Step 1:** Find the product $ac = (8)(-3) = -24$.

$\underline{-24}$	$\underline{-24}$
$-1 \cdot 24$	$-6 \cdot 4$
$-2 \cdot 12$	$-8 \cdot 3$
$-3 \cdot 8$	$-12 \cdot 2$
$-4 \cdot 6$	$-24 \cdot 1$

Step 2: List all the factors of -24 and find the pair of factors whose sum equals -2.

The numbers -6 and 4 satisfy both conditions: $(-6)(4) = -24$ and $-6 + 4 = -2$.

$= 8x^2 - 2x - 3$

Step 3: Write the middle term of the trinomial as two terms whose coefficients are the selected pair of numbers, -6 and 4.

$= 8x^2 - 6x + 4x - 3$

$= 8x^2 - 6x \mid + 4x - 3$ **Step 4:** Factor by grouping.

$= 2x(4x - 3) + 1(4x - 3)$

$= (4x - 3)(2x + 1)$

Check: $(4x - 3)(2x + 1) = 8x^2 + 4x - 6x - 3$
$= 8x^2 - 2x - 3$ ✔

example 3

Factoring a Trinomial by the Grouping Method

Factor the trinomial by the grouping method: $10x^3 - 85x^2 + 105x$

Solution:

$10x^3 - 85x^2 + 105x$ The GCF is $5x$.

$= 5x(2x^2 - 17x + 21)$ The trinomial is in the form $ax^2 + bx + c$.

$a = 2, b = -17, c = 21$ **Step 1:** Find the product $ac = (2)(21) = 42$.

Tip: Note that the positive factors of 42 do not need to be considered in this case because we are trying to form a sum of -17.

$\underline{42}$	$\underline{42}$
$1 \cdot 42$	$(-1)(-42)$
$2 \cdot 21$	$(-2)(-21)$
$3 \cdot 14$	$(-3)(-14)$
$6 \cdot 7$	$(-6)(-7)$

Step 2: List all the factors of 42 and find the pair whose sum equals -17.

The numbers -3 and -14 satisfy both conditions: $(-3)(-14) = 42$ and $-3 + (-14) = -17$.

$= 5x(2x^2 - 17x + 21)$

Step 3: Write the middle term of the trinomial as two terms whose coefficients are the selected pair of numbers, -3 and -14.

$= 5x(2x^2 - 3x - 14x + 21)$

$= 5x(2x^2 - 3x \mid - 14x + 21)$ **Step 4:** Factor by grouping.

$= 5x[x(2x - 3) - 7(2x - 3)]$

$= 5x[(2x - 3)(x - 7)]$

$= 5x(2x - 3)(x - 7)$

Tip: Notice when the GCF is removed from the original trinomial, the new trinomial has smaller coefficients. This makes the factoring process simpler because the product *ac* is smaller. It is much easier to list the factors of 42 than the factors of 1050.

Original trinomial	**With the GCF factored out**
$10x^3 - 85x^2 + 105x$	$5x(2x^2 - 17x + 21)$
$ac = (10)(105) = 1050$	$ac = (2)(21) = 42$

In most cases it is easier to factor a trinomial with a positive leading coefficient. If the leading coefficient is negative, a factor of -1 can be removed to change the sign of the leading coefficient as well as the coefficients of the remaining terms.

example 4 Factoring a Trinomial by the Grouping Method

Factor the trinomial by the grouping method: $-a^2 - 7ab + 18b^2$

Solution:

$-a^2 - 7ab + 18b^2$

$= -1(a^2 + 7ab - 18b^2)$ Factor out -1.

Step 1: Find the product $ac = -18$.

Step 2: The numbers 9 and -2 have a product of -18 and a sum of 7.

$= -1(a^2 + 9ab - 2ab - 18b^2)$ **Step 3:** Rewrite the middle term $7ab$ as $9ab - 2ab$.

$= -1(a^2 + 9ab \mid - 2ab - 18b^2)$ **Step 4:** Factor by grouping.

$= -1[a(a + 9b) - 2b(a + 9b)]$

$= -1[(a + 9b)(a - 2b)]$

<u>Check</u>: $-1[(a + 9b)(a - 2b)] = -1[a^2 - 2ab + 9ab - 18b^2]$

$= -1[a^2 + 7ab - 18b^2]$

$= -a^2 - 7ab + 18b^2$ ✔

2. Factoring Trinomials with a Leading Coefficient of 1

The grouping method is a general method to factor trinomials of the form $ax^2 + bx + c$. If the leading coefficient, a, is equal to 1, then the process can be simplified. First note that if $a = 1$, then the product $ac = 1c = c$. Therefore, in step 2, we find two integers whose product is c and whose sum is b. Then, after rewriting the middle term and factoring by grouping, we have two binomial factors whose leading terms are x and whose constant terms are the integers found in step 2.

In the next example, we will factor trinomials of the form $x^2 + bx + c$ (leading coefficient of 1) to illustrate this process.

example 5 **Factoring Trinomials with a Leading Coefficient of 1**

Factor the trinomials.

a. $w^2 + 10w + 9$ b. $x^2 - 2x - 15$

Solution:

a. $1w^2 + 10w + 9$ The trinomial is in the form $ax^2 + bx + c$, where $a = 1$.

Step 1: Because $a = 1$, the product $ac = c = 9$.

Step 2: Find two numbers whose product is 9 and whose sum is 10. The numbers are 9 and 1.

$= w^2 + 9w + 1w + 9$ **Step 3:** Rewrite the middle term as a sum of terms.

$= w(w + 9) + 1(w + 9)$ **Step 4:** Factor by grouping.

$= (w + 9)(w + 1)$ The check is left to the reader.

Tip: The constants in the binomial factors are the two integers found in step 2.

b. $1x^2 - 2x - 15$ The trinomial is in the form $ax^2 + bx + c$, where $a = 1$.

Step 1: Because $a = 1$, the product $ac = c = -15$.

$= (x - 5)(x + 3)$ **Step 2:** Find two numbers whose product is -15 and whose sum is -2. The numbers are -5 and 3. These values are the constants in the binomial factors.

3. Prime Polynomials

Note that not every trinomial is factorable by the methods presented in this text.

example 6 **Factoring a Trinomial by the Grouping Method**

Factor the trinomial by the grouping method: $2p^2 - 8p + 3$

Solution:

$2p^2 - 8p + 3$ **Step 1:** The GCF is 1.

Step 2: The product $ac = 6$.

<u>6</u>	<u>6</u>
$1 \cdot 6$	$(-1)(-6)$
$2 \cdot 3$	$(-2)(-3)$

Step 3: List the factors of 6. Notice that no pair of factors has a sum of -8. Therefore, the trinomial cannot be factored.

The trinomial $2p^2 - 8p + 3$ is prime.

section 5.2 Practice Exercises

For Exercises 1–6, factor out the GCF.

1. $8p^9 + 24p^3$
2. $5q^4 - 10q^5$
3. $9x^2y + 12xy^2 - 15x^2y^2$
4. $15ab^3 - 10a^2bc + 25b^2c^3$
5. $5x(x - 2) - 2(x - 2)$
6. $8(y + 5) + 9y(y + 5)$

For Exercises 7–10, factor by grouping.

7. $p^2 - 2pq - pq + 2q^2$
8. $2u - 10 - uv + 5v$
9. $6a^2 + 24a - 12a - 48$
10. $5b^2 + 30b - 10b - 60$
11. What is a prime polynomial?
12. How do you determine if a trinomial, $ax^2 + bx + c$, is prime?

For Exercises 13 – 20, find the pair of integers whose product and sum are given.

13. Product: 12 Sum: 13
14. Product: 12 Sum: 7
15. Product: 8 Sum: -9
16. Product: -4 Sum: -3
17. Product: -6 Sum: 5
18. Product: -18 Sum: 7
19. Product: -72 Sum: -6
20. Product: 36 Sum: -12

For Exercises 21–62, factor the trinomials using the grouping method.

21. $3x^2 + 13x + 4$
22. $2y^2 + 7y + 6$
23. $4w^2 - 9w + 2$
24. $2p^2 - 3p - 2$
25. $2m^2 + 5m - 3$
26. $6n^2 + 7n - 3$
27. $8k^2 - 6k - 9$
28. $9h^2 - 12h + 4$
29. $4k^2 - 20k + 25$
30. $16h^2 + 24h + 9$
31. $5x^2 + x + 7$
32. $4y^2 - y + 2$
33. $4p^2 + 5pq - 6q^2$
34. $6u^2 - 19uv + 10v^2$
35. $15m^2 + mn - 2n^2$
36. $12a^2 + 11ab - 5b^2$
37. $3r^2 - rs - 14s^2$
38. $3h^2 + 19hk - 14k^2$
39. $2x^2 - 13xy + y^2$
40. $3p^2 + 20pq - q^2$

41. $q^2 - 11q + 10$
42. $a^2 + 7a - 18$
43. $r^2 - 6r - 40$
44. $s^2 - 10s - 24$
45. $x^2 + 6x - 7$
46. $y^2 + 5y - 24$
47. $m^2 - 13m + 42$
48. $n^2 - 9n + 20$
49. $a^2 + 9a + 20$
50. $b^2 + 13b + 42$
51. $t^2 + 5t + 5$
52. $s^2 - 6s + 3$
53. $p^2 + 20pq + 100q^2$
54. $c^2 - 14cd + 49d^2$
55. $x^2 - xy - 42y^2$
56. $a^2 - 13ab + 40b^2$
57. $r^2 + 8rs + 15s^2$
58. $u^2 + 2uv - 15v^2$
59. $9z^2 - 21z + 10$
60. $4x^2 + 13x - 12$
61. $7y^2 + 25y + 12$
62. $20p^2 - 19p + 3$
63. Is the expression $(2x + 4)(x - 7)$ factored completely? Explain why or why not.
64. Is the expression $(3x + 1)(5x - 10)$ factored completely? Explain why or why not.

For Exercises 65–78, first factor out a common factor. Then factor the trinomials by using the grouping method.

65. $72x^2 + 18x - 2$
66. $20y^2 - 78y - 8$
67. $p^3 - 6p^2 - 27p$
68. $w^5 - 11w^4 + 28w^3$
69. $6x^2 + 20x + 14$
70. $4r^3 + 3r^2 - 10r$
71. $2p^3 - 38p^2 + 120p$
72. $4q^3 - 4q^2 - 80q$
73. $x^2y^2 + 14x^2y + 33x^2$
74. $a^2b^2 + 13ab^2 + 30b^2$
75. $-k^2 - 7k - 10$
76. $-m^2 - 15m + 34$
77. $-3n^2 - 3n + 90$
78. $-2h^2 + 28h - 90$

For Exercises 79–86, write the polynomials in descending order. Then factor the polynomials.

79. $3 - 14z + 16z^2$
80. $10w + 1 + 16w^2$
81. $b^2 + 16 - 8b$
82. $1 + q^2 - 2q$
83. $25x - 5x^2 - 30$
84. $20a - 18 - 2a^2$
85. $-6 - t + t^2$
86. $-6 + m + m^2$

Concepts

1. Trial-and-Error Method for Factoring Trinomials
2. Identifying the Signs When Using the Trial-and-Error Method
3. Factoring Trinomials with a Leading Coefficient of 1
4. Greatest Common Factor and Factoring Trinomials

section

5.3 Factoring Trinomials: Trial-and-Error Method

In Section 5.2, the grouping method was presented for factoring trinomials. In this section, we offer another method to factor trinomials, called the trial-and-error method. You and your instructor must determine which method is best for you.

1. Trial-and-Error Method for Factoring Trinomials

To understand the basis of factoring trinomials of the form $ax^2 + bx + c$, first consider the multiplication of two binomials:

Product of $2 \cdot 1$ Product of $3 \cdot 2$

$$(2x + 3)(1x + 2) = 2x^2 + \underbrace{\mathbf{4x + 3x}}_{\text{Sum of products of inner terms and outer terms}} + 6 = 2x^2 + 7x + 6$$

To factor the trinomial, $2x^2 + 7x + 6$ this operation is reversed. Hence:

Factors of 2

$$2x^2 + 7x + 6 = (\square x \quad \square)(\square x \quad \square)$$

Factors of 6

We need to fill in the blanks so that the product of the first terms in the binomials is $2x^2$ and the product of the last terms in the binomials is 6. Furthermore, the factors of $2x^2$ and 6 must be chosen so that the sum of the products of the inner terms and outer terms equals $7x$.

To produce the product $2x^2$ we might try the factors $2x$ and x within the binomials:

$$(2x \quad \square)(x \quad \square)$$

To produce a product of 6, the remaining terms in the binomials must either both be positive or both be negative. To produce a positive middle term, we will try positive factors of 6 in the remaining blanks until the correct product is found. The possibilities are $1 \cdot 6, 2 \cdot 3, 3 \cdot 2$, and $6 \cdot 1$.

$(2x + 1)(x + 6) = 2x^2 + 12x + 1x + 6 = 2x^2 + 13x + 6$ Wrong middle term

$(2x + 2)(x + 3) = 2x^2 + 6x + 2x + 6 = 2x^2 + 8x + 6$ Wrong middle term

$(2x + 3)(x + 2) = 2x^2 + 4x + 3x + 6 = 2x^2 + 7x + 6$ Correct!

$(2x + 6)(x + 1) = 2x^2 + 2x + 6x + 6 = 2x^2 + 8x + 6$ Wrong middle term

The correct factorization of $2x^2 + 7x + 6$ is $(2x + 3)(x + 2)$. ✔

As this example shows, we factor a trinomial of the form $ax^2 + bx + c$ by shuffling the factors of a and c within the binomials until the correct product is obtained. However, sometimes it is not necessary to test all the possible combinations of factors. In the previous example, the GCF of the original trinomial is 1. Therefore, any binomial factor that shares a common factor *greater than 1* does not need to be considered. In this case the possibilities $(2x + 2)(x + 3)$ and $(2x + 6)(x + 1)$ cannot work.

$$\underbrace{(2x + 2)}_{\text{Common factor of 2}}(x + 3) \qquad \underbrace{(2x + 6)}_{\text{Common factor of 2}}(x + 1).$$

The steps to factor a trinomial by the trial-and-error method are outlined in the following box.

Trial-and-Error Method to Factor $ax^2 + bx + c$

1. Factor out the GCF.
2. List all pairs of positive factors of a and pairs of positive factors of c. Consider the reverse order for either list of factors.
3. Construct two binomials of the form:

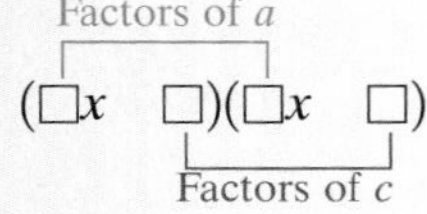

Test each combination of factors and signs until the correct product is found. If no combination of factors produces the correct product, the trinomial cannot be factored further and is called a **prime polynomial**.

Before we begin our next example, keep these two important guidelines in mind.

- For any factoring problem you encounter, always factor out the GCF from all terms first.
- To factor a trinomial, write the trinomial in the form $ax^2 + bx + c$.

example 1 **Factoring a Trinomial by the Trial-and-Error Method**

Factor the trinomial by the trial-and-error method: $10x^2 - 9x - 1$

Solution:

$10x^2 - 9x - 1$ **Step 1:** Factor out the GCF from all terms. In this case, the GCF is 1.

The trinomial is written in the form $ax^2 + bx + c$.

To factor $10x^2 - 9x - 1$, two binomials must be constructed in the form:

Factors of 10

$(\square x \quad \square)(\square x \quad \square)$

Factors of -1

Step 2: To produce the product $10x^2$, we might try $5x$ and $2x$ or $10x$ and $1x$. To produce a product of -1, we will try the factors $1(-1)$ and $-1(1)$.

Step 3: Construct all possible binomial factors using different combinations of the factors of $10x^2$ and -1.

$(5x + 1)(2x - 1) = 10x^2 - 5x + 2x - 1 = 10x^2 - 3x - 1$ Wrong middle term

$(5x - 1)(2x + 1) = 10x^2 + 5x - 2x - 1 = 10x^2 + 3x - 1$ Wrong middle term

Because the numbers 1 and -1 did not produce the correct trinomial when coupled with $5x$ and $2x$, try using $10x$ and $1x$.

$(10x - 1)(1x + 1) = 10x^2 + 10x - 1x - 1 = 10x^2 + 9x - 1$ Wrong middle term

$(10x + 1)(1x - 1) = 10x^2 - 10x + 1x - 1 = 10x^2 - 9x - 1$ Correct!

Hence $10x^2 - 9x - 1 = (10x + 1)(x - 1)$.

2. Identifying the Signs When Using the Trial-and-Error Method

In Example 1, the factors of -1 must have opposite signs to produce a negative product. Therefore, one binomial factor is a sum and one is a difference. Determining the correct signs is an important aspect of factoring trinomials. We suggest the following guidelines:

Sign Rules for the Trial-and-Error Method

Given the trinomial $ax^2 + bx + c$, $(a > 0)$ the signs can be determined as follows:

1. If c *is positive*, then the signs in the binomials must be the same (either both positive or both negative). The correct choice is determined by the middle term. If the middle term is positive, then both signs must be positive. If the middle term is negative, then both signs must be negative.

   c is positive: $20x^2 + 43x + 21$ → $(4x + 3)(5x + 7)$ — Same signs

   c is positive: $20x^2 - 43x + 21$ → $(4x - 3)(5x - 7)$ — Same signs

2. If c *is negative*, then the signs in the binomial must be different. The middle term in the trinomial determines which factor gets the positive sign and which gets the negative sign.

   c is negative: $x^2 + 3x - 28$ → $(x + 7)(x - 4)$ — Different signs

   c is negative: $x^2 - 3x - 28$ → $(x - 7)(x + 4)$ — Different signs

example 2

Factoring a Trinomial

Factor the trinomial: $8y^2 + 13y - 6$

Solution:

$8y^2 + 13y - 6$ **Step 1:** The GCF is 1.

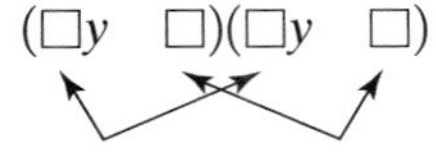

Factors of 8	Factors of 6
$1 \cdot 8$	$1 \cdot 6$
$2 \cdot 4$	$2 \cdot 3$
	$3 \cdot 2$ (reverse order)
	$6 \cdot 1$ (reverse order)

Step 2: List the positive factors of 8 and positive factors of 6. Consider the reverse order in one list of factors.

Step 3: Construct all possible binomial factors using different combinations of the factors of 8 and 6.

$(2y \quad 1)(4y \quad 6)$
$(2y \quad 2)(4y \quad 3)$
$(2y \quad 3)(4y \quad 2)$
$(2y \quad 6)(4y \quad 1)$
$(1y \quad 1)(8y \quad 6)$
$(1y \quad 3)(8y \quad 2)$

Without regard to signs, these factorizations cannot work because the terms in a binomial share a common factor greater than 1.

Test the remaining factorizations. Keep in mind that to produce a product of -6, the signs within the parentheses must be opposite (one positive and one negative). Also, the sum of the products of the inner terms and outer terms must be combined to form $13y$.

$(1y \quad 6)(8y \quad 1)$ *Incorrect.* Wrong middle term. Regardless of signs, the product of inner terms, $48y$, and the product of outer terms, $1y$, cannot be combined to form the middle term $13y$.

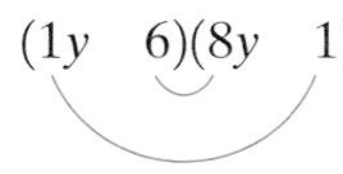

$(1y \quad 2)(8y \quad 3)$ *Correct.* The terms $16y$ and $3y$ can be combined to form the middle term $13y$, provided the signs are applied correctly. We require $+16y$ and $-3y$.

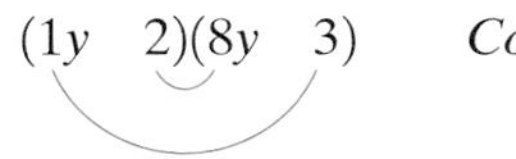

$(y + 2)(8y - 3)$

$= 8y^2 - 3y + 16y - 6$ Verify by multiplication.

$= 8y^2 + 13y - 6$

Hence, the correct factorization is $(y + 2)(8y - 3)$.

3. Factoring Trinomials with a Leading Coefficient of 1

If a trinomial has a leading coefficient of 1, the factoring process simplifies significantly. Consider the trinomial $x^2 + bx + c$. To produce a leading term of x^2, we can construct binomials of the form $(x + \quad)(x + \quad)$. The remaining terms may be satisfied by two numbers p and q whose product is c and whose sum is b:

$$(\overbrace{x + p)(x + q}^{\text{Factors of } c}) = x^2 + px + qx + pq = x^2 + \underbrace{(p + q)}_{\text{Sum} = b}x + \underbrace{pq}_{\text{Product} = c}$$

This process is demonstrated in Example 3.

example 3

Factoring a Trinomial with a Leading Coefficient of 1

Factor the trinomial: $x^2 - 10x + 16$

Solution:

$x^2 - 10x + 16$ — Factor out the GCF from all terms. In this case, the GCF is 1.

$= (x \quad)(x \quad)$ — The trinomial is written in the form $x^2 + bx + c$. To form the product x^2, use the factors x and x.

Next look for two numbers whose product is 16 and whose sum is −10. Because the middle term is negative, we will consider only the negative factors of 16.

Factors of 16	**Sum**
$-1(-16)$	$-1 + (-16) = -17$
$-2(-8)$	$-2 + (-8) = -10$
$-4(-4)$	$-4 + (-4) = -8$

The numbers are −2 and −8.

Hence $x^2 - 10x + 16 = (x - 2)(x - 8)$.

example 4

Factoring Trinomials with a Leading Coefficient of 1

Factor.

a. $t^2 + 34t + 33$ b. $c^2 - 7cd - 30d^2$

Solution:

a. $t^2 + 34t + 33$ — Factor out the GCF from all terms. In this case, the GCF is 1.

$= (t \quad)(t \quad)$

$= (t + 1)(t + 33)$

To complete the factorization, we need two numbers whose product is 33 and whose sum is 34. The numbers are 1 and 33.

b. $c^2 - 7cd - 30d^2$ — Factor out the GCF from all terms. In this case, the GCF is 1.

$= (c \quad d)(c \quad d)$

$= (c - 10d)(c + 3d)$

The presence of two variables c and d, does not change the factoring process. We will still look for two numbers whose product is −30 and whose sum is −7. The numbers are −10 and 3. These will be the coefficients on the d terms.

4. Greatest Common Factor and Factoring Trinomials

Remember that the first step in any factoring problem is to remove the GCF. By removing the GCF, the remaining terms of the trinomial will be simpler and may have smaller coefficients.

example 5 Factoring a Trinomial by the Trial-and-Error Method

Factor the trinomial by the trial-and-error method: $40x^3 - 104x^2 + 10x$

Solution:

$40x^3 - 104x^2 + 10x$

$= 2x(20x^2 - 52x + 5)$ **Step 1:** The GCF is $2x$.

$= 2x(\square x \quad \square)(\square x \quad \square)$ **Step 2:** List the factors of 20 and factors of 5. Consider the reverse order in one list of factors.

Factors of 20	Factors of 5
$1 \cdot 20$	$1 \cdot 5$
$2 \cdot 10$	$5 \cdot 1$
$4 \cdot 5$	

Step 3: Construct all possible binomial factors using different combinations of the factors of 20 and factors of 5. The signs in the parentheses must both be negative.

$= 2x(1x - 1)(20x - 5)$
$= 2x(2x - 1)(10x - 5)$
$= 2x(4x - 1)(5x - 5)$

Incorrect. The binomials contain a GCF greater than 1.

$= 2x(1x - 5)(20x - 1)$ *Incorrect.* Wrong middle term.

$2x(x - 5)(20x - 1)$
$= 2x(20x^2 - 1x - 100x + 5)$
$= 2x(20x^2 - 101x + 5)$

$= 2x(4x - 5)(5x - 1)$ *Incorrect.* Wrong middle term.

$2x(4x - 5)(5x - 1)$
$= 2x(20x^2 - 4x - 25x + 5)$
$= 2x(20x^2 - 29x + 5)$

$= 2x(2x - 5)(10x - 1)$ *Correct.* $2x(2x - 5)(10x - 1)$

$= 2x(20x^2 - 2x - 50x + 5)$
$= 2x(20x^2 - 52x + 5)$
$= 40x^3 - 104x^2 + 10x$

The correct factorization is $2x(2x - 5)(10x - 1)$.

Tip: Notice when the GCF, $2x$, is removed from the original trinomial, the new trinomial has smaller coefficients. This makes the factoring process simpler. It is easier to list the factors of 20 and 5 rather than the factors of 40 and 10.

Often it is easier to factor a trinomial when the leading coefficient is positive. If the leading coefficient is negative, consider factoring out the opposite of the GCF.

example 6 Factoring a Trinomial by the Trial-and-Error Method

Factor the trinomial by the trial-and-error method: $-w^2 - 7w + 18$

Solution:

$-w^2 - 7w + 18$

$= -1(w^2 + 7w - 18)$ — Factor out -1. The resulting trinomial has a leading coefficient of 1.

$= -1[(w \quad)(w \quad)]$ — To complete the factorization, we need two numbers whose product is -18 and whose sum is 7. The numbers are 9 and -2.

$= -1[(w + 9)(w - 2)]$

$= -(w + 9)(w - 2)$

Avoiding Mistakes

Do not forget to write the GCF as part of the final answer.

Note that not every trinomial is factorable by the methods presented here.

example 7 Factoring a Trinomial by the Trial-and-Error Method

Factor the trinomial by the trial-and-error method: $2p^2 - 8p + 3$

Solution:

$2p^2 - 8p + 3$ — **Step 1:** The GCF is 1.

$(1p \quad \square)(2p \quad \square)$ — **Step 2:** List the factors of 2 and the factors of 3.

Factors of 2	Factors of 3
$1 \cdot 2$	$1 \cdot 3$
	$3 \cdot 1$

Step 3: Construct all possible binomial factors using different combinations of the factors of 2 and 3. Because the third term in the trinomial is positive, both signs in the binomial must be the same. Because the middle term coefficient is negative, both signs will be negative.

$(p - 1)(2p - 3) = 2p^2 - 3p - 2p + 3$
$= 2p^2 - 5p + 3$ — *Incorrect.* Wrong middle term.

$(p - 3)(2p - 1) = 2p^2 - p - 6p + 3$
$= 2p^2 - 7p + 3$ — *Incorrect.* Wrong middle term.

Because none of the combinations of factors results in the correct product, we say that the trinomial $2p^2 - 8p + 3$ is prime and cannot be factored.

section 5.3 Practice Exercises

For Exercises 1–6, factor out the greatest common factor.

1. $7a^9 + 28a^3$
2. $r^4 - 9r^5$
3. $12w^2 - 4w$
4. $15x^2 + 3x$
5. $21a^2b^2 + 12ab^2 - 15a^2b$
6. $5uv^2 - 10u^2v + 25u^2v^2$

For Exercises 7–10, assume a, b, and c represent positive integers.

7. When factoring a polynomial of the form $ax^2 + bx - c$, should the signs in the binomials be both positive, both negative, or different?
8. When factoring a polynomial of the form $ax^2 - bx - c$, should the signs in the binomials be both positive, both negative, or different?
9. When factoring a polynomial of the form $ax^2 - bx + c$, should the signs in the binomials be both positive, both negative, or different?
10. When factoring a polynomial of the form $ax^2 + bx + c$, should the signs in the binomials be both positive, both negative, or different?

For Exercises 11–14, complete the factorization.

11. $x^2 + x - 56 = (x - 7)(\quad)$
12. $y^2 + y - 30 = (y - 5)(\quad)$
13. $x^2 - x - 56 = (x + 7)(\quad)$
14. $y^2 - y - 30 = (y + 5)(\quad)$
15. What is a prime polynomial?
16. How do you determine if a trinomial is prime?

For Exercises 17–58, factor the trinomial using the trial-and-error method.

17. $2y^2 - 3y - 2$
18. $2w^2 + 5w - 3$
19. $9x^2 - 12x + 4$
20. $3n^2 + 13n + 4$
21. $2a^2 + 7a + 6$
22. $8b^2 - 6b - 9$
23. $6t^2 + 7t - 3$
24. $4p^2 - 9p + 2$
25. $4m^2 - 20m + 25$
26. $16r^2 + 24r + 9$
27. $5c^2 - c + 2$
28. $7s^2 + 2s + 9$
29. $6x^2 - 19xy + 10y^2$
30. $15p^2 + pq - 2q^2$
31. $12m^2 + 11mn - 5n^2$
32. $4a^2 + 5ab - 6b^2$
33. $6r^2 + rs - 2s^2$
34. $18x^2 - 9xy - 2y^2$
35. $4s^2 - 8st + t^2$
36. $6u^2 - 10uv + 5v^2$
37. $x^2 + 7x - 18$
38. $y^2 - 6y - 40$
39. $a^2 - 10a - 24$
40. $b^2 + 6b - 7$
41. $r^2 + 5r - 24$
42. $t^2 + 20t + 100$
43. $w^2 - 14w + 49$
44. $h^2 - 11h + 10$
45. $k^2 + 5k + 4$
46. $u^2 + 9u - 22$
47. $v^2 - 4v + 1$
48. $x^2 + 5x + 2$
49. $m^2 - 13mn + 40n^2$
50. $r^2 - rs - 42s^2$
51. $a^2 + 9ab + 8b^2$
52. $y^2 - yz - 12z^2$
53. $x^2 + 9xy + 20y^2$
54. $p^2 - 13pq + 36q^2$
55. $10t^2 - 23t - 5$
56. $16n^2 + 14n + 3$
57. $14w^2 + 13w - 12$
58. $12x^2 - 16x + 5$
59. Is the expression $(3x + 6)(x - 5)$ factored completely? Explain why or why not.
60. Is the expression $(5x + 1)(4x - 12)$ factored completely? Explain why or why not.

For Exercises 61–74, first factor out the GCF. Then factor by using the trial-and-error method if possible.

61. $2m^2 - 12m - 80$
62. $3c^2 - 33c + 72$
63. $2y^5 + 13y^4 + 6y^3$
64. $3u^8 - 13u^7 + 4u^6$
65. $5d^3 + 3d^2 - 10d$
66. $3y^3 - y^2 + 12y$
67. $4b^3 - 4b^2 - 80b$
68. $2w^2 + 20w + 42$
69. $x^2y^2 - 13xy^2 + 30y^2$
70. $p^2q^2 - 14pq^2 + 33q^2$
71. $-a^2 - 15a + 34$
72. $-j^2 - 7j - 10$
73. $-2u^2 + 28u - 90$
74. $-3v^2 - 3v + 90$

For Exercises 75–82, write the polynomial in descending order. Then factor the polynomial.

75. $10x + 1 + 16x^2$
76. $k^2 + 16 - 8k$
77. $1 + c^2 - 2c$
78. $3 - 14t + 16t^2$
79. $20z - 18 - 2z^2$
80. $25t - 5t^2 - 30$
81. $42 - 13q + q^2$
82. $-5w - 24 + w^2$

---

section

5.4 Factoring Perfect Square Trinomials and the Difference of Squares

Concepts

1. **Factoring Perfect Square Trinomials**
2. **Factoring a Difference of Squares**
3. **Analyzing a Sum of Squares**
4. **Factoring Using Multiple Methods**

1. Factoring Perfect Square Trinomials

Recall from Section 4.6 that the square of a binomial always results in a perfect square trinomial:

$$(a + b)^2 = (a + b)(a + b) = a^2 + ab + ab + b^2 = a^2 + 2ab + b^2$$
$$(a - b)^2 = (a - b)(a - b) = a^2 - ab - ab + b^2 = a^2 - 2ab + b^2$$

For example, $(3x + 5)^2 = (3x)^2 + 2(3x)(5) + (5)^2 = 9x^2 + 30x + 25$

$a = 3x \quad b = 5$

To factor the trinomial $9x^2 + 30x + 25$, the grouping method or the trial-and-error method can be used. However, if we recognize that the trinomial is a perfect square trinomial, we can use one of the following patterns to reach a quick solution.

Factored Form of a Perfect Square Trinomial

$$a^2 + 2ab + b^2 = (a + b)^2$$
$$a^2 - 2ab + b^2 = (a - b)^2$$

Checking for a Perfect Square Trinomial

1. Check if the first and third terms are both perfect squares with positive coefficients.
2. If this is the case, identify a and b, and determine if the middle term equals $2ab$.

example 1 **Factoring Perfect Square Trinomials**

Factor the trinomials completely.

a. $x^2 + 14x + 49$
b. $25y^2 - 20y + 4$
c. $18c^3 - 48c^2d + 32cd^2$
d. $5w^2 + 50w + 45$

Solution:

a. $x^2 + 14x + 49$

The GCF is 1.

- The first and third terms are positive.

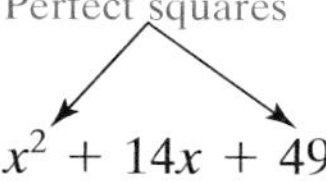

$x^2 + 14x + 49$

- The first term is a perfect square: $x^2 = (x)^2$.
- The third term is a perfect square: $49 = (7)^2$.
- The middle term is twice the product of x and 7: $14x = 2(x)(7)$

$= (x)^2 + 2(x)(7) + (7)^2$

Hence, the trinomial is in the form $a^2 + 2ab + b^2$, where $a = x$ and $b = 7$.

$= (x + 7)^2$

Factor as $(a + b)^2$.

b. $25y^2 - 20y + 4$

The GCF is 1.

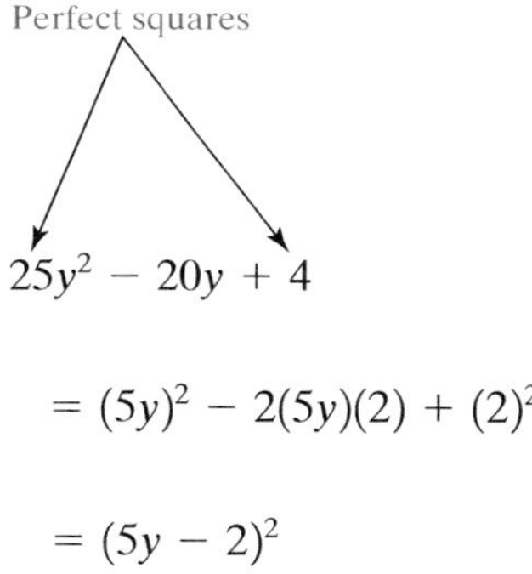

$25y^2 - 20y + 4$

- The first and third terms are positive.
- The first term is a perfect square: $25y^2 = (5y)^2$.
- The third term is a perfect square: $4 = (2)^2$.
- The middle term is twice the product of $5y$ and 2: $20y = 2(5y)(2)$

$= (5y)^2 - 2(5y)(2) + (2)^2$

$= (5y - 2)^2$

Factor as $(a - b)^2$.

c. $18c^3 - 48c^2d + 32cd^2$

$= 2c(9c^2 - 24cd + 16d^2)$

The GCF is $2c$.

Perfect squares

$= 2c(9c^2 - 24cd + 16d^2)$

- The first and third terms are positive.
- The first term is a perfect square: $9c^2 = (3c)^2$.
- The third term is a perfect square: $16d^2 = (4d)^2$.
- The middle term is twice the product of $3c$ and $4d$: $24cd = 2(3c)(4d)$

$= 2c[(3c)^2 - 2(3c)(4d) + (4d)^2]$

$= 2c(3c - 4d)^2$

Factor as $(a - b)^2$.

d. $5w^2 + 50w + 45$

$= 5(w^2 + 10w + 9)$ The GCF is 5.

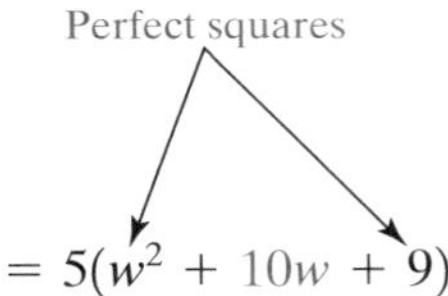

The first and third terms are perfect squares.

$$w^2 = (w)^2 \quad \text{and} \quad 9 = (3)^2$$

$= 5(w^2 + 10w + 9)$ However, the middle term is not 2 times the product of w and 3. Therefore, this is not a perfect square trinomial.

$$10w \neq 2(w)(3)$$

$= 5(w + 9)(w + 1)$ To factor, use either the grouping method or the trial-and-error method.

Tip: To help you identify a perfect square trinomial, it is recommended that you familiarize yourself with the first several perfect squares.

$1 \cdot 1 = \mathbf{2}$	$4 \cdot 4 = \mathbf{16}$	$7 \cdot 7 = \mathbf{49}$	$10 \cdot 10 = \mathbf{100}$	$13 \cdot 13 = \mathbf{169}$
$2 \cdot 2 = \mathbf{4}$	$5 \cdot 5 = \mathbf{25}$	$8 \cdot 8 = \mathbf{64}$	$11 \cdot 11 = \mathbf{121}$	$14 \cdot 14 = \mathbf{196}$
$3 \cdot 3 = \mathbf{9}$	$6 \cdot 6 = \mathbf{36}$	$9 \cdot 9 = \mathbf{81}$	$12 \cdot 12 = \mathbf{144}$	$15 \cdot 15 = \mathbf{225}$

If you do not recognize that a trinomial is a perfect square trinomial, you may still use either the trial-and-error method or the grouping method to factor the trinomial.

2. Factoring a Difference of Squares

Up to this point, we have learned several methods of factoring, including:

- Factoring out the greatest common factor from a polynomial
- Factoring a four-term polynomial by grouping
- Recognizing and factoring perfect square trinomials
- Factoring trinomials by the grouping method or by the trial-and-error method

Next, we will learn how to factor binomials that fit the pattern of a difference of squares. Recall from Section 4.6 that the product of two conjugates results in a difference of squares:

$$(a + b)(a - b) = a^2 - b^2$$

Therefore, to factor a difference of squares, the process is reversed. Identify a and b and construct the conjugate factors.

Factored Form of a Difference of Squares

$$a^2 - b^2 = (a + b)(a - b)$$

In addition to recognizing numbers that are perfect squares, it is helpful to recognize that a variable expression is a perfect square if its exponent is a multiple of 2. For example:

Perfect Squares

$x^2 = (x)^2$

$x^4 = (x^2)^2$

$x^6 = (x^3)^2$

$x^8 = (x^4)^2$

$x^{10} = (x^5)^2$

example 2

Factoring Differences of Squares

Factor the binomials.

a. $y^2 - 25$ b. $49s^2 - 4t^4$ c. $18w^2z - 2z$

Solution:

a. $y^2 - 25$ — The binomial is a difference of squares.

$= (y)^2 - (5)^2$ — Write in the form: $a^2 - b^2$, where $a = y, b = 5$.

$= (y + 5)(y - 5)$ — Factor as $(a + b)(a - b)$.

b. $49s^2 - 4t^4$ — The binomial is a difference of squares.

$= (7s)^2 - (2t^2)^2$ — Write in the form $a^2 - b^2$, where $a = 7s$ and $b = 2t^2$.

$= (7s + 2t^2)(7s - 2t^2)$ — Factor as $(a + b)(a - b)$.

c. $18w^2z - 2z$ — The GCF is $2z$.

$= 2z(9w^2 - 1)$ — $(9w^2 - 1)$ is a difference of squares.

$= 2z[(3w)^2 - (1)^2]$ — Write in the form: $a^2 - b^2$, where $a = 3w$, $b = 1$.

$= 2z(3w + 1)(3w - 1)$ — Factor as $(a + b)(a - b)$.

3. Analyzing a Sum of Squares

Suppose a and b share no common factors. Then the difference of squares $a^2 - b^2$ can be factored as $(a + b)(a - b)$. However, the sum of squares $a^2 + b^2$ cannot be factored over the real numbers. To see why, consider the expression $a^2 + b^2$. The factored form would require two binomials of the form:

$$(a \quad b)(a \quad b) \stackrel{?}{=} a^2 + b^2$$

If all possible combinations of signs are considered, none produces the correct product.

$(a + b)(a - b) = a^2 - b^2$ Wrong sign

$(a + b)(a + b) = a^2 + 2ab + b^2$ Wrong middle term

$(a - b)(a - b) = a^2 - 2ab + b^2$ Wrong middle term

After exhausting all possibilities, we see that if a and b share no common factors, then the sum of squares $a^2 + b^2$ is a prime polynomial.

4. Factoring Using Multiple Methods

Some factoring problems require more than one method of factoring. In general, when factoring a polynomial, be sure to factor completely.

example 3 **Factoring Polynomials**

Factor completely.

a. $w^4 - 16$ b. $4x^3 + 4x^2 - 25x - 25$ c. $8p^3 + 24p^2q + 18pq^2$

Solution:

a. $w^4 - 16$ The GCF is 1. $w^4 - 16$ is a difference of squares.

$= (w^2)^2 - (4)^2$ Write in the form: $a^2 - b^2$, where $a = w^2$, $b = 4$.

$= (w^2 + 4)(w^2 - 4)$ Factor as $(a + b)(a - b)$.

$= (w^2 + 4)(w + 2)(w - 2)$ Note that $w^2 - 4$ can be factored further as a difference of squares. (The binomial $w^2 + 4$ is a sum of squares and cannot be factored further.)

b. $4x^3 + 4x^2 - 25x - 25$ The GCF is 1.

$= 4x^3 + 4x^2 \mid - 25x - 25$ The polynomial has four terms. Factor by grouping.

$= 4x^2(x + 1) - 25(x + 1)$

$= (x + 1)(4x^2 - 25)$ $4x^2 - 25$ is a difference of squares.

$= (x + 1)(2x + 5)(2x - 5)$

c. $8p^3 + 24p^2q + 18pq^2$ The GCF is $2p$.

$= 2p(4p^2 + 12pq + 9q^2)$ $4p^2 + 12pq + 9q^2$ is a perfect square trinomial.

$= 2p(2p + 3q)^2$

section 5.4 Practice Exercises

For Exercises 1–10, factor the polynomials.

1. $3x^2 + x - 10$
2. $6a^2b + 3a^3b$
3. $x^2yz^2 + 6y^2z + yz$
4. $2x^2 - x - 1$
5. $12x^2 - 34x + 10$
6. $3x^2 - 3xy - 6y^2$
7. $ax + ab - 6x - 6b$
8. $2xy - 3x - 4y + 6$
9. $x^2 + 6x + 9$
10. $y^2 - 4y + 4$
11. What perfect square trinomial factors to $(2x + 3)^2$?
12. What perfect square trinomial factors to $(3k + 5)^2$?
13. What perfect square trinomial factors to $(6h - 1)^2$?
14. What perfect square trinomial factors to $(4y - 5)^2$?
15. a. Identify which trinomial is a perfect square trinomial:

    $x^2 + 4x + 4$ or $x^2 + 5x + 4$

    b. Factor both of these trinomials.
16. a. Identify which trinomial is a perfect square trinomial:

    $x^2 + 13x + 36$ or $x^2 + 12x + 36$

    b. Factor both of these trinomials.
17. a. Identify which trinomial is a perfect square trinomial:

    $4x^2 - 25x + 25$ or $4x^2 - 20x + 25$

    b. Factor both of these trinomials.
18. a. Identify which trinomial is a perfect square trinomial:

    $9x^2 + 12x + 4$ or $9x^2 + 15x + 4$

    b. Factor both of these trinomials.

For Exercises 19–40, factor the trinomials, if possible.

19. $y^2 - 10y + 25$
20. $t^2 - 16t + 64$
21. $m^2 + 6m + 9$
22. $n^2 + 18n + 81$
23. $r^2 - 2r + 36$
24. $s^2 - 4s + 100$
25. $49q^2 - 28q + 4$
26. $64y^2 - 80y + 25$
27. $9p^2 + 42p + 49$
28. $4x^2 + 36x + 81$
29. $25h^2 + 50h + 16$
30. $4w^2 - 20w + 9$
31. $16a^2 + 8ab + b^2$
32. $25m^2 + 10mn + n^2$
33. $16q^2 + 40qr + 25r^2$
34. $u^2 - 2uv + v^2$
35. $a^2 + 2ab + b^2$
36. $49h^2 - 14hk + k^2$
37. $k^2 - k + \frac{1}{4}$
38. $v^2 + \frac{2}{3}v + \frac{1}{9}$
39. $9x^2 + x + \frac{1}{36}$
40. $4y^2 - y + \frac{1}{16}$
41. What binomial factors as $(x - 5)(x + 5)$?
42. What binomial factors as $(n - 3)(n + 3)$?
43. What binomial factors as $(2w - 3)(2w + 3)$?
44. What binomial factors as $(7y - 4)(7y + 4)$?

For Exercises 45–64, factor the binomials, if possible.

45. $x^2 - 36$
46. $r^2 - 81$
47. $w^2 - 100$
48. $t^2 - 49$
49. $4a^2 - 121b^2$
50. $9x^2 - y^2$
51. $49m^2 - 16n^2$
52. $100a^2 - 49b^2$
53. $9q^2 + 16$
54. $36 + s^2$
55. $c^6 - 25$
56. $z^6 - 4$
57. $25 - 16t^2$
58. $64 - h^2$
59. $p^2 - \frac{1}{9}$
60. $q^2 - \frac{1}{36}$
61. $m^2 + \frac{100}{81}$
62. $n^2 + \frac{25}{4}$
63. $\frac{4}{9} - w^2$
64. $\frac{16}{25} - x^2$

65. a. Write a polynomial that represents the area of the shaded region in the figure.
    b. Factor the expression from part (a).

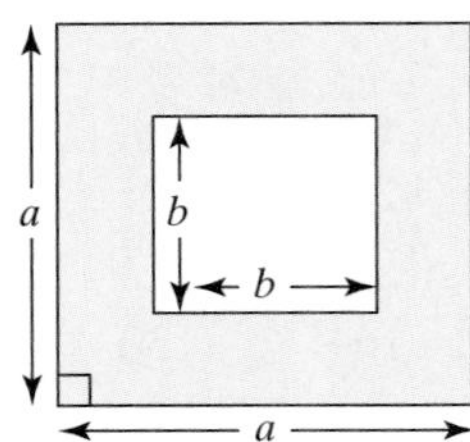

Figure for Exercise 65

66. a. Write a polynomial that represents the area of the shaded region in the figure.
    b. Factor the expression from part (a).

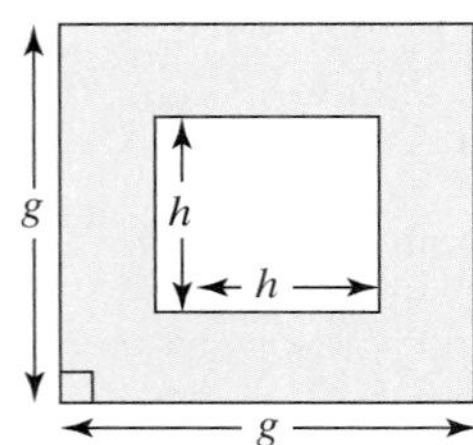

Figure for Exercise 66

For Exercises 67–88, factor the polynomials completely.

67. $3w^2 - 27$
68. $6y^2 - 6$
69. $50p^4 - 2$
70. $18q^2 - 98n^2$
71. $2x^2 + 24x + 72$
72. $3x^2y - 66xy + 363y$
73. $2t^3 - 10t^2 - 2t + 10$
74. $9a^3 + 27a^2 - 4a - 12$
75. $100y^4 + 25x^2$
76. $36a^2 + 9b^4$
77. $4a^2b - 40ab^2 + 100b^3$
78. $18u^2 + 24uv + 8v^2$
79. $2x^3 + 3x^2 - 2x - 3$
80. $3x^3 + x^2 - 12x - 4$
81. $81y^4 - 16$
82. $u^4 - 256$
83. $81k^2 + 30k + 1$
84. $9h^2 - 15h + 4$
85. $k^3 + 4k^2 - 9k - 36$
86. $w^3 - 2w^2 - 4w + 8$
87. $4m^{14} - 20m^7 + 25$
88. $9n^{12} + 24n^6 + 16$

Expanding Your Skills

For Exercises 89–100, factor the difference of squares.

89. $0.36x^2 - 0.01$
90. $0.81p^2 - 0.25q^2$
91. $\frac{1}{4}w^2 - \frac{1}{9}v^2$
92. $\frac{4}{9}c^2 - \frac{9}{16}d^2$
93. $(y - 3)^2 - 9$
94. $(x - 2)^2 - 4$
95. $(2p + 1)^2 - 36$
96. $(4q + 3)^2 - 25$
97. $16 - (t + 2)^2$
98. $81 - (a + 5)^2$
99. $100 - (2b - 5)^2$
100. $49 - (3k - 7)^2$

Concepts

1. Factoring the Sum and Difference of Cubes
2. Factoring Binomials: A Summary

section

5.5 Factoring the Sum and Difference of Cubes

1. Factoring the Sum and Difference of Cubes

In Section 5.4, you learned that a binomial $a^2 - b^2$ is a difference of squares and can be factored as $(a - b)(a + b)$. Furthermore, if a and b share no common factors, then a sum of squares $a^2 + b^2$ is not factorable over the real numbers. In this section, we will learn that both a difference of cubes, $a^3 - b^3$, and a sum of cubes $a^3 + b^3$ are factorable.

Factoring a Sum and Difference of Cubes

Sum of Cubes: $a^3 + b^3 = (a + b)(a^2 - ab + b^2)$

Difference of Cubes: $a^3 - b^3 = (a - b)(a^2 + ab + b^2)$

Multiplication can be used to confirm the formulas for factoring a sum or difference of cubes:

$$(a + b)(a^2 - ab + b^2) = a^3 - \cancel{a^2b} + \cancel{ab^2} + \cancel{a^2b} - \cancel{ab^2} + b^3 = a^3 + b^3 \checkmark$$

$$(a - b)(a^2 + ab + b^2) = a^3 + \cancel{a^2b} + \cancel{ab^2} - \cancel{a^2b} - \cancel{ab^2} - b^3 = a^3 - b^3 \checkmark$$

To help you remember the formulas for factoring a sum or difference of cubes, keep the following guidelines in mind.

- The factored form is the product of a binomial and a trinomial.
- The first and third terms in the trinomial are the squares of the terms within the binomial factor.
- Without regard to signs, the middle term in the trinomial is the product of terms in the binomial factor.

Square the first term of the binomial. Product of terms in the binomial

$$x^3 + 8 = (x)^3 + (2)^3 = (x + 2)[(x)^2 - (x)(2) + (2)^2]$$

Square the last term of the binomial.

- The sign within the binomial factor is the same as the sign of the original binomial.
- The first and third terms in the trinomial are always positive.
- The sign of the middle term in the trinomial is opposite the sign within the binomial.

Same sign Positive

$$x^3 + 8 = (x)^3 + (2)^3 = (x + 2)[(x)^2 - (x)(2) + (2)^2]$$

Opposite signs

To help you recognize a sum or difference of cubes, we recommend that you familiarize yourself with the first several perfect cubes:

Perfect Cube	Perfect Cube
$1 = (1)^3$	$216 = (6)^3$
$8 = (2)^3$	$343 = (7)^3$
$27 = (3)^3$	$512 = (8)^3$
$64 = (4)^3$	$729 = (9)^3$
$125 = (5)^3$	$1000 = (10)^3$

It is also helpful to recognize that a variable expression is a perfect cube if its exponent is a multiple of 3. For example,

Perfect Cube

$$x^3 = (x)^3$$
$$x^6 = (x^2)^3$$
$$x^9 = (x^3)^3$$
$$x^{12} = (x^4)^3$$

example 1

Factoring a Sum of Cubes

Factor: $w^3 + 64$

Solution:

$w^3 + 64$ — w^3 and 64 are perfect cubes.

$= (w)^3 + (4)^3$ — Write as $a^3 + b^3$, where $a = w, b = 4$.

$a^3 + b^3 = (a + b)(a^2 - ab + b^2)$ — Apply the formula for a sum of cubes.

$$(w)^3 + (4)^3 = (w + 4)[(w)^2 - (w)(4) + (4)^2]$$

$= (w + 4)(w^2 - 4w + 16)$ — Simplify.

Check: $(w + 4)(w^2 - 4w + 16) = w^3 - 4w^2 + 16w + 4w^2 - 16w + 64$
$= w^3 + 64$ ✔

example 2

Factoring a Difference of Cubes

Factor: $27p^3 - q^6$

Solution:

$27p^3 - q^6$ — $27p^3$ and q^6 are perfect cubes.

$(3p)^3 - (q^2)^3$ — Write as $a^3 - b^3$, where $a = 3p, b = q^2$.

$a^3 - b^3 = (a - b)(a^2 + ab + b^2)$ — Apply the formula for a difference of cubes.

$$(3p)^3 - (q^2)^3 = (3p - q^2)[(3p)^2 + (3p)(q^2) + (q^2)^2]$$

$= (3p - q^2)(9p^2 + 3pq^2 + q^4)$ — Simplify.

2. Factoring Binomials: A Summary

After removing the GCF, the next step in any factoring problem is to recognize what type of pattern it follows. Exponents that are divisible by 2 are perfect squares and those divisible by 3 are perfect cubes. The formulas for factoring binomials are summarized in the following box:

Factoring Binomials

1. Difference of Squares: $a^2 - b^2 = (a + b)(a - b)$
2. Difference of Cubes: $a^3 - b^3 = (a - b)(a^2 + ab + b^2)$
3. Sum of Cubes: $a^3 + b^3 = (a + b)(a^2 - ab + b^2)$

example 3 **Factoring Binomials**

Factor completely: a. $27y^3 + 1$ b. $m^2 - \frac{1}{4}$

c. $3y^4 - 48$ d. $z^6 - 8w^3$

Solution:

a. $27y^3 + 1$ — Sum of cubes: $27y^3 = (3y)^3$ and $1 = (1)^3$.

$= (3y)^3 + (1)^3$ — Write as $a^3 + b^3$, where $a = 3y$ and $b = 1$.

$= (3y + 1)[(3y)^2 - (3y)(1) + (1)^2]$ — Apply the formula $a^3 + b^3 = (a + b)(a^2 - ab + b^2)$.

$= (3y + 1)(9y^2 - 3y + 1)$ — Simplify.

b. $m^2 - \frac{1}{4}$ — Difference of squares

$= (m)^2 - \left(\frac{1}{2}\right)^2$ — Write as $a^2 - b^2$, where $a = m$ and $b = \frac{1}{2}$.

$= \left(m + \frac{1}{2}\right)\left(m - \frac{1}{2}\right)$ — Apply the formula $a^2 - b^2 = (a + b)(a - b)$.

c. $3y^4 - 48$

$= 3(y^4 - 16)$ — Factor out the GCF. The binomial is a difference of squares.

$= 3[(y^2)^2 - (4)^2]$ — Write as $a^2 - b^2$, where $a = y^2$ and $b = 4$.

$= 3(y^2 + 4)(y^2 - 4)$ — Apply the formula $a^2 - b^2 = (a + b)(a - b)$.

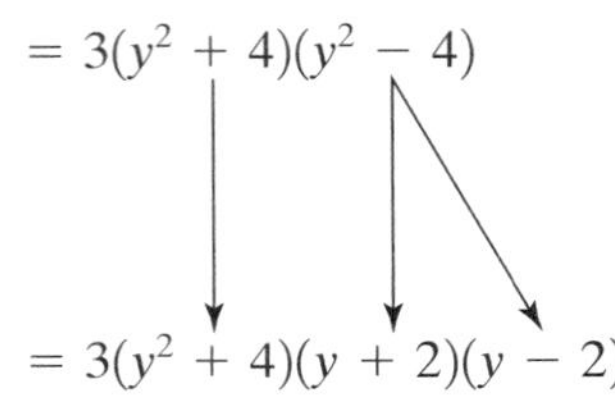

$y^2 + 4$ is a sum of squares and cannot be factored.

$= 3(y^2 + 4)(y + 2)(y - 2)$ — $y^2 - 4$ is a difference of squares and can be factored further.

d. $z^6 - 8w^3$ — Difference of cubes: $z^6 = (z^2)^3$ and $8w^3 = (2w)^3$

$= (z^2)^3 - (2w)^3$ — Write as $a^3 - b^3$, where $a = z^2$ and $b = 2w$.

$= (z^2 - 2w)[(z^2)^2 + (z^2)(2w) + (2w)^2]$ — Apply the formula $a^3 - b^3 = (a - b)(a^2 + ab + b^2)$.

$= (z^2 - 2w)(z^4 + 2z^2w + 4w^2)$ — Simplify.

Each of the factorizations in Example 3 can be checked by multiplying.

example 4 Factoring Binomials

Factor the binomial $x^6 - y^6$ as

a. A difference of cubes
b. A difference of squares

Solution:

a. $x^6 - y^6$

Difference of cubes

$= (x^2)^3 - (y^2)^3$ — Write as $a^3 - b^3$, where $a = x^2$ and $b = y^2$.

$= (x^2 - y^2)[(x^2)^2 + (x^2)(y^2) + (y^2)^2]$ — Apply the formula $a^3 - b^3 = (a - b)(a^2 + ab + b^2)$

$= (x^2 - y^2)(x^4 + x^2y^2 + y^4)$ — Factor $x^2 - y^2$ as a difference of squares.

$= (x + y)(x - y)(x^4 + x^2y^2 + y^4)$

b. $x^6 - y^6$

Difference of squares

$= (x^3)^2 - (y^3)^2$ — Write as $a^2 - b^2$, where $a = x^3$ and $b = y^3$.

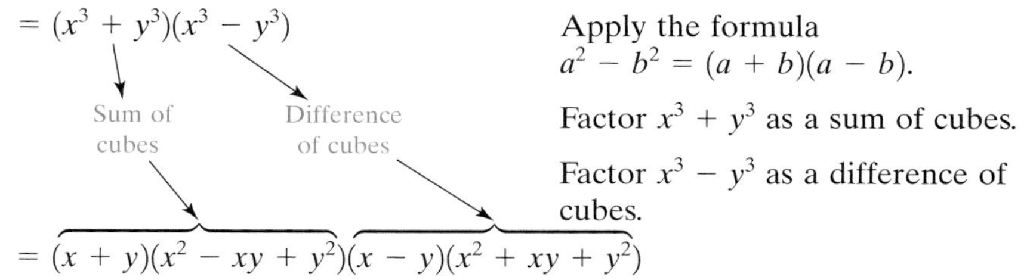

Notice that the expressions x^6 and y^6 are both perfect squares and perfect cubes because the exponents are both multiples of 2 and of 3. Consequently, $x^6 - y^6$ can be factored initially as either the difference of squares or as the difference of cubes. In such a case, it is recommended that you factor the expression as a difference of squares first because it factors more completely into polynomials of lower degree. Hence:

$$x^6 - y^6 = (x + y)(x^2 - xy + y^2)(x - y)(x^2 + xy + y^2)$$

section 5.5 Practice Exercises

1. Multiply the polynomials: $(x - y)(x^2 + xy + y^2)$
2. Multiply the polynomials: $(x + y)(x^2 - xy + y^2)$
3. Identify the expressions that are perfect cubes:

   $\{x^3, 8, 9, y^6, a^4, b^2, 3p^3, 27q^3, w^{12}, r^3s^6\}$
4. Identify the expressions that are perfect cubes:

   $\{z^9, -81, 30, 8, 6x^3, y^{15}, 27a^3, b^2, p^3q^2, -1\}$
5. How do you determine if a binomial is a sum of cubes?
6. How do you determine if a binomial is a difference of cubes?
7. From memory, write the formula to factor a sum of cubes:

   $a^3 + b^3 =$ ___________
8. From memory, write the formula to factor a difference of cubes:

   $a^3 - b^3 =$ ___________

For Exercises 9–28, factor the sums and differences of cubes.

9. $y^3 - 8$
10. $x^3 + 27$
11. $1 - p^3$
12. $q^3 + 1$
13. $w^3 + 64$
14. $8 - t^3$
15. $1000a^3 + 27$
16. $216b^3 - 125$
17. $x^3 - 1000$
18. $8y^3 - 27$
19. $64t^3 + 1$
20. $125r^3 + 1$
21. $n^3 - \dfrac{1}{8}$
22. $\dfrac{8}{27} + m^6$
23. $a^3 + b^6$
24. $u^6 - v^3$
25. $x^9 + 64y^3$
26. $125w^3 - z^9$
27. $25m^{12} + 16$
28. $36p^6 + 49q^4$
29. From memory, write the formula to factor a difference of squares.

    $a^2 - b^2 =$ ___________
30. Write a short paragraph explaining a strategy to factor binomials.

For Exercises 31–48, factor the binomials completely, if possible.

31. $x^4 - 4$
32. $b^4 - 25$
33. $a^2 + 9$
34. $w^2 + 36$
35. $t^3 + 64$
36. $u^3 + 27$

37. $g^3 - 4$

38. $h^3 - 25$

39. $4b^3 + 108$

40. $3c^3 - 24$

41. $5p^2 - 125$

42. $2q^4 - 8$

43. $\frac{1}{64} - 8h^3$

44. $\frac{1}{125} + k^6$

45. $x^4 - 16$

46. $p^4 - 81$

47. $q^6 - 64$

48. $a^6 - 1$

For Exercises 49–68, factor completely using the techniques learned in Sections 5.1–5.5.

49. $4b + 16$

50. $2a^2 - 162$

51. $y^2 + 4y + 3$

52. $6w^2 - 6w$

53. $16z^4 - 81$

54. $3t^2 + 13t + 4$

55. $5r^3 + 5$

56. $3ac + ad - 3bc - bd$

57. $7p^2 - 29p + 4$

58. $3q^2 - 9q - 12$

59. $-2x^2 + 8x - 8$

60. $18a^2 + 12a$

61. $54 - 2y^3$

62. $4t^2 - 100$

63. $4t^2 - 31t - 8$

64. $10c^2 + 10c + 10$

65. $2xw - 10x + 3yw - 15y$

66. $x^3 + 0.001$

67. $4q^2 - 9$

68. $64 + 16k + k^2$

69. What trinomial multiplied by $(x - 2)$ gives a difference of cubes?

70. What trinomial multiplied by $(p + 3)$ gives a sum of cubes?

71. Write a binomial that when multiplied by $(4x^2 - 2x + 1)$ produces a sum of cubes.

72. Write a binomial that when multiplied by $(9y^2 + 15y + 25)$ produces a difference of cubes.

Expanding Your Skills

For Exercises 73–76, factor the sum and difference of cubes.

73. $\frac{64}{125}p^3 - \frac{1}{8}q^3$

74. $\frac{1}{1000}r^3 + \frac{8}{27}s^3$

75. $a^{12} + b^{12}$

76. $a^9 - b^9$

Use Exercises 77–80, to investigate the relationship between division and factoring.

77. a. Use long division to divide $x^3 - 8$ by $(x - 2)$.
    b. Factor $x^3 - 8$.

78. a. Use long division to divide $y^3 + 27$ by $(y + 3)$.
    b. Factor $y^3 + 27$.

79. a. Use long division to divide $m^3 + 1$ by $(m + 1)$.
    b. Factor $m^3 + 1$.

80. a. Use long division to divide $n^3 - 64$ by $(n - 4)$.
    b. Factor $n^3 - 64$.

Concepts

1. A Factoring Strategy
2. Part I: General Factoring Review
3. Part II: Additional Factoring Strategies
4. Factoring Using Substitution
5. Factoring by Grouping—Rearranging Terms
6. Other Grouping Techniques for Factoring Four-Term Polynomials

section

5.6 General Factoring Summary

1. A Factoring Strategy

We now review the techniques of factoring presented thus far along with a general strategy for factoring polynomials.

Factoring Strategy

1. Factor out the greatest common factor, GCF. (Section 5.1)
2. Identify whether the polynomial has two terms, three terms, or more than three terms.
3. If the polynomial has two terms, determine if it fits the pattern for a difference of squares, difference of cubes, or sum of cubes. (Sections 5.4 or 5.5)
4. If the polynomial has three terms, check first for a perfect square trinomial (Section 5.4). Otherwise, factor the trinomial with the grouping method or the trial-and-error method. (Sections 5.2 or 5.3)
5. If the polynomial has more than three terms, try factoring by grouping. (Sections 5.1 or 5.6)
6. Be sure to factor the polynomial completely.
7. Check by multiplying.

2. Part I: General Factoring Review

example 1

Factoring Polynomials

Factor out the GCF and identify the number of terms and type of factoring pattern represented by the polynomial. Then factor the polynomial completely.

a. $abx^2 - 3ax + 5bx - 15$ b. $20y^2 - 110y - 210$
c. $4p^3 + 20p^2 + 25p$ d. $w^3 + 1000$
e. $t^3 - 25t$

Solution:

a. $abx^2 - 3ax + 5bx - 15$

$abx^2 - 3ax \mid + 5bx - 15$

$= ax(bx - 3) + 5(bx - 3)$

$= (bx - 3)(ax + 5)$

The GCF is 1. The polynomial has four terms. Therefore, factor by grouping.

Check: $(bx - 3)(ax + 5) = abx^2 + 5bx - 3ax - 15$ ✔

b. $20y^2 - 110y - 210$
$= 10(2y^2 - 11y - 21)$
$= 10(2y + 3)(y - 7)$

The GCF is 10. The polynomial has three terms. The trinomial is not a perfect square trinomial. Use either the grouping method or the trial-and-error method.

Check: $10(2y + 3)(y - 7) = 10(2y^2 - 14y + 3y - 21)$
$= 10(2y^2 - 11y - 21)$
$= 20y^2 - 110y - 210$ ✔

c. $4p^3 + 20p^2 + 25p$
$= p(4p^2 + 20p + 25)$

The GCF is p. The polynomial has three terms and is a perfect square trinomial, $a^2 + 2ab + b^2$, where $a = 2p$ and $b = 5$.

$= p(2p + 5)^2$

Apply the formula $a^2 + 2ab + b^2 = (a + b)^2$.

Check: $p(2p + 5)^2 = p[(2p)^2 + 2(2p)(5) + (5)^2]$
$= p(4p^2 + 20p + 25)$
$= 4p^3 + 20p^2 + 25p$ ✔

d. $w^3 + 1000$
$= (w)^3 + (10)^3$

The GCF is 1. The polynomial has two terms. The binomial is a sum of cubes, $a^3 + b^3$, where $a = w$ and $b = 10$.

$= (w + 10)(w^2 - 10w + 100)$

Apply the formula $a^3 + b^3 = (a + b)(a^2 - ab + b^2)$.

Check: $(w + 10)(w^2 - 10w + 100) = w^3 - \cancel{10w^2} + \cancel{100w} + \cancel{10w^2} - \cancel{100w} + 1000$
$= w^3 + 100$ ✔

e. $t^3 - 25t$
$= t(t^2 - 25)$
$= t[(t)^2 - (5)^2]$

The GCF is t. The polynomial has two terms. The binomial is a difference of squares, $a^2 - b^2$, where $a = t$ and $b = 5$.

$= t(t + 5)(t - 5)$

Apply the formula $a^2 - b^2 = (a + b)(a - b)$.

Check: $t(t + 5)(t - 5) = t(t^2 - \cancel{5t} + \cancel{5t} - 25)$
$= t(t^2 - 25)$
$= t^3 - 25t$ ✔

3. Part II: Additional Factoring Strategies

Some factoring problems may require more than one type of factoring. We also may encounter polynomials that require slight variations on the factoring techniques already learned. These are demonstrated in Examples 2–7.

example 2 **Factoring Binomials with Multiple Steps**

Factor completely. $d^4 - \frac{1}{16}$

Solution:

$d^4 - \frac{1}{16}$

$= (d^2)^2 - \left(\frac{1}{4}\right)^2$ — d^4 and $\frac{1}{16}$ are perfect squares.

$= \left(d^2 + \frac{1}{4}\right)\left(d^2 - \frac{1}{4}\right)$ — Factor as a difference of squares.

$= \left(d^2 + \frac{1}{4}\right)\left(d - \frac{1}{2}\right)\left(d + \frac{1}{2}\right)$ — The binomial $d^2 - \frac{1}{4}$ is also a difference of squares.

Avoiding Mistakes

Remember that a sum of squares such as $d^2 + \frac{1}{4}$ cannot be factored over the real numbers.

example 3 **Factoring a Trinomial Involving Fractional Coefficients**

Factor completely: $\frac{1}{9}x^2 + \frac{1}{3}x + \frac{1}{4}$

Solution:

$\frac{1}{9}x^2 + \frac{1}{3}x + \frac{1}{4}$

$= \left(\frac{1}{3}x\right)^2 + 2\left(\frac{1}{3}x\right)\left(\frac{1}{2}\right) + \left(\frac{1}{2}\right)^2$

$= \left(\frac{1}{3}x + \frac{1}{2}\right)^2$

The fractions may make this polynomial look difficult to factor. However, notice that both $\frac{1}{9}x^2$ and $\frac{1}{4}$ are perfect squares. Furthermore, the middle term $\frac{1}{3}x = 2(\frac{1}{3}x)(\frac{1}{2})$. Therefore, the trinomial is a perfect square trinomial.

4. Factoring Using Substitution

Sometimes it is convenient to use substitution to convert a polynomial into a simpler form before factoring.

example 4 **Using Substitution to Factor a Trinomial**

Factor completely: $(2x - 7)^2 - 3(2x - 7) - 40$

Solution:

$(2x - 7)^2 - 3(2x - 7) - 40$

$= u^2 - 3u - 40$ — Substitute $u = 2x - 7$. The trinomial is simpler in form.

$= (u - 8)(u + 5)$	Factor the trinomial.
$= [(2x - 7) - 8][(2x - 7) + 5]$	Reverse substitute. Replace u by $2x - 7$.
$= (2x - 7 - 8)(2x - 7 + 5)$	Simplify.
$= (2x - 15)(2x - 2)$	The second binomial has a GCF of 2.
$= (2x - 15)(2)(x - 1)$	Factor out the GCF from the second binomial.
$= 2(2x - 15)(x - 1)$	

5. Factoring by Grouping—Rearranging Terms

Sometimes it is necessary to rearrange terms when factoring by grouping.

example 5

Factoring by Grouping Where Rearranging Terms Is Necessary

Factor completely: $4x + 6pa - 8a - 3px$

Solution:

$4x + 6pa - 8a - 3px$	The GCF of all four terms is 1.
$= 4x + 6pa \mid - 8a - 3px$	Try factoring by grouping.
$= 2(2x + 3pa) - 1(8a + 3px)$	The binomial factors in each term are different.
$= 4x - 8a \mid - 3px + 6pa$	*Try rearranging the original four terms* in such a way that the first pair of coefficients is in the same ratio as the second pair of coefficients. Notice that the ratio 4 to -8 is the same as the ratio -3 to 6.
$= 4(x - 2a) - 3p(x - 2a)$	Factor out the GCF from the first pair of terms. Factor out the GCF from the second pair of terms.
$= (x - 2a)(4 - 3p)$	Factor out the common binomial factor, $(x - 2a)$.

6. Other Grouping Techniques for Factoring Four-Term Polynomials

example 6

Factoring a Four-Term Polynomial by Grouping Three Terms

Factor completely: $25w^2 + 90w + 81 - p^2$

Solution:

With a four-term polynomial, we recommend "2-by-2" grouping—that is, to group the first pair of terms and the second pair of terms. However, in this case, there is no common binomial factor shared by each pair of terms. (Even rearranging terms does not help.)

Since this polynomial is not factorable with "2-by-2" grouping, try grouping three terms. Notice that the first three terms constitute a perfect square trinomial. Hence, we will use "3-by-1" grouping.

$\underbrace{25w^2 + 90w + 81} \mid - p^2$	Group "3 by 1."
	Factor $25w^2 + 90w + 81 = (5w + 9)^2$.
$= (5w + 9)^2 - p^2$	$(5w + 9)^2 - p^2$ is a difference of squares $a^2 - b^2$, where $a = (5w + 9)$ and $b = p$.
$= [(5w + 9) + p][(5w + 9) - p]$	Factor as $a^2 - b^2 = (a + b)(a - b)$.
$= (5w + 9 + p)(5w + 9 - p)$	Simplify.

example 7 Factoring a Four-Term Polynomial by Grouping Three Terms

Factor completely: $x^2 - y^2 - 6y - 9$

Solution:

Grouping "2 by 2" will not work to factor this polynomial. However, if we factor out -1 from the last three terms, the resulting trinomial will be a perfect square trinomial.

$x^2 \mid - y^2 - 6y - 9$	Group the last three terms.
$= x^2 - 1(y^2 + 6y + 9)$	Factor out -1 from the last three terms.
$= x^2 - (y + 3)^2$	Factor the perfect square trinomial $y^2 + 6y + 9$ as $(y + 3)^2$.
	The quantity $x^2 - (y + 3)^2$ is a difference of squares, $a^2 - b^2$, where $a = x$ and $b = (y + 3)$.
$= [x - (y + 3)][x + (y + 3)]$	Factor as $a^2 - b^2 = (a + b)(a - b)$.
$= (x - y - 3)(x + y + 3)$	Apply the distributive property to clear the inner parentheses.

Avoiding Mistakes

When factoring the expression $x^2 - (y + 3)^2$ as a difference of squares, be sure to use parentheses around the quantity $(y + 3)$. This will help you remember to "distribute the negative" in the expression $[x - (y + 3)]$.

$[x - (y + 3)] = (x - y - 3)$

section 5.6 Practice Exercises

Part I:

For Exercises 1–84, factor completely using the strategy found on page 347. Identify any polynomials that are prime.

1. $x^2 - 6x - 16$
2. $a^2 - a - 42$
3. $20b^2 - 11b - 3$
4. $6y^2 - 29y + 20$
5. $100 - 9u^2$
6. $144 - x^2$
7. $p^3 - 216$
8. $125 + q^3$
9. $x^2y^3 + x^5y^2$
10. $3a^3b + 6a^2$
11. $2y - 22 + 9xy - 99x$
12. $21b + 7 - 3ab - a$
13. $w^2 - 40w + 400$
14. $z^2 + 28z + 196$
15. $x^3 - 3x^2 - 10x$
16. $x^3 - x^2 - 12x$
17. $3y^2 + 21y + 36$
18. $2w^2 + 16w + 24$
19. $p^2 + 12pq + 36q^2$
20. $m^2 - 8mn + 16n^2$
21. $x^2 + 3x + 8$
22. $x^2 - 4x + 10$
23. $2x^2 + 13x - 24$
24. $3x^2 - 17x - 6$
25. $u^2 - 25v^2$
26. $16p^2 - 1$
27. $a^3b^3 - 36ab$
28. $x^3y^3 - xy$
29. $2x^3 - 20x^2 + 18x$
30. $3x^3 - 30x^2 + 63x$
31. $-3a^2b^2 + 3ab^3 - 6ab^2$
32. $-5p^4q - 5p^3q^3 + 10p^3q$
33. $11r(s + 4) - 6(s + 4)$
34. $10a(b - 7) + 7(b - 7)$
35. $100 + t^2$
36. $81 + 4x^2$
37. $x^4 - 8x$
38. $5u - 5u^4$
39. $m^3 + 16$
40. $n^3 - 9$
41. $5xy - 3y + 15x - 9$
42. $2mn + 2m - 5n - 5$
43. $2pq - 14p - 8q + 56$
44. $3ab - 12a - 24b + 96$
45. $4x^2 - 16x + 16$
46. $5x^2 + 10x + 5$
47. $50 + p^2 - 15p$
48. $3q - 88 + q^2$
49. $24xy + 16x^2 + 9y^2$
50. $9p^2 + 4q^2 + 12pq$
51. $2x^5 + 6x^3 - 10x^4 - 30x^2$
52. $35a^2 + 14a - 15a - 6$
53. $-x^2 + 16x - 63$
54. $-x^2 - 7x + 60$
55. $6x^2 - 21x - 45$
56. $20y^2 - 14y + 2$
57. $5a^2bc^3 - 7abc^2$
58. $8a^2 - 50$
59. $t^2 + 2t - 63$
60. $b^2 + 2b - 80$
61. $ab + ay - b^2 - by$
62. $6x^3y^4 + 3x^2y^5$
63. $14u^2 - 11uv + 2v^2$
64. $9p^2 - 36pq + 4q^2$
65. $4q^2 - 8q - 6$
66. $9w^2 + 3w - 15$
67. $9m^2 + 16n^2$
68. $5b^2 - 30b + 45$
69. $6r^2 + 11r + 3$
70. $4s^2 + 4s - 15$
71. $81u^2 - 90uv + 25v^2$
72. $4x^2 + 16$
73. $2ax - 6ay + 4bx - 12by$
74. $8m^3 - 10m^2 - 3m$
75. $21x^4y + 41x^3y + 10x^2y$
76. $2m^4 - 128$
77. $8uv - 6u + 12v - 9$
78. $4t^2 - 20t + st - 5s$
79. $12x^2 - 12x + 3$
80. $p^2 + 2pq + q^2$
81. $6n^3 + 5n^2 - 4n$
82. $4k^3 + 4k^2 - 3k$
83. $64 - y^2$
84. $36b - b^3$

Part II:

For Exercises 85–112, factor completely using the strategy found on page 347 and any additional techniques of factoring illustrated in Examples 2–7.

85. $x^2(x + y) - y^2(x + y)$
86. $u^2(u - v) - v^2(u - v)$

87. $(a + 3)^4 + 6(a + 3)^5$

88. $(4 - b)^4 - 2(4 - b)^3$

89. $24(3x + 5)^3 - 30(3x + 5)^2$

90. $10(2y + 3)^2 + 15(2y + 3)^3$

91. $16p^4 - q^4$

92. $s^4t^4 - 81$

93. $y^3 + \dfrac{1}{64}$

94. $z^3 + \dfrac{1}{125}$

95. $6a^3 + a^2b - 6ab^2 - b^3$

96. $4p^3 + 12p^2q - pq^2 - 3q^3$

97. $\dfrac{1}{9}t^2 + \dfrac{1}{6}t + \dfrac{1}{16}$

98. $\dfrac{1}{25}y^2 + \dfrac{1}{5}y + \dfrac{1}{4}$

99. $x^2 + 12x + 36 - a^2$

100. $a^2 + 10a + 25 - b^2$

101. $p^2 + 2pq + q^2 - 81$

102. $m^2 - 2mn + n^2 - 9$

103. $b^2 - (x^2 + 4x + 4)$

104. $p^2 - (y^2 - 6y + 9)$

105. $4 - u^2 + 2uv - v^2$

106. $25 - a^2 - 2ab - b^2$

107. $6ax - by + 2bx - 3ay$

108. $5pq - 12 - 4q + 15p$

109. $u^6 - 64$ [*Hint*: Factor first as a difference of squares, $(u^3)^2 - (8)^2$.]

110. $1 - v^6$

111. $x^8 - 1$

112. $y^8 - 256$

Expanding Your Skills

For Exercises 113–116, factor the polynomial in part (a). Then use substitution to help factor the polynomials in parts (b) and (c).

113. a. $u^2 - 10u + 25$
 b. $x^4 - 10x^2 + 25$
 c. $(a + 1)^2 - 10(a + 1) + 25$

114. a. $u^2 + 12u + 36$
 b. $y^4 + 12y^2 + 36$
 c. $(b - 2)^2 + 12(b - 2) + 36$

115. a. $u^2 + 11u - 26$
 b. $w^6 + 11w^3 - 26$
 c. $(y - 4)^2 + 11(y - 4) - 26$

116. a. $u^2 + 17u + 30$
 b. $z^6 + 17z^3 + 30$
 c. $(x + 3)^2 + 17(x + 3) + 30$

For Exercises 117–120, use substitution to factor the expressions.

117. $(5x^2 - 1)^2 - 4(5x^2 - 1) - 5$

118. $(x^3 + 4)^2 - 10(x^3 + 4) + 24$

119. $2(3w - 5)^2 - 19(3w - 5) + 35$

120. $3(2y + 3)^2 + 23(2y + 3) - 8$

For Exercises 121–124, factor completely. Then check by multiplying.

121. $a^2 - b^2 + a + b$

122. $25c^2 - 9d^2 + 5c - 3d$

123. $5wx^3 + 5wy^3 - 2zx^3 - 2zy^3$

124. $3xu^3 - 3xv^3 - 5yu^3 + 5yv^3$

Concepts

1. Definition of a Quadratic Equation
2. Zero Product Rule
3. Solving Quadratic Equations
4. Solving Higher Degree Polynomial Equations
5. Applications of Quadratic Equations
6. Pythagorean Theorem

section

5.7 Solving Quadratic Equations Using the Zero Product Rule

1. Definition of a Quadratic Equation

In Section 2.1 we solved linear equations in one variable. These are equations of the form $ax + b = 0$ $(a \neq 0)$. A linear equation in one variable is sometimes called a first-degree polynomial equation because the highest degree of all its terms is 1. A second-degree polynomial equation in one variable is called a quadratic equation.

Definition of a Quadratic Equation in One Variable

If a, b, and c are real numbers such that $a \neq 0$, then a **quadratic equation** is an equation that can be written in the form

$$ax^2 + bx + c = 0$$

The following equations are quadratic because they can each be written in the form $ax^2 + bx + c = 0$, $(a \neq 0)$.

$$-4x^2 + 4x = 1 \qquad x(x - 2) = 3 \qquad (x - 4)(x + 4) = 9$$

$$-4x^2 + 4x - 1 = 0 \qquad x^2 - 2x = 3 \qquad x^2 - 16 = 9$$

$$x^2 - 2x - 3 = 0 \qquad x^2 - 25 = 0$$

$$x^2 + 0x - 25 = 0$$

2. Zero Product Rule

One method for solving a quadratic equation is to factor the equation and apply the zero product rule. The **zero product rule** states that if the product of two factors is zero, then one or both of its factors is zero.

Zero Product Rule

If $ab = 0$, then $a = 0$ or $b = 0$

For example, the quadratic equation $x^2 - x - 12 = 0$ can be written in factored form as $(x - 4)(x + 3) = 0$. By the zero product rule, one or both factors must be zero. Hence, either $x - 4 = 0$ or $x + 3 = 0$. Therefore, to solve the quadratic equation, set each factor equal to zero and solve for x.

$(x - 4)(x + 3) = 0$			Apply the zero product rule.
$x - 4 = 0$	or	$x + 3 = 0$	Set each factor equal to zero.
$x = 4$	or	$x = -3$	Solve each equation for x.

3. Solving Quadratic Equations

Quadratic equations, like linear equations, arise in many applications in mathematics, science, and business. The following steps summarize the factoring method for solving a quadratic equation.

Steps for Solving a Quadratic Equation by Factoring

1. Write the equation in the form: $ax^2 + bx + c = 0$.
2. Factor the equation completely.
3. Apply the zero product rule. That is, set each factor equal to zero and solve the resulting equations.

Note: The solution(s) found in step 3 may be checked by substitution into the original equation.

example 1

Solving Quadratic Equations

Solve the quadratic equations.

a. $2x^2 - 9x = 5$ b. $4x^2 + 24x = 0$ c. $5x(5x + 2) = 10x + 9$

Solution:

a. $2x^2 - 9x = 5$

$2x^2 - 9x - 5 = 0$ Write the equation in the form $ax^2 + bx + c = 0$.

$(2x + 1)(x - 5) = 0$ Factor the polynomial completely.

$2x + 1 = 0$ or $x - 5 = 0$ Set each factor equal to zero.

$2x = -1$ or $x = 5$ Solve each equation.

$x = -\frac{1}{2}$ or $x = 5$ There are two solutions, $x = -\frac{1}{2}$ and $x = 5$.

Check: $x = -\frac{1}{2}$

$$2x^2 - 9x = 5$$
$$2\left(-\frac{1}{2}\right)^2 - 9\left(-\frac{1}{2}\right) \stackrel{?}{=} 5$$
$$2\left(\frac{1}{4}\right) + \frac{9}{2} \stackrel{?}{=} 5$$
$$\frac{1}{2} + \frac{9}{2} \stackrel{?}{=} 5$$
$$\frac{10}{2} \stackrel{?}{=} 5 \checkmark$$

Check: $x = 5$

$$2x^2 - 9x = 5$$
$$2(5)^2 - 9(5) \stackrel{?}{=} 5$$
$$2(25) - 45 \stackrel{?}{=} 5$$
$$50 - 45 \stackrel{?}{=} 5 \checkmark$$

b. $4x^2 + 24x = 0$ — The equation is already in the form $ax^2 + bx + c = 0$ (Note that $c = 0$).

$4x(x + 6) = 0$ — Factor completely.

$4x = 0$ or $x + 6 = 0$ — Set each factor equal to zero.

$x = 0$ or $x = -6$ — Each solution checks in the original equation.

c. $5x(5x + 2) = 10x + 9$

$25x^2 + 10x = 10x + 9$ — Clear parentheses.

$25x^2 + 10x - 10x - 9 = 0$ — Set the equation equal to zero.

$25x^2 - 9 = 0$ — The equation is in the form $ax^2 + bx + c = 0$ (Note that $b = 0$).

$(5x - 3)(5x + 3) = 0$ — Factor completely.

$5x - 3 = 0$ or $5x + 3 = 0$ — Set each factor equal to zero.

$5x = 3$ or $5x = -3$ — Solve each equation.

$\frac{5x}{5} = \frac{3}{5}$ or $\frac{5x}{5} = \frac{-3}{5}$

$x = \frac{3}{5}$ or $x = -\frac{3}{5}$

4. Solving Higher Degree Polynomial Equations

The zero product rule can be used to solve higher degree polynomial equations provided the equations can be set to zero and written in factored form.

example 2 **Solving Higher Degree Polynomial Equations**

Solve the equations.

a. $-6(y + 3)(y - 5)(2y + 7) = 0$ b. $w^3 + 5w^2 - 9w - 45 = 0$

Solution:

a. $-6(y + 3)(y - 5)(2y + 7) = 0$ — The equation is already in factored form and equal to zero.

Set each factor equal to zero. Solve each equation for y.

$-6 \neq 0$ or $y + 3 = 0$ or $y - 5 = 0$ or $2y + 7 = 0$

No solution, $y = -3$ or $y = 5$ or $y = -\frac{7}{2}$

Notice that when the constant factor is set equal to zero, the result is a contradiction $-6 = 0$. The constant factor does not produce a solution to the equation. Therefore, the only solutions are $y = -3$, $y = 5$, and $y = -\frac{7}{2}$. Each solution can be checked in the original equation.

b. $w^3 + 5w^2 - 9w - 45 = 0$ — This is a higher degree polynomial equation.

$w^3 + 5w^2 \mid - 9w - 45 = 0$ — The equation is already set equal to zero. Now factor.

$w^2(w + 5) - 9(w + 5) = 0$

$(w + 5)(w^2 - 9) = 0$ — Because there are four terms, try factoring by grouping.

$(w + 5)(w - 3)(w + 3) = 0$ — $w^2 - 9$ is a difference of squares and can be factored further.

$w + 5 = 0$ or $w - 3 = 0$ or $w + 3 = 0$ — Set each factor equal to zero.

$w = -5$ or $w = 3$ or $w = -3$ — Solve each equation.

Each solution checks in the original equation.

5. Applications of Quadratic Equations

example 3

Using a Quadratic Equation in an Application

The base of a triangle is 3 m more than the height. The area is 35 m^2. Find the base and height of the triangle.

Solution:

Let x represent the height of the triangle.

Then $x + 3$ represents the base (Figure 5-1).

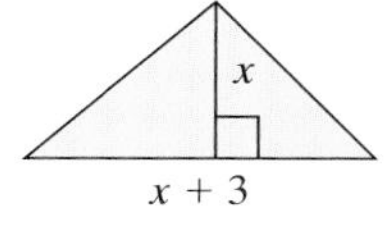

Figure 5-1

To set up an equation to solve for x, use $A = \frac{1}{2}bh$.

$$\text{Area} = \frac{1}{2}(\text{base})(\text{height})$$ Verbal equation

$$35 = \frac{1}{2}(x + 3)(x)$$ Algebraic equation

$$2 \cdot 35 = \cancel{2} \cdot \frac{1}{\cancel{2}}(x + 3)(x)$$ Multiply both sides by 2 to clear fractions.

$$70 = (x + 3)(x)$$

$70 = x^2 + 3x$	Clear parentheses.
$0 = x^2 + 3x - 70$	Write the equation in the form $ax^2 + bx + c = 0$.
$0 = (x + 10)(x - 7)$	Factor the equation.
$x + 10 = 0$ or $x - 7 = 0$	Set each factor equal to zero.
$x \not= -10$ or $x = 7$	Because x represents the height of a triangle, reject the negative solution.

The variable x represents the height of the triangle. Therefore, the height is 7 m.

The expression $x + 3$ represents the base of the triangle. Therefore, the base is 10 m.

example 4 Using Translations to Set up a Quadratic Equation

The product of two consecutive integers is 48 more than the larger integer. Find the integers.

Solution:

Let x represent the first (smaller) integer.

Then $x + 1$ represents the second (larger) integer.	Label the variables.
(First integer)(second integer) = (second integer) + 48	Verbal model
$x(x + 1) = (x + 1) + 48$	Algebraic equation
$x^2 + x = x + 49$	Simplify.
$x^2 + x - x - 49 = 0$	Set the equation equal to zero.
$x^2 - 49 = 0$	
$(x - 7)(x + 7) = 0$	Factor.
$x - 7 = 0$ or $x + 7 = 0$	Set each factor equal to zero.
$x = 7$ or $x = -7$	Solve for x.

Recall that x represents the smaller integer. Therefore, there are two possibilities for the pairs of consecutive integers.

If $x = 7$, then the larger integer is $x + 1$ or $7 + 1 = 8$.

If $x = -7$, then the larger integer is $x + 1$ or $-7 + 1 = -6$.

The integers are 7 and 8 or -7 and -6.

Tip: To check your answer in Example 4, verify that each pair of integers satisfies the requirements that the product of integers is equal to 48 more than the larger integer:

Product	Larger Integer + 48
$(7)(8) = 56$	$8 + 48 = 56$
$(-7)(-6) = 42$	$-6 + 48 = 42$

example 5

Using a Quadratic Equation in an Application

A stone is dropped off a 64-ft cliff and falls into the ocean below. The height of the stone above sea level is given by the equation

$h = -16t^2 + 64$ where h is the stone's height in feet, and t is the time in seconds.

Find the time required for the stone to hit the water.

Solution:

When the stone hits the water, its height is zero. Therefore, substitute $h = 0$ into the equation.

$h = -16t^2 + 64$ The equation is quadratic.

$0 = -16t^2 + 64$ Substitute $h = 0$.

$0 = -16(t^2 - 4)$ Factor out the GCF.

$0 = -16(t - 2)(t + 2)$ Factor as a difference of squares.

$-16 \neq 0$ or $t - 2 = 0$ or $t + 2 = 0$ Set each factor to zero.

No solution, $t = 2$ or $\cancel{t = -2}$ Solve for t.

The negative value of t is rejected because the stone cannot fall for a negative time. Therefore, the stone hits the water after 2 seconds.

In Example 5, we can analyze the path of the stone as it falls from the cliff. Compute the height values at various times between 0 and 2 sec (Table 5-1 and Table 5-2). The ordered pairs can be graphed where t is used in place of x and h is used in place of y.

Table 5-1

Time, t (sec)	Height, h (ft)
0.0	
0.5	
1.0	
1.5	
2.0	

$h = -16(0.0)^2 + 64 = 64$

$h = -16(0.5)^2 + 64 = 60$

$h = -16(1.0)^2 + 64 = 48$

$h = -16(1.5)^2 + 64 = 28$

$h = -16(2.0)^2 + 64 = 0$

Table 5-2

Time, t (sec)	Height, h (ft)
0.0	64
0.5	60
1.0	48
1.5	28
2.0	0

The graph of the height of the stone versus time is shown in Figure 5-2. From the graph, we can verify that the stone hits the water after 2 sec.

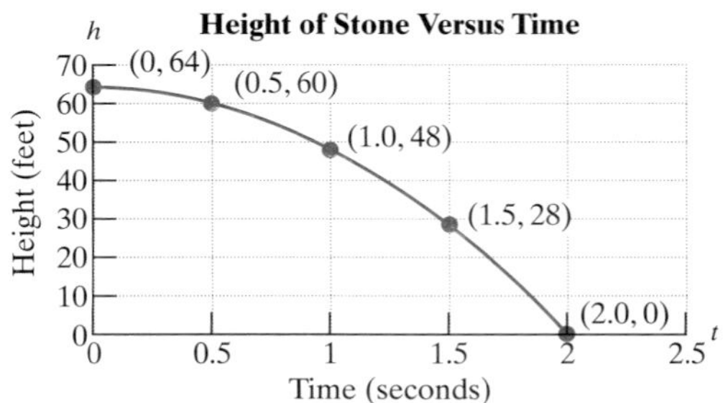

Figure 5-2

6. Pythagorean Theorem

Recall that a right triangle is a triangle that contains a 90º angle. Furthermore, the sum of the squares of the two legs (the shorter sides) of a right triangle equals the square of the hypotenuse (the longest side). This important fact is known as the Pythagorean theorem. The Pythagorean theorem is an enduring landmark of mathematical history from which many mathematical ideas have been built. Although the theorem is named after Pythagoras (sixth century B.C.E.), a Greek mathematician and philosopher, it is thought that the ancient Babylonians were familiar with the principle more than a thousand years earlier.

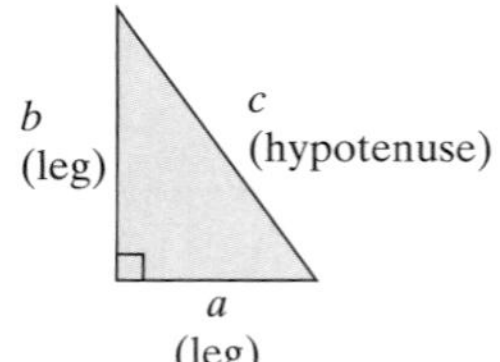

Figure 5-3

For the right triangle shown in Figure 5-3, the **Pythagorean theorem** is stated as:

$$a^2 + b^2 = c^2.$$

In this formula, a and b are the legs of the right triangle and c is the hypotenuse. Notice that the hypotenuse is the longest side of the right triangle and is opposite the 90º angle.

example 6

Applying the Pythagorean Theorem

Show that the lengths of the sides of the right triangle in the figure satisfy the Pythagorean theorem.

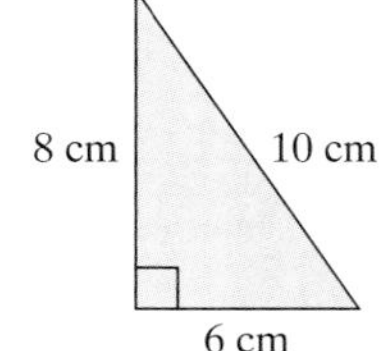

Solution:

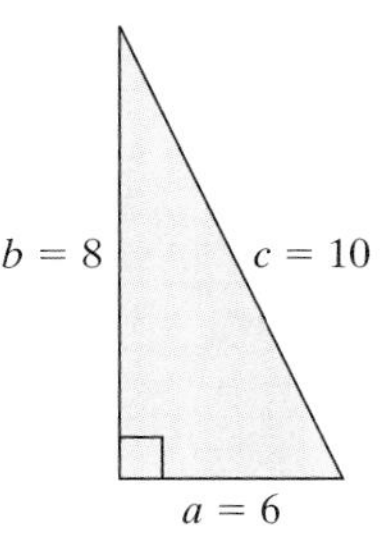

Label the triangle.

$a^2 + b^2 = c^2$ Apply the Pythagorean theorem.

$(6)^2 + (8)^2 \stackrel{?}{=} (10)^2$ $a = 6, b = 8, c = 10$

$36 + 64 \stackrel{?}{=} 100$

$100 = 100$ ✔

example 7

Using a Quadratic Equation in an Application

A 13-ft board is used as a ramp to unload furniture off a loading platform. If the distance between the top of the board and the ground is 7 ft less than the distance between the bottom of the board and the base of the platform, find both distances.

Solution:

Let x represent the distance between the bottom of the board and the base of the platform. Then $x - 7$ represents the distance between the top of the board and the ground (Figure 5-4).

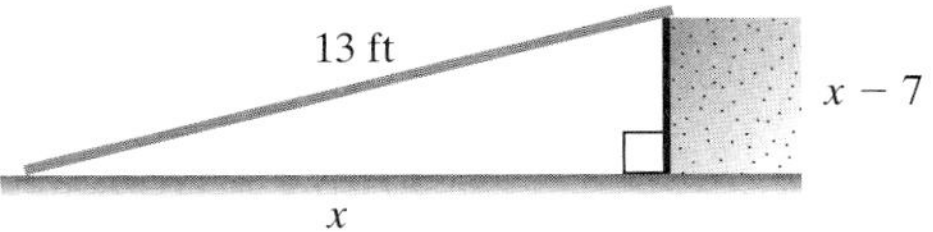

Figure 5-4

$$a^2 + b^2 = c^2$$ Pythagorean theorem

$$x^2 + (x - 7)^2 = (13)^2$$

Avoiding Mistakes

Recall that the square of a binomial results in a perfect square trinomial.

$(a - b)^2 = a^2 - 2ab + b^2$

Don't forget the middle term.

$$x^2 + [(x)^2 - 2(x)(7) + (7)^2] = 169$$

$$x^2 + x^2 - 14x + 49 = 169$$

$$2x^2 - 14x + 49 = 169$$ Combine *like* terms.

$$2x^2 - 14x + 49 - 169 = 0$$ Set the equation equal to zero.

$$2x^2 - 14x - 120 = 0$$ Write the equation in the form $ax^2 + bx + c = 0$.

$$2(x^2 - 7x - 60) = 0$$ Factor.

$$2(x - 12)(x + 5) = 0$$

$$2 \neq 0 \quad \text{or} \quad x - 12 = 0 \quad \text{or} \quad x + 5 = 0$$ Set each factor equal to zero.

$$x = 12 \quad \text{or} \quad x \neq -5$$ Solve both equations for x.

Recall that x represents the distance between the bottom of the board and the base of the platform. We reject the negative value of x because a distance cannot be negative. Therefore, the distance between the bottom of the board and the base of the platform is 12 ft. The distance between the top of the board and the ground is $x - 7 = 5$ ft.

section 5.7 Practice Exercises

For Exercises 1–10, factor completely.

1. $4x - 2 + 2bx - b$
2. $6a - 8 - 3ab + 4b$
3. $4b^2 - 44b + 120$
4. $8u^2v^2 - 4uv$
5. $16w^2 - 1$
6. $3x^2 + 10x - 8$
7. $12k + 16$
8. $3h^2 - 75$
9. $2y^2 + 3y - 44$
10. $4x^2 + 16y^2$

For Exercises 11–18, identify the polynomials as linear, quadratic, or neither.

11. $4 - 5x$
12. $5x^3 + 2$
13. $3x - 6x^2$
14. $1 - x + 2x^2$
15. $7x^4 + 8$
16. $3x + 2$
17. $6x^2 - 7x - 2$
18. $4x^2 - 1$

19. State the zero product rule.

For Exercises 20–27, solve the equations using the zero product rule.

20. $(x - 5)(x + 1) = 0$
21. $(x + 3)(x - 1) = 0$
22. $(3x - 2)(3x + 2) = 0$
23. $(2x - 7)(2x + 7) = 0$
24. $2(x - 7)(x - 7) = 0$
25. $3(x + 5)(x + 5) = 0$
26. $x(x - 4)(2x + 3) = 0$
27. $x(3x + 1)(x + 1) = 0$

28. For a quadratic equation of the form $ax^2 + bx + c = 0$, what must be done before applying the zero product rule?

For Exercises 29–40, solve the equations.

29. $p^2 - 2p - 15 = 0$
30. $y^2 - 7y - 8 = 0$
31. $z^2 + 10z - 24 = 0$
32. $w^2 - 10w + 16 = 0$
33. $2q^2 - 7q - 4 = 0$
34. $4x^2 - 11x - 3 = 0$
35. $0 = 9x^2 - 4$
36. $4a^2 - 49 = 0$
37. $2k^2 - 28k + 96 = 0$
38. $0 = 2t^2 + 20t + 50$
39. $0 = 2m^3 - 5m^2 - 12m$
40. $3n^3 + 4n^2 + n = 0$
41. What are the requirements to use the zero product rule to solve a quadratic equation or higher degree polynomial equation?

For Exercises 42–63, solve the equations.

42. $x^2 + 10x = 24$
43. $x^2 - 10x = -16$
44. $9d^2 = 4$
45. $4p^2 = 49$
46. $2(c^2 - 14c) = -96$
47. $2(q^2 + 10q) = -50$
48. $12x = 2x^3 - 5x^2$
49. $-x = 3x^3 + 4x^2$
50. $3(a^2 + 2a) = 2a^2 - 9$
51. $9(k - 1) = -4k^2$
52. $2n(n + 2) = 6$
53. $3p(p - 1) = 18$
54. $27q^2 = 9q$
55. $21w^2 = 14w$
56. $3(c^2 - 2c) = 0$
57. $2(4d^2 + d) = 0$
58. $y^3 - 3y^2 - 4y + 12 = 0$
59. $t^3 + 2t^2 - 16t - 32 = 0$
60. $(x - 1)(x + 2) = 18$
61. $(w + 5)(w - 3) = 20$
62. $(p + 2)(p + 3) = 1 - p$
63. $(k - 6)(k - 1) = -k - 2$
64. If 11 is added to the square of a number, the result is 60. Find all such numbers.
65. If a number is added to 2 times its square, the result is 36. Find all such numbers.
66. If 12 is added to 6 times a number, the result is 28 less than the square of the number. Find all such numbers.
67. The square of a number is equal to 20 more than the number. Find all such numbers.
68. The product of two consecutive odd integers is 63. Find all such integers.
69. The product of two consecutive even integers is 48. Find all such integers.
70. The sum of the squares of two consecutive integers is one more than 10 times the larger number. Find all such integers.
71. The sum of the squares of two consecutive integers is 9 less than 10 times the sum of the integers. Find all such integers.
72. The length of a rectangular room is 5 yd more than the width. If 300 yd^2 of carpeting cover the room, what are the dimensions of the room?

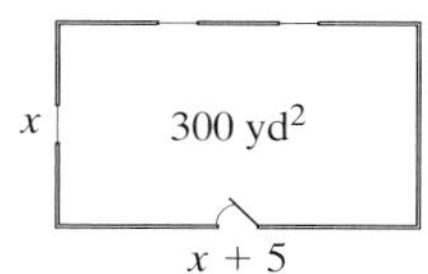

Figure for Exercise 72

73. The width of a rectangular painting is 2 in. less than the length. The area is 120 in.^{2} Find the length and width.

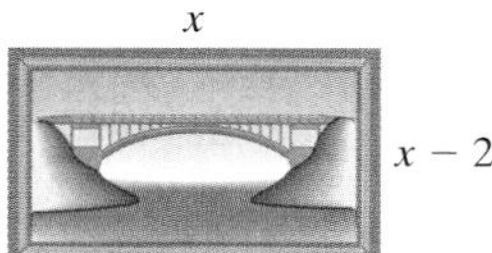

Figure for Exercise 73

74. The width of a rectangular slab of concrete is 3 m less than the length. If the area is 28 m^2,
    a. What are the dimensions of the rectangle?
    b. What is the perimeter of the rectangle?
75. The width of a rectangular picture is 7 in. less than the length. If the area of the picture is 78 in.^{2},
    a. What are the dimensions of the rectangle?
    b. What is the perimeter of the rectangle?

76. The base of a triangle is 1 ft less than twice the height. The area is 14 ft^2. Find the base and height of the triangle.

77. The height of a triangle is 5 cm less than 3 times the base. The area is 125 cm^2. Find the base and height of the triangle.

78. In a physics experiment, a ball is dropped off a 144-ft platform. The height of the ball above the ground is given by the equation

$h = -16t^2 + 144$ where h is the ball's height in feet and t is the time in seconds after the ball is dropped ($t \geq 0$).

Find the time required for the ball to hit the ground. (*Hint*: Let $h = 0$)

79. A stone is dropped off a 64-ft cliff. The height of the stone above the ground is given by the equation

$h = -16t^2 + 64$ where h is the stone's height in feet, and t is the time in seconds after the stone is dropped ($t \geq 0$).

Find the time required for the stone to hit the ground.

80. An object is shot straight up into the air from ground level with initial speed of 24 ft/sec. The height of the object (in feet) is given by the equation

$h = -16t^2 + 24t$ where t is the time in seconds after launch ($t \geq 0$).

Find the time(s) when the object is at ground level.

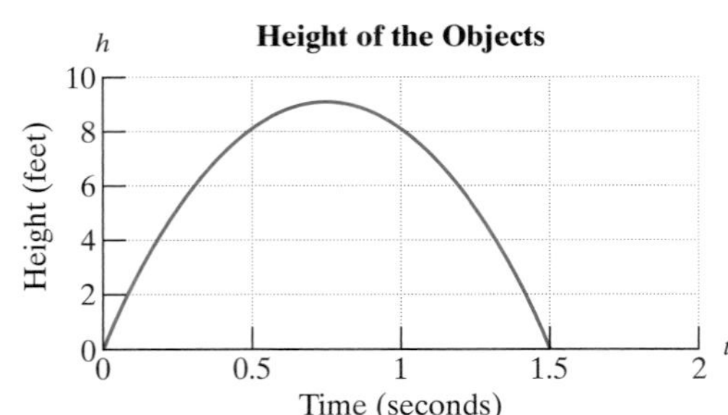

Figure for Exercise 80

81. A rocket is launched straight up into the air from the ground with initial speed of 64 ft/sec. The height of the rocket (in feet) is given by the equation

$h = -16t^2 + 64t$ where t is the time in seconds after launch ($t \geq 0$).

Find the time(s) when the rocket is at ground level.

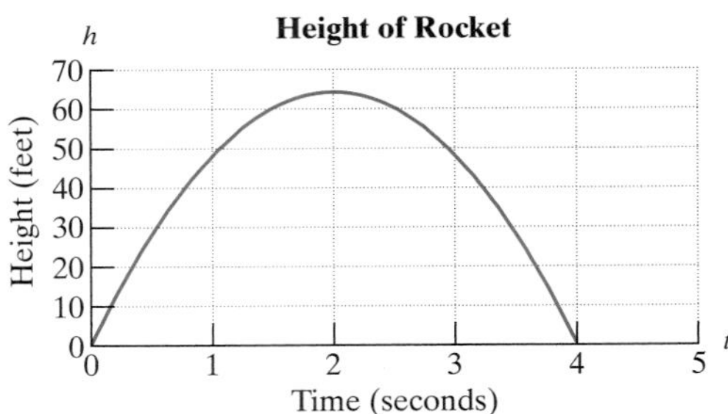

Figure for Exercise 81

82. Draw a right triangle and label the sides with the words *leg* and *hypotenuse.*

83. State the Pythagorean theorem.

For Exercises 84–87, use the Pythagorean theorem to determine whether the triangle could be a right triangle.

84. 85.

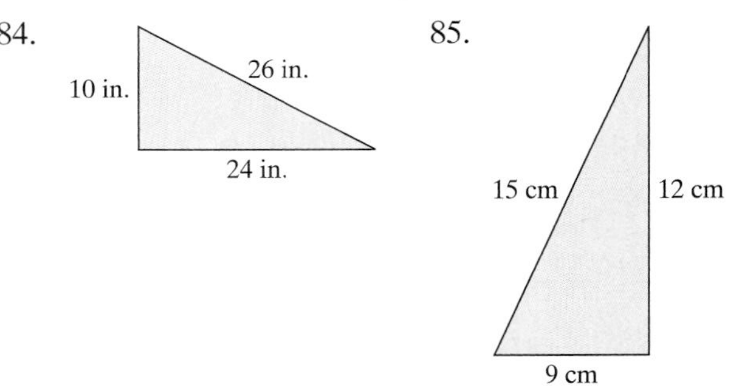

86. 87.

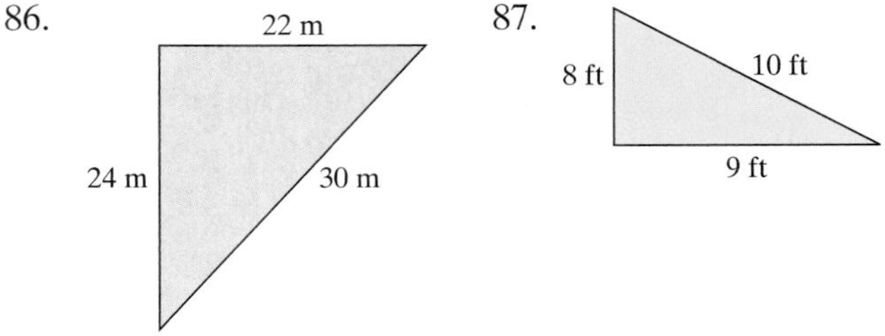

88. Darcy holds the end of a kite string 3 ft (1 yd) off the ground and wants to estimate the height of the kite. Her friend Jenna is 24 yd away from her, standing directly under the kite as shown in

the figure. If Darcy has 30 yd of string out, find the height of the kite (ignore the sag in the string).

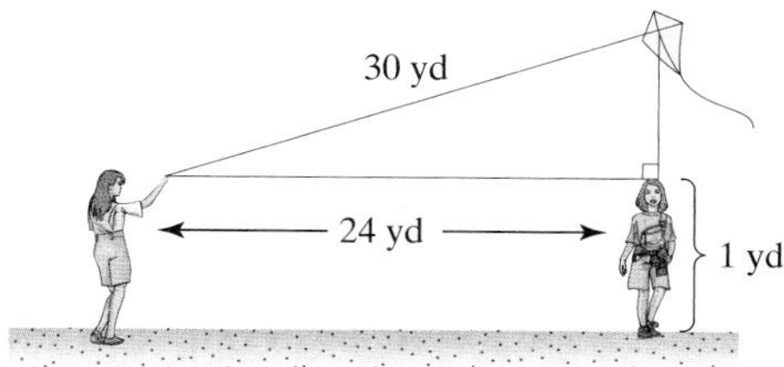

Figure for Exercise 88

89. A 17-ft ladder rests against the side of a house. The distance between the top of the ladder and the ground is 7 ft more than the distance between the base of the ladder and the bottom of the house. Find both distances.

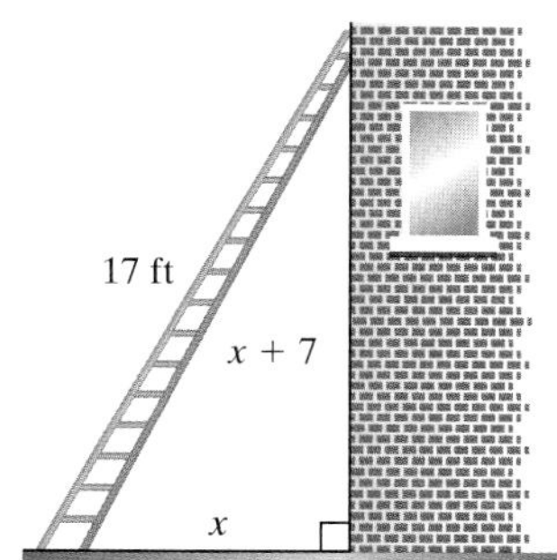

Figure for Exercise 89

90. Two boats leave a marina. One travels east, and the second travels south. After 30 min, the second boat has traveled 1 mile farther than the first boat and the distance between the boats is 5 miles. Find the distance each boat traveled.

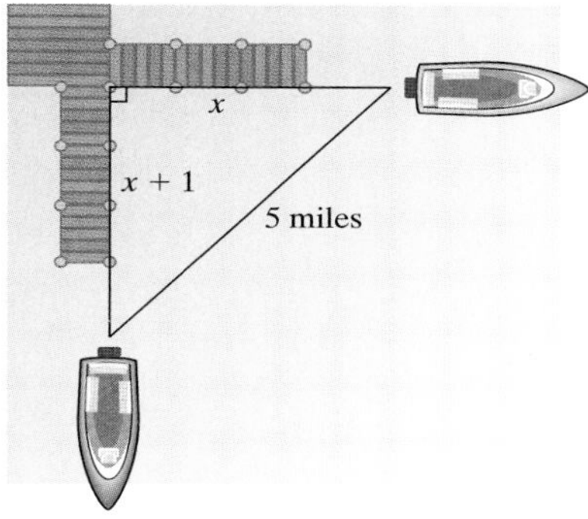

Figure for Exercise 90

91. One leg of a right triangle is 4 m less than the hypotenuse. The other leg is 2 m less than the hypotenuse. Find the length of the hypotenuse.

92. The longer leg of a right triangle is 1 cm less than twice the shorter leg. The hypotenuse is 1 cm greater than twice the shorter leg. Find the length of the shorter leg.

Expanding Your Skills

93. The formula

$$N = \frac{x(x-3)}{2}$$

gives the number of diagonals, N, for a polygon with x sides, where $x \geq 3$.

a. Find the number of diagonals for a four-sided polygon.

b. Find the number of diagonals for a five-sided polygon.

c. Find the number of sides of a polygon if the polygon has 35 diagonals.

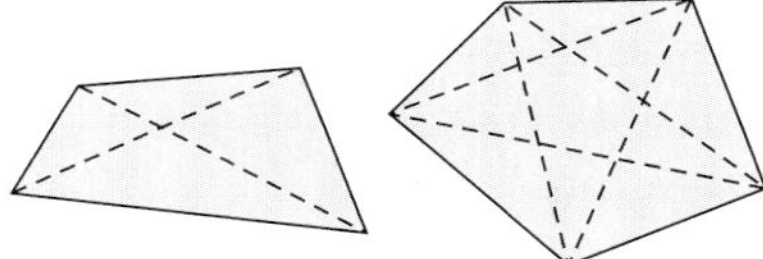

Figure for Exercise 93

94. A cardboard box is to be constructed from a square piece of cardboard by cutting out 2-in. squares from the corners and folding up the sides. If the volume of the box is 128 in.^{3}, find the original dimensions of the square piece of cardboard.

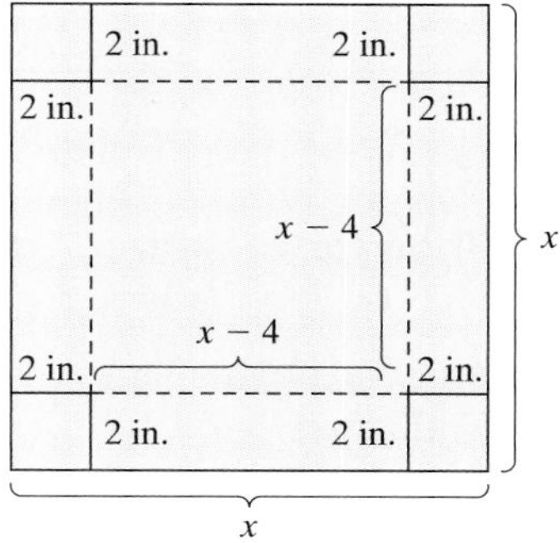

Figure for Exercise 94

For Exercises 95–98, solve the equations.

95. $(a + 2)^2 - 4(a + 2) - 21 = 0$

96. $(t - 3)^2 + 8(t - 3) + 15 = 0$

97. $2(w - 1)^2 - 7(w - 1) - 4 = 0$

98. $3(p + 4)^2 - (p + 4) - 4 = 0$

---

chapter 5

Summary

Section 5.1—Greatest Common Factor and Factoring by Grouping

Key Concepts:

The greatest common factor (GCF) is the greatest factor common to all terms of a polynomial. To factor out the GCF from a polynomial, use the distributive property.

A four-term polynomial may be factorable by grouping.

Steps to Factoring by Grouping

1. Identify and factor out the GCF from all four terms.
2. Factor out the GCF from the first pair of terms. Factor out the GCF or its opposite from the second pair of terms.
3. If the two terms share a common binomial factor, factor out the binomial factor.

Key Terms:

factoring
factoring by grouping
greatest common factor (GCF)
prime factorization

Examples:

Factor out the GCF:

$$\begin{aligned}&3x^2(a + b) - 6x(a + b)\\&= 3x(a + b)x - 3x(a + b)(2)\\&= 3x(a + b)(x - 2)\end{aligned}$$

Factor by Grouping:

$$\begin{aligned}&60xa - 30xb - 80ya + 40yb\\&= 10[6xa - 3xb - 8ya + 4yb]\\&= 10[3x(2a - b) - 4y(2a - b)]\\&= 10[(2a - b)(3x - 4y)]\\&= 10(2a - b)(3x - 4y)\end{aligned}$$

Section 5.2—Factoring Trinomials: Grouping Method

Key Concepts:

Grouping Method for Factoring Trinomials of the Form $ax^2 + bx + c$ (where $a \neq 0$)

1. Factor out the GCF from all terms.
2. Find the product ac.
3. Find two integers whose product is ac and whose sum is b. (If no pair of integers can be found, then the trinomial is prime.)

Examples:

Factor: $10y^2 + 35y - 20$
$= 5(2y^2 + 7y - 4)$

Note: $ac = (2)(-4) = -8$
$b = 7$

4. Rewrite the middle term (bx) as the sum of two terms whose coefficients are the numbers found in step 3.
5. Factor the polynomial by grouping.

Key Term:

prime polynomial

Find two integers whose product is -8 and whose sum is 7. The numbers are 8 and -1.

$$5[2y^2 + 8y - 1y - 4]$$
$$= 5[2y(y + 4) - 1(y + 4)]$$
$$= 5(y + 4)(2y - 1)$$

Section 5.3—Factoring Trinomials: Trial-and-Error Method

Key Concepts:

Trial-and-Error Method for Factoring Trinomials in the Form $ax^2 + bx + c$

1. Factor out the GCF from all terms.
2. List the pairs of factors of a and the pairs of factors of c. Consider the reverse order in either list.
3. Construct two binomials of the form

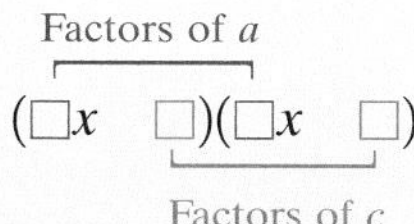

4. Test each combination of factors and signs until the product forms the correct trinomial.
5. If no combination of factors produces the correct product, then the trinomial is prime.

Key Term:

prime polynomial

Examples:

Factor: $10y^2 + 35y - 20$
$= 5(2y^2 + 7y - 4)$

The pairs of factors of 2 are: $2 \cdot 1$
The pairs of factors of -4 are:

$-1 \cdot 4$	$1(-4)$
$-2 \cdot 2$	$2(-2)$
$-4 \cdot 1$	$4(-1)$

$(2y - 2)(y + 2) = 2y^2 + 2y - 4$	No
$(2y - 4)(y + 1) = 2y^2 - 2y - 4$	No
$(2y + 1)(y - 4) = 2y^2 - 7y - 4$	No
$(2y + 2)(y - 2) = 2y^2 - 2y - 4$	No
$(2y + 4)(y - 1) = 2y^2 + 2y - 4$	No
$(2y - 1)(y + 4) = 2y^2 + 7y - 4$	Yes

The complete factorization is $5(2y - 1)(y + 4)$.

Section 5.4—Factoring Perfect Square Trinomials and the Difference of Squares

Key Concepts:

The factored form of a perfect square trinomial is the square of a binomial:

$$a^2 + 2ab + b^2 = (a + b)^2$$
$$a^2 - 2ab + b^2 = (a - b)^2$$

Difference of Squares

$$a^2 - b^2 = (a + b)(a - b)$$

Examples:

Factor: $9w^2 - 30wz + 25z^2$
$= (3w)^2 - 2(3w)(5z) + (5z)^2$
$= (3w - 5z)^2$

Factor: $25z^2 - 4y^2$
$= (5z + 2y)(5z - 2y)$

Section 5.5—Factoring the Sum and Difference of Cubes

Key Concepts:

Factoring a Difference of Cubes

$$a^3 - b^3 = (a - b)(a^2 + ab + b^2)$$

Factoring a Sum of Cubes

$$a^3 + b^3 = (a + b)(a^2 - ab + b^2)$$

Key Terms:

difference of cubes
sum of cubes

Examples:

Factor:

$$\begin{aligned} &m^3 - 64 \\ &= (m)^3 - (4)^3 \end{aligned}$$

This is a difference of cubes: $a = m$ and $b = 4$. Apply the formula:

$$\begin{aligned} m^3 - 64 &= (m)^3 - (4)^3 \\ &= (m - 4)(m^2 + 4m + 16) \end{aligned}$$

Factor:

$$\begin{aligned} &x^6 + 8y^3 \\ &= (x^2)^3 + (2y)^3 \end{aligned}$$

This is a sum of cubes: $a = x^2$ and $b = 2y$. Apply the formula:

$$\begin{aligned} x^6 + 8y^3 &= (x^2)^3 + (2y)^3 \\ &= (x^2 + 2y)(x^4 - 2x^2y + 4y^2) \end{aligned}$$

Section 5.6—General Factoring Summary

Key Concepts:

Factoring Strategy

1. Factor out the greatest common factor, GCF. (Section 5.1)
2. Identify whether the polynomial has two terms, three terms, or more than three terms.
3. If the polynomial has two terms, determine if it fits the pattern for a difference of squares, difference of cubes, or sum of cubes. (Sections 5.4 or 5.5)
4. If the polynomial has three terms, check first for a perfect square trinomial. (Section 5.4) Otherwise, factor the trinomial with the grouping method or the trial-and-error method. (Sections 5.2 or 5.3)
5. If the polynomial has more than three terms, try factoring by grouping. (Sections 5.1 or 5.6)
6. Be sure to factor the polynomial completely.
7. Check by multiplying.

Examples:

Part I: Factoring using the factoring strategy.

1. $9x^2 - 4x + 9x^3$

$= x(9x - 4 + 9x^2)$	Factor out the GCF.
$= x(9x^2 + 9x - 4)$	Descending order.
$= x(3x + 4)(3x - 1)$	Factor the trinomial.

Part II: Factoring with multiple techniques.

2. $4a^2 - 12ab + 9b^2 - c^2$

$= 4a^2 - 12ab + 9b^2 \mid - c^2$	Group 3 by 1.
$= (2a - 3b)^2 - c^2$	Perfect square trinomial.
$= (2a - 3b - c)(2a - 3b + c)$	Difference of squares.

KEY TERM:

factoring strategy

Factor completely.

3. $3w^6 - 192$

$= 3(w^6 - 64)$ Factor out the GCF.

$= 3[(w^3)^2 - (8)^2]$ Difference of squares.

$= 3(w^3 - 8)(w^3 + 8)$ Sum/difference of cubes.

$= 3(w - 2)(w^2 + 2w + 4)(w + 2)(w^2 - 2w + 4)$

SECTION 5.7—SOLVING QUADRATIC EQUATIONS USING THE ZERO PRODUCT RULE

KEY CONCEPTS:

An equation of the form: $ax^2 + bx + c = 0$, where $a \neq 0$ is a quadratic equation.

The zero product rule states that if $ab = 0$, then either $a = 0$ or $b = 0$. The zero product rule can be used to solve a quadratic equation or a higher degree polynomial equation that is factored and set to zero.

KEY TERMS:

Pythagorean theorem
quadratic equation
zero product rule

EXAMPLES:

The equation $2x^2 - 17x + 30 = 0$ is a *quadratic equation*.

Solve: $3w(w - 4)(2w + 1) = 0$

$3w = 0$ or $w - 4 = 0$ or $2w + 1 = 0$

$w = 0$ or $w = 4$ or $w = -\frac{1}{2}$

Solve: $4x^2 = 34x - 60$

$4x^2 - 34x + 60 = 0$
$2(2x^2 - 17x + 30) = 0$
$2(2x - 5)(x - 6) = 0$

$2 \neq 0$ or $2x - 5 = 0$ or $x - 6 = 0$

$x = \frac{5}{2}$ or $x = 6$

chapter 5 REVIEW EXERCISES

Section 5.1

For Exercises 1–6, identify the greatest common factor between each pair of terms.

1. 24, 18
2. $16x^2, 20x^3$
3. $15a^2b^4, 22ab^5$
4. $3(x + 5), x(x + 5)$
5. $2c^3(3c - 5), 4c(3c - 5)$
6. $-2wyz, -4xyz$

For Exercises 7–14, factor out the greatest common factor.

7. $6x^2 + 2x^3 - 8x$
8. $11w^3 - 44w^2$
9. $32y^2 - 48$
10. $5a^3 + 9a^2 + 2a$

11. $-t^2 + 5t$

12. $-6u^2 - u$

13. $3b(b + 2) - 7(b + 2)$

14. $2(5x + 9) + 8x(5x + 9)$

For Exercises 15–20, factor by grouping.

15. $7w^2 + 14w + wb + 2b$

16. $b^2 - 2b + 5b - 10$

17. $x^2 - 6x - 4x + 24$

18. $18p^2 + 12pq - 3p - 2q$

19. $60y^2 - 45y - 12y + 9$

20. $6a - 3a^2 - 2ab + a^2b$

Section 5.2

For Exercises 21–26, find a pair of integers whose product and sum are given.

21. Product: -6 sum: -5

22. Product: 12 sum: 13

23. Product: 24 sum: 11

24. Product: -60 sum: 17

25. Product: -5 sum: 4

26. Product: 15 sum: -8

For Exercises 27–40, factor the trinomial using the grouping method.

27. $3c^2 - 5c - 2$

28. $4y^2 + 13y + 3$

29. $2t^2 + 11st + 12s^2$

30. $4x^3 + 17x^2 - 15x$

31. $w^3 + 4w^2 - 5w$

32. $p^2 - 8pq + 15q^2$

33. $40v^2 + 22v - 6$

34. $40s^2 + 30s - 100$

35. $x^2 + 9x - 22$

36. $y^2 - 9y + 8$

37. $a^3b - 10a^2b^2 + 24ab^3$

38. $2z^6 + 8z^5 - 42z^4$

39. $3m + 9m^2 - 2$

40. $10 + 6p^2 + 19p$

Section 5.3

For Exercises 41–44, let a, b, and c represent positive integers.

41. When factoring a polynomial of the form $ax^2 - bx - c$, should the signs in the binomials be both positive, both negative, or different?

42. When factoring a polynomial of the form $ax^2 - bx + c$, should the signs in the binomials be both positive, both negative, or different?

43. When factoring a polynomial of the form $ax^2 + bx + c$, should the signs in the binomials be both positive, both negative, or different?

44. When factoring a polynomial of the form $ax^2 + bx - c$, should the signs in the binomials be both positive, both negative, or different?

For Exercises 45–58, factor the trinomial using the trial-and-error method.

45. $2y^2 - 5y - 12$

46. $4w^2 - 5w - 6$

47. $2p^2 - 4p - 48$

48. $3c^2 + 18c - 21$

49. $10z^2 + 29z + 10$

50. $8z^2 + 6z - 9$

51. $2p^2 - 5p + 1$

52. $5r^2 - 3r + 7$

53. $10w^2 - 60w - 270$

54. $3y^2 - 18y - 48$

55. $9c^2 - 30cd + 25d^2$

56. $121m^2 + 154mn + 49n^2$

57. $v^4 - 2v^2 - 3$

58. $x^4 + 7x^2 + 10$

Section 5.4

For Exercises 59–64, determine if the trinomial is a perfect square trinomial. If it is, factor the trinomial. If the trinomial is not a perfect square trinomial, explain why.

59. $4x^2 - 20x + 25$

60. $y^2 + 12y + 36$

61. $c^2 - 6c + 9$

62. $9b^2 + 6b + 1$

63. $t^2 + 8t + 49$

64. $k^2 - 10k + 64$

For Exercises 65–72, determine if the binomial is a difference of two squares. If it is, factor the binomial. If the binomial is not a difference of squares, explain why.

65. $a^2 - 49$

66. $d^2 - 64$

67. $h - 25$
68. $c - 9$
69. $100 - 81t^2$
70. $4 - 25k^2$
71. $x^2 + 16$
72. $y^2 + 121$

For Exercises 73–78, factor completely.

73. $2c^4 - 18$
74. $72x^2 - 2y^2$
75. $8x^2 + 24x + 18$
76. $48t^2 - 24t + 3$
77. $p^3 + 3p^2 - 16p - 48$
78. $4k - 8 - k^3 + 2k^2$

Section 5.5

79. Write the formula for factoring the sum of cubes: $a^3 + b^3$

80. Write the formula for factoring the difference of cubes: $a^3 - b^3$

For Exercises 81–88, factor the sums and differences of cubes.

81. $z^3 - w^3$
82. $r^3 + s^3$
83. $64 + a^3$
84. $125 - b^3$
85. $p^6 + 8$
86. $q^6 - \frac{1}{27}$
87. $6x^3 - 48$
88. $7y^3 + 7$

Match the polynomials in Exercises 89–91 with its factored form.

89. $a^2 - b^2$
90. $a^3 - b^3$
91. $a^3 + b^3$

i. $(a - b)^3$
ii. $(a - b)(a^2 + ab + b^2)$
iii. $(a + b)(a^2 - ab + b^2)$
iv. $(a + b)^3$
v. $(a - b)(a + b)$

For Exercises 92–97, factor the binomials completely.

92. $y^2 - 81$
93. $216w^3 - 1$
94. $4a^2 + b^2$
95. $128 + 2v^6$
96. $p^6 + 1$
97. $q^6 - 1$

For Exercises 98–107, factor completely.

98. $6y^2 - 11y - 2$
99. $3p^2 - 6p + 3$
100. $x^3 - 36x$
101. $k^2 - 13k + 42$
102. $7ac - 14ad - bc + 2bd$
103. $q^4 - 64q$
104. $8h^2 + 20$
105. $2t^2 + t + 3$
106. $m^2 - 8m$
107. $x^3 + 4x^2 - x - 4$

Section 5.6

For Exercises 108–121, factor completely using the factoring strategy found on page 347.

108. $12s^3t - 45s^2t^2 - 12st^3$
109. $5p^4q - 20q^3$
110. $4d^2(3 + d) - (3 + d)$
111. $(y - 4)^3 + 4(y - 4)^2$
112. $49x^2 + 36 - 84x$
113. $80z + 32 + 50z^2$
114. $18a^2 + 39a - 15$
115. $w^4 + w^3 - 56w^2$
116. $8n + n^4$
117. $14m^3 - 14$
118. $b^2 + 16b + 64 - 25c^2$
119. $a^2 - 6a + 9 - 16x^2$
120. $(9w + 2)^2 + 4(9w + 2) - 5$
121. $(4x + 3)^2 - 12(4x + 3) + 36$

Section 5.7

122. For which of the following equations can the zero product rule be applied directly? Explain.

$(x - 3)(2x + 1) = 0$ or
$(x - 3)(2x + 1) = 6$

For Exercises 123–136, solve the equation using the zero product rule.

123. $(4x - 1)(3x + 2) = 0$
124. $(a - 9)(2a - 1) = 0$
125. $3w(w + 3)(5w + 2) = 0$

126. $6u(u - 7)(4u - 9) = 0$

127. $7k^2 - 9k - 10 = 0$

128. $4h^2 - 23h - 6 = 0$

129. $q^2 - 144 = 0$

130. $r^2 = 25$

131. $5v^2 - v = 0$

132. $x(x - 6) = -8$

133. $36t^2 + 60t = -25$

134. $9s^2 + 12s = -4$

135. $3(y^2 + 4) = 20y$

136. $2(p^2 - 66) = -13p$

137. The base of a parallelogram is 1 ft more than twice the height. If the area is 78 ft^2 what are the base and height of the parallelogram?

138. A ball is tossed into the air from ground level with initial speed of 16 ft/sec. The height of the ball is given by the equation

$h = -16x^2 + 16x \quad (x \geq 0)$ where h is the ball's height in feet and x is the time in seconds

Find the time(s) when the ball is at ground level.

139. Using the Pythagorean theorem determine whether the triangle could be a right triangle.

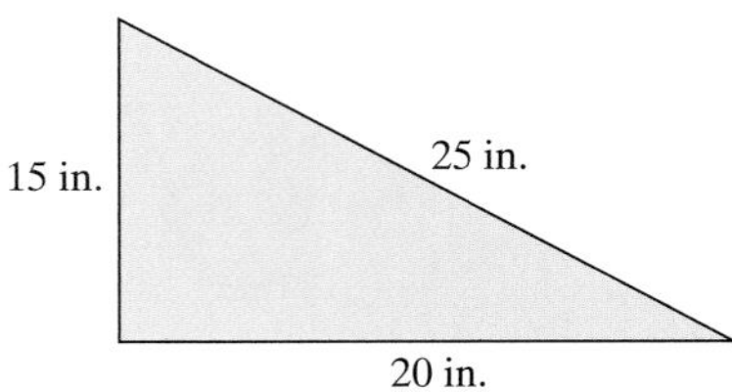

Figure for Exercise 139

140. A right triangle has one leg that is 2 ft more than the other leg. The hypotenuse is 2 ft less than twice the shorter leg. Find the lengths of all sides of the triangle.

141. If the square of a number is subtracted from 60, the result is -4. Find all such numbers.

142. The product of two consecutive integers is 44 more than 14 times their sum.

143. The base of a triangle is 1 m more than twice the height. If the area of the triangle is 18 m^2, find the base and height.

chapter 5 TEST

1. Factor out the GCF: $15x^4 - 3x + 6x^3$

2. Factor by grouping: $7a - 35 - a^2 + 5a$

3. Factor the trinomial: $6w^2 - 43w + 7$

4. Factor the difference of squares: $169 - p^2$

5. Factor the perfect square trinomial: $q^2 - 16q + 64$

6. Factor the sum of cubes: $8 + t^3$

For Exercises 7–16, factor completely.

7. $3a^2 + 27ab + 54b^2$

8. $c^4 - 1$

9. $xy - 7x + 3y - 21$

10. $49 + p^2$

11. $-10u^2 + 30u - 20$

12. $12t^2 - 75$

13. $5y^2 - 50y + 125$

14. $21q^2 + 14q$

15. $2x^3 + x^2 - 8x - 4$

16. $y^3 - 125$

17. $x^2 + 8x + 16 - y^2$

18. $r^6 - 256r^2$

19. $12a - 6ac + 2b - bc$

For Exercises 20–23, solve the equation.

20. $(2x - 3)(x + 5) = 0$

21. $x^2 - 7x = 0$

22. $x^2 - 6x = 16$

23. $x(5x + 4) = 1$

24. A tennis court has an area of 312 yd^2. If the length is 2 yd more than twice the width, find the dimensions of the court.

25. The hypotenuse of a right triangle is 2 ft less than three times the shorter leg. The longer leg is 3 ft less than three times the shorter leg. Find the length of the shorter leg.

CUMULATIVE REVIEW EXERCISES, CHAPTERS 1–5

For Exercises 1–2, simplify completely.

1. $\dfrac{|4 - 25 \div (-5) \cdot 2|}{\sqrt{8^2 + 6^2}}$

2. $-\dfrac{2}{3} - \dfrac{1}{3}(3^2 + \sqrt{81})$

3. Solve for x: $-3.5 - 2.5x = 1.5(x - 3)$

4. Solve for t: $5 - 2(t + 4) = 3t + 12$

5. Solve for y: $3x - 2y = 8$

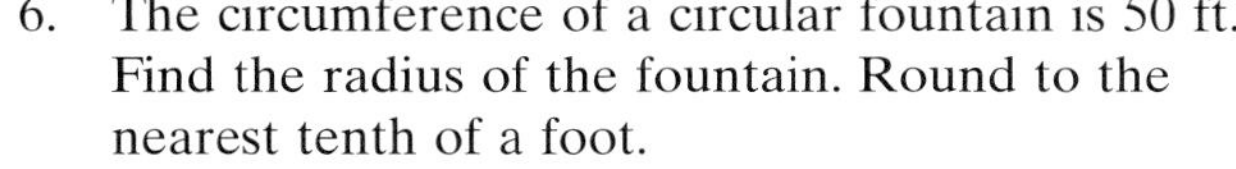

6. The circumference of a circular fountain is 50 ft. Find the radius of the fountain. Round to the nearest tenth of a foot.

7. A child's piggy bank has \$3.80 in quarters, dimes, and nickels. The number of nickels is two more than the number of quarters. The number of dimes is three less than the number of quarters. Find the number of each type of coin in the bank.

8. Solve the inequality. Graph the solution and write the solution set in interval notation.

$$-\frac{5}{12}x \leq \frac{5}{3}$$

For Exercises 9–11, perform the indicated operations.

9. $2\left(\dfrac{1}{3}y^3 - \dfrac{3}{2}y^2 - 7\right) - \left(\dfrac{2}{3}y^3 + \dfrac{1}{2}y^2 + 5y\right)$

10. $(4p^2 - 5p - 1)(2p - 3)$

11. $(2w - 7)^2$

12. Divide using long division: $(r^4 + 2r^3 - 5r + 1) \div (r - 3)$

For Exercises 13–15, simplify the expressions. Write the final answer using positive exponents only.

13. $\dfrac{c^{12}c^{-5}}{c^3}$

14. $\left(\dfrac{2a^2b^{-3}}{c}\right)^{-2}$

15. $\left(\dfrac{1}{2}\right)^0 - \left(\dfrac{1}{4}\right)^{-2}$

16. Divide. Write the final answer in scientific notation: $\dfrac{8.0 \times 10^{-3}}{5.0 \times 10^{-6}}$

For Exercises 17–24, factor completely.

17. $w^4 - 16$

18. $2ax + 10bx - 3ya - 15yb$

19. $4a^2 - 12a + 9$

20. $4x^2 - 8x - 5$

21. $y^3 - 27$

22. $p^6 + q^6$

23. $(a - 2)^2 + 5(a - 2) + 6$

24. $x^2 - (z^2 + 2z + 1)$

For Exercises 25–26, solve the equation.

25. $4x(2x - 1)(x + 5) = 0$

26. $x(x + 2) = 35$

chapter

6

Rational Expressions

The equation $d = rt$ gives the fundamental relationship among the variables: distance, rate, and time. To find the average rate of travel, we can use the rational equation:

$$r = \frac{d}{t}$$

When making a round trip, the average rate for the round trip is given by

$$r = \frac{2d}{t_1 + t_2}$$

where the distance is doubled, and t_1 and t_2 are the times for the original and return trips, respectively.

When traveling by air, the variables t_1 and t_2 may be different for each flight due to variations in wind speed. Furthermore, when traveling east to west or west to east the difference in the arrival time and departure time does not necessarily give the total travel time. Accommodations must be made for differences in time zones. For more information see the Technology Connections in MathZone at

www.mhhe.com/miller_oneill

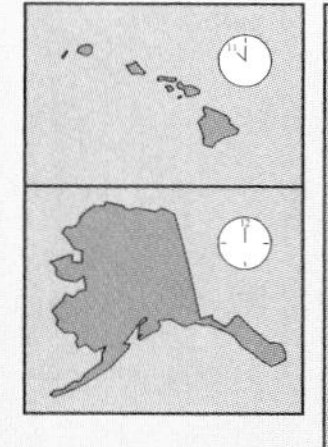

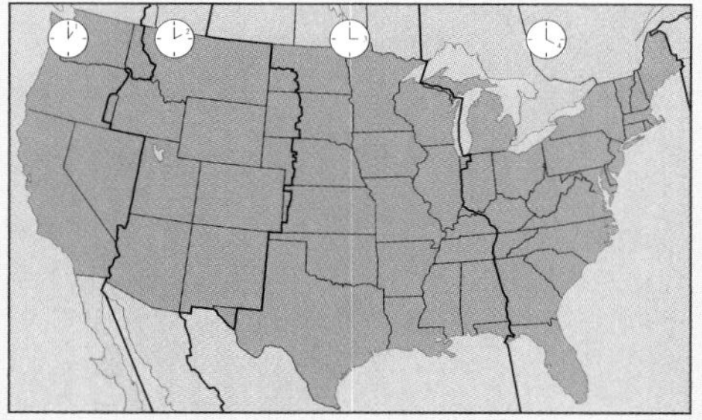

section

6.1 Introduction to Rational Expressions

Concepts

1. Definition of a Rational Expression
2. Evaluating Rational Expressions
3. Domain of a Rational Expression
4. Reducing Rational Expressions to Lowest Terms
5. Reducing a Ratio of −1

1. Definition of a Rational Expression

In Section 1.1, we defined a rational number as the ratio of two integers, $\frac{p}{q}$, where $q \neq 0$.

$$\text{Examples of rational numbers: } \frac{2}{3}, -\frac{1}{5}, 9 \text{ (because } 9 = \frac{9}{1}\text{)}$$

In a similar way, we define a **rational expression** as the ratio of two polynomials, $\frac{p}{q}$, where $q \neq 0$.

$$\text{Examples of rational expressions: } \frac{3x-6}{x^2-4}, \quad \frac{3}{4}, \quad \frac{6r^5+2r}{7}, \quad x$$

2. Evaluating Rational Expressions

example 1

Evaluating Rational Expressions

Evaluate the rational expression (if possible) for the given values of x: $\dfrac{12}{x-3}$

a. $x = 0$ b. $x = -3$ c. $x = 3$

Solution:

Substitute the variable with the given number. Then use the order of operations to evaluate.

a. $\dfrac{12}{x-3}$

$\dfrac{12}{(0)-3}$ Substitute $x = 0$.

$= -4$

b. $\dfrac{12}{x-3}$

$\dfrac{12}{(-3)-3}$ Substitute $x = -3$.

$= \dfrac{12}{-6}$

$= -2$

c. $\dfrac{12}{x-3}$

$\dfrac{12}{(3)-3}$ Substitute $x = 3$.

$= \dfrac{12}{0}$

undefined Recall that division by zero is undefined.

3. Domain of a Rational Expression

As with fractions, the denominator of a rational expression may not equal zero. If a value of the variable makes the denominator of a rational expression equal to zero, we say that the rational expression is undefined for that value of the variable. For example, the rational expression, $12/(x - 3)$ is undefined for $x = 3$. The fact that a rational expression may be defined for some values of the variable but not for others leads us to an important concept called the domain of an expression.

Informal Definition of the Domain of an Algebraic Expression

Given an algebraic expression, the **domain** of the expression consists of the real numbers that when substituted for the variable makes the expression result in a real number.

Note: For a rational expression, the domain is all real numbers except those that make the denominator zero.

According to this definition, the domain of the expression $12/(x - 3)$ is the set of real numbers *excluding* $x = 3$. This can be stated in set-builder notation as

$$\{x \mid x \neq 3\}$$

To find the domain of a rational expression we must identify and exclude the values of the variable that make the denominator zero.

Steps to Find the Domain of a Rational Expression

1. Set the denominator equal to zero and solve the resulting equation.
2. The domain is the set of real numbers *excluding* the values found in step 1.

example 2 **Finding the Domain of Rational Expressions**

Find the domain of the expressions. Write the answers in set-builder notation.

a. $\dfrac{y - 3}{2y + 7}$ b. $\dfrac{-5}{x}$ c. $\dfrac{a + 10}{a^2 - 25}$ d. $\dfrac{2x^3 + 5}{x^2 + 9}$

Solution:

a. $\dfrac{y - 3}{2y + 7}$

$2y + 7 = 0$ Set the denominator equal to zero.

$$2y = -7$$ Solve the equation.

$$\frac{2y}{2} = \frac{-7}{2}$$

$$y = -\frac{7}{2}$$ The domain is the set of real numbers except $-\frac{7}{2}$.

Domain: $\{y \mid y \neq -\frac{7}{2}\}$

b. $\dfrac{-5}{x}$

$$x = 0$$ Set the denominator equal to zero.

Domain: $\{x \mid x \neq 0\}$ The domain is the set of real numbers except 0.

c. $\dfrac{a + 10}{a^2 - 25}$

$$a^2 - 25 = 0$$ Set the denominator equal to zero. The equation is quadratic.

$$(a - 5)(a + 5) = 0$$ Factor the equation.

$$a - 5 = 0 \quad \text{or} \quad a + 5 = 0$$ Set each factor equal to zero.

$$a = 5 \quad \text{or} \quad a = -5$$ The domain is the set of real numbers except 5 and -5.

Domain: $\{a \mid a \neq 5 \text{ and } a \neq -5\}$

d. $\dfrac{2x^3 + 5}{x^2 + 9}$

Domain: $\{x \mid x \text{ is a real number}\}$

The quantity x^2 cannot be negative for any real number, x, so the denominator $x^2 + 9$ cannot equal zero. Therefore, no numbers are excluded from the domain. The domain is the set of all real numbers.

4. Reducing Rational Expressions to Lowest Terms

In many cases it is advantageous to reduce a fraction to lowest terms. The same is true for rational expressions.

The method for reducing rational expressions mirrors the process for reducing fractions. In each case, factor the numerator and denominator. Common factors in the numerator and denominator form a ratio of 1 and can be reduced.

Reducing a fraction: $\frac{21}{35} \xrightarrow{\text{factor}} \frac{3 \cdot \overset{1}{\cancel{7}}}{5 \cdot \cancel{7}} = \frac{3}{5} \cdot (1) = \frac{3}{5}$

Reducing a rational expression: $\frac{2x - 6}{x^2 - 9} \xrightarrow{\text{factor}} \frac{2\overset{1}{\cancel{(x - 3)}}}{(x + 3)\cancel{(x - 3)}} = \frac{2}{(x + 3)}(1) = \frac{2}{x + 3}$

Informally, to reduce a rational expression to lowest terms we reduce common factors whose ratio is 1. Formally, this is accomplished by applying the fundamental principle of rational expressions.

Fundamental Principle of Rational Expressions

Let p, q, and r represent polynomials. Then

$$\frac{pr}{qr} = \frac{p}{q} \quad \text{for } q \neq 0 \text{ and } r \neq 0.$$

example 3 Reducing a Rational Expression to Lowest Terms

Given the expression $\frac{2p - 14}{p^2 - 49}$

a. Factor the numerator and denominator.
b. Determine the domain of the expression and write the domain in set-builder notation.
c. Reduce the expression to lowest terms.

Solution:

a. $\frac{2p - 14}{p^2 - 49}$ Factor out the GCF in the numerator.

$= \frac{2(p - 7)}{(p + 7)(p - 7)}$ Factor the denominator as a difference of squares.

Avoiding Mistakes

The domain of a rational expression is always determined *before* reducing the expression.

b. $(p + 7)(p - 7) = 0$ To find the domain restrictions, set the denominator equal to zero. The equation is quadratic.

$p + 7 = 0$ or $p - 7 = 0$ Set each factor equal to 0.

$p = -7$ or $p = 7$ The domain is all real numbers except -7 and 7.

Domain: $\{p \mid p \neq -7, \quad p \neq 7\}$

c. $\dfrac{2(p - 7)}{(p + 7)(p - 7)}$ Reduce common factors whose ratio is 1.

$= \dfrac{2}{p + 7}$ (provided $p \neq 7$ and $p \neq -7$)

In Example 3, it is important to note that the expressions

$$\frac{2p - 14}{p^2 - 49} \quad \text{and} \quad \frac{2}{p + 7}$$

are equal for all values of p that make each expression a real number. Therefore,

$$\frac{2p - 14}{p^2 - 49} = \frac{2}{p + 7}$$

for all values of p except $p = 7$ and $p = -7$. (At $p = 7$ and $p = -7$, the original expression is undefined.) This is why the domain of an expression is always determined before the expression is reduced.

From this point forward, we will write statements of equality between two rational expressions with the assumption that they are equal for all values of the variable for which each expression is defined.

example 4

Reducing Rational Expressions to Lowest Terms

Reduce the rational expressions to lowest terms.

a. $\dfrac{18a^4}{9a^5}$ b. $\dfrac{2c - 8}{10c^2 - 80c + 160}$

Solution:

a. $\dfrac{18a^4}{9a^5}$

$= \dfrac{2 \cdot 3 \cdot 3 \cdot a \cdot a \cdot a \cdot a}{3 \cdot 3 \cdot a \cdot a \cdot a \cdot a \cdot a}$ Factor the numerator and denominator.

$= \dfrac{2 \cdot (3 \cdot 3 \cdot a \cdot a \cdot a \cdot a)}{(3 \cdot 3 \cdot a \cdot a \cdot a \cdot a) \cdot a}$ Reduce common factors.

$= \dfrac{2}{a}$

Tip: To reduce rational expressions, we can also use the property of exponents for dividing expressions of the same base. Recall, for integers, m and n, and $b \neq 0$.

$$\frac{b^m}{b^n} = b^{m-n}$$

In Example 4(a), we could have subtracted exponents to simplify the expression

$$\frac{18a^4}{9a^5} = \frac{18}{9} \cdot \frac{a^4}{a^5} = 2a^{4-5} = 2a^{-1} = \frac{2}{a}$$

Avoiding Mistakes

The fundamental principle of rational expressions indicates that common factors in the numerator and denominator may be reduced.

$$\frac{pr}{qr} = \frac{p}{q} \cdot \frac{r}{r} = \frac{p}{q}(1) = \frac{p}{q}$$

This property is based on the identity property of multiplication, so reducing or canceling applies only to *factors* (remember that factors are multiplied). Therefore, terms that are added or subtracted cannot be reduced or canceled. For example:

$$\frac{3x}{3y} = \frac{\overset{1}{\cancel{3}} \cdot x}{\cancel{3} \cdot y} = \frac{x}{y}$$

reduce common factor

however,

$\frac{x+3}{y+3}$ cannot be reduced

cannot reduce common terms

b. $\dfrac{2c - 8}{10c^2 - 80c + 160}$

$= \dfrac{2(c-4)}{10(c^2 - 8c + 16)}$ Factor out the GCF.

$= \dfrac{2(c-4)}{10(c-4)^2}$ The denominator is a perfect square trinomial.

$= \dfrac{\overset{1}{\cancel{2}}\overset{1}{\cancel{(c-4)}}}{\cancel{2} \cdot 5\cancel{(c-4)}(c-4)}$ Reduce common factors whose ratio is 1.

$= \dfrac{1}{5(c-4)}$

5. Reducing a Ratio of −1

When two factors are identical in the numerator and denominator, they form a ratio of 1 and can be reduced. Sometimes we encounter two factors that are opposites and form a ratio of −1. For example:

Reduced Form	**Details/Notes**
$\frac{-5}{5} = -1$	The ratio of a number and its opposite is −1.
$\frac{100}{-100} = -1$	The ratio of a number and its opposite is −1.
$\frac{x+7}{-x-7} = -1$	$\frac{x+7}{-x-7} = \frac{x+7}{-1(x+7)} = \frac{\overset{1}{\cancel{x+7}}}{-1\cancel{(x+7)}} = \frac{1}{-1} = -1$ (factor out −1)
$\frac{2-x}{x-2} = -1$	$\frac{2-x}{x-2} = \frac{-1(-2+x)}{x-2} = \frac{-1\overset{1}{\cancel{(x-2)}}}{\cancel{x-2}} = \frac{-1}{1} = -1$

Recognizing factors that are opposites is useful when reducing rational expressions.

example 5 Reducing Rational Expressions to Lowest Terms

Reduce the rational expressions to lowest terms.

a. $\dfrac{3c - 3d}{d - c}$ b. $\dfrac{5 - y}{y^2 - 25}$

Solution:

a. $\dfrac{3c - 3d}{d - c}$

$= \dfrac{3(c - d)}{d - c}$ Factor the numerator and denominator.

Notice that $(c - d)$ and $(d - c)$ are opposites and form a ratio of -1.

$= \dfrac{3(\overset{-1}{\cancel{c - d}})}{\cancel{d - c}}$ Details: $\dfrac{3(c - d)}{d - c} = \dfrac{3(c - d)}{-1(-d + c)} = \dfrac{3(c - d)}{-1(c - d)} = \dfrac{3}{-1}$

$= -3$

$= 3(-1)$

$= -3$

b. $\dfrac{5 - y}{y^2 - 25}$

$= \dfrac{5 - y}{(y - 5)(y + 5)}$ Factor the numerator and denominator.

Notice that $5 - y$ and $y - 5$ are opposites and form a ratio of -1.

$= \dfrac{\overset{-1}{\cancel{5 - y}}}{\cancel{(y - 5)}(y + 5)}$ Details: $\dfrac{5 - y}{(y - 5)(y + 5)} = \dfrac{-1(-5 + y)}{(y - 5)(y + 5)}$

$= \dfrac{-1(y - 5)}{(y - 5)(y + 5)} = \dfrac{-1}{y + 5}$

$= \dfrac{-1}{y + 5}$ or $\dfrac{1}{-(y + 5)}$ or $-\dfrac{1}{y + 5}$

Tip: It is important to recognize that a rational expression may be written in several equivalent forms, particularly when a negative factor is present. For example, since two numbers with opposite signs form a negative quotient, the number $-\frac{3}{4}$ can be written as $\frac{-3}{4}$ or as $\frac{3}{-4}$. The $-$ sign can be written in the numerator, in the denominator, or out in front of the fraction.

section 6.1 Practice Exercises

1. a. What is a rational number?
   b. What is a rational expression?
2. a. Write an example of a rational number. (Answers will vary.)
   b. Write an example of a rational expression. (Answers will vary.)

For Exercises 3–10, substitute the given number into the expression and simplify (if possible).

3. $\dfrac{1}{x - 6}$ let $x = -2$
4. $\dfrac{1}{x + 1}$ let $x = 4$
5. $\dfrac{w - 10}{w + 6}$ let $w = 0$
6. $\dfrac{w - 4}{2w + 8}$ let $w = 0$

7. $\dfrac{y-8}{2y^2+y-1}$ let $y = 8$

8. $\dfrac{y+3}{3y^2-25y-18}$ let $y = -3$

9. $\dfrac{(a-7)(a+1)}{(a-2)(a+5)}$ let $a = 2$

10. $\dfrac{(a+4)(a+1)}{(a-4)(a-1)}$ let $a = 1$

11. A bicyclist rides 24 miles against a wind and returns 24 miles with the same wind. His average speed for the return trip traveling with the wind is 8 mph faster than his speed going out against the wind. If x represents the bicyclist's speed going out against the wind, then the total time, t, required for the round trip is given by

$$t = \frac{24}{x} + \frac{24}{x+8}$$ where $x > 0$ and t is measured in hours.

a. Find the time required for the round trip if the cyclist rides 12 mph against the wind.

b. Find the time required for the round trip if the cyclist rides 24 mph against the wind.

12. The manufacturer of mountain bikes has a fixed cost of \$56,000, plus a variable cost of \$140 per bike. The average cost per bike, y (in dollars), is given by the equation:

$$y = \frac{56{,}000 + 140x}{x}$$ where x represents the number of bikes produced and $x \geq 0$.

a. Find the average cost per bike if the manufacturer produces 1000 bikes.

b. Find the average cost per bike if the manufacturer produces 2000 bikes.

c. Find the average cost per bike if the manufacturer produces 10,000 bikes.

For Exercises 13–18, write the domain of the expression in set-builder notation.

13. $\dfrac{5}{k+2}$

14. $\dfrac{-3}{h-4}$

15. $\dfrac{x+5}{(2x-5)(x+8)}$

16. $\dfrac{4y+1}{(3y+7)(y+3)}$

17. $\dfrac{b+12}{b^2+5b+6}$

18. $\dfrac{c-11}{c^2-5c-6}$

19. Construct a rational expression that is undefined for $x = 2$. (Answers will vary.)

20. Construct a rational expression that is undefined for $x = 5$. (Answers will vary.)

21. Construct a rational expression that is undefined for $x = -3$ and $x = 7$. (Answers will vary.)

22. Construct a rational expression that is undefined for $x = -1$ and $x = 4$. (Answers will vary.)

For Exercises 23–34, reduce the expression to lowest terms.

23. $\dfrac{7b^2}{21b}$

24. $\dfrac{15c^3}{3c^5}$

25. $\dfrac{18st^5}{12st^3}$

26. $\dfrac{20a^4b^2}{25ab^2}$

27. $\dfrac{-24x^2y^5z}{8xy^4z^3}$

28. $\dfrac{60rs^4t^2}{-12r^4s^2t^3}$

29. $\dfrac{3(y+2)}{6(y+2)}$

30. $\dfrac{8(x-1)}{4(x-1)}$

31. $\dfrac{(p-3)(p+5)}{(p+5)(p+4)}$

32. $\dfrac{(c+4)(c-1)}{(c+4)(c+2)}$

33. $\dfrac{(m+11)}{4(m+11)(m-11)}$

34. $\dfrac{(n-7)}{9(n+2)(n-7)}$

For Exercises 35–44:

a. Factor both the numerator and denominator.
b. Write the domain in set-builder notation.
c. Reduce the expression to lowest terms.

35. $\dfrac{3y + 6}{6y + 12}$

36. $\dfrac{8x - 8}{4x - 4}$

37. $\dfrac{t^2 - 1}{t + 1}$

38. $\dfrac{r^2 - 4}{r - 2}$

39. $\dfrac{7w}{21w^2 - 35w}$

40. $\dfrac{12a^2}{24a^2 - 18a}$

41. $\dfrac{9x^2 - 4}{6x + 4}$

42. $\dfrac{8b - 20}{4b^2 - 25}$

43. $\dfrac{a^2 + 3a - 10}{a^2 + a - 6}$

44. $\dfrac{t^2 + 3t - 10}{t^2 + t - 20}$

For Exercises 45–68, reduce the rational expressions to lowest terms.

45. $\dfrac{5}{20a - 25}$

46. $\dfrac{7}{14c - 21}$

47. $\dfrac{4w - 8}{w^2 - 4}$

48. $\dfrac{3x + 15}{x^2 - 25}$

49. $\dfrac{3x^2 - 6x}{9xy + 18x}$

50. $\dfrac{6p^2 + 12p}{2pq - 4p}$

51. $\dfrac{2x + 4}{x^2 - 3x - 10}$

52. $\dfrac{5z + 15}{z^2 - 4z - 21}$

53. $\dfrac{a^2 - 49}{a - 7}$

54. $\dfrac{b^2 - 64}{b - 8}$

55. $\dfrac{q^2 + 25}{q + 5}$

56. $\dfrac{r^2 + 36}{r + 6}$

57. $\dfrac{y^2 + 6y + 9}{2y^2 + y - 15}$

58. $\dfrac{h^2 + h - 6}{h^2 + 2h - 8}$

59. $\dfrac{3x^2 + 7x - 6}{x^2 + 7x + 12}$

60. $\dfrac{x^2 - 5x - 14}{2x^2 - x - 10}$

61. $\dfrac{5q^2 + 5}{q^4 - 1}$

62. $\dfrac{4t^2 + 16}{t^4 - 16}$

63. $\dfrac{ac - ad + 2bc - 2bd}{2ac + ad + 4bc + 2bd}$ (*Hint*: Factor by grouping.)

64. $\dfrac{3pr - ps - 3qr + qs}{3pr - ps + 3qr - qs}$

65. $\dfrac{49p^2 - 28pq + 4q^2}{14p - 4q}$

66. $\dfrac{3x - 3y}{a^2x - a^2y + b^2x - b^2y}$

67. $\dfrac{2x^2 - xy - 3y^2}{2x^2 - 11xy + 12y^2}$

68. $\dfrac{2c^2 + cd - d^2}{5c^2 + 3cd - 2d^2}$

69. What is the relationship between $x - 2$ and $2 - x$?

70. What is the relationship between $w + p$ and $-w - p$?

For Exercises 71–78, reduce the expression involving a ratio of -1.

71. $\dfrac{x - 5}{5 - x}$

72. $\dfrac{8 - p}{p - 8}$

73. $\dfrac{-4 - y}{4 + y}$

74. $\dfrac{z + 10}{-z - 10}$

75. $\dfrac{3y - 6}{12 - 6y}$

76. $\dfrac{4q - 4}{12 - 12q}$

77. $\dfrac{x^2 - x - 12}{16 - x^2}$

78. $\dfrac{49 - b^2}{b^2 - 10b + 21}$

79. Substitute $x = 4$ in the expressions in parts (a) and (b). Compare your answers.

a. $\dfrac{5x + 5}{x^2 - 1}$ b. $\dfrac{5}{x - 1}$

80. Substitute $x = 3$ in the expressions in parts (a) and (b). Compare your answers.

a. $\dfrac{2x^2 - 4x - 6}{2x^2 - 18}$ b. $\dfrac{x + 1}{x + 3}$

81. Substitute $x = -1$ in the expressions in parts (a) and (b). Compare your answers.

a. $\dfrac{3x^2 - 2x - 1}{6x^2 - 7x - 3}$ b. $\dfrac{x - 1}{2x - 3}$

82. Substitute $x = 4$ in the expressions in parts (a) and (b). Compare your answers.

a. $\dfrac{(x + 5)^2}{x^2 + 6x + 5}$ b. $\dfrac{x + 5}{x + 1}$

Expanding Your Skills

For Exercises 83–88, factor and reduce the expressions to lowest terms.

83. $\dfrac{w^3 - 8}{w^2 + 2w + 4}$

84. $\dfrac{y^3 + 27}{y^2 - 3y + 9}$

85. $\dfrac{z^2 - 16}{z^3 - 64}$

86. $\dfrac{x^2 - 25}{x^3 + 125}$

87. $\dfrac{5x^3 + 4x^2 - 45x - 36}{x^2 - 9}$

88. $\dfrac{x^2 - 1}{ax^3 - bx^2 - ax + b}$

Concepts

1. **Multiplication of Rational Expressions**
2. **Division of Rational Expressions**

section

6.2 Multiplication and Division of Rational Expressions

1. Multiplication of Rational Expressions

Recall from Section R.1 that to multiply fractions, we multiply the numerators and multiply the denominators. The same is true for multiplying rational expressions.

Multiplication of Rational Expressions

Let p, q, r, and s represent polynomials, such that $q \neq 0, s \neq 0$. Then,

$$\frac{p}{q} \cdot \frac{r}{s} = \frac{pr}{qs}.$$

For example,

Multiply the Fractions	**Multiply the Rational Expressions**
$\dfrac{2}{3} \cdot \dfrac{5}{7} = \dfrac{10}{21}$	$\dfrac{2x}{3y} \cdot \dfrac{5z}{7} = \dfrac{10xz}{21y}$

Sometimes it is possible to reduce a ratio of common factors to 1 *before* multiplying. To do so, we must first factor the numerators and denominators of each fraction.

$$\frac{15}{14} \cdot \frac{21}{10} = \frac{3 \cdot \overset{1}{\cancel{5}}}{2 \cdot \cancel{7}} \cdot \frac{3 \cdot \overset{1}{\cancel{7}}}{2 \cdot \cancel{5}} = \frac{9}{4}$$

The same process is also used to multiply rational expressions.

Steps to Multiply Rational Expressions

1. Factor the numerators and denominators of all rational expressions.
2. Reduce the ratios of common factors to 1.
3. Multiply the remaining factors in the numerator and multiply the remaining factors in the denominator.

example 1 **Multiplying Rational Expressions**

Multiply.

a. $\dfrac{5a^2b}{2} \cdot \dfrac{6a}{10b}$ b. $\dfrac{3c-3d}{6c} \cdot \dfrac{2}{c^2-d^2}$ c. $\dfrac{35-5x}{5x+5} \cdot \dfrac{x^2+5x+4}{x^2-49}$

Solution:

a. $\dfrac{5a^2b}{2} \cdot \dfrac{6a}{10b}$

$= \dfrac{5 \cdot a \cdot a \cdot b}{2} \cdot \dfrac{2 \cdot 3 \cdot a}{2 \cdot 5 \cdot b}$ Factor into prime factors.

$= \dfrac{\overset{1}{\cancel{5}} \cdot a \cdot a \cdot \overset{1}{\cancel{b}}}{2} \cdot \dfrac{\overset{1}{\cancel{2}} \cdot 3 \cdot a}{\cancel{2} \cdot \cancel{5} \cdot \cancel{b}}$ Reduce to lowest terms.

$= \dfrac{3a^3}{2}$ Multiply remaining factors.

b. $\dfrac{3c-3d}{6c} \cdot \dfrac{2}{c^2-d^2}$

$= \dfrac{3(c-d)}{2 \cdot 3 \cdot c} \cdot \dfrac{2}{(c-d)(c+d)}$ Factor into prime factors.

$= \dfrac{\overset{1}{\cancel{3}}\overset{1}{\cancel{(c-d)}}}{\cancel{2} \cdot \cancel{3} \cdot c} \cdot \dfrac{\overset{1}{\cancel{2}}}{\cancel{(c-d)}(c+d)}$ Reduce to lowest terms.

$= \dfrac{1}{c(c+d)}$

Avoiding Mistakes

If all factors in the numerator reduce to 1, do not forget to write the factor of 1 in the numerator.

c. $\dfrac{35-5x}{5x+5} \cdot \dfrac{x^2+5x+4}{x^2-49}$

$= \dfrac{5(7-x)}{5(x+1)} \cdot \dfrac{(x+4)(x+1)}{(x-7)(x+7)}$ Factor the numerators and denominators completely.

$= \dfrac{\overset{1}{\cancel{5}}\overset{-1}{\cancel{(7-x)}}}{\cancel{5}\cancel{(x+1)}} \cdot \dfrac{(x+4)\overset{1}{\cancel{(x+1)}}}{\cancel{(x-7)}(x+7)}$ Reduce the ratios of common factors to 1 or -1.

$= \dfrac{-1(x+4)}{x+7}$

$= \dfrac{-(x+4)}{x+7}$ or $\dfrac{x+4}{-(x+7)}$ or $-\dfrac{x+4}{x+7}$

Tip: The ratio $\dfrac{7-x}{x-7} = -1$ because $7 - x$ and $x - 7$ are opposites.

2. Division of Rational Expressions

Recall that to divide two fractions, multiply the first fraction by the reciprocal of the second.

$$\frac{21}{10} \div \frac{49}{15} \xrightarrow[\text{of the second fraction}]{\text{multiply by the reciprocal}} \frac{21}{10} \cdot \frac{15}{49} \xrightarrow{\text{factor}} \frac{3 \cdot \overset{1}{\cancel{7}}}{2 \cdot \cancel{5}} \cdot \frac{3 \cdot \overset{1}{\cancel{5}}}{\cancel{7} \cdot 7} \overset{\text{reduce}}{=} \frac{9}{14}$$

The same process is used to divide rational expressions.

Division of Rational Expressions

Let p, q, r, and s represent polynomials, such that $q \neq 0, r \neq 0, s \neq 0$. Then,

$$\frac{p}{q} \div \frac{r}{s} = \frac{p}{q} \cdot \frac{s}{r} = \frac{ps}{qr}$$

example 2 **Dividing Rational Expressions**

Divide.

a. $\dfrac{5t - 15}{2} \div \dfrac{t^2 - 9}{10}$ b. $\dfrac{p^2 - 11p + 30}{10p^2 - 250} \div \dfrac{30p - 5p^2}{2p + 4}$ c. $\dfrac{\frac{3x}{4y}}{\frac{5x}{6y}}$

Solution:

a. $\dfrac{5t - 15}{2} \div \dfrac{t^2 - 9}{10}$

$= \dfrac{5t - 15}{2} \cdot \dfrac{10}{t^2 - 9}$ Multiply the first fraction by the reciprocal of the second.

$= \dfrac{5(t - 3)}{2} \cdot \dfrac{2 \cdot 5}{(t - 3)(t + 3)}$ Factor each polynomial.

$= \dfrac{5\overset{1}{\cancel{(t - 3)}}}{\cancel{2}} \cdot \dfrac{\overset{1}{\cancel{2}} \cdot 5}{\cancel{(t - 3)}(t + 3)}$ Reduce common factors.

$= \dfrac{25}{t + 3}$

b. $\dfrac{p^2 - 11p + 30}{10p^2 - 250} \div \dfrac{30p - 5p^2}{2p + 4}$

$= \dfrac{p^2 - 11p + 30}{10p^2 - 250} \cdot \dfrac{2p + 4}{30p - 5p^2}$ Multiply the first fraction by the reciprocal of the second.

$= \dfrac{(p - 5)(p - 6)}{2 \cdot 5(p - 5)(p + 5)} \cdot \dfrac{2(p + 2)}{5p(6 - p)}$

Factor the trinomial.
$p^2 - 11p + 30 = (p - 5)(p - 6)$

Factor out the GCF.
$2p + 4 = 2(p + 2)$

Factor out the GCF. Then factor the difference of squares.
$10p^2 - 250 = 10(p^2 - 25)$
$= 2 \cdot 5(p - 5)(p + 5)$

Factor out the GCF.
$30p - 5p^2 = 5p(6 - p)$

Tip: $(p - 6)$ and $(6 - p)$ are opposites and form a ratio of -1.

$$\frac{p-6}{6-p} = \frac{p-6}{-1(-6+p)} = \frac{p-6}{-1(p-6)} = -1$$

$= \dfrac{\overset{1}{\cancel{(p - 5)}}\overset{-1}{\cancel{(p - 6)}}}{\cancel{2} \cdot 5\cancel{(p - 5)}(p + 5)} \cdot \dfrac{\overset{1}{\cancel{2}}(p + 2)}{5p\cancel{(6 - p)}}$ Reduce common factors.

$= -\dfrac{(p + 2)}{25p(p + 5)}$

c. $\dfrac{\frac{3x}{4y}}{\frac{5x}{6y}}$ ← This fraction bar denotes division (÷).

Tip: A fraction with one or more rational expressions in its numerator or denominator is called a *complex fraction.*

$$\frac{\frac{3x}{4y}}{\frac{5x}{6y}}$$

$= \dfrac{3x}{4y} \div \dfrac{5x}{6y}$

$= \dfrac{3x}{4y} \cdot \dfrac{6y}{5x}$ Multiply by the reciprocal of the second fraction.

$= \dfrac{3 \cdot \overset{1}{\cancel{x}}}{\cancel{2} \cdot 2 \cdot \cancel{y}} \cdot \dfrac{\overset{1}{\cancel{2}} \cdot 3 \cdot \overset{1}{\cancel{y}}}{5 \cdot \cancel{x}}$ Reduce common factors.

$= \dfrac{9}{10}$

section 6.2 Practice Exercises

For Exercises 1–6, write the domain in set-builder notation and reduce each rational expression.

1. $\dfrac{(x+2)(x-1)}{(x-3)(x+2)}$

2. $\dfrac{(y+6)}{(y-1)(y+6)}$

3. $\dfrac{a^2-4}{a^2-4a+4}$

4. $\dfrac{b^2+10b+25}{b^2-25}$

5. $\dfrac{12t-6}{3-6t}$

6. $\dfrac{15p-10}{8-12p}$

For Exercises 7–14, multiply or divide the fractions.

7. $\dfrac{3}{5}\cdot\dfrac{1}{2}$

8. $\dfrac{6}{7}\cdot\dfrac{5}{12}$

9. $\dfrac{3}{4}\div\dfrac{3}{8}$

10. $\dfrac{18}{5}\div\dfrac{2}{5}$

11. $6\cdot\dfrac{5}{12}$

12. $\dfrac{7}{25}\cdot 5$

13. $\dfrac{\frac{21}{4}}{\frac{7}{5}}$

14. $\dfrac{\frac{9}{2}}{\frac{3}{4}}$

For Exercises 15–50, multiply or divide as indicated.

15. $\dfrac{4x-24}{20x}\cdot\dfrac{5x}{8}$

16. $\dfrac{5a+20}{a}\cdot\dfrac{3a}{10}$

17. $\dfrac{3y+18}{y^2}\cdot\dfrac{4y}{6y+36}$

18. $\dfrac{2p-4}{6p}\cdot\dfrac{4p^2}{8p-16}$

19. $\dfrac{10}{2-a}\cdot\dfrac{a-2}{16}$

20. $\dfrac{b-3}{6}\cdot\dfrac{20}{3-b}$

21. $\dfrac{b^2-a^2}{a-b}\cdot\dfrac{a}{a^2-ab}$

22. $\dfrac{(x-y)^2}{x^2+xy}\cdot\dfrac{x}{y-x}$

23. $\dfrac{4a+12}{6a-18}\div\dfrac{3a+9}{5a-15}$

24. $\dfrac{8b-16}{3b+3}\div\dfrac{5b-10}{2b+2}$

25. $\dfrac{3x-21}{6x^2-42x}\div\dfrac{7}{12x}$

26. $\dfrac{4a^2-4a}{9a-9}\div\dfrac{5}{12a}$

27. $\dfrac{y^2+5y-36}{y^2-2y-8}\cdot\dfrac{y+2}{y-6}$

28. $\dfrac{z^2-11z+28}{z-1}\cdot\dfrac{z+1}{z^2-6z-7}$

29. $\dfrac{t^2+4t-5}{t^2+7t+10}\cdot\dfrac{t+4}{t-1}$

30. $\dfrac{p^2-3p+2}{p^2-4p+3}\cdot\dfrac{p+1}{p-2}$

31. $\dfrac{m^2-n^2}{9}\div\dfrac{3n-3m}{27m}$

32. $\dfrac{9-b^2}{15b+15}\div\dfrac{b-3}{5b}$

33. $\dfrac{3p+4q}{p^2+4pq+4q^2}\div\dfrac{4}{p+2q}$

34. $\dfrac{x^2+2xy-3y^2}{2x-y}\div\dfrac{x+3y}{5}$

35. $(w+3)\cdot\dfrac{w}{2w^2+5w-3}$

36. $\dfrac{5t+1}{5t^2-31t+6}\cdot(t-6)$

37. $\dfrac{\frac{5t-10}{12}}{\frac{4t-8}{8}}$

38. $\dfrac{\frac{6m+6}{5}}{\frac{3m+3}{10}}$

39. $\dfrac{q+1}{5q^2-28q-12}\cdot(5q+2)$

40. $(r-5)\cdot\dfrac{4r}{2r^2-7r-15}$

41. $\dfrac{2a^2+13a-24}{8a-12}\div(a+8)$

42. $\dfrac{3y^2+20y-7}{5y+35}\div(3y-1)$

43. $(5t-1)\div\dfrac{5t^2+9t-2}{3t+8}$

44. $(2q-3)\div\dfrac{2q^2+5q-12}{q-7}$

45. $\dfrac{x^2+2x-3}{x^2-3x+2}\cdot\dfrac{x^2+2x-8}{x^2+4x+3}$

46. $\dfrac{y^2+y-12}{y^2-y-20}\cdot\dfrac{y^2+y-30}{y^2-2y-3}$

47. $\dfrac{\dfrac{w^2 - 6w + 9}{8}}{\dfrac{9 - w^2}{4w + 12}}$

48. $\dfrac{\dfrac{p^2 - 6p + 8}{24}}{\dfrac{16 - p^2}{6p + 6}}$

49. $\dfrac{k^2 + 3k + 2}{k^2 + 5k + 4} \div \dfrac{k^2 + 5k + 6}{k^2 + 10k + 24}$

50. $\dfrac{4h^2 - 5h + 1}{h^2 + h - 2} \div \dfrac{6h^2 - 7h + 2}{2h^2 + 3h - 2}$

Expanding Your Skills

For Exercises 51–56, multiply or divide as indicated.

51. $\dfrac{b^3 - 3b^2 + 4b - 12}{b^4 - 16} \cdot \dfrac{3b^2 + 5b - 2}{3b^2 - 10b + 3} \div \dfrac{3}{6b - 12}$

52. $\dfrac{x^2 - 25}{3x^2 + 3xy} \cdot \dfrac{x^2 + 4x + xy + 4y}{x^2 + 9x + 20} \div \dfrac{x - 5}{x}$

53. $\dfrac{a^2 - 5a}{a^2 + 7a + 12} \div \dfrac{a^3 - 7a^2 + 10a}{a^2 + 9a + 18} \div \dfrac{a + 6}{a + 4}$

54. $\dfrac{t^2 + t - 2}{t^2 + 5t + 6} \div \dfrac{t - 1}{t} \div \dfrac{5t - 5}{t + 3}$

55. $\dfrac{p^3 - q^3}{p - q} \cdot \dfrac{p + q}{2p^2 + 2pq + 2q^2}$

56. $\dfrac{r^3 + s^3}{r - s} \div \dfrac{r^2 + 2rs + s^2}{r^2 - s^2}$

---

section

6.3 Least Common Denominator

Concepts

1. Equivalent Rational Expressions
2. Writing Equivalent Fractions
3. Least Common Denominator
4. Writing Rational Expressions with the Least Common Denominator

1. Equivalent Rational Expressions

In Sections 6.1 and 6.2, we learned how to reduce, multiply, and divide rational expressions. Our next goal is to add and subtract rational expressions. As with fractions, rational expressions may be added or subtracted only if they have the same denominator. Therefore, we must first learn how to identify a common denominator between two or more rational expressions. Then we must learn how to convert a rational expression into an **equivalent rational expression** with the indicated denominator.

Using the identity property of multiplication, we know that for $q \neq 0$ and $r \neq 0$,

$$\frac{p}{q} = \frac{p}{q} \cdot 1 = \frac{p}{q} \cdot \frac{r}{r} = \frac{pr}{qr}$$

This principle is used to convert a rational expression into an equivalent expression with a different denominator. For example, $\frac{1}{2}$ can be converted into an equivalent expression with a denominator of 12 as follows:

$$\frac{1}{2} = \frac{1}{2} \cdot \frac{6}{6} = \frac{1 \cdot 6}{2 \cdot 6} = \frac{6}{12}$$

In this example, we multiplied $\frac{1}{2}$ by a convenient form of 1. The ratio $\frac{6}{6}$ was chosen so that the product produced a new denominator of 12. Notice that multiplying $\frac{1}{2}$ by $\frac{6}{6}$ is equivalent to multiplying the numerator and denominator of the original expression by 6. In general, if the numerator and denominator of a rational expression are both multiplied by the same nonzero quantity, the value of the expression remains unchanged.

2. Writing Equivalent Fractions

example 1

Creating Equivalent Fractions

Convert each expression into an equivalent expression with the indicated denominator.

a. $\dfrac{7}{5p^2} = \dfrac{}{20p^6}$

b. $\dfrac{w}{w + 5} = \dfrac{}{w^2 + 3w - 10}$

Solution:

a. $\dfrac{7}{5p^2} = \dfrac{}{20p^6}$ Multiply the numerator and denominator of the fraction by the missing factor of $4p^4$.

$$\frac{7 \cdot 4p^4}{5p^2 \cdot 4p^4} = \frac{28p^4}{20p^6}$$

Tip: Notice that in Example 1(b) we multiplied the polynomials in the numerator but left the denominator in factored form. This convention is followed because when we add and subtract rational expressions in Section 6.4, the terms in the numerators must be combined.

b. $\dfrac{w}{w + 5} = \dfrac{}{w^2 + 3w - 10}$

$\dfrac{w}{w + 5} = \dfrac{}{(w + 5)(w - 2)}$ Factor the denominator.

$\dfrac{w}{w + 5} = \dfrac{w \cdot (w - 2)}{(w + 5) \cdot (w - 2)}$ Multiply numerator and denominator by the missing factor of $(w - 2)$.

$$= \frac{w^2 - 2w}{(w + 5)(w - 2)}$$

3. Least Common Denominator

Recall from Section R.1 that to add or subtract fractions, the fractions must have a common denominator. The same is true for rational expressions. In this section, we present a method to find the least common denominator of two rational expressions.

The **least common denominator (LCD)** of two or more rational expressions is defined as the least common multiple of the denominators. For example, consider the fractions $\frac{1}{20}$ and $\frac{1}{8}$. By inspection, you can probably see that the least common denominator is 40. To understand why, find the prime factorization of both denominators:

$$20 = 2^2 \cdot 5 \quad \text{and} \quad 8 = 2^3$$

A common multiple of 20 and 8 must be a multiple of 5, a multiple of 2^2, and a multiple of 2^3. However, any number that is a multiple of $2^3 = 8$ is automatically a multiple of $2^2 = 4$. Therefore, it is sufficient to construct the least common denominator as the product of unique prime factors, in which each factor is raised to its highest power.

$$\text{The LCD of } \frac{1}{20} \text{ and } \frac{1}{8} \text{ is } 2^3 \cdot 5 = 40.$$

Steps to Find the Least Common Denominator of Two or More Rational Expressions

1. Factor all denominators completely.
2. The LCD is the product of unique factors from the denominators, in which each factor is raised to the highest power to which it appears in any denominator.

example 2

Finding the Least Common Denominator of Rational Expressions

Find the LCD of the following sets of rational expressions.

a. $\dfrac{5}{14}; \dfrac{3}{49}; \dfrac{1}{8}$ b. $\dfrac{5}{3x^2z}; \dfrac{7}{x^5y^3}$ c. $\dfrac{a+b}{a^2-25}; \dfrac{1}{2a-10}$

d. $\dfrac{x-5}{x^2-2x}; \dfrac{1}{x^2-4x+4}$

Solution:

a. $\dfrac{5}{14}; \dfrac{3}{49}; \dfrac{1}{8}$

$= \dfrac{5}{2 \cdot 7}; \dfrac{3}{7^2}; \dfrac{1}{2^3};$ **Step 1:** Factor the denominators.

The LCD is $2^3 \cdot 7^2 = 392$. **Step 2:** The LCD is the product of unique factors, each raised to its highest power.

b. $\dfrac{5}{3x^2z}; \dfrac{7}{x^5y^3}$

$= \dfrac{5}{3x^2z}; \dfrac{7}{x^5y^3}$ **Step 1:** The denominators are already factored.

The LCD is $3x^5y^3z$. **Step 2:** The LCD is the product of unique factors, each raised to its highest power.

c. $\dfrac{a+b}{a^2-25}; \dfrac{1}{2a-10}$

$= \dfrac{a+b}{(a-5)(a+5)}; \dfrac{1}{2(a-5)}$ **Step 1:** Factor the denominators.

The LCD is $2(a-5)(a+5)$. **Step 2:** The LCD is the product of unique factors, each raised to its highest power.

d. $\dfrac{x-5}{x^2-2x}; \dfrac{1}{x^2-4x+4}$

$= \dfrac{x-5}{x(x-2)}; \dfrac{1}{(x-2)^2}$ **Step 1:** Factor the denominators.

The LCD is $x(x-2)^2$. **Step 2:** The LCD is the product of unique factors, each raised to its highest power.

4. Writing Rational Expressions with the Least Common Denominator

To add or subtract two rational expressions, the expressions must have the same denominator. Therefore, we must first practice the skill of converting each rational expression into an equivalent expression with the LCD as its denominator. The process is as follows: Identify the LCD for the two expressions. Then, multiply the numerator and denominator of each fraction by the factors from the LCD that are missing from the original denominators.

example 3 Converting to the Least Common Denominator

Find the LCD of each pair of rational expressions. Then convert each expression to an equivalent fraction with the denominator equal to the LCD.

a. $\dfrac{3}{2ab}; \dfrac{6}{5a^2}$ b. $\dfrac{4}{x+1}; \dfrac{7}{x-4}$ c. $\dfrac{w+2}{w^2-w-12}; \dfrac{1}{w^2-9}$

Solution:

a. $\dfrac{3}{2ab}; \dfrac{6}{5a^2}$ The LCD is $10a^2b$.

$\dfrac{3}{2ab} = \dfrac{3 \cdot 5a}{2ab \cdot 5a} = \dfrac{15a}{10a^2b}$ The first expression is missing the factor $5a$ from the denominator.

$\dfrac{6}{5a^2} = \dfrac{6 \cdot 2b}{5a^2 \cdot 2b} = \dfrac{12b}{10a^2b}$ The second expression is missing the factor $2b$ from the denominator.

b. $\dfrac{4}{x+1}; \dfrac{7}{x-4}$ The LCD is $(x+1)(x-4)$.

$\dfrac{4}{x+1} = \dfrac{4(x-4)}{(x+1)(x-4)} = \dfrac{4x-16}{(x+1)(x-4)}$ The first expression is missing the factor $(x-4)$ from the denominator.

$\dfrac{7}{x-4} = \dfrac{7(x+1)}{(x-4)(x+1)} = \dfrac{7x+7}{(x-4)(x+1)}$ The second expression is missing the factor $(x+1)$ from the denominator.

c. $\dfrac{w+2}{w^2-w-12};\ \dfrac{1}{w^2-9}$

To find the LCD, factor each denominator.

$\dfrac{w+2}{(w-4)(w+3)};\ \dfrac{1}{(w-3)(w+3)}$

The LCD is $(w-4)(w+3)(w-3)$.

$$\frac{w+2}{(w-4)(w+3)} = \frac{(w+2)(w-3)}{(w-4)(w+3)(w-3)}$$

The first expression is missing the factor $(w-3)$ from the denominator.

$$= \frac{w^2-w-6}{(w-4)(w+3)(w-3)}$$

$$\frac{1}{(w-3)(w+3)} = \frac{1(w-4)}{(w-3)(w+3)(w-4)}$$

The second expression is missing the factor $(w-4)$ from the denominator.

$$= \frac{w-4}{(w-3)(w+3)(w-4)}$$

example 4 Converting to the Least Common Denominator

Find the LCD of the expressions $\dfrac{3}{x-7}$ and $\dfrac{1}{7-x}$.

Solution:

Notice that the expressions $x-7$ and $7-x$ are opposites and differ by a factor of -1. Therefore, we may use either $x-7$ or $7-x$ as a common denominator. Each case is detailed in the following conversions.

Converting to the Denominator $x - 7$

$\dfrac{3}{x-7};\ \dfrac{1}{7-x}$

$$\frac{1}{7-x} = \frac{(-1)1}{(-1)(7-x)}$$

Multiply the *second* rational expression by the ratio $\frac{-1}{-1}$ to change its denominator to $x-7$.

$$= \frac{-1}{-7+x}$$

Apply the distributive property.

$$= \frac{-1}{x-7}$$

Tip: In Example 4, the expressions

$$\frac{3}{x-7} \quad \text{and} \quad \frac{1}{7-x}$$

have opposite factors in the denominators. In such a case, you do *not* need to include both factors in the LCD.

Converting to the Denominator $7 - x$

$\dfrac{3}{x-7};\ \dfrac{1}{7-x}$

$$= \frac{(-1)3}{(-1)(x-7)};$$

Multiply the *first* rational expression by the ratio $\frac{-1}{-1}$ to change its denominator to $7-x$.

$$= \frac{-3}{-x + 7} \quad \text{Apply the distributive property.}$$

$$= \frac{-3}{7 - x}$$

section 6.3 Practice Exercises

For Exercises 1–2, write the domain in set-builder notation and reduce the expression.

1. $\dfrac{3x + 3}{5x^2 - 5}$
2. $\dfrac{x + 2}{x^2 - 3x - 10}$

For Exercises 3–6, multiply or divide as indicated.

3. $\dfrac{a + 3}{a + 7} \cdot \dfrac{a^2 + 3a - 10}{a^2 + a - 6}$

4. $\dfrac{16y^2}{9y + 36} \div \dfrac{8y^3}{3y + 12}$

5. $\dfrac{6(a + 2b)}{2(a - 3b)} \cdot \dfrac{4(a + 3b)(a - 3b)}{9(a + 2b)(a - 2b)}$

6. $\dfrac{5b^2 + 6b + 1}{b^2 + 5b + 6} \div (5b + 1)$

For Exercises 7–18, fill in the blank to convert the expressions into equivalent expressions with the indicated denominator.

7. $\dfrac{6}{7} = \dfrac{\quad}{42}$
8. $\dfrac{4}{9} = \dfrac{\quad}{72}$
9. $\dfrac{2}{13} = \dfrac{\quad}{39}$
10. $\dfrac{1}{8} = \dfrac{\quad}{64}$
11. $\dfrac{3}{p^2q} = \dfrac{\quad}{5p^3q}$
12. $\dfrac{2}{3rs} = \dfrac{\quad}{18rs^3}$
13. $\dfrac{2x}{yz} = \dfrac{\quad}{6y^2z^4}$
14. $\dfrac{8a}{b^2c} = \dfrac{\quad}{2b^4c^5}$
15. $\dfrac{w + 6}{w - 7} = \dfrac{\quad}{(w - 7)(w + 2)}$
16. $\dfrac{z - 1}{z + 1} = \dfrac{\quad}{(z + 1)(z - 3)}$
17. $\dfrac{6}{x - 3} = \dfrac{\quad}{3 - x}$
18. $\dfrac{2}{a - 9} = \dfrac{\quad}{9 - a}$

19. Which of the expressions are equivalent to $-\dfrac{5}{x - 3}$?

    a. $\dfrac{-5}{x - 3}$ b. $\dfrac{5}{-x + 3}$

    c. $\dfrac{5}{3 - x}$ d. $\dfrac{5}{-(x - 3)}$

20. Which of the expressions are equivalent to $\dfrac{4 - a}{6}$?

    a. $\dfrac{a - 4}{-6}$ b. $\dfrac{a - 4}{6}$

    c. $\dfrac{-(4 - a)}{-6}$ d. $-\dfrac{a - 4}{6}$

21. Explain why the least common denominator of $\frac{1}{x^3}$, $\frac{1}{x^5}$, and $\frac{1}{x^4}$ is x^5.

22. Explain why the least common denominator of $\frac{2}{y^3}$, $\frac{9}{y^6}$, and $\frac{4}{y^5}$ is y^6.

23. Explain why the least common denominator of

    $$\frac{1}{x + 3} \quad \text{and} \quad \frac{3}{x - 2}$$

    is $(x + 3)(x - 2)$.

24. Explain why the least common denominator of

    $$\frac{7}{y - 8} \quad \text{and} \quad \frac{3}{y + 1}$$

    is $(y - 8)(y + 1)$.

25. Explain why a common denominator of

    $$\frac{b + 1}{b - 1} \quad \text{and} \quad \frac{b}{1 - b}$$

    could be either $(b - 1)$ or $(1 - b)$.

26. Explain why a common denominator of

$$\frac{1}{6-t} \quad \text{and} \quad \frac{t}{t-6}$$

could be either $(6 - t)$ or $(t - 6)$.

For Exercises 27–44, identify the LCD.

27. $\frac{4}{15}, \frac{5}{9}$

28. $\frac{7}{12}, \frac{1}{18}$

29. $\frac{3}{16}, \frac{1}{4}$

30. $\frac{1}{2}, \frac{11}{12}$

31. $\frac{1}{5}, \frac{3}{-5}$

32. $\frac{2}{3}, \frac{4}{-3}$

33. $\frac{1}{3x^2y}, \frac{8}{9xy^3}$

34. $\frac{5}{2a^4b^2}, \frac{1}{8ab^3}$

35. $\frac{6}{w^2}, \frac{7}{y}$

36. $\frac{2}{r}, \frac{3}{s^2}$

37. $\frac{p}{(p+3)(p-1)}, \frac{2}{(p+3)(p+2)}$

38. $\frac{6}{(q+4)(q-4)}, \frac{q^2}{(q+1)(q+4)}$

39. $\frac{7}{3t(t+1)}, \frac{10t}{9(t+1)^2}$

40. $\frac{13x}{15(x-1)^2}, \frac{5}{3x(x-1)}$

41. $\frac{y}{y^2-4}, \frac{3y}{y^2+5y+6}$

42. $\frac{4}{w^2-3w+2}, \frac{w}{w^2-4}$

43. $\frac{5}{3-x}, \frac{7}{x-3}$

44. $\frac{4}{x-6}, \frac{9}{6-x}$

For Exercises 45–62, find the LCD. Then convert each expression to an equivalent expression with the denominator equal to the LCD.

45. $\frac{6}{5x^2}, \frac{1}{x}$

46. $\frac{3}{y}, \frac{7}{9y^2}$

47. $\frac{4}{5x^2}, \frac{y}{6x^3}$

48. $\frac{3}{15b^2}, \frac{c}{3b^2}$

49. $\frac{5}{6a^2b}, \frac{a}{12b}$

50. $\frac{x}{15y^2}, \frac{y}{5xy}$

51. $\frac{6}{m+4}, \frac{3}{m-1}$

52. $\frac{3}{n-5}, \frac{7}{n+2}$

53. $\frac{6}{(w+3)(w-8)}, \frac{w}{(w-8)(w+1)}$

54. $\frac{t}{(t+2)(t+12)}, \frac{18}{(t-2)(t+2)}$

55. $\frac{6p}{p^2-4}, \frac{3}{p^2+4p+4}$

56. $\frac{5}{q^2-6q+9}, \frac{q}{q^2-9}$

57. $\frac{1}{a-4}, \frac{a}{4-a}$

58. $\frac{3b}{2b-5}, \frac{2b}{5-2b}$

59. $\frac{4}{x-7}, \frac{y}{14-2x}$

60. $\frac{4}{3x-15}, \frac{z}{5-x}$

61. $\frac{1}{a+b}, \frac{6}{-a-b}$

62. $\frac{p}{-q-8}, \frac{1}{q+8}$

Expanding Your Skills

For Exercises 63–66, find the LCD. Then convert each expression to an equivalent expression with the denominator equal to the LCD.

63. $\frac{z}{z^2+9z+14}, \frac{-3z}{z^2+10z+21}, \frac{5}{z^2+5z+6}$

64. $\frac{6}{w^2-3w-4}, \frac{1}{w^2+6w+5}, \frac{-9w}{w^2+w-20}$

65. $\frac{3}{p^3-8}, \frac{p}{p^2-4}, \frac{5p}{p^2+2p+4}$

66. $\frac{7}{q^3+125}, \frac{q}{q^2-25}, \frac{12}{q^2-5q+25}$

section

6.4 Addition and Subtraction of Rational Expressions

Concepts

1. Addition and Subtraction of Rational Expressions with the Same Denominator
2. Addition and Subtraction of Rational Expressions with Different Denominators
3. Translating to a Rational Expression

1. Addition and Subtraction of Rational Expressions with the Same Denominator

To add or subtract rational expressions, the expressions must have the same denominator. As with fractions, we add or subtract rational expressions with the same denominator by combining the terms in the numerator and then writing the result over the common denominator. Then, if possible, we reduce the expression to lowest terms.

Addition and Subtraction of Rational Expressions

Let p, q, and r represent polynomials where $q \neq 0$. Then,

1. $\dfrac{p}{q} + \dfrac{r}{q} = \dfrac{p + r}{q}$
2. $\dfrac{p}{q} - \dfrac{r}{q} = \dfrac{p - r}{q}$

example 1

Adding and Subtracting Rational Expressions with a Common Denominator

Add or subtract as indicated.

a. $\dfrac{2}{5p} - \dfrac{7}{5p}$ b. $\dfrac{2}{3d + 5} + \dfrac{7d}{3d + 5}$ c. $\dfrac{x^2}{x - 3} - \dfrac{-5x + 24}{x - 3}$

Solution:

a. $\dfrac{2}{5p} - \dfrac{7}{5p}$ The rational expressions have the same denominator.

$= \dfrac{2 - 7}{5p}$ Subtract the terms in the numerators and write the result over the common denominator.

$= \dfrac{-5}{5p}$

$= \dfrac{\overset{-1}{(-\cancel{5})}}{\cancel{5}p}$ Reduce to lowest terms.

$= -\dfrac{1}{p}$

b. $\dfrac{2}{3d + 5} + \dfrac{7d}{3d + 5}$ The rational expressions have the same denominator.

$= \dfrac{2 + 7d}{3d + 5}$ Add the terms in the numerators and write the result over the common denominator.

$= \dfrac{7d + 2}{3d + 5}$ Because the numerator and denominator share no common factors, the expression is in lowest terms.

c. $\dfrac{x^2}{x - 3} - \dfrac{-5x + 24}{x - 3}$ The rational expressions have the same denominator.

Subtract the terms in the numerators and write the result over the common denominator.

$= \dfrac{x^2 - (-5x + 24)}{x - 3}$

$= \dfrac{x^2 + 5x - 24}{x - 3}$ Simplify the numerator.

$= \dfrac{(x + 8)(x - 3)}{(x - 3)}$ Factor the numerator and denominator to determine if the rational expression can be reduced.

$= \dfrac{(x + 8)\overset{1}{\cancel{(x - 3)}}}{\cancel{(x - 3)}}$ Reduce common factors.

$= x + 8$

Avoiding Mistakes

When subtracting rational expressions, use parentheses to group the terms in the numerator that follow the subtraction sign. This will help you remember to apply the distributive property.

2. Addition and Subtraction of Rational Expressions with Different Denominators

To add or subtract two rational expressions with unlike denominators, we must convert the expressions to equivalent expressions with the same denominator. For example, consider adding

$$\frac{1}{10} + \frac{12}{5y}$$

The LCD is $10y$. For each expression, identify the factors from the LCD that are missing from the denominator. Then multiply the numerator and denominator of the expression by the missing factor(s).

$$\underbrace{\frac{1}{10}}_{\substack{\text{missing}\\ y}} + \underbrace{\frac{12}{5y}}_{\substack{\text{missing}\\ 2}}$$

$$= \frac{1 \cdot y}{10 \cdot y} + \frac{12 \cdot 2}{5y \cdot 2}$$

$$= \frac{y}{10y} + \frac{24}{10y}$$ The rational expressions now have the same denominators.

Avoiding Mistakes

In the expression $\frac{y + 24}{10y}$, notice that you cannot reduce the 24 and 10 because 24 is not a factor of the numerator.

$$= \frac{y + 24}{10y}$$ Add the numerators.

After successfully adding or subtracting two rational expressions, always check to see if the final answer is reduced. If necessary, factor the numerator and denominator and reduce common factors. The expression

$$\frac{y + 24}{10y}$$

is in lowest terms because the numerator and denominator do not share any common factors.

Steps to Add or Subtract Rational Expressions

1. Factor the denominators of each rational expression.
2. Identify the LCD.
3. Rewrite each rational expression as an equivalent expression with the LCD as its denominator.
4. Add or subtract the numerators and write the result over the common denominator.
5. Simplify and reduce to lowest terms.

example 2

Adding and Subtracting Rational Expressions with Unlike Denominators

Add or subtract as indicated.

a. $\frac{4}{7k} - \frac{3}{k^2}$ b. $\frac{2q - 4}{3} - \frac{q + 1}{2}$ c. $\frac{1}{x - 5} + \frac{-10}{x^2 - 25}$

Solution:

a. $\frac{4}{7k} - \frac{3}{k^2}$

Step 1: The denominators are already factored.

Step 2: The LCD is $7k^2$.

$$= \frac{4 \cdot k}{7k \cdot k} - \frac{3 \cdot 7}{k^2 \cdot 7}$$ **Step 3:** Write each expression with the LCD.

Avoiding Mistakes

Do not reduce after rewriting the fractions with the LCD. You will revert back to the original problem.

$$= \frac{4k}{7k^2} - \frac{21}{7k^2}$$

$$= \frac{4k - 21}{7k^2}$$ **Step 4:** Subtract the numerators and write the result over the LCD.

Step 5: The expression is in lowest terms because the numerator and denominator share no common factors.

b. $\dfrac{2q-4}{3} - \dfrac{q+1}{2}$

Step 1: The denominators are already factored.

Step 2: The LCD is 6.

$= \dfrac{2(2q-4)}{2 \cdot 3} - \dfrac{3(q+1)}{3 \cdot 2}$

Step 3: Write each expression with the LCD.

$= \dfrac{2(2q-4) - 3(q+1)}{6}$

Step 4: Subtract the numerators and write the result over the LCD.

$= \dfrac{4q - 8 - 3q - 3}{6}$

$= \dfrac{q-11}{6}$

Step 5: The expression is in lowest terms because the numerator and denominator share no common factors.

c. $\dfrac{1}{x-5} + \dfrac{-10}{x^2-25}$

$= \dfrac{1}{x-5} + \dfrac{-10}{(x-5)(x+5)}$

Step 1: Factor the denominators.

Step 2: The LCD is $(x-5)(x+5)$.

$= \dfrac{1(x+5)}{(x-5)(x+5)} + \dfrac{-10}{(x-5)(x+5)}$

Step 3: Write each expression with the LCD.

$= \dfrac{1(x+5) + (-10)}{(x-5)(x+5)}$

Step 4: Add the numerators and write the result over the LCD.

$= \dfrac{x+5-10}{(x-5)(x+5)}$

$= \dfrac{\overset{1}{\cancel{x-5}}}{\cancel{(x-5)}(x+5)}$

Step 5: Reduce to lowest terms.

$= \dfrac{1}{x+5}$

example 3 Adding and Subtracting Rational Expressions with Unlike Denominators

Add or subtract as indicated. $\dfrac{p+2}{p-1} - \dfrac{2}{p+6} - \dfrac{14}{p^2+5p-6}$

Solution:

$$\frac{p+2}{p-1} - \frac{2}{p+6} - \frac{14}{p^2+5p-6}$$

$$= \frac{p+2}{p-1} - \frac{2}{p+6} - \frac{14}{(p-1)(p+6)}$$

Step 1: Factor the denominators.

Step 2: The LCD is $(p-1)(p+6)$.

Step 3: Write each expression with the LCD.

$$= \frac{(p+2)(p+6)}{(p-1)(p+6)} - \frac{2(p-1)}{(p+6)(p-1)} - \frac{14}{(p-1)(p+6)}$$

$$= \frac{(p+2)(p+6) - 2(p-1) - 14}{(p-1)(p+6)}$$

Step 4: Combine the numerators and write the result over the LCD.

$$= \frac{p^2 + 6p + 2p + 12 - 2p + 2 - 14}{(p-1)(p+6)}$$

Step 5: Clear parentheses in the numerator.

$$= \frac{p^2 + 6p}{(p-1)(p+6)}$$

Combine *like* terms.

$$= \frac{p(p+6)}{(p-1)(p+6)}$$

Factor the numerator to determine if the expression is in lowest terms.

$$= \frac{p\overset{1}{\cancel{(p+6)}}}{(p-1)\cancel{(p+6)}}$$

Reduce common factors.

$$= \frac{p}{p-1}$$

When the denominator of two rational expressions are opposites, we can produce identical denominators by multiplying one of the expressions by the ratio $\frac{-1}{-1}$. This is demonstrated in Example 4.

example 4 **Adding Rational Expressions with Unlike Denominators**

Add the rational expressions. $\frac{1}{d-7} + \frac{5}{7-d}$

Solution:

$\dfrac{1}{d-7} + \dfrac{5}{7-d}$ The expressions $d - 7$ and $7 - d$ are opposites and differ by a factor of -1. Therefore, multiply the numerator and denominator of *either* expression by -1 to obtain a common denominator.

$= \dfrac{1}{d-7} + \dfrac{(-1)5}{(-1)(7-d)}$ Note that $-1(7 - d) = -7 + d$ or $d - 7$.

$= \dfrac{1}{d-7} + \dfrac{-5}{d-7}$ Simplify.

$= \dfrac{1+(-5)}{d-7}$ Add the terms in the numerators and write the result over the common denominator.

$= \dfrac{-4}{d-7}$

3. Translating to a Rational Expression

example 5

Using Rational Expressions in Translations

Translate the English phrase into a mathematical expression. Then simplify by combining the rational expressions.

The difference of the reciprocal of x and the quotient of x and 3.

Solution:

The difference of the reciprocal of x and the quotient of x and 3.

The difference of

$\left(\dfrac{1}{x}\right) - \left(\dfrac{x}{3}\right)$

the reciprocal of x — the quotient of x and 3

$\dfrac{1}{x} - \dfrac{x}{3}$ The LCD is $3x$.

$= \dfrac{3 \cdot 1}{3 \cdot x} - \dfrac{x \cdot x}{3 \cdot x}$ Write each expression over the LCD.

$= \dfrac{3 - x^2}{3x}$ Subtract the numerators.

section 6.4 Practice Exercises

1. For the rational expression

$$\frac{x^2 - 4x - 5}{x^2 - 7x + 10}$$

   a. Find the value of the expression (if possible) when $x = 0, 1, -1, 2,$ and 5.

   b. Factor the denominator and identify the domain. Write the domain in set-builder notation.

   c. Reduce the expression to lowest terms.

2. For the rational expression

$$\frac{a^2 + a - 2}{a^2 - 4a - 12}$$

   a. Find the value of the expression (if possible) when $x = 0, 1, -2, 2,$ and 6.

   b. Factor the denominator and identify the domain. Write the domain in set-builder notation.

   c. Reduce the expression to lowest terms.

For Exercises 3–4, multiply or divide as indicated.

3. $\dfrac{2b^2 - b - 3}{2b^2 - 3b - 9} \cdot \dfrac{4b - 12}{2b - 3} \div \dfrac{b^2 - 1}{4b + 6}$

4. $(t - 6) \div \dfrac{5t - 30}{6t - 1} \cdot \dfrac{10t - 25}{2t^2 - 3t - 5}$

For Exercises 5–22, add or subtract the expressions with like denominators as indicated.

5. $\dfrac{7}{8} + \dfrac{3}{8}$

6. $\dfrac{1}{3} + \dfrac{7}{3}$

7. $\dfrac{9}{16} - \dfrac{3}{16}$

8. $\dfrac{14}{15} - \dfrac{4}{15}$

9. $\dfrac{5a}{a + 2} - \dfrac{3a - 4}{a + 2}$

10. $\dfrac{2b}{b - 3} - \dfrac{b - 9}{b - 3}$

11. $\dfrac{5c}{c + 6} + \dfrac{30}{c + 6}$

12. $\dfrac{12}{2 + d} + \dfrac{6d}{2 + d}$

13. $\dfrac{5}{t - 8} - \dfrac{2t + 1}{t - 8}$

14. $\dfrac{7p + 1}{2p + 1} - \dfrac{p - 4}{2p + 1}$

15. $\dfrac{10}{3x - 7} - \dfrac{5}{3x - 7}$

16. $\dfrac{8}{2w - 1} - \dfrac{4}{2w - 1}$

17. $\dfrac{m^2}{m + 5} + \dfrac{10m + 25}{m + 5}$

18. $\dfrac{k^2}{k - 3} - \dfrac{6k - 9}{k - 3}$

19. $\dfrac{2a}{a + 3} + \dfrac{6}{a + 3}$

20. $\dfrac{5b}{b + 4} + \dfrac{20}{b + 4}$

21. $\dfrac{x^2}{x + 5} - \dfrac{25}{x + 5}$

22. $\dfrac{y^2}{y - 7} - \dfrac{49}{y - 7}$

For Exercises 23–24, find an expression that represents the perimeter of the figure (assume that $x > 0$, $y > 0$, and $t > 0$).

23.

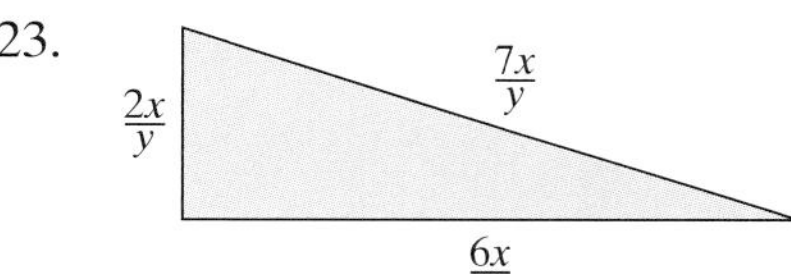

24.

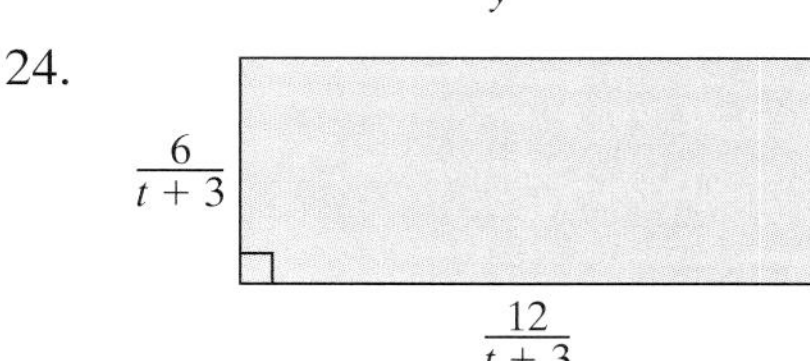

For Exercises 25–58, add or subtract the expressions with unlike denominators as indicated.

25. $\dfrac{4}{5xy^3} + \dfrac{2x}{15y^2}$

26. $\dfrac{5}{3a^2b} + \dfrac{-7}{6b^2}$

27. $\dfrac{z}{3z - 9} - \dfrac{z - 2}{z - 3}$

28. $\dfrac{3w - 8}{2w - 4} - \dfrac{w - 3}{w - 2}$

29. $\dfrac{5}{a + 1} + \dfrac{4}{3a + 3}$

30. $\dfrac{2}{c - 4} + \dfrac{1}{5c - 20}$

31. $\dfrac{k}{k^2 - 9} - \dfrac{4}{k - 3}$

32. $\dfrac{7}{h + 5} - \dfrac{2h - 3}{h^2 - 25}$

33. $\dfrac{3a - 7}{6a + 10} - \dfrac{10}{3a^2 + 5a}$

34. $\dfrac{k + 2}{8k} - \dfrac{3 - k}{12k}$

35. $\dfrac{10}{3x - 7} + \dfrac{5}{7 - 3x}$

36. $\dfrac{8}{2w - 1} + \dfrac{4}{1 - 2w}$

37. $\dfrac{6a}{a^2 - b^2} + \dfrac{2a}{a^2 + ab}$

38. $\dfrac{7x}{x^2 + 2xy + y^2} + \dfrac{3x}{x^2 + xy}$

39. $\dfrac{p}{3} - \dfrac{4p - 1}{-3}$

40. $\dfrac{r}{7} - \dfrac{r - 5}{-7}$

41. $\dfrac{4n}{n - 8} - \dfrac{2n - 1}{8 - n}$

42. $\dfrac{m}{m - 2} - \dfrac{3m + 1}{2 - m}$

43. $\dfrac{5}{x} + \dfrac{3}{x + 2}$

44. $\dfrac{6}{y - 1} + \dfrac{9}{y}$

45. $\dfrac{4w}{w^2 + 2w - 3} + \dfrac{2}{1 - w}$

46. $\dfrac{z - 23}{z^2 - z - 20} - \dfrac{2}{5 - z}$

47. $\dfrac{3a - 8}{a^2 - 5a + 6} + \dfrac{a + 2}{a^2 - 6a + 8}$

48. $\dfrac{3b + 5}{b^2 + 4b + 3} + \dfrac{-b + 5}{b^2 + 2b - 3}$

49. $\dfrac{3x}{x^2 + x - 6} + \dfrac{x}{x^2 + 5x + 6}$

50. $\dfrac{x}{x^2 + 5x + 4} - \dfrac{2x}{x^2 - 2x - 3}$

51. $\dfrac{3y}{2y^2 - y - 1} - \dfrac{4y}{2y^2 - 7y - 4}$

52. $\dfrac{5}{6y^2 - 7y - 3} + \dfrac{4y}{3y^2 + 4y + 1}$

53. $\dfrac{3}{2p - 1} - \dfrac{4p + 4}{4p^2 - 1}$

54. $\dfrac{1}{3q - 2} - \dfrac{6q + 4}{9q^2 - 4}$

55. $\dfrac{x}{x - y} - \dfrac{y}{y - x}$

56. $\dfrac{m}{m + n} - \dfrac{m}{m - n}$

57. $\dfrac{2}{a + b} + \dfrac{2}{a - b} - \dfrac{4a}{a^2 - b^2}$

58. $\dfrac{-2x}{x^2 - y^2} + \dfrac{1}{x + y} - \dfrac{1}{x - y}$

For Exercises 59–60, find an expression that represents the perimeter of the figure (assume that $x > 0$ and $t > 0$).

59. $\dfrac{2}{x + 3}$ $\dfrac{1}{x + 2}$

60. 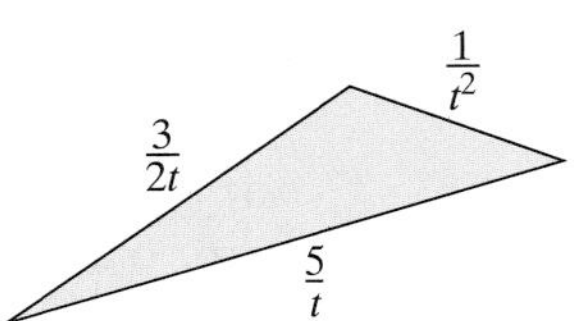

61. Let a number be represented by n. Write the reciprocal of n.

62. Write the reciprocal of the sum of a number and 6.

63. Let a number be represented by p. Write the quotient of 12 and p.

64. Write the quotient of 5 and the sum of a number and 2.

For Exercises 65–68, translate the English phrases into algebraic expressions. Then simplify by combining the rational expressions.

65. The sum of a number and the quantity seven times the reciprocal of the number.

66. The sum of a number and the quantity five times the reciprocal of the number.

67. The difference of the reciprocal of n and the quotient of 2 and n.

68. The difference of the reciprocal of m and the quotient of $3m$ and 7.

For Exercises 69–72, simplify by applying the order of operations.

69. $\left(\dfrac{2}{k + 1} + 3\right)\left(\dfrac{k + 1}{4k + 7}\right)$

70. $\left(\dfrac{p + 1}{3p + 4}\right)\left(\dfrac{1}{p + 1} + 2\right)$

71. $\left(\dfrac{1}{10a} - \dfrac{b}{10a^2}\right) \div \left(\dfrac{1}{10} - \dfrac{b}{10a}\right)$

72. $\left(\dfrac{1}{2m} + \dfrac{n}{2m^2}\right) \div \left(\dfrac{1}{4} + \dfrac{n}{4m}\right)$

Mixed Review

For Exercises 73–84, perform the indicated operation.

73. Reduce. $\dfrac{6a^2b^3}{72ab^7c}$

74. Subtract. $\dfrac{2a}{a+b} - \dfrac{b}{a-b} - \dfrac{-4ab}{a^2-b^2}$

75. Divide. $\dfrac{p^2+10pq+25q^2}{p^2+6pq+5q^2} \div \dfrac{10p+50q}{2p^2-2q^2}$

76. Add. $\dfrac{3k-8}{k-5} + \dfrac{k-12}{k-5}$

77. Reduce. $\dfrac{20x^2+10x}{4x^3+4x^2+x}$

78. Multiply.
$$\frac{w^2-81}{w^2+10w+9} \cdot \frac{w^2+w+2zw+2z}{w^2-9w+zw-9z}$$

79. Divide. $\dfrac{h^2-49}{h+1} \div \dfrac{h+7}{h^2-1}$

80. Reduce. $\dfrac{xy+7x+5y+35}{x^2+ax+5x+5a}$

81. Subtract. $\dfrac{a}{a^2-9} - \dfrac{3}{6a-18}$

82. Add. $\dfrac{4}{y^2-36} + \dfrac{2}{y^2-4y-12}$

83. Multiply. $(t^2+5t-24)\left(\dfrac{t+8}{t-3}\right)$

84. Reduce. $\dfrac{b^2+5b-14}{b-2}$

Expanding Your Skills

For Exercises 85–90, perform the indicated operations.

85. $\dfrac{-3}{w^3+27} - \dfrac{1}{w^2-9}$

86. $\dfrac{m}{m^3-1} + \dfrac{1}{(m-1)^2}$

87. $\dfrac{2p}{p^2+5p+6} - \dfrac{p+1}{p^2+2p-3} + \dfrac{3}{p^2+p-2}$

88. $\dfrac{3t}{8t^2+2t-1} - \dfrac{5t}{2t^2-9t-5} + \dfrac{2}{4t^2-21t+5}$

89. $\dfrac{3m}{m^2+3m-10} + \dfrac{5}{4-2m} - \dfrac{1}{m+5}$

90. $\dfrac{2n}{3n^2-8n-3} + \dfrac{1}{6-2n} - \dfrac{3}{3n+1}$

---

section

6.5 Complex Fractions

Concepts

1. Simplifying Complex Fractions (Method I)
2. Simplifying Complex Fractions (Method II)

1. Simplifying Complex Fractions (Method I)

A **complex fraction** is a fraction whose numerator or denominator contains one or more rational expressions. For example,

$$\frac{\frac{1}{ab}}{\frac{2}{b}} \quad \text{and} \quad \frac{1+\frac{3}{4}-\frac{1}{6}}{\frac{1}{2}+\frac{1}{3}}$$

are complex fractions.

Two methods will be presented to simplify complex fractions. The first method (Method I) follows the order of operations to simplify the numerator and denominator separately before dividing. The process is summarized as follows.

Steps to Simplify a Complex Fraction (Method I)

1. Add or subtract expressions in the numerator to form a single fraction. Add or subtract expressions in the denominator to form a single fraction.
2. Divide the rational expressions from step 1 by multiplying the numerator of the complex fraction by the reciprocal of the denominator of the complex fraction.
3. Simplify and reduce to lowest terms if possible.

example 1 **Simplifying Complex Fractions (Method I)**

Simplify the expression. $\dfrac{\frac{1}{ab}}{\frac{2}{b}}$

Solution:

Step 1: The numerator and denominator of the complex fraction are already single fractions.

$\dfrac{\frac{1}{ab}}{\frac{2}{b}}$ ← This fraction bar denotes division (÷).

$= \dfrac{1}{ab} \div \dfrac{2}{b}$

$= \dfrac{1}{ab} \cdot \dfrac{b}{2}$ **Step 2:** Multiply the numerator of the complex fraction by the reciprocal of $\frac{2}{b}$, which is $\frac{b}{2}$.

$= \dfrac{1}{a \cdot \cancel{b}} \cdot \dfrac{\overset{1}{\cancel{b}}}{2}$ **Step 3:** Reduce common factors and simplify.

$= \dfrac{1}{2a}$

Sometimes it is necessary to simplify the numerator and denominator of a complex fraction before the division can be performed. This is illustrated in Example 2.

example 2 **Simplifying Complex Fractions (Method I)**

Simplify the expression. $\dfrac{1 + \frac{3}{4} - \frac{1}{6}}{\frac{1}{2} + \frac{1}{3}}$

Solution:

$$\frac{1 + \frac{3}{4} - \frac{1}{6}}{\frac{1}{2} + \frac{1}{3}}$$

Step 1: Combine fractions in the numerator and denominator separately.

$$\frac{1 \cdot \frac{12}{12} + \frac{3}{4} \cdot \frac{3}{3} - \frac{1}{6} \cdot \frac{2}{2}}{\frac{1}{2} \cdot \frac{3}{3} + \frac{1}{3} \cdot \frac{2}{2}}$$

The LCD in the numerator is 12. The LCD in the denominator is 6.

$$= \frac{\frac{12}{12} + \frac{9}{12} - \frac{2}{12}}{\frac{3}{6} + \frac{2}{6}}$$

$$= \frac{\frac{19}{12}}{\frac{5}{6}}$$

Form a single fraction in the numerator and in the denominator.

$$= \frac{19}{\cancel{6} \cdot 2} \cdot \frac{\overset{1}{\cancel{6}}}{5}$$

Step 2: Multiply by the reciprocal of $\frac{5}{6}$, which is $\frac{6}{5}$.

$$= \frac{19}{10}$$

Step 3: Simplify and reduce to lowest terms.

example 3 Simplifying Complex Fractions (Method I)

Simplify the expression. $\dfrac{\frac{1}{x} + \frac{1}{y}}{x - \frac{y^2}{x}}$

Solution:

$$\frac{\frac{1}{x} + \frac{1}{y}}{x - \frac{y^2}{x}}$$

The LCD in the numerator is xy. The LCD in the denominator is x.

$$= \frac{\frac{1 \cdot y}{x \cdot y} + \frac{1 \cdot x}{y \cdot x}}{\frac{x \cdot x}{1 \cdot x} - \frac{y^2}{x}}$$

Rewrite the expressions using common denominators.

$$= \frac{\frac{y}{xy} + \frac{x}{xy}}{\frac{x^2}{x} - \frac{y^2}{x}}$$

$$= \frac{\frac{y + x}{xy}}{\frac{x^2 - y^2}{x}}$$ Form single fractions in the numerator and denominator.

$$= \frac{y + x}{xy} \cdot \frac{x}{x^2 - y^2}$$ Multiply by the reciprocal of the denominator.

$$= \frac{\overset{1}{\cancel{y + x}}}{\cancel{x}y} \cdot \frac{\overset{1}{\cancel{x}}}{\cancel{(x + y)}(x - y)}$$ Factor and reduce. Note that $(y + x) = (x + y)$.

$$= \frac{1}{y(x - y)}$$ Simplify.

2. Simplifying Complex Fractions (Method II)

We will now simplify the expression from Example 3 again using a second method to simplify complex fractions (Method II). Recall that multiplying the numerator and denominator of a rational expression by the same quantity does not change the value of the expression because we are multiplying by a number equivalent to 1. This is the basis for Method II.

Steps to Simplifying a Complex Fraction (Method II)

1. Multiply the numerator and denominator of the complex fraction by the LCD of *all* individual fractions within the expression.
2. Apply the distributive property and simplify the numerator and denominator.
3. Reduce to lowest terms if possible.

example 4 Simplifying a Complex Fraction (Method II)

Simplify the expression. $$\frac{\frac{1}{x} + \frac{1}{y}}{x - \frac{y^2}{x}}$$

Solution:

$$\frac{\frac{1}{x} + \frac{1}{y}}{x - \frac{y^2}{x}}$$ The LCD of the expressions: $\frac{1}{x}, \frac{1}{y}, x$, and $\frac{y^2}{x}$ is xy.

Tip: In step 1 we are multiplying the original expression by $\frac{xy}{xy}$, which equals 1.

$$= \frac{xy\left(\frac{1}{x} + \frac{1}{y}\right)}{xy\left(x - \frac{y^2}{x}\right)}$$

Step 1: Multiply numerator and denominator of the complex fraction by xy.

$$= \frac{xy \cdot \frac{1}{x} + xy \cdot \frac{1}{y}}{xy \cdot x - xy \cdot \frac{y^2}{x}}$$

Step 2: Apply the distributive property and simplify each term.

$$= \frac{y + x}{x^2y - y^3}$$

$$= \frac{y + x}{y(x^2 - y^2)}$$

Step 3: Factor completely and reduce common factors.

$$= \frac{y + x}{y(x + y)(x - y)}$$

Note that $(y + x) = (x + y)$.

$$= \frac{1}{y(x - y)}$$

example 5 Simplifying a Complex Fraction (Method II)

Simplify the expression. $\dfrac{\frac{1}{k + 1} - 1}{\frac{1}{k + 1} + 1}$

Solution:

$$\frac{\frac{1}{k + 1} - 1}{\frac{1}{k + 1} + 1}$$

The LCD of $\frac{1}{k + 1}$ and 1 is $(k + 1)$.

$$= \frac{(k + 1)\left(\frac{1}{k + 1} - 1\right)}{(k + 1)\left(\frac{1}{k + 1} + 1\right)}$$

Step 1: Multiply numerator and denominator of the complex fraction by $(k + 1)$.

$$= \frac{\cancel{(k+1)} \cdot \dfrac{1}{\cancel{(k+1)}} - (k+1) \cdot 1}{\cancel{(k+1)} \cdot \dfrac{1}{\cancel{(k+1)}} + (k+1) \cdot 1}$$

Step 2: Apply the distributive property.

$$= \frac{1 - (k+1)}{1 + (k+1)}$$

Simplify.

$$= \frac{1 - k - 1}{1 + k + 1}$$

$$= \frac{-k}{k+2}$$

Step 3: The expression is already reduced.

section 6.5 Practice Exercises

For Exercises 1–4, write the domain in set-builder notation and reduce the expression.

1. $\dfrac{(c-2)(c+3)}{(c+1)(c-2)}$

2. $\dfrac{y(2y+9)}{y^2(2y+9)}$

3. $\dfrac{6x+12}{3x^2-12}$

4. $\dfrac{a+5}{2a^2+7a-15}$

For Exercises 5–10, perform the indicated operations.

5. $\dfrac{2}{w-2} + \dfrac{3}{w}$

6. $\dfrac{6}{5} - \dfrac{3}{5k-10}$

7. $\dfrac{p^2+2p}{2p-1} \cdot \dfrac{10p^2-5p}{12p^3+24p^2}$

8. $\dfrac{x^2-2xy+y^2}{x^4-y^4} \div \dfrac{3x^2y-3xy^2}{x^2+y^2}$

9. $\left(\dfrac{1}{z} - \dfrac{1}{2z}\right) \div \left(\dfrac{1}{2} + \dfrac{1}{2z}\right)$

10. $\left(\dfrac{2}{3a^2} - \dfrac{3}{b}\right) \div \left(\dfrac{5}{ab} - 4\right)$

For Exercises 11–14, translate the English phrases into algebraic expressions. Then simplify the expressions.

11. The sum of one-half and two-thirds, all divided by five.

12. The quotient of ten and the difference of two-fifths and one-fourth.

13. The quotient of three and the sum of two-thirds and three-fourths.

14. The difference of three-fifths and one-half, all divided by four.

For Exercises 15–34, simplify the complex fractions.

15. $\dfrac{\dfrac{1}{8} + \dfrac{4}{3}}{\dfrac{1}{2} - \dfrac{5}{12}}$

16. $\dfrac{\dfrac{8}{9} - \dfrac{1}{3}}{\dfrac{7}{6} + \dfrac{1}{9}}$

17. $\dfrac{\dfrac{1}{h} + \dfrac{1}{k}}{\dfrac{1}{hk}}$

18. $\dfrac{\dfrac{1}{b} + 1}{\dfrac{1}{b}}$

19. $\dfrac{\dfrac{n+1}{n^2-9}}{\dfrac{2}{n+3}}$

20. $\dfrac{\dfrac{5}{k-5}}{\dfrac{k+1}{k^2-25}}$

21. $\dfrac{2 + \dfrac{1}{x}}{4 + \dfrac{1}{x}}$

22. $\dfrac{6 + \dfrac{6}{k}}{1 + \dfrac{1}{k}}$

23. $\dfrac{\dfrac{m}{7} - \dfrac{7}{m}}{\dfrac{1}{7} + \dfrac{1}{m}}$

24. $\dfrac{\dfrac{2}{p} + \dfrac{p}{2}}{\dfrac{p}{3} - \dfrac{3}{p}}$

25. $\dfrac{\frac{1}{5} - \frac{1}{y}}{\frac{7}{10} + \frac{1}{y^2}}$

26. $\dfrac{\frac{1}{m^2} + \frac{2}{3}}{\frac{1}{m} - \frac{5}{6}}$

27. $\dfrac{\frac{8}{a+4} + 2}{\frac{12}{a+4} - 2}$

28. $\dfrac{\frac{2}{w+1} + 3}{\frac{3}{w+1} + 4}$

29. $\dfrac{1 - \frac{4}{t^2}}{1 - \frac{2}{t} - \frac{8}{t^2}}$

30. $\dfrac{1 - \frac{9}{p^2}}{1 - \frac{1}{p} - \frac{6}{p^2}}$

31. $\dfrac{\frac{1}{z^2 - 9} + \frac{2}{z+3}}{\frac{3}{z-3}}$

32. $\dfrac{\frac{5}{w^2 - 25} - \frac{3}{w+5}}{\frac{4}{w-5}}$

33. $\dfrac{\frac{2}{x-1} + 2}{\frac{2}{x+1} - 2}$

34. $\dfrac{\frac{1}{y-3} + 1}{\frac{2}{y+3} - 1}$

35. In electronics, resistors oppose the flow of current. For two resistors in parallel, the total resistance is given by

$$R = \frac{1}{\frac{1}{R_1} + \frac{1}{R_2}}$$

a. Find the total resistance if $R_1 = 2\ \Omega$ (ohms) and $R_2 = 3\ \Omega$.

b. Find the total resistance if $R_1 = 10\ \Omega$ and $R_2 = 15\ \Omega$.

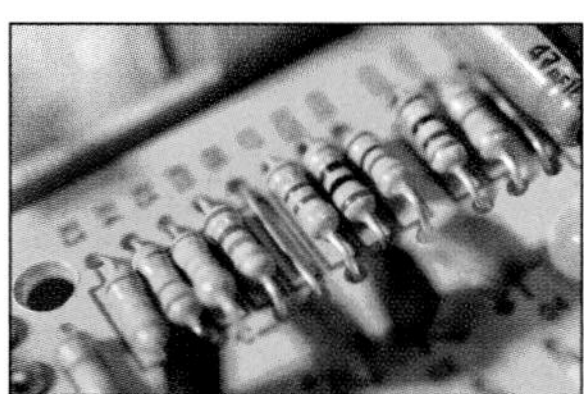

36. Suppose that Joelle makes a round trip to a location that is d miles away. If the average rate going to the location is r_1 and the average rate on the return trip is given by r_2, the average rate of the entire trip, R, is given by

$$R = \frac{2d}{\frac{d}{r_1} + \frac{d}{r_2}}$$

a. Find the average rate of a trip to a destination 30 miles away when the average rate going there was 60 mph and the average rate returning home was 45 mph. (Round to the nearest tenth of a mile per hour.)

b. Find the average rate of a trip to a destination that is 50 miles away if the driver travels at the same rates as in part (a). (Round to the nearest tenth of a mile per hour.)

c. Compare your answers from parts (a) and (b) and explain the results in the context of the problem.

Expanding Your Skills

For Exercises 37–39, simplify the complex fractions. (*Hint*: Use the order of operations and begin with the fraction on the lower right.)

37. $1 + \dfrac{1}{1+1}$

38. $1 + \dfrac{1}{1 + \frac{1}{1+1}}$

39. $1 + \dfrac{1}{1 + \frac{1}{1 + \frac{1}{1+1}}}$

Concepts

1. Definition of a Rational Equation
2. Clearing Fractions
3. Solving Rational Equations
4. Solving Formulas Involving Rational Equations

section

6.6 Rational Equations

1. Definition of a Rational Equation

Thus far we have studied two specific types of equations in one variable: linear equations and quadratic equations. Recall,

$ax + b = 0$, where $a \neq 0$, is a **linear equation**

$ax^2 + bx + c = 0$, where $a \neq 0$, is a **quadratic equation**.

We will now study another type of equation called a rational equation.

Definition of a Rational Equation

An equation with one or more rational expressions is called a **rational equation**.

The following equations are rational equations:

$$\frac{y}{2} + \frac{y}{4} = 6 \qquad \frac{1}{x} + \frac{1}{3} = \frac{5}{6} \qquad \frac{6}{t^2 - 7t + 12} + \frac{2t}{t-3} = \frac{3t}{t-4}$$

2. Clearing Fractions

To understand the process of solving a rational equation, first review the process of clearing fractions from Section 2.3.

example 1

Solving a Rational Equation

Solve. $\frac{y}{2} + \frac{y}{4} = 6$

Solution:

$\frac{y}{2} + \frac{y}{4} = 6$ The LCD of all terms in the equation is 4.

$4\left(\frac{y}{2} + \frac{y}{4}\right) = 4(6)$ Multiply both sides of the equation by 4 to clear fractions.

$4 \cdot \frac{y}{2} + 4 \cdot \frac{y}{4} = 4(6)$ Apply the distributive property.

$2y + y = 24$ Clear fractions.

$3y = 24$ Solve the resulting equation (linear).

$y = 8$

Check: $\frac{y}{2} + \frac{y}{4} = 6$

$$\frac{(8)}{2} + \frac{(8)}{4} \stackrel{?}{=} 6$$

$$4 + 2 \stackrel{?}{=} 6$$

$$6 = 6 \checkmark$$

3. Solving Rational Equations

The same process of clearing fractions is used to solve rational equations when variables are present in the denominator.

example 2 Solving a Rational Equation

Solve the equation. $\frac{1}{x} + \frac{1}{3} = \frac{5}{6}$

Solution:

$$\frac{1}{x} + \frac{1}{3} = \frac{5}{6}$$ The LCD of all expressions is $6x$.

$$6x\left(\frac{1}{x} + \frac{1}{3}\right) = 6x\left(\frac{5}{6}\right)$$ Multiply by the LCD.

$$6x \cdot \frac{1}{x} + 6x \cdot \frac{1}{3} = 6x \cdot \frac{5}{6}$$ Apply the distributive property.

$$6 + 2x = 5x$$ Clear fractions.

$$6 = 3x$$ Solve the resulting equation (linear).

$$x = 2$$

Check: $\frac{1}{x} + \frac{1}{3} = \frac{5}{6}$

$$\frac{1}{(2)} + \frac{1}{3} \stackrel{?}{=} \frac{5}{6}$$

$$\frac{3}{6} + \frac{2}{6} \stackrel{?}{=} \frac{5}{6}$$

$$\frac{5}{6} = \frac{5}{6} \checkmark$$

example 3 Solving a Rational Equation

Solve the equation. $1 + \frac{3a}{a - 2} = \frac{6}{a - 2}$

Solution:

$$1 + \frac{3a}{a-2} = \frac{6}{a-2}$$

The LCD of all expressions is $a - 2$.

$$(a-2)\left(1 + \frac{3a}{a-2}\right) = (a-2)\left(\frac{6}{a-2}\right)$$

Multiply by the LCD.

$$(a-2)1 + (a-2)\left(\frac{3a}{a-2}\right) = (a-2)\left(\frac{6}{a-2}\right)$$

Apply the distributive property.

$$a - 2 + 3a = 6$$

$$4a - 2 = 6$$

$$4a = 8$$

$$a = 2$$

Solve the resulting equation (linear).

Check: $1 + \frac{3a}{a-2} = \frac{6}{a-2}$

$$1 + \frac{3(2)}{(2)-2} \stackrel{?}{=} \frac{6}{(2)-2}$$

$$1 + \frac{6}{0} \stackrel{?}{=} \frac{6}{0}$$

The denominator is 0 when $a = 2$.

Because the value $a = 2$ makes the denominator zero in one (or more) of the rational expressions within the equation, the equation is undefined for $a = 2$. That is, $a = 2$ is not in the domain of the equation. No other potential solutions exist for the equation,

$$1 + \frac{3a}{a-2} = \frac{6}{a-2};$$

hence, it has no solution.

Examples 1–3 show that the steps to solve a rational equation mirror the process of clearing fractions from Section 2.3. However, there is one significant difference. The solutions of a rational equation must not make the denominator equal to zero for any expression within the equation. When $a = 2$ is substituted into the expression

$$\frac{3a}{a-2} \quad \text{or} \quad \frac{6}{a-2}$$

the denominator is zero and the expression is undefined. Hence, $a = 2$ cannot be a solution to the equation

$$1 + \frac{3a}{a-2} = \frac{6}{a-2}$$

The steps to solve a rational equation are summarized as follows.

Steps to Solve a Rational Equation

1. Factor the denominators of all rational expressions.
2. Identify the LCD of all expressions in the equation.
3. Multiply both sides of the equation by the LCD to clear fractions.
4. Solve the resulting equation.
5. Check potential solutions in the original equation.

example 4 **Solving Rational Equations**

Solve the equations.

a. $1 - \frac{4}{p} = -\frac{3}{p^2}$ b. $\frac{6}{t^2 - 7t + 12} + \frac{2t}{t - 3} = \frac{3t}{t - 4}$

Solution:

a. $1 - \frac{4}{p} = -\frac{3}{p^2}$ **Step 1:** The denominators are already factored.

Step 2: The LCD of all expressions is p^2.

$p^2\left(1 - \frac{4}{p}\right) = p^2\left(-\frac{3}{p^2}\right)$ **Step 3:** Multiply by the LCD.

$p^2(1) - p^2\left(\frac{4}{p}\right) = p^2\left(-\frac{3}{p^2}\right)$ Apply the distributive property.

$p^2 - 4p = -3$ **Step 4:** Solve the resulting equation. Because the resulting equation is quadratic, set the equation equal to zero and factor.

$p^2 - 4p + 3 = 0$

$(p - 3)(p - 1) = 0$

$p - 3 = 0$ or $p - 1 = 0$ Set each factor equal to zero.

$p = 3$ or $p = 1$ **Step 5:** Both solutions $p = 3$ and $p = 1$ check.

b. $\frac{6}{t^2 - 7t + 12} + \frac{2t}{t - 3} = \frac{3t}{t - 4}$

$\frac{6}{(t - 3)(t - 4)} + \frac{2t}{t - 3} = \frac{3t}{t - 4}$ **Step 1:** Factor the denominators.

Step 2: The LCD is $(t - 3)(t - 4)$.

Step 3: Multiply by the LCD on both sides.

$$(t-3)(t-4)\left(\frac{6}{(t-3)(t-4)}+\frac{2t}{t-3}\right)=(t-3)(t-4)\left(\frac{3t}{t-4}\right)$$

$$\cancel{(t-3)}\cancel{(t-4)}\left(\frac{6}{\cancel{(t-3)}\cancel{(t-4)}}\right)+\cancel{(t-3)}(t-4)\left(\frac{2t}{\cancel{t-3}}\right)=(t-3)\cancel{(t-4)}\left(\frac{3t}{\cancel{t-4}}\right)$$

$$6+2t(t-4)=3t(t-3)$$

$$6+2t^2-8t=3t^2-9t$$

Step 4: Solve the resulting equation.

$$0=3t^2-2t^2-9t+8t-6$$

$$0=t^2-t-6$$

$$0=(t-3)(t+2)$$

Because the resulting equation is quadratic, set the equation equal to zero and factor.

$t-3=0$ or $t+2=0$

$t=3$ or $t=-2$

Set each factor equal to zero.

Step 5: Check the potential solutions in the original equation.

Check: $t=3$

$t=3$ cannot be a solution to the equation because it will make the denominator zero in the original equation.

$$\frac{6}{t^2-7t+12}+\frac{2t}{t-3}=\frac{3t}{t-4}$$

$$\frac{6}{(3)^2-7(3)+12}+\frac{2(3)}{(3)-3}\stackrel{?}{=}\frac{3(3)}{(3)-4}$$

$$\frac{6}{0}+\frac{6}{0}\stackrel{?}{=}\frac{9}{-1}$$

zero in the denominator

Check: $t=-2$

$$\frac{6}{t^2-7t+12}+\frac{2t}{t-3}=\frac{3t}{t-4}$$

$$\frac{6}{(-2)^2-7(-2)+12}+\frac{2(-2)}{(-2)-3}\stackrel{?}{=}\frac{3(-2)}{(-2)-4}$$

$$\frac{6}{4+14+12}+\frac{-4}{-5}\stackrel{?}{=}\frac{-6}{-6}$$

$$\frac{6}{30}+\frac{4}{5}\stackrel{?}{=}1$$

$$\frac{1}{5}+\frac{4}{5}=1 \checkmark$$

$t=-2$ is a solution.

The only solution is $t=-2$.

Tip: $t=3$ is not a solution because it is not in the domain of the equation.

example 5 Translating to a Rational Equation

Ten times the reciprocal of a number is added to four. The result is equal to the quotient of twenty-two and the number. Find the number.

Solution:

Let x represent the number.

$$\underset{\text{is added to four}}{4} + \underset{\text{10 times the reciprocal of a number}}{10\left(\frac{1}{x}\right)} \underset{\text{the result is equal to}}{=} \underset{\text{the quotient of 22 and the number}}{\frac{22}{x}}$$

$$4 + \frac{10}{x} = \frac{22}{x}$$

Step 1: The denominators are already factored.

Step 2: The LCD is x.

$$x\left(4 + \frac{10}{x}\right) = x\left(\frac{22}{x}\right)$$

Step 3: Multiply both sides by the LCD.

$$4x + 10 = 22$$

Apply the distributive property.

$$4x = 22 - 10$$

Step 4: Solve the resulting equation (linear).

$$4x = 12$$

$x = 3$ is a potential solution.

Step 5: Substituting $x = 3$ into the original equation verifies that it is a solution.

The number is 3.

4. Solving Formulas Involving Rational Equations

A rational equation may have more than one variable. To solve for a specific variable within such a rational equation, we can still apply the principle of clearing fractions.

example 6 **Solving Formulas Involving Rational Equations**

Solve for b: $h = \dfrac{2A}{B + b}$

Algebra is case-sensitive. B and b are different variables.

Solution:

To solve for b, we must clear fractions so that b appears in the numerator.

$$h = \frac{2A}{B + b}$$ The LCD is $(B + b)$.

$$h(B + b) = \left(\frac{2A}{\cancel{B + b}}\right) \cdot \cancel{(B + b)}$$ Multiply both sides of the equation by the LCD.

$hB + hb = 2A$ — Apply the distributive property.

$hb = 2A - hB$ — Subtract hB from both sides to isolate the b term.

$\frac{\not{h}b}{\not{h}} = \frac{2A - hB}{h}$ — Divide by h.

$b = \frac{2A - hB}{h}$

Tip: The solution to Example 6 can be written in several forms. The quantity

$$\frac{2A - hB}{h}$$

can be left as a single rational expression or can be split into two fractions and reduced. Any of the following forms is a valid representation of b:

$$b = \frac{2A - hB}{h} \quad \text{or} \quad \frac{2A}{h} - \frac{hB}{h} \Rightarrow \frac{2A}{h} - B$$

example 7

Solving Formulas Involving Rational Expressions

Solve for z. $\quad y = \frac{x - z}{x + z}$

Solution:

To solve for z, we must clear fractions so that z appears in the numerator.

$y = \frac{x - z}{x + z}$ — LCD is $(x + z)$.

$y(x + z) = \left(\frac{x - z}{\cancel{x + z}}\right)(\cancel{x + z})$ — Multiply both sides of the equation by the LCD.

$yx + yz = x - z$ — Apply the distributive property.

$yz + z = x - yx$ — Add z to both sides. Subtract yx from both sides to get the terms containing z on one side of the equation.

$z(y + 1) = x - yx$ — Factor out z.

$z = \frac{x - yx}{y + 1}$ — Divide by $y + 1$ to solve for z.

section 6.6 Practice Exercises

For Exercises 1–6, perform the indicated operations.

1. $\dfrac{2}{x-3} - \dfrac{3}{x^2-x-6}$

2. $\dfrac{\dfrac{3}{a}+\dfrac{5}{2a}}{1+\dfrac{2}{a+2}}$

3. $\dfrac{t^2-5t+6}{t^2-5t-6} \div \dfrac{t^2-4}{t^2+2t+1}$

4. $\dfrac{2y}{y-3} + \dfrac{4}{y^2-9}$

5. $\dfrac{h-\dfrac{1}{h}}{\dfrac{1}{5}-\dfrac{1}{5h}}$

6. $\dfrac{w-4}{w^2-9} \cdot \dfrac{w-3}{w^2-8w+16}$

For Exercises 7–12, solve the equations by first clearing the fractions.

7. $\dfrac{1}{3}z + \dfrac{2}{3} = -2z + 10$

8. $\dfrac{5}{2} + \dfrac{1}{2}b = 5 - \dfrac{1}{3}b$

9. $\dfrac{3}{2}p + \dfrac{1}{3} = \dfrac{2p-3}{4}$

10. $\dfrac{5}{3} - \dfrac{1}{6}k = \dfrac{3k+5}{4}$

11. $\dfrac{2x-3}{4} + \dfrac{9}{10} = \dfrac{x}{5}$

12. $\dfrac{4y+2}{3} - \dfrac{7}{6} = -\dfrac{y}{6}$

For Exercises 13–16, a. identify the LCD of all of the denominators of the equations b. Solve the equations.

13. $\dfrac{1}{w} - \dfrac{1}{2} = -\dfrac{1}{4}$

14. $\dfrac{3}{z} - \dfrac{4}{5} = -\dfrac{1}{5}$

15. $\dfrac{x+1}{x^2+2x-3} = \dfrac{1}{x+3} - \dfrac{1}{x-1}$

16. $\dfrac{10}{x-2} - \dfrac{40}{x^2+x-6} = \dfrac{12}{x+3}$

For Exercises 17–42, solve the equations.

17. $\dfrac{1}{8} = \dfrac{3}{5} + \dfrac{5}{y}$

18. $\dfrac{2}{7} - \dfrac{1}{x} = \dfrac{2}{3}$

19. $\dfrac{4}{t} = \dfrac{3}{t} + \dfrac{1}{8}$

20. $\dfrac{9}{b} - \dfrac{8}{b} = \dfrac{1}{4}$

21. $\dfrac{5}{6x} + \dfrac{7}{x} = 1$

22. $\dfrac{14}{3x} - \dfrac{5}{x} = 2$

23. $1 - \dfrac{2}{y} = \dfrac{3}{y^2}$

24. $1 - \dfrac{2}{m} = \dfrac{8}{m^2}$

25. $\dfrac{a+1}{a} = 1 + \dfrac{a-2}{2a}$

26. $\dfrac{7b-4}{5b} = \dfrac{9}{5} - \dfrac{4}{b}$

27. $\dfrac{w}{5} - \dfrac{w+3}{w} = -\dfrac{3}{w}$

28. $\dfrac{t}{12} + \dfrac{t+3}{3t} = \dfrac{1}{t}$

29. $\dfrac{2}{m+3} = \dfrac{5}{4m+12} - \dfrac{3}{8}$

30. $\dfrac{2}{4n-4} - \dfrac{7}{4} = \dfrac{-3}{n-1}$

31. $\dfrac{p}{p-4} - 5 = \dfrac{4}{p-4}$

32. $\dfrac{-5}{q+5} = \dfrac{q}{q+5} + 2$

33. $\dfrac{2t}{t+2} - 2 = \dfrac{t-8}{t+2}$

34. $\dfrac{4w}{w-3} - 3 = \dfrac{3w-1}{w-3}$

35. $\dfrac{x^2-x}{x-2} = \dfrac{12}{x-2}$

36. $\dfrac{x^2+9}{x+4} = \dfrac{-10x}{x+4}$

37. $\dfrac{x^2+3x}{x-1} = \dfrac{4}{x-1}$

38. $\dfrac{2x^2-21}{2x-3} = \dfrac{-11x}{2x-3}$

39. $\dfrac{2x}{x+4} - \dfrac{8}{x-4} = \dfrac{2x^2+32}{x^2-16}$

40. $\dfrac{4x}{x+3} - \dfrac{12}{x-3} = \dfrac{4x^2+36}{x^2-9}$

41. $\dfrac{x}{x+6} = \dfrac{72}{x^2-36} + 4$

42. $\dfrac{y}{y+4} = \dfrac{32}{y^2-16} + 3$

43. The reciprocal of a number is added to three. The result is the quotient of 25 and the number. Find the number.

44. The difference of three and the reciprocal of a number is equal to the quotient of 20 and the number. Find the number.

45. If a number added to five is divided by the difference of the number and two, the result is three-fourths. Find the number.

46. If twice a number added to three is divided by the number plus one, the result is three-halves. Find the number.

For Exercises 47–66, solve for the indicated variable.

47. $K = \frac{ma}{F}$ for m

48. $K = \frac{ma}{F}$ for a

49. $K = \frac{IR}{E}$ for E

50. $K = \frac{IR}{E}$ for R

51. $I = \frac{E}{R + r}$ for R

52. $I = \frac{E}{R + r}$ for r

53. $h = \frac{2A}{B + b}$ for B

54. $\frac{C}{\pi r} = 2$ for r

55. $\frac{V}{\pi h} = r^2$ for h

56. $\frac{V}{lw} = h$ for w

57. $x = \frac{at + b}{t}$ for t

58. $\frac{T + mf}{m} = g$ for m

59. $\frac{x - y}{xy} = z$ for x

60. $\frac{w - n}{wn} = P$ for w

61. $a + b = \frac{2A}{h}$ for h

62. $1 + rt = \frac{A}{P}$ for P

63. $\frac{1}{R} = \frac{1}{R_1} + \frac{1}{R_2}$ for R

64. $\frac{b + a}{ab} = \frac{1}{f}$ for b

65. $v = \frac{s_2 - s_1}{t_2 - t_1}$ for t_2

66. $a = \frac{v_2 - v_1}{t_2 - t_1}$ for v_1

Concepts

1. Proportions
2. Applications of Proportions
3. Similar Triangles
4. Distance, Rate, Time Applications
5. "Work" Problems

section

6.7 Applications of Rational Equations and Proportions

1. Proportions

One application of rational equations is solving proportions.

Definition of Ratio and Proportion

1. The **ratio** of a to b is $\frac{a}{b}$ $(b \neq 0)$ and can also be expressed as $a{:}b$ or $a \div b$.
2. An equation that equates two ratios is called a **proportion**. Therefore, if $b \neq 0$ and $d \neq 0$, then $\frac{a}{b} = \frac{c}{d}$ is a proportion.

One method of solving a proportion is to use the technique of clearing fractions.

example 1 **Solving a Proportion**

Solve the proportion. $\frac{3}{11} = \frac{123}{w}$

Solution:

$\frac{3}{11} = \frac{123}{w}$ The LCD is $11w$.

$11w\left(\frac{3}{11}\right) = 11w\left(\frac{123}{w}\right)$ Multiply by the LCD and clear fractions.

$3w = 11 \cdot 123$ Solve the resulting equation (linear).

$3w = 1353$

$\frac{3w}{3} = \frac{1353}{3}$

$w = 451$

Check: $w = 451$

$\frac{3}{11} = \frac{123}{w}$

$\frac{3}{11} \stackrel{?}{=} \frac{123}{(451)}$ Reduce

$\frac{3}{11} = \frac{3}{11}$ ✔

Tip: The cross products of any proportion are equal. That is, for $b \neq 0$ and $d \neq 0$, the proportion $\frac{a}{b} = \frac{c}{d}$ is equivalent to $ad = bc$.

Some rational equations are proportions and can be solved by equating the cross products. Consider the proportion from Example 1:

$$\frac{3}{11} = \frac{123}{w}$$

$3 \cdot w = 11 \cdot 123$ Equate the cross products.

$3w = 1353$ Solve the resulting equation.

$\frac{3w}{3} = \frac{1353}{3}$

$w = 451$

2. Applications of Proportions

example 2

Using a Proportion in an Application

The population of Alabama in 1997 was approximately 4.2 million. At that time, Alabama had seven representatives in the U.S. House of Representatives. In 1997, North Carolina had a population of approximately 7.2 million. If representation in the House is based on population in equal proportions for each state, how many representatives would North Carolina be expected to have?

Tip: The equation from Example 2 could have been solved by first equating the cross products:

$$\frac{4.2}{7} \times \frac{7.2}{x}$$

$$4.2x = (7.2)(7)$$

$$4.2x = 50.4$$

$$x = 12$$

Solution:

Let x represent the number of representatives for North Carolina.

Set up a proportion by writing two equivalent ratios.

$$\frac{\text{Population of Alabama}}{\text{Number of representatives}} \rightarrow \frac{4.2}{7} = \frac{7.2}{x} \leftarrow \frac{\text{Population of North Carolina}}{\text{Number of representatives}}$$

$$\frac{4.2}{7} = \frac{7.2}{x}$$

$$7x \cdot \frac{4.2}{7} = 7x \cdot \frac{7.2}{x}$$ Multiply by the LCD, $7x$.

$$4.2x = (7.2)(7)$$ Solve the resulting equation (linear).

$$4.2x = 50.4$$

$$\frac{4.2x}{4.2} = \frac{50.4}{4.2}$$

$$x = 12$$ We would expect North Carolina to have had 12 representatives in 1997.

example 3 Using a Proportion in an Application

In a large lecture class in chemistry, the ratio of men to women is 3 to 1. If there are 124 more men than women, find the number of women.

Solution:

Let x represent the number of women.

Then $x + 124$ represents the number of men.

Set up a proportion by writing two equivalent ratios.

$$\frac{\text{Number of men}}{\text{Number of women}} \rightarrow \frac{3}{1} = \frac{x + 124}{x} \leftarrow \frac{\text{Number of men}}{\text{Number of women}}$$

$$x \cdot \frac{3}{1} = x \cdot \left(\frac{x + 124}{x}\right)$$ Multiply by the LCD, x.

$$3x = x + 124$$ Solve the resulting equation (linear).

$$2x = 124$$

$$x = 62$$

There are 62 women.

Because $x + 124 = 62 + 124 = 186$, there are 186 men.

3. Similar Triangles

Proportions are used in geometry with **similar triangles**. Two triangles are said to be similar if they have equal angles. In such a case, the lengths of the corresponding sides are proportional. The triangles in Figure 6-1 are similar. Therefore, the following ratios are equivalent.

$$\frac{a}{x} = \frac{b}{y} = \frac{c}{z}$$

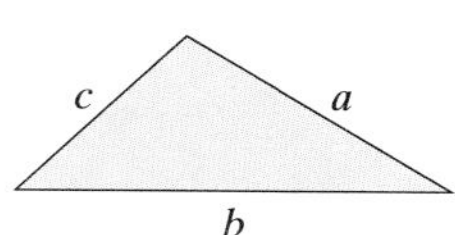

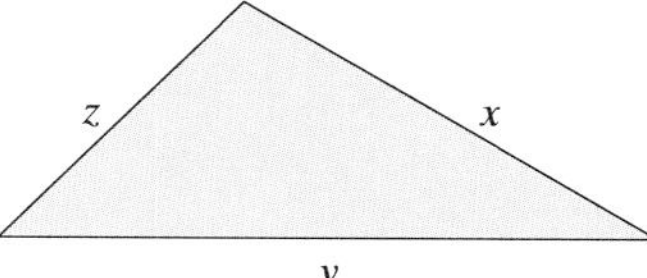

Figure 6-1

example 4 **Using Similar Triangles in an Application**

The shadow cast by a yardstick is 2 ft long. The shadow cast by a tree is 11 ft long. Find the height of the tree.

Solution:

Let x represent the height of the tree. Label the variables.

We will assume that the measurements were taken at the same time of day. Therefore, the angle of the sun is the same on both objects, and we can set up similar triangles (Figure 6-2).

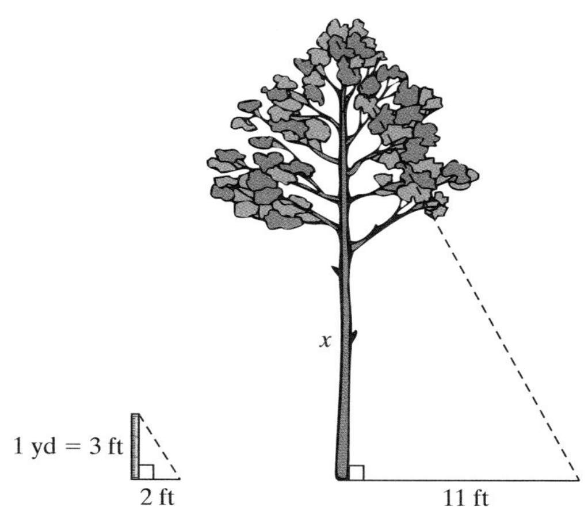

Figure 6-2

$$\boxed{\frac{\text{Height of yardstick}}{\text{Length of yardstick's shadow}}} \longrightarrow \frac{3\text{ ft}}{2\text{ ft}} = \frac{x}{11\text{ ft}} \longleftarrow \boxed{\frac{\text{Height of tree}}{\text{Length of tree's shadow}}}$$

$\frac{3}{2} = \frac{x}{11}$ Write a mathematical equation.

$3(11) = 2x$ Equate the cross products.

$33 = 2x$ Solve the equation.

$\frac{33}{2} = \frac{2x}{2}$

$16.5 = x$ Interpret the results and write the answer in words.

The tree is 16.5 ft high.

4. Distance, Rate, Time Applications

In Section 2.7, we presented applications involving the relationship among the variables distance, rate, and time. Recall that $d = rt$.

example 5 Using a Rational Equation in a Distance, Rate, Time Application

A small plane flies 440 miles with the wind from Memphis, Tennessee, to Oklahoma City, Oklahoma. In the same amount of time, the plane flies 340 miles against the wind from Oklahoma City to Little Rock, Arkansas (see Figure 6-3). If the wind speed is 30 mph, find the speed of the plane in still air.

Figure 6-3

Solution:

Let x represent the speed of the plane in still air.

Organize the given information in a chart. Note that distance, rate, and time are calculated for each event independently.

Tip: The speed with the wind is faster: $x + 30$. The speed against the wind is slower: $x - 30$.

	Distance	Rate	Time
With the Wind	440	$x + 30$	$\frac{440}{x + 30}$
Against the Wind	340	$x - 30$	$\frac{340}{x - 30}$

Because $d = rt$, $\frac{d}{r} = \frac{rt}{r}$, and $t = \frac{d}{r}$

The plane travels with the wind for the same amount of time as it travels against the wind, so we can equate the two expressions for time.

$$\begin{pmatrix}\text{Time with}\\ \text{the wind}\end{pmatrix} = \begin{pmatrix}\text{time against}\\ \text{the wind}\end{pmatrix}$$

$$\frac{440}{x+30} = \frac{340}{x-30}$$

The LCD is $(x+30)(x-30)$.

$$(x+30)(x-30)\cdot\frac{440}{x+30} = (x+30)(x-30)\cdot\frac{340}{x-30}$$

$$440(x-30) = 340(x+30)$$

$$440x - 13{,}200 = 340x + 10{,}200$$

Solve the resulting linear equation.

$$100x = 23{,}400$$

$$x = 234$$

The plane's speed in still air is 234 mph.

Tip: The equation

$$\frac{440}{x+30} = \frac{340}{x-30}$$

is a proportion. The fractions can also be cleared by equating the cross products.

$$\frac{440}{x+30} = \frac{340}{x-30}$$

$$440(x-30) = 340(x+30)$$

example 6 Using a Rational Equation in a Distance, Rate, Time Application

A motorist drives 100 miles between two cities in a bad rainstorm. For the return trip in sunny weather, she averages 10 mph faster and takes $\frac{1}{2}$ hour less time. Find the average speed of the motorist in the rainstorm and in sunny weather.

Solution:

Let x represent the motorist's speed during the rain.

Then $x + 10$ represents the speed in sunny weather.

	Distance	Rate	Time
Trip during Rainstorm	100	x	$\frac{100}{x}$
Trip during Sunny Weather	100	$x + 10$	$\frac{100}{x+10}$

Because $d = rt$, then $t = \frac{d}{r}$

Because the same distance is traveled in $\frac{1}{2}$ hr less time, the difference between the time of the trip during the rainstorm and the time during sunny weather is $\frac{1}{2}$ hr.

$$\begin{pmatrix}\text{Time during}\\ \text{the rainstorm}\end{pmatrix} - \begin{pmatrix}\text{time during}\\ \text{sunny weather}\end{pmatrix} = \left(\frac{1}{2}\text{ hr}\right)$$ Verbal model

$$\frac{100}{x} - \frac{100}{x+10} = \frac{1}{2}$$ Mathematical equation

$$2x(x+10)\left(\frac{100}{x} - \frac{100}{x+10}\right) = 2x(x+10)\left(\frac{1}{2}\right)$$ Multiply by the LCD.

$$2x(x+10)\left(\frac{100}{x}\right) - 2x(x+10)\left(\frac{100}{x+10}\right) = 2x(x+10)\left(\frac{1}{2}\right)$$ Apply the distributive property.

$$200(x+10) - 200x = x(x+10)$$ Clear fractions.

$$200x + 2000 - 200x = x^2 + 10x$$ Solve the resulting equation (quadratic).

$$2000 = x^2 + 10x$$

$$0 = x^2 + 10x - 2000$$ Set the equation equal to zero.

$$0 = (x-40)(x+50)$$ Factor.

$x = 40$ or ~~$x = -50$~~

Because a rate of speed cannot be negative, reject $x = -50$. Therefore, the speed of the motorist in the rainstorm is 40 mph. Because $x + 10 = 40 + 10 = 50$, the average speed for the return trip in sunny weather is 50 mph.

Avoiding Mistakes

The equation

$$\frac{100}{x} - \frac{100}{x+10} = \frac{1}{2}$$

is not a proportion because the left-hand side has more than one fraction. Do not try to multiply the cross products. Instead, multiply by the LCD to clear fractions.

5. "Work" Problems

Example 7 demonstrates how work rates are related to a portion of a job that can be completed in one unit of time.

example 7 **Using a Rational Equation in a Work Problem**

A new printing press can print the morning edition in 2 hours, whereas the old printer required 4 hours. How long would it take to print the morning edition if both printers were working together?

Solution:

Let x represent the time required for both printers working together to complete the job.

One method to approach this problem is to determine the portion of the job that each printer can complete in 1 hour and extend that rate to the portion of the job completed in x hours.

- The old printer can perform the job in 4 hours. Therefore, it completes $\frac{1}{4}$ of the job in 1 hour and $\frac{1}{4}x$ jobs in x hours.

- The new printer can perform the job in 2 hours. Therefore, it completes $\frac{1}{2}$ of the job in 1 hour and $\frac{1}{2}x$ jobs in x hours.

	Work Rate	Time	Portion of Job Completed
Old Printer	$\frac{1 \text{ job}}{4 \text{ hr}}$	x hours	$\frac{1}{4}x$
New Printer	$\frac{1 \text{ job}}{2 \text{ hr}}$	x hours	$\frac{1}{2}x$

The sum of the portions of the job completed by each printer must equal one whole job.

$$\begin{pmatrix}\text{Portion of job}\\ \text{completed by}\\ \text{old printer}\end{pmatrix} + \begin{pmatrix}\text{portion of job}\\ \text{completed by}\\ \text{new printer}\end{pmatrix} = \begin{pmatrix}1\\ \text{whole}\\ \text{job}\end{pmatrix}$$

$$\frac{1}{4}x + \frac{1}{2}x = 1 \qquad \text{The LCD is 4.}$$

$$4\left(\frac{1}{4}x + \frac{1}{2}x\right) = 4(1) \qquad \text{Multiply by the LCD.}$$

$$4 \cdot \frac{1}{4}x + 4 \cdot \frac{1}{2}x = 4 \cdot 1 \qquad \text{Apply the distributive property.}$$

$$x + 2x = 4 \qquad \text{Solve the resulting linear equation.}$$

$$3x = 4$$

$$x = \frac{4}{3} \quad \text{or} \quad x = 1\frac{1}{3}$$

The time required to print the morning edition using both printers is $1\frac{1}{3}$ hr.

section 6.7 Practice Exercises

For Exercises 1–8, determine whether each of the following is an equation or an expression. If it is an equation, solve it. If it is an expression, perform the indicated operation.

1. $\frac{b}{5} + 3 = 9$

2. $\frac{m}{m-1} - \frac{2}{m+3}$

3. $\frac{2}{a+5} + \frac{5}{a^2 - 25}$

4. $\frac{n}{2n+2} + \frac{5}{4n+4}$

5. $\frac{3y+6}{20} \div \frac{4y+8}{8}$

6. $\frac{z^2+z}{24} \cdot \frac{8}{z+1}$

7. $\frac{3}{p+3} = \frac{12p+19}{p^2+7p+12} - \frac{5}{p+4}$

8. $\dfrac{\frac{1}{t^2} + \frac{2}{3}}{\frac{1}{t} - \frac{5}{6}}$

For Exercises 9–18, solve the proportions.

9. $\frac{5}{3} = \frac{a}{8}$

10. $\frac{b}{14} = \frac{3}{8}$

11. $\frac{2}{1.9} = \frac{x}{38}$

12. $\frac{16}{1.3} = \frac{30}{p}$

13. $\frac{y+1}{2y} = \frac{2}{3}$

14. $\frac{w-2}{4w} = \frac{1}{6}$

15. $\dfrac{9}{2z - 1} = \dfrac{3}{z}$

16. $\dfrac{1}{t} = \dfrac{1}{4 - t}$

17. $\dfrac{8}{9a - 1} = \dfrac{5}{3a + 2}$

18. $\dfrac{4p + 1}{3} = \dfrac{2p - 5}{6}$

19. Charles' law describes the relationship between the temperature and volume of a gas held at a constant pressure.

$$\frac{V_i}{V_f} = \frac{T_i}{T_f}$$

a. Solve the equation for V_f.

b. Solve the equation for T_f.

20. The relationship between the area, height, and base of a triangle is given by the proportion

$\dfrac{A}{b} = \dfrac{h}{2}$, where A is area, b is the base, and h is the height.

a. Solve the equation for A.

b. Solve the equation for b.

21. Jennifer shot a 22 on four holes of golf. At this rate, what can she expect her score to be if she plays all 18 holes?

22. A liquid plant food is prepared by using 3 oz for each half gallon of water. At this rate, how many ounces of plant food are required for $1\frac{1}{3}$ gal of water?

23. Geoff has a garden that is 5 ft in length by 3 ft in width in the front yard. He would like a garden with the dimensions in the same proportion in the backyard. If he has a length of 8 ft available, how wide should he make the garden?

24. Cooking oatmeal requires 1 cup of water for every $\frac{1}{2}$ cup of oats. How many cups of water will be required for $\frac{3}{4}$ cup of oats?

25. A map has a scale of 75 miles/in. If two cities measure 3.5 in. apart, how many miles does this represent?

26. A map has a scale of 50 miles/in. If two cities measure 6.5 in. apart, how many miles does this represent?

27. If 12 out of 80 M&Ms are red, then how many red M&Ms would you expect to find in a bag containing 200 M&Ms?

28. In a can of mixed nuts there are approximately three peanuts out of every five nuts. If the can holds 400 nuts, how many are expected to be peanuts?

29. In a sample of ballots, it is found that 8 ballots out of 5000 are incorrectly marked and must be thrown out. How many incorrectly marked ballots would be expected in a state where 2,600,000 ballots are cast?

30. To conduct an election poll, a sample of 2000 voters is randomly selected. It is found that 100 of the people surveyed would not vote for either the Republican or Democratic presidential nominee. How many voters would not be expected to vote for the Democratic or Republican presidential nominee in a county where 800,000 votes are cast?

31. A boat travels 54 miles upstream against the current in the same amount of time it takes to travel 66 miles downstream with the current. If the boat travels 20 mph in still water, what is the speed of the current? (Use $t = \frac{d}{r}$ to complete the table.)

	Distance	Rate	Time
With the current (downstream)			
Against the current (upstream)			

Table for Exercise 31

32. A fisherman travels 9 miles downstream with the current in the same time that he travels 3 miles upstream against the current. If the speed of the current is 6 mph, what is the speed at which the fisherman travels in still water?

33. A plane flies 370 miles with the wind in the same time that it takes to fly 290 miles against the wind. If the speed of the wind is 20 mph, what is the speed of the plane in still air?

34. A plane flies 630 miles with the wind in the same time that it takes to fly 455 miles against the wind. If this plane flies at the rate of 217 mph in still air, what is the speed of the wind?

35. One motorist travels 15 mph faster than another. The faster driver can cover 360 miles in the same time as the slower driver covers 270 miles. What are the speeds of the two motorists?

36. A train can travel 325 miles in the same time an express bus can travel 200 miles. If the speed of the bus is 25 mph slower than the speed of the train, what is the speed of the train?

37. Devon can cross-country ski 5 km/hr faster than his sister Shanelle. Devon skis 45 km in the same time Shanelle skis 30 km. Find their speeds.

38. Brooke walks 2 km/hr slower than her older sister Adrianna. Brooke can walk 12 km in the same amount of time that Adriana can walk 18 km. Find their speeds.

39. If it takes a person 2 hr to paint a room, what fraction of the room would be painted in 1 hr?

40. If it takes a copier 3 hr to complete a job, what fraction of the job would be completed in 1 hr?

41. If the cold-water faucet is left on, the sink will fill in 10 min. If the hot-water faucet is left on, the sink will fill in 12 min. How long would it take the sink to fill if both faucets are left on?

42. The CUT-IT-OUT lawn mowing company consists of two people: Tina and Bill. If Tina cuts a lawn by herself, she can do it in 4 hr. If Bill cuts the same lawn himself, it takes him an hour longer than Tina. How long would it take them if they worked together?

43. A manuscript needs to be printed. One printer can do the job in 50 min and another printer can do the job in 40 min. How long would it take if both printers were used?

44. A pump can empty a small pond in 4 hr. Another more efficient pump can do the job in 3 hr. How long would it take to empty the pond if both pumps were used?

45. Tim and Al are bricklayers. Tim can construct an outdoor grill in 5 days. If Al helps Tim, they can build it in only 2 days. How long would it take Al to build the grill alone?

46. Norma is a new and inexperienced secretary. It takes her 3 hr to prepare a mailing. If her boss helps her, the mailing can be completed in 1 hr. How long would it take the boss to do the job by herself?

For Exercises 47–48, assume that the polygons are similar.

47. ΔABC is similar to ΔDEF.

a. Find the length of side $\overline{EF}$.

b. Find the length of side $\overline{DF}$.

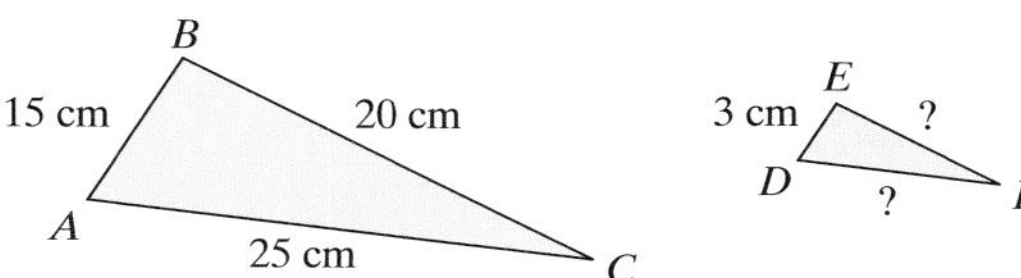

Figures for Exercise 47

48. Figure $ABCD$ is similar to Figure $EFGH$.

a. Find the length of side $\overline{EH}$.

b. Find the length of side $\overline{AB}$.

c. Find the length of side $\overline{BC}$.

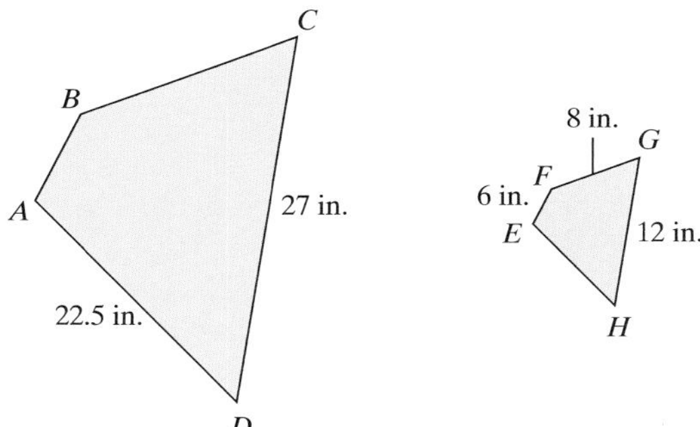

Figures for Exercise 48

For Exercises 49–50, assume that the triangles are similar. Solve for x and y.

49.

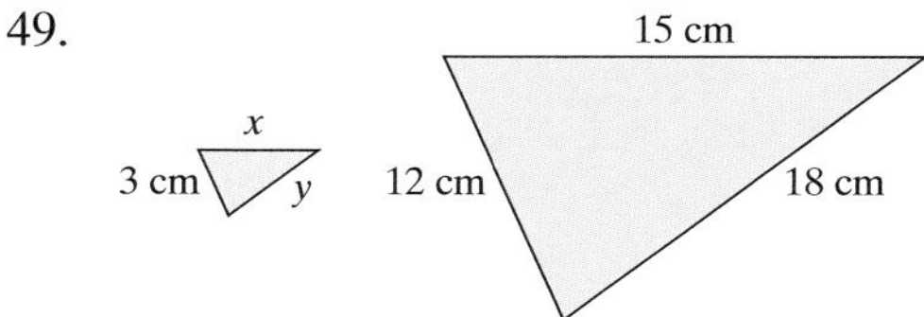

50.

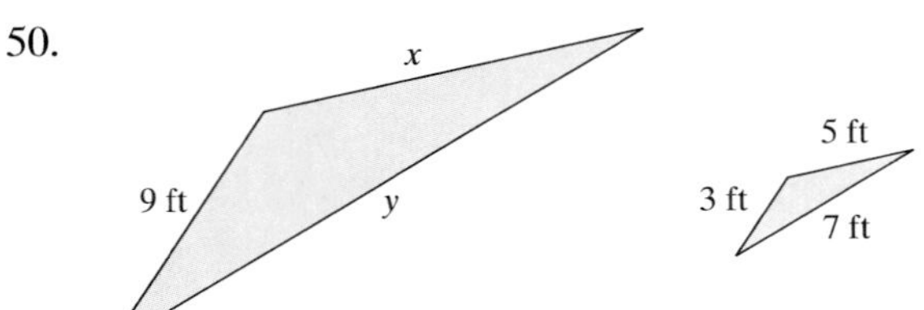

51. To estimate the height of a light pole, a mathematics student measures the length of a shadow cast by a meterstick and the length of the shadow cast by the light pole. Find the height of the light pole (see figure).

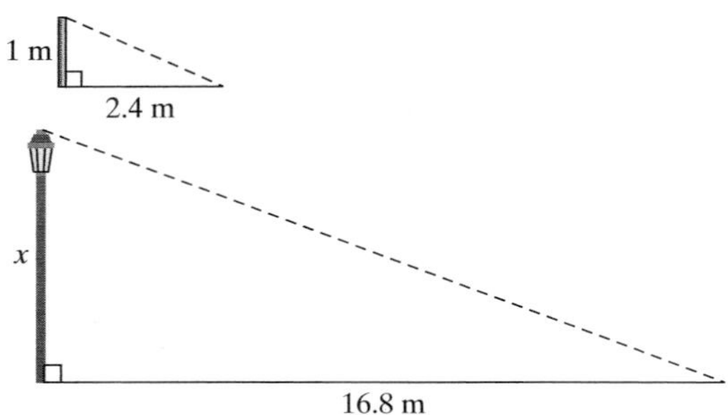

Figures for Exercise 51

52. To estimate the height of a building, a student measures the length of a shadow cast by a yardstick and the length of the shadow cast by the building (see figure). Find the height of the building.

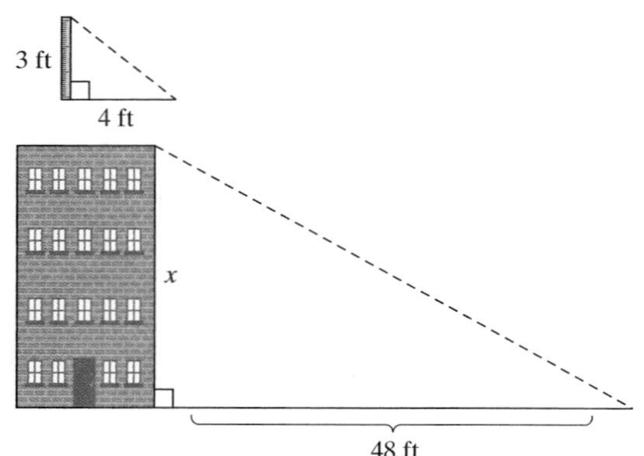

Figures for Exercise 52

53. A 6-ft tall man standing 54 ft from a light post casts an 18-ft shadow. What is the height of the light post?

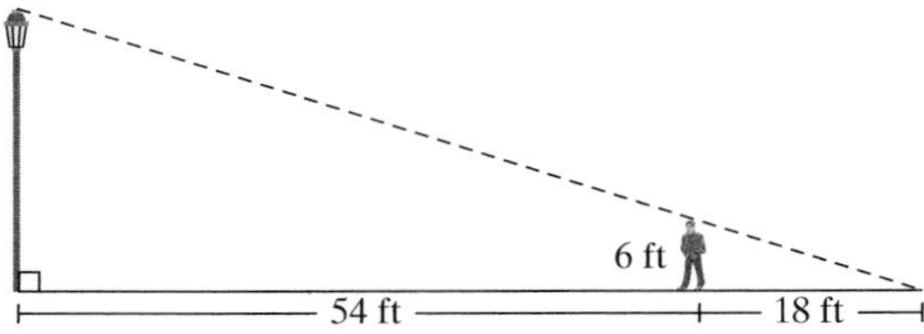

Figure for Exercise 53

54. For a science project at school, a student must measure the height of a tree. The student measures the length of the shadow of the tree and then measures the length of the shadow cast by a yardstick. Use similar triangles to find the height of the tree.

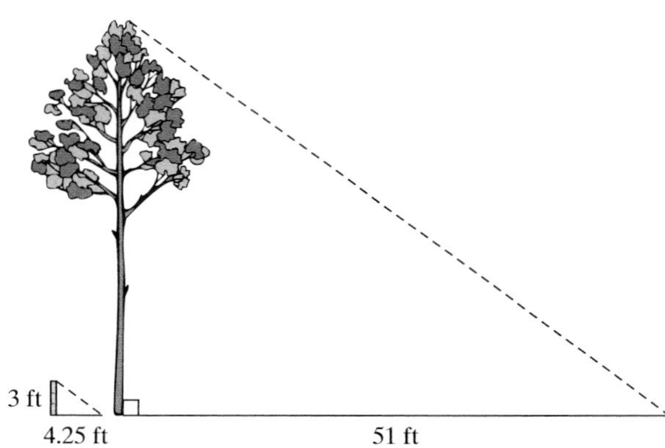

Figure for Exercise 54

chapter 6 SUMMARY

SECTION 6.1—INTRODUCTION TO RATIONAL EXPRESSIONS

KEY CONCEPTS:

A rational expression is a ratio of the form $\frac{p}{q}$, where p and q are polynomials and $q \neq 0$.

The domain* of an algebraic expression consists of the real numbers that when substituted for the variable makes the expression result in a real number.

*For a rational expression the domain is all real numbers except those that make the denominator zero.

Reducing a Rational Expression to Lowest Terms:

Factor the numerator and denominator completely and reduce factors whose ratio is equal to 1 or to -1. A rational expression written in lowest terms will still have the same restrictions on the domain as the original expression.

KEY TERMS:

domain
rational expression

EXAMPLES:

$$\frac{x+2}{x^2-5x-14}$$ is a rational expression.

Find the Domain: $\frac{x+2}{x^2-5x-14}$

$$= \frac{x+2}{(x+2)(x-7)} \qquad \text{Factor the denominator.}$$

The domain is $\{x \mid x \neq -2, x \neq 7\}$.

Reduce to lowest terms: $\frac{x+2}{x^2-5x-14}$

$$= \frac{\overset{1}{\cancel{x+2}}}{\cancel{(x+2)}(x-7)} \qquad \text{Factor and reduce.}$$

$$= \frac{1}{x-7}$$

SECTION 6.2—MULTIPLICATION AND DIVISION OF RATIONAL EXPRESSIONS

KEY CONCEPTS:

Multiplying Rational Expressions:

Factor the numerator and denominator in all expressions completely. Then reduce factors whose ratio is 1 or -1.

EXAMPLES:

Multiply: $\frac{b^2-a^2}{a^2-2ab+b^2} \cdot \frac{a^2-3ab+2b^2}{2a+2b}$

$$= \frac{\overset{-1}{\cancel{(b-a)}}\overset{1}{\cancel{(b+a)}}}{\cancel{(a-b)}\cancel{(a-b)}} \cdot \frac{(a-2b)\overset{1}{\cancel{(a-b)}}}{2\cancel{(a+b)}}$$

$$= -\frac{a-2b}{2} \quad \text{or} \quad \frac{2b-a}{2}$$

Dividing Two Rational Expressions:

Multiply the first expression by the reciprocal of the second expression. That is, for $q \neq 0$, $r \neq 0$, and $s \neq 0$,

$$\frac{p}{q} \div \frac{r}{s} = \frac{p}{q} \cdot \frac{s}{r}$$

Divide: $\dfrac{2c^2d^5}{15e^4} \div \dfrac{6c^4d^3}{20e}$

$$= \frac{2c^2d^5}{15e^4} \cdot \frac{20e}{6c^4d^3}$$

$$= \frac{40c^2d^5e}{90c^4d^3e^4}$$

$$= \frac{4d^2}{9c^2e^3}$$

Section 6.3—Least Common Denominator

Key Concepts:

Converting a Rational Expression to an Equivalent Expression with a Different Denominator:
Multiply numerator and denominator of the rational expression by the missing factors necessary to create the desired denominator.

Finding the Least Common Denominator (LCD) of Two or More Rational Expressions:

1. Factor all denominators completely.
2. The LCD is the product of unique factors from the denominators, where each factor is raised to its highest power.

Key Terms:

equivalent rational expressions
least common denominator (LCD)

Examples:

Convert $\dfrac{-3}{x-2}$ to an equivalent expression with the indicated denominator:

$$\frac{-3}{x-2} = \frac{\qquad}{5x^2 - 20}$$

$$\frac{-3}{x-2} = \frac{\qquad}{5(x^2 - 4)} \qquad \text{Factor.}$$

$$\frac{-3}{x-2} = \frac{\qquad}{5(x-2)(x+2)}$$

Multiply numerator and denominator by the missing factors from the denominator.

$$\frac{-3 \cdot 5(x+2)}{(x-2) \cdot 5(x+2)} = \frac{-15x - 30}{5(x-2)(x+2)}$$

Identify the LCD: $\dfrac{1}{8x^3y^2z}; \dfrac{5}{6xy^4}$

1. Write the denominators as a product of prime factors:

$$\frac{1}{2^3x^3y^2z}; \frac{5}{2 \cdot 3xy^4}$$

2. The LCD is $2^33x^3y^4z$ or $24x^3y^4z$

Section 6.4—Addition and Subtraction of Rational Expressions

Key Concepts:

To add or subtract rational expressions, the expressions must have the same denominator.

Steps to Add or Subtract Rational Expressions:

1. Factor the denominators of each rational expression.
2. Identify the LCD.
3. Rewrite each rational expression as an equivalent expression with the LCD as its denominator.
4. Add or subtract the numerators and write the result over the common denominator.
5. Simplify and reduce to lowest terms.

Examples:

Subtract: $\dfrac{c}{c^2 - c - 12} - \dfrac{1}{2c - 8}$

$$= \frac{c}{(c-4)(c+3)} - \frac{1}{2(c-4)}$$

The LCD is $2(c - 4)(c + 3)$.

$$= \frac{2c}{2(c-4)(c+3)} - \frac{1(c+3)}{2(c-4)(c+3)}$$

$$= \frac{2c - (c+3)}{2(c-4)(c+3)}$$

$$= \frac{2c - c - 3}{2(c-4)(c+3)} = \frac{c-3}{2(c-4)(c+3)}$$

Section 6.5—Complex Fractions

Key Concepts:

Complex fractions can be simplified by using Method I or Method II.

Method I

1. Simplify the numerator and denominator of the complex fraction separately to form a single fraction in the numerator and a single fraction in the denominator.
2. Perform the division represented by the complex fraction. (Multiply the numerator of the complex fraction by the reciprocal of the denominator of the complex fraction.)
3. Simplify and reduce to lowest terms if possible.

Examples:

Simplify Using Method I: $\dfrac{1 - \dfrac{4}{w^2}}{1 - \dfrac{1}{w} - \dfrac{6}{w^2}}$

$$\frac{1 - \dfrac{4}{w^2}}{1 - \dfrac{1}{w} - \dfrac{6}{w^2}} = \frac{\dfrac{w^2}{w^2} - \dfrac{4}{w^2}}{\dfrac{w^2}{w^2} - \dfrac{w}{w^2} - \dfrac{6}{w^2}}$$

$$= \frac{\dfrac{w^2 - 4}{w^2}}{\dfrac{w^2 - w - 6}{w^2}} = \frac{w^2 - 4}{w^2} \cdot \frac{w^2}{w^2 - w - 6}$$

$$= \frac{(w-2)\cancel{(w+2)}}{\cancel{w^2}} \cdot \frac{\cancel{w^2}}{(w-3)\cancel{(w+2)}}$$

$$= \frac{w-2}{w-3}$$

Method II

1. Multiply the numerator and denominator of the complex fraction by the LCD of all individual fractions within the expression.
2. Apply the distributive property and simplify the result.
3. Reduce to lowest terms if necessary.

KEY TERM:

complex fraction

Simplify Using Method II: $\dfrac{1 - \dfrac{4}{w^2}}{1 - \dfrac{1}{w} - \dfrac{6}{w^2}}$

$$= \frac{w^2\left(1 - \dfrac{4}{w^2}\right)}{w^2\left(1 - \dfrac{1}{w} - \dfrac{6}{w^2}\right)} = \frac{w^2 - 4}{w^2 - w - 6}$$

$$= \frac{(w-2)\cancel{(w+2)}}{(w-3)\cancel{(w+2)}} = \frac{w-2}{w-3}$$

SECTION 6.6—RATIONAL EQUATIONS

KEY CONCEPTS:

An equation with one or more rational expressions is called a rational equation.

Steps to Solve a Rational Equation

1. Factor the denominators of all rational expressions.
2. Identify the LCD of all expressions in the equation.
3. Multiply both sides of the equation by the LCD to clear fractions.
4. Solve the resulting equation.
5. Check each potential solution in the original equation.

KEY TERMS:

linear equation
quadratic equation
rational equation

EXAMPLES:

Solve: $\dfrac{1}{w} - \dfrac{1}{2w-1} = \dfrac{-2w}{2w-1}$

The LCD is $w(2w - 1)$.

$$\cancel{w}(2w-1)\frac{1}{\cancel{w}} - w\cancel{(2w-1)}\frac{1}{\cancel{2w-1}} = w\cancel{(2w-1)}\frac{-2w}{\cancel{2w-1}}$$

$$(2w-1)1 - w(1) = w(-2w)$$

$$2w - 1 - w = -2w^2 \qquad \text{Quadratic equation}$$

$$2w^2 + w - 1 = 0$$

$$(2w-1)(w+1) = 0$$

$w \not= \frac{1}{2}$ or $w = -1$

Does not check. Checks.

Solve for *I*: $q = \dfrac{VQ}{I}$

$$I \cdot q = \frac{VQ}{\cancel{I}} \cdot \cancel{I}$$

$$Iq = VQ$$

$$I = \frac{VQ}{q}$$

Section 6.7—Applications of Rational Expressions and Proportions

Key Concepts:

An equation that equates two ratios is called a proportion:

$$\frac{a}{b} = \frac{c}{d} \quad (b \neq 0, d \neq 0)$$

The cross products of a proportion are equal. $ad = bc$.

Applications of Rational Equations.

1. Applications involving $d = rt$ (Distance = rate · time)

Solve:

Two cars travel from Los Angeles to Las Vegas. One car travels an average of 8 mph faster than the other car. If the faster car travels 189 miles in the same time as the slower car travels 165 miles, what is the average speed of each car?

Let r represent the speed of the slower car.
Let $r + 8$ represent the speed of the faster car.

	Distance	Rate	Time
Slower car	165	r	$\frac{165}{r}$
Faster car	189	$r + 8$	$\frac{189}{r + 8}$

$$\frac{165}{r} = \frac{189}{r + 8}$$

$$165(r + 8) = 189r$$

$$165r + 1320 = 189r$$

$$1320 = 24r$$

$$55 = r$$

The slower car travels at 55 mph and the faster car travels $55 + 8 = 63$ mph.

Examples:

Solve:

A 90-g serving of a particular ice-cream contains 10 g of fat. How much fat does 400 g of the same ice-cream contain?

$$\frac{10 \text{ g fat}}{90 \text{ g ice-cream}} = \frac{x \text{ grams fat}}{400 \text{ g ice-cream}}$$

$$\frac{10}{90} = \frac{x}{400}$$

$$10 \cdot 400 = 90 \cdot x$$

$$4000 = 90x$$

$$x = \frac{4000}{90} = \frac{400}{9} \approx 44.4 \text{ g}$$

2. Applications involving work.

Solve:

Beth and Cecelia have a housecleaning business. Beth can clean a particular house in 5 hr by herself. Cecelia can clean the same house in 4 hr. How long would it take if they cleaned the house together?

Let x be the number of hours it takes for both Beth and Cecelia to clean the house.

Beth can clean $\frac{1}{5}$ of the house in an hour and $\frac{x}{5}$ of the house in x hours.

Cecelia can clean $\frac{1}{4}$ of the house in an hour and $\frac{x}{4}$ of the house in x hours.

Key Terms:

proportion
ratio
similar triangles

$\frac{x}{5} + \frac{x}{4} = 1$ Together they clean one whole house.

$20\left(\frac{x}{5} + \frac{x}{4}\right) = (1)20$

$4x + 5x = 20$

$9x = 20$

$x = \frac{20}{9}$

It takes $2\frac{2}{9}$ hr or $2.\overline{2}$ hr working together.

chapter 6 REVIEW EXERCISES

Section 6.1

1. For the rational expression $\frac{t - 2}{t + 9}$
   a. Evaluate the expression (if possible) for $t = 0, 1, 2, -3, -9$
   b. Write the domain of the expression in set-builder notation.

2. For the rational expression $\frac{k + 1}{k - 5}$
   a. Evaluate the expression (if possible) for $k = 0, 1, 5, -1, -2$
   b. Write the domain of the expression in set-builder notation.

3. Which of the rational expressions are equal to -1 for all values of x for which the expressions are defined?
   a. $\frac{2 - x}{x - 2}$ b. $\frac{x - 5}{x + 5}$
   c. $\frac{-x - 7}{x + 7}$ d. $\frac{x^2 - 4}{4 - x^2}$

For Exercises 4–13, write the domain in set-builder notation. Then reduce the expression to lowest terms.

4. $\frac{x - 3}{(2x - 5)(x - 3)}$ 5. $\frac{h + 7}{(3h + 1)(h + 7)}$

6. $\frac{4a^2 + 7a - 2}{a^2 - 4}$ 7. $\frac{2w^2 + 11w + 12}{w^2 - 16}$

8. $\frac{z^2 - 4z}{8 - 2z}$ 9. $\frac{15 - 3k}{2k^2 - 10k}$

10. $\frac{2b^2 + 4b - 6}{4b + 12}$ 11. $\frac{3m^2 - 12m - 15}{9m + 9}$

12. $\frac{n + 3}{n^2 + 6n + 9}$ 13. $\frac{p + 7}{p^2 + 14p + 49}$

Section 6.2

For Exercises 14–27, multiply or divide as indicated.

14. $\frac{3y^3}{3y - 6} \cdot \frac{y - 2}{y}$ 15. $\frac{2u + 10}{u} \cdot \frac{u^3}{4u + 20}$

16. $\frac{11}{v - 2} \cdot \frac{2v^2 - 8}{22}$ 17. $\frac{8}{x^2 - 25} \cdot \frac{3x + 15}{16}$

18. $\frac{4c^2 + 4c}{c^2 - 25} \div \frac{8c}{c^2 - 5c}$

19. $\frac{q^2 - 5q + 6}{2q + 4} \div \frac{2q - 6}{q + 2}$

20. $\left(\frac{-2t}{t + 1}\right)(t^2 - 4t - 5)$

21. $(s^2 - 6s + 8)\left(\frac{4s}{s - 2}\right)$

22. $\dfrac{\dfrac{a^2 + 5a + 1}{7a - 7}}{\dfrac{a^2 + 5a + 1}{a - 1}}$

23. $\dfrac{\dfrac{n^2 + n + 1}{n^2 - 4}}{\dfrac{n^2 + n + 1}{n + 2}}$

24. $\dfrac{5h^2 - 6h + 1}{h^2 - 1} \div \dfrac{16h^2 - 9}{4h^2 + 7h + 3} \cdot \dfrac{3 - 4h}{30h - 6}$

25. $\dfrac{3m - 3}{6m^2 + 18m + 12} \cdot \dfrac{2m^2 - 8}{m^2 - 3m + 2} \div \dfrac{m + 3}{m + 1}$

26. $\dfrac{x - 2}{x^2 - 3x - 18} \cdot \dfrac{6 - x}{x^2 - 4}$

27. $\dfrac{4y^2 - 1}{1 + 2y} \div \dfrac{y^2 - 4y - 5}{5 - y}$

Section 6.3

For Exercises 28–33, fill in the blank to convert each expression into an equivalent expression with the indicated denominator.

28. $\dfrac{x + 1}{x - 2} = \dfrac{\quad}{5x - 10}$

29. $\dfrac{y + 2}{y - 3} = \dfrac{\quad}{2y - 6}$

30. $\dfrac{6}{w} = \dfrac{\quad}{w^2 - 4w}$

31. $\dfrac{2}{r} = \dfrac{\quad}{r^2 + 3r}$

32. $\dfrac{s - 2}{s + 4} = \dfrac{\quad}{s^2 - 16}$

33. $\dfrac{u + 1}{u + 6} = \dfrac{\quad}{u^2 - 36}$

For Exercises 34–41, identify the LCD.

34. $\dfrac{2}{a^2bc^2}, \dfrac{5}{ab^3}$

35. $\dfrac{6x}{y^2z}, \dfrac{3}{xy^2z^4}$

36. $\dfrac{5}{p + 2}, \dfrac{p}{p - 4}$

37. $\dfrac{6}{q}, \dfrac{1}{q + 8}$

38. $\dfrac{8}{m^2 - 16}, \dfrac{7}{m^2 - m - 12}$

39. $\dfrac{6}{n^2 - 9}, \dfrac{5}{n^2 - n - 6}$

40. $\dfrac{4}{2t - 5}, \dfrac{5}{5 - 2t}$

41. $\dfrac{-2}{3k - 1}, \dfrac{6}{1 - 3k}$

42. State two possible LCDs that could be used to add the fractions.

$$\frac{7}{c - 2} + \frac{4}{2 - c}$$

43. State two possible LCDs that could be used to subtract the fractions:

$$\frac{10}{3 - x} - \frac{5}{x - 3}$$

Section 6.4

For Exercises 44–55, add or subtract as indicated.

44. $\dfrac{h + 3}{h + 1} + \dfrac{h - 1}{h + 1}$

45. $\dfrac{b - 6}{b - 2} + \dfrac{b + 2}{b - 2}$

46. $\dfrac{a^2}{a - 5} - \dfrac{25}{a - 5}$

47. $\dfrac{x^2}{x + 7} - \dfrac{49}{x + 7}$

48. $\dfrac{y}{y^2 - 81} + \dfrac{2}{9 - y}$

49. $\dfrac{3}{4 - t^2} + \dfrac{t}{2 - t}$

50. $\dfrac{4}{3m} - \dfrac{1}{m + 2}$

51. $\dfrac{5}{2r + 12} - \dfrac{1}{r}$

52. $\dfrac{4p}{p^2 + 6p + 5} - \dfrac{3p}{p^2 + 5p + 4}$

53. $\dfrac{3q}{q^2 + 7q + 10} - \dfrac{2q}{q^2 + 6q + 8}$

54. $\dfrac{1}{h} + \dfrac{h}{2h + 4} - \dfrac{2}{h^2 + 2h}$

55. $\dfrac{x}{3x + 9} - \dfrac{3}{x^2 + 3x} + \dfrac{1}{x}$

Section 6.5

For Exercises 56–63, simplify the complex fractions.

56. $\dfrac{\dfrac{a - 4}{3}}{\dfrac{a - 2}{3}}$

57. $\dfrac{\dfrac{z + 5}{z}}{\dfrac{z - 5}{3}}$

58. $\dfrac{\dfrac{2 - 3w}{2}}{\dfrac{2}{w} - 3}$

59. $\dfrac{\dfrac{2}{y} + 6}{\dfrac{3y + 1}{4}}$

60. $\dfrac{\dfrac{y}{x} - \dfrac{x}{y}}{\dfrac{1}{x} + \dfrac{1}{y}}$

61. $\dfrac{\dfrac{b}{a} - \dfrac{a}{b}}{\dfrac{1}{b} - \dfrac{1}{a}}$

62. $\dfrac{\dfrac{6}{p+2}+4}{\dfrac{8}{p+2}-4}$

63. $\dfrac{\dfrac{25}{k+5}+5}{\dfrac{5}{k+5}-5}$

Section 6.6

For Exercises 64–71, solve the equations.

64. $\dfrac{2}{x}+\dfrac{1}{2}=\dfrac{1}{4}$

65. $\dfrac{1}{y}+\dfrac{3}{4}=\dfrac{1}{4}$

66. $\dfrac{2}{h-2}+1=\dfrac{h}{h+2}$

67. $\dfrac{w}{w-1}=\dfrac{3}{w+1}+1$

68. $\dfrac{t+1}{3}-\dfrac{t-1}{6}=\dfrac{1}{6}$

69. $\dfrac{4p-4}{p^2+5p-14}+\dfrac{2}{p+7}=\dfrac{1}{p-2}$

70. $\dfrac{1}{z+2}=\dfrac{4}{z^2-4}-\dfrac{1}{z-2}$

71. $\dfrac{y+1}{y+3}=\dfrac{y^2-11y}{y^2+y-6}-\dfrac{y-3}{y-2}$

72. Four times a number is added to 5. The sum is then divided by 6. The result is $\frac{7}{2}$. Find the number.

73. Solve the formula $\dfrac{V}{h}=\dfrac{\pi r^2}{3}$ for h.

74. Solve the formula $\dfrac{A}{b}=\dfrac{h}{2}$ for b.

Section 6.7

For Exercises 75–76, solve the proportions.

75. $\dfrac{m+2}{8}=\dfrac{m}{3}$

76. $\dfrac{a-3}{4a}=\dfrac{2}{a+7}$

77. A bag of popcorn states that it contains 4 g of fat per serving. If a serving is 2 oz, how many grams of fat are in a 6-oz bag?

78. Bud goes 10 mph faster on his Harley Davidson motorcycle than Ed goes on his Honda motorcycle. If Bud travels 105 miles in the same time that Ed travels 90 miles, what are the rates of the two bikers?

79. There are two pumps set up to fill a small swimming pool. One pump takes 24 min by itself to fill the pool, but the other takes 56 min by itself. How long would it take if both pumps work together?

80. Consider the similar triangles shown below. Find the values of x and y.

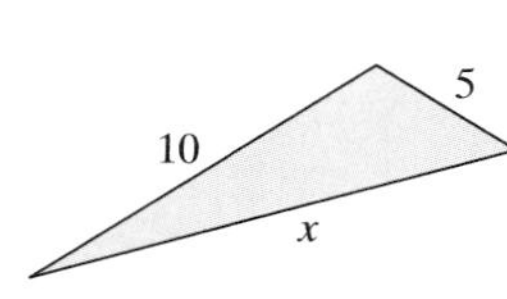

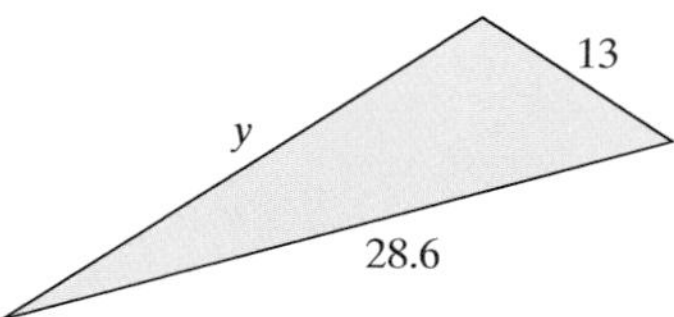

chapter 6 TEST

For Exercises 1–2,

a. Write the domain in set-builder notation.

b. Reduce the rational expression to lowest terms.

1. $\dfrac{5(x-2)(x+1)}{30(2-x)}$

2. $\dfrac{7a^2-42a}{a^3-4a^2-12a}$

3. Identify the rational expressions that are equal to -1 for all values of x for which the expression is defined.

   a. $\dfrac{x+4}{x-4}$

   b. $\dfrac{7-2x}{2x-7}$

   c. $\dfrac{9x^2+16}{-9x^2-16}$

   d. $-\dfrac{x+5}{x+5}$

For Exercises 4–9, perform the indicated operation.

4. $\dfrac{2}{y^2+4y+3}+\dfrac{1}{3y+9}$

5. $\dfrac{9-b^2}{5b+15}\div\dfrac{b-3}{b+3}$

6. $\dfrac{w^2-4w}{w^2-8w+16}\cdot\dfrac{w-4}{w^2+w}$

7. $\dfrac{t}{t-2}-\dfrac{8}{t^2-4}$

8. $\dfrac{1}{x+4}+\dfrac{2}{x^2+2x-8}+\dfrac{x}{x-2}$

9. $\dfrac{1-\dfrac{4}{m}}{m-\dfrac{16}{m}}$

For Exercises 10–13, solve the equation.

10. $\dfrac{3}{a}+\dfrac{5}{2}=\dfrac{7}{a}$

11. $\dfrac{p}{p-1}+\dfrac{1}{p}=\dfrac{p^2+1}{p^2-p}$

12. $\dfrac{3}{c-2}-\dfrac{1}{c+1}=\dfrac{7}{c^2-c-2}$

13. $\dfrac{4x}{x-4}=3+\dfrac{16}{x-4}$

14. Solve the formula $\dfrac{C}{2}=\dfrac{A}{r}$ for r.

15. If $\frac{3}{2}$ is added to the reciprocal of a number the result is $\frac{2}{5}$ times the reciprocal of that number. Find the number.

16. Solve the proportion.

$$\frac{y+7}{-4}=\frac{1}{4}$$

17. A recipe for vegetable soup calls for $\frac{1}{2}$ cup of carrots for six servings. How many cups of carrots are needed to prepare 15 servings?

18. A motorboat can travel 28 miles downstream in the same amount of time as it can travel 18 miles upstream. Find the speed of the current if the boat can travel 23 mph in still water.

19. Two printers working together can complete a job in 2 hr. If one printer requires 6 hr to do the job alone, how many hours would the second printer need to complete the job alone?

20. Consider the similar triangles shown in the figure. Find the values of a and b.

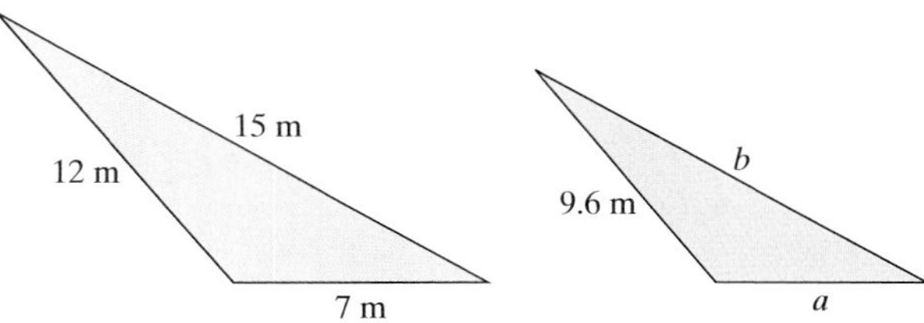

21. Find the LCD of the following pairs of rational expressions.

    a. $\dfrac{x}{3(x+3)}, \dfrac{7}{5(x+3)}$

    b. $\dfrac{-2}{3x^2y}, \dfrac{4}{xy^2}$

CUMULATIVE REVIEW EXERCISES, CHAPTERS 1–6

For Exercises 1–2, simplify completely.

1. $\left(\frac{1}{2}\right)^{-4} + 2^4$

2. $|3 - 5| + |-2 + 7|$

3. Which of the following are rational numbers and which are irrational numbers?

$$\sqrt{4}, \sqrt{5}, \sqrt{9}, \sqrt{16}, \sqrt{20}, \sqrt{49}$$

4. Solve for y: $\frac{1}{2} - \frac{3}{4}(y - 1) = \frac{5}{12}$

5. Solve the inequality. Graph the solution set and express the solution in interval notation: $-3(x - 5) - 2 < -2x + 5$

6. Complete the table.

Set-Builder Notation	Graph	Interval Notation
$\{x \mid x \geq -1\}$	⟶	
	⟶	$(-\infty, 5)$

Table for Exercise 6

7. The perimeter of a rectangular swimming pool is 104 m. The length is 1 m more than twice the width. Find the length and width.

8. The height of a triangle is 2 in. less than the base. The area is 40 in.^{2} Find the base and height of the triangle.

9. Find the values of the vertical angles.

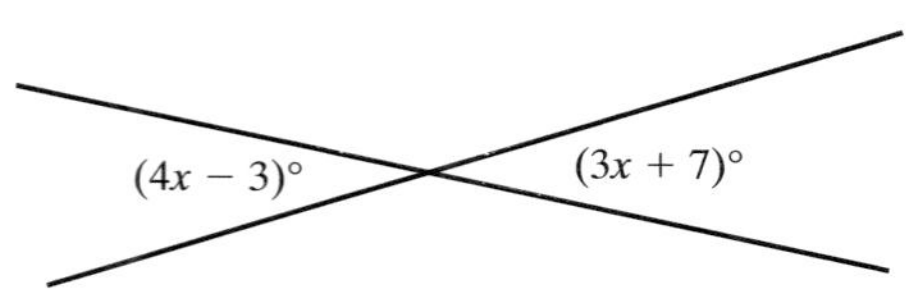

Figure for Exercise 9

10. The mass of a proton is approximately 1.66×10^{-24} g. What is the mass of 6.02×10^{23} protons?

11. Show that the lengths of the sides of the right triangle satisfy the Pythagorean theorem.

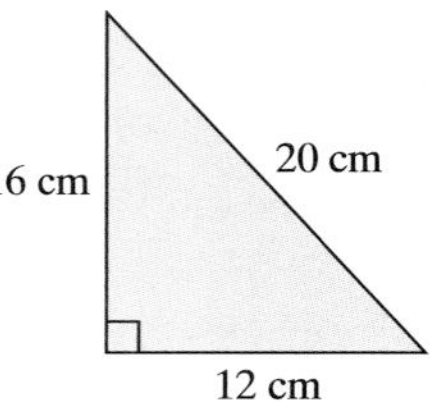

Figure for Exercise 11

For Exercises 12–13, simplify the expressions. Write the final answer with positive exponents only.

12. $\dfrac{(x^{-1})^2x^5}{x^3}$

13. $\left(\dfrac{4x^{-1}y^{-2}}{z^4}\right)^{-2}(2y^{-1}z^3)^3$

14. The length and width of a rectangle are given in terms of x.

   a. Write a polynomial that represents the perimeter of the rectangle.

   b. Write a polynomial that represents the area of the rectangle.

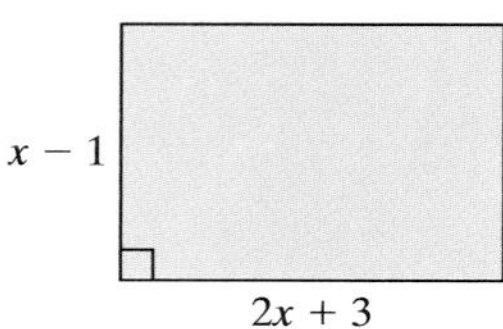

Figure for Exercise 14

15. Perform the indicated operation: $(5x - 3)^2$

16. Factor completely: $25x^2 - 30x + 9$

17. Divide.

$$\frac{8a^2b^4 - 2ab^3 + a^3b^2}{2ab^2}$$

For Exercises 18–20, factor completely.

18. $27x^2 - 75y^2$

19. $10cd + 5d - 6c - 3$

20. $x^2 - x - 20$

21. What is the domain of the expression?

$$\frac{x + 3}{(x - 5)(2x + 1)}$$

22. Reduce to lowest terms.

$$\frac{x^2 - 9}{x^2 + 8x + 15}$$

For Exercises 23–24, perform the indicated operations.

23. $\dfrac{2x - 6}{x^2 - 16} \div \dfrac{10x^2 - 90}{x^2 - x - 12}$

24. $\dfrac{x^2 - 6x}{x - 3} - \dfrac{-9}{x - 3}$

25. Simplify.

$$\frac{\frac{3}{4} - \frac{1}{x}}{\frac{1}{3x} - \frac{1}{4}}$$

26. Solve.

$$\frac{7}{y^2 - 4} = \frac{3}{y - 2} + \frac{2}{y + 2}$$

27. Solve the proportion.

$$\frac{2b - 5}{6} = \frac{4b}{7}$$

28. A small boat can sail at a rate of 18 km per hour in still water. If the boat can go 63 km downstream with the current in the same amount of time it takes to go 45 km upstream against the current, what is the speed of the current?

29. Given the equation $5x - y = 10$.
    a. Is the equation linear or nonlinear?
    b. Complete the table.

x	y
0	
	0
1	

Table for Exercise 29

    c. Graph the line passing through the points given in the table.

30. Given the equation $y = x - 25$.
    a. Is the equation linear or nonlinear?
    b. Find the y-intercept.
    c. Find the x-intercept.

31. Factor completely. $2ab + a^2 + b^2 - 16$

chapter

7

INTRODUCTION TO RELATIONS AND FUNCTIONS

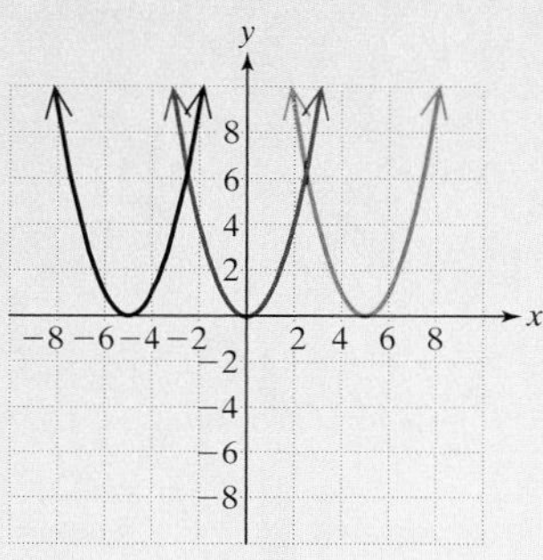

The graphical representation of numerical data can give insight to the relationship among two or more variables. Furthermore, advances in technology have made graphing equations and functions a fast and efficient process. For an equation in two variables, we can plot points by hand. For complicated functions, however, a graphing calculator or computer graphing utility is a convenient tool for analysis.

In this chapter, we investigate several basic functions and their graphs. For more information see the Technology Connections in MathZone at

www.mhhe.com/miller_oneill

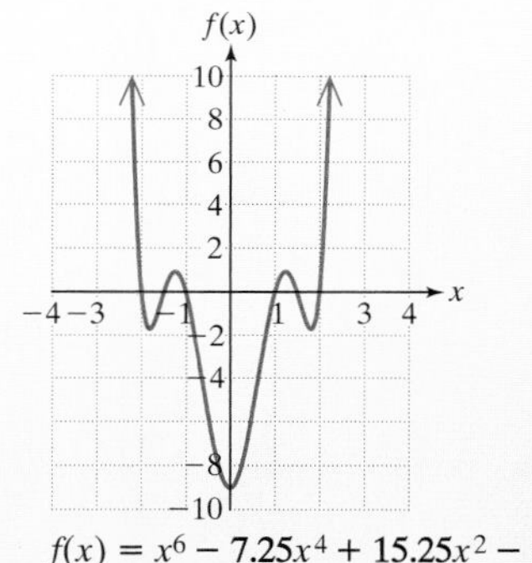

$f(x) = x^6 - 7.25x^4 + 15.25x^2 - 9$

Concepts

1. Graphs of Linear Equations in Two Variables
2. Review of Slope and Slope-Intercept Form
3. Point-Slope Formula
4. Writing an Equation of a Line Using the Point-Slope Formula
5. Writing an Equation of a Line Given Two Points
6. Writing an Equation of a Line Parallel or Perpendicular to Another Line
7. Different Forms of Linear Equations: A Summary

section

7.1 Point-Slope Formula and Review of Graphing

1. Graphs of Linear Equations in Two Variables

In Chapter 3 we learned how to plot points in a rectangular coordinate system. We also learned how to find solutions to a linear equation in two variables. The set of all solutions to such a linear equation is a line in a rectangular coordinate system. In Section 3.3, we defined an x-intercept as a point where a graph intersects the x-axis. Similarly, a y-intercept is a point where a graph intersects the y-axis. Example 1 reviews some of these important concepts learned in Chapter 3.

example 1 Finding x- and y-Intercepts and Graphing a Line

Tip: For a thorough review of Chapter 3, linear equations in two variables, please see Review C in the back of the text.

Given $2x + 4y = 4$,

a. Find the x- and y-intercepts.
b. Graph the line.

Solution:

a. To find the x-intercept, substitute $y = 0$.

$$2x + 4(0) = 4$$
$$2x = 4$$
$$x = 2$$

The x-intercept is $(2, 0)$.

To find the y-intercept substitute $x = 0$.

$$2(0) + 4y = 4$$
$$4y = 4$$
$$y = 1$$

The y-intercept is $(0, 1)$.

b. We can graph the line by plotting the points found in part (a) and graphing the line through the points. While two points are sufficient to graph a line, we can find a third solution to the equation as a check point. We can arbitrarily pick a value for x or y, substitute that value into the equation, and solve for the remaining variable. For example, letting $x = 4$ yields, $2(4) + 4y = 4$ or equivalently, $y = -1$. We have a third point $(4, -1)$ that lines up with the other two points (Figure 7-1).

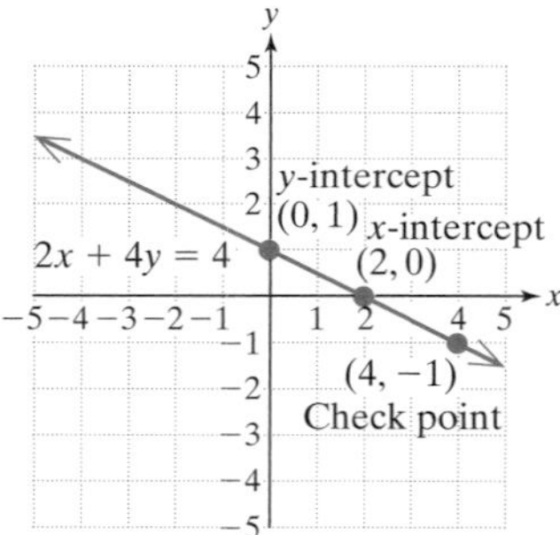

Figure 7-1

2. Review of Slope and Slope-Intercept Form

Given two points on a line (x_1, y_1) and (x_2, y_2) we can compute the slope of the line (denoted by m) from the slope formula.

$$\text{slope formula:} \quad m = \frac{y_2 - y_1}{x_2 - x_1}$$

To find the slope of the line from Example 1, we can use any two points on the line. For example, using the points $(4, -1)$ and $(2, 0)$, we have:

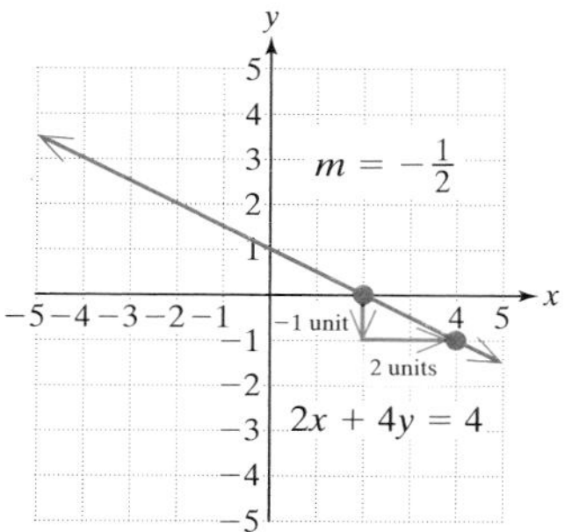

Figure 7-2

$$\overset{(x_1, y_1)}{(4, -1)} \text{ and } \overset{(x_2, y_2)}{(2, 0)} \qquad \text{Label the points.}$$

$$m = \frac{y_2 - y_1}{x_2 - x_1} = \frac{0 - (-1)}{2 - 4} = \frac{1}{-2} = -\frac{1}{2} \qquad \text{Apply the slope formula.}$$

The slope of the line from Example 1 is $-\frac{1}{2}$. The slope measures the ratio of the incremental change in y to the incremental change in x between two points on the line. In this case, a slope of $-\frac{1}{2}$ indicates that there is a change of -1 unit in the y-direction for every 2-unit change in the x-direction (Figure 7-2).

A linear equation written in the form $y = mx + b$ is said to be written in slope-intercept form. The value of m is the slope of the line and $(0, b)$ is the y-intercept.

example 2 Writing an Equation in Slope-Intercept Form

Write the equation $-4x + 3y = 6$ in slope-intercept form. Then identify the slope and y-intercept.

Solution:

$-4x + 3y = 6$	To write the form $y = mx + b$, solve the equation for y.
$3y = 4x + 6$	Add $4x$ to both sides.
$\frac{3y}{3} = \frac{4x}{3} + \frac{6}{3}$	Divide both sides by 3.
$y = \frac{4}{3}x + 2$	Slope-intercept form.

The slope is $\frac{4}{3}$ and the y-intercept is $(0, 2)$.

3. Point-Slope Formula

Given an equation of a line, the slope-intercept form is particularly helpful to identify the slope and y-intercept of the line. In this section, we present another tool called the point-slope formula that enables us to reverse this process. That is, if we know the slope of a line and any point on the line, the point-slope formula helps us construct an equation of the line. The point-slope formula can be derived from the slope formula as follows.

Suppose a line passes through a given point (x_1, y_1) and has slope m. If (x, y) is any other point on the line, then,

$$m = \frac{y - y_1}{x - x_1} \qquad \text{Apply the slope formula.}$$

$$m(x - x_1) = \frac{y - y_1}{\cancel{x - x_1}}\cancel{(x - x_1)} \qquad \text{Clear the fractions.}$$

$$m(x - x_1) = y - y_1$$

or

$$y - y_1 = m(x - x_1) \qquad \text{Point-slope formula}$$

Point-Slope Formula

The **point-slope formula** is given by

$$y - y_1 = m(x - x_1)$$

where m is the slope of the line and (x_1, y_1) is a known point on the line.

4. Writing an Equation of a Line Using the Point-Slope Formula

Example 3 illustrates the use of the point-slope formula for finding an equation of a line when a point on the line and the slope are given.

example 3 **Writing an Equation of a Line Using the Point-Slope Formula**

Use the point-slope formula to find an equation of the line having a slope of 3 and passing through the point $(-2, -4)$. Write the final answer in slope-intercept form.

Solution:

The slope of the line is given: $m = 3$.

A point on the line is given: $(x_1, y_1) = (-2, -4)$.

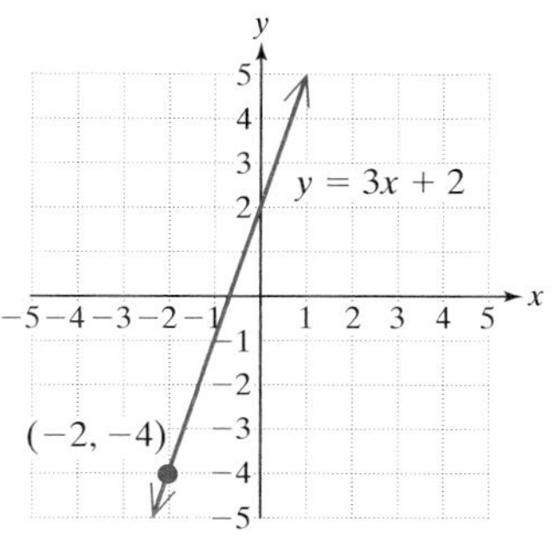

Figure 7-3

The point-slope formula:

$$y - y_1 = m(x - x_1)$$

$y - (-4) = 3[x - (-2)]$ Substitute $m = 3$, $x_1 = -2$, and $y_1 = -4$.

$y + 4 = 3(x + 2)$ Simplify. The final answer is required in slope-intercept form. Simplify the equation and solve for y.

$y + 4 = 3x + 6$ Apply the distributive property.

$y = 3x + 6 - 4$ Subtract 4 from both sides.

$y = 3x + 2$ Slope-intercept form.

The equation $y = 3x + 2$ from Example 3 is graphed in Figure 7-3. Notice that the line does indeed pass through the point $(-2, -4)$.

5. Writing an Equation of a Line Given Two Points

example 4

Writing an Equation of a Line Given Two Points

Use the point-slope formula to find an equation of the line passing through the points $(-2, 5)$ and $(4, -1)$. Write the final answer in slope-intercept form.

Solution:

Given two points on a line, the slope can be found with the slope formula.

$(-2, 5)$ and $(4, -1)$

(x_1, y_1) (x_2, y_2) Label the points.

$$m = \frac{y_2 - y_1}{x_2 - x_1} = \frac{(-1) - (5)}{(4) - (-2)} = \frac{-6}{6} = -1$$

To apply the point-slope formula, use the slope, $m = -1$ and either given point. We will choose the point $(-2, 5)$ as (x_1, y_1).

$y - y_1 = m(x - x_1)$

$y - 5 = -1[x - (-2)]$ Substitute $m = -1$, $x_1 = -2$, $y_1 = 5$.

$y - 5 = -1(x + 2)$ Simplify.

$y - 5 = -x - 2$

$y = -x + 3$ Solve for y.

Tip: The point-slope formula can be applied using either given point for (x_1, y_1). In Example 4, using the point $(4, -1)$ for (x_1, y_1) produces the same result.

$$y - y_1 = m(x - x_1)$$
$$y - (-1) = -1(x - 4)$$
$$y + 1 = -x + 4$$
$$y = -x + 3$$

6. Writing an Equation of a Line Parallel or Perpendicular to Another Line

example 5

Writing an Equation of a Line Parallel to Another Line

Use the point-slope formula to find an equation of the line passing through the point $(-1, 0)$ and parallel to the line $y = -4x + 3$. Write the final answer in slope-intercept form.

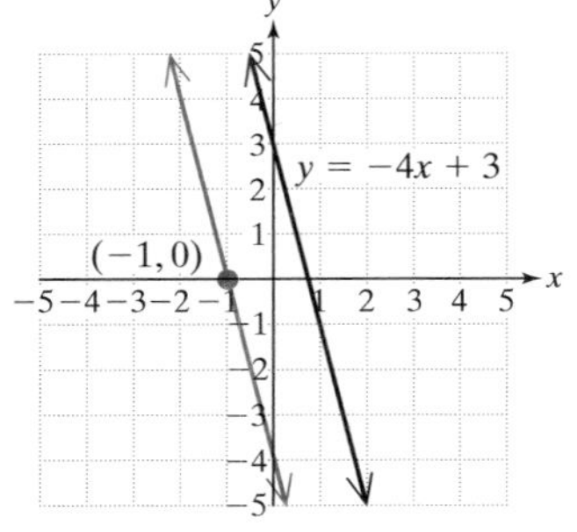

Figure 7-4

Solution:

Figure 7-4 shows the line $y = -4x + 3$ (pictured in black) and a line parallel to it (pictured in blue) that passes through the point $(-1, 0)$. The equation of the given line, $y = -4x + 3$, is written in slope-intercept form, and its slope is easily identified as -4. The line parallel to the given line must also have a slope of -4.

Apply the point-slope formula using $m = -4$ and the point $(x_1, y_1) = (-1, 0)$.

$$y - y_1 = m(x - x_1)$$
$$y - 0 = -4[x - (-1)]$$
$$y = -4(x + 1)$$
$$y = -4x - 4$$

example 6

Writing an Equation of a Line Perpendicular to Another Line

Use the point-slope formula to find an equation of the line passing through the point $(-3, 1)$ and perpendicular to the line $3x + y = -2$. Write the final answer in slope-intercept form.

Solution:

The given line can be written in slope-intercept form as $y = -3x - 2$. The slope of this line is -3. Therefore, the slope of a line perpendicular to the given line is $\frac{1}{3}$.

Apply the point-slope formula with $m = \frac{1}{3}$, and $(x_1, y_1) = (-3, 1)$.

$y - y_1 = m(x - x_1)$	Point-slope formula
$y - (1) = \frac{1}{3}[x - (-3)]$	Substitute $m = \frac{1}{3}$, $x_1 = -3$, and $y_1 = 1$.
$y - 1 = \frac{1}{3}(x + 3)$	To write the final answer in slope-intercept form, simplify the equation and solve for y.
$y - 1 = \frac{1}{3}x + 1$	Apply the distributive property.
$y = \frac{1}{3}x + 2$	Add 1 to both sides.

A graph of the perpendicular lines $y = \frac{1}{3}x + 2$ and $y = -3x - 2$ is shown in Figure 7-5. Notice that the line $y = \frac{1}{3}x + 2$ passes through the point $(-3, 1)$.

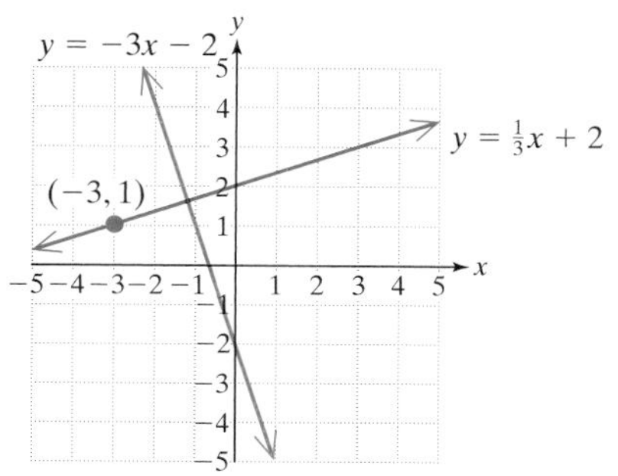

Figure 7-5

7. Different Forms of Linear Equations: A Summary

A linear equation can be written in several different forms as summarized in Table 7-1. In particular, recall from Section 3.3 that an equation written in the form $y = k$ represents a horizontal line in a rectangular coordinate system. Similarly, an equation of the form $x = k$ represents a vertical line. For example,

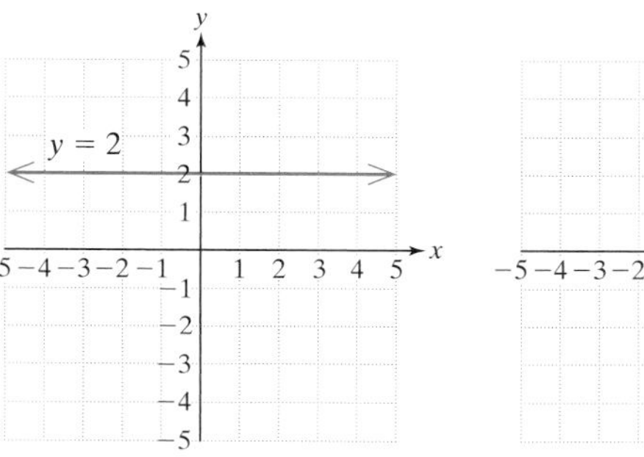

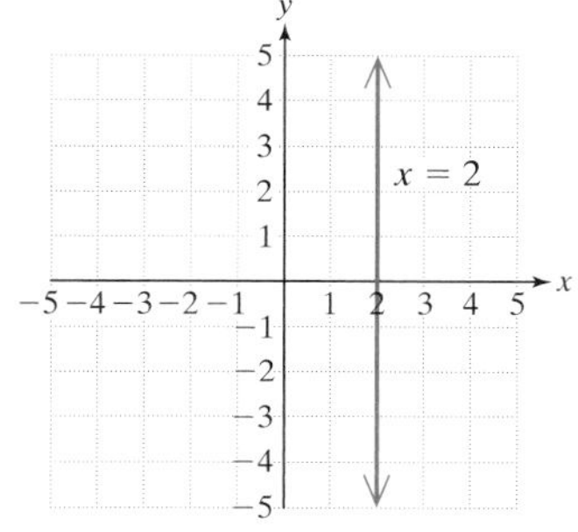

Table 7-1

Form	Example	Comments
Standard Form $ax + by = c$	$4x + 2y = 8$	a and b must not both be zero.
Horizontal Line $y = k$ (k is constant)	$y = 4$	The slope is zero and the y-intercept is $(0, k)$.
Vertical Line $x = k$ (k is constant)	$x = -1$	The slope is undefined and the x-intercept is $(k, 0)$.
Slope-Intercept Form $y = mx + b$ the slope is m y-intercept is $(0, b)$	$y = -3x + 7$ Slope $= -3$ y-intercept is $(0, 7)$	Solving a linear equation for y results in slope-intercept form. The coefficient of the x-term is the slope, and the constant defines the location of the y-intercept.
Point-Slope Formula $y - y_1 = m(x - x_1)$	$m = -3$ $(x_1, y_1) = (4, 2)$ $y - 2 = -3(x - 4)$	This formula is typically used to build an equation of a line when a point on the line is known and the slope of the line is known.

Although it is important to understand and apply slope-intercept form and the point-slope formula, they are not necessarily applicable to all problems, particularly when dealing with a horizontal or vertical line.

example 7 Writing an Equation of a Line

Find an equation of the line passing through the point $(2, -4)$ and parallel to the x-axis.

Solution:

Because the line is parallel to the x-axis, the line must be horizontal. Recall that all horizontal lines can be written in the form $y = k$, where k is a constant. A quick sketch can help find the value of the constant. See Figure 7-6.

Because the line must pass through a point whose y-coordinate is -4, then the equation of the line must be $y = -4$.

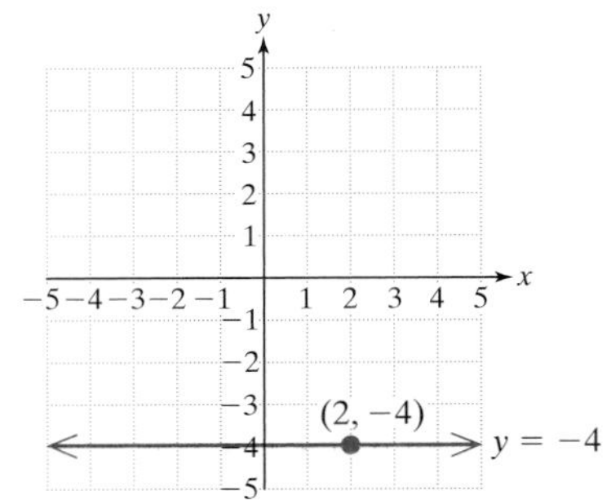

Figure 7-6

section 7.1 Practice Exercises

For Exercises 1–6, find the x- and y-intercept, and graph the line.

1. $4x + 2y = 8$
2. $5x - 2y = 5$
3. $3x - 4y = 6$
4. $3x + 5y = 15$
5. $6x - 2y = 0$
6. $15x + 10y = 0$

For Exercises 7–14, write the slope-intercept form of the line. Then identify the slope and y-intercept.

7. $4x + 2y = 8$
8. $5x - 2y = 5$
9. $9y = 5x + 3$
10. $4x = 6 - 8y$
11. $4x - 3y = 0$
12. $2y + 3x = 0$
13. $\frac{1}{3}x - \frac{2}{5}y = 1$
14. $\frac{1}{2}x + \frac{4}{7}y = 2$

For Exercises 15–20, match the form or formula on the left with its name on the right.

15. $x = k$	i. Standard form
16. $y = mx + b$	ii. Point-slope formula
17. $m = \frac{y_2 - y_1}{x_2 - x_1}$	iii. Horizontal line
18. $y - y_1 = m(x - x_1)$	iv. Vertical line
19. $y = k$	v. Slope-intercept form
20. $ax + by = c$	vi. Slope formula

For Exercises 21–24, find the slope of the line that passes through the given points.

21. $(1, -3)$ and $(2, 6)$

22. $(2, -4)$ and $(-2, 4)$

23. $(-3, -7)$ and $(4, -1)$

24. $(6, -1)$ and $(-5, 2)$

25. Given the points $(-2, 3)$ and $(4, 3)$,
    a. Graph the points and the line through the points.
    b. Find the slope of the line.
    c. Fill in the blank: The slope of a horizontal line is ________.

26. Given the points $(4, -2)$ and $(-1, -2)$,
    a. Graph the points and the line through the points.
    b. Find the slope of the line.
    c. Fill in the blank: The slope of a horizontal line is ________.

27. Given the points $(3, -1)$ and $(3, 3)$,
    a. Graph the points and the line through the points.
    b. Find the slope of the line.
    c. Fill in the blank: The slope of a vertical line is ________.

28. Given the points $(-2, 4)$ and $(-2, -1)$,
    a. Graph the points and the line through the points.
    b. Find the slope of the line.
    c. Fill in the blank: The slope of a vertical line is ________.

For Exercises 29–54, use the point-slope formula (if possible) to write an equation of the line given the following information. Write the final answer in slope-intercept form if possible.

29. The slope is 3 and the line passes through the point $(-2, 1)$.

30. The slope is -2 and the line passes through the point $(1, -5)$.

31. The slope is $\frac{1}{4}$ and the line passes through the point $(-8, 6)$.

32. The slope is $\frac{2}{5}$ and the line passes through the point $(-5, 4)$.

33. The slope is 4.1 and the line passes through the point $(5.3, -2.2)$.

34. The slope is -3.6 and the line passes through the point $(10.0, 8.2)$.

35. The slope is 0 and the line passes through the point $(3, -2)$.

36. The slope is 0 and the line passes through the point $(0, 5)$.

37. The line passes through the points $(-2, -6)$ and $(1, 0)$.

38. The line passes through the points $(-2, 5)$ and $(0, 1)$.

39. The line passes through the points $(1, -3)$ and $(-7, 2)$.

40. The line passes through the points $(0, -4)$ and $(-1, -3)$.

41. The line passes through the points $(2.2, 3.1)$ and $(12.2, -5.3)$.

42. The line passes through the points $(4.75, -2.50)$ and $(-0.25, 6.75)$.

43. The line passes through the point $(3, 1)$ and is parallel to the line $y = -4$. See the figure.

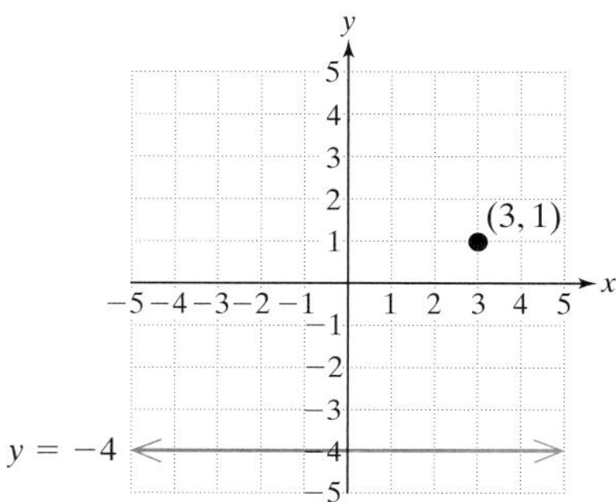

Figure for Exercise 43

44. The line passes through the point $(-1, 1)$ and is parallel to the line $y = 2$. See the figure.

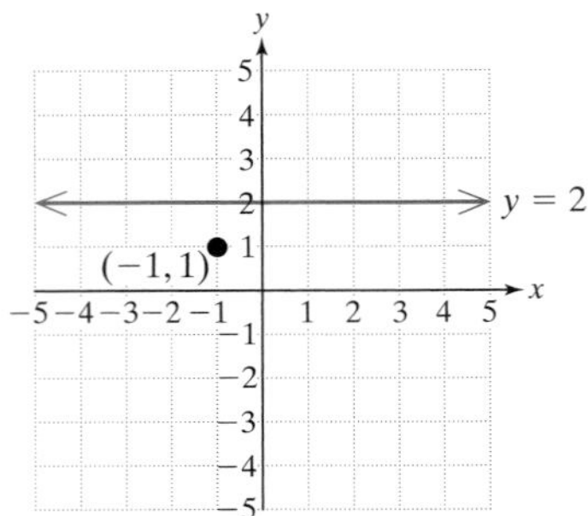

Figure for Exercise 44

45. The line passes through the point $(2, 6)$ and is perpendicular to the line $y = 1$. (*Hint*: Sketch the line first.)

46. The line passes through the point $(0, 3)$ and is perpendicular to the line $y = -5$. (*Hint*: Sketch the line first.)

47. The line passes through the point $(\frac{5}{2}, \frac{1}{2})$ and is parallel to the line $x = 4$.

48. The line passes through the point $(-6, \frac{2}{3})$ and is parallel to the line $x = -2$.

49. The line passes through the point $(2, 2)$ and is perpendicular to the line $x = 0$.

50. The line passes through the point $(5, -2)$ and is perpendicular to the line $x = 0$.

51. The slope is undefined and the line passes through the point $(-6, -3)$.

52. The slope is undefined and the line passes through the point $(2, -1)$.

53. The line passes through the points $(-4, 0)$ and $(-4, 3)$.

54. The line passes through the points $(1, 3)$ and $(1, -4)$.

55. The following table represents the median selling price, y, of new privately owned one-family houses sold in the Midwest from 1980 to 2000.

Let x represent the number of years since 1980. Let y represent price in thousands of dollars.

Year	Price in ($1000)
1980 $x = 0$	67
1985 $x = 5$	84
1990 $x = 10$	108
1995 $x = 15$	142
2000 $x = 20$	167

Table for Exercise 55

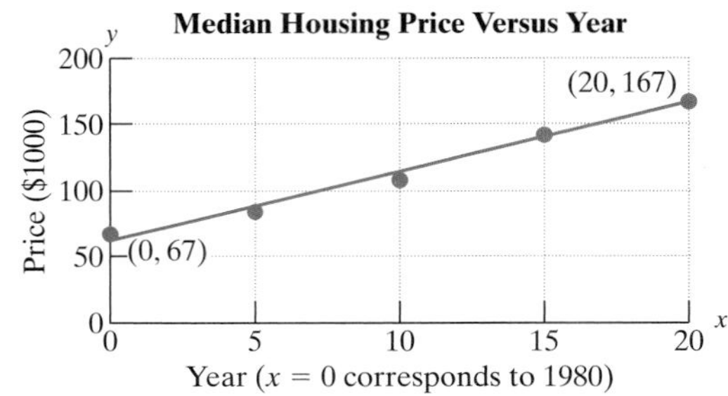

Figure for Exercise 55

Source: U.S. Bureau of Census.

a. Find the slope of the line between the points $(0, 67)$ and $(20, 167)$.

b. Find an equation of the line between the points $(0, 67)$ and $(20, 167)$. Write the answer in slope-intercept form.

c. Use the equation from part (b) to estimate the median price of a one-family house sold in the Midwest in the year 2005.

56. The following table represents the percentage of females, y, who smoked for selected years. Let x represent the number of years since 1965. Let y represent percentage of women who smoked.

Year	Percentage
1965 $x = 0$	33.9
1975 $x = 10$	32.1
1985 $x = 20$	27.9
1995 $x = 30$	23.4

Table for Exercise 56

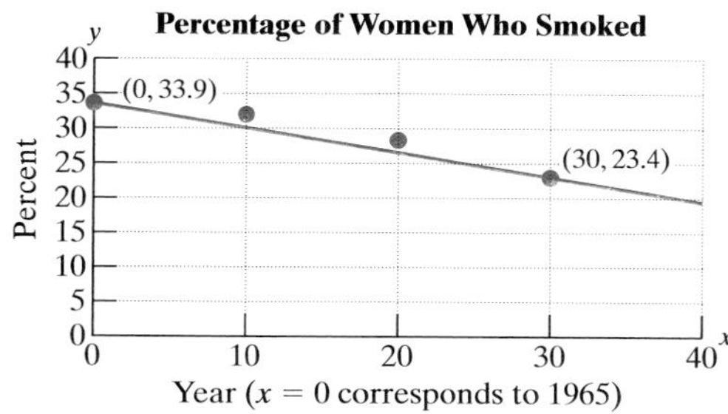

Figure for Exercise 56

Source: U.S. National Center for Health Statistics.

a. Find the slope of the line between the points (0, 33.9) and (30, 23.4).

b. Find an equation of the line between the points (0, 33.9) and (30, 23.4). Write the answer in slope-intercept form.

c. Use the equation from part (b) to estimate the percentage of women who smoked in the year 2000.

For Exercises 57–62, use the point-slope formula to write an equation of the line given the following information. Write the final answer in slope-intercept form if possible.

57. The line passes through the point $(-5, 2)$ and is perpendicular to the line $y = \frac{1}{2}x + 3$.

58. The line passes through the point $(-2, -2)$ and is perpendicular to the line $y = \frac{1}{3}x - 5$.

59. The line passes through the point $(4, 4)$ and is parallel to the line $3x - y = 6$.

60. The line passes through the point $(-1, -7)$ and is parallel to the line $5x + y = -5$.

61. The line passes through the point $(0, -6)$ and is perpendicular to the line $-5x + y = 4$.

62. The line passes through the point $(0, -8)$ and is perpendicular to the line $2x - y = 5$.

Concepts

1. Definition of Dependent and Independent Variables
2. Interpreting a Linear Equation in Two Variables
3. Writing a Linear Equation Using Observed Data Points
4. Writing a Linear Model Given a Fixed Value and a Rate of Change

section

7.2 Applications of Linear Equations

1. Definition of Dependent and Independent Variables

Linear equations can often be used to describe (or model) the relationship between two variables in a real-world event. In an *xy*-coordinate system, the variable being predicted by the mathematical equation is called the **dependent variable** (or response variable) and is represented by *y*. The variable used to make the prediction is called the **independent variable** (or predictor variable) and is represented by *x*.

2. Interpreting a Linear Equation in Two Variables

example 1 **Interpreting a Linear Equation**

The cost, y, of a speeding ticket (in dollars) is given by $y = 10x + 100$, where $x > 0$ is the number of miles per hour over the speed limit.

a. Which variable is the dependent variable?
b. Which is the independent variable?
c. What is the slope of the line?
d. Interpret the meaning of the slope in terms of cost and the number of miles per hour over the speed limit.
e. Graph the line.

Solution:

a. The dependent variable is the cost of the speeding ticket and is represented by y. The cost of the ticket *depends* on the number of miles per hour over the speed limit.
b. The independent variable is the number of miles over the speed limit and is represented by x.
c. The line is written in slope-intercept form where $m = 10$.
d. The slope $m = 10$ or $\frac{10}{1}$ indicates that there is a $10 increase in the cost of the speeding ticket for every 1 mph over the speed limit.

e.

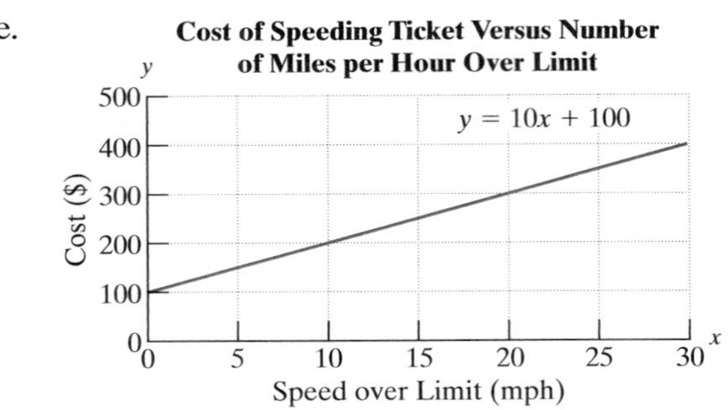

example 2 Interpreting a Linear Equation

The total number of crimes in the United States decreased from 1994 to 1997 (Figure 7-7). The decrease followed a linear trend and can be represented by the linear equation

$y = -2.6x + 45$ where y is the number of crimes measured in millions and x is the number of years since 1994.

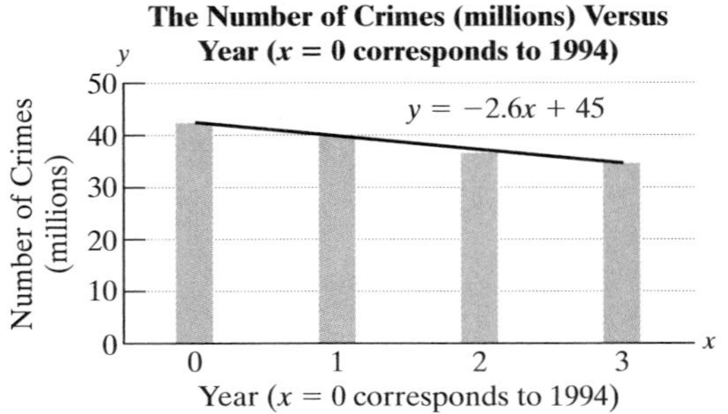

Figure 7-7

Source: U.S. Federal Bureau of Investigation.

a. Which is the dependent variable?
b. Which is the independent variable?
c. Use the equation to estimate the number of crimes in 1995.

d. What is the y-intercept of the line? Interpret the meaning of the y-intercept in terms of the number of crimes and the year.
e. What is the slope of the line? Interpret the meaning of the slope in terms of the number of crimes and the year.
f. Is it possible for this linear trend to continue indefinitely?
g. From the equation, determine the value of the x-intercept. Interpret the meaning of the x-intercept in terms of the number of crimes and the year. Realistically, is it possible for this linear trend to continue to its x-intercept?

Solution:

a. The number of crimes is the dependent variable and is represented by y.
b. The number of years since 1994 is the independent variable and is represented by x.
c. Because $x = 0$ represents the year 1994, then $x = 1$ represents 1995. Substitute $x = 1$ into the linear equation.

$$
\begin{aligned}
y &= -2.6x + 45 \\
&= -2.6(1) + 45 \qquad \text{Substitute } x = 1. \\
&= -2.6 + 45 \\
&= 42.4
\end{aligned}
$$

The number of crimes in the year 1995 was approximately 42.4 million.

d. To find the y-intercept, substitute $x = 0$.

$$
\begin{aligned}
y &= -2.6(0) + 45 \\
&= 45
\end{aligned}
$$

The y-intercept is (0, 45) and indicates that in 1994, the number of crimes was approximately 45 million.

e. From the slope-intercept form of the line, $y = -2.6x + 45$, the slope is -2.6 or equivalently $\frac{-2.6}{1}$. This indicates that the number of crimes decreased by 2.6 million per year during this time period.
f. It is not possible for the linear trend to continue indefinitely because eventually the number of crimes would reduce to a negative number.
g. To find the x-intercept, substitute $y = 0$.

$$
\begin{aligned}
y &= -2.6x + 45 \\
0 &= -2.6x + 45 \\
-45 &= -2.6x \\
\frac{-45}{-2.6} &= \frac{-2.6x}{-2.6} \\
17.3 &\approx x
\end{aligned}
$$

The x-intercept is (17.3, 0) and indicates that approximately 17.3 years after 1994 (the year 2011), the number of crimes will be 0. Although the concept of having zero crimes committed in the United States is appealing, it is not realistic. This shows that the linear trend will not continue indefinitely.

3. Writing a Linear Equation Using Observed Data Points

example 3

Writing a Linear Equation from Observed Data Points

The average amount of time per year that a person in the United States spent listening to the radio decreased between 1994 and 1999 (Figure 7-8).

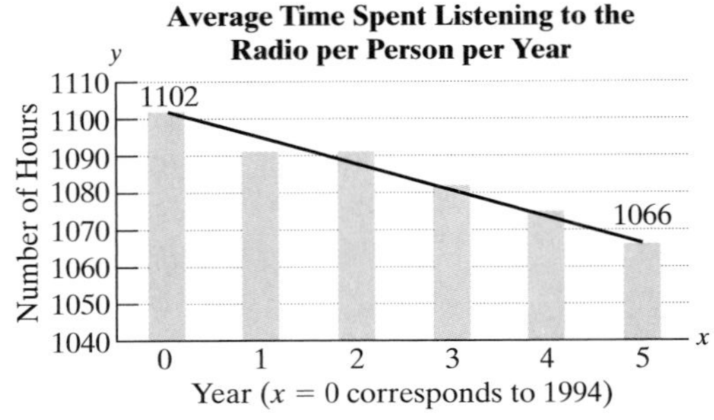

Figure 7-8

Source: Veronis, Suhler Associates, Inc., "Communication Industry Report."

Let x represent the number of years since 1994. Let y represent the average number of hours spent listening to the radio.

a. Find a linear equation that represents the average time an individual spent listening to the radio versus the number of years since 1994.
b. Use the linear equation found in part (a) to estimate the amount of time spent listening to the radio in the year 1998. Round to the nearest hour.

Solution:

a. From the graph, two data values are given. In the year 1994, ($x = 0$), the average number of hours spent listening to the radio per person was 1102. This can be written as the ordered pair (0, 1102). Similarly the ordered pair (5, 1066) indicates that 5 years later, the number of hours spent listening to the radio was 1066.

Using the points (0, 1102) and (5, 1066), find the slope of the line. Once the slope is known, either slope-intercept form or the point-slope formula can be used to find a linear equation. Because the point (0, 1102) is the y-intercept of the line, we will use slope-intercept form.

$(0, 1102)$ and $(5, 1066)$
(x_1, y_1) $\quad$ (x_2, y_2) $\qquad$ Label the points.

$$m = \frac{y_2 - y_1}{x_2 - x_1} = \frac{(1066) - (1102)}{(5) - (0)}$$ Apply the slope formula.

$$= \frac{-36}{5} \quad \text{or} \quad m = -7.2$$ The slope indicates that there has been a 7.2-hr decrease per year in the average number of hours spent listening to the radio.

With $m = -7.2$ and the y-intercept (0, 1102), we have

$$y = mx + b$$
$$y = -7.2x + 1102$$

b. The value $x = 0$ represents the year 1994. Because the year 1998 is 4 years after 1994, substitute $x = 4$ into the linear equation.

$$\begin{aligned} y &= -7.2x + 1102 \\ &= -7.2(4) + 1102 && \text{Substitute } x = 4. \\ &= -28.8 + 1102 \\ &= 1073.2 \approx 1073 && \text{Round to the nearest hour.} \end{aligned}$$

The average time spent listening to the radio per person in the United States during the year 1998 is estimated to be 1073 hr.

4. Writing a Linear Model Given a Fixed Value and a Rate of Change

Another way to look at the equation $y = mx + b$ is to identify the term mx as the variable term and the term b as the constant term. The value of the term mx will change with the value of x (this is why the slope, m, is called a rate of change). However, the term b will remain constant regardless of the value of x. With these ideas in mind, we can write a linear equation if the rate of change and the constant are known.

example 4 **Finding a Linear Equation**

A stack of posters to advertise a school play costs \$9.95 plus \$0.50 per poster at the printer.

a. Write a linear equation to compute the cost, y, of buying x posters.
b. Use the equation to compute the cost of 125 posters.

Solution:

a. The constant cost is \$9.95. The variable cost is \$0.50 per poster. If m is replaced with 0.50 and b is replaced with 9.95, the equation is

$y = 0.50x + 9.95$ where y represents the total cost of buying x posters.

b. Because x represents the number of posters, substitute $x = 125$.

$$\begin{aligned} y &= 0.50(125) + 9.95 \\ &= 62.5 + 9.95 \\ &= 72.45 \end{aligned}$$

The total cost of buying 125 posters is \$72.45.

Calculator Connections

In Example 2, the equation $y = -2.6x + 45$ was used to represent the number of crimes in the United States, y, (in millions) versus the number of years, x, since 1994. The equation is based on data between 1994 and 1997 that correspond to x-values between 0 and 3. Therefore, to graph the equation on a graphing calculator, the viewing window can be set for x between 0 and 3. Furthermore, the window must accommodate y-values up to 45 to show the y-intercept of the equation.

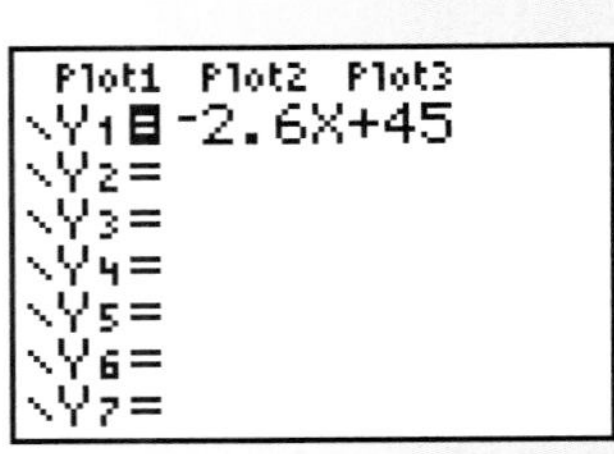

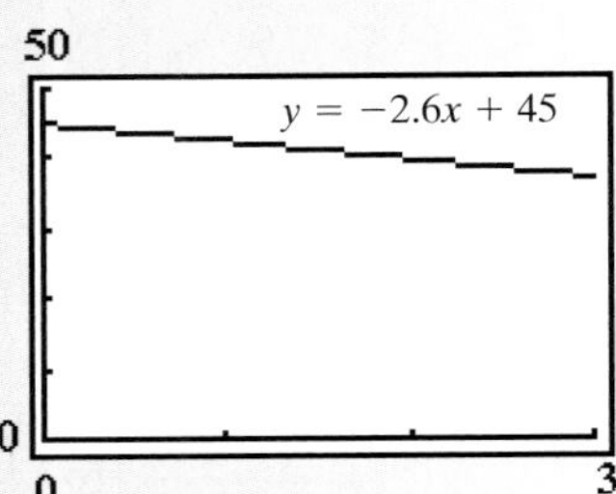

An *Eval* feature can be used to find solutions to an equation by evaluating the value of y for user-defined values of x. For example, the ordered pair (2, 39.8) indicates that in 1996, there were approximately 39.8 million crimes in the United States.

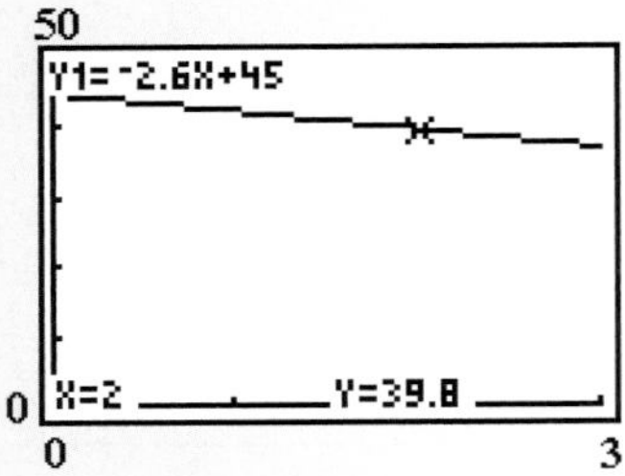

section 7.2 Practice Exercises

1. The electric bill charge for a certain utility company is \$0.095 per kilowatt-hour. The total cost, y, depends on the number of kilowatt-hours, x, according to the equation

$$y = 0.095x \quad x \geq 0$$

   a. Determine the cost of using 1000 kilowatt-hours.

   b. Determine the cost of using 2000 kilowatt-hours.

   c. What is the y-intercept of the equation? Interpret the meaning of the y-intercept in the context of this problem.

   d. What is the slope of the equation? Interpret the meaning of the slope in the context of this problem.

   e. Graph the equation.

2. The minimum hourly wage, y (in \$/hour), in the United States between 1960 and 2000 can be approximated by the equation $y = 0.10x + 0.82$, where x represents the number of years since 1960 ($x = 0$ corresponds to 1960, $x = 1$ corresponds to 1961, and so on).

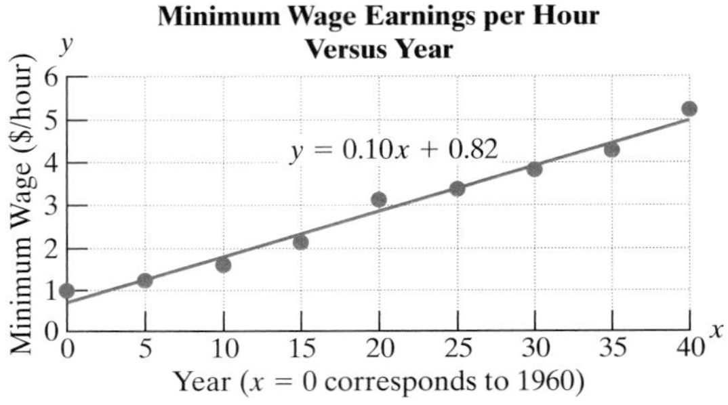

Figure for Exercise 2

a. Approximate the minimum wage in 1985.

b. Use the equation to predict the minimum wage in 2010.

c. Find the y-intercept. Interpret the meaning of the y-intercept in the context of this problem.

d. Find the slope. Interpret the meaning of the slope in the context of this problem.

3. The following graph depicts the rise in the number of jail inmates in the United States from 1987 to 1997. Two linear equations are given: one to describe the number of female inmates and one to describe the number of male inmates by year.

Let y represent the number of inmates (in thousands). Let x represent the number of years since 1987.

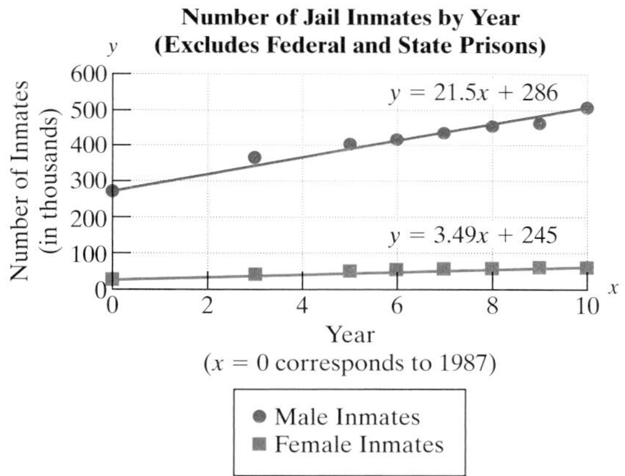

Figure for Exercise 3

Source: U.S. Bureau of Justice Statistics

a. What is the slope of the line representing the number of female inmates? Interpret the meaning of the slope in the context of this problem.

b. What is the slope of the line representing the number of male inmates? Interpret the meaning of the slope in the context of this problem.

c. Which group, males or females, has the largest slope? What does this imply about the rise in the number of male and female prisoners?

4. The following graph shows the number of points scored by Shaquille O'Neal and by Allen Iverson according to the number of minutes played for several games. Two linear equations are given: one to describe the number of points scored by O'Neal and one to describe the number of points scored by Iverson. In both equations, y represents the number of points and x represents the number of minutes played.

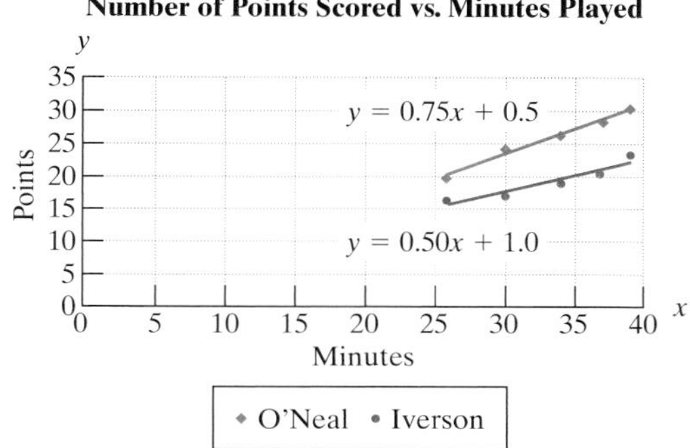

Figure for Exercise 4

a. What is the slope of the line representing the number of points scored by O'Neal? Interpret the meaning of the slope in the context of this problem.

b. What is the slope of the line representing the number of points scored by Iverson? Interpret the meaning of the slope in the context of this problem.

c. According to these linear equations, approximately how many points would each player expect to score if he played for 36 minutes? Round to the nearest point.

5. The average daily temperature in January for cities along the eastern seaboard of the United States and Canada generally decreases for cities farther north. A city's latitude in the northern

hemisphere is a measure of how far north it is on the globe.

City	x Latitude (°N)	y Average Daily Temperature, (°F)
Jacksonville, FL	30.3	52.4
Miami, FL	25.8	67.2
Atlanta, GA	33.8	41.0
Baltimore, MD	39.3	31.8
Boston, MA	42.3	28.6
Atlantic City, NJ	39.4	30.9
New York, NY	40.7	31.5
Portland, ME	43.7	20.8
Charlotte, NC	35.2	39.3
Norfolk, VA	36.9	39.1

Table for Exercise 5

The average temperature, y, (measured in degrees Fahrenheit) can be described by the equation

$y = -2.333x + 124.0$ where x is the latitude of the city.

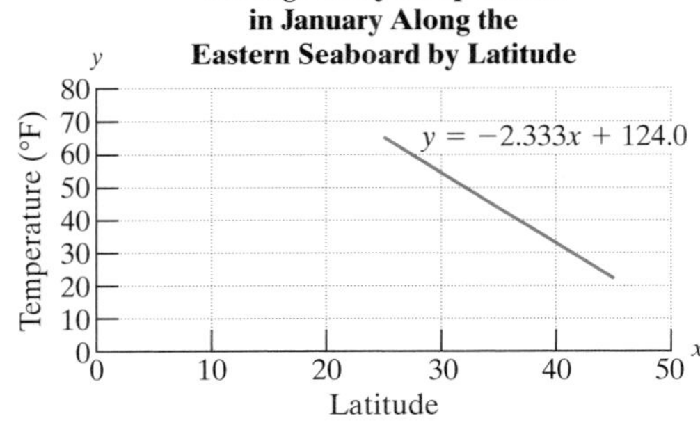

Figure for Exercise 5

Source: U.S. National Oceanic and Atmospheric Administration.

a. Which variable is the dependent variable?

b. Which variable is the independent variable?

c. Use the equation to predict the average daily temperature in January for Philadelphia, PA, whose latitude is 40.0°N. Round to one decimal place.

d. Use the equation to predict the average daily temperature in January for Edmundston, New Brunswick, Canada, whose latitude is 47.4°N. Round to one decimal place.

e. What is the slope of the line? Interpret the meaning of the slope in terms of the latitude and temperature.

f. From the equation, determine the value of the x-intercept. Round to one decimal place. Interpret the meaning of the x-intercept in terms of latitude and temperature.

6. The water bill charge for a certain utility company is \$4.20 per 1000 gallons used. The total cost, y, depends on the number of thousands of gallons of water, x, according to the equation

$$y = 4.20x \quad x \geq 0$$

a. Determine the cost of using 3000 gallons. (*Hint*: $x = 3$.)

b. Determine the cost of using 5000 gallons.

c. What is the y-intercept of the equation? Interpret the meaning of the y-intercept in the context of this problem.

d. What is the slope of the equation? Interpret the meaning of the slope in the context of this problem.

e. Graph the equation.

7. The average amount of time per year that a person in the United States spent reading newspapers decreased between 1994 and 1999.

Let x represent the number of years since 1994. Let y represent the average time (hours) spent reading newspapers.

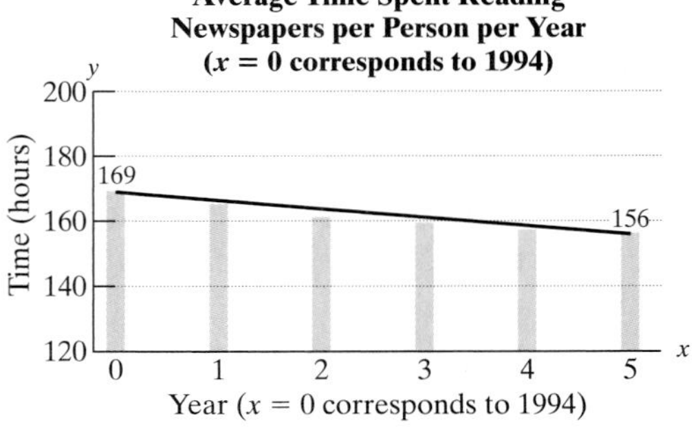

Figure for Exercise 7

Source: Veronis, Suhler & Associates Inc., "Communications Industry Report"

a. Find a linear equation that represents the time spent reading newspapers versus the year. (*Hint*: See Example 3.)

b. Use the linear equation found in part (a) to estimate the amount of time spent reading newspapers in the year 2000.

8. The average length of stay for community hospitals has been decreasing in the United States from 1980 to 1997.

Let x represent the number of years since 1980. Let y represent the average length of a hospital stay in days.

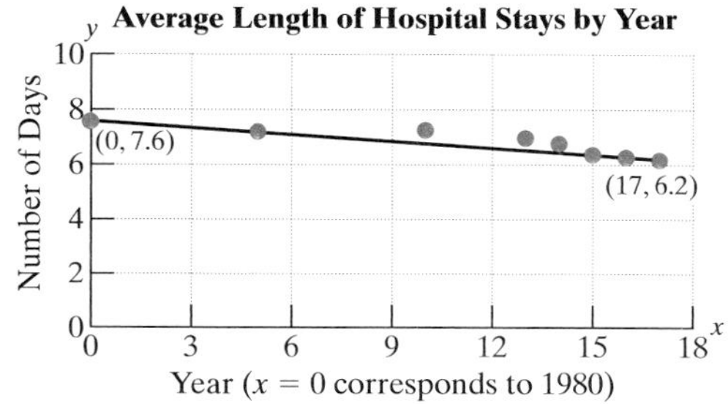

Figure for Exercise 8

Source: U.S. National Center for Health Statistics

a. Find a linear equation that relates the average length of hospital stays versus the year. (*Hint*: See Example 3.) Round the slope to three decimal places.

b. Use the linear equation found in part (a) to estimate the average length of stay in community hospitals in the year 2000.

9. The figure depicts a relationship between a person's height, y (in inches), and the length of the person's arm, x (measured in inches from shoulder to wrist).

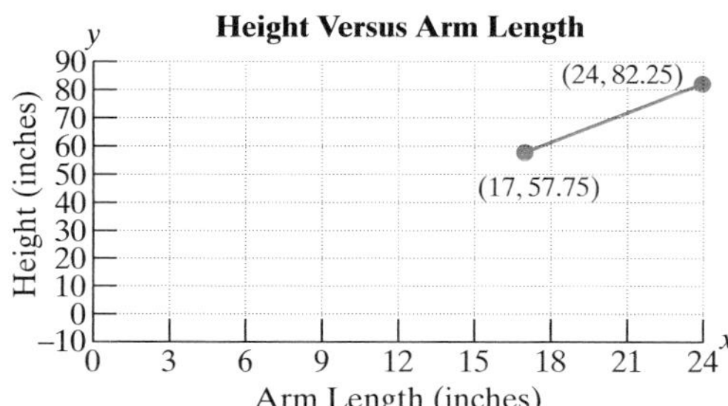

Figure for Exercise 9

a. Use the points (17, 57.75) and (24, 82.25) to find a linear equation relating height to arm length.

b. What is the slope of the line? Interpret the slope in the context of this problem.

c. Use the equation from part (a) to estimate the height of a person whose arm length is 21.5 in.

10. In a certain city, the time required to commute to work, y, (in minutes) by car is related linearly to the distance traveled, x (in miles).

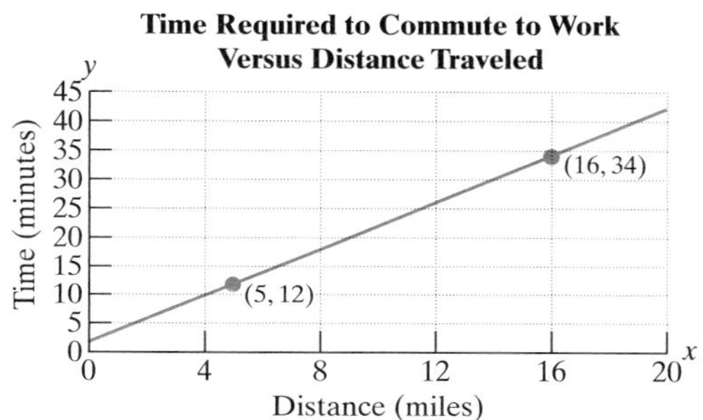

Figure for Exercise 10

a. Use the points (5, 12) and (16, 34) to find a linear equation relating the commute time to work to the distance traveled.

b. What is the slope of the line? Interpret the slope in the context of this problem.

c. Use the equation from part (a) to find the time required to commute to work for a motorist who lives 18 miles away.

11. The cost to rent a car, y, for 1 day is \$20 plus \$0.25 per mile.

a. Write a linear equation to compute the cost, y, of driving a car x miles for 1 day. (*Hint*: See Example 4)

b. Use the equation to compute the cost of driving 258 miles in the rental car.

12. A phone bill is determined each month by a \$18.95 flat fee plus \$0.08 per minute of long distance.

a. Write a linear equation to compute the monthly cost of a phone bill, y, if x minutes of long distance are used. (*Hint*: See Example 4)

b. Use the equation to compute the phone bill for a month in which 1 hr and 27 min of long distance was used.

13. A tennis instructor charges a student \$25 per lesson plus a one-time court fee of \$20.

a. Write a linear equation to compute the total cost, y, for x tennis lessons.

b. What is the total cost to a student who takes 20 tennis lessons?

14. The cost to rent a 10 ft by 10 ft storage space is \$90 per month plus a nonrefundable deposit of \$105.

a. Write a linear equation to compute the cost, y, of renting a 10 ft by 10 ft space for x months.

b. What is the cost of renting such a storage space for 1 year (12 months)?

15. A business has a fixed monthly cost of \$1200. In addition, the business has a variable cost of \$35 for each item produced.

a. Write a linear equation to compute the total cost, y, for 1 month if x items are produced.

b. Use the equation to compute the cost for 1 month if 100 items are produced.

16. An air-conditioning and heating company has a fixed monthly cost of \$5000. Furthermore, each service call costs the company \$25.

a. Write a linear equation to compute the total cost, y, for 1 month if x service calls are made.

b. Use the equation to compute the cost for 1 month if 150 service calls are made.

17. A bakery that specializes in bread rents a booth at a flea market. The daily cost to rent the booth is \$100. Each loaf of bread costs the bakery \$0.80 to produce.

a. Write a linear equation to compute the total cost, y, for 1 day if x loaves of bread are produced.

b. Use the equation to compute the cost for 1 day if 200 loaves of bread are produced.

18. A beverage company rents a booth at an art show to sell lemonade. The daily cost to rent a booth is \$35. Each lemonade costs \$0.50 to produce.

a. Write a linear equation to compute the total cost, y, for 1 day if x lemonades are produced.

b. Use the equation to compute the cost for 1 day if 350 lemonades are produced.

Graphing Calculator Exercises

For Exercises 19–22, use a graphing calculator to graph the lines on an appropriate viewing window. Evaluate the equation at the given values of x.

19. $y = -4.6x + 27.1$ at $x = 3$

20. $y = -3.6x - 42.3$ at $x = 0$

21. $y = 40x + 105$ at $x = 6$

22. $y = 20x - 65$ at $x = 8$

section

7.3 Introduction to Relations

Concepts

1. Definition of a Relation
2. Finding the Domain and Range of a Relation
3. Applications Involving Relations

1. Definition of a Relation

In Sections 7.1 and 7.2, we studied linear relationships between two variables. However, in many naturally occuring phenomena, two variables may be linked by some other type of relationship. Table 7-2 shows a correspondence between the length of a woman's femur and her height. (The femur is the large bone in the thigh attached to the knee and hip.)

Table 7-2

Length of Femur (cm) x	**Height (in.)** y		**Ordered Pair**
45.5	65.5	→	(45.5, 65.5)
48.2	68.0	→	(48.2, 68.0)
41.8	62.2	→	(41.8, 62.2)
46.0	66.0	→	(46.0, 66.0)
50.4	70.0	→	(50.4, 70.0)

Each data point from Table 7-2 may be represented as an ordered pair. In this case, the first value represents the length of a woman's femur and the second the woman's height. The set of ordered pairs: {(45.5, 65.5), (48.2, 68.0), (41.8, 62.2), (46.0, 66.0), (50.4, 70.0)} defines a relation between femur length and height.

Definition of a Relation in x and y

Any set of ordered pairs, (x, y), is called a **relation in** x **and** y. Furthermore:

- The set of first components in the ordered pairs is called the **domain of the relation**.
- The set of second components in the ordered pairs is called the **range of the relation**.

2. Finding the Domain and Range of a Relation

example 1 **Finding the Domain and Range of a Relation**

Find the domain and range of the relation linking the length of a woman's femur to her height: {(45.5, 65.5), (48.2, 68.0), (41.8, 62.2), (46.0, 66.0), (50.4, 70.0)}

Solution:

Domain: {45.5, 48.2, 41.8, 46.0, 50.4} Set of first coordinates

Range: {65.5, 68.0, 62.2, 66.0, 70.0} Set of second coordinates

The x- and y-components that constitute the ordered pairs in a relation do not need to be numerical. For example, Table 7-3 depicts 5 states in the United States and the corresponding number of representatives in the House of Representatives as of July 1998.

Table 7-3

State x	Number of Representatives y
Arizona	6
California	52
Colorado	6
Florida	23
Kansas	4

These data define a relation:

{(Arizona, 6), (California, 52), (Colorado, 6), (Florida, 23), (Kansas, 4)}

example 2 Finding the Domain and Range of a Relation

Find the domain and range of the relation:
{(Arizona, 6), (California, 52), (Colorado, 6), (Florida, 23), (Kansas, 4)}

Solution:

Domain: {Arizona, California, Colorado, Florida, Kansas}

Range: {6, 52, 23, 4} (*Note:* The element 6 is not listed twice.)

A relation may consist of a finite number of ordered pairs or an infinite number of ordered pairs. Furthermore, a relation may be defined by several different methods: by a list of ordered pairs, by a correspondence between the domain and range, by a graph, or by an equation.

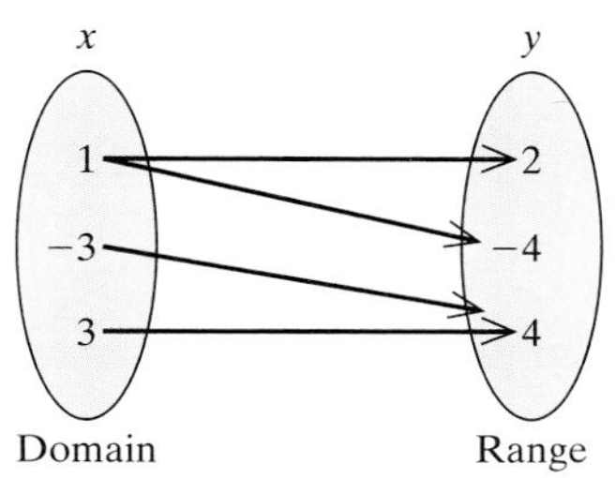

Figure 7-9

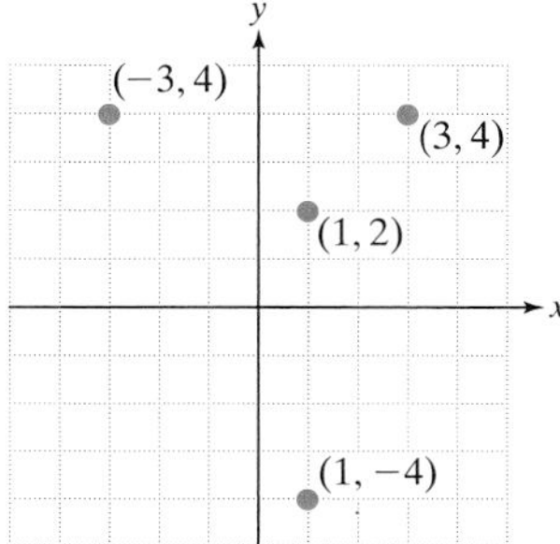

Figure 7-10

- A relation may be defined as a set of ordered pairs.
  $\{(1, 2), (-3, 4), (1, -4), (3, 4)\}$
- A relation may be defined by a correspondence (Figure 7-9). The corresponding ordered pairs are $\{(1, 2), (1, -4), (-3, 4), (3, 4)\}$.
- A relation may be defined by a graph (Figure 7-10). The corresponding ordered pairs are $\{(1, 2), (-3, 4), (1, -4), (3, 4)\}$.
- A relation may be expressed by an equation such as $x = y^2$. The solutions to this equation define an infinite set of ordered pairs of the form $\{(x, y) \mid x = y^2\}$. The solutions can also be represented by a graph in a rectangular coordinate system (Figure 7-11).

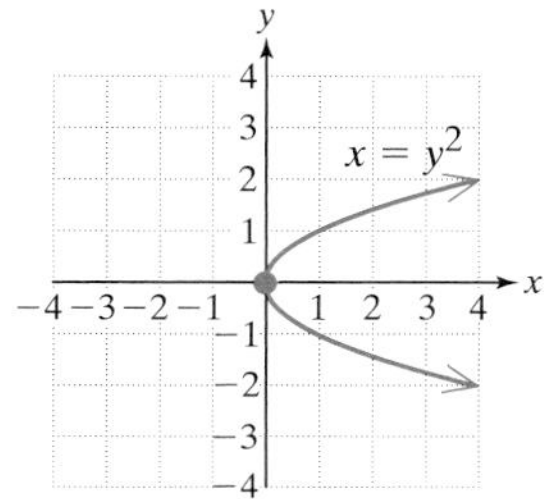

Figure 7-11

example 3 **Finding the Domain and Range of a Relation**

Find the domain and range of the following relations:

Solution:

a.

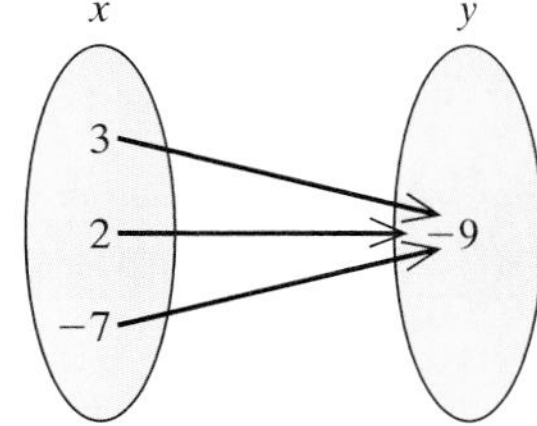

Domain: $\{3, 2, -7\}$

Range: $\{-9\}$

b.

The domain elements are the x-coordinates of the points and the range elements are the y-coordinates.

Domain: $\{-2, -1, 0, 1, 2\}$

Range: $\{-3, 0, 1\}$

c.

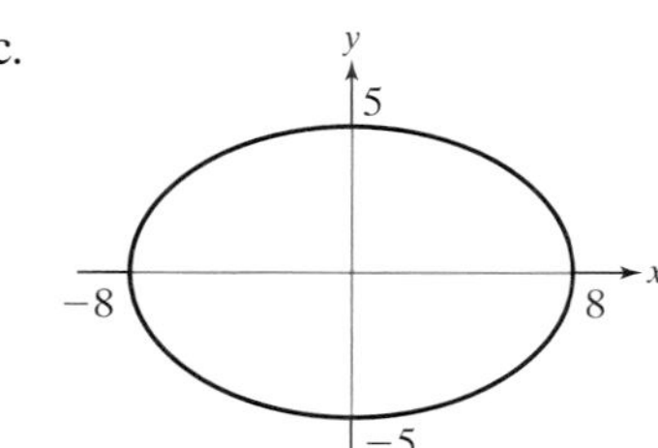

The domain consists of an infinite number of x-values extending from -8 to 8 (shown in red). The range consists of all y-values from -5 to 5 (shown in blue). Thus, the domain and range must be expressed in set-builder notation or in interval notation.

Domain: $\{x | -8 \le x \le 8\}$ or $[-8, 8]$

Range: $\{y | -5 \le y \le 5\}$ or $[-5, 5]$

d. $x = y^2$

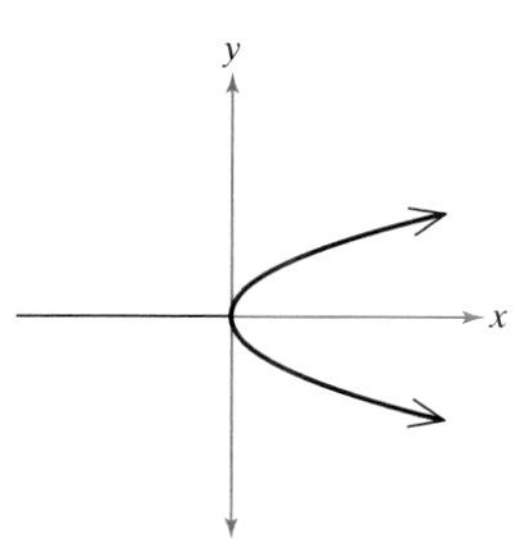

The arrows on the curve indicate that the graph extends infinitely far up and to the right and infinitely far down and to the right.

Domain: $\{x | x \ge 0\}$ or $[0, \infty)$

Range: $\{y | y \text{ is any real number}\}$ or $(-\infty, \infty)$

3. Applications Involving Relations

example 4

Analyzing a Relation

The data in Table 7-4 depict the length of a woman's femur and her corresponding height.

After collecting the data, a medical researcher finds the following linear relationship between height, y, and femur length, x:

$$y = 0.906x + 24.3 \quad 40 \le x \le 51$$

a. Find the height of a woman whose femur is 46.0 cm.

b. Find the height of a woman whose femur is 51.0 cm.

c. Why is the domain restricted to $40 \le x \le 51$?

Table 7-4

Length of Femur (cm) x	Height (in.) y
45.5	65.5
48.2	68.0
41.8	62.2
46.0	66.0
50.4	70.0

Solution:

a. $y = 0.906x + 24.3$
$= 0.906(46.0) + 24.3$ Substitute $x = 46.0$ cm.
$= 65.976$ The woman is approximately 66.0 in. tall.

b. $y = 0.906x + 24.3$
$= 0.906(51.0) + 24.3$ Substitute $x = 51.0$ cm.
$= 70.506$ The woman is approximately 70.5 in. tall.

c. The domain restricts femur length to values between 40 cm and 51 cm, inclusive. These values are within the normal lengths for an adult female and are in the proximity of the observed data (Figure 7-12).

Height of an Adult Female Based on the Length of the Femur

$y = 0.906x + 24.3$

Height (in.)

Length of Femur (cm)

Figure 7-12

section 7.3 Practice Exercises

1. Find an equation of the line passing through $(2, -7)$ having slope -3.

2. Find an equation of the line passing through $(-1, 6)$ and having slope $-\frac{1}{5}$.

3. Find an equation of the line passing through $(-1, 4)$ and $(6, -3)$.

4. Find an equation of the line passing through $(-3, 7)$ and $(-1, -5)$.

5. Find an equation of the line passing through $(-3, 4)$ and
   a. parallel to the line $y = \frac{2}{3}x + 1$.
   b. perpendicular to the line $y = \frac{2}{3}x + 1$.

6. Find an equation of the line passing through $(-4, 0)$ and
   a. parallel to the line $y = -5x - 3$.
   b. perpendicular to the line $y = -5x - 3$.

For Exercises 7–10, write each relation as a set of ordered pairs.

7.

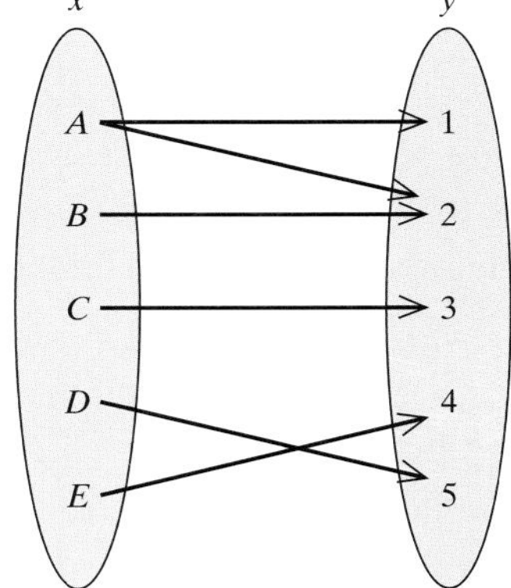

8.

State x	Year of Statehood y
Connecticut	1788
Colorado	1876
Maryland	1788
Illinois	1818
Missouri	1821

9. Reference daily intake (RDI) for proteins

Group (x)	RDI in Grams (y)
Pregnant women	60
Nursing mothers	65
Infants younger than 1 year old	14
Children from 1 to 4 years	16
Adults	50

10.

x	y
0	3
−2	$\frac{1}{2}$
5	10
−7	1
−2	8
5	1

11. List the domain and range of Exercise 7.

12. List the domain and range of Exercise 8.

13. List the domain and range of Exercise 9.

14. List the domain and range of Exercise 10.

15. a. Define a relation with four ordered pairs such that the first element of the ordered pair is the name of a friend, the second element is your friend's place of birth.

    b. State the domain and range of this relation.

16. a. Define a relation with four ordered pairs such that the first element is a state and the second element is its capital.

    b. State the domain and range of this relation.

17. a. Use a mathematical equation to define a relation whose second component, y, is 1 less than 2 times the first component, x.

    b. Sketch the relation.

    c. Write the domain and range of this relation in interval notation.

18. a. Use a mathematical equation to define a relation whose second component, y, is 3 more than the first component, x.

    b. Sketch the relation.

    c. Write the domain and range of this relation in interval notation.

For Exercises 19–32, find the domain and range of the relations. Use interval notation where appropriate.

19.

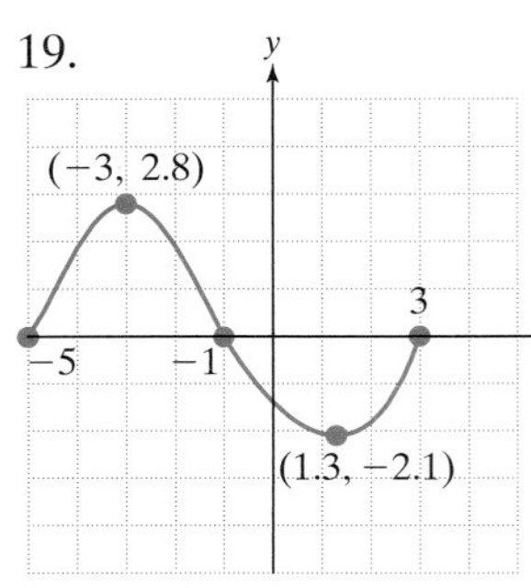

20.

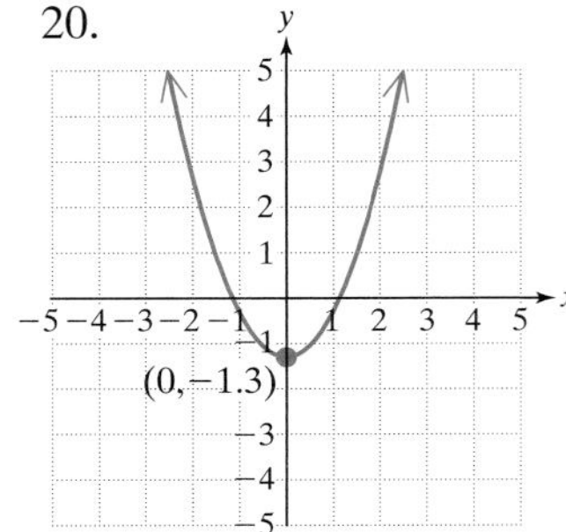

21.

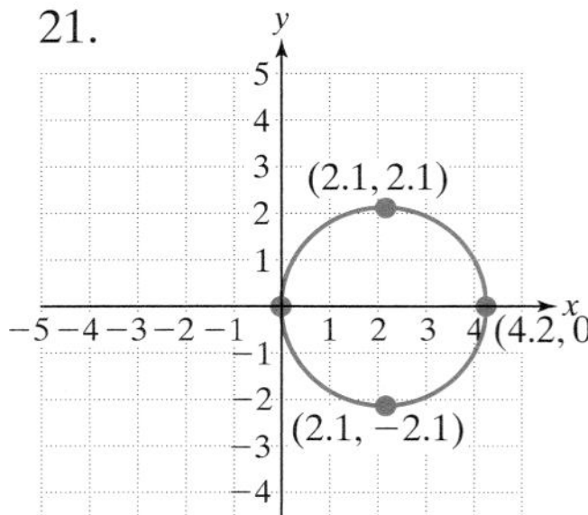

22.

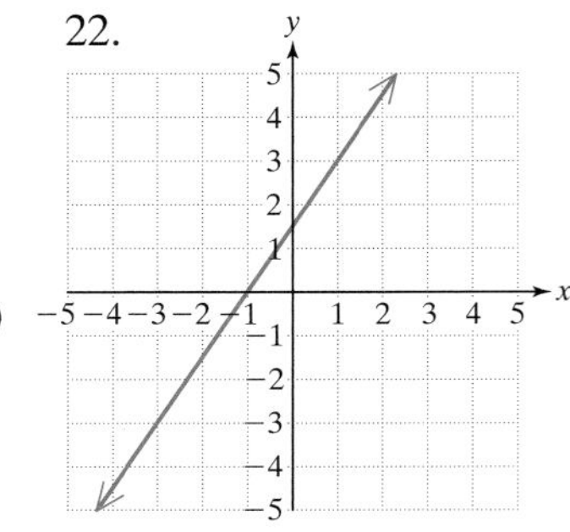

23.

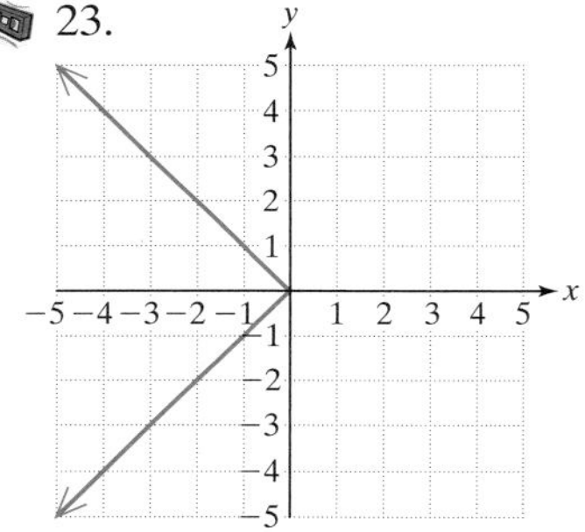

24.

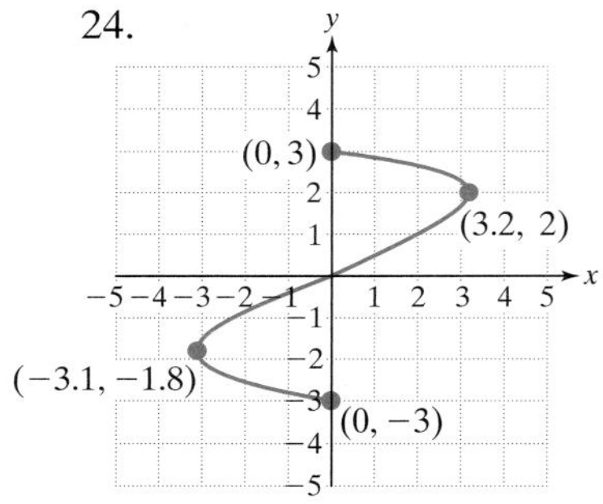

25. *Hint*: The open circle indicates that the endpoint is not included in the relation.

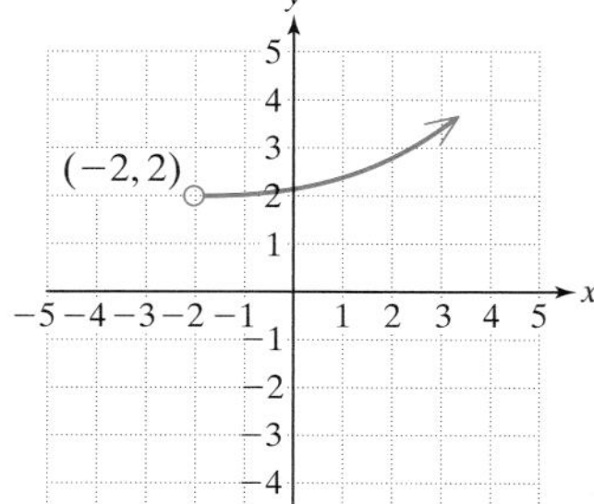

26.

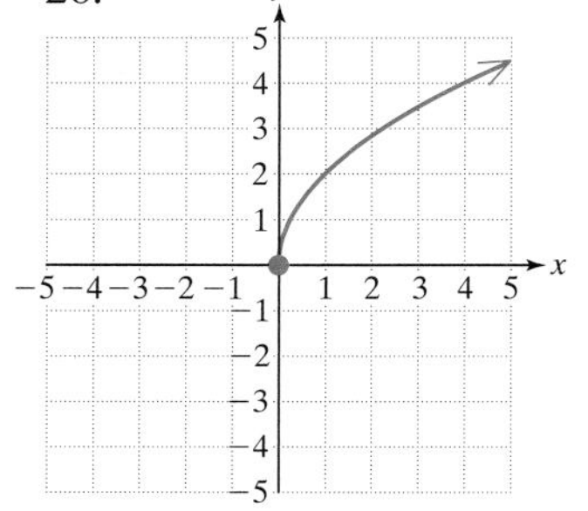

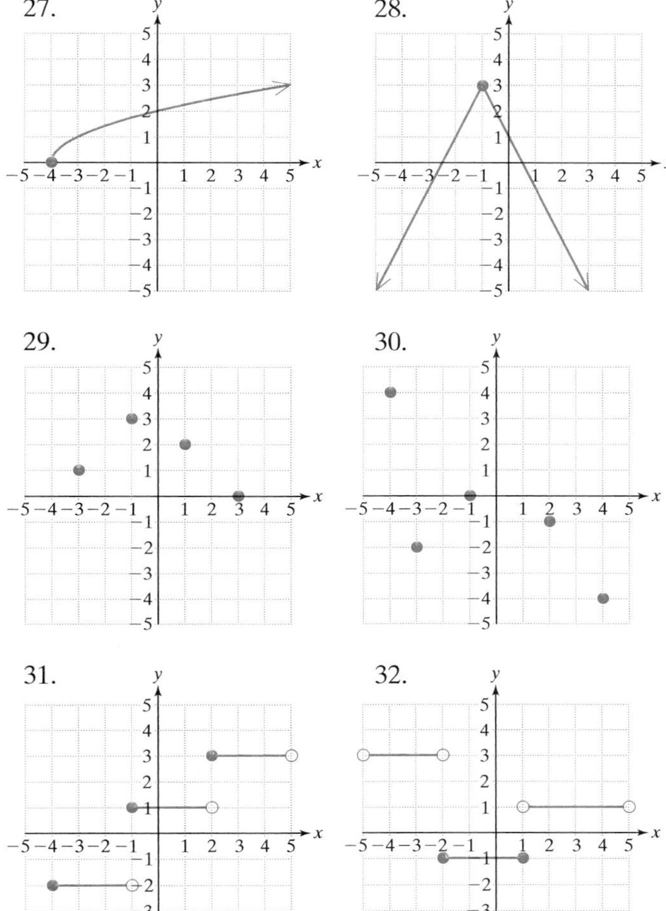

a. What is the range element corresponding to April?

b. What is the range element corresponding to June?

c. Which element in the domain corresponds to the lowest range value?

d. Complete the ordered pair: (, 2.66)

e. Complete the ordered pair: (Sept.,)

f. What is the domain of this relation?

33. The following table gives a relation between the month of the year and the average precipitation for that month for Miami, Florida.

Month x	Precipitation (in.) y	Month x	Precipitation (in.) y
Jan.	2.01	July	5.70
Feb.	2.08	Aug.	7.58
Mar.	2.39	Sept.	7.63
Apr.	2.85	Oct.	5.64
May	6.21	Nov.	2.66
June	9.33	Dec.	1.83

Source: U.S. National Oceanic and Atmospheric Administration

34. The following table gives a relation between the month of the year and the average precipitation for that month for Portland, Oregon.

Month x	Precipitation (in.) y	Month x	Precipitation (in.) y
Jan.	5.35	July	0.63
Feb.	3.85	Aug.	1.09
Mar.	3.56	Sept.	1.75
Apr.	2.39	Oct.	2.67
May	2.06	Nov.	5.34
June	1.48	Dec.	6.13

Source: U.S. National Oceanic and Atmospheric Administration

a. What is the domain?

b. What is the range?

c. Which element in the domain corresponds to the highest element in the range?

d. Complete the ordered pair: (July,)

e. Complete the ordered pair: (, 1.09)

f. Graph the ordered pairs defining this relation by plotting the month on the horizontal axis and the amount of precipitation on the vertical axis.

35. The percentage of male high school students, y, who participated in an organized physical activity for the year 1995 is approximated by $y = -12.64x + 195.22$. For this model, x represents the grade level ($9 \le x \le 12$). (*Source:* Centers for Disease Control)

Figure for Exercise 35

a. Approximate the percentage of males who participated in organized physical activity for grades 9, 10, 11, and 12, respectively.

b. Can we use this model to predict seventh-grade participation? Explain your answer.

36. As of March 1998, the world record times for selected women's track and field events are shown in the table.

Distance (m)	Time (sec)	Winner's Name and Country
100	10.49	Florence Griffith Joyner (U.S.)
200	21.34	Florence Griffith Joyner (U.S.)
400	47.60	Marita Koch (East Germany)
800	113.28	Jarmila Kratochvilova (Czechoslovakia)
1000	149.34	Maria Mutola (Mozambique)
1500	230.46	Qu Yunxia (China)

The women's world record time, y (in seconds), required to run x meters can be approximated by the relation $y = -10.78 + 0.159x$.

a. Predict the time required for a 500-m race.

b. Use this model to predict the time for a 1000-m race. Is this value exactly the same as the data value given in the table? Explain.

Expanding Your Skills

37. Write the domain and the range for the relation $y = |x|$ from the graph.

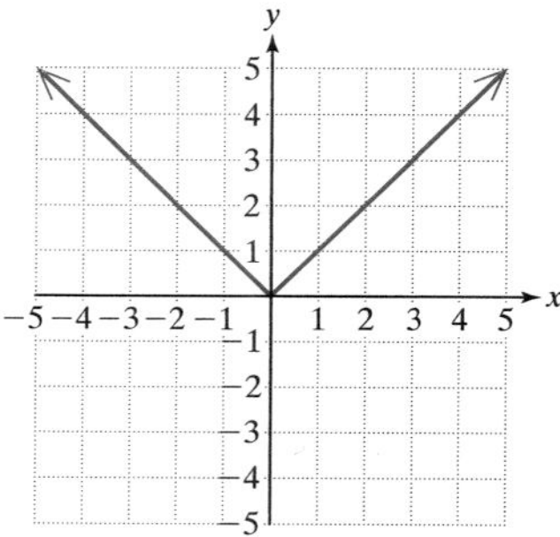

38. Write the domain and the range for the relation $y = |x - 2|$ from the graph.

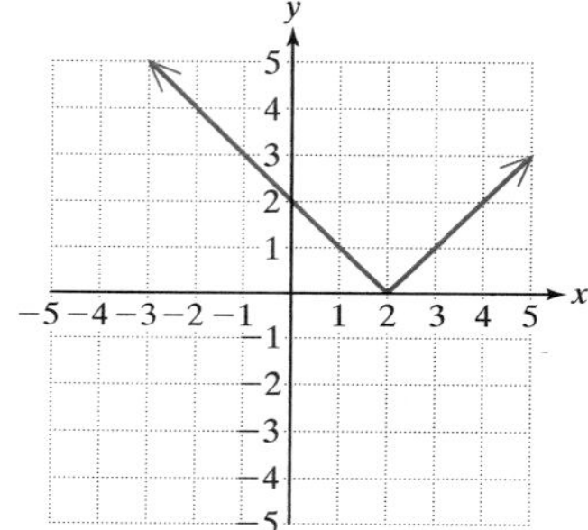

39. Write the domain and the range for the relation $y = |x + 2|$ from the graph.

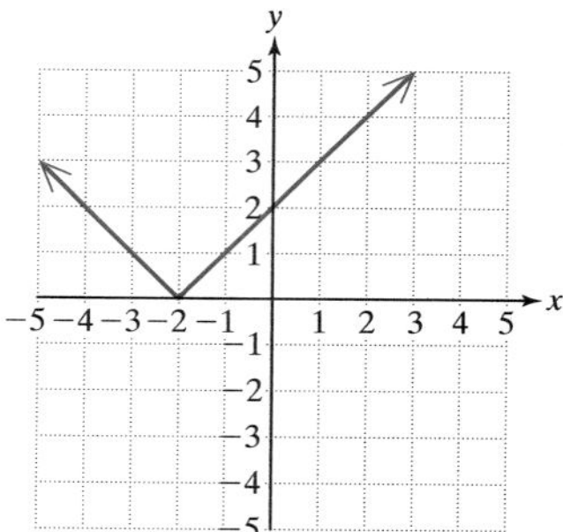

40. By comparing the domains and ranges for Exercises 37–39, what can you say about the domain and the range for the relation $y = |x + c|$, where c is any real number?

41. Write the domain and the range for the relation $y = |x| - 2$ from the graph.

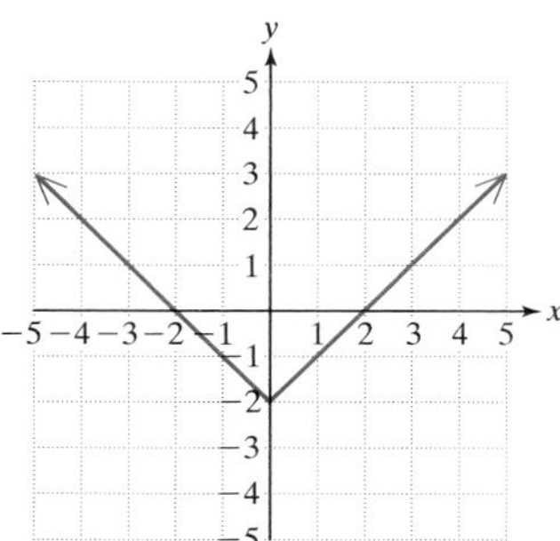

42. Write the domain and the range for the relation $y = |x| + 2$ from the graph.

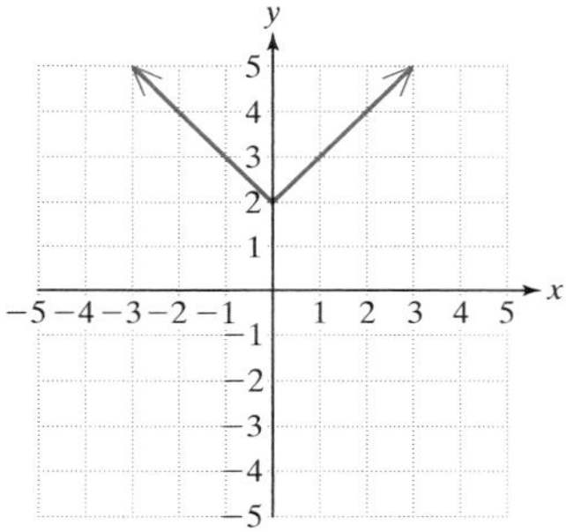

43. By comparing the domains and ranges for Exercises 37, 41–42, what can you say about the domain and the range for the relation $y = |x| + c$, where c is any real number?

Graphing Calculator Exercises

44. a. Use a graphing calculator to graph the relation $Y_1 = -12.64x + 195.22$ on a viewing window defined by $9 \le x \le 12$ and $0 \le y \le 100$.

    b. Use a *Table* feature to evaluate the relation for $x = 9, 10, 11$, and 12. Do these values agree with your solutions from Exercise 35?

45. a. Use a graphing calculator to graph the relation $Y_1 = -10.78 + 0.159x$ on a viewing window defined by $0 \le x \le 2000$ and $0 \le y \le 350$.

    b. Use a *Table* feature to evaluate the relation for $x = 500, 1000$, and 1500. Do these values agree with your solutions from Exercise 36?

Concepts

1. Definition of a Function
2. Vertical Line Test
3. Function Notation
4. Evaluating Functions
5. Finding Function Values from a Graph
6. Domain of a Function

section

7.4 Introduction to Functions

1. Definition of a Function

In this section we introduce a special type of relation called a function.

Definition of a Function

Given a relation in x and y, we say "*y is a **function** of x*" if for every element x in the domain, there corresponds exactly one element y in the range.

To understand the difference between a relation that is a function and a relation that is not a function, consider Example 1.

example 1 Determining Whether a Relation Is a Function

Determine which of the relations define y as a function of x.

a.

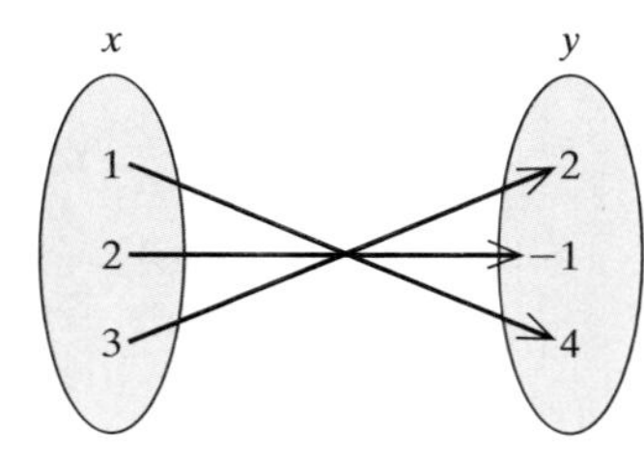

b.

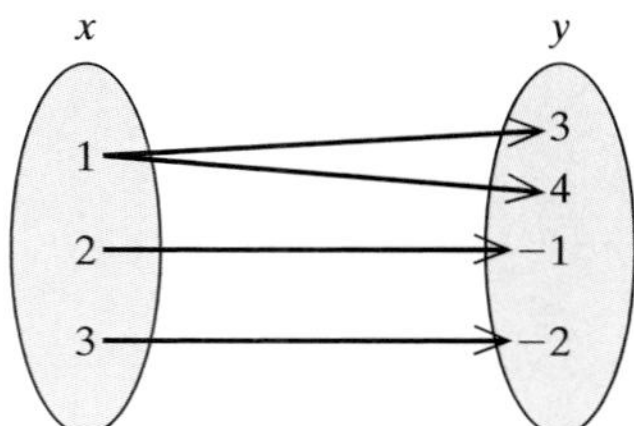

c.

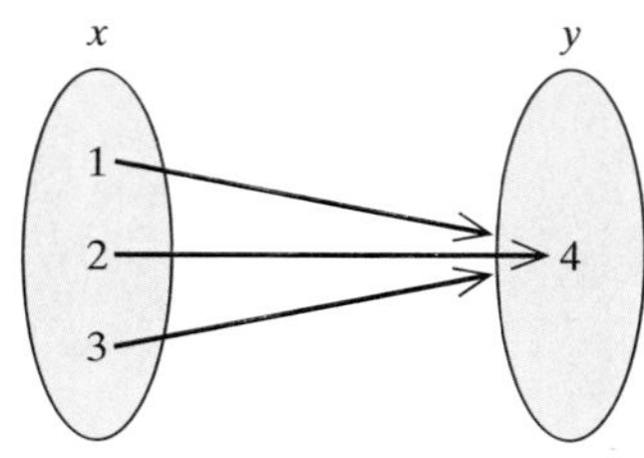

Solution:

a. This relation is defined by the set of ordered pairs: $\{(1, 4), (2, -1), (3, 2)\}$.

Notice that each x in the domain corresponds to only *one* y in the range. Therefore, this relation is a function.

When $x = 1$, there is only one possibility for y: $y = 4$

When $x = 2$, there is only one possibility for y: $y = -1$

When $x = 3$, there is only one possibility for y: $y = 2$

b. This relation is defined by the set of ordered pairs:

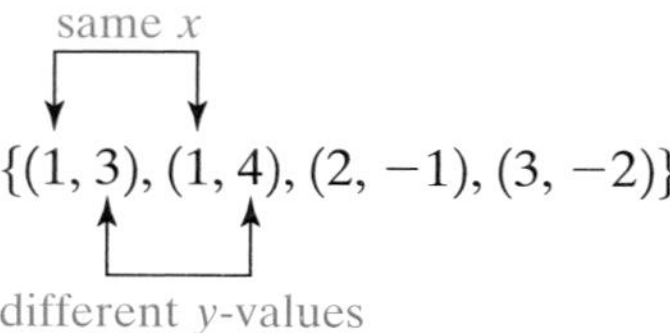

For the domain element $x = 1$, there are *two* possible range elements: $y = 3$ and $y = 4$. Therefore, this relation is *not* a function.

c. This relation is defined by the set of ordered pairs: $\{(1, 4), (2, 4), (3, 4)\}$.

When $x = 1$, there is only one possibility for y: $y = 4$

When $x = 2$, there is only one possibility for y: $y = 4$

When $x = 3$, there is only one possibility for y: $y = 4$

Because each value of x in the domain corresponds to only *one* y-value, this relation is a function.

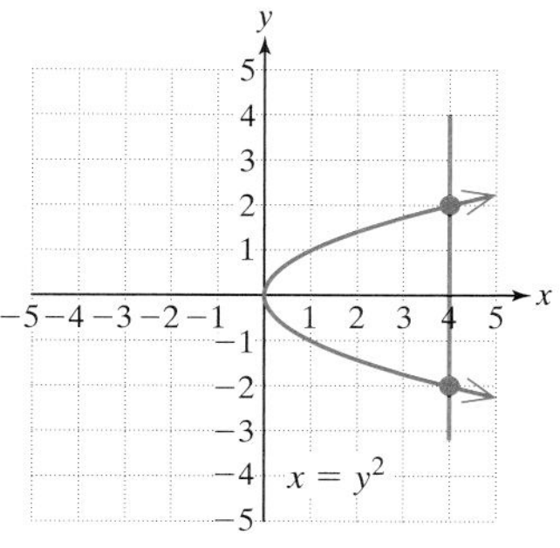

$x = y^2$ is not a function.

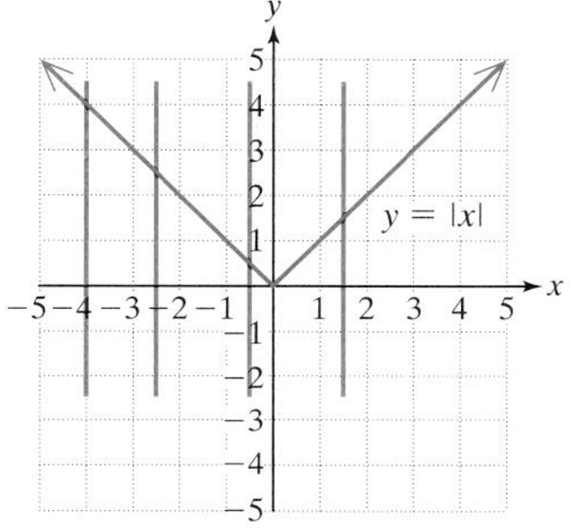

$y = |x|$ is a function.

2. Vertical Line Test

A relation that is not a function has at least one domain element, x, paired with more than one range value, y. For example, the ordered pairs (4, 2) and (4, −2) do not constitute a function. These two points are aligned vertically in the xy-plane, and a vertical line drawn through one point also intersects the other point. Thus if a vertical line drawn through a graph of a relation intersects the graph in more than one point, the relation cannot be a function. This idea is stated formally as the **vertical line test**.

The Vertical Line Test

Consider a relation defined by a set of points (x, y) in a rectangular coordinate system. Then the graph defines y as a function of x if no vertical line intersects the graph in more than one point.

The vertical line test also implies that if any vertical line drawn through the graph of a relation intersects the relation in more than one point, then the relation does *not define y as a function of x*.

The vertical line test can be demonstrated by graphing the ordered pairs from the relations in Example 1.

a. $\{(1, 4), (2, -1), (3, 2)\}$

b. $\{(1, 3), (1, 4), (2, -1), (3, -2)\}$

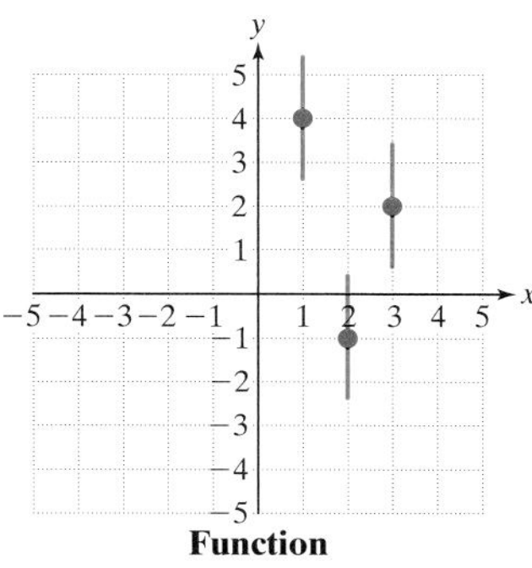

Function
No vertical line intersects more than once.

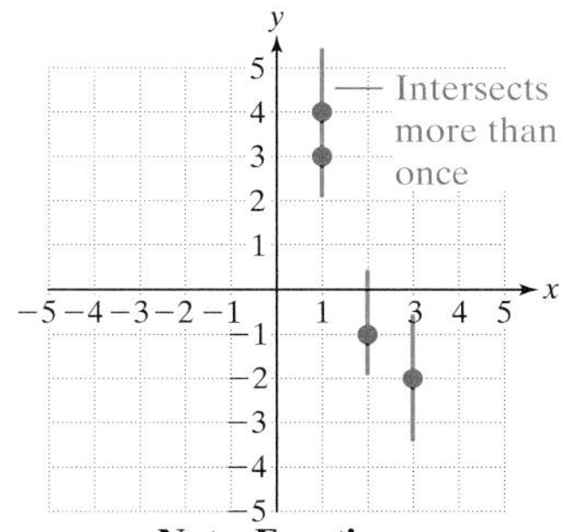

Not a Function
A vertical line intersects in more than one point.

example 2 Using the Vertical Line Test

Use the vertical line test to determine whether the following relations define y as a function of x.

a.

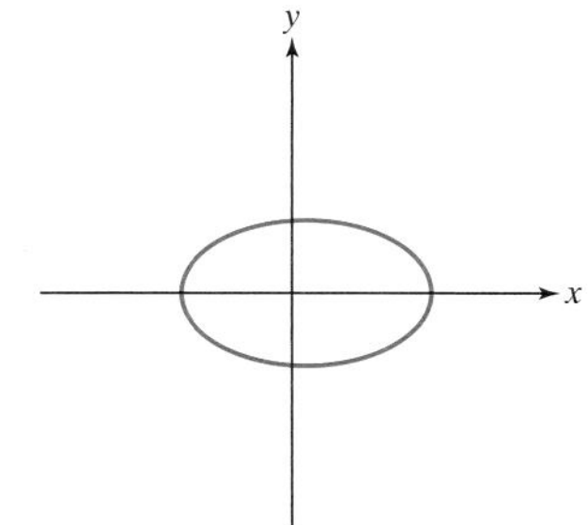

b.

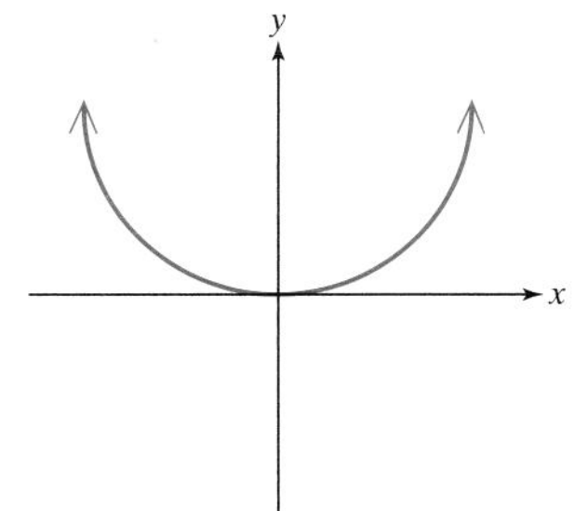

Solution:

a.

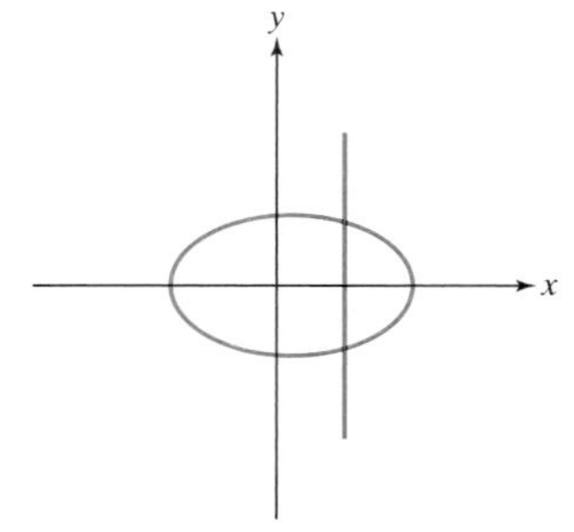

Not a Function
A vertical line intersects in more than one point.

b.

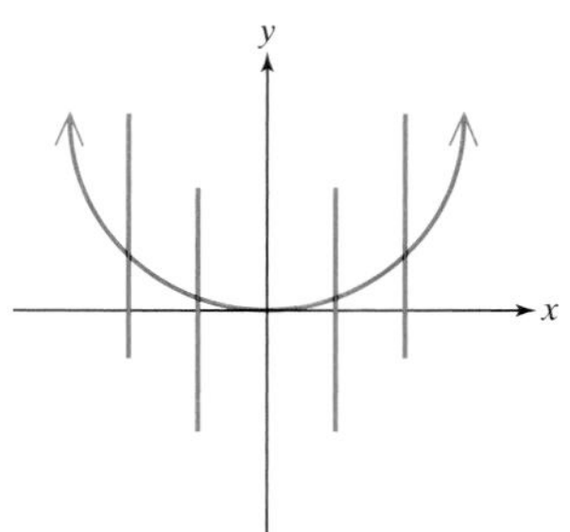

Function
No vertical line intersects in more than one point.

3. Function Notation

A function is defined as a relation with the added restriction that each value in the domain corresponds to only one value in the range. In mathematics, functions are often given by rules or equations to define the relationship between two or more variables. For example, the equation $y = 2x$ defines the set of ordered pairs such that the y-value is twice the x-value.

When a function is defined by an equation, we often use **function notation**. For example, the equation $y = 2x$ may be written in function notation as

$f(x) = 2x$, where f is the name of the function, x is an input value from the domain of the function, and $f(x)$ is the function value (or y-value) corresponding to x.

The notation $f(x)$ is read as "f of x" or "the value of the function, f, at x."

4. Evaluating Functions

A function may be evaluated at different values of x by substituting x values from the domain into the function. For example, to evaluate the function defined by $f(x) = 2x$ at $x = 5$, substitute $x = 5$ into the function.

$$f(x) = 2x$$
$$\downarrow \quad \downarrow$$
$$f(5) = 2(5)$$
$$f(5) = 10$$

Tip: The function value $f(5) = 10$ can be written as the ordered pair (5, 10).

Thus, when $x = 5$, the corresponding function value is 10. We say "f of 5 is 10" or "f at 5 is 10."

The names of functions are often given by either lowercase or uppercase letters, such as f, g, h, p, K, M, and so on.

example 3 **Evaluating a Function**

Given the function defined by $g(x) = \frac{1}{2}x - 1$, find the function values.

a. $g(0)$ b. $g(2)$ c. $g(4)$ d. $g(-2)$

Solution:

a. $g(x) = \frac{1}{2}x - 1$

$$g(0) = \frac{1}{2}(0) - 1$$
$$= 0 - 1$$
$$= -1$$

We say, "g of 0 is -1" or "g at 0 is -1." This is equivalent to the ordered pair $(0, -1)$.

b. $g(x) = \frac{1}{2}x - 1$

$$g(2) = \frac{1}{2}(2) - 1$$
$$= 1 - 1$$
$$= 0$$

We say "g of 2 is 0" or "g at 2 is 0." This is equivalent to the ordered pair $(2, 0)$.

c. $g(x) = \frac{1}{2}x - 1$

$$g(4) = \frac{1}{2}(4) - 1$$
$$= 2 - 1$$
$$= 1$$

We say "g of 4 is 1" or "g at 4 is 1." This is equivalent to the ordered pair $(4, 1)$.

d. $g(x) = \frac{1}{2}x - 1$

$$g(-2) = \frac{1}{2}(-2) - 1$$
$$= -1 - 1$$
$$= -2$$

We say "g of -2 is -2" or "g at -2 is -2." This is equivalent to the ordered pair $(-2, -2)$.

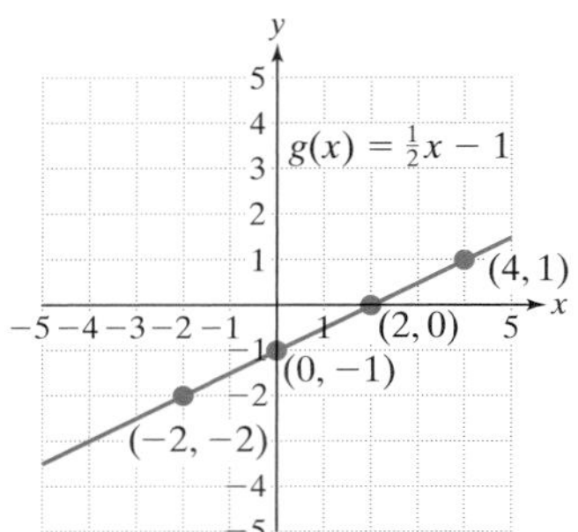

Figure 7-13

Notice that the function values $g(0)$, $g(2)$, $g(4)$, and $g(-2)$ correspond to the ordered pairs $(0, -1)$, $(2, 0)$, $(4, 1)$, and $(-2, -2)$. In the graph, these points "line up." The graph of *all* ordered pairs defined by this function is a line with a slope of $\frac{1}{2}$, and y-intercept of $(0, -1)$ (Figure 7-13). This should not be surprising because the function defined by $g(x) = \frac{1}{2}x - 1$ is equivalent to $y = \frac{1}{2}x - 1$.

Calculator Connections

The values of $g(x)$ in Example 3 can be found using a *Table* feature.

$$Y_1 = \tfrac{1}{2}x - 1$$

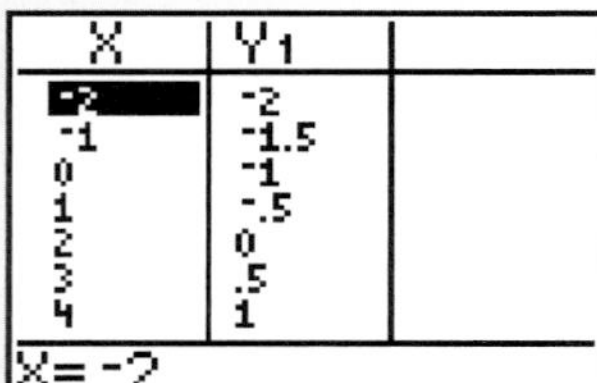

X	Y1
-2	-2
-1	-1.5
0	-1
1	-.5
2	0
3	.5
4	1

X=-2

Function values can also be evaluated by using a *Value* (or *Eval*) feature. The value of $g(4)$ is shown here.

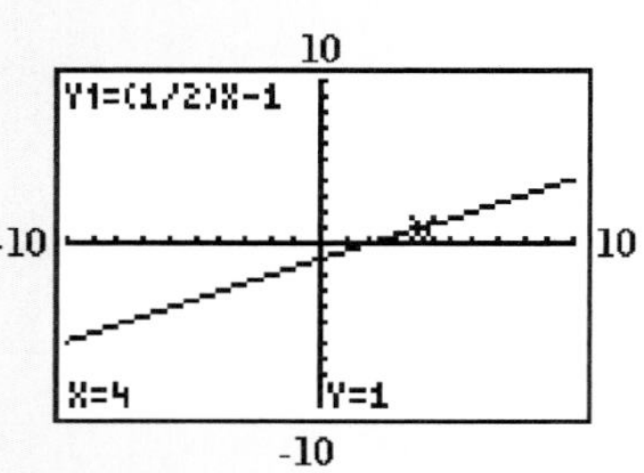

A function may be evaluated at numerical values or at algebraic expressions as shown in Example 4.

example 4 Evaluating Functions

Given the functions defined by $f(x) = x^2 - 2x$ and $g(x) = 3x + 5$, find the function values.

a. $f(t)$ b. $g(w + 4)$ c. $f(x + h)$

Solution:

a. $f(x) = x^2 - 2x$

$f(t) = (t)^2 - 2(t)$ — Substitute $x = t$ for all values of x in the function.

$= t^2 - 2t$ — Simplify.

b. $g(x) = 3x + 5$

$g(w + 4) = 3(w + 4) + 5$ — Substitute $x = w + 4$ for all values of x in the function.

$= 3w + 12 + 5$

$= 3w + 17$ — Simplify.

c. $f(x) = x^2 - 2x$

$f(x + h) = (x + h)^2 - 2(x + h)$ — Substitute the quantity $x + h$ for x.

$= x^2 + 2xh + h^2 - 2x - 2h$ — Simplify.

5. Finding Function Values from a Graph

We can find function values by looking at a graph of the function. The value $f(a)$ refers to the y-coordinate of a point with x-coordinate a.

example 5

Finding Function Values from a Graph

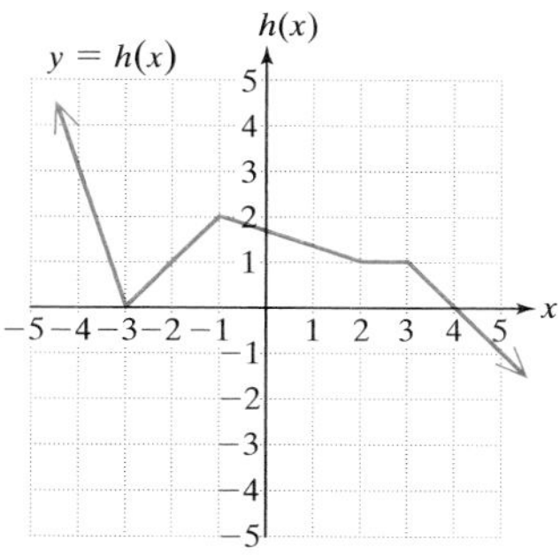

Figure 7-14

Consider the function pictured in Figure 7-14.

a. Find $h(-1)$.

b. Find $h(2)$.

c. For what value of x is $h(x) = 3$?

d. For what values of x is $h(x) = 0$?

Solution:

a. $h(-1) = 2$ This corresponds to the ordered pair $(-1, 2)$.

b. $h(2) = 1$ This corresponds to the ordered pair $(2, 1)$.

c. $h(x) = 3$ for $x = -4$ This corresponds to the ordered pair $(-4, 3)$.

d. $h(x) = 0$ for $x = -3$ and for $x = 4$. These are the ordered pairs $(-3, 0)$ and $(4, 0)$.

6. Domain of a Function

A function is a relation and it is often necessary to determine its domain and range. Consider a function defined by the equation $y = f(x)$. The **domain** of f is the set of all x-values that when substituted into the function produce a real number. The **range** of f is the set of all y-values corresponding to the values of x in the domain.

Thus far in your study of algebra we have seen one situation in which an expression will not be a real number. That is when we divide an expression by zero. A second situation is in taking a square root of a negative number. For example, $\sqrt{-4}$ is not a real number because there is no real number, b, such that $b^2 = -4$. Therefore, to find the domain of a function defined by $y = f(x)$, keep these guidelines in mind.

- Exclude values of x that make the denominator of a fraction zero.
- Exclude values of x that make the expression within a square root negative.

example 6 Finding the Domain of a Function

Find the domain of the following functions. Write the answers in interval notation.

a. $f(x) = \dfrac{x + 7}{2x - 1}$ b. $h(x) = \dfrac{x - 4}{x^2 + 9}$

c. $k(t) = \sqrt{t + 4}$ d. $g(t) = t^2 - 3t$

Solution:

a. The function will not be a real number when the denominator is zero. That is when,

$$2x - 1 = 0$$

$$2x = 1$$

$$x = \frac{1}{2}$$ The value $x = \frac{1}{2}$ must be *excluded* from the domain.

The domain of f is the set of all real numbers *excluding* $\frac{1}{2}$: $\{x \mid x \neq \frac{1}{2}\}$.

Interval notation: $\left(-\infty, \dfrac{1}{2}\right) \cup \left(\dfrac{1}{2}, \infty\right)$

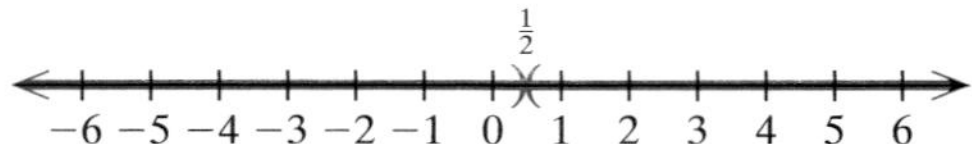

Tip: In Example 6(a), the domain of f is all real numbers except $\frac{1}{2}$. This means that any real number less than $\frac{1}{2}$, denoted $(-\infty, \frac{1}{2})$, or any real number greater than $\frac{1}{2}$, denoted $(\frac{1}{2}, \infty)$, is in the domain of f. The domain is $(-\infty, \frac{1}{2})$ or $(\frac{1}{2}, \infty)$. In interval notation, we write this as $(-\infty, \frac{1}{2}) \cup (\frac{1}{2}, \infty)$. The symbol $\cup$ is called a union symbol. Informally, the union symbol indicates that the domain is the collection (or union) of real numbers from *either* set. The formal definition of a union of two sets will be presented in Chapter 9.

b. The quantity x^2 is greater than or equal to zero for all real numbers x, and the number 9 is positive. Therefore, the sum, $x^2 + 9$, must be *positive* for all real numbers x. The denominator of $h(x) = (x - 4)/(x^2 + 9)$ will never be zero; the domain is therefore the set of all real numbers.

Interval notation: $(-\infty, \infty)$

−6 −5 −4 −3 −2 −1 0 1 2 3 4 5 6

c. The function defined by $k(t) = \sqrt{t + 4}$ will not be a real number when $t + 4$ is negative; hence the domain is the set of all t-values that make $t + 4$ *greater than or equal to zero:*

$$t + 4 \geq 0$$

$$t \geq -4$$

−6 −5 −4 −3 −2 −1 0 1 2 3 4 5 6

Interval notation: $[-4, \infty)$

d. The function defined by $g(t) = t^2 - 3t$ has no restrictions on its domain because any real number substituted for t will produce a real number. The domain is the set of all real numbers.

Interval notation: $(-\infty, \infty)$

−6 −5 −4 −3 −2 −1 0 1 2 3 4 5 6

section 7.4 Practice Exercises

For Exercises 1–4, (a) Write the relation as a set of ordered pairs. (b) Identify the domain. (c) Identify the range. (d) Is the relation a function?

1.

Parent (x)	Child (y)
Doris	Mike
Richard	Nora
Doris	Molly
Richard	Mike

2.

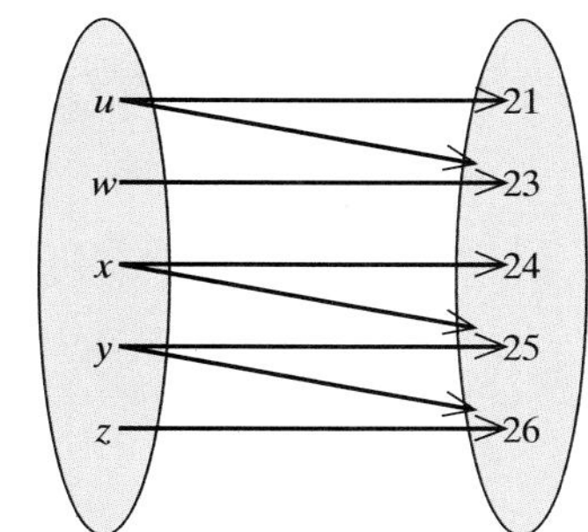

3.

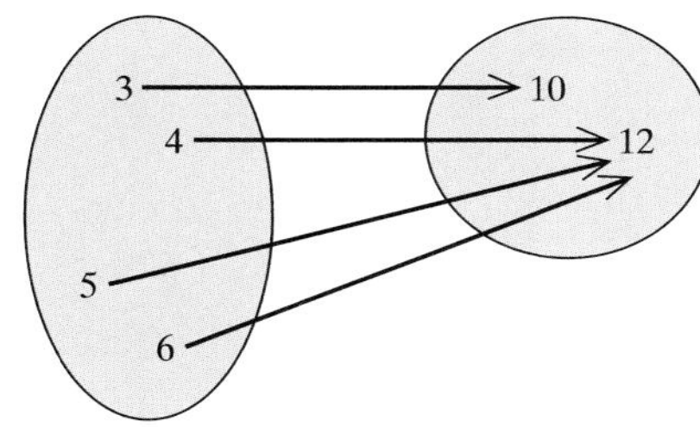

4.

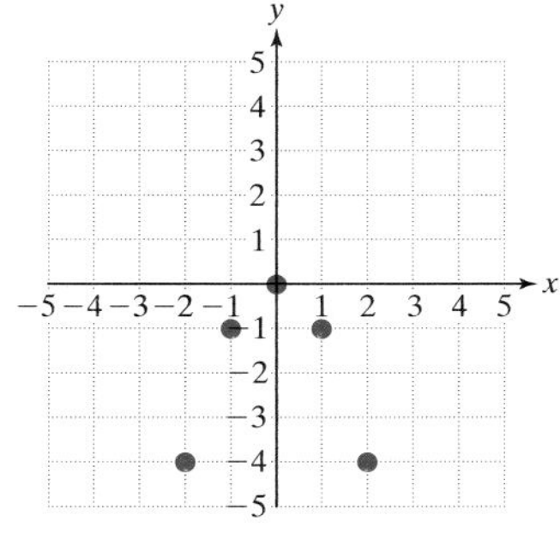

For Exercises 5–8, state the domain and range.

5.

6.

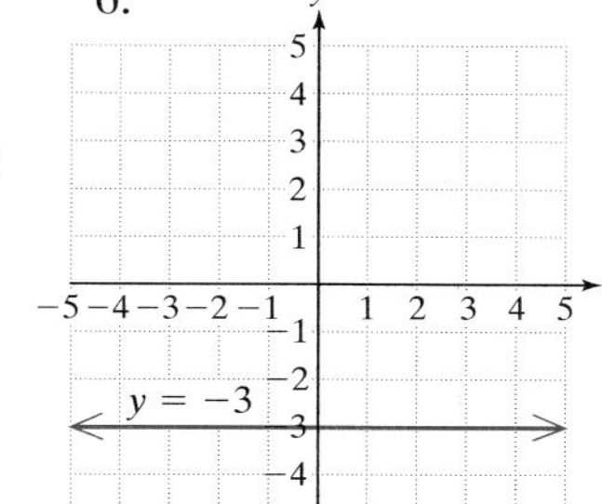

7.

8.

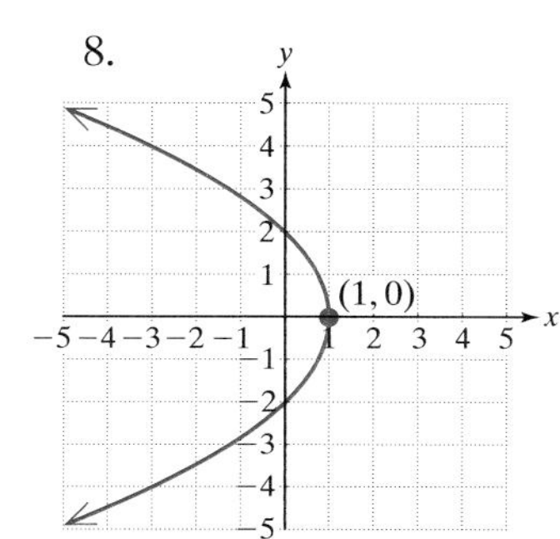

For Exercises 9–14, use the vertical line test to determine whether the relation defines y as a function of x.

9.

10.

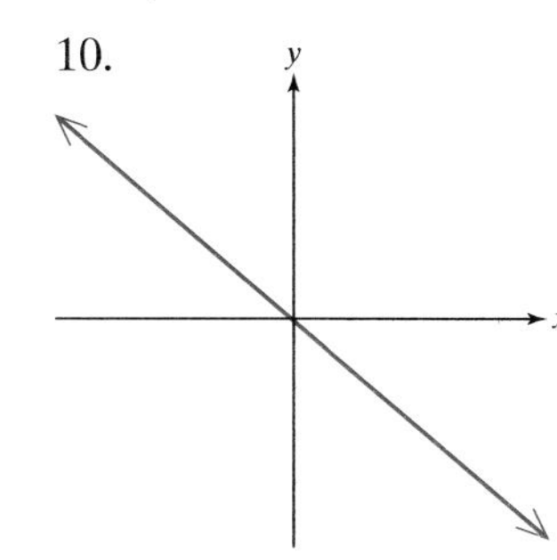

11.

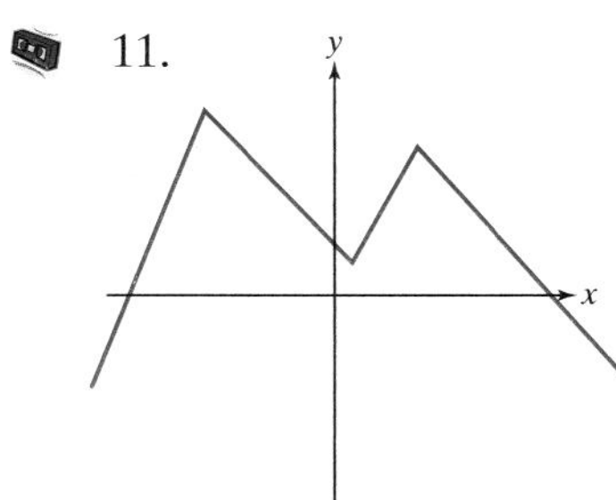

12.

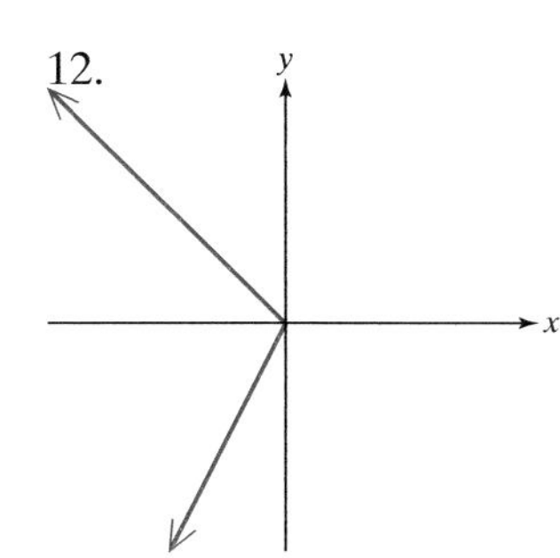

13. 14.

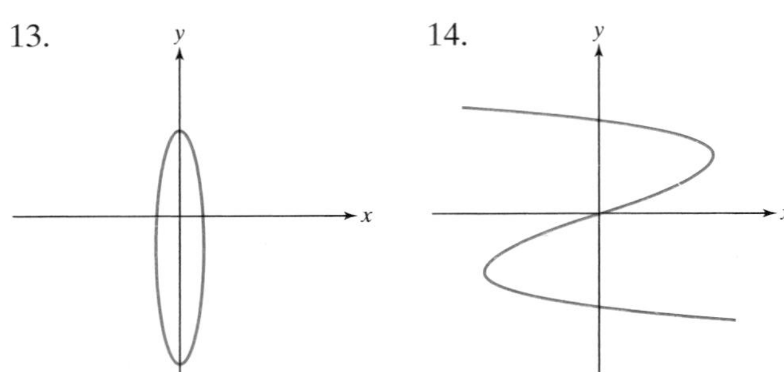

Consider the functions defined by $f(x) = 6x - 2$, $g(x) = x^2 - 4x + 1$, $h(x) = 7$ and $k(x) = |x - 2|$. For Exercises 15–48, find the function values.

15. $f(2)$
16. $g(2)$
17. $h(4)$
18. $k(2)$
19. $g(0)$
20. $h(0)$
21. $k(0)$
22. $f(0)$
23. $f(t)$
24. $g(a)$
25. $h(u)$
26. $k(v)$
27. $g(-3)$
28. $h(-5)$
29. $k(-2)$
30. $f(-6)$
31. $f(x + 1)$
32. $h(x + 1)$
33. $g(x - 2)$
34. $k(x - 3)$
35. $g(x + h)$
36. $k(x + h)$
37. $h(a + b)$
38. $f(x + h)$
39. $f(-a)$
40. $g(-b)$
41. $k(-c)$
42. $h(-x)$
43. $f\left(\frac{1}{2}\right)$
44. $g\left(\frac{1}{4}\right)$
45. $h\left(\frac{1}{7}\right)$
46. $k\left(\frac{3}{2}\right)$
47. $f(-2.8)$
48. $k(-5.4)$

Consider the functions $p = \{(\frac{1}{2}, 6), (2, -7), (1, 0), (3, 2\pi)\}$ and $q = \{(6, 4), (2, -5), (\frac{3}{4}, \frac{1}{5}), (0, 9)\}$. For Exercises 49–56, find the function values.

49. $p(2)$
50. $p(1)$
51. $p(3)$
52. $p\left(\frac{1}{2}\right)$
53. $q(2)$
54. $q\left(\frac{3}{4}\right)$
55. $q(6)$
56. $q(0)$

For Exercises 57–66, refer to the functions $y = f(x)$ and $y = g(x)$ defined as follows:

$f = \{(-3, 5), (-7, -3), (-\frac{3}{2}, 4), (1.2, 5)\}$
$g = \{(0, 6), (2, 6), (6, 0), (1, 0)\}$

57. Identify the domain of f.
58. Identify the range of f.
59. Identify the range of g.
60. Identify the domain of g.
61. For what value(s) of x is $f(x) = 5$?
62. For what value(s) of x is $f(x) = -3$?
63. For what value(s) of x is $g(x) = 0$?
64. For what value(s) of x is $g(x) = 6$?
65. Find $f(-7)$.
66. Find $g(0)$.
67. The graph of $y = f(x)$ is shown below.

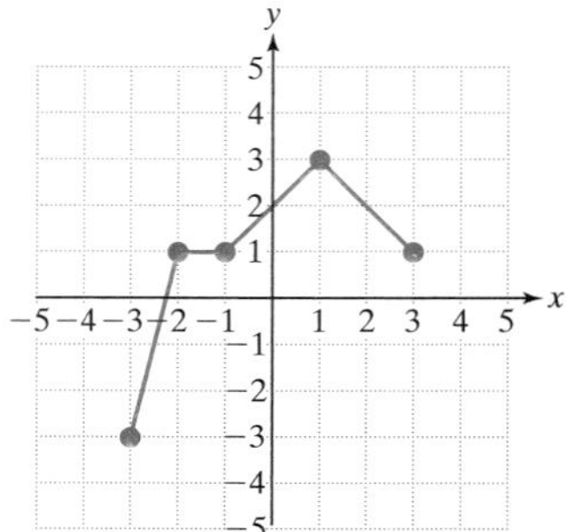

a. Evaluate $f(0)$.
b. Evaluate $f(3)$.
c. Evaluate $f(-2)$.
d. For what value(s) of x is $f(x) = -3$?
e. For what value(s) of x is $f(x) = 3$?
f. Write the domain of f.
g. Write the range of f.

68. The graph of $y = g(x)$ is shown below.

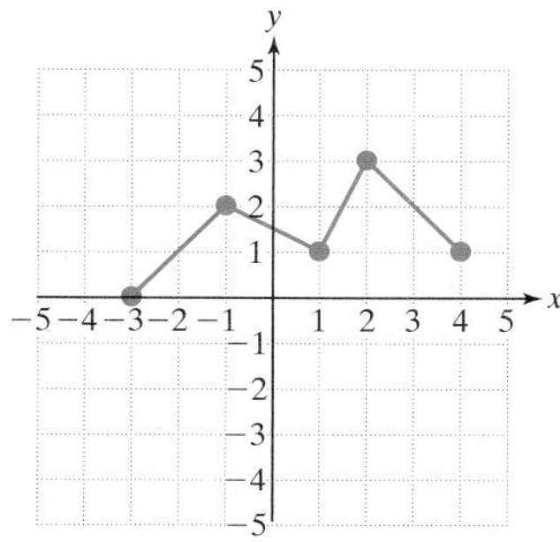

a. Evaluate $g(-1)$.
b. Evaluate $g(1)$.
c. Evaluate $g(4)$.
d. For what value(s) of x is $g(x) = 3$?
e. For what value(s) of x is $g(x) = 0$?
f. Write the domain of g.
g. Write the range of g.

69. Explain how to determine the domain of the function defined by

$$f(x) = \frac{x + 6}{x - 2}$$

For Exercises 70–75, find the domain. Write the answers in interval notation.

70. $k(x) = \dfrac{x - 3}{x + 6}$

71. $m(x) = \dfrac{x - 1}{x - 4}$

72. $f(t) = \dfrac{5}{t}$

73. $g(t) = \dfrac{t - 7}{t}$

74. $h(p) = \dfrac{p - 4}{p^2 + 1}$

75. $n(p) = \dfrac{p + 8}{p^2 + 2}$

76. Explain how to determine the domain of the function defined by $g(x) = \sqrt{x - 3}$.

For Exercises 77–82, find the domain. Write the answers in interval notation.

77. $h(t) = \sqrt{t + 7}$

78. $k(t) = \sqrt{t - 5}$

79. $f(a) = \sqrt{a - 3}$

80. $g(a) = \sqrt{a + 2}$

81. $m(x) = \sqrt{2x + 1}$

82. $n(x) = \sqrt{6x - 12}$

83. Explain how to determine the domain of the function defined by $h(x) = 2x^2 + 3$.

For Exercises 84–87, find the domain. Write the answers in interval notation.

84. $p(t) = 2t^2 + t - 1$

85. $q(t) = t^3 + t - 1$

86. $f(x) = x + 6$

87. $g(x) = 8x - \pi$

88. The height of a ball that is dropped from an 80-ft building is given by $h(t) = -16t^2 + 80$, where t is time in seconds after the ball is dropped.

a. Find $h(1)$ and $h(1.5)$
b. Interpret the meaning of the function values found in part (a).

89. A ball is dropped from a 50-m building. The height after t seconds is given by $h(t) = -4.9t^2 + 50$.

a. Find $h(1)$ and $h(1.5)$.
b. Interpret the meaning of the function values found in part (a).

90. If Alicia rides a bike at an average of 11.5 mph, the distance that she rides can be represented by $d(t) = 11.5t$, where t is the time in hours.

a. Find $d(1)$ and $d(1.5)$.
b. Interpret the meaning of the function values found in part (a).

91. If Miguel walks at an average of 5.9 km/hr, the distance that he walks can be represented by $d(t) = 5.9t$, where t is the time in hours.

a. Find $d(1)$ and $d(2)$.
b. Interpret the meaning of the function values found in part (a).

92. Brian's score on an exam is a function of the number of hours he spends studying. The function defined by $P(x) = \dfrac{100x^2}{50 + x^2}$ $(x \geq 0)$ indicates that he will achieve a score of $P\%$ if he studies for x hours.

a. Evaluate $P(0)$, $P(5)$, $P(10)$, $P(15)$, $P(20)$, and $P(25)$. (Round to one decimal place.) Interpret the results in the context of this problem.
b. Match the function values found in part (a) with the points A, B, C, D, E, and F on the graph.

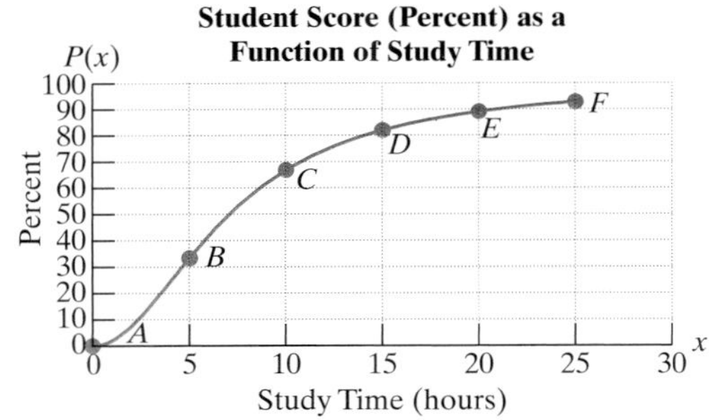

Figure for Exercise 92

93. The average number of visits to office-based physicians depends on the age of the patient according to

$N(a) = 0.0014a^2 - 0.0658a + 2.65$

where a is a patient's age in years and $N(a)$ is the average number of doctor visits per year.

a. Evaluate $N(1)$, $N(20)$, $N(40)$, and $N(75)$. (Round to one decimal place.) Interpret the results in the context of this problem.

b. Locate the function values from part (a) on the graph.

c. Based on the graph, approximately what age corresponds to the fewest doctor visits per year?

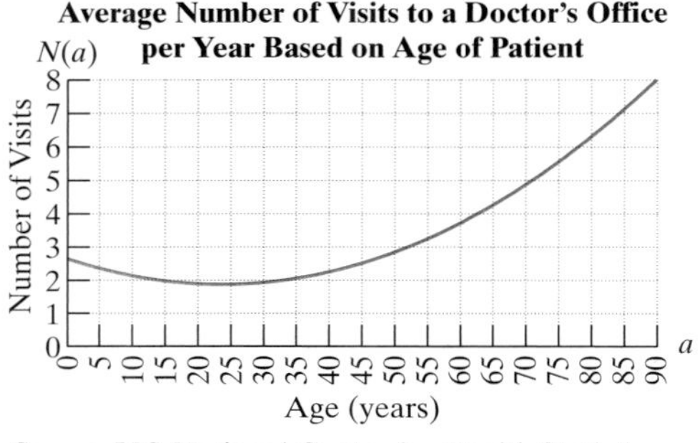

Source: U.S. National Center for Health Statistics.

Figure for Exercise 93

Expanding Your Skills

For Exercises 94–97, find the domain. Write the answers in interval notation.

94. $f(x) = \dfrac{x + 1}{3x + 1}$

95. $g(x) = \dfrac{x - 5}{6x - 2}$

96. $q(x) = \dfrac{2}{\sqrt{x + 2}}$

97. $p(x) = \dfrac{8}{\sqrt{x - 4}}$

Graphing Calculator Exercises

98. Graph $h(t) = \sqrt{t + 7}$. Use the graph to confirm the domain found in Exercise 77.

99. Graph $k(t) = \sqrt{t - 5}$. Use the graph to confirm the domain found in Exercise 78.

100. Graph $p(t) = 2t^2 + t - 1$. Use the graph to confirm the domain found in Exercise 84.

101. Graph $q(t) = t^3 + t - 1$. Use the graph to confirm the domain found in Exercise 85.

102. a. Graph $h(t) = -16t^2 + 80$ on a viewing window defined by $0 \le t \le 2$ and $0 \le y \le 100$.

b. Use the graph to approximate the function at $t = 1$ and $t = 1.5$. Use these values to support your answer to Exercise 88.

103. a. Graph $h(t) = -4.9t^2 + 50$ on a viewing window defined by $0 \le t \le 3$ and $0 \le y \le 60$.

b. Use the graph to approximate the function at $t = 1$ and $t = 1.5$. Use these values to support your answer to Exercise 89.

section

7.5 Graphs of Basic Functions

Concepts

1. Linear and Constant Functions
2. Applications of Linear Functions
3. Graphs of Basic Functions
4. Finding the x- and y-Intercepts of a Function Defined by $y = f(x)$
5. Definition of a Quadratic Function
6. Equations of Functions and Nonfunctions

1. Linear and Constant Functions

A function may be expressed as a mathematical equation that relates two or more variables. In this section, we will look at several elementary functions.

We know from Section 3.3 that an equation in the form $y = k$ is a horizontal line. In function notation, this can be written as $f(x) = k$. For example, the function defined by $f(x) = 3$ is a horizontal line as shown in Figure 7-15.

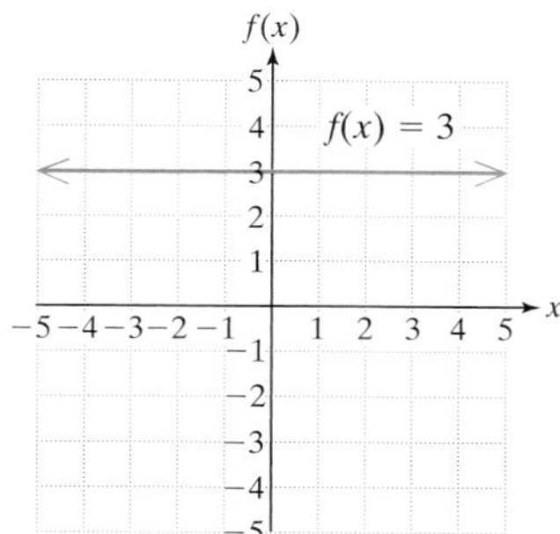

Figure 7-15

We say that a function defined by $f(x) = k$ is a constant function because for any value of x, the function value is constant.

An equation of the form $y = mx + b$ is represented graphically by a line with slope, m, and y-intercept $(0, b)$. In function notation, this may be written as $f(x) = mx + b$. A function in this form is called a linear function. For example, the function defined by $f(x) = 2x - 3$ is a linear function with slope $m = 2$ and y-intercept $(0, -3)$ (Figure 7-16).

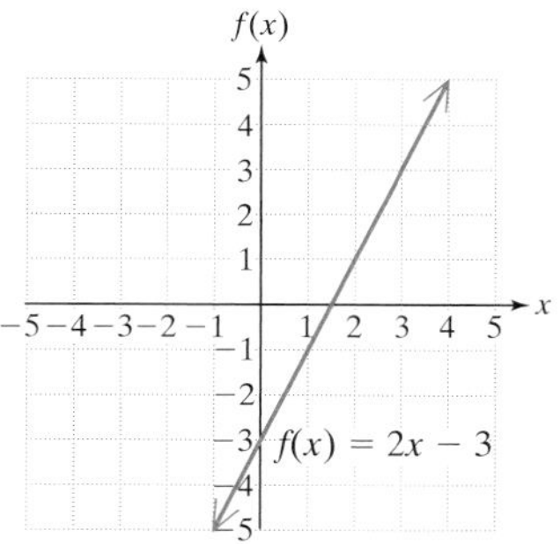

Figure 7-16

Definition of a Linear Function and a Constant Function

Let m and b represent real numbers such that $m \neq 0$; then,

A function that can be written in the form $f(x) = mx + b$ is a **linear function**.
A function that can be written in the form $f(x) = b$ is a **constant function**.

Note: The graphs of linear and constant functions are lines.

2. Applications of Linear Functions

example 1 **Solving an Application Involving a Linear Function**

The number of students receiving financial aid at a certain community college in the Midwest has grown between 1970 and 2006 according to

$$N(x) = 58x + 2050 \quad 0 \le x \le 36$$

For this function x represents the number of years since 1970 ($x = 0$ corresponds to 1970, $x = 1$ corresponds to 1971, and so on).

a. Is this function linear or nonlinear?
b. Find $N(0)$. Interpret the meaning of $N(0)$ in the context of this problem.
c. Find $N(20)$. Interpret the meaning of $N(20)$ in the context of this problem.
d. Find the year in which 2804 students received financial aid.

Solution:

a. The function $N(x) = 58x + 2050$ is linear, with $m = 58$ and $b = 2050$. The slope is 58 and the y-intercept is $(0, 2050)$.

b. $N(0) = 58(0) + 2050$ Substitute $x = 0$.
$= 2050$ This value indicates that in the year $x = 0$ (1970), 2050 students received financial aid.

c. $N(20) = 58(20) + 2050$ Substitute $x = 20$.
$= 1160 + 2050$
$= 3210$ This value indicates that in the year $x = 20$ (1990), 3210 students received financial aid.

d.
$$N(x) = 58x + 2050$$
$$2804 = 58x + 2050 \quad \text{Substitute } N(x) = 2804.$$
$$2804 - 2050 = 58x \quad \text{Solve for } x.$$
$$754 = 58x$$
$$\frac{754}{58} = \frac{58x}{58}$$
$$x = 13$$
In the year $x = 13$ (1983), 2804 students received financial aid.

The answers to this example can be confirmed from the graph of $N(x) = 58x + 2050$ (Figure 7-17).

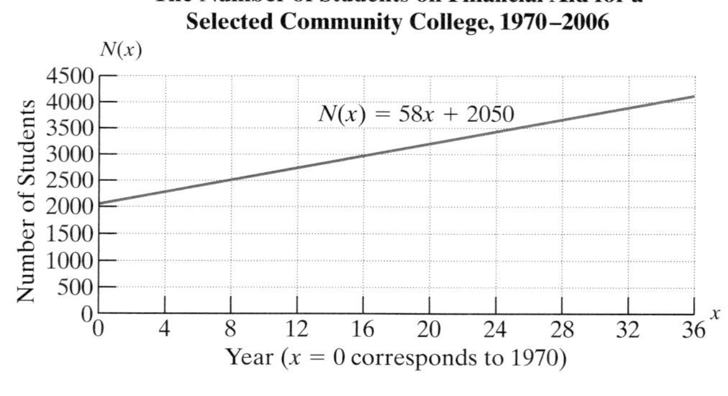

Figure 7-17

3. Graphs of Basic Functions

At this point, we are able to recognize the equations and graphs of linear and constant functions. In addition to linear and constant functions, the following equations define six basic functions that will be encountered in the study of algebra:

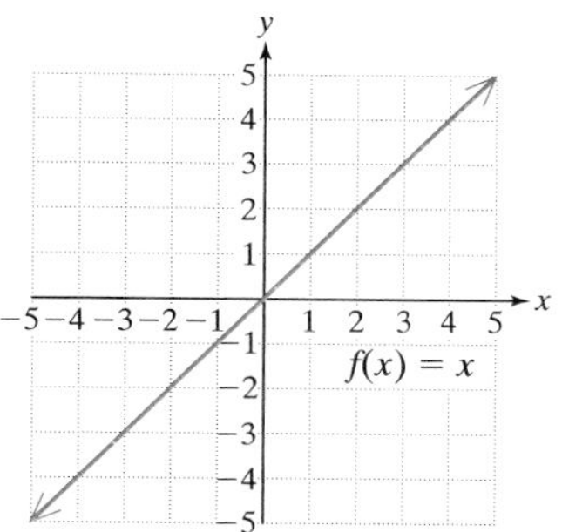

Figure 7-18

Equation		Function Notation
$y = x$		$f(x) = x$
$y = x^2$		$f(x) = x^2$
$y = x^3$	equivalent function notation →	$f(x) = x^3$
$y = \|x\|$		$f(x) = \|x\|$
$y = \sqrt{x}$		$f(x) = \sqrt{x}$
$y = \frac{1}{x}$		$f(x) = \frac{1}{x}$

The graph of the function defined by $f(x) = x$ is linear, with slope $m = 1$, and y-intercept $(0, 0)$ (Figure 7-18).

To determine the shapes of the other basic functions, we can plot several points to establish the pattern of the graph. Analyzing the equation itself may also provide insight to the domain, range, and shape of the function. To demonstrate this, we will graph $f(x) = x^2$ and $g(x) = \frac{1}{x}$.

example 2

Graphing Basic Functions

Graph the functions defined by

a. $f(x) = x^2$ b. $g(x) = \dfrac{1}{x}$

Solution:

a. The domain of the function given by $f(x) = x^2$ (or equivalently $y = x^2$) is all real numbers.

To graph the function, choose arbitrary values of x within the domain of the function. Be sure to choose values of x that are positive and values that are negative to determine the behavior of the function to the right and left of the origin (Table 7-5). The graph of $f(x) = x^2$ is shown in Figure 7-19. The function values are equated to the square of x, so $f(x)$ will always be greater than or equal to zero. Hence, the y-coordinates on the graph will never be negative. The range of the function is $\{y \mid y \geq 0\}$. The arrows on each branch of the graph imply that the pattern continues indefinitely.

Table 7-5

x	$f(x) = x^2$
0	0
1	1
2	4
3	9
−1	1
−2	4
−3	9

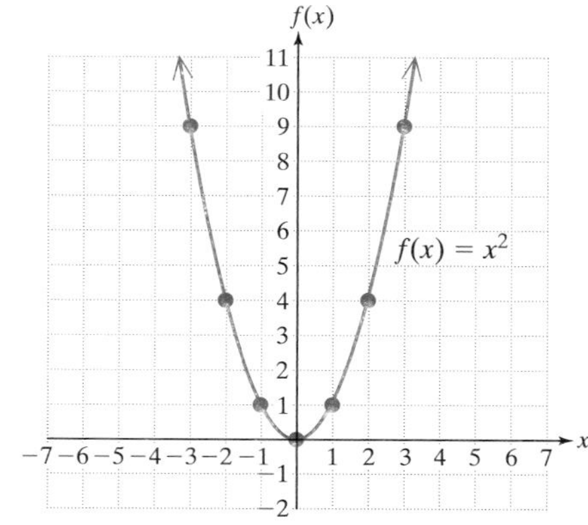

Figure 7-19

b. $g(x) = \frac{1}{x}$ Notice that $x = 0$ is not in the domain of the function. From the equation $y = \frac{1}{x}$, the y-values will be the reciprocal of the x-values. The graph defined by $g(x) = \frac{1}{x}$ is shown in Figure 7-20.

x	$g(x) = \frac{1}{x}$
1	1
2	$\frac{1}{2}$
3	$\frac{1}{3}$
-1	-1
-2	$-\frac{1}{2}$
-3	$-\frac{1}{3}$

x	$g(x) = \frac{1}{x}$
$\frac{1}{2}$	2
$\frac{1}{3}$	3
$\frac{1}{4}$	4
$-\frac{1}{2}$	-2
$-\frac{1}{3}$	-3
$-\frac{1}{4}$	-4

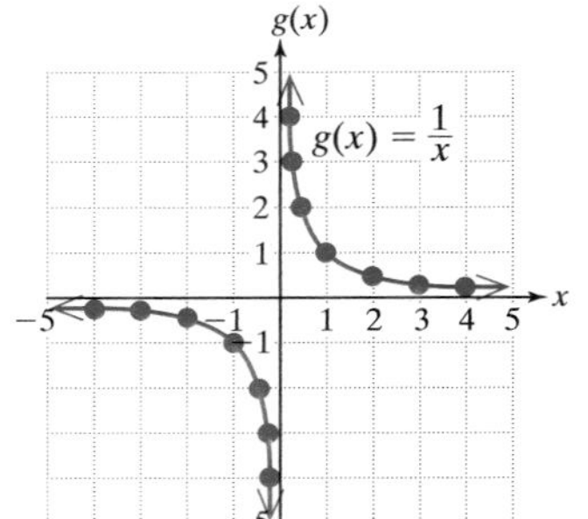

Figure 7-20

Notice that as x approaches ∞ and $-\infty$, the y-values approach zero, and the graph approaches the x-axis. In this case, the x-axis is called a **horizontal asymptote**. Similarly, the graph of the function approaches the y-axis as x gets close to zero. In this case, the y-axis is called a **vertical asymptote**.

Calculator Connections

The graphs of the functions defined by $f(x) = x^2$ and $g(x) = \frac{1}{x}$ are shown in the following calculator displays.

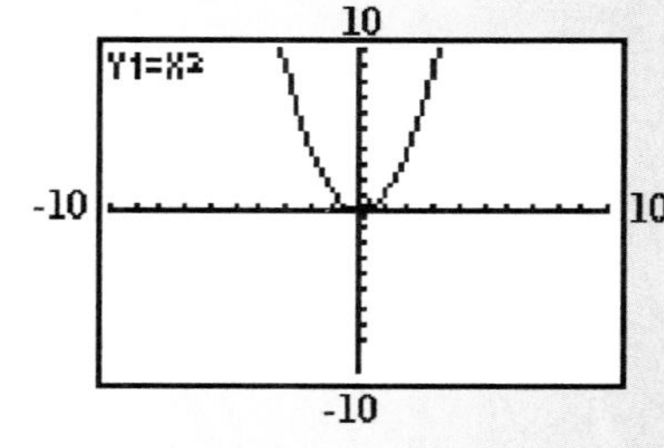

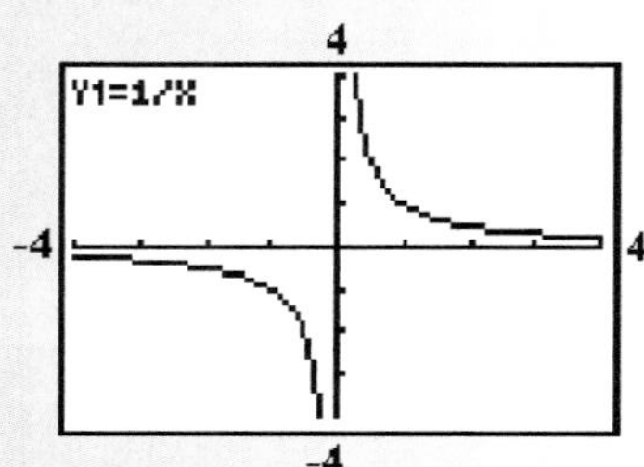

For your reference, we have provided the graphs of six basic functions in the following table.

Summary of Six Basic Functions and Their Graphs

Function	Graph	Domain and Range
1. $f(x) = x$	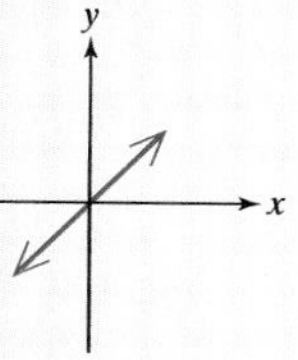	Domain $(-\infty, \infty)$ Range $(-\infty, \infty)$
2. $f(x) = x^2$	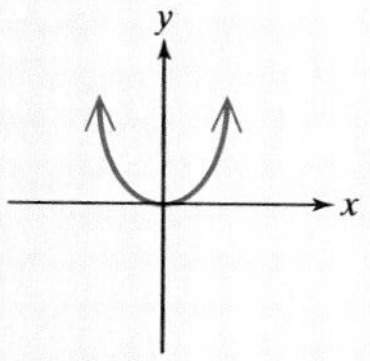	Domain $(-\infty, \infty)$ Range $[0, \infty)$
3. $f(x) = x^3$	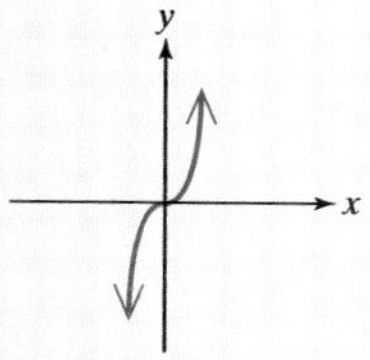	Domain $(-\infty, \infty)$ Range $(-\infty, \infty)$
4. $f(x) = \lvert x \rvert$	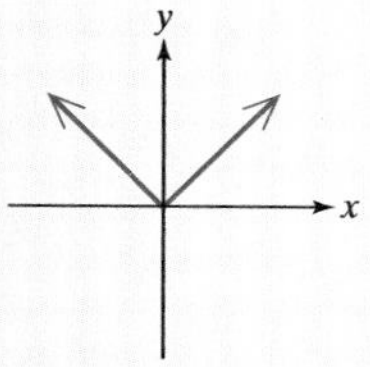	Domain $(-\infty, \infty)$ Range $[0, \infty)$
5. $f(x) = \sqrt{x}$	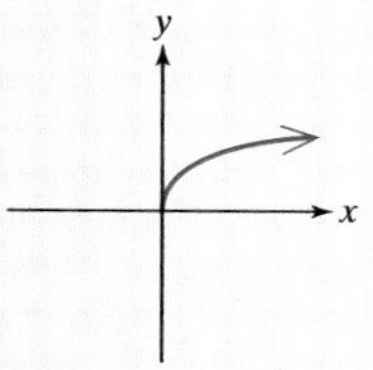	Domain $[0, \infty)$ Range $[0, \infty)$
6. $f(x) = \dfrac{1}{x}$	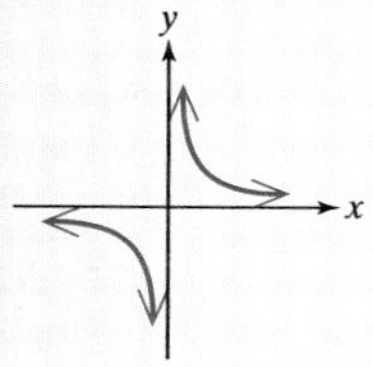	Domain $(-\infty, 0) \cup (0, \infty)$ Range $(-\infty, 0) \cup (0, \infty)$

The shapes of these six graphs will be developed in the homework exercises. These functions will be used often in the study of algebra. Therefore, we recommend that you associate an equation with its graph and commit each to memory.

4. Finding the x- and y-Intercepts of a Function Defined by $y = f(x)$

In Section 3.3, we learned that to find the x-intercept of an equation, we substitute $y = 0$ and solve for x. Using function notation, this is equivalent to finding the real solutions of the equation $f(x) = 0$. To find the y-intercept of an equation, substitute $x = 0$ and solve for y. Using function notation, this is equivalent to finding $f(0)$.

Finding the x- and y-Intercepts of a Function

Given a function defined by $y = f(x)$,

1. The ***x*-intercepts** are the real solutions to the equation $f(x) = 0$.
2. The ***y*-intercept** is given by $f(0)$.

example 3 Finding the x- and y-Intercepts of a Function

Given the function defined by $f(x) = 2x - 4$,

a. Find the x-intercept(s).
b. Find the y-intercept.
c. Graph the function.

Solution:

a. To find the x-intercept(s), find the real solutions to the equation $f(x) = 0$.

$f(x) = 2x - 4$

$0 = 2x - 4$ Substitute $f(x) = 0$.

$4 = 2x$

$2 = x$ The x-intercept is $(2, 0)$.

b. To find the y-intercept, evaluate $f(0)$.

$f(0) = 2(0) - 4$ Substitute $x = 0$.

$f(0) = -4$ The y-intercept is $(0, -4)$.

c. This function is linear, with a y-intercept of $(0, -4)$, an x-intercept of $(2, 0)$, and a slope of 2 (Figure 7-21).

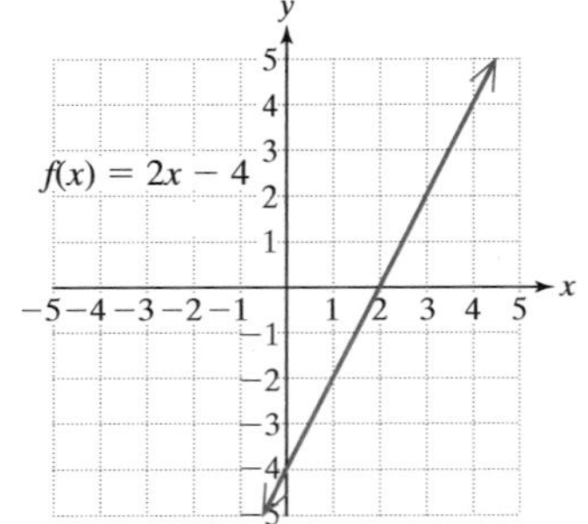

Figure 7-21

example 4

Finding the *x*- and *y*-Intercepts of a Function

Figure 7-22

For the function pictured in Figure 7-22, estimate

a. The real values of x for which $f(x) = 0$.
b. The value of $f(0)$.

Solution:

a. The real values of x for which $f(x) = 0$ are the x-intercepts of the function. For this graph, the x-intercepts are located at $x = -2$, $x = 2$, and $x = 3$.
b. The value of $f(0)$ is the value of y at $x = 0$. That is, $f(0)$ is the y-intercept. $f(0) = 6$.

example 5

Finding the *x*- and *y*-Intercepts of a Function

For the function pictured in Figure 7-23, estimate

a. The real values of x for which $f(x) = 0$.
b. The value of $f(0)$.

Solution:

a. There are no x-intercepts for this graph; therefore, there are no real values of x for which $f(x) = 0$.
b. From the graph, the value of $f(0)$ is 3.

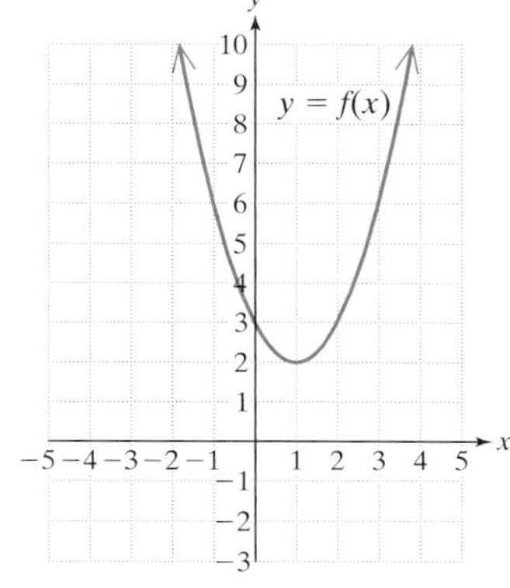

Figure 7-23

5. Definition of a Quadratic Function

In Example 2 we graphed the function defined by $f(x) = x^2$ by plotting points. This function belongs to a special category called **quadratic functions**. A quadratic function can be written in the form $f(x) = ax^2 + bx + c$, where a, b, and c are real numbers and $a \neq 0$. The graph of a quadratic function is in the shape of a **parabola**. The leading coefficient, a, determines the direction of the parabola.

If $a > 0$, then the parabola opens upward. For example, $f(x) = x^2$. The minimum point on a parabola opening upward is called the vertex.

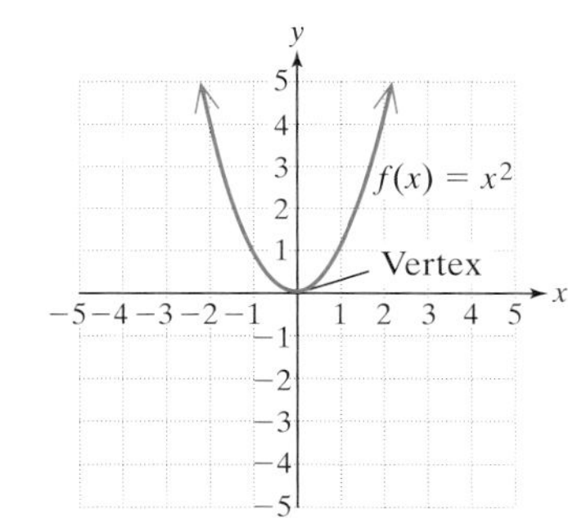

If $a < 0$, then the parabola opens downward. For example, $f(x) = -x^2$. The maximum point on a parabola opening downward is called the vertex.

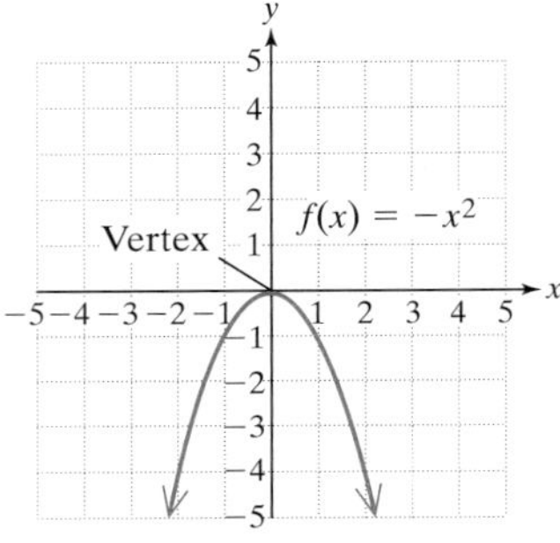

example 6 Finding the *x*- and *y*-Intercepts of a Quadratic Function

Given $f(x) = x^2 - x - 12$,

a. Find the *x*-intercept(s). b. Find the *y*-intercept.

Solution:

a. To find the *x*-intercept(s) find the real solutions to the equation $f(x) = 0$.

$f(x) = x^2 - x - 12$

$0 = x^2 - x - 12$ Substitute $f(x) = 0$. The result is a quadratic equation.

$0 = (x - 4)(x + 3)$ Factor.

$x - 4 = 0$ or $x + 3 = 0$ Set each factor equal to zero.

$x = 4$ or $x = -3$ The *x*-intercepts are $(4, 0)$ and $(-3, 0)$.

b. To find the *y*-intercept, evaluate $f(0)$.

$f(0) = (0)^2 - (0) - 12$

$f(0) = -12$ The *y*-intercept is $(0, -12)$.

Calculator Connections

In Example 6 the function defined by $f(x) = x^2 - x - 12$ is quadratic. We can use a graphing calculator to verify that its shape is a parabola. Furthermore, the graph appears to cross the *x*-axis at -3 and 4 and to cross the *y*-axis at -12, as expected.

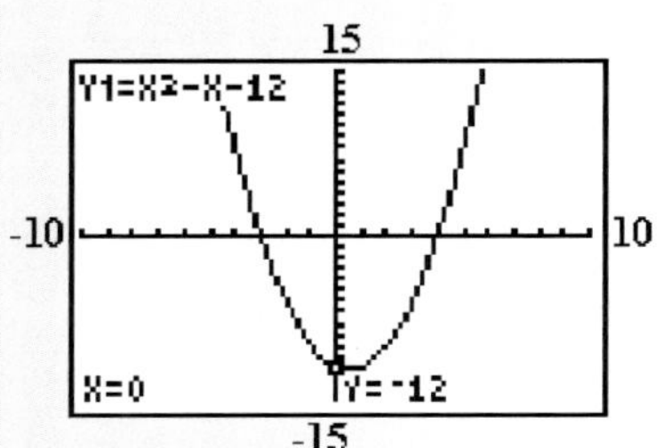

6. Equations of Functions and Nonfunctions

We have seen several equations that represent y as a function of x. In each case, notice that the y-variable is expressed in terms of x, and for every x in the domain, there is only one y-value produced by the equation. Some equations, however, do not represent y as a function of x.

example 7 Determining Whether an Equation Represents a Function

Determine whether the following equations represent y as a function of x.

a. $x = |y|$ b. $y = x^3 - 4$ c. $y = \pm\sqrt{x}$

Solution:

a. $x = |y|$

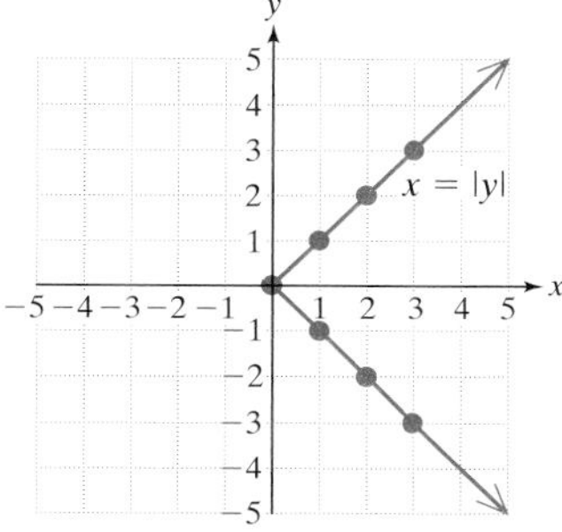

Figure 7-24

To determine if this is a function, we ask the question, "For each value of x in the domain, will there be exactly one corresponding y-value?"

If $x = 1$, we have
$$x = |y|$$
$$1 = |y|$$
$$y = 1 \quad \text{or} \quad y = -1$$

Because two values of y correspond to $x = 1$, this equation is not a function.

We can graph the equation $x = |y|$ by plotting several solutions to the equation. By the vertical line test, we can verify that this relation is not a function (Figure 7-24).

b. $y = x^3 - 4$

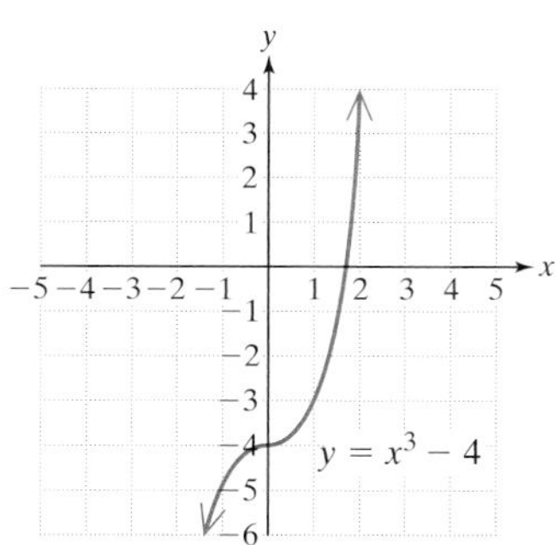

Figure 7-25

For any value of x, the cube of x is unique. Therefore, only one y-value corresponds to a given value of x. The equation is a function.

The graph of $y = x^3 - 4$ is shown in Figure 7-25. By the vertical line test, we can verify that this relation is a function.

c. $y = \pm\sqrt{x}$

The symbol, $\pm$ is read as "plus or minus." Therefore, $y = \pm\sqrt{x}$ is read as "y equals plus or minus the square root of x." For any value of $x > 0$, there will be two corresponding values of y. For example, if $x = 4$, then we have:

$$y = \pm\sqrt{x}$$
$$= \pm\sqrt{4}$$
$$= \pm 2$$

Because two values of y correspond to $x = 4$, this equation is not a function. The graph of $y = \pm\sqrt{x}$ is shown in Figure 7-26. By the vertical line test, we can verify that this relation is not a function.

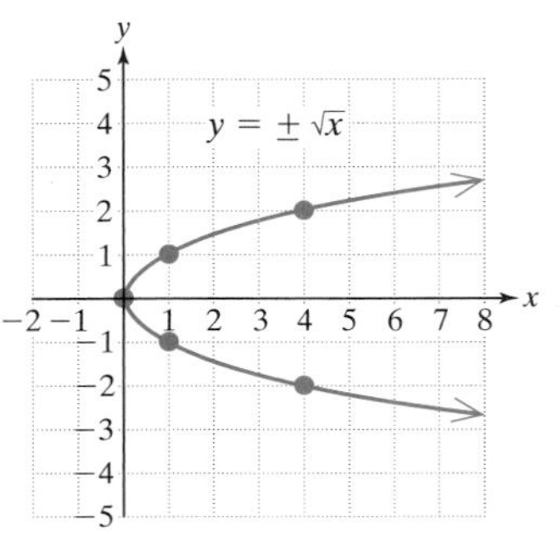

Figure 7-26

section 7.5 Practice Exercises

1. For $g = \{(6, 1), (5, 2), (4, 3), (3, 4), (2, 5), (1, 6)\}$
   a. Is this relation a function?
   b. List the elements of the domain.
   c. List the elements of the range.

2. For $f = \{(7, 3), (2, 3), (-5, 3), (9, 3)\}$
   a. Is this relation a function?
   b. List the elements of the domain.
   c. List the elements of the range.

Consider $k(x) = x^2 - 2$. For Exercises 3–10, find the function values.

3. $k(0)$
4. $k(6)$
5. $k(a)$
6. $k(a + 2)$
7. $k(-2)$
8. $k(2)$
9. $k(b - 1)$
10. $k(b)$

For Exercises 11–14, write the domain in interval notation.

11. $f(x) = x - 6$
12. $g(x) = \dfrac{x + 2}{x - 6}$
13. $h(x) = \sqrt{x - 6}$
14. $k(x) = x^2 - 6$

15. The force (measured in pounds) to stretch a certain spring x inches is given by $f(x) = 3x$.
    a. Evaluate $f(3)$ and interpret the results in the context of the problem.
    b. Evaluate $f(0)$ and interpret the results in the context of the problem.

16. The acceleration (measured in feet per second squared) of a falling object is given by $A(x) = 32$, where x is the number of seconds after the object was released.
    a. Evaluate $A(1)$ and interpret the results in the context of the problem.
    b. Evaluate $A(4)$ and interpret the results in the context of the problem.

For Exercises 17–22, sketch a graph by completing the table and plotting the points.

17. $g(x) = |x|$

x	$g(x)$
-2	
-1	
0	
1	
2	

18. $f(x) = \dfrac{1}{x}$

x	$f(x)$
-2	
-1	
$-1/2$	
$-1/4$	

x	$f(x)$
1/4	
1/2	
1	
2	

19. $h(x) = x^3$

x	$h(x)$
-2	
-1	
0	
1	
2	

20. $k(x) = x$

x	$k(x)$
-2	
-1	
0	
1	
2	

21. $p(x) = \sqrt{x}$

x	$p(x)$
0	
1	
4	
9	
16	

22. $q(x) = x^2$

x	$q(x)$
-2	
-1	
0	
1	
2	

23. For $f(x) = \sqrt{x + 4}$
   a. Write the domain of f in interval notation.
   b. Find $f(0)$, $f(5)$, $f(-3)$.

24. For $g(x) = \sqrt{x - 3}$
   a. Write the domain of g in interval notation.
   b. Find $g(19)$, $g(7)$, $g(3)$.

25. For $h(x) = -x^2 + 2$
   a. Write the domain of h in interval notation.
   b. Find $h(1)$, $h(-1)$, $h(0)$.

26. For $k(x) = x^2 + 2$
   a. Write the domain of k in interval notation.
   b. Find $k(2)$, $k(-2)$, $k(0)$.

27. For $p(x) = \dfrac{2}{x - 3}$
   a. Write the domain of p in interval notation.
   b. Find $p(0)$, $p(1)$, $p(2)$, $p(4)$, $p(5)$, $p(6)$.

28. For $q(x) = \dfrac{3}{x - 2}$
   a. Write the domain of q in interval notation.
   b. Find $q(-1)$, $q(0)$, $q(1)$, $q(3)$, $q(4)$, $q(5)$.

For Exercises 29–40, find the x- and y-intercepts.

29. $M(x) = 8x + 1$

30. $N(x) = -3x + 5$

31. $C(x) = -5x$

32. $D(x) = \dfrac{1}{2}x$

33. $A(x) = (2x + 1)(x - 5)$

34. $B(x) = (x + 3)(4x + 1)$

35. $p(x) = x^2 - 3x - 10$

36. $q(x) = x^2 - 5x + 4$

37. $g(x) = x^2 + 6x + 9$

38. $f(x) = x^2 - 9$

39. $h(x) = 4x(x - 3)(3x + 2)$

40. $k(x) = x(2x - 5)(x + 10)$

For Exercises 41–46, use the graph of the function to determine the x- and y-intercepts (if they exist).

41. $g(x) = x^3 + 1$

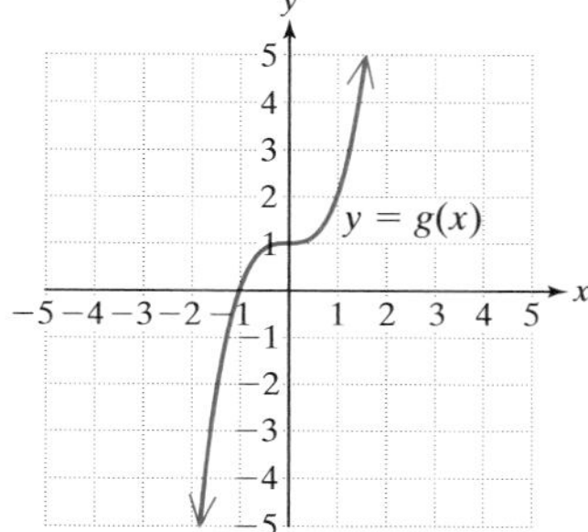

42. $h(x) = x^3 - 1$

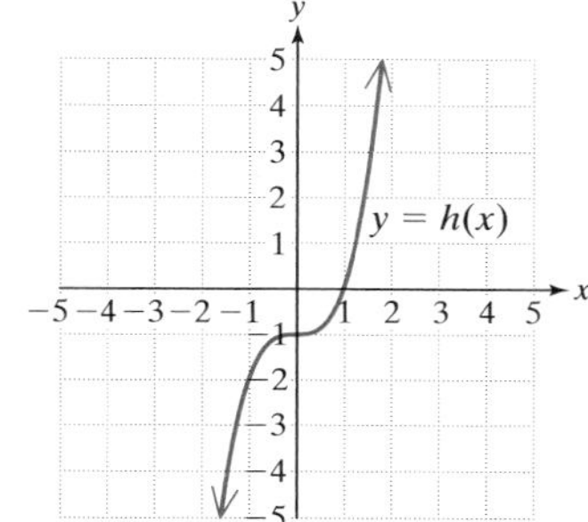

43. $f(x) = |x| - 2$

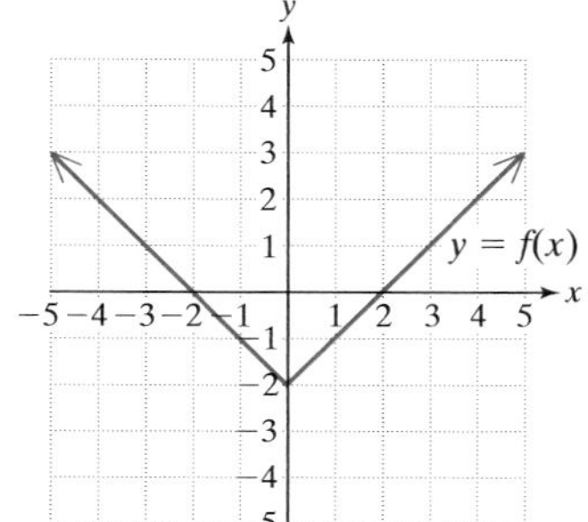

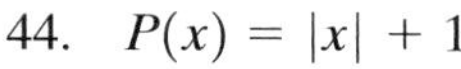
44. $P(x) = |x| + 1$

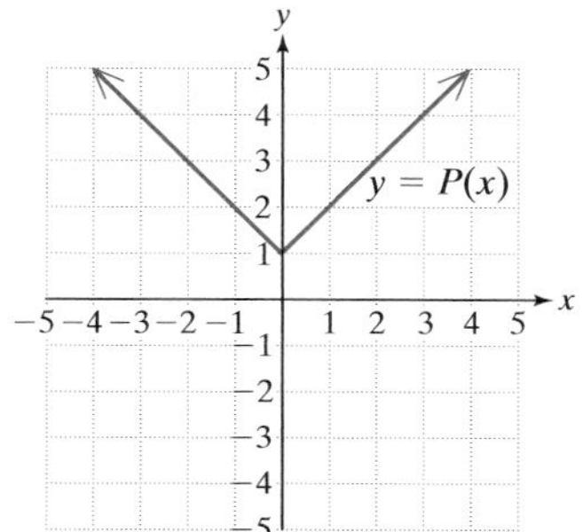

45. $r(x) = x^2 + 2$

46. $q(x) = -x^2 + 1$

47. a. Sketch the relation $y = x^2$ from memory.

b. Does $y = x^2$ define y as a function of x? Why or why not?

c. Sketch the relation $x = y^2$ by completing the table and plotting the points.

x	y
	-2
	-1
	0
	1
	2

d. Does $x = y^2$ define y as a function of x? Why or why not?

48. a. Sketch the relation $y = |x|$ from memory.

b. Does $y = |x|$ define y as a function of x? Why or why not?

c. Sketch the relation $x = |y|$ by completing the table and plotting the points.

x	y
	-2
	-1
	0
	1
	2

d. Does $x = |y|$ define y as a function of x? Why or why not?

For Exercises 49–58,

a. Identify the domain of the function.

b. Identify the y-intercept of the function.

c. Match the function with its graph by recognizing the basic shape of the function and using the results from parts (a) and (b). Plot additional points if necessary.

49. $q(x) = 2x^2$

50. $p(x) = -2x^2 + 1$

51. $h(x) = x^3 + 1$

52. $k(x) = x^3 - 2$

53. $r(x) = \sqrt{x + 1}$

54. $s(x) = \sqrt{x + 4}$

55. $f(x) = \dfrac{1}{x - 3}$

56. $g(x) = \dfrac{1}{x + 1}$

57. $k(x) = |x + 2|$

58. $h(x) = |x - 1| + 2$

i.

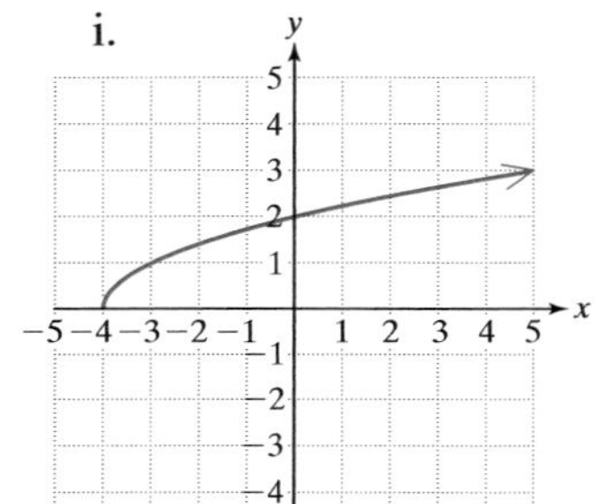

ii.

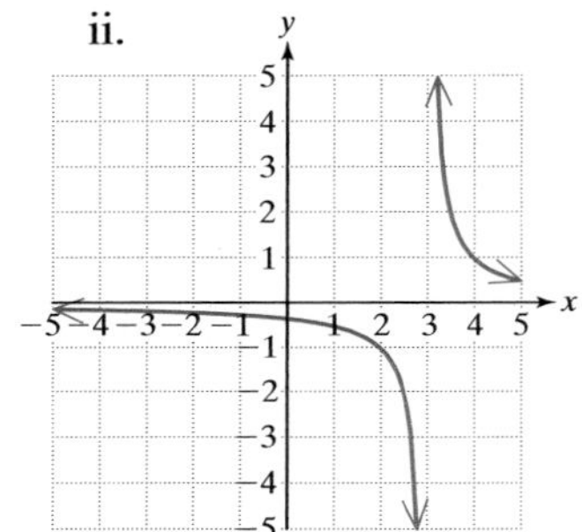

iii.

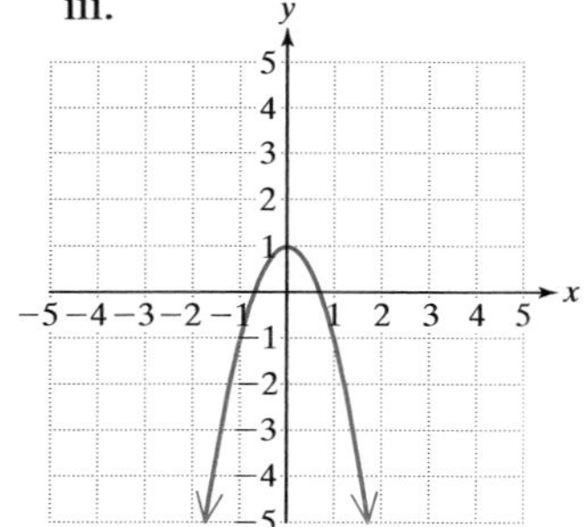

iv.

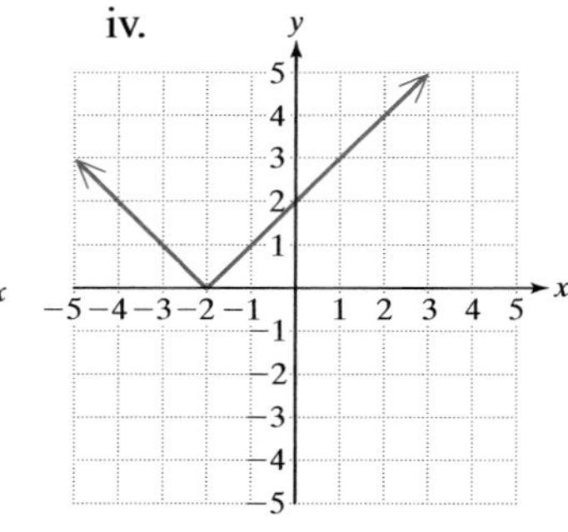

v.

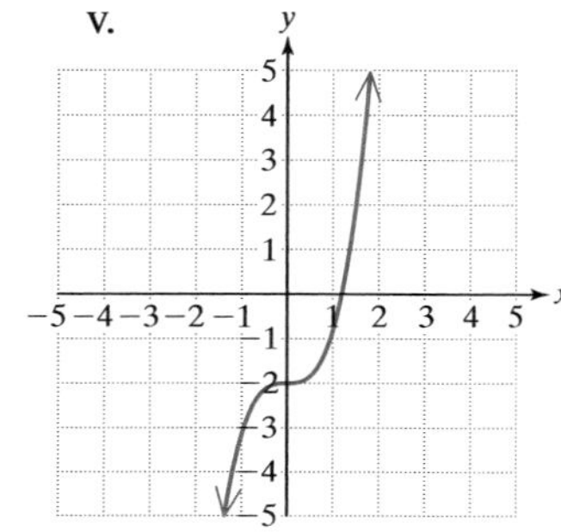

vi.

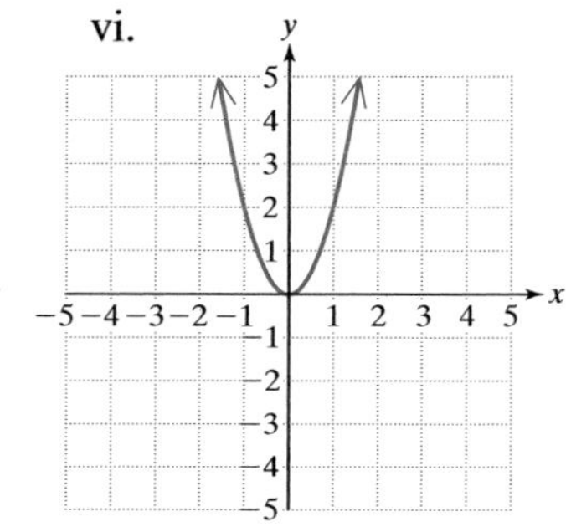

vii.

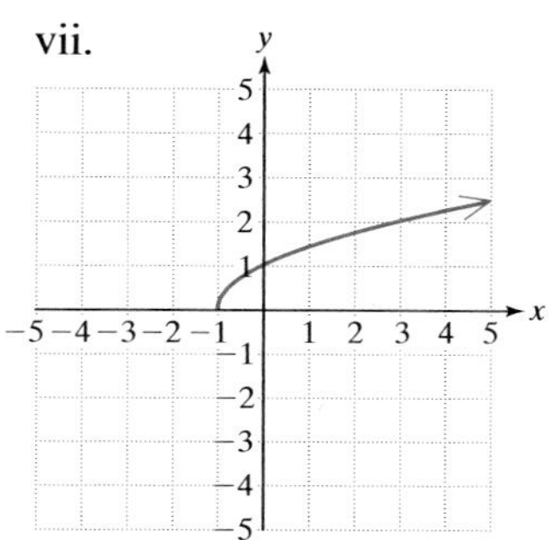

viii.

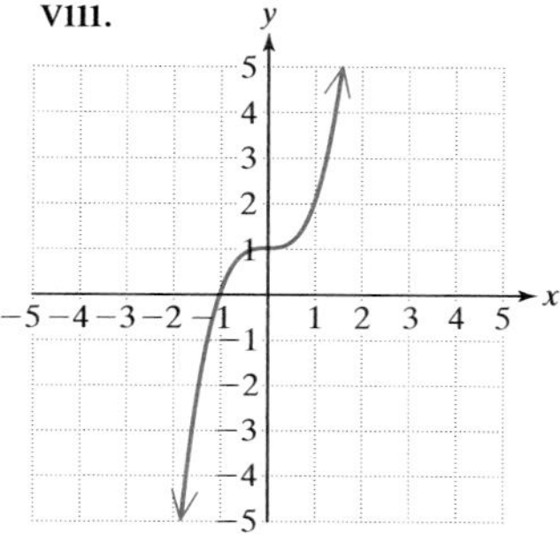

ix.

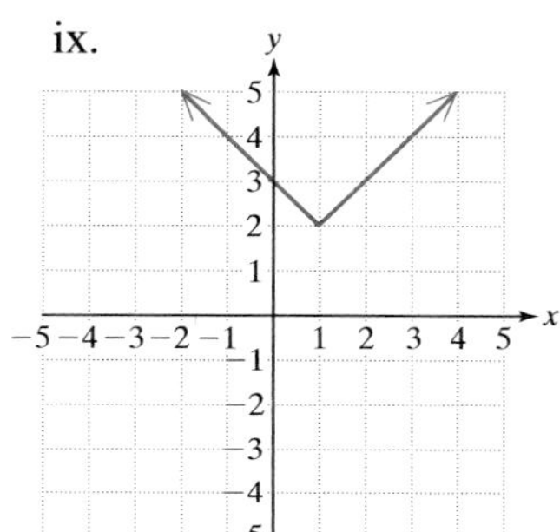

x.

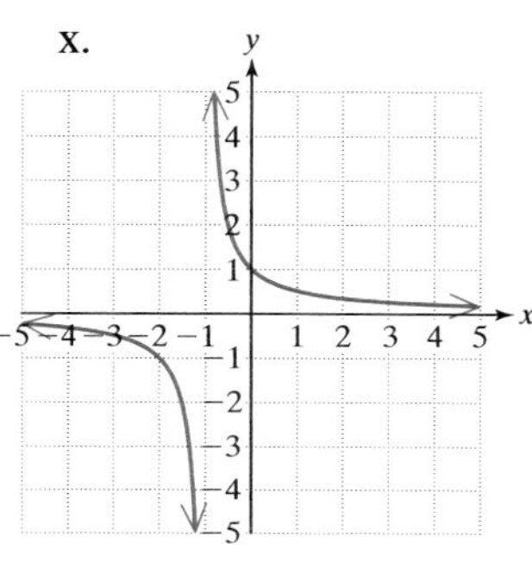

Graphing Calculator Exercises

59. Repeat Exercise 25b using a *Table* feature of your graphing calculator.

60. Repeat Exercise 26b using a *Table* feature of your graphing calculator.

61. Repeat Exercise 27b using a *Table* feature of your graphing calculator.

62. Repeat Exercise 28b using a *Table* feature of your graphing calculator.

Concepts

1. Definition of Direct and Inverse Variation
2. Translations Involving Variation Models
3. Definition of Joint Variation
4. Applications of Variation

section

7.6 Variation

1. Definition of Direct and Inverse Variation

In this section, we introduce the concept of variation. Direct and inverse variation models can show how one quantity varies in relation to another.

Definition of Direct and Inverse Variation

Let k be a nonzero constant real number. Then the following statements are equivalent:

1. y varies **directly** as x.
   y is directly proportional to x. } $y = kx$
2. y varies **inversely** as x.
   y is inversely proportional to x. } $y = \frac{k}{x}$

Note: The value of k is called the constant of variation.

For a car traveling at 30 mph, the equation $d = 30t$ indicates that the distance traveled is *directly proportional* to the time of travel. For positive values of k, when two variables are directly related, as one variable increases, the other variable will also increase. Likewise, if one variable decreases, the other will decrease. In the equation $d = 30t$, the longer the time of the trip, the greater the distance traveled. The shorter the time of the trip, the shorter the distance traveled.

For positive values of k, when two variables are *inversely related,* as one variable increases, the other will decrease, and vice versa. Consider a car traveling between Toronto and Montreal, a distance of 500 km. The time required to make the trip is inversely proportional to the speed of travel: $t = 500/r$. As the rate of speed, r, increases, the quotient $500/r$ will decrease. Hence the time will decrease. Similarly, as the rate of speed decreases, the trip will take longer.

2. Translations Involving Variation Models

The first step in using a variation model is to translate an English phrase into an equivalent mathematical equation.

example 1

Translating to a Variation Model

Translate each expression into an equivalent mathematical model.

a. The circumference of a circle varies directly as the radius.
b. At a constant temperature, the volume of a gas varies inversely as the pressure.
c. The length of time of a meeting is directly proportional to the *square* of the number of people present.

Solution:

a. Let C represent circumference and r represent radius. The variables are directly related, so use the model $C = kr$.
b. Let V represent volume and P represent pressure. Because the variables are inversely related, use the model $V = \frac{k}{P}$.
c. Let t represent time and let N be the number of people present at a meeting. Because t is directly related to N^2, use the model $t = kN^2$.

3. Definition of Joint Variation

Sometimes a variable varies directly as the product of two or more other variables. In this case, we have joint variation.

Definition of Joint Variation

Let k be a nonzero constant real number. Then the following statements are equivalent:

y varies **jointly** as w and z.
y is jointly proportional to w and z.
} $y = kwz$

example 2

Translating to a Variation Model

Translate each expression into an equivalent mathematical model.

a. y varies jointly as u and the square root of v.
b. The gravitational force of attraction between two planets varies jointly as the product of their masses and inversely as the square of the distance between them.

Solution:

a. $y = ku\sqrt{v}$

b. Let m_1 and m_2 represent the masses of the two planets. Let F represent the gravitational force of attraction and d represent the distance between the planets.

The variation model is $F = \dfrac{km_1m_2}{d^2}$.

4. Applications of Variation

Consider the variation models $y = kx$ and $y = k/x$. In either case, if values for x and y are known, we can solve for k. Once k is known, we can use the variation equation to find y if x is known, or to find x if y is known. This concept is the basis for solving many problems involving variation.

Steps to Find a Variation Model

1. Write a general variation model that relates the variables given in the problem. Let k represent the constant of variation.
2. Solve for k by substituting known values of the variables into the model from step 1.
3. Substitute the value of k into the original variation model from step 1.

example 3

Solving an Application Involving Direct Variation

The variable z varies directly as w. When w is 16, z is 56.

a. Write a variation model for this situation.
b. Solve for the constant of variation.
c. Find the value of z when w is 84.

Solution:

a. $z = kw$

b. $z = kw$

$56 = k(16)$ Substitute known values for z and w. Then solve for the unknown value of k.

$\frac{56}{16} = \frac{k(16)}{16}$ To isolate k, divide both sides by 16.

$\frac{7}{2} = k$ Reduce $\frac{56}{16}$ to $\frac{7}{2}$.

c. With the value of k known, the variation model can now be written as $z = \frac{7}{2}w$.

$z = \frac{7}{2}(84)$ To find z when $w = 84$, substitute $w = 84$ into the equation.

$z = 294$

example 4 Solving an Application Involving Direct Variation

The speed of a racing canoe in still water varies directly as the square root of the length of the canoe.

a. If a 16-ft canoe can travel 6.2 mph in still water, find a variation model that relates the speed of a canoe to its length.
b. Find the speed of a 25-ft canoe.

Solution:

a. Let s represent the speed of the canoe, and L represent the length. The general variation model is $s = k\sqrt{L}$. To solve for k, substitute the known values for s and L.

$s = k\sqrt{L}$

$6.2 = k\sqrt{16}$ Substitute $s = 6.2$ mph and $L = 16$ ft.

$6.2 = k \cdot 4$

$\frac{6.2}{4} = \frac{4k}{4}$ Solve for k.

$1.55 = k$

$s = 1.55\sqrt{L}$ Substitute $k = 1.55$ into the model $s = k\sqrt{L}$.

b. $s = 1.55\sqrt{L}$

$= 1.55\sqrt{25}$ Find the speed when $L = 25$ ft.

$= 7.75$ mph

example 5

Solving an Application Involving Inverse Variation

The loudness of sound measured in decibels varies inversely as the square of the distance between the listener and the source of the sound. If the loudness of sound is 17.92 decibels at a distance of 10 ft from a stereo speaker, what is the decibel level 20 ft from the speaker?

Solution:

Let L represent the loudness of sound in decibels and d represent the distance in feet. The inverse relationship between decibel level and the square of the distance is modeled by

$$L = \frac{k}{d^2}$$

$$17.92 = \frac{k}{(10)^2} \qquad \text{Substitute } L = 17.92 \text{ decibels and } d = 10 \text{ ft.}$$

$$17.92 = \frac{k}{100}$$

$$(17.92)100 = \frac{k}{100} \cdot 100 \qquad \text{Solve for } k \text{ (clear fractions).}$$

$$k = 1792$$

$$L = \frac{1792}{d^2} \qquad \text{Substitute } k = 1792 \text{ into the original model } L = \frac{k}{d^2}.$$

With the value of k known, we can find the value of L for any value of d.

$$L = \frac{1792}{(20)^2} \qquad \text{Find the loudness when } d = 20 \text{ ft.}$$

$$= 4.48 \text{ decibels}$$

Notice that the loudness of sound is 17.92 decibels at a distance 10 ft from the speaker. When the distance from the speaker is increased to 20 ft, the decibel level decreases to 4.48 decibels. This is consistent with an inverse relationship. For $k > 0$, as one variable is increased, the other is decreased. It also seems reasonable that the farther one moves away from the source of a sound, the softer the sound becomes.

example 6

Solving an Application Involving Joint Variation

From August 23 to August 28, 1992, Hurricane Andrew carved a path of destruction from the Caribbean to South Florida and from the Louisiana coast to the North Carolina mountains. The destructive power of a hurricane or other severe storm is related to the *square* of the wind speed. During Hurricane Andrew, the National Weather Service reported wind gusts over 180 mph. These winds were strong enough to send a piece of plywood through a tree.

The kinetic energy of an object varies jointly as the mass of the object and as the square of its velocity. During a hurricane, a 0.225-kg stone (approximately $\frac{1}{2}$ lb) traveling at 22.352 m/s (approximately 50 mph) generates a kinetic energy of 56.2 J (joules). Suppose the wind speed increases and the same stone now travels three times as fast at 67.056 m/s (150 mph). Find the kinetic energy (round to the nearest tenth of a joule).

Solution:

Let E represent the kinetic energy, let m represent the mass, and let v represent the velocity of the object. The variation model is

$$E = kmv^2$$

$$56.2 = k(0.225)(22.352)^2$$ Substitute $E = 56.2$ J, $m = 0.225$ kg, and $v = 22.352$ m/s.

$$\frac{56.2}{(0.225)(22.352)^2} = k$$

$$0.5 \approx k$$ Solve for k.

$$E = 0.5mv^2$$ Substitute $k = 0.5$ back into the original equation.

$$E = 0.5(0.225)(67.056)^2$$ To find the kinetic energy of the same stone, now traveling 3 times as fast, substitute $m = 0.225$ kg and $v = 67.056$ m/s.

$$\approx 505.9 \text{ J}$$

In Example 6, when the velocity was increased by 3 times, the kinetic energy increased by 9 times (ignoring round-off error). This factor of 9 occurs because the kinetic energy is proportional to the *square* of the velocity. When the velocity is increased by 3 times, the kinetic energy is increased by 3^2 times.

section 7.6 Practice Exercises

For Exercises 1–6, refer to the graph.

1. Find $f(1)$.
2. Find $f(-4)$.
3. Find the value(s) of x for which $f(x) = -1$.
4. Find the value(s) of x for which $f(x) = 3$.
5. Write the domain of f in interval notation.
6. Write the range of f in interval notation.

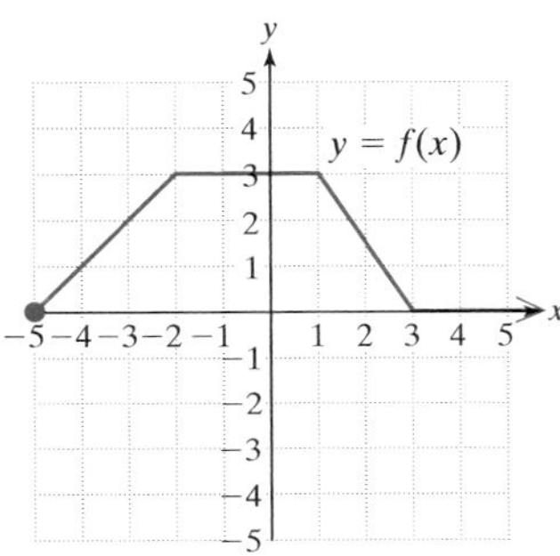

Figure for Exercises 1–6

For Exercises 7–10, determine if the relation defines y as a function of x.

7. 

8. 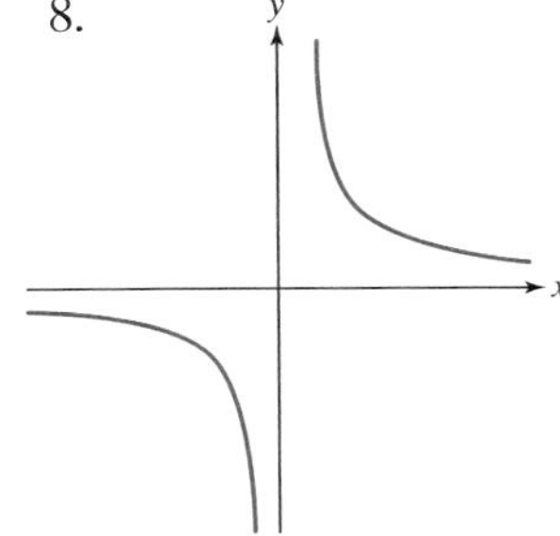

9. 

10. 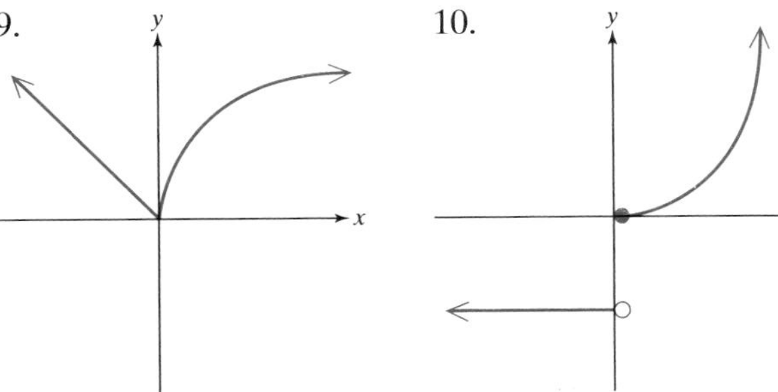

11. Suppose y varies directly as x, and $k > 0$.
    a. If x increases, then y will (increase or decrease)?
    b. If x decreases, then y will (increase or decrease)?

12. Suppose y varies inversely as x, and $k > 0$.
    a. If x increases, then y will (increase or decrease)?
    b. If x decreases, then y will (increase or decrease)?

For Exercises 13–20, write a variation model. Use k as the constant of variation.

13. T varies directly as q.
14. P varies inversely as r.
15. W varies inversely as the square of p.
16. Y varies directly as the square root of z.
17. Q is directly proportional to x and inversely proportional to the cube of y.
18. M is directly proportional to the square of p and inversely proportional to the cube of n.
19. L varies jointly as w and the square root of v.
20. X varies jointly as the square of y and w.

For Exercises 21–26, find the constant of variation, k.

21. y varies directly as x and when x is 4, y is 18.
22. m varies directly as x and when x is 8, m is 22.
23. p varies inversely as q and when q is 16, p is 32.
24. T varies inversely as x and when x is 40, T is 200.
25. y varies jointly as w and v. When w is 50 and v is 0.1, y is 8.75.
26. N varies jointly as t and p. When t is 1 and p is 7.5, N is 330.

Solve Exercises 27–32 using the steps found on page 497.

27. Z varies directly as the square of w. $Z = 14$ when $w = 4$. Find Z when $w = 8$.

28. Q varies inversely as the square of p. $Q = 4$ when $p = 3$. Find Q when $p = 2$.

29. L varies jointly as a and the square root of b. $L = 72$ when $a = 8$ and $b = 9$. Find L when $a = \frac{1}{2}$ and $b = 36$.

30. Y varies jointly as the cube of x and the square root of w. $Y = 128$ when $x = 2$ and $w = 16$. Find Y when $x = \frac{1}{2}$ and $w = 64$.

31. B varies directly as m and inversely as n. $B = 20$ when $m = 10$ and $n = 3$. Find B when $m = 15$ and $n = 12$.

32. R varies directly as s and inversely as t. $R = 14$ when $s = 2$ and $t = 9$. Find R when $s = 4$ and $t = 3$.

For Exercises 33–44, use a variation model to solve for the unknown value.

33. The amount of pollution entering the atmosphere varies directly as the number of people living in an area. If 80,000 people cause 56,800 tons of pollutants, how many tons enter the atmosphere in a city with a population of 500,000?

34. The area of a picture projected on a wall varies directly as the square of the distance from the projector to the wall. If a 10-ft distance produces a 16-ft^2 picture, what is the area of a picture produced when the projection unit is moved to a distance 20 ft from the wall?

35. The stopping distance of a car varies directly as the square of the speed of the car. If a car traveling at 40 mph has a stopping distance of 109 ft, find the stopping distance of a car that is traveling at 25 mph. (Round your answer to one decimal place.)

36. The intensity of a light source varies inversely as the square of the distance from the source. If the intensity is 48 lumens at a distance of 5 ft, what is the intensity when the distance is 8 ft?

37. The current in a wire varies directly as the voltage and inversely as the resistance. If the current is 9 A (amperes) when the voltage is 90 V (volts) and the resistance is 10 Ω (ohms), find the current when the voltage is 185 V and the resistance is 10 Ω.

38. The power in an electric circuit varies jointly as the current and the square of the resistance. If the power is 144 W (watts) when the current is 4 A and the resistance is 6 Ω, find the power when the current is 3 A and the resistance is 10 Ω.

39. The resistance of a wire varies directly as its length and inversely as the square of its diameter. A 40-ft wire 0.1 in. in diameter has a resistance of 4 Ω. What is the resistance of a 50-ft wire with a diameter of 0.20 in.?

40. The frequency of a vibrating string varies inversely as its length. A 24-in. piano string vibrates at 252 cycles/sec. What would be the frequency of an 18-in. string?

41. The weight of a medicine ball varies directly as the cube of its radius. A ball with a radius of 3 in. weighs 4.32 lb. How much would a medicine ball weigh if its radius is 5 in.?

42. The surface area of a cube varies directly as the square of the length of an edge. The surface area is 24 ft^2 when the length of an edge is 2 ft. Find the surface area of a cube with an edge that is 5 ft.

43. The strength of a wooden beam varies jointly as the width of the beam and the square of the thickness of the beam and inversely as the length of the beam. A beam that is 48 in. long, 6 in. wide, and 2 in. thick can support a load of 417 lb. Find the maximum load that can be safely supported by a board that is 12 in. wide, 72 in. long, and 4 in. thick.

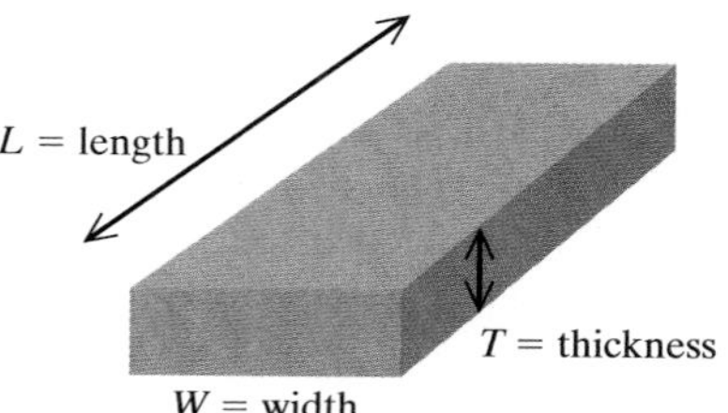

Figure for Exercise 43

44. The period of a pendulum is the length of time required to complete one swing back and forth. The period varies directly as the square root of the length of the pendulum. If it takes 1.4 sec for a 0.5-m pendulum to complete one period, what is the period of a 1.5-m pendulum? (Round your answer to two decimal places.)

Expanding Your Skills

45. The area, A, of a square varies directly as the square of the length, l, of its sides.
    a. Write a general variation model with k as the constant of variation.
    b. If the length of the sides are doubled, what effect will that have on the area?
    c. If the length of a side is tripled, what effect will that have on the area?

46. In a physics laboratory, a spring is fixed to the ceiling. With no weight attached to the end of the spring, the spring is said to be in its equilibrium position. As weights are applied to the end of the spring, the force stretches the spring a distance, d, from its equilibrium position. A student in the laboratory collects the following data:

Force, F (lb)	2	4	6	8	10
Distance, d (cm)	2.5	5.0	7.5	10.0	12.5

    a. Based on the data, do you suspect a direct relationship between force and distance or an inverse relationship?
    b. Find a variation model that describes the relationship between force and distance.

chapter 7 Summary

Section 7.1—Point-Slope Formula and Review of Graphing

Key Concepts:

The x- and y-intercepts of a graph are the points where the graph intersects the x- and y-axes, respectively.

Examples:

The x- and y-Intercepts of $-4x + y = 4$ are $(-1, 0)$ and $(0, 4)$.

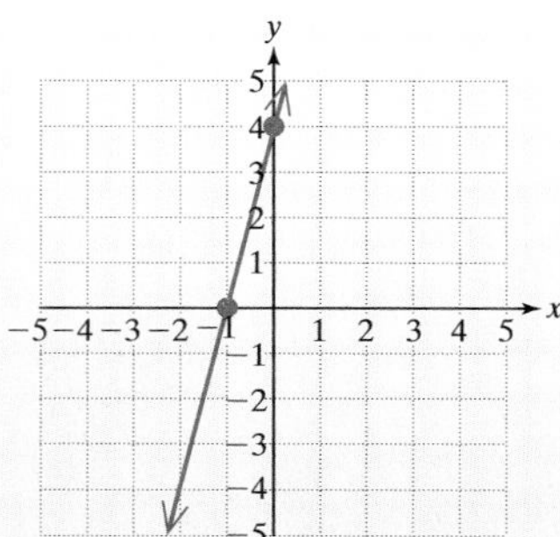

The slope of a line can be found from two points on the line, (x_1, y_1) and (x_2, y_2), using the formula

$$m = \frac{y_2 - y_1}{x_2 - x_1}$$

Find the Slope of the Line $-4x + y = 4$ Using the Points $(-1, 0)$ and $(0, 4)$.

$$m = \frac{4 - 0}{0 - (-1)} = \frac{4}{1} = 4$$

The point-slope formula is used primarily to construct an equation of a line given a point and a slope.

Equations of Lines—A Summary:

Standard form: $ax + by = c$
Horizontal line: $y = k$
Vertical line: $x = k$
Slope intercept form: $y = mx + b$
Point-slope formula: $y - y_1 = m(x - x_1)$

KEY TERM:

point-slope formula

Find an Equation of the Line Passing Through the Point (6, −4) and Having a Slope of $-\frac{1}{2}$.

Label the given information: $m = -\frac{1}{2}$ and $(6, -4) = (x_1, y_1)$

$$y - y_1 = m(x - x_1)$$
$$y - (-4) = -\tfrac{1}{2}(x - 6)$$
$$y + 4 = -\tfrac{1}{2}x + 3$$
$$y = -\tfrac{1}{2}x - 1$$

SECTION 7.2—APPLICATIONS OF LINEAR EQUATIONS

KEY CONCEPTS:

Linear equations can often be used to describe or model the relationship between variables in a real world event. In such applications, the slope may be interpreted as a rate of change.

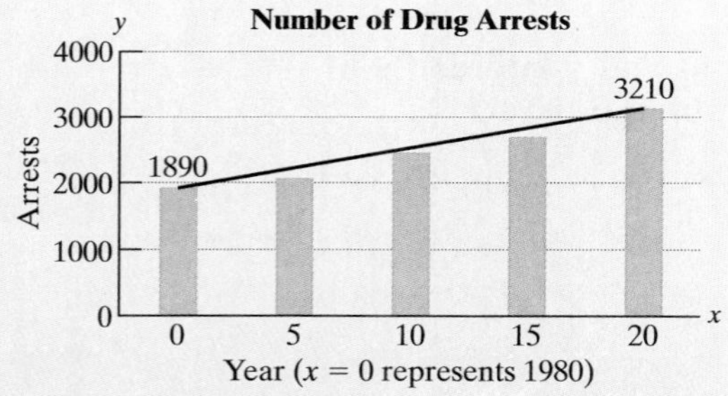

KEY TERMS:

dependent variable
independent variable

EXAMPLES:

The number of drug-related arrests for a small city has been growing approximately linearly since 1980.

Let y represent the number of drug arrests, and let x represent the number of years since 1980.

a. Use the ordered pairs (0, 1890) and (20, 3210) to find an equation of the line shown in the graph.

$$m = \frac{y_2 - y_1}{x_2 - x_1} = \frac{3210 - 1890}{20 - 0}$$
$$= \frac{1320}{20} = 66$$

The slope is 66 indicating that the number of drug arrests is increasing at a rate of 66 per year.
$m = 66$, and the y-intercept is (0,1890). Hence,

$$y = mx + b \Rightarrow y = 66x + 1890$$

b. Use the equation in part (a) to predict the number of drug-related arrests in the year 2010. (The year 2010 is 30 years after 1980. Hence $x = 30$)

$$y = 66(30) + 1890$$
$$y = 3870$$

The number of drug arrests is predicted to be 3870 by the year 2010.

Section 7.3—Introduction to Relations

Key Concepts:

Any set of ordered pairs, (x, y), is called a relation in x and y.

The domain of a relation is the set of first components in the ordered pairs in the relation. The range of a relation is the set of second components in the ordered pairs.

Key Terms:

domain of a relation
range of a relation
relation in x and y

Examples:

Find the Domain and Range of the Relation:

$\{(0, 0), (1, 1), (2, 4), (3, 9), (-1, 1), (-2, 4), (-3, 9)\}$

Domain: $\{0, 1, 2, 3, -1, -2, -3\}$

Range: $\{0, 1, 4, 9\}$

Find the Domain and Range of the Relation:

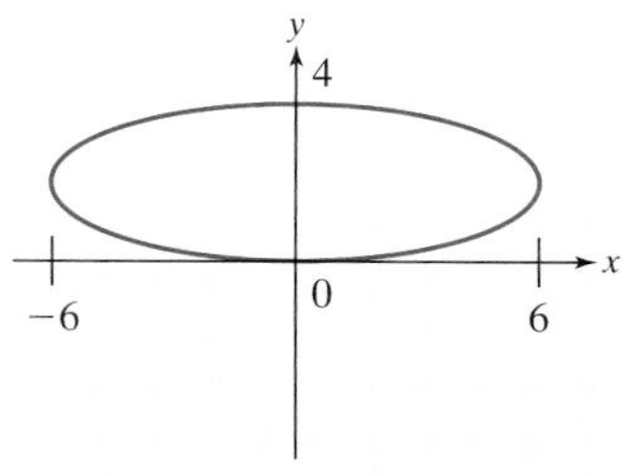

Domain: $[-6, 6]$

Range: $[0, 4]$

Section 7.4—Introduction to Functions

Key Concepts:

Given a relation in x and y, we say "y is a function of x" if for every element x in the domain, there corresponds exactly one element y in the range.

The Vertical Line Test for Functions:
Consider a relation defined by a set of points (x, y) in a rectangular coordinate system. Then the graph defines y as a function of x if no vertical line intersects the graph in more than one point.

Function Notation:
$f(x)$ is the value of the function, f, at x.

Examples:

Function $\{(1, 3), (2, 5), (6, 3)\}$

Nonfunction $\{(1, 3), (2, 5), (1, 4)\}$

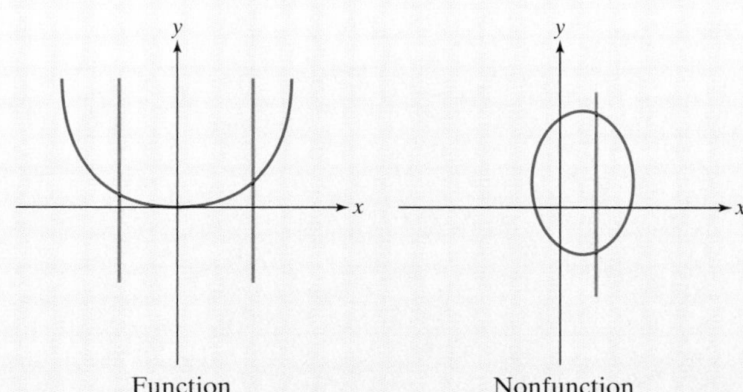

Given $f(x) = -3x^2 + 5x$, find $f(-2)$:

$$\begin{aligned} f(-2) &= -3(-2)^2 + 5(-2) \\ &= -12 - 10 \\ &= -22 \end{aligned}$$

The domain of a function defined by $y = f(x)$ is the set of x-values that when substituted into the function produces a real number. In particular,

- Exclude values of x that make the denominator of a fraction zero.
- Exclude values of x that make the expression within the square root negative.

Find the domain:

1. $f(x) = \dfrac{x+4}{(x-5)}$; $(-\infty, 5) \cup (5, \infty)$
2. $f(x) = \sqrt{x-3}$; $[3, \infty)$
3. $f(x) = 3x^2 - 5$; $(-\infty, \infty)$

Key Terms:

domain of a function
function
function notation
range of a function
vertical line test

Section 7.5—Graphs of Basic Functions

Key Concepts:

A function of the form $f(x) = mx + b$ is a linear function. Its graph is a line with slope m and y-intercept $(0, b)$.

Examples:

$f(x) = 2x - 3$

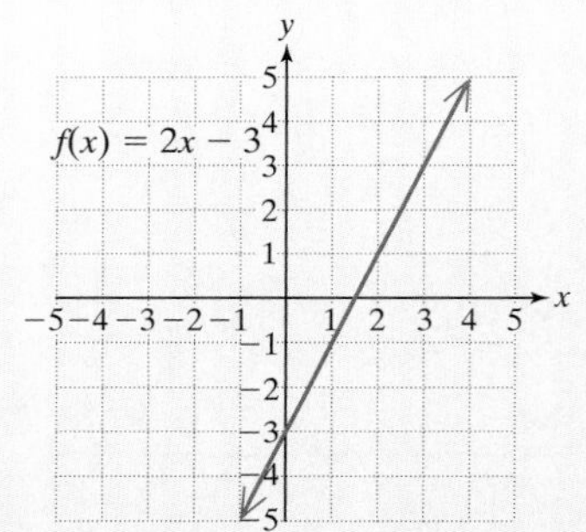

A function of the form $f(x) = k$ is a constant function. Its graph is a horizontal line.

$f(x) = 3$

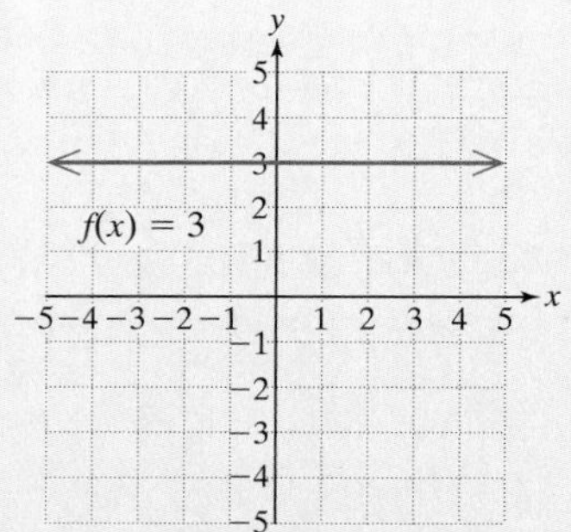

A function of the form $f(x) = ax^2 + bx + c$ $(a \neq 0)$ is a quadratic function. Its graph is a parabola.

$f(x) = x^2 - 2x - 1$

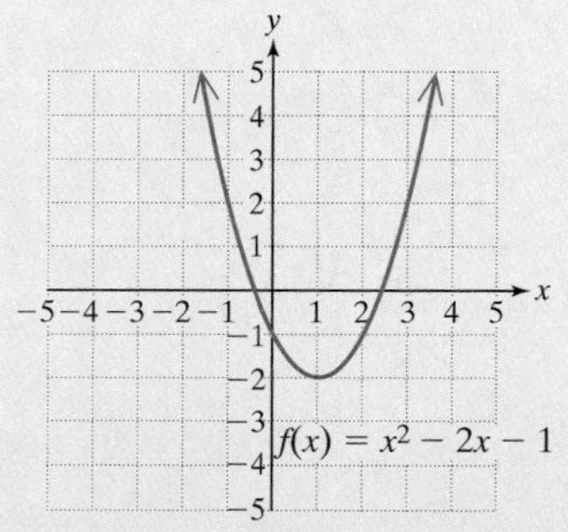

Graphs of basic functions:

$f(x) = x$ $\quad$ $f(x) = x^2$ $\quad$ $f(x) = x^3$

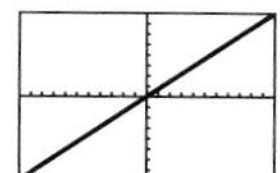

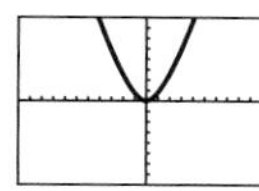

$f(x) = |x|$ $\quad$ $f(x) = \sqrt{x}$ $\quad$ $f(x) = \dfrac{1}{x}$

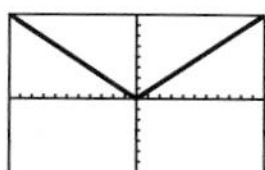

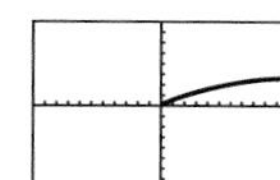

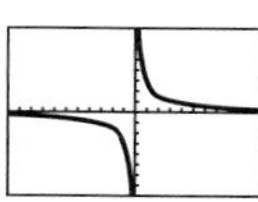

The x-intercepts of a function are determined by finding the real solutions to the equation $f(x) = 0$.

The y-intercept of a function is at $f(0)$.

Key Terms:

constant function
horizontal asymptote
linear function
parabola
quadratic function
vertical asymptote
x-intercept
y-intercept

Find the x- and y-intercepts:

$$f(x) = |x - 3| - 2$$

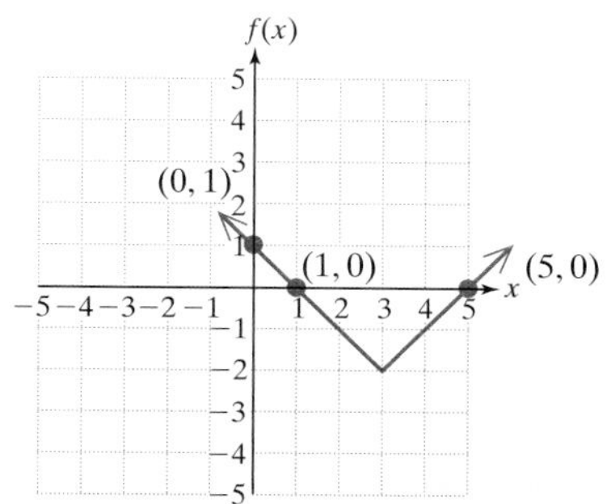

$f(x) = 0$, when $x = 1$ and $x = 5$.

Therefore, the x-intercepts are (1, 0) and (5, 0).
$f(0) = 1$.

Therefore, the y-intercept is (0, 1).

Section 7.6—Variation

Key Concepts:

Direct Variation:

y varies directly as x. $\quad y = kx$
y is directly proportional to x.

Inverse Variation:

y varies inversely as x. $\quad y = \dfrac{k}{x}$

y is inversely proportional to x.

Joint Variation:

y varies jointly as w and z. $\quad y = kwz$
y is jointly proportional to w and z.

Steps to Find a Variation Model:

1. Write a general variation model that relates the variables given in the problem. Let k represent the constant of variation.

Examples:

t varies directly as the square root of x.
$t = k\sqrt{x}$

W is inversely proportional to the cube of x.

$W = \dfrac{k}{x^3}$

y is jointly proportional to x and to the square of z.
$y = kxz^2$

C varies directly as the square root of d and inversely as t. If $C = 12$ when d is 9 and t is 6, find C if d is 16 and t is 12.

2. Solve for k by substituting known values of the variables into the model from step 1.

3. Substitute the value of k into the original variation model from step 1.

Step 1. $C = \dfrac{k\sqrt{d}}{t}$

Step 2. $12 = \dfrac{k\sqrt{9}}{6} \Rightarrow 12 = \dfrac{k \cdot 3}{6} \Rightarrow k = 24$

Step 3. $C = \dfrac{24\sqrt{d}}{t} \Rightarrow C = \dfrac{24\sqrt{16}}{12} \Rightarrow C = 8$

Key Terms:

direct variation
inverse variation
joint variation

chapter 7 REVIEW EXERCISES

Section 7.1

For Exercises 1–4,

a. Find the x- and y-intercept.
b. Find the slope (if it exists).
c. Graph the line.

1. $3x - 4y = 8$
2. $-3x = 5y$
3. $x = 4$
4. $2y = 6$

For Exercises 5–6,

a. Write the equation in slope-intercept form.
b. Identify the slope and y-intercept.

5. $4x - 2y = 8$
6. $-3x = 6y$
7. Write the point-slope formula from memory.
8. Write a formula to compute the slope of the line between the points (x_1, y_1) and (x_2, y_2).

For Exercises 9–17, write an equation of a line that satisfies the given conditions. Write the answer in slope-intercept form.

9. The slope is -6 and the line passes through $(-1, 8)$.
10. The slope is $\frac{2}{3}$ and the line passes through the point $(5, 5)$.
11. The line passes through the points $(0, -4)$ and $(8, -2)$.
12. The line passes through the points $(2, -5)$ and $(8, -5)$.
13. The line passes through the point $(8, 12)$ and is parallel to the line $y = 4x - 2$.
14. The line passes through the point $(-6, 7)$ and is parallel to the line $4x - y = 0$.
15. The line passes through the point $(5, 12)$ and is perpendicular to the line $5x + 6y = -18$.
16. The line passes through the point $(-6, 4)$ and is perpendicular to the line $2x - 3y = 6$.
17. The line has x- and y-intercepts $(-1, 0)$ and $(0, -9)$, respectively.
18. Write an equation of the line parallel to the y-axis and passing through the point $(3, -4)$.
19. Write an equation of the line perpendicular to the y-axis and passing through the point $(-2, -1)$.

Section 7.2

20. The number of reported property crimes decreased from 1994 to 1997. This decrease followed a linear trend and can be represented by the equation

    $y = -1.8x + 32.9$ where y is the number of crimes measured in millions and x is the number of years since 1994.

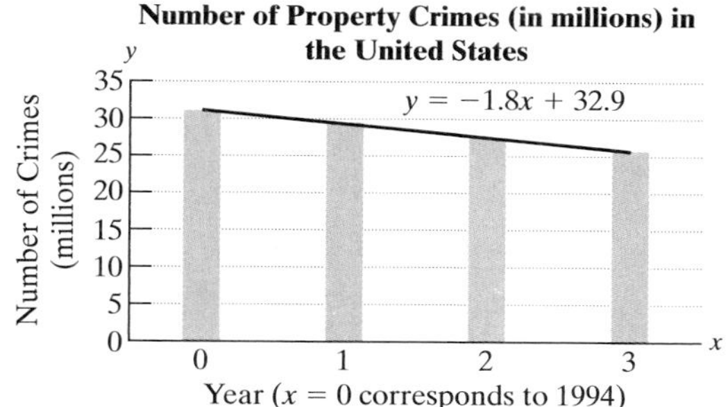

Figure for Exercise 20

Source: U.S. Federal Bureau of Investigation

a. Which variable is the dependent variable?

b. Which variable is the independent variable?

c. Use the equation to estimate the number of reported property crimes in the year 2000.

d. What is the slope of the line? Interpret the meaning of the slope in the context of the problem.

21. The number of robberies in the United States in 1994 was approximately 1.3 million. By 1997, the number dropped to approximately 0.9 million.

Let $x = 1$ represent the year 1994. Let y represent the number of robberies (in millions).

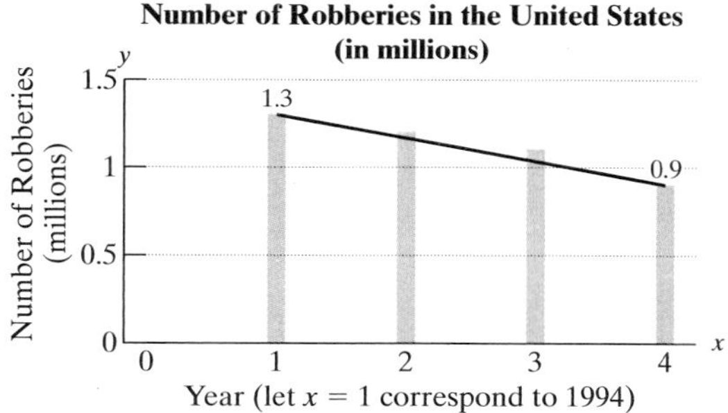

Figure for Exercise 21

Source: U.S. Federal Bureau of Investigation

a. Using the ordered pairs (1, 1.3) and (4, 0.9), find the slope of the line. Round to two decimal places.

b. Interpret the meaning of the slope in the context of this problem.

c. Write an equation that represents the number of robberies versus the year.

d. Use the equation in part (c) to approximate the number of robberies in 1998 ($x = 5$).

22. During the 1990s, consumer debt grew linearly. The amount of money (in $billions) that U.S. consumers had in outstanding automobile loans for the years 1992–1997 is expressed in the graph.

Let x represent the number of years since 1992. Let y represent the total debt in auto loans (in $billions).

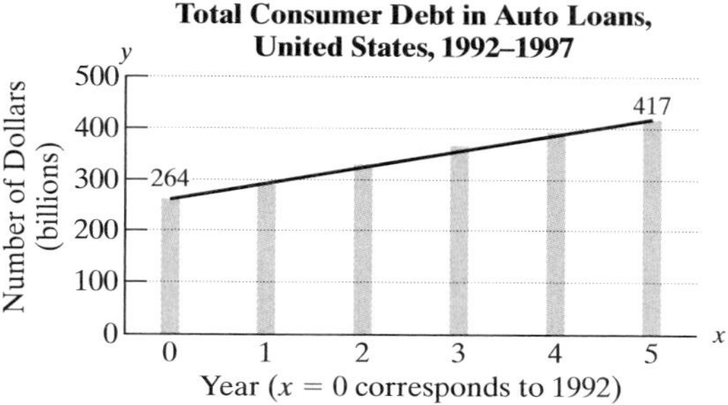

Figure for Exercise 22

Source: Federal Reserve Board.

a. Find a linear equation that represents the total debt in auto loans, y, versus the year, x.

b. Use the linear equation found in part (a) to estimate the debt in the year 2000.

23. A water purification company charges $20 per month and a $55 installation fee.

a. Write a linear equation to compute the total cost, y, of renting this system for x months.

b. Use the equation from part (a) to determine the total cost to rent the system for 9 months.

24. A small cleaning company has a fixed monthly cost of $700 and a variable cost of $8 per service call.

a. Write a linear equation to compute the total cost, y, of making x service calls in one month.

b. Use the equation from part (a) to determine the total cost of making 80 service calls.

Section 7.3

25. Write a relation with four ordered pairs for which the first element is the name of a parent and the second element is the name of the parent's child.

For Exercises 26–29, find the domain and range.

26. $\left\{\left(\frac{1}{3}, 10\right), \left(6, -\frac{1}{2}\right), \left(\frac{1}{4}, 4\right), \left(7, \frac{2}{5}\right)\right\}$

27.

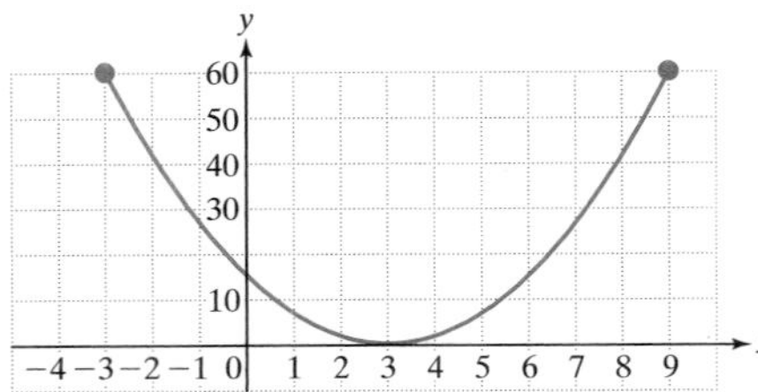

28.

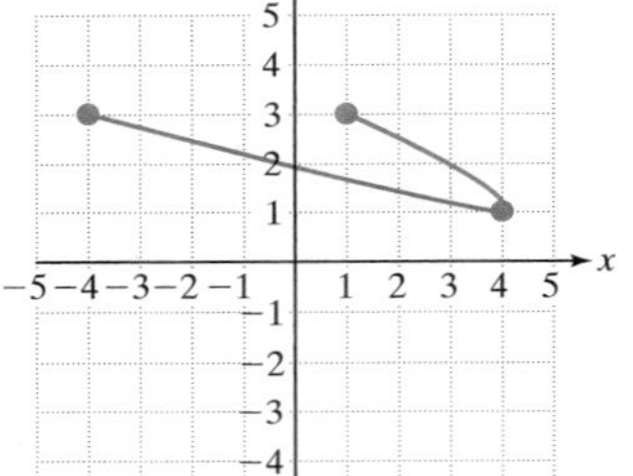

29.

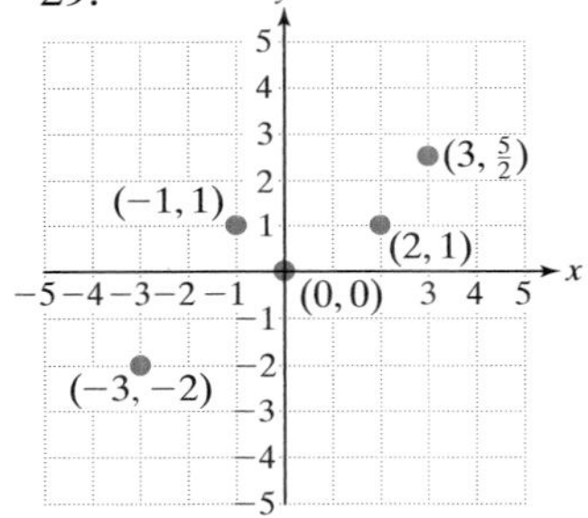

Section 7.4

30. Sketch a relation that is *not* a function. (Answers may vary.)

31. Sketch a relation that *is* a function. (Answers may vary.)

For Exercises 32–37:

a. Determine whether the relation defines y as a function of x.

b. Find the domain.

c. Find the range.

32.

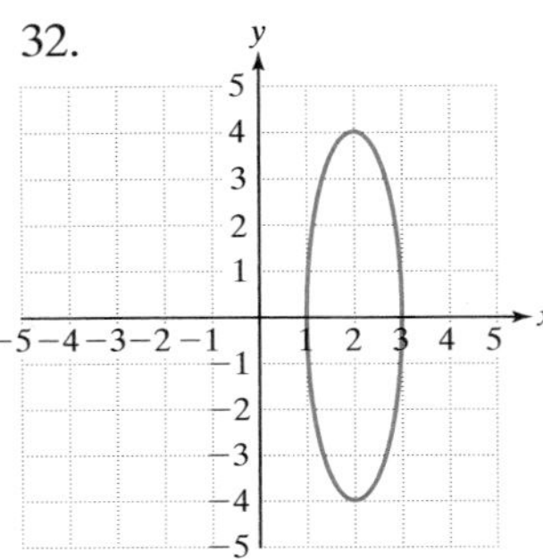

33.

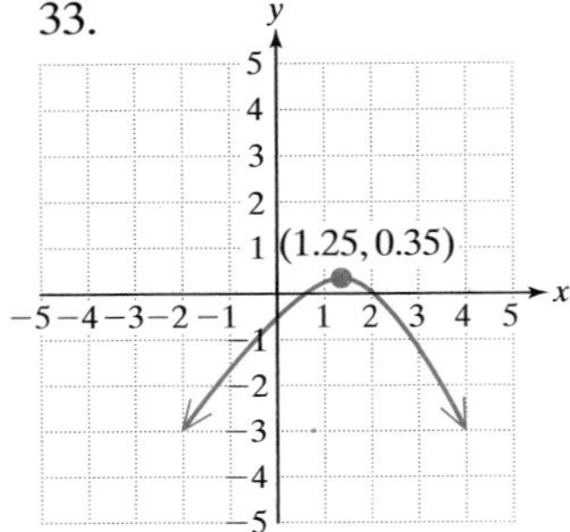

34. $\{(1, 3), (2, 3), (3, 3), (4, 3)\}$

35. $\{(0, 2), (0, 3), (4, 4), (0, 5)\}$

36.

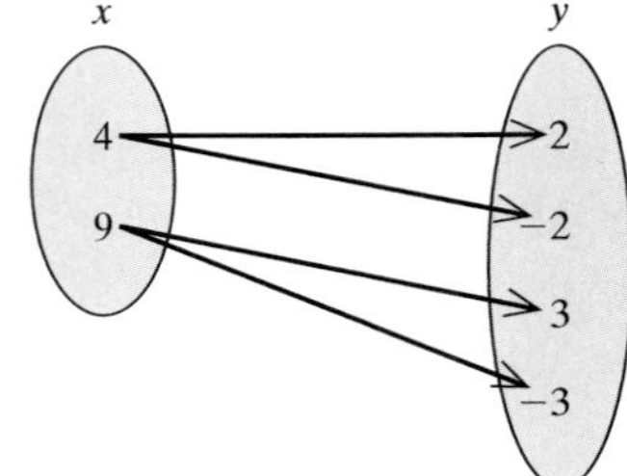

37.

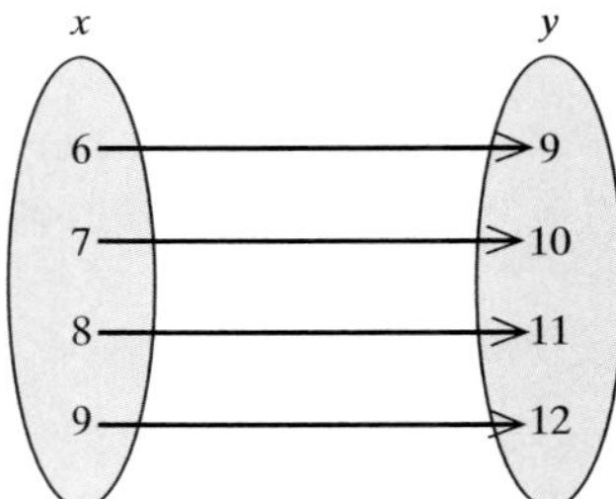

For Exercises 38–45, find the function values given $f(x) = 6x^2 - 4$.

38. $f(0)$

39. $f(1)$

40. $f(-1)$

41. $f(t)$

42. $f(b)$

43. $f(\pi)$

44. $f(\square)$

45. $f(x + h)$

For Exercises 46–49, write the domain of each function in interval notation.

46. $g(x) = 7x^3 + 1$

47. $h(x) = \dfrac{x + 10}{x - 11}$

48. $k(x) = \sqrt{x - 8}$

49. $w(x) = \sqrt{x + 2}$

50. Anita is a waitress and makes \$6/hr plus tips. Her tips average \$5 per table. In one 8-hr shift, Anita's pay can be described by $p(x) = 48 + 5x$, where x represents the number of tables she waits on. How much will Anita earn if she waits on

a. 10 tables b. 15 tables c. 20 tables

Section 7.5

For Exercises 51–56, sketch the functions from memory.

51. $h(x) = x$
52. $f(x) = x^2$
53. $g(x) = x^3$
54. $w(x) = |x|$
55. $s(x) = \sqrt{x}$
56. $r(x) = \dfrac{1}{x}$

For Exercises 57–58, sketch the functions.

57. $q(x) = 3$
58. $k(x) = 2x + 1$

For Exercises 59–64, find the x- and y-intercepts.

59. $p(x) = 4x - 7$
60. $q(x) = -2x + 9$
61. $F(x) = x^2 - 16$
62. $G(x) = x^2 + 2x - 8$
63. $r(x) = (x - 3)(x + 2)(2x - 1)$
64. $s(x) = (4x + 5)(x - 9)(x + 1)$

65. For $s(x) = (x - 2)^2$

a. Find $s(4)$, $s(3)$, $s(2)$, $s(1)$, and $s(0)$.

b. What is the domain of s?

66. For $r(x) = 2\sqrt{x - 4}$

a. Find $r(4)$, $r(5)$, and $r(8)$.

b. What is the domain of r?

67. For $h(x) = \dfrac{3}{x - 3}$

a. Find $h(-3)$, $h(-1)$, $h(0)$, $h(2)$, $h(4)$, $h(5)$, and $h(7)$.

b. What is the domain of h?

68. For $k(x) = -|x + 3|$

a. Find $k(-5)$, $k(-4)$, $k(-3)$, $k(-2)$, and $k(-1)$.

b. What is the domain of k?

69. The function defined by $b(t) = 0.7t + 4.5$ represents the per capita consumption of bottled water in the United States between 1985 and 1995. The values of $b(t)$ are measured in gallons and $t = 0$ corresponds to the year 1985. (*Source:* U.S. Department of Agriculture)

a. Evaluate $b(0)$ and $b(7)$ and interpret the results in the context of the problem.

b. What is the slope of this function? Interpret the slope in the context of the problem.

Section 7.6

70. The force applied to a spring varies directly with the distance that the spring is stretched. When 6 lb of force is applied, the spring stretches 2 ft.

a. Write a variation model using k as the constant of variation.

b. Find k.

c. How many feet will the spring stretch when 5 lb of force is applied?

71. Suppose y varies inversely with the cube of x, and $y = 32$ when $x = 2$. Find y when $x = 4$.

72. Suppose y varies jointly with x and the square root of z, and $y = 3$ when $x = 3$ and $z = 4$. Find y when $x = 8$ and $z = 9$.

73. The distance, d, that one can see to the horizon varies directly as the square root of the height above sea level. If a person 16 m above sea level can see 26.4 km, how far can a person see if she is 64 m above sea level? (Round to the nearest tenth of a kilometer.)

chapter 7 TEST

For Exercises 1–3, find the slope of the line,

1. Containing the points $(2, -3)$ and $(6, 4)$.
2. Defined by $2y = 4x + 6$.
3. Perpendicular to $y = 8x - 4$.

For Exercises 4–5, find the x- and y-intercepts.

4. $3x + 7y = 9$
5. $x = 2y$

6. Determine whether the line defined by each equation is horizontal or vertical.

 a. $4x = 2$
 b. $6y + 12 = 0$

7. Graph the line using the slope and y-intercept. $2x = 3y - 12$

8. Write an equation of the line that passes through the points $(2, 8)$, and $(4, 1)$. Write the equation in slope-intercept form.

9. Write an equation of the line that passes through the point $(2, -6)$ and is parallel to the x-axis.

10. Write an equation of the line that passes through the point $(3, 0)$ and is parallel to the line $2x + 6y = -5$.

11. Write an equation of the line that passes through the point $(-3, -1)$ and is perpendicular to the line $x + 3y = 9$.

12. Hurricane Floyd dumped rain at an average rate of $\frac{3}{4}$ in./hr on Southport, North Carolina. Further inland, in Lumberton, North Carolina, the storm dropped $\frac{1}{2}$ in. of rain per hour. The following graph depicts the total amount of rainfall (in inches) versus the time (in hours) for both locations in North Carolina.

 a. What is the slope of the line representing the rainfall for Southport?
 b. What is the slope of the line representing the rainfall for Lumberton?

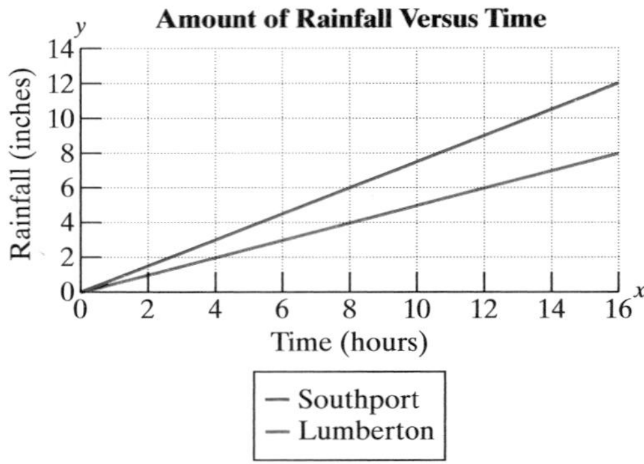

Figure for Exercise 12

13. To attend a State Fair, the cost is \$10 per person to cover exhibits and musical entertainment. There is an additional cost of \$1.50 per ride.

 a. Write an equation that gives the total cost, y, of visiting the State Fair and going on x rides.
 b. Use the equation from part (a) to determine the cost of going to the State Fair and going on 10 rides.

14. The winner of the women's 100-m freestyle swimming event in the 1912 Olympics was Fanny Durack of Australia. Her time was 82.2 sec. Since that time the winning time for this event has dropped. By the 1996 Olympics, the winning time posted by Le Jingyi of China was 54.5 sec.

 Let x represent the number of years since 1912. Let y represent the winning time in seconds for the women's 100-m freestyle event.

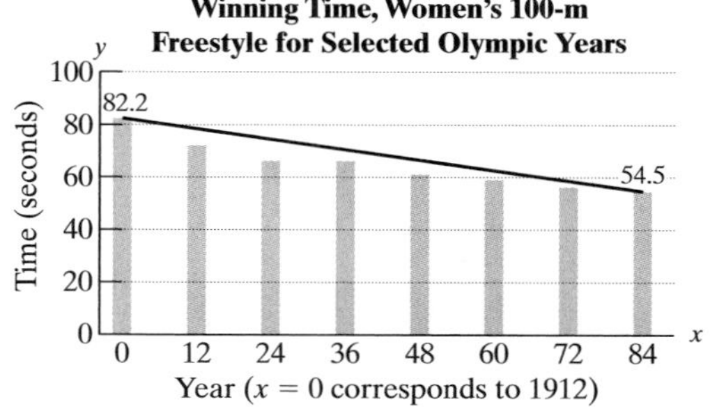

Figure for Exercise 14

a. Find the slope of the line shown in the graph. Interpret the meaning of the slope in the context of this problem. Round to two decimal places.

b. Use the slope and y-intercept to find an equation of the line that represents the winning time, y, versus the number of years, x, since 1912.

c. Use the equation from part (b) to estimate the winning time in the women's 100-m freestyle event in the year 1948.

For Exercises 15–17, (a) Determine if the relation defines y as a function of x. (b) Identify the domain. (c) Identify the range.

15. 16.

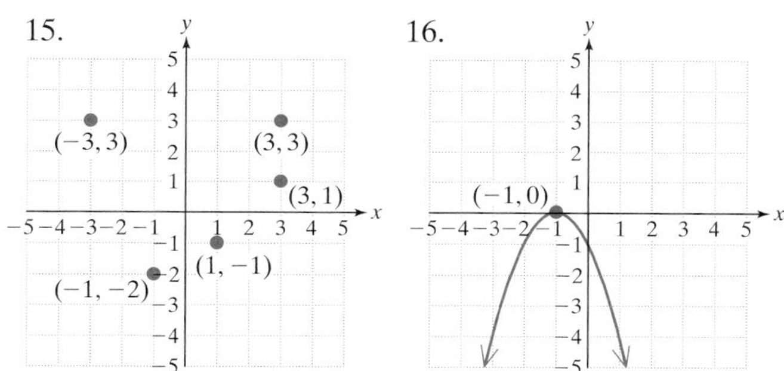

17. The percentage of mothers in the work force who have children under 18 years old is shown for selected years.

Year x	Percent y
1975	47.4%
1985	62.2%
1997	72.1%

Source: U.S. Bureau of Labor Statistics

18. Explain how to find the x- and y-intercepts of a function defined by $y = f(x)$.

19. For $f(x) = \frac{1}{3}x + 6$ and $g(x) = |x|$, find

a. $f(-3)$

b. $g(-3)$

c. $f(0) + g(6)$

d. $\dfrac{f(3)}{g(-1)}$

20. Let $k(x) = 8$

a. Find $k(0)$, $k(-2)$, and $k(15)$.

b. Write the domain of k.

c. Sketch $y = k(x)$.

d. Write the range of k.

21. For $r(x) = (x - 1)^3$

a. Find $r(-2)$, $r(-1)$, $r(0)$, $r(1)$, $r(2)$, and $r(3)$.

b. What is the domain of r?

22. Find the domain.

a. $f(x) = \dfrac{x - 5}{x + 7}$

b. $f(x) = \sqrt{x + 7}$

c. $h(x) = (x + 7)(x - 5)$

23. The function defined by $s(t) = 1.6t + 36$ approximates the per capita consumption of soft drinks in the United States between 1985 and 1995. The values of $s(t)$ are measured in gallons and $t = 0$ corresponds to the year 1985. (*Source:* U.S. Department of Agriculture)

a. Evaluate $s(0)$ and $s(7)$ and interpret the results in the context of the problem.

b. What is the slope of the function? Interpret the slope in the context of the problem.

24. For the graph of $y = f(x)$ below

a. Find $f(1)$, $f(4)$, and $f(7)$.

b. For what value(s) of x is $f(x) = 0$?

c. Write the domain of f.

d. Write the range of f.

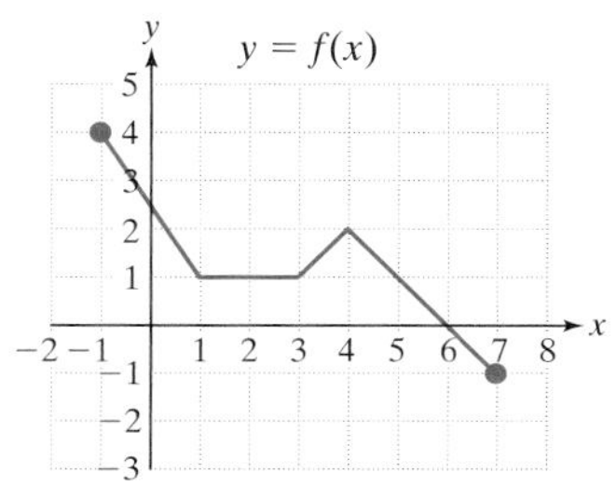

Figure for Exercise 24

25. Match the set defined in parts (a)–(c) with the appropriate interval.

a. The domain of the function given by $f(x) = \dfrac{1}{x - 3}$

b. The range of the function given by $g(x) = \sqrt{x}$

c. The domain of the function given by $h(x) = |x|$

i. $(-\infty, 0)$

ii. $(-\infty, 3) \cup (3, \infty)$

iii. $(0, \infty)$

iv. $(-\infty, \infty)$

v. $[0, \infty)$

26. The period of a pendulum varies directly as the square root of the length of the pendulum. If the period of the pendulum is 1.5 sec when the length is 2 ft, find the period when the length is 5 ft. (Round to the nearest hundredth of a second.)

CUMULATIVE REVIEW EXERCISES, CHAPTERS 1–7

1. Solve the equation.

$$\frac{1}{3}t + \frac{1}{5} = \frac{1}{10}(t - 2)$$

2. Simplify: $5 - 3(2 - \sqrt{25}) + 2 - 10 \div 5$

3. Write the inequalities in interval notation:

a. x is greater than or equal to 6.

b. x is less than 17.

c. x is between -2 and 3, inclusive.

4. Solve the inequality. Write the solution set in interval notation.

$$4 \leq -6y + 5$$

5. Determine the volume of the cone pictured here. Round your answer to the nearest whole unit.

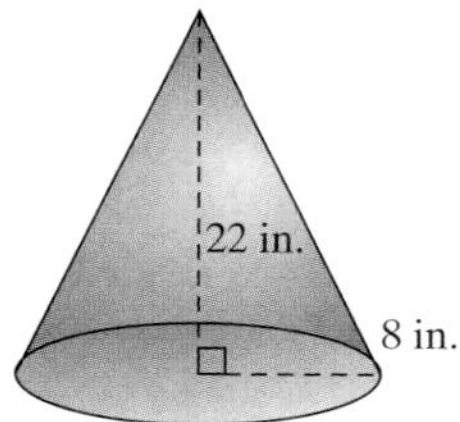

Figure for Exercise 5

6. Find an equation of the line passing through the origin and perpendicular to $3x - 4y = 1$. Write your final answer in slope-intercept form.

7. Find the pitch (slope) of the roof.
Note: Pitch is always taken as a positive slope.

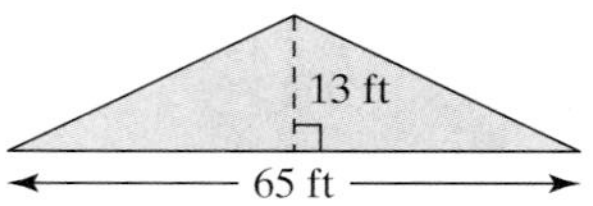

Figure for Exercise 7

8. a. Explain how to find the x- and y-intercepts of a function, $y = f(x)$.

b. Find the y-intercept of the function defined by $f(x) = 3x + 2$

c. Find the x-intercept(s) of the function defined by $f(x) = 3x + 2$

9. Find the value of each angle in the triangle.

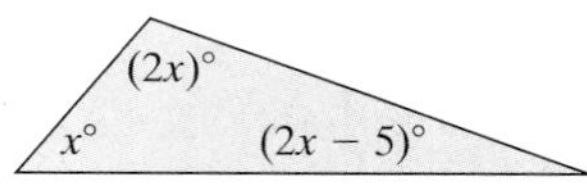

Figure for Exercise 9

10. Divide: $(x^3 + 64) \div (x + 4)$.

11. If y varies directly with x and inversely with z, and $y = 6$ when $x = 9$ and $z = \frac{1}{2}$, find y when $x = 3$ and $z = 4$.

12. Simplify the expression

$$\left(\frac{36a^{-2}b^{4}}{18b^{-6}}\right)^{-3}$$

13. Multiply the polynomials $(4b - 3)(2b^2 + 1)$.

14. Add the polynomials

$$(5a^2 + 3a - 1) + (3a^3 - 5a + 6)$$

15. Divide the polynomials $(6w^3 - 5w^2 - 2w) \div (2w^2)$.

16. Find the domain of the functions f and g.

a. $f(x) = \dfrac{x + 7}{2x - 3}$ b. $g(x) = \dfrac{x + 3}{x^2 - x - 12}$

17. Perform the indicated operations.

$$\frac{2x^2 + 11x - 21}{4x^2 - 10x + 6} \div \frac{2x^2 - 98}{x^2 - x + xa - a}$$

18. Reduce to lowest terms.

$$\frac{x^2 - 6x + 8}{20 - 5x}$$

19. Perform the indicated operations.

$$\frac{x}{x^2 + 5x - 50} - \frac{1}{x^2 - 7x + 10} + \frac{1}{x^2 + 8x - 20}$$

20. Simplify the complex fraction.

$$\frac{1 - \dfrac{49}{c^2}}{\dfrac{7}{c} + 1}$$

21. Solve the equation.

$$\frac{4y}{y + 2} - \frac{y}{y - 1} = \frac{9}{y^2 + y - 2}$$

22. Factor completely: $2x^5 - 32x$

23. Factor completely: $8x^2 + 2x - 15$

24. Solve for x: $x(x - 15) = -50$

25. The linear function defined by $N(x) = 420x + 5260$ provides a model for the number of full-time-equivalent (FTE) students attending a community college from 1988 to 2006. Assume that $x = 0$ corresponds to the year 1988.

a. Use this model to find the number of FTE students who attended the college in 1996.

b. If this linear trend continues, predict the year in which the number of FTE students will reach 14,920.

26. State the domain and range of the relation. Is the relation a function?

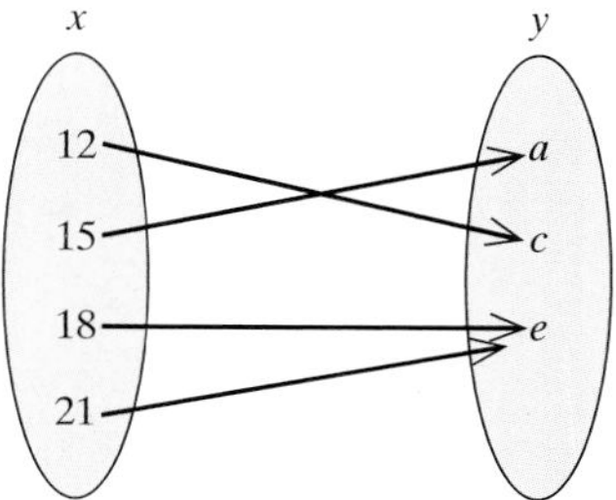

Figure for Exercise 26

27. Given $f(x) = \frac{1}{2}x - 1$ and $g(x) = 3x^2 - 2x$

a. Find $f(4)$

b. Find $g(-3)$

28. Write the domain of the functions in interval notation.

a. $f(x) = \dfrac{1}{x - 15}$ b. $g(x) = \sqrt{x - 6}$

29. Simple interest varies jointly as the interest rate and as the time the money is invested. If an investment yields \$1120 interest at 8% for 2 years, how much interest will the investment yield at 10% for 5 years?

chapter

8

Systems of Linear Equations in Two Variables

A system of linear equations can be used to solve an application where two variables are subject to two linear constraints. For example, the speed of a plane is influenced by the plane's still air speed, p, and by the speed of the wind, w. When traveling with the wind, the net speed of the plane is $p + w$. When traveling against the wind, the net speed is $p - w$.

Suppose a plane travels 1500 miles in 3 hr with a tail wind, and 1200 miles in 3 hr against a head wind. Using the relationship $d = rt$, we have two linear constraints:

$$\text{Distance} = \text{rate} \times \text{time}$$

$$1500 = (p + w)3$$

$$1200 = (p - w)3$$

Using techniques presented in this chapter, we find that $p = 450$ and $w = 50$. Therefore, the plane's still air speed is 450 mph, and the speed of the wind is 50 mph. For more information see the Technology Connections in MathZone at

www.mhhe.com/miller_oneill

section

8.1 Introduction to Systems of Linear Equations

Concepts

1. Introduction to Systems of Linear Equations in Two Variables
2. Determining Solutions to a System of Linear Equations
3. Dependent and Inconsistent Systems of Linear Equations
4. Solving Systems of Linear Equations by Graphing

1. Introduction to Systems of Linear Equations in Two Variables

Recall from Section 7.1 that a linear equation in two variables has an infinite number of solutions. The set of all solutions to a linear equation forms a line in a rectangular coordinate system. Two or more linear equations form a **system of linear equations**. For example, here are three systems of equations.

$$x - 3y = -5 \qquad y = \tfrac{1}{4}x - \tfrac{3}{4} \qquad 5a + b = 4$$
$$2x + 4y = 10 \qquad -2x + 8y = -6 \qquad -10a - 2b = 8$$

2. Determining Solutions to a System of Linear Equations

A **solution to a system of linear equations** is an ordered pair that is a solution to *each* individual linear equation.

example 1 Determining Solutions to a System of Linear Equations

Determine whether the ordered pairs are solutions to the system.

$$x + y = 4$$
$$-2x + y = -5$$

a. (3, 1) b. (0, 4)

Solution:

a. Substitute the ordered pair (3, 1) into both equations:

$$x + y = 4 \longrightarrow (3) + (1) \stackrel{?}{=} 4 \checkmark$$
$$-2x + y = -5 \longrightarrow -2(3) + (1) \stackrel{?}{=} -5 \checkmark$$

Because the ordered pair (3, 1) is a solution to each equation, it is a solution to the *system* of equations.

b. Substitute the ordered pair (0, 4) into both equations.

$$x + y = 4 \longrightarrow (0) + (4) \stackrel{?}{=} 4 \checkmark$$
$$-2x + y = -5 \longrightarrow -2(0) + (4) \stackrel{?}{=} -5 \qquad \text{False}$$

Because the ordered pair (0, 4) is not a solution to the second equation, it is *not* a solution to the system of equations.

A solution to a system of two linear equations may be interpreted graphically as a point of intersection between the two lines. Using slope-intercept form to graph the lines from Example 1, we have

$$x + y = 4 \longrightarrow y = -x + 4$$

$$-2x + y = -5 \longrightarrow y = 2x - 5$$

Notice that the lines intersect at (3, 1) (Figure 8-1).

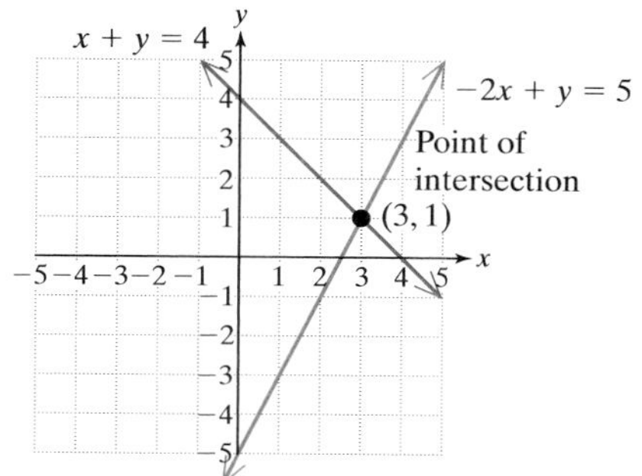

Figure 8-1

3. Dependent and Inconsistent Systems of Linear Equations

When two lines are drawn in a rectangular coordinate system, three geometric relationships are possible:

1. Two lines may intersect at *exactly one point*.
2. Two lines may intersect at *no point*. This occurs if the lines are parallel.
3. Two lines may intersect at *infinitely many points* along the line. This occurs if the equations represent the same line (the lines are coinciding).

If a system of linear equations has one or more solutions, the system is said to be **consistent**. If a linear equation has no solution, it is said to be **inconsistent**.

If two equations represent the same line, then all points along the line are solutions to the system of equations. In such a case, the system is characterized as a **dependent system**. An **independent system** is one in which the two equations represent different lines.

Solutions to Systems of Linear Equations in Two Variables

One Unique Solution	No Solution	Infinitely Many Solutions
One point of intersection	Parallel lines	Coinciding lines
System is consistent.	System is inconsistent.	System is consistent.
System is independent.	System is independent.	System is dependent.

4. Solving Systems of Linear Equations by Graphing

example 2 Solving a System of Linear Equations by Graphing

Solve the system by graphing both linear equations and finding the point(s) of intersection.

$$x - 2y = -2$$
$$-3x + 2y = 6$$

Solution:

To graph each equation, write the equation in slope-intercept form: $y = mx + b$.

$$x - 2y = -2 \qquad\qquad -3x + 2y = 6$$
$$-2y = -x - 2 \qquad\qquad 2y = 3x + 6$$
$$\frac{-2y}{-2} = \frac{-x}{-2} - \frac{2}{-2} \qquad\qquad \frac{2y}{2} = \frac{3x}{2} + \frac{6}{2}$$
$$y = \frac{1}{2}x + 1 \qquad\qquad y = \frac{3}{2}x + 3$$

From their slope-intercept forms, we see that the lines have different slopes, indicating that the lines are different and nonparallel. Therefore, the lines must intersect at exactly one point. Graph the lines to find that point (Figure 8-2).

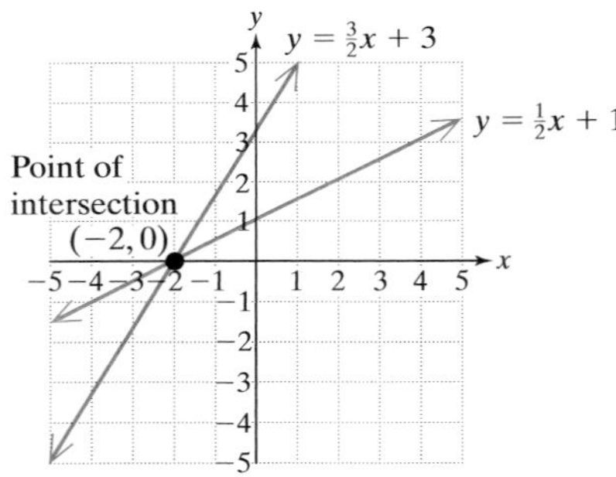

Figure 8-2

The point $(-2, 0)$ appears to be the point of intersection. This can be confirmed by substituting $x = -2$ and $y = 0$ into both equations.

$$x - 2y = -2 \longrightarrow (-2) - 2(0) \stackrel{?}{=} -2 \checkmark$$
$$-3x + 2y = 6 \longrightarrow -3(-2) + 2(0) \stackrel{?}{=} 6 \checkmark$$

The solution is $(-2, 0)$.

Tip: In Example 2, the lines could also have been graphed by using the x- and y-intercepts or by using a table of points. However, the advantage of writing the equations in slope-intercept form is that we can compare the slopes and y-intercepts of each line.

1. If the slopes differ, the lines are different and nonparallel and must cross in exactly one point.
2. If the slopes are the same and the y-intercepts are different, the lines are parallel and will not intersect.
3. If the slopes are the same and the y-intercepts are the same, the two equations represent the same line.

example 3

Solving a System of Equations by Graphing

Solve the system by graphing.

$$-x + 3y = -6$$
$$6y = 2x + 6$$

Solution:

To graph the lines, write each equation in slope-intercept form.

$$-x + 3y = -6 \qquad\qquad 6y = 2x + 6$$
$$3y = x - 6$$
$$\frac{3y}{3} = \frac{x}{3} - \frac{6}{3} \qquad\qquad \frac{6y}{6} = \frac{2x}{6} + \frac{6}{6}$$
$$y = \frac{1}{3}x - 2 \qquad\qquad y = \frac{1}{3}x + 1$$

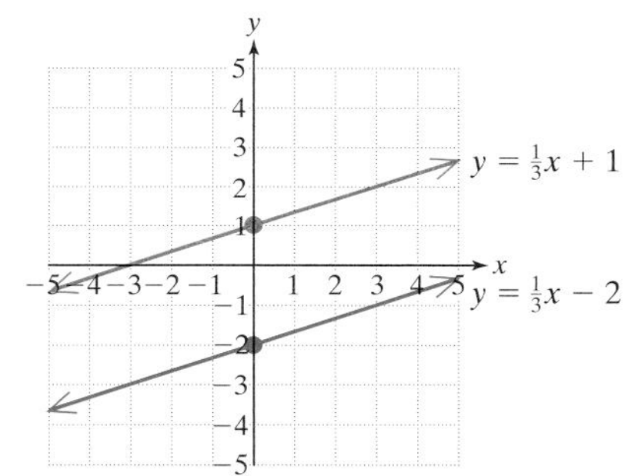

Figure 8-3

Because the lines have the same slope, but different y-intercepts, they are parallel (Figure 8-3). Two parallel lines do not intersect, which implies that the system has no solution. The system is inconsistent.

example 4

Solving a System of Linear Equations by Graphing

Solve the system by graphing.

$$x + 4y = 8$$
$$y = -\frac{1}{4}x + 2$$

Solution:

Write the first equation in slope-intercept form. The second equation is already in slope-intercept form.

$$x + 4y = 8 \qquad\qquad y = -\frac{1}{4}x + 2$$

$$4y = -x + 8$$

$$\frac{4y}{4} = \frac{-x}{4} + \frac{8}{4}$$

$$y = -\frac{1}{4}x + 2$$

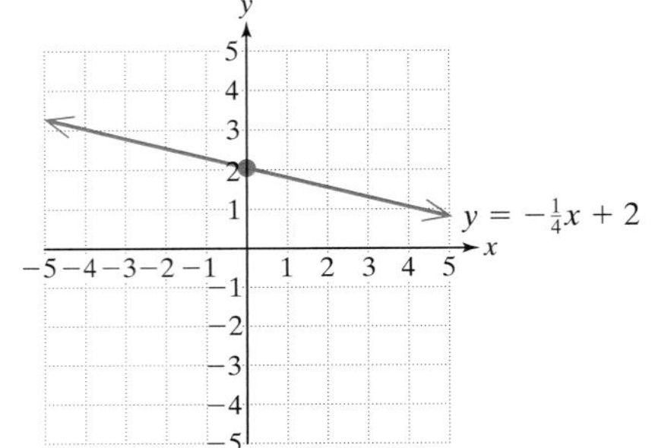

Figure 8-4

Notice that the slope-intercept forms of the two lines are identical. Therefore, the equations represent the same line (Figure 8-4). The system is dependent, and the solution to the system of equations is the set of all points on the line.

Because the ordered pairs in the solution set cannot all be listed, we can write the solution in set-builder notation. $\{(x, y) \mid y = -\frac{1}{4}x + 2\}$. This may be read as "the set of all ordered pairs, (x, y) such that the ordered pairs satisfy the equation $y = -\frac{1}{4}x + 2$."

Calculator Connections

The solution to a system of equations can be found by using either a *Trace* feature or an *Intersect* feature on a graphing calculator to find the point of intersection between two curves.

For example, consider the system:

$$-2x + y = 6$$

$$5x + y = -1$$

First graph the equations together on the same viewing window. Recall that to enter the equations into the calculator, the equations must be written with the y-variable isolated.

$$-2x + y = 6 \xrightarrow{\text{isolate } y} y = 2x + 6$$

$$5x + y = -1 \longrightarrow y = -5x - 1$$

By inspection of the graph, it appears that the solution is $(-1, 4)$. The *Trace* option on the calculator may come close to $(-1, 4)$ but may not show the exact solution (Figure 8-5). However, an *Intersect* feature on a graphing calculator may provide the exact solution (Figure 8-6). See your user's manual for further details.

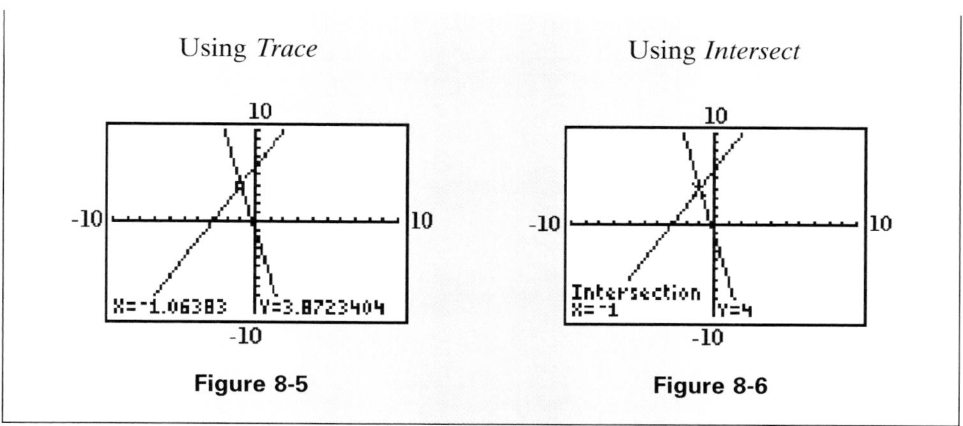

Figure 8-5

Figure 8-6

section 8.1 Practice Exercises

For Exercises 1–6, determine if the given point is a solution to the system.

1. $3x - y = 7$ $(2, -1)$
   $x - 2y = 4$

2. $x - y = 3$ $(4, 1)$
   $x + y = 5$

3. $2x - 3y = 12$ $(0, 4)$
   $3x + 4y = 12$

4. $x - 2y = 6$ $(9, -1)$
   $x + 3y = 6$

5. $3x - 6y = 9$ $\left(4, \frac{1}{2}\right)$
   $x - 2y = 3$

6. $x - y = 4$ $(6, 2)$
   $3x - 3y = 12$

7. Graph each system of equations.

   a. $y = 2x - 3$
      $y = 2x + 5$

   b. $y = 2x + 1$
      $y = 4x - 5$

   c. $y = 3x - 5$
      $y = 3x - 5$

For Exercises 8–18, determine which system of equations (a, b, or c) makes the statement true. (*Hint*: Refer to the graphs from Exercise 7.)

a. $y = 2x - 3$
   $y = 2x + 5$

b. $y = 2x + 1$
   $y = 4x - 5$

c. $y = 3x - 5$
   $y = 3x - 5$

8. The lines are parallel.
9. The lines are coinciding.
10. The lines intersect at exactly one point.
11. The system is inconsistent.
12. The system is dependent.
13. The lines have the same slope but different y-intercepts.
14. The lines have the same slope and same y-intercept.
15. The lines have different slopes.
16. The system has exactly one solution.
17. The system has infinitely many solutions.
18. The system has no solution.

For Exercises 19–22, match the graph of the system of equations with the appropriate description of the solution.

19.

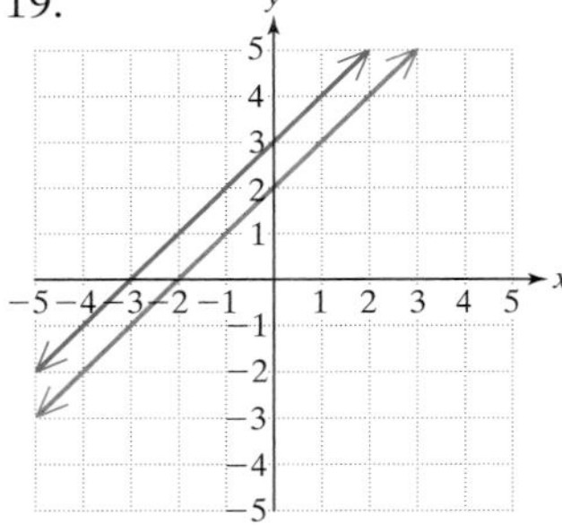

a. The solution is $(1, 3)$.

b. No solution.

c. There are infinitely many solutions.

d. The solution is $(0, 0)$.

20. 21.

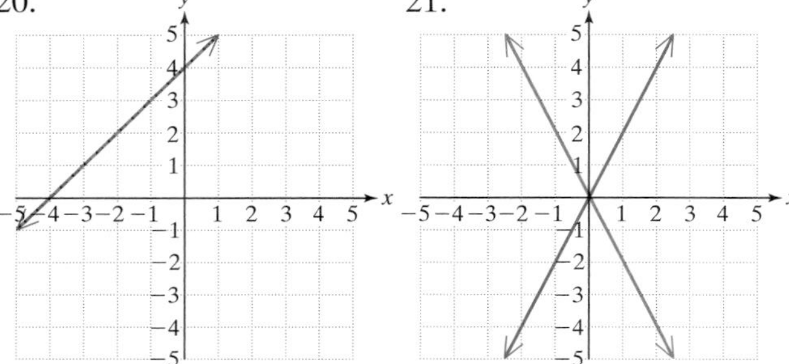

22.

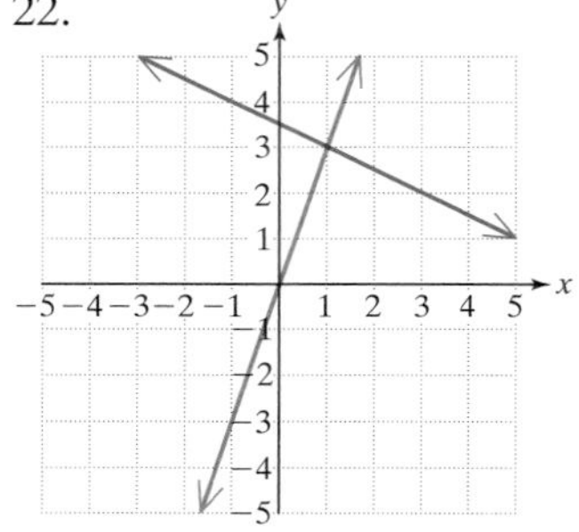

For Exercises 23–48, solve the systems by graphing. Identify each system as consistent or inconsistent. Identify each system as dependent or independent.

23. $y = -x + 4$
$y = x - 2$

24. $y = 3x + 2$
$y = 2x$

25. $2x + y = 0$
$3x + y = 1$

26. $x + y = -1$
$2x - y = -5$

27. $2x + y = 6$
$x = 1$

28. $4x + 3y = 9$
$x = 3$

29. $-6x - 3y = 0$
$4x + 2y = 4$

30. $2x - 6y = 12$
$-3x + 9y = 12$

31. $-2x + y = 3$
$6x - 3y = -9$

32. $x + 3y = 0$
$-2x - 6y = 0$

33. $y = 6$
$2x + 3y = 12$

34. $y = -2$
$x - 2y = 10$

35. $-5x + 3y = -9$
$y = \dfrac{5}{3}x - 3$

36. $4x + 2y = 6$
$y = -2x + 3$

37. $x = 4 + y$
$3y = -3x$

38. $3y = 4x$
$x - y = -1$

39. $-x + y = 3$
$4y = 4x + 6$

40. $x - y = 4$
$3y = 3x + 6$

41. $x = 4$
$2y = 4$

42. $-3x = 6$
$y = 2$

43. $2x + 3y = 8$
$-4x - 6y = 6$

44. $4x + 4y = 8$
$5x + 5y = 5$

45. $2x + y = 4$
$4x - 2y = -4$

46. $6x + 6y = 3$
$2x - y = 4$

47. $y = 0.5x + 2$
$-x + 2y = 4$

48. $3x - 4y = 6$
$-6x + 8y = -12$

49. Two tennis instructors have two different fee schedules. Owen charges \$25 per lesson plus a one-time court fee of \$20 at the tennis club. Joan charges \$30 per lesson, but does not require a court fee. The total cost, y, depends on the number of lessons, x, according to the equations

Owen: $y = 25x + 20$

Joan: $y = 30x$

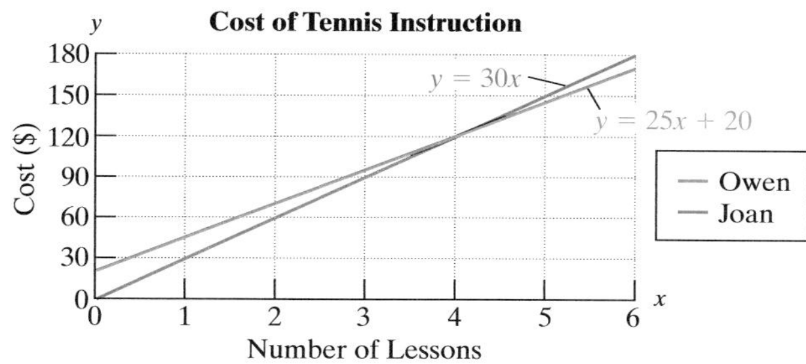

Figure for Exercise 49

From the graph, determine the number of lessons for which the total cost is the same for both instructors.

50. The cost to rent a 10 ft by 10 ft storage space is different for two different storage companies. The Storage Bin charges \$90 per month plus a nonrefundable deposit of \$120. AAA Storage charges \$110 per month with no deposit. The total cost, y, to rent a 10 ft by 10 ft space depends on the number of months, x, according to the equations

The Storage Bin: $y = 90x + 120$

AAA Storage: $y = 110x$

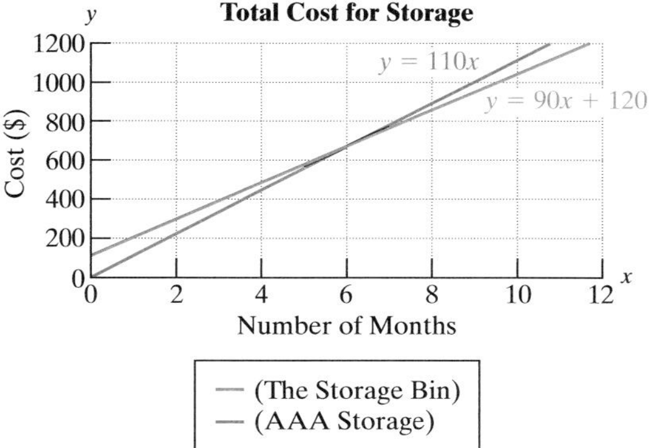

Figure for Exercise 50

From the graph, determine the number of months required for which the cost to rent space is equal for both companies.

51. The following graph depicts the number of persons in the United States living with AIDS between 1993 and 1998 by race or ethnicity. Estimate the year in which the number of black (non Hispanic) persons living with AIDS and the number of white (non Hispanic) persons living with AIDS was approximately the same.

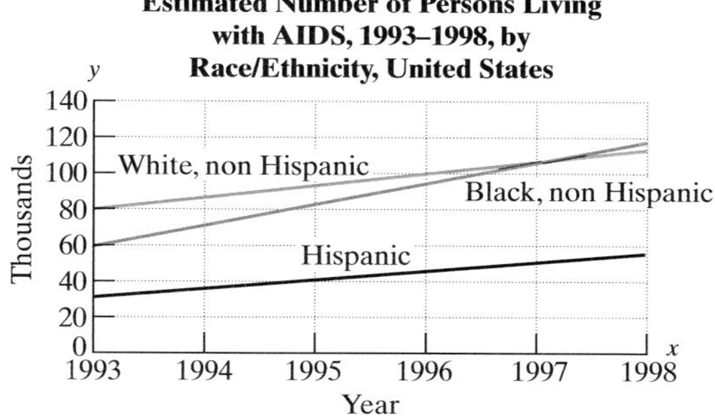

Figure for Exercise 51

Source: Centers for Disease Control and Prevention.

52. The following graph depicts the number of men and number of women in the United States living with AIDS, between 1993 and 1998. If the trends continue, will the number of women living with AIDS ever equal the number of men living with AIDS? Explain why or why not.

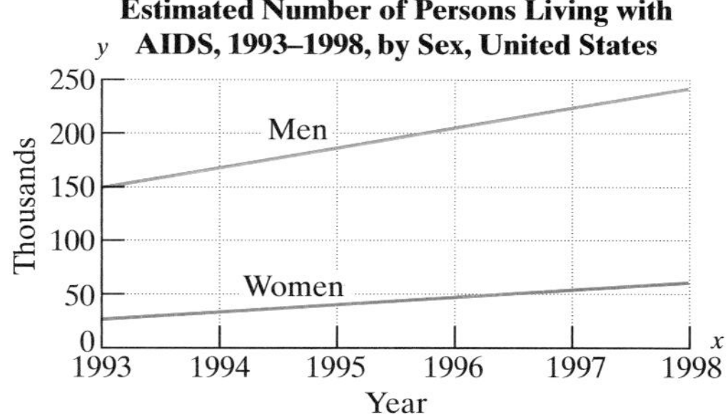

Figure for Exercise 52

Source: Centers for Disease Control and Prevention.

Expanding Your Skills

53. Write a system of linear equations whose solution is (2, 1).

54. Write a system of linear equations whose solution is (1, 4).

55. One equation in a system of linear equations is $x + y = 4$. Write a second equation such that the system will have no solution. (Answers may vary.)

56. One equation in a system of linear equations is $x - y = 3$. Write a second equation such that the system will have infinitely many solutions. (Answers may vary.)

Graphing Calculator Exercises

For Exercises 57–62, use a graphing calculator to graph each linear equation on the same viewing window. Use a *Trace* or *Intersect* feature to find the point(s) of intersection.

57. $y = 2x - 3$
$y = -4x + 9$

58. $y = -\frac{1}{2}x + 2$
$y = \frac{1}{3}x - 3$

59. $x + y = 4$ Example 1
$-2x + y = -5$

60. $x - 2y = -2$ Example 2
$-3x + 2y = 6$

61. $-x + 3y = -6$ Example 3
$6y = 2x + 6$

62. $x + 4y = 8$ Example 4
$y = -\frac{1}{4}x + 2$

Concepts

1. Introduction to the Substitution Method
2. Solving a System of Linear Equations by the Substitution Method
3. Systems of Linear Equations with No Solution or Infinitely Many Solutions
4. Solutions to Systems of Linear Equations: A Summary
5. Applications of the Substitution Method

section

8.2 Substitution Method

1. Introduction to the Substitution Method

In section 8.1 we used the graphing method to find the solution set to a system of equations. However, sometimes it is difficult to determine the solution using this method because of limitations in the accuracy of the graph. This is particularly true when the coordinates of a solution are not integer values or when the solution is a point not sufficiently close to the origin. Identifying the coordinates of the point $(\frac{3}{17}, -\frac{23}{9})$ or $(-251, 8349)$ for example, might be difficult from a graph.

In this section and the next, we will cover two algebraic methods to solve a system of equations that do not require graphing. The first method we present here is called the **substitution method**. For example, consider the following system of equations:

$$x + y = 4$$
$$-2x + y = -5$$

The first step in the substitution process is to isolate one of the variables in one of the equations. For instance, solving the first equation for x yields: $x = 4 - y$. Then, because x is equal to $4 - y$, the expression $4 - y$ may be substituted for x in the second equation. This leaves the second equation in terms of y only.

First equation: $x + y = 4 \longrightarrow x = 4 - y$

Second equation:

$$-2x + y = -5$$

$$-2(4 - y) + y = -5$$

This equation now contains only one variable.

$$-8 + 2y + y = -5$$

Solve the resulting equation.

$$-8 + 3y = -5$$

$$3y = -5 + 8$$

$$3y = 3$$

$$y = 1$$

To find x, substitute $y = 1$ back into the equation

$$x = 4 - y$$

$$x = 4 - (1)$$

$$x = 3$$

The ordered pair (3, 1) can be checked in the original equations to verify the answer.

$$x + y = 4 \longrightarrow (3) + (1) = 4 \checkmark$$

$$-2x + y = -5 \longrightarrow -2(3) + (1) = -5 \checkmark$$

The solution is (3, 1).

2. Solving a System of Linear Equations by the Substitution Method

Solving a System of Equations by the Substitution Method

1. Isolate one of the variables from one equation.
2. Substitute the quantity found in step 1 into the other equation.
3. Solve the resulting equation.
4. Substitute the value found in step 3 back into the equation in step 1 to find the value of the remaining variable.
5. Check the solution in both original equations and write the answer as an ordered pair.

example 1 **Solving a System of Linear Equations Using the Substitution Method**

Solve the system by using the substitution method.

$$3x + 5y = 17$$

$$2x - y = -6$$

Solution:

The y-variable in the second equation is the easiest variable to isolate because its coefficient is -1.

$$3x + 5y = 17$$

$$2x - y = -6 \longrightarrow -y = -2x - 6$$

$$y = 2x + 6$$

Step 1: Solve the second equation for y.

$$3x + 5(2x + 6) = 17$$

Step 2: Substitute the quantity $2x + 6$ for y in the other equation.

Avoiding Mistakes

Do not substitute $y = 2x + 6$ into the same equation from which it came. This mistake will result in an identity:

$$2x - y = -6$$
$$2x - (2x + 6) = -6$$
$$2x - 2x - 6 = -6$$
$$-6 = -6$$

$$3x + 10x + 30 = 17$$

Step 3: Solve for x.

$$13x + 30 = 17$$
$$13x = 17 - 30$$
$$13x = -13$$
$$x = -1$$

$$y = 2x + 6$$
$$y = 2(-1) + 6$$
$$y = -2 + 6$$
$$y = 4$$

Step 4: Substitute $x = -1$ into the equation $y = 2x + 6$.

$$3x + 5y = 17 \longrightarrow 3(-1) + 5(4) = 17 \checkmark$$
$$2x - y = -6 \longrightarrow 2(-1) - (4) = -6 \checkmark$$

Step 5: The ordered pair $(-1, 4)$ can be checked in the original equations to verify the answer.

The solution is $(-1, 4)$.

3. Systems of Linear Equations with No Solution or Infinitely Many Solutions

Recall from Section 8.1, that a system of linear equations may represent two parallel lines. In such a case, there is no solution to the system.

example 2 Solving a System of Linear Equations Using the Substitution Method

Solve the system by using the substitution method.

$$2x + 3y = 6$$
$$y = -\tfrac{2}{3}x + 4$$

Solution:

$$2x + 3y = 6$$

$$y = -\tfrac{2}{3}x + 4$$ **Step 1:** The variable y is already isolated in the second equation.

$$2x + 3(-\tfrac{2}{3}x + 4) = 6$$ **Step 2:** Substitute $y = -\frac{2}{3}x + 4$ from the second equation into the first equation.

$$2x - 2x + 12 = 6$$ **Step 3:** Solve the resulting equation.

$$12 = 6 \quad \text{(contradiction)}$$

The equation results in a contradiction. There are no values of x and y that will make 12 equal to 6. Therefore, there is no solution, and the system is inconsistent.

Tip: The answer to Example 2 can be verified by writing each equation in slope-intercept form and graphing the lines.

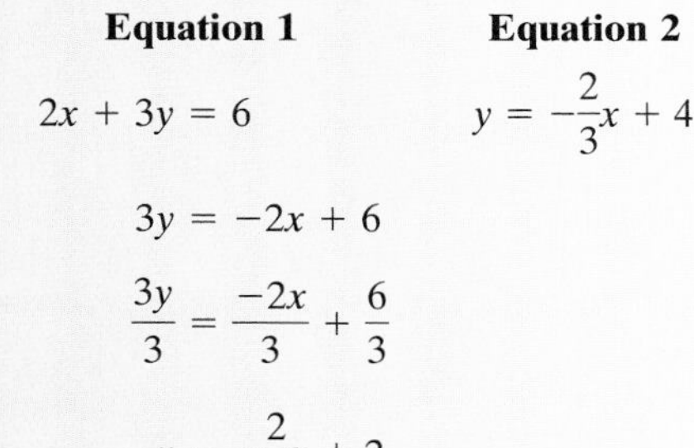

Equation 1	**Equation 2**
$2x + 3y = 6$	$y = -\frac{2}{3}x + 4$
$3y = -2x + 6$	
$\frac{3y}{3} = \frac{-2x}{3} + \frac{6}{3}$	
$y = -\frac{2}{3}x + 2$	

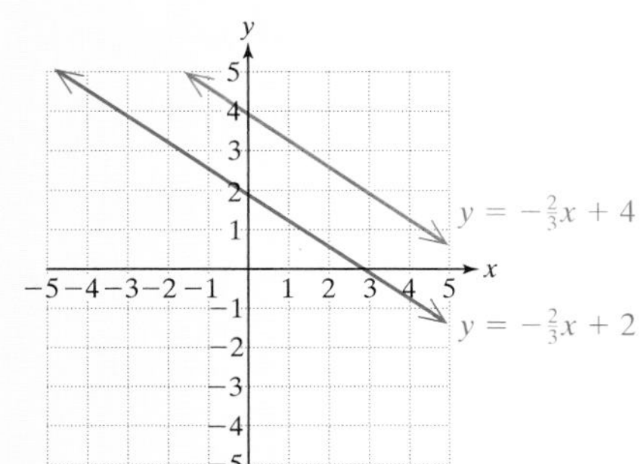

The equations indicate that the lines have the same slope but different y-intercepts. Therefore, the lines must be parallel. There is no point of intersection, indicating that the system has no solution.

Recall that a system of two linear equations may represent the same line. In such a case, the solution is the set of all points on the line.

example 3 Solving a System of Linear Equations Using Substitution

Solve the system by using the substitution method.

$$\frac{1}{2}x - \frac{1}{4}y = 1$$
$$6x - 3y = 12$$

Solution:

$$\frac{1}{2}x - \frac{1}{4}y = 1$$
$$6x - 3y = 12$$

To make the first equation easier to work with, we have the option of clearing fractions.

$$\frac{1}{2}x - \frac{1}{4}y = 1 \xrightarrow{\text{multiply by 4}} 4\left(\frac{1}{2}x\right) - 4\left(\frac{1}{4}y\right) = 4(1) \rightarrow 2x - y = 4$$

Now the system becomes:

$$2x - y = 4$$
$$6x - 3y = 12$$

The y-variable in the first equation is the easiest to isolate because its coefficient is -1.

$$2x - y = 4 \xrightarrow{\text{solve for } y} -y = -2x + 4 \rightarrow y = 2x - 4$$

Step 1: Isolate one of the variables.

$$6x - 3(2x - 4) = 12$$

Step 2: Substitute $y = 2x - 4$ from the first equation into the second equation.

$$6x - 6x + 12 = 12$$

Step 3: Solve the resulting equation.

$$12 = 12 \quad \text{(identity)}$$

Because the equation produces an identity, we know that x can be any real number. Substituting any real number into the expression $y = 2x - 4$ produces an ordered pair on the line $y = 2x - 4$. Hence the solution set to the system of equations is the set of all ordered pairs on the line $y = 2x - 4$. This can be written as $\{(x, y) \mid y = 2x - 4\}$. The system is dependent.

Tip: The solution to Example 3 can be verified by writing each equation in slope-intercept form and graphing the lines.

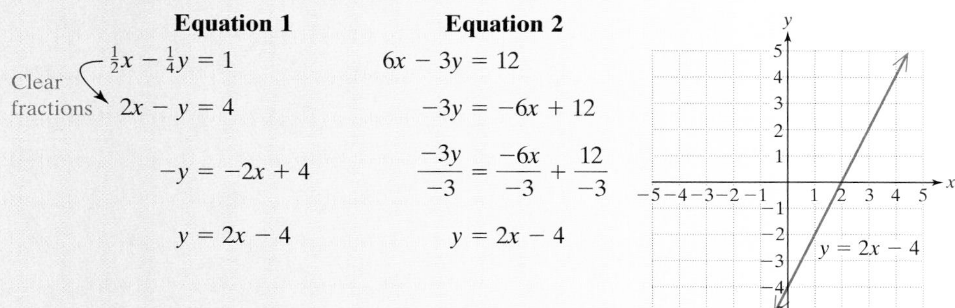

Notice that the slope-intercept forms for both equations are identical. The equations represent the same line, indicating that the system is dependent. Each point on the line is a solution to the system of equations.

4. Solutions to Systems of Linear Equations: A Summary

The following summary reviews the three different geometric relationships between two lines and the solutions to the corresponding systems of equations.

Solutions to a System of Two Linear Equations

1. The lines may intersect at one point (yielding one unique solution).
2. The lines may be parallel and intersect at no point (yielding no solution). This is detected algebraically when a contradiction (false statement) is obtained (for example: $0 = -3$ and $12 = 6$).
3. The lines may be the same and intersect at all points on the line (yielding an infinite number of solutions). This is detected algebraically when an identity is obtained (for example: $0 = 0$ and $12 = 12$).

5. Applications of the Substitution Method

example 4 **Using the Substitution Method in a Geometry Application**

Two angles are supplementary. One angle is 15° more than twice the other angle. Find the two angles.

Solution:

Let x represent one angle.
Let y represent the other angle.

The sum of supplementary angles is 180° ⟶ $x + y = 180$

One angle is 15° more than twice the other angle ⟶ $x = 2y + 15$

$x + y = 180$

$x = 2y + 15$ **Step 1:** The x-variable in the second equation is already isolated.

$(2y + 15) + y = 180$ **Step 2:** Substitute $x = 2y + 15$ from the second equation into the first equation.

$2y + 15 + y = 180$ **Step 3:** Solve the resulting equation.

$3y + 15 = 180$

$3y = 165$

$\frac{3y}{3} = \frac{165}{3}$

$y = 55$

$x = 2y + 15$ **Step 4:** Substitute $y = 55$ into the equation $x = 2y + 15$.

$x = 2(55) + 15$

$x = 110 + 15$

$x = 125$

One angle is 55° and the other is 125°.

Tip: Check that the angles 55° and 125° meet the conditions of Example 4.

- Because 55° + 125° = 180°, the angles are supplementary. ✔
- The angle 125° is 15° more than twice 55°: 125° = 2(55°) + 15°. ✔

section 8.2 Practice Exercises

For Exercises 1–6, write each pair of lines in slope-intercept form. Then identify whether the lines intersect in exactly one point or if the lines are parallel or coinciding.

1. $2x - y = 4$
   $-2y = -4x + 8$

2. $x - 2y = 5$
   $3x = 6y + 15$

3. $2x + 3y = 6$
   $x - y = 5$

4. $x - y = -1$
   $x + 2y = 4$

5. $2x = \frac{1}{2}y + 2$
   $4x - y = 13$

6. $4y = 3x$
   $3x - 4y = 15$

For Exercises 7–14, solve each system using the substitution method.

7. $3x + 2y = -3$
   $y = 2x - 12$

8. $4x - 3y = -19$
   $y = -2x + 13$

9. $x = -4y + 16$
   $3x + 5y = 20$

10. $x = -y + 3$
    $-2x + y = 6$

11. $3x + 5y = 7$
$y = -\frac{3}{5}x + 3$

12. $4x - 3y = -28$
$y = \frac{4}{3}x + 3$

13. $x = \frac{6}{5}y + 3$
$5x - 6y = 15$

14. $x = \frac{1}{2}y + 5$
$4x - 2y = 20$

15. Given the system:

$$4x - 2y = -6$$
$$3x + y = 8$$

a. Which variable from which equation is easiest to isolate and why?

b. Solve the system using the substitution method.

16. Given the system:

$$x - 5y = 2$$
$$11x + 13y = 22$$

a. Which variable from which equation is easiest to isolate and why?

b. Solve the system using the substitution method.

For Exercises 17–46, solve each system using the substitution method.

17. $4x - y = -1$
$2x + 4y = 13$

18. $5x - 3y = -2$
$10x - y = 1$

19. $x - 3y = -1$
$2x = 4y + 2$

20. $3x + y = 1$
$2y = x + 9$

21. $-2x + 5y = 5$
$x - 4y = -10$

22. $3x - 7y = -2$
$2x + y = 27$

23. $3x + 2y = -1$
$\frac{3}{2}x + y = 4$

24. $5x - 2y = 6$
$-\frac{5}{2}x + y = 5$

25. $10x - 30y = -10$
$2x - 6y = -2$

26. $3x + 6y = 6$
$-6x - 12y = -12$

27. $2x + y = 3$
$y = -7$

28. $-3x = 2y + 23$
$x = -1$

29. $x + 2y = -2$
$4x = -2y - 17$

30. $x + y = 1$
$2x - y = -2$

31. $y = -\frac{1}{2}x - 4$
$y = 4x - 13$

32. $y = \frac{2}{3}x - 3$
$y = 6x - 19$

33. $y = -2x + 1$
$y - 4 = -2(x + 3)$

34. $x = 6y + 12$
$x - 3 = 6(y + 2)$

35. $3x + 2y = 4$
$2x - 3y = -6$

36. $4x + 3y = 4$
$-2x + 5y = -2$

37. $y = 0.25x + 1$
$-x + 4y = 4$

38. $y = 0.75x - 3$
$-3x + 4y = -12$

39. $11x + 6y = 17$
$5x - 4y = 1$

40. $3x - 8y = 7$
$10x - 5y = 45$

41. $x + 2y = 4$
$4y = -2x - 8$

42. $-y = x - 6$
$2x + 2y = 4$

43. $\frac{1}{3}(2x + y) = 1$
$x + y = 4$

44. $2(x - y) = 4$
$3x + y = 10$

45. $\frac{a}{3} + \frac{b}{2} = -4$
$a - 3b = 6$

46. $a - 2b = -5$
$\frac{2a}{3} + \frac{b}{3} = 0$

For Exercises 47–56, set up a system of linear equations and solve for the indicated quantities.

47. Two numbers have a sum of 106. One number is 10 less than the other. Find the numbers.

48. Two positive numbers have a difference of 8. The larger number is 2 less than 3 times the smaller number. Find the numbers.

49. Two angles are supplementary. One angle is 15° more than 10 times the other angle. Find the measure of each angle.

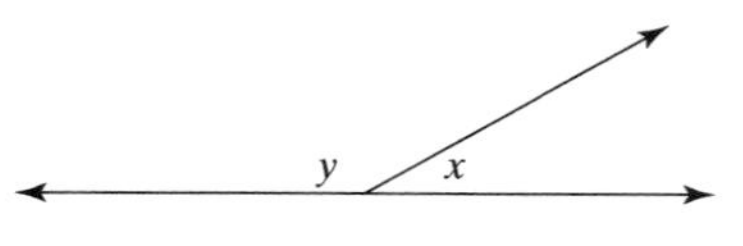

Figure for Exercise 49

50. Two angles are complementary. One angle is 1° less than 6 times the other angle. Find the measure of each angle.

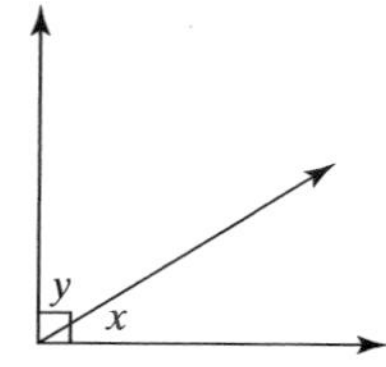

Figure for Exercise 50

51. Two angles are complementary. One angle is 10° more than 3 times the other angle. Find the measure of each angle.

52. Two angles are supplementary. One angle is 5° less than twice the other angle. Find the measure of each angle.

53. In a right triangle, one of the acute angles is 6° less than the other acute angle. Find the measure of each acute angle.

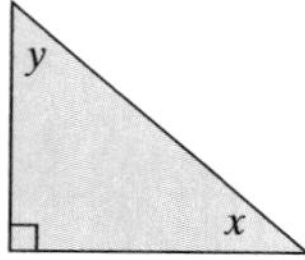

Figure for Exercise 53

54. In a right triangle, one of the acute angles is 9° less than twice the other acute angle. Find the measure of each acute angle.

55. At a ballpark the total cost of a soft drink and a hot dog is $2.50. The price of the hot dog is $1.00 more than the cost of the soft drink. Find the cost of a soft drink and the cost of a hot dog.

56. Ray played two rounds of golf for a total score of 154. If his score in the second round is 10 more than his score in the first round, find the scores for each round.

57. Two water purification systems are priced differently. Company A charges a $55 installation fee plus $20 per month. Company B charges $22.50 per month with no installation fee. The total cost, y, depends on the number of months, x, according to the following equations:

Company A: $y = 20x + 55$

Company B: $y = 22.50x$

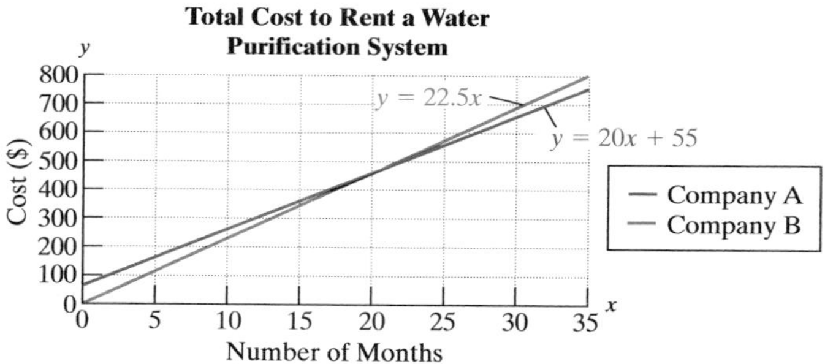

Figure for Exercise 57

a. Use the graph to determine how many months are required for the cost to rent from Company A to equal the cost to rent from Company B.

b. Use the substitution method to solve the system of equations. Interpret the answer in terms of the number of months and the total cost of renting a water purification system.

c. Which company is more expensive if the water system is rented for more than 22 months? Which is more expensive if the water system is rented for less than 22 months?

58. Two rental car companies use different fee rates. Company A charges \$20 per day and \$0.20 per mile. Company B charges 0.30 per mile with no daily fee. If a car is rented for one day, then the total cost, y, depends on the number of miles driven, x, according to the following equations:

Company A: $y = 0.20x + 20$

Company B: $y = 0.30x$

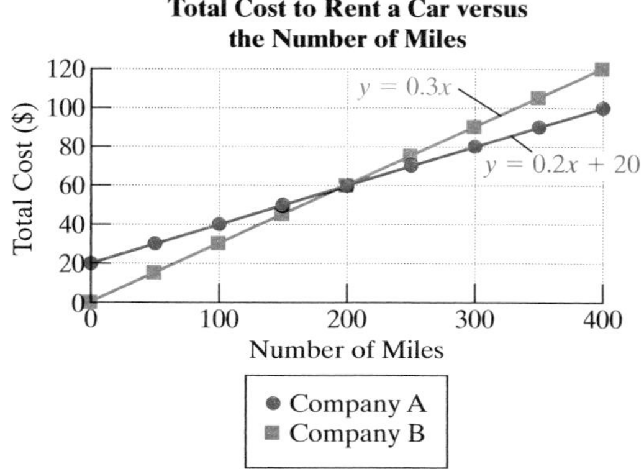

Figure for Exercise 58

a. Use the graph to determine how many miles would be required for the cost to rent from Company A to equal the cost to rent from Company B.

b. Use the substitution method to solve the system of equations. Interpret the answer in terms of the number of miles driven and the total cost.

c. Which car rental company is more expensive if a car is rented for one day and driven for more than 200 miles?

Expanding Your Skills

59. The following system of equations is dependent and has infinitely many solutions. Find three ordered pairs that are solutions to the system of equations.

$$y = 2x + 3$$
$$-4x + 2y = 6$$

60. The following system of equations is dependent and has infinitely many solutions. Find three ordered pairs that are solutions to the system of equations.

$$y = -x + 1$$
$$2x + 2y = 2$$

Concepts

1. Introduction to the Addition Method
2. Solving a System of Linear Equations by the Addition Method
3. Systems of Linear Equations with No Solution or Infinitely Many Solutions
4. Summary of Methods for Solving Linear Equations in Two Variables

section

8.3 Addition Method

1. Introduction to the Addition Method

Thus far in Chapter 8 we have used the graphing method and the substitution method to solve a system of linear equations in two variables. In this section, we present another algebraic method to solve a system of linear equations, called the **addition method** (sometimes called the elimination method). The purpose of the addition method is to eliminate one variable.

example 1 **Using the Addition Method in an Application**

The sum of two numbers is -2. The difference of the smaller number and the larger number is -6. Find the numbers.

Solution:

Let x represent the smaller number.
Let y represent the larger number.

The sum of the numbers is -2. ⟶ $x + y = -2$

The difference of the smaller and the larger is -6. ⟶ $x - y = -6$

Notice that the coefficients of the y-variables are opposites:

$$x + 1y = -2$$
$$x - 1y = -6$$

Coefficient is 1. (top equation)
Coefficient is -1. (bottom equation)

Because the coefficients of the y-variables are opposites, we can add the two equations to eliminate the y-variable.

$$\begin{aligned} x + y &= -2 \\ x - y &= -6 \\ \hline 2x \quad &= -8 \end{aligned}$$

⟵ After adding the equations, we have one equation and one variable.

$2x = -8$ Solve the resulting equation.

$x = -4$

To find the value of y, substitute $x = -4$ into *either* of the original equations.

$x + y = -2$ First equation

$(-4) + y = -2$

$y = -2 + 4$

$y = 2$

The numbers are -4 and 2.

Check that the numbers -4 and 2 meet the conditions of this problem.

1. The sum of the numbers is -2. $-4 + 2 = -2$ ✔
2. The difference of the smaller and the larger is -6. $-4 - 2 = -6$ ✔

Tip: Notice that the value $x = -4$ could have been substituted into the second equation, to obtain the same value for y.

$$x - y = -6$$
$$(-4) - y = -6$$
$$-y = -6 + 4$$
$$-y = -2$$
$$y = 2$$

2. Solving a System of Linear Equations by the Addition Method

It is important to note that the addition method works on the premise that the two equations have *opposite* values for the coefficients of one of the variables. Sometimes it is necessary to manipulate the original equations to create two coefficients that are opposites. This is accomplished by multiplying one or both equations by an appropriate constant. The process is outlined as follows.

Solving a System of Equations by the Addition Method

1. Write both equations in standard form: $ax + by = c$.
2. Clear fractions or decimals (optional).
3. Multiply one or both equations by nonzero constants to create opposite coefficients for one of the variables.
4. Add the equations from step 3 to eliminate one variable.
5. Solve for the remaining variable.
6. Substitute the known value from step 5 into one of the original equations to solve for the other variable.
7. Check the solution in both equations.

example 2

Solving a System of Linear Equations Using the Addition Method

Solve the system using the addition method.

$$3x + 5y = 17$$
$$2x - y = -6$$

Solution:

$3x + 5y = 17$ **Step 1:** Both equations are already written in standard form.

$2x - y = -6$ **Step 2:** There are no fractions or decimals.

Notice that neither the coefficients of x nor the coefficients of y are opposites. However, multiplying the second equation by 5 creates the term $-5y$ in the second equation. This is the opposite of the term $+5y$ in the first equation.

$$\begin{aligned} 3x + 5y &= 17 \\ 2x - y &= -6 \xrightarrow{\text{Multiply by 5}} \end{aligned} \qquad \begin{aligned} 3x + 5y &= 17 \\ 10x - 5y &= -30 \\ \hline 13x \quad &= -13 \end{aligned}$$

Step 3: Multiply the second equation by 5.

Step 4: Add the equations.

$$13x = -13$$
$$x = -1$$

Step 5: Solve the equation.

$$3x + 5y = 17 \quad \text{First equation}$$
$$3(-1) + 5y = 17$$
$$-3 + 5y = 17$$
$$5y = 20$$
$$y = 4$$

Step 6: Substitute the known value of x into one of the original equations.

The solution is $(-1, 4)$.

Step 7: Check the solution in both original equations.

Check:

$3x + 5y = 17 \longrightarrow 3(-1) + 5(4) \stackrel{?}{=} 17 \longrightarrow -3 + 20 = 17$ ✔

$2x - y = -6 \longrightarrow 2(-1) - (4) \stackrel{?}{=} -6 \longrightarrow -2 - 4 = -6$ ✔

example 3 Solving a System of Linear Equations Using the Addition Method

Solve the system using the addition method.

$$3(x - 10) = 7y + 11$$
$$-2(x - y) = 2x - 18$$

Solution:

Step 1: Write the equations in standard form.

$$3(x - 10) = 7y + 11 \xrightarrow{\text{Clear parentheses.}} 3x - 30 = 7y + 11 \xrightarrow{\text{Write as } ax + by = c.} 3x - 7y = 41$$
$$-2(x - y) = 2x - 18 \longrightarrow -2x + 2y = 2x - 18 \longrightarrow -4x + 2y = -18$$

Step 2: There are no fractions or decimals.

Notice that neither the coefficients of x nor the coefficients of y are opposites. However, it is possible to change the coefficients of x to 12 and -12 (notice that 12 is the LCM, of 3 and 4). This is accomplished by multiplying the first equation by 4 and the second equation by 3.

Step 3: Create opposite coefficients of x.

$$3x - 7y = 41 \xrightarrow{\text{Multiply by 4}} 12x - 28y = 164$$
$$-4x + 2y = -18 \xrightarrow[\text{Multiply by 3}]{} \underline{-12x + 6y = -54}$$
$$-22y = 110$$

Step 4: Add the equations.

$$-22y = 110$$
$$\frac{-22y}{-22} = \frac{110}{-22}$$
$$y = -5$$

Step 5: Solve the resulting equation.

$$3x - 7y = 41 \quad \text{First equation}$$
$$3x - 7(-5) = 41$$
$$3x + 35 = 41$$
$$3x = 6$$
$$x = 2$$

Step 6: Substitute the known value of y into one of the original equations.

The solution is $(2, -5)$.

Step 7: Check the solution in the original equations.

Check:

$$3(x - 10) = 7y + 11 \longrightarrow 3(2 - 10) \stackrel{?}{=} 7(-5) + 11 \longrightarrow -24 = -24 \checkmark$$
$$-2(x - y) = 2x - 18 \longrightarrow -2[2 - (-5)] \stackrel{?}{=} 2(2) - 18 \longrightarrow -14 = -14 \checkmark$$

Tip: When using the addition method, it makes no difference which variable is eliminated. In Example 3, we eliminated x. However, we could easily have eliminated y by changing the coefficients of y to -14 and 14. This would be accomplished by multiplying the first equation by 2 and the second equation by 7.

$$\begin{aligned} 3x - 7y = 41 &\xrightarrow{\text{Multiply by 2}} \quad 6x - 14y = 82 \\ -4x + 2y = -18 &\xrightarrow[\text{Multiply by 7}]{} \underline{-28x + 14y = -126} \\ & \qquad\quad -22x \qquad\quad = -44 \end{aligned}$$

Because $-22x = -44$, then $x = 2$. Substituting $x = 2$ into either original equation yields $y = -5$.

3. Systems of Linear Equations with No Solution or Infinitely Many Solutions

example 4

Solving a System of Linear Equations

Solve the system using the addition method.

$$2x - 5y = 10$$
$$\frac{1}{2}x = 1 + \frac{5}{4}y$$

Solution:

$$2x - 5y = 10 \longrightarrow 2x - 5y = 10$$
$$\frac{1}{2}x = 1 + \frac{5}{4}y \longrightarrow \frac{1}{2}x - \frac{5}{4}y = 1$$

Step 1: Write the equations in standard form.

Step 2: Multiply both sides of the second equation by 4 to clear fractions.

$$\frac{1}{2}x - \frac{5}{4}y = 1 \longrightarrow 4\left(\frac{1}{2}x - \frac{5}{4}y\right) = 4(1) \longrightarrow 2x - 5y = 4$$

Now the system becomes

$$2x - 5y = 10$$
$$2x - 5y = 4$$

To make either the x-coefficients or y-coefficients opposites, multiply either equation by -1.

$$2x - 5y = 10 \xrightarrow{\text{Multiply by } -1} -2x + 5y = -10$$
$$2x - 5y = 4 \longrightarrow 2x - 5y = 4$$
$$0 = -6$$

Step 3: Create opposite coefficients.

Step 4: Add the equations.

Because the equation results in a contradiction, there is no solution, and the system is inconsistent. Writing each line in slope-intercept form verifies that the lines are parallel (Figure 8-7).

$$2x - 5y = 10 \xrightarrow{\text{Slope-intercept form}} y = \frac{2}{5}x - 2$$

$$\frac{1}{2}x = 1 + \frac{5}{4}y \xrightarrow{\text{Slope-intercept form}} y = \frac{2}{5}x - \frac{4}{5}$$

There is no solution.

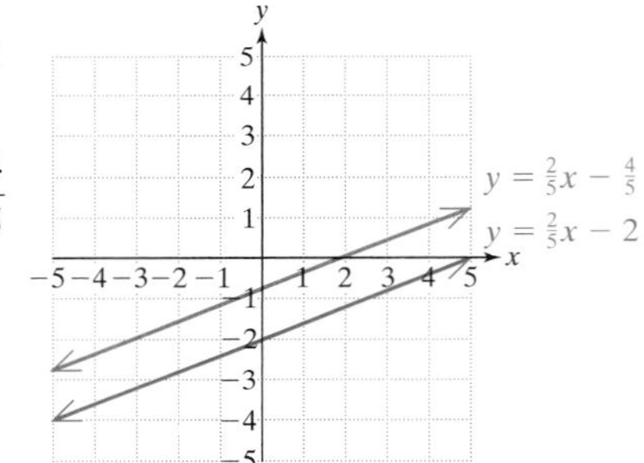

Figure 8-7

example 5

Solving a System of Linear Equations

Solve the system by the addition method.

$$3x - y = 4$$
$$2y = 6x - 8$$

Solution:

$$3x - y = 4 \longrightarrow 3x - y = 4$$

Step 1: Write the equations in standard form.

$$2y = 6x - 8 \longrightarrow -6x + 2y = -8$$

Step 2: There are no fractions or decimals.

Notice that the equations differ exactly by a factor of -2, which indicates that these two equations represent the same line. Multiply the first equation by 2 to create opposite coefficients for the variables.

$$\begin{array}{lcl} 3x - y = 4 & \xrightarrow{\text{Multiply by 2}} & 6x - 2y = 8 \\ -6x + 2y = -8 & & \underline{-6x + 2y = -8} \\ & & 0 = 0 \end{array}$$

Step 3: Create opposite coefficients.

Step 4: Add the equations.

Because the resulting equation is an identity, the original equations represent the same line. This can be confirmed by writing each equation in slope-intercept form.

$$3x - y = 4 \longrightarrow -y = -3x + 4 \longrightarrow y = 3x - 4$$

$$-6x + 2y = -8 \longrightarrow 2y = 6x - 8 \longrightarrow y = 3x - 4$$

The solution is the set of all points on the line, or equivalently, $\{(x, y) \mid y = 3x - 4\}$.

4. Summary of Methods for Solving Linear Equations in Two Variables

If no method of solving a system of linear equations is specified, you may use the method of your choice. However, we recommend the following guidelines:

1. If one of the equations is written with a variable isolated, the substitution method is a good choice. For example,

$$2x + 5y = 2 \qquad \text{or} \qquad y = \frac{1}{3}x - 2$$

$$x = y - 6 \qquad\qquad x - 6y = 9$$

2. If both equations are written in standard form, $ax + by = c$, where none of the variables has coefficients of 1 or -1, then the elimination method is a good choice.

$$4x + 5y = 12$$

$$5x + 3y = 15$$

3. If both equations are written in standard form, $ax + by = c$, and at least one variable has a coefficient of 1 or -1, then either the substitution method or the addition method is a good choice.

section 8.3 Practice Exercises

For Exercises 1–4, check to see if the given ordered pair is a solution to the system.

1. $x + y = 8$ $\quad (5, 3)$
   $y = x - 2$

2. $3x + 2y = 14$ $\quad (5, -2)$
   $5x - 2y = 29$

3. $x = y + 1$ $\quad (3, 2)$
   $-x + 2y = 0$

4. $x = 2y - 11$ $\quad(-9, 1)$
$-x + 5y = 23$

5. Given the system

$$\begin{aligned} 5x - 4y &= 1 \\ 7x - 2y &= 5 \end{aligned}$$

a. True or False: To eliminate the y-variable using the addition method, multiply the second equation by 2.

b. True or False: To eliminate the x-variable, multiply the first equation by 7 and the second equation by -5.

6. Given the system

$$\begin{aligned} 3x + 5y &= -1 \\ 9x - 8y &= -26 \end{aligned}$$

a. True or False: To eliminate the x-variable using the addition method, multiply the first equation by -3.

b. True or False: To eliminate the y-variable, multiply the first equation by 8 and the second equation by -5.

For Exercises 7 and 8,

a. Which variable, x or y, is easier to eliminate using the addition method? Explain.

b. Solve the system using the addition method.

7. $3x - 4y = 2$
$17x + y = 35$

8. $-2x + 5y = -15$
$6x - 7y = 21$

9. In solving a system of equations, suppose you get the statement $0 = 5$. How many solutions will the system have? What can you say about the graphs of these equations?

10. In solving a system of equations, suppose you get the statement $0 = 0$. How many solutions will the system have? What can you say about the graphs of these equations?

11. In solving a system of equations, suppose you get the statement $3 = 3$. How many solutions will the system have? What can you say about the graphs of these equations?

12. In solving a system of equations, suppose you get the statement $2 = -5$. How many solutions will the system have? What can you say about the graphs of these equations?

For Exercises 13–32, solve the systems using the addition method.

13. $x + 2y = 8$
$5x - 2y = 4$

14. $2x - 3y = 11$
$-4x + 3y = -19$

15. $a + b = 3$
$3a + b = 13$

16. $-2u + 6v = 10$
$-2u + v = -5$

17. $-3x + y = 1$
$-6x - 2y = -2$

18. $5p - 2q = 4$
$3p + q = 9$

19. $3x - 5y = 13$
$x - 2y = 5$

20. $7a + 2b = -1$
$3a - 4b = 19$

21. $-2x + y = -5$
$8x - 4y = 12$

22. $x - 3y = 2$
$-5x + 15y = 10$

23. $x + 2y = 2$
$-3x - 6y = -6$

24. $4x - 3y = 6$
$-12x + 9y = -18$

25. $3a + 2b = 11$
$7a - 3b = -5$

26. $4y + 5z = -2$
$5y - 3z = 16$

27. $3x - 5y = 7$
$5x - 2y = -1$

28. $4s + 3t = 9$
$3s + 4t = 12$

29. $2(x + 1) = -3y + 9$
$3x - 10 = -4y$

30. $-3(x - 2) + 7y = 5$
$5y = 2x$

31. $4x - 5y = 0$
$8(x - 1) = 10y$

32. $y = 2x + 1$
$-3(2x - y) = 0$

For Exercises 33–50, solve the system by either the addition method or the substitution method.

33. $5x - 2y = 4$
$y = -3x + 9$

34. $-x = 8y + 5$
$4x - 3y = -20$

35. $x + y = 6$
$x - y = 1$

36. $x + y = 2$
$x - y = 3$

37. $3x = 5y - 9$
$2y = 3x + 3$

38. $10x - 5 = 3y$
$2x - 3y = 1$

39. $y = -5x + 1$
$15x - 3 = -3y$

40. $4x + 5y = -2$
$8x = -10y - 4$

41. $x + 2y = 4$
$x - y = -1$

42. $-3x + y = 1$
$-6x - 2y = -2$

43. $8x - 16y = 24$
$2x - 4y = 0$

44. $y = -\frac{1}{2}x - 5$
$2x + 4y = -8$

45. $\frac{m}{2} + \frac{n}{5} = \frac{13}{10}$
$3(m - n) = m - 10$

46. $\frac{a}{4} - \frac{3b}{2} = \frac{15}{2}$
$\frac{1}{5}(a + 2b) = -2$

47. $2(p - 3q) = p + 4$
$3p + 8 = 5p - q$

48. $m - 3n = 10$
$3(m + 4n) = -12$

49. $9a - 2b = 8$
$6(3a + 1) = 4b + 22$

50. $a = 5 + 2b$
$3(a - 2b) = 15$

For Exercises 51–56, set up a system of linear equations, and solve for the indicated quantities.

51. The sum of two positive numbers is 26. Their difference is 14. Find the numbers.

52. The difference of two positive numbers is 2. The sum of the numbers is 36. Find the numbers.

53. Eight times the smaller of two numbers plus 2 times the larger number is 44. Three times the smaller number minus 2 times the larger number is zero. Find the numbers.

54. Six times the smaller of two numbers minus the larger number is −9. Ten times the smaller number plus five times the larger number is 5. Find the numbers.

55. The number of calories in a piece of cake is 20 less than 3 times the number of calories in a scoop of ice cream. Together, the cake and ice cream have 460 Calories. How many calories are in each?

56. A police force has 240 officers. If there are 116 more men than women, find the number of men and the number of women on the force.

For Exercises 57–60, use the addition method to eliminate the x-variable to solve for y. Then use the addition method to eliminate the y-variable to solve for x.

57. $2x + 3y = 6$
$x - y = 5$

58. $6x + 6y = 8$
$9x - 18y = -3$

59. $2x - 5y = 4$
$3x - 3y = 4$

60. $6x - 5y = 7$
$4x - 6y = 7$

For Exercises 61–65,

a. Which method would you choose to solve the system, the substitution method or the addition method? Explain your choice.

b. Solve the system.

61. $\begin{aligned} x &= -2y + 5 \\ 2x - 4y &= 10 \end{aligned}$

62. $\begin{aligned} 3x - 2y &= 22 \\ 5x + 2y &= 10 \end{aligned}$

63. $\begin{aligned} 3x - 6y &= 30 \\ 2x + 3y &= -22 \end{aligned}$

64. $\begin{aligned} -2x + y &= -14 \\ 4x - 2y &= 28 \end{aligned}$

65. $\begin{aligned} y &= 0.4x - 0.3 \\ -4x + 10y &= 20 \end{aligned}$

Expanding Your Skills

66. Explain why a system of linear equations cannot have exactly two solutions.

67. The solution to the following system of linear equations is $(1, 2)$. Find A and B.

$$\begin{aligned} Ax + 3y &= 8 \\ x + By &= -7 \end{aligned}$$

68. The solution to the following system of linear equations is $(-3, 4)$. Find A and B.

$$\begin{aligned} 4x + Ay &= -32 \\ Bx + 6y &= 18 \end{aligned}$$

Concepts

1. Applications Involving Cost
2. Applications Involving Principal and Interest
3. Applications Involving Mixtures
4. Applications Involving Distance, Rate, and Time

section 8.4 Applications of Linear Equations in Two Variables

1. Applications Involving Cost

In Section 2.7, we solved several applied problems by setting up a linear equation in one variable. When solving an application that involves two unknowns, sometimes it is convenient to use a system of linear equations in two variables.

example 1 **Using a System of Linear Equations Involving Cost**

At a movie theater a couple buys one large popcorn and two drinks for \$5.75. A group of teenagers buys two large popcorns and five drinks for \$13.00. Find the cost of one large popcorn and the price of one drink.

Solution:

Let x represent the cost of one large popcorn.
Let y represent the cost of one drink.

$$\begin{pmatrix}\text{Cost of 1}\\ \text{large popcorn}\end{pmatrix} + \begin{pmatrix}\text{cost of 2}\\ \text{drinks}\end{pmatrix} = \begin{pmatrix}\text{total}\\ \text{cost}\end{pmatrix} \longrightarrow x + 2y = 5.75$$

$$\begin{pmatrix}\text{Cost of 2}\\ \text{large popcorns}\end{pmatrix} + \begin{pmatrix}\text{cost of 5}\\ \text{drinks}\end{pmatrix} = \begin{pmatrix}\text{total}\\ \text{cost}\end{pmatrix} \longrightarrow 2x + 5y = 13.00$$

To solve this system, we may either use the substitution method or the addition method. We will use the substitution method by solving for x in the first equation.

$$x + 2y = 5.75 \longrightarrow x = -2y + 5.75$$
$$2x + 5y = 13.00$$

Isolate x in the first equation.

$$2(-2y + 5.75) + 5y = 13.00$$
$$-4y + 11.50 + 5y = 13.00$$
$$y + 11.50 = 13.00$$
$$y = 13.00 - 11.50$$
$$y = 1.50$$

Substitute $x = -2y + 5.75$ into the other equation.

Solve for y.

$$x = -2y + 5.75$$
$$x = -2(1.50) + 5.75$$
$$x = -3.00 + 5.75$$
$$x = 2.75$$

Substitute $y = 1.50$ into the equation $x = -2y + 5.75$.

The cost of one large popcorn is \$2.75 and the cost of one drink is \$1.50.

Check by verifying that the solutions meet the specified conditions.

1 popcorn + 2 drinks = 1(\$2.75) + 2(\$1.50) = \$5.75 ✔

2 popcorns + 5 drinks = 2(\$2.75) + 5(\$1.50) = \$13.00 ✔

2. Applications Involving Principal and Interest

In Section 2.7 we solved problems in which money is invested in two accounts at different rates. We determined how the principal was to be divided to produce a specified amount of interest. Remember, that when investing there are two sources of money: the principal amount invested and the interest earned. Recall that simple interest is computed by the formula, $I = Prt$. If the time of the investment is 1 year, we have $I = Pr$.

example 2

Using a System of Linear Equations Involving Investments

Joanne has a total of \$6000 to deposit in two accounts. One account earns 3.5% simple interest and the other earns 2.5% simple interest. If the total amount of interest at the end of 1 year is \$195, find the amount she deposited in each account.

Solution:

Let x represent the principal deposited in the 2.5% account.
Let y represent the principal deposited in the 3.5% account.

	2.5% Account	3.5% Account	Total
Principal	x	y	6000
Interest	$0.025x$	$0.035y$	195

Each row of the table yields an equation in x and y:

$$\begin{pmatrix}\text{Principal}\\ \text{invested}\\ \text{at } 2.5\%\end{pmatrix} + \begin{pmatrix}\text{principal}\\ \text{invested}\\ \text{at } 3.5\%\end{pmatrix} = \begin{pmatrix}\text{total}\\ \text{principal}\end{pmatrix} \longrightarrow x + y = 6000$$

$$\begin{pmatrix}\text{Interest}\\ \text{earned}\\ \text{at } 2.5\%\end{pmatrix} + \begin{pmatrix}\text{interest}\\ \text{earned}\\ \text{at } 3.5\%\end{pmatrix} = \begin{pmatrix}\text{total}\\ \text{interest}\end{pmatrix} \longrightarrow 0.025x + 0.035y = 195$$

We will choose the addition method to solve the system of equations. First multiply the second equation by 1000 to clear decimals.

$$\begin{aligned} x + y &= 6000 \longrightarrow & x + y &= 6000 \xrightarrow{\text{Multiply by } -25} & -25x - 25y &= -150{,}000 \\ 0.025x + 0.035y &= 195 \xrightarrow[\text{Multiply by } 1000]{} & 25x + 35y &= 195{,}000 \longrightarrow & 25x + 35y &= 195{,}000 \\ & & & & 10y &= 45{,}000 \end{aligned}$$

$$10y = 45{,}000 \qquad \text{After eliminating the } x\text{-variable, solve for } y.$$

$$\frac{10y}{10} = \frac{45{,}000}{10}$$

$$y = 4500 \qquad \text{The amount invested in the 3.5\% account is \$4500.}$$

$$x + y = 6000 \qquad \text{Substitute } y = 4500 \text{ into the equation } x + y = 6000.$$

$$x + 4500 = 6000$$

$$x = 1500 \qquad \text{The amount invested in the 2.5\% account is \$1500.}$$

Joanne deposited \$1500 in the 2.5% account and \$4500 in the 3.5% account.

To check the solution, verify that the conditions of the problem have been met.

1. The sum of \$1500 and \$4500 is \$6000 as desired. ✔
2. The interest earned on \$1500 at 2.5% is: 0.025(\$1500) = \$37.5
   The interest earned on \$4500 at 3.5% is: 0.035(\$4500) = \$157.5
   Total interest: \$195.00 ✔

3. Applications Involving Mixtures

example 3

Using a System of Linear Equations in a Mixture Application

A 10% alcohol solution is mixed with a 40% alcohol solution to produce 30 L of a 20% alcohol solution. Find the number of liters of 10% solution and the number of liters of 40% solution required for this mixture.

Solution:

Each solution contains a percentage of alcohol plus some other mixing agent such as water. Before we set up a system of equations to model this situation, it is helpful to have background understanding of the problem. In Figure 8-8, the liquid depicted in blue is pure alcohol and the liquid shown in gray is the mixing agent (such as water). Together these liquids form a solution. (Realistically the mixture may not separate as shown, but this image may be helpful for your understanding.)

Let x represent the number of liters of 10% solution.
Let y represent the number of liters of 40% solution.

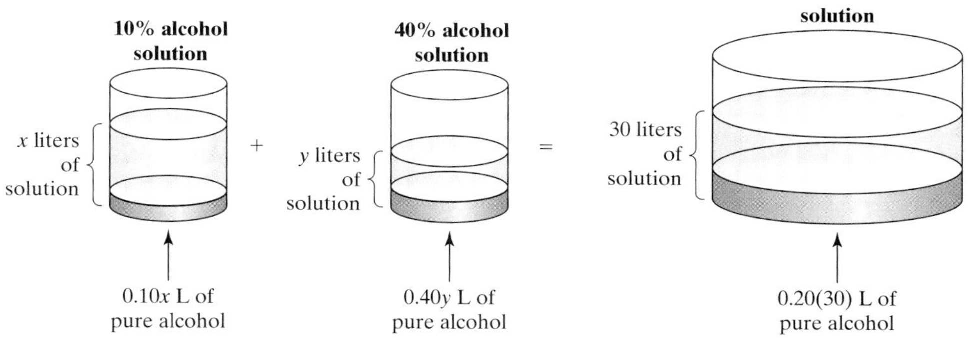

Figure 8-8

The information given in the statement of the problem can be organized in a chart.

	10% Alcohol	40% Alcohol	20% Alcohol
Number of liters of solution	x	y	30
Number of liters of pure alcohol	$0.10x$	$0.40y$	$0.20(30) = 6$

From the first row, we have

$$\begin{pmatrix}\text{Amount of}\\ \text{10\% solution}\end{pmatrix} + \begin{pmatrix}\text{amount of}\\ \text{40\% solution}\end{pmatrix} = \begin{pmatrix}\text{total amount}\\ \text{of 20\% solution}\end{pmatrix} \rightarrow x + y = 30$$

From the second row, we have

$$\begin{pmatrix}\text{Amount of}\\ \text{10\% alcohol}\\ \text{solution}\end{pmatrix} + \begin{pmatrix}\text{amount of}\\ \text{40\% alcohol}\\ \text{solution}\end{pmatrix} = \begin{pmatrix}\text{total amount of}\\ \text{alcohol 20\%}\\ \text{solution}\end{pmatrix} \longrightarrow 0.10x + 0.40y = 6$$

We will solve the system with the addition method by first clearing decimals.

$$\begin{aligned} x + y &= 30 \longrightarrow x + y = 30 \xrightarrow{\text{Multiply by } -1} -x - y = -30 \\ 0.10x + 0.40y &= 6 \xrightarrow[\text{Multiply by 10}]{} x + 4y = 60 \longrightarrow \underline{x + 4y = 60} \\ &\qquad\qquad\qquad\qquad\qquad\qquad 3y = 30 \end{aligned}$$

$3y = 30$ After eliminating the x-variable, solve for y.

$y = 10$ 10 L of 40% solution is needed.

$x + y = 30$ Substitute $y = 10$ into either of the original equations.

$x + (10) = 30$

$x = 20$ 20 L of 10% solution is needed.

10 L of 40% solution must be mixed with 20 L of 10% solution.

4. Applications Involving Distance, Rate, and Time

example 4

Using a System of Linear Equations in a Distance, Rate, Time Application

A plane travels with a tail wind from Kansas City, Missouri, to Denver, Colorado, a distance of 600 miles in 2 hours. The return trip against a head wind takes 3 hours. Find the speed of the plane in still air and find the speed of the wind.

Solution:

Let x represent the speed of the plane in still air.
Let y represent the speed of the wind.

Notice that when the plane travels with the wind, the net speed is $x + y$. When the plane travels against the wind, the net speed is $x - y$.

The information given in the problem can be organized in a chart.

	Distance	Rate	Time
With a tail wind	600	$x + y$	2
Against a head wind	600	$x - y$	3

To set up two equations in x and y, recall that $d = rt$.

From the first row, we have

$$\begin{pmatrix}\text{Distance}\\ \text{with the wind}\end{pmatrix} = \begin{pmatrix}\text{rate with}\\ \text{the wind}\end{pmatrix}\begin{pmatrix}\text{time traveled}\\ \text{with the wind}\end{pmatrix} \longrightarrow 600 = (x + y) \cdot 2$$

From the second row, we have

$$\begin{pmatrix}\text{Distance}\\ \text{against the wind}\end{pmatrix} = \begin{pmatrix}\text{rate against}\\ \text{the wind}\end{pmatrix}\begin{pmatrix}\text{time traveled}\\ \text{against the wind}\end{pmatrix} \longrightarrow 600 = (x - y) \cdot 3$$

Using the distributive property to clear parentheses, produces the following system:

$$2x + 2y = 600$$
$$3x - 3y = 600$$

The coefficients of the y-variable can be changed to 6 and -6 by multiplying the first equation by 3 and the second equation by 2.

$$\begin{aligned} 2x + 2y &= 600 \xrightarrow{\text{Multiply by 3}} & 6x + 6y &= 1800 \\ 3x - 3y &= 600 \xrightarrow[\text{Multiply by 2}]{} & 6x - 6y &= 1200 \\ & & 12x \qquad &= 3000 \end{aligned}$$

$$12x = 3000$$

$$\frac{12x}{12} = \frac{3000}{12}$$

$$x = 250$$ The speed of the plane in still air is 250 mph.

Tip: To create opposite coefficients on the y-variables, we could have divided the first equation by 2 and divided the second equation by 3:

$$\begin{aligned} 2x + 2y &= 600 \xrightarrow{\text{Divide by 2}} & x + y &= 300 \\ 3x - 3y &= 600 \xrightarrow[\text{Divide by 3}]{} & x - y &= 200 \\ & & 2x \qquad &= 500 \end{aligned}$$

$$x = 250$$

$$2x + 2y = 600$$ Substitute $x = 250$ into the first equation.

$$2(250) + 2y = 600$$

$$500 + 2y = 600$$

$$2y = 100$$

$$y = 50$$ The speed of the wind is 50 mph.

The speed of the plane in still air is 250 mph. The speed of the wind is 50 mph.

section 8.4 Practice Exercises

For Exercises 1–4, solve each system by three different methods:

a. Graphing method

b. Substitution method

c. Addition method

1. $-2x + y = 6$
   $2x + y = 2$

2. $x - y = 2$
   $x + y = 6$

3. $y = -2x + 6$
   $4x - 2y = 8$

4. $2x = y + 4$
   $4x = 2y + 8$

For Exercises 5–42, set up a system of linear equations in two variables to solve for the unknown quantities.

5. Two angles are complementary. One angle is 10° less than 9 times the other. Find the measure of each angle.

6. Two angles are supplementary. One angle is 9° more than twice the other angle. Find the measure of each angle.

7. In the 1994 Super Bowl, the Dallas Cowboys scored four more points than twice the number of points scored by the Buffalo Bills. If the total number of points scored by both teams was 43, find the number of points scored by each individual team.

8. In the 1973 Super Bowl, the Miami Dolphins scored twice as many points as the Washington Redskins. If the total number of points scored by both teams was 21, find the number of points scored by each individual team.

9. Kent bought three tapes and 2 CDs for $62.50. Demond bought one tape and four CDs for $72.50. Find the cost of one tape and the cost of one CD.

10. Tanya bought three adult tickets and one child's ticket to a movie for $23.00. Li bought two adult tickets and five children's tickets for $30.50. Find the cost of one adult ticket and the cost of one children's ticket.

11. Linda bought 100 shares of a technology stock and 200 shares of a mutual fund for $3800. Her sister, Sandie, bought 300 shares of technology stock and 50 shares of a mutual fund for $5350. Find the cost per share of the technology stock, and the cost per share of the mutual fund.

12. Two videos and three DVDs can be rented for $19.15. Four videos and one DVD can be rented for $17.35. Find the cost to rent one video and the cost to rent one DVD.

13. Shanelle invested $10,000 and at the end of 1 year she received $805 in interest. She invested part of the money in an account earning 10% simple interest and the remaining money in an account earning 7% simple interest. How much did she invest in each account?

	10% Account	7% Account	Total
Principal invested			
Interest earned			

Table for Exercise 13

14. $2000 more is invested in an account earning 12% simple interest than in an account earning 8% simple interest. If the total interest at the end of 1 year is $1240, how much was invested in each account?

15. Janise invested twice as much money in an account earning 7% simple interest as she did in an account earning 4% simple interest. If the total interest at the end of 1 year was $720, how much did Janise invest in each account?

16. Mario invested 4 times as much money in an account earning 9% simple interest as he did in an account earning 5% simple interest. If he received $1435 in total interest after 1 year, how much was invested in each account?

17. Sonia invested a total of $12,000 into two accounts paying 7.5% and 6% simple interest, respectively. If her total return at the end of the first year was $840, how much did she invest in each account?

18. Ms. Kioki divided $20,000 into two accounts paying 10% and 12% simple interest. At the end of the first year, the total interest from both accounts was $2250. Find the amount invested in each account.

19. How much 50% disinfectant solution must be mixed with a 40% disinfectant solution to produce 25 gal of a 46% disinfectant solution?

	50% Mixture	40% Mixture	46% Mixture
Amount of solution			
Amount of disinfectant			

Table for Exercise 19

20. How many gallons of 20% antifreeze solution and a 10% antifreeze solution must be mixed to obtain 40 gal of a 16% antifreeze solution?

21. How much 45% disinfectant solution must be mixed with a 30% disinfectant solution to produce 20 gal of a 39% disinfectant solution?

22. How many gallons of a 25% antifreeze solution and a 15% antifreeze solution must be mixed to obtain 15 gal of a 23% antifreeze solution?

23. A jar of face cream contains 18% moisturizer, and another is 24% moisturizer. How many ounces of each should be combined to get 12 oz of a cream that is 22% moisturizer?

24. How much pure bleach must be combined with a solution that is 4% bleach to make 12 oz of a 12% bleach solution?

25. It takes a boat 2 hr to go 16 miles downstream with the current, and 4 hr to return against the current. Find the speed of the boat in still water and the speed of the current.

	Distance	Rate	Time
Downstream			
Return			

Table for Exercise 25

26. A boat takes $1\frac{1}{2}$ hr to go 12 miles upstream against the current. It can go 24 miles downstream with the current in the same amount of time. Find the speed of the current and the speed of the boat in still water.

27. A plane can fly 800 miles with the wind in $2\frac{1}{2}$ hr. It takes the same amount of time to fly 700 miles against the wind. What is the speed of the plane in still air and the speed of the wind?

28. A plane flies 600 miles with the wind in $2\frac{1}{2}$ hr. The return trip takes $3\frac{1}{3}$ hr. What is the speed of the wind and the speed of the plane in still air?

29. Jeannie and Juan rollerblade in opposite directions. Juan averages 2 mph faster than Jeannie. If they began at the same place and ended up 20 miles apart after 2 hr, how fast did each of them travel?

30. A plane flew 720 miles in 3 hr with the wind. It would take 4 hr to travel the same distance against the wind. What is the rate of the plane in still air and the rate of wind?

31. Debi has $2.80 in a collection of dimes and nickels. The number of nickels is five more than the number of dimes. Find the number of each type of coin.

	Dimes	Nickels	Total
Number of coins			
Value of coins			

Table for Exercise 31

32. A child is collecting state quarters and new $1 coins. If she has a total of 25 coins, and the number of quarters is nine more than the number of dollar coins, how many of each type of coin does she have?

33. In the 1961–1962 NBA basketball season, Wilt Chamberlain of the Philadelphia Warriors made 2432 baskets. Some of the baskets were free-throws (worth 1 point each) and some were field goals (worth 2 points each). The number of field goals was 762 more than the number of free-throws.
    a. How many field goals did he make and how many free-throws did he make?
    b. What was the total number of points scored?
    c. If Wilt Chamberlain played 80 games during this season, what was the average number of points per game?

34. In the 1971–1972 NBA basketball season, Kareem Abdul-Jabbar of the Milwaukee Bucks made 1663 baskets. Some of the baskets were free-throws (worth 1 point each) and some were field goals (worth 2 points each). The number of field goals he scored was 151 more than twice the number of free-throws.
    a. How many field goals did he make and how many free-throws did he make?
    b. What was the total number of points scored?
    c. If Kareem Abdul-Jabbar played 81 games during this season, what was the average number of points per game?

35. A small plane can fly 350 miles with a tailwind in $1\frac{3}{4}$ hours. In the same amount of time the same plane can travel only 210 miles with a headwind. What is the speed of the plane in still air and the speed of the wind?

36. A plane takes 2 hr to travel 1000 miles with the wind. It can travel only 880 miles against the wind in the same time. Find the speed of the wind and the speed of the plane in still air.

37. At the holidays, Erica likes to sell a candy/nut mixture to her neighbors. She wants to combine candy that costs \$1.80 per pound with nuts that costs \$1.20 per pound. If Erica needs 20 lb of mixture that will sell for \$1.56 per pound, how many pounds of candy and how many pounds of nuts should she use?

38. Mary Lee's natural food store sells a combination of teas. The most popular is a mixture of a tea that sells for \$3.00 per pound with one that sells for \$4.00 per pound. If 40 lb of the mixture sells for \$3.65 per pound, how many pounds of each tea should she use?

39. A total of \$60,000 is invested in two accounts, one that earns 5.5% simple interest, and one that earns 6.5% simple interest. If the total interest at the end of 1 year is \$3750, find the amount invested in each account.

40. Jacques borrows a total of \$15,000. Part of the money is borrowed from a bank that charges 12% simple interest per year. Jacques borrows the remaining part of the money from his sister and promises to pay her 7% simple interest per year. If Jacques total interest for the year is \$1475, find the amount he borrowed from each source.

41. Miracle-Gro All-Purpose Plant Food contains 15% nitrogen. Green Light Super Bloom contains 12% nitrogen. How much Miracle-Gro and how much Green Light fertilizer must be mixed to obtain 60 oz of a mixture that is 13% nitrogen?

42. A textile manufacturer wants to combine a mixture of 20% dye with a mixture that is 50% dye to form 200 gal of a mixture that is 42.5% dye. How much of the 20% and 50% dye mixtures should he use?

43. The *demand* for a certain printer cartridge is related to the price. In general, the higher the price, x, the lower the demand, y. The *supply* for the printer cartridges is also related to price. The higher the price, the greater the incentive for the supplier to stock the item. The supply and demand for the printer cartridges depends on the price according to the equations:

$y_d = -10x + 500$ where x is the price per cartridge in dollars, and y_d is the demand measured in 1000s of cartridges.

$y_s = \frac{20}{3}x$ where x is the price per cartridge in dollars, and y_s is the supply measured in 1000s of cartridges.

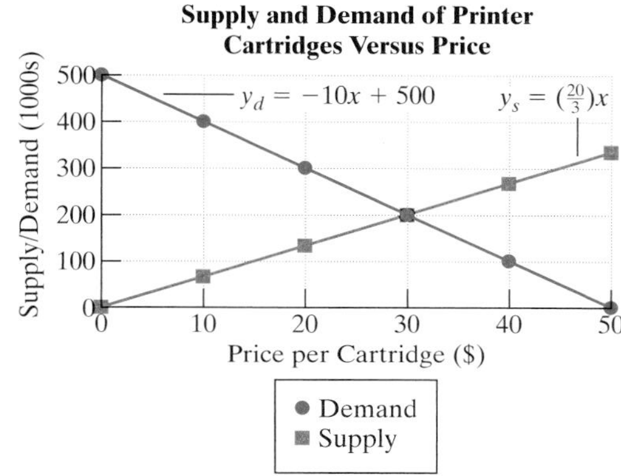

Figure for Exercise 43

Find the price at which the supply and demand are in equilibrium (supply = demand), and confirm your answer with the graph.

44. The supply and demand for a pack of note cards depends on the price according to the equations:

$y_d = -130x + 660$ where x is the price per pack in dollars and y_d is the demand in 1000s of note cards.

$y_s = 90x$ where x is the price per pack in dollars and y_s is the supply measured in 1000s of note cards.

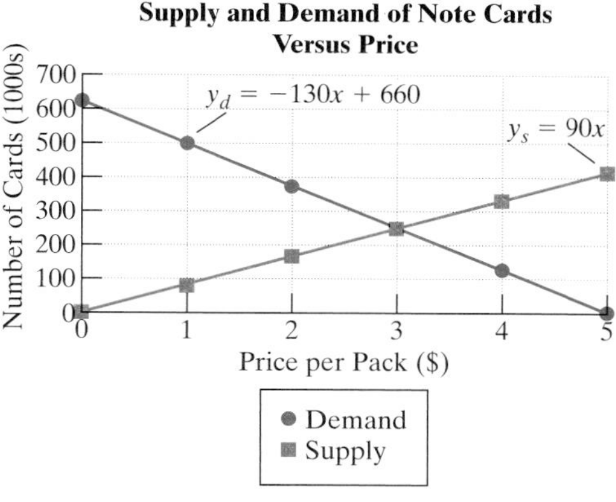

Figure for Exercise 44

Find the price at which the supply and demand are in equilibrium (supply = demand), and confirm your answer with the graph.

Expanding Your Skills

45. In a survey conducted among 500 college students, 340 said that the campus lacked adequate lighting. If $\frac{4}{5}$ of the women and $\frac{1}{2}$ of the men said that they thought the campus lacked adequate lighting, how many men and how many women were in the survey?

46. A thousand people were surveyed in southern California, and 445 said that they worked out at least three times a week. If $\frac{1}{2}$ of the women and $\frac{3}{8}$ of the men said that they worked out at least three times a week, how many men and how many women were in the survey?

47. During a 1-hour television program, there were 22 commercials. Some commercials were 15 sec, and some were 30 sec long. Find the number of 15-sec commercials and the number of 30-sec commercials if the total playing time for commercials was 9.5 min.

Concepts

1. Introduction to Linear Equations in Three Variables
2. Solutions to Systems of Linear Equations in Three Variables
3. Solving Systems of Linear Equations in Three Variables
4. Solving a Dependent System and an Inconsistent System of Linear Equations
5. Applications of Linear Equations in Three Variables

section

8.5 Systems of Linear Equations in Three Variables and Applications

1. Introduction to Linear Equations in Three Variables

In Sections 8.1–8.4, we solved systems of linear equations in two variables. In this section, we will expand the discussion to solving systems involving three variables.

A **linear equation in three variables** can be written in the form $ax + by + cz = d$, where a, b, and c are not all zero. For example, the equation $2x + 3y + z = 6$ is a linear equation in three variables. Solutions to this equation are **ordered triples** of the form (x, y, z) that satisfy the equation. Some solutions to the equation $2x + 3y + z = 6$ are:

<u>Solution</u>: <u>Check</u>:

$(1, 1, 1) \longrightarrow 2(1) + 3(1) + (1) = 6$ ✔

$(2, 0, 2) \longrightarrow 2(2) + 3(0) + (2) = 6$ ✔

$(0, 1, 3) \longrightarrow 2(0) + 3(1) + (3) = 6$ ✔

Infinitely many ordered triples serve as solutions to the equation $2x + 3y + z = 6$.

The set of all ordered triples that are solutions to a linear equation in three variables may be represented graphically by a plane in space. Figure 8-9 shows a portion of the plane $2x + 3y + x = 6$ in the first octant.

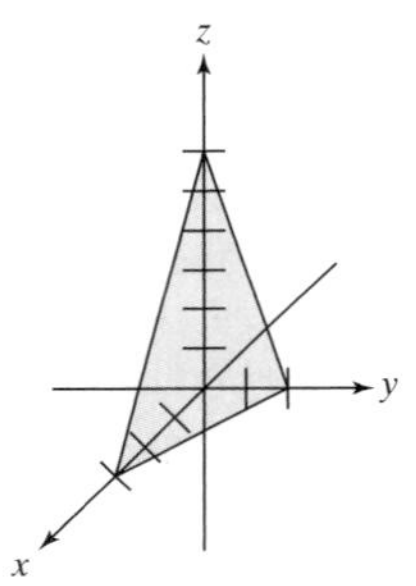

Figure 8-9

2. Solutions to Systems of Linear Equations in Three Variables

A **solution to a system of linear equations in three variables** is an ordered triple that satisfies *each* equation. Geometrically, a solution is a point of intersection of the planes represented by the equations in the system.

A system of linear equations in three variables may have *one unique solution, infinitely many solutions,* or *no solution.*

One unique solution (planes intersect at one point)

- The system is consistent.
- The system is independent.

No solution (the three planes do not all intersect)

- The system is inconsistent.
- The system is independent.

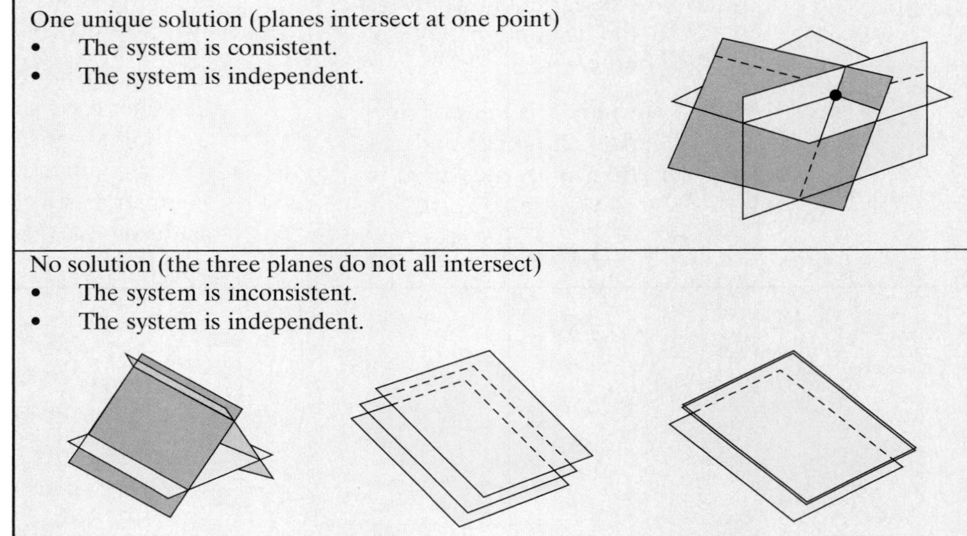

Infinitely many solutions (planes intersect at infinitely many points)

- The system is consistent.
- The system is dependent.

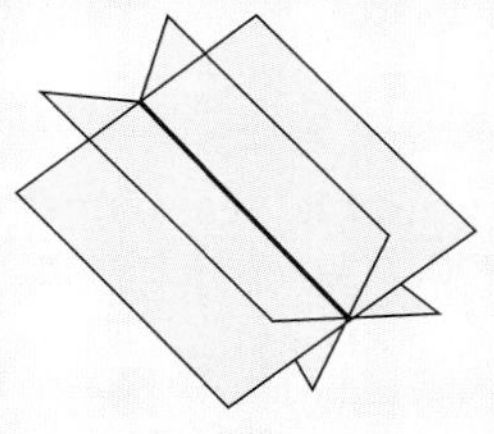
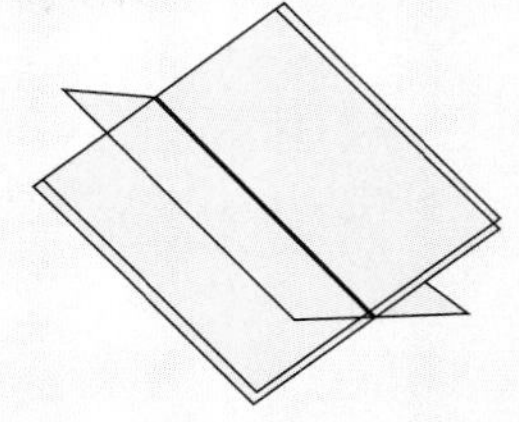
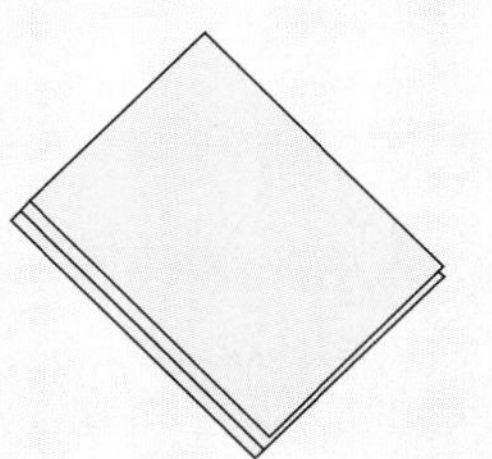

3. Solving Systems of Linear Equations in Three Variables

To solve a system involving three variables, the goal is to eliminate one variable. This reduces the system to two equations in two variables. One strategy for eliminating a variable is to pair up the original equations two at a time.

Solving a System of Three Linear Equations and Three Variables

1. Write each equation in standard form, $ax + by + cz = d$.
2. Choose a pair of equations and eliminate one of the variables using the addition method.
3. Choose a different pair of equations and eliminate the *same* variable.
4. Once steps 2 and 3 are complete, you should have two equations in two variables. Solve this system using the methods from Sections 8.2 and 8.3.
5. Substitute the values of the variables found in step 4 into any of the three original equations that contain the third variable. Solve for the third variable.
6. Check the solution in each of the original equations.

example 1 **Solving a System of Linear Equations in Three Variables**

Solve the system:

$$\begin{aligned} 2x + y - 3z &= -7 \\ 3x - 2y + z &= 11 \\ -2x - 3y - 2z &= 3 \end{aligned}$$

Solution:

[A] $2x + y - 3z = -7$

[B] $3x - 2y + z = 11$

[C] $-2x - 3y - 2z = 3$

Step 1: The equations are already in standard form.

- It is often helpful to label the equations.
- The y-variable can be easily eliminated from equations [A] and [B] and from equations [A] and [C]. This is accomplished by creating opposite coefficients for the y terms and then adding the equations.

Step 2: Eliminate the y-variable from equations [A] and [B].

$$\begin{array}{llcl} \boxed{A} & 2x + y - 3z = -7 & \xrightarrow{\text{multiply by 2}} & 4x + 2y - 6z = -14 \\ \boxed{B} & 3x - 2y + z = 11 & \longrightarrow & 3x - 2y + z = 11 \\ \hline & & & 7x - 5z = -3 \quad \boxed{D} \end{array}$$

Step 3: Eliminate the y-variable again, this time from equations [A] and [C].

$$\begin{array}{llcl} \boxed{A} & 2x + y - 3z = -7 & \xrightarrow{\text{multiply by 3}} & 6x + 3y - 9z = -21 \\ \boxed{C} & -2x - 3y - 2z = 3 & \longrightarrow & -2x - 3y - 2z = 3 \\ \hline & & & 4x - 11z = -18 \quad \boxed{E} \end{array}$$

Tip: It is important to note that in steps 2 and 3, the *same* variable is eliminated.

Step 4: Now equations [D] and [E] can be paired up to form a linear system in two variables. Solve this system.

$$\begin{array}{llcl} \boxed{D} & 7x - 5z = -3 & \xrightarrow{\text{multiply by } -4} & -28x + 20z = 12 \\ \boxed{E} & 4x - 11z = -18 & \xrightarrow{\text{multiply by 7}} & 28x - 77z = -126 \\ \hline & & & -57z = -114 \end{array}$$

$$\frac{-57z}{-57} = \frac{-114}{-57}$$

$$z = 2$$

Once one variable has been found, substitute this value into either equation in the two-variable system, that is, either equation [D] or [E].

$$\begin{aligned} \boxed{D} \quad 7x - 5z &= -3 \\ 7x - 5(2) &= -3 \qquad \text{Substitute } z = 2 \text{ into equation } \boxed{D}. \\ 7x - 10 &= -3 \\ 7x &= 7 \\ x &= 1 \end{aligned}$$

$$\begin{aligned} \boxed{A} \quad 2x + y - 3z &= -7 \\ 2(1) + y - 3(2) &= -7 \\ 2 + y - 6 &= -7 \\ y - 4 &= -7 \\ y &= -3 \end{aligned}$$

Step 5: Now that two variables are known, substitute these values (x and z) into any of the original three equations to find the remaining variable, y. Substitute $x = 1$ and $z = 2$ into equation [A].

The solution is $(1, -3, 2)$.

Step 6: Check the ordered triple in the three original equations.

Check:

$$\begin{aligned} 2x + y - 3z = -7 &\rightarrow 2(1) + (-3) - 3(2) = -7 \ \checkmark \\ 3x - 2y + z = 11 &\rightarrow 3(1) - 2(-3) + (2) = 11 \ \checkmark \\ -2x - 3y - 2z = 3 &\rightarrow -2(1) - 3(-3) - 2(2) = 3 \ \checkmark \end{aligned}$$

example 2 Applying Systems of Linear Equations in Three Variables

In a triangle, the smallest angle is 10° more than half the largest angle. The middle angle is 12° more than the smallest angle. Find the measures of each angle.

Solution:

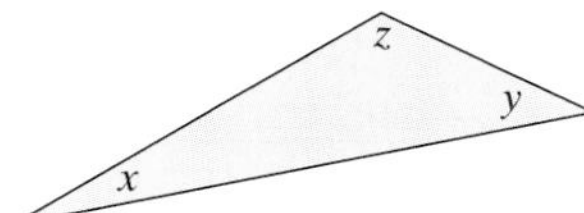

Let x represent the measure of the smallest angle.

Let y represent the measure of the middle angle.

Let z represent the measure of the largest angle.

To solve for three variables, we need to establish three independent relationships among x, y, and z.

$\boxed{A}$ $x = \frac{z}{2} + 10$	The smallest angle is 10° more than half the largest angle.
$\boxed{B}$ $y = x + 12$	The middle angle is 12° more than the smallest angle.
$\boxed{C}$ $x + y + z = 180$	The sum of the angles inscribed in a triangle is 180°.
	Clear fractions and write each equation in standard form.

Standard Form

$$\boxed{A}\quad x = \frac{z}{2} + 10 \xrightarrow{\text{multiply by 2}} 2x = z + 20 \longrightarrow 2x - z = 20$$

$$\boxed{B}\quad y = x + 12 \longrightarrow -x + y = 12$$

$$\boxed{C}\quad x + y + z = 180 \longrightarrow x + y + z = 180$$

Notice equation $\boxed{B}$ is missing the z-variable. Therefore, we can eliminate z again by pairing up equations $\boxed{A}$ and $\boxed{C}$.

$$\begin{array}{lrcl} \boxed{A} & 2x \quad\quad - z & = & 20 \\ \boxed{C} & x + y + z & = & 180 \\ \hline & 3x + y \quad\quad & = & 200 \quad \boxed{D} \end{array}$$

$$\boxed{B}\ -x + y = 12 \xrightarrow{\text{multiply by } -1} x - y = -12$$

$$\boxed{D}\ 3x + y = 200 \longrightarrow \begin{array}{rcl} 3x + y & = & 200 \\ \hline 4x & = & 188 \end{array}$$

Pair up equations $\boxed{B}$ and $\boxed{D}$ to form a system of two variables.

$$\frac{4x}{4} = \frac{188}{4}$$

Solve for x.

$$x = 47°$$

From equation $\boxed{B}$ we have: $-x + y = 12 \rightarrow -47 + y = 12 \rightarrow y = 59$

From equation $\boxed{C}$ we have: $x + y + z = 180 \rightarrow 47 + 59 + z = 180 \rightarrow z = 74$

The smallest angle is 47°, the middle angle is 59°, and the largest angle is 74°.

4. Solving a Dependent System and an Inconsistent System of Linear Equations

example 3

Solving a Dependent System of Linear Equations

Solve the system. If there is not a unique solution, label the system as either dependent or inconsistent.

$$\boxed{A}\quad 3x + y - z = 8$$
$$\boxed{B}\quad 2x - y + 2z = 3$$
$$\boxed{C}\quad x + 2y - 3z = 5$$

Solution:

The first step is to make a decision regarding the variable to eliminate. The y-variable is particularly easy to eliminate because the coefficients of y in equations $\boxed{A}$ and $\boxed{B}$ are already opposites. The y-variable can be eliminated from equations $\boxed{B}$ and $\boxed{C}$ by multiplying equation $\boxed{B}$ by 2.

$$\begin{array}{ll} \boxed{A} & 3x + y - z = 8 \\ \boxed{B} & 2x - y + 2z = 3 \\ \hline & 5x \qquad + z = 11 \quad \boxed{D} \end{array}$$

Pair up equations $\boxed{A}$ and $\boxed{B}$ to eliminate y.

$$\begin{array}{llll} \boxed{B} & 2x - y + 2z = 3 & \xrightarrow{\text{multiply by 2}} & 4x - 2y + 4z = 6 \\ \boxed{C} & x + 2y - 3z = 5 & \longrightarrow & x + 2y - 3z = 5 \\ \hline & & & 5x \qquad + z = 11 \quad \boxed{E} \end{array}$$

Pair up equations $\boxed{B}$ and $\boxed{C}$ to eliminate y.

Because equations $\boxed{D}$ and $\boxed{E}$ are equivalent equations, it appears that this is a dependent system. By eliminating variables we obtain the identity, $0 = 0$.

$$\begin{array}{llll} \boxed{D} & 5x + z = 11 & \xrightarrow{\text{multiply by } -1} & -5x - z = -11 \\ \boxed{E} & 5x + z = 11 & \longrightarrow & 5x + z = 11 \\ \hline & & & 0 = 0 \end{array}$$

The result $0 = 0$ indicates that there are infinitely many solutions and that the system is dependent.

example 4

Solving an Inconsistent System of Linear Equations

Solve the system. If there is not a unique solution, identify the system as either dependent or inconsistent.

$$\begin{aligned} 2x + 3y - 7z &= 4 \\ -4x - 6y + 14z &= 1 \\ 5x + y - 3z &= 6 \end{aligned}$$

Solution:

We will eliminate the x-variable.

$$\begin{array}{lrcl} \boxed{A} & 2x + 3y - 7z = 4 & \xrightarrow{\text{multiply by 2}} & 4x + 6y - 14z = 8 \\ \boxed{B} & -4x - 6y + 14z = 1 & \longrightarrow & -4x - 6y + 14z = 1 \\ \boxed{C} & 5x + y - 3z = 6 & & \overline{\qquad\qquad 0 = 9} \quad \text{(contradiction)} \end{array}$$

The result $0 = 9$ is a contradiction, indicating that the system has no solution. The system is inconsistent.

5. Applications of Linear Equations in Three Variables

example 5

Applying Systems of Linear Equations to Nutrition

Doctors have become increasingly concerned about the sodium intake in the American diet. Recommendations by the American Medical Association indicate that most individuals should not exceed 2400 mg of sodium per day.

Liz ate 1 slice of pizza, 1 serving of ice cream, and 1 glass of soda for a total of 1030 mg of sodium. David ate 3 slices of pizza, no ice cream, and 2 glasses of soda for a total of 2420 mg of sodium. Melinda ate 2 slices of pizza, 1 serving of ice cream, and 2 glasses of soda for a total of 1910 mg of sodium. How much sodium is in one serving of each item?

Solution:

Let x represent the sodium content of 1 slice of pizza.
Let y represent the sodium content of 1 serving of ice cream.
Let z represent the sodium content of 1 glass of soda.

From Liz's meal we have: $\boxed{A}$ $x + y + z = 1030$

From David's meal we have: $\boxed{B}$ $3x + 2z = 2420$

From Melinda's meal we have: $\boxed{C}$ $2x + y + 2z = 1910$

Equation [B] is missing the y-variable. Eliminating y from equations [A] and [C] we have

$$\begin{array}{lrcl} \text{A} & x + y + z = 1030 & \xrightarrow{\text{multiply by } -1} & -x - y - z = -1030 \\ \text{C} & 2x + y + 2z = 1910 & \longrightarrow & 2x + y + 2z = 1910 \\ & & \text{D} & x + z = 880 \end{array}$$

Solve the system formed by equations [B] and [D].

$$\begin{array}{lrcl} \text{B} & 3x + 2z = 2420 & \longrightarrow & 3x + 2z = 2420 \\ \text{D} & x + z = 880 & \xrightarrow[\text{multiply by } -2]{} & -2x - 2z = -1760 \\ & & & x = 660 \end{array}$$

From equation [D] we have: $x + z = 880 \longrightarrow 660 + z = 880 \longrightarrow z = 220$

From equation [A] we have: $x + y + z = 1030 \rightarrow 660 + y + 220 = 1030 \rightarrow y = 150$

Therefore, 1 slice of pizza has 660 mg of sodium, 1 serving of ice cream has 150 mg of sodium, and 1 glass of soda has 220 mg of sodium.

section 8.5 Practice Exercises

For Exercises 1–4, solve the systems using two methods: (a) the substitution method and (b) the addition method.

1. $3x + y = 4$
   $4x + y = 5$

2. $2x - 5y = 3$
   $-4x + 10y = 3$

3. $\frac{1}{2}x - \frac{1}{3}y = 1$
   $x - \frac{2}{3}y = 2$

4. $\frac{1}{2}x + \frac{1}{3}y = 13$
   $\frac{2}{5}x + \frac{1}{4}y = 10$

5. Two cars leave Kansas City at the same time. One travels east and one travels west. After 3 hr the cars are 369 miles apart. If one car travels 7 mph slower than the other, find the speed of each car.

6. How many solutions are possible when solving a system of three equations with three variables?

7. Which of the following points are solutions to the system?

   (2, 1, 7) (3, −10, −6) (4, 0, 2)

$$\begin{aligned} 2x - y + z &= 10 \\ 4x + 2y - 3z &= 10 \\ x - 3y + 2z &= 8 \end{aligned}$$

8. Which of the following points are solutions to the system?

   (1, 1, 3) (0, 0, 4) (4, 2, 1)

$$\begin{aligned} -3x - 3y - 6z &= -24 \\ -9x - 6y + 3z &= -45 \\ 9x + 3y - 9z &= 33 \end{aligned}$$

9. Which of the following points are solutions to the system?

   (0, 4, 3) (3, 6, 10) (3, 3, 1)

   $$\begin{aligned} x + 2y - z &= 5 \\ x - 3y + z &= -5 \\ -2x + y - z &= -4 \end{aligned}$$

10. Which of the following points are solutions to the system?

    (12, 2, −2) (4, 2, 1) (1, 1, 1)

    $$\begin{aligned} -x - y - 4z &= -6 \\ x - 3y + z &= -1 \\ 4x + y - z &= 4 \end{aligned}$$

For Exercises 11–20, solve the system of equations.

11. $$\begin{aligned} x + y + z &= 6 \\ -x + y - z &= -2 \\ 2x + 3y + z &= 11 \end{aligned}$$

12. $$\begin{aligned} x - y - z &= -11 \\ x + y - z &= 15 \\ 2x - y + z &= -9 \end{aligned}$$

13. $$\begin{aligned} -3x + y - z &= 8 \\ -4x + 2y + 3z &= -3 \\ 2x + 3y - 2z &= -1 \end{aligned}$$

14. $$\begin{aligned} 2x + 3y + 3z &= 15 \\ 3x - 6y - 6z &= -23 \\ -9x - 3y + 6z &= 8 \end{aligned}$$

15. $$\begin{aligned} 2x - y + z &= -1 \\ -3x + 2y - 2z &= 1 \\ 5x + 3y + 3z &= 16 \end{aligned}$$

16. $$\begin{aligned} 4x - 3y + 2z &= 12 \\ -3x + 2y - 3z &= -5 \\ 2x - y + 7z &= -8 \end{aligned}$$

17. $$\begin{aligned} 2x - 3y + 2z &= -1 \\ x + 2y &= -4 \\ x + z &= 1 \end{aligned}$$

18. $$\begin{aligned} x + y + z &= 2 \\ 2x - z &= 5 \\ 3y + z &= 2 \end{aligned}$$

19. $$\begin{aligned} 4x + 9y &= 8 \\ 8x + 6z &= -1 \\ 6y + 6z &= -1 \end{aligned}$$

20. $$\begin{aligned} 3x + 2z &= 11 \\ y - 7z &= 4 \\ x - 6y &= 1 \end{aligned}$$

21. A triangle has one angle that is 5° more than twice the smallest angle and the largest angle is 11° less than 3 times the smallest angle. Find the measures of the three angles.

22. The largest angle of a triangle is 4° less than 5 times the smallest angle. The middle angle is twice the smallest. Find the measures of the three angles.

23. The perimeter of a triangle is 54 cm. The longest side is equal to the sum of the other two sides. The smallest side is half as long as the middle side. Find the lengths of the three sides.

24. The perimeter of a triangle is 5 ft. The smallest side is 4 in. less than the middle side and the middle side is half the length of the largest side. What are the measures of the three sides in *inches*?

25. A movie theater charges $7 for adults, $5 for children under 17, and $4 for seniors over 60. For one showing of *Titanic* the theater sold 222 tickets and took in $1383. If twice as many adult tickets were sold as the total of children and senior tickets, how many tickets of each kind were sold?

26. Goofie Golf has 18 holes that are par 3, par 4, or par 5. Most of the holes are par 4. In fact, there are 3 times as many par 4's as par 3's. There are 3 more par 5's than par 3's. How many of each type are there?

27. Combining peanuts, pecans, and cashews makes a party mixture of nuts. If the amount of peanuts equals the amount of pecans and cashews combined, and there are twice as many cashews as pecans, how many ounces of each nut is used to make 48 oz of party mixture?

28. Souvenir hats, T-shirts, and jackets are sold at a rock concert. Three hats, two T-shirts, and one jacket cost \$140. Two hats, two T-shirts, and two jackets cost \$170. One hat, three T-shirts, and two jackets cost \$180. Find the prices of the individual items.

29. In 2002, Baylor University in Waco, Texas, had twice as many students as Vanderbilt University in Nashville, Tennessee. Pace University in New York City had 2800 more students than Vanderbilt University. If the enrollment for all three schools totaled 27,200, find the enrollment for each school.

30. Annie and Maria traveled overseas for seven days and stayed in three different hotels in three different cities: Stockholm, Sweden, Oslo, Norway, and Paris, France.

City	Number of nights	Cost/night (\$)	Tax rate
Paris, France	1	x	8%
Stockholm, Sweden	4	y	11%
Oslo, Norway	2	z	10%

The total bill for all seven nights (not including tax) was \$1040. The total tax was \$106. The nightly cost (excluding tax) to stay at the hotel in Paris was \$80 more than the nightly cost (excluding tax) to stay in Oslo. Find the cost per night for each hotel excluding tax.

For Exercises 31–40, solve the system. If there is not a unique solution, label the system as either dependent or inconsistent.

31. $$\begin{aligned} 2x + y + 3z &= 2 \\ x - y + 2z &= -4 \\ x + 3y - z &= 1 \end{aligned}$$

32. $$\begin{aligned} x + y - z &= 0 \\ 3x - 2y + 6z &= 1 \\ 7x + 3y + z &= 4 \end{aligned}$$

33. $$\begin{aligned} 6x - 2y + 2z &= 2 \\ 4x + 8y - 2z &= 5 \\ -2x - 4y + z &= -2 \end{aligned}$$

34. $$\begin{aligned} 3x + 2y + z &= 3 \\ x - 3y + z &= 4 \\ -6x - 4y - 2z &= 1 \end{aligned}$$

35. $$\begin{aligned} \tfrac{1}{2}x + \tfrac{2}{3}y \phantom{- \tfrac{1}{2}z} &= \tfrac{5}{2} \\ \tfrac{1}{5}x \phantom{+ \tfrac{2}{3}y} - \tfrac{1}{2}z &= -\tfrac{3}{10} \\ \tfrac{1}{3}y - \tfrac{1}{4}z &= \tfrac{3}{4} \end{aligned}$$

36. $$\begin{aligned} \tfrac{1}{2}x + \tfrac{1}{4}y + z &= 3 \\ \tfrac{1}{8}x + \tfrac{1}{4}y + \tfrac{1}{4}z &= \tfrac{9}{8} \\ x - y - \tfrac{2}{3}z &= \tfrac{1}{3} \end{aligned}$$

37. $$\begin{aligned} 2x + y - 3z &= -3 \\ 3x - 2y + 4z &= 1 \\ 4x + 2y - 6z &= -6 \end{aligned}$$

38. $\begin{aligned} 2x + y \quad\quad &= -3 \\ 2y + 16z &= -10 \\ -7x - 3y + 4z &= 8 \end{aligned}$

39. $\begin{aligned} -0.1y + 0.2z &= 0.2 \\ 0.1x + 0.1y + 0.1z &= 0.2 \\ -0.1x \quad\quad + 0.3z &= 0.2 \end{aligned}$

40. $\begin{aligned} 0.1x - 0.2y \quad\quad &= 0 \\ 0.3y + 0.1z &= -0.1 \\ 0.4x \quad\quad - 0.1z &= 1.2 \end{aligned}$

The systems in Exercises 41–44, are called homogeneous systems because each system has (0, 0, 0) as a solution. However if a system is dependent it will have infinitely many more solutions. For each system determine whether (0, 0, 0) is the only solution, or if the system is dependent.

41. $\begin{aligned} 2x - 4y + 8z &= 0 \\ -x - 3y + z &= 0 \\ x - 2y + 5z &= 0 \end{aligned}$

42. $\begin{aligned} 2x - 4y + z &= 0 \\ x - 3y - z &= 0 \\ 3x - y + 2z &= 0 \end{aligned}$

43. $\begin{aligned} 4x - 2y - 3z &= 0 \\ -8x - y + z &= 0 \\ 2x - y - \frac{3}{2}z &= 0 \end{aligned}$

44. $\begin{aligned} 5x + y \quad\quad &= 0 \\ 4y - z &= 0 \\ 5x + 5y - z &= 0 \end{aligned}$

Expanding Your Skills

In Section 7.1 we learned that a linear function, $y = mx + b$, can be determined from two distinct points. Likewise a quadratic model of the form $ax^2 + bx + c = 0\ (a \neq 0)$ can be used to define a parabola through three known points provided the points are not collinear. To determine the quadratic model, it is necessary to compute the values of a, b, and c. This is done by substituting the known values of x and y from each of the three points into the equation $y = ax^2 + bx + c$. The result is a system of three equations in the variables a, b, and c.

For Exercises 45–48, find the quadratic function $y = ax^2 + bx + c\ (a \neq 0)$ whose graph passes through the given points.

45. $(1, -1)$, $(0, 3)$, and $(-2, 17)$

46. $(3, 17)$, $(0, -4)$, and $(-1, -3)$

47. $(1, -5)$, $(-1, -9)$, and $(3, -17)$

48. $(-1, -10)$, $(2, -10)$, and $(4, -20)$

Concepts

section

8.6 Solving Systems of Linear Equations Using Matrices

1. Introduction to Matrices

In Sections 8.2 and 8.3, we solved systems of linear equations using the substitution method and the addition method. We now present a third method called the Gauss-Jordan method that uses matrices to solve a linear system.

A **matrix** is a rectangular array of numbers (the plural of matrix is matrices). The rows of a matrix are read horizontally and the columns of a matrix are read vertically. Every number or entry within a matrix is called an element of the matrix.

The **order of a matrix** is determined by the number of rows and number of columns. A matrix with m rows and n columns is an $m \times n$ (read as "m by n") matrix. Notice that with the order of a matrix, the number of rows is given first, followed by the number of columns.

2. The Order of a Matrix

example 1 Determining the Order of a Matrix

Determine the order of each matrix.

a. $\begin{bmatrix} 2 & -4 & 1 \\ 5 & \pi & \sqrt{7} \end{bmatrix}$ b. $\begin{bmatrix} 1.9 \\ 0 \\ 7.2 \\ -6.1 \end{bmatrix}$ c. $\begin{bmatrix} 1 & 0 & 0 \\ 0 & 1 & 0 \\ 0 & 0 & 1 \end{bmatrix}$ d. $\begin{bmatrix} a & b & c \end{bmatrix}$

Solution:

a. This matrix has two rows and three columns. Therefore, it is a 2×3 matrix.
b. This matrix has four rows and one column. Therefore, it is a 4×1 matrix. A matrix with one column is called a **column matrix**.
c. This matrix has three rows and three columns. Therefore, it is a 3×3 matrix. A matrix with the same number of rows and columns is called a **square matrix**.
d. This matrix has one row and three columns. Therefore, it is a 1×3 matrix. A matrix with one row is called a **row matrix**.

3. Augmented Matrices

A matrix can be used to represent a system of linear equations written in standard form. To do so, we extract the coefficients of the variable terms and the constants within the equation. For example, consider the system:

$$2x - y = 5$$
$$x + 2y = -5$$

The matrix **A** is called the **coefficient matrix**.

$$\mathbf{A} = \begin{bmatrix} 2 & -1 \\ 1 & 2 \end{bmatrix}$$

If we extract both the coefficients and the constants from the equations, we can construct the **augmented matrix** of the system:

$$\left[\begin{array}{rr|r} 2 & -1 & 5 \\ 1 & 2 & -5 \end{array}\right]$$

A vertical bar is inserted into an augmented matrix to designate the position of the equal signs.

example 2 Writing the Augmented Matrix of a System of Linear Equations

Write the augmented matrix for each linear system.

a. $\begin{aligned} -3x - 4y &= 3 \\ 2x + 4y &= 2 \end{aligned}$

b. $\begin{aligned} 2x \qquad - 3z &= 14 \\ 2y + z &= 2 \\ x + y \qquad &= 4 \end{aligned}$

Tip: Notice that zeros are inserted to denote the coefficient of each missing term.

Solution:

a. $\left[\begin{array}{rr|r} -3 & -4 & 3 \\ 2 & 4 & 2 \end{array}\right]$

b. $\left[\begin{array}{rrr|r} 2 & 0 & -3 & 14 \\ 0 & 2 & 1 & 2 \\ 1 & 1 & 0 & 4 \end{array}\right]$

example 3 Writing a Linear System from an Augmented Matrix

Write a system of linear equations represented by each augmented matrix.

a. $\left[\begin{array}{rr|r} 2 & -5 & -8 \\ 4 & 1 & 6 \end{array}\right]$

b. $\left[\begin{array}{rrr|r} 2 & -1 & 3 & 14 \\ 1 & 1 & -2 & -5 \\ 3 & 1 & -1 & 2 \end{array}\right]$

c. $\left[\begin{array}{rrr|r} 1 & 0 & 0 & 4 \\ 0 & 1 & 0 & -1 \\ 0 & 0 & 1 & 0 \end{array}\right]$

Solution:

a. $\begin{aligned} 2x - 5y &= -8 \\ 4x + y &= 6 \end{aligned}$

b. $\begin{aligned} 2x - y + 3z &= 14 \\ x + y - 2z &= -5 \\ 3x + y - z &= 2 \end{aligned}$

c. $\begin{aligned} x + 0y + 0z &= 4 \\ 0x + y + 0z &= -1 \\ 0x + 0y + z &= 0 \end{aligned}$ or $\begin{aligned} x &= 4 \\ y &= -1 \\ z &= 0 \end{aligned}$

4. The Gauss-Jordan Method

We know that interchanging two equations results in an equivalent system of linear equations. Interchanging two rows in an augmented matrix results in an equivalent augmented matrix. Similarly, because each row in an augmented matrix represents a linear equation, we can perform the following elementary row operations that result in an *equivalent augmented matrix*.

Elementary Row Operations

The following **elementary row operations** performed on an augmented matrix produce an equivalent augmented matrix:

1. Interchange two rows.
2. Multiply every element in a row by a nonzero real number.
3. Add a multiple of one row to another row.

5. Solving Systems of Linear Equations Using the Gauss-Jordan Method

When solving a system of linear equations by any method, the goal is to write a series of simpler but equivalent systems of equations until the solution is obvious. The **Gauss-Jordan method** uses a series of elementary row operations performed on the augmented matrix to produce a simpler augmented matrix. In particular, we want to produce an augmented matrix that has 1's along the diagonal of the matrix of coefficients and 0's for the remaining entries in the matrix of coefficients. A matrix written in this way is said to be written in **reduced row echelon form**. For example, the augmented matrix from Example 3(c) is written in reduced row echelon form.

$$\left[\begin{array}{ccc|c} 1 & 0 & 0 & 4 \\ 0 & 1 & 0 & -1 \\ 0 & 0 & 1 & 0 \end{array}\right]$$

The solution to the corresponding system of equations is easily recognized as $x = 4$, $y = -1$, and $z = 0$.

Similarly, the matrix **B** represents a solution of $x = a$ and $y = b$.

$$\mathbf{B} = \left[\begin{array}{cc|c} 1 & 0 & a \\ 0 & 1 & b \end{array}\right]$$

example 4 Solving a System of Linear Equations Using the Gauss-Jordan Method

Use the Gauss-Jordan method to solve the system:

$$\begin{aligned} 2x - y &= 5 \\ x + 2y &= -5 \end{aligned}$$

Solution:

$$\left[\begin{array}{cc|c} 2 & -1 & 5 \\ 1 & 2 & -5 \end{array}\right]$$ Write the augmented matrix.

$$\xrightarrow{R_1 \Leftrightarrow R_2} \left[\begin{array}{cc|c} 1 & 2 & -5 \\ 2 & -1 & 5 \end{array}\right]$$

Switch rows 1 and 2 to get a 1 in the upper left position.

$$\xrightarrow{-2R_1 + R_2 \Rightarrow R_2} \left[\begin{array}{cc|c} 1 & 2 & -5 \\ 0 & -5 & 15 \end{array}\right]$$

Multiply row 1 by -2 and add the result to row 2. This produces an entry of 0 below the upper left position.

$$\xrightarrow{-\frac{1}{5}R_2 \Rightarrow R_2} \left[\begin{array}{cc|c} 1 & 2 & -5 \\ 0 & 1 & -3 \end{array}\right]$$

Multiply row 2 by $-\frac{1}{5}$ to produce a 1 along the diagonal in the second row.

$$\xrightarrow{-2R_2 + R_1 \Rightarrow R_1} \left[\begin{array}{cc|c} 1 & 0 & 1 \\ 0 & 1 & -3 \end{array}\right]$$

Multiply row 2 by -2 and add the result to row 1. This produces a 0 in the first row, second column.

The matrix **C** is in reduced row echelon form. From the augmented matrix, we have, $x = 1$ and $y = -3$. The solution to the system is $(1, -3)$.

$$\mathbf{C} = \left[\begin{array}{cc|c} 1 & 0 & 1 \\ 0 & 1 & -3 \end{array}\right]$$

The order in which we manipulate the elements of an augmented matrix to produce reduced row echelon form was demonstrated in Example 4. In general, the order is as follows.

- First produce a "1" in the first row, first column. Then obtain 0's in the first column below the first row.
- Next, if possible, produce a "1" in the second row, second column, followed by 0's above and below this element.
- Next, if possible, produce a "1" in the third row, third column, followed by 0's above and below this element.
- The process continues until reduced row echelon form is obtained.

example 5

Solving a System of Linear Equations Using the Gauss-Jordan Method

Use the Gauss-Jordan method to solve the system:

$$\begin{aligned} x &= -y + 5 \\ -2x + 2z &= y - 10 \\ 3x + 6y + 7z &= 14 \end{aligned}$$

Solution:

First write each equation in the system in standard form.

$$\begin{aligned} x &= -y + 5 &\longrightarrow\quad x + y &= 5 \\ -2x + 2z &= y - 10 &\longrightarrow\quad -2x - y + 2z &= -10 \\ 3x + 6y + 7z &= 14 &\longrightarrow\quad 3x + 6y + 7z &= 14 \end{aligned}$$

$$\left[\begin{array}{ccc|c} 1 & 1 & 0 & 5 \\ -2 & -1 & 2 & -10 \\ 3 & 6 & 7 & 14 \end{array}\right]$$ Set up the augmented matrix.

$2R_1 + R_2 \Rightarrow R_2$
$-3R_1 + R_3 \Rightarrow R_3$
$$\left[\begin{array}{ccc|c} 1 & 1 & 0 & 5 \\ 0 & 1 & 2 & 0 \\ 0 & 3 & 7 & -1 \end{array}\right]$$ Multiply row 1 by 2 and add the result to row 2. Multiply row 1 by −3 and add the result to row 3.

$-1R_2 + R_1 \Rightarrow R_1$
$-3R_2 + R_3 \Rightarrow R_3$
$$\left[\begin{array}{ccc|c} 1 & 0 & -2 & 5 \\ 0 & 1 & 2 & 0 \\ 0 & 0 & 1 & -1 \end{array}\right]$$ Multiply row 2 by −1 and add the result to row 1. Multiply row 2 by −3 and add the result to row 3.

$2R_3 + R_1 \Rightarrow R_1$
$-2R_3 + R_2 \Rightarrow R_2$
$$\left[\begin{array}{ccc|c} 1 & 0 & 0 & 3 \\ 0 & 1 & 0 & 2 \\ 0 & 0 & 1 & -1 \end{array}\right]$$ Multiply row 3 by 2 and add the result to row 1. Multiply row 3 by −2 and add the result to row 2.

From the reduced row echelon form of the matrix, we have $x = 3$, $y = 2$, and $z = -1$. The solution to the system is $(3, 2, -1)$.

6. Inconsistent and Dependent Systems of Equations

It is particularly easy to recognize a dependent or inconsistent system of equations from the reduced row echelon form of an augmented matrix. This is demonstrated in Examples 6 and 7.

example 6

Solving a Dependent System of Equations Using the Gauss-Jordan Method

Use Gaussian elimination to solve the system:

$$x - 3y = 4$$
$$\frac{1}{2}x - \frac{3}{2}y = 2$$

Solution:

$$\left[\begin{array}{cc|c} 1 & -3 & 4 \\ \frac{1}{2} & -\frac{3}{2} & 2 \end{array}\right]$$ Set up the augmented matrix.

$-\frac{1}{2}R_1 + R_2 \Rightarrow R_2$
$$\left[\begin{array}{cc|c} 1 & -3 & 4 \\ 0 & 0 & 0 \end{array}\right]$$ Multiply row 1 by $-\frac{1}{2}$ and add the result to row 2.

The second row of the augmented matrix represents the equation $0 = 0$; hence, the system is dependent. From the first row of the matrix, the solution is $\{(x, y)|x - 3y = 4\}$.

example 7

Solving an Inconsistent System of Equations Using the Gauss-Jordan Method

Use the Gauss-Jordan method to solve the system:

$$2x - 5y = 10$$
$$\frac{2}{5}x - y = 7$$

Solution:

$$\left[\begin{array}{cc|c} 2 & -5 & 10 \\ \frac{2}{5} & -1 & 7 \end{array}\right]$$ Set up the augmented matrix.

$$\xrightarrow{-\frac{1}{5}R_1 + R_2 \Rightarrow R_2} \left[\begin{array}{cc|c} 2 & -5 & 10 \\ 0 & 0 & 5 \end{array}\right]$$ Multiply row 1 by $-\frac{1}{5}$ and add the result to row 2.

The second row of the augmented matrix represents the contradiction $0 = 5$; hence, the system is inconsistent. There is no solution.

Calculator Connections

Many graphing calculators have a matrix editor in which the user defines the order of the matrix and then enters the elements of the matrix. For example, the 2×3 matrix:

$$\mathbf{D} = \left[\begin{array}{cc|c} 2 & -3 & -13 \\ 3 & 1 & 8 \end{array}\right]$$

is entered as shown in the following two figures.

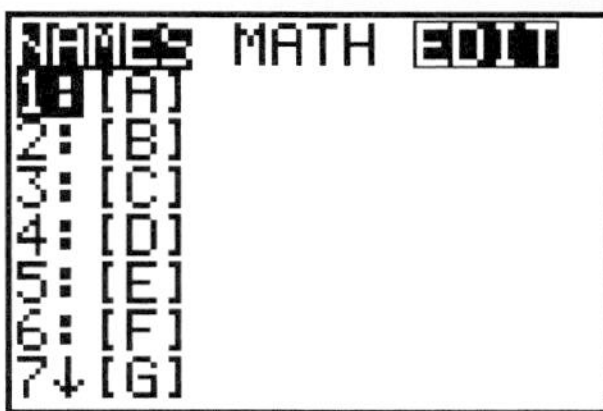

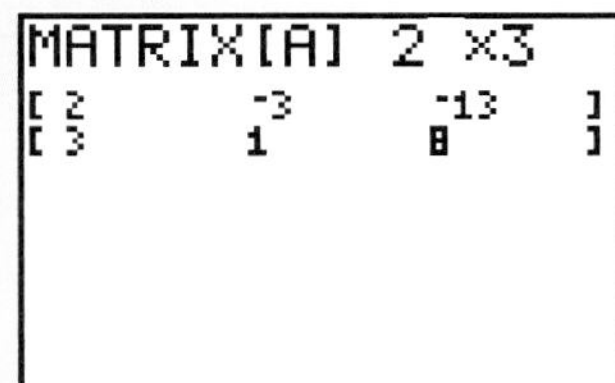

Once an augmented matrix has been entered into a graphing calculator a *rref* function can be used to transform the matrix into reduced row echelon form (see figure).

section 8.6 Practice Exercises

For Exercises 1–4, solve the system using any method.

1. $\begin{aligned} 5x + y &= 6 \\ -3x + 2y &= -1 \end{aligned}$

2. $\begin{aligned} x - 6y &= 9 \\ x + 2y &= 13 \end{aligned}$

3. $\begin{aligned} x + y - z &= 8 \\ x - 2y + z &= 3 \\ x + 3y + 2z &= 7 \end{aligned}$

4. $\begin{aligned} 2x - y + z &= -4 \\ -x + y + 3z &= -7 \\ x + 3y - 4z &= 22 \end{aligned}$

5. What is an augmented matrix?

6. What is a coefficient matrix?

7. How do you determine the order of a matrix?

8. What is a square matrix?

For Exercises 9–16, (a) Determine the order of each matrix. (b) Determine if the matrix is a row matrix, a column matrix, a square matrix, or none of these.

9. $\begin{bmatrix} 4 \\ 5 \\ -3 \\ 0 \end{bmatrix}$

10. $\begin{bmatrix} 5 \\ -1 \\ 2 \end{bmatrix}$

11. $\begin{bmatrix} -9 & 4 & 3 \\ -1 & -8 & 4 \\ 5 & 8 & 7 \end{bmatrix}$

12. $\begin{bmatrix} 3 & -9 \\ -1 & -3 \end{bmatrix}$

13. $\begin{bmatrix} 4 & -7 \end{bmatrix}$

14. $\begin{bmatrix} 0 & -8 & 11 & 5 \end{bmatrix}$

15. $\begin{bmatrix} 5 & -8.1 & 4.2 & 0 \\ 4.3 & -9 & 18 & 3 \end{bmatrix}$

16. $\begin{bmatrix} \frac{1}{3} & \frac{3}{4} & 6 \\ -2 & 1 & -\frac{7}{8} \end{bmatrix}$

For Exercises 17–24, set up the augmented matrix.

17. $\begin{aligned} x - 2y &= -1 \\ 2x + y &= -7 \end{aligned}$

18. $\begin{aligned} x - 3y &= 3 \\ 2x - 5y &= 4 \end{aligned}$

19. $\begin{aligned} -9x + 13y &= -5 \\ 7x + 5y &= 19 \end{aligned}$

20. $\begin{aligned} 3x - 2y &= 6 \\ 4x - 10y &= -3 \end{aligned}$

21. $\begin{aligned} x + y + z &= 6 \\ x - y + z &= 2 \\ x + y - z &= 0 \end{aligned}$

22. $\begin{aligned} 2x - 3y + z &= 8 \\ x + 3y + 8z &= 1 \\ 3x - y + 2z &= -1 \end{aligned}$

23. $\begin{aligned} x - 2y &= 5 - z \\ 2x + 6y + 3z &= -2 \\ 3x - y - 2z &= 1 \end{aligned}$

24. $\begin{aligned} 5x + 2z &= 17 \\ 8x - y + 6z &= 26 \\ 8x + 3y - 12z &= 24 \end{aligned}$

25. Given the matrix **E**:

$$\mathbf{E} = \left[\begin{array}{cc|c} 3 & -2 & 8 \\ 9 & -1 & 7 \end{array}\right]$$

a. What is the element in the second row and third column?

b. What is the element in the first row and second column?

26. Given the matrix **F**:

$$\mathbf{F} = \left[\begin{array}{cc|c} 1 & 8 & 0 \\ 12 & -13 & -2 \end{array}\right]$$

a. What is the element in the second row and second column?

b. What is the element in the first row and third column?

27. Given the matrix **Z**:

$$\mathbf{Z} = \left[\begin{array}{cc|c} 2 & 1 & 11 \\ 2 & -1 & 1 \end{array}\right]$$

write the matrix obtained by multiplying the elements in the first row by $\frac{1}{2}$.

28. Given the matrix **J**:

$$\mathbf{J} = \left[\begin{array}{cc|c} 1 & 1 & 7 \\ 0 & 3 & -6 \end{array}\right]$$

write the matrix obtained by multiplying the elements in the second row by $\frac{1}{3}$.

29. Given the matrix **K**:

$$\mathbf{K} = \left[\begin{array}{cc|c} 5 & 2 & 1 \\ 1 & -4 & 3 \end{array}\right]$$

write the matrix obtained by interchanging rows 1 and 2.

30. Given the matrix **L**:

$$\mathbf{L} = \left[\begin{array}{cc|c} 9 & 6 & 13 \\ -7 & 2 & 19 \end{array}\right]$$

write the matrix obtained by interchanging rows 1 and 2.

31. Given the matrix **M**:

$$\mathbf{M} = \left[\begin{array}{cc|c} 1 & 5 & 2 \\ -3 & -4 & -1 \end{array}\right]$$

write the matrix obtained by multiplying the first row by 3 and adding the result to row 2.

32. Given the matrix **N**:

$$\mathbf{N} = \left[\begin{array}{cc|c} 1 & 3 & -5 \\ -2 & 2 & 12 \end{array}\right]$$

write the matrix obtained by multiplying the first row by 2 and adding the result to row 2.

For Exercises 33–36, use the augmented matrices **A**, **B**, and **C** to answer true or false.

$$\mathbf{A} = \left[\begin{array}{cc|c} 6 & -4 & 2 \\ 5 & -2 & 7 \end{array}\right] \quad \mathbf{B} = \left[\begin{array}{cc|c} 5 & -2 & 7 \\ 6 & -4 & 2 \end{array}\right]$$

$$\mathbf{C} = \left[\begin{array}{cc|c} 1 & -\frac{2}{3} & \frac{1}{3} \\ 5 & 2 & 7 \end{array}\right]$$

33. The matrix **A** is a 3×2 matrix.

34. Matrix **B** is equivalent to matrix **A**.

35. Matrix **A** is equivalent to matrix **C**.

36. Matrix **B** is equivalent to matrix **C**.

For Exercises 37–40, write a corresponding system of equations from the augmented matrix.

37. $\left[\begin{array}{cc|c} 1 & 0 & -1 \\ 0 & 1 & -7 \end{array}\right]$

38. $\left[\begin{array}{cc|c} 1 & 0 & 0 \\ 0 & 1 & 5 \end{array}\right]$

39. $\left[\begin{array}{ccc|c} 1 & 0 & 0 & 8 \\ 0 & 1 & 0 & 0 \\ 0 & 0 & 1 & -1 \end{array}\right]$

40. $\left[\begin{array}{ccc|c} 1 & 0 & 0 & 2.7 \\ 0 & 1 & 0 & \pi \\ 0 & 0 & 1 & -1.1 \end{array}\right]$

41. What does the notation $R_2 \Leftrightarrow R_1$ mean when performing the Gauss-Jordan method?

42. What does the notation $2R_3 \Rightarrow R_3$ mean when performing the Gauss-Jordan method?

43. What does the notation $-3R_1 + R_2 \Rightarrow R_2$ mean when performing the Gauss-Jordan method?

44. What does the notation $4R_2 + R_3 \Rightarrow R_3$ mean when performing the Gauss-Jordan method?

For Exercises 45–64, solve the systems using the Gauss-Jordan method.

45. $\begin{aligned} x - 2y &= -1 \\ 2x + y &= -7 \end{aligned}$

46. $\begin{aligned} x - 3y &= 3 \\ 2x - 5y &= 4 \end{aligned}$

47. $\begin{aligned} x + 3y &= 6 \\ -4x - 9y &= 3 \end{aligned}$

48. $\begin{aligned} 2x - 3y &= -2 \\ x + 2y &= 13 \end{aligned}$

49. $\begin{aligned} x + 3y &= 3 \\ 4x + 12y &= 12 \end{aligned}$

50. $\begin{aligned} 2x + 5y &= 1 \\ -4x - 10y &= -2 \end{aligned}$

51. $\begin{aligned} x - y &= 4 \\ 2x + y &= 5 \end{aligned}$

52. $\begin{aligned} 2x - y &= 0 \\ x + y &= 3 \end{aligned}$

53. $\begin{aligned} x + 3y &= -1 \\ -3x - 6y &= 12 \end{aligned}$

54. $\begin{aligned} x + y &= 4 \\ 2x - 4y &= -4 \end{aligned}$

55. $\begin{aligned} 3x + y &= -4 \\ -6x - 2y &= 3 \end{aligned}$

56. $\begin{aligned} 2x + y &= 4 \\ 6x + 3y &= -1 \end{aligned}$

57. $\begin{aligned} x + y + z &= 6 \\ x - y + z &= 2 \\ x + y - z &= 0 \end{aligned}$

58. $\begin{aligned} 2x - 3y - 2z &= 11 \\ x + 3y + 8z &= 1 \\ 3x - y + 14z &= -2 \end{aligned}$

59. $$\begin{aligned} x - 2y &= 5 - z \\ 2x + 6y + 3z &= -10 \\ 3x - y - 2z &= 5 \end{aligned}$$

60. $$\begin{aligned} 5x - 10z &= 15 \\ x - y + 6z &= 23 \\ x + 3y - 12z &= 13 \end{aligned}$$

61. $$\begin{aligned} x + y - z &= 2 \\ 2x - y + z &= 1 \\ -x + y + z &= 2 \end{aligned}$$

62. $$\begin{aligned} x + y + z &= 6 \\ x - y - z &= -4 \\ -x + y - z &= -2 \end{aligned}$$

63. $$\begin{aligned} -x + 2y - z &= -6 \\ x - 2y + z &= -5 \\ 3x + y + 2z &= 4 \end{aligned}$$

64. $$\begin{aligned} 4x + 8y + 4z &= 9 \\ 5x + 10y + 5z &= 12 \\ x + 3y + 4z &= 10 \end{aligned}$$

Graphing Calculator Exercises

For Exercises 65–70, use the matrix features on a graphing calculator to express each augmented matrix in reduced row echelon form. Compare your results to the solution you obtained in the indicated exercise.

65. $\left[\begin{array}{cc|c} 1 & -2 & -1 \\ 2 & 1 & -7 \end{array}\right]$

Compare with Exercise 45.

66. $\left[\begin{array}{cc|c} 1 & -3 & 3 \\ 2 & -5 & 4 \end{array}\right]$

Compare with Exercise 46.

67. $\left[\begin{array}{cc|c} 1 & 3 & 3 \\ 4 & 12 & 12 \end{array}\right]$

Compare with Exercise 49.

68. $\left[\begin{array}{cc|c} 2 & 5 & 1 \\ -4 & -10 & -2 \end{array}\right]$

Compare with Exercise 50.

69. $\left[\begin{array}{ccc|c} 1 & -2 & 1 & 5 \\ 2 & 6 & 3 & -10 \\ 3 & -1 & -2 & 5 \end{array}\right]$

Compare with Exercise 59.

70. $\left[\begin{array}{ccc|c} 5 & 0 & -10 & 15 \\ 1 & -1 & 6 & 23 \\ 1 & 3 & -12 & 13 \end{array}\right]$

Compare with Exercise 60.

---

section

8.7 Determinants and Cramer's Rule

Concepts

1. Introduction to Determinants
2. Determinant of a 2 × 2 Matrix
3. Minor of an Element in a 3 × 3 Matrix
4. Determinant of a 3 × 3 Matrix
5. Cramer's Rule for a 2 × 2 System of Linear Equations
6. Cramer's Rule for a 3 × 3 System of Linear Equations
7. Inconsistent and Dependent Systems

1. Introduction to Determinants

Associated with every square matrix is a real number called the **determinant** of the matrix. A determinant of a square matrix **A**, denoted det**A**, is written by enclosing the elements of the matrix within two vertical bars. For example,

$$\text{if } \mathbf{A} = \begin{bmatrix} 2 & -1 \\ 6 & 0 \end{bmatrix}, \quad \text{then } \det\mathbf{A} = \begin{vmatrix} 2 & -1 \\ 6 & 0 \end{vmatrix}$$

$$\text{if } \mathbf{B} = \begin{bmatrix} 0 & -5 & 1 \\ 4 & 0 & \frac{1}{2} \\ -2 & 10 & 1 \end{bmatrix}, \quad \text{then } \det\mathbf{B} = \begin{vmatrix} 0 & -5 & 1 \\ 4 & 0 & \frac{1}{2} \\ -2 & 10 & 1 \end{vmatrix}$$

Determinants have many applications in mathematics, including solving systems of linear equations, finding the area of a triangle, determining whether three points are collinear, and finding an equation of a line between two points.

2. Determinant of a 2 × 2 Matrix

The determinant of a 2×2 matrix is defined as follows:

Determinant of a 2 × 2 Matrix

The determinant of the matrix $\begin{bmatrix} a & b \\ c & d \end{bmatrix}$ is the real number $ad - bc$. It is written as

$$\begin{vmatrix} a & b \\ c & d \end{vmatrix} = ad - bc$$

example 1 **Evaluating a 2 × 2 Determinant**

Evaluate the determinants.

a. $\begin{vmatrix} 6 & -2 \\ 5 & \frac{1}{3} \end{vmatrix}$ b. $\begin{vmatrix} 2 & -11 \\ 0 & 0 \end{vmatrix}$

Solution:

a. $\begin{vmatrix} 6 & -2 \\ 5 & \frac{1}{3} \end{vmatrix}$ For this determinant, $a = 6$, $b = -2$, $c = 5$, and $d = \frac{1}{3}$.

$$\begin{aligned} ad - bc &= (6)\left(\frac{1}{3}\right) - (-2)(5) \\ &= 2 + 10 \\ &= 12 \end{aligned}$$

b. $\begin{vmatrix} 2 & -11 \\ 0 & 0 \end{vmatrix}$ For this determinant, $a = 2$, $b = -11$, $c = 0$, $d = 0$.

$$\begin{aligned} ad - bc &= (2)(0) - (-11)(0) \\ &= 0 - 0 \\ &= 0 \end{aligned}$$

Tip: Example 1(b) illustrates that the value of a determinant having a row of all zeros is 0. The same is true for a determinant having a column of all zeros.

3. Minor of an Element in a 3 × 3 Matrix

To find the determinant of a 3×3 matrix, we first need to define the **minor** of an element of the matrix. For any element of a 3×3 matrix, the minor of that element is the determinant of the 2×2 matrix obtained by deleting the row and column in which the element resides. For example, consider the matrix

$$\begin{bmatrix} 5 & -1 & 6 \\ 0 & -7 & 1 \\ 4 & 2 & 6 \end{bmatrix}$$

The minor of the element 5 is found by deleting the first row and first column and then evaluating the determinant of the remaining 2×2 matrix:

$$\begin{bmatrix} 5 & -1 & 6 \\ 0 & -7 & 1 \\ 4 & 2 & 6 \end{bmatrix}$$ Now evaluate the determinant: $$\begin{vmatrix} -7 & 1 \\ 2 & 6 \end{vmatrix} = (-7)(6) - (1)(2) = -44.$$

For this matrix, the minor for the element 5 is -44.

To find the minor of the element -7, delete the second row and second column, and then evaluate the determinant of the remaining 2×2 matrix.

$$\begin{bmatrix} 5 & -1 & 6 \\ 0 & -7 & 1 \\ 4 & 2 & 6 \end{bmatrix}$$ Now evaluate the determinant: $$\begin{vmatrix} 5 & 6 \\ 4 & 6 \end{vmatrix} = (5)(6) - (6)(4) = 6.$$

For this matrix, the minor for the element -7 is 6.

example 2 Determining the Minor for Elements in a 3 × 3 matrix

Find the minor for each element in the first column of the matrix.

$$\begin{bmatrix} 3 & 4 & -1 \\ 2 & -4 & 5 \\ 0 & 1 & -6 \end{bmatrix}$$

Solution:

For 3: $$\begin{bmatrix} 3 & 4 & -1 \\ 2 & -4 & 5 \\ 0 & 1 & -6 \end{bmatrix}$$ The minor is: $$\begin{vmatrix} -4 & 5 \\ 1 & -6 \end{vmatrix} = (-4)(-6) - (5)(1) = 19$$

For 2: $$\begin{bmatrix} 3 & 4 & -1 \\ 2 & -4 & 5 \\ 0 & 1 & -6 \end{bmatrix}$$ The minor is: $$\begin{vmatrix} 4 & -1 \\ 1 & -6 \end{vmatrix} = (4)(-6) - (-1)(1) = -23$$

For 0: $$\begin{bmatrix} 3 & 4 & -1 \\ 2 & -4 & 5 \\ 0 & 1 & -6 \end{bmatrix}$$ The minor is $$\begin{vmatrix} 4 & -1 \\ -4 & 5 \end{vmatrix} = (4)(5) - (-1)(-4) = 16$$

4. Determinant of a 3 × 3 Matrix

The determinant of a 3 × 3 matrix is defined as follows.

Definition of a Determinant of a 3 × 3 Matrix

$$\begin{vmatrix} a_1 & b_1 & c_1 \\ a_2 & b_2 & c_2 \\ a_3 & b_3 & c_3 \end{vmatrix} = a_1 \cdot \begin{vmatrix} b_2 & c_2 \\ b_3 & c_3 \end{vmatrix} - a_2 \cdot \begin{vmatrix} b_1 & c_1 \\ b_3 & c_3 \end{vmatrix} + a_3 \cdot \begin{vmatrix} b_1 & c_1 \\ b_2 & c_2 \end{vmatrix}$$

From this definition, we see that the determinant of a 3 × 3 matrix can be written as

$$a_1 \cdot (\text{minor of } a_1) - a_2 \cdot (\text{minor of } a_2) + a_3 \cdot (\text{minor of } a_3)$$

Evaluating determinants in this way is called expanding minors.

example 3 **Evaluating a 3 × 3 Determinant**

Evaluate the determinant: $\begin{vmatrix} 2 & 4 & 2 \\ 1 & -3 & 0 \\ -5 & 5 & -1 \end{vmatrix}$

Solution:

$$\begin{vmatrix} 2 & 4 & 2 \\ 1 & -3 & 0 \\ -5 & 5 & -1 \end{vmatrix} = 2 \cdot \begin{vmatrix} -3 & 0 \\ 5 & -1 \end{vmatrix} - (1) \cdot \begin{vmatrix} 4 & 2 \\ 5 & -1 \end{vmatrix} + (-5) \cdot \begin{vmatrix} 4 & 2 \\ -3 & 0 \end{vmatrix}$$

$$= 2[(-3)(-1) - (0)(5)] - 1[(4)(-1) - (2)(5)] - 5[(4)(0) - (2)(-3)]$$

$$= 2(3) - 1(-14) - 5(6)$$

$$= 6 + 14 - 30$$

$$= -10$$

Tip: There is another method to determine the signs for each term of the expansion. For the a_{ij} element, multiply the term by $(-1)^{i+j}$.

Although we defined the determinant of a matrix by expanding the minors of the elements in the first column, *any row or column may be used.* However, we must choose the correct sign to apply to each term in the expansion. The following array of signs is helpful.

$$\begin{matrix} + & - & + \\ - & + & - \\ + & - & + \end{matrix}$$

The signs alternate for each row and column, beginning with + in the first row, first column.

example 4 Evaluating a 3 × 3 Determinant

Evaluate the determinant

$$\begin{vmatrix} 2 & 4 & 2 \\ 1 & -3 & 0 \\ -5 & 5 & -1 \end{vmatrix}$$

by expanding minors about the elements in the second row.

Solution:

signs obtained from the array of signs

$$\begin{vmatrix} 2 & 4 & 2 \\ 1 & -3 & 0 \\ -5 & 5 & -1 \end{vmatrix} = -(1) \cdot \begin{vmatrix} 4 & 2 \\ 5 & -1 \end{vmatrix} + (-3) \cdot \begin{vmatrix} 2 & 2 \\ -5 & -1 \end{vmatrix} - (0) \cdot \begin{vmatrix} 2 & 4 \\ -5 & 5 \end{vmatrix}$$

$$= -1[(4)(-1) - (2)(5)] - 3[(2)(-1) - (2)(-5)] - 0$$

$$= -1(-14) - 3(8)$$

$$= 14 - 24$$

$$= -10$$

Notice that the value of the determinant obtained in Examples 3 and 4 is the same.

Calculator Connections

The determinant of a matrix can be evaluated on a graphing calculator. First use the matrix editor to enter the elements of the matrix. Then use a *Det* function to evaluate the determinant. The determinant from Examples 3 and 4 is evaluated below.

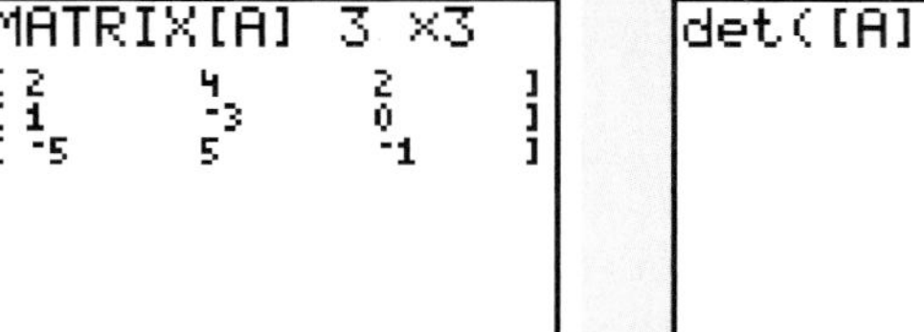

In Example 4, the third term in the expansion of minors was zero because the element 0 when multiplied by its minor is zero. To simplify the arithmetic in evaluating a determinant of a 3×3 matrix, expand about the row or column that has the most 0 elements.

5. Cramer's Rule for a 2 × 2 System of Linear Equations

In Sections 8.2, 8.3, and 8.6, we learned three methods to solve a system of linear equations: the substitution method, the addition method, and the Gauss-Jordan method. In this section, we will learn another method called **Cramer's rule** to solve a system of linear equations.

Cramer's Rule for a 2 × 2 System of Linear Equations

The solution to the system

$$a_1x + b_1y = c_1$$
$$a_2x + b_2y = c_2$$

is given by: $x = \dfrac{\mathbf{D}_x}{\mathbf{D}}$ and $y = \dfrac{\mathbf{D}_y}{\mathbf{D}}$

where $\mathbf{D} = \begin{vmatrix} a_1 & b_1 \\ a_2 & b_2 \end{vmatrix}$ (and $\mathbf{D} \neq 0$), $\mathbf{D}_x = \begin{vmatrix} c_1 & b_1 \\ c_2 & b_2 \end{vmatrix}$, $\mathbf{D}_y = \begin{vmatrix} a_1 & c_1 \\ a_2 & c_2 \end{vmatrix}$

example 5

Using Cramer's Rule to Solve a 2 × 2 System of Linear Equations

Solve the system using Cramer's rule.

$$3x - 5y = 11$$
$$-x + 3y = -5$$

Solution:

For this system: $a_1 = 3, \quad b_1 = -5, c_1 = 11$

$a_2 = -1, b_2 = 3, \quad c_2 = -5$

$$\mathbf{D} = \begin{vmatrix} 3 & -5 \\ -1 & 3 \end{vmatrix} = (3)(3) - (-5)(-1) = 9 - 5 = 4$$

$$\mathbf{D}_x = \begin{vmatrix} 11 & -5 \\ -5 & 3 \end{vmatrix} = (11)(3) - (-5)(-5) = 33 - 25 = 8$$

$$\mathbf{D}_y = \begin{vmatrix} 3 & 11 \\ -1 & -5 \end{vmatrix} = (3)(-5) - (11)(-1) = -15 + 11 = -4$$

Therefore,

$$x = \frac{\mathbf{D}_x}{\mathbf{D}} = \frac{8}{4} = 2 \qquad y = \frac{\mathbf{D}_y}{\mathbf{D}} = \frac{-4}{4} = -1$$

The solution is (2, −1). Check: $3x - 5y = 11 \longrightarrow 3(2) - 5(-1) \stackrel{?}{=} 11$ ✔

$-x + 3y = -5 \longrightarrow -(2) + 3(-1) \stackrel{?}{=} -5$ ✔

Tip: Here are some memory tips to remember Cramer's rule.

1. The determinant **D** is the determinant of the coefficients of x and y.

   coefficients of x-terms, y-terms

   $$\mathbf{D} = \begin{vmatrix} a_1 & b_1 \\ a_2 & b_2 \end{vmatrix}$$

2. The determinant $\mathbf{D}_x$ has the column of x-term coefficients replaced by c_1 and c_2.

   x-coefficients replaced by c_1 and c_2

   $$\mathbf{D}_x = \begin{vmatrix} c_1 & b_1 \\ c_2 & b_2 \end{vmatrix}$$

3. The determinant $\mathbf{D}_y$ has the column of y-term coefficients replaced by c_1 and c_2.

   y-coefficients replaced by c_1 and c_2

   $$\mathbf{D}_y = \begin{vmatrix} a_1 & c_1 \\ a_2 & c_2 \end{vmatrix}$$

It is important to note that the linear equations must be written in standard form to apply Cramer's rule.

example 6 Using Cramer's Rule to Solve a 2 × 2 System of Linear Equations

Solve the system using Cramer's rule:

$$-16y = -40x - 7$$
$$40y = 24x + 27$$

Solution:

$-16y = -40x - 7 \longrightarrow 40x - 16y = -7$ Rewrite each equation in
$40y = 24x + 27 \longrightarrow -24x + 40y = 27$ standard form.

For this system:

$$a_1 = 40, \quad b_1 = -16, c_1 = -7$$
$$a_2 = -24, b_2 = 40, \quad c_2 = 27$$

$$\mathbf{D} = \begin{vmatrix} 40 & -16 \\ -24 & 40 \end{vmatrix} = (40)(40) - (-16)(-24) = 1216$$

$$\mathbf{D}_x = \begin{vmatrix} -7 & -16 \\ 27 & 40 \end{vmatrix} = (-7)(40) - (-16)(27) = 152$$

$$\mathbf{D}_y = \begin{vmatrix} 40 & -7 \\ -24 & 27 \end{vmatrix} = (40)(27) - (-7)(-24) = 912$$

Therefore, $x = \dfrac{\mathbf{D}_x}{\mathbf{D}} = \dfrac{152}{1216} = \dfrac{1}{8}$ $\quad y = \dfrac{\mathbf{D}_y}{\mathbf{D}} = \dfrac{912}{1216} = \dfrac{3}{4}$

The solution $(\frac{1}{8}, \frac{3}{4})$ checks in the original equations.

6. Cramer's Rule for a 3 × 3 System of Linear Equations

Cramer's rule can be used to solve a 3 × 3 system of linear equations using a similar pattern of determinants.

Cramer's Rule for a 3 × 3 System of Linear Equations

The solution to the system

$$a_1x + b_1y + c_1z = d_1$$
$$a_2x + b_2y + c_2z = d_2$$
$$a_3x + b_3y + c_3z = d_3$$

is given by

$$x = \frac{\mathbf{D}_x}{\mathbf{D}}, \quad y = \frac{\mathbf{D}_y}{\mathbf{D}}, \quad \text{and} \quad z = \frac{\mathbf{D}_z}{\mathbf{D}}$$

where $\mathbf{D} = \begin{vmatrix} a_1 & b_1 & c_1 \\ a_2 & b_2 & c_2 \\ a_3 & b_3 & c_3 \end{vmatrix}$ (and $\mathbf{D} \neq 0$), $\quad \mathbf{D}_x = \begin{vmatrix} d_1 & b_1 & c_1 \\ d_2 & b_2 & c_2 \\ d_3 & b_3 & c_3 \end{vmatrix}$,

$$\mathbf{D}_y = \begin{vmatrix} a_1 & d_1 & c_1 \\ a_2 & d_2 & c_2 \\ a_3 & d_3 & c_3 \end{vmatrix}, \qquad \mathbf{D}_z = \begin{vmatrix} a_1 & b_1 & d_1 \\ a_2 & b_2 & d_2 \\ a_3 & b_3 & d_3 \end{vmatrix}$$

example 7 **Using Cramer's Rule to Solve a 3 × 3 System of Linear Equations**

Solve the system using Cramer's rule:

$$x - 2y + 4z = 3$$
$$x - 4y + 3z = -5$$
$$x + 3y - 2z = 6$$

Solution:

Tip: In Example 7, we expanded the determinants about the first column.

$$\mathbf{D} = \begin{vmatrix} 1 & -2 & 4 \\ 1 & -4 & 3 \\ 1 & 3 & -2 \end{vmatrix} = 1 \cdot \begin{vmatrix} -4 & 3 \\ 3 & -2 \end{vmatrix} - 1 \cdot \begin{vmatrix} -2 & 4 \\ 3 & -2 \end{vmatrix} + 1 \cdot \begin{vmatrix} -2 & 4 \\ -4 & 3 \end{vmatrix}$$

$$= 1(-1) - 1(-8) + 1(10)$$

$$= 17$$

$$\mathbf{D}_x = \begin{vmatrix} 3 & -2 & 4 \\ -5 & -4 & 3 \\ 6 & 3 & -2 \end{vmatrix} = 3 \cdot \begin{vmatrix} -4 & 3 \\ 3 & -2 \end{vmatrix} - (-5) \cdot \begin{vmatrix} -2 & 4 \\ 3 & -2 \end{vmatrix} + 6 \cdot \begin{vmatrix} -2 & 4 \\ -4 & 3 \end{vmatrix}$$

$$= 3(-1) + 5(-8) + 6(10)$$

$$= 17$$

$$\mathbf{D}_y = \begin{vmatrix} 1 & 3 & 4 \\ 1 & -5 & 3 \\ 1 & 6 & -2 \end{vmatrix} = 1 \cdot \begin{vmatrix} -5 & 3 \\ 6 & -2 \end{vmatrix} - 1 \cdot \begin{vmatrix} 3 & 4 \\ 6 & -2 \end{vmatrix} + 1 \cdot \begin{vmatrix} 3 & 4 \\ -5 & 3 \end{vmatrix}$$

$$= 1(-8) - 1(-30) + 1(29)$$

$$= 51$$

$$\mathbf{D}_z = \begin{vmatrix} 1 & -2 & 3 \\ 1 & -4 & -5 \\ 1 & 3 & 6 \end{vmatrix} = 1 \cdot \begin{vmatrix} -4 & -5 \\ 3 & 6 \end{vmatrix} - 1 \cdot \begin{vmatrix} -2 & 3 \\ 3 & 6 \end{vmatrix} + 1 \cdot \begin{vmatrix} -2 & 3 \\ -4 & -5 \end{vmatrix}$$

$$= 1(-9) - 1(-21) + 1(22)$$

$$= 34$$

Hence

$$x = \frac{\mathbf{D}_x}{\mathbf{D}} = \frac{17}{17} = 1 \qquad y = \frac{\mathbf{D}_y}{\mathbf{D}} = \frac{51}{17} = 3 \qquad \text{and} \qquad z = \frac{\mathbf{D}_z}{\mathbf{D}} = \frac{34}{17} = 2$$

The solution is (1, 3, 2).

Check:

$x - 2y + 4z = 3 \qquad (1) - 2(3) + 4(2) \stackrel{?}{=} 3$ ✔

$x - 4y + 3z = -5 \qquad (1) - 4(3) + 3(2) \stackrel{?}{=} -5$ ✔

$x + 3y - 2z = 6 \qquad (1) + 3(3) - 2(2) \stackrel{?}{=} 6$ ✔

Cramer's rule may seem cumbersome for solving a 3×3 system of linear equations. However, it provides convenient formulas that can be programmed into a computer or calculator to solve for x, y, and z. Cramer's rule can also be extended to solve a 4×4 system of linear equations, a 5×5 system of linear equations, and, in general, an $n \times n$ system of linear equations.

7. Inconsistent and Dependent Systems

It is important to remember that Cramer's rule does not apply if $\mathbf{D} = 0$. In such a case, the system of equations is either dependent or inconsistent, and another method must be used to analyze the system.

example 8 **Analyzing a Dependent System of Equations**

Solve the system using Cramer's rule (if possible).

$$2x - 3y = 6$$

$$-6x + 9y = -18$$

Solution:

$$\mathbf{D} = \begin{vmatrix} 2 & -3 \\ -6 & 9 \end{vmatrix} = (2)(9) - (-3)(-6) = 18 - 18 = 0$$

Because **D** = 0, Cramer's rule does not apply. Using the addition method to solve the system, we have

$$\begin{array}{rcr} 2x - 3y = 6 & \xrightarrow{\text{multiply by 3}} & 6x - 9y = 18 \\ -6x + 9y = -18 & \longrightarrow & \underline{-6x + 9y = -18} \\ & & 0 = 0 \end{array}$$ The system is dependent.

The solution is: $\{(x, y) \mid 2x - 3y = 6\}$

section 8.7 Practice Exercises

For Exercises 1–6, evaluate the determinant of the 2×2 matrix.

1. $\begin{vmatrix} -3 & 1 \\ 5 & 2 \end{vmatrix}$

2. $\begin{vmatrix} 5 & 6 \\ 4 & 8 \end{vmatrix}$

3. $\begin{vmatrix} -2 & 2 \\ -3 & -5 \end{vmatrix}$

4. $\begin{vmatrix} 5 & -1 \\ 1 & 0 \end{vmatrix}$

5. $\begin{vmatrix} \frac{1}{2} & 3 \\ -2 & 4 \end{vmatrix}$

6. $\begin{vmatrix} -3 & \frac{1}{4} \\ 8 & -2 \end{vmatrix}$

For Exercises 7–10, evaluate the minor corresponding to the given element from matrix **A**.

$$\mathbf{A} = \begin{bmatrix} 4 & -1 & 8 \\ 2 & 6 & 0 \\ -7 & 5 & 3 \end{bmatrix}$$

7. 4

8. −1

9. 2

10. 3

11. Construct the sign array for a 3×3 matrix.

12. What is the purpose of the sign array that you constructed in Exercise 11?

13. Evaluate the determinant of matrix **B** using expansion by minors.

$$\mathbf{B} = \begin{bmatrix} 0 & 1 & 2 \\ 3 & -1 & 2 \\ 3 & 2 & -2 \end{bmatrix}$$

   a. About the first column

   b. About the second row

14. Evaluate the determinant of matrix **C** using expansion by minors.

$$\mathbf{C} = \begin{bmatrix} 4 & 1 & 3 \\ 2 & -2 & 1 \\ 3 & 1 & 2 \end{bmatrix}$$

   a. About the first row

   b. About the second column

15. When evaluating the determinant of a 3×3 matrix, explain the advantage of being able to choose any row or column about which to expand minors.

For Exercises 16–21, evaluate the determinant of the 3×3 matrix.

16. $\begin{vmatrix} 8 & 2 & -4 \\ 4 & 0 & 2 \\ 3 & 0 & -1 \end{vmatrix}$

17. $\begin{vmatrix} 5 & 2 & 1 \\ 3 & -6 & 0 \\ -2 & 8 & 0 \end{vmatrix}$

18. $\begin{vmatrix} -2 & 1 & 3 \\ 1 & 4 & 4 \\ 1 & 0 & 2 \end{vmatrix}$

19. $\begin{vmatrix} 3 & 2 & 1 \\ 1 & -1 & 2 \\ 1 & 0 & 4 \end{vmatrix}$

20. $\begin{vmatrix} -5 & 4 & 2 \\ 0 & 0 & 0 \\ 3 & -1 & 5 \end{vmatrix}$

21. $\begin{vmatrix} 0 & 5 & -8 \\ 0 & -4 & 1 \\ 0 & 3 & 6 \end{vmatrix}$

For Exercises 22–27, evaluate the determinants of the matrices.

22. $\begin{vmatrix} x & 3 \\ y & -2 \end{vmatrix}$

23. $\begin{vmatrix} a & 2 \\ b & 8 \end{vmatrix}$

24. $\begin{vmatrix} a & 5 & -1 \\ b & -3 & 0 \\ c & 3 & 4 \end{vmatrix}$

25. $\begin{vmatrix} x & 0 & 3 \\ y & -2 & 6 \\ z & -1 & 1 \end{vmatrix}$

26. $\begin{vmatrix} p & 0 & q \\ r & 0 & s \\ t & 0 & u \end{vmatrix}$

27. $\begin{vmatrix} f & e & 0 \\ d & c & 0 \\ b & a & 0 \end{vmatrix}$

For Exercises 28–29, evaluate the determinants represented by $\mathbf{D}$, $\mathbf{D}_x$, and $\mathbf{D}_y$.

28. $\begin{aligned} 4x + 6y &= 9 \\ -2x + y &= 12 \end{aligned}$

29. $\begin{aligned} -3x + 8y &= -10 \\ 5x + 5y &= -13 \end{aligned}$

For Exercises 30–35, solve the system using Cramer's rule.

30. $\begin{aligned} 2x + y &= 3 \\ x - 4y &= 6 \end{aligned}$

31. $\begin{aligned} 2x - y &= -1 \\ 3x + y &= 6 \end{aligned}$

32. $\begin{aligned} x - 4y &= 8 \\ 3x + 7y &= 5 \end{aligned}$

33. $\begin{aligned} 7x + 3y &= 4 \\ 5x - 4y &= 9 \end{aligned}$

34. $\begin{aligned} 4x - 3y &= 5 \\ 2x + 5y &= 7 \end{aligned}$

35. $\begin{aligned} 2x + 3y &= 4 \\ 6x - 12y &= -5 \end{aligned}$

36. When does Cramer's rule not apply in solving a system of equations?

37. How can a system be solved if Cramer's rule does not apply?

For Exercises 38–43, solve the system of equations using Cramer's rule if possible. If not possible, use another method.

38. $\begin{aligned} 4x - 2y &= 3 \\ -2x + y &= 1 \end{aligned}$

39. $\begin{aligned} 6x - 6y &= 5 \\ x - y &= 8 \end{aligned}$

40. $\begin{aligned} 4x + y &= 0 \\ x - 7y &= 0 \end{aligned}$

41. $\begin{aligned} -3x - 2y &= 0 \\ -x + 5y &= 0 \end{aligned}$

42. $\begin{aligned} x + 5y &= 3 \\ 2x + 10y &= 6 \end{aligned}$

43. $\begin{aligned} -2x - 10y &= -4 \\ x + 5y &= 2 \end{aligned}$

For Exercises 44–49, solve for the indicated variable using Cramer's rule.

44. $\begin{aligned} 2x - y + 3z &= 9 \\ x + 4y + 4z &= 5 \\ 3x + 2y + 2z &= 5 \end{aligned}$ for x

45. $\begin{aligned} x + 2y + 3z &= 8 \\ 2x - 3y + z &= 5 \\ 3x - 4y + 2z &= 9 \end{aligned}$ for y

46. $\begin{aligned} 3x - 2y + 2z &= 5 \\ 6x + 3y - 4z &= -1 \\ 3x - y + 2z &= 4 \end{aligned}$ for z

47. $\begin{aligned} 4x + 4y - 3z &= 3 \\ 8x + 2y + 3y &= 0 \\ 4x - 4y + 6z &= -3 \end{aligned}$ for x

48. $\begin{aligned} 5x + 6z &= 5 \\ -2x + y &= -6 \\ 3y - z &= 3 \end{aligned}$ for y

49. $\begin{aligned} 8x + y &= 1 \\ 7y + z &= 0 \\ x - 3z &= -2 \end{aligned}$ for y

For Exercises 50–53, solve the system using Cramer's rule, if possible.

50. $\begin{aligned} x \qquad\qquad &= 3 \\ -x + 3y \qquad &= 3 \\ y + 2z &= 4 \end{aligned}$

51. $\begin{aligned} 4x \qquad + z &= 7 \\ y \qquad &= 2 \\ x \qquad + z &= 4 \end{aligned}$

52. $\begin{aligned} x + y + 8z &= 3 \\ 2x + y + 11z &= 4 \\ x \qquad + 3z &= 0 \end{aligned}$

53. $\begin{aligned} -8x + y + z &= 6 \\ 2x - y + z &= 3 \\ 3x \qquad - z &= 0 \end{aligned}$

Expanding Your Skills

For Exercises 54–57, solve the equation.

54. $\begin{vmatrix} 6 & x \\ 2 & -4 \end{vmatrix} = 14$

55. $\begin{vmatrix} y & -2 \\ 8 & 7 \end{vmatrix} = 30$

56. $\begin{vmatrix} 3 & 1 & 0 \\ 0 & 4 & -2 \\ 1 & 0 & w \end{vmatrix} = 10$

57. $\begin{vmatrix} -1 & 0 & 2 \\ 4 & t & 0 \\ 0 & -5 & 3 \end{vmatrix} = -4$

For Exercises 58–59, evaluate the determinant of the 4×4 matrix using expansion by minors about the first column.

58. $\begin{vmatrix} 1 & 0 & 3 & 0 \\ 0 & 1 & 2 & 4 \\ -2 & 0 & 0 & 1 \\ 4 & -1 & -2 & 0 \end{vmatrix}$

59. $\begin{vmatrix} 5 & 2 & 0 & 0 \\ 0 & 4 & -1 & 1 \\ -1 & 0 & 3 & 0 \\ 0 & -2 & 1 & 0 \end{vmatrix}$

For Exercises 60–61, refer to the following system of four variables.

$$\begin{aligned} x + y + z + w &= 0 \\ 2x \qquad - z + w &= 5 \\ 2x + y \qquad - w &= 0 \\ y + z \qquad &= -1 \end{aligned}$$

60. a. Evaluate the determinant, **D**

b. Evaluate the determinant, $\mathbf{D}_x$.

c. Solve for x by computing $\dfrac{\mathbf{D}_x}{\mathbf{D}}$

61. a. Evaluate the determinant, $\mathbf{D}_y$.

b. Solve for y by computing $\dfrac{\mathbf{D}_y}{\mathbf{D}}$

chapter 8 SUMMARY

SECTION 8.1—INTRODUCTION TO SYSTEMS OF LINEAR EQUATIONS

KEY CONCEPTS:

A system of two linear equations can be solved by graphing.

A solution to a system of linear equations is an ordered pair that satisfies each equation in the system. Graphically, this represents a point of intersection of the lines.

There may be one solution, infinitely many solutions, or no solution.

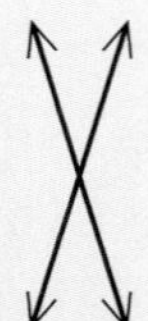

One solution
Consistent
Independent

Many solutions
Consistent
Dependent

No solution
Inconsistent
Independent

A system of equations is consistent if there is at least one solution. A system is inconsistent if there is no solution.

EXAMPLES:

Solve by graphing: $x + y = 3$
$2x - y = 0$

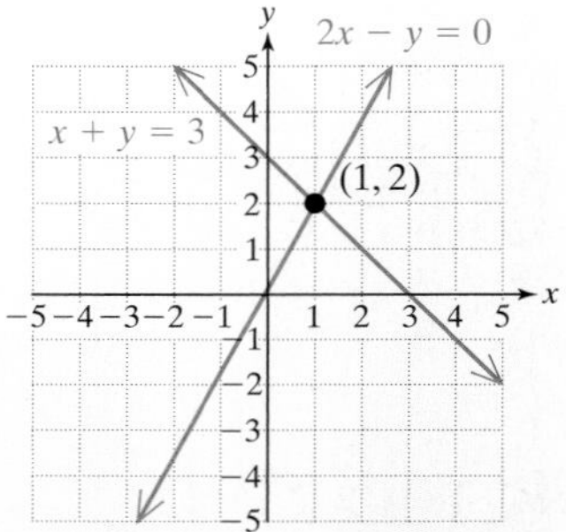

The solution is $(1, 2)$.

Solve by graphing: $3x - 2y = 2$
$-6x + 4y = 4$

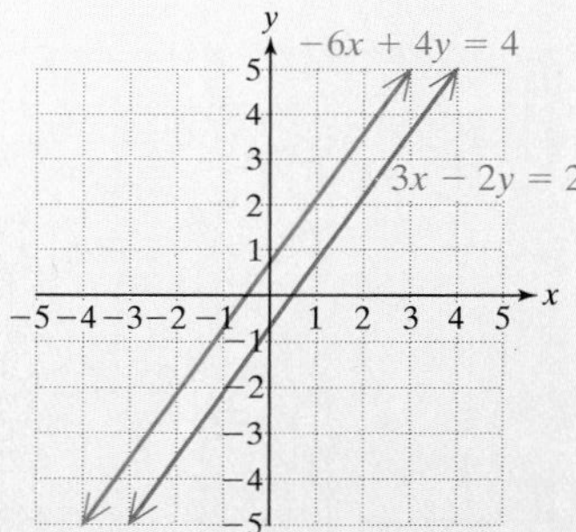

There is no solution. The system is inconsistent.

A linear system in x and y is dependent if two equations represent the same line. The solution set is the set of all points on the line.

If two linear equations represent different lines, then the system of equation is independent.

Solve by graphing: $x + 2y = 2$
$-3x - 6y = -6$

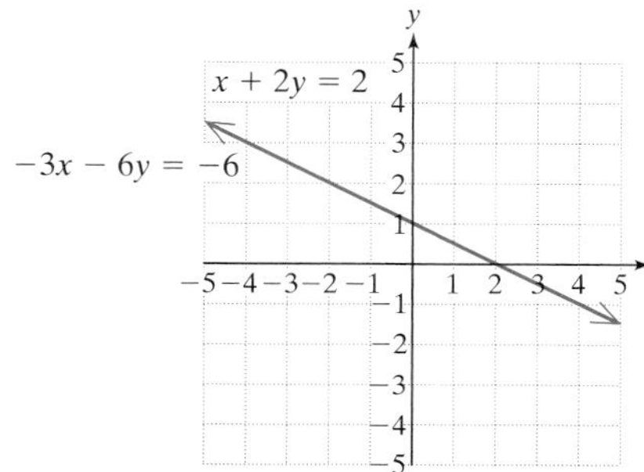

The system is dependent, and the solution set consists of all points on the line, given by $\{(x, y) | x + 2y = 2\}$

Key Terms:

consistent system
dependent system
inconsistent system
independent system
solution to a system of linear equations
system of linear equations

Section 8.2—Substitution Method

Key Concepts:

Steps to Solve a System of Equations by the Substitution Method:

1. Isolate one of the variables from one equation.
2. Substitute the quantity found in step 1 into the other equation.
3. Solve the resulting equation.
4. Substitute the value found in step 3 back into the equation in step 1 to find the remaining variable.
5. Check the solution in both original equations and write the answer as an ordered pair.

Examples:

Solve by the substitution method:

$$x + 4y = -11$$
$$3x - 2y = -5$$

Isolate x in the first equation: $x = -4y - 11$
Substitute into the second equation.

$3(-4y - 11) - 2y = -5$ Solve the equation.

$-12y - 33 - 2y = -5$

$-14y = 28$

$y = -2$

$x = -4y - 11$ Substitute $y = -2$.

$x = -4(-2) - 11$ Solve for x.

$x = -3$

The solution is $(-3, -2)$ and checks in both original equations.

An inconsistent system has no solution and is detected algebraically by a contradiction (such as $0 = 3$).

Solve by the substitution method:

$$3x + y = 4$$
$$-6x - 2y = 2$$

Isolate y in the first equation: $y = -3x + 4$. Substitute into the second equation.

$$-6x - 2(-3x + 4) = 2$$
$$-6x + 6x - 8 = 2$$
$$-8 = 2 \quad \text{Contradiction}$$

The system is inconsistent and has no solution.

If two linear equations represent the same line, the system is dependent. This is detected algebraically by an identity (such as $0 = 0$).

Key Term:

substitution method

Solve by the substitution method:

$y = x + 2$ y is already isolated.
$x - y = -2$

$x - (x + 2) = -2$ Substitute $y = x + 2$ into the second equation.

$x - x - 2 = -2$

$-2 = -2$ Identity

The system is dependent. The solution set is all points on the line $y = x + 2$ or $\{(x, y) \mid y = x + 2\}$.

Section 8.3—Addition Method

Key Concepts:

Solving a System of Linear Equations by the Addition Method:

1. Write both equations in standard form: $ax + by = c$.
2. Clear fractions or decimals (optional).
3. Multiply one or both equations by a nonzero constant to create opposite coefficients for one of the variables.
4. Add the equations to eliminate one variable.
5. Solve for the remaining variable.
6. Substitute the known value into one of the original equations to solve for the other variable.
7. Check the solution in both equations.

Key Term:

addition method

Examples:

Solve by using the addition method:

$5x = -4y - 7$ Write the first equation in standard form.
$6x - 3y = 15$

$5x + 4y = -7 \xrightarrow{\text{Multiply by 3}} 15x + 12y = -21$

$6x - 3y = 15 \xrightarrow{\text{Multiply by 4}} 24x - 12y = 60$

$$39x = 39$$
$$x = 1$$

$$5x = -4y - 7$$
$$5(1) = -4y - 7$$
$$5 = -4y - 7$$
$$12 = -4y$$
$$-3 = y$$

The solution is $(1, -3)$ and checks in both original equations.

Section 8.4—Applications of Linear Equations in Two Variables

Examples:

A riverboat travels 36 miles with the current to a marina in 2 hr. The return trip takes 3 hr against the current. Find the speed of the current and the speed of the boat in still water.

Let x represent the speed of the boat in still water.
Let y represent the speed of the current.

	Distance	Rate	Time
Against current	36	$x - y$	3
With current	36	$x + y$	2

Distance = (rate)(time)

$$36 = (x - y) \cdot 3$$
$$36 = (x + y) \cdot 2$$

$$36 = 3x - 3y \xrightarrow{\text{Multiply by 2}} 72 = 6x - 6y$$
$$36 = 2x + 2y \xrightarrow[\text{Multiply by 3}]{} 108 = 6x + 6y$$
$$180 = 12x$$
$$15 = x$$

$$36 = 2(15) + 2y$$
$$36 = 30 + 2y$$
$$6 = 2y$$
$$3 = y$$

The speed of the boat in still water is 15 mph, and the speed of the current is 3 mph.

Examples:

Diane invests \$15,000 more in an account earning 8% simple interest than in an account earning 5% simple interest. If the total interest after 1 year is \$1850, how much was invested in each account?

	8%	5%	Total
Principal	x	y	
Interest	$0.08x$	$0.05y$	1850

$$x = y + 15{,}000$$
$$0.08x + 0.05y = 1850$$

Substitute $x = y + 15{,}000$ into the second equation:

$$0.08(y + 15{,}000) + 0.05y = 1850$$
$$0.08y + 1200 + 0.05y = 1850$$
$$0.13y + 1200 = 1850$$
$$0.13y = 650$$
$$\frac{0.13y}{0.13} = \frac{650}{0.13}$$
$$y = 5000$$

$$x = y + 15{,}000$$
$$x = 5000 + 15{,}000$$
$$x = 20{,}000$$

The amount invested in the 8% account is \$20,000 and the amount invested at 5% is \$5000.

Section 8.5—Systems of Linear Equations in Three Variables and Applications

Key Concepts:

A linear equation in three variables can be written in the form: $ax + by + cz = d$, where a, b, and c are not all zero. The graph of a linear equation in three variables is a plane in space.

Examples:

A $x + 2y - z = 4$

B $3x - y + z = 5$

C $2x + 3y + 2z = 7$

A solution to a system of linear equations in three variables is an ordered triple that satisfies each equation. Graphically, a solution is a point of intersection among three planes.

A system of linear equations in three variables may have one unique solution, infinitely many solutions (dependent system), or no solution (inconsistent system).

Key Terms:

linear equation in three variables
ordered triple
solution to a system of linear equations in three variables

$$\boxed{A} \text{ and } \boxed{B} \quad \begin{aligned} x + 2y - z &= 4 \\ 3x - y + z &= 5 \\ \hline 4x + y \quad &= 9 \quad \boxed{D} \end{aligned}$$

$$2 \cdot \boxed{A} \text{ and } \boxed{C} \quad \begin{aligned} 2x + 4y - 2z &= 8 \\ 2x + 3y + 2z &= 7 \\ \hline 4x + 7y \quad &= 15 \quad \boxed{E} \end{aligned}$$

$$\boxed{D} \ 4x + y = 9 \rightarrow -4x - y = -9$$

$$\boxed{E} \ 4x + 7y = 15 \rightarrow \ \ 4x + 7y = 15$$

$$6y = 6$$

$$y = 1$$

Substitute $y = 1$ into either equation $\boxed{D}$ or $\boxed{E}$.

$$\boxed{D} \ 4x + (1) = 9$$

$$4x = 8$$

$$x = 2$$

Substitute $x = 2$ and $y = 1$ into either equation $\boxed{A}$, $\boxed{B}$, or $\boxed{C}$.

$$\boxed{A} \ (2) + 2(1) - z = 4$$

$$z = 0$$

The solution is (2, 1, 0).

Section 8.6—Solving Systems of Linear Equations Using Matrices

Key Concepts:

A matrix is a rectangular array of numbers displayed in rows and columns. Every number or entry within a matrix is called an element of the matrix.

The order of a matrix is determined by the number of rows and number of columns. A matrix with m rows and n columns is an $m \times n$ matrix.

A system of equations written in standard form can be represented by an augmented matrix consisting of the coefficients of the terms and constants of each equation in the system.

Examples:

$[1 \quad 2 \quad 5]$ is a 1×3 matrix (called a row matrix).

$\begin{bmatrix} -1 & 8 \\ 1 & 5 \end{bmatrix}$ is a 2×2 matrix (called a square matrix).

$\begin{bmatrix} 4 \\ 1 \end{bmatrix}$ is a 2×1 matrix (called a column matrix).

Write the augmented matrix for the system.

$$\begin{aligned} 4x + y &= -12 \\ x - 2y &= 6 \end{aligned} \qquad \left[\begin{array}{cc|c} 4 & 1 & -12 \\ 1 & -2 & 6 \end{array}\right]$$

The Gauss-Jordan method can be used to solve a system of equations by using the following elementary row operations on an augmented matrix.

1. Interchange two rows.
2. Multiply every element in a row by a nonzero real number.
3. Add a multiple of one row to another row.

These operations are used to write the matrix in reduced row echelon form.

$$\left[\begin{array}{cc|c} 1 & 0 & a \\ 0 & 1 & b \end{array}\right]$$

which represents the solution, $x = a$ and $y = b$.

Solve the previous system using the Gauss-Jordan method.

$R_1 \Leftrightarrow R_2$ $\left[\begin{array}{cc|c} 1 & -2 & 6 \\ 4 & 1 & -12 \end{array}\right]$

$\xrightarrow{-4R_1 + R_2 \Rightarrow R_2}$ $\left[\begin{array}{cc|c} 1 & -2 & 6 \\ 0 & 9 & -36 \end{array}\right]$

$\xrightarrow{\frac{1}{9}R_2 \Rightarrow R_2}$ $\left[\begin{array}{cc|c} 1 & -2 & 6 \\ 0 & 1 & -4 \end{array}\right]$

$\xrightarrow{2R_2 + R_1 \Rightarrow R_1}$ $\left[\begin{array}{cc|c} 1 & 0 & -2 \\ 0 & 1 & -4 \end{array}\right]$

Solution:

$x = -2$ and $y = -4$

Key Terms:

augmented matrix
coefficient matrix
column matrix
elementary row operations
Gauss-Jordan method
matrix
order of a matrix
reduced row echelon form
row matrix
square matrix

Section 8.7—Determinants and Cramer's Rule

Key Concepts:

The determinant of matrix $\mathbf{A} = \begin{bmatrix} a & b \\ c & d \end{bmatrix}$ is denoted $\det\mathbf{A} = \begin{vmatrix} a & b \\ c & d \end{vmatrix}$.

The determinant of a 2×2 matrix is defined as:

$$\begin{vmatrix} a & b \\ c & d \end{vmatrix} = ad - bc$$

The determinant of a 3×3 matrix is defined by

$$\begin{vmatrix} a_1 & b_1 & c_1 \\ a_2 & b_2 & c_2 \\ a_3 & b_3 & c_3 \end{vmatrix} = a_1 \cdot \begin{vmatrix} b_2 & c_2 \\ b_3 & c_3 \end{vmatrix} - a_2 \cdot \begin{vmatrix} b_1 & c_1 \\ b_3 & c_3 \end{vmatrix} + a_3 \cdot \begin{vmatrix} b_1 & c_1 \\ b_2 & c_2 \end{vmatrix}$$

Examples:

Find the determinant of matrix A:

$$\mathbf{A} = \begin{bmatrix} 7 & -2 \\ 3 & 2 \end{bmatrix}, \quad \det\mathbf{A} = \begin{vmatrix} 7 & -2 \\ 3 & 2 \end{vmatrix}$$

$$= 14 - (-6)$$

$$= 20$$

Find the determinant of matrix B:

$$\mathbf{B} = \begin{bmatrix} 3 & -1 & 4 \\ 0 & 5 & 1 \\ 6 & -2 & -3 \end{bmatrix}$$

$\det\mathbf{B}$

$$= 3 \cdot \begin{vmatrix} 5 & 1 \\ -2 & -3 \end{vmatrix} - 0 \cdot \begin{vmatrix} -1 & 4 \\ -2 & -3 \end{vmatrix} + 6 \cdot \begin{vmatrix} -1 & 4 \\ 5 & 1 \end{vmatrix}$$

$$= 3(-15 + 2) - 0(3 + 8) + 6(-1 - 20)$$

$$= -39 - 0 + (-126)$$

$$= -165$$

Cramer's Rule for a 2 × 2 System of Linear Equations

The solution to the system

$$a_1x + b_1y = c_1$$
$$a_2x + b_2y = c_2$$

is given by

$$x = \frac{\mathbf{D}_x}{\mathbf{D}} \quad \text{and} \quad y = \frac{\mathbf{D}_y}{\mathbf{D}}$$

where

$$\mathbf{D} = \begin{vmatrix} a_1 & b_1 \\ a_2 & b_2 \end{vmatrix} \quad (\text{and } \mathbf{D} \neq 0),$$

$$\mathbf{D}_x = \begin{vmatrix} c_1 & b_1 \\ c_2 & b_2 \end{vmatrix}, \quad \text{and} \quad \mathbf{D}_y = \begin{vmatrix} a_1 & c_1 \\ a_2 & c_2 \end{vmatrix}$$

Use Cramer's rule to solve the system.

$$2x + 8y = 0$$
$$x + 3y = 1$$

$$\mathbf{D} = \begin{vmatrix} 2 & 8 \\ 1 & 3 \end{vmatrix} = -2 \qquad \mathbf{D}_x = \begin{vmatrix} 0 & 8 \\ 1 & 3 \end{vmatrix} = -8$$

$$\mathbf{D}_y = \begin{vmatrix} 2 & 0 \\ 1 & 1 \end{vmatrix} = 2$$

Therefore,

$$x = \frac{-8}{-2} = 4, \quad y = \frac{2}{-2} = -1$$

Cramer's Rule for a 3 × 3 System of Linear Equations

The solution to the system

$$a_1x + b_1y + c_1z = d_1$$
$$a_2x + b_2y + c_2z = d_2$$
$$a_3x + b_3y + c_3z = d_3$$

is given by

$$x = \frac{\mathbf{D}_x}{\mathbf{D}}, \quad y = \frac{\mathbf{D}_y}{\mathbf{D}}, \quad \text{and} \quad z = \frac{\mathbf{D}_z}{\mathbf{D}}$$

where

$$\mathbf{D} = \begin{vmatrix} a_1 & b_1 & c_1 \\ a_2 & b_2 & c_2 \\ a_3 & b_3 & c_3 \end{vmatrix} \quad (\text{and } \mathbf{D} \neq 0),$$

$$\mathbf{D}_x = \begin{vmatrix} d_1 & b_1 & c_1 \\ d_2 & b_2 & c_2 \\ d_3 & b_3 & c_3 \end{vmatrix}, \quad \mathbf{D}_y = \begin{vmatrix} a_1 & d_1 & c_1 \\ a_2 & d_2 & c_2 \\ a_3 & d_3 & c_3 \end{vmatrix},$$

$$\text{and } \mathbf{D}_z = \begin{vmatrix} a_1 & b_1 & d_1 \\ a_2 & b_2 & d_2 \\ a_3 & b_3 & d_3 \end{vmatrix}$$

If $\mathbf{D} = 0$, use the addition method, substitution method, or Gauss-Jordan method to solve the system.

Use Cramer's rule to solve for *x*.

$$\begin{aligned} x + 3y \phantom{{}+z} &= 3 \\ 2y + z &= 3 \\ x \phantom{{}+3y} - 3z &= 2 \end{aligned}$$

$$\mathbf{D} = \begin{vmatrix} 1 & 3 & 0 \\ 0 & 2 & 1 \\ 1 & 0 & -3 \end{vmatrix} = -3$$

$$\mathbf{D}_x = \begin{vmatrix} 3 & 3 & 0 \\ 3 & 2 & 1 \\ 2 & 0 & -3 \end{vmatrix} = 15$$

Therefore,

$$x = \frac{15}{-3} = -5$$

KEY TERMS:

Cramer's rule
determinant
minor

chapter 8 REVIEW EXERCISES

Section 8.1

For Exercises 1–4, determine if the ordered pair is a solution to the system.

1. $x - 4y = -4$, $x + 2y = 8$ $(4, 2)$

2. $x - 6y = 6$, $-x + y = 4$ $(12, 1)$

3. $3x + y = 9$, $y = 3$ $(1, 3)$

4. $2x - y = 8$, $x = 2$ $(2, -4)$

For Exercises 5–10, identify whether the system represents intersecting lines, parallel lines, or coinciding lines by comparing their slopes and y-intercepts.

5. $y = -\frac{1}{2}x + 4$, $y = x - 1$

6. $y = -3x + 4$, $y = 3x + 4$

7. $y = -\frac{4}{7}x + 3$, $y = -\frac{4}{7}x - 5$

8. $y = 5x - 3$, $y = \frac{1}{5}x - 3$

9. $y = 9x - 2$, $9x - y = 2$

10. $x = -5$, $y = 2$

For Exercises 11–18, solve the systems by graphing. Identify whether the system is consistent or inconsistent. Identify whether the system is dependent or independent.

11. $y = -\frac{2}{3}x - 2$, $-x + 3y = -6$

12. $y = -2x - 1$, $2x + 3y = 5$

13. $4x = -2y + 10$, $2x + y = 5$

14. $10y = 2x - 10$, $-x + 5y = -5$

15. $6x - 3y = 9$, $y = -1$

16. $5x + y = -11$, $x = -1$

17. $x - 7y = 14$, $-2x + 14y = 14$

18. $y = -5x + 6$, $10x + 2y = 6$

19. The following graph depicts the percent of AIDS cases by race and ethnicity in the United States between 1985–1998. Estimate the year in which the percent of AIDS cases attributed to black (non Hispanic) persons equaled the percent of AIDS cases attributed to white (non Hispanic) persons.

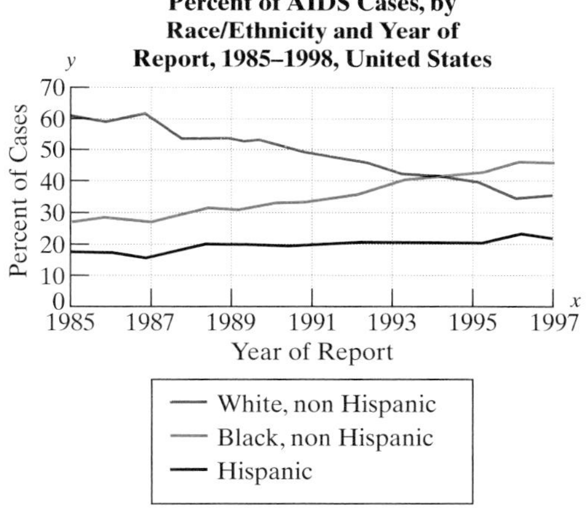

Figure for Exercise 19

Source: Centers for Disease Control and Prevention.

20. A rental car company rents a compact car for \$20 a day, plus \$0.25 per mile. A midsize car rents for \$30 a day, plus \$0.20 per mile.

The cost, y_c, to rent a compact car for one day is given by the equation:

$y_c = 0.25x + 20$ where x is the number of miles driven

The cost, y_m, to rent a midsize car for one day is given by the equation:

$y_m = 0.20x + 30$ where x is the number of miles driven

Find the number of miles at which the cost to rent either car would be the same and confirm your answer with the graph.

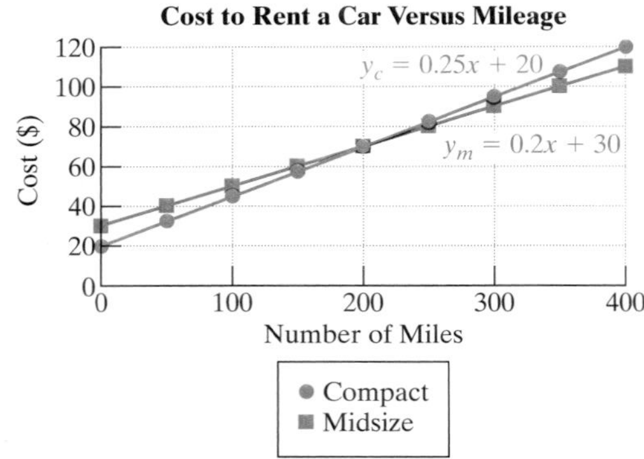

Figure for Exercise 20

Section 8.2

For Exercises 21–24, solve the systems using the substitution method.

21. $6x + y = 2$
$y = 3x - 4$

22. $2x + 3y = -5$
$x = y - 5$

23. $2x + 6y = 10$
$x = -3y + 6$

24. $4x + 2y = 4$
$y = -2x + 2$

25. Given the system:

$$x + 2y = 11$$
$$5x + 4y = 40$$

a. Which variable from which equation is easiest to isolate and why?

b. Solve the system using the substitution method.

26. Given the system:

$$4x - 3y = 9$$
$$2x + y = 12$$

a. Which variable from which equation is easiest to isolate and why?

b. Solve the system using the substitution method.

For Exercises 27–30, solve the systems using the substitution method.

27. $3x - 2y = 23$
$x + 5y = -15$

28. $x + 5y = 20$
$3x + 2y = 8$

29. $x - 3y = 9$
$5x - 15y = 45$

30. $-3x + y = 15$
$6x - 2y = 12$

31. The difference of two positive numbers is 42. The larger number is 2 more than 6 times the smaller number. Find the numbers.

32. In a right triangle, one of the acute angles is 6° less than the other acute angle. Find the measure of each acute angle.

33. During the first 13 years of his football career, Jerry Rice scored a total of 166 touchdowns. One touchdown was scored on a kickoff return and the remaining 165 were scored rushing or receiving. The number of receiving touchdowns he scored was 5 more than 15 times the number of rushing touchdowns he scored. How many receiving touchdowns and how many rushing touchdowns did he score?

34. Two angles are supplementary. One angle is 14° less than two times the other angle. Find the measure of each angle.

Section 8.3

35. Explain the process for solving a system of two equations using the addition method.

36. Given the system:

$$3x - 5y = 1$$
$$2x - y = -4$$

a. Which variable, x or y, is easier to eliminate using the addition method?

b. Solve the system using the addition method.

37. Given the system:

$$9x - 2y = 14$$
$$4x + 3y = 14$$

a. Which variable, x or y, is easier to eliminate using the addition method?

b. Solve the system using the addition method.

For Exercises 38–45, solve the systems using the addition method.

38. $2x + 3y = 1$
 $x - 2y = 4$

39. $x + 3y = 0$
 $-3x - 10y = -2$

40. $8(x + 1) = -6y + 6$
 $10x = 9y - 8$

41. $12x = 5(y + 1)$
 $5y = -1 - 4x$

42. $-4x - 6y = -2$
 $6x + 9y = 3$

43. $-8x - 4y = 16$
 $10x + 5y = 5$

44. $\frac{1}{2}x - \frac{3}{4}y = -\frac{1}{2}$
 $\frac{1}{3}x + y = -\frac{10}{3}$

45. $0.5x - 0.2y = 0.5$
 $0.4x + 0.7y = 0.4$

46. Given the system:

$$4x + 9y = -7$$
$$y = 2x - 13$$

a. Which method would you choose to solve the system, the substitution method or the addition method? Explain your choice.

b. Solve the system.

47. Given the system:

$$5x - 8y = -2$$
$$3x - y = -5$$

a. Which method would you choose to solve the system, the substitution method or the addition method? Explain your choice.

b. Solve the system.

Section 8.4

48. Miami Metrozoo charges \$8.75 for adult admission and \$5.75 for each child under 12. The total bill before tax for a school group of 60 people is \$369. How many adults and how many children were admitted?

49. Emillo invested \$20,000 and at the end of one year, he received \$1525 in interest. If he invested part of the money at 5% simple interest and the remaining money at 8% simple interest, how much did he invest in each account?

50. To produce a 16% alcohol solution, a chemist mixes a 20% alcohol solution and a 14% alcohol solution. How much 20% solution and how much 14% solution must be used to produce 15 L of a 16% alcohol solution?

51. A boat travels 80 miles downstream with the current in 4 hours and 80 miles upstream against the current in 5 hr. Find the speed of the current and the speed of the boat in still water.

52. Suzanne has a collection of new quarters and new \$1 coins. She has four more quarters than dollar coins and the total value of the coins is \$4.75. How many of each coin does Suzanne have?

53. In a recent election 5700 votes were cast and 3675 voters voted for the winning candidate. If $\frac{5}{8}$ of the women and $\frac{2}{3}$ of the men voted for the winning candidate, how many men and how many women voted?

Section 8.5

For Exercises 54–56, solve the systems of equations.

54. $5x + 5y + 5z = 30$
 $-x + y + z = 2$
 $10x + 6y - 2z = 4$

55. $5x + 3y - z = 5$
 $x + 2y + z = 6$
 $-x - 2y - z = 8$

56. $3x + 4z = 5$
 $2y + 3z = 2$
 $2x - 5y = 8$

57. The perimeter of a right triangle is 30 ft. One leg is 2 ft more than twice the shortest leg. The hypotenuse is 2 ft less than 3 times the shortest leg. Find the lengths of the sides of this triangle.

58. Determine whether the system is inconsistent or dependent.

$$\begin{aligned} x + y + z &= 4 \\ -x - 2y - 3z &= -6 \\ 2x + 4y + 6z &= 12 \end{aligned}$$

Section 8.6

For Exercises 59–62, determine the order of each matrix.

59. $\begin{bmatrix} 2 & 4 & -1 \\ 5 & 0 & -3 \\ -1 & 6 & 10 \end{bmatrix}$

60. $\begin{bmatrix} -5 & 6 \\ 9 & 2 \\ 0 & -3 \end{bmatrix}$

61. $[0 \quad 13 \quad -4 \quad 16]$

62. $\begin{bmatrix} 7 \\ 12 \\ -4 \end{bmatrix}$

For Exercises 63–64, set up the augmented matrix.

63. $\begin{aligned} x + y &= 3 \\ x - y &= -1 \end{aligned}$

64. $\begin{aligned} x - y + z &= 4 \\ 2x - y + 3z &= 8 \\ -2x + 2y - z &= -9 \end{aligned}$

For Exercises 65–66, write a corresponding system of equations from the augmented matrix.

65. $\left[\begin{array}{cc|c} 1 & 0 & 9 \\ 0 & 1 & -3 \end{array}\right]$

66. $\left[\begin{array}{ccc|c} 1 & 0 & 0 & -5 \\ 0 & 1 & 0 & 2 \\ 0 & 0 & 1 & -8 \end{array}\right]$

67. Given the matrix **A**:

$$\mathbf{A} = \left[\begin{array}{cc|c} 2 & 0 & 5 \\ 1 & 3 & -11 \end{array}\right]$$

a. What is the element in the second row and first column?

b. Write the matrix obtained by interchanging rows 1 and 2.

68. Given the matrix **D**:

$$\mathbf{D} = \left[\begin{array}{ccc|c} 1 & 2 & 0 & -3 \\ 4 & -1 & 1 & 0 \\ -3 & 2 & 2 & 5 \end{array}\right]$$

a. Write the matrix obtained by multiplying the first row by -4 and adding the result to row 2.

b. Using the matrix obtained in part (a), write the matrix obtained by multiplying the first row by 3 and adding the result to row 3.

For Exercises 69–70, solve the system by using the Gauss-Jordan method.

69. $\begin{aligned} x + y &= 3 \\ x - y &= -1 \end{aligned}$

70. $\begin{aligned} x - y + z &= 4 \\ 2x - y + 3z &= 8 \\ -2x + 2y - z &= -9 \end{aligned}$

Section 8.7

For Exercises 71–74, evaluate the determinant.

71. $\begin{vmatrix} 5 & -2 \\ 2 & -3 \end{vmatrix}$

72. $\begin{vmatrix} -6 & 1 \\ 0 & 10 \end{vmatrix}$

73. $\begin{vmatrix} \frac{1}{2} & 3 \\ 1 & 8 \end{vmatrix}$

74. $\begin{vmatrix} 9 & 3 \\ -2 & \frac{2}{3} \end{vmatrix}$

For Exercises 75–78, evaluate the minor corresponding to the given element from matrix **A**.

$$\mathbf{A} = \begin{vmatrix} 8 & 2 & 0 \\ -1 & 4 & -2 \\ 3 & -3 & 6 \end{vmatrix}$$

75. 8

76. 2

77. −2

78. 4

For Exercises 79–82, evaluate the determinant.

79. $\begin{vmatrix} 2 & 1 & 0 \\ -4 & 3 & -1 \\ 3 & 0 & 1 \end{vmatrix}$

80. $\begin{vmatrix} 1 & 0 & 2 \\ 3 & -2 & 4 \\ 0 & 1 & 1 \end{vmatrix}$

81. $\begin{vmatrix} 4 & -2 & 0 \\ 9 & 5 & 4 \\ 1 & 2 & 0 \end{vmatrix}$

82. $\begin{vmatrix} -1 & 0 & 2 \\ 5 & -2 & 6 \\ 3 & 0 & -4 \end{vmatrix}$

For Exercises 83–84, evaluate the determinants represented by $\mathbf{D}$, $\mathbf{D}_x$, and $\mathbf{D}_y$. Then use Cramer's rule to solve for x and y.

83. $\begin{aligned} 3x + 2y &= 9 \\ x - 3y &= 1 \end{aligned}$

84. $\begin{aligned} -x + 3y &= -8 \\ 2x - y &= -2 \end{aligned}$

For Exercises 85–88, solve the system using Cramer's rule.

85. $\begin{aligned} 3x - 4y &= 1 \\ 2x + 3y &= 12 \end{aligned}$

86. $\begin{aligned} 3x + 2y &= 11 \\ -x + y &= 3 \end{aligned}$

87. $\begin{aligned} 2x + 3y - z &= -7 \\ x + 3z &= 10 \\ 2y + z &= -1 \end{aligned}$

88. $\begin{aligned} 6y + 4z &= -12 \\ 3x + 3y &= 9 \\ 2x - 3z &= 10 \end{aligned}$

For Exercises 89–90, solve the system of equations using Cramer's rule if possible. If not possible, use another method.

89. $\begin{aligned} 2x - y &= 1 \\ 4x - 2y &= 2 \end{aligned}$

90. $\begin{aligned} x + y - 3z &= -1 \\ y - z &= 6 \\ -x + 2y &= 1 \end{aligned}$

chapter 8 TEST

1. Write each line in slope-intercept form. Then determine if the lines represent intersecting lines, parallel lines, or coinciding lines.

$$5x + 2y = -6$$
$$-\frac{5}{2}x - y = -3$$

2. a. Solve the system by graphing

$$y = \frac{1}{2}x + \frac{9}{2}$$
$$y = -\frac{1}{3}x + 2$$

   b. Check your answer by substituting the ordered pair in both equations.

3. The supply and demand of an item depends on the number of items produced and the price of the item. If the price is high, the demand decreases and there is a surplus in supply. Answer the following questions based on the graph provided.

   a. What is the equilibrium price (that is, for what price does supply equal demand)?

   b. How many items should be produced at the equilibrium price?

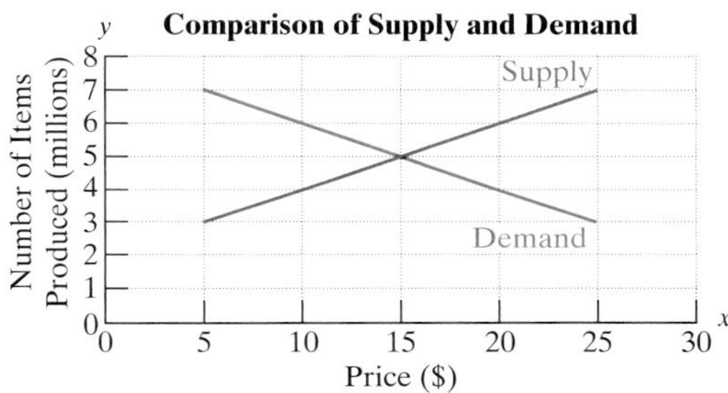

Figure for Exercise 3

4. Solve the system using the substitution method.

$$x = 5y - 2$$
$$2x + y = -4$$

5. In the 1998 WNBA season, the league's leading scorer, Cynthia Cooper from the Houston Comets scored 227 more points than her teammate Sheryl Swoopes. Together they scored a total of 1133 points. How many points did each player score?

6. Solve the system using the addition method.

$$3x - 6y = 8$$
$$2x + 3y = 3$$

7. How many milliliters of a 50% acid solution and how many milliliters of a 20% acid solution must be mixed to produce 36 mL of a 30% acid solution?

8. a. How many solutions does a system of two linear equations have if the equations represent parallel lines?

b. How many solutions does a system of two linear equations have if the equations represent coinciding lines?

c. How many solutions does a system of two linear equations have if the equations represent intersecting lines?

For Exercises 9–12, solve the systems using any method.

9. $\frac{1}{3}x + y = \frac{7}{3}$
$x = \frac{3}{2}y - 11$

10. $2(x - 6) = y$
$2x - \frac{1}{2}y = x + 5$

11. $-0.25x - 0.05y = 0.2$
$10x + 2y = -8$

12. $3(x + y) = -2y - 7$
$-3y = 10 - 4x$

13. In a right triangle, one of the acute angles is 9° less than twice the other acute angle. Find the measure of each acute angle.

14. Max has 30 coins consisting of nickels, dimes, and quarters. He has 10 nickels and the number of dimes is 1 less than twice the number of quarters.

a. How many of each type of coin does he have?

b. How much money does he have?

15. Five thousand dollars less was invested in an account earning 9% simple interest than in an account earning 11% simple interest. If the total interest at the end of 1 year is $1950, how much was invested in each account? What is the total amount invested?

16. A consumer who is tired of paying rising costs for cable television looked into the possibility of purchasing a satellite dish. The cost of a particular satellite dish is $120 plus $26 per month. The cable television is already installed and costs $34 per month. The total cost, y, for either option depends on the number of months of service, x, according to the following equations:

Satellite: $y = 26x + 120$

Cable: $y = 34x$

a. Use the graph to estimate how many months are required for the cost to use a satellite dish and the cost to use cable to be equal.

b. Solve the system of equations by any method. Interpret the meaning of the solution in terms of the number of months and the total cost.

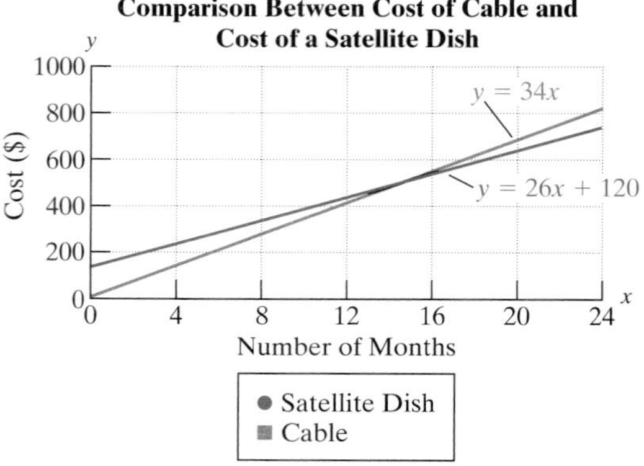

Figure for Exercise 16

17. A plane travels 880 miles in 2 hours (hr) against the wind, and 1000 miles in 2 hr with the same wind. Find the speed of the plane in still air and the speed of the wind.

18. Solve the system of equations.

$$\begin{aligned} 2x + 2y + 4z &= -6 \\ 3x + y + 2z &= 29 \\ x - y - z &= 44 \end{aligned}$$

19. Working together Joanne, Kent, and Geoff can process 504 orders per day for their business. Kent can process 20 more orders per day than Joanne can process. Geoff can process 104 fewer orders per day than Kent and Joanne combined. Find the number of orders that each person can process per day.

20. Write an example of a 3×2 matrix.

21. Given the matrix $\mathbf{A}$:

$$\mathbf{A} = \left[\begin{array}{ccc|c} 1 & 2 & 1 & -3 \\ 4 & 0 & 1 & -2 \\ -5 & -6 & 3 & 0 \end{array}\right]$$

a. Write the matrix obtained by multiplying the first row by -4 and adding the result to row 2.

b. Using the matrix obtained in part (a), write the matrix obtained by multiplying the first row by 5 and adding the result to row 3.

22. Solve the system using the Gauss-Jordan method.

$$\begin{aligned} 5x - 4y &= 34 \\ x - 2y &= 8 \end{aligned}$$

For Exercises 23–25, find the determinant of the matrix.

23. $\begin{bmatrix} 2 & -3 \\ 1 & 2 \end{bmatrix}$

24. $\begin{bmatrix} 0 & 3 \\ 2 & -4 \end{bmatrix}$

25. $\begin{bmatrix} 0 & 5 & -2 \\ 0 & 0 & 2 \\ 2 & 3 & 1 \end{bmatrix}$

26. Do all matrices have a determinant? Explain.

For Exercises 27–28, use Cramer's rule to solve for y.

27. $\begin{aligned} 6x - 5y &= 13 \\ -2x - 2y &= 9 \end{aligned}$

28. $\begin{aligned} 2x + 2z &= 2 \\ 5x + 3y &= 4 \\ 3y - 4z &= 4 \end{aligned}$

29. Solve the system: $\begin{aligned} 6x - 2y &= 0 \\ 3x - y &= 0 \end{aligned}$

CUMULATIVE REVIEW EXERCISES, CHAPTERS 1–8

1. Simplify.

$$\frac{|2 - 5| + 10 \div 2 + 3}{\sqrt{10^2 - 8^2}}$$

2. Place the numbers in the diagram: $\{-4, \pi, \sqrt{9}, 6.7, \frac{1}{3}, 0, \sqrt{2}\}$

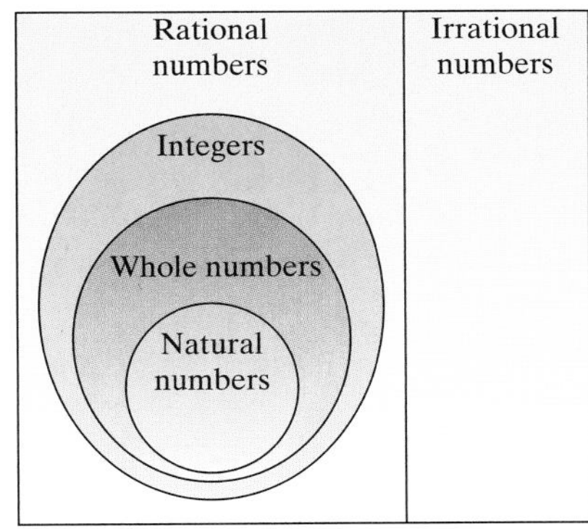

Figure for Exercise 2

3. Solve for x: $\frac{1}{3}x - \frac{3}{4} = \frac{1}{2}(x + 2)$

4. Solve for a: $-4(a + 3) + 2 = -5(a + 1) + a$

5. Solve for y: $3x - 2y = 6$

6. Solve for z. Graph the solution set and write the solution in interval notation:

$$-2(3z + 1) \leq 5(z - 3) + 10$$

7. In any triangle, what is the sum of the three inscribed angles?

8. The largest angle in a triangle is 110°. Of the remaining two angles, one is 4° less than the other angle. Find the measure of the three angles.

9. Two hikers start at opposite ends of an 18-mile trail and walk toward each other. One hiker walks predominately down hill and averages 2 mph faster than the other hiker. Find the average rate of each hiker if they meet in 3 hr.

10. Simplify. Write the final answer with positive exponents only:

    a. $(2c^2d^{-3})(5c^{-4}d^8)$ b. $\dfrac{(x^{-2})^{-3}(x^5)^2}{x^{-2}}$

11. Multiply. Write the answer in scientific notation: $(3.0 \times 10^4)(6.0 \times 10^8)$

12. Multiply: $(3y^2 + 2y - 4)(y - 2)$

13. Subtract: $(\frac{1}{2}w^2 - \frac{3}{5}w + \frac{1}{2}) - (\frac{3}{2}w^2 + \frac{1}{10}w - 2)$

14. Divide.

$$\frac{8p^3q - 4p^2q^2 + 16pq^3}{4p^2q^2}$$

15. Factor completely: $5x^2 - 125$

16. Factor completely: $5xa - 10xb - 2ya + 4yb$

17. Solve the equation: $5y(y - 3)(2y + 1) = 0$

18. Solve the equation: $t^2 - 10t = -25$

19. Given the equation: $y = x^2 - x - 12$
    a. Is the equation linear or nonlinear?
    b. Find the x- and y-intercepts.

20. Multiply.
$$\frac{20x - 30}{2x + 10} \cdot \frac{x^2 + 8x + 15}{4x^2 - 9}$$

21. Add.
$$\frac{2a}{a - 3} + \frac{28}{a^2 - 9} + \frac{6}{a + 3}$$

22. Solve for a.
$$\frac{2a}{a - 3} + \frac{28}{a^2 - 9} + \frac{6}{a + 3} = 0$$

23. Explain the difference between the process used to work Problems 21 and 22.

24. Joelle can do a job in 4 hr. Her husband Bob can do the same job in 3 hr. How long will it take them if they work together?

25. The slope of a given line is $-\frac{2}{3}$.
    a. What is the slope of a line parallel to the given line?
    b. What is the slope of a line perpendicular to the given line?

26. Find an equation of the line passing through the point $(2, -3)$ and perpendicular to the line $x - 3y = 4$. Write the final answer in slope-intercept form.

27. Sketch the following equations on the same graph.
    a. $2x + 5y = 10$
    b. $2y = 4$
    c. Find the point of intersection and check the solution in each equation.

28. Solve the system of equations using any method.
$$2x + 5y = 10$$
$$2y = 4$$

29. Graph the line $2x + y = 3$.

30. How many gallons of a 15% antifreeze solution should be mixed with a 60% antifreeze solution to produce 60 gal of a 45% antifreeze solution?

31. Set up a system of two linear equations to solve for x and y.

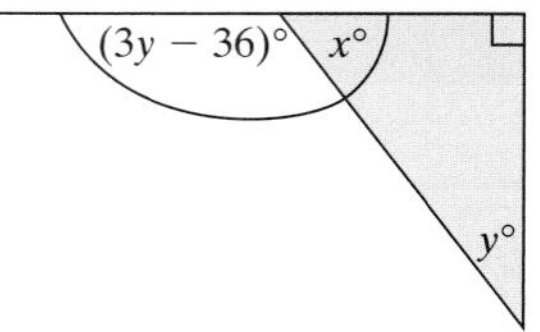

Figure for Exercise 31

32. In 1920, the average speed for the winner of the Indianapolis 500 car race was 88.6 mph. By 1990, the average speed of the winner was 186.0 mph.
    a. Find the slope of the line shown in the figure. Round to one decimal place.
    b. Interpret the meaning of the slope in the context of this problem.

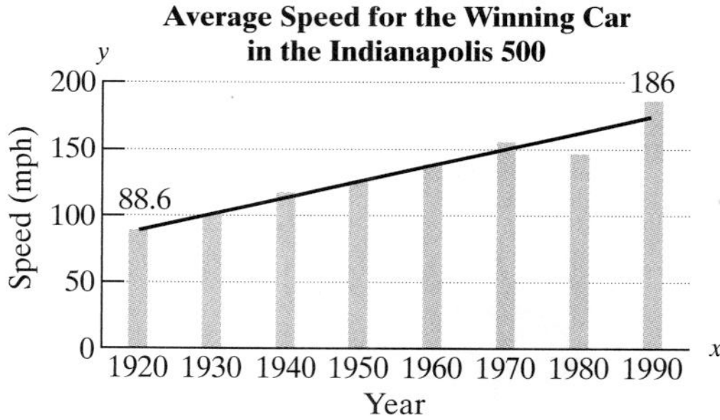

Figure for Exercise 32

33. Solve the system:
$$3x + 2y + 3z = 3$$
$$4x - 5y + 7z = 1$$
$$2x + 3y - 2z = 6$$

34. Solve the system using the Gauss-Jordan method:
$$2x - 4y = -2$$
$$4x + y = 5$$

chapter 9

More Equations and Inequalities

In the spring of 1998, the National Institutes of Health (NIH) issued new guidelines for determining whether an individual is overweight. The standard of measure is called the body mass index (BMI). Body mass index is a measure of an individual's weight in relation to the person's height according to the formula:

$$\text{BMI} = \frac{703W}{h^2}$$

where W is weight in pounds and h is height in inches. (Source: National Institutes of Health)

For more information see the Technology Connections in MathZone at

www.mhhe.com/miller_oneill

The NIH categorizes body mass indices according to the following intervals:

Body Mass Index (BMI)	Weight Status
19.5–24.9	ideal
25.0–29.9	overweight
30.0 or above	obese

Inequalities and interval notation are studied in detail in this chapter.

Concepts

1. Union and Intersection of Sets
2. Solving Compound Inequalities
3. Solving Compound Inequalities (And)
4. Solving Compound Inequalities (Or)
5. Applications of Compound Inequalities

section

9.1 Compound Inequalities

1. Union and Intersection of Sets

In Chapter 2 we graphed simple inequalities and expressed the solution set in interval notation and in set-builder notation. In this chapter, we will solve **compound inequalities** that involve the union or intersection of two or more inequalities.

A Union *B* and *A* Intersection *B*

The **union** of sets A and B, denoted $A \cup B$, is the set of elements that belong to set A or to set B or to both sets A and B.

The **intersection** of two sets A and B, denoted $A \cap B$, is the set of elements common to both A and B.

The concepts of the union and intersection of two sets are illustrated in Figure 9-1 and Figure 9-2:

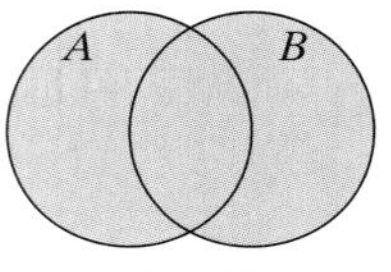

$A \cup B$
A union B
The elements in A *or* B *or* both

Figure 9-1

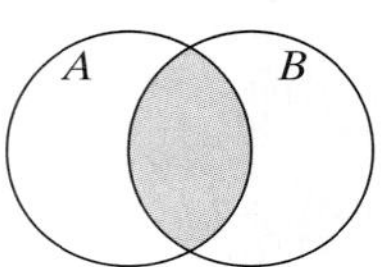

$A \cap B$
A intersection B
The elements in A *and* B

Figure 9-2

example 1 **Finding the Union and Intersection of Sets**

Given the sets: $A = \{a, b, c, d, e, f\}$ $B = \{a, c, e, g, i, k\}$ $C = \{g, h, i, j, k\}$

Find: a. $A \cup B$ b. $A \cap B$ c. $A \cap C$ d. $(A \cup B) \cap C$

Solution:

a. $A \cup B = \{a, b, c, d, e, f, g, i, k\}$

The union of A and B includes all the elements of A along with the elements of B. Notice that the elements a, c, and e are not listed twice.

b. $A \cap B = \{a, c, e\}$

The intersection of A and B includes only those elements that are common to both sets.

c. $A \cap C = \{\ \}$ (the empty set)

Because A and C share no common elements, the intersection of A and C is the empty, or null, set.

Tip: The empty set may be denoted by the symbol { } or by the symbol ∅.

d. $(A \cup B) \cap C$
$= \{a, b, c, d, e, f, g, i, k\} \cap \{g, h, i, j, k\}$
$= \{g, i, k\}$

The set C is intersected with $A \cup B$. The elements g, i, and k are common to both sets.

Tip: To confirm the solution to Example 1(d), we can set up a diagram illustrating the relationship among the sets A, B, and C.

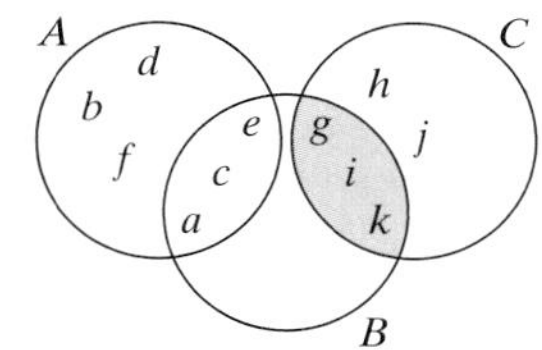

2. Solving Compound Inequalities

The solution to two inequalities joined by the word *And* is the intersection of their solution sets. The solution to two inequalities joined by the word *Or* is the union of their solution sets.

Steps to Solve a Compound Inequality

1. Solve and graph each inequality separately.
2. • If the inequalities are joined by the word *and*, find the intersection of the two solution sets.
   • If the inequalities are joined by the word *or*, find the union of the two solution sets.
3. Express the solution set in interval notation or in set-builder notation.

3. Solving Compound Inequalities (And)

example 2 **Solving Compound Inequalities (And)**

Solve the inequalities. Write the solution set in set-builder notation and in interval notation.

a. $-2x < 6$ and $x + 5 \le 7$

b. $4.4a + 3.1 < -12.3$ and $-2.8a + 9.1 < -6.3$

Avoiding Mistakes

Recall from Section 2.8 that multiplying or dividing an inequality by a negative factor reverses the direction of the inequality sign.

Solution:

a. $-2x < 6$ and $x + 5 \le 7$ — Solve each equation separately.

$\dfrac{-2x}{-2} > \dfrac{6}{-2}$ and $x \le 2$ — Reverse the first inequality sign.

$x > -3$ and $x \le 2$

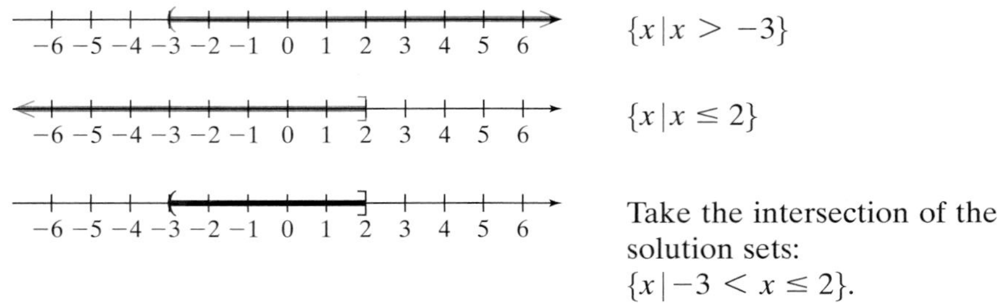

$\{x \mid x > -3\}$

$\{x \mid x \leq 2\}$

Take the intersection of the solution sets: $\{x \mid -3 < x \leq 2\}$.

The solution is $\{x \mid -3 < x \leq 2\}$ or equivalently in interval notation, $(-3, 2]$.

b. $4.4a + 3.1 < -12.3$ and $-2.8a + 9.1 < -6.3$

$4.4a < -15.4$ and $-2.8a < -15.4$ Solve each inequality separately.

$\dfrac{4.4a}{4.4} < \dfrac{-15.4}{4.4}$ and $\dfrac{-2.8a}{-2.8} > \dfrac{-15.4}{-2.8}$ Reverse the second inequality sign.

$a < -3.5$ and $a > 5.5$

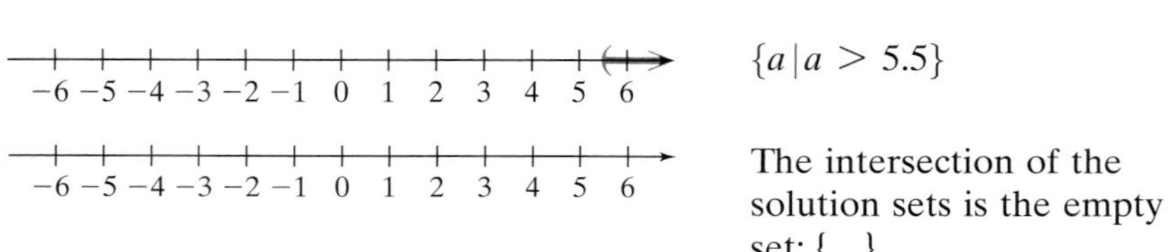

$\{a \mid a < -3.5\}$

$\{a \mid a > 5.5\}$

The intersection of the solution sets is the empty set: { }.

There are no real numbers that are simultaneously less than -3.5 and greater than 5.5. Hence, there is no solution.

In Section 2.8, we learned that the inequality $a < x < b$ is the intersection of two simultaneous conditions implied on x.

$$a < x < b$$

is equivalent to

$$a < x \quad \text{and} \quad x < b$$

example 3 **Solving Compound Inequalities (And)**

Solve the inequality: $-2 \leq -3x + 1 < 5$

Solution:

$$-2 \le -3x + 1 < 5$$

$-2 \le -3x + 1$	and	$-3x + 1 < 5$	Set up the intersection of two inequalities.
$-3 \le -3x$	and	$-3x < 4$	Solve each inequality.
$\dfrac{-3}{-3} \ge \dfrac{-3x}{-3}$	and	$\dfrac{-3x}{-3} > \dfrac{4}{-3}$	Reverse the direction of the inequality signs.
$1 \ge x$	and	$x > -\dfrac{4}{3}$	
$x \le 1$	and	$x > -\dfrac{4}{3}$	Rewrite the inequalities.

$$-\frac{4}{3} < x \le 1$$

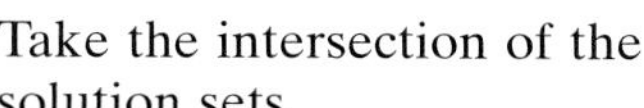

Take the intersection of the solution sets.

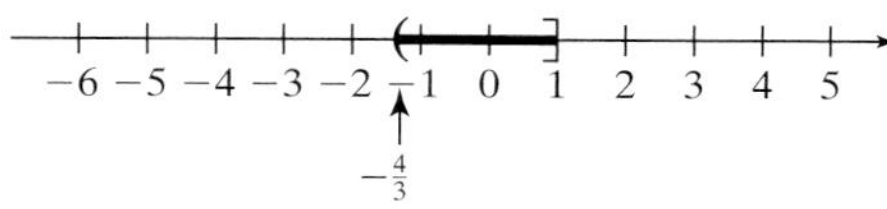

The solution is $\{x \mid -\frac{4}{3} < x \le 1\}$ or equivalently in interval notation, $(-\frac{4}{3}, 1]$.

Tip: As an alternative approach to Example 3, we can isolate the variable, x, in the "middle" portion of the inequality. Recall that the operations performed on the middle part of the inequality must also be performed on the left- and right-hand sides.

$-2 \le -3x + 1 < 5$	
$-2 - 1 \le -3x + 1 - 1 < 5 - 1$	Subtract 1 from all three parts of the inequality.
$-3 \le -3x < 4$	Simplify.
$\dfrac{-3}{-3} \ge \dfrac{-3x}{-3} > \dfrac{4}{-3}$	Divide by -3 in all three parts of the inequality. (Remember to reverse inequality signs.)
$1 \ge x > -\dfrac{4}{3}$	Simplify.
$-\dfrac{4}{3} < x \le 1$	Rewrite the inequality.

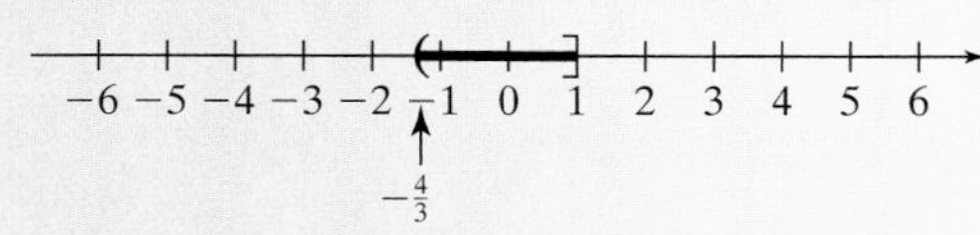

The solution is $\{x \mid -\frac{4}{3} < x \le 1\}$ or equivalently in interval notation, $(-\frac{4}{3}, 1]$.

4. Solving Compound Inequalities (Or)

example 4 Solving Compound Inequalities (Or)

Solve the inequalities. Write the solution set in set-builder notation and in interval notation.

a. $-3y - 5 > 4 \quad \text{or} \quad 4 - y < 6$
b. $4x + 3 < 16 \quad \text{or} \quad -2x < 3$

Solution:

a. $-3y - 5 > 4 \quad \text{or} \quad 4 - y < 6$

$-3y > 9 \quad \text{or} \quad -y < 2$ — Solve each inequality separately.

$\frac{-3y}{-3} < \frac{9}{-3} \quad \text{or} \quad \frac{-y}{-1} > \frac{2}{-1}$ — Reverse the inequality signs.

$y < -3 \quad \text{or} \quad y > -2$

$\{y \mid y < -3\}$

$\{y \mid y > -2\}$

Take the union of the solution sets $\{y \mid y < -3 \text{ or } y > -2\}$.

The solution is $\{y \mid y < -3 \text{ or } y > -2\}$ or equivalently in interval notation, $(-\infty, -3) \cup (-2, \infty)$.

b. $4x + 3 < 16 \quad \text{or} \quad -2x < 3$

$4x < 13 \quad \text{or} \quad x > -\frac{3}{2}$ — Solve each inequality separately.

$x < \frac{13}{4} \quad \text{or} \quad x > -\frac{3}{2}$

$\{x \mid x < \frac{13}{4}\}$

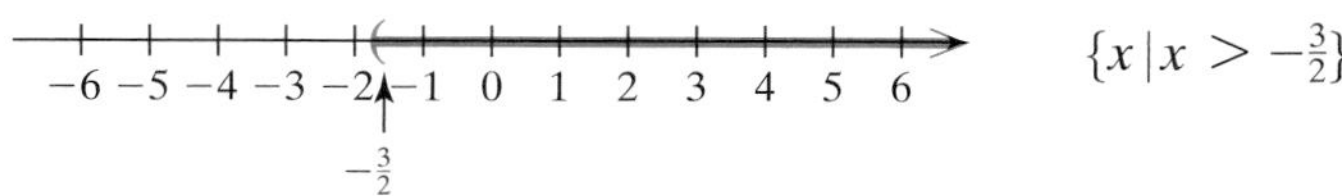

$\{x \mid x > -\frac{3}{2}\}$

Take the union of the solution sets.

The union of the solution sets is $\{x \mid x \text{ is any real number}\}$ or equivalently, $(-\infty, \infty)$.

5. Applications of Compound Inequalities

Compound inequalities are used in many applications, as shown in Examples 5 and 6.

example 5 Translating Compound Inequalities

The normal level of thyroid-stimulating hormone (TSH) for adults ranges from 0.4 μU/mL (microunits per milliliter) to 4.8 μU/mL. Let x represent the amount of TSH measured in microunits per milliliter.

a. Write an inequality representing the normal range of TSH.
b. Write a compound inequality representing abnormal TSH levels.

Solution:

a. $0.4 \le x \le 4.8$
b. $x < 0.4 \quad \text{or} \quad x > 4.8$

example 6 Translating and Solving a Compound Inequality

The sum of a number and 4 is between -5 and 12. Find all such numbers.

Solution:

Let x represent a number.

$-5 < x + 4 < 12$ Translate the inequality.

$-5 - 4 < x + 4 - 4 < 12 - 4$ Subtract 4 from all three parts of the inequality.

$-9 < x < 8$

The number may be any real number between -9 and 8: $\{x \mid -9 < x < 8\}$.

Tip: By convention for $a < b$, we will interpret the statement "x is between a and b" to exclude the endpoints a and b:

$a < x < b \quad (\text{not } a \le x \le b)$

section 9.1 Practice Exercises

For Exercises 1–6, review solving linear inequalities from Section 2.8. Write the answer in interval notation.

1. $6u + 5 > 2$
2. $-2 + 3z \le 4$
3. $-\frac{3}{4}p \le 12$
4. $-6q > -\frac{1}{3}$
5. $-1.5 < 0.1x - 8.1$
6. $4 \ge 2.6 + 7t$

For Exercises 7–8, find the intersection and union of sets as indicated. Write the answer in set notation.

7. Let $A = \{10, 20, 30, 40, 50\}$ and $B = \{5, 10, 15, 20, 25, 30, 35\}$.
   a. Find $A \cap B$.
   b. Find $A \cup B$.

8. Let $C = \{\text{a, b, c, d, e, f, g, h}\}$ and $D = \{\text{a, e, i, o, u}\}$.
   a. Find $C \cup D$.
   b. Find $C \cap D$.

9. Let $M = \{a, e, j, o, t, y\}$ and $N = \{j, k, l, m, n, o, p, q\}$.
   a. Find $M \cup N$.
   b. Find $M \cap N$.

10. Let $Q = \{2, 4, 6, 8, 10, 12\}$ and $P = \{2, 3, 4, 5, 6, 7, 8\}$.
   a. Find $Q \cup P$.
   b. Find $Q \cap P$.

11. Let $S = \{1, 2, 3, 4, 5\}$ and $T = \{6, 7, 8, 9, 10\}$.
   a. Find $S \cup T$.
   b. Find $S \cap T$.

12. Let $X = \{b, c, d, e, f\}$ and $Y = \{u, v, w, x, y, z\}$.
   a. Find $X \cup Y$.
   b. Find $X \cap Y$.

For Exercises 13–18, solve the inequalities and graph the solutions.

13. a. $4x > 8$
   b. $2 + x < 6$
   c. $4x > 8$ and $2 + x < 6$

14. a. $3x - 11 < 4$
   b. $4x + 9 > 1$
   c. $3x - 11 < 4$ and $4x + 9 > 1$

15. a. $-5 \le 3x - 4$
   b. $3x - 4 \le 8$
   c. $-5 \le 3x - 4 \le 8$

16. a. $-10 \le 3x + 2$
   b. $3x + 2 \le 17$
   c. $-10 \le 3x + 2 \le 17$

17. a. $4x - 7 < 1$
   b. $-3x + 7 > -8$
   c. $4x - 7 < 1$ and $-3x + 7 > -8$

18. a. $5 - 7x < 19$
   b. $2 - 3x < -4$
   c. $5 - 7x < 19$ and $2 - 3x < -4$

For Exercises 19–32, solve the inequality and graph the solution. Write the answer in interval notation.

19. $y - 7 \ge -9$ and $y + 2 \le 5$

20. $a + 6 > -2$ and $5a < 30$

21. $2t + 7 < 19$ and $5t + 13 > 28$

22. $5p + 2p \ge -21$ and $-9p + 3p \ge -24$

23. $0 \le 2b - 5 < 9$

24. $-6 < 3k - 9 \le 0$

25. $21k - 11 \le 6k + 19$ and $3k - 11 < -k + 7$

26. $6w - 1 > 3w - 11$ and $-3w + 7 \le 8w - 13$

27. $-1 < \frac{a}{6} \le 1$

28. $-3 \le \frac{1}{2}x < 0$

29. $-\frac{2}{3} < \frac{y - 4}{-6} < \frac{1}{3}$

30. $\frac{1}{3} > \frac{t - 4}{-3} > -2$

31. $\frac{2}{3}(2p - 1) \ge 10$ and $\frac{4}{5}(3p + 4) \ge 20$

32. $5(a + 3) + 9 < 2$ and $3(a - 2) + 6 < 10$

For Exercises 33–38, solve the inequalities and graph the solutions.

33. a. $-2x + 7 > 9$
   b. $3x + 1 < -14$
   c. $-2x + 7 > 9$ or $3x + 1 < -14$

34. a. $5 - 7x < 19$
   b. $2 - 3x < -4$
   c. $5 - 7x < 19$ or $2 - 3x < -4$

35. a. $5x + 8 \le 23$
   b. $2x - 15 \ge 1$
   c. $5x + 8 \le 23$ or $2x - 15 \ge 1$

36. a. $4x - 7 < 1$
   b. $-3x + 7 < -8$
   c. $4x - 7 < 1$ or $-3x + 7 > -8$

37. a. $-3x - 5 \le -11$
    b. $-7 \ge 5x - 2$
    c. $-3x - 5 \le -11$ or $-7 \ge 5x - 2$

38. a. $2x - 2 \ge 6$
    b. $3x - 5 \le 10$
    c. $2x - 2 \ge 6$ or $3x - 5 \le 10$

For Exercises 39–50, solve the inequality and graph the solution. Write the answer in interval notation.

39. $h + 4 < 0$ or $6h > -12$

40. $5y > 12$ or $y - 3 < -2$

41. $2y - 1 \ge 3$ or $y < -2$

42. $x < 0$ or $3x + 1 \ge 7$

43. $\frac{5}{3}v \ge 5$ or $-v - 6 > 1$

44. $\frac{3}{8}u + 1 < 0$ or $-2u \le -4$

45. $5(x - 1) \ge -5$ or $5 - x \le 11$

46. $-p + 7 \ge 10$ or $3(p - 1) \le 12$

47. $\frac{3t - 1}{10} > \frac{1}{2}$ or $\frac{3t - 1}{10} < -\frac{1}{2}$

48. $\frac{6 - x}{12} > \frac{1}{4}$ or $\frac{6 - x}{12} < -\frac{1}{6}$

49. $0.5w + 5 < 2.5w - 4$ or $0.3w \le -0.1w - 1.6$

50. $1.25a + 3 \le 0.5a - 6$ or $2.5a - 1 \ge 9 - 1.5a$

For Exercises 51–54, solve the compound inequalities and graph the solutions.

51. a. $3x - 5 < 19$ and $-2x + 3 < 23$
    b. $3x - 5 < 19$ or $-2x + 3 < 23$

52. a. $0.5(6x + 8) > 0.8x - 7$ and $4(x + 1) < 7.2$
    b. $0.5(6x + 8) > 0.8x - 7$ or $4(x + 1) < 7.2$

53. a. $8x - 4 \ge 6.4$ or $0.3(x + 6) \le -0.6$
    b. $8x - 4 \ge 6.4$ and $0.3(x + 6) \le -0.6$

54. a. $-2r + 4 \le -8$ or $3r + 5 \le 8$
    b. $-2r + 4 \le -8$ and $3r + 5 \le 8$

55. The normal number of white blood cells for human blood is between 4800 and 10,800 cells per cubic millimeter, inclusive. Let x represent the number of white blood cells per cubic millimeter (mm^3).
    a. Write an inequality representing the normal range of white blood cells per cubic millimeter.
    b. Write a compound inequality representing abnormal levels of white blood cells per cubic millimeter.

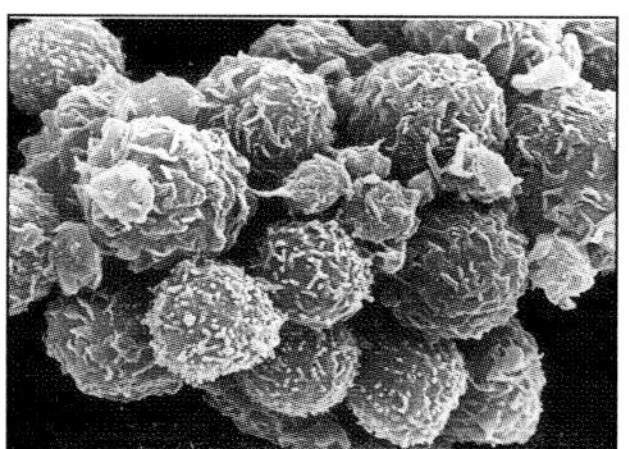

56. The normal number of platelets in human blood is between 2.0×10^5 and 3.5×10^5 platelets per cubic millimeter, inclusive. Let x represent the number of platelets per cubic millimeter.
    a. Write an inequality representing a normal platelet count per cubic millimeter.
    b. Write a compound inequality representing abnormal platelet counts per cubic millimeter.

57. Normal hemoglobin levels in human blood for adult males is between 13 and 16 g/dL (grams per deciliter), inclusive. Let x represent the level of hemoglobin measured in grams per deciliter.
    a. Write an inequality representing normal hemoglobin levels for adult males.
    b. Write a compound inequality representing abnormal levels of hemoglobin for adult males.

58. Normal hemoglobin levels in human blood for adult females is between 12 and 15 g/dL, inclusive. Let x represent the level of hemoglobin measured in grams per deciliter.
    a. Write an inequality representing normal hemoglobin levels for adult females.
    b. Write a compound inequality representing abnormal levels of hemoglobin for adult females.

59. Twice a number is between -3 and 12. Find all such numbers.

60. The difference of a number and 6 is between 0 and 8. Find all such numbers.

61. One plus twice a number is either greater than 5 or less than -1. Find all such numbers.

62. One third of a number is either less than -2 or greater than 5. Find all such numbers.

---

section

9.2 Polynomial and Rational Inequalities

Concepts

1. Solving Inequalities Graphically
2. Test Point Method
3. Solving Polynomial Inequalities Using the Test Point Method
4. Solving Rational Inequalities Using the Test Point Method
5. Inequalities with "Special Case" Solution Sets

1. Solving Inequalities Graphically

In Sections 2.8 and 9.1, we solved simple and compound linear inequalities. In this section we will solve polynomial and rational inequalities. We begin by defining a quadratic inequality.

Quadratic inequalities are inequalities that can be written in any of the following forms:

$$ax^2 + bx + c \geq 0 \qquad ax^2 + bx + c \leq 0$$

$$ax^2 + bx + c > 0 \qquad ax^2 + bx + c < 0 \quad \text{where } (a \neq 0)$$

Recall from Section 7.5 that a quadratic function defined by $f(x) = ax^2 + bx + c$ $(a \neq 0)$ is a parabola that opens upward or downward. The quadratic inequality $f(x) > 0$ or equivalently $ax^2 + bx + c > 0$ is asking the question "For what values of x is the function positive (above the x-axis)?" The inequality $f(x) < 0$ or equivalently $ax^2 + bx + c < 0$ is asking "For what values of x is the function negative (below the x-axis)?" The graph of a quadratic function can be used to answer these questions.

example 1 **Using a Graph to Solve a Quadratic Inequality**

Use the graph of $f(x) = x^2 - 6x + 8$ in Figure 9-3 to solve the inequalities:

a. $x^2 - 6x + 8 < 0$ b. $x^2 - 6x + 8 > 0$

Solution:

From Figure 9-3, we see that the graph of $f(x) = x^2 - 6x + 8$ is a parabola opening upward. The function factors as $f(x) = (x - 2)(x - 4)$. The x-intercepts are at $x = 2$ and $x = 4$, and the y-intercept is $(0, 8)$. Because the function opens up, the vertex is below the x-axis.

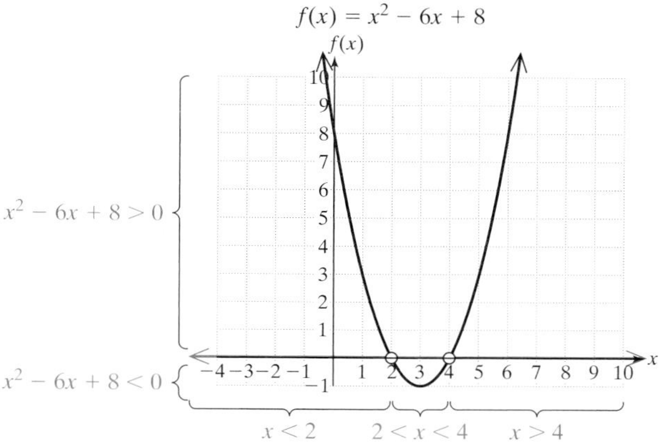

Figure 9-3

a. The solution to $x^2 - 6x + 8 < 0$ is the set of all real numbers x for which $f(x) < 0$. Graphically, this is the set of all x-values corresponding to the points where the parabola is below the x-axis (shown in red).

$$\text{Hence } x^2 - 6x + 8 < 0 \qquad \text{for} \qquad \{x \mid 2 < x < 4\}$$

b. The solution to $x^2 - 6x + 8 > 0$ is the set of x-values for which $f(x) > 0$. This is the set of x-values where the parabola is above the x-axis (shown in blue).

$$\text{Hence } x^2 - 6x + 8 > 0 \qquad \text{for} \qquad \{x \mid x < 2 \quad \text{or} \quad x > 4\}$$

Notice that the points $x = 2$ and $x = 4$ define the boundaries of the solution sets to the inequalities in Example 1. These values are the solutions to the related equation, $x^2 - 6x + 8 = 0$.

Tip: The inequalities in Example 1 are strict inequalities. Therefore, the points $x = 2$ and $x = 4$ (where $f(x) = 0$) are not included in the solution set. However, the corresponding inequalities using the symbols $\leq$ and $\geq$ do include the points where $f(x) = 0$. Hence,

The solution to $x^2 - 6x + 8 \leq 0$ is $\{x \mid 2 \leq x \leq 4\}$.

The solution to $x^2 - 6x + 8 \geq 0$ is $\{x \mid x \leq 2 \text{ or } x \geq 4\}$.

example 2 Using a Graph to Solve a Rational Inequality

Use the graph of $g(x) = \dfrac{1}{x + 1}$ in Figure 9-4 to solve the inequalities:

a. $\dfrac{1}{x + 1} < 0$

b. $\dfrac{1}{x + 1} > 0$

Solution:

a. The graph of $g(x) = \dfrac{1}{x + 1}$ shown in Figure 9-4 indicates that $g(x)$ is below the x-axis for $x < -1$ (shown in red). Therefore, the solution to $\dfrac{1}{x + 1} < 0$ is $\{x \mid x < -1\}$.

b. The graph of $g(x) = \dfrac{1}{x + 1}$ shown in Figure 9-4 indicates that $g(x)$ is above the x-axis for $x > -1$ (shown in blue). Therefore, the solution to $\dfrac{1}{x + 1} > 0$ is $\{x \mid x > -1\}$.

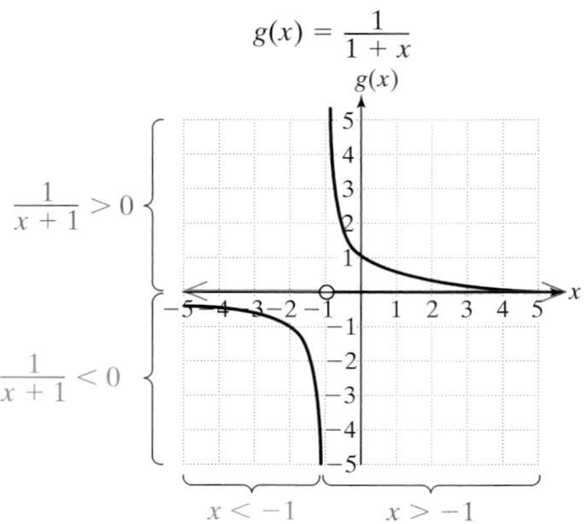

Figure 9-4

Notice that the point $x = -1$ defines the boundary of the solution sets to the inequalities in Example 2. The point $x = -1$ is a point where the inequality is undefined.

2. Test Point Method

The **boundary points** of an inequality consist of the real solutions to the related equation and the points where the inequality is undefined. Examples 1 and 2 demonstrate that the boundary points of an inequality provide the boundaries of the solution set. This is the basis of the **test point method** to solve inequalities.

Solving Inequalities Using the Test Point Method

1. Find the boundary points of the inequality.
2. Plot the boundary points on the number line. This divides the number line into regions.
3. Select a test point from each region and substitute it into the original inequality.
   - If a test point makes the original inequality true, then that region is part of the solution set.
4. Test the boundary points in the original inequality.
   - If a boundary point makes the original inequality true, then that point is part of the solution set.

3. Solving Polynomial Inequalities Using the Test Point Method

example 3 **Solving Polynomial Inequalities Using the Test Point Method**

Solve the inequalities using the test point method.

a. $2x^2 + 5x < 12$

b. $x(x - 2)(x + 4)^2(x - 4) > 0$

Solution:

a. $2x^2 + 5x < 12$

Step 1: Find the boundary points. Because polynomials are defined for all values of x, the only boundary points are the real solutions to the related equation.

$2x^2 + 5x = 12$

Solve the related equation.

$2x^2 + 5x - 12 = 0$

$(2x - 3)(x + 4) = 0$

$x = \frac{3}{2}, \; x = -4$

The boundary points are $\frac{3}{2}$ and -4.

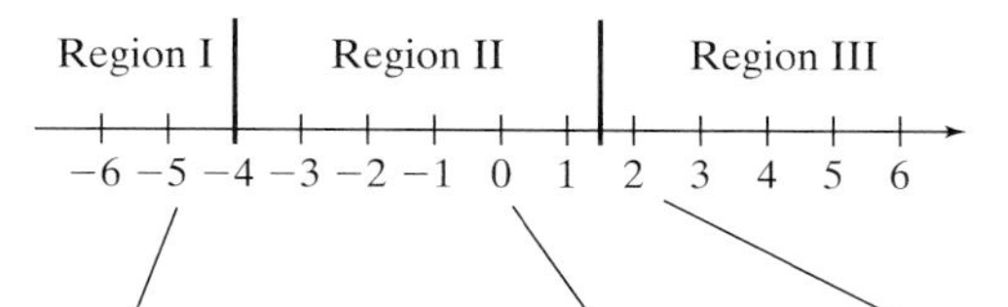

Step 2: Plot the boundary points.

Step 3: Select a test point from each region.

Test $x = -5$

$2x^2 + 5x < 12$

$2(-5)^2 + 5(-5) \stackrel{?}{<} 12$

$50 - 25 \stackrel{?}{<} 12$

$25 \stackrel{?}{<} 12$ False

Test $x = 0$

$2x^2 + 5x < 12$

$2(0)^2 + 5(0) \stackrel{?}{<} 12$

$0 \stackrel{?}{<} 12$

$0 \stackrel{?}{<} 12$ True

Test $x = 2$

$2x^2 + 5x < 12$

$2(2)^2 + 5(2) \stackrel{?}{<} 12$

$8 + 10 \stackrel{?}{<} 12$

$18 \stackrel{?}{<} 12$ False

Test $x = -4$

$$2x^2 + 5x < 12$$

$$2(-4)^2 + 5(-4) \overset{?}{<} 12$$

$$32 - 20 \overset{?}{<} 12$$

$$12 \overset{?}{<} 12 \quad \text{False}$$

Test $x = \frac{3}{2}$

$$2x^2 + 5x < 12$$

$$2\left(\frac{3}{2}\right)^2 + 5\left(\frac{3}{2}\right) \overset{?}{<} 12$$

$$2\left(\frac{9}{4}\right) + \frac{15}{2} \overset{?}{<} 12$$

$$\frac{9}{2} + \frac{15}{2} \overset{?}{<} 12$$

$$\frac{24}{2} \overset{?}{<} 12 \quad \text{False}$$

Step 4: Test the boundary points.

Tip: The strict inequality, $<$, excludes values of x for which $2x^2 + 5x = 12$. This implies that the boundary points are not included in the solution set.

Neither boundary point makes the inequality true. Therefore, the boundary points are not included in the solution set.

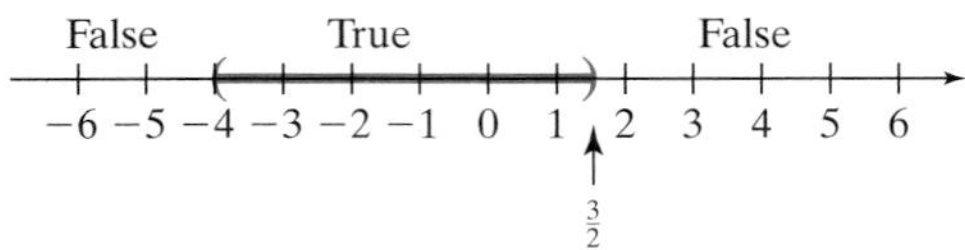

The solution is $\{x \mid -4 < x < \frac{3}{2}\}$ or equivalently in interval notation: $(-4, \frac{3}{2})$.

Calculator Connections

Graph $Y_1 = 2x^2 + 5x$ and $Y_2 = 12$. Notice that $Y_1 < Y_2$ for $\{x \mid -4 < x < \frac{3}{2}\}$.

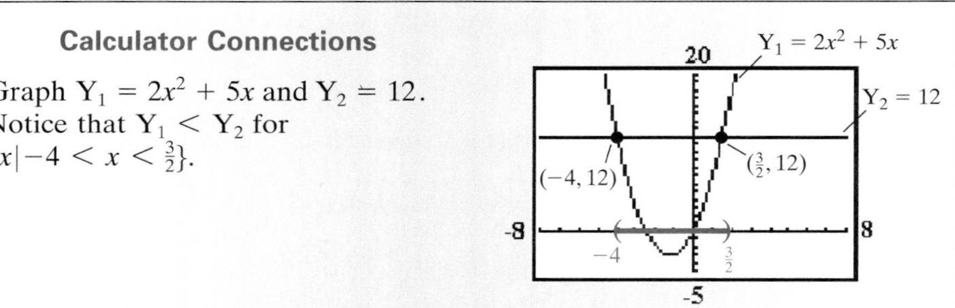

b. $x(x - 2)(x + 4)^2(x - 4) > 0$

$$x(x - 2)(x + 4)^2(x - 4) = 0$$

$$x = 0, x = 2, x = -4, x = 4$$

Step 1: Find the boundary points.

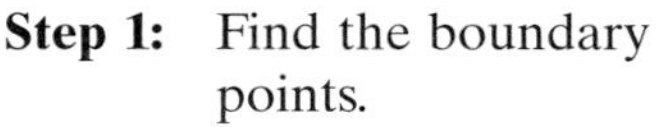

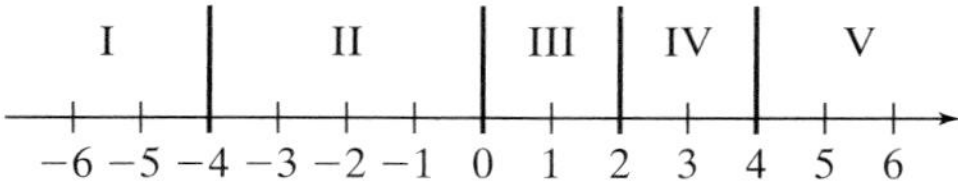

Step 2: Plot the boundary points.

Step 3: Select a test point from each region.

Test $x = -5$: $-5(-5 - 2)(-5 + 4)^2(-5 - 4) \overset{?}{>} 0$ $\quad -315 \overset{?}{>} 0$ False

Test $x = -1$: $-1(-1 - 2)(-1 + 4)^2(-1 - 4) \overset{?}{>} 0$ $\quad -135 \overset{?}{>} 0$ False

Test $x = 1$: $1(1 - 2)(1 + 4)^2(1 - 4) \overset{?}{>} 0$ $\quad 75 \overset{?}{>} 0$ True

Test $x = 3$: $3(3 - 2)(3 + 4)^2(3 - 4) \overset{?}{>} 0$ $\quad -147 \overset{?}{>} 0$ False

Test $x = 5$: $5(5 - 2)(5 + 4)^2(5 - 4) \overset{?}{>} 0$ $\quad 1215 \overset{?}{>} 0$ True

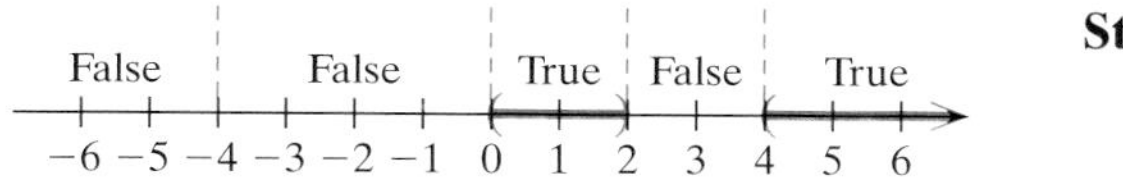

Step 4: The boundary points are not included because the inequality, $>$, is strict.

The solution is $\{x \mid 0 < x < 2 \text{ or } x > 4\}$, or equivalently in interval notation: $(0, 2) \cup (4, \infty)$.

Calculator Connections

Graph $Y_1 = x(x - 2)(x + 4)^2(x - 4)$. Y_1 is positive (above the x-axis) for $\{x \mid 0 < x < 2 \text{ or } x > 4\}$ or equivalently $(0, 2) \cup (4, \infty)$.

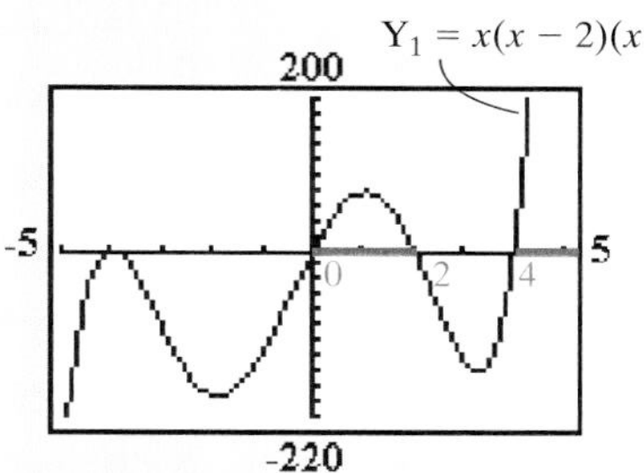

4. Solving Rational Inequalities Using the Test Point Method

The test point method can be used to solve rational inequalities. However, the solution set to a rational inequality must exclude all values of the variable that make the inequality undefined. That is, exclude all values that make the denominator equal to zero for any rational expression in the inequality.

example 4 **Solving a Rational Inequality Using the Test Point Method**

Solve the inequality using the test point method. $\dfrac{x + 2}{x - 4} \leq 3$

Solution:

$$\frac{x + 2}{x - 4} \leq 3$$

Step 1: Find the boundary points. Note that the inequality is undefined for $x = 4$. Hence $x = 4$ is automatically a boundary point. To find any other boundary points, solve the related equation.

$$\frac{x+2}{x-4} = 3 \quad \text{(related equation)}$$

$$(x-4)\left(\frac{x+2}{x-4}\right) = (x-4)(3) \qquad \text{Clear fractions.}$$

$$x + 2 = 3(x - 4) \qquad \text{Solve for } x.$$

$$x + 2 = 3x - 12$$

$$-2x = -14$$

$$x = 7$$

The solution to the related equation is $x = 7$, and the inequality is undefined for $x = 4$. Therefore, the boundary points are $x = 4$ and $x = 7$.

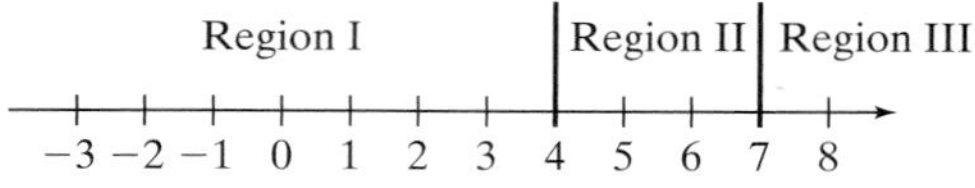

Step 2: Plot boundary points.

Step 3: Select test points.

Test $x = 0$

$$\frac{x+2}{x-4} \le 3$$

$$\frac{0+2}{0-4} \stackrel{?}{\le} 3$$

$$-\frac{1}{2} \stackrel{?}{\le} 3 \quad \text{True}$$

Test $x = 5$

$$\frac{x+2}{x-4} \le 3$$

$$\frac{5+2}{5-4} \stackrel{?}{\le} 3$$

$$\frac{7}{1} \stackrel{?}{\le} 3 \quad \text{False}$$

Test $x = 8$

$$\frac{x+2}{x-4} \le 3$$

$$\frac{8+2}{8-4} \stackrel{?}{\le} 3$$

$$\frac{10}{4} \stackrel{?}{\le} 3$$

$$\frac{5}{2} \le 3 \quad \text{or} \quad 2\frac{1}{2} \le 3 \quad \text{True}$$

Step 4: Test the boundary points.

Test $x = 4$:

$$\frac{x+2}{x-4} \le 3$$

$$\frac{4+2}{4-4} \stackrel{?}{\le} 3$$

$$\frac{6}{0} \stackrel{?}{\le} 3 \quad \text{Undefined}$$

Test $x = 7$:

$$\frac{x+2}{x-4} \le 3$$

$$\frac{7+2}{7-4} \stackrel{?}{\le} 3$$

$$\frac{9}{3} \stackrel{?}{\le} 3 \quad \text{True}$$

The boundary point $x = 4$ cannot be included in the solution set, because it is undefined in the inequality. The boundary point $x = 7$ makes the original inequality true and must be included in the solution set.

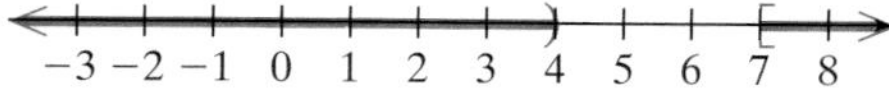

The solution is $\{x \mid x < 4 \text{ or } x \ge 7\}$ or equivalently in interval notation: $(\infty, 4) \cup [7, \infty)$.

Calculator Connections

Graph $Y_1 = \dfrac{x + 2}{x - 4}$ and $Y_2 = 3$.

Y_1 has a vertical asymptote at $x = 4$. Furthermore, $Y_1 = Y_2$ at $x = 7$. $Y_1 \leq Y_2$ (that is, Y_1 is below Y_2) for $x < 4$ and for $x \geq 7$.

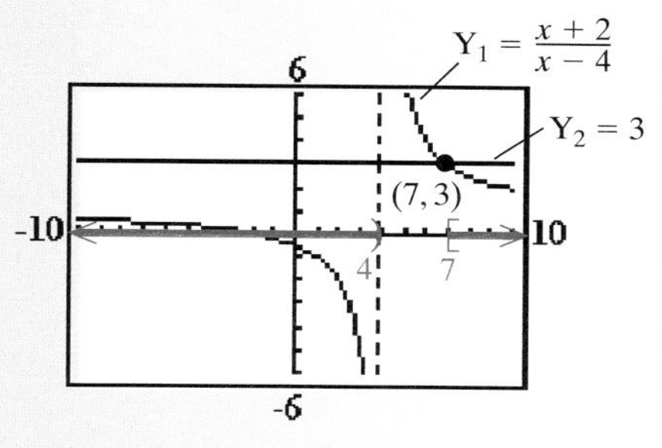

5. Inequalities with "Special Case" Solution Sets

The solution to an inequality is often one or more regions on the real number line. Sometimes, however, the solution to an inequality may be a single point on the number line, the empty set, or the set of all real numbers.

example 5 **Solving Inequalities**

Solve the inequalities.

a. $-\dfrac{16}{x^2 + 2} < 0$ b. $-\dfrac{16}{x^2 + 2} > 0$

Solution:

a. Since the expressions 16 and $x^2 + 2$ are greater than zero for all real numbers, x, then their ratio is positive for all real numbers, x. The opposite of their ratio, $-\dfrac{16}{x^2 + 2}$ will be negative for all values of x. That is, $-\dfrac{16}{x^2 + 2} < 0$ for all real numbers, x.

The solution is all real numbers, $(-\infty, \infty)$

b. Because $-\dfrac{16}{x^2 + 2} < 0$ for all real numbers, x, then there are no values of x for which $-\dfrac{16}{x^2 + 2} > 0$.

The inequality $-\dfrac{16}{x^2 + 2} > 0$ has no solution.

Calculator Connections

The graph of $Y_1 = -\frac{16}{x^2 + 2}$ is below the x-axis for all x-values on the viewing window. Therefore $-\frac{16}{x^2 + 2} < 0$ for all x on the display window. Furthermore, there are no values of x for which $-\frac{16}{x^2 + 2} \geq 0$

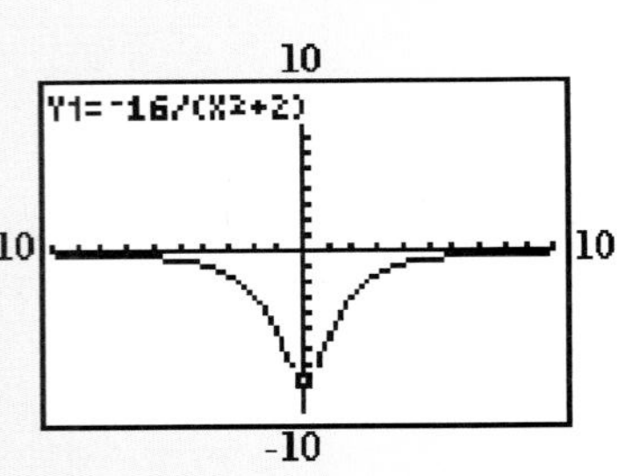

example 6 Solving Inequalities

Solve the inequalities.

a. $x^2 + 6x + 9 \geq 0$

b. $x^2 + 6x + 9 > 0$

c. $x^2 + 6x + 9 \leq 0$

d. $x^2 + 6x + 9 < 0$

Solution:

a. $x^2 + 6x + 9 \geq 0$ Notice that $x^2 + 6x + 9$ is a perfect square trinomial.

$(x + 3)^2 \geq 0$ Factor: $x^2 + 6x + 9 = (x + 3)^2$.

The quantity $(x + 3)^2$ is a perfect square and is greater than or equal to zero for all real numbers, x. The solution is all real numbers, $(-\infty, \infty)$.

True ↓ True True

−5 −4 −3 −2 −1 0 1 2 3 4 5

Tip: The graph of $f(x) = x^2 + 6x + 9$ or equivalently, $f(x) = (x + 3)^2$ is equal to zero at $x = 3$ and positive (above the x-axis) for all other values of x on its domain.

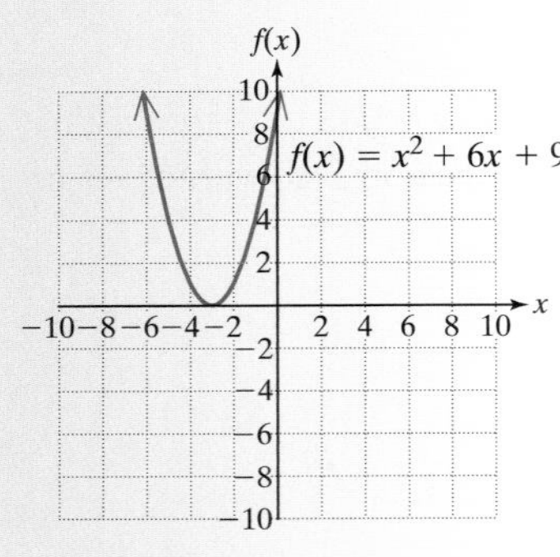

b. $x^2 + 6x + 9 > 0$

$(x + 3)^2 > 0$ This is the same inequality as in part (a) with the exception that the inequality is strict. The solution set does not include the point where $x^2 + 6x + 9 = 0$. Therefore, the boundary point $x = -3$ is *not* included in the solution set.

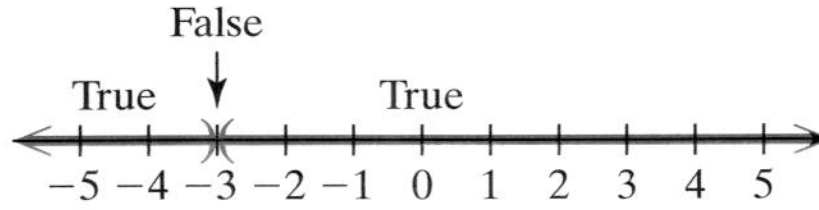

The solution set is $\{x \mid x < -3 \text{ or } x > -3\}$ or equivalently $(-\infty, -3) \cup (-3, \infty)$.

c. $x^2 + 6x + 9 \leq 0$

$(x + 3)^2 \leq 0$

A perfect square cannot be less than zero. However, $(x + 3)^2$ is equal to zero at $x = -3$. Therefore, the solution set is $\{-3\}$.

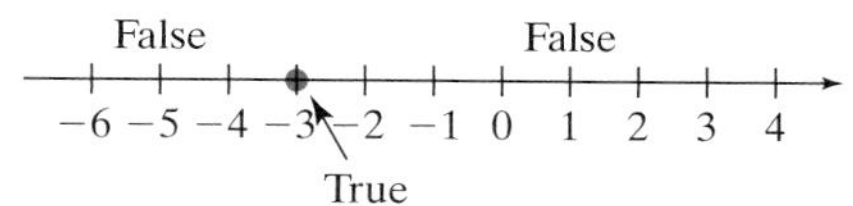

d. $x^2 + 6x + 9 < 0$

$$(x + 3)^2 < 0$$

A perfect square cannot be negative; therefore, there are no real numbers, x, such that $(x + 3)^2 < 0$. There is no solution.

section 9.2 Practice Exercises

For Exercises 1–8, solve the compound inequalities. Write the solutions in interval notation.

1. $6x - 10 > 8$ or $8x + 2 < 5$
2. $3(a - 1) + 2 > 0$ or $2a > 5a + 12$
3. $5(k - 2) > -25$ and $7(1 - k) > 7$
4. $2y + 4 \geq 10$ and $5y - 3 \leq 13$
5. $2t - 7 > 3$ and $-3t < 4$
6. $2.7c - 1.1 \leq 7$ or $9 \geq 4c$
7. $\frac{3}{2}h - 1 < 0$ or $h + 2 > \frac{7}{4}$
8. $\frac{1}{3}p + 1 \geq p$ and $p - \frac{3}{4} \geq 5$

For Exercises 9–12, estimate from the graph the intervals for which the inequality is true.

9.

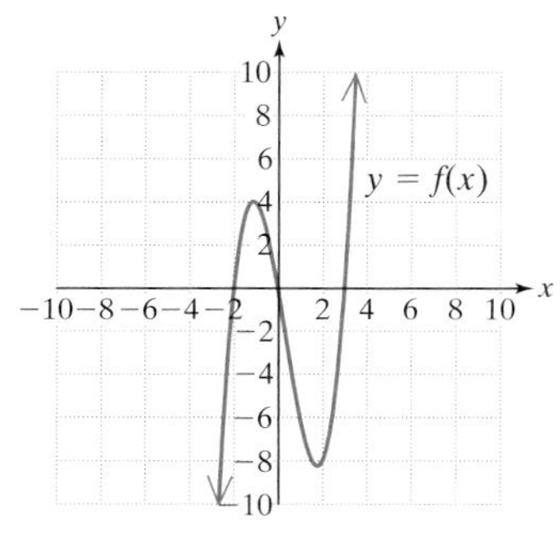

a. $f(x) > 0$ b. $f(x) < 0$
c. $f(x) \leq 0$ d. $f(x) \geq 0$

10.

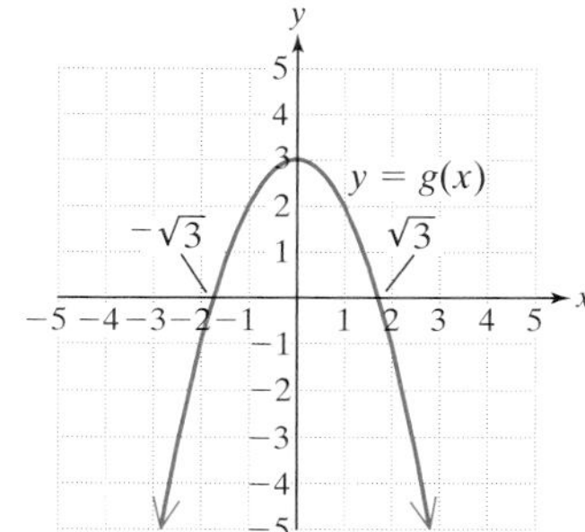

a. $g(x) < 0$ b. $g(x) > 0$
c. $g(x) \geq 0$ d. $g(x) \leq 0$

11.

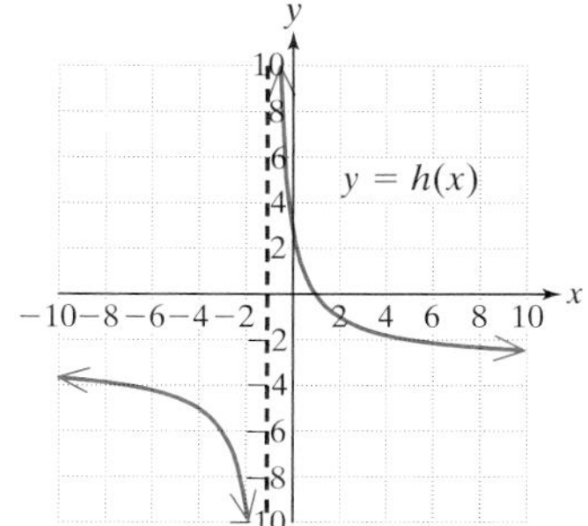

a. $h(x) \geq 0$ b. $h(x) \leq 0$
c. $h(x) < 0$ d. $h(x) > 0$

12.

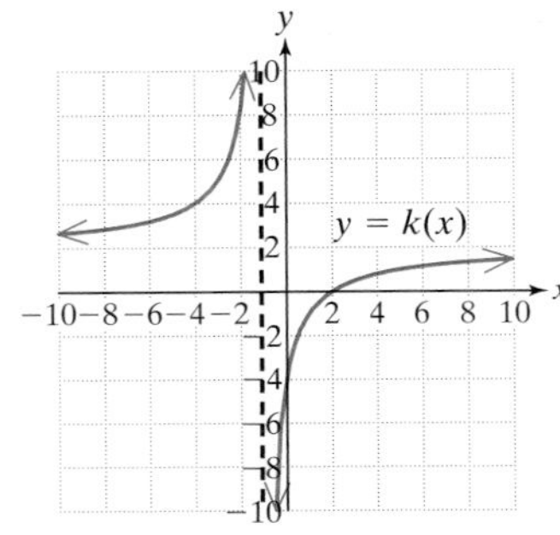

a. $k(x) \le 0$ b. $k(x) \ge 0$

c. $k(x) > 0$ d. $k(x) < 0$

For Exercises 13–20, solve the equation and related inequalities.

13. a. $3(2b - 4) - b = 5 - b$
 b. $3(2b - 4) - b < 5 - b$
 c. $3(2b - 4) - b > 5 - b$

14. a. $(a - 6) - (3a + 2) = a + 10$
 b. $(a - 6) - (3a + 2) < a + 10$
 c. $(a - 6) - (3a + 2) > a + 10$

15. a. $\frac{1}{2}y + 3 = \frac{2}{3}y$
 b. $\frac{1}{2}y + 3 \le \frac{2}{3}y$
 c. $\frac{1}{2}y + 3 \ge \frac{2}{3}y$

16. a. $-\frac{3}{2}t = \frac{1}{2}t - \frac{15}{8}$
 b. $-\frac{3}{2}t \le \frac{1}{2}t - \frac{15}{8}$
 c. $-\frac{3}{2}t \ge \frac{1}{2}t - \frac{15}{8}$

17. a. $3w(w + 4) = 10 - w$
 b. $3w(w + 4) < 10 - w$
 c. $3w(w + 4) > 10 - w$

18. a. $x^2 + 7x = 30$
 b. $x^2 + 7x < 30$
 c. $x^2 + 7x > 30$

19. a. $q^2 - 4q = 5$
 b. $q^2 - 4q \le 5$
 c. $q^2 - 4q \ge 5$

20. a. $2p(p - 2) = p + 3$
 b. $2p(p - 2) \le p + 3$
 c. $2p(p - 2) \ge p + 3$

For Exercises 21–36, solve the polynomial inequality. Write the answer in interval notation.

21. $(t - 7)(t + 1) < 0$

22. $(p - 4)(p + 2) > 0$

23. $(5y - 3)(y - 8) > 0$

24. $(2t + 5)(t - 6) < 0$

25. $a^2 - 12a \le -32$

26. $w^2 + 20w \ge -64$

27. $b^2 - 121 < 0$

28. $c^2 - 25 < 0$

29. $3p^2 - 8p - 3 \ge 0$

30. $2t^2 + 5t - 12 \le 0$

31. $2x(x - 4)(3x + 1) > 0$

32. $-y(2y - 3)(y + 3) < 0$

33. $x^3 - x^2 \le 12x$

34. $x^3 + 36 > 4x^2 + 9x$

35. $w^3 + w^2 > 4w + 4$

36. $2p^3 - 5p^2 \le 3p$

For Exercises 37–40, solve the equation and related inequalities.

37. a. $\frac{10}{x - 5} = 5$
 b. $\frac{10}{x - 5} < 5$
 c. $\frac{10}{x - 5} > 5$

38. a. $\frac{8}{a + 1} = 4$
 b. $\frac{8}{a + 1} > 4$
 c. $\frac{8}{a + 1} < 4$

39. a. $\dfrac{z+2}{z-6} = -3$
    b. $\dfrac{z+2}{z-6} \le -3$
    c. $\dfrac{z+2}{z-6} \ge -3$

40. a. $\dfrac{w-8}{w+6} = 2$
    b. $\dfrac{w-8}{w+6} \le 2$
    c. $\dfrac{w-8}{w+6} \ge 2$

For Exercises 41–52, solve the rational inequalities. Write the answer in interval notation.

41. $\dfrac{2}{x-1} \ge 0$

42. $\dfrac{-3}{x+2} \le 0$

43. $\dfrac{a+1}{a-3} < 0$

44. $\dfrac{b+4}{b-4} > 0$

45. $\dfrac{3}{2x-7} < -1$

46. $\dfrac{8}{4x+9} > 1$

47. $\dfrac{x+1}{x-5} \ge 4$

48. $\dfrac{x-2}{x+6} \le 5$

49. $\dfrac{1}{x} \le 2$

50. $\dfrac{1}{x} \ge 3$

51. $\dfrac{(x+2)^2}{x} > 0$

52. $\dfrac{(x-3)^2}{x} < 0$

For Exercises 53–64, solve the inequalities.

53. $x^2 + 10x + 25 \ge 0$

54. $x^2 + 6x + 9 < 0$

55. $x^2 + 2x + 1 < 0$

56. $x^2 + 8x + 16 \ge 0$

57. $\dfrac{x^2}{x^2+4} < 0$

58. $\dfrac{x^2}{x^2+4} \ge 0$

59. $x^4 + 3x^2 \le 0$

60. $x^4 + 2x^2 \le 0$

61. $x^2 + 4x + 4 > 0$

62. $x^2 + 12x + 36 < 0$

63. $x^2 + 4x + 4 \le 0$

64. $x^2 + 12x + 36 \ge 0$

Graphing Calculator Exercises

65. To solve the inequality

$$\frac{x}{x-2} > 0$$

enter Y_1 as $x/(x-2)$ and determine where the graph is above the x-axis. Write the solution in interval notation.

66. To solve the inequality

$$\frac{x}{x-2} < 0$$

enter Y_1 as $x/(x-2)$ and determine where the graph is below the x-axis. Write the solution in interval notation.

67. To solve the inequality $x^2 - 1 < 0$, enter Y_1 as $x^2 - 1$ and determine where the graph is below the x-axis. Write the solution in interval notation.

68. To solve the inequality $x^2 - 1 > 0$, enter Y_1 as $x^2 - 1$ and determine where the graph is above the x-axis. Write the solution in interval notation.

For Exercises 69–72, determine the solution by graphing the inequalities.

69. $x^2 + 10x + 25 \le 0$

70. $-x^2 + 10x - 25 \ge 0$

71. $\dfrac{8}{x^2+2} < 0$

72. $\dfrac{-6}{x^2+3} > 0$

section

9.3 ABSOLUTE VALUE EQUATIONS

Concepts

1. Definition of an Absolute Value Equation
2. Solving Absolute Value Equations
3. Solving Equations Having Two Absolute Values

1. Definition of an Absolute Value Equation

An equation of the form $|x| = a$ is called an **absolute value equation**. The solution includes all real numbers whose absolute value equals a. For example, the solutions to the equation, $|x| = 4$ are $x = 4$ and $x = -4$, because $|4| = 4$ and $|-4| = 4$. In Chapter 1, we introduced a geometric interpretation of $|x|$. The absolute value of a number is its distance from zero on the number line (Figure 9-5). Therefore, the solutions to the equation $|x| = 4$ are the values of x that are 4 units away from zero:

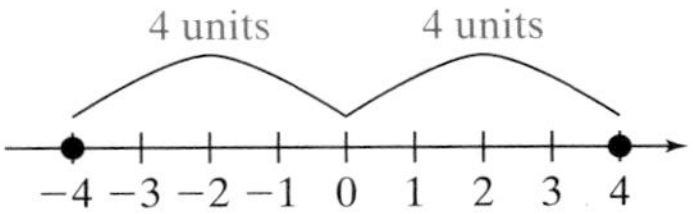

Figure 9-5

$|x| = 4$

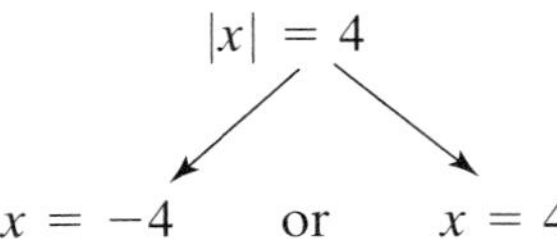

$x = -4$ or $x = 4$

2. Solving Absolute Value Equations

Solutions to Absolute Value Equations of the Form $|x| = a$

If a is a real number, then

1. If $a \geq 0$, the solutions to the equation $|x| = a$ are $x = a$ and $x = -a$.
2. If $a < 0$, there is no solution to the equation $|x| = a$.

To solve an absolute value equation of the form, $|x| = a$ $(a \geq 0)$, rewrite the equation as the union of the equations $x = a$ or $x = -a$.

example 1 **Solving Absolute Value Equations**

Solve the absolute value equations.

a. $|x| = 5$ b. $|w| - 2 = 12$ c. $|p| = 0$ d. $|x| = -6$

Solution:

a. $|x| = 5$ The equation is in the form $|x| = a$, where $a = 5$.

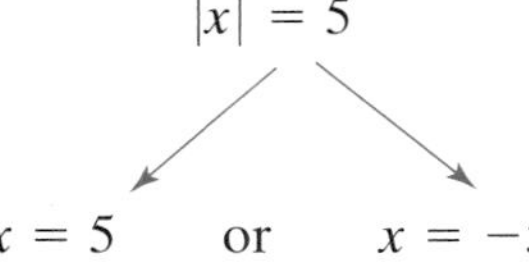

$x = 5$ or $x = -5$ Rewrite as the union of the equations $x = a$ or $x = -a$.

b. $|w| - 2 = 12$ Isolate the absolute value to write the equation in the form $|w| = a$.

$|w| = 14$

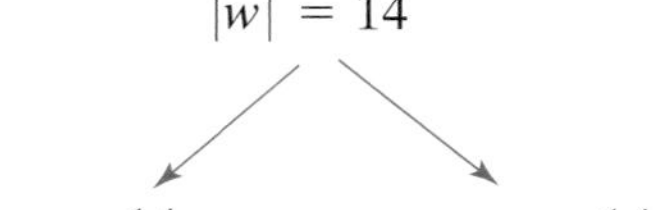

$w = 14$ or $w = -14$ Rewrite as the union of the equations $w = a$ or $w = -a$.

c. $|p| = 0$

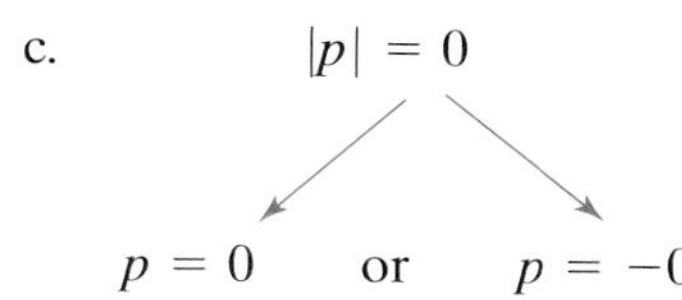

$p = 0$ or $p = -0$

Rewrite as the union of two equations. Notice that the second equation $p = -0$ is the same as the first equation. Intuitively, $p = 0$ is the only number whose absolute value equals 0.

d. $|x| = -6$

No solution

This equation is of the form $|x| = a$, but a is negative. There is no number whose absolute value is negative.

We have solved absolute value equations of the form $|x| = a$. Notice that x can represent any algebraic quantity. For example to solve the equation $|2w - 3| = 5$, we still rewrite the absolute value equation as the union of two equations. In this case, we set the quantity $2w - 3$ equal to 5 and to -5, respectively.

$$|2w - 3| = 5$$

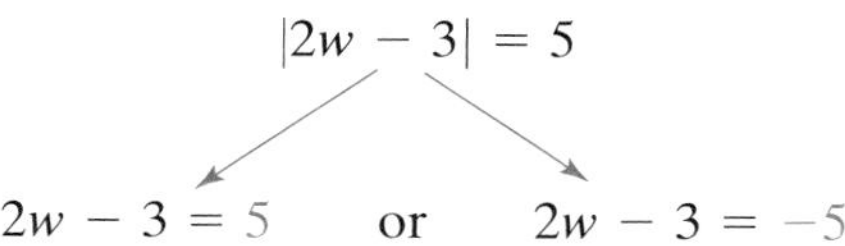

$$2w - 3 = 5 \quad \text{or} \quad 2w - 3 = -5$$

Steps to Solve an Absolute Value Equation

1. Isolate the absolute value. That is, write the equation in the form $|x| = a$, where a is a constant real number.
2. If $a < 0$, there is no solution.
3. Otherwise, if $a \geq 0$, rewrite the absolute value equation as the union of the equations $x = a$ or $x = -a$.
4. Solve the individual equations from step 3.
5. Check the answers in the original absolute value equation.

example 2

Solving Absolute Value Equations

Solve the absolute value equations:

a. $|2w - 3| = 5$ b. $|2c - 5| + 6 = 2$

Solution:

a. $|2w - 3| = 5$

The equation is already in the form $|x| = a$, where $x = 2w - 3$.

$2w - 3 = 5$ or $2w - 3 = -5$

Rewrite as the union of two equations.

$2w = 8$ or $2w = -2$

Solve each equation.

$w = 4$ or $w = -1$

The solutions are 4 and -1.

Check: $x = 4$	Check: $x = -1$	Check the solutions in the original equation.
$\lvert 2w - 3\rvert = 5$	$\lvert 2w - 3\rvert = 5$	
$\lvert 2(4) - 3\rvert \stackrel{?}{=} 5$	$\lvert 2(-1) - 3\rvert \stackrel{?}{=} 5$	
$\lvert 8 - 3\rvert \stackrel{?}{=} 5$	$\lvert -2 - 3\rvert \stackrel{?}{=} 5$	
$\lvert 5\rvert \stackrel{?}{=} 5$ ✓	$\lvert -5\rvert \stackrel{?}{=} 5$ ✓	

Calculator Connections

To confirm the answers to Example 2(a), graph $Y_1 = \text{abs}(2x - 3)$ and $Y_2 = 5$. The solutions to the equation $|2w - 3| = 5$ are the x-coordinates of the points of intersection (4, 5) and (−1, 5).

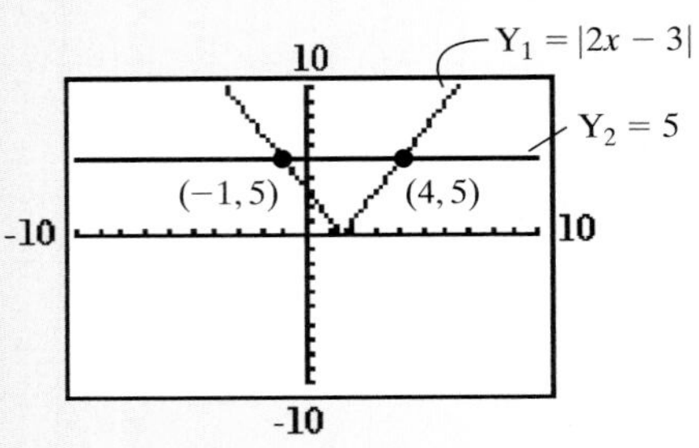

Avoiding Mistakes

Always isolate the absolute value *before* rewriting the problem as the union of two equations. In Example 2(b), if we had forgotten to isolate the absolute value first, we would have found two answers that do not check.

b. $|2c - 5| + 6 = 2$

$|2c - 5| = -4$ — Isolate the absolute value. The equation is in the form $|x| = a$, where $x = 2c - 5$ and $a = -4$. Because $a < 0$, there is no solution.

No solution — There are no numbers, x, that will make an absolute value equal to a negative number.

Calculator Connections

The graphs of $Y_1 = \text{abs}(2x - 5) + 6$ and $Y_2 = 2$ do not intersect.

Therefore, there is no solution to the equation $|2x - 5| + 6 = 2$.

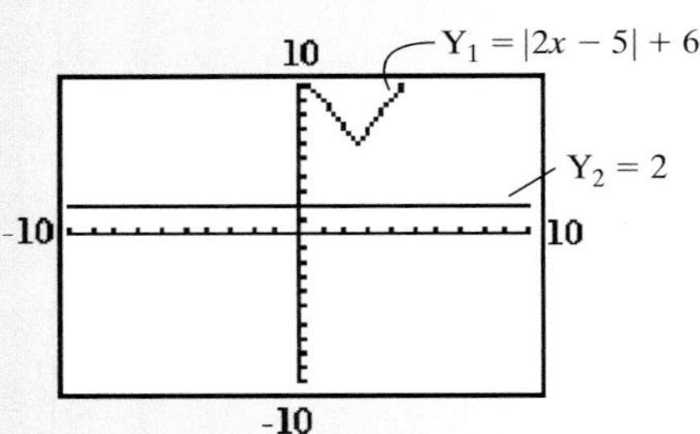

example 3 **Solving an Absolute Value Equation**

Solve the absolute value equation.

$$-2\left|\frac{2}{5}p + 3\right| - 7 = -19$$

Solution:

$$-2\left|\frac{2}{5}p + 3\right| - 7 = -19$$

$$-2\left|\frac{2}{5}p + 3\right| = -12 \qquad \text{Isolate the absolute value.}$$

$$\frac{-2\left|\frac{2}{5}p + 3\right|}{-2} = \frac{-12}{-2} \qquad \text{Divide by } -2.$$

$$\left|\frac{2}{5}p + 3\right| = 6$$

$\frac{2}{5}p + 3 = 6$	or	$\frac{2}{5}p + 3 = -6$	Rewrite as the union of two equations.
$2p + 15 = 30$	or	$2p + 15 = -30$	Multiply by 5 to clear fractions.
$2p = 15$	or	$2p = -45$	
$p = \frac{15}{2}$	or	$p = -\frac{45}{2}$	

The solutions are $\frac{15}{2}$ and $-\frac{45}{2}$. Both solutions check in the original equation.

3. Solving Equations Having Two Absolute Values

Some equations have two absolute values. The solutions to the equation $|x| = |y|$ are $x = y$ or $x = -y$. That is, if two quantities have the same absolute value, then the quantities are equal or the quantities are opposites.

Equality of Absolute Values

$|x| = |y|$ implies that $x = y$ or $x = -y$.

example 4 **Solving an Equation Having Two Absolute Values**

Solve the equations.

a. $|2w - 3| = |5w + 1|$ b. $|x - 4| = |x + 8|$

Solution:

a. $|2w - 3| = |5w + 1|$

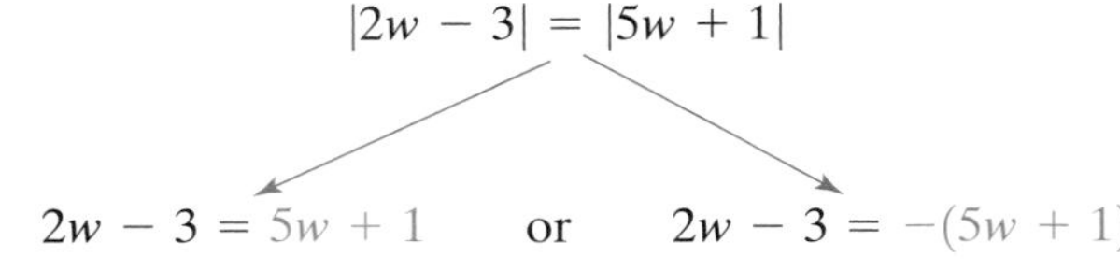

$2w - 3 = 5w + 1$	or	$2w - 3 = -(5w + 1)$	Rewrite as the union of two equations, $x = y$ or $x = -y$.
$2w - 3 = 5w + 1$	or	$2w - 3 = -5w - 1$	Apply the distributive property.
$-3w - 3 = 1$	or	$7w - 3 = -1$	Solve for w.
$-3w = 4$	or	$7w = 2$	
$w = -\frac{4}{3}$	or	$w = \frac{2}{7}$	

Avoiding Mistakes

To take the opposite of the quantity, $5w + 1$, use parentheses and apply the distributive property.

The solutions are $-\frac{4}{3}$ and $\frac{2}{7}$. Both values check in the original equation.

b. $|x - 4| = |x + 8|$

$x - 4 = x + 8$	or	$x - 4 = -(x + 8)$	Rewrite as the union of two equations $x = y$ or $x = -y$
$-4 = 8$ (contradiction)	or	$x - 4 = -x - 8$	Apply the distributive property and solve for x.
		$2x - 4 = -8$	
		$2x = -4$	
		$x = -2$	

The only solution is -2. $x = -2$ checks in the original equation.

Calculator Connections

Graph $Y_1 = \text{abs}(x - 4)$ and $Y_2 = \text{abs}(x + 8)$. There is one point of intersection at $(-2, 6)$. Therefore, the solution to $|x - 4| = |x + 8|$ is $x = -2$.

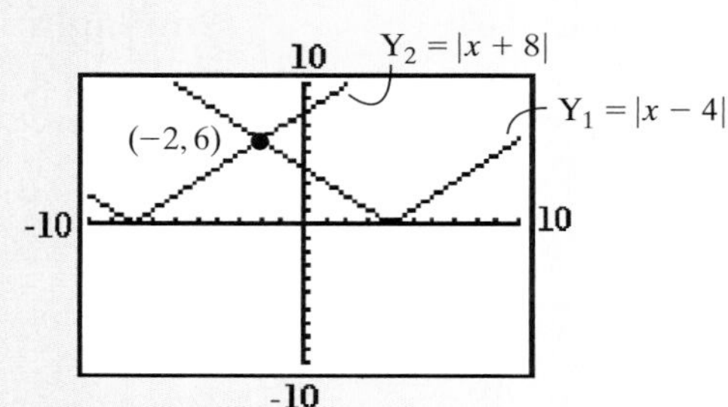

section 9.3 Practice Exercises

For Exercises 1–8, solve the inequalities. Write the answers in interval notation.

1. $3(a + 2) - 6 > 2$ and $-2(a - 3) + 14 > -3$
2. $3x - 5 \geq 7x + 3$ or $2x - 1 \leq 4x - 5$
3. $\dfrac{4}{y - 4} \geq 3$
4. $\dfrac{3}{t + 1} \leq 2$
5. The sum of a number and 6 is between -1 and 13. Find all such numbers.
6. The difference of 12 and a number is between -3 and -1. Find all such numbers.
7. $3(x - 2)(x + 4)(2x - 1) < 0$
8. $x^3 - 7x^2 - 8x > 0$

For Exercises 9–50, solve the absolute value equations.

9. $|p| = 7$
10. $|q| = 10$
11. $|x| + 5 = 11$
12. $|x| - 3 = 20$
13. $|y| = \sqrt{2}$
14. $|y| = \dfrac{5}{8}$
15. $|w| - 3 = -5$
16. $|w| + 4 = -8$
17. $|3q| = 0$
18. $|4p| = 0$
19. $\left|3x - \dfrac{1}{2}\right| = \dfrac{1}{2}$
20. $|4x + 1| = 6$
21. $|4x - 2| = |-8|$
22. $|3x + 5| = |-5|$
23. $\left|\dfrac{7z}{3} - \dfrac{1}{3}\right| + 3 = 6$
24. $\left|\dfrac{w}{2} + \dfrac{3}{2}\right| - 2 = 7$
25. $\left|\dfrac{5y + 2}{2}\right| = 6$
26. $\left|\dfrac{2t - 1}{3}\right| = 5$
27. $|0.2x - 3.5| = -5.6$
28. $|1.81 + 2x| = -2.2$
29. $|4w + 3| = |2w - 5|$
30. $|3y + 1| = |2y - 7|$
31. $|2y + 5| = |7 - 2y|$
32. $|9a + 5| = |9a - 1|$
33. $1 = -4 + \left|2 - \dfrac{1}{4}w\right|$
34. $-12 + |6 - 2x| = -6$
35. $10 = 4 + |2y + 1|$
36. $-1 = -|5x + 7|$
37. $|3b - 7| - 9 = -9$
38. $|8x + 5| = 0$
39. $|4w - 1| = |2w + 3|$
40. $|3p + 2| = |p - 4|$
41. $-2|x + 3| = 5$
42. $-3|x - 5| = 7$
43. $|6x - 9| = 0$
44. $|4k - 6| + 7 = 7$
45. $|2h - 6| = |2h + 5|$
46. $|6n - 7| = |4 - 6n|$
47. $\left|-\dfrac{1}{5} - \dfrac{1}{2}k\right| = \dfrac{9}{5}$
48. $\left|-\dfrac{1}{6} - \dfrac{2}{9}h\right| = \dfrac{1}{2}$
49. $|3.5m - 1.2| = |8.5m + 6|$
50. $|11.2n + 9| = |7.2n - 2.1|$
51. Write an absolute value equation whose solution is the set of real numbers 6 units from zero on the number line.
52. Write an absolute value equation whose solution is the set of real numbers $\frac{7}{2}$ units from zero on the number line.

53. Write an absolute value equation whose solution is the set of real numbers $\frac{4}{3}$ units from zero on the number line.

54. Write an absolute value equation whose solution is the set of real numbers 9 units from zero on the number line.

Graphing Calculator Exercises

For Exercises 55–62, enter the left side of the equation as Y_1 and enter the right side of the equation as Y_2. Then use the *Intersect* feature or *Zoom* and *Trace* to approximate the x-values where the two graphs intersect (if they intersect).

55. $|4x - 3| = 5$

56. $|x - 4| = 3$

57. $|8x + 1| + 8 = 1$

58. $|3x - 2| + 4 = 2$

59. $|x - 3| = |x + 2|$

60. $|x + 4| = |x - 2|$

61. $|2x - 1| = |-x + 3|$

62. $|3x| = |2x - 5|$

Concepts

1. Solving Absolute Value Inequalities
2. Absolute Value Inequalities with "Special Case" Solution Sets
3. Translating to an Absolute Value Expression
4. Solving Absolute Value Inequalities Using the Test Point Method

section

9.4 Absolute Value Inequalities

1. Solving Absolute Value Inequalities

In Section 9.3, we studied absolute value equations in the form $|x| = a$. In this section we will solve absolute value *inequalities*. An inequality in any of the forms $|x| < a$, $|x| \le a$, $|x| > a$, or $|x| \ge a$ is called an **absolute value inequality**.

Recall that an absolute value represents distance from zero on the real number line. Consider the following absolute value equation and inequalities.

1. $|x| = 3$

$x = 3$ or $x = -3$

Solution:

The set of all points 3 units from zero on the number line.

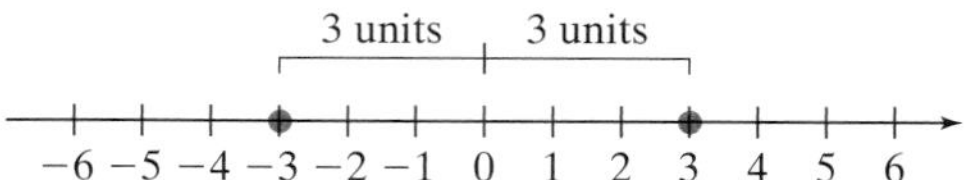

2. $|x| > 3$

$x < -3$ or $x > 3$

Solution:

The set of all points more than 3 units from zero.

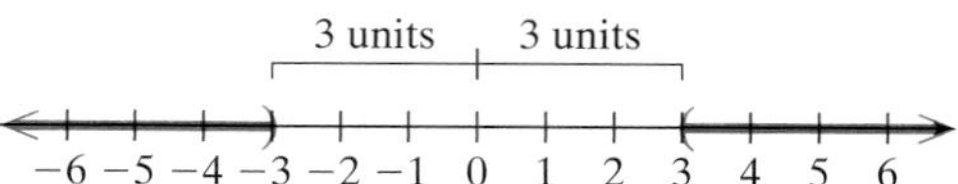

3. $|x| < 3$

$-3 < x < 3$

Solution:

The set of all points less than 3 units from zero.

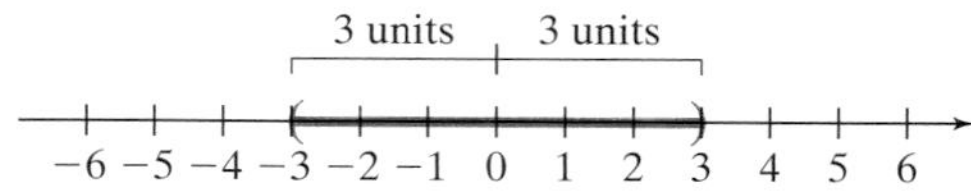

Solutions to Absolute Value Equations and Inequalities

Let a be a real number such that $a > 0$. Then,

Equation/ Inequality	Solution (equivalent form)	Graph
$\lvert x\rvert = a$	$x = -a$ or $x = a$	
$\lvert x\rvert > a$	$x < -a$ or $x > a$	
$\lvert x\rvert < a$	$-a < x < a$	

To solve an absolute value inequality, first isolate the absolute value and then rewrite the absolute value inequality in its equivalent form.

example 1 **Solving Absolute Value Inequalities**

Solve the inequalities.

a. $\lvert 3w + 1\rvert - 4 < 7$ b. $\left\lvert \frac{1}{2}t - 5\right\rvert + 1 \geq 3$

Avoiding Mistakes

Always isolate the absolute value first when solving an absolute value inequality.

Solution:

a. $\lvert 3w + 1\rvert - 4 < 7$

$\lvert 3w + 1\rvert < 11$ Isolate the absolute value first.

The inequality is in the form $\lvert x\rvert < a$, where $x = 3w + 1$.

$-11 < 3w + 1 < 11$ Rewrite in the equivalent form, $-a < x < a$.

$-12 < 3w < 10$ Solve for w.

$-4 < w < \frac{10}{3}$

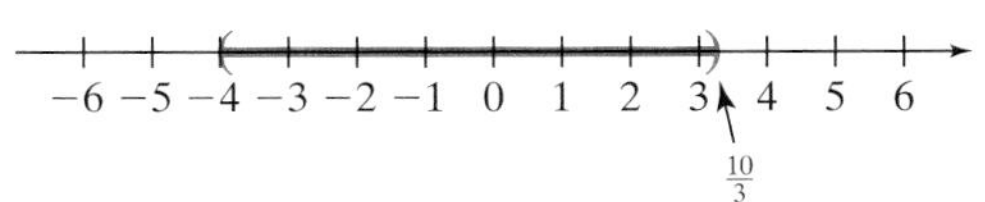

The solution is $\{w \mid -4 < w < \frac{10}{3}\}$ or equivalently in interval notation, $(-4, \frac{10}{3})$.

Calculator Connections

Graph $Y_1 = \text{abs}(3x + 1) - 4$ and $Y_2 = 7$. On the given display window, $Y_1 < Y_2$ (Y_1 is below Y_2) for $4 < x < \frac{10}{3}$.

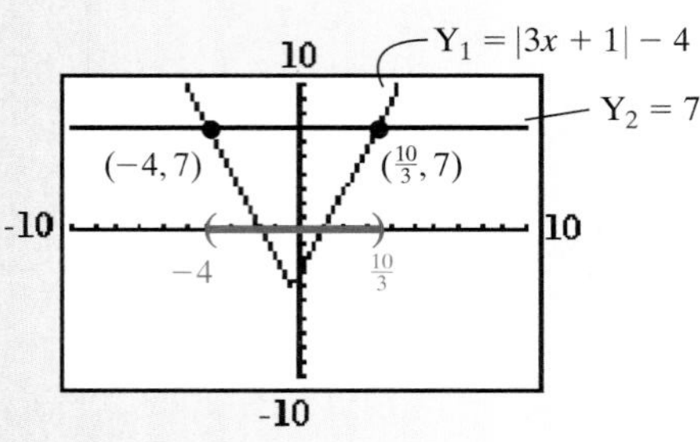

b. $\left|\frac{1}{2}t - 5\right| + 1 \geq 3$

$\left|\frac{1}{2}t - 5\right| \geq 2$ — Isolate the absolute value.

The inequality is in the form $|x| \geq a$, where $x = \frac{1}{2}t - 5$.

$\frac{1}{2}t - 5 \leq -2 \quad \text{or} \quad \frac{1}{2}t - 5 \geq 2$ — Rewrite in the equivalent form $x \leq -a$ or $x \geq a$.

$\frac{1}{2}t \leq 3 \quad \text{or} \quad \frac{1}{2}t \geq 7$ — Solve the compound inequality.

$2\left(\frac{1}{2}t\right) \leq 2(3) \quad \text{or} \quad 2\left(\frac{1}{2}t\right) \geq 2(7)$ — Clear fractions.

$t \leq 6 \quad \text{or} \quad t \geq 14$

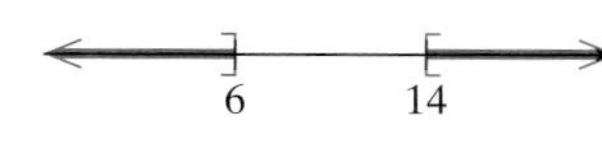

The solution is $\{t \mid t \leq 6 \text{ or } t \geq 14\}$ or equivalently in interval notation, $(-\infty, 6] \cup [14, \infty)$.

Calculator Connections

Graph $Y_1 = \text{abs}(\frac{1}{2}x - 5) + 1$ and $Y_2 = 3$. On the given display window, $Y_1 \geq Y_2$ for $x \leq 6$ or $x \geq 14$.

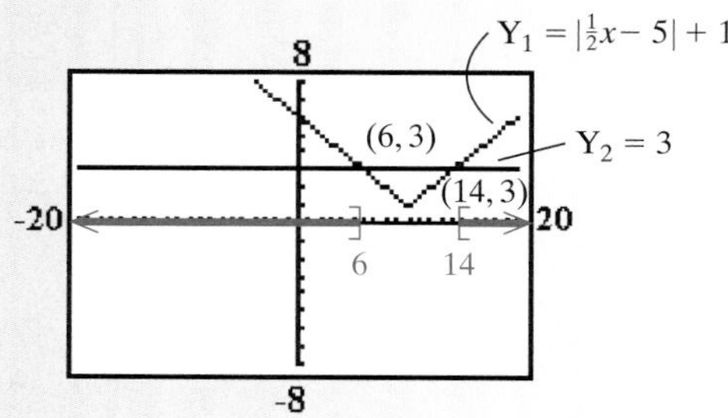

2. Absolute Value Inequalities with "Special Case" Solution Sets

Tip: By definition, the absolute value of a real number will always be nonnegative. Therefore, the absolute value of any expression will always be greater than a negative number. Similarly, an absolute value can never be less than a negative number. Let a represent a positive real number. Then,

1. The solution to the inequality $|x| > -a$ is all real numbers, $(-\infty, \infty)$.
2. There is no solution to the inequality $|x| < -a$.

example 2

Solving Absolute Value Inequalities

Solve the inequalities.

a. $|3d - 5| + 7 < 4$ b. $|3d - 5| + 7 > 4$

Solution:

a. $|3d - 5| + 7 < 4$

$|3d - 5| < -3$ Isolate the absolute value. An absolute value expression cannot be less than a negative number. Therefore, there is no solution.

No solution

b. $|3d - 5| + 7 > 4$

$|3d - 5| > -3$ Isolate the absolute value. The inequality is in the form $|x| > a$, where a is negative. An absolute value of any real number is greater than a negative number. Therefore, the solution is all real numbers.

All real numbers, $(-\infty, \infty)$

Calculator Connections

By graphing $Y_1 = \text{abs}(3x - 5) + 7$ and $Y_2 = 4$, we see that $Y_1 > Y_2$ (Y_1 is above Y_2) for all real numbers, x, on the given display window.

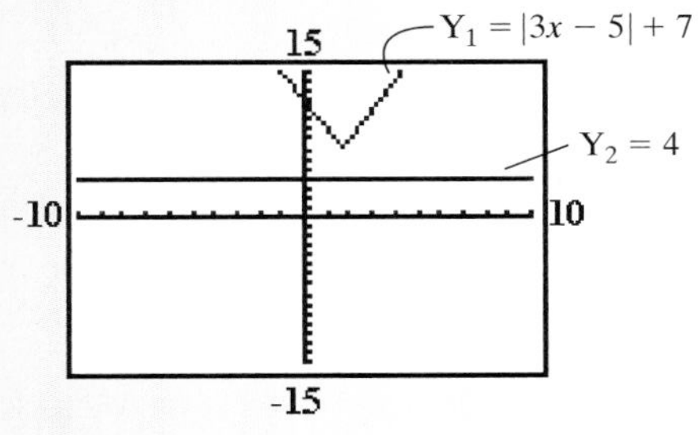

example 3 Solving Absolute Value Inequalities

Solve the inequalities.

a. $|4x + 2| \geq 0$ b. $|4x + 2| > 0$ c. $|4x + 2| < 0$

Solution:

a. $|4x + 2| \geq 0$ The absolute value is already isolated.

The absolute value of any real number is nonnegative. Therefore, the solution is all real numbers, $(-\infty, \infty)$.

b. $|4x + 2| > 0$

An absolute value will be greater than zero at all points *except where it is equal to zero*. That is, the point(s) for which $|4x + 2| = 0$ must be excluded from the solution set.

$|4x + 2| = 0$

$4x + 2 = 0$ or $4x + 2 = -0$ The second equation is the same as the first.

$4x = -2$

$x = -\dfrac{1}{2}$ Therefore, exclude $x = -\frac{1}{2}$ from the solution.

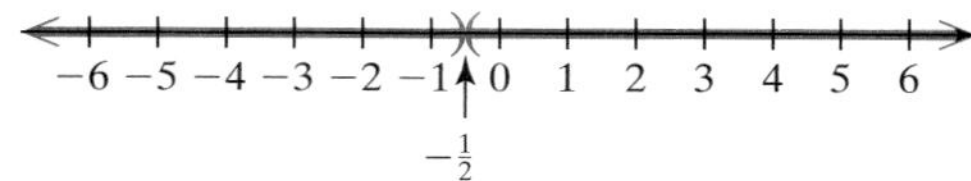

The solution is $\{x \mid x \neq -\frac{1}{2}\}$ or equivalently in interval notation, $(-\infty, -\frac{1}{2}) \cup (-\frac{1}{2}, \infty)$.

Calculator Connections

Graph $Y_1 = \text{abs}(4x + 2)$. From the graph, $Y_1 = 0$ at $x = -\frac{1}{2}$ (the x-intercept). On the given display window, $Y_1 > 0$ for $x < -\frac{1}{2}$ or $x > -\frac{1}{2}$.

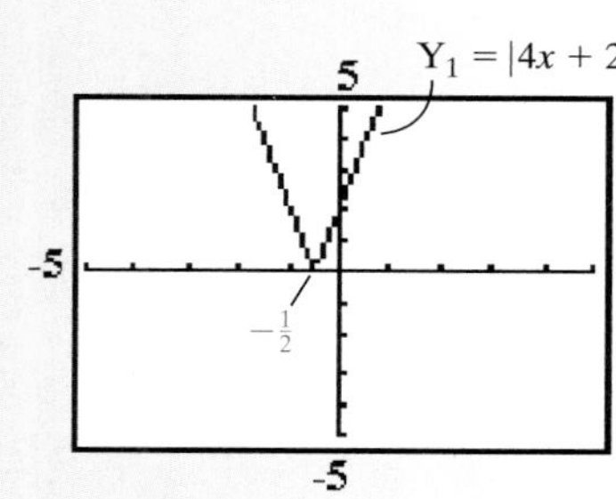

c. $|4x + 2| < 0$

The absolute value of any real number is never negative. Therefore, the inequality $|4x + 2| < 0$ has no solution.

3. Translating to an Absolute Value Expression

Absolute value expressions can be used to describe distances. The distance between c and d is given by $|c - d|$. For example, the distance between -2 and 3 on the number line is $|(-2) - 3| = 5$ as expected.

example 4 **Expressing Distances with Absolute Value**

Write an absolute value inequality to represent the following phrases.

a. All real numbers, x, whose distance from zero is greater than 5 units
b. All real numbers, x, whose distance from -7 is less than 3 units

Solution:

a. All real numbers, x, whose distance from zero is greater than 5 units

$|x - 0| > 5$ or simply $|x| > 5$

5 units 5 units

−6 −5 −4 −3 −2 −1 0 1 2 3 4 5 6

b. All real numbers, x, whose distance from -7 is less than 3 units

$|x - (-7)| < 3$ or simply $|x + 7| < 3$

3 units 3 units

−11 −10 −9 −8 −7 −6 −5 −4 −3 −2

Absolute value expressions can also be used to describe boundaries for measurement error.

example 5 **Expressing Measurement Error with Absolute Value**

Latoya measured a certain compound on a scale in the chemistry lab at school. She measured 8 g of the compound, but the scale is only accurate to the nearest tenth of a gram. Write an absolute value inequality to express an interval for the true mass, x, of the compound she measured.

Solution:

Tip: The solution to Example 5 can also be expressed as $|8.0 - x| \leq 0.1$.

Because the scale is only accurate to the nearest tenth of a gram, the true mass, x, of the compound may deviate by as much as 0.1 g above or below 8 g. This may be expressed as an absolute value inequality:

$$|x - 8.0| \leq 0.1 \qquad \text{or equivalently,} \qquad 7.9 \leq x \leq 8.1$$

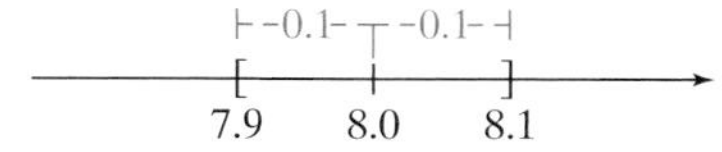

4. Solving Absolute Value Inequalities Using the Test Point Method

For each problem in Example 1, the absolute value inequality was converted to an equivalent compound inequality. However, sometimes students have difficulty setting up the appropriate compound inequality. To avoid this problem, you may want to use the test point method to solve absolute value inequalities.

Tip: The test point method was first introduced in Section 9.2. As you can see this method can be used to solve a variety of different types of inequalities.

Solving Inequalities Using the Test Point Method

1. Find the boundary points of the inequality. (Boundary points are the real solutions to the related equation and points where the inequality is undefined.)
2. Plot the boundary points on the number line. This divides the number line into regions.
3. Select a test point from each region and substitute it into the original inequality.
   - If a test point makes the original inequality true, then that region is part of the solution set.
4. Test the boundary points in the original inequality.
   - If a boundary point makes the original inequality true, then that point is part of the solution set.

To demonstrate the use of the test point method, we will repeat the absolute value inequalities from Example 1. Notice that regardless of the method used, the absolute value is always isolated *first* before any further action is taken.

example 6 — Solving Absolute Value Inequalities with the Test Point Method

Solve the inequalities using the test point method.

a. $|3w + 1| - 4 < 7$ b. $\left|\frac{1}{2}t - 5\right| + 1 \geq 3$

Solution:

a. $|3w + 1| - 4 < 7$

$|3w + 1| < 11$ Isolate the absolute value first.

$|3w + 1| = 11$ **Step 1:** Solve the related equation.

$3w + 1 = 11$ or $3w + 1 = -11$ Write as an equivalent system of two equations.

$3w = 10$ or $3w = -12$

$w = \frac{10}{3}$ or $w = -4$ These are the only boundary points.

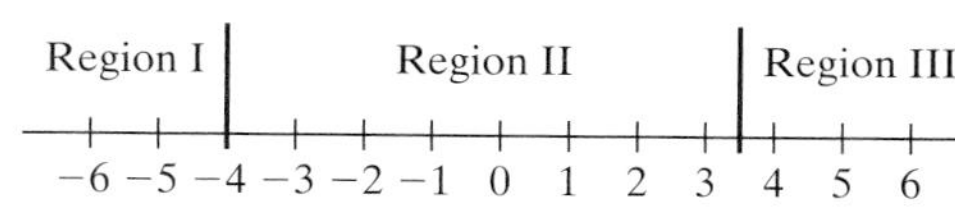

Step 2: Plot the boundary points.

Step 3: Select a test point from each region.

Test $w = -5$:

$$|3(-5) + 1| - 4 \stackrel{?}{<} 7$$
$$|-14| - 4 \stackrel{?}{<} 7$$
$$14 - 4 \stackrel{?}{<} 7$$
$$10 \stackrel{?}{<} 7 \quad \text{False}$$

Test $w = 0$:

$$|3(0) + 1| - 4 \stackrel{?}{<} 7$$
$$|1| - 4 \stackrel{?}{<} 7$$
$$-3 \stackrel{?}{<} 7 \quad \text{True}$$

Test $w = 4$:

$$|3(4) + 1| - 4 \stackrel{?}{<} 7$$
$$|13| - 4 \stackrel{?}{<} 7$$
$$13 - 4 \stackrel{?}{<} 7$$
$$9 \stackrel{?}{<} 7 \quad \text{False}$$

Step 4: The last step in the test point method is to determine whether the boundary points are part of the solution set. Because the original inequality is a strict inequality, the boundary points (where equality occurs) are not included.

False True False
−6 −5 −4 −3 −2 −1 0 1 2 3 4 5 6
$\frac{10}{3}$

The solution is $\{w \mid -4 < w < \frac{10}{3}\}$ or equivalently in interval notation, $(-4, \frac{10}{3})$.

b. $\left|\frac{1}{2}t - 5\right| + 1 \geq 3$

$$\left|\frac{1}{2}t - 5\right| \geq 2$$

Isolate the absolute value first.

$$\left|\frac{1}{2}t - 5\right| = 2$$

Step 1: Solve the related equation.

$$\frac{1}{2}t - 5 = 2 \quad \text{or} \quad \frac{1}{2}t - 5 = -2$$

Write the union of two equations.

$$\frac{1}{2}t = 7 \quad \text{or} \quad \frac{1}{2}t = 3$$

$$t = 14 \quad \text{or} \quad t = 6$$

These are the boundary points.

Region I | Region II | Region III
6 14

Step 2: Plot the boundary points.

Step 3: Select a test point from each region.

Test $t = 0$:	**Test $t = 10$:**	**Test $t = 16$:**
$\left\lvert\frac{1}{2}(0) - 5\right\rvert + 1 \overset{?}{\geq} 3$	$\left\lvert\frac{1}{2}(10) - 5\right\rvert + 1 \overset{?}{\geq} 3$	$\left\lvert\frac{1}{2}(16) - 5\right\rvert + 1 \overset{?}{\geq} 3$
$\lvert 0 - 5\rvert + 1 \overset{?}{\geq} 3$	$\lvert 5 - 5\rvert + 1 \overset{?}{\geq} 3$	$\lvert 8 - 5\rvert + 1 \overset{?}{\geq} 3$
$\lvert -5\rvert + 1 \overset{?}{\geq} 3$	$\lvert 0\rvert + 1 \overset{?}{\geq} 3$	$\lvert 3\rvert + 1 \overset{?}{\geq} 3$
$5 + 1 \overset{?}{\geq} 3$ True	$1 \overset{?}{\geq} 3$ False	$4 \overset{?}{\geq} 3$ True

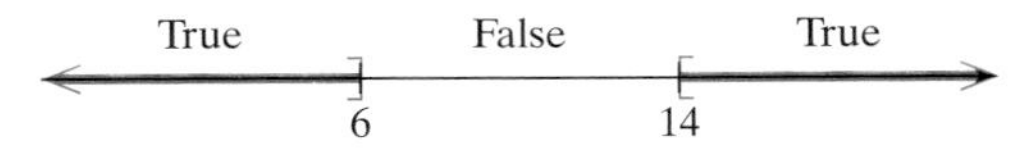

Step 4: The original inequality uses the sign $\geq$. Therefore, the boundary points (where equality occurs) must be part of the solution set.

The solution is $\{t \mid t \leq 6 \text{ or } t \geq 14\}$ or equivalently in interval notation, $(-\infty, 6] \cup [14, \infty)$.

section 9.4 PRACTICE EXERCISES

For Exercises 1–4, solve the equations.

1. $\lvert 10x - 6\rvert = -5$
2. $2 = \lvert 5 - 7x\rvert + 1$
3. $\lvert 6x\rvert = \lvert 9x + 5\rvert$
4. $\lvert 3y - 1\rvert = \lvert 3y + 4\rvert$

For Exercises 5–8, solve the inequalities and graph the solution set. Write the solution in interval notation.

5. $-15 < 3w - 6 \leq -9$
6. $5 - 2y \leq 1$ and $3y + 2 \geq 14$
7. $m - 7 \leq -5$ or $m - 7 \geq -10$
8. $3b - 2 < 7$ or $b - 2 > 4$

For Exercises 9–16, solve the equations and inequalities. For each inequality, graph the solution set and express the solution in interval notation.

9. a. $\lvert x\rvert = 5$
   b. $\lvert x\rvert > 5$
   c. $\lvert x\rvert < 5$
10. a. $\lvert a\rvert = 4$
    b. $\lvert a\rvert > 4$
    c. $\lvert a\rvert < 4$
11. a. $\lvert p\rvert = -2$
    b. $\lvert p\rvert > -2$
    c. $\lvert p\rvert < -2$
12. a. $\lvert x\rvert = -14$
    b. $\lvert x\rvert > -14$
    c. $\lvert x\rvert < -14$
13. a. $\lvert x - 3\rvert = 7$
    b. $\lvert x - 3\rvert > 7$
    c. $\lvert x - 3\rvert < 7$
14. a. $\lvert w + 2\rvert = 6$
    b. $\lvert w + 2\rvert > 6$
    c. $\lvert w + 2\rvert < 6$
15. a. $\lvert y + 1\rvert = -6$
    b. $\lvert y + 1\rvert > -6$
    c. $\lvert y + 1\rvert < -6$
16. a. $\lvert z - 4\rvert = -3$
    b. $\lvert z - 4\rvert > -3$
    c. $\lvert z - 4\rvert < -3$

For Exercises 17–40, solve the absolute value inequalities. Graph the solution set and write the solution in interval notation.

17. $\lvert x\rvert > 6$
18. $\lvert x\rvert \leq 6$
19. $\lvert t\rvert \leq 3$
20. $\lvert p\rvert > 3$
21. $\lvert y + 2\rvert \geq 0$
22. $0 \leq \lvert 7n + 2\rvert$
23. $5 \leq \lvert 2x - 1\rvert$
24. $\lvert x - 2\rvert \geq 7$
25. $\lvert k - 7\rvert < -3$
26. $\lvert h + 2\rvert < -9$

27. $\left|\frac{w-2}{3}\right| - 3 \le 1$

28. $\left|\frac{x+3}{2}\right| - 2 \ge 4$

29. $|9 - 4y| \ge 14$

30. $1 > |2m - 7|$

31. $\left|\frac{2x+1}{4}\right| < 5$

32. $\left|\frac{x-4}{5}\right| \le 7$

33. $8 < |4 - 3x| + 12$

34. $-16 < |5x - 1| - 1$

35. $5 - |2m + 1| > 5$

36. $3 - |5x + 3| > 3$

37. $|p + 5| \le 0$

38. $|y + 1| - 4 \le -4$

39. $|z - 6| + 5 \ge 5$

40. $|2c - 1| \ge 0$

For Exercises 41–44, write an absolute value inequality equivalent to the expression given.

41. All real numbers whose distance from 0 is greater than 7.

42. All real numbers whose distance from -3 is less than 4.

43. All real numbers whose distance from 2 is at most 13.

44. All real numbers whose distance from 0 is at least 6.

45. A 32-oz jug of orange juice may not contain exactly 32 oz of juice. The possibility of measurement error exists when the jug is filled in the factory. If there is a ± 0.05-oz measurement error, write an absolute value inequality representing the range of volumes, x, in which the orange juice jug may be filled.

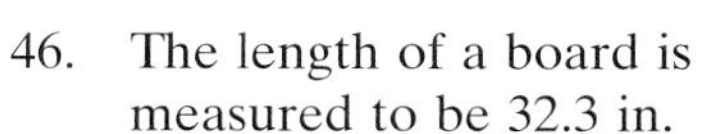

46. The length of a board is measured to be 32.3 in. There is a ± 0.2 in. measurement error. Write an absolute value inequality that represents the range for the length of the board.

47. A bag of potato chips states that its weight is $6\frac{3}{4}$ oz. There is a $\pm\frac{1}{8}$-oz measurement error. Write an absolute value inequality that represents the range for the weight, x, of the bag of chips.

48. A $\frac{7}{8}$-in. bolt varies in length by $\pm\frac{1}{16}$ in. Write an absolute value inequality that represents the range for the length of the bolt.

For Exercises 49–52, match the graph with the inequality:

49.

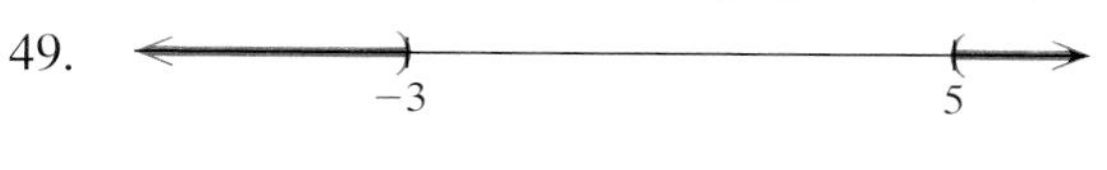

50.

51.

52.

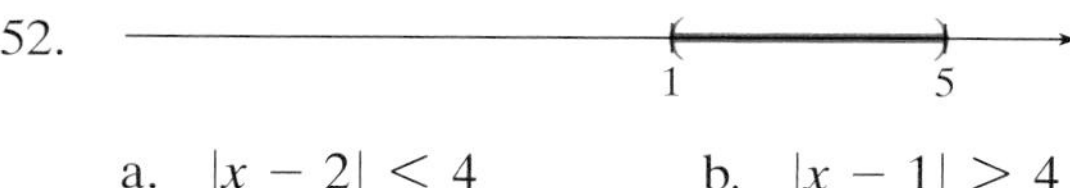

a. $|x - 2| < 4$

b. $|x - 1| > 4$

c. $|x - 3| < 2$

d. $|x - 5| > 1$

Graphing Calculator Exercises

To solve an absolute value inequality using a graphing calculator, let Y_1 equal the right side of the inequality and let Y_2 equal the left side of the inequality. Graph both Y_1 and Y_2 on a standard viewing window and use an *Intersect* feature or *Zoom* and *Trace* to approximate the intersection of the graphs. To solve $Y_1 > Y_2$, determine all x-values where the graph of Y_1 is above the graph of Y_2. To solve $Y_1 < Y_2$, determine all x-values where the graph of Y_1 is below the graph of Y_2.

For Exercises 53–62, solve the inequalities using a graphing calculator.

53. $|x + 2| > 4$

54. $|3 - x| > 6$

55. $\left|\frac{x+1}{3}\right| < 2$

56. $\left|\frac{x-1}{4}\right| < 1$

57. $|x - 5| < -3$

58. $|x + 2| < -2$

59. $|2x + 5| > -4$

60. $|1 - 2x| > -4$

61. $|6x + 1| \le 0$

62. $|3x - 4| \le 0$

Concepts

1. Introduction to Linear Inequalities in Two Variables
2. Graphing Linear Inequalities in Two Variables
3. Compound Linear Inequalities in Two Variables
4. Graphing a Feasible Region

section

9.5 Linear Inequalities in Two Variables

1. Introduction to Linear Inequalities in Two Variables

A **linear inequality in two variables** x and y is an inequality that can be written in one of the following forms: $ax + by < c$, $ax + by > c$, $ax + by \leq c$, or $ax + by \geq c$, provided a and b are not both zero.

A solution to a linear inequality in two variables is an ordered pair that makes the inequality true. For example, solutions to the inequality $x + y < 6$ are ordered pairs (x, y) such that the sum of the x- and y-coordinates is less than 6. This inequality has an infinite number of solutions, and therefore it is convenient to express the solution set as a graph.

2. Graphing Linear Inequalities in Two Variables

To graph a linear inequality in two variables, we will use the test point method. The first step is to graph the related equation. This will be a line that separates the xy-plane into three regions: (1) the region below the line, (2) the region above the line, and (3) the line itself. Then, by selecting a test point (ordered pair) from each region and substituting it into the inequality, we can determine which region(s) represents the solution set.

Test Point Method—Summary

1. Graph the related equation. The equation will be a boundary line in the xy-plane.
   - If the original inequality is a strict inequality, $<$ or $>$, then the line is not part of the solution set. Graph the boundary as a *dashed line*.
   - If the original inequality uses $\leq$ or $\geq$ then the line is part of the solution set. Graph the boundary as a *solid line*.
2. From each region above and below the line, select an ordered pair as a test point and substitute it into the original inequality.
   - If a test point makes the inequality true, then that region is part of the solution set.

example 1 **Graphing a Linear Inequality in Two Variables**

Graph the solution set of the inequality $x + y < 6$.

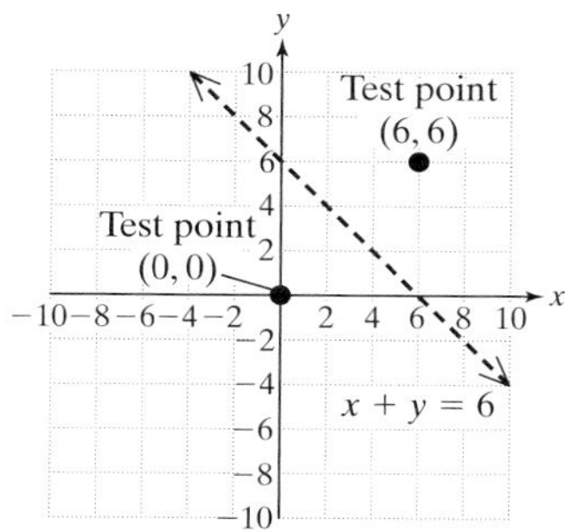

Figure 9-6

Solution:

Step 1: Graph the related equation $x + y = 6$ using a dashed line (Figure 9-6).

> **Tip:** To graph the related equation, you may either create a table of points, or you may use the slope-intercept form of the line.
>
> table of points — slope-intercept form

x	y
0	6
6	0
4	2

$$x + y = 6 \longrightarrow y = -x + 6$$

Step 2: Choose test points (ordered pairs) above and below the line and substitute the points into the original inequality.

Test Point Above: (6, 6)

$$x + y < 6$$

$$(6) + (6) \overset{?}{<} 6$$

$$12 \overset{?}{<} 6 \quad \text{False}$$

The test point (6, 6) *is not* a solution to the original inequality.

Test Point Below: (0, 0)

$$x + y < 6$$

$$(0) + (0) \overset{?}{<} 6$$

$$0 \overset{?}{<} 6 \quad \text{True}$$

The test point (0, 0) *is* a solution to the original inequality. Shade the region below the boundary. See Figure 9-7.

Figure 9-7

example 2 Graphing a Linear Inequality in Two Variables

Graph the solution set of the inequality $3x - 5y \leq 10$.

Solution:

$$3x - 5y \leq 10$$

$$3x - 5y = 10$$

$$y = \frac{3}{5}x - 2$$

Step 1: Graph the related equation to form the boundary of the solution set. (Here we use the slope-intercept form to graph the line.)

Step 2: Choose test points above and below the line.

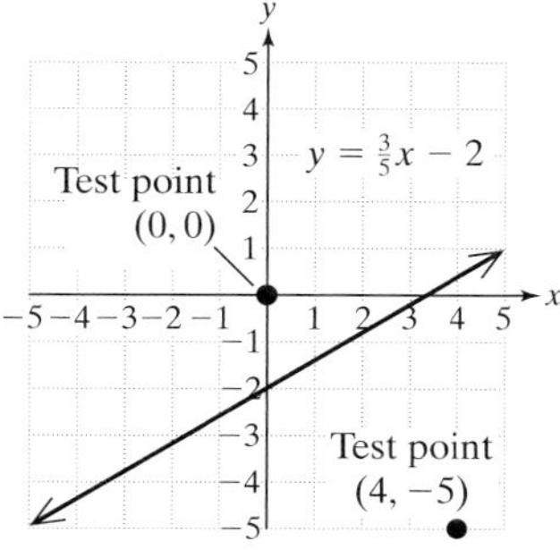

Test Point Above: (0, 0)	**Test Point Below: (4, −5)**
$3x - 5y \leq 10$	$3x - 5y \leq 10$
$3(0) - 5(0) \stackrel{?}{\leq} 10$	$3(4) - 5(-5) \stackrel{?}{\leq} 10$
$0 \stackrel{?}{\leq} 10$ True	$12 + 25 \stackrel{?}{\leq} 10$
	$37 \stackrel{?}{\leq} 10$ False

Because the test point (0, 0) above the boundary is true in the original inequality, shade the region above the line (Figure 9-8).

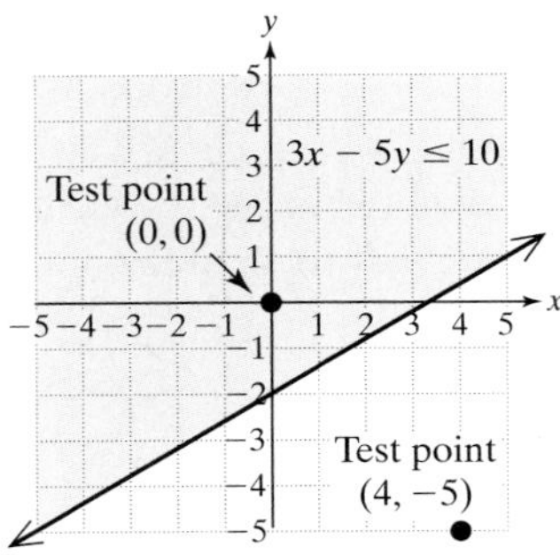

Figure 9-8

Tip: An inequality can also be graphed by first solving the inequality for *y*. Then,

- Shade below the line if the inequality is of the form $y < mx + b$ or $y \leq mx + b$.
- Shade above the line if the inequality is of the form $y > mx + b$ or $y \geq mx + b$.

From Example 2, we have:

$$3x - 5y \leq 10$$

$$-5y \leq -3x + 10$$

$$\frac{-5y}{-5} \geq \frac{-3x}{-5} + \frac{10}{-5} \qquad \text{Reverse the inequality sign.}$$

$$y \geq \frac{3}{5}x - 2 \qquad \text{Shade } \textit{above} \text{ the line.}$$

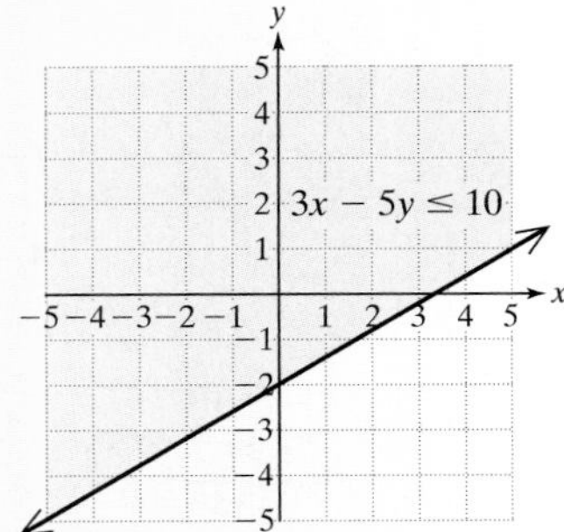

3. Compound Linear Inequalities in Two Variables

Some applications require us to find the union or intersection of two or more linear inequalities.

example 3 Graphing a Compound Linear Inequality in Two Variables

Graph the solution set of the compound inequality: $2x + y < 1$ and $2y \geq x - 4$.

Solution:

First Inequality		**Second Inequality**	
$2x + y < 1$		$2y \geq x - 4$	
$2x + y = 1$	Related equation	$2y = x - 4$	Related equation
$y = -2x + 1$	Slope-intercept form	$y = \frac{1}{2}x - 2$	Slope-intercept form

For each inequality, draw the boundary line. Then, pick test points above and below the line to determine the appropriate region to shade.

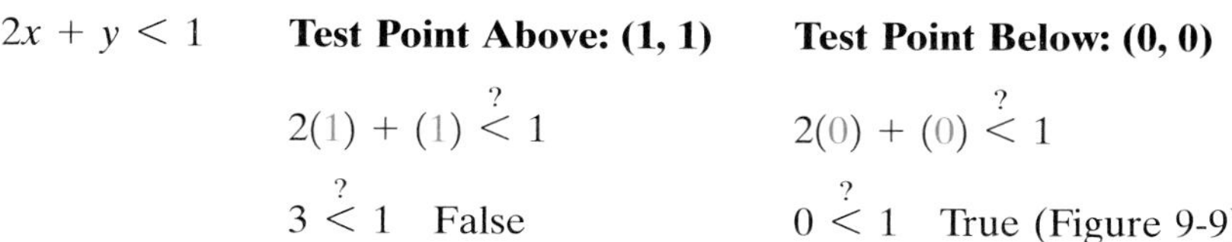

$2x + y < 1$	**Test Point Above: (1, 1)**	**Test Point Below: (0, 0)**
	$2(1) + (1) \stackrel{?}{<} 1$	$2(0) + (0) \stackrel{?}{<} 1$
	$3 \stackrel{?}{<} 1$ False	$0 \stackrel{?}{<} 1$ True (Figure 9-9)

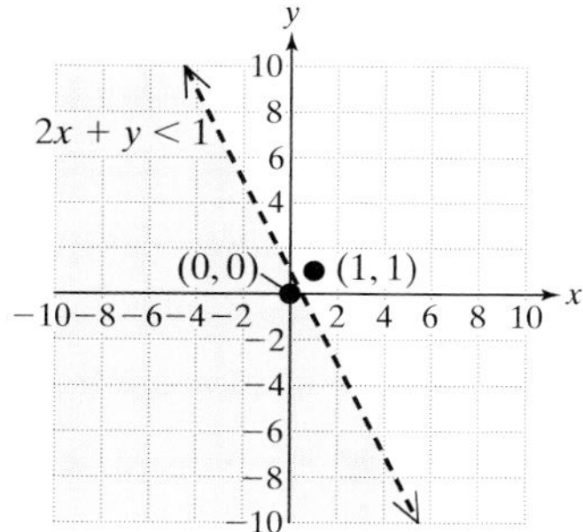

Figure 9-9

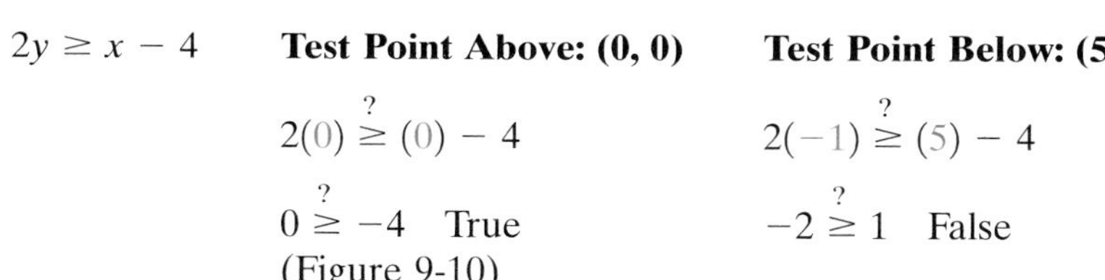

$2y \geq x - 4$	**Test Point Above: (0, 0)**	**Test Point Below: (5, −1)**
	$2(0) \stackrel{?}{\geq} (0) - 4$	$2(-1) \stackrel{?}{\geq} (5) - 4$
	$0 \stackrel{?}{\geq} -4$ True (Figure 9-10)	$-2 \stackrel{?}{\geq} 1$ False

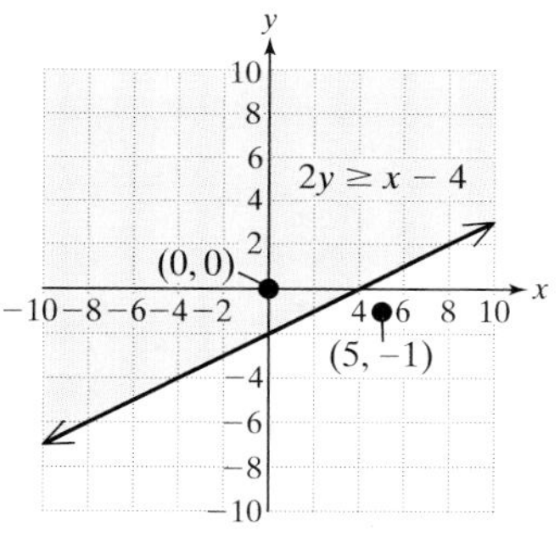

Figure 9-10

The solution to the compound inequality $2x + y < 1$ and $2y \geq x - 4$ is the intersection of the two individual solution sets. Therefore, the solution is the region of the plane below the line $y = -2x + 1$ and above the line $y = \frac{1}{2}x - 2$. This is the region shown in purple in Figure 9-11. By convention we will not display the extended boundary lines because they are not part of the solution set (see Figure 9-12).

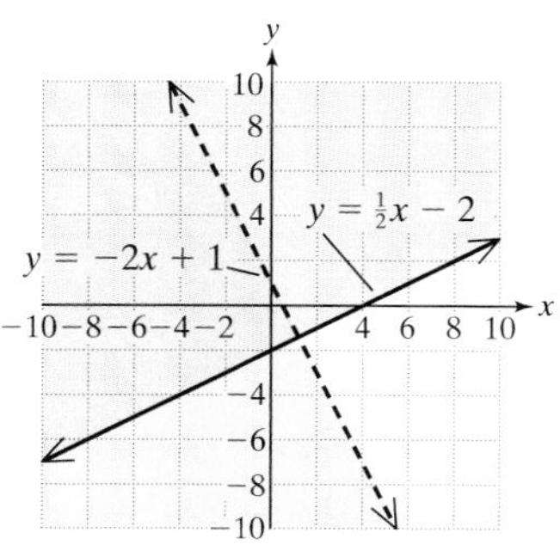

Figure 9-11

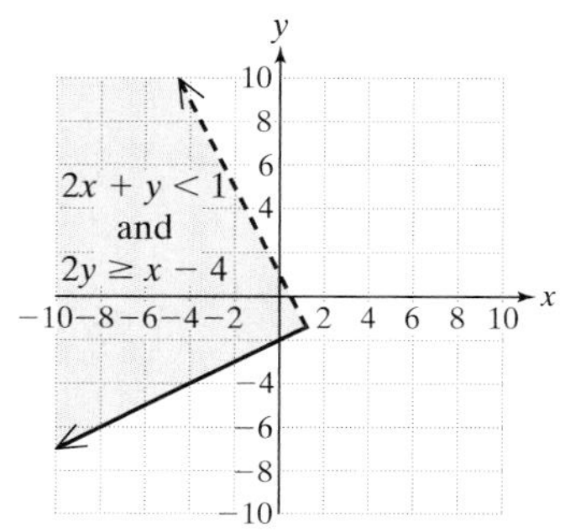

Figure 9-12

example 4 Graphing a Compound Linear Inequality in Two Variables

Graph the solution set of the inequality: $3y \geq 6$ or $y - x < 0$.

Solution:

First Inequality		**Second Inequality**	
$3y \geq 6$		$y - x < 0$	
$3y = 6$	Related equation	$y - x = 0$	Related equation
$y = 2$		$y = x$	

For each inequality, draw the boundary line. Then, pick test points above and below the line to determine the appropriate region to shade:

Figure 9-13

$3y \geq 6$ **Test Point Above: (0, 3)**

$$3(3) \overset{?}{\geq} 6$$

$$9 \overset{?}{\geq} 6 \quad \text{True (Figure 9-13)}$$

Test Point Below: (0, 1)

$$3(1) \overset{?}{\geq} 6$$

$$3 \overset{?}{\geq} 6 \quad \text{False}$$

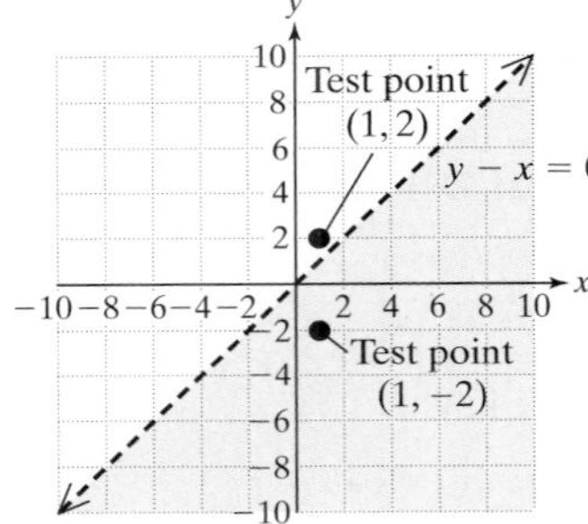

Figure 9-14

$y - x < 0$ **Test Point Above: (1, 2)**

$$(2) - (1) \overset{?}{<} 0$$

$$1 \overset{?}{<} 0 \quad \text{False}$$

Test Point Below: (1, −2)

$$(-2) - (1) \overset{?}{<} 0$$

$$-3 \overset{?}{<} 0 \quad \text{True}$$

(Figure 9-14)

The solution to the compound inequality $3y \geq 6$ or $y - x < 0$ is the union of the two solution sets (Figure 9-15).

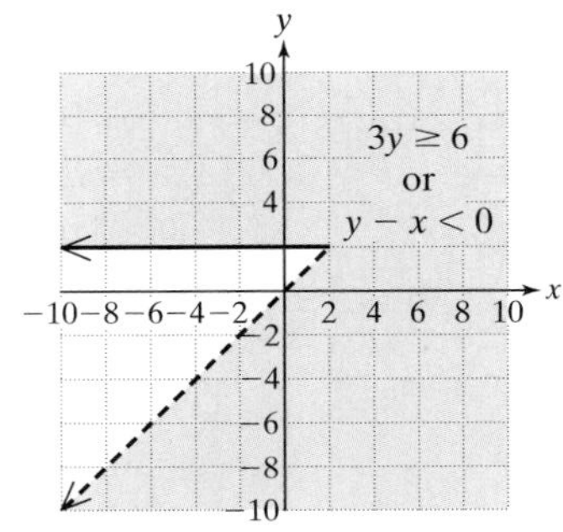

Figure 9-15

example 5 Graphing Compound Linear Inequalities

Describe the region of the plane defined by the following systems of inequalities.

a. $x > 0$ and $y < 0$ b. $x \leq 0$ and $y \geq 0$ c. $|x| \leq 4$ and $|y| \leq 4$

Solution:

a. $x > 0$ — $x > 0$ in the first and fourth quadrants.

$y < 0$ — $y < 0$ in the third and fourth quadrants.

The intersection of these inequalities is the set of points in the fourth quadrant (with the boundary excluded).

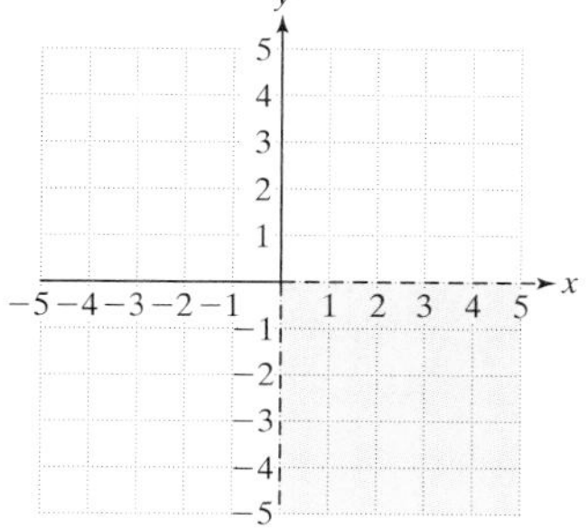

b. $x \leq 0$ — $x \leq 0$ in the second and third quadrants.

$y \geq 0$ — $y \geq 0$ in the first and second quadrants.

The intersection of these regions is the set of points in the second quadrant (with the boundary included).

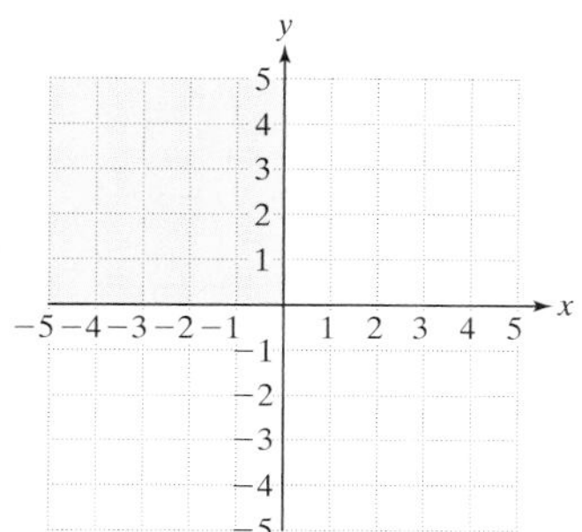

c. $|x| \leq 4$ and $|y| \leq 4$

$|x| \leq 4$ represents the set of points whose x-coordinates are between 4 and -4, inclusive.

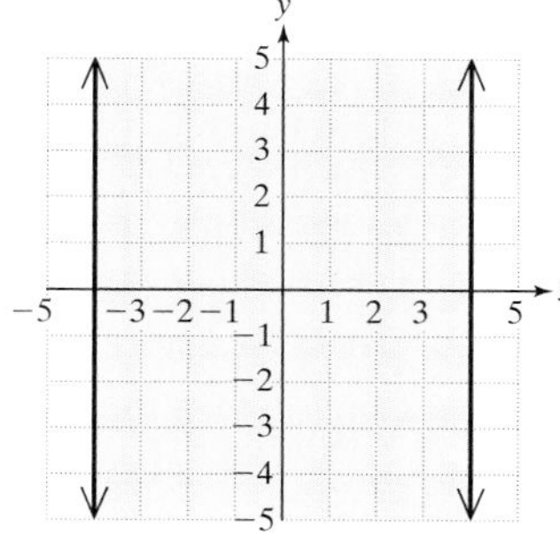

$|y| \leq 4$ represents the set of points whose y-coordinates are between 4 and -4, inclusive.

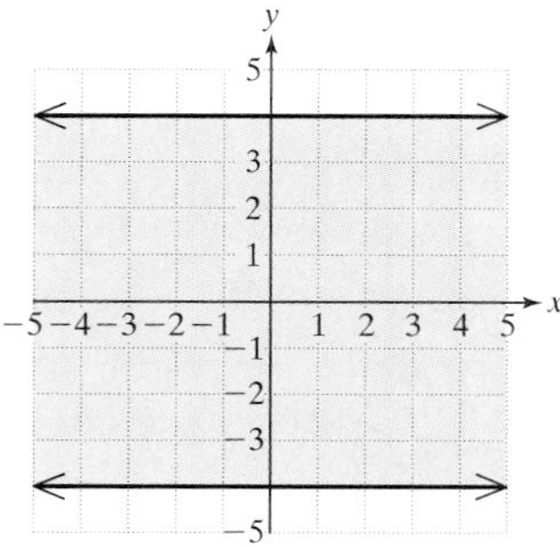

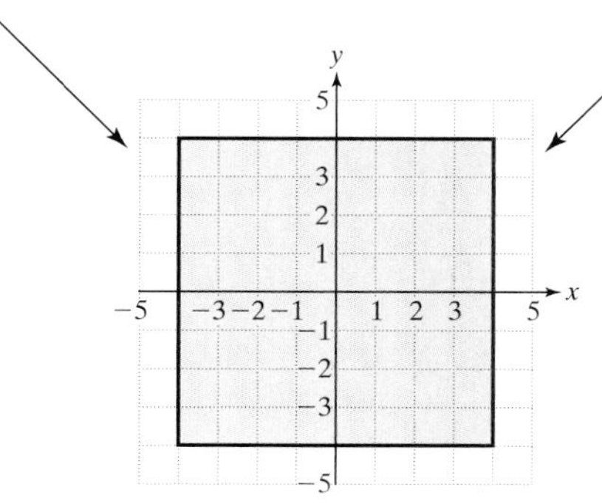

The intersection of the two inequalities represents a square.

4. Graphing a Feasible Region

When two variables are related under certain constraints, a system of linear inequalities can be used to show a region of feasible values for the variables.

example 6

Graphing a Feasible Region

Susan has two tests on Friday: one in chemistry and one in psychology. Because the two classes meet in consecutive hours, she has no study time between tests. Susan estimates that she has a maximum of 12 hr of study time before the tests, and she must divide her time between chemistry and psychology.

Let x represent the number of hours Susan spends studying chemistry.

Let y represent the number of hours Susan spends studying psychology.

a. Find a set of inequalities to describe the constraints on Susan's study time.
b. Graph the constraints to find the feasible region defining Susan's study time.

Solution:

a. Because Susan cannot study chemistry or psychology for a negative period of time, we have: $x \geq 0$ and $y \geq 0$.

Furthermore, her total time studying cannot exceed 12 hr: $x + y \leq 12$.

A system of inequalities that defines the constraints on Susan's study time is

$$x \geq 0$$

$$y \geq 0$$

$$x + y \leq 12$$

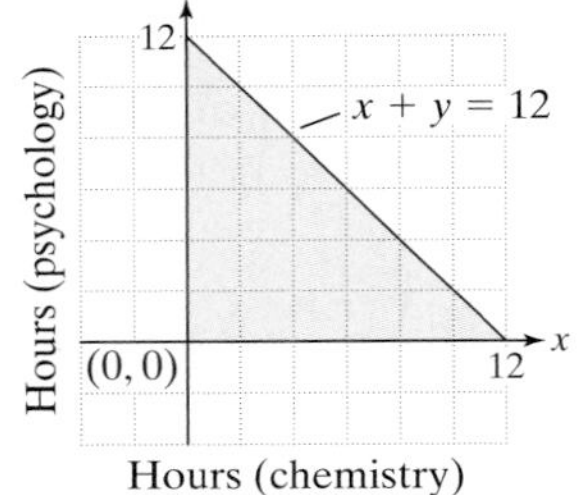

Figure 9-16

b. The first two conditions $x \geq 0$ and $y \geq 0$ represent the set of points in the first quadrant. The third condition, $x + y \leq 12$ represents the set of points below and including the line $x + y = 12$ (Figure 9-16).

Discussion:

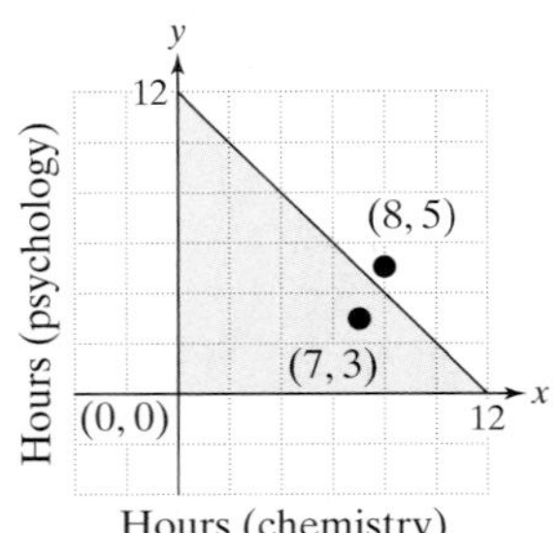

Figure 9-17

1. Refer to the feasible region drawn in Example 6(b). Is the ordered pair (8, 5) part of the feasible region?

   No. The ordered pair (8, 5) indicates that Susan spent 8 hr studying chemistry and 5 hr studying psychology. This is a total of 13 hr, which exceeds the constraint that Susan only had 12 hr to study. The point (8, 5) lies outside the feasible region, above the line $x + y = 12$ (Figure 9-17).

2. Is the ordered pair (7, 3) part of the feasible region?

   Yes. The ordered pair (7, 3) indicates that Susan spent 7 hr studying chemistry and 3 studying psychology.

   This point lies within the feasible region and satisfies all three constraints.

$$x \geq 0 \longrightarrow \quad 7 \geq 0 \quad \text{True}$$

$$y \geq 0 \longrightarrow \quad 3 \geq 0 \quad \text{True}$$

$$x + y \leq 12 \longrightarrow (7) + (3) \leq 12 \quad \text{True}$$

Notice that the ordered pair (7, 3) corresponds to a point where Susan is not making full use of the 12 hr of study time.

3. Suppose there was one additional constraint imposed on Susan's study time. She knows she needs to spend at least twice as much time studying chemistry as she does studying psychology. Graph the feasible region with this additional constraint.

   Because the time studying chemistry must be at least twice the time studying psychology, we have: $x \geq 2y$.

   This inequality may also be written as: $y \leq x/2$

   Figure 9-18 shows the feasible region with the additional constraint $y \leq x/2$.

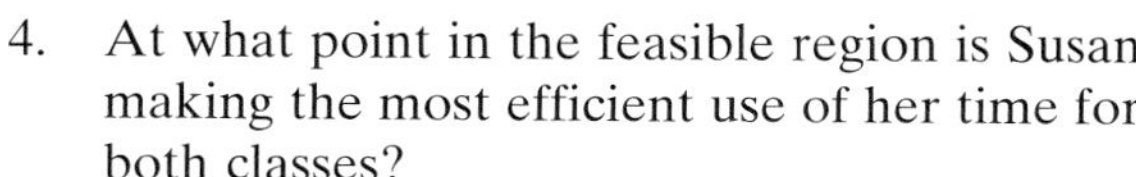

4. At what point in the feasible region is Susan making the most efficient use of her time for both classes?

   First and foremost, Susan must make use of *all* 12 hr. This occurs for points along the line $x + y = 12$. Susan will also want to study for both classes with approximately twice as much time devoted to chemistry. Therefore, Susan will be deriving the maximum benefit at the point of intersection of the line $x + y = 12$ and the line $y = x/2$.

   Using the substitution method, replace $y = x/2$ into the equation $x + y = 12$.

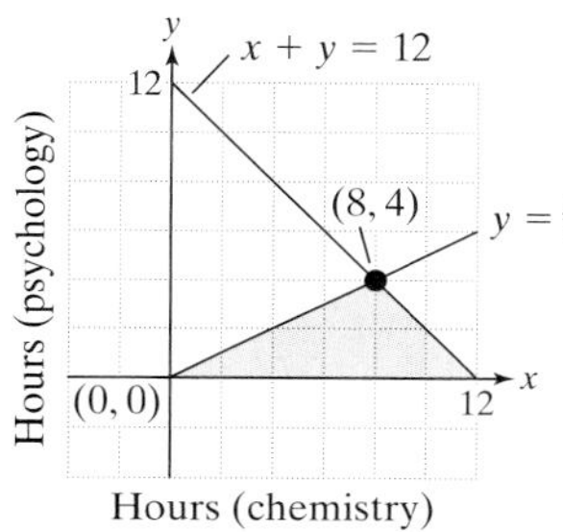

Figure 9-18

Tip: In Figure 9-18 the extended portion of each boundary line is shown for reference only and is not part of the solution set.

$$x + \frac{x}{2} = 12$$

$$2x + x = 24 \qquad \text{Clear fractions.}$$

$$3x = 24$$

$$x = 8 \qquad \text{Solve for } x.$$

$$y = \frac{(8)}{2} \qquad \text{To solve for } y \text{, substitute } x = 8 \text{ into the equation } y = x/2.$$

$$y = 4$$

Therefore Susan should spend 8 hours studying chemistry and 4 hours studying psychology.

section 9.5 Practice Exercises

1. Explain how you would solve the inequality $|x + 3| > 4$.

2. Explain how you would solve the inequality $|2x - 1| < 6$.

For Exercises 3–8, solve the inequalities.

3. $-3 < 2k - 5 < 3$

4. $\frac{1}{2} < \frac{3}{4}y < \frac{3}{5}$

5. $|6a - 1| - 4 \leq 2$

6. $|3b + 5| - 8 \le 5$
7. $|2t + 1| + 4 \ge 7$
8. $|2h - 6| - 1 \ge 3$

For Exercises 9–12, decide if the following points are solutions to the inequality.

9. $2x - y > 8$
   a. $(3, -5)$
   b. $(-1, -10)$
   c. $(4, -2)$
   d. $(0, 0)$

10. $3y + x < 5$
   a. $(-1, 7)$
   b. $(5, 0)$
   c. $(0, 0)$
   d. $(2, -3)$

11. $y \le -2$
   a. $(5, -3)$
   b. $(-4, -2)$
   c. $(0, 0)$
   d. $(3, 2)$

12. $x \ge 5$
   a. $(4, 5)$
   b. $(5, -1)$
   c. $(8, 8)$
   d. $(0, 0)$

For Exercises 13–18, decide which inequality symbol should be used. ($<, >, \ge, \le$) by looking at the graph.

13.

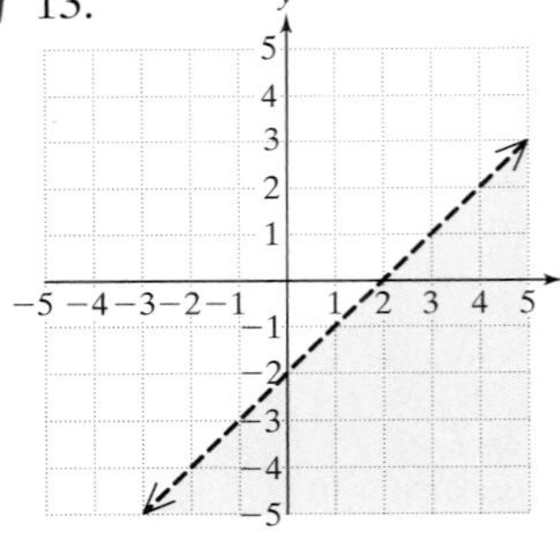

$x - y$ ____ 2

14.

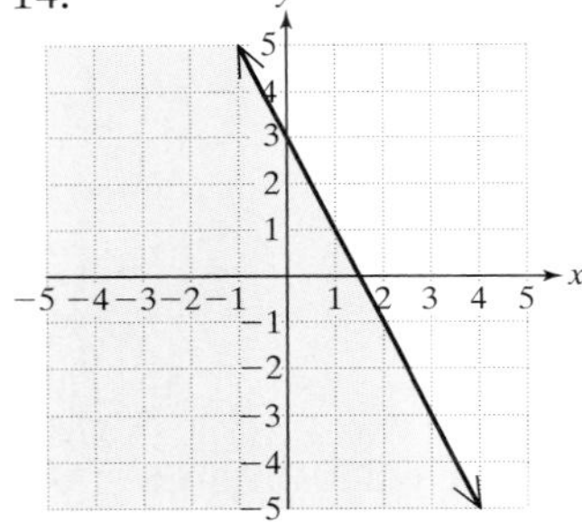

y ____ $-2x + 3$

15.

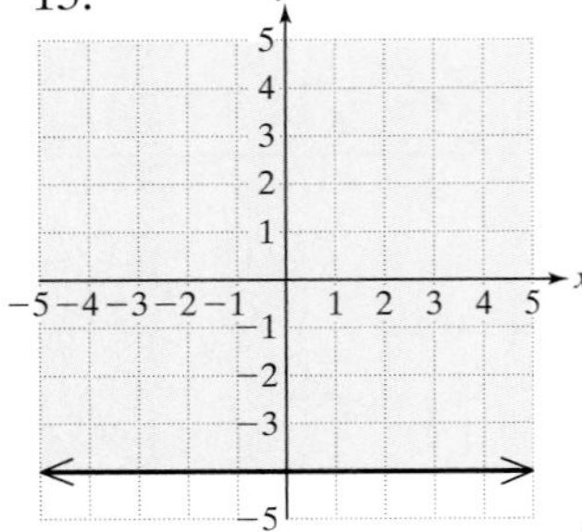

y ____ -4

16.

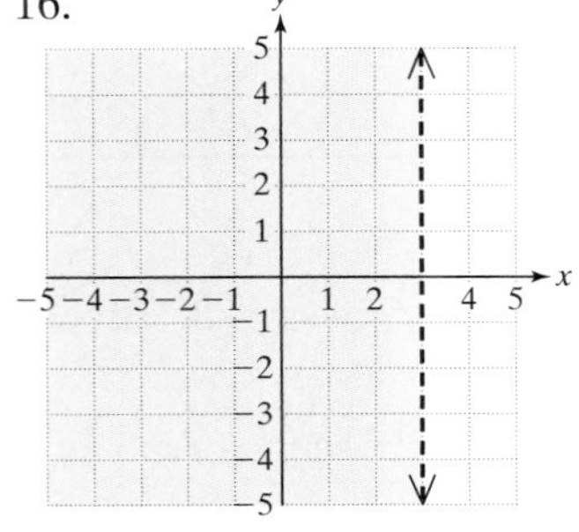

$x =$ ____ 3

17.

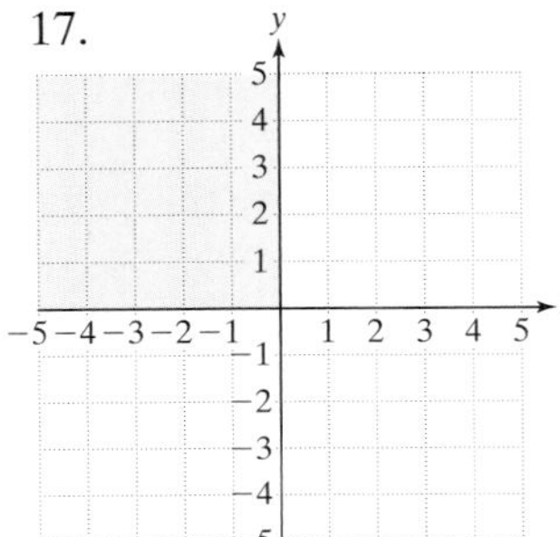

x ____ 0 and y ____ 0

18.

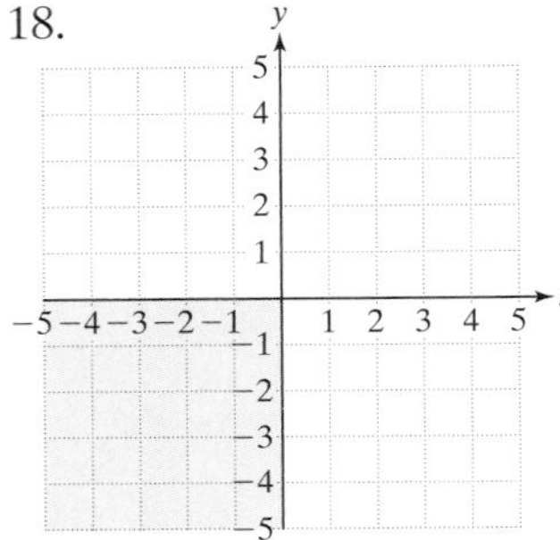

x ____ 0 and y ____ 0

For Exercises 19–38, solve the inequalities using the test point method.

19. $x - 2y > 4$
20. $x - 3y > 6$
21. $5x - 2y < 10$
22. $x - 3y < 8$
23. $2x + 6y \le 12$
24. $4x - 3y \le 12$
25. $y \ge -2$
26. $y \ge 5$
27. $4x < 5$
28. $x + 6 < 7$
29. $y \ge \frac{2}{5}x - 4$
30. $y \ge -\frac{5}{2}x - 4$
31. $y \le \frac{1}{3}x + 6$
32. $y \le -\frac{1}{4}x + 2$
33. $y > 5x$
34. $y > \frac{1}{2}x$
35. $\frac{x}{5} + \frac{y}{4} < 1$
36. $x + \frac{y}{2} \ge 2$
37. $0.1x + 0.2y \le 0.6$
38. $0.3x - 0.2y < 0.6$

For Exercises 39–54, graph the union or intersection of inequalities.

39. $y \le 4$ and $y \ge -x + 2$
40. $y < 3$ and $x + 2y < 6$
41. $2x + y < 5$ or $x > 3$
42. $x + 3y \ge 7$ or $x \le -2$
43. $x + y \le 3$ and $4x + y < 6$

44. $x + y < 4$ and $3x + y \leq 9$

45. $2x - y \leq 2$ or $2x + 3y > 6$

46. $3x + 2y > 4$ or $x - y \leq 3$

47. $x \geq 4$ and $y \leq 2$

48. $x \leq 3$ and $y \geq 4$

49. $x \leq -2$ or $y \leq 0$

50. $x \geq 0$ or $y \geq -3$

51. $x \geq 0$ and $x + y < 6$

52. $x \leq 0$ and $x + y < 2$

53. $y \leq 0$ or $x - y < -4$

54. $y \geq 0$ or $x - y > -3$

For Exercises 55–60, graph the feasible regions.

55. $x + y \leq 3$ and
$x \geq 0, y \geq 0$

56. $x - y \leq 2$ and
$x \geq 0, y \geq 0$

57. $y < \frac{1}{2}x - 3$ and
$x \leq 0, y \geq -4$

58. $y > \frac{1}{2}x - 3$ and
$x \geq -2, y \leq 0$

59. $x \geq 0, y \geq 0$
$x + y \leq 8$ and
$3x + 5y \leq 30$

60. $x \geq 0, y \geq 0$
$x + y \leq 5$ and
$x + 2y \leq 6$

61. In scheduling two drivers for delivering pizza, James needs to have at least 65 hours scheduled this week. His two drivers, Karen and Todd, are not allowed to get overtime, so each one can work at most 40 hours. Let x represent the number of hours that Karen can be scheduled and let y represent the number of hours Todd can be scheduled.

   a. Write two inequalities that express the fact that Karen and Todd cannot work a negative number of hours.

   b. Write two inequalities that express the fact that neither Karen nor Todd are allowed overtime (i.e., must have at most 40 hr).

   c. Write an inequality that expresses the fact that the total number of hours from both Karen and Todd needs to be at least 65.

   d. Graph the feasible region formed by graphing the inequalities.

62. A manufacturer produces two models of desks. Model A requires $1\frac{1}{2}$ hr to stain and finish and $1\frac{1}{4}$ hr to assemble. Model B requires 2 hr to stain and finish and $\frac{3}{4}$ hr to assemble. The total amount of time available for staining and finishing is 12 hr and for assembling is 6 hr. Let x represent the number of Model A desks, and let y represent the number of Model B desks.

   a. Write two inequalities that express the fact that the number of desks to be produced cannot be negative.

   b. Write an inequality in terms of the number of Model A and Model B desks that can be produced if the total time for staining and finishing is at most 12 hr.

   c. Write an inequality in terms of the number of Model A and Model B desks that can be produced if the total time for assembly is no more than 6 hr.

   d. Identify the feasible region formed by graphing the preceding inequalities.

chapter 9 SUMMARY

SECTION 9.1—COMPOUND INEQUALITIES

KEY CONCEPTS:

Solve two or more inequalities joined by *and* by finding the intersection of their solution sets. Solve two or more inequalities joined by *or* by finding the union of their solution sets.

Solve: $5y + 1 \geq 6$ or $2y - 5 \leq -11$

Solution: $5y \geq 5$ or $2y \leq -6$

$y \geq 1$ or $y \leq -3$

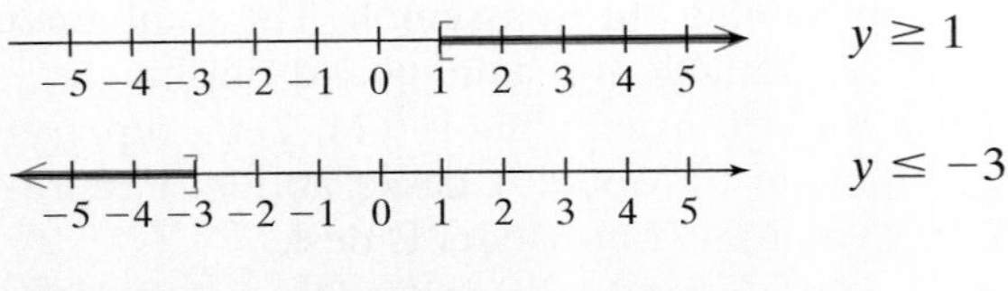

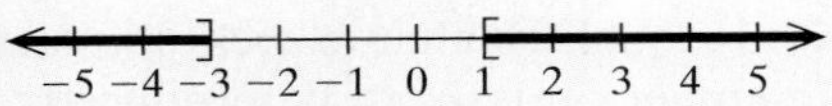

$y \geq 1$ or $y \leq -3$ or equivalently $(-\infty, -3] \cup [1, \infty)$

KEY TERMS:

and (intersection)
compound inequality
or (union)

EXAMPLES:

Solve:

$-7x + 3 \geq -11$ and $1 - x < 4.5$

Solution:

$-7x \geq -14$ and $-x < 3.5$

$x \leq 2$ and $x > -3.5$

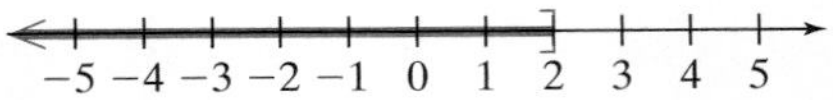

$x \leq 2$

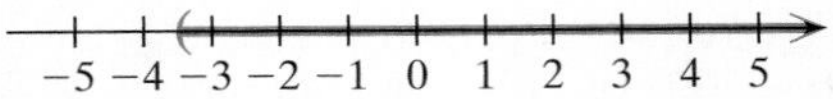

$x > -3.5$

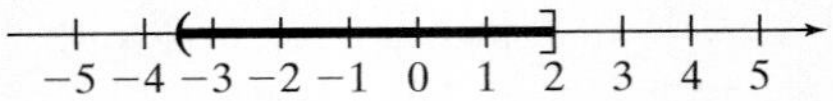

$x \leq 2$ and $x > -3.5$ or equivalently $(-3.5, 2]$

Solve: $-6 < \frac{3}{4}(x - 1) < 6$

Solution: $\frac{4}{3} \cdot -6 < \frac{4}{3} \cdot \frac{3}{4}(x - 1) < \frac{4}{3} \cdot 6$

$-8 < x - 1 < 8$

$-7 < x < 9$ or $(-7, 9)$

SECTION 9.2—POLYNOMIAL AND RATIONAL INEQUALITIES

KEY CONCEPTS:

The Test Point Method to Solve Polynomial and Rational Inequalities

1. Find the boundary points of the inequality. (Boundary points are the real solutions to the related equation and points where the inequality is undefined.)
2. Plot the boundary points on the number line. This divides the number line into regions.
3. Select a test point from each region and substitute it into the original inequality.

EXAMPLE:

Solve: $\frac{28}{2x - 3} \leq 4$

$\frac{28}{2x - 3} = 4$ Related equation

$(2x - 3) \cdot \frac{28}{2x - 3} = (2x - 3) \cdot 4$

$28 = 8x - 12$

$40 = 8x$

$x = 5$

- If a test point makes the original inequality true, then that region is part of the solution set.

4. Test the boundary points in the original inequality.

   - If a boundary point makes the original inequality true, then that point is part of the solution set.

Key Terms:

boundary points
quadratic inequality
test point method

The expression $28/(2x - 3)$ is undefined for $x = \frac{3}{2}$.

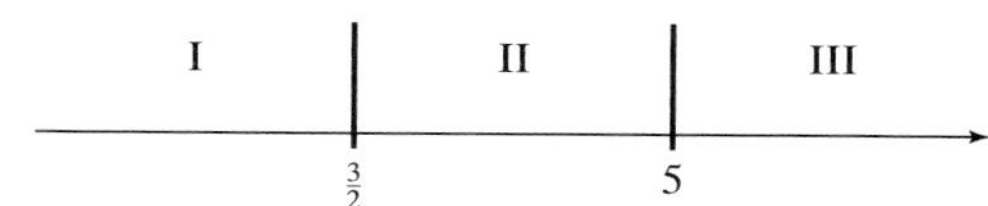

Region I: Test $x = 1$: $\frac{28}{2(1) - 3} \overset{?}{\leq} 4$ True

Region II: Test $x = 2$: $\frac{28}{2(2) - 3} \overset{?}{\leq} 4$ False

Region III: Test $x = 6$: $\frac{28}{2(6) - 3} \overset{?}{\leq} 4$ True

The boundary point $x = \frac{3}{2}$ is not included because $28/(2x - 3)$ is undefined there. The boundary $x = 5$ does check in the original inequality.

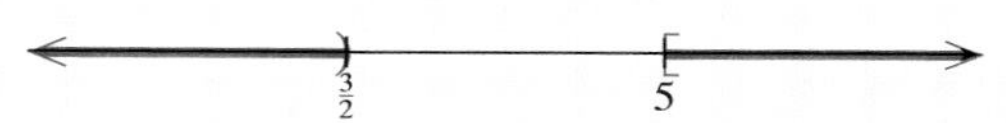

Interval notation: $(-\infty, \frac{3}{2}) \cup [5, \infty)$

Section 9.3—Absolute Value Equations

Key Concepts:

The equation $|x| = a$ is an absolute value equation. For $a \geq 0$, the solution to the equation $|x| = a$ is: $x = a$ or $x = -a$.

Steps to solve an absolute value equation

1. Isolate the absolute value to write the equation in the form $|x| = a$.
2. If $a < 0$, there is no solution.
3. Otherwise, rewrite the equation $|x| = a$ as $x = a$ or $x = -a$.
4. Solve the equations from step 3.
5. Check answers in the original equation.

The solution to the equation $|x| = |y|$ is $x = y$ or $x = -y$.

Key Term:

absolute value equation

Examples:

Solve: $|2x - 3| + 5 = 10$

$|2x - 3| = 5$

$2x - 3 = 5$ or $2x - 3 = -5$

$2x = 8$ or $2x = -2$

$x = 4$ or $x = -1$

Solve: $|x + 2| + 8 = 1$

$|x + 2| = -7$ No solution

Solve: $|2x - 1| = |x + 4|$

$2x - 1 = x + 4$ or $2x - 1 = -(x + 4)$

$x - 1 = 4$ or $2x - 1 = -x - 4$

$x = 5$ or $3x = -3$

$x = -1$

Section 9.4—Absolute Value Inequalities

Key Concepts:

Solutions to absolute value inequalities:

$$|x| > a \Rightarrow x < -a \quad \text{or} \quad x > a$$

$$|x| < a \Rightarrow -a < x < a$$

Test point method to solve inequalities

1. Find the boundary points of the inequality. (Boundary points are the real solutions to the related equation and points where the inequality is undefined.)
2. Plot the boundary points on the number line. This divides the number line into regions.
3. Select a test point from each region and substitute it into the original inequality.
   - If a test point makes the original inequality true, then that region is part of the solution set.
4. Test the boundary points in the original inequality.
   - If a boundary point makes the original inequality true, then that point is part of the solution set.

If a is *negative* ($a < 0$), then:

1. $|x| < a$ has no solution.
2. $|x| > a$ is true for all real numbers.

Key Term:

absolute value inequality

Examples:

Solve: $|5x - 2| < 12$

Solution:

$$-12 < 5x - 2 < 12$$

$$-10 < 5x < 14$$

$$-2 < x < \frac{14}{5} \quad \text{or} \quad \left(-2, \frac{14}{5}\right)$$

Solve using the test point method:

$$|x - 3| + 2 \geq 7$$

Solution: $|x - 3| \geq 5$ Isolate the absolute value.

$|x - 3| = 5$ Related equation

$x - 3 = 5$ or $x - 3 = -5$

$x = 8$ or $x = -2$ Boundary points

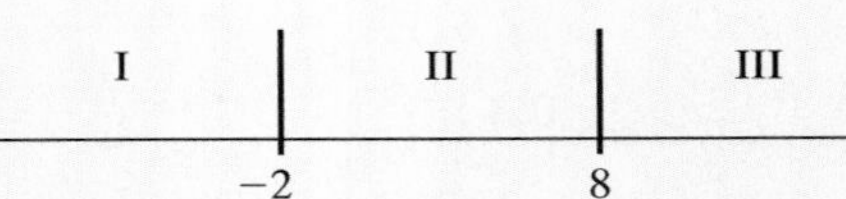

Region I:

Test $x = -3$: $|(-3) - 3| + 2 \overset{?}{\geq} 7$ True

Region II:

Test $x = 0$: $|(0) - 3| + 2 \overset{?}{\geq} 7$ False

Region III:

Test $x = 9$: $|(9) - 3| + 2 \overset{?}{\geq} 7$ True

$$(-\infty, -2] \cup [8, \infty)$$

True False True

−2 8

Section 9.5—Linear Inequalities in Two Variables

Key Concepts:

A linear inequality in two variables is an inequality of the form: $ax + by < c$, $ax + by > c$, $ax + by \leq c$, or $ax + by \geq c$, provided a and b are not both zero.

Use the test point method to solve a linear inequality in two variables. That is, graph the related equation and shade above or below the line.

If an inequality is strict ($<$, $>$) then a dashed line is used for the boundary. If the inequality contains $\leq$ or $\geq$, then a solid line is drawn.

The union or intersection of two or more linear inequalities is the union or intersection of the solution sets.

Key Term:

linear inequality in two variables

Examples:

Solve the inequality: $2x - y < 4$

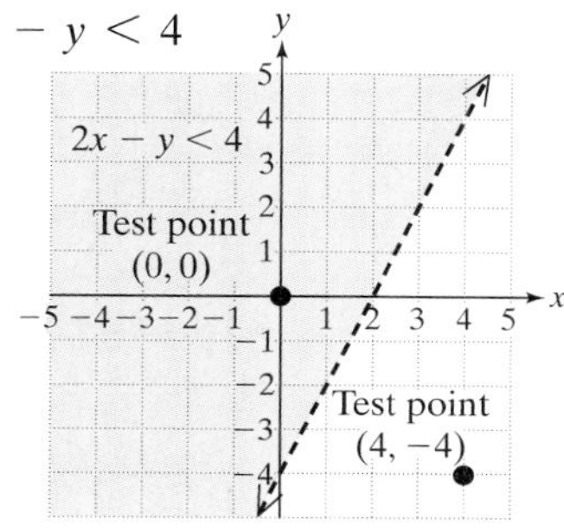

Test Points: (0, 0) and (4, −4)

$2(0) - (0) \overset{?}{<} 4$	True;	Shade above.
$2(4) - (-4) \overset{?}{<} 4$	False;	Do not shade below.

Solve:

$x < 0$ and $y \geq 2$

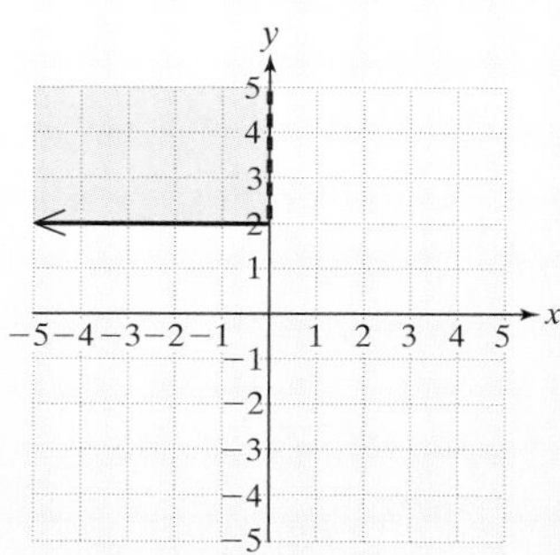

Solve:

$x < 0$ or $y \geq 2$

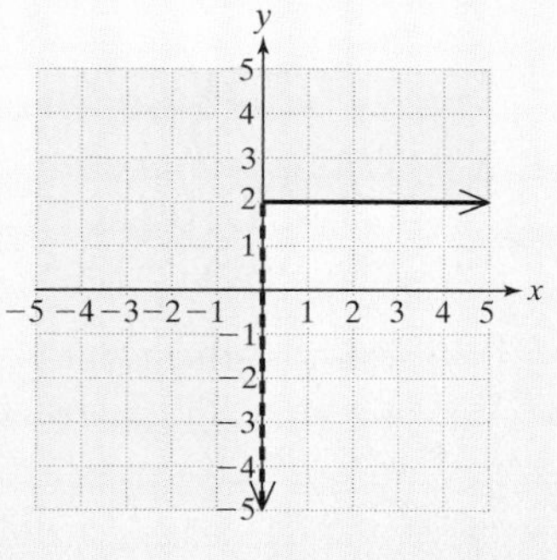

chapter 9 REVIEW EXERCISES

Section 9.1

For Exercises 1–10, solve the compound inequalities. Write the solutions in interval notation.

1. $4m > -11$ and $4m - 3 \leq 13$
2. $4n - 7 < 1$ and $7 + 3n \geq -8$
3. $-3y + 1 \geq 10$ and $-2y - 5 \leq -15$
4. $\frac{1}{2} - \frac{h}{12} \leq \frac{-7}{12}$ and $\frac{1}{2} - \frac{h}{10} > -\frac{1}{5}$
5. $\frac{2}{3}t - 3 \leq 1$ or $\frac{3}{4}t - 2 > 7$

6. $2(3x + 1) < -10$ or $3(2x - 4) \geq 0$

7. $-7 < -7(2w + 3)$ or $-2 < -4(3w - 1)$

8. $5(p + 3) + 4 > p - 1$ or $4(p - 1) + 2 > p + 8$

9. $2 \geq -(b - 2) - 5b \geq -6$

10. $-4 \leq \frac{1}{2}(x - 1) < -\frac{3}{2}$

11. The product of $\frac{1}{3}$ and the sum of a number and 3 is between −1 and 5. Find all such numbers.

12. Normal levels of total cholesterol vary according to age. For adults between 25 and 40 years old, the normal range is generally accepted to be between 140 and 225 mg/dL (milligrams per deciliter), inclusive.

    a. Write an inequality representing the normal range for total cholesterol for adults between 25 and 40 years old.

    b. Write a compound inequality representing abnormal ranges for total cholesterol for adults between 25 and 40 years old.

13. Normal levels of total cholesterol vary according to age. For adults younger than 25 years old, the normal range is generally accepted to be between 125 and 200 mg/dL, inclusive.

    a. Write an inequality representing the normal range for total cholesterol for adults younger than 25 years old.

    b. Write a compound inequality representing abnormal ranges for total cholesterol for adults younger than 25 years old.

14. In certain applications in statistics, a data value that is more than 3 standard deviations from the mean is said to be an "outlier" (a value unusually far from the average). If μ represents the mean of population, and σ represents the population standard deviation, then the inequality $|x - \mu| > 3\sigma$ can be used to test whether a data value, x, is an outlier.

    The mean height, μ, of adult men is 69.0 in. (5′9″) and the standard deviation, σ, of the height of adult men is 3.0. Determine whether the heights of the following men are outliers:

    a. Shaquille O'Neal, 7′1″ = 85 in.

    b. Charlie Ward, 6′3″ = 75 in.

    c. Elmer Fudd, 4′5″ = 53 in.

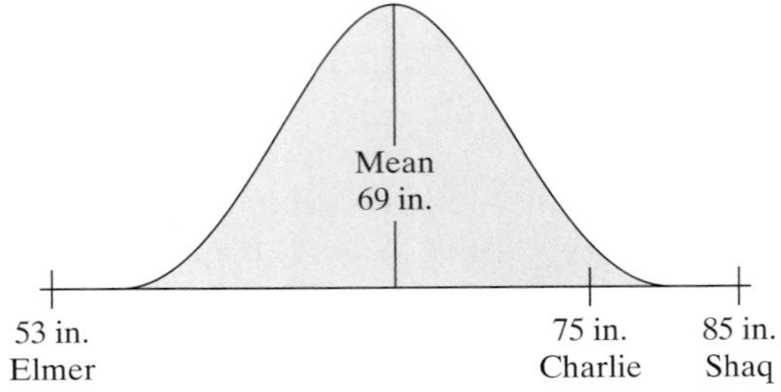

Figure for Exercise 14

15. Explain the difference between the solution sets of the following compound inequalities.

    a. $x \leq 5$ and $x \geq -2$

    b. $x \leq 5$ or $x \geq -2$

Section 9.2

16. Solve the equation and inequalities. How do your answers to parts (a), (b), and (c) relate to the graph of $f(x) = -\frac{1}{2}x - 3$?

    a. $-\frac{1}{2}x - 3 = 0$

    b. 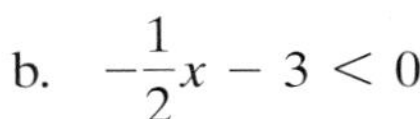 $-\frac{1}{2}x - 3 < 0$

    c. 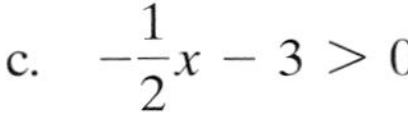 $-\frac{1}{2}x - 3 > 0$

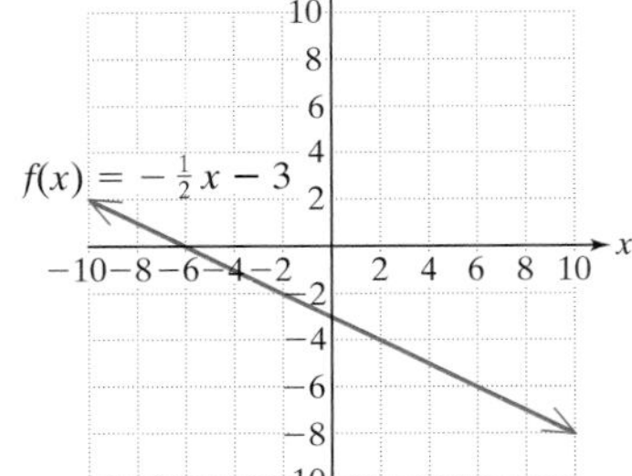

Figure for Exercise 16

17. Solve the equation and inequalities. How do your answers to parts (a), (b), and (c) relate to the graph of $g(x) = x^2 - 4$?

    a. $x^2 - 4 = 0$

    b. $x^2 - 4 < 0$

    c. $x^2 - 4 > 0$

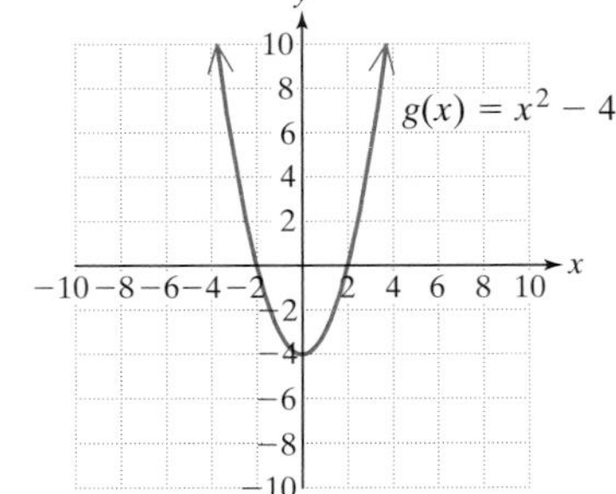

Figure for Exercise 17

18. Solve the equation and inequalities. How do your answers to parts (a), (b), (c), and (d) relate to the graph of $k(x) = 4x/(x - 2)$?

a. $\dfrac{4x}{x - 2} = 0$

b. For which values is $k(x)$ undefined?

c. $\dfrac{4x}{x - 2} \geq 0$

d. $\dfrac{4x}{x - 2} \leq 0$

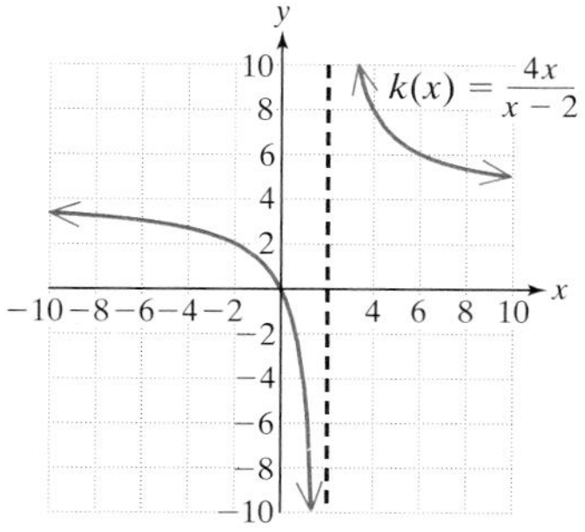

Figure for Exercise 18

For Exercises 19–30, solve the inequalities. Write the answers in interval notation.

19. $w^2 - 4w - 12 < 0$

20. $t^2 + 6t + 9 \geq 0$

21. $\dfrac{12}{x + 2} \geq 6$

22. $\dfrac{8}{p - 1} \leq -4$

23. $3y(y - 5)(y + 2) > 0$

24. $-3c(c + 2)(2c - 5) < 0$

25. $-x^2 - 4x \geq 4$

26. $y^2 + 4y > 5$

27. $\dfrac{w + 1}{w - 3} > 1$

28. $\dfrac{2a}{a + 3} < 2$

29. $t^2 + 10t + 25 \leq 0$

30. $-x^2 - 4x < 4$

Section 9.3

For Exercises 31–44, solve the absolute value equations.

31. $|x| = 10$

32. $|x| = 17$

33. $|y + 6| = \dfrac{1}{2}$

34. $|y - 3| = \dfrac{3}{4}$

35. $|8.7 - 2x| = 6.1$

36. $|5.25 - 5x| = 7.45$

37. $16 = |x + 2| + 9$

38. $5 = |x - 2| + 4$

39. $|4x - 1| + 6 = 4$

40. $|3x - 1| + 7 = 3$

41. $|7x - 3| = 0$

42. $|4x + 5| = 0$

43. $|3x - 5| = |2x + 1|$

44. $|8x + 9| = |8x - 1|$

45. Which absolute value expression represents the distance between 3 and -2 on the number line? Explain your answer.

$$|3 - (-2)|, \qquad |-2 - 3|$$

Section 9.4

46. Write the compound inequality $x < -5$ or $x > 5$ as an absolute value inequality.

47. Write the compound inequality $-4 < x < 4$ as an absolute value inequality.

48. Write an absolute value inequality that represents the solution sets graphed here:

a. 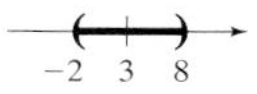

b. 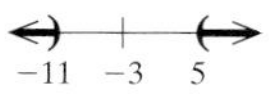

For Exercises 49–62, solve the absolute value inequalities. Graph the solution set and write the solution in interval notation.

49. $|x + 6| \geq 8$

50. $|x + 8| \leq 3$

51. $|7x - 1| > 0$

52. $|5x + 1| > 0$

53. $|3x + 4| - 6 \leq -4$

54. $|5x - 3| + 3 \leq 6$

55. $\left|\dfrac{x}{2} - 6\right| < 5$

56. $\left|\dfrac{x}{3} + 2\right| < 2$

57. $|2x - 4| + 2 > 8$

58. $|3x + 9| - 1 > 5$

59. $|5.2x - 7.8| > -13$

60. $|2.5x + 1.5| > -7$

61. $|3x - 8| < -1$

62. $|x + 5| < -4$

63. State one possible situation when an absolute value inequality will have no solution.

64. State one possible situation when an absolute value inequality will have a solution of all real numbers.

Section 9.5

For Exercises 65–74, solve the inequalities by graphing.

65. $2x + y < 5$

66. $2x + 3y \le -8$

67. $y \ge -\frac{2}{3}x + 3$

68. $y > \frac{3}{4}x - 2$

69. $x > -3$

70. $x \le 2$

71. $y < 4\frac{1}{3}$

72. $y \ge -2\frac{1}{2}$

73. $y \le 2x$

74. $y > \frac{5}{2}x$

For Exercises 75–78, solve the system of inequalities.

75. $2x - y > -2$ and $2x - y \le 2$

76. $3x + y \ge 6$ or $3x + y < -6$

77. $x \ge 0, y \ge 0$, and $y \ge -\frac{3}{2}x + 4$

78. $x \ge 0, y \ge 0$, and $y \le -\frac{2}{3}x + 4$

79. A pirate's treasure is buried on a small, uninhabited island in the eastern Caribbean. A shipwrecked sailor finds a treasure map at the base of a coconut palm tree. The treasure is buried within the intersection of three linear inequalities. The palm tree is located at the origin, and the positive y-axis is oriented due north. The scaling on the map is in 1-yd increments. Find the region where the sailor should dig for the treasure.

$$-2x + y \le 4$$
$$y \le -x + 6$$
$$y \ge 0$$

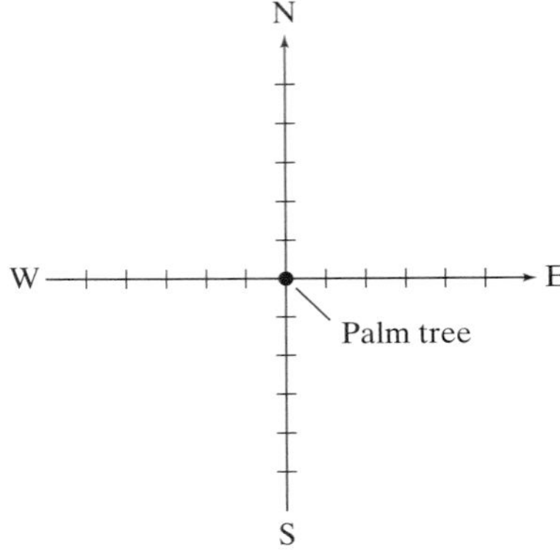

Figure for Exercise 79

80. Suppose a farmer has 100 acres of land on which to grow oranges and grapefruit. Furthermore, because of demand from his customers, he wants to plant at least four times as many acres of orange trees as grapefruit trees.

Let x represent the number of acres of orange trees.

Let y represent the number of acres of grapefruit trees.

a. Write two inequalities that express the fact that the farmer cannot use a negative number of acres to plant orange and grapefruit trees.

b. Write an inequality that expresses the fact that the total number of acres used for growing orange and grapefruit trees is at most 100.

c. Write an inequality that expresses the fact that the farmer wants to plant at least four times as many orange trees as grapefruit trees.

d. Sketch the inequalities in parts (a)–(c) to find the feasible region for the farmer's distribution of orange and grapefruit trees.

chapter 9 TEST

1. Solve the compound inequalities. Write the answers in interval notation.
   a. $-2 \le 3x - 1 \le 5$
   b. $-\frac{3}{5}x - 1 \le 8$ or $-\frac{2}{3}x \ge 16$
   c. $-2x - 3 > -3$ and $x + 3 \ge 0$
2. a. $5x + 1 \le 6$ or $2x + 4 > -6$
   b. $2x - 3 > 1$ and $x + 4 < -1$
3. The normal range in humans of the enzyme adenosine deaminase, ADA, is between 9 and 33 IU (international units), inclusive. Let x represent the ADA level in international units.
   a. Write an inequality representing the normal range for ADA.
   b. Write a compound inequality representing abnormal ranges for ADA.
   c. Patients with tuberculosis have unusually high levels of adenosine deaminase and physicians strongly suspect tuberculosis if a patient's ADA level is greater than 40 IU. Write an inequality representing ADA levels at which tuberculosis is suspected.

For Exercises 4–9, solve the polynomial and rational inequalities.

4. $\frac{2x - 1}{x - 6} \le 0$
5. $50 - 2a^2 > 0$
6. $y^3 + 3y^2 - 4y - 12 < 0$
7. $\frac{3}{w + 3} > 2$
8. $\frac{p^2}{2 + p^2} < 0$
9. $t^2 + 22t + 121 \le 0$
10. Solve the absolute value equations.
    a. $\left|\frac{1}{2}x + 3\right| = 8$
    b. $|3x + 4| = |x - 12|$
11. Solve the following equation and inequalities. How do your answers to parts (a)–(c) relate to the graph of $f(x) = |x - 3| - 4$?
    a. $|x - 3| - 4 = 0$
    b. $|x - 3| - 4 < 0$
    c. $|x - 3| - 4 > 0$

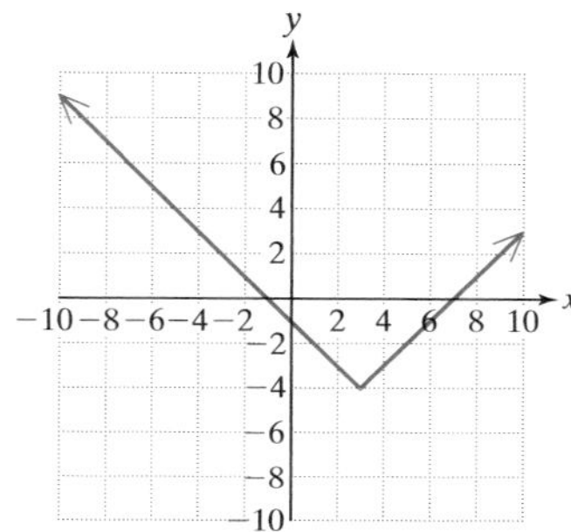

Figure for Exercise 11

For Exercises 12–15, solve the absolute value inequalities. Write the answers in interval notation.

12. $|3 - 2x| + 6 < 2$
13. $|3x - 8| > 9$
14. $|0.4x + 0.3| - 0.2 < 7$
15. $|7 - 3x| + 1 > -3$
16. The mass of a small piece of metal is measured to be 15.41 g. If the measurement error is ± 0.01 g, write an absolute value inequality that represents the possible mass, x, of the piece of metal.
17. Graph the solution to the inequality $2x - 5y \le 10$.
18. Solve the system of inequalities by graphing. $x + y < 3$ and $3x - 2y \ge -6$
19. After menopause, women are at higher risk for hip fractures as a result of low calcium. As early as their teen years, women need at least 1000 mg of calcium per day (the USDA recommended daily allowance). One 8-oz glass of skim milk contains 300 mg of calcium, and one Tums (regular strength) contains 400 mg of calcium. Let x represent the number of 8-oz glasses of milk that a woman drinks per day. Let y represent the

number of Tums tablets (regular strength) that a woman takes per day.

a. Write two inequalities that express the fact that the number of glasses of milk and the number of Tums taken each day cannot be negative.

b. Write a linear inequality in terms of x and y for which the daily calcium intake is at least 1000 mg.

c. Graph the inequalities.

CUMULATIVE REVIEW EXERCISES, CHAPTERS 1–9

1. Perform the indicated operations: $(2x - 3)(x - 4) - (x - 5)^2$

2. Solve the equation: $-9m + 3 = 2m(m - 4)$

For Exercises 3–4, solve the equation and inequalities. Write the solution to the inequalities in interval notation.

3. a. $2|3 - p| - 4 = 2$

   b. $2|3 - p| - 4 < 2$

   c. $2|3 - p| - 4 > 2$

4. a. $\left|\dfrac{y - 2}{4}\right| - 6 = -3$

   b. $\left|\dfrac{y - 2}{4}\right| - 6 < -3$

   c. $\left|\dfrac{y - 2}{4}\right| - 6 > -3$

5. Graph the inequality: $4x - y > 12$

6. The time, t, in minutes required for a rat to run through a maze depends on the number of trials, n, that the rat has practiced.

$$t(n) = \frac{3n + 15}{n + 1}; \quad n \geq 1$$

   a. Find $t(1)$, $t(50)$, and $t(500)$, and interpret the results in the context of this problem. Round to two decimal places, if necessary.

   b. Does there appear to be a limiting time in which the rat can complete the maze?

   c. How many trials are required so that the rat is able to finish the maze in under 5 min?

7. a. Solve the inequality: $2x^2 + x - 10 \geq 0$

   b. How does the answer in part (a) relate to the graph of the function 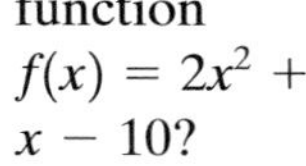 $f(x) = 2x^2 + x - 10$?

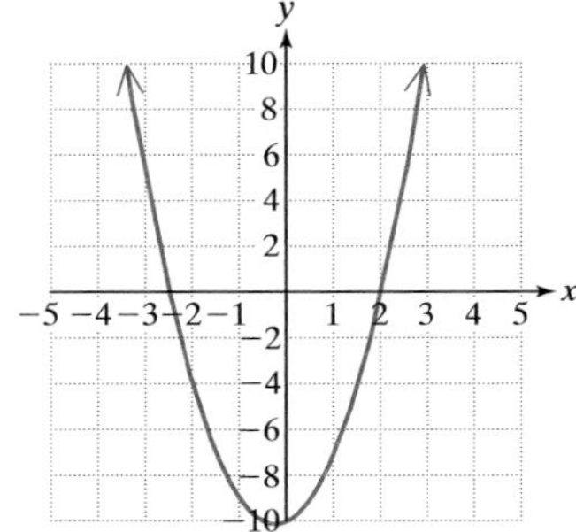

Figure for Exercise 7

8. Shade the region defined by the compound inequality: $3x + y < -2$ or $y \geq 1$.

9. Simplify the expression. $2 - 3(x - 5) + 2[4 - (2x + 6)]$

10. McDonald's corporation is the world's largest food service retailer. At the end of 1996, McDonald's operated 2.1×10^4 restaurants in over 100 countries. Worldwide sales in 1996 were nearly $\$3.18 \times 10^{10}$. Find the average sales per restaurant in 1996. Write the answer in scientific notation.

11. a. Divide the polynomials: $(2x^4 - x^3 + 5x - 7) \div (x^2 + 2x - 1)$. Identify the quotient and remainder.

    b. Check your answer by multiplying.

12. The area of a trapezoid is given by $A = \frac{1}{2}h(b_1 + b_2)$

    a. Solve for b_1.

    b. Find b_1 when $h = 4$ cm, $b_2 = 6$ cm, and $A = 32\text{ cm}^2$.

13. The speed of a car varies inversely as the time to travel a fixed distance. A car traveling the speed limit, 60 mph, travels between two points in

10 sec. How fast is a car moving if it takes only 8 sec to cover the same distance?

14. Two angles are supplementary. One angle is 9° more than twice the other angle. Find the angles.

15. Chemique invests \$3000 less in an account bearing 5% simple interest than she does in an account bearing 6.5% simple interest. At the end of one year, she earns a total \$770 in interest. Find the amount invested in each account.

16. Determine algebraically whether the lines are parallel, perpendicular, or neither:

$$\begin{aligned} 4x - 2y &= 5 \\ -3x + 6y &= 10 \end{aligned}$$

17. Find the x- and y-intercepts and slope (if they exist) of the lines. Then graph the lines.

    a. $3x + 5 = 8$ b. $\frac{1}{2}x + y = 4$

18. Find an equation of the line with slope $-\frac{2}{3}$ passing through the point $(4, -7)$. Write the final answer in slope-intercept form.

19. Solve the system of equations.

$$\begin{aligned} 3x + y &= z + 2 \\ y &= 1 - 2x \\ 3z &= -2y \end{aligned}$$

20. Identify the order of the matrices:

    a. $\begin{bmatrix} 2 & 4 & 5 \\ -1 & 0 & 1 \\ 9 & 2 & 3 \\ 3 & 0 & 1 \end{bmatrix}$ b. $\begin{bmatrix} 5 & 6 & 3 \\ 6 & 0 & -1 \\ 0 & 1 & -2 \end{bmatrix}$

21. Against a head wind, a plane can travel 6240 miles in 13 hours. The return trip flying with the same wind, the plane can fly 6240 miles in 12 hours. Find the wind speed and the speed of the plane in still air.

22. The profit that a company makes manufacturing computer desks is given by: $P(x) = -\frac{1}{5}(x - 20)(x - 650);\quad x \geq 0$, where x is the number of desks produced and $P(x)$ is the profit in dollars.

    a. Is this function constant, linear, or quadratic?

    b. Find $P(0)$ and interpret the result in the context of this problem.

    c. Find the values of x, where $P(x) = 0$. Interpret the results in the context of this problem.

23. Given $h(x) = \sqrt{50 - x}$. Find the domain of h.

24. Simplify completely.

$$\frac{x^{-1} - y^{-1}}{y^{-2} - x^{-2}}$$

25. Divide.

$$\frac{a^3 + 64}{16 - a^2} \div \frac{a^3 - 4a^2 + 16a}{a^2 - 3a - 4}$$

26. Perform the indicated operations.

$$\frac{1}{x^2 - 7x + 10} + \frac{1}{x^2 + 8x - 20}$$

chapter

10

RADICALS AND COMPLEX NUMBERS

Conditions of extreme cold can lead to frostbite, hypothermia, and even death, if proper precautions are not taken. Therefore, many weather forecasters report temperatures as well as the *wind chill factor* (WCF) to measure the effects of cold. The WCF (measured in kilocalories) is a means of quantifying the amount of heat that can be lost to the air in one hour from an exposed surface area of one square meter.

For more information see the Technology Connections in MathZone at

www.mhhe.com/miller_oneill

The value of the wind chill factor, K, can be measured by

$$K = (\sqrt{100v} - v + k)(33 - T),$$

where v is the velocity of the wind measured in meters per second (m/s), T is the temperature measured in degrees Celsius (°C), and k is a constant equal to 10.45.

This and other radical expressions are presented in this chapter.

section

10.1 Definition of an nth Root

Concepts

1. Definition of a Square Root
2. Identifying Square Roots of a Real Number
3. Definition of an nth Root
4. Identifying the Principal nth Root of a Real Number
5. Radical Functions
6. Roots of Variable Expressions
7. Applications of the Pythagorean Theorem

1. Definition of a Square Root

The reverse operation to squaring a number is to find its square roots. For example, finding a square root of 36 is equivalent to asking: "What number when squared equals 36?"

One obvious answer to this question is 6 because $(6)^2 = 36$, but -6 will also work, because $(-6)^2 = 36$.

Definition of a Square Root

b is a square root of a if $b^2 = a$.

2. Identifying Square Roots of a Real Number

example 1 **Identifying Square Roots**

Identify the square roots of the real numbers

a. 25 b. 49 c. 0 d. -9

Solution:

a. 5 is a square root of 25 because $(5)^2 = 25$
-5 is a square root of 25 because $(-5)^2 = 25$

b. 7 is a square root of 49 because $(7)^2 = 49$
-7 is a square root of 49 because $(-7)^2 = 49$

c. 0 is a square root of 0 because $(0)^2 = 0$

d. There are no real numbers that when squared will equal a negative number; therefore, there are no real-valued square roots of -9.

Tip: All positive real numbers have two real-valued square roots: one positive and one negative. Zero has only one square root, which is 0 itself. Finally, for any negative real number, there are no real-valued square roots.

Recall from Section 1.2, that the positive square root of a real number can be denoted with a **radical sign**, $\sqrt{\ }$.

Notation for Positive and Negative Square Roots

Let a represent a positive real number. Then,

1. $\sqrt{a}$ is the *positive* square root of a. The positive square root is also called the **principal square root**.
2. $-\sqrt{a}$ is the *negative* square root of a.
3. $\sqrt{0} = 0$

example 2 **Simplifying a Square Root**

Simplify the square roots.

a. $\sqrt{36}$ b. $\sqrt{\frac{4}{9}}$ c. $\sqrt{0.04}$

Solution:

a. $\sqrt{36}$ denotes the positive square root of 36.
$\sqrt{36} = 6$

b. $\sqrt{\frac{4}{9}}$ denotes the positive square root of $\frac{4}{9}$.
$\sqrt{\frac{4}{9}} = \frac{2}{3}$

c. $\sqrt{0.04}$ denotes the positive square root of 0.04.
$\sqrt{0.04} = 0.2$

Avoiding Mistakes

a. $\sqrt{36} = 6$ (not -6)

b. $\sqrt{\frac{4}{9}} = \frac{2}{3}$ $\left(\text{not } -\frac{2}{3}\right)$

c. $\sqrt{0.04} = 0.2$ (not -0.2)

Tip: On page 47 of Chapter 1 we give a list of perfect squares and their square roots.

The numbers 36, $\frac{4}{9}$ and 0.04 are **perfect squares** because their square roots are rational numbers. Radicals that cannot be simplified to rational numbers are irrational numbers. Recall that an irrational number cannot be written as a terminating or repeating decimal. For example, the symbol, $\sqrt{13}$, is used to represent the *exact* value of the square root of 13. The symbol, $\sqrt{42}$, is used to represent the *exact* value of the square root of 42. These values can be approximated by a rational number by using a calculator.

$$\sqrt{13} \approx 3.605551275 \qquad \sqrt{42} \approx 6.480740698$$

Tip: Before using a calculator to evaluate a square root, try estimating the value first.

$\sqrt{13}$ must be a number between 3 and 4 because $\sqrt{9} < \sqrt{13} < \sqrt{16}$.

$\sqrt{42}$ must be a number between 6 and 7 because $\sqrt{36} < \sqrt{42} < \sqrt{49}$.

Calculator Connections

Use a calculator to approximate the values of $\sqrt{13}$ and $\sqrt{42}$.

```
√(13)
         3.605551275
√(42)
         6.480740698
```

A negative number cannot have a real number as a square root because no real number when squared is negative. For example, $\sqrt{-25}$ is *not* a real number because there is no real number, b, for which $(b)^2 = -25$.

example 3 **Evaluating Square Roots**

Simplify the square roots if possible.

a. $\sqrt{-144}$ b. $-\sqrt{144}$ c. $\sqrt{-0.01}$

Solution:

a. $\sqrt{-144}$ is *not* a real number.

b. $-\sqrt{144}$

$= -1 \cdot \sqrt{144}$

$\downarrow \quad \downarrow$

$= -1 \cdot \quad 12$

$= -12$

c. $\sqrt{-0.01}$ is *not* a real number.

Tip: For the expression $-\sqrt{144}$, the factor of -1 is *outside* the radical.

3. Definition of an *n*th Root

Finding a square root of a number is the reverse process of squaring a number. This concept can be extended to finding a third root (called a cube root), a fourth root, and in general, an ***n*th root**.

Definition of an *n*th Root

b is an nth root of a if $b^n = a$

The radical sign $\sqrt{\ }$ is used to denote the principal square root of a number. The symbol, $\sqrt[n]{\ }$, is used to denote the principal nth root of a number. In the expression, $\sqrt[n]{a}$, n is called the **index** of the radical, and a is called the **radicand**. For a square root, the index is 2, but it is usually not written ($\sqrt[2]{a}$ is denoted simply as $\sqrt{a}$). A radical with an index of 3 is called a **cube root**, $\sqrt[3]{a}$.

Definition of $\sqrt[n]{a}$

1. If n is a positive *even* integer and $a > 0$, then $\sqrt[n]{a}$ is the principal (positive) nth root of a.
2. If n is a positive *odd* integer, then $\sqrt[n]{a}$ is the nth root of a.
3. If n is any positive integer, then $\sqrt[n]{0} = 0$.

For the purpose of simplifying radicals, it is helpful to know the following powers:

Perfect Cubes	Perfect Fourth Powers	Perfect Fifth Powers
$1^3 = 1$	$1^4 = 1$	$1^5 = 1$
$2^3 = 8$	$2^4 = 16$	$2^5 = 32$
$3^3 = 27$	$3^4 = 81$	$3^5 = 243$
$4^3 = 64$	$4^4 = 256$	$4^5 = 1024$
$5^3 = 125$	$5^4 = 625$	$5^5 = 3125$

4. Identifying the Principal *n*th Root of a Real Number

example 4

Identifying the Principal *n*th Root of a Real Number

Simplify the expressions (if possible).

a. $\sqrt{4}$ b. $\sqrt[3]{64}$ c. $\sqrt[5]{-32}$ d. $\sqrt[4]{81}$

e. $\sqrt[6]{1{,}000{,}000}$ f. $\sqrt{-100}$ g. $\sqrt[4]{-16}$

Solution:

a. $\sqrt{4} = 2$ because $(2)^2 = 4$

b. $\sqrt[3]{64} = 4$ because $(4)^3 = 64$

c. $\sqrt[5]{-32} = -2$ because $(-2)^5 = -32$

d. $\sqrt[4]{81} = 3$ because $(3)^4 = 81$

e. $\sqrt[6]{1{,}000{,}000} = 10$ because $(10)^6 = 1{,}000{,}000$

f. $\sqrt{-100}$ is not a real number. No real number when squared equals -100.

g. $\sqrt[4]{-16}$ is not a real number. No real number when raised to the fourth power equals -16.

Avoiding Mistakes

When evaluating $\sqrt[n]{a}$, where n is even, always choose the principal (positive) root. Hence:

$\sqrt[4]{81} = 3$ (not -3)

$\sqrt[6]{1{,}000{,}000} = 10$ (not -10)

Examples 4(f) and 4(g) illustrate that an *n*th root of a negative quantity is not a real number if the index is even because no real number raised to an even power is negative.

Calculator Connections

A calculator can be used to approximate *n*th roots by using the $\sqrt[x]{\ }$ function. On most calculators, the index is entered first.

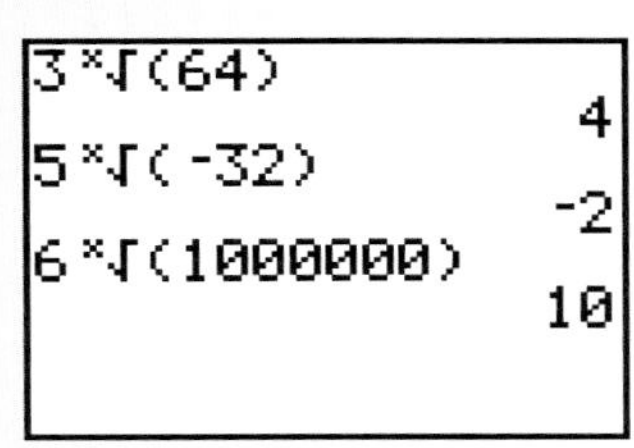

5. Radical Functions

If n is an integer greater than 1, then a function written in the form $f(x) = \sqrt[n]{x}$ is called a **radical function**. Note that if n is an even integer, then the function will be a real number only if the radicand is nonnegative. Therefore, the domain is restricted to nonnegative real numbers, or equivalently, $\{x \mid x \geq 0\}$. If n is an odd integer, then the domain is all real numbers.

example 5

Determining the Domain of Radical Functions

For each function, write the domain in interval notation.

a. $g(t) = \sqrt[4]{t - 2}$ b. $h(a) = \sqrt[3]{a - 4}$ c. $k(x) = \sqrt{3 - 5x}$

Solution:

a. $g(t) = \sqrt[4]{t - 2}$ The index is even. The radicand must be nonnegative.

$t - 2 \geq 0$ Set the radicand greater than or equal to zero.

$t \geq 2$ Solve for t.

The domain is $[2, \infty)$.

b. $h(a) = \sqrt[3]{a - 4}$ The index is odd; therefore, the domain is all real numbers.

The domain is $(-\infty, \infty)$.

c. $k(x) = \sqrt{3 - 5x}$ The index is even; therefore, the radicand must be nonnegative.

$3 - 5x \geq 0$ Set the radicand greater than or equal to zero.

$-5x \geq -3$ Solve for x.

$\dfrac{-5x}{-5} \leq \dfrac{-3}{-5}$ Reverse the inequality sign.

$x \leq \dfrac{3}{5}$

The domain is $(-\infty, \frac{3}{5}]$

Calculator Connections

The domain of the function defined by $k(x) = \sqrt{3 - 5x}$ can be confirmed from its graph.

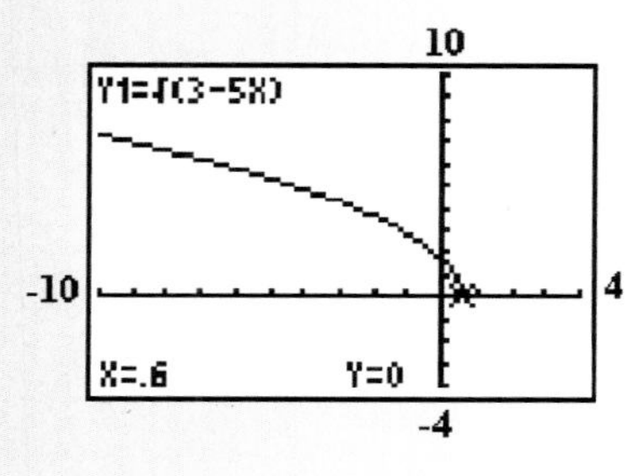

example 6 **Applying Radical Functions**

Ignoring air resistance, the velocity, v (in feet/second), of an object in free fall is a function of the distance it has fallen, x (in feet):

$$v(x) = 8\sqrt{x} \quad (x \geq 0)$$

Evaluate the function values and interpret their meaning in the context of the problem.

a. $v(25)$ b. $v(100)$

Solution:

a. $v(25) = 8\sqrt{25}$
$= 8 \cdot 5$
$= 40$ $v(25) = 40$ means that if an object has fallen 25 ft, its velocity is 40 ft/s.

b. $v(100) = 8\sqrt{100}$
$= 8 \cdot 10$
$= 80$ $v(100) = 80$ means that if an object has fallen 100 ft, its velocity is 80 ft/s.

6. Roots of Variable Expressions

Finding an nth root of a variable expression is similar to finding an nth root of a numerical expression. For roots with an even index, however, particular care must be taken to obtain a nonnegative solution.

Definition of $\sqrt[n]{a^n}$

1. If n is a positive *odd* integer, then $\sqrt[n]{a^n} = a$.
2. If n is a positive *even* integer, then $\sqrt[n]{a^n} = |a|$.

The absolute value bars are necessary for roots with an even index because the variable, a, may represent a positive quantity or a negative quantity. By using absolute value bars, $\sqrt[n]{a^n} = |a|$ is nonnegative and represents the principal nth root of a.

example 7 **Simplifying Expressions of the Form $\sqrt[n]{a^n}$**

Simplify the expressions.

a. $\sqrt[4]{(-3)^4}$ b. $\sqrt[5]{(-3)^5}$ c. $\sqrt{(x + 2)^2}$ d. $\sqrt[3]{(a + b)^3}$

Solution:

a. $\sqrt[4]{(-3)^4} = |-3| = 3$ Because this is an *even*-indexed root, absolute value bars are necessary to make the answer positive.

b. $\sqrt[5]{(-3)^5} = -3$ This is an *odd*-indexed root, so absolute value bars are not necessary.

c. $\sqrt{(x + 2)^2} = |x + 2|$ Because this is an *even*-indexed root, absolute value bars are necessary. The sign of the quantity $x + 2$ is unknown, however, $|x + 2| \geq 0$ regardless of the value of x.

d. $\sqrt[3]{(a + b)^3} = a + b$ This is an *odd*-indexed root, so absolute value bars are not necessary.

If n is an even integer, then $\sqrt[n]{a^n} = |a|$; however, if the variable a is assumed to be *nonnegative,* then the absolute value bars may be dropped. That is $\sqrt[n]{a^n} = a$ provided $a \geq 0$. In many examples and exercises, we will make the assumption that the variables within a radical expression are positive real numbers. In such a case, the absolute value bars are not needed to evaluate $\sqrt[n]{a^n}$.

It is helpful to become familiar with the patterns associated with perfect squares and perfect cubes involving variable expressions.

The following powers of x are perfect squares:

Tip: In general, any expression raised to an even power (a multiple of 2) is a perfect square.

Perfect Squares

$$(x^1)^2 = x^2$$
$$(x^2)^2 = x^4$$
$$(x^3)^2 = x^6$$
$$(x^4)^2 = x^8$$
$$\ldots$$

The following powers of x are perfect cubes:

Tip: In general, any expression raised to a power that is a multiple of 3 is a perfect cube.

Perfect Cubes

$$(x^1)^3 = x^3$$
$$(x^2)^3 = x^6$$
$$(x^3)^3 = x^9$$
$$(x^4)^3 = x^{12}$$
$$\ldots$$

These patterns may be extended to higher powers.

example 8 **Simplifying *n*th Roots**

Simplify the expressions. Assume that all variables are positive real numbers.

a. $\sqrt{y^8}$ b. $\sqrt[3]{27a^3}$ c. $\sqrt[5]{\dfrac{a^5}{b^5}}$ d. $-\sqrt[4]{\dfrac{81x^4y^8}{16}}$

Solution:

a. $\sqrt{y^8} = \sqrt{(y^4)^2} = y^4$

b. $\sqrt[3]{27a^3} = \sqrt[3]{(3a)^3} = 3a$

c. $\sqrt[5]{\dfrac{a^5}{b^5}} = \sqrt[5]{\left(\dfrac{a}{b}\right)^5} = \dfrac{a}{b}$

d. $-\sqrt[4]{\dfrac{81x^4y^8}{16}} = -\sqrt[4]{\left(\dfrac{3xy^2}{2}\right)^4} = -\dfrac{3xy^2}{2}$

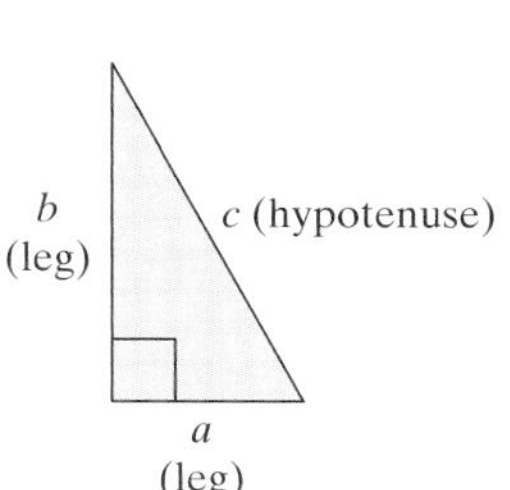

Figure 10-1

7. Applications of the Pythagorean Theorem

For the right triangle shown in Figure 10-1, the **Pythagorean theorem** is stated as: $a^2 + b^2 = c^2$. In this formula, a and b are the legs of the right triangle and c is the hypotenuse. Notice that the hypotenuse is the longest side of the right triangle and is opposite the 90° angle.

example 9 Applying the Pythagorean Theorem

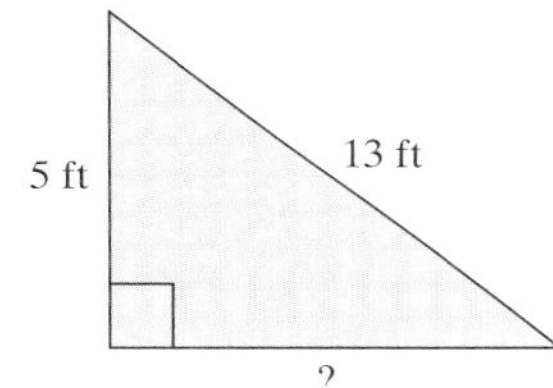

Use the Pythagorean theorem and the definition of the principal square root of a positive real number to find the length of the unknown side of the triangle shown to the left.

a = 5 ft

c = 13 ft

b = ?

Solution:

Label the sides of the triangle: $a = 5, b = ?, c = 13$.

$a^2 + b^2 = c^2$

$(5)^2 + b^2 = (13)^2$ Apply the Pythagorean theorem.

$25 + b^2 = 169$ Simplify.

$b^2 = 169 - 25$ Isolate b^2.

$b^2 = 144$ By definition, b must be one of the square roots of 144. Because b represents the length of a side of a triangle, choose the positive square root of 144.

$b = 12$

The third side is 12 ft long.

section 10.1 Practice Exercises

1. a. Find the square roots of 64.
   b. Find $\sqrt{64}$.
   c. Explain the difference between the answers in part (a) and part (b).
2. a. Find the square roots of 121.
   b. Find $\sqrt{121}$.
   c. Explain the difference between the answers in part (a) and part (b).
3. a. What is the principal square root of 81?
   b. What is the negative square root of 81?
4. a. What is the principal square root of 100?
   b. What is the negative square root of 100?

5. Using the definition of a square root, explain why $\sqrt{-36}$ is not a real number.

6. Using the definition of an *n*th root, explain why $\sqrt[4]{-36}$ is not a real number.

For Exercises 7–26, evaluate the roots without using a calculator. Identify those that are not real numbers.

7. $\sqrt{25}$
8. $-\sqrt{36}$
9. $\sqrt[3]{-27}$
10. $\sqrt{-9}$
11. $\sqrt[4]{16}$
12. $\sqrt{16}$
13. $\sqrt[3]{\dfrac{1}{8}}$
14. $\sqrt[5]{32}$
15. $\sqrt[6]{64}$
16. $\sqrt{64}$
17. $\sqrt[3]{64}$
18. $\sqrt{\dfrac{49}{100}}$
19. $\sqrt[4]{-81}$
20. $\sqrt[6]{-1}$
21. $\sqrt[5]{100{,}000}$
22. $-\sqrt[4]{625}$
23. $-\sqrt[3]{0.008}$
24. $-\sqrt{25}$
25. $-\sqrt{0.0144}$
26. $\sqrt[3]{\dfrac{27}{1000}}$

For Exercises 27–34, use a calculator to evaluate the expressions to four decimal places.

27. $\sqrt{69}$
28. $\sqrt{5798}$
29. $7\sqrt[4]{25}$
30. $-3\sqrt[3]{9}$
31. $2 + \sqrt[3]{5}$
32. $3 - 2\sqrt[4]{10}$
33. $\dfrac{3 - \sqrt{19}}{11}$
34. $\dfrac{5 + 2\sqrt{15}}{12}$

35. Use a calculator to evaluate (if possible) $h(x) = \sqrt{x - 2}$ for the given values of x (round to two decimal places). Then use interval notation to state the domain of h.

a. $h(-1)$ b. $h(0)$ c. $h(1)$ d. $h(2)$
e. $h(3)$ f. $h(4)$ g. $h(5)$ h. $h(6)$

36. Use a calculator to evaluate (if possible) $k(x) = \sqrt{x + 1}$ for the given values of x (round to two decimal places). Then use interval notation to state the domain of k.

a. $k(-4)$ b. $k(-3)$
c. $k(-2)$ d. $k(-1)$
e. $k(0)$ f. $k(1)$
g. $k(2)$ h. $k(3)$

37. Use a calculator to evaluate (if possible) $g(x) = \sqrt[3]{x - 2}$ for the given values of x (round to two decimal places). Then use interval notation to state the domain of g.

a. $g(-1)$ b. $g(0)$ c. $g(1)$ d. $g(2)$
e. $g(3)$ f. $g(4)$ g. $g(5)$ h. $g(6)$

38. Use a calculator to evaluate (if possible) $f(x) = \sqrt[3]{x + 1}$ for the given values of x (round to two decimal places). Then use interval notation to state the domain of f.

a. $f(-4)$ b. $f(-3)$
c. $f(-2)$ d. $f(-1)$
e. $f(0)$ f. $f(1)$
g. $f(2)$ h. $f(3)$

For each function defined in Exercises 39–42, write the domain in interval notation.

39. $p(x) = \sqrt{x - 1}$
40. $q(x) = \sqrt{x + 5}$
41. $R(x) = \sqrt[3]{x + 1}$
42. $T(x) = \sqrt{x - 10}$

For Exercises 43–46, match the function with the graph. Use the domain information from Exercises 39–42.

43. $p(x) = \sqrt{x - 1}$
44. $q(x) = \sqrt{x + 5}$
45. $T(x) = \sqrt{x - 10}$
46. $R(x) = \sqrt[3]{x + 1}$

a.
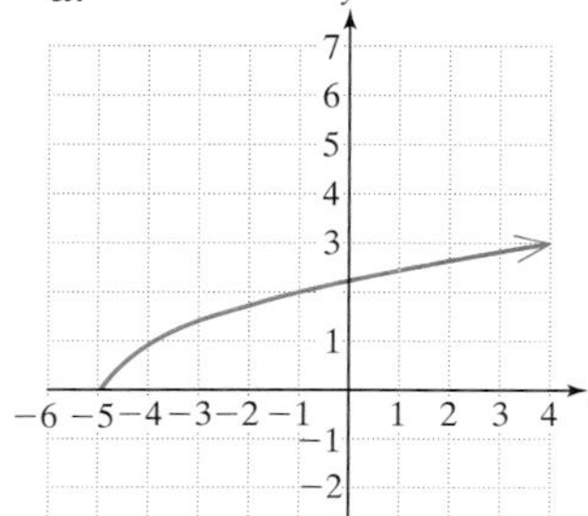

b.
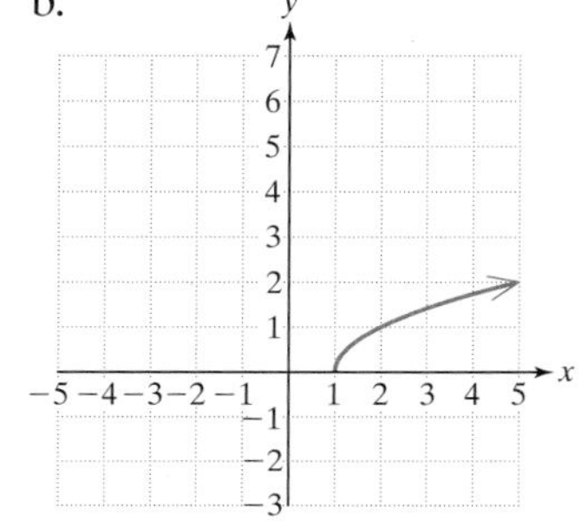

c.

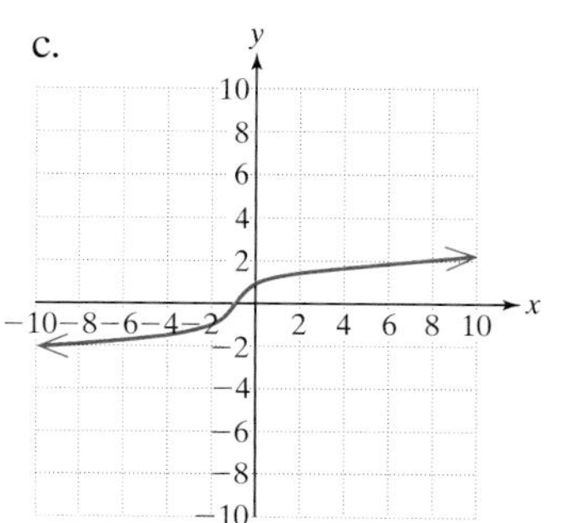

d.

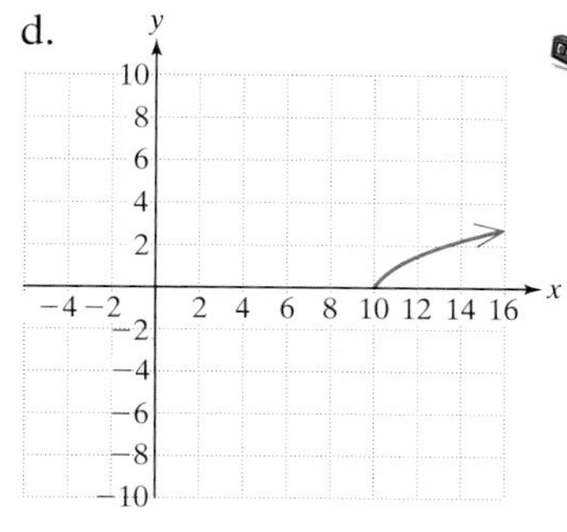

For Exercises 47–64, simplify the radical expressions.

47. $\sqrt{a^2}$

48. $\sqrt[4]{a^4}$

49. $\sqrt[3]{a^3}$

50. $\sqrt[5]{a^5}$

51. $\sqrt[6]{a^6}$

52. $\sqrt[7]{a^7}$

53. $\sqrt{x^4}$

54. $\sqrt[3]{y^{12}}$

55. $\sqrt{x^2y^4z^{10}}$

56. $\sqrt[3]{(u+v)^3}$

57. $-\sqrt[3]{\dfrac{x^3}{y^3}},\quad y \neq 0$

58. $\sqrt[4]{\dfrac{a^4}{b^8}},\quad b \neq 0$

59. $\dfrac{2}{\sqrt[4]{x^4}},\quad x \neq 0$

60. $\sqrt{(-5)^2}$

61. $\sqrt[3]{(-9)^3}$

62. $\sqrt[6]{(50)^6}$

63. $\sqrt[5]{(-2)^5}$

64. $\sqrt[10]{(-2)^{10}}$

For Exercises 65–80, simplify the expressions. Assume all variables are positive real numbers.

65. $\sqrt{x^2y^4}$

66. $\sqrt{16p^2}$

67. $\sqrt{\dfrac{a^6}{b^2}}$

68. $\sqrt{\dfrac{w^2}{z^4}}$

69. $-\sqrt{\dfrac{25}{q^2}}$

70. $-\sqrt{\dfrac{p^6}{81}}$

71. $\sqrt{9x^2y^4z^2}$

72. $\sqrt{4a^4b^2c^6}$

73. $\sqrt{\dfrac{h^2k^4}{16}}$

74. $\sqrt{\dfrac{4x^2}{y^8}}$

75. $-\sqrt[3]{\dfrac{t^3}{27}}$

76. $\sqrt[4]{\dfrac{16}{w^4}}$

77. $\sqrt[5]{32y^{10}}$

78. $\sqrt[3]{64x^6y^3}$

79. $\sqrt[6]{64p^{12}q^{18}}$

80. $\sqrt[4]{16r^{12}s^8}$

For Exercises 81–84, translate the English phrase into an algebraic expression.

81. The sum of q and the square of p.

82. The product of 11 and the cube root of x.

83. The quotient of 6 and the fourth root of x.

84. The difference of y and the square root of x.

For Exercises 85–88, translate the algebraic expression into an English phrase. Answers may vary.

85. $a^2 + \sqrt{b}$

86. $\sqrt[3]{\dfrac{x}{y}}$

87. $\dfrac{1}{(c+d)^2}$

88. $2(t + \sqrt{t})$

89. If a square has an area of 64 in.^{2}, then what are the lengths of the sides?

$s = ?$ | $A = 64$ in.^{2} | $s = ?$

90. If a square has an area of 121 m^2, then what are the lengths of the sides?

$s = ?$ | $A = 121$ m^2 | $s = ?$

91. If a square has an area of 97 in.^{2}, use a calculator to find the lengths of the sides. Round to the nearest tenth of an inch.

$A = 97$ in.^{2} | $s = ?$ | $s = ?$

92. If a square has an area of 45 cm^2, use a calculator to find the lengths of the sides. Round to the nearest tenth of a centimeter.

$A = 45$ cm^2 | $s = ?$ | $s = ?$

For Exercises 93–96, find the length of the third side of each triangle using the Pythagorean theorem.

93. 15 cm, 12 cm, ?

94.

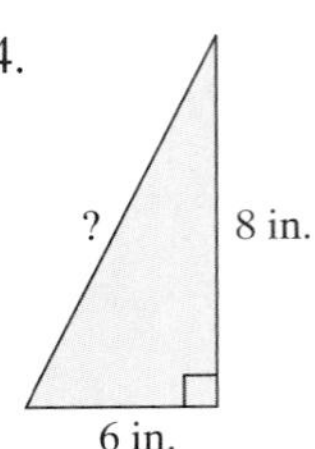

95. 5 ft, 12 ft, ?

96.

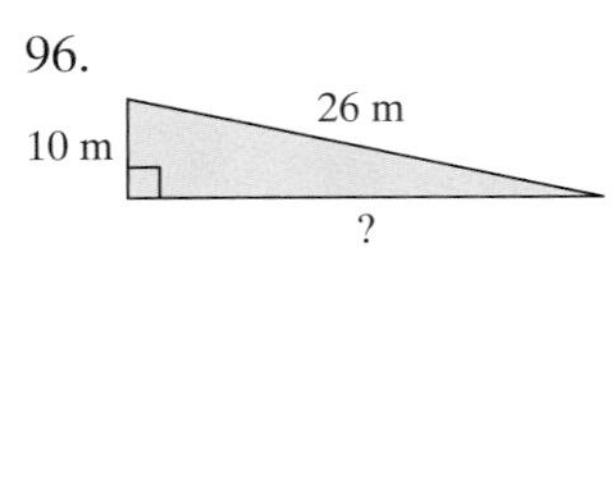

Expanding Your Skills

97. Follow the steps to provide a proof of the Pythagorean theorem.

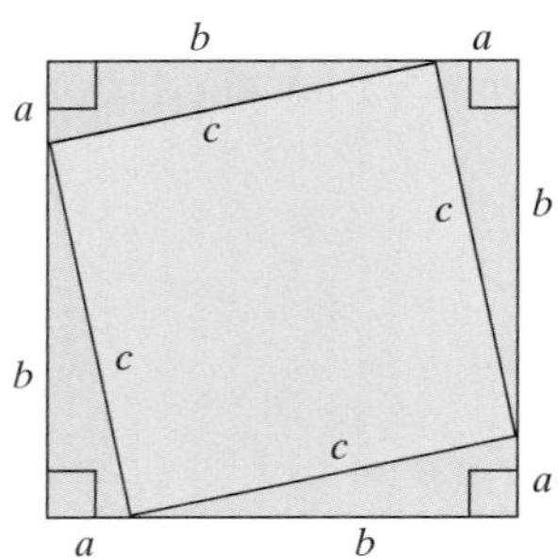

Note: The length of each side of the large outer square is $(a + b)$. Therefore, the area of the large outer square is $(a + b)^2$.

The area of the large outer square can also be found by adding the area of the inner square (pictured in yellow) plus the area of the four right triangles (pictured in blue).

Area of inner square: c^2 Area of the four right triangles: $4 \cdot (\frac{1}{2}ab)$

$\frac{1}{2}$ base · height

Now equate the two expressions representing the area of the large outer square:

$$\begin{pmatrix}\text{Area of outer}\\\text{square}\end{pmatrix} = \begin{pmatrix}\text{area of inner}\\\text{square}\end{pmatrix} + \begin{pmatrix}\text{area of the}\\\text{four right triangles}\end{pmatrix}$$

$$(a + b)^2 = c^2 + 4(\tfrac{1}{2}ab)$$

______ = ______ + ______ Clear parentheses on both sides.

______ = ______ Subtract $2ab$ from both sides.

Graphing Calculator Exercises

98. Graph $h(x) = \sqrt{x - 2}$ on the standard viewing window. Use the graph to confirm the domain found in Exercise 35.

99. Graph $k(x) = \sqrt{x + 1}$ on the standard viewing window. Use the graph to confirm the domain found in Exercise 36.

100. Graph $g(x) = \sqrt[3]{x - 2}$ on the standard viewing window. Use the graph to confirm the domain found in Exercise 37.

101. Graph $f(x) = \sqrt[3]{x + 1}$ on the standard viewing window. Use the graph to confirm the domain found in Exercise 38.

Concepts

1. Evaluating Expressions of the Form $a^{1/n}$
2. Evaluating Expressions of the Form $a^{m/n}$
3. Converting Between Rational Exponents and Radical Notation
4. Properties of Rational Exponents
5. Simplifying Expressions with Rational Exponents
6. Applications Involving Rational Exponents

section

10.2 Rational Exponents

1. Evaluating Expressions of the Form $a^{1/n}$

In Sections 4.1–4.3 the properties for simplifying expressions with integer exponents were presented. In this section, the properties are expanded to include expressions with rational exponents. We begin by defining expressions of the form: $a^{1/n}$

Definition of $a^{1/n}$

Let a be a real number, and let n be an integer such that $n > 1$. If $\sqrt[n]{a}$ is a real number, then

$$a^{1/n} = \sqrt[n]{a}$$

example 1 **Evaluating Expressions of the Form $a^{1/n}$**

Evaluate the expressions.

a. $(-8)^{1/3}$ b. $81^{1/4}$ c. $-100^{1/2}$ d. $(-100)^{1/2}$ e. $49^{1/2}$

Solution:

a. $(-8)^{1/3} = \sqrt[3]{-8} = -2$
b. $81^{1/4} = \sqrt[4]{81} = 3$
c. $-100^{1/2} = -\sqrt{100} = -10$
d. $(-100)^{1/2}$ is not a real number because $\sqrt{-100}$ is not a real number.
e. $49^{1/2} = \sqrt{49} = 7$

2. Evaluating Expressions of the Form $a^{m/n}$

If $\sqrt[n]{a}$ is a real number, then we can define an expression of the form $a^{m/n}$ in such a way that the multiplication property of exponents still holds true. For example,

$$16^{3/4} \nearrow (16^{1/4})^3 = (\sqrt[4]{16})^3 = (2)^3 = 8$$

$$16^{3/4} \searrow (16^3)^{1/4} = \sqrt[4]{16^3} = \sqrt[4]{4096} = 8$$

Definition of $a^{m/n}$

Let a be a real number, and let m and n be positive integers such that m and n share no common factors and $n > 1$. If $\sqrt[n]{a}$ is a real number, then

$$a^{m/n} = (a^{1/n})^m = (\sqrt[n]{a})^m \qquad \text{and} \qquad a^{m/n} = (a^m)^{1/n} = \sqrt[n]{a^m}$$

The rational exponent in the expression $a^{m/n}$ is essentially performing two operations. The numerator of the exponent raises the base to the mth power. The denominator takes the nth root.

example 2 **Evaluating Expressions of the Form $a^{m/n}$**

Simplify the expressions.

a. $8^{2/3}$ b. $100^{5/2}$ c. $\left(\frac{1}{25}\right)^{3/2}$

Calculator Connections

A calculator can be used to confirm the results of Example 2.

```
(8)^(2/3)
                4
(100)^(5/2)
           100000
(1/25)^(3/2)
             .008
```

Solution:

a. $8^{2/3} = (\sqrt[3]{8})^2$ Take the cube root of 8 and square the result.

$= (2)^2$ Simplify.

$= 4$

b. $100^{5/2} = (\sqrt{100})^5$ Take the square root of 100 and raise the result to the fifth power.

$= (10)^5$ Simplify.

$= 100{,}000$

c. $\left(\frac{1}{25}\right)^{3/2} = \left(\sqrt{\frac{1}{25}}\right)^3$ Take the square root of $\frac{1}{25}$ and cube the result.

$= \left(\frac{1}{5}\right)^3$ Simplify.

$= \frac{1}{125}$

3. Converting Between Rational Exponents and Radical Notation

example 3 **Using Radical Notation and Rational Exponents**

Convert each expression to radical notation.

a. $a^{3/5}$ b. $(5x^2)^{1/3}$ c. $3y^{1/4}$ $(y \geq 0)$

Solution:

a. $a^{3/5} = \sqrt[5]{a^3}$ or $\left(\sqrt[5]{a}\right)^3$

b. $(5x^2)^{1/3} = \sqrt[3]{5x^2}$

c. $3y^{1/4} = 3\sqrt[4]{y}$ Note: The coefficient, 3, is not raised to the $\frac{1}{4}$ power.

example 4 **Using Radical Notation and Rational Exponents**

Convert each expression to an equivalent expression using rational exponents. Assume that all variables represent positive real numbers.

a. $\sqrt[4]{b^3}$ b. $\sqrt{7a}$ c. $7\sqrt{a}$

Solution:

a. $\sqrt[4]{b^3} = b^{3/4}$
b. $\sqrt{7a} = (7a)^{1/2}$
c. $7\sqrt{a} = 7a^{1/2}$

4. Properties of Rational Exponents

In Sections 4.1–4.3, several properties and definitions were introduced to simplify expressions with integer exponents. These properties also apply to rational exponents.

Properties of Exponents and Definitions

Let a and b be nonzero real numbers. Let m and n be rational numbers such that a^m, a^n, and b^n are real numbers. Then,

Description	Property	Example
1. Multiplying like bases	$a^m a^n = a^{m+n}$	$x^{1/3}x^{4/3} = x^{5/3}$
2. Dividing like bases	$\dfrac{a^m}{a^n} = a^{m-n}$	$\dfrac{x^{3/5}}{x^{1/5}} = x^{2/5}$
3. The power rule	$(a^m)^n = a^{mn}$	$(2^{1/3})^{1/2} = 2^{1/6}$
4. Power of a product	$(ab)^m = a^m b^m$	$(xy)^{1/2} = x^{1/2}y^{1/2}$
5. Power of a quotient	$\left(\dfrac{a}{b}\right)^m = \dfrac{a^m}{b^m}$	$\left(\dfrac{4}{25}\right)^{1/2} = \dfrac{4^{1/2}}{25^{1/2}} = \dfrac{2}{5}$

Description	Definition	Example
1. Negative exponents	$a^{-m} = \left(\dfrac{1}{a}\right)^m = \dfrac{1}{a^m}$	$(8)^{-1/3} = \left(\dfrac{1}{8}\right)^{1/3} = \dfrac{1}{2}$
2. Zero exponent	$a^0 = 1 \quad (a \neq 0)$	$5^0 = 1$

5. Simplifying Expressions with Rational Exponents

example 5 **Simplifying Expressions with Rational Exponents**

Use the properties of exponents to simplify the expressions. Assume all variables represent positive real numbers.

a. $y^{2/5}y^{3/5}$ b. $(s^4t^8)^{1/4}$ c. $\left(\dfrac{81cd^{-2}}{3c^{-2}d^4}\right)^{1/3}$ d. $\left(\dfrac{x^{-2/3}}{y^{-1/2}}\right)^6 (x^{-1/5})^{10}$

Solution:

a. $y^{2/5}y^{3/5} = y^{(2/5)+(3/5)}$ Multiply like bases by adding exponents.

$= y^{5/5}$ Simplify.

$= y$ Reduce to lowest terms.

b. $(s^4t^8)^{1/4} = s^{4/4}t^{8/4}$ — Apply the power rule. Multiply exponents.

$= st^2$ — Reduce to lowest terms.

c. $\left(\dfrac{81cd^{-2}}{3c^{-2}d^4}\right)^{1/3} = (27c^{1-(-2)}d^{-2-4})^{1/3}$ — Simplify inside parentheses.

$= (27c^3d^{-6})^{1/3}$ — Subtract exponents.

$= \left(\dfrac{27c^3}{d^6}\right)^{1/3}$ — Simplify negative exponent.

$= \dfrac{27^{1/3}c^{3/3}}{d^{6/3}}$ — Apply the power rule. Multiply exponents.

$= \dfrac{3c}{d^2}$ — Simplify.

d. $\left(\dfrac{x^{-2/3}}{y^{-1/2}}\right)^6 (x^{-1/5})^{10} = \left(\dfrac{x^{-2/3 \cdot 6}}{y^{-1/2 \cdot 6}}\right)(x^{-1/5 \cdot 10})$ — Apply the power rule. Multiply exponents.

$= \left(\dfrac{x^{-4}}{y^{-3}}\right)(x^{-2})$ — Simplify.

$= \dfrac{x^{-4}}{y^{-3}} \cdot \dfrac{x^{-2}}{1}$

$= \dfrac{x^{-6}}{y^{-3}}$ — Multiply like bases by adding exponents.

$= \dfrac{y^3}{x^6}$ — Simplify negative exponents.

6. Applications Involving Rational Exponents

example 6

Applying Rational Exponents

Suppose P dollars in principal is invested in an account that earns interest annually. If after t years the investment grows to A dollars, then the annual rate of return, r, on the investment is given by

$$r = \left(\frac{A}{P}\right)^{1/t} - 1$$

Find the annual rate of return on \$5000 which grew to \$12,500 after 6 years.

Solution:

$$r = \left(\frac{A}{P}\right)^{1/t} - 1$$

Where $A = \$12{,}500$, $P = \$5000$ and $t = 6$. Hence,

$$= \left(\frac{12{,}500}{5000}\right)^{1/6} - 1$$

Calculator Connections

The expression

$$r = \left(\frac{12{,}500}{5000}\right)^{1/6} - 1$$

is easily evaluated on a graphing calculator.

```
(12500/5000)^(1/
6)-1
      .1649930508
```

$$= (2.5)^{1/6} - 1$$
$$\approx 1.165 - 1$$
$$\approx 0.165 \text{ or } 16.5\%$$

The annual rate of return is 16.5%.

section 10.2 Practice Exercises

For the exercises in this set, assume that all variables represent positive real numbers unless otherwise stated.

1. Given $\sqrt[3]{27}$:
   a. Identify the index
   b. Identify the radicand
2. Given $\sqrt{18}$:
   a. Identify the index
   b. Identify the radicand

For Exercises 3–10, evaluate the radicals (if possible).

3. $\sqrt{25}$
4. $\sqrt[3]{8}$
5. $\sqrt[4]{81}$
6. $(\sqrt[4]{16})^3$
7. $\sqrt{-9}$
8. $-\sqrt{9}$
9. $\sqrt[3]{(a+1)^3}$
10. $\sqrt[5]{(x+y)^5}$
11. Explain how to interpret the expression $a^{m/n}$ as a radical.
12. Explain why $(\sqrt[3]{8})^4$ is easier to evaluate than $\sqrt[3]{8^4}$.

For Exercises 13–36, simplify the expression.

13. $25^{1/2}$
14. $81^{1/2}$
15. $8^{1/3}$
16. $125^{1/3}$
17. $81^{1/4}$
18. $16^{3/4}$
19. $(-8)^{1/3}$
20. $(-9)^{1/2}$
21. $-8^{1/3}$
22. $-9^{1/2}$
23. $4^{-1/2}$
24. $121^{-1/2}$
25. $27^{-2/3}$
26. $125^{-1/3}$
27. $\dfrac{1}{36^{-1/2}}$
28. $\dfrac{1}{16^{-1/2}}$
29. $\dfrac{1}{1000^{-1/3}}$
30. $\dfrac{1}{81^{-3/4}}$
31. $\left(\dfrac{1}{8}\right)^{2/3} + \left(\dfrac{1}{4}\right)^{1/2}$
32. $\left(\dfrac{1}{8}\right)^{-2/3} + \left(\dfrac{1}{4}\right)^{-1/2}$
33. $\left(\dfrac{1}{16}\right)^{-1/4} - \left(\dfrac{1}{49}\right)^{-1/2}$
34. $\left(\dfrac{1}{16}\right)^{1/4} - \left(\dfrac{1}{49}\right)^{1/2}$
35. $\left(\dfrac{1}{4}\right)^{1/2} + \left(\dfrac{1}{64}\right)^{-1/3}$
36. $\left(\dfrac{1}{36}\right)^{1/2} + \left(\dfrac{1}{64}\right)^{-5/6}$

For Exercises 37–60, simplify the expressions using the properties of rational exponents. Write the final answer using positive exponents only.

37. $x^{1/4}x^{3/4}$
38. $2^{2/3}2^{1/3}$
39. $\dfrac{p^{5/3}}{p^{2/3}}$
40. $\dfrac{q^{5/4}}{q^{1/4}}$
41. $(y^{1/5})^{10}$
42. $(x^{1/2})^8$
43. $6^{-1/5}6^{6/5}$
44. $a^{-1/3}a^{2/3}$
45. $\dfrac{4t^{1/2}}{t^{-1/2}}$
46. $\dfrac{5s^{1/3}}{s^{-5/3}}$
47. $(a^{1/3}a^{1/4})^{12}$
48. $(x^{2/3}x^{1/2})^6$
49. $(5a^2c^{-1/2}d^{1/2})^2$
50. $(2x^{-1/3}y^2z^{5/3})^3$

51. $\left(\dfrac{x^{-2/3}}{y^{-3/4}}\right)^{12}$

52. $\left(\dfrac{m^{-1/4}}{n^{-1/2}}\right)^{-4}$

53. $\left(\dfrac{16w^{-2}z}{2wz^{-8}}\right)^{1/3}$

54. $\left(\dfrac{50p^{-1}q}{2pq^{-3}}\right)^{1/2}$

55. $(25x^2y^4z^6)^{1/2}$

56. $(8a^6b^3c^9)^{2/3}$

57. $\left(\dfrac{x^2y^{-1/3}z^{2/3}}{x^{2/3}y^{1/4}z}\right)^{12}$

58. $\left(\dfrac{a^2b^{1/2}c^{-2}}{a^{-3/4}b^0c^{1/8}}\right)^8$

59. $\left(\dfrac{x^{3m}y^{2m}}{z^{5m}}\right)^{1/m}$

60. $\left(\dfrac{a^{4n}b^{3n}}{c^n}\right)^{1/n}$

For Exercises 61–64, write the expressions in radical notation.

61. $(2x^2y)^{1/3}$

62. $(16abc)^{1/4}$

63. $\left(\dfrac{2}{y}\right)^{1/2}$

64. $(6x)^{1/2}$

For Exercises 65–68, write the expressions using rational exponents rather than radical notation.

65. $\sqrt[3]{x}$

66. $\sqrt[4]{a^2b^3}$

67. $5\sqrt{x}$

68. $\sqrt[3]{y^2}$

For Exercises 69–76, use a calculator to approximate the expressions and round to four decimal places, if necessary.

69. $9^{1/2}$

70. $125^{-1/3}$

71. $50^{-1/4}$

72. $(172)^{3/5}$

73. $\sqrt[3]{5^2}$

74. $\sqrt[4]{6^3}$

75. $\sqrt{10^3}$

76. $\sqrt[3]{16}$

77. If the area, A, of a square is known, then the length of its sides, s, can be computed by the formula: $s = A^{1/2}$

a. Compute the length of the sides of a square having an area of 100 in.[2]

b. Compute the length of the sides of a square having an area of 72 in.[2] Round your answer to the nearest 0.1 in.

78. The radius, r, of a sphere of volume, V, is given by $r = (3V/4\pi)^{1/3}$. Find the radius of a sphere having a volume of 85 in.[3] Round your answer to the nearest 0.1 in.

79. If P dollars in principal grows to A dollars after t years with annual interest, then the interest rate is given by: $r = (A/P)^{1/t} - 1$

a. In one account, \$10,000 grows to \$16,802 after 5 years. Compute the interest rate. Round your answer to a tenth of a percent.

b. In another account \$10,000 grows to \$18,000 after 7 years. Compute the interest rate. Round your answer to a tenth of a percent.

c. Which account produced a higher average yearly return?

80. Is $(a + b)^{1/2}$ the same as $a^{1/2} + b^{1/2}$? Why or why not?

Expanding Your Skills

For Exercises 81–86, write the expression as a single radical.

81. $\sqrt{\sqrt[3]{x}}$

82. $\sqrt[3]{\sqrt{x}}$

83. $\sqrt[4]{\sqrt{y}}$

84. $\sqrt{\sqrt[4]{y}}$

85. $\sqrt[5]{\sqrt[3]{w}}$

86. $\sqrt[3]{\sqrt[4]{w}}$

Concepts

1. Multiplication and Division Properties of Radicals
2. Simplified Form of a Radical
3. Simplifying Radicals Using the Multiplication Property of Radicals
4. Simplifying Radicals Using the Division Property of Radicals

section

10.3 Properties of Radicals

1. Multiplication and Division Properties of Radicals

You may have already recognized certain properties of radicals involving a product or quotient:

Multiplication and Division Properties of Radicals

Let a and b represent real numbers such that $\sqrt[n]{a}$ and $\sqrt[n]{b}$ are both real. Then,

1. $\sqrt[n]{ab} = \sqrt[n]{a} \cdot \sqrt[n]{b}$ **Multiplication property of radicals**
2. $\sqrt[n]{\dfrac{a}{b}} = \dfrac{\sqrt[n]{a}}{\sqrt[n]{b}}$ $b \neq 0$ **Division property of radicals**

Properties 1 and 2 follow from the properties of rational exponents.

$$\sqrt[n]{ab} = (ab)^{1/n} = a^{1/n}b^{1/n} = \sqrt[n]{a} \cdot \sqrt[n]{b}$$

$$\sqrt[n]{\frac{a}{b}} = \left(\frac{a}{b}\right)^{1/n} = \frac{a^{1/n}}{b^{1/n}} = \frac{\sqrt[n]{a}}{\sqrt[n]{b}}$$

The multiplication and division properties of radicals indicate that a product or quotient within a radicand can be written as a product or quotient of radicals, provided the roots are real numbers. For example,

$$\sqrt{144} = \sqrt{16} \cdot \sqrt{9}$$

$$\sqrt{\frac{25}{36}} = \frac{\sqrt{25}}{\sqrt{36}}$$

The reverse process is also true. A product or quotient of radicals can be written as a single radical provided the roots are real numbers and they have the same indices.

$$\sqrt{3} \cdot \sqrt{12} = \sqrt{36}$$

$$\frac{\sqrt[3]{8}}{\sqrt[3]{125}} = \sqrt[3]{\frac{8}{125}}$$

2. Simplified Form of a Radical

In algebra it is customary to simplify radical expressions as much as possible.

Simplified Form of a Radical

Consider any radical expression where the radicand is written as a product of prime factors. The expression is in **simplified form** if all of the following conditions are met.

1. The radicand has no factor raised to a power greater than or equal to the index.
2. The radicand does not contain a fraction.
3. There are no radicals in the denominator of a fraction.

3. Simplifying Radicals Using the Multiplication Property of Radicals

The expression $\sqrt{x^2}$ is not simplified because it fails condition 1. Because x^2 is a perfect square, $\sqrt{x^2}$ is easily simplified:

$$\sqrt{x^2} = x \quad (\text{for } x \geq 0)$$

However, how is an expression such as $\sqrt{x^9}$ simplified? This and many other radical expressions are simplified using the multiplication property of radicals. The following examples illustrate how *n*th powers can be removed from the radicands of *n*th roots.

example 1

Using the Multiplication Property to Simplify a Radical Expression

Use the multiplication property of radicals to simplify the expression $\sqrt{x^9}$. Assume $x \geq 0$.

Solution:

The expression $\sqrt{x^9}$ is equivalent to $\sqrt{x^8 \cdot x}$. By applying the multiplication property of radicals, we have

$$\sqrt{x^9} = \sqrt{x^8 \cdot x}$$

$= \sqrt{x^8} \cdot \sqrt{x}$ Apply the multiplication property of radicals.

Note that x^8 is a perfect square because $x^8 = (x^4)^2$.

$= x^4\sqrt{x}$ Simplify.

In Example 1, the expression x^9 is not a perfect square. Therefore, to simplify $\sqrt{x^9}$, it was necessary to write the expression as the product of the largest perfect square and a remaining or "left-over" factor: $\sqrt{x^9} = \sqrt{x^8 \cdot x}$. This process also applies to simplifying *n*th roots, as shown in Example 2.

example 2

Using the Multiplication Property to Simplify a Radical Expression

Use the multiplication property of radicals to simplify each expression. Assume all variables represent positive real numbers.

a. $\sqrt[4]{b^7}$ b. $\sqrt[3]{w^7z^9}$

Solution:

The goal is to rewrite each radicand as the product of the largest perfect square (perfect cube, perfect fourth power, and so on depending on the index) and a left-over factor.

a. $\sqrt[4]{b^7} = \sqrt[4]{(b^4) \cdot (b^3)}$ — b^4 is the largest perfect fourth power in the radicand.

$= \sqrt[4]{b^4} \cdot \sqrt[4]{b^3}$ — Apply the multiplication property of radicals.

$= b\sqrt[4]{b^3}$ — Simplify.

b. $\sqrt[3]{w^7z^9} = \sqrt[3]{(w^6z^9) \cdot (w)}$ — w^6z^9 is the largest perfect cube in the radicand.

$= \sqrt[3]{w^6z^9} \cdot \sqrt[3]{w}$ — Apply the multiplication property of radicals.

$= w^2z^3\sqrt[3]{w}$ — Simplify.

Each expression in Example 2 involves a radicand that is a product of variable factors. If a numerical factor is present, sometimes it is necessary to factor the coefficient before simplifying the radical.

example 3 Using the Multiplication Property to Simplify Radicals

Use the multiplication property of radicals to simplify the expressions. Assume all variables represent positive real numbers.

a. $\sqrt{56}$ b. $\sqrt[3]{40x^3y^5z^7}$

Solution:

$$\begin{array}{r|l} 2 & 56 \\ 2 & 28 \\ 2 & 14 \\ & 7 \end{array}$$

a. $\sqrt{56} = \sqrt{2^3 7}$ — Factor the radicand.

$= \sqrt{(2^2) \cdot (2 \cdot 7)}$ — 2^2 is the largest perfect square in the radicand.

$= \sqrt{2^2} \cdot \sqrt{2 \cdot 7}$ — Apply the multiplication property of radicals.

$= 2 \cdot \sqrt{14}$ — Simplify.

$= 2\sqrt{14}$

Calculator Connections

A calculator can be used to support the solution to Example 3(a). The decimal approximation for $\sqrt{56}$ and $2 \cdot \sqrt{14}$ agree for the first 10 digits. This in itself does not make $\sqrt{56} = 2 \cdot \sqrt{14}$. It is the multiplication of property of radicals that guarantees that the expressions are equal.

```
√(56)
          7.483314774
2*√(14)
          7.483314774
```

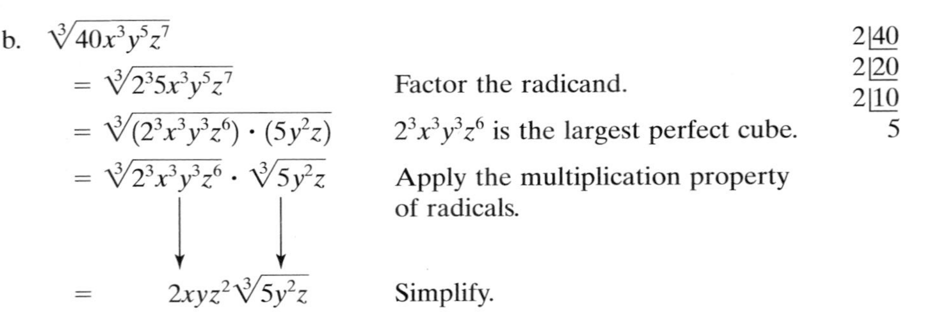

b. $\sqrt[3]{40x^3y^5z^7}$

$= \sqrt[3]{2^3 5x^3y^5z^7}$ Factor the radicand.

$= \sqrt[3]{(2^3x^3y^3z^6) \cdot (5y^2z)}$ $2^3x^3y^3z^6$ is the largest perfect cube.

$= \sqrt[3]{2^3x^3y^3z^6} \cdot \sqrt[3]{5y^2z}$ Apply the multiplication property of radicals.

$= 2xyz^2\sqrt[3]{5y^2z}$ Simplify.

$$\begin{array}{r|l} 2 & 40 \\ 2 & 20 \\ 2 & 10 \\ & 5 \end{array}$$

Avoiding Mistakes

The multiplication property of radicals allows us to simplify a product of factors within a radical. For example:

$\sqrt{x^2y^2} = \sqrt{x^2} \cdot \sqrt{y^2} = xy$
(for $x \geq 0$ and $y \geq 0$)

However, this rule does not apply to *terms* that are added or subtracted within the radical. For example,

$\sqrt{x^2 + y^2}$ and $\sqrt{x^2 - y^2}$
cannot be simplified.

4. Simplifying Radicals Using the Division Property of Radicals

The division property of radicals indicates that a radical of a quotient can be written as the quotient of the radicals and vice versa provided all roots are real numbers.

example 4

Using the Division Property to Simplify Radicals

Use the division property of radicals to simplify the expressions. Assume all variables represent positive real numbers.

a. $\sqrt{\dfrac{a^7}{a^3}}$ b. $\dfrac{\sqrt[3]{3}}{\sqrt[3]{81}}$ c. $\dfrac{7\sqrt{50}}{15}$ d. $\sqrt[4]{\dfrac{2c^5}{32cd^8}}$

Solution:

a. $\sqrt{\dfrac{a^7}{a^3}}$ The radicand contains a fraction. However, the fraction can be reduced.

$= \sqrt{a^4}$ Reduce.

$= a^2$ Simplify the radical.

b. $\dfrac{\sqrt[3]{3}}{\sqrt[3]{81}}$ The expression has a radical in the denominator.

$= \sqrt[3]{\dfrac{3}{81}}$ Because the radicands have a common factor, write the expression as a single radical and reduce (division property of radicals).

$= \sqrt[3]{\dfrac{1}{27}}$ Reduce.

$= \dfrac{1}{3}$ Simplify.

c. $\dfrac{7\sqrt{50}}{15}$ Simplify $\sqrt{50}$.

$= \dfrac{7\sqrt{5^2 \cdot 2}}{15}$ 5^2 is the largest perfect square in the radicand.

$= \dfrac{7\sqrt{5^2} \cdot \sqrt{2}}{15}$ Multiplication property of radicals.

$= \dfrac{7 \cdot 5\sqrt{2}}{15}$ Simplify the radicals.

$= \dfrac{7 \cdot \overset{1}{\cancel{5}}\sqrt{2}}{\underset{3}{\cancel{15}}}$ Reduce.

$= \dfrac{7\sqrt{2}}{3}$

Avoiding Mistakes

The division property of radicals allows us to reduce a ratio of two radicals provided they have the same index. For example:

$$\frac{\sqrt{15}}{\sqrt{5}} = \sqrt{\frac{15}{5}} = \sqrt{3}$$

However, a factor within the radicand cannot be simplified with a factor outside the radicand. For example,

$\dfrac{\sqrt{15}}{5}$ cannot be simplified.

d. $\sqrt[4]{\dfrac{2c^5}{32cd^8}}$ The radicand contains a fraction.

$= \sqrt[4]{\dfrac{c^4}{16d^8}}$ Reduce the factors in the radicand.

$= \dfrac{\sqrt[4]{c^4}}{\sqrt[4]{16d^8}}$ Apply the division property of radicals.

$= \dfrac{c}{2d^2}$ Simplify.

section 10.3 Practice Exercises

For the exercises in this set, assume that all variables represent positive real numbers unless otherwise stated.

For Exercises 1–4, simplify the expression. Write the answer with positive exponents only.

1. $(a^2b^{-4})^{1/2}\left(\dfrac{a}{b^{-3}}\right)$
2. $\left(\dfrac{p^4}{q^{-6}}\right)^{-1/2}(p^3q^{-2})$
3. $(r^{-1/4}s^{3/4})^8$
4. $(x^{1/3}y^{5/6})^{-6}$
5. Write $x^{4/7}$ in radical notation.
6. Write $y^{2/5}$ in radical notation.
7. Write $\sqrt{y^9}$ using rational exponents.
8. Write $\sqrt[3]{x^2}$ using rational exponents.
9. Approximate the expression $\dfrac{2 - \sqrt{y}}{4}$ to two decimal places for:

   a. $y = 3$ b. $y = 5$

10. Approximate the expression $\sqrt{\frac{1}{2}x + 3}$ to two decimal places for:

    a. $x = -2$ b. $x = 3$

For Exercises 11–18, use the multiplication property of radicals to multiply the expressions. Then simplify the result.

Example: $\sqrt{3x} \cdot \sqrt{3x} = \sqrt{9x^2} = 3x$

11. $\sqrt{4x} \cdot \sqrt{25x}$
12. $\sqrt{2z} \cdot \sqrt{8z}$
13. $\sqrt[3]{2x} \cdot \sqrt[3]{4x^2}$
14. $\sqrt[3]{a^2b} \cdot \sqrt[3]{ab^2}$
15. $\sqrt[4]{(x+2)^3} \cdot \sqrt[4]{(x+2)}$
16. $\sqrt[4]{8w^2} \cdot \sqrt[4]{2w^2}$
17. $\sqrt[5]{2y^2} \cdot \sqrt[5]{16y^3}$
18. $\sqrt[5]{8(t-1)^2} \cdot \sqrt[5]{4(t-1)^3}$

For Exercises 19–26, use the division property of radicals to divide the expressions. Then simplify the result.

Example: $\dfrac{\sqrt{18x}}{\sqrt{2x}} = \sqrt{\dfrac{18x}{2x}} = \sqrt{9} = 3$

19. $\dfrac{\sqrt{27y}}{\sqrt{3y}}$

20. $\dfrac{\sqrt{24r}}{\sqrt{6r}}$

21. $\dfrac{\sqrt[4]{2^5a^3b^7}}{\sqrt[4]{2a^3b^3}}$

22. $\dfrac{\sqrt[4]{8^5x^6y^3}}{\sqrt[4]{8x^2y^3}}$

23. $\dfrac{\sqrt[3]{(3x+1)^5}}{\sqrt[3]{(3x+1)^2}}$

24. $\dfrac{\sqrt[3]{(6-w)^7}}{\sqrt[3]{(6-w)^4}}$

25. $\dfrac{\sqrt{(a+b)}}{\sqrt{(a+b)^3}}$

26. $\dfrac{\sqrt{(x-3)^2}}{\sqrt{(x-3)^4}}$

For Exercises 27–46, simplify the radicals.

27. $\sqrt{28}$

28. $\sqrt{63}$

29. $\sqrt{80}$

30. $\sqrt{108}$

31. $5\sqrt{18}$

32. $2\sqrt{24}$

33. $\sqrt[3]{54}$

34. $\sqrt[3]{48}$

35. $\sqrt{25x^4y^3}$

36. $\sqrt{125p^3q^2}$

37. $\sqrt[3]{27x^2y^3z^4}$

38. $\sqrt[3]{108a^3bc^2}$

39. $\sqrt[3]{\dfrac{16a^2b}{2a^2b^4}}$

40. $\sqrt[4]{\dfrac{3s^2t^4}{10{,}000}}$

41. $\sqrt[5]{\dfrac{32x}{y^{10}}}$

42. $\sqrt[3]{\dfrac{-16j^3}{k^3}}$

43. $\dfrac{\sqrt{50x^3y}}{\sqrt{9y^4}}$

44. $\dfrac{\sqrt[3]{-27a^4}}{\sqrt[3]{8a}}$

45. $\sqrt{2^3a^{14}b^8c^{31}d^{22}}$

46. $\sqrt{7^5u^{12}v^{20}w^{65}x^{80}}$

For Exercises 47–50, write a mathematical expression for the English phrase and simplify.

47. The quotient of one and the cube root of w^6.

48. The square root of the quotient of h and 49.

49. The square root of k raised to the third power.

50. The cube root of $2x^4$.

For Exercises 51–54, find the third side of the right triangle. Write your answer as a simplified radical.

51. 10 ft, 8 ft, ?

52.

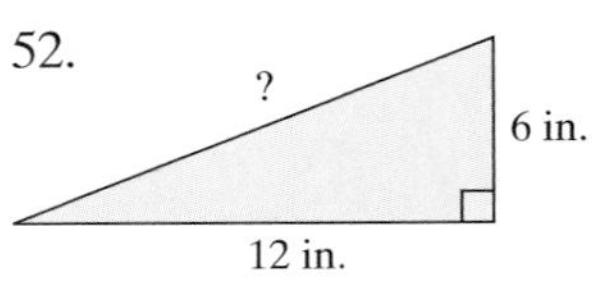

53. ?, 18 m, 12 m

54.

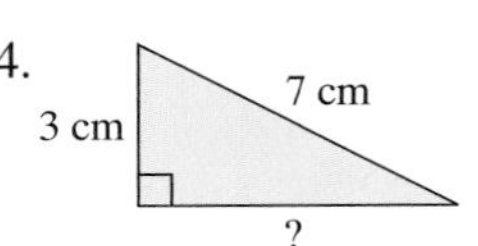

55. On a baseball diamond, the bases are 90 ft apart. Find the distance from home plate to second base. Round to one decimal place.

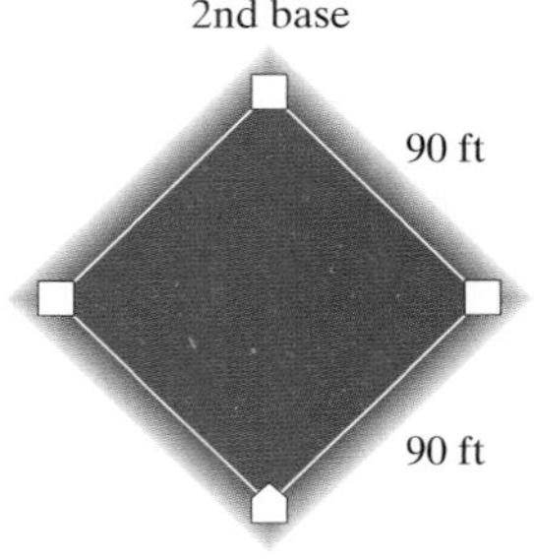

Figure for Exercise 55

56. Linda is at the beach flying a kite. The kite is directly over a sand castle 60 ft away from Linda. If 100 ft of kite string is out (ignoring any sag in the string), how high is the kite? (Assume that Linda is 5 ft tall.) See figure.

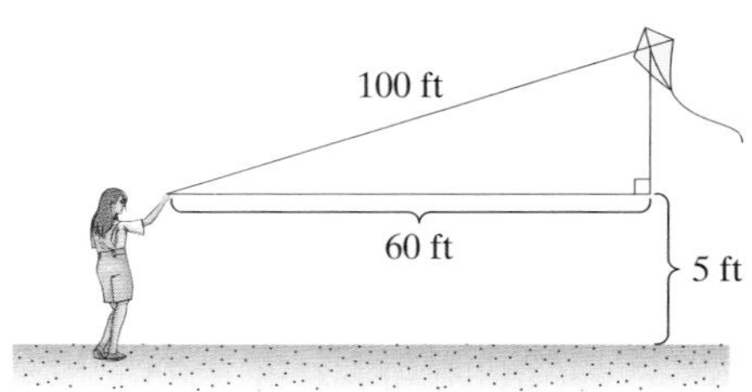

Figure for Exercise 56

Expanding Your Skills

57. Tom has to travel from town A to town C across a small mountain range. He can travel one of two routes. He can travel on a four-lane highway from A to B and then from B to C at an average

speed of 55 mph. Or he can travel on a two-lane road directly from town A to town C, but his average speed will only be 35 mph. If Tom is in a hurry, which route will take him to town C the fastest?

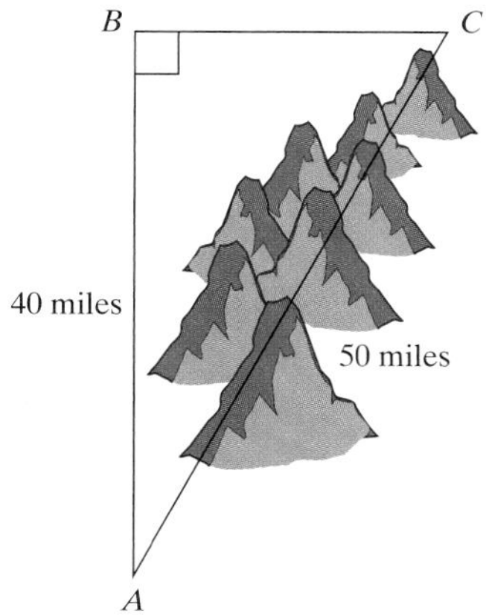

Figure for Exercise 57

58. One side of a rectangular pasture is 80 ft in length. The diagonal distance is 110 yds. If fencing costs \$3.29 per foot, how much will it cost to fence the pasture?

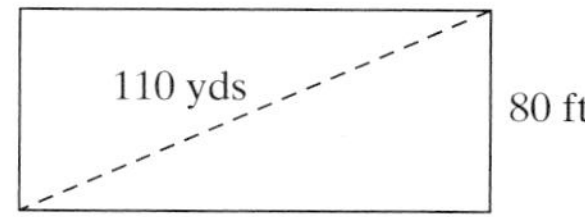

Figure for Exercise 58

Concepts

1. Definition of *Like* Radicals
2. Addition and Subtraction of Radicals
3. Recognizing *Unlike* Radicals

section

10.4 Addition and Subtraction of Radicals

1. Definition of *Like* Radicals

Definition of *Like* Radicals

Two radical expressions are said to be ***like* radicals** if their radical factors have the same index and the same radicand.

The following are pairs of *like* radicals:

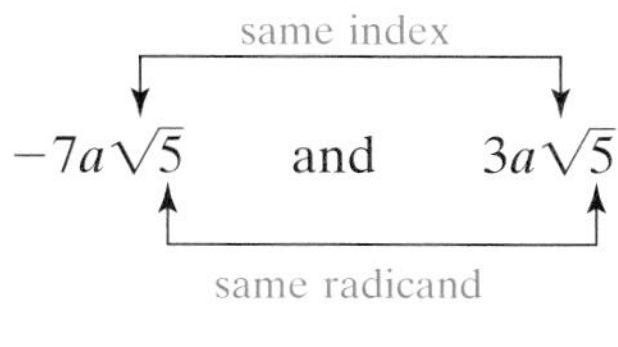

$-7a\sqrt{5}$ and $3a\sqrt{5}$ Indices and radicands are the same.

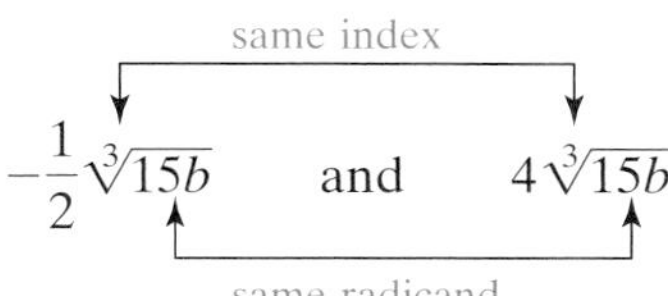

$-\frac{1}{2}\sqrt[3]{15b}$ and $4\sqrt[3]{15b}$ Indices and radicands are the same.

These pairs are not *like* radicals:

different indices

$-2\sqrt{6}$ and $13\sqrt[4]{6}$ Radicals have different indices.

$1.3cd\sqrt{3}$ and $-3.7cd\sqrt{10}$ Radicals have different radicands.

different radicands

2. Addition and Subtraction of Radicals

Expressions with radicals can be added or subtracted if they are *like* radicals. To add or subtract *like* radicals, use the distributive property. For example:

$$2\sqrt{5} + 6\sqrt{5} = (2 + 6)\sqrt{5} = 8\sqrt{5}$$

$$9\sqrt[3]{2y} - 4\sqrt[3]{2y} = (9 - 4)\sqrt[3]{2y} = 5\sqrt[3]{2y}$$

example 1

Adding and Subtracting Radicals

Add or subtract the radicals as indicated.

a. $6\sqrt{11} + 2\sqrt{11}$

b. $\sqrt{3} + \sqrt{3}$

c. $-2\sqrt[3]{ab} + 7\sqrt[3]{ab} - \sqrt[3]{ab}$

d. $\frac{1}{4}x\sqrt{3y} - \frac{3}{2}x\sqrt{3y}$

Solution:

a. $6\sqrt{11} + 2\sqrt{11}$

$= (6 + 2)\sqrt{11}$ Apply the distributive property.

$= 8\sqrt{11}$ Simplify.

b. $\sqrt{3} + \sqrt{3}$

$= 1\sqrt{3} + 1\sqrt{3}$ Note that $\sqrt{3} = 1\sqrt{3}$.

$= (1 + 1)\sqrt{3}$ Apply the distributive property.

$= 2\sqrt{3}$ Simplify.

c. $-2\sqrt[3]{ab} + 7\sqrt[3]{ab} - \sqrt[3]{ab}$

$= (-2 + 7 - 1)\sqrt[3]{ab}$ Apply the distributive property.

$= 4\sqrt[3]{ab}$ Simplify.

d. $\frac{1}{4}x\sqrt{3y} - \frac{3}{2}x\sqrt{3y}$

$= \left(\frac{1}{4} - \frac{3}{2}\right)x\sqrt{3y}$ Apply the distributive property.

Avoiding Mistakes

The process of adding *like* radicals with the distributive property is similar to adding *like* terms. The end result is that the numerical coefficients are added and the radical factor is unchanged.

$\sqrt{5} + \sqrt{5} = 1\sqrt{5} + 1\sqrt{5} = 2\sqrt{5}$ (correct)

Be careful: $\sqrt{5} + \sqrt{5} \neq \sqrt{10}$

In general: $\sqrt{x} + \sqrt{y} \neq \sqrt{x + y}$

$= \left(\frac{1}{4} - \frac{6}{4}\right)x\sqrt{3y}$ Get a common denominator.

$= -\frac{5}{4}x\sqrt{3y}$ Simplify.

Sometimes it is necessary to simplify radicals before adding or subtracting.

example 2 Adding and Subtracting Radicals

Simplify the radicals and add or subtract as indicated.

a. $3\sqrt{8} + \sqrt{2}$

b. $8\sqrt{x^3y^2} - 3y\sqrt{x^3}$

c. $\sqrt{50x^2y^5} - 13y\sqrt{2x^2y^3} + xy\sqrt{98y^3}$

Solution:

a. $3\sqrt{8} + \sqrt{2}$ The radicands are different. Try simplifying the radicals first.

$= 3 \cdot 2\sqrt{2} + \sqrt{2}$ Simplify: $\sqrt{8} = 2\sqrt{2}$

$= 6\sqrt{2} + \sqrt{2}$

$= (6 + 1)\sqrt{2}$ Apply the distributive property.

$= 7\sqrt{2}$ Simplify.

b. $8\sqrt{x^3y^2} - 3y\sqrt{x^3}$ The radicands are different. Simplify the radicals first.

$= 8xy\sqrt{x} - 3xy\sqrt{x}$ Simplify: $\sqrt{x^3y^2} = xy\sqrt{x}$ and $\sqrt{x^3} = x\sqrt{x}$

$= (8 - 3)xy\sqrt{x}$ Apply the distributive property.

$= 5xy\sqrt{x}$ Simplify.

c. $\sqrt{50x^2y^5} - 13y\sqrt{2x^2y^3} + xy\sqrt{98y^3}$

$= 5xy^2\sqrt{2y} - 13xy^2\sqrt{2y} + 7xy^2\sqrt{2y}$

Simplify each radical.

$$\begin{cases} \sqrt{50x^2y^5} = \sqrt{5^2 2x^2y^5} \\ \quad = 5xy^2\sqrt{2y} \\ -13y\sqrt{2x^2y^3} = -13xy^2\sqrt{2y} \\ xy\sqrt{98y^3} = xy\sqrt{7^2 2y^3} \\ \quad = 7xy^2\sqrt{2y} \end{cases}$$

$= (5 - 13 + 7)xy^2\sqrt{2y}$ Apply the distributive property.

$= -xy^2\sqrt{2y}$

Calculator Connections

To check the solution to Example 2(a), use a calculator to evaluate the expressions $3\sqrt{8} + \sqrt{2}$ and $7\sqrt{2}$. The decimal approximations agree to 10 digits.

```
3√(8)+√(2)
         9.899494937
7√(2)
         9.899494937
```

3. Recognizing *Unlike* Radicals

It is important to keep in mind that only *like* radicals may be added or subtracted. The next example provides extra practice for recognizing *unlike* radicals.

example 3 **Recognizing *Unlike* Radicals**

The following radicals cannot be added or subtracted to form a single radical. Explain why.

a. $\sqrt{2} + \sqrt[3]{2}$ — *Unlike* radicals. Indices are not the same.

b. $5\sqrt{3} - 2\sqrt{5}$ — *Unlike* radicals. Radicands are not the same.

c. $12\sqrt{7} - 12$ — *Unlike* radicals. One term has a radical, one does not.

d. $2\sqrt{50} - 3\sqrt{75}$ — Simplify.

$10\sqrt{2} - 15\sqrt{3}$ — *Unlike* radicals. Radicands are not the same.

section 10.4 Practice Exercises

For the exercises in this set, assume that all variables represent positive real numbers unless otherwise stated.

For Exercises 1–4, simplify the radicals.

1. $\sqrt[3]{-16s^4t^9}$
2. $-\sqrt[4]{x^7y^4}$
3. $\sqrt{3p^2} \cdot \sqrt{12p^4}$
4. $\dfrac{\sqrt[3]{7b^8}}{\sqrt[3]{56b^2}}$
5. Write the expression $(4x^2)^{3/2}$ as a radical and simplify.
6. Convert to rational exponents and simplify. $\sqrt[5]{3^5x^{15}y^{10}}$

For Exercises 7–8, simplify the expressions. Write the answer with positive exponents only.

7. $y^{2/3}y^{1/4}$
8. $(x^{1/2}y^{-3/4})^{-4}$

For Exercises 9–10, use a calculator to approximate the expressions. Round to two decimal places.

9. $(2.718)^{2/3}$
10. $\sqrt[4]{90}$
11. Explain the similarities and differences between the following pairs of expressions.
    a. $7\sqrt{5} + 4\sqrt{5}$ and $7x + 4x$
    b. $-2\sqrt{6} - 9\sqrt{3}$ and $-2x - 9y$
12. Explain the similarities and differences between the following pairs of expressions.
    a. $-4\sqrt{3} + 5\sqrt{3}$ and $-4z + 5z$
    b. $13\sqrt{7} - 6\sqrt{9}$ and $13a - 18$

For Exercises 13–30, add or subtract the radical expressions (if possible).

13. $3\sqrt{5} + 6\sqrt{5}$
14. $5\sqrt{a} + 3\sqrt{a}$
15. $3\sqrt[3]{t} - 2\sqrt[3]{t}$
16. $6\sqrt[3]{7} - 2\sqrt[3]{7}$
17. $6\sqrt{10} - \sqrt{10}$
18. $13\sqrt{11} - \sqrt{11}$
19. $\sqrt[4]{3} + 7\sqrt[4]{3} - \sqrt[4]{14}$
20. $2\sqrt{11} + 3\sqrt{13} + 5\sqrt{11}$
21. $8\sqrt{x} + 2\sqrt{y} - 6\sqrt{x}$
22. $10\sqrt{10} - 8\sqrt{10} + \sqrt{2}$
23. $\sqrt[3]{ab} + a\sqrt[3]{b}$
24. $x\sqrt[4]{y} - y\sqrt[4]{x}$
25. $\sqrt{2t} + \sqrt[3]{2t}$
26. $\sqrt[4]{5c} + \sqrt[3]{5c}$
27. $\frac{5}{6}z\sqrt[3]{6} + \frac{7}{9}z\sqrt[3]{6}$
28. $\frac{3}{4}a\sqrt[4]{b} + \frac{1}{6}a\sqrt[4]{b}$
29. $0.81x\sqrt{y} - 0.11x\sqrt{y}$

30. $7.5\sqrt{pq} - 6.3\sqrt{pq}$

31. Explain the process for adding the two radicals. $3\sqrt{2} + 7\sqrt{50}$.

32. Explain the process for adding the two radicals. $\sqrt{8} + \sqrt{32}$.

For Exercises 33–50, add or subtract the radical expressions as indicated.

33. $\sqrt{36} + \sqrt{81}$

34. $3\sqrt{80} - 5\sqrt{45}$

35. $2\sqrt{12} + \sqrt{48}$

36. $5\sqrt{32} + 2\sqrt{50}$

37. $4\sqrt{7} + \sqrt{63} - 2\sqrt{28}$

38. $8\sqrt{3} - 2\sqrt{27} + \sqrt{75}$

39. $3\sqrt{2a} - \sqrt{8a} - \sqrt{72a}$

40. $\sqrt{12t} - \sqrt{27t} + 5\sqrt{3t}$

41. $2s^2\sqrt[3]{s^2t^6} + 3t^2\sqrt[3]{8s^8}$

42. $4\sqrt[3]{x^4} - 2x\sqrt[3]{x}$

43. $7\sqrt[3]{x^4} - x\sqrt[3]{x}$

44. $6\sqrt[3]{y^{10}} - 3y^2\sqrt[3]{y^4}$

45. $5p\sqrt{20p^2} + p^2\sqrt{80}$

46. $2q\sqrt{48q^2} - \sqrt{27q^4}$

47. $\frac{3}{2}ab\sqrt{24a^3} + \frac{4}{3}\sqrt{54a^5b^2} - a^2b\sqrt{150a}$

48. $mn\sqrt{72n} + \frac{2}{3}n\sqrt{8m^2n} - \frac{5}{6}\sqrt{50m^2n^3}$

49. $x\sqrt[3]{16} - 2\sqrt[3]{27x} + \sqrt[3]{54x^3}$

50. $5\sqrt[4]{y^5} - 2y\sqrt[4]{y} + \sqrt[4]{16y^7}$

For Exercises 51–56, answer true or false. If an answer is false, explain why.

51. $\sqrt{x} + \sqrt{y} = \sqrt{x + y}$

52. $\sqrt{x} + \sqrt{x} = 2\sqrt{x}$

53. $5\sqrt[3]{x} + 2\sqrt[3]{x} = 7\sqrt[3]{x}$

54. $6\sqrt{x} + 5\sqrt[3]{x} = 11\sqrt{x}$

55. $\sqrt{y} + \sqrt{y} = \sqrt{2y}$

56. $\sqrt{c^2 + d^2} = c + d$

For Exercises 57–60, translate the English phrase into an algebraic expression. Simplify each expression if possible.

57. The sum of the square root of 48 and the square root of 12.

58. The sum of the cube root of 16 and the cube root of 2.

59. The difference of 5 times the cube root of x^6 and the square of x.

60. The sum of the cube of y and the fourth root of y^{12}.

For Exercises 61–64, write an English phrase that translates the mathematical expression. (Answers may vary.)

61. $\sqrt{18} - 5^2$

62. $4^3 - \sqrt[3]{4}$

63. $\sqrt[4]{x} + y^3$

64. $a^4 + \sqrt{a}$

Expanding Your Skills

65. a. An irregularly shaped garden is shown in the figure. All distances are expressed in yards. Find the perimeter. *Hint:* Use the Pythagorean theorem to find the length of each side. Write the final answer in radical form.

    b. Approximate your answer to two decimal places.

    c. If edging costs \$1.49 per foot and sales tax is 6%, find the total cost of edging the garden.

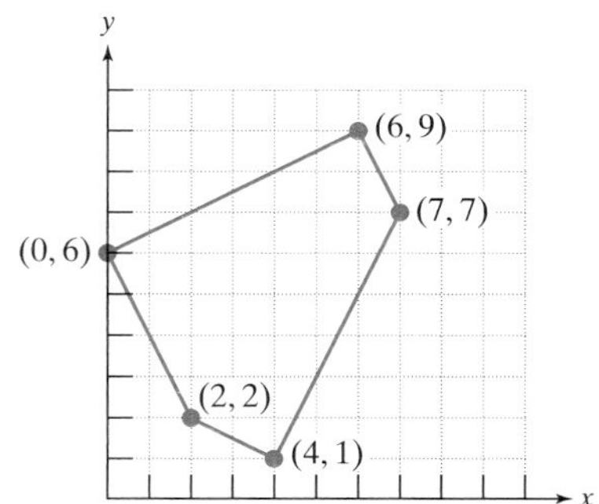

Figure for Exercise 65

66. a. An irregularly shaped garden is shown in the figure. All distances are expressed in yards. Find the perimeter. Write the final answer in radical form.

b. Approximate your answer to two decimal places.

c. If edging costs \$1.69 per foot and sales tax is 6%, find the total cost of edging the garden.

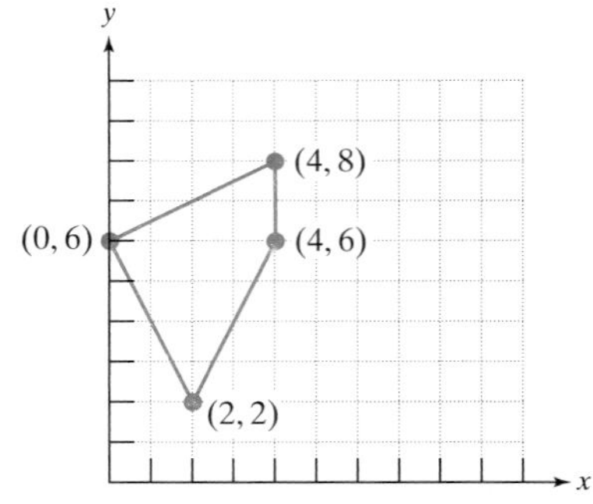

Figure for Exercise 66

Concepts

1. Multiplication Property of Radicals
2. Multiplying Radicals with Different Indices
3. Multiplying Radical Expressions Involving Multiple Terms
4. Expressions of the Form $(\sqrt[n]{a})^n$
5. Special Case Products

section

10.5 Multiplication of Radicals

1. Multiplication Property of Radicals

In this section we will learn how to multiply radicals by using the **multiplication property of radicals** first introduced in Section 10.3.

The Multiplication Property of Radicals

Let a and b represent real numbers such that $\sqrt[n]{a}$ and $\sqrt[n]{b}$ are both real. Then,

$$\sqrt[n]{ab} = \sqrt[n]{a} \cdot \sqrt[n]{b}$$

To multiply two radical expressions, we use the multiplication property of radicals along with the commutative and associative properties of multiplication.

example 1 **Multiplying Radical Expressions**

Multiply the expressions and simplify the result. Assume all variables represent positive real numbers.

a. $(3\sqrt{2})(5\sqrt{6})$ b. $(2x\sqrt{y})(-7\sqrt{xy})$ c. $(15c\sqrt[3]{cd})\left(\frac{1}{3}\sqrt[3]{cd^2}\right)$

Solution:

a. $(3\sqrt{2})(5\sqrt{6})$

$= (3 \cdot 5)(\sqrt{2} \cdot \sqrt{6})$ Commutative and associative properties of multiplication

$= 15\sqrt{12}$ Multiplication property of radicals

$= 15\sqrt{2^2 3}$

$= 15 \cdot 2\sqrt{3}$ Simplify the radical.

$= 30\sqrt{3}$

b. $(2x\sqrt{y})(-7\sqrt{xy})$

$= (2x)(-7)(\sqrt{y} \cdot \sqrt{xy})$	Commutative and associative properties of multiplication
$= -14x\sqrt{xy^2}$	Multiplication property of radicals
$= -14xy\sqrt{x}$	Simplify the radical.

c. $(15c\sqrt[3]{cd})\left(\frac{1}{3}\sqrt[3]{cd^2}\right)$

$= \left(15c \cdot \frac{1}{3}\right)(\sqrt[3]{cd} \cdot \sqrt[3]{cd^2})$	Commutative and associative properties of multiplication
$= 5c\sqrt[3]{c^2d^3}$	Multiplication property of radicals
$= 5cd\sqrt[3]{c^2}$	Simplify the radical.

2. Multiplying Radicals with Different Indices

The product of two radicals can be simplified provided the radicals have the same index. If the radicals have different indices, then we can use the properties of rational exponents to obtain a common index.

example 2 **Multiplying Radicals with Different Indices**

Multiply the expressions:

a. $\sqrt[3]{5} \cdot \sqrt[4]{5}$ b. $\sqrt[3]{7} \cdot \sqrt{2}$

Solution:

a. $\sqrt[3]{5} \cdot \sqrt[4]{5}$

$= 5^{1/3}5^{1/4}$	Rewrite each expression with rational exponents.
$= 5^{(1/3)+(1/4)}$	Because the bases are equal, we can add the exponents.
$= 5^{(4/12)+(3/12)}$	Write the fractions with a common denominator.
$= 5^{7/12}$	Simplify the exponent.
$= \sqrt[12]{5^7}$	Rewrite the expression as a radical.

b. $\sqrt[3]{7} \cdot \sqrt{2}$

$= 7^{1/3}2^{1/2}$	Rewrite each expression with rational exponents.
$= 7^{2/6}2^{3/6}$	Write the rational exponents with a common denominator.
$= (7^2 2^3)^{1/6}$	Apply the power rule of exponents.
$= \sqrt[6]{7^2 2^3}$	Rewrite the expression as a single radical.
$= \sqrt[6]{392}$	Simplify.

Avoiding Mistakes

We cannot multiply $7^{1/3} \cdot 2^{1/2}$ directly by adding exponents because the bases are different.

3. Multiplying Radical Expressions Involving Multiple Terms

When multiplying radical expressions with more than one term, we use the distributive property.

example 3 Multiplying Radical Expressions

Multiply the radical expressions. Assume all variables represent positive real numbers.

a. $3\sqrt{11}(2 + \sqrt{11})$

b. $(\sqrt{5} + 3\sqrt{2})(2\sqrt{5} - \sqrt{2})$

c. $(2\sqrt{14} + \sqrt{7})(6 - \sqrt{14} + 8\sqrt{7})$

d. $(-10a\sqrt{b} + 7b)(a\sqrt{b} + 2b)$

Solution:

a. $3\sqrt{11}(2 + \sqrt{11})$

$= 3\sqrt{11} \cdot (2) + 3\sqrt{11} \cdot \sqrt{11}$ Apply the distributive property.

$= 6\sqrt{11} + 3\sqrt{11^2}$ Multiplication property of radicals

$= 6\sqrt{11} + 3 \cdot 11$ Simplify the radical.

$= 6\sqrt{11} + 33$

b. $(\sqrt{5} + 3\sqrt{2})(2\sqrt{5} - \sqrt{2})$

$= 2\sqrt{5^2} - \sqrt{10} + 6\sqrt{10} - 3\sqrt{2^2}$ Apply the distributive property.

$= 2 \cdot 5 + 5\sqrt{10} - 3 \cdot 2$ Simplify radicals and combine *like* radicals.

$= 10 + 5\sqrt{10} - 6$

$= 4 + 5\sqrt{10}$ Combine *like* terms.

Calculator Connections

To check the solution to Example 3(b), use a calculator to evaluate the expressions $(\sqrt{5} + 3\sqrt{2})(2\sqrt{5} - \sqrt{2})$ and $4 + 5\sqrt{10}$. The decimal approximations agree to nine digits.

```
(√(5)+3√(2))*(2√
(5)-√(2))
       19.8113883
4+5√(10)
       19.8113883
```

c. $(2\sqrt{14} + \sqrt{7})(6 - \sqrt{14} + 8\sqrt{7})$

$= 12\sqrt{14} - 2\sqrt{14^2} + 16\sqrt{98} + 6\sqrt{7} - \sqrt{98} + 8\sqrt{7^2}$ Apply the distributive property.

$= 12\sqrt{14} - 2 \cdot 14 + 16 \cdot 7\sqrt{2} + 6\sqrt{7} - 7\sqrt{2} + 8 \cdot 7$ Simplify the radicals.

(*Note:* $\sqrt{98} = \sqrt{7^2 \cdot 2} = 7\sqrt{2}$)

$= 12\sqrt{14} - 28 + 112\sqrt{2} + 6\sqrt{7} - 7\sqrt{2} + 56$ Simplify.

$= 12\sqrt{14} + 105\sqrt{2} + 6\sqrt{7} + 28$ Combine *like* terms.

d. $(-10a\sqrt{b} + 7b)(a\sqrt{b} + 2b)$

$= -10a^2\sqrt{b^2} - 20ab\sqrt{b} + 7ab\sqrt{b} + 14b^2$ — Apply the distributive property.

$= -10a^2b - 13ab\sqrt{b} + 14b^2$ — Simplify and combine *like* terms.

4. Expressions of the Form $(\sqrt[n]{a})^n$

The multiplication property of radicals can be used to simplify an expression of the form $(\sqrt{a})^2$, where $a \geq 0$.

$$(\sqrt{a})^2 = \sqrt{a} \cdot \sqrt{a} = \sqrt{a^2} = a, \text{ where } a \geq 0$$

This logic can be applied to nth roots. If $\sqrt[n]{a}$ is a real number, then, $(\sqrt[n]{a})^n = a$.

example 4 **Simplifying Radical Expressions**

Simplify the expressions. Assume all variables represent positive real numbers.

a. $(\sqrt{11})^2$ b. $(\sqrt[5]{z})^5$ c. $(\sqrt[3]{pq})^3$

Solution:

a. $(\sqrt{11})^2 = 11$ b. $(\sqrt[5]{z})^5 = z$ c. $(\sqrt[3]{pq})^3 = pq$

5. Special Case Products

From Example 3, you may have noticed a similarity between multiplying radical expressions and multiplying polynomials.

Recall from Section 4.6 that the square of a binomial results in a perfect square trinomial:

$$(a + b)^2 = a^2 + 2ab + b^2$$

$$(a - b)^2 = a^2 - 2ab + b^2$$

The same patterns occur when squaring a radical expression with two terms.

example 5 **Squaring a Two-Term Radical Expression**

Square the radical expressions.

a. $(\sqrt{d} + 3)^2$ b. $(5\sqrt{y} - \sqrt{2})^2$

Solution:

a. $(\sqrt{d} + 3)^2$

This expression is in the form $(a + b)^2$, where $a = \sqrt{d}$ and $b = 3$.

$$a^2 + 2ab + b^2$$

$$= (\sqrt{d})^2 + 2(\sqrt{d})(3) + (3)^2$$

Apply the formula $(a + b)^2 = a^2 + 2ab + b^2$.

$$= d + 6\sqrt{d} + 9$$

Simplify.

Tip: The product $(\sqrt{d} + 3)^2$ can also be found using the distributive property:

$$(\sqrt{d} + 3)^2 = (\sqrt{d} + 3)(\sqrt{d} + 3) = \sqrt{d} \cdot \sqrt{d} + \sqrt{d} \cdot 3 + 3 \cdot \sqrt{d} + 3 \cdot 3$$
$$= \sqrt{d^2} + 3\sqrt{d} + 3\sqrt{d} + 9$$
$$= d + 6\sqrt{d} + 9$$

b. $(5\sqrt{y} - \sqrt{2})^2$

This expression is in the form $(a - b)^2$, where $a = 5\sqrt{y}$ and $b = \sqrt{2}$.

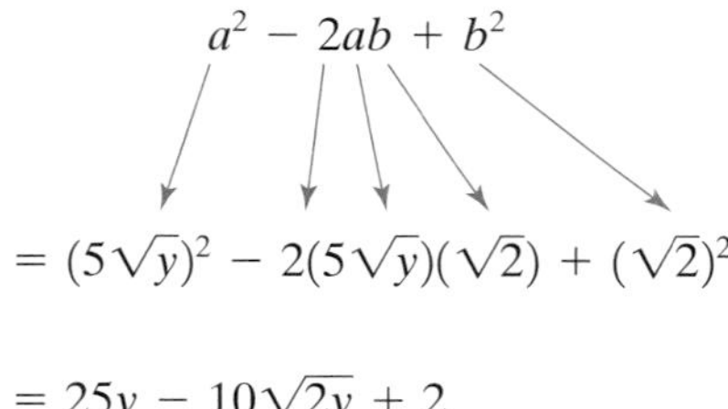

$$a^2 - 2ab + b^2$$

$$= (5\sqrt{y})^2 - 2(5\sqrt{y})(\sqrt{2}) + (\sqrt{2})^2$$

Apply the formula $(a - b)^2 = a^2 - 2ab + b^2$.

$$= 25y - 10\sqrt{2y} + 2$$

Simplify.

Recall from Section 4.6, that the product of two conjugate binomials results in a difference of squares.

$$(a + b)(a - b) = a^2 - b^2$$

The same pattern occurs when multiplying two conjugate radical expressions.

example 6

Multiplying Conjugate Radical Expressions

Multiply the radical expressions. Assume all variables represent positive real numbers.

a. $(\sqrt{3} + 2)(\sqrt{3} - 2)$

b. $\left(\frac{1}{3}\sqrt{s} - \frac{3}{4}\sqrt{t}\right)\left(\frac{1}{3}\sqrt{s} + \frac{3}{4}\sqrt{t}\right)$

Solution:

a. $(\sqrt{3} + 2)(\sqrt{3} - 2)$ — The expression is in the form $(a + b)(a - b)$, where $a = \sqrt{3}$ and $b = 2$.

$a^2 - b^2$

$= (\sqrt{3})^2 - (2)^2$ — Apply the formula $(a + b)(a - b) = a^2 - b^2$.

$= 3 - 4$ — Simplify.

$= -1$

Tip: The product $(\sqrt{3} + 2)(\sqrt{3} - 2)$ can also be found using the distributive property.

$$\begin{aligned}(\sqrt{3} + 2)(\sqrt{3} - 2) &= \sqrt{3} \cdot \sqrt{3} + \sqrt{3} \cdot (-2) + 2 \cdot \sqrt{3} + 2 \cdot (-2)\\ &= 3 - 2\sqrt{3} + 2\sqrt{3} - 4\\ &= 3 - 4\\ &= -1\end{aligned}$$

b. $\left(\frac{1}{3}\sqrt{s} - \frac{3}{4}\sqrt{t}\right)\left(\frac{1}{3}\sqrt{s} + \frac{3}{4}\sqrt{t}\right)$ — This expression is in the form $(a - b)(a + b)$, where $a = \frac{1}{3}\sqrt{s}$ and $b = \frac{3}{4}\sqrt{t}$.

$a^2 - b^2$

$= \left(\frac{1}{3}\sqrt{s}\right)^2 - \left(\frac{3}{4}\sqrt{t}\right)^2$ — Apply the formula $(a + b)(a - b) = a^2 - b^2$.

$= \frac{1}{9}s - \frac{9}{16}t$ — Simplify.

section 10.5 PRACTICE EXERCISES

For the exercises in this set, assume that all variables represent positive real numbers unless otherwise stated.

1. Given $f(x) = \sqrt{-3x + 1}$, evaluate
   a. $f(-1)$ b. $f(-5)$
2. Given $g(x) = \sqrt{5 - x}$, evaluate
   a. $g(-20)$ b. $g(-11)$

For Exercises 3–6, simplify the radicals.

3. $\sqrt[3]{(x - y)^3}$
4. $\sqrt[5]{(2 + h)^5}$
5. $\sqrt[3]{-16x^5y^6z^7}$
6. $-\sqrt{20a^2b^3c}$

For Exercises 7–12, simplify the expressions. Write the answer with positive exponents only.

7. $\dfrac{1}{9^{-1/2}}$
8. $\left(\dfrac{1}{8}\right)^{-1/3}$

9. $x^{1/3}y^{1/4}x^{-1/6}y^{1/3}$

10. $p^{1/8}q^{1/2}p^{-1/4}q^{3/2}$

11. $\dfrac{a^{2/3}}{a^{1/2}}$

12. $\dfrac{b^{1/4}}{b^{3/2}}$

Exercises 13–14, add or subtract as indicated.

13. $-2\sqrt[3]{7} + 4\sqrt[3]{7}$

14. $4\sqrt{8x^3} - x\sqrt{50x}$

For Exercises 15–40, multiply the radical expressions.

15. $\sqrt{2} \cdot \sqrt{10}$

16. $\sqrt[3]{4} \cdot \sqrt[3]{12}$

17. $\sqrt[4]{16} \cdot \sqrt[4]{64}$

18. $\sqrt{5x^3} \cdot \sqrt{10x^4}$

19. $(2\sqrt{5})(3\sqrt{7})$

20. $(4\sqrt[3]{4})(2\sqrt[3]{5})$

21. $(8a\sqrt{b})(-3\sqrt{ab})$

22. $(p\sqrt[4]{q^3})(\sqrt[4]{pq})$

23. $\sqrt{3}(4\sqrt{3} - 6)$

24. $3\sqrt{5}(2\sqrt{5} + 4)$

25. $\sqrt{2}(\sqrt{6} - \sqrt{3})$

26. $\sqrt{5}(\sqrt{3} + \sqrt{7})$

27. $-3\sqrt{x}(\sqrt{x} + 7)$

28. $-2\sqrt{y}(8 - \sqrt{y})$

29. $(\sqrt{3} + 2\sqrt{10})(4\sqrt{3} - \sqrt{10})$

30. $(8\sqrt{7} - \sqrt{5})(\sqrt{7} + 3\sqrt{5})$

31. $(\sqrt{x} + 4)(\sqrt{x} - 9)$

32. $(\sqrt{w} - 2)(\sqrt{w} - 9)$

33. $(\sqrt[3]{y} + 2)(\sqrt[3]{y} - 3)$

34. $(4 + \sqrt[5]{p})(5 + \sqrt[5]{p})$

35. $(\sqrt{a} - 3\sqrt{b})(9\sqrt{a} - \sqrt{b})$

36. $(11\sqrt{m} + 4\sqrt{n})(\sqrt{m} + \sqrt{n})$

37. $(\sqrt{7} + 3)(\sqrt{7} + \sqrt{2} - 5)$

38. $(\sqrt{2} - 6)(\sqrt{2} - \sqrt{3} - 4)$

39. $(\sqrt{p} + 2\sqrt{q})(8 + 3\sqrt{p} - \sqrt{q})$

40. $(5\sqrt{s} - \sqrt{t})(\sqrt{s} + 5 + 6\sqrt{t})$

For Exercises 41–50, multiply or divide the radicals with different indices. Write the answers in radical form and simplify.

41. $\sqrt{x} \cdot \sqrt[4]{x}$

42. $\sqrt[3]{y} \cdot \sqrt{y}$

43. $\sqrt[5]{2z} \cdot \sqrt[3]{2z}$

44. $\sqrt[3]{5w} \cdot \sqrt[4]{5w}$

45. $\sqrt[3]{p^2} \cdot \sqrt{p^3}$

46. $\sqrt[4]{q^3} \cdot \sqrt[3]{q^2}$

47. $\dfrac{\sqrt{u^3}}{\sqrt[3]{u}}$

48. $\dfrac{\sqrt{v^5}}{\sqrt[4]{v}}$

49. $\dfrac{\sqrt{(a + b)}}{\sqrt[3]{(a + b)}}$

50. $\dfrac{\sqrt[3]{(q - 1)}}{\sqrt[4]{(q - 1)}}$

51. a. Write the formula for the product of two conjugates: $(x + y)(x - y) =$

b. Multiply $(x + 5)(x - 5)$

52. a. Write the formula for squaring a binomial: $(x + y)^2 =$

b. Multiply $(x + 5)^2$

For Exercises 53–64, multiply the special products.

53. $(\sqrt{3} + x)(\sqrt{3} - x)$

54. $(y + \sqrt{6})(y - \sqrt{6})$

55. $(\sqrt{6} + \sqrt{2})(\sqrt{6} - \sqrt{2})$

56. $(\sqrt{15} + \sqrt{5})(\sqrt{15} - \sqrt{5})$

57. $(8\sqrt{x} + 2\sqrt{y})(8\sqrt{x} - 2\sqrt{y})$

58. $(4\sqrt{s} + 11\sqrt{t})(4\sqrt{s} - 11\sqrt{t})$

59. $(\sqrt{13} + 4)^2$

60. $(6 - \sqrt{11})^2$

61. $(\sqrt{p} - \sqrt{7})^2$

62. $(\sqrt{q} + \sqrt{2})^2$

63. $(\sqrt{2a} - 3\sqrt{b})^2$

64. $(\sqrt{3w} + 4\sqrt{z})^2$

For Exercises 65–72, identify each statement as true or false. If an answer is false, explain why.

65. $\sqrt{3} \cdot \sqrt{2} = \sqrt{6}$

66. $\sqrt{5} \cdot \sqrt[3]{2} = \sqrt{10}$

67. $(x - \sqrt{5})^2 = x - 5$

68. $3(2\sqrt{5x}) = 6\sqrt{5x}$

69. $5(3\sqrt{4x}) = 15\sqrt{20x}$

70. $\dfrac{\sqrt{5x}}{5} = \sqrt{x}$

71. $\dfrac{3\sqrt{x}}{3} = \sqrt{x}$

72. $(\sqrt{t} - 1)(\sqrt{t} + 1) = t - 1$

For Exercises 73–78, find the exact area.

73.

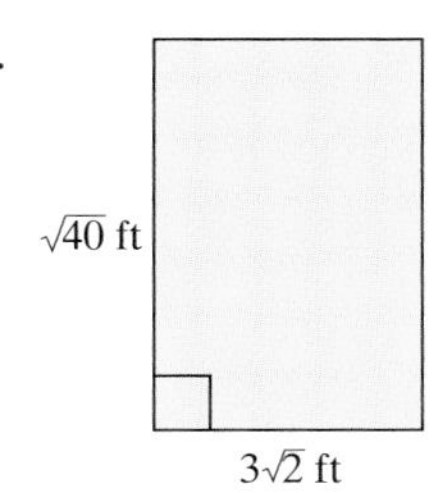

74.

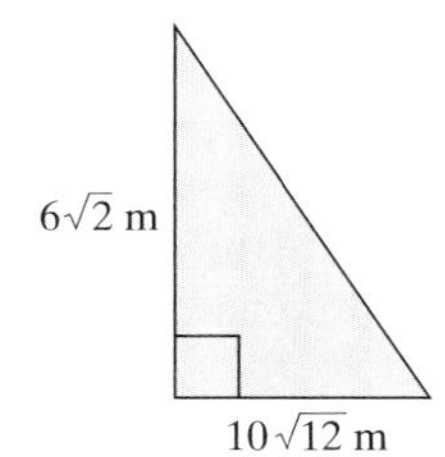

75.

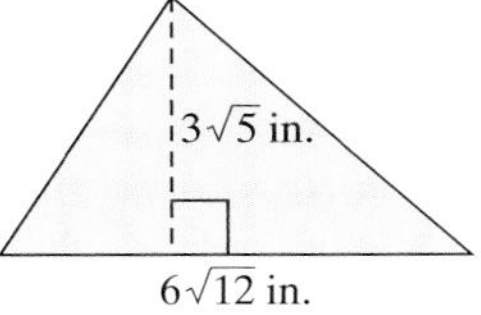

76.

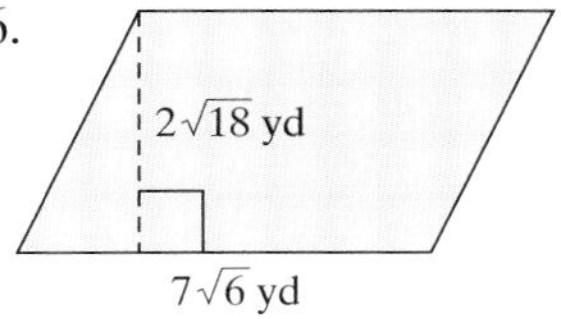

77.

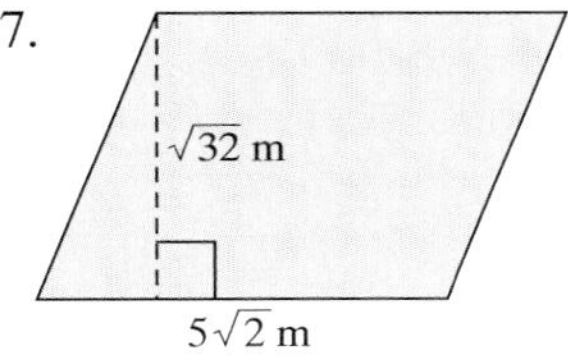

78.

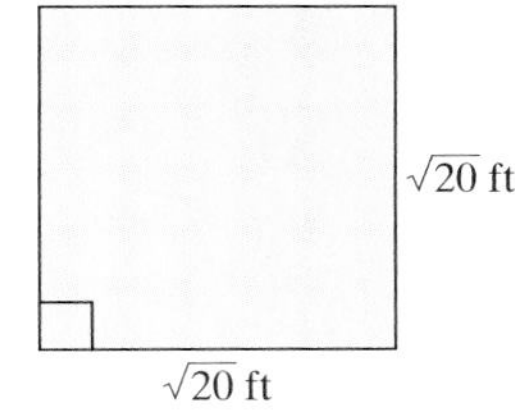

Expanding Your Skills

79. Multiply: $(\sqrt[3]{a} + \sqrt[3]{b})(\sqrt[3]{a^2} - \sqrt[3]{ab} + \sqrt[3]{b^2})$

80. Multiply: $(\sqrt[3]{a} - \sqrt[3]{b})(\sqrt[3]{a^2} + \sqrt[3]{ab} + \sqrt[3]{b^2})$

For Exercises 81–88, multiply the radicals with different indices [see Example 2(b)].

81. $\sqrt[3]{x} \cdot \sqrt[6]{y}$

82. $\sqrt{a} \cdot \sqrt[6]{b}$

83. $\sqrt[4]{8} \cdot \sqrt{3}$

84. $\sqrt{11} \cdot \sqrt[6]{2}$

85. $\sqrt[4]{6} \cdot \sqrt{2}$

86. $\sqrt[3]{10} \cdot \sqrt{3}$

87. $\sqrt[5]{p} \cdot \sqrt[3]{q}$

88. $\sqrt[6]{h} \cdot \sqrt[4]{k}$

section

10.6 Rationalization

Concepts

1. Simplified Form of a Radical
2. Rationalizing the Denominator—One Term
3. Rationalizing the Denominator—Two Terms

1. Simplified Form of a Radical

Recall that for a radical expression to be in simplified form the following three conditions must be met.

Simplified Form of a Radical

Consider any radical expression in which the radicand is written as a product of prime factors. The expression is in simplified form if all of the following conditions are met:

1. The radicand has no factor raised to a power greater than or equal to the index.
2. The radicand does not contain a fraction.
3. No radicals are in the denominator of a fraction.

The basis of the second and third conditions, which restrict radicals from the denominator of an expression, are largely historical. In some cases, removing a radical from the denominator of a fraction will create an expression that is computationally simpler. For example, we will show that

$$\frac{6}{\sqrt{3}} = 2\sqrt{3} \qquad \text{and} \qquad \frac{-2}{2+\sqrt{6}} = 2 - \sqrt{6}$$

The process to remove a radical from the denominator is called rationalizing the denominator. In this section we will consider two cases:

1. **Rationalizing the denominator (one term)**
2. **Rationalizing the denominator (two terms involving square roots)**

2. Rationalizing the Denominator—One Term

To begin the first case, recall that the *n*th root of a perfect *n*th power simplifies completely.

$$\sqrt{x^2} = x \quad x \geq 0$$
$$\sqrt[3]{x^3} = x$$
$$\sqrt[4]{x^4} = x \quad x \geq 0$$
$$\sqrt[5]{x^5} = x$$
$$\vdots$$

Therefore, to rationalize a radical expression, use the multiplication property of radicals to create an *n*th root of an *n*th power.

example 1

Rationalizing Radical Expressions

Fill in the blanks to rationalize the radical expressions. Assume all variables represent positive real numbers.

a. $\sqrt{a} \cdot \sqrt{?} = \sqrt{a^2} = a$ b. $\sqrt[3]{y} \cdot \sqrt[3]{?} = \sqrt[3]{y^3} = y$

c. $\sqrt[4]{2z^3} \cdot \sqrt[4]{?} = \sqrt[4]{2^4z^4} = 2z$

Solution:

a. $\sqrt{a} \cdot \sqrt{?} = \sqrt{a^2} = a$ What multiplied by $\sqrt{a}$ will equal $\sqrt{a^2}$?

$\sqrt{a} \cdot \sqrt{a} = \sqrt{a^2} = a$

b. $\sqrt[3]{y} \cdot \sqrt[3]{?} = \sqrt[3]{y^3} = y$ What multiplied by $\sqrt[3]{y}$ will equal $\sqrt[3]{y^3}$?

$\sqrt[3]{y} \cdot \sqrt[3]{y^2} = \sqrt[3]{y^3} = y$

c. $\sqrt[4]{2z^3} \cdot \sqrt[4]{?} = \sqrt[4]{2^4z^4} = 2z$ What multiplied by $\sqrt[4]{2z^3}$ will equal $\sqrt[4]{2^4z^4}$?

$\sqrt[4]{2z^3} \cdot \sqrt[4]{2^3z} = \sqrt[4]{2^4z^4} = 2z$

To rationalize the denominator of a radical expression, multiply the numerator and denominator by an appropriate expression to create an *n*th root of an *n*th power in the denominator.

example 2 Rationalizing the Denominator—One Term

Simplify the expression $\dfrac{5}{\sqrt[3]{a}}$, $(a \neq 0)$.

Solution:

To remove the radical from the denominator, a cube root of a perfect cube is needed in the denominator. Multiply numerator and denominator by $\sqrt[3]{a^2}$ because $\sqrt[3]{a} \cdot \sqrt[3]{a^2} = \sqrt[3]{a^3} = a$

Tip: Notice that for $a \neq 0$, the expression $\dfrac{\sqrt[3]{a^2}}{\sqrt[3]{a^2}} = 1$. Multiplying the original expression $5/\sqrt[3]{a}$ by this ratio does not change its value.

$$\frac{5}{\sqrt[3]{a}} = \frac{5}{\sqrt[3]{a}} \cdot \frac{\sqrt[3]{a^2}}{\sqrt[3]{a^2}}$$

$$= \frac{5\sqrt[3]{a^2}}{\sqrt[3]{a^3}} \qquad \text{Multiply the radicals.}$$

$$= \frac{5\sqrt[3]{a^2}}{a} \qquad \text{Simplify.}$$

example 3 Rationalizing the Denominator—One Term

Simplify the expressions. Assume all variables represent positive real numbers.

a. $\dfrac{6}{\sqrt{3}}$ b. $\sqrt{\dfrac{y^5}{7}}$ c. $\dfrac{15}{\sqrt[3]{25s}}$ d. $\dfrac{\sqrt{125p^3}}{\sqrt{5p}}$

Solution:

a. To rationalize the denominator, a square root of a perfect square is needed. Multiply numerator and denominator by $\sqrt{3}$ because $\sqrt{3} \cdot \sqrt{3} = \sqrt{3^2} = 3$.

$$\frac{6}{\sqrt{3}} = \frac{6}{\sqrt{3}} \cdot \frac{\sqrt{3}}{\sqrt{3}} \qquad \text{Rationalize the denominator.}$$

$$= \frac{6\sqrt{3}}{\sqrt{3^2}} \qquad \text{Multiply the radicals.}$$

$$= \frac{6\sqrt{3}}{3} \qquad \text{Simplify.}$$

$$= 2\sqrt{3} \qquad \text{Reduce.}$$

Calculator Connections

A calculator can be used to support the solution to a simplified radical. The calculator approximations of the expressions $6/\sqrt{3}$ and $2\sqrt{3}$ agree to 10 digits.

```
6/√(3)
         3.464101615
2√(3)
         3.464101615
```

b. $\sqrt{\dfrac{y^5}{7}}$ The radical contains an irreducible fraction.

$$= \frac{\sqrt{y^5}}{\sqrt{7}} \qquad \text{Apply the division property of radicals.}$$

$= \dfrac{y^2\sqrt{y}}{\sqrt{7}}$ Remove factors from the radical in the numerator.

$= \dfrac{y^2\sqrt{y}}{\sqrt{7}} \cdot \dfrac{\sqrt{7}}{\sqrt{7}}$ Rationalize the denominator. *Note:* $\sqrt{7} \cdot \sqrt{7} = \sqrt{7^2} = 7$.

$= \dfrac{y^2\sqrt{7y}}{\sqrt{7^2}}$

$= \dfrac{y^2\sqrt{7y}}{7}$ Simplify.

Avoiding Mistakes

A factor within a radicand cannot be simplified with a factor outside the radicand. For example, $\sqrt{7y}/7$ cannot be simplified

c. $\dfrac{15}{\sqrt[3]{25s}}$

$= \dfrac{15}{\sqrt[3]{5^2s}} \cdot \dfrac{\sqrt[3]{5s^2}}{\sqrt[3]{5s^2}}$ Because $25 = 5^2$, one additional factor of 5 is needed to form a perfect cube. Two additional factors of s are needed to make a perfect cube. Multiply numerator and denominator by $\sqrt[3]{5s^2}$.

$= \dfrac{15\sqrt[3]{5s^2}}{\sqrt[3]{5^3s^3}}$

$= \dfrac{15\sqrt[3]{5s^2}}{5s}$ Simplify the perfect cube.

$= \dfrac{\overset{3}{\cancel{15}}\sqrt[3]{5s^2}}{\underset{1}{\cancel{5}}s}$ Reduce to lowest terms.

$= \dfrac{3\sqrt[3]{5s^2}}{s}$

Tip: In the expression $\dfrac{15\sqrt[3]{s^2}}{5s}$, the factor of 15 and the factor of 5 may be reduced because both are outside the radical.

$$\frac{15\sqrt[3]{5s^2}}{5s} = \frac{15}{5} \cdot \frac{\sqrt[3]{5s^2}}{s} = \frac{3\sqrt[3]{5s^2}}{s}$$

d. $\dfrac{\sqrt{125p^3}}{\sqrt{5p}}$ Notice that the radicands in the numerator and denominator share common factors.

$= \sqrt{\dfrac{125p^3}{5p}}$ Rewrite the expression using the division property of radicals.

$= \sqrt{25p^2}$ Reduce the fraction within the radicand.

$= 5p$ Simplify the radical.

3. Rationalizing the Denominator—Two Terms

The next example demonstrates how to rationalize a two-term denominator involving square roots.

First recall from the multiplication of polynomials that the product of two conjugates results in a difference of squares.

$$(a + b)(a - b) = a^2 - b^2$$

If either a or b has a square root factor, the expression will simplify without a radical. That is, the expression is *rationalized.* For example,

$$\begin{aligned}(2 + \sqrt{6})(2 - \sqrt{6}) &= (2)^2 - (\sqrt{6})^2 \\ &= 4 - 6 \\ &= -2\end{aligned}$$

example 4 Rationalizing the Denominator—Two Terms

Simplify the expression by rationalizing the denominator. $\dfrac{-2}{2 + \sqrt{6}}$

Solution:

$\dfrac{-2}{2 + \sqrt{6}}$

$= \dfrac{(-2)}{(2 + \sqrt{6})} \cdot \dfrac{(2 - \sqrt{6})}{(2 - \sqrt{6})}$ Multiply the numerator and denominator by the conjugate of the denominator.

conjugates

$= \dfrac{-2(2 - \sqrt{6})}{(2)^2 - (\sqrt{6})^2}$ In the denominator, apply the formula $(a + b)(a - b) = a^2 - b^2$.

$= \dfrac{-2(2 - \sqrt{6})}{4 - 6}$ Simplify.

$= \dfrac{-2(2 - \sqrt{6})}{-2}$

$= \dfrac{-\cancel{2}(2 - \sqrt{6})}{-\cancel{2}}$ Reduce to lowest terms.

$= 2 - \sqrt{6}$

Calculator Connections

A calculator can be used to support the solution to a simplified radical. The calculator approximations of the expressions $-2/(2 + \sqrt{6})$ and $2 - \sqrt{6}$ agree to 10 decimal places.

```
-2/(2+√(6))
      -.4494897428
2-√(6)
      -.4494897428
```

example 5 Rationalizing the Denominator—Two Terms

Rationalize the denominator of the expression. $\dfrac{\sqrt{c} + \sqrt{d}}{\sqrt{c} - \sqrt{d}}$

Solution:

$\dfrac{\sqrt{c} + \sqrt{d}}{\sqrt{c} - \sqrt{d}}$

$= \dfrac{(\sqrt{c} + \sqrt{d})}{(\sqrt{c} - \sqrt{d})} \cdot \dfrac{(\sqrt{c} + \sqrt{d})}{(\sqrt{c} + \sqrt{d})}$ Multiply numerator and denominator by the conjugate of the denominator.

conjugates

$= \dfrac{(\sqrt{c} + \sqrt{d})^2}{(\sqrt{c})^2 - (\sqrt{d})^2}$ In the denominator apply the formula $(a + b)(a - b) = a^2 - b^2$.

$$= \frac{(\sqrt{c} + \sqrt{d})^2}{c - d}$$ Simplify.

$$= \frac{(\sqrt{c})^2 + 2\sqrt{c}\sqrt{d} + (\sqrt{d})^2}{c - d}$$ In the numerator apply the formula $(a + b)^2 = a^2 + 2ab + b^2$.

$$= \frac{c + 2\sqrt{cd} + d}{c - d}$$

section 10.6 Practice Exercises

For the exercises in this set, assume that all variables represent positive real numbers unless otherwise stated.

For Exercises 1–10, perform the indicated operations.

1. $2y\sqrt{45} + 3\sqrt{20y^2}$
2. $3x\sqrt{72x} - 9\sqrt{50x^3}$
3. $(-6\sqrt{y} + 3)(3\sqrt{y} + 1)$
4. $(\sqrt{w} + 12)(2\sqrt{w} - 4)$
5. $4\sqrt{3} + \sqrt{5} \cdot \sqrt{15}$
6. $\sqrt{7} \cdot \sqrt{21} + 2\sqrt{27}$
7. $(8 - \sqrt{t})^2$
8. $(\sqrt{p} + 4)^2$
9. $(\sqrt{2} + \sqrt{7})(\sqrt{2} - \sqrt{7})$
10. $(\sqrt{3} + 5)(\sqrt{3} - 5)$

The radical expressions in Exercises 11–18 have radicals in the denominator. Multiply the numerator and denominator by an appropriate expression to rationalize the denominator. Then simplify the result.

11. $\frac{x}{\sqrt{5}} = \frac{x}{\sqrt{5}} \cdot \frac{\sqrt{?}}{\sqrt{?}}$
12. $\frac{2}{\sqrt{x}} = \frac{2}{\sqrt{x}} \cdot \frac{\sqrt{?}}{\sqrt{?}}$
13. $\frac{7}{\sqrt[3]{x}} = \frac{7}{\sqrt[3]{x}} \cdot \frac{\sqrt[3]{?}}{\sqrt[3]{?}}$
14. $\frac{5}{\sqrt[4]{y}} = \frac{5}{\sqrt[4]{y}} \cdot \frac{\sqrt[4]{?}}{\sqrt[4]{?}}$
15. $\frac{8}{\sqrt{3z}} = \frac{8}{\sqrt{3z}} \cdot \frac{\sqrt{??}}{\sqrt{??}}$
16. $\frac{10}{\sqrt{7w}} = \frac{10}{\sqrt{7w}} \cdot \frac{\sqrt{??}}{\sqrt{??}}$
17. $\frac{1}{\sqrt[4]{2a^2}} = \frac{1}{\sqrt[4]{2a^2}} \cdot \frac{\sqrt[4]{??}}{\sqrt[4]{??}}$
18. $\frac{1}{\sqrt[3]{6b^2}} = \frac{1}{\sqrt[3]{6b^2}} \cdot \frac{\sqrt[3]{??}}{\sqrt[3]{??}}$

For Exercises 19–42, rationalize the denominator.

19. $\frac{1}{\sqrt{3}}$
20. $\frac{1}{\sqrt{7}}$
21. $\frac{10}{\sqrt{5}}$
22. $\frac{12}{\sqrt{6}}$
23. $\frac{1}{\sqrt{x}}$
24. $\frac{1}{\sqrt{z}}$
25. $\frac{6}{\sqrt{2y}}$
26. $\frac{9}{\sqrt{3t}}$
27. $\frac{-2a}{\sqrt{a}}$
28. $\frac{-7b}{\sqrt{b}}$
29. $\frac{7}{\sqrt[3]{4}}$
30. $\frac{1}{\sqrt[3]{9}}$
31. $\frac{4}{\sqrt{w^3}}$
32. $\frac{5}{\sqrt{z^3}}$
33. $\sqrt[4]{\frac{16}{3}}$
34. $\sqrt[4]{\frac{81}{8}}$

35. $\dfrac{1}{\sqrt{x^7}}$

36. $\dfrac{1}{\sqrt{y^5}}$

37. $\dfrac{2}{\sqrt{8x^5}}$

38. $\dfrac{6}{\sqrt{27t^7}}$

39. $\sqrt[3]{\dfrac{16x^3}{y}}$

40. $\sqrt{\dfrac{5}{9x}}$

41. $\dfrac{\sqrt{x^4y^5}}{\sqrt{10x}}$

42. $\sqrt[4]{\dfrac{10x^2}{15xy^3}}$

43. What is the conjugate of $\sqrt{2} - \sqrt{6}$?

44. What is the conjugate of $\sqrt{11} + \sqrt{5}$?

45. What is the conjugate of $\sqrt{x} + 23$?

46. What is the conjugate of $17 - \sqrt{y}$?

For Exercises 47–50, multiply the conjugates.

47. $(\sqrt{2} + 3)(\sqrt{2} - 3)$

48. $(4 - \sqrt{3})(4 + \sqrt{3})$

49. $(\sqrt{5} - \sqrt{2})(\sqrt{5} + \sqrt{2})$

50. $(\sqrt{3} + \sqrt{7})(\sqrt{3} - \sqrt{7})$

For Exercises 51–62, rationalize the denominators.

51. $\dfrac{4}{\sqrt{2} + 3}$

52. $\dfrac{6}{4 - \sqrt{3}}$

53. $\dfrac{1}{\sqrt{5} - \sqrt{2}}$

54. $\dfrac{1}{\sqrt{3} + \sqrt{7}}$

55. $\dfrac{\sqrt{7}}{\sqrt{3} + 2}$

56. $\dfrac{\sqrt{8}}{\sqrt{3} + 1}$

57. $\dfrac{-1}{\sqrt{p} + \sqrt{q}}$

58. $\dfrac{6}{\sqrt{a} - \sqrt{b}}$

59. $\dfrac{2\sqrt{3} + \sqrt{7}}{3\sqrt{3} - \sqrt{7}}$

60. $\dfrac{5\sqrt{2} - \sqrt{5}}{5\sqrt{2} + \sqrt{5}}$

61. $\dfrac{\sqrt{5} + 4}{2 - \sqrt{5}}$

62. $\dfrac{3 + \sqrt{2}}{\sqrt{2} - 5}$

For Exercises 63–66, translate the English phrase into an algebraic expression. Then simplify the expression.

63. Sixteen divided by the cube root of 4.

64. Twenty-one divided by the fourth root of 27.

65. Four divided by the difference of x and the square root of 2.

66. Eight divided by the sum of y and the square root of 3.

67. The approximate time, T (in seconds), for a pendulum to make one complete swing back and forth is given by

$T(x) = 2\pi\sqrt{\dfrac{x}{32}}$, where x is the length of the pendulum in feet.

a. Determine the time required for one swing of a pendulum that is 2 ft long. (Round the answer to two decimal places.)

b. Determine the time required for one swing of a pendulum that is 1 ft long. (Round the answer to two decimal places.)

c. Determine the time required for one swing of a pendulum that is $\frac{1}{2}$ ft long. (Round the answer to two decimal places.)

68. An object is dropped off a building x meters tall. The time, T, (in seconds) required for the object to hit the ground is given by

$$T(x) = \sqrt{\frac{10x}{49}}$$

a. Find the time required for the object to hit the ground if it is dropped off a 50-m building. (Round the answer to two decimal places.)

b. Find the time required for the object to hit the ground if it is dropped off a 25-m building. (Round the answer to two decimal places.)

c. Find the time required for the object to hit the ground if it is dropped off a 10-m building. (Round the answer to two decimal places.)

Expanding Your Skills

For Exercises 69–74, simplify the expression.

69. $\frac{\sqrt{6}}{2} + \frac{1}{\sqrt{6}}$

70. $\frac{1}{\sqrt{7}} + \sqrt{7}$

71. $\sqrt{15} - \sqrt{\frac{3}{5}} + \sqrt{\frac{5}{3}}$

72. $\sqrt{\frac{6}{2}} - \sqrt{12} + \sqrt{\frac{2}{6}}$

73. $\sqrt[3]{25} + \frac{3}{\sqrt[3]{5}}$

74. $\frac{1}{\sqrt[3]{4}} + \sqrt[3]{54}$

For Exercises 75–78, rationalize the numerator by multiplying both numerator and denominator by the conjugate of the numerator.

75. $\frac{\sqrt{3} + 6}{2}$

76. $\frac{\sqrt{7} - 2}{5}$

77. $\frac{\sqrt{a} - \sqrt{b}}{\sqrt{a} + \sqrt{b}}$

78. $\frac{\sqrt{p} + \sqrt{q}}{\sqrt{p} - \sqrt{q}}$

Concepts

1. Solutions to Radical Equations
2. Solving Radical Equations Involving One Radical
3. Solving Radical Equations Involving More than One Radical
4. Applications of Radical Functions

section

10.7 Radical Equations

1. Solutions to Radical Equations

An equation with one or more radicals containing a variable is called a **radical equation**. For example, $\sqrt[3]{x} = 5$ is a radical equation. Recall that $(\sqrt[n]{a})^n = a$, provided $\sqrt[n]{a}$ is a real number. The basis of solving a radical equation is to eliminate the radical by raising both sides of the equation to a power equal to the index of the radical.

To solve the equation $\sqrt[3]{x} = 5$, cube both sides of the equation.

$$\sqrt[3]{x} = 5$$
$$(\sqrt[3]{x})^3 = (5)^3$$
$$x = 125$$

By raising each side of a radical equation to a power equal to the index of the radical, a new equation is produced. It is important to note, however, that the new equation may have extraneous solutions. That is, some or all of the solutions to the new equation may *not* be solutions to the original radical equation. For this reason, it is necessary to *check all potential solutions* in the original equation. For example, consider the equation $x = 4$. By squaring both sides we produce a quadratic equation.

$$x = 4$$

square both sides

$$(x)^2 = (4)^2$$

Squaring both sides produces a quadratic equation.

$$x^2 = 16$$

Solving this equation, we find two solutions. However, the solution $x = -4$ does not check in the original equation.

$$x^2 - 16 = 0$$
$$(x - 4)(x + 4) = 0$$

$x = 4$ or $x \neq -4$ (does not check)

Steps to Solve a Radical Equation

1. Isolate the radical. If an equation has more than one radical, choose one of the radicals to isolate.
2. Raise each side of the equation to a power equal to the index of the radical.
3. Solve the resulting equation. If the equation still has a radical, repeat steps 1 and 2.

*4. Check the potential solutions in the original equation.

*Extraneous solutions can only arise when both sides of the equation are raised to an *even power.* Therefore, an equation with odd-index roots will not have an extraneous solution. However, it is still recommended that you check *all* potential solutions regardless of the type of root.

2. Solving Radical Equations Involving One Radical

example 1 **Solving Equations Containing One Radical**

Solve the equations.

a. $\sqrt{3x-2}+4=5$

b. $(w-1)^{1/3}-2=2$

c. $7=\sqrt[4]{x+3}+9$

d. $y+\sqrt{y^2+5}=7$

Solution:

a. $\sqrt{3x-2}+4=5$

$\sqrt{3x-2}=1$ Isolate the radical.

$(\sqrt{3x-2})^2=(1)^2$ Because the index is 2, square both sides.

$3x-2=1$ Simplify.

$3x=3$ Solve the resulting equation.

$x=1$

<u>Check</u>: $x=1$ Check $x=1$ as a potential solution.

$\sqrt{3x-2}+4=5$

$\sqrt{3(1)-2}+4\stackrel{?}{=}5$

$\sqrt{1}+4\stackrel{?}{=}5$

$5=5$ ✔ Therefore, $x=1$ is a solution to the original equation.

b. $(w-1)^{1/3}-2=2$ Note that $(w-1)^{1/3}=\sqrt[3]{w-1}$.

$\sqrt[3]{w-1}-2=2$

$\sqrt[3]{w-1}=4$ Isolate the radical.

$(\sqrt[3]{w-1})^3=(4)^3$ Because the index is 3, cube both sides.

$w-1=64$ Simplify.

$w=65$

Check: $w = 65$

$(w - 1)^{1/3} - 2 = 2$ — Check $w = 65$ as a potential solution.

$\sqrt[3]{65 - 1} - 2 \stackrel{?}{=} 2$

$\sqrt[3]{64} - 2 \stackrel{?}{=} 2$

$4 - 2 \stackrel{?}{=} 2$

$2 = 2$ ✔ — Therefore, $w = 65$ is a solution to the original equation.

c. $7 = \sqrt[4]{x + 3} + 9$

$-2 = \sqrt[4]{x + 3}$ — Isolate the radical.

$(-2)^4 = (\sqrt[4]{x + 3})^4$ — Because the index is 4, raise both sides to the fourth power.

$16 = x + 3$

$x = 13$ — Solve for x.

Check: $x = 13$

$7 = \sqrt[4]{x + 3} + 9$

$7 \stackrel{?}{=} \sqrt[4]{(13) + 3} + 9$

$7 \stackrel{?}{=} \sqrt[4]{16} + 9$

$7 \neq 2 + 9$ — Therefore, $x = 13$ is *not* a solution to the original equation.

The equation $7 = \sqrt[4]{x + 3} + 9$ has no solution.

Tip: After isolating the radical in Example 1(c), the equation shows a fourth root equated to a negative number:

$$-2 = \sqrt[4]{x + 3}$$

By definition, a principal fourth root of any real number must be nonnegative. Therefore, there can be no real solution to this equation.

d. $y + \sqrt{y^2 + 5} = 7$

$\sqrt{y^2 + 5} = 7 - y$ — Isolate the radical.

$(\sqrt{y^2 + 5})^2 = (7 - y)^2$ — Because the index is 2, square both sides.

Note that $(7 - y)^2 = (7 - y)(7 - y) = 49 - 14y + y^2$.

$y^2 + 5 = 49 - 14y + y^2$ — Simplify.

$5 = 49 - 14y$ — Subtract y^2 from both sides.

$-44 = -14y$ — Solve for y.

$$\frac{-44}{-14} = \frac{-14y}{-14}$$

$\frac{22}{7} = y$ — The solution checks in the original equation.

3. Solving Radical Equations Involving More than One Radical

example 2 **Solving Equations with Two Radicals**

Solve the radical equations.

$\sqrt[3]{2x - 4} = \sqrt[3]{1 - 8x}$

Solution:

$\sqrt[3]{2x - 4} = \sqrt[3]{1 - 8x}$	
$(\sqrt[3]{2x - 4})^3 = (\sqrt[3]{1 - 8x})^3$	Because the index is 3, cube both sides.
$2x - 4 = 1 - 8x$	Simplify.
$2x + 8x - 4 = 1$	Solve the resulting equation. Add $8x$ to both sides.
$10x = 5$	Combine *like* terms and add 4 to both sides.
$\frac{10x}{10} = \frac{5}{10}$	
$x = \frac{1}{2}$	The solution checks in the original equation.

Calculator Connections

The expressions on the right- and left-hand sides of the equation $\sqrt[3]{2x - 4} = \sqrt[3]{1 - 8x}$ are each functions of x. Consider the graphs of the functions:

$Y_1 = \sqrt[3]{2x - 4}$ and $Y_2 = \sqrt[3]{1 - 8x}$

The x-coordinate of the point of intersection of the two functions is the solution to the equation $\sqrt[3]{2x - 4} = \sqrt[3]{1 - 8x}$. The point of intersection can be approximated by using *Zoom* and *Trace* or by using an *Intersect* function.

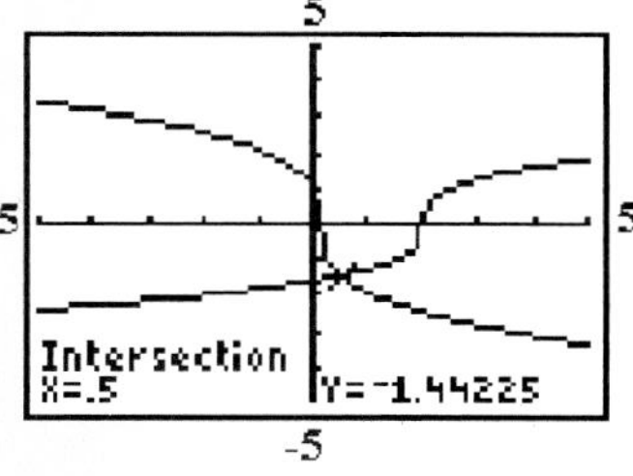

example 3 **Solving Equations with Two Radicals**

Solve the radical equation. $\sqrt{3m + 1} - \sqrt{m + 4} = 1$

Solution:

$\sqrt{3m + 1} - \sqrt{m + 4} = 1$

$\sqrt{3m + 1} = \sqrt{m + 4} + 1$ Isolate one of the radicals.

$(\sqrt{3m+1})^2 = (\sqrt{m+4}+1)^2$	Square both sides.
$3m+1 = m+4+2\sqrt{m+4}+1$	*Note:* $(\sqrt{m+4}+1)^2$ $= (\sqrt{m+4})^2 + 2(1)\sqrt{m+4} + (1)^2$ $= m+4+2\sqrt{m+4}+1$
$3m+1 = m+5+2\sqrt{m+4}$	Combine *like* terms.
$2m-4 = 2\sqrt{m+4}$	Isolate the radical again.
$m-2 = \sqrt{m+4}$	Divide both sides by 2.
$(m-2)^2 = (\sqrt{m+4})^2$	Square both sides again.
$m^2-4m+4 = m+4$	The resulting equation is quadratic.
$m^2-5m = 0$	Set the quadratic equation equal to zero.
$m(m-5) = 0$	Factor.
$m = 0$ or $m = 5$	

Check: $m = 0$

$\sqrt{3(0)+1} - \sqrt{(0)+4} \stackrel{?}{=} 1$

$\sqrt{1} - \sqrt{4} \stackrel{?}{=} 1$

$1 - 2 \neq 1$ false

Check: $m = 5$

$\sqrt{3(5)+1} - \sqrt{(5)+4} = 1$

$\sqrt{16} - \sqrt{9} \stackrel{?}{=} 1$

$4 - 3 = 1$ ✔

The solution is $m = 5$ (the value $m = 0$ does not check).

4. Applications of Radical Functions

example 4 **Analyzing a Radical Function**

On a certain surface, the speed of a car, s (in miles per hour), before the brakes were applied can be approximated from the length of its skid marks, x (in feet), by:

$$s(x) = 3.8\sqrt{x} \quad x \geq 0$$

See Figure 10-2.

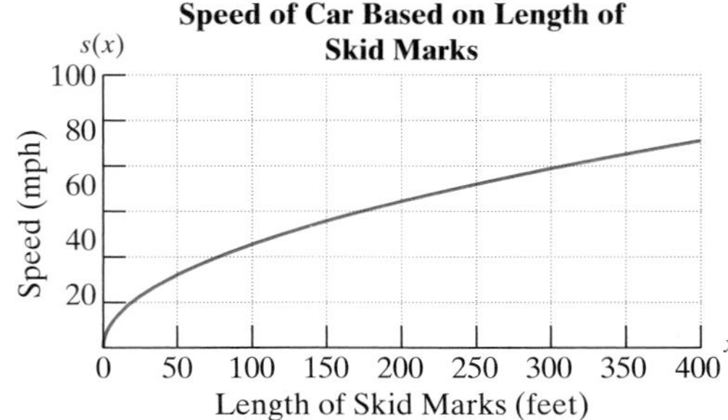

Figure 10-2

a. Find the speed of a car before the brakes were applied if its skid marks are 361 ft long.
b. How long would you expect the skid marks to be if the car had been traveling the speed limit of 50 mph? (Round to the nearest foot)

Solution:

a.

$$s(x) = 3.8\sqrt{x}$$

$$s(361) = 3.8\sqrt{361} \qquad \text{Substitute } x = 361.$$

$$= 3.8(19)$$

$$= 72.2$$

If the skid marks are 361 ft, the car was traveling approximately 72.2 mph before the brakes were applied.

b.

$$s(x) = 3.8\sqrt{x}$$

$$50 = 3.8\sqrt{x} \qquad \text{Substitute } s(x) = 50 \text{ and solve for } x.$$

$$\frac{50}{3.8} = \sqrt{x} \qquad \text{Isolate the radical.}$$

$$\left(\frac{50}{3.8}\right)^2 = x$$

$$x \approx 173$$

If the car had been going the speed limit (50 mph), then the skid marks would have been approximately 173 ft.

section 10.7 Practice Exercises

For Exercises 1–10, simplify the radical expressions, if possible. Assume all variables represent positive real numbers.

1. $\sqrt{48}$
2. $\sqrt{18w^4}$
3. $\sqrt{\frac{9w^3}{16}}$
4. $\sqrt{\frac{a^2}{3}}$
5. $\sqrt{-25}$
6. $\sqrt[3]{54c^4}$
7. $\sqrt{\frac{p^5}{q^3}}$
8. $\sqrt{\frac{x^4y^2}{16y^{10}}}$
9. $\sqrt{\frac{49}{5t^3}}$
10. $\sqrt[3]{-\frac{8}{27}}$

For Exercises 11–20, simplify each expression. Assume all radicands represent positive real numbers.

11. $(\sqrt{4x-6})^2$
12. $(\sqrt{5y+2})^2$
13. $(\sqrt[3]{9p+7})^3$
14. $(\sqrt[3]{4t+13})^3$
15. $(\sqrt{w^2+2w-17})^2$
16. $(\sqrt{6x^2-x+8})^2$
17. $(\sqrt{2x})^2$
18. $(\sqrt{5y})^2$
19. $(\sqrt[4]{7r})^4$
20. $(\sqrt[4]{3s})^4$

For Exercises 21–50, solve the equations.

21. $\sqrt{t} = 7$
22. $\sqrt{w} = 12$
23. $\sqrt{4x} = 6$
24. $\sqrt{2x} = 8$
25. $\sqrt{5y+1} = 4$
26. $\sqrt{9z-5} = 11$
27. $\sqrt[4]{2x+1} = 2$
28. $\sqrt[3]{1-3x} = 4$
29. $(2z-3)^{1/2} = 9$
30. $(8+3a)^{1/2} = 5$

31. $\sqrt[3]{x - 2} = 3$
32. $\sqrt[3]{2x - 5} = 2$
33. $(15 - w)^{1/3} = -5$
34. $(k + 18)^{1/3} = -2$
35. $\sqrt{x - 16} = -3$
36. $\sqrt{2x + 1} = -12$
37. $\sqrt[3]{x + 1} + 3 = -1$
38. $\sqrt{4y - 8} - 2 = 8$
39. $11 = 4\sqrt{3t} - 5$
40. $6 = 2\sqrt{x - 1} - 4$
41. $\sqrt{6p - 8} = p$
42. $y = \sqrt{y + 12}$
43. $2x = \sqrt{4x + 3}$
44. $2x = \sqrt{15 - 4x}$
45. $\sqrt[4]{h + 4} = \sqrt[4]{2h - 5}$
46. $\sqrt[4]{3b + 6} = \sqrt[4]{7b - 6}$
47. $\sqrt[3]{5a + 3} = \sqrt[3]{a - 13}$
48. $\sqrt[3]{k - 8} = \sqrt[3]{4k + 1}$
49. $\sqrt[4]{2x - 5} = -1$
50. $\sqrt[4]{x + 16} = -4$
51. Solve for V: $r = \sqrt[3]{\dfrac{3V}{4\pi}}$
52. Solve for V: $r = \sqrt{\dfrac{V}{h\pi}}$
53. Solve for h^2: $r = \pi\sqrt{r^2 + h^2}$
54. Solve for d: $s = 1.3\sqrt{d}$

For Exercises 55–60, use $(a + b)^2 = a^2 + 2ab + b^2$ to practice squaring a binomial.

55. $(a + 5)^2 =$
56. $(b + 7)^2 =$
57. $(5w - 4)^2 =$
58. $(2p - 3)^2 =$
59. $(\sqrt{5a} - 3)^2 =$
60. $(2 + \sqrt{b})^2 =$

For Exercises 61–86, solve the radical equations, if possible.

61. $\sqrt{a^2 + 2a + 1} = a + 5$
62. $\sqrt{b^2 - 5b - 8} = b + 7$
63. $\sqrt{25w^2 - 2w - 3} = 5w - 4$
64. $\sqrt{4p^2 - 2p + 1} = 2p - 3$
65. $\sqrt{9z^2 - z + 6} = 3z - 1$
66. $\sqrt{16x^2 - 2x - 9} = 4x + 1$
67. $\sqrt{5a - 9} = \sqrt{5a} - 3$
68. $\sqrt{8 + b} = 2 + \sqrt{b}$
69. $\sqrt{2h + 5} - \sqrt{2h} = 1$
70. $\sqrt{3k - 5} - \sqrt{3k} = -1$
71. $\sqrt{t - 9} = 3 + \sqrt{t}$
72. $\sqrt{y - 16} = \sqrt{y} + 4$
73. $\sqrt{x^2 + 3} = 6 + x$
74. $\sqrt{y^2 + 5} = 2 + y$
75. $\sqrt{3t - 7} = 2 - \sqrt{3t + 1}$
76. $\sqrt{p - 6} = \sqrt{p + 2} - 4$
77. $\sqrt{8b - 3} = 4 + \sqrt{8b + 1}$
78. $\sqrt{2w + 5} = \sqrt{2w - 1} + 3$
79. $\sqrt{p^2 + 35} = \sqrt{12p}$
80. $\sqrt{x^2 + 24} = \sqrt{11x}$
81. $\sqrt{6m + 7} = \sqrt{3m + 3} + 1$
82. $\sqrt{5w + 1} - \sqrt{3w} = 1$
83. $\sqrt{z + 1} + \sqrt{2z + 3} = 1$
84. $\sqrt{2y + 6} = \sqrt{7 - 2y} + 1$
85. $1 + \sqrt{2t + 3} + \sqrt{3t - 5} = 0$
86. $2 + \sqrt{3x + 1} + \sqrt{x - 1} = 0$
87. The time, t, in seconds it takes an object to drop d meters is given by

$$t(d) = \sqrt{\frac{d}{4.9}}$$

   a. Approximate the height of the Texas Commerce Tower in Houston, if it takes an object 7.89 sec to drop from the top. Round to the nearest meter.
   b. Approximate the height of the Shanghai World Financial Center, if it takes an object

9.69 sec to drop from the top. Round to the nearest meter.

88. If an object is dropped from an initial height h, its velocity at impact with the ground is given by

$v = \sqrt{2gh}$, where g is the acceleration due to gravity and h is the initial height.

a. Find the initial height (in feet) of an object if its velocity at impact is 44 ft/sec. (Assume that the acceleration due to gravity is $g = 32$ ft/sec^2)

b. Find the initial height (in meters) of an object if its velocity at impact is 26 m/sec. (Assume that the acceleration due to gravity is $g = 9.8$ m/sec^2) Round to the nearest tenth of a meter.

89. The airline cost for x thousand passengers to travel round trip from New York to Atlanta is given by

$C(x) = \sqrt{0.3x + 1}$, where $C(x)$ is measured in millions of dollars and $x \geq 0$.

a. Find the airline's cost for 10,000 passengers ($x = 10$) to travel from New York to Atlanta.

b. If the airline charges \$320 per passenger, find the profit made by the airline for flying 10,000 passengers from New York to Atlanta.

c. Approximate the number of passengers who traveled from New York to Atlanta if the total cost for the airline was \$4 million.

90. The time, T (in seconds), required for a pendulum to make one complete swing back and forth is approximated by

$T = 2\pi\sqrt{\frac{L}{g}}$, where g is the acceleration due to gravity and L is the length of the pendulum (in feet).

a. Find the length of a pendulum that requires 1.36 sec to make one complete swing back and forth. (Assume that the acceleration due to gravity is $g = 32$ ft/sec^2) Round to the nearest tenth of a foot.

b. Find the time required for a pendulum to complete one swing back and forth if the length of the pendulum is 4 ft. (Assume that the acceleration due to gravity is $g = 32$ ft/sec.^{2}) Round to the nearest tenth of a second.

91. a. For $x = 3$, evaluate the two expressions $\sqrt{x^2 + 4}$ and $x + 2$.

b. Are these expressions equal?

92. a. For $x = 3$, evaluate the two expressions $\sqrt{x^2 + 16}$ and $x + 4$.

b. Are these expressions equal?

Expanding Your Skills

93. The number of hours needed to cook a turkey that weighs x pounds can be approximated by

$t(x) = 0.90\sqrt[5]{x^3}$, where t is the time in hours and x is the weight of the turkey in pounds.

a. Find the weight of a turkey that cooked for 4 hr. Round to the nearest pound.

b. Find $t(18)$ and interpret the result. Round to the nearest tenth of an hour.

For Exercises 94–99, use the Pythagorean theorem to find a, b, or c.

$$a^2 + b^2 = c^2$$

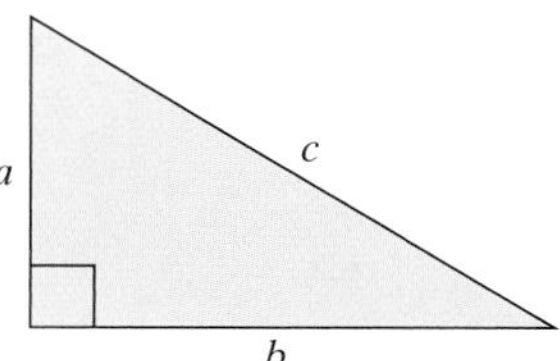

Figure for Exercises 94–99

94. Find c when $a = 6$ and $b = x$.

95. Find c when $a = k$ and $b = 9$.

96. Find b when $a = 2$ and $c = y$.

97. Find b when $a = h$ and $c = 5$.

98. Find a when $b = x$ and $c = 8$.

99. Find a when $b = 14$ and $c = k$.

For Exercises 100–103, solve the radical equations.

100. $\sqrt{y + 2 + \sqrt{y + 1}} = 1$

101. $\sqrt{4x - \sqrt{8x^2 + 1}} = 1$

102. $\sqrt[3]{p^3 + 6p^2 - 4} = p + 2$

103. $\sqrt[3]{q^3 + 9q^2 - 27} = q + 3$

Graphing Calculator Exercises

104. Refer to Exercise 23. Graph Y_1 and Y_2 on a viewing window defined by $-10 \le x \le 20$ and $-5 \le y \le 10$.

$$Y_1 = \sqrt{4x} \qquad \text{and} \qquad Y_2 = 6$$

Use an *Intersect* feature to approximate the x-coordinate of the point of intersection of the two graphs to support your solution to Exercise 23.

105. Refer to Exercise 24. Graph Y_1 and Y_2 on a viewing window defined by $-10 \le x \le 40$ and $-5 \le y \le 10$.

$$Y_1 = \sqrt{2x} \qquad \text{and} \qquad Y_2 = 8$$

Use an *Intersect* feature to approximate the x-coordinate of the point of intersection of the two graphs to support your solution to Exercise 24.

106. Graph $C(x) = \sqrt{0.3x + 1}$ on a viewing window defined by $0 \le x \le 60$ and $-2 \le y \le 6$. Use a *Table* feature to find the following function values:

$$C(10), C(20), C(30), C(40), \text{and } C(50)$$

Use these values to support your answers to Exercises 89(a) and 89(c).

107. Graph $t(x) = 0.9\sqrt[5]{x^3}$ on a viewing window defined by $0 \le x \le 30$ and $-2 \le y \le 10$. Use a *Table* feature to find the following function values:

$$t(10), t(12), t(14), t(16), t(18), \text{and } t(20)$$

Use these values to support your answers to Exercise 93.

---

section

10.8 Complex Numbers

Concepts

1. **Definition of *i***
2. **Simplifying Expressions in Terms of *i***
3. **The Powers of *i***
4. **Definition of a Complex Number**
5. **Addition, Subtraction, and Multiplication of Complex Numbers**
6. **Division of Complex Numbers**

1. Definition of *i*

In Section 10.1, we learned that there are no real-valued square roots of a negative number. For example, $\sqrt{-9}$ is not a real number because no real number when squared equals -9. However, the square roots of a negative number are defined over another set of numbers called the **imaginary numbers**. The foundation of the set of imaginary numbers is the definition of the imaginary number, i, as: $i = \sqrt{-1}$.

Definition of *i*

$$i = \sqrt{-1}$$

Note: From the definition of i, it follows that $i^2 = -1$.

2. Simplifying Expressions in Terms of *i*

Using the imaginary number i, we can define the square root of any negative real number.

Definition of $\sqrt{-b}$, $b > 0$

Let b be a real number such that $b > 0$, then $\sqrt{-b} = i\sqrt{b}$.

example 1 **Simplifying Expressions in Terms of *i***

Simplify the expressions in terms of i.

a. $\sqrt{-64}$ b. $\sqrt{-100}$ c. $\sqrt{-29}$

Solution:

a. $\sqrt{-64} = 8i$

b. $\sqrt{-100} = 10i$

c. $\sqrt{-29} = i\sqrt{29}$

Avoiding Mistakes

In an expression such as $i\sqrt{29}$ the i is usually written in front of the square root. The expression $\sqrt{29}\,i$ is also correct, but may be misinterpreted as $\sqrt{29i}$ (with i incorrectly placed under the square root).

The multiplication and division properties of radicals were presented in Sections 10.3 and 10.5 as follows:

If a and b represent real numbers such that $\sqrt[n]{a}$ and $\sqrt[n]{b}$ are both real, then

$$\sqrt[n]{ab} = \sqrt[n]{a} \cdot \sqrt[n]{b} \quad \text{and} \quad \sqrt[n]{\frac{a}{b}} = \frac{\sqrt[n]{a}}{\sqrt[n]{b}} \quad b \neq 0.$$

The conditions that $\sqrt[n]{a}$ and $\sqrt[n]{b}$ must both be real numbers prevent us from applying the multiplication and division properties of radicals for square roots with a negative radicand. Therefore, to multiply or divide radicals with a negative radicand, write the radical in terms of the imaginary number i first. This is demonstrated in the following example.

example 2 **Simplifying a Product of Expressions in Terms of *i***

Simplify the expressions.

a. $\dfrac{\sqrt{-100}}{\sqrt{-25}}$ b. $\sqrt{-25} \cdot \sqrt{-9}$

Solution:

a. $\dfrac{\sqrt{-100}}{\sqrt{-25}}$

$= \dfrac{10i}{5i}$ Simplify each radical in terms of i *before* dividing.

$= 2$ Reduce.

Avoiding Mistakes

In Example 2(b), the radical expressions were written in terms of *i* first *before* multiplying. If we had mistakenly applied the multiplication property first we would obtain an incorrect answer.

Correct: $\sqrt{-25} \cdot \sqrt{-9}$
$= (5i)(3i) = 15i^2$
$= 15(-1) = -15$ ← correct

Be careful: $\sqrt{-25} \cdot \sqrt{-9}$
$\neq \sqrt{225} = 15$ ← (incorrect answer)

b. $\sqrt{-25} \cdot \sqrt{-9}$

$= 5i \cdot 3i$	Simplify each radical in terms of *i* first *before* multiplying.
$= 15i^2$	Multiply.
$= 15(-1)$	Recall that $i^2 = -1$
$= -15$	Simplify.

example 3

Simplifying Complex Numbers

Simplify the complex numbers.

a. $\dfrac{6 + \sqrt{-18}}{9}$ b. $\dfrac{4 - \sqrt{-36}}{2}$

Solution:

a. $\dfrac{6 + \sqrt{-18}}{9} = \dfrac{6 + i\sqrt{18}}{9}$ Write the radical in terms of *i*.

$= \dfrac{6 + 3i\sqrt{2}}{9}$ Simplify $\sqrt{18} = 3\sqrt{2}$.

$= \dfrac{3(2 + i\sqrt{2})}{9}$ Factor the numerator.

$= \dfrac{\overset{1}{\cancel{3}}(2 + i\sqrt{2})}{\underset{3}{\cancel{9}}}$ Reduce the common factors.

$= \dfrac{2 + i\sqrt{2}}{3}$

b. $\dfrac{4 - \sqrt{-36}}{2} = \dfrac{4 - i\sqrt{36}}{2}$ Write the radical in terms of *i*.

$= \dfrac{4 - 6i}{2}$ Simplify $\sqrt{36} = 6$.

$= \dfrac{2(2 - 3i)}{2}$ Factor the numerator.

$= \dfrac{\overset{1}{\cancel{2}}(2 - 3i)}{\underset{1}{\cancel{2}}}$ Reduce the common factors.

$= 2 - 3i$

3. The Powers of *i*

From the definition of $i = \sqrt{-1}$, it follows that

$$
\begin{aligned}
i &= i \\
i^2 &= -1 \\
i^3 &= -i && \text{because } i^3 = i^2 \cdot i = (-1)i = -i \\
i^4 &= 1 && \text{because } i^4 = i^2 \cdot i^2 = (-1)(-1) = 1 \\
i^5 &= i && \text{because } i^5 = i^4 \cdot i = (1)i = i \\
i^6 &= -1 && \text{because } i^6 = i^4 \cdot i^2 = (1)(-1) = -1
\end{aligned}
$$

This pattern of values $i, -1, -i, 1, i, -1, -i, 1, \ldots$ continues for all subsequent powers of i. Here is a list of several powers of i.

Powers of *i*

$i^1 = i$	$i^5 = i$	$i^9 = i$
$i^2 = -1$	$i^6 = -1$	$i^{10} = -1$
$i^3 = -i$	$i^7 = -i$	$i^{11} = -i$
$i^4 = 1$	$i^8 = 1$	$i^{12} = 1$

To simplify higher powers of i, we can decompose the expression into multiples of i^4 ($i^4 = 1$) and write the remaining factors as i, i^2, or i^3.

example 4 **Simplifying Powers of *i***

Simplify the powers of i.

a. i^{13} b. i^{18} c. i^{107} d. i^{32}

Solution:

a.
$$
\begin{aligned}
i^{13} &= (i^{12}) \cdot (i) \\
&= (i^4)^3 \cdot (i) \\
&= (1)^3(i) && \text{Recall that } i^4 = 1. \\
&= i && \text{Simplify.}
\end{aligned}
$$

b.
$$
\begin{aligned}
i^{18} &= (i^{16}) \cdot (i^2) \\
&= (i^4)^4 \cdot (i^2) \\
&= (1)^4 \cdot (-1) && i^4 = 1 \text{ and } i^2 = -1 \\
&= -1 && \text{Simplify.}
\end{aligned}
$$

c.
$$
\begin{aligned}
i^{107} &= (i^{104}) \cdot (i^3) \\
&= (i^4)^{26}(i^3) \\
&= (1)^{26}(-i) && i^4 = 1 \text{ and } i^3 = -i \\
&= -i && \text{Simplify.}
\end{aligned}
$$

d. $i^{32} = (i^4)^8$

$= (1)^8$ $\quad i^4 = 1$

$= 1$ $\quad$ Simplify.

4. Definition of a Complex Number

We have already learned the definitions of the integers, rational numbers, irrational numbers, and real numbers. In this section, we define the complex numbers.

Definition of a Complex Number

A **complex number** is a number of the form $a + bi$, where a and b are real numbers and $i = \sqrt{-1}$.

Notes:

- If $b = 0$, then the complex number, $a + bi$ is a real number.
- If $b \neq 0$, then we say that $a + bi$ is an imaginary number.
- The complex number $a + bi$ is said to be written in standard form. The quantities a and b are called the **real** and **imaginary parts** (respectively) **of the complex number**.
- The complex numbers $a - bi$ and $a + bi$ are called **conjugates**.

From the definition of a complex number, it follows that all real numbers are complex numbers and all imaginary numbers are complex numbers. Figure 10-3 illustrates the relationship among the sets of numbers we have learned so far.

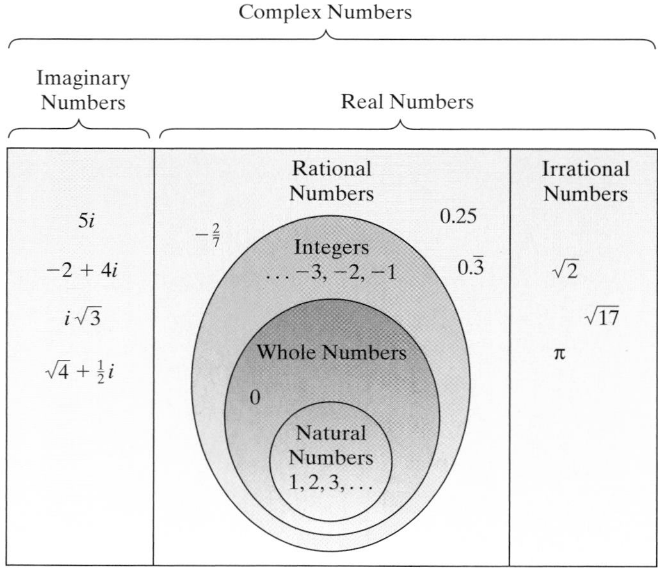

Figure 10-3

5. Addition, Subtraction, and Multiplication of Complex Numbers

The operations of addition, subtraction, and multiplication of real numbers also apply to imaginary numbers. To add or subtract complex numbers, combine the real parts and combine the imaginary parts. The commutative, associative, and distributive properties that apply to real numbers also apply to complex numbers.

example 5 **Adding, Subtracting, and Multiplying Complex Numbers**

a. Add: $(1 - 5i) + (-3 + 7i)$
b. Subtract: $(-\frac{1}{4} + \frac{3}{5}i) - (\frac{1}{2} - \frac{1}{10}i)$
c. Multiply: $(10 - 5i)(2 + 3i)$
d. Multiply: $(1.2 + 0.5i)(1.2 - 0.5i)$

Solution:

a. $(1 - 5i) + (-3 + 7i) = (1 + -3) + (-5 + 7)i$ (real parts: 1 and −3; imaginary parts: −5i and 7i) — Add real parts. Add imaginary parts.

$= -2 + 2i$ — Simplify.

b. $\left(-\frac{1}{4} + \frac{3}{5}i\right) - \left(\frac{1}{2} - \frac{1}{10}i\right) = -\frac{1}{4} + \frac{3}{5}i - \frac{1}{2} + \frac{1}{10}i$ — Apply the distributive property.

$= \left(-\frac{1}{4} - \frac{1}{2}\right) + \left(\frac{3}{5} + \frac{1}{10}\right)i$ — Add real parts. Add imaginary parts.

$= \left(-\frac{1}{4} - \frac{2}{4}\right) + \left(\frac{6}{10} + \frac{1}{10}\right)i$ — Get common denominators.

$= -\frac{3}{4} + \frac{7}{10}i$ — Simplify.

c. $(10 - 5i)(2 + 3i)$

$= (10)(2) + (10)(3i) + (-5i)(2) + (-5i)(3i)$ — Apply the distributive property.

$= 20 + 30i - 10i - 15i^2$

$= 20 + 20i - (15)(-1)$ — Recall $i^2 = -1$.

$= 20 + 20i + 15$

$= 35 + 20i$ — Write in the form $a + bi$.

d. $(1.2 + 0.5i)(1.2 - 0.5i)$

The expressions $(1.2 + 0.5i)$ and $(1.2 - 0.5i)$ are conjugates. The product is a difference of squares:

$$(a + b)(a - b) = a^2 - b^2$$

$$(1.2 + 0.5i)(1.2 - 0.5i) = (1.2)^2 - (0.5i)^2$$ Apply the formula, where $a = 1.2$ and $b = 0.5i$.

$$= 1.44 - 0.25i^2$$

$$= 1.44 - 0.25(-1)$$ Recall $i^2 = -1$.

$$= 1.44 + 0.25$$

$$= 1.69$$

Tip: The complex numbers $(1.2 + 0.5i)$ and $(1.2 - 0.5i)$ can also be multiplied by using the distributive property:

$$(1.2 + 0.5i)(1.2 - 0.5i)$$
$$= 1.44 - 0.6i + 0.6i - 0.25i^2$$
$$= 1.44 - 0.25(-1)$$
$$= 1.69$$

6. Division of Complex Numbers

The product of a complex number and its conjugate produces a real number. For example,

$$\begin{aligned}(5 + 3i)(5 - 3i) &= (5)^2 - (3i)^2 \\ &= 25 - 9i^2 \\ &= 25 - 9(-1) \\ &= 25 + 9 \\ &= 34\end{aligned}$$

To divide by a complex number, multiply the numerator and denominator by the conjugate of the denominator. This produces a real number in the denominator so that the resulting expression can be written in the form $a + bi$.

example 6

Dividing by a Complex Number

Divide the complex numbers. $\dfrac{4 - 3i}{5 + 2i}$

Solution:

$$\frac{4 - 3i}{5 + 2i}$$

Multiply the numerator and denominator by the conjugate of the denominator:

$$\frac{(4 - 3i)}{(5 + 2i)} \cdot \frac{(5 - 2i)}{(5 - 2i)} = \frac{(4)(5) + (4)(-2i) + (-3i)(5) + (-3i)(-2i)}{(5)^2 - (2i)^2}$$

$$= \frac{20 - 8i - 15i + 6i^2}{25 - 4i^2}$$ Simplify numerator and denominator.

$$= \frac{20 - 23i + 6(-1)}{25 - 4(-1)}$$ Recall $i^2 = -1$.

$$= \frac{20 - 23i - 6}{25 + 4}$$

$$= \frac{14 - 23i}{29} \qquad \text{Simplify.}$$

$$= \frac{14}{29} - \frac{23i}{29}$$

section 10.8 Practice Exercises

For Exercises 1–4, perform the indicated operations.

1. $-2\sqrt{5} - 3\sqrt{50} + \sqrt{125}$
2. $\sqrt[3]{2x}(\sqrt[3]{2x} - \sqrt[3]{4x^2})$
3. $(3 - \sqrt{x})(3 + \sqrt{x})$
4. $(\sqrt{5} + \sqrt{2})^2$

For Exercises 5–10, solve the equations.

5. $\sqrt{5y - 4} - 2 = 4$
6. $\sqrt{3w + 4} - 3 = 2$
7. $\sqrt[3]{3p + 7} - \sqrt[3]{2p - 1} = 0$
8. $\sqrt[3]{t - 5} - \sqrt[3]{2t + 1} = 0$
9. $\sqrt{36c + 15} = 6\sqrt{c} + 1$
10. $\sqrt{4a + 29} = 2\sqrt{a} + 5$
11. Define the imaginary number i.
12. Simplify i^2.
13. What is the conjugate of the complex number, $a + bi$?
14. True or False.
    a. Every real number is a complex number.
    b. Every complex number is a real number.

For Exercises 15–42, simplify the expressions:

15. $\sqrt{-144}$
16. $\sqrt{-81}$
17. $\sqrt{-3}$
18. $\sqrt{-17}$
19. $\sqrt{-20}$
20. $\sqrt{-75}$
21. $3\sqrt{-18} + 5\sqrt{-32}$
22. $5\sqrt{-45} + 3\sqrt{-80}$
23. $7\sqrt{-63} - 4\sqrt{-28}$
24. $7\sqrt{-3} - 4\sqrt{-27}$
25. $\sqrt{-7} \cdot \sqrt{-7}$
26. $\sqrt{-11} \cdot \sqrt{-11}$
27. $\sqrt{-9} \cdot \sqrt{-16}$
28. $\sqrt{-25} \cdot \sqrt{-36}$
29. $\sqrt{-15} \cdot \sqrt{-6}$
30. $\sqrt{-12} \cdot \sqrt{-50}$
31. $\dfrac{\sqrt{-50}}{\sqrt{-25}}$
32. $\dfrac{\sqrt{-27}}{\sqrt{-9}}$
33. $\dfrac{\sqrt{-90}}{\sqrt{10}}$
34. $\dfrac{\sqrt{-125}}{\sqrt{45}}$
35. $\dfrac{2 + \sqrt{-16}}{8}$
36. $\dfrac{6 - \sqrt{-4}}{4}$
37. $\dfrac{5 - \sqrt{-75}}{10}$
38. $\dfrac{8 + \sqrt{-32}}{16}$
39. $\dfrac{-6 \pm \sqrt{-72}}{6}$
40. $\dfrac{-20 \pm \sqrt{-500}}{10}$
41. $\dfrac{-8 \pm \sqrt{-48}}{4}$
42. $\dfrac{-18 \pm \sqrt{-72}}{3}$

For Exercises 43–50, add or subtract as indicated. Write the answer in the form $a + bi$.

43. $(2 - i) + (5 + 7i)$
44. $(5 - 2i) + (3 + 4i)$
45. $\left(\frac{1}{2} + \frac{2}{3}i\right) - \left(\frac{1}{5} - \frac{5}{6}i\right)$
46. $\left(\frac{11}{10} - \frac{7}{5}i\right) - \left(-\frac{2}{5} + \frac{3}{5}i\right)$

47. $(1 + 3i) + (4 - 3i)$

48. $(-2 + i) + (1 - i)$

49. $(2 + 3i) - (1 - 4i) + (-2 + 3i)$

50. $(2 + 5i) - (7 - 2i) + (-3 + 4i)$

For Exercises 51–62, simplify the powers of i.

51. i^7
52. i^{38}
53. i^{64}
54. i^{75}
55. i^{41}
56. i^{25}
57. i^{52}
58. i^0
59. i^{23}
60. i^{103}
61. i^6
62. i^{82}

For Exercises 63–74, multiply the complex numbers. Write the answer in the form $a + bi$.

63. $(8i)(3i)$
64. $(2i)(4i)$
65. $6i(1 - 3i)$
66. $-i(3 + 4i)$
67. $(2 - 10i)(3 + 2i)$
68. $(4 + 7i)(1 - i)$
69. $(-5 + 2i)(5 + 2i)$
70. $(4 - 11i)(-4 - 11i)$
71. $(4 + 5i)^2$
72. $(3 - 2i)^2$
73. $(2 + i)(3 - 2i)(4 + 3i)$
74. $(3 - i)(3 + i)(4 - i)$

For Exercises 75–78, find the conjugate. Then find the product of the number and its conjugate.

75. $1 + 3i$
76. $3 + i$
77. $4 - 3i$
78. $1 - i$

For Exercises 79–88, divide the complex numbers. Write the answer in the form $a + bi$.

79. $\dfrac{2}{1 + 3i}$
80. $\dfrac{-2}{3 + i}$
81. $\dfrac{-i}{4 - 3i}$
82. $\dfrac{3 - 3i}{1 - i}$
83. $\dfrac{5 + 2i}{5 - 2i}$
84. $\dfrac{7 + 3i}{4 - 2i}$
85. $\dfrac{3}{2i}$
86. $\dfrac{-2}{7i}$
87. $\dfrac{3}{-i}$
88. $\dfrac{-2}{-i}$

Expanding Your Skills

For Exercises 89–94, simplify the expression and write the answer in the form $a + bi$.

89. $7i^{-5}$
90. $9i^{-7}$
91. $12i^{-8}$
92. $-6i^{-12}$
93. i^{-10}
94. i^{-22}

chapter 10 Summary

Section 10.1—Definition of an nth Root

Key Concepts:	Examples:
b is an nth root of a if $b^n = a$.	2 is a square root of 4.
	-2 is a square root of 4.
	-3 is a cube root of -27.

The expression $\sqrt{a}$ represents the principal square root of a.

The expression $\sqrt[n]{a}$ represents the principal nth root of a.

$\sqrt[n]{a^n} = |a|$ if n is even.

$\sqrt[n]{a^n} = a$ if n is odd.

$\sqrt[n]{a}$ is not a real number if $a < 0$ and n is even.

$f(x) = \sqrt[n]{x}$ defines a radical function.

$\sqrt{36} = 6$ $\qquad$ $\sqrt[3]{-64} = -4$

$\sqrt[4]{(x+3)^4} = |x+3|$ $\qquad$ $\sqrt[5]{(x+3)^5} = x+3$

$\sqrt[4]{-16}$ is not a real number.

For $g(x) = \sqrt{x}$ the domain is $[0, \infty)$.

For $h(x) = \sqrt[3]{x}$ the domain is $(-\infty, \infty)$.

The Pythagorean Theorem $a^2 + b^2 = c^2$

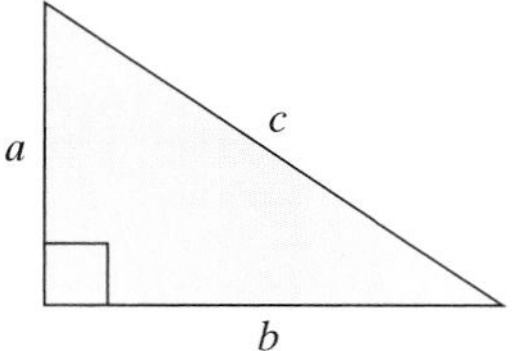

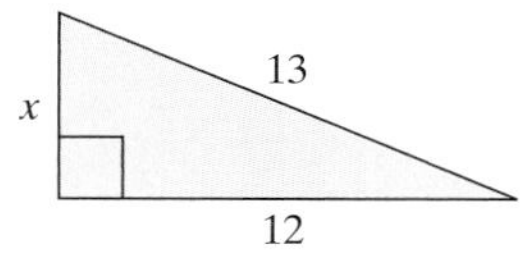

$$x^2 + 12^2 = 13^2$$
$$x^2 + 144 = 169$$
$$x^2 = 25$$
$$x = 5$$

Key Terms:

cube root	Pythagorean theorem
index	radical function
perfect square	radical sign
nth root	radicand
principal square root	

Section 10.2—Rational Exponents

Key Concepts:

Let a be a real number and n be an integer such that $n > 1$. If $\sqrt[n]{a}$ exists, then

$$a^{1/n} = \sqrt[n]{a}$$
$$a^{m/n} = (\sqrt[n]{a})^m = \sqrt[n]{a^m}$$

Examples:

$$121^{1/2} = \sqrt{121} = 11$$
$$27^{2/3} = (\sqrt[3]{27})^2 = (3)^2 = 9$$

Section 10.3—Properties of Radicals

Key Concepts:

Let a and b represent real numbers such that $\sqrt[n]{a}$ and $\sqrt[n]{b}$ are both real. Then

$\sqrt[n]{ab} = \sqrt[n]{a} \cdot \sqrt[n]{b}$ $\quad$ Multiplication property

$\sqrt[n]{\dfrac{a}{b}} = \dfrac{\sqrt[n]{a}}{\sqrt[n]{b}}$ $\quad$ Division property

Examples:

$$\sqrt{3} \cdot \sqrt{5} = \sqrt{15}$$
$$\sqrt{\frac{x}{9}} = \frac{\sqrt{x}}{\sqrt{9}} = \frac{\sqrt{x}}{3}$$

A radical expression whose radicand is written as a product of prime factors is in simplified form if all of the following conditions are met:

1. The radicand has no factor raised to a power greater than or equal to the index.
2. The radicand does not contain a fraction.
3. No radicals are in the denominator of a fraction.

KEY TERMS:

division property of radicals
multiplication property of radicals
simplified form of a radical

Simplify: $\sqrt[3]{16x^5y^7}$

$= \sqrt[3]{2^4x^5y^7}$

$= \sqrt[3]{2^3x^3y^6} \cdot \sqrt[3]{2x^2y}$

$= 2xy^2\sqrt[3]{2x^2y}$

SECTION 10.4—ADDITION AND SUBTRACTION OF RADICALS

KEY CONCEPTS:

Like radicals have radical factors with the same index and the same radicand.

Use the distributive property to add and subtract *like* radicals.

KEY TERM:

like radicals

EXAMPLES:

Perform the indicated operations:

$3x\sqrt{7} - 5x\sqrt{7} + x\sqrt{7}$

$= (3 - 5 + 1) \cdot x\sqrt{7}$

$= -x\sqrt{7}$

Subtract: $x\sqrt[4]{16x} - 3\sqrt[4]{x^5}$

$= 2x\sqrt[4]{x} - 3x\sqrt[4]{x}$

$= -x\sqrt[4]{x}$

SECTION 10.5—MULTIPLICATION OF RADICALS

KEY CONCEPTS:

The Multiplication Property of Radicals

If $\sqrt[n]{a}$ and $\sqrt[n]{b}$ are real numbers, then $\sqrt[n]{a} \cdot \sqrt[n]{b} = \sqrt[n]{ab}$

To multiply or divide radicals with different indices, convert to rational exponents and use the properties of exponents.

KEY TERM:

multiplication property of radicals

EXAMPLES:

Multiply: $3\sqrt{2}(\sqrt{2} + 5\sqrt{7} - \sqrt{6})$

$= 3\sqrt{4} + 15\sqrt{14} - 3\sqrt{12}$

$= 3 \cdot 2 + 15\sqrt{14} - 3 \cdot 2\sqrt{3}$

$= 6 + 15\sqrt{14} - 6\sqrt{3}$

Multiply: $\sqrt{p} \cdot \sqrt[5]{p^2}$

$= p^{1/2} \cdot p^{2/5}$

$= p^{5/10} \cdot p^{4/10}$

$= p^{9/10}$

$= \sqrt[10]{p^9}$

Section 10.6—Rationalization

Key Concepts:

Rationalizing a denominator with one term

Rationalizing a denominator with two terms

Key Terms:

rationalizing the denominator (one term)
rationalizing the denominator (two terms involving square roots)

Examples:

Simplify: $\dfrac{4}{\sqrt[4]{2y^3}}$

$$= \frac{4}{\sqrt[4]{2y^3}} \cdot \frac{\sqrt[4]{2^3y}}{\sqrt[4]{2^3y}} = \frac{4\sqrt[4]{8y}}{\sqrt[4]{2^4y^4}}$$

$$= \frac{4\sqrt[4]{8y}}{2y} = \frac{2\sqrt[4]{8y}}{y}$$

Rationalize the denominator:

$$\frac{\sqrt{2}}{\sqrt{x} - \sqrt{3}}$$

$$= \frac{\sqrt{2}}{(\sqrt{x} - \sqrt{3})} \cdot \frac{(\sqrt{x} + \sqrt{3})}{(\sqrt{x} + \sqrt{3})}$$

$$= \frac{\sqrt{2x} + \sqrt{6}}{x - 3}$$

Section 10.7—Radical Equations

Key Concepts:

Steps to Solve a Radical Equation

1. Isolate the radical. If an equation has more than one radical, choose one of the radicals to isolate.
2. Raise each side of the equation to a power equal to the index of the radical.
3. Solve the resulting equation. If the equation still has a radical, repeat steps 1 and 2.
4. Check the potential solutions in the original equation.

Key Term:

radical equation

Examples:

Solve:

$$\sqrt{b - 5} - \sqrt{b + 3} = 2$$

$$\sqrt{b - 5} = \sqrt{b + 3} + 2$$

$$(\sqrt{b - 5})^2 = (\sqrt{b + 3} + 2)^2$$

$$b - 5 = b + 3 + 4\sqrt{b + 3} + 4$$

$$b - 5 = b + 7 + 4\sqrt{b + 3}$$

$$-12 = 4\sqrt{b + 3}$$

$$-3 = \sqrt{b + 3}$$

$$(-3)^2 = (\sqrt{b + 3})^2$$

$$9 = b + 3$$

$$6 = b$$

Check:

$$\sqrt{6 - 5} - \sqrt{6 + 3} \stackrel{?}{=} 2$$

$$\sqrt{1} - \sqrt{9} \stackrel{?}{=} 2$$

$$1 - 3 \neq 2 \quad \text{Does not check.}$$

No solution

Section 10.8—Complex Numbers

Key Concepts:

$i = \sqrt{-1}$ and $i^2 = -1$

For a real number $b > 0$, $\sqrt{-b} = i\sqrt{b}$

A complex number is in the form $a + bi$, where a and b are real numbers. a is called the real part, and b is called the imaginary part.

To add or subtract complex numbers, combine the real parts and combine the imaginary parts.

Multiply complex numbers by using the distributive property.

Divide complex numbers by multiplying the numerator and denominator by the conjugate of the denominator.

Key Terms:

complex numbers
conjugate
i
imaginary numbers
imaginary part of a complex number
real part of a complex number

Examples:

Simplify:

$$\sqrt{-4} \cdot \sqrt{-9} = (2i)(3i) = 6i^2 = -6$$

Perform the indicated operations:

$$(3 - 5i) - (2 + i) + (3 - 2i) = 3 - 5i - 2 - i + 3 - 2i = 4 - 8i$$

Multiply:

$$(1 + 6i)(2 + 4i) = 2 + 4i + 12i + 24i^2 = 2 + 16i + 24(-1) = -22 + 16i$$

Divide: $\dfrac{3}{2 - 5i}$

$$= \frac{3}{2 - 5i} \cdot \frac{(2 + 5i)}{(2 + 5i)} = \frac{6 + 15i}{4 - 25i^2}$$

$$= \frac{6 + 15i}{29} \quad \text{or} \quad \frac{6}{29} + \frac{15}{29}i$$

chapter 10 REVIEW EXERCISES

For the exercises in this set, assume that all variables represent positive real numbers unless otherwise stated.

Section 10.1

1. True or false:
   a. The principal nth root of an even-indexed root is always positive.
   b. The principal nth root of an odd-indexed root is always positive.
2. Explain why $\sqrt{(-3)^2} = 3$ and $\sqrt{(-3)^2} \neq -3$.
3. Are the following statements true or false?
   a. $\sqrt{a^2 + b^2} = a + b$
   b. $\sqrt{(a + b)^2} = a + b$

For Exercises 4–6, simplify the radicals.

4. $\sqrt{\dfrac{50}{32}}$
5. $\sqrt[4]{625}$
6. $\sqrt{(-6)^2}$

7. Evaluate the function values for $f(x) = \sqrt{x - 1}$.

a. $f(10)$ b. $f(1)$ c. $f(8)$

d. Given $f(x) = \sqrt{x - 1}$, write the domain of f in interval notation.

8. Use a calculator to approximate the expression $-3 - 3\sqrt{5}$ to two decimal places.

9. Translate the English sentence into an algebraic expression: Four more than the quotient of the cube root of $2x$ and the fourth root of $2x$.

10. State the Pythagorean theorem and explain the theorem in your own words.

11. Use the Pythagorean theorem to find the length of the third side of the triangle.

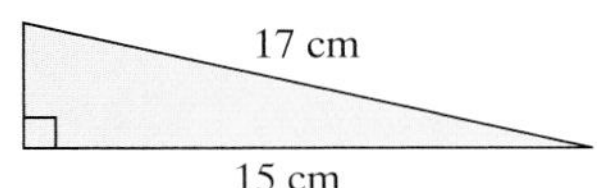

Figure for Exercise 11

12. Three sides of a right triangle are related as shown in the figure.

a. Find the lengths of the sides when $x = 3$ in.

b. Find the lengths of the sides when $x = 12$ in.

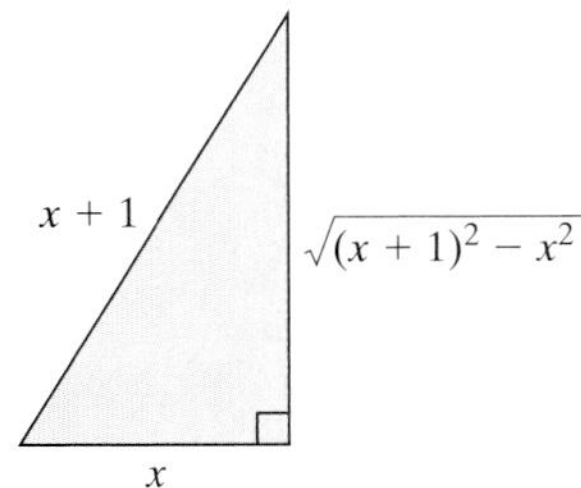

Figure for Exercise 12

Section 10.2

13. Are the properties of exponents the same for rational exponents and integer exponents? Give an example. (Answers may vary.)

14. In the expression $x^{m/n}$ what does n represent?

15. Explain the process of eliminating a negative exponent from an algebraic expression.

For Exercises 16–20, simplify the expressions. Write the answer with positive exponents only.

16. $(-125)^{1/3}$

17. $16^{-1/4}$

18. $\left(\frac{1}{16}\right)^{-3/4} - \left(\frac{1}{8}\right)^{-2/3}$

19. $(b^{1/2} \cdot b^{1/3})^{12}$

20. $\left(\frac{x^{-1/4}y^{-1/3}z^{3/4}}{2^{1/3}x^{-1/3}y^{2/3}}\right)^{-12}$

For Exercises 21–22, rewrite the expressions using rational exponents.

21. $\sqrt[4]{x^3}$

22. $\sqrt[3]{2y^2}$

For Exercises 23–25, use a calculator to approximate the expressions to four decimal places.

23. $10^{1/3}$ 24. $17.8^{2/3}$ 25. $147^{4/5}$

26. An initial investment of P dollars is made into an account in which the return is compounded quarterly. The amount in the account can be determined by

$$A = P\left(1 + \frac{r}{4}\right)^{t/3},$$

where r is the annual rate of return, and t is the time in months.

When she is 20 years old, Jenna invests \$5000 in a mutual fund that grows by an average of 11% per year. How much money does she have

a. After 6 months?
b. After 1 year?
c. At age 40?
d. At age 50?
e. At age 65?

Section 10.3

27. List the criteria for a rational expression to be simplified.

For Exercises 28–31, simplify the radicals.

28. $\sqrt{108}$

29. $\sqrt[4]{x^5yz^4}$

30. $\sqrt{5x} \cdot \sqrt{20x}$

31. $\sqrt[3]{\frac{-16x^7y^6}{z^9}}$

32. Write an English phrase that describes the following mathematical expressions: (Answers may vary.)

a. $\sqrt{\frac{2}{x}}$

b. $(x + 1)^3$

33. An engineering firm made a mistake when building a $\frac{1}{4}$-mile bridge in the Florida Keys. The bridge was made without adequate expansion joints to prevent buckling during the heat of summer. During mid-June, the bridge expanded 1.5 ft causing a vertical bulge in the middle. Calculate the height of the bulge, h, in feet. (*Note*: 1 mile = 5280 ft) Round to the nearest foot.

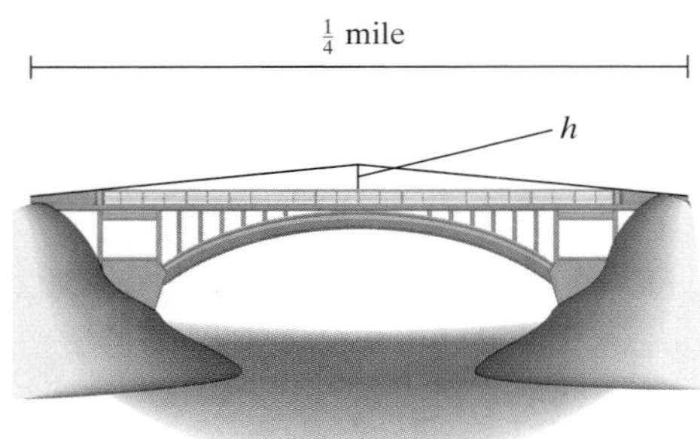

Figure for Exercise 33

Section 10.4

34. Complete the following statement: Radicals may be added or subtracted if. . . .

For Exercises 35–38, determine whether the radicals may be combined, and explain your answer.

35. $\sqrt[3]{2x} - 2\sqrt{2x}$
36. $2 + \sqrt{x}$
37. $\sqrt[4]{3xy} + 2\sqrt[4]{3xy}$
38. $-4\sqrt{32} + 7\sqrt{50}$

For Exercises 39–42, add or subtract as indicated.

39. $4\sqrt{7} - 2\sqrt{7} + 3\sqrt{7}$
40. $2\sqrt[3]{64} + 3\sqrt[3]{54} - 16$
41. $\sqrt{50} + 7\sqrt{2} - \sqrt{8}$
42. $x\sqrt[3]{16x^2} - 4\sqrt[3]{2x^5} + 5x\sqrt[3]{54x^2}$

For Exercises 43–44, answer true or false. If an answer is false, explain why.

43. $5 + 3\sqrt{x} = 8\sqrt{x}$
44. $\sqrt{y} + \sqrt{y} = \sqrt{2y}$

Section 10.5

45. Define the conjugate of a binomial expression.

46. Explain why $x - y = (\sqrt{x} + \sqrt{y})(\sqrt{x} - \sqrt{y})$.

For Exercises 47–55, multiply the radicals and simplify the answer.

47. $\sqrt{3} \cdot \sqrt{12}$
48. $\sqrt[4]{4} \cdot \sqrt[4]{8}$
49. $-2\sqrt{3}(\sqrt{3} - 3\sqrt{3})$
50. $(\sqrt{y} + 4)(\sqrt{y} - 4)$
51. $(\sqrt[3]{2x} - \sqrt[3]{4x})^2$
52. $(3\sqrt{a} - \sqrt{5})(\sqrt{a} + 2\sqrt{5})$
53. $\sqrt[3]{u} \cdot \sqrt{u^5}$
54. $\sqrt[4]{w^3} \cdot \sqrt{w}$
55. $\sqrt[3]{(a+b)} \cdot \sqrt[6]{(a+b)^5}$

Section 10.6

For Exercises 56–59, rationalize the denominator.

56. $\dfrac{-2}{\sqrt[3]{2x}}$
57. $\dfrac{2\sqrt{x} + \sqrt{5}}{\sqrt{5x}}$
58. $\dfrac{-2}{\sqrt{x} - 4}$
59. $\dfrac{2\sqrt{b} - 1}{3\sqrt{b} - 1}$

60. Translate the mathematical expression into an English phrase. (Answers may vary.)

$$\frac{\sqrt{2}}{x^2}$$

Section 10.7

Solve the radical equations in Exercises 61–68, if possible.

61. $\sqrt{2y} = 7$
62. $\sqrt{a - 6} = 5$
63. $\sqrt[3]{2w - 3} + 5 = 2$
64. $\sqrt[4]{p + 12} = \sqrt[4]{5p - 16}$
65. $\sqrt{t} + \sqrt{t - 5} = 5$
66. $\sqrt{8x + 1} = -\sqrt{x - 13}$
67. $\sqrt{2m^2 + 4} - \sqrt{9m} = 0$
68. $\sqrt{x + 2} = 1 - \sqrt{2x + 5}$

69. The velocity, v, of an ocean wave depends on the water depth, d, as the wave approaches land.

$v(d) = \sqrt{32d}$, where v is in feet/second and d is in feet.

a. Find $v(20)$ and interpret its value. Round to one decimal place.

b. Find the depth of the water at a point where a wave is traveling at 16 ft/s.

70. A Pony league baseball field is larger than a Little League diamond but still smaller than a Major League diamond. The Pony League field is a square that measures 80 ft on each side. Find the distance from home plate to second base on this field. Round to the nearest tenth of a foot.

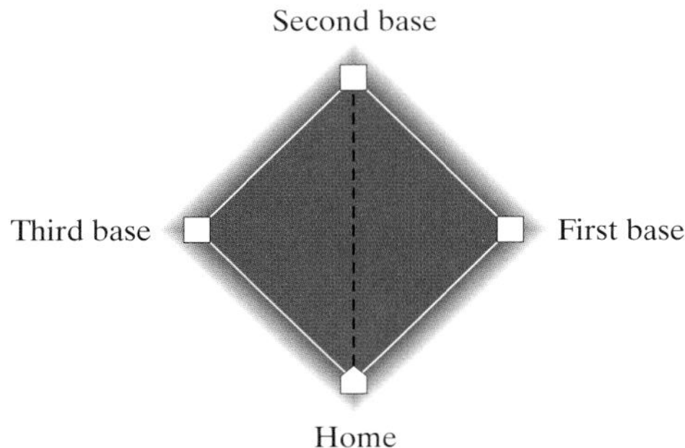

Section 10.8

71. Define a complex number.

72. Define an imaginary number.

73. Consider the following expressions.

$$\frac{3}{4 + 6i} \quad \text{and} \quad \frac{3}{\sqrt{4} + \sqrt{6}}$$

Compare the process of dividing by a complex number to the process of rationalizing the denominator.

For Exercises 74–76, rewrite the expressions in terms of i.

74. $\sqrt{-16}$

75. $-\sqrt{-5}$

76. $\sqrt{-x}$ if $x \geq 0$

For Exercises 77–81, simplify the powers of i.

77. i^{38}

78. i^{101}

79. i^{19}

80. $i^{1000} + i^{1002}$

81. $2i^{17} - 3i^{23} + 2i^{24} + 4i^{34}$

For Exercises 82–85, perform the indicated operations. Write the final answer in the form $a + bi$.

82. $(-3 + i) - (2 - 4i)$

83. $(2i + 4)(2i - 4)$

84. $(4 - 3i)(4 + 3i)$

85. $(5 - i)^2$

For Exercises 86–87, write the expressions in the form $a + bi$, and determine the real and imaginary parts.

86. $\dfrac{17 - 4i}{-4}$

87. $\dfrac{-16 - 8i}{8}$

For Exercises 88–89, divide and simplify. Write the final answer in the form $a + bi$.

88. $\dfrac{2 - i}{3 + 2i}$

89. $\dfrac{2 + i}{i^5}$

For Exercises 90–91, simplify the expression.

90. $\dfrac{-8 \pm \sqrt{-40}}{12}$

91. $\dfrac{6 \pm \sqrt{-144}}{3}$

chapter 10 TEST

1. a. What is the principal square root of 36?

   b. What is the negative square root of 36?

2. Which of the following are real numbers?

   a. $-\sqrt{100}$ b. $\sqrt{-100}$

   c. $-\sqrt[3]{1000}$ d. $\sqrt[3]{-1000}$

3. Simplify.

   a. $\sqrt[3]{y^3}$ b. $\sqrt[4]{y^4}$

For Exercises 4–11, simplify the radicals. Assume that all variables represent positive numbers.

4. $\sqrt[4]{81}$

5. $\sqrt{\dfrac{16}{9}}$

6. $\sqrt[3]{32}$

7. $\sqrt{a^4b^3c^5}$

8. $\sqrt{3x} \cdot \sqrt{6x^3}$

9. $\sqrt{\dfrac{32w^6}{3w}}$

10. $\sqrt{7y} \cdot \sqrt[5]{7y}$

11. $\dfrac{\sqrt[3]{10}}{\sqrt[4]{10}}$

12. a. Evaluate the function values $f(-8), f(-6), f(-4)$, and $f(-2)$ for $f(x) = \sqrt{-2x - 4}$.
    b. Write the domain of f in interval notation.

13. Use a calculator to evaluate $(-3 - \sqrt{5})/17$ to four decimal places.

For Exercises 14–15, simplify the expressions. Assume that all variables represent positive real numbers.

14. $-27^{1/3}$

15. $\dfrac{t^{-1} \cdot t^{1/2}}{t^{1/4}}$

16. Add or subtract as indicated:
$3\sqrt{5} + 4\sqrt{5} - 2\sqrt{20}$

17. Multiply the radicals.
    a. $3\sqrt{x}(\sqrt{2} - \sqrt{5})$
    b. $(\sqrt{2x} - 3)^2$

18. Rationalize the denominator. Assume $x > 0$.
    a. $\dfrac{-2}{\sqrt[3]{x}}$
    b. $\dfrac{\sqrt{x} + 2}{3 - \sqrt{x}}$

19. Rewrite the expressions in terms of i.
    a. $\sqrt{-8}$
    b. $2\sqrt{-16}$
    c. $\dfrac{2 \pm \sqrt{-8}}{4}$

For Exercises 20–23, perform the indicated operation and simplify completely. Write the final answer in the form $a + bi$.

20. $(3 - 5i) - (2 + 6i)$

21. $(4 + i)(8 + 2i)$

22. $\sqrt{-16} \cdot \sqrt{-49}$

23 $(10 + 3i)[(-5i + 8) - (5 - 3i)]$

24. Divide and write the final answers in the form $a + bi$.
    a. $\dfrac{-4 + i}{i^5}$
    b. $\dfrac{3 - 2i}{3 - 4i}$

25. If the volume, V, of a sphere is known, the radius of the sphere can be computed by

$$r(V) = \sqrt[3]{\frac{3V}{4\pi}}.$$

Find $r(10)$ to two decimal places. Interpret the meaning in the context of the problem.

26. A patio 20 ft wide has a slanted roof as shown in the picture. Find the length of the roof if there is an 8-in. overhang. Round the answer to the nearest foot.

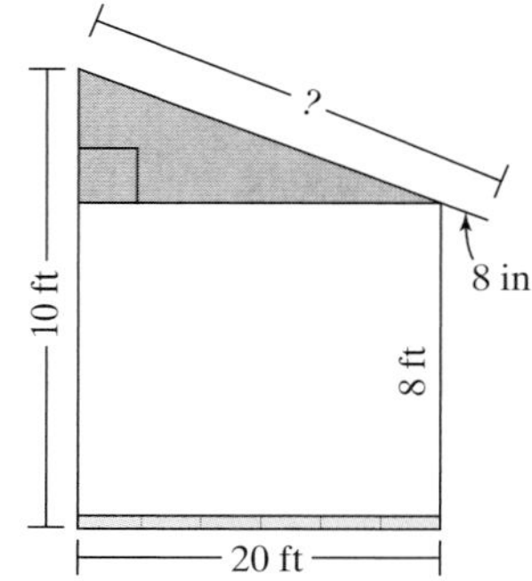

Figure for Exercise 23

For Exercises 27–29, solve the radical equations.

27. $\sqrt[3]{2x + 5} = -3$

28. $\sqrt{5x + 8} = \sqrt{5x - 1} + 1$

29. $\sqrt{t + 7} - \sqrt{2t - 3} = 2$

CUMULATIVE REVIEW EXERCISES, CHAPTERS 1–10

1. Simplify the expression:
$6^2 - 2[5 - 8(3 - 1) + 4 \div 2]$

2. Simplify the expression:
$3x - 3(-2x + 5) - 4y + 2(3x + 5) - y$

3. Solve the equation: $9(2y + 8) = 20 - (y + 5)$

4. Solve the inequality. Write the answer in interval notation. $2a - 4 < -14$

5. Write an equation of the line that is parallel to the line $2x + y = 9$ and passes through the point $(3, -1)$. Write the answer in slope-intercept form.

6. On the same coordinate system, graph the line $2x + y = 9$ and the line that you derived in Exercise 5. Verify that these two lines are indeed parallel.

7. Solve the system of equations using the addition method.

$$\begin{aligned} 2x - 3y &= 0 \\ -4x + 3y &= -1 \end{aligned}$$

8. Determine if $(2, -2, \frac{1}{2})$ is a solution to the system.

$$\begin{aligned} 2x + y - 4z &= 0 \\ x - y + 2z &= 5 \\ 3x + 2y + 2z &= 4 \end{aligned}$$

9. Write a system of linear equations from the augmented matrix. Use x, y, and z for the variables.

$$\left[\begin{array}{ccc|c} 1 & 0 & 0 & 6 \\ 0 & 1 & 0 & 3 \\ 0 & 0 & 1 & 8 \end{array}\right]$$

10. Given the function defined by $f(x) = 4x - 2$.
    a. Find $f(-2), f(0), f(4)$, and $f(\frac{1}{2})$.
    b. Write the ordered pairs that correspond to the function values in part (a).
    c. Graph the function $y = f(x)$.

11. Determine if the graph shown defines y as a function of x.

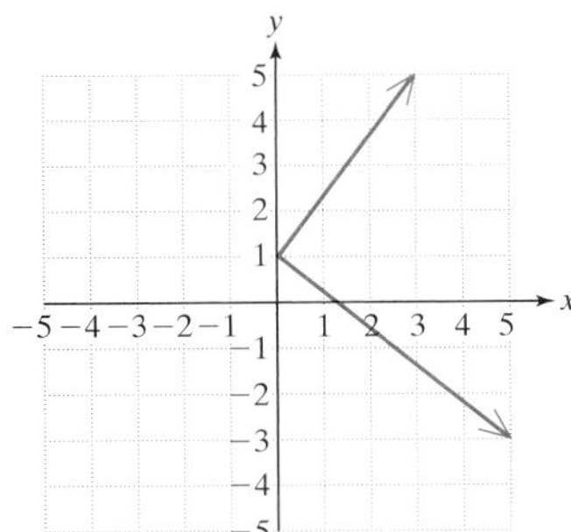

12. Simplify the expression. Write the final answer with positive exponents only.

$$\left(\frac{a^3b^{-1}c^3}{ab^{-5}c^2}\right)^2$$

13. Simplify the expression. Write the final answer with positive exponents only.

$$\left(\frac{a^{3/2}b^{-1/4}c^{1/3}}{ab^{-5/4}c^0}\right)^{12}$$

14. The escape velocity of an object is the minimum initial velocity needed to escape a planet's gravitational pull. The escape velocity, v (in meters per second), is related to the mass of the planet and the radius of the planet according to the formula:

$V = \sqrt{\frac{2GM}{R}}$, where M is the mass of the planet in kilograms; R is the radius of the planet in meters; and G is a constant equal to 6.672×10^{-11} m^3/s$^2 \cdot$ kg

Find the escape velocity of the earth whose mass is 5.97×10^{27} kg and whose radius is 6.37×10^7 m. Round to the nearest whole unit.

15. Multiply the polynomials: $(2x + 5)(x - 3)$. What is the degree of the product?

16. Perform the indicated operations and simplify: $\sqrt{3} \cdot (\sqrt{5} + \sqrt{6} + \sqrt{3})$

17. Divide: $(x^2 - x - 12) \div (x + 3)$.

18. Simplify and subtract: $\sqrt[4]{\frac{1}{16}} - \sqrt[3]{\frac{8}{27}}$.

19. Simplify: $\sqrt[3]{\frac{54c^4}{cd^3}}$.

20. Add: $4\sqrt{45b^3} + 5b\sqrt{80b}$.

21. Divide: $\frac{13i}{3 + 2i}$. Write the answer in the form $a + bi$.

22. Factor completely. $6x^2 - 31x + 18$

23. Find the x- and y-intercepts of the function defined by $y = f(x)$.

    a. $f(x) = \sqrt{2x + 1}$

    b. $f(x) = (x - 3)(2x - 5)$

24. A car travels 225 km in the same time that a truck travels 200 km. If the average speed of the car is 10 km/hr faster than the average speed of the truck, find both speeds.

25. A pump can pump water out of a pool in 12 hours. A smaller pump would take 20 hours to complete the job. How long would it take to pump the water from the pool using both pumps?

26. Perform the indicated operations and simplify.

$$\frac{-8}{y^2 - 1} + \frac{4y}{y - 1}$$

27. Perform the indicated operations and simplify.

$$\frac{2t^2 + 9t + 4}{t^2 - 16} \cdot \frac{t^2 - 4t}{2t^2 - 5t - 3}$$

28. Simplify. $\dfrac{\frac{x}{y} - \frac{y}{x}}{\frac{1}{x} - \frac{1}{y}}$

29. Factor completely. $y^6 - 8$

30. Write in standard form, $a + bi$.
    $(-3 + 5i) - (4 + 7i)$

31. Solve the inequality and write the solution in interval notation. $|2x - 7| + 5 \le 10$

32. Solve the compound inequality and write the solution in interval notation.
    $3x - 5 \le 7$ or $-2x < -12$

33. Solve the inequality and write the solution in interval notation. $x(x - 2) \le 15$

34. Rationalize the denominator. $\frac{4}{\sqrt{7} - \sqrt{5}}$

35. After a recent vacation, Jeremy and Shawn had pictures developed by a photo shop. Jeremy had 24 small (3 in. by 5 in.) photos developed and 36 midsize (4 in. by 6 in.) photos developed for \$15.36. Shawn had 48 small photos developed and 12 midsize photos developed for \$13.92. Find the unit cost to develop one of each type of photo.

QUADRATIC FUNCTIONS

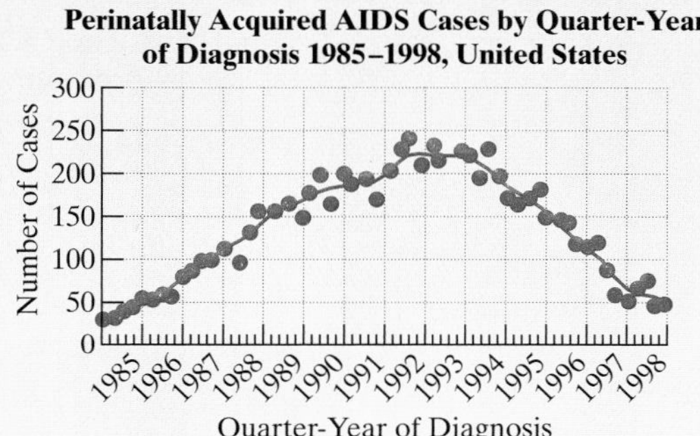

Source: Centers for Disease Control.

The Centers for Disease Control provide surveillance reports documenting the number of AIDS cases for various exposure categories. In the 1980s and 1990s, the number of children who acquired the HIV virus *perinatally* (around the time of birth) rose significantly. Then in 1992, public health organizations recommended new preventative treatments and the number of perinatally acquired AIDS cases declined.

For more information see the Technology Connections in MathZone at

www.mhhe.com/miller_oneill

Between 1985 and 1998, the number of perinatally acquired AIDS cases, N, can be approximated by the *quadratic function*:

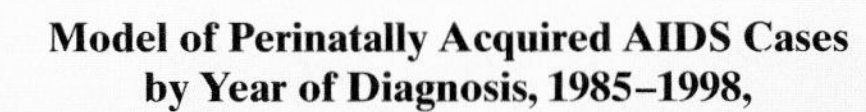

$$N(t) = -3.483t^2 + 51.64t + 35$$

where t is the number of years since 1985 and N is the number of cases.

Model of Perinatally Acquired AIDS Cases by Year of Diagnosis, 1985–1998, United States

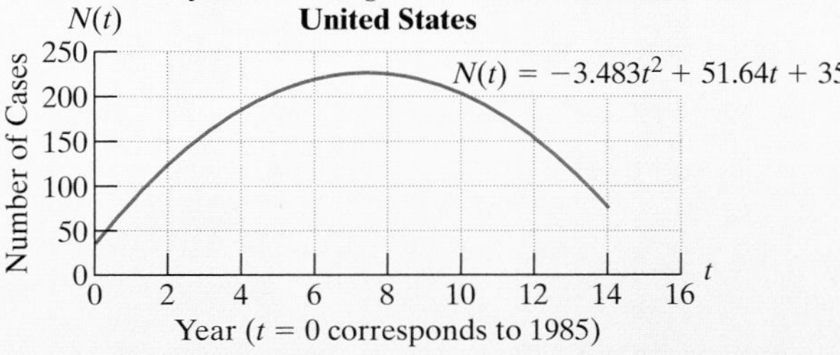

section

11.1 Square Root Property and Completing the Square

Concepts

1. Square Root Property
2. Solving Quadratic Equations Using the Square Root Property
3. Completing the Square
4. Solving Quadratic Equations by Completing the Square
5. Literal Equations

1. Square Root Property

In Section 5.7, we learned to solve quadratic equations by factoring and applying the zero product rule; however, the zero product rule can only be used if the equation is factorable. In this section and the next, we will learn two techniques to solve *all* quadratic equations, factorable and nonfactorable.

The first technique will use the **square root property**.

The Square Root Property

For any real number, k, if $x^2 = k$, then $x = \sqrt{k}$ or $x = -\sqrt{k}$

Note: The solution may also be written as $x = \pm\sqrt{k}$, read x equals plus or minus the square root of k.

2. Solving Quadratic Equations Using the Square Root Property

example 1 **Solving Quadratic Equations Using the Square Root Property**

Use the square root property to solve the equations.

a. $x^2 = 81$ b. $3x^2 + 75 = 0$ c. $(w + 3)^2 = 20$

Tip: The equation $x^2 = 81$ can also be solved by using the zero product rule.

$$x^2 = 81$$
$$x^2 - 81 = 0$$
$$(x - 9)(x + 9) = 0$$
$$x = 9 \quad \text{or} \quad x = -9$$

Solution:

a. $x^2 = 81$ The equation is in the form $x^2 = k$.

$x = \pm\sqrt{81}$ Apply the square root property.

$x = \pm 9$

The solutions are $x = 9$ and $x = -9$.

b. $3x^2 + 75 = 0$ Rewrite the equation to fit the form $x^2 = k$.

$3x^2 = -75$

$x^2 = -25$ The equation is now in the form $x^2 = k$.

$x = \pm\sqrt{-25}$ Apply the square root property.

$= \pm 5i$

Avoiding Mistakes

A common mistake is to forget the $\pm$ symbol when solving the equation $x^2 = k$

$$x = \pm\sqrt{k}$$

The solutions are $x = 5i$ and $x = -5i$.

Check: $x = 5i$	Check: $x = -5i$
$3x^2 + 75 = 0$	$3x^2 + 75 = 0$
$3(5i)^2 + 75 \stackrel{?}{=} 0$	$3(-5i)^2 + 75 \stackrel{?}{=} 0$
$3(25i^2) + 75 \stackrel{?}{=} 0$	$3(25i^2) + 75 \stackrel{?}{=} 0$
$3(-25) + 75 \stackrel{?}{=} 0$	$3(-25) + 75 \stackrel{?}{=} 0$
$-75 + 75 = 0$ ✔	$-75 + 75 = 0$ ✔

c. $(w + 3)^2 = 20$ The equation is in the form $x^2 = k$, where $x = (w + 3)$.

$w + 3 = \pm\sqrt{20}$ Apply the square root property.

$w + 3 = \pm\sqrt{2^2 \cdot 5}$ Simplify the radical.

$w + 3 = \pm 2\sqrt{5}$

$w = -3 \pm 2\sqrt{5}$ Solve for w.

The solutions are $w = -3 + 2\sqrt{5}$ and $w = -3 - 2\sqrt{5}$.

Check: $w = -3 + 2\sqrt{5}$	Check: $w = -3 - 2\sqrt{5}$
$(w + 3)^2 = 20$	$(w + 3)^2 = 20$
$(-3 + 2\sqrt{5} + 3)^2 \stackrel{?}{=} 20$	$(-3 - 2\sqrt{5} + 3)^2 \stackrel{?}{=} 20$
$(2\sqrt{5})^2 \stackrel{?}{=} 20$	$(-2\sqrt{5})^2 \stackrel{?}{=} 20$
$4 \cdot 5 \stackrel{?}{=} 20$	$4 \cdot 5 \stackrel{?}{=} 20$
$20 = 20$ ✔	$20 = 20$ ✔

3. Completing the Square

In Example 1(c), we used the square root property to solve an equation where the square of a binomial was equal to a constant.

$$(w + 3)^2 = 20$$
$$w + 3 = \pm\sqrt{20}$$
$$w = -3 \pm 2\sqrt{5}$$

In general, an equation of the form $(x - h)^2 = k$ can be solved using the square root property. Furthermore, any equation $ax^2 + bx + c = 0$ $(a \neq 0)$ can be rewritten in the form $(x - h)^2 = k$ by using a process called **completing the square**.

We begin our discussion of completing the square with some vocabulary. For a trinomial $ax^2 + bx + c$, $(a \neq 0)$, the term ax^2 is called the quadratic term. The term bx is called the linear term, and the term, c, is called the constant term.

Next, notice that the factored form of a perfect square trinomial is the square of a binomial.

Perfect Square Trinomial		**Factored Form**
$x^2 + 10x + 25$	$\longrightarrow$	$(x + 5)^2$
$t^2 - 6t + 9$	$\longrightarrow$	$(t - 3)^2$
$p^2 - 14p + 49$	$\longrightarrow$	$(p - 7)^2$

Furthermore, for a perfect square trinomial with a leading coefficient of 1, the constant term is the square of half the coefficient of the linear term. For example:

$x^2 + 10x + 25$ $\qquad$ $t^2 - 6t + 9$ $\qquad$ $p^2 - 14p + 49$

$\left[\frac{1}{2}(10)\right]^2 = (5)^2 = 25$ $\qquad$ $\left[\frac{1}{2}(-6)\right]^2 = (-3)^2 = 9$ $\qquad$ $\left[\frac{1}{2}(-14)\right]^2 = (-7)^2 = 49$

In general, an expression of the form $x^2 + bx$ will be a perfect square trinomial if the square of half the linear term coefficient, $(\frac{1}{2}b)^2$, is added to the expression.

example 2

Completing the Square

Complete the square for each expression. Then factor the expression as the square of a binomial.

a. $x^2 + 12x$ $\qquad$ b. $x^2 - 26x$ $\qquad$ c. $x^2 + 11x$ $\qquad$ d. $x^2 - \frac{4}{7}x$

Solution:

The expressions are in the form $x^2 + bx$. Add the square of half the linear term coefficient, $(\frac{1}{2}b)^2$.

a. $x^2 + 12x$

$x^2 + 12x + 36$ $\qquad$ Add $\frac{1}{2}$ of 12, squared: $[\frac{1}{2}(12)]^2 = (6)^2 = 36$.

$(x + 6)^2$ $\qquad$ Factored form

b. $x^2 - 26x$

$x^2 - 26x + 169$ $\qquad$ Add $\frac{1}{2}$ of -26, squared: $[\frac{1}{2}(-26)]^2 = (-13)^2 = 169$.

$(x - 13)^2$ $\qquad$ Factored form

c. $x^2 + 11x$

$x^2 + 11x + \frac{121}{4}$ $\qquad$ Add $\frac{1}{2}$ of 11, squared: $[\frac{1}{2}(11)]^2 = (\frac{11}{2})^2 = \frac{121}{4}$.

$\left(x + \frac{11}{2}\right)^2$ $\qquad$ Factored form

d. $x^2 - \frac{4}{7}x$

$x^2 - \frac{4}{7}x + \frac{4}{49}$ Add $\frac{1}{2}$ of $-\frac{4}{7}$, squared. $[\frac{1}{2}(-\frac{4}{7})]^2 = (-\frac{2}{7})^2 = \frac{4}{49}$

$\left(x - \frac{2}{7}\right)^2$ Factored form

4. Solving Quadratic Equations by Completing the Square

The process of completing the square can be used to write a quadratic equation $ax^2 + bx + c = 0$ $(a \neq 0)$ in the form $(x - h)^2 = k$. Then, the square root property can be used to solve the equation. The following steps outline the procedure.

Solving a Quadratic Equation in the Form $ax^2 + bx + c = 0 \quad (a \neq 0)$ by Completing the Square and Applying the Square Root Property

1. Divide both sides by a to make the leading coefficient 1.
2. Isolate the variable terms on one side of the equation.
3. Complete the square (add the square of $\frac{1}{2}$ the linear term coefficient to both sides of the equation. Then factor the resulting perfect square trinomial).
4. Apply the square root property and solve for x.

example 3

Solving Quadratic Equations by Completing the Square and Applying the Square Root Property

Solve the quadratic equations by completing the square and applying the square root property.

a. $3x^2 = 18x - 39$ b. $2x(2x - 10) = -30 + 6x$

Solution:

a. $3x^2 = 18x - 39$

$3x^2 - 18x + 39 = 0$ Write the equation in the form $ax^2 + bx + c = 0$.

$\frac{3x^2}{3} - \frac{18x}{3} + \frac{39}{3} = \frac{0}{3}$ **Step 1:** Divide both sides by the leading coefficient, 3.

$x^2 - 6x + 13 = 0$

$x^2 - 6x = -13$ **Step 2:** Isolate the variable terms on one side.

$x^2 - 6x + 9 = -13 + 9$ **Step 3:** To complete the square, add $[\frac{1}{2}(-6)]^2 = 9$ to both sides of the equation.

$(x - 3)^2 = -4$ Factor the perfect square trinomial.

$x - 3 = \pm\sqrt{-4}$ **Step 4:** Apply the square root property.

$x - 3 = \pm 2i$ Simplify the radical.

$x = 3 \pm 2i$ Solve for x.

The solutions are imaginary numbers and can be written as $x = 3 + 2i$ and $x = 3 - 2i$.

b. $2x(2x - 10) = -30 + 6x$

$4x^2 - 20x = -30 + 6x$ Clear parentheses.

$4x^2 - 26x + 30 = 0$ Write the equation in the form $ax^2 + bx + c = 0$.

$\dfrac{4x^2}{4} - \dfrac{26x}{4} + \dfrac{30}{4} = \dfrac{0}{4}$ **Step 1:** Divide both sides by the leading coefficient, 4.

$x^2 - \dfrac{13}{2}x + \dfrac{15}{2} = 0$

$x^2 - \dfrac{13}{2}x = -\dfrac{15}{2}$ **Step 2:** Isolate the variable terms on one side.

$x^2 - \dfrac{13}{2}x + \dfrac{169}{16} = -\dfrac{15}{2} + \dfrac{169}{16}$ **Step 3:** Add $[\frac{1}{2}(-\frac{13}{2})]^2 = (-\frac{13}{4})^2 = \frac{169}{16}$ to both sides.

$\left(x - \dfrac{13}{4}\right)^2 = -\dfrac{120}{16} + \dfrac{169}{16}$ Factor the perfect square trinomial. Rewrite the right-hand side with a common denominator.

$\left(x - \dfrac{13}{4}\right)^2 = \dfrac{49}{16}$

$x - \dfrac{13}{4} = \pm\sqrt{\dfrac{49}{16}}$ **Step 4:** Apply the square root property.

$x - \dfrac{13}{4} = \pm\dfrac{7}{4}$ Simplify the radical.

$x = \dfrac{13}{4} \pm \dfrac{7}{4}$

$x = \dfrac{13}{4} + \dfrac{7}{4} = \dfrac{20}{4} = 5$

$x = \dfrac{13}{4} - \dfrac{7}{4} = \dfrac{6}{4} = \dfrac{3}{2}$

The solutions are rational numbers: $x = \frac{3}{2}$ and $x = 5$. The check is left to the reader.

Tip: In general, if the solutions to a quadratic equation are rational numbers, the equation can be solved by factoring and using the zero product rule. Consider the equation from Example 3(b).

$$2x(2x - 10) = -30 + 6x$$
$$4x^2 - 20x = -30 + 6x$$
$$4x^2 - 26x + 30 = 0$$
$$2(2x^2 - 13x + 15) = 0$$
$$2(x - 5)(2x - 3) = 0$$
$$x = 5 \quad \text{or} \quad x = \frac{3}{2}$$

5. Literal Equations

example 4

Solving a Literal Equation

Ignoring air resistance, the distance, d (in meters), that an object falls in t seconds is given by the equation:

$$d = 4.9t^2, \text{ where } t \geq 0$$

a. Solve the equation for t. Do not rationalize the denominator.
b. Using the equation from part (a), determine the amount of time required for an object to fall 500 m. Round to the nearest second.

Solution:

a. $d = 4.9t^2$

$\frac{d}{4.9} = t^2$ Isolate the quadratic term. The equation is in the form $t^2 = k$.

$t = \pm\sqrt{\frac{d}{4.9}}$ Apply the square root property.

$= \sqrt{\frac{d}{4.9}}$ Because $t \geq 0$, reject the negative solution.

b. $t = \sqrt{\frac{d}{4.9}}$

$= \sqrt{\frac{500}{4.9}}$ Substitute $d = 500$.

$t \approx 10.1$

The object will require approximately 10.1 sec to fall 500 m.

section 11.1 PRACTICE EXERCISES

For Exercises 1–16, solve the equations by using the square root property.

1. $x^2 = 100$
2. $y^2 = 4$
3. $a^2 = 5$
4. $k^2 - 7 = 0$
5. $v^2 + 11 = 0$
6. $m^2 = -25$
7. $(p - 5)^2 = 9$
8. $(q + 3)^2 = 4$
9. $(x - 2)^2 = 5$
10. $(y + 3)^2 - 7 = 0$
11. $(h - 4)^2 = -8$
12. $(t + 5)^2 = -18$
13. $\left(a - \frac{1}{2}\right)^2 = \frac{3}{4}$
14. $\left(y + \frac{2}{3}\right)^2 = -\frac{5}{9}$
15. $\left(x - \frac{3}{2}\right)^2 + \frac{7}{4} = 0$
16. $\left(m + \frac{4}{5}\right)^2 - \frac{3}{25} = 0$

17. State two methods that can be used to solve the equation: $x^2 - 81 = 0$. Then solve the equation using both methods.

18. State two methods that can be used to solve the equation: $x^2 - 9 = 0$. Then solve the equation using both methods.

19. A corner shelf is to be made from a triangular piece of plywood as shown in the diagram. Find the distance, x, that the shelf will extend along the walls. Assume that the walls are at right angles. Round the answer to a tenth of a foot.

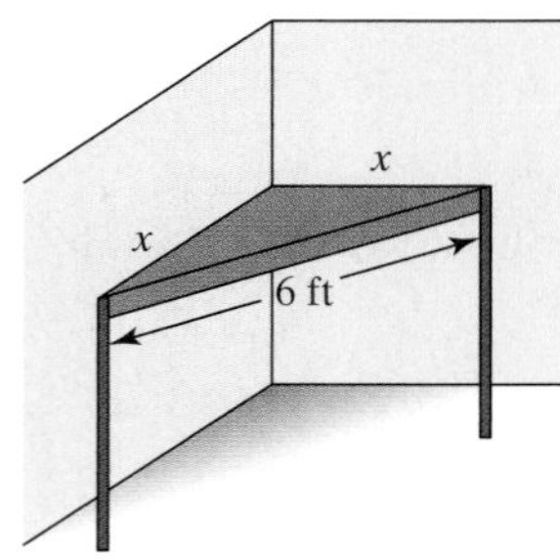

Figure for Exercise 19

20. A square has an area of 50 in.^{2} What are the lengths of the sides? Round to one decimal place.

21. The amount of money, A, in an account with an interest rate, r, compounded annually is given by

$A = P(1 + r)^t$, where P is the initial principal and t is the number of years the money is invested.

   a. If a \$10,000 investment grows to \$11,664 after 2 years, find the interest rate.

   b. If a \$6000 investment grows to \$7392.60 after 2 years, find the interest rate.

   c. Jamal wants to invest \$5000. He wants the money to grow to at least \$6500 in 2 years to cover the cost of his son's first year at college. What interest rate does Jamal need for his investment to grow to \$6500 in 2 years? Round to the nearest hundredth of a percent.

22. The volume of a box with a square bottom and a height of 4 in. is given by $V(x) = 4x^2$, where x is the length (in inches) of the sides of the bottom of the box.

   a. If the volume of the box is 289 in.^{3}, find the dimensions of the box.

   b. Are there two possible answers to part (a)? Why or why not?

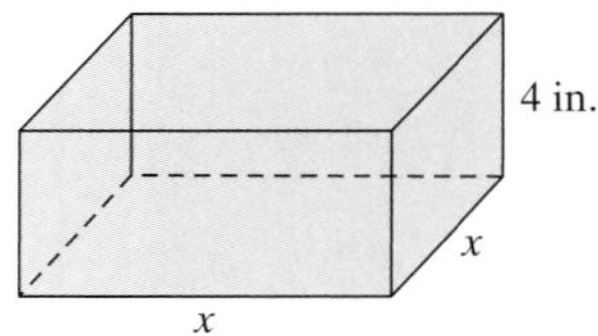

Figure for Exercise 22

For Exercises 23–28, find the value of k so that the expression is a perfect square trinomial. Then factor the trinomial.

23. $x^2 - 6x + k$
24. $x^2 + 12x + k$
25. $y^2 + 5y + k$
26. $a^2 - 7a + k$
27. $b^2 + \frac{2}{5}b + k$
28. $m^2 - \frac{2}{7}m + k$

29. Summarize the steps used in solving a quadratic equation by completing the square and applying the square root property.

30. What types of quadratic equations can be solved by completing the square and applying the square root property?

For Exercises 31–48, solve the quadratic equation by completing the square and applying the square root property.

31. $t^2 + 8t + 15 = 0$
32. $m^2 + 6m + 8 = 0$
33. $x^2 + 6x = 16$
34. $x^2 - 4x = -3$
35. $p^2 + 4p + 6 = 0$
36. $q^2 + 2q + 2 = 0$
37. $y^2 - 3y - 10 = 0$
38. $-24 = -2y^2 + 2y$
39. $2a^2 + 4a + 5 = 0$
40. $3a^2 + 6a - 7 = 0$
41. $9x^2 - 36x + 40 = 0$
42. $9y^2 - 12y + 5 = 0$
43. $p^2 - \frac{2}{5}p = \frac{2}{25}$
44. $n^2 - \frac{2}{3}n = \frac{1}{9}$
45. $(2w + 5)(w - 1) = 2$
46. $(3p - 5)(p + 1) = -3$
47. $n(n - 4) = 7$
48. $m(m + 10) = 2$

For Exercises 49–56, solve for the indicated variable.

49. $A = \pi r^2$ for r $(r > 0)$
50. $E = mc^2$ for c $(c > 0)$
51. $a^2 + b^2 + c^2 = d^2$ for a $(a > 0)$
52. $a^2 + b^2 = c^2$ for b $(b > 0)$
53. $V = \frac{1}{3}\pi r^2 h$ for r $(r > 0)$
54. $A = 6x^2$ for x $(x > 0)$
55. $s = 2\sqrt{x}$ for x
56. $A = 1.5\sqrt{x - 1}$ for x

57. A textbook company has discovered that the profit for selling its books is given by

$P(x) = -\frac{1}{8}x^2 + 5x$, where x is the number of textbooks produced (in thousands) and $P(x)$ is the corresponding profit (in thousands of dollars)

The graph of the function is shown.

a. Approximate the number of books required to make a profit of \$20,000. (*Hint:* Let $P(x) = 20$. Then complete the square to solve for x.) Round to one decimal place.

b. Why are there two answers to part (a)?

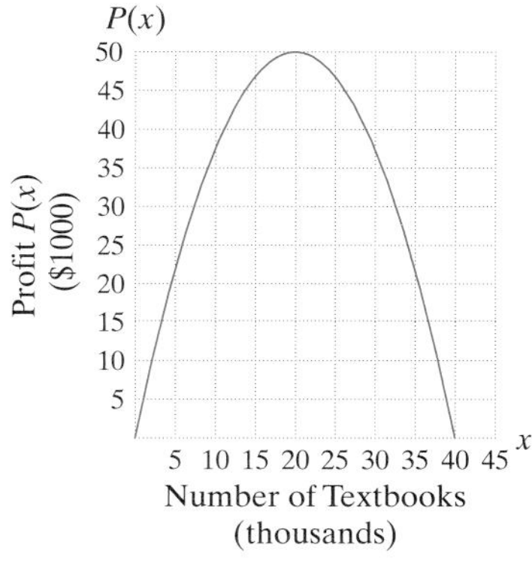

Figure for Exercise 57

58. Ignoring air resistance, the distance, d (in feet), that an object travels in free fall can be approximated by $d(t) = 16t^2$, where t is the time in seconds after the object was dropped.

a. If the CN Tower in Toronto is 1815 ft high, how long will it take an object to fall from the top of the building? Round to one decimal place.

b. If the Renaissance Tower in Dallas is 886 ft high, how long will it take an object to fall from the top of the building? Round to one decimal place.

section

11.2 Quadratic Formula

Concepts

1. Derivation of the Quadratic Formula
2. Solving Quadratic Equations Using the Quadratic Formula
3. Using the Quadratic Formula in Applications
4. Discriminant
5. Review of the Methods to Solve a Quadratic Equation

1. Derivation of the Quadratic Formula

If we solve a general quadratic equation $ax^2 + bx + c = 0\ (a \neq 0)$ by completing the square and using the square root property, the result is a formula that gives the solutions for x in terms of a, b, and c.

$ax^2 + bx + c = 0$	Begin with a quadratic equation in standard form.
$x^2 + \frac{b}{a}x + \frac{c}{a} = 0$	Divide by the leading coefficient.
$x^2 + \frac{b}{a}x = -\frac{c}{a}$	Isolate the terms containing x.
$x^2 + \frac{b}{a}x + \left(\frac{1}{2} \cdot \frac{b}{a}\right)^2 = \left(\frac{1}{2} \cdot \frac{b}{a}\right)^2 - \frac{c}{a}$	Add the square of $\frac{1}{2}$ the linear term coefficient to both sides of the equation.
$\left(x + \frac{b}{2a}\right)^2 = \frac{b^2}{4a^2} - \frac{c}{a}$	Factor the left side as a perfect square.
$\left(x + \frac{b}{2a}\right)^2 = \frac{b^2 - 4ac}{4a^2}$	Combine fractions on the right side by getting a common denominator.
$x + \frac{b}{2a} = \pm\sqrt{\frac{b^2 - 4ac}{4a^2}}$	Apply the square root property.
$x + \frac{b}{2a} = \frac{\pm\sqrt{b^2 - 4ac}}{2a}$	Simplify the denominator.
$x = -\frac{b}{2a} \pm \frac{\sqrt{b^2 - 4ac}}{2a}$	Subtract $\frac{b}{2a}$ from both sides.
$= \frac{-b \pm \sqrt{b^2 - 4ac}}{2a}$	Combine fractions.

The solution to the equation, $ax^2 + bx + c = 0$, for x in terms of the coefficients a, b, and c, is given by the **quadratic formula**.

The Quadratic Formula

For any quadratic equation of the form $ax^2 + bx + c = 0\ (a \neq 0)$ the solutions are

$$x = \frac{-b \pm \sqrt{b^2 - 4ac}}{2a}$$

2. Solving Quadratic Equations Using the Quadratic Formula

example 1

Solving a Quadratic Equation Using the Quadratic Formula

Solve the quadratic equation using the quadratic formula.

$$3x^2 + 8x = -5$$

Solution:

$$3x^2 + 8x = -5$$

$$3x^2 + 8x + 5 = 0$$ Write the equation in the form $ax^2 + bx + c = 0$.

$$a = 3, b = 8, c = 5$$ Identify a, b, and c.

$$x = \frac{-(8) \pm \sqrt{(8)^2 - 4(3)(5)}}{2(3)}$$ Apply the quadratic formula.

$$= \frac{-8 \pm \sqrt{64 - 60}}{6}$$ Simplify.

$$= \frac{-8 \pm \sqrt{4}}{6}$$

$$= \frac{-8 \pm 2}{6}$$

There are two rational solutions.

$$x = \frac{-8 + 2}{6} = \frac{-6}{6} = -1$$

$$x = \frac{-8 - 2}{6} = \frac{-10}{6} = -\frac{5}{3}$$

Both solutions check in the original equation.

Tip: Because the solutions to the equation $3x^2 + 8x = -5$ are rational numbers, the equation could have been solved by factoring and using the zero product rule.

$$3x^2 + 8x = -5$$

$$3x^2 + 8x + 5 = 0$$

$$(3x + 5)(x + 1) = 0$$

$$x = -\tfrac{5}{3} \quad \text{or} \quad x = -1$$

example 2 Solving a Quadratic Equation Using the Quadratic Formula

Solve the quadratic equation using the quadratic formula.

$$x(x + 7) + 4 = 0$$

Solution:

$x(x + 7) + 4 = 0$

$x^2 + 7x + 4 = 0$ Write the equation in the form $ax^2 + bx + c = 0$.

$a = 1, b = 7, c = 4$ Identify a, b, and c.

$x = \dfrac{-(7) \pm \sqrt{(7)^2 - 4(1)(4)}}{2(1)}$ Apply the quadratic formula.

$= \dfrac{-7 \pm \sqrt{49 - 16}}{2}$ Simplify.

$= \dfrac{-7 \pm \sqrt{33}}{2}$ The solutions are irrational numbers.

The solutions can be written as

$$x = \frac{-7 + \sqrt{33}}{2} \approx -0.628 \quad \text{and} \quad x = \frac{-7 - \sqrt{33}}{2} \approx -6.372$$

Calculator Connections

Consider the equation $x(x + 7) + 4 = 0$ from Example 2. In standard form, this equation is written as $x^2 + 7x + 4 = 0$. Using the quadratic formula, we have

$$x = \frac{-(7) \pm \sqrt{(7)^2 - 4(1)(4)}}{2(1)}$$

A calculator can be used to apply the quadratic formula directly; however, each solution must be entered separately. The solution can be checked on the calculator by using the *Ans* variable. This contains the result of the calculator's most recent computation.

$$x = \frac{-(7) + \sqrt{(7)^2 - 4(1)(4)}}{2(1)} \approx -0.6277186767$$

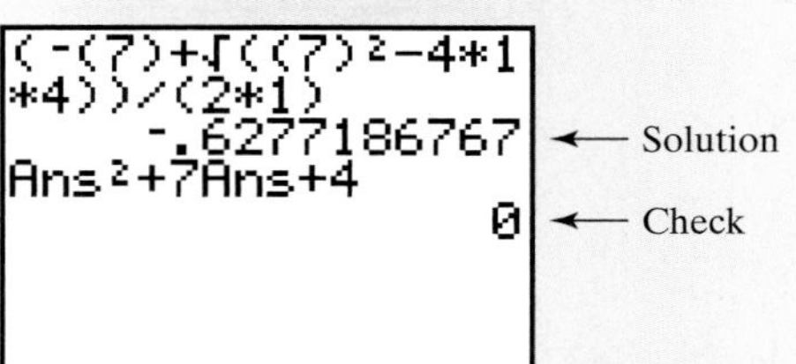

$$x = \frac{-(7) - \sqrt{(7)^2 - 4(1)(4)}}{2(1)} \approx -6.372281323$$

```
(-(7)-√((7)²-4*1
*4))/(2*1)
            -6.372281323   ← Solution
Ans²+7Ans+4
                       0   ← Check
```

3. Using the Quadratic Formula in Applications

example 3

Using the Quadratic Formula in an Application

A delivery truck travels south from Hartselle, Alabama, to Birmingham, Alabama, along Interstate 65. The truck then heads east to Atlanta, Georgia, along Interstate 20. The distance from Birmingham to Atlanta is 8 miles less than twice the distance from Hartselle to Birmingham. If the straight line distance from Hartselle to Atlanta is 165 miles, find the distance from Hartselle to Birmingham and from Birmingham to Atlanta. (Round the answers to the nearest mile.)

Solution:

The motorist travels due south and then due east. The three cities therefore form the vertices of a right triangle (Figure 11-1).

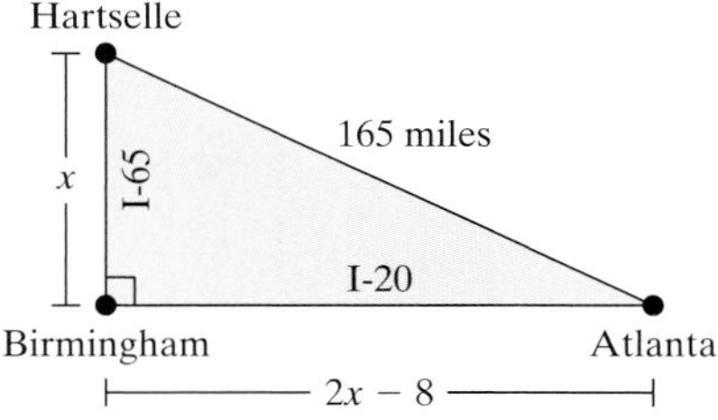

Figure 11-1

Let x represent the distance between Hartselle and Birmingham.

Then $2x - 8$ represents the distance between Birmingham and Atlanta.

Use the Pythagorean theorem to establish a relationship among the three sides of the triangle.

$$(x)^2 + (2x - 8)^2 = (165)^2$$

$$x^2 + 4x^2 - 32x + 64 = 27{,}225$$

$$5x^2 - 32x - 27{,}161 = 0$$ Write the equation in the form $ax^2 + bx + c = 0$.

$a = 5, b = -32, c = -27{,}161$ Identify a, b, and c.

$$x = \frac{-(-32) \pm \sqrt{(-32)^2 - 4(5)(-27{,}161)}}{2(5)}$$ Apply the quadratic formula.

$$= \frac{32 \pm \sqrt{1024 + 543{,}220}}{10}$$ Simplify.

$$= \frac{32 \pm \sqrt{544,244}}{10}$$

$$x = \frac{32 + \sqrt{544,244}}{10} \approx 76.97 \text{ miles} \quad \text{or}$$

$$x = \frac{32 - \sqrt{544,244}}{10} \approx -70.57 \text{ miles}$$ We reject the negative distance.

Recall that x represents the distance from Hartselle to Birmingham; therefore, to the nearest mile, the distance between Hartselle and Birmingham is 77 miles.

The distance between Birmingham and Atlanta is $2x - 8 = 2(77) - 8 =$ 146 miles.

example 4 Analyzing a Quadratic Function

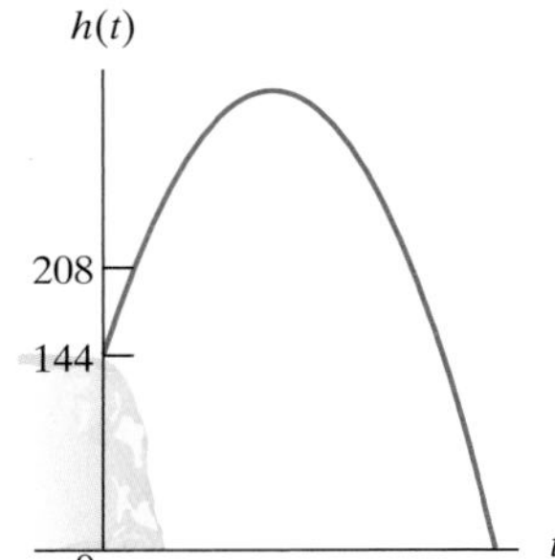

Figure 11-2

A model rocket is launched straight up from the side of a 144-ft cliff (Figure 11-2). The initial velocity is 112 ft/sec. The height of the rocket, $h(t)$, is given by

$$h(t) = -16t^2 + 112t + 144,$$ where $h(t)$ is measured in feet and t is the time in seconds.

Find the time(s) at which the rocket is 208 ft above the ground.

Solution:

$$h(t) = -16t^2 + 112t + 144$$

$$208 = -16t^2 + 112t + 144$$ Substitute 208 for $h(t)$.

$$16t^2 - 112t + 64 = 0$$ The equation is quadratic.

$$\frac{16t^2}{16} - \frac{112t}{16} + \frac{64}{16} = \frac{0}{16}$$ Divide by 16. This makes the coefficients smaller and less cumbersome to apply the quadratic formula.

$$t^2 - 7t + 4 = 0$$ The equation is not factorable. Apply the quadratic formula.

$$t = \frac{-(-7) \pm \sqrt{(-7)^2 - 4(1)(4)}}{2(1)}$$ Let $a = 1$, $b = -7$, and $c = 4$.

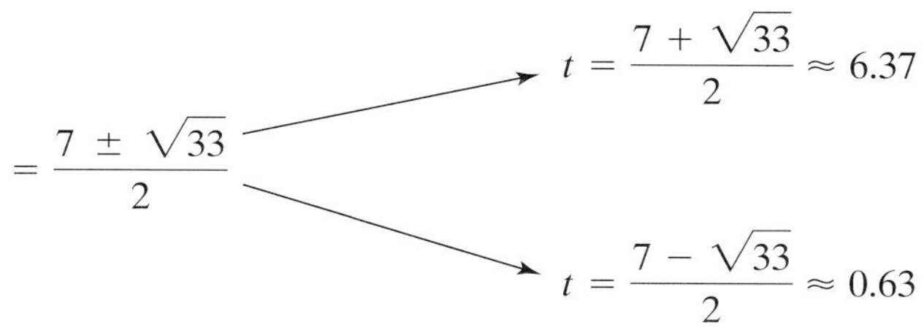

$$= \frac{7 \pm \sqrt{33}}{2} \quad \begin{cases} t = \dfrac{7 + \sqrt{33}}{2} \approx 6.37 \\[2ex] t = \dfrac{7 - \sqrt{33}}{2} \approx 0.63 \end{cases}$$

The rocket will reach a height of 208 ft after approximately 0.63 sec (on the way up) and after 6.37 sec (on the way down).

example 5 Finding the *x*- and *y*-Intercepts of a Quadratic Function

Given $f(x) = \frac{1}{4}x^2 + \frac{1}{4}x + \frac{1}{2}$,

a. Find the y-intercept.
b. Find the x-intercept(s).

Solution:

a. The y-intercept is the value of $f(0) = \frac{1}{4}(0)^2 + \frac{1}{4}(0) + \frac{1}{2}$

$$= \frac{1}{2}$$

The y-intercept is $(0, \frac{1}{2})$.

b. The x-intercepts are the real solutions to the equation $f(x) = 0$. In this case, we have

$$f(x) = \frac{1}{4}x^2 + \frac{1}{4}x + \frac{1}{2} = 0$$

$$4\left(\frac{1}{4}x^2 + \frac{1}{4}x + \frac{1}{2}\right) = 4(0)$$ Multiply by 4 to clear fractions.

$$x^2 + x + 2 = 0$$ The equation is in the form $ax^2 + bx + c = 0$, where $a = 1$, $b = 1$, and $c = 2$.

$$x = \frac{-(1) \pm \sqrt{(1)^2 - 4(1)(2)}}{2(1)}$$ Apply the quadratic formula.

$$= \frac{-1 \pm \sqrt{-7}}{2}$$ Simplify.

$$= -\frac{1}{2} \pm \frac{\sqrt{7}}{2}i$$

These solutions are *imaginary numbers*. Because there are no real solutions to the equation $f(x) = 0$, the function has no x-intercepts. The function is graphed in Figure 11-3.

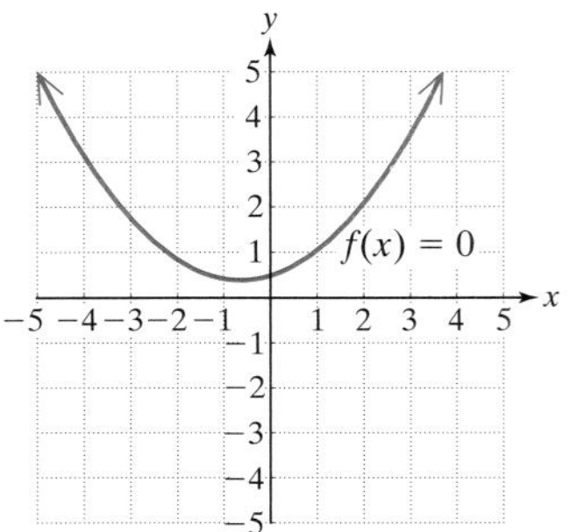

Figure 11-3

4. Discriminant

From Examples 1, 2, and 5, we see that the solutions to a quadratic equation may be rational, irrational, or imaginary numbers. The *number* and the *type* of solutions can be determined by noting the value of the square root term in the quadratic formula. The radicand of the square root, $b^2 - 4ac$, is called the discriminant.

Using the Discriminant to Determine the Number and Type of Solutions of a Quadratic Equation

Consider the equation, $ax^2 + bx + c = 0$, where a, b, and c are rational numbers and $a \neq 0$. The expression $b^2 - 4ac$, is called the **discriminant**. Furthermore,

1. If $b^2 - 4ac > 0$ then there will be two real solutions. Moreover,
   a. If $b^2 - 4ac$ is a perfect square, the solutions will be rational numbers.
   b. If $b^2 - 4ac$ is not a perfect square, the solutions will be irrational numbers.
2. If $b^2 - 4ac < 0$ then there will be two imaginary solutions.
3. If $b^2 - 4ac = 0$ then there will be one rational solution.

example 6

Using the Discriminant

Use the discriminant to determine the type and number of solutions for each equation.

a. $2x^2 - 5x + 9 = 0$

b. $3x^2 = -x + 2$

c. $-2x(2x - 3) = -1$

d. $3.6x^2 = -1.2x - 0.1$

Solution:

For each equation, first write the equation in standard form, $ax^2 + bx + c = 0$. Then determine the discriminant.

	Equation	Discriminant	Solution Type and Number
a.	$2x^2 - 5x + 9 = 0$	$b^2 - 4ac$ $= (-5)^2 - 4(2)(9)$ $= 25 - 72$ $= -47$	Because $-47 < 0$, there will be two imaginary solutions.
b.	$3x^2 = -x + 2$ $3x^2 + x - 2 = 0$	$b^2 - 4ac$ $= (1)^2 - 4(3)(-2)$ $= 1 - (-24)$ $= 25$	Because $25 > 0$ and 25 is a perfect square, there will be two rational solutions.
c.	$-2x(2x - 3) = -1$ $-4x^2 + 6x = -1$ $-4x^2 + 6x + 1 = 0$	$b^2 - 4ac$ $= (6)^2 - 4(-4)(1)$ $= 36 - (-16)$ $= 52$	Because $52 > 0$, but 52 is *not* a perfect square, there will be two irrational solutions.
d.	$3.6x^2 = -1.2x - 0.1$ $36x^2 = -12x - 1$ $36x^2 + 12x + 1 = 0$	Clear decimals first. $b^2 - 4ac$ $= (12)^2 - 4(36)(1)$ $= 144 - 144$ $= 0$	Because the discriminant equals 0, there will be only one rational solution.

With the discriminant we can determine the number of real-valued solutions to the equation $ax^2 + bx + c = 0$, and thus the number of x-intercepts to the function $f(x) = ax^2 + bx + c$. The following illustrations show the graphical interpretation of the three cases of the discriminant.

$f(x) = x^2 - 4x + 3$

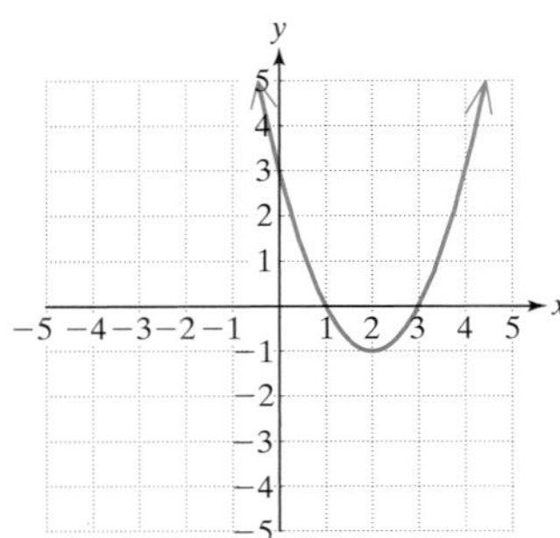

Use $x^2 - 4x + 3 = 0$ to find the value of the discriminant.

$$b^2 - 4ac = (-4)^2 - 4(1)(3) = 4$$

Since $b^2 - 4ac = 4$,
$\mathbf{b^2 - 4ac > 0}$
There are two real solutions and two x-intercepts (1, 0) and (3, 0).

$f(x) = x^2 - x + 1$

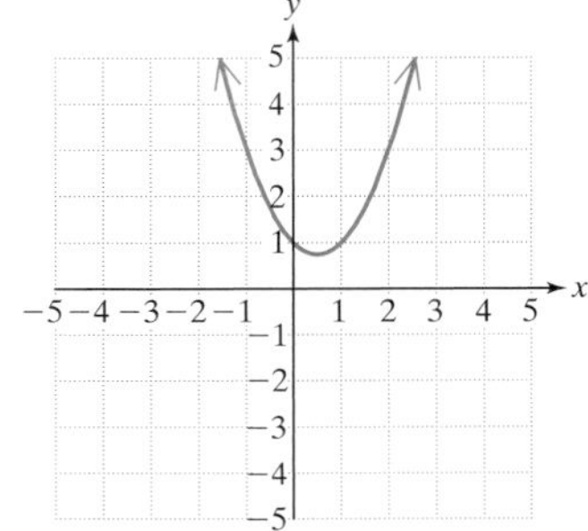

Use $x^2 - x + 1 = 0$ to find the value of the discriminant.

$$b^2 - 4ac = (-1)^2 - 4(1)(1) = -3$$

Since $b^2 - 4ac = -3$,
$\mathbf{b^2 - 4ac < 0}$
There are no real solutions and no x-intercepts.

$f(x) = x^2 - 2x + 1$

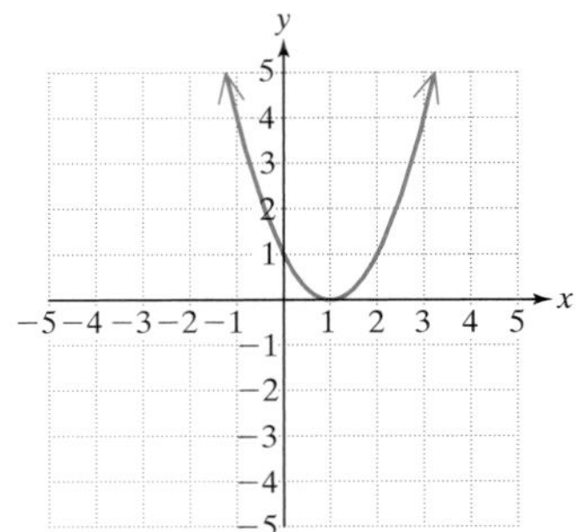

Use $x^2 - 2x + 1 = 0$ to find the value of the discriminant.

$$b^2 - 4ac = (-2)^2 - 4(1)(1) = 0$$

$\mathbf{b^2 - 4ac = 0}$
There is one real solution and one x-intercept (1, 0).

5. Review of the Methods to Solve a Quadratic Equation

Three methods have been presented to solve quadratic equations.

Methods to Solve a Quadratic Equation

- Factor and use the zero product rule (Section 5.7).
- Use the square root property. Complete the square if necessary (Section 11.1).
- Use the quadratic formula (Section 11.2).

Using the zero product rule is the simplest method, but it only works if you can factor the equation. The square root property and the quadratic formula can be used to solve any quadratic equation. Before solving a quadratic equation, take a minute to analyze it first. Each problem must be evaluated individually before choosing the most efficient method to find its solutions.

example 7 **Solving Quadratic Equations Using Any Method**

Solve the quadratic equations using any method.

a. $(x + 3)^2 + x^2 - 9x = 8$ b. $\dfrac{x^2}{2} + \dfrac{5}{2} = -x$ c. $(p - 2)^2 - 11 = 0$

Solution:

a. $(x + 3)^2 + x^2 - 9x = 8$

$x^2 + 6x + 9 + x^2 - 9x - 8 = 0$ Clear parentheses and write the equation in the form $ax^2 + bx + c = 0$.

$2x^2 - 3x + 1 = 0$ This equation is factorable.

$(2x - 1)(x - 1) = 0$ Factor.

$2x - 1 = 0 \quad \text{or} \quad x - 1 = 0$ Apply the zero product rule.

$x = \frac{1}{2} \quad \text{or} \quad x = 1$ Solve for x.

This equation could have been solved using any of the three methods but factoring was the most efficient method.

b. $\frac{x^2}{2} + \frac{5}{2} = -x$ Clear fractions and write the equation in the form $ax^2 + bx + c = 0$.

$x^2 + 5 = -2x$

$x^2 + 2x + 5 = 0$ This equation does not factor. Use the quadratic formula.

$a = 1, b = 2, c = 5$ Identify a, b, and c.

$x = \frac{-(2) \pm \sqrt{(2)^2 - 4(1)(5)}}{2(1)}$ Apply the quadratic formula.

$= \frac{-2 \pm \sqrt{-16}}{2}$ Simplify.

$= \frac{-2 \pm 4i}{2}$ Simplify the radical.

$= -1 \pm 2i$ Reduce the expression.

$$\frac{-2 \pm 4i}{2} = \frac{\cancel{2}(-1 \pm 2i)}{\cancel{2}}$$

This equation could also have been solved by completing the square and applying the square root property.

c. $(p - 2)^2 - 11 = 0$

$(p - 2)^2 = 11$ The equation is in the form $x^2 = k$, where $x = (p - 2)$.

$p - 2 = \pm\sqrt{11}$ Apply the square root property.

$p = 2 \pm \sqrt{11}$ Solve for p.

This problem could have been solved by the quadratic formula but that would have involved clearing parentheses and collecting *like* terms first.

Calculator Connections

The real solutions to a quadratic equation $ax^2 + bx + c = 0$ can be interpreted graphically as the x-intercepts of the corresponding function, $f(x) = ax^2 + bx + c$. Consequently, we can use the graph of f to approximate the real solutions to the equation $ax^2 + bx + c = 0$.

For example, the real solutions to the equation $-x^2 - 3x + 6 = 0$ can be approximated by using the *Zero* or *Root* feature on a graphing calculator. From the graphs we have $x \approx 1.3722813$ and $x \approx -4.372281$.

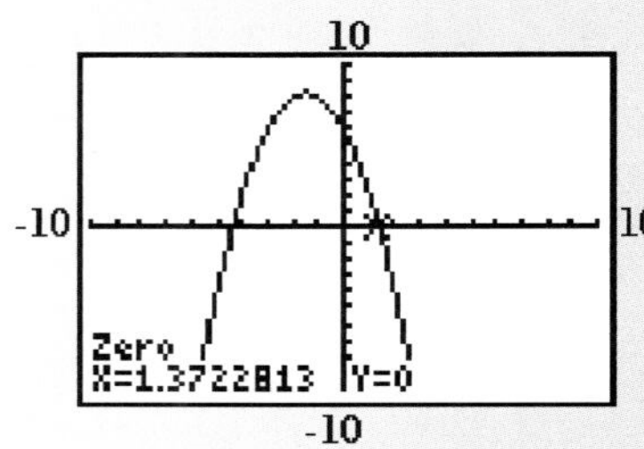

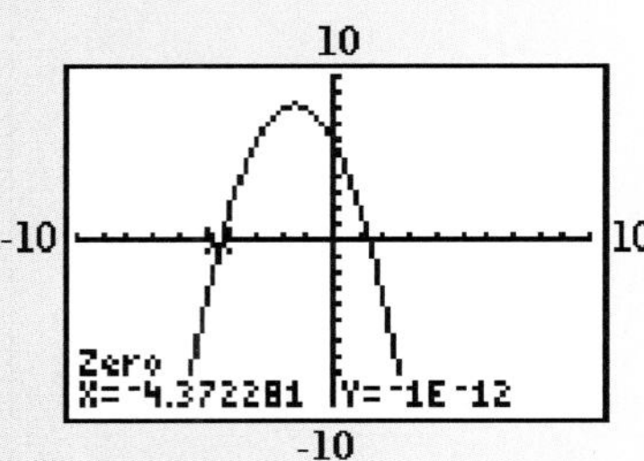

section 11.2 Practice Exercises

For Exercises 1–8, either factor the polynomial or solve the polynomial equation as the directions indicate.

1. Solve: $(x + 5)^2 = 49$
2. Solve: $16 = (2x - 3)^2$
3. Factor: $x^3 - 1$
4. Factor: $y^4 - \dfrac{1}{81}$
5. Solve: $x^3 - 2x^2 - 9x + 18 = 0$
6. Solve: $k^3 + 2k^2 - 3k = 0$
7. Factor: $8uv - 6u + 12v - 9$
8. Factor: $4t^2 - 20t + 25 - s^2$

For Exercises 9–12, simplify the expressions.

9. $\dfrac{16 - \sqrt{640}}{4}$
10. $\dfrac{18 + \sqrt{180}}{3}$
11. $\dfrac{14 - \sqrt{-147}}{7}$
12. $\dfrac{10 - \sqrt{-175}}{5}$

For Exercises 13–20, write the equation in the form $ax^2 + bx + c = 0, a > 0$, and identify a, b, and c.

13. $x^2 + 2x = -1$
14. $12y - 9 = 4y^2$
15. $19m^2 = 8m$
16. $2n - 5n^2 = 0$
17. $5p^2 - 21 = 0$
18. $3k^2 = 7$
19. $4n(n - 2) - 5n(n - 1) = 4$
20. $(2x + 1)(x - 3) = -9$

For Exercises 21–28, find the discriminant, $b^2 - 4ac$, and determine the number and type of solutions to the indicated exercise. Choose from two rational solutions, one rational solution, two irrational solutions, or two imaginary solutions.

21. Exercise 13
22. Exercise 14
23. Exercise 15
24. Exercise 16
25. Exercise 17
26. Exercise 18
27. Exercise 19
28. Exercise 20

For Exercises 29–34, (a) find the discriminant and determine the number of solutions to the equation and (b) determine the number of x-intercepts to the related function.

29. a. $3x^2 - 3x - 1 = 0$
    b. $f(x) = 3x^2 - 3x - 1$

30. a. $x^2 - 10x - 3 = 0$
    b. $g(x) = x^2 - 10x - 3$

31. a. $4x^2 + 4x + 1 = 0$
    b. $h(x) = 4x^2 + 4x + 1$

32. a. $x^2 - 5x + \frac{25}{4} = 0$
    b. $k(x) = x^2 - 5x + \frac{25}{4}$

33. a. $2x^2 - 6x + 5 = 0$
    b. $p(x) = 2x^2 - 6x + 5$

34. a. $-3x^2 + x - 5 = 0$
    b. $q(x) = -3x^2 + x - 5$

35. Describe the circumstances in which factoring can be used as a method for solving a quadratic equation.

36. Describe the circumstances in which the square root property can be used as a method for solving a quadratic equation.

37. Describe the circumstances in which the quadratic formula can be used as a method for solving a quadratic equation.

38. Write the quadratic formula from memory.

For Exercises 39–62, solve the equation using the quadratic formula.

39. $a^2 + 11a - 12 = 0$
40. $5b^2 - 14b - 3 = 0$
41. $9y^2 - 2y + 5 = 0$
42. $2t^2 + 3t - 7 = 0$
43. $12p^2 - 4p + 5 = 0$
44. $5n^2 - 4n + 6 = 0$
45. $z^2 = 2z + 35$
46. $12x^2 - 5x = 2$
47. $a^2 + 3a = 8$
48. $k^2 + 4 = 6k$
49. $25x^2 - 20x + 4 = 0$
50. $9y^2 = -12y - 4$
51. $w^2 - 6w + 14 = 0$
52. $2m^2 + 3m = -2$
53. $(x + 2)(x - 3) = 1$
54. $3y(y + 1) - 7y(y + 2) = 6$
55. $\frac{1}{2}y^2 + \frac{2}{3} = -\frac{2}{3}y$
(*Hint:* Clear the fractions first.)
56. $\frac{2}{3}p^2 - \frac{1}{6}p + \frac{1}{2} = 0$
57. $\frac{1}{5}h^2 + h + \frac{3}{5} = 0$
58. $\frac{1}{4}w^2 + \frac{7}{4}w + 1 = 0$
59. $0.01x^2 + 0.06x + 0.08 = 0$
(*Hint:* Clear the decimals first.)
60. $0.5y^2 - 0.7y + 0.2 = 0$
61. $0.3t^2 + 0.7t - 0.5 = 0$
62. $0.01x^2 + 0.04x - 0.07 = 0$

63. a. Factor: $x^3 - 27$.
    b. Use the zero product rule and the quadratic formula to solve $x^3 - 27 = 0$. There should be three solutions (one real and two imaginary).

64. a. Factor: $64x^3 + 1$.
    b. Use the zero product rule and the quadratic formula to solve $64x^3 + 1 = 0$. There should be three solutions (one real and two imaginary).

65. a. Factor: $3x^3 - 6x^2 + 6x$.
    b. Use the zero product rule and the quadratic formula to solve $3x^3 - 6x^2 + 6x = 0$. There should be three complex solutions.

66. a. Factor: $5x^3 + 5x^2 + 10x$.
    b. Use the zero product rule and the quadratic formula to solve $5x^3 + 5x^2 + 10x = 0$. There should be three complex solutions.

67. The volume of a cube is 27 ft^3. Find the lengths of the sides.

Figure for Exercise 67

68. The volume of a rectangular box is 64 ft^3. If the width is three times longer than the height, and the length is nine times longer than the height, find the dimensions of the box.

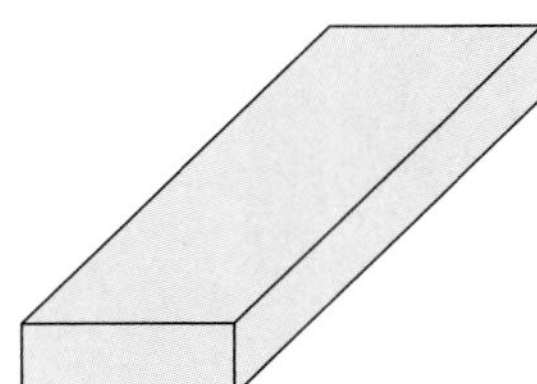

Figure for Exercise 68

For Exercises 69–76, solve the quadratic equations using any method.

69. $a^2 + 3a + 4 = 0$

70. $4z^2 + 7z = 0$

71. $x^2 - 2 = 0$

72. $b^2 + 7 = 0$

73. $4y^2 + 8y - 5 = 0$

74. $k^2 - k + 8 = 0$

75. $\left(x + \frac{1}{2}\right)^2 + 4 = 0$

76. $(2y + 3)^2 = 9$

77. The braking distance, d (in feet), of a car going v miles per hour is given by

$$d(v) = \frac{v^2}{20} + v \quad v \geq 0$$

a. How fast would a car be traveling if its braking distance is 150 ft? Round to the nearest mile per hour.

b. How fast would a car be traveling if its braking distance is 100 ft? Round to the nearest mile per hour.

78. The number of lawyers, N, in the United States from the year 1951 through 1989 can be approximated by $N(t) = 1060t^2 - 7976t + 202{,}209$ where t represents the number of years after 1951 (*Source:* Datapedia of the United States).

a. Approximate the number of lawyers in the United States in the year 1978. Round to the nearest thousand.

b. In what year after 1951 did the number of lawyers in the United States hit 400,000. (Round to the nearest year.)

c. In what year after 1951 did the number of lawyers hit a half-million? (Round to the nearest year.)

d. If this trend continues, predict the number of lawyers in the United States in the year 2010. Round to the nearest thousand.

79. The hypotenuse of a right triangle is 10.2 m long. One leg is 2.1 m shorter than the other leg. Find the lengths of the legs. Round to one decimal place.

80. The hypotenuse of a right triangle is 17 ft long. One leg is 3.4 ft longer than the other leg. Find the lengths of the legs.

81. The number of farms in the United States increased between 1890 and 1920 and then began a downward trend.

The number of farms, N, can be approximated as a function of time by $N(t) = -1.43t^2 + 94.56t + 4825$, where $t = 0$ corresponds to the year 1890 and N is measured in thousands of farms.

a. Use the function to approximate the number of farms in the year 1930. Round to the nearest whole unit.

b. Why is the function value slightly different from the actual data value of 6295?

c. Approximate the years when the number of farms in the United States was 5,000,000 ($N = 5000$ thousands).

Expanding Your Skills

82. An artist has been commissioned to make a stained glass window in the shape of a regular octagon. The octagon must fit inside an 18-in. square space. See the figure.

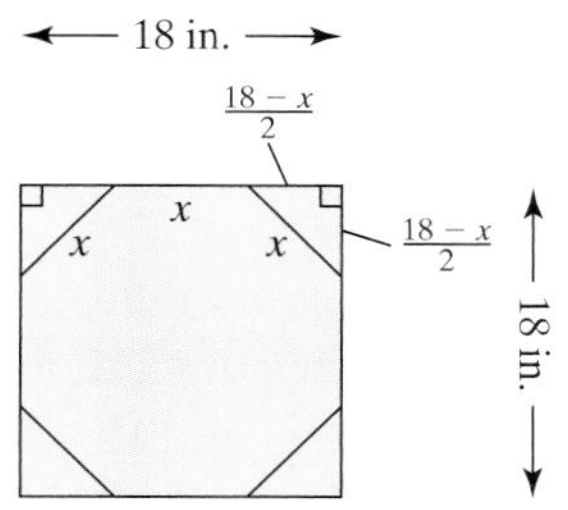

Figure for Exercise 82

a. Let x represent the length of each side of the octagon. Verify that the legs of the small triangles formed by the corners of the square can be expressed as $\frac{18 - x}{2}$.

b. Use the Pythagorean theorem to set up an equation in terms of x that represents the relationship between the legs of the triangle and the hypotenuse.

c. Simplify the equation by clearing parentheses and clearing fractions.

d. Solve the resulting quadratic equation by using the quadratic formula. Use a calculator and round your answers to the nearest tenth of an inch.

e. There are two solutions for x. Which one is appropriate and why?

Graphing Calculator Exercises

83. Graph $Y_1 = x^3 - 27$. Compare the x-intercepts with the solutions to the equation $x^3 - 27 = 0$ found in Exercise 63.

84. Graph $Y_1 = 64x^3 + 1$. Compare the x-intercepts with the solutions to the equation $64x^3 + 1 = 0$ found in Exercise 64.

85. Graph $Y_1 = 3x^3 - 6x^2 + 6x$. Compare the x-intercepts with the solutions to the equation $3x^3 - 6x^2 + 6x = 0$ found in Exercise 65.

86. Graph $Y_1 = 5x^3 + 5x^2 + 10x$. Compare the x-intercepts with the solutions to the equation $5x^3 + 5x^2 + 10x = 0$ found in Exercise 66.

87. The recent population, P (in thousands), of Ecuador can be approximated by $P(t) = 1.12t^2 + 204.4t + 6697$, where $t = 0$ corresponds to the year 1974.

a. Approximate the number of people in Ecuador in the year 1980.

b. If this trend continues, approximate the number of people in Ecuador in the year 2010.

c. In what year after 1974 did the population of Ecuador reach 10 million? Round to the nearest year. (*Hint:* 10 million equals 10,000 thousands.)

d. Use a graphing calculator to graph the function, P on the window $0 \leq x \leq 20$, $4000 \leq y \leq 12{,}000$. Use a *Trace* feature to approximate the year when the population in Ecuador was 10 million (10,000 thousands).

88. The recent population, P (in thousands), of New Zealand can be approximated by $P(t) = 0.089t^2 + 25.7t + 3601$, where $t = 0$ corresponds to the year 1995.

a. Approximate the number of people in New Zealand in the year 1999.

b. If this trend continues, approximate the number of people in New Zealand in the year 2005.

c. In what year after 1995 did the population of New Zealand reach 3.8 million? Round to the nearest year. (*Hint*: 3.8 million equals 3800 thousands.)

d. Use a graphing calculator to graph the function, $P(t)$ on the window $0 \leq x \leq 50$, $3000 \leq y \leq 5000$. Use the *Trace* feature to determine the year when the population in New Zealand was 3.8 million.

Concepts

1. Equations Reducible to a Quadratic
2. Solving Equations Using Substitution

section
11.3 Equations in Quadratic Form

1. Equations Reducible to a Quadratic

We have learned to solve a variety of different types of equations, including linear, quadratic, rational, radical, and polynomial equations. Sometimes, however, it is necessary to use a quadratic equation as a tool to solve other types of equations. For instance, the equation in Example 1 is a radical equation that reduces to a quadratic equation after squaring both sides.

example 1

Solving an Equation in Quadratic Form

Solve the equation. $x - \sqrt{x} - 12 = 0$

Solution:

$x - \sqrt{x} - 12 = 0$	This is a radical equation.
$x - 12 = \sqrt{x}$	Isolate the radical.
$(x - 12)^2 = (\sqrt{x})^2$	Square both sides.
$x^2 - 24x + 144 = x$	The resulting equation is quadratic.
$x^2 - 25x + 144 = 0$	Write the equation in the form $ax^2 + bx + c = 0$.
$(x - 9)(x - 16) = 0$	The equation is factorable.
$x = 9$ or $x = 16$	Apply the zero product rule.
Check: $x = 9$	Check: $x = 16$
$x - \sqrt{x} - 12 = 0$	$x - \sqrt{x} - 12 = 0$
$(9) - \sqrt{9} - 12 \stackrel{?}{=} 0$	$(16) - \sqrt{16} - 12 \stackrel{?}{=} 0$
$9 - 3 - 12 \neq 0$	$16 - 4 - 12 = 0$ ✓

$x = 16$ is the only solution. ($x = 9$ does not check.)

Avoiding Mistakes

Recall that if we raise both sides of a radical equation to an even power, the potential solutions must be checked in the original equation.

example 2

Solving an Equation Quadratic in Form

Solve the equation. $w^4 - 81 = 0$

Solution:

This equation is a higher order polynomial equation.

$$w^4 - 81 = 0$$

$$(w^2 - 9)(w^2 + 9) = 0 \qquad \text{The equation is factorable.}$$

$$(w + 3)(w - 3)(w^2 + 9) = 0$$

$w + 3 = 0$ or	$w - 3 = 0$ or	$w^2 + 9 = 0$	Apply the zero product rule.
$w = -3$ or	$w = 3$ or	$w^2 = -9$	Solve for w.
		$w = \pm\sqrt{-9}$	
		$w = \pm 3i$	

The solutions are $w = 3$, $w = -3$, $w = 3i$, and $w = -3i$.

2. Solving Equations Using Substitution

In this section, we will see that some equations that are not quadratic can be manipulated to appear as **equations in quadratic form** by using substitution.

example 3

Solving an Equation Quadratic in Form

Solve the equation. $(2x^2 - 5)^2 - 16(2x^2 - 5) + 39 = 0$

Solution:

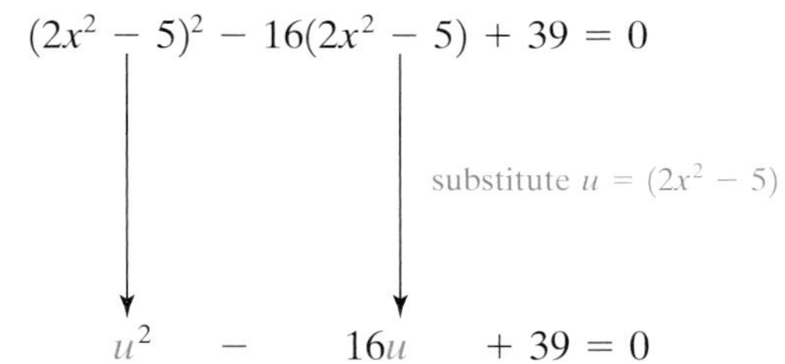

$(2x^2 - 5)^2 - 16(2x^2 - 5) + 39 = 0$

substitute $u = (2x^2 - 5)$

Notice this equation is a trinomial. If the substitution $u = (2x^2 - 5)$ is made, the equation becomes quadratic in the variable u.

$u^2 - 16u + 39 = 0$ — The equation is in the form $au^2 + bu + c = 0$.

$(u - 13)(u - 3) = 0$ — The equation is factorable.

$u = 13$ or $u = 3$ — Apply the zero product rule.

reverse substitute

$2x^2 - 5 = 13$ or $2x^2 - 5 = 3$

$2x^2 = 18$ or $2x^2 = 8$

$x^2 = 9$ or $x^2 = 4$ — Write the equations in the form $x^2 = k$.

$x = \pm\sqrt{9}$ or $x = \pm\sqrt{4}$ — Apply the square root property.

$x = \pm 3$ or $x = \pm 2$

The solutions are $x = 3$, $x = -3$, $x = 2$, $x = -2$. — Substituting these values in the original equation verifies that these are all valid solutions.

Avoiding Mistakes

When using substitution, it is critical to reverse substitute to solve the equation in terms of the original variable.

example 4 Solving an Equation Quadratic in Form

Solve the equation. $p^{2/3} - 2p^{1/3} = 8$

Solution:

$p^{2/3} - 2p^{1/3} = 8$

$p^{2/3} - 2p^{1/3} - 8 = 0$ Set the equation equal to zero.

$(p^{1/3})^2 - 2(p^{1/3}) - 8 = 0$ Make the substitution $u = p^{1/3}$.

substitute $u = p^{1/3}$

$u^2 - 2u - 8 = 0$ Then, the equation is in the form $au^2 + bu + c = 0$.

$(u - 4)(u + 2) = 0$ The equation is factorable.

$u = 4$ or $u = -2$ Apply the zero product rule.

reverse substitute

$p^{1/3} = 4$ or $p^{1/3} = -2$

$\sqrt[3]{p} = 4$ or $\sqrt[3]{p} = -2$ The equations are radical equations.

$(\sqrt[3]{p})^3 = (4)^3$ or $(\sqrt[3]{p})^3 = (-2)^3$ Cube both sides.

$p = 64$ or $p = -8$

Check: $p = 64$

$p^{2/3} - 2p^{1/3} = 8$

$(64)^{2/3} - 2(64)^{1/3} \stackrel{?}{=} 8$

$16 - 2(4) \stackrel{?}{=} 8$

$8 = 8$ ✓

Check: $p = -8$

$p^{2/3} - 2p^{1/3} = 8$

$(-8)^{2/3} - 2(-8)^{1/3} \stackrel{?}{=} 8$

$4 - 2(-2) \stackrel{?}{=} 8$

$4 + 4 = 8$ ✓

The solutions are $p = 64$ and $p = -8$.

example 5 Solving a Quadratic Equation Using Substitution

Solve the equation $(t - 5)^2 - 4(t - 5) + 13 = 0$ by using the substitution $u = t - 5$.

Solution:

$(t - 5)^2 - 4(t - 5) + 13 = 0$ This equation is quadratic; however, we can make it a simpler quadratic equation by letting $u = (t - 5)$.

substitute $u = t - 5$

$u^2 - 4u + 13 = 0$ This equation does not factor.

$$u = \frac{-(-4) \pm \sqrt{(-4)^2 - 4(1)(13)}}{2(1)}$$ Apply the quadratic formula: $a = 1$, $b = -4$, $c = 13$.

$$= \frac{4 \pm \sqrt{16 - 52}}{2}$$

$$= \frac{4 \pm \sqrt{-36}}{2}$$

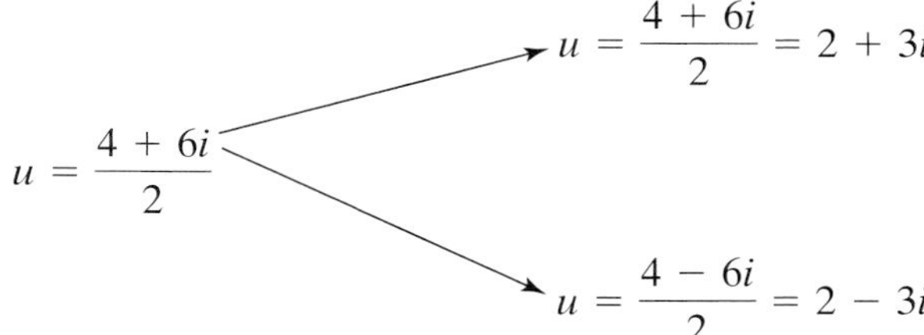

$$u = \frac{4 + 6i}{2} = 2 + 3i$$

$$u = \frac{4 - 6i}{2} = 2 - 3i$$

$u = 2 + 3i$ or $u = 2 - 3i$

reverse substitute

$t - 5 = 2 + 3i$ or $t - 5 = 2 - 3i$

$t = 7 + 3i$ or $t = 7 - 3i$ Both values check in the original equation.

The solutions are $t = 7 + 3i$ and $t = 7 - 3i$.

section 11.3 Practice Exercises

For Exercises 1–16, solve the equation. For the equations involving radicals, be sure that you check all solutions in the original equation.

1. $y + 6\sqrt{y} = 16$
2. $p - 8\sqrt{p} = -15$
3. $2x + 3\sqrt{x} - 2 = 0$
4. $3t + 5\sqrt{t} - 2 = 0$
5. $\sqrt{4b + 1} - \sqrt{b - 2} = 3$
6. $\sqrt{6a + 7} - \sqrt{3a + 3} = 1$
7. $\sqrt{w - 6} + 3 = \sqrt{w + 9}$
8. $\sqrt{z + 15} - \sqrt{2z + 7} = 1$
9. $x^4 - 16 = 0$
10. $t^4 - 625 = 0$
11. $m^4 - 81 = 0$
12. $n^4 - 256 = 0$
13. $a^3 + 8 = 0$
14. $b^3 - 1 = 0$
15. $5p^3 - 5 = 0$
16. $2t^3 + 54 = 0$
17. a. Solve the quadratic equation by factoring: $u^2 + 10u + 24 = 0$
    b. Solve the equation using substitution: $(y^2 + 5y)^2 + 10(y^2 + 5y) + 24 = 0$
18. a. Solve the quadratic equation by factoring: $u^2 - 2u - 35 = 0$
    b. Solve the equation using substitution: $(w^2 - 6w)^2 - 2(w^2 - 6w) - 35 = 0$
19. a. Solve the quadratic equation by factoring: $u^2 - 2u - 24 = 0$
    b. Solve the equation using substitution: $(x^2 - 5x)^2 - 2(x^2 - 5x) - 24 = 0$

20. a. Solve the quadratic equation by factoring: $u^2 - 4u + 3 = 0$

    b. Solve the equation using substitution: $(2p^2 + p)^2 - 4(2p^2 + p) + 3 = 0$

For Exercises 21–34, solve using substitution.

21. $(4x + 5)^2 + 3(4x + 5) + 2 = 0$

22. $2(5x + 3)^2 - (5x + 3) - 28 = 0$

23. $16\left(\frac{x + 6}{4}\right)^2 + 8\left(\frac{x + 6}{4}\right) + 1 = 0$

24. $9\left(\frac{x + 3}{2}\right)^2 - 6\left(\frac{x + 3}{2}\right) + 1 = 0$

25. $(x^2 - 2x)^2 + 2(x^2 - 2x) = 3$

26. $(x^2 + x)^2 - 8(x^2 + x) = -12$

27. $x^4 - 13x^2 + 36 = 0$

28. $y^4 - 5y^2 + 4 = 0$

29. $x^6 - 9x^3 + 8 = 0$

30. $x^6 - 26x^3 - 27 = 0$

31. $m^{2/3} - m^{1/3} - 6 = 0$

32. $2n^{2/3} + 7n^{1/3} - 15 = 0$

33. $2t^{2/5} + 7t^{1/5} + 3 = 0$

34. $p^{2/5} + p^{1/5} - 2 = 0$

Expanding Your Skills

35. Solve $x^2 - 4 = 0$ three ways:
    a. By factoring
    b. By using the square root property
    c. By using the quadratic formula

36. Solve $9x^2 - 16 = 0$ three ways:
    a. By factoring
    b. By using the square root property
    c. By using the quadratic formula

For Exercises 37–40, solve the equation. *Hint*: Factor by grouping first.

37. $a^3 + 16a - a^2 - 16 = 0$

38. $b^3 + 9b - b^2 - 9 = 0$

39. $x^3 + 5x - 4x^2 - 20 = 0$

40. $y^3 + 8y - 3y^2 - 24 = 0$

Graphing Calculator Exercises

41. a. Solve the equation $x^4 + 4x^2 + 4 = 0$.
    b. How many solutions are real and how many solutions are imaginary?
    c. How many x-intercepts do you anticipate for the function defined by $y = x^4 + 4x^2 + 4$?
    d. Graph $Y_1 = x^4 + 4x^2 + 4$ on a standard viewing window.

42. a. Solve the equation $x^4 - 2x^2 + 1 = 0$.
    b. How many solutions are real and how many solutions are imaginary?
    c. How many x-intercepts do you anticipate for the function defined by $y = x^4 - 2x^2 + 1$?
    d. Graph $Y_1 = x^4 - 2x^2 + 1$ on a standard viewing window.

43. a. Solve the equation $x^4 - x^3 - 6x^2 = 0$.
    b. How many solutions are real and how many solutions are imaginary?
    c. How many x-intercepts do you anticipate for the function defined by $y = x^4 - x^3 - 6x^2$?
    d. Graph $Y_1 = x^4 - x^3 - 6x^2$ on a standard viewing window.

44. a. Solve the equation $x^4 - 10x^2 + 9 = 0$.
    b. How many solutions are real and how many solutions are imaginary?
    c. How many x-intercepts do you anticipate for the function defined by $y = x^4 - 10x^2 + 9$?
    d. Graph $Y_1 = x^4 - 10x^2 + 9$ on a standard viewing window.

section

11.4 GRAPHS OF QUADRATIC FUNCTIONS

Concepts

1. Quadratic Functions of the Form $f(x) = x^2 + k$
2. Quadratic Functions of the Form $f(x) = (x - h)^2$
3. Quadratic Functions of the Form $f(x) = ax^2$ $(a \neq 0)$
4. Quadratic Functions of the Form $f(x) = a(x - h)^2 + k$ $(a \neq 0)$

In Section 7.5, we defined a quadratic function as a function of the form $f(x) = ax^2 + bx + c$ $(a \neq 0)$. We also learned that the graph of a quadratic function is a parabola. The parabola opens up if $a > 0$ (Figures 11-4 and 11-5), and opens down if $a < 0$ (Figure 11-6). If a parabola opens up, the **vertex** is the lowest point on the graph. If a parabola opens down, the *vertex* is the highest point on the graph. The **axis of symmetry** is the vertical line that passes through the vertex.

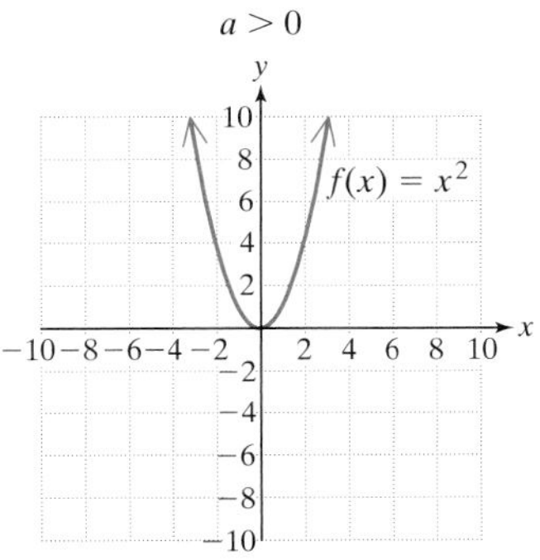

Figure 11-4

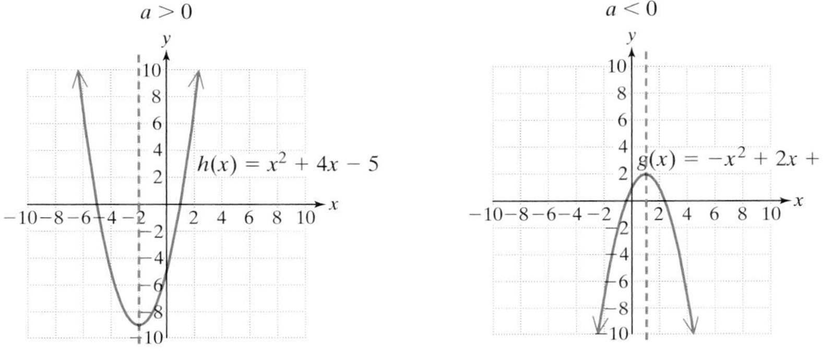

Figure 11-5

Figure 11-6

1. Quadratic Functions of the Form $f(x) = x^2 + k$

One technique for graphing a function is to plot a sufficient number of points on the function until the general shape and defining characteristics can be determined. Then sketch a curve through the points.

example 1 Graphing Quadratic Functions in the Form $f(x) = x^2 + k$

Graph the functions f, g, and h on the same coordinate system.

$$f(x) = x^2 \qquad g(x) = x^2 + 1 \qquad h(x) = x^2 - 2$$

Solution:

Several function values for f, g, and h are shown in Table 11-1 for selected values of x. The corresponding graphs are pictured in Figure 11-7.

g(x) = x² + 1
f(x) = x²
h(x) = x² − 2

Figure 11-7

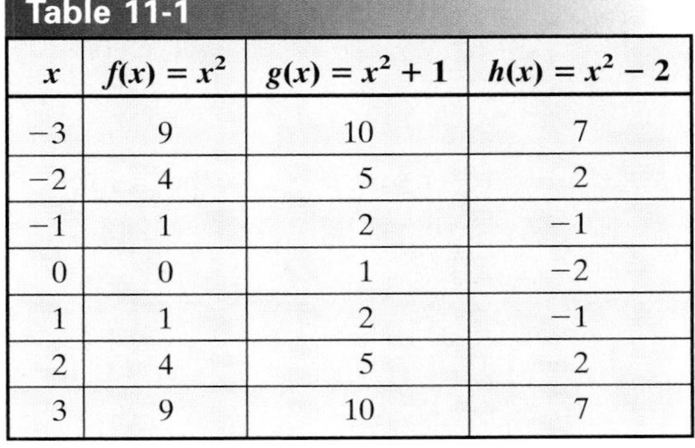

Table 11-1

x	$f(x) = x^2$	$g(x) = x^2 + 1$	$h(x) = x^2 - 2$
−3	9	10	7
−2	4	5	2
−1	1	2	−1
0	0	1	−2
1	1	2	−1
2	4	5	2
3	9	10	7

Notice that the graphs of $g(x) = x^2 + 1$ and $h(x) = x^2 - 2$ have the same shape as $f(x) = x^2$. However, the y-values of g are 1 more than the y-values of f. Hence the graph of $g(x) = x^2 + 1$ is the same as the graph of $f(x) = x^2$ shifted *up* 1 unit. Likewise the y-values of h are 2 less than those of f. The graph of $h(x) = x^2 - 2$ is the same as the graph of $f(x) = x^2$ shifted *down* 2 units.

The functions in Example 1 illustrate the following properties of quadratic functions of the form $f(x) = x^2 + k$.

Graphs of $f(x) = x^2 + k$

If $k > 0$, then the graph of $f(x) = x^2 + k$ is the same as the graph of $y = x^2$ shifted *up* k units.

If $k < 0$, then the graph of $f(x) = x^2 + k$ is the same as the graph of $y = x^2$ shifted *down* $|k|$ units.

Calculator Connections

Try experimenting with a graphing calculator by graphing functions of the form $y = x^2 + k$ for several values of k.

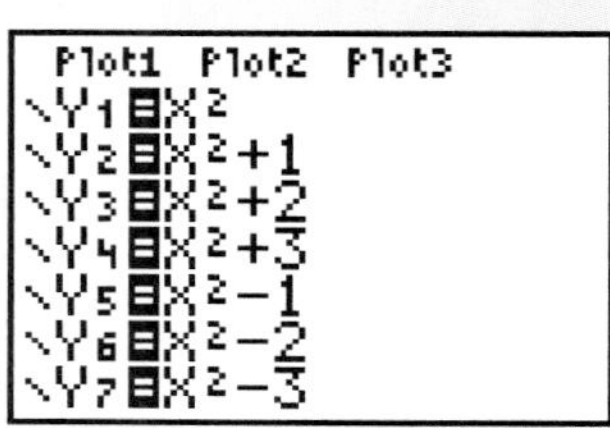

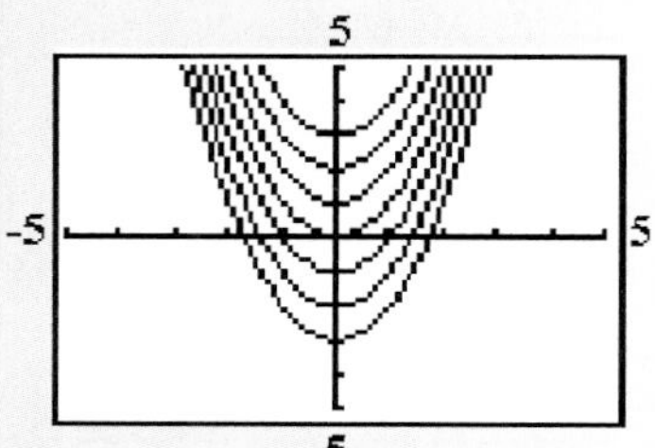

example 2 Graphing Quadratic Functions of the Form $f(x) = x^2 + k$

Sketch the functions defined by.

a. $m(x) = x^2 - 4$

b. $n(x) = x^2 + \frac{7}{2}$

Solution:

a. $m(x) = x^2 - 4$

$m(x) = x^2 + (-4)$

Because $k = -4$, the graph is obtained by shifting the graph of $y = x^2$ down $|-4|$ units (Figure 11-8).

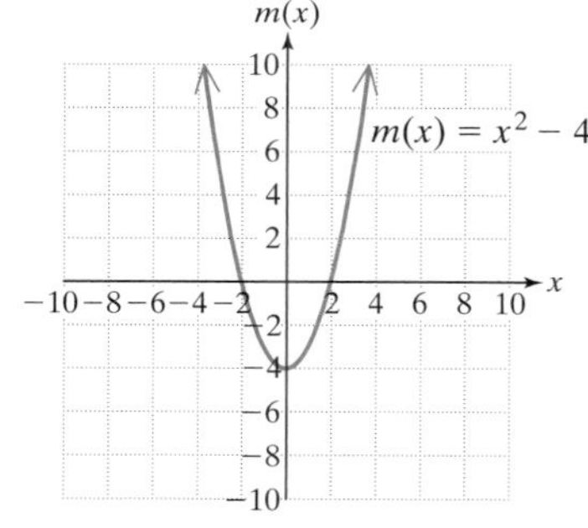

Figure 11-8

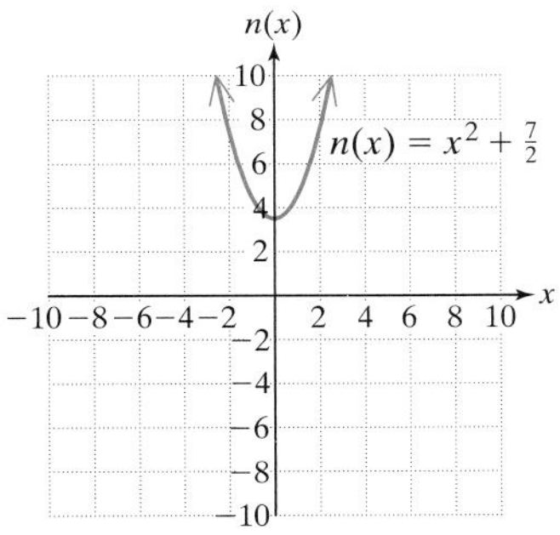

Figure 11-9

b. $n(x) = x^2 + \dfrac{7}{2}$

Because $k = \frac{7}{2}$, the graph is obtained by shifting the graph of $y = x^2$ up $\frac{7}{2}$ units (Figure 11-9).

2. Quadratic Functions of the Form $f(x) = (x - h)^2$

The graph of $f(x) = x^2 + k$ represents a vertical shift (up or down) of the function $y = x^2$. Example 3 shows that functions of the form $f(x) = (x - h)^2$ represent a horizontal shift (left or right) of the function $y = x^2$.

example 3 **Graphing Quadratic Functions of the Form $f(x) = (x - h)^2$**

Graph the functions f, g, and h on the same coordinate system.

$$f(x) = x^2 \qquad g(x) = (x + 1)^2 \qquad h(x) = (x - 2)^2$$

Solution:

Several function values for f, g, and h are shown in Table 11-2 for selected values of x. The corresponding graphs are pictured in Figure 11-10.

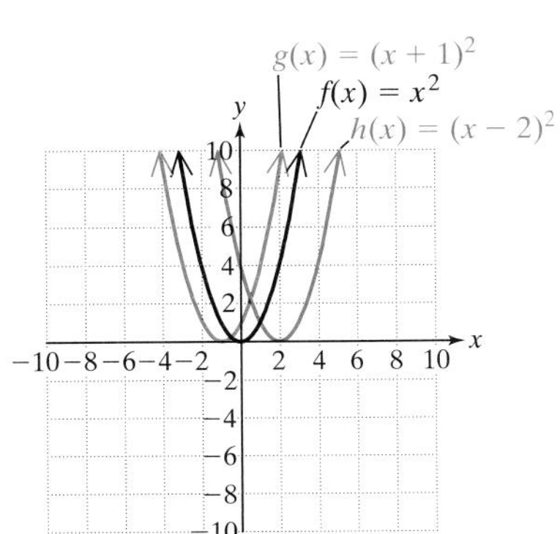

Figure 11-10

Table 11-2

x	$f(x) = x^2$	$g(x) = (x + 1)^2$	$h(x) = (x - 2)^2$
−4	16	9	36
−3	9	4	25
−2	4	1	16
−1	1	0	9
0	0	1	4
1	1	4	1
2	4	9	0
3	9	16	1
4	16	25	4
5	25	36	9

Example 3 illustrates the following properties of quadratic functions of the form $f(x) = (x - h)^2$.

Graphs of $f(x) = (x - h)^2$

If $h > 0$, then the graph of $f(x) = (x - h)^2$ is the same as the graph of $y = x^2$ shifted h units to the *right*.

If $h < 0$, then the graph of $f(x) = (x - h)^2$ is the same as the graph of $y = x^2$ shifted $|h|$ units to the *left*.

From Example 3 we have

$$h(x) = (x - 2)^2 \qquad \text{and} \qquad g(x) = [x - (-1)]^2$$

$y = x^2$ shifted 2 units to the right

$y = x^2$ shifted $|-1|$ unit to the left

example 4

Graphing Functions of the Form $f(x) = (x - h)^2$

Sketch the functions p and q.

a. $p(x) = (x - 7)^2$

b. $q(x) = (x + 1.6)^2$

Solution:

a. $p(x) = (x - 7)^2$

Because $h = 7 > 0$, shift the graph of $y = x^2$ to the *right* 7 units (Figure 11-11).

b. $q(x) = (x + 1.6)^2$

$q(x) = [x - (-1.6)]^2$

Because $h = -1.6 < 0$, shift the graph of $y = x^2$ to the *left* 1.6 units (Figure 11-12).

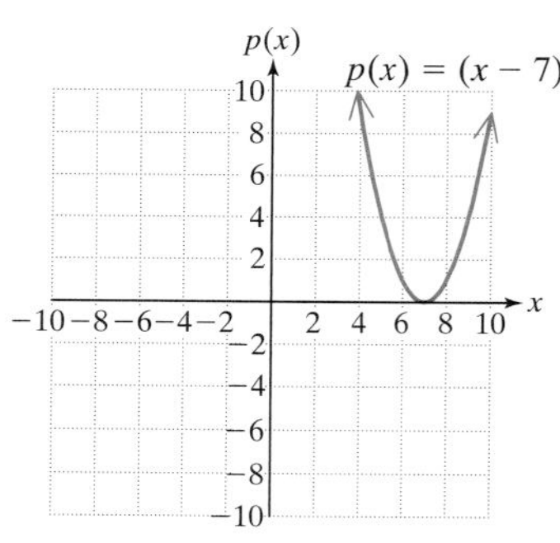

Figure 11-11

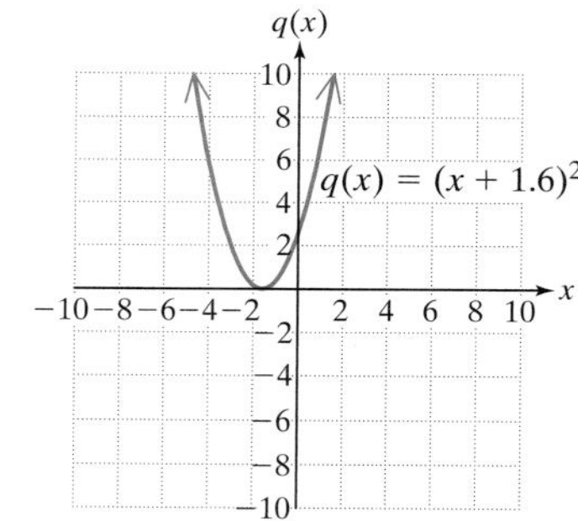

Figure 11-12

3. Quadratic Functions of the Form $f(x) = ax^2 \quad (a \neq 0)$

Examples 5 and 6 investigate functions of the form $f(x) = ax^2 \quad (a \neq 0)$.

example 5

Graphing Functions of the Form $f(x) = ax^2 \quad (a \neq 0)$

Graph the functions f, g, and h on the same coordinate system.

$$f(x) = x^2 \qquad g(x) = 2x^2 \qquad h(x) = \frac{1}{2}x^2$$

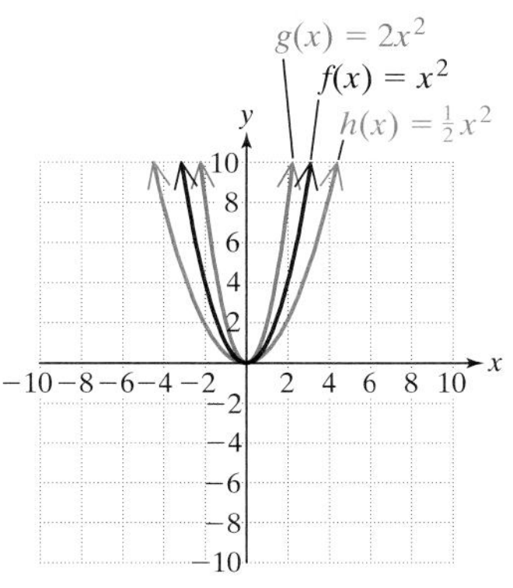

Figure 11-13

Solution:

Several function values for f, g, and h are shown in Table 11-3 for selected values of x. The corresponding graphs are pictured in Figure 11-13.

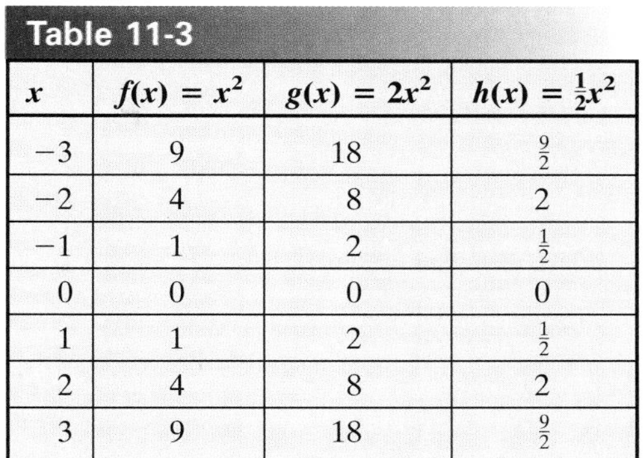

Table 11-3

x	$f(x) = x^2$	$g(x) = 2x^2$	$h(x) = \frac{1}{2}x^2$
−3	9	18	$\frac{9}{2}$
−2	4	8	2
−1	1	2	$\frac{1}{2}$
0	0	0	0
1	1	2	$\frac{1}{2}$
2	4	8	2
3	9	18	$\frac{9}{2}$

In Example 5, the function values defined by $g(x) = 2x^2$ are twice those of $f(x) = x^2$. The graph of $g(x) = 2x^2$ is the same as is the graph of $f(x) = x^2$ *stretched vertically* by a factor of 2 [the graph appears narrower than $f(x) = x^2$].

In Example 5, the function values defined by $h(x) = \frac{1}{2}x^2$ are one half those of $f(x) = x^2$. The graph of $h(x) = \frac{1}{2}x^2$ is the same as the graph of $f(x) = x^2$ *shrunk vertically* by a factor of $\frac{1}{2}$ [the graph appears wider than $f(x) = x^2$].

example 6 Graphing Functions of the Form $f(x) = ax^2 \quad (a \neq 0)$

Graph the functions f, g, and h on the same coordinate system.

$$f(x) = -x^2 \qquad g(x) = -3x^2 \qquad h(x) = -\frac{1}{3}x^2$$

Solution:

Several function values for f, g, and h are shown in Table 11-4 for selected values of x. The corresponding graphs are pictured in Figure 11-14.

Table 11-4

x	$f(x) = -x^2$	$g(x) = -3x^2$	$h(x) = -\frac{1}{3}x^2$
−3	−9	−27	−3
−2	−4	−12	$-\frac{4}{3}$
−1	−1	−3	$-\frac{1}{3}$
0	0	0	0
1	−1	−3	$-\frac{1}{3}$
2	−4	−12	$-\frac{4}{3}$
3	−9	−27	−3

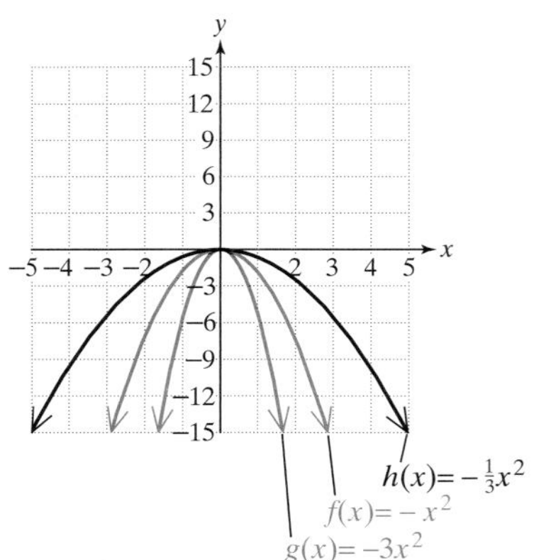

Figure 11-14

Example 6 illustrates that if the coefficient of the square term is negative, the parabola opens down. The graph of $g(x) = -3x^2$ is the same as the graph of $f(x) = -x^2$ with a *vertical stretch* by a factor of $|-3|$. The graph of $h(x) = -\frac{1}{3}x^2$ is the same as the graph of $f(x) = -x^2$ with a *vertical shrink* by a factor of $|-\frac{1}{3}|$.

Graphs of $f(x) = ax^2 \quad (a \neq 0)$

1. If $a > 0$, the parabola opens up. Furthermore,
   - If $0 < a < 1$, then the graph of $f(x) = ax^2$ is the same as the graph of $y = x^2$ with a *vertical shrink* by a factor of a.
   - If $a > 1$, then the graph of $f(x) = ax^2$ is the same as the graph of $y = x^2$ with a *vertical stretch* by a factor of a.
2. If $a < 0$, the parabola opens down. Furthermore,
   - If $0 < |a| < 1$, then the graph of $f(x) = ax^2$ is the same as the graph of $y = -x^2$ with a *vertical shrink* by a factor of $|a|$.
   - If $|a| > 1$, then the graph of $f(x) = ax^2$ is the same as the graph of $y = -x^2$ with a *vertical stretch* by a factor of $|a|$.

4. Quadratic Functions of the Form $f(x) = a(x - h)^2 + k$ $(a \neq 0)$

We can summarize our findings from Examples 1–6 by graphing functions of the form $f(x) = a(x - h)^2 + k \quad (a \neq 0)$.

The graph of $y = x^2$ has its vertex at the origin, $(0, 0)$. The graph of $f(x) = a(x - h)^2 + k$ is the same as the graph of $y = x^2$ shifted to the right or left h units and shifted up or down k units. Therefore, the vertex is shifted from $(0, 0)$ to (h, k). The axis of symmetry is the vertical line through the vertex. Hence the axis of symmetry must be the line $x = h$.

Graphs of $f(x) = a(x - h)^2 + k \quad (a \neq 0)$

1. The vertex is located at (h, k).
2. The axis of symmetry is the line $x = h$.
3. If $a > 0$, the parabola opens up, and k is the **minimum value** of the function.
4. If $a < 0$, the parabola opens down, and k is the **maximum value** of the function.

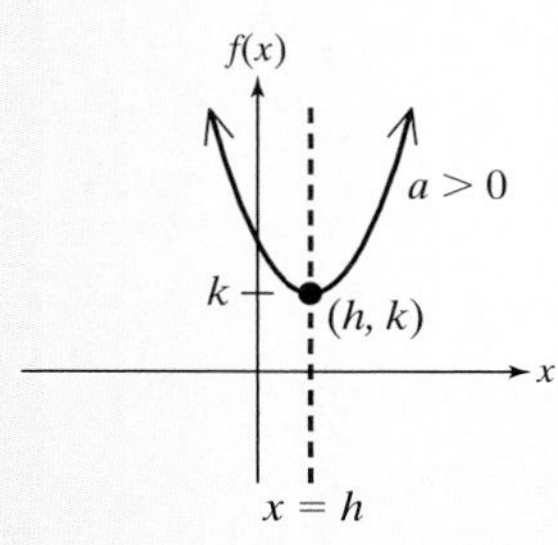

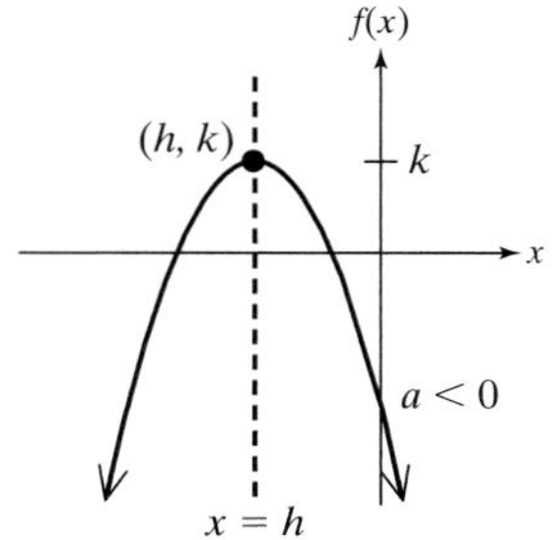

example 7

Graphing a Function of the Form $f(x) = a(x - h)^2 + k \quad (a \neq 0)$

Given the function defined by

$$f(x) = 2(x - 3)^2 + 4$$

a. Identify the vertex.
b. Sketch the function.
c. Identify the axis of symmetry.
d. Identify the maximum or minimum value of the function.

Solution:

a. $f(x) = 2(x - 3)^2 + 4$
The function is in the form $f(x) = a(x - h)^2 + k$, where $a = 2$, $h = 3$, and $k = 4$. Therefore, the vertex is at $(3, 4)$.

b. The graph of f is the same as the graph of $y = x^2$ shifted to the right 3 units, shifted up 4 units, and stretched vertically by a factor of 2 (Figure 11-15).

c. The axis of symmetry is the line $x = 3$.

d. Because $a > 0$, the function opens up. Therefore, the minimum function value is 4. Notice that the minimum value is the minimum y-value on the graph.

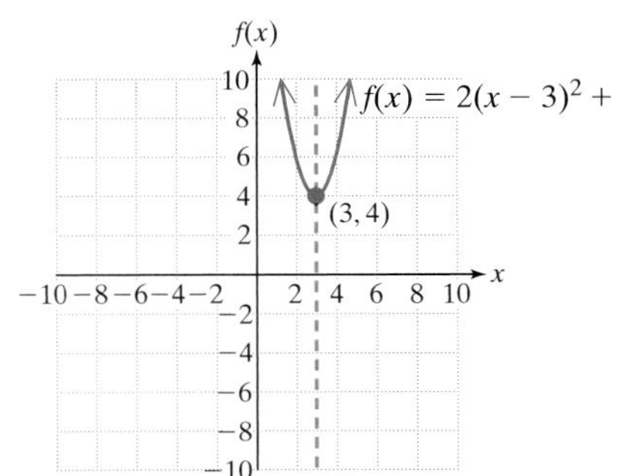

Figure 11-15

example 8 Graphing a Function of the Form $f(x) = a(x - h)^2 + k \quad (a \neq 0)$

Given the function defined by

$$g(x) = -(x + 2)^2 - \frac{7}{4}$$

a. Identify the vertex.
b. Sketch the function.
c. Identify the axis of symmetry.
d. Identify the maximum or minimum value of the function.

Solution:

a. $g(x) = -(x + 2)^2 - \dfrac{7}{4}$

$$= -1[x - (-2)]^2 + \left(-\frac{7}{4}\right)$$

The function is in the form $g(x) = a(x - h)^2 + k$, where $a = -1$, $h = -2$, and $k = -\frac{7}{4}$. Therefore, the vertex is at $(-2, -\frac{7}{4})$.

b. The graph of g is the same as the graph of $y = x^2$ shifted to the left 2 units, shifted down $\frac{7}{4}$ units, and opening down (Figure 11-16).

c. The axis of symmetry is the line $x = -2$.

d. The parabola opens down, so the maximum function value is $-\frac{7}{4}$.

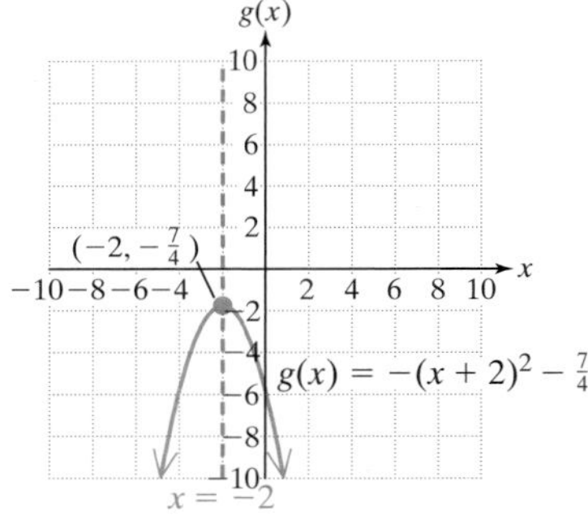

Figure 11-16

section 11.4 Practice Exercises

1. Describe the variation in the graphs of functions of the form $f(x) = x^2 + k$.

For Exercises 2–11, graph the functions.

2. $g(x) = x^2 + 1$
3. $f(x) = x^2 + 2$
4. $p(x) = x^2 - 3$
5. $q(x) = x^2 - 4$
6. $T(x) = x^2 + \dfrac{3}{4}$
7. $S(x) = x^2 + \dfrac{3}{2}$
8. $M(x) = x^2 - \dfrac{5}{4}$
9. $n(x) = x^2 - \dfrac{1}{3}$
10. $P(x) = x^2 + \dfrac{1}{2}$
11. $Q(x) = x^2 + \dfrac{1}{4}$

12. Describe the variation in the graphs of functions of the form $f(x) = (x - h)^2$.

For Exercises 13–22, graph the functions.

13. $r(x) = (x + 1)^2$
14. $h(x) = (x + 2)^2$
15. $k(x) = (x - 3)^2$
16. $L(x) = (x - 4)^2$
17. $A(x) = \left(x + \dfrac{3}{4}\right)^2$
18. $r(x) = \left(x + \dfrac{3}{2}\right)^2$
19. $W(x) = \left(x - \dfrac{5}{4}\right)^2$
20. $V(x) = \left(x - \dfrac{1}{3}\right)^2$
21. $M(x) = \left(x + \dfrac{1}{2}\right)^2$
22. $N(x) = \left(x + \dfrac{1}{4}\right)^2$

23. Describe the variation in the graphs of functions of the form $f(x) = ax^2$, where $a \neq 0$.

24. How do you determine whether the graph of a function defined by $h(x) = ax^2 + bx + c \quad (a \neq 0)$ opens up or down?

For Exercises 25–32, match the function with its graph.

25. $f(x) = -\frac{1}{4}x^2$

26. $g(x) = (x + 3)^2$

27. $k(x) = (x - 3)^2$

28. $h(x) = \frac{1}{4}x^2$

29. $t(x) = x^2 + 2$

30. $m(x) = x^2 - 4$

31. $n(x) = -(x - 2)^2 + 3$

32. $p(x) = (x + 1)^2 - 3$

a.

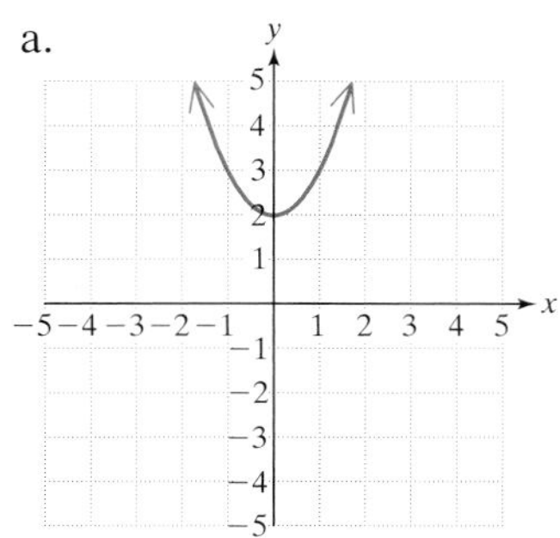

b.

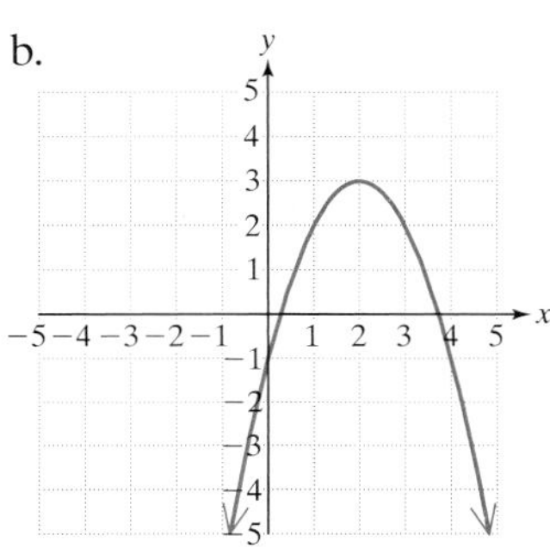

c.

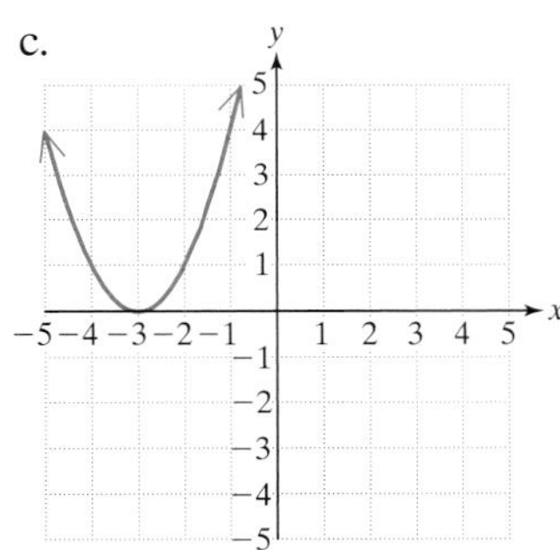

d.

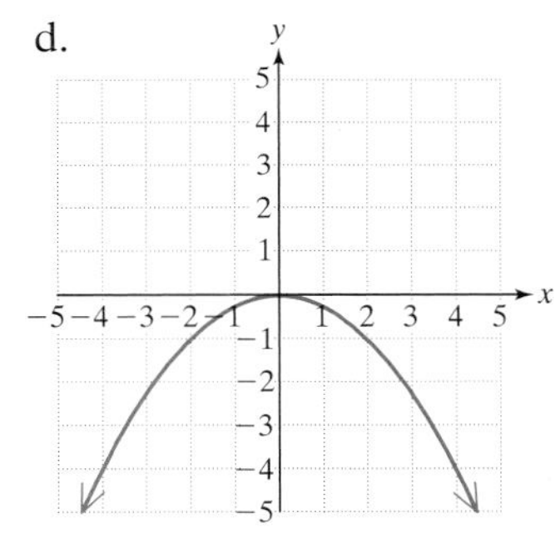

e.

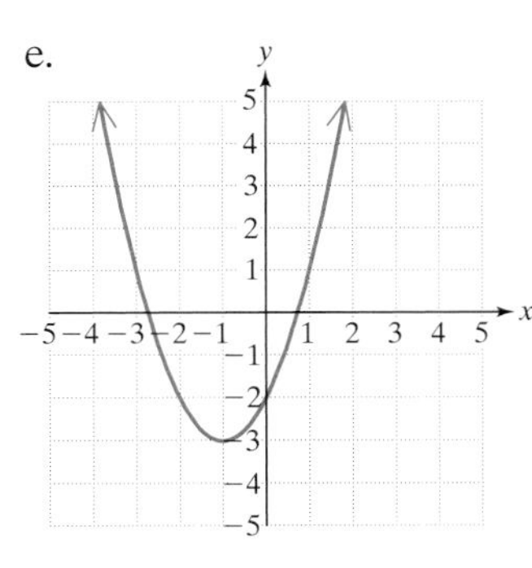

f.

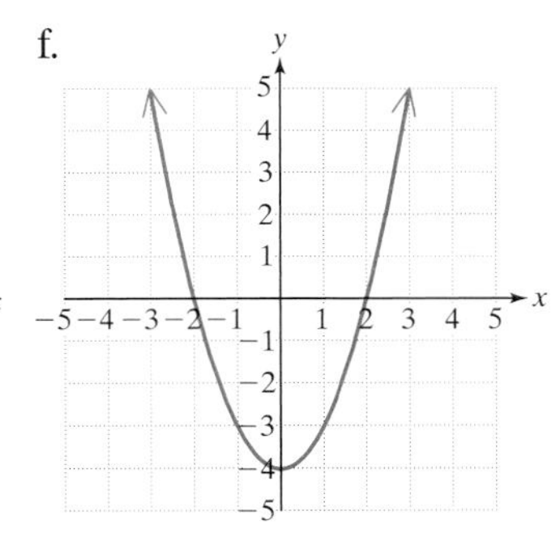

g. h.

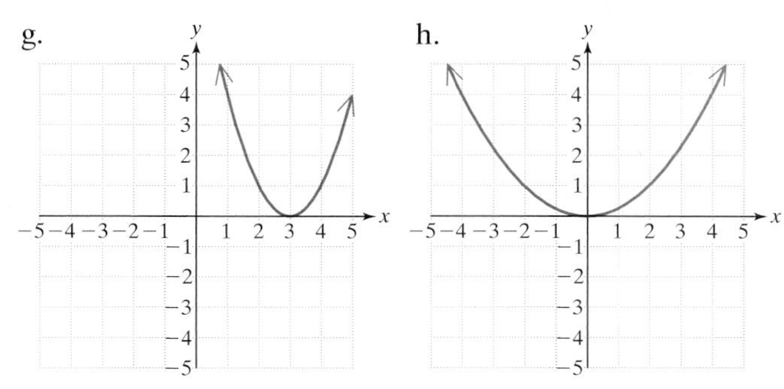

For Exercises 33–42, graph the parabola and the axis of symmetry. Label the vertex and the axis of symmetry.

33. $y = (x - 3)^2 + 2$

34. $y = (x - 2)^2 + 3$

35. $y = (x + 1)^2 - 3$

36. $y = (x + 3)^2 - 1$

37. $y = -(x - 4)^2 - 2$

38. $y = -(x - 2)^2 - 4$

39. $y = -(x + 3)^2 + 3$

40. $y = -(x + 2)^2 + 2$

41. $y = (x + 1)^2 + 1$

42. $y = (x - 4)^2 - 4$

For Exercises 43–52, without graphing the quadratic function, identify the vertex and determine if it is a maximum point or a minimum point. Then, write the maximum or minimum value.

43. $f(x) = 4(x - 6)^2 - 9$

44. $g(x) = 3(x - 4)^2 - 7$

45. $p(x) = -\frac{2}{5}(x - 2)^2 + 5$

46. $h(x) = -\frac{3}{7}(x - 5)^2 + 10$

47. $k(x) = \frac{1}{2}(x + 8)^2 - 3$

48. $m(x) = \frac{2}{9}(x + 11)^2 - 2$

49. $n(x) = -6\left(x + \frac{3}{4}\right)^2 + \frac{21}{4}$

50. $q(x) = -4\left(x + \frac{5}{6}\right)^2 + \frac{1}{6}$

51. $A(x) = 2(x - 7)^2 - \frac{3}{2}$

52. $B(x) = 5(x - 3)^2 - \frac{1}{4}$

53. True or False: The function defined by $g(x) = -5x^2$ has a maximum value but no minimum value.

54. True or False: The function defined by $f(x) = 2(x - 5)^2$ has a maximum value but no minimum value.

55. True or False: If the vertex $(-2, 8)$ represents a minimum point, then the minimum value is -2.

56. True or False: If the vertex $(-2, 8)$ represents a maximum point, then the maximum value is 8.

57. A suspension bridge is 120 ft long. Its supporting cable hangs in a shape that resembles a parabola. The function defined by $H(x) = \frac{1}{90}(x - 60)^2 + 30$ (where $0 \le x \le 120$) approximates the height of the supporting cable a distance of x feet from the end of the bridge (see figure).
    a. What is the location of the vertex of the parabolic cable?
    b. What is the minimum height of the cable?
    c. How high are the towers at either end of the supporting cable?

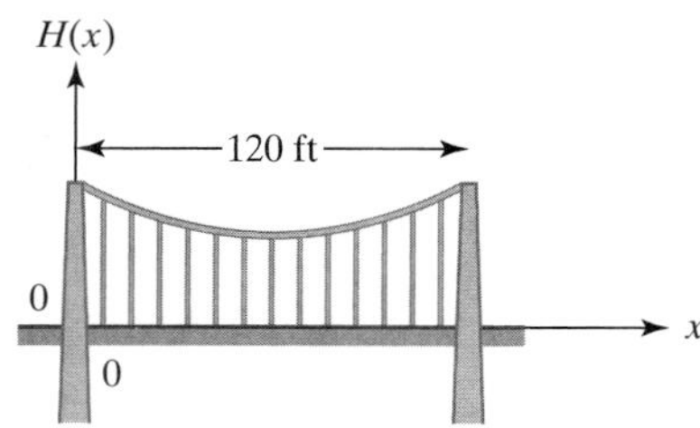

Figure for Exercise 57

58. A 50-m bridge over a crevasse is supported by a parabolic arch. The function defined by $f(x) = -0.16(x - 25)^2 + 100$ (where $0 \le x \le 50$) approximates the height of the supporting arch x meters from the end of the bridge (see figure).
    a. What is the location of the vertex of the arch?
    b. What is the maximum height of the arch?

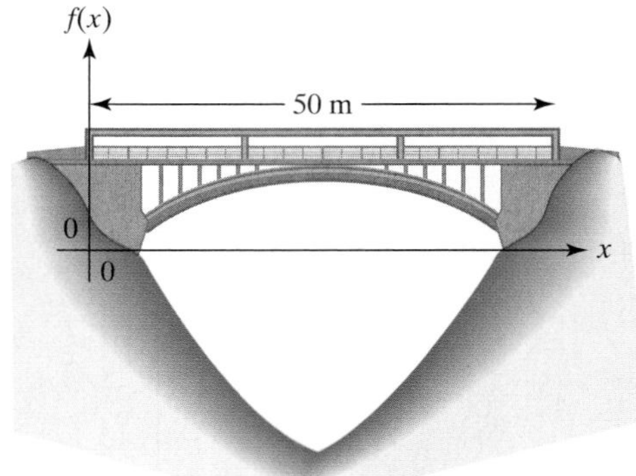

Figure for Exercise 58

Graphing Calculator Exercises

For Exercises 59–64, verify the maximum and minimum points found in Exercises 43–46, 49, and 50, by graphing each function on the calculator.

59. $Y_1 = 4(x - 6)^2 - 9$ (Exercise 43)

60. $Y_1 = 3(x - 4)^2 - 7$ (Exercise 44)

61. $Y_1 = -\frac{2}{5}(x - 2)^2 + 5$ (Exercise 45)

62. $Y_1 = -\frac{3}{7}(x - 5)^2 + 10$ (Exercise 46)

63. $Y_1 = -6\left(x + \frac{3}{4}\right)^2 + \frac{21}{4}$ (Exercise 49)

64. $Y_1 = -4\left(x + \frac{5}{6}\right)^2 + \frac{1}{6}$ (Exercise 50)

section

11.5 Applications of Quadratic Functions

Concepts

1. Writing a Quadratic Function in the Form $f(x) = a(x - h)^2 + k$
2. Vertex Formula
3. Determining the Vertex and Intercepts of a Quadratic Function
4. Vertex of a Parabola—Applications

1. Writing a Quadratic Function in the Form $f(x) = a(x - h)^2 + k$

The graph of a quadratic function is a parabola, and if the function is written in the form $f(x) = a(x - h)^2 + k \quad (a \neq 0)$, then the vertex is at (h, k). A quadratic function can be written in the form $f(x) = a(x - h)^2 + k \quad (a \neq 0)$ by completing the square. The process is similar to the steps outlined in Section 11.1 except that all algebraic manipulation is performed on the right-hand side of the function.

example 1

Writing a Quadratic Function in the Form $f(x) = a(x - h)^2 + k \quad (a \neq 0)$

Given $f(x) = x^2 + 8x + 13$

a. Write the function in the form $f(x) = a(x - h)^2 + k$.
b. Identify the vertex, the axis of symmetry, and the minimum function value.

Solution:

a. $f(x) = x^2 + 8x + 13$

$= 1(x^2 + 8x) + 13$ — Rather than dividing by the leading coefficient on both sides, we will factor out the leading coefficient from the variable terms on the right-hand side.

$= 1(x^2 + 8x \qquad) + 13$ — Next, complete the square on the expression within the parentheses: $[\frac{1}{2}(8)]^2 = 16$.

$= 1(x^2 + 8x + 16 - 16) + 13$ — Rather than adding 16 to both sides of the function, we will *add and subtract 16* within the parentheses on the right-hand side. This has the effect of adding 0 to the right-hand side.

$= 1(x^2 + 8x + 16) - 16 + 13$ — Use the associative property of addition to regroup terms and isolate the perfect square trinomial within the parentheses.

$= (x + 4)^2 - 3$ — Factor and simplify.

b. $f(x) = (x + 4)^2 - 3$

The vertex is at $(-4, -3)$.

The axis of symmetry is $x = -4$.

Because $a > 0$, the parabola opens up.

The minimum value is -3 (Figure 11-17).

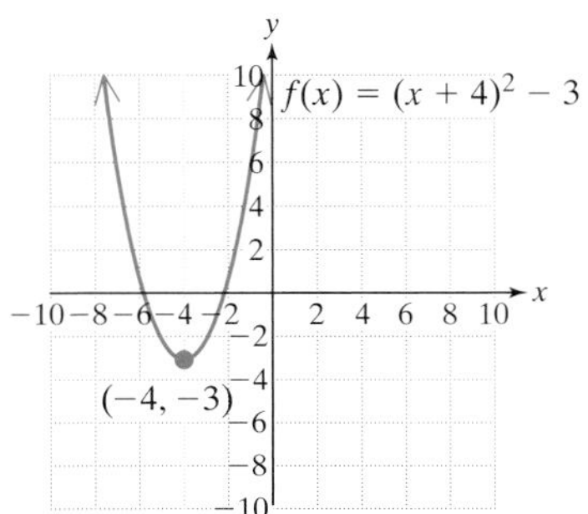

Figure 11-17

example 2 Analyzing a Quadratic Function

Given $f(x) = -2x^2 + 12x - 16$

a. Write the function in the form $f(x) = a(x - h)^2 + k$.
b. Find the vertex, axis of symmetry, and maximum function value.
c. Find the x- and y-intercepts.
d. Sketch the function.

Solution:

a. $f(x) = -2x^2 + 12x - 16$ — To find the vertex, write the function in the form $f(x) = a(x - h)^2 + k$.

Avoiding Mistakes

Do not factor out the leading coefficient from the constant term.

$= -2(x^2 - 6x \quad) - 16$ — If the leading coefficient is not 1, factor the coefficient from the variable terms.

$= -2(x^2 - 6x + 9 - 9) - 16$ — Add and subtract the quantity $[\frac{1}{2}(-6)]^2 = 9$ within the parentheses.

$= -2(x^2 - 6x + 9) + (-2)(-9) - 16$ — To remove the term -9 from the parentheses, we must first apply the distributive property. When -9 is removed from the parentheses, it carries with it a factor of -2.

$= -2(x - 3)^2 + 18 - 16$ — Factor and simplify.

$= -2(x - 3)^2 + 2$

b. $f(x) = -2(x - 3)^2 + 2$
The vertex is $(3, 2)$.

The line of symmetry is $x = 3$. Because $a < 0$, the parabola opens down and the maximum value is 2.

c. The y-intercept is given by $f(0) = -2(0)^2 + 12(0) - 16 = -16$.
The y-intercept is $(0, -16)$.

To find the x-intercept(s), find the real solutions to the equation $f(x) = 0$.

$f(x) = -2x^2 + 12x - 16$

$0 = -2x^2 + 12x - 16$ — Substitute $f(x) = 0$.

$0 = -2(x^2 - 6x + 8)$ — Factor.

$0 = -2(x - 4)(x - 2)$

$x = 4 \quad \text{or} \quad x = 2$

The x-intercepts are $(4, 0)$ and $(2, 0)$.

d. Using the information from parts (a)–(c), sketch the graph (Figure 11-18).

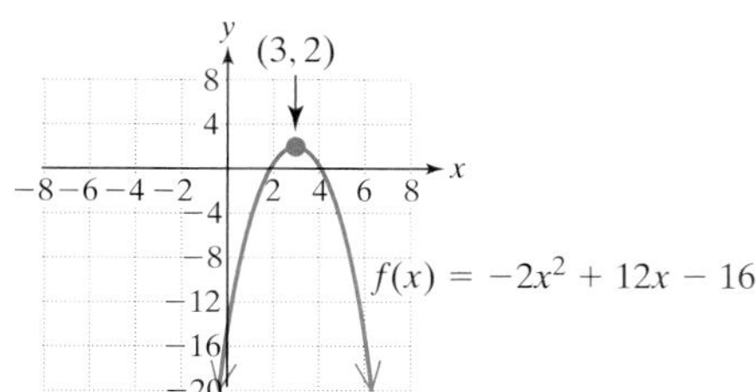

Figure 11-18

2. Vertex Formula

Completing the square and writing a quadratic function in the form $f(x) = a(x - h)^2 + k \quad (a \neq 0)$ is one method to find the vertex of a parabola. Another method is to use the vertex formula. The **vertex formula** can be derived by completing the square on the function defined by $f(x) = ax^2 + bx + c \quad (a \neq 0)$.

$$f(x) = ax^2 + bx + c \quad (a \neq 0)$$

$$= a\left(x^2 + \frac{b}{a}x \qquad\qquad \right) + c$$ Factor a from the variable terms.

$$= a\left(x^2 + \frac{b}{a}x + \frac{b^2}{4a^2} - \frac{b^2}{4a^2}\right) + c$$ Add and subtract $[\frac{1}{2}(b/a)]^2 = b^2/4a^2$ within the parentheses.

$$= a\left(x^2 + \frac{b}{a}x + \frac{b^2}{4a^2}\right) + (a)\left(-\frac{b^2}{4a^2}\right) + c$$ Apply the distributive property and remove the term $-b^2/4a^2$ from the parentheses.

$$= a\left(x + \frac{b}{2a}\right)^2 - \frac{b^2}{4a} + c$$ Factor the trinomial and simplify.

$$= a\left(x + \frac{b}{2a}\right)^2 + c - \frac{b^2}{4a}$$ Apply the commutative property of addition to reverse the last two terms.

$$= a\left(x + \frac{b}{2a}\right)^2 + \frac{4ac}{4a} - \frac{b^2}{4a}$$ Obtain a common denominator.

$$= a\left(x + \frac{b}{2a}\right)^2 + \frac{4ac - b^2}{4a}$$

$$= a\left[x - \left(-\frac{b}{2a}\right)\right]^2 + \frac{4ac - b^2}{4a}$$

$$f(x) = a(x \;-\; h)^2 \;+\; k$$

The function is in the form $f(x) = a(x - h)^2 + k$, where

$$h = \frac{-b}{2a} \quad \text{and} \quad k = \frac{4ac - b^2}{4a}$$

Hence, the vertex is at

$$\left(\frac{-b}{2a}, \frac{4ac - b^2}{4a}\right)$$

Although the y-coordinate of the vertex is given as $(4ac - b^2)/4a$, it is usually easier to determine the x-coordinate of the vertex first and then find y by evaluating the function at $x = -b/2a$.

The Vertex Formula

For $f(x) = ax^2 + bx + c \quad (a \neq 0)$, the vertex is given by

$$\left(\frac{-b}{2a}, \frac{4ac - b^2}{4a}\right) \quad \text{or} \quad \left(\frac{-b}{2a}, f\left(-\frac{b}{2a}\right)\right)$$

3. Determining the Vertex and Intercepts of a Quadratic Function

example 3 **Determining the Vertex and Intercepts of a Quadratic Function**

Given $h(x) = x^2 - 2x + 5$

a. Use the vertex formula to find the vertex.
b. Find the x- and y-intercepts.
c. Sketch the function.

Solution:

a. $h(x) = x^2 - 2x + 5$

$a = 1, b = -2, c = 5$ Identify a, b, and c.

The x-coordinate of the vertex is $\dfrac{-b}{2a} = \dfrac{-(-2)}{2(1)} = 1$.

The y-coordinate of the vertex is $h(1) = (1)^2 - 2(1) + 5 = 4$.

The vertex is $(1, 4)$.

b. The y-intercept is given by $h(0) = (0)^2 - 2(0) + 5 = 5$.

The y-intercept is $(0, 5)$.

To find the x-intercept(s), find the real solutions to the equation $h(x) = 0$.

$$h(x) = x^2 - 2x + 5$$

$$0 = x^2 - 2x + 5$$ This quadratic equation is not factorable. Apply the quadratic formula: $a = 1, b = -2, c = 5$.

$$x = \frac{-(-2) \pm \sqrt{(-2)^2 - 4(1)(5)}}{2(1)}$$

$$= \frac{2 \pm \sqrt{4 - 20}}{2(1)}$$

$$= \frac{2 \pm \sqrt{-16}}{2}$$

$$= \frac{2 \pm 4i}{2}$$

$$= 1 \pm 2i$$

The solutions to the equation $h(x) = 0$ are not real numbers. Therefore, there are no x-intercepts.

c.

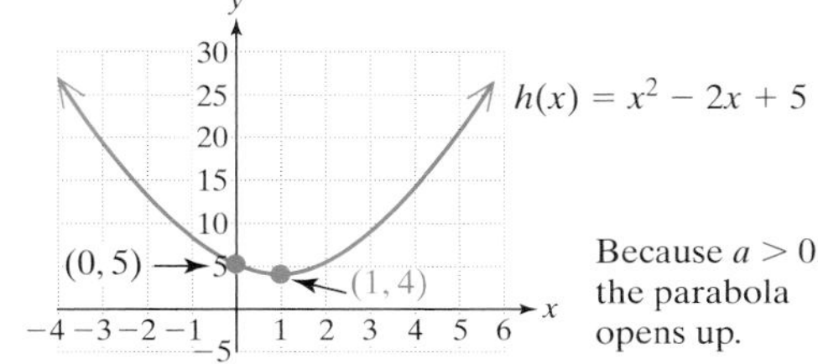

Because $a > 0$, the parabola opens up.

Figure 11-19

Tip: Recall from Section 11.2 that the discriminant for the quadratic equation $ax^2 + bx + c = 0$ can be used to determine the number of x-intercepts of the function $f(x) = ax^2 + bx + c$. The location of the vertex and the direction that a parabola opens can also be used to determine the number of x-intercepts.

For example, given $h(x) = x^2 - 2x + 5$, the vertex $(1, 4)$ is *above* the x-axis. Furthermore, because $a > 0$, the parabola opens upward. Therefore, it is not possible for the function h to cross the x-axis (Figure 11-19).

4. Vertex of a Parabola—Applications

example 4 **Analyzing a Quadratic Function**

The crew from Extravaganza Entertainment launches fireworks at an angle of 60° from the horizontal. The height, h, of one particular type of display can be approximated by the following function:

$$h(t) = -16t^2 + 128\sqrt{3}t,$$ where h is measured in feet and t is measured in seconds.

a. How long will it take the fireworks to reach their maximum height? Round to the nearest second.
b. Find the maximum height. Round to the nearest foot.

Solution:

$h(t) = -16t^2 + 128\sqrt{3}t$ — This parabola opens downward; therefore, the maximum height of the fireworks will occur at the vertex of the parabola.

$a = -16, b = 128\sqrt{3}, c = 0$ — Identify a, b, and c, and apply the vertex formula.

The x-coordinate of the vertex is

$$\frac{-b}{2a} = \frac{-128\sqrt{3}}{2(-16)} = \frac{-128\sqrt{3}}{-32} \approx 6.9$$

The y-coordinate of the vertex is approximately

$$h(6.9) = -16(6.9)^2 + 128\sqrt{3}(6.9) = 768$$

The vertex is (6.9, 768).

a. The fireworks will reach their maximum height in 6.9 sec.
b. The maximum height is 768 ft.

Calculator Connections

Some graphing calculators have *Minimum* and *Maximum* features that enable the user to approximate the minimum and maximum values of a function. Otherwise, *Zoom* and *Trace* can be used.

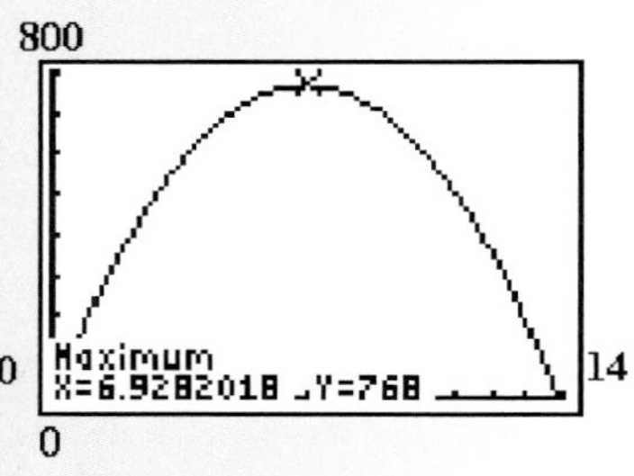

example 5

Analyzing a Quadratic Function

In business, profit is defined as the difference of total revenue and total cost. That is,

$P(x) = R(x) - C(x)$, where x represents the number of units sold.

Suppose a concession stand at the Daytona International Speedway sells a hot dog/drink combo for \$4. Then the revenue function is defined as $R(x) = 4x$. It is also known that the cost to operate the stand is $C(x) = 0.001x^2 + 1.4x + 400$.

a. Find the profit function, P, in terms of x.
b. Find the x-intercepts of the profit function (to the nearest whole unit) and interpret the meaning of the x-intercepts in the context of this problem.

c. Find the y-intercept of the profit function and interpret its meaning in the context of this problem.
d. Find the vertex of the profit function and interpret its meaning in the context of this problem.
e. Sketch the profit function.

Solution:

a.
$$\begin{aligned} P(x) &= R(x) - C(x) \\ &= 4x - (0.001x^2 + 1.4x + 400) \\ &= 4x - 0.001x^2 - 1.4x - 400 \\ &= -0.001x^2 + 2.6x - 400 \end{aligned}$$

b. $P(x) = -0.001x^2 + 2.6x - 400$

The x-intercepts are the real solutions of the equation $P(x) = 0$.

$$0 = -0.001x^2 + 2.6x - 400$$

$$x = \frac{-(2.6) \pm \sqrt{(2.6)^2 - 4(-0.001)(-400)}}{2(-0.001)}$$

Use the quadratic formula.

$$= \frac{-2.6 \pm \sqrt{6.76 - 1.6}}{-0.002}$$

$$= \frac{-2.6 \pm \sqrt{5.16}}{-0.002}$$

$$x = \frac{-2.6 + \sqrt{5.16}}{-0.002} \approx 164 \quad \text{or} \quad x = \frac{-2.6 - \sqrt{5.16}}{-0.002} \approx 2436$$

The x-intercepts are (164, 0) and (2436, 0). The x-intercepts represent the break-even points where profit is zero. At these points, the revenue and cost are the same. That is, they are in equilibrium.

c. The y-intercept is $P(0) = -0.001(0)^2 + 2.6(0) - 400 = -400$.

The y-intercept is at the point $(0, -400)$. The y-intercept indicates that if no hot dog/drink combos are sold, the vendor has a \$400 loss.

d. The x-coordinate of the vertex is

$$\frac{-b}{2a} = \frac{-2.6}{2(-0.001)} = 1300$$

The y-coordinate of the vertex is $P(1300) = -0.001(1300)^2 + 2.6(1300) - 400 = 1290$.

The vertex is (1300, 1290).

A maximum profit of \$1290 is obtained when the vendor sells 1300 hot dog/drink combos.

e. Using the information from parts (a)–(d), sketch the profit function (Figure 11-20).

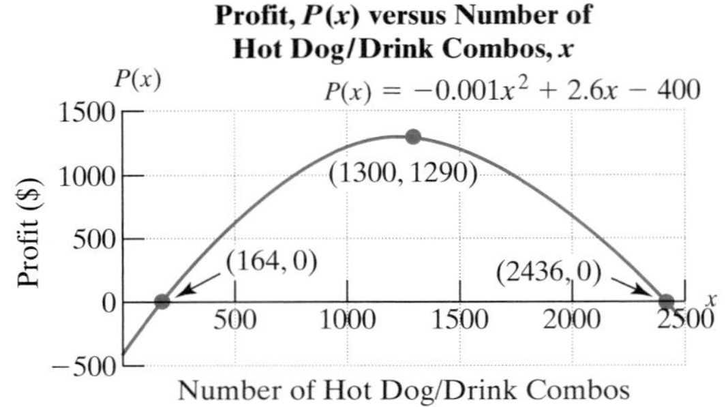

Figure 11-20

section 11.5 Practice Exercises

1. How does the graph of $f(x) = -2x^2$ compare with the graph of $y = x^2$?

2. How does the graph of $p(x) = \frac{1}{4}x^2$ compare with the graph of $y = x^2$?

3. How does the graph of $Q(x) = x^2 - \frac{8}{3}$ compare with the graph of $y = x^2$?

4. How does the graph of $r(x) = x^2 + 7$ compare with the graph of $y = x^2$?

5. How does the graph of $s(x) = (x - 4)^2$ compare with the graph of $y = x^2$?

6. How does the graph of $t(x) = (x + 10)^2$ compare with the graph of $y = x^2$?

For Exercises 7–14, find the value of k to complete the square.

7. $x^2 - 8x + k$
8. $x^2 + 4x + k$
9. $y^2 + 7y + k$
10. $a^2 - a + k$
11. $b^2 + \frac{2}{9}b + k$
12. $m^2 - \frac{2}{7}m + k$
13. $t^2 - \frac{1}{3}t + k$
14. $p^2 + \frac{1}{4}p + k$

For Exercises 15–24, write the function in the form $f(x) = a(x - h)^2 + k$ by completing the square. Then identify the vertex.

15. $g(x) = x^2 - 8x + 5$
16. $h(x) = x^2 + 4x + 5$
17. $n(x) = 2x^2 + 12x + 13$
18. $f(x) = 4x^2 + 16x + 19$
19. $p(x) = -3x^2 + 6x - 5$
20. $q(x) = -2x^2 + 12x - 11$
21. $k(x) = x^2 + 7x - 10$
22. $m(x) = x^2 - x - 8$
23. $f(x) = x^2 + 8x - 1$
24. $g(x) = x^2 + 5x - 2$

For Exercises 25–34, find the vertex by using the vertex formula.

25. $Q(x) = x^2 - 4x + 7$
26. $T(x) = x^2 - 8x + 17$
27. $r(x) = -3x^2 - 6x - 5$
28. $s(x) = -2x^2 - 12x - 19$
29. $N(x) = x^2 + 8x + 1$
30. $M(x) = x^2 + 6x - 5$
31. $m(x) = \frac{1}{2}x^2 + x + \frac{5}{2}$
32. $n(x) = \frac{1}{2}x^2 + 2x + 3$
33. $k(x) = -x^2 + 2x + 2$
34. $h(x) = -x^2 + 4x - 3$

For Exercises 35–40,

a. Find the vertex.
b. Find the y-intercept.
c. Find the x-intercept(s), if they exist.
d. Use this information to graph the function.

35. $y = x^2 + 9x + 8$

36. $y = x^2 + 7x + 10$

37. $y = 2x^2 - 2x + 4$

38. $y = 2x^2 - 12x + 19$

39. $y = -x^2 + 3x - \frac{9}{4}$

40. $y = -x^2 - \frac{3}{2}x - \frac{9}{16}$

41. The pressure, x, in an automobile tire can affect its wear. Both overinflated and underinflated tires can lead to poor performance and poor mileage. For one particular tire, the function, P, represents the number of miles that a tire lasts (in thousands) for a given pressure, x.

$P(x) = -0.857x^2 + 56.1x - 880$, where x is the tire pressure in pounds per square inch (psi).

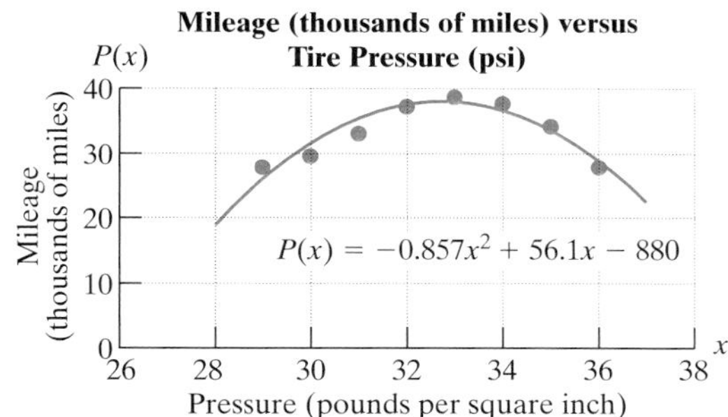

Figure for Exercise 41

a. Use the function to approximate the number of miles a set of tires will get for a pressure of 28 psi. Round to one decimal place.

b. Find the tire pressure that will yield the maximum mileage. Round to the nearest pound per square inch.

42. A baseball player tosses a ball and the height of the ball (in feet) can be approximated by $y(x) = -0.011x^2 + 0.577x + 5$, where x is the horizontal position of the ball measured in feet from the origin.

a. What is the height of the ball when its horizontal position is 20 ft from the origin?

b. At what horizontal position will the ball reach its highest point? Round to the nearest foot.

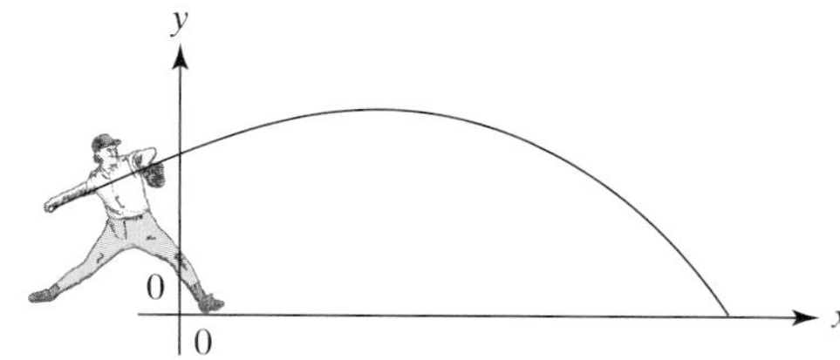

Figure for Exercise 42

43. For a fund-raising activity, a charitable organization produces cookbooks to sell in the community. The profit, P (in dollars), depends on the number of cookbooks produced, x, according to

$$P(x) = -\frac{1}{50}x^2 + 12x - 550, \text{ where } x \geq 0$$

a. How much profit is made when 100 cookbooks are produced?

b. How much profit is made when 150 cookbooks are produced?

c. How many cookbooks must be produced for the organization to break even?

d. Find the vertex.

e. Sketch the function.

f. How many cookbooks must be produced to maximize profit? What is the maximum profit?

44. A jewelry maker sells bracelets at art shows. The profit, P (in dollars), depends on the number of bracelets produced, x, according to

$$P(x) = -\frac{1}{10}x^2 + 42x - 1260, \text{ where } x \geq 0$$

a. How much profit does the jeweler make when 10 bracelets are produced?

b. How much profit does the jeweler make when 50 bracelets are produced?

c. How many bracelets must be produced for the jeweler to break even? Round to the nearest whole unit.

d. Find the vertex.

e. Sketch the function.

f. How many bracelets must be produced to maximize profit? What is the maximum profit?

Expanding Your Skills

45. A farmer wants to fence a rectangular corral adjacent to the side of a barn; however, she has only 200 ft of fencing and wants to enclose the largest possible area. See figure.

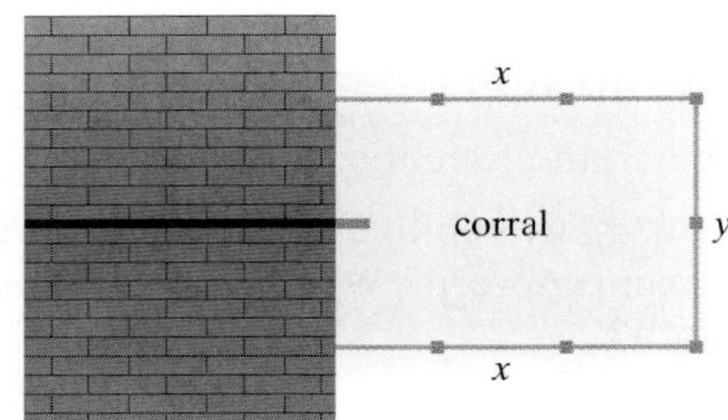

Figure for Exercise 45

a. If x represents the length of the corral, and y represents the width, explain why the dimensions of the corral are subject to the constraint: $2x + y = 200$.

b. The area of the corral is given by $A = xy$. Use the constraint equation from part (a) to express A as a function in terms of x, where $0 < x < 100$.

c. Use the function from part (b) to find the dimensions of the corral that will yield the maximum area. [*Hint*: Find the vertex of the function from part (b).]

46. A veterinarian wants to construct two equal-sized pens of maximum area out of 240 ft of fencing. See figure.

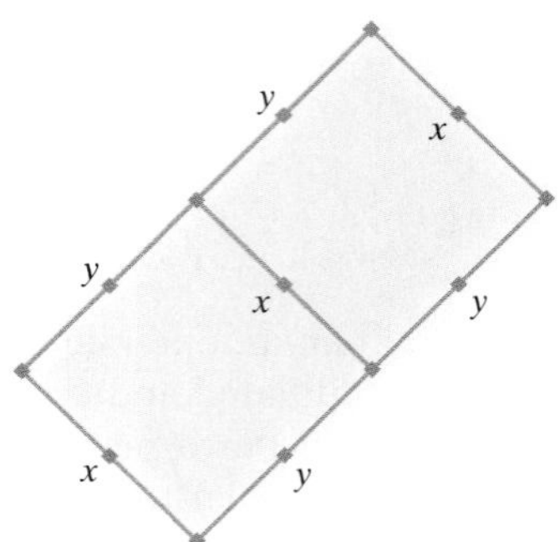

Figure for Exercise 46

a. If x represents the length of the pens and y represents the width of each pen, explain why the dimensions of the pens are subject to the constraint: $3x + 4y = 240$.

b. The area of each individual pen is given by $A = xy$. Use the constraint equation from part (a) to express A as a function in terms of x, where $0 < x < 80$.

c. Use the function from part (b) to find the dimensions of an individual pen that will yield the maximum area. [*Hint*: Find the vertex of the function from part (b).]

Graphing Calculator Exercises

For Exercises 47–52, graph the function in Exercises 35–40 on a graphing calculator. Use the *Root* or *Zero* function to approximate the x-intercepts. Use the *Max* or *Min* feature or *Zoom* and *Trace* to approximate the vertex.

47. $Y_1 = x^2 + 9x + 8$ (Exercise 35)

48. $Y_1 = x^2 + 7x + 10$ (Exercise 36)

49. $Y_1 = 2x^2 - 2x + 4$ (Exercise 37)

50. $Y_1 = 2x^2 - 12x + 19$ (Exercise 38)

51. $Y_1 = -x^2 + 3x - \frac{9}{4}$ (Exercise 39)

52. $Y_1 = -x^2 - \frac{3}{2}x - \frac{9}{16}$ (Exercise 40)

chapter 11 SUMMARY

SECTION 11.1—SQUARE ROOT PROPERTY AND COMPLETING THE SQUARE

KEY CONCEPTS:

The square root property states that if

$$x^2 = k, \text{ then } x = \pm\sqrt{k}.$$

The square root property can be used to solve quadratic equations in the form:

$$(x - h)^2 = k$$

Steps to solve a quadratic equation in the form $ax^2 + bx + c = 0$ $(a \neq 0)$ by completing the square and applying the square root property:

1. Divide both sides by a to make the leading coefficient 1.
2. Isolate the variable terms on one side of the equation.
3. Complete the square: Add the square of one-half the linear term coefficient to both sides of the equation. Then factor the resulting perfect square trinomial.
4. Apply the square root property and solve for x.

KEY TERMS:

completing the square
square root property

EXAMPLES:

Solve:

$$(x - 5)^2 = -13$$

$$x - 5 = \pm\sqrt{-13} \quad \text{(square root property)}$$

$$x = 5 \pm i\sqrt{13}$$

Solve the equation:

$$2x^2 - 6x - 5 = 0$$

$$\frac{2x^2}{2} - \frac{6x}{2} - \frac{5}{2} = \frac{0}{2}$$

$$x^2 - 3x = \frac{5}{2}$$

Note: $[\frac{1}{2} \cdot (-3)]^2 = \frac{9}{4}$

$$x^2 - 3x + \frac{9}{4} = \frac{5}{2} + \frac{9}{4}$$

$$\left(x - \frac{3}{2}\right)^2 = \frac{19}{4}$$

$$x - \frac{3}{2} = \pm\sqrt{\frac{19}{4}}$$

$$x = \frac{3}{2} \pm \frac{\sqrt{19}}{2}$$

$$x = \frac{3 \pm \sqrt{19}}{2}$$

Section 11.2—Quadratic Formula

Key Concepts:

The solutions to a quadratic equation $ax^2 + bx + c = 0$ $(a \neq 0)$ are given by the quadratic formula:

$$x = \frac{-b \pm \sqrt{b^2 - 4ac}}{2a}$$

The discriminant of a quadratic equation $ax^2 + bx + c = 0$ is $b^2 - 4ac$. If a, b, and c are rational numbers, then:

1. If $b^2 - 4ac > 0$, then there will be two real solutions. Moreover,
   a. If $b^2 - 4ac$ is a perfect square, the solutions will be rational numbers.
   b. If $b^2 - 4ac$ is not a perfect square, the solutions will be irrational numbers.
2. If $b^2 - 4ac < 0$, then there will be two imaginary solutions.
3. If $b^2 - 4ac = 0$, then there will be one rational solution.

Three methods to solve a quadratic equation are

1. Factoring
2. Completing the square and applying the square root property
3. Using the quadratic formula

Key Terms:

discriminant
quadratic formula

Examples:

Solve the equation:

$$0.03x^2 - 0.02x + 0.04 = 0$$

$$3x^2 - 2x + 4 = 0 \quad \text{(multiply by 100)}$$

$$a = 3, b = -2, c = 4$$

$$x = \frac{-(-2) \pm \sqrt{(-2)^2 - 4(3)(4)}}{2(3)}$$

$$= \frac{2 \pm \sqrt{4 - 48}}{6}$$

$$= \frac{2 \pm \sqrt{-44}}{6}$$

$$= \frac{2 \pm 2i\sqrt{11}}{6}$$

$$= \frac{1 \pm i\sqrt{11}}{3}$$

Section 11.3—Equations in Quadratic Form

Key Concepts:

Equations may reduce to a quadratic equation.

Substitution may be used to solve equations that are in quadratic form.

Key Term:

equations in quadratic form

Examples:

Solve: $x^{2/3} - x^{1/3} - 12 = 0$

Let $u = x^{1/3}$

$$u^2 - u - 12 = 0$$

$$(u - 4)(u + 3) = 0$$

$u = 4$ or $u = -3$

$x^{1/3} = 4$ or $x^{1/3} = -3$ Reverse substitute.

$x = 64$ or $x = -27$ Cube both sides.

Section 11.4—Graphs of Quadratic Functions

Key Concepts:

A quadratic function of the form $f(x) = x^2 + k$ shifts the graph of $y = x^2$ up k units if $k > 0$, and down $|k|$ units if $k < 0$.

A quadratic function of the form $f(x) = (x - h)^2$ shifts the graph of $y = x^2$ right h units if $h > 0$, and left $|h|$ units if $h < 0$.

A quadratic function of the form $f(x) = ax^2$ is a parabola that opens up when $a > 0$ and opens down when $a < 0$. If $|a| > 1$ the graph of $y = x^2$ is stretched vertically by a factor of $|a|$. If $0 < |a| < 1$ the graph of $y = x^2$ is shrunk vertically by a factor of $|a|$.

Examples:

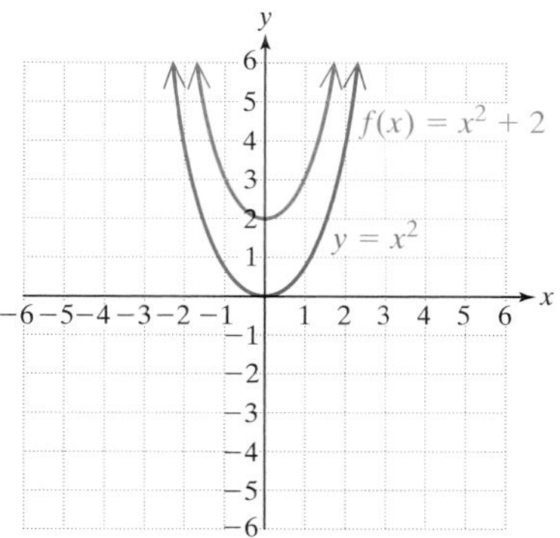

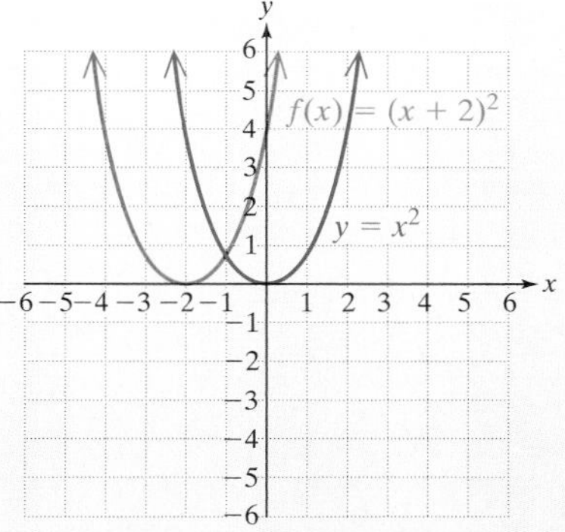

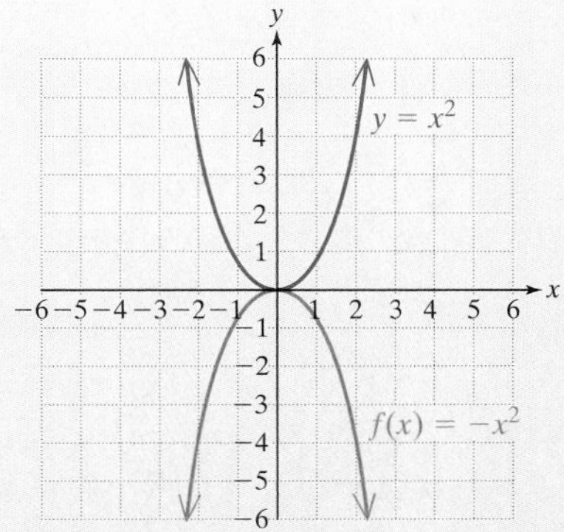

A quadratic function of the form $f(x) = a(x - h)^2 + k$ has vertex (h, k). If $a > 0$ the vertex represents the minimum point. If $a < 0$ the vertex represents the maximum point.

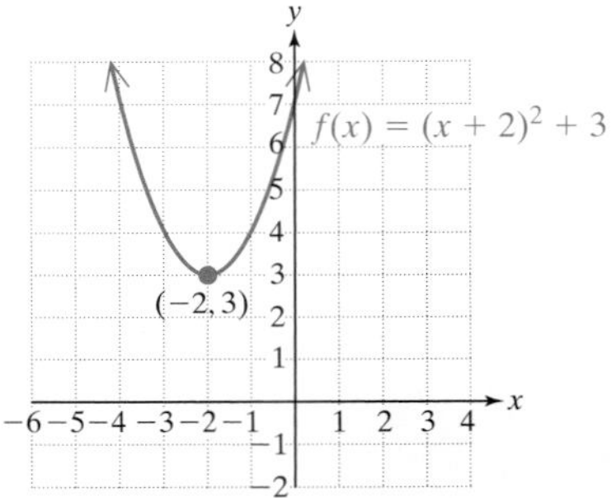

Key Terms:

axis of symmetry
maximum value
minimum value
vertex

Section 11.5—Applications of Quadratic Functions

Key Concepts:

Completing the square is used to write a quadratic function $f(x) = ax^2 + bx + c \quad (a \neq 0)$ in the form $f(x) = a(x - h)^2 + k$ for the purpose of identifying the vertex, (h, k).

The vertex formula finds the vertex of a quadratic function $f(x) = ax^2 + bx + c \quad (a \neq 0)$. The vertex is

$$\left(\frac{-b}{2a}, \frac{4ac - b^2}{4a}\right) \quad \text{or} \quad \left(\frac{-b}{2a}, f\left(\frac{-b}{2a}\right)\right)$$

Key Term:

vertex formula

Examples:

Find the vertex:

$$\begin{aligned} f(x) &= 3x^2 + 6x + 11 \\ &= 3(x^2 + 2x \qquad\quad) + 11 \\ &= 3(x^2 + 2x + 1 - 1) + 11 \\ &= 3(x^2 + 2x + 1) - 3 + 11 \\ &= 3(x + 1)^2 + 8 \\ &= 3[x - (-1)]^2 + 8 \end{aligned}$$

The vertex is $(-1, 8)$. Because $a = 3 > 0$, $(-1, 8)$ is a minimum point.

Find the vertex:

$$f(x) = -5x^2 + 4x - 1 \qquad a = -5, b = 4, c = -1$$

$$x = \frac{-b}{2a}$$

$$x = \frac{-4}{2(-5)} = \frac{2}{5}$$

$$f\left(\frac{2}{5}\right) = -5\left(\frac{2}{5}\right)^2 + 4\left(\frac{2}{5}\right) - 1 = -\frac{1}{5}$$

The vertex is $(\frac{2}{5}, -\frac{1}{5})$. Because $a = -5 < 0$, $(\frac{2}{5}, -\frac{1}{5})$ is a maximum point.

chapter 11 REVIEW EXERCISES

Section 11.1

For Exercises 1–8, solve the equations using the square root property.

1. $x^2 = 5$
2. $2y^2 = -8$
3. $a^2 = -81$
4. $3b^2 = -19$
5. $(x - 2)^2 = 72$
6. $(2x - 5)^2 = -9$
7. $(3y - 1)^2 = 3$
8. $3(m - 4)^2 = 15$
9. The length of each side of an equilateral triangle is 10 in. Find the height of the triangle. Round the answer to the nearest tenth of an inch.

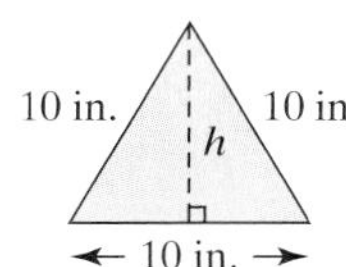

Figure for Exercise 9

10. Use the square root property to find the length of the sides of a square whose area is 81 in.^{2}
11. Use the square root property to find the length of the sides of a square whose area is 150 in.^{2} Round the answer to the nearest tenth of an inch.

For Exercises 12–15, find the value of k so that the expression is a perfect square trinomial. Then factor the trinomial.

12. $x^2 + 16x + k$
13. $x^2 - 9x + k$
14. $y^2 + \frac{1}{2}y + k$
15. $z^2 - \frac{2}{5}z + k$

For Exercises 16–21, solve the equation by completing the square and applying the square root property.

16. $w^2 + 4w + 13 = 0$
17. $4y^2 - 12y + 13 = 0$
18. $3x^2 + 2x = 1$
19. $b^2 + \frac{7}{2}b = 2$
20. $2x^2 = 12x + 6$
21. $-t^2 + 8t - 25 = 0$

Section 11.2

22. Explain how the discriminant can determine the type and number of solutions to a quadratic equation with rational coefficients.

For Exercises 23–28, determine the type (rational, irrational, or imaginary) and number of solutions for the equations using the discriminant.

23. $x^2 - 5x = -6$
24. $2y^2 = -3y$
25. $z^2 + 23 = 17z$
26. $a^2 + a + 1 = 0$
27. $10b + 1 = -25b^2$
28. $3x^2 + 15 = 0$

For Exercises 29–38, solve the equations by using the quadratic formula.

29. $y^2 - 4y + 1 = 0$
30. $m^2 - 5m + 25 = 0$
31. $6a^2 - 7a - 10 = 0$
32. $3x^2 - 10x + 8 = 0$
33. $b^2 - \frac{4}{25} = \frac{3}{5}b$
34. $k^2 + 0.4k = 0.05$
35. $32 + 4x - x^2 = 0$
36. $8y - y^2 = 0$
37. $5x^2 - 20 = 0$
38. $14 = a(a - 5)$
39. The landing distance that a certain plane will travel on a runway is determined by the initial landing speed at the instant the plane touches down. The function D relates landing distance in feet to initial landing speed, s:

$$D(s) = \frac{1}{10}s^2 - 3s + 22, \text{ where } s \text{ is in feet per second.}$$

   a. Find the landing distance for a plane traveling 150 ft/sec at touchdown.
   b. If the landing speed is too fast, the pilot may run out of runway. If the speed is too slow, the plane may stall. Find the maximum initial landing speed of a plane for a runway that is 1000 ft long. Round to one decimal place.

c. Convert the landing speed you found in part (b), into miles per hour. Note that 1 mile = 5280 feet, and 1 hour = 3600 seconds. Round to the nearest mile per hour.

40. The recent population, P (in thousands), of Kenya can be approximated by: $P(t) = 4.62t^2 + 564.6t + 13{,}128$, where t is the number of years since 1974.

a. Approximate the number of people in Kenya in the year 1990.

b. If this trend continues, approximate the number of people in Kenya in the year 2025.

c. What is the y-intercept of this function and what does it mean in the context of this problem?

d. In what year after 1974 will the population of Kenya reach 50 million? (*Hint*: 50 million equals 50,000 thousands)

Section 11.3

For Exercises 41–48, solve the equations using substitution, if necessary.

41. $x - 4\sqrt{x} - 21 = 0$

42. $n - 6\sqrt{n} + 8 = 0$

43. $y^4 - 11y^2 + 18 = 0$

44. $2m^4 - m^2 - 3 = 0$

45. $t^{2/5} + t^{1/5} - 6 = 0$

46. $p^{2/5} - 3p^{1/5} + 2 = 0$

47. $\sqrt{4a - 3} - \sqrt{8a + 1} = -2$

48. $\sqrt{2b - 5} - \sqrt{b - 2} = 2$

Section 11.4

For Exercises 49–56, graph the functions.

49. $g(x) = x^2 - 5$

50. $f(x) = x^2 + 3$

51. $h(x) = (x - 5)^2$

52. $k(x) = (x + 3)^2$

53. $m(x) = -2x^2$

54. $n(x) = -4x^2$

55. $p(x) = -2(x - 5)^2 - 5$

56. $q(x) = -4(x + 3)^2 + 3$

For Exercises 57–58, identify the vertex of the parabola and determine if it is a maximum or a minimum point. Then write the maximum or the minimum value.

57. $t(x) = \frac{1}{3}(x - 4)^2 + \frac{5}{3}$

58. $s(x) = -\frac{5}{7}(x - 1)^2 - \frac{1}{7}$

For Exercises 59–60, identify the axis of symmetry.

59. $a(x) = -\frac{3}{2}\left(x + \frac{2}{11}\right)^2 - \frac{4}{13}$

60. $w(x) = -\frac{4}{3}\left(x - \frac{3}{16}\right)^2 + \frac{2}{9}$

Section 11.5

For Exercises 61–64, write the function in the form $f(x) = a(x - h)^2 + k$ by completing the square. Then identify the vertex.

61. $z(x) = x^2 - 6x + 7$

62. $b(x) = x^2 - 4x - 44$

63. $p(x) = -5x^2 - 10x - 13$

64. $q(x) = -3x^2 - 24x - 54$

For Exercises 65–68, find the vertex of each function by using the vertex formula.

65. $f(x) = -2x^2 + 4x - 17$

66. $g(x) = -4x^2 - 8x + 3$

67. $m(x) = 3x^2 - 3x + 11$

68. $n(x) = 3x^2 + 2x - 7$

69. Tetanus bacillus bacterium is cultured to produce tetanus toxin used in an inactive form for the tetanus vaccine. The amount of toxin produced per batch increases with time and then becomes unstable. The yield of toxin (in grams) as a function of time (in hours) can be approximated by the following function.

$$y(t) = -\frac{1}{1152}t^2 + \frac{1}{12}t \qquad 0 \le t \le 96$$

Figure for Exercise 69

a. Use the function to approximate the yield of toxin after 12 hours, 36 hours, 48 hours, and 60 hours. Verify your answer from the graph.

b. Find the maximum yield of toxin and the time required to produce the maximum yield.

70. The bacterium *Pseudomonas aeruginosa* is cultured with an initial population of 10^4 active organisms. The population of active bacteria increases up to a point, and then due to a limited food supply and an increase of waste products, the population of living organisms decreases.

Over the first 48 hours, the population, P, can be approximated by the following function.

$$P(t) = -1718.75t^2 + 82{,}500t + 10{,}000 \quad \text{where } 0 \le t \le 48$$

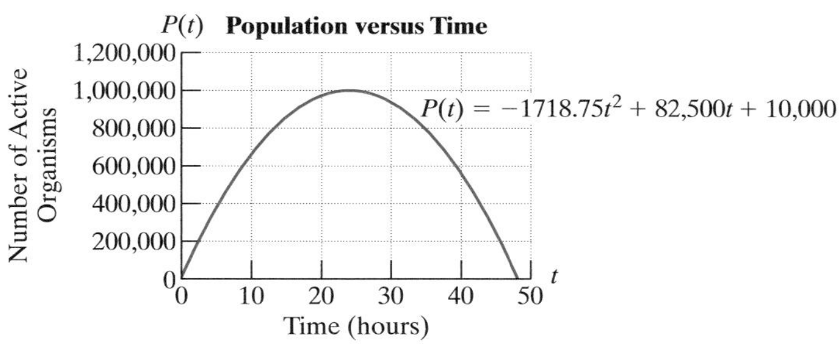

Figure for Exercise 70

a. Use the function to approximate the number of active bacteria after 6 hours, 20 hours, and 30 hours. Round to the nearest thousand.

b. Find the time required for the population to reach its maximum value. What is the maximum number of active bacteria?

chapter 11 TEST

For Exercises 1–3, solve the equation using the square root property.

1. $(x + 3)^2 = 25$

2. $(p - 2)^2 = 12$

3. $(m + 1)^2 = -1$

4. Find the value of k so that the expression is a perfect square trinomial. Then factor the trinomial: $d^2 + 7d + k$.

For Exercises 5–6, solve the equation by completing the square and applying the square root property.

5. $2x^2 + 12x - 36 = 0$

6. $2x^2 = 3x - 7$

For Exercises 7–8:

a. Write the equation in standard form, $ax^2 + bx + c = 0$.

b. Identify a, b, and c.

c. Find the discriminant.

d. Determine the number and type (rational, irrational, or imaginary) of solutions.

7. $x^2 - 3x = -12$

8. $y(y - 2) = -1$

For Exercises 9–10, solve the equation using the quadratic formula.

9. $3x^2 - 4x + 1 = 0$

10. $x^2 + 7x + 11 = 0$

11. The base of a triangle is 3 ft less than twice the height. The area of the triangle is 14 ft^2. Find the base and the height. Round the answers to the nearest tenth of a foot.

12. A circular garden has an area of approximately 450 ft^2. Find the radius. Round the answer to the nearest tenth of a foot.

For Exercises 13–14, solve the equation using substitution if necessary.

13. $x - \sqrt{x} - 6 = 0$

14. $y^{2/3} + 2y^{1/3} = 8$

For Exercises 15–18, find the x- and y-intercepts of the function. Then match the function with its graph.

15. $f(x) = x^2 - 6x + 8$

16. $k(x) = x^3 + 4x^2 - 9x - 36$

17. $p(x) = -2x^2 - 8x - 6$

18. $q(x) = x^3 - x^2 - 12x$

a.

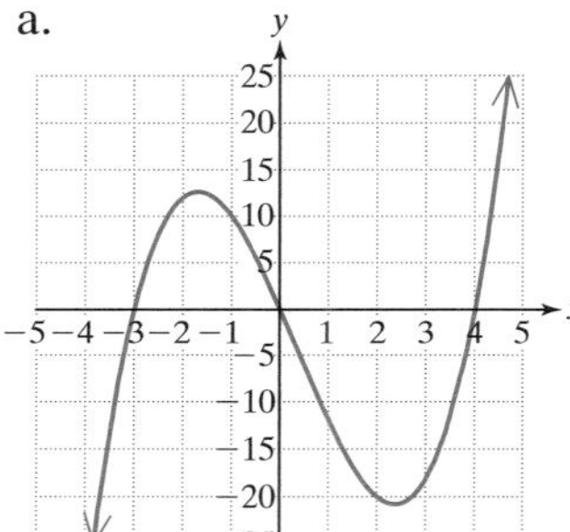

b.

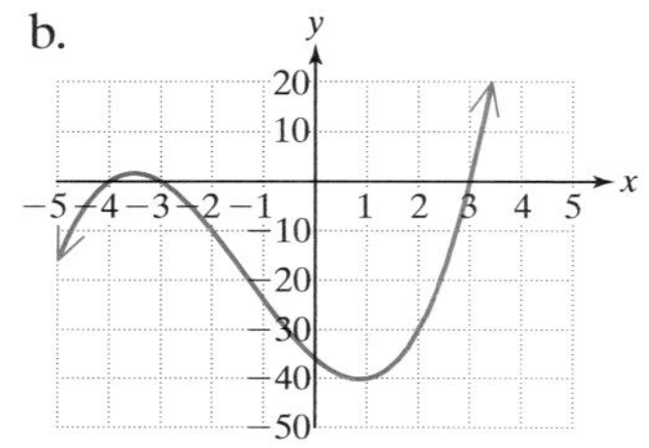

c.

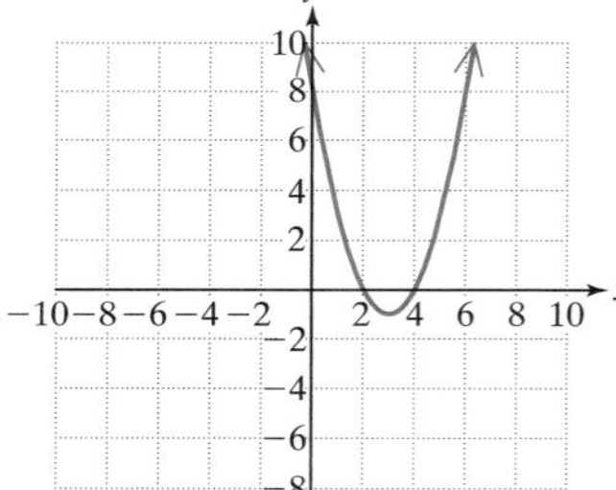

d.

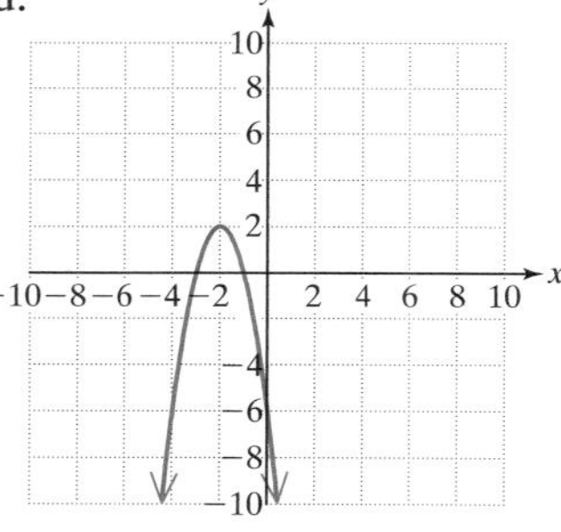

19. A child launches a toy rocket from the ground. The height of the rocket, h, can be determined by its horizontal distance from the launch pad, x, by:

$$h(x) = -\frac{x^2}{256} + x, \text{ where } x \text{ and } h \text{ are in feet and } x \geq 0 \text{ and } h \geq 0.$$

How many feet from the launch pad will the rocket hit the ground?

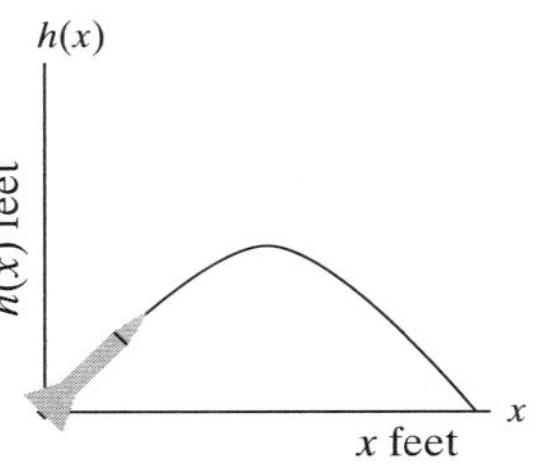

Figure for Exercise 19

20. The recent population, P (in millions), of India can be approximated by: $P(t) = 0.136t^2 + 12.6t + 607.7$, where $t = 0$ corresponds to the year 1974.

    a. Approximate the number of people in India in the year 1978. (Round to the nearest million.)

    b. If this trend continues approximate the number of people in India in the year 2010.

    c. In what year after 1974 did the population of India reach 700 million? (Round to the nearest year.)

    d. Approximate the year in which the population of India reached 1 billion (1000 million). (Round to the nearest year.)

21. Explain the relationship between the graphs of $y = x^2$ and $y = x^2 - 2$.

22. Explain the relationship between the graphs of $y = x^2$ and $y = (x + 3)^2$.

23. Explain the relationship between the graphs of $y = 4x^2$ and $y = -4x^2$.

24. Given the function defined by $f(x) = -(x - 4)^2 + 2$.
    a. Identify the vertex of the parabola.
    b. Does this parabola open up or down?
    c. Does the vertex represent the maximum or minimum point of the function?
    d. What is the maximum or minimum value of the function f?
    e. What is the axis of symmetry for this parabola?

25. For the function defined by $g(x) = 2x^2 - 20x + 51$, find the vertex using two methods:
    a. Complete the square to write g in the form $g(x) = a(x - h)^2 + k$. Identify the vertex.
    b. Use the vertex formula to find the vertex.

26. A farmer has 400 feet of fencing with which to enclose a rectangular field. The field is situated such that one of its sides is adjacent to a river and requires no fencing. The area of the field (in square feet) can be modeled by

$$A(x) = -\frac{x^2}{2} + 200x$$

where x is the length of the side parallel to the river (measured in feet).

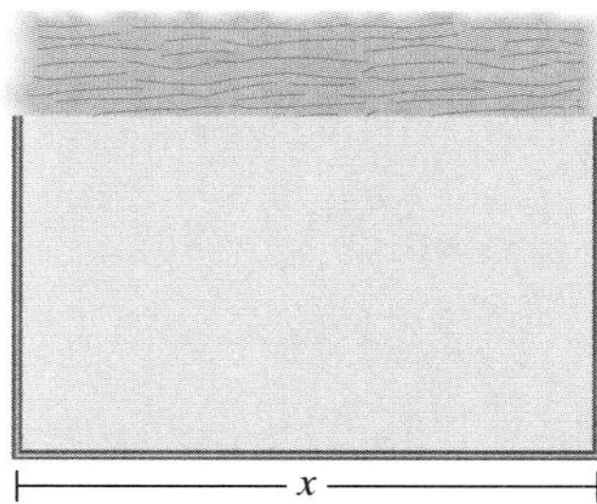

Figure for Exercise 26

Use the function to determine the maximum area that can be enclosed.

CUMULATIVE REVIEW EXERCISES, CHAPTERS 1–11

1. Given $A = \{2, 4, 6, 8, 10\}$ and $B = \{2, 8, 12, 16\}$
    a. Find $A \cup B$ b. $A \cap B$

2. Perform the indicated operations and simplify: $(2x^2 - 5) - (x + 3)(5x - 2)$

3. Simplify completely: $4^0 - \left(\frac{1}{2}\right)^{-3} - 81^{1/2}$

4. Perform the indicated operations. Write the answer in scientific notation:

$$\frac{(3.0 \times 10^{12})(6.0 \times 10^{-3})}{(2.5 \times 10^{-4})}$$

5. a. Factor completely: $x^3 + 2x^2 - 9x - 18$
    b. Divide using long division. Identify the quotient and remainder.

$$(x^3 + 2x^2 - 9x - 18) \div (x - 3)$$

6. Multiply: $(\sqrt[3]{x} + \sqrt[3]{2})(\sqrt[3]{x^2} - \sqrt[3]{2x} + \sqrt[3]{4})$

7. Simplify: $\frac{4}{\sqrt{2x}}$

8. Jacques invests a total of \$10,000 in two mutual funds. After 1 year, one fund produced 12% growth, and the other lost 3%. Find the amount invested in each fund if the total investment grew by \$900.

9. Solve the system of equations:

$$\frac{1}{9}x - \frac{1}{3}y = -\frac{13}{9}$$
$$x - \frac{1}{2}y = \frac{9}{2}$$

10. An object is fired straight up into the air from an initial height of 384 ft with an initial velocity of 160 ft/sec. The height, h, in feet is given by:

$h(t) = -16t^2 + 160t + 384$, where t is the time in seconds.

a. Find the height of the object after 3 sec.

b. Find the height of the object after 7 sec.

c. Find the time required for the object to hit the ground.

11. Solve the equation: $(x - 3)^2 + 16 = 0$.

12. Solve the equation: $2x^2 + 5x - 1 = 0$.

13. What number would have to be added to the quantity $x^2 + 10x$ to make it a perfect square trinomial?

14. Factor completely: $2x^3 + 250$.

15. Graph the line $3x - 5y = 10$.

16. a. Find the x-intercepts of the function defined by $g(x) = 2x^2 - 9x + 10$.

    b. What is the y-intercept of $y = g(x)$?

17. The poverty threshold for four-person families in 1960 was \$3022. The poverty threshold for four-person families in 1990 was \$13,359. Let y represent the poverty threshold, and let x represent the year, where $x = 0$ corresponds to 1960. (*Source:* U.S. Bureau of the Census)

    a. Plot the ordered pairs (0, 3022) and (30, 13,359).

    b. Find a linear model that represents the poverty threshold y, as a function of the year, x. Round the slope to the nearest whole number.

    c. Use the model you found in part (b) to estimate the poverty level in 1980.

18. Michael Jordan was the NBA leading scorer for 10 of 12 seasons between 1987 and 1998. In his 1998 season, he scored a total of 2357 points consisting of 1-point free throws, 2-point field goals, and 3-point field goals. He scored 286 more 2-point shots than he did free throws. The number of 3-point shots was 821 less than the number of 2-point shots. Determine the number of free throws, 2-point shots, and 3-point shots scored by Michael Jordan during his 1998 season.

19. Explain why this relation is *not* a function.

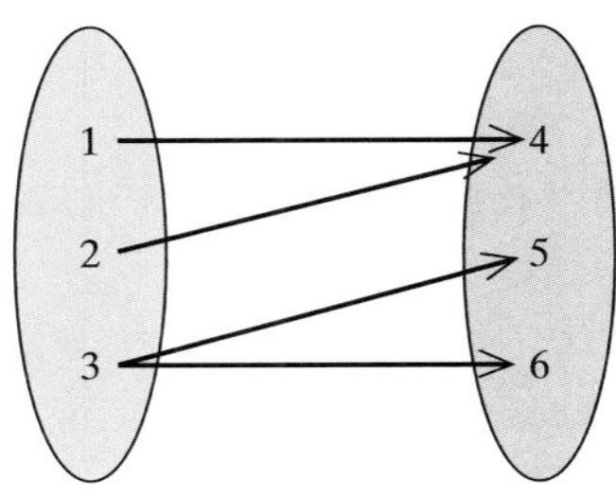

20. Graph the function defined by $f(x) = \dfrac{1}{x}$.

21. The quantity y varies directly as x and inversely as z. If $y = 15$ when $x = 50$ and $z = 10$, find y when $x = 65$ and $z = 5$.

22. The total number of flights, F (including passenger flights and cargo flights), at a large airport can be approximated by $F(x) = 300{,}000 + 0.008x$, where x is the number of passengers.

    a. Is this function linear, quadratic, constant, or other?

    b. What is the y-intercept and interpret its meaning in the context of this problem.

    c. What is the slope of the function and what does the slope mean in the context of this problem?

23. Given the function defined by $g(x) = \sqrt{2 - x}$, find the function values (if they exist) over the set of real numbers:

    a. $g(-7)$ b. $g(0)$ c. $g(3)$

24. Let $m(x) = \sqrt{x + 4}$, and $n(x) = x^2 + 2$, find:

    a. the domain of m. b. the domain of n.

25. Consider the function $y = f(x)$ graphed here. Find:

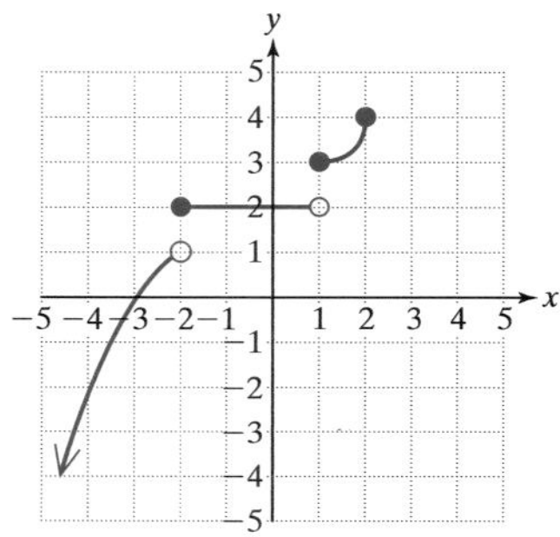

a. The domain b. The range

c. $f(-2)$ d. $f(1)$ e. $f(0)$

f. For what value(s) of x is $f(x) = 0$?

For Exercises 26–30, solve the inequality and write the solution in interval notation.

26. $2x - 3 \le x + 5$ and $2x + 1 \ge -3$

27. $2x - 3 \le x + 5$ or $2x + 1 \ge -3$

28. $\left|\frac{1}{2}x + 5\right| + 7 < 1$

29. $2x^2 + 8x - 10 \ge 0$

30. $\dfrac{2x + 5}{x - 3} \le 0$

31. Graph the solution set: $3x - 2y < 6$

32. Solve: $\sqrt{8x + 5} = \sqrt{2x} + 2$

33. Solve for f: $\dfrac{1}{p} + \dfrac{1}{q} = \dfrac{1}{f}$

34. Solve: $\dfrac{15}{t^2 - 2t - 8} = \dfrac{1}{t - 4} + \dfrac{2}{t + 2}$

35. Simplify: $\dfrac{y - \dfrac{4}{y - 3}}{y - 4}$

36. Given the function defined by $f(x) = 2(x - 3)^2 + 1$

a. Write the coordinates of the vertex.

b. Does the graph of the function open upward or downward?

c. Write the coordinates of the y-intercept.

d. Find the x-intercepts (if possible).

e. Sketch the function.

37. Use the method of completing the square to solve the equation.

$$x^2 - 16x + 2 = 0$$

38. Use the method of completing the square to find the vertex of the parabola. Check your answer using the vertex formula.

$$f(x) = x^2 - 16x + 2$$

chapter

Exponential and Logarithmic Functions

12

In the year 1999, the population of Mexico was approximately 100 million and the population of Japan was 126 million. Although the population of Mexico was less than that of Japan in 1999, its growth rate is higher. The population of Mexico is growing at a rate of 2.02% per year, whereas the population of Japan is growing at a rate of 0.24% per year.

For t representing the number of years since 1999, the population (in millions) of each country can be modeled by an exponential function:

Mexico: $M(t) = 100(1.0202)^t$

Japan: $J(t) = 126(1.0024)^t$

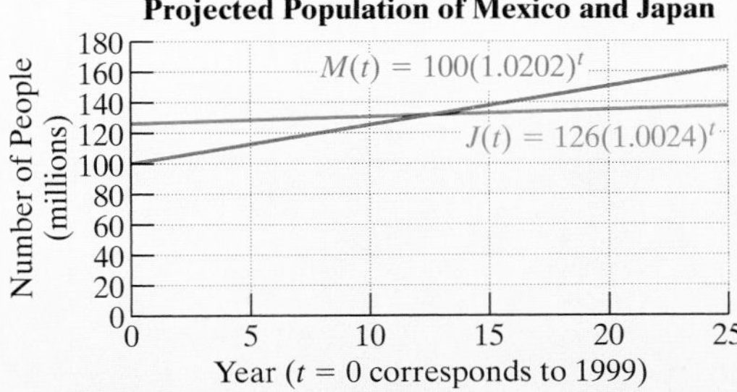

From the graphs of $M(t) = 100(1.0202)^t$ and $J(t) = 126(1.0024)^t$, we see that after approximately 13 years, the population of Mexico will overtake the population of Japan.

For more information about population statistics see the Technology Connections in MathZone at

www.mhhe.com/miller_oneill

Concepts

1. Algebra of Functions
2. Composition of Functions
3. Multiple Operations on Functions

section

12.1 Algebra and Composition of Functions

1. Algebra of Functions

Addition, subtraction, multiplication, and division can be used to create a new function from two or more functions.

Sum, Difference, Product, and Quotient of Functions

Given two functions, f and g, the functions $f + g$, $f - g$, $f \cdot g$, and $\frac{f}{g}$ are defined as

$$(f + g)(x) = f(x) + g(x)$$

$$(f - g)(x) = f(x) - g(x)$$

$$(f \cdot g)(x) = f(x) \cdot g(x)$$

$$\left(\frac{f}{g}\right)(x) = \frac{f(x)}{g(x)} \qquad \text{provided } g(x) \neq 0$$

For example, suppose $f(x) = |x|$ and $g(x) = 3$. Taking the sum of the functions produces a new function denoted by $(f + g)$. In this case, $(f + g)(x) = |x| + 3$. Graphically, the y-values of the function $(f + g)$ are given by the sum of the corresponding y-values of f and g. This is depicted in Figure 12-1. The function $(f + g)$ appears in red. In particular, notice that $(f + g)(2) = f(2) + g(2) = 2 + 3 = 5$.

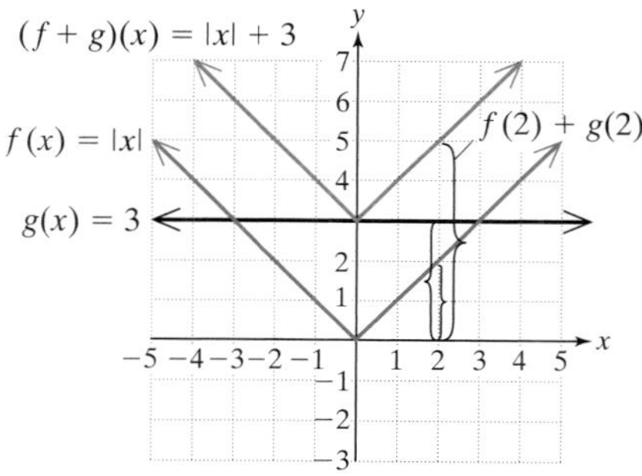

Figure 12-1

example 1

Adding, Subtracting, and Multiplying Functions

Given the functions defined by $g(x) = 4x$, $h(x) = x^2 - 3x$, and $k(x) = \sqrt{x - 2}$,

a. Find $(g + h)(x)$ and write the domain of $(g + h)$ in interval notation.

b. Find $(h - g)(x)$ and write the domain of $(h - g)$ in interval notation.

c. Find $(g \cdot k)(x)$ and write the domain of $(g \cdot k)$ in interval notation.

Solution:

a. $(g + h)(x) = g(x) + h(x)$
$= (4x) + (x^2 - 3x)$
$= 4x + x^2 - 3x$
$= x^2 + x$ The domain is all real numbers, $(-\infty, \infty)$.

b. $(h - g)(x) = h(x) - g(x)$
$= (x^2 - 3x) - (4x)$
$= x^2 - 3x - 4x$
$= x^2 - 7x$ The domain is all real numbers, $(-\infty, \infty)$.

c. $(g \cdot k)(x) = g(x) \cdot k(x)$
$= (4x)(\sqrt{x - 2})$
$= 4x\sqrt{x - 2}$ The domain is $[2, \infty)$.

example 2

Dividing Functions

Given the functions defined by $h(x) = x^2 - 3x$ and $k(x) = \sqrt{x - 2}$, find $\left(\frac{k}{h}\right)(x)$ and write the domain of $\left(\frac{k}{h}\right)$ in interval notation.

Solution:

$$\left(\frac{k}{h}\right)(x) = \frac{\sqrt{x - 2}}{x^2 - 3x}.$$

To find the domain we must consider the restrictions on x imposed by the square root and by the fraction.

- From the numerator we have $x - 2 \geq 0$, or equivalently, $x \geq 2$.
- From the denominator we have $x^2 - 3x \neq 0$, or equivalently, $x(x - 3) \neq 0$. Hence, $x \neq 3$ and $x \neq 0$.

Thus, the domain of $\frac{k}{h}$ is the set of real numbers greater than or equal to 2, but not equal to 3 or 0. This is shown graphically in Figure 12-2.

The domain is $[2, 3) \cup (3, \infty)$.

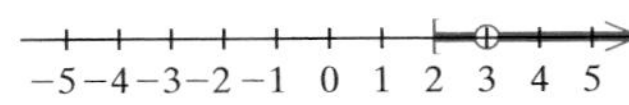

Figure 12-2

2. Composition of Functions

Composition of Functions

The **composition** of f and g, denoted $f \circ g$, is defined by the rule:

$(f \circ g)(x) = f(g(x))$ provided that $g(x)$ is in the domain of f.*

The composition of g and f, denoted $g \circ f$, is defined by the rule:

$(g \circ f)(x) = g(f(x))$ provided that $f(x)$ is in the domain of g.†

*$f \circ g$ is also read as "f compose g."
†$g \circ f$ is also read as "g compose f."

For example, given $f(x) = 2x - 3$ and $g(x) = x + 5$, we have

$$(f \circ g)(x) = f(g(x))$$
$$= f(x + 5)$$ Substitute $g(x) = x + 5$ into the function f.
$$= 2(x + 5) - 3$$
$$= 2x + 10 - 3$$
$$= 2x + 7$$

In this composition, the function g is the innermost operation and acts on x first. Then the output value of function g becomes the domain element of the function f as shown in the figure.

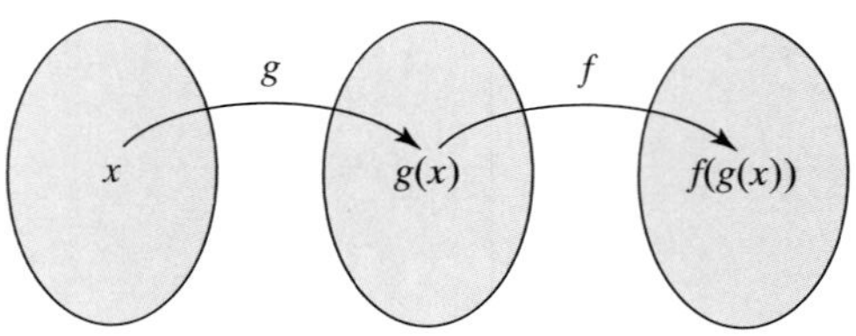

example 3

Composing Functions

Given the functions defined by $f(x) = x - 5$, $g(x) = x^2$, and $n(x) = \sqrt{x + 2}$,

a. Find $(f \circ g)(x)$ and write the domain of $(f \circ g)$ in interval notation.

b. Find $(g \circ f)(x)$ and write the domain of $(g \circ f)$ in interval notation.

c. Find $(n \circ f)(x)$ and write the domain of $(n \circ f)$ in interval notation.

Solution:

a. $(f \circ g)(x) = f(g(x))$

$= f(x^2)$ Evaluate the function f at x^2.

$= (x^2) - 5$ Replace x by x^2 in the function, f.

$= x^2 - 5$ The domain is all real numbers, $(-\infty, \infty)$.

b. $(g \circ f)(x) = g(f(x))$

$= g(x - 5)$ Evaluate the function, g, at $x - 5$.

$= (x - 5)^2$ Replace x by the quantity $(x - 5)$ in function g.

$= x^2 - 10x + 25$ The domain is all real numbers, $(-\infty, \infty)$.

c. $(n \circ f)(x) = n(f(x))$

$= n(x - 5)$ Evaluate the function, n, at $x - 5$.

$= \sqrt{(x - 5) + 2}$ Replace x by the quantity $(x - 5)$ in function n.

$= \sqrt{x - 3}$ The domain is $[3, \infty)$.

Tip: Examples 3(a) and 3(b) illustrate that the order in which two functions are composed may result in different functions. That is, $f \circ g$ does not necessarily equal $g \circ f$.

3. Multiple Operations on Functions

example 4 **Combining Functions**

Given the functions defined by $f(x) = x - 7$ and $h(x) = 2x^3$, find the function values if possible.

a. $(f \cdot h)(3)$ b. $\left(\frac{h}{f}\right)(7)$ c. $(h \circ f)(2)$

Solution:

a. $(f \cdot h)(3) = f(3) \cdot h(3)$

$= (3 - 7) \cdot 2(3)^3$

$= (-4) \cdot 2(27)$

$= -216$

Avoiding Mistakes

If you had tried evaluating the function $\frac{h}{f}$ at $x = 7$, the denominator would be zero and the function is undefined.

$$\frac{h(7)}{f(7)} = \frac{2(7)^3}{7 - 7}$$

b. The function $\frac{h}{f}$ has restrictions on its domain. $\left(\frac{h}{f}\right)(x) = \frac{h(x)}{f(x)} = \frac{2x^3}{x - 7}$.

Therefore, $x = 7$ is not in the domain and $\left(\frac{h}{f}\right)(7) = \frac{h(7)}{f(7)}$ is undefined.

c. $(h \circ f)(2) = h(f(2))$ Evaluate $f(2)$ first. $f(2) = 2 - 7 = -5$.

$= h(-5)$ Substitute the result into the function h.

$= 2(-5)^3$

$= 2(-125)$

$= -250$

example 5 **Finding Function Values from a Graph**

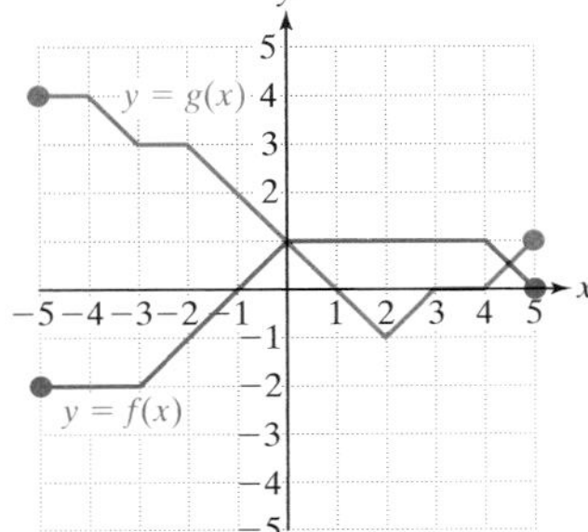

For the functions f and g pictured at left, find the function values if possible.

a. $(f - g)(-3)$ b. $\left(\frac{g}{f}\right)(5)$ c. $(f \circ g)(4)$

Solution:

a. $(f - g)(-3) = f(-3) - g(-3)$ Evaluate the difference of $f(-3)$ and $g(-3)$.

$= -2 - (3)$ Estimate function values from the graph.

$= -5$

b. $\left(\frac{g}{f}\right)(5) = \frac{g(5)}{f(5)}$ Evaluate the quotient of $g(5)$ and $f(5)$.

$= \frac{1}{0}$ (undefined)

The function $\frac{g}{f}$ is undefined at 5 because the denominator is zero.

c. $(f \circ g)(4) = f(g(4))$ — From the red graph, find the value of $g(4)$ first.

$= f(0)$ — From the blue graph, find the value of f at $x = 0$.

$= 1$

section 12.1 PRACTICE EXERCISES

For Exercises 1–18, refer to the functions defined below.

$f(x) = x + 4 \qquad g(x) = 2x^2 + 4x$

$h(x) = \sqrt{x - 1} \qquad k(x) = \dfrac{1}{x}$

Find the indicated functions. Write the domain in interval notation.

1. $(f + g)(x)$
2. $(f - g)(x)$
3. $(g - f)(x)$
4. $(f + h)(x)$
5. $(f \cdot h)(x)$
6. $(h \cdot k)(x)$
7. $(g \cdot f)(x)$
8. $(f \cdot k)(x)$
9. $\left(\dfrac{h}{f}\right)(x)$
10. $\left(\dfrac{g}{f}\right)(x)$
11. $\left(\dfrac{f}{g}\right)(x)$
12. $\left(\dfrac{f}{h}\right)(x)$
13. $(f \circ g)(x)$
14. $(g \circ f)(x)$
15. $(f \circ k)(x)$
16. $(k \circ f)(x)$
17. $(k \circ h)(x)$
18. $(h \circ k)(x)$
19. Based on your answers to Exercises 13 and 14 is it true in general that $(f \circ g)(x) = (g \circ f)(x)$?
20. Based on your answers to Exercises 15 and 16, is it true in general that $(f \circ k)(x) = (k \circ f)(x)$?

For Exercises 21–34, refer to the functions defined below.

$m(x) = x^3 \qquad n(x) = x - 3$

$r(x) = \sqrt{x + 4} \qquad p(x) = \dfrac{1}{x + 2}$

Find the function values if possible.

21. $(m \cdot r)(0)$
22. $(n \cdot p)(0)$
23. $(m + r)(-4)$
24. $(n - m)(4)$
25. $(r \circ n)(3)$
26. $(n \circ r)(5)$
27. $(p \circ m)(-1)$
28. $(m \circ n)(5)$
29. $(m \circ p)(2)$
30. $(r \circ m)(2)$
31. $(r + p)(-3)$
32. $(n + p)(-2)$
33. $(m \circ p)(-2)$
34. $(r \circ m)(-2)$

For Exercises 35–48, approximate the function values from the graph if possible.

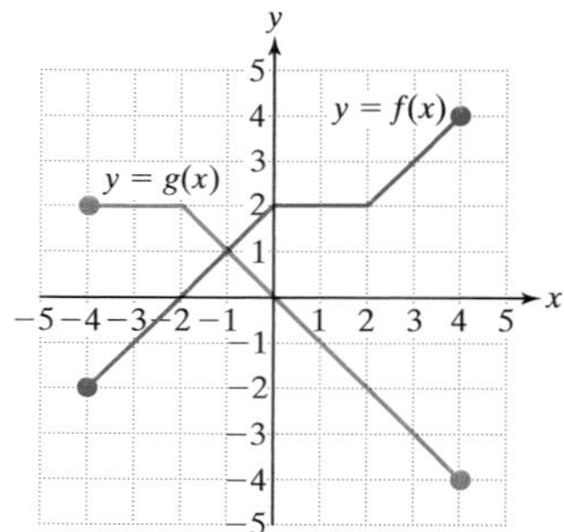

Figure for Exercises 35–48

35. $(f + g)(2)$
36. $(g - f)(3)$
37. $(f \cdot g)(-1)$
38. $(g \cdot f)(-4)$
39. $\left(\dfrac{g}{f}\right)(0)$
40. $\left(\dfrac{f}{g}\right)(-2)$
41. $\left(\dfrac{f}{g}\right)(0)$
42. $\left(\dfrac{g}{f}\right)(-2)$

43. $(g \circ f)(-1)$
44. $(f \circ g)(0)$
45. $(f \circ g)(-4)$
46. $(g \circ f)(-4)$
47. $(g \circ g)(2)$
48. $(f \circ f)(-2)$

49. The cost in dollars of producing x toy cars is $C(x) = 2.2x + 1$. The revenue received is $R(x) = 5.98x$. To calculate profit, subtract the cost from the revenue.
   a. Write and simplify a function P that represents profit in terms of x.
   b. Find the profit of producing 50 toy cars.

50. The cost in dollars of producing x lawn chairs is $C(x) = 2.5x + 10.1$. The revenue for selling x chairs is $R(x) = 6.99x$. To calculate profit, subtract the cost from the revenue.
   a. Write and simplify a function P that represents profit in terms of x.
   b. Find the profit of producing 100 lawn chairs.

51. The functions defined by $D(t) = 0.925t + 13.083$ and $R(t) = 0.725t + 8.683$ approximate the amount of child support (in billions of dollars) that was due and the amount of child support actually received in the United States between 1985 and 1989. In each case, $t = 0$ corresponds to 1985.
   a. Find the function F defined by $F(t) = D(t) - R(t)$. What does F represent in the context of this problem?
   b. Find $F(0)$, $F(2)$, and $F(4)$. What do these function values represent in the context of this problem?

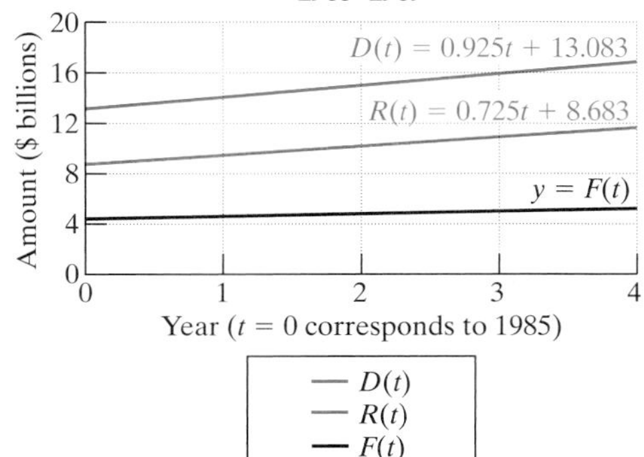

Figure for Exercise 51

52. If t represents the number of years after 1900, then the rural and urban populations in the South (United States) between 1900 and 1970 can be approximated by

$$r(t) = -3.497t^2 + 266.2t + 20{,}220$$

$t = 0$ corresponds to 1900 and $r(t)$ represents the rural population in thousands.

$$u(t) = 0.0566t^3 + 0.952t^2 + 177.8t + 4593$$

$t = 0$ corresponds to 1900 and $u(t)$ represents the urban population in thousands.

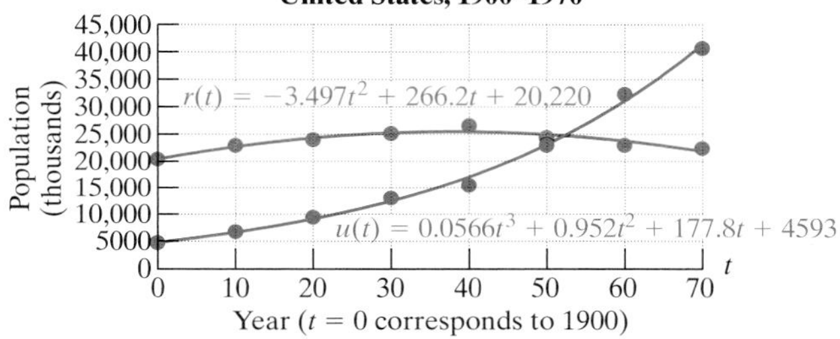

Source: Historical Abstract of the United States.

Figure for Exercise 52

   a. Find the function T defined by $T(t) = r(t) + u(t)$. What does the function T represent in the context of this problem?
   b. Use the function T to approximate the total population in the South for the year 1940.

53. Joe rides a bicycle at a constant rate of 80 revolutions per minute (that is, the wheels on his bike complete a complete circle 80 times per minute). Therefore, the total number of revolutions, r, is given by $r(t) = 80t$, where t is the time in minutes. For each revolution of the wheels of the bike, he travels approximately 7 feet. Therefore, the total distance he travels, D, depends on the total number of revolutions, r, according to the function $D(r) = 7r$.
   a. Find $(D \circ r)(t)$ and interpret its meaning in the context of this problem.
   b. Find Joe's total distance in feet after 10 minutes.

54. The area, A, of a square is given by the function, $a(x) = x^2$, where x is the length of the sides of the square. If carpeting costs \$9.95 per square yard, then the cost, C, to carpet a square room is given by $C(a) = 9.95a$, where a is the area of the room in square yards.

a. Find $(C \circ a)(x)$ and interpret its meaning in the context of this problem.

b. Find the cost to carpet a square room if its floor dimensions are 15 yd. by 15 yd.

---

section

12.2 Inverse Functions

Concepts

1. Definition of the Inverse of a Function
2. Definition of a One-to-One Function
3. Horizontal Line Test
4. Finding an Equation of the Inverse of a Function
5. Composition of a Function and Its Inverse
6. Restricting the Domain of a Function

1. Definition of the Inverse of a Function

Recall from Section 7.4 that a function is a set of ordered pairs (x, y), such that for every element x in the domain, there corresponds exactly one element y in the range. For example, the function, f, relates the weight of a package of deli meat, x, to its cost, y.

$$f = \{(1, 2.99), (1.5, 4.49), (4, 11.96)\}$$

That is, 1 lb of meat sells for \$2.99, 1.5 lb sells for \$4.49, and 4 lb sells for \$11.96. Now suppose we create a new function in which the values of x and y are interchanged. The new function, called the **inverse of *f*** (denoted f^{-1}) relates the price of meat, x, to its weight, y.

$$f^{-1} = \{(2.99, 1), (4.49, 1.5), (11.96, 4)\}$$

Avoiding Mistakes

f^{-1} denotes the inverse of a function. The "-1" does not represent an exponent.

Notice that interchanging the x- and y-values has the following outcome. The domain of f is the same as the range of f^{-1}, and the range of f is the domain of f^{-1}.

2. Definition of a One-to-One Function

A necessary condition for a function f to have an inverse function is that no two ordered pairs in f may have different x-coordinates and the same y-coordinate. A function that satisfies this condition is called a **one-to-one function**. The function relating the weight of a package of meat to its price is a one-to-one function. However, consider the function g defined by

$$g = \{(1, 4), (2, 3), (-2, 4)\}$$

same y

different x

This function is not one-to-one because the range element 4 has two different x-coordinates, 1 and -2. Interchanging the x- and y-values produces a relation that violates the definition of a function.

$$\{(4, 1), (3, 2), (4, -2)\}$$

same x

different y

This relation is not a function because for $x = 4$ there are two different y-values, $y = 1$ and $y = -2$.

3. Horizontal Line Test

In Section 7.4, you learned the vertical line test to determine visually if a graph represents a function. Similarly, we use a **horizontal line test** to determine whether a function is one-to-one.

Horizontal Line Test

Consider a function defined by a set of points (x, y) in a rectangular coordinate system. The graph of the ordered pairs defines y as a *one-to-one* function of x if no horizontal line intersects the graph in more than one point.

To understand the horizontal line test, consider the functions f and g.

$f = \{(1, 2.99), (1.5, 4.49), (4, 11.96)\}$ $\quad g = \{(1, 4), (2, 3), (-2, 4)\}$

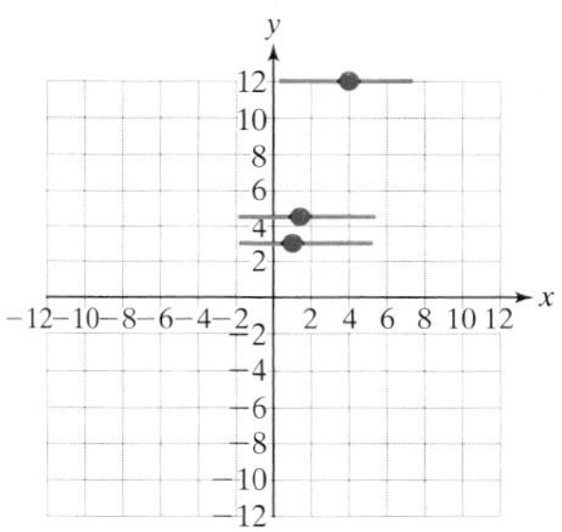

This function is one-to-one.
No horizontal line intersects more than once.

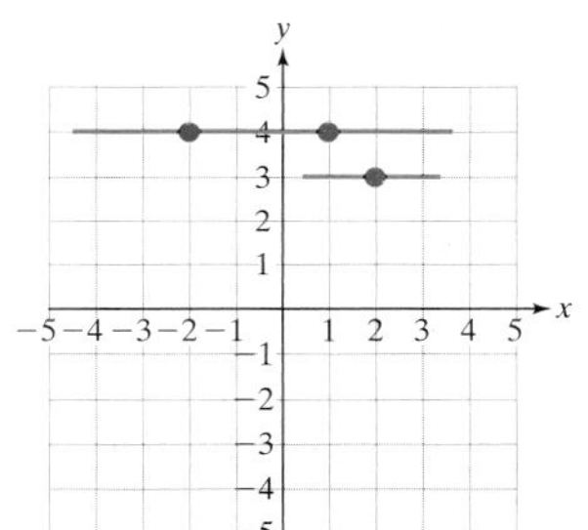

This function is *not* one-to-one.
A horizontal line intersects more than once.

example 1 **Identifying One-to-One Functions**

Determine whether the function is one-to-one.

a.

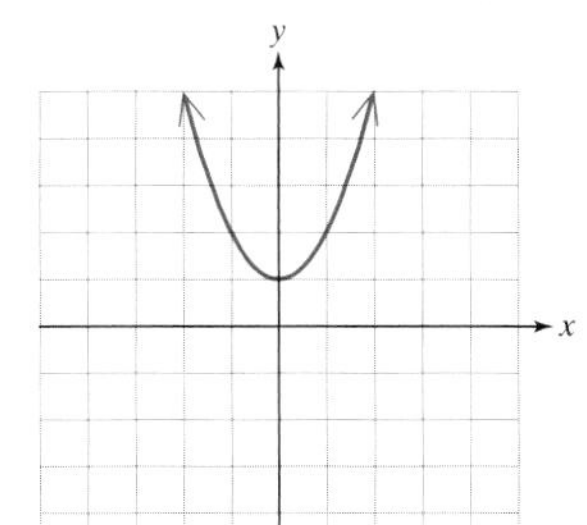

b.

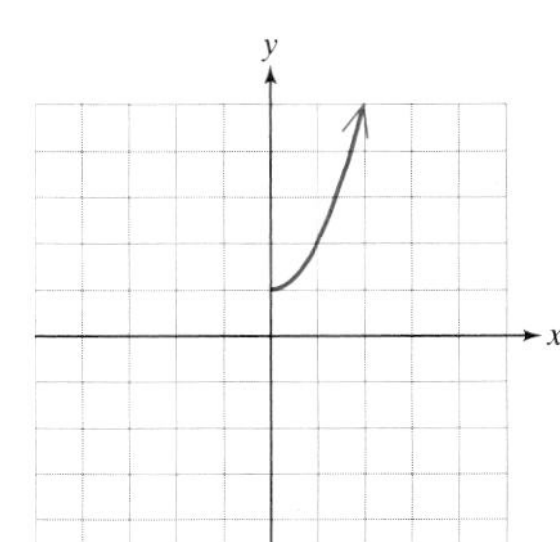

Solution:

a. Function is not one-to-one. A horizontal line intersects in more than one point.

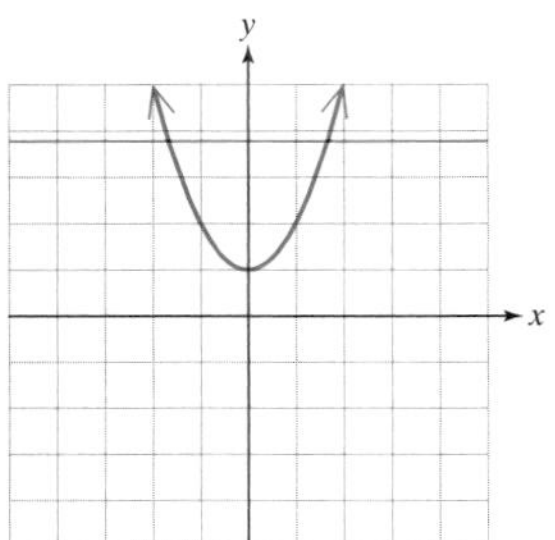

b. Function is one-to-one. No horizontal line intersects more than once.

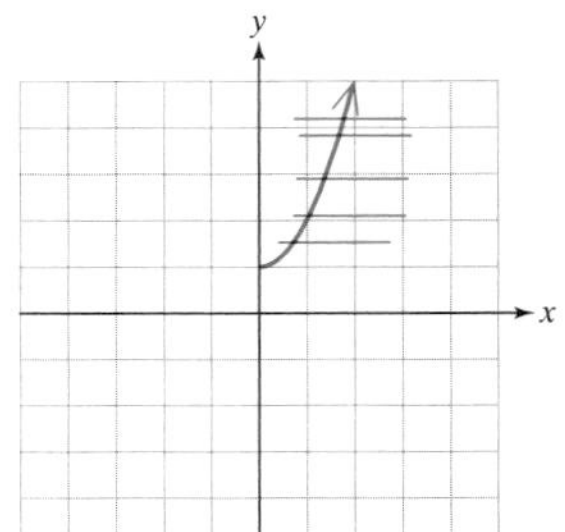

4. Finding an Equation of the Inverse of a Function

Another way to view the construction of the inverse of a function is to find a function that performs the inverse operations in the reverse order. For example, the function defined by $f(x) = 2x + 1$ multiplies x by 2 and then adds 1. Therefore, the inverse function must *subtract* 1 from x and *divide* by 2. We have

$$f^{-1}(x) = \frac{x - 1}{2}$$

To facilitate the process of finding an equation of the inverse of a one-to-one function, we offer the following steps.

Finding an Equation of an Inverse of a Function

For a one-to-one function defined by $y = f(x)$, the equation of the inverse can be found as follows:

1. Replace $f(x)$ by y.
2. Interchange x and y.
3. Solve for y.
4. Replace y by $f^{-1}(x)$.

example 2

Finding an Equation of the Inverse of a Function

Find the inverse: $f(x) = 2x + 1$

Solution:

Foremost, we know the graph of f is a nonvertical line. Therefore, $f(x) = 2x + 1$ defines a one-to-one function. To find the inverse we have

$y = 2x + 1$	**Step 1:**	Replace $f(x)$ by y.
$x = 2y + 1$	**Step 2:**	Interchange x and y.

$x - 1 = 2y$	**Step 3:**	Solve for y. Subtract 1 from both sides.
$\frac{x-1}{2} = y$		Divide both sides by 2.
$f^{-1}(x) = \frac{x-1}{2}$	**Step 4:**	Replace y by $f^{-1}(x)$.

The key step in determining the equation of the inverse of a function is interchanging x and y. By so doing, a point (a, b) on f corresponds to a point (b, a) on f^{-1}. For this reason, the graph of f and f^{-1} are symmetric with respect to the line $y = x$ (Figure 12-3).

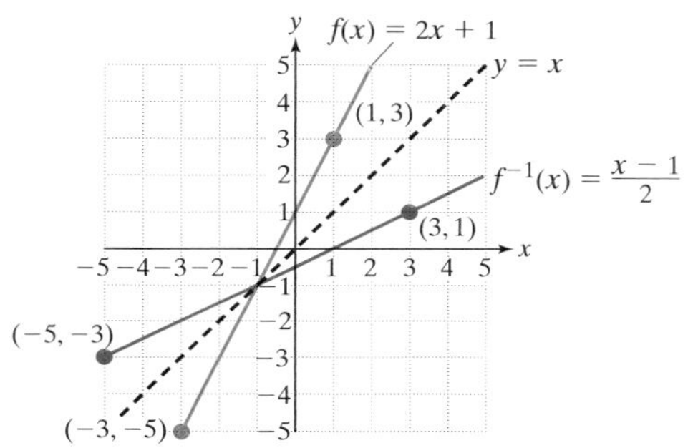

Figure 12-3

example 3 Finding an Equation of the Inverse of a Function

Find the inverse of the one-to-one function: $g(x) = \sqrt[3]{5x} - 4$.

Solution:

$y = \sqrt[3]{5x} - 4$	**Step 1:**	Replace $g(x)$ by y.
$x = \sqrt[3]{5y} - 4$	**Step 2:**	Interchange x and y.
$x + 4 = \sqrt[3]{5y}$	**Step 3:**	Solve for y. Add 4 to both sides.
$(x + 4)^3 = (\sqrt[3]{5y})^3$		To eliminate the cube root, cube both sides.
$(x + 4)^3 = 5y$		Simplify the right side.
$\frac{(x + 4)^3}{5} = y$		Divide both sides by 5.
$g^{-1}(x) = \frac{(x + 4)^3}{5}$	**Step 4:**	Replace y by $g^{-1}(x)$.

The graphs of g and g^{-1} are shown in Figure 12-4. Once again we see that the graphs of a function and its inverse are symmetric with respect to the line $y = x$.

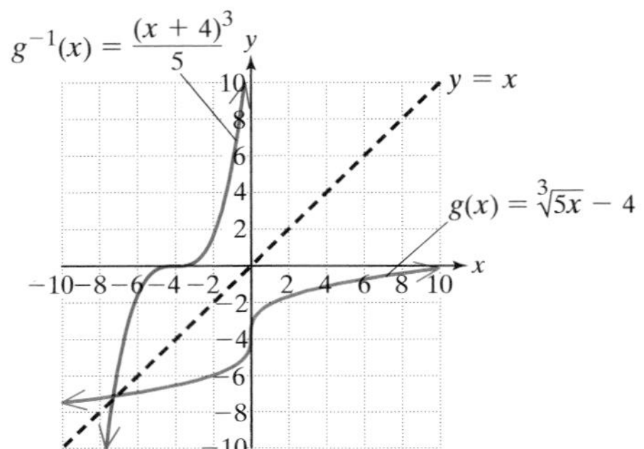

Figure 12-4

5. Composition of a Function and Its Inverse

An important relationship between a function and its inverse is shown in Figure 12-5.

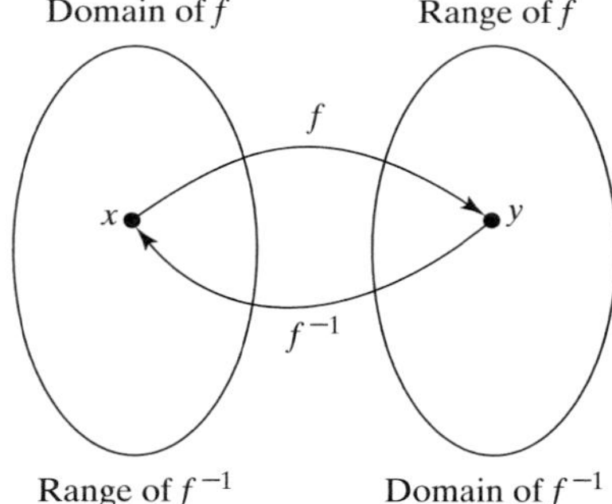

Figure 12-5

Recall that the domain of f is the range of f^{-1} and the range of f is the domain of f^{-1}. The operations performed by f are reversed by f^{-1}. This leads to a formal definition of an inverse of a function.

Definition of the Inverse of a Function

If f is a one-to-one function, then g is the inverse of f if and only if

$$(f \circ g)(x) = x \quad \text{for all } x \text{ in the domain of } g$$

and

$$(g \circ f)(x) = x \quad \text{for all } x \text{ in the domain of } f$$

example 4

Composing a Function with Its Inverse

Show that the functions are inverses.

$$h(x) = 2x + 1 \qquad \text{and} \qquad k(x) = \frac{x - 1}{2}$$

Solution:

To show that the functions h and k are inverses we need to confirm that $(h \circ k)(x) = x$ and $(k \circ h)(x) = x$.

$$\begin{aligned}(h \circ k)(x) &= h(k(x)) = 2(k(x)) + 1 \\ &= \cancel{2}\left(\frac{x - 1}{\cancel{2}}\right) + 1 \\ &= x - 1 + 1 \\ &= x \checkmark \qquad (h \circ k)(x) = x \text{ as desired.}\end{aligned}$$

$$\begin{aligned}(k \circ h)(x) = k(h(x)) &= \frac{(h(x)) - 1}{2} \\ &= \frac{(2x + 1) - 1}{2} \\ &= \frac{2x + \cancel{1} - \cancel{1}}{2} \\ &= \frac{2x}{2} \\ &= x \checkmark \qquad (k \circ h)(x) = x \text{ as desired.}\end{aligned}$$

The functions h and k are inverses because $(h \circ k)(x) = x$ and $(k \circ h)(x) = x$, for all real numbers, x.

6. Restricting the Domain of a Function

For a function that is not one-to-one, sometimes we can restrict its domain to create a new function that is one-to-one. This is demonstrated in the next example.

example 5

Finding the Equation of an Inverse of a Function with a Restricted Domain

Given the function defined by $m(x) = x^2 + 4$ for $x \geq 0$, find an equation defining m^{-1}.

Solution:

From Section 11.4, we know that $y = x^2 + 4$ is a parabola with the y-intercept at (0, 4) (Figure 12-6). The graph represents a function that is not one-to-one. However, with the restriction on the domain $x \geq 0$, the graph of $m(x) = x^2 + 4$, $x \geq 0$ consists of only the "right" branch of the parabola (Figure 12-7). This *is* a one-to-one function.

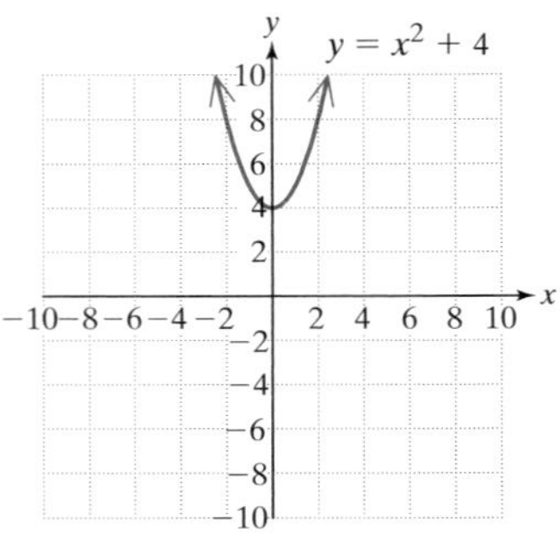

Figure 12-6

Figure 12-7

To find the inverse, we have

$y = x^2 + 4; \quad x \geq 0$	**Step 1:** Replace $m(x)$ by y.
$x = y^2 + 4; \quad y \geq 0$	**Step 2:** Interchange x and y. Notice that the restriction $x \geq 0$ becomes $y \geq 0$.
$x - 4 = y^2; \quad y \geq 0$	**Step 3:** Solve for y. Subtract 4 from both sides.
$\sqrt{x - 4} = y; \quad y \geq 0$	Apply the square root property. Notice that we obtain the *positive* square root of $x - 4$ because of the restriction $y \geq 0$.
$m^{-1}(x) = \sqrt{x - 4}; \quad x \geq 4$	**Step 4:** Replace y by $m^{-1}(x)$. Notice that the domain of m^{-1} has the same values as the range of m.

Figure 12-8 shows the graphs of m and m^{-1}.

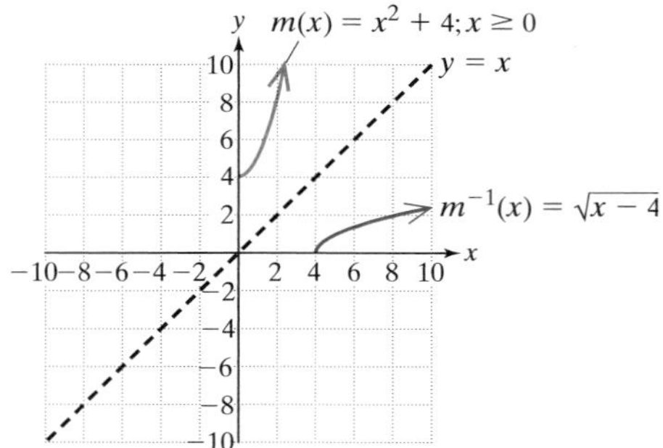

Figure 12-8

section 12.2 Practice Exercises

For Exercises 1–6, determine if the relation is a function by using the vertical line test. (See Section 7.4.)

1.

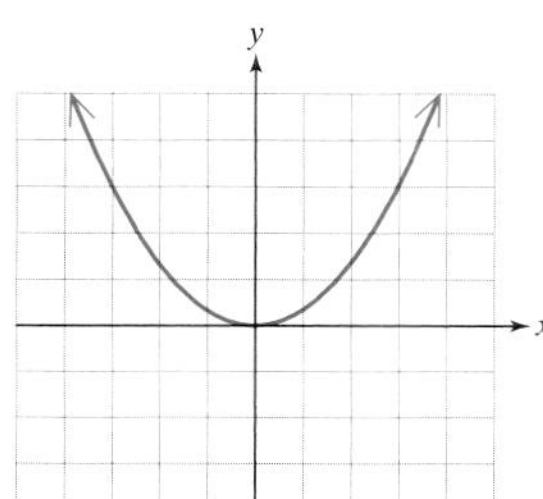

2.

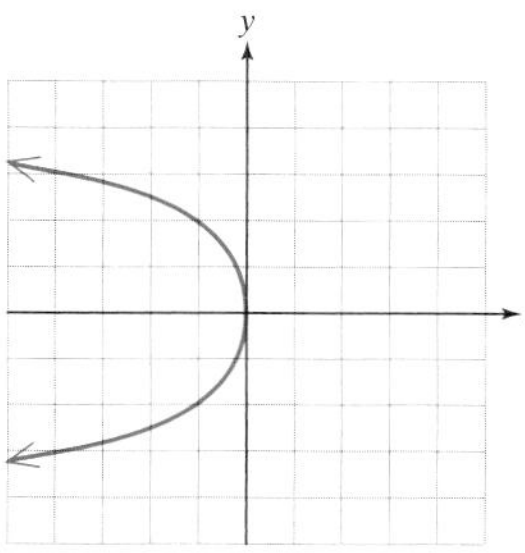

3.

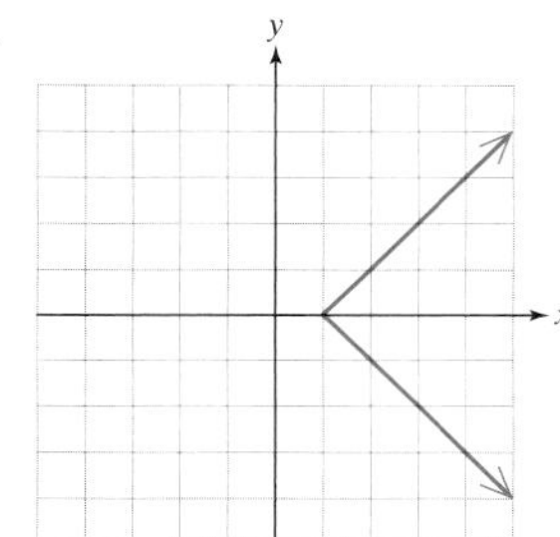

4.

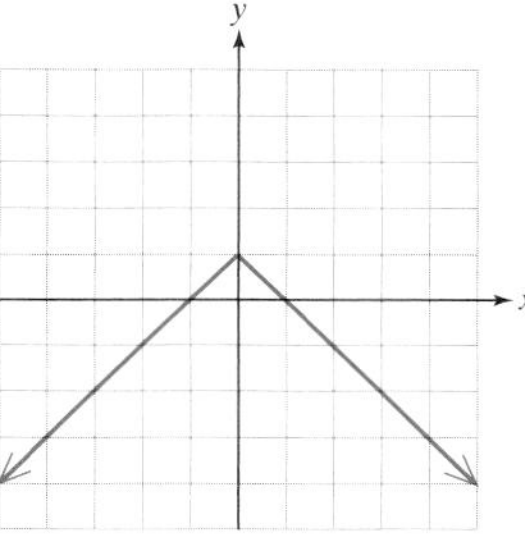

5.

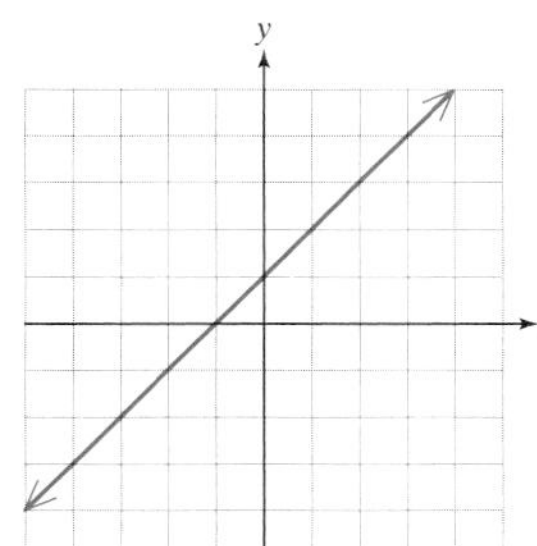

6.

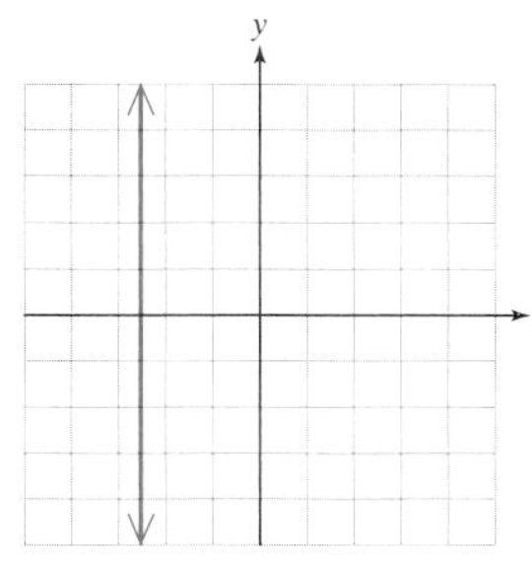

For Exercises 7–12, determine if the function is one-to-one by using the horizontal line test.

7.

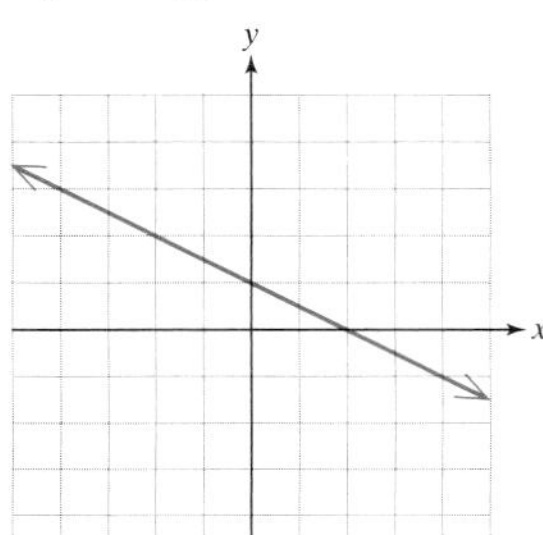

8.

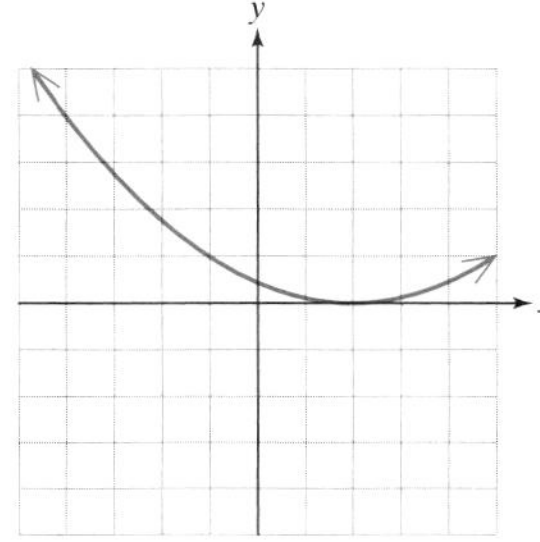

9.

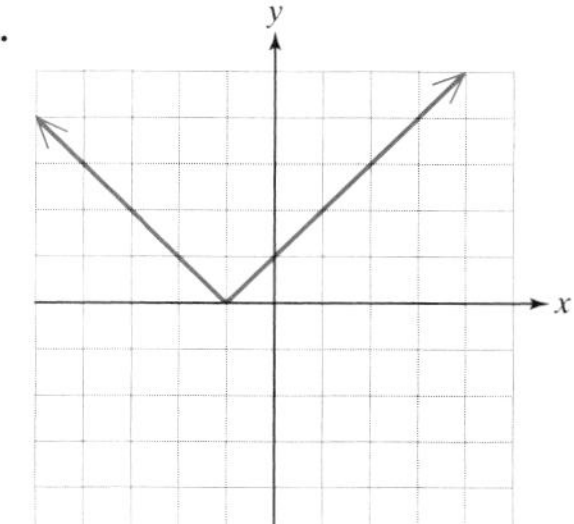

10.

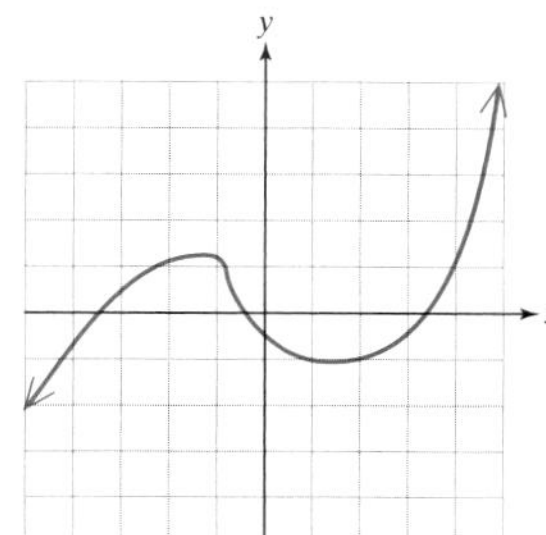

11.

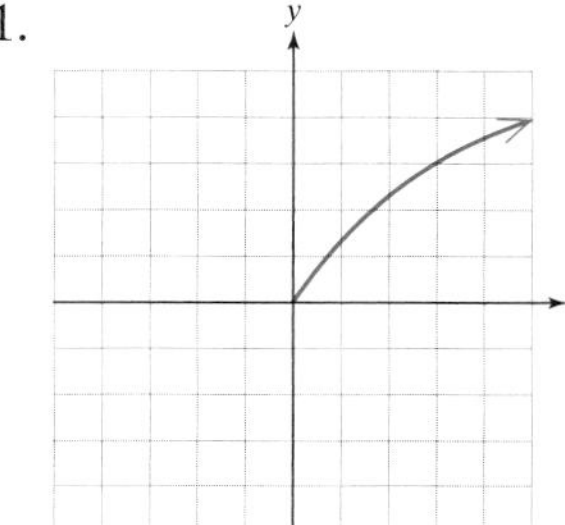

12.

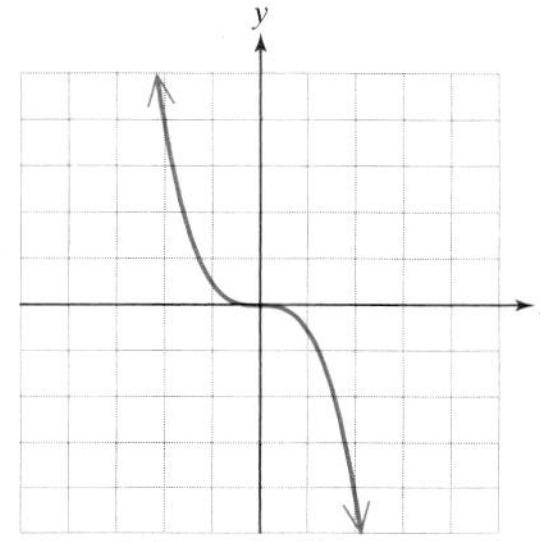

For Exercises 13–16, write the inverse function for each function as defined.

13. $g = \{(3, 5), (8, 1), (-3, 9), (0, 2)\}$

14. $f = \{(-6, 2), (-9, 0), (-2, -1), (3, 4)\}$

15. $r = \{(a, 3), (b, 6), (c, 9)\}$

16. $s = \{(-1, x), (-2, y), (-3, z)\}$

17. The table relates a state (x) to the number of representatives in the House of Representatives (y) in the year 2000. Does this relation define a one-to-one function? If so, write a function defining the inverse as a set of ordered pairs.

State (x)	Number of Representatives (y)
Colorado	6
California	52
Texas	30
Connecticut	6
Pennsylvania	21

Table for Exercise 17

18. The table relates a city, (x) to its average January temperature (y). Does this relation define a one-to-one function? If so, write a function defining the inverse as a set of ordered pairs.

City (x)	Temperature (°C) (y)
Gainesville, FL	13.6
Keene, NH	−6.0
Wooster, OH	−4.0
Rock Springs, WY	−6.0
Lafayette, LA	10.9

Table for Exercise 18

For Exercises 19–28, write an equation of the inverse for each one-to-one function as defined.

19. $h(x) = x + 4$
20. $k(x) = x - 3$
21. $m(x) = \frac{1}{3}x - 2$
22. $n(x) = 4x + 2$
23. $p(x) = -x + 10$
24. $q(x) = -x - \frac{2}{3}$
25. $f(x) = x^3$
26. $g(x) = \sqrt[3]{x}$
27. $g(x) = \sqrt[3]{2x - 1}$
28. $f(x) = x^3 - 4$

29. The function defined by $f(x) = 0.3048x$ converts a length of x feet into $f(x)$ meters.
    a. Find the equivalent length in meters for a 4-ft board and a 50-ft wire.
    b. Find an equation defining $y = f^{-1}(x)$.
    c. Use the inverse function from part (b) to find the equivalent length in feet for a 1500-m race in track and field. Round to the nearest tenth of a foot.

30. The function defined by $s(x) = 1.47x$ converts a speed of x miles per hour to $s(x)$ feet per second.
    a. Find the equivalent speed in feet per second for a car traveling 60 mph.
    b. Find an equation defining $y = s^{-1}(x)$.
    c. Use the inverse function from part (b) to find the equivalent speed in miles per hour for a train traveling 132 ft/sec. Round to the nearest tenth.

For Exercises 31–37, answer true or false.

31. The function defined by $y = 2$ has an inverse function defined by $x = 2$.
32. The domain of any one-to-one function is the same as the domain of its inverse.
33. All linear functions have an inverse function.
34. The function defined by $g(x) = |x|$ is one-to-one.
35. The function defined by $k(x) = x^2$ is one-to-one.
36. The function defined by $h(x) = |x|$ for $x \geq 0$ is one-to-one.
37. The function defined by $L(x) = x^2$ for $x \geq 0$ is one-to-one.
38. Explain how the domain and range of a one-to-one function and its inverse are related.
39. If $(0, b)$ is the y-intercept of a one-to-one function, what is the x-intercept of its inverse?
40. If $(a, 0)$ is the x-intercept of a one-to-one function, what is the y-intercept of its inverse?
41. Can you think of any function that is its own inverse?
42. a. What are the domain and range of the function defined by $f(x) = \sqrt{x - 1}$?
    b. What are the domain and range of the function defined by $f^{-1}(x) = x^2 + 1; x \geq 0$?
43. a. What are the domain and range of the function defined by $g(x) = x^2 - 4; x \leq 0$?
    b. What are the domain and range of the function defined by $g^{-1}(x) = -\sqrt{x + 4}$?

For Exercises 44–47, given the graph of $y = f(x)$,

a. State the domain of f.
b. State the range of f.
c. State the domain of f^{-1}.
d. State the range of f^{-1}.
e. Graph the function defined by $y = f^{-1}(x)$. The line $y = x$ is provided for your reference.

44.

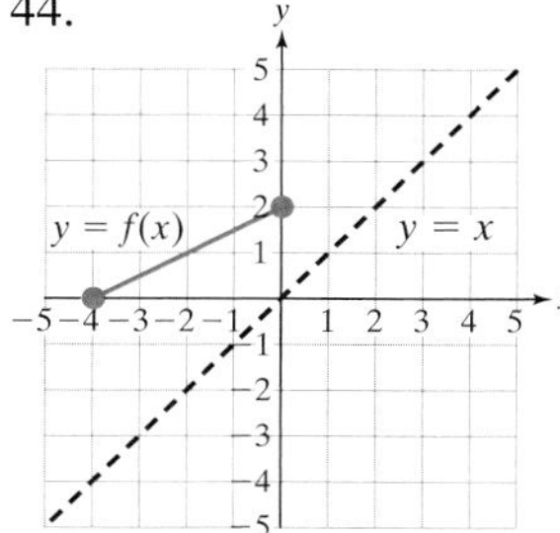

45.

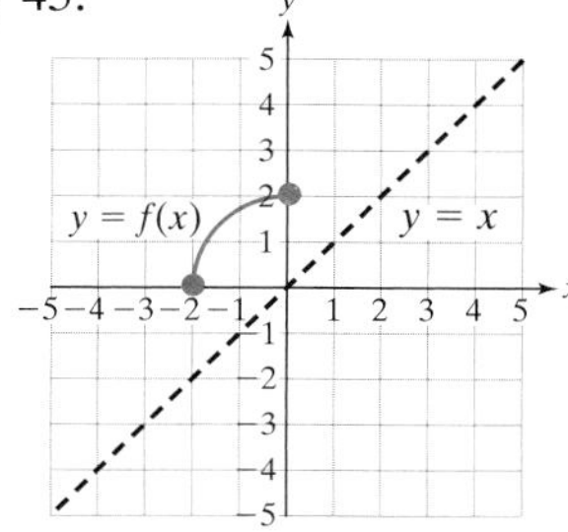

46.

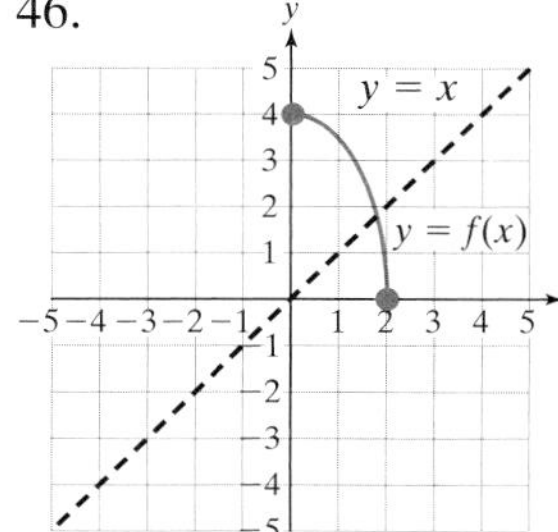

47.

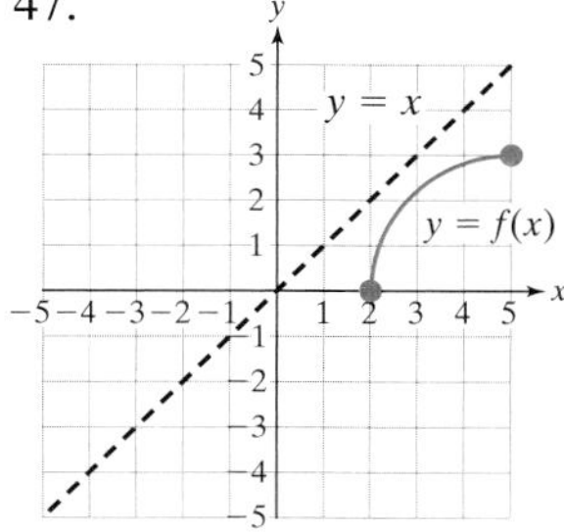

Expanding Your Skills

For Exercises 48–59, write an equation of the inverse of the one-to-one function.

48. $f(x) = \dfrac{x-1}{x+1}, \quad x \neq -1$

49. $p(x) = \dfrac{3-x}{x+3}, \quad x \neq -3$

50. $t(x) = \dfrac{2}{x-1}, \quad x \neq 1$

51. $w(x) = \dfrac{4}{x+2}, \quad x \neq -2$

52. $g(x) = x^2 + 9, \quad x \geq 0$

53. $m(x) = x^2 - 1, \quad x \geq 0$

54. $n(x) = x^2 + 9, \quad x \leq 0$

55. $g(x) = x^2 - 1, \quad x \leq 0$

56. $q(x) = \sqrt{x+4}$

57. $v(x) = \sqrt{x+16}$

58. $z(x) = -\sqrt{x+4}$

59. $u(x) = -\sqrt{x+16}$

Graphing Calculator Exercises

For Exercises 60–63, use a graphing calculator to graph each function on the standard viewing window defined by $-10 \leq x \leq 10$ and $-10 \leq y \leq 10$. Use the graph of the function to determine if the function is one-to-one on the interval $-10 \leq x \leq 10$. If the function is one-to-one, find its inverse and graph both functions on the standard viewing window.

60. $f(x) = \sqrt[3]{x+5}$

61. $k(x) = x^3 - 4$

62. $g(x) = 0.5x^3 - 2$

63. $m(x) = 3x - 4$

Concepts

1. Definition of an Exponential Function
2. Approximating Exponential Expressions with a Calculator
3. Graphs of Exponential Functions
4. Applications of Exponential Functions—Radioactive Decay
5. Applications of Exponential Functions—Population Growth

section

12.3 Exponential Functions

1. Definition of an Exponential Function

The concept of a function was first introduced in Section 7.4. Since then we have learned to recognize several categories of functions, including constant, linear, rational, radical, and quadratic functions. In this section and Section 12.4, we will define two new types of functions called exponential and logarithmic functions.

To introduce the concept of an exponential function, consider the following salary plans for a new job. Plan A pays \$1 million for a month's work. Plan B starts with a 1¢ signing bonus, and every day beginning with day 1 the salary is doubled.

At first glance, the million-dollar plan appears to be more favorable. Look, however, at Table 12-1, which shows the daily payments for 30 days under plan B.

Table 12-1

Day	Payment	Day	Payment	Day	Payment
1	2¢	11	$20.48	21	$20,971.52
2	4¢	12	$40.96	22	$41,943.04
3	8¢	13	$81.92	23	$83,886.08
4	16¢	14	$163.84	24	$167,772.16
5	32¢	15	$327.68	25	$335,544.32
6	64¢	16	$655.36	26	$671,088.64
7	$1.28	17	$1310.72	27	$1,342,177.28
8	$2.56	18	$2621.44	28	$2,684,354.56
9	$5.12	19	$5242.88	29	$5,368,709.12
10	$10.24	20	$10,485.76	30	$10,737,418.24

Notice that the salary on the 30th day for plan B is over $10 million. Taking the sum of the payments, the total salary for the 30-day period is $21,474,836.46.

The daily salary for plan B can be represented by the function, $y = 2^x$, where x is the number of days on the job, and y is the salary (in cents) for that day. An interesting feature of this function is that for every positive 1-unit change in x, the y-value doubles. The function $y = 2^x$ is called an exponential function.

Definition of an Exponential Function

Let b be any real number such that $b > 0$ and $b \neq 1$. Then a function of the form $y = b^x$ is called an **exponential function**.

An exponential function is easily recognized as a function with a constant base and variable exponent. Notice that the base of an exponential function must be a positive real number not equal to 1.

2. Approximating Exponential Expressions with a Calculator

Up to this point, we have evaluated exponential expressions with integer exponents and with rational exponents. For example, $4^3 = 64$ and $4^{1/2} = \sqrt{4} = 2$. However,

Calculator Connections

On a graphing calculator, use the ^ key to approximate an expression with an irrational exponent.

```
4^π
        77.88023365
```

how do we evaluate an exponential expression with an irrational exponent such as 4^{π}? In such a case, the exponent is a nonterminating and nonrepeating decimal. The value of 4^{π} can be thought of as the limiting value of a sequence of approximations using rational exponents:

$$4^{3.14} \approx 77.7084726$$
$$4^{3.141} \approx 77.81627412$$
$$4^{3.1415} \approx 77.87023095$$
$$\ldots$$
$$4^{\pi} \approx 77.88023365$$

An exponential expression can be evaluated at all rational numbers and at all irrational numbers. Hence, the domain of an exponential function is all real numbers.

example 1

Approximating Exponential Expressions with a Calculator

Approximate the expressions. Round the answers to four decimal places.

a. $8^{\sqrt{3}}$ b. $5^{-\sqrt{17}}$ c. $\sqrt{10}^{\sqrt{2}}$

Solution:

a. $8^{\sqrt{3}} \approx 36.6604$
b. $5^{-\sqrt{17}} \approx 0.0013$
c. $\sqrt{10}^{\sqrt{2}} \approx 5.0946$

Calculator Connections

The expressions from Example 1 can be approximated on a graphing calculator as shown.

```
8^√(3)
        36.66044576
5^-√(17)
        .00131242
(√(10))^√(2)
        5.09456117
```

3. Graphs of Exponential Functions

The functions defined by $f(x) = 2^x$, $g(x) = 3^x$, $h(x) = 5^x$, and $k(x) = (\frac{1}{2})^x$ are all examples of exponential functions. Example 2 illustrates the two general shapes of exponential functions.

example 2

Graphing an Exponential Function

Graph the functions f and g.

a. $f(x) = 2^x$ b. $g(x) = (\frac{1}{2})^x$

Solution:

Table 12-2 shows several function values, $f(x)$ and $g(x)$, for both positive and negative values of x. The graph is shown in Figure 12-9.

Table 12-2

x	$f(x) = 2^x$	$g(x) = (\frac{1}{2})^x$
-4	$\frac{1}{16}$	16
-3	$\frac{1}{8}$	8
-2	$\frac{1}{4}$	4
-1	$\frac{1}{2}$	2
0	1	1
1	2	$\frac{1}{2}$
2	4	$\frac{1}{4}$
3	8	$\frac{1}{8}$
4	16	$\frac{1}{16}$

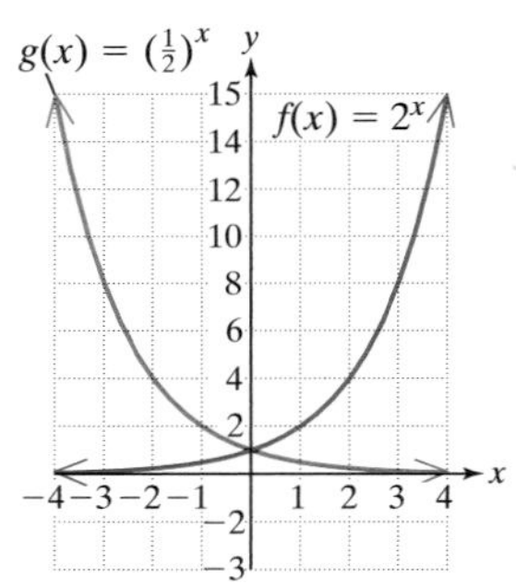

Figure 12-9

The graphs in Figure 12-9 illustrate several important features of exponential functions.

Graphs of $f(x) = b^x$

The graph of an exponential function defined by $f(x) = b^x$ ($b > 0$ and $b \neq 1$) has the following properties.

1. If $b > 1$, f is an *increasing* exponential function (sometimes called an **exponential growth function**).

   If $0 < b < 1$, f is a *decreasing* exponential function (sometimes called an **exponential decay function**).
2. The domain is the set of all real numbers.
3. The range is $(0, \infty)$.
4. The x-axis is a horizontal asymptote.
5. The function passes through the point $(0, 1)$ because $f(0) = b^0 = 1$.

These properties state that the graph of an exponential function is an increasing function if the base is greater than 1. Furthermore, the base affects its "steepness." Consider the graphs of $f(x) = 2^x$, $h(x) = 3^x$, and $k(x) = 5^x$ (Figure 12-10). For every positive 1-unit change in x, $f(x) = 2^x$ increases by 2 times, $h(x) = 3^x$ increases by 3 times, and $k(x) = 5^x$ increases by 5 times (Table 12-3).

Table 12-3

x	$f(x) = 2^x$	$h(x) = 3^x$	$k(x) = 5^x$
-3	$\frac{1}{8}$	$\frac{1}{27}$	$\frac{1}{125}$
-2	$\frac{1}{4}$	$\frac{1}{9}$	$\frac{1}{25}$
-1	$\frac{1}{2}$	$\frac{1}{3}$	$\frac{1}{5}$
0	1	1	1
1	2	3	5
2	4	9	25
3	8	27	125

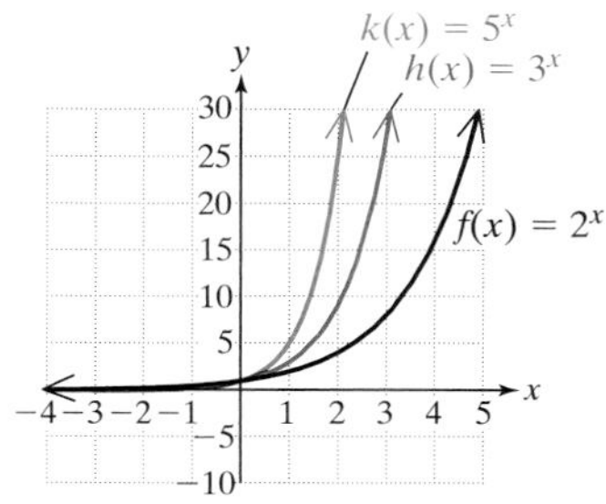

Figure 12-10

The graph of an exponential function is a *decreasing function* if the base is between 0 and 1. Consider the graphs of $g(x) = (\frac{1}{2})^x$, $m(x) = (\frac{1}{3})^x$, and $n(x) = (\frac{1}{5})^x$ (Table 12-4 and Figure 12-11).

Table 12-4

x	$g(x) = (\frac{1}{2})^x$	$m(x) = (\frac{1}{3})^x$	$n(x) = (\frac{1}{5})^x$
-3	8	27	125
-2	4	9	25
-1	2	3	5
0	1	1	1
1	$\frac{1}{2}$	$\frac{1}{3}$	$\frac{1}{5}$
2	$\frac{1}{4}$	$\frac{1}{9}$	$\frac{1}{25}$
3	$\frac{1}{8}$	$\frac{1}{27}$	$\frac{1}{125}$

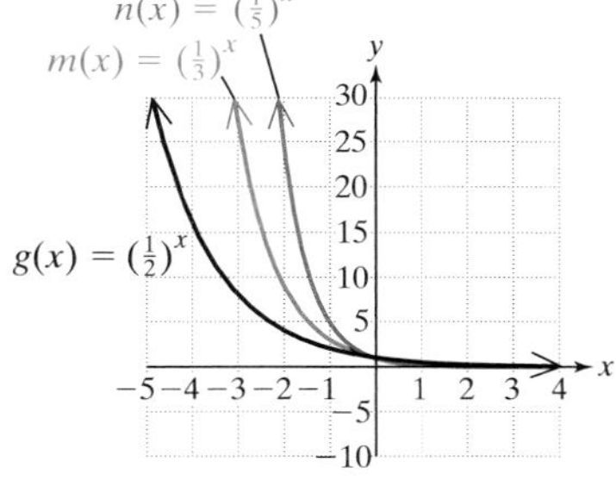

Figure 12-11

4. Applications of Exponential Functions—Radioactive Decay

Exponential growth and decay can be found in a variety of real-world phenomena. For example,

- Population growth can often be modeled by an exponential function.
- The growth of an investment under compound interest grows exponentially.
- The mass of a radioactive substance decreases exponentially with time.
- The temperature of a cup of coffee decreases exponentially as it approaches room temperature.

example 3 Applying an Exponential Function

Marie and Pierre Curie

A substance that undergoes radioactive decay is said to be radioactive. The **half-life** of a radioactive substance is the amount of time it takes for one half of the original amount of the substance to change into something else. That is, after each half-life the amount of the original substance decreases by half.

In 1898, Marie Curie discovered the highly radioactive element, radium. She shared the 1903 Nobel Prize in physics for her research on radioactivity and was awarded the 1911 Nobel Prize in chemistry for her discovery of radium and polonium. Radium-226 (an isotope of radium) has a half-life of 1620 years and decays into radon-222 (a radioactive gas).

In a sample originally having 1 g of radium-226, the amount of radium-226 present after t years is given by

$A(t) = (\frac{1}{2})^{t/1620}$, where A is the amount of radium in grams, and t is the time in years.

a. How much radium-226 will be present after 1620 years?
b. How much radium-226 will be present after 3240 years?
c. How much radium-226 will be present after 4860 years?

Solution:

a. $$\begin{aligned} A(1620) &= \left(\frac{1}{2}\right)^{1620/1620} \\ &= \left(\frac{1}{2}\right)^{1} \\ &= 0.5 \end{aligned}$$

After 1620 years (1 half-life), 0.5 g remains.

b. $$\begin{aligned} A(3240) &= \left(\frac{1}{2}\right)^{3240/1620} \\ &= \left(\frac{1}{2}\right)^{2} \\ &= 0.25 \end{aligned}$$

After 3240 years (2 half-lifes), the amount of the original substance is reduced by half, two times: 0.25 g remains.

c. $$\begin{aligned} A(4860) &= \left(\frac{1}{2}\right)^{4860/1620} \\ &= \left(\frac{1}{2}\right)^{3} \\ &= 0.125 \end{aligned}$$

After 4860 years (3 half-lifes), the amount of the original substance is reduced by half, three times: 0.125 g remains.

5. Applications of Exponential Functions—Population Growth

Exponential functions are often used to model population growth. Suppose the initial value of a population at some time $t = 0$ is P_0. If the rate of increase of a population is r, then after 1, 2, and 3 years, the new population can be found as follows:

After 1 year:
$$\begin{pmatrix}\text{Total}\\ \text{population}\end{pmatrix} = \begin{pmatrix}\text{initial}\\ \text{population}\end{pmatrix} + \begin{pmatrix}\text{increase in}\\ \text{population}\end{pmatrix}$$
$$= P_0 + P_0 r$$
$$= P_0(1 + r) \qquad \text{Factor out } P_0.$$

After 2 years:
$$\begin{pmatrix}\text{Total}\\ \text{population}\end{pmatrix} = \begin{pmatrix}\text{population}\\ \text{after 1 year}\end{pmatrix} + \begin{pmatrix}\text{increase in}\\ \text{population}\end{pmatrix}$$
$$= P_0(1 + r) + P_0(1 + r)r$$
$$= P_0(1 + r)1 + P_0(1 + r)r$$
$$= P_0(1 + r)(1 + r) \qquad \text{Factor out } P_0(1 + r).$$
$$= P_0(1 + r)^2$$

After 3 years:
$$\begin{pmatrix}\text{Total}\\ \text{population}\end{pmatrix} = \begin{pmatrix}\text{population}\\ \text{after 2 years}\end{pmatrix} + \begin{pmatrix}\text{increase in}\\ \text{population}\end{pmatrix}$$
$$= P_0(1 + r)^2 + P_0(1 + r)^2 r$$
$$= P_0(1 + r)^2 1 + P_0(1 + r)^2 r$$
$$= P_0(1 + r)^2(1 + r) \qquad \text{Factor out } P_0(1 + r)^2.$$
$$= P_0(1 + r)^3$$

This pattern continues, and after t years, the population $P(t)$ is given by

$$P(t) = P_0(1 + r)^t$$

example 4 **Applying an Exponential Function**

The population of the Bahamas in 1998 was estimated at 280,000 with an annual rate of increase of 1.39%.

a. Find a mathematical model that relates the population of the Bahamas as a function of the number of years after 1998.
b. If the annual rate of increase remains the same, use this model to predict the population of the Bahamas in the year 2010. Round to the nearest thousand.

Solution:

a. The initial population is $P_0 = 280{,}000$ and the rate of increase is $r = 1.39\%$.

$P(t) = P_0(1 + r)^t$ — Substitute $P_0 = 280{,}000$ and $r = 0.0139$.

$= 280{,}000(1 + 0.0139)^t$

$= 280{,}000(1.0139)^t$ — Where $t = 0$ corresponds to the year 1998

b. Because the initial population ($t = 0$) corresponds to the year 1998, we use $t = 12$ to find the population in the year 2010.

$P(12) = 280{,}000(1.0139)^{12}$
$\approx 330{,}000$

section 12.3 Practice Exercises

For Exercises 1–8, evaluate the expression without the use of a calculator.

1. 5^2
2. 2^{-3}
3. 10^{-3}
4. 3^4
5. $36^{1/2}$
6. $27^{1/3}$
7. $16^{3/4}$
8. $8^{2/3}$

For Exercises 9–16, evaluate the expression using a calculator. Round to four decimal places.

9. $5^{1.1}$
10. $2^{\sqrt{3}}$
11. 10^{π}
12. $3^{4.8}$
13. $36^{-\sqrt{2}}$
14. $27^{-0.5126}$
15. $16^{-0.04}$
16. $8^{-0.61}$

17. Solve for x.
    a. $3^x = 9$
    b. $3^x = 27$
    c. Approximate the solution for x in $3^x = 11$.

18. Solve for x.
    a. $5^x = 125$
    b. $5^x = 625$
    c. Approximate the solution for x in $5^x = 130$.

19. Solve for x.
    a. $2^x = 16$
    b. $2^x = 32$
    c. Approximate the solution for x in $2^x = 30$.

20. Solve for x.
    a. $4^x = 16$
    b. $4^x = 64$
    c. Approximate the solution for x in $4^x = 20$.

21. For $f(x) = (\frac{1}{5})^x$ find $f(0), f(1), f(2), f(-1)$, and $f(-2)$.

22. For $g(x) = (\frac{2}{3})^x$ find $g(0), g(1), g(2), g(-1)$, and $g(-2)$.

23. For $h(x) = \pi^x$ use a calculator to find $h(0), h(1), h(-1), h(\sqrt{2})$, and $h(\pi)$. Round to two decimal places.

24. For $k(x) = (\sqrt{5})^x$ use a calculator to find $k(0), k(1), k(-1), k(\pi)$, and $k(\sqrt{2})$. Round to two decimal places.

25. For $r(x) = 3^{x+2}$ find $r(0), r(1), r(2), r(-1), r(-2)$, and $r(-3)$.

26. For $s(x) = 2^{2x-1}$ find $s(0), s(1), s(2), s(-1)$, and $s(-2)$.

27. How do you determine whether the graph of $f(x) = b^x$ is increasing or decreasing?

28. For $f(x) = b^x, (b > 0, b \neq 1)$, find $f(0)$.

Graph the functions defined in Exercises 29–36. Plot at least three points for each function.

29. $f(x) = 4^x$
30. $g(x) = 6^x$

31. $m(x) = \left(\frac{1}{8}\right)^x$

32. $n(x) = \left(\frac{1}{3}\right)^x$

33. $h(x) = 2^{x+1}$

34. $k(x) = 5^{x-1}$

35. $g(x) = 5^{-x}$

36. $f(x) = 2^{-x}$

37. Suppose \$1000 is initially invested in an account and the value of the account grows exponentially. If the investment doubles in 7 years, then the amount in the account t years after the initial investment is given by

$A(t) = 1000(2)^{t/7}$, where t is expressed in years and $A(t)$ is the amount in the account.

a. Find the amount in the account after 5 years.

b. Find the amount in the account after 10 years.

c. Find $A(0)$ and $A(7)$ and interpret the answers in the context of the problem.

38. Suppose \$1500 is initially invested in an account and the value of the account grows exponentially. If the investment doubles in 8 years, then the amount in the account t years after the initial investment is given by

$A(t) = 1500(2)^{t/8}$, where t is expressed in years and $A(t)$ is the amount in the account.

a. Find the amount in the account after 5 years.

b. Find the amount in the account after 10 years.

c. Find $A(0)$ and $A(8)$ and interpret the answers in the context of the problem.

39. The population of Ireland in 1998 was estimated at 3,600,000, with an annual rate of increase of 0.36%. The population of Singapore in 1998 was estimated at 3,500,000, with an annual rate of increase of 1.20%.

a. Write a mathematical model that describes the population of Ireland, $I(t)$, as a function of the number of years, t, after 1998. (See Example 4.)

b. Write a mathematical model that describes the population of Singapore, $S(t)$, as a function of the number of years, t, after 1998.

c. The populations of the two countries were very nearly the same in 1998. However, Singapore has a higher rate of increase. Examine how the rate of increase affects population over time by predicting the populations of Ireland and Singapore 20, 40, and 60 years after 1998. Round to the nearest hundred thousand.

d. Although Singapore had fewer people in 1998, it also has a higher growth rate. What is the effect of the growth rate when comparing the populations of two countries?

e. The population of Singapore is growing more rapidly than that of Ireland. Furthermore, the land area in Singapore is only about 1/100 that of Ireland. What conclusion can you make about *population density* of the two countries?

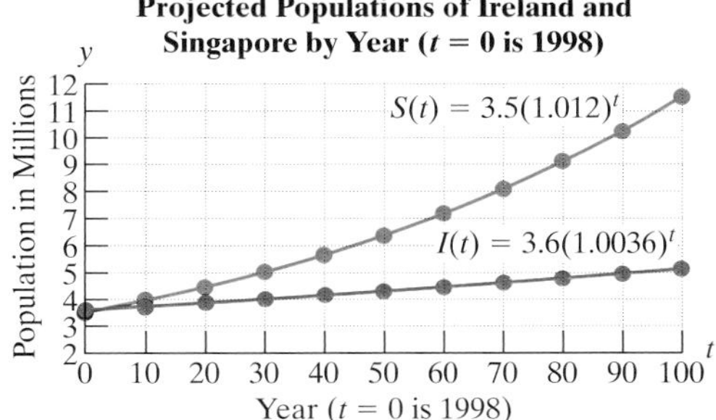

Figure for Exercise 39

40. The population of Pakistan in 1998 was estimated at 135,000,000, with an annual rate of increase of 2.20%. The population of Brazil in 1998 was estimated at 170,000,000, with an annual rate of increase of 1.24%.

a. Write a mathematical model that describes the population of Pakistan as a function of the number of years, t, after 1998. (See Example 4.)

b. Write a mathematical model that describes the population of Brazil as a function of the number of years, t, after 1998.

c. Use your models from parts (a) and (b) to predict the populations of Pakistan and Brazil 10, 20, and 30 years after 1998. Round to the nearest million.

d. Although Pakistan had approximately 35 million fewer people than Brazil in 1998, it also has a higher growth rate. What is the effect of growth rate when comparing the populations of two countries?

Graphing Calculator Exercises

For Exercises 41–48, graph the functions on your calculator to support your solutions to the indicated exercises.

41. $f(x) = 4^x$ (see Exercise 29)

42. $g(x) = 6^x$ (see Exercise 30)

43. $m(x) = (\frac{1}{8})^x$ (see Exercise 31)

44. $n(x) = (\frac{1}{3})^x$ (see Exercise 32)

45. $h(x) = 2^{x+1}$ (see Exercise 33)

46. $k(x) = 5^{x-1}$ (see Exercise 34)

47. $g(x) = 5^{-x}$ (see Exercise 35)

48. $f(x) = 2^{-x}$ (see Exercise 36)

49. The function defined by $A(x) = 1000(2)^{x/7}$ represents the total amount, A, in an account x years after an initial investment of \$1000.
    a. Graph $y = A(x)$ on the window where $0 \le x \le 25$ and $0 \le y \le 10{,}000$.
    b. Use *Zoom* and *Trace* to approximate the times required for the account to reach \$2000, \$4000, and \$8000.

50. The function defined by $A(x) = 1500(2)^{x/8}$ represents the total amount, A, in an account x years after the initial investment of \$1500.
    a. Graph $y = A(x)$ on the window where $0 \le x \le 40$ and $0 \le y \le 25{,}000$.
    b. Use *Zoom* and *Trace* to approximate the times required for the account to reach \$3000, \$6000, and \$12,000.

---

section

12.4 Logarithmic Functions

Concepts

1. Definition of a Logarithmic Function
2. Converting Between Logarithmic Form and Exponential Form
3. Evaluating Logarithmic Expressions
4. The Common Logarithmic Function
5. Applications of the Common Logarithmic Function
6. Graphs of Logarithmic Functions
7. Determining the Domain of Logarithmic Functions

1. Definition of a Logarithmic Function

Consider the following equations in which the variable is located in the exponent of an expression. In some cases the solution can be found by inspection because the constant on the right-hand side of the equation is a perfect power of the base.

Equation		**Solution**
$5^x = 5$	⟶	$x = 1$
$5^x = 20$	⟶	$x = ?$
$5^x = 25$	⟶	$x = 2$
$5^x = 60$	⟶	$x = ?$
$5^x = 125$	⟶	$x = 3$

The equation $5^x = 20$ cannot be solved by inspection. However, we might suspect that x is between 1 and 2. Similarly, the solution to the equation $5^x = 60$ is between 2 and 3. To solve an exponential equation for an unknown exponent we must use a new type of function called a logarithmic function.

Definition of a Logarithmic Function

If x and b are positive real numbers such that $b \neq 1$, then $y = \log_b(x)$ is called the **logarithmic function** with base b and

$$y = \log_b(x) \text{ is equivalent to } b^y = x.$$

Note: In the expression, $y = \log_b(x)$, y is called the **logarithm**, b is called the **base**, and x is called the **argument**.

The expression $y = \log_b(x)$ is equivalent to $b^y = x$ and indicates that *the logarithm, y, is the exponent to which b must be raised to obtain x*. The expression $y = \log_b(x)$ is called the logarithmic form of the equation, and the expression $b^y = x$ is called the exponential form of the equation.

The definition of a logarithmic function suggests a close relationship with an exponential function of the same base. In fact, a logarithmic function is the inverse of the corresponding exponential function. For example, the following steps find the inverse of the exponential function defined by $f(x) = b^x$.

$f(x) = b^x$	
$y = b^x$	Replace $f(x)$ by y.
$x = b^y$	Interchange x and y.
$y = \log_b x$	Solve for y using the definition of a logarithmic function.
$f^{-1}(x) = \log_b x$	Replace y by $f^{-1}(x)$.

2. Converting Between Logarithmic Form and Exponential Form

Because the concept of a logarithm is new and unfamiliar, it may be helpful to rewrite a logarithm in its equivalent exponential form.

example 1 Converting from Logarithmic Form to Exponential Form

Rewrite the logarithmic equations in exponential form.

a. $\log_2(32) = 5$ b. $\log_{10}\left(\frac{1}{1000}\right) = -3$ c. $\log_5(1) = 0$

Solution:

	Logarithmic Form		**Exponential Form**
a.	$\log_2(32) = 5$	$\Leftrightarrow$	$2^5 = 32$
b.	$\log_{10}\left(\frac{1}{1000}\right) = -3$	$\Leftrightarrow$	$10^{-3} = \frac{1}{1000}$
c.	$\log_5(1) = 0$	$\Leftrightarrow$	$5^0 = 1$

Tip: To understand the meaning of a logarithmic function intuitively, consider the function defined by $y = \log_3(x)$. If we evaluate the function at $x = 9$, $x = 27$, and $x = 81$, we have

$y = \log_3(9)$ or equivalently $3^y = 9$ ← The value of the logarithm, y, is the exponent to which 3 is raised to produce 9.
$\log_3(9) = 2$ because $3^2 = 9$

$y = \log_3(27)$ or equivalently $3^y = 27$ ← The value of the logarithm, y, is the exponent to which 3 is raised to produce 27.
$\log_3(27) = 3$ because $3^3 = 27$

$y = \log_3(81)$ or equivalently $3^y = 81$ ← The value of the logarithm, y, is the exponent to which 3 is raised to produce 81.
$\log_3(81) = 4$ because $3^4 = 81$

3. Evaluating Logarithmic Expressions

example 2

Evaluating Logarithmic Expressions

Evaluate the logarithmic expressions.

a. $\log_{10}(10{,}000)$ b. $\log_5\left(\frac{1}{125}\right)$ c. $\log_{(1/2)}\left(\frac{1}{8}\right)$

d. $\log_b(b)$ e. $\log_c(c^7)$ f. $\log_3 \sqrt[4]{3}$

Solution:

a. $\log_{10}(10{,}000)$ is the exponent to which 10 must be raised to obtain 10,000.

$y = \log_{10}(10{,}000)$ Let y represent the value of the logarithm.

$10^y = 10{,}000$ Rewrite the expression in exponential form.

$y = 4$

Therefore, $\log_{10}(10{,}000) = 4$.

b. $\log_5(\frac{1}{125})$ is the exponent to which 5 must be raised to obtain $\frac{1}{125}$.

$y = \log_5\left(\frac{1}{125}\right)$ Let y represent the value of the logarithm.

$5^y = \frac{1}{125}$ Rewrite the expression in exponential form.

$y = -3$

Therefore, $\log_5(\frac{1}{125}) = -3$.

c. $\log_{(1/2)}(\frac{1}{8})$ is the exponent to which $\frac{1}{2}$ must be raised to obtain $\frac{1}{8}$.

$y = \log_{(1/2)}\left(\frac{1}{8}\right)$ Let y represent the value of the logarithm.

$\left(\frac{1}{2}\right)^y = \frac{1}{8}$ Rewrite the expression in exponential form.

$y = 3$

Therefore, $\log_{(1/2)}(\frac{1}{8}) = 3$.

d. $\log_b(b)$ is the exponent to which b must be raised to obtain b.

$y = \log_b(b)$ Let y represent the value of the logarithm.

$b^y = b$ Rewrite the expression in exponential form.

$y = 1$

Therefore, $\log_b(b) = 1$.

e. $\log_c(c^7)$ is the exponent to which c must be raised to obtain c^7.

$y = \log_c(c^7)$ Let y represent the value of the logarithm.

$c^y = c^7$ Rewrite the expression in exponential form.

$y = 7$

Therefore, $\log_c(c^7) = 7$.

f. $\log_3 \sqrt[4]{3} = \log_3 3^{1/4}$ is the exponent to which 3 must be raised to obtain $3^{1/4}$.

$y = \log_3 3^{1/4}$ Let y represent the value of the logarithm.

$3^y = 3^{1/4}$ Rewrite the expression in exponential form.

$y = \frac{1}{4}$

Therefore, $\log_3 \sqrt[4]{3} = \frac{1}{4}$.

4. The Common Logarithmic Function

The logarithmic function with base 10 is called the **common logarithmic function** and is denoted by $y = \log(x)$. Notice that the base is not explicitly written but is understood to be 10. That is, $y = \log_{10}(x)$ is written simply as $y = \log(x)$.

Calculator Connections

On most calculators, the log key is used to compute logarithms with base 10. For example, we know the expression $\log(1{,}000{,}000) = 6$ because $10^6 = 1{,}000{,}000$. Use the log key to show this result on a calculator.

```
log(1000000)
               6
```

example 3 Evaluating Common Logarithms on a Calculator

Evaluate the common logarithms. Round the answers to four decimal places.

a. $\log(420)$ b. $\log(8.2 \times 10^9)$ c. $\log(0.0002)$

Solution:

a. $\log(420) \approx 2.6232$
b. $\log(8.2 \times 10^9) \approx 9.9138$
c. $\log(0.0002) \approx -3.6990$

Calculator Connections

The expressions from Example 3 can be approximated on a graphing calculator as shown.

```
log(420)
          2.62324929
log(8.2E9)
          9.913813852
log(0.0002)
         -3.698970004
```

5. Applications of the Common Logarithmic Function

example 4 Applying Logarithmic Functions to a Memory Model

One method of measuring a student's retention of material after taking a course is to retest the student at specified time intervals after the course has been completed. A student's score on a calculus test t months after completing a course in calculus is approximated by

$S(t) = 85 - 25\log(t + 1)$, where t is the time in months after completing the course, and $S(t)$ is the student's score as a percent.

a. What was the student's score at the time the course was completed ($t = 0$)?
b. What was the student's score after 2 months?
c. What was the student's score after 1 year?

Solution:

a. $S(t) = 85 - 25\log(t + 1)$

$S(0) = 85 - 25\log(0 + 1)$ Substitute $t = 0$.

$= 85 - 25\log(1)$ $\log(1) = 0$ because $10^0 = 1$.

$= 85 - 25(0)$

$= 85 - 0$

$= 85$ The student's score at the time the course was completed was 85%.

b. $S(t) = 85 - 25\log(t + 1)$

$S(2) = 85 - 25\log(2 + 1)$

$= 85 - 25\log(3)$ Use a calculator to approximate log(3).

≈ 73.1 The student's score dropped to 73.1%.

c. $S(t) = 85 - 25\log(t + 1)$

$S(12) = 85 - 25\log(12 + 1)$ 1 year = 12 months. Substitute $t = 12$.

$= 85 - 25\log(13)$ Use a calculator to approximate log(13).

≈ 57.2 The student's score dropped to 57.2%.

6. Graphs of Logarithmic Functions

In Section 12.3 we studied the graphs of exponential functions. In this section, we will graph logarithmic functions. First, recall that $f(x) = \log_b(x)$ is the inverse of $g(x) = b^x$. Therefore, the graph of $y = f(x)$ is symmetric to the graph of $y = g(x)$ about the line $y = x$, as shown in Figures 12-12 and 12-13.

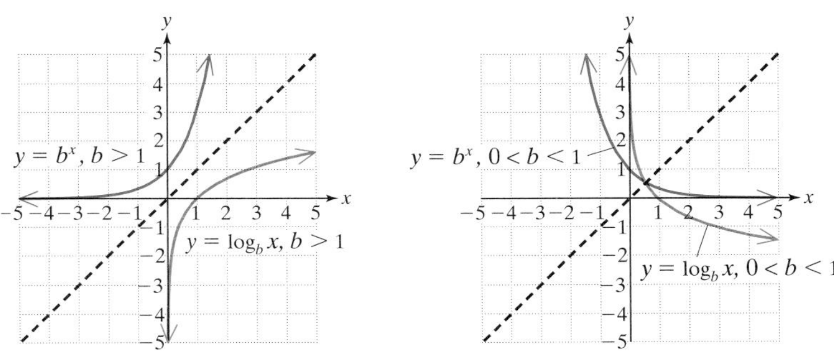

Figure 12-12 **Figure 12-13**

From Figures 12-12 and 12-13, we see that the range of $y = b^x$ is the set of positive real numbers. As expected, the domain of its inverse, the logarithmic function $y = \log_b(x)$, is also the set of positive real numbers. Therefore, the **domain of the logarithmic function** $y = \log_b(x)$ is the set of positive real numbers.

example 5 **Graphing Logarithmic Functions**

Graph the functions and compare the graphs to examine the effect of the base on the shape of the logarithmic function.

a. $y = \log_2(x)$ b. $y = \log(x)$

Solution:

We can write each equation in its equivalent exponential form and create a table of values (Table 12-5). To simplify the calculations, choose integer values of y and then solve for x.

$$y = \log_2(x) \text{ or } 2^y = x \qquad y = \log(x) \text{ or } 10^y = x$$

Choose values for y

Solve for x

Table 12-5

$x = 2^y$	$x = 10^y$	y
$\frac{1}{8}$	$\frac{1}{1000}$	-3
$\frac{1}{4}$	$\frac{1}{100}$	-2
$\frac{1}{2}$	$\frac{1}{10}$	-1
1	1	0
2	10	1
4	100	2
8	1000	3

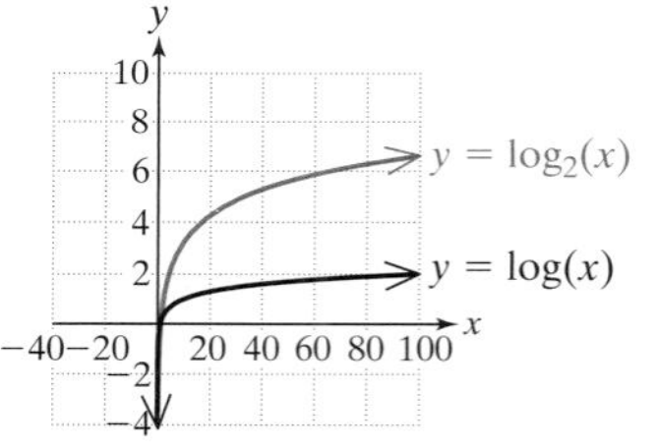

Figure 12-14

The graphs of $y = \log_2(x)$ and $y = \log(x)$ are shown in Figure 12-14. Both graphs exhibit the same general behavior, and the steepness of the curve is affected by the base. The function $y = \log(x)$ requires a 10-fold increase in x to increase the y-value by 1 unit. The function $y = \log_2(x)$ requires a twofold increase in x to increase the y-value by 1 unit. In addition, the following characteristics are true for both graphs.

- The domain is the set of real numbers, x, such that $x > 0$.
- The range is the set of real numbers.
- The y-axis is a vertical asymptote.
- Both graphs pass through the point $(1, 0)$.

Example 5 illustrates that a logarithmic function with base, $b > 1$ is an increasing function. In Example 6, we see that if the base, b, is between 0 and 1, the function decreases over its entire domain.

example 6 Graphing a Logarithmic Function

Graph $y = \log_{1/4}(x)$.

Solution:

$y = \log_{1/4}(x)$ can be written in exponential form as $(\frac{1}{4})^y = x$. By choosing several values for y, we can solve for x and plot the corresponding points (Table 12-6).

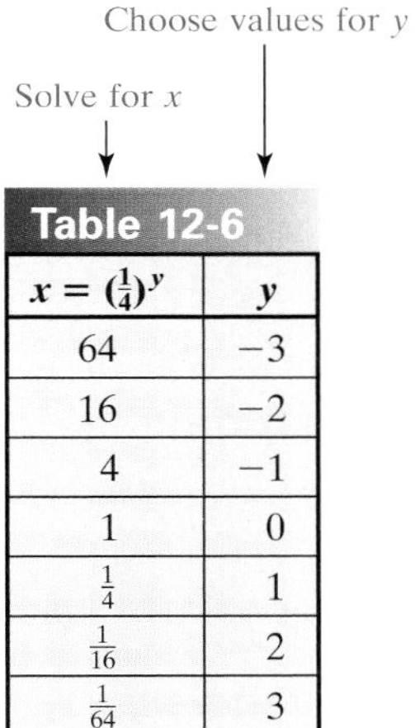
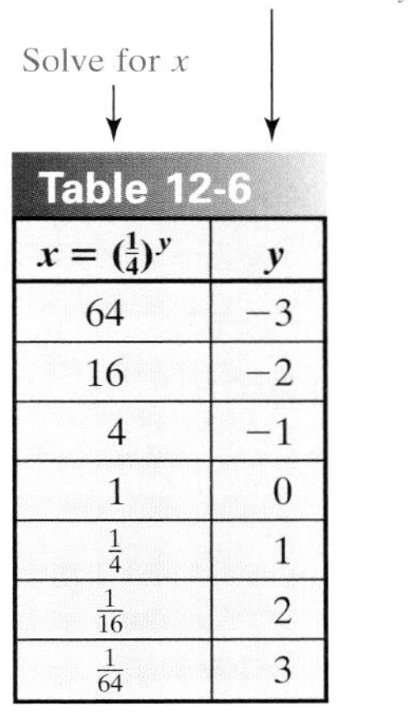

Table 12-6

$x = (\frac{1}{4})^y$	y
64	−3
16	−2
4	−1
1	0
$\frac{1}{4}$	1
$\frac{1}{16}$	2
$\frac{1}{64}$	3

Figure 12-15

The expression $y = \log_{1/4}(x)$ defines a decreasing logarithmic function (Figure 12-15). Notice that the vertical asymptote, domain, and range are the same for both increasing and decreasing logarithmic functions.

Graphs of Exponential and Logarithmic Functions—A Summary

Exponential Functions

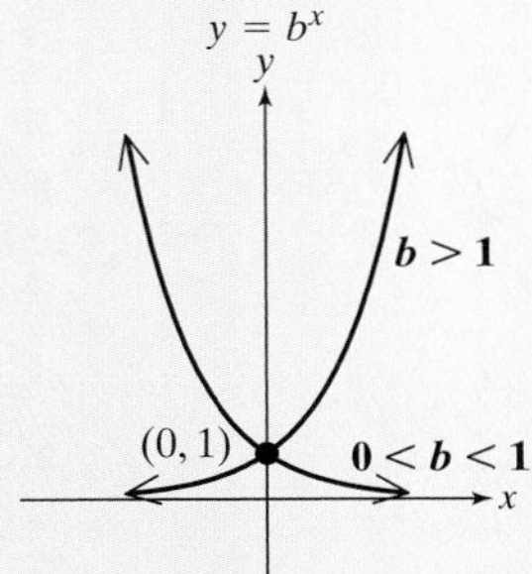

Domain: $(-\infty, \infty)$
Range: $(0, \infty)$
Horizontal asymptote: $y = 0$
Passes through (0, 1)
If $b > 1$, the function is increasing.
If $0 < b < 1$, the function is decreasing.

Logarithmic Functions

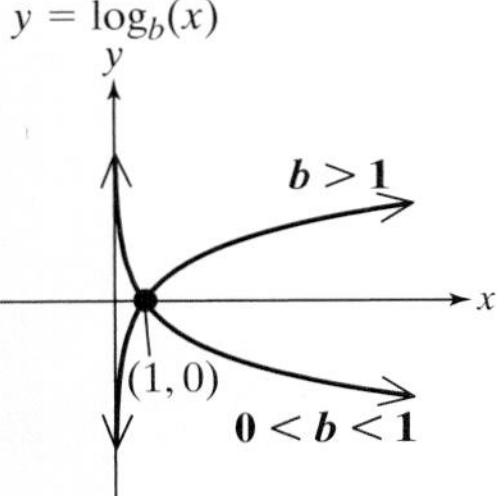

Domain: $(0, \infty)$
Range: $(-\infty, \infty)$
Vertical asymptote: $x = 0$
Passes through (1, 0)
If $b > 1$, the function is increasing.
If $0 < b < 1$, the function is decreasing.

Notice that the roles of x and y are interchanged for the functions $y = b^x$ and $b^y = x$. Therefore, it is not surprising that the domain and range are reversed between exponential and logarithmic functions. Moreover, an exponential function passes through $(0, 1)$, whereas a logarithmic function passes through $(1, 0)$. An exponential function has a horizontal asymptote at $y = 0$, whereas a logarithmic function has a vertical asymptote at $x = 0$. With the roles of x and y interchanged, the exponential function, base b, and the logarithmic function, base b, are *inverses* of each other.

7. Determining the Domain of Logarithmic Functions

example 7

Identifying the Domain of a Logarithmic Function

Find the domain of the functions.

a. $f(x) = \log(4 - x)$

b. $g(x) = \log(2x + 6)$

Solution:

The domain of the function $y = \log_b(x)$ is the set of all positive real numbers. That is, the argument, x, must be greater than zero: $x > 0$.

a. $f(x) = \log(4 - x)$ The argument is $4 - x$.

$4 - x > 0$ The argument of the logarithm must be greater than zero.

$-x > -4$ Solve for x.

$x < 4$ Divide by -1 and reverse the inequality sign.

The domain is $\{x \mid x < 4\}$ or equivalently $(-\infty, 4)$.

b. $g(x) = \log(2x + 6)$ The argument is $2x + 6$.

$2x + 6 > 0$ The argument of the logarithm must be greater than zero.

$2x > -6$ Solve for x.

$x > -3$

The domain is $\{x \mid x > -3\}$ or equivalently $(-3, \infty)$.

Calculator Connections

The graphs of $Y_1 = \log(4 - x)$ and $Y_2 = \log(2x + 6)$ are shown here and can be used to confirm the solutions to Example 7. Notice that each function has a vertical asymptote at the value of x where the argument equals zero.

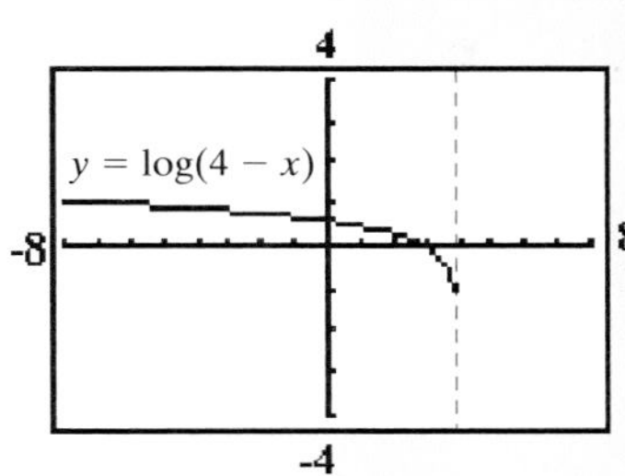

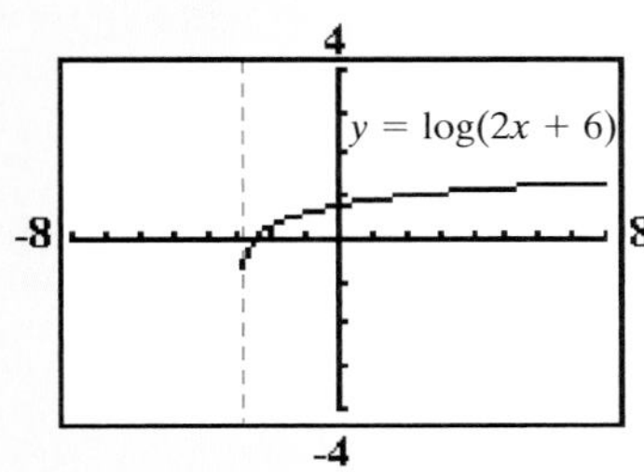

section 12.4 PRACTICE EXERCISES

1. For which graph of $y = b^x$ is $0 < b < 1$?

i.

ii.

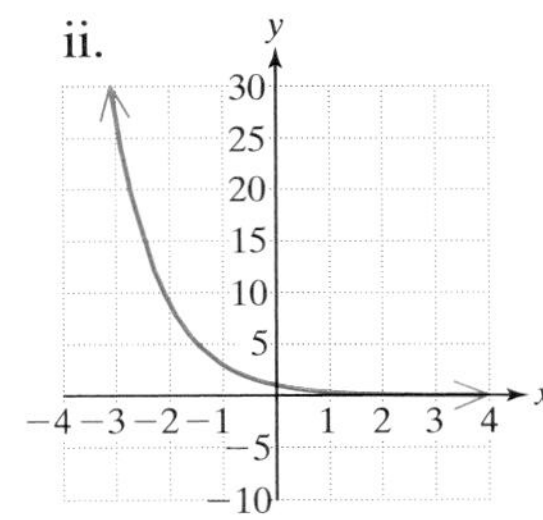

2. Let $f(x) = 6^x$.
   a. Find $f(-2), f(-1), f(0), f(1)$, and $f(2)$.
   b. Graph $y = f(x)$ by plotting the points found in part (a).

3. Let $g(x) = 3^x$.
   a. Find $g(-2), g(-1), g(0), g(1)$, and $g(2)$.
   b. Graph $y = g(x)$ by plotting the points found in part (a).

4. Let $r(x) = (\frac{3}{4})^x$.
   a. Find $r(-2), r(-1), r(0), r(1)$, and $r(2)$.
   b. Graph $y = r(x)$ by plotting the points found in part (a).

5. Let $s(x) = (\frac{2}{5})^x$.
   a. Find $s(-2), s(-1), s(0), s(1)$, and $s(2)$.
   b. Graph $y = s(x)$ by plotting the points found in part (a).

6. For the expression $y = \log_b(x)$, identify the base, the argument, and the logarithm.

7. Rewrite the expression in exponential form: $y = \log_b(x)$.

For Exercises 8–19, write the expression in logarithmic form.

8. $3^x = 81$
9. $10^3 = 1000$
10. $5^2 = 25$
11. $8^{1/3} = 2$
12. $7^{-1} = \frac{1}{7}$
13. $8^{-2} = \frac{1}{64}$
14. $b^x = y$
15. $b^y = x$
16. $e^x = y$
17. $e^y = x$
18. $K^n = p$
19. $H^m = q$

For Exercises 20–31, write the expression in exponential form.

20. $\log_5 625 = 4$
21. $\log_{125} 25 = \frac{2}{3}$

22. $\log_{10} 0.0001 = -4$
23. $\log_{25}\left(\frac{1}{5}\right) = -\frac{1}{2}$
24. $\log_6 36 = 2$
25. $\log_2 128 = 7$
26. $\log_b 15 = x$
27. $\log_b 82 = y$
28. $\log_3 5 = x$
29. $\log_2 7 = x$
30. $\log_4 x = 10$
31. $\log_{1/2} x = 6$

For Exercises 32–52, find the logarithms without the use of a calculator.

32. $\log_7 49$
33. $\log_3 27$
34. $\log_{10} 0.1$
35. $\log_2\left(\frac{1}{16}\right)$
36. $\log_{16} 4$
37. $\log_8 2$
38. $\log_6 1$
39. $\log_8 8$
40. $\log_3 3^5$
41. $\log_9 9^3$
42. $\log_{10} 10$
43. $\log_7 1$
44. $\log(10)$
45. $\log(100)$
46. $\log(1000)$
47. $\log(10{,}000)$
48. $\log(1.0 \times 10^6)$
49. $\log(0.1)$
50. $\log(0.01)$
51. $\log(0.001)$
52. $\log(1.0 \times 10^{-6})$

For Exercises 53–64, use a calculator to approximate the logarithms. Round to 4 decimal places.

53. $\log 6$
54. $\log 18$
55. $\log \pi$
56. $\log \frac{1}{8}$
57. $\log\left(\frac{1}{32}\right)$
58. $\log(\sqrt{5})$
59. $\log(0.0054)$
60. $\log(0.0000062)$
61. $\log(3.4 \times 10^5)$
62. $\log(4.78 \times 10^9)$
63. $\log(3.8 \times 10^{-8})$
64. $\log(2.77 \times 10^{-4})$

65. Given that $\log 10 = 1$ and $\log 100 = 2$,
    a. Estimate $\log 93$.
    b. Estimate $\log 12$.
    c. Evaluate the logarithms in parts (a) and (b) on a calculator and compare to your estimates.

66. Given that $\log \frac{1}{10} = -1$ and $\log 1 = 0$,
    a. Estimate $\log \frac{9}{10}$.
    b. Estimate $\log \frac{1}{5}$.
    c. Evaluate the logarithms in parts (a) and (b) on a calculator and compare to your estimates.

67. Let $f(x) = \log_4 x$
    a. Find the values of $f(\frac{1}{64}), f(\frac{1}{16}), f(\frac{1}{4}), f(1), f(4), f(16)$, and $f(64)$.
    b. Graph $y = f(x)$ by plotting the points found in part (a).

68. Let $g(x) = \log_2 x$
    a. Find the values of $g(\frac{1}{8}), g(\frac{1}{4}), g(\frac{1}{2}), g(1), g(2), g(4)$, and $g(8)$.
    b. Graph $y = g(x)$ by plotting the points found in part (a).

Graph the logarithmic functions in Exercises 69–72 by writing the function in exponential form and making a table of points (see Examples 5 and 6).

69. $y = \log_3 x$
70. $y = \log_5 x$
71. $y = \log_{1/2} x$
72. $y = \log_{1/3} x$
73. State the domain for the function in Exercise 69.
74. State the domain for the function in Exercise 70.
75. State the domain for the function in Exercise 71.
76. State the domain for the function in Exercise 72.

For Exercises 77–82, find the domain of the function and express the domain in interval notation.

77. $y = \log_7(x - 5)$
78. $y = \log_3(2x + 1)$
79. $y = \log_3(x + 1.2)$
80. $y = \log\left(x - \frac{1}{2}\right)$
81. $y = \log(x^2)$
82. $y = \log(x^2 + 1)$

83. A graduate student in education is doing research to compare the effectiveness of two different teaching techniques designed to teach vocabulary to sixth-graders. The first group of students (group 1) was taught with method I, in which the students worked individually to complete the assignments in a workbook. The second group (group 2) was taught with method II, in which the students worked cooperatively in groups of four to complete the assignments in the same workbook.

None of the students knew any of the vocabulary words before the study began. After completing the assignments in the workbook, the students were then tested on the vocabulary at 1-month intervals to assess how much material they had retained over time. The students' average score t months after completing the assignments are given by the following functions:

Method I: $S_1(t) = 91 - 30\log(t + 1)$, where t is the time in months and $S_1(t)$ is the average score of students in group 1.

Method II: $S_2(t) = 88 - 15\log(t + 1)$, where t is the time in months and $S_2(t)$ is the average score of students in group 2.

a. Complete the table to find the average scores for each group of students after the indicated number of months. Round to one decimal place.

t (months)	0	1	2	6	12	24
$S_1(t)$						
$S_2(t)$						

Table for Exercise 83

b. Based on the table of values, what were the average scores for each group immediately after completion of the assigned material $(t = 0)$?

c. Based on the table of values, which teaching method helped students retain the material better for a long period of time?

84. Generally, the more money a company spends on advertising, the higher the sales. Let a represent the amount of money spent on advertising (in \$100s). Then the amount of money in sales, $S(a)$, (in \$1000s) is given by

$$S(a) = 10 + 20\log(a + 1), \text{ where } a \geq 0.$$

a. The value of $S(1) \approx 16.0$, which means that if \$100 is spent on advertising, \$16,000 is returned in sales. Find the values of $S(11)$, $S(21)$, and $S(31)$. Round to one decimal place. Interpret the meaning of each function value in the context of the problem.

b. The graph of $y = S(a)$ is shown here. Use the graph and your answers from part (a) to explain why the money spent in advertising becomes less "efficient" as it is used in larger amounts.

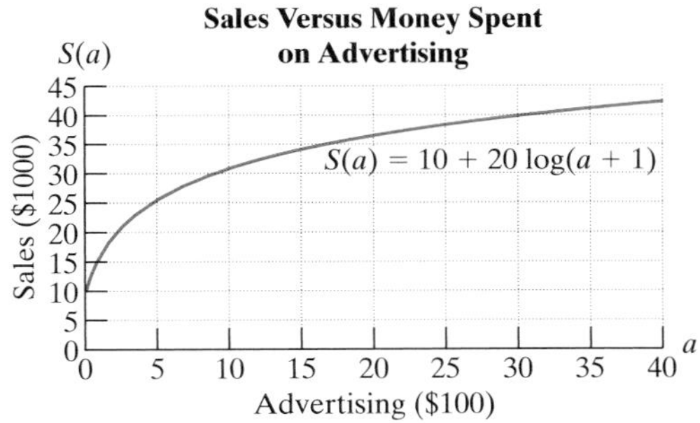

Figure for Exercise 84

85. The pH (hydrogen potential) of a solution is defined as

$\text{pH} = -\log[H^+]$, where $[H^+]$ represents the concentration of hydrogen ions in moles per liter (mol/L).

The pH scale ranges from 0 to 14. The midpoint of this range, 7, represents a neutral solution. Values below 7 are progressively more acidic, and values above 7 are progressively more alkaline. For example, the pH of orange juice is roughly 3.5. Based on the equation $\text{pH} = -\log[H^+]$, a 1-unit change in pH means a 10-fold change in hydrogen ion concentration.

a. Normal rain has a pH of 5.6. However, in some areas of the northeast United States the rainwater is more acidic. What is the pH of a rain sample for which the concentration of hydrogen ions is 0.0002 mol/L?

b. Find the pH of household ammonia if the concentration of hydrogen ions is 1.0×10^{-11} mol/L.

Graphing Calculator Exercises

For Exercises 86–91, graph the function on an appropriate viewing window. From the graph, identify the domain of the function and the location of the vertical asymptote.

86. $y = \log(x + 6)$
87. $y = \log(2x + 4)$
88. $y = \log(0.5x - 1)$
89. $y = \log(x + 8)$
90. $y = \log(2 - x)$
91. $y = \log(3 - x)$

section

12.5 Properties of Logarithms

Concepts

1. Properties of Logarithms
2. Expanded Logarithmic Expressions
3. Single Logarithmic Expressions

1. Properties of Logarithms

You have already been exposed to certain properties of logarithms that follow directly from the definition. Recall

$$y = \log_b(x) \text{ is equivalent to } b^y = x \quad \text{for } x > 0, b > 0, \text{ and } b \neq 1.$$

The following properties follow directly from the definition.

$\log_b(1) = 0$ Property 1

$\log_b(b) = 1$ Property 2

$\log_b(b^p) = p$ Property 3

$b^{\log_b(x)} = x$ Property 4

example 1

Applying the Properties of Logarithms to Simplify Expressions

Use the properties of logarithms to simplify the expressions. Assume that all variable expressions within the logarithms represent positive real numbers.

a. $\log_8(8) + \log_8(1)$ b. $10^{\log(x+2)}$ c. $\log_{1/2}\left(\frac{1}{2}\right)^x$

Solution:

a. $\log_8(8) + \log_8(1)$

$= 1 + 0$ Properties 2 and 1

$= 1$

b. $10^{\log(x+2)} = x + 2$ Property 4

c. $\log_{1/2}\left(\frac{1}{2}\right)^x = x$ Property 3

Three additional properties are useful when simplifying logarithmic expressions. The first is the product property for logarithms.

Product Property for Logarithms

Let b, x, and y be positive real numbers where $b \neq 1$. Then,

$$\log_b(xy) = \log_b(x) + \log_b(y)$$

The logarithm of a product equals the sum of the logarithms of the factors.

Proof:

Let $M = \log_b(x)$, which implies $b^M = x$.

Let $N = \log_b(y)$, which implies $b^N = y$.

Then $xy = b^M b^N = b^{M+N}$

Writing the expression $xy = b^{M+N}$ in logarithmic form, we have

$$\log_b(xy) = M + N$$

$$\log_b(xy) = \log_b(x) + \log_b(y) \checkmark$$

To demonstrate the product property for logarithms, simplify the following expressions using the order of operations.

$$\log_3(3 \cdot 9) \stackrel{?}{=} \log_3(3) + \log_3(9)$$

$$\log_3(27) \stackrel{?}{=} 1 + 2$$

$$3 = 3 \checkmark$$

Quotient Property for Logarithms

Let b, x, and y be positive real numbers where $b \neq 1$. Then,

$$\log_b\left(\frac{x}{y}\right) = \log_b(x) - \log_b(y)$$

The logarithm of a quotient equals the difference of the logarithms of the numerator and denominator.

The proof of the quotient property for logarithms is similar to the proof of the product property and is omitted here. To demonstrate the quotient property for logarithms, simplify the following expressions using the order of operations.

$$\log\left(\frac{1{,}000{,}000}{100}\right) \stackrel{?}{=} \log(1{,}000{,}000) - \log(100)$$

$$\log(10{,}000) \stackrel{?}{=} 6 - 2$$

$$4 = 4 \checkmark$$

Power Property for Logarithms

Let b and x be positive real numbers where $b \neq 1$. Let p be any real number. Then,

$$\log_b(x^p) = p\log_b(x)$$

Proof:

Let $M = \log_b(x)$, which implies $b^M = x$.

Raise both sides to the p power: $(b^M)^p = (x)^p$, or equivalently $b^{Mp} = (x^p)$

Write the expression $b^{Mp} = (x^p)$ in logarithmic form: $\log_b(x^p) = Mp = pM$, or equivalently: $\log_b(x^p) = p\log_b(x)$ ✓

To demonstrate the power property for logarithms, simplify the following expressions using the order of operations.

$$\log_4(4^2) \stackrel{?}{=} 2\log_4(4)$$
$$2 \stackrel{?}{=} 2 \cdot 1$$
$$2 = 2 \checkmark$$

The properties of logarithms are summarized in the box.

Properties of Logarithms

Let b, x, and y be positive real numbers where $b \neq 1$, and let p be a real number. Then the following **properties of logarithms** are true.

1. $\log_b(1) = 0$
2. $\log_b(b) = 1$
3. $\log_b(b^p) = p$
4. $b^{\log_b(x)} = x$
5. $\log_b(xy) = \log_b(x) + \log_b(y)$ — Product property for logarithms
6. $\log_b\left(\dfrac{x}{y}\right) = \log_b(x) - \log_b(y)$ — Quotient property for logarithms
7. $\log_b(x^p) = p\log_b(x)$ — Power property for logarithms

2. Expanded Logarithmic Expressions

In many applications, it is advantageous to expand a logarithm into a sum or difference of simpler logarithms.

example 2 **Writing a Logarithmic Expression in Expanded Form**

Write the expressions as the sum or difference of logarithms of x, y, and z. Assume all variable expressions within the logarithms represent positive real numbers.

a. $\log_3\left(\frac{xy^3}{z^2}\right)$ b. $\log\left(\frac{\sqrt{x+y}}{10}\right)$ c. $\log_b\sqrt[5]{\frac{x^4}{yz^3}}$

Solution:

a. $\log_3\left(\frac{xy^3}{z^2}\right)$

$= \log_3(xy^3) - \log_3(z^2)$	Quotient property for logarithms (property 6)
$= [\log_3(x) + \log_3(y^3)] - \log_3(z^2)$	Product property for logarithms (property 5)
$= \log_3(x) + 3\log_3(y) - 2\log_3(z)$	Power property for logarithms (property 7)

b. $\log\left(\frac{\sqrt{x+y}}{10}\right)$

$= \log\sqrt{x+y} - \log(10)$	Quotient property for logarithms (property 6)
$= \log(x+y)^{1/2} - 1$	Write $\sqrt{x+y}$ as $(x+y)^{1/2}$ and simplify $\log(10) = 1$.
$= \frac{1}{2}\log(x+y) - 1$	Power property for logarithms (property 7)

c. $\log_b\sqrt[5]{\frac{x^4}{yz^3}}$

$= \log_b\left(\frac{x^4}{yz^3}\right)^{1/5}$	Write $\sqrt[5]{\frac{x^4}{yz^3}}$ as $\left(\frac{x^4}{yz^3}\right)^{1/5}$.
$= \frac{1}{5}\log_b\left(\frac{x^4}{yz^3}\right)$	Power property for logarithms (property 7)
$= \frac{1}{5}(\log_b x^4 - \log_b(yz^3))$	Quotient property for logarithms (property 6)
$= \frac{1}{5}(\log_b x^4 - [\log_b y + \log_b z^3])$	Product property for logarithms (property 5)
$= \frac{1}{5}(\log_b x^4 - \log_b y - \log_b z^3)$	Distributive property
$= \frac{1}{5}(4\log_b x - \log_b y - 3\log_b z)$	Power property for logarithms (property 7)

or $\frac{4}{5}\log_b x - \frac{1}{5}\log_b y - \frac{3}{5}\log_b z$

3. Single Logarithmic Expressions

In some applications it is necessary to write a sum or difference of logarithms as a single logarithm.

example 3 Writing a Sum or Difference of Logarithms as a Single Logarithm

Rewrite each expression as a single logarithm, and simplify the result if possible. Assume all variable expressions within the logarithms represent positive real numbers.

a. $\log_2 560 - \log_2 7 - \log_2 5$

b. $2 \log x - \frac{1}{2} \log y + 3 \log z$

c. $\frac{1}{2}[\log_5(x^2 - y^2) - \log_5(x + y)]$

Solution:

a. $\log_2 560 - \log_2 7 - \log_2 5$

$= \log_2 560 - (\log_2 7 + \log_2 5)$ Factor out -1 from the last two terms.

$= \log_2 560 - \log_2(7 \cdot 5)$ Product property for logarithms (property 5)

$= \log_2\left(\frac{560}{7 \cdot 5}\right)$ Quotient property for logarithms (property 6)

$= \log_2(16)$ Simplify inside parentheses.

$= 4$

b. $2 \log x - \frac{1}{2} \log y + 3 \log z$

$= \log x^2 - \log y^{1/2} + \log z^3$ Power property for logarithms (property 7)

$= \log x^2 + \log z^3 - \log y^{1/2}$ Group terms with positive coefficients.

$= \log(x^2z^3) - \log y^{1/2}$ Product property for logarithms (property 5)

$= \log\left(\frac{x^2z^3}{y^{1/2}}\right)$ or $\log\left(\frac{x^2z^3}{\sqrt{y}}\right)$ Quotient property for logarithms (property 6)

c. $\frac{1}{2}[\log_5(x^2 - y^2) - \log_5(x + y)]$

$= \frac{1}{2}\log_5\left(\frac{x^2 - y^2}{x + y}\right)$ Quotient property for logarithms (property 6)

$$= \frac{1}{2}\log_5\left[\frac{(x+y)(x-y)}{x+y}\right]$$ Factor and simplify within the parentheses.

$$= \frac{1}{2}\log_5(x-y)$$

$$= \log_5(x-y)^{1/2} \text{ or } \log_5\sqrt{x-y}$$ Power property for logarithms (property 7)

It is important to note that the properties of logarithms may be used to write a single logarithm as a sum or difference of logarithms. Furthermore, the properties may be used to write a sum or difference of logarithms as a single logarithm. In either case, these operations may change the domain.

For example, consider the function $y = \log_b(x^2)$. Using the power property for logarithms we have $y = 2\log_b(x)$. Consider the domain of each function:

$y = \log_b(x^2)$ Domain: $\{x \mid x \neq 0\}$

$y = 2\log_b(x)$ Domain: $\{x \mid x > 0\}$

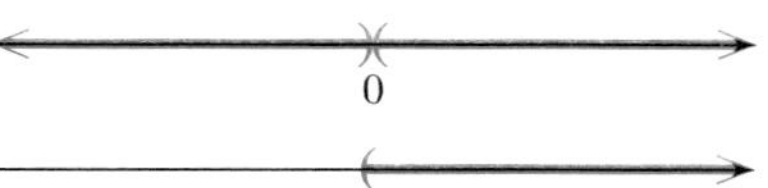

These two functions are equivalent only for values of x in the intersection of the two domains. That is for $x > 0$.

section 12.5 Practice Exercises

For Exercises 1–4, find the values of the logarithmic and exponential expressions without using a calculator.

1. 8^{-2}
2. $\log 10{,}000$
3. $\log_2 32$
4. 6^{-1}

For Exercises 5–8, approximate the values of the logarithmic and exponential expressions by using a calculator.

5. $(\sqrt{2})^{\pi}$
6. $\log 8$
7. $\log 27$
8. $\pi^{\sqrt{2}}$

For Exercises 9–12, match the function with the appropriate graph.

9. $f(x) = 4^x$
10. $q(x) = \left(\frac{1}{5}\right)^x$
11. $h(x) = \log_5 x$
12. $k(x) = \log_{1/3} x$

a. b.

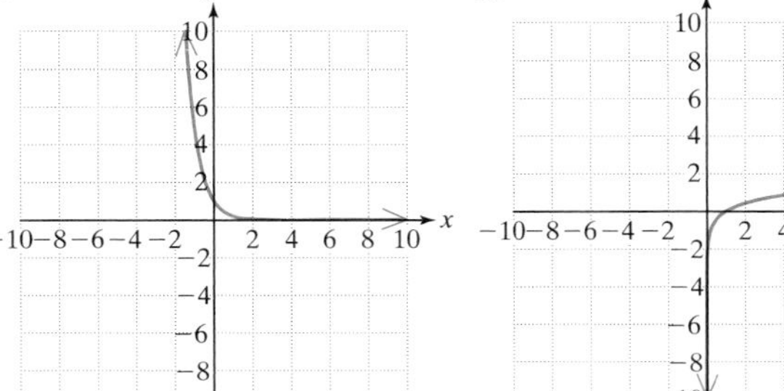

c. d.

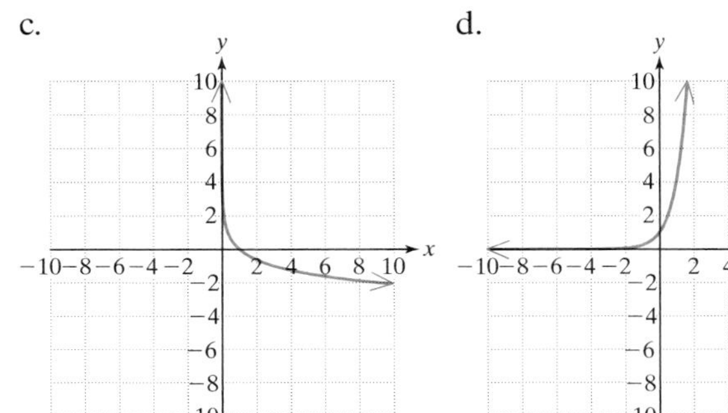

13. Property 1 of logarithms states that $\log_b 1 = 0$. Write an example of this property.

14. Property 2 of logarithms states that $\log_b b = 1$. Write an example of this property.

15. Property 3 of logarithms states that $\log_b(b^n) = n$. An example is $\log_6 6^3 = 3$. Write another example of this property.

16. Property 4 of logarithms states that $b^{\log_b(x)} = x$. An example is $10^{\log_{10} 5} = 5$. Write another example of this property.

For Exercises 17–28, evaluate each expression.

17. $\log_3 3$ 18. $\log 10$ 19. $\log_5 5^4$

20. $\log_4 4^5$ 21. $6^{\log_6 11}$ 22. $7^{\log_7 2}$

23. $\log 10^3$ 24. $\log_6 6^3$ 25. $\log_3 1$

26. $\log_8 1$ 27. $10^{\log 9}$ 28. $8^{\log_8 5}$

For Exercises 29–32, compare the expressions by approximating their values on a calculator. Which two expressions appear to be equivalent?

29. a. $\log(3 \cdot 5)$
b. $\log 3 \cdot \log 5$
c. $\log 3 + \log 5$

30. a. $\log\left(\frac{6}{5}\right)$
b. $\frac{\log 6}{\log 5}$
c. $\log 6 - \log 5$

31. a. $\log(20^2)$
b. $(\log 20)^2$
c. $2 \log 20$

32. a. $\log \sqrt{4}$
b. $\frac{1}{2}\log 4$
c. $\sqrt{\log 4}$

For Exercises 33–42, expand into sums and differences of logarithms.

33. $\log_3\left(\frac{x}{5}\right)$ 34. $\log_2\left(\frac{y}{z}\right)$

35. $\log(2x)$ 36. $\log_6(xyz)$

37. $\log_{10}(x^4)$ 38. $\log_7(z^{1/3})$

39. $\log_4\left(\frac{ab}{c}\right)$ 40. $\log_2\left(\frac{x}{yz}\right)$

41. $\log_b\left(\frac{\sqrt{xy}}{z^3w}\right)$ 42. $\log\left(\frac{a \cdot \sqrt[3]{b}}{cd^2}\right)$

For Exercises 43–50, write the expressions as a single logarithm.

43. $\log C + \log A + \log B + \log I + \log N$

44. $\log x + \log y - \log z$

45. $2 \log_3 x - 3 \log_3 y + \log_3 z$

46. $\log_5 a - \frac{1}{2}\log_5 b - 3 \log_5 c$

47. $\log_b x - 3 \log_b x + 4 \log_b x$

48. $2 \log_3 z + \log_3 z - \frac{1}{2}\log_3 z$

49. $5 \log_8 a - \log_8 1 + \log_8 8$

50. $\log_2 2 + 2 \log_2 b - \log_2 1$

51. The intensity of sound waves is measured in decibels and is calculated by the formula

$B = 10 \log\left(\frac{I}{I_0}\right)$, where I_0 is the minimum detectable decibel level.

a. Expand this formula using the properties of logarithms.
b. Let $I_0 = 10^{-16}$ W/cm^2 and simplify.

52. The Richter scale is used to measure the intensity of an earthquake and is calculated by the formula

$R = \log\left(\frac{I}{I_0}\right)$, where I_0 is the minimum level detectable by a seismograph.

a. Expand this formula using the properties of logarithms.
b. Let $I_0 = 1$ and simplify.

53. It is difficult to compare the magnitudes (brightness) and luminosities (total energy radiated) of stars because of the vast differences in their distances from the earth. Stars that are closer might appear brighter because of their proximity to the earth, not because they necessarily radiate more

energy. The absolute magnitude of a star is a measure of its brightness if the star were located a distance of 10 parsecs (approximately 1.9×10^{13} miles) from the earth. The luminosity of a star, L, is related to its absolute magnitude, M, by the following formula:

$M = 4.71 + 2.5\log(3.9 \times 10^{26}) - 2.5\log(L)$, where M is the absolute magnitude of the star, and L is the luminosity of the star measured in watts (W).

a. Show that the formula can be written as

$$M = 4.71 + 2.5\log\left(\frac{3.9 \times 10^{26}}{L}\right)$$

b. The luminosity of the sun is 3.9×10^{26} W. Find the absolute magnitude of the sun.

c. The luminosity of the "nearby" star, Sirius, is 8.2×10^{27} W. Find its absolute magnitude.

Graphing Calculator Exercises

54. a. Graph $Y_1 = \log(x - 1)^2$ in *Dot Mode* and state its domain.

    b. Graph $Y_2 = 2\log(x - 1)$ in *Dot Mode* and state its domain.

    c. For what values of x are the expressions $\log(x - 1)^2$ and $2\log(x - 1)$ equivalent?

55. a. Graph $Y_1 = \log(x^2)$ in *Dot Mode* and state its domain.

    b. Graph $Y_2 = 2\log(x)$ in *Dot Mode* and state its domain.

    c. For what values of x are the expressions $\log(x^2)$ and $2\log(x)$ equivalent?

Concepts

1. Definition of the Irrational Number *e*
2. Graph of $f(x) = e^x$
3. Computing Compound Interest
4. The Natural Logarithmic Function
5. Properties of the Natural Logarithmic Function
6. Simplifying Logarithmic Expressions
7. Change-of-Base Formula
8. Applications of the Natural Logarithmic Function

section

12.6 The Irrational Number *e*

1. Definition of the Irrational Number *e*

The exponential function base 10 is particularly easy to work with because integral powers of 10 represent different place positions in the base 10 numbering system. In this section, we introduce another important exponential function whose base is an irrational number called *e*.

Consider the expression $(1 + \frac{1}{x})^x$. The value of the expression for increasingly large values of x approaches a constant (Table 12-7).

Table 12-7

x	$\left(1 + \frac{1}{x}\right)^x$
100	2.70481382942
1000	2.71692393224
10,000	2.71814592683
100,000	2.71826823717
1,000,000	2.71828046932
1,000,000,000	2.71828182710

As x approaches infinity, the expression $(1 + \frac{1}{x})^x$ approaches a constant value that we call ***e***. From Table 12-7, this value is approximately 2.718281828.

$$e \approx 2.718281828$$

The value of e is an irrational number (a nonterminating, nonrepeating decimal) and is a universal constant like π.

2. Graph of $f(x) = e^x$

example 1

Graphing $f(x) = e^x$

Graph the function defined by $f(x) = e^x$.

Solution:

Because the base of the function is greater than 1 ($e \approx 2.718281828$), the graph is an increasing exponential function. We can use a calculator to evaluate $f(x) = e^x$ at several values of x.

Practice using your calculator by evaluating e^x for the following values of x. If you are using your calculator correctly, your answers should match those found in Table 12-8. Values are rounded to three decimal places. The corresponding graph of $f(x) = e^x$ is shown in Figure 12-16.

Calculator Connections

```
e^(1)
         2.718281828
e^(2)
         7.389056099
e^(-1)
         .3678794412
```

Table 12-8

x	$f(x) = e^x$
−3	0.050
−2	0.135
−1	0.368
0	1.000
1	2.718
2	7.389
3	20.086

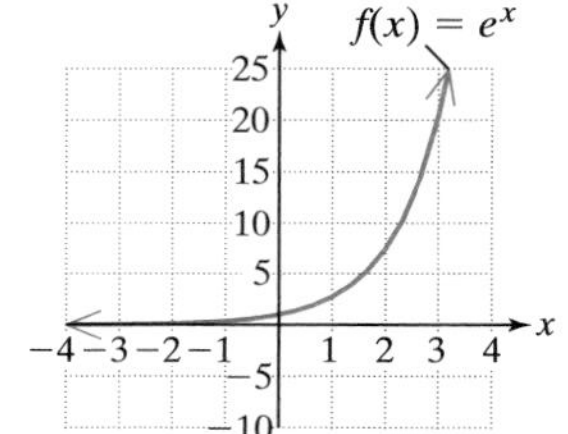

Figure 12-16

3. Computing Compound Interest

One particularly interesting application of exponential functions is in computing compound interest.

1. If the number of compounding periods per year is finite, then the amount in an account, A, is given by

   $A(t) = P\left(1 + \frac{r}{n}\right)^{nt}$, where P is the initial principal, r is the annual interest rate, n is the number of times compounded per year, and t is the time in years that the money is invested.

2. If the number of compound periods per year is infinite, then interest is said to be **compounded continuously**. In such a case, the amount, A, in an account is given by

   $A(t) = Pe^{rt}$, where P is the initial principal, r is the annual interest rate, t is the time in years that the money is invested.

example 2 Computing the Balance on an Account

Suppose \$5000 is invested in an account earning 6.5% interest. Find the balance in the account after 10 years under the following compounding options.

a. Compounded annually
b. Compounded quarterly
c. Compounded monthly
d. Compounded daily
e. Compounded continuously

Solution:

Compounding Option	n value	Formula	Result
annually	$n = 1$	$A(10) = 5000\left(1 + \frac{0.065}{1}\right)^{(1)(10)}$	\$9385.69
quarterly	$n = 4$	$A(10) = 5000\left(1 + \frac{0.065}{4}\right)^{(4)(10)}$	\$9527.79
monthly	$n = 12$	$A(10) = 5000\left(1 + \frac{0.065}{12}\right)^{(12)(10)}$	\$9560.92
daily	$n = 365$	$A(10) = 5000\left(1 + \frac{0.065}{365}\right)^{(365)(10)}$	\$9577.15
continuously	not applicable	$A(10) = 5000e^{(0.065)(10)}$	\$9577.70

Notice that there is a \$191.46 difference in the account balance between annual compounding and daily compounding. However, the difference between compounding daily and compounding continuously is small: \$0.55. As n gets infinitely large, the function defined by

$$A(t) = P\left(1 + \frac{r}{n}\right)^{nt}$$

converges to $A(t) = Pe^{rt}$.

4. The Natural Logarithmic Function

Recall that the common logarithmic function $y = \log(x)$ has a base of 10. Another important logarithmic function is called the **natural logarithmic function**. The natural logarithmic function has a base of e and is written as $y = \ln(x)$. That is,

$$y = \ln(x) = \log_e(x)$$

example 3

Graphing $y = \ln(x)$

Graph $y = \ln(x)$.

Solution:

Because the base of the function $y = \ln(x)$ is base e, and $e > 1$, then the graph is an increasing logarithmic function. We can use a calculator to find specific points on the graph of $y = \ln(x)$ by using the [ln] key.

Practice using your calculator by evaluating $\ln(x)$ for the following values of x. If you are using your calculator correctly, your answers should match those found in Table 12-9. Values are rounded to three decimal places. The corresponding graph of $y = \ln(x)$ is shown in Figure 12-17.

Calculator Connections

```
ln(1)
                0
ln(2)
      .6931471806
ln(3)
      1.098612289
```

Table 12-9

x	$\ln(x)$
1	0.000
2	0.693
3	1.099
4	1.386
5	1.609
6	1.792
7	1.946

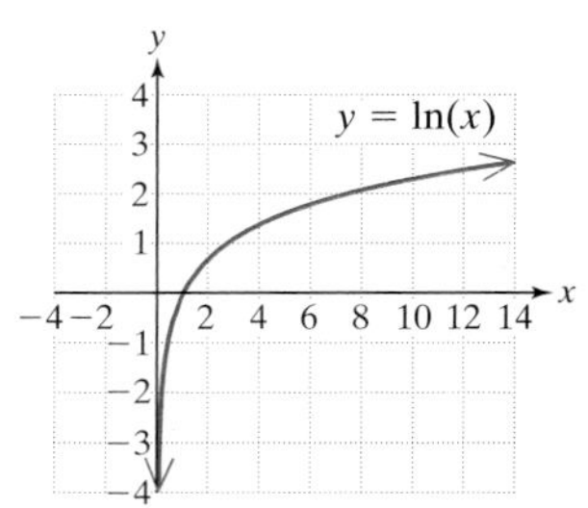

Figure 12-17

5. Properties of the Natural Logarithmic Function

The properties of logarithms stated in Section 12.5 are also true for natural logarithms.

Properties of the Natural Logarithmic Function

Let x and y be positive real numbers, and let p be a real number. Then the following properties are true.

1. $\ln(1) = 0$	5. $\ln(xy) = \ln(x) + \ln(y)$	Product property for logarithms
2. $\ln(e) = 1$	6. $\ln\left(\frac{x}{y}\right) = \ln(x) - \ln(y)$	Quotient property for logarithms
3. $\ln(e^p) = p$	7. $\ln(x^p) = p\ln(x)$	Power property for logarithms
4. $e^{\ln(x)} = x$		

6. Simplifying Logarithmic Expressions

example 4 Simplifying Expressions with Natural Logarithms

Simplify the expressions. Assume that all variable expressions within the logarithms represent positive real numbers.

a. $\ln(e)$ b. $\ln(1)$ c. $\ln e^{(x+1)}$ d. $e^{\ln(x+1)}$

Solution:

a. $\ln(e) = 1$ Property 2

b. $\ln(1) = 0$ Property 1

c. $\ln e^{(x+1)} = x + 1$ Property 3

d. $e^{\ln(x+1)} = x + 1$ Property 4

example 5 Writing a Sum or Difference of Natural Logarithms as a Single Logarithm

Write the expression as a single logarithm. Assume that all variable expressions within the logarithms represent positive real numbers.

$$\ln(x^2 - 9) - \ln(x - 3) - 2\ln(x)$$

Solution:

$\ln(x^2 - 9) - \ln(x - 3) - 2\ln(x)$

$= \ln(x^2 - 9) - \ln(x - 3) - \ln(x^2)$ Power property for logarithms (property 7)

$= \ln(x^2 - 9) - [\ln(x - 3) + \ln(x^2)]$ Factor out -1 from the last two terms.

$= \ln(x^2 - 9) - \ln[(x - 3)x^2]$ Product property for logarithms (property 5)

$= \ln\left(\dfrac{x^2 - 9}{(x - 3)x^2}\right)$ Quotient property for logarithms (property 6)

$= \ln\left(\dfrac{(x - 3)(x + 3)}{(x - 3)x^2}\right)$ Factor.

$= \ln\left(\dfrac{x + 3}{x^2}\right)$ provided $x \neq 3$ Simplify.

example 6 **Writing a Logarithmic Expression in Expanded Form**

Write the expression

$$\ln\left(\frac{e}{x^2\sqrt{y}}\right)$$

as a sum or difference of logarithms of x and y. Assume all variable expressions within the logarithm represents positive real numbers.

Solution:

$\ln\left(\frac{e}{x^2\sqrt{y}}\right)$

$= \ln e - \ln(x^2\sqrt{y})$	Quotient property for logarithms (property 6)
$= \ln e - (\ln x^2 + \ln\sqrt{y})$	Product property for logarithms (property 5)
$= 1 - \ln x^2 - \ln y^{1/2}$	Distributive property. Also simplify $\ln e = 1$.
$= 1 - 2\ln x - \frac{1}{2}\ln y$	Power property for logarithms (property 7)

7. Change-of-Base Formula

A calculator can be used to approximate the value of a logarithm with a base of 10 or a base of e by using the [log] key or the [ln] key, respectively. However, to use a calculator to evaluate a logarithmic expression with a base other than 10 or e, we must use the **change-of-base formula**.

Change-of-Base Formula

Let a and b be positive real numbers such that $a \neq 1$ and $b \neq 1$. Then for any positive real number x,

$$\log_b(x) = \frac{\log_a(x)}{\log_a(b)}$$

Proof:

Let $M = \log_b(x)$, which implies that $b^M = x$

Now, take the logarithm, base a, on both sides: $\log_a(b^M) = \log_a(x)$

Apply the power property for logarithms: $M \cdot \log_a(b) = \log_a(x)$

Divide both sides by $\log_a(b)$: $\frac{M \cdot \cancel{\log_a(b)}}{\cancel{\log_a(b)}} = \frac{\log_a(x)}{\log_a(b)}$

$$M = \frac{\log_a(x)}{\log_a(b)}$$

Because $M = \log_b(x)$, we have: $\log_b(x) = \frac{\log_a(x)}{\log_a(b)}$ ✓

The change-of-base formula converts a logarithm of one base to a ratio of logarithms of a different base. For the sake of using a calculator, we often apply the change-of-base formula with base 10 or base *e*.

example 7

Using the Change-of-Base Formula

a. Use the change-of-base formula to evaluate $\log_4(80)$ by using base 10. (Round the result to three decimal places.)

b. Use the change-of-base formula to evaluate $\log_4(80)$ by using base *e*. (Round the result to three decimal places.)

Solution:

a. $\log_4(80) = \dfrac{\log_{10}(80)}{\log_{10}(4)} = \dfrac{\log(80)}{\log(4)} \approx \dfrac{1.903089987}{0.6020599913} \approx 3.161$

b. $\log_4(80) = \dfrac{\log_e(80)}{\log_e(4)} = \dfrac{\ln(80)}{\ln(4)} \approx \dfrac{4.382026635}{1.386294361} \approx 3.161$

To check the result, we see that $4^{3.161} \approx 80$.

Calculator Connections

The change-of-base formula can be used to graph logarithmic functions with bases other than 10 or *e*. For example, to graph $Y_1 = \log_2(x)$ we can enter the function as either

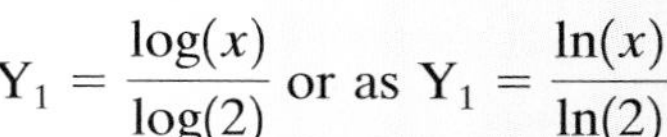

$$Y_1 = \frac{\log(x)}{\log(2)} \text{ or as } Y_1 = \frac{\ln(x)}{\ln(2)}$$

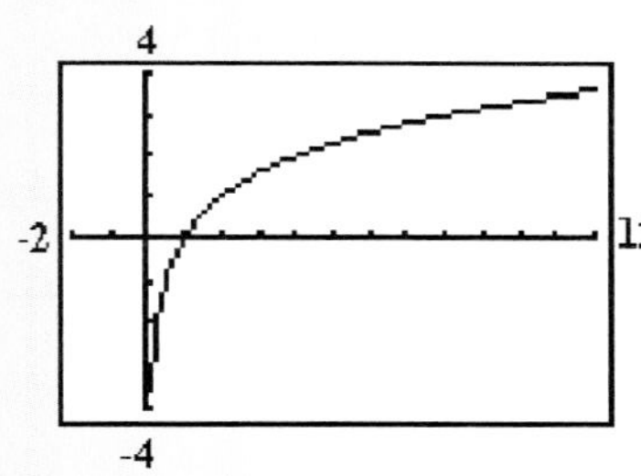

8. Applications of the Natural Logarithmic Function

example 8

Applying the Natural Logarithm Function to Radioactive Decay

Plant and animal tissue contain both carbon-12 and carbon-14. Carbon-12 is a stable form of carbon, whereas carbon-14 is a radioactive isotope with a half-life of approximately 5730 years. While a plant or animal is living, it takes in carbon from the atmosphere either through photosynthesis or through its food. The ratio of carbon-14 to carbon-12 in a living organism is constant and is the same as the ratio found in the atmosphere.

When a plant or animal dies, it no longer ingests carbon from the atmosphere. The amount of stable carbon-12 remains unchanged from the time of death, but the carbon-14 begins to decay. Because the rate of decay is constant, a tissue sample can be dated by comparing the percentage of carbon-14 still present to the percentage of carbon-14 assumed to be in its original living state.

The age of a tissue sample is a function of the percentage of carbon-14 still present in the organism according to the following model:

$A(p) = \dfrac{\ln(p)}{-0.000121}$, where A is the age in years and p is the percentage (in decimal form) of carbon-14 still present.

a. Find the age of a bone that has 72% of its initial carbon-14.
b. Find the age of the Iceman, a body uncovered in the mountains of northern Italy in 1991. Samples of his hair revealed that 51.4% of the original carbon-14 was present after his death.

Solution:

a. $A(p) = \dfrac{\ln(p)}{-0.000121}$

$A(0.72) = \dfrac{\ln(0.72)}{-0.000121}$ Substitute 0.72 for p.

≈ 2715 years

b. $A(p) = \dfrac{\ln(p)}{-0.000121}$

$A(0.514) = \dfrac{\ln(0.514)}{-0.000121}$ Substitute 0.514 for p.

≈ 5500 years The body of the Iceman is approximately 5500 years old.

section 12.6 Practice Exercises

For Exercises 1–4, fill out the tables and graph the functions. For Exercises 3 and 4, round to two decimal places.

1. $f(x) = \left(\dfrac{3}{2}\right)^x$

x	$f(x)$
−3	
−2	
−1	
0	
1	
2	
3	

2. $g(x) = \left(\dfrac{1}{5}\right)^x$

x	$g(x)$
−3	
−2	
−1	
0	
1	
2	
3	

3. $q(x) = \log(x + 1)$

x	$q(x)$
−0.75	
−0.50	
−0.25	
0	
1	
2	

4. $r(x) = \log x$

x	$r(x)$
0.25	
0.50	
0.75	
1	
2	
3	

For Exercises 5–8, write the expression as a sum or difference of $\ln a$, $\ln b$, and $\ln c$.

5. $\ln\left(\dfrac{a^4\sqrt{b}}{c}\right)$

6. $\ln\left(\dfrac{\sqrt{ab}}{c^3}\right)$

7. $\ln\left(\dfrac{ab}{c^2}\right)^{1/5}$

8. $\ln\sqrt{2ab}$

For Exercises 9–12, write the expression as a single logarithm.

9. $2 \ln a - \ln b - \frac{1}{3} \ln c$

10. $-\ln x + 3 \ln y - \ln z$

11. $4 \ln x - 3 \ln y - \ln z$

12. $\frac{1}{2} \ln c + \ln a - 2 \ln b$

13. a. Graph $f(x) = e^x$
    b. Identify the domain and range of f.
    c. Graph $g(x) = \ln x$
    d. Identify the domain and range of g.

14. a. Graph $f(x) = 10^x$
    b. Identify the domain and range of f.
    c. Graph $g(x) = \log x$
    d. Identify the domain and range of g.

For Exercises 15–18, complete the table and graph the function. Round values to two decimal places if necessary. Identify the domain and range.

15. $y = e^{x+1}$

x	y
−4	
−3	
−2	
−1	
0	
1	
2	

16. $y = e^{x+2}$

x	y
−5	
−4	
−3	
−2	
−1	
0	
1	

17. $y = \ln(x - 2)$

x	y
2.25	
2.50	
2.75	
3	
4	
5	
6	

18. $y = \ln(x - 1)$

x	y
1.25	
1.50	
1.75	
2	
3	
4	
5	

19. a. Evaluate $\log_6(200)$ by computing $\log(200)/\log(6)$ to four decimal places.
    b. Evaluate $\log_6(200)$ by computing $\ln(200)/\ln(6)$ to four decimal places.
    c. How do your answers to parts (a) and (b) compare?

20. a. Evaluate $\log_8(120)$ by computing $\log(120)/\log(8)$ to four decimal places.
    b. Evaluate $\log_8(120)$ by computing $\ln(200)/\ln(8)$ to four decimal places.
    c. How do your answers to parts (a) and (b) compare?

For Exercises 21–32, use the change-of-base formula to approximate the logarithms to four decimal places. Check your answers by using the exponential key on your calculator. (See Example 7.)

21. $\log_2 7$

22. $\log_3 5$

23. $\log_8 24$

24. $\log_4 17$

25. $\log_8 0.012$

26. $\log_7 0.251$

27. $\log_9 1$

28. $\log_2\left(\frac{1}{5}\right)$

29. $\log_4\left(\frac{1}{100}\right)$

30. $\log_5 0.0025$

31. $\log_7 0.0006$

32. $\log_2 0.24$

33. Given that $\log_3 9 = 2$ and $\log_3 27 = 3$,
    a. Estimate $\log_3 15$
    b. Estimate $\log_3 25$
    c. Evaluate the logarithms in parts (a) and (b) on a calculator and compare to your estimates.

34. Given that $\log_5 1 = 0$ and $\log_5 5 = 1$,
    a. Estimate $\log_5 2$
    b. Estimate $\log_5 4$
    c. Evaluate the logarithms in parts (a) and (b) on a calculator and compare to your estimates.

35. Given that $\log_6 6 = 1$ and $\log_6 36 = 2$,
    a. Estimate $\log_6 10$
    b. Estimate $\log_6 30$
    c. Evaluate the logarithms in parts (a) and (b) on a calculator and compare to your estimates.

36. Given that $\log_4 4 = 1$ and $\log_4 16 = 2$,
    a. Estimate $\log_4 6$
    b. Estimate $\log_4 12$
    c. Evaluate the logarithms in parts (a) and (b) on a calculator and compare to your estimates.

37. On August 31, 1854, an epidemic of cholera was discovered in London, England, resulting from a contaminated community water pump at Broad Street. By the end of September more than 600 citizens who drank water from the pump had died.

    The cumulative number of deaths from cholera in the 1854 London epidemic can be approximated by

    $D(t) = 91 + 160 \ln(t + 1)$, where t is the number of days after the start of the epidemic ($t = 0$ corresponds to September 1, 1854).

    a. Approximate total number of deaths as of September 1 ($t = 0$).
    b. Approximate total number of deaths as of September 5, September 10, and September 20.

Under continuous compounding, the amount of time, t, in years required for an investment to double is a function of the interest rate, r:

$$t = \frac{\ln(2)}{r}$$

Use the formula for Exercises 38–39.

38. a. If you invest \$5000, how long will it take the investment to reach \$10,000 if the interest rate is 4.5%? Round to one decimal place.
    b. If you invest \$5000, how long will it take the investment to reach \$10,000 if the interest rate is 10%? Round to one decimal place.
    c. Using the doubling time found in part (b), how long would it take a \$5000 investment to reach \$20,000 if the interest rate is 10%?

39. a. If you invest \$3000, how long will it take the investment to reach \$6000 if the interest rate is 5.5%? Round to one decimal place.
    b. If you invest \$3000, how long will it take the investment to reach \$6000 if the interest rate is 8%? Round to one decimal place.
    c. Using the doubling time found in part (b), how long would it take a \$3000 investment to reach \$12,000 if the interest rate is 8%?

In Exercises 40–45 use the model

$$A(t) = P\left(1 + \frac{r}{n}\right)^{nt}$$

for interest compounded n times per year. Use the model $A(t) = Pe^{rt}$ for interest compounded continuously. **How does interest rate affect an investment?**

40. Suppose an investor deposits \$10,000 in a certificate of deposit for 5 years for which the interest is compounded monthly. Find the total amount of money in the account for the following interest rates. Compare your answers and comment on the effect of interest rate on an investment.
    a. $r = 4.0\%$
    b. $r = 6.0\%$
    c. $r = 8.0\%$
    d. $r = 9.5\%$

41. Suppose an investor deposits \$5000 in a certificate of deposit for 8 years for which the interest is compounded quarterly. Find the total amount of money in the account for the following interest rates. Compare your answers and comment on the effect of interest rate on an investment.
    a. $r = 4.5\%$
    b. $r = 5.5\%$
    c. $r = 7.0\%$
    d. $r = 9.0\%$

How does the number of compounding periods affect an investment?

42. Suppose an investor deposits \$8000 in a savings account for 10 years at 4.5% interest. Find the total amount of money in the account for the following compounding options. Compare your answers. How does the number of compounding periods per year affect the total investment?

a. Compounded annually
b. Compounded quarterly
c. Compounded monthly
d. Compounded daily
e. Compounded continuously

43. Suppose an investor deposits \$15,000 in a savings account for 8 years at 5.0% interest. Find the total amount of money in the account for the following compounding options. Compare your answers. How does the number of compound periods per year affect the total investment?
    a. Compounded annually
    b. Compounded quarterly
    c. Compounded monthly
    d. Compounded daily
    e. Compounded continuously

How does the length of time money is invested affect the account value?

44. Suppose an investor deposits \$5000 in an account bearing 6.5% interest compounded continuously. Find the total amount in the account for the following time periods.
    a. 5 years b. 10 years
    c. 15 years d. 20 years
    e. 30 years

45. Suppose an investor deposits \$10,000 in an account bearing 6.0% interest compounded continuously. Find the total amount in the account for the following time periods.
    a. 5 years b. 10 years
    c. 15 years d. 20 years
    e. 30 years

Graphing Calculator Exercises

46. a. Graph the function defined by $f(x) = \log_3(x)$ by graphing $Y_1 = \log(x)/\log(3)$.
    b. Graph the function defined by $f(x) = \log_3(x)$ by graphing $Y_2 = \ln(x)/\ln(3)$.
    c. Does it appear that $Y_1 = Y_2$?

47. a. Graph the function defined by $f(x) = \log_7(x)$ by graphing $Y_1 = \log(x)/\log(7)$.
    b. Graph the function defined by $f(x) = \log_7(x)$ by graphing $Y_2 = \ln(x)/\ln(7)$.
    c. Does it appear that $Y_1 = Y_2$?

48. a. Graph the function defined by $f(x) = \log_{1/3}(x)$ by graphing $Y_1 = \log(x)/\log(\frac{1}{3})$.
    b. Graph the function defined by $f(x) = \log_{1/3}(x)$ by graphing $Y_2 = \ln(x)/\ln(\frac{1}{3})$.
    c. Does it appear that $Y_1 = Y_2$?

49. a. Graph the function defined by $f(x) = \log_{1/7}(x)$ by graphing $Y_1 = \log(x)/\log(\frac{1}{7})$.
    b. Graph the function defined by $f(x) = \log_{1/7}(x)$ by graphing $Y_2 = \ln(x)/\ln(\frac{1}{7})$.
    c. Does it appear that $Y_1 = Y_2$?

For Exercises 50–55, graph the functions on your calculator.

50. Graph $s(x) = \log_{1/2}(x)$
51. Graph $w(x) = \log_{1/3}(x)$
52. Graph $y = e^{x-2}$
53. Graph $y = e^{x-1}$
54. Graph $y = \ln(x + 1)$
55. Graph $y = \ln(x + 2)$

section

12.7 Exponential and Logarithmic Equations

Concepts

1. Solving Logarithmic Equations
2. Solving Logarithmic Equations in Quadratic Form
3. Solving Exponential Equations
4. Applications of Exponential Equations—Population Growth
5. Applications of Exponential Equations—Radioactive Decay

1. Solving Logarithmic Equations

Equations containing one or more logarithms are called **logarithmic equations**. For example,

$$\log_4 x = 1 - \log_4(x - 3) \qquad \text{and} \qquad \ln(x + 2) + \ln(x - 1) = \ln(9x - 17)$$

are logarithmic equations. To solve logarithmic equations of first degree, use the following guidelines.

Guidelines to Solve Logarithmic Equations

1. Isolate the logarithms on one side of the equation.
2. Write a sum or difference of logarithms as a single logarithm.
3. Rewrite the equation in step 2 in exponential form.
4. Solve the resulting equation from step 3.
5. Check all solutions to verify that they are within the domain of the logarithmic expressions in the original equation.

example 1 Solving a Logarithmic Equation

Solve the equation. $\log_4 x = 1 - \log_4(x - 3)$

Solution:

$$\log_4 x = 1 - \log_4(x - 3)$$

$$\log_4 x + \log_4(x - 3) = 1$$ Isolate the logarithms on one side of the equation.

$$\log_4[x(x - 3)] = 1$$ Write as a single logarithm.

$$\log_4(x^2 - 3x) = 1$$ Simplify inside the parentheses.

$$4^1 = x^2 - 3x$$ Write the equation in exponential form.

$$x^2 - 3x - 4 = 0$$ The resulting equation is quadratic.

$$(x - 4)(x + 1) = 0$$ Factor.

$$x = 4 \quad \text{or} \quad x = -1$$ Apply the zero product rule.

Notice that $x = -1$ is *not* a solution because $\log_4(x)$ is not defined at $x = -1$. However, $\log_4(x)$ and $\log_4(x - 3)$ are both defined at $x = 4$. We can substitute $x = 4$ into the original equation to show that it checks.

Tip: The equation from Example 1 involved the logarithmic functions $y = \log_4(x)$ and $y = \log_4(x - 3)$. The domains of these functions are $\{x \mid x > 0\}$ and $\{x \mid x > 3\}$, respectively. Therefore, the solutions to the equation are restricted to x values in the intersection of these two sets. That is, $\{x \mid x > 3\}$. The solution $x = 4$ satisfies this requirement, whereas $x = -1$ does not.

Check: $x = 4$

$$\log_4 x = 1 - \log_4(x - 3)$$
$$\log_4(4) \stackrel{?}{=} 1 - \log_4(4 - 3)$$
$$1 \stackrel{?}{=} 1 - \log_4(1)$$
$$1 = 1 - 0 \checkmark$$

The solution is $x = 4$.

example 2 Solving Logarithmic Equations

Solve the equations.

a. $\ln(x + 2) + \ln(x - 1) = \ln(9x - 17)$

b. $\log(x + 300) = 3.7$

Solution:

a. $$\ln(x + 2) + \ln(x - 1) = \ln(9x - 17)$$

$$\ln(x + 2) + \ln(x - 1) - \ln(9x - 17) = 0$$ Isolate the logarithms on one side.

$$\ln\left[\frac{(x + 2)(x - 1)}{9x - 17}\right] = 0$$ Write as a single logarithm.

$$e^0 = \frac{(x + 2)(x - 1)}{9x - 17}$$ Write the equation in exponential form.

$$1 = \frac{(x + 2)(x - 1)}{9x - 17}$$ Simplify.

$$(1) \cdot (9x - 17) = \left(\frac{(x + 2)(x - 1)}{\cancel{9x - 17}}\right) \cdot \cancel{(9x - 17)}$$ Multiply by the LCD.

$$9x - 17 = (x + 2)(x - 1)$$ The equation is quadratic.

$$9x - 17 = x^2 + x - 2$$
$$0 = x^2 - 8x + 15$$
$$0 = (x - 5)(x - 3)$$
$$x = 5 \quad \text{or} \quad x = 3$$

The solutions $x = 5$ and $x = 3$ are both within the domain of the logarithmic functions in the original equation. Both solutions check.

b. $\log(x + 300) = 3.7$ — The equation has a single logarithm that is already isolated.

$$10^{3.7} = x + 300$$ Write the equation in exponential form.

$$10^{3.7} - 300 = x$$ Solve for x.

$$x = 10^{3.7} - 300$$

$$\approx 4711.87$$

The solution $x = 10^{3.7} - 300$ checks in the original equation.

2. Solving Logarithmic Equations in Quadratic Form

example 3

Solving a Logarithmic Equation that Is in Quadratic Form

Solve the equation. $(\ln x)^2 - 7\ln x + 12 = 0$

Solution:

$$(\ln x)^2 - 7\ln x + 12 = 0$$ By letting $u = \ln x$, we see that the equation is in quadratic form.

$$u^2 - 7u + 12 = 0$$

$$(u - 4)(u - 3) = 0$$

$u = 4$ or $u = 3$ — Reverse substitute.

$\ln x = 4$ or $\ln x = 3$

$x = e^4$ or $x = e^3$ — Write the logarithmic equations in exponential form.

$x \approx 54.6$ or $x \approx 20.1$

Check: $x = e^4$

$$(\ln x)^2 - 7\ln x + 12 = 0$$

$$[\ln(e^4)]^2 - 7\ln(e^4) + 12 \stackrel{?}{=} 0$$

$$(4)^2 - 7 \cdot 4 + 12 \stackrel{?}{=} 0$$

$$16 - 28 + 12 \stackrel{?}{=} 0$$

$$-12 + 12 = 0 \checkmark$$

Check: $x = e^3$

$$(\ln x)^2 - 7\ln x + 12 = 0$$

$$[\ln(e^3)]^2 - 7\ln(e^3) + 12 \stackrel{?}{=} 0$$

$$(3)^2 - 7 \cdot 3 + 12 \stackrel{?}{=} 0$$

$$9 - 21 + 12 \stackrel{?}{=} 0$$

$$-12 + 12 = 0 \checkmark$$

Both solutions check.

Calculator Connections

To support the solution to Example 3 graph $Y_1 = (\ln x)^2 - 7\ln(x) + 12$. The values of x for which $Y_1 = 0$ can be approximated by using a *Zoom* and *Trace* feature.

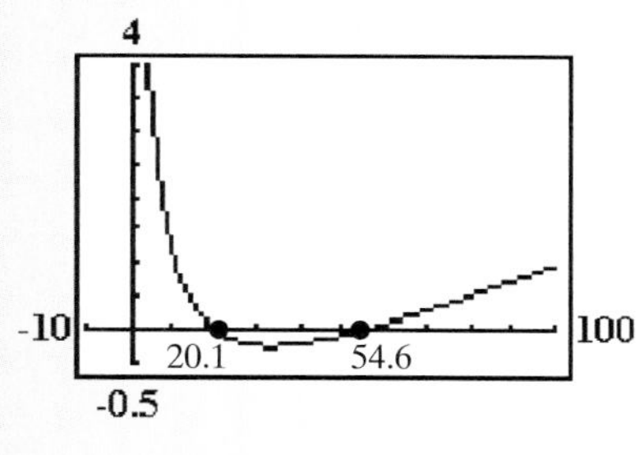

3. Solving Exponential Equations

An equation with one or more exponential expressions is called an **exponential equation**. The following property is often useful in solving exponential equations.

Equivalence of Exponential Expressions

Let x, y, and b be real numbers such that $b > 0$ and $b \neq 1$. Then

$$b^x = b^y \quad \text{implies} \quad x = y$$

The equivalence of exponential expressions indicates that if two exponential expressions of the same base are equal, their exponents must be equal.

example 4 **Solving Exponential Equations**

Solve the equations.

a. $4^{2x-9} = 64$ b. $(2^x)^{x+3} = \dfrac{1}{4}$

Solution:

a. $4^{2x-9} = 64$

$4^{2x-9} = 4^3$ Write both sides with a common base.

$2x - 9 = 3$ If $b^x = b^y$, then $x = y$.

$2x = 12$ Solve for x.

$x = 6$

To check, substitute $x = 6$ into the original equation.

$$4^{2(6)-9} \stackrel{?}{=} 64$$

$$4^{12-9} \stackrel{?}{=} 64$$

$$4^3 = 64 \checkmark$$

b. $(2^x)^{x+3} = \frac{1}{4}$

$2^{x^2+3x} = 2^{-2}$ Apply the multiplication property of exponents. Write both sides of the equation with a common base.

$x^2 + 3x = -2$ If $b^x = b^y$, then $x = y$.

$x^2 + 3x + 2 = 0$ The resulting equation is quadratic.

$(x + 2)(x + 1) = 0$ Solve for x.

$x = -2$ or $x = -1$

Check: $x = -2$

$(2^{-2})^{(-2)+3} \stackrel{?}{=} \frac{1}{4}$

$(2^{-2})^1 \stackrel{?}{=} \frac{1}{4}$

$2^{-2} = \frac{1}{4}$ ✓

Check: $x = -1$

$(2^{-1})^{(-1)+3} \stackrel{?}{=} \frac{1}{4}$

$(2^{-1})^2 \stackrel{?}{=} \frac{1}{4}$

$2^{-2} = \frac{1}{4}$ ✓

Both solutions check.

example 5

Solving an Exponential Equation

Solve the equation: $4^x = 25$.

Solution:

Because 25 cannot be written as an integral power of 4, we cannot immediately use the property that if $b^x = b^y$, then $x = y$. Instead we can rewrite the equation in its corresponding logarithmic form to solve for x.

$4^x = 25$

$x = \log_4(25)$ Write the equation in logarithmic form.

$= \frac{\ln(25)}{\ln(4)} \approx 2.322$ Apply the change-of-base formula.

Calculator Connections

Graph $Y_1 = 4^x$ and $Y_2 = 25$.

An *Intersect* feature can be used to find the x-coordinate where $Y_1 = Y_2$.

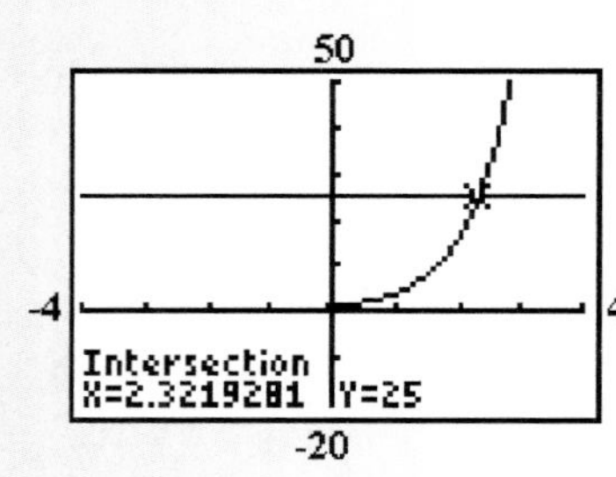

The same result can be reached by taking a logarithm of any base on both sides of the equation. Then by applying the power property of logarithms, the unknown exponent can be written as a factor.

$$4^x = 25$$

$$\log 4^x = \log 25$$ Take the common logarithm of both sides.

$$x \log 4 = \log 25$$ Apply the power property of logarithms to express the exponent as a factor. This is now a linear equation in x.

$$\frac{x \log 4}{\log 4} = \frac{\log 25}{\log 4}$$ Solve for x.

$$x = \frac{\log(25)}{\log(4)} \approx 2.322$$

Guidelines to Solve Exponential Equations

1. Isolate one of the exponential expressions in the equation.
2. Take a logarithm on both sides of the equation. (The natural logarithm function or the common logarithm function is often used so that the final answer can be approximated with a calculator.)
3. Use the power property of logarithms (property 7) to write exponents as factors. Recall: $\log_b(M^p) = p \log_b(M)$
4. Solve the resulting equation from step 3.

example 6

Solving Exponential Equations by Taking a Logarithm on Both Sides

Solve the equations.

a. $2^{x+3} = 7^x$ b. $e^{-3.6x} = 9.74$

Solution:

a.

$$2^{x+3} = 7^x$$

$$\ln 2^{(x+3)} = \ln 7^x$$ Take the natural logarithm of both sides.

$$(x + 3)\ln 2 = x \ln 7$$ Express the exponents as factors.

$$x(\ln 2) + 3(\ln 2) = x \ln 7$$ Apply the distributive property.

$$x(\ln 2) - x(\ln 7) = -3 \ln 2$$ Collect x-terms on one side.

$$x(\ln 2 - \ln 7) = -3 \ln 2$$ Factor out x.

$$\frac{x(\ln 2 - \ln 7)}{(\ln 2 - \ln 7)} = \frac{-3 \ln 2}{(\ln 2 - \ln 7)} \qquad \text{Solve for } x.$$

$$x = \frac{-3 \ln 2}{\ln 2 - \ln 7} \approx 1.660$$

Tip: The exponential equation $2^{x+3} = 7^x$ could have been solved by taking a logarithm of *any* base on both sides of the equation. For example, using base 10 yields:

$$\log 2^{x+3} = \log 7^x$$

$$(x + 3) \log 2 = x \log 7 \qquad \text{Apply the power property for logarithms.}$$

$$x \log 2 + 3 \log 2 = x \log 7 \qquad \text{Apply the distributive property.}$$

$$x \log 2 - x \log 7 = -3 \log 2 \qquad \text{Collect } x\text{-terms on one side of the equation.}$$

$$x(\log 2 - \log 7) = -3 \log 2 \qquad \text{Factor out } x.$$

$$x = \frac{-3 \log 2}{\log 2 - \log 7} \approx 1.660$$

b.

$$e^{-3.6x} = 9.74$$

$$\ln e^{-3.6x} = \ln 9.74$$

The exponential expression has a base of e, so it is convenient to take the natural logarithm of both sides.

$$(-3.6x)\ln e = \ln 9.74 \qquad \text{Express the exponent as a factor.}$$

$$-3.6x = \ln 9.74 \qquad \text{Simplify (recall that } \ln e = 1).$$

$$x = \frac{\ln 9.74}{-3.6} \approx -0.632$$

4. Applications of Exponential Equations—Population Growth

example 7 **Applying an Exponential Function to World Population**

The population of the world was estimated to have reached 6 billion in October 1999. The population growth rate for the world is estimated to be 1.4%. (*Source*: *Information Please Almanac*, 1999) Therefore,

$$P(t) = 6(1.014)^t$$

represents the world population in billions as a function of the number of years after October 1999 ($t = 0$ represents October 1999).

a. Use the function to estimate the world population in October of 2005 and in October of 2010.
b. Use the function to estimate the amount of time after October 1999 required for the world population to reach 12 billion.

Solution:

a.

$P(t) = 6(1.014)^t$

$P(6) = 6(1.014)^6$ — The year 2005 corresponds to $t = 6$.

≈ 6.5 — In 2005, the world's population will be approximately 6.5 billion.

$P(11) = 6(1.014)^{11}$ — The year 2010 corresponds to $t = 11$.

≈ 7.0 — In 2010, the world's population will be approximately 7.0 billion.

b.

$P(t) = 6(1.014)^t$

$12 = 6(1.014)^t$ — Substitute $P(t) = 12$ and solve for t.

$\frac{12}{6} = \frac{6(1.014)^t}{6}$ — Isolate the exponential expression on one side of the equation.

$2 = 1.014^t$

$\ln 2 = \ln 1.014^t$ — Take the natural logarithm of both sides.

$\ln 2 = t \ln 1.014$ — Express the exponent as a factor.

$\frac{\ln 2}{\ln 1.014} = \frac{t \ln 1.014}{\ln 1.014}$ — Solve for t.

$t = \frac{\ln 2}{\ln 1.014} \approx 50$ — The population will reach 12 billion (double the October 1999 value) approximately 50 years after 1999.

Note: It has taken thousands of years for the world's population to reach 6 billion. However, with a growth rate of 1.4%, it will take only 50 years to gain an additional 6 billion.

5. Applications of Exponential Equations—Radioactive Decay

example 8 Applying an Exponential Equation to Radioactive Decay

On Friday, April 25, 1986, a nuclear accident occurred at the Chernobyl nuclear reactor, resulting in radioactive contaminates being released into the atmosphere. The most hazardous isotopes released in this accident were ^{137}Cs (cesium-137),

^{131}I (iodine-131), and ^{90}Sr (strontium-90). People living close to Chernobyl (in the Ukraine) were at risk of radiation exposure from inhalation, from absorption through the skin, and from food contamination. Years after the incident, scientists have seen an increase in the incidence of thyroid disease among children living in the contaminated areas. Because iodine is readily absorbed in the thyroid gland, scientists suspect that radiation from iodine-131 is the cause.

The half-life of radioactive iodine (^{131}I) is 8.04 days. If 10 g of I-131 is initially present, then the amount of radioactive iodine still present after t days is approximated by:

$$A(t) = 10e^{-0.0862t}, \text{ where } t \text{ is the time in days.}$$

a. Use the model to approximate the amount of ^{131}I still present after 2 weeks. Round to the nearest 0.1 g.
b. How long will it take for the amount of ^{131}I to decay to 0.5 g? Round to the nearest 0.1 day.

Solution:

a. $A(t) = 10e^{-0.0862t}$

$A(14) = 10e^{-0.0862(14)}$ Substitute $t = 14$ (2 weeks).

$= 3.0$ g

b. $A(t) = 10e^{-0.0862t}$

$0.5 = 10e^{-0.0862t}$ Substitute $A = 0.5$.

$\dfrac{0.5}{10} = \dfrac{10e^{-0.0862t}}{10}$ Isolate the exponential expression.

$0.05 = e^{-0.0862t}$

$\ln(0.05) = \ln(e^{-0.0862t})$ Take the natural logarithm of both sides.

$\ln(0.05) = -0.0862t$ The resulting equation is linear.

$\dfrac{\ln(0.05)}{-0.0862} = \dfrac{-0.0862t}{-0.0862}$ Solve for t.

$t = \dfrac{\ln(0.05)}{-0.0862}$

≈ 34.8 days

Note: Radioactive iodine (^{131}I) is used in medicine in appropriate dosages to treat patients with hyperactive (overactive) thyroids. Because iodine is readily absorbed in the thyroid gland, the radiation is localized and will reduce the size of the thyroid while minimizing damage to surrounding tissues.

section 12.7 Practice Exercises

1. a. Graph $f(x) = e^x$.
   b. Write the domain and range in interval notation.
2. a. Graph $g(x) = 3^x$.
   b. Write the domain and range in interval notation.
3. a. Graph $h(x) = \ln(x)$.
   b. What is the vertical asymptote?
   c. Write the domain and range in interval notation.
4. a. Graph $k(x) = \log(x)$.
   b. What is the vertical asymptote?
   c. Write the domain and range in interval notation.

For Exercises 5–8, write the expression as a single logarithm.

5. $\log_b(x - 1) + \log_b(x + 2)$
6. $\log_b(x) + \log_b(2x + 3)$
7. $\log_b(x) - \log_b(1 - x)$
8. $\log_b(x + 2) - \log_b(3x - 5)$

For Exercises 9–14, identify the location of the vertical asymptote. Determine the domain of the function and write the answer in interval notation.

9. $y = \ln(x - 5)$
10. $y = \ln(x - 10)$
11. $y = \log(x + 2)$
12. $y = \log(x + 3)$
13. $y = \log_5(2x + 1)$
14. $y = \log_3(3x - 1)$

For Exercises 15–24, solve the exponential equation using the property that $b^x = b^y$ implies $x = y$, for $b > 0$ and $b \neq 1$.

15. $5^x = 625$
16. $3^x = 81$
17. $2^{-x} = 64$
18. $6^{-x} = 216$
19. $36^x = 6$
20. $343^x = 7$
21. $4^{2x-1} = 64$
22. $5^{3x-1} = 125$
23. $81^{3x-4} = \frac{1}{243}$
24. $4^{2x-7} = \frac{1}{128}$

For Exercises 25–36, solve the exponential equations by taking a logarithm of both sides. (Round the answers to three decimal places.)

25. $8^a = 21$
26. $6^y = 39$
27. $e^x = 8.1254$
28. $e^x = 0.3151$
29. $10^t = 0.0138$
30. $10^p = 16.8125$
31. $e^{0.07h} = 15$
32. $e^{0.03k} = 4$
33. $32e^{0.04m} = 128$
34. $8e^{0.05n} = 160$
35. $3^{x+1} = 5^x$
36. $2^{x-1} = 7^x$
37. Suppose \$5000 is invested at 7% interest compounded continuously. How long will it take for the investment to grow to \$10,000? Use the model $A(t) = Pe^{rt}$ and round to the nearest tenth of a year.
38. Suppose \$2000 is invested at 10% interest compounded continuously. How long will it take for the investment to triple? Use the model $A(t) = Pe^{rt}$ and round to the nearest year.
39. Phosphorus-32 (^{32}P) has a half-life of approximately 14 days. If 10 g of ^{32}P is present initially, then the amount, A, of phosphorus-32 still present after t days is given by $A(t) = 10(0.5)^{t/14}$.
    a. Find the amount of phosphorus-32 still present after 5 days. Round to the nearest tenth of a gram.
    b. Find the amount of time necessary for the amount of ^{32}P to decay to 4 g. Round to the nearest tenth of a day.
40. Polonium-210 (^{210}Po) has a half-life of approximately 138.6 days. If 4 g of ^{210}Po is present initially, then the amount, A, of polonium-210 still present after t days is given by $A(t) = 4e^{-0.005t}$.

a. Find the amount of polonium-138 still present after 50 days. Round to the nearest tenth of a gram.

b. Find the amount of time necessary for the amount of ^{32}P to decay to 0.5 g. Round to the nearest tenth of a day.

41. The population of China can be modeled by the function

$P(t) = 1237(1.0095)^t$, where $P(t)$ is in millions and t is the number of years since 1998.

a. Using this model, what was the population in the year 2002?

b. Predict the population in the year 2012.

c. If this growth rate continues, in what year will the population reach 2 billion people (2 billion is 2000 million)?

42. The population of Delhi, India, can be modeled by the function

$P(t) = 9817(1.031)^t$, where $P(t)$ is in thousands and t is the number of years since 2001.

a. Using this model, predict the population in the year 2010.

b. If this growth rate continues, in what year will the population reach 15 million (15 million is 15,000 thousand)?

43. The growth of a certain bacteria in a culture is given by the model

$A(t) = 500e^{0.0277t}$, where $A(t)$ is the number of bacteria and t is time in minutes.

a. What is the initial number of bacteria?

b. What is the population after 10 minutes?

c. How long will it take for the population to double (that is, reach 1000)?

44. The population of the bacteria *Salmonella typhimurium* is given by the model

$A(t) = 300e^{0.01733t}$, where $A(t)$ is the number of bacteria and t is time in minutes.

a. What is the initial number of bacteria?

b. What is the population after 10 minutes?

c. How long will it take for the population to double?

45. Suppose you save \$10,000 from working an extra job. Rather than spending the money, you decide to save the money for retirement by investing in a mutual fund that averages 12% per year. How long will it take for this money to grow to \$1,000,000? Use the model $A(t) = Pe^{rt}$ and round to the nearest tenth of a year.

46. The model $A = Pe^{rt}$ is used to compute the total amount of money in an account after t years at an interest rate, r, compounded continuously. P is the initial principal. Find the amount of time required for the investment to double as a function of the interest rate. (*Hint*: Substitute $A = 2P$ and solve for t.)

Solve the logarithmic equations in Exercises 47–72.

47. $\log_3 x = 2$

48. $\log_4 x = 9$

49. $\log p = 42$

50. $\log q = \frac{1}{2}$

51. $\ln x = 0.08$

52. $\ln x = 19$

53. $\log_x 25 = 2 \quad (x > 0)$

54. $\log_x 100 = 2 \quad (x > 0)$

55. $\log_b 10{,}000 = 4 \quad (b > 0)$

56. $\log_b e^3 = 3 \quad (b > 0)$

57. $\log_y 5 = \frac{1}{2} \quad (y > 0)$

58. $\log_b 8 = \frac{1}{2} \quad (b > 0)$

59. $\log_4(c + 5) = 3$

60. $\log_5(a - 4) = 2$

61. $\log_5(4y + 1) = 1$

62. $\log_6(5t - 2) = 1$

63. $\log_3 k + \log_3(2k + 3) = 2$

64. $\log_2(h - 1) + \log_2(h + 1) = 3$

65. $\log(x + 2) = \log(3x - 6)$

66. $\log x = \log(1 - x)$

67. $\log_5(3t + 2) - \log_5 t = \log_5 4$

68. $\log(6y - 7) + \log y = \log 5$

69. $\log(4m) = \log 2 + \log(m - 3)$

70. $\log(-h) + \log 3 = \log(2h - 15)$

71. $(\log_2 x)^2 - 12 \log_2 x = -32$

72. $(\log_3 x)^2 - \log_3 x^2 = 3$

73. The isotope of plutonium of mass 238 (written ^{238}Pu) is used to make thermoelectric power sources for spacecraft. The heat and electric power derived from such units have made the Voyager, Gallileo, and Cassini missions to the outer reaches of our solar system possible. The half-life of ^{238}Pu is 87.7 years.

Suppose a hypothetical space probe was launched in the year 2002 with 2.0 kg of ^{238}Pu. Then the amount of ^{238}Pu available to power the spacecraft decays over time according to

$P(t) = 2e^{-0.0079t}$, where $t \geq 0$ is the time in years, and $P(t)$ is the amount of plutonium still present (in kilograms).

a. Suppose the space probe is due to arrive at Pluto in the year 2045. How much plutonium will remain when the spacecraft reaches Pluto? Round to two decimal places.

b. If 1.5 kg of ^{238}Pu is required to power the spacecraft's data transmitter, will there be enough power in the year 2045 for us to receive close-up images of Pluto?

74. ^{99m}Tc is a radionuclide of technetium that is widely used in nuclear medicine. Although its half-life is only 6 hr, the isotope is continuously produced via the decay of its longer-lived parent, ^{99}Mo (molybdenum-99) whose half-life is approximately 3 days. ^{99}Mo generators (or "cows") are sold to hospitals in which the ^{99m}Tc can be "milked" as needed over a period of a few weeks. Once separated from its parent, the ^{99m}Tc may be chemically incorporated into a variety of imaging agents, each of which is designed to be taken up by a specific target organ within the body. Special cameras, sensitive to the gamma rays emitted by the technetium, are then used to record a "picture" (similar in appearance to an x-ray film) of the selected organ.

Suppose a technician prepares a sample of ^{99m}Tc-pyrophosphate to image the heart of a patient suspected of having had a mild heart attack. If the injection contains 10 mCi (millicuries) of ^{99m}Tc at 1:00 P.M., then the amount of technetium still present is given by

$T(t) = 10e^{-0.1155t}$, where $t > 0$ represents the time in hours after 1:00 P.M. and $T(t)$ represents the amount of ^{99m}Tc (in millicuries) still present.

a. How many millicuries of ^{99m}Tc will remain at 4:20 P.M. when the image is recorded? Round to the nearest tenth of a millicurie.

b. How long will it take for the radioactive level of the ^{99m}Tc to reach 2 mCi? Round to the nearest tenth of an hour.

75. The magnitude of an earthquake (the amount of seismic energy released at the hypocenter of the earthquake) is measured on the Richter scale. The Richter scale value, R, is determined by the formula

$R = \log\left(\frac{I}{I_0}\right)$, where I is the intensity of the earthquake and I_0 is the minimum measurable intensity of an earthquake. (I_0 is a "zero level" quake—one that is barely detected by a seismograph.)

a. Compare the Richter scale values of earthquakes that are (i) 100,000 times (10^5 times) more intense than I_0 and (ii) 1,000,000 times (10^6 times) more intense than I_0.

b. On October 17, 1989, an earthquake measuring 7.1 on the Richter scale occurred in the Loma Prieta area in the Santa Cruz Mountains. The quake devastated parts of San Francisco and Oakland, California, bringing 63 deaths and over 3700 injuries. Determine how many times more intense this earthquake was than a zero-level quake.

Graphing Calculator Exercises

76. The amount of money a company receives from sales is related to the money spent on advertising according to

$S(x) = 400 + 250\log(x) \quad x \geq 1$, where $S(x)$ is the amount in sales (in \$1000s) and x is the amount spent on advertising (in \$1000s).

a. The value of $S(1) = 400$ means that if \$1000 is spent on advertising, the total sales will be \$400,000.

   i. Find the total sales for this company if \$11,000 is spent on advertising.

   ii. Find the total sales for this company if \$21,000 is spent on advertising.

   iii. Find the total sales for this company if \$31,000 is spent on advertising.

b. Graph the function $y = S(x)$ on a window where $0 \leq x \leq 40$ and $0 \leq y \leq 1000$. Using the graph and your answers from part (a), describe the relationship between total sales and the money spent on advertising. As the money spent on advertising is increased what happens to the rate of increase of total sales?

c. How many advertising dollars are required for the total sales to reach \$1,000,000? That is, for what value of x will $S(x) = 1000$?

77. Graph $Y_1 = 8\text{^}x$ and $Y_2 = 21$ on a window where $0 \leq x \leq 5$ and $0 \leq y \leq 40$. Use the graph and an *Intersect* feature or *Zoom* and *Trace* to support your answer to Exercise 25.

78. Graph $Y_1 = 6\text{^}x$ and $Y_2 = 39$ on a window where $0 \leq x \leq 5$ and $0 \leq y \leq 50$. Use the graph and an *Intersect* feature or *Zoom* and *Trace* to support your answer to Exercise 26.

79. Graph $Y_1 = \log_3(x)$ (use the change-of-base formula) and $Y_2 = 2$ on a window where $0 \leq x \leq 40$ and $-4 \leq y \leq 4$. Use the graph and an *Intersect* feature or *Zoom* and *Trace* to support your answer to Exercise 47.

80. Graph $Y_1 = \log_4(x)$ (use the change-of-base formula) and $Y_2 = 9$ on a window where $0 \leq x \leq 1{,}000{,}000$ and $-2 \leq y \leq 12$. Use the graph and an *Intersect* feature or *Zoom* and *Trace* to support your answer to Exercise 48.

---

chapter 12 Summary

Section 12.1—Algebra and Composition of Functions

Key Concepts:

The Algebra of Functions:
Given two functions, f and g, the functions $f + g, f - g, f \cdot g$, and $\frac{f}{g}$ are defined as:

$$(f + g)(x) = f(x) + g(x)$$
$$(f - g)(x) = f(x) - g(x)$$
$$(f \cdot g)(x) = f(x) \cdot g(x)$$
$$\left(\frac{f}{g}\right)(x) = \frac{f(x)}{g(x)} \quad \text{provided } g(x) \neq 0$$

Examples:

Let $g(x) = 5x + 1$ and $h(x) = x^3$. Find:

1. $(g + h)(3) = g(3) + h(3) = 16 + 27 = 43$
2. $(g \cdot h)(-1) = g(-1) \cdot h(-1) = (-4) \cdot (-1) = 4$
3. $(g - h)(x) = 5x + 1 - x^3$
4. $\left(\frac{g}{h}\right)(x) = \frac{5x + 1}{x^3}$

Composition of Functions:
The composition of f and g, denoted $f \circ g$, is defined by the rule:

$(f \circ g)(x) = f(g(x))$ provided that $g(x)$ is in the domain of f.

The composition of g and f, denoted $g \circ f$, is defined by the rule:

$(g \circ f)(x) = g(f(x))$ provided that $f(x)$ is in the domain of g.

Find $(f \circ g)(x)$ and $(g \circ f)(x)$ given the functions defined by $f(x) = 4x + 3$ and $g(x) = 7x$.

$$\begin{aligned}(f \circ g)(x) &= f(g(x))\\ &= 4(g(x)) + 3\\ &= 4(7x) + 3\\ &= 28x + 3\end{aligned}$$

$$\begin{aligned}(g \circ f)(x) &= g(f(x))\\ &= 7(f(x))\\ &= 7(4x + 3)\\ &= 28x + 21\end{aligned}$$

Key Term

composition of functions

Section 12.2—Inverse Functions

Key Concepts:

Horizontal Line Test
Consider a function defined by a set of points (x, y) in a rectangular coordinate system. Then the graph defines y as a one-to-one function of x if no horizontal line intersects the graph in more than one point.

Examples:

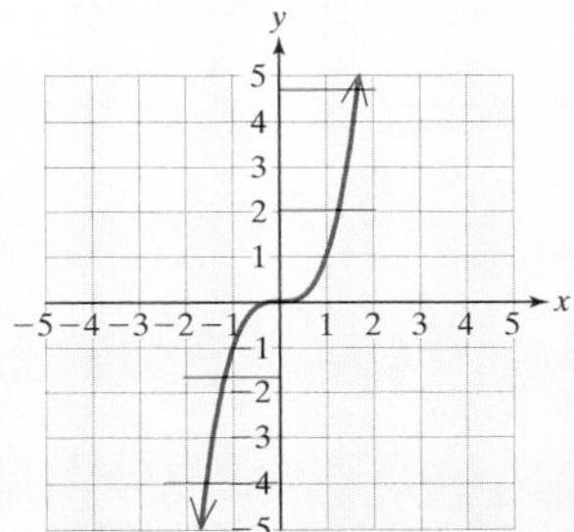

The function is one-to-one because it passes the horizontal line test.

Finding an Equation of the Inverse of a Function
For a one-to-one function defined by $y = f(x)$, the equation of the inverse can be found as follows:

1. Replace $f(x)$ by y.
2. Interchange x and y.
3. Solve for y.
4. Replace y by $f^{-1}(x)$.

Find the inverse of the one-to-one function defined by $f(x) = 3 - x^3$.

1. $y = 3 - x^3$
2. $x = 3 - y^3$
3. $x - 3 = -y^3$

$-x + 3 = y^3$

$\sqrt[3]{-x + 3} = y$

The graphs defined by $y = f(x)$ and $y = f^{-1}(x)$ are symmetric with respect to the line $y = x$.

4. $f^{-1}(x) = \sqrt[3]{-x + 3}$

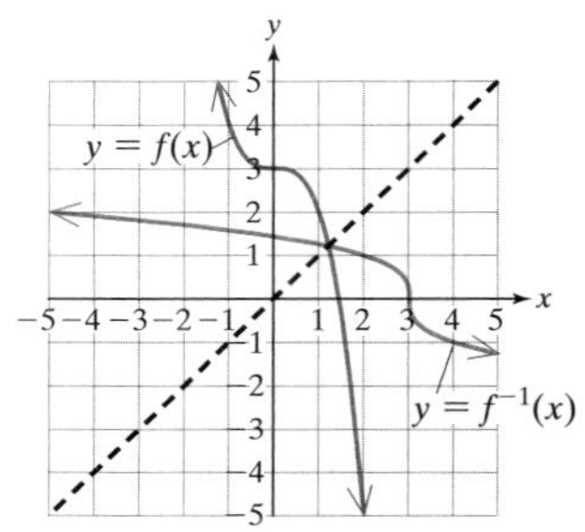

Definition of the Inverse of a Function
If f is a one-to-one function, then g is the inverse of f if and only if $(f \circ g)(x) = x$ for all x in the domain of g, and $(g \circ f)(x) = x$ for all x in the domain of f.

Verify that the functions defined by $f(x) = x - 1$ and $g(x) = x + 1$ are inverses.

$$(f \circ g)(x) = f(x + 1) = (x + 1) - 1 = x$$
$$(g \circ f)(x) = g(x - 1) = (x - 1) + 1 = x$$

Key Terms:

horizontal line test
inverse of f
one-to-one function

Section 12.3—Exponential Functions

Key Concepts:

A function $y = b^x (b > 0, b \neq 1)$ is an exponential function.

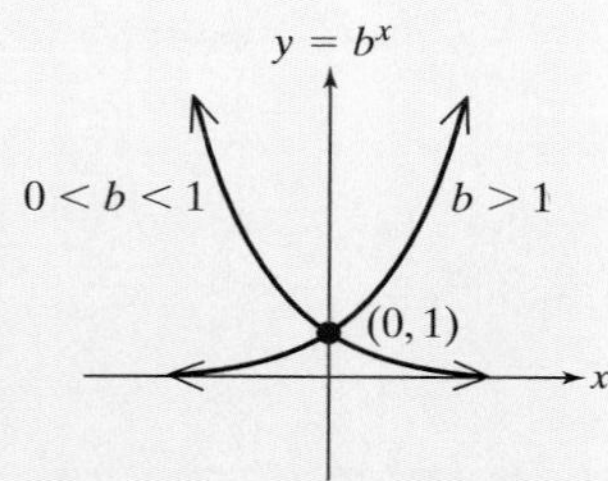

The domain is $(-\infty, \infty)$.
The range is $(0, \infty)$
The line $y = 0$ is a horizontal asymptote.
The y-intercept is $(0, 1)$.

Key Terms:

exponential decay function
exponential function
exponential growth function
half-life

Examples:

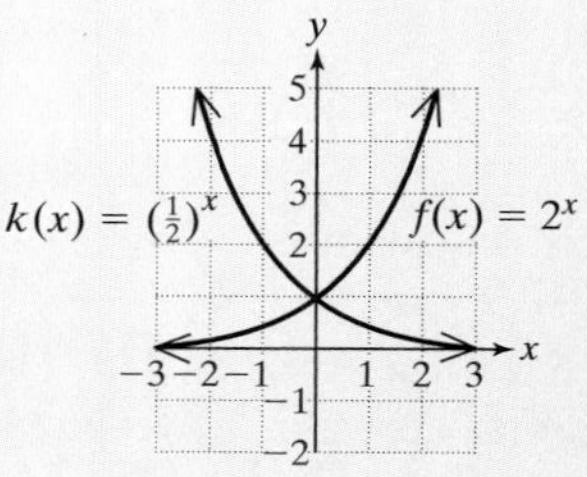

Section 12.4—Logarithmic Functions

Key Concepts:

The function $y = \log_b(x)$ is a logarithmic function.

$$y = \log_b(x) \Leftrightarrow b^y = x \quad (x > 0, b > 0, b \neq 1)$$

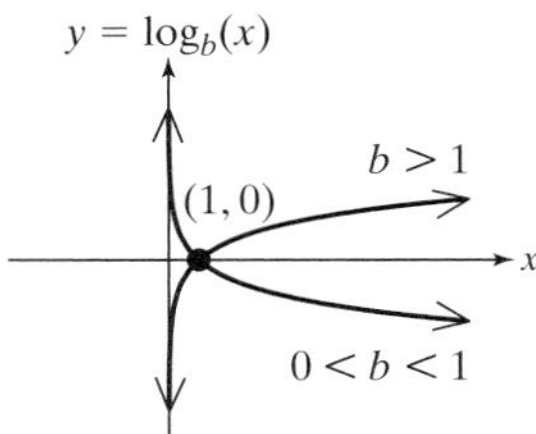

For $y = \log_b(x)$, the domain is $(0, \infty)$.
The range is $(-\infty, \infty)$.
The line $x = 0$ is a vertical asymptote.
The x-intercept is $(1, 0)$.

$y = \log(x)$ is the common logarithmic function (base 10).

Key Terms:

argument
base
common logarithmic function
domain of a logarithmic function
logarithm
logarithmic function

Examples:

$\log_4(64) = 3$ because $4^3 = 64$.

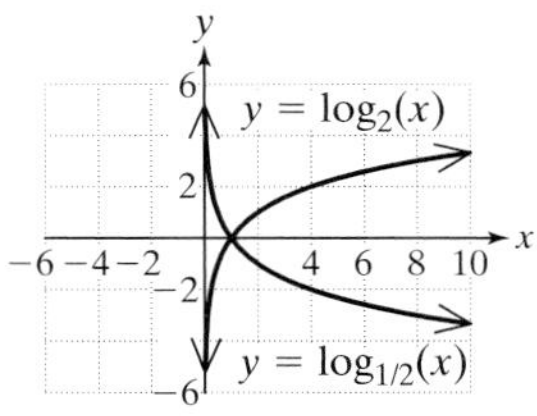

$\log(10{,}000) = 4$ because $10^4 = 10{,}000$.

Section 12.5—Properties of Logarithms

Key Concepts:

Let b, x, and y be positive real numbers where $b \neq 1$, and let p be a real number. Then the following properties are true.

1. $\log_b(1) = 0$
2. $\log_b(b) = 1$
3. $\log_b(b^p) = p$
4. $b^{\log_b(x)} = x$
5. $\log_b(xy) = \log_b(x) + \log_b(y)$
6. $\log_b\left(\frac{x}{y}\right) = \log_b(x) - \log_b(y)$
7. $\log_b(x^p) = p\log_b(x)$

Examples:

1. $\log_5(1) = 0$
2. $\log_6(6) = 1$
3. $\log_4(4^7) = 7$
4. $2^{\log_2(5)} = 5$
5. $\log(5x) = \log(5) + \log(x)$
6. $\log_7\left(\frac{z}{10}\right) = \log_7(z) - \log_7(10)$
7. $\log x^5 = 5 \log x$

The properties of logarithms can be used to write multiple logarithms as a single logarithm.

Write as a single logarithm:

$$\log x - \frac{1}{2}\log y - 3\log z$$
$$= \log x - (\log y^{1/2} + \log z^3)$$
$$= \log x - \log(\sqrt{y}z^3)$$
$$= \log\left(\frac{x}{\sqrt{y}z^3}\right)$$

The properties of logarithms can be used to write a single logarithm as a sum or difference of logarithms.

Expand into sums or differences of logarithms:

$$\log\sqrt[3]{\frac{x}{y^2}}$$
$$= \frac{1}{3}\log\left(\frac{x}{y^2}\right)$$
$$= \frac{1}{3}(\log x - \log y^2)$$
$$= \frac{1}{3}(\log x - 2\log y)$$
$$= \frac{1}{3}\log x - \frac{2}{3}\log y$$

KEY TERM:

properties of logarithms

SECTION 12.6—THE IRRATIONAL NUMBER e

KEY CONCEPTS:

As x becomes infinitely large, the expression, $\left(1 + \frac{1}{x}\right)^x$ approaches the irrational number, e, where $e \approx 2.718281$.

The balance of an account earning compound interest, n times per year is given by

$$A(t) = P\left(1 + \frac{r}{n}\right)^{nt}$$

where P = principal, r = annual interest rate, t = time in years, and n = number of compound periods per year.

The balance of an account earning interest compounded continuously is given by:

$$A(t) = Pe^{rt}$$

EXAMPLES:

Find the account balance for \$8000 invested for 10 years at 7% compounded quarterly:

$$P = 8000, t = 10, r = 0.07, n = 4$$
$$A(10) = 8000\left(1 + \frac{0.07}{4}\right)^{(4)(10)}$$
$$A(10) = \$16{,}012.78$$

Find the account balance for the same investment compounded continuously:

$$P = 8000, t = 10, r = 0.07$$
$$A(t) = 8000e^{(0.07)(10)}$$
$$= \$16{,}110.02$$

The function $y = e^x$ is the exponential function with base e.

The natural logarithm function, $y = \ln(x)$, is the logarithm function with base e.

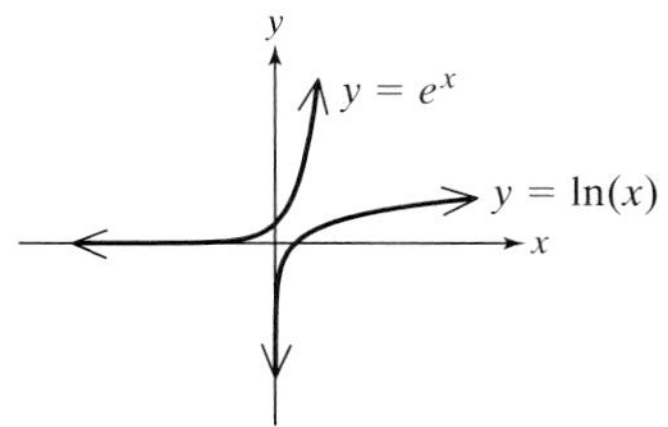

Use a calculator to approximate the value of the expressions.

$$e^{7.5} \approx 1808.04 \qquad e^{-\pi} \approx 0.0432$$

$$\ln(107) \approx 4.6728$$

$$\ln\left(\frac{1}{\sqrt{2}}\right) \approx -0.3466$$

Change-of-Base Formula:

$$\log_b(x) = \frac{\log_a(x)}{\log_a(b)} \qquad a > 0, a \neq 1, b > 0, b \neq 1$$

$$\log_3(59) = \frac{\log(59)}{\log(3)} \approx 3.7115$$

Key Terms:

change-of-base formula
continuous compounding
e
natural logarithmic function

Section 12.7—Exponential and Logarithmic Equations

Key Concepts:

Guidelines to Solve Logarithmic Equations

1. Isolate the logarithms on one side of the equation.
2. Write a sum or difference of logarithms as a single logarithm.
3. Rewrite the equation in step 2 in exponential form.
4. Solve the resulting equation from step 3.
5. Check all solutions to verify that they are within the domain of the logarithmic expressions in the equation.

Examples:

Solve:

$$\log(3x - 1) + 1 = \log(2x + 1)$$

Step 1: $\log(3x - 1) - \log(2x + 1) = -1$

Step 2: $\log\left(\frac{3x - 1}{2x + 1}\right) = -1$

Step 3: $10^{-1} = \frac{3x - 1}{2x + 1}$

Step 4: $\frac{1}{10} = \frac{3x - 1}{2x + 1}$

$$2x + 1 = 10(3x - 1)$$

$$2x + 1 = 30x - 10$$

$$-28x = -11$$

$$x = \frac{11}{28}$$

The equivalence of exponential expressions can be used to solve exponential equations.

If $b^x = b^y$ then $x = y$

Step 5: $x = \frac{11}{28}$ Checks in original equation.

Solve: $5^{2x} = 125$

$5^{2x} = 5^3$ implies that $2x = 3$

$x = \frac{3}{2}$

Guidelines to Solve Exponential Equations

1. Isolate one of the exponential expressions in the equation.
2. Take a logarithm of both sides of the equation.
3. Use the power property of logarithms to write exponents as factors.
4. Solve the resulting equation from step 3.

Solve: $4^{x+1} - 2 = 1055$

Step 1: $4^{x+1} = 1057$

Step 2: $\ln(4^{x+1}) = \ln(1057)$

Step 3: $(x + 1)\ln 4 = \ln 1057$

Step 4: $x + 1 = \frac{\ln 1057}{\ln 4}$

$x = \frac{\ln 1057}{\ln 4} - 1 \approx 4.023$

KEY TERMS:

exponential equations
logarithmic equations

chapter 12 REVIEW EXERCISES

Section 12.1

For Exercises 1–8, refer to the functions defined here.

$f(x) = x - 7 \quad g(x) = -2x^3 - 8x$

$m(x) = \sqrt{x} \quad n(x) = \frac{1}{x - 2}$

Find the indicated function values. Write the domain in interval notation.

1. $(f - g)(x)$
2. $(f + g)(x)$
3. $(f \cdot n)(x)$
4. $(f \cdot m)(x)$
5. $\left(\frac{f}{g}\right)(x)$
6. $\left(\frac{g}{f}\right)(x)$
7. $(m \circ f)(x)$
8. $(n \circ f)(x)$

For Exercises 9–12, refer to the functions defined for Exercises 1–8. Find the function values if possible.

9. $(m \circ g)(-2)$
10. $(n \circ g)(-1)$
11. $(f \circ g)(4)$
12. $(g \circ f)(8)$

13. Given $f(x) = 2x + 1$ and $g(x) = x^2$,
   a. Find $(g \circ f)(x)$.
   b. Find $(f \circ g)(x)$.
   c. Based on your answers to part (a) is $f \circ g$ equal to $g \circ f$?

For Exercises 14–19, refer to the graph. Approximate the function values if possible.

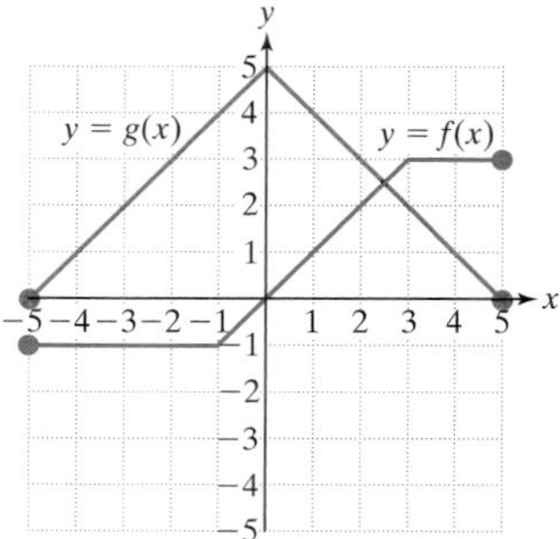

14. $\left(\frac{f}{g}\right)(1)$
15. $(f \cdot g)(-2)$

16. $(f + g)(-4)$

17. $(f - g)(2)$

18. $(g \circ f)(-3)$

19. $(f \circ g)(4)$

20. The following graph depicts the per capita consumption of refined sugar (cane and beet) and the per capita consumption of corn sweeteners as a function of the year from 1980 to 1996.

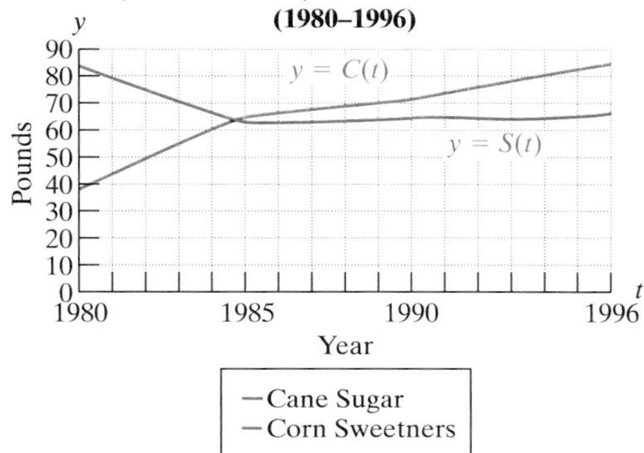

Source: U.S. Department of Agriculture.

Figure for Exercise 20

Let the function C represent the amount of corn sweeteners consumed per capita, and let the function S represent the amount of refined sugar (cane and beet) consumed per capita. Both functions measure sugar consumption in pounds. The variable t represents the year where $1980 \le t \le 1996$.

Suppose the function G is defined as $G(t) = C(t) + S(t)$. What does the function G represent?

Section 12.2

For Exercises 21–24, determine if the function is one-to-one by using the horizontal line test.

21.

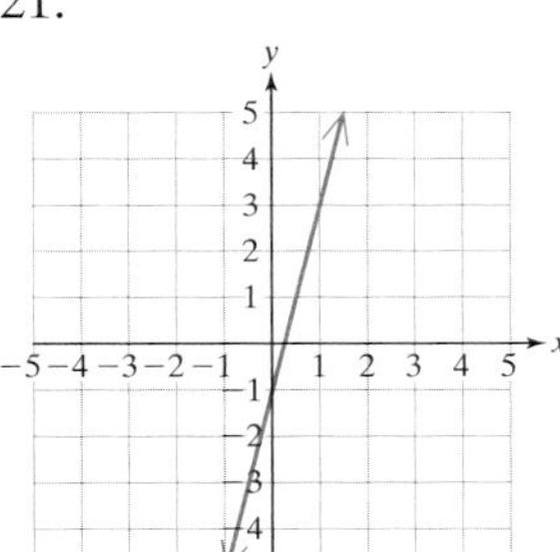

22.

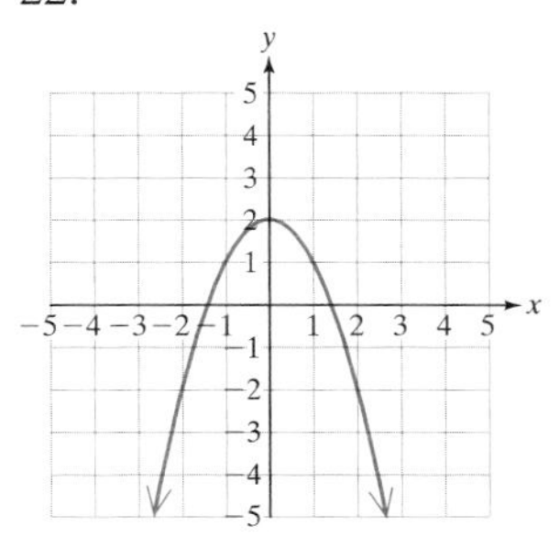

23.

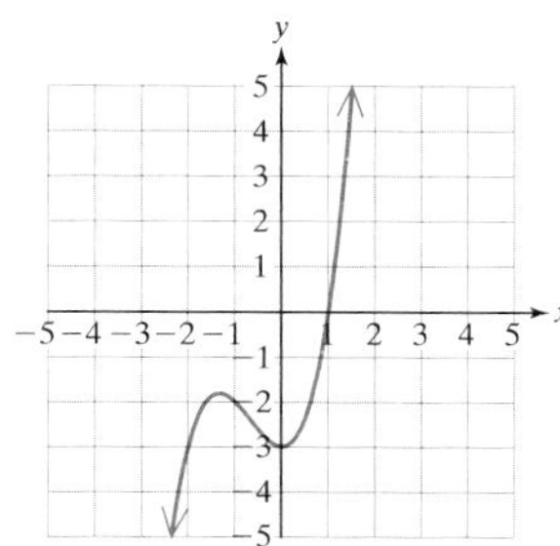

24.

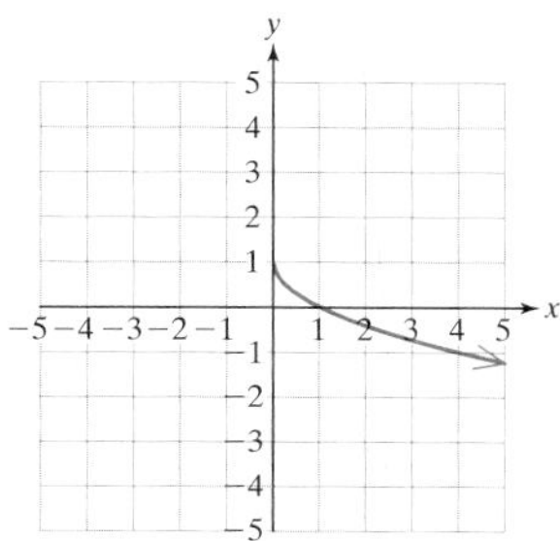

For Exercises 25–32, write the inverse for each one-to-one function.

25. $\{(3, 5), (2, 9), (0, -1), (4, 1)\}$

26. $\{(2, 0), (-1, 7), (1, 1), (0, 4)\}$

27. $p(x) = 3 - 4x$

28. $q(x) = \frac{3}{4}x - 2$

29. $g(x) = \sqrt[5]{x} + 3$

30. $f(x) = (x - 1)^3$

31. $m(x) = \frac{x - 2}{x + 2}$

32. $n(x) = \frac{4}{x - 2}$

33. Verify that the functions defined by $f(x) = 5x - 2$ and $g(x) = \frac{1}{5}x + \frac{2}{5}$ are inverses by showing that $(f \circ g)(x) = x$ and $(g \circ f)(x) = x$.

34. Graph the functions p and p^{-1} from Exercise 27 on the same coordinate axes. What can you say about the relationship between these two graphs?

35. a. What are the domain and range of the function defined by $h(x) = \sqrt{x + 1}$?

b. What are the domain and range of the function defined by $k(x) = x^2 - 1, x \ge 0$?

36. Determine the inverse of the function $p(x) = \sqrt{x} + 2$.

Section 12.3

For Exercises 37–44, evaluate the exponential expressions. Use a calculator and round to three decimal places, if necessary.

37. 4^5

38. 6^{-2}

39. $8^{1/3}$

40. $\left(\frac{1}{100}\right)^{-1/2}$

41. 2^{π}

42. $5^{\sqrt{3}}$

43. $(\sqrt{7})^{1/2}$

44. $\left(\frac{3}{4}\right)^{4/3}$

For Exercises 45–48, graph the functions.

45. $f(x) = 3^x$

46. $g(x) = \left(\frac{1}{4}\right)^x$

47. $h(x) = 5^{-x}$

48. $k(x) = \left(\frac{2}{5}\right)^{-x}$

49. a. Does the graph of $y = b^x, b > 0, b \neq 1$, have a vertical or a horizontal asymptote?
    b. Write an equation of the asymptote.

50. Background radiation is radiation that we are exposed to from naturally occurring sources including the soil, the foods we eat, and the sun. Background radiation varies depending on where we live. A typical background radiation level is 150 mrem (millirems) per year. (A rem is a measure of energy produced from radiation.) Suppose a substance emits 30,000 mrem per year and has a half-life of 5 years. The function defined by

$A(t) = 30{,}000\left(\frac{1}{2}\right)^{t/5}$ gives the radiation level (in millirems) of this substance after t years.

    a. What is the radiation level after 5 years?
    b. What is the radiation level after 15 years?
    c. Will the radiation level of this substance be below the background level of 150 mrem after 50 years?

Section 12.4

For Exercises 51–58, evaluate the logarithms without using a calculator.

51. $\log_3 \frac{1}{27}$

52. $\log_5 1$

53. $\log_7 7$

54. $\log_2 2^8$

55. $\log_2 16$

56. $\log_3 81$

57. $\log(100{,}000)$

58. $\log_8\left(\frac{1}{8}\right)$

For Exercises 59–60, graph the logarithmic functions.

59. $q(x) = \log_3 x$

60. $r(x) = \log_{1/2} x$

61. a. Does the graph of $y = \log_b x$ have a vertical or a horizontal asymptote?
    b. Write an equation of the asymptote.

62. Acidity of a substance is measured by its pH. The pH can be calculated by the formula $\text{pH} = -\log[\text{H}^+]$ where $[\text{H}^+]$ is the hydrogen ion concentration.
    a. What is the pH of a fruit with a hydrogen ion concentration of 0.00316 mol/L? Round to one decimal place.
    b. What is the pH of an antacid tablet with $[\text{H}^+] = 3.16 \times 10^{-10}$? Round to one decimal place.

Section 12.5

For Exercises 63–66, evaluate the logarithms without using a calculator.

63. $\log_8 8$

64. $\log_{11}(11^6)$

65. $\log_{1/2} 1$

66. $12^{\log_{12} 7}$

67. Complete the properties. Assume x, y, and b are positive real numbers such that $b \neq 1$.
    a. $\log_b(xy) =$
    b. $\log_b x - \log_b y =$
    c. $\log_b x^p =$

For Exercises 68–77, expand the logarithms to be in terms of $\log_b 2$, $\log_b 3$, and $\log_b 5$. Then find the values of the logarithms given that $\log_b 2 \approx 0.693$, $\log_b 3 \approx 1.099$, and $\log_b 5 \approx 1.609$.

68. $\log_b 6$

69. $\log_b 4$

70. $\log_b 12$

71. $\log_b 25$

72. $\log_b 81$

73. $\log_b 30$

74. $\log_b\left(\frac{5}{2}\right)$

75. $\log_b\left(\frac{25}{3}\right)$

76. $\log_b(10^6)$

77. $\log_b(2^{12})$

For Exercises 78–81, write the logarithmic expressions as a single logarithm.

78. $\frac{1}{4}(\log_b y - 4\log_b z + 3\log_b x)$

79. $\frac{1}{2}\log_3 a + \frac{1}{2}\log_3 b - 2\log_3 c - 4\log_3 d$

80. $\log 540 - 3\log 3 - 2\log 2$

81. $-\log_4 18 + \log_4 6 + \log_4 3 - \log_4 1$

Section 12.6

For Exercises 82–93, use a calculator to approximate the expressions to four decimal places.

82. e^5

83. $e^{\sqrt{7}}$

84. $32\,e^{0.008}$

85. $58\,e^{-0.0125}$

86. $\ln 6$

87. $\ln\left(\frac{1}{9}\right)$

88. $\ln 1$

89. $\ln 0.0162$

90. $\log 200$

91. $\log(0.0058)$

92. $\log 22$

93. $\log e^3$

For Exercises 94–99, use the change-of-base formula to approximate the logarithms to four decimal places.

94. $\log_2 10$

95. $\log_9 80$

96. $\log_{1/2}(20)$

97. $\log_{1/3}(100)$

98. $\log_5(0.26)$

99. $\log_4(0.0062)$

100. An investor wants to deposit \$20,000 in an account for 10 years at 5.25% interest. Compare the amount she would have if her money were invested with the following different compounding options. Use

$$A(t) = P\left(1 + \frac{r}{n}\right)^{nt}$$

for interest compounded n times per year and $A(t) = Pe^{rt}$ for interest compounded continuously.

a. Compounded annually

b. Compounded quarterly

c. Compounded monthly

d. Compounded continuously

101. To measure a student's retention of material at the end of a course, researchers give the student a test on the material every month for 24 months after the course is over. The student's average score t months after completing the course is given by

$S(t) = 75e^{-0.5t} + 20$, where t is the time in months, and S is the test score.

a. Find $S(0)$ and interpret the result.

b. Find $S(6)$ and interpret the result.

c. Find $S(12)$ and interpret the result.

d. The graph of $y = S(t)$ is shown here. Does it appear that the student's average score is approaching a limiting value? Explain.

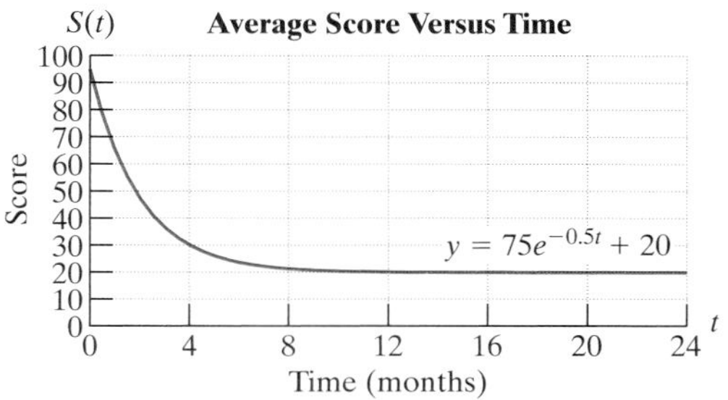

Figure for Exercise 101

102. A lake is stocked with 1000 fish and the fish population can be modeled by

$p(t) = \dfrac{20{,}000}{2 + 18e^{-t/4}}$, where p is the fish population and t is the time in months after the lake was initially stocked.

a. Find $p(0)$ and interpret the result.

b. Find $p(12)$ and interpret the result.

c. Does it appear that the fish population is approaching a limiting value? Explain.

d. Use the graph to approximate the fish population after 8 months.

Figure for Exercise 102

Section 12.7

For Exercises 103–110, identify the domain. Write the answer in interval notation.

103. $f(x) = e^x$

104. $g(x) = e^{x+6}$

105. $h(x) = e^{x-3}$

106. $k(x) = \ln x$

107. $q(x) = \ln(x + 5)$

108. $p(x) = \ln(x - 7)$

109. $r(x) = \ln(3x - 4)$

110. $w(x) = \ln(5 - x)$

Solve the logarithmic equations in Exercises 111–118. If necessary, round to two decimal places.

111. $\log_5 x = 3$

112. $\log_7 x = -2$

113. $\log_6 y = 3$

114. $\log_3 y = \frac{1}{12}$

115. $\log(2w - 1) = 3$

116. $\log_2(3w + 5) = 5$

117. $\log p - 1 = -\log(p - 3)$

118. $\log_4(2 + t) - 3 = \log_4(3 - 5t)$

Solve the exponential equations in Exercises 119–126. If necessary, round to four decimal places.

119. $4^{3x+5} = 16$

120. $5^{7x} = 625$

121. $4^a = 21$

122. $5^a = 18$

123. $e^{-x} = 0.1$

124. $e^{-2x} = 0.06$

125. $10^{2n} = 1512$

126. $10^{-3m} = \frac{1}{821}$

127. Radioactive iodine (^{131}I) is used to treat patients with a hyperactive (overactive) thyroid. Patients with this condition may have symptoms that include rapid weight loss, heart palpitations, and high blood pressure. Because iodine is readily absorbed in the thyroid gland, the radiation is localized and will reduce the size of the thyroid while minimizing damage to surrounding tissues. The half-life of radioactive iodine is 8.04 days. If a patient is given an initial dose of 2 μg, then the amount of iodine in the body after t days is approximated by:

$A(t) = 2e^{-0.0862t}$, where t is the time in days and $A(t)$ is the amount (in micrograms) of ^{131}I remaining.

a. How much radioactive iodine is present after a week? Round to two decimal places.

b. How much radioactive iodine is present after 30 days? Round to two decimal places.

c. How long will it take for the level of radioactive iodine to reach 0.5 μg?

128. The growth of a certain bacteria in a culture is given by the model $A(t) = 150e^{0.007t}$, where $A(t)$ is the number of bacteria and t is time in minutes. Let $t = 0$ correspond to the initial number of bacteria.

a. What is the initial number of bacteria?

b. What is the population after $\frac{1}{2}$ hour?

c. How long will it take for the population to double?

129. The value of a car depreciates with time according to

$V(t) = 15{,}000e^{-0.15t}$, where $V(t)$ is the value in dollars and t is the time in years.

a. Find $V(0)$ and interpret the result in the context of the problem.

b. Find $V(10)$ and interpret the result in the context of the problem. Round to the nearest dollar.

c. Find the time required for the value of the car to drop to \$5000. Round to the nearest tenth of a year.

d. The graph of $y = V(t)$ is shown here. Does it appear that the value of the car is approaching a limiting value? Explain.

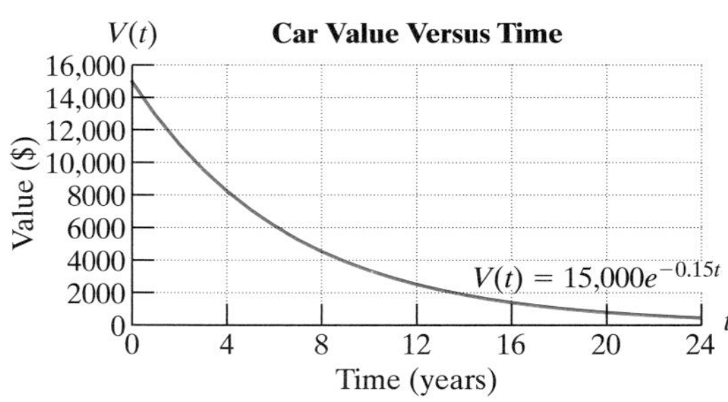

Figure for Exercise 129

chapter 12 TEST

For Exercises 1–8, refer to these functions.

$f(x) = x - 4 \qquad g(x) = \sqrt{x + 2} \qquad h(x) = \dfrac{1}{x}$

Find the function values if possible.

1. 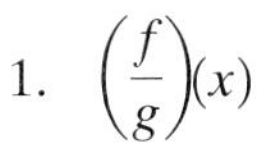 $\left(\dfrac{f}{g}\right)(x)$
2. $(h \cdot g)(x)$
3. $(g \circ f)(x)$
4. $(h \circ f)(x)$
5. $(f - g)(7)$
6. $(h + f)(2)$
7. $(h \circ g)(14)$
8. $(g \circ f)(0)$
9. If $f(x) = x - 4$ and $g(x) = \sqrt{x + 2}$, write the domain of the function, $\frac{g}{f}$.
10. 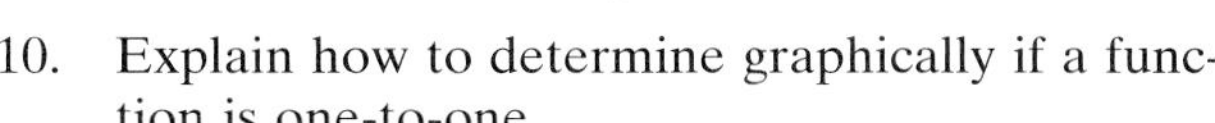 Explain how to determine graphically if a function is one-to-one.
11. Which of the functions is one-to-one?

a.

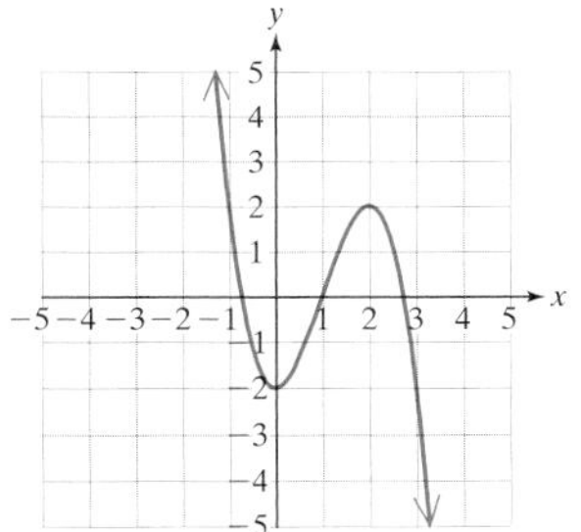

b.

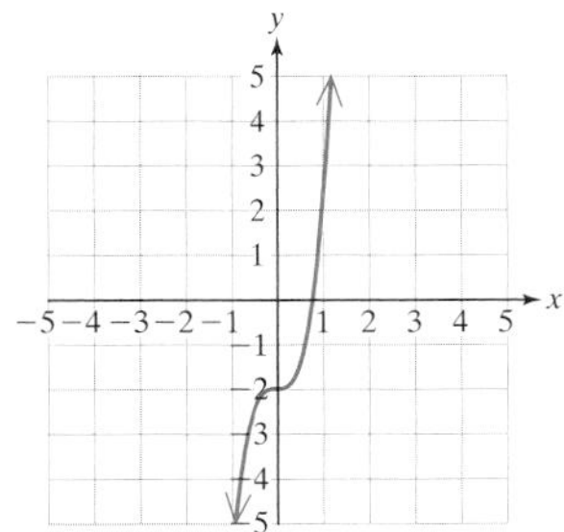

12. Write an equation of the inverse of the one-to-one function defined by $f(x) = \frac{1}{4}x + 3$.
13. Write an equation of the inverse of the function defined by $g(x) = (x - 1)^2, x \geq 1$.
14. Given the graph of the function $y = p(x)$, graph its inverse, $p^{-1}(x)$.

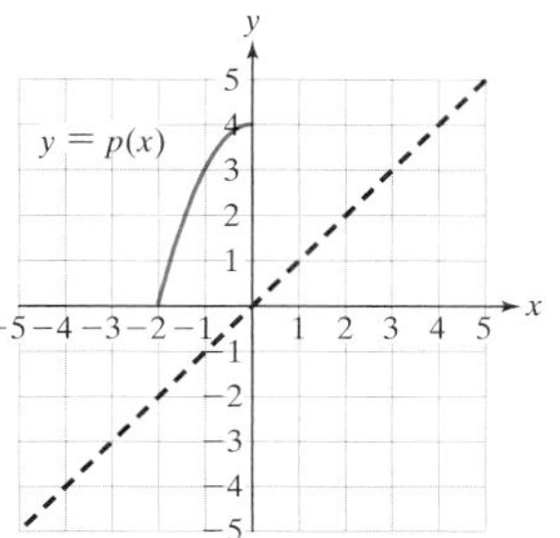

15. Use a calculator to approximate the expression to four decimal places.

a. $10^{2/3}$ b. $3^{\sqrt{10}}$ c. 8^{π}

16. Graph $f(x) = 4^{x-1}$.
17. a. Write in logarithmic form: $16^{3/4} = 8$

b. Write in exponential form: $\log_x 31 = 5$

18. Graph $g(x) = \log_3 x$.
19. Complete the change-of-base formula: $\log_b n =$ ______
20. Use a calculator to approximate the expression to four decimal places:

a. $\log 21$ b. $\log_4 13$ c. $\log_{1/2} 6$

21. Using the properties of logarithms, expand and simplify. Assume all variables represent positive real numbers.

a. $-\log_3\left(\dfrac{3}{9x}\right)$ b. $\log\left(\dfrac{1}{10^5}\right)$

22. Write as a single logarithm. Assume all variables represent positive real numbers.

a. $\dfrac{1}{2}\log_b x + 3\log_b y$ b. $\log a - 4\log a$

23. Use a calculator to approximate the expression to four decimal places if necessary:

a. $e^{1/2}$ b. e^{-3} c. $\ln\left(\dfrac{1}{3}\right)$ d. $\ln e$

24. Identify the graphs as $y = e^x$ or $y = \ln x$.

a.

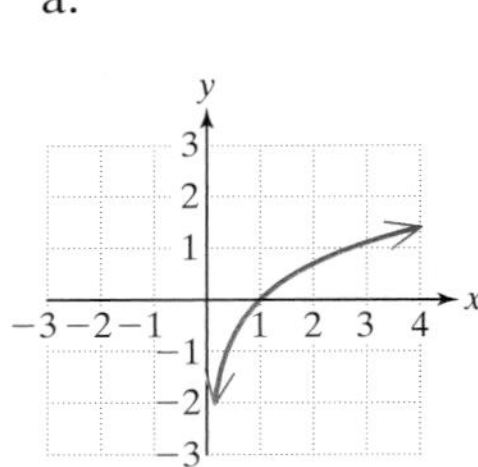

b.

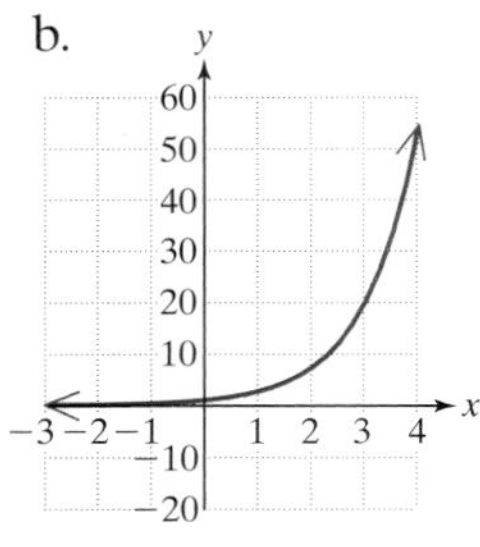

25. Researchers found that t months after taking a course, students remembered $p\%$ of the material according to

$p(t) = 92 - 20\ln(t + 1)$, where $0 \le t \le 24$ is the time in months.

a. Find $p(4)$ and interpret the results.

b. Find $p(12)$ and interpret the results.

c. Find $p(0)$ and interpret the results.

26. The population of New York City has a 2% growth rate and can be modeled by the function $P(t) = 8008(1.02)^t$, where $P(t)$ is in thousands and t is in years ($t = 0$ corresponds to the year 2000).

a. Using this model, predict the population in the year 2010.

b. If this growth rate continues, in what year will the population reach 12 million (12 million is 12,000 thousand)?

27. A certain bacterial culture grows according to

$$P(t) = \frac{1{,}500{,}000}{1 + 5000e^{-0.8t}},$$ where P is the population of the bacteria and t is the time in hours.

a. Find $P(0)$ and interpret the result. Round to the nearest whole number.

b. How many bacteria will be present after 6 hr?

c. How many bacteria will be present after 12 hr?

d. How many bacteria will be present after 18 hr?

e. From the graph does it appear that the population of bacteria is reaching a limiting value? Explain.

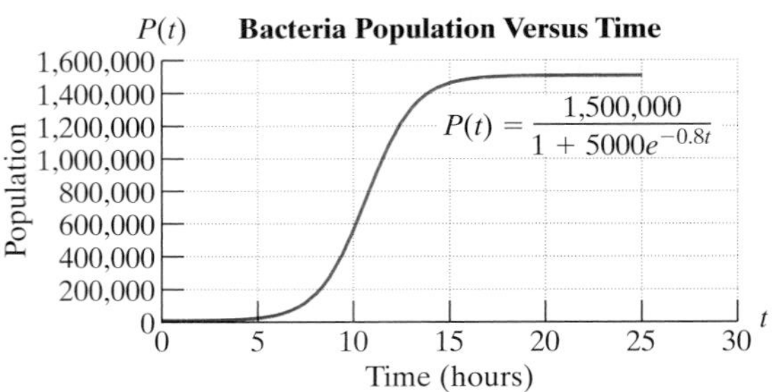

Figure for Exercise 27

For Exercises 28–35, solve the exponential and logarithmic equations. If necessary, round to three decimal places.

28. $\log x + \log(x - 21) = 2$

29. $\log_{1/2} x = -5$

30. $\ln(x + 7) = 2.4$

31. $3^{x+4} = \frac{1}{27}$

32. $4^x = 50$

33. $e^{2.4x} = 250$

34. Atmospheric pressure, P, decreases exponentially with altitude, x according to

$P(x) = 760e^{-0.000122x}$, where $P(x)$ is the pressure measured in millimeters of mercury (mm Hg) and x is the altitude measured in meters.

a. Find $P(2500)$ and interpret the result. Round to one decimal place.

b. Find the pressure at sea level.

c. Find the altitude at which the pressure is 633 mm Hg.

35. Use the formula $A(t) = Pe^{rt}$ to compute the value of an investment under continuous compounding.

a. If \$2000 is invested at 7.5% compounded continuously, find the value of the investment after 5 years.

b. How long will it take the investment to double? Round to two decimal places.

CUMULATIVE REVIEW EXERCISES, CHAPTERS 1–12

1. Simplify completely.

$$\frac{8 - 4 \cdot 2^2 + 15 \div 5}{|-3 + 7|}$$

2. Divide.

$$\frac{-8p^2 + 4p^3 + 6p^5}{8p^2}$$

3. Divide. $(t^4 - 13t^2 + 36) \div (t - 2)$. Identify the quotient and remainder.

4. Simplify. $\sqrt{x^2 - 6x + 9}$

5. Simplify. $\dfrac{4}{\sqrt[3]{40}}$

6. Find the length of the missing side.

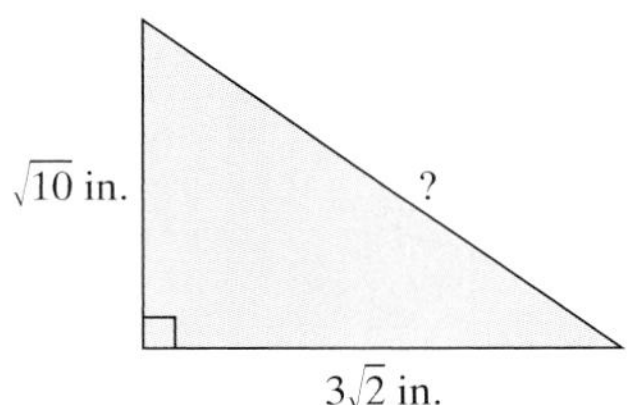

Figure for Exercise 6

7. Simplify. Write the answer with positive exponents only.

$$\frac{2^{2/5}c^{-1/4}d^{1/5}}{2^{-8/5}c^{3/4}d^{1/10}}$$

8. Find the area of the rectangle.

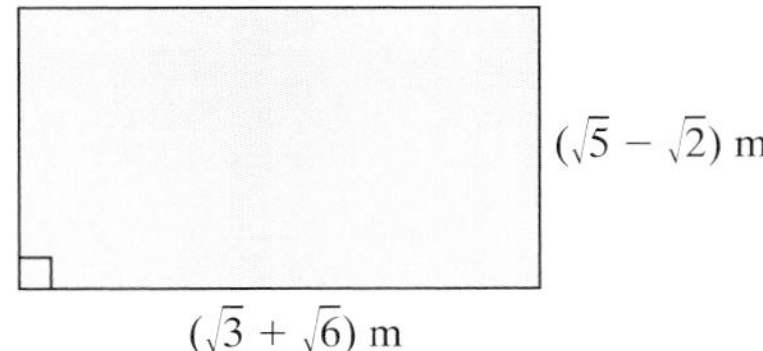

Figure for Exercise 8

9. Perform the indicated operation.

$$\frac{4 - 3i}{2 + 5i}$$

10. Find the measure of each angle in the right triangle.

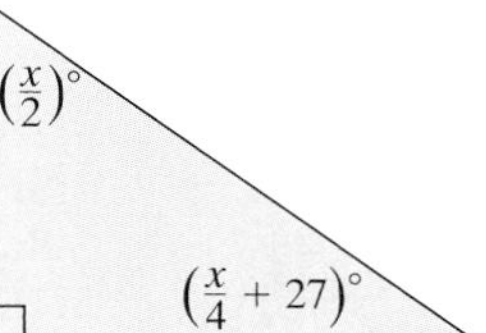

Figure for Exercise 10

11. Find the positive slope of the sides of a pyramid with a square base 66 ft on a side and height of 22 ft.

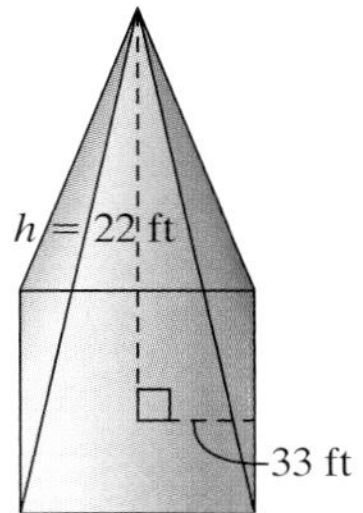

Figure for Exercise 11

12. Solve for x: $2x(x - 7) = x - 18$

13. How many liters of pure alcohol must be mixed with 8 L of 20% alcohol to bring the concentration up to 50% alcohol?

14. Bank robbers leave the scene of a crime and head north through winding mountain roads to their hideaway. Their average rate of speed is 40 mph. The police leave 6 min later in pursuit. If the police car averages 50 mph traveling the same route, how long will it take the police to catch the bank robbers?

15. Solve the system using the Gauss-Jordan method:

$$\begin{aligned} 5x + 10y &= 25 \\ -2x + 6y &= -20 \end{aligned}$$

16. Solve for w: $-2[w - 3(w + 1)] = 4 - 7(w + 3)$

17. Solve for x: $ax - c = bx + d$

18. Solve for t: $s = \frac{1}{2}gt^2, t \geq 0$

19. Solve for T: $\sqrt{1 - kT} = \dfrac{V_0}{V}$

20. Find the x-intercepts of the function defined by $f(x) = |x - 5| - 2$.

21. Let $f(t) = 6$, $g(t) = -5t$, and $h(t) = 2t^2$, find

    a. $(fg)(t)$ b. $(g \circ h)(t)$ c. $(h - g)(t)$

22. Solve for q: $|2q - 5| = |2q + 5|$

23. a. Find an equation of the line parallel to the y-axis and passing through the point (2, 6).

    b. Find an equation of the line perpendicular to the y-axis and passing through the point (2, 6).

    c. Find an equation of the line perpendicular to the line $2x + y = 4$ and passing through the point (2, 6). Write the answer in slope-intercept form.

24. The number of inmates in U.S. state and federal prisons has increased with time between 1990 and 1995. See the following table.

Year	x	Number of Inmates y
1990	0	1,148,000
1991	1	1,129,000
1992	2	1,295,000
1993	3	1,369,000
1994	4	1,478,000
1995	5	1,585,000

Source: U.S. Department of Justice

Table for Exercise 24

    a. Let x represent the year, where $x = 0$ corresponds to 1990. Let y represent the number of inmates. Plot the ordered pairs.

    b. Use the ordered pairs (0, 1,148,000) and (4, 1,478,000) to find a linear equation to model the number of inmates as a function of the year after 1990. Write the equation in slope-intercept form.

    c. What is the slope of the line from part (b)? What does the slope mean in the context of this problem?

    d. Use the equation from part (b) to estimate the prison population in the year 2000.

25. The smallest angle in a triangle is half the largest angle. The smallest angle is 20° less than the middle angle. Find the measure of all three angles.

26. Solve the system.

$$\frac{1}{2}x - \frac{1}{4}y = 1$$
$$-2x + y = -4$$

27. Match the function with the appropriate graph.

i.	$f(x) = \ln(x)$	ii.	$g(x) = 3^x$
iii.	$h(x) = x^2$	iv.	$k(x) = -2x - 3$
v.	$L(x) = \lvert x \rvert$	vi.	$m(x) = \sqrt{x}$
vii.	$n(x) = \sqrt[3]{x}$	viii.	$p(x) = x^3$
ix.	$q(x) = \dfrac{1}{x}$	x.	$r(x) = x$

a.

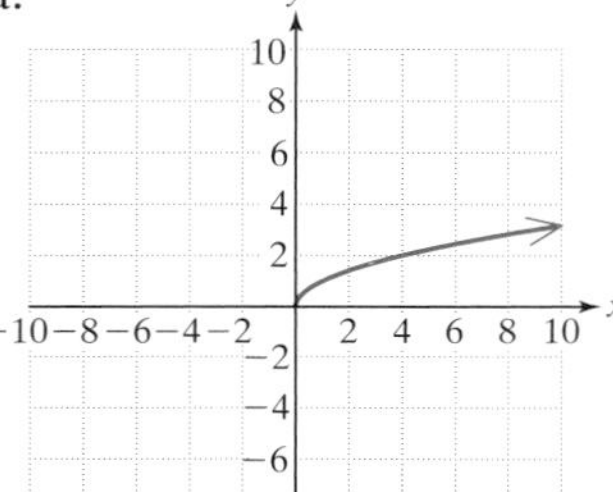

b.

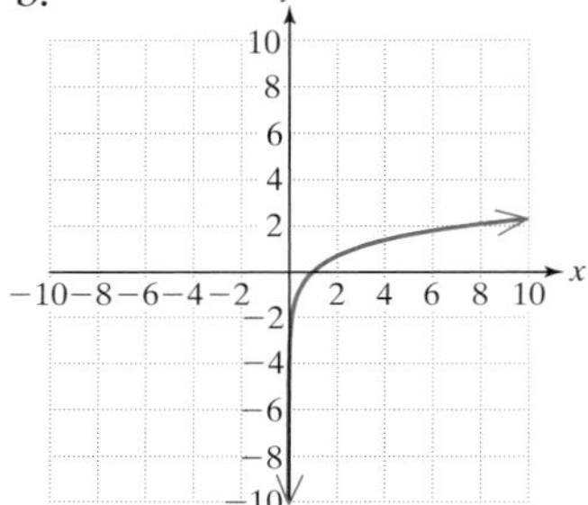

c.

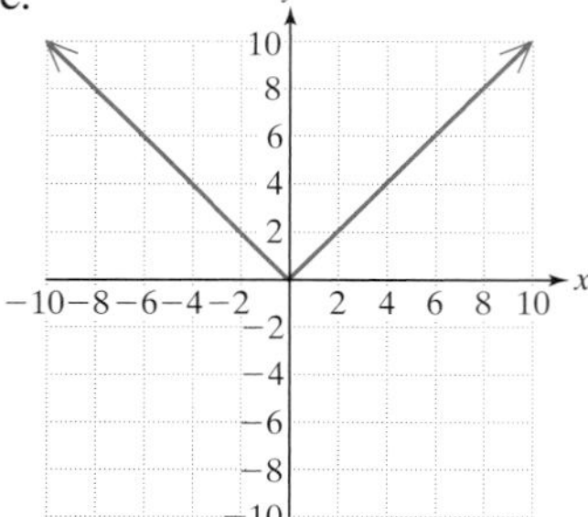

d.

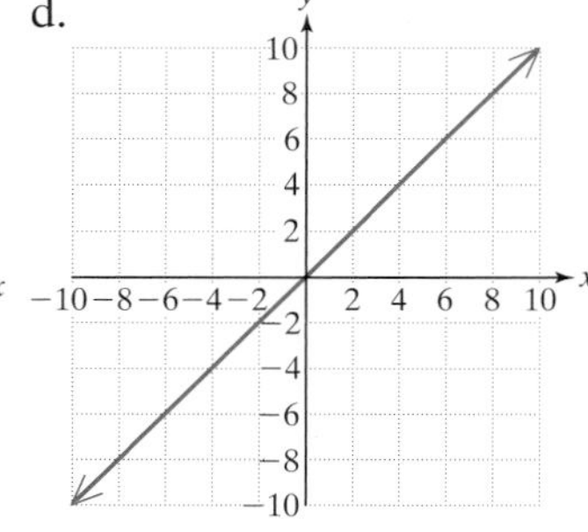

e.

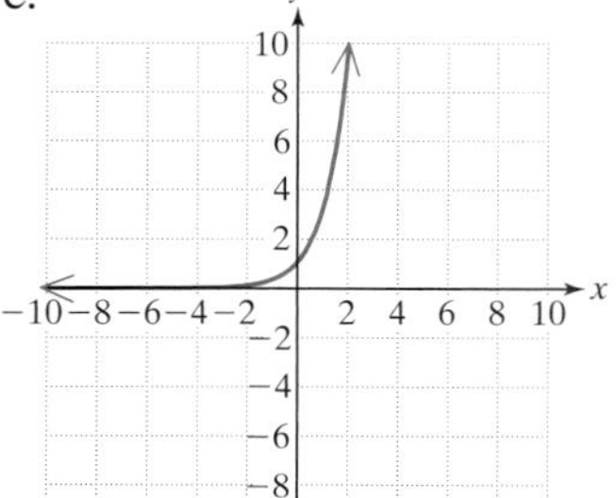

f.

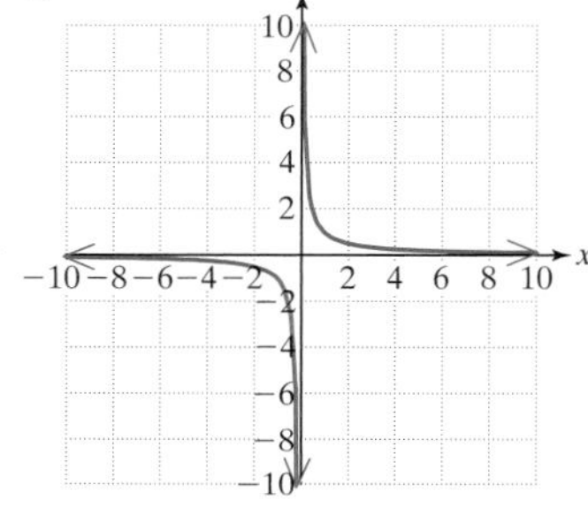

g.

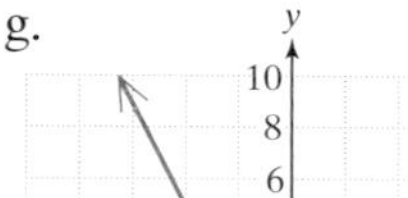

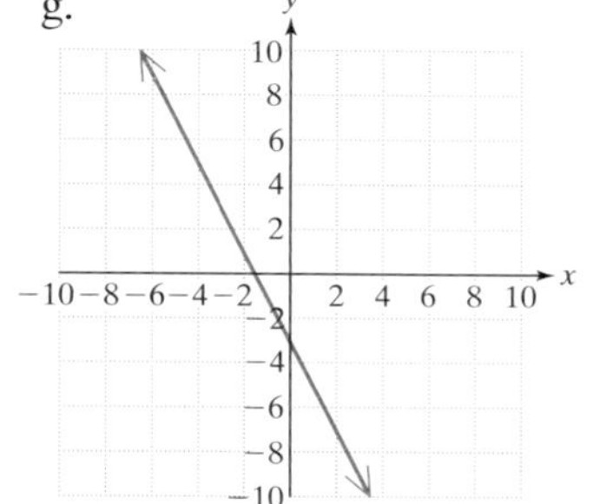

h.

i.

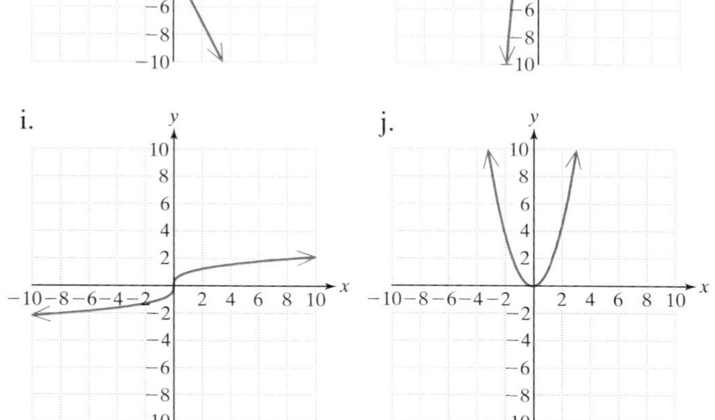

j.

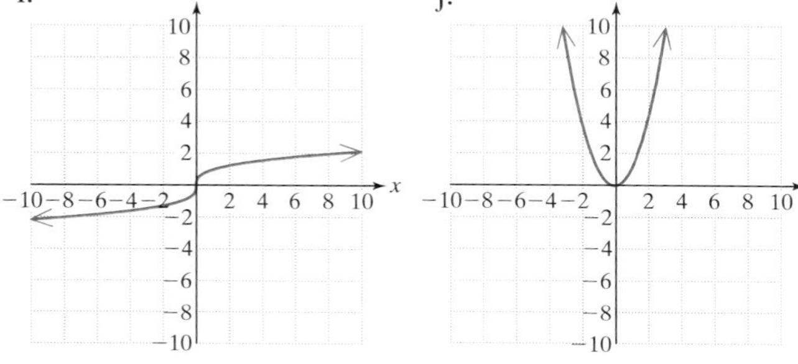

28. Find the domain: $f(x) = \sqrt{2x - 1}$.

29. Find $f^{-1}(x)$ given $f(x) = 5x - \frac{2}{3}$

30. The volume of a gas varies directly as its temperature and inversely with pressure. At a temperature of 100 kelvins and a pressure of 10 N/m^2 (Newtons per square meter), the gas occupies a volume of 30 m^3. Find the volume at a temperature of 200 kelvins and pressure of 15 N/m^2.

31. Perform the indicated operations.

$$\frac{5x - 10}{x^2 - 4x + 4} \div \frac{5x^2 - 125}{25 - 5x} \cdot \frac{x^3 + 125}{10x + 5}$$

32. Perform the indicated operations.

$$\frac{x}{x - y} + \frac{y}{y - x} + x$$

33. Given the equation

$$\frac{2}{x - 4} = \frac{5}{x + 2}$$

a. Are there any restrictions on x for which the rational expressions are undefined?

b. Solve the equation.

c. Solve the related inequality.

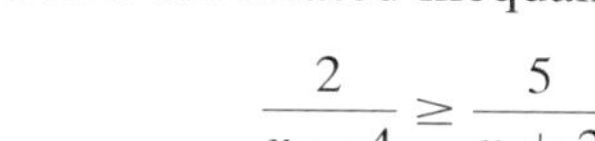

$$\frac{2}{x - 4} \geq \frac{5}{x + 2}$$

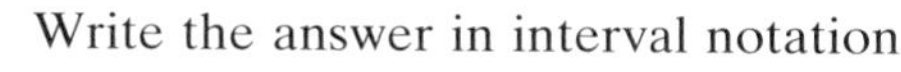

Write the answer in interval notation.

34. Two more than 3 times the reciprocal of a number is $\frac{5}{4}$ less than the number. Find all such numbers.

35. Solve the equation: $\sqrt{-x} = x + 6$

36. Solve the inequality: $2|x - 3| + 1 > -7$. Write the answer in interval notation.

37. Four million *Escherichia coli* (*E. coli*) bacteria are present in a laboratory culture. An antibacterial agent is introduced and the population of bacteria, P, decreases by half every 6 hr according to

$$P(t) = 4{,}000{,}000\left(\frac{1}{2}\right)^{t/6} \quad t \geq 0, \text{ where } t \text{ is the time in hours.}$$

a. Find the population of bacteria after 6, 12, 18, 24, and 30 hr.

b. Sketch a graph of $y = P(t)$ based on the function values found in part (a).

c. Predict the time required for the population to decrease to 15,625 bacteria.

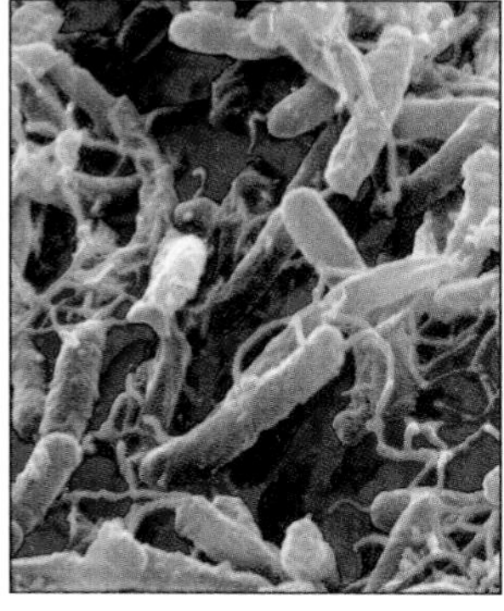

38. Evaluate the expressions without a calculator.

a. $\log_7 49$

b. $\log_4\left(\frac{1}{64}\right)$

c. $\log(1{,}000{,}000)$

d. $\ln(e^3)$

39. Use a calculator to approximate the expressions to four decimal places.

a. $\pi^{4.7}$ b. e^{π}

c. $(\sqrt{2})^{-5}$ d. $\log(5362)$

e. $\ln(0.67)$ f. $\log_4(37)$

40. Solve the equation: $5^2 = 125^x$

41. Solve the equation: $e^x = 100$

42. Solve the equation: $\log_3(x + 6) - 3 = -\log_3(x)$

43. Write the following expression as a single logarithm: $\frac{1}{2}\log(z) - 2\log(x) - 3\log(y)$

44. Write the following expression as a sum or difference of logarithms:

$$\ln\sqrt[3]{\frac{x^2}{y}}$$

chapter

13

Conic Sections and Nonlinear Systems

The vertical position, y, (in meters) of an object can be modeled by the *quadratic function*:

$$y(t) = \frac{1}{2}gt^2 + v_0t + y_0$$

where g is the acceleration due to gravity (on the earth, $g \approx -9.8$ m/sec^2), t is the time (in seconds), v_0 is the initial velocity (in meters per second), and y_0 is the initial height (in meters).

A diver jumps off a springboard with an initial velocity of 6 m/sec. The height of the diver's center of mass is given by

$$y(t) = -4.9t^2 + 6t + 3.8$$

The maximum height the diver's center of mass is approximately 5.6 m, which is the y-coordinate of the vertex of the graph.

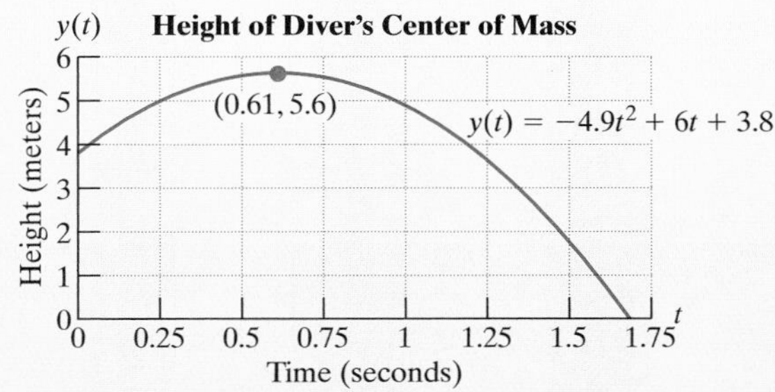

To graph this and other quadratic functions, see the Technology Connections in MathZone at

www.mhhe.com/miller_oneill

section

13.1 Distance Formula and Circles

Concepts

1. Distance Formula
2. Finding the Distance Between Two Points
3. Circles
4. Graphing a Circle
5. Writing an Equation of a Circle
6. The Midpoint Formula
7. Applications of the Midpoint Formula

1. Distance Formula

Suppose we are given two points (x_1, y_1) and (x_2, y_2) in a rectangular coordinate system. The distance between the two points can be found using the Pythagorean theorem (Figure 13-1).

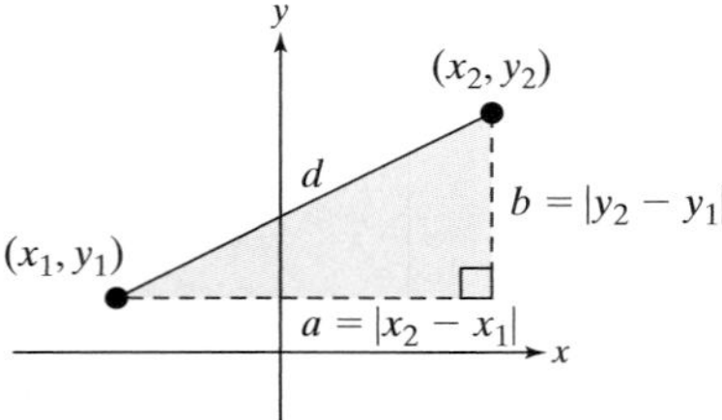

Figure 13-1

First draw a right triangle with the distance, d, as the hypotenuse. The length of the horizontal leg, a, is $|x_2 - x_1|$, and the length of the vertical leg, b, is $|y_2 - y_1|$. From the Pythagorean theorem we have

$$d^2 = a^2 + b^2 \qquad \text{Pythagorean theorem}$$
$$= (x_2 - x_1)^2 + (y_2 - y_1)^2$$
$$d = \pm\sqrt{(x_2 - x_1)^2 + (y_2 - y_1)^2}$$
$$= \sqrt{(x_2 - x_1)^2 + (y_2 - y_1)^2} \qquad \text{Because distance is positive, reject the negative value.}$$

The Distance Formula

The distance, d, between the points (x_1, y_1) and (x_2, y_2) is

$$d = \sqrt{(x_2 - x_1)^2 + (y_2 - y_1)^2}$$

2. Finding the Distance Between Two Points

example 1 **Finding the Distance Between Two Points**

Find the distance between the points $(-2, 3)$ and $(4, -1)$ (Figure 13-2).

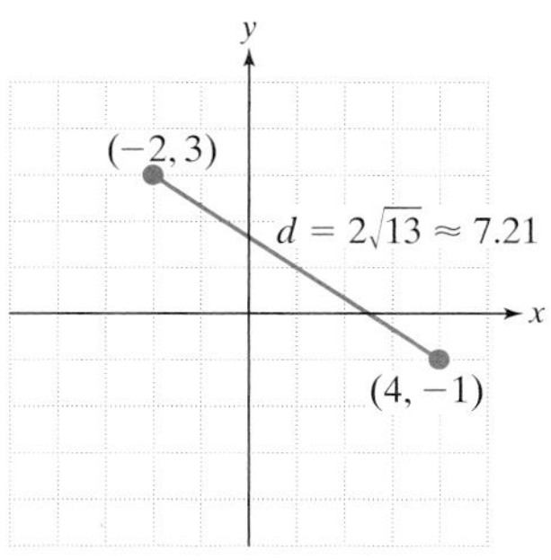

Figure 13-2

Solution:

$(-2, 3)$ and $(4, -1)$

(x_1, y_1) and (x_2, y_2) Label the points.

$$d = \sqrt{(x_2 - x_1)^2 + (y_2 - y_1)^2}$$

$$= \sqrt{[(4) - (-2)]^2 + [(-1) - (3)]^2}$$ Apply the distance formula.

$$= \sqrt{(6)^2 + (-4)^2}$$

$$= \sqrt{36 + 16}$$

$$= \sqrt{52}$$

$$= \sqrt{2^2 \cdot 13}$$

$$= 2\sqrt{13}$$

Tip: The order in which the points are labeled does not affect the result of the distance formula. For example, if the points in Example 1 had been labeled in reverse, the distance formula still yields the same result:

$(-2, 3)$ and $(4, -1)$

(x_2, y_2) and (x_1, y_1)

$$d = \sqrt{(x_2 - x_1)^2 + (y_2 - y_1)^2}$$

$$= \sqrt{[(-2) - (4)]^2 + [(3) - (-1)]^2}$$

$$= \sqrt{(-6)^2 + (4)^2}$$

$$= \sqrt{36 + 16}$$

$$= \sqrt{52}$$

$$= 2\sqrt{13}$$

3. Circles

A **circle** is defined as the set of all points in a plane that are equidistant from a fixed point called the **center**. The fixed distance from the center is called the **radius** and is denoted by r, where $r > 0$.

Suppose a circle is centered at the point (h, k) and has radius, r (Figure 13-3). The distance formula can be used to derive an equation of the circle.

Let (x, y) be any arbitrary point on the circle. Then by definition, the distance between (h, k) and (x, y) must be r.

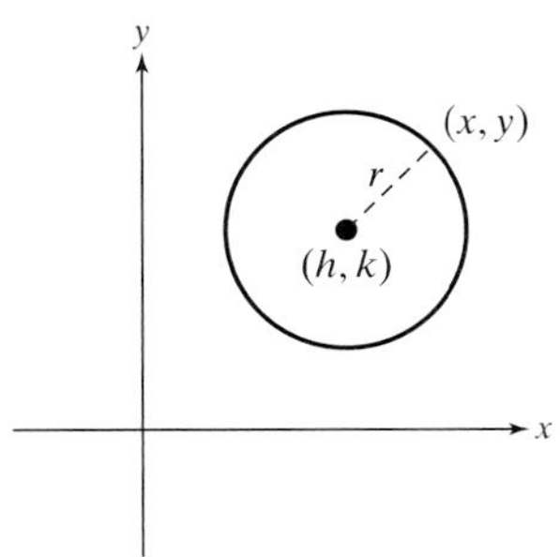

Figure 13-3

$$\sqrt{(x - h)^2 + (y - k)^2} = r$$

$$(x - h)^2 + (y - k)^2 = r^2$$ Square both sides.

Standard Equation of a Circle

The **standard equation of a circle**, centered at (h, k) with radius r is given by:

$$(x - h)^2 + (y - k)^2 = r^2, \text{ where } r > 0.$$

Note: For a circle centered at the origin, $(0, 0)$, then $h = 0$ and $k = 0$, and the equation simplifies to $x^2 + y^2 = r^2$.

4. Graphing a Circle

example 2

Graphing a Circle

Find the center and radius of each circle. Then graph the circle.

a. $(x - 3)^2 + (y + 4)^2 = 36$ b. $x^2 + \left(y - \frac{10}{3}\right)^2 = \frac{25}{9}$ c. $x^2 + y^2 = 10$

Solution:

a. $(x - 3)^2 + (y + 4)^2 = 36$

$(x - 3)^2 + [y - (-4)]^2 = (6)^2$

$h = 3, k = -4, \text{ and } r = 6$

The center is at $(3, -4)$ and the radius is 6 (Figure 13-4).

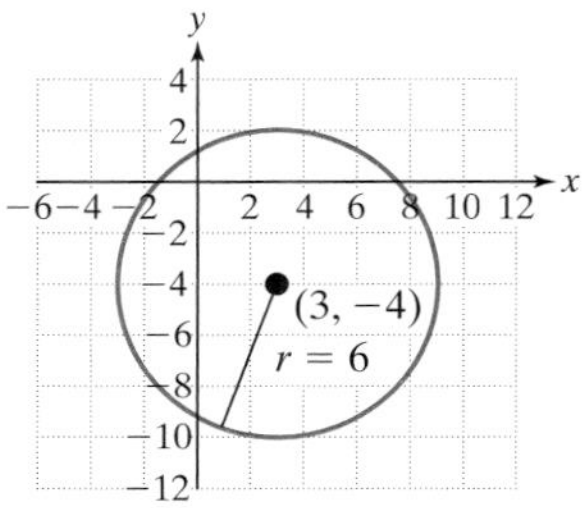

Figure 13-4

Calculator Connections

Graphing calculators are designed to graph *functions*, in which y is written in terms of x. A circle is not a function. However, it can be graphed as the union of two functions, one representing the top semicircle, and the other representing the bottom semicircle.

Solving for y in Example 2(a), we have

$(x - 3)^2 + (y + 4)^2 = 36$

$(y + 4)^2 = 36 - (x - 3)^2$

$y + 4 = \pm\sqrt{36 - (x - 3)^2}$

$y = -4 \pm \sqrt{36 - (x - 3)^2}$

Graph these functions as Y_1 and Y_2 using a square viewing window.

$Y_1 = -4 + \sqrt{36 - (x - 3)^2}$

$Y_2 = -4 - \sqrt{36 - (x - 3)^2}$

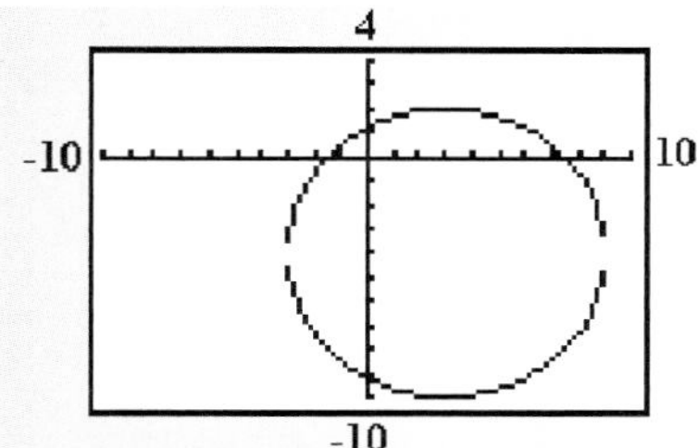

Notice that the image from the calculator does not show the upper and lower semicircles connecting at their endpoints, when in fact the semicircles should "hook up." This is due to the calculator's limited resolution.

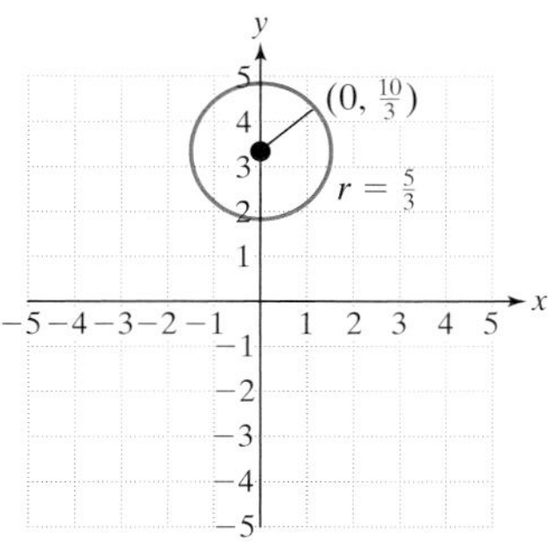

Figure 13-5

b. $x^2 + \left(y - \dfrac{10}{3}\right)^2 = \dfrac{25}{9}$

$$(x - 0)^2 + \left(y - \frac{10}{3}\right)^2 = \left(\frac{5}{3}\right)^2$$

The center is $(0, \frac{10}{3})$ and the radius is $\frac{5}{3}$ (Figure 13-5).

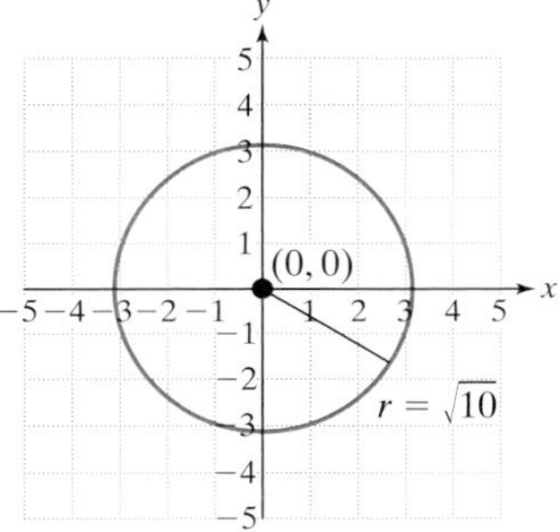

Figure 13-6

c. $x^2 + y^2 = 10$

$$(x - 0)^2 + (y - 0)^2 = (\sqrt{10})^2$$

The center is (0, 0) and the radius is $\sqrt{10} \approx 3.16$ (Figure 13-6).

5. Writing an Equation of a Circle

example 3

Writing an Equation of a Circle

Write an equation of the circle shown in Figure 13-7.

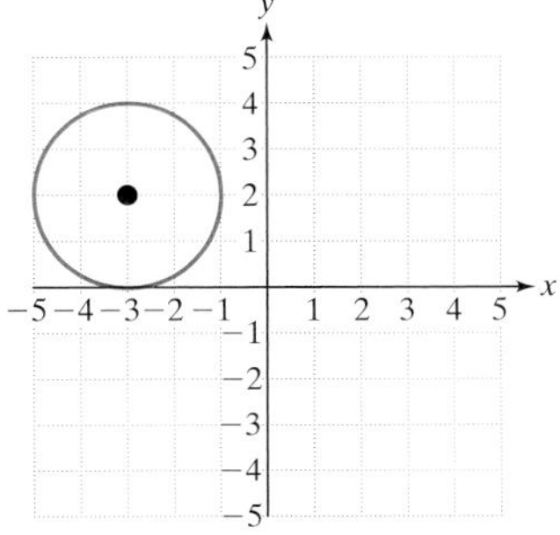

Figure 13-7

Solution:

The center is $(-3, 2)$; therefore, $h = -3$ and $k = 2$.
From the graph the radius is $r = 2$.

$$(x - h)^2 + (y - k)^2 = r^2$$

$$[x - (-3)]^2 + (y - 2)^2 = (2)^2$$

$$(x + 3)^2 + (y - 2)^2 = 4$$

Sometimes it is necessary to complete the square to write an equation of a circle in standard form.

example 4

Writing the Equation of a Circle in the Form $(x - h)^2 + (y - k)^2 = r^2$

Identify the center and radius of the circle given by the equation: $x^2 + y^2 + 2x - 16y + 61 = 0$.

Solution:

$$x^2 + y^2 + 2x - 16y + 61 = 0$$

To identify the center and radius, write the equation in the form $(x - h)^2 + (y - k)^2 = r^2$.

$$(x^2 + 2x \quad) + (y^2 - 16y \quad) = -61$$

Group the x-terms and group the y-terms. Move the constant to the right-hand side.

$$(x^2 + 2x + 1) + (y^2 - 16y + 64) = -61 + 1 + 64$$

- Complete the square on x. Add $[\frac{1}{2}(2)]^2 = 1$ on both sides of the equation.
- Complete the square on y. Add $[\frac{1}{2}(-16)]^2 = 64$ on both sides of the equation.

$$(x + 1)^2 + (y - 8)^2 = 4$$

Factor and simplify.

$$[x - (-1)]^2 + (y - 8)^2 = 2^2$$

The center is $(-1, 8)$ and the radius is 2.

6. The Midpoint Formula

Tip: The midpoint of a line segment is found by taking the *average* of the x-coordinates and the *average* of the y-coordinates of the endpoints.

Consider the line segment defined by the points (x_1, y_1) and (x_2, y_2). The midpoint of the line segment is given by the formula:

Midpoint formula: $\left(\dfrac{x_1 + x_2}{2}, \dfrac{y_1 + y_2}{2}\right)$

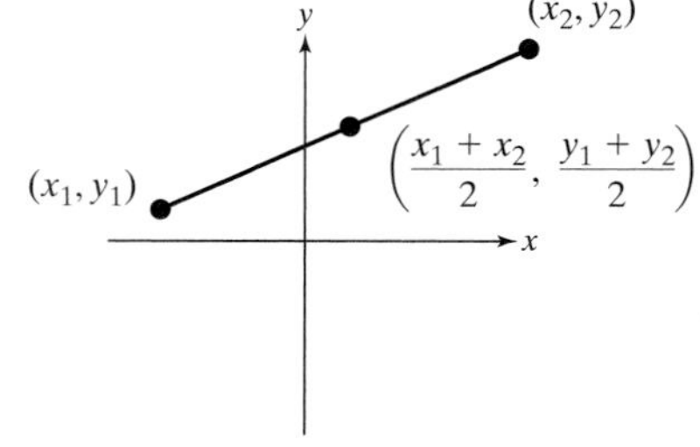

7. Applications of the Midpoint Formula

example 5

Applying the Midpoint Formula

A map of a national park is created so that the ranger station is at the origin of a rectangular grid. Two hikers are located at positions $(2, 3)$ and $(-5, -2)$ with respect to the ranger station where all units are in miles. The hikers would like to meet at a point halfway between them (Figure 13-8), but they are too far apart to communicate their positions to each other via radio. However, the hikers are both within radio range of the ranger station. If the ranger station relays each hiker's position to the other, at what point on the map should the hikers meet?

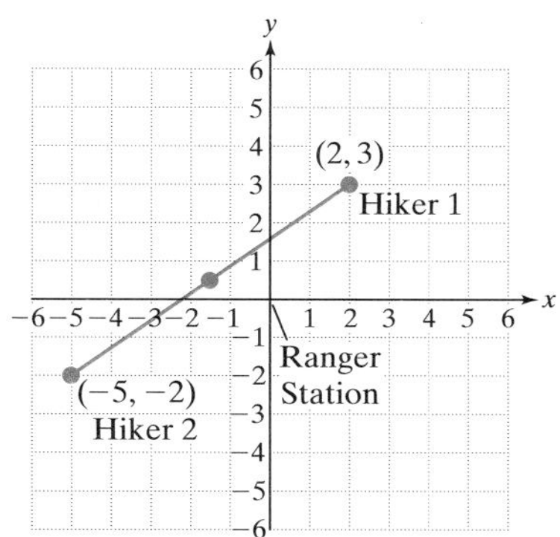

Figure 13-8

Solution:

To find the halfway point on the line segment between the two hikers, apply the midpoint formula:

$(2, 3)$ and $(-5, -2)$

(x_1, y_1) and (x_2, y_2) Label the points.

$$\left(\frac{x_1 + x_2}{2}, \frac{y_1 + y_2}{2}\right)$$

$$\left(\frac{2 + (-5)}{2}, \frac{3 + (-2)}{2}\right) \quad \text{Apply the midpoint formula.}$$

$$\left(-\frac{3}{2}, \frac{1}{2}\right)$$

The halfway point between the hikers is located at $(-\frac{3}{2}, \frac{1}{2})$ or $(-1.5, 0.5)$.

section 13.1 Practice Exercises

For Exercises 1–12, use the distance formula to find the distance between the two points.

1. $(-2, 7)$ and $(3, -9)$
2. $(1, 10)$ and $(-2, 4)$
3. $(0, 5)$ and $(-3, 8)$
4. $(6, 7)$ and $(3, 2)$
5. $\left(\frac{2}{3}, \frac{1}{5}\right)$ and $\left(-\frac{5}{6}, \frac{3}{10}\right)$
6. $\left(-\frac{1}{2}, \frac{5}{8}\right)$ and $\left(-\frac{3}{2}, \frac{1}{4}\right)$
7. $(4, 13)$ and $(4, -6)$
8. $(-2, 5)$ and $(-2, 9)$
9. $(8, -6)$ and $(-2, -6)$
10. $(7, 2)$ and $(15, 2)$
11. $(3\sqrt{5}, 2\sqrt{7})$ and $(-\sqrt{5}, -3\sqrt{7})$
12. $(4\sqrt{6}, -2\sqrt{2})$ and $(2\sqrt{6}, \sqrt{2})$
13. Explain how to find the distance between 5 and −7 on the y-axis.

14. Explain how to find the distance between 15 and -37 on the x-axis.

15. Find a value of y such that the distance between the points $(4, 7)$ and $(-4, y)$ is 10 units.

16. Find a value of x such that the distance between the points $(-4, -2)$ and $(x, 3)$ is 13 units.

17. Find a value of x such that the distance between the points $(x, 2)$ and $(4, -1)$ is 5 units.

18. Find a value of y such that the distance between the points $(-5, 2)$ and $(3, y)$ is 10 units.

For Exercises 19–22, determine if the three points define the vertices of a right triangle.

19. $(-3, 2), (-2, -4)$, and $(3, 3)$

20. $(1, -2), (-2, 4)$, and $(7, 1)$

21. $(-3, -2), (4, -3)$, and $(1, 5)$

22. $(1, 4), (5, 3)$, and $(2, 0)$

For Exercises 23–38, identify the center and radius of the circle and then graph the circle. Complete the square if necessary.

23. $(x - 4)^2 + (y + 2)^2 = 9$

24. $(x - 3)^2 + (y + 1)^2 = 16$

25. $(x + 1)^2 + (y + 1)^2 = 1$

26. $(x - 4)^2 + (y - 4)^2 = 4$

27. $x^2 + (y - 5)^2 = 25$

28. $(x + 1)^2 + y^2 = 1$

29. $(x - 3)^2 + y^2 = 8$

30. $x^2 + (y + 2)^2 = 20$

31. $x^2 + y^2 = 6$

32. $x^2 + y^2 = 15$

33. $\left(x + \frac{4}{5}\right)^2 + y^2 = \frac{64}{25}$

34. $x^2 + \left(y - \frac{5}{2}\right)^2 = \frac{9}{4}$

35. $x^2 + y^2 - 2x - 6y - 26 = 0$

36. $x^2 + y^2 + 4x - 8y + 16 = 0$

37. $x^2 + y^2 + 6y + \frac{65}{9} = 0$

38. $x^2 + y^2 + 12x + \frac{143}{4} = 0$

For Exercises 39–44, write an equation that represents the graph of the circle.

39.

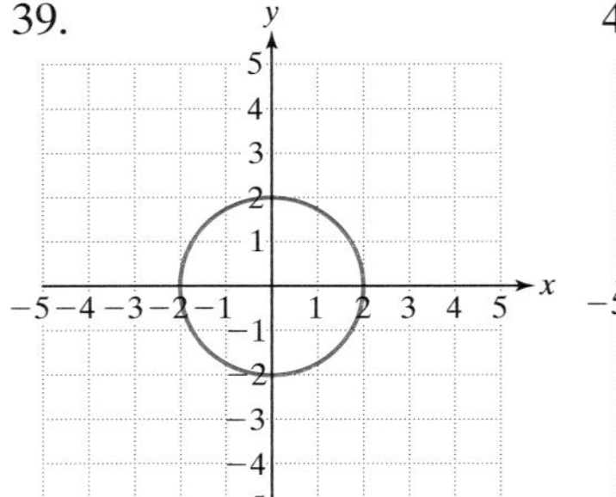

40.

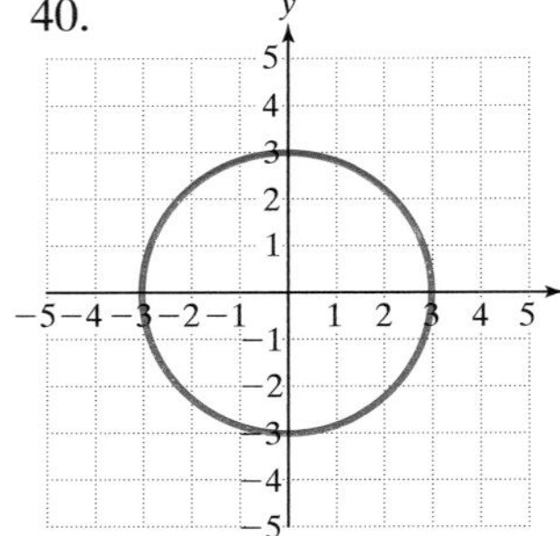

41.

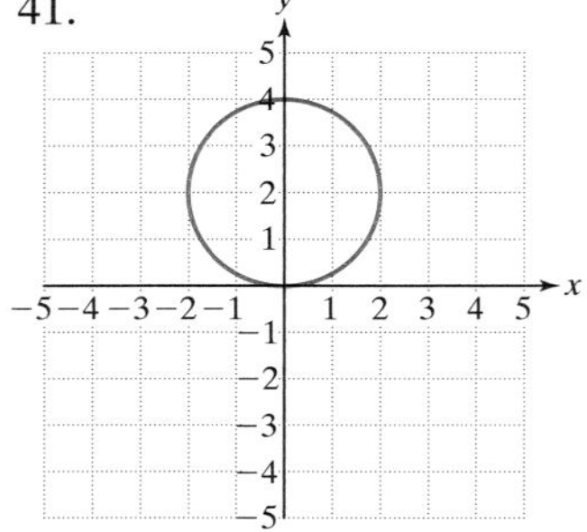

42.

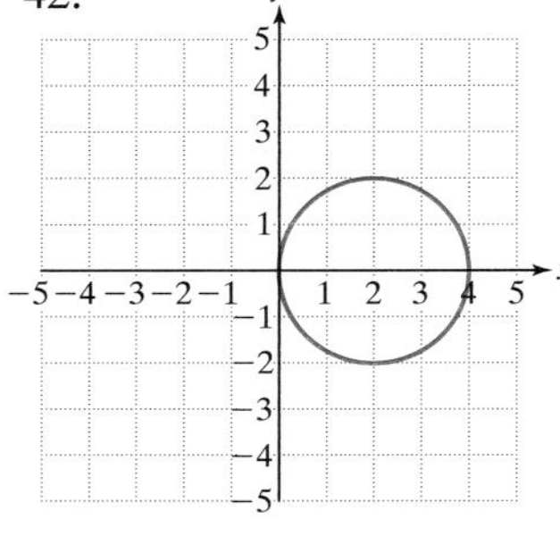

43.

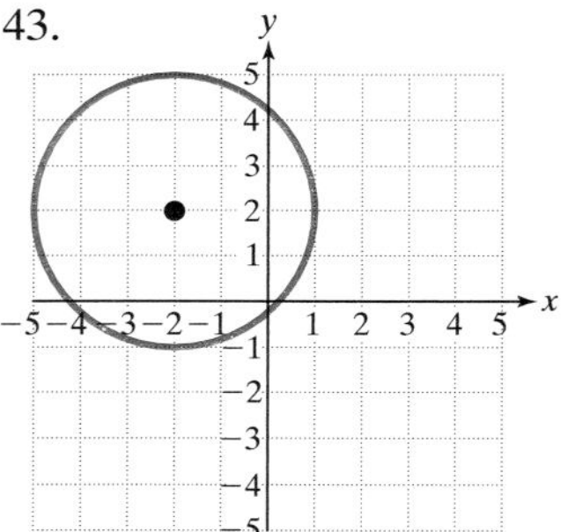

44.

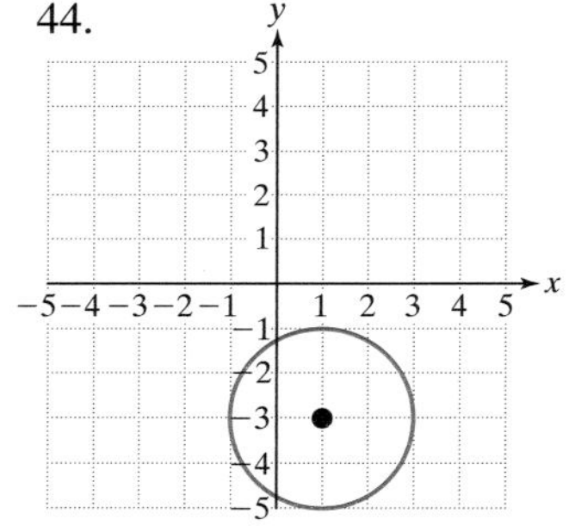

45. Write an equation of a circle centered at the origin with radius 7 m.

46. Write an equation of a circle centered at the origin with a radius of 12 m.

47. Write an equation of a circle centered at $(-3, -4)$ with a diameter of 12 ft.

48. Write an equation of a circle centered at $(5, -1)$ with a diameter of 8 ft.

For Exercises 49–52, find the midpoint of the line segment. Check your answers by plotting the midpoint on the graph.

49.
50.

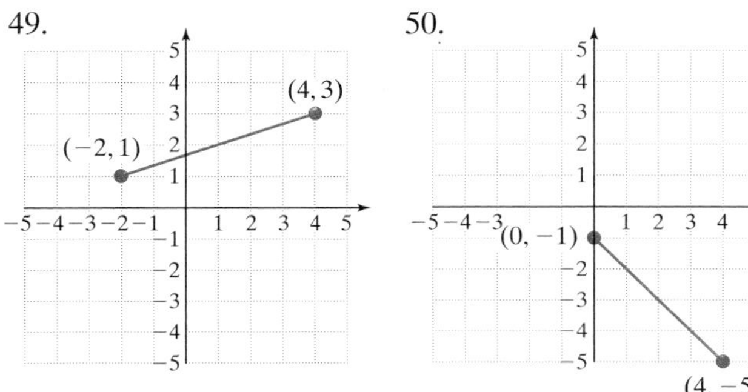

51.
52.

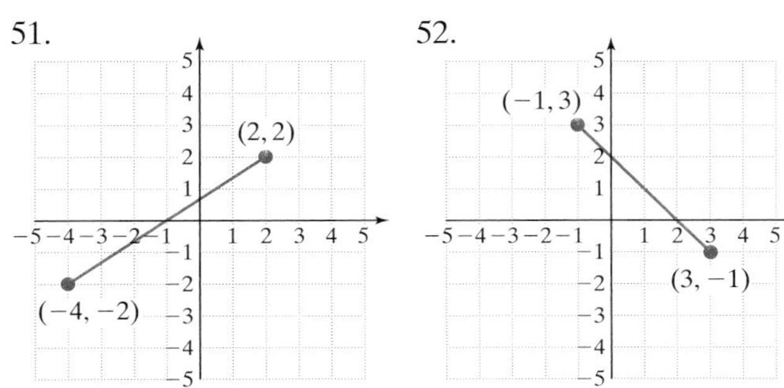

For Exercises 53–60, find the midpoint of the line segment between the two given points.

53. (4, 0) and (−6, 12)
54. (−7, 2) and (−3, −2)
55. (−3, 8) and (3, −2)
56. (0, 5) and (4, −5)
57. (5, 2) and (−6, 1)
58. (−9, 3) and (0, −4)
59. (−2.4, −3.1) and (1.6, 1.1)
60. (0.8, 5.3) and (−4.2, 7.1)

61. Two courier trucks leave their warehouse to make deliveries. One travels 20 miles north and 30 miles east. The other truck travels 5 miles south and 50 miles east. If the two drivers want to meet for lunch at a restaurant at a point halfway between them, where should they meet relative to the warehouse? (*Hint*: Label the warehouse as the origin and find the coordinates of the restaurant. See figure.)

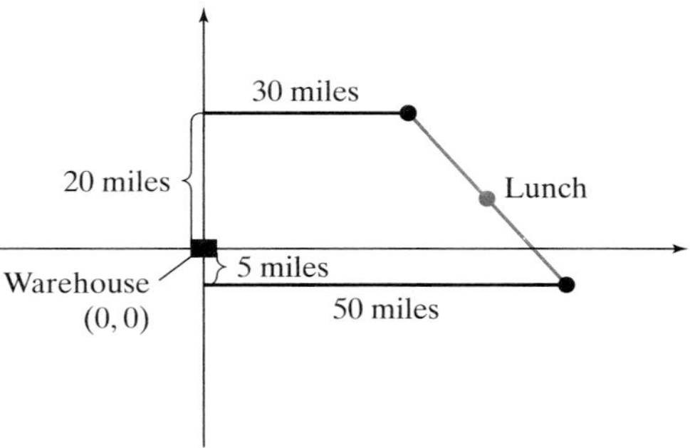

Figure for Exercise 61

62. A map of a hiking area is drawn so that the visitors' center is at the origin of a rectangular grid. Two hikers are located at positions (−1, 1) and (−3, −2) with respect to the visitors' center where all units are in miles. A campground is located exactly halfway between the hikers. What are the coordinates of the campground?

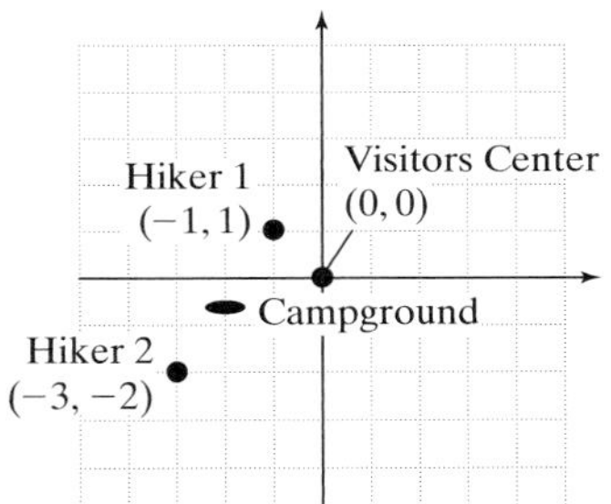

Figure for Exercise 62

Expanding Your Skills

63. Write an equation of a circle whose center is at (4, 4) and is tangent to the x- and y-axes. (*Hint*: Sketch the circle first.)

64. Write an equation of a circle whose center is at (−3, 3) and is tangent to the x- and y-axes. (*Hint*: Sketch the circle first.)

65. Write an equation of a circle whose center is at (1, 1) and that passes through the point (−4, 3).

66. Write an equation of a circle whose center is at (−3, −1) and that passes through the point (5, −2).

Graphing Calculator Exercises

For Exercises 67–72, graph the circles from the indicated exercise on a square viewing window and approximate the center and the radius from the graph.

67. $(x - 4)^2 + (y + 2)^2 = 9$ (Exercise 23)

68. $(x - 3)^2 + (y + 1)^2 = 16$ (Exercise 24)

69. $x^2 + (y - 5)^2 = 25$ (Exercise 27)

70. $(x + 1)^2 + y^2 = 1$ (Exercise 28)

71. $x^2 + y^2 = 6$ (Exercise 31)

72. $x^2 + y^2 = 15$ (Exercise 32)

Concepts

1. Introduction to Conic Sections
2. Definition of a Parabola
3. Standard Form of the Equation of a Parabola (Vertical Axis of Symmetry)
4. Finding the Focus, Directrix, and Vertex of a Parabola
5. Completing the Square to Obtain Standard Form
6. Standard Form of the Equation of a Parabola (Horizontal Axis of Symmetry)

section

13.2 More on the Parabola

1. Introduction to Conic Sections

Recall that the graph of a quadratic equation, $y = ax^2 + bx + c$ $(a \neq 0)$, is a parabola. In Section 13.1 we learned that the graph of $(x - h)^2 + (y - k)^2 = r^2$ is a circle. These and two other types of figures called ellipses and hyperbolas are called **conic sections**. Conic sections derive their name because each is the intersection of a plane and a double-napped cone (Figure 13-9).

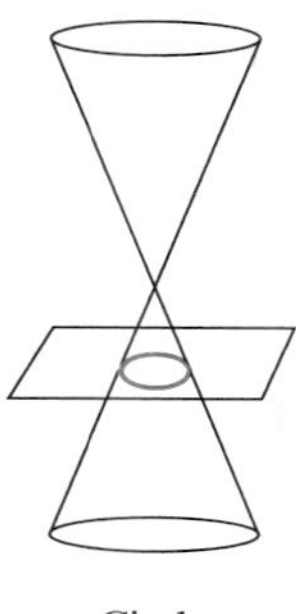

Circle

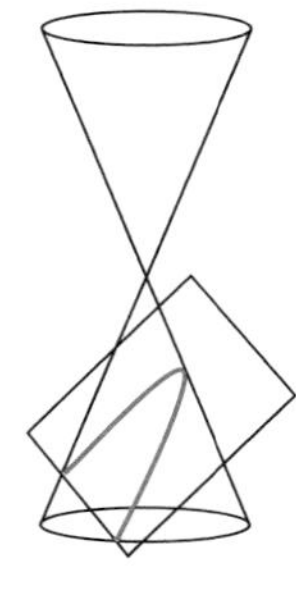

Parabola

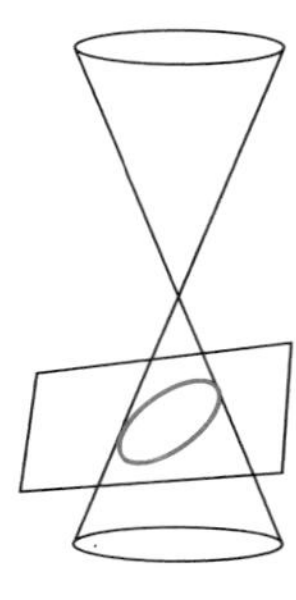

Ellipse

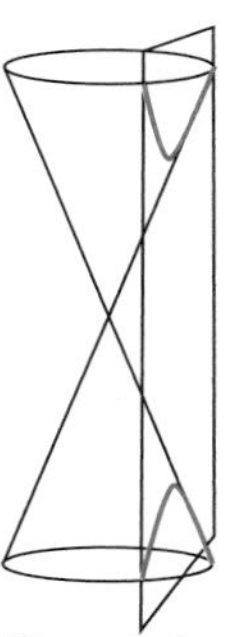

Hyperbola

Figure 13-9

2. Definition of a Parabola

The points that make up the graph of a parabola may be defined by the following geometric approach.

Definition of a Parabola

A **parabola** is the set of points in a plane that are equidistant (the same distance) from a fixed line (called the **directrix**) and a fixed point (called the **focus**) not on the directrix.

Note: The distance between a point and a line is taken to be the shortest distance. This is measured along a line perpendicular to the given line and through the given point.

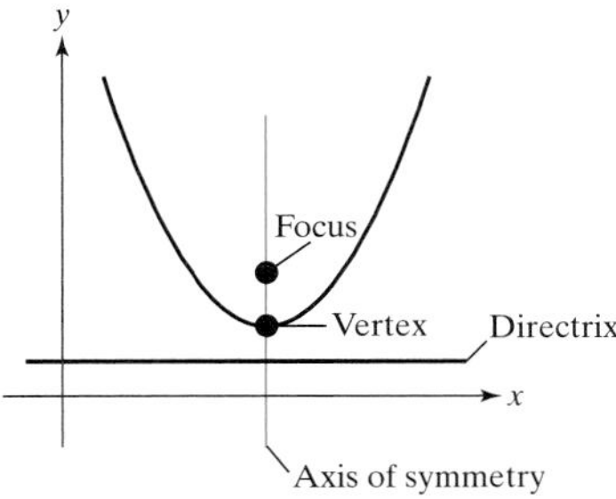

Figure 13-10

The **axis of symmetry** of the parabola passes through the focus and is perpendicular to the directrix. Furthermore, the **vertex** of the parabola is located on the axis of symmetry at the midpoint between the focus and directrix (Figure 13-10).

Suppose that the vertex of a parabola is located at $(0, 0)$ and that the axis of symmetry is the y-axis. If the focus is located at the point $(0, p)$ where $p \geq 0$, then the directrix must have the equation $y = -p$ (Figure 13-11). If the point (x, y) represents any point on the parabola, then by definition, (x, y) must be equidistant from the focus and the directrix.

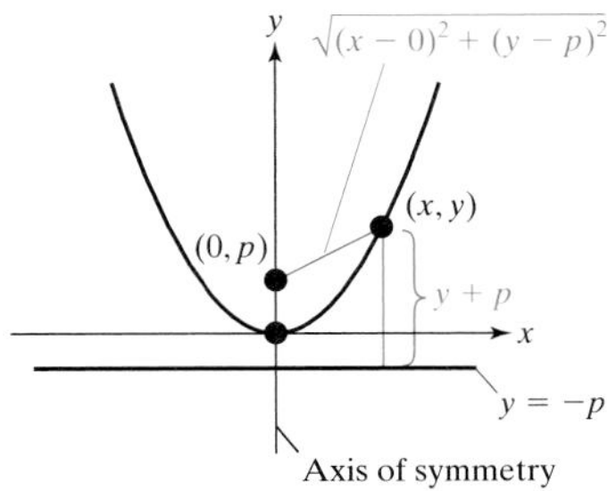

Figure 13-11

Using the distance formula, the distance between (x, y) and the focus $(0, p)$ is $\sqrt{(x - 0)^2 + (y - p)^2}$.

The distance between (x, y) and the directrix is $y + p$.
Equating the distances, we have

$\sqrt{(x - 0)^2 + (y - p)^2} = y + p$	
$(x - 0)^2 + (y - p)^2 = (y + p)^2$	Square both sides.
$x^2 + y^2 - 2py + p^2 = y^2 + 2py + p^2$	Expand the binomials.
$x^2 - 2py = 2py$	Subtract y^2 and p^2 from both sides.
$x^2 = 4py$	Add $2py$ to both sides.
$y = \frac{1}{4p}x^2$	Solve for y.

By letting $a = \frac{1}{4p}$, we have the familiar equation $y = ax^2$. If the vertex of the parabola is located at the point (h, k), then the graph of $y = ax^2$ would be shifted $|h|$ units to the left or right and $|k|$ units up or down. The corresponding equation is

$$y = a(x - h)^2 + k, \quad \text{where } a \neq 0 \text{ and } a = \frac{1}{4p}$$

3. Standard Form of the Equation of a Parabola (Vertical Axis of Symmetry)

Standard Form of the Equation of a Parabola (Vertical Axis of Symmetry)

The standard form of the equation of a parabola with vertex (h, k) and vertical axis of symmetry is

$$y = a(x - h)^2 + k \quad \text{where } a \neq 0 \text{ and } a = \frac{1}{4p}$$

The focus is located at $(h, k + p)$.
The equation of the directrix is $y = k - p$.
If $a > 0$, then the parabola opens upward, and if $a < 0$, the parabola opens downward.

4. Finding the Focus, Directrix, and Vertex of a Parabola

example 1

Determining the Focus, Directrix, and Vertex of a Parabola

Given the equation of the parabola, $y = (x + 3)^2 - 2$

a. Determine the coordinates of the vertex.

b. Determine the coordinates of the focus and the equation of the directrix.

c. Graph the parabola.

Solution:

a. The equation can be written in the form

$$y = a(x - h)^2 + k$$
$$y = 1(x + 3)^2 - 2$$
$$= 1[x - (-3)]^2 + (-2)$$

Therefore, $h = -3$, $k = -2$, and $a = 1$.
The vertex is at (h, k), or $(-3, -2)$.

b. $a = \dfrac{1}{4p} \Rightarrow 1 = \dfrac{1}{4p}$

Solving for p, we have: $4p = 1$ or $p = \frac{1}{4}$.

The focus is located at $(h, k + p)$ or $[-3, -2 + (\frac{1}{4})] = (-3, -\frac{7}{4})$.

The directrix is $y = k - p$ or $y = -2 - (\frac{1}{4})$ or equivalently, $y = -\frac{9}{4}$.

Tip: Notice that the location of the focus is "inside" the curve of the parabola and $|p|$ units from the vertex along the axis of symmetry.

c.

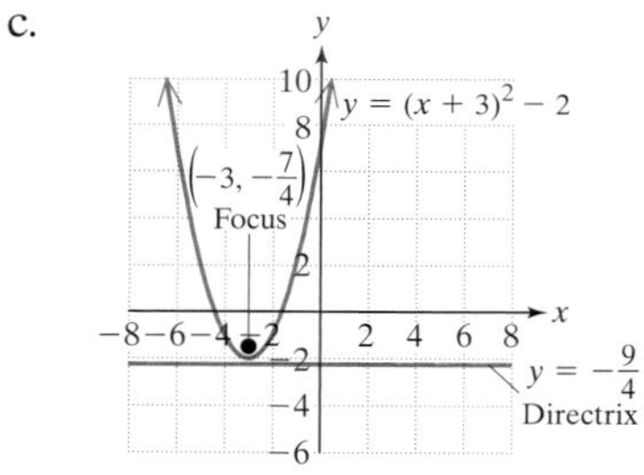

5. Completing the Square to Obtain Standard Form

To write the equation of a parabola in standard form, sometimes it is necessary to complete the square.

example 2

Determining the Focus, Directrix, and Vertex of a Parabola

Given the equation of the parabola, $y = -2x^2 + 4x + 1$

a. Write the equation in standard form, $y = a(x - h)^2 + k$.

b. Identify the vertex, focus, and directrix.

Avoiding Mistakes

It is important to understand the difference between an equation of a circle and an equation of a parabola. A circle will have square terms for both x and y with equal coefficients. A parabola will be of second degree for one variable and first degree for the other variable. For example,

Equations of Circles

$$x^2 + y^2 + 2x - 16y + 61 = 0$$

$$(x - 2)^2 + (y + 3)^2 = 4$$

Equations of Parabolas

$$x^2 + 2x + y - 3 = 0$$

$$y = 3(x - 2)^2 + 1$$

Solution:

a. Recall from Section 11.5 that we can complete the square to write the equation in the form

$y = a(x - h)^2 + k$

$= -2x^2 + 4x + 1$

$= -2(x^2 - 2x) + 1$ — Factor out -2 from the variable terms.

$= -2(x^2 - 2x + 1 - 1) + 1$ — Add and subtract the quantity $[\frac{1}{2}(-2)]^2 = 1$.

$= -2(x^2 - 2x + 1) + (-1)(-2) + 1$ — Remove the -1 term from within the parentheses by first applying the distributive property. When -1 is removed from the parentheses, it carries with it the factor of -2 from outside the parentheses.

$= -2(x - 1)^2 + 2 + 1$

$= -2(x - 1)^2 + 3$

The equation is in the form $y = a(x - h)^2 + k$, where $a = -2$, $h = 1$, and $k = 3$.

b. The vertex is $(1, 3)$.

Because $a = -2$, then $-2 = \frac{1}{4p}$; hence $-8p = 1$ and $p = -\frac{1}{8}$.

The focus is $(h, k + p)$: $[1, 3 + (-\frac{1}{8})]$, or $(1, \frac{23}{8})$.

The directrix is $y = k - p$: $y = 3 - (-\frac{1}{8})$, or $y = \frac{25}{8}$ (Figure 13-12).

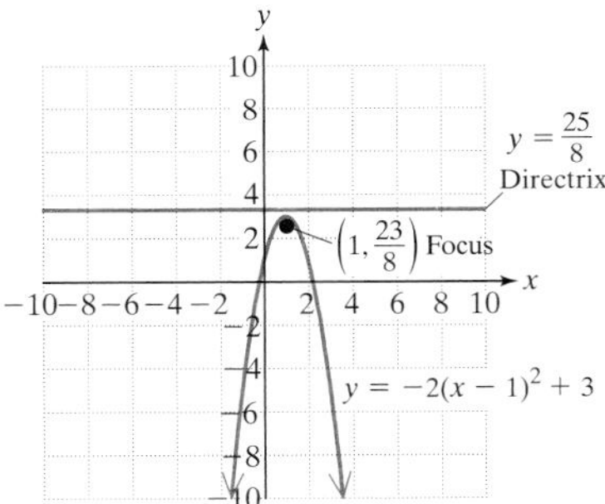

Figure 13-12

6. Standard Form of the Equation of a Parabola (Horizontal Axis of Symmetry)

We have seen that the graph of a parabola $y = a(x - h)^2 + k$ opens upward if $a > 0$ and downward if $a < 0$. A parabola can also open to the left or right. In such a case, the "roles" of x and y are essentially interchanged in the equation. Thus, the graph of $x = a(y - k)^2 + h$ opens to the right if $a > 0$ and to the left if $a < 0$.

Standard Form of the Equation of a Parabola (Horizontal Axis of Symmetry)

The standard form of the equation of a parabola with vertex (h, k) and horizontal axis of symmetry is

$$x = a(y - k)^2 + h, \text{ where } a \neq 0 \text{ and } a = \frac{1}{4p}$$

The focus is located at $(h + p, k)$.
The equation of the directrix is $x = h - p$.
If $a > 0$, then the parabola opens to the right and if $a < 0$, the parabola opens to the left.

example 3 **Determining the Focus, Directrix, and Vertex of a Parabola**

Given the equation of the parabola, $x = -\frac{1}{12}(y - 4)^2 + 2$

a. Determine the coordinates of the vertex.

b. Determine the coordinates of the focus and the equation of the directrix.

c. Graph the parabola.

Solution:

a. The equation is written in the form

$$x = a(y - k)^2 + h$$

$$x = -\frac{1}{12}(y - 4)^2 + 2$$

Therefore, $a = -\frac{1}{12}$, $h = 2$, $k = 4$.

The vertex is $(2, 4)$.

b. Because $a = -\dfrac{1}{12}$, then $-\dfrac{1}{12} = \dfrac{1}{4p}$

Solving for p, we have $-4p = 12$, or $p = -3$.

The focus is $(h + p, k)$: $(2 + (-3),\ 4)$, or $(-1, 4)$.

The directrix is $x = h - p$: $x = 2 - (-3)$, or $x = 5$.

c.

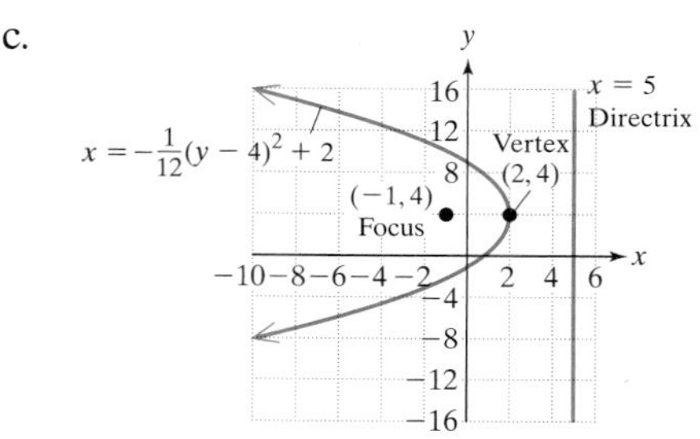

section 13.2 Practice Exercises

1. Explain how to determine whether a parabola opens upward, downward, left, or right.

2. Explain how to determine whether a parabola has a vertical or horizontal axis of symmetry.

For Exercises 3–12, use the equation of the parabola to first determine whether the axis of symmetry is vertical or horizontal. Then determine if the parabola opens upward, downward, left, or right.

3. $y = (x - 2)^2 + 3$

4. $y = (x - 4)^2 + 2$

5. $y = -2(x + 1)^2 - 4$

6. $y = -3(x + 2)^2 - 1$

7. $x = y^2 + 4$

8. $x = y^2 - 2$

9. $x = -(y + 3)^2 + 2$

10. $x = -2(y - 1)^2 - 3$

11. $y = -2x^2 - 5$

12. $y = -x^2 + 3$

13. Explain how to determine the coordinates of the focus of a parabola whose equation is written in the form $y = a(x - h)^2 + k$.

14. Explain how to determine the equation of the directrix of a parabola whose equation is written in the form $y = a(x - h)^2 + k$.

15. Explain how to determine the equation of the directrix of a parabola whose equation is written in the form $x = a(y - k)^2 + h$.

16. Explain how to determine the coordinates of the focus of a parabola whose equation is written in the form $x = a(y - k)^2 + h$.

17. True or False: The directrix of a parabola is perpendicular to its axis of symmetry.

18. True or False: A parabola never crosses its directrix.

19. True or False: The vertex of a parabola lies on the directrix.

20. True or False: The focus of a parabola lies on the directrix.

21. True or False: The focus of a parabola lies on the axis of symmetry.

22. True or False: The vertex of a parabola lies on the axis of symmetry.

For Exercises 23–36, use the equation of the parabola to determine the coordinates of the vertex and focus and the equation of the directrix. Then use this information to graph the parabola.

23. $y = \frac{1}{2}x^2$

24. $y = \frac{1}{4}x^2$

25. $y = -4x^2$

26. $y = -2x^2$

27. $y = (x - 3)^2$

28. $y = (x + 2)^2$

29. $x = y^2$

30. $x = 3y^2$

31. $x = -2(y + 1)^2$

32. $x = -(y - 2)^2$

33. $y = \frac{1}{8}(x + 2)^2 - 5$

34. $y = \frac{1}{2}(x - 1)^2 + 1$

35. $x = -\frac{1}{12}(y - 3)^2 - 3$

36. $x = -\frac{3}{4}(y + 2)^2 + 1$

For Exercises 37–44, find the value of k to complete the square. (See Section 11.1.)

37. $x^2 - 4x + k$

38. $x^2 + 6x + k$

39. $y^2 + 2y + k$

40. $y^2 - 8y + k$

41. $x^2 + x + k$

42. $x^2 - 3x + k$

43. $y^2 - 5y + k$

44. $y^2 - y + k$

For Exercises 45–54, write the equation in standard form: $y = a(x - h)^2 + k$ or $x = a(y - k)^2 + h$. Then identify the vertex, focus, and directrix of the parabola. (See Example 2.)

45. $y = x^2 - 4x + 3$
46. $y = x^2 + 6x - 2$
47. $x = y^2 + 2y + 6$
48. $x = y^2 - 8y + 3$
49. $y = -2x^2 + 8x$
50. $y = -3x^2 - 6x$
51. $y = x^2 - 3x + 2$
52. $y = x^2 + x - 4$
53. $x = -2y^2 + 16y + 1$
54. $x = -3y^2 - 6y + 7$

55. A reflecting telescope has a mirror with cross section in the shape of a parabola. A parabolic mirror has the property that incoming rays of light are reflected from the surface of the mirror to the focus.

Suppose a reflecting telescope has a parabolic mirror 6 m in diameter. With the vertex of the parabola located at the origin, the mirror traces the curve defined by

$$y = \frac{1}{50}x^2 \quad -3 \le x \le 3$$

Find the coordinates of the focus of the mirror.

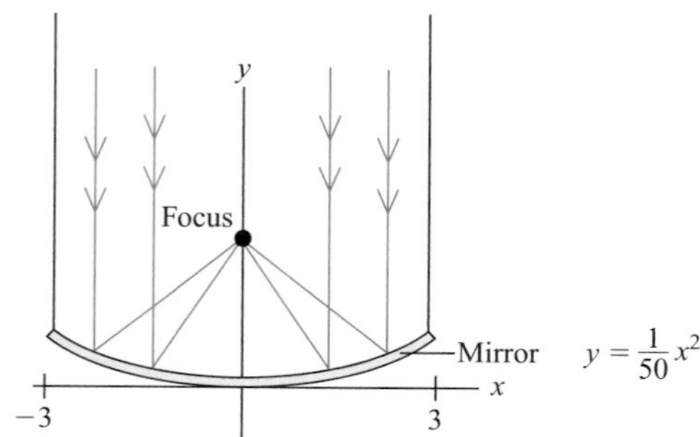

Figure for Exercise 55

56. A radio telescope has a parabolic dish 500 ft in diameter. The antenna of the telescope is located at the focus. With the vertex of the parabola located at the origin, the parabola traces the curve

$$y = 0.0016x^2 \quad -250 \le x \le 250$$

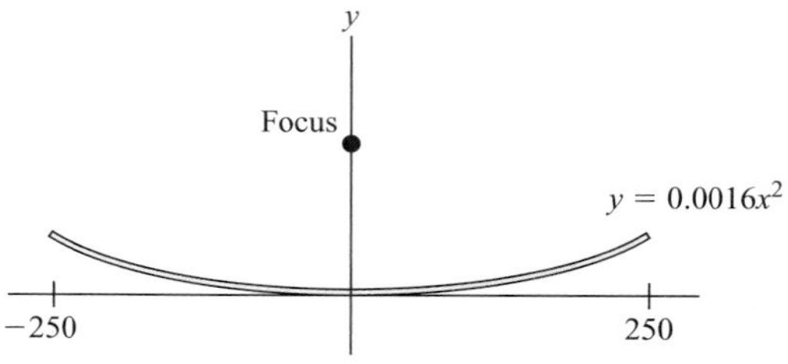

Figure for Exercise 56

Find the coordinates of the focus.

Expanding Your Skills

For Exercises 57–68, determine if the parabola opens upward, downward, left, or right given the following information. (*Hint:* Sketch the information.)

57. Directrix is $y = 3$; focus is $(0, 0)$
58. Directrix is $y = -2$; focus is $(0, 0)$
59. Directrix is $x = -1$; focus is $(1, -2)$
60. Directrix is $x = 3$; focus is $(1, 3)$
61. Directrix is $y = 0$; vertex is $(-5, -3)$
62. Directrix is $y = -4$; vertex is $(2, -3)$
63. Directrix is $x = 0$; vertex is $(-6, 6)$
64. Directrix is $x = -2$; vertex is $(-1, 0)$
65. Vertex is $(-2, 3)$; focus is $(1, 3)$
66. Vertex is $(1, 4)$; focus is $(1, -1)$
67. Vertex is $(-3, -2)$; focus is $(-3, 2)$
68. Vertex is $(4, 1)$; focus is $(-2, 1)$

section

13.3 Ellipse and Hyperbola

Concepts

1. Definition of an Ellipse
2. Standard Forms of an Equation of an Ellipse
3. Graphing an Ellipse
4. Definition of a Hyperbola
5. Standard Forms of an Equation of a Hyperbola
6. Graphing a Hyperbola

1. Definition of an Ellipse

In this section we will study the two remaining conic sections: the ellipse and the hyperbola. An **ellipse** is the set of all points (x, y) such that the sum of the distance between (x, y) and two distinct points is a constant (Figure 13-13). The fixed points are called the **foci** (plural of focus) of the ellipse.

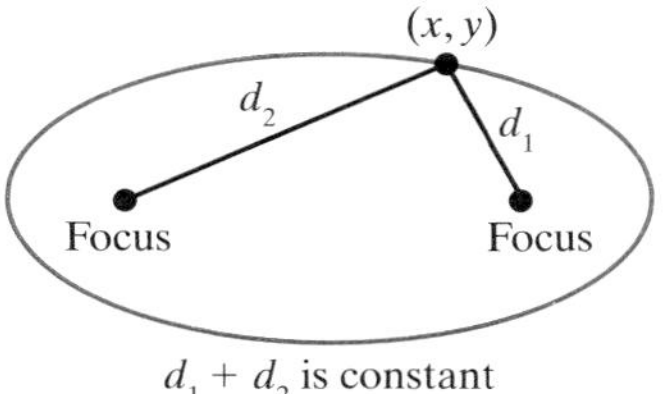

Figure 13-13

The **vertices** of the ellipse are the points where the ellipse intersects the line through the foci. The line segment connecting the vertices is called the **major axis** of the ellipse. The midpoint of the major axis is the **center** of the ellipse. The line perpendicular to the major axis and passing through the center intersects the ellipse in two points called **covertices**. The line segment defined by these points is called the **minor axis** (Figure 13-14).

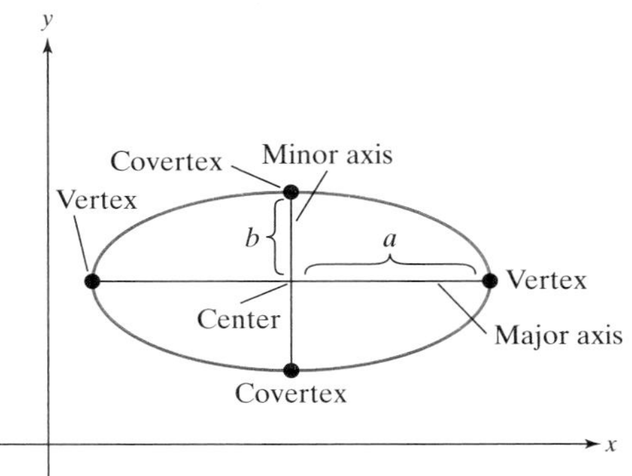

Figure 13-14

It is customary to denote the length of the major axis as $2a$ and the length of the minor axis as $2b$. The length of the major axis is longer than the length of the minor axis; therefore, $a > b$ (see Figure 13-14).

2. Standard Forms of an Equation of an Ellipse

The definition of an ellipse and the distance formula can be used to find an equation of an ellipse. The standard forms of an equation of an ellipse are shown here.

Standard Forms of an Equation of an Ellipse

Let a and b represent positive real numbers such that $a > b$.

Horizontal major axis:

The standard form of an equation of an ellipse with a *horizontal major axis* and center at (h, k) is given by

$$\frac{(x-h)^2}{a^2} + \frac{(y-k)^2}{b^2} = 1$$

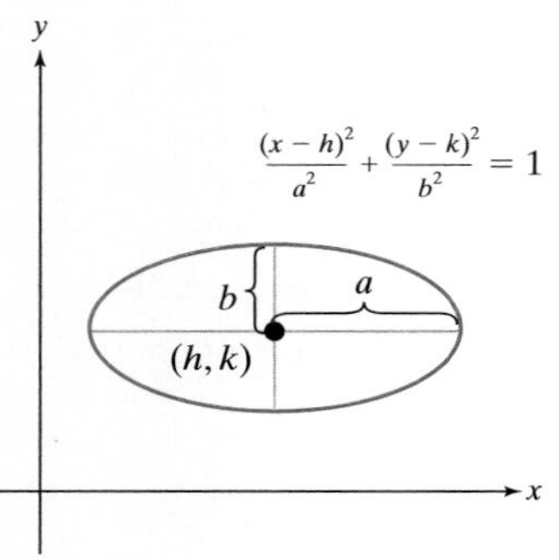

Vertical major axis:

The standard form of an equation of an ellipse with a *vertical major axis* and center at (h, k) is given by

$$\frac{(x-h)^2}{b^2} + \frac{(y-k)^2}{a^2} = 1$$

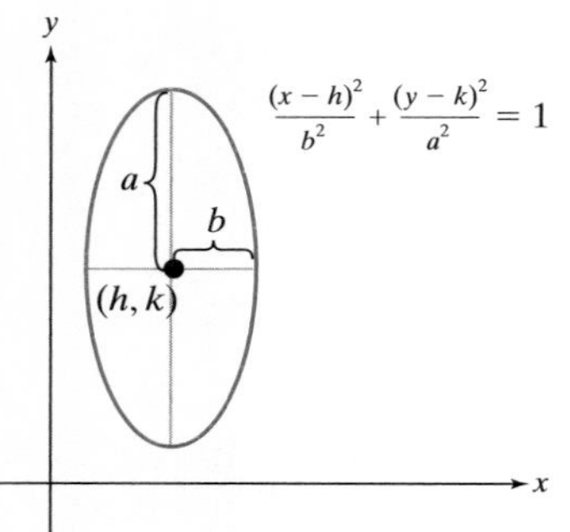

3. Graphing an Ellipse

example 1

Graphing an Ellipse

Given the equation of the ellipse

$$\frac{(x-3)^2}{16} + \frac{(y+1)^2}{81} = 1$$

a. Determine the coordinates of the center.

b. Determine whether the major axis is horizontal or vertical.

c. Graph the ellipse.

d. Determine the coordinates of the vertices and covertices.

Solution:

a. The equation can be written as

$$\frac{(x-3)^2}{(4)^2} + \frac{[y-(-1)]^2}{(9)^2} = 1$$

Therefore the center is $(3, -1)$.

b. $a > b$. Therefore, $a = 9$ and $b = 4$.

The major axis is vertical.

c.

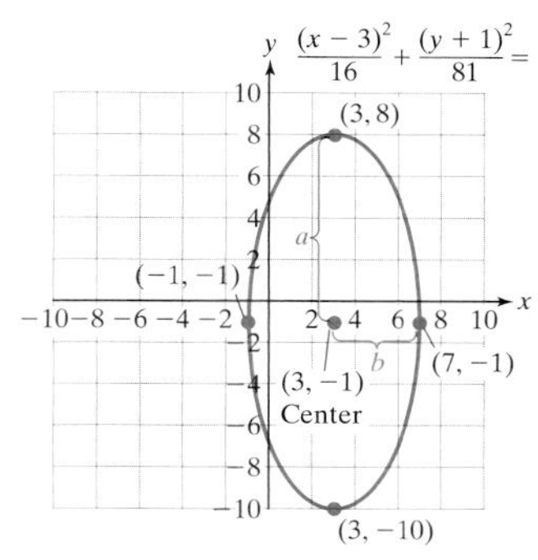

d. The vertices are located 9 units from the center and define the endpoints of the major axis. The coordinates of the vertices are $(3, -10)$ and $(3, 8)$. The covertices are located 4 units from the center and define the endpoints of the minor axis. The coordinates of the covertices are $(-1, -1)$ and $(7, -1)$.

example 2

Graphing an Ellipse

Given the equation of the ellipse $4x^2 + 9y^2 = 36$.

a. Determine the coordinates of the center.

b. Determine whether the major axis is horizontal or vertical.

c. Graph the ellipse.

d. Determine the coordinates of the vertices and covertices.

Solution:

a. To write the equation $4x^2 + 9y^2 = 36$ in standard form, begin by dividing both sides by 36. This will make the constant term equal to 1.

$$\frac{4x^2}{36} + \frac{9y^2}{36} = \frac{36}{36}$$

$$\frac{x^2}{9} + \frac{y^2}{4} = 1$$

The equation can be written as $\frac{(x-0)^2}{(3)^2} + \frac{(y-0)^2}{(2)^2} = 1$. Therefore the center is $(0, 0)$.

Tip: In a standard equation of an ellipse, the constant term must be 1.

$$\frac{(x-h)^2}{a^2} + \frac{(y-k)^2}{b^2} = 1$$

must be 1

Therefore, in Example 2, we divided both sides of the equation by 36.

Tip: For an ellipse written in standard form, the direction of the major axis is determined by the term that has a greater value in the denominator. In Example 2, the denominator of the x-term is 9, whereas the denominator of the y-term is 4. Therefore, the major axis is in the x-direction (horizontal).

b. $a > b$; therefore, $a = 3$ and $b = 2$.

The major axis is horizontal.

c.

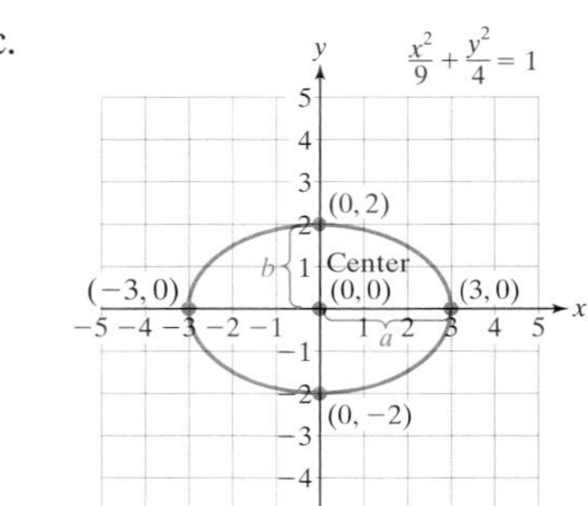

d. The vertices are located 3 units from the center and define the endpoints of the major axis. The coordinates of the vertices are $(-3, 0)$ and $(3, 0)$. The co-vertices are located 2 units from the center and define the endpoints of the minor axis. The coordinates of the covertices are $(0, -2)$ and $(0, 2)$.

4. Definition of a Hyperbola

A **hyperbola** is the set of all points (x, y) such that the *difference* of the distances between (x, y) and two distinct points is a constant (Figure 13-15). The fixed points are called the **foci** of the hyperbola.

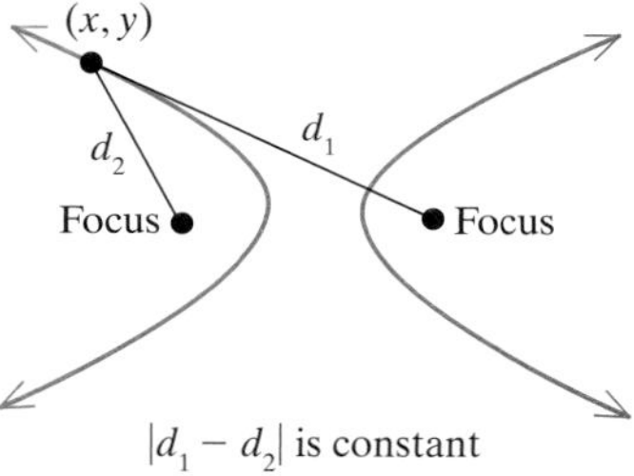

Figure 13-15

The line through the foci intersects the hyperbola at two points called the **vertices** of the hyperbola. The line segment connecting the vertices is called the **transverse axis**. The midpoint of this line segment is the **center** of the hyperbola. The **conjugate axis** passes through the center of the hyperbola and is perpendicular to the transverse axis.

A hyperbola has two branches, and each branch approaches a pair of intersecting lines called **asymptotes**. The asymptotes are not actually part of the hyperbola but may be used as an aid in sketching a hyperbola. The asymptotes of a hyperbola are the lines passing through the diagonals of a fundamental rectangle (Figure 13-16).

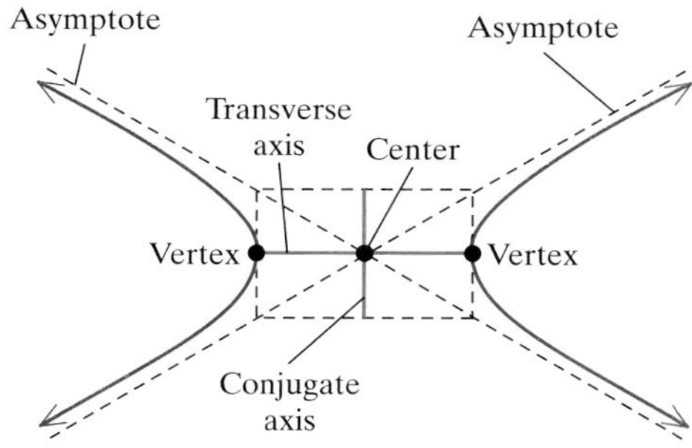

Figure 13-16

The dimensions of the fundamental rectangle are the lengths of the transverse axis and conjugate axis. Furthermore, it is customary to denote the length of the transverse axis by $2a$ and the length of the conjugate axis by $2b$.

5. Standard Forms of an Equation of a Hyperbola

The definition of a hyperbola and the distance formula can be used to find an equation of a hyperbola. The standard forms of an equation of a hyperbola with a horizontal or vertical transverse axis are shown here.

Standard Forms of an Equation of a Hyperbola

Let a and b represent positive real numbers.

Horizontal transverse axis:

The standard form of an equation of a hyperbola with a *horizontal transverse axis* and center at (h, k) is given by

$$\frac{(x-h)^2}{a^2} - \frac{(y-k)^2}{b^2} = 1$$

Vertical transverse axis:

The standard form of an equation of a hyperbola with a *vertical transverse axis* and center at (h, k) is given by

$$\frac{(y-k)^2}{a^2} - \frac{(x-h)^2}{b^2} = 1$$

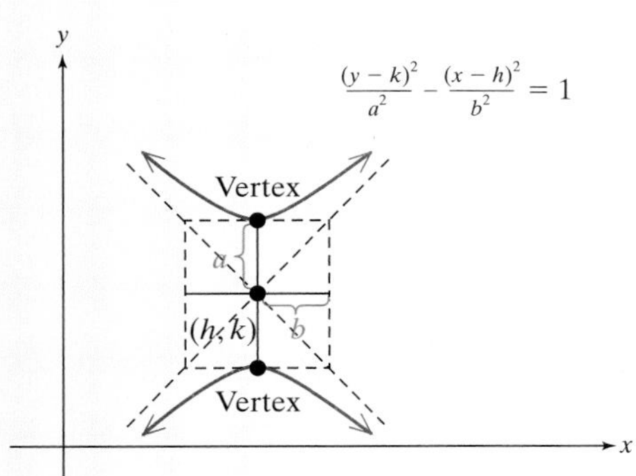

6. Graphing a Hyperbola

Given an equation of a hyperbola in standard form, the term with a positive coefficient determines the orientation of the transverse axis (either horizontal or vertical). To sketch a hyperbola, first determine the location of the center and the values of a and b. Then draw the fundamental rectangle, asymptotes, and vertices of the hyperbola. Use these features as an aid to sketch the curve. This is demonstrated in Examples 3 and 4.

example 3

Graphing a Hyperbola

Given the equation of the hyperbola

$$\frac{x^2}{36} - \frac{y^2}{9} = 1$$

a. Determine the coordinates of the center.

b. Determine whether the transverse axis is horizontal or vertical.

c. Graph the hyperbola.

d. Determine the coordinates of the vertices.

e. Determine the equations of the asymptotes.

Solution:

a. The equation can be written as

$$\frac{(x - 0)^2}{(6)^2} - \frac{(y - 0)^2}{(3)^2} = 1$$

Therefore, the center is (0, 0).

b. Because the positive coefficient is on the x-term, the transverse axis is horizontal and $a = 6$. Thus $b = 3$.

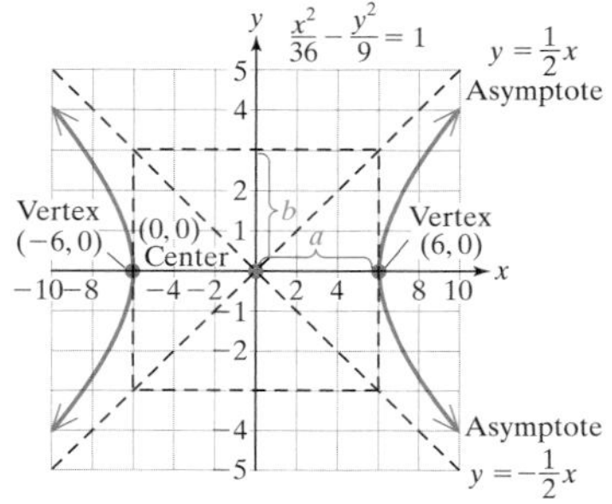

d. The vertices are located 6 units from the center and define the endpoints of the transverse axis. The coordinates of the vertices are $(-6, 0)$ and $(6, 0)$.

e. The asymptotes must pass through the diagonals of the fundamental rectangle; therefore, one asymptote must pass through $(-6, -3)$ and $(6, 3)$. The other must pass through $(-6, 3)$ and $(6, -3)$. Finding the slope of each line and applying the point-slope formula, we have

Tip: The slopes of the asymptotes of a hyperbola are opposites of each other.

Line through $(-6, -3)$ and $(6, 3)$

$$m = \frac{3 - (-3)}{6 - (-6)} = \frac{6}{12} = \frac{1}{2}$$

$$y - y_1 = m(x - x_1)$$

$$y - (-3) = \frac{1}{2}(x - (-6))$$

$$y + 3 = \frac{1}{2}x + 3$$

$$y = \frac{1}{2}x$$

Line through $(-6, 3)$ and $(6, -3)$

$$m = \frac{-3 - (3)}{6 - (-6)} = \frac{-6}{12} = -\frac{1}{2}$$

$$y - y_1 = m(x - x_1)$$

$$y - 3 = -\frac{1}{2}(x - (-6))$$

$$y - 3 = -\frac{1}{2}x - 3$$

$$y = -\frac{1}{2}x$$

example 4 Graphing a Hyperbola

Given the equation of the hyperbola $(y - 2)^2 - 4(x + 4)^2 = 16$

a. Determine the coordinates of the center.

b. Determine whether the transverse axis is horizontal or vertical.

c. Graph the hyperbola.

d. Determine the coordinates of the vertices.

Solution:

a. By dividing $(y - 2)^2 - 4(x + 4)^2 = 16$ by 16 we have

$$\frac{(y-2)^2}{16} - \frac{(x+4)^2}{4} = 1 \Rightarrow \frac{(y-2)^2}{(4)^2} - \frac{[x-(-4)]^2}{(2)^2} = 1$$

Therefore the center is $(-4, 2)$.

b. Because the positive coefficient is on the y-term, the transverse axis is vertical and $a = 4$. Thus $b = 2$.

c.

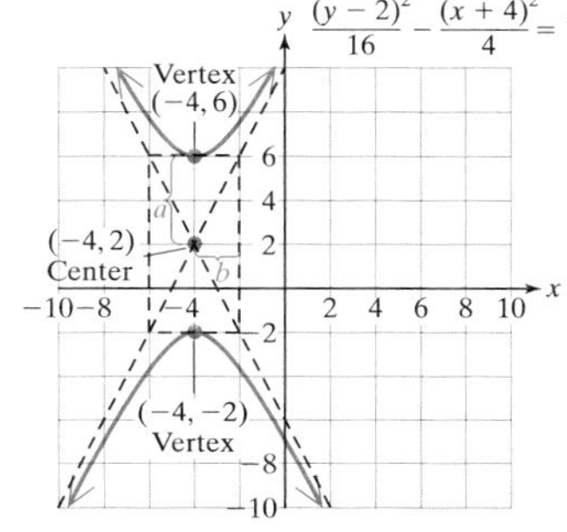

d. The vertices are located 4 units from the center and define the endpoints of the transverse axis. The coordinates of the vertices are $(-4, 6)$ and $(-4, -2)$.

section 13.3 Practice Exercises

1. Draw an ellipse with a vertical major axis and label the following:
   a. The center
   b. The vertices
   c. The covertices
   d. The major axis
   e. The minor axis

2. Draw an ellipse with a horizontal major axis and label the following:
   a. The center
   b. The vertices
   c. The covertices
   d. The major axis
   e. The minor axis

For Exercises 3–10, use the equation of the ellipse to identify the center and determine whether the major axis is horizontal or vertical.

3. $\frac{x^2}{4} + \frac{y^2}{12} = 1$

4. $\frac{x^2}{9} + \frac{y^2}{14} = 1$

5. $\frac{(x-2)^2}{9} + \frac{(y-5)^2}{4} = 1$

6. $\frac{(x-6)^2}{16} + \frac{(y-2)^2}{4} = 1$

7. $5(x+2)^2 + (y+4)^2 = 10$

8. $\frac{(x+1)^2}{5} + \frac{(y+8)^2}{8} = 1$

9. $\frac{(x+4)^2}{7} + \frac{(y-1)^2}{1} = 1$

10. $(x-4)^2 + 3(y+4)^2 = 9$

For Exercises 11–20, graph the ellipse and identify the center, vertices, and covertices.

11. $\frac{x^2}{4} + \frac{y^2}{9} = 1$

12. $\frac{x^2}{16} + \frac{y^2}{25} = 1$

13. $\frac{x^2}{25} + \frac{y^2}{16} = 1$

14. $\frac{x^2}{81} + \frac{y^2}{36} = 1$

15. $\frac{x^2}{16} + \frac{(y-2)^2}{4} = 1$

16. $\frac{(x-2)^2}{9} + \frac{y^2}{4} = 1$

17. $\frac{(x+1)^2}{1} + \frac{(y+3)^2}{36} = 1$

18. $\frac{(x+4)^2}{16} + \frac{(y+2)^2}{25} = 1$

19. $\frac{(x-3)^2}{49} + \frac{(y+1)^2}{25} = 1$

20. $\frac{(x+3)^2}{4} + \frac{(y-1)^2}{1} = 1$

21. Sonya wants to cut an elliptical rug from a rectangular rug to avoid a stain made by the family dog. She places two tacks along the center horizontal line. Then she ties the ends of a slack piece of rope to each tack. With the rope pulled tight, she traces out a curve. Use the definition of an ellipse to explain why this method will produce an elliptical shape curve.

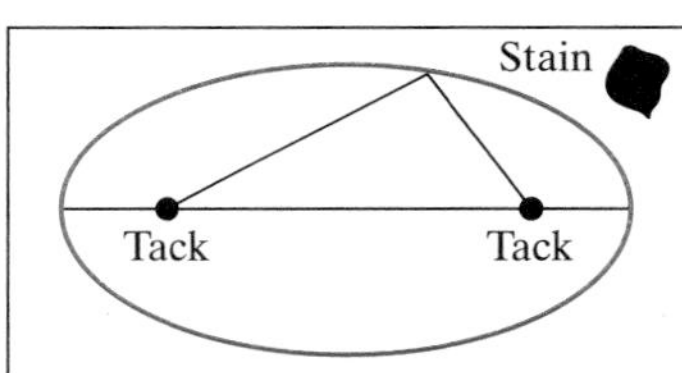

Figure for Exercise 21

22. An arch for a tunnel is in the shape of a semi-ellipse. The distance between vertices is 120 ft, and the height to the top of the arch is 50 ft. Find the height of the arch 10 ft from the center. Round to the nearest foot.

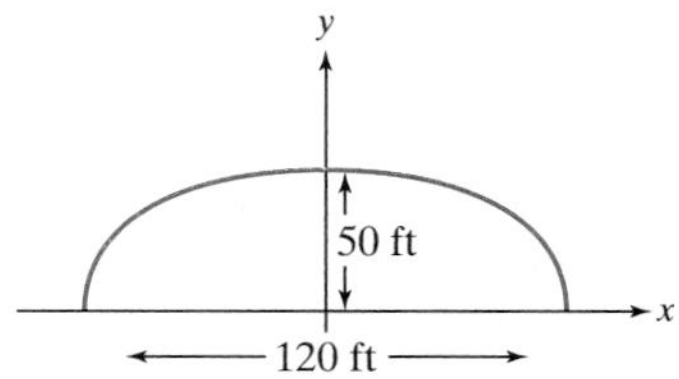

Figure for Exercise 22

23. A bridge over a gorge is supported by an arch in the shape of a semiellipse. The length of the bridge is 400 ft, and the height is 100 ft. Find the height of the arch 50 ft from the center. Round to the nearest foot.

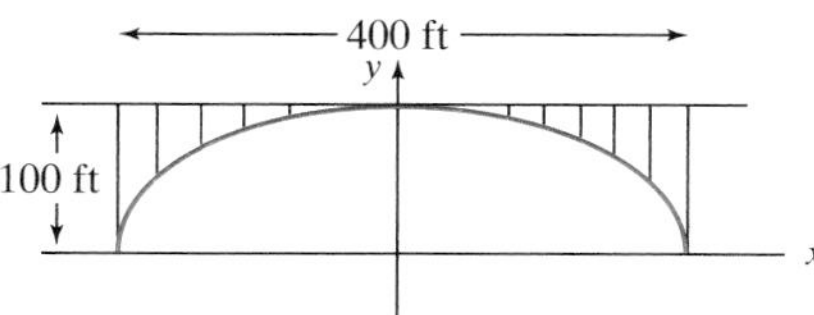

Figure for Exercise 23

24. Draw a hyperbola with a vertical transverse axis and label the following:
    a. The center
    b. The vertices
    c. The asymptotes
    d. The transverse axis
    e. The conjugate axis

25. Draw a hyperbola with a horizontal transverse axis and label the following:
    a. The center
    b. The vertices
    c. The asymptotes
    d. The transverse axis
    e. The conjugate axis

For Exercises 26–33, identify the center of the hyperbola and determine whether the transverse axis is horizontal or vertical.

26. $\dfrac{x^2}{14} - \dfrac{y^2}{8} = 1$

27. $\dfrac{x^2}{10} - \dfrac{y^2}{4} = 1$

28. $\dfrac{y^2}{6} - \dfrac{x^2}{16} = 1$

29. $\dfrac{y^2}{9} - \dfrac{x^2}{25} = 1$

30. $\dfrac{(x-2)^2}{12} - \dfrac{(y+4)^2}{9} = 1$

31. $(x+1)^2 - 6(y+8)^2 = 24$

32. $(y-5)^2 - (x-2)^2 = 9$

33. $\dfrac{(y+4)^2}{12} - \dfrac{(x+1)^2}{12} = 1$

For Exercises 34–43, graph the hyperbola and identify the center and vertices.

34. $\dfrac{x^2}{25} - \dfrac{y^2}{16} = 1$

35. $\dfrac{x^2}{9} - \dfrac{y^2}{36} = 1$

36. $\dfrac{y^2}{4} - \dfrac{x^2}{4} = 1$

37. $\dfrac{y^2}{9} - \dfrac{x^2}{9} = 1$

38. $\dfrac{(x-1)^2}{25} - \dfrac{(y-1)^2}{4} = 1$

39. $\dfrac{(x-2)^2}{36} - \dfrac{(y-2)^2}{25} = 1$

40. $\dfrac{(y+3)^2}{1} - \dfrac{x^2}{16} = 1$

41. $\dfrac{y^2}{4} - \dfrac{(x+4)^2}{9} = 1$

42. $\dfrac{(x+5)^2}{49} - \dfrac{(y+1)^2}{9} = 1$

43. $\dfrac{(x+6)^2}{25} - \dfrac{(y+2)^2}{1} = 1$

44. Explain how to find the equations of the asymptotes of a hyperbola written in the form

$$\frac{(x-h)^2}{a^2} - \frac{(y-k)^2}{b^2} = 1$$

For Exercises 45–48, find the equations of the asymptotes of the hyperbola.

45. $\dfrac{x^2}{16} - \dfrac{y^2}{16} = 1$

46. $\dfrac{x^2}{9} - \dfrac{y^2}{1} = 1$

47. $\dfrac{(x-2)^2}{36} - \dfrac{(y+4)^2}{9} = 1$

48. $\dfrac{(x-1)^2}{25} - \dfrac{(y-1)^2}{4} = 1$

49. Explain how to tell the difference between the equations of a hyperbola and an ellipse.

For Exercises 50–57, determine if the equation represents an ellipse or a hyperbola.

50. $\dfrac{x^2}{6} - \dfrac{y^2}{10} = 1$

51. $\dfrac{x^2}{14} + \dfrac{y^2}{2} = 1$

52. $\dfrac{(y-4)^2}{4} + \dfrac{(x+3)^2}{16} = 1$

53. $\dfrac{(x+1)^2}{5} + \dfrac{y^2}{5} = 1$

54. $4x^2 + y^2 = 16$

55. $-3x^2 - 4y^2 = -36$

56. $-y^2 + 2x^2 = -10$

57. $x^2 - y^2 = -1$

Expanding Your Skills

The foci of an ellipse are located $|c|$ units from the center along the major axis, where $c^2 = a^2 - b^2$.

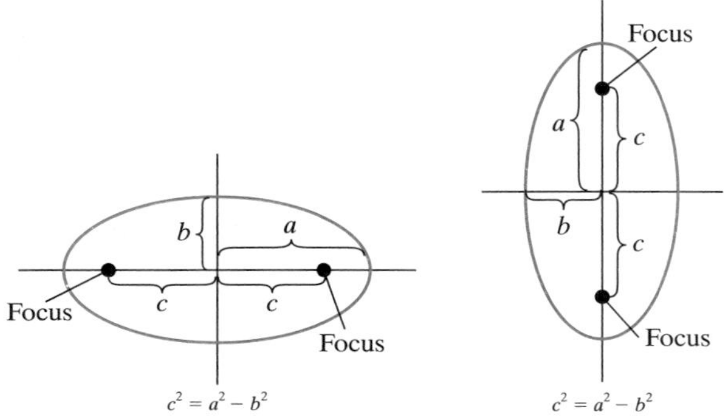

For Exercises 58–61, find the coordinates of the foci of the ellipse.

58. $\dfrac{x^2}{25} + \dfrac{y^2}{16} = 1$

59. $\dfrac{x^2}{9} + \dfrac{y^2}{25} = 1$

60. $\dfrac{x^2}{64} + \dfrac{y^2}{100} = 1$

61. $\dfrac{x^2}{100} + \dfrac{y^2}{36} = 1$

62. The orbit of Venus around the sun can be approximated by an ellipse with an equation

$$\frac{x^2}{(113.93)^2} + \frac{y^2}{(113.34)^2} = 1$$

where the sun is located at one of the foci and all distances are in millions of kilometers.

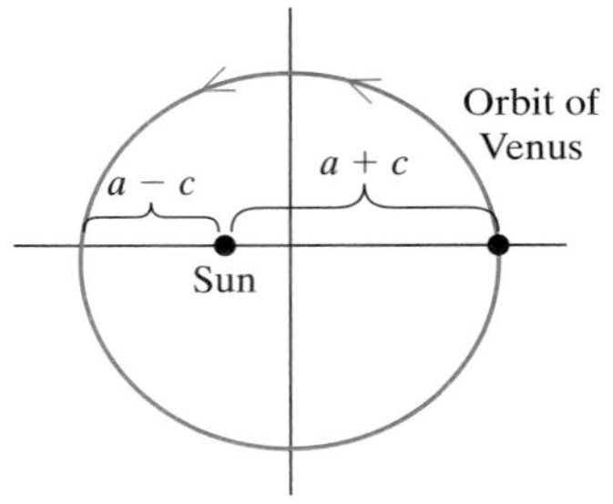

Figure for Exercise 62

a. Find the coordinates of the foci. Round to the nearest hundredth.

b. Find the greatest distance from Venus to the sun (this distance is called the apogee).

c. Find the least distance from Venus to the sun (this distance is called the perigee).

The foci of a hyperbola are located $|c|$ units from the center along the line defined by the transverse axis, where $c^2 = a^2 + b^2$.

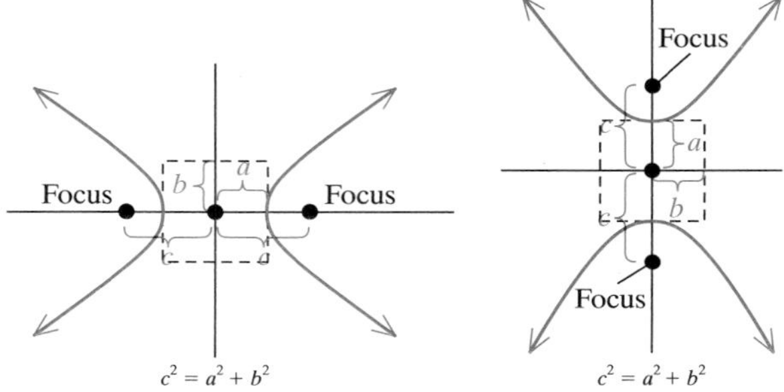

For Exercises 63–66, find the coordinates of the foci of the hyperbola.

63. $\dfrac{x^2}{16} - \dfrac{y^2}{9} = 1$

64. $\dfrac{y^2}{36} - \dfrac{x^2}{64} = 1$

65. $\dfrac{y^2}{144} - \dfrac{x^2}{25} = 1$

66. $\dfrac{x^2}{64} - \dfrac{y^2}{36} = 1$

section

13.4 Nonlinear Systems of Equations in Two Variables

Concepts

1. **Solving Nonlinear Systems of Equations by the Substitution Method**
2. **Solving Nonlinear Systems of Equations by the Addition Method**

1. Solving Nonlinear Systems of Equations by the Substitution Method

Recall that a linear equation in two variables x and y is an equation that can be written in the form $ax + by = c$, where a and b are not both zero. In Sections 8.1–8.3, we solved systems of linear equations in two variables using the graphing method, the substitution method, and the addition method. In this section, we will solve *nonlinear* systems of equations using the same methods. A **nonlinear system of equations** is a system in which at least one of the equations is nonlinear.

Graphing the equations in a nonlinear system helps to determine the number of solutions and to approximate the coordinates of the solutions. The substitution method is used most often to solve a nonlinear system of equations analytically.

example 1 **Solving a Nonlinear System of Equations by the Substitution Method**

Given the system

$$x - 7y = -25$$
$$x^2 + y^2 = 25$$

a. Solve the system by graphing.

b. Solve the system by the substitution method.

Solution:

a. $x - 7y = -25$ is a line (the slope-intercept form is $y = \frac{1}{7}x + \frac{25}{7}$).

$x^2 + y^2 = 25$ is a circle centered at the origin with radius 5.

From Figure 13-17, we appear to have two solutions at $(-4, 3)$ and $(3, 4)$.

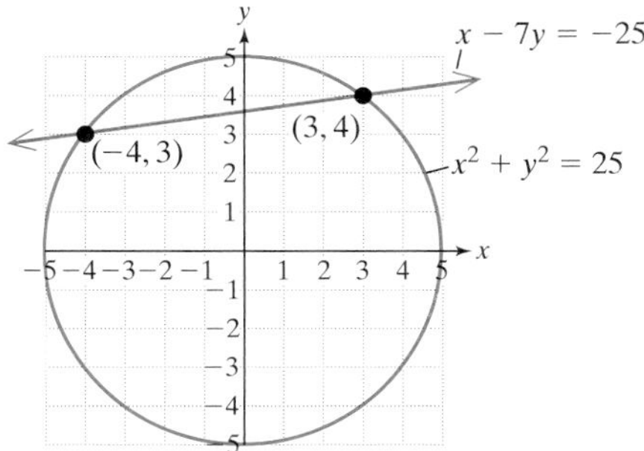

Figure 13-17

b. To use the substitution method, isolate one of the variables from one of the equations. We will solve for x in the first equation.

A $\quad x - 7y = -25 \xrightarrow{\text{solve for } x} x = 7y - 25$

B $\quad x^2 + y^2 = 25$

B $\quad (7y - 25)^2 + y^2 = 25$ Substitute $(7y - 25)$ for x in the second equation.

$49y^2 - 350y + 625 + y^2 = 25$ The resulting equation is quadratic in y.

$50y^2 - 350y + 600 = 0$ Set the equation equal to zero.

$50(y^2 - 7y + 12) = 0$ Factor.

$50(y - 3)(y - 4) = 0$

$y = 3 \quad \text{or} \quad y = 4$

For each value of y, find the corresponding x value from the equation $x = 7y - 25$.

$y = 3$: $x = 7(3) - 25 = -4$ The solution point is $(-4, 3)$.

$y = 4$: $x = 7(4) - 25 = 3$ The solution point is $(3, 4)$. (See Figure 13-17.)

example 2 Solving a Nonlinear System by the Substitution Method

Given the system

$$y = \sqrt{x}$$
$$x^2 + y^2 = 20$$

a. Sketch the equations.

b. Solve the system by the substitution method.

Solution:

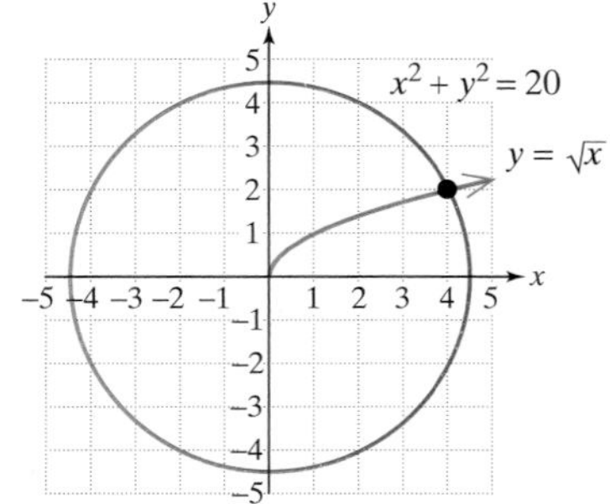

Figure 13-18

a. $y = \sqrt{x}$ is one of the six basic functions graphed in Section 7.5.

$x^2 + y^2 = 20$ is a circle centered at the origin with radius $\sqrt{20} \approx 4.5$.

From Figure 13-18, we see that there is one solution.

b. To use the substitution method, we will substitute $y = \sqrt{x}$ into equation B.

A $\quad y = \sqrt{x}$

B $\quad x^2 + y^2 = 20$

B $\quad x^2 + (\sqrt{x})^2 = 20$ Substitute $y = \sqrt{x}$ into the second equation.

$x^2 + x = 20$

$x^2 + x - 20 = 0$ Set the second equation equal to zero.

$(x + 5)(x - 4) = 0$ Factor.

~~$x = -5$~~ or $x = 4$ Reject $x = -5$ because it is not in the domain of $y = \sqrt{x}$.

Substitute $x = 4$ into the equation $y = \sqrt{x}$.

If $x = 4$, then $y = \sqrt{4} = 2$ The solution point is $(4, 2)$.

Calculator Connections

Graph the equations from Example 2 to confirm your solution to the system of equations. Use an *Intersect* feature or *Zoom* and *Trace* to approximate the point of intersection. Recall that the circle must be entered into the calculator as two functions:

$$Y_1 = \sqrt{20 - x^2}$$
$$Y_2 = -\sqrt{20 - x^2}$$
$$Y_3 = \sqrt{x}$$

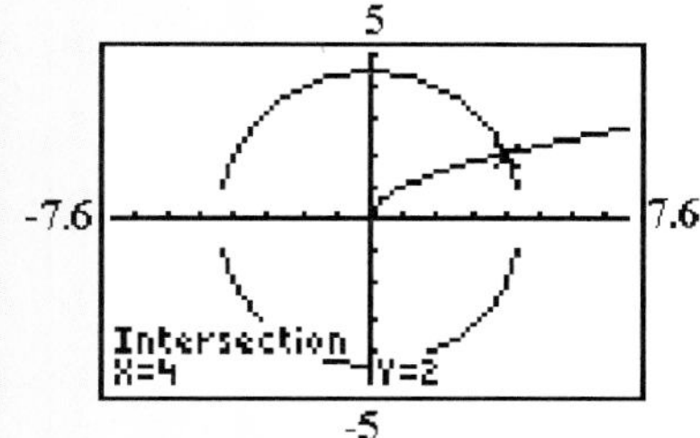

example 3 Solving a Nonlinear System Using the Substitution Method

Solve the system using the substitution method.

$$y = \sqrt[3]{x}$$
$$y = x$$

Solution:

A $y = \sqrt[3]{x}$

B $y = x$

$\sqrt[3]{x} = x$ Because y is isolated in both equations, we can equate the expressions for y.

$(\sqrt[3]{x})^3 = (x)^3$ To solve the radical equation, raise both sides to the third power.

$x = x^3$ This is a third-degree polynomial equation.

$0 = x^3 - x$ Set the equation equal to zero.

$0 = x(x^2 - 1)$ Factor out the GCF.

$0 = x(x + 1)(x - 1)$ Factor completely.

$x = 0$ or $x = -1$ or $x = 1$

For each value of x, find the corresponding y-value from either original equation. We will use equation $\boxed{B}$: $y = x$.

If $x = 0$, then $y = 0$ The solution point is (0, 0).

If $x = -1$, then $y = -1$ The solution point is $(-1, -1)$.

If $x = 1$, then $y = 1$ The solution point is (1, 1).

Calculator Connections

Graph the equations $y = \sqrt[3]{x}$ and $y = x$ to support the solutions to Example 3.

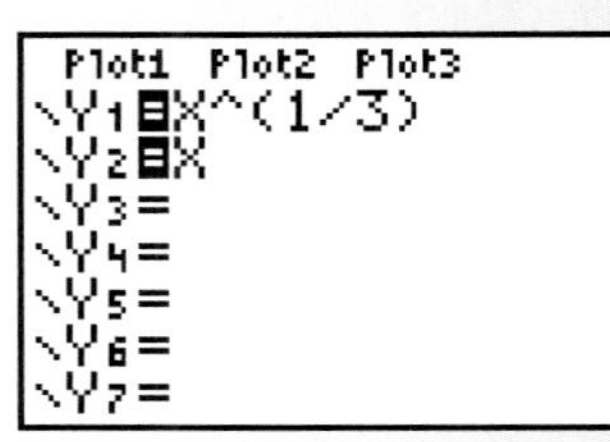

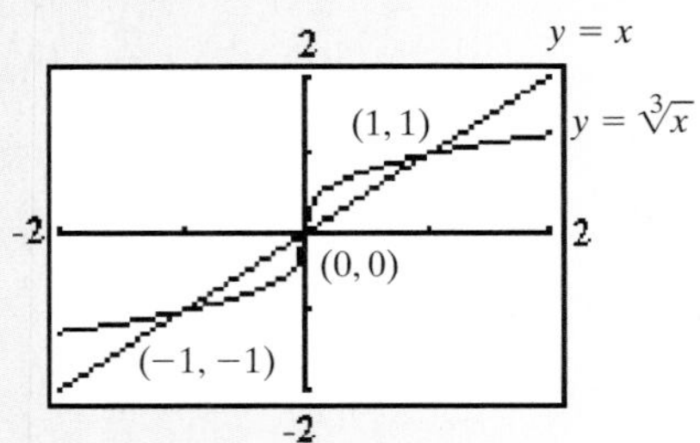

2. Solving Nonlinear Systems of Equations by the Addition Method

The substitution method is used most often to solve a system of nonlinear equations. In some situations, however, the addition method offers an efficient means of finding a solution. Example 4 demonstrates that we can eliminate a variable from both equations provided the terms containing that variable are *like* terms.

example 4 Solving a Nonlinear System of Equations by the Addition Method

Solve the system

$$2x^2 + y^2 = 17$$
$$x^2 + 2y^2 = 22$$

Solution:

$\boxed{A}$ $2x^2 + y^2 = 17$ Notice that the y^2 terms are *like* in each equation.

$\boxed{B}$ $x^2 + 2y^2 = 22$ To eliminate the y^2 terms, multiply the first equation by -2.

Tip: In Example 4, the x^2 terms are also *like* in both equations. We could have eliminated the x^2 terms by multiplying equation $\boxed{B}$ by -2.

$\boxed{A}$ $2x^2 + y^2 = 17 \xrightarrow{\text{multiply by } -2} -4x^2 - 2y^2 = -34$

$\boxed{B}$ $x^2 + 2y^2 = 22 \longrightarrow x^2 + 2y^2 = 22$

$$-3x^2 = -12 \quad \text{Eliminate the } y^2 \text{ term.}$$

$$\frac{-3x^2}{-3} = \frac{-12}{-3}$$

$$x^2 = 4$$

$$x = \pm 2$$

Substitute each value of x into one of the original equations to solve for y. We will use equation $\boxed{A}$: $2x^2 + y^2 = 17$.

$x = 2$: $\boxed{A}$ $2(2)^2 + y^2 = 17$

$$8 + y^2 = 17$$

$$y^2 = 9$$

$y = \pm 3$ The solution points are $(2, 3)$ and $(2, -3)$.

$x = -2$: $\boxed{A}$ $2(-2)^2 + y^2 = 17$

$$8 + y^2 = 17$$

$$y^2 = 9$$

$y = \pm 3$ The solution points are $(-2, 3)$ and $(-2, -3)$.

Calculator Connections

The solutions to Example 4 can be checked from the graphs of the equations.

For the equation $2x^2 + y^2 = 17$, we have: $y = \pm\sqrt{17 - 2x^2}$

For the equation $x^2 + 2y^2 = 22$, we have: $y = \pm\sqrt{\dfrac{22 - x^2}{2}}$

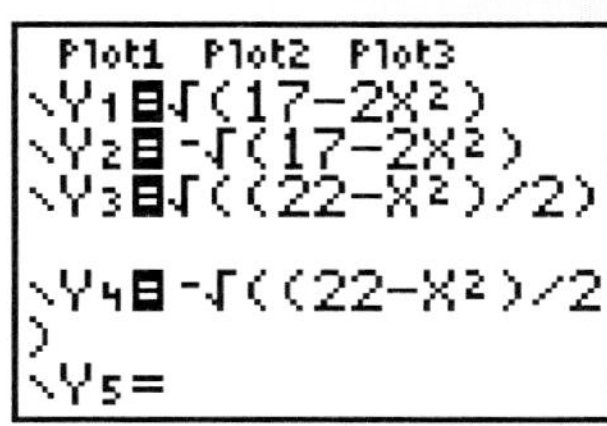

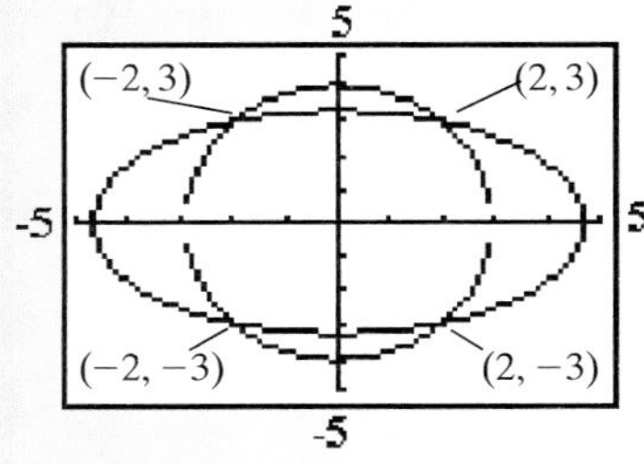

Tip: It is important to note that the addition method can only be used if two equations share a pair of *like* terms. The substitution method is effective in solving a wider range of systems of equations. The system in Example 4 could also have been solved using substitution.

A $\quad 2x^2 + y^2 = 17 \xrightarrow{\text{solve for } y^2} y^2 = 17 - 2x^2$

B $\quad x^2 + 2y^2 = 22$

B $\quad x^2 + 2(17 - 2x^2) = 22$

$x^2 + 34 - 4x^2 = 22$

$-3x^2 = -12$

$x^2 = 4$

$x = \pm 2$

$x = 2$: $\quad y^2 = 17 - 2(2)^2$

$y^2 = 9$

$y = \pm 3$ The solutions are (2, 3) and (2, −3).

$x = -2$: $\quad y^2 = 17 - 2(-2)^2$

$y^2 = 9$

$y = \pm 3$ The solutions are (−2, 3) and (−2, −3).

section 13.4 Practice Exercises

1. Write the distance formula between two points (x_1, y_1) and (x_2, y_2) from memory.

2. Find the distance between the two points $(8, -1)$ and $(1, -8)$.

3. Write an equation representing the set of all points 2 units from the point $(-1, 1)$.

4. Write an equation representing the set of all points 8 units from the point $(-5, 3)$.

For Exercises 5–12, determine if the equation represents a parabola, circle, ellipse, or hyperbola.

5. $x^2 + y^2 = 15$

6. $\dfrac{x^2}{4} - \dfrac{y^2}{2} = 1$

7. $y = (x - 6)^2 + 4$

8. $\dfrac{(x + 1)^2}{2} + \dfrac{(y + 1)^2}{5} = 1$

9. $\dfrac{(y - 1)^2}{3} - \dfrac{(x + 2)^2}{3} = 1$

10. $3x^2 + 3y^2 = 1$

11. $\dfrac{x^2}{9} + \dfrac{y^2}{12} = 1$

12. $x = (y + 2)^2 - 5$

For Exercises 13–20, use sketches to explain.

13. How many points of intersection are possible between a line and a parabola?

14. How many points of intersection are possible between a line and a circle?

15. How many points of intersection are possible between two distinct circles?

16. How many points of intersection are possible between two distinct parabolas of the form $y = ax^2 + bx + c, a \neq 0$?

17. How many points of intersection are possible between a circle and a parabola?

18. How many points of intersection are possible between two distinct lines?

19. How many points of intersection are possible with an ellipse and a hyperbola?

20. How many points of intersection are possible with an ellipse and a parabola?

For Exercises 21–26, sketch each system of equations. Then solve the system by the substitution method.

21. $y = x + 3$
$x^2 + y = 9$

22. $y = x - 2$
$x^2 + y = 4$

23. $x^2 + y^2 = 1$
$y = x + 1$

24. $x^2 + y^2 = 25$
$y = 2x$

25. $x^2 + y^2 = 6$
$y = x^2$

26. $x^2 + y^2 = 12$
$y = x^2$

For Exercises 27–34, solve the system by the substitution method.

27. $x^2 + y^2 = 20$
$y = \sqrt{x}$

28. $x^2 + y^2 = 30$
$y = \sqrt{x}$

29. $y = x^2$
$y = -\sqrt{x}$

30. $y = -x^2$
$y = -\sqrt{x}$

31. $y = x^2$
$y = (x - 3)^2$

32. $y = (x + 4)^2$
$y = x^2$

33. $y = x^2 + 6x$
$y = 4x$

34. $y = 3x^2 - 6x$
$y = 3x$

For Exercises 35–48, solve the system of nonlinear equations by the addition method.

35. $x^2 + y^2 = 13$
$x^2 - y^2 = 5$

36. $4x^2 - y^2 = 4$
$4x^2 + y^2 = 4$

37. $9x^2 + 4y^2 = 36$
$x^2 + y^2 = 9$

38. $x^2 + y^2 = 4$
$2x^2 + y^2 = 8$

39. $3x^2 + 4y^2 = 16$
$2x^2 - 3y^2 = 5$

40. $2x^2 - 5y^2 = -2$
$3x^2 + 2y^2 = 35$

41. $y = x^2 - 2$
$y = -x^2 + 2$

42. $y = x^2$
$y = -x^2 + 8$

43. $\dfrac{x^2}{4} + \dfrac{y^2}{9} = 1$
$x^2 + y^2 = 4$

44. $\dfrac{x^2}{16} + \dfrac{y^2}{4} = 1$
$x^2 + y^2 = 4$

45. $x^2 + 6y^2 = 9$
$\dfrac{x^2}{9} + \dfrac{y^2}{12} = 1$

46. $\dfrac{x^2}{10} + \dfrac{y^2}{10} = 1$
$2x^2 + y^2 = 11$

47. $x^2 - xy = -4$
$2x^2 - xy = 12$

48. $x^2 - xy = 3$
$2x^2 + xy = 6$

Expanding Your Skills

49. The sum of two numbers is 7. The sum of the squares of the numbers is 25. Find the numbers.

50. The sum of the squares of two numbers is 100. The sum of the numbers is 2. Find the numbers.

51. The sum of the squares of two numbers is 32. The difference of the squares of the numbers is 18. Find the numbers.

52. The sum of the squares of two numbers is 24. The difference of the squares of the numbers is 8. Find the numbers.

Graphing Calculator Exercises

For Exercises 53–56, use the *Intersect* feature or *Zoom* and *Trace* to approximate the solutions to the system.

53. $y = x + 3$ (Exercise 21)
$x^2 + y = 9$

54. $y = x - 2$ (Exercise 22)
$x^2 + y = 4$

55. $y = x^2$ (Exercise 29)
$y = -\sqrt{x}$

56. $y = -x^2$ (Exercise 30)
$y = -\sqrt{x}$

For Exercises 57–58, graph the system on a square viewing window. What can be said about the solution to the system?

57. $x^2 + y^2 = 4$
$y = x^2 + 3$

58. $x^2 + y^2 = 16$
$y = -x^2 - 5$

section

13.5 Nonlinear Inequalities and Systems of Inequalities

Concepts

1. Nonlinear Inequalities in Two Variables
2. Systems of Nonlinear Inequalities in Two Variables

1. Nonlinear Inequalities in Two Variables

In Section 9.5 we graphed the solution sets to linear inequalities in two variables, such as $y \le 2x + 1$. This involved graphing the related equation (a line in the xy-plane) and then shading the appropriate region above or below the line. See Figure 13-19. In this section, we will employ the same technique to solve nonlinear inequalities in two variables.

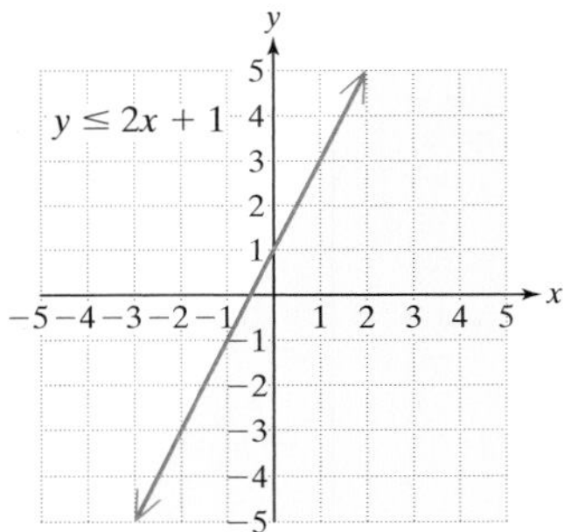

Figure 13-19

example 1 Graphing a Nonlinear Inequality in Two Variables

Graph the solution set of the inequality $x^2 + y^2 < 16$.

Solution:

The related equation $x^2 + y^2 = 16$ is a circle of radius 4, centered at the origin. Graph the related equation using a dashed curve because the points satisfying the equation $x^2 + y^2 = 16$ are not part of the solution to the strict inequality, $x^2 + y^2 < 16$. See Figure 13-20.

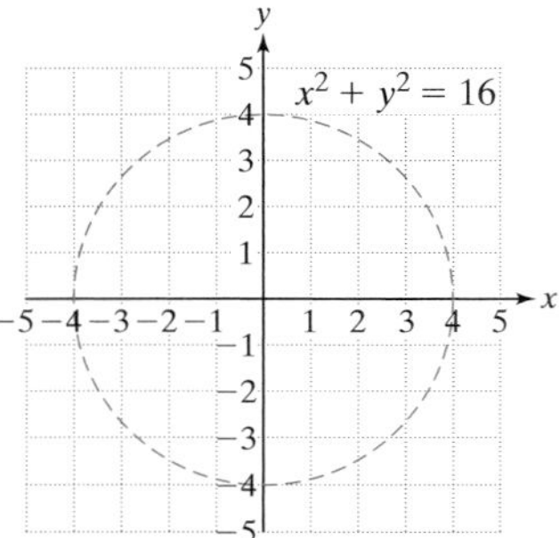

Figure 13-20

Notice that the dashed curve separates the xy-plane into two regions, one "inside" the circle, the other "outside" the circle. Select a test point from each region and test the point in the original inequality.

Test Point "Inside": (0, 0)	**Test Point "Outside": (4, 4)**
$x^2 + y^2 < 16$	$x^2 + y^2 < 16$
$(0)^2 + (0)^2 \overset{?}{<} 16$	$(4)^2 + (4)^2 \overset{?}{<} 16$
$0 \overset{?}{<} 16$ True	$32 \overset{?}{<} 16$ False

The inequality $x^2 + y^2 < 16$ is true at the test point $(0, 0)$. Therefore, the solution set is the region "inside" the circle. See Figure 13-21.

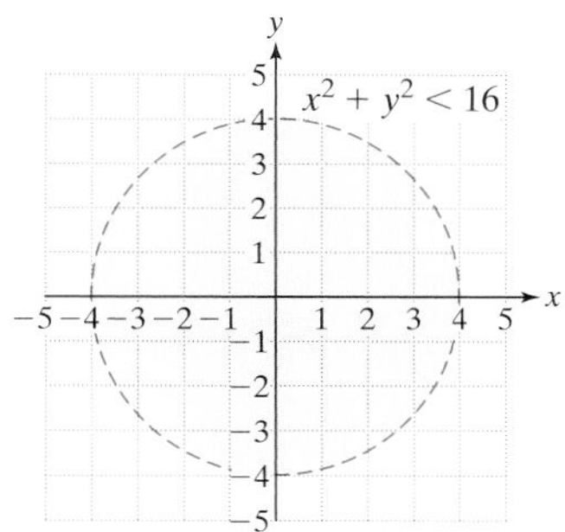

Figure 13-21

example 2 Graphing a Nonlinear Inequality in Two Variables

Graph the solution set of the inequality $9y^2 \geq 36 + 4x^2$.

Solution:

First graph the related equation, $9y^2 = 36 + 4x^2$. Notice that the equation can be written in the standard form of a hyperbola.

$$9y^2 = 36 + 4x^2$$

$$9y^2 - 4x^2 = 36 \qquad \text{Subtract } 4x^2 \text{ from both sides.}$$

$$\frac{y^2}{4} - \frac{x^2}{9} = 1 \qquad \text{Divide both sides by 36.}$$

Graph the hyperbola as a solid curve, because the original inequality includes equality. See Figure 13-22.

Figure 13-22

The hyperbola divides the xy-plane into three regions, a region above the upper branch, a region between the branches, and a region below the lower branch. Select a test point from each region.

$$9y^2 \geq 36 + 4x^2$$

Test: (0, 3)	**Test: (0, 0)**	**Test: (0, −3)**
$9(3)^2 \overset{?}{\geq} 36 + 4(0)^2$	$9(0)^2 \overset{?}{\geq} 36 + 4(0)^2$	$9(-3)^2 \overset{?}{\geq} 36 + 4(0)^2$
$81 \overset{?}{\geq} 36$ True	$0 \overset{?}{\geq} 36$ False	$81 \overset{?}{\geq} 36$ True

Shade the regions above the top branch and below the bottom branch of the hyperbola. See Figure 13-23.

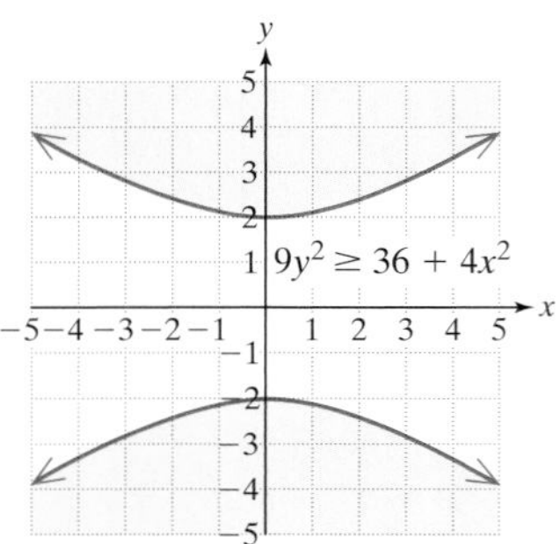

Figure 13-23

2. Systems of Nonlinear Inequalities in Two Variables

In Section 13.4 we solved systems of nonlinear equations in two variables. The solution set for such a system is the set of ordered pairs that satisfy both equations simultaneously. We will now solve systems of nonlinear inequalities in two variables. Similarly, the solution set is the set of all ordered pairs that simultaneously satisfy each inequality. To solve a system of inequalities, we will graph the solution to each individual inequality and then take the intersection of the solution sets.

example 3 **Graphing a System of Nonlinear Inequalities in Two Variables**

Graph the solution set.

$$y \geq e^x$$
$$y \leq -x^2 + 4$$

Solution:

The solution to $y \geq e^x$ is the set of points on and above the curve $y = e^x$. See Figure 13-24. The solution to $y \leq -x^2 + 4$ is the set of points on and below the parabola $y = -x^2 + 4$. See Figure 13-25.

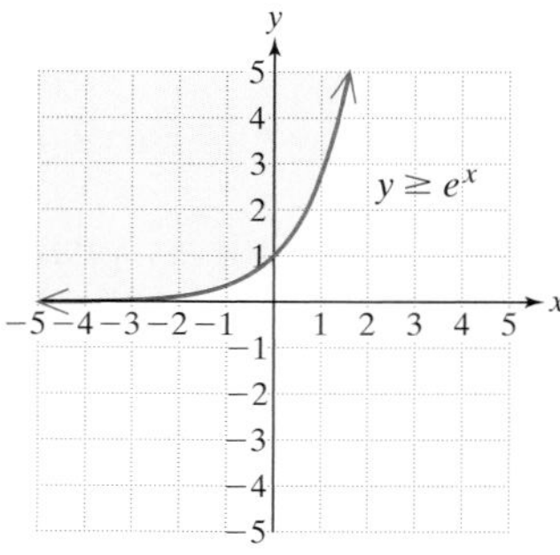

Figure 13-24

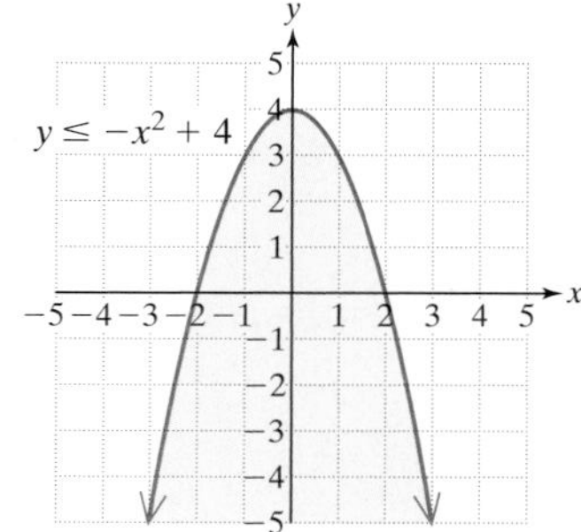

Figure 13-25

The solution to the system of inequalities is the intersection of the solution sets of the individual inequalities. See Figure 13-26.

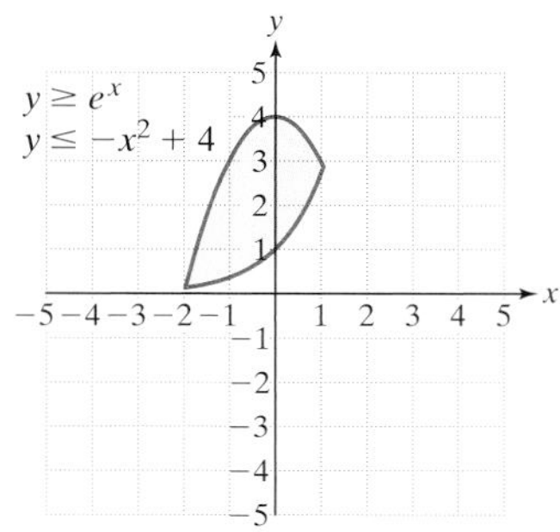

Figure 13-26

section 13.5 Practice Exercises

For Exercises 1–11, match the equation with its graph.

1. $y = 4x - 1$
2. $y = -4x^2$
3. $y = e^x$
4. $y = x^3$
5. $\dfrac{x^2}{4} + \dfrac{y^2}{9} = 1$
6. $\dfrac{x^2}{4} - \dfrac{y^2}{9} = 1$
7. $y = \left(\dfrac{2}{3}\right)^x$
8. $y = \dfrac{1}{x}$
9. $y = \log_2(x)$
10. $y = \sqrt{x}$
11. $(x + 2)^2 + (y - 1)^2 = 4$

a.

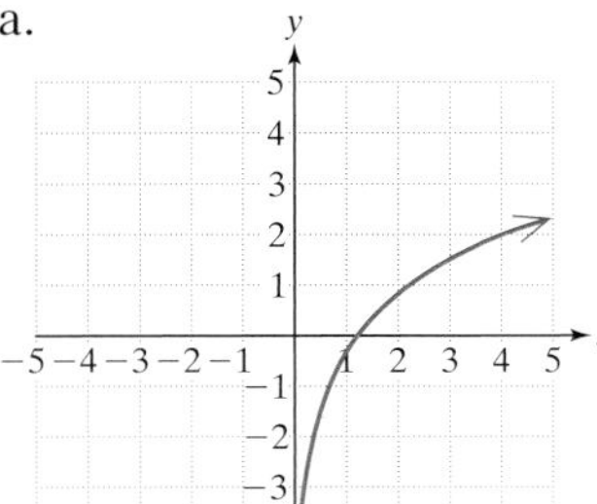

b.

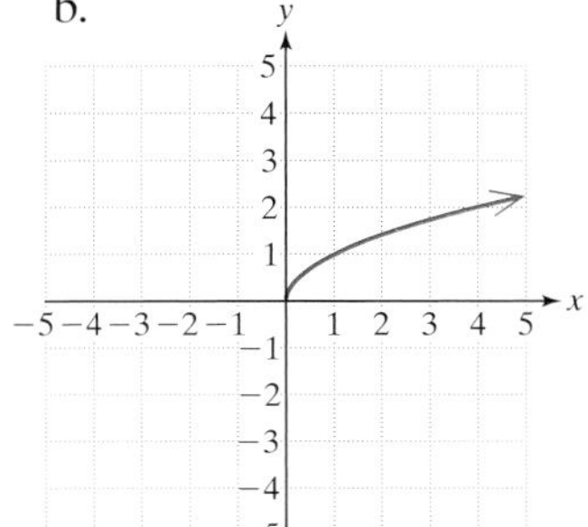

c.

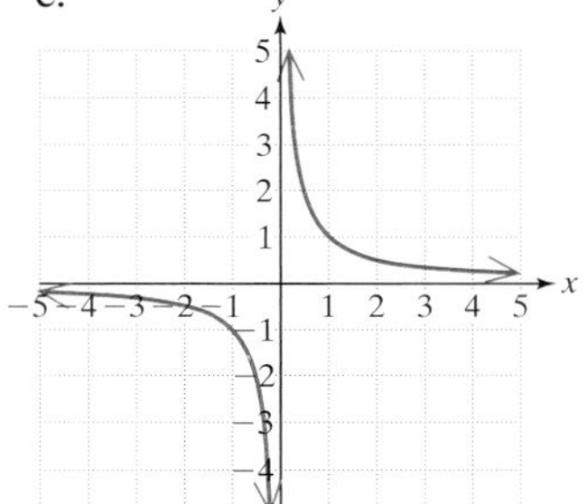

d.

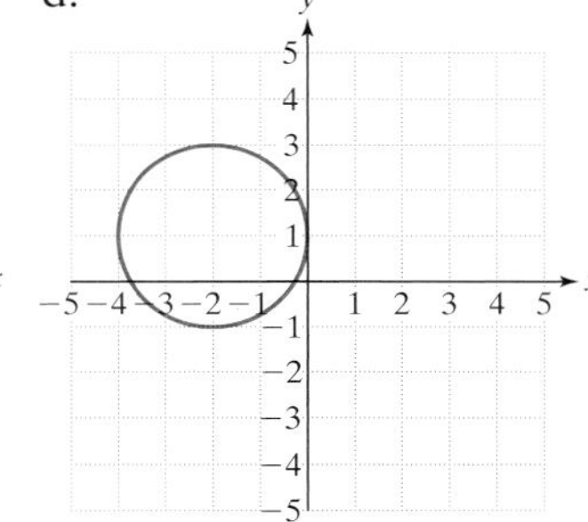

e.

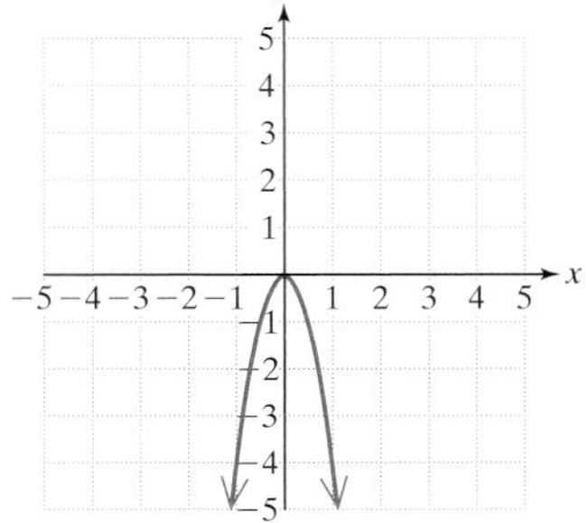

f.

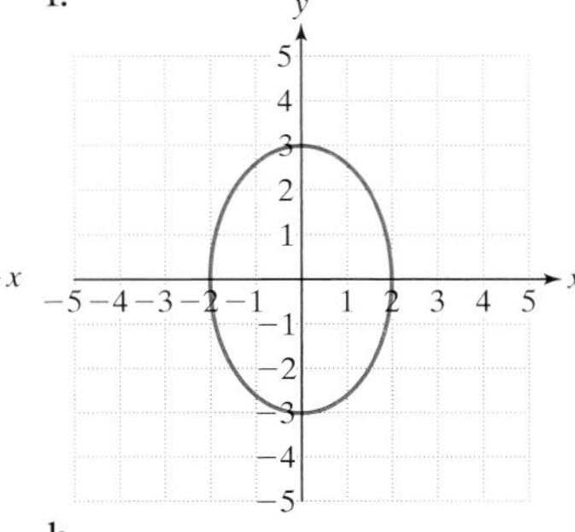

g.

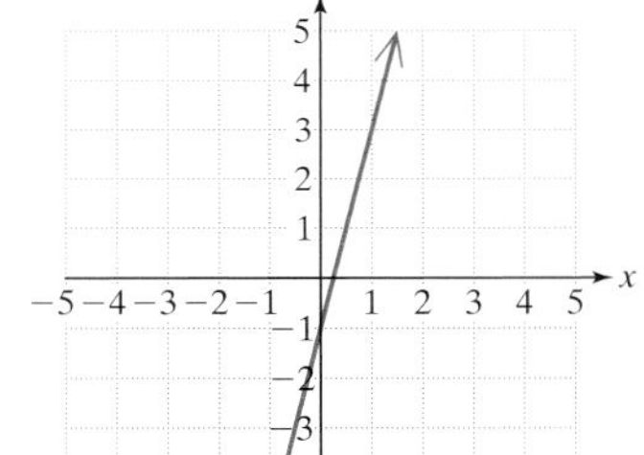

h.

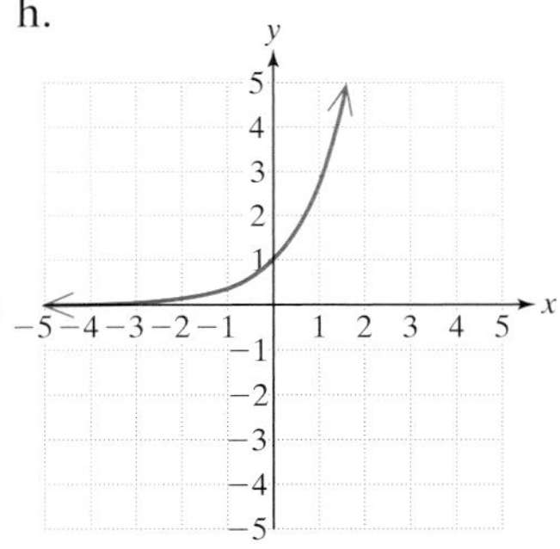

i.

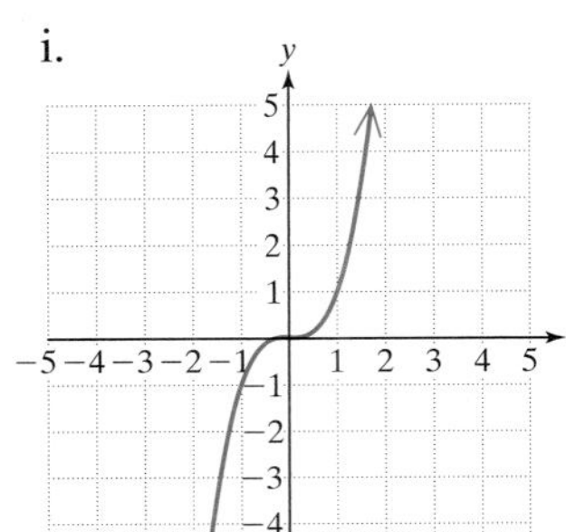

j.

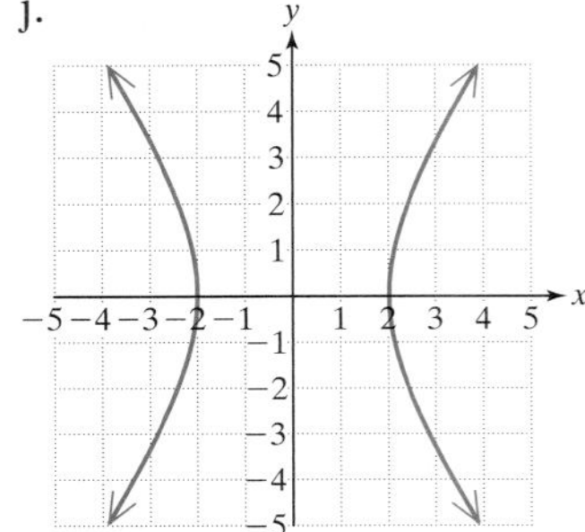

k.

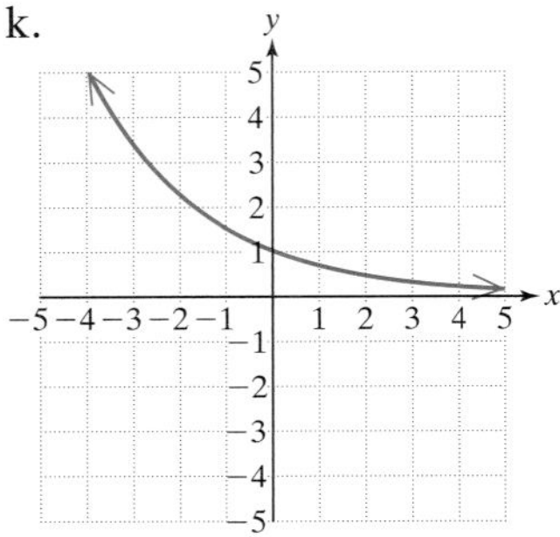

12. True or false? The point (1, 1) satisfies the inequality $-x^2 + y^3 > 1$.

13. True or false? The point (4, −2) satisfies the inequality $4x^2 - 2x + 1 + y^2 < 3$.

14. True or false? The point (5, 4) satisfies the system of inequalities:

$$\frac{x^2}{36} + \frac{y^2}{25} < 1$$
$$x^2 + y^2 \geq 4$$

15. True or false? The point (1, −2) satisfies the system of inequalities:

$$y < x^2$$
$$y > x^2 - 4$$

16. True or false? The point (−3, 5) satisfies the system of inequalities:

$$(x + 3)^2 + (y - 5)^2 \leq 2$$
$$y > x^2$$

17. a. Graph the solution set: $x^2 + y^2 \leq 9$.
    b. Describe the solution set for the inequality $x^2 + y^2 \geq 9$.
    c. Describe the solution set of the equation $x^2 + y^2 = 9$.

18. a. Graph the solution set: $\frac{x^2}{4} + \frac{y^2}{9} \geq 1$.
    b. Describe the solution set for the inequality $\frac{x^2}{4} + \frac{y^2}{9} \leq 1$.
    c. Describe the solution set for the equation $\frac{x^2}{4} + \frac{y^2}{9} = 1$.

19. a. Graph the solution set: $y \geq x^2 + 1$.
    b. How would the solution change for the strict inequality $y > x^2 + 1$?

20. a. Graph the solution set: $\frac{x^2}{4} - \frac{y^2}{9} \leq 1$.
    b. How would the solution change for the strict inequality $\frac{x^2}{4} - \frac{y^2}{9} < 1$?

For Exercises 21–36, graph the solution set.

21. $2x + y \geq 1$
22. $3x + 2y \geq 6$
23. $x \leq y^2$
24. $y \leq -x^2$
25. $(x - 1)^2 + (y + 2)^2 > 9$
26. $(x + 1)^2 + (y - 4)^2 > 1$
27. $x^2 + y^2 + 2x - 24 < 0$
28. $x^2 + y^2 - 6y < 0$
29. $9x^2 - y^2 > 9$
30. $y^2 - 4x^2 \leq 4$
31. $x^2 + 16y^2 \leq 16$
32. $4x^2 + y^2 \leq 4$
33. $y \leq \ln x$
34. $y \leq \log x$
35. $y > 5^x$
36. $y \geq \left(\frac{1}{3}\right)^x$

For Exercises 37–50, graph the solution set to the system of inequalities.

37. $y \leq \sqrt{x}$, $x \geq 1$
38. $y \geq \sqrt{x}$, $x \geq 0$
39. $\frac{x^2}{36} + \frac{y^2}{25} < 1$, $x^2 + y^2 \geq 4$
40. $x^2 - y^2 \geq 1$, $x \leq 0$
41. $y < x^2$, $y > x^2 - 4$
42. $y^2 - x^2 \geq 1$, $y \geq 0$

43. $y \leq \frac{1}{x}$
$y \geq 0$
$y \leq x$

44. $y \geq x^3$
$y \leq 8$
$x \geq 0$

45. $x^2 + y^2 \geq 25$
$x^2 + y^2 \leq 9$

46. $\frac{x^2}{4} + \frac{y^2}{25} \geq 1$
$x^2 + \frac{y^2}{4} \leq 1$

47. $x < -(y-1)^2 + 3$
$x + y > 2$

48. $x > (y-2)^2 + 1$
$x - y < 1$

49. $x^2 + y^2 \leq 25$
$y \leq \frac{4}{3}x$
$y \geq -\frac{4}{3}x$

50. $y \leq e^x$
$y \geq 1$
$x \leq 2$

Expanding Your Skills

For Exercises 51–54, graph the compound inequalities.

51. a. $x^2 + y^2 \leq 36$ and $x + y \geq 0$
b. $x^2 + y^2 \leq 36$ or $x + y \geq 0$

52. a. $y \leq -x^2 + 4$ and $y \geq x^2 - 4$
b. $y \leq -x^2 + 4$ or $y \geq x^2 - 4$

53. a. $y + 1 \geq x^2$ and $y + 1 \leq -x^2$
b. $y + 1 \geq x^2$ or $y + 1 \leq -x^2$

54. a. $(x+2)^2 + (y+3)^2 \leq 4$ and $x \geq y^2$
b. $(x+2)^2 + (y+3)^2 \leq 4$ or $x \geq y^2$

chapter 13 Summary

Section 13.1—Distance Formula and Circles

Key Concepts:	Examples:
The distance between two points (x_1, y_1) and (x_2, y_2) is $d = \sqrt{(x_2 - x_1)^2 + (y_2 - y_1)^2}$	**Find the distance between two points:** $(5, -2)$ and $(-1, -6)$ $d = \sqrt{(-1-5)^2 + [-6-(-2)]^2}$ $= \sqrt{(-6)^2 + (-4)^2}$ $= \sqrt{36 + 16}$ $= \sqrt{52} = 2\sqrt{13}$
The standard equation of a circle with center (h, k) and radius r is $(x-h)^2 + (y-k)^2 = r^2$	**Find the center and radius of the circle:** $x^2 + y^2 - 8x + 6y = 0$ $(x^2 - 8x + 16) + (y^2 + 6y + 9) = 16 + 9$ $(x-4)^2 + (y+3)^2 = 25$ The center is $(4, -3)$ and the radius is 5.

The midpoint between two points is found using the formula:

$$\left(\frac{x_1 + x_2}{2}, \frac{y_1 + y_2}{2}\right)$$

Find the midpoint between (−3, 1) and (5, 7)

$$\left(\frac{-3 + 5}{2}, \frac{1 + 7}{2}\right) = (1, 4)$$

Key Terms:

center of a circle
circle
midpoint formula
radius
standard equation of a circle

Section 13.2—More on the Parabola

Key Concepts:

A parabola is the set of points in a plane that are equidistant (the same distance) from a fixed line (called the directrix) and a fixed point (called the focus) not on the directrix.

The standard form of an equation of a parabola with vertex (h, k) and vertical axis of symmetry is

$$y = a(x - h)^2 + k, \text{ where } a = \frac{1}{4p}$$

The focus is located at $(h, k + p)$.

The equation of the directrix is $y = k - p$.

If $a > 0$, then the parabola opens upward, and if $a < 0$, the parabola opens downward.

The standard form of an equation of a parabola with vertex (h, k) and horizontal axis of symmetry is

$$x = a(y - k)^2 + h, \text{ where } a = \frac{1}{4p}$$

The focus is located at $(h + p, k)$.

The equation of the directrix is $x = h - p$.

If $a > 0$, then the parabola opens to the right, and if $a < 0$, the parabola opens to the left.

Key Terms:

axis of symmetry
conic sections
directrix
focus
parabola
vertex

Examples:

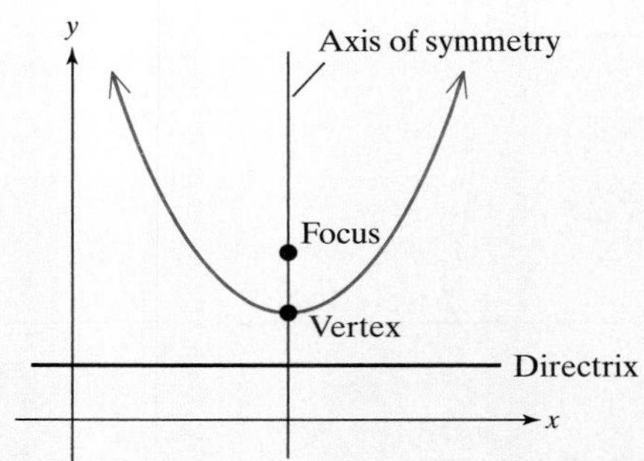

Given the parabola, $x = -\frac{1}{4}y^2 + 1$, **determine the coordinates of the vertex and the focus and write an equation of the directrix.**

$$x = -\frac{1}{4}(y - 0)^2 + 1$$

The vertex is $(1, 0)$.

$-\frac{1}{4} = \frac{1}{4p}$, therefore $p = -1$.

The focus is $(0, 0)$.

The directrix is $x = 2$.

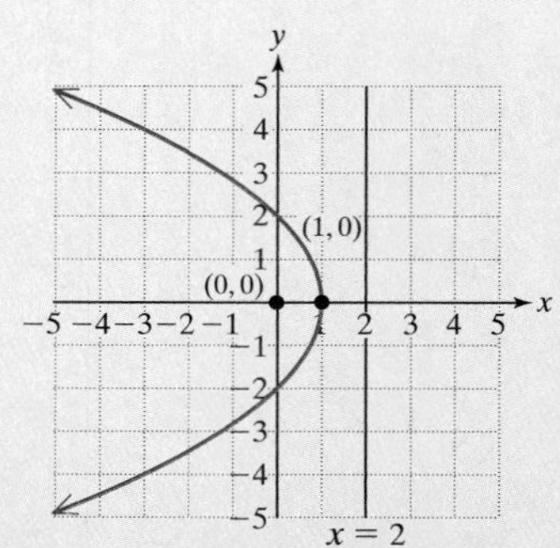

SECTION 13.3—ELLIPSE AND HYPERBOLA

KEY CONCEPTS:

An ellipse is the set of all points (x, y) such that the sum of the distances between (x, y) and two distinct points is a constant. The fixed points are called the foci of the ellipse.

Standard Forms of an Equation of an Ellipse
Let a and b represent positive real numbers such that $a > b$.

Horizontal Major Axis:
The standard form of an equation of an ellipse with a horizontal major axis and center at (h, k) is given by

$$\frac{(x-h)^2}{a^2} + \frac{(y-k)^2}{b^2} = 1$$

Vertical Major Axis:
The standard form of an equation of an ellipse with a vertical major axis and center at (h, k) is given by

$$\frac{(x-h)^2}{b^2} + \frac{(y-k)^2}{a^2} = 1$$

A hyperbola is the set of all points (x, y) such that the difference of the distances between (x, y) and two distinct points is a constant. The fixed points are called the foci of the hyperbola.

Standard Forms of an Equation of a Hyperbola
Let a and b represent positive real numbers.

Horizontal Transverse Axis:
The standard form of an equation of a hyperbola with a horizontal transverse axis and center at (h, k) is given by

$$\frac{(x-h)^2}{a^2} - \frac{(y-k)^2}{b^2} = 1$$

EXAMPLES:

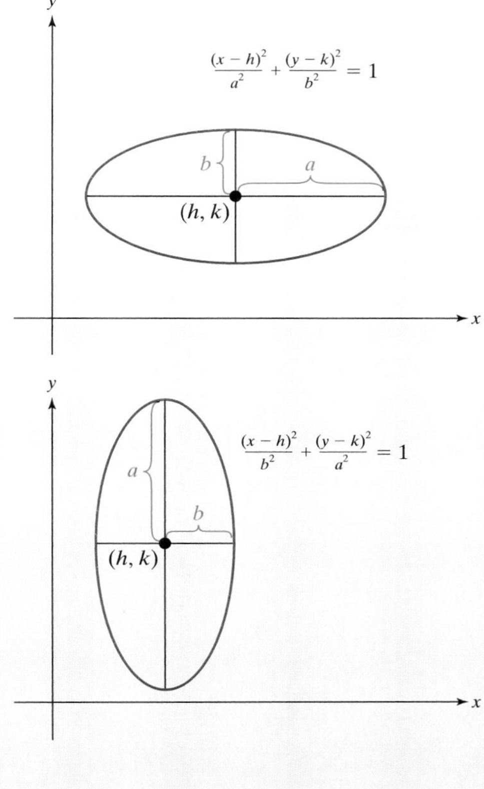

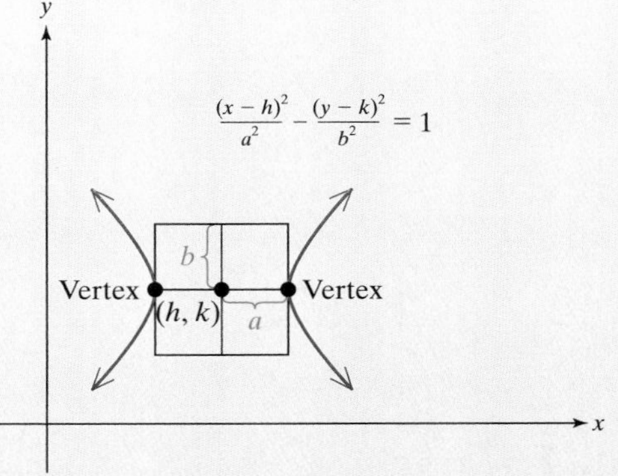

Vertical Transverse Axis:
The standard form of an equation of a hyperbola with a vertical transverse axis and center at (h, k) is given by

$$\frac{(y-k)^2}{a^2} - \frac{(x-h)^2}{b^2} = 1$$

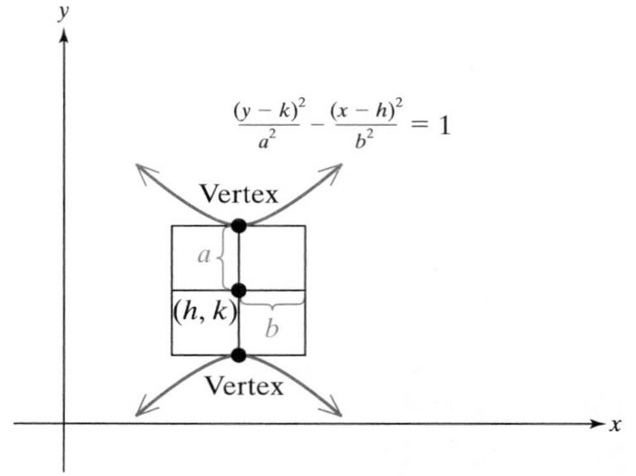

Key Terms:

asymptotes
center
conjugate axis
covertices
ellipse
foci
hyperbola
major axis
minor axis
transverse axis
vertices

Section 13.4—Nonlinear Systems of Equations in Two Variables

Key Concepts:

A nonlinear system of equations can be solved by graphing or by the substitution method.

$$2x^2 + y^2 = 15$$
$$x^2 - y = 0$$

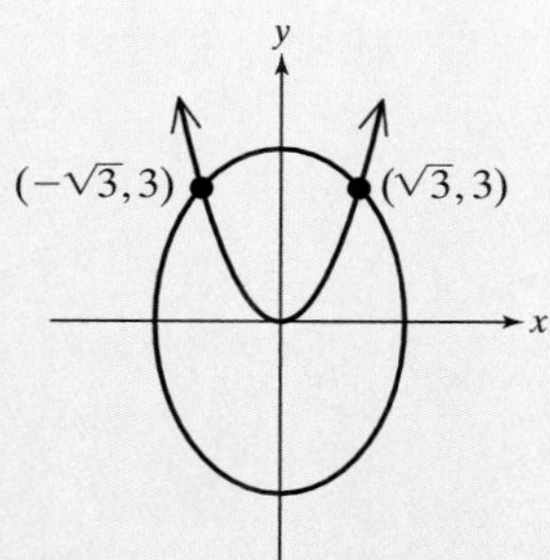

Examples:

Solve the nonlinear system:

A $\quad 2x^2 + y^2 = 15$

B $\quad x^2 - y = 0 \qquad$ Solve for y: $y = x^2$

A $\quad 2x^2 + (x^2)^2 = 15 \qquad$ Substitute in first equation.

$$2x^2 + x^4 = 15$$
$$x^4 + 2x^2 - 15 = 0$$
$$(x^2 + 5)(x^2 - 3) = 0$$

$x^2 + 5 = 0 \quad$ or $\quad x^2 - 3 = 0$

$\cancel{x^2 = -5} \quad$ or $\quad x^2 = 3$

$x = \pm\sqrt{3}$

If $x = \sqrt{3}$, then $y = (\sqrt{3})^2 = 3$.

If $x = -\sqrt{3}$, then $y = (-\sqrt{3})^2 = 3$.

Points of intersection are: $(\sqrt{3}, 3)$ and $(-\sqrt{3}, 3)$.

A nonlinear system may also be solved using the addition method when the equations share *like* terms.

Solve the nonlinear system:

$$\begin{aligned} 2x^2 + y^2 &= 4 \xrightarrow{\text{mult. by } -5} & -10x^2 - 5y^2 &= -20 \\ 3x^2 + 5y^2 &= 13 \longrightarrow & 3x^2 + 5y^2 &= 13 \\ & & -7x^2 \quad &= -7 \end{aligned}$$

$$\frac{-7x^2}{-7} = \frac{-7}{-7}$$

$$x^2 = 1 \longrightarrow x = \pm 1$$

If $x = 1$, $2(1)^2 + y^2 = 4$.

$$y^2 = 2$$

$$y = \pm\sqrt{2}$$

If $x = -1$, $2(-1)^2 + y^2 = 4$.

$$y^2 = 2$$

$$y = \pm\sqrt{2}$$

The points of intersection are

$(1, \sqrt{2}), (1, -\sqrt{2}), (-1, \sqrt{2})$, and $(-1, -\sqrt{2})$

KEY TERM:

nonlinear system of equations

SECTION 13.5—NONLINEAR INEQUALITIES AND SYSTEMS OF INEQUALITIES

KEY CONCEPTS:

Graph a nonlinear inequality using the test point method. That is, graph the related equation. Then choose test points in each region to determine where the inequality is true.

EXAMPLES:

Graph the inequalities.

1. $y \geq x^2$

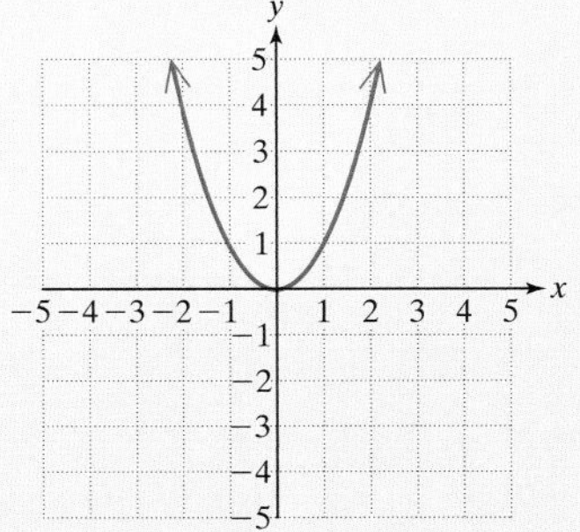

2. $x^2 + y^2 < 4$

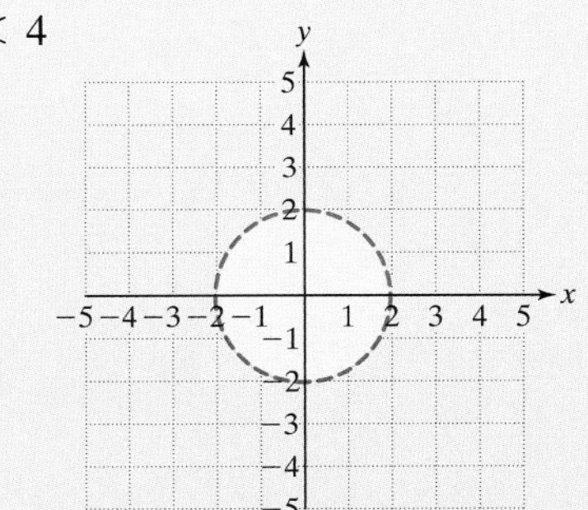

Graph a system of nonlinear inequalities by finding the intersection of the solution sets. That is, graph the solution set for each individual inequality, then take the intersection.

Graph the sytem of inequalities:

$x \geq 0, y \geq x^2$ and $x^2 + y^2 < 4$

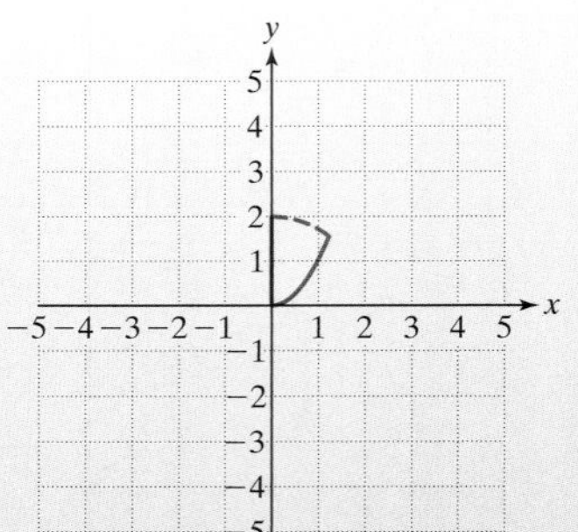

chapter 13 REVIEW EXERCISES

Section 13.1

For Exercises 1–4, find the distance between the two points using the distance formula.

1. $(-6, 3)$ and $(0, 1)$
2. $(4, 13)$ and $(-1, 5)$
3. Find x such that $(x, 5)$ is 5 units from $(2, 9)$.
4. Find x such that $(-3, 4)$ is 3 units from $(x, 1)$.

Points are said to be collinear if they all lie on the same line. If three points are collinear then the distance between the outer most points will equal the sum of the distances between the middle point and each of the outer points. For Exercises 5–6, determine if the three points are collinear.

5. $(-2, -3), (1, 3)$, and $(5, 11)$
6. $(-2, 11), (0, 5)$, and $(4, -7)$

For Exercises 7–10, find the center and the radius of the circle.

7. $(x - 12)^2 + (y - 3)^2 = 16$
8. $(x + 7)^2 + (y - 5)^2 = 81$
9. $(x + 3)^2 + (y + 8)^2 = 20$
10. $(x - 1)^2 + (y + 6)^2 = 32$

11. A stained glass window is in the shape of a circle with a 16-in. diameter. Find an equation of the circle relative to the origin for each of the following graphs.

a.

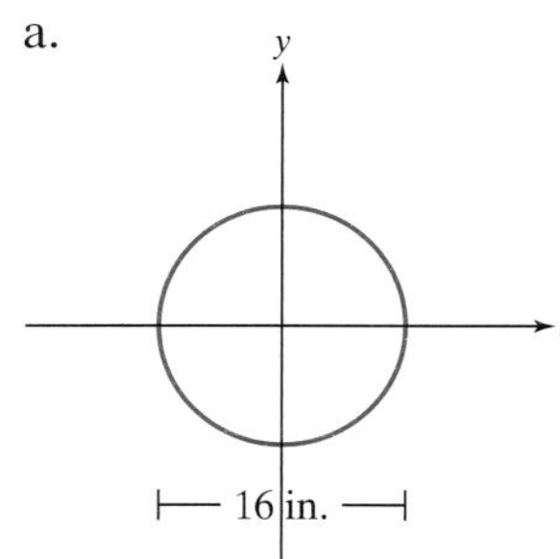

b.

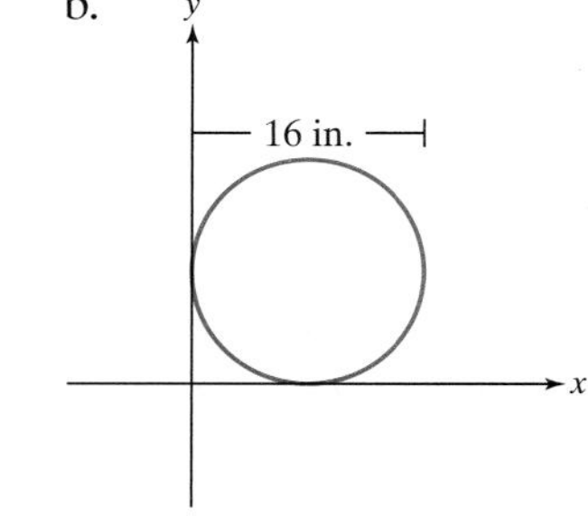

For Exercises 12–15, write the equation of the circle in standard form by completing the square.

12. $x^2 + y^2 + 12x - 10y + 51 = 0$
13. $x^2 + y^2 + 4x + 16y + 60 = 0$
14. $x^2 + y^2 - x - 4y + \frac{1}{4} = 0$
15. $x^2 + y^2 - 6x - \frac{2}{3}y + \frac{1}{9} = 0$
16. Write an equation of a circle with center at the origin and a diameter of 7 m.
17. Write an equation of a circle with center at $(0, 2)$ and a diameter of 6 m.

18. Find the midpoint of the line segment between the two points $(-13, 12)$ and $(4, -18)$.

19. Find the midpoint of the line segment between the two points $(1.2, -3.7)$ and $(-4.1, -8.3)$.

Section 13.2

For Exercises 20–23, determine whether the axis of symmetry is vertical or horizontal and if the parabola opens upward, downward, left, or right.

20. $y = -2(x - 3)^2 + 2$

21. $x = 3(y - 9)^2 + 1$

22. $x = -(y + 4)^2 - 8$

23. $y = (x + 3)^2 - 10$

For Exercises 24–27, determine the coordinates of the vertex and focus and the equation of the directrix. Then use this information to graph the parabola.

24. $x = \frac{1}{8}(y - 1)^2$

25. $y = (x + 2)^2$

26. $y = -\frac{1}{4}x^2$

27. $x = 2y^2 - 1$

For Exercises 28–31, write the equation in standard form: $y = a(x - h)^2 + k$ or $x = a(y - k)^2 + h$. Then identify the vertex, focus, and directrix.

28. $y = x^2 - 6x + 5$

29. $x = y^2 + 4y + 2$

30. $x = -4y^2 + 4y$

31. $y = -2x^2 - 2x$

Section 13.3

32. For the ellipse shown in the graph, label the center, foci, vertices, covertices, major axis, and minor axis.

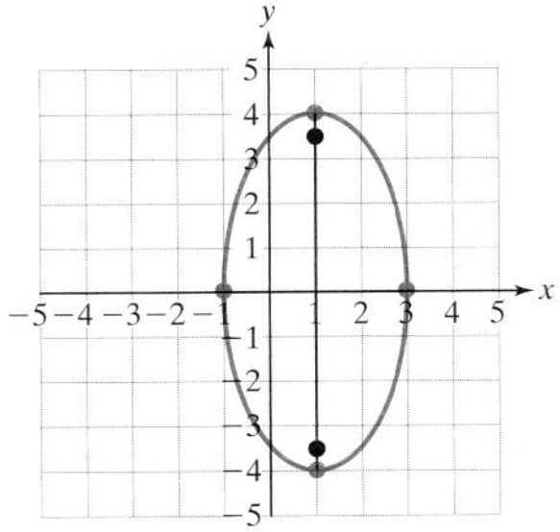

Figure for Exercise 32

For Exercises 33–36, identify the center of the ellipse and determine whether the major axis is horizontal or vertical.

33. $\frac{x^2}{9} + \frac{y^2}{25} = 1$

34. $\frac{x^2}{12} + \frac{y^2}{8} = 1$

35. $\frac{(x - 1)^2}{36} + \frac{y^2}{9} = 1$

36. $\frac{x^2}{1} + \frac{(y + 2)^2}{6} = 1$

37. For the hyperbola pictured, label the center, foci, vertices, and transverse axis. Draw the asymptotes.

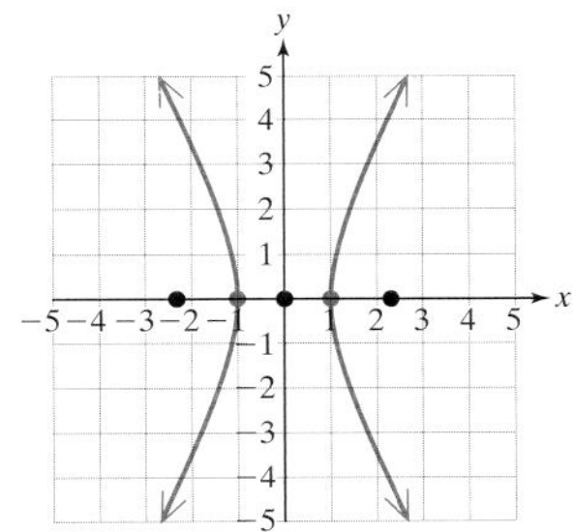

Figure for Exercise 37

For Exercises 38–41, identify the center of the hyperbola and determine whether the transverse axis is horizontal or vertical.

38. $\frac{x^2}{12} - \frac{y^2}{16} = 1$

39. $\frac{y^2}{9} - \frac{x^2}{9} = 1$

40. $\frac{(y - 3)^2}{24} - \frac{(x + 2)^2}{10} = 1$

41. $\frac{(x + 1)^2}{6} - \frac{y^2}{16} = 1$

For Exercises 42–45, identify as an ellipse or a hyperbola. Then graph, labeling the center and the vertices.

42. $\frac{x^2}{4} - \frac{y^2}{9} = 1$

43. $\frac{x^2}{16} + \frac{y^2}{9} = 1$

44. $\frac{x^2}{4} + \frac{(y+2)^2}{1} = 1$

45. $\frac{(y-1)^2}{1} - \frac{(x-1)^2}{16} = 1$

Section 13.4

For Exercises 46–49,

a. Identify each equation as a line, a parabola, a circle, an ellipse, or a hyperbola.

b. Graph both equations on the same coordinate system.

c. Solve the system analytically and verify the answers from the graph.

46. $3x + 2y = 10$
$y = x^2 - 5$

47. $4x + 2y = 10$
$y = x^2 - 10$

48. $\frac{x^2}{9} + \frac{y^2}{9} = 1$
$2x + y = 3$

49. $x^2 + y^2 = 16$
$x - 2y = 8$

For Exercises 50–55, solve the system of nonlinear equations using either the substitution method or the addition method.

50. $x^2 + 2y^2 = 8$
$2x - y = 2$

51. $x^2 + 4y^2 = 29$
$x - y = -4$

52. $x - y = 4$
$y^2 = 2x$

53. $y = x^2$
$6x^2 - y^2 = 8$

54. $x^2 + y^2 = 10$
$x^2 + 9y^2 = 18$

55. $x^2 + y^2 = 61$
$x^2 - y^2 = 11$

Section 13.5

For Exercises 56–63, graph the solution set to the inequality.

56. $3x + y \le 4$

57. $x - 2y \ge -2$

58. $\frac{x^2}{16} + \frac{y^2}{81} < 1$

59. $\frac{x^2}{25} + \frac{y^2}{4} > 1$

60. $(x-3)^2 + (y+1)^2 \ge 9$

61. $(x+2)^2 + (y+1)^2 \le 4$

62. $y > (x-1)^2$

63. $y > x^2 - 1$

For Exercises 64–67, graph the solution set to the system of nonlinear inequalities.

64. $y \ge 4^x$
$x^2 + y^2 \le 4$

65. $y \le 3^x$
$x^2 + y^2 \le 9$

66. $\frac{x^2}{4} - y^2 \le 1$
$x^2 + y^2 \le 9$

67. $\frac{y^2}{9} - x^2 \ge 1$
$x^2 + (y-2)^2 \le 4$

---

chapter 13 TEST

1. Determine the coordinates of the vertex and focus and write an equation of the directrix.

$$x = -\frac{1}{4}(y+6)^2 + 3$$

2. Write the equation in standard form, $y = a(x-h)^2 + k$, and graph.

$$y = x^2 - 4x + 5$$

3. Use the distance formula to find the distance between the two points (5, 19) and (−2, 13).

4. Determine if the three points, (3, 4), (−1, 1), and (6, 0) are vertices of a right triangle.

5. Determine the center and radius of the circle.

$$\left(x - \frac{5}{6}\right)^2 + \left(y + \frac{1}{3}\right)^2 = \frac{25}{49}$$

6. Determine the center and radius of the circle. $x^2 + y^2 - 4y - 5 = 0$.

7. Let (0, 4) be the center of a circle that passes through the point (−2, 5).

   a. What is the radius of the circle?

   b. Write the equation of the circle in standard form.

8. Find the midpoint of the line segment between the two points $(21, -15)$ and $(5, 32)$.

9. Identify the center of the ellipse and determine whether the major axis is horizontal or vertical.

$$\frac{(x-3)^2}{8} + \frac{y^2}{14} = 1$$

10. Graph the hyperbola and identify the center and vertices.

$$\frac{y^2}{1} - \frac{x^2}{4} = 1$$

11. Solve the systems and identify the correct graph of the equations:

a. $16x^2 + 9y^2 = 144$
   $4x - 3y = -12$

b. $x^2 + 4y^2 = 4$
   $4x - 3y = -12$

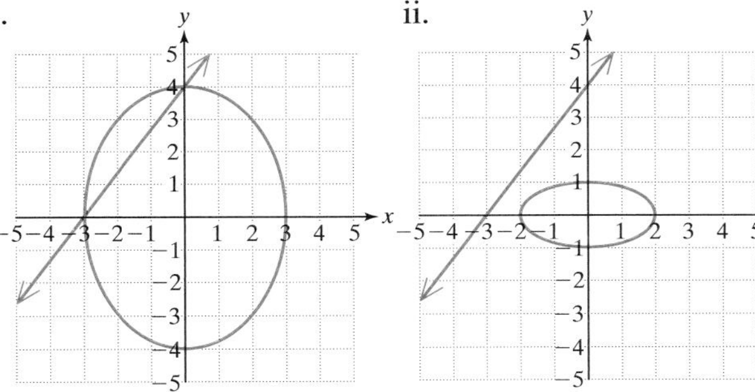

12. Describe the circumstances in which a nonlinear system of equations can be solved using the addition method.

13. Solve the system using either the substitution method or the addition method.

$$25x^2 + 4y^2 = 100$$
$$25x^2 - 4y^2 = 100$$

For Exercises 14–17, graph the solution set.

14. $x \le y^2 + 1$

15. $y \ge -\frac{1}{3}x + 1$

16. $x \le y^2 + 1$
    $y \ge -\frac{1}{3}x + 1$

17. $y \le \sqrt{x}$
    $y > x - 2$
    $x \ge 0$

18. Graph the ellipse $\frac{(x+2)^2}{9} + \frac{(y+1)^2}{4} = 1$, and label the center, the vertices, and covertices.

CUMULATIVE REVIEW EXERCISES, CHAPTERS 1–13

1. Solve the equation:
   $5(2y - 1) = 2y - 4 + 8y - 1$

2. Solve the inequality. Graph the solution and write the solution in interval notation.

$$4(x-1) + 2 > 3x + 8 - 2x$$

3. The product of two integers is 150. If one integer is 5 less than twice the other, find the integers.

4. For $5y - 3x - 15 = 0$:

   a. Find the x- and y-intercepts.

   b. Find the slope.

   c. Graph the line.

5. The amount spent by the Philippines on arms imports each year between 1985 and 1989 is linear as shown in the graph (see figure). Let x represent the year, where $x = 0$ corresponds to 1985. Let y represent the amount spent on arms imports (in millions of dollars).

a. Use any two data points to find the slope of the line.

b. Find an equation of the line. Write the answer in slope-intercept form.

c. Use the linear model found in part (b) to approximate the amount spent on arms imports by the Philippines in the year 1990.

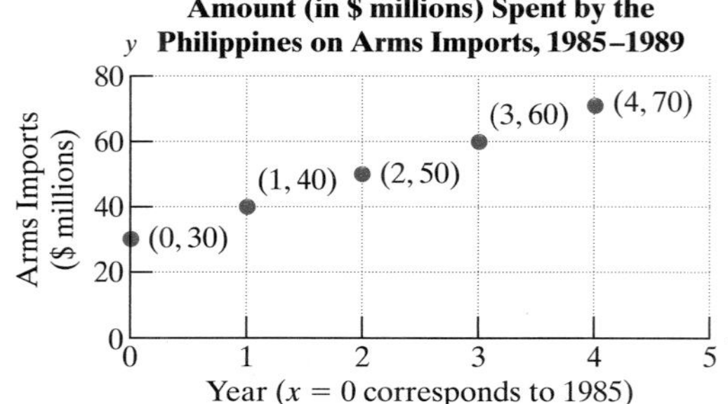

Source: Statistical Abstract of the World.

Figure for Exercise 5

6. A collection of dimes and quarters has a total value of \$2.45. If there are 17 coins, how many of each type are there?

7. Solve the system.

$$\begin{aligned} x + y \qquad &= -1 \\ 2x \qquad - z &= 3 \\ y + 2z &= -1 \end{aligned}$$

8. a. Given the matrix $\mathbf{A} = \left[\begin{array}{cc|c} 1 & -2 & -8 \\ 0 & 3 & 6 \end{array}\right]$, write the matrix obtained by multiplying the elements in the second row by $\frac{1}{3}$.

   b. Using the matrix obtained from part (a), write the matrix obtained by multiplying the second row by 2 and adding the result to the first row.

9. Solve the following system using Cramer's rule.

$$\begin{aligned} 4x - 2y &= 7 \\ -3x + 5y &= 0 \end{aligned}$$

10. For $f(x) = 3x - x^2 - 12$, find the function values: $f(0), f(-1), f(2)$, and $f(4)$.

11. For $g = \{(2, 5), (8, -1), (3, 0), (-5, 5)\}$ find the function values: $g(2), g(8), g(3)$, and $g(-5)$.

12. The quantity z varies jointly as y and as the square of x. If z is 80 when x is 5 and y is 2, find z when $x = 2$ and $y = 5$.

13. a. Find the value of the expression $x^3 + x^2 + x + 1$ for $x = -2$.

    b. Factor the expression $x^3 + x^2 + x + 1$ and find the value when x is -2.

    c. Compare the values for parts (a) and (b).

14. Solve the radical equations.

    a. $\sqrt{2x - 5} = -3$

    b. $\sqrt[3]{2x - 5} = -3$

15. Perform the indicated operations with complex numbers.

    a. $6i(4 + 5i)$ b. $\dfrac{3}{4 - 5i}$

16. An automobile starts from rest and accelerates at a constant rate for 10 sec. The distance, $d(t)$, in feet traveled by the car is given by

    $d(t) = 4.4t^2$ where $0 \le t \le 10$ is the time in seconds.

    a. How far has the car traveled after 2, 3, and 4 sec, respectively?

    b. How long will it take for the car to travel 281.6 ft?

17. Solve the equation $125w^3 + 1 = 0$ by factoring and using the quadratic formula. (*Hint:* You will find one real solution and two imaginary solutions.)

18. Solve the rational equation.

$$\frac{x}{x + 2} - \frac{3}{x - 1} = \frac{1}{x^2 + x - 2}$$

19. Solve the inequality and write the answer in interval notation.

$$|x - 9| - 3 < 7$$

20. Write the expression in logarithmic form: $8^{5/3} = 32$.

21. Find the vertex of $f(x) = x^2 + 10x - 11$ by completing the square.

22. Graph the quadratic function defined by $g(x) = -x^2 - 2x + 3$.
    a. Label the x-intercepts.
    b. Label the y-intercept.
    c. Label the vertex.

23. Write an equation representing the set of all points 4 units from the point (0, 5).

24. Can a circle and a parabola intersect in only one point? Explain.

25. Solve the system of nonlinear equations.

$$x^2 + y^2 = 16$$
$$y = -x^2 - 4$$

26. Graph the solution set:

$$y^2 - x^2 < 1$$

27. Graph the solution set to this system:

$$y \geq \left(\frac{1}{2}\right)^x$$
$$x \leq 0$$

28. Factor completely:

$$x^2 - y^2 - 6x - 6y$$

chapter

14

Sequences, Series, Counting, and Probability

To supplement their revenue income, many states have instituted games of chance called lotteries. In Pennsylvania for example, the game Match 6 Lotto calls for the player to select a group of 6 numbers taken from 1–49. If a person matches three, four, five, or all six numbers, the player wins a cash prize. The amount of the prize depends on the number of numbers matched, the amount of money collected by the state in ticket sales, and the number of other players who also won.

The state makes money regardless of the number of winners. The payoffs are computed so that the total revenue from ticket sales exceeds the total payout in cash prizes.

In this chapter, we will learn how to compute the number of possible outcomes in such a game of chance and the probability of winning. For more information about lotteries and expected payoffs, see the Technology Connections in MathZone at

www.mhhe.com/miller_oneill

section

14.1 Sequences and Series

Concepts

1. Definition of Finite and Infinite Sequences
2. Listing the Terms of a Sequence
3. Finding a Formula for the *n*th Term of a Sequence
4. Applications of Sequences
5. Definition of a Series
6. Summation Notation
7. Writing a Sum Using Sigma Notation

1. Definition of Finite and Infinite Sequences

In everyday life we think of a sequence as a set of events or items with some order or pattern. In mathematics, a sequence is a list of terms that correspond with the set of positive integers. For example, the sequence

$$1, 4, 9, 16, 25$$

represents the squares of the first five positive integers. This sequence has a finite number of terms and is called a *finite sequence*. The sequence

$$1, 4, 9, 16, 25, \ldots$$

represents the squares of *all* positive integers. This sequence continues indefinitely and is called an *infinite sequence*.

Because the terms in a sequence are related to the set of positive integers, we give a formal definition of finite and infinite sequences using the language of functions.

Definition of Finite and Infinite Sequences

An **infinite sequence** is a function whose domain is the set of positive integers.

A **finite sequence** is a function whose domain is the set of the first n positive integers.

For any positive integer n (the independent variable), the value of the sequence (the dependent variable) is denoted by $\boldsymbol{a_n}$ (read as "a sub-n"). The values a_1, a_2, $a_3, \ldots$ are called the **terms of the sequence**. The expression a_n defines the ***n*th term** (or general term) **of the sequence**.

2. Listing the Terms of a Sequence

example 1

Listing the Terms of a Sequence

List the terms of the following sequences:

a. $a_n = 3n^2 - 4; \quad 1 \le n \le 4$

b. $a_n = 3 \cdot 2^n$

Solution:

a. The domain is restricted to the first four positive integers, indicating that the sequence is finite.

n	a_n
1	$3(1)^2 - 4 = -1$
2	$3(2)^2 - 4 = 8$
3	$3(3)^2 - 4 = 23$
4	$3(4)^2 - 4 = 44$

The sequence is $-1, 8, 23, 44$.

b. The sequence $a_n = 3 \cdot 2^n$ has no restrictions on its domain; therefore, it is an infinite sequence.

n	a_n
1	$3 \cdot 2^1 = 6$
2	$3 \cdot 2^2 = 12$
3	$3 \cdot 2^3 = 24$
4	$3 \cdot 2^4 = 48$
. . .	

The sequence is $6, 12, 24, 48, \ldots$

Calculator Connections

If the nth term of a sequence is known, a *Seq* function on a graphing calculator can quickly display a list of terms. It is necessary to enter the formula for the nth term, the independent variable, the starting value for n, the ending value for n, and the increment for n.

Seq(a_n, n, begin, end, increment)

The first four terms of the sequence $a_n = 3 \cdot 2^n$ from Example 1(b) are given by:

```
seq(3*2^n,n,1,4,
1)
       {6 12 24 48}
```

Sometimes the terms of a sequence may have alternating signs. Such a sequence is called an **alternating sequence**.

example 2 Listing the Terms of an Alternating Sequence

List the first four terms of each alternating sequence.

a. $a_n = (-1)^n \cdot \dfrac{1}{n}$

b. $a_n = (-1)^{n+1} \cdot \left(\dfrac{2}{3}\right)^n$

Tip: Notice that the factor $(-1)^n$ makes the even-numbered terms positive and the odd-numbered terms negative.

Solution:

a.

n	a_n
1	$(-1)^1 \cdot \frac{1}{1} = -1$
2	$(-1)^2 \cdot \frac{1}{2} = \frac{1}{2}$
3	$(-1)^3 \cdot \frac{1}{3} = -\frac{1}{3}$
4	$(-1)^4 \cdot \frac{1}{4} = \frac{1}{4}$

The first four terms are $-1, \frac{1}{2}, -\frac{1}{3}, \frac{1}{4}$.

Tip: Notice that the factor $(-1)^{n+1}$ makes the odd-numbered terms positive and the even-numbered terms negative.

b.

n	a_n
1	$(-1)^{1+1} \cdot \left(\frac{2}{3}\right)^1 = (-1)^2 \cdot \left(\frac{2}{3}\right) = \frac{2}{3}$
2	$(-1)^{2+1} \cdot \left(\frac{2}{3}\right)^2 = (-1)^3 \cdot \left(\frac{4}{9}\right) = -\frac{4}{9}$
3	$(-1)^{3+1} \cdot \left(\frac{2}{3}\right)^3 = (-1)^4 \cdot \left(\frac{8}{27}\right) = \frac{8}{27}$
4	$(-1)^{4+1} \cdot \left(\frac{2}{3}\right)^4 = (-1)^5 \cdot \left(\frac{16}{81}\right) = -\frac{16}{81}$

The first four terms are $\frac{2}{3}, -\frac{4}{9}, \frac{8}{27}, -\frac{16}{81}$.

3. Finding a Formula for the *n*th Term of a Sequence

In Examples 1 and 2, we were given the formula for the *n*th term of a sequence and asked to list several terms of the sequence. We now consider the reverse process. Given several terms of the sequence, we will find a formula for the *n*th term. To do so, look for a pattern that establishes each term as a function of the term number.

example 3 Finding the *n*th Term of a Sequence

Find a formula for the *n*th term of the sequence:

a. $\frac{1}{2}, \frac{2}{3}, \frac{3}{4}, \frac{4}{5}, \ldots$ b. $-2, 4, -6, 8, -10, \ldots$ c. $\frac{1}{2}, \frac{1}{4}, \frac{1}{8}, \frac{1}{16}, \ldots$

Solution:

a. For each term in the sequence, the numerator is equal to the term number, and the denominator is equal to one more than the term number. Therefore, the *n*th term may be given by

$$a_n = \frac{n}{n+1}$$

b. The odd-numbered terms are negative and the even-numbered terms are positive. The factor $(-1)^n$ will produce the required alternation of signs. The numbers $2, 4, 6, 8, 10, \ldots$ are equal to $2(1), 2(2), 2(3), 2(4), 2(5), \ldots$ Therefore, the nth term may be given by

$$a_n = (-1)^n \cdot 2n$$

c. The denominators are consecutive powers of 2. The sequence can be written as

$$\frac{1}{2^1}, \frac{1}{2^2}, \frac{1}{2^3}, \frac{1}{2^4}, \ldots$$

Therefore, the nth term may be given by

$$a_n = \frac{1}{2^n}$$

Tip: The first few terms of a sequence are not sufficient to define the sequence uniquely. For example, consider the sequence:

$$\frac{1}{2}, \frac{1}{4}, \frac{1}{8}, \ldots$$

The following formulas both produce the first three terms, but differ at the fourth term:

$$a_n = \frac{1}{2^n} \quad \text{and} \quad b_n = \frac{1}{n^2 - n + 2}$$

$$a_4 = \frac{1}{16} \quad \text{whereas} \quad b_4 = \frac{1}{14}$$

To define a sequence uniquely, the nth term must be provided.

4. Applications of Sequences

example 4 **Using a Sequence in an Application**

A child drops a ball from a height of 4 ft. With each bounce, the ball rebounds to 50% of its height. Write a sequence whose terms represent the heights from which the ball falls (begin with the initial height from which the ball was dropped).

Solution:

The ball first drops 4 ft and then rebounds to a new height of 0.50(4 ft) = 2 ft. Similarly, it falls from 2 ft and rebounds 0.50(2 ft) = 1 ft. Repeating this process, we have

$$4, 2, 1, \frac{1}{2}, \frac{1}{4}, \frac{1}{8}, \ldots$$

The nth term can be represented by $a_n = 4 \cdot (0.50)^{n-1}$.

5. Definition of a Series

In many mathematical applications it is important to find the sum of the terms of a sequence. For example, suppose that the yearly interest earned in an account over a 4-year period is given by the sequence

$$\$250, \quad \$265, \quad \$278.25, \quad \$292.16$$

The sum of the terms gives the total interest earned

$$\$250 + \$265 + \$278.25 + \$292.16 = \$1085.41$$

By adding the terms of a sequence, we obtain a series.

Definition of a Series

The indicated sum of the terms of a sequence is called a **series**.

Calculator Connections

The sum of a finite series may be found on a graphing calculator if the nth term of the corresponding sequence is known. The calculator takes the sum of a sequence by using the *Sum* and *Seq* functions.

```
sum(seq(n^3,n,1,
4,1)
                100
```

As with a sequence, a series may be a finite or infinite sum of terms.

6. Summation Notation

A convenient notation used to denote the sum of a set of terms is called **sigma notation**. The Greek letter Σ (sigma) is used to indicate sums. For example, the first four terms of the sequence defined by $a_n = n^3$ is denoted by

$$\sum_{n=1}^{4} n^3$$

This is read as "the sum from n equals 1 to 4 of n^3," and is simplified as

$$\begin{aligned}\sum_{n=1}^{4} n^3 &= (1)^3 + (2)^3 + (3)^3 + (4)^3 \\ &= 1 + 8 + 27 + 64 \\ &= 100\end{aligned}$$

In this example, the letter n is called the **index of summation**. Many times, the letters i, j, and k are also used for the index of summation.

example 5 **Finding a Sum from Sigma Notation**

Find the sum $\quad \sum_{i=1}^{3} (-1)^{i+1} \cdot (3i + 4).$

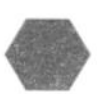

Avoiding Mistakes

The letter i is used as a variable for the index of summation. In this context, it does not represent an imaginary number.

Solution:

$$\sum_{i=1}^{3}(-1)^{i+1}\cdot(3i+4) = (-1)^{1+1}\cdot[3(1)+4] + (-1)^{2+1}\cdot[3(2)+4] + (-1)^{3+1}\cdot[3(3)+4]$$

$$= (-1)^2\cdot(7) + (-1)^3\cdot(10) + (-1)^4\cdot(13)$$

$$= 7 - 10 + 13$$

$$= 10$$

7. Writing a Sum Using Sigma Notation

example 6 Converting to Summation Notation

Write the series in summation notation.

a. $\frac{2}{3}+\frac{4}{9}+\frac{8}{27}+\frac{16}{81}$ Use n as the index of summation.

b. $1-\frac{1}{2}+\frac{1}{3}-\frac{1}{4}+\frac{1}{5}-\frac{1}{6}$ Use j as the index of summation.

Solution:

a. The sum can be written as

$$\left(\frac{2}{3}\right)^1+\left(\frac{2}{3}\right)^2+\left(\frac{2}{3}\right)^3+\left(\frac{2}{3}\right)^4$$

Taking n from 1 to 4, we have

$$\sum_{n=1}^{4}\left(\frac{2}{3}\right)^n$$

b. The even-numbered terms are negative. The factor $(-1)^{j+1}$ is negative for even values of j. Therefore, the series can be written as

$$\sum_{j=1}^{6}(-1)^{j+1}\cdot\frac{1}{j}$$

section 14.1 Practice Exercises

For Exercises 1–16, list the terms of each sequence.

1. $a_n = 3n + 1, \quad 1 \le n \le 4$
2. $a_n = -2n + 3, \quad 1 \le n \le 4$
3. $a_n = \sqrt{n+2}, \quad 1 \le n \le 4$
4. $a_n = \sqrt{n-1}, \quad 1 \le n \le 4$
5. $a_n = \frac{3}{n}, \quad 1 \le n \le 5$
6. $a_n = \frac{n}{n+2}, \quad 1 \le n \le 5$

7. $a_n = (-1)^n \dfrac{n+1}{n+2}, \quad 1 \le n \le 4$

8. $a_n = (-1)^n \dfrac{n-1}{n+2}, \quad 1 \le n \le 4$

9. $a_n = (-1)^{n+1}(n^2 - 1), \quad 1 \le n \le 4$

10. $a_n = (-1)^{n+1}(n^2), \quad 1 \le n \le 4$

11. $a_n = 3 - \dfrac{1}{n}, \quad 2 \le n \le 5$

12. $a_n = 2 - \dfrac{1}{n+1}, \quad 2 \le n \le 5$

13. $a_n = n^2 - n, \quad 1 \le n \le 4$

14. $a_n = n(n^2 - 1), \quad 1 \le n \le 4$

15. $a_n = (-1)^n 3^n, \quad 1 \le n \le 4$

16. $a_n = (-1)^n n, \quad 1 \le n \le 4$

17. If the nth term of a sequence is $a_n = (-1)^n \cdot n^2$, which terms are positive and which are negative?

18. If the nth term of a sequence is $a_n = (-1)^{n-1} \cdot \frac{1}{n}$, which terms are positive and which are negative?

19. Edmond borrowed \$500. To pay off the loan, he agreed to pay 2% of the balance plus \$50 each month. Write a sequence representing the amount Edmond will pay each month for the next 6 months. Round each term to the nearest cent.

20. Janice deposited \$1000 in a savings account that pays 3% interest compounded annually. Write a sequence representing the amount Janice receives in interest each year for the first 4 years. Round each term to the nearest cent.

21. A certain bacteria culture doubles its size each day. If there are 25,000 bacteria on the first day, write a sequence representing the population each day for the first week (7 days).

22. A radioactive chemical decays by half of its amount each week. If there is 16 g of the chemical in week 1, write a sequence representing the amount present each week for 2 months (8 weeks).

For Exercises 23–34, find a formula for the nth term of the sequence. Answers may vary.

23. $2, 4, 6, 8, \ldots$

24. $3, 6, 9, 12, \ldots$

25. $1, 3, 5, 7, \ldots$

26. $3, 5, 7, 9, \ldots$

27. $1, \dfrac{1}{4}, \dfrac{1}{9}, \dfrac{1}{16}, \ldots$

28. $\dfrac{1}{2}, \dfrac{2}{3}, \dfrac{3}{4}, \dfrac{4}{5}, \ldots$

29. $1, -1, 1, -1, \ldots$

30. $-1, 1 -1, 1, \ldots$

31. $-2, 4, -8, 16, \ldots$

32. $3, -9, 27, -81, \ldots$

33. $\dfrac{3}{5}, \dfrac{3}{25}, \dfrac{3}{125}, \dfrac{3}{625}, \ldots$

34. $\dfrac{1}{4}, \dfrac{1}{16}, \dfrac{1}{64}, \dfrac{1}{256}, \ldots$

35. What is the difference between a sequence and a series?

36. What is the index of a summation?

For Exercises 37–52, find the sums.

37. $\displaystyle\sum_{i=1}^{4} (3i^2)$

38. $\displaystyle\sum_{i=1}^{4} (2i^2)$

39. $\displaystyle\sum_{j=0}^{4} \left(\frac{1}{2}\right)^j$

40. $\displaystyle\sum_{j=0}^{4} \left(\frac{1}{3}\right)^j$

41. $\displaystyle\sum_{i=1}^{6} 5$

42. $\displaystyle\sum_{i=1}^{7} 3$

43. $\displaystyle\sum_{j=1}^{4} (-1)^j(5j)$

44. $\displaystyle\sum_{j=1}^{4} (-1)^j(4j)$

45. $\displaystyle\sum_{i=1}^{4} \frac{i+1}{i}$

46. $\displaystyle\sum_{i=2}^{5} \frac{i-1}{i}$

47. $\displaystyle\sum_{j=1}^{3} (j+1)(j+2)$

48. $\displaystyle\sum_{j=1}^{3} j(j+2)$

49. $\displaystyle\sum_{k=1}^{7} (-1)^k$

50. $\displaystyle\sum_{k=0}^{5} (-1)^{k+1}$

51. $\displaystyle\sum_{k=1}^{5} k^2$

52. $\displaystyle\sum_{k=1}^{5} 2^k$

For Exercises 53–62, write the series in summation notation.

53. $1 + 2 + 3 + 4 + 5 + 6$

54. $1 - 2 + 3 - 4 + 5 - 6$

55. $4 + 4 + 4 + 4 + 4$

56. $8 + 8 + 8 + 8 + 8$

57. $4 + 8 + 12 + 16 + 20$

58. $3 + 6 + 9 + 12 + 15$

59. $\frac{1}{3} - \frac{1}{9} + \frac{1}{27} - \frac{1}{81}$

60. $\frac{1}{2} - \frac{1}{4} + \frac{1}{8} - \frac{1}{16}$

61. $x + x^2 + x^3 + x^4 + x^5$

62. $y^2 + y^4 + y^6 + y^8 + y^{10}$

63. A certain plant will grow $1\frac{1}{2}$ in. each day for 1 week (7 days). If the plant begins with a height of 1 in., the height of the plant can be represented as a series. For example, after 1 day the plant will be $1 + 1\frac{1}{2}$ in. tall. Write out a series to represent the height of the plant each day for 1 week (7 days). Then write the series in summation notation. Finally, determine the height after 1 week.

64. A company produces a product and sells 5000 units. The company's research department found that for each year, 10% of the units cease to operate.
   a. Write out a sum representing how many of the original 5000 units will become inoperative in the first 3 years.
   b. How many of the original 5000 units will still be operational at the end of the 3 years?

Summation notation is used extensively in statistics. The sample mean (average), $\bar{x}$, of a set of n values $x_1, x_2, x_3, \ldots, x_n$ is given by

$$\bar{x} = \frac{1}{n}\sum_{i=1}^{n} x_i$$

Use this formula for Exercises 65–66.

65. Find the mean number of grams of protein for a sample of five energy bars: 10, 15, 12, 18, 22

66. Find the mean age of a sample of viewers who regularly watch CNN on television: 29, 56, 62, 39, 58, 74.

The sample standard deviation, s, of a set of n values $x_1, x_2, x_3, \ldots, x_n$ is given by:

$$s = \sqrt{\frac{n \cdot \sum_{i=1}^{n} x_i^2 - \left(\sum_{i=1}^{n} x_i\right)^2}{n(n-1)}}$$

Use this formula for Exercises 67–68.

67. Find the standard deviation for the number of grams of protein for a sample of five energy bars: 10, 15, 12, 18, 22. Round to one decimal place.

68. Find the standard deviation of a sample of viewers who regularly watch CNN on television: 29, 56, 62, 39, 58, 74. Round to one decimal place.

Expanding Your Skills

Some sequences are defined by a recursion formula, which defines each term of a sequence in terms of one or more of its preceding terms. For example, if $a_1 = 5$ and $a_n = 2a_{n-1} + 1$ for $n > 1$, then the terms of the sequence are 5, 11, 23, 47, . . . In this case, each term after the first is one more than twice the term before it.

For Exercises 69–72, list the first five terms of the sequence.

69. $a_1 = -3, a_n = a_{n-1} + 5 \quad \text{for } n > 1$

70. $a_1 = 10, a_n = a_{n-1} - 3 \quad \text{for } n > 1$

71. $a_1 = 5, a_n = 4a_{n-1} + 1 \quad \text{for } n > 1$

72. $a_1 = -2, a_n = -3a_{n-1} + 4 \quad \text{for } n > 1$

73. A famous sequence in mathematics is called the Fibonacci sequence, named after the Italian mathematician Leonardo Fibonacci of the 13th century. The Fibonacci sequence is defined by

$$a_1 = 1$$
$$a_2 = 1$$
$$a_n = a_{n-1} + a_{n-2} \quad \text{for } n > 2$$

This definition implies that beginning with the third term, each term is the sum of the preceding two terms. Write out the first 10 terms of the Fibonacci sequence.

Graphing Calculator Exercises

For Exercises 74–77, use a graphing calculator to list the first five terms of the sequence.

74. $a_n = 1 + 3n$

75. $a_n = \dfrac{8}{n}$

76. $a_n = (-1)^n\left(\dfrac{7}{8}\right)^n$

77. $a_n = (-1)^n (n^2)$

For Exercises 78–81, use the *Sum* and *Seq* functions on a graphing calculator to find the sum.

78. $\sum_{n=1}^{20} (1 + 3n)$

79. $\sum_{n=1}^{4} \dfrac{8}{n}$

80. $\sum_{n=1}^{6} (-1)^n\left(\dfrac{1}{2}\right)^n$

81. $\sum_{n=1}^{100} (-1)^n (n^2)$

---

section

14.2 Arithmetic and Geometric Sequences and Series

Concepts

1. Definition of an Arithmetic Sequence
2. Writing the *n*th Term of an Arithmetic Sequence
3. Finding a Specific Term of an Arithmetic Sequence
4. Definition of an Arithmetic Series
5. Definition of a Geometric Sequence
6. Definition of a Geometric Series
7. Infinite Geometric Series

1. Definition of an Arithmetic Sequence

In this section we will study two special types of sequences and series. The first is called an arithmetic sequence. For example:

$$4, 7, 10, 13, 16, \ldots$$

This sequence is an arithmetic sequence. Note the characteristic that each successive term after the first is a fixed value more than the previous term (in this case the terms differ by 3).

Definition of an Arithmetic Sequence

An **arithmetic sequence** is a sequence in which the difference between consecutive terms is constant.

The fixed difference between a term and its predecessor is called the **common difference** and is denoted by the letter d. The common difference is the difference between a term and its predecessor. That is,

$$d = a_{n+1} - a_n$$

Furthermore, if a_1 is the first term then

$a_2 = a_1 + d$ is the second term,

$a_3 = a_1 + 2d$ is the third term,

$a_4 = a_1 + 3d$ is the fourth term and so on.

. . .

In general, $a_n = a_1 + (n - 1)d$

*n*th Term of an Arithmetic Sequence

The nth term of an arithmetic sequence is given by

$$a_n = a_1 + (n - 1)d$$

where a_1 is the first term of the sequence and d is the common difference.

2. Writing the *n*th Term of an Arithmetic Sequence

example 1

Writing the *n*th Term of an Arithmetic Sequence

Write the nth term of the sequence: $9, 2, -5, -12, \ldots$

Solution:

$9, \quad 2, \quad -5, \quad -12, \ldots$

$-7 \quad -7 \quad -7$

The common difference can be found by subtracting a term from its predecessor: $2 - 9 = -7$

With $a_1 = 9$ and $d = -7$, we have

$$\begin{aligned} a_n &= 9 + (n - 1)(-7) \\ &= 9 - 7n + 7 \\ &= -7n + 16 \end{aligned}$$

In Example 1, the common difference between terms is -7. Accordingly, each term of the sequence *decreases* by 7.

3. Finding a Specific Term of an Arithmetic Sequence

The formula $a_n = a_1 + (n - 1)d$ contains four variables: $a_n, a_1, n,$ and d. Consequently, if we know the value of three of the four variables, we can solve for the fourth.

example 2

Finding a Specified Term of an Arithmetic Sequence

Find the ninth term of the arithmetic sequence in which $a_1 = -4$ and $a_{22} = 164$.

Solution:

To find the value of the ninth term, a_9, we need to determine the value of d. To find d, substitute $a_1 = -4$, $n = 22$, and $a_{22} = 164$ into the formula $a_n = a_1 + (n - 1)d$.

$$\begin{aligned} a_n &= a_1 + (n - 1)d \\ 164 &= -4 + (22 - 1)d \\ 164 &= -4 + 21d \\ 168 &= 21d \\ d &= 8 \end{aligned}$$

Therefore,

$$\begin{aligned} a_9 &= -4 + (9 - 1)(8) \\ &= -4 + (8)(8) \\ &= 60 \end{aligned}$$

example 3

Finding the Number of Terms in an Arithmetic Sequence

Find the number of terms of the sequence 7, 3, −1, −5, . . . , −113.

Solution:

To find the number of terms, n, we can substitute $a_1 = 7$, $d = -4$, and $a_n = -113$ into the formula for the nth term.

$$\begin{aligned} a_n &= a_1 + (n - 1)d \\ -113 &= 7 + (n - 1)(-4) \\ -113 &= 7 - 4n + 4 \\ -113 &= 11 - 4n \\ -124 &= -4n \\ n &= 31 \end{aligned}$$

4. Definition of an Arithmetic Series

The indicated sum of an arithmetic sequence is called an **arithmetic series**. For example, the series

$$3 + 7 + 11 + 15 + 19 + 23$$

is an arithmetic series because the common difference between terms is constant (4). Adding the terms in a lengthy sum is cumbersome, so we offer the following "short-cut," which is developed here. Let S represent the sum of the terms in the series.

$S = 3 + 7 + 11 + 15 + 19 + 23$	Add the terms in ascending order.
$S = 23 + 19 + 15 + 11 + 7 + 3$	Add the terms in descending order.
$2S = 26 + 26 + 26 + 26 + 26 + 26$	Adding the two series produces 6 terms of 26.

$$2S = 6 \cdot 26$$

$$S = \frac{6 \cdot 26}{2}$$

$$= 78$$

By adding the terms in ascending and descending order, we double the sum, but create a pattern that is easily added. This is true in general. To find the sum, S_n, of the first n terms of the arithmetic series $a_1 + a_2 + a_3 + \cdots + a_n$, we have

$$S_n = a_1 + (a_1 + d) + (a_1 + 2d) + \cdots + a_n \quad \text{Ascending order}$$

$$S_n = a_n + (a_n - d) + (a_n - 2d) + \cdots + a_1 \quad \text{Descending order}$$

$$2S_n = (a_1 + a_n) + (a_1 + a_n) + (a_1 + a_n) + \cdots + (a_1 + a_n)$$

$$2S_n = n(a_1 + a_n)$$

$$S_n = \frac{n}{2}(a_1 + a_n)$$

Sum of an Arithmetic Series

The sum, S_n, of the first n terms of an arithmetic series is given by

$$S_n = \frac{n}{2}(a_1 + a_n)$$

where a_1 is the first term of the series and a_n is the nth term of the series.

example 4 **Finding the Sum of an Arithmetic Series**

Find the sum of the series

$$\sum_{i=1}^{25} (2i + 3)$$

Solution:

In this series, $n = 25$. Furthermore, $a_1 = 2(1) + 3 = 5$ and $a_{25} = 2(25) + 3 = 53$. Therefore,

$$S_{25} = \frac{25}{2}(5 + 53) \quad \text{Apply the formula } S_n = \frac{n}{2}(a_1 + a_n).$$

$$= \frac{25}{2}(58) \quad \text{Simplify.}$$

$$= 725$$

Avoiding Mistakes

When applying the sum formula $S_n = \frac{n}{2}(a_1 + a_n)$ to find the sum $\sum_{i=1}^{n} a_n$, the index of summation, i, must begin at 1.

example 5

Finding the Sum of an Arithmetic Series

Find the sum of the series: $-3 + (-5) + (-7) + \cdots + (-127)$

Solution:

For this series, $a_1 = -3$ and $a_n = -127$. However, to determine the sum we first need to find the value of n. The difference between the second term and its predecessor is $-5 - (-3) = -2$. Thus, $d = -2$. We have

$-127 = -3 + (n - 1)(-2)$	Apply the formula $a_n = a_1 + (n - 1)d$.
$-127 = -3 - 2n + 2$	Apply the distributive property.
$-127 = -1 - 2n$	Combine *like* terms.
$-126 = -2n$	Solve for n.
$n = 63$	

Using $n = 63$, $a_1 = -3$ and $a_{63} = -127$, we have

$$S_n = \frac{n}{2}(a_1 + a_n) = \frac{63}{2}[-3 + (-127)] = \frac{63}{2}(-130) = -4095$$

5. Definition of a Geometric Sequence

The sequence 2, 4, 8, 16, 32, . . . is not an arithmetic sequence because the difference between terms is not constant. However, a different pattern exists. Notice that each term after the first is 2 times the preceding term. This sequence is called a geometric sequence.

Definition of a Geometric Sequence

A **geometric sequence** is a sequence in which each term after the first term is a constant multiple of the preceding term.

The constant multiple between a term and its predecessor is called the **common ratio** and is denoted by r. The common ratio is found by dividing a term by the preceding term. That is,

$$r = \frac{a_{n+1}}{a_n}$$

For the sequence 2, 4, 8, 16, 32, . . . we have $r = \frac{4}{2} = \frac{8}{4} = \frac{16}{8} = \frac{32}{16} = 2$.

If a_1 denotes the first term of a geometric sequence, then

$a_2 = a_1 r$	is the second term.
$a_3 = a_1 r^2$	is the third term.
$a_4 = a_1 r^3$	is the fourth term, and so on.
$\ldots$	

This pattern gives $a_n = a_1 r^{n-1}$

*n*th Term of a Geometric Sequence

The *n*th term of a geometric sequence is given by

$$a_n = a_1 r^{n-1}$$

where a_1 is the first term and r is the common ratio.

example 6

Finding the *n*th Term of a Geometric Sequence

Find the *n*th term of the sequence.

a. $-1, 4, -16, 64, \ldots$

b. $12, 8, \frac{16}{3}, \frac{32}{9}, \ldots$

Solution:

a. The common ratio is found by dividing any term (after the first) by its predecessor.

$$r = \frac{4}{-1} = -4$$

With $r = -4$ and $a_1 = -1$, we have $a_n = -1(-4)^{n-1}$.

b. The common ratio is $r = \frac{8}{12} = \frac{2}{3}$. With $a_1 = 12$ and $r = \frac{2}{3}$, we have

$$a_n = 12\left(\frac{2}{3}\right)^{n-1}$$

The formula $a_n = a_1 r^{n-1}$ contains the variables a_n, a_1, n, and r. If we know the value of three of the four variables, we can find the fourth.

example 7

Finding a Specified Term of a Geometric Sequence

Given $a_n = 6(\frac{1}{2})^{n-1}$ find a_5.

Solution:

$$a_n = 6\left(\frac{1}{2}\right)^{n-1}$$

$$a_5 = 6\left(\frac{1}{2}\right)^{5-1} = 6\left(\frac{1}{2}\right)^4 = 6\left(\frac{1}{16}\right) = \frac{3}{8}$$

example 8 **Finding a Specified Term of a Geometric Sequence**

Find the first term of the geometric sequence where $a_5 = -162$ and $r = 3$.

Solution:

$-162 = a_1(3)^{5-1}$ Substitute $a_5 = -162$, $n = 5$, and $r = 3$ into the formula $a_n = a_1r^{n-1}$.

$-162 = a_1(3)^4$ Simplify and solve for a_1.

$-162 = a_1(81)$

$a_1 = -2$

6. Definition of a Geometric Series

The indicated sum of a geometric sequence is called a **geometric series.** For example

$$1 + 3 + 9 + 27 + 81 + 243$$

is a geometric series. To find the sum, consider the following procedure. Let S represent the sum

$$S = 1 + 3 + 9 + 27 + 81 + 243$$

Now multiply S by the common ratio, which in this case is 3.

$$3S = 3 + 9 + 27 + 81 + 243 + 729$$

Then

$$3S - S = (3 + 9 + 27 + 81 + 243 + 729) - (1 + 3 + 9 + 27 + 81 + 243)$$

$$2S = 3 + 9 + 27 + 81 + 243 + 729 - 1 - 3 - 9 - 27 - 81 - 243$$

$2S = 729 - 1$ The terms in red form a sum of zero.

$$2S = 728$$

$$S = 364$$

A similar procedure can be used to find the sum S_n of the first n terms of any geometric series. Subtract rS_n from S_n.

$$S_n = a_1 + a_1r + a_1r^2 + a_1r^3 + \cdots + a_1r^{n-1}$$

$$rS_n = a_1r + a_1r^2 + a_1r^3 + a_1r^4 + \cdots + a_1r^n$$

$$S_n - rS_n = (a_1 - a_1r) + (a_1r - a_1r^2) + (a_1r^2 - a_1r^3) + (a_1r^3 - a_1r^4) + \cdots + (a_1r^{n-1} - a_1r^n)$$

$S_n - rS_n = a_1 - a_1r^n$ The terms in red form a sum of zero.

$S_n(1 - r) = a_1(1 - r^n)$ Factor each side of the equation.

$S_n = \dfrac{a_1(1 - r^n)}{1 - r}$ Divide by $(1 - r)$.

Sum of a Geometric Series

The sum, S_n, of the first n terms of a geometric series, $\sum_{i=1}^{n}(a_1r^{i-1})$, is given by

$$S_n = \frac{a_1(1 - r^n)}{1 - r}$$

where a_1 is the first term of the series and r is the common ratio and $r \neq 1$.

example 9 **Finding the Sum of a Geometric Series**

Find the sum of the series

$$4 + 2 + 1 + \frac{1}{2} + \frac{1}{4} + \frac{1}{8}$$

Solution:

$$a_1 = 4, \quad r = \frac{1}{2}, \text{ and } n = 6$$

$$S_n = \frac{a_1(1 - r^n)}{1 - r} = \frac{4[1 - (\frac{1}{2})^6]}{1 - \frac{1}{2}} = \frac{4(1 - \frac{1}{64})}{\frac{1}{2}} = 8\left(\frac{63}{64}\right) = \frac{63}{8}$$

example 10 **Finding the Sum of a Geometric Series**

Find the sum of the series: $5 + 10 + 20 + \cdots + 5120$.

Solution:

The common ratio is 2, and $a_1 = 5$. The nth term of the sequence can be written as $a_n = 5(2)^{n-1}$. To find the value of n, substitute 5120 for a_n.

$5120 = 5(2)^{n-1}$

$\frac{5120}{5} = \frac{5(2)^{n-1}}{5}$ Divide both sides by 5.

$1024 = 2^{n-1}$ To solve the exponential equation, write each side as a power of 2.

$2^{10} = 2^{n-1}$

$10 = n - 1$ From Section 12.7, we know that if $b^x = b^y$, then $x = y$.

$n = 11$

With $a_1 = 5$, $r = 2$, and $n = 11$, we have

$$S_n = \frac{a_1(1 - r^n)}{1 - r} = \frac{5(1 - 2^{11})}{1 - 2} = \frac{5(1 - 2048)}{-1} = \frac{5(-2047)}{-1} = 10{,}235$$

7. Infinite Geometric Series

Consider a geometric series where $|r| < 1$. For increasing values of n, r^n decreases. For example,

$$\left(\frac{1}{2}\right)^5 = 0.03125, \qquad \left(\frac{1}{2}\right)^{10} \approx 0.00097656, \qquad \left(\frac{1}{2}\right)^{15} \approx 0.00003052$$

For $|r| < 1$, r^n approaches 0 as n gets larger and larger. As n approaches infinity, the sum

$$S = \frac{a_1(1 - r^n)}{1 - r} \qquad \text{approaches} \qquad \frac{a_1(1 - 0)}{1 - r} = \frac{a_1}{1 - r}$$

Sum of an Infinite Geometric Series

Given an infinite geometric series $a_1 + a_1r + a_1r^2 + \cdots$, with $|r| < 1$, the sum S of all terms in the series is given by

$$S = \frac{a_1}{1 - r}$$

Note: If $|r| \geq 1$, then the sum does not exist.

example 11 **Finding the Sum of an Infinite Geometric Series**

Find the sum of the series

$$1 + \frac{1}{3} + \frac{1}{9} + \frac{1}{27} + \frac{1}{81} + \cdots$$

Solution:

This is a geometric series with $a_1 = 1$ and $r = \frac{1}{3}$. Because $|r| = |\frac{1}{3}| < 1$, we have

$$S = \frac{a_1}{1 - r} = \frac{1}{1 - \frac{1}{3}} = \frac{1}{\frac{2}{3}} = \frac{3}{2}$$

The sum is $\frac{3}{2}$.

example 12 **Using Geometric Series in a Physics Application**

A child drops a ball from a height of 4 ft. With each bounce, the ball rebounds to 50% of its original height. Determine the total distance traveled by the ball.

Solution:

The heights from which the ball drops are given by the sequence

$$4, 2, 1, \frac{1}{2}, \frac{1}{4}, \ldots$$

After the ball falls from its initial height of 4 ft, the distance traveled for every bounce thereafter is doubled (the ball travels up and down). Therefore, the total distance traveled is given by the series

$$4 + 2 \cdot 2 + 2 \cdot 1 + 2 \cdot \frac{1}{2} + 2 \cdot \frac{1}{4} + \cdots \quad \text{or equivalently}$$

$$4 + 4 + 2 + 1 + \frac{1}{2} + \cdots$$

The series $4 + 2 + 1 + \frac{1}{2} + \cdots$ is an infinite geometric series with $a_1 = 4$ and $r = \frac{1}{2}$.

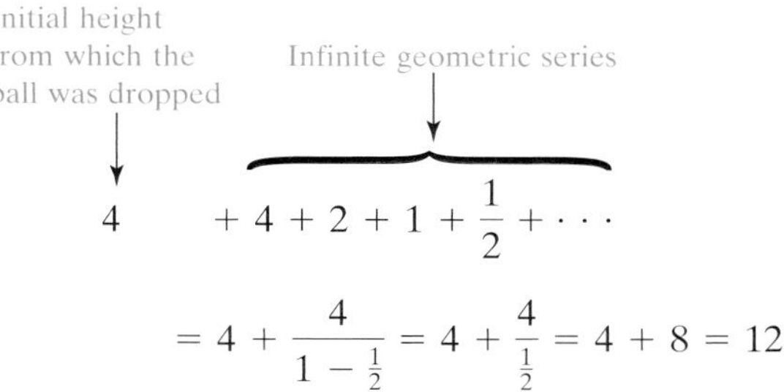

The ball traveled a total of 12 ft.

section 14.2 Practice Exercises

1. Explain how to determine if a sequence is arithmetic.

For Exercises 2–7, find the common difference, d, for each arithmetic sequence.

2. $1, 3, 5, 7, 9, \ldots$
3. $2, 8, 14, 20, 26, \ldots$
4. $6, 3, 0, -3, -6, \ldots$
5. $8, 3, -2, -7, -12, \ldots$
6. $-7, -9, -11, -13, -15, \ldots$
7. $-15, -11, -7, -3, 1, \ldots$

For Exercises 8–13, write the first five terms of the arithmetic sequence.

8. $a_1 = 3, d = 5$
9. $a_1 = -3, d = 2$
10. $a_1 = 2, d = \frac{1}{2}$
11. $a_1 = 0, d = \frac{1}{3}$
12. $a_1 = 2, d = -4$
13. $a_1 = 10, d = -6$

For Exercises 14–23, write the nth term of the sequence.

14. $0, 5, 10, 15, 20, \ldots$
15. $7, 12, 17, 22, 27, \ldots$
16. $-2, -4, -6, -8, -10, \ldots$
17. $1, -3, -7, -11, -15, \ldots$
18. $2, \frac{5}{2}, 3, \frac{7}{2}, 4, \ldots$
19. $1, \frac{4}{3}, \frac{5}{3}, 2, \frac{7}{3}, \ldots$
20. $21, 17, 13, 9, 5, \ldots$
21. $9, 6, 3, 0, -3, \ldots$
22. $-8, -2, 4, 10, 16, \ldots$
23. $-9, -1, 7, 15, 23, \ldots$

For Exercises 24–31, find the indicated term of each sequence.

24. Find the eighth term given $a_1 = -6$ and $d = -3$.
25. Find the sixth term given $a_1 = -3$ and $d = 4$.

26. Find the 12th term given $a_1 = -8$ and $d = -2$.

27. Find the ninth term given $a_1 = -1$ and $d = 6$.

28. Find the seventh term given $a_1 = 0$ and $d = -5$.

29. Find the 10th term given $a_1 = 1$ and $d = 5$.

30. Find the sixth term given $a_1 = 3$ and $d = 3$.

31. Find the 11th term given $a_1 = 12$ and $d = -6$.

For Exercises 32–39, find the number of terms, n, of each sequence.

32. $2, 0, -2, \ldots, -56$

33. $8, 13, 18, \ldots, 98$

34. $1, -3, -7, \ldots, -67$

35. $1, 5, 9, \ldots, 85$

36. $1, \frac{3}{4}, \frac{1}{2}, \ldots, -4$

37. $2, \frac{5}{2}, 3, \ldots, 13$

38. $-\frac{5}{3}, -1, -\frac{1}{3}, \ldots, 12\frac{1}{3}$

39. $\frac{13}{3}, \frac{19}{3}, \frac{25}{3}, \ldots, \frac{73}{3}$

40. If the third and fourth terms of an arithmetic sequence are 18 and 21, what are the first and second terms?

41. If the third and fourth terms of an arithmetic sequence are -8 and -11, what are the first and second terms?

42. Explain the difference between an arithmetic sequence and an arithmetic series.

For Exercises 43–56, find the sum of the arithmetic series.

43. $\sum_{i=1}^{20} (3i + 2)$

44. $\sum_{i=1}^{15} (2i - 3)$

45. $\sum_{i=1}^{20} (i + 4)$

46. $\sum_{i=1}^{25} (i - 3)$

47. $\sum_{j=1}^{10} (4 - j)$

48. $\sum_{j=1}^{10} (6 - j)$

49. $\sum_{j=1}^{15} \left(\frac{2}{3}j + 1\right)$

50. $\sum_{j=1}^{15} \left(\frac{1}{2}j - 2\right)$

51. $4 + 8 + 12 + \cdots + 84$

52. $4 + 9 + 14 + \cdots + 49$

53. $6 + 8 + 10 + \cdots + 34$

54. $-4 + (-3) + (-2) + \cdots + 12$

55. $-3 + (-7) + (-11) + \cdots + (-39)$

56. $2 + 5 + 8 + \cdots + 53$

57. Find the sum of the first 100 positive integers.

58. Find the sum of the first 50 positive even integers.

59. Find the sum of the first 50 positive odd integers.

60. Find the sum of the first 20 positive multiples of 5.

61. The seating in a certain theater is arranged so that there are 30 seats in row 1, 32 in row 2, 34 in row 3, and so on. If there are 20 rows, how many total seats are there? What is the total revenue if the average ticket price is \$15 per seat and the theater is sold out?

62. A triangular array of dominos has one domino in the first row, two dominos in the second row, three dominos in the third row and so on. If there are 15 rows, how many dominos are in the array?

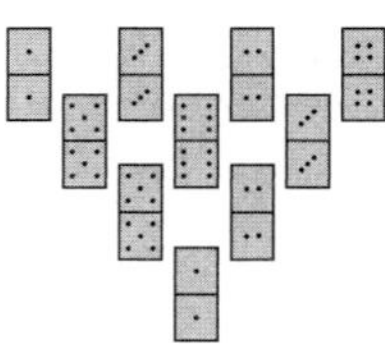

Figure for Exercise 62

63. Explain how to determine if a sequence is geometric.

For Exercises 64–69, determine the common ratio, r, for the geometric sequence.

64. $5, 10, 20, 40, \ldots$

65. $-2, -1, -\frac{1}{2}, -\frac{1}{4}, \ldots$

66. $8, -2, \frac{1}{2}, -\frac{1}{8}, \ldots$

67. $4, -12, 36, -108, \ldots$

68. $3, -6, 12, -24, \ldots$

69. $1, 4, 16, 64, \ldots$

For Exercises 70–75, write the first five terms of the geometric sequence.

70. $a_1 = -3, r = -2$

71. $a_1 = -4, r = -1$

72. $a_1 = 6, r = \frac{1}{2}$

73. $a_1 = 8, r = \frac{1}{4}$

74. $a_1 = -1, r = 6$

75. $a_1 = 2, r = -3$

For Exercises 76–81, find the nth term of each geometric sequence.

76. 3, 12, 48, 192, . . .

77. 2, 6, 18, 54, . . .

78. −5, 15, −45, 135, . . .

79. −6, 12, −24, 48, . . .

80. $\frac{1}{2}$, 2, 8, 32, . . .

81. $\frac{16}{3}$, 4, 3, $\frac{9}{4}$, . . .

For Exercises 82–91, find the indicated term of each sequence.

82. Given $a_n = 2(\frac{1}{2})^{n-1}$, find a_8.

83. Given $a_n = -3(\frac{1}{2})^{n-1}$, find a_8.

84. Given $a_n = 4(-\frac{3}{2})^{n-1}$, find a_6.

85. Given $a_n = 6(-\frac{1}{3})^{n-1}$, find a_6.

86. Given $a_n = -3(2)^{n-1}$, find a_5.

87. Given $a_n = 5(3)^{n-1}$, find a_4.

88. Given $a_5 = -\frac{16}{9}$ and $r = -\frac{2}{3}$, find a_1.

89. Given $a_6 = \frac{5}{16}$ and $r = -\frac{1}{2}$, find a_1.

90. Given $a_7 = 8$ and $r = 2$, find a_1.

91. Given $a_6 = 27$ and $r = 3$, find a_1.

92. If the second and third terms of a geometric sequence are 16 and 64, what is the first term?

93. If the second and third terms of a geometric sequence are $\frac{1}{3}$ and $\frac{1}{9}$, what is the first term?

94. Explain the difference between a geometric sequence and a geometric series.

For Exercises 95–104, find the sum of the geometric series.

95. $10 + 2 + \frac{2}{5} + \frac{2}{25} + \frac{2}{125}$

96. $1 + 3 + 9 + 27 + 81 + 243$

97. $-2 + 1 + \left(-\frac{1}{2}\right) + \frac{1}{4} + \left(-\frac{1}{8}\right)$

98. $\frac{1}{4} + (-1) + 4 + (-16) + 64$

99. $12 + 16 + \frac{64}{3} + \frac{256}{9} + \frac{1024}{27}$

100. $9 + 6 + 4 + \frac{8}{3} + \frac{16}{9}$

101. $1 + \frac{2}{3} + \frac{4}{9} + \cdots + \frac{64}{729}$

102. $\frac{8}{3} + 2 + \frac{3}{2} + \cdots + \frac{243}{512}$

103. $-4 + 8 + (-16) + \cdots + (-256)$

104. $-\frac{7}{3} + 7 + (-21) + \cdots + (-1701)$

105. A deposit of \$1000 is made in an account that earns 5% interest compounded annually. The balance in the account after n years is given by

$$a_n = 1000(1.05)^n \quad \text{for } n \geq 1$$

a. List the first four terms of the sequence. Round to the nearest cent.

b. Find the balance after 10 years, 20 years, and 40 years by computing a_{10}, a_{20}, and a_{40}. Round to the nearest cent.

106. A home purchased for \$125,000 increases by 4% of its value each year. The value of the home after n years is given by

$$a_n = 125{,}000(1.04)^n \quad \text{for } n \geq 1$$

a. List the first four terms of the sequence. Round to the nearest dollar.

b. Find the value of the home after 5 years, 10 years, and 20 years by computing a_5, a_{10}, and a_{20}. Round to the nearest dollar.

For Exercises 107–112, first find the common ratio, r. Then determine the sum of the infinite series if it exists.

107. $1 + \frac{1}{6} + \frac{1}{36} + \frac{1}{216} + \cdots$

108. $-2 + \left(-\frac{1}{2}\right) + \left(-\frac{1}{8}\right) + \left(-\frac{1}{32}\right) + \cdots$

109. $-3 + 1 + \left(-\frac{1}{3}\right) + \frac{1}{9} + \cdots$

110. $\frac{1}{2} + \left(-\frac{1}{10}\right) + \frac{1}{50} + \left(-\frac{1}{250}\right) + \cdots$

111. $\frac{2}{3} + (-1) + \frac{3}{2} + \left(-\frac{9}{4}\right) + \cdots$

112. $3 + 5 + \frac{25}{3} + \frac{125}{9} + \cdots$

113. Suppose \$200 million is spent by tourists at a certain resort town. Further suppose that 75% of the revenue is respent in the community, and then respent over and over again, each time at a rate of 75%. The series

$$200 + 200(0.75) + 200(0.75)^2 + 200(0.75)^3 + \cdots$$

gives the total amount spent (and respent) in the community. Find the sum of the infinite series.

114. A bungee jumper jumps off a platform and stretches the cord 80 ft before rebounding upward. Each successive bounce stretches the cord 60% of its previous length. The total vertical distance traveled is given by

$$80 + 2(0.60)(80) + 2(0.60)^2(80) + 2(0.60)^3(80) + \cdots$$

Ignoring the first term, the series is an infinite geometric series. Compute the total vertical distance traveled. See Example 12.

115. A ball drops from a height of 4 ft. With each bounce, the ball rebounds to $\frac{3}{4}$ of its height. The total vertical distance traveled is given by

$$4 + 2\left(\frac{3}{4}\right)(4) + 2\left(\frac{3}{4}\right)^2(4) + 2\left(\frac{3}{4}\right)^3(4) + \cdots$$

Ignoring the first term, the series is an infinite geometric series. Compute the total vertical distance traveled. See Example 12.

116. The repeating decimal number $0.\overline{2}$ can be written as an infinite geometric series by:

$$\frac{2}{10} + \frac{2}{100} + \frac{2}{1000} + \cdots$$

a. What is a_1?

b. What is r?

c. Find the sum of the series.

117. The repeating decimal number $0.\overline{7}$ can be written as an infinite geometric series by

$$\frac{7}{10} + \frac{7}{100} + \frac{7}{1000} + \cdots$$

a. What is a_1?

b. What is r?

c. Find the sum of the series.

For Exercises 118–131, determine if the sequence is arithmetic, geometric, or neither. If the sequence is arithmetic, find d. If the sequence is geometric, find r.

118. $5, -\frac{5}{2}, \frac{5}{4}, -\frac{5}{8}, \ldots$

119. $1, -\frac{3}{2}, \frac{9}{4}, -\frac{27}{8}, \ldots$

120. $\frac{1}{2}, 1, \frac{3}{2}, 2, \frac{5}{2}, \ldots$

121. $-\frac{1}{3}, \frac{1}{3}, 1, \frac{5}{3}, \ldots$

122. $-2, -4, -8, -16, -32, \ldots$

123. $2, 6, 18, 54, 162, \ldots$

124. $-2, -4, -6, -8, -10, \ldots$

125. $2, 6, 10, 14, 18, 22, \ldots$

126. $-2, -4, -7, -11, -16, \ldots$

127. $2, 6, 11, 17, 24, \ldots$

128. $1, 3, 1, 3, 1, \ldots$

129. $0, 1, 0, 1, 0, 1, \ldots$

130. $2, -2, 2, -2, 2, \ldots$

131. $-1, 1, -1, 1, -1, \ldots$

Expanding Your Skills

132. The yearly salary for job A is \$30,000 initially with an annual raise of \$2000 per year. The yearly salary for job B is \$29,000 initially with an annual raise of 6% per year.

a. Find the total earnings for job A over 20 years (there will be 19 raises). Is this an arithmetic or geometric series?

b. For job B, what is the amount of the raise after 1 year?

c. Find the total earnings for job B over 20 years. Round to the nearest dollar. Is this an arithmetic or geometric series?

d. What is the difference in total salary earned over 20 years between job A and job B.

133. a. Brook has a job that pays $40,000 the first year. She receives a 4% raise each year. Find the sum of her yearly salaries over a 20-year period. Round to the nearest dollar.

b. Chamille has a job that pays $40,000 the first year. She receives a 4.5% raise each year. Find the sum of her yearly salaries over a 20-year period. Round to the nearest dollar.

c. Chamille's raise each year was 0.5% higher than Brook's raise. How much more total income did Chamille receive than Brook over 20 years?

Graphing Calculator Exercises

134. Use the *Seq* function on a graphing calculator to list the first four terms of the sequence $a_n = 1000(1.05)^n$. Verify your answer to Exercise 105(a).

135. Use the *Seq* function on a graphing calculator to list the first four terms of the sequence $a_n = 125{,}000(1.04)^n$. Verify your answer to Exercise 106(a).

136. Use the *Sum* and *Seq* functions on a graphing calculator to confirm your answer to Exercise 61.

137. Use the *Sum* and *Seq* functions on a graphing calculator to confirm your answer to Exercise 62.

138. Use the *Sum* and *Seq* functions on a graphing calculator to find the sum of the first 10 terms and the first 20 terms of the series given below. Then compute the exact value of the infinite geometric series.

$$10 + 10(0.6) + 10(0.6)^2 + 10(0.6)^3 + \cdots$$

139. Use the *Sum* and *Seq* functions on a graphing calculator to find the sum of the first 10 terms and the first 20 terms of the series given below. Then compute the exact value of the infinite geometric series.

$$4 + 2 + 1 + \frac{1}{2} + \cdots$$

section

14.3 Binomial Expansions

Concepts

1. Introduction to Binomial Expansions
2. Determining the Coefficients of a Binomial Expansion Using Pascal's Triangle
3. Determining the Coefficients of a Binomial Expansion Using Factorial Notation
4. The Binomial Theorem
5. Finding a Specific Term of a Binomial Expansion

1. Introduction to Binomial Expansions

In Section 4.6 we learned how to square a binomial.

$$(a + b)^2 = a^2 + 2ab + b^2$$

The expression $a^2 + 2ab + b^2$ is called the **binomial expansion** of $(a + b)^2$. To expand $(a + b)^3$, we can find the product $(a + b)(a + b)^2$.

$$\begin{aligned}(a + b)(a + b)^2 &= (a + b)(a^2 + 2ab + b^2)\\ &= a^3 + 2a^2b + ab^2 + a^2b + 2ab^2 + b^3\\ &= a^3 + 3a^2b + 3ab^2 + b^3\end{aligned}$$

Similarly, to expand $(a + b)^4$, we can multiply $(a + b)$ by $(a + b)^3$. Using this method, we can expand several powers of $(a + b)$ to find the following pattern:

$$(a + b)^0 = 1$$
$$(a + b)^1 = a + b$$
$$(a + b)^2 = a^2 + 2ab + b^2$$
$$(a + b)^3 = a^3 + 3a^2b + 3ab^2 + b^3$$
$$(a + b)^4 = a^4 + 4a^3b + 6a^2b^2 + 4ab^3 + b^4$$
$$(a + b)^5 = a^5 + 5a^4b + 10a^3b^2 + 10a^2b^3 + 5ab^4 + b^5$$

Notice that the exponent on a decreases from left to right while the exponents on b increase from left to right. Also observe that for each term, the sum of the exponents is equal to the exponent to which $(a + b)$ is raised. Finally, notice that the number of terms in the expansion is exactly one more than the power to which $(a + b)$ is raised. For example, the expansion of $(a + b)^4$ has five terms and the expansion of $(a + b)^5$ has six terms.

With these guidelines in mind, we know that $(a + b)^6$ will contain seven terms involving

$$a^6, \quad a^5b, \quad a^4b^2, \quad a^3b^3, \quad a^2b^4, \quad ab^5, \quad b^6$$

We can complete the expansion of $(a + b)^6$ if we can determine the correct coefficients of each term.

2. Determining the Coefficients of a Binomial Expansion Using Pascal's Triangle

If we write the coefficients for several expansions of $(a + b)^n$, where $n \geq 0$, we have a triangular array of numbers.

$(a + b)^0 = 1$	1
$(a + b)^1 = 1a + 1b$	1 1
$(a + b)^2 = 1a^2 + 2ab + 1b^2$	1 2 1
$(a + b)^3 = 1a^3 + 3a^2b + 3ab^2 + 1b^3$	1 3 3 1
$(a + b)^4 = 1a^4 + 4a^3b + 6a^2b^2 + 4ab^3 + 1b^4$	1 4 6 4 1
$(a + b)^5 = 1a^5 + 5a^4b + 10a^3b^2 + 10a^2b^3 + 5ab^4 + 1b^5$	1 5 10 10 5 1

Each row begins and ends with a 1 and each entry in between is the sum of the two entries from the line above. For example, in the last row, $5 = 1 + 4$, $10 = 4 + 6$, and so on. This triangular array of coefficients for a binomial expansion is called **Pascal's triangle**, named after French mathematician Blaise Pascal (1623–1662).

1
1 1
1 2 1
1 3 3 1
1 4 6 4 1
1 5 10 10 5 1
1 6 15 20 15 6 1

Using the pattern shown in Pascal's triangle, the coefficients corresponding to $(a + b)^6$ would be 1, 6, 15, 20, 15, 6, 1. Inserting the coefficients, the sum becomes

$$(a + b)^6 = 1a^6 + 6a^5b + 15a^4b^2 + 20a^3b^3 + 15a^2b^4 + 6ab^5 + 1b^6$$

3. Determining the Coefficients of a Binomial Expansion Using Factorial Notation

Although Pascal's triangle provides an easy method to find the coefficients of $(a + b)^n$, it is impractical for large values of n. A more efficient method to find the coefficients of a binomial expansion involves **factorial notation**.

Definition of n!

Let n be a positive integer. Then $n!$ (read as "n factorial") is defined as the product of integers from 1 through n. That is,

$$n! = n(n - 1)(n - 2) \cdot \cdot \cdot (2)(1)$$

Note: For $n = 0$, then $n! = 0! = 1$.

example 1 Evaluating Factorial Notation

Evaluate the expressions.

a. 4!   b. 10!   c. 0!

Solution:

a. $4! = 4 \cdot 3 \cdot 2 \cdot 1 = 24$

b. $10! = 10 \cdot 9 \cdot 8 \cdot 7 \cdot 6 \cdot 5 \cdot 4 \cdot 3 \cdot 2 \cdot 1 = 3{,}628{,}800$

c. $0! = 1$ by definition

Calculator Connections

Most calculators have a [!] function. Try evaluating the expressions from Example 1 on a calculator:

```
4!
              24
10!
         3628800
0!
               1
```

Sometimes factorial notation is used with other operations such as multiplication and division.

example 2 Operations with Factorials

Evaluate the expressions.

a. $\dfrac{4!}{4! \cdot 0!}$ b. $\dfrac{4!}{3! \cdot 1!}$ c. $\dfrac{4!}{2! \cdot 2!}$ d. $\dfrac{4!}{1! \cdot 3!}$ e. $\dfrac{4!}{0! \cdot 4!}$

Solution:

a. $\dfrac{4!}{4! \cdot 0!} = \dfrac{(4 \cdot 3 \cdot 2 \cdot 1)}{(4 \cdot 3 \cdot 2 \cdot 1) \cdot (1)} = \dfrac{\cancel{4 \cdot 3 \cdot 2 \cdot 1}}{\cancel{4 \cdot 3 \cdot 2 \cdot 1} \cdot 1} = 1$

b. $\dfrac{4!}{3! \cdot 1!} = \dfrac{(4 \cdot 3 \cdot 2 \cdot 1)}{(3 \cdot 2 \cdot 1) \cdot (1)} = \dfrac{4 \cdot \cancel{3 \cdot 2 \cdot 1}}{\cancel{3 \cdot 2 \cdot 1} \cdot 1} = 4$

c. $\dfrac{4!}{2! \cdot 2!} = \dfrac{(4 \cdot 3 \cdot 2 \cdot 1)}{(2 \cdot 1) \cdot (2 \cdot 1)} = \dfrac{4 \cdot 3 \cdot \cancel{2 \cdot 1}}{2 \cdot 1 \cdot \cancel{2 \cdot 1}} = 6$

d. $\dfrac{4!}{1! \cdot 3!} = \dfrac{(4 \cdot 3 \cdot 2 \cdot 1)}{(1) \cdot (3 \cdot 2 \cdot 1)} = \dfrac{4 \cdot \cancel{3 \cdot 2 \cdot 1}}{1 \cdot \cancel{3 \cdot 2 \cdot 1}} = 4$

e. $\dfrac{4!}{0! \cdot 4!} = \dfrac{(4 \cdot 3 \cdot 2 \cdot 1)}{(1) \cdot (4 \cdot 3 \cdot 2 \cdot 1)} = \dfrac{\cancel{4 \cdot 3 \cdot 2 \cdot 1}}{1 \cdot \cancel{4 \cdot 3 \cdot 2 \cdot 1}} = 1$

Calculator Connections

To evaluate the expressions from Example 2 on a calculator, use parentheses around the factors in the denominator:

```
4!/(4!*0!)
                  1
4!/(3!*1!)
                  4
4!/(2!*2!)
                  6
```

4. The Binomial Theorem

Notice from Example 2 that the values of

$$\frac{4!}{4! \cdot 0!}, \quad \frac{4!}{3! \cdot 1!}, \quad \frac{4!}{2! \cdot 2!}, \quad \frac{4!}{1! \cdot 3!}, \quad \text{and} \quad \frac{4!}{0! \cdot 4!}$$

correspond to the values 1, 4, 6, 4, 1, which are the coefficients for the expansion of $(a + b)^4$. Generalizing this pattern, the coefficients for the terms in the expansion of $(a + b)^n$ are given by

$\dfrac{n!}{r! \cdot (n - r)!}$ where r corresponds to the exponent on the factor of a and $(n - r)$ corresponds to the exponent on the factor of b.

Using this formula for the coefficients in a binomial expansion results in the **binomial theorem**.

The Binomial Theorem

For any positive integer, n,

$$(a + b)^n = \frac{n!}{n! \cdot 0!}a^n + \frac{n!}{(n - 1)! \cdot 1!}a^{(n-1)}b + \frac{n!}{(n - 2)! \cdot 2!}a^{(n-2)}b^2 + \cdots + \frac{n!}{0! \cdot n!}b^n$$

example 3 **Applying the Binomial Theorem**

Write out the first three terms of the expansion of $(a + b)^{10}$.

Solution:

The first three terms of $(a + b)^{10}$ are $\frac{10!}{10! \cdot 0!}a^{10} + \frac{10!}{9! \cdot 1!}a^9b + \frac{10!}{8! \cdot 2!}a^8b^2$

$$= \frac{\cancel{10!}}{\cancel{10!} \cdot 1}a^{10} + \frac{10 \cdot \cancel{9!}}{\cancel{9!} \cdot 1!}a^9b + \frac{10 \cdot 9 \cdot \cancel{8!}}{\cancel{8!} \cdot 2 \cdot 1}a^8b^2$$

$$= a^{10} + 10a^9b + 45a^8b^2$$

example 4 **Applying the Binomial Theorem**

Use the binomial theorem to find the expansion of $(3x^2 - 5y)^4$.

Solution:

Write $(3x^2 - 5y)^4$ as $[(3x^2) + (-5y)]^4$. In this case, the expressions $3x^2$ and $-5y$ may be substituted for a and b in the expansion of $(a + b)^4$.

$$(a + b)^4 = \frac{4!}{4! \cdot 0!}a^4 + \frac{4!}{3! \cdot 1!}a^3b + \frac{4!}{2! \cdot 2!}a^2b^2 + \frac{4!}{1! \cdot 3!}ab^3 + \frac{4!}{0! \cdot 4!}b^4$$

$$= \frac{4!}{4! \cdot 0!}(3x^2)^4 + \frac{4!}{3! \cdot 1!}(3x^2)^3(-5y) + \frac{4!}{2! \cdot 2!}(3x^2)^2(-5y)^2 + \frac{4!}{1! \cdot 3!}(3x^2)(-5y)^3 + \frac{4!}{0! \cdot 4!}(-5y)^4$$

$$= 1 \cdot (81x^8) + 4 \cdot (27x^6)(-5y) + 6 \cdot (9x^4)(25y^2) + 4 \cdot (3x^2)(-125y^3) + 1 \cdot (625y^4)$$

$$= 81x^8 - 540x^6y + 1350x^4y^2 - 1500x^2y^3 + 625y^4$$

Tip: The values of $\frac{n!}{r!(n-r)!}$ can also be found using Pascal's Triangle.

1
1 1
1 2 1
1 3 3 1
1 4 6 4 1

5. Finding a Specific Term of a Binomial Expansion

The binomial theorem may also be used to find a specific term in a binomial expansion.

example 5 **Finding a Specific Term in a Binomial Expansion**

Find the fourth term of the expansion $(a + b)^{13}$.

Solution:

There are 14 terms in the expansion of $(a + b)^{13}$. The first term will have variable factors $a^{13}b^0$. The second term will have variable factors $a^{12}b^1$. The third term will have $a^{11}b^2$, and the fourth term will have $a^{10}b^3$. Hence the fourth term is

$$\frac{13!}{10! \cdot 3!}a^{10}b^3 = 286a^{10}b^3$$

From Example 5, we see that for the kth term in the expansion $(a + b)^n$, where k is an integer greater than zero, the variable factors are $a^{n-(k-1)}$ and b^{k-1}. Therefore, to find the kth term of $(a + b)^n$ we can make the following generalization.

Finding a Specific Term in a Binomial Expansion

Let n and k be positive integers such that $k \le n$. Then the kth term in the expansion of $(a + b)^n$ is given by

$$\frac{n!}{[n - (k - 1)]! \cdot (k - 1)!} \cdot a^{n-(k-1)} \cdot b^{k-1}$$

example 6

Finding a Specific Term in a Binomial Expansion

Find the sixth term of $(p^3 + 2w)^8$.

Solution:

Apply the formula

$$\frac{n!}{[n - (k - 1)]! \cdot (k - 1)!} \cdot a^{n-(k-1)} \cdot b^{k-1} \quad \text{with } n = 8, k = 6, a = p^3, \text{ and } b = 2w$$

$$\frac{8!}{[8 - (6 - 1)]! \cdot (6 - 1)!} \cdot (p^3)^{8-(6-1)} \cdot (2w)^{6-1}$$

$$= \frac{8!}{[8 - (5)]! \cdot (5)!} \cdot (p^3)^{8-(5)} \cdot (2w)^5$$

$$= \frac{8!}{(3)! \cdot (5)!} \cdot (p^3)^3 \cdot (2w)^5$$

$$= 56 \cdot (p^9) \cdot (32w^5)$$

$$= 1792p^9w^5$$

section 14.3 Practice Exercises

For Exercises 1–6, expand the binomials. Use Pascal's triangle to find the coefficients.

1. $(x + y)^4$
2. $(a + b)^3$
3. $(4 + p)^3$
4. $(1 + g)^4$
5. $(a^2 + b)^6$
6. $(p + q^2)^7$

For Exercises 7–8, rewrite each binomial of the form $(a - b)^n$ as $[a + (-b)]^n$. Then expand the binomials. Use Pascal's triangle to find the coefficients.

7. $(p^2 - w)^3$
8. $(5 - u^3)^4$
9. For $a > 0$ and $b > 0$, what happens to the signs of the terms when expanding the binomial $(a - b)^n$ compared with $(a + b)^n$?

For Exercises 10–13, evaluate the expression.

10. $5!$
11. $3!$
12. $0!$
13. $1!$
14. True or False: $0! \neq 1!$
15. True or False: $n!$ is defined for negative integers.
16. True or False: $n! = n$ for $n = 1$ and 2.
17. Show that $9! = 9 \cdot 8!$
18. Show that $6! = 6 \cdot 5!$

For Exercises 19–26, evaluate the expression.

19. $\dfrac{8!}{4!}$
20. $\dfrac{7!}{5!}$
21. $\dfrac{3!}{0!}$
22. $\dfrac{4!}{0!}$
23. $\dfrac{8!}{3!\,5!}$
24. $\dfrac{6!}{2!\,4!}$
25. $\dfrac{4!}{0!\,4!}$
26. $\dfrac{6!}{0!\,6!}$

For Exercises 27–36, use the binomial theorem to expand the binomials.

27. $(s + t)^6$
28. $(h + k)^4$
29. $(b - 3)^3$
30. $(c - 2)^5$
31. $(2x + y)^4$
32. $(p + 3q)^3$
33. $(c^2 - d)^7$
34. $(u - v^3)^6$
35. $\left(\dfrac{a}{2} - b\right)^5$
36. $\left(\dfrac{s}{3} + t\right)^5$

For Exercises 37–40, find the first three terms of the expansion.

37. $(m - n)^{11}$
38. $(p - q)^9$
39. $(u^2 - v)^{12}$
40. $(r - s^2)^8$
41. How many terms are in the expansion of $(a + b)^8$?
42. How many terms are in the expansion of $(x + y)^{13}$?

For Exercises 43–48, find the indicated term of the binomial expansion.

43. $(m - n)^{11}$; sixth term
44. $(p - q)^9$; fourth term
45. $(u^2 - v)^{12}$; fifth term
46. $(r - s^2)^8$; sixth term
47. $(5f + g)^9$; 10th term
48. $(4m + n)^{10}$; 11th term

Concepts

1. Fundamental Principle of Counting
2. Permutations
3. Combinations
4. Comparing Permutations and Combinations

section

14.4 Fundamentals of Counting

1. Fundamental Principle of Counting

Suppose a child makes an ice cream sundae with one scoop of ice cream plus a syrup. Suppose there are three choices of ice cream (vanilla, chocolate, and strawberry) and two choices of syrup (fudge and caramel). For each of the three ice cream choices there are two possible syrups, yielding $3 \cdot 2 = 6$ possible sundae combinations (Figure 14-1).

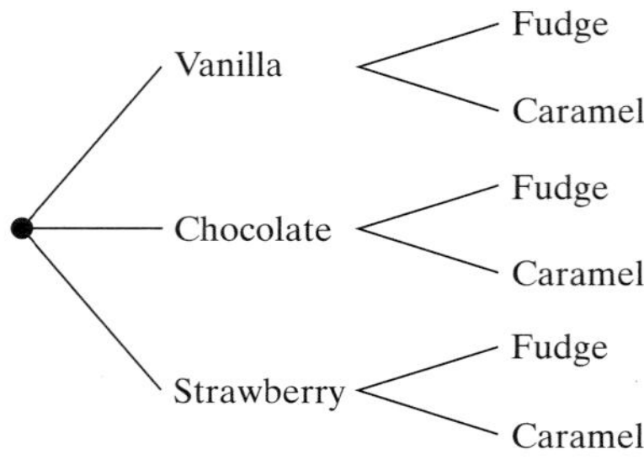

Figure 14-1

This example illustrates the **fundamental principle of counting**.

Fundamental Principle of Counting

If one event can occur in m different ways and a second event can occur in n different ways, then the sequence of both events can occur in $m \cdot n$ different ways.

The fundamental principle of counting can be extended to more than two events as demonstrated in the next example.

example 1

Applying the Fundamental Principle of Counting

Suppose Denisha visits Jimmy G's, a Cajun restaurant in Houston, Texas. She opts for a combo-meal in which she may choose from 12 different entrées, 4 different soups, and 3 different desserts. How many different dinners can she choose if she selects one item from each category?

Solution:

Applying the fundamental principle of counting, we have 144 different dinners.

$$12 \cdot 4 \cdot 3 = 144$$

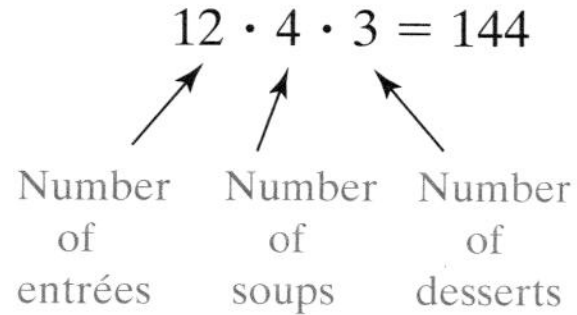

Sometimes the events in a sequence depend on a preceding event as shown in Example 2.

example 2

Applying the Fundamental Principle of Counting

Maximus has five different photos that he wants to arrange on a shelf. In how many different ways can he arrange the five photos?

Solution:

Think of the photo arrangement as five different slots on the shelf (Figure 14-2).

Figure 14-2

The first slot can have any of the five pictures. However, once the first picture is in place, there are only four available for the second slot. Similarly, once the first two pictures are in place, there are only three remaining, and so on. The total number of photo arrangements is

$$\underline{\quad 5 \quad} \cdot \underline{\quad 4 \quad} \cdot \underline{\quad 3 \quad} \cdot \underline{\quad 2 \quad} \cdot \underline{\quad 1 \quad} = 120.$$

The solution to Example 2 could have been written using factorial notation.

$$5 \cdot 4 \cdot 3 \cdot 2 \cdot 1 = 5!$$

Let n be a positive integer. Recall from Section 14.3, that $n!$ (read as "n factorial") is defined as the product of integers from 1 to n. That is,

$$n! = n(n-1)(n-2) \cdot \cdot \cdot (1) \text{ and by definition, } 0! = 1.$$

example 3

Applying the Fundamental Principle of Counting

Suppose eight horses run in a race. How many first-, second-, third-place arrangements are possible assuming no ties?

Solution:

Any of the eight horses can come in first place. That leaves seven possibilities for second place, and six possibilities for third. Therefore, there are $8 \cdot 7 \cdot 6 = 336$ possible first-, second-, third-place arrangements.

2. Permutations

The scenario presented in Example 3 has three different horses selected in a specified order from a group of eight horses. Each of the 336 arrangements is called a permutation of eight horses taken three at a time. In general, an ordered arrangement of r different items selected from n different items is called a **permutation** of n items taken r at a time. The number of all such arrangements is given by the following formula.

Permutation Rule

If ${}_nP_r$ represents the number of permutations of n items taken r at a time, then

$$ {}_nP_r = \frac{n!}{(n-r)!} $$

Other commonly used notations for the number of permutations of n items taken r at a time are $P(n, r)$ and P^n_r.

From Example 3, we found the number of ways three different horses could be selected from eight horses in a race in a specified order. This value is equivalent to ${}_8P_3$.

$$ {}_8P_3 = \frac{8!}{(8-3)!} = \frac{8!}{5!} = \frac{8 \cdot 7 \cdot 6 \cdot \cancel{5 \cdot 4 \cdot 3 \cdot 2 \cdot 1}}{\cancel{5 \cdot 4 \cdot 3 \cdot 2 \cdot 1}} = 8 \cdot 7 \cdot 6 = 336 $$

This is consistent with the result of Example 3.

example 4

Computing Permutations

Compute: a. ${}_{12}P_4$ b. ${}_7P_7$

Solution:

a. $$ {}_{12}P_4 = \frac{12!}{(12-4)!} = \frac{12!}{8!} = \frac{12 \cdot 11 \cdot 10 \cdot 9 \cdot \cancel{8 \cdot 7 \cdot 6 \cdot 5 \cdot 4 \cdot 3 \cdot 2 \cdot 1}}{\cancel{8 \cdot 7 \cdot 6 \cdot 5 \cdot 4 \cdot 3 \cdot 2 \cdot 1}} $$

$$ = 12 \cdot 11 \cdot 10 \cdot 9 $$

$$ = 11{,}880 $$

This result implies that there are 11,880 ways to select 4 items from a group of 12 items where the 4 items are selected in a specified order.

Avoiding Mistakes

Remember that by definition, $0! = 1$.

b. $$ {}_7P_7 = \frac{7!}{(7-7)!} = \frac{7!}{0!} = \frac{7!}{1} = 7! = 5040 $$

This result implies that there are 5040 ways to select seven out of seven items in a specific order.

Example 5 will apply the permutation formula to solve a counting problem.

example 5 Applying Permutations to a Counting Problem

From a group of nine students, four are to be selected to receive cash prizes of \$100, \$50, \$25, and \$10, respectively. In how many ways can the four students be selected?

Tip: The order of selection is important in Example 5, because the four prizes are different. For example, suppose person A is awarded the \$100 prize and B is awarded the \$50 prize. It would be a different outcome if B got the \$100 prize and A received the \$50 prize.

Solution:

Because four different students are to be selected from a group of nine in a specified order, we have ${}_9P_4$.

$$ {}_9P_4 = \frac{9!}{(9-4)!} = \frac{9!}{5!} = \frac{9 \cdot 8 \cdot 7 \cdot 6 \cdot \cancel{5 \cdot 4 \cdot 3 \cdot 2 \cdot 1}}{\cancel{5 \cdot 4 \cdot 3 \cdot 2 \cdot 1}} = 9 \cdot 8 \cdot 7 \cdot 6 = 3024 $$

There are 3024 ways to pick four different students from a group of nine in a specific order.

3. Combinations

In Example 5, we saw that a permutation defines a selection of four distinct items in a specified order from nine items. Now we will compare an event in which the order is important to one in which the order is not important.

Suppose that from a group of four members of student government we want to select two to serve as president and vice-president. If we label the students as A, B, C, and D, we can list all possible selections. The first student listed is designated as president and the second student listed is designated as vice-president. There are 12 possible outcomes shown here.

AB	AC	AD	BC	BD	CD	12 permutations
BA	CA	DA	CB	DB	DC	

These outcomes represent the set of all permutations of four students taken two at a time. Therefore, the number of president/vice-president selections can also be found by computing

$$ {}_4P_2 = \frac{4!}{(4-2)!} = \frac{4!}{2!} = \frac{4 \cdot 3 \cdot \cancel{2 \cdot 1}}{\cancel{2 \cdot 1}} = 12 $$

Now suppose that we want to choose two students from the group of four to form a safety committee. In this case, there is no specific label assigned to the two students selected (in other words, we are not assigning roles to the students such as president or vice-president). In some sense, the two people selected are "equals." They are not holding distinguishable positions. In such a case, we can see that the order of selection is not important. Hence, selecting A first followed by B forms the same committee as selecting B first followed by A. We do not want to count the outcomes AB and BA twice. Therefore, there are only six possible groups as shown here.

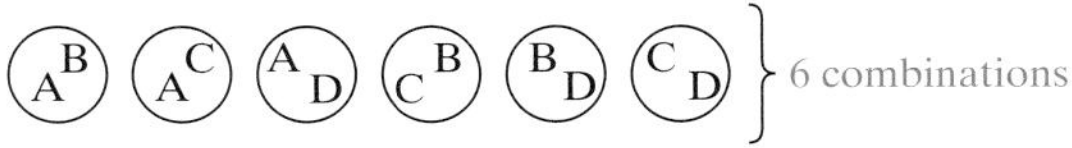

Tip: The number of combinations of n items taken r at a time can also be found by computing the number of permutations and dividing the result by $r!$. The factor of $r!$ divides out repeated combinations that result from order within the permutations. Therefore, the combination formula

$${}_nC_r = \frac{n!}{(n-r)! \cdot r!}$$ can also be written as $${}_nC_r = \frac{{}_nP_r}{r!}$$

In this situation, the order of selection is not important. In general, an *unordered* selection of r different items taken from n different items is called a **combination** of n items taken r at a time. The number of all such groupings is given by the following formula.

Combination Rule

If ${}_nC_r$ represents the number of combinations of n items taken r at a time, then

$${}_nC_r = \frac{n!}{(n-r)! \cdot r!}$$

To count the number of ways two students can be selected from a group of four without regard to order, we compute ${}_4C_2$.

$${}_4C_2 = \frac{4!}{(4-2)! \cdot 2!} = \frac{4!}{2! \cdot 2!} = \frac{4 \cdot 3 \cdot 2 \cdot 1}{(2 \cdot 1) \cdot (2 \cdot 1)} = 6$$

Notice that the combination formula differs from the permutation formula by a factor of $r!$ in the denominator. In this example, there is an additional factor of 2! in the denominator. This factor divides out the repeated arrangements within the groups of two that result from order. In other words, the factor of 2! divides the total list of permutations by 2 to divide out the repeated arrangements such as AB and BA, AC and CA, and so on.

example 6 Computing Combinations

Calculator Connections

Most scientific and graphing calculators have a permutation function and a combination function.

```
4 nPr 2
            12
4 nCr 2
             6
12 nCr 12
             1
```

Compute:

a. ${}_{10}C_2$

b. ${}_{12}C_{12}$

Solution:

a. $${}_{10}C_2 = \frac{10!}{(10-2)! \cdot 2!} = \frac{10!}{8! \cdot 2!} = \frac{10 \cdot 9 \cdot \cancel{8!}}{\cancel{8!} \cdot 2 \cdot 1} = 45$$

This result implies that there are 45 ways to select 2 items from a group of 10 when the order of selection is not considered.

b. $${}_{12}C_{12} = \frac{12!}{(12-12)! \cdot 12!} = \frac{12!}{0! \cdot 12!} = \frac{\cancel{12!}}{1 \cdot \cancel{12!}} = 1$$

This result implies that there is one way to select 12 items from a group of 12 without regard to order.

Tip: You might have noticed that the formula to compute ${}_nC_r$ follows the same format as the formula used to compute the coefficients of a binomial expansion (see Section 14.3).

example 7 **Applying Combinations to a Counting Problem**

Sara picked out 15 different CDs that she likes equally well. However, she only has enough money to purchase four. In how many ways can she select 4 CDs from the group of 15?

Solution:

In this example, there is no implied order. Therefore, the number of ways she can select 4 CDs from 15 CDs is given by ${}_{15}C_4$.

$$ {}_{15}C_4 = \frac{15!}{(15-4)! \cdot 4!} = \frac{15!}{11! \cdot 4!} = \frac{15 \cdot 14 \cdot 13 \cdot 12 \cdot \cancel{11!}}{\cancel{11!} \cdot 4 \cdot 3 \cdot 2 \cdot 1} = 1365 $$

4. Comparing Permutations and Combinations

In this section three different rules for counting have been presented: the fundamental principle of counting, the permutation rule, and the combination rule. Perhaps the most difficult part of solving a counting problem is to select the most appropriate rule to apply.

- The fundamental principle of counting can always be applied, but it is not always the most convenient method.
- The permutation rule can be applied when r different items are selected from a group of n different items in a specified order.
- The combination rule can be applied when r different items are selected from a group of n different items in no specific order.

section 14.4 Practice Exercises

For Exercises 1–4, evaluate the factorial expressions.

1. 6!
2. 8!
3. 0!
4. 1!
5. Evaluate ${}_{10}P_3$ and interpret its meaning.
6. Evaluate ${}_{7}P_2$ and interpret its meaning.
7. Evaluate ${}_{10}C_3$ and interpret its meaning.
8. Evaluate ${}_{7}C_2$ and interpret its meaning.

For Exercises 9–20, compute the permutation or combination.

9. ${}_{12}P_9$
10. ${}_{6}P_4$
11. ${}_{12}C_9$
12. ${}_{6}C_4$
13. ${}_{7}P_1$
14. ${}_{5}P_1$
15. ${}_{7}C_1$
16. ${}_{5}C_1$
17. ${}_{8}P_8$
18. ${}_{4}P_4$
19. ${}_{8}C_8$
20. ${}_{4}C_4$
21. Given the set of elements {A, B, C},
    a. List all permutations of two elements.
    b. List all combinations of two elements.
22. Three people must be selected from a group of 10 people.
    a. Explain the difference between permutations of three people versus combinations of three people.
    b. How many permutations of three can be found from the group of 10?
    c. How many combinations of three can be found from the group of 10?

For Exercises 23–28, apply the fundamental principle of counting.

23. How many different ways can six people be seated in a row?

24. In how many ways can 10 children line up to leave the classroom?

25. A computer password consists of six digits. How many six-digit passwords are possible?

26. A computer filename can consist of eight characters, where a character is any one of the 26 letters of the alphabet or any of the 10 digits. How many computer filenames are possible?

27. At a certain hospital, the dinner menu consists of four choices of entrée, three choices of salad, eight choices of beverage, and six choices of dessert. How many different meals can be formed if a patient chooses one item from each category?

28. Mr. Dehili must dress for an important business meeting. He can choose his outfit from three different suits, six different shirts, and 12 different ties. Assuming that Mr. Dehili has no regard for color combinations, how many different outfits can he form, given that he picks one item from each category?

For Exercises 29–36, use the permutation or combination rule.

29. How many first-, second-, and third-place finishes are possible in a dog race containing 10 dogs?

30. In how many ways can a judge award blue, red, and yellow ribbons to 3 films at a film festival if there are 12 films entered in the contest?

31. Heather wants to invite 11 girls from her fifth-grade class to a slumber party. However, her mother will only allow her to invite six people. In how many ways can she invite 6 girls out of the 11 to her party?

32. How many different five-member committees can be formed from the 100 U.S. senators?

33. A book club offers five books for $9.99 as an introductory offer. If you can choose from a list of 10 books, in how many ways can you make your selection of 5?

34. A basketball coach must pick 5 players from a roster of 12 to start a game. In how many ways can he choose his starting five players assuming that all players have equal ability?

35. Suppose there are eight employees who work at a chain coffee shop in Chicago. The manager wants to select two employees to work at two new shops (shop A and shop B) to help train new employees. If one of the selected employees is to work in shop A and the other is to work in shop B, how many possible selections can the manager make?

36. A disc jockey has five songs that he must play in a half-hour period. How many different ways can he arrange the five songs?

For Exercises 37–54, use an appropriate rule of counting.

37. Most radio stations that were licensed after 1927 have four call letters starting with K or W such as WROD. Assuming no repetitions of letters, how many four-letter sets are possible?

38. A lock on a school locker consists of three different numbers taken from 1 to 39 in a specific order. How many three-number codes are possible?

39. In how many ways can a book buyer select 4 books from a list of 10 different books?

40. How many samples of size 6 can be selected from a population of 30 members?

41. Three men and three women have reserved six seats in a row at a concert. In how many ways can they arrange themselves if the men and women are to alternate seats and a man must sit in the first seat?

42. Three men and three women have reserved six seats in a row for a play. In how many ways can they arrange themselves if the men must all sit together and the women must all sit together?

43. A musician plans to perform nine selections. In how many ways can she arrange the musical selections?

44. From a pool of 12 candidates, the offices of president, vice-president, treasurer, and secretary must be filled. In how many different ways can the offices be filled?

45. A committee is to be formed from a collection of 10 men and eight women. How many committees can be made consisting of exactly three men and one woman?

46. A committee is to be formed from a collection of 10 men and eight women. How many committees of four can be made consisting of two men and two women?

47. If a fair coin is flipped three times, how many different sequences of heads and tails can be formed?

48. If a couple plans to have four children, how many different gender sequences can be formed?

49. From a jury pool of 40 people, 12 are to be selected. In how many different ways can a jury of 12 be selected?

50. To play the Georgia lottery, a person must choose 6 numbers (in any order) from a list of 49 numbers. How many different combinations of six numbers are possible?

51. There are six Republicans and four Democrats sitting in a room.
    a. In how many different ways can a committee of two be selected?
    b. In how many different ways can a committee of two be selected if both people must be Republican?
    c. In how many different ways can a committee of two be selected if both people must be Democrat?
    d. In how many different ways can a committee of two be selected if one person from each party is selected?

52. In how many ways can a 10-question true or false test be answered assuming that a student answers all questions?

53. At a pizza place a customer can order a pizza with or without any of the following options: pepperoni, sausage, mushrooms, peppers, onions, olives, or anchovies. How many different pizzas can be formed?

54. Jeff wants to give his five most valuable baseball cards to his three brothers. In how many ways can he make the distribution?

Expanding Your Skills

55. In how many ways can the batting order be determined for a co-ed softball team with five women and four men if
    a. There are no restrictions.
    b. The first and last batters must be women.
    c. The men must all bat after the women.

56. In how many ways can a group of six men and five women be lined up for a photograph if
    a. There are no restrictions.
    b. The men and women must alternate.
    c. The men must stand to the left of the women.
    d. There must be a man on each end.

57. a. List all the different sequences of letters that can be made from the word *pot.*
    b. List all the different sequences of letters that can be made from the word *tot.*

58. a. List all the different sequences that can be made from the letters in the word *desk*.

    b. List all of the different sequences that can be made from the letters in the word *door*.

59. Given the set of numbers {2, 3, 4, 5, 6, 7, 8}

    a. How many different three-digit numbers can be formed?

    b. How many different three-digit numbers can be formed if the number cannot have any repeated digits?

    c. How many different three-digit numbers can be formed if the number is to be even and repetition of digits is allowed?

60. Given the set of numbers {1, 2, 3, 4, 5, 6, 7}

    a. How many different four-digit numbers can be formed?

    b. How many different four-digit numbers can be formed if the number cannot have any repeated digits?

    c. How many different four-digit numbers can be formed if the number is to be divisible by 5 and repetition of digits is allowed?

For Exercises 61–64, use the following description of a standard deck of 52 cards. A standard deck of cards has four suits (clubs, diamonds, hearts, and spades). Each suit has 13 cards. The hearts and diamonds are red cards, and the clubs and spades are black cards. Assume that five cards are selected from a standard deck. In how many ways can the following occur?

61. All five cards are black.

62. All five cards are hearts.

63. There are three diamonds and two clubs.

64. There are more red cards than black cards.

Concepts

1. Basic Definitions
2. Probability of an Event
3. Estimating Probabilities from Observed Data
4. Events Expressed as Alternatives

section

14.5 Introduction to Probability

1. Basic Definitions

The study of probability provides a mathematical means to measure the likelihood of an event occurring. It is of particular interest because of its application to everyday events.

- The National Cancer Institute estimates that a woman has a 1 in 8 chance of developing breast cancer in her lifetime.
- The probability of winning the California Fantasy Five lottery grand prize is

$$\frac{1}{575{,}757}$$

- Genetic DNA analysis can be used to determine the risk that a child will be born with cystic fibrosis. If both parents test positive, the probability is 25% that a child will be born with the disease.

To begin our discussion, we must first understand some basic definitions.

An activity with observable outcomes is called an **experiment**. Each repetition of an experiment is called a **trial**. The result of a trial is called an **outcome** of the experiment. The set of all possible outcomes of an experiment is called the **sample space**, S, of the experiment.

For example, if a single die is rolled, the sample space is {1, 2, 3, 4, 5, 6}. If a coin is tossed, the outcomes are heads (H) or tails (T). The sample space is {H, T}.

Tip: The word *die* is the singular of the word *dice.* Therefore, we may roll a pair of dice, but we roll a single die.

Any subset of a sample space is called an **event**. For example, if we define event, E_1, as the event that a number greater than 4 is rolled on a die, then, $E_1 = \{5, 6\}$. If event E_2 is the event that a coin lands as a head, then $E_2 = \{\text{H}\}$.

2. Probability of an Event

The number of elements in a sample space is denoted by $n(S)$. The number of elements in the sample space that are also in event E is denoted by $n(E)$. The notation $P(E)$ denotes the probability of event E, defined as follows.

Probability of Event, *E*

In a sample space, S, of equally likely outcomes, the **probability of *E*** is given by

$$P(E) = \frac{n(E)}{n(S)}$$

For the event, E_1, of rolling a number greater than 4 on a die, we have $E_1 = \{5, 6\}$ and $S = \{1, 2, 3, 4, 5, 6\}$. Then

$$P(E_1) = \frac{n(E_1)}{n(S)} = \frac{2}{6} = \frac{1}{3}$$

For the event, E_2, that a coin will land as a head, we have $E_2 = \{\text{H}\}$ and $S = \{\text{H}, \text{T}\}$. Then

$$P(E_2) = \frac{n(E_2)}{n(S)} = \frac{1}{2}$$

The value of a probability can be written as a fraction, as a decimal, or as a percent. Therefore, $P(E_2) = \frac{1}{2}$, or 0.5 or 50%. In words, this means that theoretically we expect half (50%) of the outcomes to land as a head. This does not necessarily mean that exactly one out of every two coin tosses will land as a head. Instead, it is a ratio that we expect experimental observations to approach after a large number of trials. For example, if we flip a coin 1000 times, we might get 493 heads for a ratio of $\frac{493}{1000}$, or 0.493. This value is close to the theoretical value of 0.500.

example 1

Computing Probabilities

A box contains two red, four white, and one blue marble. Suppose one marble is selected at random. Find the probability of selecting

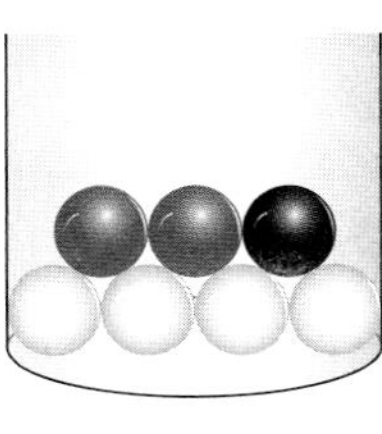

a. A red marble.
b. A white marble.
c. A green marble.

Solution:

Denote the sample space as $S = \{\text{R}_1, \text{R}_2, \text{W}_1, \text{W}_2, \text{W}_3, \text{W}_4, \text{B}_1\}$

a. Let event R represent the event of selecting a red marble. Because there are two red marbles, then $R = \{\text{R}_1, \text{R}_2\}$. Then

$$P(R) = \frac{n(R)}{n(S)} = \frac{2}{7}$$

b. Let W represent the event of selecting a white marble. Because there are four white marbles, then $W = \{W_1, W_2, W_3, W_4\}$. Then $P(W) = \frac{4}{7}$.
c. Let G represent the event of selecting a green marble. Because there are no green marbles in the box, then G equals the empty set. $G = \{\ \}$. Then $P(G) = \frac{0}{7} = 0$.

From Example 1(c), the event of selecting a green marble is impossible. The probability of an **impossible event** is 0. An event that is certain to happen is called a **certain event**. Its probability is 1. For example, if a die is tossed, the probability that the die will land as a number between 0 and 7 is certain to happen. Any of the six outcomes 1, 2, 3, 4, 5, 6 will satisfy the event. Therefore, the probability of rolling a number between 0 and 7 is $\frac{6}{6} = 1$. In general, for any event E,

$$0 \leq P(E) \leq 1$$

The counting rules learned in Section 14.4 can be helpful in determining the number of elements in an event and in a sample space. This is illustrated in Example 2.

example 2

Applying the Counting Rules to Probability

Suppose a group of politicians consists of nine men and five women. If three people are selected at random to form a committee, what is the probability that all are women?

Solution:

Let W represent the event that a committee of three women is selected.

Let S represent the sample space consisting of all possible committees of three with no restrictions.

There are 14 people available to form a committee of 3. If no restrictions are imposed, the number of possible committees of three is given by ${}_{14}C_3 = 364$. The number of possible committees of three selected from the group of women is given by ${}_5C_3 = 10$. Therefore, the probability of selecting a committee that consists of all women is given by

$$P(W) = \frac{n(W)}{n(S)} = \frac{{}_5C_3}{{}_{14}C_3} = \frac{10}{364} = \frac{5}{182}$$

3. Estimating Probabilities from Observed Data

We were able to compute the probabilities in Examples 1 and 2 because the sample space was known. Sometimes we need to collect information from a series of repeated trials to help us estimate probabilities.

example 3

Estimating Probabilities from Observed Data

In a carnival game, Erin will win a prize if she can toss a ring around the neck of a bowling pin. After observing 200 players that had gone before her, she learns that 15 players won a prize. Based on this observation, what is the probability of winning a prize?

Solution:

In this situation, we may think of the outcomes of the 200 trials as the sample space. In this case, 15 trials came out as wins. Therefore, if W represents the event of a win, then

$$P(W) = \frac{15}{200} = \frac{3}{40}, \text{ or } 0.075$$

example 4

Determining Probabilities from a Graph

The data in the pie chart (Figure 14-3) categorizes the number of accidental deaths in the United States for the year 2001.

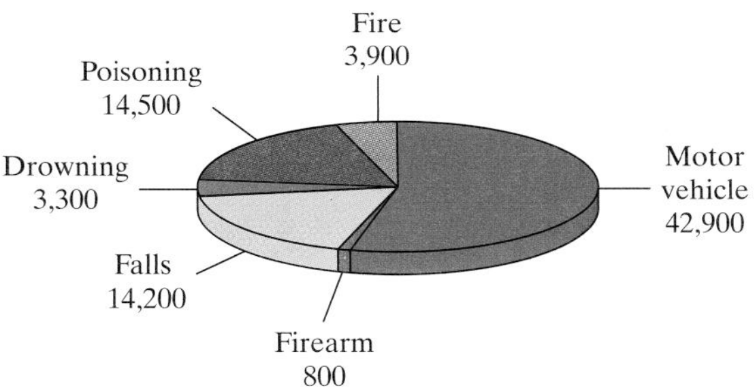

Figure 14-3

a. Based on the chart, what is the probability that an accidental death is caused by fire?
b. What is the probability that an accidental death is caused by a motor vehicle?

Solution:

First note that the total number of accidental deaths depicted in the graph is 79,600.

a. Let F represent the event that the death is caused by fire.

$$\text{Then } P(F) = \frac{3900}{79{,}600} = \frac{39}{796} \text{ or about } 4.9\%.$$

b. Let M represent the event that the death was caused by a motor vehicle.

$$\text{Then } P(M) = \frac{42{,}900}{79{,}600} = \frac{429}{796} \text{ or about } 53.9\%.$$

4. Events Expressed as Alternatives

A compound event in probability involves two or more events. If two events are expressed as alternatives, they can be considered as a compound event often joined by the word *or.*

example 5

Finding the Probability of Two or More Events

The safety and security department at a certain college asked a sample of 265 students to respond to the following question.

"Do you think that the College has adequate lighting on campus at night?"

The table shows the results of the survey by gender and response.

	Yes	No	No Opinion	Total
Male	92	7	4	103
Female	36	102	24	162
Total	128	109	28	265

If one student is selected at random from the group, find the probability that

a. The student answered yes or had no opinion.
b. The student answered no or was female.

Solution:

a. Let Y be the event that a student answered yes, and let Z represent the event that the student had no opinion. There are 128 people who answered "yes" and 28 who had "no opinion." The total number of unique elements in the event Y or Z is 156. Therefore,

$$P(Y \text{ or } Z) = \frac{156}{265}$$

b. Let N be the event that a student answered no, and let F be the event that a student is female. The events N and F "overlap." That is, some people who answered "no" are also "female." We must be careful not to count any element in the sample space twice. There are 109 people who answered "no" (some of whom are female). Counting 109 "no's" and the remaining females, we have a total of $109 + 36 + 24 = 169$ unduplicated individuals who answered "no" or who are "female." Therefore,

$$P(N \text{ or } F) = \frac{169}{265}$$

section 14.5 PRACTICE EXERCISES

1. In how many ways can 4 songs be selected from 10 different songs without regard to order?

2. Samira wants her father to buy her six different toys, but he only has enough money to buy two. In how many ways can Samira's father buy two toys from a group of six?

3. In how many different orders can the five musical notes A, B, C, D, and E be played?

4. In how many ways can Mr. Bishop rank three movies from a group of seven?

5. Which of the values can represent the probability of an event?

   a. 0.5 b. $\frac{2}{3}$ c. $-\frac{7}{5}$ d. 1.00

   e. 150% f. 3.7 g. 3.7% h. 0.92

6. Which of the values can represent the probability of an event?

   a. 1.62 b. $-\frac{2}{7}$ c. 0.00 d. 1.00

   e. 200% f. 4.5 g. 4.5% h. 0.87

7. If a single die is rolled, what is the probability that it will come up as a number less than 4?

8. If a single die is rolled, what is the probability that it will come up as a number greater than 5?

9. The final exam in a course in contemporary science resulted in the following distribution:

Grade	A	B	C	D	F
Number of students	8	15	21	10	5

   a. What is the probability that a student selected at random received an "A" in the course?

   b. What is the probability that a student did not pass the course if a passing grade is a "C" or better?

10. A sample of students is taken from a physical education class at a 4-year college. The distribution is given in the table. If one student is selected at random, find the probability that

   a. The student is a sophomore.

   b. The student is not a senior.

Class	Number of students
Freshman	15
Sophomore	11
Junior	6
Senior	3

11. The tardy record for a group of 2nd graders for one school year is given in the table. If one student is picked at random, find the probability that the student was late

   a. Exactly 3 days.

   b. Between 1 and 5 days, inclusive.

   c. At least 4 days.

   d. More than 5 days or fewer than 2 days.

Number of days late	Number of students
0	4
1	2
2	14
3	10
4	16
5	18
6	10
7	6

12. The data in the pie chart categorizes the method by which workers in a college town commute to work. If one working member of the community is selected at random, find the probability that the individual

   a. Commutes by bicycle.

   b. Does not use public transportation.

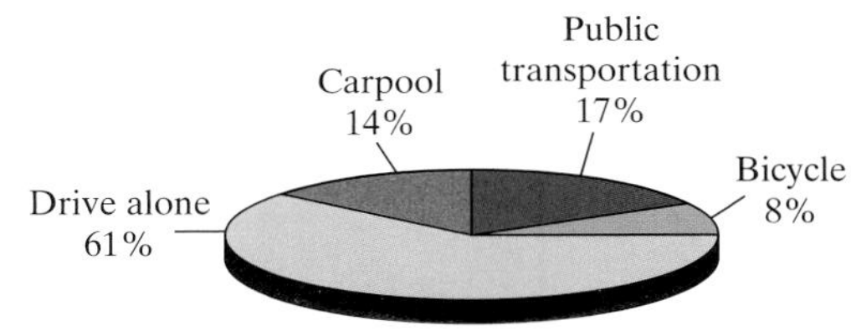

Figure for Exercise 12

13. In a group of eight students, three are female and five are male. If a committee of two is to be selected at random, what is the probability that
    a. Both members are female?
    b. Both members are male?

14. In a group of 10 students at a community college, 6 are freshmen and 4 are sophomores. If a committee of two is selected at random, what is the probability that
    a. Both members are sophomores?
    b. Both members are freshmen?

15. In the California Fantasy Five lottery a player wins the grand prize if the player picks the winning 5 numbers (in any order) out of 39 numbers.
    a. What is the probability that a player will win the grand prize?
    b. What is the probability that a player will not win the grand prize?
    c. Joanne thinks that the probability of winning the grand prize in the lottery is 0.50, because according to her "There's a 50–50 chance of winning, because you either win or you lose." What is wrong with Joanne's logic?

16. In the Florida lottery, a player wins the grand prize if the player picks the winning 6 numbers (in any order) out of 53 numbers.
    a. What is the probability that a player will win the grand prize?
    b. What is the probability that a player will not win the grand prize?

For Exercises 17–21 use the following table categorizing a sample of smokers and nonsmokers according to their blood pressure levels.

	Normal Blood Pressure	Elevated Blood Pressure	Total
Smokers	32	18	50
Nonsmokers	71	9	80
Total	103	27	130

If one person from the sample is selected at random, find the probability that

17. The person has elevated blood pressure.
18. The person is a nonsmoker.
19. The person does not have normal blood pressure.
20. The person is a smoker or has normal blood pressure.
21. The person is a nonsmoker or has normal blood pressure.

For Exercises 22–26 use the following table categorizing a sample of alcoholics and nonalcoholics according to their cholesterol level.

	Elevated Cholesterol	Normal Cholesterol	Total
Alcoholic	120	30	150
Nonalcoholic	60	240	300
Total	180	270	450

If one person from the sample is selected at random, find the probability that

22. The person is a nonalcoholic.
23. The person has elevated cholesterol.
24. The person does not have elevated cholesterol.
25. The person is an alcoholic or has elevated cholesterol.
26. The person is a nonalcoholic or has normal cholesterol.

A standard deck of cards has 52 cards divided into four suits: clubs; diamonds; hearts; and spades. Each suit has 13 cards consisting of an ace, a king, a queen, a jack, and cards numbered from 2 to 10. Assume that one card is selected from a standard deck. For Exercises 27–38 find the probability of selecting the indicated card.

27. A heart?
28. A face card (face cards are jacks, queens, and kings)?
29. A red card (hearts and diamonds are red)?
30. A red card or a jack?
31. A heart or a 6?

32. A spade or a heart?

33. A club?

34. A 5 or a 10?

35. A black card (clubs and spades are black cards)?

36. A black card or an ace?

37. A diamond or a 2?

38. A club or a diamond?

Expanding Your Skills

39. The firefighters in one county in the United States consist of 600 members: 120 women and 480 men. Over a five year period, 160 firefighters were promoted. The table shows the breakdown of promotions for male and female firefighters.

	Promoted	Not Promoted	Total
Male	140	340	480
Female	20	100	120
	160	440	600

If one firefighter is selected at random, what is the probability that

a. The firefighter is a female?

b. The firefighter was promoted?

c. The firefighter is male or was not promoted?

d. The firefighter was promoted, given that he is male?

e. The firefighter was promoted, given that she is female?

f. Write the probabilities in parts (d) and (e) in decimal form (round to three decimal places). What does the difference in probabilities mean in the context of this problem?

40. The following table depicts the grade distribution for a college algebra class based on age and grades.

	A	B	C	F	Total
17–26 years	150	220	370	180	920
27–36 years	140	200	220	90	650
37–46 years	150	120	60	40	370
Total	440	540	650	310	1940

If one student is selected at random from the group, find the probability that

a. The student received an "A" in the course.

b. The student was in the 27–36 age group.

c. The student was in the 37–46 age group or received a "B" in the course.

d. The student received an "A" in the course, given that the student is in the 37–46 age group.

e. The student received an "A" in the course, given that the student is in the 17–26 age group.

f. Write the probabilities in parts (d) and (e) in decimal form (round to three decimal places). What does the difference in probabilities mean in the context of this problem?

chapter 14 SUMMARY

SECTION 14.1—SEQUENCES AND SERIES

KEY CONCEPTS:

An infinite sequence is a function whose domain is the set of positive integers.

A finite sequence is a function whose domain is the set of the first n positive integers.

The expression a_n defines the nth term (or general term) of a sequence.

Sometimes the terms of a sequence may have alternating signs. Such a sequence is called an alternating sequence.

The indicated sum of the terms of a sequence is called a series. The Greek letter Σ (sigma) is used to indicate sums.

KEY TERMS:

a_n
alternating sequence
finite sequence
index of summation
infinite sequence
nth term of a sequence
series
sigma notation
terms of a sequence

EXAMPLES:

1, 4, 9, 16, 25, . . . is an infinite sequence.

1, 4, 9, 16, 25 is a finite sequence.

The nth term of the sequence 1, 4, 9, 16, 25, . . . is given by $a_n = n^2$.

Write the terms of the alternating sequence

$$a_n = (-1)^n(3n) \quad 1 \le n \le 6.$$

Solution: $-3, 6, -9, 12, -15, 18$

Evaluate the sum. $\sum_{i=1}^{4} 5i$

Solution: $\sum_{i=1}^{4} 5i = 5(1) + 5(2) + 5(3) + 5(4)$

$= 5 + 10 + 15 + 20$

$= 50$

SECTION 14.2—ARITHMETIC AND GEOMETRIC SEQUENCES AND SERIES

KEY CONCEPTS:

An arithmetic sequence is a sequence in which the difference between consecutive terms is constant.

The nth term of an arithmetic sequence is given by $a_n = a_1 + (n - 1)d$, where a_1 is the first term and d is the common difference between a term and its predecessor.

The indicated sum of an arithmetic sequence is called an arithmetic series. The sum, S_n, of the first n terms of an arithmetic series is given by

$$S_n = \frac{n}{2}(a_1 + a_n)$$

where a_1 is the first term and a_n is the nth term of the series.

EXAMPLES:

The following sequence is arithmetic:

$$-4, -2, 0, 2, 4, 6, 8, 10, \ldots$$

The nth term of the preceding sequence can be written as: $a_n = -4 + (n - 1)(2)$

Find the sum of the series.

$$\sum_{i=1}^{30} [5 + (i - 1)(2)]$$

Solution: $a_1 = 5, n = 30, d = 2, a_{30} = 63$
Therefore,

$$S_{30} = \tfrac{30}{2}(5 + 63) = 1020$$

A geometric sequence is a sequence in which each term after the first term is a constant multiple of the preceding term.

The following sequence is geometric:

$$50, 30, 18, \frac{54}{5}, \frac{162}{25}, \ldots$$

The nth term of a geometric sequence is given by $a_n = a_1 r^{n-1}$, where a_1 is the first term and r is the common ratio.

The nth term of the preceding sequence can be written as $a_n = 50\left(\frac{3}{5}\right)^{n-1}$.

The indicated sum of a geometric sequence is called a geometric series. The sum, S_n, of the first n terms of a geometric series is given by:

$$S_n = \frac{a_1(1 - r^n)}{1 - r}$$

where a_1 is the first term of the series and r is the common ratio and $r \neq 1$.

Find the sum of the series.

$$\sum_{i=1}^{6} 48\left(\frac{1}{2}\right)^{i-1}$$

Solution: $a_1 = 48$, $r = \frac{1}{2}$, $n = 6$. Hence,

$$S_n = \frac{48[1 - (\frac{1}{2})^6]}{1 - \frac{1}{2}} = \frac{189}{2}$$

Given an infinite geometric series

$$a_1 + a_1 r + a_1 r^2 + \cdots, \text{ with } |r| < 1$$

the sum S of all terms in the series is given by

$$S = \frac{a_1}{1 - r}$$

Note: If $|r| \geq 1$, then the sum does not exist.

Find the sum of the infinite series.

$$4 - 2 + 1 - \frac{1}{2} + \cdots$$

Solution: $a_1 = 4$, $r = -\frac{1}{2}$.
Hence

$$S = \frac{4}{1 - (-\frac{1}{2})} = \frac{8}{3}$$

Key Terms:

arithmetic sequence
arithmetic series
common difference
common ratio
geometric sequence
geometric series

Section 14.3—Binomial Expansions

Key Concepts:

The expression $n!$ (read as "n factorial") is defined as the product of integers from 1 to n. That is,

$$n! = n(n - 1)(n - 2) \cdots (2)(1)$$

Note: $0! = 1$.

For a positive integer, n, the expression $(a + b)^n$ can be expanded using the binomial theorem.

$$(a + b)^n = \frac{n!}{n! \cdot 0!}a^n + \frac{n!}{(n - 1)! \cdot 1!}a^{(n-1)}b + \frac{n!}{(n - 2)! \cdot 2!}a^{(n-2)}b^2 + \cdots + \frac{n!}{0! \cdot n!}b^n$$

The coefficients of the expansion may also be found using Pascal's triangle.

Examples:

$$6! = (6)(5)(4)(3)(2)(1) = 720$$

Expand: $(x + 2y)^4$

Solution:
Using the fifth row of Pascal's triangle for the coefficients, we have

$$\begin{aligned} &(x + 2y)^4 \\ &= 1(x)^4 + 4(x)^3(2y)^1 + 6(x)^2(2y)^2 + 4(x)^1(2y)^3 + 1(2y)^4 \\ &= x^4 + 8x^3y + 24x^2y^2 + 32xy^3 + 16y^4 \end{aligned}$$

Pascal's triangle:

$$\begin{array}{ccccccccc} & & & & 1 & & & & \\ & & & 1 & & 1 & & & \\ & & 1 & & 2 & & 1 & & \\ & 1 & & 3 & & 3 & & 1 & \\ 1 & & 4 & & 6 & & 4 & & 1 \end{array}$$

(and so on . . .)

Let n and k be positive integers such that $k \leq n$. Then the kth-term in the expansion of $(a + b)^n$ is given by

$$\frac{n!}{[n-(k-1)]! \cdot (k-1)!} \cdot a^{n-(k-1)} \cdot b^{k-1}$$

KEY TERMS:

binomial expansion
binomial theorem
factorial notation
Pascal's triangle

Find the third term of $(3s + t^2)^7$.

Solution: $n = 7$, $k = 3$, $a = 3s$, and $b = t^2$. The third term is given by

$$\frac{7!}{[7-(3-1)]! \cdot (3-1)!} \cdot (3s)^{7-(3-1)} \cdot (t^2)^{3-1}$$

$$= \frac{7!}{5! \cdot 2!}(3s)^5(t^2)^2$$

$$= 21(3s)^5(t^2)^2$$

$$= 5103s^5t^4$$

SECTION 14.4—FUNDAMENTALS OF COUNTING

KEY CONCEPTS:

Fundamental Principle of Counting
If one event can occur in m different ways and a second event can occur in n different ways, then the sequence of both events can occur in $m \cdot n$ different ways.

An ordered arrangement of r different items selected from n different items is called a permutation of n items taken r at a time. The number of such permutations is given by

$$_nP_r = \frac{n!}{(n-r)!}$$

An unordered selection of r different items taken from n different items is called a combination of n items taken r at a time. The number of such combinations is given by

$$_nC_r = \frac{n!}{(n-r)! \cdot r!}$$

KEY TERMS:

combination
fundamental principle of counting
permutation

EXAMPLES:

A menu has four choices for salad, six main dishes, and three desserts. How many different meals are available if one item is selected from each category?

Solution: $4 \cdot 6 \cdot 3 = 72$

Suppose there are 12 students in a class and 3 are to be selected to be president, vice-president, and treasurer of the class. How many different arrangements are possible?

Solution: $_{12}P_3 = \frac{12!}{(12-3)!} = \frac{12!}{9!} = 1320$

Suppose there are 12 students in a class, and 3 are to be selected to represent the class at a fund raiser. How many different combinations are possible?

Solution: $_{12}C_3 = \frac{12!}{(12-3)! \cdot 3!} = \frac{12!}{9! \cdot 3!}$

$= 220$

Section 14.5—Introduction to Probability

Key Concepts:

The probability of an event, E, is given by

$$P(E) = \frac{n(E)}{n(S)}$$

where $n(E)$ is the number of elements in the event common to the sample space, and $n(S)$ is the number of elements in the sample space.

Key Terms:

certain event
event
experiment
impossible event
outcome
probability of an event, E
sample space
trial

Examples:

A group of single men and single women who own pets are categorized in the table according to the type of pet they own.

	Cat	Dog	Bird	Total
Male	20	45	4	69
Female	35	30	6	71
Total	55	75	10	140

If one person is selected at random from this group, find the probability that the person

a. owns a cat.
b. owns a dog or a bird.
c. is male or owns a dog.

Solution:

a. $P(C) = \frac{55}{140} = \frac{11}{28}$

b. $P(D \text{ or } B) = \frac{85}{140} = \frac{17}{28}$

c. $P(M \text{ or } D) = \frac{99}{140}$

chapter 14 REVIEW EXERCISES

Section 14.1

1. List the terms of the sequence $a_n = (\frac{3}{4})^n$ for $1 \le n \le 3$

2. Find a formula for the nth term of the sequence: $\frac{2}{3}, \frac{4}{9}, \frac{8}{27}, \frac{16}{81}, \ldots$

3. Find a formula for the nth term of the sequence: $1, 4, 7, 10, 13, \ldots$

4. Find the sum:

$$\sum_{i=1}^{5} 4i^2$$

5. Find the sum:

$$\sum_{k=1}^{4} (-1)^k(2k)$$

6. Write the series in summation notation: $2 - 4 + 6 - 8 + 10 - 12 + 14$

7. Write the series in summation notation: $x^3 + x^6 + x^9 + x^{12}$

8. A radioactive substance decays with a half-life of 1 week. This means that each week the substance will lose half of its mass from the week before. If there is 10 g of the substance initially, write a sequence that represents the amount present at the end of each week for the next 5 weeks.

9. In Miami, a speeding ticket costs \$100 plus \$10 for each mile per hour over the speed limit.
   a. Write the nth term of a sequence describing the cost of a speeding ticket for a car traveling n mph over the speed limit.
   b. Write out the first five terms of the sequence.
   c. Find the 27th term and interpret its meaning in the context of this problem.

Section 14.2

10. Find the common difference for the arithmetic sequence: $10, 6, 2, -2, -6, \ldots$
11. Find the common difference for the arithmetic sequence: $\frac{1}{2}, \frac{1}{6}, -\frac{1}{6}, -\frac{1}{2}, -\frac{5}{6}, \ldots$
12. Write the first five terms of the arithmetic sequence in which $a_1 = 3$ and $d = 4$.
13. Write the nth term of the sequence: $1, 6, 11, 16, \ldots$
14. Write the nth term of the sequence: $2, \frac{2}{3}, \frac{2}{9}, \frac{2}{27}, \ldots$
15. Find the 10th term of the arithmetic sequence in which $a_1 = 7$ and $d = -4$.
16. Find the number of terms in the arithmetic sequence: $10, 12, 14, \ldots 94$.
17. Find the sum of the series:
$$\sum_{i=1}^{50} (i - 4)$$
18. Find the sum of the series: $-4 + (-7) + (-10) + \cdots + (-67)$.
19. Determine the common ratio for the geometric sequence: $24, -12, 6, -3, \ldots$
20. Write the first five terms of the geometric sequence in which $a_1 = 8$ and $r = \frac{1}{2}$.
21. Find the nth term of the geometric sequence: $4, 6, 9, \frac{27}{2}, \frac{81}{4}, \ldots$
22. Find a_8 for the sequence $a_n = -4\left(\frac{3}{2}\right)^{n-1}$.
23. Find the sum of the series:
$$\sum_{k=1}^{5} -4\left(\frac{3}{4}\right)^{k-1}$$
24. Find the sum of the series:
$$9 + 3 + 1 + \frac{1}{3} + \frac{1}{9} + \frac{1}{27} + \frac{1}{81} + \frac{1}{243}$$
25. Find the common ratio, r, then determine the sum of the infinite series, if it exists.
$$2 + \frac{1}{3} + \frac{1}{18} + \frac{1}{108} + \cdots$$
26. Suppose \$1000 is invested at 8% interest compounded annually.
   a. What will the value of the investment be at the end of the first year? At the end of the second year?
   b. Write a formula for the nth term of the sequence describing the value of the investment after n years. Is this sequence arithmetic or geometric?

Section 14.3

27. Expand the binomial. Use Pascal's triangle to find the coefficients. $(x^2 + 4)^5$
28. Evaluate the expression.
$$\frac{8!}{6!}$$
29. Evaluate the expression.
$$\frac{7!}{3! \cdot 4!}$$
30. Use the binomial theorem to expand the binomial: $(c - 3d)^4$.
31. Find the first three terms of the binomial expansion: $(a + 2b)^{11}$
32. Find the eighth term of the binomial expansion: $(5x - y^2)^{10}$.
33. Find the middle term of the binomial expansion: $(a + 2b)^6$.

Section 14.4

34. A pasta bar allows customers to "build" their own dinner. They may choose from five different kinds of pasta, six different meats, 16 different vegetables, and eight sauces. If a customer chooses one item from each category, how many different dinners are possible?

35. If a couple plans to have five children, how many different gender sequences can be formed?

36. A company sells a certain automobile in eight different colors with either automatic or standard transmission. In addition, the car comes in a deluxe, standard, or economy model. How many automobiles would a dealer have to stock if the dealer wanted to have one car of every color, model, and transmission option?

37. In how many ways can a woman pick a computer password if the password must contain four letters and no letter can be repeated?

38. Jonas found five books at a bookstore that he thinks he would like to read. If he can only select two of the books, how many different choices of two books does he have?

39. A committee is to be formed from a collection of 10 men and eight women. How many committees of four can be made consisting of three women and one man?

40. Seven horses are in a race. How many different first- and second-place arrangements are possible?

Section 14.5

41. In 1997, Florida International University conducted a survey of 1200 randomly selected Cuban Americans to measure attitudes regarding Cuban and U.S. political issues. One question regarded the timeframe for a major political change in Cuba. The results of the survey are shown in the following table. If one survey respondent is selected at random, what is the probability that:

When do you think major political change will occur in Cuba?	Percentage response (%)
2–5 years	26.9
6–10 years	16.2
More than 10 years	12.7
Never	18.6
No opinion	25.6

a. The person thought that major political change will never occur in Cuba?

b. The person thought that major political change would occur in more than 5 years?

c. The person had no opinion.

42. The following table describes the smoking habits of a group of asthma sufferers.

	Nonsmoker	Occasional smoker	Regular smoker	Heavy smoker	Total
Men	382	37	60	34	513
Women	403	31	74	37	545
Total	785	68	134	71	1058

If one of the 1058 people is randomly selected, find the probability that the person is

a. A heavy smoker.

b. A woman.

c. Not a heavy smoker.

d. An occasional smoker or a regular smoker.

e. A nonsmoker or a man.

f. A woman or a regular smoker.

chapter 14 TEST

1. Write the terms of the sequence. Is the sequence arithmetic or geometric?

$$a_n = -3\left(\frac{2}{3}\right)^{n-1} \quad \text{for } 1 \le n \le 4.$$

2. Write the nth term of the sequence. Is the sequence arithmetic or geometric?

$$-6, -\frac{11}{2}, -5, -\frac{9}{2}, \ldots$$

3. Find the sum:

$$\sum_{i=1}^{7} 5(-1)^{i-1}$$

4. Study the following figures and the corresponding expression for the area of each region. Write the nth term of a sequence representing the area of the region.

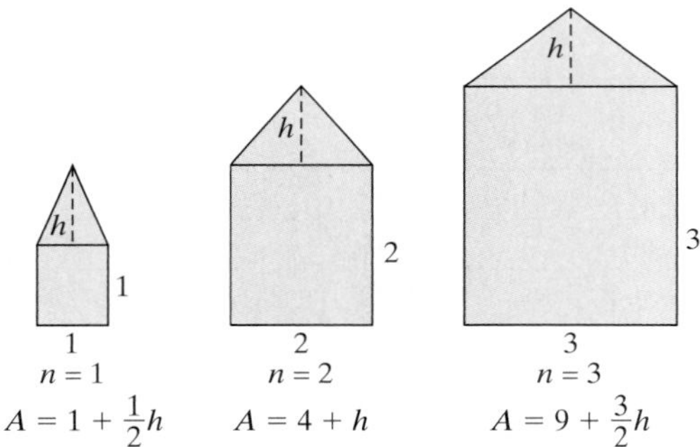

Figure for Exercise 4

5. Write the sum in summation notation.

$$\frac{4}{1} + \frac{5}{2} + \frac{6}{3} + \frac{7}{4} + \frac{8}{5} + \frac{9}{6} + \frac{10}{7}$$

6. Find the 33rd term of the arithmetic sequence in which $a_1 = -7$ and $d = -16$.

7. Find the sum of the series.

$$\sum_{i=1}^{45} [7 + (i - 1)2]$$

8. a. Find the number of terms of the sequence: 6, 10, 14, 18, 22, . . . , 210

   b. Find the sum of the series: $6 + 10 + 14 + 18 + 22 + \cdots + 210$

9. Find the fifth term of the sequence defined by $a_n = \frac{1}{3}(2)^n$.

10. Find the first term of the geometric sequence in which $a_4 = \frac{5}{81}$ and $r = \frac{1}{3}$.

11. Find the sum: $4 + (-8) + 16 + (-32) + \cdots + (-2048)$

12. Find the sum.

$$\sum_{n=1}^{6} 5\left(\frac{3}{2}\right)^{n-1}$$

13. Suppose a student smokes one pack of cigarettes per day at a cost of \$3.00 per pack.

    a. If the student quits smoking and puts the \$3.00 saved per day in a jar, how much money will the student have saved at the end of 1 year? (Ignore leap years.)

    b. If the student takes the savings from part (a) and invests the money in a bond fund paying 6% annual interest, how much will this money be worth 1 year later?

    c. Suppose that the student reinvests the principal and interest each year so that effectively the money is compounded annually. The sequence defined by $a_n = 1095(1.06)^n$ represents the value of the investment after n years. Find the value of the investment after 29 years. (Round to two decimal places.)

    d. Now suppose the student saves \$1095 per year, each year for 30 years, by not buying cigarettes. At the end of each year, the student invests the year's savings at 6% interest compounded annually. The table shows the amount of money that the student has at the end of each year.

Year	Amount
End of first year	\$1095
End of second year	\$1095(1.06) + \$1095
End of third year	$\$1095(1.06)^2 + \$1095(1.06) + \$1095$
. . .	. . .
End of 30th year	$\$1095(1.06)^{29} + \$1095(1.06)^{28} + \cdots + \$1095$

Notice that the first year's savings of \$1095 will be compounded 29 times. The second year's savings will be compounded 28 times and so on. Evaluate the sum

$$\sum_{i=1}^{30} 1095(1.06)^{n-1}$$

to two decimal places and interpret this value in the context of the problem.

14. Find the infinite sum, if possible: $2 + \frac{2}{3} + \frac{2}{9} + \frac{2}{27} + \cdots$

15. Write the expansion of $(a + b)^4$. Use Pascal's triangle to determine the coefficients of the expansion.

16. Expand $(3y - 2x^2)^4$.

17. Find the third term of the expansion of $(w + 3z)^7$.

18. Three airlines fly from Atlanta to Philadelphia. If you can choose from any of the three airlines and can travel either first class or business class, how many different travel options do you have for a one-way trip between the two cities?

19. At an art show, 25 artists display their work. In how many ways can the judges choose the first-, second-, and third-place winners?

20. If a fair coin is flipped two times, how many different sequences of heads and tails can be formed?

21. To play the Colorado lottery game, Lotto, a person must choose 6 numbers in any order from a list of 42 numbers. How many different combinations of 6 numbers are possible?

22. A survey of 2225 people in a community revealed the following distribution by type of housing. If one of the survey respondents is randomly selected, find the probability of selecting someone who
    a. Lives in a mobile home.
    b. Does not live in a house, mobile home, or apartment.

Type of Housing	Number of People
House	690
Mobile home	387
Apartment	543
Other	605

23. The following table depicts a sample of 50 people organized by educational level and gender. If one person is selected at random from the group, find the probability that
    a. The person is a college graduate.
    b. The person is female.
    c. The person is male or a college graduate.

	College Graduate	Not a College Graduate	Total
Male	8	16	24
Female	12	14	26
Total	20	30	50

CUMULATIVE REVIEW EXERCISES, CHAPTERS 1–14

1. Simplify.

$$\frac{-2 - 3\sqrt{25 - 16}}{\sqrt{8^2 + 6^2}}$$

2. Simplify: $-3\{4 - 2[5 + 2(3 - x)] - 5x\}$

3. Solve: $-4(2x - 5) - 2(3x + 1) = -10 + 7x$

4. \$10,000 is invested in two accounts, one earning 8% simple interest and one earning 5% simple interest. If the total interest at the end of 1 year is \$566.75, find the amount invested in each account.

5. Two complementary angles are such that the measure of one is 6° more than three times the other. Find the measure of each angle.

6. Solve the inequality. Write the final answer in interval notation. $-3(x - 4) - x > -8$

7. Find the x- and y-intercept and graph the line: $y = 3x$

8. Find the slope and y-intercept of the line.
$$-\frac{x}{2} + \frac{y}{2} = 1$$

9. Find an equation of a line passing through (2, 3) and (−3, 1). Write the final answer in slope-intercept form.

10. Find an equation of a line passing through (−2, 4) and perpendicular to the line $x + 3y = 4$.

11. Solve the system.
$$\begin{aligned} 2x + 3y - 4z &= 17 \\ x - 2y + z &= -5 \\ 3x + 4y - 3z &= 21 \end{aligned}$$

12. Solve the system.
$$\begin{aligned} y &= \frac{1}{3}x + 4 \\ -x + 3y &= 6 \end{aligned}$$

13. Solve the system using Gauss-Jordan elimination.
$$\begin{aligned} 2x - 4y &= -10 \\ -5x + y &= 16 \end{aligned}$$

14. Evaluate the determinant.
$$\begin{vmatrix} -6 & 3 \\ -1 & -4 \end{vmatrix}$$

15. Solve for y using Cramer's rule.
$$\begin{aligned} 2x - 3y \quad &= 12 \\ -x + 4y + z &= 2 \\ y + 3z &= 6 \end{aligned}$$

16. State the domain and range of the relation: {(4, 2), (3, −6), (2, 2), (4, 5), (4, 1)}

17. The wavelength, w, of a radio wave varies inversely as its frequency, f. A wave with a length of 300 m has a frequency of 1200 kHz (kilohertz). Find the frequency of a 450-m wave.

18. Simplify.
$$\left(\frac{3x^3y^{-4}}{x^4y^2}\right)^{-2}$$

19. Simplify. Write the answer in scientific notation.
$$\frac{(2.0 \times 10^3)(3.2 \times 10^{-7})}{1.6 \times 10^{-9}}$$

20. Add or subtract as indicated.
$$(3x^3 - 2x - 4) - (4x^3 + 2x^2 + 5) + (-8x^2 - 5x - 3)$$

21. Multiply and simplify: $[(x + 2y) - 3][(x + 2y) + 3]$

22. Divide: $(3x^4 - 2x^2 + 5x - 3) \div (x^2 + 2)$

23. Simplify. Assume that x can be any real number. $\sqrt[4]{x^4}$

24. Find the domain: $g(x) = \sqrt{2x - 6}$

25. Simplify: $8^{-2/3}$

26. Simplify.
$$\left(\frac{x^{1/3}y^{5/6}}{x^{-1/3}y^{1/6}}\right)^{12}$$

27. Simplify: $\sqrt[3]{16x^5y^{12}}$

28. Rationalize the denominator.
$$\frac{2}{\sqrt[3]{4x}}$$

29. Solve: $\sqrt[3]{4x + 2} + 5 = 3$

30. Simplify: $3i^4 - 2i^2 + 5i - 1$

31. Divide and write the answer in the form $a + bi$.
$$\frac{-6}{3 - 7i}$$

32. Factor completely: $6x^2 + 19x - 36$

33. Factor completely: $8x^6 + y^3$

34. Solve: $4x^2 - 9 = 0$

35. Solve by completing the square and applying the square root property: $2x^2 + 12x = 10$

36. Solve by using the quadratic formula: $2x(x - 5) + 7x = -2$

37. Solve: $x^4 - x^2 - 12 = 0$

38. The number of diagonals, N, of a polygon having k sides (where k is an integer and $k \geq 3$) is given by

$$N = \frac{k^2 - 3k}{2}$$

   a. Find the number of diagonals for a four-sided polygon.

   b. Find the number of sides for a polygon having 209 diagonals.

39. Write the domain in interval notation.

$$f(x) = \frac{1}{x^2 - 49}$$

40. Divide.

$$\frac{2x^2 + 10x - 12}{x^2 - 16} \div \frac{5x + 30}{x^2 - 3x - 4}$$

41. Simplify.

$$\frac{\frac{1}{y} - \frac{y}{x^2}}{\frac{1}{x} + \frac{1}{y}}$$

42. Solve.

$$\frac{x + 5}{x - 2} = \frac{5}{x + 2} + \frac{28}{x^2 - 4}$$

43. Solve.

   a. $-3x + 2 \leq 4$ or $2x + 3 \leq -5$

   b. $-3x + 2 \leq 4$ and $2x + 3 \leq -5$

44. Solve: $|2x - 3| = |x + 5|$

45. Solve and write the solution in interval notation: $|x - 5| + 3 < 10$

46. Solve and write the solution in interval notation.

$$\frac{2}{x - 4} + 1 \leq 0$$

47. Graph the inequality: $2x + y \geq 3$

48. Find the inverse.

$$f(x) = \frac{3x - 1}{2}$$

For Exercises 49–50, state the domain and range and graph the function.

49. $f(x) = e^x$

50. $g(x) = \log_2 x$

51. Evaluate: $\log_2 8$

52. Evaluate: $\ln(e^5)$

53. Write as a single logarithm: $3 \ln x - \ln y + \frac{1}{2} \ln z$

54. Marc invests \$5000 for his son's college education. At the end of 18 years, Marc hopes to have \$20,000. At what interest rate (compounded continuously) must Marc average to reach his goal? (*Hint:* Use $A = Pe^{rt}$.) Round to the nearest tenth of a percent.

55. Solve: $\log(x + 2) + \log(x - 1) = 1$

For Exercises 56–57, state whether the sequence is arithmetic, geometric, or neither, and then find the nth term of the sequence.

56. $-16, -21, -26, -31, -36, \ldots$

57. $\frac{3}{4}, \frac{3}{2}, 3, 6, 12, \ldots$

58. Find the sum of the series: $\sum_{i=1}^{6} 3\left(\frac{1}{5}\right)^{i-1}$

59. Find the sum of the series: $-5 + (-2) + 1 + 4 + 7 + \cdots + 88$

60. Expand the binomial: $(3 - 2t^2)^4$

61. To play the Arizona lottery, a person must choose 6 numbers in any order from a list of 41 numbers. How many different combinations of 6 numbers are possible?

62. There are six front-row seats in a classroom. In how many ways can 6 students out of 15 students be arranged in the front-row seats?

63. The students in a psychology class are categorized by gender and year of study as shown in the table.

	Freshman	Sophomore	Junior	Senior	Total
Male	16	15	8	1	40
Female	12	15	3	4	34
Total	28	30	11	5	74

If one student is selected at random, find the probability that

a. The student is female.

b. The student is not a freshman.

c. The student is a sophomore or a male.

d. The student is a sophomore or a junior.

64. Given the parabola defined by $y = \frac{1}{4}(x - 2)^2 - 4$

a. Find the coordinates of the vertex.

b. Find the y-intercept.

c. Find the x-intercept(s) if any exist.

d. Write an equation for the directrix of the parabola.

e. Sketch the parabola and the features found in parts (a)–(d).

65. Identify the center and radius of the circle: $(x - 3)^2 + (y + 4)^2 = 35$

66. Graph the ellipse.

$$\frac{x^2}{9} + \frac{y^2}{4} = 1$$

67. Graph the hyperbola.

$$\frac{y^2}{16} - \frac{x^2}{9} = 1$$

68. Solve the system.

$$x^2 + y^2 = 100$$
$$3y = 4x$$

69. Solve the system of equations.

$$x^2 + y^2 = 20$$
$$y = x^2$$

70. Graph the solution set of the system of inequalities.

$$x^2 + y^2 \le 20$$
$$y \ge x^2$$

Beginning Algebra Review

Concepts

1. Sets of Real Numbers
2. Symbols and Mathematical Language
3. Operations on Real Numbers
4. Exponents and Square Roots
5. Order of Operations
6. Properties of Real Numbers
7. Simplifying Expressions

review

A Set of Real Numbers

1. Sets of Real Numbers

the real number line, p. 34

The numbers we are familiar with in everyday life comprise the **set of real numbers**. Every real number corresponds to a unique point on a number line.

For example:

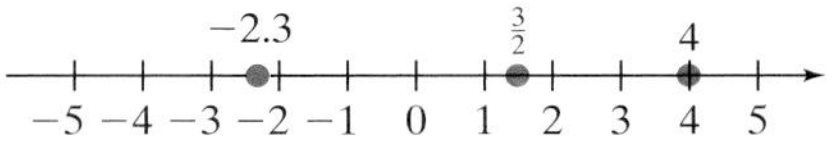

subsets of real numbers, p. 35

Several subsets of the real numbers are given below. We use the symbols { } to enclose the elements of a set.

Set	**Definition**
Natural numbers	$\{1, 2, 3, \ldots\}$
Whole numbers	$\{0, 1, 2, 3, \ldots\}$
Integers	$\{\ldots, -3, -2, -1, 0, 1, 2, 3, \ldots\}$
Rational numbers	the set of numbers of the form $\frac{p}{q}$ where p and q are integers and $q \neq 0$.
Irrational numbers	the set of real numbers that are not rational

identifying rational numbers, p. 35

Rational numbers can be written as a ratio of two integers such as $\frac{1}{3}$ or $\frac{3}{4}$. When expressed in decimal form a rational number is either a repeating decimal or a terminating decimal such as $0.\overline{3}$ or 0.75. The decimal form of an irrational number is non-repeating and non-terminating. Examples of irrational numbers are π and $\sqrt{2}$.

2. Symbols and Mathematical Language

In mathematics, the symbol $<$ (meaning "is less than") and the symbol $>$ (meaning "is greater than") are used to express inequalities. These and other inequality symbols are summarized in Table A-1.

inequality symbols, p. 37

Table A-1

Inequality	**In Words**	**Example**
$a < b$	a is less than b	$5 < 7$
$a > b$	a is greater than b	$8 > 3$
$a \leq b$	a is less than or equal to b	$8 \leq 9$
$a \geq b$	a is greater than or equal to b	$3 \geq 3$
$a < x < b$	x is between a and b	$3 < 4 < 5$
$a \neq b$	a is not equal to b	$4 \neq 6$

The operations of addition, subtraction, multiplication, and division can be denoted several ways. These along with other common operations on real numbers are summarized in (Table A-2).

common mathematical translations, p. 44

Table A-2

Operation	Symbols	Translation/Example
Addition	$a + b$	"the **sum** of a and b" "the sum of 8 and -4" $\Rightarrow 8 + (-4)$
Subtraction	$a - b$	"the **difference** of a and b" "the difference of -8 and 2" $\Rightarrow -8 - 2$
Multiplication	$a \times b, a \cdot b, a(b),$ $(a)b, (a)(b), ab$	"the **product** of a and b" "the product of 3 and -2" $\Rightarrow 3(-2)$
Division	$a \div b, \frac{a}{b}, a/b, b\overline{)a}$	"the **quotient** of a and b" "the quotient of 10 and 5" $\Rightarrow \frac{10}{5}$
Absolute value	$\|a\|$	"the **absolute value** of a" "the absolute value of -3" $\Rightarrow \|-3\|$
Opposite	$-a$	"the **opposite** of a" "the opposite of 7" $\Rightarrow -(7)$
Reciprocal	$\frac{1}{a} (a \neq 0)$	"the **reciprocal** of a" "the reciprocal of 5" $\Rightarrow \frac{1}{5}$
Square root	$\sqrt{a}$	"the **square root** of a" "the square root of 64" $\Rightarrow \sqrt{64}$

3. Operations on Real Numbers

absolute value and opposite, pp. 39–40

The **absolute value** of a real number, a, denoted $|a|$, is its distance from 0 on the number line. Two numbers that are the same distance from 0 but on opposite sides of the number line are called **opposites**. The opposite of a is denoted $-a$.

example 1 **Finding Opposites and Absolute Value**

a. Find the opposite of 5.

b. Find the opposite of $-\frac{1}{2}$.

c. Evaluate $|-6|$.

d. Evaluate $|2.7|$.

e. Evaluate $-|-8|$.

Solution:

a. The opposite of 5 is -5.

b. The opposite of $-\frac{1}{2}$ is $\frac{1}{2}$.

c. $|-6| = 6$

d. $|2.7| = 2.7$

e. $-|-8| = -8$ Take the opposite of the absolute value of -8.

The rules for adding, subtracting, multiplying, and dividing real numbers are given as follows.

adding real numbers, p. 55

Addition of Two Real Numbers	
• To add two numbers with the *same sign:* Add the absolute values of the numbers and apply the common sign to the sum.	example: $-3 + (-7) = -(3 + 7) = -10$
• To add two numbers with *unlike signs:* Subtract the smaller absolute value from the larger absolute value. Then apply the sign of the number having the larger absolute value.	examples: $10 + (-4) = 10 - 4 = 6$ $-8 + 5 = -(8 - 5) = -3$

subtracting real numbers, p. 59

Subtraction of Two Real Numbers	
Add the opposite of the second number to the first number. In symbols: $a - b = a + (-b)$	examples: $8 - (-4) = 8 + (4) = 12$ $-3 - 6 = -3 + (-6) = -9$

multiplying and dividing real numbers, pp. 67–69

Multiplication and Division of Two Real Numbers	
• The product or quotient of two real numbers with the *same* sign is positive.	example: $-20 \div (-4) = 5$
• The product or quotient of two real numbers with *unlike* signs is negative.	example: $\dfrac{30}{-2} = -15$
Notes: • The product of any real number and 0 is 0. • The quotient of 0 and any nonzero real number is 0. • The quotient of a real number and 0 is undefined.	example: $5 \cdot 0 = 0$ example: $0 \div 6 = 0$ example: $6 \div 0$ is undefined

4. Exponents and Square Roots

exponents, p. 45

We can use **exponents** to represent repeated multiplication. For example, the product

$$3 \cdot 3 \cdot 3 \cdot 3 \cdot 3 \quad \text{can be written as} \quad 3^5.$$

(exponent: 5; base: 3)

The expression 3^5 is written in exponential form. The exponent (or **power**) is 5 and represents the number of times the **base**, 3, is multiplied. The expression 3^5 is read as "three to the fifth power." In general,

$$x^n = \underbrace{x \cdot x \cdot x \cdot \cdots \cdot x}_{\text{the factor } x \text{ occurs } n \text{ times}}$$

square roots, p. 46

The number b is a **square root** of a if $b^2 = a$. A **radical sign**, $\sqrt{\ }$, is used to denote the principle (or positive) square root of a nonnegative real number.

example 2 Evaluating Expressions Containing Exponents and Square Roots

Evaluate. a. 2^4 b. $\sqrt{36}$

Solution:

a. $2^4 = 2 \cdot 2 \cdot 2 \cdot 2 = 16$

b. $\sqrt{36} = 6$ because $6^2 = 36$.

5. Order of Operations

Algebraic expressions often have more than one operation. In such a case, it is important to follow the order of operations.

order of operations, p. 47

Order of Operations

1. Simplify expressions within parentheses and other grouping symbols first. These include absolute value bars, fraction bars, and radicals. If imbedded parentheses are present, start with the innermost parenthesis.
2. Evaluate expressions involving exponents and radicals.
3. Perform multiplication or division in the order that they occur from left to right.
4. Perform addition or subtraction in the order that they occur from left to right.

example 3 **Applying the Order of Operations**

Simplify. $50 \div [15 - 2(4 + 6) + 15] \cdot 2^3$

Solution:

$50 \div [15 - 2(4 + 6) + 15] \cdot 2^3$

$= 50 \div [15 - 2(10) + 15] \cdot 2^3$	Simplify inside the inner parentheses.
$= 50 \div [15 - 20 + 15] \cdot 2^3$	Simplify inside [ ]. Within [ ], we perform multiplication before addition or subtraction.
$= 50 \div [-5 + 15] \cdot 2^3$	Within [ ], subtract and add from left to right.
$= 50 \div [10] \cdot 2^3$	Simplify within [ ].
$= 50 \div [10] \cdot 8$	Evaluate the exponential expressions.
$= 5 \cdot 8$	Divide and multiply from left to right.
$= 40$	Multiply.

example 4 **Applying the Order of Operations within a Formula**

Given $x = -3$, evaluate the expressions. a. x^2 b. $-x^2$

Solution:

a. x^2

$= (\ \ )^2$	Use parentheses in place of the variable.
$= (-3)^2$	Substitute -3 for x.
$= 9$	Evaluate expressions with exponents.

b. $-x^2$

$= -(\ \ )^2$	Use parentheses in place of the variable.
$= -(-3)^2$	Substitute -3 for x.
$= -(9)$	Simplify $(-3)^2$ first. Note that $(-3)^2 = 9$.
$= -9$	Now take the opposite of 9.

Tip: The expression $-(-3)^2$ is equivalent to $-1(-3)^2$. The order of operations indicates that we should square -3 first, and then multiply the result by -1.

6. Properties of Real Numbers

Several properties of real numbers are reviewed in Table A-3.

commutative property, p. 76

associative property, p. 78

distributive property of multiplication over addition, p. 80

identity property, p. 79

inverse property, p. 80

Table A-3

Property Name	Algebraic Representation	Example	Description/Notes
Commutative property of addition	$a + b = b + a$	$5 + 3 = 3 + 5$	The order in which two real numbers are added or multiplied does not affect the result.
Commutative property of multiplication	$a \cdot b = b \cdot a$	$(5)(3) = (3)(5)$	
Associative property of addition	$(a + b) + c = a + (b + c)$	$(2 + 3) + 7 = 2 + (3 + 7)$	The manner in which two real numbers are grouped under addition or multiplication does not affect the result.
Associative property of multiplication	$(a \cdot b)c = a(b \cdot c)$	$(2 \cdot 3)7 = 2(3 \cdot 7)$	
Distributive property of multiplication over addition	$a(b + c) = ab + ac$	$3(5 + 2) = 3 \cdot 5 + 3 \cdot 2$	A factor outside the parentheses is multiplied by each term inside the parentheses.
Identity property of addition	0 is the identity element for addition because $a + 0 = 0 + a = a$	$5 + 0 = 0 + 5 = 5$	Any number added to the identity element, 0, will remain unchanged.
Identity property of multiplication	1 is the identity element for multiplication because $a \cdot 1 = 1 \cdot a = a$	$5 \cdot 1 = 1 \cdot 5 = 5$	Any number multiplied by the identity element, 1, will remain unchanged.
Inverse property of addition	a and $(-a)$ are additive inverses because $a + (-a) = 0$ and $(-a) + a = 0$	$3 + (-3) = 0$	The sum of a number and its additive inverse (**opposite**) is the identity element, 0.
Inverse property of multiplication	a and $\frac{1}{a}$ are multiplicative inverses because $a \cdot \frac{1}{a} = 1$ and $\frac{1}{a} \cdot a = 1$ (provided $a \neq 0$)	$5 \cdot \frac{1}{5} = 1$	The product of a number and its multiplicative inverse (**reciprocal**) is the identity element, 1.

7. Simplifying Expressions

combining like terms, p. 82

We can simplify expressions containing variables by adding or subtracting *like* terms. Recall that *like* terms have the same variables and the corresponding variables are raised to the same powers. To combine *like* terms, we use the distributive property. The same result may also be obtained by adding or subtracting the coefficients of each term and leaving the variable factors unchanged.

$$5x + 2x - 3x = 4x$$

example 5 Simplifying Expressions

Simplify. $4x - 3(2x - 8) - 1 + 5x$

Solution:

$4x - 3(2x - 8) - 1 + 5x$

$= 4x - 6x + 24 - 1 + 5x$ Apply the distributive property.

$= 4x - 6x + 5x + 24 - 1$ Arrange *like* terms together.

$= 3x + 23$ Combine *like* terms.

review A PRACTICE EXERCISES

For Exercises 1–8, refer to sets
$A = \{-6, -\sqrt{5}, -\frac{7}{8}, -0.\overline{3}, 0, \frac{4}{9}, \pi, 7, 10.4\}$ and
$B = \{-2, -\pi, -\frac{1}{2}, 0, 1, \sqrt{3}, \frac{5}{4}, 8\}$

1. List all rational numbers in set A.
2. List all irrational numbers in set A.
3. List all integers in set A.
4. List all whole numbers in set A.
5. List all natural numbers in set B.
6. List all integers in set B.
7. List all irrational numbers in set B.
8. List all rational numbers in set B.

For Exercises 9–16, write the expression in words.

9. $5 > -2$
10. $8 \leq 9$
11. $-3 \geq -4$
12. $\frac{1}{2} > -\frac{1}{2}$
13. $6 < 10 < 12$
14. $-1 < 0 < 1$
15. $2 \neq -5$
16. $\frac{2}{3} \neq -\frac{2}{3}$

For Exercises 17–26, translate the English phrase into an algebraic expression.

17. The product of 6 and 3
18. The difference of 7 and 4
19. The quotient of 20 and 5
20. The sum of 2 and 12
21. The absolute value of -5

22. The absolute value of -13

23. The sum of 2 and the absolute value of -8.

24. The difference of 10 and the absolute value of -1

25. The product of 4 and the reciprocal of 3

26. The quotient of 8 and the square root of 16

For Exercises 27–46, add or subtract as indicated.

27. $8 + (-10)$

28. $23 + (-9)$

29. $51 - 32$

30. $12 - 15$

31. $0 + 16$

32. $26 + 0$

33. $-13 - 20$

34. $-5 - 30$

35. $-2.1 + (-2.2)$

36. $-4.0 + (-0.5)$

37. $-19 - (-3)$

38. $-7 - (-28)$

39. $0 - 19$

40. $0 - 41$

41. $5.2 - (-0.4)$

42. $0.03 - (-1.2)$

43. $-\frac{3}{4} - \frac{3}{8}$

44. $-\frac{5}{2} - \frac{1}{6}$

45. $-\frac{7}{3} - \left(-\frac{1}{9}\right)$

46. $-\frac{7}{10} - \left(-\frac{4}{5}\right)$

For Exercises 47–66, multiply or divide as indicated.

47. $3(-9)$

48. $(-8)(2)$

49. $(-5)(-7)$

50. $(-11)(-3)$

51. $0 \cdot 32$

52. $0 \cdot (-12)$

53. $(2.05)(-4.2)$

54. $(-10.7)(3.4)$

55. $\left(-\frac{5}{9}\right) \cdot \left(-\frac{3}{2}\right)$

56. $\left(-\frac{1}{6}\right) \cdot \left(-\frac{12}{7}\right)$

57. $-56 \div (-8)$

58. $-51 \div (-3)$

59. $-78 \div 13$

60. $112 \div (-16)$

61. $0 \div (-5)$

62. $0 \div 12$

63. $3.1 \div 0$

64. $-1.4 \div 0$

65. $\left(-\frac{7}{8}\right) \div \left(-\frac{7}{4}\right)$

66. $\left(-\frac{3}{11}\right) \div \left(\frac{21}{22}\right)$

For Exercises 67–80, simplify using the order of operations.

67. $|-14| - |-5|$

68. $\sqrt{81} - |-4|$

69. $10 \div 5 \cdot 4$

70. $12 \div 2 - 3$

71. $6 + 42 \div 7 - 10$

72. $21 - 3 \cdot 5 + 6$

73. $(8 - 5)^2 - (2 + 3)^2$

74. $16 - (10 - 3)^2 + 2 \cdot 7$

75. $3^3 - 2 + 5(3 - 7)$

76. $6 - 4^2 + (8 \div 4 + 5)$

77. $12 - \sqrt{25} + 4^2 \div 2$

78. $\sqrt{81} - \sqrt{16} \div 4 + 18$

79. $\dfrac{6 + 8 - (-2)}{-40 \div 8 - 1}$

80. $\dfrac{\sqrt{32 \div 2} + 7}{-52 + (7 \cdot 4 + 2)}$

For Exercises 81–84, evaluate the expressions given $a = 3$, $b = -5$, $c = 9$, and $d = -2$.

81. $a^2 - b^2$

82. $c \div a - b$

83. $b + \sqrt{c} \cdot d$

84. $-a^2 + c$

85. The area of a trapezoid is given by the formula $A = \frac{1}{2}(b_1 + b_2)h$. Find the area of the given trapezoid.

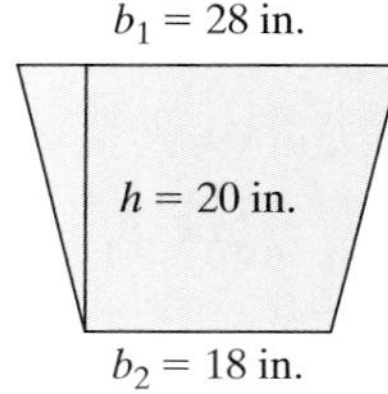

86. The surface area of a rectangular solid with a square base is given by $S = 4ab + 2a^2$. Find the surface area of the given rectangular solid.

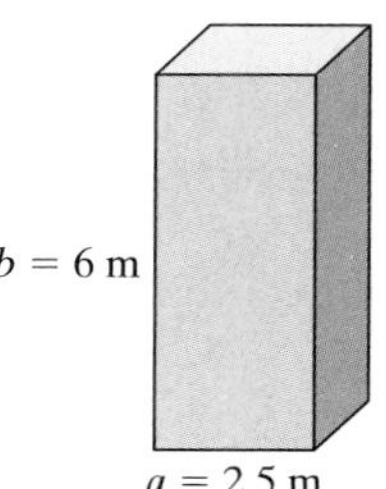

For Exercises 87–92, identify the property that is shown. Choose from the commutative property of addition, commutative property of multiplication, associative property of addition, associative property of multiplication, or distributive property of multiplication over addition.

87. $4(8 \cdot 14) = (4 \cdot 8)(14)$

88. $(7 + 4) + 6 = 7 + (4 + 6)$

89. $-6(3 + 9) = -6(3) + -6(9)$

90. $9 \cdot 2 = 2 \cdot 9$

91. $5 + 15 = 15 + 5$

92. $9(11 + 4) = 9(11) + 9(4)$

93. What is the identity element for multiplication? Use it in an example.

94. What is the identity element for addition? Use it in an example.

95. a. What is another name for multiplicative inverse?

    b. Write the multiplicative inverse of 4.

96. a. What is another name for the additive inverse?

    b. Write the additive inverse of 6.

For Exercises 97–104, simplify the expression by applying the distributive property.

97. $5(a + 9)$

98. $3(x + 12)$

99. $-7(y - z)$

100. $-2(a - b)$

101. $\frac{1}{3}(6b + 9c - 15)$

102. $\frac{1}{5}(10h - 20k - 5)$

103. $-(2x - 3y + 8)$

104. $-(-3p - q + 9)$

For Exercises 105–112, simplify the expression by clearing parentheses and combining like terms.

105. $5a + 2(a + 7)$

106. $6b - 3(b + 4)$

107. $8y - 3(y + 2) - 19$

108. $9k - 3(k - 4) + 8$

109. $4w + 9u - (5w - 3u)$

110. $-5m + 2(m + 4n) - 7n$

111. $2g - 4[2 - 3(g - 4h) - 2]$

112. $3p + 8[q - (p - 4) - 3p]$

---

review

B LINEAR EQUATIONS IN ONE VARIABLE

Concepts

1. Solving Linear Equations in One Variable
2. Problem Solving
3. Linear Inequalities

1. Solving Linear Equations in One Variable

An equation that can be written in the form $ax + b = 0$, where $a \neq 0$, is called a **linear equation in one variable**. The steps for solving a linear equation in one variable are given next.

solving linear equations, p. 117

Steps for Solving a Linear Equation in One Variable

1. Consider clearing fractions or decimals (if any are present) by multiplying both sides of the equation by a common denominator of all terms.
2. Simplify both sides of the equation by clearing parentheses and combining *like* terms.
3. Use the addition and subtraction properties of equality to collect the variable terms on one side of the equation.
4. Use the addition and subtraction properties of equality to collect the constant terms on the other side of the equation.
5. Use the multiplication and division properties of equality to make the coefficient of the variable term equal to 1.
6. Check your answer.

example 1 **Solving Linear Equations in One Variable**

Solve. $\frac{1}{3}x + \frac{1}{4}x = x + 5$

Solution:

$\frac{1}{3}x + \frac{1}{4}x = x + 5$ The LCD (least common denominator) is 12.

$12\left(\frac{1}{3}x + \frac{1}{4}x\right) = 12(x + 5)$ Multiply both sides by 12.

$\frac{12}{1}\left(\frac{1}{3}x\right) + \frac{12}{1}\left(\frac{1}{4}x\right) = \frac{12}{1}(x) + \frac{12}{1}(5)$ Apply the distributive property.

$4x + 3x = 12x + 60$

$7x = 12x + 60$ Combine *like* terms.

$-5x = 60$ Subtract $12x$ from both sides.

$x = -12$ Divide both sides by -5.

The answer can be checked by substituting $x = -12$ back into the original equation and verifying that the left-hand side equals the right-hand side.

2. Problem Solving

Solving word problems in mathematics takes practice and often a little patience. We also offer the following guidelines for problem solving.

problem-solving flowchart, p. 123

Problem-Solving Guidelines

Step 1: Read the problem carefully.
Step 2: Assign labels to unknown quantities.
Step 3: Develop an equation in words.
Step 4: Write a mathematical equation.
Step 5: Solve the equation.
Step 6: Interpret the results and write the answer in words.

example 2

Solving an Application Involving Principal and Interest

Kadriana took out two student loans for a total of \$8500. One loan was for 5% simple interest and the other was for 8% simple interest. If the total interest after one year was \$530, find the amount borrowed at each rate.

Solution:

interest problems, p. 149

We have two unknown values. We arbitrarily let x represent the amount borrowed at 5%. Then the remaining principal, $8500 - x$, is the amount borrowed at 8%.

Step 1: Read the problem.

	5% Loan	8% Loan	Total
Principal	x	$8500 - x$	
Interest	$0.05x$	$0.08(8500 - x)$	530

Step 2: Label the variables. A chart may help to organize the information given in the problem.

$$\begin{pmatrix}\text{Interest from}\\ \text{5\% loan}\end{pmatrix} + \begin{pmatrix}\text{Interest from}\\ \text{8\% loan}\end{pmatrix} = \begin{pmatrix}\text{Total}\\ \text{interest}\end{pmatrix}$$

Step 3: Verbal equation.

$$0.05x + 0.08(8500 - x) = 530$$

Step 4: Mathematical equation.

$$0.05x + 680 - 0.08x = 530$$

$$-0.03x + 680 = 530$$

Step 5: Solve the equation. Combine *like* terms.

$$-0.03x = -150$$

Subtract 680 from both sides.

$$\frac{-0.03x}{-0.03} = \frac{-150}{-0.03}$$

Divide both sides by -0.03.

$$x = 5000$$

Step 6: Write the answer in words.

The amount borrowed at 5% is \$5000. The amount borrowed at 8% is given by $8500 - x = 8500 - 5000 = 3500$. Thus, \$3500 was borrowed at 8%.

example 3 Solving an Application Involving Distance, Rate, and Time

distance, rate, time problems, p. 151

Two cars leave a rest area at 12:00 P.M. One car travels north on Interstate 25 through Colorado and the other travels south. The southbound car travels 10 mph slower than the northbound car. After 2 hours, the cars are 260 miles apart. How fast is the northbound car traveling?

Solution: **Step 1:** Read the problem.

Let x represent the speed of the northbound car. **Step 2:** Label the variables.

	Distance	Rate	Time
Northbound Car	$2(x)$	x	2
Southbound Car	$2(x - 10)$	$x - 10$	2

$d = rt$ distance = (rate)(time)

$$\begin{pmatrix}\text{Distance by}\\ \text{northbound car}\end{pmatrix} + \begin{pmatrix}\text{distance by}\\ \text{southbound car}\end{pmatrix} = \begin{pmatrix}\text{total}\\ \text{distance}\end{pmatrix}$$ **Step 3:** Verbal equation.

$$2x + 2(x - 10) = 260$$ **Step 4:** Mathematical equation.

$$2x + 2x - 20 = 260$$ **Step 5:** Solve the equation.

$$4x - 20 = 260$$

$$4x = 280$$

$$x = 70$$

The northbound car travels at 70 mph. **Step 6:** Interpret the answer.

3. Linear Inequalities

A linear inequality is a mathematical sentence written with one of the symbols, $<, >, \leq$, or $\geq$. The solution to an inequality can be visualized graphically on a number line or can be written in interval notation. For any real numbers a and b, Table A-4 summarizes the solution sets for five general inequalities.

interval notation, p. 160

Table A-4

Inequality	Graph	Interval Notation
$x > a$	(at a, shaded to the right	(a, ∞)
$x \geq a$	[at a, shaded to the right	$[a, \infty)$
$x < a$	) at a, shaded to the left	$(-\infty, a)$
$x \leq a$	] at a, shaded to the left	$(-\infty, a]$
$a < x < b$	(at a,) at b, shaded between	(a, b)

The use of a parenthesis (or) indicates that an endpoint is not included in the solution. The use of a square bracket [or] indicates that an endpoint *is* included in the solution. Note that a parenthesis is always used for infinity.

The steps to solve a linear inequality mirror those to solve a linear equation. **However, if an inequality is multiplied or divided by a negative number, then the direction of the inequality sign must be reversed.** This is demonstrated in the next example.

example 4 Solving a Linear Inequality

solving linear inequalities, p. 161

Solve the inequality. Graph the solution and write the solution in interval notation.

$$-4(x-3)+1 \le x+23$$

Solution:

$-4(x-3)+1 \le x+23$	
$-4x+12+1 \le x+23$	Apply the distributive property.
$-4x+13 \le x+23$	Combine *like* terms.
$-4x-x+13 \le x-x+23$	Subtract x from both sides.
$-5x+13 \le 23$	
$-5x+13-13 \le 23-13$	Subtract 13 from both sides.
$-5x \le 10$	
$\dfrac{-5x}{-5} \ge \dfrac{10}{-5}$	Divide both sides by -5. Reverse the direction of the inequality sign.
$x \ge -2$	

Graph: −5 −4 −3 −2 −1 0 1 2 3 4 5

Interval notation: $[-2, \infty)$

review B PRACTICE EXERCISES

For Exercises 1–30, solve the equation.

1. $4x = -84$
2. $-3y = 72$
3. $\dfrac{t}{5} = 12$
4. $\dfrac{r}{3} = 13$
5. $6 + p = -14$
6. $-5 + q = -23$
7. $7x + 12 = 33$
8. $-3x - 4 = -1$
9. $4(x + 5) = 14$
10. $5(y - 2) = -22$
11. $3b + 25 = 5b - 21$
12. $4c - 9 = 6c + 17$
13. $3w + 2(w - 7) = 7(w - 1)$
14. $x + 5(x - 4) = 3(x - 8) - 5$
15. $\dfrac{1}{5}y - \dfrac{3}{10}y = y - \dfrac{2}{5}y + 1$
16. $\dfrac{2}{3}q + 2 - \dfrac{1}{9}q = \dfrac{5}{9}q + \dfrac{4}{3}q$

17. $\frac{5}{6}(x + 2) = \frac{1}{3}x - \frac{3}{2}$

18. $\frac{1}{4}(6 - p) = \frac{3}{8}(2 + 3p)$

19. $0.2a + 6 = 1.8a - 2.8$

20. $0.3(v - 3) = 1.4v - 3.1$

21. $0.72t - 1.18 = 0.28(4 - t)$

22. $0.08(100 - 73x) = 1.2(1 - 5x)$

23. $-4(x + 1) - 2x = -2(3x + 2) - x$

24. $5(z + 3) - 2z + 6(1 - z) = 4 - 2(1 + z)$

25. $11 - 5(y + 3) = -2[2y - 3(1 - y)]$

26. $-4p + 2[p - 4(2 - p)] = 4(p - 5)$

27. $-8m - 2[4 - 3(m + 1)] = 4(3 - m) + 4$

28. $10x - 5[1 + 3(1 - 2x)] = [4(x - 3) + 27] + x$

29. $4 - 6[2y - 2(y + 3)] = 7 - y$

30. $-3(x + 4) + 2x + 12 = -6(x - 5)$

For Exercises 31–40, solve.

31. Five times the product of 8 and a number is 20. What is the number?

32. One half of the total of 6 times a number and 20 is 4. What is the number?

33. The sum of three consecutive integers is -57. Find the integers.

34. The sum of three consecutive odd integers is 111. Find the integers.

35. If Casey invested \$2500 in an account and earned \$112.50 in interest the first year, what was the simple interest rate?

36. Roberto deposited \$4050 in an account that made \$202.50 simple interest the first year. Find the simple interest rate.

37. Ketul made a total investment of \$3500 into two accounts earning 4.2% simple interest and 3.8% simple interest. If his total interest after one year was \$141.40, how much was invested in each account?

38. Katlin borrowed \$1800, part from her parents and part from a bank. The bank charged 8% interest but her parents only charged 2%. If she paid off the loan in one year and paid a total of \$102 interest, how much did she borrow from her parents?

39. A car and a truck leave home at the same time traveling in opposite directions. The truck travels 4 mph slower than the car. If the distance between the two vehicles is 372 miles after 3 hours, find the speed of each vehicle.

40. Juan and Peter both leave a shopping center at the same time going in opposite directions. Juan is on his bike and travels 3 mph faster than Peter who is on his skateboard. After 1.5 hours they are 19.5 miles apart. How fast does Peter travel.

For Exercises 41–56, solve the inequality. Graph the solution set and write the set in interval notation.

41. $3 - x > 5$

42. $4 - 2d < 6$

43. $4y + 6 \geq -18$

44. $3r - 13 \leq 27$

45. $8m - 15 < 9m - 13$

46. $6n + 23 > -3n - 13$

47. $1 - 5(2t - 7) < 38t$

48. $9r + 3(r - 14) \geq 6$

49. $-5 < 4p + 2 \leq 6$

50. $-8 \leq 7 + 5k \leq 17$

51. $2 > -y + 1 > -6$

52. $0 \geq -3x \geq -10$

53. $\frac{1}{7}h + 2 > \frac{5}{14}h - \frac{1}{2}$

54. $\frac{2}{3}b - \frac{1}{6} \leq 3b + \frac{5}{6}$

55. $0.4w + 3.1(w - 1) \leq -3.8$

56. $2.1x - 2(x + 5.1) > -8.2$

review

C LINEAR EQUATIONS IN TWO VARIABLES

Concepts

1. Plotting Ordered Pairs
2. Graphing Linear Equations in Two Variables
3. x- and y-Intercepts
4. Horizontal and Vertical Lines
5. Slope of a Line
6. Slope-Intercept Form

1. Plotting Ordered Pairs

rectangular coordinate system, p. 185

In Figure A-1, we review some important features of a rectangular coordinate system. The x- and y-axes are perpendicular lines that cross at a common point called the **origin**. The arrows indicate the positive direction for the x- and y-axes. The four regions in the graph are called **quadrants**. Each point in a rectangular coordinate system is represented by an **ordered pair**. Figure A-2 shows the location of the following ordered pairs: $(4, 1)$, $(-2, 5)$, $(5, -2)$, $(-2, 1)$, $(0, 4)$, and $(-\frac{7}{2}, -2)$. For example, to plot the point represented by $(-2, 5)$ move -2 units in the x-direction (left) and 5 units in the y-direction (up) and draw a dot.

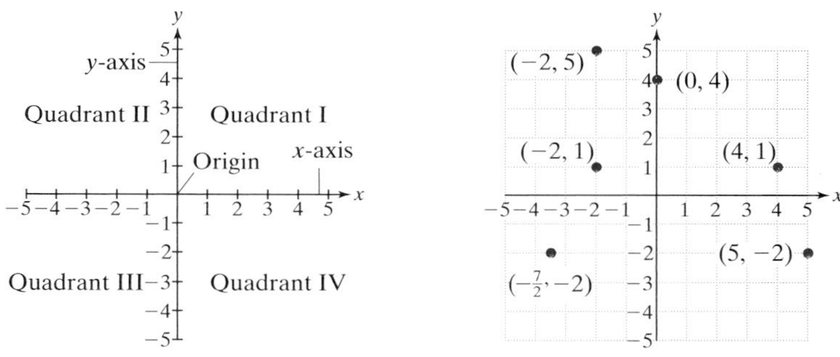

Figure A-1 **Figure A-2**

2. Graphing Linear Equations in Two Variables

linear equations in two variables, p. 193

An equation that can be written in the form $ax + by = c$ (where a and b are not both 0) is called a linear equation in two variables. The equation $3x + y = 5$ is a linear equation in two variables. A solution to such an equation is an ordered pair (x, y) that when substituted into the equation makes the equation a true statement. For example, the ordered pair $(-1, 8)$ is a solution to the equation $3x + y = 5$ because $3(-1) + 8 = 5$. A linear equation in two variables has infinitely many solutions that when plotted form a line in a rectangular coordinate system. To find a solution to a linear equation in two variables, we substitute any real number for one of the variables in the equation and then solve for the other variable.

example 1 **Graphing a Linear Equation in Two Variables**

Complete the ordered pairs to form solutions to the equation $3x + y = 5$. Then graph the equation.

$(1, \ )$ $(\ , -4)$ $(0, \ )$ $(\ , 0)$

solutions to linear equations in two variables, p. 194

Solution:

The ordered pairs can also be represented in tabular form as follows.

x	y
1	
	−4
0	
	0

$3x + y = 5$

$3(1) + y = 5$

$3x + (-4) = 5$

$3(0) + y = 5$

$3x + (0) = 5$

x	y	Ordered pair
1	2	(1, 2)
3	−4	(3, −4)
0	5	(0, 5)
$\frac{5}{3}$	0	$(\frac{5}{3}, 0)$

graphing linear equations in two variables, p. 194

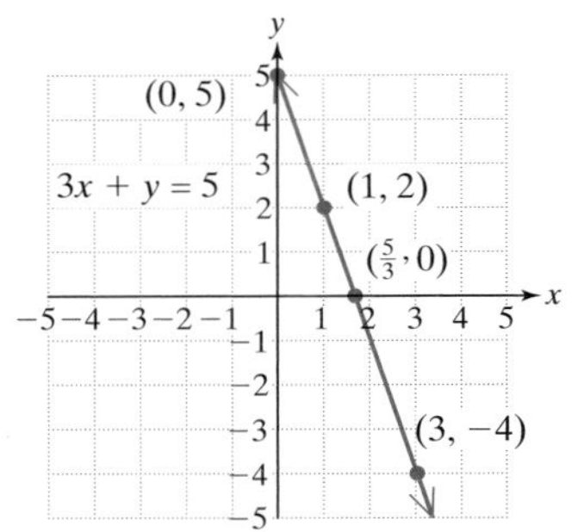

Figure A-3

The ordered pairs are graphed in Figure A-3. Notice that the points all "line up." The line drawn through the points represents all solutions to the equation. Therefore, we say the line represents a graph of the equation.

x- and y-intercepts, p. 202

3. *x*- and *y*-Intercepts

In Example 1, the points $(\frac{5}{3}, 0)$ and $(0, 5)$ lie on the x-axis and y-axis, respectively. The point $(\frac{5}{3}, 0)$ is called an x-intercept and the point $(0, 5)$ is called a y-intercept. In general, the x- and y-intercepts of a graph are points where the graph intersects the x- and y-axes, respectively. Notice that the x-intercept $(\frac{5}{3}, 0)$ has a y-coordinate of 0. The y-intercept $(0, 5)$ has an x-coordinate of 0.

Finding the *x*- and *y*-Intercepts of a Linear Equation in Two Variables

Consider an equation in the variables x and y.

1. To find the x-intercept substitute $y = 0$ into the equation and solve for x.
2. To find the y-intercept substitute $x = 0$ into the equation and solve for y.

example 2

Finding *x*- and *y*-Intercepts

Given $7x - 3y = 6$, find the x- and y-intercepts.

Solution:

To find the x-intercept, substitute zero for y:

$$7x - 3(0) = 6$$
$$7x = 6$$
$$x = \frac{6}{7}$$

To find the y-intercept, substitute zero for x:

$$7(0) - 3y = 6$$
$$-3y = 6$$
$$y = -2$$

The x-intercept is $(\frac{6}{7}, 0)$ and the y-intercept is $(0, -2)$.

4. Horizontal and Vertical Lines

horizontal and vertical lines, p. 207

In Section 3.3, we learned that an equation of the form $x = k$ is a vertical line in a rectangular coordinate system. An equation of the form $y = k$ is a horizontal line.

example 3

Graphing Horizontal and Vertical Lines

graphing horizontal and vertical lines, p. 208

Graph the lines. a. $x = 4$ b. $-2y = 5$

Solution:

a. The equation $x = 4$ is in the form $x = k$. Therefore, the graph of the equation is a vertical line that passes through the x-axis at 4.

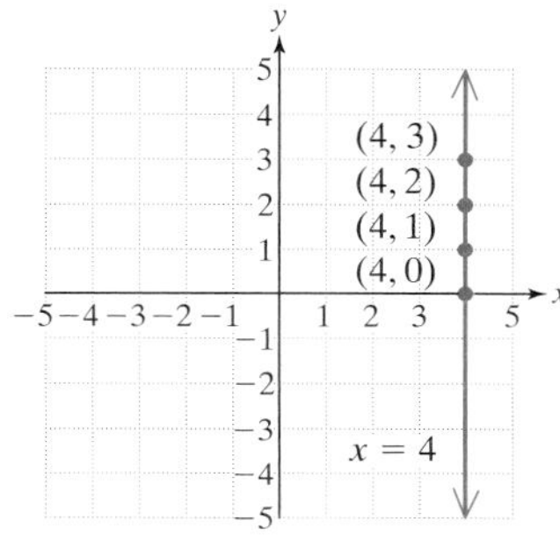

The solutions to the equation $x = 4$ are ordered pairs whose x-coordinate is 4 and whose y-coordinate has no restriction. Several such solutions are (4, 0), (4, 1), (4, 2), (4, 3), and so on.

b. The equation $-2y = 5$ is equivalent to $y = -\frac{5}{2}$. The equation is in the form $y = k$. Therefore, the graph of the equation is a horizontal line passing through the y-axis at $-\frac{5}{2}$.

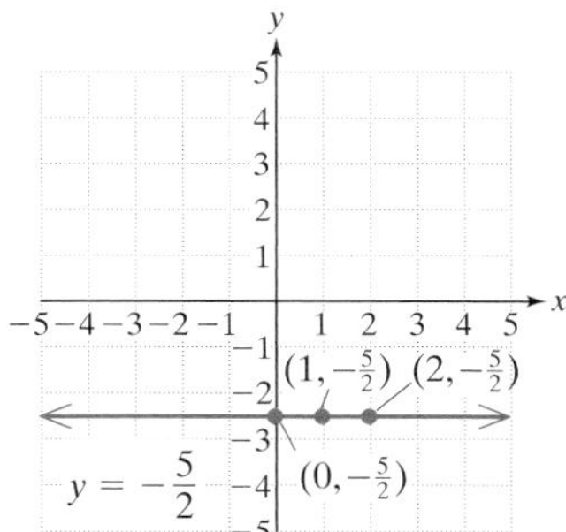

The solutions to the equation $y = -\frac{5}{2}$ are ordered pairs whose y-coordinate is $-\frac{5}{2}$ and whose x-coordinate has no restriction. Several such solutions are $(0, -\frac{5}{2})$, $(1, -\frac{5}{2})$, $(2, -\frac{5}{2})$, and so on.

5. Slope of a Line

slope of a line, p. 214

The slope of a line (often denoted by m) is a measure of the line's "steepness." It measures the ratio of an incremental change in y to an incremental change in x between two points on the line.

The slope is $\frac{3}{2}$.

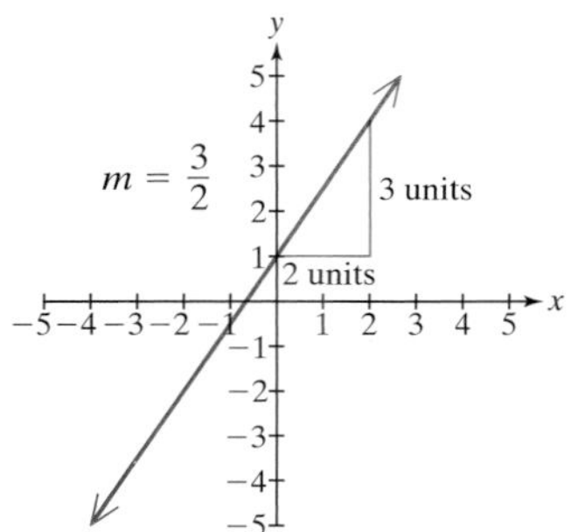

The slope is 4.

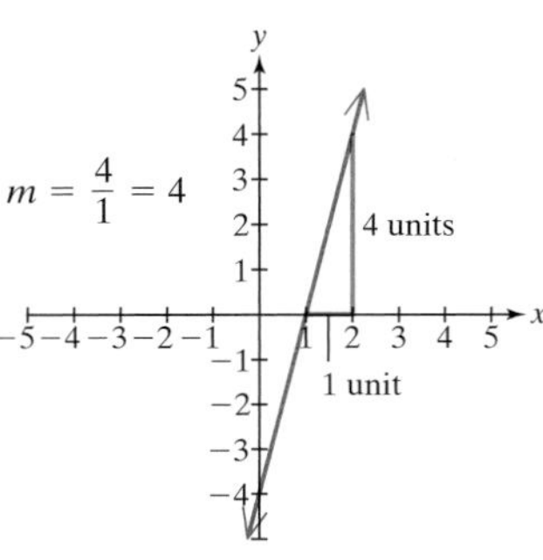

positive, negative, zero, and undefined slopes, p. 216

The slope of a line may be positive, negative, zero, or undefined.

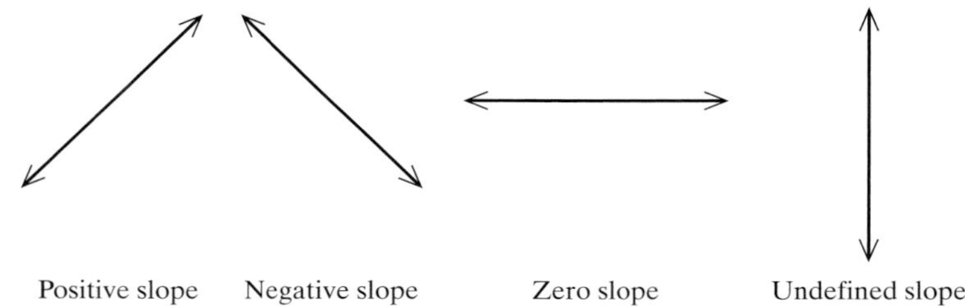

To find the slope of a line given two points, (x_1, y_1) and (x_2, y_2), we use the slope formula:

$$m = \frac{y_2 - y_1}{x_2 - x_1}.$$

example 4

Finding the Slope Given Two Points on a Line

slope formula, p. 215

Find the slope of the line between $(-3, -4)$ and $(7, -1)$.

Solution:

(x_1, y_1) (x_2, y_2)
$(-3, -4)$ $(7, -1)$ Label the points.

$$m = \frac{y_2 - y_1}{x_2 - x_1} = \frac{-1 - (-4)}{7 - (-3)} = \frac{3}{10}$$ The slope is $\frac{3}{10}$.

parallel and perpendicular lines, p. 218

The slopes of two lines can be used to determine whether the lines are parallel or perpendicular.

- Two lines that are parallel have the same slope. That is, $m_1 = m_2$.
- The slopes of perpendicular lines are related such that one slope is the opposite of the reciprocal of the other.

  That is, $m_1 = -\frac{1}{m_2}$.

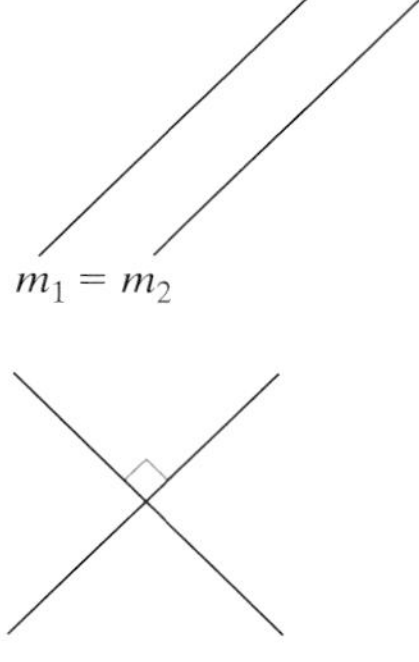

6. Slope-Intercept Form

A linear equation in two variables written in the form $y = mx + b$ is said to be in slope-intercept form. In such a case, m is the slope, and $(0, b)$ is the y-intercept. To write a linear equation in slope-intercept form, solve the equation for y.

example 5 Writing an Equation in Slope-Intercept Form

slope-intercept form of a line, p. 226

Given $2x + 5y = 20$,

a. Write the equation in slope-intercept form.
b. Identify the slope and y-intercept.
c. Graph the line using the slope and y-intercept.

Solution:

a.

$2x + 5y = 20$	To write in the form $y = mx + b$, solve for y.
$5y = -2x + 20$	Subtract $2x$ from both sides.
$\frac{5y}{5} = \frac{-2x}{5} + \frac{20}{5}$	Divide by 5.
$y = -\frac{2}{5}x + 4$	The equation is in slope-intercept form.

graphing a line using the slope and y-intercept, p. 227

b. The slope is $-\frac{2}{5}$ and the y-intercept is $(0, 4)$.
c. To graph the line using the slope and y-intercept, first plot the y-intercept. Then we can use the slope to find a second point on the line. For example,

The slope can be interpreted as $\frac{-2}{5}$, which indicates that we move two units in the negative y-direction (down) and 5 units in the positive x-direction (right).

The slope can also be interpreted as $\frac{2}{-5}$ which indicates that we move two units in the positive y direction (up) and 5 units in the negative x-direction (left). See Figure A-4.

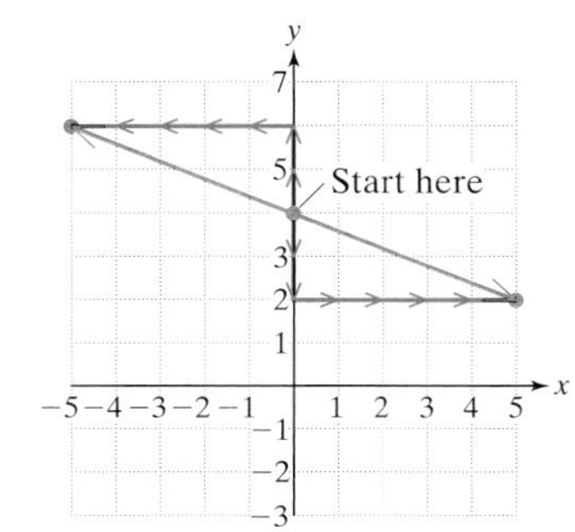

Figure A-4

example 6 Determining Whether Lines are Parallel, Perpendicular, or Neither

comparing slopes of lines, p. 229

Determine whether the lines l_1 and l_2 are parallel, perpendicular, or neither.

l_1: $2x - 3y = 12$ $\qquad$ l_2: $6x + 4y = 4$

Solution:

Write each line in slope-intercept form and compare the slopes.

l_1: $2x - 3y = 12$

$$-3y = -2x + 12$$

$$\frac{-3y}{-3} = \frac{-2x}{-3} + \frac{12}{-3}$$

$$y = \frac{2}{3}x - 4$$

l_2: $6x + 4y = 4$

$$4y = -6x + 4$$

$$\frac{4y}{4} = \frac{-6x}{4} + \frac{4}{4}$$

$$y = -\frac{3}{2}x + 1$$

The slope of l_1 is $\frac{2}{3}$. $\qquad$ The slope of l_2 is $-\frac{3}{2}$.

The slope of l_1 is the opposite of the reciprocal of the slope of l_2. Therefore, the lines are perpendicular.

review C Practice Exercises

For Exercises 1–8, plot the points on a rectangular coordinate system.

1. $(2, 4)$
2. $(4, 2)$
3. $(4, -3)$
4. $(0, 2)$
5. $(-3, 0)$
6. $(0, 0)$
7. $\left(-\frac{3}{2}, -2\right)$
8. $(-1, 3)$

For Exercises 9–16, identify the quadrant in which each point is located.

9. $(20, -6)$
10. $(-3, -3)$
11. $\left(-13, -\frac{7}{8}\right)$
12. $\left(\frac{12}{7}, 5\right)$
13. $(\pi, 14)$
14. $(7.8, -42)$
15. $(-3.8, 6.2)$
16. $(-7.8, 42)$

For Exercises 17–22, complete the table and graph the line.

17. $x + y = 4$

x	y
3	
	-1
0	

18. $3x - y = 6$

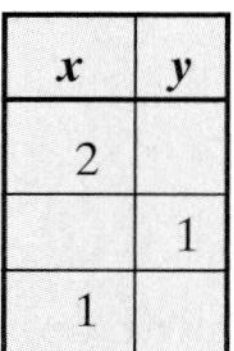

x	y
2	
	1
1	

19. $y = 5x$

x	y
0	
$\frac{2}{5}$	
-1	

20. $y = -2x + 3$

x	y
1	
$\frac{1}{2}$	
-1	

21. $y = \frac{2}{3}x + 1$

x	y
3	
−3	
0	

22. $y = -\frac{4}{3}x + 2$

x	y
3	
0	
−1	

For Exercises 23–34, find the x- and y-intercepts.

23. $3x - 2y = 18$
24. $x - 4y = 16$
25. $5x + y = -15$
26. $-6x - y = 6$
27. $\frac{1}{2}x + y = 2$
28. $x - \frac{1}{3}y = -4$
29. $-3x + 4y = 0$
30. $6x - 5y = 0$
31. $y = -\frac{2}{3}x - 3$
32. $y = \frac{3}{5}x + 1$
33. $y = 6x - 5$
34. $y = -2x - 3$

For Exercises 35–40, write the equation in the form $x = k$ or $y = k$ and identify the line as horizontal or vertical.

35. $2x = 6$
36. $5y = 10$
37. $-3y + 1 = 2$
38. $7x - 1 = 3$
39. $5x = 0$
40. $\frac{2}{3}y = 0$

For Exercises 41–50, write the equation in slope-intercept form (if possible). Then identify the slope and the y-intercept (if they exist).

41. $4x - 5y = 6$
42. $2x - 5y = 11$
43. $-6x + 2y = 3$
44. $-x + 3y = 4$
45. $5x - 1 = 4$
46. $2y - 3 = 0$
47. $x + 4y = 0$
48. $8x + 2y = 0$
49. $y + 4 = 8$
50. $x - 3 = 5$

For Exercises 51–56, graph the line given the slope and y-intercept.

51. Slope is $\frac{1}{2}$, y-intercept is $(0, -2)$
52. Slope is -3, y-intercept is $(0, 1)$
53. Slope is $-\frac{3}{2}$, y-intercept is $(0, 3)$
54. Slope is $\frac{5}{3}$, y-intercept is $(0, -1)$
55. Slope is 4, y-intercept is $(0, -2)$
56. Slope is 2, y-intercept is $(0, -3)$

For Exercises 57–68, graph the line.

57. $2x - 3y = 12$
58. $2x + 4y = -8$
59. $x - 2y = -4$
60. $6x - y = 4$
61. $4x + y = 0$
62. $-x - 3y = 0$
63. $y = -\frac{2}{3}x + 1$
64. $y = \frac{1}{4}x - 1$
65. $-2x + 1 = 5$
66. $4y - 2 = 6$
67. $y = 3x - 3$
68. $y = -4x - 4$

For Exercises 69–76, determine if the lines are parallel, perpendicular, or neither.

69. l_1: $-2x + y = 1$
l_2: $2x - y = 3$
70. l_1: $y = 3x + 6$
l_2: $-x + 3y = -3$
71. l_1: $-2x + 5y = 5$
l_2: $5x - 2y = -2$
72. l_1: $-7x + y = 1$
l_2: $-7x = 0$
73. l_1: $x + y = 7$
l_2: $x - y = -2$
74. l_1: $y = -\frac{1}{4}x + 2$
l_2: $2x + 8y = 24$
75. l_1: $y = 6$
l_2: $y = -2$
76. l_1: $x = 3$
l_2: $y = 0$

review

D Polynomials and Properties of Exponents

Concepts

1. Properties of Exponents
2. Scientific Notation
3. Addition and Subtraction of Polynomials
4. Multiplication of Polynomials
5. Division of Polynomials

1. Properties of Exponents

Recall that expressions with exponents can be simplified using the following definitions and properties.

Properties and Definitions Involving Exponents

Assume that a and b are real numbers such that $a \neq 0, b \neq 0$ and m and n are integers.

Property/Definition		Example
Multiplication of like bases:	$b^m b^n = b^{m+n}$	$x^3x^5 = x^{3+5} = x^8$
Division of like bases:	$\dfrac{b^m}{b^n} = b^{m-n}$	$\dfrac{y^7}{y^3} = y^{7-3} = y^4$
The power rule:	$(b^m)^n = b^{m \cdot n}$	$(5^4)^2 = 5^{4 \cdot 2} = 5^8$
Power of a product:	$(ab)^m = a^m b^m$	$(5x)^2 = 5^2x^2 = 25x^2$
Power of a quotient:	$\left(\dfrac{a}{b}\right)^m = \dfrac{a^m}{b^m}$	$\left(\dfrac{x}{4}\right)^3 = \dfrac{x^3}{4^3} = \dfrac{x^3}{64}$
Definition of b^0:	$b^0 = 1$	$5^0 = 1$
Definition of b^{-n}:	$b^{-n} = \left(\dfrac{1}{b}\right)^n = \dfrac{1}{b^n}$	$x^{-6} = \left(\dfrac{1}{x}\right)^6 = \dfrac{1}{x^6}$

multiplying and dividing common bases, p. 248

power rule for exponents, p. 255

definition of b^0 and b^{-n}, p. 260

These properties and definitions may be combined to simplify exponential expressions.

example 1 **Simplifying Exponential Expressions**

Simplify. a. $\dfrac{x^4x^6}{x^3}$ b. $(2y^3)^4$ c. $\left(\dfrac{pw^2}{2z^5}\right)^3$

d. $\dfrac{4x^6z^{-3}}{y^{-2}}$ e. $\left(\dfrac{1}{2}\right)^{-3} + \left(\dfrac{1}{4}\right)^0$

Solution:

a. $\dfrac{x^4x^6}{x^3} = \dfrac{x^{4+6}}{x^3} = \dfrac{x^{10}}{x^3} = x^{10-3} = x^7$

b. $(2y^3)^4 = 2^4(y^3)^4 = 2^4y^{3 \cdot 4} = 16y^{12}$

c. $\left(\dfrac{pw^2}{2z^5}\right)^3 = \dfrac{p^3w^6}{2^3\,z^{15}} = \dfrac{p^3w^6}{8z^{15}}$

d. $\dfrac{4x^6z^{-3}}{y^{-2}} = \dfrac{4x^6y^2}{z^3}$

e. $\left(\dfrac{1}{2}\right)^{-3} + \left(\dfrac{1}{4}\right)^0 = \left(\dfrac{2}{1}\right)^3 + 1 = 8 + 1 = 9$

2. Scientific Notation

Recall that scientific notation is a convenient method to write very large or very small numbers. A number is in scientific notation if it is written in the form $a \times 10^n$, where $1 \le |a| < 10$ and n is an integer. For example,

Number	Equivalent Form	Scientific Notation
4200	4.2×1000	4.2×10^3
0.008	$8.0 \times 0.001 = 8.0 \times \dfrac{1}{1000} = 8.0 \times \dfrac{1}{10^3}$	8.0×10^{-3}

scientific notation, p. 267

Writing a number in scientific notation can be done easily by moving the decimal point to the left or right and then multiplying the resulting number by the appropriate power of 10.

$730{,}000 = 7.3 \times 10^5$ $\qquad$ $0.000046 = 4.6 \times 10^{-5}$

example 2 Multiplying and Dividing Numbers in Scientific Notation

multiplying and dividing numbers in scientific notation, p. 269

Perform the indicated operations and write the answer in scientific notation.

$$\frac{(6.0 \times 10^3)(9.0 \times 10^5)}{2.0 \times 10^{-2}}$$

Solution:

$\dfrac{(6.0 \times 10^3)(9.0 \times 10^5)}{2.0 \times 10^{-2}} = \dfrac{(6.0)(9.0) \times 10^3 10^5}{2.0 \times 10^{-2}}$ Multiply the numerators.

$= \dfrac{54.0 \times 10^8}{2.0 \times 10^{-2}}$ Simplify.

$= \left(\dfrac{54.0}{2.0}\right) \times \dfrac{10^8}{10^{-2}}$ Divide.

$= 27.0 \times 10^{8-(-2)}$ Simplify.

$= 27.0 \times 10^{10}$ This number is not in proper scientific notation.

$= (2.7 \times 10^1) \times 10^{10}$ Write 27.0 as 2.7×10^1.

$= 2.7 \times 10^{11}$

3. Addition and Subtraction of Polynomials

To add or subtract polynomials, we add or subtract *like* terms.

example 3

adding and subtracting polynomials, pp. 276–278

Adding and Subtracting Polynomials

Perform the indicated operations.

$$\left(2x^4 - 5x^2 - \frac{3}{4}x\right) + (7x^2 + 13) - \left(8x^4 + x^3 + \frac{1}{4}x\right)$$

Solution:

$$\left(2x^4 - 5x^2 - \frac{3}{4}x\right) + (7x^2 + 13) - \left(8x^4 + x^3 + \frac{1}{4}x\right)$$

$= 2x^4 - 5x^2 - \frac{3}{4}x + 7x^2 + 13 - 8x^4 - x^3 - \frac{1}{4}x$ Apply the distributive property.

$= 2x^4 - 8x^4 - x^3 - 5x^2 + 7x^2 - \frac{3}{4}x - \frac{1}{4}x + 13$ Collect *like* terms together.

$= -6x^4 - x^3 + 2x^2 - \frac{4}{4}x + 13$ Add like terms.

$= -6x^4 - x^3 + 2x^2 - x + 13$ Simplify.

4. Multiplication of Polynomials

To multiply polynomials, multiply each term in the first polynomial by each term in the second and combine *like* terms.

example 4

multiplying polynomials, p. 283

Multiplying Polynomials

Multiply. a. $(2y + 5)(6y^2 - 4y + 3)$ b. $(2x - 3y)(5x + 7y)$

Solution:

a. $(2y + 5)(6y^2 - 4y + 3)$

$= 2y(6y^2) + 2y(-4y) + 2y(3) + 5(6y^2) + 5(-4y) + 5(3)$ Multiply.

$= 12y^3 - 8y^2 + 6y + 30y^2 - 20y + 15$ Simplify.

$= 12y^3 + 22y^2 - 14y + 15$ Combine *like* terms.

b. $(2x - 3y)(5x + 7y)$

$= 2x(5x) + 2x(7y) + (-3y)(5x) + (-3y)(7y)$ Multiply.

$= 10x^2 + 14xy - 15xy - 21y^2$ Simplify.

$= 10x^2 - xy - 21y^2$ Combine *like* terms.

When squaring a binomial such as $(a + b)^2$, we can write the expression as $(a + b)(a + b)$. Proceeding as in Example 4, we have $(a + b)(a + b) = a^2 + ab + ab + b^2 = a^2 + 2ab + b^2$. The resulting trinomial is a **perfect square trinomial**.

To multiply the product of conjugates such as $(a + b)(a - b)$, we have $a^2 - ab + ab - b^2 = a^2 - b^2$. This result is called a **difference of squares**.

These general results are summarized in Table A-5.

special case products, p. 286

Table A-5

Product	Example
The square of a binomial results in a perfect square trinomial. 1. $(a + b)^2 = a^2 + 2ab + b^2$ 2. $(a - b)^2 = a^2 - 2ab + b^2$	1. $(2x + 5y)^2 = (2x)^2 + 2(2x)(5y) + (5y)^2$ $= 4x^2 + 20xy + 25y^2$ 2. $(4m - 3n)^2 = (4m)^2 - 2(4m)(3n) + (3n)^2$ $= 16m^2 - 24mn + 9n^2$
The product of conjugate factors results in a difference of squares. 3. $(a + b)(a - b) = a^2 - b^2$	3. $(7w + 3)(7w - 3) = (7w)^2 - (3)^2$ $= 49w^2 - 9$

5. Division of Polynomials

Division of polynomials is presented in two separate cases.

1. To divide a polynomial by a monomial divisor (that is, when the divisor has only one term), we use the properties:

$$\frac{a+b}{c} = \frac{a}{c} + \frac{b}{c} \quad \text{and} \quad \frac{a-b}{c} = \frac{a}{c} - \frac{b}{c}.$$

2. If the divisor has more than one term, we use long division.

example 5

division by a monomial, p. 291

Dividing a Polynomial by a Monomial

Divide. $(15xy^2 - 5x^2y^3 + 10x^3y^6) \div (5x^2y^2)$

Solution:

We can write the expression as $\dfrac{15xy^2 - 5x^2y^3 + 10x^3y^6}{5x^2y^2}$

$= \dfrac{15xy^2}{5x^2y^2} - \dfrac{5x^2y^3}{5x^2y^2} + \dfrac{10x^3y^6}{5x^2y^2}$ Divide each term in the numerator by $5x^2y^2$.

$= \dfrac{3}{x} - y + 2xy^4$ Simplify each term.

example 6

Using Long Division to Divide Polynomials

Divide. $(8x^3 + 14x^2 - 7) \div (x + 3)$

long division, p. 292

Solution:

Write the polynomials in long division format. The term $0x$ is inserted in the dividend as a place holder for the missing power of x.

$$\begin{array}{r|l} & 8x^2 \\ \hline x + 3 & 8x^3 + 14x^2 + 0x - 7 \end{array}$$

Divide the leading term of the dividend by the leading term of the divisor $(8x^3/x)$. The result, $8x^2$, is the first term in the quotient.

$$\begin{array}{r|l} & 8x^2 \\ \hline x + 3 & 8x^3 + 14x^2 + 0x - 7 \\ & -(8x^3 + 24x^2) \end{array}$$

Multiply $8x^2$ by the divisor and subtract the result.

$$\begin{array}{r|l} & 8x^2 \\ \hline x + 3 & 8x^3 + 14x^2 + 0x - 7 \\ & \underline{-8x^3 - 24x^2} \quad \downarrow \\ & \quad -10x^2 + 0x \end{array}$$

Subtract the quantity $8x^3 + 24x^2$. To do this we can add the opposite.

Bring down the next column and repeat the process.

$$\begin{array}{r|l} & 8x^2 - 10x \\ \hline x + 3 & 8x^3 + 14x^2 + 0x - 7 \\ & \underline{-8x^3 - 24x^2} \\ & \quad -10x^2 + 0x \\ & \quad -(-10x - 30x) \end{array}$$

$$\begin{array}{r|l} & 8x^2 - 10x + 30 \\ \hline x + 3 & 8x^3 + 14x^2 + 0x - 7 \\ & \underline{-8x^3 - 24x^2} \\ & \quad -10x^2 + 0x \\ & \quad \underline{10x + 30x} \quad \downarrow \\ & \qquad 30x - 7 \\ & \qquad -(30x + 90) \end{array}$$

$$\begin{array}{r|l} & 8x^2 - 10x + 30 \\ \hline x + 3 & 8x^3 + 14x^2 + 0x - 7 \\ & \underline{-8x^3 - 24x^2} \\ & \quad -10x^2 + 0x \\ & \quad \underline{10x + 30x} \\ & \qquad 30x - 7 \\ & \qquad \underline{-30x - 90} \\ & \qquad\qquad -97 \end{array}$$

Therefore, $(8x^3 + 14x^2 - 7) \div (x + 3) = 8x^2 - 10x + 30 + \dfrac{-97}{x + 3}$.

review D PRACTICE EXERCISES

Assume that all variables represent positive numbers.

For Exercises 1–8, simplify and state the property used. Choose from multiplication of like bases, division of like bases, power rule, power of a product, or power of a quotient.

1. $6^2 \cdot 6^5$
2. $(x^3)^4$
3. $\dfrac{y^{10}}{y^7}$
4. $(10p)^5$
5. $\left(\dfrac{w}{4}\right)^4$
6. $x^{10} \cdot x^3 \cdot x$
7. $(q^5)^6$
8. $\dfrac{z^4}{z^4}$

For Exercises 9–28, simplify. Write the answers with positive exponents.

9. 3^0
10. 5^{-1}
11. $\left(\dfrac{1}{2}\right)^{-1}$
12. $(-6)^0$
13. $(-5)^{-2}$
14. $\left(\dfrac{2}{3}\right)^{-2}$
15. $6^{-1} + (-4)^0$
16. $(-2)^{-2} + 8^{-1}$
17. $\dfrac{r^2 \cdot r^6}{r^{10}}$
18. $\dfrac{t^3 \cdot t^7}{t^{13}}$
19. $\dfrac{b^3}{b^4 \cdot b^{-3}}$
20. $\dfrac{c^{-2}}{c^{-5} \cdot c^2}$
21. $(3x^2)^3$
22. $(-7y^4)^2$
23. $\dfrac{6w^2z}{2z^{-1}}$
24. $\dfrac{-8m^4n^{-2}}{4n^3}$
25. $\left(\dfrac{a^2b^{-1}}{c^3d^{-4}}\right)^{-2}$
26. $\left(\dfrac{p^{-3}q}{r^2s^{-1}}\right)^{-3}$
27. $(5^0w^2z) \cdot (w^{-3}z^{-4})$
28. $(4p^{-3}q) \cdot (p^{-2}q^4)$

For Exercises 29–36, write the number in scientific notation.

29. 3,050,000
30. 82,500,000
31. 0.0000251
32. 0.0038
33. There were 89,600,000 people who watched the Super Bowl on Sunday, February 1, 2004.
34. Mont Blanc, a mountain in the Alps, is about 15,800 ft high.
35. A single blood cell is approximately 0.00039 in. long.
36. The shortest flash of light recorded was 0.000 000 000 000 000 69 seconds.

For Exercises 37–44, perform the operation. Write the answer in scientific notation.

37. $(2.4 \times 10^1)(1.5 \times 10^3)$
38. $(8 \times 10^{-5})(1.2 \times 10^9)$
39. $\dfrac{6.3 \times 10^{14}}{3.0 \times 10^{-2}}$
40. $\dfrac{9.6 \times 10^{-12}}{4.0 \times 10^3}$
41. $(3.0 \times 10^{-5})(7.3 \times 10^2)$
42. $(2.1 \times 10^{13})(5.0 \times 10^{-2})$
43. $\dfrac{(4.0 \times 10^3)(1.2 \times 10^{-4})}{(6.0 \times 10^{10})}$
44. $\dfrac{(2.8 \times 10^2)(4.0 \times 10^4)}{(7.0 \times 10^{-6})}$

For Exercises 45–50, add or subtract as indicated.

45. $(5p^3 + 2p - 3) + (8p^3 - 4p^2 + 14)$
46. $(2m^2 - 3m + 9) - (-3m^2 + 2m - 8)$
47. $\left(10n^2 - \dfrac{3}{8}n + 2\right) - \left(11n^2 + \dfrac{5}{8}n - 6\right)$
48. $(a^2 - 5a + 10) + \left(-3a^3 + \dfrac{1}{2}a^2 - 13\right)$
49. $(-u^2v + 6u^2 - 2uv^2) + (11u^2 + 7uv^2) - (9u^2v + u^2 + 3uv^2)$
50. $(8a^2b - 5ab + 12ab^2) - (4a^2b - 6ab - ab^2) + (10a^2b + 3ab^2)$

For Exercises 51–66, multiply the polynomials.

51. $4p(p^3 - 5p^2 + 2p + 8)$
52. $5mn(-3m^2 + 6mn + n^2)$

53. $(3x + y)(-2x - 5y)$

54. $(4u - 7v)(3u - v)$

55. $(8b + 4)(2b - 3)$

56. $(-5a + 2)(5a - 3)$

57. $(x - 3y)(x^2 - 3xy + 5y^2)$

58. $(3p - 5q)(p - 2pq + q)$

59. $(3h - 8)(3h + 8)$

60. $(k - 5)(k + 5)$

61. $(4x - 5)^2$

62. $(3y + 8)^2$

63. $\left(\frac{1}{3}t^2 - 9\right)\left(\frac{1}{3}t^2 + 9\right)$

64. $\left(ab - \frac{3}{4}\right)\left(ab + \frac{3}{4}\right)$

65. $(0.2x^2 - 3)^2$

66. $(1.2y^2 + 4)^2$

For Exercises 67–76, divide the polynomials.

67. $(6a^3b^2 - 18a^2b^2 + 3ab^2 - 9ab) \div (3ab)$

68. $(8x^4y^4 + 4x^3y^2 - 12x^2y^4 + 4x^2y^2) \div (4x^2y^2)$

69. $(x^3 - 2x^2 - 4x + 33) \div (x + 3)$

70. $(2y^3 - 7y^2 + y + 10) \div (y - 2)$

71. $(2p^3 - 17p^2 + 39p - 10) \div (p - 5)$

72. $(t^3 - t^2 - 10t + 4) \div (t + 3)$

73. $(10m^5 - 16m^4 + 8m^3 + 8m^2 - 2m) \div (-2m)$

74. $(-25b^5 + 10b^4 - 5b^3 - 35b^2) \div (-5b^2)$

75. $(8a^3 - 9) \div (2a - 3)$

76. $(32k^3 + 5) \div (2k - 1)$

review

E FACTORING POLYNOMIALS AND SOLVING QUADRATIC EQUATIONS

Concepts

1. **Factoring Polynomials**
2. **Solving Quadratic Equations Using the Zero Product Rule**

1. Factoring Polynomials

The steps to factor a polynomial are outlined in detail in Section 5.6. Recall that the first step in factoring is to factor out the greatest common factor, GCF. Then we proceed with an appropriate technique of factoring based on whether the polynomial has two terms (a binomial), three terms (a trinomial), or four terms.

example 1 **Factoring Binomials**

Factor completely. a. $81w^4 - 1$ b. $2y^3 + 128$

Solution:

factoring a difference of squares, p. 336

When factoring a binomial, determine if it fits any of the following patterns.

$a^2 - b^2 = (a + b)(a - b)$ Difference of squares (Section 5.4)

$a^3 - b^3 = (a - b)(a^2 + ab + b^2)$ Difference of cubes (Section 5.5)

$a^3 + b^3 = (a + b)(a^2 - ab + b^2)$ Sum of cubes (Section 5.5)

Avoiding Mistakes

The binomial $9x^2 - 1$ is a difference of squares and may be factored as $(3x + 1)(3x - 1)$. However, the binomial $9x^2 + 1$ is a sum of squares and may not be factored over the real numbers.

a. $81w^4 - 1$ The GCF is 1.

$= (9x^2)^2 - (1)^2$ This is a difference of squares, $a^2 - b^2$, where $a = 9x^2$ and $b = 1$.

$= (9x^2 + 1)(9x^2 - 1)$ Apply the formula $a^2 - b^2 = (a + b)(a - b)$.

$= (9x^2 + 1)(3x + 1)(3x - 1)$ The factor $9x^2 - 1$ factors further as a difference of squares. $9x^2 - 1 = (3x + 1)(3x - 1)$.

factoring a sum or difference of cubes, p. 340

b. $2y^3 + 128$ — The GCF is 2.

$= 2(y^3 + 64)$ — Factor out 2.

$= 2[(y)^3 + (4)^3]$ — The resulting binomial is a sum of cubes, $a^3 + b^3$, where $a = y$ and $b = 4$.

$= 2(y + 4)(y^2 - 4y + 16)$ — Apply the formula $a^3 + b^3 = (a + b)(a^2 - ab + b^2)$

Tip: Recall that a factoring problem may be checked by multiplication.

$$\begin{aligned} 2(y + 4)(y^2 - 4y + 16) &= (2y + 8)(y^2 - 4y + 16) \\ &= 2y(y^2) + 2y(-4y) + 2y(16) + 8(y^2) + 8(-4y) + 8(16) \\ &= 2y^3 - 8y^2 + 32y + 8y^2 - 32y + 128 \\ &= 2y^3 + 128 \end{aligned}$$

example 2 Factoring Trinomials

Factor. a. $49w^2 - 28w + 4$ b. $5x^4y - 8x^3y + 3x^2y$

Solution:

Check first to see if the trinomial is a perfect square trinomial, in which case factor as the square of a binomial.

Perfect square trinomials $\begin{cases} a^2 + 2ab + b^2 = (a + b)^2 \\ a^2 - 2ab + b^2 = (a - b)^2 \end{cases}$

Otherwise factor the trinomial using the trial-and-error method (Section 5.3) or the grouping method (Section 5.2).

factoring a perfect square trinomial, p. 334

a. $49w^2 - 28w + 4$ — The GCF is 1.

$= (7w)^2 - 2(7w)(2) + (2)^2$ — The polynomial is a perfect square trinomial, $a^2 - 2ab + b^2$ where $a = 7w$ and $b = 2$.

$= (7w - 2)^2$ — Apply the formula $a^2 - 2ab + b^2 = (a - b)^2$.

factoring trinomials with the trial-and-error method, p. 325

b. $5x^4y - 8x^3y + 3x^2y$ — The GCF is x^2y.

$= x^2y(5x^2 - 8x + 3)$ — Factor out x^2y. The factored form of $5x^2 - 8x + 3$ must be a product of binomials of the form:

factoring trinomials with the grouping method, p. 318

factors of 5

$(\Box x - \Box)(\Box x - \Box)$

factors of 3

The signs within parentheses must both be negative for a product of 3 and a sum of $-8x$.

Trying all possible combinations we have:

$(5x - 1)(x - 3) = 5x^2 - 15x - x + 3$

$= 5x^2 - 16x + 3$ Wrong middle term.

$(5x - 3)(x - 1) = 5x^2 - 5x - 3x + 3$

$= 5x^2 - 8x + 3$ Correct!

$= x^2y(5x - 3)(x - 1)$ Factored completely.

example 3 Factoring a Four-Term Polynomial

factoring a four-term polynomial by grouping, p. 314

Factor completely. a. $6ab + 8ax + 15bx + 20x^2$ b. $w^2 + 16w + 64 - z^2$

Solution:

This polynomial has 4 terms. Therefore, try factoring by grouping.

a. $6ab + 8ax + 15bx + 20x^2$ The GCF is 1.

$= 6ab + 8ax \mid + 15bx + 20x^2$ Group the first pair of terms and the second pair of terms.

$= 2a(3b + 4x) + 5x(3b + 4x)$ Factor out the GCF from each pair. Note that the two resulting terms share a common binomial factor.

$= (3b + 4x)(2a + 5x)$ Factor out the common binomial factor.

b. $w^2 + 16w + 64 - z^2$

factoring a four-term polynomial (3 terms with 1 term), p. 350

$= w^2 + 16w + 64 \mid - z^2$ The first three terms are a perfect square trinomial. Group three terms with one term.

$= (w + 8)^2 - z^2$ Factor the first three terms. The result is a difference of squares, $a^2 - b^2$, where $a = (w + 8)$ and $b = z$.

$= [(w + 8) + z][(w + 8) - z]$ Apply the formula $a^2 - b^2 = (a + b)(a - b)$.

$= (w + 8 + z)(w + 8 - z)$

2. Solving Quadratic Equations Using the Zero Product Rule

definition of a quadratic equation, p. 354

A quadratic equation is an equation of the form $ax^2 + bx + c = 0$, where $a \neq 0$. To solve a quadratic equation in this form, we factor the expression $ax^2 + bx + c$ and set each factor equal to zero.

example 4 Solving Quadratic Equations

solving a quadratic equation by factoring, p. 355

Solve. a. $(2x - 1)(x - 3) = 0$ b. $10x^2 + 50x = 140$

Solution:

a. $(2x - 1)(x - 3) = 0$ — The equation is already in factored form and set equal to zero.

$2x - 1 = 0 \quad \text{or} \quad x - 3 = 0$ — Set each factor equal to zero.

$2x = 1 \quad \text{or} \quad x = 3$ — Solve each equation.

$x = \frac{1}{2} \quad \text{or} \quad x = 3$ — The solutions are $x = \frac{1}{2}$ and $x = 3$.

b. $10x^2 + 50x = 140$

$10x^2 + 50x - 140 = 0$ — Write the equation in the form $ax^2 + bx + c = 0$.

$10(x^2 + 5x - 14) = 0$ — Factor out the GCF.

$10(x - 2)(x + 7) = 0$ — Factor the trinomial.

$10 \neq 0 \quad \text{or} \quad x - 2 = 0 \quad \text{or} \quad x + 7 = 0$ — Set each factor equal to zero.

$x = 2 \quad \text{or} \quad x = -7$ — The solutions are $x = 2$ and $x = -7$.

review E PRACTICE EXERCISES

1. Write the formula to factor the difference of squares, $a^2 - b^2$.
2. Write the formula to factor the difference of cubes, $a^3 - b^3$.
3. Write the formula to factor the sum of cubes, $a^3 + b^3$.
4. Is it possible to factor the sum of squares $a^2 + b^2$ over the real numbers?

For Exercises 5–50,

a. Identify the category in which the polynomial best fits. Choose from

- difference of squares
- sum of squares
- difference of cubes
- sum of cubes
- trinomial (perfect square trinomial)
- trinomial (nonperfect square trinomial)
- four terms—grouping
- none of these

Hint: (You may have to factor out the greatest common factor first to categorize the polynomial.)

b. Factor the polynomial completely.

5. $t^2 - 100$
6. $4m^2 - 49n^2$
7. $y^3 + 27$
8. $x^3 + 1$
9. $d^2 + 3d - 28$
10. $c^2 + 5c - 24$
11. $x^2 - 12x + 36$
12. $p^2 + 16p + 64$
13. $2ax^2 - 5ax + 2bx - 5b$
14. $8x^2 - 4bx + 2ax - ab$
15. $10y^2 + 3y - 4$
16. $12z^2 + 11z + 2$
17. $10p^2 - 640$
18. $50a^2 - 72$
19. $z^4 - 64z$
20. $t^4 - 8t$

21. $b^3 - 4b^2 - 45b$

22. $y^3 - 14y^2 + 40y$

23. $9w^2 + 24wx + 16x^2$

24. $4k^2 - 20kp + 25p^2$

25. $60x^2 - 20x + 30ax - 10a$

26. $50x^2 - 200x + 10cx - 40c$

27. $x^3 + 4x^2 - 9x - 36$

28. $m^3 + 5m^2 - 4m - 20$

29. $w^4 - 16$

30. $k^4 - 81$

31. $t^6 - 8$

32. $p^6 + 27$

33. $8p^2 - 22p + 5$

34. $9m^2 - 3m - 20$

35. $36y^2 - 12y + 1$

36. $9a^2 + 42a + 49$

37. $x^2 + 4x + 4 - y^2$

38. $m^2 - 6m + 9 - n^2$

39. $2x^2 + 50$

40. $4y^2 + 64$

41. $12r^2s^2 + 7rs^2 - 10s^2$

42. $7z^2w^2 - 10zw^2 - 8w^2$

43. $x^2 + 8xy - 33y^2$

44. $s^2 - 9st - 36t^2$

45. $m^6 + n^3$

46. $a^3 - b^6$

47. $x^2(a + b) - x(a + b) - 12(a + b)$

48. $y^2(y + 2) + 10y(y + 2) + 9(y + 2)$

49. $x^2 - 4x$

50. $y^2 - 9y$

For Exercises 51–64, solve the equation.

51. $(2x + 5)(x - 3) = 0$

52. $(4x - 1)(x + 6) = 0$

53. $x(x - 5) = 0$

54. $y(y + 7) = 0$

55. $5(w + 3) = 0$

56. $-2(t - 10) = 0$

57. $z^2 - 2z - 8 = 0$

58. $p^2 - 15p - 16 = 0$

59. $6x^2 - 7x = 5$

60. $4x^2 = x + 14$

61. $2x(x - 10) = -9x - 12$

62. $x(x - 12) + 5 = -30$

63. $w^3 - w^2 - w + 1 = 0$

64. $z^3 - 4z^2 - 16z + 64 = 0$

---

review

F RATIONAL EXPRESSIONS

Concepts

1. Definition of a Rational Expression
2. Reducing Rational Expressions to Lowest Terms
3. Multiplying and Dividing Rational Expressions
4. Adding and Subtracting Rational Expressions
5. Simplifying Complex Fractions
6. Solving Rational Equations

definition of a rational expression, p. 376

1. Definition of a Rational Expression

A rational expression is an expression of the form $\frac{p}{q}$, where p and q are polynomials. The following are rational expressions.

$$\frac{x^2 + x - 4}{x - 2}, \quad \frac{1}{x^3}, \quad \frac{2}{5}$$

The domain of a rational expression is the set of all real numbers except those that make the denominator zero. For example, the domain of

$\frac{x^2 + x - 4}{x - 2}$ Is all real numbers except 2, $\{x|x \neq 2\}$.

$\frac{1}{x^3}$ Is all real numbers except 0, $\{x|x \neq 0\}$.

$\frac{2}{5}$ Is all real numbers, $\{x|x \text{ is any real number}\}$.

reducing rational expressions to lowest terms, p. 378

2. Reducing Rational Expressions to Lowest Terms

The expressions $\frac{6}{12}$, $\frac{2}{4}$, and $\frac{1}{2}$ are all equivalent. However, the value $\frac{1}{2}$ is in lowest terms. To reduce a rational expression to lowest terms, first factor the numerator and denominator. Then reduce the common factors that form a ratio of 1.

example 1

Reducing a Rational Expression to Lowest Terms

Reduce to lowest terms. $\dfrac{x^2 + 10x + 9}{x^2 - 1}$

Solution:

$$\frac{x^2 + 10x + 9}{x^2 - 1} = \frac{(x + 1)(x + 9)}{(x + 1)(x - 1)} \quad \text{Factor numerator and denominator.}$$

$$= \frac{\overset{1}{\cancel{(x + 1)}}(x + 9)}{\cancel{(x + 1)}(x - 1)} \quad \text{Reduce } (x + 1)/(x + 1) \text{ to 1.}$$

$$= \frac{x + 9}{x - 1} \quad (\text{where } x \neq 1 \text{ and } x \neq -1)$$

In Example 1, it is important to realize that the original restrictions on the variable $x \neq 1$ and $x \neq -1$, are still in effect even after the expression is reduced to lowest terms. That is,

$$\frac{x^2 + 10x + 9}{x^2 - 1} = \frac{x + 9}{x - 1} \quad \text{for } x \neq 1 \text{ and } x \neq -1.$$

multiplying and dividing rational expressions, p. 385

3. Multiplying and Dividing Rational Expressions

To multiply two rational expressions, factor the numerator and denominator of both expressions. Then reduce common factors that form a ratio of 1.

example 2

Multiplying Rational Expressions

Multiply and simplify the result. $\dfrac{2x - 4}{4x + 4} \cdot \dfrac{2x^2 - 7x - 4}{x^2 - 6x + 8}$

Solution:

$$\frac{2x - 4}{4x + 4} \cdot \frac{2x^2 - 7x - 4}{x^2 - 6x + 8} = \frac{2(x - 2)}{4(x + 1)} \cdot \frac{(2x + 1)(x - 4)}{(x - 4)(x - 2)} \quad \text{Factor.}$$

$$= \frac{\overset{1}{\cancel{2}}\overset{1}{\cancel{(x - 2)}}}{\cancel{2} \cdot 2(x + 1)} \cdot \frac{(2x + 1)\overset{1}{\cancel{(x - 4)}}}{\cancel{(x - 4)}\cancel{(x - 2)}} \quad \text{Reduce ratios of 1.}$$

$$= \frac{2x + 1}{2(x + 1)}$$

Tip: In the expression $\dfrac{2x + 1}{2(x + 1)}$, do not try to "cancel" the 2's. The 2 in the numerator is not a factor of the numerator.

To divide two rational expressions, multiply the first expression by the reciprocal of the second expression.

example 3

Dividing Rational Expressions

Divide and simplify. $\dfrac{-x^2 + x}{-x^2} \div \dfrac{3x^2 + 2x - 5}{5x}$

Solution:

$$\frac{-x^2 + x}{-x^2} \div \frac{3x^2 + 2x - 5}{5x} = \frac{-x^2 + x}{-x^2} \cdot \frac{5x}{3x^2 + 2x - 5}$$

Multiply the first expression by the reciprocal of the second.

$$= \frac{-x(x - 1)}{-x \cdot x} \cdot \frac{5x}{(3x + 5)(x - 1)}$$

Factor.

$$= \frac{\overset{1}{\cancel{(-x)}}\overset{1}{\cancel{(x - 1)}}}{\cancel{(-x)} \cdot x} \cdot \frac{5\overset{1}{\cancel{x}}}{(3x + 5)\cancel{(x - 1)}}$$

Reduce ratios of 1.

$$= \frac{5}{3x + 5}$$

4. Adding and Subtracting Rational Expressions

To add or subtract a rational expression it is necessary that the expressions have a common denominator. To add or subtract expressions with a common denominator add or subtract the terms in the numerators and write the result over the common denominator. Then reduce the expression to lowest terms if possible.

example 4

adding and subtracting rational expressions with the same denominator, p. 397

Subtracting Rational Expressions with a Common Denominator

Subtract. $\dfrac{x^2 - 2x}{x - 5} - \dfrac{8x - 25}{x - 5}$

Solution:

$$\frac{x^2 - 2x}{x - 5} - \frac{8x - 25}{x - 5} = \frac{x^2 - 2x - (8x - 25)}{x - 5}$$

Subtract terms in the numerator. Use parentheses to help you remember to subtract both terms in the numerator of the second expression.

$$= \frac{x^2 - 2x - 8x + 25}{x - 5}$$

Apply the distributive property.

$$= \frac{x^2 - 10x + 25}{x - 5}$$

Combine *like* terms.

$$= \frac{\cancel{(x - 5)}(x - 5)}{\cancel{x - 5}}$$

Factor and reduce to lowest terms.

$$= x - 5$$

To add or subtract two rational expressions that do not have a common denominator, it is necessary to convert each expression to an equivalent rational expression with a common denominator. The steps below outline the process to find the least common denominator of two or more rational expressions.

finding the LCD, p. 391

Steps to find the LCD of Two or More Rational Expressions

1. Factor all denominators completely.
2. The LCD is the product of unique factors from the denominators, where each factor is raised to the highest power to which it appears in any denominator.

For example, the LCD of the expressions $\frac{2}{5x^2y^4}$ and $\frac{1}{7x^3y}$ is given by the product:

$5^1 \cdot 7^1 \cdot x^3 \cdot y^4 = 35x^3y^4$.

To convert a rational expression to an equivalent expression with a common denominator, we multiply the numerator and denominator by the factors in the LCD that are missing from the denominator of the original expression. This is mathematically feasible because we are multiplying the original expression by 1. For example, we convert $\frac{2}{5x^2y^4}$ and $\frac{1}{7x^3y}$ to equivalent expressions with a denominator of $35x^3y^4$ as follows.

$$\frac{2}{5x^2y^4} = \frac{2 \cdot 7x}{5x^2y^4 \cdot 7x} = \frac{14x}{35x^3y^4} \qquad \text{Note that } \frac{7x}{7x} = 1.$$

$$\frac{1}{7x^3y} = \frac{1 \cdot 5y^3}{7x^3y \cdot 5y^3} = \frac{5y^3}{35x^3y^4} \qquad \text{Note that } \frac{5y^3}{5y^3} = 1.$$

example 5 Subtracting Two Rational Expressions with Different Denominators

adding and subtracting rational expressions with different denominators, p. 398

Subtract the rational expressions. $\frac{3x}{x^2 + 7x + 10} - \frac{2x}{x^2 + 6x + 8}$

Solution:

$$\frac{3x}{x^2 + 7x + 10} - \frac{2x}{x^2 + 6x + 8}$$

$$= \frac{3x}{(x + 2)(x + 5)} - \frac{2x}{(x + 4)(x + 2)}$$

Factor the denominators. The LCD is $(x + 2)(x + 5)(x + 4)$.

$$= \frac{3x(x + 4)}{(x + 2)(x + 5)(x + 4)} - \frac{2x(x + 5)}{(x + 4)(x + 2)(x + 5)}$$

Convert each expression to an equivalent expression with the LCD.

$$= \frac{3x(x + 4) - 2x(x + 5)}{(x + 2)(x + 5)(x + 4)}$$

Subtract the numerators.

$$= \frac{3x^2 + 12x - 2x^2 - 10x}{(x + 2)(x + 5)(x + 4)}$$ Apply the distributive property.

$$= \frac{x^2 + 2x}{(x + 2)(x + 5)(x + 4)}$$ Combine *like* terms.

$$= \frac{x\overset{1}{\cancel{(x + 2)}}}{\cancel{(x + 2)}(x + 5)(x + 4)}$$ Factor and reduce to lowest terms.

$$= \frac{x}{(x + 5)(x + 4)}$$

5. Simplifying Complex Fractions

A complex fraction is a fraction whose numerator or denominator contains one or more rational expressions. For example:

$$\frac{1 - \dfrac{4}{3y}}{y - \dfrac{16}{9y}}$$

One method to simplify such an expression is to multiply the numerator and denominator of the complex fraction by the LCD of all four individual fractions. This is demonstrated in Example 6.

example 6 **Simplifying a Complex Fraction**

complex fractions, p. 405

Simplify. $\dfrac{1 - \dfrac{4}{3y}}{y - \dfrac{16}{9y}}$

Solution:

$$\frac{1 - \dfrac{4}{3y}}{y - \dfrac{16}{9y}}$$ The LCD of the expressions $\frac{1}{1}, \frac{4}{3y}, \frac{y}{1}$, and $\frac{16}{9y}$ is $9y$.

$$\frac{1 - \dfrac{4}{3y}}{y - \dfrac{16}{9y}} = \frac{9y\left(1 - \dfrac{4}{3y}\right)}{9y\left(y - \dfrac{16}{9y}\right)} = \frac{(9y) \cdot 1 - (9y) \cdot \dfrac{4}{3y}}{(9y) \cdot y - (9y) \cdot \dfrac{16}{9y}}$$ Apply the distributive property.

$$= \frac{9y - 12}{9y^2 - 16}$$ Simplify.

$$= \frac{3\overset{1}{\cancel{(3y - 4)}}}{\cancel{(3y - 4)}(3y + 4)}$$ Factor and reduce to lowest terms.

$$= \frac{3}{3y + 4}$$

The expression in Example 6 could also have been simplified by using the order of operations to combine the expressions in the numerator and denominator separately and then dividing the resulting expressions.

6. Solving Rational Equations

solving rational equations, p. 412

An equation with one or more rational expressions is called a rational equation. To solve a rational equation, we offer the following steps.

Steps to Solve a Rational Equation

1. Factor the denominators of all rational expressions.
2. Identify the LCD of all expressions in the equation.
3. Multiply both sides of the equation by the LCD.
4. Solve the resulting equation.
5. Check potential solutions in the original equation.

example 7 **Solving a Rational Equation**

Solve. $\frac{4}{y - 4} + \frac{y}{2} = \frac{y}{y - 4}$

Solution:

$$\frac{4}{y - 4} + \frac{y}{2} = \frac{y}{y - 4}$$

Step 1: The denominators are already factored.

Step 2: The LCD is $2(y - 4)$.

$$2(y - 4)\left(\frac{4}{y - 4} + \frac{y}{2}\right) = 2(y - 4)\left(\frac{y}{y - 4}\right)$$

Step 3: Multiply by the LCD.

$$\frac{2(y - 4)}{1} \cdot \left(\frac{4}{y - 4}\right) + \frac{2(y - 4)}{1} \cdot \frac{y}{2} = \frac{2(y - 4)}{1} \cdot \left(\frac{y}{y - 4}\right)$$

Apply the distributive property.

$$2 \cdot 4 + (y - 4) \cdot y = 2 \cdot y$$

Step 4: Solve the resulting equation.

$$8 + y^2 - 4y = 2y$$

Apply the distributive property to clear parentheses.

$$y^2 - 4y - 2y + 8 = 0$$

The equation is quadratic. Set the equation equal to zero.

Tip: Before solving a rational equation you might consider noting any values of the variable for which the equation is undefined. In Example 7, the equation $\frac{4}{y - 4} + \frac{y}{2} = \frac{y}{y - 4}$ is not defined for $y = 4$ because this value makes the denominator zero in the first and third rational expressions. If one of your potential solutions matches a restricted value, then you automatically know that this value cannot be a solution to the original equation.

$$y^2 - 6y + 8 = 0$$

$$(y - 4)(y - 2) = 0 \qquad \text{Factor.}$$

$$y - 4 = 0 \quad \text{or} \quad y - 2 = 0 \qquad \text{Set each factor equal to zero.}$$

$$y = 4 \quad \text{or} \quad y = 2$$

Step 5: Check each potential solution.

Check $y = 4$:

$$\frac{4}{(4) - 4} + \frac{(4)}{2} \stackrel{?}{=} \frac{(4)}{(4) - 4}$$

$$\frac{4}{0} + \frac{4}{2} \stackrel{?}{=} \frac{4}{0} \text{ (Undefined)}$$

Check $y = 2$:

$$\frac{4}{(2) - 4} + \frac{(2)}{2} \stackrel{?}{=} \frac{(2)}{(2) - 4}$$

$$\frac{4}{-2} + 1 \stackrel{?}{=} \frac{2}{-2}$$

$$-2 + 1 = -1 \checkmark$$

The solution is $y = 2$. The value $y = 4$ does not check because it makes the denominator zero in one or more of the rational expressions.

review F PRACTICE EXERCISES

For Exercises 1–6, determine domain of the expression.

1. $\dfrac{x + 2}{x + 4}$

2. $\dfrac{x - 5}{x - 1}$

3. $\dfrac{2x}{25x^2 - 9}$

4. $\dfrac{5y}{4y^2 - 49}$

5. $\dfrac{t - 7}{8}$

6. $\dfrac{x + 2}{5}$

For Exercises 7–12, reduce the expression to lowest terms.

7. $\dfrac{3x^4y^7}{12xy^8}$

8. $\dfrac{21ab^5}{7a^3b^4}$

9. $\dfrac{t^2 - 4}{2t - 4}$

10. $\dfrac{m^2 - 9}{3m - 9}$

11. $\dfrac{2y^2 + 5y - 12}{2y^2 + y - 6}$

12. $\dfrac{2x^2 + 20x + 48}{4x^2 - 144}$

For Exercises 13–22, multiply or divide as indicated and reduce to lowest terms.

13. $\dfrac{2x}{15y^2} \cdot \dfrac{3y^5}{4x^2}$

14. $\dfrac{4s^2t}{8t^3} \cdot \dfrac{2t^5}{6s^4}$

15. $\dfrac{5y - 15}{10y + 40} \cdot \dfrac{2y^2 + y - 28}{2y^2 - 13y + 21}$

16. $\dfrac{x^2 - 36}{x^2 - 4x - 12} \cdot \dfrac{4x + 8}{4x - 24}$

17. $\dfrac{a^2b^3}{5c^2} \div \dfrac{ab}{15c^3}$

18. $\dfrac{m^3}{14n^5} \div \dfrac{m^7}{21n}$

19. $\dfrac{p^2 - 36}{2p - 4} \div \dfrac{2p + 12}{p^2 - 2p}$

20. $\dfrac{y^2 + y}{y^2 - y - 12} \div \dfrac{5y + 5}{y^2 - 5y + 4}$

21. $\dfrac{t^2 + 7t}{3 - t} \div \dfrac{t^2 - 49}{t^2 - 3t}$

22. $\dfrac{x^2 - 2x}{2 - x} \div \dfrac{x^2 + 5x}{x^2 - 25}$

For Exercises 23–26, find the LCD for each pair of expressions.

23. $\dfrac{1}{4x^3y^7}, \dfrac{1}{8xy^{10}}$

24. $\dfrac{1}{16ab^4}, \dfrac{1}{24a^2b^2}$

25. $\dfrac{1}{x^2 - x - 12}, \dfrac{1}{x^2 - 9}$

26. $\dfrac{1}{x^2 + 10x + 9}, \dfrac{1}{x^2 - 81}$

For Exercises 27–40, add or subtract as indicated.

27. $\dfrac{1}{2x} - \dfrac{5}{6x}$

28. $\dfrac{7}{15t} - \dfrac{4}{5t}$

29. $\dfrac{1}{2x^3y} + \dfrac{5}{4xy^2}$

30. $\dfrac{1}{3a^2b^4} + \dfrac{7}{9a^2b}$

31. $\dfrac{x^2}{x-7} - \dfrac{14x-49}{x-7}$

32. $\dfrac{z^2}{z-10} - \dfrac{20z-100}{z-10}$

33. $\dfrac{7x}{4x-2} + \dfrac{5x}{2x-1}$

34. $\dfrac{-8}{5x+4} + \dfrac{7}{10x+8}$

35. $\dfrac{x}{x^2+x-12} - \dfrac{2}{x^2+3x-4}$

36. $\dfrac{4y}{y^2-4y+4} + \dfrac{3}{y^2-7y+10}$

37. $\dfrac{-3}{y-2} + \dfrac{2y+11}{y^2+y-6} - \dfrac{2}{y+3}$

38. $\dfrac{-2}{x+2} + \dfrac{x}{x-1} + \dfrac{3x-6}{x^2+x-2}$

39. $\dfrac{5}{x-3} - \dfrac{x+2}{x-3}$

40. $\dfrac{3x}{x+2} - \dfrac{4x+2}{x+2}$

For Exercises 41–48, simplify the complex fractions.

41. $\dfrac{\dfrac{3}{a} + \dfrac{4}{b}}{\dfrac{7}{a} - \dfrac{1}{b}}$

42. $\dfrac{\dfrac{2}{p} - \dfrac{4}{q}}{\dfrac{2}{p} + \dfrac{1}{q}}$

43. $\dfrac{8x - \dfrac{1}{2x}}{1 - \dfrac{1}{4x}}$

44. $\dfrac{\dfrac{1}{2} - \dfrac{1}{2t}}{1 - \dfrac{1}{t^2}}$

45. $\dfrac{\dfrac{u^2-v^2}{uv}}{\dfrac{u+v}{uv}}$

46. $\dfrac{\dfrac{a^2-b^2}{ab}}{\dfrac{2a-2b}{ab}}$

47. $\dfrac{\dfrac{1}{y-5}}{\dfrac{2}{y+5} + \dfrac{1}{y^2-25}}$

48. $\dfrac{\dfrac{3}{x+1}}{\dfrac{2}{x^2-1} + \dfrac{1}{x-1}}$

For Exercises 49–56, solve the equation.

49. $\dfrac{1}{8} - \dfrac{3}{4x} = \dfrac{1}{2x}$

50. $\dfrac{3}{10} - \dfrac{4}{5x} = \dfrac{1}{x}$

51. $\dfrac{5}{x-3} = \dfrac{2}{x-3}$

52. $\dfrac{6}{x+2} = \dfrac{4}{x+2}$

53. $1 + \dfrac{6}{x} = -\dfrac{8}{x^2}$

54. $1 = \dfrac{8}{x} - \dfrac{15}{x^2}$

55. $\dfrac{4x+11}{x^2-4} - \dfrac{5}{x+2} = \dfrac{2x}{x^2-4}$

56. $\dfrac{3x}{x+1} - 2 = \dfrac{12}{x^2-1}$

Appendix A

section

A.1 Synthetic Division

Concepts

1. Introduction to Synthetic Division
2. Using Synthetic Division to Divide Polynomials
3. Choosing Synthetic Division or Long Division

1. Introduction to Synthetic Division

In Section 4.7 we introduced the process of long division to divide two polynomials. In this section, we will learn another technique, called **synthetic division**, to divide two polynomials. Synthetic division may be used when dividing a polynomial by a first-degree divisor of the form, $x - r$, where r is a constant. Synthetic division is considered a "short cut" because it uses the coefficients of the divisor and dividend without writing the variables.

Consider dividing the polynomials $(3x^2 - 14x - 10) \div (x - 2)$.

$$\begin{array}{r} 3x - 8 \\ x - 2 \overline{\big) 3x^2 - 14x - 10} \\ -(3x^2 - 6x) \\ \hline -8x - 10 \\ -(-8x + 16) \\ \hline -26 \end{array}$$

First note that the divisor, $x - 2$, is in the form $x - r$, where $r = 2$. Hence synthetic division can also be used to find the quotient and remainder.

$$\begin{array}{r|rrr} 2 & 3 & -14 & -10 \\ & & & \\ \hline & 3 & & \end{array}$$

Step 1: Write the value of r in a box.

Step 2: Write the coefficients of the dividend to the right of the box.

Step 3: Skip a line and draw a horizontal line below the list of coefficients.

Step 4: Bring down the leading coefficient from the dividend and write it below the line.

$$\begin{array}{r|rrr} 2 & 3 & -14 & -10 \\ & & 6 & \\ \hline & 3 & -8 & \end{array}$$

Step 5: Multiply the value of r by the number below the line $(2 \times 3 = 6)$. Write the result in the next column above the line.

Step 6: Add the numbers in the column above the line $(-14 + 6)$ and write the result below the line.

Repeat Steps 5 and 6 until all columns have been completed.

Step 7: To get the final result, we use the numbers below the line. The number in the last column is the remainder. The other numbers are the coefficients of the quotient.

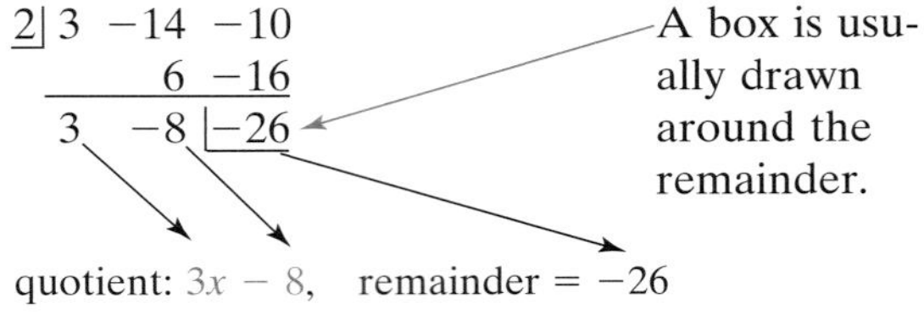

The degree of the quotient will always be one less than that of the dividend. Because the dividend is a second-degree polynomial, the quotient will be a first-degree polynomial. In this case, the quotient is $3x - 8$ and the remainder is -26.

2. Using Synthetic Division to Divide Polynomials

example 1

Using Synthetic Division to Divide Polynomials

Divide the polynomials $(5x + 4x^3 - 6 + x^4) \div (x + 3)$ using synthetic division.

Solution:

As with long division the terms of the dividend and divisor should be written in descending order. Furthermore, missing powers must be accounted for using place holders (shown here in bold).

Hence,

$$5x + 4x^3 - 6 + x^4$$
$$= x^4 + 4x^3 + \mathbf{0x^2} + 5x - 6$$

To use synthetic division, the divisor must be in the form $(x - r)$. The divisor, $x + 3$, can be written as $x - (-3)$. Hence, $r = -3$.

Step 1: Write the value of r in a box.

Step 2: Write the coefficients of the dividend to the right of the box.

$$\begin{array}{r|rrrrr} -3 & 1 & 4 & 0 & 5 & -6 \\ & & & & & \\ \hline & 1 & & & & \end{array}$$

Step 3: Skip a line and draw a horizontal line below the list of coefficients.

Step 4: Bring down the leading coefficient from the dividend and write it below the line.

Step 5: Multiply the value of r by the number below the line $(-3 \times 1 = -3)$. Write the result in the next column above the line.

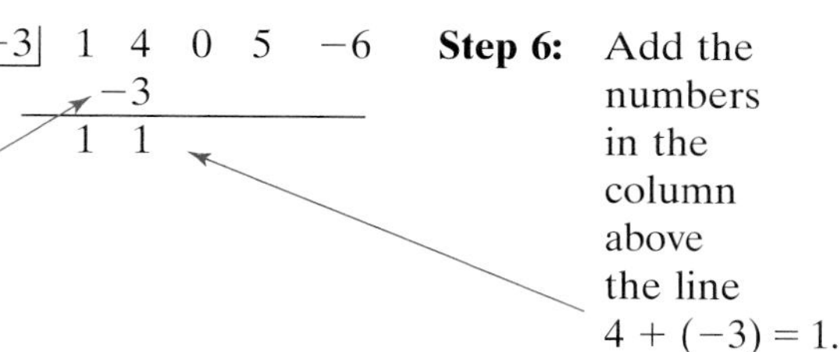

Step 6: Add the numbers in the column above the line $4 + (-3) = 1$.

Repeat Steps 5 and 6:

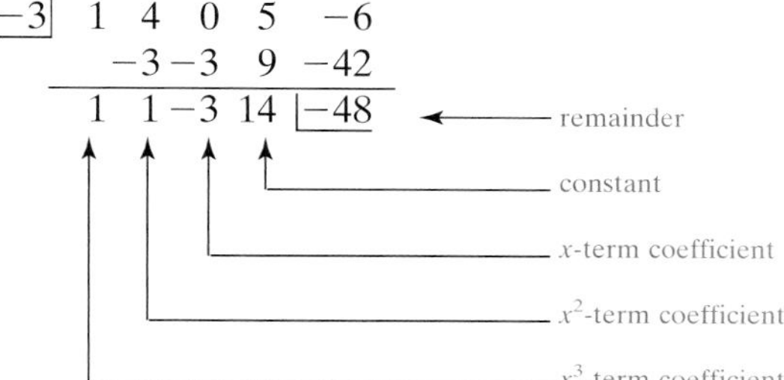

The quotient is

$x^3 + x^2 - 3x + 14$.

The remainder is -48.

Tip: It is interesting to compare the long division process to the synthetic division process. For Example 1, long division is shown on the left, and synthetic division is shown on the right. Notice that the same pattern of coefficients used in long division appears in the synthetic division process.

$$\begin{array}{r} x^3 + x^2 - 3x + 14 \\ x+3\overline{)x^4 + 4x^3 + 0x^2 + 5x - 6} \\ -(x^4 + 3x^3) \\ x^3 + 0x^2 \\ -(x^3 + 3x^2) \\ -3x^2 + 5x \\ -(-3x^2 - 9x) \\ 14x - 6 \\ -(14x + 42) \\ -48 \end{array}$$

$$\begin{array}{r|rrrrr} -3 & 1 & 4 & 0 & 5 & -6 \\ & & -3 & -3 & 9 & -42 \\ \hline & 1 & 1 & -3 & 14 & \boxed{-48} \\ & x^3 & x^2 & x & \text{constant} & \text{remainder} \end{array}$$

Quotient: $x^3 + x^2 - 3x + 14$
Remainder: -48

example 2 Using Synthetic Division to Divide Polynomials

Divide the polynomials using synthetic division. Identify the quotient and remainder.

a. $(2m^7 - 3m^5 + 4m^4 - m + 8) \div (m + 2)$ b. $(p^4 - 81) \div (p - 3)$

Solution:

a. Insert place holders (bold) for missing powers of m.

$(2m^7 - 3m^5 + 4m^4 - m + 8) \div (m + 2)$

$(2m^7 + \mathbf{0m^6} - 3m^5 + 4m^4 + \mathbf{0m^3} + \mathbf{0m^2} - m + 8) \div (m + 2)$

Because $m + 2$ can be written as $m - (-2)$, $r = -2$

$\underline{-2}\rvert$	2	0	−3	4	0	0	−1	8
		−4	8	−10	12	−24	48	−94
	2	−4	5	−6	12	−24	47	−86

Quotient: $2m^6 - 4m^5 + 5m^4 - 6m^3 + 12m^2 - 24m + 47$
Remainder: -86

The quotient is one degree less than dividend.

b. $(p^4 - 81) \div (p - 3)$

$(p^4 + \mathbf{0p^3} + \mathbf{0p^2} + \mathbf{0p} - 81) \div (p - 3)$

Insert place holders (bold) for missing powers of p.

$3\rvert$	1	0	0	0	−81
		3	9	27	81
	1	3	9	27	0

Quotient: $p^3 + 3p^2 + 9p + 27$
Remainder: 0

3. Choosing Synthetic Division or Long Division

Synthetic division is a time-saving shortcut. It is important to remember, however, that we will only use synthetic division if the divisor is in the form $(x - r)$, where r is a constant.

example 3 Choosing Long Division or Synthetic Division

Determine whether synthetic division can be used to divide the following polynomials.

a. $(x^3 - 4x^2 + 5) \div (x - 7)$
b. $(w^8 - w^2) \div (w^2 - 3)$
c. $\left(x^2 - 4x + \dfrac{1}{2}\right) \div (2x + 3)$

Solution:

a. Yes. The divisor $(x - 7)$ is in the form $(x - r)$, where $r = 7$. Either long division or synthetic division may be used.
b. No. $(w^2 - 3)$ is a second-degree divisor instead of first degree. Long division must be used.
c. No. The leading coefficient in the divisor is not 1. That is, $(2x + 3)$ is not in the form $(x - r)$. Long division will be used.

section A.1 PRACTICE EXERCISES

1. Explain the conditions under which you may use synthetic division to divide polynomials.

2. Can synthetic division be used to divide $(4x^4 + 3x^3 - 7x + 9)$ by $(2x + 5)$? Explain why or why not.

3. Can synthetic division be used to divide $(6y^5 - 3y^2 + 2y - 14)$ by $(y^2 - 3)$? Explain why or why not.

4. Can synthetic division be used to divide $(3y^4 - y + 1)$ by $(y - 5)$? Explain why or why not.

5. The following table represents the result of a synthetic division.

5	1	−2	−4	3
		5	15	55
	1	3	11	58

Using x as the variable,

a. Identify the divisor.

b. Identify the quotient.

c. Identify the remainder.

6. The following table represents the result of a synthetic division.

−2	2	3	0	−1	6
		−4	2	−4	10
	2	−1	2	−5	16

Using x as the variable,

a. Identify the divisor.

b. Identify the quotient.

c. Identify the remainder.

For Exercises 7–28, divide using synthetic division. Check your answer by multiplication.

7. $(x^2 - 2x - 48) \div (x - 8)$

8. $(x^2 - 4x - 12) \div (x - 6)$

9. $(t^2 - 3t - 4) \div (t + 1)$

10. $(h^2 + 7h + 12) \div (h + 3)$

11. $(5y^2 + 5y + 1) \div (y - 1)$

12. $(3w^2 + w - 5) \div (w + 2)$

13. $(x^2 + 11x + 28) \div (x + 4)$

14. $(x^3 - 7x^2 + 13x - 6) \div (x - 2)$

15. $(3y^3 + 7y^2 - 4y + 3) \div (y + 3)$

16. $(z^3 - 2z^2 + 2z - 5) \div (z + 3)$

17. $(x^3 - 3x^2 + 4) \div (x - 2)$

18. $(3y^4 - 25y^2 - 18) \div (y - 3)$

19. $(m^3 + 27) \div (m + 3)$

20. $(n^4 - 16) \div (n - 2)$

21. $(a^4 - 2a^2 - 8) \div (a + 2)$

22. $(b^4 - 6b^2 - 27) \div (b + 3)$

23. $(3p^4 - 11p^3 - 14p - 10) \div (p - 4)$

24. $(t^4 + 3t^3 - 3t + 4) \div (t + 2)$

25. $(4w^4 - w^2 + 6w - 3) \div \left(w - \frac{1}{2}\right)$

26. $(-12y^4 - 5y^3 - y^2 + y + 3) \div \left(y + \frac{3}{4}\right)$

27. $(8x^3 - 2x^2 + 1) \div \left(x + \frac{1}{4}\right)$

28. $(3x^3 + 4x^2 - x - 4) \div \left(x - \frac{2}{3}\right)$

29. Given the polynomial function defined by $T(x) = 4x^3 + 10x^2 - 8x - 20$,

a. Evaluate $T(-4)$.

b. Divide $(4x^3 + 10x^2 - 8x - 20) \div (x + 4)$.

c. Compare the value found in part (a) to the remainder found in part (b).

30. Given the polynomial function defined by $M(x) = -3x^3 - 12x^2 + 5x - 8$,
    a. Evaluate $M(-6)$.
    b. Divide $(-3x^3 - 12x^2 + 5x - 8) \div (x + 6)$.
    c. Compare the value found in part (a) to the remainder found in part (b).

31. Based on your solutions to Exercises 29–30, make a conjecture about the relationship between the value of a polynomial function, $P(x)$ at $x = r$ and the value of the remainder of $P(x) \div (x - r)$.

32. a. Use synthetic division to divide $(7x^2 - 16x + 9) \div (x - 1)$.
    b. Based on your solution to part (a), is $x - 1$ a *factor* of $7x^2 - 16x + 9$?

33. a. Use synthetic division to divide $(8x^2 + 13x + 5) \div (x + 1)$.
    b. Based on your solution to part (a), is $x + 1$ a *factor* of $8x^2 + 13x + 5$?

Concepts

1. Mean
2. Median
3. Mode
4. Weighted Mean

section

A.2 Mean, Median, and Mode

1. Mean

Mathematics is used in a wide variety of fields to analyze **data** (information). When given a list of numerical data, it is often helpful to obtain a single number that represents the central value of the data. In this section, we introduce three such values called the mean, median, and mode. The first calculation we present is the mean (or average) of a list of data values.

Mean

The **mean** (or average) of a set of numbers is the sum of the numbers divided by the number of values.

example 1 **Finding the Mean of a Data Set**

At a certain gas station, the weekly price for one gallon of regular unleaded gasoline is given for an 8-week period.

\$2.59	\$2.64	\$2.61	\$2.74
\$2.57	\$2.65	\$2.67	\$2.73

Find the mean price of regular unleaded gasoline from this gas station for the 8-week period.

Solution:

Divide the sum of the data by the number of values:

$$\text{mean} = \frac{2.59 + 2.57 + 2.64 + 2.65 + 2.61 + 2.67 + 2.74 + 2.73}{8}$$

$$= \frac{21.2}{8}$$

$$= 2.65 \qquad \text{Divide.}$$

The mean (or average) price of gas for this time period was $2.65.

example 2 **Finding the Mean of a Data Set**

Radcliffe received the following scores on his first five algebra tests: 92, 96, 85, 89, 90. On his sixth test, he neglected his studies and earned a score of 22.

a. Find the mean of Radcliffe's first five test scores.

b. Find the mean of all six scores.

c. Suppose that to earn an A in the course, a student must obtain an average test score of 90 or better. To earn a B, the student must score between an 80 and 90. The student will receive a C if the average is below 80. What affect did the low score have on Radcliffe's average and on his grade?

Solution:

a. $\text{mean of first five scores} = \frac{92 + 96 + 85 + 89 + 90}{5}$

$$= 90.4$$

b. $\text{mean of all six scores} = \frac{92 + 96 + 85 + 89 + 90 + 22}{6}$

$$= 79.0$$

c. The low score of 22 dropped Radcliffe's average dramatically. By not studying for the last test, his grade dropped from an A to a C.

2. Median

In Example 2, we saw that the mean test score was greatly affected by the presence of the low score of 22. If we want to de-emphasize extreme values (unusually high or low numbers) in a list of numbers, we may choose to compute the median. The median is the "middle" number in an ordered list of numbers.

Median

To compute the **median** of a list of numbers, first arrange the numbers in order from least to greatest.

- If the number of data values in the list is *odd*, then the median is the middle number in the list.
- If the number of data values is *even*, there is no single middle number. Therefore, the median is the mean (average) of the two middle numbers in the list.

To understand how to compute the median, consider Radcliffe's first five test scores from Example 2, arranged in order.

First five test scores:	85, 89, 90, 92, 96	Arrange the data in order.
The median is 90.		Because there are 5 data values (an *odd* number), the median is the middle number.

Now consider all 6 of Radcliffe's test scores.

First six test scores:	22, 85, 89, 90, 92, 96	Arrange the data in order. There are 6 data values (an *even* number). The median is the average of the two middle numbers.
The median is 89.5.	$\dfrac{89 + 90}{2} = 89.5$	

The mean of Radcliffe's six tests is 79, whereas the median is 89.5. The median was not significantly influenced by the presence of the low score of 22. This example shows that the median is a good choice for a central value when the data list has unusually high or low data values.

example 3 Finding the Median of a Data Set

The birth weights of a sample of 10 babies born at Massachusetts General Hospital in Boston are given below.

5.9 lb	6.8 lb	8.2 lb	7.6 lb	7.3 lb
7.1 lb	8.0 lb	9.3 lb	8.5 lb	6.8 lb

Find the median weight.

Solution:

First arrange the numbers in order from least to greatest:

5.9 6.8 6.8 7.1 7.3 7.6 8.0 8.2 8.5 9.3

$$\text{median} = \frac{7.3 + 7.6}{2} = 7.45$$

There are 10 data values (an *even* number). Therefore, the median is the average of the middle two numbers. The median weight is 7.45 pounds.

3. Mode

The last representative value for a list of data is called the mode.

Mode

The **mode** of a set of data is the value (or values) that occurs most often. If each value occurs the same number of times, then there is no mode.

example 4 Finding the Mode of a Data Set

Find the mode of the weights of babies from Example 3.

Solution:

5.9 lb	6.8 lb	8.2 lb	7.6 lb	7.3 lb
7.1 lb	8.0 lb	9.3 lb	8.5 lb	6.8 lb

The data value 6.8 lb appears the most often. Therefore, the modal weight of the babies is 6.8 pounds.

example 5 Finding the Mode of a Data Set

Find the mode for the list of high temperatures (in Fahrenheit) for a 2-week period in July in Asheville, North Carolina.

78	75	81	84	83	73	86
76	82	77	74	71	88	87

Solution:

No data value occurs most often. There is no mode for this set of data.

4. Weighted Mean

Sometimes data values in a list appear multiple times. In such a case, we can use the fact that multiplication represents repeated addition to compute a weighted mean. In Example 6, each data value is "weighted" by the number of times it appears in the list.

example 6

Computing a Weighted Mean

Donations are made to a certain charitable organization in increments of \$35, \$50, \$100, and \$200 as shown in Figure A-5. Find the mean amount donated.

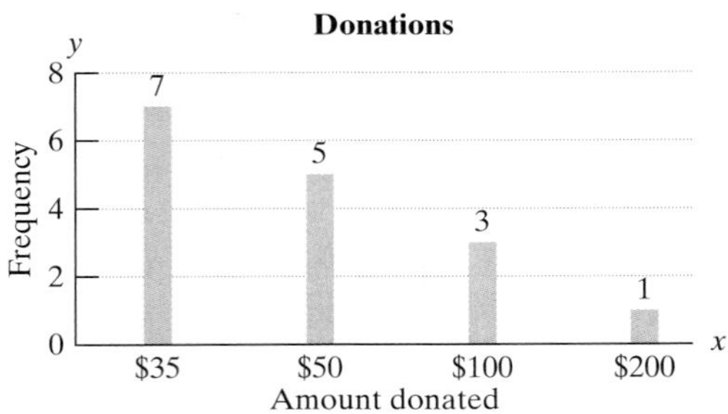

Figure A-5

Solution:

Notice that there are 16 data values represented in the graph:

7 of these: \$35, \$35, \$35, \$35, \$35, \$35, \$35

5 of these: \$50, \$50, \$50, \$50, \$50

3 of these: \$100, \$100, \$100

1 of these: \$200

The data value \$35 occurs seven times. Rather than adding \$35 seven times, we can find the sum by multiplying \$35(7) = \$245. Similarly, the value \$50 occurs 5 times for a sum of \$50(5) = \$250, and so on. To find the sum of all data values, we multiply each data value by the number of times it occurs (its frequency) and add the results. This process can be organized easily in a table.

Amount Donated (\$)	Frequency	Product (\$)
35	7	(35)(7) = 245
50	5	(50)(5) = 250
100	3	(100)(3) = 300
200	1	(200)(1) = 200
Total:	**16**	**995**

Total frequency 16 ← total number of donations made

Total product 995 ← sum of all data values

The mean is the sum of all data values divided by the total number of donations made (total frequency).

$$\text{mean} = \frac{995}{16} = 62.1875$$

The mean amount donated is $62.19.

section A.2 Practice Exercises

For Exercises 1–6, find the mean, median, and the mode.

1. 6, 5, 10, 4, 5
2. 5, 7, 4, 2, 7
3. 3, 5, 7, 4, 7, 2, 9, 3
4. 7, 4, 5, 0, 2, 2, 4, 8
5. 10, 13, 18, 20, 15, 11
6. 22, 14, 12, 16, 15, 17
7. A group of five people from the office went on a diet and exercise program. After three months, the following weight losses were recorded: 10, 14, 12, 8, 21. Find the mean and median weight loss.
8. A student received grades of 85, 92, 86, 98, and 99. Find her mean and median grade.
9. A small class has students of the following ages: 18, 20, 17, 36, 45, 20. Find the mean and median age for this class.
10. A college ice hockey team played its last 10 games. The goals scored for each game were as follows: 3, 4, 1, 0, 2, 2, 4, 1, 1, 0. Find the mean and median number of goals scored.
11. A survey of car wash prices produced the following list: $5, $7, $6, $7, $10, $8, $7. What is the most common price (mode)?
12. When asked the number of children in a family, the following list was formed: 3, 2, 1, 3, 2, 4, 6, 1, 3, 3. What is the most common number of children (mode) in a family from this survey?

The mean and median can be computed if the data are numerical. However, the mode can be found with numerical or nonnumerical data. For Exercises 13–14, find the mode.

13. An observer notes the type of vehicles in a mall parking lot.

SUV Pickup truck
Pickup truck Car
Pickup truck Car
Car SUV
Car SUV
Car Car

14. An observer notes the color of vehicles in a mall parking lot.

red white
white red
white brown
black blue
yellow black
black white

15. The prices for an oil change for a car are $25, $16, $22, $20, $25, $22, $22. What is the modal price?
16. The number of pets in a survey of families is 0, 1, 0, 2, 3, 1, 1, 3, 0, 1. What is the mode of this list?
17. The recent populations of the five most populous cities in the world are given in the table. Find the mean.

City	Population (millions)
Seoul, South Korea	10.2
São Paulo, Brazil	10.0
Bombay (Mumbai), India	9.9
Jakarta, Indonesia	9.1
Moscow, Russia	8.4

18. The recent populations of the five most populous countries are given in the table. Find the mean.

Country	Population (millions)
China	1,237
India	1,001
United States	273
Indonesia	216
Brazil	172

19. The given graph depicts the tuition for the year 2001–2002 for five leading private universities. Find the mean yearly tuition.

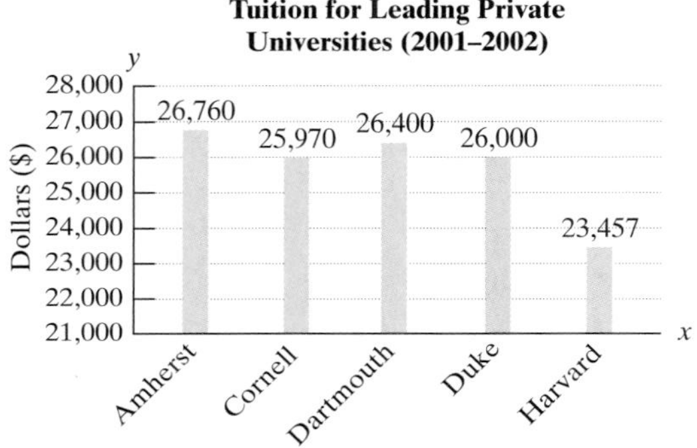

Figure for Exercise 19

20. The given graph depicts the in-state tuition for five state universities. Find the mean yearly cost.

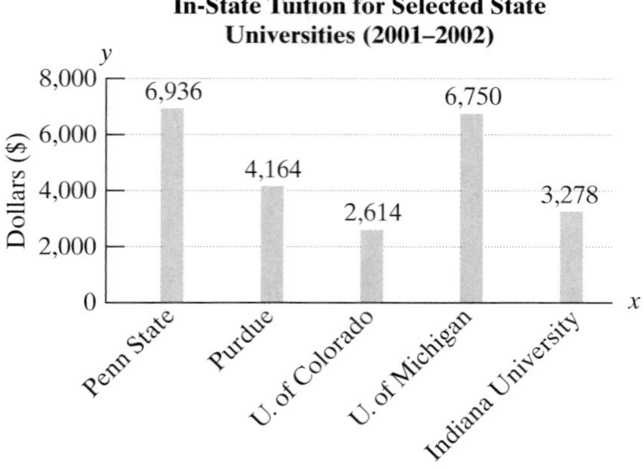

Figure for Exercise 20

21. The ages of several top women tennis players and several top women golfers are given for July 2003. Find the mean age of each group. For which group is the average age older?

Golfers Age (yrs)	Tennis Player Age (yrs)
25	27
40	20
24	27
43	21
38	28
31	27
41	21
38	23
21	
40	

22. The Center for Great Apes in Wauchula, Florida, provides sanctuary to chimpanzees and orangutans rescued from dire living conditions. The ages and weights for nine chimpanzees at the Center for Great Apes in July 2003 are given in the table.

Chimpanzee	Age (yr)	Weight (lb)
Knuckles	4	40
Noelle	8	105
Brooks	8	110
Kenya	10	120
Grub	12	145
Roger	24	140
Chipper	28	130
Toddy	29	110
Butch	30	135

a. Find the mean, median, and modal age.

b. Find the mean and median weight of all nine chimpanzees listed.

c. Find the mean and median weight of the eight chimpanzees, excluding 4-year-old Knuckles. How do the mean and median weights differ by excluding Knuckles' weight?

At most colleges and universities, weighted means are used to compute students' grade point averages (GPAs). At one college, the grades A–F are assigned numerical values as follows:

A = 4.0 C = 2.0

B+ = 3.5 D+ = 1.5

B = 3.0 D = 1.0

C+ = 2.5 F = 0.0

Grade point average is a weighted mean where the "weights" for each grade are the number of credit hours for that class. Use this information to answer Exercises 23–24.

23. Compute the GPA for the following grades. Round to the nearest hundredth.

Course	Grade	Number of Credit Hours (weights)
General Psychology	B+	3
Beginning Algebra	A	4
Student Success	A	1
Freshman English	B	3

24. Compute the GPA for the following grades. Round to the nearest hundredth.

Course	Grade	Number of Credit Hours (weights)
Intermediate Algebra	B	4
Theater	C	1
Music Appreciation	A	3
World History	D	5

25. There are 60 children at a daycare center. Their ages are given in the table. Compute the mean age. Round to the nearest year.

Age	Number of Children
2	5
3	15
4	22
5	16
6	2

26. A police officer writes 20 tickets for drivers speeding in a 30 mph zone. The speeds are recorded in the table. Compute the mean speed.

Speed (mph)	Frequency
44	1
45	3
48	4
50	6
52	3
54	2
55	1

STUDENT ANSWER APPENDIX

CHAPTER R

Section R.1 Practice Exercises, pp. 12–16

1. Numerator: 7; denominator: 8; proper **3.** Numerator: 9; denominator: 5; improper **5.** Numerator: 6; denominator: 6; improper **7.** Numerator: 12; denominator: 1; improper **9.** $\frac{3}{4}$ **11.** $\frac{4}{3}$ **13.** $\frac{1}{6}$ **15.** $\frac{2}{2}$ **17.** $\frac{5}{2}$ or $2\frac{1}{2}$ **19.** $\frac{6}{2}$ or 3 **21.** The set of whole numbers includes the number 0 and the set of natural numbers does not. **23.** For example: $\frac{2}{4}$ **25.** Prime **27.** Composite **29.** Composite **31.** Prime **33.** $2 \times 2 \times 3 \times 3$ **35.** $2 \times 3 \times 7$ **37.** $2 \times 5 \times 11$ **39.** $3 \times 3 \times 3 \times 5$ **41.** $\frac{1}{5}$ **43.** $\frac{3}{8}$ **45.** $\frac{7}{8}$ **47.** $\frac{3}{4}$ **49.** $\frac{5}{8}$ **51.** $\frac{3}{4}$ **53.** False: When adding or subtracting fractions it is necessary to have a common denominator. **55.** $1\frac{2}{9}$ **57.** $3\frac{1}{2}$ **59.** $\frac{27}{5}$ **61.** $\frac{15}{8}$ **63.** $\frac{4}{3}$ or $1\frac{1}{3}$ **65.** $\frac{2}{3}$ **67.** $\frac{9}{2}$ or $4\frac{1}{2}$ **69.** 46 **71.** $\frac{14}{5}$ or $2\frac{4}{5}$ **73.** $\frac{11}{54}$ **75.** \$300 **77.** 4 eggs **79.** 8 jars **81.** 12 servings **83.** 12 batches **85.** 24 **87.** 40 **89.** 90 **91.** $\frac{3}{7}$ **93.** $\frac{1}{2}$ **95.** $\frac{7}{8}$ **97.** $\frac{3}{40}$ **99.** $\frac{3}{26}$ **101.** $\frac{23}{36}$ **103.** $\frac{11}{10}$ **105.** $\frac{19}{48}$ **107.** $\frac{7}{2}$ or $3\frac{1}{2}$ **109.** $\frac{23}{24}$ **111.** $\frac{59}{12}$ or $4\frac{11}{12}$ **113.** $\frac{1}{8}$ **115.** $5\frac{7}{12}$ ft **117.** $1\frac{7}{12}$ cups **119.** $\frac{2}{3}$ mi **121.** $1\frac{1}{4}$ hours **123.** $244\frac{1}{2}$ yd **125.** Approximately 2 **127.** Approximately 8

Section R.2 Practice Exercises, pp. 24–30

1. b, e, i **3.** 108 cm; 704 cm^2 **5.** 1 ft; 0.0625 ft^2 **7.** $11\frac{1}{2}$ in. **9.** 31.4 ft **11.** a, f, g **13.** 0.0004 m^2 **15.** 40 mi^2 **17.** 132.665 cm^2 **19.** 66 in.^{2} **21.** 6 km^2 **23.** 75.36 ft^3 **25.** 39 in.^{3} **27.** 3052.08 in.^{3} **29.** 113.04 cm^3 **31.** 32.768 ft^3 **33.** 82 ft **35.** 36 in.^{2} **37.** 15.2464 cm^2 **39. a.** \$0.31/ft^2 **b.** \$129 **41.** Perimeter **43. a.** 50.24 in.^{2} **b.** 113.04 in.^{2} **c.** One 12-in. pizza **45.** 289.3824 cm^3 **47.** True **49.** True **51.** True **53.** True **55.** 45° **57.** Not possible **59.** For example: 100°, 80° **61. a.** $\angle 1$ and $\angle 3$, $\angle 2$ and $\angle 4$ **b.** $\angle 1$ and $\angle 2$, $\angle 2$ and $\angle 3$, $\angle 3$ and $\angle 4$, $\angle 1$ and $\angle 4$ **c.** $m(\angle 1) = 100°$, $m(\angle 2) = 80°$, $m(\angle 3) = 100°$ **63.** 57° **65.** 78° **67.** 60° **69.** 20° **71.** 147° **73.** 58° **75.** 135° **77.** 45° **79.** 7 **81.** 1 **83.** 1 **85.** 5 **87.** $m(\angle a) = 45°$, $m(\angle b) = 135°$, $m(\angle c) = 45°$, $m(\angle d) = 135°$, $m(\angle e) = 45°$, $m(\angle f) = 135°$, $m(\angle g) = 45°$ **89.** Scalene **91.** Isosceles **93.** No, a 90° angle plus an angle greater than 90° would make the sum of the angles greater than 180°. **95.** 40° **97.** 37° **99.** $m(\angle a) = 80°$, $m(\angle b) = 80°$, $m(\angle c) = 100°$, $m(\angle d) = 100°$, $m(\angle e) = 65°$, $m(\angle f) = 115°$, $m(\angle g) = 115°$, $m(\angle h) = 35°$, $m(\angle i) = 145°$, $m(\angle j) = 145°$ **101.** $m(\angle a) = 70°$, $m(\angle b) = 65°$, $m(\angle c) = 65°$, $m(\angle d) = 110°$, $m(\angle e) = 70°$, $m(\angle f) = 110°$, $m(\angle g) = 115°$, $m(\angle h) = 115°$, $m(\angle i) = 65°$, $m(\angle j) = 70°$, $m(\angle k) = 65°$

CHAPTER 1

Section 1.1 Practice Exercises, pp. 42–44

1.

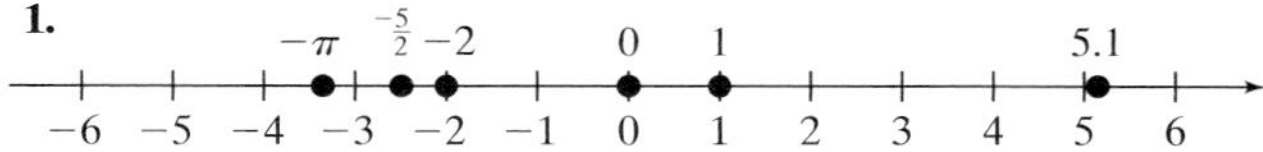

3. a **5.** b **7.** a **9.** a **11.** c **13.** a **15.** a **17.** a **19.** b **21.** c **23.** 0.29, 3.8, $\frac{1}{9}$, $\frac{1}{3}$, $\frac{1}{8}$, $\frac{1}{5}$, 5, 2, -0.125, -3.24, -3, -6, $\frac{7}{20}$, $\frac{5}{8}$, $0.\overline{2}$, $0.\overline{6}$ **25.** $\{1, 2, 3, 4, \ldots\}$ **27.** The set of all real numbers that are not rational **29.** The set of all numbers that includes all rational and irrational numbers **31.** For example: $\pi, -\sqrt{2}, \sqrt{3}$ **33.** For example: $-4, -1, 0$ **35.** For example: $-2, \frac{1}{2}, 0$ **37.** Yes **39.** $\frac{0}{5}, 1$

41. $\sqrt{11}, \sqrt{7}$

43.

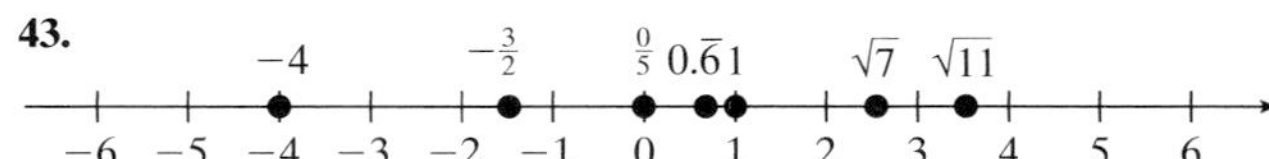

45.

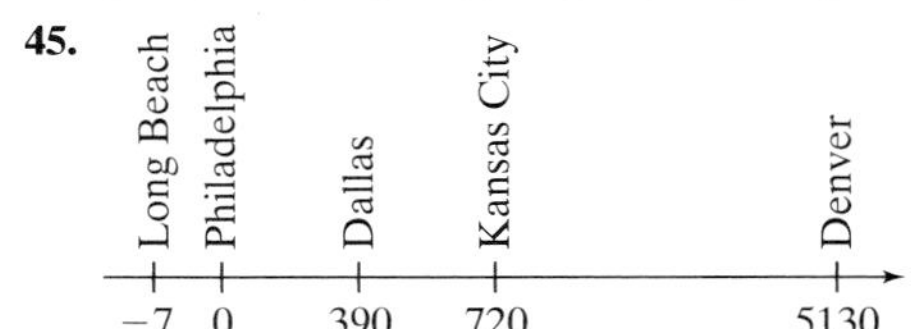

a. > **b.** < **c.** < **d.** <

47.

Webb Fukushima Mallon Kane Inkster

-7 -6 -3 0 2

a. < **b.** > **c.** > **d.** >

49. $-2, 2$ **51.** 2.5, 2.5 **53.** $\frac{1}{3}, \frac{1}{3}$ **55.** $-\frac{1}{9}, \frac{1}{9}$ **57.** 5.1 **59.** 7 **61.** False, $|m|$ is never negative. **63.** True **65.** False **67.** True **69.** False **71.** False **73.** False **75.** False **77.** True **79.** False **81.** True **83.** True **85.** True **87.** For all $a \geq 0$

Section 1.2 Practice Exercises, pp. 50–53

1.

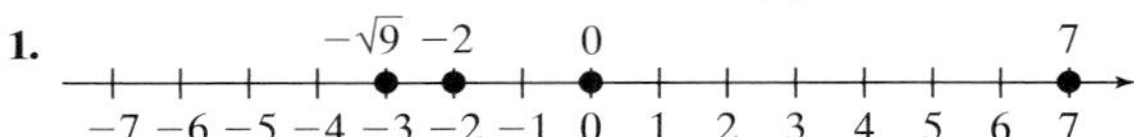

3. Number line from −7 to 7 with points plotted at $\sqrt{5}$ and π.

−7 −6 −5 −4 −3 −2 −1 0 1 2 3 4 5 6 7

5. a. $<$ **b.** $>$ **c.** $>$ **d.** $<$ **7.** $\left(\frac{1}{6}\right)^4$ **9.** a^3b^2 **11.** $(5c)^5$ **13.** $8yx^6$ **15.** $x \cdot x \cdot x$ **17.** $2b \cdot 2b \cdot 2b$ **19.** $10 \cdot y \cdot y \cdot y \cdot y \cdot y$ **21.** $2 \cdot w \cdot z \cdot z$ **23.** 25 **25.** $\frac{1}{49}$ **27.** 0.015625 **29.** 64 **31.** 9 **33.** 2 **35.** 10 **37.** 4 **39.** 20 **41.** 60 **43.** 8 **45.** 0 **47.** $\frac{7}{8}$ **49.** 45 **51.** 16 **53.** 15 **55.** 19 **57.** 3 **59.** $\frac{5}{12}$ **61.** $\frac{5}{2}$ **63.** 39 **65.** 26 **67.** 15 **69.** 3 **71.** 5 **73.** 9 **75.** $\frac{2}{3}$ **77.** 17 **79.** 57,600 ft^2 **81.** 21 ft^2 **83. a.** x **b.** Yes, 1 **85.** $3x$ **87.** $\frac{x}{7}$ or $x \div 7$ **89.** $2 - a$ **91.** $2y + x$ **93.** $4(x + 12)$ **95.** $21 - 2x$ **97.** $t - 14$ **99.** The sum of 5 and r **101.** The difference of s and 14 **103.** The quotient of 5 and the product of 2 and p **105.** One more than the product of 7 and x **107.** 5, squared **109.** The square root of 5 **111.** 7, cubed **113.** The sum of 2 and the square of x **115.** The sum of 3 and the square root of r

117. a. $36 \div 4 \cdot 3 = 9 \cdot 3 = 27$ Division must be performed before multiplication.

b. $36 - 4 + 3 = 32 + 3 = 35$ Subtraction must be performed before addition.

119. This is acceptable, provided division and multiplication are performed in order from left to right, and subtraction and addition are performed in order from left to right. **121.** 2 **123.** 1

125.
```
(4+6)/(8-3)
                2
110-5*(2+1)-4
               91
100-2*(5-3)^3
               84
```
127–129.
```
3+(4-1)²
               12
(12-6+1)²
               49
3*8-√(32+2²)
               18
```
131.
```
√(18-2)
                4
(4*3-3*3)^3
               27
(20-3²)/(26-2²)
               .5
```

Section 1.3 Practice Exercises, pp. 57–59

1. Rational **3.** Rational **5.** Irrational **7.** Rational **9.** $>$ **11.** $>$ **13.** $>$ **15.** 3 **17.** −3 **19.** −17 **21.** 7 **23.** −19 **25.** −23 **27.** −5 **29.** −3 **31.** 0 **33.** 0 **35.** −5 **37.** −3 **39.** 0 **41.** −23 **43.** −6 **45.** −3 **47.** $-5 + 13 + (-11)$; $-3°$ **49.** $3 + (-5) + 14$; 12-yd gain **51.** To add two numbers with different signs, subtract the smaller absolute value from the larger absolute value and apply the sign of the number with the larger absolute value. **53.** 21.3 **55.** $-\frac{3}{14}$ **57.** −2.4 or $-\frac{12}{5}$ **59.** $\frac{1}{4}$ or 0.25 **61.** 0 **63.** $-\frac{7}{8}$ **65.** $\frac{11}{9}$ **67.** −23.08 **69.** 494.686 **71.** −0.002117 **73. a.** $52.23 + (-52.95)$ **b.** Yes **75. a.** $100 + 200 + (-500) + 300 + 100 + (-200)$ **b.** \$0 **77.** −1 **79.** 10 **81.** 5 **83.** $-6 + (-10)$; −16 **85.** $-3 + 8$; 5 **87.** $-21 + 17$; −4 **89.** $3(-14 + 20)$; 18 **91.** $(-7 + (-2)) + 5$; −4

Section 1.4 Practice Exercises, pp. 64–66

1. For example: 0, 1, 2, 3, 4 **3.** For example: $-\sqrt{2}$, $\sqrt{3}$, $\sqrt{5}$, -2π, π **5.** For example: 1, 2, 3, 4, 5 **7.** $\sqrt{6}$ **9.** $-7 + 10$ **11.** −3 **13.** −12 **15.** 4 **17.** 3 **19.** −2 **21.** 8 **23.** −8 **25.** 2 **27.** 6 **29.** 40 **31.** −40 **33.** −6 **35.** −20 **37.** −24 **39.** 25 **41.** −5 **43.** $-\frac{3}{2}$ **45.** $\frac{41}{24}$ **47.** $\frac{2}{5}$ **49.** $-\frac{2}{3}$ **51.** 9.2 **53.** −5.72 **55.** −10 **57.** −14 **59.** −51 **61.** −173.188 **63.** 3.243 **65.** $6 - (-7)$; 13 **67.** $3 - 18$; −15 **69.** $-5 - (-11)$; 6 **71.** $-1 - (-13)$; 12 **73.** $-32 - 20$; −52 **75.** 13 **77.** −9 **79.** $-\frac{11}{30}$ **81.** $\frac{1}{18}$ **83.** 5 **85.** −25 **87.** −25 **89.** 19,881 m **91.** $200 + 400 + 600 + 800 - 1000$; \$1000 **93.** 152° **95.** −7 **97.** 5 **99.** 5 **101.** 3 **103.** −2 **105.** −11 **107.** 2

109–111.
```
-8+(-5)
              -13
4+(-5)+(-1)
               -2
627-(-84)
              711
```
113.
```
-0.06-0.12
             -.18
-3.2+(-14.5)
            -17.7
-472+(-518)
             -990
```
115.
```
-12-9+4
              -17
209-108+(-63)
               38
```

Section 1.5 Practice Exercises, pp. 74–76

1. True **3.** False **5.** True **7.** False **9.** $4 + 4 + 4 + 4 + 4$ **11.** $(-2) + (-2) + (-2)$ **13.** $(-2)(-7) = 14$ **15.** $-5 \cdot 0 = 0$ **17.** No number multiplied by zero equals 6. **19.** 6 **21.** −6 **23.** −6 **25.** 6 **27.** 8 **29.** −8 **31.** −8 **33.** 8 **35.** 0 **37.** Undefined **39.** 0 **41.** 0 **43.** 2 **45.** $-\frac{1}{5}$ **47.** $-\frac{3}{2}$ **49.** $\frac{3}{10}$ **51.** −2 **53.** −7.912 **55.** 0.092 **57.** −6 **59.** 2.1 **61.** 9 **63.** −9 **65.** $-\frac{8}{27}$ **67.** 0.0016 **69.** −0.0016 **71.** −3 **73.** 3 **75.** −7 **77.** 7 **79.** −30 **81.** 96 **83.** 2 **85.** $-\frac{4}{33}$ **87.** $-\frac{4}{7}$ **89.** −1 **91.** $-2(3) + 3$; loss of \$3 **93.** −29 **95.** 48 **97.** −14.28 **99.** 340 **101.** $-\frac{10}{9}$ **103.** $\frac{14}{9}$ **105.** −24 **107.** $-\frac{1}{20}$ **109.** No, parentheses are required around the quantity $5x$; $10/(5x)$ **111.** $-3.75(0.3)$; −1.125 **113.** $\frac{16}{5} \div \left(-\frac{8}{9}\right)$; $-\frac{18}{5}$ **115.** $-0.4 + 6(-0.42)$; −2.92 **117.** $-\frac{1}{4} - 6\left(-\frac{1}{3}\right)$; $\frac{7}{4}$ **119. a.** −10 **b.** 24 **c.** In part (a) we subtract; in part (b) we multiply. **121.** −23 **123.** 12 **125.** $\frac{9}{7}$ **127.** Undefined **129.** −2 **131.** −6 **133.** −1 **135.** 12 **137.** −40 **139.** $\frac{7}{2}$ **141.** $-\frac{5}{3}$ **143.** For $x = 2$, 10; for $x = -2$, 10

145–147.
```
-6(5)
              -30
-5.2/2.6
               -2
(-5)(-5)(-5)(-5)
              625
```
149.
```
(-5)^4
              625
-5^4
             -625
-2.4²
            -5.76
```
151. **153.**

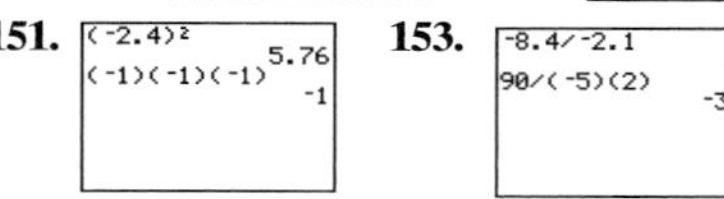

Section 1.6 Practice Exercises, pp. 85–87

1. −8 **3.** $-\frac{9}{2}$ or −4.5 **5.** $\frac{1}{3}$ **7.** $-\frac{4}{45}$ **9.** Reciprocal **11.** 0 **13.** b **15.** i **17.** g **19.** d **21.** h **23.** $30x + 6$ **25.** $-2a - 16$

27. $15c - 3d$ **29.** $-7y + 14$ **31.** $\frac{1}{3}m - 1$ **33.** $\frac{3}{2} + 3s$
35. $-\frac{2}{3}x + 4$ **37.** $-2p - 10$ **39.** $3w + 5z$ **41.** $4x + 8y - 4z$
43. $6w - x + 3y$ **45.** $4(90 + 2)$; 368 **47.** $4(900 + 2)$; 3608
49. term: $3xy$, coefficient: 3;
term: $-6x^2$, coefficient: -6;
term: y, coefficient: 1;
term: -17, coefficient: -17
51. term: x^4, coefficient: 1;
term: $-10xy$, coefficient: -10;
term: 12, coefficient: 12;
term: $-y$, coefficient: -1
53. The exponents on x are different. **55.** The variables are the same and raised to the same power. **57.** For example: $5y$, $-2y$, y
59. $-17k + 23$ **61.** $21x + 7y$ **63.** $-\frac{1}{2}a - 4b$ **65.** $-5.3z - 9.2$
67. $-6x + 22$ **69.** $4w$ **71.** $-3x + 17$ **73.** $10t - 44$ **75.** -18
77. $-2t + 7$ **79.** $51a - 27$ **81.** -6 **83.** $4q - \frac{1}{3}$ **85.** $6n$
87. $2x + 18$ **89.** $32.33z - 30.81$ **91.** Associative property of multiplication **93.** Associative property of addition **95.** Distributive property of multiplication over addition **97.** Identity property of addition **99.** Identity property of multiplication **101.** Inverse property of addition **103.** Equivalent **105.** Not equivalent. The terms are not *like* terms and cannot be combined. **107.** Not equivalent; subtraction is not commutative. **109.** Equivalent **111.** $14\frac{2}{7} + (2\frac{1}{3} + \frac{2}{3})$ is easier. **113. a.** 55 **b.** 210

Chapter 1 Review Exercises, pp. 92–94

1. a. 7, 1 **b.** 7, -4, 0, 1 **c.** 7, 0, 1 **d.** 7, $\frac{1}{3}$, -4, 0, $-0.\overline{2}$, 1
e. $-\sqrt{3}$, π **f.** 7, $\frac{1}{3}$, -4, 0, $-\sqrt{3}$, $-0.\overline{2}$, π, 1 **3.** 6 **5.** 0 **7.** False
9. True **11.** True **13.** True **15.** $\frac{7}{y}$ or $7 \div y$ **17.** $a - 5$
19. $13z - 7$ **21.** $3y + 12$ **23.** $2p - 5$ **25.** 225 **27.** $\frac{1}{16}$ **29.** $\frac{27}{8}$
31. 11 **33.** 10 **35.** 4 **37.** -17 **39.** $-\frac{5}{22}$ **41.** $-\frac{27}{10}$ **43.** -4.28
45. 8 **47.** When a and b are both negative or when a and b have different signs and the number with the larger absolute value is negative **49.** -12 **51.** -1 **53.** $-\frac{29}{18}$ **55.** -1.2 **57.** -10.2
59. $\frac{10}{3}$ **61.** -1 **63.** $-7 - (-18)$; 11 **65.** $7 - 13$; -6
67. $(6 + (-12)) - 21$; -27 **69.** -170 **71.** -2 **73.** $-\frac{1}{6}$ **75.** 0
77. 0 **79.** 2.25 **81.** $-\frac{3}{2}$ **83.** -30 **85.** $\frac{1}{4}$ **87.** -2 **89.** 17
91. $-\frac{7}{120}$ **93.** -2 **95.** -23 **97.** 70.6 **99.** True **101.** True
103. True **105.** For example: $2 + 3 = 3 + 2$ **107.** For example: $5 + (-5) = 0$ **109.** For example: $5 \cdot 2 = 2 \cdot 5$ **111.** For example: $3 \cdot \frac{1}{3} = 1$ **113.** $5x - 2y = 5x + (-2y)$, then use the commutative property of addition. **115.** $3y$, $10x$, -12, xy **117. a.** $8a - b - 10$ **b.** $-7p - 11q + 16$ **119.** $p - 2$ **121.** $-14q - 1$ **123.** $4x + 24$

Chapter 1 Test, p. 95

1. Rational, all repeating decimals are rational numbers.
2.
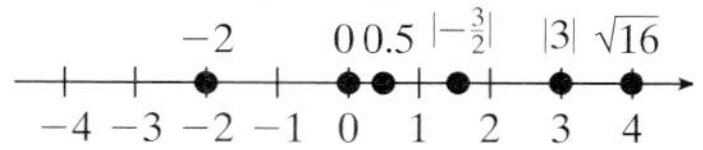

3. a. False **b.** True **c.** True **d.** True **4. a.** $(4x)(4x)(4x)$ **b.** $4 \cdot x \cdot x \cdot x$ **5. a.** Twice the difference of a and b **b.** The difference of twice a and b **6.** $\frac{\sqrt{c}}{d^2}$ or $\sqrt{c} \div d^2$ **7.** 6 **8.** 28 **9.** $-\frac{7}{8}$
10. 4.66 **11.** -32 **12.** -12 **13.** Undefined **14.** -28 **15.** 0
16. 96 **17.** $\frac{2}{3}$ **18.** -8 **19.** 9 **20.** $\frac{1}{3}$ **21.** $-\frac{3}{5}$
22. a. Commutative property of multiplication **b.** Identity property of addition **c.** Associative property of addition **d.** Inverse property of multiplication **e.** Associative property of multiplication
23. $-6k - 8$ **24.** $-4p - 23$ **25.** $4p - \frac{4}{3}$ **26.** $-3.72x + 13.12$
27. -51 **28.** 9 **29.** 576 **30.** -21 **31.** $12 - (-4)$; 16
32. $6 - 8$; -2

CHAPTER 2

Section 2.1 Practice Exercises, pp. 106–107

1. Expression **3.** Equation **5.** Equation **7.** Expression **9.** Substitute the value into the equation and determine if the right-hand side is equal to the left-hand side. **11.** No **13.** Yes **15. a.** No **b.** Yes **c.** No **d.** Yes **17.** $x = -1$ **19.** $q = 20$
21. $m = -17$ **23.** $y = -16$ **25.** $z = \frac{11}{2}$ or $5\frac{1}{2}$ **27.** $x = -2$
29. $a = 1.3$ **31.** $c = 0$ **33.** $c = -2.13$ **35.** $t = -3.2675$
37. $-8 + x = 42$, $x = 50$ **39.** $x - (-6) = 18$, $x = 12$
41. $x + \frac{5}{8} = \frac{13}{8}$, $x = 1$ **43.** $x = 9$ **45.** $p = -4$ **47.** $y = 0$
49. $y = -15$ **51.** $t = -\frac{4}{5}$ **53.** $a = -10$ **55.** $b = 4$ **57.** $x = 41$
59. $p = -127$ **61.** $y = -2.6$ **63.** $x \cdot 7 = -63$ or $7x = -63$; $x = -9$ **65.** $\frac{x}{12} = \frac{1}{3}$; $x = 4$ **67.** $a = 10$ **69.** $x = -\frac{1}{9}$
71. $h = -12$ **73.** $t = \frac{22}{3}$ or $7\frac{1}{3}$ **75.** $r = -36$ **77.** $k = 16$
79. $k = 2$ **81.** $q = -\frac{7}{4}$ or $-1\frac{3}{4}$ **83.** $q = 11$ **85.** $d = -36$
87. $z = \frac{7}{2}$ or $3\frac{1}{2}$ **89.** $y = 4$ **91.** $y = 3.6$ **93.** $y = 0.4084$ **95.** 14 cm
97. 4.04 ft **99.** $1\frac{1}{3}$ hr **101.** 8 mph

Section 2.2 Practice Exercises, pp. 113–115

1. $-7t - 6$ **3.** $-5z + 2$ **5.** $10p - 10$ **7.** $20a + 15$ **9.** To simplify an expression, clear parentheses and combine *like* terms. To solve an equation, use the addition, subtraction, multiplication, and division properties of equality to isolate the variable. **11.** $y = -3$
13. $z = -5$ **15.** $b = \frac{59}{5}$ or $11\frac{4}{5}$ **17.** $h = \frac{1}{20}$ **19.** First use the addition property, then the division property. **21.** $x = -3$ **23.** $w = 2$
25. $q = -\frac{3}{4}$ **27.** $n = -2$ **29.** $b = \frac{21}{8}$ or $2\frac{5}{8}$ **31.** $a = -2$

33. $r = \frac{3}{7}$ **35.** $v = 16$ **37.** $u = 1$ **39.** $b = 34$ **41.** $x = 2$

43. $t = 4$ **45.** $x = -1$ **47.** $t = -10$ **49.** $k = \frac{1}{4}$ **51.** $w = 22$

53. $p = 8$ **55.** $y = 1$ **57.** $k = \frac{3}{7}$ **59.** $z = 5$ **61.** $m = -0.65$

63. No solution **65.** Contradiction; no solution **67.** Conditional equation; $x = -15$ **69.** Identity; all real numbers **71.** Identity; all real numbers **73.** Conditional equation; $x = 0$ **75.** $a = 15$ **77.** $a = 4$ **79.** Contradiction

Section 2.3 Practice Exercises, pp. 121–122

1. $x = -\frac{3}{5}$ **3.** $m = 46$ **5.** $b = 22$ **7.** No solution **9.** 18, 36

11. 100; 1000; 10,000 **13.** $x = 4$ **15.** $y = -\frac{19}{2}$ **17.** $q = -\frac{15}{4}$

19. $w = 8$ **21.** $m = 3$ **23.** $s = 15$ **25.** No solution **27.** All real numbers **29.** $x = 5$ **31.** $w = 2$ **33.** The number is -16.

35. The number is $\frac{9}{5}$. **37.** The number is $\frac{7}{16}$. **39.** $y = 6$ **41.** $w = 3$

43. All real numbers **45.** $x = 67$ **47.** $p = 90$ **49.** $y = -2$ **51.** $b = 7$ **53.** $y = -8$ **55.** $a = 20$ **57.** $x = 0$ **59.** $h = 3$

61. $x = \frac{8}{3}$ or $2\frac{2}{3}$ **63.** No solution **65.** $w = \frac{25}{2}$ or $12\frac{1}{2}$ **67.** All real numbers **69.** $a = -6$ **71.** $h = \frac{1}{3}$ **73.** $t = -6$ **75.** $a = -1$

77. $b = 2$

Section 2.4 Practice Exercises, pp. 129–131

1. $x + 16 = -31; x = -47$ **3.** $x - 6 = -3; x = 3$ **5.** $x - 16 = -1; x = 15$ **7.** Contradiction; no solution **9.** Conditional equation; $y = -7$ **11.** Identity; all real numbers **13.** The number is -3. **15.** The number is 11. **17.** The number is 5. **19.** The number is -5. **21.** The number is 10. **23.** The number is 9. **25.** There were 165 Republicans and 269 Democrats. **27.** The lengths of the pieces are 33 cm and 53 cm. **29.** The Congo River is 4370 km long, and the Nile River is 6825 km. **31. a.** $x + 1, x + 2$ **b.** $x - 1, x - 2$ **33. a.** $x + 2, x + 4$ **b.** $x - 2, x - 4$ **35.** The page numbers are 470 and 471. **37.** The numbers are 17, 19, and 21. **39.** 42, 43, and 44 **41.** The sides are 13 in., 14 in., and 15 in. **43.** The sides are 14 in., 15 in., 16 in., 17 in., and 18 in. **45.** The area of New Guinea is 792,500 km^2. **47.** The area of Africa is 30,065,000 km^2. The area of Asia is 44,579,000 km^2.

Section 2.5 Practice Exercises, pp. 136–138

1. -69 and -71 **3.** $a = 40$ **5.** 0.57 **7.** 1.35 **9.** 69% **11.** 0.6% **13.** 12.5% **15.** 85% **17.** 0.75 **19.** 310.8 **21.** 885 **23.** 2200 **25.** 50.9% **27.** 30.9% **29.** \$80.20 **31.** \$16.00 **33.** 5% **35.** \$29.96 **37.** \$4562.50 **39.** \$3879 **41.** 10% **43.** \$420 **45.** \$1200 **47.** 6% **49. a.** \$330 **b.** \$2330 **51.** \$645 **53.** \$60

Section 2.6 Practice Exercises, pp. 143–147

1. $y = -5$ **3.** $x = 0$ **5.** $y = -2$ **7.** $a = P - b - c$

9. $y = x + z$ **11.** $q = p - 250$ **13.** $t = \frac{d}{r}$ **15.** $t = \frac{PV}{nr}$

17. $x = 5 + y$ **19.** $y = -3x - 19$

21. $y = \frac{-2x + 6}{3}$ or $y = -\frac{2}{3}x + 2$

23. $x = \frac{y + 9}{-2}$ or $x = -\frac{1}{2}y - \frac{9}{2}$

25. $y = \frac{-4x + 12}{-3}$ or $y = \frac{4}{3}x - 4$

27. $y = \frac{-ax + c}{b}$ or $y = -\frac{a}{b}x + \frac{c}{b}$

29. $t = \frac{A - P}{Pr}$ or $t = \frac{A}{Pr} - \frac{1}{r}$ **31.** $c = \frac{a - 2b}{2}$ or $c = \frac{a}{2} - b$

33. $y = 2Q - x$ **35.** $a = MS$ **37.** $R = \frac{P}{I^2}$ **39.** The length is 7 ft and the width is 5 ft. **41.** The length is 195 m and the width is 100 m. **43.** The measures of the angles are 30°, 60°, and 90°. **45.** The measures of the angles are 42°, 54°, and 84°. **47.** $x = 17$; the measures of the angles are 34° and 56°. **49.** Adjacent **S**upplementary angles form a **S**traight angle. The words *supplementary* and *straight* both begin with the same letter. **51.** The angles are 30° and 60°. **53.** The angles are 45° and 135°. **55.** The angles are 34.8° and 145.2°. **57.** $y = 40$; the vertical angles measure 146°. **59. a.** $A = bh$ **b.** $b = \frac{A}{h}$ **c.** The base is 8 m. **61. a.** $P = 4s$ **b.** $s = \frac{P}{4}$ **c.** The length of each side at the bottom of the pyramid is 230.4 m.

63. a. $V = lwh$ **b.** $h = \frac{V}{lw}$ **c.** The height of the box is 2 ft.

65. $r^3 = \frac{3V}{4\pi}$ **67.** $h = \frac{3V}{lw}$ **69. a.** 1256 ft^2 **b.** 10,048 ft^3 **71. a.** 28 m^2 **b.** 14 m^2 **c.** The area of the triangle is one-half the area of the parallelogram.

73. 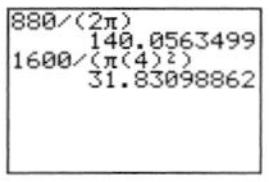

140.056 **75.** 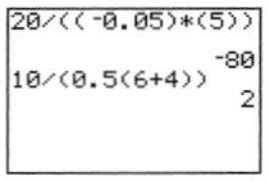

-80

Section 2.7 Practice Exercises, pp. 154–157

1. $x = 2$ **3.** $y = \frac{-4x + 20}{5}$ or $y = -\frac{4}{5}x + 4$ **5.** $a = 0.5$

7. $r = \frac{1}{4}$ **9. a.** \$1200 **b.** $6x$ **c.** $6(750 - x)$ or $4500 - 6x$

11. 53 tickets were sold at \$3 and 28 tickets were sold at \$2. **13.** 112 \$2 tickets and 96 \$1 tickets **15. a.** 2 lb **b.** $0.10x$ **c.** $0.10(x + 3) = 0.10x + 0.30$ **17.** Ten pounds of coffee sold at \$12 per pound and 40 lb of coffee sold at \$8 per pound. **19.** 2 lb of raisins and 4 lb of granola **21.** 2.5 lb of black tea and 1.5 lb of orange pekoe tea **23. a.** \$300 **b.** $0.06x$ **c.** $0.06(20{,}000 - x)$ or $1200 - 0.06x$ **25.** There is \$400 in the 8% account and \$1200 in the 6% account. **27.** \$8500 at 8% and \$4000 at 12% **29.** \$4500 at 5% and \$9000 at 6% **31.** \$5000 should be invested in the 6% account and \$3750 should be invested in the 8% account. **33. a.** 300 miles **b.** $5x$ **c.** $5(x + 12)$ or $5x + 60$ **35.** The car travels 5 hr. **37.** The slower car travels at 46 mph and the faster car travels at 50 mph. **39.** The Cessna's speed is 110 mph and the Piper Cub's speed is 120 mph. **41.** She hikes 2 mph to the lake. **43.** The rates of the boats are 20 mph and 40 mph. **45. a.** \$3.50 **b.** $0.05x$ **c.** $0.05(30 + x)$ or $1.5 + 0.05x$ **47.** Jean-Paul has 3 quarters and 9 dimes. **49.** She receives four \$10 bills and eight \$20 bills.

Section 2.8 Practice Exercises, pp. 167–170

1. a. $x + 13$ **b.** $-7x - 11$ **c.** $x = -3$ **3.** The United States scored $11\frac{1}{2}$ points, and Europe scored $16\frac{1}{2}$ points.

5. 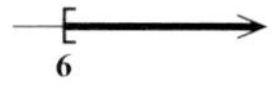

$[6, \infty)$ **7.** 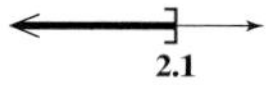

$(-\infty, 2.1]$

9. $(-2, 7]$ **11.** $\left\{x \mid x > \frac{3}{4}\right\}$ $\left(\frac{3}{4}, \infty\right)$

13. $\{x \mid -1 < x < 8\}$ $(-1, 8)$ **15.** $\{x \mid x < -14\}$ $(-\infty, -14)$
17. $\{x \mid x \geq 18\}$ **19.** $\{x \mid x < -0.6\}$

21. $\{x \mid -3.5 \leq x < 7.1\}$

23. a. $x = 3$ **b.** $x > 3$ **25. a.** $p = 13$ **b.** $p \leq 13$

27. a. $c = -3$ **b.** $c < -3$ **29. a.** $z = -\frac{3}{2}$ **b.** $z \geq -\frac{3}{2}$

31. No **33.** Yes
35. $(-\infty, 1]$ **37.** $(10, \infty)$
39. $(3, \infty)$ **41.** $(-\infty, 8]$
43. $(2, \infty)$ **45.** $(-\infty, -2)$
47. $[14, \infty)$ **49.** $[-24, \infty)$
51. $[-3, 3)$ **53.** $\left(0, \frac{5}{2}\right)$
55. $(10, 12)$ **57.** $[-1, 4)$
59. $[90, \infty)$ **61.** $(-9, \infty)$
63. $\left[-\frac{15}{2}, \infty\right)$ **65.** $\left[-\frac{1}{3}, \infty\right)$
67. $(-3, \infty)$ **69.** $(-\infty, 7)$
71. $\left(-\infty, \frac{15}{4}\right]$ **73.** $(-3, \infty)$

75. a. A [93, 100]; B+ [89, 93); B [84, 89); C+ [80, 84); C [75, 80); F [0, 75) **b.** B **c.** C **77.** $s \geq 110$ **79.** $t > 90$ **81.** $h \leq 2$ **83.** $t \geq 100$ **85.** $d \leq 10$ **87.** $2 < h < 5$ **89.** More than 10.2 in. of rain is needed. **91. a.** \$1539 **b.** 200 birdhouses cost \$1440. It is cheaper to purchase 200 birdhouses because the discount is greater. **93. a.** $2.00x > 75 + 0.17x$ **b.** $x > 41$; profit occurs when more than 41 lemonades are sold. **95.** $[13, \infty)$

97. $[-4, \infty)$ **99.** $(14.5, \infty)$

Chapter 2 Review Exercises, pp. 177–180

1. a. Equation **b.** Expression **c.** Equation **d.** Equation **3.** b, d

5. $a = -8$ **7.** $k = \frac{21}{4}$ or $5\frac{1}{4}$ **9.** $x = -\frac{21}{5}$ or $-4\frac{1}{5}$

11. $k = -\frac{10}{7}$ or $-1\frac{3}{7}$ **13.** The number is 60. **15.** The number is −8.

17. $d = 1$ **19.** $c = \frac{9}{4}$ or $2\frac{1}{4}$ **21.** $b = -3$ **23.** $p = \frac{3}{4}$ **25.** $a = 0$

27. $b = \frac{3}{8}$ **29.** A contradiction has no solution and an identity is true for all real numbers. **31.** $x = 6$ **33.** $z = -3$ **35.** $p = -10$

37. $t = \frac{5}{3}$ **39.** $q = 2.5$ **41.** $a = -4.2$ **43.** $x = -312$

45. No solution **47.** All real numbers **49.** $w = -3$ **51.** The number is 18. **53.** The number is 30. **55.** The number is −7. **57.** 66, 68, 70 **59.** The sides are 25 in., 26 in., and 27 in. **61.** The minimum salary was \$30,000 in 1980. **63.** 23.8 **65.** 12.5% **67.** 160 **69.** The sale price is \$26.39. **71.** She sold \$1700. **73. a.** \$840 **b.** \$3840 **75.** $K = C + 273$ **77.** $s = \frac{P}{4}$ **79.** $y = \frac{-2 - 2x}{5}$

81. The width is 5 ft. **83.** The angles are 32° and 58°. **85.** Peggy invested \$3000 in savings and \$1500 in stocks. **87.** She buys 10 ice cream on a stick and 14 Popsicles. **89.** The Millers travel 55 mph, and the O'Neills travel 50 mph. **91. a.** \$637 **b.** Three hundred plants cost \$1410 and 295 plants cost \$1416. Buying 300 plants gives a greater discount. **93.** $\left(-\frac{1}{3}, \infty\right)$

95. $\left(-\infty, -\frac{14}{5}\right]$ **97.** $(-\infty, 34.5)$

99. $\left(-\infty, \frac{5}{2}\right]$ **101.** $[-6, 5]$

103. a. $1.50x > 33 + 0.4x$ **b.** $x > 30$; a profit is realized if more than 30 hot dogs are sold.

Chapter 2 Test, pp. 180–181

1. $t = -16$ **2.** $p = 12$ **3.** $t = -\frac{16}{9}$ **4.** $x = \frac{7}{3}$ **5.** $p = 15$

6. $d = \frac{13}{4}$ **7.** $x = \frac{20}{21}$ **8.** No solution **9.** $y = -13$ **10.** $c = -47$

11. All real numbers **12.** $y = -3x - 4$ **13.** $r = \frac{C}{2\pi}$

14. $z = \frac{t - 4x - 4y}{4}$ or $z = \frac{t}{4} - x - y$ **15.** The sides are 61 in., 62 in., 63 in., 64 in., and 65 in. **16.** The basketball tickets were \$36.32, and the hockey tickets were \$40.64. **17.** The cost was \$82.00. **18.** They invested \$3200 in the 8% account and \$4000 in the 10% account. **19.** The angles are 32° and 58°. **20.** Their speeds were 50 mph and 55 mph. **21.** 8000

22. a. $(-\infty, 0)$ **b.** $[-2, 5)$

23. $(-2, \infty)$ **24.** $(-\infty, -4]$

25. $[-5, 1]$ **26.** More than 26.5 in. is required.

Cumulative Review Exercises, Chapters 1–2, pp. 181–182

1. −4 **2.** $\frac{1}{2}$ **3.** −2.75 **4.** −7 **5.** $-\frac{5}{12}$ **6.** 16 **7.** 4 **8.** −3.5

9. Whole numbers 2; Integers 2, −2; Rational numbers 2, −2, $0.\overline{7}$, 0.25, $\frac{3}{5}$; Irrational numbers $\sqrt{7}$, π **10. a.** $\frac{1}{3} \cdot \frac{3}{4}; \frac{1}{4}$ **b.** $\sqrt{5^2 - 9}; 4$

11. a. $-\frac{5}{2}$ **b.** $\frac{1}{3}$ **c.** Not possible **12.** $-7x^2y, 4xy, -6$ **13.** 1

14. $9x + 13$ **15.** $x = -7.2$ **16.** All real numbers **17.** The numbers are 77 and 79. **18.** The cost before tax was \$350.00. **19.** The height is $6\frac{5}{6}$ cm. **20.** Rachael invests \$3600 in the 4% account and \$7200 in the 9% account. **21.** Her average running rate is 8 mph and her average biking rate is 20 mph.

22. $(-2, \infty)$ **23.** −10

CHAPTER 3

Section 3.1 Practice Exercises, pp. 189–192

1.

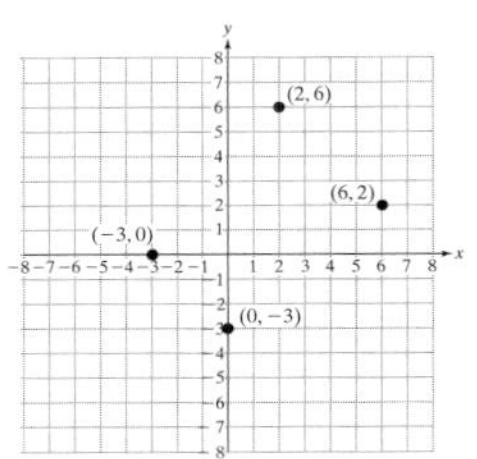

3.

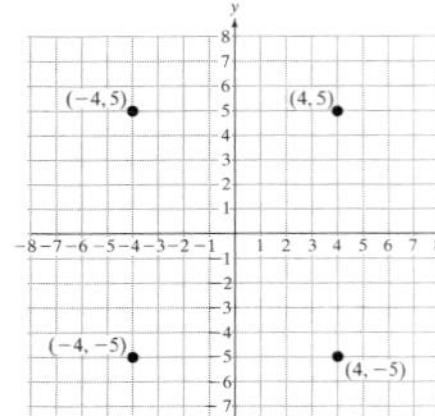

5.

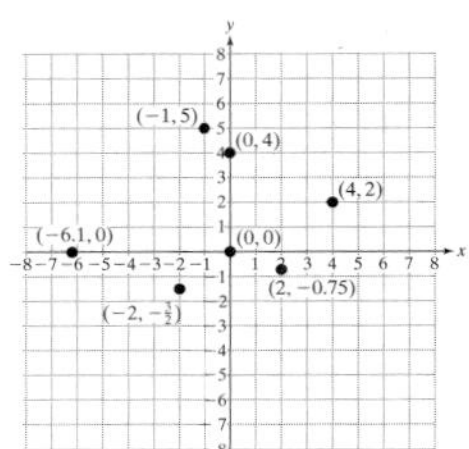

7. IV **9.** II **11.** III **13.** I **15.** $(0, -5)$ lies on the y-axis. **17.** $(\frac{7}{8}, 0)$ is located on the x-axis. **19.** $A(-4, 2)$, $B(\frac{1}{2}, 4)$, $C(3, -4)$, $D(-3, -4)$, $E(0, -3)$, $F(5, 0)$

21. a. (250, 225), (175, 193), (315, 330), (220, 209), (450, 570), (400, 480), (190, 185); (250, 225) 250 people produce \$225 in popcorn sales.

b.

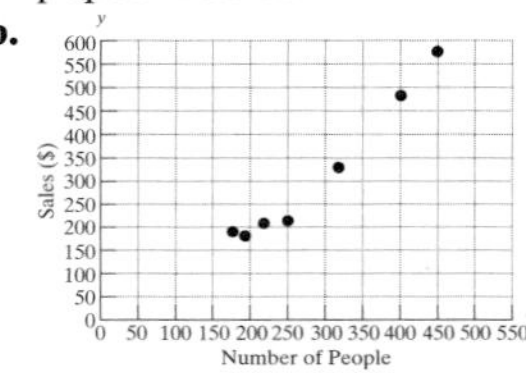

23. a. (10, 332000) In 1710 the population of the U.S. colonies was 332,000.

b.

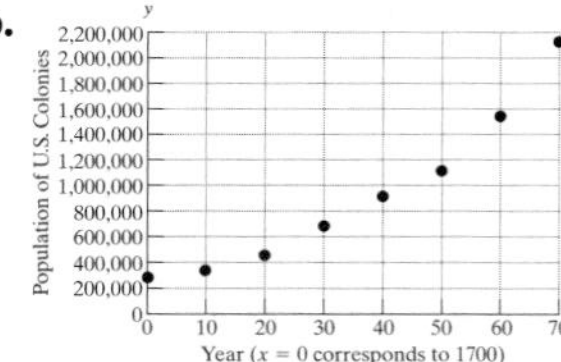

25. a. $(1, -10.2)$, $(2, -9.0)$, $(3, -2.5)$, $(4, 5.7)$, $(5, 13.0)$, $(6, 18.3)$, $(7, 20.9)$, $(8, 19.6)$, $(9, 14.8)$, $(10, 8.7)$, $(11, 2.0)$, $(12, -6.9)$.

b.

Temperature (°C)

Month

27. a. month 10 **b.** 30 **c.** between months 3 and 5 and between months 10 and 12 **d.** months 8 and 9 **e.** month 3 **f.** 80

29. a. (1, 89.25) On day 1 the closing price per share was \$89.25.
(2, 92.50)
(3, 91.25)
(4, 93.00)
(5, 90.25)
b. \$1.75
c. −\$2.75

31. a. $A(400, 200)$, $B(200, -150)$, $C(-300, -200)$, $D(-300, 250)$, $E(0, 450)$ **b.** 450 m

Section 3.2 Practice Exercises, pp. 200–202

1. Yes **3.** Yes **5.** No

7.

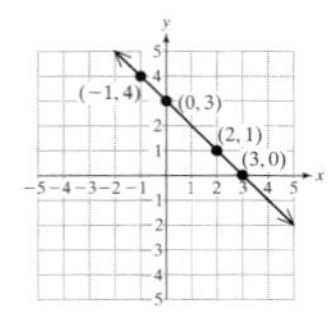

9.

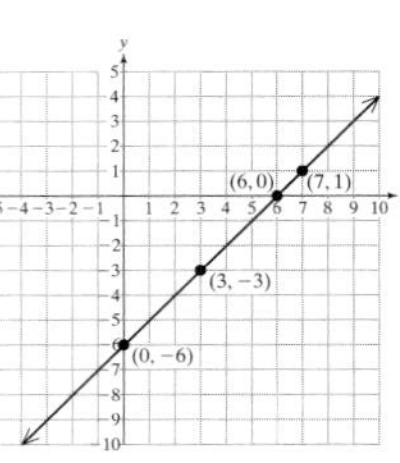

11.

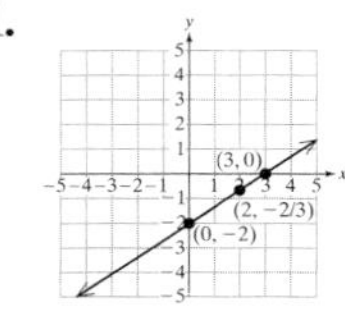

13.

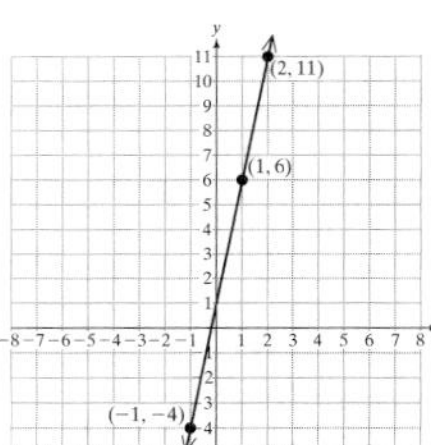

15.

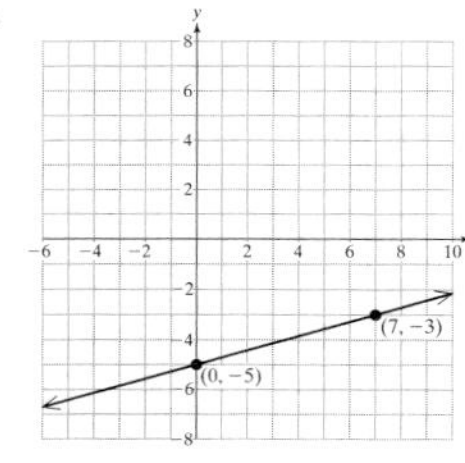

17.

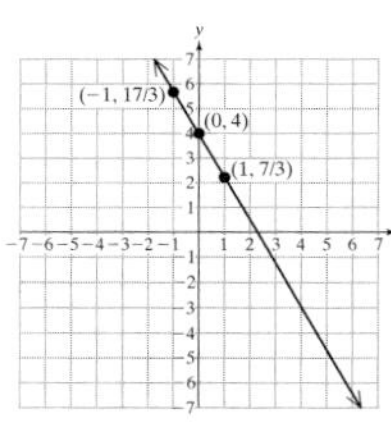

19.

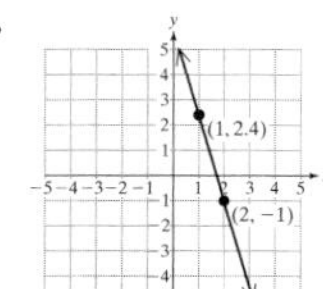

21.

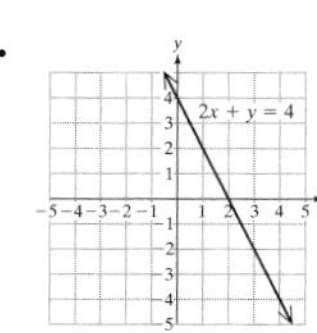

23.

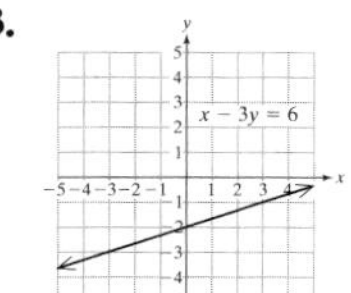

25.

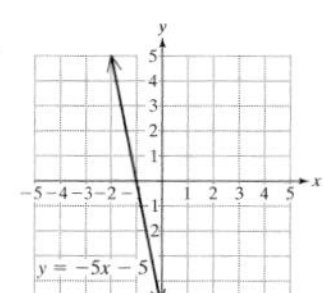

27.

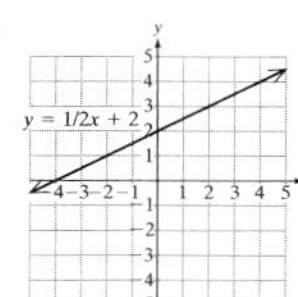

29.

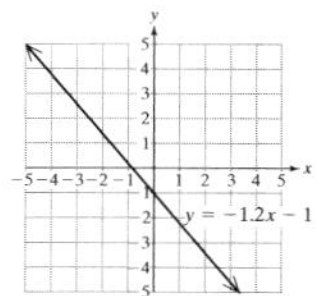

31. a. $y = 17.95$ **b.** $x = 145$ **c.** (55, 17.95) Collecting 55 lb of cans yields \$17.95. (145, 80.05) Collecting 145 lb of cans yields \$80.05.

d.

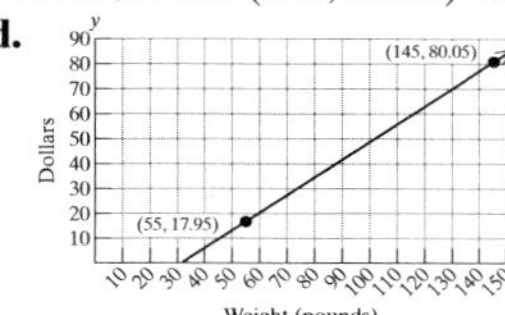

33. a. $y = 10{,}068$ **b.** $x = 3$ **c.** (1, 10068) One year after purchase the value of the car is \$10,068. (3, 7006) Three years after purchase the value of the car is \$7006. **35. a.** \$9926 million **b.** \$13,334 million **c.** 1997 **d.** 1999

37.

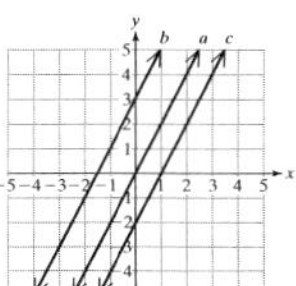

39. a.

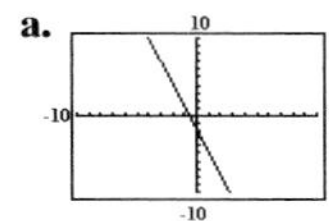

b.

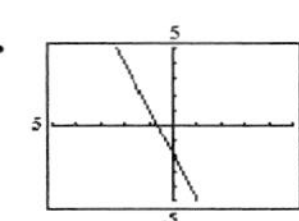

41. a.

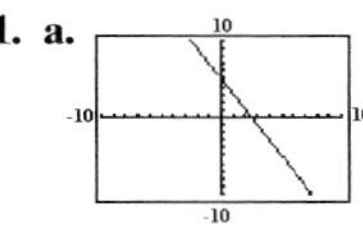

b.

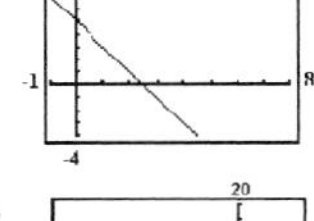

43. a.

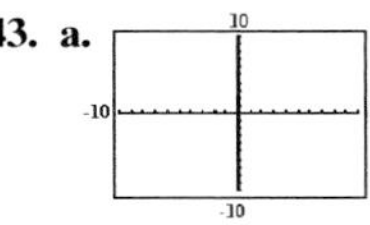

b.

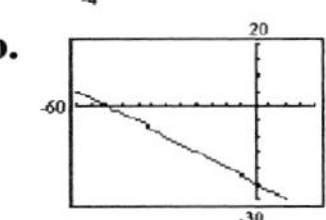

Section 3.3 Practice Exercises, pp. 210–213

1. II **3.** III **5. a.** Yes **b.** No **c.** Yes **d.** Yes **7.** An x-intercept is a point $(a, 0)$ where a graph intersects the x-axis. **9.** Substitute $x = 0$, and solve for y. **11.** x-intercept: (0, 0); y-intercepts: (0, 0)(0, −3) **13.** x-intercept: (−2, 0); y-intercepts: (0, 4)(0, −3) **15.** x-intercepts: (2, 0)(−2, 0); y-intercept: (0, 2) **17.** x-intercept: (4, 0); y-intercept: (0, −2) **19.** x-intercept: (−2, 0); y-intercept: (0, 10) **21.** x-intercept: (2, 0); y-intercept: (0, −2) **23.** x-intercept: (0, 0); y-intercept: (0, 0) **25.** x-intercept: (6, 0); y-intercept: (0, 3)

27. x-intercept: $\left(\frac{9}{4}, 0\right)$;

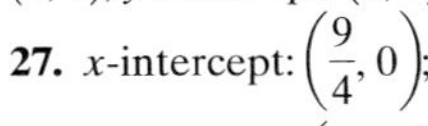

y-intercept: $\left(0, -\frac{9}{7}\right)$

29. a. False, $x = 3$ is vertical. **a.** True

31. Vertical **33.** Horizontal

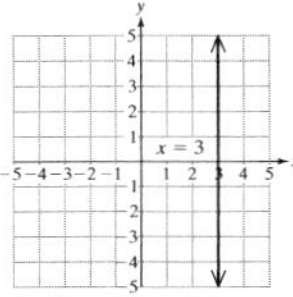

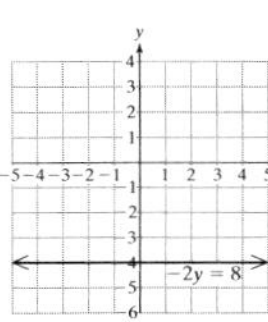

35. Vertical

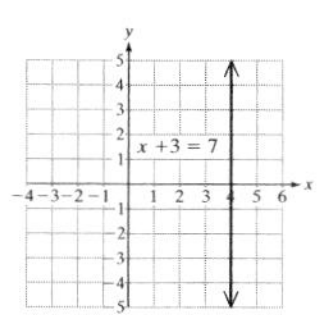

37. Horizontal

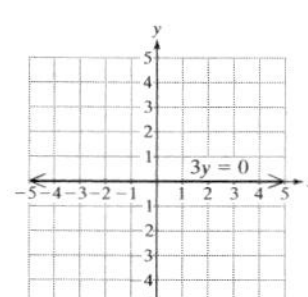

39. Vertical

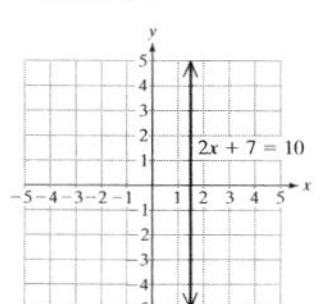

41. Horizontal

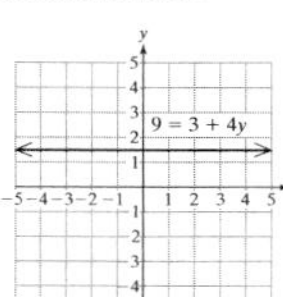

43. A horizontal line may not have an x-intercept. A vertical line may not have a y-intercept. **45.** y-axis **47.** a, b, c

49. x-intercept: (−9, 0); y-intercept: (0, 3)

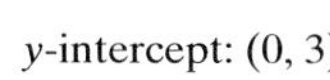

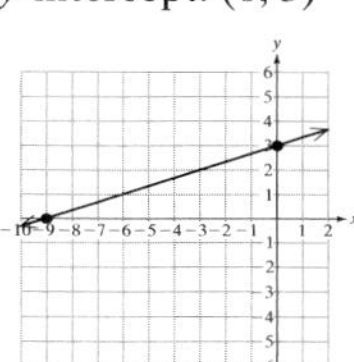

51. x-intercept: $\left(\frac{8}{3}, 0\right)$; y-intercept: (0, 2)

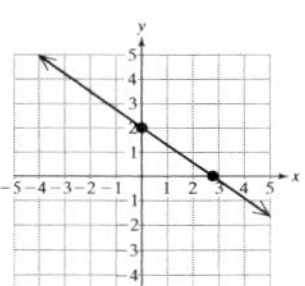

53. x-intercept: (−4, 0); y-intercept: (0, 8)

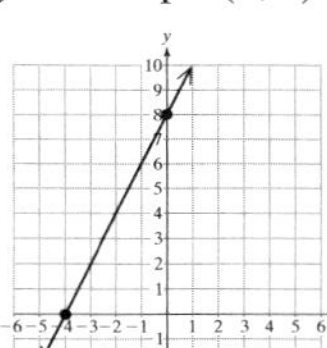

55. x-intercept: (1, 0); y-intercept: none

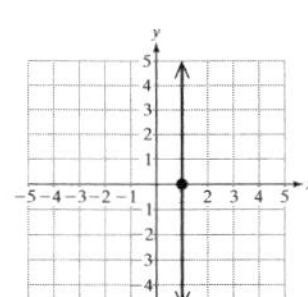

57. x-intercept: none; y-intercept: (0, −2)

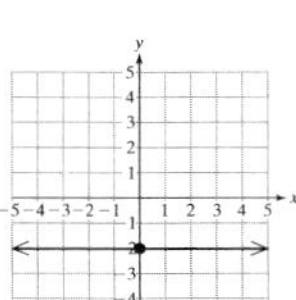

59. x-intercept: $\left(\frac{5}{4}, 0\right)$; y-intercept: none

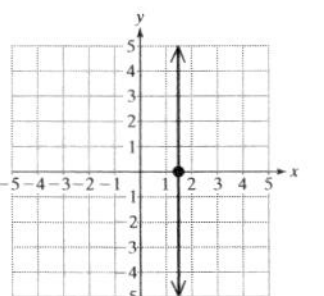

61. x-intercept: (10, 0); y-intercept: (0, 5)

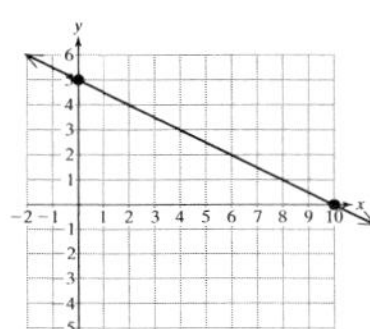

63. x-intercept: (−2, 0); y-intercept: (0, −1.5)

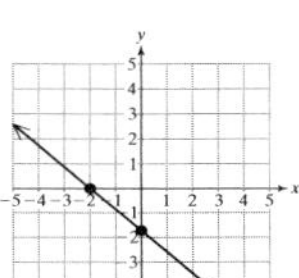

65. x-intercept: (0, 0); y-intercept: (0, 0)

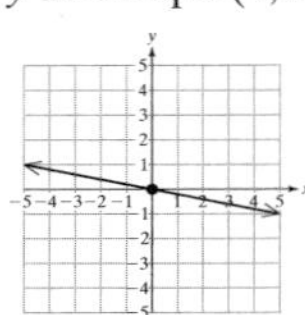

67. x-intercept: none; y-intercept: (0, −4)

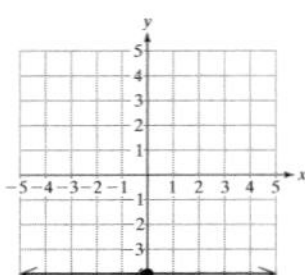

69. x-intercept: (1, 0); y-intercept: none

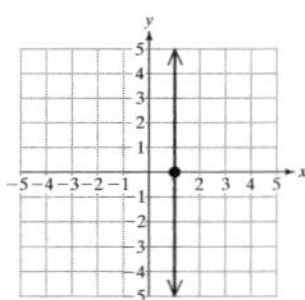

71. a. \$400 **b.** \$200 **c.** (0, 800) If the stereo lasts 0 years (it breaks near the time of purchase) the store will provide a full refund of \$800. **d.** (2, 0) If the stereo breaks after 2 years the store pays nothing. **73. a.** 16,000 tickets **b.** 4000 tickets **c.** (0, 20000) If tickets cost \$0 (free), the promoter will "sell" 20,000 tickets. **d.** (50, 0) If tickets cost \$50, the promoter will sell 0 tickets. **75.** $x = 0$ **77.** $x = 4$ **79.** $y = 5$ **81.** x-intercept: (2, 0); y-intercept: (0, −4)

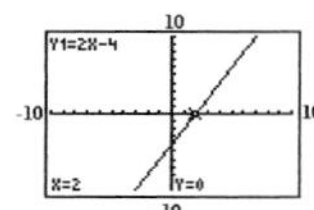

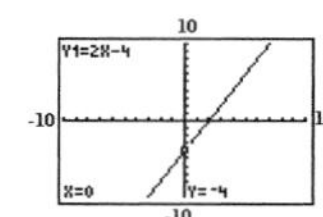

83. x-intercept: (2, 0); y-intercept: $\left(0, \frac{3}{2}\right)$

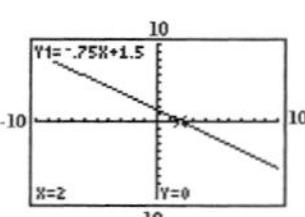

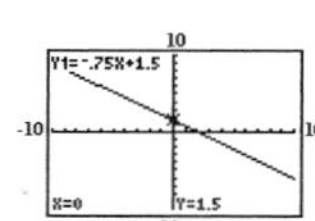

85. x-intercept: (15, 0); y-intercept: (0, −15)

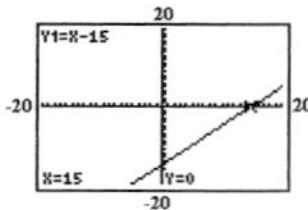

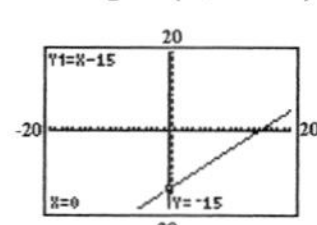

Section 3.4 Practice Exercises, pp. 220–225

1. x-intercept: (6, 0); y-intercept: (0, −2)

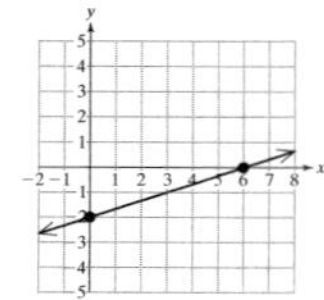

3. x-intercept: (7, 0); y-intercept: none

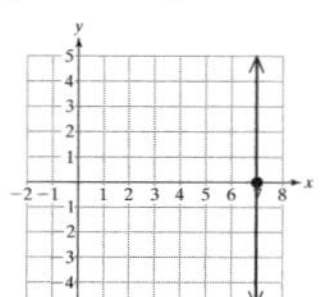

5. x-intercept: none; y-intercept: $(0, \frac{3}{2})$

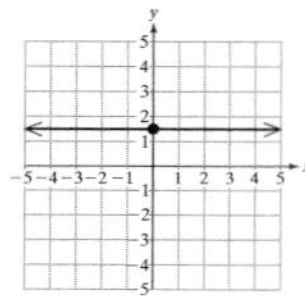

7. x-intercept: (0, 0); y-intercept: (0, 0)

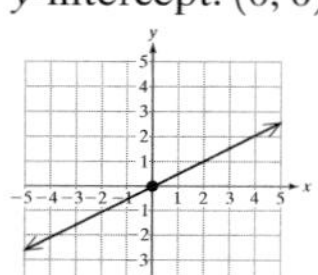

9. undefined **11.** positive **13.** Negative **15.** Zero **17.** Undefined **19.** Positive **21.** $\frac{5}{3}$ **23.** −3 **25.** Zero **27.** Undefined **29.** $\frac{28}{5}$ **31.** $-\frac{31}{35}$ **33.** −0.45 **35.** −1.6375 **37.** −1833.3 or $-\frac{5500}{3}$ **39.** $\frac{3}{4}$ **41.** −1 **43.** Zero **45. a.** $\frac{765}{16}$ **b.** The number of male inmates increased by 765 thousand in 16 years, or approximately 47.8 thousand inmates per year. **47. a.** $m = 450$. The median income for women in the United States increased by \$450/year. **b.** No, the rate of increase in women's median income is less than the rate of increase in men's median income.

49.

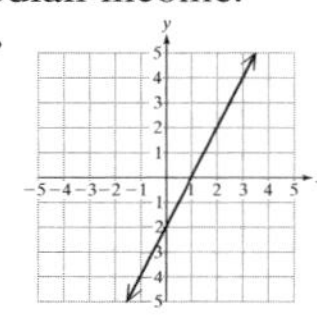

51.

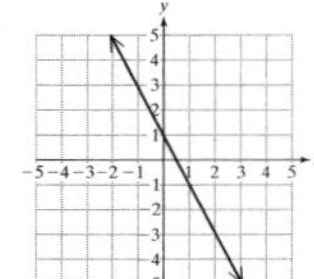

53.

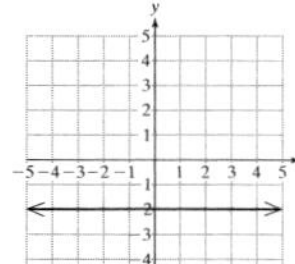

55.

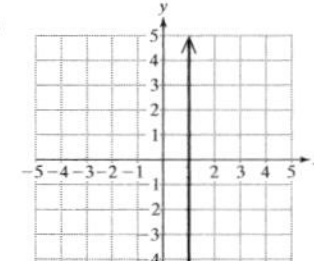

57.

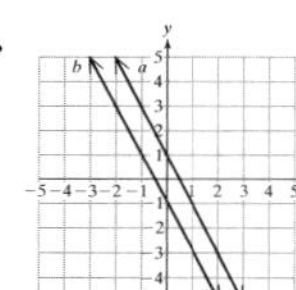

59.

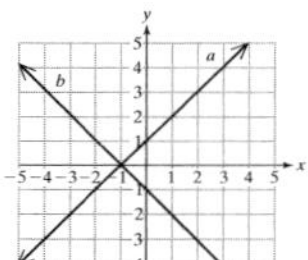

61.

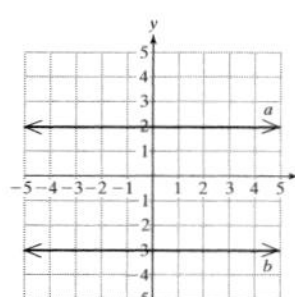

63. a. $\frac{2}{3}$ **b.** $-\frac{3}{2}$ **65. a.** Undefined **b.** 0 **67.** l_1: $m = -2$, l_2: $m = -2$; parallel **69.** l_1: $m = 1$, l_2: $m = -1$; perpendicular **71.** l_1: $m = 2$, l_2: $m = \frac{1}{2}$; neither **73.** l_1: $m = 0$, l_2: $m = 0$; parallel **75.** $\frac{9}{2}$ ft or $4\frac{1}{2}$ ft **77. a.** \$230.00 **b.** \$241.50 **c.** \$253.00 **d.** $m = 11.5$; Jorge's pay increases \$11.50 for each additional hour worked. **79.** $\frac{2t}{2c + d}$ or $\frac{-2t}{-2c - d}$ **81.** $\left(0, \frac{c}{b}\right)$ **83.** For example: (1, 5)

Section 3.5 Practice Exercises, pp. 231–234

1. x-intercept: (10, 0); y-intercept: (0, −2) **3.** x-intercept: none; y-intercept: (0, −3) **5.** x-intercept: (0, 0); y-intercept: (0, 0) **7.** x-intercept: (−5, 0); y-intercept: none **9.** $y = \frac{2}{5}x - \frac{4}{5}$; $m = \frac{2}{5}$; y-intercept: $\left(0, -\frac{4}{5}\right)$ **11.** $y = 3x - 5$; $m = 3$; y-intercept (0, −5) **13.** $y = -x + 6$; $m = -1$; y-intercept (0, 6) **15.** $x = 2$; Cannot be written in slope-intercept form; undefined slope; no y-intercept **17.** $y = -\frac{1}{4}$; $m = 0$; y-intercept $\left(0, -\frac{1}{4}\right)$ **19.** $y = \frac{2}{3}x$; $m = \frac{2}{3}$; y-intercept: (0, 0)

21.

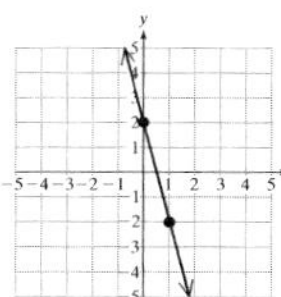

23.

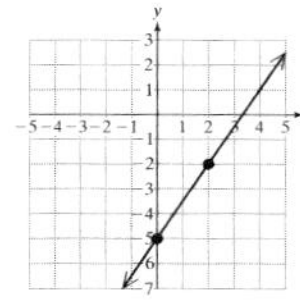

25. $y = \frac{1}{2}x - 3$

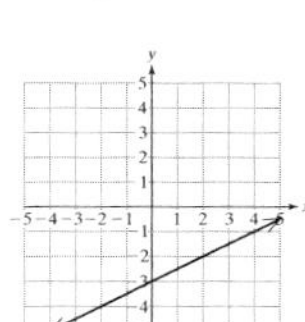

27. $y = -2x + 9$

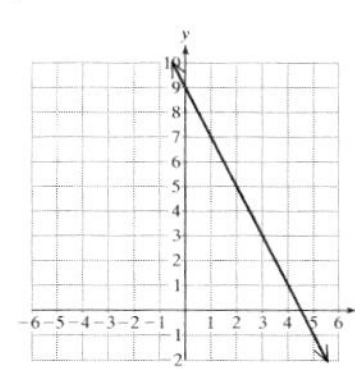

29. $y = -2x - \frac{1}{3}$

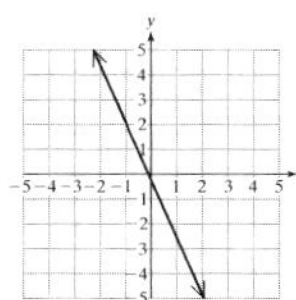

31. $y = -\frac{1}{2}x + \frac{3}{2}$

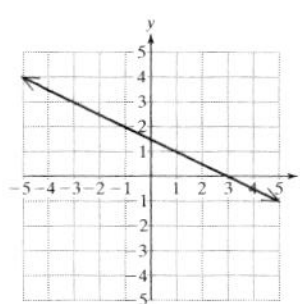

33. $y = -x$

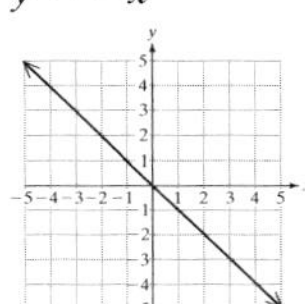

35. $y = 0.4x - 0.2$

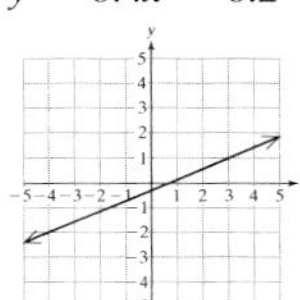

37. $y = \frac{9}{5}x$

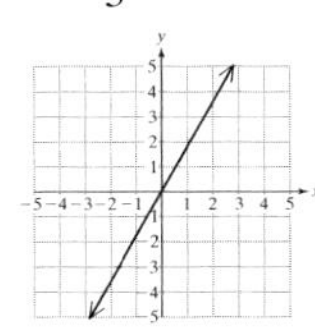

39. $y = -\frac{2}{3}$

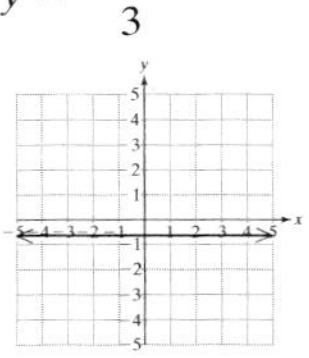

41. $x = 2$

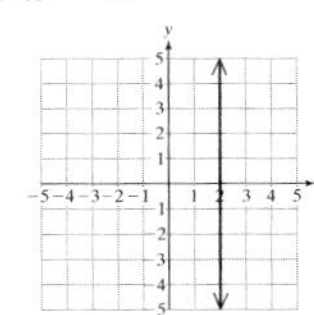

43. $y = -2x + 2$

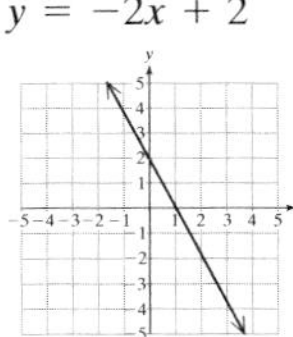

45. Perpendicular **47.** Parallel **49.** Neither **51.** Perpendicular **53.** Neither **55.** Parallel **57.** Perpendicular **59.** Parallel **61.** Perpendicular **63.** Neither **65.** Parallel **67.** $y = -\frac{1}{3}x + 2$

69. $y = 5x$ **71.** $y = 6x - 2$ **73. a.** $m = 49.95$. The cost increases $49.95 per day. **b.** (0, 31.95). The cost to rent the car for 0 days is $31.95. **c.** $381.60 **75. a.** $75 **b.** 2.5; It costs $2.50 per mile to tow a car. **c.** (0, 45); Towing a car 0 miles will cost $45. In other words, there is a $45 fee for the tow truck to respond to a call.

77.

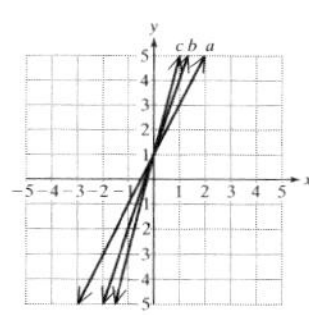

79.

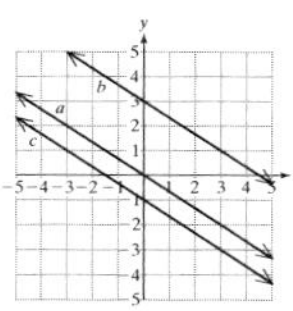

81. $y = -\frac{a}{b}x + \frac{c}{b}$; The slope is $-\frac{a}{b}$. **83.** $m = -\frac{6}{7}$ **85.** $m = \frac{11}{8}$

87. Parallel

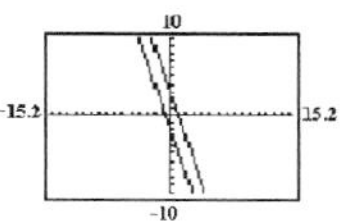

89. The lines may appear parallel; however, they are not parallel because the slopes are different.

Chapter 3 Review Exercises, pp. 238–241

1.

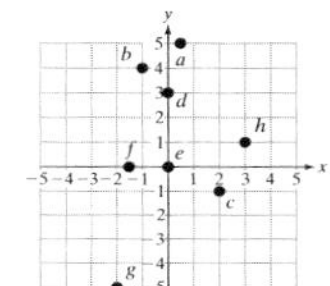

3. III **5.** IV **7.** IV **9.** x-axis **11. a.** (1, 26.25), (2, 28.50), (3, 28.00), (4, 27.00), (5, 24.75); On day 1, the price was $26.25. **b.** day 2 **c.** $2.25 **13.** No **15.** yes

17.

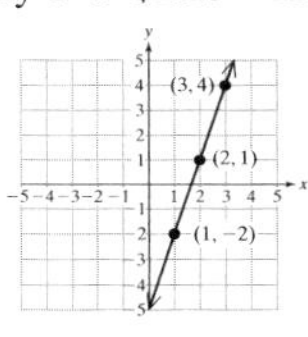

19.

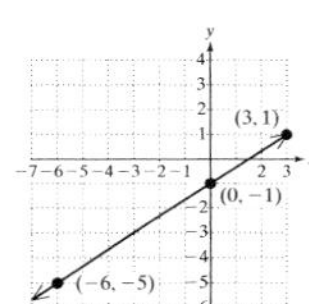

21.

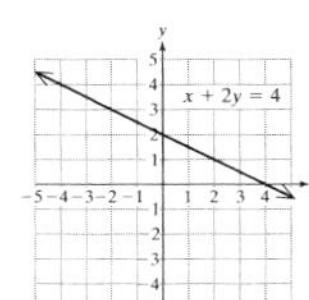

23.

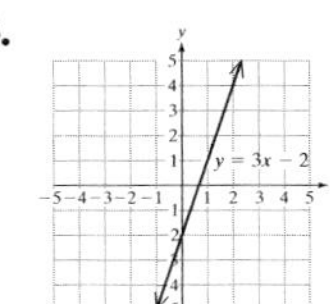

25. a. $19.10 **b.** 7.70, 11.55, 15.40, 19.25, 23.10 **c.** (5.0, 7.70), (7.5, 11.55), (10.0, 15.40), (12.5, 19.25), (15.0, 23.10) **d.** 6.5 gal **e.** 17 gal

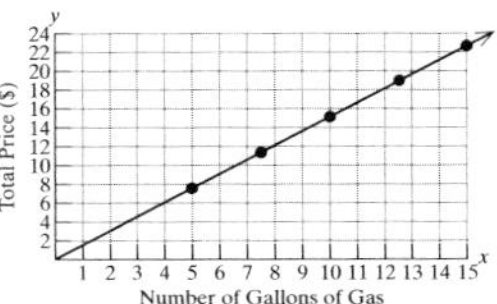

27. Vertical;

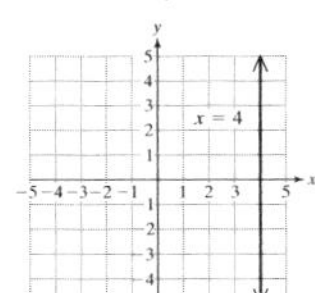

29. Horizontal;

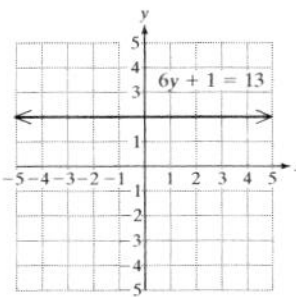

31. x-intercept: $(-3, 0)$; y-intercept: $\left(0, \frac{3}{2}\right)$ **33.** x-intercept: (0, 0); y-intercept: (0, 0) **35.** x-intercept: none; y-intercept: $(0, -4)$

37. x-intercept: $\left(-\frac{5}{2}, 0\right)$; y-intercept: none **39. a.** (0, 30,000)
b. At the time of purchase ($n = 0$), the value of the car is \$30,000.
c. \$15,000
41. For example:

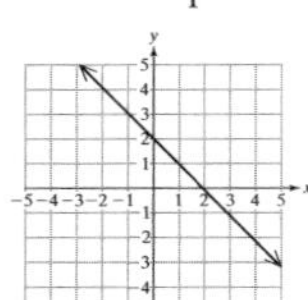

43. For example:

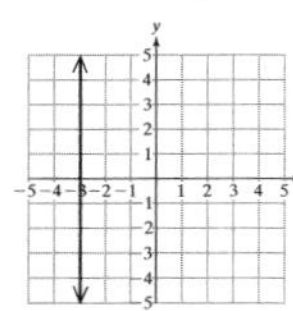

45. −2 **47.** 8 **49.** 0 **51. a.** 0 **b.** Undefined **53.** $m_1 = 8$ $m_2 = 8$; parallel **55.** $m_1 =$ undefined $m_2 = 0$; perpendicular

57. $y = -\frac{3}{4}x + 3$; $m = -\frac{3}{4}$; y-intercept: (0, 3)

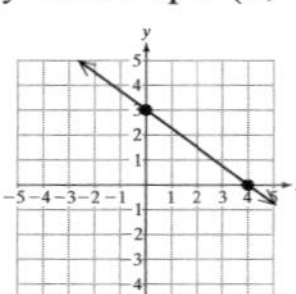

59. $y = 3x - 4$; $m = 3$; y-intercept: (0, −4)

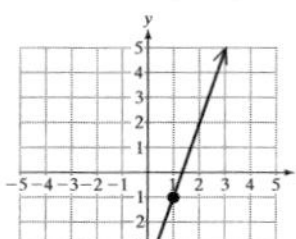

61. $y = \frac{12}{5}$; $m = 0$; y-intercept: $\left(0, \frac{12}{5}\right)$

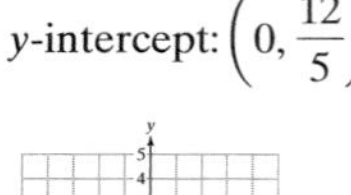

63. $y = x$; $m = 1$; y-intercept: (0, 0)

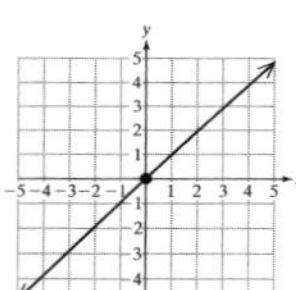

65. Perpendicular **67.** Parallel **69.** $y = 2$

Chapter 3 Test, pp. 241–242

1. a. II **b.** IV **c.** III **2.** 0 **3.** 0
4. a. (5, 46) At age 5 the boy's height was 46 in.
(7, 50)
(9, 55)
(11, 60)
b.

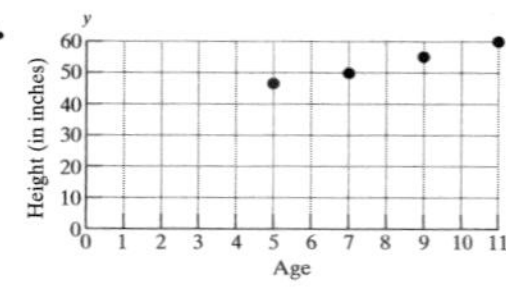

c. 57.5 in. **d.** No, his height will maximize in his teen years. No.
5. a. No **b.** Yes **c.** Yes **d.** Yes
6.

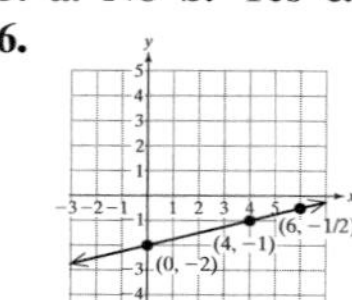

7. a. 202 beats per minute **b.** (20, 200) (30, 190) (40, 180) (50, 170) (60, 160)

8.

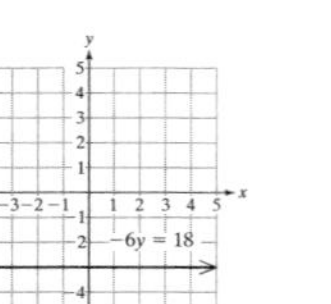

9.

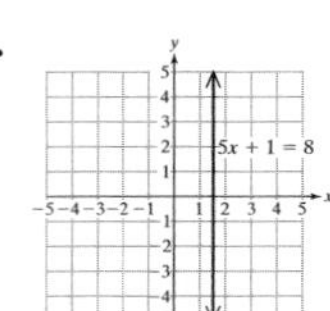

10. x-intercept: $\left(-\frac{3}{2}, 0\right)$; y-intercept: (0, 2) **11.** $\frac{2}{5}$ **12. a.** $\frac{1}{3}$ **b.** $\frac{4}{3}$

13. a. $-\frac{1}{4}$ **b.** 4 **14. a.** Undefined **b.** 0

15. x-intercept: $\left(-\frac{1}{4}, 0\right)$; y-intercept: (0, 2)

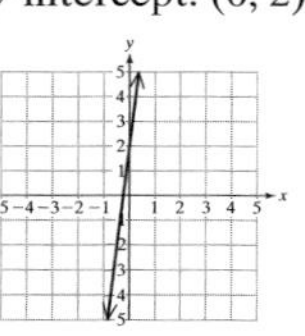

16. x-intercept: (0, 0); y-intercept: (0, 0)

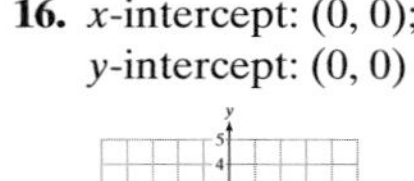

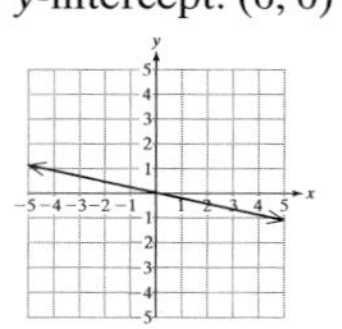

17. x-intercept: (3, 0); y-intercept: none

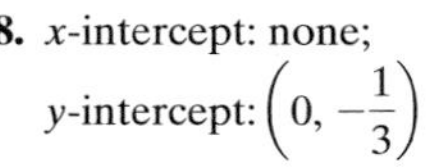

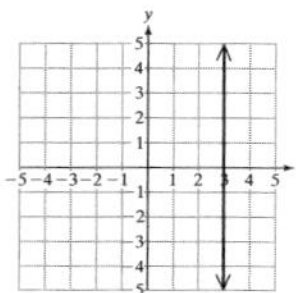

18. x-intercept: none; y-intercept: $\left(0, -\frac{1}{3}\right)$

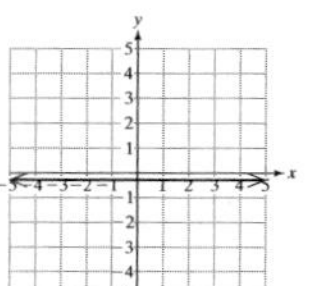

19. Perpendicular **20.** $y = \frac{1}{4}x + \frac{1}{2}$

21. slope: $-\frac{2}{3}$; y-intercept: (0, −3)

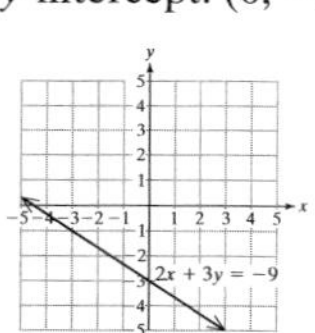

22. slope: $\frac{4}{5}$; y-intercept: (0, 0)

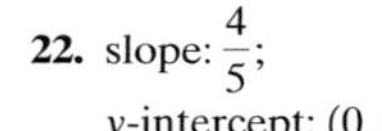

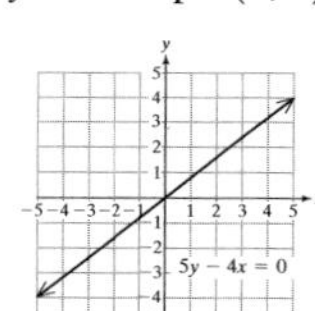

23. a. \$610 **b.** \$2200 **c.** y-intercept: (0, 400); If Girard has \$0 in sales (he sells no merchandise) his salary will be \$400. **d.** slope: 0.15; The slope represents Girard's commission rate. That is, he makes \$0.15 in income for every \$1 sold in merchandise.

Cumulative Review Exercises, Chapters 1–3, pp. 242–243

1. a. Rational **b.** Rational **c.** Irrational **d.** Rational **2. a.** $\frac{2}{3}$; $\frac{2}{3}$
b. −5.3; 5.3 **3.** 69 **4.** −13 **5.** 18 **6.** $\frac{3}{4} \div -\frac{7}{8}$; $-\frac{6}{7}$
7. (−2.1)(−6); 12.6 **8.** The associative property of addition
9. $x = 4$ **10.** $m = 5$ **11.** $y = -\frac{9}{2}$ **12.** $z = -2$ **13.** 9241 mi^2
14. $a = \frac{c - b}{3}$

15.

16. $A(-1, 1), B(3, 0), C(-4, -2), D(1, -3)$ **17.** x-intercept $(-2, 0)$; y-intercept $(0, 1)$ **18.** $(1, 3), (13, -3)$ **19.** $y = -\frac{3}{2}x - 6$; slope: $-\frac{3}{2}$; y-intercept: $(0, -6)$ **20.** $y = \frac{1}{2}x - 5$ **21.** $2x + 3 = 5$ can be written as $x = 1$, which represents a vertical line. A vertical line of the form $x = k\ (k \neq 0)$ has an x-intercept of $(k, 0)$ and no y-intercept. **22.** $-6x + 2y = 0$ is of the form $ax + by = 0$ and has only one intercept, $(0, 0)$, that is both the x- and y-intercept.

CHAPTER 4

Section 4.1 Practice Exercises, pp. 253–255

1. Base: r; exponent: 4 **3.** Base: 5; exponent: 2 **5.** Base: -4; exponent: 8 **7.** Base: x; exponent: 1 **9.** y **11.** x **13.** No; $-5^2 = -25$ and $(-5)^2 = 25$ **15.** Yes; $-2^5 = -32$ and $(-2)^5 = -32$ **17.** $\left(\frac{1}{2}\right)^3 = \frac{1}{8}$ and $\frac{1}{2^3} = \frac{1}{8}$ **19.** $\left(\frac{3}{10}\right)^2 = \frac{9}{100}$ and $(0.3)^2 = 0.09$ **21. a.** $(x \cdot x \cdot x \cdot x)(x \cdot x \cdot x) = x^7$ **b.** $(5 \cdot 5 \cdot 5 \cdot 5)(5 \cdot 5 \cdot 5) = 5^7$ **23.** z^8 **25.** a^9 **27.** 4^{14} **29.** 9^5 **31.** c^{14} **33. a.** $\frac{p \cdot p \cdot p \cdot p \cdot p \cdot p \cdot p \cdot p}{p \cdot p \cdot p} = p^5$ **b.** $\frac{8 \cdot 8 \cdot 8 \cdot 8 \cdot 8 \cdot 8 \cdot 8 \cdot 8}{8 \cdot 8 \cdot 8} = 8^5$ **35.** x^2 **37.** a^9 **39.** 7^7 **41.** 5^7 **43.** y **45.** h^4 **47.** x^5 **49.** 7^7 **51.** x^6 **53.** 10^9 **55.** 6^{10} **57.** z^4 **59.** $40a^5b^5$ **61.** $13r^8s^5$ **63.** $16m^{20}n^{10}$ **65.** $2cd^4$ **67.** $\frac{x^2y^2}{4}$ **69.** $\frac{25hjk^4}{12}$ **71.** 201 in.^{2} **73.** 113 cm^3 **75.** 34,560 in.^{2}, or 240 ft^2 **77.** \$5724.50 **79.** \$4764.06 **81.** x^{2n+1} **83.** p^{2m+3} **85.** z **87.** r^3

89–91.

```
(1.06)^5
        1.338225578
(1.02)^40
        2.208039664
5000(1.06)^5
        6691.127888
```

93.

```
2000(1.02)^40
        4416.079327
3000(1+.06)^2
             3370.8
```

Section 4.2 Practice Exercises, pp. 258–260

1. 4^9 **3.** a^{20} **5.** d^9 **7.** 7^6 **9.** When multiplying expressions with the same base, add the exponents. When raising an expression with an exponent to a power, multiply the exponents. **11.** 5^{12} **13.** 12^6 **15.** y^{14} **17.** w^{25} **19.** a^{36} **21.** y^{14} **23. a.** x^2 **b.** x^{12} **c.** x^{35} **25. a.** $\frac{y^8}{z^{12}}$ **b.** y^8z^{12} **27.** $\frac{8}{27}$ **29.** $\frac{1}{16}$ **31.** $\frac{x^5}{y^5}$ **33.** $\frac{1}{t^4}$ **35.** $81a^4$ **37.** $-27a^3b^3c^3$ **39.** $3u^3$ **41.** $6x^3y^3$ **43.** a^2b^6 **45.** $(4m)$ **47.** $(2a + b)^2$ **49.** $216\,u^6v^{12}$ **51.** $5x^8y^4$ **53.** $\frac{1024}{r^5s^{20}}$ **55.** $\frac{243p^5}{q^{15}}$ **57.** y^{14} **59.** x^{31} **61.** $a^{26}b^{18}$ **63.** $25a^{18}b^6$ **65.** $3x^{12}y^4$ **67.** $\frac{4}{9}c^2d^6$ **69.** $\frac{c^{27}d^{31}}{2}$ **71.** x^{2m} **73.** $125a^{6n}$ **75.** $\frac{m^{2b}}{n^{3b}}$ **77.** $\frac{3^na^{3n}}{5^nb^{4n}}$ **79.** They are both equal to 2^6. **81.** $2^{(2^4)} = 2^{16}$; $(2^2)^4 = 2^8$; $2^{(2^4)}$ is greater than $(2^2)^4$.

Section 4.3 Practice Exercises, pp. 265–266

1. b^{11} **3.** x^4 **5.** 9^{11} **7.** $7776a^5b^{15}c^{10}$ **9.** $\frac{s^6t^{15}}{64}$ **11. a.** 1 **b.** 1 **13.** 1 **15.** 1 **17.** -1 **19.** 1 **21.** 1 **23.** -7 **25.** a **27. a.** $\frac{1}{t^5}$ **b.** $\frac{1}{t^5}$ **29.** $\frac{x^4}{x^{-6}} = x^{4-(-6)} = x^{10}$ **31.** $2a^{-3} = 2 \cdot \frac{1}{a^3} = \frac{2}{a^3}$ **33.** $\frac{343}{8}$ **35.** 25 **37.** $\frac{1}{a^3}$ **39.** $\frac{1}{12}$ **41.** $\frac{1}{16b^2}$ **43.** $\frac{6}{x^2}$ **45.** $\frac{1}{w^6}$ **47.** $\frac{1}{x^4}$ **49.** 1 **51.** y^4 **53.** $\frac{n^{27}}{m^{18}}$ **55.** $\frac{81k^{24}}{j^{20}}$ **57.** $\frac{1}{p^6}$ **59.** $\frac{1}{r^3}$ **61.** $\frac{1}{7^7}$ **63.** $\frac{1}{a^4b^6}$ **65.** $\frac{1}{w^{21}}$ **67.** $\frac{-16y^4}{z^2}$ **69.** $-\frac{a^{12}}{6}$ **71.** $80c^{21}d^{24}$ **73.** $\frac{p^{27}}{8}$ **75.** $\frac{2d^8}{c}$ **77.** 3 **79.** $\frac{b^9}{2^{15}}$ **81.** $\frac{16y^4}{81x^4}$ **83.** $3a^7b^5$ **85.** $\frac{y^4}{x^8}$ **87.** $\frac{1}{t^2}$ **89.** $\frac{8w^6x^9}{27}$ **91.** $\frac{q^3s}{r^2t^5}$ **93.** $\frac{1}{y^{13}}$ **95.** $-\frac{1}{8a^{18}b^6}$ **97.** $\frac{k^8}{5h^6}$ **99.** $\frac{9}{20}$ **101.** $\frac{9}{10}$ **103.** $\frac{26}{81}$

Section 4.4 Practice Exercises, pp. 272–274

1. $\frac{1}{a}$ **3.** $\frac{1}{10}$ **5.** $\frac{1}{x^3}$ **7.** $\frac{1}{10^3}$ **9.** z^{10} **11.** 10^{10} **13.** Move the decimal point between 2 and 3 and multiply by 10^{-10}; 2.3×10^{-10}. **15.** 6.8×10^7 gal; 1.0×10^2 miles **17.** 4.2×10^8 **19.** 8×10^{-6} **21.** 1.7×10^{-24} g **23.** 1.4115999×10^8 shares **25.** Move the decimal point nine places to the left; 0.000 000 0031 **27.** 0.00005 **29.** 2800 **31.** 0. 000 000 000 001 g **33.** 1600 calories and 2800 calories **35.** 5.0×10^4 **37.** 3.6×10^{11} **39.** 2.2×10^4 **41.** 2.25×10^{-13} **43.** 3.2×10^{14} **45.** 2.432×10^{-10} **47.** 3.0×10^{13} **49.** 6.0×10^5 **51.** 1.38×10^1 **53.** 5.0×10^{-14} **55.** 3.75 in. **57.** Approximately \$1714 per commercial **59. a.** 6.5×10^7 **b.** 2.3725×10^{10} days **c.** 5.694×10^{11} hr **d.** 2.04984×10^{15} sec **61. a.** 6.75×10^5 workers **b.** 4.5×10^4 workers **63.** Approximately $\$5.66 \times 10^{10}$

65.

```
(5.2E6)*(4.6E-3)
           2.392E4
(2.19E-8)*(7.84E
-4)
       1.71696E-11
```

67.

```
(4.76E-5)/(2.38E
9)
             2E-14
(8.5E4)/(4.0E-1)
           2.125E5
```

69.

```
((5.0E-12)*(6.4E
-5))/((1.6E-8)*(
4.0E2))
            5E-11
```

Section 4.5 Practice Exercises, pp. 280–282

1. $4p^2$ **3.** $12y^6$ **5.** $\frac{1}{8^5}$ **7.** 3.0×10^7 is scientific notation in which 10 is raised to the seventh power. 3^7 is not scientific notation and 3 is being raised to the seventh power. **9.** $-7x^4 + 7x^2 + 9x + 6$ **11.** Binomial; 10; 2 **13.** Monomial; 6; 2 **15.** Trinomial; -1; 4 **17.** Trinomial; 12; 4 **19.** Monomial; 5; 3 **21.** Binomial; 1; 4 **23.** The exponents on the x-factors are different. **25.** $35x^2y$ **27.** $10y$ **29.** $8b^2 - 9$ **31.** $4y^2 + y - 9$ **33.** $4a - 8c$ **35.** $a - \frac{1}{2}b - 2$ **37.** $\frac{4}{3}z^2 - \frac{5}{3}$ **39.** $7.9t^3 - 3.4t^2 + 6t - 4.2$ **41.** $4y^3 + 2y^2 + 2$ **43. a.** 134 ft, 114 ft, 86 ft **b.** 150 ft **45.** $-4h + 5$ **47.** $2m^2 - 3m + 15$ **49.** $-3v^3 - 5v^2 - 10v - 22$ **51.** $9t^4 + 8t + 39$ **53.** $-8a^3b^2$ **55.** $-53x^3$ **57.** $-5a - 3$ **59.** $16k + 9$ **61.** $2s + 14$ **63.** $3t^2 - 4t - 3$ **65.** $-2r - 3s + 3t$ **67.** $\frac{3}{4}x + \frac{1}{3}y - \frac{3}{10}$ **69.** $-\frac{2}{3}h^2 + \frac{3}{5}h - \frac{5}{2}$ **71.** $2.4x^4 - 3.1x^2 - 4.4x - 6.7$ **73.** $3a^2 - 3a + 5$ **75.** $-3x^3 - 2x^2 + 11x - 31$ **77.** $4b^3 + 12b^2 - 5b - 12$ **79.** $9ab^2 - 3ab + 16a^2b$ **81.** $4z^5 + z^4 + 9z^3 - 3z - 2$ **83.** $2x^4 + 11x^3 - 3x^2 + 8x - 4$ **85.** $-2w^2 - 7w + 18$ **87.** $-p^2q - 4pq^2 + 3pq$ **89.** For example: $x^3 + 6$ **91.** For example: $8x^5$ **93.** For example: $-6x^2 + 2x + 5$

Section 4.6 Practice Exercises, pp. 289–291

1. $9x$ **3.** $20x^2$ **5.** $-7a^3b$ **7.** $10a^6b^2$ **9.** $4c^2 - c$ **11.** $-4c^3$ **13.** 9.89×10^{12} **15.** -1.953×10^{-23} **17.** $32x$ **19.** $-50z$ **21.** $4x^{13}$ **23.** $-12m^9n^8$ **25.** $16p^2q^2 - 24p^2q + 40pq^2$ **27.** $-4k^3 + 52k^2 + 24k$ **29.** $-45p^3q - 15p^4q^3 + 30pq^2$ **31.** $y^2 - y - 90$ **33.** $m^2 - 14m + 24$ **35.** $p^2 - p - 2$ **37.** $w^2 + 11w + 24$ **39.** $p^2 - 14p + 33$ **41.** $12x^2 + 28x - 5$ **43.** $8a^2 - 22a + 9$ **45.** $9t^2 - 18t - 7$ **47.** $3x^2 + 28x + 32$ **49.** $5s^3 + 8s^2 - 7s - 6$ **51.** $27w^3 - 8$ **53.** $9a^2 - 16b^2$ **55.** $81k^2 - 36$ **57.** $\frac{1}{4} - t^2$ **59.** $u^6 - 25v^2$ **61.** $a^2 + 2ab + b^2$ **63.** $x^2 - 2xy + y^2$ **65.** $4c^2 + 20c + 25$ **67.** $9t^4 - 24st^2 + 16s^2$ **69. a.** 36 **b.** 20 **c.** $(a + b)^2 \neq a^2 + b^2$ in general **71.** $4x^2 - 25$ **73.** $16p^2 + 40p + 25$ **75.** $49x^2 - y^2$ **77.** $25s^2 + 30st + 9t^2$ **79.** $21x^2 - 65xy + 24y^2$ **81.** $2t^2 + \frac{26}{3}t + 8$ **83.** $5z^3 + 23z^2 + 7z - 3$ **85.** $\frac{1}{9}m^2 - \frac{2}{3}mn + n^2$ **87.** $42w^3 - 84w^2$ **89.** $16y^2 - 65.61$ **91.** $21c^4 + 4c^2 - 32$ **93.** $9.61x^2 + 27.9x + 20.25$ **95.** $k^3 - 12k^2 + 48k - 64$ **97.** $15a^5 - 6a^2$ **99.** $27p^3 - 135p^2 + 225p - 125$ **101.** $6a^3 + 22a^2 - 40a$ **103.** $2x^3 - 13x^2 + 17x + 12$ **105.** $2x - 7$

Section 4.7 Practice Exercises, pp. 297–298

1. $6z^5 - 10z^4 - 4z^3 - z^2 - 6$ **3.** $10x^2 - 29xy - 3y^2$ **5.** $4w^6 + 20w^3 + 25$ **7.** $\frac{49}{64}w^2 - 1$ **9.** Use long division when the divisor is a polynomial with two or more terms. **11. a.** $5t^2 + 6t$ **13.** $3a^2 + 2a - 7$ **15.** $x^2 + 4x - 1$ **17.** $3p^2 - p$ **19.** $1 + \frac{2}{m}$ **21.** $-2y^2 + y - 3$ **23.** $x^2 - 6x - \frac{1}{4} + \frac{2}{x}$ **25.** $a - 1 + \frac{b}{a}$ **27.** $3t - 1 + \frac{3}{2t} - \frac{1}{2t^2} + \frac{2}{t^3}$ **29. a.** $z + 2 + \frac{1}{z + 5}$ **31.** $t + 3$ **33.** $7b + 4$ **35.** $k - 6$ **37.** $2p^2 + 3p - 4$ **39.** $k - 2 + \frac{-4}{k + 1}$ **41.** $2x^2 - x + 6 + \frac{2}{2x - 3}$ **43.** $a - 3 + \frac{18}{a + 3}$ **45.** $w^2 + 5w - 2 + \frac{1}{w^2 - 3}$ **47.** $n^2 + n - 6$ **49.** $x - 1 + \frac{-8}{5x^2 + 5x + 1}$ **51.** To check, multiply $(x - 2)(x^2 + 4) = x^3 - 2x^2 + 4x - 8$, which does not equal $x^3 - 8$. **53.** Monomial division; $3a^2 + 4a$ **55.** Long division; $p + 2$ **57.** Long division; $t^3 - 2t^2 + 5t - 10 + \frac{4}{t + 2}$ **59.** Long division; $w^2 + 3 + \frac{1}{w^2 - 2}$ **61.** Long division; $n^2 + 4n + 16$ **63.** Monomial division; $-3r + 4 - \frac{3}{r^2}$ **65.** $x + 1$ **67.** $x^3 + x^2 + x + 1$ **69.** $x + 1 + \frac{1}{x - 1}$ **71.** $x^3 + x^2 + x + 1 + \frac{1}{x - 1}$

Chapter 4 Review Exercises, pp. 302–305

1. Base: 5; exponent: 3 **3.** Base: -2; exponent: 0 **5. a.** 36 **b.** 36 **c.** -36 **7.** 5^{13} **9.** x^9 **11.** 10^3 **13.** b^8 **15.** k **17.** 2^8 **19.** Exponents are added only when multiplying factors with the same base. In such a case, the base does not change. **21.** \$7146.10 **23.** 7^{12} **25.** p^{18} **27.** $\frac{a^2}{b^2}$ **29.** $\frac{5^2}{c^4d^{10}}$ **31.** $2^4a^4b^8$ **33.** $\frac{-3^3x^9}{5^3y^6z^3}$ **35.** a^{11} **37.** $4h^{14}$ **39.** $\frac{x^6y^2}{4}$ **41.** 1 **43.** 1 **45.** 2 **47.** $\frac{1}{z^5}$ **49.** $\frac{1}{36a^2}$ **51.** $\frac{17}{16}$ **53.** $\frac{1}{t^8}$ **55.** $\frac{2y^7}{x^6}$ **57.** $\frac{n^{16}}{16m^8}$ **59.** $\frac{k^{21}}{5}$ **61.** 6 **63. a.** $\$5.9148 \times 10^{12}$ **b.** 4.2×10^{-3} in. **c.** 1.66241×10^8 km^2 **65.** 9.43×10^5 **67.** 2.5×10^8 **69.** $\approx 9.5367 \times 10^{13}$. This number has too many digits to fit on most calculator displays. **71. a.** $\approx 5.84 \times 10^8$ miles **b.** $\approx 6.67 \times 10^4$ mph **73. a.** Trinomial **b.** 4 **c.** 7 **75.** $7x - 3$ **77.** $14a^2 - 2a - 6$ **79.** $\frac{15}{2}x^3 + \frac{1}{4}x^2 + \frac{1}{2}x + 2$ **81.** $-2x^2 - 9x - 6$ **83.** For example: $-5x^2 + 2x - 4$ **85.** $6w + 6$ **87.** $18a^8b^4$ **89.** $-2x^3 - 10x^2 + 6x$ **91.** $20t^2 + 3t - 2$ **93.** $2a^2 + 4a - 30$ **95.** $b^2 - 8b + 16$ **97.** $-2w^3 - 5w^2 - 5w + 4$ **99.** $\frac{1}{9}r^8 - s^4$ **101.** $2h^5 + h^4 - h^3 + h^2 - h + 3$ **103.** $4y^2 - 2y$ **105.** $-3x^2 + 2x - 1$ **107.** $x + 2$ **109.** $p - 3 + \frac{5}{2p + 7}$ **111.** $b^2 + 5b + 25$ **113.** $y^2 - 4y + 2 + \frac{9y - 4}{y^2 + 3}$ **115.** $2x^2 - 3x + 2 + \frac{1}{3x + 2}$ **117.** $t^2 - 3t + 1 + \frac{-2t - 6}{3t^2 + t + 1}$

Chapter 4 Test, pp. 305–306

1. $\frac{(3 \cdot 3 \cdot 3 \cdot 3) \cdot (3 \cdot 3 \cdot 3)}{3 \cdot 3 \cdot 3 \cdot 3 \cdot 3 \cdot 3} = 3$ **2.** 9^6 **3.** q^8 **4.** $27a^6b^3$
5. $\frac{16x^4}{y^{12}}$ **6.** 1 **7.** $\frac{1}{c^3}$ **8.** 14 **9.** $49s^{18}t$ **10.** $\frac{4}{b^{12}}$ **11.** $\frac{16a^{12}}{9b^6}$
12. a. 4.3×10^{10} **b.** 0.000 0056 **13. a.** $2.4192 \times 10^8 \text{ m}^3$ **b.** $8.83008 \times 10^{10} \text{ m}^3$ **14.** $5x^3 - 7x^2 + 4x + 11$ **a.** 3 **b.** 5
15. $24w^2 - 3w - 4$ **16.** $-10x^5 - 2x^4 + 30x^3$ **17.** $8a^2 - 10a + 3$
18. $4y^3 - 25y^2 + 37y - 15$ **19.** $4 - 9b^2$ **20.** $25z^2 - 60z + 36$
21. Perimeter: $12x - 2$; area: $5x^2 - 13x - 6$
22. a. $-3x^6 + \frac{x^4}{4} - 2x$ **b.** $2y - 7$

Cumulative Review Exercises, Chapters 1–4, pp. 306–307

1. $-\frac{35}{2}$ **2.** 4 **3.** 4 **4.** $5^2 - \sqrt{4}$; 23 **5.** $-7, \frac{0}{4}, 2, 0.8, \sqrt{100}$
6. $x = \frac{28}{3}$ **7.** No solution **8.** Quadrant III **9.** y-axis **10.** The measures are 31°, 54°, 95°. **11.** He sold $12,000 worth of merchandise.
12. The dimensions are 24 ft × 120 ft. **13. a.** 12 in. **b.** 19.5 in. **c.** 5.5 hr **d.**

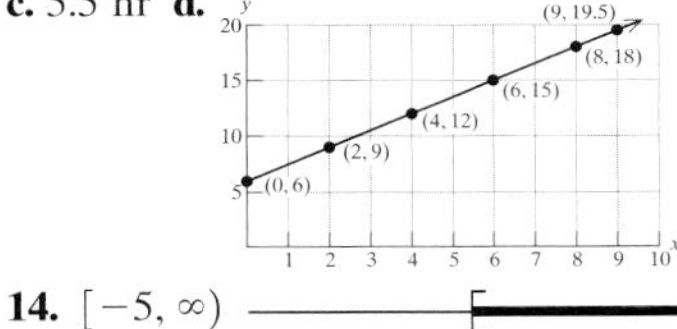

14. $[-5, \infty)$ (number line shaded from -5 to the right) **15.** $5x^2 - 9x - 15$

16. $-2y^2 - 13yz - 15z^2$ **17.** $16t^2 - 24t + 9$ **18.** $\frac{4}{25}a^2 - \frac{1}{9}$
19. $-x^2 + x - 17$ **20.** $-4a^3b^2 + 2ab - 1$
21. $4m^2 + 8m + 11 + \frac{24}{m - 2}$ **22.** $\frac{1}{x}$ **23.** $\frac{c^2}{16d^4}$ **24.** $\frac{2b^3}{a^2}$
25. a. 4.071×10^8 **b.** 4.071×10^{-6} **26. a.** 0.000 389 **b.** 4,500,000,000,000 **27.** 2.788×10^{-2}

CHAPTER 5

Section 5.1 Practice Exercises, pp. 316–318

1. 7 **3.** 6 **5.** ab **7.** $4w^2z$ **9.** $(x - y)$ **11.** $7(3x + 1)$
13. a. $3x - 6y$ **b.** $3(x - 2y)$ **15.** $4(p + 3)$ **17.** $5c(c - 2)$
19. $x^3(x^2 + 1)$ **21.** $t(t^3 - 4)$ **23.** $2ab(1 + 2a^2)$ **25.** $19x^2y(2 - y^3)$
27. $7pq^2(6p^2 + 2 - p^3q^2)$ **29.** $t^2(t^3 + 2rt - 3t^2 + 4r^2)$
31. $(a + 6)(13 - 4b)$ **33.** $(w^2 - 2)(8v + 1)$ **35.** $7x(x + 3)^2$
37. $(z - 1)(13z^2 - 19z + 5)$ **39. a.** $-2x(x^2 + 2x - 4)$
b. $2x(-x^2 - 2x + 4)$ **41.** $-1(8t^2 + 9t + 2)$ **43.** $-1(4y^3 - 5y + 7)$
45. $-15p^2(p + 2)$ **47.** $-q(q^3 - 2q + 9)$ **49.** $-1(7x + 6y + 2z)$
51. $-1(2c + 5)(3 + 4c)$ **53.** $(2a - b)(4a + 3c)$ **55.** $(q + p)(3 + r)$
57. $(2x + 1)(3x + 2)$ **59.** $(t + 3)(2t - 5)$ **61.** $(3y - 1)(2y - 3)$
63. $(b + 1)(b^3 - 4)$ **65.** $(j^2 + 5)(3k + 1)$ **67.** $(2x^6 + 1)(7w^6 - 1)$
69. $5x(x^2 + y^2)(3x + 2y)$ **71.** $4b(a - b)(x - 1)$ **73.** $6t(t - 3)(s - t^2)$
75. $P = 2(l + w)$ **77.** $S = 2\pi r(r + h)$ **79.** $\frac{1}{7}(x^2 + 3x - 5)$
81. $\frac{1}{4}(5w^2 + 3w + 9)$ **83.** $\frac{1}{12}(z^2 + 4z + 6)$ **85.** $\frac{1}{6}(5q^2 + 2q - 12)$
87. For example: $6x^2 + 9x$ **89.** For example: $16p^4q^2 + 8p^3q - 4p^2q$

Section 5.2 Practice Exercises, pp. 324–325

1. $8p^3(p^6 + 3)$ **3.** $3xy(3x + 4y - 5xy)$ **5.** $(x - 2)(5x - 2)$
7. $(p - 2q)(p - q)$ **9.** $6(a + 4)(a - 2)$ **11.** A polynomial that cannot be factored is prime. **13.** 12, 1 **15.** $-8, -1$ **17.** $-1, 6$
19. $-12, 6$ **21.** $(x + 4)(3x + 1)$ **23.** $(w - 2)(4w - 1)$
25. $(m + 3)(2m - 1)$ **27.** $(4k + 3)(2k - 3)$ **29.** $(2k - 5)^2$
31. Prime **33.** $(p + 2q)(4p - 3q)$ **35.** $(5m + 2n)(3m - n)$
37. $(r + 2s)(3r - 7s)$ **39.** Prime **41.** $(q - 10)(q - 1)$
43. $(r + 4)(r - 10)$ **45.** $(x + 7)(x - 1)$ **47.** $(m - 6)(m - 7)$
49. $(a + 5)(a + 4)$ **51.** Prime **53.** $(p + 10q)^2$ **55.** $(x - 7y)(x + 6y)$
57. $(r + 5s)(r + 3s)$ **59.** $(3z - 5)(3z - 2)$ **61.** $(7y + 4)(y + 3)$
63. No, $(2x + 4)$ contains a common factor of 2.
65. $2(12x - 1)(3x + 1)$ **67.** $p(p + 3)(p - 9)$ **69.** $2(3x + 7)(x + 1)$
71. $2p(p - 15)(p - 4)$ **73.** $x^2(y + 3)(y + 11)$ **75.** $-1(k + 2)(k + 5)$
77. $-3(n + 6)(n - 5)$ **79.** $(2z - 1)(8z - 3)$ **81.** $(b - 4)^2$
83. $-5(x - 2)(x - 3)$ **85.** $(t - 3)(t + 2)$

Section 5.3 Practice Exercises, pp. 333–334

1. $7a^3(a^6 + 4)$ **3.** $4w(3w - 1)$ **5.** $3ab(7ab + 4b - 5a)$
7. Different **9.** Both negative **11.** $(x + 8)$ **13.** $(x - 8)$
15. A polynomial that cannot be factored is prime.
17. $(2y + 1)(y - 2)$ **19.** $(3x - 2)^2$ **21.** $(2a + 3)(a + 2)$
23. $(2t + 3)(3t - 1)$ **25.** $(2m - 5)^2$ **27.** Prime
29. $(2x - 5y)(3x - 2y)$ **31.** $(4m + 5n)(3m - n)$
33. $(3r + 2s)(2r - s)$ **35.** Prime **37.** $(x + 9)(x - 2)$
39. $(a - 12)(a + 2)$ **41.** $(r + 8)(r - 3)$ **43.** $(w - 7)^2$
45. $(k + 4)(k + 1)$ **47.** Prime **49.** $(m - 8n)(m - 5n)$
51. $(a + 8b)(a + b)$ **53.** $(x + 5y)(x + 4y)$ **55.** $(2t - 5)(5t + 1)$
57. $(7w - 4)(2w + 3)$ **59.** No, $(3x + 6)$ has a common factor of 3.
61. $2(m + 4)(m - 10)$ **63.** $y^3(2y + 1)(y + 6)$ **65.** $d(5d^2 + 3d - 10)$
67. $4b(b - 5)(b + 4)$ **69.** $y^2(x - 3)(x - 10)$ **71.** $-1(a + 17)(a - 2)$
73. $-2(u - 5)(u - 9)$ **75.** $(8x + 1)(2x + 1)$ **77.** $(c - 1)^2$
79. $-2(z - 9)(z - 1)$ **81.** $(q - 7)(q - 6)$

Section 5.4 Practice Exercises, pp. 339–340

1. $(3x - 5)(x + 2)$ **3.** $yz(x^2z + 6y + 1)$ **5.** $2(3x - 1)(2x - 5)$
7. $(x + b)(a - 6)$ **9.** $(x + 3)^2$ **11.** $4x^2 + 12x + 9$
13. $36h^2 - 12h + 1$ **15. a.** $x^2 + 4x + 4$ **b.** $(x + 2)^2$; $(x + 4)(x + 1)$
17. a. $4x^2 - 20x + 25$ **b.** $(x - 5)(4x - 5)$; $(2x - 5)^2$ **19.** $(y - 5)^2$
21. $(m + 3)^2$ **23.** Prime **25.** $(7q - 2)^2$ **27.** $(3p + 7)^2$
29. $(5h + 2)(5h + 8)$ **31.** $(4a + b)^2$ **33.** $(4q + 5r)^2$ **35.** $(a + b)^2$
37. $\left(k - \frac{1}{2}\right)^2$ **39.** $\left(3x + \frac{1}{6}\right)^2$ **41.** $x^2 - 25$ **43.** $4w^2 - 9$
45. $(x - 6)(x + 6)$ **47.** $(w - 10)(w + 10)$ **49.** $(2a - 11b)(2a + 11b)$
51. $(7m - 4n)(7m + 4n)$ **53.** Prime **55.** $(c^3 - 5)(c^3 + 5)$
57. $(5 - 4t)(5 + 4t)$ **59.** $\left(p - \frac{1}{3}\right)\left(p + \frac{1}{3}\right)$ **61.** Prime
63. $\left(\frac{2}{3} - w\right)\left(\frac{2}{3} + w\right)$ **65. a.** $a^2 - b^2$ **b.** $(a - b)(a + b)$
67. $3(w - 3)(w + 3)$ **69.** $2(5p^2 - 1)(5p^2 + 1)$ **71.** $2(x + 6)^2$
73. $2(t - 5)(t - 1)(t + 1)$ **75.** $25(4y^4 + x^2)$ **77.** $4b(a - 5b)^2$
79. $(2x + 3)(x - 1)(x + 1)$ **81.** $(3y - 2)(3y + 2)(9y^2 + 4)$
83. $(27k + 1)(3k + 1)$ **85.** $(k + 4)(k - 3)(k + 3)$ **87.** $(2m^7 - 5)^2$
89. $(0.6x - 0.1)(0.6x + 0.1)$ **91.** $\left(\frac{1}{2}w - \frac{1}{3}v\right)\left(\frac{1}{2}w + \frac{1}{3}v\right)$

93. $y(y - 6)$ **95.** $(2p - 5)(2p + 7)$ **97.** $(-t + 2)(t + 6)$ or $-1(t - 2)(t + 6)$ **99.** $(-2b + 15)(2b + 5)$ or $-1(2b - 15)(2b + 5)$

Section 5.5 Practice Exercises, pp. 345–346

1. $x^3 - y^3$ **3.** $x^3, 8, y^6, 27q^3, w^{12}, r^3s^6$ **5.** If the binomial is of the form $a^3 + b^3$. **7.** $(a + b)(a^2 - ab + b^2)$ **9.** $(y - 2)(y^2 + 2y + 4)$ **11.** $(1 - p)(1 + p + p^2)$ **13.** $(w + 4)(w^2 - 4w + 16)$ **15.** $(10a + 3)(100a^2 - 30a + 9)$ **17.** $(x - 10)(x^2 + 10x + 100)$ **19.** $(4t + 1)(16t^2 - 4t + 1)$ **21.** $\left(n - \frac{1}{2}\right)\left(n^2 + \frac{1}{2}n + \frac{1}{4}\right)$ **23.** $(a + b^2)(a^2 - ab^2 + b^4)$ **25.** $(x^3 + 4y)(x^6 - 4x^3y + 16y^2)$ **27.** Prime **29.** $(a - b)(a + b)$ **31.** $(x^2 - 2)(x^2 + 2)$ **33.** Prime **35.** $(t + 4)(t^2 - 4t + 16)$ **37.** Prime **39.** $4(b + 3)(b^2 - 3b + 9)$ **41.** $5(p - 5)(p + 5)$ **43.** $(\frac{1}{4} - 2h)(\frac{1}{16} + \frac{1}{2}h + 4h^2)$ **45.** $(x - 2)(x + 2)(x^2 + 4)$ **47.** $(q - 2)(q^2 + 2q + 4)(q + 2)(q^2 - 2q + 4)$ **49.** $4(b + 4)$ **51.** $(y + 3)(y + 1)$ **53.** $(2z + 3)(2z - 3)(4z^2 + 9)$ **55.** $5(r + 1)(r^2 - r + 1)$ **57.** $(7p - 1)(p - 4)$ **59.** $-2(x - 2)^2$ **61.** $2(3 - y)(9 + 3y + y^2)$ **63.** $(4t + 1)(t - 8)$ **65.** $(w - 5)(2x + 3y)$ **67.** $(2q - 3)(2q + 3)$ **69.** $x^2 + 2x + 4$ **71.** $2x + 1$ **73.** $\left(\frac{4}{5}p - \frac{1}{2}q\right)\left(\frac{16}{25}p^2 + \frac{2}{5}pq + \frac{1}{4}q^2\right)$ **75.** $(a^4 + b^4)(a^8 - a^4b^4 + b^8)$ **77. a.** The quotient is $x^2 + 2x + 4$. **b.** $(x - 2)(x^2 + 2x + 4)$ **79. a.** The quotient is $m^2 - m + 1$. **b.** $(m + 1)(m^2 - m + 1)$

Section 5.6 Practice Exercises, pp. 352–353

1. $(x - 8)(x + 2)$ **3.** $(5b + 1)(4b - 3)$ **5.** $(10 - 3u)(10 + 3u)$ **7.** $(p - 6)(p^2 + 6p + 36)$ **9.** $x^2y^2(y + x^3)$ **11.** $(2 + 9x)(y - 11)$ **13.** $(w - 20)^2$ **15.** $x(x - 5)(x + 2)$ **17.** $3(y + 3)(y + 4)$ **19.** $(p + 6q)^2$ **21.** Prime **23.** $(2x - 3)(x + 8)$ **25.** $(u - 5v)(u + 5v)$ **27.** $ab(ab - 6)(ab + 6)$ **29.** $2x(x - 1)(x - 9)$ **31.** $-3ab^2(a - b + 2)$ **33.** $(11r - 6)(s + 4)$ **35.** Prime **37.** $x(x - 2)(x^2 + 2x + 4)$ **39.** Prime **41.** $(y + 3)(5x - 3)$ **43.** $2(p - 4)(q - 7)$ **45.** $4(x - 2)^2$ **47.** $(p - 5)(p - 10)$ **49.** $(4x + 3y)^2$ **51.** $2x^2(x - 5)(x^2 + 3)$ **53.** $-(x - 7)(x - 9)$ **55.** $3(2x + 3)(x - 5)$ **57.** $abc^2(5ac - 7)$ **59.** $(t + 9)(t - 7)$ **61.** $(b + y)(a - b)$ **63.** $(7u - 2v)(2u - v)$ **65.** $2(2q^2 - 4q - 3)$ **67.** Prime **69.** $(3r + 1)(2r + 3)$ **71.** $(9u - 5v)^2$ **73.** $2(x - 3y)(a + 2b)$ **75.** $x^2y(3x + 5)(7x + 2)$ **77.** $(4v - 3)(2u + 3)$ **79.** $3(2x - 1)^2$ **81.** $n(2n - 1)(3n + 4)$ **83.** $(8 - y)(8 + y)$ **85.** $(x - y)(x + y)^2$ **87.** $(a + 3)^4(6a + 19)$ **89.** $18(3x + 5)^2(4x + 5)$ **91.** $(4p^2 + q^2)(2p - q)(2p + q)$ **93.** $\left(y + \frac{1}{4}\right)\left(y^2 - \frac{1}{4}y + \frac{1}{16}\right)$ **95.** $(a + b)(a - b)(6a + b)$ **97.** $\left(\frac{1}{3}t + \frac{1}{4}\right)^2$ **99.** $(x + 6 - a)(x + 6 + a)$ **101.** $(p + q - 9)(p + q + 9)$ **103.** $(b - x - 2)(b + x + 2)$ **105.** $(2 + u - v)(2 - u + v)$ **107.** $(3a + b)(2x - y)$ **109.** $(u - 2)(u + 2)(u^2 + 2u + 4)(u^2 - 2u + 4)$ **111.** $(x^4 + 1)(x^2 + 1)(x + 1)(x - 1)$ **113. a.** $(u - 5)^2$ **b.** $(x^2 - 5)^2$ **c.** $(a - 4)^2$ **115. a.** $(u + 13)(u - 2)$ **b.** $(w^3 + 13)(w^3 - 2)$ **c.** $(y + 9)(y - 6)$ **117.** $5x^2(5x^2 - 6)$ **119.** $9(w - 4)(2w - 5)$ **121.** $(a + b)(a - b + 1)$ **123.** $(x + y)(x^2 - xy + y^2)(5w - 2z)$

Section 5.7 Practice Exercises, pp. 362–366

1. $(2x - 1)(2 + b)$ **3.** $4(b - 5)(b - 6)$ **5.** $(4w - 1)(4w + 1)$ **7.** $4(3k + 4)$ **9.** $(2y + 11)(y - 4)$ **11.** Linear **13.** Quadratic **15.** Neither **17.** Quadratic **19.** If $ab = 0$, then $a = 0$ or $b = 0$. **21.** $x = -3, x = 1$ **23.** $x = \frac{7}{2}, x = -\frac{7}{2}$ **25.** $x = -5$ **27.** $x = 0, x = -\frac{1}{3}, x = -1$ **29.** $p = 5, p = -3$ **31.** $z = -12, z = 2$ **33.** $q = 4, q = -\frac{1}{2}$ **35.** $x = \frac{2}{3}, x = -\frac{2}{3}$ **37.** $k = 6, k = 8$ **39.** $m = 0, m = -\frac{3}{2}, m = 4$ **41.** The equation must have one side equal to zero and the other side factored completely. **43.** $x = 8, x = 2$ **45.** $p = \frac{7}{2}, p = -\frac{7}{2}$ **47.** $q = -5$ **49.** $x = 0, x = -\frac{1}{3}, x = -1$ **51.** $k = \frac{3}{4}, k = -3$ **53.** $p = 3, p = -2$ **55.** $w = 0, w = \frac{2}{3}$ **57.** $d = 0, d = -\frac{1}{4}$ **59.** $t = -2, t = 4, t = -4$ **61.** $w = -7, w = 5$ **63.** $k = 4, k = 2$ **65.** The numbers are $-\frac{9}{2}$ and 4. **67.** The numbers are 5 and −4. **69.** The numbers are 6 and 8 or −8 and −6. **71.** The numbers are 0 and 1 or 9 and 10. **73.** The painting has length 12 in. and width 10 in. **75. a.** The picture is 13 in. by 6 in. **b.** 38 in. **77.** The base is 10 cm and the height is 25 cm. **79.** 2 sec **81.** 0 sec and 4 sec **83.** Given a right triangle with legs a and b and hypotenuse c, then $a^2 + b^2 = c^2$. **85.** Yes **87.** No **89.** The bottom of the ladder is 8 ft from the house. The distance from the top of the ladder to the ground is 15 ft. **91.** 10 m **93. a.** Two diagonals **b.** Five diagonals **c.** Ten sides **95.** $a = 5, a = -5$ **97.** $w = \frac{1}{2}, w = 5$

Chapter 5 Review Exercises, pp. 369–372

1. 6 **3.** ab^4 **5.** $2c(3c - 5)$ **7.** $2x(3x + x^2 - 4)$ **9.** $16(2y^2 - 3)$ **11.** $t(-t + 5)$ or $-t(t - 5)$ **13.** $(b + 2)(3b - 7)$ **15.** $(w + 2)(7w + b)$ **17.** $(x - 6)(x - 4)$ **19.** $3(4y - 3)(5y - 1)$ **21.** $-6, 1$ **23.** $8, 3$ **25.** $5, -1$ **27.** $(c - 2)(3c + 1)$ **29.** $(t + 4s)(2t + 3s)$ **31.** $w(w + 5)(w - 1)$ **33.** $2(4v + 3)(5v - 1)$ **35.** $(x - 2)(x + 11)$ **37.** $ab(a - 6b)(a - 4b)$ **39.** $(3m - 1)(3m + 2)$ **41.** Different **43.** Both positive **45.** $(2y + 3)(y - 4)$ **47.** $2(p - 6)(p + 4)$ **49.** $(2z + 5)(5z + 2)$ **51.** Prime **53.** $10(w - 9)(w + 3)$ **55.** $(3c - 5d)^2$ **57.** $(v^2 + 1)(v^2 - 3)$ **59.** $(2x - 5)^2$ **61.** $(c - 3)^2$ **63.** Not a perfect square trinomial. The middle term does not equal $2(t)(7)$. **65.** $(a - 7)(a + 7)$ **67.** Not a difference of squares. h is not a perfect square. **69.** $(10 - 9t)(10 + 9t)$ **71.** This is a sum of squares. **73.** $2(c^2 - 3)(c^2 + 3)$ **75.** $2(2x + 3)^2$ **77.** $(p + 3)(p - 4)(p + 4)$ **79.** $(a + b)(a^2 - ab + b^2)$ **81.** $(z - w)(z^2 + zw + w^2)$ **83.** $(4 + a)(16 - 4a + a^2)$ **85.** $(p^2 + 2)(p^4 - 2p^2 + 4)$ **87.** $6(x - 2)(x^2 + 2x + 4)$ **89.** v. **91.** iii. **93.** $(6w - 1)(36w^2 + 6w + 1)$ **95.** $2(4 + v^2)(16 - 4v^2 + v^4)$ **97.** $(q - 1)(q^2 + q + 1)(q + 1)(q^2 - q + 1)$ **99.** $3(p - 1)^2$ **101.** $(k - 7)(k - 6)$ **103.** $q(q - 4)(q^2 + 4q + 16)$ **105.** Prime **107.** $(x + 4)(x + 1)(x - 1)$ **109.** $5q(p^2 - 2q)(p^2 + 2q)$ **111.** $y(y - 4)^2$ **113.** $2(5z + 4)^2$ **115.** $w^2(w + 8)(w - 7)$ **117.** $14(m - 1)(m^2 + m + 1)$ **119.** $(a - 3 - 4x)(a - 3 + 4x)$ **121.** $(4x - 3)^2$ **123.** $x = \frac{1}{4}, x = -\frac{2}{3}$ **125.** $w = 0, w = -3, w = -\frac{2}{5}$ **127.** $k = -\frac{5}{7}, k = 2$ **129.** $q = 12, q = -12$ **131.** $v = 0, v = \frac{1}{5}$ **133.** $t = -\frac{5}{6}$ **135.** $y = \frac{2}{3}, y = 6$ **137.** The height is 6 ft, and the base is 13 ft. **139.** Yes **141.** The numbers are −8 and 8. **143.** The height is 4 m and the base is 9 m.

Chapter 5 Test, pp. 372–373

1. $3x(5x^3 - 1 + 2x^2)$ **2.** $(a - 5)(7 - a)$ **3.** $(6w - 1)(w - 7)$
4. $(13 - p)(13 + p)$ **5.** $(q - 8)^2$ **6.** $(2 + t)(4 - 2t + t^2)$
7. $3(a + 6b)(a + 3b)$ **8.** $(c - 1)(c + 1)(c^2 + 1)$ **9.** $(y - 7)(x + 3)$
10. Prime **11.** $-10(u - 2)(u - 1)$ **12.** $3(2t - 5)(2t + 5)$
13. $5(y - 5)^2$ **14.** $7q(3q + 2)$ **15.** $(2x + 1)(x - 2)(x + 2)$
16. $(y - 5)(y^2 + 5y + 25)$ **17.** $(x + 4 - y)(x + 4 + y)$
18. $r^2(r^2 + 16)(r - 4)(r + 4)$ **19.** $(2 - c)(6a + b)$
20. $x = \frac{3}{2}, x = -5$ **21.** $x = 0, x = 7$ **22.** $x = 8, x = -2$
23. $x = \frac{1}{5}, x = -1$ **24.** The tennis court is 12 yd by 26 yd.
25. The shorter leg is 5 ft.

Cumulative Review Exercises, Chapters 1–5, p. 373

1. $\frac{7}{5}$ **2.** $-\frac{20}{3}$ **3.** $x = 0.25$ **4.** $t = -3$ **5.** $y = \frac{8 - 3x}{-2}$ or $y = \frac{3x - 8}{2}$ **6.** The radius is 8.0 ft. **7.** There are 10 quarters, 12 nickels, and 7 dimes. **8.** 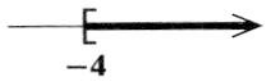

$[-4, \infty)$
9. $-\frac{7}{2}y^2 - 5y - 14$ **10.** $8p^3 - 22p^2 + 13p + 3$ **11.** $4w^2 - 28w + 49$
12. $r^3 + 5r^2 + 15r + 40 + \frac{121}{r - 3}$ **13.** c^4 **14.** $\frac{b^6c^2}{4a^4}$ **15.** -15
16. 1.6×10^3 **17.** $(w - 2)(w + 2)(w^2 + 4)$ **18.** $(a + 5b)(2x - 3y)$
19. $(2a - 3)^2$ **20.** $(2x - 5)(2x + 1)$ **21.** $(y - 3)(y^2 + 3y + 9)$
22. $(p^2 + q^2)(p^4 - p^2q^2 + q^4)$ **23.** $a(a + 1)$
24. $(x - z - 1)(x + z + 1)$ **25.** $x = 0, x = \frac{1}{2}, x = -5$
26. $x = -7, x = 5$

CHAPTER 6

Section 6.1 Practice Exercises, pp. 382–385

1. a. A number $\left(\frac{p}{q}\right)$, where p and q are integers and $q \neq 0$.
b. An expression $\left(\frac{p}{q}\right)$, where p and q are polynomials and $q \neq 0$.
3. $-\frac{1}{8}$ **5.** $-\frac{5}{3}$ **7.** 0 **9.** Undefined **11. a.** $3\frac{1}{5}$ hr or 3.2 hr **b.** $1\frac{3}{4}$ hr or 1.75 hr **13.** $\{k \mid k \neq -2\}$ **15.** $\{x \mid x \neq \frac{5}{2}, x \neq -8\}$
17. $\{b \mid b \neq -2, b \neq -3\}$ **19.** For example: $\frac{1}{x - 2}$ **21.** For example: $\frac{1}{(x + 3)(x - 7)}$ **23.** $\frac{b}{3}$ **25.** $\frac{3}{2}t^2$ **27.** $-\frac{3xy}{z^2}$ **29.** $\frac{1}{2}$
31. $\frac{p - 3}{p + 4}$ **33.** $\frac{1}{4(m - 11)}$ **35. a.** $\frac{3(y + 2)}{6(y + 2)}$ **b.** $\{y \mid y \neq -2\}$ **c.** $\frac{1}{2}$
37. a. $\frac{(t - 1)(t + 1)}{t + 1}$ **b.** $\{t \mid t \neq -1\}$ **c.** $t - 1$
39. a. $\frac{7w}{7w(3w - 5)}$ **b.** $\left\{w \mid w \neq 0, w \neq \frac{5}{3}\right\}$ **c.** $\frac{1}{3w - 5}$
41. a. $\frac{(3x - 2)(3x + 2)}{2(3x + 2)}$ **b.** $\left\{x \mid x \neq -\frac{2}{3}\right\}$ **c.** $\frac{3x - 2}{2}$
43. a. $\frac{(a + 5)(a - 2)}{(a + 3)(a - 2)}$ **b.** $\{a \mid a \neq -3, a \neq 2\}$ **c.** $\frac{a + 5}{a + 3}$
45. $\frac{1}{4a - 5}$ **47.** $\frac{4}{w + 2}$ **49.** $\frac{x - 2}{3(y + 2)}$ **51.** $\frac{2}{x - 5}$ **53.** $a + 7$
55. Not reducible **57.** $\frac{y + 3}{2y - 5}$ **59.** $\frac{3x - 2}{x + 4}$ **61.** $\frac{5}{(q + 1)(q - 1)}$
63. $\frac{c - d}{2c + d}$ **65.** $\frac{7p - 2q}{2}$ **67.** $\frac{x + y}{x - 4y}$ **69.** They are opposites.
71. -1 **73.** -1 **75.** $-\frac{1}{2}$ **77.** $-\frac{x + 3}{4 + x}$ **79. a.** $\frac{5}{3}$ **b.** $\frac{5}{3}$ **81. a.** $\frac{2}{5}$ **b.** $\frac{2}{5}$ **83.** $w - 2$ **85.** $\frac{z + 4}{z^2 + 4z + 16}$ **87.** $5x + 4$

Section 6.2 Practice Exercises, pp. 389–390

1. $\{x \mid x \neq 3, x \neq -2\}, \frac{x - 1}{x - 3}$ **3.** $\{a \mid a \neq 2\}, \frac{a + 2}{a - 2}$
5. $\left\{t \mid t \neq \frac{1}{2}\right\}, -2$ **7.** $\frac{3}{10}$ **9.** 2 **11.** $\frac{5}{2}$ **13.** $\frac{15}{4}$ **15.** $\frac{x - 6}{8}$
17. $\frac{2}{y}$ **19.** $-\frac{5}{8}$ **21.** $-\frac{b + a}{a - b}$ **23.** $\frac{10}{9}$ **25.** $\frac{6}{7}$ **27.** $\frac{y + 9}{y - 6}$
29. $\frac{t + 4}{t + 2}$ **31.** $-m(m + n)$ **33.** $\frac{3p + 4q}{4(p + 2q)}$ **35.** $\frac{w}{2w - 1}$ **37.** $\frac{5}{6}$
39. $\frac{q + 1}{q - 6}$ **41.** $\frac{1}{4}$ **43.** $\frac{3t + 8}{t + 2}$ **45.** $\frac{x + 4}{x + 1}$ **47.** $-\frac{w - 3}{2}$ **49.** $\frac{k + 6}{k + 3}$
51. 2 **53.** $\frac{1}{a - 2}$ **55.** $\frac{p + q}{2}$

Section 6.3 Practice Exercises, pp. 395–396

1. $\{x \mid x \neq 1, x \neq -1\}; \frac{3}{5(x - 1)}$ **3.** $\frac{a + 5}{a + 7}$ **5.** $\frac{4(a + 3b)}{3(a - 2b)}$ **7.** 36
9. 6 **11.** $15p$ **13.** $12xyz^3$ **15.** $w^2 + 8w + 12$ **17.** -6 **19.** a, b, c, d **21.** x^5 is the lowest power of x that has x^3, x^5, x^4 as factors.
23. The product of unique factors is $(x + 3)(x - 2)$. **25.** Because $(b - 1)$ and $(1 - b)$ are opposites, they differ by a factor of -1.
27. 45 **29.** 16 **31.** 5 or -5 **33.** $9x^2y^3$ **35.** w^2y
37. $(p + 3)(p - 1)(p + 2)$ **39.** $9t(t + 1)^2$ **41.** $(y - 2)(y + 2)(y + 3)$
43. $3 - x$ or $x - 3$ **45.** $\frac{6}{5x^2}, \frac{5x}{5x^2}$ **47.** $\frac{24x}{30x^3}, \frac{5y}{30x^3}$ **49.** $\frac{10}{12a^2b}, \frac{a^3}{12a^2b}$
51. $\frac{6m - 6}{(m + 4)(m - 1)}, \frac{3m + 12}{(m + 4)(m - 1)}$
53. $\frac{6w + 6}{(w + 3)(w - 8)(w + 1)}, \frac{w^2 + 3w}{(w + 3)(w - 8)(w + 1)}$
55. $\frac{6p^2 + 12p}{(p - 2)(p + 2)^2}, \frac{3p - 6}{(p - 2)(p + 2)^2}$
57. $\frac{1}{a - 4}, \frac{-a}{a - 4}$ or $\frac{-1}{4 - a}, \frac{a}{4 - a}$
59. $\frac{8}{2(x - 7)}, \frac{-y}{2(x - 7)}$ or $\frac{-8}{2(7 - x)}, \frac{y}{2(7 - x)}$
61. $\frac{1}{a + b}, \frac{-6}{a + b}$ or $\frac{-1}{-a - b}, \frac{6}{-a - b}$ **63.** $\frac{z^2 + 3z}{(z + 2)(z + 7)(z + 3)}, \frac{-3z^2 - 6z}{(z + 2)(z + 7)(z + 3)}, \frac{5z + 35}{(z + 2)(z + 7)(z + 3)}$
65. $\frac{3p + 6}{(p - 2)(p^2 + 2p + 4)(p + 2)}, \frac{p^3 + 2p^2 + 4p}{(p - 2)(p^2 + 2p + 4)(p + 2)}, \frac{5p^3 - 20p}{(p - 2)(p^2 + 2p + 4)(p + 2)}$

Section 6.4 Practice Exercises, pp. 403–405

1. a. $-\frac{1}{2}$, -2, 0, undefined, undefined **b.** $(x-5)(x-2)$; $\{x \mid x \neq 5, x \neq 2\}$ **c.** $\frac{x+1}{x-2}$ **3.** $\frac{8}{b-1}$ **5.** $\frac{5}{4}$ **7.** $\frac{3}{8}$ **9.** 2 **11.** 5 **13.** $\frac{-2(t-2)}{t-8}$ **15.** $\frac{5}{3x-7}$ **17.** $m+5$ **19.** 2 **21.** $x-5$ **23.** $\frac{15x}{y}$ **25.** $\frac{2(6+x^2y)}{15xy^3}$ **27.** $-\frac{2}{3}$ **29.** $\frac{19}{3(a+1)}$ **31.** $\frac{-3(k+4)}{(k-3)(k+3)}$ **33.** $\frac{a-4}{2a}$ **35.** $\frac{5}{3x-7}$ or $\frac{-5}{7-3x}$ **37.** $\frac{2(4a-b)}{(a+b)(a-b)}$ **39.** $\frac{5p-1}{3}$ or $\frac{-5p+1}{-3}$ **41.** $\frac{6n-1}{n-8}$ or $\frac{-6n+1}{8-n}$ **43.** $\frac{2(4x+5)}{x(x+2)}$ **45.** $\frac{2(w-3)}{(w+3)(w-1)}$ **47.** $\frac{4a-13}{(a-3)(a-4)}$ **49.** $\frac{4x(x+1)}{(x+3)(x-2)(x+2)}$ **51.** $\frac{-y(y+8)}{(2y+1)(y-1)(y-4)}$ **53.** $\frac{1}{2p+1}$ **55.** $\frac{x+y}{x-y}$ or $\frac{-x-y}{y-x}$ **57.** 0 **59.** $\frac{2(3x+7)}{(x+3)(x+2)}$ **61.** $\frac{1}{n}$ **63.** $\frac{12}{p}$ **65.** $n+\left(7\cdot\frac{1}{n}\right)$; $\frac{n^2+7}{n}$ **67.** $\frac{1}{n}-\frac{2}{n}$; $-\frac{1}{n}$ **69.** $\frac{3k+5}{4k+7}$ **71.** $\frac{1}{a}$ **73.** $\frac{a}{12b^4c}$ **75.** $\frac{p-q}{5}$ **77.** $\frac{10}{2x+1}$ **79.** $(h-7)(h-1)$ **81.** $\frac{1}{2(a+3)}$ **83.** $(t+8)^2$ **85.** $\frac{-w^2}{(w+3)(w-3)(w^2-3w+9)}$ **87.** $\frac{p^2-2p+7}{(p+2)(p+3)(p-1)}$ **89.** $\frac{-m-21}{2(m+5)(m-2)}$ or $\frac{m+21}{2(m+5)(2-m)}$

Section 6.5 Practice Exercises, pp. 410–411

1. $\{c \mid c \neq -1, c \neq 2\}$; $\frac{c+3}{c+1}$ **3.** $\{x \mid x \neq 2, x \neq -2\}$; $\frac{2}{x-2}$ **5.** $\frac{5w-6}{w(w-2)}$ **7.** $\frac{5}{12}$ **9.** $\frac{1}{z+1}$ **11.** $\frac{\frac{1}{2}+\frac{2}{3}}{5}$; $\frac{7}{30}$ **13.** $\frac{3}{\frac{2}{3}+\frac{3}{4}}$; $\frac{36}{17}$ **15.** $\frac{35}{2}$ **17.** $k+h$ **19.** $\frac{n+1}{2(n-3)}$ **21.** $\frac{2x+1}{4x+1}$ **23.** $m-7$ **25.** $\frac{2y(y-5)}{7y^2+10}$ **27.** $-\frac{a+8}{a-2}$ or $\frac{a+8}{2-a}$ **29.** $\frac{t-2}{t-4}$ **31.** $\frac{2z-5}{3(z+3)}$ **33.** $-\frac{x+1}{x-1}$ or $\frac{x+1}{1-x}$ **35. a.** $\frac{6}{5}\,\Omega$ **b.** $6\,\Omega$ **37.** $\frac{3}{2}$ **39.** $\frac{8}{5}$

Section 6.6 Practice Exercises, pp. 419–420

1. $\frac{2x+1}{(x-3)(x+2)}$ **3.** $\frac{(t-3)(t+1)}{(t-6)(t+2)}$ **5.** $5(h+1)$ **7.** $z=4$ **9.** $p=-\frac{13}{12}$ **11.** $x=-\frac{1}{2}$ **13. a.** $4w$ **b.** $w=4$ **15. a.** $(x+3)(x-1)$ **b.** $x=-5$ **17.** $y=-\frac{200}{19}$ **19.** $t=8$ **21.** $x=\frac{47}{6}$ **23.** $y=3, y=-1$ **25.** $a=4$ **27.** $w=5$; ($w=0$ does not check.) **29.** $m=-5$ **31.** No solution; ($p=4$ does not check.) **33.** $t=4$ **35.** $x=4, x=-3$ **37.** $x=-4$; ($x=1$ does not check) **39.** No solution; ($x=-4$ does not check.) **41.** $x=4$; ($x=-6$ does not check.) **43.** The number is 8. **45.** The number is -26. **47.** $m=\frac{FK}{a}$ **49.** $E=\frac{IR}{K}$ **51.** $R=\frac{E-Ir}{I}$ or $R=\frac{E}{I}-r$ **53.** $B=\frac{2A-hb}{h}$ or $B=\frac{2A}{h}-b$ **55.** $h=\frac{V}{r^2\pi}$ **57.** $t=\frac{b}{x-a}$ or $t=\frac{-b}{a-x}$ **59.** $x=\frac{y}{1-yz}$ or $x=\frac{-y}{yz-1}$ **61.** $h=\frac{2A}{a+b}$ **63.** $R=\frac{R_1R_2}{R_2+R_1}$ **65.** $t_2=\frac{s_2-s_1+vt_1}{v}$ or $t_2=\frac{s_2-s_1}{v}+t_1$

Section 6.7 Practice Exercises, pp. 427–430

1. Equation; $b=30$ **3.** Expression; $\frac{2a-5}{(a+5)(a-5)}$ **5.** Expression; $\frac{3}{10}$ **7.** Equation; $p=2$ **9.** $a=\frac{40}{3}$ **11.** $x=40$ **13.** $y=3$ **15.** $z=-1$ **17.** $a=1$ **19. a.** $V_f=\frac{V_iT_f}{T_i}$ **b.** $T_f=\frac{T_iV_f}{V_i}$ **21.** 99 **23.** 4.8 ft **25.** 262.5 miles **27.** 30 red M&Ms **29.** 4160 incorrectly marked ballots are expected. **31.** The speed of the current is 2 mph. **33.** The plane travels 165 mph in still air. **35.** The speeds are 45 mph and 60 mph. **37.** Shanelle skis 10 km/hr and Devon skis 15 km/hr. **39.** $\frac{1}{2}$ of the room **41.** $5\frac{5}{11}$ (or $5.\overline{45}$) minutes **43.** $22\frac{2}{9}$ (or $22.\overline{2}$) min **45.** $3\frac{1}{3}$ (or $3.\overline{3}$) days **47. a.** 4 cm **b.** 5 cm **49.** $x=3.75$ cm; $y=4.5$ cm **51.** The height of the pole is 7 m. **53.** The light post is 24 ft high.

Chapter 6 Review Exercises, pp. 436–438

1. a. $-\frac{2}{9}$, $-\frac{1}{10}$, 0, $-\frac{5}{6}$, undefined **b.** $\{t \mid t \neq -9\}$ **3.** a, c, d **5.** $\left\{h \mid h \neq -\frac{1}{3}, h \neq -7\right\}$; $\frac{1}{3h+1}$ **7.** $\{w \mid w \neq 4, w \neq -4\}$; $\frac{2w+3}{w-4}$ **9.** $\{k \mid k \neq 0, k \neq 5\}$; $-\frac{3}{2k}$ **11.** $\{m \mid m \neq -1\}$; $\frac{m-5}{3}$ **13.** $\{p \mid p \neq -7\}$; $\frac{1}{p+7}$ **15.** $\frac{u^2}{2}$ **17.** $\frac{3}{2(x-5)}$ **19.** $\frac{q-2}{4}$ **21.** $4s(s-4)$ **23.** $\frac{1}{n-2}$ **25.** $\frac{1}{m+3}$ **27.** $-\frac{2y-1}{y+1}$ **29.** $2y+4$ **31.** $2r+6$ **33.** u^2-5u-6 **35.** xy^2z^4 **37.** $q(q+8)$ **39.** $(n-3)(n+3)(n+2)$ **41.** $3k-1$ or $1-3k$ **43.** $3-x$ or $x-3$ **45.** 2 **47.** $x-7$ **49.** $\frac{t^2+2t+3}{(2-t)(2+t)}$ **51.** $\frac{3(r-4)}{2r(r+6)}$ **53.** $\frac{q}{(q+5)(q+4)}$ **55.** $\frac{1}{3}$ **57.** $\frac{3(z+5)}{z(z-5)}$ **59.** $\frac{8}{y}$ **61.** $-(b+a)$ **63.** $-\frac{k+10}{k+4}$ **65.** $y=-2$ **67.** $w=2$ **69.** $p=3$ **71.** $y=-11, y=1$ **73.** $h=\frac{3V}{\pi r^2}$ **75.** $m=\frac{6}{5}$ **77.** 12 g **79.** Together the pumps would fill the pool in 16.8 min.

Chapter 6 Test, p. 439

1. a. $\{x \mid x \neq 2\}$ **b.** $-\frac{x+1}{6}$ **2. a.** $\{a \mid a \neq 0, a \neq 6, a \neq -2\}$ **b.** $\frac{7}{a+2}$ **3.** b, c, d **4.** $\frac{y+7}{3(y+3)(y+1)}$ **5.** $-\frac{b+3}{5}$ **6.** $\frac{1}{w+1}$

7. $\dfrac{t+4}{t+2}$ **8.** $\dfrac{x(x+5)}{(x+4)(x-2)}$ **9.** $\dfrac{1}{m+4}$ **10.** $a = \dfrac{8}{5}$ **11.** $p = 2$
12. $c = 1$ **13.** No solution. ($x = 4$ does not check) **14.** $r = \dfrac{2A}{C}$
15. The number is $-\dfrac{2}{5}$. **16.** $y = -8$ **17.** $1\frac{1}{4}$ (1.25) cups of carrots
18. The speed of the current is 5 mph. **19.** It would take the second printer 3 hr to do the job working alone. **20.** $a = 5.6$ m, $b = 12$ m
21. a. $15(x+3)$ **b.** $3x^2y^2$

Cumulative Review Exercises, Chapters 1–6, pp. 440–441

1. 32 **2.** 7 **3.** Rational: $\sqrt{4}, \sqrt{9}, \sqrt{16}, \sqrt{49}$; irrational: $\sqrt{5}, \sqrt{20}$
4. $y = \dfrac{10}{9}$ **5.** $(8, \infty)$

6.

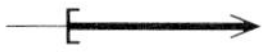

$[-1, \infty)$; $\{x \mid x < 5\}$

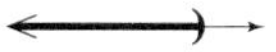

7. The width is 17 m and the length is 35 m. **8.** The base is 10 in. and the height is 8 in. **9.** $x = 10$; the angles are 37°. **10.** 9.9932×10^{-1} g or 0.99932 g **11.**
$$12^2 + 16^2 = 20^2$$
$$144 + 256 = 400$$
$$400 = 400 \checkmark$$

12. 1 **13.** $\dfrac{x^2yz^{17}}{2}$ **14. a.** $6x + 4$ **b.** $2x^2 + x - 3$
15. $25x^2 - 30x + 9$ **16.** $(5x-3)^2$ **17.** $4ab^2 - b + \dfrac{a^2}{2}$
18. $3(3x-5y)(3x+5y)$ **19.** $(2c+1)(5d-3)$ **20.** $(x-5)(x+4)$
21. $\left\{x \mid x \neq 5, x \neq -\dfrac{1}{2}\right\}$ **22.** $\dfrac{x-3}{x+5}$ **23.** $\dfrac{1}{5(x+4)}$ **24.** $x - 3$
25. −3 **26.** $y = 1$ **27.** $b = -\dfrac{7}{2}$ **28.** The speed of the current is 3 mph. **29. a.** Linear **b.**

x	y
0	−10
2	0
1	−5

c.

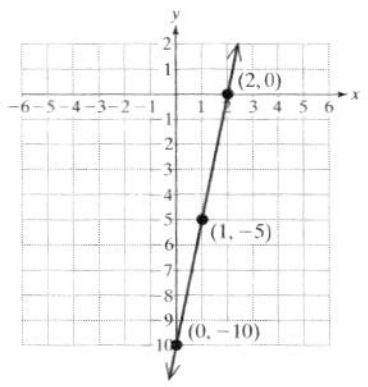

30. a. linear **b.** $(0, -25)$ **c.** $(25, 0)$ **31.** $(a+b-4)(a+b+4)$

CHAPTER 7

Section 7.1 Practice Exercises, pp. 450–453

1. x-intercept: (2, 0); y-intercept: (0, 4)

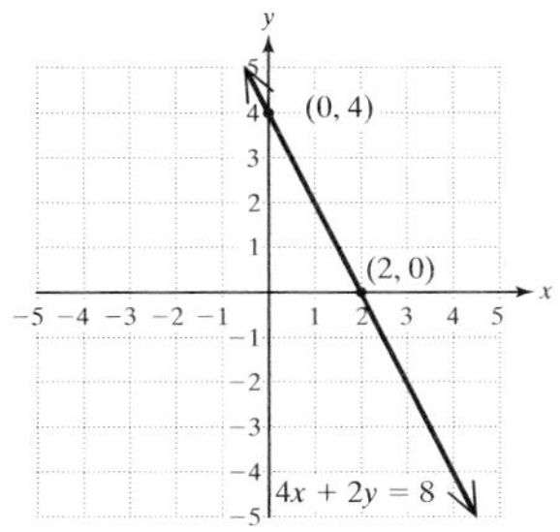

3. x-intercept: (2, 0); y-intercept: $\left(0, -\dfrac{3}{2}\right)$

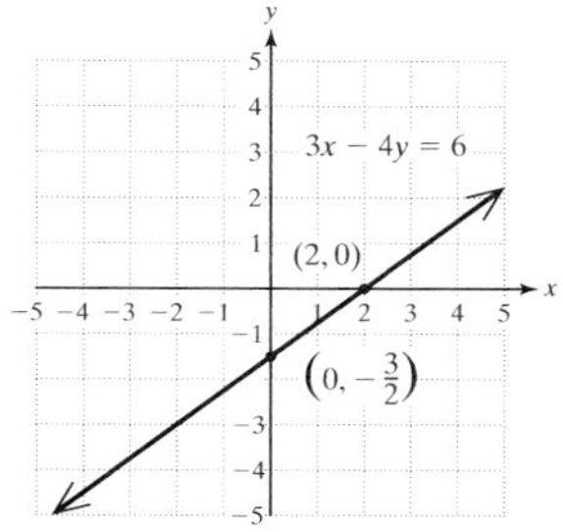

5. x-intercept: (0, 0); y-intercept: (0, 0)

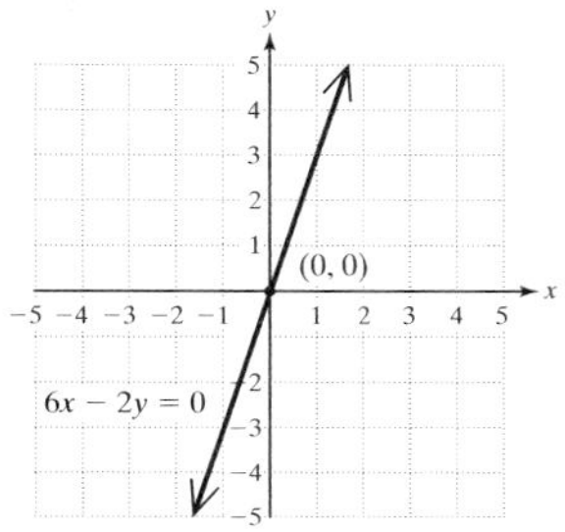

7. $y = -2x + 4$; slope: −2; y-intercept: (0, 4)
9. $y = \dfrac{5}{9}x + \dfrac{1}{3}$; slope: $\dfrac{5}{9}$; y-intercept: $\left(0, \dfrac{1}{3}\right)$
11. $y = \dfrac{4}{3}x$; slope: $\dfrac{4}{3}$; y-intercept: (0, 0)
13. $y = \dfrac{5}{6}x - \dfrac{5}{2}$; slope: $\dfrac{5}{6}$; y-intercept: $\left(0, -\dfrac{5}{2}\right)$
15. iv **17.** vi **19.** iii **21.** 9 **23.** $\dfrac{6}{7}$
25. a.

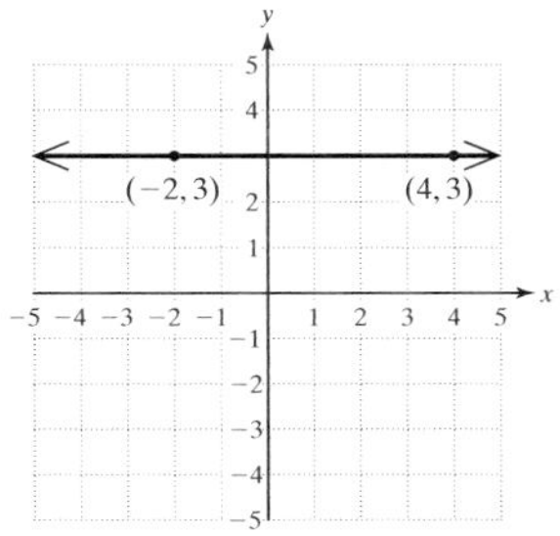

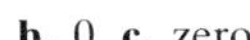

b. 0 **c.** zero

27. a.

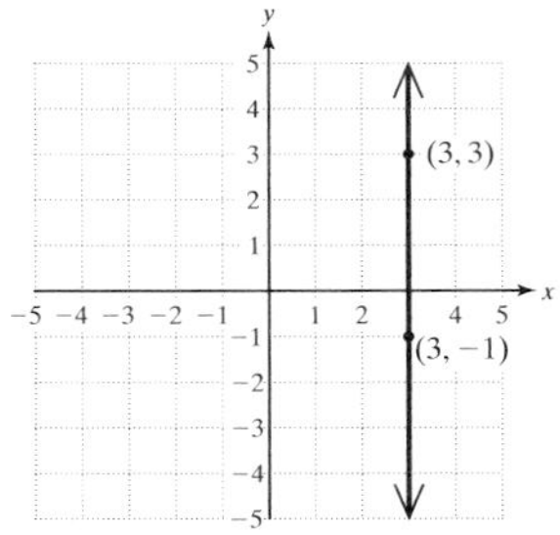

b. undefined **c.** undefined

29. $y = 3x + 7$ **31.** $y = \dfrac{1}{4}x + 8$ **33.** $y = 4.1x - 23.93$
35. $y = -2$ **37.** $y = 2x - 2$ **39.** $y = -\dfrac{5}{8}x - \dfrac{19}{8}$

41. $y = -0.84x + 4.948$ **43.** $y = 1$ **45.** $x = 2$ **47.** $x = \frac{5}{2}$ **49.** $y = 2$ **51.** $x = -6$ **53.** $x = -4$ **55. a.** 5 **b.** $y = 5x + 67$ **c.** The median price of a one-family house in 2005 would be \$192,000. **57.** $y = -2x - 8$ **59.** $y = 3x - 8$ **61.** $y = -\frac{1}{5}x - 6$

Section 7.2 Practice Exercises, pp. 458–462

1. a. \$95 **b.** \$190 **c.** (0, 0). For 0 kilowatt-hours used, the cost is \$0. **d.** $m = 0.095$. The cost increases by \$0.095 for each kilowatt-hour used. **e.**

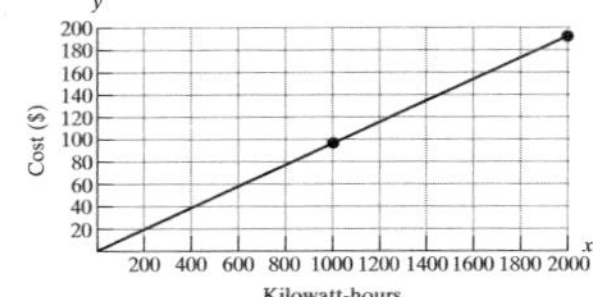

3. a. $m = 3.49$. The number of female inmates increased by 3.49 thousand per year between 1987 and 1997. **b.** $m = 21.5$. The number of male inmates increased by 21.5 thousand per year between 1987 and 1997. **c.** Males. The number of male inmates is increasing at a faster rate than the number of female inmates. **5. a.** y, temperature **b.** x, latitude **c.** 30.7° **d.** 13.4° **e.** $m = -2.333$. The average temperature in January decreases 2.333° per 1° of latitude. **f.** (53.2, 0). At 53.2° latitude, the average temperature in January is 0°.
7. a. $y = -2.6x + 169$ **b.** 153.4 hr **9. a.** $y = 3.5x - 1.75$ **b.** $m = 3.5$. For each additional inch in length of a person's arm, the person's height increases by 3.5 in. **c.** 73.5 in. or 6 ft $1\frac{1}{2}$ in.
11. a. $y = 0.25x + 20$ **b.** \$84.50 **13. a.** $y = 25x + 20$ **b.** \$520.00 **15. a.** $y = 35x + 1200$ **b.** \$4700.00 **17. a.** $y = 0.8x + 100$ **b.** \$260.00 **19.** 13.3 **21.** 345

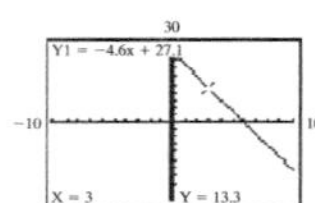

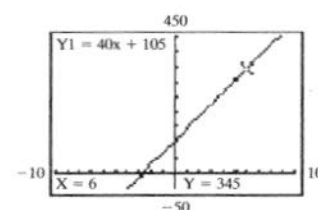

Section 7.3 Practice Exercises, pp. 467–471

1. $y = -3x - 1$ **3.** $y = -x + 3$ **5. a.** $y = \frac{2}{3}x + 6$ **b.** $y = -\frac{3}{2}x - \frac{1}{2}$ **7.** $\{(A, 1)(A, 2)(B, 2)(C, 3)(D, 5)(E, 4)\}$
9. {(Pregnant women, 60) (Nursing mothers, 65)(Infants under 1 year, 14)(Children 1–4 years, 16)(Adults, 50)} **11.** Domain $\{A, B, C, D, E\}$; range {1, 2, 3, 4, 5} **13.** Domain {Pregnant women, nursing mothers, infants under 1, children 1–4, adults}; range {60, 65, 14, 16, 50}
15. a. For example: {(Julie, New York)(Peggy, Florida)(Stephen, Kansas)(Pat, New York)} **b.** Domain {Julie, Peggy, Stephen, Pat}; range {New York, Florida, Kansas}
17. a. $y = 2x - 1$ **b.**

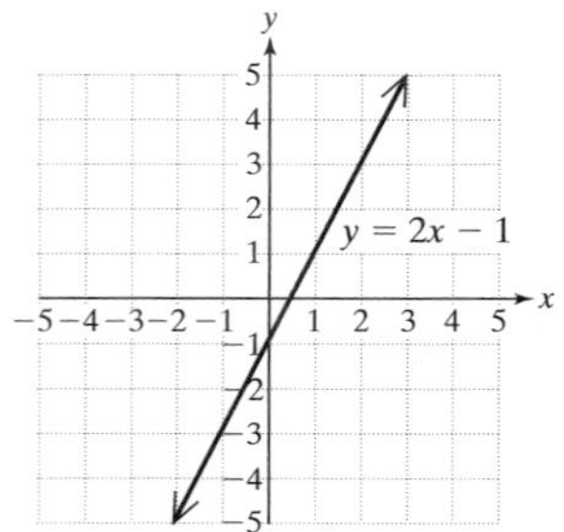

c. Domain: $(-\infty, \infty)$; range: $(-\infty, \infty)$ **19.** Domain $[-5, 3]$; range $[-2.1, 2.8]$ **21.** Domain $[0, 4.2]$; range $[-2.1, 2.1]$ **23.** Domain $(-\infty, 0]$; range $(-\infty, \infty)$ **25.** Domain $(-2, \infty)$; range $(2, \infty)$
27. Domain $[-4, \infty)$; range $[0, \infty)$ **29.** Domain $\{-3, -1, 1, 3\}$; range $\{0, 1, 2, 3\}$ **31.** Domain $[-4, 5)$; range $\{-2, 1, 3\}$ **33. a.** 2.85 **b.** 9.33 **c.** Dec. **d.** Nov. **e.** 7.63 **f.** {Jan, Feb, Mar, Apr, May, June, July, Aug, Sept, Oct, Nov, Dec} **35. a.** 81.46%, 68.82%, 56.18%, 43.54% **b.** No, 7 is not in the domain. **37.** Domain $(-\infty, \infty)$; range $[0, \infty)$ **39.** Domain $(-\infty, \infty)$; range $[0, \infty)$ **41.** Domain $(-\infty, \infty)$; range $[-2, \infty)$ **43.** The domain $(-\infty, \infty)$ and the range $[c, \infty)$ will be the same for all values of c.
45. a.

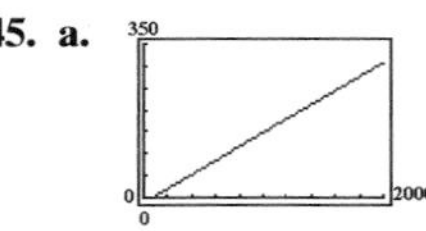

b.

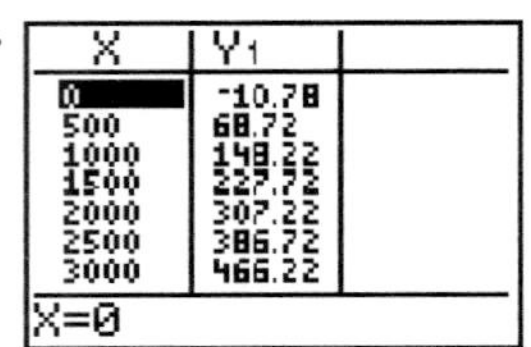

X	Y1
0	-10.78
500	68.72
1000	148.22
1500	227.72
2000	307.22
2500	386.72
3000	466.22

X=0

Section 7.4 Practice Exercises, pp. 479–482

1. a. {(Doris, Mike)(Richard, Nora)(Doris, Molly)(Richard, Mike)} **b.** {Doris, Richard} **c.** {Mike, Nora, Molly} **d.** Not a function
3. a. {(3, 10)(4, 12)(5, 12)(6, 12)} **b.** {3, 4, 5, 6} **c.** {10, 12} **d.** Is a function **5.** Domain $[0, 4]$; range $[1, 4]$ **7.** Domain $\{-4\}$; range $(-\infty, \infty)$ **9.** Not a function **11.** Function **13.** Not a function
15. 10 **17.** 7 **19.** 1 **21.** 2 **23.** $6t - 2$ **25.** 7 **27.** 22 **29.** 4
31. $6x + 4$ **33.** $x^2 - 8x + 13$ **35.** $x^2 + 2xh + h^2 - 4x - 4h + 1$
37. 7 **39.** $-6a - 2$ **41.** $|-c - 2|$ **43.** 1 **45.** 7 **47.** -18.8
49. -7 **51.** 2π **53.** -5 **55.** 4 **57.** $\{-3, -7, -\frac{3}{2}, 1.2\}$ **59.** {6, 0}
61. -3 and 1.2 **63.** 6 and 1 **65.** -3 **67. a.** 2 **b.** 1 **c.** 1 **d.** $x = -3$ **e.** $x = 1$ **f.** $[-3, 3]$ **g.** $[-3, 3]$ **69.** The domain is all real numbers that do not make the denominator zero. Domain: $\{x \mid x \neq 2\}$ or $(-\infty, 2) \cup (2, \infty)$ **71.** $(-\infty, 4) \cup (4, \infty)$
73. $(-\infty, 0) \cup (0, \infty)$ **75.** $(-\infty, \infty)$ **77.** $[-7, \infty)$ **79.** $[3, \infty)$
81. $\left[-\frac{1}{2}, \infty\right)$ **83.** The domain of a polynomial is all real numbers.
85. $(-\infty, \infty)$ **87.** $(-\infty, \infty)$ **89. a.** 45.1, 38.975 **b.** After 1 sec, the height of the ball is 45.1 m. After 1.5 sec, the height of the ball is 38.975 m. **91. a.** 5.9, 11.8 **b.** After 1 hr, the distance is 5.9 km. After 2 hr, the distance is 11.8 km. **93. a.** $N(1) = 2.6$. A person 1 year old averages 2.6 visits per year to a doctor. $N(20) = 1.9$. A person 20 years old averages 1.9 visits per year to a doctor. $N(40) = 2.3$. A person 40 years old averages 2.3 visits per year to a doctor. $N(75) = 5.6$. A person 75 years old averages 5.6 visits per year to a doctor. **c.** About 24 years old **95.** $\left(-\infty, \frac{1}{3}\right) \cup \left(\frac{1}{3}, \infty\right)$
97. $(4, \infty)$ **99.**

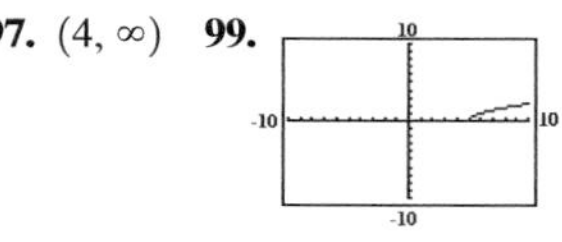

101.

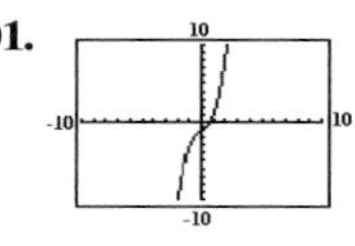

103. a.

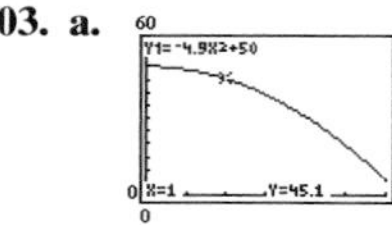

b. $h(1) = 45.1$; $h(1.5) = 38.975$

Section 7.5 Practice Exercises, pp. 492–495

1. a. Yes **b.** {6, 5, 4, 3, 2, 1} **c.** {1, 2, 3, 4, 5, 6} **3.** -2 **5.** $a^2 - 2$ **7.** 2 **9.** $b^2 - 2b - 1$ **11.** $(-\infty, \infty)$ **13.** $[6, \infty)$ **15. a.** $f(3) = 9$;

It takes 9 lb to stretch the spring 3 in. **b** $f(0) = 0$; It takes 0 lb to stretch the spring 0 in.

17.

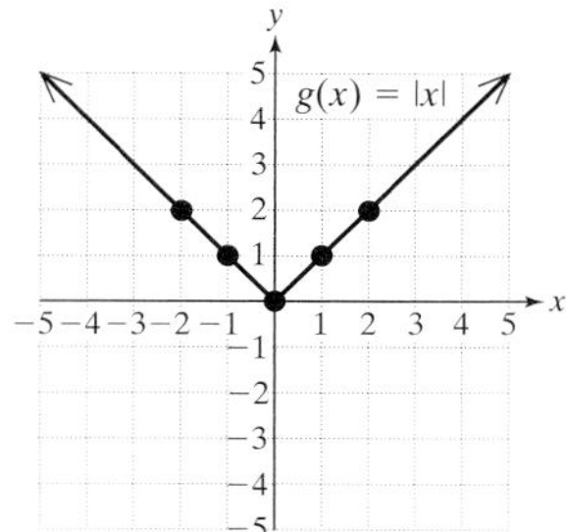

19.

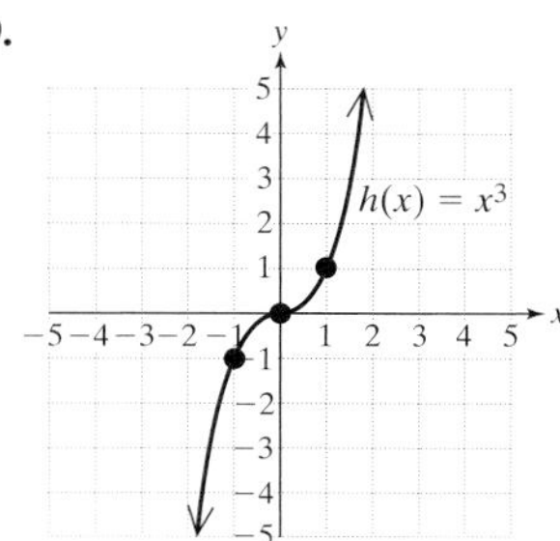

21.

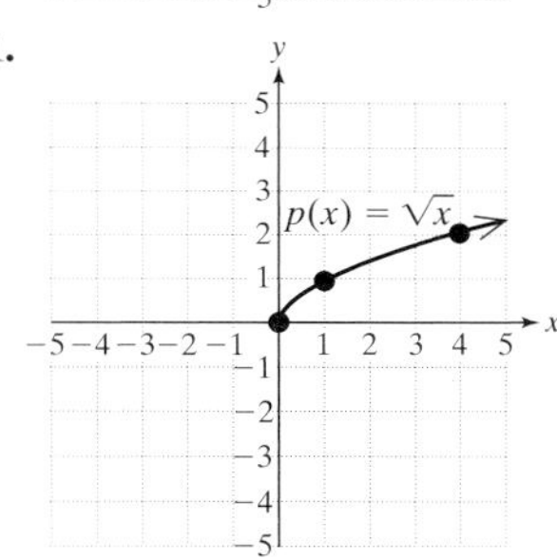

23. a. $[-4, \infty)$ **b.** $f(0) = 2, f(5) = 3, f(-3) = 1$ **25. a.** $(-\infty, \infty)$ **b.** $h(1) = 1, h(-1) = 1, h(0) = 2$ **27. a.** $(-\infty, 3) \cup (3, \infty)$ **b.** $p(0) = -\frac{2}{3}, p(1) = -1, p(2) = -2, p(4) = 2, p(5) = 1, p(6) = \frac{2}{3}$

29. x-intercept: $\left(-\frac{1}{8}, 0\right)$; y-intercept: $(0, 1)$ **31.** x-intercept: $(0, 0)$; y-intercept: $(0, 0)$ **33.** x-intercepts: $(5, 0), \left(-\frac{1}{2}, 0\right)$; y-intercept: $(0, -5)$ **35.** x-intercepts: $(-2, 0), (5, 0)$; y-intercept: $(0, -10)$ **37.** x-intercept: $(-3, 0)$; y-intercept: $(0, 9)$ **39.** x-intercepts: $(0, 0), (3, 0), \left(-\frac{2}{3}, 0\right)$; y-intercept: $(0, 0)$ **41.** x-intercept $(-1, 0)$; y-intercept $(0, 1)$ **43.** x-intercepts $(-2,0), (2,0)$; y-intercept $(0, -2)$ **45.** x-intercept–none; y-intercept $(0, 2)$

47. a.

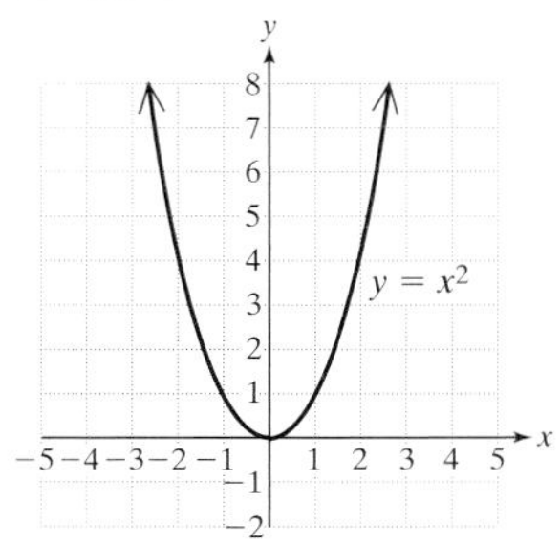

b. $y = x^2$ is a function. It passes the vertical line test.

c.

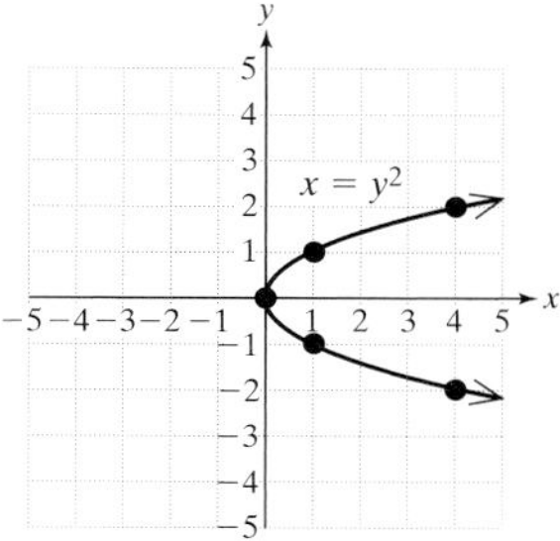

d. $x = y^2$ is not a function. The graph does not pass the vertical line test. **49. a.** $(-\infty, \infty)$ **b.** $(0, 0)$ **c.** vi **51. a.** $(-\infty, \infty)$ **b.** $(0, 1)$ **c.** viii **53. a.** $[-1, \infty)$ **b.** $(0, 1)$ **c.** vii **55. a.** $(-\infty, 3) \cup (3, \infty)$ **b.** $\left(0, -\frac{1}{3}\right)$ **c.** ii **57. a.** $(-\infty, \infty)$ **b.** $(0, 2)$ **c.** iv

59. **61.**

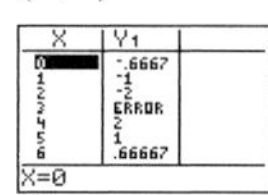

Section 7.6 Practice Exercises, pp. 501–503

1. 3 **3.** None **5.** $[-5, \infty)$ **7.** Not a function **9.** Function **11. a.** Increase **b.** Decrease **13.** $T = kq$ **15.** $W = \frac{k}{p^2}$ **17.** $Q = \frac{kx}{y^3}$ **19.** $L = kw\sqrt{v}$ **21.** $k = \frac{9}{2}$ **23.** $k = 512$ **25.** $k = 1.75$ **27.** $Z = 56$ **29.** $L = 9$ **31.** $B = \frac{15}{2}$ **33.** 355,000 tons **35.** 42.6 ft **37.** 18.5 A **39.** 1.25 Ω **41.** 20 lb **43.** 2224 lb **45. a.** $A = kl^2$ **b.** The area will increase by 4 times. **c.** The area will increase by 9 times.

Chapter 7 Review Exercises, pp. 508–511

1. a. x-intercept: $(\frac{8}{3}, 0)$; y-intercept: $(0, -2)$ **b.** $m = \frac{3}{4}$

c.

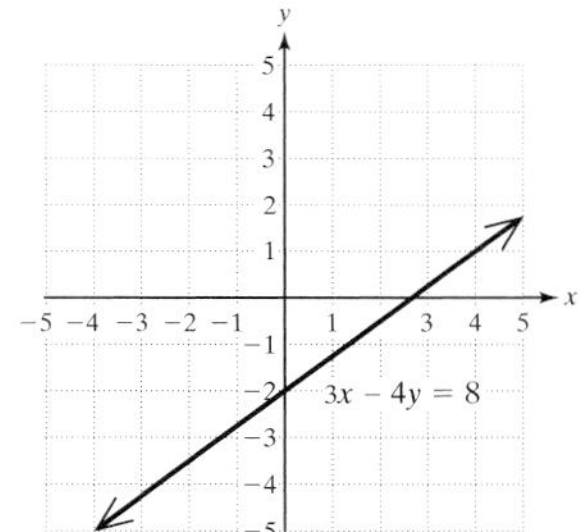

3. a. x-intercept: $(4, 0)$; y-intercept: none **b.** slope is undefined

c.

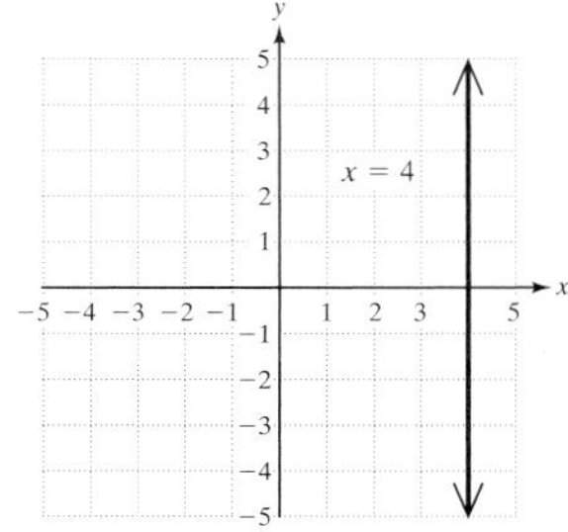

5. a. $y = 2x - 4$; slope is 2; **b.** y-intercept is $(0, -4)$

7. $y - y_1 = m(x - x_1)$ **9.** $y = -6x + 2$ **11.** $y = \frac{1}{4}x - 4$

13. $y = 4x - 20$ **15.** $y = \frac{6}{5}x + 6$ **17.** $y = -9x - 9$ **19.** $y = -1$

21. a. $m = -0.13$ **b.** The number of robberies decreased by an average of 0.13 million per year between 1994 and 1999. **c.** $y = -0.13x + 1.43$ **d.** 0.78 million robberies **23. a.** $y = 20x + 55$ **b.** $235 **25.** For example: {(Peggy, Kent)(Charlie, Laura)(Tom, Matt)(Tom, Chris)} **27.** Domain $[-3, 9]$; range $[0, 60]$ **29.** Domain $\{-3, -1, 0, 2, 3\}$; range $\left\{-2, 1, 0, \frac{5}{2}\right\}$

31. For example:

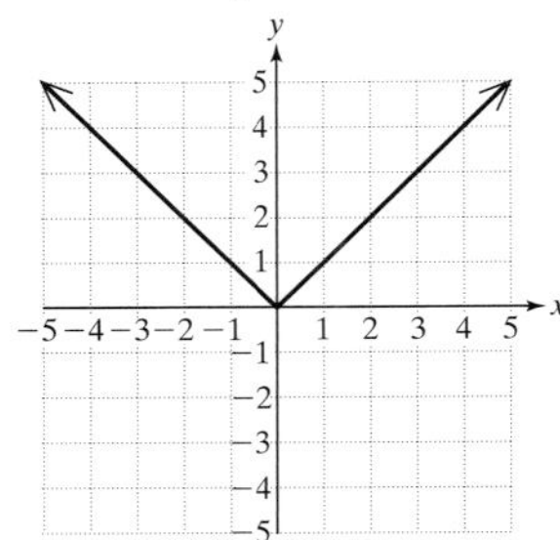

33. a. Function **b.** $(-\infty, \infty)$ **c.** $(-\infty, 0.35)$ **35. a.** Not a function **b.** $\{0, 4\}$ **c.** $\{2, 3, 4, 5\}$ **37. a.** Function **b.** $\{6, 7, 8, 9\}$ **c.** $\{9, 10, 11, 12\}$ **39.** 2 **41.** $6t^2 - 4$ **43.** $6\pi^2 - 4$ **45.** $6x^2 + 12xh + 6h^2 - 4$ **47.** $(-\infty, 11) \cup (11, \infty)$ **49.** $[-2, \infty)$

51.

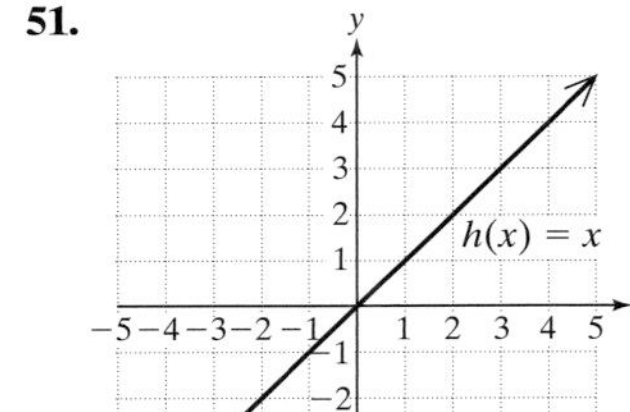

53.

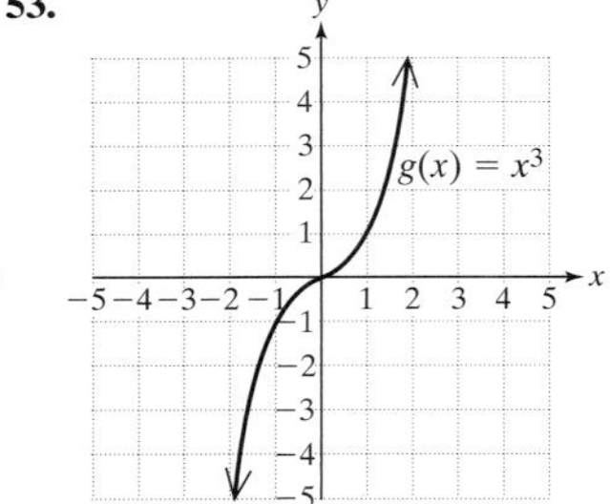

55.

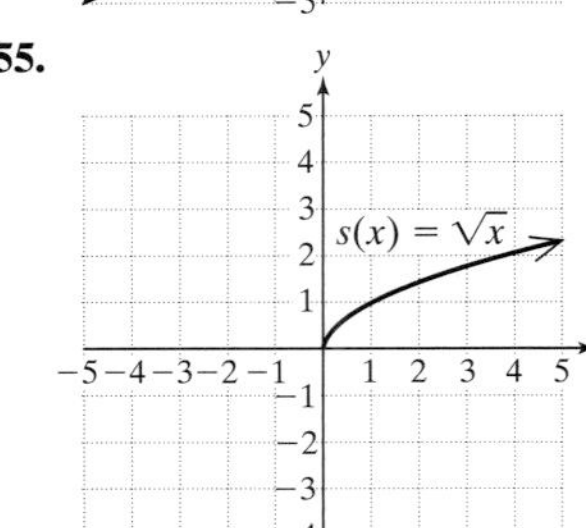

57.

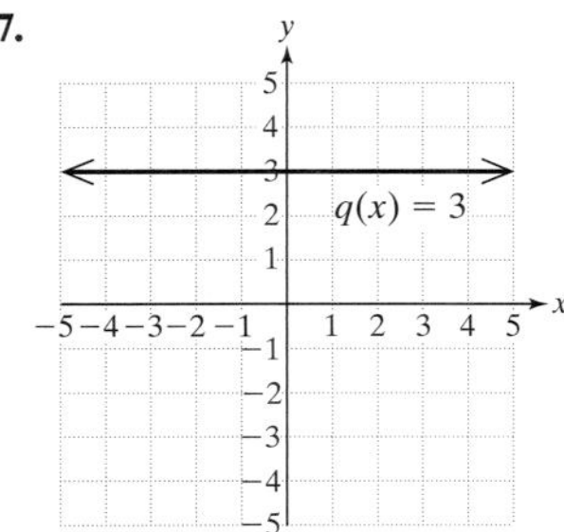

59. x-intercept: $\left(\frac{7}{4}, 0\right)$; y-intercept: $(0, -7)$ **61.** x-intercepts: $(4, 0)$, $(-4, 0)$; y-intercept: $(0, -16)$ **63.** x-intercepts: $(3, 0)$, $(-2, 0)$, $\left(\frac{1}{2}, 0\right)$; y-intercept: $(0, 6)$

65. a. $s(4) = 4, s(3) = 1, s(2) = 0, s(1) = 1, s(0) = 4$ **b.** $(-\infty, \infty)$

67. a. $h(-3) = -\frac{1}{2}, h(-1) = -\frac{3}{4}, h(0) = -1, h(2) = -3, h(4) = 3, h(5) = \frac{3}{2}, h(7) = \frac{3}{4}$ **b.** $(-\infty, 3) \cup (3, \infty)$ **69. a.** $b(0) = 4.5$. In 1985 consumption was 4.5 gal of bottled water per capita. $b(7) = 9.4$. In 1992 consumption was 9.4 gal of bottled water per capita. **b.** $m = 0.7$. Consumption increased by 0.7 gal/year. **71.** $y = 4$ **73.** 52.8 km

Chapter 7 Test, pp. 512–514

1. $\frac{7}{4}$ **2.** 2 **3.** $-\frac{1}{8}$ **4.** x-intercept: $(3, 0)$; y-intercept: $\left(0, \frac{9}{7}\right)$ **5.** x-intercept: $(0, 0)$; y-intercept: $(0, 0)$ **6. a.** Vertical **b.** Horizontal

7.

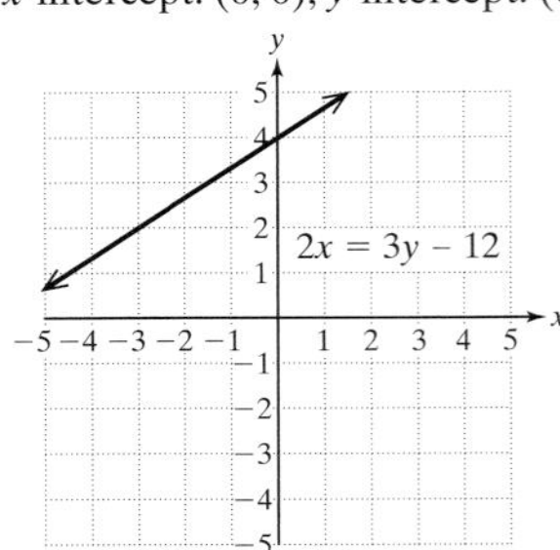

8. $y = -\frac{7}{2}x + 15$ **9.** $y = -6$ **10.** $y = -\frac{1}{3}x + 1$ **11.** $y = 3x + 8$

12. a. $\frac{3}{4}$ **b.** $\frac{1}{2}$ **13. a.** $y = 1.5x + 10$ **b.** $25 **14. a.** $m = -0.33$. The winning time has dropped by an average of 0.33 sec/year between 1912 and 1996. **b.** $y = -0.33x + 82.2$ **c.** 70.32 sec **15. a.** Not a function **b.** $\{-3, -1, 1, 3\}$ **c.** $\{3, -2, -1, 1\}$ **16. a.** Function **b.** $(-\infty, \infty)$ **c.** $(-\infty, 0]$ **17. a.** Function **b.** {1975, 1985, 1997} **c.** {47.4%, 62.2%, 72.1%} **18.** To find the x-intercept(s), solve for the real solutions of the equation $f(x) = 0$. To find the y-intercept, find $f(0)$. **19. a.** 5 **b.** 3 **c.** 12 **d.** 7 **20. a.** $k(0) = 8, k(-2) = 8, k(15) = 8$ **b.** $(-\infty, \infty)$ **c.** **d.** $\{8\}$

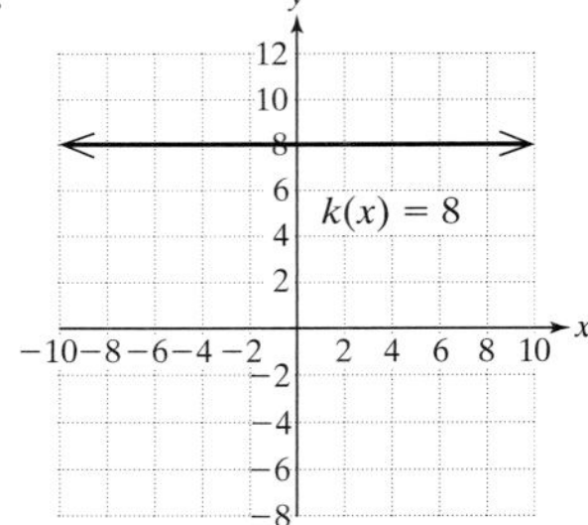

21. a. $r(-2) = -27, r(-1) = -8, r(0) = -1, r(1) = 0, r(2) = 1, r(3) = 8$ **b.** $(-\infty, \infty)$ **22. a.** $(-\infty, -7) \cup (-7, \infty)$ **b.** $[-7, \infty)$ **c.** $(-\infty, \infty)$ **23. a.** $s(0) = 36$. In 1985 the per capita consumption was 36 gal; $s(7) = 47.2$. In 1992 the per capita consumption was 47.2 gal **b.** $m = 1.6$. Increase of 1.6 gal/year **24. a.** $f(1) = 1, f(4) = 2, f(7) = -1$ **b.** $x = 6$ **c.** $[-1, 7]$ **d.** $[-1, 4]$ **25. a.** ii **b.** v **c.** iv **26.** 2.37 sec

Cumulative Review Exercises, Chapters 1–7, pp. 514–515

1. $t = -\frac{12}{7}$ **2.** 14 **3. a.** $[6, \infty)$ **b.** $(-\infty, 17)$ **c.** $[-2, 3]$ **4.** $\left(-\infty, \frac{1}{6}\right]$ **5.** 1474 in.^{3} **6.** $y = -\frac{4}{3}x$ **7.** $m = \frac{2}{5}$ **8. a.** To find the x-intercepts, let $f(x) = 0$ and solve for x. To find the y-intercept, find $f(0)$. **b.** $(0, 2)$ **c.** $\left(-\frac{2}{3}, 0\right)$ **9.** 37°, 74°, 69° **10.** $x^2 - 4x + 16$ **11.** $\frac{1}{4}$ **12.** $\frac{a^6}{8b^{30}}$ **13.** $8b^3 - 6b^2 + 4b - 3$ **14.** $3a^3 + 5a^2 - 2a + 5$

15. $3w - \frac{5}{2} - \frac{1}{w}$ **16. a.** $\left(-\infty, \frac{3}{2}\right) \cup \left(\frac{3}{2}, \infty\right)$
b. $(-\infty, -3) \cup (-3, 4) \cup (4, \infty)$ **17.** $\frac{x + a}{4(x - 7)}$
18. $\frac{-(x - 2)}{5}$ or $\frac{2 - x}{5}$ **19.** $\frac{x + 3}{(x + 10)(x - 2)}$ **20.** $\frac{c - 7}{c}$
21. $y = 3, y = -1$ **22.** $2x(x^2 + 4)(x - 2)(x + 2)$
23. $(2x + 3)(4x - 5)$ **24.** $x = 5, x = 10$ **25. a.** 8620 students
b. The year 2011 **26.** Domain $\{12, 15, 18, 21\}$; range $\{a, c, e\}$, yes
27. a. 1 **b.** 33 **28. a.** $(-\infty, 15) \cup (15, \infty)$ **b.** $[6, \infty)$ **29.** \$3500

CHAPTER 8

Section 8.1 Practice Exercises, pp. 523–526

1. Yes **3.** No **5.** Yes

7. a.

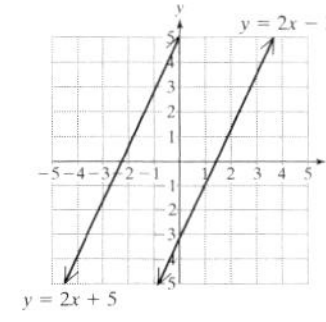

b.

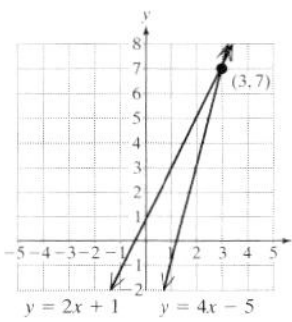

c.

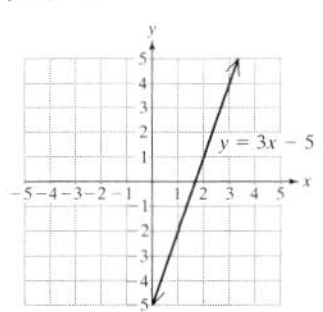

9. c **11.** a **13.** a **15.** b **17.** c **19.** b **21.** d

23. (3, 1)
Consistent; independent

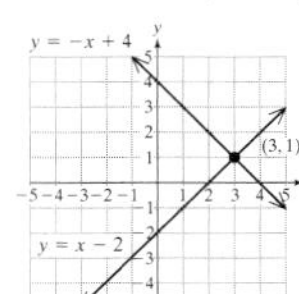

25. (1, −2)
Consistent; independent

27. (1, 4)
Consistent; independent

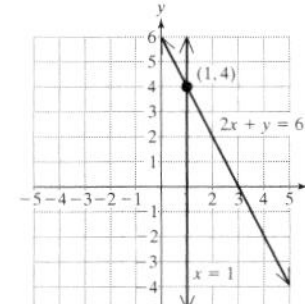

29. No solution
Inconsistent; independent

31. Infinitely many solutions
$\{(x, y) | y = 2x + 3\}$;
Consistent; dependent

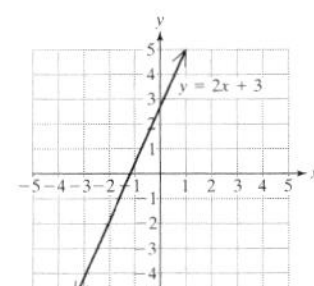

33. (−3, 6)
Consistent; independent

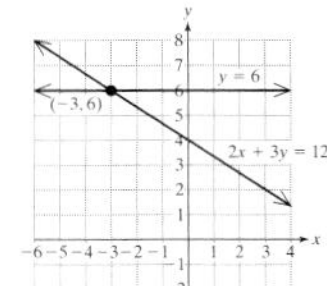

35. Infinitely many solutions
$\{(x, y) | y = \frac{5}{3}x - 3\}$;
Consistent; dependent

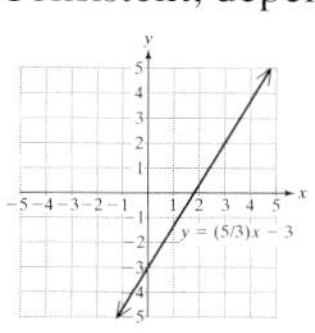

37. (2, −2)
Consistent; independent

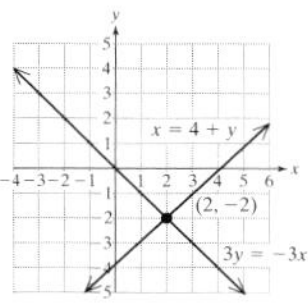

39. No solution
Inconsistent; independent

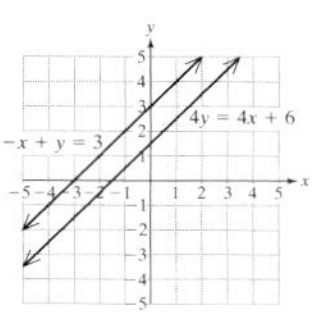

41. (4, 2)
Consistent; independent

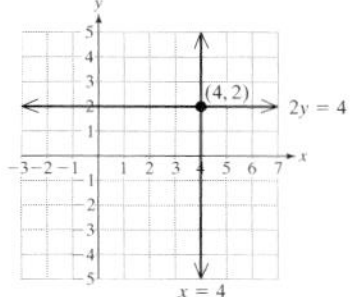

43. No solution
Inconsistent; independent

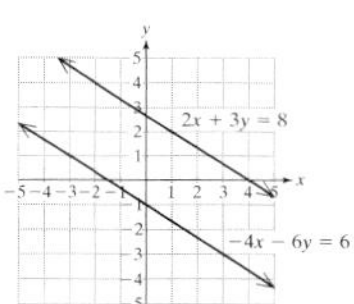

45. $\left(\frac{1}{2}, 3\right)$
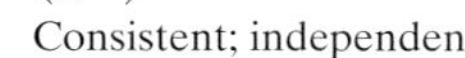
Consistent; independent

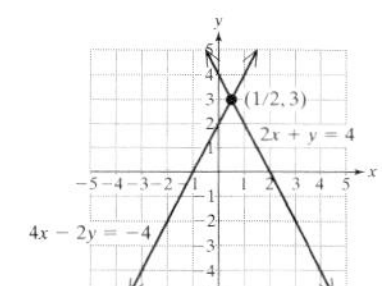

47. Infinitely many solutions
$\{(x, y) | y = 0.5x + 2\}$;
Consistent; dependent

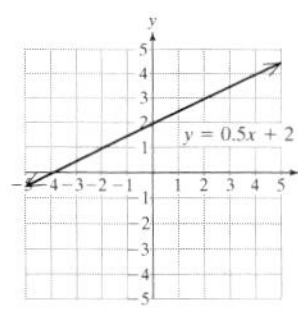

49. 4 lessons will cost \$120 for each instructor. **51.** 1997 **53.** For example, $4x + y = 9$, $-2x - y = -5$ **55.** For example, $2x + 2y = 1$

57. (2, 1)

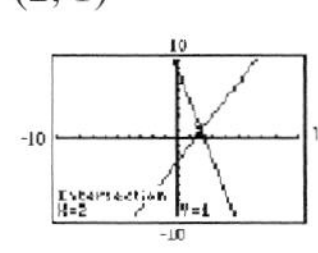

59. (3, 1)

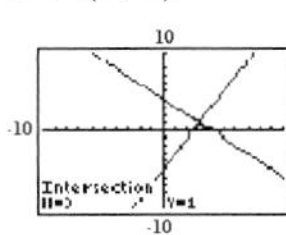

61. No solution

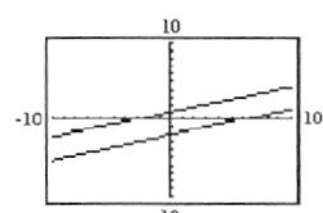

Section 8.2 Practice Exercises, pp. 532–535

1. $y = 2x - 4$ $y = 2x - 4$; coinciding lines **3.** $y = -\frac{2}{3}x + 2$
$y = x - 5$; intersecting lines **5.** $y = 4x - 4$
$y = 4x - 13$; parallel lines **7.** (3, −6) **9.** (0, 4) **11.** No solution
13. Infinitely many solutions $\{(x, y) | x = \frac{6}{5}y + 3\}$ **15. a.** y in the second equation is easiest to solve for because its coefficient is 1.
b. (1, 5) **17.** $\left(\frac{1}{2}, 3\right)$ **19.** (5, 2) **21.** (10, 5) **23.** No solution
25. Infinitely many solutions $\{(x, y) | y = \frac{1}{3}x + \frac{1}{3}\}$ **27.** (5, −7)
29. $\left(-5, \frac{3}{2}\right)$ **31.** (2, −5) **33.** No solution **35.** (0, 2) **37.** Infinitely many solutions $\{(x, y) | y = 0.25x + 1\}$ **39.** (1, 1) **41.** No solution
43. (−1, 5) **45.** (−6, −4) **47.** The numbers are 48 and 58.

49. The angles are 165° and 15°. **51.** The angles are 70° and 20°. **53.** The angles are 42° and 48°. **55.** A hot dog costs \$1.75 and a drink costs \$0.75. **57. a.** 22 months **b.** (22, 495); After renting a system for 22 months, both Company A and Company B will charge \$495. **c.** Company B is more expensive than Company A for more than 22 months. Company A is more expensive than Company B for less than 22 months. **59.** For example, (0, 3)(1, 5)(−1, 1)

Section 8.3 Practice Exercises, pp. 541–544

1. Yes **3.** No **5. a.** False, multiply by −2. **b.** True **7. a.** y would be easier. **b.** (2, 1) **9.** The system will have no solution. The lines are parallel. **11.** There are infinitely many solutions. The lines coincide. **13.** (2, 3) **15.** (5, −2) **17.** (0, 1) **19.** (1, −2) **21.** No solution **23.** Infinitely many solutions $\{(x, y) \mid y = -\frac{1}{2}x + 1\}$ **25.** (1, 4) **27.** (−1, −2) **29.** (2, 1) **31.** No solution **33.** (2, 3) **35.** $\left(\frac{7}{2}, \frac{5}{2}\right)$ **37.** $\left(\frac{1}{3}, 2\right)$ **39.** Infinitely many solutions $\{(x, y) \mid y = -5x + 1\}$ **41.** $\left(\frac{2}{3}, \frac{5}{3}\right)$ **43.** No solution **45.** (1, 4) **47.** (4, 0) **49.** Infinitely many solutions $\{(a, b) \mid b = \frac{9}{2}a - 4\}$ **51.** The numbers are 20 and 6. **53.** The numbers are 4 and 6. **55.** The cake has 340 Calories, and the ice cream has 120 Calories. **57.** $\left(\frac{21}{5}, -\frac{4}{5}\right)$ **59.** $\left(\frac{8}{9}, -\frac{4}{9}\right)$ **61.** (5, 0) **63.** (−2, −6) **65.** No solution **67.** $A = 2, B = -4$

Section 8.4 Practice Exercises, pp. 550–553

1. (−1, 4) **3.** $\left(\frac{5}{2}, 1\right)$ **5.** The angles are 80° and 10°. **7.** Dallas scored 30 points, and Buffalo scored 13 points. **9.** Tapes are \$10.50 each, and CDs are \$15.50 each. **11.** Technology stock costs \$16 per share, and the mutual fund costs \$11 per share. **13.** Shanelle invested \$3500 in the 10% account and \$6500 in the 7% account. **15.** Janise invested \$8000 in the 7% account and \$4000 in the 4% account. **17.** \$8000 in the 7.5% account; \$4000 in the 6% account **19.** 15 gallons of the 50% mixture should be mixed with 10 gal of the 40% mixture. **21.** 12 gal of the 45% disinfectant solution should be mixed with 8 gal of the 30% disinfectant solution. **23.** 4 oz of 18% moisturizer; 8 oz of 24% moisturizer **25.** The speed of the boat in still water is 6 mph, and the speed of the current is 2 mph. **27.** The speed of the plane in still air is 300 mph, and the wind is 20 mph. **29.** Juan 6 mph; Jeannie 4 mph **31.** There are 17 dimes and 22 nickels. **33. a.** 835 free-throws and 1597 field goals **b.** 4029 points **c.** Approximately 50 points per game **35.** The speed of the plane in still air is 160 mph, and the wind is 40 mph. **37.** 12 pounds of candy should be mixed with 8 lb of nuts. **39.** \$15,000 is invested in the 5.5% account, and \$45,000 is invested in the 6.5% account. **41.** 20 oz of Miracle-Gro should be mixed with 40 oz of Green Light. **43.** \$30 **45.** There were 300 women and 200 men in the survey. **47.** There are six 15-sec commercials and sixteen 30-sec commercials.

Section 8.5 Practice Exercises, pp. 560–563

1. (1, 1) **3.** $\left\{(x, y) \,\middle|\, x - \frac{2}{3}y = 2\right\}$ **5.** 65 mph, 58 mph **7.** (4, 0, 2) is a solution. **9.** None **11.** (1, 2, 3) **13.** (−2, −1, −3) **15.** (−1, 3, 4) **17.** (−6, 1, 7) **19.** $\left(\frac{1}{2}, \frac{2}{3}, -\frac{5}{6}\right)$ **21.** 67°, 82°, 31° **23.** 9 cm, 18 cm, 27 cm **25.** 148 adult tickets, 51 children's tickets, 23 senior tickets **27.** 24 oz peanuts, 8 oz pecans, 16 oz cashews **29.** Vanderbilt: 6100; Baylor: 12,200; Pace: 8900 **31.** (−9, 5, 5) **33.** No solution; inconsistent **35.** (1, 3, 1) **37.** Dependent **39.** (1, 0, 1) **41.** (0, 0, 0) **43.** Dependent **45.** $y = x^2 - 5x + 3$ **47.** $y = -2x^2 + 2x - 5$

Section 8.6 Practice Exercises, pp. 570–572

1. (1, 1) **3.** (6, 1, −1) **5.** An augmented matrix is one constructed from the coefficients of the variable terms and the constants. **7.** The order is the number of rows by the number of columns. **9.** 4 × 1, column matrix **11.** 3 × 3, square matrix **13.** 1 × 2, row matrix **15.** 2 × 4, none of these **17.** $\left[\begin{array}{cc|c} 1 & -2 & -1 \\ 2 & 1 & -7 \end{array}\right]$ **19.** $\left[\begin{array}{cc|c} -9 & 13 & -5 \\ 7 & 5 & 19 \end{array}\right]$ **21.** $\left[\begin{array}{ccc|c} 1 & 1 & 1 & 6 \\ 1 & -1 & 1 & 2 \\ 1 & 1 & -1 & 0 \end{array}\right]$ **23.** $\left[\begin{array}{ccc|c} 1 & -2 & 1 & 5 \\ 2 & 6 & 3 & -2 \\ 3 & -1 & -2 & 1 \end{array}\right]$ **25. a.** 7 **b.** −2 **27.** $\left[\begin{array}{cc|c} 1 & \frac{1}{2} & \frac{11}{2} \\ 2 & -1 & 1 \end{array}\right]$ **29.** $\left[\begin{array}{cc|c} 1 & -4 & 3 \\ 5 & 2 & 1 \end{array}\right]$ **31.** $\left[\begin{array}{cc|c} 1 & 5 & 2 \\ 0 & 11 & 5 \end{array}\right]$ **33.** False **35.** False **37.** $\begin{array}{l} x = -1 \\ y = -7 \end{array}$ **39.** $\begin{array}{l} x = 8 \\ y = 0 \\ z = -1 \end{array}$ **41.** Interchange rows 1 and 2. **43.** Multiply row 1 by −3 and add to row 2. Replace row 2 with the result. **45.** (−3, −1) **47.** (−21, 9) **49.** Dependent **51.** (3, −1) **53.** (−10, 3) **55.** No solution **57.** (1, 2, 3) **59.** (1, −2, 0) **61.** (1, 2, 1) **63.** No solution **65.** $\left[\begin{array}{cc|c} 1 & 0 & -3 \\ 0 & 1 & -1 \end{array}\right]$ **67.** $\left[\begin{array}{cc|c} 1 & 3 & 3 \\ 0 & 0 & 0 \end{array}\right]$ Dependent **69.** $\left[\begin{array}{ccc|c} 1 & 0 & 0 & 1 \\ 0 & 1 & 0 & -2 \\ 0 & 0 & 1 & 0 \end{array}\right]$

Section 8.7 Practice Exercises, pp. 581–583

1. −11 **3.** 16 **5.** 8 **7.** 18 **9.** −43 **11.** $\begin{bmatrix} + & - & + \\ - & + & - \\ + & - & + \end{bmatrix}$ **13. a.** 30 **b.** 30 **15.** Choosing the row or column with the most zero elements simplifies the arithmetic when evaluating a determinant. **17.** 12 **19.** −15 **21.** 0 **23.** $8a - 2b$ **25.** $4x - 3y + 6z$ **27.** 0 **29.** $\mathbf{D} = -55$; $\mathbf{D}_x = 54$; $\mathbf{D}_y = 89$ **31.** (1, 3) **33.** (1, −1) **35.** $\left(\frac{11}{14}, \frac{17}{21}\right)$ **37.** Elimination method, substitution method, or the Gauss-Jordan method will determine if a system is inconsistent or dependent. **39.** No solution (inconsistent) **41.** (0, 0) **43.** $\{(x, y) \mid x + 5y = 2\}$ (dependent) **45.** $y = -2$ **47.** $x = \frac{1}{4}$ **49.** $y = -\frac{17}{167}$ **51.** (1, 2, 3) **53.** Cramer's rule does not apply. **55.** $y = 2$ **57.** $t = -12$ **59.** 32 **61. a.** 8 **b.** $y = 4$

Chapter 8 Review Exercises, pp. 591–595

1. Yes **3.** No **5.** Intersecting lines **7.** Parallel lines
9. Coinciding lines

11. $(0, -2)$
Consistent; independent

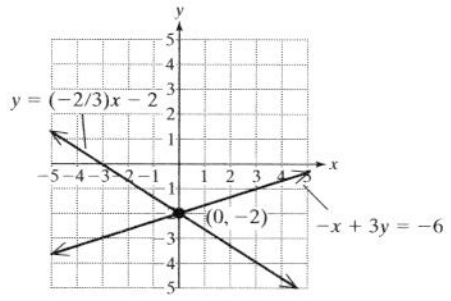

13. Infinitely many solutions
$\{(x, y) | y = -2x + 5\}$;
Consistent; dependent

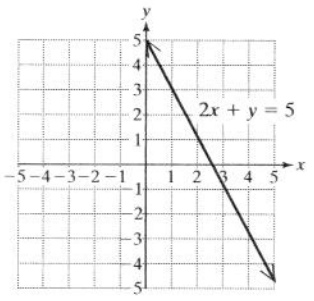

15. $(1, -1)$
Consistent; independent

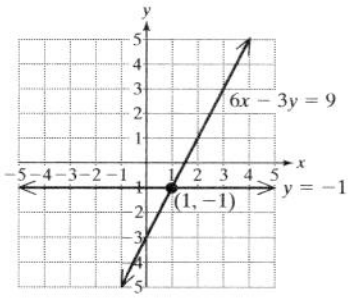

17. No solution
Inconsistent; independent

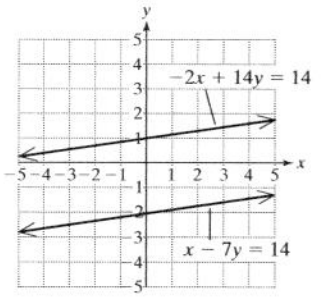

19. 1994 **21.** $\left(\frac{2}{3}, -2\right)$ **23.** No solution **25. a.** x in the first equation is easiest to solve for because its coefficient is 1. **b.** $\left(6, \frac{5}{2}\right)$
27. $(5, -4)$ **29.** Infinitely many solutions $\{(x, y) | y = \frac{1}{3}x - 3\}$
31. The numbers are 50 and 8. **33.** He scored 155 receiving touchdowns and 10 touchdowns rushing.
35.
1. Write both equations in standard form.
2. Multiply one or both equations by a constant to create opposite coefficients for one of the variables.
3. Add the equations to eliminate the variable.
4. Solve for the remaining variable.
5. Substitute the known variable into an original equation to solve for the other variable.

37. a. Answers may vary. **b.** $(2, 2)$ **39.** $(-6, 2)$ **41.** $\left(\frac{1}{4}, -\frac{2}{5}\right)$
43. No solution **45.** $(1, 0)$ **47. a.** Answers may vary. **b.** $(-2, -1)$
49. Emillo invested \$2500 in the 5% account and \$17,500 in the 8% account. **51.** The speed of the boat is 18 mph, and that of the stream is 2 mph. **53.** 3000 women and 2700 men **55.** No solution
57. 5 ft, 12 ft, 13 ft **59.** 3×3 **61.** 1×4 **63.** $\left[\begin{array}{cc|c} 1 & 1 & 3 \\ 1 & -1 & -1 \end{array}\right]$
65. $x = 9$, $y = -3$ **67. a.** 1 **b.** $\left[\begin{array}{cc|c} 1 & 3 & -11 \\ 2 & 0 & 5 \end{array}\right]$ **69.** $(1, 2)$ **71.** -11
73. 1 **75.** 18 **77.** -30 **79.** 7 **81.** -40
83. $\mathbf{D} = -11$ $\quad x = \frac{29}{11}$
$\mathbf{D}_x = -29$
$\mathbf{D}_y = -6$ $\quad y = \frac{6}{11}$
85. $(3, 2)$ **87.** $(1, -2, 3)$ **89.** $\{(x, y) | y = 2x - 1\}$

Chapter 8 Test, pp. 595–597

1. Parallel lines **2. a.** $(-3, 3)$

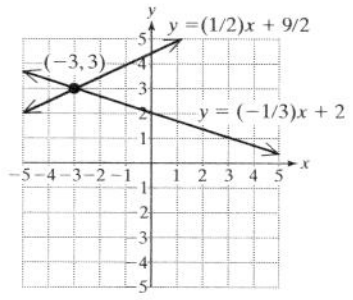

3. a. \$15 **b.** 5,000,000 items **4.** $(-2, 0)$ **5.** Cooper had 680 points, and Swoopes had 453 points. **6.** $\left(2, -\frac{1}{3}\right)$ **7.** 12 milliliters of the 50% acid solution should be mixed with 24 mL of the 20% solution. **8. a.** No solution **b.** Infinitely many solutions **c.** One solution **9.** $(-5, 4)$ **10.** No solution **11.** Infinitely many solutions $\{(x, y) | y = -5x - 4\}$ **12.** $(1, -2)$ **13.** The angles are 57° and 33°. **14. a.** 13 dimes, 7 quarters, 10 nickels **b.** \$3.55
15. \$7000 was invested at 9%, \$12,000 was invested at 11%, totaling \$19,000. **16. a.** 15 months **b.** For 15 months of service the cost is \$510 for either cable or a satellite dish. **17.** The plane travels 470 mph in still air, and the wind is 30 mph. **18.** $(16, -37, 9)$
19. Joanne: 142 orders; Kent: 162 orders; Geoff: 200 orders **20.** For example: $\begin{bmatrix} 2 & 1 \\ 0 & -4 \\ 2.6 & 7 \end{bmatrix}$ **21. a.** $\left[\begin{array}{ccc|c} 1 & 2 & 1 & -3 \\ 0 & -8 & -3 & 10 \\ -5 & -6 & 3 & 0 \end{array}\right]$
b. $\left[\begin{array}{ccc|c} 1 & 2 & 1 & -3 \\ 0 & -8 & -3 & 10 \\ 0 & 4 & 8 & -15 \end{array}\right]$ **22.** $(6, -1)$ **23.** 7 **24.** -6 **25.** 20
26. No. The matrix must be square. **27.** $y = -\frac{40}{11}$ **28.** $y = 8$
29. $\{(x, y) | y = 3x\}$

Cumulative Review Exercises, Chapters 1–8, pp. 597–598

1. $\frac{11}{6}$ **2.** Natural numbers $\sqrt{9}$; Whole numbers $\sqrt{9}$, 0; Integers $\sqrt{9}$, 0, -4; Rational numbers $\sqrt{9}$, 0, -4, 6.7, $\frac{1}{3}$; Irrational numbers π, $\sqrt{2}$ **3.** $x = -\frac{21}{2}$ **4.** No solution **5.** $y = \frac{3}{2}x - 3$
6. $\left[\frac{3}{11}, \infty\right)$

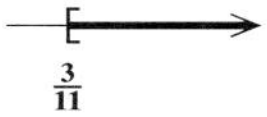

7. 180° **8.** The angles are 37°, 33°, 110°. **9.** The rates of the hikers are 2 mph and 4 mph.
10. a. $\frac{10d^5}{c^2}$ **b.** x^{18} **11.** 1.8×10^{13} **12.** $3y^3 - 4y^2 - 8y + 8$
13. $-w^2 - \frac{7}{10}w + \frac{5}{2}$ **14.** $\frac{2p}{q} - 1 + \frac{4q}{p}$ **15.** $5(x - 5)(x + 5)$
16. $(a - 2b)(5x - 2y)$ **17.** $y = 0, y = 3, y = -\frac{1}{2}$ **18.** $t = 5$
19. a. Nonlinear **b.** x-intercepts: $(4, 0)(-3, 0)$; y-intercept: $(0, -12)$
20. $\frac{5(x + 3)}{2x + 3}$ **21.** $\frac{2(a + 5)(a + 1)}{(a + 3)(a - 3)}$ **22.** $a = -5, a = -1$
23. In Problem 21 you must change the fractions to equivalent fractions with a common denominator. In Problem 22 you must clear the denominators. **24.** It would take them $1\frac{5}{7}$ hr. **25. a.** $-\frac{2}{3}$ **b.** $\frac{3}{2}$
26. $y = -3x + 3$

27. a. b.

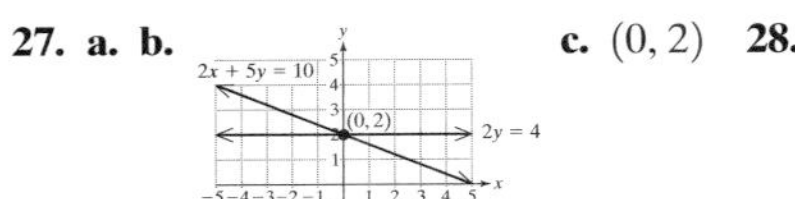

c. (0, 2) **28.** (0, 2)

29.

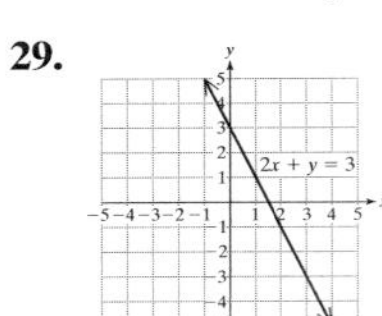

30. 20 gal of the 15% solution should be mixed with 40 gal of the 60% solution. **31.** x is 27°; y is 63° **32. a.** 1.4 **b.** Between 1920 and 1990 the winning speed in the Indianapolis 500 increased on average by 1.4 mph per year. **33.** $(2, 0, -1)$ **34.** $(1, 1)$

CHAPTER 9

Section 9.1 Practice Exercises, pp. 605–608

1. $\left(-\frac{1}{2}, \infty\right)$ **3.** $[-16, \infty)$ **5.** $(66, \infty)$

7. a. $\{10, 20, 30\}$ **b.** $\{5, 10, 15, 20, 25, 30, 35, 40, 50\}$

9. a. $\{a, e, j, k, l, m, n, o, p, q, t, y\}$ **b.** $\{j, o\}$

11. a. $\{1, 2, 3, 4, 5, 6, 7, 8, 9, 10\}$ **b.** { }, the empty set

13. a. $x > 2$ **b.** $x < 4$ **c.** $2 < x < 4$ **15. a.** $x \geq -\frac{1}{3}$

b. $x \leq 4$ **c.** $-\frac{1}{3} \leq x \leq 4$

17. a. $x < 2$ **b.** $x < 5$ **c.** $x < 2$ **19.** $[-2, 3]$

21. $(3, 6)$ **23.** $\left[\frac{5}{2}, 7\right)$

25. $(-\infty, 2]$ **27.** $(-6, 6]$

29. $(2, 8)$ **31.** $[8, \infty)$

33. a. $x < -1$ **b.** $x < -5$ **c.** $x < -1$ **35. a.** $x \leq 3$

b. $x \geq 8$ **c.** $x \leq 3$ or $x \geq 8$

37. a. $x \geq 2$ **b.** $x \leq -1$ **c.** $x \geq 2$ or $x \leq -1$ **39.** $(-\infty, -4) \cup (-2, \infty)$

41. $(-\infty, -2) \cup [2, \infty)$ **43.** $(-\infty, -7) \cup [3, \infty)$

45. $[-6, \infty)$ **47.** $\left(-\infty, -\frac{4}{3}\right) \cup (2, \infty)$

49. $(-\infty, -4] \cup (4.5, \infty)$ **51. a.** $-10 < x < 8$ **b.** All real numbers

53. a. $x \geq 1.3$ or $x \leq -8$ **b.** No solution

55. a. $4800 \leq x \leq 10{,}800$ **b.** $x < 4800$ or $x > 10{,}800$

57. a. $13 \leq x \leq 16$ **b.** $x < 13$ or $x > 16$

59. All real numbers between $-\frac{3}{2}$ and 6

61. All real numbers greater than 2 or less than -1

Section 9.2 Practice Exercises, pp. 617–619

1. $\left(-\infty, \frac{3}{8}\right) \cup (3, \infty)$ **3.** $(-3, 0)$ **5.** $(5, \infty)$ **7.** All real numbers

9. a. $(-2, 0) \cup (3, \infty)$ **b.** $(-\infty, -2) \cup (0, 3)$ **c.** $(-\infty, -2] \cup [0, 3]$ **d.** $[-2, 0] \cup [3, \infty)$ **11. a.** $(-1, 1]$ **b.** $(-\infty, -1) \cup [1, \infty)$ **c.** $(-\infty, -1) \cup (1, \infty)$ **d.** $(-1, 1)$ **13. a.** $b = \frac{17}{6}$ **b.** $\left(-\infty, \frac{17}{6}\right)$ **c.** $\left(\frac{17}{6}, \infty\right)$ **15. a.** $y = 18$ **b.** $[18, \infty)$ **c.** $(-\infty, 18]$

17. a. $w = \frac{2}{3}, w = -5$ **b.** $\left(-5, \frac{2}{3}\right)$ **c.** $(-\infty, -5) \cup \left(\frac{2}{3}, \infty\right)$

19. a. $q = 5, q = -1$ **b.** $[-1, 5]$ **c.** $(-\infty, -1] \cup [5, \infty)$

21. $(-1, 7)$ **23.** $\left(-\infty, \frac{3}{5}\right) \cup (8, \infty)$ **25.** $[4, 8]$ **27.** $(-11, 11)$

29. $\left(-\infty, -\frac{1}{3}\right] \cup [3, \infty)$ **31.** $\left(-\frac{1}{3}, 0\right) \cup (4, \infty)$

33. $(-\infty, -3] \cup [0, 4]$ **35.** $(-2, -1) \cup (2, \infty)$ **37. a.** $x = 7$ **b.** $(-\infty, 5) \cup (7, \infty)$ **c.** $(5, 7)$ **39. a.** $z = 4$ **b.** $[4, 6)$ **c.** $(-\infty, 4] \cup (6, \infty)$ **41.** $(1, \infty)$ **43.** $(-1, 3)$ **45.** $\left(2, \frac{7}{2}\right)$ **47.** $(5, 7]$

49. $(-\infty, 0) \cup \left[\frac{1}{2}, \infty\right)$ **51.** $(0, \infty)$ **53.** All real numbers

55. No solution **57.** No solution **59.** $\{0\}$

61. $(-\infty, -2) \cup (-2, \infty)$ **63.** $\{-2\}$

65. $(-\infty, 0) \cup (2, \infty)$

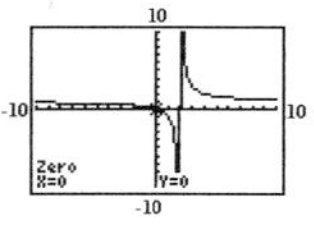

67. $(-1, 1)$

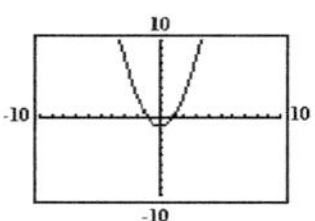

69. $\{-5\}$

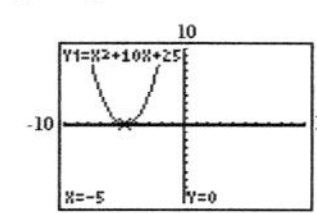

71. No solution

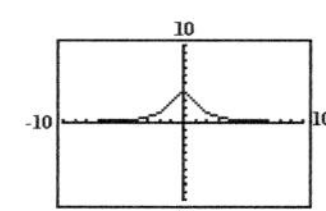

Section 9.3 Practice Exercises, pp. 625–626

1. $\left(\frac{2}{3}, \frac{23}{2}\right)$ **3.** $\left(4, \frac{16}{3}\right]$ **5.** All real numbers between -7 and 7 **7.** $(-\infty, -4) \cup \left(\frac{1}{2}, 2\right)$ **9.** $p = 7, p = -7$ **11.** $x = 6, x = -6$ **13.** $y = \sqrt{2}, y = -\sqrt{2}$ **15.** No solution **17.** $q = 0$ **19.** $x = \frac{1}{3}, x = 0$ **21.** $x = \frac{5}{2}, x = -\frac{3}{2}$ **23.** $z = \frac{10}{7}, z = -\frac{8}{7}$ **25.** $y = 2, y = -\frac{14}{5}$ **27.** No solution **29.** $w = -4, w = \frac{1}{3}$ **31.** $y = \frac{1}{2}$ **33.** $w = -12, w = 28$ **35.** $y = \frac{5}{2}, y = -\frac{7}{2}$ **37.** $b = \frac{7}{3}$ **39.** $w = 2, w = -\frac{1}{3}$ **41.** No solution **43.** $x = \frac{3}{2}$ **45.** $h = \frac{1}{4}$ **47.** $k = -4, k = \frac{16}{5}$ **49.** $m = -1.44, m = -0.4$ **51.** $|x| = 6$ **53.** $|x| = \frac{4}{3}$ **55.** $x = 2, x = -\frac{1}{2}$

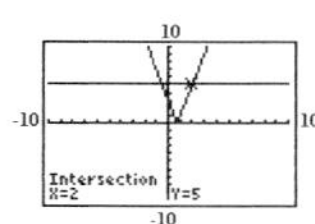
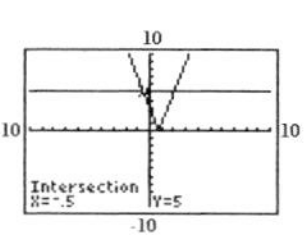

57. No solution **59.** $x = \frac{1}{2}$

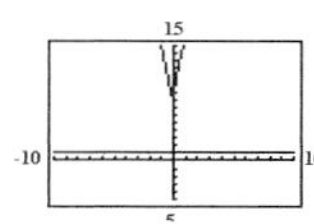
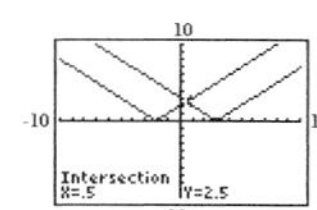

61. $x = \frac{4}{3}, x = -2$

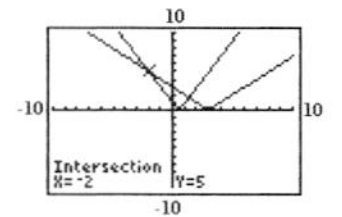
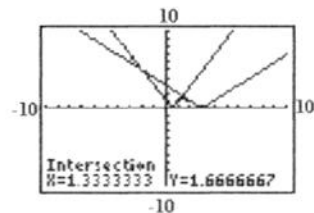

Section 9.4 Practice Exercises, pp. 634–635

1. No solution **3.** $x = -\frac{5}{3}, x = -\frac{1}{3}$ **5.** $(-3, -1]$ **7.** All real numbers **9. a.** $x = -5, x = 5$ **b.** $(-\infty, -5) \cup (5, \infty)$ **c.** $(-5, 5)$ **11. a.** No solution **b.** $(-\infty, \infty)$ **c.** No solution **13. a.** $x = 10, x = -4$ **b.** $(-\infty, -4) \cup (10, \infty)$ **c.** $(-4, 10)$ **15. a.** No solution **b.** $(-\infty, \infty)$ **c.** No solution **17.** $(-\infty, -6) \cup (6, \infty)$ **19.** $[-3, 3]$ **21.** $(-\infty, \infty)$ **23.** $(-\infty, -2] \cup [3, \infty)$ **25.** No solution **27.** $[-10, 14]$ **29.** $\left(-\infty, -\frac{5}{4}\right] \cup \left[\frac{23}{4}, \infty\right)$ **31.** $\left(-\frac{21}{2}, \frac{19}{2}\right)$ **33.** $(-\infty, \infty)$ **35.** No solution **37.** $\{-5\}$ **39.** $(-\infty, \infty)$ **41.** $|x| > 7$ **43.** $|x - 2| \le 13$ **45.** $|x - 32| \le 0.05$ **47.** $|x - 6\frac{3}{4}| \le \frac{1}{8}$ **49.** b **51.** a **53.** $(-\infty, -6) \cup (2, \infty)$

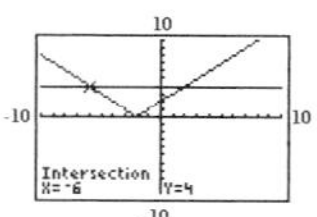
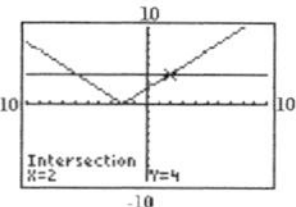

55. $(-7, 5)$

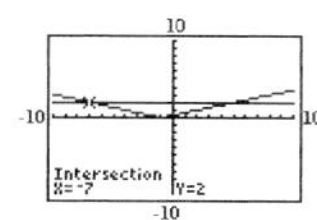
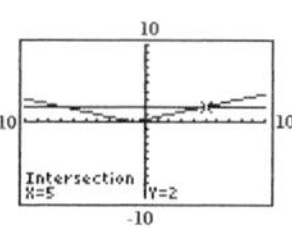

57. No solution **59.** $(-\infty, \infty)$ **61.** $x = -\frac{1}{6}$

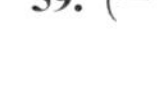
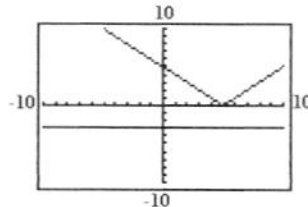
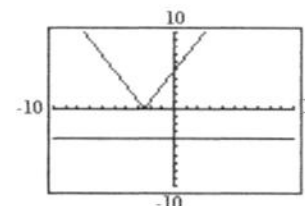
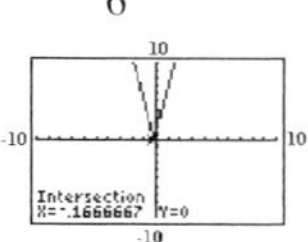

Section 9.5 Practice Exercises, pp. 643–645

1. Solve the inequality: $x + 3 < -4$ or $x + 3 > 4$ or use the test point method. **3.** $(1, 4)$ **5.** $\left[-\frac{5}{6}, \frac{7}{6}\right]$ **7.** $(-\infty, -2] \cup [1, \infty)$ **9. a.** Yes **b.** No **c.** Yes **d.** No **11. a.** Yes **b.** Yes **c.** No **d.** No **13.** $>$ **15.** $\ge$ **17.** $\le, \ge$

19.

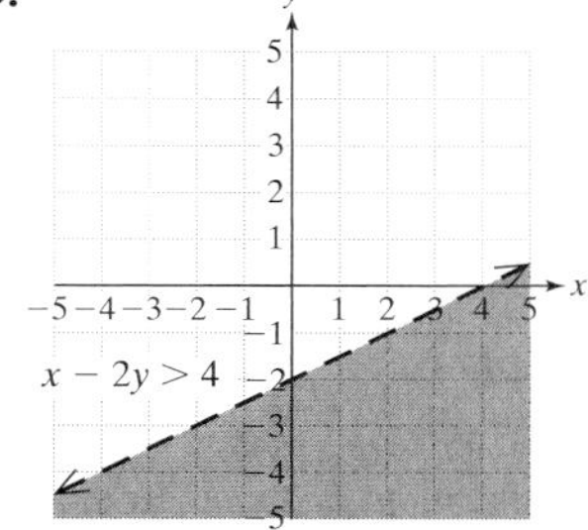

21.

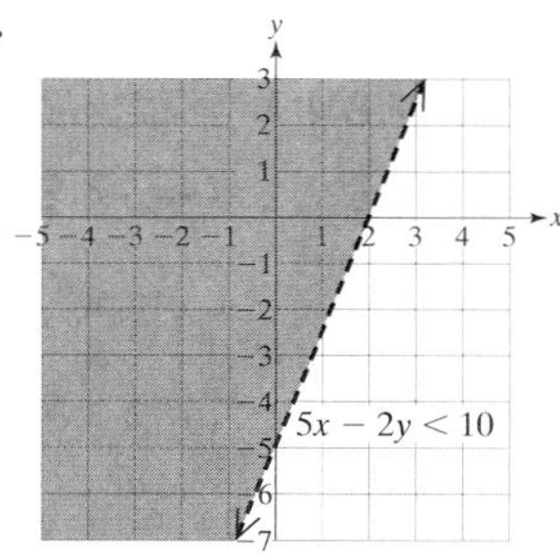

23.

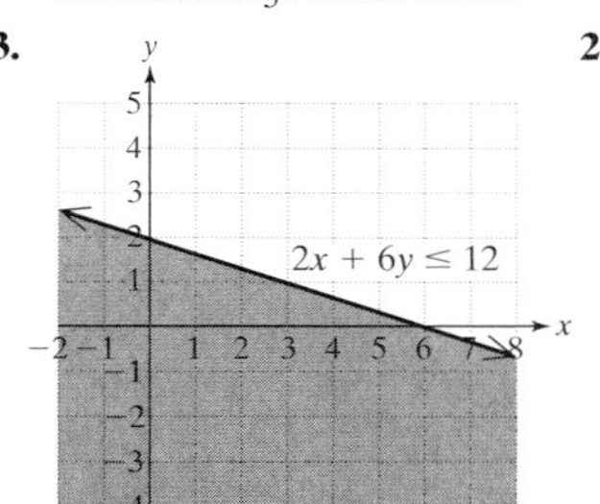

25.

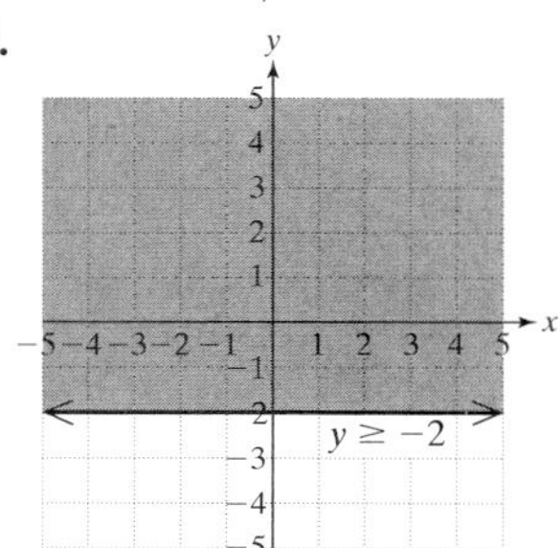

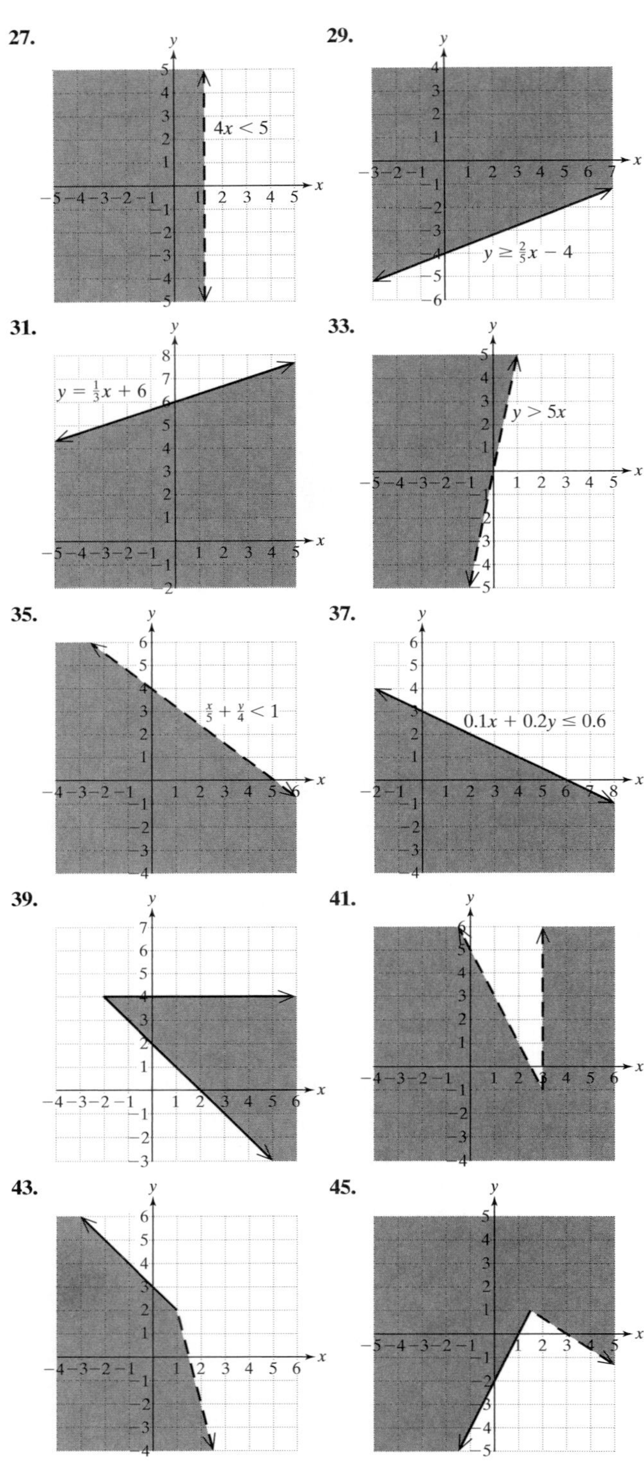

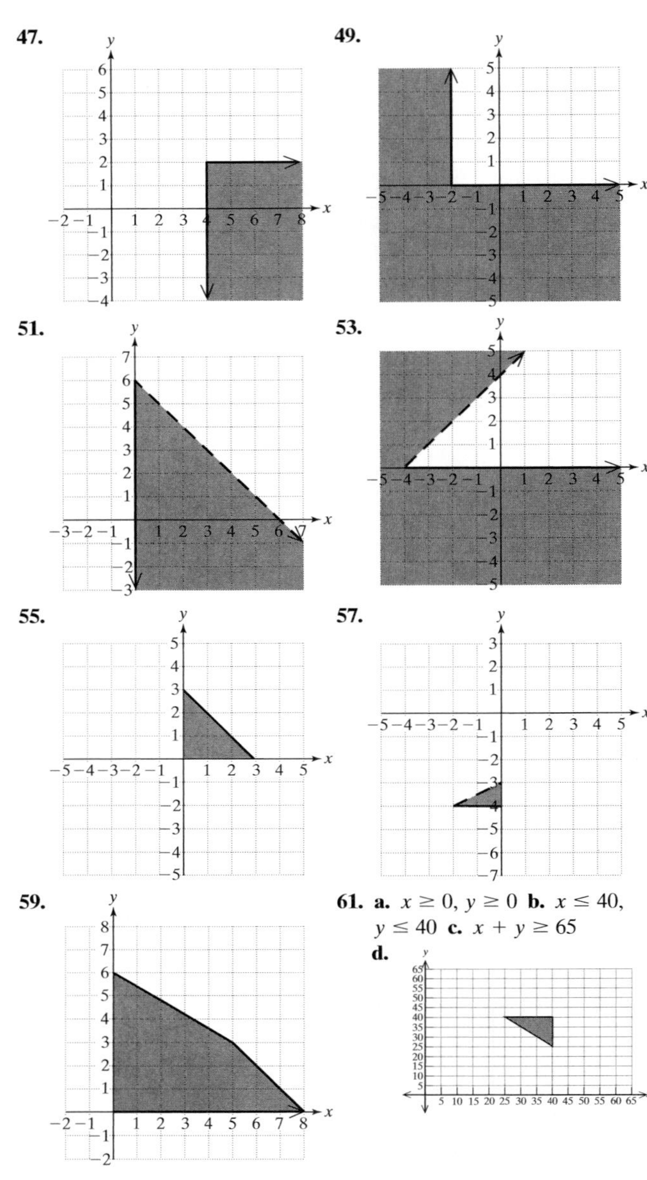

61. a. $x \geq 0, y \geq 0$ **b.** $x \leq 40$, $y \leq 40$ **c.** $x + y \geq 65$
d.

Chapter 9 Review Exercises, pp. 649–652

1. $\left(-\dfrac{11}{4}, 4\right]$ **3.** No solution **5.** $(-\infty, 6] \cup (12, \infty)$ **7.** $\left(-\infty, \dfrac{1}{2}\right)$

9. $\left[0, \dfrac{4}{3}\right]$ **11.** All real numbers between -6 and 12

13. a. $125 \leq x \leq 200$ **b.** $x < 125$ or $x > 200$
15. a. The solution is the intersection of the two inequalities. Answer: $-2 \leq x \leq 5$. **b.** The solution is the union of the two inequalities. Answer: all real numbers. **17. a.** $x = -2, x = 2$; $(-2, 0)(2, 0)$ are the x-intercepts. **b.** $-2 < x < 2$; On the interval $(-2, 2)$ the graph is below the x-axis. **c.** $x < -2$ or $x > 2$; On the interval $(-\infty, -2)$ and $(2, \infty)$ the graph is above the x-axis. **19.** $(-2, 6)$ **21.** $(-2, 0]$
23. $(-2, 0) \cup (5, \infty)$ **25.** $\{-2\}$ **27.** $(3, \infty)$ **29.** $\{-5\}$

31. $x = 10, x = -10$ **33.** $y = -\frac{11}{2}, y = -\frac{13}{2}$ **35.** $x = 1.3, x = 7.4$

37. $x = 5, x = -9$ **39.** No solution **41.** $x = \frac{3}{7}$ **43.** $x = 6, x = \frac{4}{5}$

45. Both expressions give the distance between 3 and −2.

47. $|x| < 4$ **49.** $(-\infty, -14] \cup [2, \infty)$ −14 2

51. $\left(-\infty, \frac{1}{7}\right) \cup \left(\frac{1}{7}, \infty\right)$ $\frac{1}{7}$

53. $\left[-2, -\frac{2}{3}\right]$ −2 $-\frac{2}{3}$ **55.** $(2, 22)$ 2 22

57. $(-\infty, -1) \cup (5, \infty)$ −1 5

59. $(-\infty, \infty)$ **61.** No solution

63. If an absolute value is less than a negative number there will be no solution.

65.

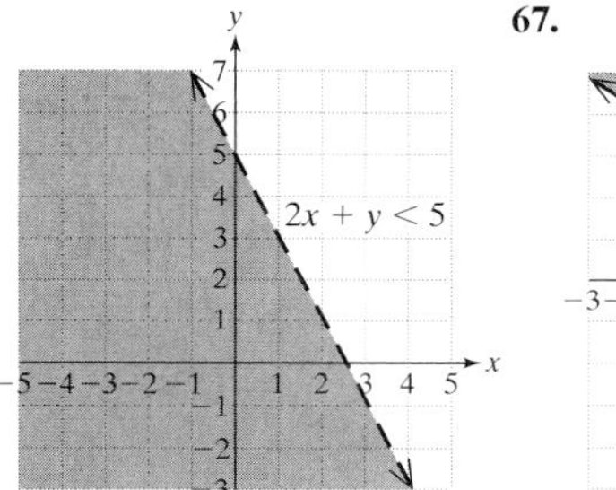

67.

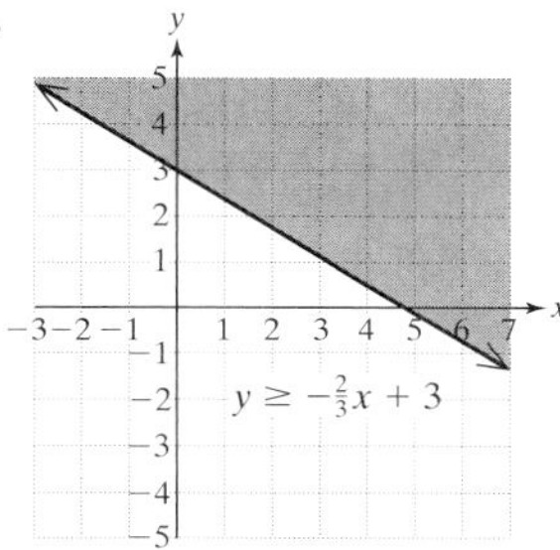

69.

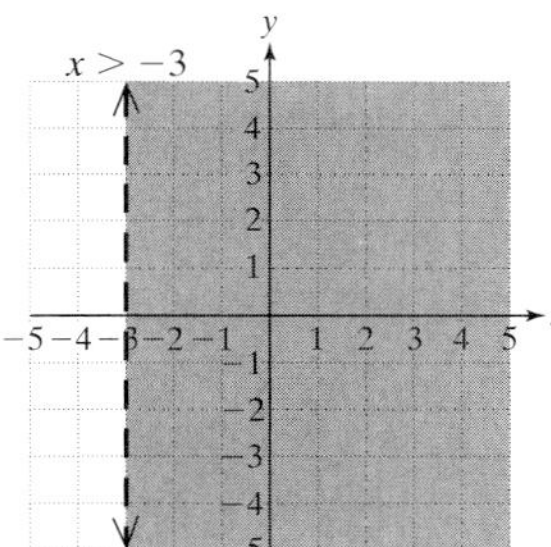

71.

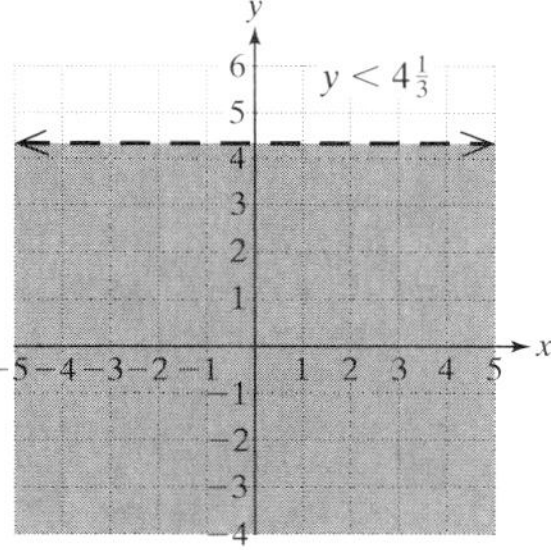

73.

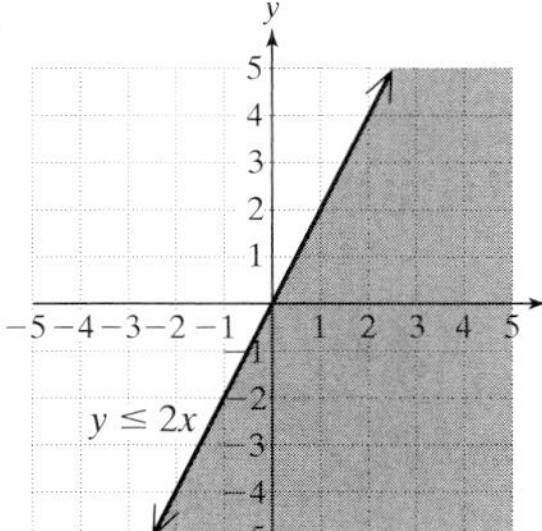

75.

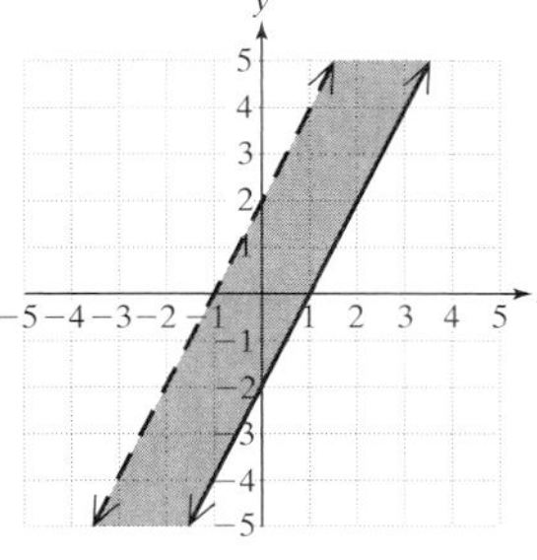

77.

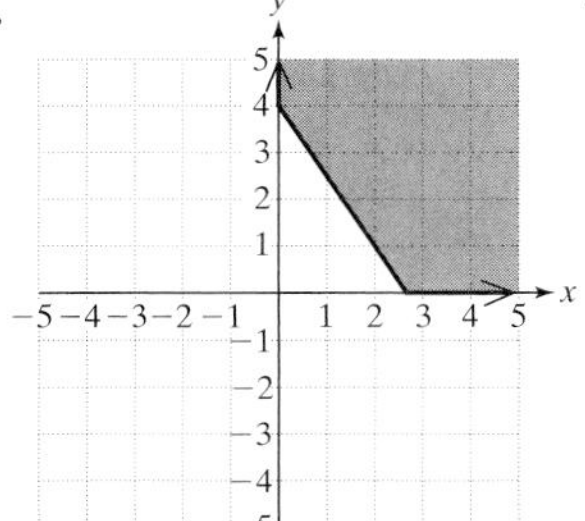

79.

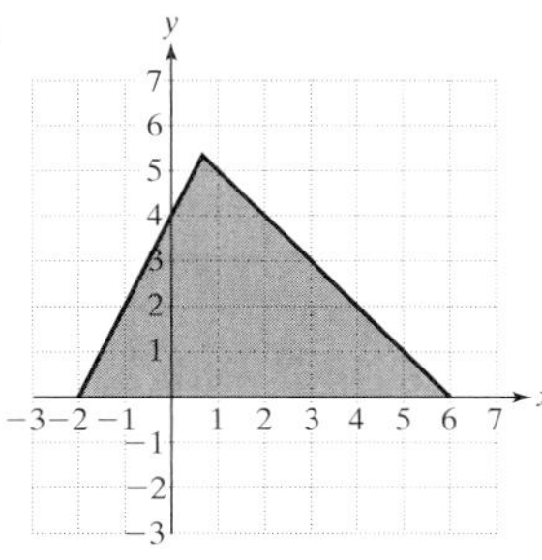

Chapter 9 Test, pp. 653–654

1. a. $\left[-\frac{1}{3}, 2\right]$ **b.** $(-\infty, -24] \cup [-15, \infty)$ **c.** $[-3, 0)$ **2. a.** $(-\infty, \infty)$ **b.** no solution **3. a.** $9 \le x \le 33$ **b.** $x < 9$ or $x > 33$ **c.** $x > 40$ **4.** $\left[\frac{1}{2}, 6\right)$ **5.** $(-5, 5)$ **6.** $(-\infty, -3) \cup (-2, 2)$ **7.** $\left(-3, -\frac{3}{2}\right)$ **8.** No solution **9.** $\{-11\}$ **10. a.** $x = 10, x = -22$ **b.** $x = -8, x = 2$ **11. a.** $x = 7, x = -1$; $(7, 0)(-1, 0)$ are the x-intercepts. **b.** $-1 < x < 7$; On the interval $(-1, 7)$ the graph below the x-axis. **c.** $x < -1$ or $x > 7$; On the intervals $(-\infty, -1)$ and $(7, \infty)$ the graph is above the x-axis. **12.** No solution **13.** $\left(-\infty, -\frac{1}{3}\right) \cup \left(\frac{17}{3}, \infty\right)$ **14.** $(-18.75, 17.25)$ **15.** $(-\infty, \infty)$ **16.** $|x - 15.41| \le 0.01$

17.

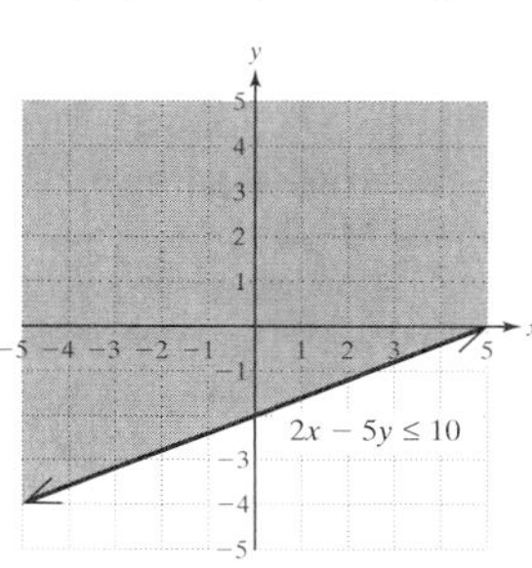

18.

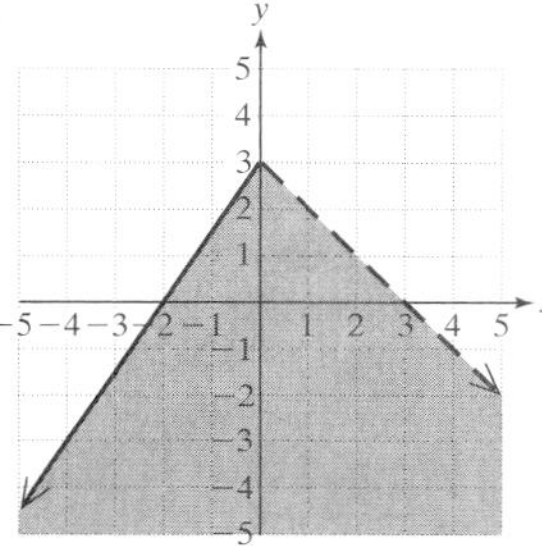

19. a. $x \ge 0, y \ge 0$ **b.** $300x + 400y \ge 1000$
c.

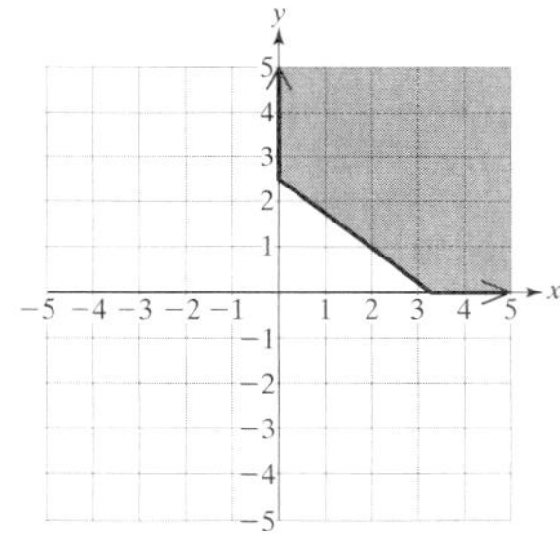

Cumulative Review Exercises, Chapters 1–9, pp. 654–655

1. $x^2 - x - 13$ **2.** $m = -\frac{3}{2}, m = 1$ **3. a.** $p = 0, p = 6$ **b.** $(0, 6)$ **c.** $(-\infty, 0) \cup (6, \infty)$ **4. a.** $y = 14, y = -10$ **b.** $(-10, 14)$ **c.** $(-\infty, -10) \cup (14, \infty)$

5.

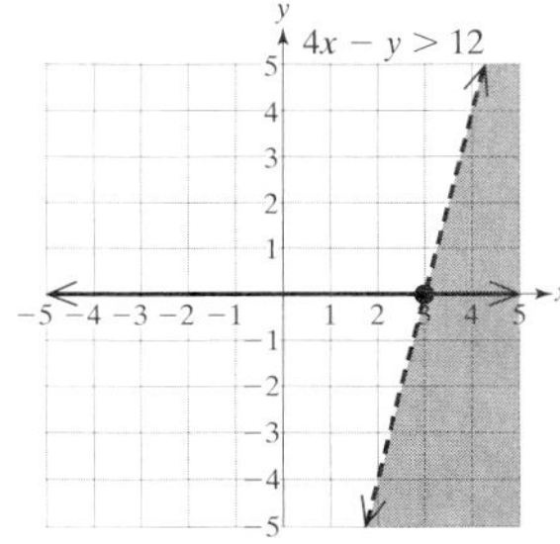

6. a. $t(1) = 9$. After one trial the rat requires 9 min to complete the maze. $t(50) = 3.24$. After 50 trials the rat requires 3.24 min to complete the maze. $t(500) = 3.02$. After 500 trials the rat requires 3.02 min to complete the maze. **b.** The limiting time is 3 min **c.** Over 5 trials **7. a.** $\left(-\infty, -\frac{5}{2}\right] \cup [2, \infty)$ **b.** On these intervals, the graph is on or above the x-axis (greater than or equal to 0).

8.

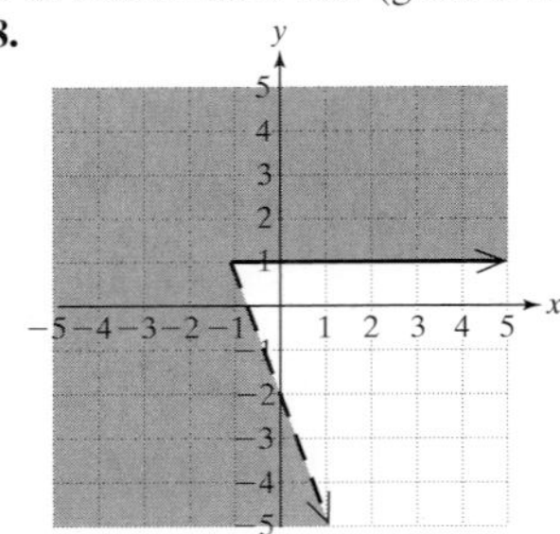

9. $-7x + 13$ **10.** approximately $\$1.5 \times 10^6$ per restaurant
11. a. Quotient: $2x^2 - 5x + 12$; remainder: $-24x + 5$
12. a. $b_1 = \frac{2A - hb_2}{h}$ or $b_1 = \frac{2A}{h} - b_2$ **b.** 10 cm **13.** 75 mph
14. 57°, 123° **15.** \$8000 in the 6.5% account; \$5000 in the 5% account **16.** Neither
17. a. x-intercept $(1, 0)$; no y-intercept; undefined slope **b.** x-intercept $(8, 0)$; y-intercept $(0, 4)$; slope $-\frac{1}{2}$

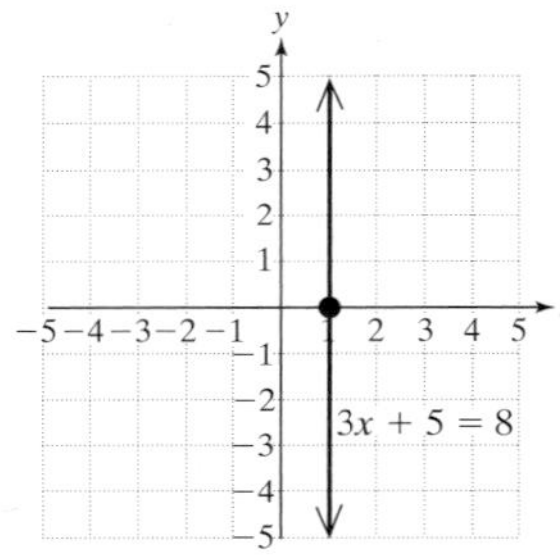

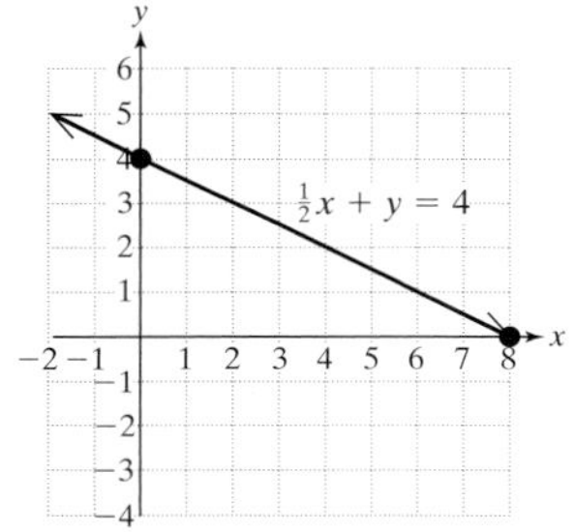

18. $y = -\frac{2}{3}x - \frac{13}{3}$ **19.** $(-1, 3, -2)$ **20. a.** 4×3 **b.** 3×3
21. Plane: 500 mph Wind: 20 mph **22. a.** Quadratic
b. $P(0) = -2600$; The company will lose \$2600 if no desks are produced. **c.** $P(x) = 0$ for $x = 20$ and $x = 650$; The company will break even (\$0 profit) when 20 desks or 650 desks are produced.
23. $\{x|x \le 50\}$ **24.** $-\frac{xy}{x + y}$ **25.** $-\frac{a + 1}{a}$
26. $\frac{2x + 5}{(x - 5)(x - 2)(x + 10)}$

CHAPTER 10

Section 10.1 Practice Exercises, pp. 665–668

1. a. 8, −8 **b.** 8 **c.** There are two square roots for every positive number. $\sqrt{64}$ identifies the positive square root. **3. a.** 9 **b.** −9
5. There is no real number b such that $b^2 = -36$. **7.** 5 **9.** −3
11. 2 **13.** $\frac{1}{2}$ **15.** 2 **17.** 4 **19.** Not a real number **21.** 10
23. −0.2 **25.** −0.12 **27.** 8.3066 **29.** 15.6525 **31.** 3.7100
33. −0.1235 **35. a.** Not a real number **b.** Not a real number **c.** Not a real number **d.** 0 **e.** 1 **f.** 1.41 **g.** 1.73 **h.** 2 Domain: $[2, \infty)$ **37. a.** −1.44 **b.** −1.26 **c.** −1 **d.** 0 **e.** 1 **f.** 1.26 **g.** 1.44 **h.** 1.59 Domain: $(-\infty, \infty)$ **39.** $[1, \infty)$
41. $(-\infty, \infty)$ **43.** b **45.** d **47.** $|a|$ **49.** a **51.** $|a|$ **53.** x^2
55. $|xz^5|y^2$ **57.** $-\frac{x}{y}$ **59.** $\frac{2}{|x|}$ **61.** −9 **63.** −2 **65.** xy^2 **67.** $\frac{a^3}{b}$
69. $-\frac{5}{q}$ **71.** $3xy^2z$ **73.** $\frac{hk^2}{4}$ **75.** $-\frac{t}{3}$ **77.** $2y^2$ **79.** $2p^2q^3$
81. $q + p^2$ **83.** $\frac{6}{\sqrt[4]{x}}$ **85.** The sum of the square of a and the square root of b. **87.** The quotient of 1 and the square of the quantity c plus d. **89.** 8 in. **91.** 9.8 in. **93.** 9 cm
95. 13 ft **97.** $a^2 + 2ab + b^2 = c^2 + 2ab$; $a^2 + b^2 = c^2$
99.

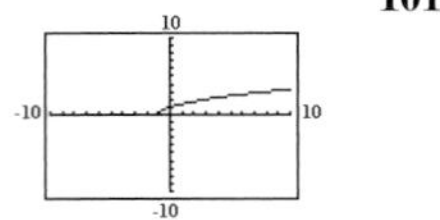

101.

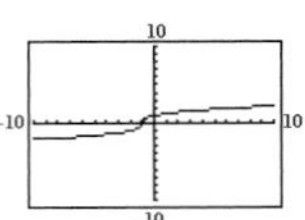

Section 10.2 Practice Exercises, pp. 673–674

1. a. 3 **b.** 27 **3.** 5 **5.** 3 **7.** Not a real number **9.** $a + 1$
11. $a^{m/n} = \sqrt[n]{a^m}$; The numerator of the exponent represents the power of the base. The denominator of the exponent represents the index of the radical. **13.** 5 **15.** 2 **17.** 3 **19.** −2 **21.** −2 **23.** $\frac{1}{2}$ **25.** $\frac{1}{9}$
27. 6 **29.** 10 **31.** $\frac{3}{4}$ **33.** −5 **35.** $\frac{9}{2}$ **37.** x **39.** p **41.** y^2
43. 6 **45.** $4t$ **47.** a^7 **49.** $\frac{25a^4d}{c}$ **51.** $\frac{y^9}{x^8}$ **53.** $\frac{2z^3}{w}$ **55.** $5xy^2z^3$
57. $\frac{x^{16}}{y^7z^4}$ **59.** $\frac{x^3y^2}{z^5}$ **61.** $\sqrt[3]{2x^2y}$ **63.** $\sqrt{\frac{2}{y}}$ **65.** $x^{1/3}$ **67.** $5x^{1/2}$
69. 3 **71.** 0.3761 **73.** 2.9240 **75.** 31.6228 **77. a.** 10 in. **b.** 8.5 in.
79. a. 10.9% **b.** 8.8% **c.** The account in part (a). **81.** $\sqrt[6]{x}$ **83.** $\sqrt[8]{y}$
85. $\sqrt[15]{w}$

Section 10.3 Practice Exercises, pp. 679–681

1. a^2b **3.** $\frac{s^6}{r^2}$ **5.** $\sqrt[7]{x^4}$ **7.** $y^{9/2}$ **9. a.** 0.07 **b.** −0.06 **11.** $10x$
13. $2x$ **15.** $x + 2$ **17.** $2y$ **19.** 3 **21.** $2b$ **23.** $3x + 1$ **25.** $\frac{1}{a + b}$
27. $2\sqrt{7}$ **29.** $4\sqrt{5}$ **31.** $15\sqrt{2}$ **33.** $3\sqrt[3]{2}$ **35.** $5x^2y\sqrt{y}$
37. $3yz\sqrt[3]{x^2z}$ **39.** $\frac{2}{b}$ **41.** $\frac{2\sqrt[5]{x}}{y^2}$ **43.** $\frac{5x\sqrt{2xy}}{3y^2}$ **45.** $2a^7b^4c^{15}d^{11}\sqrt{2c}$
47. $\frac{1}{\sqrt[3]{w^6}} = \frac{1}{w^2}$ **49.** $\sqrt{k^3} = k\sqrt{k}$ **51.** $2\sqrt{41}$ ft **53.** $6\sqrt{5}$ m
55. 127.3 ft **57.** The path from A to B and B to C is faster.

Section 10.4 Practice Exercises, pp. 684–686

1. $-2st^3\sqrt[3]{2s}$ **3.** $6p^3$ **5.** $\sqrt{(4x^2)^3} = 8x^3$ **7.** $y^{11/12}$ **9.** 1.95
11. a. Both expressions can be simplified using the distributive property. $7\sqrt{5} + 4\sqrt{5} = (7 + 4)\sqrt{5} = 11\sqrt{5}$; $7x + 4x = (7 + 4)x = 11x$ **b.** Neither expression can be simplified

because they do not contain *like* terms or *like* radicals. **13.** $9\sqrt{5}$ **15.** $\sqrt[3]{t}$ **17.** $5\sqrt{10}$ **19.** $8\sqrt[4]{3} - \sqrt[4]{14}$ **21.** $2\sqrt{x} + 2\sqrt{y}$ **23.** Cannot be simplified further **25.** Cannot be simplified further **27.** $\frac{29}{18}z\sqrt[3]{6}$ **29.** $0.70x\sqrt{y}$ **31.** Simplify each radical: $3\sqrt{2} + 35\sqrt{2}$. Then add *like* radicals: $38\sqrt{2}$ **33.** 15 **35.** $8\sqrt{3}$ **37.** $3\sqrt{7}$ **39.** $-5\sqrt{2a}$ **41.** $8s^2t^2\sqrt[3]{s^2}$ **43.** $6x\sqrt[3]{x}$ **45.** $14p^2\sqrt{5}$ **47.** $2a^2b\sqrt{6a}$ **49.** $5x\sqrt[3]{2} - 6\sqrt[3]{x}$ **51.** False For example: $\sqrt{9} + \sqrt{16} \neq \sqrt{9 + 16}$; $7 \neq 5$ **53.** True **55.** False $\sqrt{y} + \sqrt{y} = 2\sqrt{y} \neq \sqrt{2y}$ **57.** $\sqrt{48} + \sqrt{12}$; $6\sqrt{3}$ **59.** $5\sqrt[3]{x^6} - x^2$; $4x^2$ **61.** The difference of the square root of 18 and the square of 5 **63.** The sum of the fourth root of x and the cube of y **65. a.** $10\sqrt{5}$ yards **b.** 22.36 yards **c.** $105.95

Section 10.5 Practice Exercises, pp. 691–693

1. a. 2 **b.** 4 **3.** $x - y$ **5.** $-2xy^2z^2\sqrt[3]{2x^2z}$ **7.** 3 **9.** $x^{1/6}y^{7/12}$ **11.** $a^{1/6}$ **13.** $2\sqrt[3]{7}$ **15.** $2\sqrt{5}$ **17.** $4\sqrt[4]{4}$ **19.** $6\sqrt{35}$ **21.** $-24ab\sqrt{a}$ **23.** $12 - 6\sqrt{3}$ **25.** $2\sqrt{3} - \sqrt{6}$ **27.** $-3x - 21\sqrt{x}$ **29.** $-8 + 7\sqrt{30}$ **31.** $x - 5\sqrt{x} - 36$ **33.** $\sqrt[3]{y^2} - \sqrt[3]{y} - 6$ **35.** $9a - 28\sqrt{ab} + 3b$ **37.** $-8 + \sqrt{14} - 2\sqrt{7} + 3\sqrt{2}$ **39.** $8\sqrt{p} + 3p + 5\sqrt{pq} + 16\sqrt{q} - 2q$ **41.** $\sqrt[4]{x^3}$ **43.** $\sqrt[15]{(2z)^8}$ **45.** $p^2\sqrt[6]{p}$ **47.** $u\sqrt[6]{u}$ **49.** $\sqrt[6]{(a + b)}$ **51. a.** $x^2 - y^2$ **b.** $x^2 - 25$ **53.** $3 - x^2$ **55.** 4 **57.** $64x - 4y$ **59.** $29 + 8\sqrt{13}$ **61.** $p - 2\sqrt{7p} + 7$ **63.** $2a - 6\sqrt{2ab} + 9b$ **65.** True **67.** False; $(x - \sqrt{5})^2 = x^2 - 2x\sqrt{5} + 5$ **69.** False; 5 is multiplied only with the 3. **71.** True **73.** $12\sqrt{5}$ ft^2 **75.** $18\sqrt{15}$ in.^{2} **77.** 40 m^2 **79.** $a + b$ **81.** $\sqrt[6]{x^2y}$ **83.** $\sqrt[6]{2^3 \cdot 3^2}$ or $\sqrt[6]{72}$ **85.** $\sqrt[4]{3 \cdot 2^3}$ or $\sqrt[4]{24}$ **87.** $\sqrt[15]{p^3q^5}$

Section 10.6 Practice Exercises, pp. 698–700

1. $12y\sqrt{5}$ **3.** $-18y + 3\sqrt{y} + 3$ **5.** $9\sqrt{3}$ **7.** $64 - 16\sqrt{t} + t$ **9.** -5 **11.** $\frac{\sqrt{5}}{\sqrt{5}}; \frac{x\sqrt{5}}{5}$ **13.** $\frac{\sqrt[3]{x^2}}{\sqrt[3]{x^2}}; \frac{7\sqrt[3]{x^2}}{x}$ **15.** $\frac{\sqrt{3z}}{\sqrt{3z}}; \frac{8\sqrt{3z}}{3z}$ **17.** $\frac{\sqrt[4]{2^3a^2}}{\sqrt[4]{2^3a^2}}; \frac{\sqrt[4]{8a^2}}{2a}$ **19.** $\frac{\sqrt{3}}{3}$ **21.** $2\sqrt{5}$ **23.** $\frac{\sqrt{x}}{x}$ **25.** $\frac{3\sqrt{2y}}{y}$ **27.** $-2\sqrt{a}$ **29.** $\frac{7\sqrt[3]{2}}{2}$ **31.** $\frac{4\sqrt{w}}{w^2}$ **33.** $\frac{2\sqrt[4]{27}}{3}$ **35.** $\frac{\sqrt{x}}{x^4}$ **37.** $\frac{\sqrt{2x}}{2x^3}$ **39.** $\frac{2x\sqrt[3]{2y^2}}{y}$ **41.** $\frac{xy^2\sqrt{10xy}}{10}$ **43.** $\sqrt{2} + \sqrt{6}$ **45.** $\sqrt{x} - 23$ **47.** -7 **49.** 3 **51.** $\frac{4\sqrt{2} - 12}{-7}$ or $\frac{-4\sqrt{2} + 12}{7}$ **53.** $\frac{\sqrt{5} + \sqrt{2}}{3}$ **55.** $-\sqrt{21} + 2\sqrt{7}$ **57.** $\frac{-\sqrt{p} + \sqrt{q}}{p - q}$ **59.** $\frac{5 + \sqrt{21}}{4}$ **61.** $-6\sqrt{5} - 13$ **63.** $\frac{16}{\sqrt[3]{4}}; 8\sqrt[3]{2}$ **65.** $\frac{4}{x - \sqrt{2}}; \frac{4x + 4\sqrt{2}}{x^2 - 2}$ **67. a.** 1.57 s **b.** 1.11 s **c.** 0.79 s **69.** $\frac{2\sqrt{6}}{3}$ **71.** $\frac{17\sqrt{15}}{15}$ **73.** $\frac{8\sqrt[3]{25}}{5}$ **75.** $\frac{-33}{2\sqrt{3} - 12}$ **77.** $\frac{a - b}{a + 2\sqrt{ab} + b}$

Section 10.7 Practice Exercises, pp. 705–708

1. $4\sqrt{3}$ **3.** $\frac{3w\sqrt{w}}{4}$ **5.** Not a real number **7.** $\frac{p^2\sqrt{pq}}{q^2}$ **9.** $\frac{7\sqrt{5t}}{5t^2}$ **11.** $4x - 6$ **13.** $9p + 7$ **15.** $w^2 + 2w - 17$ **17.** $2x$ **19.** $7r$ **21.** $t = 49$ **23.** $x = 9$ **25.** $y = 3$ **27.** $x = \frac{15}{2}$ **29.** $z = 42$ **31.** $x = 29$ **33.** $w = 140$ **35.** No solution **37.** $x = -65$ **39.** $t = \frac{16}{3}$ **41.** $p = 4, p = 2$ **43.** $x = \frac{3}{2}$ $\left(x = -\frac{1}{2}\text{ does not check}\right)$ **45.** $h = 9$ **47.** $a = -4$ **49.** No solution **51.** $V = \frac{4\pi r^3}{3}$ **53.** $h^2 = \frac{r^2 - \pi^2r^2}{\pi^2}$ **55.** $a^2 + 10a + 25$ **57.** $25w^2 - 40w + 16$ **59.** $5a - 6\sqrt{5a} + 9$ **61.** $a = -3$ **63.** No solution **65.** No solution **67.** $a = \frac{9}{5}$ **69.** $h = 2$ **71.** No solution **73.** $x = -\frac{11}{4}$ **75.** No solution **77.** No solution **79.** $p = 7, p = 5$ **81.** $m = \frac{1}{3}, m = -1$ **83.** $z = -1$ ($z = 3$ does not check) **85.** No solution ($t = 3, t = 23$ do not check) **87. a.** 305 m **b.** 460 m **89. a.** $2 million **b.** $1.2 million **c.** 50,000 passengers **91. a.** $\sqrt{13}$ versus 5 **b.** Not equal **93. a.** 12 lb **b.** $t(18) = 5.1$. An 18-lb turkey will take about 5.1 hr to cook. **95.** $c = \sqrt{k^2 + 81}$ **97.** $b = \sqrt{25 - h^2}$ **99.** $a = \sqrt{k^2 - 196}$ **101.** $x = 1$ ($x = 0$ does not check) **103.** $q = -2$

105. **107.**

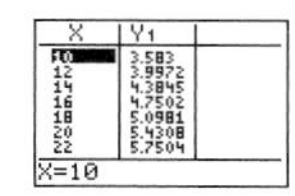

Section 10.8 Practice Exercises, pp. 715–716

1. $3\sqrt{5} - 15\sqrt{2}$ **3.** $9 - x$ **5.** $y = 8$ **7.** $p = -8$ **9.** $c = \frac{49}{36}$ **11.** $i = \sqrt{-1}$ **13.** $a - bi$ **15.** $12i$ **17.** $i\sqrt{3}$ **19.** $2i\sqrt{5}$ **21.** $29i\sqrt{2}$ **23.** $13i\sqrt{7}$ **25.** -7 **27.** -12 **29.** $-3\sqrt{10}$ **31.** $\sqrt{2}$ **33.** $3i$ **35.** $\frac{1 + 2i}{4}$ **37.** $\frac{1 - i\sqrt{3}}{2}$ **39.** $-1 \pm i\sqrt{2}$ **41.** $-2 \pm i\sqrt{3}$ **43.** $7 + 6i$ **45.** $\frac{3}{10} + \frac{3}{2}i$ **47.** $5 + 0i$ **49.** $-1 + 10i$ **51.** $-i$ **53.** 1 **55.** i **57.** 1 **59.** $-i$ **61.** -1 **63.** $-24 + 0i$ **65.** $18 + 6i$ **67.** $26 - 26i$ **69.** $-29 + 0i$ **71.** $-9 + 40i$ **73.** $35 + 20i$ **75.** $1 - 3i$; 10 **77.** $4 + 3i$; 25 **79.** $\frac{1}{5} - \frac{3}{5}i$ **81.** $\frac{3}{25} - \frac{4}{25}i$ **83.** $\frac{21}{29} + \frac{20}{29}i$ **85.** $0 - \frac{3}{2}i$ **87.** $0 + 3i$ **89.** $0 - 7i$ **91.** $12 + 0i$ **93.** $-1 + 0i$

Chapter 10 Review Exercises, pp. 720–723

1. a. False; $\sqrt{0} = 0$ is not positive. **b.** False; $\sqrt[3]{-8} = -2$ **3. a.** False **b.** True **5.** 5 **7. a.** 3 **b.** 0 **c.** $\sqrt{7}$ **d.** $[1, \infty)$ **9.** $\frac{\sqrt[3]{2x}}{\sqrt[4]{2x}} + 4$ **11.** 8 cm **13.** Yes, provided the expressions are well defined. For example: $x^5 \cdot x^3 = x^8$ and $x^{1/5} \cdot x^{1/3} = x^{8/15}$

15. Take the reciprocal of the base and change the exponent to positive. **17.** $\frac{1}{2}$ **19.** b^{10} **21.** $x^{3/4}$ **23.** 2.1544 **25.** 54.1819
27. 1. Factors of the radicand must have powers less than the index 2. No fractions in the radicand 3. No radical in the denominator of a fraction. **29.** $xz\sqrt[4]{xy}$ **31.** $\frac{-2x^2y^2\sqrt[3]{2x}}{z^3}$ **33.** 31 ft **35.** Cannot be combined; The indices are different. **37.** Can be combined: $3\sqrt[4]{3xy}$
39. $5\sqrt{7}$ **41.** $10\sqrt{2}$ **43.** False; 5 and $3\sqrt{x}$ are not *like* radicals.
45. $a + b$ and $a - b$ are conjugates. **47.** 6 **49.** 12
51. $\sqrt[3]{4x^2} - 4\sqrt[3]{x^2} + 2\sqrt[3]{2x^2}$ **53.** $u^2\sqrt[6]{u^5}$ **55.** $(a + b)\sqrt[6]{(a + b)}$
57. $\frac{2x\sqrt{5} + 5\sqrt{x}}{5x}$ **59.** $\frac{6b - \sqrt{b} - 1}{9b - 1}$ **61.** $y = \frac{49}{2}$ **63.** $w = -12$
65. $t = 9$ **67.** $m = \frac{1}{2}, m = 4$ **69. a.** 25.3 ft/s; When the water depth is 20 ft, a wave travels about 25.3 ft/s. **b.** 8 ft **71.** $a + bi$, where a and b are real numbers and $i = \sqrt{-1}$. **73.** In each case we simplify the expression by multiplying the numerator and denominator by the conjugate of the denominator. **75.** $-i\sqrt{5}$ **77.** -1
79. $-i$ **81.** $-2 + 5i$ **83.** $-20 + 0i$ **85.** $24 - 10i$ **87.** $-2 - i$, Real part: -2, Imaginary part: -1 **89.** $1 - 2i$ **91.** $2 \pm 4i$

Chapter 10 Test, pp. 723–724

1. a. 6 **b.** -6 **2. a.** Real **b.** Not real **c.** Real **d.** Real
3. a. y **b.** $|y|$ **4.** 3 **5.** $\frac{4}{3}$ **6.** $2\sqrt[3]{4}$ **7.** $a^2bc^2\sqrt{bc}$ **8.** $3x^2\sqrt{2}$
9. $\frac{4w^2\sqrt{6w}}{3}$ **10.** $\sqrt[10]{(7y)^7}$ **11.** $\sqrt[12]{10}$ **12. a.** $f(-8) = 2\sqrt{3}$; $f(-6) = 2\sqrt{2}$; $f(-4) = 2$; $f(-2) = 0$ **b.** $(-\infty, -2]$ **13.** -0.3080
14. -3 **15.** $\frac{1}{t^{3/4}}$ **16.** $3\sqrt{5}$ **17. a.** $3\sqrt{2x} - 3\sqrt{5x}$
b. $2x - 6\sqrt{2x} + 9$ **18. a.** $\frac{-2\sqrt[3]{x^2}}{x}$ **b.** $\frac{x + 6 + 5\sqrt{x}}{9 - x}$ **19. a.** $2i\sqrt{2}$
b. $8i$ **c.** $\frac{1 \pm i\sqrt{2}}{2}$ **20.** $1 - 11i$ **21.** $30 + 16i$ **22.** -28
23. $36 - 11i$ **24. a.** $1 + 4i$ **b.** $\frac{17}{25} + \frac{6}{25}i$ **25.** $r(10) = 1.34$; the radius of a sphere of volume 10 cubic units is 1.34 units. **26.** 21 ft
27. $x = -16$ **28.** $x = \frac{17}{5}$ **29.** $t = 2$ ($t = 42$ does not check)

Cumulative Review Exercises, Chapters 1–10, pp. 724–726

1. 54 **2.** $15x - 5y - 5$ **3.** $y = -3$ **4.** $(-\infty, -5)$
5. $y = -2x + 5$
6.

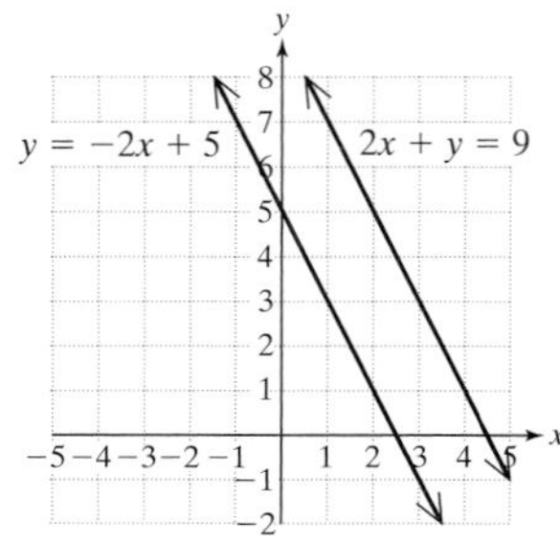

7. $\left(\frac{1}{2}, \frac{1}{3}\right)$ **8.** $\left(2, -2, \frac{1}{2}\right)$ is not a solution. **9.** $x = 6, y = 3, z = 8$
10. a. $f(-2) = -10$; $f(0) = -2$; $f(4) = 14$; $f\left(\frac{1}{2}\right) = 0$
b. $(-2, -10)$ $(0, -2)$ $(4, 14)$ $\left(\frac{1}{2}, 0\right)$
c.

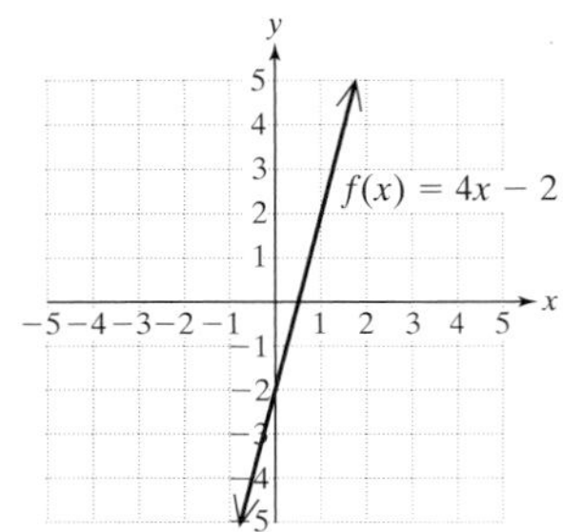

11. Not a function **12.** $a^4b^8c^2$ **13.** $a^6b^{12}c^4$ **14.** 111,831 m/s
15. $2x^2 - x - 15$; Second degree **16.** $\sqrt{15} + 3\sqrt{2} + 3$ **17.** $x - 4$
18. $-\frac{1}{6}$ **19.** $\frac{3c\sqrt[3]{2}}{d}$ **20.** $32b\sqrt{5b}$ **21.** $2 + 3i$
22. $(2x - 9)(3x - 2)$ **23. a.** x-intercept: $(-\frac{1}{2}, 0)$; y-intercept: $(0, 1)$ **b.** x-intercepts: $(3, 0)$, $(\frac{5}{2}, 0)$; y-intercept $(0, 15)$ **24.** The car travels 90 km/hr and the truck travels 80 km/hr. **25.** 7.5 hours
26. $\frac{4(y + 2)}{y + 1}$ **27.** $\frac{t}{t - 3}$ **28.** $-(x + y)$ **29.** $(y^2 - 2)(y^4 + 2y^2 + 4)$
30. $-7 - 2i$ **31.** $[1, 6]$ **32.** $(-\infty, 4] \cup (6, \infty)$ **33.** $[-3, 5]$
34. $2\sqrt{7} + 2\sqrt{5}$ **35.** Small: \$0.22 per photo; midsize: \$0.28 per photo

CHAPTER 11

Section 11.1 Practice Exercises, pp. 734–735

1. $x = \pm 10$ **3.** $a = \pm\sqrt{5}$ **5.** $v = \pm i\sqrt{11}$ **7.** $p = 8, p = 2$
9. $x = 2 \pm \sqrt{5}$ **11.** $h = 4 \pm 2i\sqrt{2}$ **13.** $a = \frac{1}{2} \pm \frac{\sqrt{3}}{2}$
15. $x = \frac{3}{2} \pm \frac{i\sqrt{7}}{2}$ **17.** 1. Factoring and applying the zero product rule. 2. Applying the square root property. $x = \pm 9$ **19.** 4.2 ft
21. a. 8% **b.** 11% **c.** 14.02% **23.** $k = 9$; $(x - 3)^2$
25. $k = \frac{25}{4}$; $\left(y + \frac{5}{2}\right)^2$ **27.** $k = \frac{1}{25}$; $\left(b + \frac{1}{5}\right)^2$
29. 1. Write equation in the form $ax^2 + bx + c = 0$. 2. Divide each term by a. 3. Isolate the variable terms. 4. Complete the square and factor. 5. Apply the square root property.
31. $t = -3, t = -5$ **33.** $x = 2, x = -8$ **35.** $p = -2 \pm i\sqrt{2}$
37. $y = 5, y = -2$ **39.** $a = -1 \pm \frac{i\sqrt{6}}{2}$ **41.** $x = 2 \pm \frac{2}{3}i$
43. $p = \frac{1}{5} \pm \frac{\sqrt{3}}{5}$ **45.** $w = -\frac{3}{4} \pm \frac{\sqrt{65}}{4}$ **47.** $n = 2 \pm \sqrt{11}$
49. $r = \sqrt{\frac{A}{\pi}}$ or $r = \frac{\sqrt{A\pi}}{\pi}$ **51.** $a = \sqrt{d^2 - b^2 - c^2}$
53. $r = \sqrt{\frac{3V}{\pi h}}$ or $r = \frac{\sqrt{3V\pi h}}{\pi h}$ **55.** $x = \frac{s^2}{4}$

57. a. 4.5 thousand textbooks or 35.5 thousand textbooks **b.** Profit increases to a point as more books are produced. Beyond that point, the market is "flooded," and profit decreases. Hence there are 2 points at which the profit is \$20,000. Producing 4.5 thousand books makes the same profit using fewer resources than producing 35.5 thousand books.

Section 11.2 Practice Exercises, pp. 746–749

1. $x = 2, x = -12$ **3.** $(x - 1)(x^2 + x + 1)$
5. $x = 2, x = -3, x = 3$ **7.** $(4v - 3)(2u + 3)$ **9.** $4 - 2\sqrt{10}$
11. $2 - i\sqrt{3}$ **13.** $x^2 + 2x + 1 = 0; a = 1, b = 2, c = 1$
15. $19m^2 - 8m + 0 = 0; a = 19, b = -8, c = 0$
17. $5p^2 + 0p - 21 = 0; a = 5, b = 0, c = -21$
19. $n^2 + 3n + 4 = 0; a = 1, b = 3, c = 4$
21. 0; One rational solution **23.** 64; Two rational solutions
25. 420; Two irrational solutions **27.** -7; Two imaginary solutions
29. a. Discriminant = 21; two real solutions **b.** Two x-intercepts
31. a. Discriminant = 0; one real solution **b.** One x-intercept
33. a. Discriminant = -4; two imaginary solutions **b.** No x-intercepts
35. If the equation is factorable **37.** Any quadratic equation written in the form $ax^2 + bx + c = 0$ **39.** $a = -12, a = 1$
41. $y = \dfrac{1 \pm 2i\sqrt{11}}{9}$ **43.** $p = \dfrac{1 \pm i\sqrt{14}}{6}$ **45.** $z = 7, z = -5$
47. $a = \dfrac{-3 \pm \sqrt{41}}{2}$ **49.** $x = \dfrac{2}{5}$ **51.** $w = 3 \pm i\sqrt{5}$
53. $x = \dfrac{1 \pm \sqrt{29}}{2}$ **55.** $y = \dfrac{-2 \pm 2i\sqrt{2}}{3}$ **57.** $h = \dfrac{-5 \pm \sqrt{13}}{2}$
59. $x = -2$ **61.** $t = \dfrac{-7 \pm \sqrt{109}}{6}$ **63. a.** $(x - 3)(x^2 + 3x + 9)$
$x = -4$
b. $x = 3, x = \dfrac{-3 \pm 3i\sqrt{3}}{2}$ **65. a.** $3x(x^2 - 2x + 2)$
b. $x = 0, x = 1 \pm i$ **67.** 3 ft **69.** $a = \dfrac{-3 \pm i\sqrt{7}}{2}$ **71.** $x = \pm\sqrt{2}$
73. $y = -\dfrac{5}{2}, y = \dfrac{1}{2}$ **75.** $x = -\dfrac{1}{2} \pm 2i$ **77. a.** 46 mph **b.** 36 mph
79. 8.2 m and 6.1 m **81. a.** 6319 thousand farms **b.** $N(t)$ gives approximate values. **c.** 1892, 1954
83.

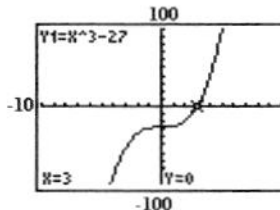

85.

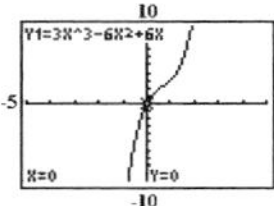

87. a. 7964 thousand **b.** 15,507 thousand **c.** 1989
d.

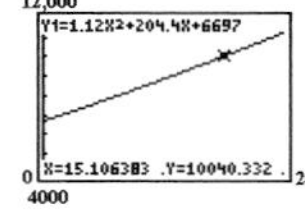

Section 11.3 Practice Exercises, pp. 753–754

1. $y = 4$ **3.** $x = \dfrac{1}{4}$ **5.** $b = 6, b = 2$ **7.** $w = 7$
9. $x = \pm 2, x = \pm 2i$ **11.** $m = \pm 3, m = \pm 3i$
13. $a = -2, a = 1 \pm i\sqrt{3}$ **15.** $p = 1, p = \dfrac{-1 \pm i\sqrt{3}}{2}$
17. a. $u = -4, u = -6$ **b.** $y = -4, y = -1, y = -2, y = -3$
19. a. $u = 6, u = -4$ **b.** $x = 6, x = -1, x = 4, x = 1$
21. $x = -\dfrac{7}{4}, x = -\dfrac{3}{2}$ **23.** $x = -7$ **25.** $x = 1 \pm i\sqrt{2}, x = 1 \pm \sqrt{2}$
27. $x = 3, x = -3, x = 2, x = -2$
29. $x = 2, x = 1, x = -1 \pm i\sqrt{3}, x = \dfrac{-1 \pm i\sqrt{3}}{2}$
31. $m = 27, m = -8$ **33.** $t = -\dfrac{1}{32}, t = -243$ **35.** $x = \pm 2$
37. $a = \pm 4i, a = 1$ **39.** $x = 4, x = \pm i\sqrt{5}$ **41. a.** $x = \pm i\sqrt{2}$
b. Two imaginary solutions; zero real solutions **c.** No x-intercepts
d. **43. a.** $x = 0, x = 3, x = -2$
b. Three real solutions; zero imaginary solutions **c.** Three x-intercepts
d.

Section 11.4 Practice Exercises, pp. 762–764

1. The value of k shifts the graph of $y = x^2$ vertically.
3.

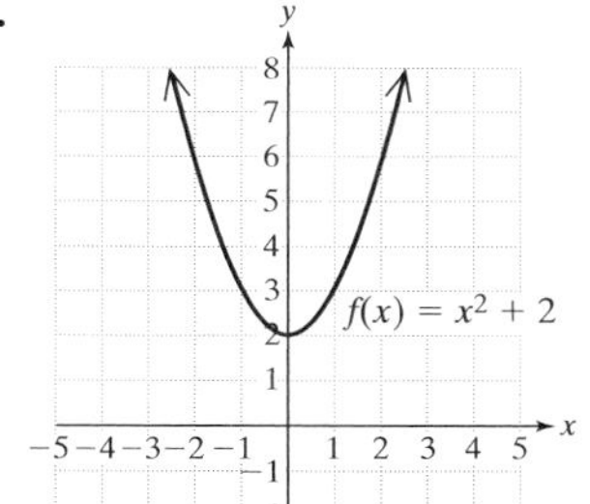

5.

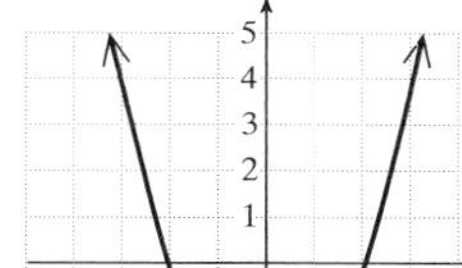
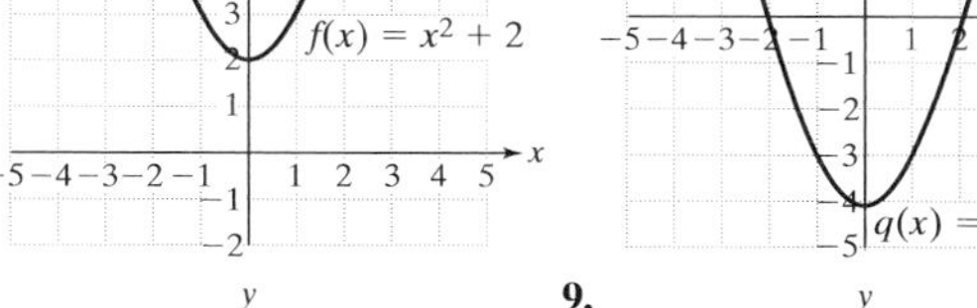

7.

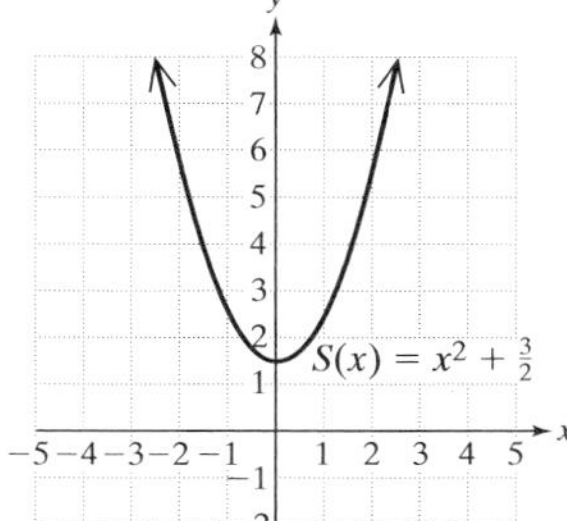

9.

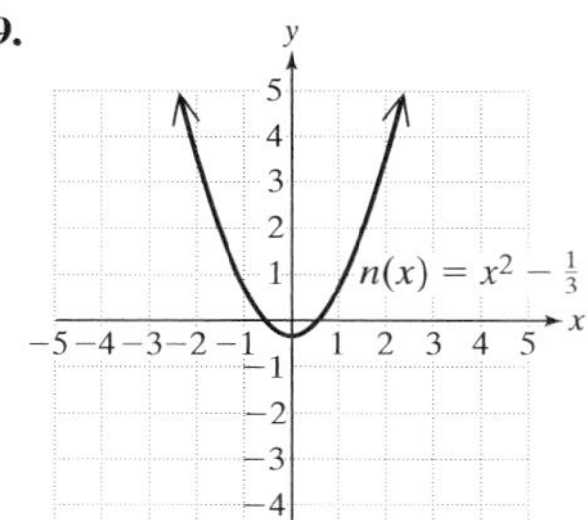

11.

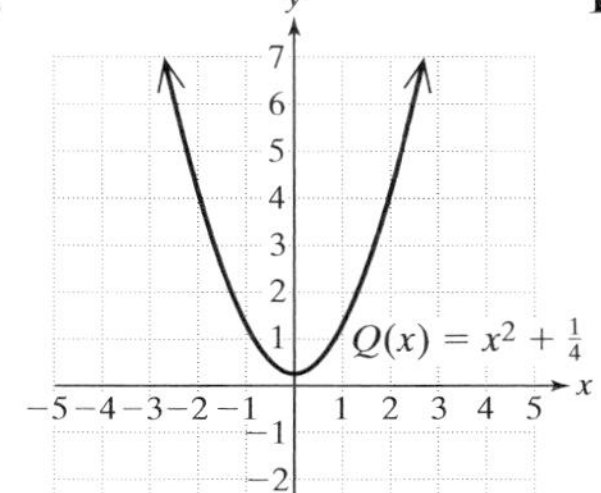

13.

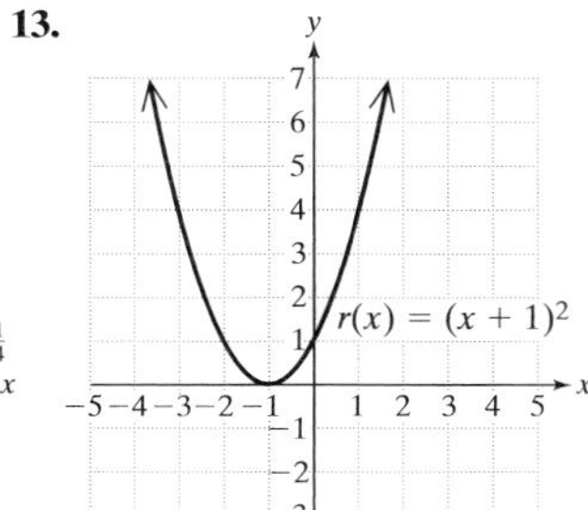

15.

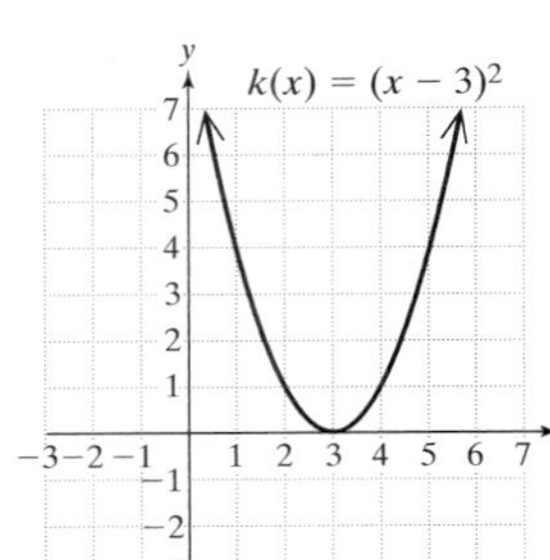

17.

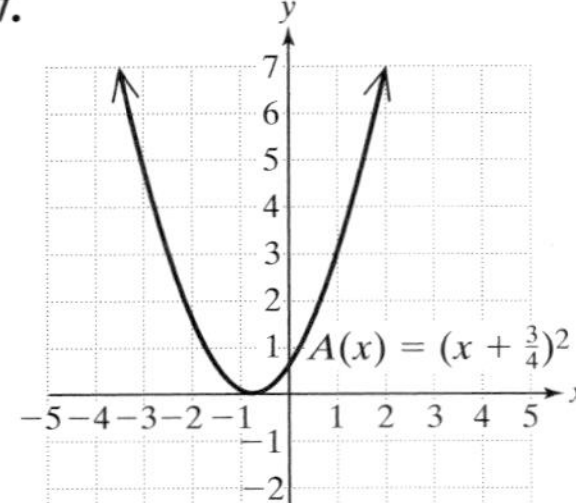

19.

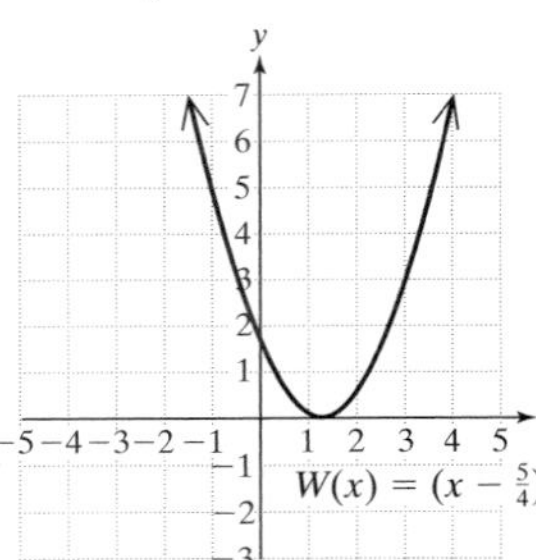

21.

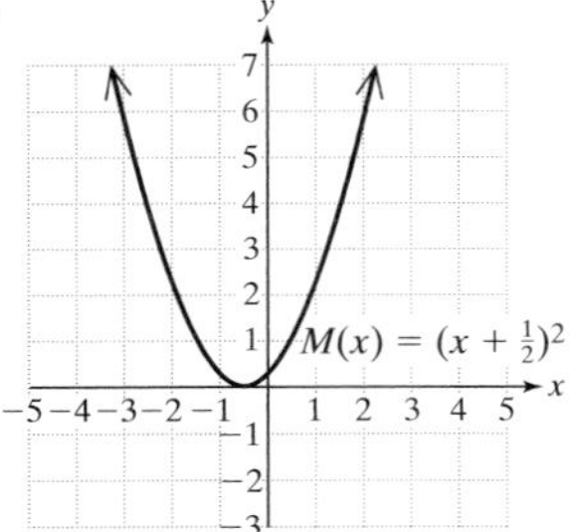

23. The value of a vertically stretches or shrinks the graph of $y = x^2$. **25.** d **27.** g **29.** a **31.** b

33.

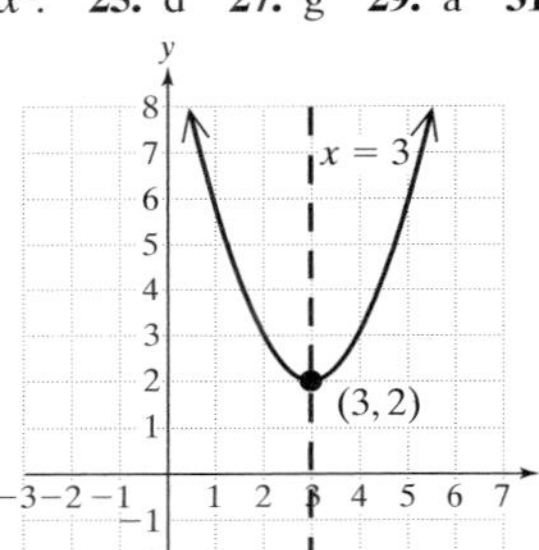

35.

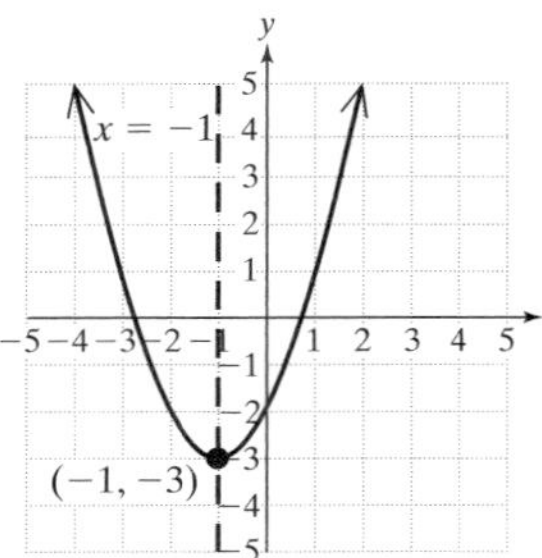

37.

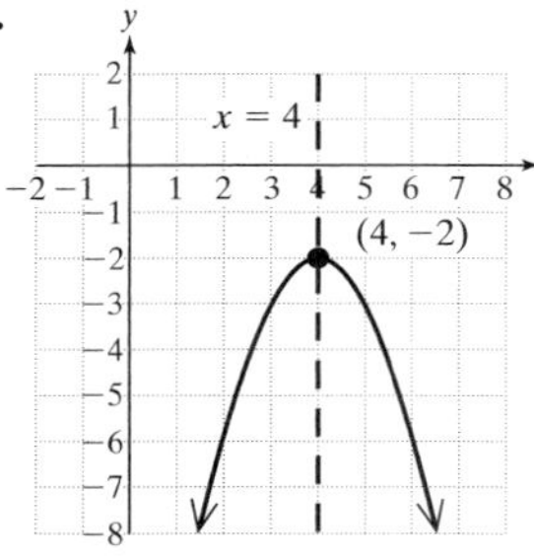

39.

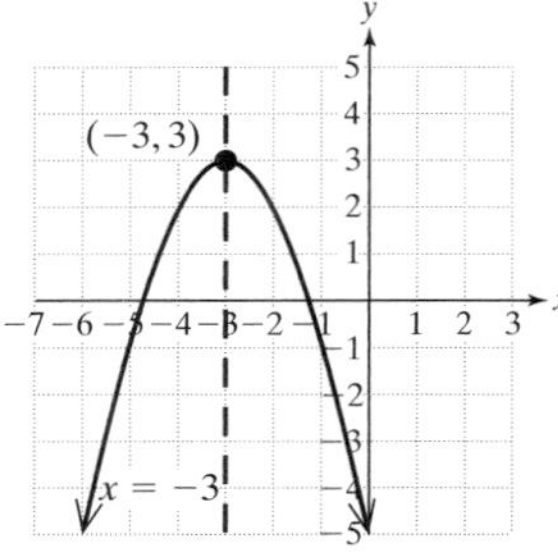

41.

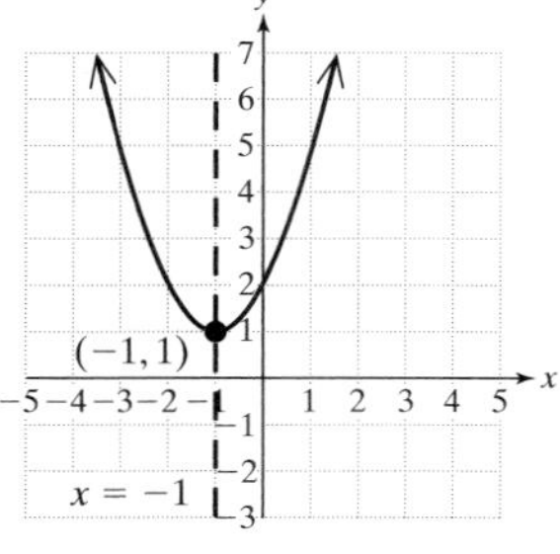

43. Vertex $(6, -9)$ minimum point; minimum value: -9
45. Vertex $(2, 5)$; maximum point; maximum value: 5
47. Vertex $(-8, -3)$; minimum point; minimum value: -3
49. Vertex $\left(-\frac{3}{4}, \frac{21}{4}\right)$; maximum point; maximum value: $\frac{21}{4}$
51. Vertex $\left(7, -\frac{3}{2}\right)$; minimum point; minimum value: $-\frac{3}{2}$
53. True **55.** False **57. a.** $(60, 30)$ **b.** 30 ft **c.** 70 ft

59.

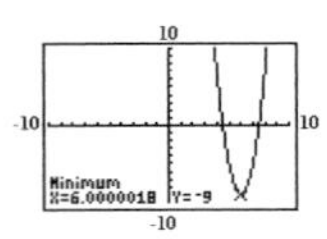

61.

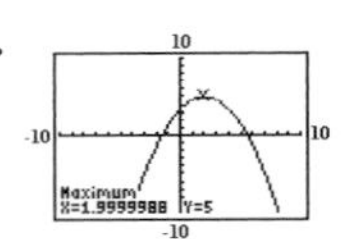

63.

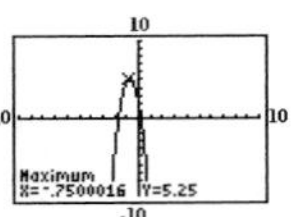

Section 11.5 Practice Exercises, pp. 772–774

1. $f(x)$ is the graph of $y = x^2$ but opens downward and is stretched vertically by a factor of 2. **3.** $Q(x)$ is the graph of $y = x^2$ shifted down $\frac{8}{3}$ units. **5.** $s(x)$ is the graph of $y = x^2$ shifted to the right 4 units. **7.** 16 **9.** $\frac{49}{4}$ **11.** $\frac{1}{81}$ **13.** $\frac{1}{36}$
15. $g(x) = (x - 4)^2 - 11; (4, -11)$
17. $n(x) = 2(x + 3)^2 - 5; (-3, -5)$
19. $p(x) = -3(x - 1)^2 - 2; (1, -2)$
21. $k(x) = \left(x + \frac{7}{2}\right)^2 - \frac{89}{4}; \left(-\frac{7}{2}, -\frac{89}{4}\right)$
23. $f(x) = (x + 4)^2 - 17; (-4, -17)$ **25.** $(2, 3)$ **27.** $(-1, -2)$ **29.** $(-4, -15)$ **31.** $(-1, 2)$ **33.** $(1, 3)$ **35. a.** $\left(-\frac{9}{2}, -\frac{49}{4}\right)$ **b.** $(0, 8)$ **c.** $(-8, 0)(-1, 0)$ **d.**

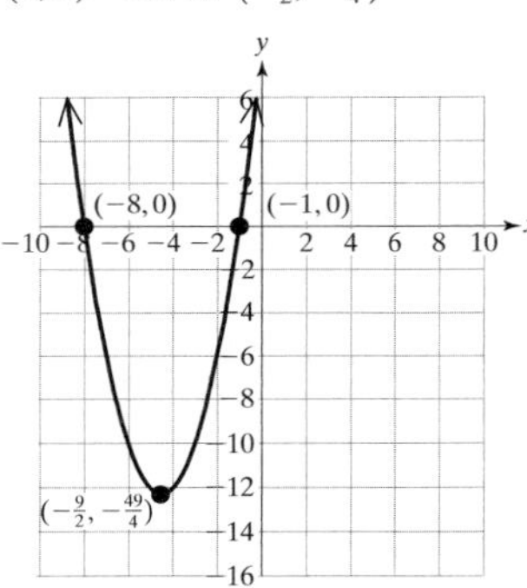

37. a. $\left(\frac{1}{2}, \frac{7}{2}\right)$ **b.** $(0, 4)$ **c.** No x-intercepts
d.

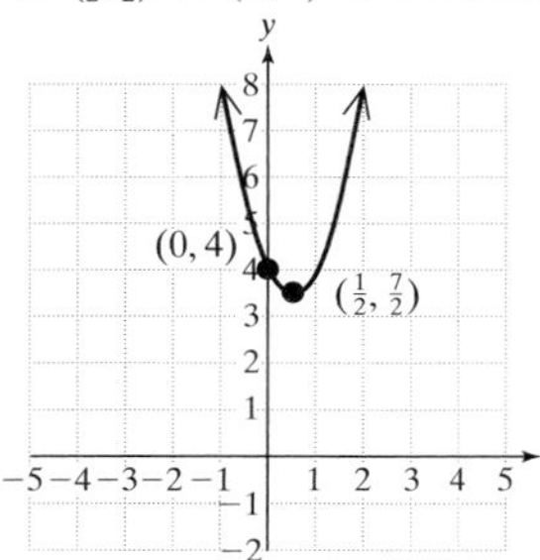

39. a. $(\frac{3}{2}, 0)$ **b.** $(0, -\frac{9}{4})$ **c.** $(\frac{3}{2}, 0)$
d.

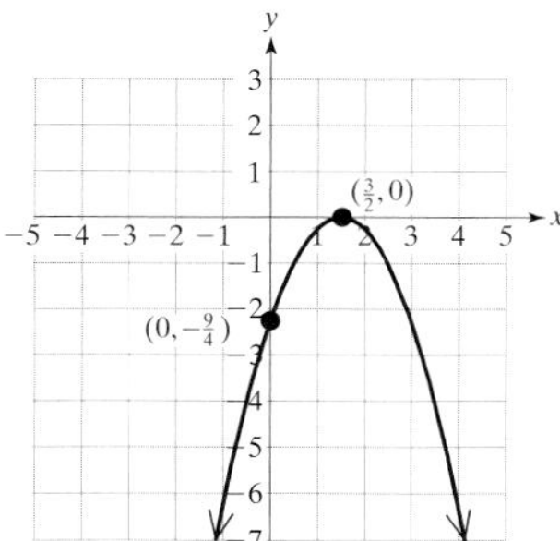

41. a. 18.9 thousand miles (18,900) **b.** 33 psi
43. a. \$450 **b.** \$800 **c.** 50 books or 550 books **d.** (300, 1250)
e.

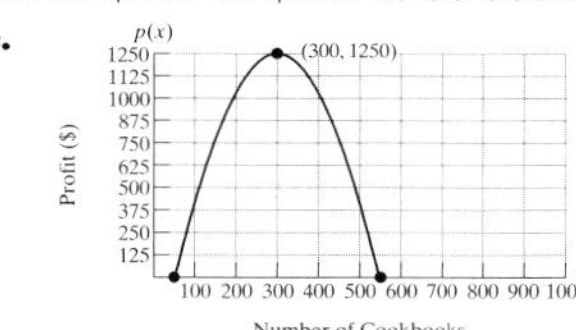

f. 300 books produced will yield a maximum profit of \$1250.
45. a. The sum of the three sides must equal the total amount of fencing. **b.** $A = x(200 - 2x)$ **c.** 50 ft by 100 ft

47.

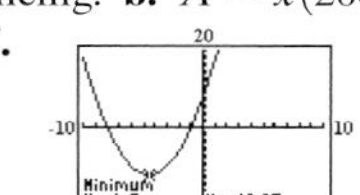

49.

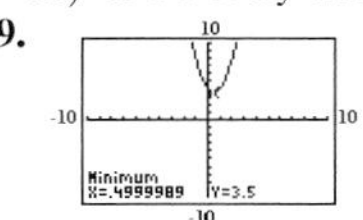

51.

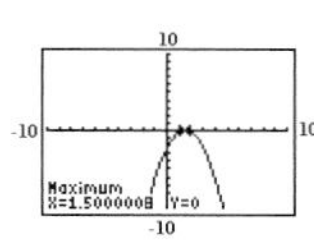

Chapter 11 Review Exercises, pp. 779–781

1. $x = \pm\sqrt{5}$ **3.** $a = \pm 9i$ **5.** $x = 2 \pm 6\sqrt{2}$ **7.** $y = \dfrac{1 \pm \sqrt{3}}{3}$

9. 8.7 in. **11.** 12.2 in. **13.** $k = \dfrac{81}{4}; \left(x - \dfrac{9}{2}\right)^2$

15. $k = \dfrac{1}{25}; \left(z - \dfrac{1}{5}\right)^2$ **17.** $y = \dfrac{3}{2} \pm i$ **19.** $b = \dfrac{1}{2}, b = -4$

21. $t = 4 \pm 3i$ **23.** Two rational solutions **25.** Two irrational solutions **27.** One rational solution **29.** $y = 2 \pm \sqrt{3}$

31. $a = 2, a = -\dfrac{5}{6}$ **33.** $b = \dfrac{4}{5}, b = -\dfrac{1}{5}$ **35.** $x = 8, x = -4$

37. $x = 2, x = -2$ **39. a.** 1822 ft **b.** 115 ft/sec **c.** 78 mph
41. $x = 49$ **43.** $y = \pm 3, y = \pm\sqrt{2}$ **45.** $t = -243, t = 32$
47. $a = 3, a = 1$

49.

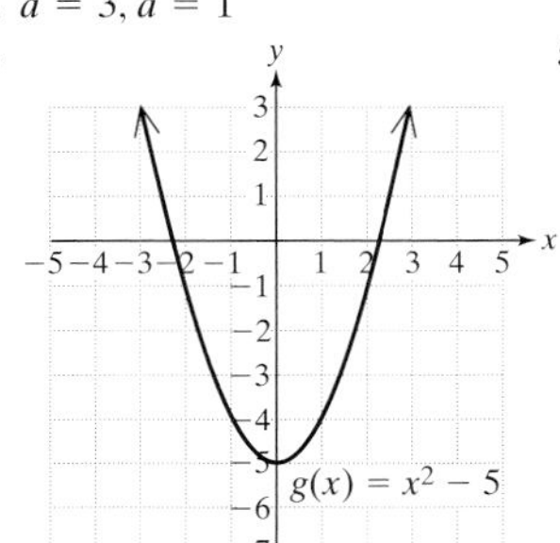

51.

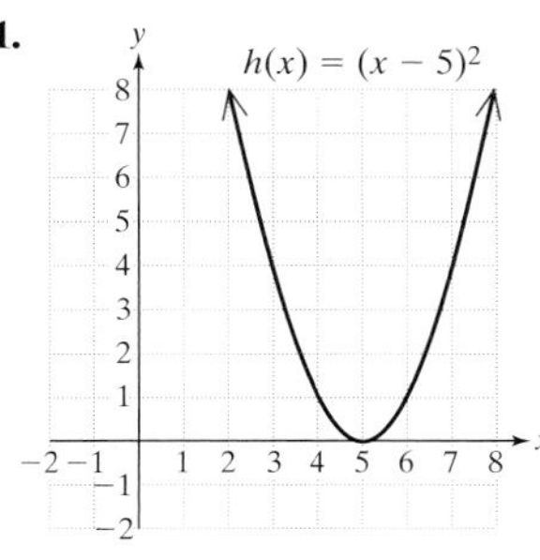

53. **55.**

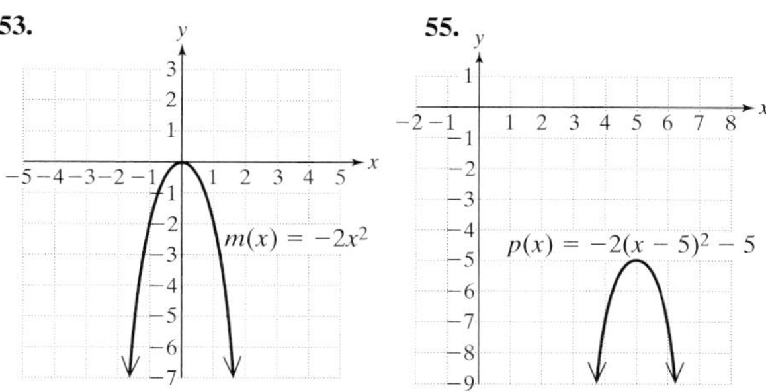

57. $\left(4, \dfrac{5}{3}\right)$ is the minimum point. The minimum value is $\dfrac{5}{3}$.

59. $x = -\dfrac{2}{11}$ **61.** $z(x) = (x - 3)^2 - 2; (3, -2)$

63. $p(x) = -5(x + 1)^2 - 8; (-1, -8)$ **65.** $(1, -15)$ **67.** $\left(\dfrac{1}{2}, \dfrac{41}{4}\right)$

69. a. 0.875 grams; 1.875 grams; 2 grams; 1.875 grams **b.** 2 grams after 48 hours

Chapter 11 Test, pp. 781–783

1. $x = 2, x = -8$ **2.** $p = 2 \pm 2\sqrt{3}$ **3.** $m = -1 \pm i$
4. $k = \dfrac{49}{4}; \left(d + \dfrac{7}{2}\right)^2$ **5.** $x = -3 \pm 3\sqrt{3}$ **6.** $x = \dfrac{3 \pm i\sqrt{47}}{4}$
7. a. $x^2 - 3x + 12 = 0$ **b.** $a = 1, b = -3, c = 12$ **c.** -39 **d.** Two imaginary solutions **8. a.** $y^2 - 2y + 1 = 0$ **b.** $a = 1, b = -2, c = 1$ **c.** 0 **d.** One rational solution
9. $x = 1, x = \dfrac{1}{3}$ **10.** $x = \dfrac{-7 \pm \sqrt{5}}{2}$ **11.** Height 4.6 ft; base 6.2 ft
12. Radius 12.0 ft **13.** $x = 9$ **14.** $y = 8, y = -64$
15. $(4, 0)(2, 0)(0, 8)$; c **16.** $(-4, 0)(3, 0)(-3, 0)(0, -36)$; b
17. $(-3, 0)(-1, 0)(0, -6)$; d **18.** $(4, 0)(-3, 0)(0, 0)$; a **19.** 256 ft
20. a. 660 million **b.** 1238 million **c.** 1981 **d.** 1999
21. $y = x^2 - 2$ is the graph of $y = x^2$ shifted down 2 units.
22. $y = (x + 3)^2$ is the graph of $y = x^2$ shifted 3 units to the left.
23. $y = -4x^2$ is the graph of $y = 4x^2$ opening downward instead of upward. **24. a.** (4, 2) **b.** Down **c.** Maximum point **d.** 2 **e.** $x = 4$ **25. a.** $2(x - 5)^2 + 1; (5, 1)$ **b.** (5, 1) **26.** 20,000 ft^2

Cumulative Review Exercises, Chapters 1–11, pp. 783–785

1. a. {2, 4, 6, 8, 10, 12, 16} **b.** {2, 8} **2.** $-3x^2 - 13x + 1$
3. -16 **4.** 7.2×10^{13} **5. a.** $(x + 2)(x + 3)(x - 3)$
b. Quotient: $x^2 + 5x + 6$; no remainder **6.** $x + 2$ **7.** $\dfrac{2\sqrt{2x}}{x}$
8. \$8000 in the 12% account; \$2000 in the 3% account **9.** (8, 7)
10. a. 720 ft **b.** 720 ft **c.** 12 sec **11.** $x = 3 \pm 4i$
12. $x = \dfrac{-5 \pm \sqrt{33}}{4}$ **13.** 25 **14.** $2(x + 5)(x^2 - 5x + 25)$

15.

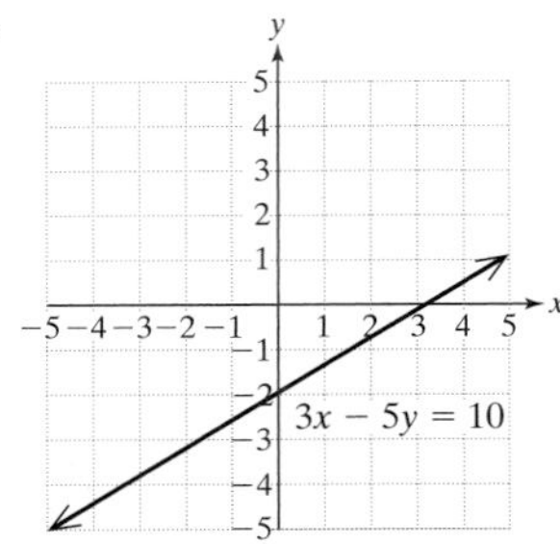

16. a. $\left(\frac{5}{2}, 0\right)(2, 0)$ **b.** (0, 10)

17. a.

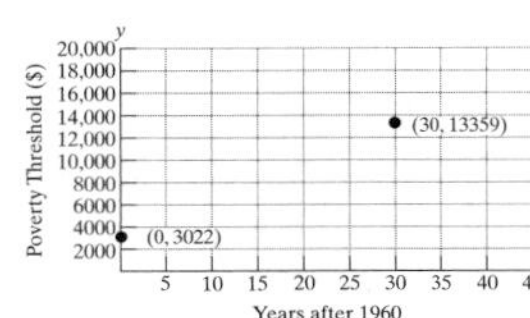

b. $y = 345x + 3022$ **c.** $9922

18. free throws: 565 2-pt. shots: 851 3-pt. shots: 30
19. The domain element 3 has more than one corresponding range element.
20.

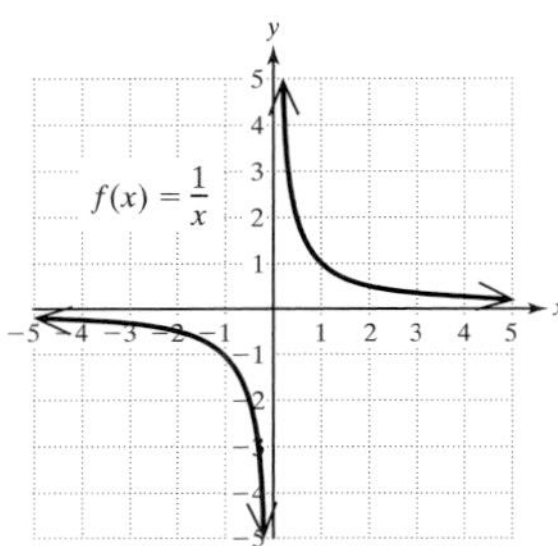

21. 39 **22. a.** Linear **b.** (0, 300000). If there are no passengers, the airport runs 300,000 flights per year. **c.** $m = 0.008$ or $m = \frac{8}{1000}$. There are eight additional flights per 1000 passengers.
23. a. 3 **b.** $\sqrt{2}$ **c.** Not a real number **24. a.** $[-4, \infty)$ **b.** $(-\infty, \infty)$
25. a. $(-\infty, 2]$ **b.** $(-\infty, 1) \cup \{2\} \cup [3, 4]$ **c.** 2 **d.** 3 **e.** 2 **f.** $x = -3$
26. $[-2, 8]$ **27.** $(-\infty, \infty)$ **28.** No solution **29.** $(-\infty, -5] \cup [1, \infty)$
30. $\left[-\frac{5}{2}, 3\right)$ **31.**

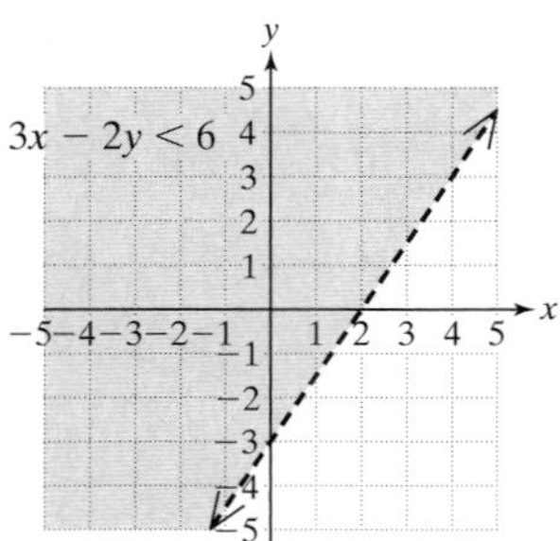

32. $x = \frac{1}{18}, x = \frac{1}{2}$ **33.** $f = \frac{pq}{p+q}$ **34.** $t = 7$ **35.** $\frac{y+1}{y-3}$
36. a. (3, 1) **b.** Upward **c.** (0, 19) **d.** No x-intercept

e.

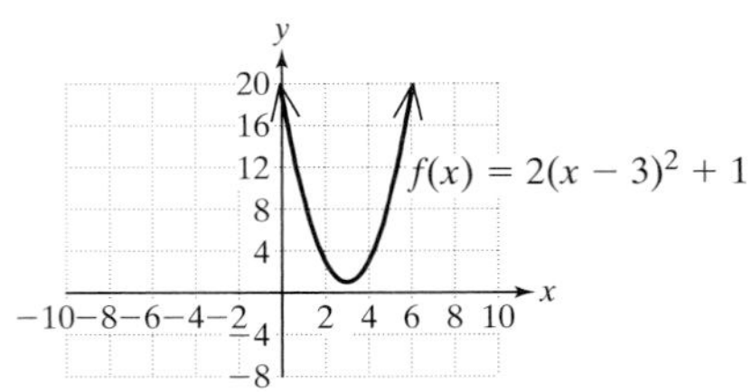

37. $x = 8 \pm \sqrt{62}$ **38.** Vertex: (8, −62)

CHAPTER 12

Section 12.1 Practice Exercises, pp. 792–794

1. $(f + g)(x) = 2x^2 + 5x + 4; (-\infty, \infty)$
3. $(g - f)(x) = 2x^2 + 3x - 4; (-\infty, \infty)$
5. $(f \cdot h)(x) = (x + 4)\sqrt{x - 1};\ [1, \infty)$
7. $(g \cdot f)(x) = 2x^3 + 12x^2 + 16x; (-\infty, \infty)$
9. $\left(\frac{h}{f}\right)(x) = \frac{\sqrt{x-1}}{x+4}; [1, \infty)$
11. $\left(\frac{f}{g}\right)(x) = \frac{x+4}{2x^2+4x}; (-\infty, -2) \cup (-2, 0) \cup (0, \infty)$
13. $(f \circ g)(x) = 2x^2 + 4x + 4;\ (-\infty, \infty)$
15. $(f \circ k)(x) = \frac{1}{x} + 4;\ (-\infty, 0) \cup (0, \infty)$
17. $(k \circ h)(x) = \frac{1}{\sqrt{x-1}};\ (1, \infty)$ **19.** No **21.** 0 **23.** −64 **25.** 2
27. 1 **29.** $\frac{1}{64}$ **31.** 0 **33.** Undefined **35.** 0 **37.** 1 **39.** 0
41. Undefined **43.** −1 **45.** 2 **47.** 2 **49. a.** $P(x) = 3.78x - 1$ **b.** $188 **51. a.** $F(t) = 0.2t + 4.4$; F represents the outstanding child support due. **b.** $F(0) = 4.4$ means that in 1985, 4.4 billion dollars of child support was not paid. $F(2) = 4.8$ means that in 1987, 4.8 billion dollars of child support was not paid. $F(4) = 5.2$ means that in 1989, 5.2 billion dollars of child support was not paid. **53. a.** $(D \circ r)(t) = 560t$; This function represents the total distance Joe travels as a function of time. **b.** 5600 ft

Section 12.2 Practice Exercises, pp. 801–803

1. Yes **3.** No **5.** Yes **7.** Yes **9.** No **11.** Yes
13. $g^{-1} = \{(5, 3), (1, 8), (9, -3), (2, 0)\}$
15. $r^{-1} = \{(3, a), (6, b), (9, c)\}$ **17.** The function is not one-to-one.
19. $h^{-1}(x) = x - 4$ **21.** $m^{-1}(x) = 3(x + 2)$ **23.** $p^{-1}(x) = -x + 10$
25. $f^{-1}(x) = \sqrt[3]{x}$ **27.** $g^{-1}(x) = \frac{x^3+1}{2}$ **29. a.** 1.2192 m, 15.24 m
b. $f^{-1}(x) = \frac{x}{0.3048}$ **c.** 4921.3 ft **31.** False **33.** True **35.** False
37. True **39.** $(b, 0)$ **41.** For example: $f(x) = x$
43. a. Domain $\{x \mid x \le 0\}$, range $\{y \mid y \ge -4\}$ **b.** Domain $\{x \mid x \ge -4\}$, range $\{y \mid y \le 0\}$ **45. a.** $\{x \mid -2 \le x \le 0\}$ **b.** $\{y \mid 0 \le y \le 2\}$ **c.** $\{x \mid 0 \le x \le 2\}$ **d.** $\{y \mid -2 \le y \le 0\}$

e.

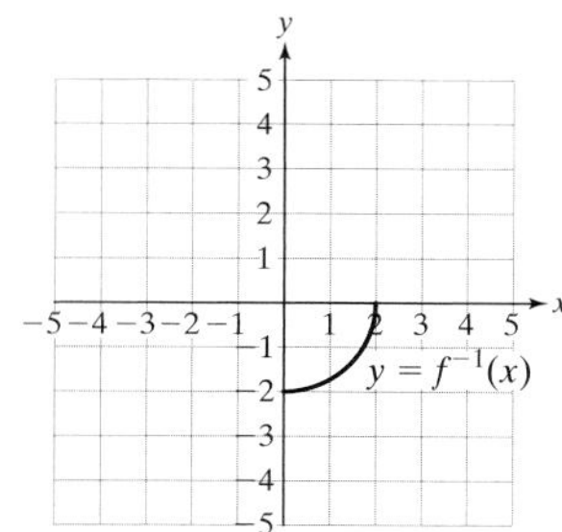

47. a. $\{x \mid 2 \le x \le 5\}$ **b.** $\{y \mid 0 \le y \le 3\}$ **c.** $\{x \mid 0 \le x \le 3\}$ **d.** $\{y \mid 2 \le y \le 5\}$ **e.**

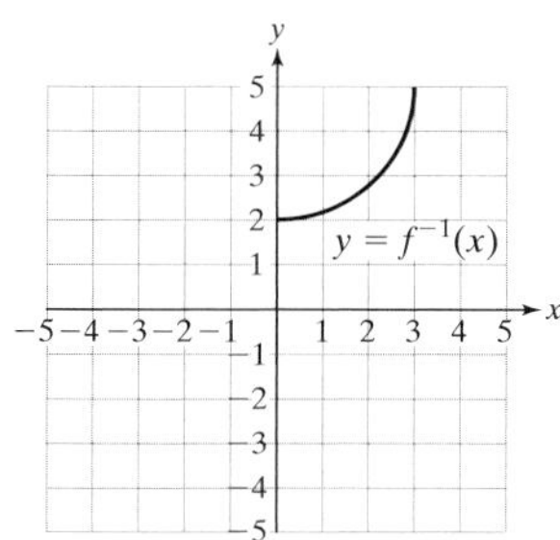

49. $p^{-1}(x) = \dfrac{3 - 3x}{x + 1}$, $x \ne -1$ **51.** $w^{-1}(x) = \dfrac{4 - 2x}{x}$, $x \ne 0$
53. $m^{-1}(x) = \sqrt{x + 1}$ **55.** $g^{-1}(x) = -\sqrt{x + 1}$
57. $v^{-1}(x) = x^2 - 16$, $x \ge 0$ **59.** $u^{-1}(x) = x^2 - 16$, $x \le 0$

61.

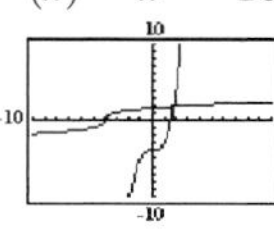

$k^{-1}(x) = \sqrt[3]{x + 4}$

63.

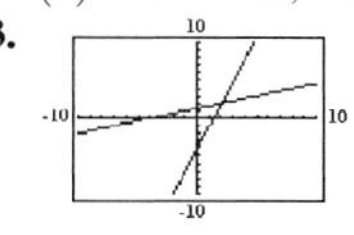

$m^{-1}(x) = \dfrac{x + 4}{3}$

Section 12.3 Practice Exercises, pp. 810–812

1. 25 **3.** $\dfrac{1}{1000}$ **5.** 6 **7.** 8 **9.** 5.8731 **11.** 1385.4557 **13.** 0.0063 **15.** 0.8950 **17. a.** 2 **b.** 3 **c.** Between 2 and 3, closer to 2 **19. a.** 4 **b.** 5 **c.** Between 4 and 5, closer to 5

21. $f(0) = 1$, $f(1) = \dfrac{1}{5}$, $f(2) = \dfrac{1}{25}$, $f(-1) = 5$, $f(-2) = 25$

23. $h(0) = 1$, $h(1) = \pi \approx 3.14$, $h(-1) = 0.32$, $h(\sqrt{2}) = 5.05$, $h(\pi) = 36.46$

25. $r(0) = 9$, $r(1) = 27$, $r(2) = 81$, $r(-1) = 3$, $r(-2) = 1$, $r(-3) = \dfrac{1}{3}$

27. If $b > 1$, the graph is increasing. If $0 < b < 1$, the graph is decreasing.

29.

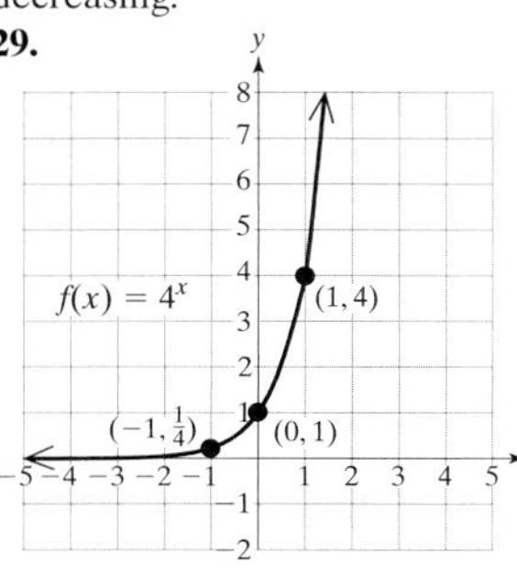

31.

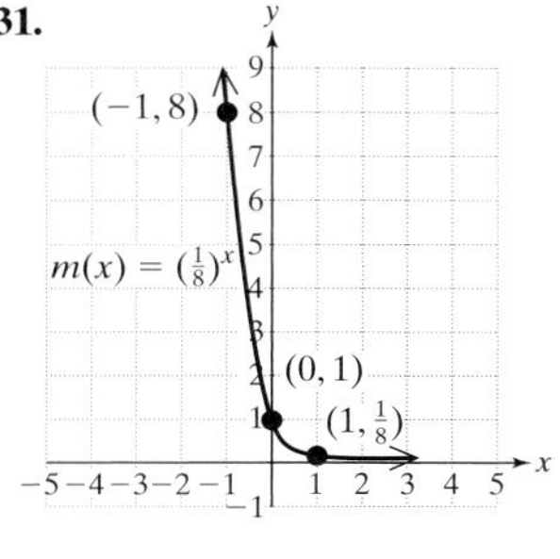

33.

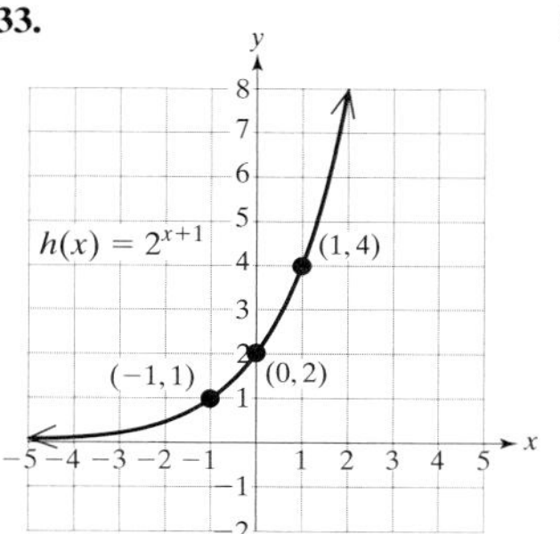

35.

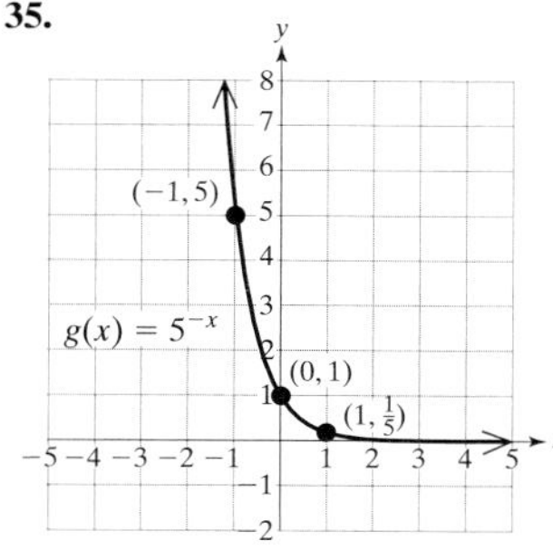

37. a. \$1640.67 **b.** \$2691.80 **c.** $A(0) = 1000$. The initial amount of the investment is \$1000. $A(7) = 2000$. The amount of the investment doubles in 7 years.

39. a. $I(t) = 3{,}600{,}000(1.0036)^t$ **b.** $S(t) = 3{,}500{,}000(1.012)^t$
c. $I(20) = 3{,}900{,}000$ $S(20) = 4{,}400{,}000$
$I(40) = 4{,}200{,}000$ $S(40) = 5{,}600{,}000$
$I(60) = 4{,}500{,}000$ $S(60) = 7{,}200{,}000$
d. Because Singapore has a higher growth rate, the population of Singapore will eventually overtake the population of Ireland. **e.** The population density (number of people per square mile) is more than 100 times as large for Singapore as for Ireland.

41.

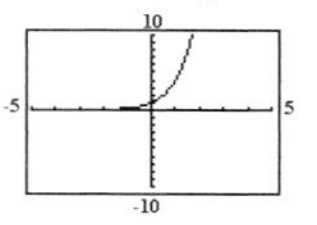

43.

45

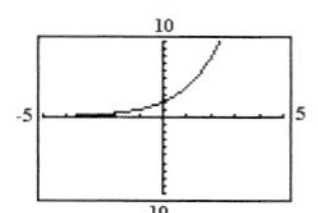

47

49. a.

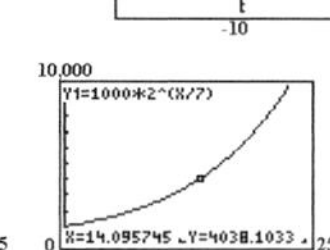

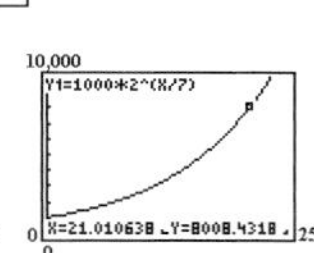

b. 7, 14, 21

Section 12.4 Practice Exercises, pp. 821–824

1. ii **3. a.** $g(-2) = \dfrac{1}{9}$, $g(-1) = \dfrac{1}{3}$, $g(0) = 1$, $g(1) = 3$, $g(2) = 9$
b.

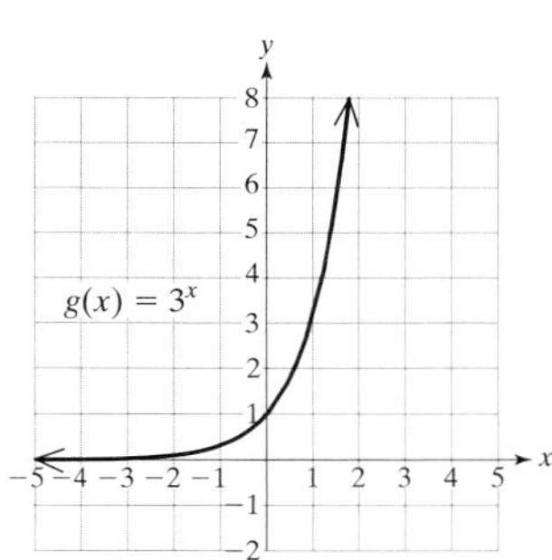

5. a. $s(-2) = \dfrac{25}{4}$, $s(-1) = \dfrac{5}{2}$, $s(0) = 1$, $s(1) = \dfrac{2}{5}$, $s(2) = \dfrac{4}{25}$

b.

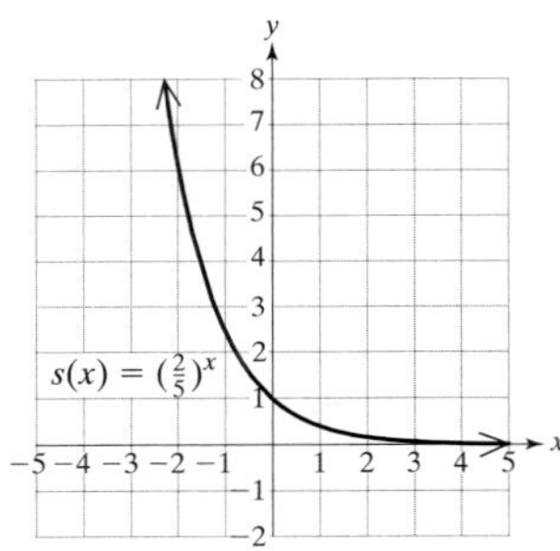

7. $b^y = x$ **9.** $\log_{10}(1000) = 3$ **11.** $\log_8(2) = \frac{1}{3}$

13. $\log_8\left(\frac{1}{64}\right) = -2$ **15.** $\log_b(x) = y$ **17.** $\log_e(x) = y$

19. $\log_H(q) = m$ **21.** $125^{2/3} = 25$ **23.** $25^{-1/2} = \frac{1}{5}$ **25.** $2^7 = 128$

27. $b^y = 82$ **29.** $2^x = 7$ **31.** $\left(\frac{1}{2}\right)^6 = x$ **33.** 3 **35.** −4 **37.** $\frac{1}{3}$

39. 1 **41.** 3 **43.** 0 **45.** 2 **47.** 4 **49.** −1 **51.** −3 **53.** 0.7782
55. 0.4971 **57.** −1.5051 **59** −2.2676 **61.** 5.5315 **63.** −7.4202
65. a. Slightly less than 2 **b.** Slightly more than 1
c. log 93 = 1.9685, log 12 = 1.0792

67. a. $f\left(\frac{1}{64}\right) = -3, f\left(\frac{1}{16}\right) = -2, f\left(\frac{1}{4}\right) = -1, f(1) = 0,$
$f(4) = 1, f(16) = 2, f(64) = 3$

b.

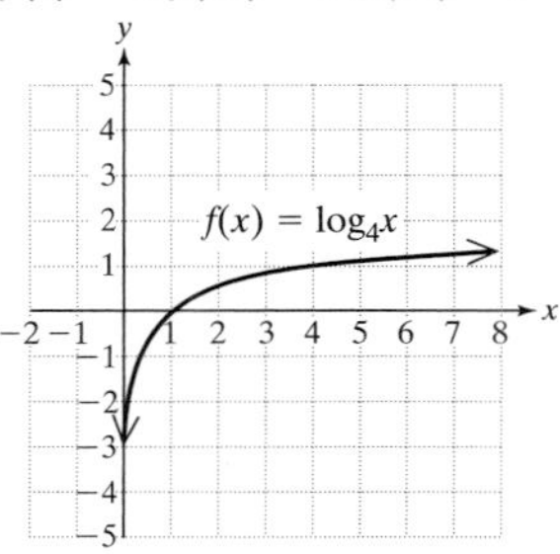

69. $3^y = x$

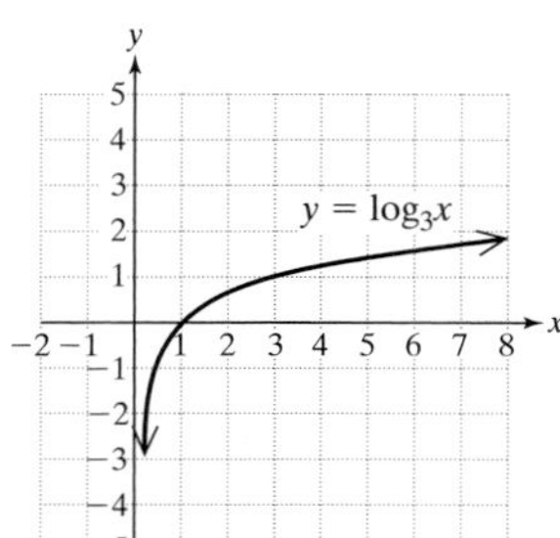

x	y
$\frac{1}{9}$	−2
$\frac{1}{3}$	−1
1	0
3	1
9	2

71. $\left(\frac{1}{2}\right)^y = x$

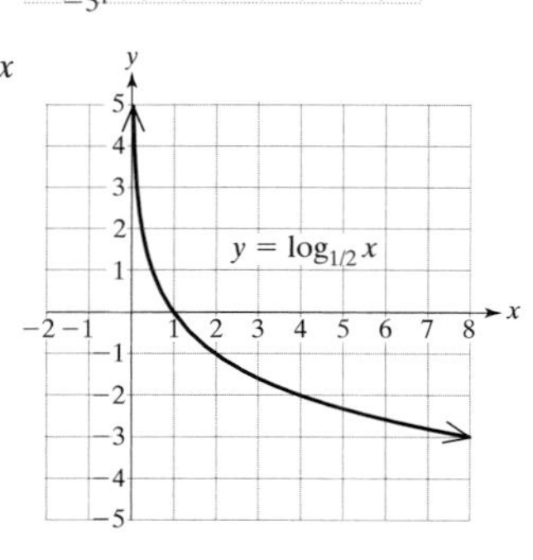

x	y
4	−2
2	−1
1	0
$\frac{1}{2}$	1
$\frac{1}{4}$	2

73. $\{x \mid x > 0\}$ **75.** $\{x \mid x > 0\}$ **77.** $(5, \infty)$ **79.** $(-1.2, \infty)$
81. $(-\infty, 0) \cup (0, \infty)$

83. a.

t (months)	0	1	2	6	12	24
$S_1(t)$	91	82.0	76.7	65.6	57.6	49.1
$S_2(t)$	88	83.5	80.8	75.3	71.3	67.0

b. Group 1: 91 Group 2: 88 **c.** Method II **85. a.** ≈3.7 **b.** 11
87. Domain: $(-2, \infty)$; asymptote: $x = -2$

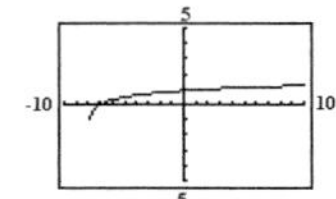

89. Domain: $(-8, \infty)$; asymptote: $x = -8$

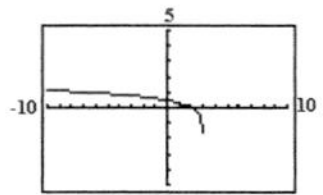

91. Domain: $(-\infty, 3)$; asymptote: $x = 3$

Section 12.5 Practice Exercises, pp. 829–831

1. $\frac{1}{64}$ **3.** 5 **5.** 2.9707 **7.** 1.4314 **9.** d **11.** b
13. For example: $\log_{10} 1 = 0$ **15.** For example: $\log_4 4^2 = 2$ **17.** 1
19. 4 **21.** 11 **23.** 3 **25.** 0 **27.** 9 **29.** Expressions a and c are equivalent. **31.** Expressions a and c are equivalent.
33. $\log_3 x - \log_3 5$ **35.** $\log 2 + \log x$ **37.** $4 \log_{10} x$
39. $\log_4 a + \log_4 b - \log_4 c$
41. $\frac{1}{2}\log_b x + \log_b y - 3\log_b z - \log_b w$ **43.** $\log(CABIN)$
45. $\log_3\left(\frac{x^2 z}{y^3}\right)$ **47.** $\log_b(x^2)$ **49.** $\log_8(8a^5)$ or $\log_8 a^5 + 1$
51. a. $B = 10 \log I - 10 \log I_0$ **b.** $10 \log I + 160$ **53. b.** 4.71
c. ≈1.4
55. Domain: $(-\infty, 0) \cup (0, \infty)$ Domain: $(0, \infty)$

a.

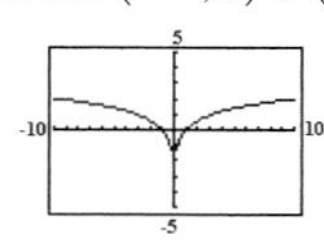

b.

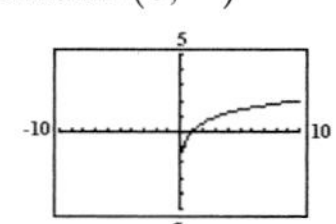

c. They are equivalent for all x in the intersection of their domains, $(0, \infty)$.

Section 12.6 Practice Exercises, pp. 838–841

1.

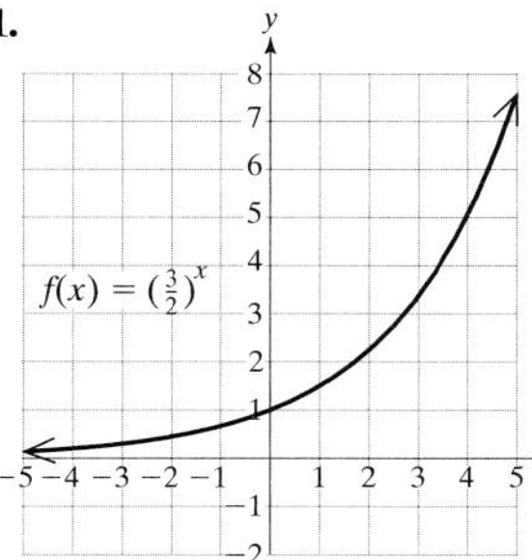

3.

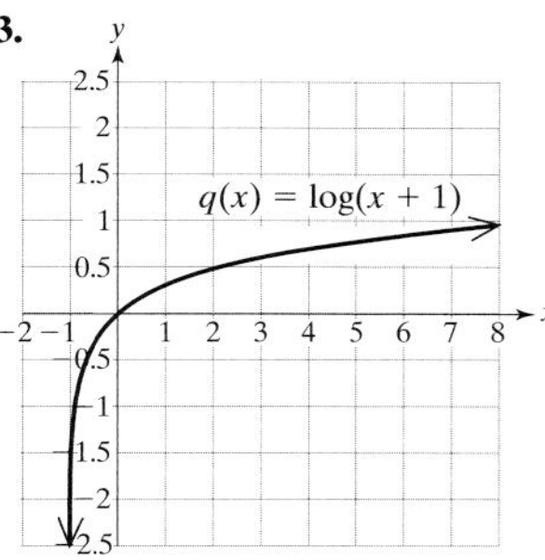

5. $4 \ln a + \frac{1}{2} \ln b - \ln c$ **7.** $\frac{1}{5} \ln a + \frac{1}{5} \ln b - \frac{2}{5} \ln c$ **9.** $\ln\left(\frac{a^2}{b\sqrt[3]{c}}\right)$

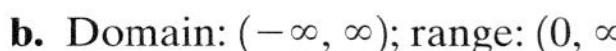

11. $\ln\left(\frac{x^4}{y^3 z}\right)$

13. a.

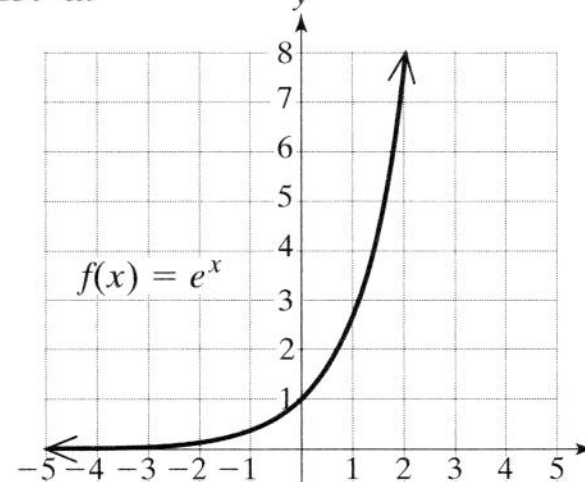

b. Domain: $(-\infty, \infty)$; range: $(0, \infty)$

c.

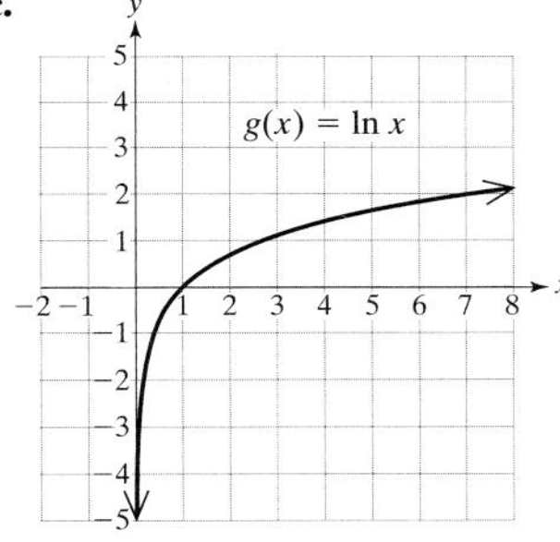

d. Domain: $(0, \infty)$; range: $(-\infty, \infty)$

15. Domain: $(-\infty, \infty)$; range: $(0, \infty)$

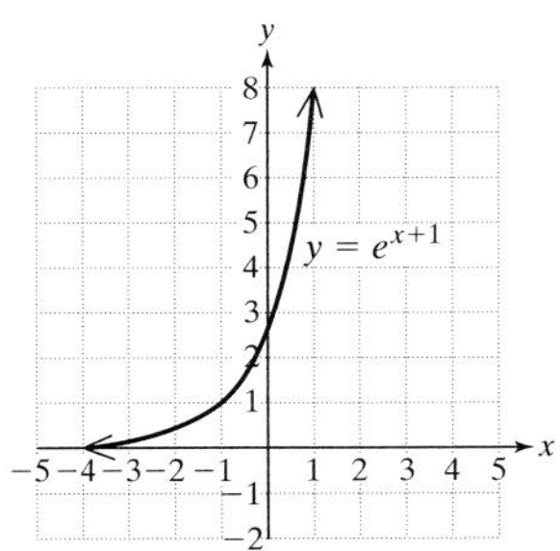

0.05, 0.14, 0.37, 1, 2.72, 7.39, 20.09

17. Domain: $(2, \infty)$; range: $(-\infty, \infty)$

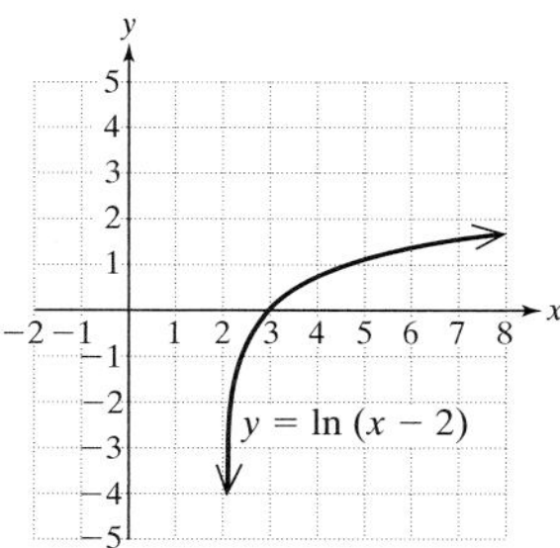

−1.39, −0.69, −0.29, 0, 0.69, 1.10, 1.39

19. a. 2.9570 **b.** 2.9570 **c.** They are the same. **21.** 2.8074 **23.** 1.5283 **25.** −2.1269 **27.** 0 **29.** −3.3219 **31.** −3.8124 **33. a.** Between 2 and 3 **b.** Slightly less than 3 **c.** $\log_3 15 = 2.4650$, $\log_3 25 = 2.9299$ **35. a.** Slightly more than 1 **b.** Slightly less than 2 **c.** $\log_6 10 = 1.2851$, $\log_6 30 = 1.8982$ **37. a.** 91 deaths **b.** Sept. 5: 349 deaths, Sept. 10: 459 deaths, Sept. 20: 570 deaths **39. a.** 12.6 years **b.** 8.7 years **c.** 17.4 years **41. a.** \$7152.26 **b.** \$7740.30 **c.** \$8711.07 **d.** \$10,190.52 An investment grows more rapidly at higher interest rates. **43. a.** \$22,161.83 **b.** \$22,321.96 **c.** \$22,358.78 **d.** \$22,376.76 **e.** \$22,377.37 More money is earned at a greater number of compound periods per year. **45. a.** \$13,498.59 **b.** \$18,221.19 **c.** \$24,596.03 **d.** \$33,201.17 **e.** \$60,496.47 More money is earned over a longer period of time.

47. a. b.

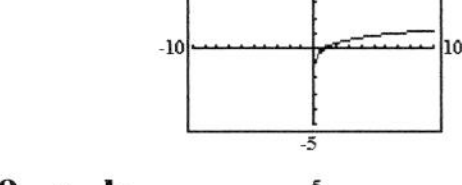

c. They appear to be the same.

49. a. b.

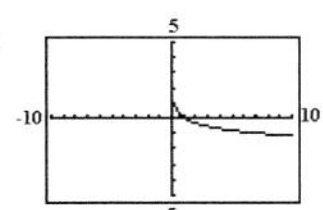

c. They appear to be the same.

51. **53.**

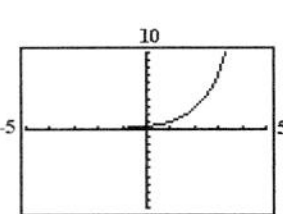

55.

Section 12.7 Practice Exercises, pp. 851–854

1. a.

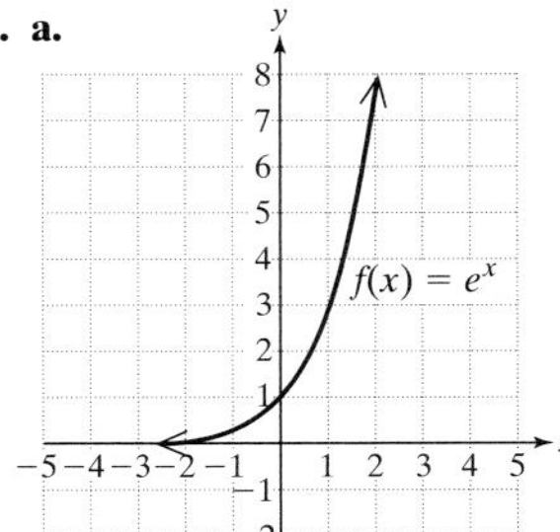

b. Domain: $(-\infty, \infty)$; range: $(0, \infty)$

3. a.

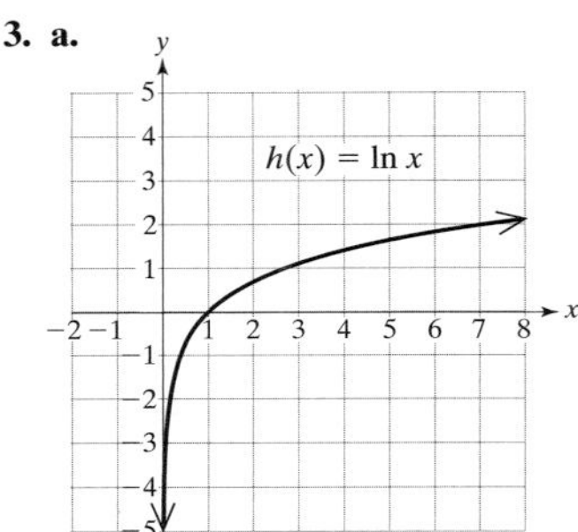

b. $x = 0$ **c.** Domain: $(0, \infty)$; range: $(-\infty, \infty)$ **5.** $\log_b[(x - 1)(x + 2)]$

7. $\log_b\left(\frac{x}{1 - x}\right)$ **9.** $x = 5$; domain: $(5, \infty)$

11. $x = -2$; domain: $(-2, \infty)$ **13.** $x = -\frac{1}{2}$; domain: $\left(-\frac{1}{2}, \infty\right)$

15. $x = 4$ **17.** $x = -6$ **19.** $x = \frac{1}{2}$ **21.** $x = 2$ **23.** $x = \frac{11}{12}$

25. $a = \frac{\ln 21}{\ln 8} \approx 1.464$ **27.** $x = \ln 8.1254 \approx 2.095$

29. $t = \log 0.0138 \approx -1.860$ **31.** $h = \frac{\ln 15}{0.07} \approx 38.686$

33. $m = \frac{\ln 4}{0.04} \approx 34.657$ **35.** $x = \frac{\ln 3}{\ln 5 - \ln 3} \approx 2.151$ **37.** 9.9 years **39. a.** 7.8 g **b.** 18.5 days **41. a.** ≈ 1285 million (or 1,285,000,000) people **b.** ≈ 1412 million (or 1,412,000,000) people **c.** The year 2049 ($t \approx 50.8$) **43. a.** 500 bacteria **b.** ≈ 660 **c.** ≈ 25 minutes **45.** 38.4 years **47.** $x = 9$ **49.** $p = 10^{42}$ **51.** $x = e^{0.08} \approx 1.083$ **53.** $x = 5$ **55.** $b = 10$ **57.** $y = 25$ **59.** $c = 59$ **61.** $y = 1$

63. $k = \frac{3}{2}$ **65.** $x = 4$ **67.** $t = 2$ **69.** No solution

71. $x = 16, x = 256$ **73. a.** 1.42 kg **b.** No **75. a.** i) 5 ii) 6 **b.** $10^{7.1}$ times ($\approx 12{,}590{,}000$) more intense

77.

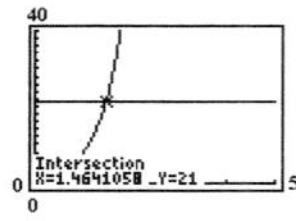

79.

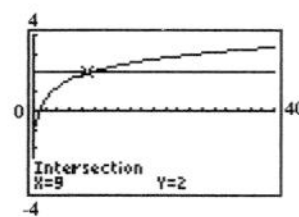

Chapter 12 Review Exercises, pp. 860–864

1. $(f - g)(x) = 2x^3 + 9x - 7; (-\infty, \infty)$

3. $(f \cdot n)(x) = \frac{x - 7}{x - 2}; (-\infty, 2)\cup(2, \infty)$

5. $\left(\frac{f}{g}\right)(x) = \frac{x - 7}{-2x^3 - 8x}; (-\infty, 0)\cup(0, \infty)$

7. $(m \circ f)(x) = \sqrt{x - 7}; [7, \infty)$ **9.** $\sqrt{32}$ or $4\sqrt{2}$ **11.** -167 **13. a.** $(2x + 1)^2$ or $4x^2 + 4x + 1$ **b.** $2x^2 + 1$ **c.** No, $f\circ g \neq g\circ f$ **15.** -3 **17.** -1 **19.** 1 **21.** Yes **23.** No

25. $\{(5, 3), (9, 2), (-1, 0), (1, 4)\}$ **27.** $p^{-1}(x) = -\frac{1}{4}x + \frac{3}{4}$

29. $g^{-1}(x) = (x - 3)^5$ **31.** $m^{-1}(x) = \frac{-2x - 2}{x - 1}$

33. $(f \circ g)(x) = 5\left(\frac{1}{5}x + \frac{2}{5}\right) - 2 = x + 2 - 2 = x$

$(g \circ f)(x) = \frac{1}{5}(5x - 2) + \frac{2}{5} = x - \frac{2}{5} + \frac{2}{5} = x$

35. a. Domain: $\{x \mid x \geq -1\}$; Range: $\{y \mid y \geq 0\}$ **b.** Domain: $\{x \mid x \geq 0\}$; Range: $\{y \mid y \geq -1\}$ **37.** 1024 **39.** 2 **41.** 8.825 **43.** 1.627

45.

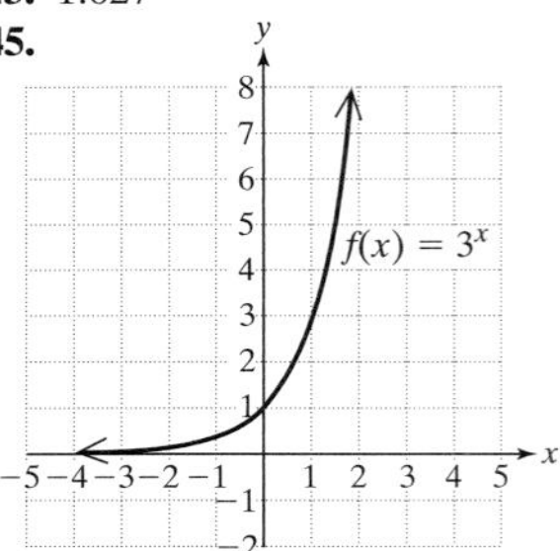

47.

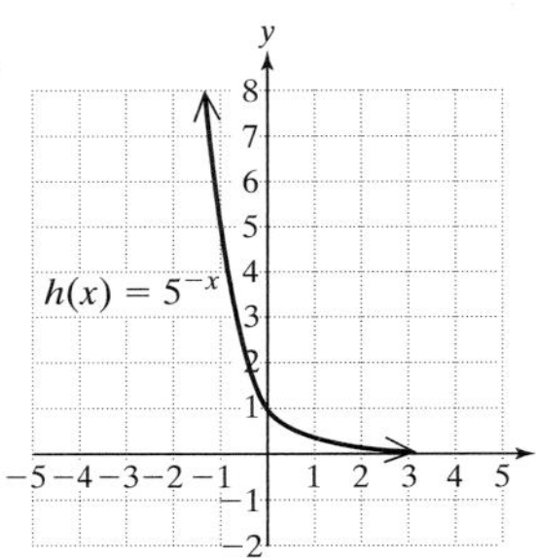

49. a. Horizontal **b.** $y = 0$ **51.** -3 **53.** 1 **55.** 4 **57.** 5

59.

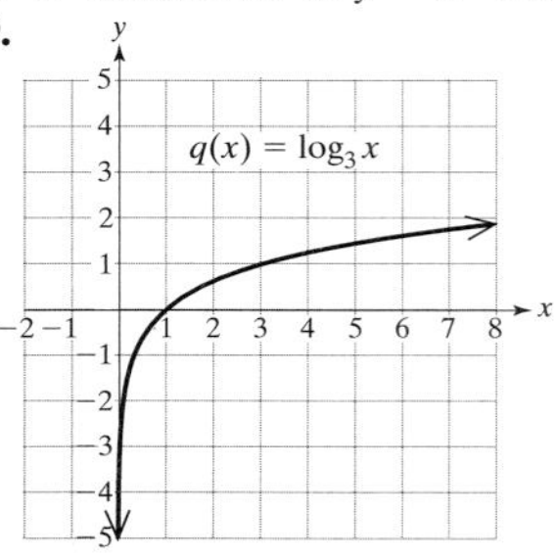

61. a. Vertical asymptote **b.** $x = 0$ **63.** 1 **65.** 0

67. a. $\log_b x + \log_b y$ **b.** $\log_b\left(\frac{x}{y}\right)$ **c.** $p \log_b x$ **69.** 1.386 **71.** 3.218

73. 3.401 **75.** 2.119 **77.** 8.316 **79.** $\log_3\left(\frac{\sqrt{ab}}{c^2d^4}\right)$ **81.** 0

83. 14.0940 **85.** 57.2795 **87.** -2.1972 **89.** -4.1227 **91.** -2.2366 **93.** 1.3029 **95.** 1.9943 **97.** -4.1918 **99.** -3.6668 **101. a.** $S(0) = 95$; the student's score is 95 at the end of the course. **b.** $S(6) \approx 23.7$; the student's score is 23.7 after 6 months. **c.** $S(12) \approx 20.2$; the student's score is 20.2 after 1 year. **d.** The limiting value is 20. **103.** $(-\infty, \infty)$ **105.** $(-\infty, \infty)$ **107.** $(-5, \infty)$

109. $\left(\frac{4}{3}, \infty\right)$ **111.** $x = 125$ **113.** $y = 216$ **115.** $w = \frac{1001}{2} = 500.5$

117. $p = 5$ **119.** $x = -1$ **121.** $a = \frac{\ln 21}{\ln 4} \approx 2.1962$

123. $x = -\ln 0.1 \approx 2.3026$ **125.** $n = \frac{\log 1512}{2} \approx 1.5898$

127. a. 1.09 μg **b.** 0.15 μg **c.** 16.08 days **129. a.** $V(0) = 15{,}000$; the initial value of the car is \$15,000. **b.** $V(10) = 3347$; the value of the car after 10 years is \$3347. **c.** 7.3 years **d.** The limiting value is 0.

Chapter 12 Test, pp. 865–866

1. $\left(\frac{f}{g}\right)(x) = \frac{x - 4}{\sqrt{x + 2}}$ **2.** $(h \cdot g)(x) = \frac{\sqrt{x + 2}}{x}$

3. $(g \circ f)(x) = \sqrt{x - 2}$ **4.** $(h \circ f)(x) = \frac{1}{x - 4}$ **5.** 0 **6.** $-\frac{3}{2}$ **7.** $\frac{1}{4}$

8. Undefined **9.** $\left(\frac{g}{f}\right)(x) = \frac{\sqrt{x + 2}}{x - 4}$; Domain: $[-2, 4)\cup(4, \infty)$

10. A function is one-to-one if it passes the horizontal line test. **11.** b **12.** $f^{-1}(x) = 4x - 12$ **13.** $g^{-1}(x) = \sqrt{x} + 1$

14.

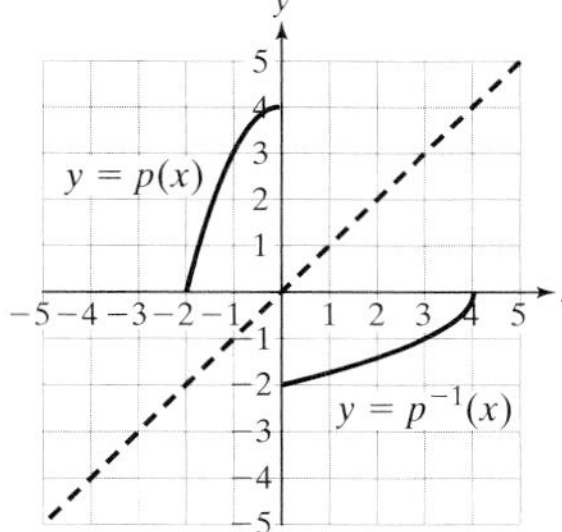

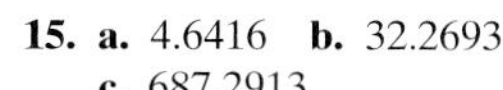
15. a. 4.6416 **b.** 32.2693 **c.** 687.2913

16.

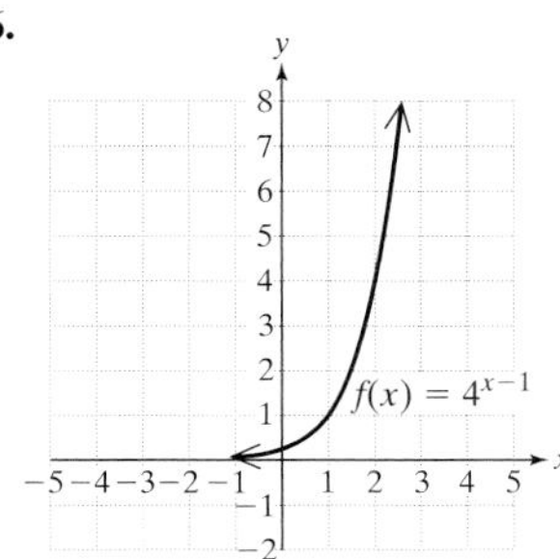

17. a. $\log_{16} 8 = \frac{3}{4}$ **b.** $x^5 = 31$

18.

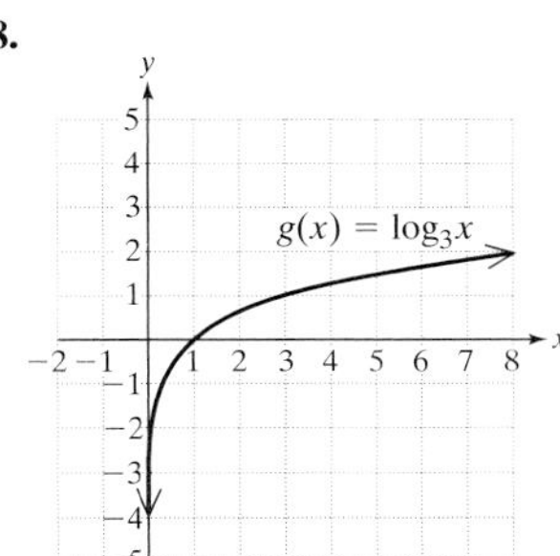

19. $\frac{\log_a n}{\log_a b}$ **20. a.** 1.3222 **b.** 1.8502 **c.** −2.5850 **21. a.** $1 + \log_3 x$ **b.** −5 **22. a.** $\log_b(\sqrt{x}\, y^3)$ **b.** $\log \frac{1}{a^3}$ or $-\log a^3$ **23. a.** 1.6487 **b.** 0.0498 **c.** −1.0986 **d.** 1 **24. a.** $y = \ln x$ **b.** $y = e^x$
25. a. $p(4) \approx 59.8$; 59.8% of the material is retained after 4 months. **b.** $p(12) \approx 40.7$; 40.7% of the material is retained after 1 year. **c.** $p(0) = 92$; 92% of the material is retained at the end of the course. **26. a.** 9762 thousand (or 9,762,000) people **b.** The year 2021 ($t \approx 20.4$) **27. a.** $P(0) = 300$; there are 300 bacteria initially. **b.** 35,588 bacteria **c.** 1,120,537 bacteria **d.** 1,495,831 bacteria **e.** The limiting amount appears to be 1,500,000. **28.** $x = 25$
29. $x = 32$ **30.** $x \approx 4.023$ **31.** $x = -7$ **32.** $x \approx 2.822$
33. $x \approx 2.301$ **34. a.** $p(2500) = 560.2$; At 2500 m the atmospheric pressure is 560.2 mm Hg. **b.** 760 mm Hg **c.** 1498.8 m
35. a. \$2909.98 **b.** 9.24 years to double

Cumulative Review Exercises, Chapters 1–12, pp. 867–870

1. $-\frac{5}{4}$ **2.** $-1 + \frac{p}{2} + \frac{3p^3}{4}$ **3.** Quotient: $t^3 + 2t^2 - 9t - 18$; remainder: 0 **4.** $|x - 3|$ **5.** $\frac{2\sqrt[3]{25}}{5}$ **6.** $2\sqrt{7}$ in. **7.** $\frac{4d^{1/10}}{c}$

8. $(\sqrt{15} - \sqrt{6} + \sqrt{30} - 2\sqrt{3})\ \text{m}^2$ **9.** $-\frac{7}{29} - \frac{26}{29}i$ **10.** 42°, 48°
11. $m = \frac{2}{3}$ **12.** $x = 6, x = \frac{3}{2}$ **13.** 4.8 L **14.** 24 min **15.** (7, −1)
16. $w = -\frac{23}{11}$ **17.** $x = \frac{c + d}{a - b}$ **18.** $t = \frac{\sqrt{2sg}}{g}$
19. $T = \frac{1 - \left(\frac{V_0}{V}\right)^2}{k}$ or $\frac{V^2 - V_0^2}{kV^2}$ **20.** (7, 0), (3, 0)
21. a. $-30t$ **b.** $-10t^2$ **c.** $2t^2 + 5t$ **22.** $q = 0$ **23. a.** $x = 2$ **b.** $y = 6$ **c.** $y = \frac{1}{2}x + 5$

24. a.

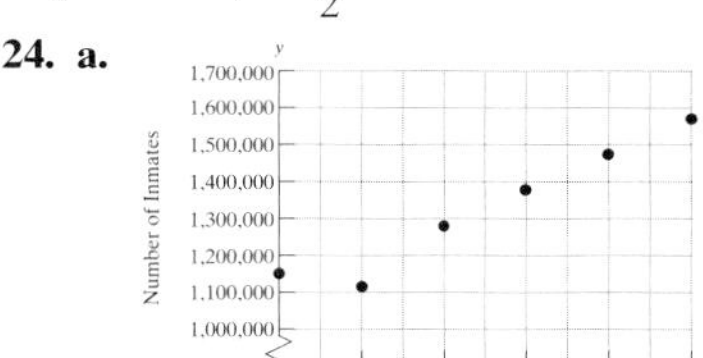

b. $y = 82{,}500x + 1{,}148{,}000$ **c.** $m = 82{,}500$; There is an increase of 82,500 inmates per year. **d.** 1,973,000 inmates

25. 40°, 80°, 60° **26.** $\{(x, y) \mid -2x + y = -4\}$ **27. a.** vi **b.** i **c.** v **d.** x **e.** ii **f.** ix **g.** iv **h.** viii **i.** vii **j.** iii **28.** $\left[\frac{1}{2}, \infty\right)$
29. $f^{-1}(x) = \frac{1}{5}x + \frac{2}{15}$ **30.** 40 m³ **31.** $\frac{-x^2 + 5x - 25}{(2x + 1)(x - 2)}$ **32.** $1 + x$
33. a. Yes; $x \neq 4, x \neq -2$ **b.** $x = 8$ **c.** $(-\infty, -2) \cup (4, 8]$
34. The numbers are $-\frac{3}{4}$, 4 **35.** $x = -4$ **36.** $(-\infty, \infty)$
37. a. $P(6) = 2{,}000{,}000$, $P(12) = 1{,}000{,}000$, $P(18) = 500{,}000$, $P(24) = 250{,}000$, $P(30) = 125{,}000$
b.

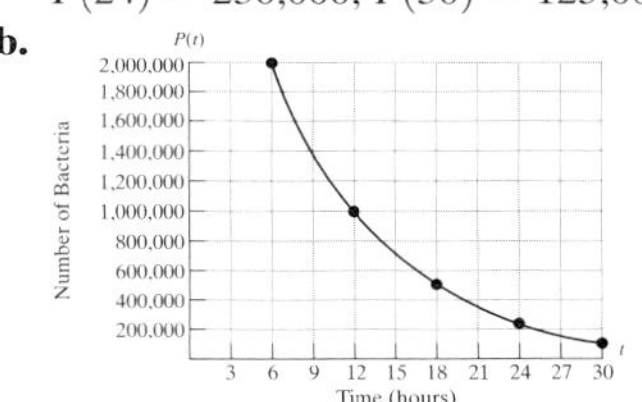

c. 48 hr

38. a. 2 **b.** −3. **c.** 6 **d.** 3 **39. a.** 217.0723 **b.** 23.1407 **c.** 0.1768 **d.** 3.7293 **e.** –0.4005 **f.** 2.6047 **40.** $x = \frac{2}{3}$
41. $x = \ln 100 \approx 4.6052$ **42.** $x = 3$ **43.** $\log\left(\frac{\sqrt{z}}{x^2y^3}\right)$
44. $\frac{2}{3}\ln x - \frac{1}{3}\ln y$

CHAPTER 13

Section 13.1 Practice Exercises, pp. 877–880

1. $\sqrt{281}$ **3.** $3\sqrt{2}$ **5.** $\frac{\sqrt{226}}{10}$ **7.** 19 **9.** 10 **11.** $\sqrt{255}$
13. Subtract 5 and −7. This becomes $5 - (-7) = 12$. **15.** $y = 13$ or $y = 1$ **17.** $x = 0$ or $x = 8$ **19.** Yes **21.** No

23. Center $(4, -2)$; $r = 3$;

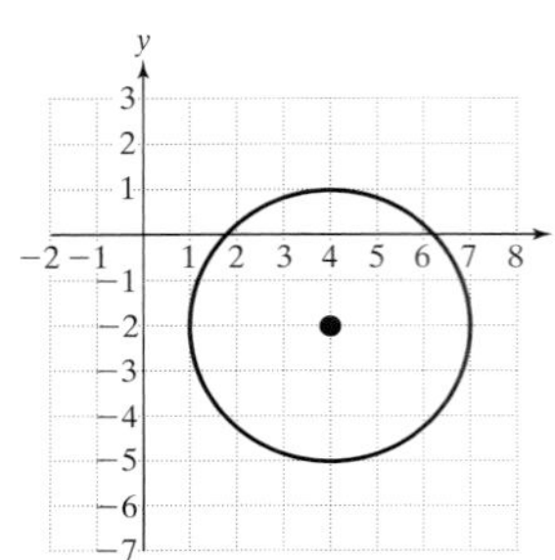

25. Center $(-1, -1)$; $r = 1$;

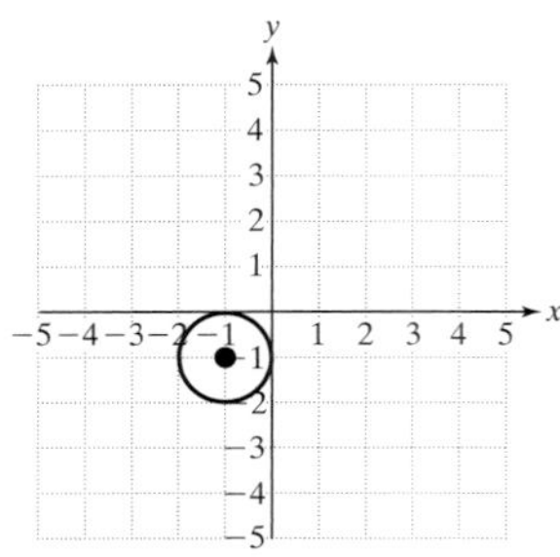

27. Center $(0, 5)$; $r = 5$;

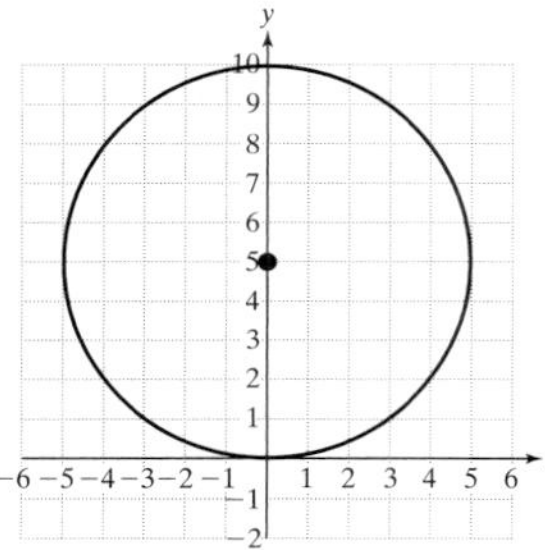

29. Center $(3, 0)$; $r = 2\sqrt{2}$;

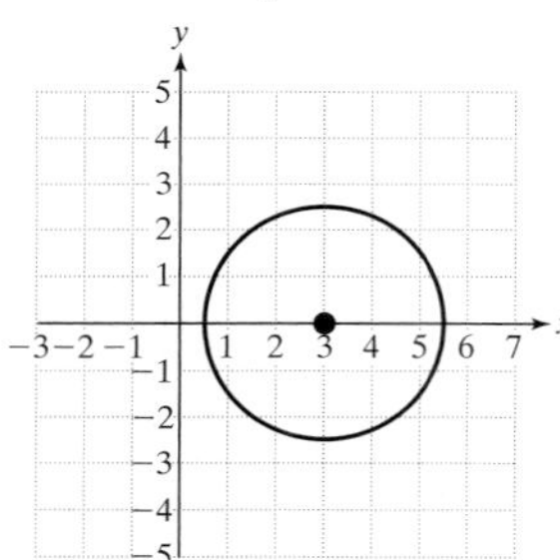

31. Center $(0, 0)$; $r = \sqrt{6}$;

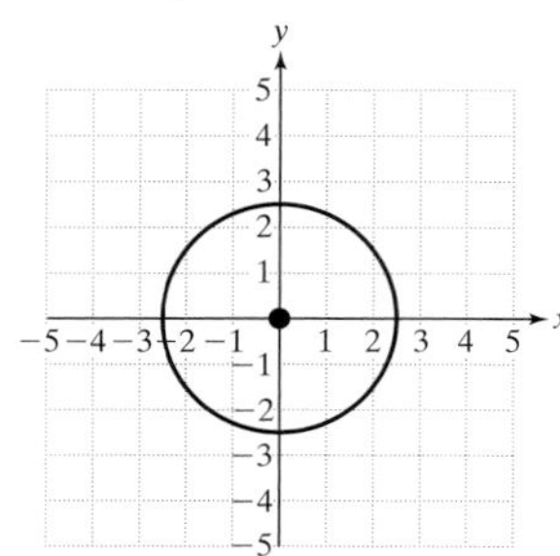

33. Center $\left(-\frac{4}{5}, 0\right)$; $r = \frac{8}{5}$;

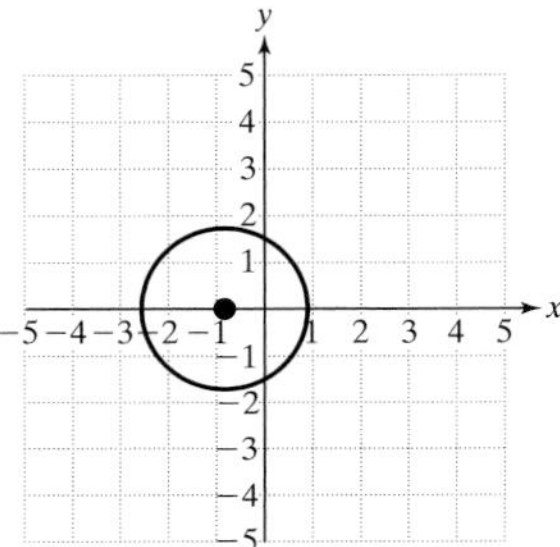

35. Center $(1, 3)$; $r = 6$;

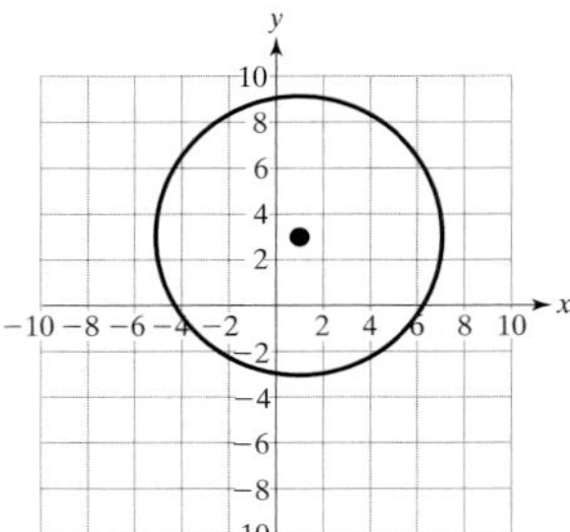

37. Center $(0, -3)$; $r = \frac{4}{3}$;

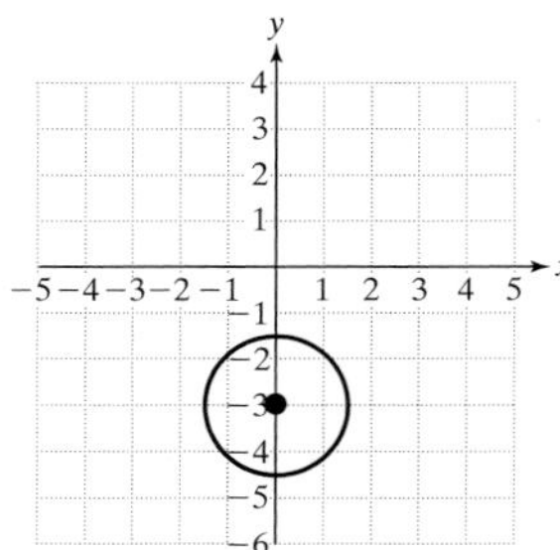

39. $x^2 + y^2 = 4$ **41.** $x^2 + (y - 2)^2 = 4$ **43.** $(x + 2)^2 + (y - 2)^2 = 9$ **45.** $x^2 + y^2 = 49$ **47.** $(x + 3)^2 + (y + 4)^2 = 36$ **49.** $(1, 2)$

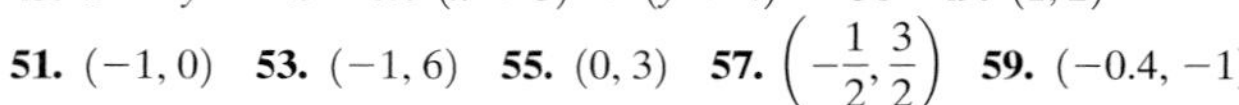

51. $(-1, 0)$ **53.** $(-1, 6)$ **55.** $(0, 3)$ **57.** $\left(-\frac{1}{2}, \frac{3}{2}\right)$ **59.** $(-0.4, -1)$

61. $(40, 7\frac{1}{2})$; they should meet 40 miles east, $7\frac{1}{2}$ miles north of the warehouse.

63. $(x - 4)^2 + (y - 4)^2 = 16$

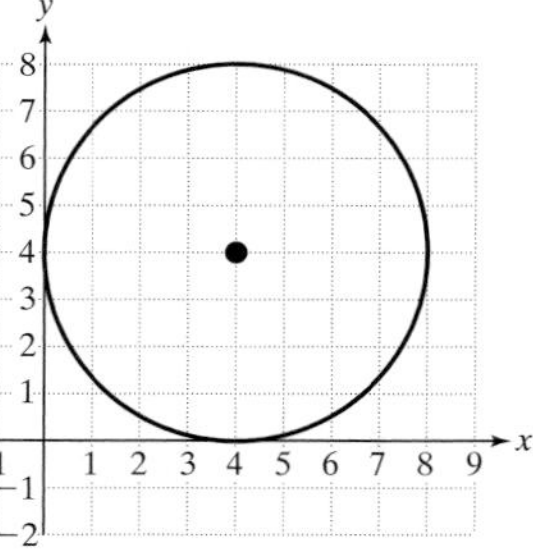

65. $(x - 1)^2 + (y - 1)^2 = 29$

67.

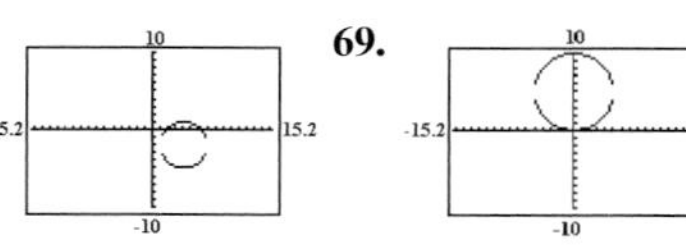

69.

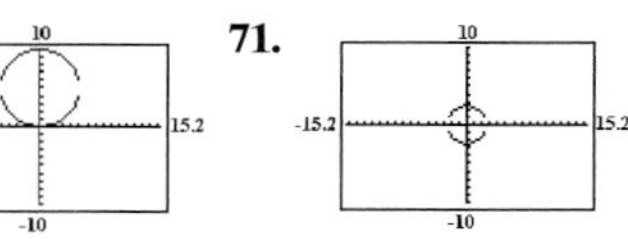

71.

Section 13.2 Practice Exercises, pp. 885–886

1. For a parabola whose equation is written in the form $y = a(x-h)^2 + k$, if $a > 0$ the parabola opens upward, if $a < 0$ the parabola opens downward. For a parabola written in the form $x = a(y-k)^2 + h$, if $a > 0$ the parabola opens right, if $a < 0$ the parabola opens left. **3.** Vertical axis of symmetry; opens upward **5.** Vertical axis of symmetry: opens downward **7.** Horizontal axis of symmetry; opens right **9.** Horizontal axis of symmetry; opens left **11.** Vertical axis of symmetry; opens downward **13.** The focus is $(h, k+p)$, where $p = \frac{1}{4a}$. **15.** The directrix is the line $x = h - p$, where $p = \frac{1}{4a}$. **17.** True **19.** False **21.** True

23.

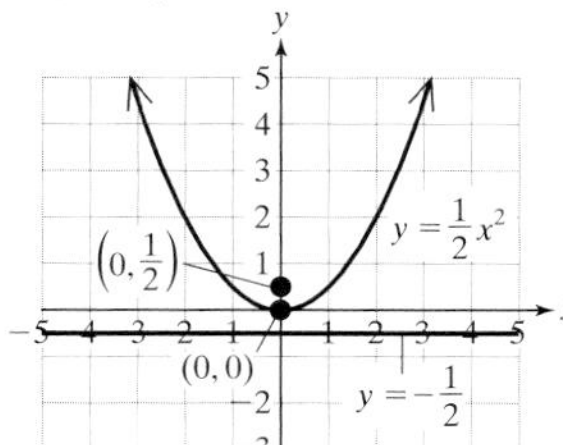

25.

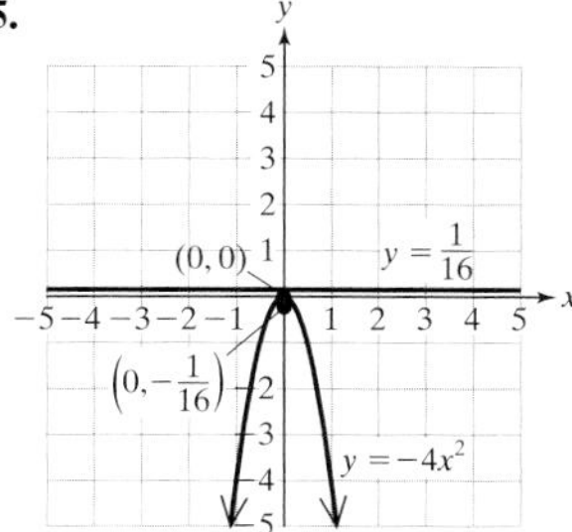

27.

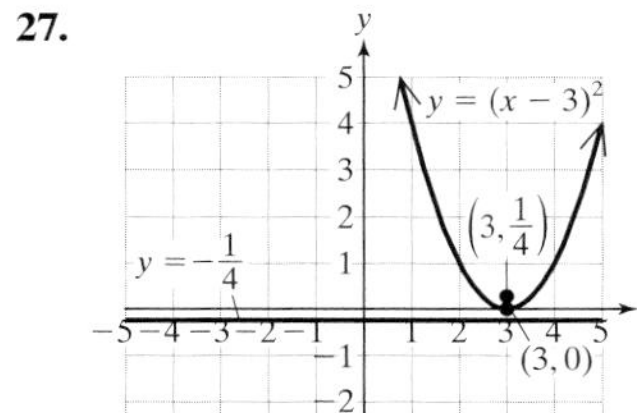

29.

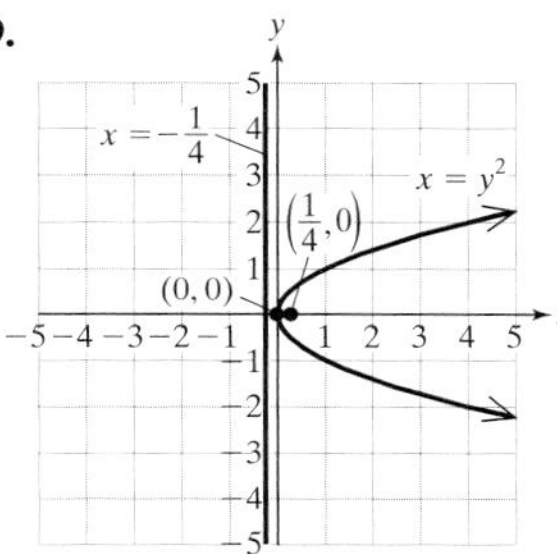

31.

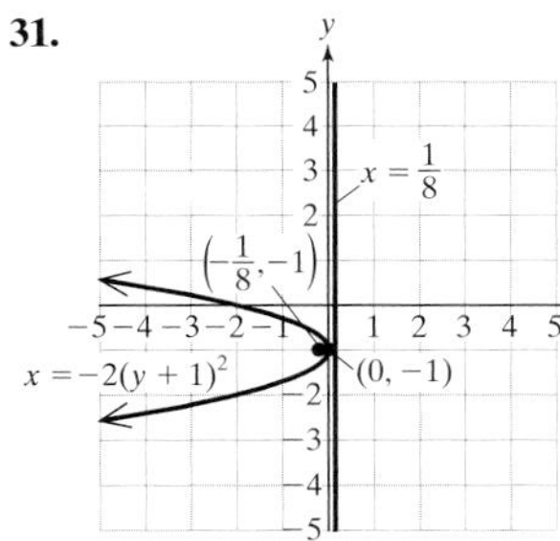

33.

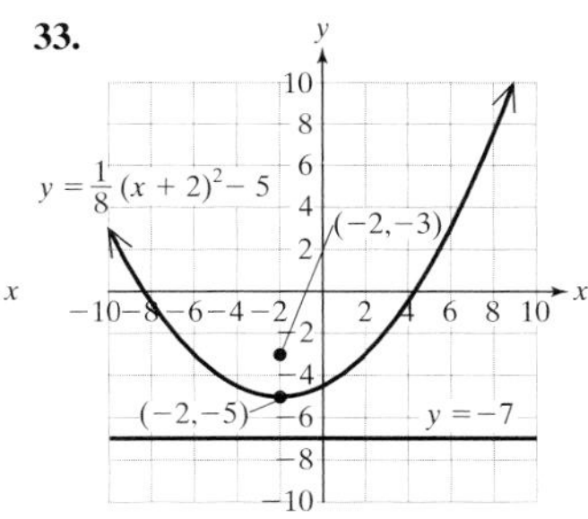

35.

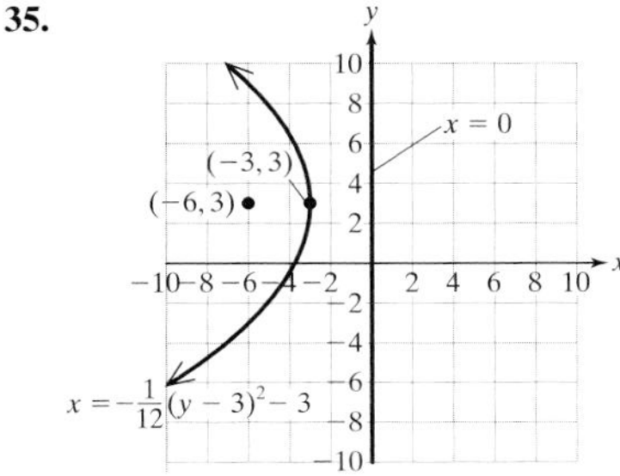

37. 4 **39.** 1 **41.** $\frac{1}{4}$ **43.** $\frac{25}{4}$ **45.** $y = (x-2)^2 - 1$; vertex $(2, -1)$; focus $(2, -\frac{3}{4})$; directrix $y = -\frac{5}{4}$ **47.** $x = (y+1)^2 + 5$; vertex $(5, -1)$; focus $(\frac{21}{4}, -1)$; directrix $x = \frac{19}{4}$ **49.** $y = -2(x-2)^2 + 8$; vertex $(2, 8)$; focus $(2, \frac{63}{8})$; directrix $y = \frac{65}{8}$ **51.** $y = (x - \frac{3}{2})^2 - \frac{1}{4}$; vertex $(\frac{3}{2}, -\frac{1}{4})$; focus $(\frac{3}{2}, 0)$; directrix $y = -\frac{1}{2}$ **53.** $x = -2(y-4)^2 + 33$; vertex $(33, 4)$; focus $(\frac{263}{8}, 4)$; directrix $x = \frac{265}{8}$ **55.** Focus $(0, \frac{25}{2})$ **57.** Opens downward **59.** Opens right **61.** Opens downward **63.** Opens left **65.** Opens right **67.** Opens upward

Section 13.3 Practice Exercises, pp. 894–896

1.

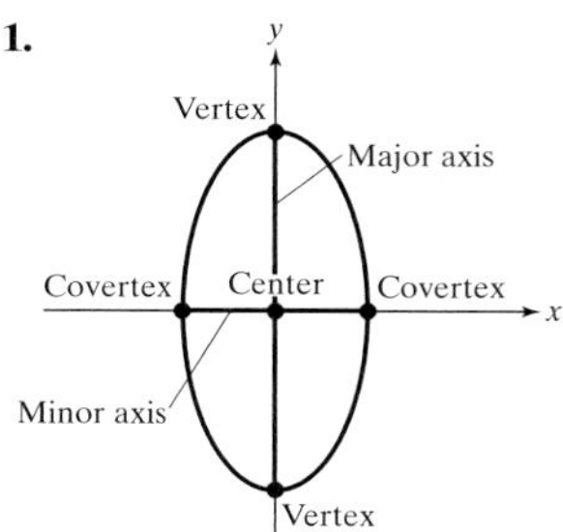

3. (0, 0); vertical **5.** (2, 5); horizontal **7.** (−2, −4); vertical **9.** (−4, 1); horizontal

11.

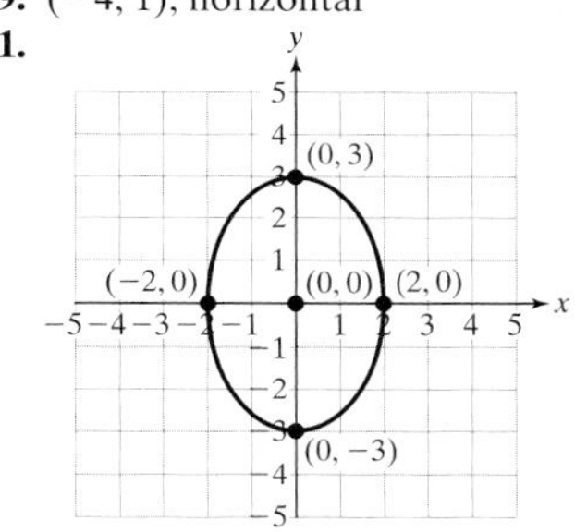

13.

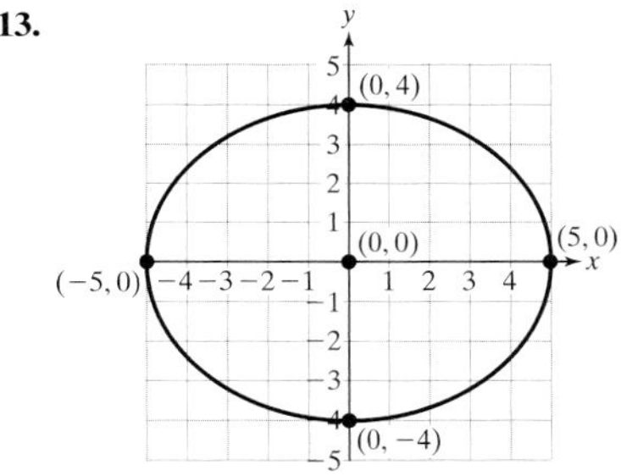

15. **17.**

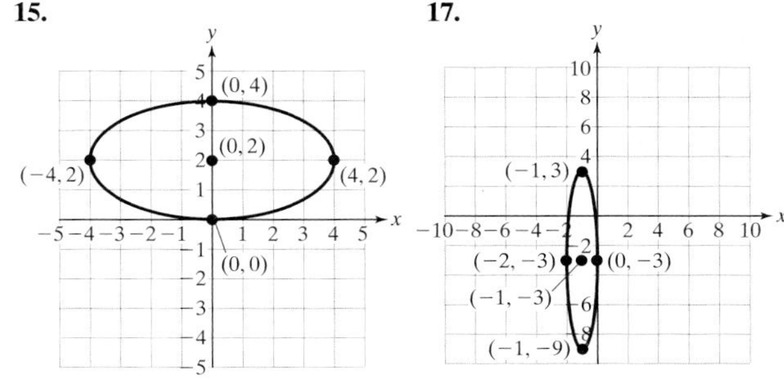

19.

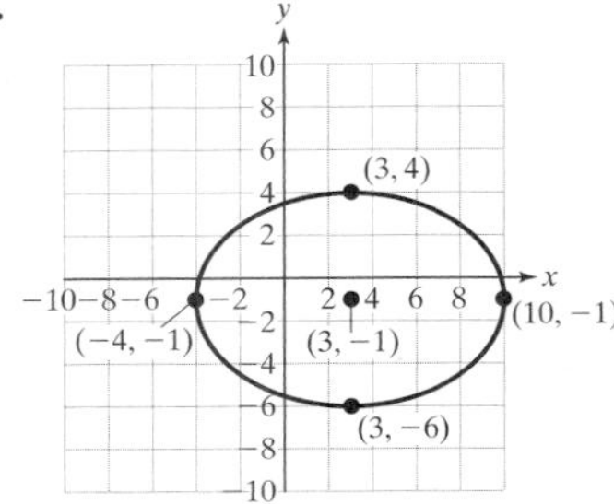

21. The length of the string is constant and the two tacks are fixed points. The sum of the distances from each tack to a point on the curve is constant, therefore an elliptical curve is traced. **23.** 97 ft

25.

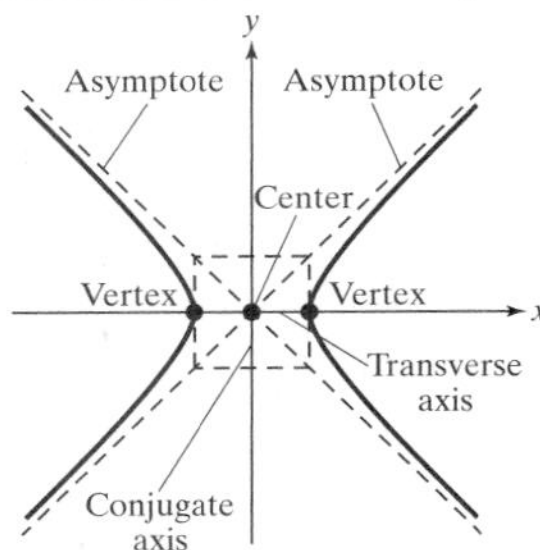

27. (0, 0); horizontal **29.** (0, 0); vertical **31.** (−1, −8); horizontal
33. (−1, −4); vertical

35.

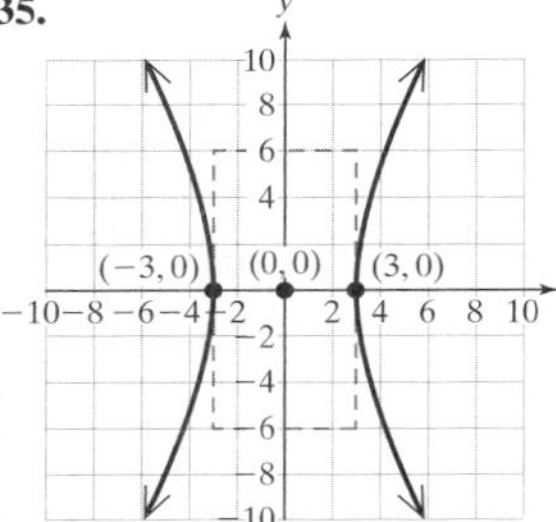

37.

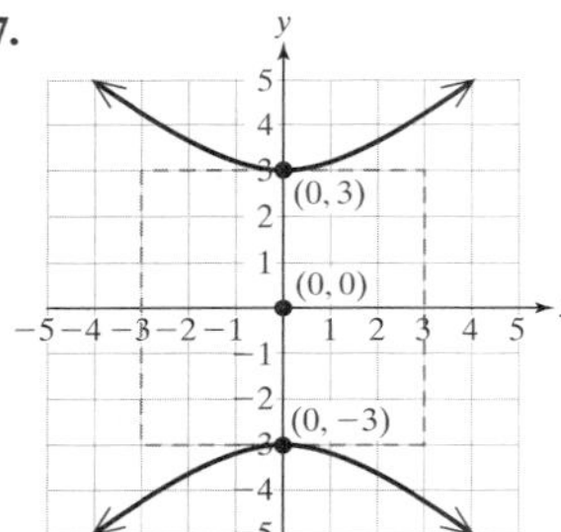

39.

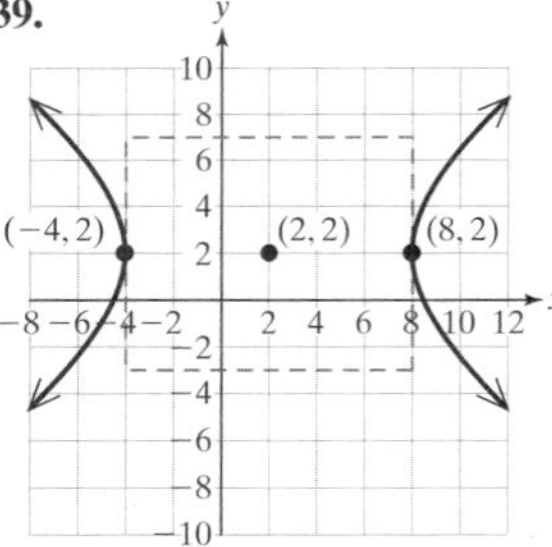

41.

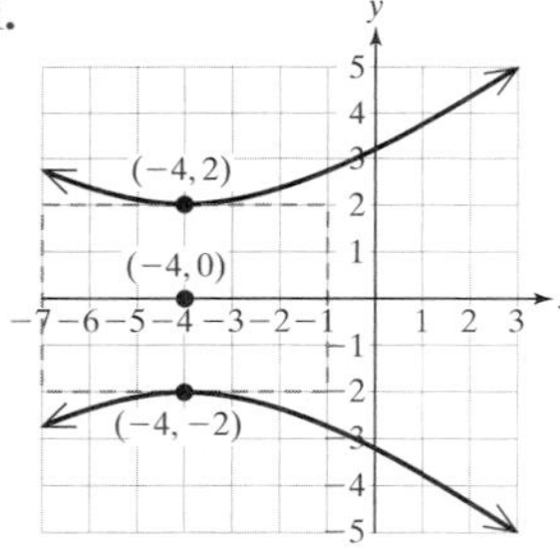

43.

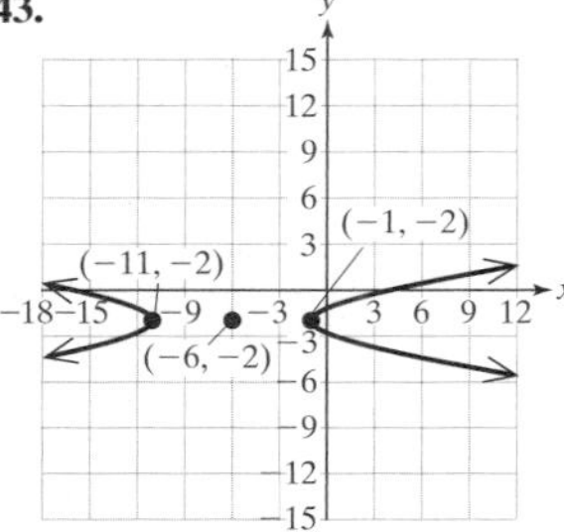

45. $y = x, y = -x$ **47.** $y = \frac{1}{2}x - 5,\ y = -\frac{1}{2}x - 3$

49. An ellipse can be written in the form

$$\frac{(x-h)^2}{a^2} + \frac{(y-k)^2}{b^2} = 1 \quad \text{or} \quad \frac{(x-h)^2}{b^2} + \frac{(y-k)^2}{a^2} = 1$$

whereas a hyperbola can be written in the form

$$\frac{(x-h)^2}{a^2} - \frac{(y-k)^2}{b^2} = 1 \quad \text{or} \quad \frac{(y-k)^2}{a^2} - \frac{(x-h)^2}{b^2} = 1$$

51. Ellipse **53.** Hyperbola **55.** Ellipse **57.** Hyperbola
59. (0, 4), (0, −4) **61.** (8, 0), (−8, 0) **63.** (5, 0), (−5, 0)
65. (0, 13), (0, −13)

Section 13.4 Practice Exercises, pp. 902–903

1. $d = \sqrt{(x_2 - x_1)^2 + (y_2 - y_1)^2}$ **3.** $(x + 1)^2 + (y - 1)^2 = 4$
5. Circle **7.** Parabola **9.** Hyperbola **11.** Ellipse
13. Zero, one, or two **15.** Zero, one, or two **17.** Zero, one, two, three, or four **19.** Zero, one, two, three, or four
21. (−3, 0)(2, 5)

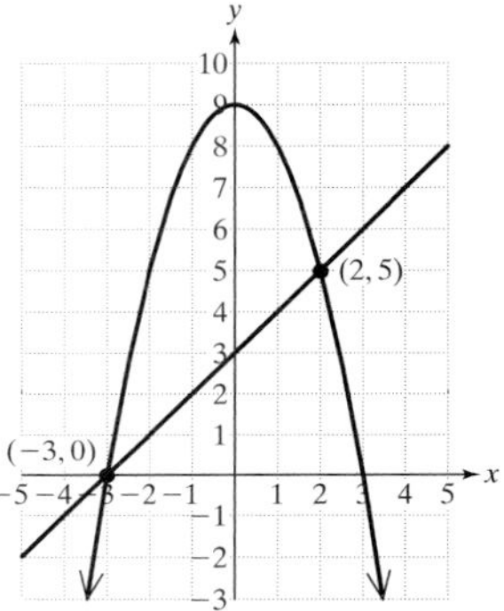

23. (0, 1)(−1, 0)

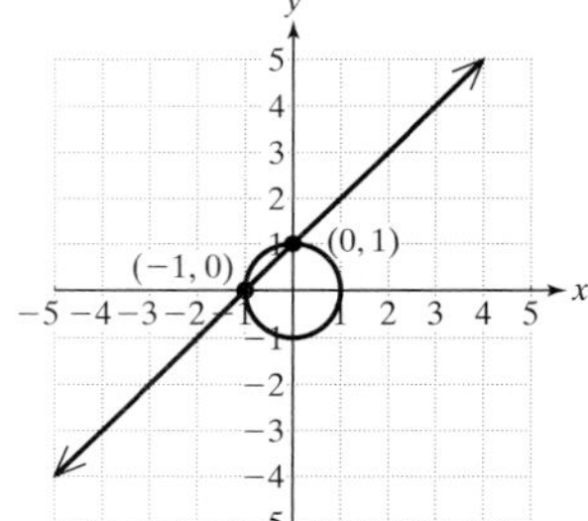

25. $(\sqrt{2}, 2)(-\sqrt{2}, 2)$

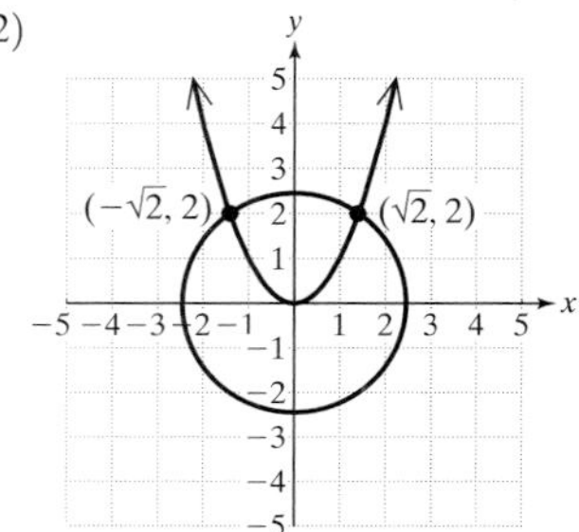

27. (4, 2) **29.** (0, 0) **31.** $\left(\frac{3}{2}, \frac{9}{4}\right)$ **33.** (0, 0), (−2, −8)
35. (3, 2)(3, −2) (−3, 2)(−3, −2) **37.** (0, 3)(0, −3)
39. (2, 1)(−2, 1) (2, −1)(−2, −1) **41.** $(\sqrt{2}, 0)(-\sqrt{2}, 0)$
43. (2, 0)(−2, 0) **45.** (3, 0), (−3, 0) **47.** (4, 5), (−4, −5)
49. 3 and 4 **51.** 5 and $\sqrt{7}$, −5 and $\sqrt{7}$, 5 and $-\sqrt{7}$, or −5 and $-\sqrt{7}$

53.

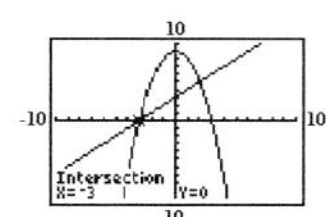

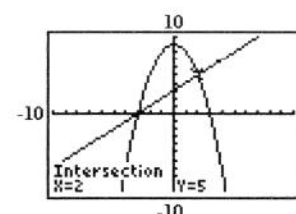

55.

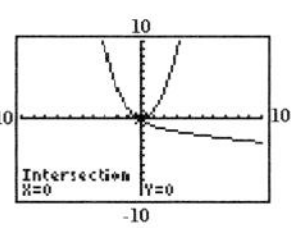

57.

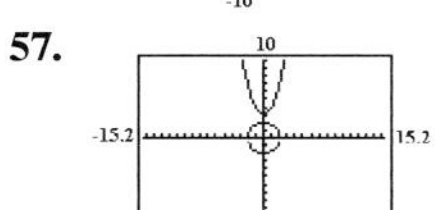

No solution

Section 13.5 Practice Exercises, pp. 907–909

1. g **3.** h **5.** f **7.** k **9.** a **11.** d **13.** False **15.** True

17. a.

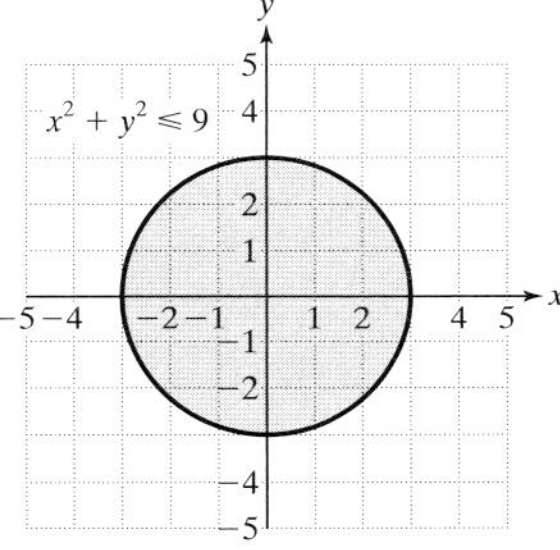

b. The set of points on and "outside" the circle $x^2 + y^2 = 9$.
c. The set of points on the circle $x^2 + y^2 = 9$.

19. a.

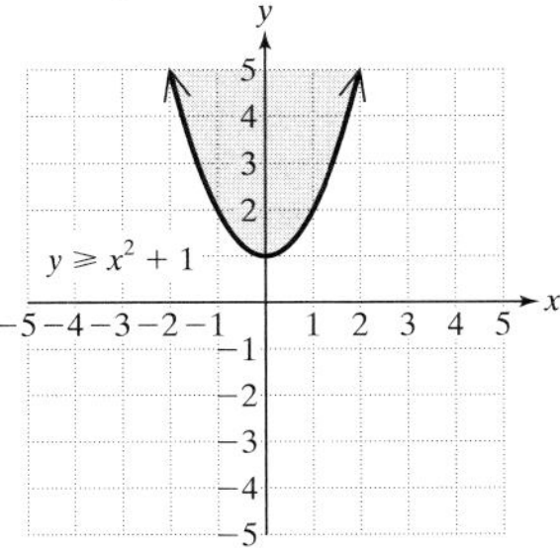

b. The parabola, $y = x^2 + 1$ would be drawn as a dashed curve.

21.

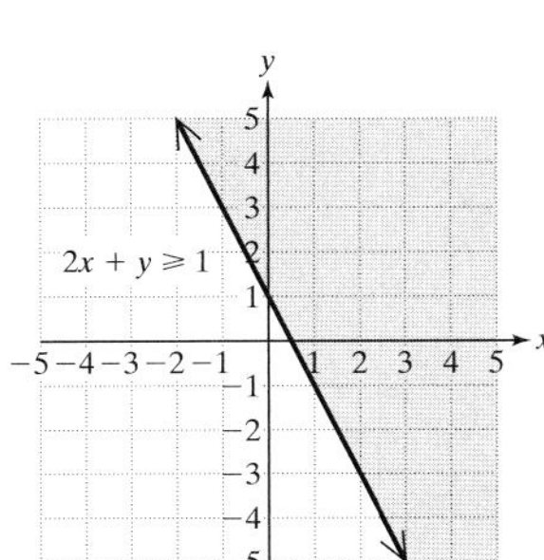

23.

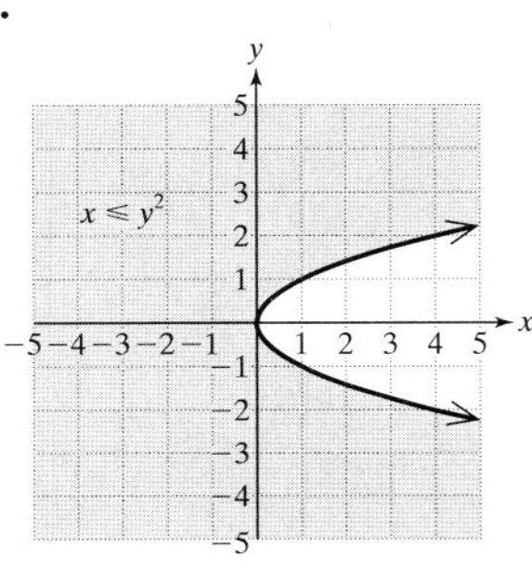

25.

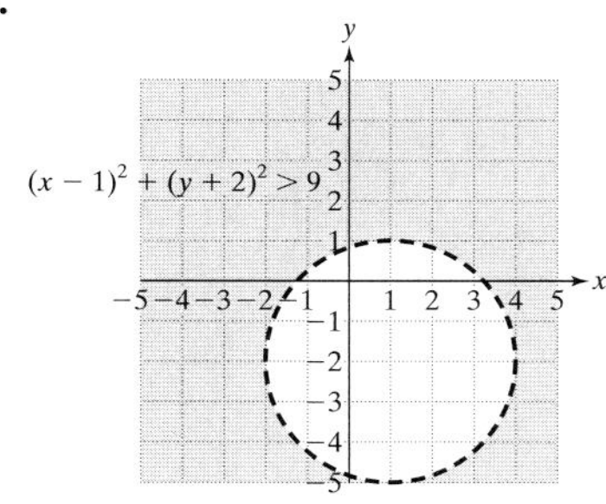

27.

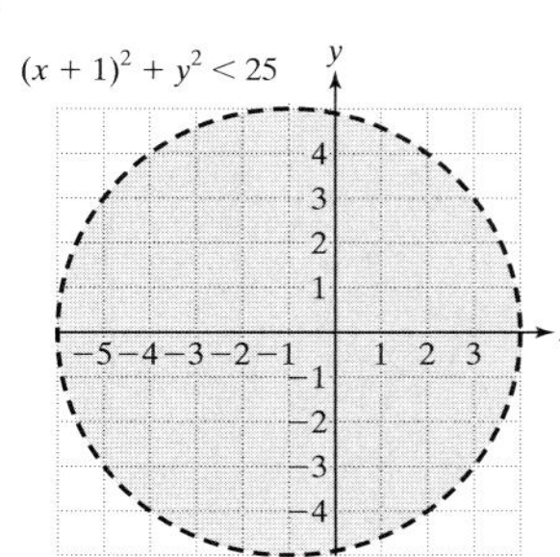

29.

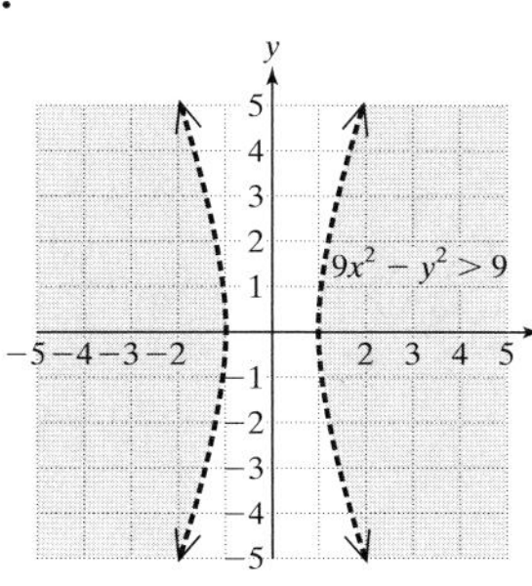

31.

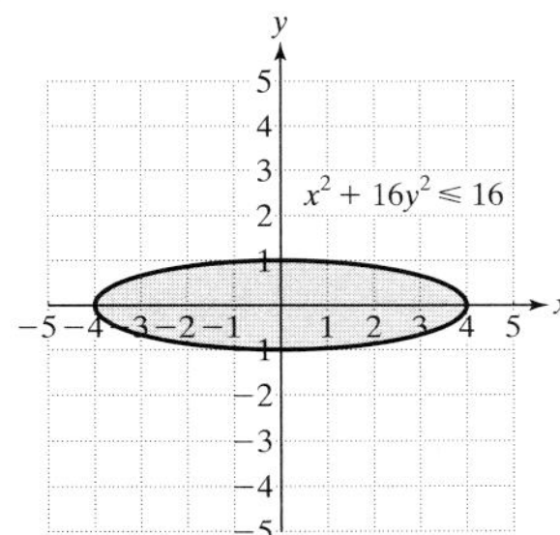

33.

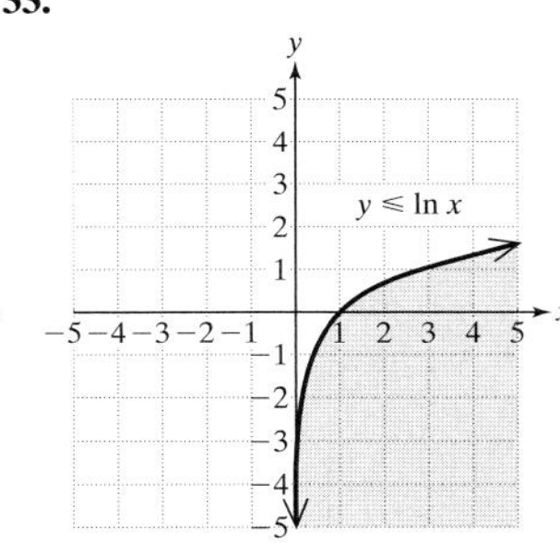

35.

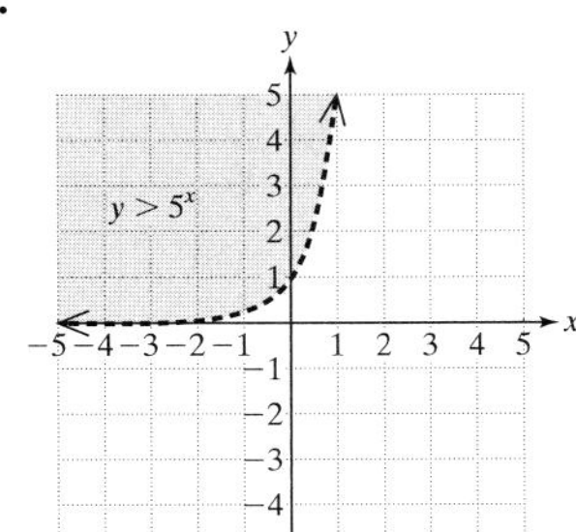

37.

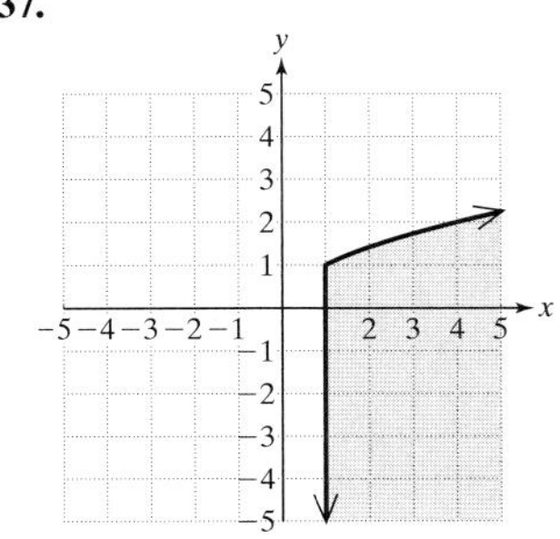

39. **41.**

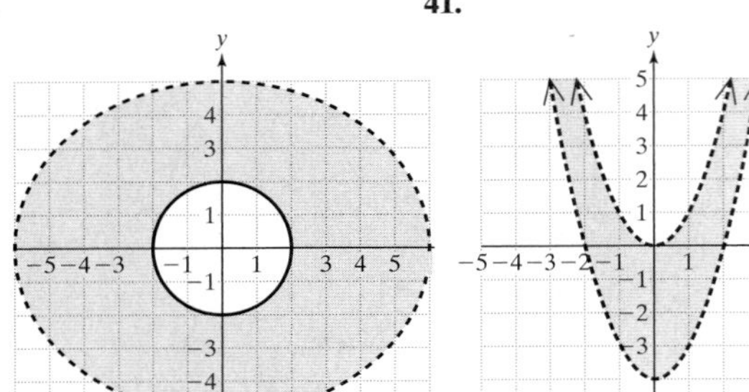

43.

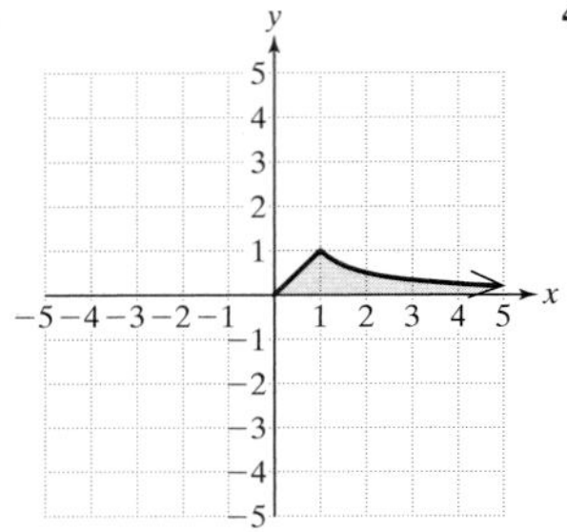

45. No solution

47. **49.**

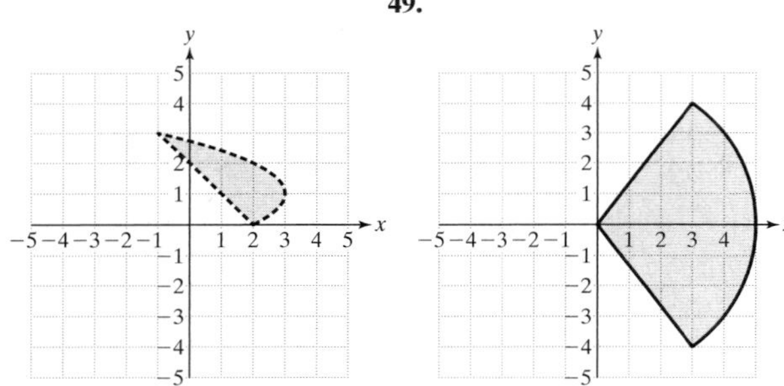

51. a.

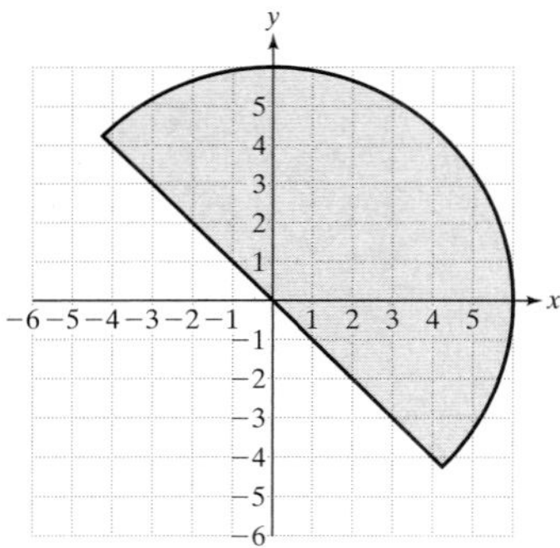

b.

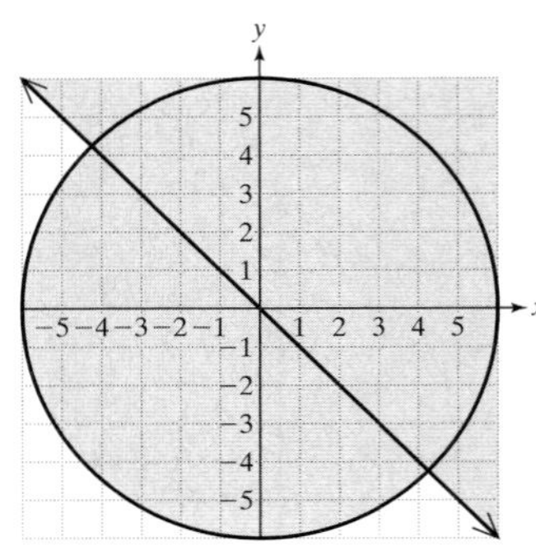

53. a. **b.**

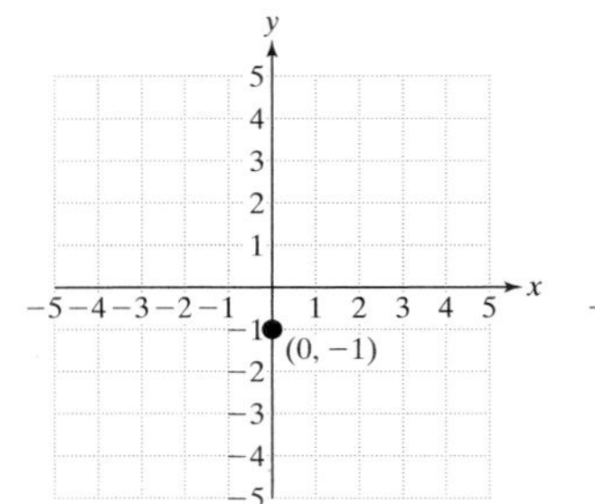

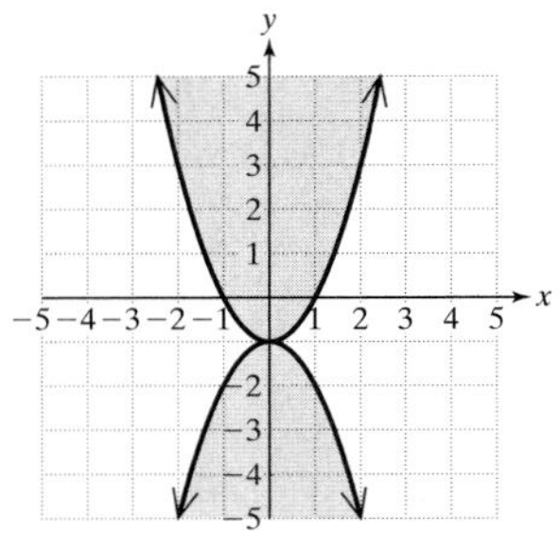

Chapter 13 Review Exercises, pp. 914–916

1. $2\sqrt{10}$ **3.** $x = 5$ or $x = -1$ **5.** Collinear **7.** Center (12, 3); $r = 4$ **9.** Center $(-3, -8)$; $r = 2\sqrt{5}$ **11. a.** $x^2 + y^2 = 64$ **b.** $(x - 8)^2 + (y - 8)^2 = 64$ **13.** $(x + 2)^2 + (y + 8)^2 = 8$ **15.** $(x - 3)^2 + \left(y - \frac{1}{3}\right)^2 = 9$ **17.** $x^2 + (y - 2)^2 = 9$ **19.** $(-1.45, -6)$ **21.** Horizontal axis of symmetry; parabola opens right **23.** Vertical axis of symmetry; parabola opens upward **25.** Vertex: $(-2, 0)$; focus: $(-2, \frac{1}{4})$; Directrix: $y = -\frac{1}{4}$

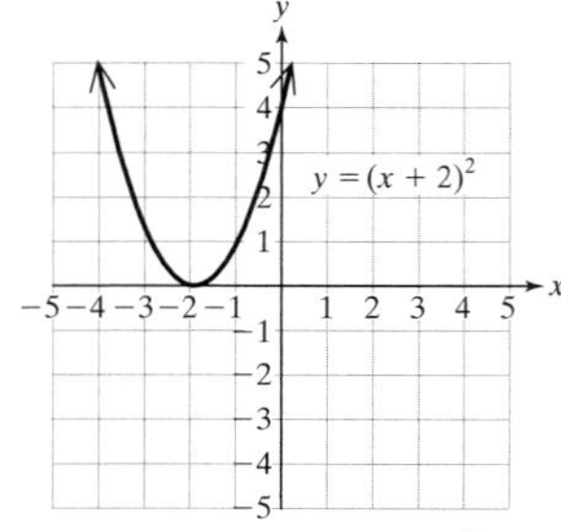

27. Vertex: $(-1, 0)$; focus: $(-\frac{7}{8}, 0)$; Directrix: $x = -\frac{9}{8}$

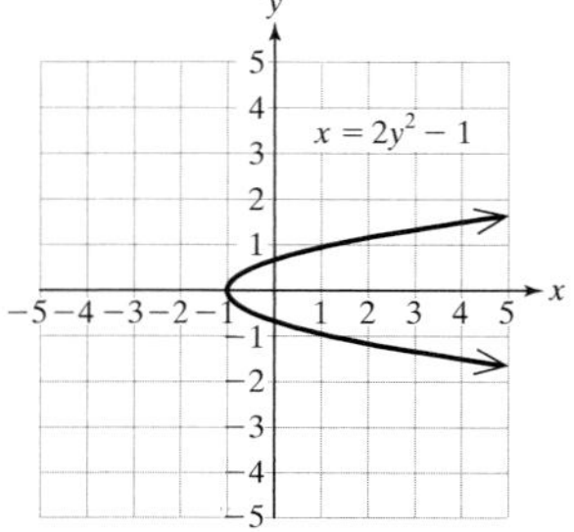

29. $x = (y + 2)^2 - 2$; vertex: $(-2, -2)$; Focus: $(-\frac{7}{4}, -2)$; directrix: $x = -\frac{9}{4}$ **31.** $y = -2(x + \frac{1}{2})^2 + \frac{1}{2}$; vertex: $(-\frac{1}{2}, \frac{1}{2})$; Focus: $(-\frac{1}{2}, \frac{3}{8})$; directrix: $y = \frac{5}{8}$ **33.** Center: (0, 0); major axis is vertical
35. Center: (1, 0); major axis is horizontal
37.

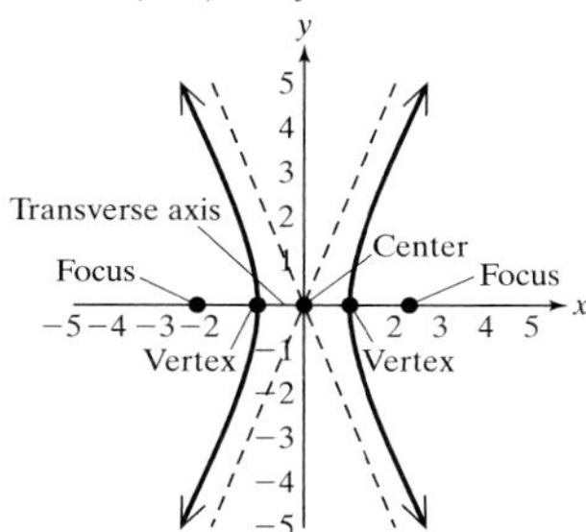

39. Center: (0, 0); transverse axis is vertical. **41.** Center: $(-1, 0)$; transverse axis is horizontal.

43. Ellipse

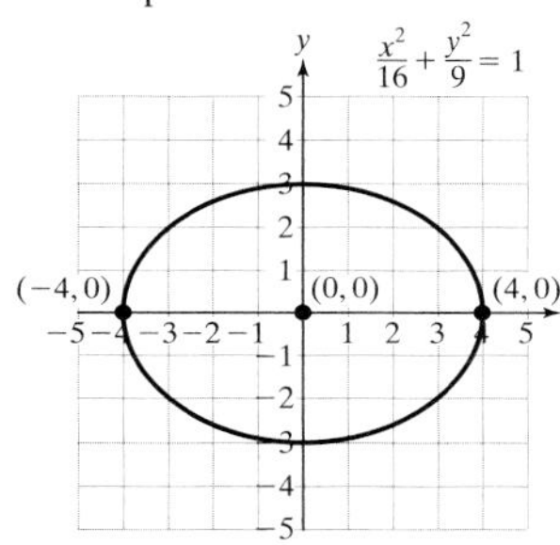

45. Hyperbola

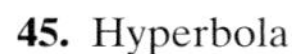

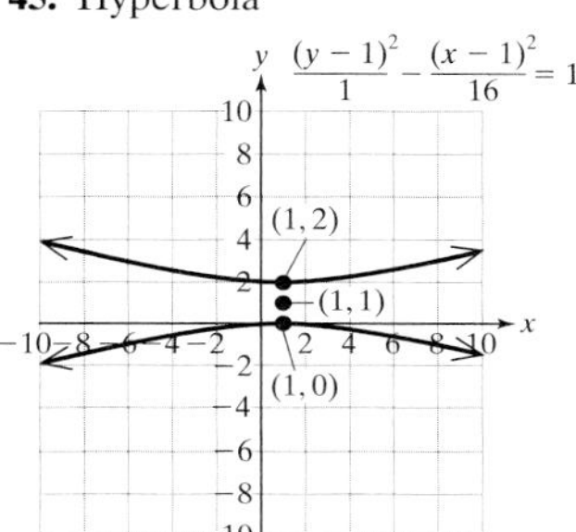

47. a. Line and parabola
c. $(-5, 15)(3, -1)$
b.

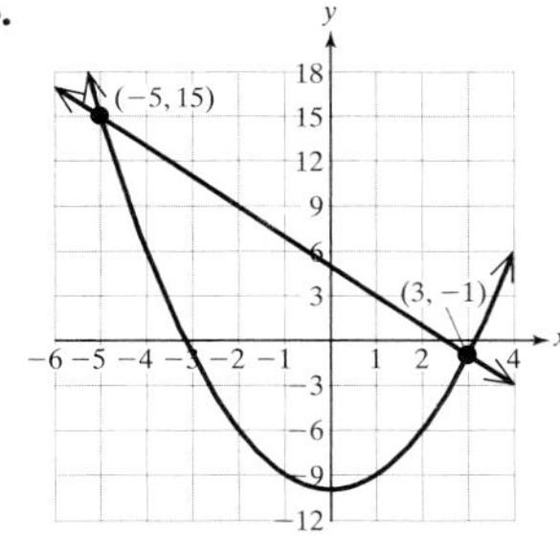

49. a. Circle and line
c. $(0, -4)\left(\frac{16}{5}, -\frac{12}{5}\right)$
b.

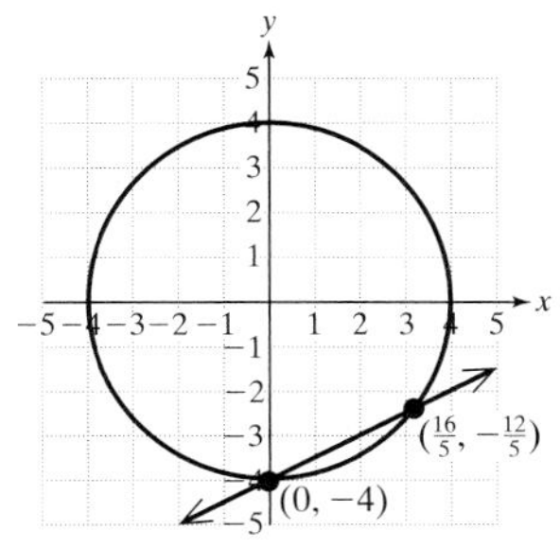

51. $\left(-\frac{7}{5}, \frac{13}{5}\right)(-5, -1)$ **53.** $(2, 4)(-2, 4)$ $(\sqrt{2}, 2)(-\sqrt{2}, 2)$
55. $(6, 5)(-6, 5)$ $(6, -5)(-6, -5)$

57.

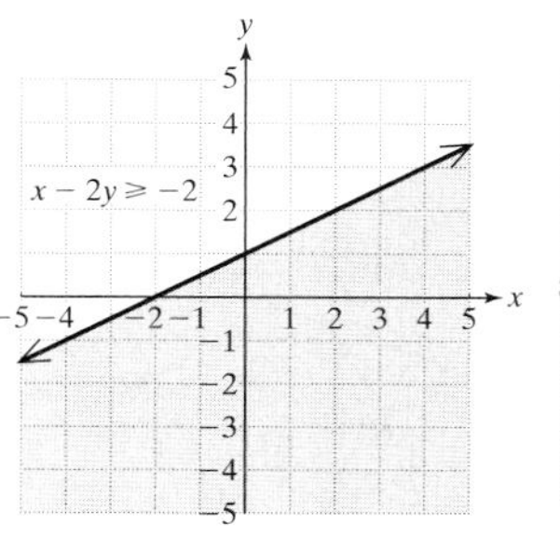

59.

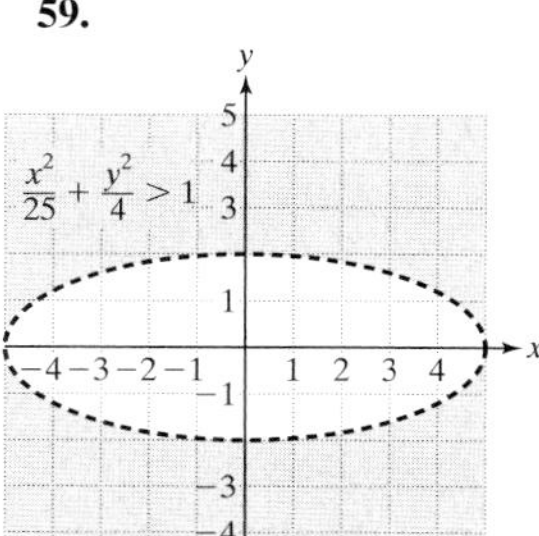

61.

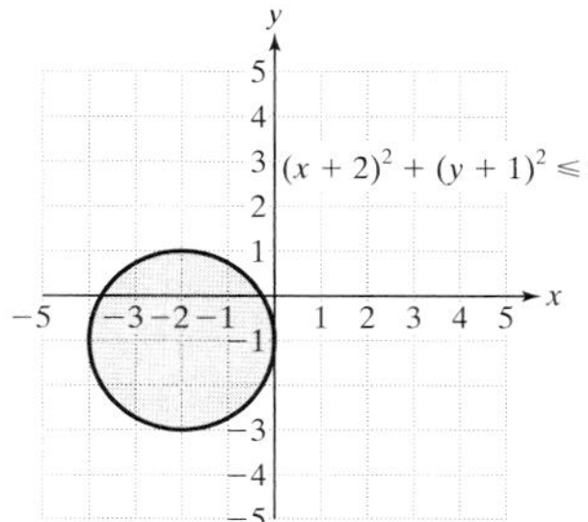

63.

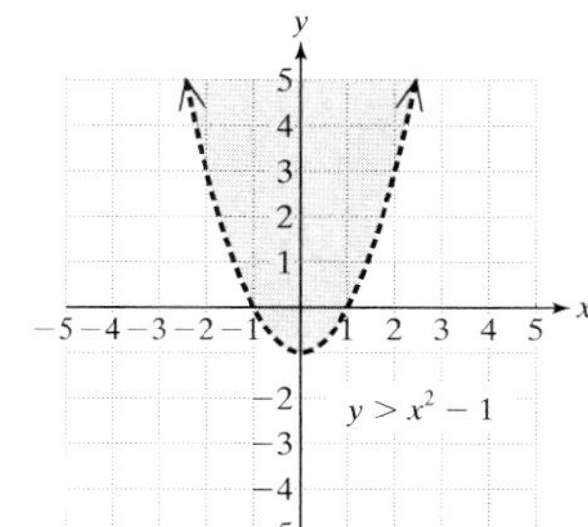

65. **67.**

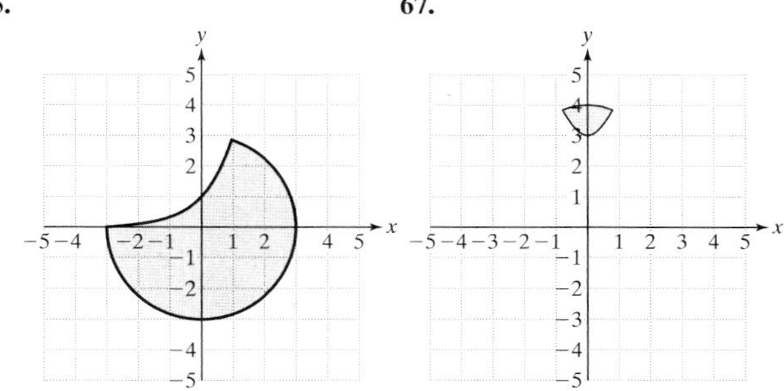

Chapter 13 Test, pp. 916–917

1. Vertex: $(3, -6)$; focus: $(2, -6)$; directrix: $x = 4$
2. $y = (x - 2)^2 + 1$

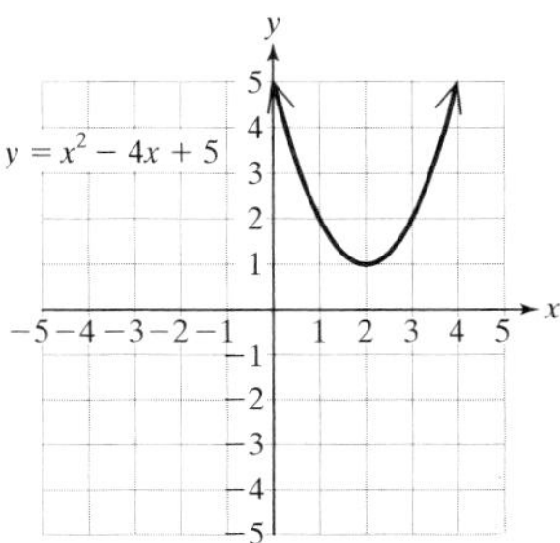

3. $\sqrt{85}$ **4.** Yes, they form a right triangle. **5.** $\left(\frac{5}{6}, -\frac{1}{3}\right)$; $r = \frac{5}{7}$
6. $(0, 2)$; $r = 3$ **7. a.** $\sqrt{5}$ **b.** $x^2 + (y - 4)^2 = 5$ **8.** $\left(13, \frac{17}{2}\right)$
9. Center: (3, 0); major axis is vertical.

10. Center: (0, 0); vertices: (0, 1), (0, −1)

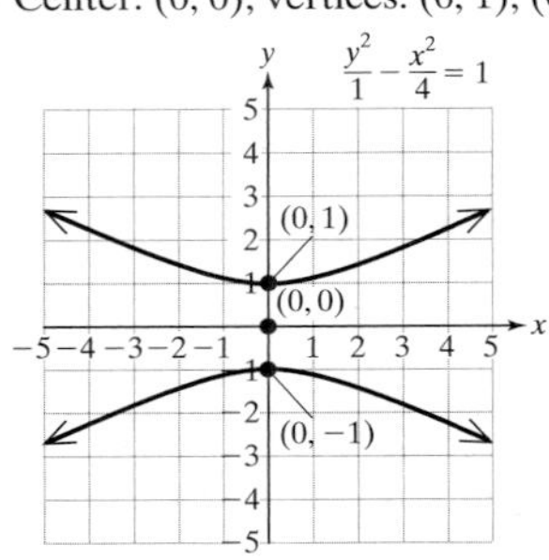

11. a. (−3, 0)(0, 4); i **b.** No solution; ii **12.** The addition method can be used if the equations have corresponding *like* terms.
13. (2, 0)(−2, 0)
14.

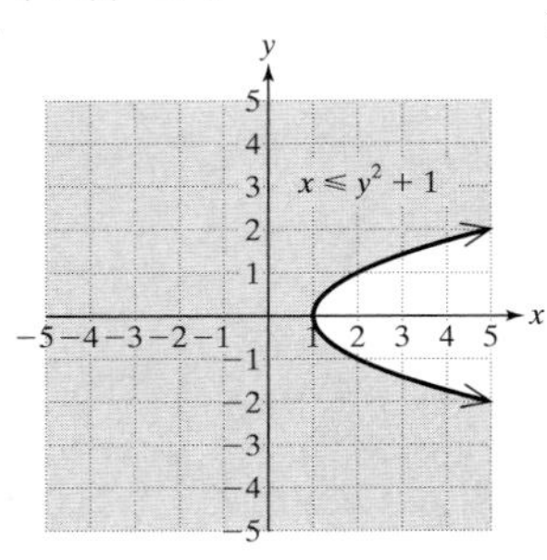

15.

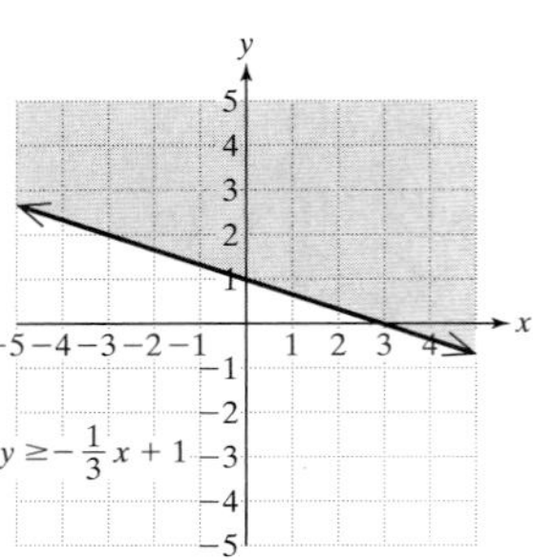

16.

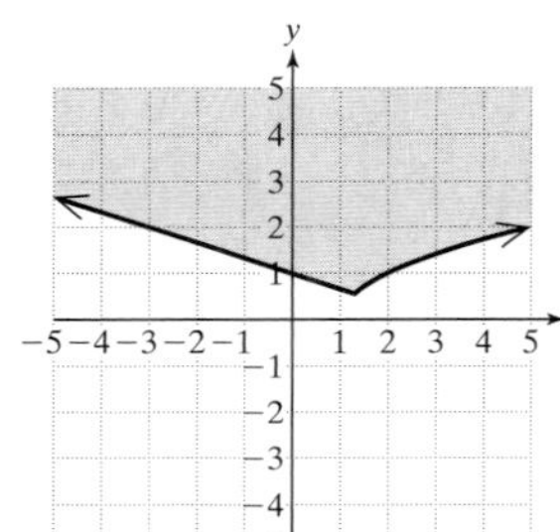

17.

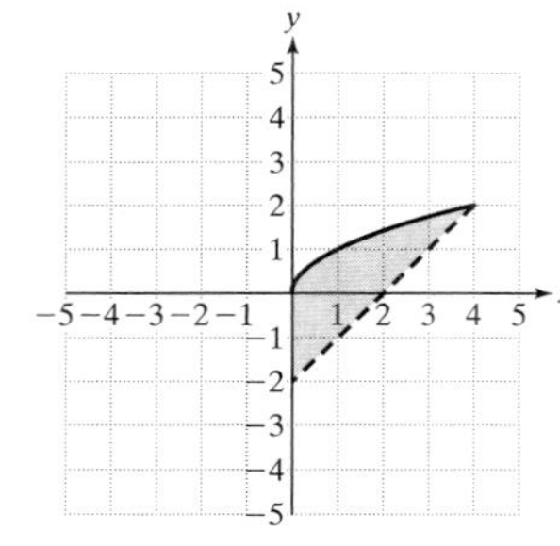

18.

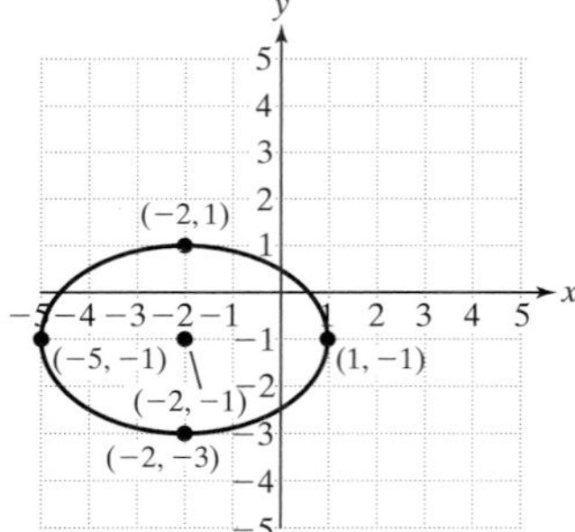

Cumulative Review Exercises, Chapters 1–13, pp. 917–919

1. All real numbers **2.** $\left(\frac{10}{3}, \infty\right)$ **3.** 10, 15

4. a. (−5, 0)(0, 3) **b.** $m = \frac{3}{5}$ **c.**

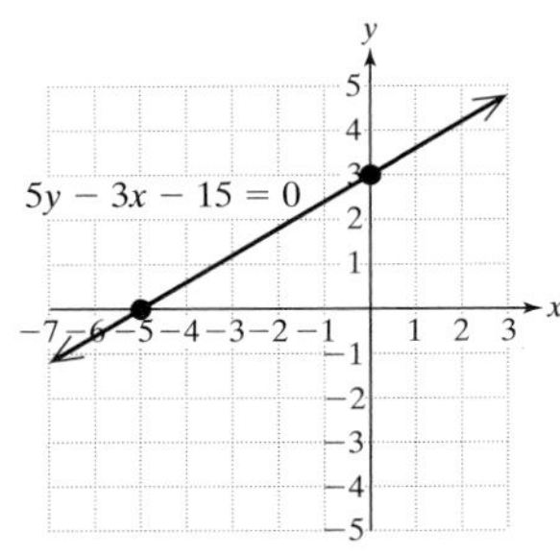

5. a. $m = 10$ **b.** $y = 10x + 30$ **c.** $80 million
6. 12 dimes, 5 quarters **7.** (2, −3, 1)
8. a. $\left[\begin{array}{cc|c} 1 & -2 & -8 \\ 0 & 1 & 2 \end{array}\right]$ **b.** $\left[\begin{array}{cc|c} 1 & 0 & -4 \\ 0 & 1 & 2 \end{array}\right]$ **9.** $(\frac{5}{2}, \frac{3}{2})$
10. $f(0) = -12; f(-1) = -16; f(2) = -10; f(4) = -16$
11. $g(2) = 5; g(8) = -1; g(3) = 0; g(-5) = 5$
12. 32 **13. a.** −5 **b.** $(x + 1)(x^2 + 1)$; −5 **c.** They are the same.
14. a. No solution **b.** $x = -11$ **15. a.** $-30 + 24i$
b. $\frac{12}{41} + \frac{15}{41}i$ **16. a.** 17.6 ft; 39.6 ft; 70.4 ft
b. 8 sec **17.** $w = -\frac{1}{5}$; $w = \frac{1}{10} \pm \frac{i\sqrt{3}}{10}$ **18.** $x = 2 \pm \sqrt{11}$
19. (−1, 19) **20.** $\log_8 32 = \frac{5}{3}$ **21.** (−5, −36) **22. a.** (−3, 0)(1, 0)
b. (0, 3) **c.** (−1, 4)

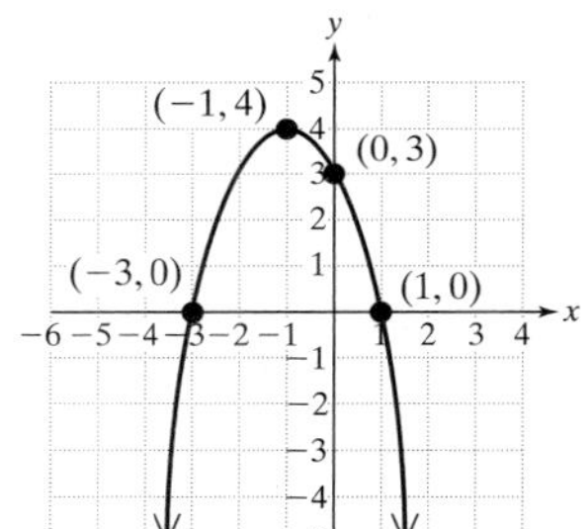

23. $x^2 + (y - 5)^2 = 16$ **24.** Yes, the circle can be tangent to the parabola. **25.** (0, −4)
26.

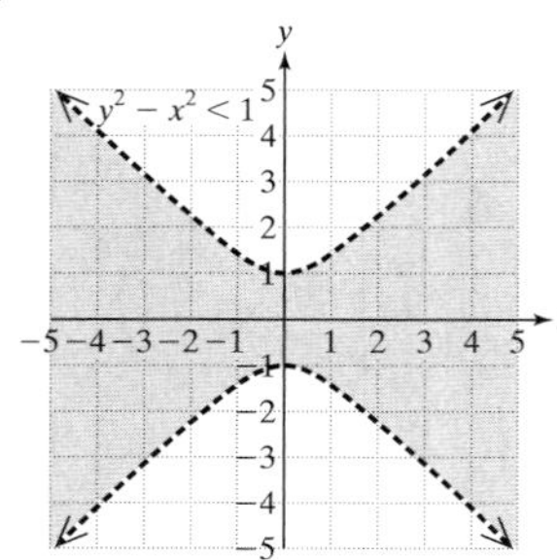

27.

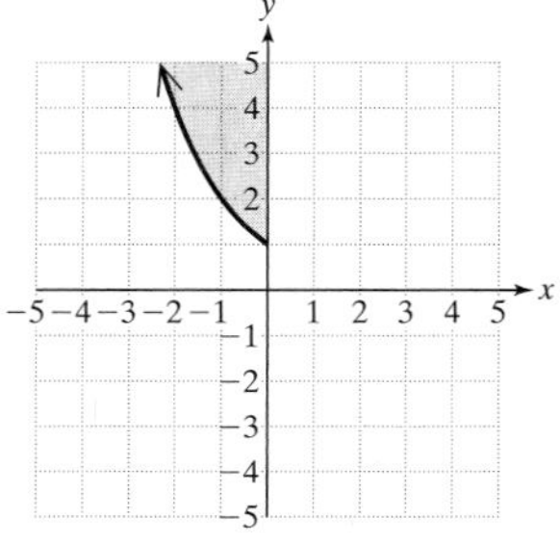

28. $(x + y)(x - y - 6)$

CHAPTER 14

Section 14.1 Practice Exercises, pp. 927–930

1. 4, 7, 10, 13 **3.** $\sqrt{3}, 2, \sqrt{5}, \sqrt{6}$ **5.** $3, \frac{3}{2}, 1, \frac{3}{4}, \frac{3}{5}$ **7.** $-\frac{2}{3}, \frac{3}{4}, -\frac{4}{5}, \frac{5}{6}$ **9.** 0, −3, 8, −15 **11.** $\frac{5}{2}, \frac{8}{3}, \frac{11}{4}, \frac{14}{5}$ **13.** 0, 2, 6, 12 **15.** −3, 9, −27, 81 **17.** When n is odd, the term is negative. When n is even, the term is positive. **19.** \$60, \$58.80, \$57.62, \$56.47, \$55.34, \$54.24 **21.** 25,000; 50,000; 100,000; 200,000; 400,000; 800,000; 1,600,000 **23.** $a_n = 2n$ **25.** $a_n = 2n - 1$ **27.** $a_n = \frac{1}{n^2}$ **29.** $a_n = (-1)^{n+1}$ **31.** $a_n = (-1)^n 2^n$ **33.** $a_n = \frac{3}{5^n}$ **35.** A sequence is an ordered list of terms. A series is the sum of the terms of a sequence. **37.** 90 **39.** $\frac{31}{16}$ **41.** 30 **43.** 10 **45.** $\frac{73}{12}$ **47.** 38 **49.** −1 **51.** 55 **53.** $\sum_{n=1}^{6} n$ **55.** $\sum_{i=1}^{5} 4$ **57.** $\sum_{j=1}^{5} 4j$ **59.** $\sum_{k=1}^{4} (-1)^{k+1}\frac{1}{3^k}$ **61.** $\sum_{n=1}^{5} x^n$

63. $1 + 1\frac{1}{2} + 1\frac{1}{2} + 1\frac{1}{2} + 1\frac{1}{2} + 1\frac{1}{2} + 1\frac{1}{2} + 1\frac{1}{2}$; $1 + \sum_{i=1}^{7} 1\frac{1}{2}$; $11\frac{1}{2}$ in.

65. 15.4 g **67.** 4.8 g **69.** −3, 2, 7, 12, 17 **71.** 5, 21, 85, 341, 1365 **73.** 1, 1, 2, 3, 5, 8, 13, 21, 34, 55

75.
```
seq(8/n,n,1,5,1)
{8 4 2.66666666...
Ans▸Frac
 {8 4 8/3 2 8/5}
```
77.
```
seq((-1)^n*n²,n,
1,5,1)
{-1 4 -9 16 -25}
```
79.
```
sum(seq(8/n,n,1,
4,1)
         16.66666667
Ans▸Frac
                50/3
```
81.
```
sum(seq((-1)^n*n
²,n,1,100,1)
                5050
```

Section 14.2 Practice Exercises, pp. 939–943

1. A sequence is arithmetic if the difference between any term and the preceding term is constant. **3.** 6 **5.** −5 **7.** 4 **9.** −3, −1, 1, 3, 5 **11.** $0, \frac{1}{3}, \frac{2}{3}, 1, \frac{4}{3}$ **13.** 10, 4, −2, −8, −14 **15.** $a_n = 2 + 5n$ **17.** $a_n = 5 - 4n$ **19.** $a_n = \frac{2}{3} + \frac{1}{3}n$ **21.** $a_n = 12 - 3n$ **23.** $a_n = -17 + 8n$ **25.** 17 **27.** 47 **29.** 46 **31.** −48 **33.** 19 **35.** 22 **37.** 23 **39.** 11 **41.** $a_1 = -2, a_2 = -5$ **43.** 670 **45.** 290 **47.** −15 **49.** 95 **51.** 924 **53.** 300 **55.** −210 **57.** 5050 **59.** 2500 **61.** 980 seats; \$14,700 **63.** A sequence is geometric if the ratio between a term and the preceding term is constant. **65.** $\frac{1}{2}$ **67.** −3 **69.** 4 **71.** −4, 4, −4, 4, −4 **73.** $8, 2, \frac{1}{2}, \frac{1}{8}, \frac{1}{32}$ **75.** 2, −6, 18, −54, 162 **77.** $a_n = 2(3)^{n-1}$ **79.** $a_n = -6(-2)^{n-1}$ **81.** $a_n = \frac{16}{3}\left(\frac{3}{4}\right)^{n-1}$ **83.** $-\frac{3}{128}$ **85.** $-\frac{2}{81}$ **87.** 135 **89.** −10 **91.** $\frac{1}{9}$ **93.** 1 **95.** $\frac{1562}{125}$ **97.** $-\frac{11}{8}$ **99.** $\frac{3124}{27}$ **101.** $\frac{2059}{729}$ **103.** −172 **105. a.** \$1050.00, \$1102.50, \$1157.63, \$1215.51 **b.** a_{10} = \$1628.89; a_{20} = \$2653.30; a_{40} = \$7039.99 **107.** $r = \frac{1}{6}; \frac{6}{5}$ **109.** $r = -\frac{1}{3}; -\frac{9}{4}$ **111.** $r = -\frac{3}{2}$; sum does not exist **113.** \$800 million **115.** 28 ft **117. a.** $\frac{7}{10}$ **b.** $\frac{1}{10}$ **c.** $\frac{7}{9}$ **119.** Geometric, $r = -\frac{3}{2}$ **121.** Arithmetic, $d = \frac{2}{3}$ **123.** Geometric, $r = 3$ **125.** Arithmetic, $d = 4$ **127.** Neither **129.** Neither **131.** Geometric, $r = -1$ **133. a.** \$1,191,123 **b.** \$1,254,857 **c.** \$63,734

135.
```
seq(125000*1.04^
n,n,1,4,1)
{130000 135200 ...
```
137.
```
sum(seq(n,n,1,15
,1)
                 120
```
139.
```
sum(seq(4*(1/2)^
(n-1),n,1,10,1)
           7.9921875
sum(seq(4*(1/2)^
(n-1),n,1,20,1)
         7.999992371
```

Section 14.3 Practice Exercises, pp. 948–949

1. $x^4 + 4x^3y + 6x^2y^2 + 4xy^3 + y^4$ **3.** $64 + 48p + 12p^2 + p^3$ **5.** $a^{12} + 6a^{10}b + 15a^8b^2 + 20a^6b^3 + 15a^4b^4 + 6a^2b^5 + b^6$ **7.** $p^6 - 3p^4w + 3p^2w^2 - w^3$ **9.** The signs alternate on the terms of the expression $(a - b)^n$. The signs for the expression $(a + b)^n$ are all positive. **11.** 6 **13.** 1 **15.** False **17.** $9! = 9 \cdot (8 \cdot 7 \cdot 6 \cdot 5 \cdot 4 \cdot 3 \cdot 2 \cdot 1) = 9 \cdot 8!$ **19.** 1680 **21.** 6 **23.** 56 **25.** 1 **27.** $s^6 + 6s^5t + 15s^4t^2 + 20s^3t^3 + 15s^2t^4 + 6st^5 + t^6$ **29.** $b^3 - 9b^2 + 27b - 27$ **31.** $16x^4 + 32x^3y + 24x^2y^2 + 8xy^3 + y^4$ **33.** $c^{14} - 7c^{12}d + 21c^{10}d^2 - 35c^8d^3 + 35c^6d^4 - 21c^4d^5 + 7c^2d^6 - d^7$ **35.** $\frac{1}{32}a^5 - \frac{5}{16}a^4b + \frac{5}{4}a^3b^2 - \frac{5}{2}a^2b^3 + \frac{5}{2}ab^4 - b^5$ **37.** $m^{11} - 11m^{10}n + 55m^9n^2$ **39.** $u^{24} - 12u^{22}v + 66u^{20}v^2$ **41.** Nine terms **43.** $-462m^6n^5$ **45.** $495u^{16}v^4$ **47.** g^9

Section 14.4 Practice Exercises, pp. 955–958

1. 720 **3.** 1 **5.** 720; There are 720 ways in which 3 items can be selected from 10 in a specified order. **7.** 120; There are 120 ways in which 3 items can be selected from 10 in *no* specific order. **9.** 79,833,600 **11.** 220 **13.** 7 **15.** 7 **17.** 40,320 **19.** 1 **21. a.** AB, BA, AC, CA, BC, CB **b.** AB, AC, BC **23.** 6! = 720 **25.** 10^6 = 1,000,000 **27.** $4 \cdot 3 \cdot 8 \cdot 6 = 576$ **29.** ${}_{10}P_3 = 720$ **31.** ${}_{11}C_6 = 462$ **33.** ${}_{10}C_5 = 252$ **35.** ${}_8P_2 = 56$ **37.** $2 \cdot 25 \cdot 24 \cdot 23 = 27{,}600$ **39.** ${}_{10}C_4 = 210$ **41.** $3 \cdot 3 \cdot 2 \cdot 2 \cdot 1 \cdot 1 = 36$ **43.** ${}_9P_9 = 362{,}880$ **45.** ${}_{10}C_3 \cdot {}_8C_1 = 960$ **47.** $2^3 = 8$ **49.** ${}_{40}C_{12}$ or 5,586,853,480 **51. a.** ${}_{10}C_2 = 45$ **b.** ${}_6C_2 = 15$ **c.** ${}_4C_2 = 6$ **d.** ${}_6C_1 \cdot {}_4C_1 = 24$ **53.** $2^7 = 128$ **55. a.** 9! = 362,880 **b.** $5 \cdot 4 \cdot 7! = 100{,}800$ **c.** $5! \cdot 4! = 2880$ **57. a.** pot, pto, opt, otp, top, tpo **b.** tot, tto, ott **59. a.** $7^3 = 343$ **b.** $7 \cdot 6 \cdot 5 = 210$ **c.** $7 \cdot 7 \cdot 4 = 196$ **61.** ${}_{26}C_5 = 65{,}780$ **63.** ${}_{13}C_3 \cdot {}_{13}C_2 = 22{,}308$

Section 14.5 Practice Exercises, pp. 963–965

1. ${}_{10}C_4 = 210$ **3.** 5! = 120 **5.** a, b, d, g, h **7.** $\frac{3}{6} = \frac{1}{2}$ **9. a.** $\frac{8}{59}$ **b.** $\frac{15}{59}$ **11. a.** $\frac{10}{80} = \frac{1}{8}$ **b.** $\frac{60}{80} = \frac{3}{4}$ **c.** $\frac{50}{80} = \frac{5}{8}$ **d.** $\frac{22}{80} = \frac{11}{40}$ **13. a.** $\frac{{}_3C_2}{{}_8C_2} = \frac{3}{28}$ **b.** $\frac{{}_5C_2}{{}_8C_2} = \frac{5}{14}$ **15. a.** $\frac{1}{{}_{39}C_5} = \frac{1}{575{,}757}$ **b.** $\frac{575{,}756}{575{,}757}$ **c.** The events of winning the grand prize and losing are not equally likely events. **17.** $\frac{27}{130}$ **19.** $\frac{27}{130}$ **21.** $\frac{112}{130} = \frac{56}{65}$ **23.** $\frac{180}{450} = \frac{2}{5}$ **25.** $\frac{210}{450} = \frac{7}{15}$

27. $\frac{13}{52}=\frac{1}{4}$ **29.** $\frac{26}{52}=\frac{1}{2}$ **31.** $\frac{16}{52}=\frac{4}{13}$ **33.** $\frac{13}{52}=\frac{1}{4}$ **35.** $\frac{26}{52}=\frac{1}{2}$ **37.** $\frac{16}{52}=\frac{4}{13}$ **39. a.** $\frac{120}{600}=\frac{1}{5}$ **b.** $\frac{160}{600}=\frac{4}{15}$ **c.** $\frac{580}{600}=\frac{29}{30}$ **d.** $\frac{140}{480}=\frac{7}{24}$ **e.** $\frac{20}{120}=\frac{1}{6}$ **f.** $\frac{7}{24}\approx 0.292$; $\frac{1}{6}\approx 0.167$; male firefighters are more likely to be promoted.

Chapter 14 Review Exercises, pp. 969–971

1. $\frac{3}{4}, \frac{9}{16}, \frac{27}{64}$ **3.** $a_n = 1 + (n-1)(3)$ **5.** 4 **7.** $\sum_{k=1}^{4} x^{3k}$

9. a. $a_n = 100 + 10n$ **b.** 110, 120, 130, 140, 150 **c.** $a_{27} = 370$; The cost of a speeding ticket is \$370 for a driver traveling 27 mph over the speed limit. **11.** $-\frac{1}{3}$ **13.** $a_n = 1 + (n-1)(5)$

15. $a_{10} = -29$ **17.** 1075 **19.** $-\frac{1}{2}$ **21.** $a_n = 4\left(\frac{3}{2}\right)^{n-1}$ **23.** $-\frac{781}{64}$

25. $r = \frac{1}{6}$; $S = \frac{12}{5}$ **27.** $x^{10} + 20x^8 + 160x^6 + 640x^4 + 1280x^2 + 1024$ **29.** 35 **31.** $a^{11} + 22a^{10}b + 220a^9b^2$ **33.** $160a^3b^3$ **35.** $2^5 = 32$ **37.** ${}_{26}P_4 = 358{,}800$ **39.** ${}_8C_3 \cdot {}_{10}C_1 = 560$ **41. a.** 0.186 **b.** 0.289 **c.** 0.256

Chapter 14 Test, pp. 972–973

1. $-3, -2, -\frac{4}{3}, -\frac{8}{9}$; geometric

2. $a_n = -6 + (n-1)\left(\frac{1}{2}\right)$; arithmetic **3.** 5 **4.** $a_n = n^2 + \left(\frac{n}{2}\right)h$

5. $\sum_{i=1}^{7} \frac{3+i}{i}$ **6.** $a_{33} = -519$ **7.** 2295 **8. a.** 52 **b.** 5616

9. $a_5 = \frac{32}{3}$ **10.** $a_1 = \frac{5}{3}$ **11.** -1364 **12.** $\frac{3325}{32}$ **13. a.** \$1095 **b.** \$1160.70 **c.** \$5933.13 **d.** 86,568.71. By saving money instead of buying cigarettes, the total savings plus interest amounts to \$86,568.71 **14.** 3 **15.** $a^4 + 4a^3b + 6a^2b^2 + 4ab^3 + b^4$ **16.** $81y^4 - 216y^3x^2 + 216y^2x^4 - 96yx^6 + 16x^8$ **17.** $189w^5z^2$ **18.** $3 \cdot 2 = 6$

19. ${}_{25}P_3 = 13{,}800$ **20.** $2 \cdot 2 = 4$ **21.** ${}_{42}C_6 = 5{,}245{,}786$ **22. a.** $\frac{387}{2225}$ **b.** $\frac{605}{2225}=\frac{121}{445}$ **23. a.** $\frac{20}{50}=\frac{2}{5}$ **b.** $\frac{26}{50}=\frac{13}{25}$ **c.** $\frac{36}{50}=\frac{18}{25}$

Cumulative Review Exercises, Chapters 1–14, pp. 973–976

1. $-\frac{11}{10}$ **2.** $3x + 54$ **3.** $x = \frac{4}{3}$ **4.** \$2225 is invested at 8% and \$7775 is invested at 5%. **5.** The angles are 69° and 21°. **6.** $(-\infty, 5)$ **7.** x-intercept: (0, 0); y-intercept: (0, 0);

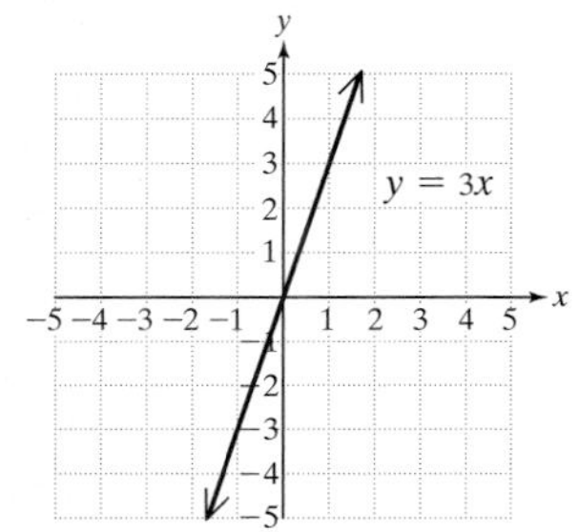

8. Slope: 1; y-intercept: (0, 2) **9.** $y = \frac{2}{5}x + \frac{11}{5}$ **10.** $y = 3x + 10$

11. (2, 3, −1) **12.** No solution **13.** (−3, 1) **14.** 27 **15.** $y = \frac{36}{13}$

16. Domain: {4, 3, 2}; range: {2, −6, 5, 1} **17.** 800 kHz **18.** $\frac{x^2y^{12}}{9}$

19. 4.0×10^5 **20.** $-x^3 - 10x^2 - 7x - 12$ **21.** $x^2 + 4xy + 4y^2 - 9$

22. $3x^2 - 8 + \frac{5x+13}{x^2+2}$ **23.** $|x|$ **24.** $\{x \mid x \geq 3\}$ **25.** $\frac{1}{4}$ **26.** x^8y^8

27. $2xy^4\sqrt[3]{2x^2}$ **28.** $\frac{\sqrt[3]{2x^2}}{x}$ **29.** $x = -\frac{5}{2}$ **30.** $4 + 5i$

31. $-\frac{9}{29} - \frac{21}{29}i$ **32.** $(2x + 9)(3x - 4)$

33. $(2x^2 + y)(4x^4 - 2x^2y + y^2)$ **34.** $x = \pm\frac{3}{2}$ **35.** $x = -3 \pm \sqrt{14}$

36. $x = \frac{3 \pm i\sqrt{7}}{4}$ **37.** $x = \pm 2, x = \pm i\sqrt{3}$ **38. a.** 2 **b.** 22 sides

39. $(-\infty, -7) \cup (-7, 7) \cup (7, \infty)$ **40.** $\frac{2(x-1)(x+1)}{5(x+4)}$ **41.** $\frac{x-y}{x}$

42. $x = -4$ ($x = 2$ does not check) **43. a.** $(-\infty, -4] \cup [-\frac{2}{3}, \infty)$ **b.** No solution **44.** $x = 8, x = -\frac{2}{3}$ **45.** (−2, 12) **46.** [2, 4)

47.

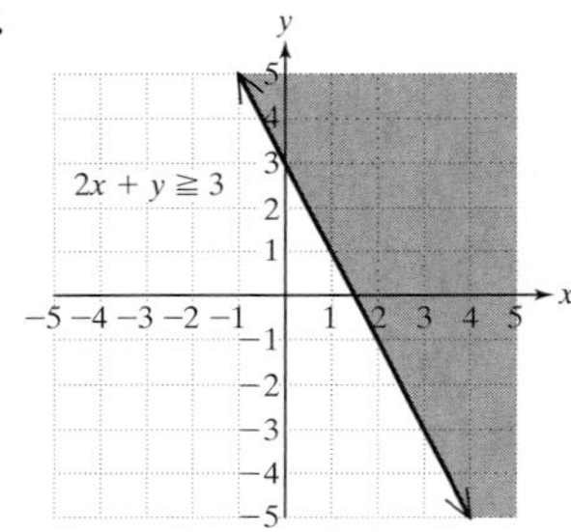

48. $f^{-1}(x) = \frac{2x+1}{3}$

49. Domain: $(-\infty, \infty)$; range: $(0, \infty)$

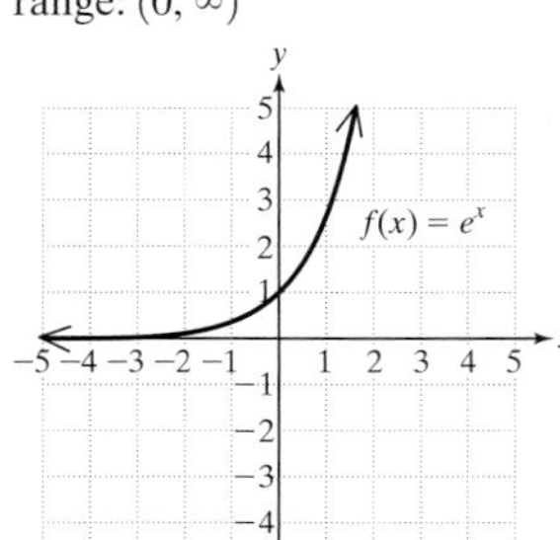

50. Domain: $(0, \infty)$; range: $(-\infty, \infty)$

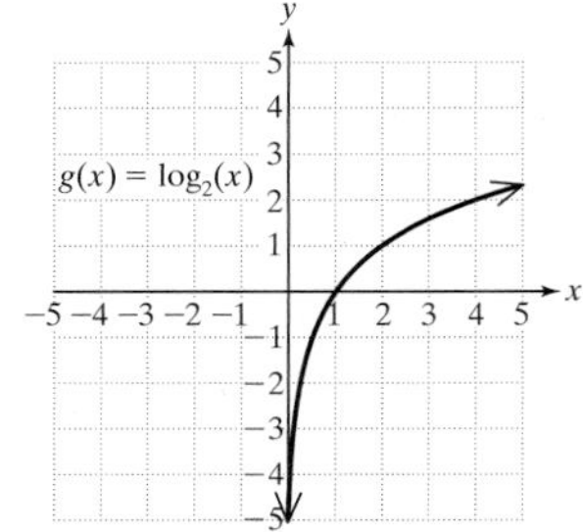

51. 3 **52.** 5 **53.** $\ln\left(\frac{x^3\sqrt{z}}{y}\right)$ **54.** 7.7% **55.** $x = 3$ ($x = -4$ does not check) **56.** Arithmetic; $a_n = -16 + (n-1)(-5)$

57. Geometric; $a_n = \frac{3}{4}(2)^{n-1}$ **58.** $\frac{11{,}718}{3125} = 3.74976$ **59.** 1328

60. $81 - 216t^2 + 216t^4 - 96t^6 + 16t^8$ **61.** ${}_{41}C_6 = 4{,}496{,}388$

62. ${}_{15}P_6 = 3{,}603{,}600$ **63. a.** $\frac{34}{74}=\frac{17}{37}$ **b.** $\frac{46}{74}=\frac{23}{37}$ **c.** $\frac{55}{74}$ **d.** $\frac{41}{74}$

64. a. (2, −4) **b.** (0, −3) **c.** (−2, 0) and (6, 0) **d.** $y = -5$ **e.**

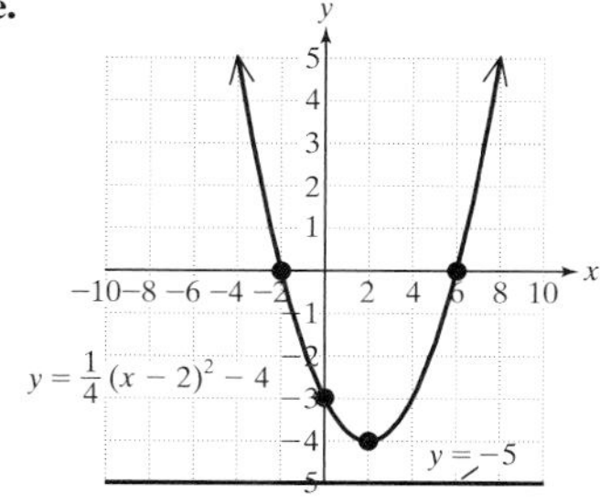

65. Center: (3, −4); radius: $\sqrt{35}$

66.

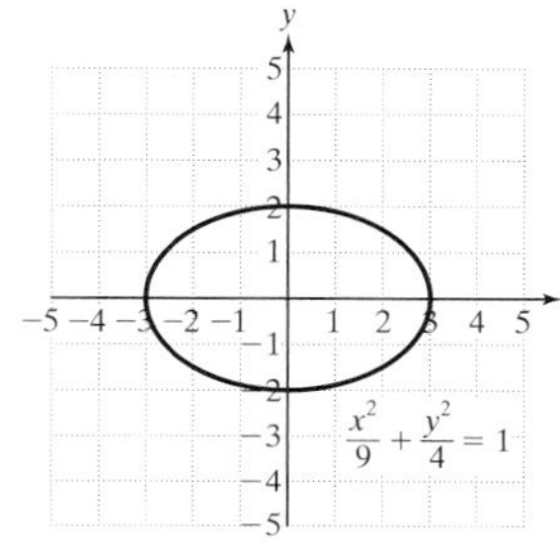

67.

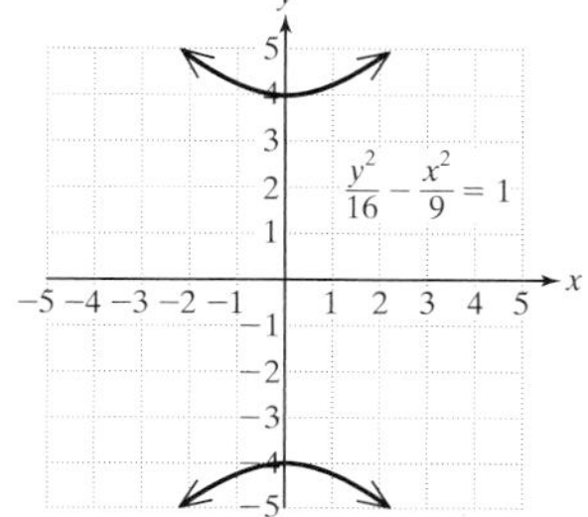

68. (6, 8) and (−6, −8) **69.** (2, 4), (−2, 4)

70.

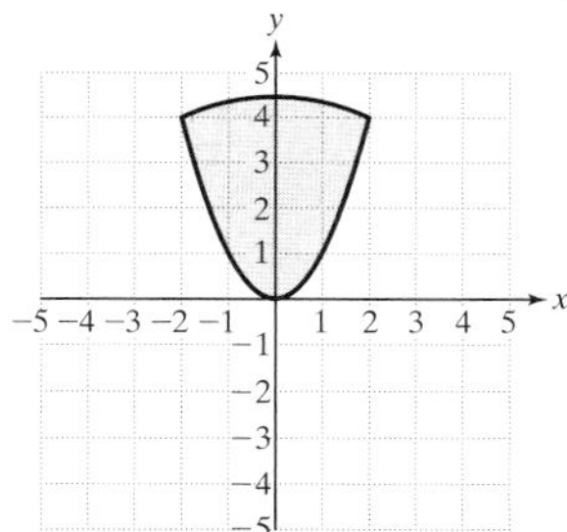

REVIEWS

Review A Practice Exercises, pp. 983–985

1. $\left\{-6, -\frac{7}{8}, -0.\overline{3}, 0, \frac{4}{9}, 7, 10.4\right\}$ **3.** {−6, 0, 7} **5.** {1, 8} **7.** $\{-\pi, \sqrt{3}\}$ **9.** 5 is greater than −2 **11.** −3 is greater than or equal to −4 **13.** 10 is between 6 and 12 **15.** 2 is not equal to −5 **17.** $6 \cdot 3$ **19.** $20 \div 5$ **21.** $|-5|$ **23.** $2 + |-8|$ **25.** $4 \cdot \frac{1}{3}$ **27.** −2 **29.** 19 **31.** 16 **33.** −33 **35.** −4.3 **37.** −16 **39.** −19 **41.** 5.6 **43.** $-\frac{9}{8}$ **45.** $-\frac{20}{9}$ **47.** −27 **49.** 35 **51.** 0 **53.** −8.61 **55.** $\frac{5}{6}$ **57.** 7 **59.** −6 **61.** 0 **63.** Undefined **65.** $\frac{1}{2}$ **67.** 9 **69.** 8 **71.** 2 **73.** −16 **75.** 5 **77.** 15 **79.** $-\frac{8}{3}$ **81.** −16 **83.** −11 **85.** 460 in.^{2} **87.** Associative property of multiplication **89.** Distributive property of multiplication over addition **91.** Commutative property of addition **93.** 1; for example 1(3) = 3 **95. a.** reciprocal **b.** $\frac{1}{4}$ **97.** $5a + 45$ **99.** $-7y + 7z$ **101.** $2b + 3c - 5$ **103.** $-2x + 3y - 8$ **105.** $7a + 14$ **107.** $5y - 25$ **109.** $-w + 12u$ **111.** $14g - 48h$

Review B Practice Exercises, pp. 989–990

1. $x = -21$ **3.** $t = 60$ **5.** $p = -20$ **7.** $x = 3$ **9.** $x = -\frac{3}{2}$ **11.** $b = 23$ **13.** $w = -\frac{7}{2}$ **15.** $y = -\frac{10}{7}$ **17.** $x = -\frac{19}{3}$ **19.** $a = 5.5$ **21.** $t = 2.3$ **23.** $x = 0$ **25.** $y = 2$ **27.** $m = 9$ **29.** $y = -33$ **31.** $\frac{1}{2}$ **33.** −20, −19, −18 **35.** 4.5% **37.** $2100 in the 4.2% account, $1400 in the 3.8% account **39.** Car: 64 mph; truck: 60 mph

41. $(-\infty, -2)$

43. $[-6, \infty)$

45. $(-2, \infty)$

47.

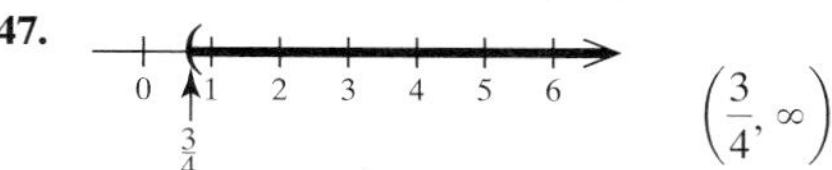

$\left(\frac{3}{4}, \infty\right)$

49. $\left(-\frac{7}{4}, 1\right]$

51. $(-1, 7)$

53. $\left(-\infty, \frac{35}{3}\right)$

55. $(-\infty, -0.2]$

Review C Practice Exercises, pp. 996–997

1–8.

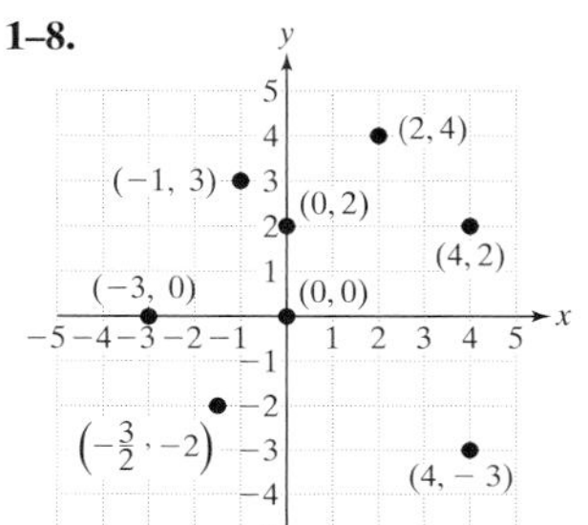

9. IV **11.** III **13.** I **15.** II

17.

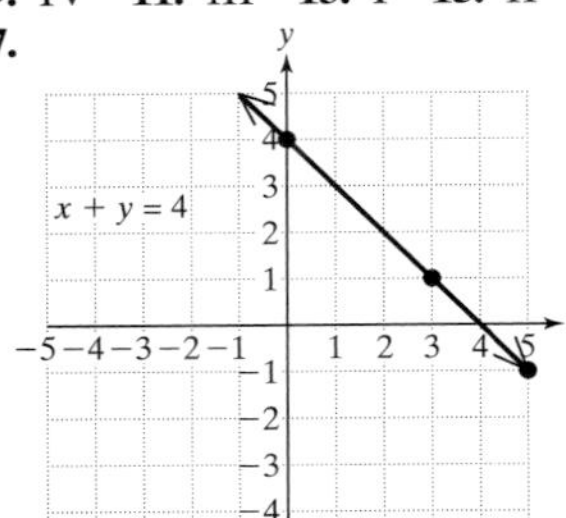

19.

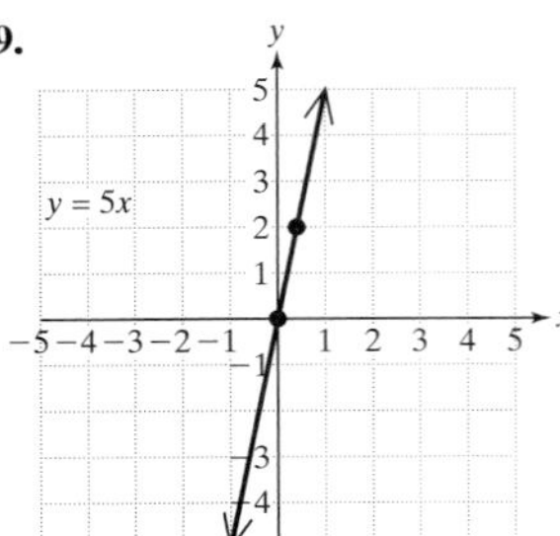

21.

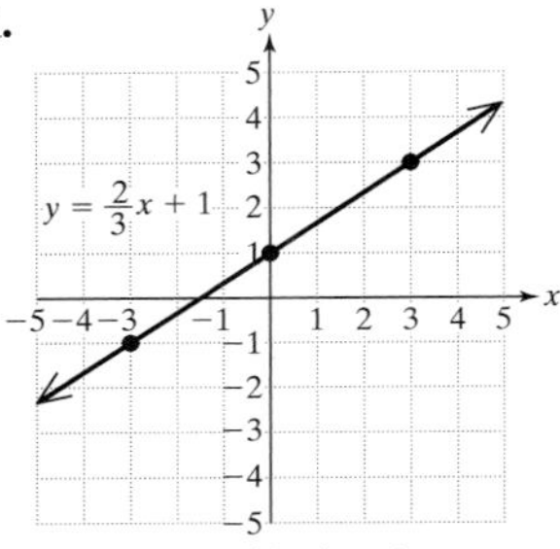

23. x-intercept $(6, 0)$; y-intercept $(0, -9)$ **25.** x-intercept $(-3, 0)$; y-intercept $(0, -15)$ **27.** x-intercept $(4, 0)$; y-intercept $(0, 2)$ **29.** x-intercept $(0, 0)$; y-intercept $(0, 0)$ **31.** x-intercept $\left(-\frac{9}{2}, 0\right)$; y-intercept $(0, -3)$ **33.** x-intercept $\left(\frac{5}{6}, 0\right)$; y-intercept $(0, -5)$

35. $x = 3$; vertical **37.** $y = -\frac{1}{3}$; horizontal **39.** $x = 0$; vertical

41. $y = \frac{4}{5}x - \frac{6}{5}$; slope is $\frac{4}{5}$; y-intercept $\left(0, -\frac{6}{5}\right)$ **43.** $y = 3x + \frac{3}{2}$; slope is 3; y-intercept $\left(0, \frac{3}{2}\right)$ **45.** $x = 1$; undefined slope; no y-intercept **47.** $y = -\frac{1}{4}x$; slope is $-\frac{1}{4}$; y-intercept $(0, 0)$

49. $y = 4$; slope is 0; y-intercept is $(0, 4)$

51.

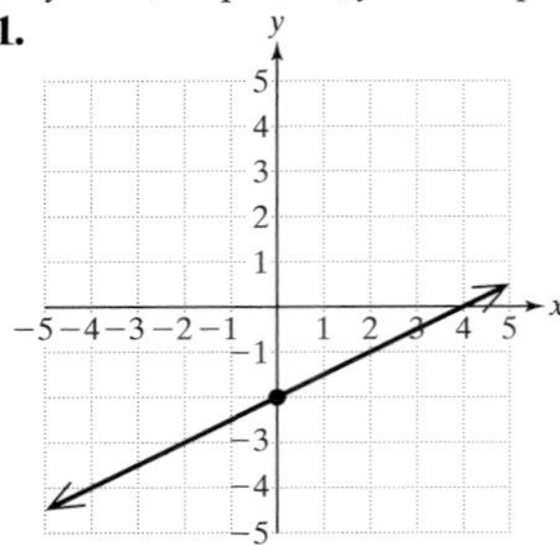

53.

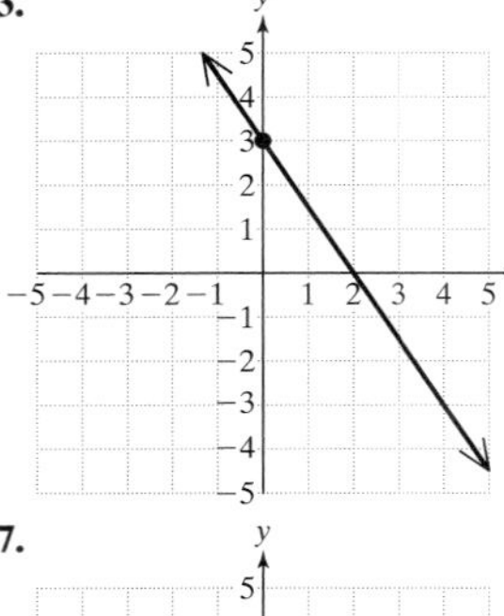

55.

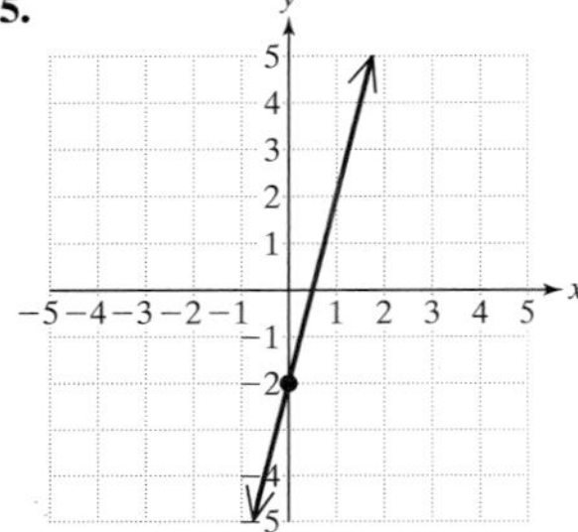

57.

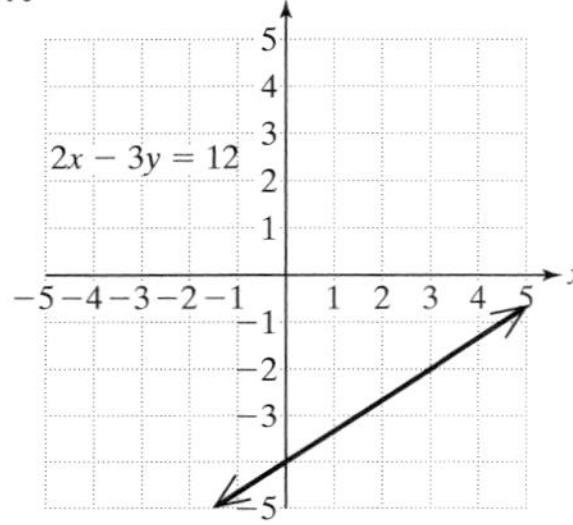

59.

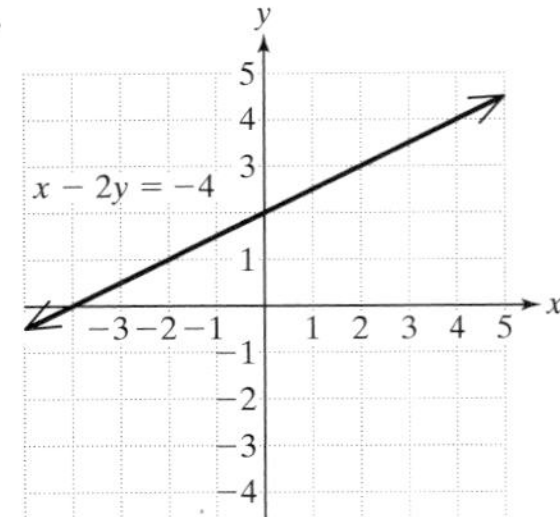

61.

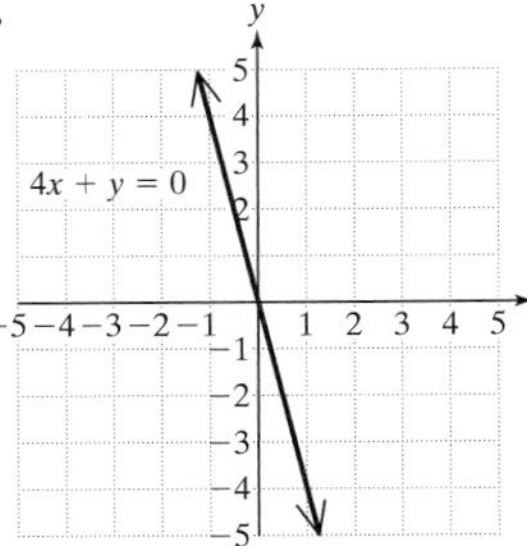

63.

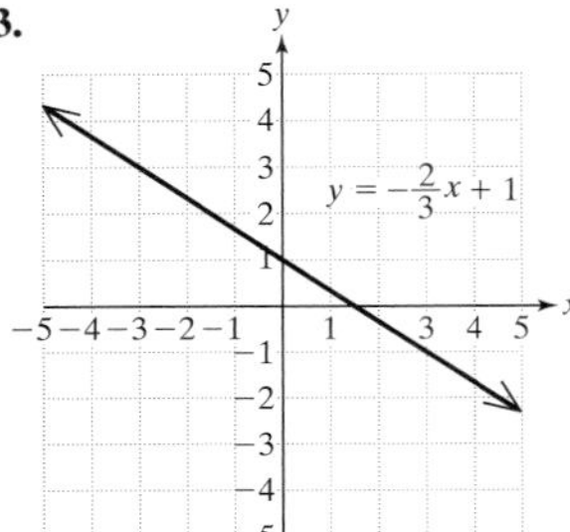

65.

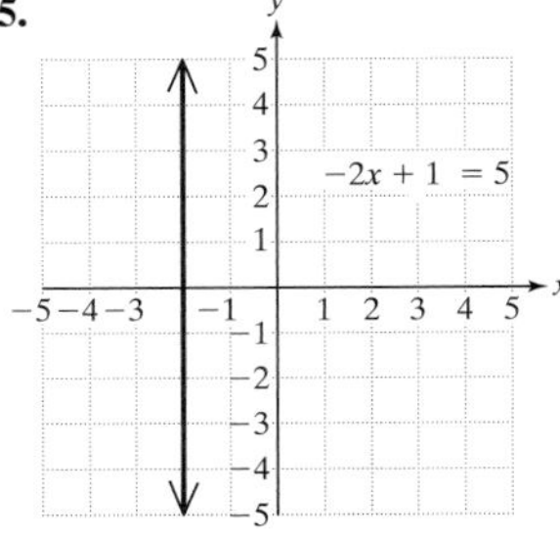

67.

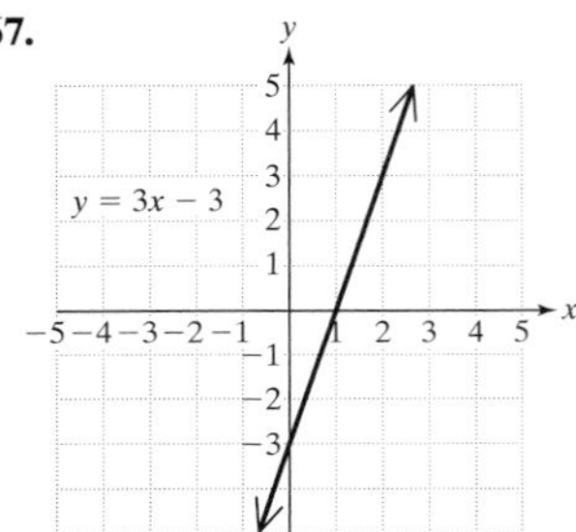

69. Parallel **71.** Neither **73.** Perpendicular **75.** Parallel

Review D Practice Exercises, pp. 1003–1004

1. 6^7; multiplication of like bases **3.** y^3, division of like bases **5.** $\frac{w^4}{256}$; power of a quotient **7.** q^{30}; power rule **9.** 1 **11.** 2 **13.** $\frac{1}{25}$ **15.** $\frac{7}{6}$ **17.** $\frac{1}{r^2}$ **19.** b^2 **21.** $27x^6$ **23.** $3w^2z^2$ **25.** $\frac{b^2c^6}{a^4d^8}$ **27.** $\frac{1}{wz^3}$ **29.** 3.05×10^6 **31.** 2.51×10^{-5} **33.** 8.96×10^7 **35.** 3.9×10^{-4} **37.** 3.6×10^4 **39.** 2.1×10^{16} **41.** 2.19×10^{-2} **43.** 8.0×10^{-12} **45.** $13p^3 - 4p^2 + 2p + 11$ **47.** $-n^2 - n + 8$ **49.** $-10u^2v + 16u^2 + 2uv^2$ **51.** $4p^4 - 20p^3 + 8p^2 + 32p$ **53.** $-6x^2 - 17xy - 5y^2$ **55.** $16b^2 - 16b - 12$ **57.** $x^3 - 6x^2y + 14xy^2 - 15y^3$ **59.** $9h^2 - 64$ **61.** $16x^2 - 40x + 25$ **63.** $\frac{1}{9}t^4 - 81$ **65.** $0.04x^4 - 1.2x^2 + 9$ **67.** $2a^2b - 6ab + b - 3$ **69.** $x^2 - 5x + 11$ **71.** $2p^2 - 7p + 4 + \frac{10}{p - 5}$ **73.** $-5m^4 + 8m^3 - 4m^2 - 4m + 1$ **75.** $4a^2 + 6a + 9 + \frac{18}{2a - 3}$

Review E Practice Exercises, pp. 1007–1008

1. $a^2 - b^2 = (a + b)(a - b)$ **3.** $a^3 + b^3 = (a + b)(a^2 - ab + b^2)$ **5. a.** Difference of squares **b.** $(t - 10)(t + 10)$ **7. a.** Sum of cubes **b.** $(y + 3)(y^2 - 3y + 9)$ **9. a.** Trinomial (nonperfect square trinomial) **b.** $(d - 4)(d + 7)$ **11. a.** Trinomial (perfect square trinomial) **b.** $(x - 6)^2$ **13. a.** Four terms—grouping **b.** $(ax + b)(2x - 5)$ **15. a.** Trinomial (nonperfect square trinomial) **b.** $(2y - 1)(5y + 4)$ **17. a.** Difference of squares **b.** $10(p - 8)(p + 8)$ **19. a.** Difference of cubes **b.** $z(z - 4)(z^2 + 4z + 16)$ **21. a.** Trinomial (nonperfect square trinomial) **b.** $b(b + 5)(b - 9)$ **23. a.** Trinomial (perfect square trinomial) **b.** $(3w + 4x)^2$ **25. a.** Four terms—grouping **b.** $10(2x + a)(3x - 1)$ **27. a.** Four terms—grouping **b.** $(x - 3)(x + 3)(x + 4)$ **29. a.** Difference of squares **b.** $(w^2 + 4)(w - 2)(w + 2)$ **31. a.** Difference of cubes **b.** $(t^2 - 2)(t^4 + 2t^2 + 4)$ **33. a.** Trinomial (nonperfect square trinomial) **b.** $(4p - 1)(2p - 5)$ **35. a.** Trinomial (perfect square trinomial) **b.** $(6y - 1)^2$ **37. a.** Four terms—grouping (three terms by one term) **b.** $(x + 2 - y)(x + 2 + y)$ **39. a.** Sum of squares **b.** $2(x^2 + 25)$ **41. a.** Trinomial (nonperfect square trinomial) **b.** $s^2(3r - 2)(4r + 5)$ **43. a.** Trinomial (nonperfect square trinomial) **b.** $(x - 3y)(x + 11y)$ **45. a.** Sum of cubes **b.** $(m^2 + n)(m^4 - m^2n + n^2)$ **47. a.** Trinomial (nonperfect square trinomial) **b.** $(a + b)(x - 4)(x + 3)$ **49. a.** None of these **b.** $x(x - 4)$ **51.** $x = -\frac{5}{2}$ or $x = 3$ **53.** $x = 0$ or $x = 5$ **55.** $w = -3$ **57.** $z = 4$ or $z = -2$ **59.** $x = \frac{5}{3}$ or $x = -\frac{1}{2}$ **61.** $x = \frac{3}{2}$ or $x = 4$ **63.** $w = 1$ or $w = -1$

Review F Practice Exercises, pp. 1014–1015

1. $\{x \mid x \neq -4\}$ **3.** $\left\{x \mid x \neq \frac{3}{5} \text{ and } x \neq -\frac{3}{5}\right\}$ **5.** $\{t \mid t \text{ is any real number}\}$ **7.** $\frac{x^3}{4y}$ **9.** $\frac{t + 2}{2}$ **11.** $\frac{y + 4}{y + 2}$ **13.** $\frac{y^3}{10x}$ **15.** $\frac{1}{2}$ **17.** $3ab^2c$ **19.** $\frac{p(p - 6)}{4}$ **21.** $-\frac{t^2}{t - 7}$ **23.** $8x^3y^{10}$ **25.** $(x - 4)(x + 3)(x - 3)$ **27.** $-\frac{1}{3x}$ **29.** $\frac{5x^2 + 2y}{4x^3y^2}$ **31.** $x - 7$ **33.** $\frac{17x}{2(2x - 1)}$ **35.** $\frac{x^2 - 3x + 6}{(x + 4)(x - 3)(x - 1)}$ **37.** $\frac{-3}{y + 3}$ **39.** -1 **41.** $\frac{3b + 4a}{7b - a}$ **43.** $2(4x + 1)$ **45.** $u - v$ **47.** $\frac{y + 5}{2y - 9}$ **49.** $x = 10$ **51.** No solution **53.** $x = -2, x = -4$ **55.** $x = 7$

APPENDIX A

Section A.1 Practice Exercises, pp. 1021–1022

1. The divisor must be of the form $x - r$. **3.** No, the divisor must be of the form $x - r$. **5. a.** $x - 5$ **b.** $x^2 + 3x + 11$ **c.** 58 **7.** $x + 6$ **9.** $t - 4$ **11.** $5y + 10 + \frac{11}{y - 1}$ **13.** $x + 7$ **15.** $3y^2 - 2y + 2 + \frac{-3}{y + 3}$ **17.** $x^2 - x - 2$ **19.** $m^2 - 3m + 9$ **21.** $a^3 - 2a^2 + 2a - 4$ **23.** $3p^3 + p^2 + 4p + 2 + \frac{-2}{p - 4}$ **25.** $4w^3 + 2w^2 + 6$ **27.** $8x^2 - 4x + 1 + \frac{\frac{3}{4}}{x + \frac{1}{4}}$ **29. a.** -84 **b.** $4x^2 - 6x + 16$; remainder -84 **31.** $P(r)$ equals the remainder of $P(x) \div (x - r)$. **33. a.** $8x + 5$ **b.** Yes

Section A.2 Practice Exercises, pp. 1027–1029

1. Mean: 6; median: 5; mode: 5 **3.** Mean: 5; median: 4.5; modes: 3, 7 **5.** Mean: 14.5; median: 14; mode: none **7.** Mean: 13 pounds; median: 12 pounds **9.** Mean: 26; median: 20 **11.** $7 **13.** car **15.** $22 **17.** 9.52 million **19.** $25,717.40 **21.** Mean age for golfers: 34.1 years; Mean age for tennis players: 24.25 years. On the average, women golfers are older **23.** 3.59 **25.** 4 years old

Applications Index

Biology/Health/Life Sciences

Business

CHEMISTRY

CONSTRUCTION

CONSUMER APPLICATIONS

DISTANCE/SPEED/TIME

ENVIRONMENT

INVESTMENT

POLITICS

SCHOOL

SCIENCE

SPORTS

STATISTICS/DEMOGRAPHICS

INDEX

D

Q

R

S

T

U

V

W

X

Properties of Real Numbers

Commutative Property of Addition	$a + b = b + a$
Commutative Property of Multiplication	$ab = ba$
Associative Property of Addition	$(a + b) + c = a + (b + c)$
Associative Property of Multiplication	$(ab)c = a(bc)$
Distributive Property of Multiplication over Addition	$a(b + c) = ab + ac$
Identity Property of Addition	0 is the **identity element for addition** because $a + 0 = 0 + a = a$
Identity Property of Multiplication	1 is the **identity element for multiplication** because $a \cdot 1 = 1 \cdot a = a$
Inverse Property of Addition	a and $(-a)$ are **additive inverses** because $a + (-a) = 0$ and $(-a) + a = 0$
Inverse Property of Multiplication	a and $\frac{1}{a}$ are **multiplicative inverses** because $a \cdot \frac{1}{a} = 1$ and $\frac{1}{a} \cdot a = 1$ (provided $a \neq 0$)

Sets of Real Numbers

Natural numbers: $\{1, 2, 3, \ldots\}$

Whole numbers: $\{0, 1, 2, 3, \ldots\}$

Integers: $\{\ldots -3, -2, -1, 0, 1, 2, 3, \ldots\}$

Rational numbers: $\{\frac{p}{q} | p$ and q are integers and q does not equal $0\}$

Irrational numbers: $\{x | x$ is a real number that is not rational$\}$

Application Formulas

Sales tax = (cost of merchandise)(tax rate)

Commission = (dollars in sales)(commission rate)

Simple interest = (principal)(rate)(time): $I = Prt$

Distance = (rate)(time): $d = rt$

Compound interest: $A(t) = P\left(1 + \frac{r}{n}\right)^{nt}$

Continuous compound interest: $A(t) = Pe^{rt}$,

where $A(t)$ = balance of account after t years,
P = principal, r = annual interest rate, t = time in years,
n = number of compound periods per year

Proportions

An equation that equates two ratios is called a proportion:

$$\frac{a}{b} = \frac{c}{d} \quad (b \neq 0, d \neq 0)$$

The cross products are equal: $ad = bc$.

Exponential Functions

A function defined by $y = b^x$ $(b > 0, b \neq 1)$ is an exponential function.

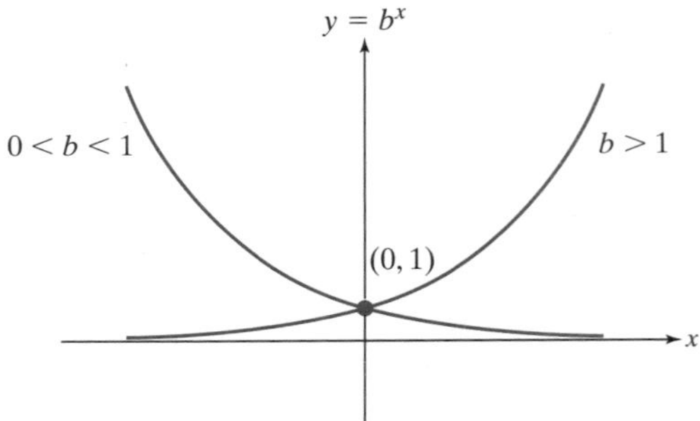

Logarithmic Functions

A function defined by $y = \log_b(x)$ is a logarithmic function.

$y = \log_b(x) \Leftrightarrow b^y = x \quad (x > 0, b > 0, b \neq 1)$

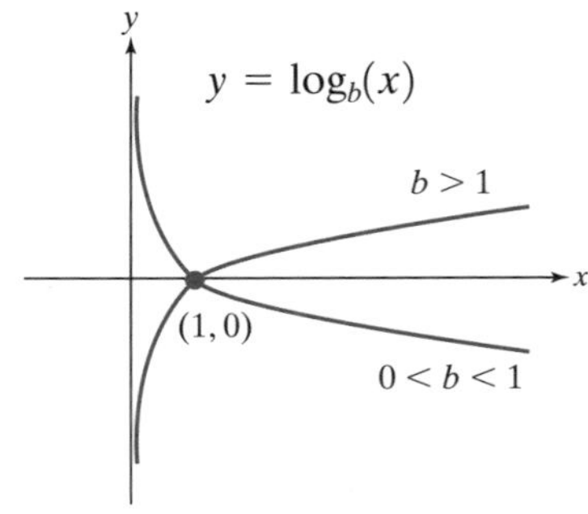

Linear Equations and Slope

The slope, m, of a line between two distinct points (x_1, y_1) and (x_2, y_2):

$$m = \frac{y_2 - y_1}{x_2 - x_1}, \quad x_2 - x_1 \neq 0$$

Standard form: $ax + by = c$, a and b are not both zero

Horizontal line: $y = k$

Vertical line: $x = k$

Slope intercept form: $y = mx + b$

Point-slope formula: $y - y_1 = m(x - x_1)$

Midpoint Formula

Given two points (x_1, y_1) and (x_2, y_2), the midpoint is

$$\left(\frac{x_1 + x_2}{2}, \frac{y_1 + y_2}{2}\right)$$

Angles

Two angles are **complementary** if the sum of their measures is 90°.

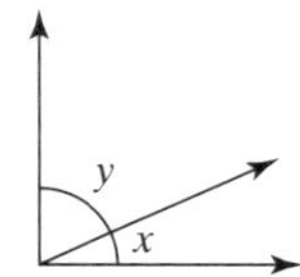

Two angles are **supplementary** if the sum of their measures is 180°.

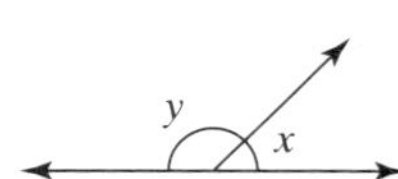

In the figure below, $\angle a$ and $\angle c$ are vertical angles and $\angle b$ and $\angle d$ are vertical angles. The measures of vertical angles are equal.

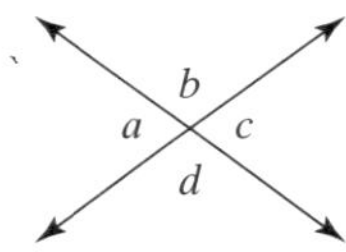

The sum of the measures of the angles of a triangle is 180°.

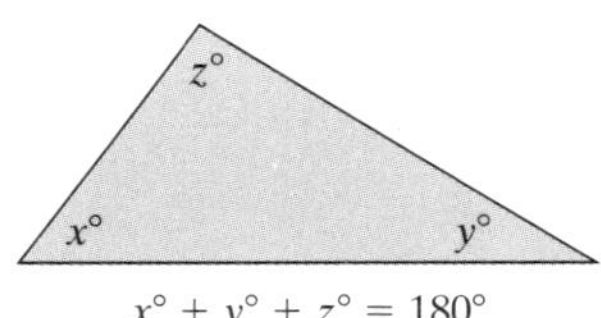

$x° + y° + z° = 180°$

Properties and Definitions of Exponents

Let a and b $(b \neq 0)$ represent real numbers and m and n represent positive integers.

$$b^m b^n = b^{m+n}; \quad \frac{b^m}{b^n} = b^{m-n}; \quad (b^m)^n = b^{mn};$$

$$(ab)^m = a^m b^m; \quad \left(\frac{a}{b}\right)^m = \frac{a^m}{b^m}; \quad b^0 = 1; \quad b^{-n} = \left(\frac{1}{b}\right)^n$$

Difference of Squares:

$a^2 - b^2 = (a + b)(a - b)$

Difference of Cubes:

$a^3 - b^3 = (a - b)(a^2 + ab + b^2)$

Sum of Cubes:

$a^3 + b^3 = (a + b)(a^2 - ab + b^2)$

Perfect Square Trinomials:

$a^2 + 2ab + b^2 = (a + b)^2$

$a^2 - 2ab + b^2 = (a - b)^2$

The Quadratic Formula

The solutions to $ax^2 + bx + c = 0 \quad (a \neq 0)$ are given by

$$x = \frac{-b \pm \sqrt{b^2 - 4ac}}{2a}$$

The Vertex Formula

For $f(x) = ax^2 + bx + c$ $(a \neq 0)$, the vertex is

$$\left(\frac{-b}{2a}, \frac{4ac - b^2}{4a}\right) \quad \text{or} \quad \left(\frac{-b}{2a}, f\left(\frac{-b}{2a}\right)\right)$$

The Distance Formula

The distance between two points (x_1, y_1) and (x_2, y_2) is $d = \sqrt{(x_2 - x_1)^2 + (y_2 - y_1)^2}$

The Standard Form of a Circle

$(x - h)^2 + (y - k)^2 = r^2$ with center (h, k) and radius r